The Birds of Africa
Volume V

Edited by

EMIL K. URBAN
Department of Biology, Augusta State University,
Augusta, Georgia, USA

C. HILARY FRY
Department of Zoology, University of Aberdeen,
Aberdeen, UK

STUART KEITH
Department of Ornithology, American Museum of Natural History,
New York, New York, USA

Colour Plates by Martin Woodcock

Line Drawings by Ian Willis
Acoustic References by Claude Chappuis
Bibliography and Index by Lois L. Urban

ACADEMIC PRESS
San Diego · London · Boston · New York · Sydney · Tokyo · Toronto

This book is printed on acid-free paper.

Copyright © 1997 by ACADEMIC PRESS

All rights reserved.
No part of this publication may be reproduced or transmitted in any form or by any means, electronic or mechanical, including photocopy, recording, or any information storage and retrieval system, without permission in writing from the publisher.

Academic Press
525 B Street, Suite 1900, San Diego, California 92101–4495, USA
http://www.apnet.com

Academic Press Limited
24–28 Oval Road, London NW1 7DX, UK
http://www.hbuk.co.uk/ap/

ISBN 0-12-137305-3

A catalogue record for this book is available from the British Library

97 98 99 00 01 02 EB 9 8 7 6 5 4 3 2 1

Editorial and production services by Moira Fisher, Ringwood, Hampshire
Maps produced by Hardlines, Charlbury, Oxford
Typeset by Phoenix Photosetting, Chatham, Kent
Colour plates printed in the UK by George Over Ltd, Rugby
Printed and bound in the UK by Mackays of Chatham PLC, Chatham, Kent

ACKNOWLEDGEMENTS

As in previous volumes we are extremely grateful to our authors for the tremendous amount of time and effort spent in researching their species in the field, museum and library and in writing accounts to exacting standards; and for their patience and goodwill as we edited their accounts. We express our sincere thanks and appreciation to them.

During the preparation of species accounts authors have drawn heavily upon the special knowledge of numerous colleagues who have given them access to unpublished and prepublished works, their valuable time to referee the accounts and also their hospitality. We particularly appreciate all those who sent us unpublished field notes from recent trips; their contributions have helped make this volume more complete. For all their help we take great pleasure in thanking: J. S. Ash, the late D. R. Aspinwall, T. Aversa, G. C. Backhurst, N. and L. Baker, M. Bellatreche, L. Bennun, S. Bensch, A. Berruti, L. Birch, C. G. R. Bowden, the late R. K. Brooke, A. Brosset, T. M. Butynski, M. Carswell, C. Carter, the late R. T. Chapin, C. Chappuis, R. A. Cheke, P. Christy, P. A. Clancey, B. D. Colahan, P. R. Colston, N. J. Cordeiro, G. Cowles, H. Q. P. Crick, A. J. Cruickshank, P. Davidson, W. R. J. Dean, J.-P. Decoux, A. Devez, R. J. Dowsett, F. Dowsett-Lemaire, R. J. Douthwaite, C. Dranzoa, M. Dyer, C. Erard, J. and H. Eriksen, T. Evans, the late J. Farrand Jr, L. D. C. Fishpool, K. L. Garrett, M. E. Gartshore, J. P. Gee, C. M. Gichuki, D. Goodall, A. A. Green, L. G. Grimes, D. Halleux, D. B. Hanmer, D. Hasselquist, P. A. R. Hockey, C. W. Hustler, M. P. S. Irwin, G. Jarry, B. L. Johnson, E. D. H. Johnson, J. Kalina, D. Kelly, A. C. Kemp, R. Kettle, P. C. Lack, B. Lamarche, J.-M. Lernould, M. Louette, A. J. Manson, R. Martin, A. N. B. Masterson, G. Mauersberger, the late J. Mendelsohn, R. Medland, J. Miskell, G. J. Morel, D. C. Moyer, P. J. Mundy, R. de Naurois, G. Nikolaus, T. B. Oatley, S. L. Olson, U. Olsson, P. E. Osborne, R. K. Panza, D. A. Parkes, K. C. Parkes, J. R. Paxton, H. Penry, the late A. Prigogine, R. Quantrill, P. A. Rodewald, D. Sargent, K. Schulze-Hagen, L. L. Short, N. J. Skinner, K. de Smet, D. W. Snow, E. H. Stickney, R. Stjernstedt, S. N. Stuart, L. Svensson, W. Tarboton, P. B. Taylor, M. Thévenot, B. J. Tigar, M. A. Traylor, D. A. Turner, A. Tye, G. D. Underhill, J.-P. Vande weghe, J. D. R. Vernon, R. Vernon, J. Vincent, Y.-M. de Viviès, M. P. Walters. Special thanks are due to D. A. Zimmerman for refereeing the *Cisticola* texts, and to D. J. Pearson for checking the *Cisticola* maps.

Again we are greatly indebted to the many museums holding African avian collections. For study facilities and specimen loans we thank the Trustees and staff of: The American Museum of Natural History, Department of Ornithology (and the Frank M. Chapman Memorial Fund for financial support for C. Erard); Natural History Museum (London), Department of Ornithology (Tring); Carnegie Museum of Natural History; Durban Natural History Museum; Field Museum of Natural History, Bird Division; Institut Royal des Sciences Naturelles (Brussels); Los Angeles County Museum, Bird Division; Musée National d'Histoire Naturelle (Paris); Musée Royale de l'Afrique Centrale (Tervuren); Museum of Comparative Zoology (Harvard); National Museums of Kenya; National Museum of Zimbabwe; Naturhistorisches Museum Wien; Naturhistoriska Riksmuseet (Stockholm); Peabody Museum of Natural History (New Haven); Philadelphia Academy of Natural Sciences; Royal Ontario Museum; Smithsonian Institution, US National Museum of Natural History, Department of Ornithology; Staatliches Museum für Naturkunde (Stuttgart); Transvaal Museum, Bird Division; Zoologishes Forschungsinstitut und Museum Alexander Koenig (Bonn); Zoologisches Museum (Berlin); and Zoologisk Museum (Copenhagen). For copious assistance with literature we (and particularly Hilary Fry) thank L. Birch and the Edward Grey Institute of Field Ornithology, Oxford University; also J. Hinshaw, The Josselyn Van Tyne Memorial Library, Wilson Ornithological Society, University of Michigan. We are especially grateful to Aberdeen University Library and Augusta State University Reese Library for inter-library loan services. For supplying bird sound recordings we are grateful to R. Ranft and the British Library of Wildlife Sounds; G. F. Budney and the Library of Natural Sounds, Cornell University; Percy FitzPatrick Institute of African Ornithology; the Transvaal Museum Bird Department; and most especially C. Chappuis and F. Dowsett-Lemaire. We owe special thanks to K. L. Garrett and P. I. LaFollete (Los Angeles County Museum) and L. B. Kiff (Western Foundation of Vertebrate Zoology) for providing printouts of their specimen collections and to J. S. Ash, R. J. Dowsett, G. Nikolaus, S. P. Rodwell and S. J. R. Rumsey for giving us substantial amounts of unpublished weight data. We acknowledge with thanks R. D. Wassenaar, Euring Data Bank and Netherlands Institute Ecology for providing European-African ringing data. We would especially like to acknowledge P. R. Colston of the Natural History Museum, Tring, who has willingly supplied authors with information on numerous occasions. L. G. Grimes provided valuable editorial support in researching literature from the Edward Grey Institute Library in Oxford.

The following sources were used in the preparation of line drawings: photo by P. W. Atkinson in *Bird Conservation International* 1991, **1** (*Amaurocichla bocagei*); drawings in G. L. Bates, *Ibis* 1927 (flycatcher bills); drawings by F. M. Benson in A. W. Vincent, *Ibis* 1947, **89** (*Apalis* nests); drawings in *The Birds of the Western Palearctic* 1992, **VI**, with kind permission of Oxford University Press (*Cisticola juncidis, Scotocerca inquieta*); photo in J. P. Chapin, *The Birds of the Belgian Congo* 1953 (*Schistolais leucopogon*); photo in R. T. Chapin, *Revue de Zoologie Africaine* 1978, **92** (*Muscicapa lendu*); photos by A. Devez in A. Brosset, *La Vie dans la Forêt Équatoriale* 1976 (several nests); drawings in C. Erard, *Écologie et Comportement des Gobe-mouches (Aves: Muscicapinae, Platysteirinae, Monarchinae) du Nord-Est du Gabon* 1987 (flycatcher and batis behaviour); photo by J. Eriksen and H. Eriksen (*Hippolais languida*); sketch by

M. Gartshore in *An Avifaunal Survey of Taï National Park, Ivory Coast* 1989 (*Melaenornis annamarulae*); drawings by D. Goodall in D. A. Parkes, *Honeyguide* 1993, **39** (*Sylvietta* nests); photos in P. J. Ginn, W. G. McIlleron and P. le S. Milstein, *The Complete Book of Southern African Birds* 1989 (several nests and portraits); drawings in T. Harris and G. Arnott, *Shrikes of Southern Africa* 1988 (*Batis capensis, Lanioturdus torquatus*); photo by Eric Hosking (*Prinia gracilis*); sketch in P. Kunkel *Zeitschrift für Tierpsychologie* 1974, **34** (*Eminia lepida*); photo by I. Makatsch in *British Birds* 1965, **58** (*Ficedula parva*); photos in C. W. Mackworth-Praed and C. H. B. Grant, *African Handbook of Birds* **I** (2) 1960, **II** (2) 1963 and **III** (2) 1973 (many nests); photos by G. L. Maclean (*Bradypterus baboecala, Eremomela icteropygialis*); skins in Natural History Museum (London) (heads, tails); sketch by R. de Naurois (*Prinia molleri*); photo in K. Newman, *Bird Life in Southern Africa,* 1979 (*Prinia flavicans*); photos by W. Nicholl (*Sylvietta rufescens*) and C. J. Uys (*Parisoma layardi*) in P. J. Ginn, W. G. McIlleron and P. le S. Milstein (eds), *The Complete Book of Southern African Birds* 1989; photo by R. Nöller in *Bokmakierie* 1975, **27** (*Batis pririt*); photo by P. Petit in *Alauda* 1993, **61** (*Cisticola juncidis*); photos in V. G. L. van Someren, *Days with Birds* 1956 (nests); and drawings reproduced from L. Svensson, *Identification Guide to European Passerines* 4th ed 1992, with kind permission of the author (wings).

Colour plates are the work of Martin Woodcock, and line illustrations are mainly by Ian Willis with some by Hilary Fry (*Amaurochichla bocagei, Scotocerca inquieta, Ficedula parva*), Lars Svensson (wings) and Martin Woodcock (warbler and flycatcher bills and heads). The artists, like the authors, have accommodated all referees' and editors' demands, and we are particularly grateful to them for their skill and ready cooperation. Emil Urban, as '*primus inter pares*' editor for Volume 5, wants to thank especially Jane Millward, Lois Urban and colleagues at Augusta State University for all their help in coordinating the entire work and in preparing much of the front and back matter. Stuart Keith is extremely grateful to M. G. Fugate for providing accommodation during his visits to New York, and Emil Urban to J. and M. L. Schmidt for doing likewise during his visits to Washington, D.C.

Again it is our pleasure to thank C. Chappuis for preparing the acoustic references, Lois Urban for preparing the bibliography and indexing the book, our wives Lois Urban, Kathie Fry and Sallyann Keith for their ever-present support, encouragement and understanding, and Andrew Richford of Academic Press (London) Ltd., who continues to be the driving force behind this project.

January 1997

Emil K. Urban
C. Hilary Fry
Stuart Keith

CONTENTS

ACKNOWLEDGEMENTS v

LIST OF PLATES xi

INTRODUCTION. xii

ORDER PASSERIFORMES

    Turdidae, thrushes (C. H. Fry, S. Keith, P. C. Lack, R. de Naurois, A. Prigogine and E. K. Urban) 1

    Sylviidae, Old World warblers (C. Erard, C. H. Fry, L. G. Grimes, M. P. S. Irwin, S. Keith, P. C. Lack, D. J. Pearson and A. Tye) 57

    Muscicapidae, flycatchers (C. Erard, C. H. Fry and D. J. Pearson) 432

    Monarchidae, paradise-flycatchers and monarchs (C. Erard, S. Keith and R. de Naurois) . . 508

    Platysteiridae, shrike-flycatchers, wattle-eyes and batises (C. Erard and C. H. Fry) . . 548

BIBLIOGRAPHY

    1. General and Regional References 607

    2. Family References 616

    3. Acoustic References 639

ERRATA TO PREVIOUS VOLUMES 643

INDEXES

    1. Scientific Names 645

    2. English Names 662

    3. French Names 667

Authorship by family

TURDIDAE

C. H. Fry: *Monticola.*
S. Keith and A. Prigogine: *Zoothera.*
P. C. Lack: *Turdus tephronotus.*
R. de Naurois and C. H. Fry: *Turdus olivaceofuscus.*
E. K. Urban: *Psophocichla, Turdus* (remaining 9 species).

SYLVIIDAE

C. Erard: *Hyliota violacea.*
C. H. Fry: *Cettia, Bathmocercus, Schoenicola, Cisticola juncidis, Regulus, Sylvia leucomelaena, S. nana, S. rueppelli, S. melanocephala, S. melanothorax, S. mystacea, S. conspicillata, S. deserticola, S. undata, S. sarda, Parisoma, Hyliota flavigaster, H. australis, H. usambarae.*
L. G. Grimes, C. H. Fry, and S. Keith: *Bradypterus, Melocichla, Sphenoeacus, Incana, Heliolais, Urolais, Spiloptila, Drymocichla, Phyllolais, Urorhipis, Poliolais, Camaroptera, Calamonastes, Eminia, Hypergerus.*
M. P. S. Irwin: *Cisticola rufilatus, C. subruficapillus, C. pipiens, C. fulvicapillus, C. angusticaudus, C. melanurus, C. textrix, C. dambo, Phragmacia, Prinia* (except *P. molleri* and *P. gracilis*), *Oreophilais, Schistolais, Apalis, Malcorus, Artisornis.*
M. P. S. Irwin and C. H. Fry: *Prinia gracilis.*
S. Keith: *Hemitesia, Chloropeta, Cisticola chubbi, C. hunteri, C. nigriloris* (and **Field Characters** and **Voice** for all *Cisticola* species), *Scotocerca, Euryptila, Graueria, Macrosphenus, Hylia.*
P. C. Lack: *Eremomela, Sylvietta.*
R. de Naurois and C. H. Fry: *Prinia molleri, Amaurocichla.*
D. Pearson: *Locustella, Acrocephalus, Hippolais, Phylloscopus, Sylvia* (remaining 7 species).
A. Tye: *Cisticola* (remaining 31 species).

MUSCICAPIDAE

C. Erard: *Fraseria, Melaenornis edolioides, M. pallidus, M. microrhynchus, Muscicapa striata, M. caerulescens, M. cassini, M. olivascens, M. epulata, M. sethsmithi, M. comitata, M. infuscata, Myioparus.*
C. H. Fry: *Melaenornis* (remaining 9 species), *Empidornis, Muscicapa* (remaining 7 species), *Stenostira.*
D. Pearson: *Ficedula.*

MONARCHIDAE

C. Erard: *Erythrocercus mccallii, Elminia, Trochocercus, Terpsiphone* (except *T. atrochalybeia*).
R. de Naurois and C. H. Fry: *Terpsiphone atrochalybeia.*
S. Keith: *Erythrocercus holochlorus, E. livingstonei.*

PLATYSTEIRIDAE

C. Erard: *Megabyas, Bias, Dyaphorophyia, Platysteira.*
C. Erard and C. H. Fry: *Batis senegalensis, B. poensis, B. minima, B. ituriensis.*
C. H. Fry: *Batis* (remaining 12 species), *Lanioturdus.*

Species authorship by contributor

C. Erard: *Hyliota violacea, Fraseria, Melaenornis edolioides, M. pallidus, M. microrhynchus, Muscicapa striata, M. caerulescens, M. cassini, M. olivascens, M. epulata, M. sethsmithi, M. comitata, M. infuscata, Myioparus, Eythrocercus mccallii, Elminia, Trochocercus, Terpsiphone* (except *T. atrochalybeia*), *Megabyas, Bias, Dyaphorophyia, Platysteira.*

C. Erard and C. H. Fry: *Batis senegalensis, B. poensis, B. minima, B. ituriensis.*

C. H. Fry: *Monticola, Cettia, Bathmocercus, Schoenicola, Cisticola juncidis, Regulus, Sylvia leucomelaena, S. nana, S. rueppelli, S. melanocephala, S. melanothorax, S. mystacea, S. conspicillata, S. deserticola, S. undata, S. sarda, Parisoma, Hyliota* (except *H. violacea*), *Melaenornis* (remaining 9 species), *Empidornis, Muscicapa* (remaining 7 species), *Stenostira, Batis* (remaining 12 species), *Lanioturdus.*

L. G. Grimes, C. H. Fry and S. Keith: *Bradypterus, Melocichla, Sphenoeacus, Incana, Heliolais, Urolais, Spiloptila, Drymocichla, Phyllolais, Urorhipis, Poliolais, Camaroptera, Calamonastes, Eminia, Hypergerus.*

M. P. S. Irwin: *Cisticola rufilatus, C. subruficapillus, C. pipiens, C. fulvicapillus, C. angusticaudus, C. melanurus, C. textrix, C. dambo, Phragmacia, Prinia* (except *P. molleri* and *P. gracilis*), *Oreophilais, Schistolais, Apalis, Malcorus, Artisornis.*

M. P. S. Irwin and C. H. Fry: *Prinia gracilis.*

S. Keith: *Hemitesia, Chloropeta, Cisticola chubbi, C. hunteri, C. nigriloris* (and **Field Characters** and **Voice** for all *Cisticola* species), *Scotocerca, Euryptila, Graueria, Macrosphenus, Hylia, Erythrocercus* (remaining 2 species).

S. Keith and A. Prigogine: *Zoothera.*

P. C. Lack: *Turdus tephronotus, Eremomela, Sylvietta.*

R. de Naurois and C. H. Fry: *Turdus olivaceofuscus, Prinia molleri, Amaurocichla, Terpsiphone atrochalybeia.*

D. Pearson: *Locustella, Acrocephalus, Hippolais, Phylloscopus, Sylvia* (remaining 7 species), *Ficedula.*

A. Tye: *Cisticola* (remaining 31 species).

E. K. Urban: *Psophocichla, Turdus* (remaining 9 species).

LIST OF PLATES

Plate	Facing Page	Plate	Facing Page
1 Rock-Thrushes	32	17 Warblers (*Camaroptera* and others)	320
2 Ground-Thrushes	33	18 Warblers (*Eremomela*)	321
3 Thrushes (*Turdus*)	48	19 Warblers (*Sylvietta*)	336
4 Thrushes (*Turdus*)	49	20 Warblers (*Macrosphenus* and others)	337
5 Warblers (*Bradypterus, Cettia*)	96	21 Warblers (*Phylloscopus, Regulus*)	400
6 Warblers (*Bathmocercus* and others)	97	22 Warblers (*Parisoma, Sylvia*)	401
7 Warblers (*Acrocephalus*)	112	23 Warblers (*Sylvia*)	416
8 Warblers (*Chloropeta, Hippolais*)	113	24 Warblers (*Sylvia*)	417
9 Warblers (*Cisticola*)	160	25 Flycatchers (*Melaenornis, Empidornis*)	480
10 Warblers (*Cisticola*)	161	26 Flycatchers (*Muscicapa, Fraseria*)	481
11 Warblers (*Cisticola*)	176	27 Flycatchers (*Muscicapa* and others)	496
12 Warblers (*Cisticola*)	177	28 Flycatchers (*Elminia* and others)	497
13 Warblers (*Cisticola, Incana*)	240	29 Flycatchers (*Trochocercus, Terpsiphone*)	544
14 Warblers (*Prinia* and others)	241	30 Wattle-eyes (*Platysteira, Dyaphorophyia*)	545
15 Warblers (*Apalis*)	256	31 Batises	560
16 Warblers (*Apalis*)	257	32 Batises	561

INTRODUCTION

This volume includes the four genera of thrushes (Turdidae) not dealt with in Volume IV (*Monticola, Zoothera, Psophocichla* and *Turdus*) and the families Sylviidae, Muscicapidae, Monarchidae and Platysteiridae.

The periodical literature has been reviewed comprehensively up to late 1995, and reference has also been made to some 1996 publications. Noteworthy contributions to African ornithology since Volume IV appeared, in 1992, are the landmark works 'Checklist of Birds of the Afrotropical and Malagasy Regions', by R.J. Dowsett and A.D. Forbes-Watson (1993), and its companion volume 'A Contribution to the Distribution and Taxonomy of Afrotropical and Malagasy Birds' by R.J. Dowsett and F. Dowsett-Lemaire (1993a). Other major contributions include the sixth edition of 'Roberts' Birds of Southern Africa' (Maclean 1993) and the second of 'The Birds of Nigeria' (Elgood *et al.* 1994), the 'Bird Atlas of Botswana' (Penry 1994), and checklists for Lesotho (Bonde 1993) and Bioko (Perez de Val *et al.* 1994). The final four volumes of 'The Birds of the Western Palearctic' (Cramp 1992 and Cramp and Perrins 1993, 1994, 1994) have also appeared. Welcome events have been the formation of the Avian Demography Unit in the University of Cape Town and of the African Bird Club, and the publication of their respective journals *Bird Numbers* and the *ABC Bulletin*.

The only stylistic novelty in this Volume is that in the maps breeding ranges are shown in red. Figure 1 shows the shading conventions.

Thanks to the suggestion of M.P.S. Irwin, we present detailed maps below of Northwest and West Africa (Fig. 3), Northeast and East Africa (Fig. 4), and Central and Southern Africa (Fig. 5). These show hard-to-find localities, including recently-gazetted national parks and forest and game reserves, and forest islands, wetlands and other threatened habitats which have been featured in recent ICBP and IUCN publications (Collar and Stuart 1988, Stuart and Adams 1990, Anon. 1991, Bibby *et al.* 1992, Sayer *et al.* 1992). For other localities, the reader should refer to gazetteers and maps in references given for many of the nations in Figures 3–5; complete details of these references are in the General/Regional References in the Bibliography of this volume.

Political change in Africa has resulted in some new regional names. We usually use the new names, but this may not be feasible when the sources cited refer to former political entities whose boundaries are not congruous with the new ones. Data may be given for 'Ethiopia' without distinguishing between Ethiopia and Eritrea, and data for Senegal and Gambia may be lumped under 'Senegambia'. We continue to use the old names for South African provinces because these are the areas referred to in standard works such as Maclean (1993). Figure 2 updates the political map of Africa.

For French names we follow Devillers and Ouellet (1993). Our use of English names is discussed in the introduction to Volume II; for this volume, Sibley and Monroe (1990), Short *et al.* (1990) and Dowsett and Forbes-Watson (1993) have been fruitful sources.

Systematic Treatment

For higher taxa we generally follow the traditional classification of Campbell and Lack (1985), although we raise wattle-eyes and relatives to family level, the Platysteiridae, and place it after the Monarchidae. We consider *Lanioturdus* to be a platysteirid close to *Batis*. *Achaetops pycnopygius* appears on Plate 6 but we are transposing its text to Volume VI, where it will appear *incertae sedis* in Timaliidae (babblers). At the genus, species and subspecies levels authors have made their own decisions after discussions, often lengthy, with the Editors. No single classification is followed, but the new Afrotropical list of Dowsett and Forbes-Watson (1993) is proving to be a very valuable guideline.

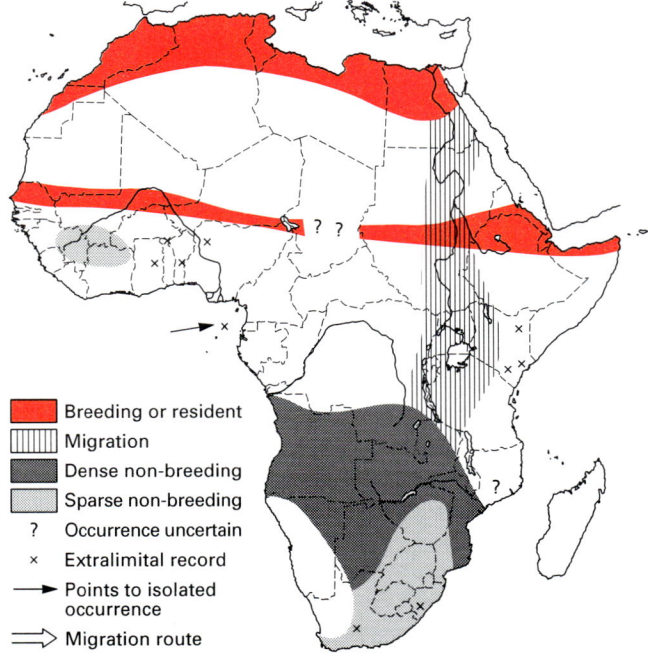

Fig. 1. Shading, symbols and arrows used on maps.

Fig. 2. Political map.

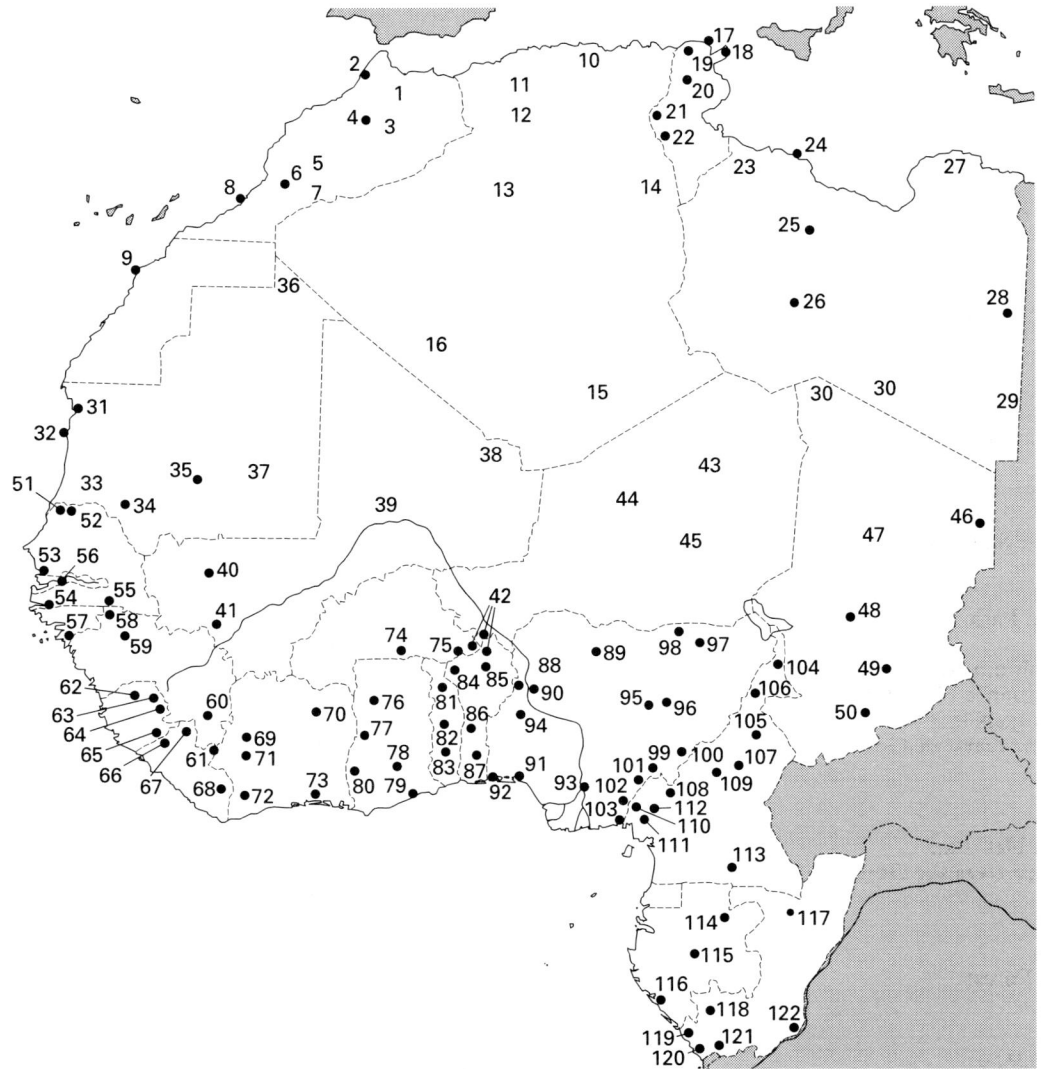

Fig. 3. Some localities, many mentioned in text, in Northwest and West Africa.

Morocco
1. Rif Mts
2. Merja Zerga
3. Middle Atlas
4. Central Plateau
5. High Atlas
6. Souss
7. Anti-Atlas
8. Cap Draa

For a map see Thévenot et al. (1982).

Western Sahara
9. Cap Boüjdour

Algeria
10. Djurdjura Mts
11. Tell-Atlas
12. Sahara-Atlas
13. Grand Erg Occidental
14. Grand Erg Oriental
15. Hoggar Mts; Atakor Nat. Park
16. Tanezrouft

For a gazetteer see Ledant et al. (1981).

Tunisia
17. L. Ichkeul
18. Cap Bon Peninsula
19. Kroumirie forests
20. Djebel Jugurtha
21. Metlaoui Mts
22. Chott el Djerid

For a map see Thomsen and Jacobsen (1979).

Libya
23. Jebel Nafusa
24. Wadi Turghat
25. Jafra Oasis
26. Sebha
27. Jebel Akhdar
28. Kufra Oasis
29. Jebel Awainat
30. Tibesti

For maps see Bundy (1976).

Mauritania
31. Banc d'Arguin Nat. Park
32. Cap Timirist
33. Aftout-es-Sahel
34. Lac d'Aleg
35. Rkiz Massif
36. Erg d'Iguidi
37. Aouker Depression

For maps and a gazetteer see Gee (1984), Lamarche (1988).

Mali
38. Adrar des Iforhas
39. Niger Delta inundation zone
40. Boucle du Baoulé Nat. Park
41. Mandingues Mts

For map and a gazetteer see Lamarche (1980).

Niger
42. W Nat. Park
43. Ténéré Desert
44. Aïr Mts
45. Termit Massif

For a gazetteer see Giraudoux et al. (1988).

Chad
30. Tibesti
46. Ennedi Massif
47. Ouadi Rimé-Ouadi Achim Faunal Reserve
48. L. Fitri
49. Zakouma Nat. Park
50. Manda Nat. Park

For a gazetteer see Salvan (1969).

Senegal
51. Djoudj Nat. Park
52. L. Guier
53. Delta du Saloum Nat. Park
54. Basse-Casamance Nat. Park
55. Niokolo-Koba Nat. Park

For maps and gazetteers see Gore (1990), Morel and Morel (1990).

Gambia
56. Kiang West Nat. Park

For maps and gazetteers see Gore (1990), Morel and Morel (1990).

Guinea-Bissau
57. Cantanhez Forest

Guinea
58. Badiar Nat. Park
59. Fouta Djallon Plateau; Mt Loura
60. Ziama Massif
61. Mt Nimba

For maps and a gazetteer see Demey (1995) and Hayman et al. (1995).

Sierra Leone
62. Sula Hills
63. Loma Mts
64. Tingi Mts
65. Gola Hills

Liberia
66. Lofa Mano Nat. Park
67. Wologizi Mts
61. Mt Nimba
68. Sapo Nat. Park

For a map see Gatter (1988).

Ivory Coast (Cote d'Ivoire)
61. Mt Nimba
69. Mt Sangbe
70. Comoé Nat. Park
71. Mt Pekoé
72. Taï Nat. Park; N'Zo Forest Reserve
73. Yapo Forest

For a gazetteer see Thiollay (1985).

Burkina Faso
74. Po Nat. Park
75. Aril Nat. Park
42. W Nat. Park

For maps and a gazetteer see Thonnerieux (1998b), Holyoak and Seddon (1989).

Ghana
76. Mole Nat. Park
77. Bui Nat. Park
78. Digya Nat. Park
79. Shai Hills
80. Bia Nat. Park

For maps and a gazetteer see Grimes (1987).

Togo
81. Keran Nat. Park
82. Fazao-Malfakassa Nat. Park; Fazao Mts
83. Togo Mts

For maps and a gazetteer see Cheke and Walsh (1996).

Benin
42. W Nat. Park
84. Pendjari Nat. Park
85. Atakora Mts
86. Kouffe Mts
87. Lama Forest

For maps and a gazetteer see Brunel (1958) and Claffey (1995).

Nigeria
88. Borgu Game Reserve
89. Funtua
90. L. Kainji Nat. Park
91. Omo Forest Reserve
92. Badagri
93. Anambara Creek
94. Old Oyo Nat. Park
95. Nindam Forest Reserve
96. Yankari Nat. Park
97. Hadejia-Nguru Wetlands
98. Bulatura Oases
99. Mambilla Plateau; Gotel Mts
100. Gashaka-Gumti Nat. Park
101. Obudu Plateau
102. Stubbs Creek Game Reserve
103. Oban Hills; Cross River Nat. Park

For maps and a gazetteer see Elgood et al. (1994).

Cameroon
104. Waza Nat. Park
105. Benoué Nat. Park
106. Mandara Highlands
107. Adamawa Plateau
108. Mt Oku
109. Tchabel Mbabo
110. Korup Nat. Park
111. Rumpi Hills
112. Mt Manenguba; Mt Nlonako; Mt Kupé
113. Dja Game Reserve

For a gazetteer see Louette (1981).

Gabon
114. Makokou; M'Passa; Ivindo Basin; Mt Sassamongo
115. Lopé-Okanda Nat. Park; Mt Iboundji
116. Sétté-Cama Reserve

For a map see Rand et al. (1959).

Congo
117. Odzala Nat. Park
118. Mt Fouari
119. Conkouati Faunal Reserve
120. Kouilou Basin
121. Mayombe
122. Léfini Faunal Reserve

For a gazetteer see Dowsett (1991).

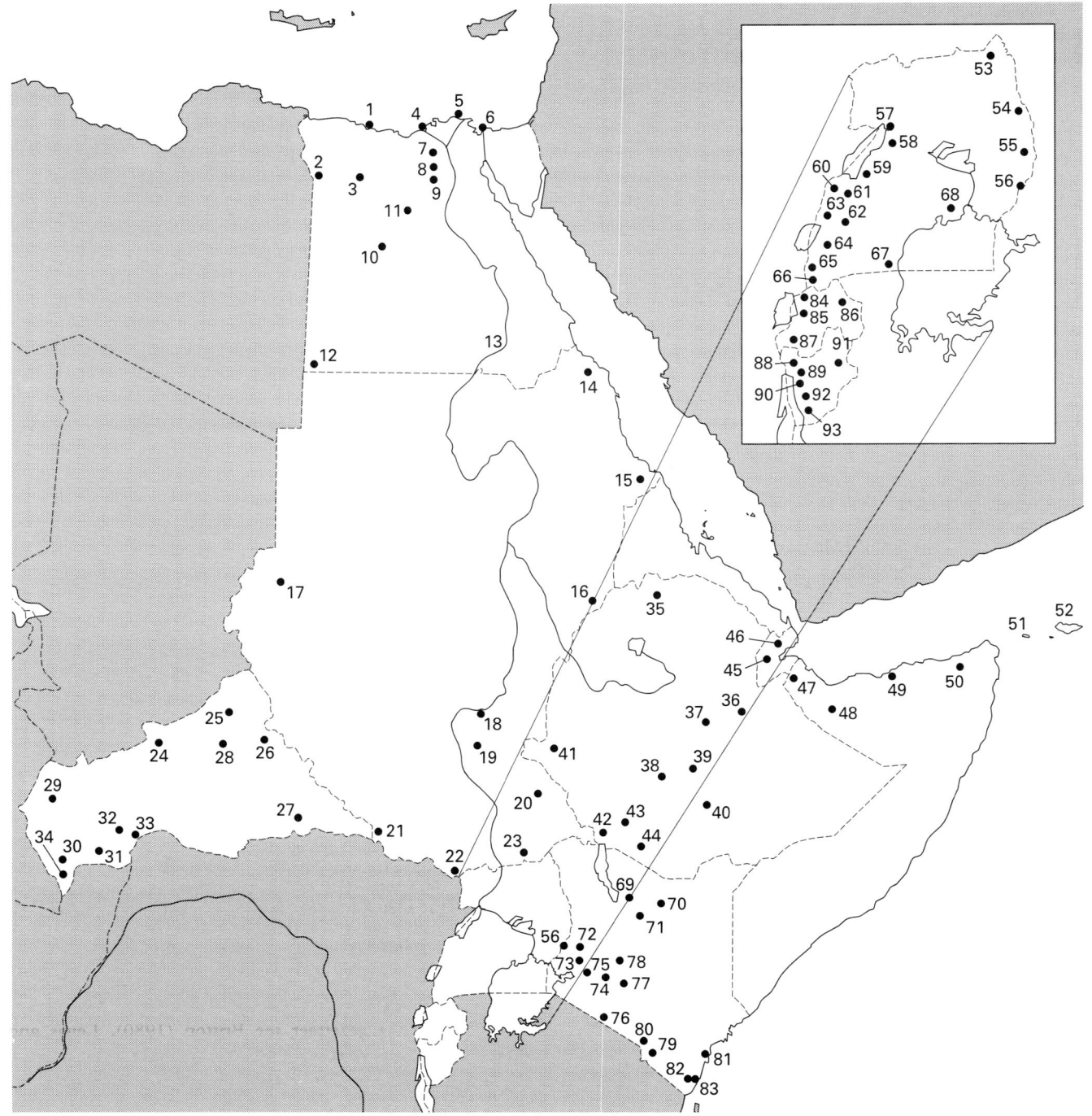

Fig. 4. Some localities, many mentioned in text, in Northeast and East Africa.

Egypt
1. Matruh
2. Siwa Oasis
3. Qattara Depression
4. L. Maryut
5. L. Burullus
6. L. Manzala
7. Wadi el Natrun
8. L. Qarun
9. Wadi el Rayan
10. Farafra Oasis
11. Bahariya Oasis
12. Gebel Uweinat
13. L. Nasser

For a gazetteer see Goodman and Meininger (1989).

Sudan
14. Gebel Elba
15. Red Sea Hills
16. Dinder Nat. Park
17. Jebel Marra Massif
18. Malakal
19. Jonglei Canal
20. Boma Nat. Park
21. Bengengai Forest
22. Aloma Plateau
23. Imatong Mts; Didinga Mts; Dongotona Mts; Lotti and Talanga forests; Mt Kinyeti

For a gazetteer see Nikolaus (1987).

Central African Republic
24. Bamingui-Bangoran Nat. Park
25. Manovo-Gounda-St. Floris Nat. Park
26. Dar Challa Massif and André Félix Nat. Park
27. Ouossi R. (Baroua)
28. Bongo Massif
29. Adamoua Massif
30. Dzanga-Sangha Forest
31. Mbaéré-Bodingué-Ngota Forest
32. Lobaye Prefecture
33. Bangui
34. Haute Sangha Prefecture

For a map see Carroll (1988).

Eritrea
For a gazetteer see Smith (1957).

Ethiopia
35. Simien Mts and Mt Ras Dejen
36. Chercher Mts and Harar
37. Awash Nat. Park
38. Abijata-Shala Lakes Nat. Park
39. Arsi Mts and Mt Chilalo
40. Bale Mts Nat. Park; Mt Batu; Harenna Forest
41. Gambela
42. Omo and Mago Nat. Parks
43. Nechisar Nat. Park
44. Yabelo

For a gazetteer see Urban and Brown (1971).

Djibouti
45. Forêt du Day
46. Mabla Mts

For maps see Welch and Welch (1984a) and Laurent (1990).

Somalia
47. Zeila Wildlife Reserve
48. Wagger Mts and Gaan Libaah Forest
49. Daalo Forest
50. Ahl Mescat Mts and Warsangeli Escarpment

For gazetteers see Ash and Miskell (1983), Clarke (1985) and Douthwaite and Miskell (1991).

Socotra and Abd-el-Kuri
51. Abd-el-Kuri
52. Socotra

For a map see Ripley and Bond (1966).

Uganda
53. Kidepo Nat. Park
54. Mt Moroto
55. Mt Kadam
56. Mt Elgon
57. Murchison Falls Nat. Park
58. Budongo Forest
59. Bugoma Forest
60. Bwamba (Semliki) Forest
61. Itwara Forest
62. Kibale Forest
63. Rwenzori Mts and Forest
64. Kasyoha-Kitomi Forest
65. Maramagambo and Kalinzu forests
66. Bwindi–Impenetrable Forest and Nat. Park
67. Malabigambo Forest
68. Mabira Forest

For maps and a gazetteer see Britton (1980), Short et al. (1990).

Kenya
56. Mt Elgon
69. Mt Kulal
70. Mt Marsabit
71. Ndoto Mts
72. Cherangani Hills
73. Kakamega and Nandi forests
74. L. Nakuru
75. Mau Forest
76. Nguruman Hills
77. Aberdare Mts
78. Laikipia Plateau
79. Taita Hills
80. Chyulu Hills
81. Arabuko-Sokoke Forest
82. Shimba Hills
83. Diani Forest

For maps and a gazetteer see Britton (1980), Lewis and Pomeroy (1989), Short et al. (1990).

Rwanda
84. Volcanoes Nat. Park
85. Gishwati and Mukura forests
86. Akagera Nat. Park
87. Nyungwe and Cyamudongo forests

Burundi
88. Ruzizi Nat. Park
89. Kibira Nat. Park; Mt Teza; Ijenda Forest
90. Mt Hema
91. Ruvubu Nat. Park
92. Mt Bururi and Bururi Forest
93. Rumonga-Vyanda and Karehe forests

For a map see Gaugris et al. (1981).

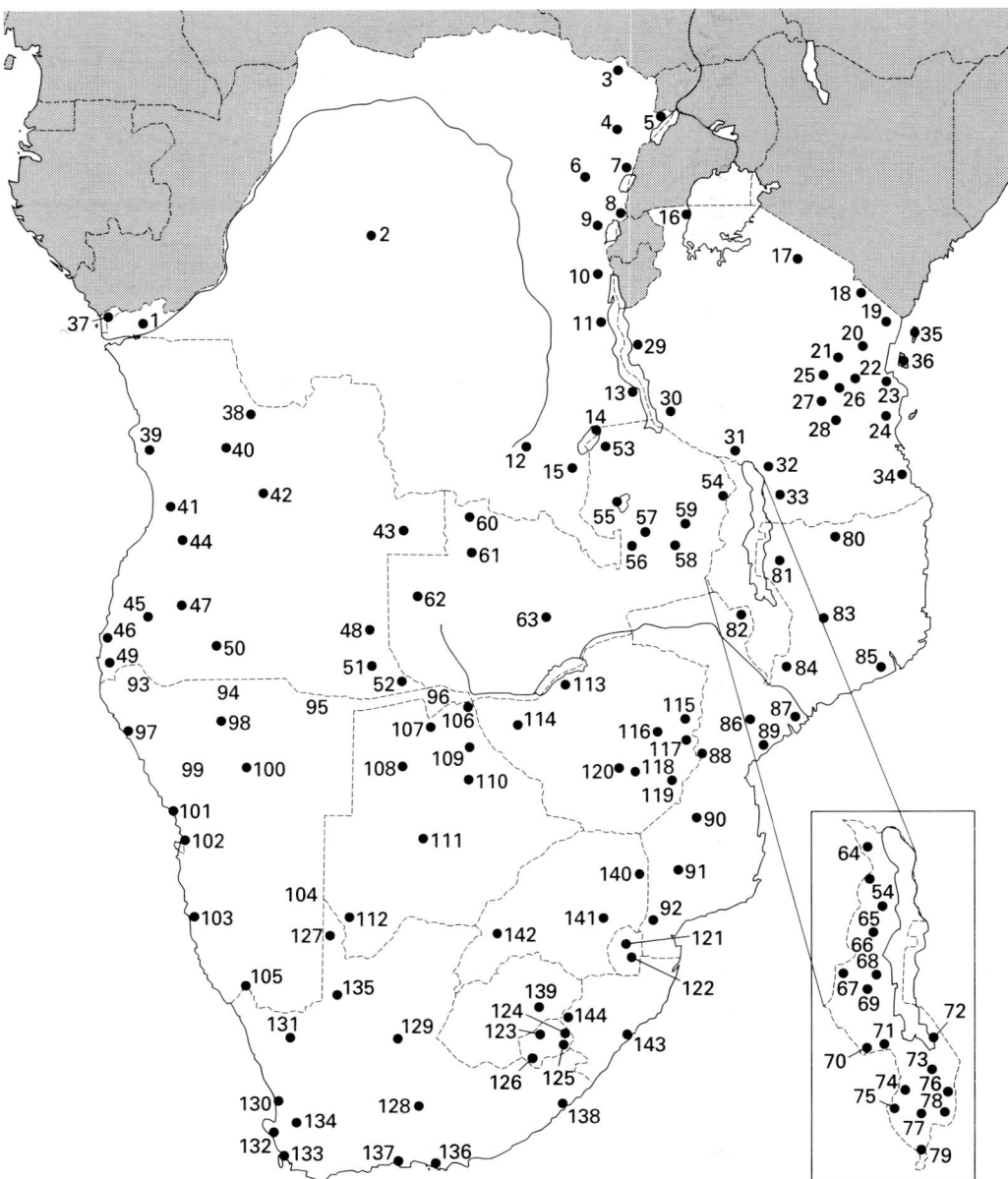

Fig. 5. Some localities, many mentioned in text, in Central and Southern Africa.

Zaïre
1. Luki Forest
2. Salonga Nat. Park
3. Garamba Nat. Park
4. Ituri Forest
5. Lendu Plateau
6. Maïko Nat. Park
7. Virunga Nat. Park; Rwenzori and Lubero forests
8. Virunga Volcanoes
9. Kahuzi-Biega Forest and Nat. Park
10. Itombwe Forest and Mts; Mt Mohi
11. Mt Kabobo
12. Upemba Nat. Park; Mulumbe Mts; Mt Kibara
13. Marungu Mts; Mitumba Mts; Mt Lusale
14. L. Mweru
15. Kundelungu Nat. Park

For a map and a gazetteer see Chapin (1954), Lippens and Wille (1976).

Tanzania
16. Minziro Forest
17. Crater Highlands
18. N and S Pare Mts
19. W and E Usambara Mts
20. Nguru Mts.
21. Kiboriani and Ukaguru Mts
22. Uluguru Mts; Lukwangule Plateau; Kinole, Bagilo and Kimboza forests
23. Pugu Hills
24. Kirengoma Forest
25. Rubeho Mts
26. Uvidunda Mts
27. Udzungwa Mts; Mwanihana and Magombera forests
28. Mahenge Forest
29. Kungwe-Mahari Mts
30. Ufipa Plateau
31. Poroto Mts and Mt Rungwe
32. Mdando Forest
33. Matengo Highlands; Livingstone Mts
34. Rondo and Nyangamara forests
35. Ngezi, Mwilu Mkuu and Ras Kiuyu forests
36. Jozani and Muyuni forests

For a gazetteer see Britton (1980), Short et al. (1990).

Angola
37. Cabinda
38. Milando Special Reserve
39. Kisama Nat. Park
40. Kangandala Nat. Park
41. Gabela and Amboim forests and Angolan Escarpment
42. Luando Integral Nature Reserve
43. Kameia Nat. Park
44. Bailundo Highlands; Mt Soque, Mt Moco, Mombolo Plateau
45. Serra de Chela
46. Moçâmedes Partial Reserve
47. Bikuar Nat. Park
48. Mavinga Partial Reserve
49. Iona Nat. Park
50. Mupa Nat. Park
51. Cuando-Cabango Coutadas Controlled Hunting Area
52. Luiana Partial Reserve

For a map and gazetteers see Traylor (1963), da Rosa Pinto (1983), Dean et al. (1988).

Zambia
53. Lusenga Plain Nat. Park
54. Nyika Plateau; Mt Makutu; Mafinga Mts
55. L. Bangweulu
56. Kasanka Nat. Park
57. Lavushi Mande Nat. Park
58. Luangwa Nat. Park
59. Muchinga Mts and Escarpment
60. Mwinilunga
61. West Lunga Nat. Park
62. Liuwa Plain Nat. Park
63. Kafue Flats; Blue Lagoon and Lochinvar Nat. Parks

For a gazetteer see Benson et al. (1971).

Malaŵi
64. Misuku Hills; Igembe Hills; Musisi Hill
54. Nyika Plateau; Mafinga Mts; Jembya Plateau
65. North Viphya Plateau; Kaningina Hills
66. South Viphya Plateau; Kuwilwe Hill
67. Kasungu Nat. Park
68. Nkhota-kota Game Reserve
69. Chipata Mts; Ntchisi Mts
70. Dzalanyama Range
71. Chongoni Mts; Dedza Mts
72. Namizimu Hills; Mt Mangochi
73. Chikala Hill; Malosa and Zomba Mts
74. Kirk Range
75. Thambani and Zobue Hills
76. L. Chilwa
77. Chiradzulu Mts; Hills around Blantyre; Mt Thyolo
78. Mt Mulanje
79. Malaŵi Hills

For a gazetteer see Benson and Benson (1977).

Mozambique
80. Niassa Game Reserve
81. Njesi Plateau; Mt Unango
82. Mt Domue
83. Mt Namuli
84. Mt Chiperone
85. Gile Game Reserve
86. Gorongoza Nat. Park; Mt Gorongoza
87. Marromeu Game Reserve
88. Chimanimani Mts; Mt Binga
89. Sofala coastal forests; Inhamitanga Forest
90. Zinave Nat. Park
91. Banhine Nat. Park
92. Lebombo Mts

Namibia
93. Kaokoland
94. Owambo (Ovamboland) and Cuvelai Drainage
95. Kavango
96. Caprivi East
97. Skeleton Coast Park
98. Etosha Nat. Park
99. Damaraland
100. Waterberg Plateau Park
101. Swakopmund
102. Walvis Bay
103. Namib-Naukluft Park
104. Namaland
105. Fish River Canyon

For a map see Winterbottom (1971a).

Botswana
106. Chobe Nat. Park
107. Moremi Wildlife Reserve
108. L. Ngami
109. Nxai Pan Nat. Park
110. Makgadikgadi Pans Nat. Park
111. Central Kalahari and Kutse Game Reserves
112. Gemsbok Nat. Park and Mabuasehube Game Reserve

For a gazetteer see Penry (1994).

Zimbabwe
88. Chimanimani Mts.; Melsetter Highlands
113. L. Kariba
114. Hwange (Wankie) Nat. Park
115. Inyanga (Nyanga) Mts
116. Wedza Mts
117. Vumba Mts; Banti Forest Reserve; Stapleford Forest Reserve
118. Great Zimbabwe Nat. Monument
119. Chirinda Forest
120. Mt Buhwa

For a gazetteer see Irwin (1981).

Swaziland
121. Malolotja Nature Reserve
122. Mbabane

Lesotho
123. Maseru
124. Mokhotlong
125. Sehlabathebe Nat. Park
126. Senqu Valley

For a gazetteer see Bonde (1993).

South Africa
127. Kalahari Gemsbok Nat. Park
128. Karoo Nat. Park
129. 'Die Bos' Nat. Reserve, Prieska
130. Lambert's Bay
131. Goegap Provincial Nat. Reserve
132. Langebaan; West Coast Nat. Park
133. Helderberg and Rondevlei Nat. Reserves
134. Cedarberg Wilderness Area
135. Augrabies Nat. Park
136. Tsitsikamma Coastal Nat. Park
137. Knysna
138. Dwesa Nature Reserve (Transkei)
139. Golden Gate Highlands Nat. Park
140. Kruger Nat. Park
141. Wakkerstroom Reserve
142. Barberspan Nat. Reserve
143. Ngoye Forest
144. Giant's Castle Game Reserve

For maps see Cyrus and Robson (1980), Earlé and Grobler (1987), Tarboton et al. (1987), Hockey et al. (1989), Harrison (1995).

Genus *Monticola* Boie

Thrushes of mountains, rocky terrain and woodland in Afrotropical, Palearctic and Oriental regions: ranging in size from that of chats such as redstarts (*Phoenicurus*) to typical thrushes (*Turdus*). Bill slender, straight, hook-tipped; nares basal, round, not concealed; wings rather long, P10 variable in size but always less than half length of P9; tail short and square-ended. Sexes differ markedly, except in *M. rufocinereus*. ♂♂ rather alike, blue-grey with orange bellies and tails, *M. saxatilis* with white back, *M. gularis* and *M. cinclorhynchus* (Asia) with blackish back and white in wings; western races of *M. solitarius* all blue. ♀♀ are variously mottled brown, and incubating birds can be highly cryptic. Bulky cup nest, in rock crevice, tree cavity, or beneath rock overhang.

13 species: 4 endemic to Africa, 1 to Africa and SW Arabia, 2 in NW Africa, S Europe and Asia (both wintering in Africa), 3 in Madagascar and 3 in Himalayas and E Asia. Ranges of 10 are parapatric or allopatric (*M. explorator*, *M. brevipes*, *M. angolensis* and *M. rufocinereus* in subsaharan Africa, *M. sharpei*, *M. bensoni* and *M. imerinus* in Madagascar, *M. saxatilis* from W Mediterranean to E Asia, and *M. rufiventris* and *M. gularis* in E Asia). All African populations except *M. solitarius* were treated by Meinertzhagen (1951) as a single species; and most of the 10 are generally regarded as composing a superspecies (Hall and Moreau 1970, Harrison 1982, Cramp 1988). ♂ plumages are quite alike but ♀ plumages are dissimilar from each other, subtly or strikingly so; moreover, habitats and foraging behaviours of African rock-thrushes differ importantly, and so we consider that the 7 species dealt with are independent, not comprising any superspecies.

Monticola may be distantly related with redstarts *Phoenicurus* (like them, *M. saxatilis* and *M. rufocinereus* quiver the tails). Placed by Sibley and Monroe (1990) in Muscicapidae: Turdinae, in sequence *Neocossyphus*, *Monticola*, *Myiophonus*, *Zoothera*, *Sialia*, *Myadestes*, *Catharus*, *Turdus* and *Alethe*.

Monticola rupestris (Vieillot). Cape Rock-Thrush. Monticole rocar.

Plate 1 (Opp. p. 32)

Turdus rupestris Vieillot, 1818. Nouv. Dict. d'Hist. Nat. 20, p. 281; Table Mountain near Cape Town.

Monticola rupestris

Range and Status. Endemic resident, S and E South Africa, Lesotho, Swaziland, and just into Botswana and Mozambique. Locally common, from Cape Town to Swaziland and N Transvaal. Cape Province: in SW Cape mainly in mountains but occurs down to sea level; occurs north to De Aar district where widespread in mountains and occasional in grassland and gardens; widespread in Cape Province east of De Aar. Natal: widespread on rocky hill sides; near coast confined to gorges; uncommon east of 31°E, absent north of St Lucia. Orange Free State: locally common in mountains in east, in Warden–Harrismith–Clarens areas; isolated populations between Bloemfontein, Ladybrand and Wepener, near Zastron, around Bethulie, and between Bloemhof and Hoopstad. Transvaal: widespread in Magaliesberg, Waterberg, Soutpansberg, Magato Mts, Sekukuneland, on Escarpment, and in other mountains; absent west of 27°E; in E and NE, absent from areas below 1000 m. Lesotho, uncommon lowlands, foothills and upland river valleys. Botswana, only between Kanye and Mafeking. Mozambique, a few in S Lebombo Mts on Swaziland border.

Description. ADULT ♂: forehead, chin and throat blue-grey; lores blackish; crown, nape, hindneck, sides of neck, cheeks, ear-coverts and throat dark blue-grey; blue-grey head sharply demarcated all round. Mantle, scapulars and upper back dark russet; feathers dark brown, broadly fringed chestnut; lower back, rump and uppertail-coverts bright chestnut. Tail: T1 dark brown, T2–T5 bright chestnut with some blackish along edge of outer vane near tip, T6 chestnut with outer vane mainly blackish; underside of tail dull orange. Underparts below throat bright rufous, richest on breast, slightly paler towards undertail-coverts. Wings blackish brown, primaries narrowly edged buff proximal to emargination point, secondaries edged and tipped buff, tertials more broadly edged and tipped rufous, all coverts fringed bright rufous-cinnamon. Underwing-coverts and axillaries orange; underside of flight feathers shiny dark grey. Bill black; eyes very dark brown; legs and feet black. ADULT ♀: nape, hindneck and sides of neck greyish brown; forehead, crown, mantle and scapulars greyish brown, each feather blackish near shaft (forehead and forecrown look more blackish than grey-brown); rump, uppertail-coverts and tail like ♂. Lores, malar region, cheeks and ear-coverts freckled pale buff and dark brown. Chin pale buff in centre, freckled grey-brown, black and pale rufous at sides; throat buff-rufous in centre, mainly grey-brown at sides, freckled; remaining underparts like ♂, but feathers pale-tipped and with narrow, wavy, blackish subterminal bar. Wing like ♂ but feathers edged buff rather than rufous. SIZE: wing, ♂ (n = 21) 110–119 (113), ♀ (n = 12) 106–115 (109); tail, ♂♀ (n = 33) 72–88, ♂ (n = 6) 78·5–86 (80·8), ♀ (n = 5) 76–78·5 (77·3); bill, ♂♀ (n = 33) 21–25; tarsus, ♂♀ (n = 33) 28–31. WEIGHT: 1 ♂ 60·5, 2 ♀♀ 60, 63·7 (Maclean 1993).

IMMATURE: above rufescent dark brown, head and neck heavily spotted with pale buff; mantle, back and scapulars barred pale rufous and black, rump rufous, wing feathers fringed and tipped rufous; breast mottled rufous-buff and blackish, belly pale rufous, lightly barred blackish.

NESTLING: not described.

Field Characters. Length 21–22 cm; largest rock-thrush in southern Africa. Differs from sympatric Sentinel Rock-Thrush *M. explorator* and Short-toed Rock-Thrush *M. brevipes* in more *Turdus*-like stance; ♂ has blue-grey head sharply cut off from mottled brown back and rufous breast; back blue-grey in the other 2, and in Sentinel grey extends onto breast. ♀ has rich orange-rufous underparts, like ♀ Short-toed but richer, and further differs in streaked back and freckled throat (back plain and throat white in ♀ Short-toed); ♀ Sentinel has mottled brown and white underparts.

Voice. Tape-recorded (20, 75, 91, F, WALK, WAT). Song, given by ♂ and ♀, a phrase of varied, clear mellow whistles, repeated several times after short pauses. One song had 5 phrases in 28 s. Phrase is 2–5 s long, with bold, sweeping whistles interspersed with 'ch' and high-pitched 'see' sounds, and sometimes ending with short trill. Pauses are 1–2 s long. All phrases in one song bout differ in duration and number of included notes: 'weet leeo-o pee'p see peeu chewee trrr' or 'tsee-tsee-tseet-chee-chweeeoo' or 'chireewoo chirri wee roo'. Alarm a harsh, sharp, grating 'charr'; also (♂ alerting incubating ♀) a wispy, mournful whistle. See also under *Breeding Habits*.

General Habits. Inhabits steep, rocky, boulder-strewn mountain sides, cliffs in mountains and by lowland rivers, sea-cliffs, gorges, quarries, scree slopes with scattered bushes and small trees, isolated old buildings, and human habitation; occasionally in open grassland and in lightly wooded areas, particularly if burnt; fynbos, at flowering aloes; gardens in winter (De Aar). Usually in pairs, but often seen solitarily; family parties in summer. Wary, but can become quite tame around habitations; very wary when breeding. Pairs keep year-round to territory; sit motionless for long periods, hence easily overlooked. Perches on cliff-face, tree or boulder, with bill somewhat inclined and body either attenuated or a bit pot-bellied. ♂ has favourite song perches. Often flicks wings after alighting. Forages by hopping over burnt ground, around rocks or on tussocky grass; sometimes feeds in trees and at aloe flowers. Sedentary.

Food. Insects, spiders, millipedes, centipedes, molluscs, sometimes frogs; also fruits, seeds and nectar (Maclean 1993).

Breeding Habits. Monogamous, solitary breeder. Aggressively territorial; adjacent pairs of birds chase each other in fast, swooping flight, calling excitedly with garbled whistles and mimicry of other birds. At perch calling bird lifts head, inclines bill, and sings with body erect and tail fanned (C. J. Vernon *in* Ginn *et al.* 1989).

NEST: a bulky, untidy structure made of coarse dry grasses on a platform of rootlets, twigs and soil, and sometimes pieces of dead aloe leaves; with a shallow cup-shaped depression in centre, lined with fine rootlets and a little hair. Int. diam. 100, depth 50. Sited about 4 m above ground, on ledge beneath rock overhang, wedged into rock crevice or nook in wall of building, on beams in old building, or built into leaf rosette of aloe.

EGGS: 2–4; av. (23 clutches) 3·1. Pale blue or cream, immaculate, or sparingly spotted with rust, or indistinctly and streakily freckled all over with pale yellowish brown and grey. SIZE: (n = 41) 25·5–29·8 × 19·1–20·7 (27·3 × 19·7).

LAYING DATES: South Africa, Sept–Feb.

INCUBATION: when person enters territory, ♂ alerts incubating ♀ with special alarm call; ♀ slips off nest undetected and will not return until ♂ stops calling.

DEVELOPMENT AND CARE OF YOUNG: young fed by ♂ and ♀.

BREEDING SUCCESS/SURVIVAL: sometimes parasitized by Red-chested Cuckoo *Cuculus solitarius*.

References
Farkas, T. (1962a).
Ginn, P. J. *et al.* (1989).

Monticola explorator (Vieillot). Sentinel Rock-Thrush. Monticole espion.

Turdus explorator Vieillot, 1818. Nouv. Dict. d'Hist. Nat. 20, p. 260; Cape of Good Hope.

Plate 1
(Opp. p. 32)

Monticola explorator

Range and Status. Endemic resident, temperate-zone and montane grassland in South Africa, Lesotho, Swaziland and extreme S Mozambique. In South Africa a common breeding resident above 1200 m with some birds wintering down to about 600 m and a few to near sea level in S Cape. Ranges from SW Cape (Cape Peninsula) to E Orange Free State (in winter birds occurring west of 27°E and south of 28°S), discontinuously in E highveld of Transvaal above 2000 m, and over 3000 m on Maluti Plateau, from Wakkerstroom to north of Dullstroom, where a locally common resident (in winter birds occurring west to Potchefstroom and north to Mooketsi), and W Natal and Lebombo Mts in NE Zululand (a few birds wintering down to Natal coast). Lesotho, common higher elevations above 2500 m, rare lowlands, absent below 2000 m. Mozambique, non-breeding visitor from Lesotho in June–Aug on Mt Meponduine in the Lebombo Mts. In optimum habitat reaches density of a pair every 500 m (W. R. Tarboton *in* Ginn *et al.* 1989).

Description. *M. e. explorator*: range of species except Lebombo Mts, Swaziland. ADULT ♂: forehead, crown, nape, hindneck, sides of neck, mantle, scapulars, back, lores, ear-coverts, chin, throat and upper breast all rather uniform grey-blue, darkest on lores (which are blackish in some birds) and back, palest on chin. Rump and uppertail-coverts bright orange. Tail: T1 blackish brown, T2 bright rufous with blackish corners to both webs, T3–T5 bright rufous with blackish corner to outer web, T6 rufous with black streak 18 mm long at end of outer web. Underside of tail orange. Lower breast bright orange-rufous, sharply demarcated from blue-grey upper breast. Remaining underparts, including underwing-coverts and axillaries, orange, paling backwards. Wings blackish, primaries narrowly edged pale grey proximal to emargination point, secondaries narrowly edged and tipped pale grey, tertials more broadly edged and tipped pale grey or buff; larger wing-coverts blackish, edged blue-grey; smaller ones wholly grey-blue. Bill black, long, slender, often hook-tipped; eyes dark brown; legs and feet black. ADULT ♀: forehead dark brown, forecrown blackish brown, hindcrown to back dark greyish brown, rump and tail like ♂; wings like ♂ but feathers edged buff, not blue-grey; lesser coverts fringed blue-grey but not wholly blue-grey; lores and narrow orbital ring of featherlets pale buff; cheeks and ear-coverts greyish brown, streaked dark brown; chin and throat with broad pale buff stripe down centre, and mottled at sides (mainly grey-brown with pale buff stripes, each stripe with narrow blackish borders); upper breast the same but without broad buff stripe in centre; lower breast light orange-buff, freckled with dark brown and pale buff; rest of underparts creamy light orange. SIZE: (5 ♂♂, 5 ♀♀) wing, ♂ 100–104 (102), ♀ 95–101 (97·8); tail, ♂ 57–63 (58·6), ♀ 54–58 (55·8); bill, ♂ 24–26 (24·6), ♀ 24–25 (24·6); tarsus, ♂ 33–35 (34·2), ♀ 33–34 (33·4) (P. Colston, pers. comm.); (16 ♂♀) wing, 97–108 (102); tail, 56–65; bill, 19–22·5; tarsus, 32–35 (Maclean 1993).

IMMATURE: upperparts mainly dull brown, profusely spotted with pale buff on head and hindneck, less profusely so on back; rump and tail like adult but colour duller; wings like adult ♀ but duller; cheeks and throat warm buff, mottled with brown; breast mid-brown, heavily but diffusely streaked with pale buff; upper belly warm buff with small brown scallops; lower belly and flanks plain buff with light orange wash; undertail-coverts orange.

NESTLING: at about 5–7 days when eyes opening, nestling has pink skin, grey in nuchal tract and where feather papillae are forming; quite long but sparse greyish down on crown; gape yellowish white, inside of mouth yellow at edges shading to bright orange on tongue, palate and gullet.

M. e. tenebriformis Clancey: Lebombo Mts, Swaziland. ♂ with crown, nape and mantle darker grey-blue, and ear-coverts, sides of neck, throat and breast darker and more intensely blue, than in nominate subspecies; underparts, wings and tail also somewhat darker.

Field Characters. Length 18 cm. A rock-thrush of upland grass and pasture with rock outcrops; tail rather short. ♂ distinguished by rather uniform blue-grey head, breast, back and wing-coverts; ♂ Cape and Short-toed Rock-Thrushes *M. rupestris* and *M. brevipes* have rufous breasts; Cape Rock-Thrush has mottled brown back and Short-toed Rock-Thrush has whitish forehead and eyebrow contrasting with dark face. ♀ differs from others in mottled brown and white underparts with only a trace of orange. (Overlaps Short-toed Rock-Thrush in SW and E Orange Free State and, marginally, in SW-central Transvaal and N Swaziland; sympatric with Cape Rock-Thrush.)

Voice. Tape-recorded (91, F, LEM). Song of ♂ a bold, lively but brief warble; duration 3 s. Song repeated several times with little variation, sometimes after only 6 s pause. Very like songs of some *Turdus* thrushes. Song consists of about a dozen whistles run into each other, some sweeping, some see-sawing, usually with 1–2 chattering trills in the middle: 'tew-tchui-see-saw-chui-tutututu-tchui-tews-tew' or 'teeu-teeu-teeu-preet-trrrr-peety-prrrr-peu' or 'worr, chilli-chilli, worr, treeo, treeo, worr'. Not very melodious, mainly at 1 pitch, but a fine, stirring song nonetheless. ♀ call, between successive songs of ♂, a trill of *c*. 6 notes, duration *c*. 0·5 s, descending scale somewhat. Alarm, a soft chattering 'e-e-e-ee-ee-ee-e'.

General Habits. Breeds in elevated open grassland and sheep pasture with scattered rocky outcrops and ridges; bare, windswept grassy mountain sides; rolling grasslands mainly above 1200 m; stony or boulder-strewn hill sides, and occasionally human habitations and (at Kangwane) mine dumps. Shuns trees. In winter some birds move down to 600 m or below to rocky slopes, recently burnt veld and felled plantations. Chiefly solitary; on breeding grounds ♂ and ♀ sometimes together but often forage well apart. Commonly perches on stone, rock or termite mound to survey surroundings, then flies or jumps to ground to forage by hopping about amongst grass tufts, sometimes bounding over a large clump. Often flicks wings after alighting. Active and conspicuous in its open habitat, but rather shy and difficult to watch foraging except at a distance, and often lost to view behind rocks. Habitually resorts to rocks if it senses danger (W. R. Tarboton *in* Ginn *et al.* 1989); perches in attenuated, upright posture, with bill a little inclined; but at other times can look plump and round. Sings from elevated rock perch. When agitated makes nervous twitches and little jumps on rock.

Most birds appear to be paired year-round residents in a large territory; however, many single birds can occur anywhere in highveld grassland in May–July, occupying a burnt area for a few weeks then moving away. Also a partial vertical migrant to low altitudes, in May–Aug.

Food. Field observation suggests insects 'such as ants, beetles, moths, caterpillars and . . . pupae' (Ginn *et al.* 1989); ants especially, also fruit and seeds (Maclean 1993). Items brought to nest include millipedes (Broekhuysen 1965).

Breeding Habits. Monogamous, territorial, solitary breeder.

NEST: a broad-based truncated-cone-shaped structure, rather large and untidy, with cup-shaped depression in centre; foundation made of grass, thin twigs and roots, walls of grass stems, shreds, moss and lichen, rim and lining of fine grasses and rootlets laid down circularly. Sited typically on earth or bare rock littered with dead vegetation, in acute angle between ground and overhanging rock; also under a low shrub beneath rock overhang, in a rock crevice, under a stone or boulder, and under leaning tuft of dense grass on open grassy hill side. Nest built by ♀.

EGGS: 2–4, usually 3. Immaculate sky-blue or greenish blue; occasionally spotted with red-brown. SIZE: (n = 18) 24·2–28·5 × 18·3–20·4 (26·5 × 19·1).

LAYING DATES: Transvaal, Oct–Dec (Oct 6, Nov 6, Dec 2 clutches); Natal, Sept–Dec.

INCUBATION: mainly or entirely by ♀. Period: 13–14 days.

DEVELOPMENT AND CARE OF YOUNG: young fed by ♀ and to lesser extent by ♂; nestling period, 16–18 days.

Reference
Ginn, P. J. *et al.* (1989).

Plate 1   *Monticola brevipes* (Waterhouse). **Short-toed Rock-Thrush.** Monticole à doigts courts.
(Opp. p. 32)

Petrocincla brevipes Waterhouse, 1838. *In* Alexander, Exped. Int. Africa 2, p. 263; Gamsberg, near Windhoek.

Range and Status. Endemic resident, W Angola to South Africa and Swaziland. Angola, common in woods near Iona Posto and Cambeno, Iona Nat. Park; also in semi-desert coastal plain of Benguela, and along Mossamedes Escarpment; 1 record Gambos, SW Huila. Namibia, common in mountainous country and isolated hills; probably more widespread than shown in map, perhaps with a continuous population from Angola to Cape. South Africa: N Karoo; flat SW Transvaal highveld and Magaliesberg, east to Pretoria; in Orange Free State mainly in SW, with a few records in N and an isolated population east of 29°E. SE Botswana, fairly common north to Molepolole and west to hills E of Jwaneng. Swaziland, Siteki and Lebombo Mts.

Description. *M. b. brevipes* Waterhouse (includes '*leucocapilla*' and '*kaokensis*'): Angola, Namibia, Botswana, Cape Prov. east to *c*. 23°E. ADULT ♂: forehead, forecrown and superciliary area greyish white or pale grey, darkening backwards to blue-grey on hindcrown, hindneck, mantle, back and scapulars; in worn birds whole of crown, nape, hindneck, and even upper mantle are greyish white, sharply darkening to blue-grey on lower mantle. Rump and uppertail-coverts bright orange. Tail: T1 black (base orange where overlain by coverts); T2–T6 bright orange, with small black mark at tip of outer web; underside of tail orange. Lores, area under eye, and ear-coverts blackish grey. Chin pale blue-grey, merging to darker blue-grey on throat and sides of neck (same shade as mantle). Breast to undertail-coverts, also underwing-coverts and axillaries, bright rufous-orange, darkest on breast, paling a little further back. Primaries blackish brown, narrowly edged blue-grey proximal to emargination point (P5–P8 emarginated), inner primaries narrowly tipped whitish; secondaries blackish brown, fringed blue-grey and narrowly tipped whitish; tertials broadly and loosely fringed blue-grey, and tipped buff; all upperwing-coverts blackish brown, broadly fringed and tipped blue-grey, lesser coverts appearing entirely blue-grey. Bill black; eyes dark brown; legs and feet black. ADULT ♀: forehead, crown, nape, mantle, back, scapulars and

Monticola brevipes

Field Characters. Length 18 cm. Like most rock-thrushes, a rather chunky, sleek-plumaged bird of rocky outcrops and mountain sides, rather long-billed, large-headed and short-tailed. Somewhat shorter-legged than typical thrushes and most chats. Usually seen solitarily; tends to perch with bill a little inclined. ♂ has blue-grey head, throat, back and wing-coverts, but differs from Cape and Sentinel Rock-Thrushes *M. rupestris* and *M. explorator* in having forehead and eyebrow and sometimes entire top of head very pale grey to whitish, contrasting with dark face. ♀ has orange underparts, like ♀ Cape Rock-Thrush, but is smaller and overall paler, throat with much white and back plain, not streaked. (Parapatric with Miombo Rock-Thrush *M. angolensis* and overlaps Sentinel in SW and extreme E Orange Free State, in Magaliesberg, Transvaal, and in N Swaziland; widely sympatric with Cape Rock-Thrush.)

Voice. Tape-recorded (91). ♂ song like that of Sentinel Rock-Thrush but without the trills, a lively warbling mixture of whistles, somewhat monotonous but rich and pleasant, containing occasional slurred and scratchy notes; not easily imitable by person. Typically lasts *c*. 4 s and repeated after 1–2 s pause; but sometimes 3–4 songs given in succession without pauses, in 12–15 s, each slightly different. Said to imitate other bird species.

General Habits. Inhabits koppies, escarpments, inselbergs, ruined stone buildings, quarries, grassy slopes with scattered rocky outcrops, river valleys with scattered bushes and low trees; and occasionally house tops in rural villages. Solitary or in pairs; not easy to approach; when taking up a winter territory around a settlement, tolerates human activities close at hand but tends to keep out of sight. Freely perches in trees (and nests in fig trees) but forages mainly on ground among rocks, also on flat roof tops. Foraging is chat-like or thrush-like, bird hopping quickly then standing still for a few s, looking and listening, then making a quick lunge.

Food. Insects, scorpions, seeds (Maclean 1993); once a small gecko, fed to nestlings.

Breeding Habits. Solitary nester, monogamous, territorial.

NEST: bulky cup made of dry grass and roots, lined with fine grass and rootlets. In Windhoek area, Namibia, nests always sited amongst dead roots of rock fig tree climbing against vertical rock surface; elsewhere, in hollows beneath rocks (C. F. Clinning *in* Ginn *et al.* 1989).

EGGS: 2–3, usually 3. Immaculate pale blue or greenish blue. SIZE: (n = 7) 21·1–24·2 × 17·1–18·3 (22·7 × 18·3).

LAYING DATES: Namibia, Nov–Mar; Botswana, Aug–Dec; Transvaal, Oct (Tarboton *et al.* 1987) or Sept–Jan (Maclean 1993).

References
Clancey, P. A. (1968).
Farkas, T. (1966, 1979).

upperwing-coverts brownish grey, uniform and without mottling, but forehead sometimes suffused with rufous; rump, uppertail-coverts and tail like ♂. Wings like ♂ (much greyer than wings of ♀ *M. explorator*) but feather fringes buffy rather than blue-grey. Lores brownish; ear-coverts brown, mottled buff; chin and throat white in mid-line forming white stripe, mottled with brown at edges and towards breast; sides of chin and throat whitish, heavily mottled buff and brown; breast rufous, the feathers inconspicuously tipped white, breast usually more or less heavily mottled with grey-brown; belly paler rufous, not mottled; flanks and undertail-coverts pale orange. SIZE: wing, ♂ (n = 7) 100–108 (102), ♀ (n = 6) 96–102 (98·5); tail, ♂ (n = 5) 55–64·5 (61·0), ♀ (n = 5) 54·5–65 (61·9); bill, ♂ (n = 5) 22·5–27·2 (25·1), ♀ (n = 5) 23·8–25 (24·3); tarsus, ♂ (n = 5) 24–27 (25·2), ♀ (n = 5) 23·5–26 (25·2).

IMMATURE: ♂ with blue-grey mantle; crown and mantle heavily spotted with black. ♀ with upperparts mainly dull brown, profusely spotted with pale buff on head and hindneck, less profusely so on back. ♂ and ♀ rump and tail like adult but colour duller; wings like adult ♀ but duller; cheeks, throat and breast buff with orange wash, heavily marked with dark brown crescents; ♀ upper belly warm buff with small brown scallops; lower belly and flanks plain buff with light orange wash; undertail-coverts orange. ♂ underparts more heavily scalloped, except for bright orange patch at side of breast which is free of scallops.

NESTLING: not described.

M. b. pretoriae Gunning and Roberts: South Africa and Swaziland, from Griqualand West to Orange Free State, highveld of Transvaal north to about Tzaneen, and Lebombo Mts (Swaziland). Intergrades with nominate race in Kuruman District, N Cape Province. In breeding plumage like *brevipes* but crown and mantle paler, the grey less bluish and more brownish; in worn plumage crown and nape remain grey (not turning whitish as in nominate race). Smaller: wing, ♂ 100–105.

TAXONOMIC NOTE: Farkas (1962b, 1966) showed that nominate *brevipes* is seasonally dimorphic and that *pretoriae* is not. On that basis he has argued (1966, 1979) that the latter is a separate species. We agree with Clancey (1968) that the 2 are best treated conspecifically.

Plate 1
(Opp. p. 32)

Monticola angolensis Sousa. Miombo Rock-Thrush. Monticole angolais.

Monticola angolensis Sousa, 1888. J. Lisbõa 12, p. 225; Caconda, Angola.

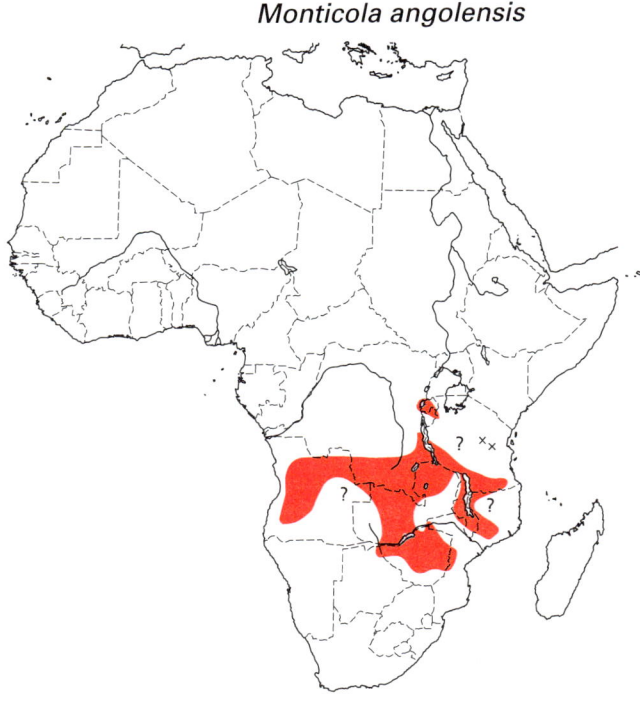

Monticola angolensis

Range and Status. Endemic resident, miombo woodland in S-central Africa. Zaïre, widespread about 1500 m in Shaba, north near L. Tanganyika to about Katenga; also in plains around L. Kivu. SW Rwanda, NE Burundi, and adjacently in Tanzania at Kibondo. Fairly common in SW and S Tanzania from Kigoma to L. Rukwa and Songea, east to Nandembo and Nachingwea; records from Dodoma and near Kilosa; locally numerous in Songea in Mar–Sept. Angola, locally common throughout central plateau from central Huila and N Bihé, to S Cuanza Sul, central Malanje, N Lunda and NE Moxico. Zambia, quite common throughout, but absent from W Balovale, Barotse (scarce in Mankoya Province), middle Zambezi valley and Luangwa valley; up to 1725 m on Nyika Plateau. Malaŵi, widespread between 900 and 1550 m, ascending to 2000 m on Mafinga Mts and 2150 m on Nyika Plateau; absent south of Blantyre. Zimbabwe from Zambezi to central plateau, south to E Matobo Hills, Mts Buhwa and Emberengwa, and Igar; formerly considered common and widespread over whole of central plateau (Irwin 1981), but now patchy and local there, and rarely common (Tree 1994, 1995b); ranges up to 1500 m in E highlands (Inyanga to Chipinga). Botswana, sparse in NE. Mozambique, fairly common in W Manica e Sofala, between Inyamadzi and Buzi R.; probably more widespread in N Mozambique than shown.

Distribution within miombo woodland is localized. Pairs occur year after year in some areas but bird is absent from apparently perfectly suitable intervening woods; local distribution may be determined by nest-hole availability (A. J. S. Weaving *in* Ginn *et al.* 1989).

Description. *M. a. angolensis* Sousa (including '*niassae*'): range of species except S Zambia, Zimbabwe, W Mozambique and W Malaŵi. ADULT ♂: whole of head, neck, throat, mantle, back, scapulars and upper wing-coverts blue-grey, all except chin and throat very variably mottled with black. Blue-grey rather uniform, palest on forehead, chin and centre of throat, darkest on wing-coverts; lores and sometimes cheeks dark grey to blackish. Black mottling usually consists of scattered small spots on hindcrown, nape, hindneck, mantle, back and scapulars, generally heaviest on hindneck and upper mantle; spots sometimes in rows on hindcrown and nape, forming irregular streaks. Some birds have no spots at all on top of head or upperwing-coverts. At other extreme many birds have large black spots mottling whole of head top and coalescing irregularly on hindneck, mantle, back and scapulars so that these parts look more black than blue-grey; such birds have black crescents and bars on all upperwing-coverts, and even rump and tail can be tiger-banded. Rump and uppertail-coverts bright rufous-orange, plain or lightly or heavily barred black. Tail: T1 brown or blackish, usually with narrow black-bordered orange shaft-streak, widening proximally to orange feather base; T1 sometimes with *c*. 10 wavy black bands; T2–T6 bright rufous-orange, T2–T5 with long black line at end of outer webs and blackish marks at end of inner webs, and T6 with broad black edge to whole of outer web and blackish mark near end of inner web; underside of tail orange. Blue-grey throat sharply demarcated from orange underparts. Breast rich cinnamon or dark orange, colour paling gradually to light orange-buff on belly, vent and undertail-coverts. Flanks, thighs, underwing-coverts and axillaries bright orange. All flight feathers blackish brown (underside shiny grey-brown), primaries narrowly edged blue-grey proximal to emargination point (P5–P8 emarginated), secondaries fringed blue-grey and tipped buff, tertials with outer webs broadly and loosely fringed blue-grey, mottled with wavy black bars, and tipped with long pale buff crescent; primary coverts black-brown with blue-grey outer edges; remaining coverts blue-grey, generally with black shaft-streak and large crossbow-shaped subterminal black bar. Bill black, broader and deeper than *M. explorator*; eyes dark brown; legs and feet black. ADULT ♀: variable, birds in high-rainfall areas often very dark and richly patterned ('*niassae*'); upperparts variegated, greyish or buffy, heavily mottled and barred with black (or blackish, marked with buff-grey), rather like Northern Wryneck *Jynx torquilla*; rump tiger-banded black and rufous. Tail mainly chestnut, T1 greyish brown, all feathers heavily marked with black, either in wavy bands or in several diagonal stripes on inner web and 1–2 longitudinal lines on outer web. Throat region distinctive: blackish cheeks, pale buff moustachial stripe, irregular black malar stripe, and broad pale buff stripe down centre of chin and throat. Breast rich orange, feathers narrowly fringed blackish; flanks orange, with narrow blackish bars and buffy blotches; belly and undertail-coverts cream with light orange wash; thighs orange with *c*. 5 thin wavy black bars. Sometimes only sides of breast are rufous, and remaining underparts mainly buffy. Soft parts like ♂. SIZE: wing, ♂ (n = 10, Angola) 99–107 (103), (n = 13, elsewhere) 93–100 (98·5), ♀ (n = 10, Angola) 98–102 (100), (n = 5, elsewhere) 90–99 (94·0); tail, ♂ (n = 10) 54–63 (60·1), ♀ (n = 5) 58–63 (60·2); bill, ♂ (n = 5) 21–23·5 (22·9), ♀ (n = 5) 22·5–24 (22·8); tarsus, ♂ (n = 3) 26–26·5 (26·3), ♀ (n = 3) 25–26·5 (25·3). WEIGHT: unsexed (n = 2) 44, 44·7.

IMMATURE: like adult ♀, even more variegated by profuse buff spotting; sides of neck, cheeks, chin, throat, breast and flanks buff, heavily scaled with black, with patches of orange in centre and at sides of breast; belly, vent, thighs and undertail-coverts cream-buff, unmarked.

NESTLING: not described.

M. a. hylophila Clancey: Zimbabwe, S Zambia, W Malaŵi and W Mozambique. ♂ with breast, sides and flanks much paler than in nominate race, giving less contrast between chest and rest of underparts; belly and undertail-coverts white. Uppertail-coverts paler than in *angolensis*; T6 with dusky mark at tip of inner vane. ♀ like ♀ *angolensis*.

Field Characters. Length 18 cm. A shy and easily overlooked, rather poorly known rock-thrush endemic to miombo woodland, where it is the only *Monticola* sp. Makes close approach to Short-toed Rock-Thrush *M. brevipes* in W Angola where, however, Miombo Rock-Thrush is on escarpment and Short-toed Rock-Thrush below it on littoral plains. ♂ readily recognized by pale grey-blue head and upperparts, back and wings lightly or heavily spotted with black, and orange breast and tail. Belly white in Zimbabwe, orange elsewhere. ♀ is also variable, with upperparts handsomely mottled and variegated; identified by orange tail (central feathers dark brown) and orange on sides of breast (often whole breast, and flanks, orange), and by conspicuous, blackish, malar stripe, flanked by white throat and whitish moustachial stripe. Larger Kurrichane Thrush *Turdus libonyanus* has similar throat stripes and underparts, but has plain upperparts and orange bill. Miombo Rock-Thrush's eyes are larger than in other rock-thrushes, and in both sexes large, black eyes add to distinctive appearance. For distinction from Mountain Rock-Thrush *M. saxatilis*, see that species.

Voice. Tape-recorded (86, 91, B, F, CART, CHA, FAR). Song of ♂ lasts 30 s or more, a sweet, fluty, rather haunting series of measured phrases, each 2 s long, separated by 2 s pause, with 7 phrases in 27 s. Every phrase starts with sweeping, melodious whistle 'wheeeouu', most phrases end with 1–2 bisyllabic whistles, and a few phrases include a scratchy note 'tchr'. Successive phrases differ slightly: 'wheoou'u'u seoo hetchr (pause), heooo heo hehe (pause), wheoou'u'u seeti seeti (pause), heeiou'u tch situ-situ (pause) . . .'. Also rendered 'peee peeu, pweetweet'. Not loud. Call a fluty bisyllabic whistle, the second syllable higher pitched. Alarm a chatter. Said to mimic other birds.

General Habits. Practically confined to miombo (*Brachystegia*) and mutemwa or gusu (*Baikiaea*) woodlands in hilly plateau country; often on Kalahari Sand, also in Mountain 'Acacia' *Brachystegia glaucescens* woodland; and at edges of *Eucalyptus* plantations 5 km from nearest *Brachystegia* (Malaŵi, Zimbabwe). Solitary or in pairs. Rather quiet, shy and unassertive, but when breeding quite bold and tolerant of a person. Behaves like a *Turdus* thrush; forages mainly on ground, in leaf litter, poking under logs and bark; perches freely in trees, sings from canopy. When alarmed, flies up into tree, perches motionless for a few s, then flies away (Maclean 1993). Sedentary in Zambia and Zimbabwe; but in Malaŵi occurs at Bana, W coast of L. Malaŵi, where unlikely to breed, only in June–Oct.

Food. Ants and termites, also beetles, crickets, moths, spiders, centipedes, and small worm-lizards (Ginn *et al.* 1989).

Breeding Habits. Monogamous, territorial, solitary nester. Territory seems to be maintained for several years, and same nest-site sometimes used more than once (A. J. S. Weaving *in* Ginn *et al.* 1989).

NEST: shallow to quite deep saucer-shaped depression in circularly interlaced mass of fine or wiry grass and rootlets, lined with fine dry grass and creeper tendrils, on bulky and untidy base made of twigs and coarse grass. Some nests lack foundation or have only flimsy one. Sited generally in hollow in jagged stump of large vertical branch or trunk of *Uapaca kirkiana* or similar tree; sometimes on a branch or in shallow cavity in trunk; 1–3 m above ground.

EGGS: 2–4, Malaŵi mean (11 clutches) 2·7. Turquoise or bluish white, immaculate or sparsely speckled with reddish brown. SIZE: (n = 18) 20·4–26 × 16·8–19 (23·8 × 17·8).

LAYING DATES: Angola, Sept; Zambia, Sept (3), Oct (1), Dec (1); Malaŵi, Sept (2), Oct (9), Nov (1), Dec (1); Zimbabwe, Aug–Dec (Aug 7, Sept 46, Oct 32, Nov 14, Dec 5 clutches).

INCUBATION: by ♀ and ♂ alternately. Dead-leaf colour and pattern of ♀'s back is highly cryptic when she is incubating. Period: 13·5–15 days (Vernon 1968, Kirkpatrick 1993).

DEVELOPMENT AND CARE OF YOUNG: young fed by ♂ and ♀. Fledging period: (n = 2) 18·5–20 days (Vernon 1968).

BREEDING SUCCESS/SURVIVAL: 1 pair raised 3 broods in 3 months and 23 young in same place in 3 years (Vernon 1968).

Monticola rufocinereus (Rüppell). Little Rock-Thrush. Monticole rougequeue.

Saxicola rufocinerea Rüppell, 1837. N. Wirbelt. Vög., p. 76; Simen, Abyssinia.

Plate 1
(Opp. p. 32)

Range and Status. Africa and SW Arabia. Resident, perhaps a partial migrant. Ethiopia, frequent to common and widespread between 1000 and 3100 m; in Eritrea very common at about 1400 m (in Yemen, > 50 birds seen, at 1700–2500 m, in 2 months: Cornwallis and Porter 1982). Sudan, in extreme S in Imatong and Didinga Mts, Nov–Mar only, at 2000 m. Somalia, locally common on high ground north of 9°30′ N.

Monticola rufocinereus

Kenya, uncommon and local, between 1400 m and 2500 m: Mts Kulal, Nyiru, Ololokwe and Elgon, and Tugen hills to Narok, Nandi, Nakuru and Naivasha; records from S end L. Turkana. Uganda, Kidepo Valley Nat. Park, Mt Kakamari, Mt Moroto and Sebei. Tanzania, only at Longido, 25 km from Kenya border.

Description. *M. r. rufocinereus* (only subspecies in Africa). ADULT ♂: forehead above bill blue-grey, colour extending above lores and eye as short superciliary stripe; crown, nape, hindneck, sides of neck, mantle, scapulars, back, greater, median and lesser coverts, uniform olive-grey; rump and uppertail-coverts bright orange. Tail: T1 blackish with orange shaft and shaft-streak broadening proximally into orange feather-base; T2–T6 orange, with tips of both webs blackish in well-demarcated V-shaped patch 5 mm deep in centre with arms of V *c*. 13 mm long, or on outer web of T6 *c*. 27 mm long. Underside of tail orange, with dusky edges. Lores black; cheeks, ear coverts, chin, throat and breast uniform blue-grey, colour sharply demarcated from orange underparts; belly, flanks, undertail-coverts, underwing-coverts and axillaries orange, vent cream-buff, thighs grey-brown. Wing black-brown, P6–P8 emarginated, primaries and secondaries narrowly and tertials broadly edged olive-grey, greater and median primary-coverts narrowly edged olive-grey, alula black, remaining coverts olive-grey, underside of flight feathers shiny dark brown. Bill black, slender; eyes dark brown; legs and feet black, soles grey. ADULT ♀: ♂-plumaged; differs in being olivaceous or greyish dark brown above where ♂ is olive-grey, superciliary stripe whitish, chin and throat paler than ♂, olive-grey rather than blue-grey, lower breast faintly banded whitish; not as distinct a demarcation between breast and belly colours as in ♂. SIZE: (6 ♂♂, 6 ♀♀) wing, ♂ 85·5–88 (87·1), ♀ 80–85 (82·0); tail, ♂ 62·5–65 (64·3), ♀ 54–59 (57·2); bill, ♂ 19–20·5 (20·0), ♀ 18–21 (20·2); tarsus, ♂ 24–27 (25·2), ♀ 22·5–26 (25·1). WEIGHT: (Kenya) ♂ (n = 5) 21–27 (24·25), ♀ (n = 2) 20, 27.

IMMATURE: not described. Immature of Arabian *M. r. sclateri* (adult *sclateri* and *rufocinereus* very alike) is like adult ♀, head and back heavily spotted pale buff, underparts except lower belly and undertail-coverts buff, heavily scaled with blackish brown, with orange patches in breast (very like immature Miombo Rock-Thrush *M. angolensis*).

Field Characters. Length 15–16 cm. Mountains of NE and E Africa; arboreal. The smallest rock-thrush, almost chat-sized. ♂ has blue-grey head and breast, grey-brown back and orange belly and tail; ♀ similar but duller. Quivers tail on alighting, which distinguishes it from all other birds in range except redstarts and Mountain Rock-thrush *M. saxatilis*; latter is considerably larger (19 cm) and has white patch on back (♂) or speckled and scalloped plumage (♀ and non-breeding ♂). Slightly larger than redstarts, with stockier shape, shorter tail and longer bill. Adult superficially like adult ♂ of Asian race of Black Redstart *Phoenicurus ochruros phoenicuroides*, which has rufous belly but differs in black face and breast. A particularly dingy ♀ in poor light might possibly be mistaken for a young Black Redstart or Common Redstart *P. phoenicurus*, but Little Rock-Thrush can be told from all redstarts in all plumages by distinctive wheatear-like tail pattern, an inverted black 'T' on an orange tail, on underside appearing as a black terminal band.

Voice. Unknown.

General Habits. In Kenya and Uganda inhabits grass-strewn rocky escarpments without bushes, forested ravines with or without rocks, forest clearings, cultivation with dead trees, and isolated patches of thicket in open bushland or grassland (Britton 1980a); sometimes around buildings (e.g. on Mt Kakamari, Uganda, and Nanyuki, Kenya); clumps of mixed scrub and nettles in old kraals; in Sudan and Ethiopia, open slopes with scattered trees near edges of forest; in Eritrea, upland scrub, light woodland on steep rocky slopes, stands of juniper and euphorbia; in Somalia, inhabits low open copses and thickets, keeping largely to high branches of trees, also native cultivation with dead trees, and rocks and boulders where there are no dead trees; in Darass Forest, N Somalia, lives in clearings and perches in lower branches and dead stumps of juniper trees (Archer and Godman 1961).

In pairs; tame and confiding, at least in breeding season. At perch habitually shivers tail, like Mountain Rock-Thrush. Said to use dead trees as commanding positions from which to scan ground, and to drop onto insect prey on ground; also said to hawk for flying insects from top of trees, like a flycatcher; and to be crepuscular (Jackson and Sclater 1938).

Mainly resident; any movements that there are in Kenya are not at all understood; recorded in S Sudan only Nov–Mar; in Yemen a partial vertical migrant, from 800–3000 m in summer down to 600 m or less in winter (Brooks *et al*. 1987).

Food. Insects, including grasshoppers and beetles (Kenya); fruits of *Rosa abyssinica* and *Olea chrysophylla*, a 1-cm beetle and a 3-cm caterpillar (Yemen).

Breeding Habits. Solitary nester. Probably monogamous and territorial: sexual chasing by 4 birds in rapid flight through foliage of trees and over open ground noted in Yemen, Sept, and song and territorial defence, Nov (Phillips 1982, Brooks *et al.* 1987).

NEST: the only nest described was 1·25 m above ground in cleft in trunk of tamarind tree *Tamarindus indica*, at elevation of 1200 m. Nests on buildings at Nanyuki, Kenya. Both ♂ and ♀ attend at nest.

EGGS: 4 (1 clutch). Turquoise, freckled with light red-brown around broad end. SIZE: (n = 4) all 21 × 16.

LAYING DATES: Somalia, Apr–May. BREEDING INDICATIONS: Ethiopia, Feb–Mar, May, Sept–Oct; Kenya, Dec, Mar, June.

Reference
Archer, G. and Godman, E. M. (1961).

Monticola saxatilis (Linnaeus). Mountain Rock-Thrush; Rock Thrush. Monticole merle-de-roche.

Turdus saxatilis Linnaeus, 1766. Syst. Nat. (12th ed.), p. 294; Switzerland.

Plate 1
(Opp. p. 32)

Range and Status. NW Africa, S Europe from Spain to Turkey, Transcaspia, and Asia east to 115°E and north to L. Baikal. Winters in subsaharan Africa (and <2% in N Africa and S Arabia). Vagrant Seychelles.

In NW Africa sparse breeding summer visitor and passage migrant. Morocco, recent breeding records from Oukaimeden, Jebel Hebri and Jebel Lakraa, all between 1750 and 2600 m; other sight records between 1600 and 2700 m. Algeria, breeds above 1900 m in Aurès (up to 2300 m on Jebel Chelia), Oursenis Mts and Djurdjura and at 1500–1800 m at Babor; passage migrant in Mar–May and from mid-Sept to Oct, on the coast, in northern mountains and N Sahara; in winter, known only from Hoggar Mts in S Algeria.

East of Algeria a passage migrant, mainly in spring. Tunisia, breeding not proven; a rare but regular migrant from late Mar to early May, mainly in oases and at Cap Bon; no autumn migration. Libya: in Tripolitania common Mar to early May, once Sept, twice Jan; in Cyrenaica regular near coast Mar–May, none recorded in autumn; in desert found sparsely in Mar–May (Wau en Namus, Sebha, El Hammam, Jalo, Kufra, Serir). Egypt, fairly common spring migrant, mid-Mar to late Apr (a few late Feb and early May); rare in autumn (*c*. 10 records); once in winter (Faiyum, Jan).

Main wintering grounds are from E Chad to highland Ethiopia, Uganda, Kenya and N and E Tanzania. Chad, in winter common above 700 m in sahelian and soudanian zones east of 19°E; juveniles arrive in mid-Oct (sometimes Sept), main passage of adults in late Oct, when 40 birds per ha on summit of Tondou; *c*. 12 pairs winter on Kilingen (825 m, near Abéché); most leave in Jan, latest early Mar. Winters in Kapka, Ouaddai and Jebel Mara (Ennedi Mts), E Chad and W Sudan (Newby 1980, Hogg *et al.* 1984). Sudan, uncommon in winter in rocky hills and woodland mainly south of Boro R., Wau, Amadi and Boma; common spring migrant further north. Central African Republic, known only from Manovo–Gounda–Saint Floris Nat. Park; Zaïre, only in NE where frequent southeast of Niangara (3° 42′N, 27° 53′E). Ethiopia, frequent to common, mainly above 1200 m, in thorn and broad-leaved woodland and grassland, mid-Sept to late Apr. Uganda, frequent and widespread in NE, uncommon south of Teso, Acholi and Murchison Falls Nat. Park. Kenya, common or locally common winter visitor Nov–Mar and regular passage migrant in Oct at L. Turkana and Mar–Apr on coast; ranges up to 3000 m but keeps mainly below 1500 m. Tanzania, winters in N, centre and E, west to central Tabora and Mpanda, south to Njombe and Mikindani, east to coast, with 4 records from Zanzibar; in autumn reaches Serengeti Nat. Park in Nov; quits S and central Tanzania in Feb (latest, early Mar); passage in N Tanzania Nov–Dec and Mar to early Apr.

Rare to locally not infrequent to west of Chad. Mauritania, records at Kaédi, Moudjéria and Iwik and 4 at Nouakchott (Sept–Oct, Apr). Senegal and Gambia, 9 records with several in Feb–Apr on coast at Popenguine (Morel *et al.* 1983). Mali, *c*. 10 records Kassa, Baoulé, Sikasso; several in hills north of

Sigimindji. Guinea-Bissau, once at Bissau. Guinea, 4 records Mt Nimba, 2 elsewhere. Sierra Leone, sparse, hills in NE (Tembikunda). Liberia, regular on Mt Nimba at 1250–1400 m. Ivory Coast, 3, Mt Nimba, Comoé Nat. Park and Ferkéssédougou. Ghana, uncommon and sporadic in N, mainly Bolgatanga; 1 record at Ejura. Niger, records from Aoudéras, Tebernit and Aïr. Nigeria, uncommon between 10° and 12°N, rare elsewhere (records at Sokoto, Kano, Igbetti, also Tatara in Benue Province); at Zaria, 24 records in 3 winters, evenly spread from Nov to mid-Mar; small numbers recorded in Falgore, Borgu, Plateau Province over 900 m; no records east of 10°E. Cameroon, records at Yagoua, Koum and Genderu; regular at Waza.

Zambia, vagrant, Ngitwa and Mbala, Dec. Rwanda, rare, Akagera. Djibouti, Arta, Oct, Garrab, Mar. Somalia, locally common in NW and south of 4°N in autumn and spring; Sept–Nov and Mar–May; 3 records in Jan and Feb.

Description. ADULT ♂ (breeding): forehead, crown, nape, hindneck, sides of neck, mantle, ear-coverts, chin and throat pale greyish blue; scapulars dark bluish brown; back white; rump bluish dark brown; uppertail-coverts bright orange. Tail: T1 dark grey-brown with orange shaft, rest of tail orange, outer web of outer feathers edged with dark brown near tip. Breast orange, sharply demarcated from throat; belly, flanks, thighs, vent and undertail-coverts orange, with some whitish feather-tips. Wings blackish brown, the primaries, secondaries, tertials and greater coverts pale-fringed and all coverts with dark bluish wash. Underside of flight-feathers grey, underwing-coverts and axillaries deep rufous-orange. Bill black; eyes dark brown; legs and feet brown-black. ADULT ♂ (non-breeding): the grey-blue and dark brown feathers have broad buff tip and narrow black subterminal band; back mottled brown (not white); rufous feathers in underparts with broad whitish tips. ADULT ♀ (breeding and non-breeding): forehead, crown, nape and sides of neck grey, variegated with very pale grey and dark blackish scallops (each feather has whitish tip and narrow blackish subterminal line). Mantle, back and rump dark bluish grey, the feathers with 2–3 mm broad buff tip and blackish subterminal line, giving effect of bluish grey ground heavily spotted with buff. Ear-coverts striped dark brown and pale rufous. Chin and throat white or whitish in centre, bluish or dark brownish at sides, where heavily marked with dark scallops; breast light orange, heavily scalloped with blackish; belly light orange, washed creamy white. Wings more variegated than in breeding ♂: tertials and most flight feathers broadly tipped whitish or pale buffy. SIZE: wing, ♂ (n = 21) 119–129 (124), ♀ (n = 10) 115–126 (120); tail, ♂ (n = 24) 56–65 (60·8), ♀ (n = 12) 56–67 (62·0); bill, ♂ (n = 30) 23·2–27·2 (25·0), ♀ (n = 13) 23·3–27 (25·0); tarsus, ♂ (n = 26) 26·8–30·1 (28·3), ♀ (n = 13) 27·1–29·6 (28·2). WEIGHT: (Kenya) ♂ (n = 7, Oct–Mar) 42–50 (45·9), ♀ (n = 7, Oct–Mar) 44–59 (49·9), ♂♀ (n = 78, Nov–Dec) 39–57 (48·0); (Nigeria) ♂♀ (n = 4, Dec–Apr) 47–72 (60·0). Afghanistan and Kazakhstan, May–Aug, ♂ (n = 53) 40–65 (49·5), ♀ (n = 23) 42–65 (52·2) (Cramp 1988).

IMMATURE: quite like non-breeding adult ♂, but buffier overall: tail and uppertail-coverts orange like adult, but rest of upperparts, head and neck buff, each feather narrowly tipped dark brown; underparts buff, warmest on breast, scaly, each feather blackish-tipped.

NESTLING: mouth yellow or orange, no tongue spots; gape flanges pale yellow; down long, dark bluish grey.

Field Characters. A thrush-sized but chat-like bird, sleek, rather short-tailed, with wings nearly reaching tail-tip; solitary; breeding habitat is rocky mountains, but inhabits savanna woodland in winter. Often quivers tail briefly on alighting. ♂ in full breeding plumage readily distinguished by blue-grey head and mantle, white patch on back, blackish wings and bright orange tail and underparts. However, most birds seen south of the Sahara (both sexes) are in nondescript brown winter plumage, spotted and scalloped, with just a rusty wash on breast and flanks, the only distinctive character being the black-centred orange tail. Winter birds overlap in E Africa with Little Rock-Thrush *M. rufocinereus*, which is smaller and shorter-winged, with blue-grey extending to breast and no white patch on back, and retains same (unspotted) plumage all year; however, scaly immature might cause confusion; best identified by black tips to all tail feathers. Winter birds occasionally wander into range of Miombo Rock-Thrush *M. angolensis* in S Tanzania/Zambia. Blue-grey ♂ Miombo Rock-Thrush has black scaling on upperparts, no white on back; ♀ Miombo more difficult, similarly brown and scaly above, but below has orange breast and flanks with only a little scaling, and black and white striped throat; scaly immature has orange patches at sides of breast.

Voice. Tape-recorded (53–62, 73, 93, B, C, F). Song, by ♂ and ♀ (mainly ♂), a measured series of 3–6 mellow phrases separated by pauses of 2 to several s; each phrase consists of 6–12 notes, mainly sweeping, fluty whistles with a few grating 'tschk' notes interspersed and with considerable mimicry of other bird species. Phrase lasts 1·5–4·5 s; whole song is euphonious but not very melodious. About 23 Palearctic birds known to be mimicked; main one is Chaffinch *Fringilla coelebs* whose mimicked song 'chip chip chip chip tell r-r-r-r tu tu wechoo tu wechoo' can comprise one complete phrase in Mountain Rock-Thrush's song. Same phrase may be repeated, but usually all phrases in a song bout differ. ♂ sings mainly at perch, sometimes in flight; aerial song adds to one phrase of warbled perch-song with medley of mimicry, twittering, rattling and gurgling. When disturbed, or in territorial confrontation, ♂ sings loud, low-pitched, continuous song. Contact call, single 'tak', sometimes 'tak-tak-tak' accompanied by tail-flicking. Alarm, quiet plaintive whistle recalling Eurasian Bullfinch *Pyrrhula pyrrhula*; fright, emphatic 'tschak-tschak-tschak-tschak'. Rather silent on wintering grounds, but sometimes subsings (Kenya).

General Habits. Breeding habitat is sunny, rocky hill sides with scattered shrubs, rocky heaths, crags, farmland with dry-stone walls and old stone buildings; mountain ravines and high, rock-strewn valleys with stunted tree growth, at 1500–2600 m (possibly 1250–3000 m). In winter inhabits grassland and open bouldery moorland, cliffs, villages and buildings in Eritrea, degraded savanna woodland in Nigeria with gulley-erosion and scrub, bare mined ground with much exposed rock on Mt Nimba (Liberia), lightly bushed grassland with living and dead trees, rocks, a few buildings and burnt areas (E Africa); also, increasingly, building complexes such as industrial estates, sewage farms and airport buildings (Nairobi, Kenya). On migration occurs in even greater variety of habitats, including sea-cliffs in Senegal and lowland acacia scrub in Eritrea.

Solitary and in pairs; sometimes in small, loose flocks on migration. Rather shy, and furtive when breeding. Bird disturbed on open ground makes for cover of trees (Sierra Leone). Unobtrusive; often sits still; hence easily overlooked. In fact can be tame and confiding in winter quarters: in a noisy Nairobi factory yard or aircraft hangar during a lull a bird can be heard singing, hidden only 2 m away (Backhurst 1987). Perches on stony ground, rocks, buildings, sometimes wires and quite freely on trees. Perches looking slim and upright, bill often pointing a little upward. Gait a long-paced hop; nimble over rocky ground. Flight low, rather floating; plummets down crags and steep hill sides. Often quivers tail on alighting, much like Common Redstart *Phoenicurus phoenicurus*. Forages by scanning from elevated perch on rock or tree, then dropping or flying down to ground, seizing and eating prey, and then returning to its perch or hopping or running around for a while to search for more prey. Of 116 items in Kenya, 96% were taken from ground; in 13 out of 27 sorties to the ground bird took multiple items before returning to perch (Cramp 1988). Occasionally makes brief aerial sortie to catch flying insect; takes fruit direct from tree; and eats insects on dung. Hammers at and shakes larger items (beetle, lizard). ♂ and ♀ may defend separate territories in winter (Tanzania: Sinclair 1978). Chases Blue Rock-Thrushes *M. solitarius* off territory (Sierra Leone: G. D. Field *in* Cramp 1988). Courtship once noted in Tanzania: ♀ accepted several mulberries *Morus* offered by ♂, Mar (Moreau 1943).

Migratory; whole population winters in Africa. Chinese birds must migrate 7500 km each way. Migrates at night, singly or in small, loose groups, often with Blue Rock-Thrushes. In autumn young birds migrate slightly earlier than adults; in spring ♂♂ precede ♀♀ (Morocco, Egypt). Winters in much greater abundance east of Nigeria than west of it. In autumn, scarce throughout N Africa, suggesting that S European birds overfly it; appears to cross N Africa on broad front, often making Saharan landfalls at oases and on rocky high ground. On spring passage common on NE African coast (Libya to Gulf of Suez), less so in Maghreb. Areas in subsaharan Africa where common on migration, in both autumn and spring, all lie northeast of main wintering areas, suggesting main migration route is along northeast–southwest axis: there may well be major undetected movement across Red Sea (Cramp 1988). In central Sudan west of about 35°E, markedly commoner on autumn than on spring migration (Hogg *et al.* 1984).

Marked passage through Cyprus and slight one through N Egypt from late Aug to late Sept. Immigration into eastern Africa starts in mid-Sept with main passage in late Oct, and some movement across Kenyan/Tanzanian border (e.g. Ngulia) until late Dec. Winters in Tanzania Nov–Mar.

South of about 15°N, main spring emigration is in last half of Mar; passage continues well into Apr (E African coast, and Ethiopia), with stragglers in early May (Eritrea). North of 15°N main migration is in first week Apr (E Sudan) and first half Apr (N Egypt). Situation to northeast of E Africa is instructive: Oman has marked passage in early Mar lessening to mid-Apr, but much heavier one in autumn, from mid-Sept to mid-Oct, particularly last week Sept.

Food. Invertebrates, mainly large insects; some fruit. Kenya, 7 birds contained beetles, one small grasshoppers, and 6 'insects'; others contained Orthoptera, small snails and a few berries. Zaïre, 4 birds contained insects including ants, also a millipede and fruits. Liberia (Mt Nimba), 1 bird contained alate black ants. Tanzania, eats mulberries *Morus*. In Europe beetles, Orthoptera and caterpillars are staples; also eats Odonata, earwigs, flies, wasps, bees, spiders, centipedes, millipedes, earthworms, small snails, a few lizards and frogs, and fruits of *Spiraea*, *Prunus*, *Sorbus*, *Vitis*, *Ribes*, *Viburnum* and *Sambucus*.

Breeding Habits. Monogamous, territorial, solitary breeder. Territory size 8·0–12·7 ha (Italy); pairs 200–800 m apart. Neighbouring pairs not very aggressive and freely intrude into each other's territory. May defend territory against a Northern Wheatear *Oenanthe oenanthe*. Territory maintained by ♂ in song-flight: ♂ takes off suddenly, flies low then ascends steeply with slow, powerful wing-beats, tail fanned, begins to sing during ascent, flutters down singing with mimicry, then plummets. Threatens by snapping bill and lifting 1 wing. Courting ♂ runs towards ♀ with head held high, uttering subsong. He stops, moves head on stretched neck from side to side in slow, snake-like way for 15–20 s, and moves sideways with tripping gait, singing all the while; pair may then copulate. ♂ also has bowing display: he bows deeply, raising and quivering fanned tail.

NEST: moderately neat rather flat cup of coarse grass, rootlets and moss, lined with moss and fine rootlets. Sited beneath rock overhang, in horizontal crevice in rock face or wall, under boulder on steeply sloping ground, and sometimes in tree-hole. ♀ selects site, accompanied by ♂ who often sings. ♀ builds nest. Last year's site often used.

EGGS: 4–6, usually 4–5. Subelliptical, smooth, glossy; pale blue, immaculate or with faint rufous speckles around broad end. SIZE: (n = 100, Europe) 22·5–29·0 × 16·5–21·0 (26·0 × 18·2). WEIGHT: av. *c.* 4·95 g. Double-brooded.

LAYING DATES: NW Africa, late Apr to June.

INCUBATION: by ♀ only; she scarcely leaves nest for 3 days before eggs hatch, subsings to attract ♂'s attention, and is fed on nest by ♂; ♂ known to remove her faeces. Period: 14–15 days.

DEVELOPMENT AND CARE OF YOUNG: young cared for by ♀ and ♂, mainly ♀. At first ♀ broods young; ♂ gives her food items which she passes to young; later ♂ (and ♀) feed young directly. Both parents remove faecal sacs. Young leave nest unable to fly at 14–15 days, dispersing widely, hiding under rocks and bushes. When they can fly they re-group and are fed and guarded by both parents. ♂ guards young for 3–4 weeks after they have left nest.

References
Beven, G. and England, M. D. (1969).
Cramp, S. (1988).
Schmidt, E. and Farkas, T. (1974).

Plate 1
(Opp. p. 32)

Monticola solitarius (Linnaeus). **Blue Rock-Thrush. Monticole merle-bleu.**

Turdus solitarius Linnaeus, 1758. Syst. Nat. (10th ed.), p. 170; Italy.

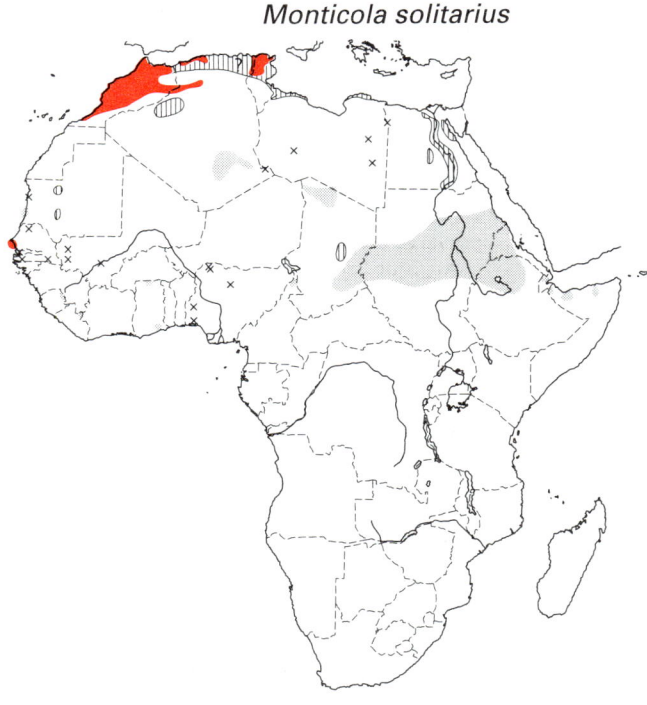

Range and Status. Mediterranean basin east continuously to Japan. Breeding latitudes in W about 30–45°N but in E breeds south to 20°N (China) and to equator (Sumatra). Resident and partial migrant in Mediterranean basin (NW Africa; Portugal to Yugoslavian states and SW Turkey); summer visitor, central Turkey to Kirgizia; resident and/or full migrant in China and Japan. Migrants winter in Arabia, India, Indochina and Indonesia, and south to about 8°N in Africa. Vagrant N Europe.

Widespread breeding resident and vertical or partial migrant in Morocco, quite common, from Atlantic coast at 28°N to Strait of Gibraltar and Mediterranean coast east to Saidia (abundant on sea-cliffs at Fnidek, Cabo Negro, between Oued Lao and Bou Ahmed, Juan en Nich, El Jebha, Cala Iris, Al Hoceima, Cap des Trois Fourches and Ras Kebdana), in plains and up to 2700 m in all Moroccan mountains. Commoner on plains west and north of mountains in winter than summer. Algeria, resident in Mts des Ksours, and near coast on Massif de Dahra and at Oran where common; uncommon in winter, when occurs on passage in Saharan oases (Beni-Abbès, Timimoun) and in Tassili and Hoggar Mts. Has been suspected to breed in Hoggar, where winter birds seem very fat (Niethammer 1963). Tunisia, scarce resident on rocky coasts between Tabarka and Cap Bon and Enfida, and breeds on crags and in ancient ruins in mountains south to Bou Hedma and Gafsa.

A July record near Tripoli; but not proved to breed in Libya, where a common winter visitor to coast Sept–Apr (commoner in Tripolitania than Cyrenaica); desert records Jan–May at Ghat, Sebha, Jaghbub, Kufra and Serir. May have bred in Egypt (60 years ago said to be present all year at Giza; bird carrying food, Cairo, Apr 1981); otherwise in Egypt a fairly common passage migrant and winter visitor mid-Sept to late Apr (exceptionally mid-May), in Nile valley, Eastern Desert, Gulf of Suez coast, and Western Desert oases. Probably breeds Senegal (see below).

Winters in Hoggar Mts (Algeria), Tibesti highlands (Chad), eastern Africa from 19°E in Chad through Darfur and hills in central Sudan to N Ethiopia, Eritrea, and N Somalia. Chad, abundant Oct–Feb at Abéché, Kilingen, Tondou and Goz Beida; on passage Sept–Oct through E Wadi Rime – Wadi Achim. Sudan, sparse or locally quite common winter visitor on high ground throughout central latitudes. Ethiopia, frequent in NW, status poorly known; common (commoner than Mountain Rock-Thrush *M. saxatilis*) in Eritrea. Djibouti, widespread but uncommon in Forêt du Day and Mabla Mts; also Arta. Somalia, frequent on Bihendula hills (30 km south of Berbera) at 600 m; also Somadu, Lower Sheikh and Golis foothills, Sheikh and Gidil on Mt Wagar; and Warsangeli escarpment, on passage (?) at 1200 m. Probably more widespread in N Somalia than shown.

Overwintering populations less numerous and more localized in W Africa. Mauritania, regular in small numbers in Nouakchott, Oct–Mar; also on passage Cap Timirist and in Adrar and Tagant areas. Senegal, a few dozens winter along coast from Cap Vert to Gorée and Popenguine, Nov to early Apr. In 1983 and 1984 pairs at Popenguine and Cap de Naze were singing, defending territories, and carrying materials; 5 ♂♂ and a juvenile present in July, and birds seem certain to have bred there (Rouchouse 1985, Morel and Morel 1990). Well-marked spring passage at Popenguine. Only 3 other sight records in Senegal, and *c*. 4 in Gambia (Jan–Mar) and 3 in Mali (Jan–Feb). Sierra Leone, up to 5 seen daily at 1500 m on Lomas Mts, and 3 on summit of Tingi Mts at 1700 m, outnumbering Mountain Rock-Thrushes there (Field 1973). Liberia/Guinea/Ivory Coast, regular since 1960s on mine workings on Mt Nimba. Ghana, regular but uncommon on Bongo hills near Bolgatanga; 4 stragglers at Mampong quarry, S Ghana, Nov–Mar. Nigeria, 5 records, Nov–Mar, mainly on inselbergs: Sokoto, Zaria, Igbetti, Shasha. Cameroon, regular but scarce, Waza and (border with Chad) Ndjaména.

Description. *M. s. solitarius* (Linnaeus): NW Africa to Caucasus; winters in NW, W and NE Africa. ADULT ♂: wings and tail blackish brown with blue tinge, rest of plumage greyish blue. Blue brightest on crown, nape and mantle. Lores and ear-coverts duskier. Lower breast, belly and flanks indistinctly banded dusky. Outer webs of all tail feathers, tertials, P9–P5, and all upperwing-coverts except alula, narrowly fringed pale blue. Underside of tail and remiges shiny black; underwing-coverts black with blue tinge; axillaries blue-grey. In first autumn freshly moulted ♂ has plumage as above, but with buffy feather-tips forming small, narrow crescents on head and back and quite large and copious ones on underparts. Bill black; eyes dark brown or brown-black; legs and feet brown-black. ADULT ♀: forehead, crown, nape, sides of neck and mantle dark grey-brown, most feathers narrowly tipped white and with narrow dark grey subterminal band; scapulars, back and rump similar but with bluish wash; uppertail-coverts brown with pale tips and blackish subterminal band. Tail like that of ♂ or slightly browner. Lores buffy; ear-coverts dark grey; chin and throat buff, scaled with dark bluish grey; breast grey with buff spots, like throat; lower breast-feathers white-tipped and with dark grey subterminal mark; belly, flanks and undertail-coverts dark grey, banded with whitish or pale buff and with wavy blackish lines. SIZE: wing, ♂ (n = 17) 123–133 (128), ♂ (n = 12) 120–128 (124); tail, ♂ (n = 13) 79–92 (85·2); ♀ (n = 13) 79–88 (85·4); bill to skull, ♂ (n = 17) 28·5–32·7 (30·4), ♀ (n = 13) 28·2–31·8 (29·7); tarsus, ♂ (n = 12) 28·8–31·2 (30·0), ♀ (n = 13) 27·8–31·1 (29·7). WEIGHT: 1 ♂ (Nigeria, Mar) 70·5; 1 ♀ (S Algeria, Jan) 69, 1 ♀ (SE Morocco, May) 58; unsexed (n = 26, Malta) 50·5–63·5 (57·0).

IMMATURE: warm brown; crown, mantle and underparts spotted with buff and scaled with dark brown crescents (each feather with buff centre and dark tip). Tail blackish. Wings blackish, lesser and median coverts with white tip and black subterminal bar, tertials with buff-white fringe.

NESTLING: down black; on head and upperparts only (Cramp 1988).

M. s. longirostris (Blyth): NE Iraq, Iran, Turkmenistan, Afghanistan. Winters in E Africa, west to about 20°E. ♀ and juvenile have underparts paler, greyer and less barred than in nominate ♀. Smaller. SIZE: wing, ♂ (n = 14) 118–127 (123), ♀ (n = 9) 114–119 (117); bill, ♂ (n = 13) 26·7–30·1 (28·3), ♀ (n = 9) 26·6–28·5 (27·7). WEIGHT: (Afghanistan, July–Aug) 2 ♂♂ 50, 53, 1 ♀ 51.

Field Characters. Length 20 cm. A solitary, shy bird of precipitous cliffs and smooth-topped inselbergs, difficult to observe, yet distinctive. Adult ♂ blue, with quite long black bill, black wings and tail. Adult ♀ and immature birds somewhat nondescript, dark, rather uniform, with slaty or bluish grey upperparts and dark-barred brown underparts; throat buffier and scaly. Characteristic stance: stands still for long periods with head up, tail and wing-tips nearly touching ground, and bill pointing up at angle of *c*. 20°. Buff-spotted brown juvenile could be confused with young Eurasian Blackbird *Turdus merula* but is scaly, with spotted mantle and barred wing-coverts (both streaked in Eurasian Blackbird).

Voice. Tape-recorded (53–62, 73, 93, B, C, F). Song of ♂ a deliberate, loud, melodious, fluty or piping whistling, recalling Eurasian Blackbird. ♀ occasionally sings. Song consists of several short phrases, each 1·3–2·2 s long and followed by pause of 0·5 s or longer. Phrase has 5–8 notes: mainly clear whistles, sometimes short fast churrs, e.g. 'tju-sri tjurr-titi'; sometimes song includes mimicry, of e.g. Willow Warbler *Phylloscopus trochilus* and Cirl Bunting *Emberiza cirlus*. Successive phrases differ slightly; different ♂♂ have individual songs. Sings Jan to mid-May, with resurgence in Aug–Nov (Malta). Sometimes sings on wintering grounds (e.g. sings Eritrea Dec–Feb). Both sexes use subsong, on breeding and wintering grounds: a pleasant, quiet warbling, somewhat like Common Starling *Sturnus vulgaris*. Characteristic call, a liquid, soft, penetrating whistle, 'uit-uit'. Contact call a deep 'tak-tak'. Alarm, 'tchuc-tchuc', like Eurasian Blackbird, also 'pee', given very high-pitched as a warning.

General Habits. Inhabits precipitous mountain crags and valleys, sea-cliffs and rocky coasts, quarries, ruins, isolated buildings like churches in farmed valleys, also buildings (preferably flat-roofed) in hamlets and suburbs. In Morocco breeds mainly at 2000–2700 m, descending to valleys, plains, and lowland towns, olive groves and gardens in winter; also breeds commonly on sea-cliffs. On passage, uses almost any terrain including oases and coastal desert wadis thinly fringed with bushes (Eritrea). In winter, mountain tops with vertical rock faces and some woody vegetation; inselbergs in rainforest and savanna zones; quarries; mine tailings (Mt Nimba, heavily deforested and mined for iron ore), rocky coasts (NW Africa, Senegal, Eritrea), and buildings.

In pairs when breeding, solitary in winter. ♂ and ♀ defend separate territories in winter, excluding conspecifics, wheatears *Oenanthe* and bulbuls *Pycnonotus*. Chased off by Mountain Rock-Thrush *M. saxatilis* (Sierra Leone). Very shy. Sings from eminence such as tree or TV aerial; perches otherwise on lower vantage point, often the very edge of top of precipice or 5 cm in from edge of flat roof so that from below person can see only bird's head. Stance upright, sleek, bill looking long and usually pointing markedly upward, wings slightly drooping and wing and tail-tips almost touching ground. When disturbed, drops quietly and disappears, or reappears on similar perch some distance away; or swoops fast down cliff-face with wings only half extended. Very good at keeping itself out of human sight, vanishing from view behind or under boulders, on broken ground, or structures on tops of buildings.

Forages by keeping watch from low vantage point and dropping from perch onto prey, also by hopping and running on ground and by making short aerial chases. Aerial flycatching, with sorties of 7–30 m, may be an important technique (Spain: de los Santos *et al.* 1987). Takes fruits from ground or direct from plant. Beats large insect against ground to immobilize it; decapitates snakes and centipedes. Roosts at night in rock crevice or under eaves of houses. Waterbathes; sunbathes with bill pointing up to sun, or prostrates body with wings and tail outspread. In heat of day sometimes seeks out shade under rocks and bushes.

Resident, vertical migrant and intercontinental migrant. Abundances of the 2 races wintering in eastern Africa are about the same, so most birds probably originate in Turkey to Iran. In absence of ringing returns, origin of W African wintering populations is unknown, but probably S Europe rather than N Africa. N Africa also receives winter migrants, from S Europe. Migrates at night, singly or in small loose flocks, sometimes with Mountain Rock-Thrushes. For timings, see *Range and Status*.

Food. Invertebrates (mainly insects), small vertebrates (up to 10 g), seeds and fruits. In Europe, grasshoppers, crickets, flies, ants, beetles, spiders, snails, worms, lizards, snakes, frogs; ivy, olive, fig, vine, hawthorn, *Myrtus*, *Daphne*, *Viburnum*, *Ephedra*. In Tunisia, locusts, beetles and lizards. Lizards and caterpillars commonly fed to young (Spain).

Breeding Habits. Solitary breeder, monogamous, territorial. Aggressive, chasing away buntings, chats and other thrushes. Territory, of at least 2·5 ha, may need to include perpendicular cliffs totalling 0·5 ha of cliff-face (Cramp 1988). In Sicily at least 15 pairs in *c*. 5 km². 2 pairs with nests sometimes only 200 m apart. Song-flight starts from rock or dead tree with upward glide, bird spreading tail and wings for lift; swoops down at angle, drops almost vertically, or flutters down like pipit *Anthus* (**A**), then glides up to perch and drops softly onto it. On alighting, also as prelude to copulation, makes asymmetric wing-flicking movements, wing-tips describing more or less circular paths. Other advertising and courtship behaviours described (Cramp 1988).

A

NEST: shallow cup of coarse dry grasses, moss and rootlets, loosely constructed and quite bulky; lined with fine, soft grass and rootlets with occasional feathers and plant down. Cup diam. 100–110 and depth 70. Built by ♀. Sited beneath overhanging rock, in crevice in cliff, wall of old building, quarry or cave; sometimes in tree-hole or horizontal drainage pipe. Most nests in Spain 2·3–4·8 m above level ground; one nest 120 m up precipice. Nest may be refurbished and used in successive years.

EGGS: 3–6, Europe av. (37 clutches) 4·54. Subelliptical, smooth and quite glossy. Very pale blue or turquoise, immaculate, or with fine rufous or brown speckles and mottling, mainly at broad end. SIZE: (n = 125, Europe) 25·4–30·4 × 18·3–21·2 (27·6 × 20·1). WEIGHT: (n = 17) 4·2–6·2 (5·4). Double-brooded.

LAYING DATES: NW Africa, Apr–May.

INCUBATION: mainly or entirely by ♀. Begins with last egg (captive bird). Period: 12–15 days.

DEVELOPMENT AND CARE OF YOUNG: young brooded for 3–4 days. Cared for by ♀ and ♂; at 1 nest ♂ visited young twice as much as ♀. Faeces removed by both parents, and often deposited at regular site, e.g. electricity pylon (Cramp 1988). Fledging period *c*. 18 days. After leaving nest young depend on parents for *c*. 2 weeks.

References
Beven, G. (1968).
Cramp, S. (1988).
Morel, G. J. *et al*. (1983).
Rouchouse, C. (1985).

Genus *Zoothera* Vigors

A group of primitive terrestrial thrushes, similar to *Turdus* but wing and tail shorter, wing more rounded, legs and feet strong and pale in colour. Differs principally in underwing pattern, having basal portion of inner webs of all secondaries and most primaries pale, usually white, contrasting with dark colour of rest of feathers; axillaries pale with dark terminal half, underwing-coverts dark with pale terminal half. In underwing of *Turdus*, flight feathers lack stripe, axillaries and underwing-coverts are more uniform (although often a different colour from rest of wing, e.g. rufous in *T. olivaceus*: see photographs of underwing of *Zoothera* and *Turdus* in Irwin 1984).

All African *Zoothera*, and many elsewhere, have 2 rows of white spots on upperwing-coverts; 3 African species have black patches across face and ear-coverts in the adult, and all have a similar double or single patch in the immature. 6 African species have orange underparts and are closely related; the seventh, *Z. guttata*, has spotted underparts and longer tail, and is the only migratory one; it may be most closely related to the similar-looking *Z. spiloptera* of Sri Lanka (Irwin 1984, 1992).

We agree with Irwin (1984) that the ground-thrushes are a more primitive group than *Turdus* and part of an older radiation, and should therefore be placed in a separate genus. This treatment is also recommended by Clancey (1991).

Mainly Old World tropics. 36 living and 1 extinct species listed for the world by Sibley and Monroe (1990), of which 2 African spp. (*Z. kibalensis* and *Z. tanganjicae*) are not recognized here. 7 species in Africa, 1 superspecies (*Z. gurneyi/Z. crossleyi*).

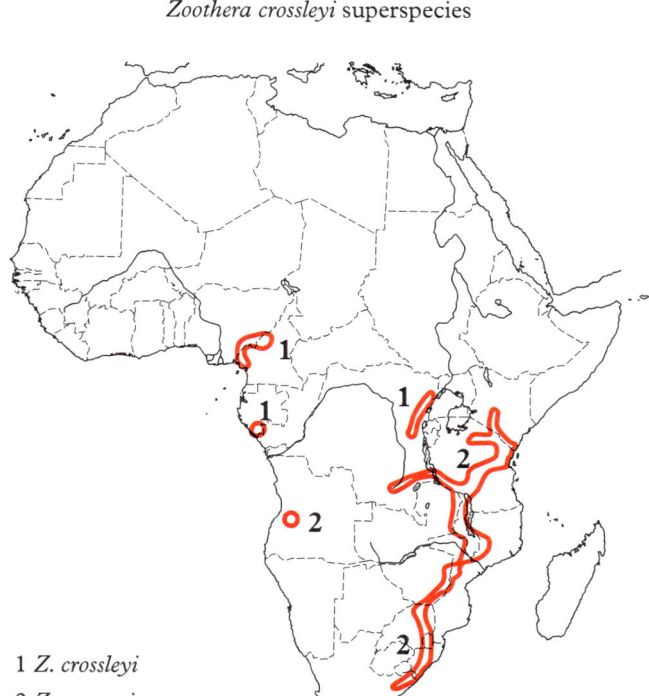

Zoothera crossleyi superspecies

1 *Z. crossleyi*
2 *Z. gurneyi*

Zoothera oberlaenderi. Oberländer's Ground-Thrush. Grive d'Oberlaender.

Plate 2
(Opp. p. 33)

Geocichla gurneyi oberlaenderi Sassi, 1914. Anz. k. Akad. Wiss. Wien, Math.-Naturwiss. 51, p. 310; between Beni and Mawambi, NE Congo.

Range and Status. Endemic resident. NE Zaïre in Ituri (Bondo Mabe, Semliki valley, between Beni and Mawambi) and Kivu (around Kamituga); W Uganda in Bwamba and Bwindi-Impenetrable Forests (Keith and Garrett 1994).

Listed as rare in ICBP/IUCN Red Data Book. May no longer exist in Bwamba Forest, which has been largely destroyed. Forest clearance is also taking place around Beni, and clearance is threatened around Kamituga, which has become a mining centre (Collar and Stuart 1985).

Description. ADULT ♂: forehead, crown, cheeks, ear-coverts and sides of neck rich rufous-brown, becoming brighter on nape and hindneck. Narrow loral line buff, incomplete eye-ring white, interrupted by black above and below eye; indistinct blackish smudges below eye and on ear-coverts. Mantle, back, scapulars and tertials warm brown, less rufous than head; rump and uppertail-coverts rufous. Tail dark brown, outer feathers sometimes with narrow whitish tips. Chin buff, throat, breast and flanks deep orange-rufous, faint moustachial streak sometimes present; belly white, undertail-coverts white or buff. Primaries and secondaries dark brown edged reddish brown, greater and median coverts blackish brown with large white tips; axillaries white with greyish tips, underwing-coverts brown edged white, broad white bar across underside of secondaries and all primaries except outermost. Bill black, base of lower mandible paler; bare skin behind eye dark grey; eyes dark brown; legs and feet pinkish white. Sexes alike. SIZE: (4 ♂♂, 8 ♀♀) wing, ♂ 98–103 (101), ♀ 95–102 (98·0); tail, ♂ 60–68 (63·8), ♀ 60–67 (62·9); bill, ♂ 19–22 (20·6), ♀ 18–21 (19·6); tarsus, ♂ 27–28 (27·5), ♀ 27–30

Zoothera oberlaenderi

(28·4). WEIGHT: (Uganda) ♂ (n = 3) 41–44 (42·0), ♀ (n = 3) 43–48 (av.?).

IMMATURE: crown darker than adult, feathers of crown and mantle with buff shaft-streaks, mantle sometimes with olivaceous wash; pronounced rufous nuchal collar; face rufous, black patch below eye and smaller one on ear-coverts, separated by rufous zone (in adults, black patch below eye is indistinct, black on ear-coverts vestigial or absent); blackish moustachial streak present; breast and upper belly pale orange densely spotted blackish brown, flanks pale orange, belly and undertail-coverts white; bill and eyes dark brown; legs and feet cream.

NESTLING: unknown.

Field Characters. A small ground-thrush confined to E Zaïre and W Uganda, with entire head rufous, and orange-rufous underparts. Lacks the pale face with contrasting black patches of Grey Ground-Thrush *Z. princei* and Black-eared Ground-Thrush *Z. cameronensis*, and further distinguished from them by bright colour of underparts, which are grey-brown in Grey Ground-Thrush and dull orange in Black-eared Ground-Thrush. The similar-looking race *tanganjicae* of Abyssinian Ground-Thrush *Z. piaggiae* normally lives at higher altitudes than Oberländer's Ground-Thrush but may overlap in a few places; it also has an all-rufous head and no face markings but is much larger and has a different song.

Voice. Tape-recorded (32, 56). Song, a series of phrases of 2–12 mellow, fluty notes, each phrase different and given once only, not repeated as in Orange Ground-Thrush *Z. gurneyi* and Abyssinian Ground-Thrush. Variation seems endless; 'for sheer inventiveness, this has to be one of the world's greatest singers' (Keith and Gunn 1971). Notes are rich and warbled, with tone and quality recalling Eurasian Blackbird *Turdus merula*.

General Habits. Inhabits primary lowland and transitional forest, at 700–1300 m in northern part of range, 1080–1420 m in south, reaching 1620 m in Bwindi-Impenetrable Forest. In Bwamba Forest lives in tall stands of ironwood *Cynometra alexandri*, where ground is fairly open, avoiding dense brush and tangled secondary growth at edge of stands. When disturbed flies up to branches in mid-stratum, 6–9 m above ground, from which level it also sings (Keith 1968). Solitary or in pairs. Local movements unlikely.

Food. Invertebrates: slugs; insects, including caterpillars and other larvae.

Breeding Habits. Largely unknown; no nests or eggs found. ♀ with enlarged ovary Uganda, July; fledglings under parental care Zaïre, Mar and Sept. These dates suggest breeding during rains and at end of dry season.

References
Prigogine, A. (1980, 1985b).

Plate 2
(Opp. p. 33)

Zoothera cameronensis (Sharpe). Black-eared Ground-Thrush. Grive du Cameroun.

Geocichla cameronensis Sharpe, 1905. Ibis, p. 472; Efulen, Cameroon.

Range and Status. Endemic resident. Cameroon in coastal forests south of *c.* 4°N, and inland to Sakbayeme and Nsimalen, also Korup Nat. Park; NE Gabon; NE Zaïre in Ituri; forests of W Uganda (Budongo, Bugoma, Kibale). Rare; *Z. c. kibalensis* known from only 2 specimens.

Description. *Z. c. cameronensis* (Sharpe): Cameroon, Gabon. ADULT ♂: forehead, crown and nape dark brown; lores, cheeks and ear-coverts pale orange-buff, with black patch below eye and another on ear-coverts, separated by pale orange-buff; eye-ring buff, broken by black above and below. Mantle, back, scapulars and tertials russet-brown, rump and uppertail-coverts rufous-chestnut; tail dark reddish brown. Chin pale

Zoothera cameronensis

orange-buff, grading to orange-chestnut on throat, black moustachial streak generally present; rest of underparts orange-chestnut, centre of belly whitish, undertail-coverts white tipped orange-chestnut. Primaries and secondaries dark brown edged pale rufous-brown; greater and median coverts blackish with large white terminal spots; axillaries white tipped dark grey, underwing-coverts dark grey tipped white, broad white bar across underside of secondaries and all but outermost primaries. Bill black, slightly paler at base of lower mandible; eyes dark brown; legs and feet whitish. Sexes alike. SIZE: (4 ♂♂, 4 ♀♀) wing, ♂ 94–102 (97·6), ♀ 98–100 (98·0); tail, ♂ 58–68 (62·0), ♀ 57–64 (60·4); bill, ♂ 18–20 (19·1), ♀ 17–19 (17·9); tarsus, ♂ 27–29 (27·9), ♀ 26–28 (27·4). WEIGHT: ♂♀ (n = 5, Gabon) 42–47 (44).

IMMATURE: like adult but crown, nape and mantle chestnut-brown with whitish buff shaft-streaks, contrasting with rufous rump and uppertail-coverts; chin buff becoming pale rufous on throat; breast, flanks and belly lightly spotted dark brown, undertail-coverts pale rufous; secondaries with dark rufous tips; white wing-spots present even on very young birds.

NESTLING: unknown.

Z. c. graueri (Sassi): Zaïre and Uganda. Crown, mantle and back greyish brown; broken eye-ring buffy white; underparts much paler, less chestnut, with more white on centre of belly and pronounced streaks on upper breast, undertail-coverts buffy white. Somewhat larger: (6 ♂♂, 3 ♀♀) wing, ♂ 103–106 (105), ♀ 96–102 (100); tail, ♂ 63–67 (65·3), ♀ 60–64 (62·2); bill, ♂ 18–20 (19·3), ♀ 18–19 (18·5); tarsus, ♂ 28–30 (29·4), ♀ 27–29 (28·5). WEIGHT: ♂ (n = 3, Uganda) 46–52 (48·0), 2♀♀ 45, 45.

IMMATURE: crown dark brown washed with grey, feathers with pale shaft-streaks, nuchal collar rufous, mantle and back with rufous shaft-streaks, rump and uppertail-coverts dark rufous; tail brownish rufous. Chin and throat white becoming buff on upper breast; hint of moustachial streak; breast pale rufous spotted dark brown, belly pale rufous, centre white washed with rufous, flanks and undertail-coverts rufous.

Z. c. kibalensis (Prigogine): Kibale Forest, Uganda. Known only from 2 adult ♂♂. Crown and upperparts more rufous, belly whiter; bill strongly compressed at tip, basal half of lower mandible grey, bare skin behind eye dark grey. Largest race: (2 ♂♂) wing 109, 112; tail 69, 73; bill 21, 21; tarsus 30·5, 33; weight 55, 57.

TAXONOMIC NOTE: the 2 birds from Kibale Forest, originally considered to be *Z. cameronensis* by Friedmann and Williams (1968), were later elevated to a full species, *Z. kibalensis*, by Prigogine (1978b). This classification was questioned by Payne (1981) and Vuilleumier and Mayr (1987), and rejected by Dowsett and Dowsett-Lemaire (1993b), who consider it to be *Z. cameronensis*, 'perhaps subspecifically distinct'. We agree with the latter. Morphological differences (colour, size) between '*kibalensis*' and *Z. c. graueri* are only at the subspecies level. Prigogine (1978) argued that '*kibalensis*' is a montane forest bird, whereas *Z. cameronensis* is confined to lowland forest. The distinction is not so neat. Kibale, at 1650–1700 m, is a transitional forest, with lowland as well as montane elements; on the other hand, *Z. cameronensis graueri* is found in W Uganda in Bugoma and Budongo forests, up to 1370 m; these are lowland forests but with a few transitional elements, differing in character from the pure lowland forests of the Semliki, where *Z. c. graueri* also occurs. In any case, a number of taxa occur in both lowland and montane forests without differentiating into species.

Field Characters. A rare bird of lowland forests in central Africa, shy and very difficult to see. Heavy bill and short tail give it a chunky appearance. Occurs with Grey Ground-Thrush *Z. princei*, which has similar black marks on pale face but is somewhat larger, with longer bill and tail; best distinguished from Grey Ground-Thrush by dull orange underparts. In E Zaïre may also occur with Oberländer's Ground-Thrush *Z. oberlaenderi*, which has deep orange-chestnut head, throat and breast, without pale face or contrasting black face marks.

Voice. Unknown.

General Habits. Inhabits lowland primary forest, in E up to 1700 m. In Gabon found near forest edge, once also on island in river, in low vine tangles. Solitary or in pairs. Scratches on ground for food.

Food. Insects, including beetles, ants and cockroaches; small snails.

Breeding Habits. Almost unknown; nest and eggs not described.

LAYING DATES: Gabon (♂♂ with enlarged testes Dec–Jan, immature June); probably nests during short dry season and long rains (Brosset and Erard 1986); Zaïre (♀ with enlarged ovary June, fledglings and immatures Oct); Uganda (♀ in breeding condition May, fledgling May).

References
Brosset, A. and Erard, C. (1977).
Friedmann, H. and Williams, J. G. (1968).
Prigogine, A. (1978b, 1985b).

Plate 2
(Opp. p. 33)

Zoothera princei (Sharpe). Grey Ground-Thrush. Grive olivâtre.

Chamaetylas princei Sharpe, 1873. Proc. Zool. Soc. London, p. 625; Denkera, Gold Coast.

Range and Status. Endemic resident. Liberia, south and east from Lofa County and Mt Nimba through Grand Gedeh County (Jouadi, Dugbe R.) into Ivory Coast (Taï Nat. Park, Abidjan and Yapo Forest north to Lamto and Comoé Nat. Park); S Ghana (only 2 records but probably more widespread: Grimes 1987); Nigeria (Umuagwu); S Cameroon (coast inland to Yokodouma, also Korup Nat. Park and Mt Kupé); NE Gabon; NE Zaïre in Ituri; W Uganda (Bugoma, Budongo and Bwamba: Dranzoa 1994). Generally rare but perhaps more common than supposed since new populations recently discovered in Liberia (Dickerman *et al.* 1994), Ivory Coast, Cameroon, Gabon and Uganda. Density in Gabon, 1 pair/12 ha.

Zoothera princei

Description. *Z. p. batesi* (Sharpe): Cameroon to Uganda. ADULT ♂: forehead, crown and nape dark brown tinged olive, somewhat paler on hindneck, sides of neck light greyish brown, lores, cheeks and ear-coverts buffy white, with black patch below eye and another behind ear-coverts; eye-ring white broken by black above and below eye. Mantle and upper back brown with olive wash, grading to brown on lower back, scapulars and tertials and rufous on rump and uppertail-coverts; tail feathers dark reddish brown, outermost narrowly tipped white, shafts maroon above, white below. Chin white with buff tinge, becoming brownish buff on throat; moustachial streak black; breast light greyish brown with indistinct dark streaks; flanks greyish brown, belly pale grey-brown, undertail-coverts dirty white. Primaries and secondaries dark brown edged pale brown, with broad white bar across centre of underside except on outer 2 primaries; greater and median coverts blackish brown with large white terminal spots, axillaries white tipped blackish, underwing-coverts dark greyish brown with white tips, bend of wing white. Bill blackish brown, base of lower mandible horn, gape yellowish white; bare skin behind eye grey or flesh; eyes brown to dark brown; legs and feet pinkish flesh tinged grey. ADULT ♀: like ♂ but gape greyish, base of lower mandible dark, bare skin behind eye less extensive. SIZE: (5 ♂♂, 2 ♀♀) wing, ♂ 102–110 (108), ♀ 102, 110; tail, ♂ 68–73 (71·2), ♀ 70, 79; bill, ♂ 19–22 (20·8), ♀ 20, 22; tarsus, ♂ 30–33 (31·6), ♀ 30, 32. WEIGHT: unsexed (n = 2, Gabon) 59, 61.

IMMATURE: head and mantle pale brown heavily streaked pale russet, rump and uppertail-coverts rufous, contrasting with mantle; underparts tinged rufous, breast rufous-brown lightly spotted dark brown, belly buffy white.

NESTLING: unknown.

Z. p. princei (Sharpe): Liberia to Ghana. Head and mantle with grey rather than olive wash. Larger: (4 ♂♂, 4 ♀♀) wing, ♂ 112–117 (114), ♀ 106–112 (109); tail, ♂ 76–78 (77·3), ♀ 71–77 (74·9); bill, ♂ 20–24 (22·6), ♀ 19–21 (20·0); tarsus, ♂ 34–37 (35·8), ♀ 32–35 (33·4). WEIGHT: (Liberia) ♂ (n = 7) 59–69 (63·1), ♀ (n = 7) 59–83 (67·7).

IMMATURE: head and mantle brownish russet heavily streaked pale rufous, 2 face patches present; chin white, throat buff, becoming progressively more rufous on rest of underparts; breast spotted dark brown, some dark spots also on belly.

Field Characters. A predominantly brown or grey-brown ground-thrush without the orange underparts characteristic of other species. Readily identified as a ground-thrush by black-and-white face pattern, white wing-bars and white stripe on underside of wing. In Cameroon, local race *saturatus* of African Thrush *Turdus pelios* is rather similar in colour but larger and greyer with yellow bill, plain face, no wing-bars, orange underwing. Forest Scrub-Robin *Cercotrichas leucosticta* has white spots on wing but is smaller and darker, with white supercilium, grey breast-band and conspicuous white spots in tail. Brown-chested Alethe *Alethe poliocephala* is another small dark thrush of the forest floor but lacks wing and face markings (except pale eyebrow), white throat contrasts with dark breast-band, shorter tail gives different shape.

Voice. Tape-recorded (ERA). Song unknown. Contact call, frequently given by members of pair, a long, high-pitched, rolling trill 'tsssrrr', drawn out at the end, reminiscent of calls of White-browed Forest-Flycatcher *Fraseria cinerascens* and Ashy Flycatcher *Muscicapa caerulescens* (Brosset and Erard 1977). Alarm call (not described) once heard when squirrel approached nest (Brosset and Erard 1986).

General Habits. Inhabits lowland primary forest, riverine forest and moist second-growth. Frequents dense shrubby undergrowth and vine tangles, sometimes where there is a carpet of herbaceous plants, e.g. Marantaceae. Up to 550 m in Liberia (Mt Nimba), 610 m in Cameroon (Efulen).

Lives in pairs on territory. Forages by hopping along on ground like thrush *Turdus*. Turns over dead leaves and searches among the leaf litter, digs up moss to look for worms; excavates earth accumulated on dead stumps and fallen trunks and branches; makes holes in rotten logs with blows from its bill. Seen to join mixed flocks of insectivores, sometimes around army ants. Shy and difficult to observe; at the least disturbance runs off or flies away low over the ground. Sedentary, at least in NE Gabon.

Food. Invertebrates: insects, including beetles and their larvae; spiders, worms, millipedes, snails. Also a few small frogs.

Breeding Habits. Monogamous; territorial.

NEST: bulky, open cup of twigs, dead leaves and plant stems, placed on thick base of vegetation and dead leaves picked from the ground, lined with rootlets and plant fibres; 1 nest (Gabon) ext. diam. 180, int. diam. 100 × 90, ext. depth 140, int. depth 45; situated 1·5–3 m above ground in fork in centre of crown of small isolated tree in understorey; highly visible, but easily confused with the bunches of dead leaves accumulated by squirrels in tree forks (which sometimes form the base for a nest) (Brosset and Erard 1986).

EGGS: 1–3, Gabon av. (9 clutches) 2·2. Broad oval, once truncated at large end (Serle 1957); smooth, glossy; turquoise blue to emerald green, spotted and blotched with red-brown over lilac undermarkings. SIZE: 22·8–25 × 18–19·6 (24·3 × 18·6).

LAYING DATES: Liberia, June, Aug; Nigeria, Aug; Gabon, Oct–Apr; Zaïre, June.

INCUBATION: by ♀ only.

DEVELOPMENT AND CARE OF YOUNG: ♀ broods young for a few days after hatching, while ♂ brings food. Young fed more frequently than other forest species: 1 ♂ brought food every 10–15 min for 6 h, another 41 times in 6 h (A. Devez *in* Brosset and Erard 1986); latter ♂ brought only long slender worms. Young develop rapidly and leave nest when *c.* 12 days old.

BREEDING SUCCESS/SURVIVAL: of 13 nests followed, 11 were destroyed by predators.

References
Brosset, A. and Erard, C. (1976, 1977, 1986).
Prigogine, A. (1985b).

Zoothera crossleyi (Sharpe). Crossley's Ground-Thrush. Grive de Crossley.

Plate 2
(Opp. p. 33)

Turdus crossleyi Sharpe, 1871. Proc. Zool. Soc. London, p. 607; Cameroon Mt.

Forms a superspecies with *Z. gurneyi*.

Range and Status. Endemic resident. SE Nigeria (Obudu and Mambilla plateaux, Gotel Mts); Cameroon highlands (Rumpi Hills, Mt Kupé, Mt Nlonako, Mt Cameroon); very small population in S Congo (Mayombe); NE Zaïre in Semliki valley and on W slopes of Itombwe highlands around Kamituga.

Rare. In Cameroon, widely distributed in highland forests, probably not immediately threatened (Stuart and Jensen 1986). Congo population confined to small patch of forest threatened by encroaching banana plantations (Dowsett 1989).

Description. *Z. c. crossleyi* (Sharpe): Nigeria, Cameroon, Congo. ADULT ♂: forehead and crown chestnut-brown, hindcollar, sides of neck and posterior ear-coverts rufous; small patch of pale orange-buff at base of upper mandible; blackish 'mask' beginning on lores and preorbital region, forming large black patch under eye and continuing onto anterior ear-coverts; eye-ring white, broken by black above and below eye. Mantle, upper back, scapulars and tertials reddish brown with olive wash, lower back, rump and uppertail-coverts more rufous; tail dark reddish brown, rarely with narrow whitish tips to outer feathers. Chin, throat, breast, upper belly and flanks brownish orange, some black spots on chin and upper throat, lower belly and undertail-coverts white. Primaries and secondaries dark brown edged buff, median coverts blackish brown with large white terminal spots; greater coverts similar but with white confined to outer web; axillaries white with greyish tips, underwing-coverts brown edged white, bend of wing white; broad white bar across centre of underside of secondaries and all primaries except outermost. Bill black; eyes brown; legs and feet flesh to pinkish white. Sexes alike. SIZE: (8 ♂♂, 6 ♀♀) wing, ♂ 105–114 (110), ♀ 107–116 (110); tail, ♂ 69–82 (74·6), ♀ 72–80 (75·5); bill, ♂ 23–25 (24·3), ♀ 22·5–25 (23·4); tarsus, ♂ 32–37 (34·7), ♀ 35–37·5 (36·0). Unsexed (n = 6, Mt Kupé): wing 112–118 (114·8) (C. G. R. Bowden, pers. comm.). WEIGHT: (Cameroon) 1 ♂ 70, 1 ♀ 80, unsexed (n = 12) 63–82 (71·5).

IMMATURE: crown and mantle with buff shaft streaks; rufous nuchal collar very pronounced; black patch below eye, behind which is a rufous patch. Chin and throat blackish becoming buff towards breast; breast spotted blackish brown, centre of belly buff with some dark spots, flanks rufous, lightly spotted. Upper mandible greyish, lower mandible buff becoming darker towards tip.

NESTLING: unknown.

Z. c. pilettei (Schouteden): E Zaïre. Crown and upperparts olive-brown, nuchal collar scarcely visible; tail brown without reddish tinge, outer feathers usually with whitish tips; underparts paler. Size similar but bill shorter: ♂ (n = 12) 20–23 (22·0), ♀ (n = 2) 20, 22.

IMMATURE: head and upperparts streaked rufous-buff, rufous nuchal collar pronounced. Lores black, supraloral line buff; blackish brown patch below eye, ear-coverts reddish buff, chin and throat buff, with hint of moustachial streak; breast and upper belly light rufous spotted dark brown, undertail-coverts white; tips to greater coverts white on outer feathers, becoming pale rufous inwardly.

JUVENILE (under parental care): similar to immature but crown and upperparts dark russet-brown lightly streaked rufous-buff, lores dark russet-brown, chin and throat whiter, belly dull rufous, undertail-coverts pale rufous.

TAXONOMIC NOTE: variously treated as a distinct species, e.g. by Chapin (1953a), Ripley (1964) Prigogine (1977, 1985b), and as a race of Orange Ground-Thrush *Z. gurneyi* (White 1962, Hall and Moreau 1970, Dowsett-Lemaire and Dowsett 1989b). The last authors (Dowsett and Dowsett-Lemaire 1993b) later reversed their opinion, noting that plumage differences, especially the distinctive face pattern of *crossleyi*, are greater than between other taxa of *Zoothera* accepted as good species, and believing that these outweigh the vocal similarity between *gurneyi* and *crossleyi*, given that songs of clearly divergent taxa of *Zoothera* are also indistinguishable.

Field Characters. A shy and rare ground-thrush with a restricted range; lives in forests at low and middle elevations. Large, with orange underparts and much black on face; crown rufous in W, olive-brown in E; white eye-ring narrow, broken by black above and below; stance very upright. Song resembles that of its close relative Orange Ground-Thrush *Z. gurneyi*; sings from high perch, hidden in foliage. Oberländer's Ground-Thrush *Z. oberlaenderi*, Grey Ground-Thrush *Z. princei* and Black-eared Ground-Thrush *Z. cameronensis* are all much smaller; Oberländer's has rufous head but overlaps only with brown-headed race of Crossley's, and has almost no black on face; Grey has no orange in plumage, Black-eared has dull orange underparts, both have vertical black streaks across pale face. Abyssinian Ground-Thrush *Z. piaggiae* is same size but lacks black on face, has broad, unbroken white eye-ring.

Voice. Tape-recorded (LEM). Song, whistled phrases of 7–10 rich, mellow notes; takes a phrase and repeats it but with minor variations, consecutive phrases not quite identical. Songs tape-recorded by F. Dowsett-Lemaire in Nigeria all start with 2 low notes, second higher than first, and work their way up the scale, rather in the manner of North American Hermit Thrush *Catharus guttatus* or Swainson's Thrush *C. ustulatus*; last few notes high and thin, intervening ones slurred, e.g. 'hor-her-heewo-chichiwo-tsitsi'. Quality similar to Orange Ground-Thrush but composition less inventive, more structured.

General Habits. Inhabits mature primary forest at middle elevations, 1000–2300 m in Cameroon, 500–600 m in Congo (Col de Bamba), 960–1850 m in Zaïre. On Mt Cameroon, bird mist-netted at 650 m on 24 Dec was retrapped at 200 m on 20 Jan, indicating altitudinal migration. Confined to wetter parts of forest; favours ravines. Lives low down, in understorey, on logs or on ground. Solitary or in pairs; uncommon and shy, difficult to see. Stance very upright. Sings both from ground and 15–25 m up in tree, hidden in foliage.

Food. Mainly insects; also some seeds.

Breeding Habits. Breeding behaviour, nest and eggs unknown.

BREEDING INDICATIONS: Nigeria (breeding condition Apr–June); Cameroon (breeding condition Apr, June, Aug, during rainy season), and singing also increases during rains; Zaïre, Oct (breeding condition Aug–Nov), indicating breeding at end of dry season and beginning of rains.

References
Prigogine, A. (1965, 1980b, 1985b).
Stuart, S. N. and Jensen, F. P. (1986).

Plate 2
(Opp. p. 33)

Zoothera gurneyi (Hartlaub). Orange Ground-Thrush. Grive de Gurney.

Turdus gurneyi Hartlaub, 1864. Ibis, p. 350; Pietermaritzburg, Natal.

Forms a superspecies with *Z. crossleyi*.

Range and Status. Endemic resident and partial local migrant. Kenya on SE slopes of Mt Kenya (Irangi, Chuka and Meru Forests), Kikuyu Escarpment, Chyulu and Taita Hills, Mrima Hill; widespread in highlands of E Tanzania from Ngorongoro, Oldeani, Mbulu, Mt Meru, N Pare Mts and E Usambaras to Njombe, Mt Rungwe and Isoka, also in coastal forest at Kiono (6° 10'S, 38° 35'E) and once in Pugu Hills (Baker and Baker 1992); also in SW on Ufipa Plateau. Angola (Mt Moco); SE Zaïre (Upemba Nat. Park); Nyika Plateau (NE Zambia/N Malaŵi); S Malaŵi below 1370 m (common); mountains of Mozambique

Zoothera gurneyi

in N, along Zimbabwe border from headwaters of Pungwe R. south to Chimanimani Mts and Lusitu R., and Mt Gorongoza; Zimbabwe in E highlands from Inyanga to Melsetter, but absent Chipinga Uplands; Escarpment region of Transvaal from Soutpansberg to Swaziland border; presumably Swaziland; interior of Natal from W Zululand south into E Cape (E Griqualand, headwaters of Buffalo R., Umtata, Stutterheim, Hogsback, wintering in coastal forests of Transkei). Scarce to locally common. Density in Udzungwa Mts, Tanzania, 4·6 pairs/10 ha (Moyer and Lovett in press), on Nyika Plateau 2–5 pairs/10 ha, in E Transvaal 4 pairs/4·5 ha.

Description. *Z. g. otomitra* (Reichenow): Tanzania, N Malaŵi, Zambia, Zaïre, Angola. ADULT ♂: forehead and crown dark brown-olive, somewhat greyer in northern populations; lores orange-rufous; eye-ring white, broken above and below by black; cheeks and ear-coverts black with intervening buff patch; indistinct moustachial streak sometimes present. Upperparts including scapulars and wing-coverts dark olive-brown with gingery or rufous wash, rump and uppertail-coverts greenish rufous; tail feathers dark brown variably tinged rufous, with narrow white tips to outer feathers when fresh. Chin to upper belly and flanks brownish orange, flanks washed olive, lower belly and undertail-coverts pure white. Primaries and secondaries blackish brown edged pale reddish buff, greater and median coverts brownish black with large terminal white spots; axillaries white tipped dark olive-brown, underwing-coverts dark olive-brown tipped white; broad white bar across centre of underside of secondaries and all but outer 2 primaries. Bill black, inside of mouth pink; bare skin behind eye grey; eyes brown to dark brown; legs and feet pale lilac-flesh. Sexes alike. SIZE: (29 ♂♂, 25 ♀♀) wing, ♂ 105–109 (111), ♀ 105–113 (109); tail, ♂ 74–85 (79·1), ♀ 73–85 (77·8), longer in Angolan birds (5 ♂♀ 82–87 (85·0)); bill, ♂ 21–24 (22·3), ♀ 20–25 (22·2); tarsus, ♂ 33–38 (35·4), ♀ 33–37 (35·1). WEIGHT: (Tanzania) ♂ (n = 20) 54–72 (60·4), ♀ (n = 20) 54–76 (62·0); (Zambia) ♂♀ (n = 32) av. 57·2; (Zaïre) 1 ♀ 52.

IMMATURE: like adult but top of head, upperparts and wings streaked buff, face and chin buff, trace of blackish patch below eye, faint moustachial streak; underparts spotted blackish, breast pale rufous, undertail-coverts buffy white. At later stage streaks confined to head and nape, lores, chin and throat reddish buff, breast and flanks brownish orange, belly and undertail-coverts buffy white, black streaks on face pronounced, white wing-spots present.

JUVENILE (under parental care; *Z. g. gurneyi*): head dark brown with some whitish shaft-streaks, rufous hindcollar, mantle with buff shaft-streaks; chin whitish buff, throat buff, breast buff heavily spotted black, centre of belly white, with a few spots on upper belly, flanks buff with blackish brown spots.

NESTLING: hatches naked; skin pink.

Z. g. gurneyi (Hartlaub): South Africa except Transvaal, Swaziland. Lacks greyish wash on head; orange underparts of ♂ generally darker than ♀. Wing same size but tail longer: (3 ♂♂, 4 ♀♀) wing, ♂ 113–114 (113), ♀ 108–112 (110); tail, ♂ (n = 3) 83–90 (87·0), ♀ 85–88 (86·8). WEIGHT: (southern Africa) ♂ (n = 54) 44·5–64·5 (54·4), ♀ (n = 56) 48·5–70·2 (54·9).

Z. g. disruptans (Clancey): central Malaŵi to Mozambique, Zimbabwe and Transvaal. Smaller and somewhat lighter than nominate; measurements smaller than *otomitra* but weight about the same. SIZE: (8 ♂♂, 10 ♀♀) wing, ♂ 100–109 (104), ♀ 101–110 (106); tail, ♂ 73–79 (76·2), ♀ 74–82 (77·4); bill, ♂ 21–23 (21·9), ♀ 20–24 (22·1); tarsus, ♂ 32–35 (33·8), ♀ 32–36 (33·2). WEIGHT: 1 ♂ (Zimbabwe) 55; ♂♀ (n = 22, Transvaal) 59–70 (63·7). Intergrades with both *gurneyi* and *otomitra*.

Z. g. raineyi (Mearns) (including '*chyulu*'): SE Kenya (Taita and Chyulu Hills). Mantle and underparts paler than *otomitra*. Minor colour and size differences between *raineyi* and '*chyulu*' do not justify separation of the latter. WEIGHT: 1 ♂ 56.

Z. g. chuka (van Someren): Mt Kenya and Kikuyu Escarpment. Larger than other races, with longer and stronger bill. SIZE: (17 ♂♂, 6 ♀♀) wing, ♂ 117–124 (120), ♀ 117–122 (120); tail, ♂ 83–88 (84·9), ♀ 82–87 (85·0); bill, ♂ 24–28 (25·3), ♀ 24–26 (24·9); tarsus, ♂ 34–38 (35·9), ♀ 34–36 (35·3). WEIGHT: 2 ♂♂ 79, 82; 2 ♀♀ 68, 69.

Field Characters. The orange ground-thrush of southern Africa, with orange breast and flanks, dark bill, broken white eye-ring, black face patches, 2 rows of white spots on black wing-coverts and broad white bar on underside of wing; bars on both wing surfaces conspicuous in flight. Spotted immature already has white wing-bars. In central Kenya (Mt Kenya, Kikuyu Escarpment) meets Abyssinian Ground-Thrush *Z. piaggiae*, which is smaller, with shorter bill, rufous forehead, brown cap and no black marks on face; songs of these 2 are similar, although that of Orange Ground-Thrush is rather slower and richer. Olive Thrush *Turdus olivaceus*, present in same forests, also has orange on underparts but in different pattern (grey-brown breast and orange belly instead of orange breast and white belly); it lacks markings on face and wings and has yellow bill, orange underwing, different song.

Voice. Tape-recorded (32, 56, 58, 86, 88, 91, B, F, LEM). Song in Kenya a series of phrases of 5–18 rich, fluty, whistles, each phrase typically repeated 3–5 times before singer switches to next one; phrases very variable, no 2 exactly alike; song may continue for several min. Whistle sometimes has human quality. After most phrases singer adds a few barely audible, high-pitched, whispered notes. Song slower, mellower and more relaxed than that of Abyssinian Ground-Thrush. Southern African birds *Z. g. gurneyi* have rather different song, less structured and repetitive, more rambling, with sweet liquid quality of Brown Scrub-Robin *Cercotrichas signata*, for which it can be mistaken. Call, a hissing trill, 'tsirrt'; also a brief call at dusk, 'ti-tue-tue-too-wee-to' and a chuckle when flushed from ground (van Someren 1939); bird carrying food gave buzzy 'bzie-e-e-ie'.

General Habits. Inhabits montane primary forest, including very small patches; in South Africa primarily in *Podocarpus* forests of mistbelt. Exceptionally extends into riverine forest and even scrub by streams (Ngorongoro Crater, Tanzania); generally avoids second growth and edge habitats. In mistbelt frequents depths of moister parts of forest, often on steep slopes near streams, but on Kikuyu Escarpment (Kenya) all birds netted were on sloping ground away from water, none near streams (P. B. Taylor, pers. comm.). Requires carpet of leaf litter. Nominate *gurneyi* prefers places with little undergrowth, but in Chyulu Hills (Kenya), *raineyi* inhabits damp undergrowth; it is most abundant in areas where *Connophyringia* (Apocynaceae) is plentiful, since it feeds on Dipterous larvae in the fallen fruits of this tree (van Someren 1939). Requires the moist soil of indigenous forest, typically near streams, and does not enter plantations of pine, acacia or eucalyptus where soil is drier (South Africa: Earlé and Oatley 1983). Altitudinal range 1830–2300 m in central Kenya, 1370–2140 m in Chyulu Hills, 1600–2500 m in most of Tanzania and NE Zambia, but 910–1220 m at Amani, 1050–1900 m on Udzungwas, up to 3500 m on Mt Meru, down to 920 m on Ngurus. Range in Malaŵi 1450–2350 m in N, 1200–2200 m south of 14°S; in Zimbabwe, 1590–2140 m; south of Zimbabwe, below 1220 m. Population in SE Zaïre confined to gallery forest in Upemba highlands, 1250–1750 m; on Mt Moco, Angola, in montane forest at 1800–2400 m. Sympatric with Abyssinian Ground-Thrush in Irangi Forest, Mt Kenya, 1970 m, and Kikuyu Escarpment, 2200–2300 m.

Solitary or in pairs. Lives at lowest levels of forest, on ground and in understorey, although sings from perch up to 15 m high. Forages on ground, hopping along with wings depressed and tail slightly raised, scratching among dry leaves and other forest debris, turning over rotting vegetation with quick lateral flick of bill; occasionally follows ant swarms; 1 in Malaŵi was seen making a few sallies for prey near swarm, but it was much more attracted to fallen fruits of fig trees (Willis 1985). Associates with mole-rats *Cryptomys*, taking advantage of their mound-building activities which flush litter-dwelling invertebrates from cover (Earlé and Oatley 1983, cf. Dean and MacDonald 1981). Shy and elusive but at the same time inquisitive; will approach to within 5 m of stationary observer. Partly crepuscular; sings mainly in early morning and at dusk (Maclean 1993). Scratches head over wing. Sometimes competes for food with larger Olive Thrush, to which it gives way in aggressive encounters (Earlé and Oatley 1983).

Mainly sedentary but partial altitudinal migrant. Noted in coastal Kenya at Mrima Hill (270 m); Tanzania at Amani Sigi (450 m) and Magombera (300 m) Forests, and once in Pugu Hills (22 May 1988); Malaŵi on Mt Mlanje at 1000 m, Zomba at 910 m and below, and 550 m on Yembe Hill; Mozambique at foot of Mt Gorongoza; Zimbabwe at Haroni–Lusitu confluence, 350 m. Birds also move locally in search of the moist soil necessary for their invertebrate prey.

Food. Chiefly invertebrates, especially earthworms, also millipedes, slugs and other molluscs; insects, including beetles, crickets, dipterans, caterpillars and other larvae, cocoons. Small amphibians. Some berries, seeds from fruit and other vegetable matter, but frugivory seems rare in South Africa (Earlé and Oatley 1983); diet in Kenya 20% fruit (van Someren 1939). Young given only earthworms (South Africa: Earlé and Oatley 1983).

Breeding Habits. Monogamous; territorial. Mean distance to nest on adjoining territory 70 m. Of 7 territorial adults, 4 held same territory for 2 seasons (Dowsett 1985). Unmated birds sometimes present in breeding population (Dowsett-Lemaire 1985).

NEST: bulky cup of twigs, roots, dead leaves, fern fronds and dead moss, with green moss on rim, lined with black rootlets, exceptionally with green leaves (*gurneyi*), or of green moss lined with fern rootlets, black fibres of dead moss and ferns (*disruptans*); ext. diam. 180–200, int. diam. 30–85, int. depth 45–50 (n = 5, southern Africa); placed 1–4 m above ground among vines near tree trunk, on stump or horizontal branch or in fork of sapling or bush in understorey (where sometimes conspicuous), often near footpath or over stream, in thick tangle of creepers and *Piper*, or in crown of tree fern, once in recessed ledge in earthen wall of gully sheltered by overhanging bank.

EGGS: 1–3, laid at 1 day intervals; mean (23 clutches) 2·0. Oval; glossy; turquoise or blue spotted brown to red-brown over lilac blotches, usually most heavily at large end; exceptionally unspotted. SIZE: (n = 18, southern Africa) 24–31·2 × 19·2–21·8 (28·4 × 20·5).

LAYING DATES: Kenya, Jan, May (breeding condition Apr); Tanzania, Aug, Dec; Malaŵi, Oct–Jan; Mozambique, Nov–Dec (breeding condition Sept); Zimbabwe, Nov–Jan, peaking in Dec; Transvaal, Oct–Nov; Natal, Oct–Jan. Breeds after beginning of rains.

INCUBATION: begins when 2nd egg laid. Eggs hatch on consecutive days, or on same day in morning and afternoon. Incubating and brooding birds sit very tight and can be touched. Period: 15 days.

DEVELOPMENT AND CARE OF YOUNG: weight of 2 chicks soon after hatching (day 2?) 7 and 8 g; day 3, 11 and 13; day 4, 15 and 18; day 6, 24 and 28; day 7, 30 and 33; day 8, 34 and 38; day 10, 42 and 43; day 11, 46 and 47; day 13, 47 and 49 (see Earlé and Oatley 1983,

Fig. 4). 32 days after bird left nest its breast feathers were orange like adult.

Young brooded until almost fully fledged. Frequent brooding of young may result in higher success rate by protecting them from cool and wet conditions characteristic of forests during breeding season; during period of exceptionally heavy rain, 2 chicks *c.* 8 days old were found dead in nest, having apparently died of chilling and exposure (Earlé and Oatley 1983). Nestling period 14–19 days; chicks leave nest of their own accord, remain under parental care for several months.

BREEDING SUCCESS/SURVIVAL: from 12 eggs in 6 clutches, Transvaal, 3 young were reared, i.e. success rate was 25%; annual adult survival rate in a 2-year period was 85·7% (Earlé and Oatley 1983). 1 ringed bird recaptured after 5 years 70 days (Zimbabwe, Bvumba: Harwin *et al.* 1994).

References
Clancey, P. A. (1955b).
Earlé, R. A. and Oatley, T. B. (1983).
Prigogine, A. (1985b).
van Someren, V. G. L. (1939).
Taylor, P. B. and Taylor, C. A. (1988).

Zoothera piaggiae (Bouvier). Abyssinian Ground-Thrush. Grive de Piaggia.

Plate 2 (Opp. p. 33)

Turdus piaggiae Bouvier, 1877. Bull. Soc. Zool. France 2, p. 456; Uganda (M'tesa's country) (= L. Tana, N Ethiopia, *vide* Chapin 1953, p. 579).

Range and Status. Endemic resident. Ethiopia (W Highlands north to *c.* 10°N, southern SE Highlands, S Ethiopia); S Sudan in Imatong and Dongotona Mts and Boma Hills (a ground-thrush assigned to this species was seen in Bangangai Forest in SW (Hillman and Hillman 1986), but the unusually low altitude (500 m) and the possibility of confusion with one of the lowland species make identification uncertain). Mountains of Albertine Rift in Zaïre/W Uganda (Rwenzoris, Mt Tshiaberimu northwest of L. Edward (T. M. Butynski, pers. comm.) Bwindi-Impenetrable Forest, Virunga volcanoes, Itombwe, Mt Kabobo); highlands of W Rwanda and W Burundi south to Bururi Forest. NE Uganda at Mt Moroto; Mt Elgon (Kenya/Uganda); isolated mountains in N Kenya (Kulal, Nyiru, Marsabit, Uaraguess), and more widespread in central highlands, west of Rift from Cheranganis to *c.* 1°S and Nguruman Forest, east of Rift on Aberdares, Kikuyu Escarpment, Mt Kenya and Nyambenis. N Tanzania in Loliondo and Magaidu Forests and on Mt Kilimanjaro. Scarce to locally common. Density in primary forest 0·6 birds/ha, in coniferous plantation 0·007 birds/ha (Kinale, Kenya: Carlson 1986).

Zoothera piaggiae

Description. *Z. p. piaggiae* (Bouvier): E Zaïre (Kahuzi and Itombwe Mts above 1900 m), Rwenzori Mts (Zaïre/Uganda), Ethiopia, SE Sudan (Boma Hills), E Uganda, N and W Kenya. ADULT ♂: forehead and lores dark rufous, complete eye-ring white; centre of crown to ear-coverts, sides of neck and rest of upperparts including scapulars and upperwing-coverts olive-brown, variably rufous on rump and uppertail-coverts; tail-feathers dark reddish brown, outer pair (and sometimes others) narrowly tipped white, mainly on inner web, when fresh. Chin and throat orange-rufous, no moustachial streak; breast orange, flanks and upper belly paler orange, sides of breast and sometimes flanks washed olive, lower belly and undertail-coverts white, thighs brown. Primaries and secondaries dark brown edged pale reddish brown; greater and median coverts blackish brown tipped with large white 'teardrop'; broad white stripe across centre of underside of secondaries and all but outer 2 primaries; axillaries white tipped brownish grey, underwing-coverts brownish grey tipped white. Bill black; bare skin behind eye greyish; eyes dark brown; legs and feet whitish flesh to pale pinkish or pale brown. Sexes alike. SIZE: (38 ♂♂, 28 ♀♀) wing, ♂ 96–109 (103), ♀ 95–108 (102); tail, ♂ 77–90 (83·2), ♀ 76–91 (82·5); bill, ♂ 20–24 (21·5), ♀ 20–23 (21·5); tarsus, ♂ 33–37 (34·3), ♀ 31–37 (34·4). WEIGHT: (Ethiopia, Uganda, Kenya) ♂ (n = 25) 45–58 (50·3), ♀ (n = 14) 43–65 (52·4).

IMMATURE: upperparts dark brown streaked buff; blackish band across cheeks; blackish mark interrupting white eye-ring above and below eye; throat buff, with black moustachial streak; upper breast buff heavily spotted black, grading to orange-rufous on lower breast; belly and flanks buff with blackish brown spots; undertail-coverts rufous to buff.

NESTLING: unknown.

Z. p. kilimensis (Neumann): Kenya east of Rift and Mt

Kilimanjaro. Top of head and upperparts somewhat colder and darker, more olive; throat and breast darker, more rufous. WEIGHT: (Kenya) ♂ (n = 11) 42–52 (47·4), ♀ (n = 9) 45–60 (52·6), ♂♀ (n = 40) 48–63 (54·7). Kikuyu Escarpment (Kenya), 3 laying ♀♀ 62·7–69·3 (66·0), 4 non-breeding ♀♀ 52·5–59·4 (55·9), 59 unsexed adults 48·3–63·4 (54·5), 6 unsexed immatures 51·1–57·4 (53·8); maximum weight variation in 24 h period: adult, from 52·4 at 08·40 h to 57·7 at 18·25 h (= gain of 10%); immature, from 57·4 at 17·25 h to 53·7 at 10·15 h (= overnight loss of 6·5%) (P. B. Taylor, pers. comm.).

IMMATURE: less streaked above, spots on underparts dark brown, less blackish, undertail-coverts buff to whitish buff.

Z. p. hadii (Macdonald): Sudan on Imatong and Dongotona Mts. Like nominate but upperparts darker, especially mantle, tail dark brown; primaries and secondaries darker (dark brownish grey); lesser coverts blackish. SIZE: slightly smaller: wing, ♂ (n = 6) 98–100 (98·8), ♀ (n = 4) 95–99 (97·5). WEIGHT: ♂ (n = 5) 43–50 (45·8), ♀ (n = 3) 45–51 (47·8).

Z. p. ruwenzorii Prigogine: Rwenzori Mts of Uganda/Zaïre. Rufous of forehead extends further onto top of head, almost to hindcrown; orange of underparts darker and more rufous than nominate, extending further down onto centre of belly; white eye-ring of adults usually has clear rose feathers where immatures have black mark. SIZE: slightly larger than nominate: (4 ♂♂, 4 ♀♀) wing, ♂ 101–108 (106), ♀ 100–109 (104); tail, ♂ 84–90 (87·5), ♀ 82–90 (85·6). WEIGHT: 1 ♂ (Uganda) 54.

Z. p. tanganjicae (Sassi): SW Uganda (Bwindi-Impenetrable Forest), Virunga volcanoes (Uganda/Zaïre), Rwanda, N Burundi; also Zaïre in Mt Tshiaberimu, Kahuzi and Itombwe Mts below 2040 m, and Mt Kabobo. Differs from all other races in having entire head from forehead to hindneck and sides of neck uniform rufous, somewhat darker on lores, cheeks and ear-coverts.

Z. p. rowei (Grant and Mackworth-Praed): N Tanzania in Loliondo and Magaidu forests. More olive above than other races, rump and uppertail-coverts paler; orange of underparts duller and paler.

TAXONOMIC NOTE: delineation of races in this species is complicated by considerable variation within populations in back colour, extent of rufous on crown, and extent, shade and intensity of orange on underparts. Further, in 3 places in the Albertine Rift (Rwenzoris, Mt Kahuzi, Itombwe) there are apparently altitudinal races; nominate *piaggiae* occupies higher levels and other races lower levels. Prigogine (1984b) described *Z. p. ruwenzorii* for the Rwenzori Mts, but there is a specimen in the American Museum of Natural History from Bugongo Ridge, 2745 m, on the Rwenzoris which matches nominate *Z. p. piaggiae* from Itombwe and differs from *Z. p. ruwenzorii* from lower levels. Similarly, on Mt Kahuzi and the Itombwe, *Z. p. piaggiae* occurs at higher levels than *Z. p. tanganjicae*; they overlap in a limited zone, but this could be due to altitudinal movement. Until overlapping breeding is proven, we prefer to follow Dowsett-Lemaire and Dowsett (1990) and Short *et al.* (1990), *contra* Prigogine (1977), in considering *tanganjicae* a race of *piaggiae* rather than a distinct species. Morphological differences are only at the megasubspecies level.

Field Characters. A montane ground-thrush of NE Africa, common in Kenya highlands, with olive-brown upperparts, reddish orange underparts, conspicuous white eye-ring and 2 white wing-bars; in flight shows broad white bar on underside of wing. Highly vocal in breeding season, when it sings from mid-levels in forest, but at other times very hard to see as it feeds silently on the forest floor. Race *tanganjicae* (Albertine Rift) has entire head rufous, without olive on back of head and neck. On Mt Kenya and Kikuyu Escarpment (race *kilimensis*) overlaps with Orange Ground-Thrush *Z. gurneyi chuka*, which is larger, with longer and stronger bill, dark grey top of head, black face markings and more white on underparts, orange reaching only to lower breast. Their songs are similar and should be distinguished with care, although that of Abyssinian Ground-Thrush is rather faster and less rich.

Voice. Tape-recorded (32, 53, C, GREG, LEM). Song, phrases of 3–12 whistled notes, repeated after pause of 2–4 s, given in series lasting several min. Song-phrases very variable; each is repeated 2–4 times, often with minor variations, before a new one is chosen. Quality sweet and tuneful but somewhat thinner and less mellow than Orange Ground-Thrush. Songs of nominate race on Mt Kulal (N Kenya) rather shorter and faster than those of *Z. p. tanganjicae* in Rwanda, typically 4–6 notes, usually ending with a tuneless 'chuck' or 'pseet'; songs in Rwanda longer and more rambling, less structured. Does not respond to playback of song of Orange Ground-Thrush (P. B. Taylor, pers. comm.). Alarm, a shrill, musical rattle; young birds give high-pitched 'seep'.

General Habits. Inhabits primary montane forest of various types, from tall trees with closed canopy, extensive mid-stratum and clear forest floor to small scattered trees with broken canopy, much undergrowth and even short-grass areas. Usually not present in secondary or cutover forest; exceptionally occurs in pine plantations (Kenya: Carlson 1986). At lower elevations occurs in lush undergrowth with tree ferns, balsams and brambles *Rubus*, sometimes along streams; at higher levels enters mixed forest and bamboo and even pure stands of bamboo, and extends onto moorlands in patches of trees, e.g. *Hagenia*, *Erica*, *Philippia*. In Ethiopia, found in olive/juniper/*Podocarpus* forest; on Mt Kilimanjaro, principal tree species include *Macaranga kilimanjarica*, *Agauria salicifolia*, *Myrica kilimanjarica*, *Xymalos monospora*, *Ocotea usambarensis*, also *Podocarpus milanjianus* above 2100 m, and in Afroalpine zone *Hagenia abyssinica* (Moreau 1936); on other mountains also *Dombeya* and *Juniperus*. Most trees, especially junipers, are draped with lichens *Usnea* spp. and moss; the many herbaceous plants in undergrowth include *Mimulopsis*, *Berberis* and Compositae. On Kilimanjaro occurs especially where forest floor is densely covered with mosses and lichens (Cordeiro 1994). Generally avoids open spaces, but sometimes emerges to feed on tracks and in low cover at forest edge (Taylor and Taylor 1988).

Normal altitudinal range *c.* 2000–3000 m (1700–3600 m on Mt Kilimanjaro, up to 3300 m on Mt Kenya). Occurs below 2000 m on some isolated mountains (Mt Kulal 1800–1860 m, Mt Marsabit 1310–1560 m), and in E Zaïre *Z. p. tanganjicae* descends to 1520 m. Normally at higher altitudes than Orange Ground-Thrush, but sympatric with it at Irangi Forest, Mt Kenya (1970 m) and Kikuyu Escarpment (2200–2300 m).

Solitary or in pairs. Lives in densest parts of forest, mainly on ground but also in lower branches of trees, in dense liana patches and among tops of tree ferns. Forages on ground among moss, lichens and leaf litter, turning over dead leaves for invertebrates, and on short-grassed forest tracks for worms; seasonally joins bulbuls in low fruiting bushes, especially *Pauridintha paucinervis* (P. B. Taylor, pers. comm.). Follows ant columns. Appears not to be in competition with Orange Ground-Thrush, which is larger, with longer and heavier bill, nor generally with Olive Thrush *Turdus olivaceus*, although dominated by the latter on Mt Kenya (P. B. Taylor, pers. comm.). Active from *c*. 45 min after dawn to dusk, with break from 11·30 to 13·15 h (Mt Kenya: P. B. Taylor, pers. comm.); maximum weight increase in 1 day (10·15–17·25 h) *c*. 7%. Normally shy and elusive, but can become tame, feeding in the open within a few m of observer, even entering a house (Mt Kenya: P. B. Taylor, pers. comm.). Sings from lower and middle strata.

Adults undergo complete moult after breeding. Primary moult duration not known but bird captured in worn plumage in Nov was in fresh plumage by Feb (P. B. Taylor, pers. comm.). Primary moult in Kenya June–Aug and Feb, suggesting at least 2 breeding seasons; body moult Feb–Mar, July–Aug and Nov–Dec (P. B. Taylor, pers. comm.); wide spread of moult over much of year suggests protracted breeding season.

No local movements certainly recorded, although circumstantial evidence suggests they may take place at least in E Zaïre (see *Taxonomic Note*).

Food. Invertebrates: millipedes, worms, snails and other small molluscs, including gastropods; insects, including eggs and larvae (grasshoppers, beetles, caterpillars); also berries and other small fruits and fruit pulp, seeds.

Breeding Habits. Monogamous; courtship unknown. Sings from mid-stratum during breeding season.

NEST: built mostly of moss, with lining of fern stems and rootlets; int. diam. 75–80, int. depth 35–40; placed *c*. 1·5 m above ground in horizontal fork in moss-covered branch of small tree. Structure and site resemble those of Orange Ground-Thrush in Chyulu Hills, Kenya (van Someren and van Someren 1949); nest on Rwenzori said to resemble that of Eurasian Blackbird *Turdus merula* (Ogilvie-Grant 1910). Nest in Uganda entirely of loose green moss, well hidden in dense upper foliage of *Brillintasia*, 5 m above ground, 1 m from bole of *Maesia* tree 3 m from forest track (T. Butynski, J. Dubois, I. Francis, N. Penford and K. Swinnerton, nest record card).

EGGS: 2. Oval; somewhat glossy; pale greenish blue to bluish green, marked with blotches and spots of chestnut to maroon, sometimes with purplish grey undermarkings; some have complete ring of spots round large end. SIZE: (n = 4) 27–28 × 19·5–20 (27·4 × 19·9).

LAYING DATES: Ethiopia, Feb–May, Sept; Sudan, May; Zaïre (♂♂ with enlarged testes, Apr, ♀ with egg in oviduct Oct); Rwanda (♂♂ with enlarged testes Jan, nest-building Apr, Oct, immatures under parental care May–June, Nov–Dec); Uganda, Mar, July (nest-building Nov); Kenya, Mar–June, probably Nov–Dec. Breeds mainly during rains, but 2 Rwenzori records from beginning of short dry season.

DEVELOPMENT AND CARE OF YOUNG: fledged young remain under parental care for *c*. 3 months.

BREEDING SUCCESS/SURVIVAL: data very meagre; 3 adults each tending single fledged young from clutch of 2 suggest survival rate of *c*. 50%.

References
Prigogine, A. (1977, 1985b).
van Someren, V. G. L. and van Someren, G. R. C. (1949).
Taylor, P. B. and Taylor, C. A. (1988).

Zoothera guttata (Vigors). Spotted Ground-Thrush. Grive tachetée.

Plate 2
(Opp. p. 33)

Turdus guttatus Vigors, 1831. Proc. Zool. Soc. London, p. 92; Natal.

Range and Status. Endemic resident and intra-African migrant. S Sudan (resident in Lotti Forest, Imatong Mts). Non-breeding visitor to coastal Kenya and NE Tanzania from Lamu to Pugu Hills; breeding grounds of this population unknown, possibly S Tanzania/Mozambique; birds in Rondo Forest Reserve, S Tanzania, Nov, had brood patches which increased over time, making this a possible breeding site (Holsten *et al.* 1991). SE Zaïre (1 specimen, Upemba Nat. Park). Small breeding population (30–40 pairs) in S Malawi east of Rift on Mts Chiradzulu, Soche, Thyolo and Mulanje, 1200–1500 m. In South Africa breeds in Natal (Ndumu Game Reserve, Ngoye Forest, Isipingo, Oribi Gorge) and Transkei (Nquleni Forest); E Cape populations appear to migrate to Natal in winter (Maclean 1993); occurs west to E London area, including 1 at Yellowsands, Kwelera R. mouth, Jan (Brieschke 1989).

Generally rare but locally fairly common at a very few localities, e.g. Gede, Kenya, where population density estimated at 2·9 birds/ha, giving total of 113 individuals in the 39 ha forest area; density in nearby Sokoke Forest probably much lower (Bennun 1985). Classed as rare in IUCN/ICBP Red Data Book (Collar and Stuart 1985); threatened by destruction of its forest habitat, and in South Africa also by low breeding success, availability of food and mortality during migration (Anon 1993).

Zoothera guttata

Description. *Z. g. guttata* (Vigors): South Africa, south of lower Tugela R. ADULT ♂: forehead and crown dark olivaceous brown, hindcollar a little paler; pre-orbital region, eye-ring, malar streak and patch on ear-coverts white, eye-ring interrupted by black above and below eye; dark spot in front of eye, black patch on face below eye and another behind ear-coverts. Upperparts, including scapulars, olive-brown, a little darker than hindcollar (more olivaceous in birds breeding in KwaZulu), some feathers in middle of mantle with whitish shafts; tail-feathers olive-brown, outermost pair (T6) with large triangular white patch, mainly on inner web, extending *c.* 25 mm in from tip, T5 with small (5 mm) triangular white tip, T4 with just a trace of white at tip. Chin and throat white, latter sometimes with a few small blackish spots; conspicuous black moustachial streak; rest of underparts white, tinged buff on breast and flanks, with large, drop-shaped blackish brown spots, those on flanks sometimes joining to form irregular bars; centre of belly and undertail-coverts unspotted, faintly tinged buff (birds breeding in KwaZulu have belly and flanks generally whiter, with larger blackish spots). Primaries and secondaries greyish brown edged pale buff, lesser coverts olive-brown with indistinct buff spots at tip, median coverts black broadly tipped with triangular white spots, greater coverts similar but with olive-brown outer webs, smaller white spots, alula with brown-buff outer web, grey-brown inner web, greater primary coverts blackish brown with ochre to buff median patch; axillaries white tipped olive-brown, under-wing-coverts blackish brown tipped white, bend of wing white, broad white stripe across centre of underside of secondaries and all but outer 2 primaries. Bill brownish black, base of lower mandible buffy white to pale grey-brown or yellowish; small patch of bare skin behind eye pink to pinkish grey or purplish blue; eyes brown to dark brown; legs and feet whitish flesh to pale pinkish. Sexes alike. SIZE: (22 ♂♂, 18 ♀♀) wing, ♂ 114–124 (119), ♀ 110–120 (117); tail, ♂ 80–97 (86·6), ♀ 82–90 (86·3); bill, ♂ 23–26 (24·4), ♀ 23–26 (23·3); tarsus, ♂ 30–33 (32·0), ♀ 31–34 (32·1). WEIGHT: (Pondoland) 1 ♂ 72, (Natal) 1 ♂ 78, (KwaZulu) 1 ♂ 72.

IMMATURE: crown and mantle streaked and spotted buff; russet nuchal collar; spotting on underparts finer, undertail-coverts washed buff; some rusty spots on wing-coverts.

NESTLING: unknown.

Z. g. fischeri (Hellmayr): Kenya, Tanzania. Upperparts paler, spots on underparts smaller and sparser. Upper mandible blackish brown becoming pale brown at tip, tomium and gape pale flesh, lower mandible flesh becoming pale greyish brown on distal third. Smaller, especially tail: (6 ♂♂, 2 ♀♀) wing, ♂ 114–117 (116), ♀ 112, 116, ♂♀ (n = 17) wing 108–119 (113); tail, ♂ 78–84 (80·3), ♀ 75, 75; bill, ♂ 21–24 (22·6), ♀ 23, 24; tarsus, ♂ 28–32 (30·4), ♀ 29, 30. WEIGHT: (Kenya) ♂ (n = 6) 45–60 (55·3), 1 ♀ 65, ♂♀ (n = 25) 52–65 (57·7), subadult unsexed (n = 4) 52–56 (53·2).

IMMATURE: pale buff streaks on crown and mantle, and more pronounced on wing-coverts; underparts washed buffy ochre, undertail-coverts washed buff.

Z. g. belcheri (Benson): Malaŵi; Zululand north of lower Tugela R. Belly and flanks more intense white than in nominate race, not tinged buff, spotting more intense black. SIZE: (1 ♂, 1 ♀) wing, ♂ 118, ♀ 119; tail, ♂ 84, ♀ 86; bill, ♀ 23; tarsus, ♂ 32, ♀ 30.

Z. g. maxis (Nikolaus): Sudan. Upperparts darker and browner than other races; primary and secondary coverts blackish brown, contrasting with mantle; band of yellowish white spots across centre of greater primary coverts, largest on outer 3 feathers and becoming smaller inwardly; undertail-coverts orange-buff. SIZE: (1 subadult ♀) wing 108, tail 88, bill 24, tarsus 31; weight 62.

Z. g. lippensi Prigogine and Louette: Zaïre. Differs from all other races by olive-grey upperparts; undertail-coverts orange-buff. SIZE: (1 ♀) wing, 112, tail 89, bill 24, tarsus 30.

Field Characters. A brown thrush with spotted underparts, superficially resembling Song Thrush *Turdus philomelos* but with the tell-tale white wing-spots and underwing bar of a *Zoothera*. Brown plumage blends perfectly with leaf litter, and this combined with habit of standing still for long periods makes it extremely hard to find. Same size as Groundscraper Thrush *Psophocichla litsitsirupa*, but latter has greyer upperparts, ochre underwing, different shape (longer bill, shorter tail) and more upright stance, and lacks white wing-spots; the 2 are not likely to occur together as Spotted Ground-Thrush is confined to forest, Groundscraper Thrush to open country outside forest. Might be confused with spotted immature of Orange Ground-Thrush *Zoothera gurneyi* or Olive Thrush *Turdus olivaceus*, but these have underparts partly or wholly orange with smudgy brown spots or bars, rather than pure white or buff with large black spots, and immature Olive Thrush lacks white wing-spots.

Voice. Tape-recorded (88, 91, F, HAAG, KEI). (a) *Z. g. guttata*: typical *Zoothera* song with tonal quality somewhere between Oberländer's Ground-Thrush *Z. oberlaenderi* and Orange Ground-Thrush. Repeated short, variable phrases of 3–6 sweet, melodic notes, with emphasis on the first note, 'teeyoo-tu-tyoodu', 'hoo-eeyer', 'swee-toot-toodle', 'preeu-pree-pree-swee'; sweet notes sometimes followed by an 'afterthought', a softer, higher-pitched bubbling trill or cadence. Song of Orange Ground-Thrush has same flavour but phrases much longer and more drawn-out. (b) *Z. g. fischeri*: song recorded by S. Keith in coastal Kenya (Gede) very different, quiet, dry, not very musical, with chuckling quality; usually a rolling, almost continuous series of notes not broken into phrases, but sometimes short phrases distinguishable, e.g. 'tcheew tu tu wee-u tuwi', or a rolling 'teerrlu toorrli teerrlu'. This may well have been a subsong rather than a true song, especially since birds at Gede are non-breeding migrants. Call, a thin, high-pitched 'pssssss...' lasting *c*. 0·5 s. Birds at Diani (Kenya) sometimes responded to tape-recording of nominate race but with different calls. Foraging bird gives scarcely audible 'tsee-tsee', like call of Brown Scrub-Robin *Cercotrichas signata*.

General Habits. Inhabits forest of various types: in Sudan, intermediate forest at 1220 m; in Zaïre, gallery forest near streams at 1750 m; in E Africa, low-altitude moist evergreen forest, especially coral rag forest with nearly complete canopy cover, deep shade, extensive moist, thick leaf litter and relatively sparse undergrowth. Likes to have nearby areas of low, thick undergrowth with dead wood and vine tangles for cover when threatened. In Sokoke Forest (Kenya) occurs in *Afzelia* forest and dry *Cynometra* thicket; in Malaŵi, evergreen forest at 1370–1530 m. In South Africa, coastal evergreen forest, up to 150 m, especially in narrow strips along streams and bases of nearby steep slopes, also coastal dune forest at Mtunzini; migrants occur further inland, in moist bushland and thickets, occasionally in shady gardens.

Solitary or in pairs; sometimes in small parties on migration. Site fidelity marked on wintering grounds at Gede, Kenya, where birds appeared to have home ranges (Bennun 1985, 1987). 3 of 9 birds retrapped in 1986 had been ringed 3 years earlier; 2 returned to the same home ranges they had previously occupied, the 3rd had been a subadult 3 years earlier and possibly had not established a clear home range. Home ranges partly overlap; size of 2 calculated as 1360 m^2 and 1347 m^2. No territorial interactions were observed.

Inconspicuous and mainly silent; frequents forest floor and lower branches of leafy trees; searches for food on rotting logs and scratches among leaves on ground in deep shade; makes a few rapid hops then plunges bill repeatedly into leaf litter. Shows no interest in swarming ants, even in its home range. Solitary feeder; does not join bird parties. Seen to displace Red-capped Robin-Chat *Cossypha natalensis* in foraging area (Bennun 1987). When disturbed remains motionless, sometimes for as long as 6 min, then runs away or flies off low over ground or up into tree. Agile in flight, twisting and turning rapidly through trees (Chiazzari 1952). Can be fairly tame: migrant in South Africa watched feeding in public park at distance of only 2 m; another in Kenya exploring leaf litter in forest close to hotel, unperturbed by people passing a few m away (Irvine and Irvine 1988). Sings at dawn and dusk from upper branches of tree.

Annual post-breeding moult ends Mar–Apr in nominate *guttata*. Wing-moult was complete in 5 *fischeri* by end July, and well advanced in 4 others, but timing of moult apparently varies. 1 bird was in early moult by May 5, suggesting moult begins soon after arrival on wintering grounds, but another already had fresh primaries by May 3, indicating moult was complete (Bennun 1985).

In South Africa, breeds mainly in Pondoland and Transkei, where present late Sept–Apr, and starts moving north in Mar to winter on Natal coast. However, movements not completely understood. Present mainly Apr–Aug in Durban area, but there are also 'summer' specimens from there (15 Jan, 26 Mar); 'winter' specimens from East London (E Cape) in June and Aug suggest there may not be a complete exodus from the breeding grounds (Quickelberge 1969). Small population in KwaZulu breeds in Ngoye Forest and moves locally in May and June. *Z. g. fischeri* is present in E Africa late Mar to late Nov, mainly May–Oct, once 15 Dec (Backhurst 1992). Bird in Upemba Nat. Park, Zaïre (*Z. g. lippensi*) may have been migrant (Prigogine and Louette 1984). Migrates at night, and nocturnal migrants collide with buildings, e.g. in Durban.

Food. Land molluscs; worms, large (up to 8 cm) and small millipedes *Prionopetalum*; insects and their larvae, including termites and ants; some seeds and fruit pulp.

Breeding Habits. Monogamous; territorial.

NEST: (nominate race) heavy bowl of mud, small twigs, leaves, roots, grasses and moss, lined with feathers, fine plant fibres and leaves of creepers, placed 2–3 m above ground in low forest tree, often on top of horizontal branch next to trunk; in Ngoye Forest favours *Garcinia gerrardii* (Maclean 1993); also among lianas, or in bush festooned with creepers; (*belcheri*) bulky oval cup of dark tendrils, on base of thick tendrils, roots and dead leaves, unlined but with some moss on rim, diam. 70 × 50, depth 38, placed 2 m above ground in small tree (Belcher 1930).

EGGS: 2–3, av. 2·5. Oval; greenish blue, heavily blotched with dark red-brown and greenish brown; like those of Olive Thrush but markings somewhat darker. SIZE: (n = 7, South Africa) 22·3–28 × 17·6–20·3 (25·5 × 19·2); (n = ?, Malaŵi) av. 30 × 19.

LAYING DATES: Malaŵi, Nov; South Africa, Oct–Feb.

BREEDING SUCCESS/SURVIVAL: in South Africa, survival of nestlings depends on availability of earthworms, which form 95% of their diet; years of low rainfall cause a decline in earthworm numbers; also, nestlings are preyed upon by snakes, especially the Boomslang, and by African Harrier-Hawk *Polyboroides typus* (Anon. 1993). From a study of 10 active nests at Dhlinza Forest, Zululand, in 1988, it was estimated that heavy predation resulted in a failure rate of 60–70% (H. Chittenden, *in* Clancey 1992–1993). Ringing recaptures at Gede, Kenya, imply minimum annual adult survival rate of 67%, which is among the higher values recorded for African passerines (Bennun 1987).

References
Bennun, L. A. (1985, 1987).
Chiazzari, W. L. (1952).
Clancey, P. A. (1985).
Clancey, P. A. (1992–1993).
Collar, N. J. and Stuart, S. N. (1985).

Genus *Psophocichla* Cabanis

An open savanna woodland thrush, probably linking the primitive *Zoothera* ground-thrushes with the advanced *Turdus* thrushes. Has the black patches on cheeks and ear-coverts found in many *Zoothera* spp. but lacks the characteristic *Zoothera* underwing pattern (white stripe across centre of dark underwing); instead underwing is a plain yellow-buff, like that of Song Thrush *Turdus philomelos*. Wing proportionately longer and more pointed than *Zoothera*. Similar to *Turdus*, but tail short and bill larger, longer and broader at base. Underparts white and heavily marked with drop-like spotting as in *Z. guttata* and *T. viscivorus*.

Has habit of flicking wings at intervals, displaying underwing markings. Tends to be more social than *Zoothera*, occurring in small groups for much of the year.

Variously placed in *Geokichla* (Chapin 1953a), *Zoothera* (Irwin 1984) and *Turdus* (Ripley 1964, Hall and Moreau 1970). We agree with Irwin (1984) that the Groundscraper Thrush is more closely related to *Zoothera* than to *Turdus*, and that the resemblance to *T. viscivorus* is superficial and due to convergence (see Milstein 1968). We believe, however, it is sufficiently distinct to warrant its own genus, and we follow Short *et al.* (1990) and Dowsett and Forbes-Watson (1993) in placing it in the monotypic genus *Psophocichla*. We cannot endorse the suggestion by Hall and Moreau (1970) that it forms a superspecies with *T. philomelos*.

Endemic; monotypic.

Plate 2
(Opp. p. 33)

Psophocichla litsitsirupa (Smith). Groundscraper Thrush. Merle litsitsirupa.

Merula litsitsirupa A. Smith, 1836. Rep. Exped. Centr. Africa, p. 45; 'country between the Orange and the Tropic' (= Zeerust, W Transvaal, South Africa). (See Cole (1984) for correct spelling of *litsitsirupa* rather than *litsipsirupa*.)

Range and Status. Endemic resident in 2 areas: (1) Eritrea (plateau over 2100 m, common) and Ethiopia (W and SE highlands, rarely below 1800 m, common); and (2) Central and southern Africa from Angola (mainly highlands from the provinces of Cuanza Sul and S Malanje south to Huila and Cunene and east to Lunda Sul, Moxico, Cuando Cubango and Zambian border, frequent to common), Zambia (widespread; common in lighter woodlands of SW and W, uncommon to frequent in heavier woodlands in N; absent Luangwa and Middle Zambezi valleys), SE Zaïre (Shaba and S Kivu Provinces north to Baraka, Dogodo, Kabinda and Kabalo and east to W shore of L. Tanganyika, frequent), and SW Tanzania (Ufipa and Udzungwa Mts to Southern Highlands and Livingstone Mts, uncommon) to Malaŵi (west of Rift at 1000–1600 m; Dedza and Ntcheu Districts, frequent; Viphya Plateau, uncommon), N and NE Namibia (Kaokoland, Ovambo and Damaraland south to Naukluft Mts, and east to Okavango R. valley and Caprivi Strip, uncommon to frequent), Botswana (throughout except SW dunelands, common), Zimbabwe (widespread, frequent to common, but absent Middle Zambezi valley and E highlands above 1600 m), SW Mozambique (upper Limpopo R. to Maputo, uncommon), Swaziland, and South Africa: N Cape (N Gordonia and N Kuruman districts); NW and E Orange Free State south to *c.* 30°S, uncommon; Transvaal (throughout but mainly in bushveld and lowveld: frequent to common) and Natal (west of coastal belt and south to *c.* 30° 45'S: frequent).

Description. *P. l. litsitsirupa* (Smith): central Namibia, Botswana (except NW), SW and S Zambia, Zimbabwe, South Africa and Swaziland. ADULT ♂: forehead to rump pale slate-grey, some feathers with dark brown shafts. Tail feathers greyish brown, outer tips of T6–T5 narrowly fringed buff. Lores, ear-coverts and cheeks white; orbital ring white with black spot above and below eye; bold black streak over front part of ear-coverts and adjacent cheek, and another over rear fringe of ear-coverts. Chin and centre of throat white, unspotted; sides of throat, neck and foreneck white spotted black. Breast, flanks and upper belly white, washed with buff and densely speckled with drop-shaped black spots; lower belly and undertail-coverts white with buff tinge, and sometimes with a few black spots. Primaries and secondaries dark brown, outer webs edged greyish, basal half of inner webs yellow-buff; tertials and upperwing-coverts warm olive-brown, except that feathers of median and secondary uppercoverts are fringed greyish buff; axillaries and underwing-coverts yellow-buff. Bill black, basal half of lower mandible yellow; eyes chestnut to brown; legs and feet yellowish flesh. Sexes alike. SIZE: (10 ♂♂, 10 ♀♀) wing, ♂ 121–139 (127), ♀ 120–130 (124); tail, ♂ 67–77 (70·5), ♀ 55–77 (63·1); bill, ♂ 21–26 (22·8), ♀ 21–26 (23·2); tarsus, ♂ 27–34 (31·2), ♀ 30–34 (32·0). WEIGHT: ♂ (n = 13, Namibia, Zambia, Botswana) 70–83 (75·8); ♀ (n = 7, Namibia, Botswana) 71–84·4 (76·1); unsexed (n = 19, Botswana, Zambia, South Africa) 67–82·2 (74·0).

IMMATURE: like adult but head, mantle and back sparingly speckled with off-white; black spots on throat, breast and belly smaller; lesser median and greater upperwing-coverts conspicuously tipped with buffy white.

NESTLING: not described.

P. l. pauciguttatus Clancey: N Namibia, S Angola and NW Botswana. Upperparts paler grey, underparts whiter, less buff and less spotted than nominate race.

P. l. stierlingi (Reichenow): central and E Angola, N and E Zambia, SE Zaïre, SW Mozambique, Malawi, and SW Tanzania. Upperparts and underparts more ashy grey than nominate race; bill shorter: 1 ♂ 21, 1 ♀ 22.

P. l. simensis (Rüppell): Eritrea and Ethiopia. Like *litsitsirupa* but upperparts more brown tinged, underparts buffier.

Field Characters. A medium-sized, rather thick-set thrush with brown or greyish upperparts, contrasting black face markings, boldly spotted underparts and short tail. In flight, shows conspicuous yellow-buff panel on underside of wing. Stands upright, flicking its wings open at intervals to display the underwing colour. Not unlike Spotted Ground-Thrush *Zoothera guttata* which has 2 bold white wing-bars, white stripe across dark underwing, different shape (shorter bill, longer tail) and forest habitat. In Eritrea overlaps with Song Thrush *Turdus philomelos* which has shorter bill, longer tail, more horizontal stance and lacks black face marks.

Voice. Tape-recorded (33, 74, 88, 91, B, F). Song, a variety of short phrases, typically 4–8 notes, combining whistles with drier grating or clicking notes, 'tseeoo-tsitsi-choy', 'choy-chichi-choy', 'wheee-pu-chichichi-chit', 'pachaw, chee-puchewy, pachaw', 'chacha-wheeyo-pu-chee-cho', 'chyoo-cheepa-chyoo-cheepa'; also rendered e.g. 'pray-do pray-do now-then now-then' (Liversidge 1991), or 'sweet sweet wip-wip,' (Maclean 1993). Rather unmusical compared to *Turdus* thrushes and certainly inferior to *Zoothera*; the whistles are not rich or melodious. Said to imitate other birds (Maclean 1993). Call, an onomatopoeic version of its Setswana name, rendered 'li-tsi-tsi-rufa' or 'tsi-tsi-tsi-rufa'; also several sharp clicks or a chuckling note when alarmed (Liversidge 1991).

General Habits. Lives in a variety of habitats including moorland, grassland, open woodland, secondary growth, cultivated lands and gardens. Rarely found in forest, where replaced by Kurrichane Thrush *Turdus libonyanus* or Olive Thrush *T. olivaceus*. In Eritrea and Ethiopia lives mainly in grassland and stands of *Hagenia* and bamboo, and moorlands with *Hypericum*, heath and grass at 1800–4100 m. In Tanzania occurs in miombo and acacia habitats at 900–1900 m. In Zambia prefers lighter woodlands, especially on Kalahari sands; also gardens and miombo. In Botswana inhabits any woodland or tree and bush savanna with short grass cover or bare ground; prefers open areas with scattered clumps of bushes and tall trees. In Zimbabwe occurs mainly in miombo woods, and in Mozambique dry acacia savanna and dry mopane woodland. In Angola likes miombo, especially between 500 and 1000 m; in Namibia acacia, thick riverine growth including tall mopane, grassland and dry savanna with shrubby mopane, *Terminalia* and euphorbias; and in South Africa found mainly in bushveld.

Solitary, in pairs, or in small parties; more social than other thrushes, 'associating in what appear to be family parties for the greater part of the year' (Irwin 1984). Often shy although less so near human habitation. Runs quickly, then stops suddenly in an upright position, sometimes flicking 1 wing at a time. Perches on prominent spot on ground or in top of tree; from there often flies far out into open field to feed (Ethiopia). Flight is strong, powerful and direct; dips in flight like Mistle Thrush *T. viscivorus* (Desfayes 1975). When disturbed, flies to perch and calls 'li-tsi-tsi-rufa'. Typically one of the last birds to call after dark and the first at dawn. Forages by hopping on ground like *Turdus* thrush, scraping among fallen leaves and decaying vegetation. Feeds in more open areas than does Kurrichane Thrush. Sometimes forages over burnt area 1–24 h after a fire; keeps near fire front, hawking or pouncing on prey (Dean 1987).

Post-nuptial moult begins late Oct to early Nov (Zambia: Traylor 1965a).

Mainly resident. SW Tanzanian records (Njombe, Mbeya) are all Aug–Nov, suggesting some movement. In Zambia some withdraw south during rains (Dec–Feb) and return Mar–July; others appear to depart in Aug and return in Oct (when commonest: Aspinwall 1981). In Zimbabwe (Zimbabwe Ruins) seldom seen June–Aug (Vernon 1977) and in South Africa (Kruger Nat. Park) present all year but commoner Dec–Feb (Kemp 1976). 2 ringed birds in Bulawayo, Zimbabwe recovered in same place after 9 months (Elliott and Jarvis 1973) and 2 years (Irwin 1981).

Food. Insects and their larvae including dipterans, termites, grasshoppers, beetles; also spiders, isopods, slugs, earthworms and skinks.

Breeding Habits. Monogamous; courtship behaviour not described.

NEST: bulky, open cup of twigs, roots, tendrils, seeds and sometimes wool, cotton, string and dung, lined with tiny rootlets and grass; outside sometimes with lichens; no mud. Int. diam. 70–90, int. depth 40–70. Placed 1·5–7 m above ground, sometimes over water, often in a fork but also at end of horizontal branch. Parents swoop at intruders when defending nest and perch nearby, uttering alarm calls. Often nests in association with Fork-tailed Drongo *Dicrurus adsimilis*, 4 m apart in same tree or in another tree up to 45 m away; 5 out of 7 nests in Transvaal and 4 out of 9 in Namibia were associated with drongo nests (probably to take advantage of drongo's pugnacity: Tarboton and Clinning 1977).

EGGS: 1–4, usually 2–3; Ethiopia, Angola, Botswana, Malaŵi and Zimbabwe (9 clutches) 1–3 (2·7); Zambia (14 clutches) 2–3 (2·5); southern Africa (55 clutches) 2–4 (2·7). Oval; somewhat glossy; pale creamy blue to greenish blue, marked with some blotches of lilac and numerous blotches and spots of red-brown especially towards the blunt end. SIZE: (n = 142) 24·7–31·9 × 18–22·4 (27·5 × 20·3); *P. l. simensis* (n = 8) 27–29 × 20–21 (28·5 × 20·6), *P. l. litsitsirupa* (n = 12) 26·6–29·4 × 19·7–21·3 (28·1 × 20·7), *P. l. pauciguttatus* (n = 12) 25·9–28·7 × 19·6–21·1 (27·2 × 20·5), *P. l. stierlingi* (n = 7) 26·2–31·3 × 17·8–21·7 (27·8 × 19·1).

LAYING DATES: Eritrea, June–Aug; Ethiopia, Mar–July (before and after first part of main rains); Angola, Sept–Oct; Zaïre, July–Sept; Tanzania, Oct–Dec (before and during early rains); Malaŵi, Sept–Oct; Zambia, Aug–Oct, Jan (Aug 5, Sept 8, Oct 6, Jan 2 clutches); Namibia, Oct–Mar; Botswana, Aug–Mar (Aug 4, Sept 4, Oct 9, Nov 7, Dec 4, Jan 2, Feb 2, Mar 1 clutches); Zimbabwe, Aug–Jan (Aug 16, Sept 117, Oct 146, Nov 78, Dec 22, Jan 2 clutches); South Africa: Transvaal, Aug–Jan (Aug 3, Sept 9, Oct 25, Nov 26, Dec 6, Jan 2 clutches), Natal, Sept–Nov.

DEVELOPMENT AND CARE OF YOUNG: during hatch of termites both parents alternately fed young every 20–30 s for up to 20 min (Anon. 1971). Parents sometimes bring skinks for young (Broadley 1974); they remove faecal sacs. 4 adults fed a brood of 3 young 4 times, Zimbabwe (Hezekia 1987).

References
Aspinwall, D. R. (1981).
Cole, D. (1984).
Tarboton, W. R. and Clinning, C. F. (1977).

Genus *Turdus* Linnaeus

Medium to large thrushes. Bill stout, fairly short, about length of or slightly shorter than head, and distinctly curved; tip of upper mandible compressed, notched and decurved. Nostril oval or round, partly covered by a membrane and partly overhung by frontal feathers. Rictal and nasal bristles well developed; loral and chin feathers with distinct bristle-like points. Legs stout and strong; feet strong, middle toe without claw, and at least two-thirds as long as tarsus; tarsus in front with sheath, subdivided only at base. Wing long and pointed, longer than tail; 10 primaries, outermost P10 minute; P8, P7 and P6 sinuated on outer web, P8 and P7 usually longest. Flight feathers of underwing uniform, axillaries and underwing-coverts often a different colour from rest of wing. Tail usually with 12 feathers, all about same length. Plumage diverse but sexes of African species often alike; young spotted at least on breast. Nests deep open cups; eggs spotted.

Run and/or hop. Feed mainly on invertebrates on ground, also on berries in trees. Most species use bill to flick aside leaf litter; some also scratch backwards with 1 or both feet.

Thought to be the most advanced genus in its family. *Zoothera* (= *Geokichla*) sometimes included, but these are distinct thrushes (q.v.), and we follow most authors including Irwin (1984) in separating *Zoothera* from *Turdus*.

Almost cosmopolitan; 59–62 spp. 11 in Africa: 5 endemic, 2 (*T. viscivorus*, *T. merula*) resident in N Africa which are also winter visitors from Europe, 1 (*T. torquatus*) a Palearctic visitor possibly resident in N Africa, and 3 Palearctic winter visitors. 1 superspecies in Africa (*T. olivaceus/T. pelios/T. libonyanus/T. tephronotus*), with an allospecies in SW Arabia (*T. menachensis*). While there is considerable geographical overlap between species, especially between *olivaceus* and the other 3, they remain segregated ecologically with *olivaceus* typically in forest, *pelios* in open deciduous woodland, *libonyanus* in miombo woodland and *tephronotus* in coastal and semi-arid bush and scrub (Keith and Urban 1992). No hybridization is known. Because of the extremely limited on-the-ground overlap between them they are essentially allopatric and we consider them allospecies of a single superspecies. Alternatively, because of its geographic overlap with these other species, *olivaceus* could form a separate superspecies with *menachensis*. *T. olivaceofuscus* is best placed in a separate species group with *T. bewsheri* of the Comoros; we consider them older than, and not particularly closely related to, present-day mainland birds (Keith and Urban 1992).

Turdus olivaceus superspecies

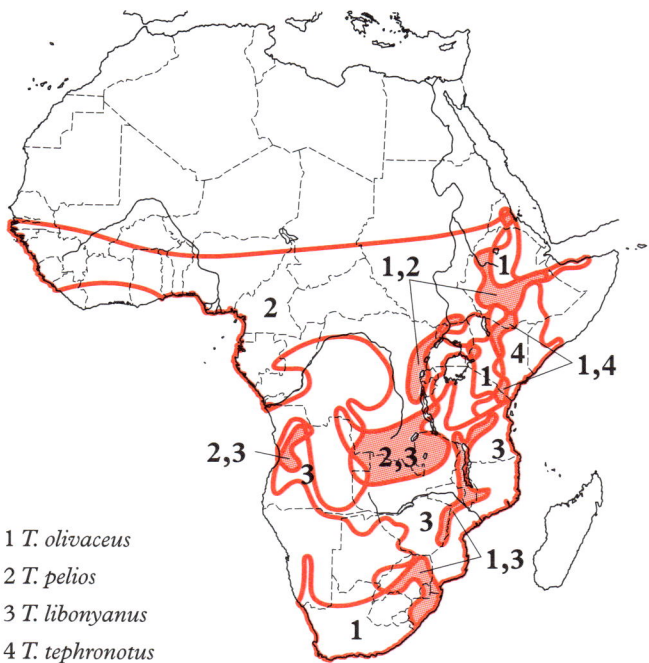

1 *T. olivaceus*
2 *T. pelios*
3 *T. libonyanus*
4 *T. tephronotus*

Turdus olivaceofuscus Hartlaub. Gulf of Guinea Thrush. Merle de São Tomé.

Plate 4
(Opp. p. 49)

Turdus olivaceofuscus Hartlaub, 1852. Abh. Geb. Naturw. Hamburg, 2, p. 49; São Tomé.

Range and Status. Endemic resident, São Tomé and Principe islands. São Tomé subspecies is common in well-vegetated areas at all elevations up to 1500–1600 m, with pairs only a few hundred m apart. However, described also as occurring at low densities, so population must be fairly small (Atkinson *et al.* 1991). Principe subspecies, described in 1901, was always very uncommon, restricted to SW coast and precipitous slopes of SW mountains. 4 specimens were taken by J. Correira in 1925; it has not been seen since despite extensive searching (de Naurois 1984c) and may be extinct.

Description. *T. o. olivaceofuscus* Hartlaub: São Tomé. ADULT ♂: forehead, crown, nape, hindneck, mantle, back, rump, upperwing-coverts and tail brown, washed with dark olive-green; tail somewhat darker than neck. Primaries and secondaries blackish brown, chin dirty white finely mottled with brownish dots, throat whitish with 5–6 long blackish brown dashed lines; breast and belly pale fulvous-brown with dark brown bars up to 10 mm long on flanks and sides of belly; undertail-coverts whitish, with brown markings. Underwing-coverts rusty. Bill horn; eyes light brown or yellow-brown; legs dark brown; claws blackish brown. Sexes alike. SIZE: (14 ♂♂, 7 ♀♀) wing, ♂ 123–136 (130), ♀ 122–128 (126); tail, ♂ 88–97 (92), ♀ 84–94 (89); bill, ♂ 25·5–29·0 (25·8), ♀ 21–28 (25·8); tarsus, ♂ 39–43 (40·8), ♀ 38–41 (39·0). WEIGHT: ♂♀ (n = 6) 77–92 (87).

IMMATURE: like adult but duller, and bars on underparts less distinct.

NESTLING: not described.

T. o. xanthorhynchus Salvadori: Principe. Like nominate race but underparts whiter, bars on sides of belly broader and darker, underwing-coverts pale ochre, bill bright yellow, legs pale brown. Smaller: 1 unsexed, wing 120, bill 23, tarsus 38·5.

Plate 1

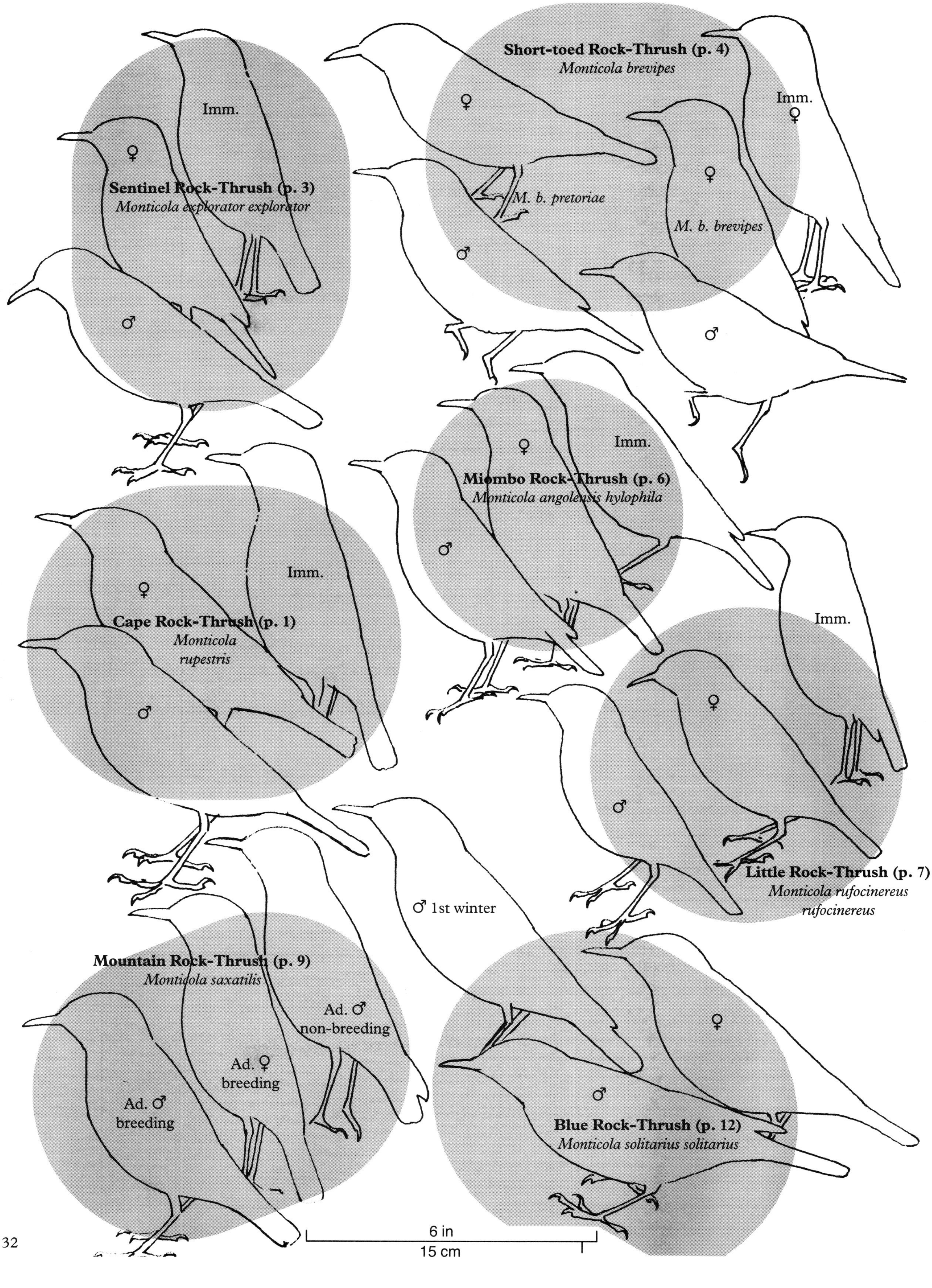
Sentinel Rock-Thrush (p. 3)
Monticola explorator explorator
Imm.
♀
♂
Short-toed Rock-Thrush (p. 4)
Monticola brevipes
♀
M. b. pretoriae
♂
Imm.
♀
♀
M. b. brevipes
♂
Miombo Rock-Thrush (p. 6)
Monticola angolensis hylophila
♀
Imm.
♂
Cape Rock-Thrush (p. 1)
Monticola rupestris
♀
Imm.
♂
Imm.
♀
♂
Little Rock-Thrush (p. 7)
Monticola rufocinereus rufocinereus
♂ 1st winter
Mountain Rock-Thrush (p. 9)
Monticola saxatilis
Ad. ♂ non-breeding
Ad. ♀ breeding
Ad. ♂ breeding
♀
♂
Blue Rock-Thrush (p. 12)
Monticola solitarius solitarius
6 in
15 cm

Plate 2

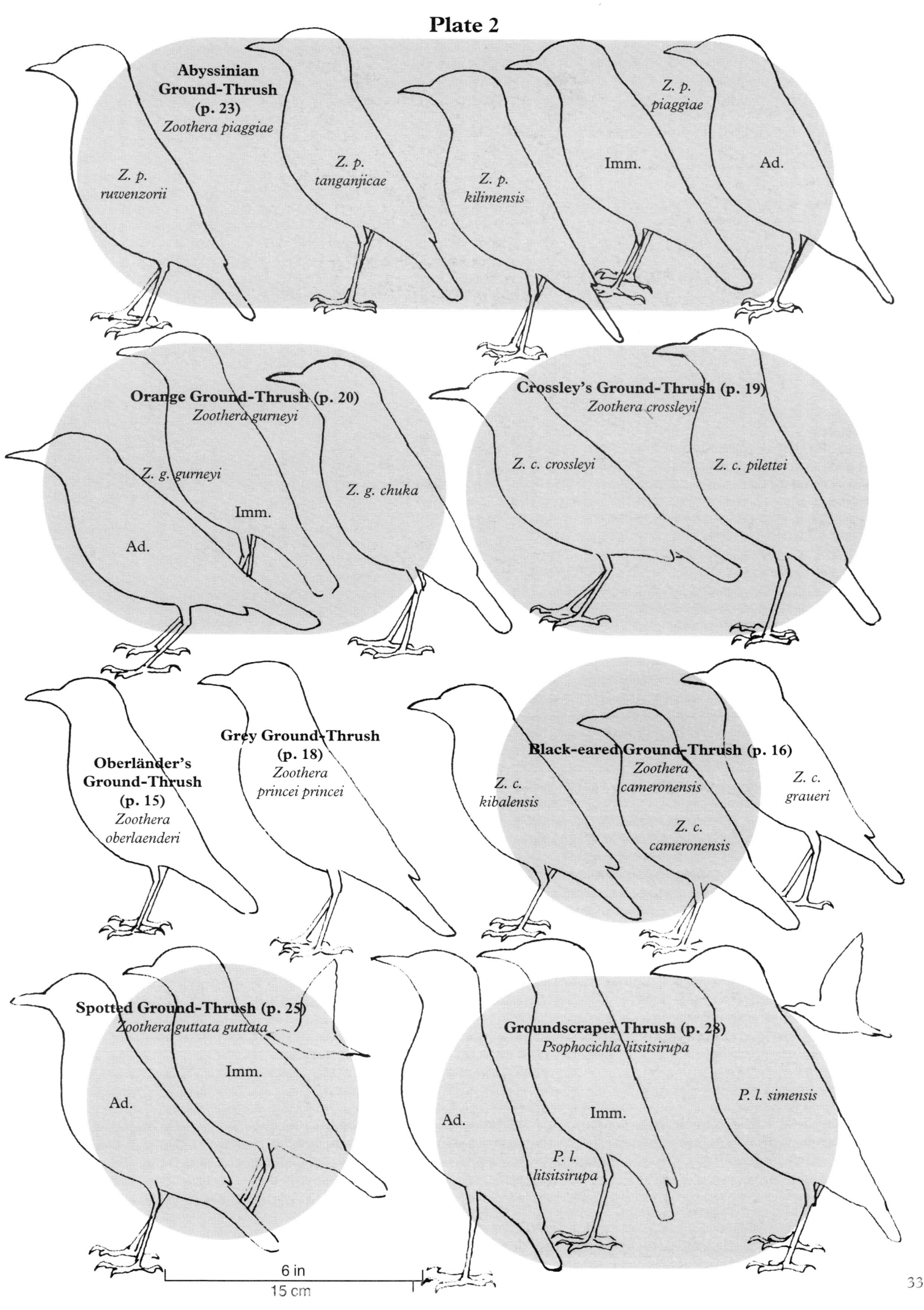

Field Characters. The only thrush in São Tomé and Principe; all brown; a bit like ♀ Eurasian Blackbird *T. merula* but with barring on underparts; bill brown (São Tomé) or bright yellow (Principe).

Voice. Tape-recorded (B, ALEX, JOPJ, TYE). Song, 3–5 mellow, slurred whistles followed by several soft, high-pitched hisses or trills, 'tyoo-tooeeyoo-teeyoo-tsss-tsss', 'twee-tooeeyoo-teewoh-tooee-too-tsss-trrr-trrr-tsss'; most songs include the characteristic up-and-down slurred 'tooeeyoo'. Delivery is unhurried and notes are measured, with a clear space between each, not run together. Phrases may be given singly, separated by a brief pause, or more enthusiastically strung together in a continuum. One version, with alternating low and high notes, 'twoo-twee-twoo-twee-tsss-tsss', is reminiscent of some phrases of Olive Thrush *T. olivaceus* from Kenya (see Keith and Gunn 1971). In any case, this song is of a very different type from that of African Thrush *T. pelios* which inhabits the opposite mainland and Bioko (see Keith and Urban 1992). Alarm, a low 'chup' or 'chupchup', like muffled Eurasian Blackbird.

General Habits. Inhabits dense lowland primary forest, cocoa plantations with *Erythrina* shade-trees and (in north of São Tomé) savanna woodland; rarely away from cover of tall trees. From sea level to *c*. 1600 m. Shy and elusive, but sometimes permits close approach. Habits typical of its genus: forages on ground, hopping and running among leaf litter and where earth contains decaying wood. Perches mainly at *c*. 2 m above ground, but up to 5 m in tall trees. When disturbed flies off rapidly, close to ground, disappearing into dense undergrowth; 'makes a peculiar noise with its wings, by shaking or vibrating them' (Snow 1950). May be partly crepuscular (Jones and Tye 1988).

Food. Invertebrates and fruits. Former include earthworms, Hemiptera, Lepidoptera (caterpillars), Coleoptera (Nitidulidae, Curculionidae) and snails.

Breeding Habits. Monogamous, territorial, solitary nester.
NEST: bulky cup, mainly of mixture of dried or rotten vegetable matter and mud, coated outside with dead leaves, moss and interlocked twigs, and lined with skeletonized leaves and dry grass stems. Ext. diam. 130–170, int. diam. 90–110, int. depth 60–80. Placed 0·5–4 m high in dense vegetation, on horizontal branch or in fork of cocoa tree or bamboo; occasionally placed up to 12 m high in tree, several m out from trunk on a bough.
EGGS: 2. Pointed ovals; slightly glossy; blue-green, sparsely flecked and dotted with dark reddish or violaceous brown. SIZE: (n = ?, São Tomé) 27·0–33·1 × 20·0–23·1 (31·5 × 21·8).
LAYING DATES: São Tomé, Aug-Dec, mainly Oct–Dec.
BREEDING SUCCESS/SURVIVAL: very many clutches and broods destroyed by brown rats, which are abundant on São Tomé.

References
Atkinson, P. *et al.* (1991).
de Naurois, R. (1984c).

Plate 3
(Opp. p. 48)

Turdus olivaceus **Linnaeus. Olive Thrush. Merle olivâtre.**

Turdus olivaceus Linnaeus, 1766. Syst. Nat. (12th ed.), 1, p. 292; Cape of Good Hope.

Forms a superspecies with *T. pelios*, *T. libonyanus* and *T. tephronotus* (and *T. menachensis* of SW Arabia).

Range and Status. Endemic resident, E and southern Africa, typically in highland forests. Eritrea (plateau and E escarpment down to 1050 m, abundant at lower and uncommon at higher elevations), Ethiopia (W and SE highlands), S Sudan (Imatong Mts, above 1600 m, common), N Somalia (mountains, Haraf to Wagger Range in W, Daalo Forest to Warsangeli Escarpment in E, locally common), Uganda (Mts Lonyili, Morongole, Moroto and Elgon in N and E, Rwenzoris to Bwindi-Impenetrable Forest in W and SW, frequent to common), Kenya (highlands west of *c*. 38° 30′E and north to Mts Elgon, Loima and Kulal; Ndoto Mts and Marsabit, common; single record Moyale), Tanzania (Bukoba and Minziro Forest in NW, highlands from Loliondo to Mt Kilimanjaro in N, and Usambara, Nguru, Uluguru and Udzungwa Mts to Southern Highlands, common), Zaïre (highlands from S Virunga Nat. Park and E Ituri forests south through Mitumba Mts and E Kivu Province to N Shaba Province at Bigogo and Mt Kabobo, 1520–4510 m, common), Rwanda (mainly highlands in W from Volcanoes Nat. Park and Gishwati Forest to Nyungwe Forest, common), Burundi (Teza-Nyungwe range to Bururi Forest, above 1750 m, common to abundant), NE Zambia (Mafinga Mts and Nyika Plateau, frequent) and Malaŵi (in N, Viphya Mts and Nyika Plateau, common; in S, Dedza, Zomba, Thyolo, Mulanje districts, uncommon to frequent; absent from central regions). Zimbabwe (Inyanga Highlands, Chimanimani Mts and Chipinga Uplands along Mozambique border above 900 m, frequent to common; also Honde valley, 750 m, and the Lusitu–Haroni confluence, 350 m, uncommon; once Theydon), Mozambique (in N, Namuli Mts, frequent to common; in S, central Chimanimani Mts, headwaters of Pungwe R. and Mt Gorongoza, uncommon; and extreme south Delagoa (= Maputo), vagrant), Namibia

Turdus olivaceus

(lower Orange R. and Fish R. in Namaqualand, also Naukluft Mts, rare to uncommon), Botswana (Gaborone and Lobatse areas, common; once northwest of Molepolole), South Africa (Cape Province north to Orange R., common; throughout Orange Free State highlands, common, but uncommon in drier W parts; Transvaal, common at high elevations along escarpment, rare at lower elevations in bushveld and lowveld; Natal, common above 760 m, absent NE coastal region); W Swaziland; and W Lesotho (Maseru area, common; also Senqu and Quthing valleys). Not northwest of L. Victoria (cf. Hall and Moreau 1970).

Description. *T. o. olivaceus* Linnaeus: SW Cape Province especially sea-facing sides of southern mountain ranges east to East London district. ADULT ♂: forehead, crown, lores, cheeks and ear-coverts medium brown; rest of upperparts from crown to uppertail-coverts olivaceous slate-brown; feathers of upperparts indistinctly edged with darker brown. Tail rufous-brown. Chin and sides of throat dusky white, with medium-sized brown streaks; centre of lower throat greyish without streaks; sides of neck and lower side of throat washed with some orange-buff. Upper breast greyish brown or tawny; lower breast to belly tawny-orange; flanks olivaceous slate-brown. Undertail-coverts dusky white with tawny-orange bases and brown marks at tips. Primaries and secondaries rufous-brown; upperwing-coverts olivaceous slate-brown; axillaries and underwing-coverts rich reddish orange. Bill orange or yellow, upper mandible often dusky, lower mandible yellow, inside of mouth orange; eyes brown; legs and feet yellow-ochre. Sexes alike. SIZE: (6 ♂♂, 7 ♀♀) wing, ♂ 108–138 (128), ♀ 119–128 (123); tail, ♂ 75–91 (83), ♀ 75–92 (84); bill, ♂ 19–24 (21·6), ♀ 20–24 (21·4); tarsus, ♂ 30–35 (30·0), ♀ 29–33 (30·5). WEIGHT: ♂ (n = 19, South Africa) 60–82 (72); ♀ (n = 4, South Africa) 66–80 (71); unsexed (n = 22, SW Cape) 60–85 (74·5), (n = 179, South Africa) 60–98 (78·1).

IMMATURE: like adult but breast, flanks and belly spotted dark brown.

NESTLING: skin orange to flesh-brown; head and back with some short tufts of greyish yellow down, tipped rusty.

T. o. pondoensis (Reichenow): E Cape and Natal from East London along coast and lower midlands to Ngoye Forest, Mtunzini and L. St Lucia (perhaps to Delagoa, Mozambique). Upperparts, tail and wings more olive tinged than in nominate race; upper breast more orange, less greyish, and lower breast and belly brighter.

T. o. smithi (Bonaparte): S Namibia, N Cape south to Little Karoo, Orange Free State, SW Transvaal highlands, W Lesotho and SE Botswana. Greyer, paler and duller than nominate race, without olive tinge; dark streaks on throat finer and lighter brown, orange of breast and belly paler.

T. o. transvaalensis (Roberts): N and E Transvaal, W Swaziland. Upperparts darker and more olivaceous than *culminans*; underparts like *culminans* but orange less reddish or brownish, more ochreous.

T. o. culminans Clancey: Natal: Drakensberg east to Nkandhla, Qudeni and Ngome forests. Like *pondoensis*, but without olive wash on back, wings and tail; underparts darker, breast duskier, and orange on lower belly browner, less bright.

T. o. swynnertoni (Bannerman): montane areas of E Zimbabwe and adjacent Mozambique. Like nominate race but upperparts richer and darker brown; cheeks, ear-coverts and lores darker brown, throat buffier, streaking heavier, breast and belly paler, upper mandible darker. Smaller: wing ♂♀ (n = 8) 111–118 (114). WEIGHT: (Zimbabwe) ♂ (n = 74) 59–77 (66), ♀ (n = 78) 57–79 (66).

T. o. milanjensis (Shelley): S Malaŵi and Mozambique, mountains from Dedza south and east to Namuli. Like nominate race but upperparts slightly darker, throat buffier with bold black streaking; belly paler and duller; upper mandible darker.

T. o. nyikae (Reichenow): Tanzania (Ngurus and Ulugurus to Mt. Rungwe), N Malaŵi (Viphya Mts, Nyika Plateau and Mafinga Mts); NE Zambia. Darker below than *milanjensis*, very little white on throat, breast and underparts brown, rusty wash on flanks only, centre of belly whitish. WEIGHT: unsexed (n = 3, Tanzania) 61–70·5 (64·9) (J. O. Svendsen, pers. comm.).

T. o. bambusicola Neumann: highlands of Burundi, Rwanda, SW Uganda, NW Tanzania and E Zaïre (Kivu) north and west of L. Tanganyika. Like nominate race but throat and upper breast more extensively ashy brown, lower belly whiter. Upperparts more olive than *abyssinicus*. WEIGHT: (Bwindi-Impenetrable Forest, Uganda) ♂ (n = 9) 62–70 (65·7), ♀ (n = 4) 52–69 (62·9).

T. o. baraka (Sharpe): E Zaïre (mountains of S Virunga Nat. Park and Rwenzoris); also W Uganda in Rwenzoris. Like nominate race but lower breast and belly more tawny, throat paler brown. Lower breast and flanks darker, richer chestnut than *abyssinicus*.

T. o. abyssinicus (Gmelin) (including '*fuscatus*' and '*mwaki*'): highlands of Ethiopia, SE Sudan, N Uganda and Kenya south to Chyulu Hills, and N Tanzania at Loliondo. Like nominate race but upperparts dark brownish grey, not olivaceous slate-brown; forehead, crown, lores and ear-coverts richer brown. Chin and throat more olive-grey, less dusky white; streaks finer and paler. Upper breast brown with more extensive greyish olive; lower belly white. WEIGHT: (Kenya) ♂ (n = 23) 50–76 (60·7), ♀ (n = 23) 49–87 (62·1); unsexed (n = 30, Addis Ababa, Ethiopia) 55–81 (66·6).

T. o. deckeni (Cabanis): N Tanzania from Longido and Ketumbeine to Monduli and Mt Kilimanjaro. Like nominate race but upperparts richer, darker brown; throat greyish brown with streaks finer and paler. Upper breast more grey, less brown; lower breast and belly duller, darker orange, almost brown with orange tinge.

T. o. roehli (Reichenow): NE Tanzania including Pare and Usambara Mts. Like *deckeni* but head darker and greyer.

Lower breast and belly white, like *helleri*; orange restricted to flanks. WEIGHT: ♂ (n = 4) 66·5–75·5 (71·5), 1 ♀ 75·4.

T. o. helleri (Mearns): SE Kenya in Taita Hills and Mt Kasigau; reported vagrant on Mt Kilimanjaro probably an error (Cordeiro 1994). Distinct; differs from nominate race in having forehead to nape, cheeks and ear-coverts black, throat dark grey with fine black streaks; upper breast dark grey, sides of lower breast and flanks rich tawny-orange, middle of lower breast and belly white.

T. o. oldeani Sclater and Moreau: N Tanzania at Embagai, Elanairobi, Ngorongoro, Oldeani, Mt Meru, Mbulu, Mt Lolkissale, Mt Hanang and Ufiome Forest. Upperparts darker and greyer than nominate race, underparts almost completely grey with a faint wash of orange on belly, flanks greyish brown, underwing-coverts tawny.

T. o. ludoviciae (Phillips): mountains of N Somalia. General colour slate-grey, paler on belly; undertail-coverts dark grey tipped pale grey; lores, cheeks and ear-coverts black; feathers of chin and throat black edged with pale grey; breast streaked grey and black; no orange in plumage except on underwing-coverts; bill light yellow.

TAXONOMIC NOTE: there has been much confusion about the taxonomy of the various 'olive' thrushes, including separation of *olivaceus* and *abyssinicus* as distinct species. We follow here the revision of the group by Keith and Urban (1992), and consider all forms from Eritrea and Somalia to South Africa as a single species.

T. o. ludoviciae, 'Somali Blackbird', often considered a separate species (e.g. Ripley 1964), is the darkest race, with black head and breast and grey underparts without orange. The amount of orange on the breast, flanks and belly varies greatly within *olivaceus*. Lack of orange on the underparts probably cannot be used to characterize the species; *T. o. oldeani*, for example, also has almost no orange on its underparts. On the other hand, *T. o. ludoviciae* has the orange underwing-coverts characteristic of all *olivaceus* races, and does not cock its tail on alighting (Archer and Godman 1961), which link it with *olivaceus*, and not with the Eurasian Blackbird *T. merula*, as suggested by Hall and Moreau (1970).

T. o. helleri, 'Taita Thrush', with its distinctive black head, has also been considered a separate species (e.g. Ripley 1964). However, *T. o. roehli*, which lies geographically between *helleri* and *nyikae*, is also intermediate between them in plumage, having a dark head and white breast and belly, suggesting *helleri* is another melanistic race of *T. olivaceus*. It is a very silent bird; its voice could prove to be distinctive. Recent recordings attributed to it are referable to either Orange Ground-Thrush *Zoothera gurneyi* or Rüppell's Robin-Chat *Cossypha semirufa*.

T. olivaceus and *T. pelios* were frequently confused until it was found they are readily separable by voice; *pelios*, for example, was considered to be just a race of *olivaceus* by Ripley (1964). All birds from W and W-central Africa are now known to be *pelios*, including the races *nigrilorum* and *poensis*, which had been ascribed to *olivaceus* by Hall and Moreau (1970); intergrades between *nigrilorum* and *T. p. saturatus* are known (Eisentraut 1970) and their songs are similar (Stuart and Jensen 1986).

Field Characters. A medium to fairly large thrush, typically with dark brown upperparts, uniformly streaked throat, reddish orange underparts with a variable amount of brown on breast, and orange bill and legs. Very variable, with especially distinctive races in Somalia (*ludoviciae*, very dark overall with black head and breast and grey underparts without orange), and Taita Hills, SE Kenya (*helleri*, with black head and breast and white belly). Best distinguished from other members of the superspecies by voice and habitat. Not found together with African Bare-eyed Thrush *T. tephronotus*, which would be told by broad yellow eye-ring, but does meet very similar African Thrush *T. pelios*. Where *T. o. abyssinicus* and *T. p. centralis* meet in the same habitat in central Kenya (Laikipia Plateau), African Thrush can be told by its longer, leaner, long-legged look, and its overall duller colour, whereas Olive Thrush appears chunkier, shorter-legged and brighter (L. L. Short, pers. comm. and photo). Bills and facial patterns differ. African Thrush has longer, yellow-orange bill and grey-brown lores concolorous with crown and forehead, dull orange eye-ring not contrasting with surrounding feathers, and light orange-brown iris; Olive Thrush has reddish orange bill and bright orange eye-ring contrasting with black pre-orbital region continuing as ring outside eye-ring, and dark red-brown eye (looks black). The net effect is that the eye-ring is prominent in Olive Thrush and the iris is prominent in African Thrush (L. L. Short, pers. comm.). Flanks and underwing are deeper orange in Olive Thrush (not easy to see in the field); throat streaking is slightly more pronounced in African Thrush. Where it meets Kurrichane Thrush *T. libonyanus* in southern Africa, Olive Thrush has yellow bill, plain face (no black lores or orange eye-ring), inconspicuously striped throat and various shades of orange on flanks and belly; Kurrichane has orange bill and patterned face, with white eyebrow, orange eye-ring, dark lores and pronounced black malar stripe beside unstreaked white throat; flanks orange but belly white.

Voice. Tape-recorded (10, 14, 32, 75, 86, 88, 91, B, F). Song in E Africa, phrases of 5–8 notes, typically with alternating high and low notes, usually followed by some short tuneless notes or a hissing trill, often barely audible: 'tyoo-tee-tyoo-tee . . . titi'; 'tee-tyoo-tee-tyoo-tee . . . tssss'; 'tee-too-tee-tay-tyoo-tee . . . tiddly'. Phrases last 2·5–3·5 s, but sometimes are run together into a continous song of 10 s or more, and lose the up–down quality of single phrases. Delivery is unhurried, measured, notes well separated; tone is clear and pleasant but not particularly tuneful, lacking the full rich quality of ground-thrushes *Zoothera*. Song in southern Africa is more hurried and lacks the deliberate quality of E African birds, and phrases are shorter, often only 2 s or less. Notes are run together and have a more rounded, warbling quality, and there may be no 'afterthought' at the end: 'teewo-teewo-tyui-psss-tyui'; 'weeyo-weeyo-widdlyo-pss-trrt'; 'waydee-waydee-wippatur-chew'. Some variations retain the up–down quality of E African birds, e.g. 'trootee-trootee-trootee-trootee, treetrrroo' (Newman 1989). Call, given on ground or in flight, a sharp 'chink', 'wheet', 'chook' or thin 'tsey'. Sometimes mimics songs and calls of other birds (Vernon 1973); once a ♂ made a 'perfect' imitation of Cape White-eye *Zosterops pallidus* and Orange-breasted Sunbird *Nectarinia violacea* (Steyn 1985).

General Habits. Inhabits forest of all types including primary, secondary, gallery and riverine forests and their edges and clearings; mainly in highlands, but reaches lowlands in S. In parts of its range also occupies woodland, bushveld, and man-made habitats such as exotic plantations, lawns and gardens. In Eritrea inhabits upland scrub, *Combretum* and wet woodlands receiving >760 mm annual rainfall; rarely in eucalyptus woods; in Somalia prefers *Juniperus procera* forest and neighbouring open areas; in Kenya 7 times more common in primary than in coniferous forest (Carlson 1986). Where it meets African Thrush on Laikipia Plateau, prefers moist, dense forest, woods and thickets while African Thrush inhabits drier areas, but both occupy degraded forest and woods (L. L. Short, pers. comm.). Altitudinal range in Ethiopia 900–3200 m (mainly over 1500 m); in Tanzania *c*. 950–2300 m (but mainly 1100–1830 m); in Malaŵi 1700–2450 m north of 14°S, 1600–2200 m south of 14°S (Dowsett-Lemaire 1989), usually 1 pair/ha of forest, typically in drier areas than Orange Ground-Thrush. In Burundi also found on moors with ferns and heather. In Namibia prefers riverine bush. In South Africa inhabits evergreen forests in E, while elsewhere, especially where Kurrichane Thrush does not occur, riverine bush, exotic plantations, gardens, parks and orchards. In SW Cape race *olivaceus* prefers well-wooded areas, dense riparian vegetation and gardens while race *smithi* occurs in more arid areas along rivers and streams. In E Cape commoner in W than in E, varying from *c*. 0·96 pairs/ha of woodland at Kirstenbosch to 0·02 pairs/ha at Alexandria (Cody 1983). In Transvaal, Natal and W Swaziland keeps above 800 m while Kurrichane Thrush usually occurs below this level. In Lesotho below 2000 m along rivers with bush.

Solitary, in pairs and groups of 3–4; occasionally 12 or so in roosts, and up to 30 congregate at a fruiting juniper. Garden fish pond regularly used as social gathering place in late afternoon and evening; up to 12 individuals observed together, bathing and interacting, including half-hearted chases (Douglas 1995). Shy except when near or in man-made habitats. Sometimes regarded as a pest due to fondness for fruit. Hops; perches on ground and in trees; flight strong and powerful. Occasionally dustbathes in sand, fanning tail, spreading wings and resting on ground (Winterbottom 1966). One of the first birds to sing at dawn and last to call at dusk; seldom sings outside breeding season. Likewise, forages actively in early dawn and late dusk, and comes to water when nearly dark (L. L. Short, pers. comm.). Forages usually on ground, scratching among the dead leaves; flicks leaves with bill and scratches backward with 1 foot. Takes insects from ground and soft fruit from trees, sometimes fallen fruit; occasionally picks insects and lizards from branches (Dowsett-Lemaire 1983). Sometimes cracks snail by beating it against a knot of a root (Sclater and Moreau 1933). Follows, sometimes in mixed flocks, safari ants (Dorylinae) and feeds on prey flushed by them (Willis 1985); can remain with them for 22–175 min, hopping over logs, stopping to toss leaves by run-and-peck, catching insects and occasionally picking up fruits. Also feeds on emerging termites.

Known to moult Dec–Mar, Nyika Plateau, Malaŵi (Dowsett and Dowsett-Lemaire 1984).

Sedentary in most of its range, e.g. in Malaŵi, where 11 ringed birds in 10 years ranged from only 0 to 1000 m (Dowsett 1985). Some dispersal after breeding season to lower elevations (Malaŵi, Apr–Sept; Zimbabwe, winter months); somewhat migratory in Namibia (Naukluft Mts, July–Sept, 400 km north of known range, and also in winter months along Kuiseb R., 130 km northwest of Naukluft Mts: Boyer and Bridgeford 1988) and South Africa (Barberspan, May–Sept; and Natal, coastal regions).

Food. Insects (beetles, moths, grasshoppers, mantids, glow-worms, caterpillars, pupae and grubs) and fruits (lianas, *Cotoneaster, Olea africana, Celtis, Rapanea, Afrocrania volkensis, Myrica salicifolia, Polyscias fulva, Syzygium guineense afromontanum, Clidemia*, mulberry, date and strawberry). Also earthworms, snails, slugs, small bivalves, spiders, small lizards (including chameleons), fish, nestling birds, seeds and flower buds. Rwanda: fruit of *Bridelia, Ilex, Polyscias, Rytigynia, Scheffleria, Trema, Urera* (Dowsett-Lemaire 1990).

Breeding Habits. Monogamous, terrestrial. In SW Cape av. territory (n = 34) 16 ha/pair; birds defend area of radius 6 m around nest (Winterbottom 1966). In Malaŵi, forest patch of 6–12 ha can be occupied by 2 pairs; sometimes 1 pair in patch of only 0·5 ha (Dowsett-Lemaire 1989). Maintains territory up to 6 years (Dowsett 1985).

Sometimes at beginning of breeding season up to 5 birds gather, then with wings fanned out and tail opened, 'call and jump about' (Liversidge 1991). ♂, displaying to ♀, cocks tail straight up, fans it and beats outstretched wings. Defends territory by chasing intruder along ground with tail fanned, wings spread out and down, head thrust forward and crown and orange belly feathers fluffed out; sometimes 2 birds fight with bills and claws interlocked. Occasionally chases intruders in and out of trees and bushes. Typically 1 bird of the pair defends nest and territory: ♂ definitely defends nest; role of ♀ not recorded. Regularly seen attacking its reflection in a window pane (Kenya: S. Keith, pers. comm.). Breeding ♂ sings from branch or ground, usually early in morning and again late afternoon and early evening. Said to duet (Winterbottom 1966).

NEST: large cup or bowl, rather untidy but strongly constructed; built of small twigs, strips of bark, rootlets, stems, coarse grasses, bits of lichens, dry leaves, bracken leaves, moss (some nests made of moss with a few bracken leaves: T. M. Butynski, pers. comm.) and sometimes strips of white plastic; bound together with mud and lined with fine grass and rootlets tightly interwoven. Ext. diam. 140–150, int. diam. 80–90, ext. depth 70–80, int. depth 40–50. 1·5–23 m, usually 2–5 m above ground, on top of thick foliage or concealed in fork of bush or tree usually against trunk; sometimes on side of house, sheltered by eaves of roof.

Constructed in *c*. 10 days, largely by ♀; building usually ceases from middle of morning to early afternoon (van Someren 1956).

EGGS: 1–4, usually 2–3; Kenya 1–3 (11 clutches, av. 2·0), Malaŵi 2, Mozambique 1–3 (8 clutches, av. 1·88), South Africa 1–4 (95 clutches, av. 2·1). Laid at 1 day intervals. Oval, somewhat blunt at large end; smooth, slightly glossy; bluish to light green, blotched, spotted and streaked with yellowish brown and red-brown; sometimes prominent pale purple marks at broad end. SIZE: (n = 114, southern Africa) 25–34·1 × 18.9–23·5 (29·3 × 21·6), race *olivaceus* (n = 18) 28–31·8 × 20·6–23·6 (29·9 × 21·6), race *pondoensis* (n = 5) 29·5–34 × 21·2–23·6 (30·7 × 22·3), race *abyssinicus* (n = 5) 26–29 × 20 (27·4 × 20·0). Sometimes double-brooded (van Someren 1956, Dowsett–Lemaire 1989).

LAYING DATES: Eritrea, July–Aug; Ethiopia, Jan, Mar–Aug, Dec; Sudan, Jan–Apr; E Africa: Kenya, Jan–July, Oct–Dec; Uganda, Feb–Mar, June–July, Nov; Tanzania, Aug; Region A, Mar–Apr, Oct–Dec; Region B, Dec; Region C, Jan, Apr–May; Region D, Jan–Dec, peak Mar–June, at lower altitudes mainly Mar–June (long rains), higher altitudes in dry and moderately wet months (Mar–June 33, July–Feb 24 clutches); Zaïre, July–Aug, Oct–Nov (enlarged gonads Feb–June, breeding condition Dec); Rwanda, Mar–May (4 clutches), Oct–Dec (5 clutches), Malaŵi, Sept–Dec, peak Oct (Sept 4, Oct 15, Nov 9, Dec 3 clutches); Zambia, Sept–Mar; Zimbabwe, Sept–Jan, peak Nov–Dec (16 out of 24 clutches); Angola, Sept–Dec; Mozambique, Nov–Jan; Botswana, Oct–Nov; South Africa: Transvaal, Aug–Mar (mainly Sept–Nov), Natal, Sept–Jan; W Cape, all months (Jan–July 45, Aug–Dec 114 clutches) with major peak Oct, minor one Feb, fewest nests Dec–Jan when relative humidity lowest and May–July when rainfall heaviest.

INCUBATION: begins with 1st egg; by ♀ (during 12 h 20 min of observation, ♀ at nest 78% of time). Occasionally ♂ flies to nest and stands on its edge for less than 1 min, once 4 min, while ♀ incubates (Winterbottom 1966). ♀ away from nest for <30 s–49 min. Period: 14 days (SW Cape), 14–15 days (Kenya).

DEVELOPMENT AND CARE OF YOUNG: young naked for first 4 days when flight feathers show; spotted plumage by day 14; day 16, leave nest but can barely fly; day 30, size of adult.

Young brooded almost constantly for first 2 days; ♀ broods until midday when ♂ relieves her; ♂ then broods until she returns. Very small young fed by ♀; ♂ usually passes food to ♀ who then feeds young. Young fed 3–13, av. 6·6, times/h. Young sometimes scratch among dead leaves for food while waiting for parents to bring food to them. Remain with adults up to 2 months.

BREEDING SUCCESS/SURVIVAL: of 15 adults ringed Malaŵi, av. annual mortality 48% over 10 years (Dowsett 1985); Kenya, 8 pairs each with 1 fledgling, 1 pair with 2 (Bennun 1989). Parasitized by Red-chested Cuckoo *Cuculus solitarius*; and nesting birds attacked by Wahlberg's Eagle *Aquila wahlbergi* (Ethiopia: Vittery 1978), White-browed Coucal *Centropus superciliosus* (Kenya: Ng'weno 1986) and Laughing Dove *Streptopelia senegalensis* (SW Cape: Winterbottom 1966). Oldest known birds at least 10 years Kenya, 7 and 4½ years South Africa, 5⅓ years Zimbabwe, 5½ years Malaŵi, and 5½ years Ethiopia (Whitelaw 1983, Dowsett 1985, Ng'weno 1986, Tyler 1987, Harwin *et al.* 1994).

References
Keith, S. and Urban, E. K. (1992).
van Someren, V. G. L. (1956).
Winterbottom, M. G. (1966).

Plate 3
(Opp. p. 48)

Turdus pelios Bonaparte. African Thrush. Merle africain.

Turdus pelios Bonaparte, 1851. Consp. Av. 1 (1850), p. 273; 'ex Asia centrali' (Fazoglu = Fazughli, Sudan, Rensch 1923, J. Orn., 71, p. 99).

Forms a superspecies with *T. olivaceus*, *T. libonyanus* and *T. tephronotus* (and *T. menachensis* of SW Arabia).

Range and Status. Endemic resident and partial intra-African migrant, widespread, frequent to common, in W, central and eastern Africa. In W Africa from Senegal (south of 14° 30′N, 1 record Richard-Toll), Gambia (less common in interior) and Mali (occasionally north to 13° 30′N) through Guinea-Bissau, Guinea (Conakry, Gaoual, Massif de Fouta-Djalon, Kankan, Mt Nimba), Sierra Leone, Liberia (Mt Nimba), Ivory Coast and Ghana (outside rain forest, most common in N guinean zone), Togo, Burkina Faso and Benin (in Arli and Pendjari and 'W' Nat. Parks, north to Ouagadougou, south to Save), Nigeria (less common north to Borgu, Kano and Maiduguri), Niger (in SW at 'W' Nat. Park, Korogoungou and Makalondi north to about Filingué), Chad (SW at N'Djamena and Sarh, also Abéché) and Central African Republic (Birao and Bamingui-Bangoran Nat. Park in N, Lobaye Préfecture and Haute Sangha Préfecture including Bayauga in SW); Cameroon (less common in dry N, not recorded in SE), Bioko, Rio Muni, Gabon, Congo (not recorded N), and Angola (Cabinda south along escarpment but not on arid coast, Cuanza Norte, Cuanza Sul, N Huila and Ulge and Malanje Provinces; also E Moxico Province at Luau R. and Nana Candundo). In Zaïre (absent S Equator, N Upper Zaïre, W and N West Kasai, N East Kasai and SW Shaba Provinces) and Zambia (west of

Turdus pelios

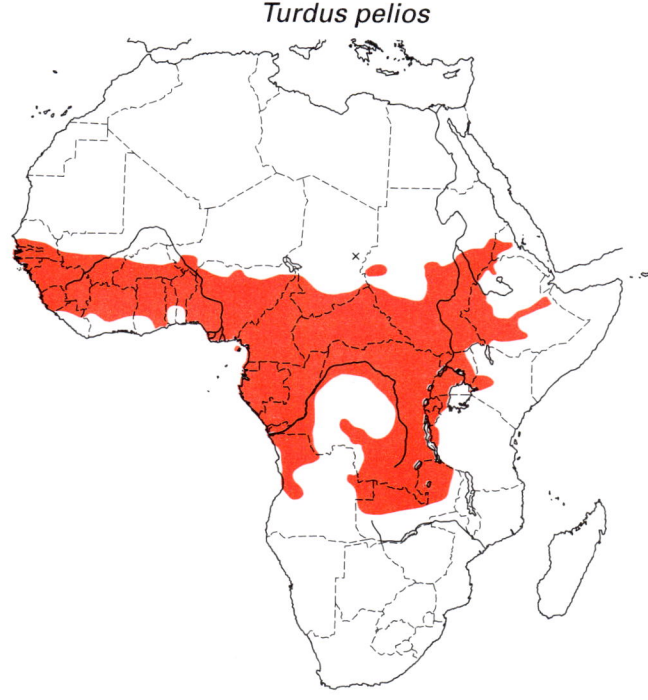

Luangwa Rift in Luapula, Western and North Western Provinces from Mweru Marsh south and east to Muchinga Escarpment, and west and south to Balovale and S Kasiji R.). In eastern Africa in Sudan (north and west to Darfur, Kordofan at El Obeid; also north and east to *c*. 17°N along Eritrean border; uncommon Darfur), N and W Eritrea (below 1500 m), W and S Ethiopia (below 1500 m); also Rift valley from L. Langano and Shashamanne southward and Jijiga area in E; Kenya (Mt Elgon, Kerio valley and Karissia Hills south through Laikipia Plateau to L. Nakuru; and in W south to 0° 30′S), and Uganda (mainly below 1600 m; not NE in parts of Karamoja) to Rwanda, Burundi and NW and W Tanzania (south to Tanda Mbuga, Sumbawanga district).

Description. *T. p. pelios* Bonaparte: Eritrea, N, central and E Ethiopia, central Sudan, Chad, N Central African Republic, E Cameroon. ADULT ♂: upperparts from forehead to upper-tail-coverts ashy grey; feathers of forehead to crown with indistinct black-brown edgings. Lores dark brown; cheeks and ear-coverts lighter brown, feathers with fine light brown shafts; sides of neck, especially near base, washed with light buff to off-white. Tail greyish brown. Chin and throat off-white, streaked with ashy brown; upper breast ashy brown; sides of lower breast and flanks with pale tawny-orange; centre of lower breast, belly and undertail-coverts white. Wings greyish brown, axillaries and underwing-coverts tawny-orange. Bill yellowish brown to yellow; eyes dark brown; legs and feet pale brown to olive-brown. Sexes alike. SIZE: (10♂♂, 10♀♀) wing, ♂ 108–118 (113), ♀ 103–118 (111); tail, ♂ 69–88 (78·6), ♀ 70–90 (81·1); bill, ♂ 17–20 (18·2), ♀ 16–19 (17·6); tarsus, ♂ 26–32 (28·3), ♀ 27–29 (27·4). WEIGHT: unsexed (n = 6, Chad) 51–66 (59).

IMMATURE: like adult but spotted ashy grey on breast and upper flanks; some buff above and behind eye; feather tips of upperwing-coverts tawny.

NESTLING: down dull buff-brown.

T. p. centralis Reichenow: S and SW Ethiopia, S Sudan, Uganda, Kenya, NW Tanzania, NE and N Zaïre, S Central African Republic and E Congo. Like nominate race, but with slightly darker upperparts and upper breast; axillaries and underwing-coverts paler tawny-orange. Bill yellow-orange, eye light orange-brown. WEIGHT: (Uganda, Kenya) ♂ (n = 3) 62–69 (64·7) ♀ (n = 6) 58–65 (61·5); unsexed (n = 9, Uganda) 65–66 (av. ?).

T. p. graueri (Neumann): W Tanzania, Burundi, Rwanda, E Zaïre (north end of L. Tanganyika to Shaba Province). Like nominate race, but upperparts more brownish olive, lower breast and flanks more tawny-orange; upper belly with some tawny-orange; lower belly white.

T. p. stormsi (Hartlaub): Zambia, E Angola, SE Zaïre (Shaba). Darkest and most richly coloured form; upperparts dark brownish olive; streaks on throat larger and darker than nominate race. Breast brighter tawny-orange than *graueri*; flanks and belly even brighter tawny-orange; white restricted to centre of lower belly. Large. WEIGHT: 1 ♂ (Zambia) 80.

T. p. bocagei (Cabanis): W and NW Angola, W Zaïre. Like nominate race, but ashy brown on upper breast darker and more extensive; sides of lower breast and flanks with more tawny-orange; centre of lower breast and belly with less white, but more white than *stormsi* or *graueri*.

T. p. saturatus (Cabanis): Ghana (not in N) to Cameroon (except E, and Mt Cameroon) to W Congo and Gabon. Like nominate race, but upperparts darker, throat darker ashy brown with dark brown streaks; upper breast ashy grey; sides of upper breast and flanks with less extensive orange; centre of lower breast, belly and undertail-coverts dirtier white to ashy grey; axillaries and underwing-coverts paler tawny-orange. WEIGHT: (Cameroon) ♂ (n = 7) 55–70 (63·2), ♀ (n = 4), 62–69 (65·7); unsexed (n = 5, Nigeria) 56·7–79 (62.8).

T. p. nigrilorum (Reichenow): Cameroon: Mt Cameroon above 500 m. Hybridizes with *saturatus* below 500 m; hybrids known from Mt Cameroon between 370 and 610 m (but up to 1070 m), and from Bamenda Highlands. Breast and flanks grey-brown, belly white, no orange except on underwing-coverts. WEIGHT: ♂ (n = 3) 58–64 (60·7); ♀ (n = 4) 61–72 (66·3); unsexed (n = 50) 55–70 (63·1).

T. p. poensis (Alexander): mountains of Bioko. Like *nigrilorum* but upperparts, breast and throat streaking paler brown.

T. p. chiguancoides (Seebohm): Senegal, Gambia, Guinea, Sierre Leone, Liberia, N Ghana. Pale. Like nominate race above; chin and throat pale ashy brown with indistinct darker brown streaks; upper breast pale ashy grey; sides of lower breast and flanks with slight orange wash; centre of lower breast, belly and undertail-coverts off-white to light ashy grey; axillaries and underwing-coverts pale tawny-orange. WEIGHT: (Ghana) ♀ (n = 6) 57·9–65·8 (60·5); unsexed (n = 16) 57–71·5 (66·1).

TAXONOMIC NOTE: We follow Keith and Urban (1992) in considering *nigrilorum* and the very similar *poensis* to be races of *pelios* rather than of *olivaceus*; see note under *T. olivaceus*, p. 36.

Field Characters. A fairly large thrush, with a 'washed out' appearance; in open deciduous wooded habitats. Most races with light-grey upperparts and breast, orangish flanks and white belly; variable streaking on throat, yellow bill and orange underwing. W African *chiguancoides* is palest race with only small amount of orange on flanks; *nigrilorum* (Mt Cameroon) and *poensis* (Bioko) are browner without orange on flanks. Readily distinguished from both Olive Thrush *T. olivaceus* and Kurrichane Thrush *T. libonyanus* by characteristic repetitive song, and is usually not in same habitat.

Where it meets Olive Thrush in Kenya, African Thrush has a longer, leaner, long-legged look and is overall duller, while Olive Thrush appears chunkier, shorter-legged and brighter; for further differences see under Olive Thrush. Race *stormsi* (Zambia), with dark upperparts and extensive orange on breast and belly, looks like an Olive Thrush and until recently was thought to be a race of *T. olivaceus*; it has, however, the African Thrush song and habitat. Kurrichane Thrush has patterned face, with white eyebrow, orange eye-ring, dark lores and pronounced black malar stripe beside unstreaked white throat. In Cameroon, plain race *saturatus* meets Grey Ground-Thrush *Zoothera princei*, which is smaller and darker, with blackish (not yellow) bill, black stripes on face, white wing-bars and white stripe on underwing.

Voice. Tape-recorded (10, 53, C, ERA, GRI, MOR). Song, a long, melodious series of simple repeated phrases or single notes delivered in a continuous stream for 35–40 s, sometimes 1 or several minutes. Most frequent are 2-note phrases with first note higher, 'wee-poo' or its reverse, 'poo-lee', and variations thereon, sometimes extending to 3 notes, interspersed with a few single notes or a dry trill, 'wee-poo, wee-poo, poo-lee, poo-lee, wee-pyoo-wee, wee-pyoo-wee, pee-wit, pee-wit, peewit, chit, chit, chit, woo-yee-poor, woo-yee-poor, wee-poo, wee-poo, pyoolee, pyoo-lee, chew, chew, chew, woo-lee-pee, woo-lee-pee …'. Each phrase is repeated 2 or 3 times, sometimes 4 or 5. General effect is like Song Thrush *T. philomelos* but a little faster. Mimics other species (Goodman and Goodman 1985; L. L. Short, pers. comm.). Call, a hard 'chuk', often lengthened into a dry trill, 'chukukukukuk …'; alarm, the typical *Turdus* 'creaking hinge' note, high, thin and downslurred. Call of young, a hissing trill.

General Habits. Inhabits open deciduous woodland, often associated with watercourses, also forest edges and clearings, gardens, parks, orchards, farmland, dense coastal bush, savanna and evergreen woodlands, secondary forest and occasionally mature primary forest. In Sierra Leone also in copses above 1440 m and upper limit of forest. In Nigeria lives from sea level to 2060 m, in riparian woodland and thickets in guinea zone in N and in a wide variety of habitats in S. In Cameroon occurs in dense shrubs, gardens, villages and secondary forest, from sea level to 1500 m; on Mt Cameroon inhabits montane forest, also secondary shrubbery, clearings, cultivated areas and copses in ravines from 500 to 3000 m (mainly above 1300 m). In Bioko, lives in forest, cultivation and human settlements.

Where it meets Olive Thrush in Zaïre, Uganda and Kenya, prefers wooded lowlands, typically below *c.* 1660 m, while Olive Thrush is mainly in montane forests above 1700 m. Where they meet in central Kenya (Laikipia Plateau), African Thrush is in drier woodland while Olive Thrush is in moist dense forest, woods and thickets; however, they occur together in degraded forest and woods, and African Thrush can occur even in moist riverine cover (L. L. Short, pers. comm.). Where it meets Kurrichane Thrush in Tanzania, Zambia and Angola, prefers heavier evergreen and riparian woodland while Kurrichane Thrush usually occupies drier woodland and savanna. Where it overlaps with Olive Thrush in Sudan, lives in woodland while Olive Thrush tends to remain in forest and secondary growth above 1600 m. In Ethiopia overlaps with Olive Thrush and Bare-eyed Thrush *T. tephronotus*, but their ecological separation remains unknown.

Solitary, in pairs, sometimes in flocks of up to 15–20. Shy, but less so around settlements and cultivation. Frequents bush stratum, perching on small trees and bushes. Sings typically before dawn and often all day until dark; sings all year but less so at height of dry season. Forages most actively in early dawn and late dusk, and comes to water in near-dark (L. L. Short, pers. comm.). Forages on ground, turning over leaf litter. Often hops, stands bolt upright, then thrusts bill into soil to obtain food. Sometimes gathers in flocks in trees to eat fruit and berries. Occasionally follows safari ants (Dorylinae), when it may fly up 4–6 m to catch arthropods that they flush (Willis 1986b). Thought to crack *Pila* snails against stones (anvils) like Song Thrush (Burkina Faso: Walsh and Walsh 1983).

Mainly sedentary (e.g. Gabon bird ringed M'Passa caught same area 32 months later: Brosset and Erard 1986). Partially migratory in northern part of range in Sahel, probably associated with wet/dry seasons; in Gambia, some dispersal in dry season; in Mali, some movement northward to 13° 30′ in wet season; in Ghana, numbers augmented at Mole Nat. Park in wet season; in Nigeria, numbers increase in wet season in N in Zaria and Kano, present only Sept–Oct in Serti, and in S flocks present in dry season in Ibadan; in Chad most common in July in wet season; and in Sudan (Darfur) possible summer/wet season visitor. Local wanderers in Ethiopia (L. Langano, occasional transients noted in dry season, juveniles left after 6 months: Beals 1970).

Food. Insects including moth pupae, harvester ants, termites, locusts, beetles; millipedes, snails, earthworms, small fish; seeds, acacia flowers, berries (*Rauwolfia*, *Loranthus*) and fruits (*Azadirachta indica*, figs, papaya). Of 12 stomachs (Zaïre), 10 contained fruits, berries or seeds of fruits, 3 had several small red peppers, 5 had beetle larvae, caterpillars and termites and 2 contained millipedes (Chapin 1953a).

Breeding Habits. Monogamous. Territorial; in Gabon usually 3–5 pairs/ha, but some territories up to 2–3 ha. ♂ has several song posts in territory; sings all year but frequency increases before beginning of wet season (Brosset and Erard 1986).

NEST: large cup, neatly built of mud, moss, grass, stems, leaves, lined with moss, grass, rootlets. Ext. diam. *c.* 140, int. diam. 80–90, int. depth 50–55. 1–9 m above ground in fork on bough of tree (*Harungana*, *Dracaena*, *Musanga*, *Erythrina*, *Acacia*, *Ficus*, *Cassia*), palm (*Borassus*), shrub (*Veronia*), clump of lianas, or once on top of tent peg. Sometimes uses part of its own old nest (Beals 1970), or nest of mannikin *Lonchura* or Laughing Dove *Streptopelia senegalensis* (Chapin 1953a, Grimes 1972, Brosset and Erard 1986). Constructed in 14 days (Lynes 1925), by ♀.

EGGS: 1–4, usually 2–3; 42 clutches W Africa and 5 E Africa averaged 2·4 eggs. Laid at 1 day intervals. Oval, somewhat blunt at larger end, sometimes almost spherical; smooth, slightly glossy; pale bluish green to pale bluish white, rarely warm cream colour, with coarse and fine rufous spots all over but more so at blunt end. SIZE: (n = 35, Nigeria) 22·5–30 × 19–20·4 (25·9 × 20·2), (n = 13, W Africa) 26·3–30·7 × 18·7–21.3 (27·1 × 19·9), (n = 3, Uganda) av. 26·5–27 × 19·5–20. Sometimes double-brooded (Bannerman 1936).

LAYING DATES: Senegal, June–Aug, Oct; Gambia, June–Nov; Sierra Leone (singing and enlarged gonads May); Ghana, Apr–Aug (nestlings and fledglings Feb); Togo, June; Mali, June, Sept; Burkina Faso (singing Mar, Aug, juvenile Aug); Central African Republic (probable breeding June, Aug); Nigeria, Apr–Sept (singing Feb–Mar, Oct); Cameroon, Jan, Mar–Sept, Dec but largely in wet season; Bioko, Nov–Mar (enlarged gonads and singing Sept–Oct); Congo (breeding condition, Feb); Gabon, Aug–Oct (singing Nov–Jan, carrying food Jan); Angola, Sept–Dec (singing Feb–Apr); Zaïre, Jan–Dec; Zambia, Feb, Sept, Nov–Dec (enlarged gonads Jan, Mar, Aug, Oct); Sudan, Jan–Aug; Eritrea (young Aug); Ethiopia, Apr–July (breeding possible Mar); Kenya, Apr; Uganda, Jan–Dec (Jan 12, Feb 13, Mar 52, Apr 17, May 17, June 3, July 3, Aug 3, Sept 4, Oct 8, Nov 5, Dec 12 clutches).

INCUBATION: starts after last egg laid; by ♀ only.

DEVELOPMENT AND CARE OF YOUNG: both parents rear young.

BREEDING SUCCESS/SURVIVAL: African Pied Hornbill *Tockus fasciatus* took a clutch, Gabon (Brosset and Erard 1986).

References
Brosset, A. and Erard, C. (1986).
Stuart, S. N. and Jensen, F. P. (1986).

Turdus libonyanus (Smith). Kurrichane Thrush. Merle kurrichane.

Plate 3 (Opp. p. 48)

Merula libonyanus Smith, 1836. Rep. Exped. Centr. Africa, p. 45; near Kurrichane, W Transvaal.

Forms a superspecies with *T. olivaceus*, *T. pelios* and *T. tephronotus* (and *T. menachensis* of SW Arabia).

Range and Status. Endemic resident, central and SE Africa. Angola (Malange, Cuanza Sul and Lunda to W Huila and Moxico, common), S central and SE Zaïre (S Kasai and Shaba Provinces north to Moba, common), SE Burundi (Kumoso between Giofi and Kinyinya, common), and Tanzania (common in SE, uncommon in N and W, north in W to Kibondo, Tabora and Mwanza, also in Minziro F. R. and in E to Amani and Tanga). Throughout Zambia (common), Malaŵi (common along W shore L. Malaŵi and Thyolo and Nsanje districts, uncommon elsewhere), Zimbabwe (widespread and usually common but uncommon in semi-arid savannas of major river valleys, W Matabeleland outside of Kalahari Sands, and Chimanimani Mts and Inyanga Highlands; absent much of Limpopo R. valley away from water) and Mozambique (locally common south of Zambezi R. but uncommon lower Zambezi and Nampula Province); NE Namibia (Kavango in E Owambo and Caprivi Strip, common E Caprivi, uncommon to frequent elsewhere); N and E Botswana (Sepopa, Gumare and Okavango delta to Francistown and Gaborone, locally frequent to common). Swaziland and South Africa: N Cape (Mafeking), Transvaal (throughout: common lowveld and bushveld, uncommon Kruger Nat. Park, rare highveld), Natal (south to Transkei, locally common coastal and midlands, rare above 950 m) and Orange Free State (rare, 5 records). Single record Lesotho (Bonde 1993).

Turdus libonyanus

Description. *T. l. libonyanus* (Smith): Botswana, South Africa (N Cape, Transvaal), Swaziland (except SE). ADULT ♂: forehead to back brownish slate-grey; rump and uppertail-coverts greyer, with little brown. Tail brownish slate-grey. Lower cheeks off-white; indistinct buff-orange superciliary stripe; eye-ring orange; cheeks and ear-coverts slate-grey, ear-coverts finely streaked off-white. Chin, sides of neck and throat off-white; sides of throat with several rows of pronounced dark brown to black-brown streaks; centre of throat not streaked. Upper breast ashy grey; breast, flanks and belly pale rufous-orange, lower breast and belly pale cream in midline, undertail-coverts pale cream. Primaries, secondaries and upperwing-coverts brown to slate-grey, greater and median primary coverts slatier; axillaries and underwing-coverts pale tawny-orange. Bill bright orange; eyes brown; legs and feet chrome-yellow. Sexes alike. SIZE: (10 ♂♂, 10 ♀♀): wing, ♂ 113–120 (115), ♀ 106–119 (111); tail, ♂ 69–83 (77·3), ♀ 68–81 (74·1); bill, ♂ 16–21 (18·8), ♀ 17–23 (19·3); tarsus, ♂ 25–33 (27·8), ♀ 25–29 (26·6). WEIGHT: (southern Africa) ♂ (n = 4) 46·2–64·5 (55·6), ♀ (n = 3) 50·8–59·3 (55·3) unsexed (n = 65) 51–70 (60·6).
IMMATURE: like adult but sides of throat and breast with pronounced black-brown spots, a few on flanks; upperwing-coverts with buff tips.
NESTLING: not described.
T. l. peripheris Clancey: Natal, SE Swaziland and S Mozambique (Maputo district). Like nominate race, but upperparts, wings and tail darker brown; upper breast darker ashy brown, sides of lower breast and flanks with more extensive and richer rufous-orange, centre of belly with less cream.
T. l. verreauxi (Bocage) (including '*chobiensis*'): S central Zaïre, Angola, Namibia, W Zambia, W Zimbabwe and N Botswana (southeast to Francistown). Like nominate race, but upper breast greyer, less ashy; sides of lower breast and flanks with less extensive and paler rufous-orange.
T. l. tropicalis (Peters): Mozambique (north of about Limpopo R.), Zimbabwe (except W), Zambia (except W), Malaŵi, SE Zaïre, Burundi and Tanzania. Like nominate but upperparts slightly greyer; breast and flanks with more rufous-orange than *verreauxi*. WEIGHT: 1 ♂ (Mozambique) 68·6; (Zimbabwe) ♂ (n = 5) 46·2–75·5 (59·6), ♀ (n = 5) 50·8–80 (61·1), unsexed (n = 46) 52·2–69·3 (60·2); unsexed (n = 9, Zambia) 51–69 (57·0).

Field Characters. A medium-sized thrush of dry open woodland, with grey-brown upperparts, whitish belly, orange flanks, bright orange bill, white throat unstreaked in centre but with prominent black malar stripes at sides; at close range shows pale buffy supercilium and orange eye-ring. Best told from similar-looking congeners Olive Thrush *T. olivaceus* and African Thrush *T. pelios* by voice and habitat. Sympatric races of Olive Thrush are darker and have yellow bill, plain face with inconspicuously streaked throat and no malar stripes, and orange (not white) belly. African Thrush similarly has unmarked face and lightly streaked throat, and sympatric race *stormsi* in Zambia and surrounding areas has entirely orange underparts. Bare-eyed Thrush *T. tephronotus* easily distinguished by broad yellow eye-ring.

Voice. Tape-recorded (12, 58, 86, 88, 91, B, C, F). Song, a series of phrases with a pause of 1–2 s between each, in manner of Olive Thrush but quite unlike continuous delivery of African Thrush. Phrases very variable, some rich and warbling, others thinner but still sweet, and frequently ending in a high, tuneless trill or hiss, like Olive Thrush: 'wordy, wordy, pss-pss', 'wordy, wordy, tirrrr', 'peeoo, peeoo, psst', 'woody-woody-woody', 'tyeoo-weet-weet', 'wooyee-wooyee-tssst', 'way-cheeyo-werchick', 'way-peeyo-tertlili', or longer 'weeyo-wer-cheep-weeyo-weeyo-weeyo'. Phrase may be repeated, but usually with slight variation. Sometimes mimics other species (Vernon 1973). Call, a double 'pss-chew', 'pss-chewi' or 'tsi-tseeoo'.

General Habits. Prefers miombo woodland but also inhabits acacia woodland, open bushland, wooded rocky hill sides, riverine bush, forest fringes, gardens, parks, cultivated land, areas around houses, and exotic plantations including eucalyptus stands. In Natal also found in euphorbia forest on steep hill sides. Tends to avoid forest, arid savanna and pure stands of mopane or mopane/*Combretum*, although occurs in mopane in Angola (Traylor 1965a), Botswana (Brewster 1991) and Namibia (Koen 1988, Brown 1993) where miombo is absent. Range overlaps Olive, African, and Bare-eyed Thrushes, but Kurrichane usually ecologically separated from them. Hybrids unknown. In Burundi and NW Tanzania occupies miombo savanna woodland (African Thrush in degraded savanna and cultivation, Olive Thrush in dense forest and galleries). Where it overlaps with Bare-eyed Thrush in Tanzania, occupies miombo while the Bare-eyed inhabits coastal and semi-arid bush and scrub. In Tanzania ranges from sea level to 1900 m. In Malaŵi occurs mainly at 900–1500 m in woodlands while Olive Thrush occurs in forests above 1400 m; in Viphya Mts where Olive Thrush absent, inhabits forests up to 1700 m. Overlaps with Olive and African Thrushes in Zambia but is ecologically separated from them, inhabiting miombo woodland and savanna rather than evergreen forest (Olive Thrush) or riverine woodland (African Thrush); occurs up to 1740 m in Zambia. In Zimbabwe in woodland, savanna, man-made habitats, and edges of forest up to 2200 m but remains separated from Olive Thrush which tends to be in forest. In Transvaal mainly in bushveld and lowveld, sometimes side-by-side with Olive Thrush (e.g. Pretoria: Tarboton *et al.* 1987).

Solitary or in pairs. Generally shy but becomes tame in man-made habitats, and sometimes forages around houses and safari camps for crumbs and soft fruits. Perches on ground and in trees and shrubs. When disturbed, flies off with 2-note call (see *Voice*), usually landing in tree. Flight strong, powerful and direct. Sometimes bathes in wet foliage of trees, flying into foliage without perching and fluttering its wings as if taking a bath in a shallow pool (Beasley 1993). ♂ sings typically after rain in spring and summer, in early morning (sometimes before dawn), late afternoon and evening, and in moonlight. Forages mainly on ground, by running or hopping, stopping, listening, then pecking and tossing litter aside to expose prey. Very occasionally flies almost vertically to hawk prey in the air, then returns to ground (Campbell 1973). Sometimes pecks at hanging avocado pears. Once noted attempting to steal a *Lycosa* spider from wasp *Hemipepsis*, which stung bird on head (Wilson 1978).

Sedentary in most of its range, e.g. ringed birds remained in same locality for 3½ years in Zimbabwe (Irwin 1981) and 3 years in Transvaal (Elliot and Jarvis 1973). Some birds disperse into SE Botswana Dec–Mar (Beesley and Irving 1976), and after breeding, move in 'winter' from miombo woodland to mixed woodland in Zimbabwe (Vernon 1977).

Food. Insects including beetles, grubs, caterpillars (including those of the mopane emperor moth *Imbrasia belina*: Styles 1995), grasshoppers, locusts, crickets; also spiders, millipedes, earthworms, molluscs, lizards; seeds and fruits of avocado pears, *Ficus* and *Rhus* spp.

Breeding Habits. Monogamous, territorial. Territory size *c*. 2 ha (Natal: Chittenden 1982; in Zimbabwe 46 pairs in 100 ha of miombo woodland); pairs can nest within 22–36 m of each other (Steyn and Brooke 1973). Sings much of day, even at midday during breeding season; courtship not described.

NEST: large cup or bowl, rather untidy, built of twigs, stems, grass and roots; also cobweb, cotton and bits of plaster, paper and rags, held together with some mud, usually lined with thin layer of mud (5 of 7 nests: Chittenden 1982), rootlets, tendrils and dry grass. Ext. diam. 140–150, int. diam. 65–80, ext. depth 70–80, int. depth 35–50. 1–10 m above ground, usually 3–4 m (n = 33, 1·5–6·7, av. 3·7 m: Chittenden 1982), in fork of large tree usually against trunk or in clump of mistletoe or tree orchid, hollow lip of tall vertical pipe in garden (da Rosa Pinto and Lamm 1953), and house gutter choked with debris (Steyn 1965). Constructed in 1–2 days, by ♀; building begins often immediately after or during light rain, or in early morning when nest material is damp, and lasts most of day. During each collecting trip ♀ gathers large beakfuls of material, some of which dangles beyond her tail when she returns to nest. Occasionally builds nest on old nest of Laughing Dove *Streptopelia senegalensis* or Fiscal Shrike *Lanius collaris* (Ginn *et al.* 1989). Sometimes nest used for second brood, at other times new nest is constructed, often near the first. ♂ and ♀ roost together for 3–6 days after nest completed and before egg-laying begins.

EGGS: 1–4, usually 3; Malaŵi 1–3 (1 C/1, 13 C/2, 57 C/3), Zambia 2–3 (5 C/2, 11 C/3), Mozambique 2–4 (3 C/2, 7 C/3, 2 C/4), Zimbabwe 3 (3 clutches), southern Africa 1–4 (n = 363 clutches, av. 2·9), South Africa: Natal 1–4 (2 C/1, 12 C/2, 59 C/3, 3 C/4). Laid at 1–2 day intervals in Zimbabwe; at 1 day intervals in Natal: often between 08.15 and 09.15 h (7/13 eggs). Oval; pale green to pale blue, finely speckled and spotted with light red-brown to yellowish brown. SIZE: (n = 146, southern Africa) 22·3–30 × 17–21·3 (26·5 × 19·3), race *tropicalis* (n = 14) 23·5–27 × 17–20·1 (26·3 × 19·7). Sometimes double and occasionally triple-brooded; once 1 ♀ apparently laid 7 clutches (Chittenden 1982).

LAYING DATES: Tanzania, Sept–Nov; Zaïre, Sept–Dec; Angola, Aug, Oct–Nov; Zambia, Aug–Dec, Mar–Apr (mainly Sept–Nov, rare Mar–Apr); Malaŵi, Aug–Dec, peak Oct (Aug 1, Sept 14, Oct 46, Nov 14, Dec 8 clutches); Zimbabwe, Aug–Mar, peak Sept–Nov (Aug 9, Sept 385, Oct 536, Nov 341, Dec 65, Jan 5, Feb 2, Mar 1 clutches); Mozambique, Sept–Dec (enlarged gonads Mar); Namibia, Oct–Jan; Botswana, Sept–Feb; South Africa: Transvaal, Sept–Feb, peak Oct–Nov (Sept 9, Oct 40, Nov 39, Dec 15, Jan 4, Feb 3 clutches), Natal, Aug–Jan, peak Oct–Nov (Aug 1, Sept 13, Oct 30, Nov 28, Dec 11, Jan 1 clutches).

INCUBATION: begins evening before last egg laid (Chittenden 1982); by ♀ only; ♂ often roosts near by at night. Period: 12–14 days.

DEVELOPMENT AND CARE OF YOUNG: weight: (n = 9, Natal) 1·5 h after hatching 4·8–5·6 (5·2). Young brooded mainly by ♀. Both parents feed young; sometimes ♂ feeds young of 1st brood while ♀ incubates 2nd brood. Small young fed small winged insects and small worms, half-grown young fed grasshoppers and large worms. In Mozambique from 15·45 to 16·55 h, brood fed once every 8·5 min. Both parents remove faecal sacs from edge of nest and eat them. ♂ sometimes feeds brooding ♀ rather than small young. Young leave nest after 15 or 16 days.

BREEDING SUCCESS/SURVIVAL: Botswana, 2 nests each produced 2 fledglings; Transvaal, 2 nests produced 3 and 2 fledglings; Natal, in 7 nests no chicks fledged; Zimbabwe, of 7 nests, 5 failed due to predators. Parasitized by Red-chested Cuckoo *Cuculus solitarius* (Pitman 1961). Young taken by Shikra *Accipiter badius* and Pied Crow *Corvus albus*, and eggs probably taken by boomslang *Dispholidus typus* and thick-tailed galago *Galago crassicaudatus* (Parnell 1974, Chittenden 1982). Nests highly susceptible to predation by monkeys and baboons, to heavy rainfall, and probably to anti-tsetse fly applications of many years earlier (Macdonald and Birkenstock 1980). Oldest ringed birds at least 8 years (Botswana: Ginn 1982) and 3½ years (Zimbabwe: Irwin 1981). A bird in Kruger Nat. Park with malformed bill 86 mm long lived only 4 months (Whyte 1993).

References
Chittenden, H. (1982).
Clancey, P. A. (1965b).
Keith, S. and Urban, E. K. (1992).

44 TURDIDAE

Plate 4
(Opp. p. 49)

Turdus tephronotus **Cabanis. Bare-eyed Thrush. Merle cendré.**

Turdus tephronotus Cabanis, 1878. J. Orn., p. 205; Ndi, Taita District, Kenya.

Forms a superspecies with *T. olivaceus*, *T. pelios* and *T. libonyanus* (and *T. menachensis* of SW Arabia)

Turdus tephronotus

Range and Status. Endemic resident, in drier parts of E Africa. Patchy and local in S and SE Ethiopia; common but rather local in E half of Kenya west to E wall of Rift valley (L. Turkana, Wajir, Meru, Thika R., Kitui and Tsavo); local and uncommon in S Somalia west of 46°E in semi-arid areas, and less common still and perhaps absent in the most arid parts of W Somalia and the E interior of Kenya. Common in Mkomazi area of N Tanzania. Occurs south at least to Chalinze (100 km north of Dar es Salaam) and southwest to Dodoma; also in Dar es Salaam. 2 records from Selous Game Reserve (Howell and Msuya 1979).

Commonest in coastal lowlands of Kenya. Occurs up to 2000 m in Ethiopia.

Description. ADULT ♂: crown, face, lores, nape, mantle, back, rump, uppertail-coverts and tail ashy grey, sometimes with faint brownish tinge; chin and throat white with very dark brown streaks; breast pale ashy grey; flanks and belly orange with white patch around cloaca; undertail-coverts orange. Primaries ashy grey, with basal third and edge of middle third of inner webs orange; secondaries, tertials and all wing-coverts ashy grey; underwing-coverts orange. Bill orange or orange-brown; bare skin around eyes chrome yellow; eyes brown; legs and feet orange. Sexes alike. SIZE: (15 ♂♂, 6 ♀♀) wing, ♂ 105–113 (108), ♀ 105–114 (107); tail, ♂ 82–95 (88·5), ♀ 86–97 (89·3); bill, ♂ 21–26 (24·0), ♀ 21·5–25 (23·0); tarsus, ♂ 28–31·5 (30·4), ♀ 27·5–31·5 (29·8). WEIGHT: (Kenya, Mar–June) ♂ (n = 6) 45–55 (48·7), ♀ (n = 3) 48–55 (51·0), unsexed (n = 3) 46–55 (49·2).

IMMATURE: like adult but with dark brown spots on breast and upper belly.
NESTLING: not described.

Field Characters. A greyish thrush with distinctive yellow skin around eye; largely allopatric with similar species and occurs in drier habitats. Distinguished from Kurrichane Thrush *T. libonyanus* and African Thrush *T. pelios* by greyer upperparts and orange belly, and from Olive Thrush *T. olivaceus* by white, not plain brownish, throat with dark streaks, and greyer upperparts.

Voice. Tape-recorded (B, C, F, CHA, NOR, ROC). Song of typical *Turdus* quality, rather short and fluty, and not as varied as in some other species. Typically, 2 strong fluty whistles followed by 2–3 weaker and faster notes, either higher or lower pitched. Call, a loud, liquid rattling or often bubbling series of 4–5 notes; and a soft whinnying call, often from ground. Alarm note soft, sibilant and ventriloquial, usually uttered from ground.

General Habits. Inhabits thorn scrub, thickets and fairly thick bushland, especially with some rocks, mainly in semi-arid lowland areas; also orchards and other cultivated areas with thick vegetation. In Tsavo East Nat. Park (Kenya) and Mkomazi Game Reserve (Tanzania) almost restricted to, but quite common in, *Commiphora* woodland thicket (Lack 1985). At edges of range prefers vicinity of watercourses.

A wild, shy bird, solitary or in pairs. Behaviourally a typical thrush, feeding predominantly on ground, usually under or close to thick bushes, tossing leaves and litter about. When alarmed, flies quietly to dense cover. Sings usually from top of trees. Closely resembles Eurasian Blackbird *Turdus merula* in all habits except that it keeps to thick cover.

Sedentary.

Food. Mainly insects: especially caterpillars (Fuggles-Couchman and Elliott 1946) and beetles; also flies. Also fruits, berries and seeds. 4 stomachs contained insects, 1 stomach fruits, 1 berries and 1 seeds.

Breeding Habits. Monogamous and territorial.
NEST: open cup of twigs, grasses, roots and dead leaves placed in tree or bush, usually well hidden.
EGGS: 2–3; pale bluish green with grey or violet undermarkings, spotted with dark rufous at large end. SIZE: (n = 3, Somalia) 25–27 × 19–19·5 (25·8 × 19·2).
LAYING DATES: E Africa: Kenya, Mar, Nov; Region D, May, Nov, Dec; Region E, Apr–June; Somalia, May. Mainly in rains.

Reference
Lack, P. C. (1985).

Turdus philomelos Brehm. Song Thrush. Grive musicienne.

Turdus philomelos Brehm, 1831. Handb. Naturgesch. Vög. Deutschl., p. 382; central Germany.

Plate 4
(Opp. p. 49)

Range and Status. W Europe to Siberia; Asia Minor to N Iran. Winters S Europe, N Africa and SW Asia.

Winter visitor to N Africa from Morocco to Egypt; vagrant south to *c.* 11°N. In Morocco, common to abundant west and north of High Atlas Mts, uncommon to rare south and east of them to fringes of Sahara desert, south to about Tiznit and Goulimine; Algeria, common north of Saharien Atlas, uncommon in S, south to Ghardaïa, also in Sahara at Tadjemout and Arak; in Tunisia, frequent north of Gafsa and Gabès especially along N coast, rare in S in oases. Common in coastal belt of Libya from Tripolitania to Cyrenaica, uncommon inland (Jebel Nafusa), vagrant in interior at Sebha; in Egypt, common along N coast, Nile delta and Suez Canal area, sometimes in flocks of 250–300, uncommon to frequent Red Sea coast, Wadi el Natrun, Farafra Oasis and Nile valley south to Sudan.

South of Sahara, in Mauritania (several records, Cap Timirist, Cansado, Nouadhibou, Nouakchott and at 21° 18′N, 7° 45′W); Senegal (1, Richard-Toll); Mali (1, Araouane); Central African Republic (1, Manovo-Gounda-St Floris Nat. Park); Chad (1, Abéché); Sudan (frequent to common along Red Sea coast, uncommon Nile valley south to Khartoum, also around Jebel Marra and in N Kordofan Province); Eritrea (frequent along coastal areas in most winters; very common in winter of 1952–53: Smith 1957); and Djibouti (rare to uncommon Forêt du Day, Wadi Ambouli and Djibouti City).

Common on passage in N Morocco in Oct–Dec and Feb–Apr; also High Atlas and Rabat and Casablanca region, Algeria (Saharien Atlas, along coast Oct–Dec, Feb), Egypt (especially along Gulf of Suez and Red Sea coast), Sudan (locally common Erkowit) and Eritrea (autumn and spring).

Description. *T. p. philomelos* Brehm: W Europe to central Siberia; winters S Europe, SW Asia and N Africa. ADULT ♂: forehead and crown warm brown; nape, hindcrown and sides of neck brown. Inconspicuous line from base of bill to over eye pale buff; lores blackish brown; eye-ring pale buff; cheeks and ear-coverts buff-brown, heavily streaked blackish brown, particularly on posterior ear-coverts. Mantle and back grey-brown with a slight rufescent tinge; rump brown tinged olive. Uppertail-coverts and upperside of tail warm brown; underside of tail grey-brown. Malar stripe blackish brown, speckled with some pale buff; indistinct moustachial stripe pale cream; chin and throat pale buff with a few lateral blackish brown spots. Upper breast pale buff with larger round blackish brown spots, lower breast and belly white with some blackish brown spots. Primaries, secondaries, scapulars and upperwing-coverts brown, primaries tinged warm brown; median and greater coverts with yellowish buff tips; underside of flight feathers grey-brown, underwing-coverts and axillaries orange-buff. Bill blackish brown, base of lower mandible yellow; eyes dark brown; legs and feet pale flesh colour. Sexes alike. SIZE: (10 ♂♂, 10 ♀♀) wing, ♂ 112–119 (116), ♀ 110–120 (114); tail, ♂ 78–92 (85·7), ♀ 81–86 (82·8); bill, ♂ 15–17 (15·9), ♀ 15–17 (16·1); tarsus, ♂ 28–32 (29·4), ♀ 27–32 (29·4). WEIGHT: unsexed (n = 1, Chad) 77, (n = 6, Egypt) 49·8–63·5 (54·6), (n = 32, Israel) av. 70·9.

IMMATURE: like adult but upperparts with more warm brown tinge, feathers of head, mantle and back with dark brown tips and buff medial streaks; white of underparts from chin to belly washed with buff, brown spots smaller and buffier brown, less blackish brown; more yellow-buff streaks and spots on forewing, especially median and greater coverts.

T. p. clarkei Hartert: W Palearctic; winters S Europe and Algeria. Like nominate race but upperparts warmer brown; underparts more heavily suffused with yellow-buff, breast often appearing darker. Underwing brighter, rich yellow-olive.

T. p. hebridensis (Clarke): Outer Hebrides and Scotland; winters England, once Algeria (Vaurie 1959). Like nominate race but upperparts darker and colder, spots on underparts blacker and bolder.

Field Characters. A medium-sized thrush with upright stance, sandy brown upperparts, boldly spotted underparts with buff wash on breast and flanks, indistinct face pattern and orange-buff underwings. Smaller than Mistle Thrush *T. viscivorus* and browner, with smaller spots, shorter tail and without pale edges to wing-feathers. Redwing *T. iliacus* is similar at rest, when closed wing may conceal rufous flanks, but shows distinct whitish supercilium and streaked rather than spotted breast; easily distinguished in flight by rufous flanks and underwings. In Eritrea overlaps with Groundscraper Thrush *Psophocichla litsitsirupa* which has longer bill, shorter tail and black face marks.

Voice. Tape-recorded (53, 62, 73, 89, B, C, F). Territorial song, clear, flute-like, repetitive, far-carrying, lacking melodious quality of many other thrushes. Given in series of phrases; individual birds use 100–200 (av. 130) different phrases (Cramp 1988). Each phrase repeated 3–4 times, then a short pause, then a new series of phrases. Rarely sings in Africa (see below). Subsong, especially when interacting with other ♂♂, a low twittering warble. Calls, 'sipp', 'tic', or 'seep' in flight; sharp explosive 'tchuk-tchuk' in alarm; rapid succession of 'tschi' when excited; 'kschri kschri' in distress; and 'siih' when predators fly over. (For further details see Cramp 1988.)

General Habits. Occurs near open areas in bush, scrub, and woods with undergrowth. In Morocco inhabits mainly low and tall scrub and trees; also in orchards and vineyards, euphorbia, *Argon* bush, *Salicornia*, *Ilex* woods, and forests up to 2700 m. In Tunisia, deciduous forest and olive plantations. In Sudan occurs mainly in arid open bush up to 1050 m; in Eritrea, in acacia bush along coast, sometimes in high moorland.

Solitary or in pairs; small flocks when feeding and roosting and flocks of up to 300 when migrating. Agile; flight powerful, fairly fast (48 km/h), without the marked undulations of larger thrushes. Roosts gregariously, often in thick cover. Runs briskly; hops, then pauses and stands upright, sometimes with head cocked to one side. Flicks wings and tail when excited. Tends to forage more under bushes and trees and less in open fields than other *Turdus* thrushes. Alert and shy; retreats quickly into cover. Feeds close to cover on ground, moving litter with sideways sweep of bill; usually 3–8 movements at a time, occasionally almost continuously for up to 1·5 min (Cramp 1988). Uses stones, rocks and roads as anvils to smash snails; flicks out snail's body and eats it. Sometimes steals earthworms from conspecifics (Cramp 1988). Singing recorded once in Africa, Ifrane, Morocco, Apr.

Population wintering in Morocco, Algeria, Tunisia and Libya largely from N and central Europe; migrates on broad front across Europe and Mediterranean to NW Africa; may cross Mediterranean Sea to NE Africa and along NW Egyptian coast. Present Morocco Sept–May (earliest 5 Sept, latest 12 May); most movement mid-Oct to late Nov, and Feb to end Mar; commoner some years than others. Present Algeria mid-Oct to late Apr; Libya, earliest 28 Oct, latest 28 Mar; Egypt, early Oct to late Apr, most Dec to mid-Mar with fluctuating numbers that suggest some local or long distance movements; Sudan, Oct–Mar; Eritrea, late Nov to late Mar; Djibouti, Dec–Mar; Mauritania, Nov, Jan, Feb; Senegal, Nov; Mali, Dec; Chad, Mar.

Birds recovered Morocco, ringed Britain, Belgium, Luxembourg, Denmark, Germany, Norway, Poland, Czechoslovakia, Switzerland, France and Italy; recovered Algeria, ringed Britain, Luxembourg, Germany, Finland, Russia, Poland, Czechoslovakia, France, and Italy; recovered Tunisia, ringed Germany, France and Italy; 29 of 32 NW Africa recoveries from Algeria. Bird ringed Egypt (Bahig, Mar 1967) recovered same locality 3 years later.

Food. In Europe invertebrates (grasshoppers, earwigs, bugs, lacewings, scorpion flies, butterflies, moths, flies, beetles, spiders, harvestmen, mites, isopods, sandhoppers, millipedes, centipedes, snails and slugs, and earthworms), also lizards, shrews, bird faeces, once an egg of its own species, fruits, berries and seeds. In Cordoba, Spain, winter, mostly fruits; Negev desert, Israel, large numbers of snails (Cramp 1988). In Egypt, desert snails (*Eremina desertorum*) (Goodman and Watson 1984).

References
Cramp, S. (1988).
Melde, F. and Melde, M. (1991).
Simms, E. (1978).

Plate 4
(Opp. p. 49)

Turdus iliacus Linnaeus. Redwing. Grive mauvis.

Turdus iliacus Linnaeus, 1766. Syst. Nat. (12th ed.), p. 192; Sweden.

Range and Status. Iceland and N Europe to central Siberia (L. Baikal, Lena and Kolyma). Winters central and S Europe, N Africa, Asia Minor, Iran and Turkestan.

Winter visitor to NW Africa, a few to Egypt; numbers fluctuate markedly from year to year. In Morocco, west and north of High Atlas, uncommon east and south to Goulimine and Cap Draa; Algeria, frequent along coastal plain south to Saharien Atlas, rare N Sahara (Ghardaïa); N Tunisia, uncommon and irregular, mainly near coast, south to about Sousse; NW Libya, rare to uncommon in coastal zone (Sabratha, Wadi Turghat) south to Jefren Escarpment; Egypt, uncommon and irregular in Nile delta and coast west to Burg el Arab and south to Cairo, Giza, Wadi el Natrun, and Aïn Sukhna area.

Passage occurs on N Morocco coast and Tangier Peninsula in Nov and Feb–Mar; also as far south as 29°N at Goulimine, so some birds possibly winter in Western Sahara.

Description. *T. i. iliacus* Linnaeus: Europe to Siberia; only race in Africa. ADULT ♂: upperparts from forehead to back olive-brown, feathers of crown with indistinct dark centres, rump and uppertail-coverts slightly more olivaceous. Tail-feathers dark brown, tip of outer web of T6 buffy. Sides of neck white shading to buff at rear; broad pale buff supercilium from bill to behind eye, sometimes interrupted over eye; lores, ear-coverts and cheeks sooty brown, ear-coverts with fine pale shaft-streaks. Chin pale buff to off-white with small black-brown streaks; throat pale buff to off-white with broader and larger black-brown streaks, longest on sides of throat; indistinct sooty brown malar stripe. Breast pale buff to cream-white with obscure dark brown streaks; flanks bright rufous

Turdus iliacus

Eurasian Blackbird *T. merula*; far less loud than other Palearctic thrushes. Subsong, a babbling warble interspersed with fluty notes. Calls, 'tsseep', 'seez', 'seeih' or 'seeip' given in flight, often during night migration; an abrupt 'tchop' when feeding or roosting; and a 'chit-tuck' or 'trrrt trrt trrrt' in alarm (Cramp 1988).

General Habits. In winter inhabits woods, forest, hedges, bushes, fields, pasture, gardens and parks. In Morocco, also cedar forest and olives up to 750 m on High Plateaux and in High Atlas, where reported in trees next to snow-covered fields.

Gregarious outside breeding season, sometimes in large flocks of up to several hundred. Mixes freely with Song Thrush, Mistle Thrush *T. viscivorus*, Fieldfare *T. pilaris* and Ring Ouzel *T. torquatus*. Restless and often shy when migrating. Flight strong, fairly fast (35–50 km/h), slightly undulating and with alternate closing of wings at short intervals. At night roosts gregariously outside breeding season, hundreds gathering at traditional sites in trees, thick shrubs and hedgerows; 2–3 individuals at a time enter roost; often roosts with other thrushes and Common Starling *Sturnus vulgaris*. Feeds on ground, in trees and shrubs. When feeding on ground, tends to be in open areas. Runs or hops usually 1–5 times, then stops, scans ground and pecks at surface item; or hesitates, cocks head to one side, sometimes takes a short step backwards, then stabs downward with bill at prey in ground. Sometimes moves litter with sideways sweep of bill and at same time may displace litter with a backward jump on both feet. Takes small items mainly from soil surface (Britain: Tye 1981). Uses stones and rocks as anvil to smash snails. Several hundred birds sometimes gather in fruit trees to feed. Often gives subsong in winter but rarely full song (see *Voice*).

Migrates from N Europe to winter mainly in S Europe, some crossing Mediterranean to Africa. Present in Morocco Oct–Mar, with most movement mid-Oct to mid-Nov and late Feb to mid-Mar (latest 26 Mar). Present in Algeria and Tunisia Nov–Mar; Libya, Jan–Feb; Egypt, late Oct to late April (earliest 30 Oct, latest 23 Apr).

Birds recovered Morocco, ringed Finland and Netherlands; recovered Algeria, ringed Britain, Netherlands, Norway, Sweden, Finland and Germany.

with some dark brown streaks and spots; belly white, streaked olive-brown at sides; undertail-coverts white with some olive-brown spots mainly at sides. Primaries and secondaries dark olive-brown with outer edges paler; upperwing-coverts dark olive-brown, outer webs of greater upperwing-coverts with faint buff tip; outer webs of tertials with small whitish tip; axillaries and underwing-coverts rufous. Bill black-brown, base of lower mandible and cutting edge of upper mandible yellow; eyes dark brown; legs and feet flesh-pink. Sexes alike. SIZE: (10 ♂♂, 10 ♀♀) wing, ♂ 111–124 (117), ♀ 111–119 (114); tail, ♂ 75–87 (81·7), ♀ 73–82 (78·4); bill, ♂ 15–17 (16·0), ♀ 15–17 (16·1); tarsus, ♂ 22–28 (25·7), ♀ 24–28 (25·6). WEIGHT: unrecorded in Africa; in Europe, unsexed (n = 143, Norway, autumn) 50–84 (68·4), (n = 53, Germany, on migration) 47–77 (58) (Cramp 1988).

IMMATURE: like adult but mantle, scapulars and back spotted, each feather with pale buff shaft-streak and sooty tip; breast, flanks and belly with sooty spots (not streaks), largest on breast and flanks; outer webs of tertials and greater upperwing-coverts with buff tips. Axillaries and flanks paler rufous.

Field Characters. A small thrush with dark upperparts, streaked and speckled underparts, told by bold creamy eye-stripe and moustachial stripe framing dark ear-coverts; also by conspicuous rufous flanks (not always visible in perched birds) and underwing. Similar-sized Song Thrush *T. philomelos* told by plain face without pale stripes, no rufous on flanks and in flight by orange buff underwing.

Voice. Tape-recorded (62, 73, 89, 93, B). Song variable, usually 4–6 fairly loud, flute-like phrases, often descending, then followed by a prolonged, low, quieter warble; sometimes a throaty chuckle at end recalling

Food. In Africa nothing recorded except fruits of olive trees. In Europe invertebrates (mayflies, dragonflies, crickets, bugs, moths, butterflies, flies, sawflies, ants, beetles, spiders, sandhoppers, millipedes, small crabs, snails, slugs, bivalves, earthworms and marine worms); also seeds of conifers and seeds and fruits of flowering plants.

References
Cramp, S. (1988).
Simms, E. (1978).

Plate 3

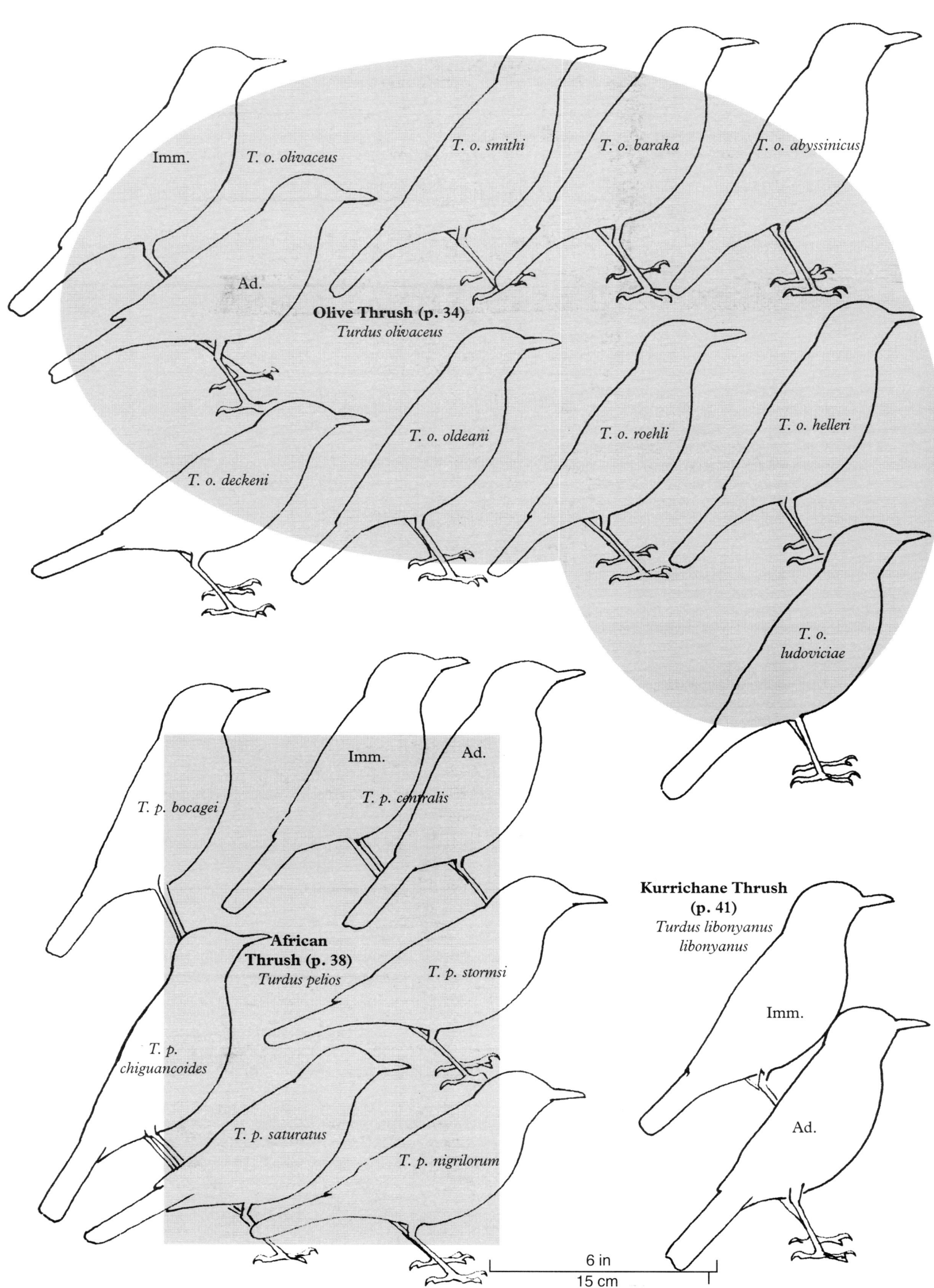

Plate 4

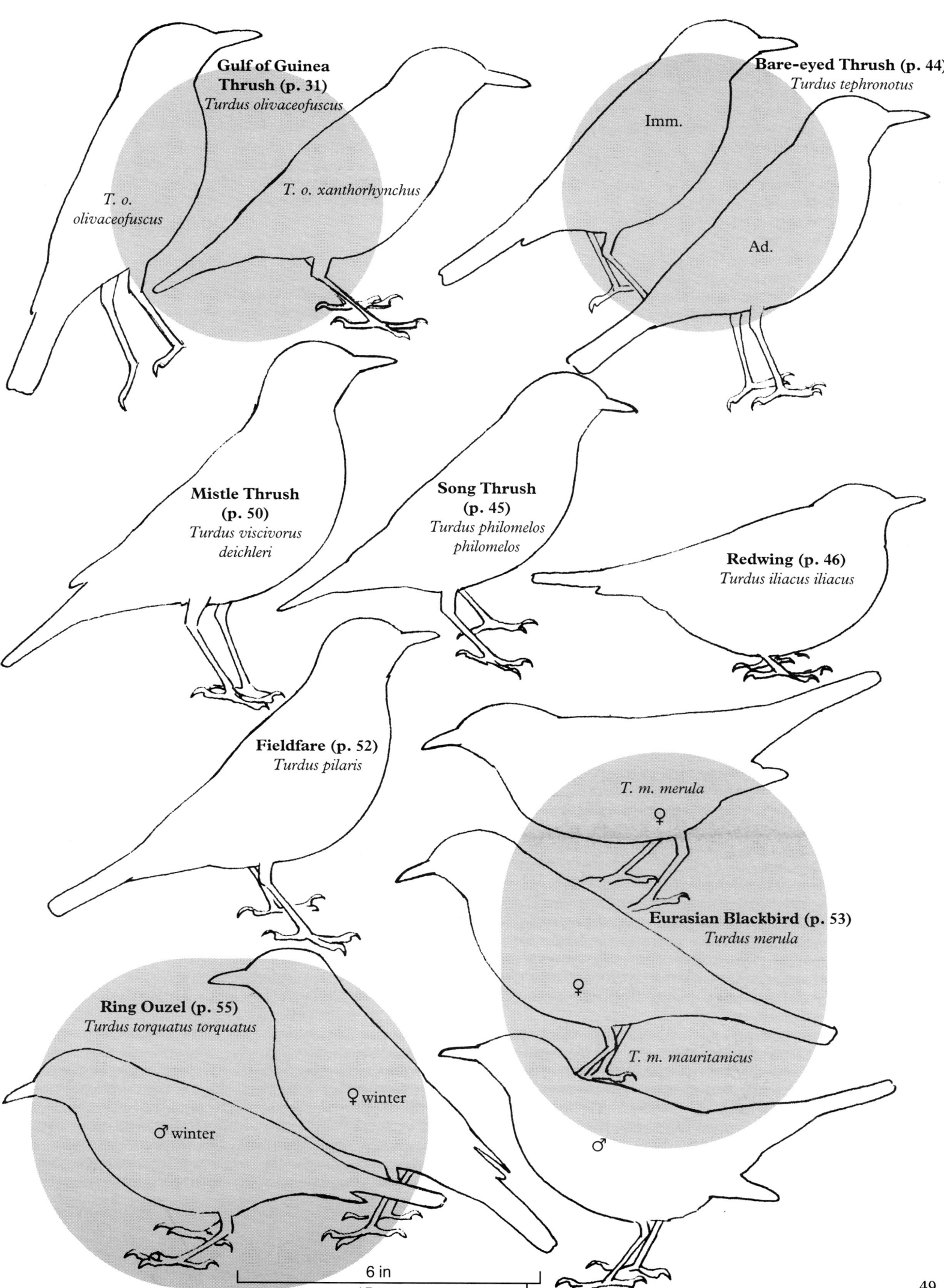

Plate 4
(Opp. p. 49)

Turdus viscivorus Linnaeus. Mistle Thrush. Grive draine.

Turdus viscivorus Linnaeus, 1758. Syst. Nat. (10th ed.), p. 168; England.

Range and Status. Europe to central Siberia, Mediterranean basin, NW Africa, Asia Minor, Afghanistan, W Himalayas and W China; winters in southern part of breeding region.

Resident in NW Africa; winter visitor to NW Africa and a few to NE Africa. Breeds in Morocco, in Tangier Peninsula, Rharb region, Middle Atlas, Central Plateau and High Atlas; Algeria, in Saharien Atlas (Djelfa and to west) and mountains along Tunisia border (Tebessa); and Tunisia, in mountains along Algeria border and central plateau. Uncommon to frequent in all 3 countries.

Winter visitor Morocco to Egypt. In Morocco, uncommon in south to *c.* 31°N; common in some years along Mediterranean coast and Strait of Gibraltar (flocks of up to 150 Dayet Afourgah, Dec). Algeria, present in winter (Mayaud 1988). Tunisia, uncommon to frequent south to Gabès. Libya, uncommon and irregular, sometimes frequent, on Tripolitanian coastal belt (Tripoli, Wadi Rami, Wadi Turghat) and Jefren Escarpment, vagrant Tobruk area. Egypt, uncommon, but when European winters harsh, common Nile delta, Alexandria, Giza, NW coast (Matruh) and Suez.

On passage Morocco across Strait of Gibraltar (large flocks late Oct), Tunisia along coast (flocks of 15–20 at Cap Bon, end Oct) and Libya on Tripolitanian coast (occasionally flocks of up to 30, mid-Feb), and Cyrenaican coast (small numbers 1959–1961 but none since: Bundy 1976).

Description. *T. v. viscivorus* Linnaeus: Europe to China; winters in southern part of range including Africa. ADULT ♂: upperparts from forecrown to uppertail-coverts greyish olive-brown, back and rump with golden brown wash, feathers of uppertail-coverts edged whitish. Tail-feathers grey-brown, T6 to T4 with varying amount of white on tips. Lores and feathers in front of and below eye greyish white; distinct eye-ring buff-white; ear-coverts, cheeks and sides of neck pale buff with brown spots, rear border of ear-coverts and patch below rear corner of eye blackish; chin and throat buff with small sooty brown spots. Breast with many large, wedge-shaped sooty-brown spots, upper breast golden buff, lower breast whitish; belly whitish, with some large round sooty brown spots; undertail-coverts whitish with golden buff wash and without spots. Primaries, secondaries, and upperwing-coverts grey-brown, edges of outer webs pale buff but greater coverts with buff edges and greater primary coverts black-brown with olive-brown outer web edges; axillaries and underwing-coverts whitish. Bill dark horn-brown, base of lower mandible yellow; eyes brown; legs and feet yellow-brown. Sexes alike. SIZE: (10 ♂♂, 10 ♀♀) wing, ♂ 143–156 (152), ♀ 142–154 (147); tail, ♂ 96–115 (106), ♀ 95–110 (104); bill, ♂ 17–21 (18·7), ♀ 18–21 (19·1); tarsus, ♂ 26–33 (29·2), ♀ 27–32 (30·3). WEIGHT: unrecorded in Africa; in Britain, unsexed (n = 15, Sept–Nov) 117–139 (127), (n = 5, Mar) 117–148 (131).

IMMATURE: like adult but head paler, with fine brown bars and spots; feathers of mantle with pale buff centres and blackish tips; lores, chin and throat almost white; chin and throat without spots, underparts deeper buff with smaller spots.

NESTLING: covered with fairly long buffish white down.

T. v. deichleri Erlanger: Morocco, Algeria, Tunisia; also Corsica and Sardinia. Paler and greyer than nominate race; bill more slender.

Turdus viscivorus

Field Characters. A large thrush with greyish upperparts, creamy white underparts with bold black spots, white underwing and white corners of tail obvious in flight. Much larger than Song Thrush *T. philomelos* which is less bulky, browner above, with shorter tail, smaller spots on breast and orange-buff underwing. Closes wings during flight like Fieldfare *T. pilaris* but Fieldfare has greyish head, chestnut back and rusty breast. Often draws attention to itself with rattling flight call.

Voice. Tape-recorded (CHA – 62, 73, 89, 93, B). In Europe, song, 3–6 (up to 12) loud, far-carrying rather monotonous phrases; usually a short pause between each phrase; same phrase may be given 2–3 times in succession. Recalls Eurasian Blackbird *T. merula* in tone and sluggish Song Thrush in form (Jonsson 1993). Has a small song repertoire (8–20 variations: Ince and Slater 1985). Subsong, harsh, rambling notes with warbles, sometimes a faint version of full song with rapid warbling passages. Rarely mimics other species. Flight call and alarm, loud rattling 'khrr' or 'rrr', like drawing a piece of wood over coarse teeth of a comb; also in flight, 'churr'. Recognition call, 'tuc, tuc'. (For further details see Cramp 1988.)

General Habits. In N Africa favours cedar, pine, oak and juniper forests above 600 m, usually 1500–2700 m; also palm groves, gardens, lawns and fields; avoids semi-arid and arid habitats. In Europe, inhabits stands of trees with open grassland and scrub; often in fields in winter; avoids dense forest and treeless areas.

Occurs singly, in pairs, family groups and loose parties of up to about 12; more gregarious on migration, with flocks of 150 or more. Often mixed with Fieldfares but less often with Redwings *T. iliacus* and Ring Ouzels *T. torquatus*. Wary and shy. When perching, flicks tail and wings in excitement, giving 'tuc, tuc' call. Flight strong and direct; closes wings but flight does not undulate; flies higher than most thrushes, up to 30 m above ground. Roosts in trees and hedges, in groups of up to 40 but usually less. Sometimes bird or pair maintains feeding territory in winter, defending area with radius of up to 100 m around tree. Forages on ground in open, also in bushes and trees; runs or hops, stops and stands upright with head held up, tail down and wings often drooped, then pecks at food; rarely moves litter with sideways sweeps of bill. Sometimes flies 15 m up in air to capture insect or take berries from tree (does not hover). Often sings when other *Turdus* thrushes do not, especially in poor weather such as light rain.

NW Africa race mainly sedentary, although birds move locally (Tunisia: Mayaud 1988). Origin of European birds wintering in N Africa unknown except for a bird ringed in Austria, recovered in Morocco. Migrants present Morocco mid-Oct to mid-Feb; in Tunisia earliest date 31 Oct; Libya, late Oct to mid-Feb; and Egypt mid-Oct to early Mar.

Food. Unrecorded in Africa. In Europe, grasshoppers, earwigs, bugs, moths, butterflies, flies, ants, beetles, spiders, millipedes, snails, slugs, earthworms, occasional young birds; fruits and seeds of conifers and flowering plants, moss and fungi. In Israel eats many snails. In Britain (1924–27), 55% of diet invertebrates, 45% plant material (Cramp 1988). Of total weight of contents of 68 stomachs (SE Spain), 71·4% was vegetable matter with a predominance of fruits such as olives and grapes; of total weight of animal parts, 63·7% were beetles and 63·2% were invertebrates that live on top of the ground (Perez-Gonzalez and Soler 1990).

Breeding Habits. Monogamous; territorial, in some places somewhat colonial. Territory size in Morocco 5·2 pairs/km^2; in Europe 0·6 ha to 'several acres' (Cramp 1988). ♂ and ♀ often chase each other as bond forms. During pause in chase, ♀ opens and shivers wings while ♂ spreads wings and fans tail, revealing the white tips. ♂ sometimes feeds ♀ just before copulation. Pair-bond established 5–7 days after arrival in territory. ♂ defends territory by singing from high perch, often top of tree; sings before daylight, up to 1000 songs/h. In song-flight ♂ flies for several hundred m, singing constantly. ♂, bold and aggressive, chases away other ♂♂ and other larger bird species. ♀ sometimes assists ♂ in defending territory.

NEST: large cup, built of grass, plant stems, roots and mosses bound together with mud. Ext. diam. 150–240, int. diam. 86–140, ext. depth 84–110, int. depth 55–72. Placed 2–10 m from ground in fork of tree, in Morocco cedar *Cedrus atlantica*, aleppo pine *Pinus halepensis*, holm oak *Quercus ilex* and juniper.

EGGS: 3–4, sometimes 2. Morocco, 2 C/2, 15 C/3, 16 C/4; Algeria and Tunisia, 14 C/3, 19 C/4. Laid at 1 day intervals. Subelliptical; smooth, glossy; bluish green to pale buff, spotted and blotched with shades of red-brown to purple; sometimes prominent marks at broad end. SIZE: (n = 250, Europe) 26·5–34 × 20·2–24 (30·2 × 22·3). Sometimes double-brooded.

LAYING DATES: Morocco, Algeria and Tunisia, Mar–June (extremes 28 Mar–18 June).

INCUBATION: begins when clutch complete, by ♀, occasionally by ♂ for short periods (up to 6 min). ♂ may feed ♀ when she incubates. ♀ leaves nest 8–10 times daily, for up to 20 min at a time. Period: 12–15 days.

DEVELOPMENT AND CARE OF YOUNG: on days 14–16, young leave nest when almost able to fly; they can fly by day 20; about days 30–35, sometimes follow parents to feeding grounds, begging for food; independent by about day 35. Young of 1st and 2nd broods often join up to form single family (Cramp 1988).

Young brooded by ♀ or occasionally ♂ for 4–6 days; thereafter brooded only in bad weather. Both parents feed young; ♂ looks after them when ♀ begins 2nd clutch. Parents swallow faecal sacs or carry them off. Both defend nest by chasing predators with loud rattling alarm call or attacking them.

BREEDING SUCCESS/SURVIVAL: no information for Africa; in Britain 40% of 435 nests produced at least 1 fledged young; annual adult mortality 48% with mortality of 1-year-old fledged young 62%. Adult ringed bird in Europe lived 9·5 years.

References
Cramp, S. (1988).
Simms, E. (1978).

Plate 4
(Opp. p. 49)

Turdus pilaris Linnaeus. Fieldfare. Grive litorne.

Turdus pilaris Linnaeus, 1758. Syst. Nat. (10th ed.), p. 168; Sweden.

Range and Status. S Greenland, N and central Europe (south to Austria) east to Siberia (L. Baikal); winters throughout Europe, to central Asia and NW India; irregularly in N Africa.

Irruptive Palearctic winter visitor to N Africa, in most years uncommon or absent, but in some years common, due to harsh winters in Europe. In Morocco uncommon (N coast, Tangier Peninsula: Cap Sartel, Charf-el Akab, Rif) but some years common (occasionally large numbers cross Strait of Gibraltar); Algeria, uncommon (Lion Mts near Oran, slope of Mt Zaccor, L. Oubeira *c.* 60 km east of Annala); Tunisia, most years uncommon in coastal regions (Maklar, Sousse, El Djem, Bahiret el Bibane) but common 1940–41 (south to Sousse); Libya, uncommon (Tarhuna, Brace, Al Adem), absent 1968–69, but common 1965–66 and 1969–70 (Wadi Kamm, Wadi Turghat, Jefren Escarpment); and Egypt, uncommon, but some years common and other years absent (most records Nile delta south to Cairo, west to Alexandria; others, lower Nile south to Sohag and between Luxor and Dendera, N coast Matruh, Wadi el Natrun, Bahariya, L. Qarun, and Suez).

Turdus pilaris

Description. ADULT ♂: forehead, crown, nape, hind neck, sides of neck and ear-coverts bluish grey, slightly tinged brown; feathers of forehead and crown with brownish blue streaks. Mantle and scapulars rufous-chestnut with black mottling; back, rump and uppertail-coverts bluish grey tinged brown. Tail brownish black; tips of inner webs of 2 outer tail-feathers with whitish margin. Indistinct grey-white supercilium; lores and patch under eye black; cheeks grey; chin cream-white; throat orange-buff boldly marked with brown-black scaly spots. Upper breast and flanks orange-buff boldly marked with brown-black scaly spots; flanks with indistinct bluish grey markings; lower breast, belly and undertail-coverts with indistinct bluish grey markings. Primaries and secondaries black; outer edges of primaries pale grey, of secondaries cinnamon-brown. Upperwing-coverts rufous-chestnut with black mottling except greater coverts blackish brown, edged rufous and tipped grey; axillaries and underwing-coverts white. Bill black-brown; eyes dark brown; legs and feet dark brown. ADULT ♀: like ♂ but duller; head, nape, hindneck, sides of neck and ear-coverts tinged more brown; forehead and crown with fewer streaks; underparts paler, less orange-buff and less heavily streaked. SIZE: (19 ♂♂, 10 ♀♀): wing, ♂ 133–148 (143), ♀ 137–147 (141); tail, ♂ 90–110 (99·6), ♀ 91–107 (98·1); bill, ♂ 16–19 (17·4); ♀ 16–19 (17·7); tarsus, ♂ 30–33 (31·4), ♀ 29–33 (31·4). WEIGHT: unrecorded in Africa; in England, Nov–Dec ♂ (n = 15) 105–132 (115), ♀ (n = 17) 101–141 (117).

IMMATURE: like adult but upperparts browner, sometimes olive-brown; mantle feathers with pale buff central streaks; chin and throat whitish without spots; upper breast and flanks with rounded, not scaly spots; upperwing-coverts with whitish fringes.

Field Characters. A medium-sized thrush, between Song Thrush *T. philomelos* and Mistle Thrush *T. viscivorus* in size; easily told from congeners by combination of bluish grey head and rump and rufous-chestnut back. Often draws attention to itself, especially in flight, by characteristic 'chack chack' call. Similar to Mistle Thrush in flight with white underwing, but Mistle Thrush larger and longer-tailed, with bolder spots and rattling call.

Voice. Tape-recorded (62, 73, 89, 93, B). Song, an almost tuneless medley of chuckles, squeaks and whistles, weak and feeble, suggesting poor song of Eurasian Blackbird *T. merula* but with chattering notes; rendered 'took-took-cherri-weeoo' or 'took-took-took-cherri-wee-chee' (Simms 1978). Has song repertoire (26–28 variations) larger than Mistle Thrush, Ring Ouzel *T. torquatus* and Redwing *T. iliacus* but smaller than Eurasian Blackbird and Song Thrush (Ince and Slater 1985). Subsong, like full song but fainter and more guttural and warbling. Commonest call, a clear raucous 'chack-chack' or 'chak-chak-chak' or 'chacker-chack-chack', used as contact when feeding and in flight. Other calls: a 'huee huit' when migrating; soft 'quok' when ground predator about; deep resonant 'wu wu' in threat; and agitated 'tjetjetjetje' in alarm. Also bill-snapping sounds when threatening conspecifics and other species (Cramp 1988).

General Habits. Inhabits boreal forests, mixed pine and birch woods, forest edges and clearings, parks and gardens when breeding; in non-breeding period tends to occupy open country more than forests.

Gregarious, in flocks of up to 300 or more in harsh winters. Often forms loose flocks scattered over fields;

flies into trees if disturbed, rarely escaping into undergrowth. Occasionally solitary or in pairs. Wary in non-breeding season; noisy. Flight direct; wing-beats alternating with brief glides with wings closed; not undulating as Mistle Thrush; travels up to c. 48 km/h and at heights 100 m above ground. Flocks of up to several hundred. Roosts communally in conifers, hedgerows, shrubs, reeds, sometimes only 1 m from ground, once (Britain) in mixed group of up to 15,000–20,000 with Redwings (60% Fieldfare, 40% Redwing: Simms 1978).

Some individuals defend feeding territories during non-breeding season. Forages on ground, also in trees and bushes; runs along ground in short bursts or hops, stops, scans ground, then pecks at prey on surface or in soil; very occasionally bill-sweeps loose litter to one side with bill. Takes more in-soil than soil-surface items (63% in-soil: Carlson and Moreno 1986). Sometimes takes aerial insects, flying high in air to catch them; feeds on berries and fruits in trees. Bird feeding in mixed-thrush flock (Redwings, Song Thrushes), threatens in horizontal posture, with plumage ruffled, flicking and shaking fanned tail, twitching slightly open wings and raising bill almost vertically; then lowers head and tail and rushes at rival (Cramp 1988). In spring, migrating ♂ commonly subsings, and both sexes regularly give 'chack' call.

From breeding areas in N Europe migrates south and west to winter in S Europe, although some cross Mediterranean. Flocks smaller in autumn than spring, when several thousand can pass by in a matter of hours. Present Morocco mid-Oct to mid-Feb; Algeria, latest 20 Apr; Tunisia, present Jan; Libya mid-Oct to early Mar with some movement mid-Nov to early Dec; Egypt, most mid-Nov to mid-Feb (earliest 1 Nov, latest 14 Mar).

Food. In Africa unrecorded. In Europe, dragonflies, crickets, bugs, scorpion flies, caterpillars, flies, ants, beetles, spiders, harvestmen, millipedes, snails, slugs, earthworms, leeches, fish up to 7 cm long (Cramp 1988), fruits, berries and seeds of conifers and flowering plants, shoots and buds. In Sweden, earthworms commonest prey (Carlson and Moreno 1986); in Britain (Sept–May), 37·5% insects, 14·5% earthworms, 4·5% slugs, 2·5% other invertebrates, 36% fruits and seeds, 5% other plant materials (Cramp 1988).

References
Cramp, S. (1988).
Lübcke, W. and Furrer, R. (1985).
Simms, E. (1978).

Turdus merula Linnaeus. Eurasian Blackbird. Merle noir.

Turdus Merula Linnaeus, 1758. Syst. Nat. (10th ed.), p. 170; Sweden.

Plate 4
(Opp. p. 49)

Range and Status. N Africa and Europe across Asia to China; resident and partial migrant with some wintering in S Europe, N Africa and Middle East.

Resident race in N Africa (Morocco to Tunisia; also Egypt). European races visit NW Africa and Egypt. In Morocco resident south to about Draa, Agdzi, Tafilalet, Erfoud and Zagora, common; N Algeria, including Tell, south to Saharien Atlas (Figuig, Aïn Sefra, Djelfa, Biskra), uncommon (Ledant *et al.* 1981), not observed in recent years (Mayaud 1988); Tunisia, south to about Kasserine and along coast to El Djem, common in N and E, less common in SW. First known to breed in Egypt's N Sinai in 1975 (El Arish) and by 1984 locally common resident there (El Arish, Rafa area). In mid-1980s common resident in S and central Nile delta, and in 1993 nesting south to Cairo (Gezira) and Fayoum Oasis (Evans and Dijkstra 1993).

Winter visitor to Morocco, evidently throughout range of resident race (south to about Draa and Zagora), scattering of records most years, occasionally flocks of up to 200 in autumn. Algeria, N coastal belt south to Aïn Sefra and Biskra, uncommon; Tunisia,

coastal areas, common but irregular; Libya, regular near coast south to Jebel Nafusa Escarpment (mainly immature ♂♂: Bundy 1976), common Tripolitania, frequent Cyrenaica, vagrant in desert, Serir (twice: Jan, Feb); Egypt, N coast, Nile delta and valley south to Cairo and Suez Canal area; in some years commoner than in others, rare in desert oases (Siwa, Qattara Depression, El Maghra and Wadi el Natrun). Sudan, vagrant Port Sudan 1912.

On passage Morocco (Tangier, commonest 1st week Apr), Algeria (N coast at Reghaïa Oct–Nov), Tunisia (autumn only), and Egypt (N coast at Bahig, mid-Feb to mid-Apr).

Description. *T. m. merula* Linnaeus: Europe; winters throughout African range. ADULT ♂: entirely sooty black. Bill yellow to orange-yellow; eye-ring orange-yellow; eyes dark brown; legs and feet dark brown. ADULT ♀: upperparts from head to uppertail-coverts rufescent to brown-black. Tail feathers dark brown to black. Ear-coverts, cheeks and sides of neck brown-black, ear-coverts finely streaked off-white; chin and throat pale brown to black-brown, streaked with pale grey. Breast rufous-brown with triangular dark brown spots; belly and undertail-coverts blackish brown. Bill dark brown to yellow; eye-ring yellow; eyes dark brown; legs and feet dark brown. SIZE: (10 ♂♂, 10 ♀♀) wing, ♂ 121–132 (126), ♀ 120–128 (122); tail, ♂ 83–110 (99·9), ♀ 86–108 (95·5); bill, ♂ 20–23 (20·6), ♀ 19–21 (19·8); tarsus, ♂ 28–32 (30·4), ♀ 24–33 (29·9). WEIGHT: 1 ♂ (Egypt) 60·4; ♀ (n = 2, Egypt) 61·5, 61·5; ♂♀ (n = 32, Israel) av. 82·8 (see Cramp 1988 for European data).
IMMATURE: ♂ like adult ♂, but body-feathers tinged rufous; head to mantle with rufous streaks; underparts rufous-buff; feathers of breast often edged and tipped brown, giving heavily spotted appearance; throat pale; indistinct pale buff malar streak; flight feathers brownish, innermost greater coverts with prominent rufous streaks; bill blackish. ♀ like adult ♀, but upperparts paler brown, mantle with paler shaft-streaks, indistinct pale buff malar streak; underparts more rufous and more spotted; bill dark brown.
NESTLING: sparsely covered with dark grey or pale buffish grey down, eyes closed.

T. m. mauritanicus Hartert: Morocco, Algeria and Tunisia. ♂ like nominate race but glossier, ♀ darker and greyer than nominate ♀, without rufous; heavier-billed than nominate race.

T. m. aterrimus (Madarász): E Europe and SW Asia (N Iran); winter visitor to Egypt (Cramp 1988 where no specific localities given; race not included in Goodman and Meininger 1989). Like nominate race but ♂ duller black, ♀ duller and greyer; also smaller and weaker-billed.

T. m. syriacus Hemprich and Ehrenberg: SW Asia east to S Iran, vagrant Libya (1, Tobruk) and Egypt (Matruh, Burg el Arab, Bahig); also intermediates between *merula* and *syriacus* from Matruh, Bahig and Nile delta). Racial identity of Nile delta breeding population unknown (Goodman and Watson 1984); birds intermediate in colour between *merula* and *syriacus* (Goodman and Meininger 1989). (A small population in Sinai seems to be *syriacus*.) Like nominate race, but ♂ slatier, ♀ greyer, less rufous below; throat with fewer streaks.

Field Characters. A medium-sized thrush distinguished from all other Palearctic *Turdus* spp. by its black or dark brown plumage. ♂ is jet black with yellow bill and yellow eye-ring, ♀ dark brown above, paler below with indistinct dark mottling but without distinct bold spots of Song Thrush *T. philomelos* and Mistle Thrush *T. viscivorus*. Tail rather long and rounded; in flight wing-tips appear rounded.

Voice. Tape-recorded (CHA – 62, 73, 89, 93, B). Song, one of the world's finest, a series of rich flute-like whistles. Sometimes ends with feeble chuckling notes. Given by ♂; ♀ has a subdued, strained version during courtship. Richer, mellower and lower-pitched than Song Thrush, with almost no repetition. Has repertoire of 20–51 songs, fewer than Song Thrush but more than other Palearctic *Turdus* thrushes (Ince and Slater 1985). Subsong a subdued, continuous warbling, sometimes with alarm notes interspersed. Mild alarm (as when cat spotted at distance) 'chook' or 'tuk-tuk', often delivered from bottom of hedge. Fright, a loud, somewhat explosive rattle, 'tuk tuk tuk tchook-tch-WEE tchWEE cheWEEcheeWEE cheWEEcheWEE tchook chook'. Characteristic going-to-roost call, same as mild alarm. Also 'chuk chuk chuk' in flight, 'ziep' as warning. Sometimes mimics; also snaps bill (see also Cramp 1988).

General Habits. Inhabits broad-leaved and coniferous woods and forests, spending much time in undergrowth near open areas; also hedgerows, scrub, citrus groves, eucalyptus, palm groves, olive plantations, farmlands, gardens and parks. In Morocco up to 2300 m in High Atlas, rarely to 2700 m, in *Oleo-Lentisetum* stands, *Calenduleta-Junipericum* forest and oak forest; in Egypt, oases, gardens and mango and guava plantations. Absent in areas with less than 300 mm rainfall per year except in oases.

Solitary or in pairs. Seasonally territorial, some individuals and pairs maintaining territories all year. Feed singly or in twos or threes, sometimes in groups of up to 10–20. Roosts singly, in small groups or occasionally up to several hundred (Cramp 1988) and once 2000 (England: Simms 1978). Migrate in loose flocks of up to 10–20, occasionally up to 200. Skulking, readily moves to cover. Flight agile, direct, rapid; sometimes glides or swoops for 1–3 m in open spaces; easily flies into and passes through dense cover. Upon landing, usually flits tail up vertically; then drops it slowly and droops wings. Frequently flicks wings and cocks tail. Runs or hops, pauses, then runs and hops again. Stands less upright than other thrushes with tail held horizontally or slightly raised. Feeds mainly on ground, also often on berries in trees and bushes. Turns leaf litter with sideways sweep of bill, sometimes scratching litter with a backward jump on 1 or both feet (Clark 1983). Pulls earthworms from ground. Rarely breaks snails on anvil, and steals food from other birds (including snails extracted from shells by Song Thrush). Sings commonly in breeding season, rarely otherwise, but subsong can be given at any time of year.

Population breeding in N Africa is evidently sedentary; some birds south of breeding range in Morocco and Tunisia suggest possible wandering after nesting. European and Asian birds winter in Morocco from early Nov to first week Apr; in Algeria from Oct–Nov to Mar; in Libya mainly from Oct (earliest 3 Oct) to Apr (latest

20 Apr); in Egypt from late Oct to mid-Apr (earliest mid-Sept). Birds ringed in Germany, Poland and Belgium recovered in Algeria; 1 bird ringed Morocco (Cap Bon) recovered Italy 2 years later, and 1 ringed Egypt, recovered Cyprus 3 years later.

Food. Unrecorded in Africa. In Europe mainly insects, earthworms and fruits (springtails, mayflies, damselflies, grasshoppers, earwigs, cockroaches, bugs, lacewings, moths, butterflies, flies, ants, wasps, bees, beetles, spiders, harvestmen, woodlice, snails, slugs, earthworms, small fish and frogs; also mammalian faeces; fruits and seeds of conifers and broad-leaved trees).

Breeding Habits. No data for Africa; in Europe monogamous, pair remaining together for life. Territorial; population density 1 pair/km^2 (open country) to 200–300 pairs/km^2 (suburban habitats, Britain, Germany) with nests sometimes 10 m but usually 20–30 m or more apart. Both birds defend territory, ♂ fighting ♂♂, ♀ fighting ♀♀. Up to 6 birds involved in territorial chases. When ♀ first moves into ♂ territory, ♂ chases her from it, with head stretched up, crown-feathers raised, tail fanned, wings drooped and back hunched. Courtship begins when ♂, with head stretched out and tail fanned, bows his head, then turns around or jumps up in front of ♀. He may fly with slow wing-beats from perch to perch, and may touch ♀'s bill. ♂ sings in flight or from high perch, often a house roof top or TV aerial. Once pair formation is completed, ♀ becomes dominant member. At copulation, 1–5 days before egg-laying, ♀ first raises head with bill pointing almost vertically and places tail almost horizontally; she then walks to ♂ with wings trembling.

NEST: large bulky cup, built of grasses and small twigs, packed with mud, and lined with fine grass and stems. Mud hardens, and nest often lasts a year or more even when regularly rained on. Average nest measurements (n = >500): ext. diam. 152 (120–260), int. diam. 98 (75–150), ext. depth 124 (50–300), int. depth 67 (40–140). Built 0·5–15 (1·9) m above ground in fork of small tree or bush; sometimes on wall of house or well; rarely on ground. ♂ and ♀ look for nest sites; ♀ decides on it. Constructed in 2–14 (av. about 6) days, mainly by ♀; egg-laying starts c. 2 days after nest completed.

EGGS: 2–6 (usually 3–4). Morocco, 1 C/2, 25 C/3, 49 C/4, 6 C/5, 1 C/6. Laid at 1 day intervals. Oval, somewhat blunt at large end; smooth, glossy; bluish green to grey-green, marked with red-brown spots concentrated at blunt end. SIZE: (n = 250, race *merula*) 25–34·5 × 18·5–23·5 (29·3 × 21·4). 2–3 (sometimes 4) broods per year (Europe).

LAYING DATES: Morocco, Mar–June (Mar 4, Apr 7, May 5, June 5 clutches), (carrying nest material Feb); Algeria, Mar–July; Tunisia, Apr–Sept; Egypt, Mar–July.

INCUBATION: begins with laying of penultimate egg, or sometimes when clutch complete; by ♀ who spends 90% of day and all night on nest. When ♀ away, ♂ guards nest but does not incubate. Period: 10–19 days (av. 13, n = 184, central Europe).

DEVELOPMENT AND CARE OF YOUNG: on day 7, first juvenile feathers appear; by day 8 eyes open; by days 9–10 well-feathered; on days 13–14 (end of week 2) young leave nest but remain in cover; at end of week 3 move into open; and by end of week 5 feed on their own (Snow 1987). Breed when 1 year old.

Young brooded by ♀ almost constantly for first 3–4 days, then decreasingly until brooding only at night when young 8 or more days old. Only ♀ feeds young for first few days, then both parents feed them. Both swallow faecal sacs or carry them away. Known rarely to soak breast-feathers in water, then fly to young who drink from them (Cramp 1988). When ♀ starts second clutch, ♂ takes over feeding of first brood. Once, a second ♂ assisted a pair in raising young (N coast France: Erard 1990b). Young fed earthworms or insects if earthworms not available. Parents protect young by giving warning calls, sitting tight, or by striking at intruder; rarely they perform distraction display by running to and fro in a crouched position.

BREEDING/SUCCESS SURVIVAL: no data for Africa. Of 1428 nests (Britain) 1 young hatched in 56% and at least 1 young fledged in 41%; of 6664 eggs laid in 1601 nests (Czechoslovakia) 35·7% lost before fledging (Cramp 1988). Hazards for survival in W Palearctic include predators, desertion as a result of disturbance, unseasonable weather, nest collapse and starvation. Oldest ringed bird in Europe 20 years (see also Cramp 1988).

References
Cramp, S. (1988).
Pikula, J. and Beklova, M. (1983).
Simms, E. (1978).
Stephen, B. (1985).

Turdus torquatus Linnaeus. Ring Ouzel. Merle à plastron.

Turdus torquatus Linnaeus, 1758. Syst. Nat. (10th ed.), p. 170; Sweden.

Plate 4
(Opp. p. 49)

Range and Status. Europe east to Armenia, Azerbaijan, W Turkmenistan and N Iran; winters in S of range and south to Mediterranean, N Africa, Sinai and S Iran.

Winter visitor to NW Africa and Nile delta; possibly breeds in Algeria. Morocco, mainly Middle and High Atlas, common, but also in oases south to Goulimine, frequent to locally common; recorded on passage

Turdus torquatus

Tangier Peninsula Oct–Nov, Mar–Apr; Algeria in Tell region, High Plateau, Aurès and Saharien Atlas, common, south to Beni Abbès and Ghardaïa; Tunisia in N and central mountains south to Gafsa and Gabès, uncommon to frequent; Libya in NW Tripolitania south to Jebel Nafusa, uncommon, vagrant Fezzan (Hon, Brak); and Egypt on N coast west to Bahig, Nile delta and river south to Cairo, and Suez Canal area, uncommon. Vagrant Mauritania (2 records Nouakchott) and Sudan (Dongola, Khartoum).

Following is evidence for breeding in Algeria: 1 pair giving alarm call Aurès (Rhouf), Apr 1977, 2 ♂♂ fighting Djurdjura Apr 1980, ♀ present Djurjura, 10 June 1977 (Ledant et al. 1981, Mayaud 1988); 'adults seen feeding young by D. Vallet' Djurdjura Nat. Park (date?) and 'several observations of breeding (Djurdjura) reported by Ochando and de Smet' in the 1980s (E. D. H. Johnson, pers. comm.).

Description. *T. t. torquatus* Linnaeus: N and W Europe; winters south to Morocco and Algeria (and probably this race in Mauritania). ADULT ♂: mainly sooty black; breast with broad whitish band, feathers with indistinct brown edges. Feathers of mantle to rump, and uppertail-coverts with dark brown edges, those of underparts from chin to belly with white edges; undertail-coverts with broader white edges and some white shaft-streaks. Tail and wing-feathers sooty black; outer webs of primaries and secondaries with fine white edges; underwing-coverts dirty white, underside of flight feathers greyish. Bill yellow, tip of upper mandible blackish; eyes dark brown; legs and feet dark brown. ADULT ♀: like ♂ but entire body more sooty brown, feathers of upperparts with brown edges; feathers of underparts with broad white edges; white upper breast-band less extensive. SIZE: (10 ♂♂, 10 ♀♀) wing, ♂ 137–143 (140), ♀ 135–143 (138); tail, ♂ 89–102 (94·8), ♀ 87–98 (92·5); bill, ♂ 18–22 (19·6), ♀ 18–21 (19·9); tarsus, ♂ 28–34 (30·8), ♀ 31–35 (31·9). WEIGHT: no data for Africa; unsexed (n = 34, Heligoland) 92–138 (111), (n = 5, Norway) 95–120 (av. ?) (and see Cramp 1988).

IMMATURE: similar to ♀ but most body-feathers rusty-brown, feathers of chin and throat with white and buffy edges; upper breast with only slight indication of white band; feathers of undertail-coverts with broad white to buffy edges; edges of flight feathers more cream, less white.

NESTLING: covered with fairly long buff down.

T. t. alpestris (Brehm): central, S and E Europe; winters south to Morocco, Algeria, Tunisia, Libya, Egypt and Sudan. Like nominate race but body-feathers with much broader pale edges, giving overall paler appearance, white edges to wing-feathers more extensive.

Field Characters. Medium-sized, similar to Eurasian Blackbird *T. merula* in black or sooty brown colour but readily distinguished from it by scaly plumage (in all except breeding ♂ of nominate race), silvery wing-patch, and breast-band, pure white in breeding ♂, white obscured by dark scales in ♀ and winter ♂. First winter ♀ lacks pale band, and immature is spotty rather than scaly. Race *alpestris* paler and more scaly than nominate *torquatus* in all plumages. Calls distinctive.

Voice. Tape-recorded (62, 73, 93, B). Song, 2–4 simple, fluty, piping notes, rendered 'pi-ree pi-ree pi-ree' or 'pee pee pee' (Cramp 1988), and repeated several times with a brief pause between phrases. Each ♂ has limited repertoire, sometimes only 2–4 song types; often repeats one several times, shifts to next one, then eventually returns to original song. Subsong, a series of piping notes interspersed with chuckling. Calls include characteristic loud rattling 'tac-tac-tac', harsher 'chrechrechrechra' given in alarm, 'tjuck' or 'tjuck-uck-uck' when on ground or taking off, chuckling rattle rendered 'tchook-tchook-tchook' when migrating or in flocks; also 'dcharr', 'tschirh' or 'zrrp' in flight (Cramp 1988).

General Habits. Inhabits dry and rugged slopes of mountains from 1200 to 2700 m, with conifers, often near open areas. Also inhabits small trees and shrubs on cliffs and sides of ravines, sometimes on open broken ground, and occasionally visits palm groves in oases.

Solitary or in groups of 2 or 3, but often in flocks when feeding, roosting or migrating. Gathers sometimes in trees in feeding flocks with other thrushes, especially Song Thrush *T. philomelos*, Redwing *T. iliacus* and Mistle Thrush *T. viscivorus*. Migrates singly, or ♂♂ and ♀♀ together in flocks of 5–10, sometimes up to 50. The shyest N African thrush. Flight rapid, direct, with little undulation. When landing, cocks tail like the Eurasian Blackbird. Feeds equally on ground and in tree. On ground, runs, pauses, runs again, then pecks at food; does not move leaf-litter sideways with bill. Sometimes sings when migrating in spring but not on winter grounds.

Wintering birds arrive in N Africa mainly in Nov (with some in late Sept or Oct) and leave in Mar to mid-Apr. Extreme dates, Morocco 5 Oct–26 Apr, Libya 30 Sept–1 Mar, and Egypt 25 Oct–20 Mar. Race *torquatus* ringed Germany and Britain recovered Morocco and Algeria, race *alpestris* ringed Switzerland and Italy recovered Morocco.

Food. In Africa nothing recorded, except in Algeria in winter mainly berries of junipers (*Juniperus oxicedrus* and *J. phoenicea*). In Europe, in spring and early summer eats adult and larval insects and earthworms, for rest of year eats mainly fruits. In Sierra Nevada, SE Spain, eats mainly berries of *J. communis*, also, by frequency, 5–42% beetles, millipedes, grasshoppers, flies, ants and insect larvae (Zamora 1990). Also eats earwigs, bugs, moths, butterflies, caterpillars, flies, grubs, sawflies, ants, beetles, spiders, millipedes, small snails, slugs, earthworms, lizards, salamanders, and fruits of bramble, strawberries, cherries and olives (Cramp 1988).

Breeding Habits. Monogamous; territorial, with nests 160–200 m and up to 3 km apart. ♂ apparently arrives on territory before ♀. In pair formation ♂ and ♀ walk on ground with heads erect, ♀ advances, ♂ moves up to her, she moves forward and he catches up repeatedly. Both may also flutter up into the air facing each other, then return to ground where they continue walking. ♂ sometimes displays white breast to ♀ and raises tail; he sings from exposed, elevated perch or while flying up into the air. ♂ (and probably ♀) bold and aggressive on territory. ♂ regularly flies after others and will pursue them on ground with tail raised and head pointed upward, displaying white on breast.

NEST: bulky cup of grass, stems, moss and leaves mixed with mud and lined with fine grass. Ext. diam. 155–205, int. diam. 93–113, ext. depth 95–200, int. depth 50–70. Usually on ground, sometimes in shrub or tree 1–16 m (3·5 m) from ground. Both sexes build nest.

EGGS: 4–5 (3–6), laid one per day. Sub-elliptical; smooth, glossy; pale blue to bluish green, marked with small red-brown spots. SIZE: (*torquatus*) (n = 100) 26·3–33·5 × 19·6–23·4 (30·2 × 21·5). 1–2 broods.

LAYING DATES: Algeria evidently Apr (if at all).

INCUBATION: begins with penultimate or sometimes last egg, by ♀, with ♂ guarding nest but not incubating when ♀ away. ♂ sometimes feeds ♀ on nest. Period: 13–14 days.

DEVELOPMENT AND CARE OF YOUNG: young leave nest on days 14 or 15, barely able to fly; independent at about 26–27 days. ♀ broods young most of day and night until days 3–5; by days 10–14 only at night. Both parents feed young, but only ♂ feeds them while ♀ incubating 2nd clutch. Breeds at 1 year.

Reference
Cramp, S. (1988).

Family SYLVIIDAE: Old World warblers

A large family of diverse small songbirds living in leafy vegetation, closely related to Turdidae (thrushes), Muscicapidae (Old World flycatchers) and probably to Timaliidae (babblers).

Small songbirds; sexes generally of similar size and plumage. Bill thin (or fairly stout in some species), straight, of medium length or occasionally long. Nasal and rictal bristles usually vestigial, sometimes strong. Wing-tips short and rounded in non-migratory species, longer and more pointed in migratory ones; 10 primaries, P10 short in non-migratory species, tiny in migratory species. Tail short to medium length, sometimes long; square or rounded but in some warblers graduated or fan-shaped; usually of 12 feathers, but 10 in e.g. *Prinia*. Tarsus scutellate, sometimes booted. Feet usually weak, but quite strong in e.g. *Locustella* and *Acrocephalus*. Plumage plain, usually greenish in foliage-inhabiting species; upperparts often profusely streaked in grass- and reed-living species. Tail often white-sided or with patterned tip on corners. Nestling naked but with scanty down in some species. Some Palearctic species wintering in Africa undergo complete moult before reaching their winter quarters, others have one complete moult on their breeding grounds and a second on their wintering grounds.

Mainly arboreal, inhabiting woods, scrub, reedbeds and grass; some species feed partly on the ground. Eat chiefly insects; also other small invertebrates, small fruits (often berries), seeds and other vegetable matter. Pick prey from vegetation, bird sometimes hovering briefly to do so; some make flycatching sallies. Sustained flight rather weak in non-migratory species, stronger in migratory birds; flight often rapid, straight, direct and agile. Migrate by night, feed by day. ♂♂ often perform aerial displays accompanied by song and/or wing-noises (e.g. *Cisticola*, *Prinia*). Hop; a few species also run. Songs well developed and varied. Typically not social in non-breeding season, feeding and roosting solitarily, but some sylviids join mixed-species flocks when feeding. Usually monogamous but some polygynous. Both sexes (mainly ♀) build nest which varies from single cup to complex domed structure.

Old World warblers were formerly united with thrushes, flycatchers and babblers as a single, huge, amorphous family. Sibley and Monroe (1990), following the DNA-DNA results of Sibley and Ahlquist (1990), separated Old World warblers into 3 families: Regulidae (kinglets), Cisticolidae (African warblers) and Sylviidae. The last was divided into 4 subfamilies: Acrocephalinae, Megalurinae, Garrulacinae (laughing thrushes) and Sylviinae (*Sylvia*, babblers and wrentits). Further, they interposed bulbuls (Pycnonotidae) and *Hypocolius* (Hypocoliidae) between kinglets and cisticolas, and white-eyes (Zosteropidae) between cisticolas and 'Sylviidae'. Dowsett and Forbes-Watson (1993) recognized only the family Sylviidae which they divided into Megalurinae, Acrocephalinae, Sylviinae and Cisticolinae. Watson *et al.* (1986) also treated Old World warblers as a single family but did not recognize subfamilies, and nor do we; in the main we follow their arrangement for Palearctic genera, but opinions have changed recently about many Afrotropical genera, so our sequence here is our own.

We have transferred two of the sylviid taxa of Watson *et al.* (1986) elsewhere: we treat *Stenostira* as a flycatcher and *Achaetops pycnopygius* as a babbler. The text for *Achaetops* (*see* Plate 6, this volume) will appear in Volume VI. Too little is known about relationships among the 44 African genera to produce any kind of evolutionary tree. Our linear sequence is based mainly on recent published papers or comments; details may be found in the generic diagnoses. One notable divergence from Watson *et al.* (1986) is that we split several prinias off from *Prinia* into monotypic genera, such as *Heliolais* and *Oreophilais*. We agree with Chapin (1948) that the closest relative of *Hemitesia* is probably the Asian *Tesia*, so, like *Tesia* in Watson *et al.* (1986), we have placed it at the beginning. We have 'clumped' certain genera which seem to form natural groups, like the African tailorbirds *Artisornis, Poliolais, Camaroptera, Calamonastes* and *Euryptila* (see Fry 1976). *Eremomela* and *Sylvietta* belong together, but how they relate to other genera is uncertain. There remain many thorny questions. Why is *Bathmocercus*, a denizen of tropical forest undergrowth, traditionally placed among the grassland and Palearctic genera at the beginning of the list, none of which it resembles at all? Where does *Drymocichla* belong? Is *Macrosphenus* a warbler, a babbler or a bulbul (all have been suggested)? It remains a fertile field for molecular and other studies.

358 species (Watson *et al.* 1986), 389 (Sibley and Monroe 1990); mainly Afrotropical, Palearctic and Oriental. 204 species in Africa, of which 160 are endemic, 23 are non-endemic breeders (most also Palearctic migrants), and 21 are Palearctic winter visitors. Of the 44 genera in Africa no less than 32 are endemic.

Genus *Hemitesia* Chapin

Head broad, bill flattened, wings rounded, tail very short, feet large. Head pattern distinct: top and sides black with broad grey crown-stripe and narrow white supercilium. Originally placed in *Sylvietta* but its head and bill are much broader and feet far larger; the only similarity between the two is the short tail. Chapin (1948a) showed that it resembles the Asian genus *Tesia*, from which it differs only in relatively longer wing (P4 and P5 longest) and slightly shorter tail (tail-feathers extending 12–13 mm beyond uppertail-coverts, whereas tail of *Tesia* is even shorter, completely hidden beneath coverts); he proposed the monotypic genus *Hemitesia* for it. Pending anatomical and behavioural research we agree with him and with Hall and Moreau (1970) that it is closer to *Tesia* than to any African genus.

Endemic; monotypic.

Plate 20
(Opp. p. 337)

Hemitesia neumanni **(Rothschild). Neumann's Warbler. Crombec de Neumann.**

Sylvietta neumanni Rothschild, 1908. Bull. Br. Orn. Club 23, p. 42; forest west of L. Tanganyika, Belgian Congo, 2000 m.

Range and Status. Endemic resident. E Zaïre in highlands from west of L. Edward south to Mt Kabobo; SW Uganda in Bwindi-Impenetrable Forest (fairly common at 1630 m); W Rwanda (Nyungwe Forest). Rare to fairly common; density in Nyungwe Forest < 5 pairs/km^2 (Dowsett-Lemaire 1990).

Description. ADULT ♂: broad black stripe curving outwards from base of bill along sides of crown and broadening onto sides of nape; broad brownish grey stripe from forehead to centre of nape; width and extent of stripes variable: central grey stripe may reach base of bill, or black stripes may meet across base of bill and on nape. Supraloral line brownish grey to dull greenish or off-white, continuing as white line from

Hemitesia neumanni

Field Characters. A rare warbler confined to a few mountain forests in E Zaïre and W Uganda. Head mainly black with long white supercilium; back green, throat and breast yellow, tail very short. Large head and short tail give different shape from Green Hylia *Hylia prasina*, the only species with which it is likely to be confused, and its weak little song is very different from the loud 'pee-pee' of Green Hylia. Hylia also has pale supercilium and black line through eye but lacks yellow on underparts and has normal-length tail.

Voice. Tape-recorded (B, CART, LEM). Song loud and explosive, brief (1 s) phrases of 2–3 high-pitched, pure notes, often slurred, 'tee-tiyoo-tee', 'tee-tyer-tyoo', 'tyoowi-tyee', 'tee-teeyoo-tyoowi', 'tay-tiyoo-tay'; given at intervals of *c.* 5 s.

General Habits. Inhabits primary montane and transition forest and gallery forest, at 1200–2250 m in Zaïre, 1500–2130 m in Uganda, and from below 2000 m up to 2350 m in Rwanda. Frequents thick undergrowth near rivers, humid valley bottoms, and flat swampy forest with ground cover of shrubs, e.g. *Mimulopsis*, *Alchornea* and Rubiaceae; in Rwanda also forested slopes of sheltered valleys, where it occurs in low ground cover of creepers *Sericostachys* in open forest, and enters closed forest in dense patches of *Mimulopsis* (Dowsett-Lemaire 1990). Forages low down, on ground and in vegetation at height of 1–2 m, also sings from that level. Occurs singly or in pairs; joins mixed-species flocks. When disturbed remains very quiet, without moving; very hard to see.

Food. Small insects, especially beetles.

Breeding Habits. Almost unknown. Territories in Rwanda, defined by position of singing ♂, appear quite large; one was estimated at 10 ha (Dowsett-Lemaire 1990). Birds in breeding condition, Zaïre, Apr–June, Aug–Oct; singing output increases in Rwanda in Oct, and ♀ taken in Jan was about to lay.

above eye back onto sides of hindneck, bordered below by black line from base of bill through and behind eye. Upperparts olive-green, tail dark brown and very short. Cheeks, ear-coverts, sides of neck, chin, throat and breast yellow with variable admixture of olive on ear-coverts, sides of neck and breast, sometimes across centre of breast; flanks and thighs olive-green, lower breast and belly white, undertail-coverts yellow. Wing-feathers dark brown edged olive-green, underside of inner webs of primaries and secondaries edged pale brown, axillaries and underwing-coverts olive-green becoming pale yellow on bend of wing. Upper mandible brownish black, lower mandible and gape yellowish to pinkish buff with dark brown area at sides of distal half; eyes dark brown to reddish brown; legs and feet brownish to light brownish olive or flesh, claws light grey. SIZE: wing, ♂ (n = 5) 56–64 (60·2), ♀ (n = 2) 59, 63; tail, ♂ (n = 4) 27–32 (30·0), ♀ (n = 2) 29, 29; bill to skull, ♂ (n = 4) 14–15 (14·3), ♀ 14, 14, to feathers, 1 ♂ 11; tarsus, ♂ (n = 5) 19·5–25 (22·8), ♀ (n = 2) 25, 25. WEIGHT: 2 ♂♂ 14·5, 17; 3 ♀♀ 13·5–14·5 (14·2).

IMMATURE and NESTLING: unknown.

Genus *Cettia* Bonaparte

Chestnut- and olive-brown warblers, mostly with distinct superciliary stripe and white underparts, often a yellowish suffusion, living in forest undergrowth, damp thickets and scrub. Plumage soft, contour feathers long, especially undertail-coverts. Wings not long; rounded; P10 broad, half length of P9. Tail of 10 broad, soft feathers, medium length or very short, square or strongly graduated. Sexes alike. Nest a deep cup; eggs of some species uniform brick-red.

15 species (Sibley and Monroe 1990), S and E Asia to Solomon Is. 1, *C. cetti*, ranges from W China to Europe and N Africa. Superspecific affinities have not been studied; the genus contains several insular endemics (Palau I., Tanimbar I., Taiwan, Philippines, Solomons).

60 SYLVIIDAE

Plate 5
(Opp. p. 96)

Cettia cetti (Temminck). Cetti's Warbler. Bouscarle de Cetti.

Sylvia cetti Temminck, 1820. Man. d'Orn. (2nd ed.) 1, p. 194; Sardinia.

Cettia cetti

Range and Status. W Palearctic: NW Africa, Spain to Turkey, Iraq to L. Balkhash; France and S England. Mainly resident west of 30°E and mainly migratory east of 30°E. Has colonized much of W Europe north of 45°N in last 30 years; but decimated in cold winters.

Resident and winter visitor in NW Africa. Morocco, common resident at low altitudes, sea level to 1800 m (occasionally 2100 m), in Middle Atlas and High Atlas, but in any one valley prone to disappear in some years; also occurs in pre-desert areas, e.g. Errachidia. Algeria, quite widespread breeding resident on N and S slopes of Tell Atlas Mts; in autumn spreads to Hauts Plateaux (Boughzoul) and N oases (Touggourt, Ghardaïa); once in spring at Laghouat. Tunisia, scarce resident breeder in N; scarce winter visitor from Sept to Apr, south to Gabès; birds ringed Cap Bon, Apr, recovered there May, thought to be residents. Egypt, sight records Sharm el Sheikh (Sinai) Mar 1979 and near Suez Mar 1982; since Cetti's Warbler is summer visitor to much of Turkey and a winter visitor at Eilat, Israel, further Egyptian records are anticipated.

2 birds mist-netted Kano, Nigeria, Apr 1964 (Elgood *et al.* 1966), possibly misidentified (Moreau 1972).

Density: 1 pair/50 m of hedge (Portugal); 58 breeding pairs/km^2 (Italy).

Description *C. c. cetti* Temminck: NW Africa, Europe (only race in Africa, though Egyptian vagrants probably *C. c. orientalis*). ADULT ♂: forehead to uppertail-coverts, also scapulars, wings and tail, dark rufous-brown, mantle tinged olive, rump tinged cinnamon, flight feathers of wing and tail dark brown with rufous-brown outer edges. Lores dusky brown, ear-coverts brown-grey, upper ones mottled with rufous, lower ones with pale grey. Long, quite well-defined superciliary stripe, from nostril nearly to nape. Sides of neck grey. Chin and throat off-white, breast pale brownish grey, belly off-white, vent and flanks olive-brown, thighs brown, undertail-coverts olive-brown with pale tips. Underside of flight feathers and tail-feathers, underwing-coverts and axillaries, whitish brown-grey. Bill dark brown, cutting edge of upper mandible horn-brown, lower mandible with yellow-horn base; eyes dark brown; legs and feet flesh-brown. Sexes alike. SIZE: (Algeria, S Europe) wing, ♂ (n = 35) 58–66 (61·9), ♀ (n = 13) 52–57 (54·0); tail, ♂ (n = 31) 54–63 (58·5), ♀ (n = 13) 48–56 (51·3); bill to skull, ♂ (n = 33) 13·6–15·6 (14·4), ♀ (n = 13) 12·8–14·4 (13·7); tarsus, ♂ (n = 26) 20·7–23·6 (22·1), ♀ (n = 13) 18·4–20·2 (19·4). WEIGHT: (Balearic Is) ♂ (n = 72) 10·1–15·5 (12·8), ♀ (n = 63) 8·3–12·5 (9·9).

IMMATURE: like adult but duller, less rufous, browns greyer, whites creamier.

NESTLING: upperparts of hatchling with scant long black down.

Field Characters. A warbler with dark rufescent brown upperparts and whitish underparts that skulks in woody ditches, best identified by its loud, abrupt song with explosive first note, given freely all year from cover. Upperparts unstreaked, tail broad and rounded, often cocked or flirted, long straight white supercilium, pale-tipped undertail-coverts. Nightingale *Luscinia megarhynchos* and Sprosser *L. luscinia* have brighter rufous tails contrasting with brown backs. Plumage similar to Savi's Warbler *Locustella luscinioides*, which is distinguished by pale lores (dusky in Cetti's Warbler), short post-orbital eye-stripe and browner underparts; Savi's Warbler and plain-backed reed warblers *Acrocephalus* best told by voice and habitat; reed warblers also have brown or buffy (not greyish) flanks, shorter, less conspicuous eye-stripe, narrower tail.

Voice. Tape-recorded (62, 73, 93, B). ♂ song, 'one of the most distinctive of West Palearctic warblers' (Cramp 1992), is abrupt outburst of staccato notes 'CHEE, cheeweecho-weecho-weechoo-chew' or 'TCHI tchitchirititchitchirititchi' or 'TI tipitipitipi tipi tipi', slightly variable geographically. Song lasts 2·5–5·0 s (usually 3 s) and is audible at 300 m. Rarely, ♀ sings. Contact call a staccato 'chip'. Common call (function?), a stuttering 'chick-ich-ich-ich-ich'. Alarm (with human intruder) 'tsuk tsuk tsuk'; high-intensity alarm a rattle, like a Winter Wren *Troglodytes troglodytes*.

General Habits. Favours thick, damp cover of hedges next to water, ditches, bracken and coarse grass mixed with woody perennials like blackthorn *Prunus spinosa*, broom *Sarothamnus*, honeysuckle *Lonicera*, bramble *Rubus* and willow *Salix*. Prefers substrate of bare earth to one of standing water – avoids reedbeds except near coast. Skulks in dense woody growth within 1·5 m of ground, foraging solitarily or in pairs, creeping and darting to pick insects from bare twigs and from under leaves. Feeds on ground by hopping, like Hedge

Prunella modularis; grips rotten wood with strong feet and tugs out insect larvae. Nervously flicks wings and tail; when alarmed by person, cocks tail high, droops wings, lowers head. ♂ sings from cover, in upright posture, head thrown back, tail depressed. Sings all day, most of year, sometimes at night; after dawn gives 1–2 songs per min, sings much less frequently towards noon.

Mainly sedentary. In NW Africa populations evidently crash in cold weather, as they do in Europe. Some vertical migration in High Atlas, Morocco, where in winter some birds occur on N fringes of Sahara; more widespread in E Morocco in winter than in summer; passage recorded at Defilia (Smith 1968) although not at Tafilalet. No evidence of movement in Algeria. In Tunisia, small numbers winter at Gabès (Castan 1963).

Food. Large variety of small insects, including aquatic ones and their larvae; some spiders, snails and seeds (Cramp 1992).

Breeding Habits. Solitary breeder, aggressively territorial, monogamous or (usually) successively polygynous, a ♂ with 2 or more ♀♀. Courtship behaviour not well known, but is mainly vocal, with wing- and tail-flicking (Cramp 1992).

NEST: untidy cup made of stems and roots, with bulky base composed of stems and leaves; lined with reed flowers, feathers and hair. Sited in thick twiggy vegetation with nettles *Urtica* and reeds *Phragmites*, 30–45 cm above ground. Height (n = 6 nests) 81–104 (91·5), width 84–91(87), int. diam. 49–54 (51), cup depth 47–56 (51). Built by ♀.

EGGS: (n = 18 clutches) 4–5 (av. 4·8: W Europe). Subelliptical, glossy, uniform immaculate dark reddish brown, sometimes grey-brown. SIZE: (n = 178, Europe) 16·5–19·8 × 12·45 – 14·5 (18·0 × 13·9). WEIGHT: *c.* 1·8.

LAYING DATES: Morocco, Apr–May.

INCUBATION: by ♀ in spells of av. 11.8 min with absences av. 6·3 min. Period: 16–17 days.

DEVELOPMENT AND CARE OF YOUNG: young brooded and fed by ♀. Nestling period: 14–16 days. After leaving nest, brood fed by parent(s) for 15 days or more; young independent at 30 days, disperse at 35 days.

BREEDING SUCCESS/SURVIVAL: 40 out of 41 1st-clutch eggs hatched (England). Drastic declines after severe winters, e.g. 60 territories in Netherlands in 1978 reduced to 10 in 1979. Longevity: one bird 7 years 4 months.

Reference
Cramp, S. (1992).

Genus *Bradypterus* Swainson

Dark brown warblers with little morphological distinction; plumage plain, paler below than above, some species spotted or streaked on throat and breast, one with white in wing; rather short-winged; several species with tail feathers broad and marginally disintegrated. Tail with 12 feathers, 10 in *B. lopezi*, 10 or 12 in *B. cinnamomeus*. Nestlings of some species with distinctive mouth spots. Best characterized biologically: live close to ground in dense vegetation of swamps, bamboo, forest edge or interior of montane or lowland forests; several forage partly on ground; distinctive, often far-carrying songs; open cup nests bound only loosely to vegetation.

Usually placed in Acrocephalinae; *Bradypterus* seems closest to *Cettia*, on basis of voice and plumage. Within African *Bradypterus* there has been a wide diversity of treatment, both of species limits and racial affiliation. Decisions were made from study of museum skins, without reference to voice, and the result was confusion. Voice is paramount for the understanding of this genus. Recently Dowsett and Forbes-Watson (1993) rearranged the genus, based at least partly on voice; we adopt the same approach but take it a few steps further. We use the following sequence:

1. The central African swamp group – *baboecala, carpalis, graueri, grandis*. This subdivides into (a) the 'bouncing ball' group, *baboecala/carpalis*. These 2 have similar accelerating songs that have been likened to the sound made by a bouncing ball. This suggests a recent common ancestry, but since they are sympatric they cannot be allospecies, and are best linked in a species group; (b) *graueri* and *grandis* have a trill with some introductory notes, reminiscent of *Cisticola robustus*; *graueri* is sympatric with *carpalis*, and we agree with Vande weghe (1983) and Dowsett and Dowsett-Lemaire (1993b) that the 2 are 'not closely related', *contra* Watson *et al.* (1986). *B. grandis* is extremely similar in plumage and voice to *graueri* and may even be conspecific with it (F. Dowsett-Lemaire, pers. comm.). Pending further research, we place the 2 in a superspecies.
2. *B. alfredi* has a dry, monotonous song and lives in a wide variety of habitats, including marshes. It may be a link between the swamp group and the forest-edge group. On the basis of skins it has almost universally been regarded as close to *B. sylvaticus*, and placed in the same superspecies by Watson *et al.* (1986), Short *et al.* (1990) and Sibley and Monroe (1990), but this relationship has quite rightly not been accepted by Dowsett and Forbes-

Watson (1993). The songs of *alfredi* and *sylvaticus* are totally different, and in any case it seems unlikely that 2 such sedentary birds living thousands of miles apart only recently diverged from a common ancestor, i.e. are allospecies.

3. The forest and forest edge group – *lopezi*, *cinnamomeus*, *barratti*. These 3 are overall very similar in plumage but each varies throughout its range. This has produced extensive interspecific morphological overlap and has left museum workers uncertain as to which race should be assigned to which species. Watson *et al.* (1986), for instance, subsumed all races of *lopezi* under *barratti*, except one with cinnamon plumage, *bangwaensis*, which he assigned to *cinnamomeus*. A study of their voices suggests a quite different arrangement.

B. lopezi is a widespread species with a remarkably consistent song throughout its range, a monotonous series of notes that increase in volume as they progress. There is much individual variation in song, and each bird may sing several types, but songs from Tanzania sound essentially the same as those from Cameroon. The song of *bangwaensis* is clearly within the range of *lopezi*, as admitted by Dowsett-Lemaire and Dowsett (1989c), but they elevated *bangwaensis* to specific rank and allied it with *cinnamomeus* on the basis of its call note. Calls of *Bradypterus* are not well studied, and we believe song more important; we consider *bangwaensis* a race of *lopezi*. It would also be a biogeographic oddity if a single race of *cinnamomeus* occurred among a cluster of races of *lopezi* when the nearest race of *cinnamomeus* is 1000 miles to the east, in the Albertine Rift.

The song of *B. cinnamomeus* is quite different from that of *lopezi*, one or more ringing introductory notes followed by a chatter. The 2 species can be heard together on Keith and Gunn (1971). The song of *barratti* has the same form as *cinnamomeus* but is a little drier and faster; *barratti* is probably the result of an earlier incursion of *cinnamomeus* into southern Africa, and we agree with Dowsett and Forbes-Watson (1993) that the 2 should be grouped in a superspecies.

4. The 2 South African endemics, *sylvaticus* and *victorini*, are best placed at the end of the list as independents. Their songs are so different from any other African *Bradypterus* that it is not possible to say what their closest relatives are. *B. sylvaticus* has a sharp, accelerating trill reminiscent of a Wood Warbler *Phylloscopus sibilatrix* (cf. *alfredi*, above); *victorini* has a rollicking little song more like that of the Cape Grass-Warbler *Sphenoeacus afer* than any *Bradypterus*, and we cannot accept Watson *et al.*'s (1986) suggestion that it forms a superspecies with *cinnamomeus*.

About 18 species: 10 in Africa and 8 in Asia (of which 3 in Palearctic).

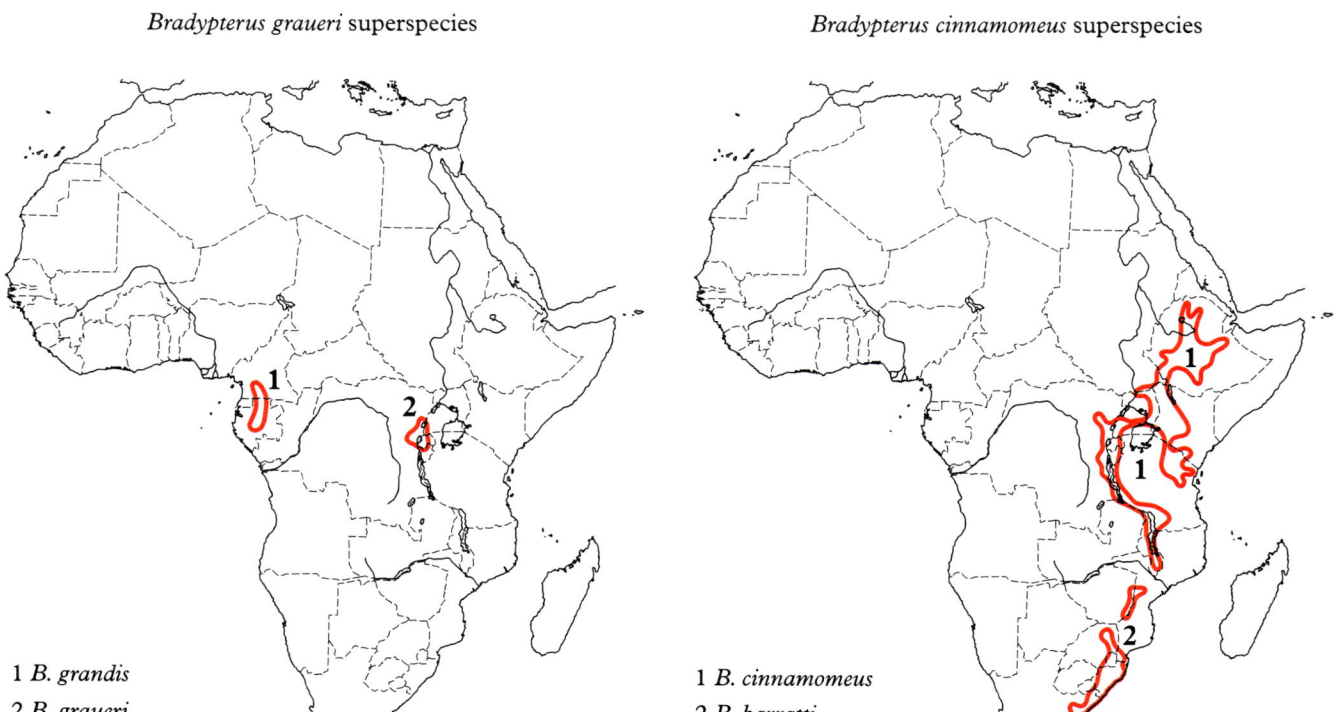

Bradypterus graueri superspecies

1 *B. grandis*
2 *B. graueri*

Bradypterus cinnamomeus superspecies

1 *B. cinnamomeus*
2 *B. barratti*

Bradypterus baboecala (Vieillot). Little Rush-Warbler; African Sedge Warbler. Bouscarle caqueteuse.

Plate 5
(Opp. p. 96)

Sylvia baboecala Vieillot, 1817. Nouv. Dict. d'Hist. Nat., nouv. éd., 11, p. 172; Knysna District, southern Cape (ex Levaillant).

Range and Status. Endemic, widespread resident. Distribution fragmented in northern tropics where generally uncommon and highly localized; common and widespread throughout southern range. Record of a singing bird in a swamp at Ferkessedougou, Ivory Coast (Thiollay 1985) requires substantiation. Nigeria, small populations Onitsha and Kano, and heard singing at Obubra on Cross R. Cameroon (coastal forest, Mt Kupé, Buea, and from Nyong R. and Efulen east to Akomolinga and possibly in N); rare Chad (L. Chad, Baga-Kawa). Sudan, known only from localities in SW and SE and from L. No and Tonga, also Juba area, on Nile, but probably overlooked in Sudd. Locally common in Ethiopia and probably more widespread than shown: W Highlands north to L. Tana, and records from Harrar, Aletta, Dangila and Challa. Common and widespread in NE, E and SE Zaïre: Faradje, Nzoro, Uele District and south to Kibali R., Semliki valley between Karebumba and Kabiabo, Kivu (Katana, Butembo, Miki), Katanga to L. Ngami, Itombwe District at Kitoga, Luiko, Kikenge and Walungu. Local and uncommon in Uganda, in NE (Kidepo valley), SW (L. Bunyoni, L. Mutanda) and SE (Bukedi and around base of Mt Elgon); also in Burundi (lowlands and Bututsi plateau), Rwanda (Nyungwe, Mkingo, Muhera and R. Akanyaru). More widespread and locally common in SE Kenya (coastal lowlands from Tana R. south to Vanga region and Tanzanian border and inland at L. Jipe) and, disjunctly, in W and central regions from base of Mt Elgon, Cherangani Hills and Laikipia south to L. Victoria basin (Siayu, L. Kanyaboli) and through highlands (Kavirondo, Nandi, Nakuru, Naivasha, Nairobi) to Mara Game Res. and Nguruman Hills. Locally common NW Tanzania (Rwandan border, Kagera and Lugusye swamps and Kibondo) and in N central and E Tanzania (Arusha Nat. Park, Mbulu and E Usambaras and along coastal lowlands to Mikindani west to Songea) and SW highlands (Umalila and Ufipa Plateau). Zambia, almost throughout, but rare in middle Zambezi valley and absent from Luangwa valley. Uncommon W Angola (Chicuma, central Malanje in Cangandala Nat. Park, W and N Huila, extreme E Benguela, reaching Huambo and Bié); also in E (Macondo district, Moxico) and SE. Widespread and common Malaŵi, Zimbabwe (Matabeleland, Midlands, Mashonaland, up to 1750 m in E Highlands, Mana Pools on Middle Zambezi in NW and Sabi valley in SE). NE Namibia (Shorobe, and along Kwando, Kavango and Sterksruit rivers in Caprivi Strip). Mozambique, local but widespread throughout, occurs in small numbers along Zambezi near Mopeia, uncommon in Furancungo District in Tete Province, occurs on Lagão de Ura near Beira; many records south of Save R., numerous on Inhatouco R. near Panda. Sparse to locally fairly common in N Botswana (Okavango

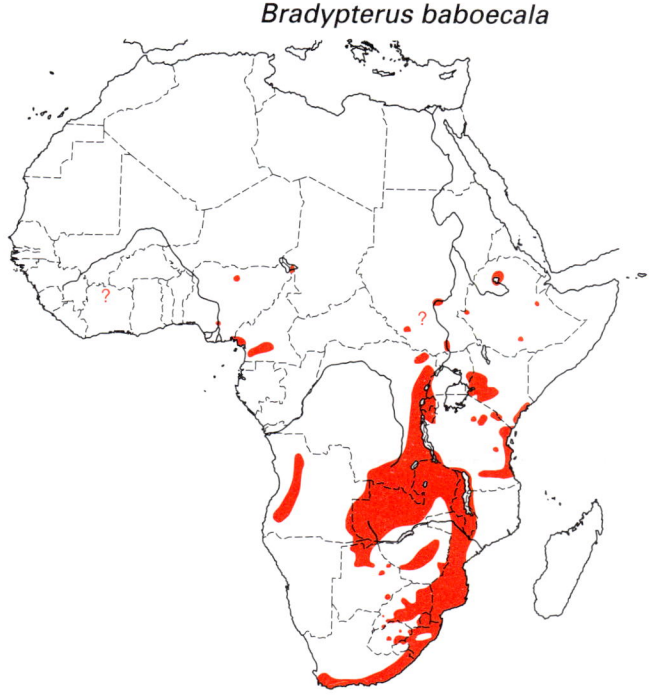

Bradypterus baboecala

Swamp, Linyauti and Chobe river systems), and regular at 4 sites near Gaborone in SE; also recorded in E at Nata, Orapa and Francistown. Widespread in Transvaal, locally common, commonest in Witwatersrand; occurs at Mataffin, Mtsoli R., Waterberg and (rare) Kangwane in Mswati District, also (rare) in Kruger Nat. Park. Swaziland. In Orange Free State occurs in E and NE, common in Memela area and along Vaal R., elsewhere uncommon. Zululand, Natal, Transkei coast (Port St Johns, Ntafufu), E Griqualand and westwards to SW and S Cape. In Lesotho, known only from Collett Dam.

Description. *B. b. baboecala* (Vieillot): southern South Africa east to Great Kei R. ADULT ♂: whole of head and upperparts dark olive-brown, lower back, rump and uppertail-coverts tinged red-brown. Tail strongly graduated: T1 16·3–19·6 (17·7, n = 5) longer than T6; feathers broad and rounded, dark brown with faint underlying narrow, dark transverse bars all along (T1 with 18 bars *c*. 2 mm apart), edges of outer webs red-brown, tips of T1, T2 and T3 olive-buff. Lores and spot behind eye dusky, cheeks and ear-coverts buff-brown faintly streaked white, supercilium buff-white. Throat off-white washed at the sides with buff, lower throat faintly streaked with fuscous. Rest of underparts yellowish white, undertail-coverts washed with olive or rusty buff. Primaries and secondaries dark brown, upperwing-coverts margined on outer edge with red-brown; axillaries and underwing-coverts buff. Bill blackish horn, lower mandible flesh to yellowish at base; eyes brown; legs and feet flesh colour. Sexes alike. SIZE: (7 ♂♂, 10 ♀♀) wing, ♂ 57–61 (59·2), ♀ 56–61 (58·4); tail, ♂ 61–67 (63·6), ♀ 57–69 (62·5); bill to feathers, ♂ 12–14 (13·1), ♀ 12–14 (12·9); culmen, ♂♀, (n = 126, southern Africa), 11–19 (15·6) (Maclean 1993); tarsus, ♂ 19–22

(20·4), ♀ 20–22 (20·8). WEIGHT: unsexed (n = 83) 11·2–17 (14·0) (Maclean 1993).

IMMATURE: like adult but underparts with strong yellowish wash.

NESTLING: 3 tongue spots, 1 at tip, 1 on each side of centre (Maclean and Vernon 1976).

B. b. tongensis Roberts (including '*moreaui*'): SE Kenya, Tanzania (probably except SW), E and S Zambia, Malawi, Mozambique, extreme E Zimbabwe in Zambezi valley from Tete to Chobe R. junction where it merges with *msiri* and E South Africa (E and NE Transvaal, E Swaziland, Zululand, coastal and lower Midland Natal, Transkei coast). Similar to nominate race but upperparts and tail more rufescent, less streaking on lower throat and breast; underparts pale yellow, sides and flanks more yellowish, undertail-coverts paler. Smaller and shorter-tailed: (9 ♂♂, 6 ♀♀) wing, ♂ 53–58 (56·3), ♀ 55–65 (58·3); tail, ♂ 54–65 (60·2), ♀ 50–61 (54·5). WEIGHT: (Tanzania) ♂ (n = 3) 15·2–16 (15·6), ♀ (n = 2) 15·5, 17·0.

B. b. transvaalensis Roberts: Zimbabwe (central plateau), W Swaziland, Transvaal highveld, Orange Free State, lowlands of Lesotho, W Natal and adjacent Zululand above c. 1370 m and E Griqualand. Similar to nominate race but markedly whiter below, sides, flanks and undertail-coverts lighter, less olive-brown. Streaking on lower and central part of throat more prominent than in *tongensis*. Tail longer than tail of *tongensis*. SIZE: unsexed (n = 20) wing, 55–61·5 (58·5), tail, 62–69·5 (65·1), bill to skull, 15–18 (16·2).

B. b. msiri Neave (including '*bedfordi*'): E Angola, W and N Zambia, SE Zaïre (Shaba) and probably SW Tanzania, NE Botswana (Okavango and Botletle R. and Caprivi Strip) where it merges with *tongensis*. Darker brown than *tongensis* and less reddish, whiter below with darker flanks and undertail-coverts; streaks on lower throat less distinct; rump and uppertail-coverts darker than *transvaalensis*. A semi-albino once recorded (Benson 1961). SIZE: (unsexed, n = 20) wing 54·5–61 (57·9), tail 60–71·5 (66·3).

B. b. benguellensis Bannerman: W Angola. Like *centralis* and *msiri* but darker and wing longer: ♂ (n = 3) 63–65 (64·3), ♀ (n = 2) 58, 63.

B. b. elgonensis Madarász: highlands of W and central Kenya, SE Uganda and possibly NE Uganda. Upperside a uniform deep rich reddish brown, spots on foreneck distinct. SIZE: wing (unsexed, n = 8, L. Kanyaboli, W Kenya) 53–55 (54·3), (unsexed, n = 19, L. Naivasha) 54–67 (56·7); tail, ♂ (n = 2) 60, 62. WEIGHT: ♂♀ (n = 7, L. Kanyaboli) 11·7–15 (12·9); ♂♀ (n = 19, L. Naivasha) 10·4–13·5 (11·85).

B. b. centralis Neumann: Nigeria, S Cameroon, NE Zaïre, Burundi, Rwanda, SW Uganda. Similar to *tongensis* but upperparts dark olive-brown, throat well streaked. SIZE: (18 ♂♂, 2 ♀♀) wing, ♂ 52–57 (54·7), ♀ 54–55 (54·5); tail, ♂ 50–62 (57·5), ♀ 59–62 (60·5).

B. b. abyssinicus (Blundell and Lovat): Ethiopia. Similar to *centralis* but upperparts lighter olive-brown, wash on underparts more olive-tawny than in *sudanensis*. SIZE: wing, ♂ (n = 6) 52–60 (55·8), ♀ (n = 2) 55–57 (56·0); tail, ♂ (n = 4) 47–66 (59·5), ♀ (n = 2) 55–59 (57·0); bill to feathers, ♂ (n = 6) 12–13 (12·8), ♀ (n = 2) 12–13 (12·5).

B. b. sudanensis Grant and Mackworth-Praed: S Sudan. Like *abyssinicus* but smaller. SIZE: (5 ♂♂, 3 ♀♀) wing, ♂ 52–54 (52·8), ♀ 50–53 (51·7); tail, ♂ 54–63 (58·0), ♀ 52–63 (55·4).

B. b. chadensis Bannerman: W Chad. Similar to *abyssinicus* but larger and upperparts rufous-brown, no spotting or streaks on throat; underparts, flanks, sides and undertail-coverts browner than *centralis*.

TAXONOMIC NOTE: variation, studied adequately only in southern Africa (Clancey 1976), is clinal. Taxonomy of northern populations requires revision; in the meantime we follow Watson *et al.* 1986.

Field Characters. Medium-sized, distinguished from other marsh warblers by dark brown upperparts, broad rounded tail and dark spots or streaks on throat (faint or absent in some races). Pale supercilium; throat and underparts whitish; rump tinged rusty in some populations. Secretive and difficult to observe, but its presence easily detected by distinctive accelerating song. Barratt's Warbler *B. barratti* is similar but greyer below, and lives in different habitat, has ringing song; Greater Swamp-Warbler *Acrocephalus rufescens* is also dark above but lacks pale eye-stripe and is larger and heavier, with long bill, orange gape, greyish underparts, chuckling song. Lesser Swamp-Warbler *A. gracilirostris* and African Reed-Warbler *A. baeticatus* are lighter and browner and have more melodious songs. Song similar to White-winged Swamp-Warbler *B. carpalis*, but where they occur together (e.g. Rwanda) they occur in different habitats (Vande weghe 1981). Grauer's Swamp-Warbler *B. graueri* is similar and occurs in similar habitat in Rwanda but at higher altitudes (Vande weghe 1983).

Voice. Tape-recorded (11, 57, 75, 86, 91, B, F, WAT). Song of ♂ a series of loud, sharp chirps 'tirrup', repeated 12–15 times, slowly at first then accelerating, trailing away somewhat at the end: 'tirrup tirrup tirrup tirrup tirrup trip trip trip trip trip-trip-trip-trip-trip-trip'. Overall effect is like a bouncing ball or stick held against spokes of accelerating wheel. Form similar to song of White-winged Swamp-Warbler which, however, adds a trill at the end. Song lasts 1·8–12·5 s (av. 4·9 s, n = 22); 'tirrup' note often repeated 4–5 times without full song developing, and sometimes continues at steady rate for up to 60 s. Alarm, a catlike 'meew'; contact or alarm call of ♀, a sad, bleating, repeated 'pee-et' (D. Watts, pers. comm.).

General Habits. Inhabits tall reed beds and rushes by more or less permanent water; occurs along estuaries, rivers and streams, often abundantly, particularly in South Africa, and quick to exploit man-made swamp habitats; favours reeds *Phragmites* and reedmace *Typha* vegetation in southern Africa and *Papyrus* elsewhere; occurs also in dry stands of elephant grass. In Kenya ranges entirely within 500 mm rainfall areas; occurs below 500 m in semi-arid and subhumid coastal areas, and up to 2300 m in highlands. Solitary or in pairs or sometimes small groups. Secretive, but may be coaxed into view by playback of song or by imitating distress call. In South Africa sings all year, mainly when cool; elsewhere sings mainly in breeding season. Often sings in middle of night. Keeps in lower strata of vegetation: in one marsh with many warbler species, netted most frequently low down near edge (Tyler 1991). Mainly sedentary, but local seasonal movements occur in South Africa when water levels change; one ringed bird caught at same site 29 months later. Moves locally and extensively in Zimbabwe, particularly in winter months, using transitory habitats as well as more permanent ones. In Kenya small influx to L. Baringo, Feb–Apr, and sudden appearance at man-made dams, suggest local movements. Occurs in dry habitats at Buea,

Cameroon, during Dec (the dry season). Song is used in territorial display and several birds often sing simultaneously.

Food. Small insects, ant eggs.

Breeding Habits. Monogamous, territorial, loosely colonial. Bird often, but not always, begins a flight display in the middle or towards the end of its song, and gives 4–5 loud wing beats as it flutters and hovers with tail pointed downwards, in amongst the vegetation.

NEST: a neat, deep, tightly formed cup made of dry reed, weed or coarse grass which straggles out untidily below, lined with finer fibres or rootlets (**A**); sited usually over water, in base of sedge tussock or between blades of fallen sheaf of dead rushes, usually < 70 cm above ground. Never bound to reeds; nest comes away from base easily when lifted. Ext. diam. 76, int. diam. 51, ext. depth 76, int. depth 38 (Benson and Benson 1947).

EGGS: usually 2–3, av. 2·4 in southern Africa (n = 9), occasionally 1 egg (Zimbabwe, Malawi). Colour varies from pinkish white or greenish to pale cream, patterned with small pink, red and brown freckles and spots, with smattering of small purple-grey and mauve marks that form a well-defined girdle around large end. SIZE (n = 7, southern Africa) 18·5–20·4 × 13·5–13·9 (19·2 × 13·7).

LAYING DATES: Nigeria (fluffy juv. late May); W Cameroon, June–Aug; Chad, Apr–May; Ethiopia, Aug–Sept; E Africa: Region D, Apr, June, Aug, Nov; Zaïre, Apr–Oct, (gonads enlarged Apr–Oct); Malawi, Mar, May; Zambia, Nov–Mar (Nov 1, Dec 2, Jan 2, Feb 3, Mar 1 clutches), (gonads enlarged Mar, Apr,

A

Aug, Sept); Zimbabwe, Nov–Mar, (Nov 1, Dec 2, Jan 2, Feb 3, Mar 1 clutches); SW Cape, Aug–Dec (season later and more protracted in N), Transvaal, Sept–Mar (Sept 1, Oct 3, Nov 1, Dec 6, Jan 5, Feb 2, Mar 2 clutches), Natal, Sept–Mar.

INCUBATION: period 12–14 days.

DEVELOPMENT AND CARE OF YOUNG: nestling period 12–13 days. Young fed by both parents.

BREEDING SUCCESS/SURVIVAL: parasitized by Diederik Cuckoo *Chrysococcyx caprius* (Squelch and Safe-Squelch 1994).

References
Britton, P. L. (1978b).
Clancey, P. A. (1976).

Bradypterus carpalis Chapin. White-winged Swamp-Warbler; White-winged Warbler. Bouscarle à ailes blanches.

Plate 5
(Opp. p. 96)

Bradypterus carpalis Chapin, 1916. Bull. Amer. Mus. Nat. Hist. 35, p. 27, fig. 4; Faradje, upper Uele District, Belgian Congo.

Range and Status. Endemic resident. Locally common or abundant in E Zaïre, Uganda, W Kenya, Rwanda, Burundi and NW Tanzania. Zaïre: occurs Faradje and north of Nzoro in NE Uele District, along Shari R. between Kilo and Nizi in E Ituri, and Lwiro in Kivu District. Uganda: in S occurs in Kampala area at Namulonge and (formerly) in swamps along Sezibwa R.; in SW occurs at edge of Kibale Forest Res., in Kazinga Channel in Queen Elizabeth Park, also 70 km southwest of Mbarara and west of Kabale. W Kenya: at Ukwala, in Yala Swamp, L. Kanyaboli, Kisumu, Kendu Bay and at Waturi on south side of Kavirondo Gulf. Rwanda: in Mulindi valley and Rugezi swamp in N, abundant throughout the Akagera–Nyabarongo–Akanyaru river system, where papyrus swamps cover 50,000–60,000 ha, in central and E; used to inhabit papyrus swamps in delta of Kamiranzovu R. until they were drained in 1975; a few may survive in adjacent elephant grass *Pennisetum purpureum* (Nov 1976: Vande weghe 1981). E and central Burundi: occurs all along Akagera R. in Bugesera, and along Akanyaru R. in Nyamushwaga, Ndurumu, Ruvubu, Kayongozi and Nyakijanda valleys. Also in swamps of Malagarazi R., particularly in Nyamabuye, Muyovozi, Lugoma and Lumpungu valleys. NW Tanzania: occurs in swamps in middle Akagera basin, Ruvubu Valley and Malagarazi basin along Burundi border (Vande weghe 1981). Density at Kadenge, Kenya: 4–21 birds per ha (estimates depend on method used: Britton 1978b).

Bradypterus carpalis

Description. ADULT ♂: entire upperparts brownish black, crown and forehead slightly darker, rump slightly paler, tail-feathers faintly barred darker throughout, underlying bands distinct. Tail has 12 feathers (Vande weghe 1983), not 10 (Chapin 1916). Faint whitish superciliary stripe from near nostril; ear-coverts and lores blackish brown, shafts of feathers whitish; cheeks white, each feather broadly tipped brown giving blotched appearance. Chin, throat and middle of breast white, sides of breast washed greyish yellowish brown; throat and upper breast heavily marked with large, roughly triangular spots of blackish brown; central line of belly white washed buff, flanks white washed greyish yellowish brown; undertail-coverts dark grey-brown with broad white tips; thighs dark brown, some feathers with white flecks. Primaries and secondaries dark brown, most feathers brownish white along edge of inner web. Lesser coverts white with concealed blackish brown bases, median coverts fuscous, tipped with white, forming a white wing-bar, greater coverts blackish brown bordered on outer webs with lighter brown; underwing-coverts white blotched brown, leading edge of wing white. Bill mostly black, but lower mandible with bluish grey base; eyes dark or light brown; feet brown or flesh. ADULT ♀: very similar to ♂, but lesser coverts dusky-centred, not white, undertail-coverts a little browner than in ♂; smaller. SIZE: wing, ♂ (n = 17) 70–74 (71·4), ♀ (n = 14) 61–69 (68·4); tail, ♂ (n = 11) 69–78·5 (av.?), ♂ (n = 5) 66–74 (70·0); bill to skull, ♂ (n = 11) 18–19·5 (av.?), ♀ (n = 5) 17–18·5 (17·6); tarsus, ♂ (n = 9) 27–30, ♀ (n = 4) 26–29 (♂, ♀ av.?). WEIGHT: ♂ (n = 23, Kadenge, Kenya) 18·9–26·2 (23·5), (n = 1, Lwiro, Zaïre, Dec) 27; ♀ (n = 18, Kadenge, Kenya) 19·0–23·0 (21·3), (n = 1, Lwiro, Zaïre, Dec) 25.

IMMATURE: one (probably ♀) had throat and breast cream, not white as in adult, covered with indistinct dusky markings, remaining underparts dusky brown, darker on flanks, sides of breast, belly and thighs; centre of breast and belly not white; upperparts slightly darker than in adult, edges of lesser coverts creamy white, primaries and median upperwing-coverts without pale edges of adult; outer webs of greater coverts edged brown, tips of median coverts marginally brown (Britton 1970b).

Field Characters. A large, skulking warbler of papyrus swamps, with very dark brown upperparts and flanks, triangular black markings on white breast; readily distinguished from other swamp warblers by whitish wing-bar and white or creamy shoulder patch, conspicuous in flight. Accelerating song similar in quality and structure to Little Rush-Warbler *B. baboecala*, but changes in pitch after introductory notes and ends with a trill.

Voice. Tape-recorded (FISP, GREG, McVIC, LEM). Song a succession of loud dry notes, starting slowly, then becoming lower in pitch and accelerating (like a bouncing ball), and changing abruptly at the end into a pea-whistle-like trill 'chirip, chirip, chuuw chuuw chuuw-chuw-chuw-chuw-chuw-chuw-preee-eeeeee'. Mean duration 6·2 s (range 5·3–6·7 s, n = 5). Song is often followed by 4–5 loud, explosive wing beats. Song audible over a range of several hundred m (Chapin 1953a, Vande weghe 1983).

General Habits. Inhabits papyrus swamps. May move to elephant grass and broad-leaved sedges when swamp is drained (Vande weghe 1983). In Rwanda (Rugezi swamp) inhabits mixed patches of papyrus and reeds (mainly *Miscanthidium*). In SW Uganda occurs from 1460 to 1890 m, in Zaïre at 1580 m, in Rwanda from 1290 to 2050 m and in Kenya from 1100 to 1300 m where annual rainfall is > 1000 mm (Lewis and Pomeroy 1989). Very secretive; reluctant even to cross rides through papyrus, made for mist-nets. Captured bird, when released, seeks any available cover even when papyrus is within 30 m range (Britton 1978b). Creeps with small jerky movements, much like a mouse, in lower layers of papyrus, among dead and rotting vegetation, with body and tail close to vegetation. In Rwanda, White-winged and Grauer's Swamp-Warbler *B. graueri* are allopatric, separated by both altitude and habitat, except in Rugezi swamp where they occur together (Vande weghe 1983).

Food. Insects.

Breeding Habits. Makes short display flight in the more open middle layer of papyrus bed, immediately following song.

BREEDING INDICATIONS: Zaïre, ♂ with gonads enlarged mid-Dec, end Feb, Apr, Sept; ♀ with partially enlarged ovary mid-Dec; egg in oviduct mid-Sept. Kenya, ♂ with enlarged gonads late May; pair displaying Kendu Bay early Apr; pair with dependent young (little skull ossification) early June.

References
Britton, P. L. (1970b, 1978b).
Chapin, R. T. (1973).
Vande weghe, J.-P. (1981, 1983).

Bradypterus graueri Neumann. Grauer's Swamp-Warbler; Grauer's Rush-Warbler. Bouscarle de Grauer.

Plate 5
(Opp. p. 96)

Bradypterus graueri Neumann, 1908. Bull. Br. Orn. Club, 21, 56; Western Kivu Volcanoes, Belgian Congo.

Forms a superspecies with *B. grandis*.

Range and Status. Endemic resident, probably sedentary. Locally common in highland swamps in E Zaïre, SW Uganda, Rwanda and N Burundi. E Zaïre: west of L. Edward near Lubero and north of Alimbongo, W Kivu volcanoes, west of L. Kivu at Mumba and Nyawaronga in mountains above Kalehe and in Kahuzi swamp, in Kahuzi-Biega Nat. Park. SW Uganda: Mubwindi and Ruhizha swamps in Bwindi-Impenetrable Forest. Rwanda: in N at Tshava, Kitabi and Rwasenkoko in Rugezi swamp, western Kivu Volcanoes below Mt Sabinyo in marshes between Virunga Volcanoes, and in SW at Kamiranzovu and Mukohole swamps in Nyungwe (Rugege) Forest. N Burundi: Rwegura, and between Teza and Rwandan border. Common, but total population size probably small due to restricted montane habitat, which is threatened by human pressure. In 1984, population estimated to be *c.* 3000 pairs in Kamiranzovu swamp (*c.* 900 ha) in Rwanda, but only *c.* 100 pairs in the whole of Burundi (J.-P. Vande weghe, *in* Collar and Stuart 1985).

Some of its swamps in danger of being drained (Rugezi Swamp, 8500 ha, is threatened by tea planting), but others have been conserved (Kahuzi Swamp in E Zaïre) or are likely to be conserved (swamps in Nyungwe Forest), or have had threat of drainage removed (Kamiranzovu Swamp, where gold extraction was planned). Some swamps are isolated and surrounded by forest, and birds probably reach these micro-habitats by following streams hidden by dense overhanging forest undergrowth (Friedmann and Williams 1970b), others are isolated and far from forest areas (e.g. in N Rwanda); yet others appear suitable (e.g in Itombwe Mts) but bird is absent.

Description. ADULT ♂: forehead, crown, hindneck, mantle and back blackish brown, rump, uppertail-coverts and tail of 12 feathers darker; tail has distinct underlying darker bands uniformly spaced throughout length. Supercilium from base of nostril to beyond eye white, lores and ear-coverts dark brown, shafts of ear-coverts whitish; cheeks whitish blotched pale brown. Chin, throat and breast white, with large arrow-shaped dark brown blotches, sides of belly, flanks, thighs and undertail-coverts pale brown, rest of underparts white. Primaries and secondaries dark brown, outer webs edged paler, basal part of inner web whitish, coverts similar but edged brown; secondaries with underlying darker bands uniformly spaced, as in tail, but less distinct. Under surfaces of wings paler brown, underwing-coverts buffish white, mottled with darker blotches. Upper mandible brownish black, lower mandible greyish; eyes brown to dark brown; legs and feet pale brown or pale flesh brown. Sexes similar, but ♀ has spots small, confined to breast, and throat almost pure white. SIZE: wing, ♂ (n = 7) 56–60·5 (58·0), ♀ (n = 5) 56–58·8 (57·5); tail, ♂ (n = 7) 71–75 (72·9), ♀ (n = 5) 65·7–68·8 (67·4); bill to skull, ♂ (n = 7) 15–16·4 (15·7), ♀ (n = 5) 14·5–15·3 (14·9); tarsus, ♀ (n = 1) 23·5. WEIGHT: (Uganda, Mar) ♂ (n = 9) 16–19 (17·3), ♀ (n = 6) 15–18 (16·7).

IMMATURE: ♂ similar to adult but spots on breast narrower, lower breast and underparts mainly white with pale yellow wash, flank feathers cinnamon-brown.

Field Characters. A medium-sized dark brown bird with white supercilium, ♂ with heavy spotting on throat and upper breast, ♀ with smaller spots on breast. Tail long and steeply graduated, but often looks frayed and worn. Often skulks, but unusually visible for a *Bradypterus* warbler and will sit on high, exposed perch, particularly when disturbed or in song. Song distinctive. Plumage similar to Little Rush-Warbler *B. baboecala*, which has darker brown upperparts, with the white superciliary stripe thinner and shorter; Little Rush-Warbler is also smaller and its song is quite different.

Voice. Tape-recorded (B, GUL, LEM). Song a rapid trill, preceded by 1–2 loud guttural notes, 'tchew' or 'chew', and occasionally followed by similar notes, 'tchew-tchew-trrrrrrrr' or 'tchew-tchew-trrrrrrrr-tchew'. Mean duration of trill 0·7 s (range 0·61–0·79 s, n = 7); occasionally only 3 or 4 'tchew' notes are sung, without the trill. Song lasts 1–1·5 s. Pairs duet, one calling 'tchew-tchew-tchew . . .' while the other adds an excited chattering call (Bennun 1986b, Dowsett-Lemaire 1990). In display, calls are repeated without pause: 'chup-chup-chup-trrr, chup-chup-chup-trrr' (Bennun 1986b). Alarm and contact notes not described.

General Habits. Inhabits montane forest swamps at altitude 1950–2600 m, and also swamps outside forest. Occurs in wide variety of swamps, some having short grasses only, some with a predominance of medium-sized sedges, some with long grass and others with mainly dense scrubby vegetation (Vande weghe 1983). Not shy; when disturbed flies low with fluttering flight for a few m giving short rattling trill, then drops down into cover. Sings from tops of stems and twigs, then makes short display flight with snapping wings low over vegetation; very vocal and impossible to miss if present in a swamp. Feeds in the open, near ground and sometimes on floating vegetation, when it uses tail to balance. Outside breeding season moves in groups of 10–12 birds (Vande weghe 1983). All the same, more localized than Little Rush-Warbler; the 2 seem to displace one another ecologically. In Nyungwe Forest, Rwanda, Grauer's occurs at 1950 m and Little Rush at 2500 m; in Itombwe, E Zaïre, Little Rush lives at 2600 m and Grauer's is absent (Vande weghe 1983). In Rwanda Grauer's and White-winged Swamp-Warbler *B. carpalis* are largely allopatric but occur together in Rugezi swamp, which is mainly covered with tall reeds *Miscanthidium violaceum* with sedges *Cyperus latifolius* at its edge; there Grauer's is dominant and White-winged is confined to large area of papyrus near swamp outlet. In transition zone between papyrus and *Miscanthidium*, both warblers sing within metres of each other, without any observed interaction (Vande weghe 1983).

Food. Small beetles, caterpillars, spiders and small seeds.

Breeding Habits. Monagamous, territorial. Territories of isolated pairs at Kitabi, Rwanda, 0·1–0·5 ha in extent (Dowsett-Lemaire 1990). In sexual display bird perches near top of sedge, rapidly flutters wings above back and calls constantly. Display may last *c.* 90 s, but stops abruptly when bird gives chase to probable mate alighting in sedge a few m away (Bennun 1986b). No nests yet found.

BREEDING INDICATIONS: E Zaïre, ♀ with brood patch mid-Mar, gonads enlarged Mar, immature Nyawarongo, end Feb and Alimbongo, early Oct.

References
Ash, J. S. *et al.* (1991).
Vande weghe, J.-P. (1983).

Plate 5
(Opp. p. 96)

Bradypterus grandis Ogilvie-Grant. Ja River Scrub-Warbler. Bouscarle géante.

Bradypterus grandis, Ogilvie-Grant, 1917. Ibis, p. 78; Bitye, Ja (= Dja) River, Southern Cameroons.

Forms a superspecies with *B. graueri*.

Range and Status. Endemic resident. S Cameroon, known from Bitye (600 m) on Dja R., but not found during the ECOFAC study of the birds of the Réserve du Dja in 1993. Gabon, found at Mimongo (800 m) and M'Bigou (700 m) in the Massif du Chaillu, and 3 skeletons in the Field Museum in Chicago are from unknown localities. Bird caught at Makokou airport in NE was probably a vagrant (A. Brosset *in* Collar and Stuart 1985). Recently discovered at Réserve de la Lopé in central Gabon 0°10'S, 11°30'E (Christy 1994). Rare, but possibly more widespread than map shows; skulking behaviour and dense swamp habitat make it extremely difficult to observe. Not a forest bird so not threatened by forest clearance unless this is accompanied by drainage of adjacent swamps.

Bradypterus grandis

Description. ADULT ♂: entire head and upperparts dark brown, rump and uppertail-coverts with faint rufous wash. Tail long and graduated, brown with definite but indistinct narrow blackish brown bars uniformly spaced throughout its length. Supercilium brownish white, lores brown, ear-coverts blackish, chin and throat white with some blackish spots, breast brownish buff, cheeks and middle of upper breast distinctly marked with narrow black streaks; sides of body, flanks and undertail-coverts dark olive-brown to dusky, centre line of lower breast and belly white; thighs light brown. Primaries and secondaries dark brown, outer webs paler more rufous-brown, lesser and median coverts dark brown edged brownish buff; underwing-coverts and axillaries light brown. Bill black, lower mandible grey beneath; eyes brown, legs and feet grey. Sexes alike (?). SIZE: wing, ♂ (n = 3) 64–66·5 (65·5), 1 unsexed 65; tail, ♂ (n = 3) 72–77 (75·0), 1 unsexed 76; bill to skull, ♂ (n = 2) 17, 17, 1 unsexed 17; bill to feathers, ♂ (n = 1) 14; tarsus, ♂ (n = 3) 24–26 (25·3), 1 unsexed 26.

IMMATURE: upperparts of a ♂ lighter than adult, throat and upper breast moderately spotted with blackish, remainder of underparts white; flanks washed brown, but no brownish breast band; lesser coverts edged white.

Field Characters. A rather large warbler with dark brown upperparts, flanks and undertail-coverts and contrasting white throat; breast buffy brown with dark streaks; tail long, graduated and broad, feathers with pointed tips. Secretive; keeps near ground in dense growth. Cannot be confused with any other species in Gabon, but closely-related Little Rush-Warbler *B. baboecala* is present in Cameroon, although whether the 2 occur in the same marshes is not yet known. Little Rush-Warbler is smaller, but no direct field comparison has been made; their songs are similar.

Voice. Tape-recorded (CHR). Song, loud and repetitive, given for long periods during the breeding season, consists of 4 separate introductory notes followed by a trill, rendered 'psuit-psuit-psuit-its-struuuuuuu' (Christy and Clarke 1994). When excited, also gives a guttural 'oua-oua-oua'. Sonagrams of the songs of *B. graueri* and *B. grandis* made by F. Dowsett-Lemaire (pers. comm.) show they are similar in form but yet different enough to represent a strong dialect, if not a different species. Song said by Beatty (*in* Rand *et al.* 1959) to be beautiful and distinctive, but while distinctive it is certainly not beautiful, so this remark may have referred to a different species.

General Habits. Principal habitat appears to be marshes with tall Cyperaceae, especially *Rhynchospora corymbosa*, lying between forest (including gallery forest) and savanna; may also be expected in similar marshes within forest or bordering rivers that flow through forest. Also occurs in tall elephant grass *Pennisetum* and low, dense growth in abandoned plantations. Does not necessarily require extensive marshes: bird in Gabon inhabited patch only 50 m × 12 m (Christy 1994). Feet very strong, characteristic of a species which moves about in a dense environment with short hops from stem to stem, rather than flying any distance (J. Kingdon *in* Christy 1994). Responds well to song playback, otherwise almost impossible to see; flies heavily and noisily over tops of sedges, in partly upright position, with tail spread, sometimes giving some notes of the song, and lands on exposed perch (Christy and Clarke 1994).

Food. Unknown.

Breeding Habits. Almost unknown. During display flight over territory makes a curious wing-snapping noise, which may come from an area under the wings of bare or sparsely-feathered skin; the long and dense flank feathers may help protect the bird as it plunges down into the rough vegetation at the end of its flight (J. Kingdon *in* Christy 1994).

BREEDING INDICATIONS: 2 birds taken in mid-June were probably young.

References
Christy, P. (1994).
Christy, P. and Clarke, W. (1994).
Collar, N. J. and Stuart, S. N. (1985).
Rand, A. L. *et al.* (1959).

Bradypterus alfredi Hartlaub. Bamboo Warbler. Bouscarle des bambous.

Plate 5 (Opp. p. 96)

Bradypterus alfredi Hartlaub, 1890. J. Orn. 38, p. 152; Njangalo (= Nyangabo), NE Congo Free State.

Range and Status. Endemic resident, rare and local, from W Ethiopia to NW Zambia. W Ethiopia: known from 3 sites, Didessa (Shoa Province), Bulcha Forest (Sidamo Province, 360 km southeast of Didessa) and Gambela (200 km southwest of Didessa). S Sudan: only at Gilo (4°N, 32° 51′E), where frequent. E Zaïre: from plateau west of L. Albert and Rwenzoris south to mountains northwest of Baraka, Bunia in Ituri District, Nyangabo (S Lendu plateau), and Lwiro and Dipidi in Upemba Nat. Park; possibly occurs in SE Zaïre (Chapin 1953a). W Uganda: Bugoma Forest and Mubuku Valley. W Tanzania: Mt Kungwe and Mt Mahari. An isolated population in NW Mwinilunga District, Zambia: Isombo, Salujinga, Mundwiji Plain and Mwinilunga. Everywhere rare and localized.

Bradypterus alfredi

Description. *B. a. kungwensis* Moreau: W Tanzania, NW Zambia. ADULT ♂: upperparts from forehead to rump and uppertail-coverts blackish olivaceous-brown; 12 tail-feathers, which are broad, dark brown, with faint traces of darker transverse barring throughout length, inner edge of most feathers brownish white. Faint pale buff supercilium, from nostril to just beyond eye. Chin and throat whitish, more olive at sides, breast brownish olive, central part of lower breast and belly dirty white, sides of belly and flanks dark olive-grey. Undertail-coverts whitish grey, barred brown, central ones brown with broad white margins. Lores dark brown, ear-coverts and cheeks blackish brown, feather-shafts whitish. Primaries and secondaries blackish brown, much paler underneath, upperwing-coverts blackish brown edged brown, underwing-coverts whitish mottled olive-grey, axillaries white. Upper mandible black, lower mandible slate grey; eyes dark brown; feet brown. Sexes alike, but ♀ warmer brown above than ♂. SIZE: wing, ♂ (n = 4) 55–60 (58·0), ♀ (n = 4) 56–62 (59·0); tail, ♂ (n = 3) 53, 53, 41 (the last well worn), ♀ (n = 3) 51, 55, 57; bill to feathers, 1 ♂ 14, ♀ (n = 2) 13, 14; bill to skull, ♂ (n = 3) 16, 17, 17·5, ♀ (n = 2) 16, 16·5; tarsus, ♂ (n = 2) 24, 24, ♀ (n = 3) 24, 25, 26 (Prigogine 1980c). WEIGHT: not known.

IMMATURE: ♂ has yellowish wash on underparts.

B. a. alfredi Hartlaub: W Ethiopia to E Zaïre. Similar to *kungwensis* but upperparts more rufous, flanks paler, edges of undertail-coverts whiter. Immature paler grey below, undertail-coverts buff. SIZE: (1 ♀) wing 65, tail 61, bill to skull 16·5, tarsus 24; wing, unsexed (n = 3, Sudan) 58–61 (60·0) (Nikolaus 1982).

Field Characters. A medium-sized warbler with dark brown upperparts, rounded tail, pale supercilium, whitish throat and central line of belly, rest of underparts olive-brown. Difficult to observe as it moves through ground vegetation; and even in the hand easy to mistake for other *Bradypterus* warblers, e.g. Cinnamon Bracken Warbler *B. cinnamomeus* (which is more rufous above and below). Song distinctive. Best recognized by short ticking call, like Common Redstart *Phoenicurus phoenicurus* (Nikolaus 1982).

Voice. Tape-recorded (B, F, ASP, CART). Song of ♂ a monotonously repeated, hard, not melodious, 2-syllabled note, 'tchuu-ka', the 'ka' being staccato: 'tchuu-ka, tchuu-ka, tchuu-ka, tchuu-ka, tchuu-ka, tchuu-ka . . .'. Song often consists of 20–40 'tchuu-ka's, sometimes up to 60. Song lasts 5–20 s (av. 11·6 s, n = 8); sung at rate of 2·7 'tchuu-ka's per s (range 2·5–3·0, n = 5). Volume is constant. Possible alarm, a repeated short 'whitt' (C. Carter, pers. comm.); doubtless the Redstart-like ticking (see above).

General Habits. A rare warbler occurring in a wide variety of habitats, mainly bamboo, also long grass and dense cover in and outside forest, at altitudes of 1200–2500 m in Zaïre, 1200–1600 m in W Uganda, 1800–2300 m in W Tanzania and 525–1260 m in Ethiopia. Keeps to ground stratum, mainly below 1·5 m. In Zaïre and Ethiopia occurs in complex mosaic of riverine marshland, *Combretum/Terminalia* savanna, crops, mixed deciduous woodland; near villages. Sometimes lives up to 50 km away from bamboo stands (Ash 1977). In Sudan found in grassy secondary growth, and is easier to find after fire has destroyed most low vegetation (Nikolaus 1982). In Zambia inhabits ground stratum of moist evergreen forest, sometimes alongside Cinnamon Bracken Warbler, but the 2 are usually mutually exclusive (Benson *et al.* 1971).

Food. Insects.

Breeding Habits. Not known.

References
Ash, J. S. (1977).
Prigogine, A. (1980c).
van den Elzen, R. and Konig, C. (1983).

Plate 5
(Opp. p. 96)

Bradypterus lopezi (Alexander). Evergreen-Forest Warbler. Bouscarle de Lopes.

Phlexis lopesi Alexander, 1903. Bull. Br. Orn. Club 13, p. 48; Moka, Fernando Po.

Range and Status. Endemic, localized resident and partial altitudinal migrant. Occurs E Nigeria, Bioko and W Cameroon; W Angola; and E and S-central Africa; at *c.* 1700–3000 m in W and E Africa and 900–2500 m in southern Africa. Locally very common E Nigeria (Obudu Plateau, Gotel Mts at Gangirwal, Chappel

Bradypterus lopezi

Wadi, Leinde Fadali), Bioko (Moka valley and N highlands) and Cameroon (Mt Oku, Banso Mts, Kumbo, in forest patches at Sabga Pass, between Mbengwi and Tinachong, Bamenda Highlands (L. Bambulue), Bamboutos Mts, and south to Bangwa on Bamileke Plateau, Mt Manenguba and Mt Cameroon, but not on Rumpi Hills or Mt Kupé). Locally common in Angola (S Cuanza Sul and Huambo, Mt Soque, Mt Moco, Mombolo, Benguela Plateau).

Frequent in E and SE Zaïre (Rwenzoris, L. Edward at Lutunguru, Kivu volcanoes, Itombwe district, north of L. Tanganyika, Upemba Nat. Park); local and uncommon in SW Uganda (Rwenzoris, Bwindi-Impenetrable Forest), commoner in W and central Kenya (Nandi, Mau and Molo Forests to Aberdares, Irangi Forest on Mt Kenya and Limuru) and in SE (Chyulu Hills, Taita Hills, Taveta on Lumi R.). Local and generally uncommon in Rwanda (Mt Muzimu, Kamiranzovu, Uwinka within Nyungwe Forest, Burunga, Lulenga, Bugoie) but more common Gisovu area; rare in S and SE (Schouteden 1966, Dowsett-Lemaire 1990). Common in NE Tanzania (Oldeani to Arusha Nat. Park and Mt Kilimanjaro, N Pare Mts, E and W Usambaras), and in E, SW and S Tanzania (Mbisi Forest on Ufipa Plateau, Ilembo, Tukuyu, Mbeya, Mt Rungwe, Njombe, Songea, Ukaguru, Nguru, Uluguru and Udzungwa Mts). Locally common Zambia in NW (Mwinilunga, Solwezi and Ndola regions), N (Luapula, Mbala), and NE (border with NW Malaŵi). Widespread in N Malaŵi (Misuku, Nyika Mts in NW, Uzumara, Chimaliro, Nythungwa, Chamambo and Kawandama forests in Viphya); uncommon at Misuku in N, rare at Ntchisi and local in E Viphya; common and more widespread in S Malaŵi (Dzalanyama Mts, Chongoni, Mlunduni, Dedza, Chirobwe, Zomba and Malosa Mts, and Kirk Range), but less common in drier forests of Mangochi and Lisau. Locally common in N Mozambique (Namuli, Unango, Mt Chiperone).

Densities of 3–6 pairs per 10 ha at 1700–2300 m on Nyika plateau, Malaŵi, and in some Viphya forests (Uzumara, Kawandama and central S Viphya); in large forests 1–4 breeding pairs per 4 ha, av. 1 pair per 2 ha (39 pairs in 78·7 ha, Dowsett-Lemaire 1983); forest patches as small as 0·9 ha can hold a breeding pair. Absence from Mafinga-Jembya-Musisi forest, Malaŵi, not understood (Dowsett-Lemaire 1989). Populations on Mts Manenguba and Oku possibly threatened by habitat destruction, elsewhere not threatened.

Description. *B. l. lopezi* (Alexander): Bioko. ADULT ♂: whole of upperparts from forehead to tail dark chocolate-brown with russet wash, head a little darker, feathers of forehead and crown with paler centres; rump-feathers loose and long, broadly edged tawny brown; 10 tail-feathers, greatly frayed when old, shafts blackish. Lores, and ear-coverts brown, cheeks mottled cinnamon-brown which merges into colour of mantle, each feather tipped brown with whitish shaft; superciliary stripe cinnamon, extending from nostril to well beyond eye; indistinct brown stripe through eye. Chin and throat cinnamon, breast and sides of body tawny brown, flanks and thighs darker, inside of thighs paler, more olive; centre of belly cinnamon, becoming more buffish on lower belly; undertail-coverts dark tawny-brown. Primaries brown, outer web of each edged russet, inner web edged slightly lighter; secondaries and upperwing-coverts brown, broadly edged russet, secondaries with some faint underlying dark bands, marginal coverts cinnamon giving pale leading edge to wing; underwing-coverts grey-brown, some tipped tawny giving mottled appearance, axillaries tawny brown. Bill black, lower mandible horn colour; eyes brown; legs brownish flesh. Sexes alike. SIZE: wing, ♂ (n = 7) 57–60 (58·5), ♀ (n = 5) 55–59 (57·0); tail, ♂ (n = 7) 54–58 (57·0), ♀ (n = 5) 55–60 (56·5); bill to feathers, ♂ (n = 5) 11–13 (11·5); tarsus, ♂ (n = 5) 20–23 (21·4).

WEIGHT: (Bioko) unsexed (n = 12) 16–20 (17·7).

IMMATURE: upperparts like adult but chin, throat and belly dull olive-yellow, breast dark olive, flanks, thighs and undertail-coverts olive-brown, webs of tail narrow. Bill blackish above, yellowish below, tipped dusky; iris and feet grey brown.

NESTLING: black spot at tip and one on either side at back of tongue.

B. l. camerunensis Alexander (including '*youngi*'): Mt Cameroon. Similar to *lopezi* above but brown of underparts with only faint cinnamon wash, throat whitish (pale brown in some birds), mottled brown; lower breast and belly whitish with suggestion of streaking on sides of belly in some birds; superciliary stripe pale brown.

WEIGHT: (Mt Cameroon) ♂ (n = 10) 17–19·5 (18·3), ♀ (n = 2) 14, 16, unsexed (n = 2) 16, 19.

B. l. granti (Benson): Malaŵi (south of Nyika) and Mt Chiperone, N Mozambique. Similar to *camerunensis* but underparts washed more strongly cinnamon, pattern of paler colours the same. WEIGHT: (*granti* and *usambarae*, Tanzania, Malaŵi, n = 29) unsexed 12·0–19·9 (16·9).

B. l. mariae (including '*altumi*', '*sjöstedti*' and '*mitoni*') van Someren: Kenya (except Taita Hills), NE Tanzania. Like *camerunensis* but darker, entire upperparts brownish olive, superciliary stripe greyish white; underparts olive-brown, some feathers on upper breast with darker centres (like diffuse streaks); sides, flanks and thighs darker and chin, throat and centre line of belly whiter than in *camerunensis*. SIZE: (10 ♂♂, 6 ♀♀) wing, ♂ 59–64 (62·5), ♀ 58–60 (59·2); tail, ♂ 58–67 (63·4), ♀ 45–60 (55·7). WEIGHT: (Udzungwa Mts, Tanzania) unsexed (n = 21) 15·1–20·8 (18·9).

B. l. usambarae Reichenow (including '*roehli*'): Taita Hills, Kenya, E and SW Tanzania, Zambia (Nyika Plateau, Tukuyu, Ilembo), Malaŵi north of Nyika Plateau and N Mozambique at Unango. Similar to *mariae* but underparts browner, less olive, white on belly more extensive; also like *camerunensis* but lacks cinnamon wash. WEIGHT: ♂ (n = 4) 20–21 (20·2); ♀ (n = 4) 16·1–18·4 (17·3).

B. l. boultoni Chapin: Angola. Similar to *usambarae* but upperparts lighter brown, some tail-feathers with darker underlying bars uniformly spaced along them. Chin and upper throat whitish buff, lower throat and breast pale antique-brown faintly streaked dark brown, lower breast and upper belly tawny cinnamon along centre line, sides darker and merging into burnt umber of flanks; centre line of belly whitish. Larger: wing, unsexed (n = 14) 64–68 (66·1); tail, unsexed (n = 14) 65–73 (68·7).

B. l. ufipae (Grant and Mackworth-Praed): SW Tanzania (Ufipa plateau, Mbisi forest), SE Zaïre (Marungu Plateau), N Zambia. Like *boultoni* above but area of white on belly more extensive, breast tawny brown and tail-feathers barred uniformly darker; undertail-coverts tawny cinnamon. Differs from *bangwaensis* in having dark olive (not dark tawny brown) sides. WEIGHT: unsexed (n = 17) 17–24 (20·2); (Upemba, Zaïre) ♂ (n = 2) 22, 22, ♀ (n = 5) 18–23 (20·8).

B. l. bangwaensis (including '*castaneus*' and '*manengubae*'): E Nigeria, Cameroon except Mt Cameroon. Similar to *ufipae* but much richer tawny cinnamon below, sides dark tawny brown; white on belly more extensive and extending on to lower breast; thighs dark on outside, whitish on inside; upperparts rustier than in other races. Although the underparts of the two *manengubae* skins are fulvous brown, they are probably only aberrant forms of *bangwaensis*. WEIGHT: (Cameroon) ♂ (n = 5) 18–27 (21·0).

B. l. barakae Sharpe: Zaïre (except Marungu Plateau), Rwanda, SW Uganda. Closest to *bangwaensis* but upperparts rich reddish dark brown, throat to lower breast rich tawny cinnamon; centre line of belly pale. WEIGHT: (Kigezi District, Uganda) ♂ (n = 6) 16–18 (17·1), ♀ (n = 2) 15, 17, 1 unsexed 18·5.

TAXONOMIC NOTE: formerly treated as 2 species, *B. lopezi* and *B. mariae*. Populations vary in plumage colour, but the songs of *camerunensis*, *mariae*, *ufipae*, *usambarae*, *bangwaensis* and *lopezi* are similar (agreeing with descriptions of *barakae* and *boultoni*) and each race responds to played-back songs of others. Moreover they all have similar tail/wing length ratios (racial averages 0·95–1·01) (Dowsett and Stjernstedt 1979, Dowsett-Lemaire and Dowsett 1989c). See also Taxonomic Note in *B. cinnamomeus*.

Field Characters. A dingy warbler of forest undergrowth, dark brown above, paler below, with whitish throat and supercilium; in some races upperparts washed russet, underparts cinnamon. Tail-feathers narrow, often frayed. Easily located by contact call (Eisentraut 1973). Best identified by loud, monotonous song and by habitat – no other *Bradypterus* lives in forest interior, although Cinnamon Bracken Warbler *B. cinnamomeus* and Bamboo Warbler *B. alfredi* live on forest edge; both have very different songs. Does not overlap with Barratt's Warbler *B. barratti*.

Voice. Tape-recorded (32, 57, 86, CART, CHA, GREG, GRI, LEM, McVIC). Song of ♂ loud, sometimes explosive, a series of 3–10 (6–8) single or double notes on same pitch that increase in volume, giving the impression the bird is moving towards you, 'chee-chee-chee-chee-CHEE-CHEE-CHEE', 'chitrip-chitrip-chi-trip-chitrip-CHITRIP-CHITRIP'. Individual variation seems almost endless: 'tyoo-tyoo-tyoo . . .', 'chip-chip-chip . . .' 'chwee-chwee-chwee . . .', 'set-set-set', or double 'cheewa-cheewa-cheewa . . .', 'choopi-choopi-choopi . . .', 'weecha-weecha-weecha . . .', 'tillick-tillick-tillick . . .', 'wiji-wiji-wiji . . .'; nevertheless, pattern and structure remain constant throughout range (Dowsett-Lemaire and Dowsett 1989c). Each individual often has several song types. Sometimes a duet, ♀ joining in towards end of song with some high-pitched whistles, 'wheee-ooo', which may continue for a short time after the ♂ stops. Frequency range of each note 1·8–6 kHz; each note has 1 or 2 peaks (sonagrams, Grimes 1976a, Dowsett and Stjernstedt 1979). One song of *bangwaensis* was more complex (Dowsett-Lemaire and Dowsett 1989c).

Number of notes per s is 3–7·5 in Zambia, Malaŵi and Tanzania (n = 13); 3–11 on Mt Cameroon (n = 17); and 3–6·5 at Obudu and Gangirwal in Gotel Mts, Nigeria. Song duration is 2–5 s in Zambia, Malaŵi and Tanzania (n = 31); 2·3–3·5 s on Mt Cameroon (n = 18); 5–10 s at Gangirwal (n = 7); and 4–4·25 s at Obudu (n = 6). One song in Rwanda lasted 13 s (Dowsett-Lemaire and Dowsett 1990). Time intervals between songs range from 4 to 10 s (single note song) and 8 to 12 s (double note song).

Alarm call of *lopezi* a 'yoop-yoop-ipp' (Eisentraut 1973) and a loud 'chick' (Stuart and Jensen 1986), of *mariae* 1–2 loud, clicking 'tchic, tchi-tchic', of *camerunensis* similar but usually 3–8 together. Other calls: a vigorous 'weet-weet, weet-weet, weet-weet', a vibrant 'pirr', and a prolonged, harsh 'churr' (*bangwaensis*) (Serle 1965); contact call a distinctive 'cheep' repeated 3–4 times (Eisentraut 1973). Call of *bangwaensis*, when either alarmed or excited, a short rattle 'krrrr' (frequency range 1–6 kHz – sonagrams in Dowsett 1989).

General Habits. Inhabits dense undergrowth inside montane forest; gullies near streams (Rwanda) and bracken-briar, Malaŵi, where a < 1 ha patch of forest, mainly a tangle of *Mimulopsis sulmsii*, once held a pair; in W Africa, rank herbage along streams and plantations of spaced trees, tall thick grass, scrub, bracken, *Dissotis–Hypericum* thickets, brambles at high elevations, undergrowth of *Podocarpus* forests, and bamboo thickets growing under the canopy (Stuart and Jensen 1986, Wilson 1987). Readily adapts to habitat changes; in Dedza, Malaŵi, inhabits pine plantations with plenty of ground cover. Altitudinal ranges: 1600–2300 m in Nigeria; 800–2150 on Mt Cameroon, where abundance increases with altitude; 1900–2200 on Mt Manenguba; < 2000–2950 on Mt Oku where abundant only above 2400 m; 1700–2600 m elsewhere in Bamenda highlands; 1500–3100 m in Kenya (in 1000+ mm annual rainfall areas, except forest islands of Chyulu and Taita Hills); up to 2800 m on Mt Muzimu, Rwanda, where usually not below 2100 m except very locally in west (e.g. Kamiranzovu, 1950–2000 m); 2300–2400 m near Uwinka, Rwanda (where pairs scattered and isolated); generally at 1700–2250 m in Zaïre but down to *c*. 1500 m at Musangakye; 900–2500 m in Tanzania; 900–2500 m in N Malaŵi, commonest at 1700–2300 m (or lower in Ntchisi and on Viphya); 1150–2450 m in S Malaŵi but at 950 m on S Mulanje.

Keeps within 2 m of ground. Secretive, inquisitive, but difficult to locate and observe as it clambers among low thickets of shrubs and creepers in deep shade, even when drawn close to observer by song playback. Flight short and ungainly. Usually in pairs, occasionally in family parties. Gleans from leaves and bark in dense shrubby understorey, picking off insects where soaking wet lianas and beard moss descend some way down trunk (van Someren 1939); also feeds on ground and at ant swarms (Cordeiro 1994). On Nyika Plateau, Malaŵi, dominates Cinnamon Bracken Warbler; interactions twice observed near edge of a 2-ha forest, ♂♂ counter singing within a few m of each other; if Evergreen-Forest Warbler is absent from patch of forest Cinnamon Bracken Warbler occurs in its interior but otherwise remains at the edge (Dowsett-Lemaire 1983). On Mt Kilimanjaro, Tanzania, occurs at forest edge and interior, but scarce in thin and shrubby forest where Cinnamon Bracken Warbler often occurs; absent from bracken areas and rare at elevations where grasses are dominant (Cordeiro 1994).

Mainly resident, but some altitudinal migration; in Kibungo Forest, Tanzania once recorded at 250 m in cool season (June) (Stuart and Jensen 1985). Moves down Mt Mulanje, Malaŵi to 600–700 m in cool season (Mar–Aug); once down to *c*. 550 m (June); deserts highlands of Zambia in Apr–May (Dowsett-Lemaire 1889); may move down mountains in Cameroon.

Food. Fly larvae, small beetles, ants, caterpillars, insect eggs, and small snails.

Breeding Habits. Monogamous, solitary, territorial nester.

NEST: one was compactly made of leaf skeletons mixed with bits of lichen and other plants, the cup lined with hard, dry plant threads; another was thickly compacted with dry grass stalks and blades and some leaves, the cup lined with thin, long plant fibres (Eisentraut 1963, 1973); a third consisted of blades of grass and reeds (van Someren 1919). Int. diam. 50–60, int. depth *c*. 40. Placed at upper edge of a 3 m depression in thick undergrowth at the base of a tree-fern, or well hidden in grass tussock some 20–30 cm above ground, or in clump of reeds or willow-like shrub.

EGGS: 2–3. Whitish pink, speckled and streaked with violet-lilac and liver-brown, or (Bioko) covered with dull brownish red and brownish grey speckles which merge into dark band at blunt end. SIZE: (Bioko, n = 2) 20·6 × 15·2, 20·7 × 16·0.

LAYING DATES: Cameroon, Oct–Nov, occasionally Mar–Apr (immatures, Sept, Oct, pair feeding young, mid-Jan, brood patch, Jan, breeding condition, Aug, adult carrying food, Aug: Wilson 1987); Bioko, Oct–Nov; Zaïre, Mt Kivu volcanoes, Apr (Kivu, juv., early Mar, Itombwe, gonads enlarged, Mar–May, July); Angola, Oct; Tanzania, Mt Kilimanjaro, Dec (2 clutches); Malaŵi and Malaŵi/Zambia border, Sept–Jan (Sept 1, Oct 4, Nov 9, Dec 9, Jan 1 clutches) (end of dry season, continuing into rainy season: Dowsett-Lemaire 1989).

References
Dowsett, R. J. and Dowsett-Lemaire, F. (1980, 1984).
Dowsett, R. J. and Stjernstedt, R. (1979).
Eisentraut, M. (1973).

Bradypterus cinnamomeus (Rüppell). Cinnamon Bracken Warbler. Bouscarle cannelle.

Plate 5
(Opp. p. 96)

Sylvia? (Salicaria) cinnamomea Rüppell, 1840. Neue Wirbelthiere Fauna Abyssinien, Vögel, p. 111, pl. 42, fig. 1; Entschetqab, Semien Province, N Abyssinia.

Forms a superspecies with *B. barratti*.

Range and Status. Endemic, resident, mountains of Ethiopia, S Sudan, Uganda, Kenya, E and SE Zaïre, Rwanda, Burundi, Tanzania, Zambia and Malaŵi, at 1700–4100 m. Common in isolated pockets in W and SE Highlands of Ethiopia; SE Sudan (Imatong and Dongotona Mts only, where very common); mountains along Zaïre/Uganda border, N and S Kivu District and NW of L. Tanganyika and Marungu highlands in SE Zaïre; Uganda in NE (Mt Moroto, Mt Morongole) and SW (Rwenzori Mts, Ankole, Kigezi); Kenya from Mt Elgon, Mt Nyiru, Aberdare Mts and Mt Kenya south to Mau Forest, Trans-Mara Forest in SW, Nguruman Hills, Nakuru, Kikuyu and Chyulu Hills; Rwanda (Astrida, Nyungwe Forest, Lulenge, Burungu, Nyamuzinga, Rwabeya, Mt Sabinyo, Mt Bisoke, Mt Karisimbi, Mt Muhavura), Burundi (abundant from N border with Rwanda to Ijenda and Teza in S), Tanzania (from Crater Highlands and Mt Kilimanjaro east and south to Usambaras and Ngurus, Ufipa Plateau in SW, and Njombe to Mbeya and Ilembo in S). Locally common N Malaŵi (Nyika Plateau and border of Zambia only) and S Malaŵi on Mulanje Mt, some 550 km further south (Dowsett-Lemaire 1989).

Bradypterus cinnamomeus

Description. *B. c. cinnamomeus* (Rüppell) (including '*rufoflavidus*', '*pallidior*', '*salvadorii*', '*chyuluensis*' and '*macdonaldi*'): Ethiopia, Kenya, Uganda, E Zaïre (except Rwenzori Mts), Uganda, Burundi, Rwanda and N Tanzania (north of *nyassae*). ADULT ♂: forehead to crown dark brown, hindcrown to back lighter, rump and uppertail-coverts reddish brown, rump-feathers in some birds tipped cinnamon, tail-feathers up to 15 mm broad, more rufous than upperparts, with faint underlying dark barring (T1 with at least 19 bars spaced *c.* 2 mm apart). 10 or 12 tail-feathers (Chapin 1953a). Lores blackish, ear-coverts brown, cheeks brownish cinnamon, feathers with whitish shafts, merging into cinnamon on sides of throat; supercilium from nostril to beyond eye cinnamon, stripe through eye blackish. Chin and throat buffish white, chin and sides of throat washed cinnamon, upper breast, sides of body and flanks rich tawny cinnamon, each feather with cinnamon tip; flank-feathers long and fluffy; lower breast and belly white, vent buffish, undertail-coverts and outer thighs tawny cinnamon, inside thighs mottled buffish. Primaries, secondaries and upperwing-coverts dark brown, outer webs of primaries paler and edged rufous-cinnamon, secondaries and upperwing-coverts edged rufous-cinnamon, secondaries with faint underlying dark barring; marginals pale buff giving pale leading edge to wing; underwing-coverts brown, some edged and mottled cinnamon, axillaries pale cinnamon. Bill dark horn; eyes hazel; legs and feet brown. Sexes similar, but ♀ with head greyer, supercilium buffy and sides of face without cinnamon wash. SIZE: (8 ♂♂, 6 ♀♀) wing, ♂ 57–64 (59·3), ♀ 57–60 (58·0); tail, ♂ 55–70 (62·1), ♀ 55–67 (60·5); bill to feathers, ♂ 11–12 (11·7), ♀ 11–12 (11·5); tarsus, ♂ 20–22 (20·7), ♀ 20–22 (21·2). WEIGHT: ♂ (n = 9, Mt Nyiru, Kenya) 15–25 (19·0), (n = 23, Kenya) 15–20 (17·8), (n = 4, Ghera Region, Ethiopia) 17–19 (18·2), (n = 16, Uganda) 15–20 (18·3); ♀ (n = 7, Kenya) 10–25 (17·0), (n = 16, Uganda) 14·5–25 (16·9).

IMMATURE: browner above and below, lacking rufous-brown of adult; chin pale yellow, supercilium, throat and belly olive-yellow, breast olive-brown with dark blotches and washed cinnamon in some birds, sides of body and flanks olive-brown; older birds have yellowish wash confined to belly. Bill black, yellow at base of lower mandible, legs yellowish flesh-brown.

NESTLING: tongue yellow with black spot on each side at back.

B. c. cavei Macdonald: SE Sudan. Similar to *cinnamomeus* but upperparts, including head and tail, darker and deeper brown, less rufous; underparts much more tawny cinnamon, throat with cinnamon wash, smaller area of white on lower breast and belly.

B. c. mildbreadi Reichenow: Rwenzori Mts, Uganda/Zaïre. Very similar to *cinnamomeus*, but richer reddish above; not well differentiated in plumage, but a little larger than nominate race. SIZE: ♂ (n = 10), wing 59–65 (61·9), tail 56–72 (65·9). WEIGHT: (Uganda) 1 ♂ 18; 3 ♀♀ 18–20 (18·7).

B. c. nyassae Shelley: NE and SW Tanzania, SE Zaïre, NE Zambia and Malaŵi. Like *cinnamomeus* but duller; breast and sides of body buffish brown, flanks and undertail-coverts with cinnamon wash, throat and belly buffish white, supercilium buff. SIZE: (7 ♂♂, 9 ♀♀) wing, ♂ 57–65 (61·7), ♀ 57–63 (60·6); tail, ♂ 64–76 (71·1), ♀ 40–71 (62·8). WEIGHT: unsexed (n = 20) 16·3–22 (18·5); (Udzungwa Mts, Tanzania) ♂ (n = 5) 15·2–18·5 (16·8), ♀ (n = 1) 16·6.

TAXONOMIC NOTE: *B. cinnamomeus* is so similar morphologically to the sympatric *B. lopezi* that they have been widely confused. *B. cinnamomeus* has wider rectrices: T1 when fresh is *c.* 15 mm wide, at one-third of its length from tip, *vs.* 10 mm in *B. lopezi*. In *B. cinnamomeus* tail/wing ratios are: nominate *cinnamomeus* (n = 2) 1·15, *nyassae* (n = 21) 1·11–1·31 (1·18) and *mildbreadi* (n = 10) 1·01–1·16 (1·07). But *cavei* is only (n = 10) 0·94–1·15 (1·03) and so may belong to *B. lopezi* which has a ratio of 1·0 (Dowsett-Lemaire and Dowsett 1989c). Songs of *B. cinnamomeus* and *B. lopezi* differ; pending study of its song we retain *cavei* in *B. cinnamomeus*. The song of *B. cinnamomeus* is much like that of *B. barratti* (Dowsett and Stjernstedt 1979, Dowsett and Dowsett-Lemaire 1980) and for that reason we treat them as a superspecies (*contra* Hall and Moreau 1970, who made *B. cinnamomeus* a superspecies with *B. victorini*).

Field Characters. A medium-sized warbler, above dark brown with rufous wash, particularly on tail; pale buff-brown supercilium extending well beyond eyes. Throat whitish, breast and sides of body rich cinnamon, rest of underparts white, undertail-coverts tawny-cinnamon. Tail long and slightly graduated. Occurs alongside Evergreen-Forest Warbler *B. lopezi* in parts of Uganda, Zaïre, Kenya, Tanzania and Malaŵi; the 2 are ecologically separate, *B. lopezi* inhabiting forest interior and *B. cinnamomeus* the forest edge and other non-forest habitats; very alike in appearance but with quite different songs: *B. lopezi* has short crescendo song, without introductory whistles; *B. cinnamomeus* has loud, ringing 2-part song, see *Voice*. Hybrids may occur on Nyika Plateau, Malaŵi, as one recorded song is a mixture of both song types (I. Sinclair, pers. comm.).

Voice. Tape-recorded (32, 57, 86, B, C, MANN, SIN). Song of ♂ loud and ringing, with many variations; begins with 1–5 high-pitched slurred notes followed by a rapid repetition of a single short note: 'tee tee tee tew tsu-tsu-tsu-tsu-tsu . . .', 'chee woy woy chichichichichi', or 'wee chyoo-chyoo-chyoo-chyoo'. One song recorded at Mbulu, N Tanzania, ends with a slurred note in which pitch rises; this may have been contributed by a ♀. Song lasts 0·6–2·0 s (av. 1·4 s (n = 19), Molo, Kenya; av. 1·1 s (n = 40), Mt Kenya; av. 0·7 s (n = 14), Mbulu, Tanzania). In Malaŵi, a song given in response to playback of own song lasted for 4 s. The faster notes are usually given at rate of 10–20 per s, but may be as few as 4 per s. Volume of song is constant throughout. A second bird sometimes accompanies in duet with 3 or 4 high-pitched slurred whistles, 'tsee-too-way-wee', and may continue after first bird has stopped. Pitched at 3–4 kHz, but up to 6 kHz (Dowsett and Stjernstedt 1979). Intervals between start of 19 consecutive songs 4·7–10·8 (6·6) s. Alarm call a low pitched rattle 'trrr' (N Malaŵi) or 'brrrr brrrr', also a higher pitched squeal 'schreep' (Mt Kenya). Contact call in Nyungwe Forest, Rwanda, a loud, sharp 'pîe', not heard in other E African populations (Dowsett and Stjernstedt 1979, Dowsett-Lemaire 1990).

General Habits. Occurs in undergrowth of dense shrubbery in fairly open montane forest, bamboo and forest edge, penetrating small patches of montane forest with herbaceous understorey (where Evergreen-Forest Warbler absent: Dowsett-Lemaire 1983). Also in thick cover in plantations with spaced trees, tall thick grass, bracken briar, *Hypericum*–*Erica* scrub, edges of marshes, overgrown gardens, brambles and almost pure bracken *Pteridium aquilinum* with some grasses. Not dependent on forest. Breeds in shrub 'island' as small as 0·1 ha (Dowsett-Lemaire 1983). In SE Sudan occurs above 1800 m (Nikolaus 1989). In Kenya mainly between 2000–4000 m in subhumid–humid zone (rainfall areas of > 500 mm) which includes the forest islands of Mt Nyiru and Chyulu Hills (Lewis and Pomeroy 1989). In Nyungwe Forest, Rwanda, occurs usually above 2000 m but descends to 1950 m at edge of Kamiranzovu and to 1800 m in Shava valley, absent from the more exposed Tangaro Valley and Bureyeye Hills at 1700–2100 m. Penetrates forest only under rather broken canopy (Dowsett-Lemaire 1990). In Zaïre occurs in dense herbaceous growth in *Hagenia* woods at 3350 m and in grasses and *Alchemilla* in the alpine zone (Chapin 1953a), and at 1520–4100 m in Kivu District. On W slopes of Rwenzori Mts occurs at 2100–2750 m, but not in heath zone; on E slopes reaches 3050 m where there are moist valley bottoms with ground cover. On Mt Kilimanjaro, Tanzania, occurs in scrub areas with bracken, and in heath zone (Cordeiro 1994). In Malaŵi lives in montane scrub and bracken on forest edges at 1300–2800 m.

Usually solitary or in pairs. Secretive, keeps to ground cover; movements when foraging mouse-like; does not join mixed bird parties. Generally located by song; responds positively to playback of song, which often allows a brief glimpse of bird.

Food. Insects, many tiny, including beetles, ants and caterpillars.

Breeding Habits. Monogamous, solitary, territorial nester.
NEST: bulky deep cup of dry grass, leaves, and occasionally fern fronds, covered outside with plant down and feathers and neatly lined with hair-like grasses and plant down. One nest placed in tangle against trunk of *Senecio* sp., another in side of rank tuft of grass in thick growth on edge of a stream 140 m from patch of evergreen forest, and a third was attached to brambles *c.* 30 cm above ground (Chapin 1953a).
EGGS: 2–3, rather rounded; smooth; dull white or pinkish white, fairly well speckled and blotched with purplish brown on underlying pale grey blotches, heavily marked at large end. Single brooded (in Malaŵi).
SIZE: (n = 2) 20·7 × 16·3, 19·9 × 16·3.
LAYING DATES: Ethiopia, May, Oct (juvs early Sept); Sudan, May (juvs early Feb, Mar and end June); E Africa: Region A, Oct (breeding condition mid-June, Nov, juvs early Sept); Region B, (juvs early Mar); Region D, Mt Kilimanjaro (1750–2800 m), Aug–Nov (Aug 1, Oct 1, Nov 1 clutches), elsewhere Sept–Feb (Sept 1, Jan 2, Feb 1) (juvs early Aug, end Sept, end Dec); Itombwe, Zaïre, Feb–June; Nyika Plateau, Zambia/Malaŵi, Nov–Feb (Nov 3, Dec 9, Jan 2, Feb 1).

References
Chapin, J. P. (1953a).
Dowsett, R. J. and Dowsett-Lemaire, F. (1984).
Dowsett, R. J. and Stjernstedt, R. (1979).
Dowsett-Lemaire, F. and Dowsett, R. J. (1989c).

Bradypterus barratti Sharpe. Barratt's Warbler. Bouscarle des fourrés.

Plate 5 (Opp. p. 96)

Bradypterus barratti Sharpe, 1876. Ibis, p. 53; Macamac = Mac Mac Forest Reserve, Pilgrim's rest, Lyndenberg District, E Transvaal.

Forms a superspecies with *B. cinnamomeus*.

Range and Status. Endemic resident and partial altitudinal migrant. Locally common in E Zimbabwe (Inyanga Highlands, Vumba Highlands, Chimanimani Mts and Melsetter, absent from Chipinga Uplands) and adjacent S Mozambique (Espungabera north to the Vumba and headwaters of Pungwe R.), also on Mt Gorongoza and Lebombo Range north to Namaacha, Tukuyu on Liwiru R. Not uncommon Swaziland. In South Africa range discontinuous and status uncertain; locally common in Transvaal but restricted to escarpment region and patches of relict forests in SE Highveld (Giant's Castle, Monks Cowl, Woodbush, Tzaneen, Tobacco Dam, Mataffin; scarce Kangwane (Mswati District); absent from Kruger Nat. Park); scarce

Bradypterus barratti

Orange Free State (Golden Gate Nat. Park, Kranskop on Natal border, Mt Pelaan, Groothock, Van Reenen, Verkykerskop); widespread in Natal and E Cape Province (in W and N from Weza/Ingeli north to Wakkerstroom and inland at Kloof, Hillcrest, Richmond, Dargle, Karkloof, Pietermaritzburg, also Cathedral Peak southwest of Ladysmith), Zululand (Eshowe and Nkandhla north to Ngome and Lebombo Range); common Transkei (Ingeli and Sulenkama forests; occurs Tanti forest) and E Cape Province (Pirie, Alexandria Forest) west to Great Fish River. Early record from border of E Lesotho with Natal (Bonde 1993), recent one there and 5 others in S, W-central and N Lesotho (P.E. Osborne and B.J. Tigar, pers. comm.).

Description. *B. b. barratti* Sharpe: SW Mozambique, E Transvaal, Swaziland, N Zululand and N Natal. ADULT ♂: entire upperparts chocolate-brown, rump and tail with rufous wash; 10 tail-feathers; they have faint underlying dark bars uniformly spaced (T1 with 18 bars *c.* 2 mm apart). Chin and throat buffish white, throat faintly streaked dark brown, feathers with diffuse dark centres, breast and sides of neck greyish olive with buffish wash, breast streaked more distinctly dark brown; sides of body greyish olive, flanks, thighs and undertail-coverts brownish olive, flanks in some birds faintly and sparsely streaked darker; centre of lower breast and belly buffish white with some greyish streaks. Indistinct supercilium greyish buff, extending from nostril to beyond eye, lores, cheeks and ear-coverts dark brown, shafts of ear-coverts whitish. Primaries rufous-brown, outer webs of primaries edged cinnamon-brown, inner webs edged paler, secondaries and upperwing-coverts rufous-brown, feathers edged cinnamon-brown, secondaries with faint underlying dark bands; marginal coverts buffish giving indistinct pale leading edge to wing; axillaries and underwing-coverts ashy brown, the latter mottled darker brown. Bill black; eyes dark brown, hazel, or pale brown; legs and toes dark brownish flesh. Sexes alike.

SIZE: (unsexed; all races) wing (n = 43) 59·5–67 (63·4); tail (n = 36) 61–74 (66·9); bill to feathers (n = 46) 11–16·9 (13·9); tarsus (n = 45) 21–25·5 (23·2). WEIGHT: unsexed (n = 36, southern Africa) 16·1–22·2 (18·8).

IMMATURE: smaller than adult; pattern of markings as adult, but above more olive, below yellowish; in older birds yellow confined to centre of belly, breast-feathers with olive-brown centres. Plumage of juv. similar to juv. *B. sylvaticus* (Berruti *et al.* 1993).

NESTLING: at *c.* 10 days, forehead grey owing to numerous pin feathers, long wispy brown down on hindcrown, rest of plumage like adult but greyer, mouth dark orange, gape pale yellow (photo in Ginn *et al.* 1989).

B. b. godfreyi (Roberts) (including '*wilsoni*'): Cape Province, Natal (but in Zululand only low-lying parts). Similar to *barratti* but upperparts, including wings and tail, darker; underparts greyer, streaking less distinct and confined to throat. SIZE: wing, unsexed (n = 17) 62–67 (64·6); tail, unsexed (n = 17) 66–81 (72·6). WEIGHT: ♂ (n = 1, Balgowan, Natal, Feb) 19·5, ♀ (n = 2, Natal) 17·5, 18·5, unsexed (n = 1, Natal), 18·3.

B. b. cathkinensis Vincent (including '*major*' and '*lysis*'): Drakensberg range and its main outliers from E Griqualand, north-east to Natal/Transvaal border, Orange Free State. Similar to *barratti* but olivaceous above, underparts, including flanks and undertail-coverts, lighter; streaking confined to throat and upper breast.

B. b. priesti Benson: E Zimbabwe and adjacent Mozambique. Similar to *barratti* but streaking on throat does not extend onto breast and flanks, centre of chin whiter, underparts more extensively white, contrasting with dark flanks. SIZE: (6♂♂, 3♀♀) wing, ♂ 61–65 (62·5), ♀ 59–61 (60·3); tail, ♂ 62–69 (66·2), ♀ 64–67 (65·0). WEIGHT: (Inyanga, Zimbabwe) ♂ (n = 2) 18·6, 19·8; ♀ (n = 1) 19·1.

Field Characters. 15–16 cm. An active, medium-sized warbler with long rounded tail; dark chocolate-brown above, greyish white below with faint streaks on breast and sometimes on flanks. Usually secretive, but not at Inyanga, Zimbabwe, where its harsh alarm calls draw attention to it (Mees 1970). Range overlaps that of Little Rush-Warbler *B. baboecala* but they live in different habitats (Barratt's never enters marshes). Overlaps in winter on coast of E South Africa with Knysna Warbler *B. sylvaticus*, which is extremely similar even in the hand, but Barratt's is redder above, especially on tail, and more heavily streaked below, with a whiter chin (Berruti *et al.* 1993). The two are best distinguished by song.

Voice. Tape-recorded (58, 88, 91, F, CART, JOJ, WAT). Song begins slowly and quietly with 2–3 (rarely 4) high-pitched, thin notes, which merge into a louder liquid trill of lower pitch, even in tempo and volume, and ends abruptly: 'tseep, tseep, tirp tchi-tchi-tchi-tchi-tchi-tchi-tchi'. Song lasts 1·5–4·5 s (av. 3·3 s, n = 50, Giant's Castle, Transvaal; av 2·7 s, n = 24, Woodbush, E Transvaal; 2·5 s, n = 11, Zimbabwe; av. 2·9, n = 10, Vumba, Zimbabwe). Speed of trill varies. Intervals between starts of 41 songs of one bird were 2·6–8·5 s (av. 4·9 s) and of 9 songs of another 6·6–11·6 s (av. 8·07 s). Contact call a soft, slowly repeated 'dritt', 'tuc' or 'trrr', alarm call a low 'churr churr churr' or 'churrt-churrt', or 'tr.r.r.r . . . tar . . . r.r.r' (Mees 1970). Does not duet (Oatley 1969). Reacts well to playback.

General Habits. Occurs in dense bush, tangled scrub along streams in temperate interior forests of coastal E Cape Province, and in gullies in upland mist forest generally above 1370 m in Drakensberg Range and its outliers. In Natal breeds in scrub and glades in and at edge of evergreen forest. Favoured habitat in Lesotho is dense *Buddleia salviifolia* scrub 1–3 m tall, often near water (P.E. Osborne and B.J. Tigar, pers. comm.). Breeds usually above 1500 m in Zimbabwe (2200 m in Inyanga Highlands), occurring in scrub and heathlands, particularly of *Philippia*, in rank vegetation at edges of montane forest and occasionally in clearings within forest, along paths and stands of tree-ferns. Usually solitary. Not as secretive as other *Bradypterus* species. Forages in vegetation close to ground, or on ground; runs mouse-like over ground and does not hop. Quite plentiful in pockets of forests; calls frequently even when other bird species are silent.

At onset of cool weather in Apr, abandons highlands of Zimbabwe and occurs then in herbaceous tangles at edges of lowland evergreen forest, reaching Honde Valley (900 m) in July, and Haroni–Lusitu confluence (350 m) in June; greater part of the Zimbabwe population probably moves east into Mozambique lowlands (Irwin 1981); only present in Chinanimani Mts Sept–Jan (Beasley 1995). In Transkei 'probably breeds inland and winters along coast' (Quickelberge 1989). Some birds move from uplands of Natal and Drakensberg to Natal coast in winter, e.g. Nahoon and Quinera estuaries, also reaching Transkei; those from Amatola Mts reach East London (Bonza Bay) where 88% of records occur between Apr and Sept (Vernon 1989, Johnson and Maclean 1994).

Food. Insects, including crickets.

Breeding Habits. Monogamous, solitary, territorial nester.

NEST: bulky cup of leaves, interwoven with stems and grasses, placed low down in tangled scrub, often in centre, and close to ground.

EGGS: 2 (n = 3). Pinkish white, thinly speckled with yellowish brown and blue-grey. SIZE: (n = 8) 19·6–21·4 × 14·5–16·5 (20·5 × 15·5).

LAYING DATES: Zimbabwe, Oct–Dec (Oct 2, Nov 3, Dec 2 clutches); South Africa, late Nov and late Dec (Transvaal), Sept–Nov (Natal).

References
Berruti, A. *et al.* (1993).
Brooke, R. K. (1966).
Clancey, P. A. (1955b).

Bradypterus sylvaticus Sundevall. Knysna Warbler. Bouscarle de Knysna. Plate 5 (Opp. p. 96)

Bradypterus sylvaticus Sundevall, 1860. In Grill, K. Sv. Vet.-Akad. Handl., ser. 2, 2, No. 10 (1858), p. 30; Knysna, southern Cape.

Range and Status. Endemic, highly localized and rare resident, South Africa; locally dispersive. Rare in SW Cape, confined to areas of indigenous forest: SE slopes of Table Mt between Tokai and Newlands, Kirstenbosch Bot. Garden (Cape Town), Riviersonderend and along Buffelijags R., Oubos (Caledon district, 130 km east of Cape Town), Grootvadersbosch (Heidelberg district), Garcia State Forest Res., near Riversdale, and Knysna. In E Cape occurs in coastal forests (mouth of Ntafufu R., Umpambinyoni, Lombazi R., Dwesa, Port St Johns). Status in Natal uncertain, has occurred Wentworth, the Bluff, Durban and Umhlanga Rocks; only one record since 1966 ('probable', Cyrus and Robson 1980), but may still occur in Oribi Gorge, S Natal. Formerly thought to be a winter visitor to Natal (Clancey 1964, 1980b), but present evidence suggests that these were relict populations, now extinct through habitat destruction. Western populations do not migrate (Berruti *et al.* 1993). Rare and local; vulnerable to habitat destruction, but tolerates invasion by such aliens as *Rubus* brambles and *Lantana*. Many populations isolated. In SW Cape estimated <

Bradypterus sylvaticus

200 pairs in 3 isolated forest patches – Oubos, Grootvadersbosch, Garcia State Forest Res. (Martin et al. 1982, Hockey et al. 1989). Total population in 10,000s rather than 100,000s (Siegfried 1992, Berruti et al. 1993). Requires conservation.

Description. *B. s. sylvaticus* Sundevall: Cape Peninsula to Port Elizabeth, eastern limits uncertain. ADULT ♂: whole of upperparts warm dark brown, wings and tail darker; 12 tail-feathers, which are broad, blackish with brown fringes and faint underlying darker bands uniformly spaced throughout. Head, lores and ear-coverts brown, shafts of ear-coverts whitish, cheeks brown, mottled with buff, merging into olive-brown of throat; faint, hardly discernible, supercilium. Chin and throat whitish, faintly mottled and streaked olive-brown; breast more uniform olive-brown, sides of belly the same, flanks with browner wash; mid-line of belly whitish olive; undertail-coverts rusty brown, barred and tipped whitish grey; lower thighs brownish olive, upper thighs paler. Primaries, secondaries and upperwing-coverts blackish brown, outer web of primaries edged paler brown, underwing-coverts whitish grey, mottled brown; leading edge of wing buffish white; axillaries dull yellowish. Bill dark horn, lower mandible paler and yellowish near base; mouth yellow; eyes dark brown; legs and feet pale flesh, olive-brown or flesh-brown. Sexes alike, but throat of ♀ whiter. SIZE: (8 ♂♀) wing, 58–62 (60·1); tail, 58–62 (59·6); bill to feathers, 11·8–13·9 (12·7); tarsus, 16·7–18·6 (17·6) (Berruti et al. 1993).

IMMATURE: like adult but face and underparts strongly washed with yellowish, and streaks on throat and breast much more prominent. Plumage evidently similar to juv. Barratt's Warbler *B. barratti* (Berruti et al. 1993).

NESTLING: skin pink, mouth orange; 5 mouth spots, one at tip of tongue and 2 at sides, and 2 smaller ones on edge of upper mandible (Pringle 1977). Nearly fledged young have upperparts blue-grey, underparts yellow-olive, bill and legs horn-coloured.

B. s. pondoensis Haagner: E Cape, Transkei and Natal. Upperparts generally darker than in *sylvaticus*, wings and tail more olive-brown, breast more dusky, streaks on throat do not extend onto upper breast; sides of body and flanks darker and washed with yellow. Same size as nominate race (Clancey 1955a, 1985).

Field Characters. A small, elusive warbler, uniform brown above with no hint of rufous, paler and greyer below with faint streaks on lower throat and upper breast. Tail broad, slightly fan-shaped and rounded, but shorter and more square than other *Bradypterus* spp. Habitat different to Little Rush-Warbler *B. baboecala*, but in winter occurs with very similar Barratt's Warbler *B. barratti*, from which best told by song. The two can be drawn from cover by 'pishing', but are very difficult to tell apart unless throat is seen (white in Barratt's, dusky in Knysna); Knysna is slightly smaller, with less heavy tail, less heavily streaked breast. Both have same churring alarm call, but Knysna also has a loud 'peeet' or 'oeeit' call (Newman 1992).

Voice. Tape-recorded (75, 91, C, F, CART, HRS, PRIN, WAT). Warbling song of ♂ starts abruptly and consists of about 14 very loud, thin, high-pitched notes increasing in tempo and volume until almost a trill, then switches to a lower bubbling trill or reel, 'wit wit wit wit wit twit-twit-twit-twit-trrrrrrrrrrrrrrrrrrrrrrr', reminiscent of Wood Warbler *Phylloscopus sibilatrix*. Song lasts 4·2–9·0 s (av. 6·8 s, n = 46); intervals between starts of 17 successive songs 9·5–23 s (13·0 s). Alarm a repetitive 'prrrit prrrit prrrit-prrrit' (D. Watts, pers. comm), and, near nest, a short, often repeated, croaking or growling 'tjorr-tjorr' (Hofmeyer et al. 1961, Pringle 1977). Contact call a low, repeated and slurred 'brrit', 'churr churr', a soft 'trr-up' or a loud 'peeeit' or 'peet', which is usually answered by another bird (Newman 1992); when feeding bird utters soft 'prrup' note. Contact call of fledglings a high-pitched 'tseet-tseet'.

General Habits. Restricted to dense, gloomy tangles of undergrowth and forest debris in densest parts of isolated indigenous evergreen forests, often near streams, also in bramble thickets. Possibly in Cape macchia (Clancey 1985: 352). Creeps secretively through thick, low matted vegetation; feeds almost entirely on ground, moving at an extraordinarily slow walk looking hunched up and mouse-like; also occurs in mid and low strata (Newman 1992). To disturb the humus the bird either scratches, or else crouches, flutters wings and wags tail as if dust-bathing, then searches the exposed area, much like an Olive Thrush *Turdus olivaceus* (Pringle 1977). Vocal all year in East London but less so after breeding (Berruti et al. 1993), sings from Aug to Dec in SW Cape (Clancey 1985: 352); sings from within thicket but occasionally from exposed perch; ventriloquial. During singing, yellow mouth is noticeable; head held high and white throat puffed out during the trill sequence (T. Harris, pers. comm.). Responds vigorously to song playback 30–50 m from nest and comes within 3 m of loudspeaker; does not respond to playback near nest. Emerges from thicket in response to sibilant alarm notes of another bird. Usually sings no nearer nest-site than 50–100 m (but once much closer when a Southern Boubou *Laniarius ferrugineus* was near nest). No response shown to Cape Robin-Chat *Cossypha caffra* or Cape Batis *Batis capensis* when near nest.

Resident, but disperses seasonally, e.g. near East London occurs all year on Nahoon estuary but only a winter visitor to Quinera estuary (Berruti et al. 1993). At Bonza Bay, however, occurs all months and probably resident; isolated ♂ singing in a garden (5 Nov) may have come from Newland Forest or followed river from Kirstenbosch (Steyn 1994).

Food. Invertebrates: grasshoppers, insect larvae, spiders, slugs, worms.

Breeding habits. Monogamous, territorial, solitary nester.

NEST: a relatively large, bulky, bowl-shaped structure loosely constructed in layers at bottom but becoming more interwoven on inside with small neat cup, built into living plants (such as irises) or in debris, well concealed in tangled vegetation, often underneath a canopy of leaves, 0·50–1·20 m above ground. Foundation of one nest mainly layered oak leaves interspersed with strands of dry grass and dry bramble leaves, the inside interwoven with coarse and fine strands of grass, thin grass-like vegetation and plant fibres, cup lined with a thin layer of fine vegetable fibres. Outer bowl of nest (in thickly matted branches of fallen pine tree) was composed of layered rings of dead pine needles and damp decaying leaves, inner cup neatly formed from finer stems, needles and dead leaves. At nest placed in bramble thicket, only one bird built during a 45 min period, although mate was close at hand; material was collected within 18 m of nest and bird always approached nest on the ground. Nest size (n = 3): ext. width 115–140, ext. depth 75–140, cup width 40–50, cup depth 30–50.

EGGS: 2–3. Ground colour pinkish white, dusted all over with small red spots, with larger red speckles, sparsely but evenly distributed. SIZE: (n = 3) 19·6–20·0 × 14·8–15·0 (19·8 × 14·9).

LAYING DATES: Sept (1 clutch), Oct (1).

INCUBATION: by one parent only; period c. 12–13 days.

DEVELOPMENT AND CARE OF YOUNG: both adults feed young and remove faeces either by swallowing them or flying away with them. Food for nestlings collected within 20 m of nest and birds usually approach nest along ground, very infrequently flying. At 6–7 days each nestling given c. 3 feeds per h and at 10 days c. 8 feeds. Nestling period 12–14 days (Hofmeyer et al. 1961, Pringle 1977). Birds show little fear of person near nest or when they are foraging. In distraction display, body held close to ground, tail fanned to one side and wings occasionally held loosely to body and fluttered, and whole body quivers as if bird injured or disabled (Hofmeyer et al. 1961).

References
Berruti, A. et al. (1993).
Clancey, P. A. (1955a).
Hofmeyer, J. H. et al. (1961).
Pringle, J. S. (1977).

Bradypterus victorini Sundevall. Victorin's Warbler. Bouscarle de Victorin.

Plate 5
(Opp. p. 96)

Bradypterus victorini Sundevall, 1860. In Grill, K. Sv. Vet.-Akad. Handl., ser. 2, 2, No. 10 (1858), p. 29; Knysna, southern Cape.

Range and Status. Endemic resident, winter rainfall region of SW and S Cape Province, South Africa. Ranges from coast to about 75 km inland, from N Cedarberg Mts (Citrusdal) and E side of False Bay to E slopes of Tsitsikama hills, Uitenhage, Gamtoos R., Baviaanskloof and Witteklip Mt (20 km west of Port Elizabeth). Mainly on seaward slopes of mountains, locally common in wet areas, commonest in high rainfall areas, scarce on drier north-facing slopes of E mountains; at least 10 birds in 4 ha in some places.

Description. ADULT ♂: forehead dark greyish brown, rest of head to hindneck dark brown, rest of upperparts dark reddish brown, feathers of back and rump very loose and long, base of feathers very dark grey. Tail dark reddish brown, feathers narrow, long and crossed throughout their length with underlying indistinct paler bars of equal thickness and spacing. Ear-coverts, cheeks and lores dark grey-brown, faint buffish brown supercilium from nostril to above eye. Chin, throat, breast and sides of belly tawny-cinnamon, central part of belly lightly streaked pale buff, undertail-coverts and thighs tawny, flank-feathers grey at base becoming dark tawny in middle and light tawny at tip. Primaries, secondaries and upperwing-coverts dark reddish brown, underwing-coverts and underside of wing brown. Bill greyish brown, paler below, base flesh colour; eyes reddish brown to orange; legs and feet brownish flesh to greyish brown. Sexes similar, but upperparts and underparts of ♀ duller. SIZE: (unsexed, n = 6) wing, 50–55 (53·0); tail, 64–72 (69·0), bill to feathers, 11–12 (11·5); tarsus, 19–22 (20·4). WEIGHT: 1 unsexed, 10.

IMMATURE: on leaving nest upperparts dark ash-grey, wings with some brown on coverts; tail, only a few mm long, brownish; breast and chin light rufous; bill dark grey with yellow tip, gape yellow; iris dark grey; tarsus dark flesh, underside of feet yellowish, claws grey (MacLeod et al. 1958); later, above more rufous, below paler than adult.

NESTLING: skin pink, mouth orange; first plumage blackish above, greyish black below streaked with white (Frandsen 1982; photo in Ginn et al. 1989).

Field Characters. The most striking feature of this bird as it skulks in underbrush is orange-yellow eye set in greyish face. Appears dark brown from above, but on song perch reveals bright cinnamon underparts. Tail long, wedge-shaped, often flicked up and down. Song a little bit like that of Cape Grass-Warbler *Sphenoeacus afer*, with which it often occurs, but rollicking and repetitive rather than a formless jumble. More often heard than seen; an excellent ventriloquist (Frandsen 1982).

Voice. Tape-recorded (75, 88, 91, B, C, F, CART, HRS, WAT). Song unique among African *Bradypterus*, lively and bubbling, a repetition of a phrase lasting 0·5–1·2 s, consisting of lilting clear and sibilant notes of varying pitch: 'wee-tee-wititi, wee-tee-wititi . . .' or 'tip twiddy tsip tsip tsee weet'. Sings 2–7 phrases, usually 4–5 times, in 4·0–7·4 s (av. 5·5 s, n = 37); intervals between start of 30 successive songs 9–29 (19) s. Often aborts song after one partial phrase. Song starts slowly, speeds up, and often ends with a series of grating rattles ('sucking notes': Maclean 1993) '. . . tip twiddy tsip tsip tsee weet tchii tchii tchii tchii tchii tchii'. Duration of the rattle varies from 0·5 to 9·7 s (av. 3·2 s, n = 28). Subsong, given in winter (July), consists of 2–3 poorly developed phrases. Alarm at nest, a sharp, churring 'chip chip chip'; when disturbed bird uses a single loud 'purr-r-' (MacLeod 1946). Fledglings out of nest use 2 call notes, a piercing 'teep' and a rasping 'tjorrr' (much like alarm call of Cape Sparrow *Passer melanurus*: MacLeod *et al.* 1958).

A

General Habits. Confined to thickets and rank vegetation, especially sedges along mountain streams, in Cape macchia vegetation on moist ground with thick grass, in rushes and ferns, stands of proteas and heath along coast, also undergrowth at forest edge. Favours wet and misty mountain slopes. In Langerberg, near Swellendam, occurs in rank waist-high rushes and herbs where cedar trees are regenerating and in vegetation on boggy watercourses. Usually solitary; forages on ground in amongst grass and shrubs like Cape Rockjumper *Chaetops frenatus*, and scuttles over it at great speed like a mouse. Sings from top of bush or rock and from mid stratum of coastal forest edge (not from within thicket like other *Bradypterus* species: Clancey 1985), and often close to nest after feeding nestlings. Usually very tame, creeping within 1 m or so of sitting observer. At one nest an adult (♀?) continued to feed nestlings while observers were less than 1 m away; in contrast its mate was timid, and displacement-fed and sang nearby.

Food. Insects, possibly other small invertebrates.

Breeding Habits. Monogamous, territorial, solitary nester.

NEST: bulky, deep cup, consisting of roughly woven grass base and sides of dead leaves and bark, lined with finer grass and plant down (**A**). In one nest under construction, walls included a number of round dry leaves placed vertically on their edges which were later hidden by lining. Nest placed on ground or within 30 cm of it, well hidden in dense grass or clump of vegetation; one was close to mountain stream. Both sexes build. Material collected within 4–5 m of nest (MacLeod 1946).

EGGS: 2–3. Pinkish white or white, spotted with red all over, spots concentrated at large end; almost an exact miniature of those of Fiscal Shrike *Lanius collaris* (MacLeod 1946). SIZE: (n = 2) 20·8 × 15·3, 21 × 15.

LAYING DATES: Sept–Nov (Sept 2, Oct 1 clutches).

INCUBATION: period 21 days.

DEVELOPMENT AND CARE OF YOUNG: both parents feed young; at one nest one ♀ (?) brought 6–8 food items per h, with occasional long gaps. Birds sometimes fly in with food from quite a distance and approach nest in a series of short relays. On first landing they almost tumble into the grass or bush, or sometimes land on a rock only to move a short distance towards the nest, either on the ground or from twig to twig. The birds then emerge briefly on some small twig or other exposed perch and fly another relay towards the nest before tumbling into the grass once again; flight said to be bumble-bee-like, and rather like Cape Rockjumper; flies with tail spread (MacLeod *et al.* 1958). After a number of relays, depending on distance from nest, birds land within a few m of nest and final movement towards it is a mouse-like scuttle, usually right on the ground.

Reference
MacLeod, J.G.R. *et al.* (1958).

Genus *Bathmocercus* Reichenbach

Forest ground-cover warblers, ♂♂ black and rufous, ♀♀ olivaceous dark grey. Bill stout, culmen ridged, nostrils partly covered by forward-pointing featherlets, 2 well-developed rictal bristles; throat skin blue; wings very short and rounded, primaries strongly decurved towards tips, flight weak; tail of 10 narrow feathers, steeply graduated, unpatterned, tail-feathers rather decomposed, becoming ragged with wear; legs robust, claws very sharp. At least 1 species sings in duet. '*Scepomycter*' *winifredae* accords with this diagnosis and we therefore include it in *Bathmocercus*.

Endemic. 3 species, the W African *B. cerviniventris*, central African *B. rufus* and Tanzanian *B. winifredae*. The first 2 have often been treated conspecifically. We separate them following Louette (1976) and Chappuis (1976b), and regard them as a superspecies.

Bathmocercus rufus superspecies

1 *B. cerviniventris*
2 *B. rufus*

Bathmocercus winifredae (Moreau). Mrs Moreau's Warbler. Bathmocerque de Winifred.

Artisornis winifredae Moreau, 1938. Bull. Br. Orn. Club 58, p. 139; Uluguru.

Plate 6
(Opp. p. 97)

Range and Status. Endemic resident, E Tanzanian montane forests. Ukaguru Mts, common 1500–1650 m (Ukagurus have not been explored ornithologically above 1650 m). Uluguru Mts, widespread between 1350 and 2350 m, uncommon below and common above 1650 m. Udzungwa Mts, known only from Mwanihana Forest on E escarpment, where fairly common 1300–1700 m; however, less common and more local than in Ukagurus or Ulugurus (Jensen and Brøgger-Jensen 1992). 'Rare' status; vulnerable to forest clearance; parts of all 3 forests have already been exploited (Collar and Stuart 1985).

Description. ADULT ♂: forehead, crown, nape, lores, cheeks, ear-coverts, chin, throat and breast (except at sides) rufous-chestnut; hindneck, sides of neck, mantle, scapulars, back, rump and uppertail-coverts uniform mid to dark brown, with olivaceous tinge; tail of 10 narrow, dark brown feathers, with olivaceous brown edges to outer vanes. Sides of breast buffy grey-brown; centre of upper belly buff; rest of belly, flanks, thighs, vent and undertail-coverts buffy greyish brown with olive tinge. Wing-feathers blackish brown, primaries and greater and median primary-coverts narrowly edged olivaceous brown, secondaries, tertials, greater, median and lesser coverts edged paler olivaceous brown. Bill black; eyes dark; legs and feet dark. Sexes similar, but ♀ has head and breast

SYLVIIDAE

Bathmocercus winifredae

pale rufous rather than chestnut. SIZE: (2 ♂♂, 1♀) wing, ♂ 57, 59·5, ♀ 54·5; tail, ♂ 52·5, 53, ♀ 51; bill to skull, ♂ 17·1, 17·4, ♀ 15; tarsus, ♂ 21, 22, ♀ 22. WEIGHT: (Morogoro, Tanzania) 2 ♂♂ 20·0, 20·0, 1 ♀ 11·0.

IMMATURE: entire upperparts olive; ear-coverts, sides of face, chin, throat and breast pale tawny rufous, rest of underparts pale olive.

NESTLING: unknown.

Field Characters. An olivaceous brown warbler of thick undergrowth in highland forest of E Tanzania; rufous head and breast preclude confusion with any other bird. African Tailorbird *Artisornis metopias* is olive-brown above with chestnut cap, but has white underparts. Keeps near ground. Shy. Best told by its whistling duet.

Voice. Tape-recorded (57, B, STJ). Song has elements very like songs of Black-faced and Black-headed Rufous Warblers *B. rufus* and *B. cerviniventris*: a clear whistle on one pitch, repeated evenly, with short pauses, 6 times in 13 s, recalling Grey-headed Bush-Shrike *Malaconotus blanchoti*. Evident duet, a high-pitched whistle 'eeeee' (by ♂?) alternating without pause with a melodic 'dewy-dewy-dewy-dewy' (by ♀?); also 'eeeee' alternating or overlapping with lower-pitched 'wheeewooo'.

General Habits. Inhabits natural clearings in forest at 1350–2350 m, where undergrowth is very dense; often near streams; occurs in wet and some dry forests, but unaccountably absent from some dry forests in Udzungwa Mts (Collar and Stuart 1985). Also areas with plenty of thick undergrowth and creeper-covered trees, where timber has been felled and removed (Turner 1977). In pairs; constantly on the move, keeping to dense undergrowth, foraging close to ground or on it, occasionally moving up to 6 m above ground. Extremely shy, and difficult to locate except by call.

Food. Eats a wide variety of insects and other invertebrates, especially weevils Curculionidae (Collar and Stuart 1985), 'beetles' and earwigs.

Breeding Habits. Nest and eggs undescribed, but breeding season is Oct–Mar.

Reference
Collar, N. J. and Stuart, S. N. (1985).

Plate 6
(Opp. p. 97)

Bathmocercus rufus Reichenow. **Black-faced Rufous Warbler.** Bathmocerque à face noire.

Bathmocercus rufus Reichenow, 1895. Orn. Monatsber., p. 113; Yaoundé.

Forms a superspecies with *B. cerviniventris*.

Range and Status. Endemic resident, equatorial forests, in 2 main populations (Cameroon/Gabon and E Zaïre/Uganda) and small outliers in Sudan and Kenya. Cameroon, rather local and uncommon, between 370 and 1400 m, in SW to west of Mt Kupé and in SE. Gabon, widespread and quite common in Ivindo basin in NE; records from SW; a difficult bird to find, and may be more widespread than map indicates. Central African Republic, uncommon in Dzanga Reserves, probably also Lobaye Préfecture. Zaïre, frequent and widespread from about Buta to eastern border between Simliki valley and Kiliza (3° 42'S in SE Kivu Province); common in Itombwe, 930–1920 m. Uganda, common and wide-ranging at 700–1800 m in Bwamba, Kibale, Bwindi-Impenetrable and Malabigambo Forests, Kifu and Mabira. S Sudan, rather local but not uncommon in Imatong Mts up to 2400 m; also found breeding near Sudan/Uganda/Zaïre border point. W Kenya, locally common in highlands at 1700–2800 m, at 8 localities from Elgon to Kericho. Tanzania, Bukoba only (W shore of L. Victoria).

Description. *B. r. vulpinus* Reichenow (includes 'jacksoni'): E Zaïre, Sudan, Uganda, Kenya and Tanzania. ADULT ♂: forehead, lores, cheeks, ear-coverts, chin and throat black; extent of black on forehead individually somewhat variable. Crown to scapulars, back, rump and uppertail-coverts, also tail and wings, rich russet-brown or bright foxy red-brown. Tail of 10 feathers, graduated; underside russet-brown. Skin of throat bright blue, bare or only sparsely feathered at sides. Breast black, confluent with throat, with sides of breast russet; black

Bathmocercus rufus

patch narrows, continuing onto upper belly as irregular black line; lower belly pale grey in centre, belly olivaceous grey at sides, some russet colour extending onto sides of upper belly. Flanks, thighs, vent and undertail-coverts olivaceous grey to bluish slate-grey. Wing-feathers blackish brown, all with such broad foxy russet fringes that closed wing looks uniformly rufous. Bill black; eyes dark red-brown or dark red; throat skin bright blue; legs and feet purplish grey or bluish grey. ADULT ♀: black parts like ♂, but upperparts olivaceous grey where ♂ is rufous. Sides of neck, sides of breast, sides of upper belly and forepart of flanks, pale bluish grey; rest of underparts greyish buff. SIZE: (10 ♂♂, 10 ♀♀, Zaïre) wing, ♂ 53–58 (56·2), ♀ 52·5–56 (54·3); tail, ♂ 51·5–55 (53·5), ♀ 46·5–51 (49·0); bill, ♂ 12·5–13·5 (13·0), ♀ 12·5–14 (13·1); tarsus, ♂ 24·5–26 (25·2), ♀ 23–26 (25·1). WEIGHT: (E Africa) ♂ (n = 54) 10–21 (16·6), ♀ (n = 25) 7–20 (14·3), 1 juv. 16; (Kenya) unsexed (n = 32) av. 16·9 (Mann 1985).

IMMATURE: wings and tail olivaceous rufous, less bright than in adult; rest of plumage dark olive-grey, breast and belly paler in the midline; lower mandible yellow.

NESTLING: not described.

B. r. rufus Reichenow: Cameroon, Gabon; Central African Republic population probably of this race. ♂ slightly darker and less orange-rufous above and on sides of breast. ♀ lacks the pale area on lower breast.

Field Characters. A thickset warbler of forest undergrowth, in pairs, near ground. ♂ bright foxy red, with black face and long V-shaped patch from chin to upper belly, rest of underparts grey; ♀ olive-grey where ♂ is foxy, the black face/breast-patch bordered by pale grey. Cocks tail. Distinctive voice. Unlike any other bird, although on poor view in undergrowth ♀ might be mistaken for one of the Masked Apalises *A. binotata/A. personata*, q.v.

Voice. Tape-recorded (57, F, ERA, HOR, KEI, STJ). Song of ♂, 15–20 or more very high pitched, thin, penetrating whistles, 'eeeee', all on one pitch, whistle lasting 1·0–1·2 s, followed by pause of 0·4–0·5 s; one song of 13 whistles in 22 s, another of 16 in 23 s; each 'eeeee' note slightly louder at abrupt beginning and end than in middle or increasing in volume to abrupt end; whistles uttered in even tempo. Sometimes sings 'for a considerable period' (Chapin 1953a). ♂ song has substantial carrying power. Song of ♀: shorter, less high-pitched whistles, in even tempo; one song had 50 whistles in 21 s. Duet a measured 'seee-oooo-ee' the 'seee' and 'ee' by ♂ and 'oooo' by ♀; no pause between the 3 notes; ♂'s voice very thin and reedy, ♀'s lower-pitched and euphonic; in recording, 6 'seee-oooo-ee' duets given in 16 s at even rate with short pauses between them. Call of ♀, whilst ♂ singing 'eeeee', a short scolding trill, 'trr-trr, trrt trrt trrt r r'.

General Habits. Inhabits thick undergrowth with herbaceous vegetation and luxuriant bushes on boggy ground in forest, particularly old cultivated land, abandoned and thickly overgrown with impenetrable bushes, vines, sedges and young trees. Occurs typically also along damp edges of clearings and roadways through forest. Lowlands (up to 1400 m) in W Africa, but up to 2400 m in E Africa. Lives in pairs, low down in dense vegetation, often feeding on damp ground; usually near stream. Sometimes in small parties; does not flock with other species. Once seen attending an ant swarm (Kakamega, Kenya). Skulking and shy. Cocks tail forward over back.

Sings all year; ♂ and ♀ have different voices, often combined in duet. Singing ♂ stands high on its legs, straightens closed tail, and rhythmically raises half-opened wings. When bird sings, bright blue throat skin may prove to show at bare sides of throat, as in camaropteras and some other warblers. In breeding season ♂ and ♀ both roost in nest at night and have been readily caught, taken with the nest (Cameroon and Ituri).

Food. Small insects, including beetles, caterpillars, eggs and ants; also snails and small millipedes.

Breeding Habits. Solitary breeder. Pairs are territorial (Brosset and Erard 1986).

NEST: bulky, globular, with a side entrance, made of dry leaves and petioles, lined with fine grasses including flowering heads of *Panicum* and a few feathers. Sited 50–100 cm above ground in small bush in thick ground layer of vegetation. Ext. diam. 90 and height 60, int. diam. 50 and depth 40.

EGGS: 2. Matt; white, either immaculate or with large black and grey blotches (Prigogine 1972). SIZE: *c*. 18 × 13.

LAYING DATES: Zaïre: N Ituri, Mar, July (Oct), Itombwe, Mar–May, Dec–Jan; E Africa: Region A, June, Dec; Region B, Apr, June; Kenya, Kakamega, (nest-building Aug); Uganda, Kibale (nest-building Feb).

References
Chapin, J. P. (1953a).
Chappuis, C. (1976b).

Plate 6
(Opp. p. 97)

Bathmocercus cerviniventris (Sharpe). **Black-headed Rufous Warbler.**
Bathmocerque à capuchon.

Apalis cerviniventris Sharpe, 1877. Proc. Zool. Soc. Lond., p. 22; Gold Coast.

Forms a superspecies with *B. rufus*.

Range and Status. Endemic resident, W African rain forest from Sierra Leone to Ghana. Range fragmented. Sierra Leone, fairly common in streamside vegetation in lowland, gallery, and highland forest (Collar and Stuart 1985); not uncommon in Nimmini Mts, Konno district, also Sandaru and probably Kankordu. Guinea, 1 record, Seredon and locally common Ziama Massif. Liberia, frequent on Mt Nimba. Ivory Coast, Mt Nimba and Sipilou in W, and Taï, Gagnoa and Lamto in S. Ghana, known only from 2 birds collected last century at Denkere and probably near Dompoasi.

Bathmocercus cerviniventris

Description. ADULT ♂: head and neck entirely black, including chin and throat and sides of neck. Mantle, scapulars, back, rump and uppertail-coverts dark rufescent brown. Tail dark brown, strongly graduated, of 10 narrow feathers, shafts black. Central part of upper breast black, confluent with black throat, but sides bright orange-chestnut; lower breast bright orange, raggedly black in midline; belly, thighs, vent and undertail-coverts fulvous or buffy brown; flanks fulvous, washed chestnut. Wing-feathers blackish brown, primaries and greater and median primary-coverts narrowly edged fulvous, secondaries, tertials, greater, median and lesser coverts edged rufous-brown. Bill black; skin of throat bright blue (Bannerman 1939); eyes chestnut; legs and feet blue-grey, legs powerful, claws very sharp. ADULT ♀: upperparts like ♂; chin and sides of throat off-white, centre of throat black, confluent with breast; breast like ♂ but sides rich rufous rather than bright chestnut; rest of underparts like ♂ (1 specimen known, Mt Nimba). Another specimen, the unsexed type of the species, may be ♀ (not adult?: Louette 1976): like ♂, but black of head duller, chin and throat pale buff, upper breast with black patch in centre, remaining underparts fulvous-brown (no orange band around the black breast). SIZE: (Liberia) unsexed (n = 7) wing, av. 58.6 ± 1.5, 1 imm. ♀ 56; tail, av. 49.6 ± 2.4, 1 imm. ♀ 45; bill, av. 16.1 ± 0.4, 1 imm. ♀ 16; tarsus, ♂ (n = 3) 24–26, 1 ♂ 23; (all other skins, ♂ ♀, n = 7, Ghana, Sierra Leone, Liberia, Guinea, Ivory Coast) wing 51–58 (54.8), tail 45–54 (50.8), bill 14–15.5 (14.9), tarsus 22.5–24.5 (23.8). WEIGHT: (Liberia) unsexed (n = 7) av. 15.8 ± 2.1, 1 imm. ♀ 15.7.

NESTLING AND IMMATURE: 1 bird, just able to fly, collected at Kankordu, Sierra Leone, June 1930, its parents unseen, is probably but not certainly of this species: crown sooty, rest of upperparts brown, throat and breast-feathers downy, dull greenish brown, rest of upperparts fulvous, wing and tail quills dark brown. Immature throat probably almost white (Louette 1976); a moulting immature ♀, Liberia, Feb, has black throat feathers replacing white ones.

Field Characters. A vocal, tail-cocking warbler of damp, ferny undergrowth near streams in W African rain forests; in pairs, foraging near ground. ♀, see above; ♂ distinctive, with *Ploceus* weaver-like black head, black breast bordered by bright chestnut band, and buffy belly. Blackish tail steeply graduated, often cocked high over back. Wings short, flight weak. Distinctive voice.

Voice. Tape-recorded (57, B, C, F, CHA, KEI). ♂ song a loud, variable, quite euphonic whistle of 3 syllables (sometimes nearer 2 or 4), 'whee-ee oo', 'whee-ery deee', 'tiu, tiu-whee', 'wit-ui dee' or 'wheeeeee-heeeee', repeated rhythmically about 6 times, at rate of 6 in 16 s, with a short pause after each whistle. Emphasis is even, or first and last parts of each whistle louder than middle, or first part loudest. Song is strongly reminiscent of Oriole Warbler *Hypergerus atriceps*. Call of ♀, whilst ♂ singing, a scolding 'trr-trr, trrt trrt trrt r r . . .', with harsher notes recalling African Paradise-Flycatcher *Terpsiphone viridis*.

General Habits. Inhabits thick undergrowth of tall primary and secondary lowland and highland rain forest, also gallery forest; typically in damp hollows by edges of sluggish small stream, with dense thickets of bracken, saw-edged grass, *Selaginella* club-moss, and Marantaceae (Walker 1939); on Ziama Massif (Guinea), humid, open sites near old clearings within mature forest (Halleux 1994). Always in pairs, working their way through undergrowth, near the ground. Robust legs and very sharp claws suggest that climbing may be an important means of locomotion. Cocks tail over back. Vocal.

Food. Insects, including small beetles, grasshoppers and a mantis nymph; also small spiders.

Breeding Habits. Unknown.
BREEDING INDICATIONS: Sierra Leone (bird just fledged, very probably of this species, late June, points to May laying); Liberia (immature Feb).

References
Bannerman, D. A. (1939).
Chappuis, C. (1976b).
Louette, M. (1976).

Genus *Melocichla* Hartlaub

A large and bulky warbler, with powerful, deep curved bill and prominent rictal bristles, short rounded wings and long, broad, rounded tail of 12 feathers, T5 and T6 with cinnamon patch. Upperparts brown and unstreaked, underparts cinnamon, throat white with black malar stripe. Habitat rank grass; nest a rough, loose cup placed in grass tussock.

Monotypic, endemic. Hall and Moreau (1970) not only merged both *Melocichla* and *Achaetops* into *Sphenoeacus* but placed them all in a 'superspecies'. Subsequent authors have not accepted this close a relationship, although the 3 genera are placed next to one another in sequence by Watson *et al.* (1986). More recently, *Achaetops* has been considered a babbler (Olson 1990, Dowsett and Forbes–Watson 1993). We find the morphological differences between *Melocichla* and *Sphenoeacus* great enough to warrant generic separation, and the distinctive accelerating song of *Melocichla* does not suggest to us either *Sphenoeacus* or *Achaetops*, as has been claimed.

Melocichla mentalis (Fraser). Moustached Grass-Warbler. Mélocichle à moustaches.

Drymoica mentalis Fraser, 1843. Proc. Zool. Soc. Lond. p.16; Accra, Gold Coast.

Plate 6
(Opp. p. 97)

Range and Status. Endemic, resident and locally possibly a partial migrant. In S Mauritania a scarce wet season visitor to Guidimaka; uncommon resident S Senegal (Badi, Kédougou, Niokola-Kobe), Guinea-Bissau, Guinea (north of Fouta-Djalon mountains, also near Mt Nimba, but abundant in Macenta area in SE), Sierra Leone, Liberia (Mt Nimba). In S Mali rare; probably both resident and intra-African migrant (Banankourou, Baoulé, Bamako, Sélingué, Kangaba), also rare SE Burkina Faso. Uncommon or locally common in savannas north of lowland forest in Ivory Coast and Ghana and south of forest zone in drier coastal strip; widespread throughout Togo, Benin and Nigeria (north to Zaria and Aliya, reaches to the coast at Badagri and Bonny); widespread in Cameroon but rare in far N, and in S Chad (Irene, Sarh). Common in Central African Republic (Shari and Ubangi rivers, Bozoum, Lobaye Préfecture, Bangui area); common in N Zaïre, E Zaïre to shores of L. Tanganyika, S and SE Zaïre, also in pockets of derived savanna within forest zone (Kwamouth, Kinshasa) and Gabon; Rwanda (Astrida, R. Akanyaru, Rwinkwavu); Burundi (Bujumbura, Mosso, Kininya, L. Nyanza, Ruzizi valley); locally common Sudan (southwest of line Raga-

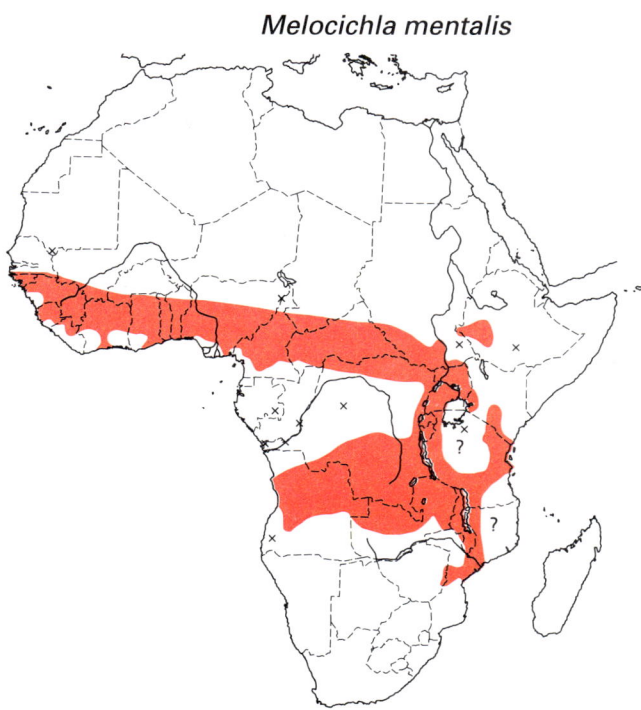
Melocichla mentalis

Wau-Bor-Imatong Mts, south of *c*. 9° N, Bahr el Ghazal) and Ethiopia (W, and W Highlands up to 2000 m; sight record from east of Rift at 2650 m on Wendo-Negele road, 15 km south of Agere Salem (Robertson 1995). Locally common but sparsely distributed in Uganda except in NE; Kenya (Mt Elgon south to S Nyanza in W and uplands in E from Meru to Taita Hills, rare Tsavo E Nat. Park); Tanzania (in W from Minziro Forest Reserve and Ngara south to Ufipa Plateau, in NE from Arusha Nat. Park, Mt Kilimanjaro and Pangani R. south to coast at Dar es Salaam, inland to Njombe, Mahenge and Utete; 1 recent record Serengeti Nat. Park, Aug 1985 (Stronach 1990)). Uncommon and local N and central Angola, Zambia (Luapula, Northern, Central, Western and Northwestern Provinces, also Eastern Province from Luanga valley south to Chipata); fairly common throughout Malaŵi; local and uncommon E highlands of Zimbabwe at Inyanga at 2250 m, Honde valley, Botanical Garden, Vumba (Beasley 1988), Chimanimani Mts, Mt Selinda and Melsetter. N Mozambique (along Malaŵi border south to coastal areas), inland to E-central Mozambique (near Zimbabwe border). (Reported sighting in Transvaal not now accepted: Tarboton *et al.* 1987.)

Description. *M. m. mentalis* (Fraser): (including '*grandis*', '*meridionalis*', '*adamauae*'): W Africa east to Central African Republic, N and S Zaïre, Gabon, Congo, Angola, NW and N Zambia, N Malaŵi. ADULT ♂: forehead russet, shading into olive-brown on crown, mantle and scapulars, rump russet, uppertail-coverts cinnamon. Rump feathers fluffy. Tail broad and rounded at tip, dusky brown, T5 and T6 with pale cinnamon terminal patch; freshly moulted tail feathers have pale terminal patch on underside. Lores and superciliary stripe white, the latter with blackish border below, eye-ring whitish; ear-coverts olive-brown with cinnamon shafts. Rictal bristles prominent. Chin, throat and cheeks white with distinct black malar stripe; breast cinnamon-buff, variable in shade, becoming paler on belly, tawny-cinnamon on flanks and tawny on undertail-coverts; thighs pale cinnamon. Upperwing-coverts, primaries and secondaries olive-brown, outer edge of primaries and secondaries russet (barely discernible on worn secondaries); underwing-coverts and axillaries pale cinnamon. Upper mandible black, lower whitish; eyes pale yellow or creamy white; feet and legs blue grey or whitish brown. Sexes alike. SIZE: (17 ♂♂, 16 ♀♀) wing, ♂ 72–81 (76·8), ♀ 70–80 (72·8); tail, ♂ 81–99 (91·2), ♀ 76–95 (85.2); bill to feathers, ♂ 15–18 (16·5), ♀ 14–17 (15·9); tarsus, ♂ 26–30 (28·5), ♀ 25–30 (27·6). WEIGHT: ♂♀ (n = 12, Ghana) 30–37 (33·2), (n = 7, Uganda) 29–40 (34·1).

IMMATURE: similar to adult but forehead olive-brown (rather than russet), lores darker and mantle more russet, breast mottled brown, and malar stripe absent; flight feathers with broader russet edges.

NESTLING: bill brownish black, corners and interior of mouth yellow, 2 black spots on back of tongue; eyes blackish, shading to grey on outer rim; legs and feet dull dark brown, lighter on rear of tarsometatarsus, yellowish grey beneath toes. Featherless body dusky red, quills slate blue.

M. m. amaurourus (Pelzeln) (including '*atricauda*', '*chyulu*', '*granviki*'): S Sudan east to SW Ethiopia and south through central East Africa to NW, N and central Zambia. Similar to *mentalis* but consistently more dusky brown above; ear-coverts with whitish shafts giving stripy appearance. Birds in Uele District, Zaïre, probably intermediate between *mentalis* and *amaurourus* (Chapin 1953a). WEIGHT: 1 ♂ (Kenya) 39.

M. m. orientalis (Sharpe): E Kenya, E and S Tanzania, S Malaŵi and adjoining Zambia, Zimbabwe and Mozambique. Like nominate race but upperparts light sandy brown. WEIGHT: ♂ (n = 2) 32·9, 38.

M. m. incanus Diesselhorst: NE Tanzania (Momella, Mt Meru at 1800 m). An 'island' population with greyer upperparts than neighbouring populations. Similar to *orientalis* but crown and mantle grey-brown and forehead much paler reddish brown. Only 5 specimens known, all distinctly grey above, race may range more widely (Diesselhorst 1959).

M. m. luangwae Benson: Zambia (plateau of Eastern Province: Luangwa R. in Mpika and Lundazi districts east to Fort Jameson and south to Jumbe), one Malaŵi bird (Hewe R. at Katumbi) closer to *luangwae* than *orientalis* (Benson and Benson 1977). Differs from *mentalis*, *amauroura* and *orientalis* in cold, more greyish tone of upperparts, paler underparts, and paler russet of edges of wing-feathers. SIZE: wing, ♂♀ (n = 6) 78–83 (79·8).

Field Characters. A large, heavy-looking warbler of rank grass, with plain brown upperparts, unstreaked underparts and conspicuous black moustache stripe at sides of white throat. Facial pattern, with white eye-ring and eyebrow and chestnut ear-coverts, similar to Cape Grass-Warbler *Sphenoeacus afer* except that latter has 2 black malar stripes, but tail is broader, rounded and blackish, and rest of plumage quite different; Cape Grass-Warbler is a streaked bird, with a longer, looser rusty tail. Songs are very different. Broad-tailed Warbler *Schoenicola p. brevirostris* has broad dark tail but is smaller and has plain face and no moustachial stripe, very short stubby bill and distinctive song of single metallic notes.

Voice. Tape-recorded (5, 68, 86, 91, B, C, F, GRI). Song melodious, cheerful, with much variation, begins quietly and hesitantly, becomes louder and bolder, and ends in a flurry of notes: 'tip-tip-tip-tip-tip-tip-twiddle-iddle-eee'; 'chip-chip-chip-chee-choodly-twor'; 'chip-chop-chip-chep-chee-tiddly-oh, tiddi-er'. Av. duration (Ghana, breeding season) 2·6 s (n = 19, range 1·23–3·63 s), av. intervals between start of consecutive songs 10 s (n = 10, range 6·8–11·8 s). Alarm note a loud 'cheeep', or 'tchaaa', repeated 3–4 times per s and lasting up to 15 s or more: 'cheeep-cheeep-cheeep-cheeep . . .' or 'tchaaa-tchaaa-tchaaa-tchaaa . . .', an explosive edition of Sedge Warbler's *Acrocephalus scirpaceus* alarm; also a 'tukk' like a wheatear.

General Habits. Inhabits rank grass and coarse vegetation with scattered bushes and trees in areas of high and medium rainfall; usually near water and in valley bottoms; avoids forest edge and forest clearings within forest blocks. Local in rank grass with bushes and trees near water or moist valley bottoms, absent from forest and more arid savannas. Widespread in guinea and derived savannas of W, E and S-central Africa. In SE Kenya, occurs at 500–2500 m where annual rainfall is 250–1100 mm and also in more arid areas; in SW Kenya occurs at 1000–2000 m, with rainfall > 1000 mm. In Zimbabwe occurs from 750 m (Honde valley) to 2200 m (Inyanga Downs).

Solitary or in pairs; a ground feeder. Rather retiring and skulking, and difficult to flush; when disturbed drops quickly into cover. Flight in open heavy and laboured. ♂ sings from exposed perch on grass stem or shrub, at all times of year but mainly in breeding season. When alarmed flirts tail.

Mainly sedentary, but descends Malaŵi Hill (920 m) in July and Aug (Benson and Benson 1977), and may be a partial migrant at N limits in W Africa (occurs Guidimaka, S Mauritania, only in June–Oct, and at Serti, Nigeria, only in wet season).

Food. Insects: mantids, grasshoppers, small beetles.

Breeding Habits. Monogamous, territorial.

NEST: a cup, often shallow, made of long blades of dry coarse grass and tendrils, outer structure rough and loose, and lined with grass, rootlets, and fine stems, sited in grass tussock 1–2·5 m high and placed up to 0·3 m above ground. One was made from broad dry strips of millet and bound on outside with a little cobweb (Vincent 1948); ext. diam. 90–100, ext. depth 63, depth of cup 50.

EGGS: 2–3, usually 2. Slightly glossed; pinkish white, marbled with red-brown, a wreath of heavy markings encircling large end. SIZE: (n = 11, Zaïre and southern Africa) 21·5–23·5 × 15·3–16·5 (22·4 × 15·9).

LAYING DATES: Sierra Leone, July, Nov; Mali, June–Aug, occasionally Sept; Nigeria, Apr–July, Sept; N Zaïre, Mar–Dec, SW Zaïre, Nov–Apr (complete moult occurs May–June); Rwanda, Nov–May; E Africa: Region B, Jan (1), Apr (2), May (2), June (6), Nov (1), Dec (2) clutches; Region D, Feb (1), Mar (3), Apr (1), May (4), June (1), Aug (1), Nov (1) clutches; Zambia, Oct, Nov, mid-Jan (Chipata) (adults feeding young mid-Dec and mid-Jan, fledgling late Mar); Malaŵi (adults carrying food and agitated late Dec); Zimbabwe (moult of tail-feathers complete May). Season ill-defined but throughout range eggs are laid in the main wet season.

Reference
Beasley, A. (1988).

Genus *Sphenoeacus* Strickland

A large warbler with long, narrow, ragged tail often kept closed. Upperparts streaked black on brown, underparts buffy, black malar stripe. Lives in grassy habitats and macchia scrub; nest a deep cup placed in grass tuft or low in bush.

Monotypic, endemic to southern Africa. Head and face markings remarkably similar to *Melocichla*, with which it has been linked in a superspecies by Hall and Moreau (1970), but *Sphenoeacus* has streaked upperparts and a very different tail, and we follow recent authors in retaining it in a separate genus. For further notes, see under *Melocichla*.

Sphenoeacus afer (Gmelin). Cape Grass-Warbler; Grassbird. Sphénoèque du Cap.

Muscicapa afra Gmelin, 1789. Syst. Nat. 1, pt. 2, p. 940; Cape of Good Hope, South Africa.

Plate 6
(Opp. p. 97)

Range and Status. Endemic resident, southern Africa. Common SW Cape (where widespread in south, scarcer north of Worcester), scarcer in E Cape, uncommon Lesotho (Maseru-Roma area, also Morija and Sengu R.), locally common Natal and W Zululand (high hills and Lebombo Range) but absent from lowlands in E, fairly common E Orange Free State (mountainous areas from about Harrismith south to Golden Gate) but vagrant in N (records few and widely scattered), W Swaziland and possibly E Swaziland (Lebombo hills), locally common Transvaal (most widespread in Escarpment region, SE Highveld and Central area west to Rustenburg and Magaliesberg); Zimbabwe (Inyanga Highlands south to Melsetter District and Chimanimani Mts); Mozambique (Vumba, Chimanimani Mts, Pungwe R. to Espungabera; also Namaacha on Mt Meponduine in extreme south). Locally common, becoming scarce when habitat is destroyed through either bush encroachment or degradation by poor stock farming methods; absent from areas invaded by dense stands of Australian acacias, around East London confined to areas where habitat is protected.

Sphenoeacus afer

Description. *S. a. afer* (Gmelin): SW Cape south of about 31° 45′ S and S Cape east to about Humansdorp and Gamtoos R.; enters Little Karoo at Oudtshoorn. ADULT ♂: forehead, crown and nape dark red-brown, crown and hind-crown streaked with black; sides of neck, mantle and upper-wing-coverts black, feathers edged with pale buff, some washed with dark brown. Rump and uppertail-coverts tawny with some black streaks on rump. Tail tawny with dusky brown shaft-streaks, strongly graduated, each feather sharply pointed. Lores and superciliary stripe buffy white, eye-ring buffy white with black outer ring, ear-coverts russet; 2 black malar stripes. Underparts cinnamon buff, paler on throat; sides and front of breast mottled blackish brown, flanks and undertail-coverts with broad blackish brown shaft-streaks. Primaries, secondaries and scapulars dark blackish brown, entirely or partially margined with tawny. Bill blackish horn, base paler; eyes brown; legs and feet brownish green. Sexes alike. SIZE: wing, ♂ (n = 10) 67–74 (69·7), ♀ (n = 10) 61–69 (66·6); tail, ♂ (n = 10) 92·5–109 (101), ♀ (n = 10) 94–102 (98·5); bill to feathers, ♂♀ (n = ?) 15–17 (16·4); tarsus, ♂♀ (n = ?) 22–24 (22·7).

IMMATURE: much like adult but head and mantle lighter brown, feathers with dark blackish shafts, underparts paler than adult and only sparsely streaked on flanks.

NESTLING: not described.

S. a. intermedius Shelley: E Cape Province, east of range of *afer*, from Port Elizabeth to Transkei, Natal (E Griqualand and Alfred County); probably this race in Lesotho. Similar to nominate *afer* but crown and nape paler, mantle without dark brown wash, rump, uppertail-coverts and outer vane of tail-feathers more dark brown than tawny, no streaks on uppertail-coverts. Underparts much more uniform cinnamon, mottling on breast confined to sides only, none on belly or undertail-coverts; brown shaft-streaks confined to lower sides and flanks. Larger: (10 ♂♂, 5 ♀♀) wing, ♂ 69–75 (72·2), ♀ 67–70 (67·5); tail, ♂ 95–114 (106), ♀ 90–104 (94·9).

S. a. natalensis Shelley: Natal and Zululand, W Swaziland, Orange Free State, N Lesotho, and Transvaal (west to Magaliesberg, Swartruggens and Rustenberg and north to Zoutpansberg). Like *intermedius* but lacks dark shaft-streaks on lower sides and flanks. Larger than *intermedius*: (10 ♂♂, 10 ♀♀) wing, ♂ 74–80 (77·2), ♀ 69–75 (71·8); tail, ♂ 98–115 (104), ♀ 90–103 (95·1).

S. a. excisus Clancey: E Zimbabwe and adjacently in Mozambique; generally above 1200 m. In fresh plumage underparts much paler than in *natalensis*, flanks duller and more olivaceous, and streaks on upperparts finer. Tail slightly darker and shorter. ♀ had dark red eyes and pinkish grey legs (Mees 1970).

WEIGHT: ♂ (n = 2, Zimbabwe) 32·3, 28·3; ♀ (n = 4, Zimbabwe) 28·3–33·7 (31.4); sex? (n = 1, Mozambique) 27·7; unsexed (n = ?, South Africa) 27–32·3.

Field Characters. A large warbler, far larger than any cisticola, with reddish crown and ear-coverts, heavily striped upperparts and buffy underparts, streaked on sides except in race *natalensis*. Long ragged-looking tawny tail is diagnostic; the feathers are opened wide when bird climbs through grasses but otherwise tail is kept tightly closed into a point. Double black malar stripes are also a good field mark. A skulker except in early morning when it perches on grass stem to sun itself or sing. May just overlap in Zimbabwe/Mozambique with similar-sized Moustached Grass-Warbler *Melocichla mentalis* which is unstreaked above and below and has blackish, rounded tail, single malar streak and different song.

Voice. Tape-recorded (58, 75, 88, 91, B, F, CHA, WALK). Song a rapid medley of jumbled musical notes (none staccato): 'chu-tee-ree chwiddly-chwiddly-chitty-chitty-chitty chi chi chreeeeeeeu'. Usually begins quietly, becomes louder, and ends in drawn-out trill. Song duration may vary geographically: SW Cape Coast (n = 9) av. 4.0 s; Kirstenbosch, Cape (n = 22) av. 3·21 s; Sani Pass, Natal (n = 27) av. 2.16 s; El Mirador, Natal (n = 8) av. 2·89 s; Giants Castle, Natal (n = 8) av. 2·88 s, (n = 9) av. 2·78 s; Hill Crest, Natal (n = 16) av. 1·80 s; N Transvaal (n = 9) av. 1·80 s. Interval between starts of consecutive song bursts (range 5–10 s), and quality and intensity of trill at end of song, also vary geographically. Duetting once recorded; loud song of 2nd bird (♀?) consists of 2 similar phrases 0·95 s long and a final phrase 0.7 s long (A. Walker, pers. comm.): 'chee-ee-wichee-wichee chee-ee-wichee-wichee, chee-ee-wi'. Alarm, a cat-like 'meeew' rising in pitch, duration 0·8 s, repeated every 1·4 s; also a 'tear, tear'.

General Habits. Occurs along streams and rivers, inhabiting rank grass, macchia scrub in lowland coastal areas, short montane grassland with scattered *Protea* or other bushes, fynbos on inland hills, and forest edge. Found in Cape Province and Natal at sea level but in north of range it occurs at higher altitudes, e.g. 2100–2300 m on Drakensberg escarpment in Natal (Clancey 1964). In Zimbabwe usually above 1500 m but occasionally down to 1200 m.

Generally singly or in pairs. Shy and skulking except while singing when it usually perches on exposed grass head or branch of shrub; rarely if ever uses trees. Feeds on ground under vegetation. Flight laboured, clumsy and short; flops down into cover; prefers to creep into cover than fly.

Sedentary.

Food. Mainly insects (larvae, mainly caterpillars), occasionally grass seeds (da Rosa Pinto and Lamm 1953), and possibly guava fruits (Priest 1935).

Breeding Habits. Monogamous, territorial.

NEST: deep cup of coarse dry leaves and reedy grass blades, neatly and compactly lined with finer grass or fibres, concealed in grass tuft, low bush, or in weedy growth among saplings, within 0·1–0·46 m of ground; ext. diam. 83, depth 76, depth of cup 50 (Vincent 1948a). ♀ alone builds.

EGGS: 2–3, usually 2 in Zimbabwe but 3 in Natal and Zululand. Dull greyish white, spotted with brown and grey; spotting denser near large end. In NE of range eggs said to be larger with greenish purple suffusion (Mackworth-Praed and Grant 1963). SIZE: (n = 52, South Africa) 20–24 × 14·8–17 (22 × 15·8).

LAYING DATES: mainly in wet season. SW Cape, July–Dec (mainly Aug and Sept); N Lesotho, Jan; Natal, Oct–Feb (mainly Dec); Transvaal, Sept–Mar, 21 of 55 clutches in Oct (Tarboton *et al.* 1987); Mozambique and Zimbabwe, Oct–Feb (Mar and Apr in 1913, a year of prolonged rainfall).

INCUBATION: by both sexes? Period: 14–17 days.

DEVELOPMENT AND CARE OF YOUNG: nestlings fed by both ♂ and ♀. Period: 14–16 days.

BREEDING SUCCESS/SURVIVAL: 1 bird recaptured after 3 years 137 days.

References
Clancey, P. A. (1973).
Maclean, G. L. (1993).

Genus *Schoenicola* Blyth

A grass-dwelling warbler with short, strong, slightly decurved bill, 2 strong rictal bristles, powerful feet, and strongly graduated tail composed of 12 broad feathers, black with pale tips; uppertail-coverts long and lax; undertail-coverts very long, lax, pale-tipped. Tail slightly pleated or keeled (V-shaped in cross section) (Chapin 1953a). Several distal tendons of tibia ossified.

1 species; widespread in subsaharan Africa and SW India (and may have occurred formerly in Sri Lanka). African and Indian populations often treated as allospecies (*S. brevirostris* and *S. platyura* respectively), but differences are trivial.

Placed by Sibley and Monroe (1990) in Subfamily Megalurinae, with *Megalurus*, *Amphilais* (Madagascar), and 8 other genera of Oriental/Australasian grass warblers.

Schoenicola platyura (Jerdon). Broad-tailed Warbler; Fan-tailed Grassbird. Graminicole à queue large.

Plate 6
(Opp. p. 97)

Thimalia platyura Jerdon, 1844. Madras J. Litt. Sci. 13, p. 170; Goodaloor, Neilgherry Hills, India.

Range and Status. Africa south of Sahara, and SW India.

Resident, widespread in S-central Africa but with highly fragmented, localized range at higher latitudes (10°N to South Africa). Migratory south of Zambezi R. Sierra Leone, common on Birwa Plateau (Tingi Mts); occurs adjacently in Guinea, also on Mt Nimba in Guinea. Nigeria, locally frequent in SE: Mt Kolukoshon at 1400 m and on Obudu and Mambilla Plateaux. Cameroon, known from Nkongsamba, Manenguba, Bamenda, Yoko, Lépopomo and Akonolinga. Sudan, 2 records in Bahr el Ghazal and Bahr el Jebel and 3 in Nagishot/Didinga, Lotti Forest and Gilo. Ethiopia, a few records in W Highlands only. In Uganda and Kenya widespread in L. Victoria basin, locally common at 800–2000 m especially in Kenya; north in W Uganda to Bunyoro and Acholi. Tanzania, as mapped; in NE, west to Crater Highlands at Mbulumbulu. E, S and W Zaïre; Congo. NE Angola, locally frequent, through plateau from N Huila to Cuanza Norte and Malanje east to Zaïre and Zambia borders. Zambia, widespread north of line from Kabompo to Lusaka to Isoka; sparse south of Kabompo–Lusaka line, south to Kabompo/Zambezi confluence and Kafue R. in N Mazabuka; east of Lusaka–Isoka line, known only at Chipata and Nyika Plateau at 2150 m (Dowsett 1979); absent from Luangwa valley. Malaŵi, sparse and local throughout, above 1100 m, south to Blantyre. Zimbabwe, Mozambique, Transvaal, Lesotho, Natal and E Cape, sparse and local, sometimes common, in narrow corridor from Harare and Inyanga Highlands (up to 1800

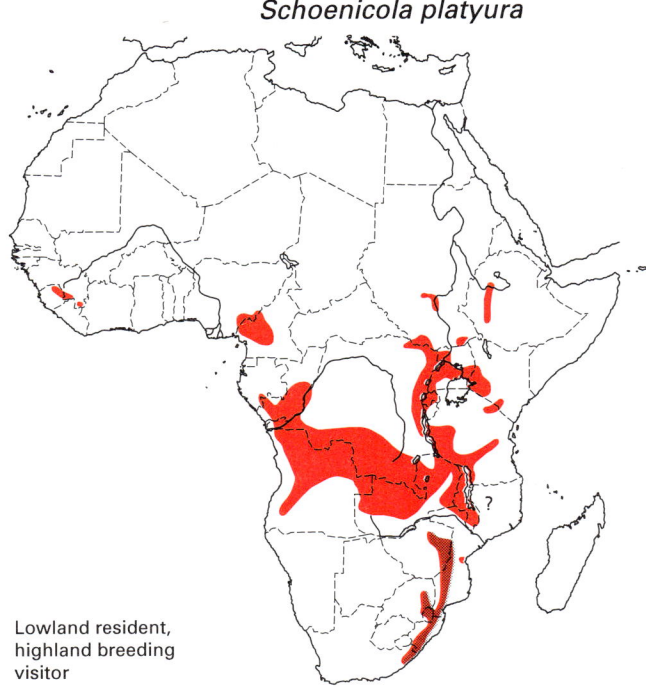

Schoenicola platyura

Lowland resident, highland breeding visitor

m) to Transkei, mainly resident below 1000 m but a breeding visitor, Nov–Apr, at greater elevations (e.g. Zimbabwean plateau, most of Transvaal, Lesotho). Also Beira (Mozambique).

Numbers fluctuate: breeding birds frequent one year, absent the next if ground too dry (Tarboton *et al.* 1987). Habitat vulnerable to burning, damming, commercial afforestation, and trampling by stock (Allan *et al.* 1988); not endangered in South Africa but at risk and thought to be decreasing there.

Description. *S. p. brevirostris* Sundevall: Malaŵi to South Africa. ADULT ♂: forehead, crown, nape, hindneck and sides of neck greyish brown, merging into warmer, ochreous or sometimes rufescent brown on mantle, scapulars and back. Warm mid-brown merges to rufescent dark brown on rump and to rufescent blackish brown on uppertail-coverts, which are lax and extend to about half length of tail. Central tail feathers very dark brown, somewhat glossy, with about 10 narrow bars, which are structural (in degree of gloss) rather than pigmentary. T2 to T6 progressively shorter, T6 being little longer than undertail-coverts; T2 to T6 have well-demarcated grey or grey-brown tips, sometimes also outer fringes, 4 mm wide, paler and more distinct on underside of tail (which is otherwise matt black) than on upperside. All tail feathers soft and broad; whole tail is broad at the base, and rounded and graduated. Ear-coverts and area under eye grey-brown, minutely shaft-streaked with white. Chin, throat and belly silky white or greyish white, buffy towards ear-coverts, sides of neck and flanks. Breast pale buff, richer towards shoulders. Flanks warm buff. Undertail-coverts ochreous grey-brown or blackish brown, lax, rather long, lying towards side of tail-base rather than near midline, difficult to distinguish from T6. Wing short and rounded, primaries and secondaries dark brown, narrowly fringed sandy or rufescent brown; tertials narrowly fringed warm or rufescent mid-brown. Undersides of primaries and secondaries glossy silvery grey; underwing-coverts white, washed buff. Lores pale buff. Upper mandible black with light horn cutting edges; lower mandible very pale grey; eyes light brown; legs and feet pale greyish flesh. Sexes alike. SIZE: (7 ♂♂, 3 ♀♀) wing, ♂ 60–63 (61·6), ♀ 56–62 (59); tail, ♂♀ 69–81; bill, ♂♀ 10–12.5; tarsus, ♂♀ 19–20. WEIGHT: (*S. p. alexinae*) ♂ (n = 5, Uganda) 14–17 (15·7).

IMMATURE: upperparts like adult but top of head rufescent brown, concolorous with mantle; chin, throat, breast and belly pale yellow, feathers soft and fluffy; washed with buff at sides of breast and flanks, but underparts more sharply demarcated from upperparts, along sides, than in adult. Tail shorter (very short at fledging) and less broad.

NESTLING: hatchling not described; at about 1 week young have mouth deep yellow with 3 tongue-spots, 1 apical and 2 lateral (photo of young in Allan *et al.* 1988); plumage of half-fledged young markedly paler than adult.

S. p. alexinae Heuglin: north and west of Malaŵi, to Ethiopia and Guinea. Darker and more olive-brown above than *brevirostris*; flanks and undertail-coverts less buffy/ochreous. Intergrades with *brevirostris*.

Field Characters. Characteristic of marshy highland ground; easily seen on misty mornings perched singly on rank grass heads, but once it goes to cover very difficult to flush. A small-headed, short-billed plain warbler, brown above, whitish below, with long, broad, black tail, jerked up and down as bird flies but carried mainly drooping, as if too heavy. Tail is half length of bird, strongly graduated, not fanned, all but central feathers tipped whitish. Undertail-coverts long, droopy, dark with whitish tips, conspicuous as bird perches. Distinctive jerky flight, low over grasses; equally distinctive voice. Seen from behind when perched on grass head it can resemble a miniature coucal *Centropus* (M. Woodcock, pers. comm.).

Voice. Tape-recorded (68, 75, 86, 88, 91, B, C, F, CHA, McVIC, MOY). Song a short, rather weak but far-carrying, high-pitched, metallic, ringing, whistle: 'sweep', 'tseenk', 'tch', 'peee', 'zink', or 'twink', repeated 2–4 times at intervals of 1·6–1·8 s. Call, 'jur-jur-jur', generally 3-syllabled; also gives a quick, harsh rattle, a clear 'chiyyink, chiyyink' (Sierra Leone) and a wheezy 'tzzzt-tzzt'. Only ♂ said to sing, and only call of ♀ an occasional harsh 'chick' (Chapin 1953a). Alarm a single rasping note at measured intervals, uttered persistently (Vincent 1935).

General Habits. Habitat, long rank grasses, tussocky sedges and tangled reeds in damp or marshy areas or near rivers or dams; boggy drainage lines; swampy, short-grass dambos; rough cover on open mountainsides; water-logged, open, tufty, coarse, boggy, montane grassland; reeds and bushes fringing mountain streams; tall *Themeda* grassland with *Veronia* scrub, above 1550 m (NE Tanzania); *Pennisetum* prairie at 600–1200 m in Itombwe (E Zaïre); *Hyparrhenia* grassland and *Imperata cylindrica* grassland along watercourses, in Transvaal.

Solitary (or if in pairs, commonly only 1 bird seen at a time). After drizzle, or in early morning when grass is wet with dew, sits on top of tall grass stem in sunshine as if drying out, with tail hanging down. Sometimes several can be seen, perching solitarily, spaced out, on grass heads. Also perches on grass head to sing; most vocal (and visible) in wet, misty weather. Rather shy;

can be readily flushed once, but then takes alarm, flies low over grass tops and dives down into grass, when almost impossible to flush again without a dog (Vincent 1935). Flight rather feeble, hesitating, butterfly-like, tail-heavy, tail drooping or jerking up and down; but can fly strongly and fast when really alarmed. Most active at sunrise and after rain, singing and cruising around in wide circle or spiral 6–25 m above ground, slowly, although wings beat quickly making a loud burring or snapping noise. When perched often flicks wings and twitches tail. Once seen flying repeatedly straight up into air for *c*. 6 m, dropping back into grass each time (Bannerman 1939). Feeds only on or near ground (Benson and Benson 1977).

Resident north of Zambezi (Dowsett 1979, for Zambia). In highland Zimbabwe a breeding summer visitor Nov–Apr (occasionally May); in Transvaal present above 1000 m Sept–May; all records Natal Sept–Apr except near Pietermaritzburg, where also July–Aug.

Food. Small insects including tiny beetles, grasshoppers and hairless caterpillars.

Breeding Habits. Solitary breeder, evidently territorial; mating system not studied. Circular song-flight over breeding marsh (see above) presumably has territorial/courtship function. 'Beautiful to watch as they plane steeply down a hillside in display flight, the long broad tail rising and falling rhythmically' (Serle 1950a).

NEST: bulky bowl-shaped structure of coarse, broad-leaved, raffia-like dry grass, not woven or only roughly interwoven, unlined or lined flimsily with a few blades of wiry dry grass, built deep down in large dry tuft of coarse grass standing in *c*. 15 cm of stagnant water, or at least on wet, recently waterlogged earth. Nest extremely well concealed, built into the grass stems but not attached; if nest removed by person it invariably falls apart. Base of nest up to 30 cm above water; or rests on earth caught up in base of grass clump, 5 cm above general wet ground level. Ext. diam. 65–90 (1 nest 65 × 70), ext. depth 65, 69 and *c*. 150, int. diam. 45–60, int. depth 25–50 (n = 3, and 1 nest 25 × 50).

EGGS: 2–3 (av. 2·3, n = 9 clutches Malaŵi to Natal). Laid at 24 h intervals. White or pale cream, faintly and finely speckled with red-brown, mainly in ring at broad end, where some underlying grey or lilac clouding.

SIZE: (n = 4, Natal) 17·9–18·9 × 13·5–13·9 (18·4 × 13·7), (n = 4, Malaŵi) 18·0–18·5 × 12·5–13·8 (18·1 × 13·1).

LAYING DATES: Sierra Leone (breeding condition July); Kenya (Kakamega), July (egg laid by netted bird, late May; young nestlings, mid-Aug); E Africa: Region B, Feb, Mar, June; Region D, Jan, May, Oct; Zaïre, Dec (display, Uele District, Mar–Oct); Zambia, Jan, Feb; Malaŵi, Dec–Mar; Mozambique, Mar; Zimbabwe, Nov–Feb; Transvaal, Dec–Feb (and nest-building Mar).

INCUBATION: when approached by person incubating bird scolds, or departs from nest silently, unobserved. Period unknown.

DEVELOPMENT AND CARE OF YOUNG: at 1 nest only presumed ♂ fed young, at another both sexes did.

References
Allan, D. G. *et al.* (1988).
Took, J. M. E. (1959).
Vincent, J. (1935).

Genus *Locustella* Kaup

Medium-sized, slim-bodied warblers with attenuated heads. Bill fine and pointed; rictal bristles minute; legs slender, toes long; tail broad and rounded with 12 tail-feathers; undertail-coverts long, reaching almost to tail-tip; wing long, sharply or bluntly pointed. Olive or warm brown, plain or streaked above; usually streaked on breast, and streaked or barred on undertail-coverts; supercilium poorly defined. Sexes alike. Generally secretive birds of low, rank herbage, thickets, dense grass and reedbeds, spending much time on ground. Give sharp ticking calls and insect-like reeling or pulsating songs.

7 species, all breeding in Palearctic. 3 in Africa, all Palearctic migrants, but 1 also breeds in NW Africa. *Locustella certhiola* sight record Sidi-Moussa, Morocco, Sept 1984 by G. Balança, not accepted by M. Thévenot (pers. comm.) for Centrale Ornithologique Marocaine.

Plate 6
(Opp. p. 97)

Locustella naevia (Boddaert). **Grasshopper Warbler. Locustelle tachetée.**

Motacilla naevia Boddaert, 1783. Table Planches Enlum., p. 35, *ex* Daubenton pl. 581, fig. 3 and Brisson, Ornithologie, 3, p. 389; Italy.

Locustella naevia

Range and Status. Europe from about 45°N to S Baltic, and through Russia and Kazakhstan to about 100°E; also in Caucasus; winters in Africa and S Asia.

Palearctic migrant. Distribution in Africa patchy; status poorly understood. Winters in tropical W Africa: locally frequent to common W Mauritania and NW Senegal (Djoudj), Guinea (Sierra Leone border, Mt Nimba foothills) and S Mali (margins of Niger inundation zone); occasional Gambia (Feb), Sierra Leone (Dec and Mar–Apr), Liberia (Jan) and N Ghana (Nov–Dec). A few winter north of Sahara in W and N Morocco. On autumn passage, frequent to common in W Morocco and W Mauritania, rare in Tunisia. In spring, locally frequent to common N Senegal, W Mauritania, Morocco and NW Algeria (Beni-Abbès); also scattered records N and E Algeria, and Libya (Sebha and Tobruk). In E Africa, uncommon central and NW Ethiopia (Koka, Awash valley, L. Tana, Sept–Nov and Feb–Apr); rare on passage Sudan coast (Khor Arba'at, Aug–Sept) and Egypt (Suez, Sept; Kom Ombo, Apr); vagrant S Kenya (Loita hills, June).

Description. *L. n. naevia* (Boddaert): breeds in Europe, east in Russia to about 50°E; occurs in N and W Africa. ADULT ♂: upperparts typically olive-brown with broad blackish streaks, uppertail-coverts plain and more rufous. Pale supercilium short and indistinct; lores and ear-coverts olive-brown, speckled yellowish buff. Tail warm brown. Underparts whitish or pale yellowish, sides of throat and breast, flanks and band across upper breast buff or yellowish buff; lower throat usually spotted and flanks streaked dark brown. Undertail-coverts buff, faintly streaked dark brown. Ground colour variable; some birds more buffish yellow below and tinged more greenish above. Strength of streaking above and spotting on breast also varies; some birds are only faintly marked below. Flight feathers dark brown, narrowly edged warm brown; tertials and upperwing-coverts warm brown; axillaries and underwing-coverts pale buff. Wing bluntly pointed; P8 longest, P7 0·5–3 shorter, P6 3–5, P5 5–8, P1 12–16; P9 0·5–4, between P6 and P8; P10 minute; P8 emarginated. Bill dark brown above, pale yellowish brown below; eyes brown; legs and feet pinkish flesh. Sexes alike. SIZE: (10 ♂♂, 10 ♀♀, Europe) wing, ♂ 62–65 (63·6), ♀ 61–64 (62·5); tail, ♂ 52–58 (55·6), ♀ 49–57 (53·7); bill, ♂ 13–15 (14·3), ♀ 13–15 (14·2); tarsus, ♂ 19–21 (19·7), ♀ 18–20 (19·3). WEIGHT: unsexed (n = 43, Senegal, Dec–Feb) 10·9–19·6 (12·9), (n = 13, Senegal, Mar–Apr) 12·1–20·5 (16·0), (n = 5, N Ghana, Nov–Dec) 12–14·8 (13·2).

Complete moult in Africa, though some birds begin in Europe.

IMMATURE: 1st winter bird like adult, but flight feathers still unworn in Sept–Oct.

L. n. obscurior Buturlin: breeds Caucasus; probably winters NE Africa. More olive than nominate race above with blacker, more contrasting streaking. Size similar.

L. n. straminea Seebohm: breeds Siberia and W central Asia; occurs Sudan coast, Ethiopia and Kenya. Greyer above than nominate race, paler below; streaking more contrasting. Slightly smaller: wing, ♂ (n = 10, Turkestan) 55–62 (58·7). WEIGHT: unsexed (n = 11, Ethiopia, Sept–Apr) 10·2–15·4 (12·4).

Field Characters. A small slim tapered-looking warbler, usually skulking in grassy cover or low bushes. Streaked olive-brown above and whitish or pale yellowish below with spots across lower throat and fine streaks on long undertail-coverts. Lacks distinct supercilium. Wing-tip pointed and tail rounded. Legs pinkish. The only streaked *Locustella* in Africa; streaked *Acrocephalus* warblers have bold head patterns and squarer tails, and Sedge Warbler *A. schoenobaenus* has greyish legs. *Cisticola* spp. have thicker bills, short rounded wings and short undertail-coverts. Insect-like reeling song, higher pitched than that of Savi's Warbler *L. luscinioides*, is highly distinctive.

Voice. Tape-recorded (62, 73, 89, 93, B). Song, a high-pitched, insect-like ventriloquial reel, with a tinkling quality, which can carry over 300 m. Volume and tone seem to change as bird turns its head. Continuous singing lasts from a few s up to many min. Contact-alarm call, a short hard 'tic'.

General Habits. Frequents bushy herbaceous cover or dense grass, including undergrowth in acacia and open deciduous woodland; commonly near water; in Mali in *Mimosa pigra* on margins of flooded ground; in Guinea (Mt Nimba foothills) in clearings recolonized by 2 m high grasses *Andropogon* and *Hyparrhenia* with a scattering of shrubs (Brosset 1984). Usually at low altitude, but occurs above 2000 m in Ethiopia and Kenya.

Solitary in winter, but sometimes in parties on migration. Skulks and creeps among low tangled cover. Hops and runs on ground. Forages from leaves and plant stems, and among ground debris; occasionally takes insects in flight (Cramp 1992). Flight flitting and low; usually dives quickly into cover. Sings in migration stopover stations in Morocco (Dec, Apr: Thévenot 1982); also occasionally in Senegal (Feb) and Sierra Leone (Mar).

Migration occurrences patchy; scarcity in many areas suggests long unbroken flights. Ringing evidence and passage records indicate that nominate birds head towards Iberia in autumn and along W African coast (those from E Europe migrating southwest). Passage through Morocco is mainly Sept–Oct, through SW Mauritania Aug–Nov; arrives in Mali from Sept. Northward migration starts early, but is protracted. Passage through Senegal occurs early Mar to mid-May, at Gibralter in late Mar to late May. Widespread records in Algeria to mid-May suggest that many European birds use a more easterly spring return route. Recoveries from Britain to Algeria (1, Feb) and Senegal (1, Jan), from Germany to Morocco (1, Oct) and from Sweden to Mauritania (1, Oct). 2 birds ringed Senegal recovered Britain.

Food. Mainly insects; also spiders and small molluscs. No information from Africa.

References
Brosset, A. (1984).
Cramp, S. (1992).

Locustella fluviatilis (Wolf). Eurasian River Warbler. Locustelle fluviatile.

Plate 6
(Opp. p. 97)

Sylvia fluviatilis Wolf, 1810. *In* Meyer and Wolf, Taschenb. I, p. 229; Austria.

Range and Status. Central and E Europe, and east through Russia between 50° and 60°N to Irtysh R. at about 75°E; winters in Africa.

Palearctic migrant. Winters sparsely in E and SE Africa, but probably widely overlooked and winter range poorly defined. During late Jan to early Mar, small numbers locally in SE Kenya (Kitui, Kibwezi, Amboseli, Bissel); locally frequent to common in S Malaŵi and adjacent NW Mozambique, Zambia (north to Mwinilunga and the Copperbelt, especially Middle Zambezi), and NE Botswana (Kasane Forest Reserve); occasional records further south, in Zimbabwe, E Botswana (Francistown) and N South Africa (Transvaal to *c.* 26°S). On southward passage, occurs in Egypt (uncommon, Bahariya, Bahig, L. Maryut); NE Sudan (frequent, Suakin, Khor Arba'at); central Ethiopia (few records, Rift valley); S Somalia (once, Jilib); and central and SE Kenya (locally frequent to common Samburu and Meru to Naivasha, Ukambani, Tsavo and Taita-Taveta); probably only vagrant in Uganda (once L. Edward, Nov). On northward passage, occurs N Tanzania (once, Arusha); central and SE Kenya (uncommon, Nairobi, Tsavo); Ethiopia (twice Koka, Jijiga) and E Egypt (rare, El Maadi, Bahig). A series of late Mar to early Apr ringing records NW Algeria (Beni-Abbès, Hassi-Messaoud) and SE Morocco (once, Defilia) (Depuy and Johnson 1967, Dowsett 1972) suggests an undiscovered small W African wintering population.

Locustella fluviatilis

Description. ADULT ♂: entire upperparts, tertials and upper-wing-coverts dark olive-brown, tinged greenish. Lores and ear-coverts olive-brown streaked with yellowish buff; supercilium indistinct, tinged yellowish buff. Tail dark brown. Chin, throat and centre of breast washed yellowish buff, chin and upper throat finely spotted, lower throat and breast broadly streaked dark brown; sides of breast and flanks olive-brown; centre of belly whitish; undertail-coverts pale olive-brown broadly tipped white. Primaries and secondaries dark brown, narrowly edged pale olive; axillaries and underwing-coverts olive-brown and whitish. Wing pointed; P9 longest (sometimes = P8), P8 0–2 shorter, P7 3–6, P6 6–8, P5 9–11, P1 18–21; P10 minute; no emargination. Bill horn above, pale flesh below; eyes hazel to dark brown; legs and feet flesh pink. ADULT ♀: like ♂, but slightly more greyish, less greenish above, supercilium almost absent, breast streaking less extensive. SIZE: (10 ♂♂, 10 ♀♀, Europe) wing, ♂ 72–78 (75·4), ♀ 72–78 (75·2); tail, ♂ 57–61 (59·0), ♀ 56–60 (58·0); bill, ♂ 15–17 (16·0), ♀ 15–16 (15·5); tarsus, ♂ 20–23 (21·6), ♀ 20–22 (21·2). WEIGHT: unsexed (n = 66, NE Sudan,

Aug–Oct) 11–18 (14·3); Ethiopia (n = 9, Sept–Nov) 14·1–19·8 (17·0), (n = 1, Apr) 22·3; Kenya, (n = 500, Nov–Jan) 13·3–21·2 (16·1), (n = 20, Mar–Apr) 15–19·6 (16·9); (n = 18, Zambia and Malaŵi, Dec–Mar) 15·5–19·7 (17·4).

Partial post-nuptial moult in NE Africa, Sept–Nov; complete moult in southern Africa, Jan–Mar.

IMMATURE: 1st year usually like adult, but some are rustier above and more yellow on face and underparts; wing-coverts and flight feather edgings usually rustier.

Field Characters. A medium-sized warbler which skulks and creeps close to ground. Upperparts unstreaked dark brown, tail broad and rounded, throat and breast streaked. Head rather plain, with faint supercilium. Undertail-coverts bulky, reaching to tip of tail, and prominently barred. Legs pinkish. Distinguished from similar Savi's Warbler *L. luscinioides* by more conspicuous markings on breast and undertail-coverts; also by colder tone above and yellowish tinge to face and throat. Song distinctive.

Voice. Tape-recorded (86, 91, F, HRS, PEAR – 62, 73, B). Song, a high-pitched, rather mechanical, throbbing 'dzi-dzi-dzi-dzi-dzi-dzi- . . .', in burst lasting from a few s to over half a min; resembles stridulation of bush cricket or display call of certain bishops *Euplectes*; individual 'dzi' notes, delivered at *c*. 7 per s, are clearly separated. Contact-alarm call, a sharp explosive 'phit', sometimes singly but often repeated persistently, 1·5–4 times per s.

General Habits. Frequents dense green bush and thicket, rank herbage and woodland undergrowth, at low to medium altitude. In Kenya, winters up to 1700 m in sites with tall grass, acacia scrub and thickets of *Premna*, *Grewia* and *Combretum*. In NE Botswana, in understorey of *Baphia* and *Bauhinia* scrub interspersed with long grasses, in more open parts of Rhodesian teak *Baikiaea plurijuga* woodlands (Herremans 1994). Utilizes wider range of moist habitat than does Savi's Warbler.

Skulks in low cover and usually located by calls, but is inquisitive, and may emerge into top of bush when excited, especially at sunrise and sunset. Creeps among twigs with body typically horizontal and tapered-looking. Runs on ground through grassy vegetation. Most food taken directly from ground or plants (Cramp 1992). Drops to ground when disturbed. Flies reluctantly, straight and low, with broad tail spread, usually dropping back into cover after a few m. Solitary in Africa. Birds occupy territories on southward passage stopover and in winter quarters, apparently maintained by repeated 'phit' call. In NE Botswana, densities of 3–7 birds per ha. in teak woodland understorey (Herremans 1994). In Zambia, a moulting bird retrapped Feb at place where ringed 6 weeks earlier (Tucker 1978). Sings frequently on southward passage in Kenya and in winter quarters.

Adults moult body plumage and outer primaries in NE Africa, Sept–Nov (Pearson and Backhurst 1983). Undergoes a 2nd, complete and quite rapid moult in winter quarters (all birds) from late Jan.

Known only from a few areas in Africa, and migrations still somewhat obscure, so it may make long unbroken flights. Birds from Europe head southeast in autumn but few make landfall in E Mediterranean (Cramp 1992). Enters Africa east of 30°E and apparently follows narrow route through NE Sudan, central (and probably W) Ethiopia and central and SE Kenya. Migration initially rapid, with main passage at Sudan coast from mid-Aug to early Oct. The few Ethiopian records are early Sept to mid-Nov, but birds presumably remain there 2–3 months, for earliest Kenya record is 2 Nov. Kenya passage is mainly east of highlands, from mid-Nov to early Jan, with concentrations in Kibwezi/Tsavo area where *c*. 200 ringed annually after attraction to lights at Ngulia Lodge; many of these birds are markedly fat (Pearson and Backhurst 1976). 1 record Tanzania (Kilimanjaro, Dec: Cordeiro 1994). Arrives Zambia and Malaŵi from mid-Dec. Birds in Zambia in late Dec and Jan apparently still moving south (Dowsett 1972). Remains at most wintering sites until late Mar to early Apr; birds still moulting Kenya until early Apr, and singing and territorial in Botswana up to 5 Apr. Northward passage through Kenya is inconspicuous, mainly mid to late Apr, again east of highlands; most birds appear to overfly, but 15 once trapped in a day at Ngulia (Pearson 1980). Absence of spring records in Ethiopia suggests localized and undiscovered staging areas, or passage through Somalia further east. Few records in E Egypt, mid-Apr to mid-May, and sparse spring passage through Sinai and Eilat, Israel.

Food. No information from Africa. In Palearctic, mainly insects; also spiders, small ticks, millipedes and small molluscs.

References
Cramp, S. (1992).
Dowsett, R. J. (1972).
Herremans, H. (1994).
Pearson, D. J. and Backhurst, G. C. (1976, 1983).

Plate 6 *Locustella luscinioides* (Savi). **Savi's Warbler. Locustelle luscinioïde.**
(Opp. p. 97)
Sylvia luscinioides Savi, 1824. N. Giorn. Litter. 7, p. 341; Pisa, Italy.

Range and Status. NW Africa; Europe north to Denmark and Estonia, east to S Urals; Turkey and Jordan to Transcaspia, Kazakhstan and W Mongolia; winters in Africa.

Locustella luscinioides

Palearctic migrant. Breeds Morocco (Bas Loukkos, Middle Atlas, Moulouya, perhaps Oued Smir), N Algeria (Oranie, near Algiers, Grand Kabilie, Lac Tonga) and probably N Tunisia (L. Ichkeul); local and uncommon. Winters in scattered areas south of Sahara (but perhaps widely overlooked). Uncommon to frequent in Senegal (Lower Senegal and Casamance valleys), S Mali (Bamako, Parc du Baoulé, Diaka R., and probably in Niger delta north to 17°N) and N Ghana (Navrongo, Vea dam); rare N Nigeria (Kano, L. Chad). In the east, uncommon W Sudan (Darfur) and S Sudan (Nile valley near Juba), but locally common in Ethiopia (in and west of the Rift, south to Abiata). Rare SE Kenya (twice Ngulia, Dec). Winters occasionally north of Sahara, in Algeria and apparently Egypt (several Feb records). Scarce on southward passage in Morocco, Algeria, Egypt and SW Mauritania, but locally common in N, central and W Sudan (Red Sea coast, Nile valley, Darfur) and in Eritrea. On northward passage, frequent to common locally in Morocco, Algeria and Egypt, but uncommon in Tunisia and rare Libya; occasional at Saharan oases.

Description. *L. l. luscinioides* (Savi): NW Africa and Europe east to Crimea; occurs W Africa and probably Sudan. ADULT ♂: entire upperparts, upperwing-coverts and tertials dark brown, tinged rufous. Supercilium buffish, short and indistinct; lores and ear-coverts pale brown, speckled buff. Tail dark rufous-brown. Underparts whitish, sides of breast and flanks suffused deep buff; undertail-coverts deep buff with paler tips; sometimes a few grey-brown spots form gorget across base of throat. Primaries and secondaries dark brown with narrow rufous fringes; axillaries and underwing-coverts buff. Wing pointed; P9 longest, P8 0·5–3 shorter, P7 4–6, P6 6–9, P5 9–11, P1 17–19; P10 minute; no primaries emarginated. Bill dark horn above, pale below; eyes brown; legs and feet pale brown or pinkish. Sexes alike. SIZE: (10 ♂♂, 10 ♀♀, Europe) wing, ♂ 69–73 (70·3), ♀ 65–72 (69·2); tail, ♂ 57–62 (58·8), ♀ 55–59 (57·2); bill, ♂ 15–17 (16·0), ♀ 15–17 (16·0); tarsus, ♂ 20–22 (20·8), ♀ 20–21 (20·6). WEIGHT: unsexed, Senegal (n = 87, Oct–Feb) 11·2–20·5 (13·8), (n = 41, Mar–Apr) 13–23·4 (17·2); (n = 13, Ghana, Nov–Dec) 12·2–15 (14·5).

Complete moult in Europe July–Sept or in Africa, Oct–Feb; but some birds begin in Europe, suspend, then finish in Africa.

IMMATURE: juvenile and 1st winter bird like adult but flight feathers still unworn in Sept–Oct.

NESTLING: covered with sparse grey or brown down above. Mouth deep yellow; gape flanges deep yellow; 3 indistinct tongue spots.

L. l. sarmatica Kazakov: E Ukraine and Sea of Azov to lower Volga and S Urals; winters in Sudan and Ethiopia. Upperparts and side of body slightly less dark, more olive-brown than in nominate race.

L. l. fusca (Severtzov): Turkey and Jordan to central Asia; winters in Sudan and Ethiopia, recorded Kenya. Less brown on upperparts, more olive-grey or sandy than *luscinioides* or *sarmatica*, and whiter below, with flanks and undertail-coverts paler. WEIGHT: E Sudan, this race and *sarmatica*, unsexed (n = 14, Aug–Sept) 11·5–18 (12·9); Ethiopia, this race and *sarmatica*, unsexed (n = 33, Sept–Feb) 12–19·1 (14·6), (n = 63, Mar–Apr) 12·6–25·2 (17·1).

Field Characters. A smallish warbler with fine pointed bill, unstreaked dark brown upperparts, full rounded tail and long bulky-looking undertail-coverts. Whitish below with buff sides; legs pinkish. Usually on or near ground among swampy vegetation. Resembles Eurasian River Warbler *L. fluviatilis* in plumage and structure, but warmer toned, with fainter barring on undertail-coverts, and spots absent on underparts or restricted to gorget on lower throat. Distinguished from Eurasian Reed-Warbler *Acrocephalus scirpaceus* by habits and voice; also by rounder tail, longer undertail-coverts, paler legs and less distinct supercilium.

Voice. Tape-recorded (62, 73, 89, 93, B). Song, a mechanical reeling, faster, lower-pitched and less insect-like than that of Grasshopper Warbler *L. naevia*, but often carrying over 300 m; preceded by a series of 'twik, twik' notes, audible at close range; often in short bursts of less than 20 s, but sometimes sustained for over 2 min. Contact-alarm call, 'tchik' or 'pink', singly or repeated, sometimes as a rapid chatter.

General Habits. More confined to wet places than the other 2 African *Locustella*. Breeds in marshes of reeds, reedmace *Typha* or sedges standing in shallow fresh or brackish water. Winters in reedbeds, *Typha*, marsh and swamp vegetation, rank grass, rice fields, sugar cane and occasionally garden shrubbery; in N Ethiopia, occurs in dense *Salvadora persica* thickets by permanent springs in acacia bush (Smith 1953).

Usually solitary, creeping in dense low cover and difficult to observe. Returned to same site in Ethiopia in 2 successive years. Often hops and runs on ground. Picks food from low stems, also from mud and water surface (Cramp 1992). When singing, may appear in full view near top of reed stem. Sings in winter quarters in Senegal (Dec–Feb), and once Ghana (Jan: Walsh and Grimes 1981); also on spring passage.

Plate 5

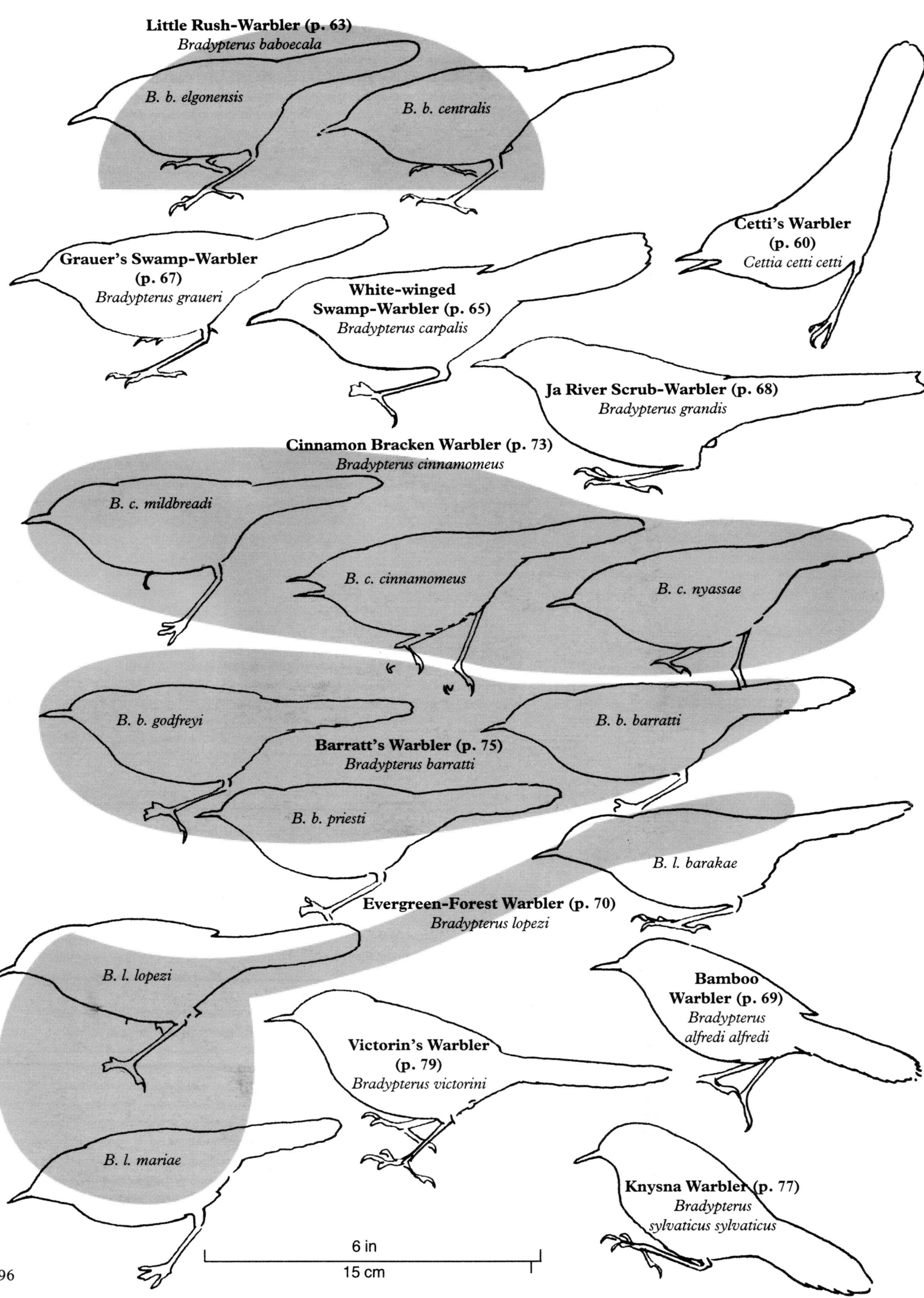

Plate 6

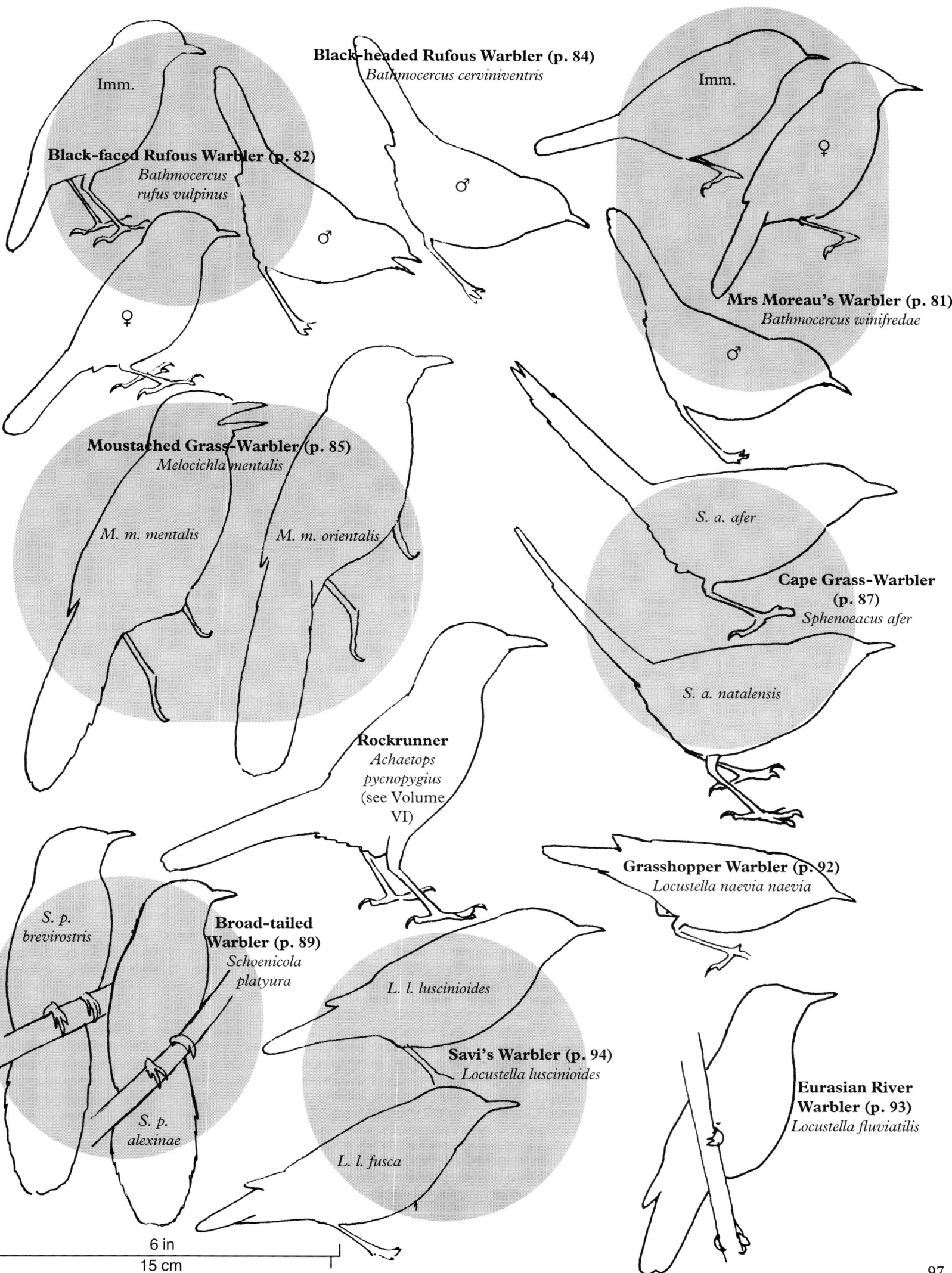

Many W European birds begin adult moult on breeding grounds, with unusual primary sequence initiated in middle of tract (Thomas 1977). Some finish there, others suspend and finish in W Africa, Oct–Feb. Eastern races usually undergo adult wing-moult in Africa, although some renew a few inner primaries before autumn migration.

W European birds head south to southwest in autumn, central and E European birds southeast, and Asian *fusca* southwest. Probable migratory divide in Europe supported by paucity of records in central N Africa. European breeding grounds vacated in Aug to mid-Sept, E European and Russian ones in late Aug to early Oct, and Moroccan ones in Oct. Passage through N Africa is from mid-Aug to late Oct, but scarcity in autumn in Mediterranean area suggests long unbroken flights to tropics. Recorded Nigeria in Sept, but mainly Nov onwards; in Senegal and S Mali from mid-Nov. In Sudan, main autumn passage on Red Sea coast is late Aug to early Oct; birds present and moulting Darfur from Oct (Lynes 1925). Earliest record in Ethiopia, late Aug.

Spring movement begins early, during Feb. Most remain Senegal to Mar, latest to mid-Apr. Passage in NW Africa more marked in spring than in autumn, with main movement through Algeria early Mar to mid-May. Returns to Moroccan breeding sites in Mar. Pre-migratory weights in Senegal approximately double usual winter weight (S. J. R. Rumsey, pers. comm.) suggest some birds make long uninterrupted flights back to Europe. Most birds remain in S Sudan to Mar, and in Ethiopia to late Mar or Apr (latest 4 May). Strong passage in Egypt late Mar to mid-May. A bird ringed France recovered Niger; 2 birds ringed Spain recovered Senegal (Jan, Feb).

Food. No information from Africa. In Europe, mainly insects, including mayflies, adult and larval damselflies and dragonflies, small grasshoppers, bugs, adult and larval Lepidoptera, adult and larval Diptera, and beetles; also spiders, small worms and molluscs.

Breeding Habits. Few African data; well studied in Europe. Monogamous. Solitary and territorial; territory 0·08–0·8 ha (Europe); serves for pair formation and nesting; occupied by ♂ immediately on arrival and defended vigorously against conspecifics. ♂ has several song-posts within territory. Sings from part way up reed stem or from bush, with body upright, tail pointing vertically down and head held back with bill almost vertical (Cramp 1992); sings with bill open, head turning from side to side; mainly in mornings and evenings, but also at night. Song period Apr–July (Europe). Prolonged singing is typical of unpaired ♂. Paired birds sing briefly during nest-building and incubation, but song-length often increases about late June, prior to 2nd brood. Pairing and nest-building take *c.* 10 days. Displaying ♂ chases ♀, sometimes in flight; spreads and flaps wings; and raises and lowers fanned tail (Cramp 1992).

NEST: deep cup, loosely built from dead sedges, reeds and aquatic plants, lined with finer leaves and plant fibres. Generally well concealed, resting low down in reeds, sedges or grasses, a few cm to over 50 cm above shallow water or damp ground; often partially covered. Built by ♀ or by both sexes.

EGGS: in Europe 3–6 (usually 4–5), laid at daily intervals. Glossy; white, often densely speckled with brown, grey-brown or purplish, mainly at broad end. SIZE: (n = 100, Europe) 17–22 × 13–15. Often double-brooded (Europe).

LAYING DATES: Algeria, probably late Apr–May.

INCUBATION: by both sexes, mainly ♀. Period: 10–12 days.

DEVELOPMENT AND CARE OF YOUNG: young cared for and fed by both parents, mainly by ♂; remain in nest 11–15 days.

BREEDING SUCCESS/SURVIVAL: of 73 nests, Poland, 44% lost by predation; loss more frequent on lower-sited nests over shallow water (Pikulski 1986).

Reference
Cramp, S. (1992).

Genus *Acrocephalus* Naumann

Medium-sized or large warblers with slim bodies and somewhat attenuated heads. Bill long and pointed; rictal bristles few, usually distinct and fairly strong; legs and feet strong; tail of 12 feathers, slightly or moderately rounded; wing bluntly pointed, in some cases rounded, with outermost primary typically very small. Most species plain rich olive or warm brown above and buff or whitish below; some with streaked upperparts; supercilium distinct, in some broad and conspicuous. Sexes alike. Head profiles illustrated: (**A**) *A. griseldis*, (**B**) *A. arundinaceus*, (**C**) *A. stentoreus*.

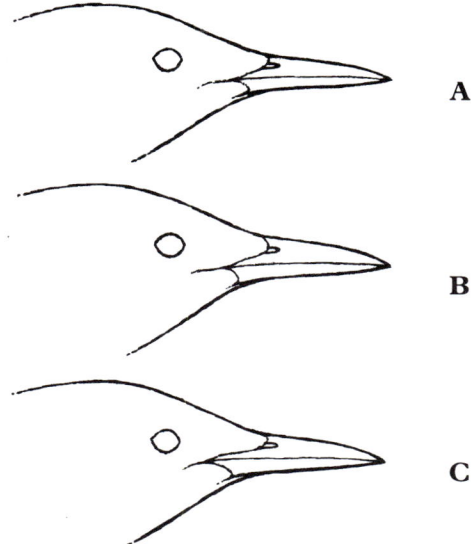

Inhabit thick cover, often near water: reedbeds, marshes, leafy thicket and herbaceous undergrowth. Nest open, usually suspended on upright plant stems, often over water. Vocally active, with chattering, grating, typically rhythmic songs, and churring or chacking calls.

33 species, all in Old World. 11 in Africa, of which 2 are endemic residents south of Sahara, 1 is a resident and local migrant breeding also in Arabia, 1 is a Palearctic species resident in NE Africa and 7 are Palearctic migrants (2 also breeding NW Africa).

The 2 swamp-warblers *A. rufescens* and *A. gracilirostris*, formerly placed in *Calamocichla*, have short rounded wings with well-developed outermost primary, long rounded tails, very strong dark legs and feet, and rather flattened base to lower mandible. Their low chuckling songs are different from those of most *Acrocephalus* spp. With some reservations, however, we follow White (1952) and later authors in sinking *Calamocichla* in *Acrocephalus*. Following Parker and Harrison (1963) and most subsequent authors we also include *melanopogon* (formerly in the monotypic genus *Lusciniola*) in *Acrocephalus*. We recognize one superspecies, *A. scirpaceus/A. baeticatus*.

Acrocephalus scirpaceus superspecies

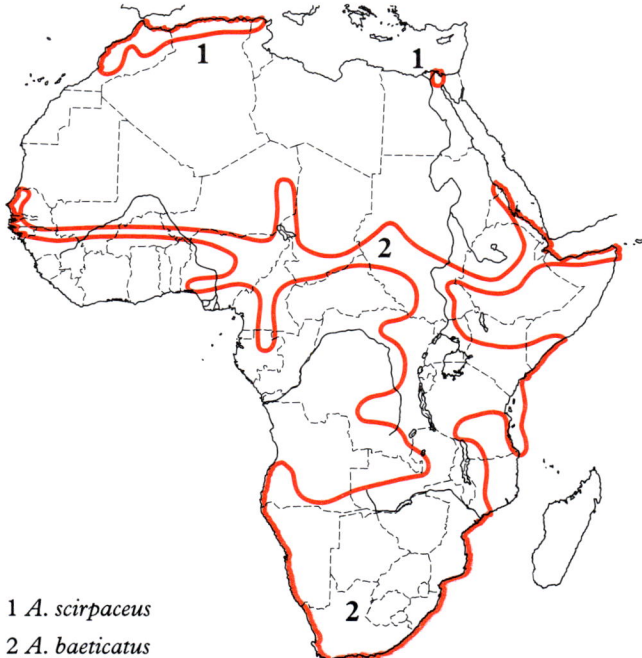

1 *A. scirpaceus*
2 *A. baeticatus*

Plate 7
(Opp. p. 112)

Acrocephalus melanopogon (Temminck). Moustached Warbler. Lusciniole à moustaches.

Sylvia melanopogon Temminck, 1823. *In* Temminck and Laugier, Planches Col. livr. 41, pl. 245, fig. 2; Roman Campania.

Range and Status. NW Africa, S Europe east to Black Sea, and SW Asia; partial migrant, wintering in Mediterranean basin, SW Asia, Arabia and India.

Palearctic visitor and scarce local resident. In Morocco, breeds Loukkos marshes (where apparently sedentary) and possibly also lower Moulouya valley near Berkane and mouth of Wadi Massa; increased numbers in late autumn perhaps due to immigration; little evidence of passage through Gibraltar, but recorded occasionally in Moroccan non-breeding areas. In Algeria, several winter Jan–Apr at Reghaïa. A few breed in Tunisia at L. Ichkeul (and formerly bred Cap Bon), but otherwise rare. Scarce but regular on passage or wintering in Egypt: several Aug–Nov and Feb–May records from Suez Canal corridor and Nile delta, and once, Feb, near L. Qarun. No confirmation of occurrence south of Sahara. Postulated wintering L. Chad (Grote 1928) and L. Albert (Schouteden 1957) is unlikely, and reports of passage birds in Libya (Jan 1960: Bundy 1976) lack supporting evidence.

Description. *A. m. melanopogon* (Temminck): NW Africa and S Europe; only subspecies in Africa. ADULT ♂: top of head black, feathers fringed chestnut-brown; mantle and scapulars chestnut-brown streaked black; back, rump and uppertail-coverts uniform chestnut-brown. Tail-feathers blackish brown, fringed chestnut-brown. Supercilium white and prominent, broad behind eye and ending squarely on hindcrown; lores and ear-coverts blackish brown; nape and sides of neck chestnut-brown. Chin, throat and centre of belly white; breast, flanks and undertail-coverts chestnut-buff, breast often with fine brown streaks. Flight feathers blackish brown, edged chestnut-brown, broadly so on tertials. Greater and median coverts as tertials; lesser coverts blackish; underwing-coverts and axillaries white. Wing rounded; P6 and P7 longest, P5 0·5–2 shorter; P4 3–4, P1 7–10; P8 0·5–2 shorter, P9 5–8, between P2 and P4; P10 small, about one-third of length of P9; P6 to P8 emarginated. Bill dark horn, pale at base of lower mandible; eyes rich brown; legs and feet greyish or dark brown. Sexes alike. SIZE: (10 ♂♂, 10 ♀♀, N Africa and Europe) wing, ♂ 55–62 (57·7), ♀ 53–59 (56·4); tail, ♂ 48–54 (51·6), ♀ 47–54 (50·7); bill, ♂ 14–16 (15·0), ♀ 13–16 (15.0); tarsus, ♂ 19–22 (20·4), ♀ 19–21 (20·1). WEIGHT: unsexed (n = 33, Hungary, July–Sept) 9–13 (10·9).

Complete moult in breeding area July–Oct.

IMMATURE: juvenile like adult but top of head blacker, underparts paler, upperparts less rufous; upper breast with pale greyish spots.

NESTLING: naked; mouth bright yellow; 2 tongue spots.

Field Characters. A small dark reed-warbler with heavily streaked brown upperparts, plain rufous rump and prominent supercilium. Resembles a small Sedge Warbler *A. schoenobaenus*, but darker and more rufous above, with almost unmarked rufous nape and sides of neck. Head more strongly patterned, with crown blacker and eye-stripe darker; supercilium whiter, extending well back and square-ended. Also distinguished from Sedge Warbler by voice, more graduated tail (**A**), shorter, rounder wing, and in hand by longer P10 (**B**).

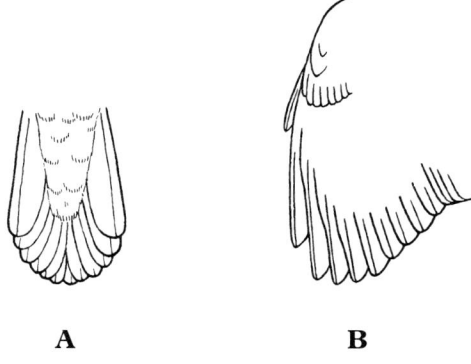

A   B

Acrocephalus melanopogon

Voice. Tape-recorded (62, 73, 93, B). Song, a prolonged scratchy medley recalling Sedge Warbler, but softer and more musical, and lacks harsh chattering; includes characteristic, fluty 'du-du-du-du' introductory notes. Contact calls, a soft 't-trrrt' and a louder 'tuc', the latter repeated as a churring alarm.

General Habits. Occurs along ditches and lake margins, mainly in reedbeds. Breeds in dense stands of reed, *Typha* and sedges.

Essentially solitary although it may congregate at food sources. Skulking but not shy; usually sings from concealment, but may allow close approach. Forages low down, picking prey from vegetation near water surface. Characteristically cocks and spreads tail. Flight less fluent, more flitting, than Sedge Warbler, with tail held straight, not drooping (Cramp 1992).

Vacates northern parts of range, leaving late (Sept–Nov) and returning early (Mar to early Apr). Occupies Mediterranean wintering areas from Oct to early Mar.

Food. Mainly small beetles and other tiny insects, including bugs, Hymenoptera, Diptera and larvae; also small spiders and water snails.

Breeding Habits. Monogamous. Solitary and territorial; territory 0·3–0·5 ha (Austria); used for courtship, nesting and most foraging; defended aggressively by both sexes. ♂ has several song-posts within territory; sings from reed-top or other exposed perch with tail held vertically down; does not have song-flight (Cramp 1992).

NEST: deep untidy cup of loosely woven leaves and stems of aquatic plants lined with fine plant material and feathers; sometimes with partial roof; suspended from several vertical stems by loops, c. 30–60 cm above water; built by ♀.

EGGS: in Europe 3–5 (6); white or greyish, finely speckled light olive. SIZE: (n = 90, Europe) 16–19 × 12–14 (18·0 × 13·2).

LAYING DATES: in Europe, Apr to early May.

INCUBATION: by both sexes. Period: 14–15 days.

DEVELOPMENT AND CARE OF YOUNG: young fed by both parents; remain in nest c. 12 days.

Reference
Cramp, S. (1992).

Acrocephalus paludicola (Vieillot). Aquatic Warbler. Phragmite aquatique.

Plate 7 (Opp. p. 112)

Sylvia paludicola Vieillot, 1817. Nouv. Dict. d'Hist. Nat., nouv. éd. II, p. 202; Lorraine and Picardy.

Range and Status. E Germany, Poland, Slovakia and Hungary, and probably former Soviet Union to c. 60°E; also W Siberia (R. Ob); winters in Africa.

Palearctic migrant. Status in Africa poorly known. Winters sparsely in Senegal (Senegal delta, Dec–Mar); Mali (3 records Dec, Gossi, Mare Takadji, Bamako); and N Ghana (once Nov, Navrongo: Hedenström *et al.* 1990). On autumn passage, formerly frequent NE Morocco (Ras el Ma and Moulouya estuary: Brosset 1956); now rare Morocco (3 recent records, Tangier, Fes, and off Dakhla), coastal Algeria (Reghaïa), coastal Mauritania and Tunisia; 2 records N Mali (Arawan). On spring passage, uncommon Morocco (20 widely scattered records) and coastal Algeria, rare Mauritania and Tunisia (Mayaud 1990, Cramp 1992, J. D. R. Vernon, pers. comm.). A few sight records Egypt but occurrence there requires confirmation (Goodman and Meininger 1989).

Acrocephalus paludicola

Description. ADULT ♂: upperparts olive- or sandy-buff, heavily streaked with black on mantle and scapulars, rather less so on tawny rump and uppertail-coverts. Tail-feathers dark brown with broad pale fringes. Forehead tawny-buff; line down centre of crown yellowish buff; broad line along side of crown blackish; well-defined long supercilium yellowish buff; lores and ear-coverts yellowish brown streaked blackish. Chin and centre of belly whitish; sides of throat and breast, and flanks, yellowish to tawny buff with narrow brown streaks; undertail-coverts yellowish buff. Primaries, secondaries and primary coverts blackish brown, edged paler; tertials and greater and median coverts blackish, broadly fringed and tipped yellowish or tawny-brown; lesser coverts greyish brown; axillaries and underwing-coverts white with dusky centres. Wing bluntly pointed; P8 longest (often = P9); P7 2–3 shorter, P6 5–6, P5 9–8, P1 16–17; P9 0–2, longer than P7; P10 minute; P8 emarginated (**A**). Bill blackish brown, lower mandible flesh with darker tip; eyes brown; legs and feet yellowish or pinkish brown. Sexes alike. SIZE: (10 ♂♂, 10 ♀♀, Europe): wing, ♂ 59–64 (62·5), ♀ 59–64 (61·3); tail, ♂ 44–50 (47·4), ♀ 47–49 (48·0); bill, ♂ 13–15 (14·0), ♀ 13–15 (13·8); tarsus, ♂ 20–21 (20·4), ♀ 20–21 (20·4). WEIGHT: unsexed (n = 21, France, Aug) 10·5–14·5 (12·0), (n = 4, Senegal, Mar) 10–16 (11·9), (n = 1, Ghana, Nov) 10·1.

Apparently moults wing and tail fully in winter quarters.

IMMATURE: 1st year bird like adult but in unworn plumage Aug–Sept, generally sandier buff above and below.

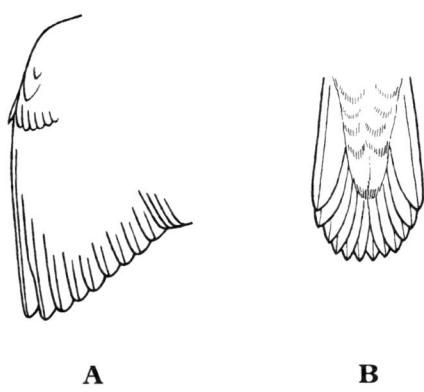

A   B

Field Characters. A small, slim buffish warbler with streaked upperparts. Head longitudinally striped, with blackish on sides of crown and with broad yellowish or creamy supercilia and central crown-stripe. More strongly patterned above than Sedge Warbler *A. schoenobaenus*, long black stripes on centre of mantle contrasting with pale bands along scapulars, and continuing as streaks on rump. Lacks discrete rufous rump; usually has distinct fine streaks on flanks. Head more boldly marked than in Sedge, and longer looking with shorter bill. Tail narrower with pointed feather-tips (**B**), and legs usually paler.

Voice. Tape-recorded (62, 73, 93, B, CHA). No information from Africa. Song recalls Sedge Warbler but is shorter, less musical and varied, notes slower and softer; comprises a few distinct phrases with discernible pauses between, 'trr-trr-ju-ju, trrr-du-du-du, trrt-di-di . . .', lasting a few s in all. Calls lower than Sedge Warbler, a hard 'chat', a soft 'trrt-trrt', and a hard 'tek'.

General Habits. On passage frequents waterside habitat, both on coasts and inland; especially by rivers and along ditches, among sedges, reed patches, reedmace *Typha* and bushes (Cramp 1992); in *Salicornia* in NE Morocco. In tropical Africa, found in reedbeds and flooded grassland, and on shallow lake edges.

Apparently solitary. More secretive than Sedge Warbler, and confined mainly to dense low vegetation. Forages usually close above ground or water, creeping like a *Locustella* warbler (Cramp 1992). Scurries rapidly on ground. At times climbs to feed higher up in bushes. Alert bird may adopt characteristic slim upright posture with neck stretched (Schulze-Hagen 1989). Flight lighter, more darting, than Sedge Warbler, though tail often depressed as in that species (Cramp 1992).

Migrations poorly understood. Winter quarters of European populations probably in tropical W Africa. Many birds apparently head west to southwest in autumn to staging areas in W Europe, then south through Iberia. Passage records in Morocco and W Mauritania are Sept to early Oct, with Oct birds also in Algeria and Tunisia and 2 in Nov in N Mali. Spring passage more protracted, with birds in Morocco and occasionally Algeria and Tunisia from early Feb to early May. Few spring records from W Europe, suggesting return to breeding areas by more direct route.

Food. No information from Africa. In Europe, mainly insects; also spiders and small snails. Tends to take larger items than do other small *Acrocephalus* spp. (Schulze-Hagen *et al.* 1989).

References
Cramp, S. (1992).
Schulze-Hagen, K. (1989).

Plate 7
(Opp. p. 112)

Acrocephalus schoenobaenus (Linnaeus). **Sedge Warbler. Phragmite des joncs.**

Motacilla schoenobaenus Linnaeus, 1758. Syst. Nat. (10th ed.), p. 184; Sweden.

Range and Status. Europe from *c.* 45°N to Arctic, Turkey and Caucasus, and through Russia and N Kazakhstan to W Siberia; winters in Africa.

Palearctic migrant. Winters from SW Mauritania, Senegal, Mali (north to Niger delta at 17°N), S Niger, central Chad, central Sudan (north to Khartoum), W and central Ethiopia, W, central and SE Kenya and S Somalia, south to Angola, NE Namibia, N and E Botswana and South Africa (Transvaal, Orange Free State, Natal, E Cape); local and confined to the vicinity of water. Common to very abundant in W African wetlands north of 9–10°N from Senegal to L. Chad; scarce Liberia, Ivory Coast and near coast from S Nigeria to Angola. In E and S, common to abundant by lakes and marshes from Uganda and W Kenya to E and S Zaïre, Zambia, Malaŵi, Zimbabwe, NW Botswana and Transvaal. Occasional birds in Egypt Dec-Jan are prob- ably wintering. On autumn passage in N Africa, small numbers in Morocco; scarce Algeria, Tunisia and Libya; locally common to abundant in Egypt (especially Nile valley, also Eastern and Western Desert oases). In spring, locally common to abundant throughout N Africa, including Saharan oases. South of Sahara, locally abundant on southward passage in Senegal, Chad, and N, central and W Sudan (L. Nasser, Nile valley, Darfur); rare N Somalia; frequent to common locally in Uganda and Kenya. On northward passage, locally very abundant in Kenya, Ethiopia and N Somalia, and also N Nigeria, N Burkina Faso and N Senegal.

Numbers breeding in Britain and N Europe since the late 1960s appear to have been limited mainly by West African rainfall (Peach *et al.* 1991).

Acrocephalus schoenobaenus

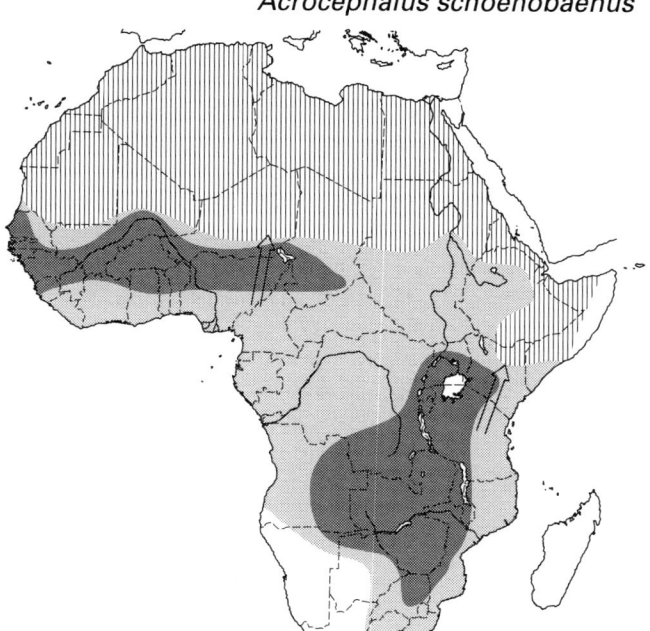

Description. ADULT ♂: top of head, mantle and scapulars olive-brown with broad blackish brown streaking; back, rump and uppertail-coverts uniform tawny. Tail dark brown with pale edging. Supercilium broad and creamy; lores dusky; cheeks and ear-coverts yellowish brown; chin, throat and centre of belly creamy white; breast and undertail-coverts pale yellowish buff, flanks same but darker. Flight feathers blackish brown, edged pale brown, broadly so on tertials; primary coverts blackish with narrow pale edging; rest of upperwing-coverts same with broad pale brown edges and tips; axillaries and underwing-coverts whitish with dusky centres. Wing bluntly pointed; P8 longest (rarely = P9), P7 1·5–4 shorter, P6 4–6, P5 7–9, P1 14–17; P9 0–2, usually longer than P7; P8 emarginated (**A**). Bill blackish brown, base of lower mandible yellowish flesh; eyes brown; legs and feet greyish brown. Sexes alike. SIZE: (10 ♂♂, 10 ♀♀, Europe and Siberia) wing, ♂ 66–70 (67·5), ♀ 64–71 (65·9); tail (**B**), ♂ 49–52 (50·8), ♀ 46–54 (48·8); bill, ♂ 14–16 (15·2), ♀ 15–16 (15·2); tarsus, ♂ 20–23 (21·6), ♀ 20–23 (21·3). A cline of increasing size from southwest to northeast: wing, ♂ (n = 10, Britain) 62–65, (n = 6, Siberia and Arabia) 68–70. WEIGHT: unsexed, Senegal (n = 79, Oct–Nov) 8·7–12 (12·5), (n = 129, Dec–Feb) 9–13·2 (10·3), (n = 126, Mar–Apr) 8·9–17 (11·0); N Ghana (n = 167, Nov–Dec) 9·1–13·2 (10·5); N Nigeria (n = 300, Mar–May) 8·2–20; Uganda (n = 100, Dec–Feb) 8·8–13·8 (11·5), (n = 138, Mar–May) 9·6–18·6 (12·5); Kenya (n = 189, Oct–Feb) 8·9–14·3 (11·5), (n = 1515, Mar–May) 8–21·5 (11·9); Zambia (n = 59, Nov–Apr) 9–15·4 (11·2); Transvaal, South Africa (n = 95, Dec–Feb) 8·1–14·7 (11·5), (n = 64, Mar–Apr) 10·5–14·7 (12·0); Morocco (n = 91, Mar–May) 7·8–12·2 (9·5).

Moults completely in Africa, in N tropics Sept–Nov or further south Jan–Mar.

IMMATURE: 1st winter bird similar to adult but generally more yellowish brown, breast usually speckled dark brown; crown sometimes with pale central stripe.

Field Characters. A small streaked warbler with a conspicuous broad creamy supercilium, dark crown, and almost plain rufous-tinged rump contrasting with well-marked olive-brown back. Buffish white below, some young birds with a few spots on breast. Legs greyish. Calls and song distinctive. For separation in N Africa from much scarcer Moustached Warbler *A. melanopogon* and Aquatic Warbler *A. paludicola* see under those species.

Voice. Tape-recorded (86, 88, 91, B, C, F). Song, loud and vigorous; a sustained medley of chattering, buzzing phrases, interrupted by clear musical passages, 'chit-chit-tuk-tuk-tuk-chit-terwee-terwee-tit-tit-twee-twee-tit-it-it-it-cherwee . . .' (Simms 1985); includes mimicry. Faster, less rhythmic, notes less grating than in Eurasian Reed-Warbler *A. scirpaceus*; more buzzy, less fluent and liquid than in Marsh Warbler *A. palustris*. Contact-alarm notes, a buzzing 'tsrrr' and a short frequently repeated 'tuc'.

General Habits. In winter inhabits mainly waterside and emergent vegetation, often in more open situations and more seasonal habitats than other members of the genus. Reeds, sedges, reedmace *Typha* and sometimes papyrus, by lakes, dams and rivers; wet grassland; rice fields; rank herbage and bushes or low tree canopy beside water; small acacias and scrub along ditches. Often up to 2500 m in E Africa. On passage occurs also in drier bush and thicket, cultivation and even desert scrub.

Mainly solitary in Africa, but also commonly in loose groups. Occurs locally in large wintering or passage concentrations. Thus, scores per day caught for ringing in N Senegal (Djoudj) in Feb–Mar, N Nigeria (L. Chad) in Mar–Apr and Kenya (L. Nakuru) in Apr to early May. Frequently remains attached to winter territory for several weeks. In Uganda, of 15 birds caught and then released 40 km away, 4 returned to capture site within 7 days (Pearson 1972). Return of ringed birds in successive years demonstrated in Senegal, Nigeria, Zimbabwe and Zambia; in Uganda, half of wintering birds ringed one year estimated to have returned to same place the next (Pearson 1972). Typically less sedentary in wintering area, however, than other *Acrocephalus* spp., moving in response to changing water levels and temporary food sources such as lakefly swarms. Forages low down, often in thick cover close to water where detected by buzzy call. Hops through vegetation and on ground, and sidles inquisitively up plant stems (Cramp 1992). Gleans mainly

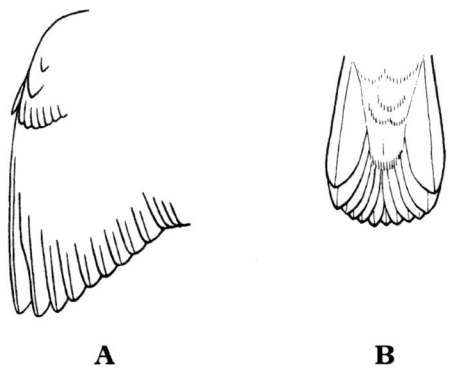

A **B**

slow-moving or stationary prey from aquatic plants, foliage and twigs of bushes and low canopy, and occasionally from mud surface; sometimes leaps to catch flying insects. Flight typically low and direct over short distances, with tail spread and slightly depressed (Cramp 1992).

Most birds from N and NW Europe move south to southwest in autumn, but some recoveries from Finland and Poland are south southeast. Autumn scarcity in Morocco–Libya suggests that most birds fly direct to tropics from Europe. Most British birds probably begin long-distance flights from north of 45°N (Cramp 1992). Movement through Africa widespread and protracted, with passage in Morocco Sept–Nov (peak mid-Sept to mid-Oct), Chad Sept–Dec. Occurs Senegal from late Aug, but main arrival lower Senegal R. not until Nov. Arrives N Nigeria and N Ghana from Sept, mainly in Oct. Passage through Egypt and N and central Sudan mid-Aug to mid-Oct, arrival Ethiopia from mid-Sept. Earliest birds reach E African equator in late Sept and Zambia in mid-Oct, but main arrival in Zambia and further south is in Nov–Dec; light passage near E African equator suggests that low latitudes are largely overflown. Arrival of wintering birds in Kenya, Uganda and Rwanda is mainly in Dec–Jan. Most birds wintering in W Africa and E Africa south to c. 10°S moult during autumn in N tropics, but most wintering in southern Africa moult there after arrival.

Some birds remain in southern Africa until Apr (late date, Zimbabwe, 5 May: C.J. Pollard *in* Tree 1995), but most wintering birds leave Uganda and Kenya between late Mar and mid-Apr. Heavy passage occurs through Kenya mid-Apr to early May, Ethiopia mid to late Apr and N Somalia May; this suggests that many birds use a more easterly route in spring than in autumn. In W Africa, most leave lower Senegal R. by early Mar, although passage continues to early May, with occasional records to June. Main passage L. Chad (Nigeria) is from late Mar to early May. Spring passage in N Africa is heavier and more widespread than in autumn; first birds in late Feb, but main movement Morocco–Algeria late Mar to late Apr, Egypt late Mar to early May. High pre-migratory weights and ringing recovery patterns suggest that long unbroken flights are commonplace between widely separated staging areas. Spring birds approximately double the typical winter weight have been noted in Nigeria, Uganda and Kenya. At a Kenya site, birds gained on average 0·64 g per day in late April (Pearson *et al.* 1979).

Birds ringed Europe recovered N Africa as follows: Britain to Morocco (21, autumn and spring) and Algeria (2, spring); France (2), Holland (1) and Belgium (1) to Morocco; Sweden (1) and Czechoslovakia (1) to Tunisia; and Finland to Egypt (2, autumn). Birds ringed Morocco (spring) recovered Britain (4) and Portugal (1), and 1 ringed Tunisia (spring) recovered Russia (29°E). South of Sahara, >80 recoveries in Senegal from Britain (>50), France/Channel Is (23), Belgium (5), Holland (1) and Sweden (1); and 23 in Mali from Britain (9), Holland (2), Germany (1), Norway (1), Sweden (9) and Finland (1). Other recoveries Britain to Sierra Leone (1, Mar), Liberia (1, Feb), Ghana (3, Dec, Feb, May) and Burkina Faso (2, Oct, Apr); from France to Ivory Coast (1, Mar); and from Finland to Congo (1, Dec), Zaïre (2, Jan, Feb) and Zambia (1). Birds ringed Senegal (1) and Mali (1) recovered Britain, 1 ringed Nigeria (Apr) recovered Russia (43°E), 1 Sudan (Sept) recovered Turkey and 1 Kenya (Apr) recovered Russia (48°E). It appears that most birds wintering from Senegal to Ghana are from NW Europe, while those wintering further east and in southern Africa are from east of the Baltic.

Food. Insects and their larvae; also spiders, small slugs and worms and some plant material. Spring birds fattening at L. Chad ate mainly midges (Diptera), also Odonata, Hemiptera, Neuroptera, Lepidoptera, Hymenoptera, Coleoptera and spiders; also flowers and fruit of *Salvadora*, sedge seeds and other plant items (Fry *et al.* 1970).

References
Cramp, S. (1992).
Fry, C. H. *et al.* (1970).
Peach, W. *et al.* (1991).
Pearson, D. J. *et al.* (1979).

Plate 7
(Opp. p. 112)

Acrocephalus scirpaceus (Hermann). Eurasian Reed-Warbler. Rousserolle effarvate.

Turdus scirpaceus Hermann, 1804. Obs. Zool., p. 202; Alsace.

Forms a superspecies with *A. baeticatus*.

Range and Status. NW Africa; Europe north to S Baltic and east to Volga river; Turkey, Levant, Caucasus and W Iran; and N Caspian area to N Afghanistan and Kazakhstan to c. 85°E; winters in Africa.

Palearctic migrant. A locally common breeding summer visitor Morocco (NW coast, NE coast, Middle Atlas, oases near Marrakech); N Algeria (coastal marshes near Oran, Algiers and El Kala, Boughzoul,

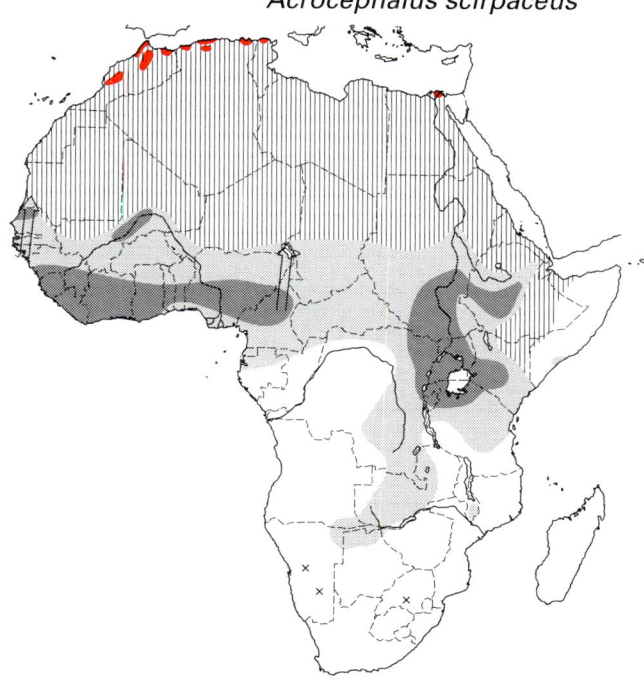

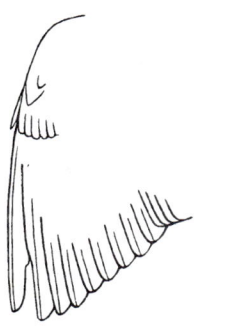

and oases of Bordj Saada and perhaps El Goléa); and N Tunisia (L. Ichkeul). Has also bred recently Egypt (L. Manzala and probably other lakes in Nile delta). Common and widespread on autumn passage in Morocco and Egypt, but scarce elsewhere in N Africa. On spring passage, common to abundant in Morocco, N Algeria and Egypt, and frequent in Libya. Occurs on both migrations in Saharan oases.

Winters from Senegal, Mali (north to Niger delta), SW Niger, Nigeria, S Chad, Sudan (north to Khartoum) and Ethiopia (in and west of Rift, north to c. 12°N), to Gulf of Guinea coast, Cameroon, N Gabon, N and E Zaïre, Rwanda, Burundi, Uganda, Kenya (except arid NE), S Somalia (Webi Shabeelle) and Tanzania (south to Iringa), with small numbers to S Zaïre (Shaba), central and S Zambia, S Malaŵi, N Botswana (Okavango) and NE Namibia (Kavango, Bushmanland); rarely to central Namibia (Windhoek, Keetmanshoop) and South Africa (Transvaal: Klip R., Vanderbijlpark, Olifantsvlei). A few remain north of Sahara in Morocco and Egypt. In W Africa, abundant in Niger delta at Lakes Horo and Aougoundou (Lamarche 1981), and N Senegal at Djoudj, but winters mainly in more humid savannas south of 10°N; locally abundant S Guinea, common N Liberia, Ivory Coast and S Ghana, widespread S Nigeria and Cameroon. Locally common to abundant E Africa, with major wintering concentrations in Ethiopian Rift valley, along Nile, and near shores of L. Victoria and lakes of W Ugandan Rift. Common to abundant locally on autumn passage in Mauritania, Senegal, S Mali, central Chad and Sudan (Darfur, Nile valley, Red Sea coast). On spring passage, common to very abundant locally Senegal, S Mali, Burkina Faso, N Nigeria, central Chad, central Sudan, central Kenya and Ethiopia.

Description. *A. s. scirpaceus* (Hermann): NW Africa and Europe east to Volga R.; occurs throughout African non-breeding range of species. ADULT ♂: entire upperparts warm olive-brown, rump and uppertail-coverts more rufous. Tail dark brown. Short whitish supercilium narrow and rather poorly defined; lores and ear-coverts brown; underparts whitish, suffused with pale buff on breast and undertail-coverts, and darker rusty-buff on flanks. Flight feathers dark brown with paler fringes, broader on tertials; upperwing-coverts like mantle; axillaries and underwing-coverts buffish white. Wing bluntly pointed; P8 longest, P7 1–3 shorter, P6 3–6, P5 6–8, P1 13–16; P9 1–3, between P6 and P8; P10 minute; P8 emarginated; notch on inner web of P9 at or below tip of P2 on closed wing (**A**). Bill dark brown, lower mandible mainly yellowish or flesh; eyes medium brown to bright reddish brown; legs and feet dark greyish brown to pale horn. Sexes alike. SIZE: (10 ♂♂, 10 ♀♀, Europe) wing, ♂ 64–70 (67·5), ♀ 63–67 (64·8); tail, ♂ 52–55 (54·0), ♀ 50–54 (51·7); bill, ♂ 16–18 (17·6), ♀ 15–18 (17·3); tarsus, ♂ 21–24 (22·8), ♀ 21–23 (21·9). W European birds are smaller than those of central and E Europe: wing, ♂ (n = 22, Britain) 62–68 (65·4), (n = 5, Iberia) 63–65 (64·8). WEIGHT: unsexed, Morocco (n = 145, autumn) 8–16·6 (11·9), (n = 15, Mar–May) 8·1–10·3 (9·0); Mauritania (n = 13, autumn) 8·7–13 (10·4); Senegal (n = 56, Oct–Nov) 8·3–12·9 (10·2), (n = 104, Dec–Feb) 7·6–12·3 (9·6), (n = 146, Apr) 7·4–15·7 (10·1); N Ghana (n = 36, Nov-Dec) 8·2–13·8 (9·4); N Nigeria (n = 30, autumn) 8·6–12·5 (10·4), (n = 608, spring) 8·0–22·3 (11.9).

Moults completely in Africa, in N tropics in Oct–Nov or further south in Dec–Mar.

IMMATURE: juvenile brighter and rustier than adult. 1st winter slightly rustier than adult above and deeper buff below; flight feathers fresh-looking until Oct; legs dark greyish brown, usually tinged greenish; notch on P9 inner web may fall as high as tip of P4.

NESTLING: naked; mouth orange-yellow; gape flanges yellow; 2 black tongue spots.

A. s. fuscus (Hemprich and Ehrenberg): E Mediterranean, and Caspian area to Kazakhstan; winters NE, E and central Africa, west to Zaïre and south to Zambia. Paler, more olive-brown above and whiter below than nominate race, usually tinged greyish on head and hindneck. Warm tinge confined to rump and uppertail-coverts; sometimes lacking, and some birds wholly greyish olive-brown above. Slightly larger: ♂ (n = 10) wing, 66–72 (68·3); bill, 17–19 (18·1). WEIGHT: unsexed, Kenya and Uganda (n = 96, Nov–Feb) 9·1–15·5 (10·9), (n = 173, Mar–May) 9·2–17·1 (11·6); Ethiopia (n = 202, Jan–Mar, mainly this race) 8·1–17·6 (11·0), (n = 234, Apr–May) 8·9–23·5 (12·8).

Field Characters. A small unstreaked warbler with a longish bill, olive-brown or warm brown above and whitish below with buff flanks. Creamy supercilium moderately pronounced, extending to just behind eye. Crown usually looks peaked. Very like Marsh Warbler *A. palustris*, and in the absence of diagnostic vocalizations often impossible to distinguish from this species in the field. Upperparts of nominate race warmer and darker than in Marsh, but those of paler, more olive *fuscus* sometimes identical to Marsh. Underparts of Eurasian Reed-Warbler lack yellowish tinge of Marsh, and those of *fuscus* are especially white, but this is often difficult to appreciate. Head typically more pointed looking than in Marsh, with flatter forehead and slightly longer, narrower bill. In the hand, adults and many 1st year birds are separable from Marsh using position of ninth primary notch (**A**), and bill measurements are useful (Schulze-Hagen and Barthel 1993). Slightly larger than African Reed-Warbler *A. baeticatus*, with more pointed wing (**A**). Paler and less brightly coloured, especially in E and central Africa. Smaller and paler than Lesser Swamp-Warbler *A. gracilirostris*, with more pointed wing and browner legs.

Voice. Tape-recorded (57, 91 – 62, 73, B, C, F). Song, a strident rhythmic series of grating and squeaky notes, each repeated 2–3 times: 'chup-chup-chir-chir-churric-churric-whit-whit-whit-churric-churric . . .', in bursts lasting up to 20 s (occasionally 60 s); slower, less chattering than Sedge Warbler *A. schoenobaenus*; much less mimetic than Marsh Warbler, more nasal and grating, with slower, steadier tempo. In winter quarters, song like that on breeding grounds but quieter. Contact calls, a low grating 'churr' and a conversational 'chrr-chrr'; threat or alarm call a harsher longer 'tcharr'; a sharp 'chek' is less common.

General Habits. Breeds mainly in *Phragmites* reedbeds, but often forages in adjacent herbaceous vegetation, scrub, low trees or willows. In winter quarters and on migration frequents tall grass, reeds and green thicket; often along river courses and near lake shores, but also away from water in secondary bush, acacia and *Lantana* scrub, forest edge and garden hedges; commonly prefers drier sites to marshy habitat in same area. Ranges typically below 1400 m, but up to 2000 m in Ethiopia and Kenya.

Usually solitary or in loose groups; sometimes in considerable wintering concentrations, with 'tens per ha' Guinea (Brosset 1984) and *c.* 40 per ha locally at L. Victoria, Uganda (Pearson 1972). Wintering birds use overlapping feeding territories which they occupy for up to 5 months. A high proportion returned to same Uganda site in successive years. In Malaŵi, birds retrapped up to 3 years after ringing (Hanmer 1982a). Forages mainly at middle height in reeds and within bushes and thickets where difficult to observe. Takes prey from vegetation, sometimes by leaping or hovering, but also flycatches and snaps aerial insects (Cramp 1992). Climbs easily up and down reeds with feet on separate stems. Frequently adopts horizontal posture. Flight fast and confident, usually low and direct (Cramp 1992). Sings commonly in Africa from Nov onwards, typically throughout the morning.

A migratory divide in Europe. Birds from W Europe, Scandinavia, Poland and Germany head southwest through Iberia and Morocco; those from E central Europe head southeast through the Balkans and Egypt (Dowsett-Lemaire and Dowsett 1987a). N African passage lasts from early Aug to early Nov, but limited practically to Morocco (mainly Sept–Oct) and Egypt (late Aug to mid-Oct), with few birds in Algeria. Passage through Senegal mainly mid-Sept to mid-Nov, with arrival in Mali in Sept–Oct and strong movement in Chad in late Sept to early Nov. Birds reach W Africa south of 10°N in Nov, most already in fresh plumage, having presumably moulted in stopover areas nearer S edge of Sahara. Departs from wintering sites mainly late Mar to early Apr. Strong passage Senegal Mar–May or early June, with peak early Apr, and in N Nigeria late Feb to mid-May (peak late Apr), with many birds commonly 50–80% above mean winter weight (Aidley and Wilkinson 1987a); also Chad with heavy passage by late Mar, latest bird 2 June. Spring passage conspicuous across whole of N Africa with birds commonly at Saharan oases, indicating migration on a broader front than in autumn. Occurs from Morocco to Tunisia mainly Apr–May, and in Egypt from late Mar to mid-May.

Many birds ringed Europe recovered N Africa, and these emphasize the migratory divide. From Britain, over 90 to Morocco (mainly autumn) and 2 to Algeria (spring); from France, Holland, Belgium and Germany, over 70 to Morocco and 4 to Algeria; 2 from Denmark, 1 from Sweden and 1 from Poland to Morocco; and 1 from Finland to Algeria. From E Austria, by contrast, 1 to E Libya and 14 to Egypt, while 1 from Hungary to Egypt, and birds from Czechoslovakia to both Egypt (1) and Algeria (1). Birds ringed Morocco found Britain (7) and Austria (1).

Also many recoveries W Africa, as follows: Britain to Mauritania (7), Senegal (29, mostly autumn and spring), Guinea-Bissau (1), Mali (1), Ivory Coast (1), Burkina Faso (1) and Ghana (1); France and Belgium to Senegal (8), Mali (4) Ivory Coast (3) and Nigeria (1); Spain and Portugal to Senegal (9); Holland, Germany and Switzerland to Mauritania (1), Sierra Leone (1), Ghana (2) and Togo (2); Sweden to Liberia (1), Mali (1), Ivory Coast (1) and Ghana (1); Czechoslovakia to Guinea (1); and Austria to Nigeria (1) and Cameroon (1). Also, 1 bird ringed Senegal and 2 ringed Guinea-Bissau recovered in Britain. Thus, birds from W and N Europe appear to winter across W Africa east to Nigeria. Regular passage through NE Nigeria and Chad involves birds of unknown origin, probably from N Mediterranean or central and E Europe.

In E Africa, *fuscus* winters alongside equally long-winged birds apparently of nominate race; *fuscus* predominates in most areas. Passage on Sudan coast is in late Aug to mid-Oct, but arrival in Uganda, Kenya and further south is not until late Oct–Jan. Main E African movement is apparently through Uganda and W Kenya. Many birds reaching E African equator have already moulted, but about half of those wintering in Uganda, and most in southern Africa moult after arrival. Moult in Uganda is mainly Dec–Mar, individuals taking *c.* 70 days (Pearson 1973). Birds remain in wintering sites in Zambia until late Mar to early Apr, and in Uganda to early to mid-Apr. Northward passage conspicuous in central Kenya in Apr to early May, and Ethiopia in mid-

Apr to late May. Only 1 E African recovery from Palearctic ringing (Cyprus to N Sudan). E African ringed birds recovered abroad: ringed Sudan (Oct) to Iran (1), Ethiopia (Apr) to Kuwait (1), and Kenya (Apr) to Saudi Arabia (1) and Russia (Astrakhan) (1).

Food. Insects and their larvae, spiders, some small snails, occasionally fruit. In Europe, mainly beetles, Hymenoptera, Diptera and spiders. At L. Chad, Nigeria, flowers and fruits of *Salvadora persica* eaten before migration, also sedge seeds and cortical strips; insects mainly flies then beetles, some bugs and hymenopterans, and a few damselflies, butterflies and neuropterans (Fry *et al.* 1970). In Zambia, mosquitos are main prey, also winged termites when available (Cramp 1992). In Uganda, only insects eaten even when birds fattening in spring (Fogden 1972).

Breeding Habits. Well studied Europe. Monogamous; sometimes bigamous; often paired with same mate in successive years (Catchpole 1974). Loosely colonial, often with neighbourhood groups of <10 to >100 pairs. Strictly territorial. Territories small, mean values from European studies 306–439 m^2 (Cramp 1992); often >1000 pairs of birds per km^2 of reedbed. From arrival ♂ sings in long bursts, typically perched near top of reed with throat ruffled and conspicuous. Uses 2–3 main song posts. After pairing bird sings less, mainly at dawn and in evening, but singing may increase again before 2nd brood (Cramp 1992). Territorial intrusion elicits short bursts from ♂, and sometimes also from ♀.

NEST: deep cup of dried grasses, reed leaves and sedge, woven onto 2–10 reed stems 30–110 cm above water; outer diam. *c.* 80 mm, inner diam. *c.* 50, cup depth *c.* 55; built by ♀ in about 7 days.

EGGS: NW Africa 3–4 (Europe 3–5). Pale green or greenish white, densely blotched or mottled greenish brown, grey and olive. SIZE: 16–21 × 12–15.

LAYING DATES: NW Africa, late Apr to early June; Egypt, May. Often double-brooded in Europe.

INCUBATION: by both sexes, mainly by ♀, starting once clutch complete. Period: 9–12 days.

DEVELOPMENT AND CARE OF YOUNG: young fed by both parents; remain in nest 9–13 days.

BREEDING SUCCESS/SURVIVAL: of 2700 eggs, Britain, 69% hatched and 89% of these produced fledged young (overall success 61%). 70% of all clutches and 85% of 2nd brood clutches produced fledged young. Of 496 clutches, Germany, overall success 55% (Cramp 1992). Legs of 2 spring migrants (out of 391) at L. Chad (Nigeria) grossly damaged by mite *Knemidokoptes jamaicensis*, debilitating if not fatal (Fry *et al.* 1969).

References
Brown, P. E. and Davies, M. G. (1949).
Cramp, S. (1992).
Dowsett-Lemaire, F. and Dowsett, R. J. (1987a).
Pearson, D. J. (1973).

Acrocephalus baeticatus (Vieillot). African Reed-Warbler. Rousserolle africaine.

Sylvia baeticatus Vieillot, 1817. Nouv. Dict. d'Hist. Nat. 11, p. 195; South Africa (= Knysna).

Plate 7
(Opp. p. 112)

Forms a superspecies with *A. scirpaceus*.

Range and Status. Africa; also locally on W Arabian coast (Yanbu).

Resident and intra-African migrant. Distribution local and fragmented in W, N central and NE Africa: Senegal (L. Guier, Casamance river); Gambia; S Mali (Bamako); NE Niger (L. Arragai); Nigeria (Kano, L. Chad, Gashaka-Gumti, Ibadan, Serti); S Cameroon (Akonolinga); N Gabon; S Chad (N'Djamena, Sarh); W Sudan (Darfur); W Ethiopia (Gambela); central Ethiopia (Rift valley); and S Somalia (Dannow); also in mangrove in Sudan (Suakin), Eritrea and N Somalia. Locally frequent to common, but numbers generally small. In E Africa widespread in S Sudan (Nile valley); W and S Uganda; W, central and SE Kenya; N, W and central Tanzania; Rwanda, Burundi and E Zaïre borders; N Zaïre (Buta); S Zaïre (Kasai); and in Tanzanian mangroves (Pemba and Zanzibar to Mikandani). Widespread and more numerous in southern Africa, where locally common from SW Angola (Moçâmedes, N Huila), Zambia (except NW), Malaŵi and Mozambique (Tete and Zambesia) south to Cape. Apparently resident in many areas, but in South Africa mainly a breeding visitor, also probably in W Sudan (Darfur) and on Sudan coast. Common summer visitor to lowlands and foothills of W Lesotho (P. Osborne and B. Tigar, pers. comm.).

Acrocephalus baeticatus

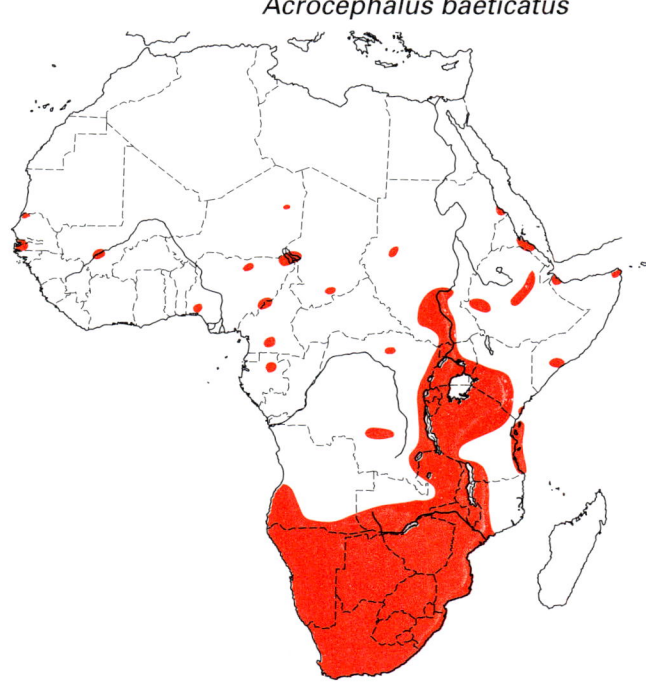

A

Description. *A. b. baeticatus* (Vieillot): South Africa (S and E Cape to Natal and Transvaal), NE and N Botswana, locally in Zimbabwe; non-breeding visitor to lowlands of S Mozambique. ADULT ♂: entire upperparts warm brown, more rufous on rump and uppertail-coverts. Indistinct supercilium buffish; lores and ear-coverts brown. Tail dark brown. Underparts whitish, suffused with buff on breast and undertail-coverts, rich buff on flanks. Flight feathers dark brown with paler warm brown fringes, especially on tertials. Upperwing-coverts like mantle; underwing-coverts and axillaries pale buff. Wing rounded; P6 to P8 longest, P5 1–2 shorter, P1 9–11; P9 4–5, between P3 and P5; P10 very small; P7 to P8 emarginated (**A**). Bill dark brown above, yellowish below; eyes light brown; legs and feet dark horn to flesh-brown. Sexes alike. SIZE: (10 ♂♂, 10 ♀♀) wing, ♂ 58–62 (60·2), ♀ 58–61 (59·5); tail, ♂ 41–47 (44·5), ♀ 40–45 (42·7); bill, ♂ 16–19 (17·3), ♀ 16–18 (16·8); tarsus, ♂ 22–24 (22·6), ♀ 21–24 (22·3). WEIGHT: unsexed (n = 387, Transvaal, South Africa) 7–15.5 (10·2).

IMMATURE: juvenile somewhat brighter and rustier than adult.

NESTLING: unknown.

A. b. hallae White: South Africa (N and W Cape) to W and SW Botswana, Namibia, SW Zambia and (probably this race) SW Angola. Slightly paler than nominate race; more olive-brown above and whiter below.

A. b. cinnamomeus Reichenow: S and E Zambia, Malawi and Mozambique to E Africa, S Somalia, W and central Ethiopia and S Sudan, and west to Chad, Nigeria, Niger, and S Senegal. Brighter, more rusty above than nominate race. Smaller: wing, ♂ (n = 10, central and E Africa) 53–56 (54·6). P9 averages shorter, between P1 and P4; P1 only 6–9 shorter than wing-point. Southern African birds ('*fraterculus*') a little darker. L. Chad birds ('*hopsoni*') browner than typical *cinnamomeus*; wing longer, 56–59 (n = 5, unsexed), with P1 9–11 shorter than wing-point, and P9 between P4 and P5. WEIGHT: unsexed (n = 8, Kenya) 7·4–8·6 (8·1).

A. b. guiersi Colston and Morel: N Senegal; restricted to *Typha*. Colder brown above than *cinnamomeus*, and greyer, especially on head; whiter below; slightly larger: wing, ♂ (n = 10) 56–60 (57·9).

A. b. suahelicus Grote: coastal Tanzania and probably coastal Mozambique and Natal; associated with mangroves. Slightly darker rufous brown above than nominate race, and deeper buff on breast and flanks. Similar in size: wing, ♂ (n = 9) 56–61 (59·4).

A. b. avicenniae Ash, Pearson, Nikolaus and Colston: coastal Sudan, Eritrea and NW Somalia; associated with mangroves. Paler olive-brown above, tinged rusty on rump and uppertail-coverts; creamy white below with little buff on flanks. Wing longer than in *cinnamomeus*: ♂ (n = 10) 57–61 (59·0), P9 usually between P4 and P5, occasionally longer than P5. WEIGHT: unsexed (n = 20, Sudan) 7–10·5 (8·0).

TAXONOMIC NOTE: Clancey (1975a) recognized 2 species of Afrotropical reed-warblers, *A. cinnamomeus* (central, E and N tropical Africa) and *A. baeticatus* (S areas and E African coast), based on the supposed separation by the small *cinnamomeus* of 2 *baeticatus suahelicus* populations (Upper Zambezi and E coast). But existence of *suahelicus* on the Zambezi is doubtful; moreover, some W African populations, clearly allied to *cinnamomeus*, have the slightly more pointed wing of southern *baeticatus*. We therefore submerge *cinnamomeus* in *A. baeticatus*. Clancey (1994) argues that the large Upper Zambezi birds are non-breeding migrants, but still proposes on other grounds that *A. cinnamomeus*, including *suahelicus*, is specifically separable from *A. baeticatus*.

Affinities of some W African and Red Sea populations immediately south of the Sahara and on the Red Sea coast have been debated (Smith 1964, Ash *et al.* 1989b). On the basis of voice, coloration, bill and foot structure, and nest structure, *A. baeticatus* appears closely allied to *A. scirpaceus* (not to the Palearctic *A. dumetorum* as proposed by Fry *et al.* 1974). Songs of *baeticatus* and *scirpaceus* are so similar that the 2 could be regarded as conspecific (Dowsett-Lemaire and Dowsett 1987b). However, in view of their different migratory behaviour and associated wing structure difference, we retain them as separate members of a superspecies.

Field Characters. A small unstreaked *Acrocephalus*, warm brown or olive-brown above, with buff flanks, whitish throat and underparts, and a narrow but noticeable pale supercilium. Very similar to Eurasian Reed-Warbler *A. scirpaceus*, but wing shorter and rounder (cf. *A. scirpaceus*, **A**); tips of exposed primaries form only about one-fifth of the length of the closed wing (between a quarter and a third in Eurasian Reed). Central and E African birds distinguished by more brightly coloured plumage, but those of W Africa, the Red Sea coast and southern Africa resemble Eurasian Reed-Warbler more closely. Much smaller, paler and sleeker than Lesser Swamp-Warbler *A. gracilirostris* and Greater Swamp-Warbler *A. rufescens*, with browner, more slender legs.

Voice. Tape-recorded (11, 22, 38, 57, B, C, F, CHA, LEM, MOR). Song, a sustained rhythmic series of repeated grating and squeaky notes: 'chir-chir-churric-churric- . . .'; more subdued in moulting birds in non-breeding season. Songs of *cinnamomeus*, nominate *baeticatus* and *A. scirpaceus* all the same (Dowsett-Lemaire and Dowsett 1987b). Contact-alarm note, a short 'churr'.

General Habits. Breeds in wet or moist habitat: edges of *Phragmites* reedbeds, seasonal floodplains, beds of reedmace *Typha* and sedge *Cyperus* over mud or water, edges of papyrus swamp, rank vegetation along ditches, stands of tall grass and scrub along river banks; in Senegambia, in sugar cane fields and along irrigation canals; sometimes in moist thickets away from water. Occurs in drier places when not breeding, e.g. riverside thickets of acacia, *Lantana*, *Rhus* in S Zambia (Dowsett-Lemaire and Dowsett 1987b). NE and E African coastal races occur mainly in mangroves, though birds on Pemba I. noted singing 10 m up in large trees. Occurs up to 1900 m in Kenya, but further south not above 1500 m.

Solitary, in pairs or small groups. Secretive, foraging mainly at mid-height or low down among reeds, sedges and papyrus, or within bushes, hedges and low tree canopy where it is difficult to see. Carriage typically horizontal. Gleans from vegetation and snaps aerial prey. Tends to feed nearer bases of reeds than Lesser Swamp-Warbler. Flies reluctantly, usually low and for short distances.

Some populations migratory. Most South African birds move north in May, returning in Aug. At Darfur, arrives to breed in Sept, while leaves Sudan coast in Aug after breeding, moulting and putting on fat. In Zambia, non-breeding birds present Ndola May–Nov, and Livingstone July–Sept. In Zimbabwe a non-resident form, *hallae*?, has been detected at Umwindisidale and L. Chivero in Sept–Dec (Tree 1995a). Movements of *cinnamomeus* probably mainly local, but birds occasionally attracted to lights at Ngulia, Kenya.

Food. Insects.

Breeding Habits. Monogamous. Territorial, but several pairs may form neighbourhood group. Often feeds in dry herbaceous growth and low trees outside territory. ♂ typically sings from concealment.

NEST: deep cup of grass and reeds lined with fine grass; bound to 4–6 upright stems of reeds, sedges or grasses; in South Africa to drooping branches of willow (Maclean 1993); usually at a height of 0·6–1·5 m; on Sudan coast, a deep cup woven to fork in slender mangrove shoot, c. 20 cm above mud.

EGGS: 2–3, rarely 4, laid at daily intervals; South Africa mean (23 clutches) 2·7, Senegal mean (6 clutches) 2·7. Pale greenish or bluish white, blotched and mottled olive-brown and slate-grey. SIZE: (n = 82, South Africa) 16–19 × 12–15 (17·6 × 13·2).

LAYING DATES: mainly during rains: South Africa, Sept–Feb; Botswana, Oct–Mar; Zimbabwe, Nov–Mar; Zambia, Feb; E Africa: Region C, Jan; Senegal, June–July; W Sudan (Darfur) Sept–Oct.

INCUBATION: period 12–14 days (South Africa).

DEVELOPMENT AND CARE OF YOUNG: young remain in nest about 14 days.

References
Ash, J. S. *et al.* (1989b).
Clancey, P. A. (1975a).
Dowsett-Lemaire, F. and Dowsett, R. J. (1987b).

Acrocephalus palustris (Bechstein). Marsh Warbler. Rousserolle verderolle.

Plate 7 (Opp. p. 112)

Sylvia palustris Bechstein, 1798. Latham's Allg. Uebersicht der Vögel, iii, p. 545; Thuringia, Germany.

Range and Status. NW and central Europe, and east between about 40° and 60°N to Urals and W Caspian; winters in Africa.

Palearctic migrant. Winters mainly SE Africa: locally in Kenya east of the highlands (Isiolo and Meru south to Bissel, Amboseli and Tsavo); and from S Tanzania (Iringa), SE Zaïre (S Shaba), Zambia (except NW) and central and S Malaŵi to Zimbabwe, Mozambique (north to Zambesia), N and E Botswana, Namibia (35 km east of Windhoek), Swaziland and South Africa (Transvaal, Orange Free State, Natal, E and S Cape west to Sedgefield and twice in SW Cape: Martin and Martin 1993, Fraser and McMahon 1995). Locally common to abundant in much of wintering range, but scarce at N limits in Kenya in dry years. On southward passage, frequent to common in E Egypt (Nile corridor, Red Sea coast); locally very abundant NE Sudan (Red Sea coast, Red Sea hills, small oases); common central Sudan along Nile but scarce in W (Darfur); frequent to common in Ethiopia in and west of Rift. Common to very abundant central and SE Kenya, and locally N Kenya east of L. Turkana; uncommon S Somalia (Juba valley), and rare Kenya west of Rift, Uganda (Kampala, Masindi) and Rwanda. Pronounced southward migration Zambia, but passage status in central and S Tanzania and N Mozambique poorly known. On north-

Acrocephalus palustris

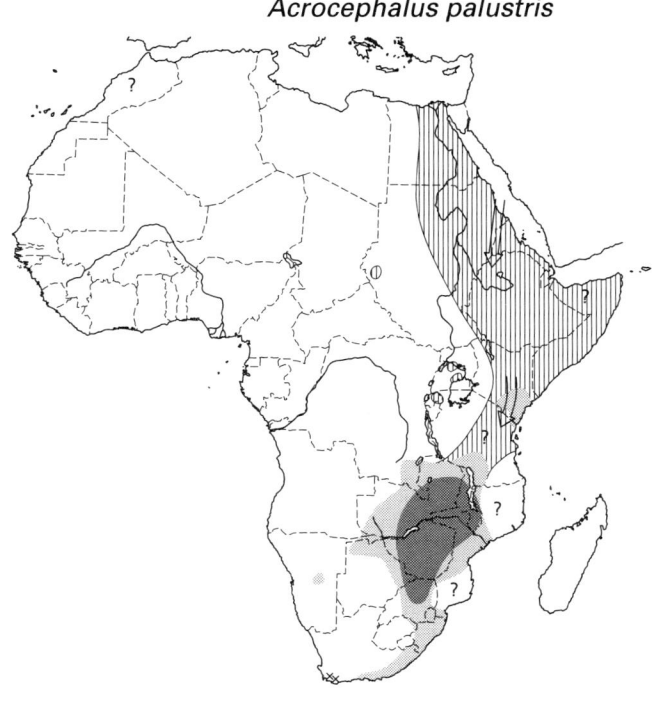

ward migration, frequent to common in NE Tanzania and SE Kenya including coastal areas; locally common N Somalia, but few records Ethiopia (Rift valley, Jijiga) and rare Sudan (once Red Sea); scarce Egypt (6 records Bahig, once Western Desert). Possibly vagrant Morocco (9 records, Feb (1), Apr (6) and Sept (2) but none convincingly substantiated).

Landfalls of tens or hundreds of thousands on S Oman coast, 2nd week May, evidently from Somalia, have traditionally provided harvest for local people. Newly-arrived migrants are netted in bushes using nylon fish-nets. Probably tens of thousands, perhaps 100,000, killed each May; sometimes sold in fish markets. Methods no longer being traditional, the practice could soon pose a major threat to the Marsh Warbler population concerned.

Description. ADULT ♂: entire upperparts olive-brown, tinged greenish. Tail blackish brown. Short supercilium and narrow eye-ring creamy; lores, cheeks and ear-coverts olive-brown; below, breast, flanks and undertail-coverts suffused yellowish buff, chin, throat and centre of belly creamy white. Flight feathers blackish brown, fringed olive or greenish olive, broadly on tertials; wing-coverts dark brown broadly fringed and tipped olive-brown; underwing-coverts and axillaries pale yellowish buff. Wing bluntly pointed; P8 longest (sometimes = P9), P7 1–3 shorter, P6 3–6, P5 6–8, P1 14–17; P9 0–2, usually longer than P7; P10 minute; P8 emarginated; notch on inner web of P9 falls between tips of P2 and P5 in closed wing. Bill dark horn, lower mandible paler yellowish or flesh; eyes olive-brown; legs and feet usually pinkish brown or pale horn. Sexes alike. SIZE: (10 ♂♂, 10 ♀♀, Europe and Russia) wing, ♂ 66–70 (69·0), ♀ 67–70 (68·3); tail, ♂ 49–56 (53·5), ♀ 51–57 (53·3); bill, ♂ 15–17 (16·0), ♀ 15–17 (16·4); tarsus, ♂ 21–23 (22·2), ♀ 21–24 (22·5). WEIGHT: unsexed, E Sudan (n = 220, Aug–Sept) 7·5–13 (10·1); Kenya (n = 1200, Nov–Jan) 9–17·4 (11·5), (n = 77, Apr) 9·1–13·2 (11·2); Somalia (n = 4, May) 14·9–16·5 (16·0); Zambia (n = 159, Nov–Feb) 8·6–13·7 (11·0), (n = 33, Mar–Apr) 10·1–15·8 (11·9); Malawi (n = 53, Nov–Feb) 9·6–14·2 (11·4), (n = 30, Mar–Apr) 10·1–15·5 (11·9); Zimbabwe (n = 111, Dec–Feb) 8·7–13·9 (11·2), (n = 18, Mar–Apr) 9·8–13 (11·6); Transvaal, South Africa (n = 47, Dec–Feb) 10–15 (12·1), (n = 12, Mar) 11·6–16 (12·3).

Complete moult in Africa Jan–Mar; post-nuptial body moult in NE Africa Sept–Oct.

IMMATURE: 1st winter bird before moulting is like adult but often tinged bronzy or warm brown on wings, rump and uppertail-coverts, sometimes on whole upperparts; legs usually darker, often tinged greenish.

Field Characters. A small unstreaked warbler, olive-brown above and creamy below with paler throat often contrasting. Short supercilium and eye-ring well-marked. Legs typically pale in adults, but often dark brown in young birds. Separation from Eurasian Reed-Warbler *A. scirpaceus* is usually difficult in the field; paler looking, more olive or greenish above than nominate (European) Reed-Warbler; closely similar to eastern Reed-Warbler (*fuscus*) but usually cooler-toned on rump. Differs from both races of Eurasian Reed in having yellow-buff tinge to underparts, and brighter edges to fresh tertials and larger wing-coverts, contrasting more with dark feather-centres, but these details often difficult to appreciate. Young Marsh may have distinctive warm golden or bronzy tinge above in Aug–Sept, especially on wings. Head rather less attenuated-looking than in Eurasian Reed, crown more rounded and bill slightly shorter and broader. In the hand, separable from adult (but not usually from 1st winter) Eurasian Reed using position of notch on P9 (**A**) (cf. *A. scirpaceus*, **A**); bill shape also useful (see Schülze-Hagen and Barthel 1993). Less warmly coloured than African Reed-Warbler *A. baeticatus*, and longer-winged, with closed wing-point forming almost one-third of total wing-length. Voice useful in identification; full song distinctive.

A

Voice. Tape-recorded (75, 86, 88, 91, F). Adult song, a prolonged, rapidly delivered warbling chatter, including clear liquid trills, sweet high-pitched notes and a characteristic harsh 'zi-chay'; combines a remarkable variety of mimicked sounds into lively fluent sequences (Dowsett-Lemaire 1981). Song may be almost entirely imitative: at least 99 European bird species and 113 African species copied as well as mechanical sounds, repertoire being learned by young birds during first 8 months (Dowsett-Lemaire 1979, 1981). Lacks sustained underlying rhythm of Reed-Warbler song; sweeter and mostly higher pitched. Juvenile song a subdued chatter, lacking mimicry. Contact-alarm calls, a slightly nasal grating 'cherr' or 't-cherrr', not unlike call of Sedge Warbler *A. schoenobaenus* but less buzzy, and a short dry *Hippolais*-like 'tek' or 'tchrek'; also a chattering alarm, 'chre-chre-chre-chre- . . .'.

General Habits. Winter habitat is leafy thickets with herbaceous undergrowth, often along river banks and overgrown stream beds, below 1600 m. Occurs locally in gardens and hedges, bushy forest edge and patches of tall grass. Passage habitat more varied, including dry thornbush and coastal mangroves. Ranges up to 2200 m on migration.

Mainly solitary in Africa, though commonly in loose groups and at times scores together on migration. Tends to form local wintering pockets, within which individuals establish loose territories. Up to 25 singing birds per ha in Zambia, where overlapping territories of colour-ringed birds averaged *c.* 800 m^2, and 47% of ringed birds returned essentially to previous year's territory (Kelsey 1989, Cramp 1992). 1 bird retrapped 6 years after initial capture Malaŵi (Hanmer 1982a). Site fidelity rarely recorded in Kenya passage birds. Often concealed within thick cover, but not especially shy. Readily flies fast and low to next bush when disturbed. Gait and carriage much like Eurasian Reed-Warbler, but movements rather more deliberate, less agile. Feeds mainly by gleaning, typically in herbaceous shrub layer, but also in low tree canopy; in Zambia, forages among coarse grasses, picking insects from underside of blades. Usually takes fewer flying insects than Eurasian Reed-Warbler (Cramp 1992). In Tsavo, Kenya, 91% of food taken from leaves, 4% from air, 3% from ground (Lack 1985). Song infrequent and subdued on passage Kenya Nov–Dec, but full imitative song heard increasingly in wintering areas Feb to early Apr. Juvenile song noted only up to Feb (Dowsett-Lemaire 1981).

Populations breeding in NW and central Europe migrate initially southeast. A few birds enter Africa through Egypt, but main influx occurs across Red Sea coast of Sudan and probably N Ethiopia. Birds follow narrow route through W (and perhaps central) Ethiopia, and later through Kenya east of the highlands, W and E breeding populations apparently migrating together. Recorded in Egypt from 7 July, but main passage from mid-Aug to early Sept. Earliest bird Sudan coast 27 July, and main migration there from mid-Aug to mid-Sept, with adults about 10 days before young birds. Recorded in Ethiopia from mid-Aug to mid-Dec, but birds apparently stay there for about 3 months, and earliest Kenya records are late Oct. Passage peaks in SE Kenya in mid-Nov to mid-Dec and continues to early Jan. Often the most abundant migrant in Kibwezi/Tsavo area where over 47,000 ringed at Ngulia 1969–1991 after attraction to lights (Backhurst and Pearson 1993). Arrives in Zambia and Malaŵi from last week Nov, but mainly in Dec, and passage continues to Jan.

Adults moult body plumage in NE Africa before arrival in Kenya. All birds moult completely in winter quarters, usually Jan–Mar, finishing shortly before migration; individual duration *c.* 60 days (Zambia). Main departure southern Africa late Mar to mid-Apr (latest Malaŵi 23 Apr, Zambia 29 Apr). Spring passage through E Africa less heavy and protracted than autumn, but with more birds along Tanzania and Kenya coasts; migration through SE Kenya peaks in mid-Apr (latest 5 May). Absent from Sudan in spring, and few records from central Ethiopia. Main route probably lies further east, through N Somalia, where passage recorded late Apr to mid-May and some birds very fat. Records and ringing recoveries suggest more easterly migration in spring than autumn across Arabia. Heavy passage of fat birds occurs through Oman early to mid-May (J. S. Ash, pers. comm., C. H. Fry, pers. comm.). However, main staging areas NE Africa remain unknown.

Birds ringed Europe recovered/controlled as follows: Belgium to N Sudan (1, Sept), Kenya (6, Nov–Dec) and Mozambique (1, Dec); Germany to Egypt (1, Sept), N Sudan (1, Sept) and Kenya (3, Nov–Dec); Czech Republic to Kenya (4, Nov–Dec) and Malaŵi (3, Dec, Feb); Sweden to Kenya (1, Nov); Finland to South Africa (1, Transvaal, Dec); and Cyprus to N Sudan (1, Aug). A bird ringed Oman (May) was recovered Kenya the following Nov. Birds ringed Kenya Nov–Dec recovered/controlled as follows: 1 France, 1 Belgium, 1 Switzerland, 1 Denmark, 3 Germany, 2 Czech Republic, 1 Slovenia, 3 Russia (2 Dagestan, 1 St Petersburg), 4 Saudi Arabia (1 spring, 3 autumn), 5 Oman (4 spring, 1 autumn), 2 Malaŵi (Dec), 2 Zambia (Jan), 1 Zimbabwe (Dec) and 1 SE Zaïre (Apr).

Food. Mainly insects, spiders, some snails, rarely plant material. No detailed information for Africa, but eats mosquitos and termites in Zambia.

References
Cramp, S. (1992).
Dowsett-Lemaire, F. and Dowsett, R. J. (1987a).
Kelsey, M. G. (1989).
Pearson, D. J. (1982).

Plate 7

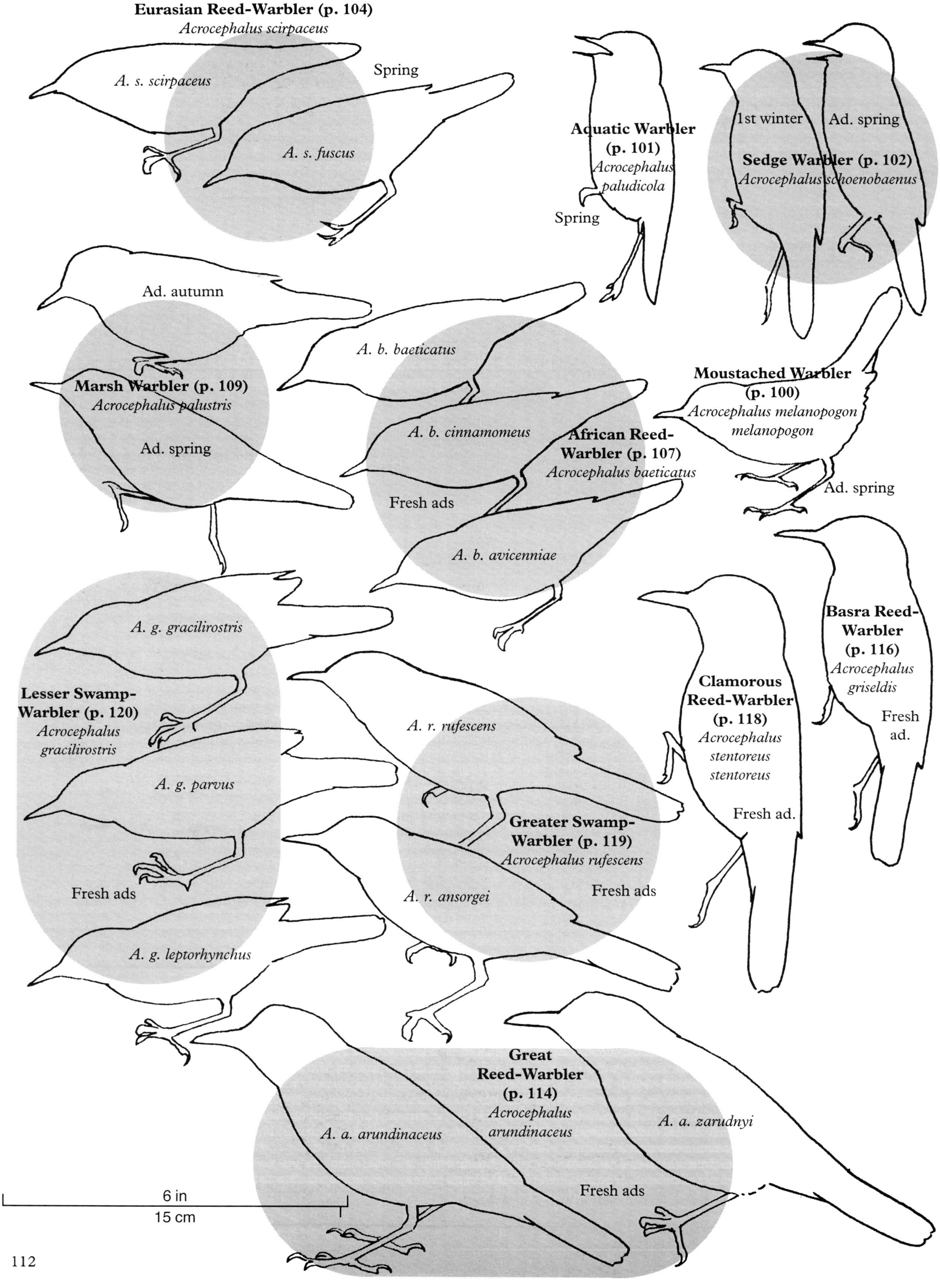

Eurasian Reed-Warbler (p. 104)
Acrocephalus scirpaceus

A. s. scirpaceus — Spring

A. s. fuscus

Aquatic Warbler (p. 101)
Acrocephalus paludicola
Spring

Sedge Warbler (p. 102)
Acrocephalus schoenobaenus
1st winter — Ad. spring

Marsh Warbler (p. 109)
Acrocephalus palustris
Ad. autumn
Ad. spring

A. b. baeticatus

A. b. cinnamomeus

African Reed-Warbler (p. 107)
Acrocephalus baeticatus
Fresh ads

A. b. avicenniae

Moustached Warbler (p. 100)
Acrocephalus melanopogon melanopogon
Ad. spring

Lesser Swamp-Warbler (p. 120)
Acrocephalus gracilirostris

A. g. gracilirostris

A. g. parvus

Fresh ads

A. r. rufescens

Greater Swamp-Warbler (p. 119)
Acrocephalus rufescens
Fresh ads

A. r. ansorgei

A. g. leptorhynchus

Clamorous Reed-Warbler (p. 118)
Acrocephalus stentoreus stentoreus
Fresh ad.

Basra Reed-Warbler (p. 116)
Acrocephalus griseldis
Fresh ad.

Great Reed-Warbler (p. 114)
Acrocephalus arundinaceus

A. a. arundinaceus

A. a. zarudnyi
Fresh ads

6 in
15 cm

Plate 8

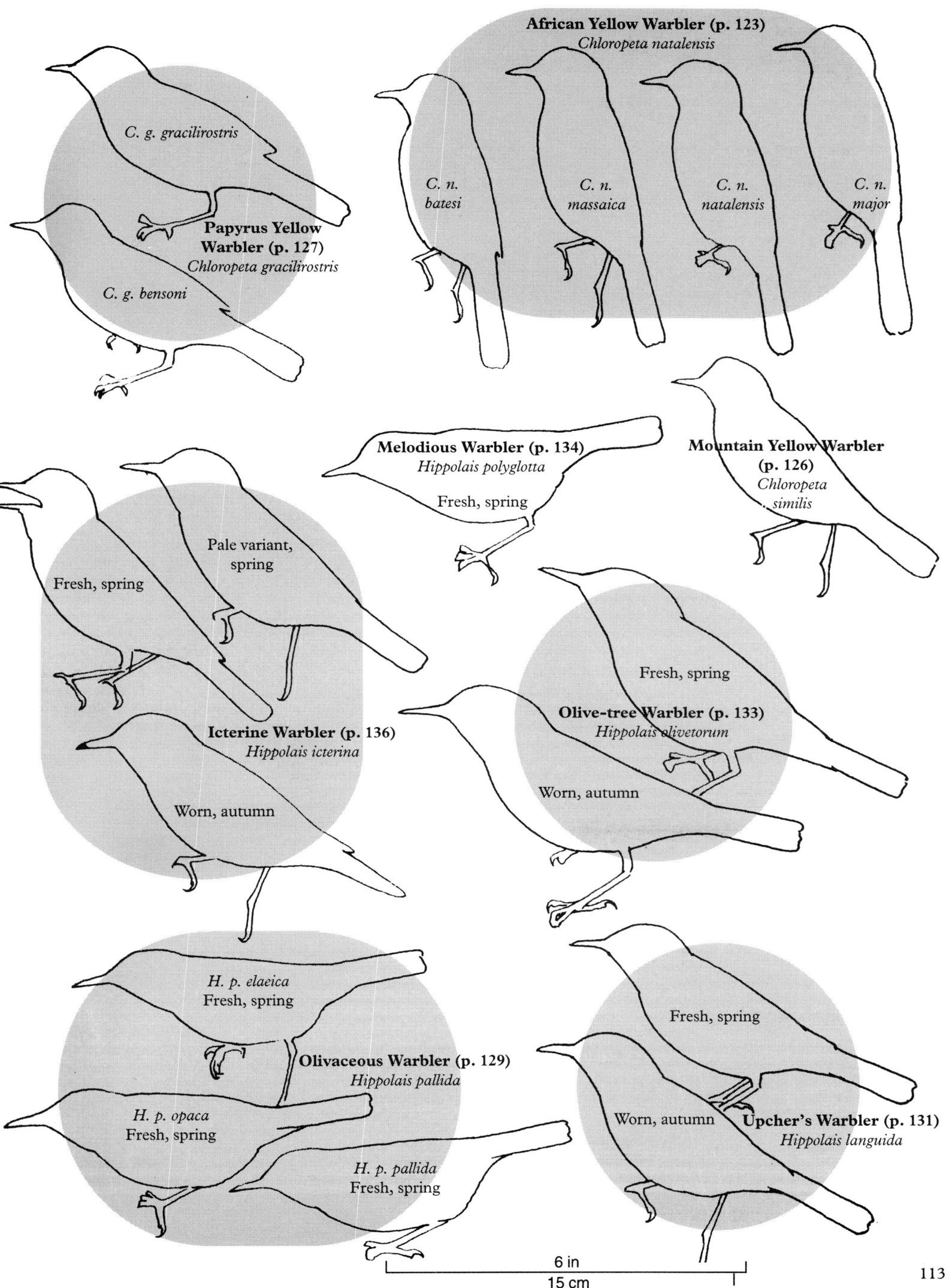

Plate 7
(Opp. p. 112)

Acrocephalus arundinaceus (Linnaeus). Great Reed-Warbler. Rousserolle turdoïde.

Turdus arundinaceus Linnaeus, 1758. Syst. Nat. (10th ed.), p. 170; Danzig.

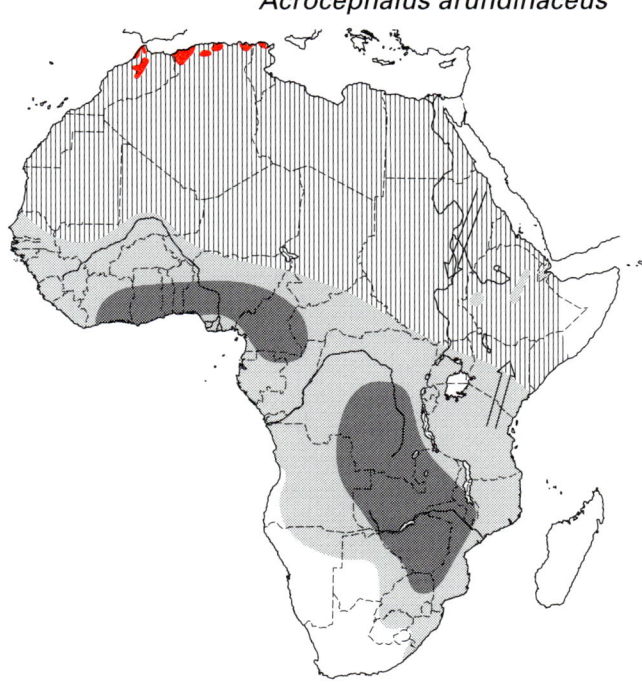

Acrocephalus arundinaceus

Range and Status. NW Africa, continental Europe north to *c.* 60°N, Turkey to Caucasus and N Iran, S Russia to *c.* 85°E, and E Caspian and Kazakhstan to NW China; winters Africa.

Palearctic migrant. Breeding summer visitor locally and in small numbers in N Morocco (coastal marshes near Larache and Tetouan, Fes, Middle Atlas up to 1600 m), N Algeria (Oranie, Reghaïa, Dar el Beida, Boughzoul, near El Kala), and N Tunisia (L. Ichkeul). Scarce autumn passage migrant in Morocco and Algeria, rare in Tunisia and Libya, but frequent Egypt. On spring passage frequent to common along whole N African coast, occurring also in N African oases.

Winters south of Sahara, from Senegal, S Mali, SW Niger, Nigeria, S Chad, Central African Republic, Zaïre, S and W Uganda and W and S Kenya south to Angola, NE Namibia, N and E Botswana and E South Africa (Transvaal, Orange Free State, Natal, E Cape); 2 records W Lesotho, Jan 1989 (P. Osborne and B. Tigar, pers. comm.); also W and central Ethiopia to *c.* 12°N, and Djibouti. Main wintering areas in W Africa are south of 9–10°N; locally common from Ivory Coast to Cameroon, though scarce further west. Uncommon in Sahel and soudanian belts, but a few winter on lower Senegal delta and on inland delta of R. Niger in Mali. Small numbers winter in Ethiopia and E Africa where decidedly local. Many winter south of equator, where species is common through much of range from Zaïre south to Transvaal. On southward passage and stopovers, frequent in Mali, N Ghana and N Nigeria, but rare in Senegal and Gambia; few records in Chad; locally common NE, central and W Sudan (Red Sea coast, Nile valley, Darfur), but scarce E Africa. On northward passage, locally common to abundant E Tanzania, central and E Kenya and Ethiopia, and a few records Somalia, but scarce in Sudan; small spring numbers W African Sahel, but commoner in Sierra Leone and Senegal.

Description. *A. a. arundinaceus* (Linnaeus): NW Africa, Europe to W Siberia, and Turkey to N Iran; occurs throughout African non-breeding range of species. ADULT ♂: entire upperparts warm olive-brown, more strongly rufous on rump and uppertail-coverts. Tail-feathers brown with narrow pale tips. Supercilium well-marked, creamy; narrow eye-ring creamy white; lores, cheeks and ear-coverts brown. Underparts suffused with warm buff, especially on flanks; chin, throat and belly whiter, lower throat with pale brown shaft-streaks. Flight feathers dark brown with narrow paler edges; upperwing-coverts same as upperparts; axillaries and underwing-coverts creamy to warm buff. Wing-tip bluntly pointed; P8 longest (occasionally = P9), P7 2–4 shorter, P6 5–8, P5 8–13, P1 21–26; P9 0–2, usually longer than P7; P10 minute; P8 emarginated, sometimes also tip of P7. Bill dark brown above, pinkish flesh below with dark tip; eyes olive-brown to reddish brown; legs and feet pale brown or yellowish brown. ADULT ♀: whitish more restricted below, and throat-streaks less pronounced. SIZE: (10 ♂♂, 10 ♀♀, Europe and W Africa) wing, ♂ 93–99 (95·5), ♀ 89–96 (91·3); tail, ♂ 74–81 (77·5), ♀ 73–80 (76·1); bill, ♂ 23–25 (24·1), ♀ 20–24 (22·9); tarsus, ♂ 29–31 (30·0), ♀ 27–29 (28·4). WEIGHT: unsexed, Morocco (n = 12, Mar–May) 22–31 (27·2); N Ghana (n = 38, Oct–Dec) 24–36 (29·3); N Nigeria (n = 6, Oct–Dec) 27–33 (29·5), (n = 5, Mar–May) 29–38 (34·7); Ethiopia (this race and *zarudnyi*), (n = 22, Jan–Feb) 24–40 (31·1), (n = 49, Apr–May) 24–47 (31·6); Kenya/Uganda (this race and *zarudnyi*), (n = 39, Nov–Feb) 23–38 (28·6), (n = 103, Mar–May) 21–52 (31·6); Malaŵi (n = 230, Nov–Feb) 23–37 (29), (n = 36, Mar–Apr) 23–38 (32); South Africa (Transvaal). n = 34, Dec–Feb) 25·1–36 (30·3), (n = 16, Mar) 28·5–39 (33·8).

Moults completely in Africa, most birds in N tropics in Oct–Nov, but some in southern Africa in Jan–Mar.

IMMATURE: juvenile and 1st winter like adult but rustier above, strongly suffused with buff below and lacking throat-streaks; eyes dark brown; legs usually darker. Flight feather-tips unworn in Aug–Sept.

NESTLING: hatches naked. Mouth yellow or orange-yellow; gape flanges pale yellow; 2 black tongue spots.

A. a. zarudnyi Hartert: Khirgiz steppes and E Caspian to NW China border; winters Africa south to Natal and west to Zaïre. Paler throughout, more olive above than nominate race, sometimes with greyish tinge; whiter below, with buff more confined to sides.

Field Characters. A large, heavy-looking warbler with longish tail and thick thrush-like bill. Plain olive-brown above, usually warmly tinged, and pale below with buff flanks and faint streaks on throat. Conspicuous creamy supercilium. Legs usually pale brown. Larger, more heavily built than Basra Reed-Warbler *A. griseldis*, with much stouter bill (see generic diagnosis) and warmer, buffier coloration. Similar in size and colour to

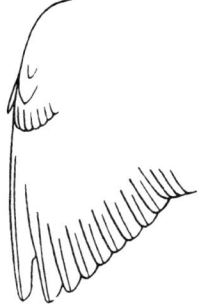

Clamorous Reed-Warbler *A. stentoreus* but wing longer and more pointed (**A**) (cf. *A. stentoreus*, **A**); tail squarer, bill thicker and face pattern bolder. Song distinctive.

Voice. Tape-recorded (57, 75, 86, 88, 91, B, C, F, PEAR). Advertising song (see *Breeding Habits*) loud, far-carrying and rhythmic, a sequence of varied harsh grating and creaky notes, each characteristically repeated 3–4 times; 'krerk-krerk-krerk-kri-kri-kri-kruk-kruk-karra-karra-karra-...' lasting 3–20 s; of same general form as song of Eurasian Reed-Warbler *A. scirpaceus*, but much louder, more guttural. Bird sings for up to 20 min with pauses of only 1–3 s between sequences. Territorial song (see *Breeding Habits*) shorter and quieter; a few harsh grating notes lasting 1–2 s. Song in winter quarters like advertising song but rather less powerful, usually sustained for a few s only. Contact-alarm calls, a harsh 'krack', a chattering alarm 'kaa-kaa-kaaa' and a low croaking 'crr-crr-' given near nest and with fledged young.

General Habits. Breeds in beds of reed *Phragmites* or sometimes reedmace *Typha*, standing in shallow fresh or brackish water; prefers strong tall reeds with thick stems, especially near open water. In winter inhabits reeds, *Typha*, beds of *Cyperus*, tall grass, green thickets, crops such as maize and sugar cane; shrubbery and garden hedges; often on river and lake edges, but also commonly away from water. Occurs up to 1800 m (Kenya).

In winter quarters solitary and territorial; in Gabon, territories are contiguous, one of *c.* 600 m^2, 2 others of tens of m^2 (Brosset 1971). Birds return to same site each winter (demonstrated in Ghana, Zaïre, Uganda, Zambia, Malaŵi and Zimbabwe); at a site in Malaŵi ringed birds remained up to 3 months and returned for up to 7 winters after first capture (Hanmer 1986). Bolder, less secretive than Eurasian Reed-Warbler or Basra Reed-Warbler, less confined to thickets. Feeds low down, hopping heavily near base of vegetation; may leap to catch flying insects and sometimes pounces to ground; in breeding season takes dragonfly nymphs, fish fry and newly emergent adult insects from or near water surface (Cramp 1992). Also forages in hedges and small trees such as acacias, sometimes perching on open sprays. Flies more freely than most *Acrocephalus* spp., low and direct, with longish tail slightly spread. Sings loudly and persistently in winter quarters from late Nov to early Apr.

Ringing recoveries indicate that most birds from European breeding grounds initially head south in autumn, but central Asian *zarudnyi* must migrate southwest. The paucity of autumn records N Africa suggests most birds fly non-stop to tropics. Southward passage through Egypt is in late Aug to late Sept, on Sudan coast in late Aug to mid-Oct, and Darfur and central Sudan in Sept to early Nov. NW African breeding sites are vacated by Oct. Main arrival in Mali is in Sept–Oct, in N Nigeria Oct, and in Ethiopia late Sept–Oct. Most birds, destined to go further south, stay for about 2 months in the N tropics, and many moult completely there. Thus, adult and 1st year birds wintering in S Ghana, S Nigeria, central Africa, Kenya, Uganda, Zambia and Malaŵi arrive in late Nov–Dec in fresh plumage. In contrast, most birds wintering in Zimbabwe and South Africa moult after arrival (Pearson 1975). The sparseness of autumn passage through S Sudan, Uganda and Kenya points to E Africa being overflown by birds wintering to the south.

The main departure from southern Africa is in Mar to early Apr. Many birds apparently return by a more easterly route, with strong passage through Kenya (Apr) and Ethiopia (mid-Apr to early May). W African wintering sites are vacated mainly by late Mar to early Apr, but some birds remain in Nigeria and Ghana until May, and in Senegal to late May or early June. Birds seem to embark on the desert crossing from low latitudes. Some in Kenya in Apr are double lean winter weight. Spring passage through N Africa, from mid-Mar to mid- or late May, is much heavier and more widespread than in autumn.

Ringing recoveries N Africa from Spain to Morocco (1, Apr), Holland to Algeria (1, Nov), and Austria to Algeria (2, spring, and 1, autumn) and Tunisia (1, Apr). Birds ringed Algeria (May) found Belgium (2), ringed Tunisia (May) found Italy (2), Yugoslavia (1) and Czechoslovakia (1). Recoveries south of Sahara from France to Ivory Coast (1, Mar); Germany to Ghana (1, Jan); Austria to Ghana (1, Feb), Benin (1, Jan) and Nigeria (1, Jan); and Sweden to Ivory Coast (1, Nov) and Chad (1, Dec). Thus, W, central and S European birds appear to winter in W Africa. Nominate birds occurring in E and central Africa (typically outnumbering *zarudnyi*) are presumably from east of the Baltic.

Food. Mainly insects and their larvae, especially beetles and Odonata; some spiders, snails, small vertebrates; fruit and berries outside breeding season. In Zambia, Lepidoptera, grasshoppers, dragonflies, mayflies, winged termites, small frogs; *Salvadora persica* berries in spring, L. Chad (Nigeria).

Breeding Habits. Well-studied in Europe. Usually monogamous, but frequently polygynous. In Poland 12–14% of ♂♂ are polygynous and in Sweden *c.* 40% are polygynous with 2–4 ♀♀ each (Bensch and Hasselquist 1991). Solitary and territorial but tends to form neighbourhood groups near good feeding sites; in Germany, territories of 400–500 m^2, or smaller if contiguous (Cramp 1992). ♀♀ of polygynous ♂♂ live in same territory. Territory established by ♂, which arrives first and seeks to attract mate with loud advertising song from top of tall reed or from bush or tree. ♂ may use several song posts on edge of territory. Short territorial song is used by paired and unpaired ♂♂ in territorial defence (Cramp 1992). ♀ usually appears in ♂'s territory about 10 days after first song, and nest-building then starts within a week. During courtship ♂ uses territorial song and the 2 birds climb and fly about excitedly; ♂ sometimes displays in flight, with slow wing-beats and glides low over reeds; shivers wings, and may fly up to hover over ♀ with tail spread and crown-feathers ruffled before dropping to reeds (Cramp 1992).

NEST: deep cylindrical cup or coarsely woven reed and plant stems and fibres, attached to several growing reed stems, 0·5–2 m above water; outer diam. c. 90, inner diam. c. 70, cup depth c. 70; built by ♀.

EGGS: usually 4 in NW Africa (4–6 in Europe); bluish or greenish, boldly marked with umber and black. SIZE: (n = 100, Europe) 21–25 × 15–17 (22·6 × 16·2).

LAYING DATES: NW Africa May–June; usually single-brooded.

INCUBATION: by ♀, starting once clutch is complete. Period: 14–15 days.

DEVELOPMENT AND CARE OF YOUNG: young cared for and fed by both parents; brooded by ♀ for first 6–7 days; remain in nest 12–14 days; become independent 12–14 days after fledging.

BREEDING SUCCESS/SURVIVAL: in 390 clutches, Germany, 60% of eggs hatched and 73% of these fledged: overall success 44%, with average 2·0 young fledged per nest (Beier 1981).

References
Bensch, S. and Hasselquist, D. (1991).
Cramp, S. (1992).
de Roo, A. and Deheegher, J. (1969).
Pearson, D. J. (1975).

Plate 7
(Opp. p. 112)

Acrocephalus griseldis (Hartlaub). Basra Reed-Warbler. Rousserolle d'Irak.

Calamoherpe griseldis Hartlaub, 1981. Abh. Nat. Ver. Bremen 12, p. 7; Nguru, Kilosa, Tanganyika Territory.

Acrocephalus griseldis

Range and Status. Iraq on Euphrates R. and Tigris R.; winters in Africa.

Palearctic visitor. Migrates to eastern Africa, but distribution there poorly known. Winters regularly in SE Kenya (N Uaso Nyiro and middle and lower Tana to coast, Tsavo and Mtito Andei; frequent to locally common, often abundant on lower Tana flood-plain) and S Malaŵi (lower Shire valley, locally common); probably also S Somalia (lower Webi Shabeelle), NE Tanzania (south to Dar es Salaam and inland to Kilosa) and SE Tanzania (around Mikindani). Other Jan–Feb records from central Kenya (Naivasha), central Mozambique (lower Zambezi), Malaŵi (twice Lilongwe) and South Africa (once Transvaal, twice Natal: Hockey 1995). On southward passage, regular in NE Sudan (frequent Red Sea coast, once Atbara), Ethiopian Rift valley (locally common), central and SE Kenya (locally common, 935 ringed Ngulia 1969–1991: Backhurst and Pearson 1993) and E Tanzania; recorded also S Sudan (Juba, Aug), W Ethiopia (Gambela, Oct), S Somalia (lower Juba R., Nov) and Uganda (Kampala, Serere, Nov). Scarce on northward migration, with a few late Mar–Apr records from coastal Tanzania, coastal and SE Kenya and central Kenya (once Baringo).

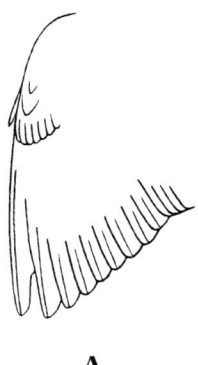

A

Description. ADULT ♂: entire upperparts including wing-coverts cold olive-brown. Tail blackish brown. Distinct supercilium and narrow eye-ring whitish; lores, cheeks and ear-coverts olive-brown. Underparts white, sides of breast and flanks suffused buff. Flight feathers blackish brown, edged olivaceous; axillaries and underwing-coverts pale buff. Wing bluntly pointed; P8 longest, P7 1–2 shorter, P6 4–6, P5 7–9, P1 17–20; P9 0·5–1·5, between P7 and P8; P10 minute; P8 emarginated (**A**). Bill dark brown above, mainly pinkish below; eyes olive-brown to reddish brown; legs and feet dark grey or brownish grey. Sexes alike. SIZE: (5 ♂♂, 4 ♀♀ Iraq and E Africa): wing, ♂ 82–85 (83·0), ♀ 78–84 (81·0); tail, ♂ 62–67 (64·6), ♀ 58–64 (62·0); bill, ♂ 20–24 (22·4), ♀ 22–23 (22·3); tarsus, ♂ 25–26 (25·8), ♀ 24–26 (24·8). WEIGHT: unsexed, Kenya (n = 440, Nov–Jan) 13–24 (16·7), (n = 4, Apr) 20·1–29·3 (23·4); Malaŵi (n = 149, Nov–Apr) 15–25·5 (18·5).

Moults completely in Africa, most in Ethiopia Sept–Nov, some in southern Africa Jan–Mar.

IMMATURE: 1st winter bird in Sept greyer than adult with buff fringes to feathers of upperparts and pale buff edges to wing-coverts. Legs dark grey.

TAXONOMIC NOTE: often regarded as a race of *A. arundinaceus*, but treated by several recent authors (e.g. Voous 1960, Pearson and Backhurst 1976, Benson and Benson 1977, Clancey 1980b) as a full species. It is quite different from *arundinaceus* in size, bill shape, leg colour and voice, and its adult and immature plumages are distinct (Pearson and Backhurst 1988); moreover the 2 taxa have different winter habitat preferences.

Field Characters. A slim, medium-sized warbler, cold olive-brown above and white below, with distinctive long slender bill; whitish supercilium narrow but conspicuous. Smaller than Great Reed-Warbler, with much thinner bill (see generic diagnosis), and also differs in lack of throat-streaks, grey leg colour and voice. Larger than Eurasian Reed-Warbler *A. scirpaceus* and Marsh Warbler *A. palustris*, with longer bill, more prominent supercilium and longer, darker-looking tail; call harsher.

Voice. Tape-recorded (McVIC, PEAR). Song in Africa of similar form to that of Eurasian Reed-Warbler and Great Reed-Warbler, but tempo much slower, less rhythmic; a loose sequence of low throaty notes, 'chrk-chri-chrk-chchuk-chrk-churrik . . .', *c.* 1·5 notes per s, sometimes sustained for several min; lacks volume and vigorous grating quality of Great Reed-Warbler song. Contact call, a harsh 'chaarr', stronger and more nasal than similar call of Eurasian Reed-Warbler.

General Habits. Frequents dense green thickets and moist undergrowth, rank grass, sedges along ditches, and bushes over flood-pans; less partial than Great Reed-Warbler to more open tall grass and crops. The most numerous wintering *Acrocephalus* sp. along the lower Tana, Kenya, where it occurs in *Terminalia* thickets by water and in 1–2 m high saltbush *Sueda monoeca* over seasonal flood. Regular wintering sites are all in hot country below 1000 m.

Solitary, but often in small loose groups; 10–20 per ha in saltbush in the Tana delta (Pearson *et al.* 1978). Individual occupies territory in Kenya (also Malaŵi) for up to 3 months. Fidelity to wintering site shown by several birds ringed at Nchalo, Malaŵi, where 1 returned 8 years after first capture (Hanmer 1986). Secretive and shy; forages mainly within thick bushes, often low over water, hopping among leafy cover; sometimes takes insects from mud. Typical posture horizontal and elongated. Flight fluent; lighter, more agile than Great Reed-Warbler; usually low, diving quickly into cover. Quiet song or subsong frequently heard in Kenya Jan–Mar, throughout morning and in late afternoon.

Migrates south-southwest across Arabia from Iraq breeding grounds, and occurs on Sudan coast from late Aug to mid-Oct. Recorded Ethiopia from late Aug to early Dec (mainly Sept), but most birds must remain there about 3 months, since earliest record Kenya is 28 Oct, and main passage at Ngulia is from mid-Nov to early Jan. Reaches Kenya wintering sites from early Dec, Malaŵi from mid-Dec (earliest, 27 Nov). Most birds have complete autumn moult in Ethiopian stopover area, but some Ngulia passage birds and Malaŵi wintering arrivals are still unmoulted or incompletely moulted (Hanmer 1979, Pearson 1982). Wintering sites are occupied until late Mar to early Apr; latest Malaŵi bird, 12 Apr. Paucity of spring passage records suggests long unbroken return flights, and 2 Kenya birds had very high pre-migratory weights (Pearson *et al.* 1978). Latest Kenya record 18 Apr. Bird ringed Ethiopia (Dec) recovered Mozambique the following Feb; 1 ringed Kenya (Dec) recovered Saudi Arabia (Aug).

Food. Little information; mainly insects.

References
Hanmer, D. B. (1979).
Pearson, D. J. (1982).
Pearson, D. J. and Backhurst, G. C. (1988).

Acrocephalus stentoreus (Hemprich and Ehrenberg). Clamorous Reed-Warbler. Rousserolle stentor.

Curruca stentorea Hemprich and Ehrenberg, 1833. Symb. Phys. Avium, fol. bb; Damietta, Egypt.

Acrocephalus stentoreus

Range and Status. NE Africa, Levant, Jordan, Arabia, and Iran to Kazakhstan and Mongolia and to N India, Sri Lanka, S China, Philippines, Indonesia, New Guinea and Australia. African and S Asian birds mainly resident, but central Asian birds winter in Arabian Gulf and India, and some other populations partially migratory.

Resident Egypt: abundant Nile delta (tens of thousands of pairs), Nile valley south to Aswan, Suez canal area, Faiyum and Wadi el Rayan; probably also breeds Dakhla and Bahariya oases in Western Desert; spreading in Nile valley following completion of Aswan dam and creation of new lakes and reedbeds, and also in Wadi el Rayan since flooding and reedbed establishment in early 1980s. Also locally common, presumed resident, in mangroves on coasts of Sudan and Eritrea and islands off NW and possibly NE Somalia.

Density of 10–20 singing birds per ha, Nile delta.

Description. *A. s. stentoreus* (Hemprich and Ehrenberg): Egypt, Levant and Jordan. ADULT ♂: entire upperparts warm olive-brown, inclining to rufous on rump and uppertail-coverts. Tail brown. Supercilium whitish; lores, cheeks and ear-coverts olive-brown; chin and throat white, sometimes with a few dark shaft-streaks; rest of underparts suffused with warm buff. Flight feathers dark brown, edged paler; tertials and wing coverts warm olive-brown. Wing bluntly pointed; P7 and P8 longest, P6 1–2 shorter, P5 4–5, P4 6–8, P1 12–14; P9 4–6 shorter, usually between P5 and P6; P10 minute; P7 to P8 emarginated, sometimes also P6. Bill horn above, pinkish flesh tipped horn below; mouth bright orange-red; eyes pale brown; legs and feet dark grey to dark horn, soles pale greenish. Sexes alike. SIZE: (10 ♂♂, 10 ♀♀) wing, ♂ 80–84 (81·8), ♀ 74–80 (77·5); tail, ♂ 75–80 (77·5), ♀ 72–77 (75·0); bill, ♂ 25–27 (26·3), ♀ 25–26 (25·4); tarsus, ♂ 27–29 (27·7), ♀ 27–29 (27·8). WEIGHT: unsexed (n = 13, Israel, Sept–Apr) 21·5–28·5 (24·2).

Complete moult Sept–Oct.

IMMATURE: juvenile more rusty-buff than adult.

NESTLING: naked when hatched.

A. s. brunnescens (Jerdon): Arabia, and Iran to central Asia and India; birds of coastal Sudan, Eritrea and NW Somalia perhaps belong here. Less warmly tinged above, paler and whiter below. WEIGHT: unsexed (n = 14, Sudan, Aug–Sept) 21–29 (24·5).

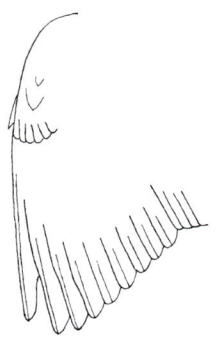

A

Field Characters. A large, rather slender, long-billed warbler, warm brown or olive-brown above and buff or whitish below, with greyish or dark brown legs. Close in size to Great Reed-Warbler *A. arundinaceus* but generally slimmer, with shorter, rounder wing (**A**); appears more laboured and long-tailed in flight. Face pattern less distinct than in Great Reed-Warbler, and usually lacks streaks on throat. Larger and more warmly coloured than Basra Reed-Warbler *A. griseldis*, with heavier bill (see generic diagnosis) and shorter wing. Song distinctive.

Voice. Tape-recorded (92, B, CHA, HOL). Song combines harsh grating and chattering sounds together with sweeter notes and loud squeaks. Recalls Great Reed-Warbler, but generally less raucous, more melodious and higher-pitched; varied phrases 2–8 s long are separated by short pauses; phrases consist mainly of 3–4 repetitions of same unit, 'kerri-kerri-kerri, kureek-kureek-kureek, chi-kerruchi-kerruchi-kerruchi . . .'. Contact-alarm notes, a loud 'chack' like call of Great Reed-Warbler and a low 'churr'.

General Habits. Frequents reeds *Phragmites*, reedmace *Typha*, papyrus and bushes by water; also sugar cane and other crops. Abundant in reedbeds in Nile delta. Apparently confined to mangroves on Red Sea and N Somalia coasts.

Skulking but not shy. Forages mainly low down within reeds or bushes, near or over water, but may sing from exposed perch. Takes food from water surface and vegetation, and by hopping about on ground near water. Sometimes feeds in trees near reedbeds, when movements among foliage noticeably slow and deliberate.

Nominate race mainly sedentary but disperses locally outside breeding season (Cramp 1992). Red Sea birds probably also sedentary.

Food. Insects and aquatic larvae, spiders, small snails and slugs, small frogs, seeds. In Egypt, stomachs contained mantises, Lepidoptera larvae and spiders (Lynes 1912).

Breeding Habits. Monogamous; sometimes polygynous. Solitary; territorial but often concentrated in neighbourhood groups; in Nile delta 10–20 ♂♂ singing per ha (Meininger *et al.* 1986); at least 20 singing in May along 1 km of mangroves on Saad el Din island, Somalia (Ash 1983).

NEST: deep cylindrical cup of stems and reeds, lined with finer material, attached to mangrove or reed stems, usually low over water; outer diam. 110–130, inner diam. *c.* 60, cup depth *c.* 60; built mainly by ♀.

EGGS: 3–6, pale blue-green with bold blackish markings. SIZE: (n = 3, Asia) 22–25 × 16–18.

LAYING DATES: Egypt, Mar to late June.

INCUBATION: by ♀ alone. Period: 13–15 days.

DEVELOPMENT AND CARE OF YOUNG: young cared for and fed by both parents; brooded at first by ♀; fledging period 11–13 days, post-fledging care for several days more.

References
Cramp, S. (1992).
Goodman, S. and Meininger, P. L. (1989).

Acrocephalus rufescens (Sharpe and Bouvier). Greater Swamp-Warbler. Rousserolle des cannes.

Plate 7
(Opp. p. 112)

Calamocichla rufescens Sharpe and Bouvier, 1876. Bull. Soc. Zool. France I, p. 30; Landana, Cabinda.

Range and Status. Endemic resident. Distribution local and fragmented, mainly in extensive swamps, from Senegal to Sudan and south to Botswana. Rather scarce to locally common NW Senegal (Richard Toll to Mboro); Ghana (SE coast, Mole); Togo (Lomé); S Nigeria to S Cameroon and Bioko; N Nigeria (Kano); L. Chad; N Central African Republic; NW Zaïre (Ubangi R.); NE Zaïre (Uelle R., border areas south to Bukavu) to S Sudan (Nile north to Malakal), W and S Uganda, Rwanda and W Kenya (Nyanza); NW Angola (Cabinda, Cuanza Norte, W Malanje); SE Zaïre (L. Upemba); Zambia (Luapula and Northern Provinces east to Sumba and Isangara, Kafue, Upper Zambezi); N Botswana (Okavango) and W Zimbabwe (Zambezi R. between Kazungulu and Katombora rapids). Density of *c.* 100 pairs in 9·0 ha of papyrus swamps on *c.* 9 km of Zambezi R. in W Zimbabwe (a further 4 km of that stretch of river lacks papyrus: Hustler 1995).

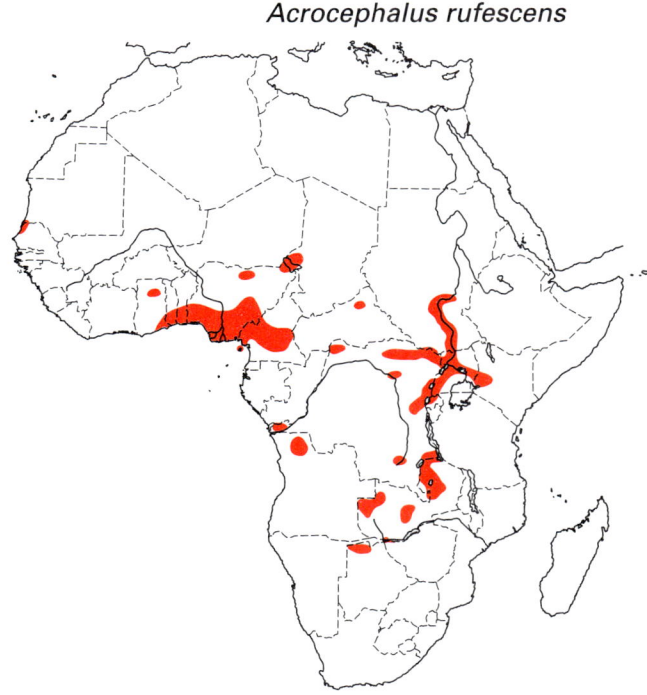

Acrocephalus rufescens

Description. *A. r. rufescens* (Sharpe and Bouvier): Ghana, Nigeria, Cameroon, Bioko, Cabinda and NW Zaïre. ADULT ♂: upperparts dark, cold olive-brown, head rather darker and greyer, rump paler. Tail-feathers dark brown. Supercilium poorly defined, greyish. Underparts greyish white, suffused with grey-brown on breast and flanks. Flight feathers and larger upperwing-coverts dark brown, edged olive-brown; lesser coverts same as mantle; underwing-coverts and axillaries creamy white. Wing rounded; P5, P6 and P7 longest, P4 *c.* 2 shorter; P9 *c.* 8 shorter; P10 about half length of P9. Bill black above, yellowish brown below with dark tip; eyes hazel to red-brown; legs and feet dark grey. Sexes alike. SIZE: (10 ♂♂, 10 ♀♀) wing, ♂ 71–77 (74·4), ♀ 68–74 (71·3); tail, ♂ 69–74 (71·3), ♀ 67–72 (68·2); bill, ♂ 21–24 (22·4), ♀ 21–22 (21·8); tarsus, ♂ 28–31 (29·4), ♀ 25–30 (28·2). WEIGHT: 1 ♂ (Ghana) 23·2; unsexed (n = 6, Nigeria) mean 19·8, s.d. 1·6, (n = 2, Ghana) 17·9, 19·6.

IMMATURE: juvenile like adult.
NESTLING: mouth yellow.

A. r. senegalensis Colston and Morel: Senegal. Paler above than nominate race, greyer on head; whiter below.

A. r. chadensis (Alexander): L. Chad. Slightly paler above than nominate race, whiter below.

A. r. ansorgei (Hartert) (including '*niloticus*' and '*foxi*'): S Sudan, Uganda, Rwanda and adjacent Zaïre, W Kenya, Zambia, N Botswana and NW Angola. Larger and darker than nominate race, greyer below with dark shaft-streaks on throat. SIZE: wing, ♂ (n = 12) 76–85 (80·5). WEIGHT: unsexed (n = 81, Kenya) 18–29·3 (22·3).

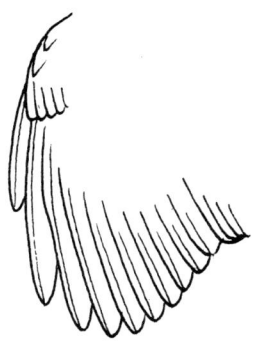

A

Field Characters. A large, robust reed warbler with a long bill and strong greyish black legs and feet. Dark olive-brown above and greyish below, with whiter throat and belly. Distinguished from similar Lesser Swamp-Warbler *A. gracilirostris* by being rather larger, less warmly coloured and greyer below, the flanks never rufous; supercilium less distinct. Darker than Great Reed-Warbler *A. arundinaceus*, with much stouter, blacker legs, shorter, rounded wing with long P10 (**A**), relatively longer tail and quite different voice.

Voice. Tape-recorded (13, 57, 86, 91, B, F, GR1). Song loud and throaty, rather ventriloquial (van Someren 1956); varied phrases of 1–2 s are separated by short pauses, each consisting of 3–5 repetitions of low chuckling notes, 'krup-krr-krr-krr, kikweu-kikweu-kikweu-kikweu, kieru-kwee-kwee-kwee . . .'. Contact-alarm note, a harsh 'kweeok' or 'kierok', sometimes a longer 'klieu-rali-wok' (van Someren 1956); anxiety note 'klok, klok'. Responds vigorously to playback (Hustler 1995).

General Habits. Usually occurs in vegetation emergent from water. In W Africa, inhabits reeds, papyrus, wet elephant grass and sugar cane. In E and southern Africa, restricted to papyrus in swamps and on lake edges. Where it occurs with Lesser Swamp-Warbler the 2 tend to be segregated, Greater Swamp-Warbler in the interior of papyrus, Lesser on the edges and utilizing other vegetation (Britton 1978b, Hustler 1995).

Skulks low down, foraging singly or in pairs. Frequently calls, but may be difficult to see. Climbs easily up and down vertical stems, tail often cocked; hops from one stem to another. Descends to base of papyrus clumps to pick insects off stem sheaths, and flies out to feed on lily-pads (van Someren 1956). Flight rather weak, with rapid wing-beats, from one reed or papyrus clump to the next.

Food. Insects, small frogs. In Kenya, young fed beetle larvae, noctuid larvae, damsel flies, a moth and other aquatic insects (van Someren 1956).

Breeding Habits. Territorial. Probably monogamous. Pair occupies territory of c. 625 m^2 (Hustler 1995).

NEST: deep cup of reed, papyrus or grass strips, lined with finer strips of vegetation and sometimes a few feathers; some strips (from papyrus leaves?) can be very long, making nest resemble a weaver's *Ploceus*; slung between upright stems 2–2·5 m above surface of deep water, or built within papyrus *Cyperus papyrus* inflorescence on shorter stem below papyrus canopy (van Someren 1956, Hustler 1995, nest photo de Naurois 1985). 2 nests measured: ext. diam. 81, 83, int. diam. 49, 52 × 61, int. depth 45, 48, floor thickness 45, 60.

EGGS: 2–3. Laid evidently at daily intervals. Greyish white or pale greenish with brown, grey-brown or reddish spots, throughout or mainly at broad end. SIZE: 20–21 × 15–16.

LAYING DATES: Senegal, May-July; Nigeria, Apr–Oct; Kenya, Mar–Apr, July, Oct-Dec; Zambia, Oct–Feb; Zimbabwe, Feb; Botswana Sept, Nov–Dec.

INCUBATION: bird sits very tight. Eggs hatch at 1 day intervals. Period: at least 14 days (Hustler 1995).

DEVELOPMENT AND CARE OF YOUNG: young fed in nest by both parents.

References
Hustler, K. (1995).
de Naurois, R. (1985).
van Someren, V. G. L. (1956).

Plate 7
(Opp. p. 112)

Acrocephalus gracilirostris (Hartlaub). Lesser Swamp-Warbler. Rousserolle à bec fin.

Calamoherpe gracilirostris Hartlaub, 1864. In Gurney, Ibis, p. 348; Natal.

Range and Status. Endemic, resident. Ranges from S Sudan and Ethiopia to the Cape, with apparently isolated populations in Nigeria, Chad and Central African Republic. Locally common N Nigeria (Kano); local and poorly known W Chad (L. Chad, Sounta–Bahr Azoum confluence) and N Central African Republic (Manovo-Gounda-Saint Floris Nat. Park); locally common but fragmented in Ethiopia (Gambela, W Highlands north to L. Tana, Rift valley, Juba R.) and S Somalia (Webi Shabeelle, lower Juba). Local and uncommon central and S Sudan (S Kordofan, upper Nile valley). Locally common to abundant W and S Uganda, Rwanda,

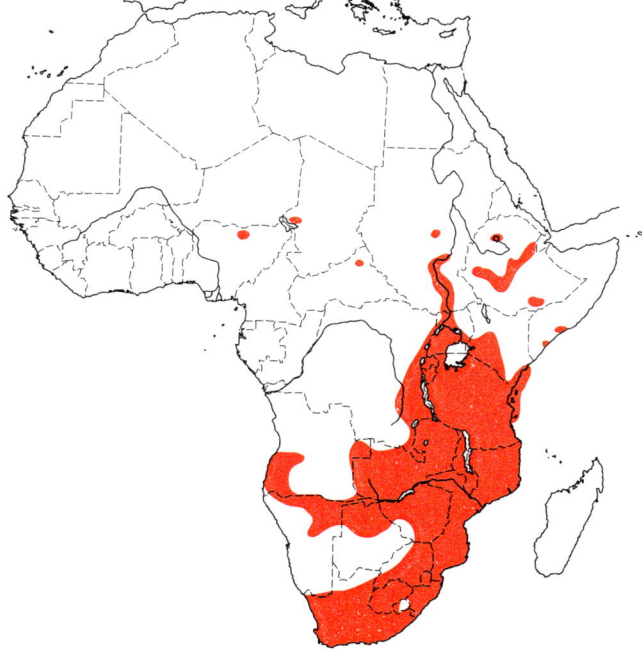

Burundi and adjacent E Zaïre; SW, central and SE Kenya; Tanzania; and SE Zaïre (Shaba) to Zambia, E Angola (L. Dililo and Macondo), Malaŵi, Mozambique, SW Angola (coastal plain to Benguela), NE Namibia, N Botswana, Zimbabwe, South Africa and Swaziland. Scarce in lowlands of NW Lesotho (P. Osborne and B. Tigar, pers. comm.).

Description. *A. g. gracilirostris* (Hartlaub) (including '*zuluensis*'): South Africa (Cape, interior of Natal, Orange Free State, Transvaal highlands), Lesotho, W Swaziland, S Mozambique and SE Zimbabwe. ADULT ♂: upperparts warm brown, richest on rump, darker and greyer on head and hindneck. Tail-feathers dark brown, edged buffish brown. Supercilium whitish; lores and ear-coverts greyish brown with stripe behind eye slightly darker. Chin to breast whitish; flanks, belly and undertail-coverts pale buffish brown. Flight feathers and larger upperwing-coverts dark brown, edged buffish brown; lesser coverts warm brown; underwing-coverts and axillaries buffish white. Wing rounded; P6 longest; P5 and P7 *c.* 1 mm shorter, P4 2–3 shorter; P9 8–10 shorter; P10 about half length of P9 (**A**). Bill blackish horn, base of lower mandible yellowish; eyes brown; legs and feet dark greenish or bluish horn. Sexes alike. SIZE: (10 ♂♂, 10 ♀♀) wing, ♂ 72–80 (75·7), ♀ 72–75 (73·9); tail, ♂ 67–73 (70·3), ♀ 63–73 (69·1); bill, ♂ 20–22 (20·9), ♀ 19–22 (20·6); tarsus, ♂ 26–29 (27·8), ♀ 25–28 (26·4). WEIGHT: unsexed (n = 29, South Africa) 11·3–20·4 (16·7).

IMMATURE: juvenile like adult.

NESTLING: not described.

A. g. cunenensis (Hartert): SW Angola, N Namibia, N Botswana, SW Zambia and W Zimbabwe. Whiter below than nominate race, whitish extending to belly and flanks; less warm, more olive-brown above. Slightly smaller: wing, ♂ (n = 4) 72–74, ♀ (n = 2) 68, 70.

A. g. winterbottomi (White): N and NW Zambia to E Angola and SW Tanzania. Greyer below, with no buff on flanks or undertail-coverts, slight streaking on throat. Darker, greyer above than *cunenensis*, size similar.

A. g. leptorhynchus (Reichenow): South Africa (coastal Natal and E Transvaal), and E Swaziland to E Zimbabwe, Mozambique, Malaŵi, SE Zambia, E and central Tanzania, SE Zaïre, SE Kenya, S Somalia and E Ethiopia. Like nominate race, but brighter, more reddish brown above. Smaller: wing, ♂ (n = 10) 63–68 (65·5).

A. g. parvus (Fischer and Reichenow): Kenya highlands, N and NW Tanzania, Rwanda, Burundi, SW Ethiopia. Differs from nominate race in having upperparts dark olive-brown, head more uniform with mantle. Below, whitish chin and throat contrast with greyish breast, belly and undertail-coverts; flanks browner. Size similar to nominate race: wing, ♂ (n = 5) 70–76 (72.6). WEIGHT: unsexed (n = 221, Kenya highlands) 11·9–21·2 (16·5).

A. g. jacksoni (Neumann): S Sudan, W Kenya, Uganda and adjacent Zaïre. Similar to *parvus* but smaller, and slightly greyer. SIZE: wing, ♂ (n = 7) 64–70 (65·8). WEIGHT: unsexed (n = 9, W Kenya) 12·5–16·2 (14·4).

A. g. tsanae (Bannerman): NW Ethiopia (L. Tana). Like *parvus* but darker grey below, less white on throat.

A. g. neglectus (Alexander): W Chad. Paler than *parvus*, buffier below. Smaller: wing, ♂ (n = 2) 67, 69, ♀ (n = 5) 62–69 (64·8).

Field Characters. A medium-sized to rather large warbler with a long bill, rounded wing and strong, dark legs and feet. Darkish brown above, and whitish below with brown or rufous flanks. Birds of the Nile valley, Uganda and the E African and Ethiopian highlands are more olive above, and greyer below with contrasting white throat. All races are more richly coloured than Greater Swamp-Warbler *A. rufescens*, with whiter underparts, more conspicuous supercilium, and rather smaller bill. Short wing helps to separate it from Palearctic Great Reed-Warbler *A. arundinaceus* and Basra Reed-Warbler *A. griseldis*.

Voice. Tape-recorded (57, 75, 86, 88, 91, B, C, F). Song rather thrush- or bulbul-like; similar to that of Greater Swamp-Warbler but more melodious, less harsh and throaty. Consists of loud, chuckling notes repeated 3–4 times to form varied phrases of about 1 s duration, separated by short pauses: 'klieru-klieru-klieru, klikliu-klikliu-klikliu-klee, kliew-klikliklikli . . .'. Contact-alarm notes, loud 'kierok', low 'chuck' and chattering 'chuk-uk-uk-k-k'.

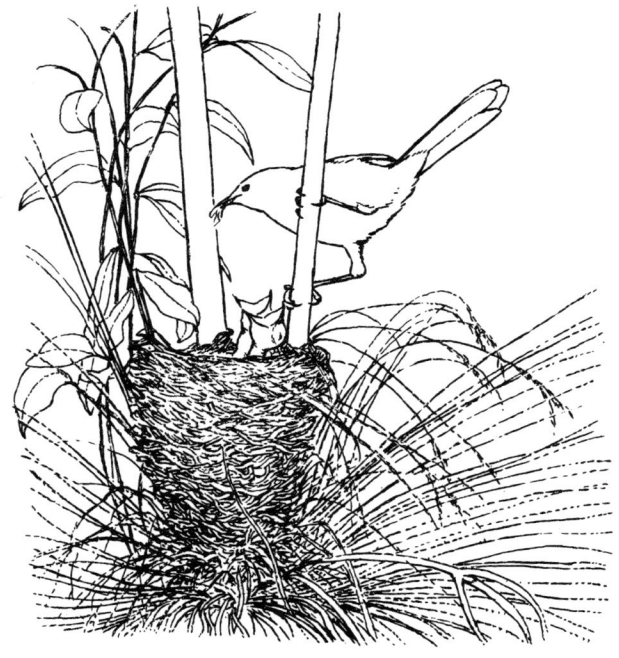

B

General Habits. Frequents beds of tall reeds, reedmace *Typha*, sedges and papyrus, on lake shores and edges of swamps, dams, rivers and estuaries; also rank grass, low shrubbery and thick foliage by water. Where present alongside Greater Swamp-Warbler, as in W Kenya, S Sudan and N Zambia, found mainly on swamp edges and rocky islands, utilizing vegetation other than papyrus. Occurs up to 2400 m in Kenya and Ethiopia, but only up to 1500 m in southern Africa.

Usually solitary, though often in pairs and small groups. Forages mainly within reed or sedgebeds, at mid-level or low down above water. Moves easily among vertical stems. Inquisitive, and emerges on edges of reed or sedgebeds, picking food from low stems, water surface or floating vegetation. Occasionally feeds in papyrus heads (Tyler 1991). Flight rather weak, low and fluttery, from one reed clump to the next. Sings frequently, usually from within cover 1–2 m above water. Several singing together in a reedbed produce a melodious chuckling chorus.

Food. Insects; small frogs.

Breeding Habits. Territorial, but several pairs may nest within a small area. Probably monogamous.
NEST: deep conical cup of dry reed strips, coarse grass, and sometimes water weed, lined with fine strips, fine grass and sometimes a few feathers (**B**); built around vertical stems, or in thick vegetation or shrubs; in South Africa, across fork of small flooded wattle tree; usually 0·4–1·8 m above water; ext. diam. *c.* 100, cup diam. *c.* 50, cup depth *c.* 85.
EGGS: 2–3, occasionally 4; South Africa mean (30 clutches) 2·3. White or greenish white, closely freckled with black, dark grey, pale brown and ashy grey. SIZE: (n = 31, South Africa) 16–21 × 13–15 (18·7 × 14·2).
LAYING DATES: Kenya and Tanzania, Mar–Dec; Malaŵi, Jan–Aug; Zambia, Feb; Zimbabwe, Aug–May (mainly Nov–Mar); Botswana, Aug, Nov, Feb; South Africa, Aug–Feb; mainly in the rains.

Not illustrated

[*Acrocephalus aedon* (Pallas). Thick-billed Warbler. Rousserolle à gros bec.

Muscicapa Aëdon Pallas, 1776. Reise d. versch. Prov. Russ. Reichs., 3, p. 695; Dauria.

Range and Status. E Palearctic: Transbaykalia to NW Mongolia and S Siberia. Winters from India to Indochina and S China. Vagrant to Britain (twice) and Egypt.
Sight record, St Katherine Monastery, S Sinai, Nov 1991 (Grieve 1992). (Although outside Africa as here delimited, this record is included to draw attention to an easily-overlooked 'great reed warbler' which might occur again in NE Africa.)

Reference
Grieve, A. (1992)]

Genus *Chloropeta* Smith

A genus with characters of both flycatchers and warblers. Flycatcher characters are broad, flat bill and well-developed rictal bristles; warbler characteristics are long legs, slim build and warbler shape, 2 black tongue spots of nestling, unspotted juvenile plumage, nest, foraging behaviour (mainly gleaning, although sometimes makes aerial sallies); and vocalizations are distinctly acrocephaline. Wings short and rounded, reaching just beyond base of tail; tail long and rounded, with 10 feathers in *natalensis* and *gracilirostris*, 12 in *similis*. Plumage olive above, yellow below. Decidely acrocephaline: voices, foraging behaviour and nests recall *Acrocephalus* spp. and habitats, plumages and bill shapes recall *Hippolais* spp.; the Afrotropical *Chloropeta* seems to have the same relationship to Palearctic *Hippolais* as have '*Calamocichla*' to *Acrocephalus*, '*Seicercus*' to *Phylloscopus* and '*Parisoma*' to *Sylvia* (C.H. Fry, pers. comm.).

C. gracilirostris has narrower bill (**A**, *C. g. gracilirostris*, **B**, *C. g. bensoni*) and larger feet than the other 2 species. *C. similis* has a wide bill (**C**: see measurements in Keith and Vernon 1966), which, together with swamp habitat, caused Grant and Mackworth-Praed (1940) to place it in a monotypic genus *Calamonastides*, close to *Calamocaetor* [*sic*] (= *Calamocichla*, now *Acrocephalus*). On the other hand, vocal characters and olive and yellow plumage suggest it should be retained in *Chloropeta* (Keith and Vernon 1966), as is done by Hall and Moreau (1970) and Traylor (*in* Watson *et al.* 1986).

Endemic; 3 species. Geographical and ecological overlap and difference in number of tail-feathers prevent *C. natalensis* and *C. similis* from being placed in a superspecies, yet they are close enough to be considered members of a species group; *C. gracilirostris* is too divergent to be included in the same species group, as was done by Hall and Moreau (1970).

 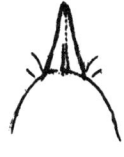 

A   B   C

Chloropeta natalensis Smith. African Yellow Warbler. Chloropète jaune.

Chloropeta natalensis A. Smith, 1847. Illustr. Zool. South Africa, Aves, pl. 112, fig. 2 (and text); near Port Natal (= Durban), Natal.

Plate 8 (Opp. p. 113)

Range and Status. Endemic resident and local migrant. Highlands of W and SE Ethiopia (frequent); highlands of E Nigeria (Obudu and Mambilla plateaux), and probable sight record Mbaakon (7° 12′N, 9°E); S Cameroon lowlands, north and east to Djokong and Mieri, and north in highlands to Bamenda and Mt Oku; S Gabon (Mbigou, Tchibanga); SW Central African Republic; Zaïre around edge of forest zone, south in E to Itombwe and Marungu Plateau (but absent Mt Kabobo); extreme S Sudan (rare); Rwanda, NE Burundi (Kigamba); S and W Uganda from Rwenzoris, Kigezi and S Ankole north to Mubende, Lango and Jinja; Mt Elgon (Kenya/Uganda) and W and central Kenya from Saiwa swamp, Maralal and Mt Kenya south to S Nyanza and Nairobi region, also Chyulus and Taita Hills; W Tanzania in Bukoba District and from Kasulu to Ufipa Plateau and southern highlands, and in E from Kilimanjaro and Usambaras to Iringa, Songea and Mahenge. Angola (Cabinda, E highlands from S Cuanza Norte to Huila, and central plateau); Zambia except SW and Luangwa valley, continuing up Zambezi R. valley to E Caprivi (Namibia); Malawi above 1000 m; highlands of Mozambique north of Zambezi, and along Zimbabwe border from headwaters of Pungwe R. south to Spungabera, and Mt Gorongoza, wintering sparingly on coast of Manica e Sofala and Sul do Save. In Zimbabwe in E highlands from Inyanga to Chipinga, and on Mashonaland Plateau north to Zambezi escarpment, east to Miami, south to Bikita and Great Zimbabwe, moving to lower levels in cold season; wanderers (pair) at Aisleby Sewage Works, Bulawayo, Mar (Tree 1991). In Transvaal, locally common along base of escarpment, scarce elsewhere, mainly along rivers; Swaziland, Natal and Zululand, and E Cape Province to King William's Town; once at Amanzi near Uitenhage. Widespread, locally common, but generally rather uncommon; scarce now in E Cape (Vernon 1993).

Chloropeta natalensis

Description. *C. n. massaica* Fischer and Reichenow: SE Sudan and Ethiopia to E Zaïre, Uganda, Kenya and S Tanzania. ADULT ♂: top of head blackish brown, crown-feathers often raised to form slight crest; lores and incomplete narrow eye-ring yellow, ear-coverts, sides of neck and upperparts, including scapulars, yellowish olive-brown, somewhat paler and yellower on rump and uppertail-coverts. Tail-feathers blackish brown with narrow yellowish margins and tips, outermost pair overall paler, with broader yellow margins. 10 tail-feathers. Entire underparts bright rich yellow, olive-brown wash on sides of breast and flanks. Wing-feathers blackish brown edged pale yellow, inner margins of underside of primaries and secondaries pale sulphur yellow, axillaries, underwing-coverts and bend of wing bright yellow. Upper mandible dark brown to dark horn, lower mandible flesh to pinkish buff, dull ochreous yellow, or pale yellow with bluish veins; inside of mouth reddish orange to pinkish orange, rictal bristles black; eyes brown to dark brown, reddish brown or greyish brown; legs and feet blackish to dark slate, grey, lead grey and dark greenish grey, claws black. Sexes alike except ♀ slightly duller. SIZE: (10 ♂♂, 10 ♀♀) wing, ♂ 58–65 (63·1), ♀ 57–63 (60·0); tail, ♂ 57–63 (59·7), ♀ 52–59·5 (55·9); bill, ♂ 15–17 (15·7), ♀ 14–16·5 (15·5); tarsus, ♂ 21–22 (21·4), ♀ 19·5–23 (21·2). WEIGHT: 1 ♂ 11; ♀ (n = 4) 10–11.5 (10·6). (N.B. Weights given in Sclater and Moreau (1933) of 1 ♂ 23, 1 ♀ 24 must be in error.)

IMMATURE: top of head dull dark brown; upperparts brown washed cinnamon, rump, uppertail-coverts and edges of wing- and tail-feathers cinnamon; underparts, axillaries and underwing-coverts yellow-ochre, yellower on throat, whiter on abdomen. Tongue spotted.

NESTLING: dark brown, slightly yellower on belly; gape yellow, inside of mouth and tongue deeper yellow or orange, 2 black spots at base of tongue.

C. n. batesi Sharpe: Nigeria to N Zaïre and SW Sudan. Doubtfully distinct; very like *massaica* but top of head browner, less black, contrasting less with back. WEIGHT: (Cameroon) 2 ♀♀ 11, 12.

C. n. major Hartert: Gabon and Angola to S Zaïre and N Zambia. Somewhat larger, upperparts paler and greener, crown concolorous with back. WEIGHT: 2 ♂♂ 12, 13; unsexed (n = 6) 10–15 (12·8).

C. n. natalensis Smith: South Africa north to S Zambia and S Tanzania. Darker and duller than *major*, upperparts more like *massaica*; head contrasts somewhat with back but less than *massaica*. WEIGHT: 2 ♂♂ 11·9, 12.8; unsexed (n = 68) 9–13·6 (11·7).

Field Characters. Widespread but shy and uncommon, easy to overlook; keeps low in vegetation and emerges only to sing. Olive-green to olive-brown above, crown either blackish, contrasting with back (*massaica*), concolorous with back (*major*) or an intermediate shade (other races). Underparts bright rich yellow. Shaped rather like reed-warbler *Acrocephalus*, and feeds like warbler. Generally at lower elevations than Mountain Yellow Warbler *C. similis* but overlaps with it at 1800–2200 m, and sometimes found in same habitat. For differences see that species. Palearctic Willow Warbler *Phylloscopus trochilus* and Icterine Warbler *Hippolais icterina* also have yellow underparts but are paler and more washed out, duller above, more greyish olive, without contrasting cap, and underparts pale sulphur, not orange-yellow; Willow Warbler is also smaller.

Voice. Tape-recorded (20, 58, 75, 88, 91, B, C, F). Pleasant little song is a series of brief (*c.* 2 s) phrases, each consisting of a slow, rippling trill of clear, often ringing notes introduced by 2–4 short, dry chips, 'chip-chip-chipa-rililililili', 'chip-chip-titit-reep-reep-reep-reep', 'trip-trip-trelelelelelel', or reversed, long notes followed by short ones; clear notes of trill often have a rather throaty undertone, and are reminiscent of Lesser Swamp-Warbler *Acrocephalus gracilirostris*; trill has also been likened to that of paradise-flycatcher *Terpsiphone*. Each phrase is repeated several times and then changed; repertoire is considerable. It has been claimed that ♂ and ♀ duet, ♂ giving short warble instantly answered by harsh note from ♀ (Beatty *in* Rand *et al.* 1959), but this is not usual. Alarm, sharp chip or harsh churr; soft trill given in courtship display.

General Habits. Inhabits a wide variety of rank vegetation: tall grass, bracken-briar, willows and tangled growth along streams, nettle beds, fallow land and waste ground overgrown with *Triumfetta* and other shrubs, dense herbaceous growth at forest edge, grassland near forest; in Gabon, gallery forest and undergrowth in old plantations; in Zaïre, forest clearings, often with *Pennisetum*, but towards edge, not in deep forest; in E Africa, bushy country close to water (swamps, lakes, rivers), edges of reedbeds, reedy patches in open country, rank gardens; in southern Africa, sheltered stream gullies choked with dense vegetation where they flow out of evergreen forest patches, swampy vegetation below dam walls, scattered scrub and bushes, locally into thornveld (Natal); in Transvaal, rank growth of forbs and weeds (*Vernonia amydalina*, *Solanum* spp.: Tarboton *et al.* 1987); in winter, coastal scrub. Broad altitudinal range, lowlands to lower montane levels, up to 2150 m in Cameroon, 2600 m in Ethiopia, 2100 m in Zimbabwe; in E Africa 500–2300 m, in Malaŵi 900–2450 m.

Solitary or in pairs except after breeding, when in small parties with fledged young. Does not join mixed-bird parties, but 1 briefly joined ants swarming in weeds and on fallen tree (Tanzania: Willis 1986a). Forages mainly like warbler, creeping about in low vegetation, sidling up and down stems like reed-warbler *Acrocephalus*, snapping up prey it flushes, often with short flutters; also gleans leaves in trees like *Phylloscopus*, once observed at top of 7 m high acacia (Madge 1972). Sometimes makes sallies after flying insects from exposed perch, like flycatcher. Shy, more often heard than seen; on approach of man tends to drop into vegetation and creep away, reappearing at some distance. Sings year-round, often from exposed perch on top of bush. Bird in Zimbabwe sang while perched on nearly vertical twig, hopped up twig between each burst of song; after reaching the top it flew further down and repeated the process (Brooke 1954), once interrupting song to take a small flying insect on the wing.

Moves to lower levels during cold season in southern Africa. Birds below 1000 m in Malaŵi probably non-breeders (Benson and Benson 1977), likewise records on Zambezi R. at Tete (Mozambique) and Chirundu (Zimbabwe). Moves east from Zimbabwe highlands to lower levels, e.g. Haroni-Lusitu, and possibly joins birds from interior Mozambique on Mozambique coast; moves from higher parts of interior South Africa to coastal lowlands. In Zambia changes habitats seasonally; bird ringed by river in dry season retrapped 1 km from water in wet season, in rank grass and weeds; latter habitat is burned in dry season, when birds presumably restricted to waterside habitat (Bowen 1983).

Food. Insects, including Coleoptera and Lepidoptera and their larvae. Young fed small larvae of mantids, locusts and Lepidoptera, and a few moths and small winged termites.

Breeding Habits. Monogamous; territorial. Defends territory by force, driving off intruder with much singing, crest-raising and other dislays. ♂♂ on adjacent territories countersing. In courtship display, ♂ sits on prominent twig and sings loudly, with throat vibrating visibly, shivering wings, raising and lowering crest, then flies 2–3 m up in air and drops down in a series of jerky glides to where ♀ is sitting. ♀ flies off and zigzags through bushes with ♂ in pursuit; ♂ perches, raises and spreads tail, then repeats flight, and alights near ♀, giving soft trill (van Someren 1956).

NEST: usually a deep cup, often bulky, compactly built of broad blades of grass or reeds, rootlets and other fibres, sometimes strips of maize blades and awn of maize ears, bound with cobwebs, sometimes with stringy, ragged bits of material hanging from the underside (**A**); one had complete sloughed skin of whip-snake 0·75 m long wrapped around outside of nest and held in

A

place by grass blades (Jackson and Sclater 1938); with lining or inner cup of fine grass leaves and fluffy seed-heads, stems of *Thalictrum rhynchocarpum* and other plants, maize awn and hair; ext. diam. 70–80, int. diam. 45–65, ext. depth 70–90, int. depth 40–65; overall resembles nest of Sedge Warbler *Acrocephalus schoenobaenus*. Situated 0·3–1·3 m above ground in double or triple upright fork, with foundation material pushed down into cleft and sides attached to surrounding stalks, or slung between upright stems, sometimes incorporating one or more of them into the wall of the nest, i.e. built around them; in low bush, bracken, reeds, sedges or herbaceous plants, e.g. *Leonotis*, hibiscus, heliotrope, often near stream or swamp, or even over water on crossed reed stems; sometimes well hidden, sometimes not, and often placed below gap in cover so that light and air can enter; in Cameroon on cultivated ground in weed *Triumfetta* or tangled, even thorny bush. 1 nest in Zimbabwe was shallow, thin-walled cup of fine grass and fibres, with back wall extended upwards and attached to horizontal fork of *Acacia sieberiana*, 1·5 m above ground, hung like nest of Long-billed Crombec *Sylvietta rufescens*; ext. diam. 38 × 63, ext. depth, rear 88, front 25; from a short distance it resembled old nest of Common Bulbul *Pycnonotus barbatus* tilted on its side (Vernon 1962).

EGGS: usually 2, sometimes 1 or 3, exceptionally 4; Southern Africa av. (11 clutches) 1·4. Wide to narrow oval; smooth, unglossed or with slight gloss; white to cream or pale pink, unmarked or lightly speckled or streaked with red-brown, maroon or yellow-brown and occasionally blackish brown over mauve or lilac-grey, mainly around large end. SIZE: (tropical Africa, n = 8) 16·5–18·5 × 12·8–14 (17·3 × 13·4); (southern Africa, n = 15) 17–19 × 12·7–14·1 (18·4 × 13·5). Double-brooded, at least in Kenya.

LAYING DATES: Cameroon, Apr, June–July, Oct; Zaïre, Apr–Sept (Uele), Dec–Mar (during rains) in S (juv. Oct), Mar, June and Dec in Itombwe; Central African Republic (breeding condition May); Sudan, June; E

Africa: Uganda, July; Kenya, June; Region A, May; Region C, Dec; Region D, Mar–June; Zambia, Dec; Malaŵi, Jan, Oct–Nov; Mozambique, Oct–Mar; Zimbabwe, Sept–Feb (Sept 2, Oct 8, Nov 8, Dec 22, Jan 8, Feb 7 clutches) (also imm. Apr); Transvaal, Nov–Dec, Feb; Natal, Oct–Feb.

INCUBATION: mainly by ♀, but ♂ takes over for short spells. Period: 12 days. ♀ sits close; when disturbed, vanishes into bush, then perches on nearby tree branch and gives anxiety call. ♂ soon appears and tries to distract observer by calling and making short flights. ♀ soon creeps back into nest when observer leaves.

DEVELOPMENT AND CARE OF YOUNG: feathers grow fast after first few days; ♂ helps with brooding when chicks are small. Young fed by both parents (mainly by ♀), who are attentive and fearless; at 1 nest ♀ brought food every 10 min; ♂ sings between feeding bouts.

Nestling period: 14–16 days. Fledged young remain with parents for 1 month unless there is a 2nd brood.

BREEDING SUCCESS/SURVIVAL: in nest with 2 eggs, 1 young hatched, in another with 3 eggs, 2 young hatched. Young preyed on by ground predators such as mongooses; shrikes also take eggs and newly-hatched young. Full brood seldom reaches maturity (van Someren 1956). 1 bird ringed as adult recaptured after 3 years 330 days (Harwin *et al.* 1994).

References
Bates, G. L. (1911).
van Someren, V. G. L. (1956)
Vincent, A. W. (1947).

Plate 8
(Opp. p. 113)

Chloropeta similis **Richmond. Mountain Yellow Warbler. Chloropète de montagne.**

Chloropeta similis Richmond, 1897. Auk 14, p. 163; Mt Kilimanjaro, Tanganyika, 10,000 ft.

Range and Status. Endemic resident. Extreme SE Sudan in Imatong and Dongatona Mts above 1800 m (common), and nearby NE Uganda at Mt Morongole. Mountains either side of the Albertine Rift in E Zaïre, W Uganda, Rwanda and Burundi, from Rwenzoris south through Kivu to Itombwe and Mt Kabobo, and through Bwindi-Impenetrable Forest and volcano region to Mt Heha (Burundi). Mt Elgon (Kenya/Uganda) and highlands of W and central Kenya from Cheranganis, Mt Nyiru and Mt Kenya south to Trans-Mara (Mau) Forest, Kinangop Plateau and Nguruman Hills; N Tanzania on Mt Kilimanjaro and Crater Highlands south to Mbulu; E Tanzania from Nguru, Ukaguru and Uluguru Mts to Udzungwa Mts and Njombe; Nyika Plateau (Zambia/Malaŵi). Locally common; density on Nyika Plateau 1–2 pairs/ha (Dowsett-Lemaire 1989).

Description. ADULT ♂: top of head, upperparts, ear-coverts, sides of neck and line from bill through eye olive-green, somewhat paler and yellower on rump and uppertail-coverts; narrow yellow loral line. Tail-feathers olive-brown edged yellow-green, shaft-streaks maroon above, whitish below. 12 tail-feathers. Entire underparts from chin to undertail-coverts bright rich yellow. Wing-feathers olive-brown edged yellow-green, inner webs of underside of primaries and secondaries margined yellowish white, axillaries, underwing-coverts and bend of wing rich yellow. Upper mandible dark brown, lower mandible yellowish, especially at base, or salmon pink with buff tip and dark brown at sides towards tip (Chapin 1953a); gape and inside of mouth light reddish orange, rictal bristles black; eyes hazel, brown or dark brown; legs and feet slate-grey to greenish grey. Sexes alike. SIZE: (10 ♂♂, 10 ♀♀) wing, ♂ 57–61 (59·1), ♀ 56–60 (57·6); tail, ♂ 54–59 (56·2), ♀ 52–57 (54·0); bill, ♂ 15–16 (15·6), ♀ 15–16 (15·6); tarsus, ♂ 20–23·5 (22·3), ♀ 21–23 (22·3). WEIGHT: ♂ (n = 14) 9–15 (10·9); ♀ (n = 9) 10–13 (11·3).

IMMATURE: upperparts browner, less green, with slight buffy wash, especially on rump; underparts paler yellow with buffy wash.

NESTLING: naked young has black spot on each side of tongue.

TAXONOMIC NOTE: some incipient tendencies towards geographic differentiation in this species were noted by Parkes (1987), but we agree with him and with Traylor (*in* Watson *et al.* 1986) that it should be treated as monospecific.

Field Characters. Similar to African Yellow Warbler *C. natalensis* (and once thought to be a race of it). Has a much more musical song and is more of a mountain bird, but they overlap at 1800–2200 m and may even occupy the same habitat, e.g. bracken-briar on Nyika Plateau (Keith and Vernon 1966). Upperparts uniform olive-green, without contrast between head and back. Race *major* of *C. natalensis* shares this character but is allopatric; sympatric races all have dark caps. Palearctic Willow Warbler *Phylloscopus trochilus* might occur in same habitat but is smaller and much paler yellow, and its call in winter is a soft 'hoo-eet'; similar-sized Icterine Warbler *Hippolais icterina* is duller and paler and occurs in drier lowland habitat.

Voice. Tape-recorded (32, 56, B, C, F, HOR, STJ). Song remarkable for its pure, sweet notes, vocal range and inventiveness; rightly presented as one of the most beautiful bird songs of the world by Gunn and Gulledge (1977). Warbled notes alternate with trills in a pleasant medley lasting 4–7 (5) s; highest notes can be a little sharp, lowest have a nasal quality reminiscent of *Acrocephalus*, but most are clear and brilliant. Individual on Mt Kenya usually began its song with 2 long, pure notes, first high, next low, 'peeee, poooo . . .' and continued with a variety of warbles and trills. Songs may continue in series for 5 min or more, with intervals of *c.* 3 s between each; no two are the same. Alarm, 'cha-cha-cha'.

General Habits. Occurs in a variety of montane habitats, including primary forest, especially in rank growth with bracken at edges or in gaps or marshy clearings; in Rwanda (Nyungwe) in clearings with *Sericostachys*; second growth, bamboo, heath scrub, herbaceous undergrowth in *Hagenia* woods, secondary brush mixed with giant lobelias, bracken-briar, tangled undergrowth along streams, swampy valleys, overgrown fallow land in cutover areas. Altitudinal range in Zaïre 1850–3700 m, in Rwanda over 2100 m, in E Africa 1800–3400 m, on Nyika Plateau (Zambia/Malaŵi) 2000–2450 m. On Nyika Plateau overlaps with African Yellow Warbler in narrow band at *c.* 2000–2100 m in same habitat (bracken-briar, rank growth along streams), and the two once nested 50 m apart (Keith and Vernon 1966); on Mt Kilimanjaro both occur at 1900 m but only Mountain Yellow Warbler occurs inside or at edge of forest, African Yellow Warbler in bush below forest (Cordeiro 1994).

Occurs usually in pairs, sometimes singly. Forages like warbler, gleaning insects from leaves and branches with short hops and sallies; usually fairly low down, but up to 18 m; in Rwanda (Nyungwe) in lianas around isolated trees in the open (Dowsett-Lemaire 1990). On Nyika Plateau, in tall submontane forest (trees 20–25 m high) prefers edges and mid-stratum thickets in clearings; in low-canopy montane forest (8–15 m) feeds at all levels, to tree tops; also in forest regrowth 1–4 m high (Dowsett-Lemaire 1989). Not sociable; only occasionally joins mixed-species flocks. Once seen to follow ant raid in bracken and heath, foraging while giving alarm call (it had large fledged young in the area), but on other days ignored ants and the thrushes feeding around them (Willis 1986a).

Food. Insects; in Itombwe, only Diptera (Prigogine 1971).

Breeding Habits. Territorial. Of 7 territories, Malaŵi, 3 were held for at least 2 years, 4 for at least 3 years (Dowsett 1985).

NEST: bulkier than that of African Yellow Warbler, a thick-walled cup of broad grass blades, leaves and seed-heads of grasses *Panicum* spp. mixed with fine grass, feathers, coarse and fine fern ramenta and moss (e.g. *Brachythecium impucatum*), bound on outside with cobwebs; included in nest near rest-house was non-vegetable fibre like fine green nylon, probably from human clothing (Keith and Vernon 1966); lined with feathers, hair and fine branches of *Thalictrum rhynchocarpum* and other plant fibres; ext. diam. 100–120, int. diam. 40, ext. depth 100; placed 0·7–1·3 m above ground in fork of herb, low shrub or thorny bush, often near stream.

EGGS: usually 2, sometimes 1. Ovate, rounded, elliptical; smooth but not glossy; pale pink or cream with a few tawny and ash-grey spots mainly around or on large end. SIZE: (n = 6) 17·9–19 × 13·3–14·2 (18·2 × 13·7).

LAYING DATES: Sudan, Apr, Oct; Zaïre, Apr–June (Kivu), also Oct (Itombwe); E Africa: Tanzania, Sept, Dec; Region D, July–Aug, Nov; Zambia/Malaŵi, Dec–Feb.

INCUBATION: pair 'seemed to be taking turns to incubate the eggs' (Cordeiro 1994).

Reference
Keith, S. and Vernon, C. J. (1966).

Chloropeta gracilirostris Ogilvie-Grant. Papyrus Yellow Warbler. Chloropète aquatique.

Chloropeta gracilirostris Ogilvie-Grant, 1906. Bull. Br. Orn. Club 19, p. 33; Muhokya, near L. George, Uganda.

Plate 8
(Opp. p. 113)

Range and Status. Endemic resident, restricted to papyrus in central Africa, in 2 disjunct populations. Nominate race recorded from Zaïre (Kabare, shore of L. Edward, and Kibga, south of Mt Visoke, Virunga Volcanoes); Rwanda around L. Lubondo and L. Bulera in N and nearby Rugezi and Mulindi swamps, and fur-

Chloropeta gracilirostris

ther south in Akanyaru swamp and Kagogo and Kibaya valleys, but not on lakes of Akagera Nat. Park; Burundi near Karuzi and in Ndurumu valley, and Burundi–Tanzania on both sides of Ruvubu R. where it flows along the border; SW Uganda (Lakes Bunyonyi, Mutanda, Edward and George; sight record from S end of L. Albert is unconfirmed); W Kenya at L. Kanyaboli, Kisumu and Kendu Bay. *C. g. bensoni* known only from mouth of Luapula R. in Zambia (doubtless in adjacent Zaïre). Locally common but vulnerable due to restricted range and specialized habitat; classed as rare in IUCN/ICBP Red Data Book; threatened by draining of swamps for agriculture and cutting of papyrus for fuel.

Description. *C. g. gracilirostris* Ogilvie-Grant: Zaïre to Kenya and Burundi. ADULT ♂: top of head to hindneck and upper mantle, lores, cheeks, ear-coverts and sides of neck greenish olive; lower mantle and back olive-brown, becoming tawny on rump and uppertail-coverts; tail-feathers dark reddish brown edged tawny. 10 tail-feathers. Underparts yellow, palest on chin and throat, breast lightly washed with ochre, flanks and undertail-coverts tawny-ochre. Wing-feathers dark brown edged olive to tawny, underside of inner webs of primaries and secondaries edged whitish, axillaries, underwing-coverts and bend of wing ochre. In worn plumage upperparts become duller, and there is some greyish frosting on feathers of head and upper mantle. Upper mandible blackish brown, lower mandible bright pinkish buff; gape dull orange, tongue and inside of mouth rich reddish orange, rictal bristles black; eyes reddish brown, pale rufous or bright orange-rufous; legs and feet black to lead-grey tinged bluish, soles light buffy grey, feet large. Sexes alike. SIZE: wing, ♂ (n = 4) 62–64 (63·0), 1 ♀ 57, unsexed (n = 9) 59–65 (61·6 ± 2·35); tail, ♂ (n = 3) 56–66 (60·8), 1 ♂ 59; bill to skull, ♂ (n = 3) 15·5–16 (15·8), 1 ♀ 16; bill to feathers, 1 ♂ 14; tarsus, ♂ (n = 4) 21–24·5 (23·3), 1 ♀ 23·5. WEIGHT: unsexed (n = 13) 10–12·5 (10·8 ± 0·68).

IMMATURE: like adult but more tawny below.
NESTLING: unknown.

C. g. bensoni Amadon: Zambia (see above). Lacks all tawny and ochre coloration. Upperparts darker olive-green becoming browner on rump; sides of breast and flanks with olive-brown wash, undertail-coverts yellow. Upper mandible brownish black, lower mandible whitish; legs and feet black. Bill wider (see generic diagnosis). SIZE: (1 ♂, 1 ♀) wing, ♂ 57, ♀ 55; tail, ♂ 58, ♀ 55; bill, ♂ (to feathers) 13, ♀ (to skull) 14·5; tarsus, ♂ 20, ♀ 21.

IMMATURE: like adult but more ochreous.

Field Characters. Easily distinguished from other swamp warblers by greenish upperparts, yellow underparts, liquid song. Other plain warblers in same habitat include Greater and Lesser Swamp-Warblers *Acrocephalus rufescens* and *A. gracilirostris* (large, with grey-brown upperparts, raucous voices), African Reed-Warbler *A. baeticatus* (same size but brown above, with grating and squeaky song) and Little Rush-Warbler *Bradypterus baboecala* (same size, dark brown above, song a dry, accelerating trill). Skulking; best located by call.

Voice. Tape-recorded (F, KEI). Song, a series of short (0·7–0·8 s) 2- or 3-note phrases, uttered at irregular intervals, 'to-tslo-wee', 'tee-tschlee-wo', 'tslo-tschlee-wo', 'tschlee-ow'; has a tendency to repeat one phrase a number of times before switching to the next, which it likewise repeats. Song is liquid but rather weak, with a plaintive, sibilant quality.

General Habits. Principal habitat papyrus; secondarily also reeds. In Rwanda confined to papyrus in drier parts of range, also occurring in reeds in wetter areas; in Uganda occurs in thick fringe of reeds around L. Bunyonyi; on L. Mweru (Zambia/Zaïre) inhabits papyrus with outer fringe of reeds. In Rwanda, most plentiful in swamps at 1850–2050 m, with mean annual rainfall of 1050–1200 mm; less common at lower and slightly drier levels (1350 m, rainfall 1000–1050 mm); absent from apparently suitable areas in Rwanda, Tanzania and SE Zaïre, probably because annual rainfall there < 1000 mm; may be in competition with African Reed-Warbler, which prefers lower and drier swamps (Vande weghe 1981).

Occurs singly or in pairs. Forages at all levels in papyrus; tends to feed over water. Hops from stem to stem like reed warbler *Acrocephalus*. A seasonal and erratic singer; has no apparent song perch but sings as it moves about in low vegetation.

Food. Tiny insects.

Breeding Habits. Unknown. Bird in breeding condition, Kenya, June.

References
Britton, P. L. (1978b).
Collar, N. C. and Stuart, S. N. (1985).
Keith, G. S. and Vernon, C. J. (1966).
Vande weghe, J.-P. (1981).

Genus *Hippolais* Baldenstein

Slim, medium-sized warblers with low, sloping foreheads. Bill long and strong, usually rather wide; rictal bristles short; legs strong with tarsus scutellated in front; tail square or almost so; wing long, bluntly pointed, with outermost feather minute. Feet smaller than in *Acrocephalus*, undertail-coverts shorter, plumage looser. Plain greyish or greenish brown above; some species yellow below. Tail in most edged and tipped whitish; some species with pronounced pale wing-panel. Sexes alike. Head profiles illustrated: (**A**) *H. olivetorum*, (**B**) *H. icterina*, (**C**) *H. languida*.

Inhabit woodland and bushy growth; more arboreal than *Acrocephalus*, generally preferring drier habitats. Produce chattering warbling songs and hard chacking call notes.

6 species, all in Old World. 5 in Africa, all Palearctic migrants, but one also breeds in NW Africa and another breeds in NW and NE Africa and in Sahara.

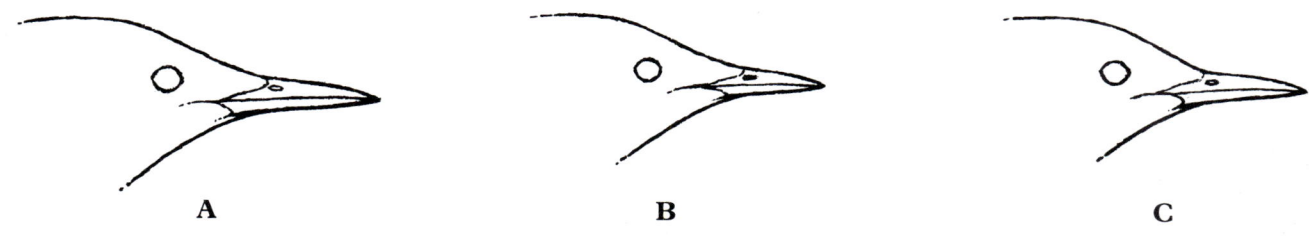

A B C

Hippolais pallida (Hemprich and Ehrenberg). Olivaceous Warbler. Hypolaïs pâle.

Plate 8 (Opp. p. 113)

Curruca pallida Hemprich and Ehrenberg, 1833. Symb. Phys. Avium, fol. bb; the Nile in Egypt and Nubia.

Range and Status. Africa south to about 12°N, Iberia, SE Europe, Near East, and Turkey to W and S Caspian, W Afghanistan, and S Kazakhstan to about 70°E. Palearctic populations winter in Africa.

3 populations in Africa: (1) Iberian and NW African race *opaca* is breeding summer visitor, widespread and common to abundant Morocco (south to Anti-Atlas), local in N Algeria (near Oran, Atlas Saharien, NE coast south to Batna); also Tunisia (south to Gafsa and Gabès) and W Libya (NW Tripolitania). Common on passage in W Saharan oases; frequent spring migrant E Libya (Cyrenaica), rare in W Egypt (Matruh). Winters across Sahel from S Mauritania and Senegal to Niger (occasional in Chad), south to N Guinea, S Mali, Burkina Faso, N Ghana, central Nigeria and N Cameroon; common to abundant. Rarely south to Ivory Coast (Lamto), S Ghana (Accra, Achimota) and S Nigeria (Lagos and Sapele). Frequently oversummers in Sahel but breeding not yet demonstrated.

(2) SE European and SW Asian race *elaeica* winters from S Chad (rarely NE Nigeria), central and NE Sudan and Eritrea south to Central African Republic (Lobaye, probably this race), NE Zaïre, Uganda, Rwanda, Kenya, Somalia (to 46°E) and NE Tanzania (to Dodoma and Dar es Salaam); locally common to abundant. Rarely SW Tanzania (Rukwa). Common passage migrant Egypt.

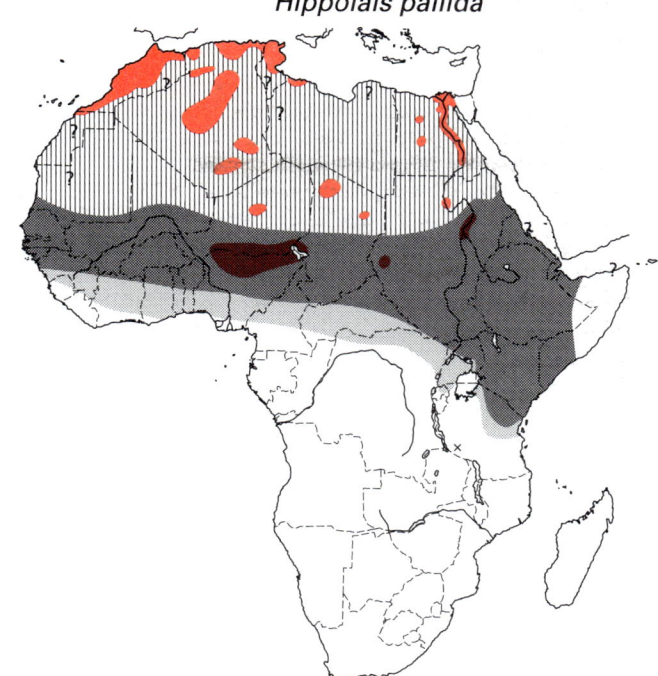

(3) Endemic small Saharan and NE African races (*reiseri*, *laeneni* and *pallida*) breed at oases in Algeria (Laghouat and Biskra south to Aïn Saleh, Ahaggar and Tassili d'Ajjer) and probably S Morocco (Tafilalet), Mauritania, S Tunisia and W Libya (Fezzan); Egypt (Nile delta and valley, Suez corridor, Wadi el Natrun, Faiyum, Western Desert oases); Niger (Aïr, S borders); Chad (Tibesti, Bol, Fada); N Nigeria (Sokoto to L. Chad); W Sudan (Darfur); N Sudan (Nile valley); and probably Eritrea and NE Somalia coasts. Abundant Egypt, frequent to locally common elsewhere. Nominate *pallida* vacates N Egypt and migrates to Sudan south to 10°N. Algerian *reiseri* is resident in Ahaggar, but a breeding visitor in the north; *laeneni* is a breeding visitor in N Niger but is resident in Nigeria. Saharan birds are frequent in N Senegal, and occasional in S Mali (Bamako) and Burkina Faso (Ouagadougou), but their status is unclear.

Description. *H. p. opaca* Cabanis: Iberia, and Morocco to NW Libya; winters W Africa east to Chad. ADULT ♂: entire upperparts greyish olive. Short supercilium buffish white; lores buffish white, ear-coverts pale brown. Tail dark brown, T6 and T5 tipped and distally edged whitish. Underparts creamy white, sides of breast and flanks washed with pale brown. Flight feathers and upperwing-coverts dark brown, broadly fringed and tipped greyish olive; axillaries and underwing-coverts creamy white. Wing rather rounded; P7 and P8 longest (rarely = P6); P6 0–2 shorter, P5 3–5, P4 4–8, P1 10–13; P9 6–9, usually between P3 and P4; P10 very small, 4–10 longer than primary coverts; P6 to P8 emarginated. Bill dark brown above, yellowish brown or flesh below; eyes brown; legs and feet pale brown, tinged grey. Sexes alike. SIZE: (10 ♂♂, 10 ♀♀) wing, ♂ 68–72 (70.6), ♀ 65–71 (68.0); tail, ♂ 53–58 (55.3), ♀ 51–57 (52.9); bill, ♂ 17–20 (18.5), ♀ 17–19 (18.1); tarsus, ♂ 21–24 (22.6), ♀ 22–24 (22.6). WEIGHT: unsexed, Senegal (n = 84, Oct–Nov) 10.4–14.5 (11.8), (n = 15, Dec–Feb) 9.2–11.9 (10.8), (n = 33, Mar–Apr) 9.4–16.8 (12.3); N Ghana (n = 11, Nov–Dec) 9.9–11.8 (11.0); Nigeria (n = 51, autumn) 8.3–13.1 (10.7), (n = 7, spring) 8.5–12.9 (11.3).

Complete moult in Africa, Oct–Jan.

IMMATURE: juvenile and 1st winter like adult, but flight feathers only slightly worn until Oct.

NESTLING: naked. Bill yellowish with dark tip above; gape yellow; 2 black tongue spots.

H. p. elaeica (Lindermeyer): SE Europe and SW Asia; winters in NE and E Africa. Smaller than *opaca*, bill rather shorter and narrower at base; less olive, colder grey-brown above, whiter below. SIZE: (10 ♂♂) wing 65–71 (67.8), bill 15–17 (16.2). WEIGHT: unsexed (n = 82, E Sudan, Aug–Oct) 7–10.5 (8.9); Ethiopia (n = 70, Jan–Mar) 8–12.3 (9.9), (n = 6, Apr–May) 8.3–10.3 (9.2); Kenya (n = 309, Nov–Jan) 7–10.9 (9.0), (n = 18, Mar–Apr) 8.7–10.8 (9.7). Wing slightly less rounded; P9 usually equals P4 or between P4 and P5; P10 3–7 longer than primary coverts (**A**).

H. p. pallida (Hemprich and Ehrenberg): Egypt; winters in S Egypt, Sudan and Ethiopia. Similar to *elaeica* but browner and slightly smaller: wing, ♂ (n = 10) 62–68 (64.4).

H. p. reiseri Hilgert: Algerian Sahara and perhaps S Tunisia, S Morocco, Mauritania and Libya (Fezzan). Paler and sandier than *opaca* or *elaeica*. Similar in size to *elaeica*: wing, ♂ (n = 8) 64–68 (66.6).

H. p. laeneni Niethammer: Niger, Chad, Nigeria and (perhaps this race) N Sudan. Paler and greyer than *reiseri*, smaller: wing, ♂ (n = 6) 60–66. WEIGHT: (Niger, June) ♂ (n = 9) 8.6–10.3 (9.4), ♀ (n = 4) 8.4–9.9 (9.3).

Field Characters. A smallish warbler (Sahara, E Africa) or medium-sized one (W Africa) with a rather long bill. Olivaceous or greyish brown above and whitish below, with short pale supercilium and whitish tips to fresh outer tail-feathers. Lacks pale wing-panel. Rather short-winged for a *Hippolais*, the exposed wing-point only a quarter of length of closed wing. Legs pinkish or brownish grey. In W Africa, *opaca* distinguished from Melodious Warbler *H. polyglotta* by lack of yellow, and by longer bill, accentuated by flat forecrown. In E Africa, *elaeica* distinguished from similar Upcher's Warbler *H. languida* by smaller size, browner upperparts, slightly shorter bill and tail, lack of wing-panel, and voice; in hand, P10 rather longer (**A**). Separated from Eurasian Reed-Warbler *Acrocephalus scirpaceus* and Marsh Warbler *A. palustris* by squarer tail, shorter undertail-coverts and shorter exposed wing-point; also by 'bare' looking lores and greyer toned plumage.

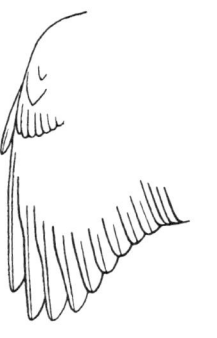

A

Voice. Tape-recorded (57, GREG, HOR, McVIC, MOR, PEAR). Song, a vigorous scratchy warble, chattering and repetitive, with underlying rhythm recalling Eurasian Reed-Warbler; duration a few s to a minute or more. Contact-alarm calls, a persistent repeated hard 'tak . . . tak . . . tak . . .'; also 'trrt . . . trrt . . . trrt . . .'; and a series of sharp clicking sounds given, for instance, when bird disturbed at nest.

General Habits. On breeding grounds in N Africa frequents olive groves, orchards, gardens, tamarisk thickets and palm oases; often occurs in trees in built-up areas. Commonly up to 1200 m in High Atlas, and occasionally over 2000 m. Winters mainly in arid or semi-arid country. In W Africa inhabits tall dense stands of *Acacia nilotica*, big trees, gardens and riparian woodland. In E Africa occurs in open woodland, thornbush and dry thicket, especially large acacias and *Balanites*; up to 1700 m in Kenya.

Usually solitary in winter quarters, but forms loose feeding groups and associates with mixed bird parties. In Sudan, occurs with Common Chiffchaffs *Phylloscopus collybita* and Lesser Whitethroats *Sylvia curruca* in acacias (Hogg *et al.* 1984). Some remain sedentary for weeks, giving regular and persistent song from same post, even through hot hours of day. Fidelity to wintering site in Senegal, Ethiopia and Kenya shown by capture of birds ringed 1–4 years previously (Cramp 1992).

In Tsavo, Kenya, peak winter count was of at least 14 birds per 10 ha (Lack 1985); in Senegal 6–8 per ha in *A. nilotica* woodland throughout most of year (Morel and Roux 1966). Restless but not shy. Forages with deliberate movements in open canopy and bush tops, picking insects from vegetation. More arboreal in E Africa than Upcher's Warbler. In Tsavo, Kenya, average foraging height 5·7 m; 86% of prey taken from leaves, 9% from twigs and stems, 1% from the air (Lack 1985). Carriage typically horizontal. Calls persistently and flicks tail. When excited raises feathers to produce peak at centre of crown.

Race *opaca* passes through Morocco in late July–Oct (peak in 2nd half Aug); vacates NW African breeding grounds during mid-Aug to Oct. Southward movement through Africa masked by overlap with *reiseri* and *laeneni*: arrives in Mali Aug–Sept, N Nigeria from end Aug. Most birds leave wintering areas in mid-Apr, but some remain in Mali until May, and a few oversummer in the Sahel. Spring passage in Mauritania occurs Apr–June, and in N Africa from late Mar to early June. Returns to N African breeding sites between mid-Apr and early May. Bird ringed Spain recovered Morocco in Sept.

H. p. elaeica migrates mainly east of Mediterranean in autumn: common in Levant but few in N Egypt. Autumn passage on Sudan coast mainly from early Aug to mid-Sept, in Chad (probably this race) in Sept–Oct; arrival in Ethiopia occurs in Aug–Sept. First birds reach Uganda and Kenya about mid-Oct, but most arrive in Nov–Dec, when some are already partly or fully moulted. Main spring departure from E Africa late Mar to mid-Apr, but some birds remain in Ethiopia until mid-May and in Sudan until late May (occasionally June). Main passage in Egypt, mid-Mar to mid-May. Nominate *pallida* quits breeding areas in N Egypt in Aug–Sept, returning in late Mar to early Apr. Races *reiseri* and *laeneni* are partially migratory, vacating areas further north in Sahara after breeding (Cramp 1992).

Food. Small arthropods; occasionally plant material. Birds wintering Ghana, Senegal and Gambia ate small beetles, flies, ants and spiders (Wink 1981, Cramp 1992). In Niger, June, *laeneni* fed on beetles, Lepidoptera, grasshoppers, termites, larval Diptera, spiders and plant material (Fairon 1975); in N Nigeria (L. Chad) ate small insects and *Salvadora* berries (Fry et al. 1970). Food of nestlings in N Morocco, June–July, includes dragonflies, adult Lepidoptera and spiders; in NW Libya fledglings ate green caterpillars (Cramp 1992).

Breeding Habits. Monogamous. Territorial but tends to form loose breeding concentrations. In Egypt, 4 pairs in *c.* 2·4 ha (Simmons 1954); territory used for pair formation, nesting and rearing young. In Egypt, ♂ arrives several days before ♀; sings from perch near top of bush, usually within foliage, and builds rough nests (Simmons 1952). ♂ in threat display sang with tail fanned and wings shivered and lowered. In courtship, ♂ pursued ♀, crouched with ruffled body-feathers and wings drooped, flicked wings and tail erratically and gave bursts of subsong; ♀ responded with less intense version of same display (Simmons 1952). Song decreases after pairing and reaches lowest level during feeding of young.

NEST: strong cup of grass stems, fine twigs, sedges and hair, woven to plant stems, lined with fine fibres, hair or feathers; in outer twigs of shrubs or small trees; in Egypt often in creepers, hedges, small palms, garden plants, usually 0·3–2 m above ground; in NW Africa usually in tamarisk *c.* 3 m high, or in citrus or olive groves; built by ♂ only (Egypt) or both sexes; ext. diam. 75–90, int. diam. *c.* 45, cup depth *c.* 45.

EGGS: 2–4 (usually 3) Egypt; 4–5 Tunisia; 2–5 Morocco and Algeria, mean (n = 47 clutches) 3·3; 2 in 1 clutch L. Chad (*laeneni*). Pale greyish white, but sometimes pinkish or lilac with a few black speckles; laid at daily intervals. SIZE: 17–20 × 12–15 (n = 17, *opaca*); 18–19 × 14 (n = 3, *reiseri*).

LAYING DATES: Egypt (*pallida*) Apr; Morocco and N Algeria (*opaca*) early May to early July; S Algeria (*reiseri*) May; L. Chad (*laeneni*) Mar–June.

INCUBATION: usually by ♀ who is occasionally fed by ♂. Period: 10–12 days (Egypt).

DEVELOPMENT AND CARE OF YOUNG: young brooded at first by ♀; fed by both parents; remain in nest 11–15 days.

References
Cramp, S. (1992).
Simmons, K. E. L. (1952).

Hippolais languida (Hemprich and Ehrenberg). Upcher's Warbler. Hypolaïs d'Upcher.

Curruca languida Hemprich and Ehrenberg, 1833. Symb. Phys. Avium, fol. cc; Syria.

Plate 8
(Opp. p. 113)

Range and Status. Near East, SE Turkey, Azerbaijan, Iran, W Pakistan and N Afghanistan to the Kyzyl Kum and S Kazakhstan (to *c.* 80°E); winters in E Africa. Palearctic migrant. Winters in S Somalia (south of *c.* 3°N), Kenya (north and east of the highlands, but absent from coast), S Ethiopia (west of Dolo), E

Hippolais languida

A

Uganda (Teso, Bukedi) and NE Tanzania (Moshi and Amani); locally common to abundant in S Somalia and SE Kenya, frequent or uncommon elsewhere. On southward passage, common in N Somalia, Eritrea, central and E Ethiopia and Djibouti; rare on Sudan coast. On northward passage, common in N Somalia, scarce in central Ethiopia and Eritrea.

Description. ADULT ♂: upperparts pale brownish grey. Supercilium, extending to behind eye, and narrow eye-ring whitish; lores, ear-coverts and sides of neck pale brownish grey. Tail blackish, T6 narrowly edged white and broadly tipped white on inner web, T5 narrowly tipped white. Underparts whitish, sides of breast, flanks and undertail-coverts suffused with pale greyish buff. Flight feathers blackish brown, primaries narrowly edged grey, secondaries and tertials edged white when fresh to form a noticeable pale panel on closed wing. Upperwing-coverts pale brownish grey; underwing-coverts and axillaries pale buffish white. Wing-tip bluntly pointed; P7 and P8 longest; P6 0·5–2 shorter, P5 2–6, P4 6–9, P1 14–17; P9 2–7, between P5 and P6 (sometimes P4 and P5); P10 minute, 4 shorter to 2 longer than primary coverts; P6 to P8 emarginated (**A**). Bill dark brown above, pinkish flesh below; eyes dark reddish brown; legs and feet light brown or pinkish brown, tinged grey. Sexes alike. SIZE: (10 ♂♂, 10 ♀♀, Asia) wing, ♂ 75–78 (76·4), ♀ 73–79 (75·4); tail, ♂ 60–64 (61·9), ♀ 59–63 (61·2); bill, ♂ 17–20 (18·3), ♀ 17–19 (18·0); tarsus, ♂ 22–23 (22·4), ♀ 22–23 (22·7). WEIGHT: unsexed, Kenya (n = 241, Nov–Jan) 9·9–17 (12·0), (n = 13, Mar–Apr) 12·1–13·6 (13·0).

Complete moult in Africa, most birds starting Sept–Nov and finishing Dec–Jan.

IMMATURE: 1st winter like adult, but flight feathers remain relatively unworn until Sept–Oct.

Field Characters. A medium-sized warbler with long bill, pale greyish above and rather clean white below, with white supercilium extending to behind eye. In spring (and probably all year) its continuous, emphatic tail-waving distinguishes it, even at a distance or in silhouette, from all other warblers except Ménétries's Warbler *Sylvia mystacea* and perhaps Arabian Warbler *S. leucomelaena*; also flicks wings and flags one wing (Fry 1990 and pers. comm.). Tail quite long and bulky looking; dark when freshly moulted and with white outer feather tips and narrow white outer edge. Wings dark when freshly moulted, with paler panel on secondaries. Bill pinkish below. Legs greyish. Slightly larger than Olivaceous Warbler *H. pallida* of the race *elaeica* (with which it occurs in Africa), and has longer, fuller tail and rather longer bill. Upperparts greyer than in Olivaceous, contrasting more with darker wings; also distinguished by voice and by characteristic tail movements (see below). Smaller than Olive-tree Warbler *H. olivetorum* and paler grey, with longer supercilium and paler cheeks; bill and legs more slender. In the hand, wing-tip slightly less rounded than in Olivaceous, with shorter P10, but less pointed than in Olive-tree Warbler.

Voice. Tape-recorded (F, GREG, McVIC, PEAR – 92, 94). Song, a fluent warbling with a sweet quality suggesting a quiet *Sylvia* warbler; often sustained for over 20 s. Contact call, a repeated 'tuc', likened to knocking two stones together (Shirihai 1987); less hard and persistent than similar call of Olivaceous Warbler; also a low churring alarm.

General Habits. Winters in scattered bush and thicket and in *Commiphora* or low acacia woodland, often with little leaf; mainly in hot arid to semi-arid country below 1200 m.

Usually dispersed in ones and twos although groups of 5–10 occur on passage. Apparently territorial in late winter when aggressive chasing occurs in Kenya. Forages unobtrusively within bushes or low trees, hopping among dense or open thorn branches and twigs, typically with rather horizontal carriage. Gleans arthropods from fine leaves of acacias. Extends legs and peers under foliage looking for prey. Makes occasional flycatching sallies. Searches all levels and sometimes descends briefly to ground, pouncing on prey from low perch, or hopping like a chat (Fry 1990). In Tsavo, Kenya, average foraging height 4·2 m; 60% of prey taken from leaves, 37% from twigs or stems, 3% in the air (Lack 1985). When excited raises feathers to give peak in centre of crown. Performs continuous and deliberate movements of the tail, depressing it through an arc of c. 60° then raising it back to body axis, fanning it slightly and usually swinging it to one side or the other on downward wag. Rate of tail movement not constant, but averages about once per s. Flicks wings frequently; in Oman, sometimes raises one wing sideways or upward (**B**) (Fry 1990). Rather shy, and flies to next bush when approached within a few m. Flight low and direct with tail held straight. Sings on southward passage in Sept–Oct, and also frequently in winter quarters in Feb–Mar.

Birds move south or southwest across Arabia in autumn, then through Somalia and Ethiopia along and east of Rift. Passage in Eritrea from mid-Aug to early Oct, in N Somalia from mid-Aug; main movement through central Ethiopia late Sept to early Oct, with last birds late Oct. Arrival in S Somalia and Kenya is not until Nov–Dec; most reach wintering sites in Tsavo, SE Kenya, in late Dec–Jan; birds thus remain for about 2 months in NE Africa. Most begin wing-moult there and finish later further south. Present in SE Kenya and S

B

Somalia wintering sites until late Mar or early Apr. Spring passage in central Ethiopia, mid-Mar to early Apr, much lighter than autumn passage. Last birds Kenya are in late Apr, in Somalia in early May.

Food. Chiefly invertebrates; no details.

References
Cramp, S. (1992).
Fry, C. H. (1990).
Shirihai, H. (1987).

Hippolais olivetorum (Strickland). Olive-tree Warbler. Hypolaïs des oliviers.

Salicaria olivetorum Strickland, 1837. *In* Gould's Bds. Europe 2, pl. 107; Zante, Ionian Is.

Plate 8
(Opp. p. 113)

Range and Status. Balkans, Crete, and W and S Turkey; winters in Africa.

Palearctic migrant. Main wintering area southern Africa, from S Zambia (Upper Zambezi valley) to W and S Zimbabwe, E and S Botswana and N South Africa (N Transvaal); frequent to locally common; occasional Namibia (Windhoek) and South Africa south to Johannesburg and N Natal. Also winters at scattered sites in E Africa: N Kenya (Baringo, Isiolo, Ololokwe), SE Kenya (Tsavo East, Galana Ranch, Mombasa), central Tanzania (Dodoma) and probably SW Tanzania (Rukwa); local and uncommon. Birds Eritrea (Jan, Massawa: Zedlitz 1910–11) and Egypt (Feb and early Mar: Goodman and Meininger 1989) were probably overwintering. On southward passage occurs in Egypt (mainly NE, uncommon), N Sudan (Khartoum, Wad Medani, Suakin, apparently rare), Eritrea (once, Aruba) and Djibouti (once); locally frequent in Kenya north and east of highlands (once in west at Usengi), SW Tanzania (Rukwa), central and S Zambia, and S Malawi. Vagrant west to Niger (once Aug, N Aïr) and N Nigeria (once Oct, Kano). Remarkably few records on northward passage: N Tanzania (once, Arusha), N, E and coastal Kenya (uncommon), N Somalia (once, Alula), Eritrea (once, Asmara), Egypt (once, Cairo) and Libya (once, Serir). Also records from SE Egypt in late June and N Sudan in early July, but status of these birds unclear.

Hippolais olivetorum

Description. ADULT ♂: upperparts brownish grey; supercilium from bill to eye and narrow eye-ring whitish, lores grey; ear-coverts brownish grey. Tail dark brown, T6 whitish on outer web, T6 and T5 with white crescent at tip. Underparts whitish suffused with pale creamy on breast and with grey on flanks. Flight feathers dark brown, secondaries and tertials edged whitish, producing a conspicuous pale wing-panel; upperwing-coverts brown, greaters edged and tipped whitish; underwing-coverts and axillaries creamy white. Wing bluntly pointed; P8 longest (sometimes = P7), P7 0–3 shorter, P6 5–7, P5 7–11, P1 20–24; P9 1·5–4, between P6 and P7; P10 minute, 3–8 shorter than primary coverts; P7 and P8 emarginated. Bill dark horn above, orange-yellow below with dusky tip; eyes dark brown; legs and feet dull bluish grey. Sexes alike. SIZE: (10 ♂♂, 5 ♀♀, SE Europe and Africa) wing, ♂ 85–89 (87·3), ♀ 83–86 (84·4); tail, ♂ 63–70 (67·2), ♀ 64–68 (66·2); bill, ♂ 19–21 (20·2), ♀ 20–21 (20·3); tarsus, ♂ 23–24 (23·5), ♀♀ all 23. WEIGHT: unsexed, Kenya (n = 270, Nov–Dec) 13·7–23·1 (17·5), (n = 2, Apr) 17·1, 16·2; Malaŵi and Zambia (n = 4, Nov–Mar) 15·5–17·9 (16·6).

Complete moult in Africa Dec–Mar.

IMMATURE: 1st winter like adult, but flight feathers unworn until Oct.

Field Characters. A distinctive, large, greyish warbler with rather flat forehead, long strong bill, strong grey legs and long wings. Exposed wing-point about one-third of total length of closed wing. Brownish grey above and dusky white below; wings noticeably browner than back, with pale panel prominent in fresh plumage. Tail-feather tips and outer tail edge whitish. Larger than Upcher's Warbler *H. languida*, with darker grey upperparts, darker cheeks, and stouter bill (see generic diagnosis) with orange-yellow (rather than pinkish) lower mandible; supercilium shorter, extending back only to eye. In the hand, wing-tip more pointed than in Upcher's Warbler, with smaller P10 and larger P9. Close in size to Barred Warbler *Sylvia nisoria*, but Olive-tree Warbler has longer bill, shorter tail and eyes always dark.

Voice. Tape-recorded (75, 88, 91, F, CHA, GREG, McVIC). Song a deep, rich medley with an underlying rhythm, slower in delivery than in other *Hippolais*; includes croaks and squawks reminiscent of Great Reed-Warbler *Acrocephalus arundinaceus*, but less harsh and more musical; sustained for 3–15 s. Contact call, a deep hard 'chack'; gives a loud nasal 'chaarr' in alarm.

General Habits. Frequents open dry bush and woodland, especially acacia thornbush. Usually singly. Territorial in winter quarters: 1 ringed at Pretoria site late Jan retrapped 42 days later (Cramp 1992); 1 sang at same Tanzanian site up to 43 days and returned there in 3 successive years (Harpum 1978a). In winter quarters commonly associates with Red-backed Shrike *Lanius collurio*. Unobtrusive but often quite approachable. Less skulking in Africa than Upcher's Warbler, foraging more in open canopy and among outer twigs of bushes and small trees; also in low scrub and on ground. Hops and clambers rather heavily. Frequently flicks tail. Flight from one bush to another strong and buoyant. Song heard in Djibouti in Oct, and occasionally uttered in autumn in Kenya; more frequent and sustained in winter quarters in Feb–Mar.

Migrations through Africa poorly documented. The few autumn records in Egypt and Sudan are mid-Aug to early Oct. Most birds presumably spend Sept–Oct in Ethiopia, although there are no records there at this time. Passage regular through Kenya, north and east of highlands, from early Nov (earliest 2 Nov) to early Dec; over 500 ringed to 1992 at Ngulia, nearly all in Nov, many with high weights suggesting that they make long onward flights. Occurs SW Tanzania in Oct–Jan. Reaches Zambia and Malaŵi from late Nov and Botswana from mid-Nov. Southern African wintering area occupied mainly from Dec to late Mar. A few birds remain in Botswana until early Apr, and Malaŵi until mid-Apr. Sparse northward passage through Zambia and Malaŵi, and through E and N Kenya in early and mid-April. Rare in spring in NE Africa: single records Libya (Apr), Egypt (Apr), Eritrea (May) and N Somalia (May).

Food. Mainly invertebrates; little information.

Reference
Cramp. S. (1992).

Plate 8
(Opp. p. 113)

Hippolais polyglotta (Vieillot). Melodious Warbler. Hypolaïs polyglotte.

Sylvia polyglotta Vieillot, 1817. Nouv. Dict. d'Hist. Nat., nouv. éd. II, p. 200; France.

Range and Status. SW Europe and NW Africa; winters in W Africa.

Palearctic migrant. Breeds commonly N Morocco (south to Meknès and Fes, Middle Atlas, High Atlas foothills), N Algeria (the Tell, Aurès Mts, near Aïn Sefra, Djebel Amour) and frequently NW Tunisia (east to L. Ichkeul, south to Kef). Winters in W Africa; from Gambia, S Senegal, Guinea, S Mali (Bamako) and Burkina Faso (south of 11°N) to Sierra Leone, Liberia, Ivory Coast and Ghana, and east to Togo, Nigeria (north to Kano) and NW Cameroon; locally common to abundant south of 8–9°N and east to S Nigeria. On passage, frequent to common in Morocco and Algeria east to Tassili d'Ajjer (mainly in spring); rare in Tunisia and NW Libya; frequent to common in S Mali (both seasons), S Mauritania and Senegal (mainly autumn), N Nigeria (both seasons) and central Burkina Faso (spring); scarce S Niger (May, Sept); 3 records for Central African Republic, Mar (Germain and Cornet 1994).

Hippolais polyglotta

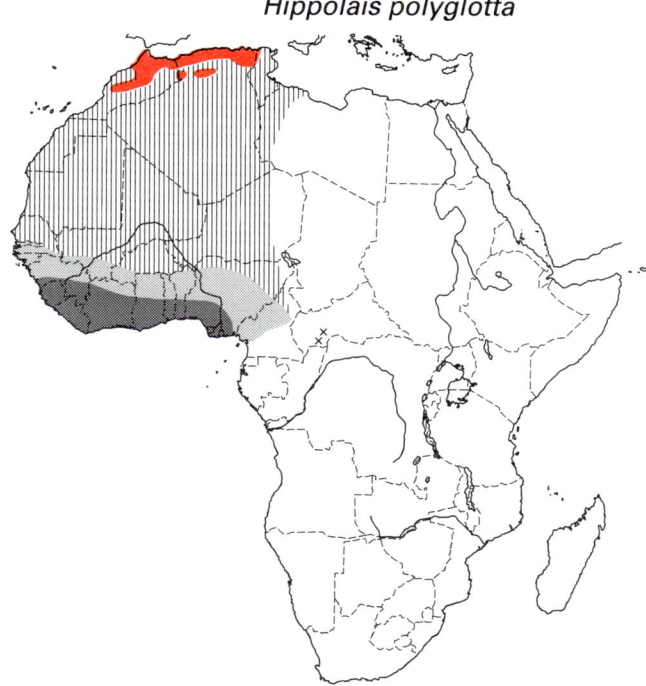

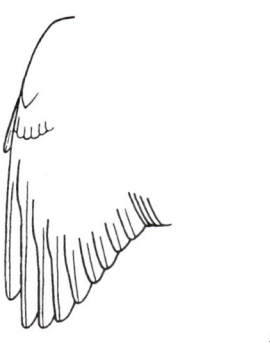

A

Description. ADULT ♂: upperparts greenish brown. Short supercilium yellow; lores and ear-coverts olive-yellow. Tail dark brown. Entire underparts yellow, sides of breast and flanks tinged brownish. Occasional pale variant has sides of face and underparts pale yellow or almost white. Flight feathers and upperwing-coverts dark brown, edged buffish; underwing-coverts, axillaries and bend of wing yellow. Wing-tip rounded; P7 and P8 longest; P6 0·5–1·5 shorter, P5 2–5, P4 5–7, P1 13–15; P9 4–10, shorter than P5; P10 very small, 3–8 longer than primary coverts; P6 to P8 emarginated. Bill dark brown above, yellowish flesh below; eyes dark brown; legs and feet brownish horn or grey. Sexes alike. SIZE: (10 ♂♂, 10 ♀♀, SW Europe) wing, ♂ 63–69 (66·0), ♀ 61–66 (63·5); tail, ♂ 50–53 (51·5), ♀ 49–52 (50·6); bill, ♂ 15–17 (16·0), ♀ 14–17 (15·8); tarsus, ♂ 20–21 (20·5), ♀ 20–21 (20·4). WEIGHT: unsexed, Senegal (n = 7, Oct–Nov) 8·5–12 (10·3), (n = 41, Mar–Apr) 8·1–13·9 (11·1); N Ghana (n = 15, Oct–Dec) 8·5–11 (10·0); Nigeria (n = 8, Dec–Apr) 8·4–14·6 (11·3).

Complete moult in Africa, Oct–Jan.

IMMATURE: juvenile and 1st winter same as adult but browner above, usually paler below; flight feathers fresh until Oct.

NESTLING: naked. Mouth deep yellow; gape flanges pale yellow; 2 dark tongue spots.

Field Characters. A rather small, compact warbler, greenish brown above and rich yellow below. Short yellow supercilium contrasts with dark crown, less so with ear-coverts. Lores pale, giving typical *Hippolais* 'barefaced' appearance. Slightly smaller than Icterine Warbler *H. icterina* with shorter, more rounded wing (**A**); on closed wing, exposed primary tips are about one-quarter of total wing-length, primaries not reaching end of uppertail-coverts. Lacks pale wing-panel even in fresh plumage, and legs usually brownish grey (rather than blue-grey). Pale variants and birds in worn plumage can be confused with Olivaceous Warbler *H. pallida*, but bill shorter and head rounder; even palest Melodious Warblers show some yellow on throat and breast.

Voice. Tape-recorded (57, GR1 – 62, 73, 93, B). Song, a succession of brisk chattering warbling phrases of several s duration. Often introduced by a repeated 'twi-twi-twi . . .' and phrases separated by a sparrow-like 't-t-t-t- . . .'; lacks the power and musical variety of Icterine Warbler song, and delivery more hurried. Main contact-alarm calls, a distinctive loud 'chuk' and churring 'trr-trr', both recalling House Sparrow *Passer domesticus*.

General Habits. Breeds in bushy woodland and thick scrub; in NW Africa in tamarisks, parks, gardens and wooded hill sides up to 1800 m. Winters in savanna woodland, forest edge and clearings, secondary growth, dense bushes and gardens, and mangrove; prefers less fragmented habitat than Icterine Warbler, and avoids acacia.

Usually solitary, but once 12 in a tree (Gambia, Jan: Cawkell and Moreau 1963). Largely territorial in winter quarters. Density of tens per ha in Guinea (Nimba: Brosset 1984). Forages within bushes or tree canopy, usually below 3–4 m, gleaning insects from foliage and also snapping up aerial prey. More skulking, less dashing than Icterine Warbler, flight more fluttery (Cramp 1992). Carriage typically rather horizontal. Foraging movements heavy as in other *Hippolais*, and frequently flicks tail. Often sings persistently in winter quarters, from Dec onwards.

Most European birds head southwest in autumn and enter Africa via S Iberia. They continue south together with NW African populations, apparently keeping near the W African coast. Main passage in Morocco Aug to early Sept, with stragglers to mid-Oct; in Mauritania Aug–Oct; in Senegal mid-Aug to early Nov; and in Mali Aug–Oct. Common and moulting in N Nigeria (Kano) in late Sept to late Oct, but most move on during Nov (Aidley and Wilkinson 1987a). Arrives in main guinea savanna wintering area in mid-Oct to Nov. Spring movement begins in Senegal and Mauritania in late Feb to Mar, but most birds are still on wintering grounds until early Apr, some until May. Returns to NW African breeding sites in Apr to early May. Spring passage through Morocco and Algeria is heavier than in autumn, mainly late Apr to mid-May. Migrants occur in both seasons in Saharan oases. Bird ringed France recovered Morocco (Apr); 1 ringed Morocco (May) recovered France.

Food. Adult and larval insects, especially small flies and aphids; spiders; some fruit prior to migration.

Breeding Habits. Monogamous; probably sometimes polygynous. Territorial, but may form neighbourhood groups. Territory size in Europe ranges from < 0.36 to c. 3 ha (Cramp 1992). In Morocco, pairs scattered (Brosset 1961); 1.5 pairs per km² in woodland, 2·8 pairs per km² in maquis (Thévenot 1982). Territory serves for courtship, nesting and some feeding; established by ♂, who arrives a few days before ♀, and defended by song of ♂ to about July. ♂ usually has several song posts, and also sings in 'butterfly-flight' with wings outspread and beating quickly and regularly (Cramp 1992). Neighbouring ♂♂ confront each other, singing from prominent perches with crown-feathers raised.

NEST: deep cup of grasses and plant stems lined with fine grass, seeds, plant down and hair; outer diam. c. 80, inner diam. c. 50, cup depth c. 40; in fork in shrub, usually 0·4–1·3 m from ground, sometimes in small olive tree; built by ♀, accompanied and watched by ♂.

EGGS: 3–5. Pinkish with black spots and faint dark hairlines. SIZE: (n = 100, Europe) 16–20 × 12–15 (17·8 × 13·3).

LAYING DATES: Morocco, early June.

INCUBATION: by ♀ alone; starts when clutch complete. Period: 12–13 days.

DEVELOPMENT AND CARE OF YOUNG: young brooded for c. 5 days by ♀; fed by both parents; remain in nest for about 12 days; fed by parents for about 9 days after fledging.

BREEDING SUCCESS/SURVIVAL: of 125 eggs in 28 nests in France, 66% hatched and 79% of the young fledged (overall success thus 47%: Cramp 1992).

Reference
Cramp, S. (1992).

Plate 8
(Opp. p. 113)

Hippolais icterina (Vieillot). **Icterine Warbler. Hypolaïs ictérine.**

Sylvia icterina Vieillot, 1817. Nouv. Dict. d'Hist. Nat. 11, p. 194; Nancy, France.

Range and Status. Central, N and E Europe east to about 90°E in W Siberia; also N Iran. Winters in Africa.

Palearctic migrant. Winters mainly south of equator, from E and S Zaïre, Rwanda, Burundi, S Uganda, SW Kenya and W Tanzania (to c. 35°E) to Angola, Zambia, Malaŵi, central and S Mozambique, Zimbabwe, N and NE Namibia, Botswana, Swaziland and South Africa (Transvaal, N Cape, Natal, rarely E Cape, once in Cape of Good Hope Nature Reserve: Fraser 1986); scarce Uganda and Kenya, but locally common to abundant from central Zambia, S Malaŵi and interior of Angola to Botswana and N Transvaal. North of equator, presumed wintering birds reported in S Mali (Dec–Feb), Ivory Coast (once Dec), S Nigeria (once Feb), and Egypt (once Jan, once Feb).

On passage N Africa, rare in autumn in Tunisia, uncommon in Libya and Egypt; in spring, rare in Morocco (12 records, mainly in E), occasional NW Algeria, common from E Algeria to Egypt. South of Sahara, common locally on southward passage in E Nigeria and Chad; frequent N Zaïre, Rwanda and W Uganda; occasional NE and central Sudan, Ethiopia and Kenya. Locally abundant on both passages in Zambia. On northward migration, locally common in Nigeria; uncommon in Uganda, Kenya and Ethiopia, rare Somalia. West of main migrations, recorded in Mali (Sahel and Niger delta; autumn), Togo (once Nov, once Mar), Liberia (once Mar) and Ghana (twice Apr).

Description. ADULT ♂: upperparts greenish brown; supercilium pale yellow; lores yellowish, ear-coverts yellow-olive. Tail dark brown, outer feathers narrowly edged yellowish. Entire underparts lemon yellow, sides of breast and flanks tinged brownish. Flight feathers and primary coverts blackish brown,

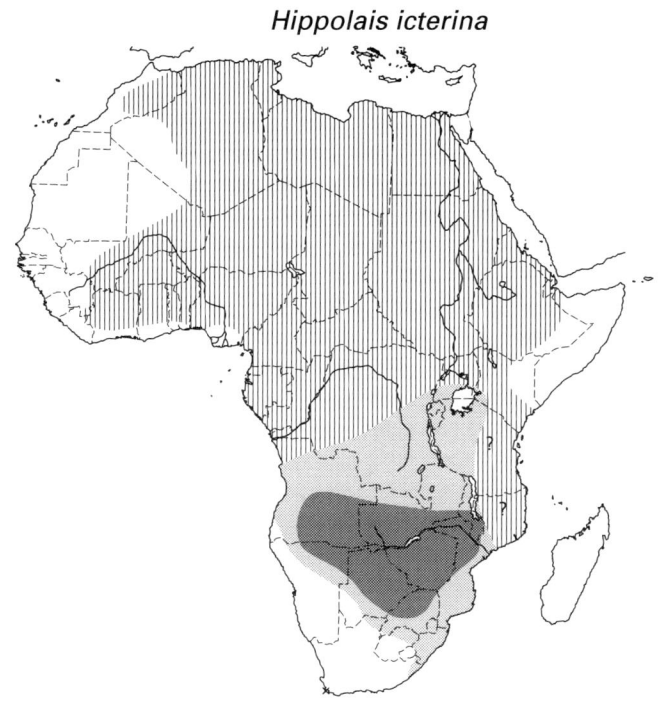
Hippolais icterina

primaries and primary coverts narrowly edged olive, secondaries and tertials more broadly edged golden yellow; greater and median coverts dark brown, fringed yellowish; lesser coverts olive-brown. Bright edging to secondaries, tertials and greater coverts forms noticeable pale panel on closed wing. Axillaries, underwing-coverts and bend of wing pale yellow. Wing-tip bluntly pointed; P8 longest (occasionally = P7), P7 0–3 shorter, P6 3–6, P5 7–11, P1 20–25; P9 1·5–5, between

P6 and P7 (rarely equal to or shorter than P6); P10 minute, 3 mm shorter to 3 mm longer than primary coverts; P7 and P8 emarginated, and sometimes tip of P6. Bill dark brown above, yellowish flesh below; eyes dark brown or olive-brown; legs and feet bluish grey. Sexes alike. A variant form has upperparts browner or greyer, underparts paler yellow, belly and undertail-coverts almost white. SIZE: (10 ♂♂, 10 ♀♀, Europe) wing, ♂ 78–82 (79·8), ♀ 74–79 (77·0); tail, ♂ 52–58 (54·8), ♀ 51–56 (54·0); bill, ♂ 17–18 (17·4), ♀ 16–18 (16·7); tarsus, ♂ 20–22 (21·1), ♀ 20–22 (21·0). WEIGHT: unsexed, N Nigeria (n = 11, autumn) 9·1–12·9 (11·4), (n = 65, spring) 10·4–22·8 (16·2); (n = 24, central Nigeria, Apr–May) 10–18·5 (15·5); (n = 6, Kenya, Nov–Dec) 9·4–17·8 (12·3); (n = 14, South Africa (Transvaal)) 12–18·6 (14·3).

Complete moult in Africa Dec–Feb.

IMMATURE: 1st winter bird same as adult, except that flight feathers are fresh, and pale wing-panel is often noticeable until Oct–Nov.

Field Characters. A medium-sized warbler, greenish olive above and lemon yellow below, with rather flat crown and long wings. Yellow supercilium contrasts with olive crown and ear-coverts, but pale yellow on lores and below eye gives bare-faced look. Fresh-plumaged bird shows conspicuous yellowish wing panel. Rather broad bill has pinkish base to lower mandible. Distinguished from very similar Melodious Warbler *H. polyglotta* by wing-panel (if present), blue-grey (not grey-brown) legs, and longer, more pointed wing (**A**); exposed wing-tip almost one-third of total closed wing length, and primary tips reach end of uppertail-coverts. Voice also distinctive. Some birds are greyer, with little green or yellow in plumage; these can be confused with Olivaceous Warbler *H. pallida*, especially western *opaca*, but have shorter bill, longer wing and bluer legs, and never show white on tail-feather tips. Olive-tree Warbler *H. olivetorum* has long wings like Icterine and often shows pale wing-panel, but is larger, with longer bill, and is never yellow. Yellow-tinged migrant *Phylloscopus* warblers are smaller than Icterine with slender legs, fine bills and dark streak through lores.

Voice. Tape-recorded (57, 86, 88, 91, F, WALK). Song, a vigorous sustained medley, in which rich musical passages are interspersed with harsh chattering and strident sounds; includes much repetition and mimicry, and at times recalls Marsh Warbler *Acrocephalus palustris*; frequently includes a characteristic 3-syllabled 'di-de-roi'. Contact-alarm, 'tec' or 'tseck'.

General Habits. Mainly inhabits open woodland with scattered trees and bushes, especially acacia. In N Africa occurs in palm groves, gardens, orchards and scattered desert scrub; in southern Africa in miombo (at least on passage).

Usually solitary, although small groups may occur on spring migration. Territorial in winter quarters; individuals sometimes remain in same garden for months and return in successive years. Forages actively among foliage, often over 10 m from ground but also low down. Takes insects from leaves and twigs while perched or fluttering and flies out to catch aerial prey

A

(Cramp 1992). Carriage often rather upright. Crown-feathers raised when excited to give obvious peak above eye. Frequently flicks tail. Flight buoyant and agile, at times recalling that of Spotted Flycatcher *Muscicapa striata*. Sings frequently in winter quarters from Nov onwards, usually hidden in foliage near top of tree.

Birds from NW and central Europe move south southeast. Autumn passage apparently concentrated across E Mediterranean and through Chad; however, only small numbers occur in NE Africa. Arrival in Africa is early, with passage between Tunisia and Egypt in Aug–Oct, Sudan and Ethiopian coasts late Aug to mid-Sept, Chad late Aug to mid-Oct and NE Nigeria Sept–Oct. Birds move on quickly to southern Africa, most using a route west of L. Victoria. Passage is conspicuous in Rwanda Oct to early Nov, and the species is regular then in W Uganda. Occurs in coastal Eritrea early to mid-Sept, and occasionally in Kenya Oct to early Dec. Arrives in Zaïre from Sept, Zambia Oct–Nov, Zimbabwe, Botswana and South Africa from Oct but mainly in Nov. Northward movement begins in Zambia in late Feb, but most birds remain in wintering sites to late Mar. Pronounced passage through Zimbabwe and Zambia late Mar to early Apr. Northbound migrants occur west to W Nigeria and E Algeria: a strong passage at Zaria, Kano and L. Chad Apr to mid-May (peak late Apr). The few spring records in Kenya and Ethiopia are early to mid-Apr. Migration across N Africa is mainly late Apr to mid-May, much heavier than in autumn with more birds apparently crossing W Mediterranean. Recoveries from Italy to Tunisia (1), Belgium to Ghana (1, Apr), Finland to Congo (1, Apr) and Tunisia to SW Zaïre (1). 1 ringed Tunisia (May) recovered Poland.

Spring migrants in Nigeria move at night, when some killed by striking wires *c*. 20 m high (Fry 1970).

Food. No information in Africa; in Europe insects and their larvae, especially Diptera, adult and larval Lepidoptera, aphids and small beetles; also spiders; occasional berries.

Reference
Cramp, S. (1992).

Genus *Cisticola* Kaup

Medium-heavy to tiny warblers, ♂♂ same size as ♀♀ or up to 20% larger; bill stout and markedly decurved to short, slender and slightly curved; bill comparatively heavy, perhaps adapted for terrestrial foraging. Rictal bristles small except in 3 largest species; wings short and blunt, P3–P6 emarginated, P1 notched in 3 species and very short (P1 < half of P2) in smallest species, tail of 12 feathers, graduated (rounded in smallest species), shorter in breeding than in non-breeding plumage, in some species considerably so. Plumage mainly brown above, streaked or unstreaked, crown and wing-patch often rufous, pale to rich buff below; tail usually of 'spotted fan' pattern, with subterminal dark band and pale tip, but spotting obscure in some species and almost absent in a few (plain black in *melanurus*); underside of tail paler, making spots more evident. Two moults a year in most species, post-nuptial (complete) and pre-nuptial (partial); in equatorial regions, most populations have only a single annual moult; in many species, breeding and non-breeding plumages differ; sexes alike except some ♂♂ brighter in breeding plumage; palate black in breeding ♂♂, pink in most breeding ♀♀ (but brown in *galactotes*, *hunteri* and *ruficeps*) and in non-breeding ♂♂. Juveniles often with sulphur-yellow suffusion on underparts.

Inhabit grassland, wooded and bushy savanna, scrub and marshes; none truly adapted to desert conditions. Mainly sedentary. Generally remain within cover; largely terrestrial and low-level foragers, creeping within vegetation and running on ground. Breeding ♂♂ sing conspicuously from exposed perch or in flight. Songs and calls distinctive. Often several species sympatric; they then show marked lack of interspecific interaction. Best identified by combination of major plumage characters, voice, display and habitat. Nest usually enclosed, typically a ball with side-top entrance or oval with entrance at top, or stitched within large living leaves. Egg colour highly variable.

45 species; 42 confined to Africa, one (*juncidis*) in Africa, Eurasia and Australia, one (*cherina*) in Madagascar and one (*exilis*) in S India to Australia. Relationships within *Cisticola* remain unclear; while some natural groups can be identified, the links between these groups are often obscure. We recognize the following superspecies: the widely-accepted *C. woosnami*/*C. anonymus*/*C. bulliens*, *C. hunteri*/*C. chubbi*/*C. nigriloris*, *C. rufus*/*C. troglotytes*, and *C. juncidis*/*C. haesitatus*, also *C. fulvicapillus*/*C. angusticaudus* which have been considered conspecific, and *C. brunnescens*/*C. cinnamomeus* which are here proposed as separate species.

C. erythrops and *C. cantans* are closely related (placed in a superspecies by Hall and Moreau 1970), but we agree with Short *et al.* (1990) that there is too much sympatry for them to be considered allospecies. *C. lateralis* is closely related to the *C. woosnami* superspecies (perhaps originally the W African counterpart of it) but its range at least partly overlaps with those of all 3 members. *C. chiniana* and *C. bodessa* are sibling species with considerable geographical overlap. *C. njombe* has been linked with this species-pair (Hall and Moreau 1970) but it may not be closely related to them (Short *et al.* 1990); the relationships of *C. njombe* and *C. aberrans* are uncertain. *C. subruficapillus*, *C. rufilatus*, *C. cinereolus* and the enigmatic *C. restrictus* may form a lowland, dry-country species group, while *C. lais* may be a highland offshoot of it or may be closest to *C. subruficapillus*. The heavily streaked, dark-backed marshland species *C. galactotes*, *C. carruthersi*, *C. pipiens* and *C. tinniens* have similar songs but are widely sympatric, and form a species group. *C. robustus* and *C. aberdare* are sibling species; the other large, highland species *C. natalensis* may be linked with them, although its voice is different. *C. ruficeps* and *C. dorsti* are sibling species ('sibling species' are lookalikes with behavioural [voice, habitat] differences but, unlike members of a superspecies, they are essentially sympatric. Should *C. ruficeps* and *C. dorsti* prove to be allopatric they would be better treated as a superspecies.) *C. nanus* was associated with *ruficeps* in a superspecies by Hall and Moreau (1970) but their dissimilar plumages and size argue against close relationship. The *C. fulvicapillus* superspecies may be close to the *C. rufus* superspecies; all are small, plain-backed, bush and wooded savanna species, with suggestive parapatry. The more heavily-marked, widespread *C. brachypterus* is also close to the *rufus* superspecies and was even included in it by Dowsett and Forbes-Watson (1993) although sympatry is extensive and plumage differences greater. The aberrant *C. melanurus* is perhaps also closest to this group of species, while *C. nanus* appears to link this plain-backed group with the *ruficeps-dorsti* pair. Finally, the 11 tiny species with streaked backs and aerial display flights ('cloudscrapers') seem the most recently radiated, some having spread out of Africa, with *C. juncidis* still expanding its range in Europe. Extralimital *C. exilis* and *C. cherina* seem closest to the *juncidis* superspecies and *C. aridulus* may be an arid-zone offshoot of this group. Among the remaining 6 species, *C. textrix*, *C. eximius* and *C. dambo* may be very close and are largely parapatric, while these 3, along with the *brunnescens* superspecies and *C. ayresii*, form a species group.

Size range illustrated by (**A**) *C. textrix*, (**B**) *C. cantans*, (**C**) *C. chiniana*, (**D**) *C. robustus* and (**E**) *C. natalensis*.

A

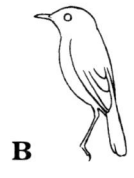
B

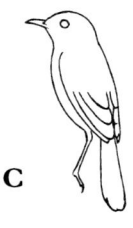
C

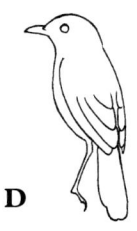
D

E

Cisticola woosnami superspecies

1 *C. woosnami*
2 *C. anonymus*
3 *C. bulliens*

Cisticola chubbi superspecies

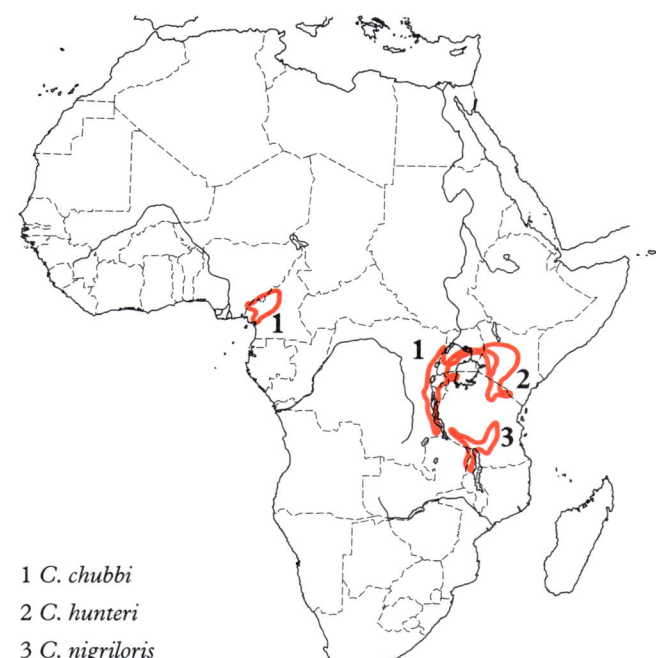

1 *C. chubbi*
2 *C. hunteri*
3 *C. nigriloris*

Cisticola rufus superspecies

1 *C. rufus*
2 *C. troglodytes*

Cisticola fulvicapillus superspecies

1 *C. angusticaudus*
2 *C. fulvicapillus*

140 SYLVIIDAE

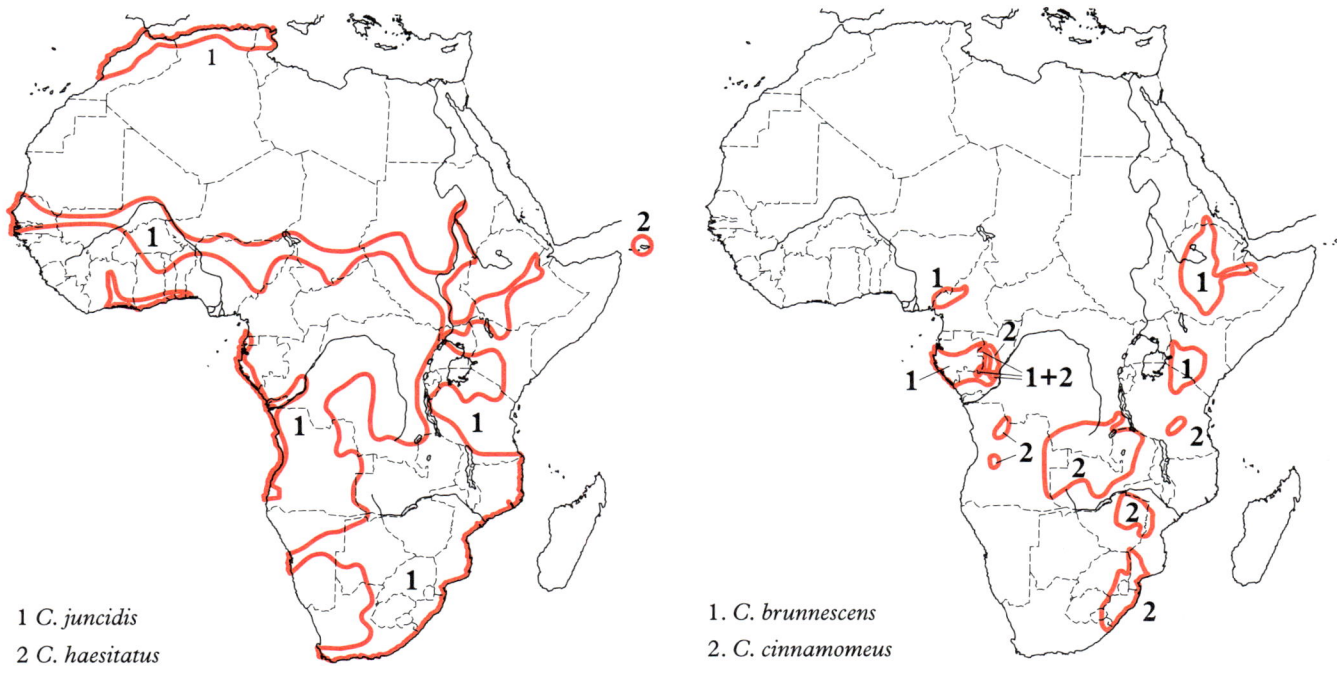

Cisticola juncidis superspecies

1 *C. juncidis*
2 *C. haesitatus*

Cisticola brunnescens superspecies

1. *C. brunnescens*
2. *C. cinnamomeus*

Plate 9
(Opp. p. 160)

Cisticola erythrops **(Hartlaub). Red-faced Cisticola. Cisticole à face rousse.**

Drymoeca erythrops Hartlaub, 1857. System. der Orn. Westafrika's, p. 58; between Liberia and Nigeria (= Calabar).

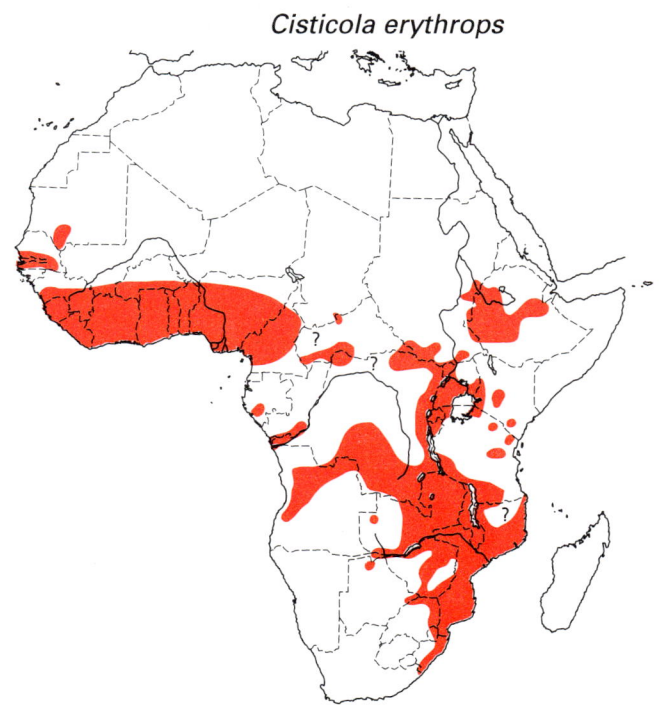

Cisticola erythrops

Range and Status. Endemic, mostly resident; in places local migrant. Rare visitor to S Mauritania June–Nov, reaching 17°N during rains. Locally common resident in lower river region of Gambia; uncommon S Senegal. Widespread SW Guinea (Demey 1995), and common Macenta Préfecture (Halleux 1994). Frequent to common resident Sierra Leone, Liberia, Ivory Coast, Ghana, Togo. Common Mali south of 13°N, perhaps local migrant. Occurs S Burkina Faso and common June–July in Arli-Pendjari Nat. Parks (Benin-Burkina Faso). Rare central Nigeria (Zaria, Kaduna, Kafanchan, Jos Plateau, Falgore G.R.), locally common in S. Common W and S Cameroon, also Adamawa and S Benue Plain in N. Rare in S Central African Republic (frequent near Bangui) and in Bamingui-Bangoran Nat. Park. Frequent in SE Sudan, where resident in E, local migrant (non-breeding visitor present Jan–May) in S. Locally frequent to abundant resident in central and SW Ethiopia, 300–2400 m. Local in savannas of Gabon. Local S Congo, recorded Cabinda and elsewhere on lower Congo R. Frequent to common in N, E and SE Zaïre, Rwanda and Burundi. Common Kenya in SW and Taita Hills (900–2300 m), Uganda and Tanzania (0–2300 m), perhaps commonest Uganda (Jackson and Sclater 1938). Common 600–1600 m in Malaŵi (breeding up to at least 1200 m). Occurs in

Zambia east of 27°E, with separate population at c. 14°10'S, 23°35'E around Kabompo tributaries of Zambezi R. Locally frequent to common in N and central Angola (NW Huila District to S Cuanza Norte, W Malange and N Lunda), Caprivi Strip (Namibia) and adjacent N Botswana (Chobe river system and Moremi region of Okavango). Occurs on Zimbabwe plateau (commoner in E, local in dry areas). In Mozambique, frequent on central plateau, common to abundant in lowlands. Locally frequent to common in lowveld of Transvaal, adjacent E Botswana (Limpopo valley), lowland Natal; occurs E Swaziland.

Description. *C. e. erythrops* (Hartlaub): Gambia to Central African Republic, south to mouth of Congo R. Dress seasonal except in Cameroon, where perennial dress is like non-breeding plumage elsewhere. ADULT ♂ (breeding): forehead, lores, supercilium, ear-coverts and sides of neck rufous; crown to nape olivaceous brown; rest of upperparts dark grey-brown, tinged olivaceous; tail-feathers dull brown with dark brown subterminal band (less obvious on central pair) and grey-brown tips; entire underparts rich buff, paler on chin, throat and centre of belly, and washed olive-grey on flanks; primaries and secondaries brown, edged rufous-buff; tertials and wing-coverts olivaceous grey-brown; underwing-coverts rufous-buff; axillaries olivaceous buff. Upper mandible very dark brown to black, lower grey or horn; eyes orange-brown; legs and feet pinkish brown. ADULT ♂ (non-breeding): crown and nape browner; rest of upperparts more olivaceous; underparts richer rufous-buff. Sexes alike. SIZE: (15♂♂, 12♀♀) wing, ♂ 55–61 (58·7), ♀ 51–55 (53·2); tail, ♂ 53–64 (57·1), ♀47–60 (50·9); bill, ♂ 16–18 (17·3), ♀ 15–17 (16·3); tarsus, ♂ 22–24 (23·5), ♀ 20–23 (21·7). WEIGHT: (Mt Nimba, Liberia) ♂ (n = 8) 11·0–15·9 (13·6), ♀ (n = 4) 11·8–14·2 (12·9); (Ghana) ♀ (n = 3) 12·1–14·1 (13·1), unsexed (n = 3) 13–15 (14·0); Nigeria, 1 unsexed 13·1.

IMMATURE (*C. e. sylvia*): browner than adult, with broader tail-feathers.

NESTLING (*C. e. sylvia*): hatches naked, pinkish, with yellow bill and black spot on each side of tongue.

C. e. sylvia (Reichenow): NE Zaïre and S Sudan to central Tanzania and Kasai area of S Zaïre. Same as Cameroon population of nominate race, but averages slightly paler and larger. SIZE: (8♂♂, 5♀♀) wing, ♂ 59–65 (61·7), ♀ 56–58 (57·2); tail, ♂ 54–63 (58·0), ♀ 49–57 (51·6); bill, ♂ 16–17 (16·2), ♀ 15–16 (15·8); tarsus, ♂ 23–27 (24·5), ♀ 22–25 (23·2). WEIGHT: (Kenya, Uganda) ♂ (n = 10) 12–18·2 (15·3), ♀ (n = 9) 10–16 (13·4).

C. e. nyasa Lynes (including '*arcana*' and '*elusa*'): S Tanzania, extreme SE Zaïre, to N Botswana and Natal. Breeding dress like *sylvia*. In non-breeding dress, upperparts tinged more tawny yellow; forehead and cheeks redder; breast and flanks richer rufous-buff. SIZE: (10♂♂, 12♀♀) wing, ♂ 53–59 (56·2), ♀ 48–55 (50·7); tail, ♂ breeding (n = 5) 45–60 (52·2), non-breeding (n = 5) 50–60 (55·4), ♀ breeding (n = 7) 42–57 (47·4), non-breeding (n = 5) 49–54 (51·8); bill, ♂ 15–17 (16·2), ♀ 14–16 (14·7); Mozambique birds in lower part of this range. WEIGHT: (Zaïre) ♂ (n = 8) 11–16 (14·2), ♀ (n = 2) 11, 12; (Mozambique) 1♂ 13·3, 1♀ 11·9; (Malaŵi) ♂ (n = 7) 14·1–17·1 (15·7), ♀ (n = 8) 11·9–14·0 (13·1); (Zambia) 4 unsexed 11–15·3 (13·6); (Zimbabwe) ♂ (n = 29) 13·7–17·8 (15·7), ♀ (n = 47) 11·2–15·9 (13·9).

C. e. lepe Lynes: Angola. Seasonal dress like corresponding plumages of nominate race but upperparts slightly paler (breeding) or less yellowish (non-breeding) and underparts much less rufous (both plumages). Breeding dress same as *sylvia*.

C. e. pyrrhomitrus (Reichenow): Ethiopia, SE Sudan. Like nominate race but, in both breeding and non-breeding plumages, head-top yellower, rest of upperparts slightly paler and more olive-tinged. Both plumages slightly richer-coloured than *sylvia*, especially on underparts.

C. e. niloticus (Madarasz): upper Blue Nile, Sudan. Like corresponding plumages of *pyrrhomitrus* but breeding dress paler; non-breeding dress paler and upperparts yellower. Smaller. SIZE: (2♂♂, 2♀♀) wing, ♂ 53, 55, ♀ 51, 51; tail, ♂ 56, 57, ♀ 47, 55; bill, ♂ 15, 17, ♀ 15, 16; tarsus, ♂ 21, 22, ♀ 20, 21.

TAXONOMIC NOTE: Dowsett and Prigogine (1974) treated *C. lepe* as a distinct species because of apparent sympatry with *C. erythrops* in SE Zaïre, but they later considered the case to need further investigation (Dowsett and Dowsett-Lemaire 1993b).

Field Characters. A plain cisticola of damp hollows and waterside vegetation with distinctive red face and bright reddish buff underparts. Upperparts unstreaked; crown, back and wings uniform grey-brown, without any red. Southern African birds are less richly coloured, especially in the breeding season. Most likely confused with Singing Cisticola *C. cantans* (upperparts also unstreaked, but has rufous cap and wing-patch contrasting with grey-brown back; whiter below; no red on face). Some songs and calls of the 2 species are quite similar in tone, and singing birds should not be identified by voice alone until one is completely familiar with their repertoires. Where sympatric with Singing Cisticola, often separated by habitat, with Singing in higher, drier sites.

Voice. Tape-recorded (7, 10, 45, 87, 88, 91, B, C, F, KEI). Song a duet. Primary singer, presumably ♂, has varied repertoire of loud, often ringing notes; these are typically a series of repeated single or double notes on same pitch: upslurred 'weep-weep-weep . . .', downslurred 'peeuw-peeuw-peeuw . . .', hard 'pee-pee-pee . . .', dry, prinia-like 'jik-jik-jik . . .', sharp double 'chip-weep, chip-weep, chip-weep . . .', grating 'zee-tsik, zee-tsik, zee-tsik . . .'. Some songs are a medley of single notes and trills at varying speeds: 'wawawawawa, pee-pee-pee-pee, churrrrrr, pipipipipi'; common in southern Africa is a series of ringing notes decreasing in pitch, 'tee-tee-tee-tee-tay-tay-toy-toy-toy-toy . . .'; common in E Africa is a series of buzzy, rather nasal notes ascending the scale and getting louder, 'jee-jee-jee-chee-chee-cheer-cheer-CHEER-CHEER'. Second bird accompanies with low rasping or clicking notes, 'jidit', 'zititit', 'clikikikik', which may either be synchronized with primary call, 'peeuw-jidit, peeuw-jidit, peeuw-jidit . . .' or come in at irregular intervals. Thin 'seeep-seeep-seeep', uttered from tall perch, is probably an alarm (Lewis 1981–1982); alarm in Zimbabwe, a 'thin, drawn out plaintive 'teeeeeeee, teeeeeeee' with an appreciable pause between syllables, very similar to alarm of Tawny-flanked Prinia *Prinia subflava*, which often occurs in same habitat' (Masterson 1992).

General Habits. Dwells in rich growth of rank grass, reedbeds and shrubs with scattered trees by water, including rivers, estuaries, lakes and pools; also in grass and stunted bushes in marshy forest or on other damp ground away from water. In some countries (e.g. Sierra Leone, Nigeria, Gabon, Zaïre and Tanzania) also found in gardens, farmland, airfields, dense secondary bush, broad-leaved tall-grass savanna or open places in forest, often far from water, although in drier parts more or less restricted to reedbeds during the dry season. In Ethiopia also inhabits olive–*Podocarpus*–juniper forest and lowland subtropical humid forest (Urban and Brown 1971). In SE Zaïre mainly in wooded savanna, especially *Acacia hockii*, also on fallow land at the bushy, regenerative stage (Ruwet 1965). In Angola, grasslands and vegetation along rivers and streams, especially cane fields (da Rosa Pinto 1970).

Usually solitary or in pairs; small groups after breeding season. Noisy but shy; keeps to cover, except ♂ in breeding season. Feeds on and near ground.

Food. Insects, including caterpillars up to 3 cm long (Sclater and Moreau 1933), ants, beetles, bluebottles (Calliphoridae) and alate termites. Young fed on moths, caterpillars, grasshopper nymphs, mantids (van Someren 1956).

Breeding Habits. ♂ sings from cover or exposed on low bush or other vegetation; when disturbed, drops into cover. In slow, butterfly-like, song-flight over territory, bird circled at height of 6 m, uttering loud version of 'chip-pwi' call, with chuckling 'chicker-chucker-chucker' at the end (Lewis 1981–1982).

NEST: usually oval with entrance at side-top, occasionally an open cup; of a little coarse, dry grass (and occasionally other materials, e.g. banana 'bark' shreds), thinly lined with vegetable down; built between 2–3 broad, green leaves (**A**) which are 'stitched' together with silk (actually, rivetted: thread is made into lump on one side of leaf, pushed through leaf surfaces and flattened into lump on other side) to form a cup, with one leaf over the top; well-hidden among dense grass or soft, bushy, wet-season vegetation (*c.* 2 m high), 50–150 cm above ground, often at edge of stream or marsh; in Kenya and Tanzania, commonly nests in old farms in a *Solanum* sp. whose coarse leaves stitch well (Sclater and Moreau 1933), or in other similar large-leaved herbs. Built by both sexes (van Someren 1956).

EGGS: 2–4; Gambia 3; Nigeria (1 clutch) 3; Cameroon 2–4; Uganda (1 complete clutch) 1; Zaïre (8 clutches) 2–3; Malaŵi 2 (n = 15), 3 (19), 4 (1); Zimbabawe (2 clutches) 3, 4; Natal 2 (n = 1), 3 (2), 4 (1). Elongate to rounded oval; moderately glossy; usually pale greenish or blue-green, finely but thickly spotted or blotched with brownish reds, and sometimes with underlying purplish greys, spots either evenly distributed or with ring of bolder marks at big end. Occasionally immaculate or with merely faint freckling of rust in ring at big end; others white, stippled rusty red. SIZE: (n = 1, Ghana) 16·1 × 12·4; (n = 40, Cameroon) 16·5–18·5 × 12·5–13·5; (n = 53,

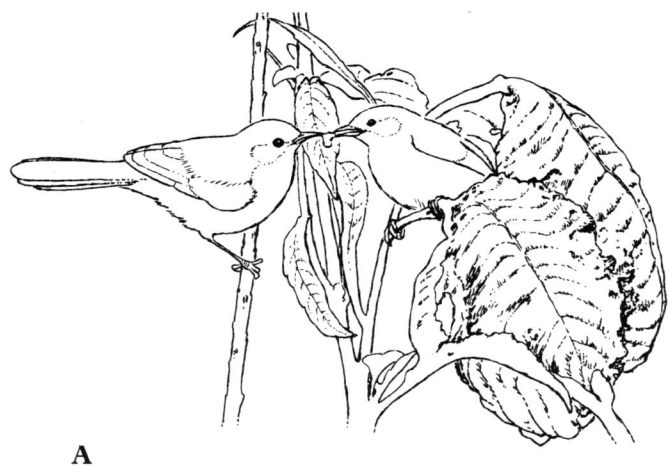

A

Cameroon) av. 17·3 × 12·9; (n = 13, S Sudan, Uganda) av. 17·2 × 13·0; (n = 1, W Uganda) 19 × 13; (Kenya) av. 17 × 12; (Zaïre) 17·2–18·7 × 13–13·4; (n = 5, S Zaïre) 17·5–18·5 × 12·9–13·1 (18·1 × 13·0); (n = 26, Malaŵi) av. 17·3 × 12·7; (n = 3, Natal) av. 18·8 × 13·7; (n = 45, southern Africa) 16–18·8 × 11·6–13·5 (17·5 × 12·6).

LAYING DATES: Gambia, May–June; Sierra Leone (courting June); Liberia, July, Sept; Ghana, Mar–July; Togo, July; Nigeria, June, Oct, Dec; Cameroon, May, July–Dec; Central African Republic (breeding condition Sept–Oct); Sudan, June; Ethiopia, Sept; Rwanda, May, Oct–Dec; S Zaïre, Dec–Apr, E Zaïre, Mar–Apr, June, Oct, possibly year-round; Uele (N Zaïre), July–Sept. E Africa: Kenya (Ngong), Mar–July; W Uganda, Nov; Kenya, Uganda two broods per year; Region A, May (1 record); Region B, Mar–July, Nov–Dec; Region D, Mar–July (peak Apr), Oct–Dec; breeding virtually confined to long rains or just after. Malaŵi, Zambia Dec–Apr, peak Feb; Angola highlands, Jan–Feb, probably to June; Zimbabwe, Oct–Mar, peak Dec–Feb (57 records: Irwin 1981); Botswana, Dec; Transvaal, Dec–Feb, Apr; Natal, Nov, Jan.

INCUBATION: mainly by ♀ which continues to add down; later in period, which lasts 12–16 days, sits for greater proportion of daylight hours than earlier. Incubating ♀ often fed by ♂.

DEVELOPMENT AND CARE OF YOUNG: ♀ broods nestlings during first 2 days, while ♂ brings food, giving it to ♀ if she is there (**A**); faeces initially swallowed, after 1–2 days carried off; *c.* 1 visit every 10 min.; large insects mandibulated before being presented to young; if prey still too large, adult swallows it; adult approaches nest by hopping through vegetation but flies away with faeces (van Someren 1956). Young in nest 14–16 days. Fledglings drop to low cover or ground when adults give alarm call.

BREEDING SUCCESS/SURVIVAL: parasitized by Parasitic Weaver *Anomalospiza imberbis*. One adult recaptured after 4 years 3 months (Manson 1985). Host of *Haemoproteus wenyoni* (Peirce 1984).

References
Chapin, J. P. (1953a).
Lynes, H. (1930).
van Someren, V. G. L. (1956).
Vincent, A. W. (1948a).

Cisticola cantans (Heuglin). Singing Cisticola. Cisticole chanteuse.

Drymoeca cantans. Heuglin, 1869. Ibis ser. 2, vol. 5, p. 96; Tigré & Semien (= Gondar), Ethiopia.

Plate 9
(Opp. p. 160)

Range and Status. Endemic resident. Frequent Gambia and S Senegal, locally common near coast. Common Mali south of 14°30′N. Recorded N Sierra Leone, and S Guinea at Conakry, Kindia, Kankan, Beyla. Rare Liberia (Mt Nimba). Common, locally abundant, Ivory Coast, S Ghana, Togo; frequent N Ghana, S Burkina Faso and soudan zone of extreme SW Niger (Sept, Nov–Mar). Recorded Benin. Common in savannas of central and NE Nigeria. In Cameroon from border of forest zone to at least *c.* 11°N. Present SW Chad (Moundou), Apr–May. Throughout much of N and central Central African Republic. Locally frequent Sudan south of *c.* 13°N. Locally frequent Ethiopia above 1200 m, common 2000–2500 m but not in dry highlands. In Eritrea common to abundant above 1200 m. Common Zaïre in N and E in area 6–7°S, 19–25°E, and in Rwanda and Burundi; mainly lowlands but up to 1700 m in E Zaïre, Rwanda, Burundi. Locally common to abundant Uganda, Kenya (SW from Elgon to central highlands and Taita Hills; Huri Hills) and Tanzania, up to 2500 m (mainly highlands above 1000 m). Local E Zambia (Nyika Plateau up to 2200 m). Occurs 600–1800 m in Malaŵi (up to 2150 m on Mt Dedza). Local E Zimbabwe west to Shurugwi; locally abundant Bvumba highlands and in adjacent Mozambique on central plateau and Mt Gorongoza, up to 1220 m.

Description. *C. c. swanzii* (Sharpe): Gambia, Sierra Leone to S Nigeria. ADULT ♂ (breeding): forehead, crown, nape and face tawny brown (lores and ear-coverts slightly buffier); rest of upperparts including scapulars greyish brown, tail tawny brown with subterminal dark brown bar and buffish brown tip. Entire underparts buff, palest (often near-white) on chin, throat and centre of belly, richest on flanks and undertail-coverts; flanks and axillaries washed with grey; primaries, secondaries and tertials dull brown, fringed on both webs with rusty brown; wing-coverts dull brown fringed tawny; underwing-coverts buff. Bill dark olive-green or dark brown to black, basal two-thirds of lower mandible bluish grey to pale horn; eyes orange-brown; feet pale pinkish brown. ADULT ♂ (non-breeding): head top richer, more chestnut; rest of upperparts and tail more tawny; underparts average richer buff. Sexes alike. SIZE: (10 ♂♂, 10 ♀♀) wing, ♂ 52–58 (54·7), ♀ 48–52 (49·6); tail, ♂ breeding (n = 5) 45–50 (47·2), non-breeding (n = 5) 52–55 (53·2), ♀ breeding (n = 5) 41–45 (43·2), non-breeding (n = 5) 40–48 (45·2); bill, ♂ 15–18 (16·7), ♀ 15–17 (15·6); tarsus, ♂ 19–22 (20·8), ♀ 19–22 (20·4). WEIGHT: Ghana, 1 unsexed 13·0; Nigeria, unsexed (n = 34) 9·7–14·0 (11·7).

IMMATURE: like non-breeding adult but underparts paler, washed lemon-yellow.

NESTLING: unknown.

C. c. concolor (Heuglin): N Nigeria to Sudan; probably also this race in Mali. Like *swanzii* except upperparts paler, with more contrast between head and back; underparts brighter buff, especially on flanks; differences less evident in non-breeding plumage.

C. c. adamauae (Reichenow): Cameroon, Congo, NW Zaïre. Perennial dress, like breeding *swanzii* but upperparts darker; head top darker and more rufous (more like non-breeding *swanzii* in this respect).

C. c. belli (Ogilvie-Grant): Central African Republic to NE Zaïre, Uganda and NW Tanzania. Perennial dress, like *adamauae* but head top richer rufous.

C. c. cantans (Heuglin): S Ethiopia and S Eritrea, except dry SE. Seasonal dress. In breeding plumage like *belli*. Non-breeding dress similarly coloured but upperparts with broad streaks of very dark brown (feather centres).

C. c. pictipennis (Madarasz): Kenya, N Tanzania south to Nguru Mts. Perennial dress like *belli* but averages larger than it and all previous races. SIZE: wing, ♂ (n = 6) 58–63 (60·8), ♀ (n = 5) 55–61 (57·6); tail, ♂ (n = 6) 54–61 (58·7), ♀ (n = 5) 54–61 (56·2); bill, ♂ (n = 5) 15–16 (15·6), ♀ (n = 5) 14–16 (15·0); tarsus, ♂ (n = 6) 24–25 (24·5), ♀ (n = 4) 23–25 (23·7). WEIGHT: (Kenya) ♂ (n = 7) 12–17 (15·0), 1 ♀ 12, 1 immature ♂ 13, 1 immature ♀ 14.

C. c. muenzneri (Reichenow): S Tanzania (from Uluguru Mts south) to Zimbabwe. Breeding dress like *belli* and *pictipennis* but mantle and back slightly darker, browner. Non-breeding dress like non-breeding *swanzii* but brighter: strongly washed yellow-brown on upperparts; crown brighter orange-brown than both non-breeding *swanzii* and breeding *muenzneri*. Small. SIZE: wing, ♂ (n = 10) 53–56 (54·5), ♀ (n = 10) 48–55 (51·3); tail, ♂ breeding (n = 5) 43–51 (46·8), non-breeding (n = 5) 52–56 (54·8), ♀ breeding (n = 5) 41–50 (45·6), non-breeding (n = 5) 45–54 (48·8); bill, ♂ (n = 10) 14–15 (14·6), ♀ (n = 8) 13–15 (14·4); tarsus, ♂ (n = 10) 21–23 (22·2), ♀ (n = 10) 21–23 (21·7). WEIGHT: (Malaŵi) unsexed (n = 6) 12·2–12·9 (12·6); (Zimbabwe) 1 ♂ 14·5, 1 ♀ 10·4.

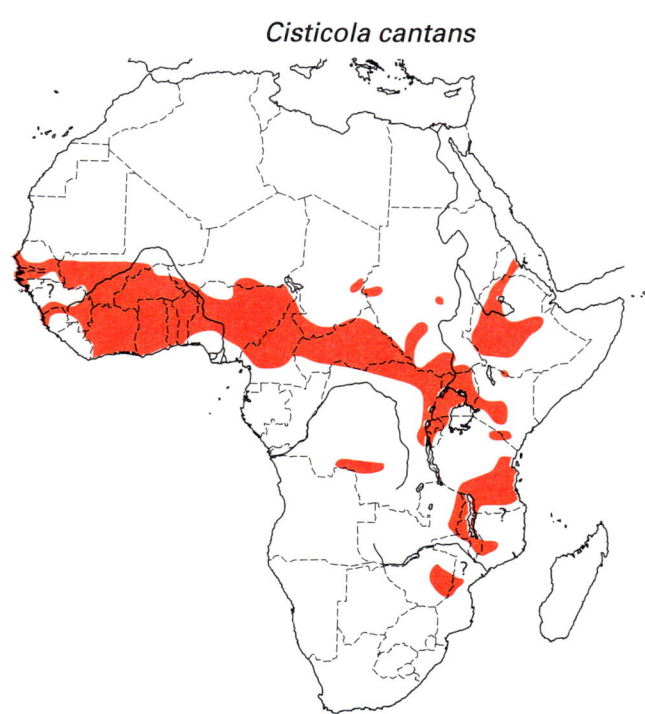
Cisticola cantans

Field Characters. A common and familiar bird, with a loud, arresting song. Upperparts unstreaked; red-brown crown and reddish wing-patch contrast with grey-brown back (crown, back and wings concolorous in Red-faced Cisticola *C. erythrops*). Otherwise similar to Red-faced but underparts whiter, no red on face (face pale buff with a black subloral spot more noticeable in ♂ (Zimmerman *et al.* 1996)). Song and calls of the 2 species have different form but similar ringing tone, and should be distinguished with care. Singing Cisticola lives in drier habitat than Red-faced and is less inclined to skulk.

Voice. Tape-recorded (10, 45, 87, 91, B, C, F, KEI, ZIM). Song, 2–3 loud, ringing, metallic notes, 'wee-tyer', 'kowee-chup', 'ploo-wee', 'plu-sip', 'plotchit-tee', 'pitch-chew', 'tsoo-chew-wip', 'trick-or-treat'; very variable but always with same tone. Sometimes duets, second bird giving either high and squeaky or (usually) low grating notes. An abrupt 't'yaw' said to be 'anger call' (North and McChesney 1964); also has low churr. Alarm, a prolonged burst of rapid chattering, similar to call of Roberts's Prinia (= Briar Warbler, *Oreophilais robertsi*) (Masterson 1992).

General Habits. Usually associated with tangled undergrowth, including thick, low scrub with long grass and herbage, savanna with thick bushes and rank undergrowth, secondary bush and overgrown cultivation, woodland with dense ground cover (including olive/*Podocarpus*/juniper forest and subtropical humid forest in Ethiopia), grass and sedge with scattered bushes, bracken–briar, *Philippia* heath; occupies these habitats in the open, or close to evergreen forest edge, and often along streams or lake shores. In Sierra Leone and Nigeria on grass savanna with scattered shrubs and trees. In Nairobi has adapted to gardens. In places occurs together with Red-faced Cisticola.

Perches on bushes more often than grass; normally stays in cover. Flits and hops about with wings drooped, tail cocked and wagging sideways and vertically. In pairs and family parties, which have small foraging range. Mostly forages close to ground, often in grass clumps. ♂ defends territory; ♀ creeps silently away when disturbed.

Food. Insects, especially small caterpillars and beetles, also small moths, small grasshoppers, crickets, flies and their larvae, ants; spiders. Once a skipper butterfly given to young (van Someren 1956).

Breeding Habits. ♂ sings from within or on top of bush, or from low branch of small tree.

NEST: flimsy, usually domed; ext. depth av. 140, ext. diam. 75, depth of cup from lip 50; occasionally an open cup. Made of dry grass with fine grass seed-heads or bark fibre (**A**); lined, especially in lower part, with plant down; sewn with caterpillar or spider silk (or occasionally with dry grass-fibre) between 2–3 broad leaves, often with one sewn over top; occasionally in single large leaf with the sides drawn and sewn together;

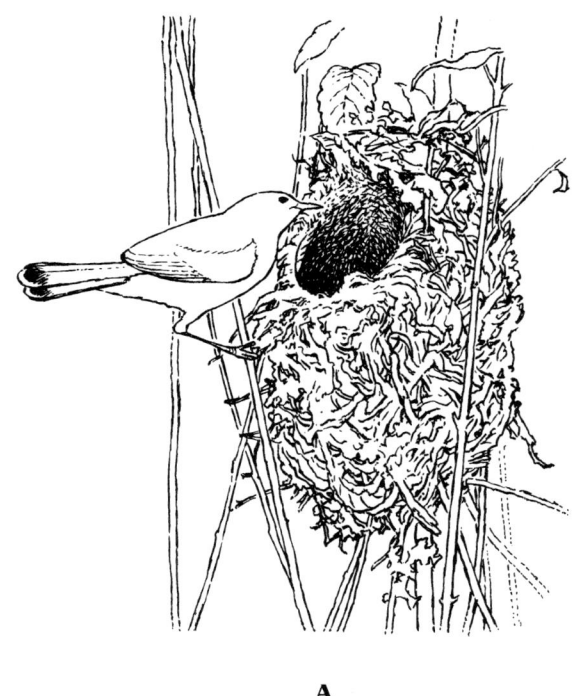

A

placed 50–100 cm (or less) above ground, well-hidden in a bushy plant (favours small-leaved aromatic shrubs in NW Ethiopia; cassava in Nigeria).

EGGS: Gambia 3; Ghana (2 clutches) 3; Nigeria (1 clutch) 3; Ethiopia (5 clutches) 2–3 (2·2); Kenya 2–4 (usually 3); Zaïre 3–4, Malaŵi (32 clutches) 2–3 (2·5); Zimbabwe 2–4; S Mozambique up to 4. Slightly glossy; very variable in colour: white or cream to pink, or very pale blue, or greenish blue, with freckles or larger spots of red-brown and underlying greys or ashy purples, often in ring at big end but sometimes evenly spread; or immaculate pale blue; or pale green or blue with obscure pale lilac mottling or small pink or purple spots at big end; or creamy brown, closely spotted with red-brown and underlying lilac. (Some of the eggs on which this description is based may be of Pin-tailed Whydah *Vidua macroura*.) SIZE: (n = 14, Ghana, Nigeria) 16–17·7 × 11·5–12·5 (17·0 × 12·0); (n = 1, Ethiopia) 17 × 12·7; (n = 15, Kenya) 16·4–18·8 × 12·1–13·6 (17·3 × 12·6); (n = 2, NE Zaïre) 17·9 × 13·9, 19·2 × 14·7; (n = 41, Malaŵi, Zimbabwe) av. 16·9 × 12·4; (n = 61, southern Africa) 15·9–19·5 × 11·9–13·1 (17·0 × 12·3).

LAYING DATES: Gambia, Feb, Apr–May; Guinea, Oct; Sierra Leone (courting May–June); Ghana (copulation Mar, singing Mar–Oct); Togo, May–Aug; N Nigeria, June–Aug; Ethiopia, June–Nov (rains, double-brooded); Eritrea, June; Rwanda, Feb; NE Zaïre, Apr–July (rains); E Zaïre, Mar. E Africa: Kenya, Nov–July; Tanzania, Feb; Region C, Jan; Region D, Jan–July (peak Apr–May), Oct–Dec. Zambia, Jan; Zimbabwe, Nov–Apr, peak Dec–Jan (46 records); Malaŵi, Dec–Apr (peak Dec–Jan); Mozambique, Sept, late Nov to late Apr.

INCUBATION: by ♀, which sits tight after first day; period 12–14 days. ♀ continues to add lining during incubation; ♂ often brings down to sitting ♀.

DEVELOPMENT AND CARE OF YOUNG: ♂ drives *Prinia* spp. from area within 20 m of nest. Young fed by both parents: at one nest with 2 young, ♀ visited 8 times, ♂ 5 times, in 30 min (van Someren 1956). Young in nest *c*. 16 days; leave when hardly able to fly.

BREEDING SUCCESS/SURVIVAL: may re-nest twice in season if first 2 clutches predated. Parasitized by Pin-tailed Whydah and Klaas's Cuckoo *Chrysococcyx klaas*; latter may destroy one or more of host eggs or may leave them, in which case young cuckoo ejects host young. ♂ ringed as adult recaptured after 2 years, 262 days (Harwin *et al*. 1994).

References
Chapin, J. P. (1953a).
Cheesman, R. E. and Sclater, W. L. (1935).
Lynes, H. (1930).
van Someren, V. G. L. (1956).

Cisticola lateralis (Fraser). Whistling Cisticola. Cisticole siffleuse.

Drymoica lateralis Fraser, 1843. Proc. Zool. Soc. Lond., p. 16; Cape Palmas, Liberia.

Plate 9
(Opp. p. 160)

Range and Status. Endemic resident. Uncommon Gambia, mostly along lower river; 1 record MacCarthy Is. Common Casamance and Niokolo-Koba Nat. Park (S Senegal). One record Mali (Kangaba, Nov). Recorded Guinea-Bissau. Guinea in SW coastal area, north to Boffa, and in SE. Frequent Sierra Leone. Common Liberia (up to 1300 m on Mt Nimba). Abundant Ivory Coast. Locally frequent central and S Ghana and Togo, and in band across S Nigeria, also at Kagoro on Jos Plateau and in extreme E on Chappal Waddi. In Cameroon, from Adamawa Plateau south to forest edge. Local S Central African Republic, rare Bamingui-Bangoran Nat. Park. Occurs across N Zaïre. Frequent S and SW Sudan. Common N Uganda (but absent Karamoja) at 900–1900 m, southeast to Kakamega (SW Kenya). Rare Burundi (only between Rumonge and Kigwena) and nearby in E Zaïre. Occurs in SE Zaïre at 20–28°E, south of 4°S, and adjacent extreme NW Zambia (Mwinilunga), where uncommon. Locally common in N Angola (NE Malange, Lunda and extreme NE Moxico Districts). Common Cabinda, W Zaïre and adjacent coastal Angola, along Congo R. inland to 1°S. Rare central and S Congo and Gabon.

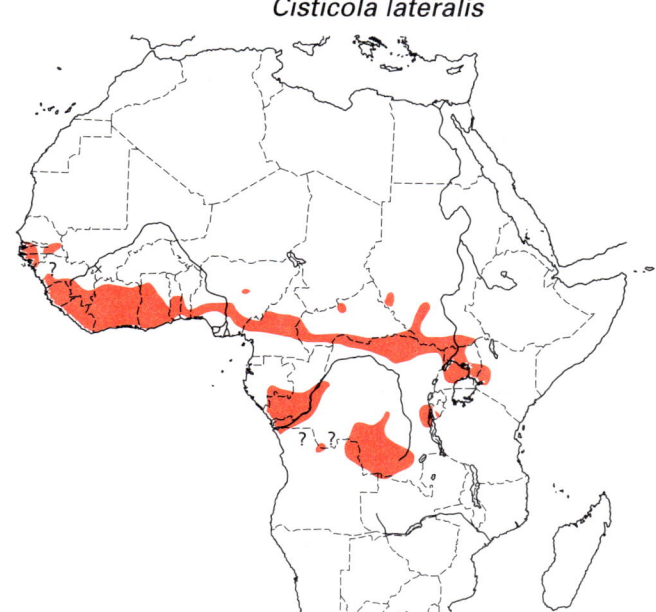

Description. *C. l. lateralis* (Fraser): Gambia to Cameroon highlands. ADULT ♂ (breeding): entire upperparts, face and inner wing-coverts dark sooty brown to brown (clinal: browner in E); tail-feathers like upperparts or faintly tinged reddish, tipped buff and with subterminal darker brown band; underparts buff, paler (near-white) on throat and centre of belly, richer buff or greyer on flanks, and with grey or grey-buff patch on sides of breast; primaries, secondaries, tertials and outer wing-coverts like upperparts, fringed rufous; under-wing-coverts buff. Upper mandible dark brown to black, lower pinkish grey; eyes orange-brown; legs and feet pinkish brown. ADULT ♂ (non-breeding): upperparts, tail and wings rufous brown; underparts richer buff, not grey. Sexes alike. One melanistic individual collected Ivory Coast. SIZE: wing, ♂ (n = 10) 64–69 (66·6), ♀ (n = 9) 51–55 (53·7); tail, ♂ (n = 10) 51–60 (56·4), ♀ (n = 9) 43–46 (44·2); bill, ♂ (n = 9) 15–17 (16·2), ♀ (n = 8) 14–16 (14·9); tarsus, ♂ (n = 10) 23–24 (23·6), ♀ (n = 9) 20–22 (20·8). WEIGHT: (Mt Nimba, Liberia) ♂ (n = 2) 18, 23·4, ♀ (n = 3) 10·1–14·0 (12·4); (Ghana) unsexed (n = 14) 12·0–22·0 (17·1).

IMMATURE: like non-breeding adult but throat to breast at first washed sulphur; bill paler.

NESTLING: unknown.

C. l. antinorii (Heuglin): Central African Republic to W Kenya; racial attribution of population around north end of L. Tanganyika unknown. In breeding dress, upperparts, wings and tail lighter, browner, less sooty; wing edgings less rufous; underparts average slightly buffier. Non-breeding upperparts slightly paler, less reddish, more tawny. WEIGHT: (Uganda) 1 ♂ 19, 1 ♀ 19.

C. l. modesta (Bocage) (including '*vincenti*'): Gabon to N Angola, S Zaïre. In breeding dress, head to nape greyer, less brown than previous races; upperparts, tail and wings less rufous and more tawny than nominate race, more tawny than *antinorii*; much richer rufous-buff on flanks and undertail-coverts (richest in E: '*vincenti*'). In non-breeding dress, entire upperparts less rufous and more tawny than nominate, richer, more tawny than *antinorii*; flanks and undertail-coverts richer rufous-buff. Slightly smaller. SIZE: wing, ♂ (n = 7) 61–64 (62·9), ♀ (n = 6) 51–52 (51·5); tail, ♂ (breeding, n = 4) 51–54 (52·2), (non-breeding, n = 3) 57–59 (58·0), ♀ (breeding, n = 2) 44, 48, (non-breeding, n = 3) 44–50 (48·0); bill, ♂ (n = 7)

14–16 (15·3), ♀ (n = 6) 13–15 (14·0); tarsus, ♂ (n = 7) 22–24 (23·4), ♀ (n = 6) 20–21 (20·8).

TAXONOMIC NOTE: *C. l. modesta* was previously suppressed in *antinorii* by Chapin (1953b), when he erected *vincenti* for birds from N Angola to S Zaïre; however, specimens from the Congo type locality of *modesta* agree with '*vincenti*' more than with *antinorii* (see Lynes 1930), so '*vincenti*' becomes a synonym of *modesta*.

Field Characters. Plain, with stout bill, pale lores and whistled song. Plumage quite variable but always unstreaked, from dark sooty grey or brown with contrasting reddish wing-patch in W Africa, to medium brown without wing-patch in Uganda. Young birds washed yellowish below. Head always concolorous with back; black tailspots large and conspicuous, sometimes even on central feathers. More of a forest bird than other cisticolas, in clearings and on edge, also in damp woodland and rich savanna. Same-sized Trilling Cisticola *C. woosnami* looks extremely similar, although redder on head and with less bold spots on tail, which help distinguish it at close range; best told by trilling song and drier habitat.

Voice. Tape-recorded (45, B, C, F, FOR, GRI, KEI, MAC). Usual song, in both W Africa and Uganda, is a series of 7–8 identical whistled notes, all on the same pitch but with a diagnostic rhythm: brief pause after first note, the rest together without pause but first 2 somewhat slower, 'whew, whew-whew-whewhewhe-whew'. Notes are liquid and pleasant but not ringing or melodious. Many individual variations, with changes in pitch and differing notes; nearly always with characteristic pause after first 1 or 2 notes; the remaining notes are often run together into a continuous warble, e.g. in Liberia (KEI) 'pwee, pwee-cho, purpurtitichyow', or descending the scale, 'pee, pee, direediryediroo'. In NW Zambia (KEI) has a 3-part song, 'chew-chew-willilli-wyoo', hurried, definite and powerful, first 2 notes not unlike first notes of Rattling Cisticola *C. chiniana*; song sometimes has a preamble of a few dry notes, 'cher-didit'. Intervals between songs typically 2–5 s. Songs vary individually but each bird repeats its own song without variation. One bird put his buzzy scold into the form of the song, 'bzee, bzeezeezeezee', then gave a combination 'bzee, whewhewhewhewhew' (Chappuis 1974). Calls include a loud 'pyew' and a hard, dry chatter.

General Habits. Inhabits rank, soft vegetation on grassy, open hillsides dotted with small bushes and bushy ant-heaps, tall-grass savannas with bushes and trees, small clumps of trees, light woodland often with thick scrub undergrowth, bush and scrub fringes at forest edge, forest clearings, well-wooded watercourses, farmland (including fields of upland rice) with scattered tall trees and low thickets of second-growth.

Usually in parties of 4–5 (pair with young). Noisy, especially when disturbed. ♂ uses elevated perches, others stay lower, in cover. Sings in or close to cover, hidden in tree-crown or in open from high tree top, bush top or telephone wire. Sits on song perch hour after hour, with occasional perch changes; quickly drops to cover if disturbed.

Food. Arthropods, including Orthoptera, Coleoptera (including Cicindelidae), Heteroptera, alate Isoptera, spiders.

Breeding Habits.
NEST: oval, with side entrance; of dry grass and spider web, lined with plant down; placed *c.* 30 cm above ground in low bush amid grass; sometimes bound to vegetation with web.

EGGS: W Africa 2–3; Angola, Zaïre 2. In W Africa greenish blue, speckled brown; in Angola, white to bluish grey, strongly marked with bold spots and small blotches of reds and underlying greys; in NE Zaïre plain, pale turquoise-blue. SIZE: (n = 2, Sierra Leone) 18·2–18·8 × 13–13·2; (n = 6, Ivory Coast) 16·3–17·2 × 12·9–13·0 (16·9 × 13·2); (n = 2, NE Zaïre) av. 15·5 × 12·9; (n = 7, Uganda) 15·3–19·0 × 12·7–13·7 (17·6 × 13·4); (n = 2, Angola) av. 17·5 × 11·8.

LAYING DATES: Gambia, June–July; S Senegal, June–Oct; Guinea (singing Nov); Sierra Leone, Mar–June, Aug; Liberia, June; Ghana, May, July; Nigeria, Apr–Aug; Sudan, July; NE Zaïre in rains (May–Nov); W Zaïre, Sept onward (rains); Angola, Nov–Feb. E Africa: Kenya (juv. Sept); Region A, Mar, June; Region B, Mar, Apr, June, Sept, Oct; Zambia, Feb, Oct (immatures late Apr).

DEVELOPMENT AND CARE OF YOUNG: fledged young tended by both parents. Adult, if disturbed when with fledged young, creeps about inside bushes, remaining in cover, uttering alarm notes.

References
Chapin, J. P. (1953a).
Jackson, F. J. and Sclater, W. L. (1938).
Lynes, H. (1930).

Plate 9
(Opp. p. 160)

***Cisticola woosnami* Ogilvie-Grant. Trilling Cisticola. Cisticole de Woosnam.**

Cisticola woosnami Ogilvie-Grant, 1908. Bull. Br. Orn. Club, vol. 21, p. 72; SE Ruwenzori (= Mokia), Uganda.

Forms a superspecies with *C. anonymus* and *C. bulliens*.

Range and Status. Endemic resident, E Africa. Common to abundant in S and SW Uganda, SW Kenya, N and W Tanzania; absent from Kenya between Mt Elgon and S Nyanza and from N Tanzania high-

Cisticola woosnami

lands above 750 m. Frequent in Rwanda, Burundi and Zaïre, in E (east of 29°E, between 1°N and 2°S) and SE. Local and uncommon in N half of Zambia. Common Malaŵi above 1600 m, north of *c.* 12°S.

Description. *C. w. woosnami* Ogilvie-Grant: Uganda, NE Zaïre and Burundi, east across N Tanzania (south to Iringa). ADULT ♂: entire upperparts dark reddish brown, slightly redder on head-top; sides of head grey-brown; tail-feathers dark reddish brown, tipped buff, with subterminal very dark brown band; chin, throat and centre of belly white; centre of breast and undertail-coverts washed buff; sides of breast and flanks buffish grey; primaries, secondaries, tertials and wing-coverts dark brown, fringed reddish brown; underwing-coverts buff. Upper mandible dark brown to black, lower light brown to grey; eyes light brown; legs and feet pinkish brown. ADULT ♀: like ♂ but underparts richer buff, less grey. SIZE: (10♂♂, 10♀♀) wing, ♂ 64–68 (66·5), ♀ 53–56 (54·5); tail, ♂ 49–58 (54·2), ♀ 42–48 (45·2); bill, ♂ 15–16 (15·3), ♀ 13–14 (13·9); tarsus, ♂ 21–24 (23·0), ♀ 20–22 (21·0). WEIGHT: (Uganda) 1♀ 13; (Kenya) 1♀ 14.

IMMATURE: like adult ♀ but upperparts more rufous; underparts still richer buff.

JUVENILE: like immature but underparts washed with sulphur.

NESTLING: unknown.

C. w. lufira Lynes: Kigoma and Rukwa (SW Tanzania) southwards. Slight seasonal change in plumage. Breeding dress like nominate race but slightly redder on crown and wings; richer buff on underparts (differences barely perceptible, especially in ♀). In non-breeding dress, upperparts slightly redder still, a rich, warm brown; underparts richer buff (in ♀, flanks almost brown).

Field Characters. Plain, with stout bill; ♂ medium large, ♀ much smaller. Brown back grades into somewhat redder crown, wings have slightly rusty tone; immatures foxy red above and pale yellowish below. Extremely similar to Whistling Cisticola *C. lateralis* but lives in drier habitat (open woods, bushed grassland), and song is very different. Whistling has bolder tail-spots and lacks red on crown.

Voice. Tape-recorded (10, 45, 87, B, C, F, PAR). Song a pure, liquid, penetrating, ringing trill lasting *c.* 2·5–3 s, gradually increasing in intensity; often ventriloqual; typically all on one pitch but sometimes ascending. Rapid; individual notes barely discernible. Call, a single harsh 'chip' or 'chik' or double 'chit-tup' or 'tsee-tuk', sometimes running into a dry trill, 'chit-trrrrrrrrr'.

General Habits. Inhabits bushy, wooded (Uganda, W Tanzania) or treeless (E Tanzania) grassland, including rocky hillsides, *Acacia* savanna, *Brachystegia* woodland with bushy undergrowth, stunted woodland, thick riparian bush and grass, overgrown cultivation, second-growth at forest edge. Usually in or near open groves of trees with grass beneath.

Usually in noisy and conspicuous groups of 4–5; often high in trees or bush tops. When flushed, flies to lower branches of trees.

Food. Insects, including Orthoptera, small beetles, larvae, other winged insects.

Breeding Habits. Sings from cover in low bush or exposed on high tree. Sits on song-perch for hours. Will sing in heavy rain (Sclater and Moreau 1933). Drops into cover silently when disturbed.

NEST: a ball or upright ellipse, ext. depth *c.* 8–10 cm, ext. width 6–9 cm, with side-top entrance 3–4 cm diam; of broad, dry grass blades, bound with spider web and enwrapped with green grass, lined with finer grass and inner lining of plant down (thicker at bottom); placed 7–15 cm above ground in grass and shrubs *c.* 60 cm high.

EGGS: Tanzania (2 clutches) 2–3; Zaïre/Zambia 2–3. Slightly glossy; variable in colour: unspotted blue; or white, thinly spotted with red-brown and lilac, mostly at big end and occasionally with odd marks of dark brown or grey; or pale blue to turquoise-blue, finely and evenly marked with scattered dots and streaks of pinkish brown, dull red, reddish purple, or occasionally yellow-brown, often with underlying greys or violets; or pale blue, thinly spotted with black, dark grey and dark brown, more at big end, often in ring. SIZE: (n = 2, Burundi) 20·2 × 13·3, 20·3 × 13·2; (n = 5, Tanzania) 17·0–19·0 × 12·3–13·5 (18·0 × 12·8); (n = 9, S Zaïre) 16–19 × 12·4–13·1 (17·6 × 12·7); (n = 2, Zaïre/Zambia) av. 17·5 × 13·0.

LAYING DATES: Rwanda, Apr, May, Oct–Dec; E Zaïre, probably Mar–May; S Zaïre, Dec–Jan. E Africa: Tanzania, Jan–Mar, July; Region B, Apr; Region C, Dec–Mar; Region D, Nov–Jan, Mar–Apr, July; in Region C, a well-defined rains breeder. Zambia, Jan–Feb, Apr; Malaŵi (sings Nov to mid-Mar, juvs Apr–June).

DEVELOPMENT AND CARE OF YOUNG: when disturbed with fledged young, parents keep inside bushes, giving alarm calls.

References
Chapin, J. P. (1953a).
Lynes, H. (1930).
Vincent, A. W. (1948a).

Plate 9
(Opp. p. 160)

Cisticola anonymus (von Müller). Chattering Cisticola. Cisticole babillarde.

Drymoeca anonyma von Müller, 1855. J. Orn. 3, p. 197; W Africa (Nun R., S Nigeria).

Forms a superspecies with *C. bulliens* and *C. woosnami*.

Range and Status. Endemic resident. Birds referred to this species locally frequent SE Sierra Leone (Allport *et al.* 1989, G. D. Field, pers. comm.). Common to abundant in lowland forest zone from S Nigeria (near Lagos) through extreme SW Central African Republic, to N and central Zaïre south to *c.* 2°S in E and border with Angola. Common W Gabon; 26 adults in 15 ha clearing at M'Passa in 1977, reduced to 8–12 in 1983–85 after bush growth (Brosset and Erard 1986). Local S Congo (Pointe-Noire); uncommon Cabinda and recorded Cuanza Norte in Angola. From coastal lowlands, up to at least 1900 m in S Cameroon.

Cisticola anonymus

Description. ADULT ♂: forehead, crown, nape and hindneck brown; lores buff; rest of upperparts dull olivaceous brown; tail-feathers brown with subterminal dark brown bar and buff tip. Entire underparts and underwing yellowish white or buffy white, palest on chin, throat and belly and washed with grey on flanks, underwing, sides of breast and sides of neck; primaries, secondaries and tertials brown with narrow paler fringes on outer webs and broad pale fringes on basal two-thirds of inner webs; wing-coverts brown, fringed olivaceous brown. Bill black, central area of lower mandible grey; eyes light brown; feet pale brown. Sexes alike. SIZE: wing, ♂ (n = 10) 60–63 (61·4), ♀ (n = 10) 52–56 (53·9); tail, ♂ (n = 10) 50–58 (53·6), ♀ (n = 10) 45–50 (47·0); bill, ♂ (n = 8) 15–17 (16·0), ♀ (n = 9) 14–16 (14·9); tarsus, ♂ (n = 10) 22–24 (22·8), ♀ (n = 10) 21–22 (21·6). WEIGHT: (Cameroon) ♂ (n = 4) 15–16 (15·5), ♀ (n = 2) 12, 13.

IMMATURE: like adult except chin, throat and breast pale lemon-yellow; less contrast between head-top and back (more uniform olivaceous brown, crown duller than adult, back brighter than adult).

NESTLING: feathered nestling like immature.

Field Characters. Medium-small, with unstreaked upperparts; rusty cap contrasts with greyish back. Chattering song reminiscent of Rattling Cisticola *C. chiniana*. Confined to clearings in lowland forest belt from Sierra Leone to Zaïre, where usually the only cisticola present. At fringes of range overlaps in habitat with larger Whistling Cisticola *C. lateralis* (browner, no red on head, whistled song) and 'Brown-backed' Cisticola *C. chubbi discolor* (much larger, reddish brown back, black lores, loud duetting song). Common around villages; attracts attention with its loud calls.

Voice. Tape-recorded (45, 87, B, C, F, CART, ERA, LEM). Song very variable but similar in form to that of Rattling Cisticola, i.e. some introductory notes followed by a trill: 'ji-ji-ji-jerrr' or 'chi-chi-wrreee'. The trill is 0·5–1·0 s long, and varies from liquid and bubbly to dry and tuneless; sometimes it is slow enough for syllables to be discernible: 'chi-chi-wiwiwiwiwi'. The loud, chattering call, a string of grating, buzzy notes, is often interspersed between these short songs, to produce a medley of calls and trills.

General Habits. Inhabits high grass, bushy grassland or low second-growth in forest clearings and open places in oil-palm and old rubber plantations; also forest edge, open cultivation and regrowth, path-sides, edges of villages, airfields, bushes along sea-shore. In Sierra Leone mainly in swamps, also marshy fields. Rapidly colonizes forest clearance, via paths and streams at least, but no firm evidence for flying over forest canopy; on old farmland and other clearings, disappears as bush develops over the years.

Usually in pairs or (family?) parties, remaining all year on territory, within which it breeds and does most foraging. In Gabon, territories always included much *Paspalum* grass, with bushy areas (used as cover from danger) and a few isolated trees (song-posts) (Gowthorpe 1977). Territories of larger groups with more bushes were larger: *c.* 9000 m² for 2 birds; *c.* 12,000 m² (3 birds); *c.* 30,000 m² (4). Boundaries fixed and territory defended by male(s) of the group, using song, although ♀ may help defend near nest. Group sizes in one study area: solitary ♂ (1); pairs (3); pair plus extra ♀ (4); pair plus extra ♂ (1); 2 ♂♂ with 2 ♀♀ (1).

Runs, walks or hops on ground; perches on grass stalks and bushes. Searches lawns after rain. Cocks tail. Calls much from cover; sings from a shrub, high grass-head, telegraph wire or other elevated perch. Curious, will approach observer, calling and flicking wings.

Food. Insects; exploits termite flights. In Gabon, items given to young included grasshoppers (80–95%), caterpillars (1–11%), flies, crickets, butterflies (0–5%), other (2–4%); grasshoppers mostly of one green species; prey size was initially 3–4 mm, finally 40–50 mm, with little variation between 5 broods (Gowthorpe 1977).

Breeding Habits. Information mainly from Gowthorpe (1977). Monogamous, group-territorial, helpers at some nests. ♂ sings from grass stalk or bush top, occasionally from tree or in cruising flight c. 6 m above ground. Courtship dance: ♂ bounces in air (vertical bounces c. 15 cm), c. 30 cm above ♀ for c. 1 min, while singing in time with bounces (Rand et al. 1959). ♀ sometimes strays outside own territory; ♂ sometimes tolerates ♂ trespassers. In groups with extra ♀, only one breeds; same ♀ builds replacement nests. In groups with extra ♂, one is dominant and displays more. Helpers are in adult plumage and thought to be offspring of previous year. ♀ chooses nest site and often builds near edge of territory (once outside). Aggression increases during incubation and is high during hatching and fledging. Territory area decreases slightly during incubation. Alarm calls given near nest in response to human, coucal, sparrow. Aggressive to other species only near nest.

NEST: oval, domed, with side-top opening. Outside built first, of dry, broad grass blades, sometimes anchored by gossamer; lined with fine grass stems and heads and occasionally rootlets; inner lining of vegetable down (in Zaïre down of *Funtumia* seed pods), may be added after eggs laid, during incubation, ending up a thick layer on floor. Incubating birds often seen carrying down to nest. Nest well-concealed in rank vegetation 20–60 cm above ground: in heart of grass tussock or edge of larger patch, or lodged between weed stalks. In plantations may be comparatively exposed, 0·5–1 m up, attached to young palm frond or between old frond and trunk. Built by ♀ (once ♂ helped) in 1–2 days (except lining).

EGGS: Cameroon 2 (sometimes 3); Gabon 1 (n = 2), 2 (4), 3 (43); Zaïre 2(–3). Ovate to obtuse-ovate; smooth, matt or slightly glossy; various shades of blue or turquoise-blue, occasionally pale green, with sparse spots or blotches of purplish brown, rufous-brown or light brown and with ashy violet shell-marks; marks often clustered at big end. SIZE: (n = 17, Cameroon) 15·7–18 × 12–13·2 (16·7 × 12·7); (Gabon) av. 16·4 × 12·5; (E Zaïre) 17·3–18 × 12·4–13·1.

LAYING DATES: W Cameroon, all months, peaks May and Oct (beginning and end of rains); Gabon (73 nests), Jan–Feb (46), Mar (3), May (3), June (5), July (4), Sept (6), Oct (5), Dec (1) (peak in short dry season); Congo, Nov; S Zaïre, N Angola, Oct–Nov, probably throughout year; E Zaïre, Jan, Aug–Oct.

INCUBATION: starts after 3rd egg laid; by ♀; occupies c. 60% of daylight hours; period 13–15 days (all hatch same day). ♂ defends territory and warns ♀ of danger; joins ♀ when she leaves nest.

DEVELOPMENT AND CARE OF YOUNG: chicks brooded for 4–5 days after hatching and after heavy rain when older. Fed by ♀ (once ♂ brought 3 prey items during period in which ♀ brought 580). ♀ feeds young for 10–15 min, bringing prey every 2–3 min, then 'rests' for 10–15 min, on nest at first, then with ♂. Faeces are carried off. Extra ♀♀ help only with feeding young and bring 3–5% of food. Supernumerary ♂ does not feed young. Fledglings fed by ♀.

BREEDING SUCCESS/SURVIVAL: groups consisting of parents and newly-fledged young usually include 2 fledglings. 13 of 18 (72%) Gabon nests predated, at night. Predators apparently terrestrial mammals and snakes. One nest attacked by ants: ♀ picked them off for 20 min but 3 chicks eventually eaten. Predation quickly followed by replacement nest (5–6 days between destruction and laying), up to 4 replacements per pair.

References
Gowthorpe, P. (1977, 1978).
Lynes, H. (1930).
Serle, W. (1949b).

Cisticola bulliens Lynes. Bubbling Cisticola. Cisticole murmure.

Plate 9
(Opp. p. 160)

Cisticola bulliens Lynes, 1930. Ibis ser. 12, vol. 6 suppl., p. 315–321; Lobito Bay, Angola.

Forms a superspecies with *C. anonymus* and *C. woosnami*.

Range and Status. Endemic resident in Angola and lower Congo Valley. Common from Moçâmedes north along coast to mouth of Congo R. and Landana (Cabinda). Common inland in valleys of Cuanza and other rivers and at least to Matadi on Congo R.; less common away from rivers. Local on escarpment in NW Huila District. 3 records of birds seen and heard in 1981 near Mouila, Gabon (Y.-M. de Yiviès, pers. comm.).

Description. *C. b. septentrionalis* Tye: Cabinda, lower Congo R., N Angola south to about Gabela. ADULT ♂ (breeding): forehead, crown and nape tawny brown, lores and ear-coverts slightly paler; rest of upperparts including scapulars dull greyish brown, faintly mottled darker brown; tail tawny brown with subterminal dark brown bar (less marked on central pair) and buffish brown tip. Chin, throat and centre of belly silvery white; rest of underparts pale yellowish buff: richer or washed grey on flanks and axillaries; thighs deep buff; primaries, secondaries, tertials and wing-coverts dull brown, fringed buffish

Cisticola bulliens

or tawny brown; underwing-coverts buff. Upper mandible dark grey to dark brown, lower mandible paler grey, often with dark tip; eyes brown; feet pale pinkish brown. ADULT ♂ (non-breeding): richer-coloured, especially on upperparts (plumage not worn or faded); head-top dark reddish brown. Sexes alike. SIZE: wing, ♂ (n = 14) 61–68 (64·6), ♀ (n = 5) 51–56 (54·2); tail, ♂ breeding (n = 8) 53–58 (55·0), non-breeding (n = 6) 57–65 (60·8), ♀ breeding (n = 3) 43–50 (46·0), non-breeding (n = 2) 52, 57; bill, ♂ (n = 14) 16–18 (16·8), ♀ (n = 5) 14–15 (14·8); tarsus, ♂ (n = 13) 24–27 (24·5), ♀ (n = 5) 21–22 (21·6).

IMMATURE: like non-breeding adult but upperparts richer brown; tail and wings paler; underparts at first washed with lemon-yellow, grey of breast and flanks replaced by richer buff.

NESTLING: unknown.

C. b. bulliens Lynes: S Angola (Benguela). Head-top paler, yellower, tawny brown (difference clearest in non-breeding plumage); also in non-breeding dress, rest of upperparts paler, greyer brown. SIZE: wing, ♂ (n = 11) 62–66 (63·7), ♀ (n = 11) 51–55 (53·1); tail, ♂ breeding (n = 6) 54–61 (55·8), non-breeding (n = 5) 58–63 (60·6), ♀ breeding (n = 6) 46–50 (47·5), non-breeding (n = 5) 51–53 (52·0); bill, ♂ (n = 11) 15–17 (16·0), ♀ (n = 11) 14–15 (14·7); tarsus, ♂ (n = 11) 24–25 (24·4), ♀ (n = 10) 21–23 (22·1).

Field Characters. Very plain; dull grey-brown with tawny wash on head; looks like a bleached version of Chattering Cisticola *C. anonymus*. Somewhat redder in non-breeding season. Distinctive bubbling song, given from tall post. Restricted to W Angola, extreme SW Zaïre and Gabon, where it lives in coarse grass in salt marshes on coast and in dry bush country inland. Further told from other cisticolas in its range and habitat by unstreaked upperparts.

Voice. Not tape-recorded. Song fairly musical, rippling and bubbling, preceded by 3 short introductory notes, 'di-di-di-drrrreee', 'twee-twee-twee-trrrrrrrr'; tone and tempo vary.

General Habits. Occurs near coast in brackish meadows thickly covered with coarse grass up to 2 m high and dotted with bushes, inland from mangrove zone; also bushy sisal fields. Further inland found in dry savanna with similar grasses, or denser bush and palm groves.

Food. Insects, including beetles.

Breeding Habits. ♂ sings from bush top or telegraph wire at daybreak and more or less throughout the day, with bill wide open showing black mouth, tail spread and flicked but wings not moved. May give bubbling part of song in flight, but no real aerial display. When alarmed, ♂ flies towards ♀ calling.

NEST: a ball; placed 30–100 cm above ground, usually in high grass, occasionally in thorn bush.

EGGS: 3 (4 clutches). Slightly glossy; white to pale or medium blue or greenish blue, blotched or finely spotted with red, dark purple, chocolate or rust-brown, marks sometimes concentrated near big end. SIZE: (n = 17, Angola) av. 16·3 × 12·1.

LAYING DATES: Oct–July (irregular, peak Feb–Apr).

References
Chapin, J. P. (1953a).
Lynes, H. (1930).

Plate 9 (Opp. p. 160)

Cisticola chubbi Sharpe. Chubb's Cisticola. Cisticole de Chubb.

Cisticola chubbi Sharpe, 1892. Ibis, p. 157; Kimangtichi (= Mangiki), Mt Elgon, Kenya.

Forms a superspecies with *C. hunteri* and *C. nigriloris*.

Range and Status. Endemic resident. Common in highlands of SE Nigeria (Gotel Mts, Chappal Hendu, Obudu and Mambilla plateaux, Leinde Fadali) and W Cameroon south through Bamenda and all highlands except Rumpi Hills to Mt Cameroon. Widespread in highlands of E Zaïre from Lendu Plateau and Rwenzoris to Mt Kabobo and Marungu Plateau; volcano region of NW Rwanda and south in highlands of W Rwanda and W Burundi to Bururi Forest and over into W Tanzania at Manyovu; widespread in highlands of SW Uganda

Cisticola chubbi

north to Kibale Forest, and an isolated population at Mubende (Boma Hill); Mt Elgon (Kenya/Uganda) and W Kenya from Kapenguria to Kakamega, Nandi Escarpment, Eldama Ravine and Molo; NW Tanzania at Karagwe (1° 30′S, 31° 00′E) (not Bukoba: Keith 1993). Common to abundant; adaptable, range expanding as forest is cleared; in Cameroon probably not immediately threatened (Stuart 1986).

Description. *C. c. chubbi* Sharpe: Zaïre except Marungu Plateau, Rwanda, Burundi, Uganda, Kenya. ADULT ♂: forehead to hindneck light rufous, palest on forehead; rest of upperparts umber-brown, unstreaked, uppertail-coverts rather paler and warmer. Tail graduated, producing fan shape, feathers umber-brown, central pair with vestige of black subterminal spot and slightly paler tip, rest with black subterminal band *c.* 4 mm wide (variable: range 3–5), and 3–5 mm pale rufous tip; underside of tail paler, making black bands more conspicuous. Pre-orbital region black; ear-coverts pale reddish brown. Chin and throat white; breast creamy white in centre, greyish at sides; belly white, flanks buffy brown; undertail-coverts buff, thighs pale rufous. Wings dark brown, feathers variably edged brown to pale rufous-brown on outer webs, underside of flight feathers margined buff to reddish buff on inner webs, axillaries and underwing-coverts reddish buff. Bill straight and fairly stout, black, base of lower mandible sometimes grey; inside of mouth black except for bluish grey underside of tongue; eyes hazel, light brown, warm brown or deep brown-red, rim of eyelids light brown; legs and feet pinkish buff, pinkish flesh or brownish flesh, claws brown. Sexes alike. SIZE: (10 ♂♂, 10 ♀♀) wing, ♂ 60–64 (62·8), ♀ 56–59 (57·5); tail, ♂ 60–64 (61·9), ♀ 52–57·5 (55·4); bill, ♂ 15·5–17 (16·0), ♀ 14·5–16 (15·0); tarsus, ♂ 25–26·5 (25·5), ♀ 23–25 (24·2). WEIGHT: (Kenya/Uganda) ♂ (n = 24) 12–21 (17·5), ♀ (n = 22) 11–19 (14·7).

IMMATURE: upperparts more rufous-brown than adult, contrasting only a little with head; rump and uppertail-coverts paler rufous, like head; tail rufous-brown, like back, almost uniform, with only a hint of black band and rufous tip. Below, sides of breast with brownish wash, less grey, buff of flanks and undertail-coverts brighter, more rusty than adult, and extending to lower belly. Bill yellow-ochre, culmen dark brown, gape yellow-ochre; eyes grey, grey-brown or light brown; legs and feet pale flesh.

NESTLING: bill yellowish, culmen brown; inside of mouth yellow except for 2 black tongue spots; eyes grey-brown; feet flesh.

C. c. marungensis Chapin: Marungu Plateau, SE Zaïre. Upperparts and tail rather warmer brown; dark band on tail duller, less contrasting; entire underparts washed with buff, flanks and undertail-coverts rich buff. Tail slightly longer.

C. c. adametzi Reichenow: Nigeria, Cameroon except Mt Cameroon. Differs from *chubbi* mainly in colour of underparts: chin, throat and belly rich buff, breast brownish grey, washed buff in centre, flanks and undertail-coverts warmer, smoky buff, thighs rufous-cinnamon. SIZE: (7 ♂♂, 4 ♀♀) wing, ♂ 60–62 (61·0), ♀ 55–57 (56·0); tail, ♂ 55–60 (57·3), ♀ 51–54 (52·3); bill, ♂ 14–16 (15·3), ♀ 14–15 (14·5); tarsus, ♂ 25–27 (26·2), ♀ 24–25 (24·6). WEIGHT: 2 ♂♂ 20, 20.

C. c. discolor Sjöstedt: Mt Cameroon. Differs from other races only in tail being almost plain, dark spots obscure, almost absent. WEIGHT: ♂ (n = 3) 17–18 (17·3) ♀ (n = 7) 12–19 (15.1).

TAXONOMIC NOTE: populations in Cameroon and Nigeria (*adametzi*, *discolor*) were treated by Lynes (1930), Grimes (1976a) and Sibley and Monroe (1990) as a separate species, *C. discolor*. However, Chappuis (1974) pointed out the great similarity between the songs of *discolor* and *C. chubbi*, and believed the birds to be conspecific. Experimental playback of songs of *chubbi* to *discolor* produced both positive and negative responses. We follow Traylor (*in* Watson *et al.* 1986) and Dowsett and Forbes-Watson (1993) in treating the *discolor* group as races of *C. chubbi*. It is remarkable that in spite of presumed geographic isolation of E and W populations for thousands of years the song has remained constant; another instance of 'la très grande stabilité dans le temps des caractères acoustiques par rapport aux autres caractères génétiques' (Chappuis 1974).

Field Characters. A medium-large cisticola of middle-altitude scrub and grass in woodland and along forest edge, with rufous cap, black lores, unstreaked brown upperparts, wings only slightly warmer brown than back, not contrasting, underparts either white with grey sides (eastern populations) or buff and grey (Cameroon). Easily identified by loud duet. Just meets Hunter's Cisticola *C. hunteri* at around 2500 m on Mt Elgon; Hunter's Cisticola lives at higher elevations and has very different duet; it can also be told by smaller size, longer tail, darker plumage, red on head darker and less bright, contrasting less with back. Similar species in east are Singing Cisticola *C. cantans* (rusty cap and unstreaked back, lives in forest edge scrub, but is smaller, with chestnut wing-patch contrasting with grey back, buffy underparts, song of short phrases of 2–3 notes); Trilling Cisticola *C. woosnami* (same size, unstreaked upperparts, but lacks black lores and rusty cap, lives in lower altitude dry wooded grassland, trills from top of tree or bush); and Whistling Cisticola *C. lateralis* (unstreaked upperparts and overlapping habitat, but lacks rufous crown and black lores, gives measured song from tall perch). In W, it is the only cisticola in most of its montane habitat, but at lowest levels might be confused with smaller Chattering Cisticola *C. anonymus*, which has red-brown crown and unstreaked grey back but no black lores, lives in lowland forest clearings and has chattering song.

Voice. Tape-recorded (45, 87, B, C, F, GRI, HOR, KEI, LEM); see also sonograms in Grimes (1976a) and Thorpe (1972). Song a precisely antiphonal duet. Principal song phrase consists of 3–4 ringing notes, 'twee-hweeyo', 'tswee-tee-woy', 'see-tee-tweet', 'which-cherry', 'see-which-cherry'; this is accompanied by a dry chatter. The phrase 'twee-hweeyo' (Kakamega, Kenya) is accompanied, in addition to the chatter, by a sharp 'tick' which coincides precisely with 'twee' and 'yo' (S. Keith tape). Principal song in *discolor* group very similar, 'tsyoo-weep', 'tweety-woy', 'weety-choy', 'switch-a-bee', with same ringing tone, but accompanist gives 1–3 high thin notes, 'wee-tsit', 'tsee-tswee', 'tsi-wee-tsit', 'tssswit', rather than a chatter. Alarm, a short chatter. Snaps wings in display.

General Habits. A bird of wooded highlands. Principal habitat is bushes and scrub along borders of forest, usually with grass nearby; grassy forest clearings, wooded streams, edges of gallery forest; also bracken-briar with some admixture of grass, cutover areas with dense roadside bush, elephant grass in some places; enters giant lobelia zone in Nyungwe Forest (Rwanda). In Cameroon, dense shrubby undergrowth of forest clearings sometimes 1–2 km from nearest grass, brushy second growth, shrubs and herbage in ravines and sheltered hollows, less often in intermediate zone of shrubs, bracken and grass between forest and farmland; at lower altitudes, more grassy terrain, abandoned cultivation, eucalyptus plantations. On Mt Elgon found along streams in forest and in scrub by road (2400 m), in bamboo and thick scrub (2450 m), and overlaps with Hunter's Cisticola at 2550 m in large glades by rocks in mixed bamboo/*Podocarpus* forest (Britton and Sugg 1973). Intermediate and montane forest, 850-3000 m in Cameroon, 1200–3350 m in Zaïre/E Africa.

Singly or in pairs, often in family groups of 4–5. Does not associate with other species. Noisy and conspicuous when singing; at other times slips about in thick cover, foraging among tangled vegetation, only seen when it flies from one bush to another. 1 briefly followed ant swarm in weeds and bushes of large clearing by road in forest, Rwanda (Willis 1986a). Duetting takes place throughout the year; besides maintaining the pair bond, it may help birds to keep in contact in long, dense grass (Thorpe 1972). Duetting birds sing from leafless bush, grass stem, or other visible perch (in Cameroon prefers cocoa bush or banana plant to grass). When one bird starts to sing its mate flies to join it; duetting not heard when birds were out of sight of each other (Grimes 1976a). They sit close together, facing each other, with tails spread, and bob up and down, bowing to one another, swaying sideways with a quick, jerky motion, and turning around. Leader repeats the song theme up to 10 or even 20 times, partner countersings, and often one or more family members join in with shorter notes. Often fearless of man, especially when singing or nesting; 1 pair nesting by hut entered and left nest unconcernedly, often perching on railing of verandah within arm's length of observer.

Moult in W Kenya took *c.* 159 days (Mann 1985).

Food. Insects.

Breeding Habits. Monogamous; territorial. Duetting intensifies during breeding season, and adjacent pairs define territorial boundaries by countersinging regularly from same stations every day (Cameroon: Grimes 1976a). Displays at trespassers and responds to song playback by duetting, often with loud wing snaps, and one bird (♂?) also bobs up and down.

NEST: spherical or oval, large and untidy, with circular side entrance *c.* 50 diam. near top. Built mainly of grass, with other materials matching surroundings, e.g. dry roots, leaves, lichens; lined with plant down, including thistles and groundsel, fine grass and vegetable fibres, also feathers; ext. diam. 100–210, ext. depth 140–280; placed 0·5–2·0 m above ground in tuft of grass, clump of herbs or in low tree close to trunk, well concealed; 1 was built between 2 leaves sewn together, with a 3rd leaf over the top as a screen; 1 was situated 1.6 m from steps of mountain hut.

EGGS: 2–3, sometimes 4; av. (12 clutches) 2·5. Elongated to regular oval; smooth, somewhat glossy; bluish white or pale blue to pale bluish or greyish green or beige, spotted with pale red, red-brown, purplish brown or pink over lilac and ash-grey, mainly at large end. SIZE: (n = 10) 17·7–19 × 12–14 (18·4 × 13·3).

LAYING DATES: Nigeria, Nov; Cameroon, Mar–Apr, Aug, Nov–Dec (juveniles Mar–Apr, June); Zaïre, Jan, Apr–May, July (juveniles Oct), mainly during rains; E Africa: Region A, Apr–May, July, Sept; Region B, Dec–Jan, Apr–May; season ill-defined, not confined to rains.

INCUBATION: probably by ♀. Incubating bird usually very shy, leaving nest long before human observer reaches it, although one near hut allowed observer to pass within 0·3 m; also, when presumed ♂ sang, incubating ♀ left nest to join him (Young 1946).

BREEDING SUCCESS/SURVIVAL: in 1 nest, 2 eggs were cracked and nest was invaded by ants (Bennun 1989).

References
Chappuis, C. (1974).
Grimes, L. G. (1976a).
Stuart, S. N. (1986a).
Thorpe, W. H. (1972).

Cisticola hunteri Shelley. Hunter's Cisticola. Cisticole de Hunter.

Cisticola hunteri Shelley, 1889. Proc. Zool. Soc. Lond., p. 364; Mt Kilimanjaro.

Forms a superspecies with *C. chubbi* and *C. nigriloris*.

Range and Status. Endemic resident. Mt Elgon (Kenya/Uganda), and widespread in W and central Kenya and N Tanzania from Cherangani Mts, Laikipia, Mt Uraguess, Mt Kenya and Nyambenis to Loita Hills, Ngong and Nairobi region in Kenya, and in Tanzania from Loliondo to Crater Highlands (to Oldeani), Mt Meru and Mt Kilimanjaro. Common to abundant.

Cisticola hunteri

Description. (including '*hunteri*', '*masaba*', '*hypernephela*', '*prinioides*'). ADULT ♂: forehead to hindneck dark reddish brown with indistinct streaks, amount of streaking and degree of redness varying considerably; rest of upperparts rather dark brown with variable reddish wash, all except rump with variable amount of dark streaking. Tail-feathers dark brown with light brown to pale rufous edges and tips, all except central pair with black subterminal spot 3–5 mm wide; underside of tail paler, spots more contrasting. Lores dark brown; ear-coverts light brown to pale rufous with pale shaft-streaks. Underparts dull creamy white, palest on throat, sides of breast, flanks and undertail-coverts greyish brown, thighs brownish rufous. Wings dark brown, with narrow light brown to pale rufous margins to outer webs of primaries and secondaries, inner and outer webs and tips of tertials and wing-coverts; axillaries, underwing-coverts and margins of inner webs of underside of primaries and secondaries pale reddish buff; bend of wing noticeably pale creamy. Albinos recorded. Bill black to brownish black, keel sometimes paler; eyes brown to dark brown; legs and feet greyish brown to dark flesh. Sexes alike. SIZE: (10 ♂♂, 10 ♀♀) wing, ♂ 59–65 (61·9), ♀ 55–62 (58·9); tail, ♂ 62–71 (66·2), ♀ 58–68 (62·4); bill, ♂ 14–16 (14·9), ♀ 14–16 (15·0); tarsus, ♂ 24–27 (25·0), ♀ 23–25 (24·4). WEIGHT: ♂ (n = 13) 14–19 (16·2); ♀ (n = 11) 12–18 (14·8).

IMMATURE: less heavily patterned above than adult, somewhat more rufous, whiter below, some with slight yellowish tinge; black on lores reduced or absent. Bill yellow-ochre; eyes pale brown; legs and feet pale flesh.

NESTLING: unknown.

TAXONOMIC NOTE: birds from *c.* 2900–3050 m and above on Mts Elgon, Kenya and Kilimanjaro (but not Mt Meru) are overall duller and darker (darker brown above and grey below); birds in SE part of range (especially Mt Kilimanjaro) are more heavily streaked than those from NW (especially Mt Elgon), which are almost plain. Variation is both clinal and individual, and we agree with White (1960b) that it is not practicable to recognize any races.

Field Characters. A duetting and chorus-singing bird of high altitudes, where it is generally the only cisticola present. Long-tailed and dark, with blackish lores and obscurely streaked crown and back which can appear plain. Overlaps with Chubb's Cisticola *C. chubbi* only in narrow zone on Mt Elgon; when not singing its very different song, Chubb's Cisticola separable by larger size, shorter tail, paler and brighter rufous cap contrasting with completely unstreaked back.

Voice. Tape-recorded (10, 13, 45, 46, 87, B, C, F, KEI). Song a remarkable duet, repeated many times in succession. In Kenya highlands, one bird (probably the ♀: Todt 1970) produces a long shivering trill which rises and falls in waves, while its mate gives a sweet whistle which also rises and falls, 'too-it', 'see-too-it', 'see-here', 'chorry-cheer', 'tooi-tooi-weet'. In N Tanzania (Ngorongoro) trill is shorter and drier and notes of 2nd bird less sweet, 'chee-chee-chee', 'kowit' or a brief 'chit-chit' or 'chup'. Timing is very precise; the whistle comes in at exactly the same note of the trill each time (see sonograms in Todt 1970 and Thorpe 1972). The whistled component is often melodious but can also be dry and clipped, a single 'chip' or 'chichichi', and the trilled component may not rise and fall but remain more or less on 1 pitch. In trios, the 3rd bird is a (presumed) ♂, and he uses his own whistle, not copying that of the other ♂; in quartets (counter-duetting between neighbouring pairs), songs are temporally co-ordinated (Todt 1970). Birds foraging in thick bush said to give a 'more throaty warble: "whi-chu-wou-itthri, rirrrrr", and mate replies "chii-cheedle-lu-u-u-whit"' (van Someren 1956). Alarm, 'tsi-tsi-tsi'.

General Habits. Inhabits forest edge and clearings, dense shrubby undergrowth, grass tangles, fringes of damp hollows and marshes, bushy hill sides, streamside vegetation, gardens; at higher altitudes, clearings in bamboo zone, moorlands with tree heaths and lobelias, at highest point with only sparse cover. 1550–4400 m, seldom below 1700 m; on Mt Elgon not below 2500 m (overlaps with Chubb's Cisticola at 2500–2550 m).

Occurs singly but more often in pairs or family groups; flocks of 6–10 noted on Kilimanjaro (King 1973). Site fidelity sometimes marked; pairs observed for periods of 10 months and 2 years respectively were never observed >30 m from centre of range, usually less (Sessions 1966). Associates with Moorland Chat *Cercomela sordida*; they apparently compete for insects (King 1973). Birds forage in pairs and groups, calling and whistling as they hop from twig to twig, wagging raised tails. Singing continues all day, including dawn and dusk, and all year, and is one of the most characteristic sounds of the Kenya highlands. Often climbs to top of shrub or bush to sing, but also sings from inside cover. A pure albino duetted successfully with a pure brown partner, birds even touching each other as they sang (Sessions 1966). Probable biological functions of duetting are: (1) synchronization of sexual activity of paired birds; (2) maintenance of pair bond; (3) facilitating recognition and contact between paired birds; (4) territorial defence (Todt 1970).

Food. Small insects, including beetles, young mantids, nymphal grasshoppers, caterpillars, geometer and other moths, and butterflies (including a lycaenid).

Breeding Habits. Territorial. Courtship song noisy, birds approachable, almost oblivious of observer. After duet performed within cover, birds fly up and meet on perch, almost touching one another, and continue to duet with excited movements, bobbing, turning to and fro, beating wings and twirling half-spread tail. Group singing by 3, 4 or 5 individuals takes place within cover. In response to playback of own duet, birds of pair (1) make vocal contact; (2) aggressively approach loudspeaker; (3) give loud counter-duet, using same motif that is being played back to them. During communal singing, playback causes only 1 pair to emerge from undergrowth, the pair whose duet is being played back; this pair then sings from exposed perch.

NEST: large, stout ball with circular entrance near the top (once with 2 entrances, 1 in front and 1 at the side: A. Williams, nest record card), the entrance sometimes covered; made mainly of dry grass blades, with a few plant stems, leaf ribs and fibres, profusely lined with plant down, also with some fine grass and feathers; strands of grass sometimes hang from the entrance, causing the nest to resemble a mass of leaves and debris; ext. diam. (n = 1) 120. Placed 0·1–1·3 m above ground in grass tussock, among sedges or soft-leaved plants, or in small shrub or bush, once at base of fence post; base and sides supported by surrounding vegetation, adjacent grass blades sometimes bent down and woven into top of dome. Once used old nest of Cape Bishop *Euplectes capensis*, adding thick lining of plant down and other soft material to the original structure, which was 0.3 m above ground in lush grass (A. Williams, nest record card).

EGGS: 2 (11 clutches) or 3 (10 clutches), rarely 4, once 5 (Sessions 1966). White, pale blue, green or pink, with small to bold spots and freckles of brown, red-brown, blackish brown or maroon, either mainly at large end or evenly distributed. SIZE: (n = ?) 16–17·5 × 11–12·5. Double-brooded.

LAYING DATES: E Africa: Kenya, Mar, May–June, Oct, possibly all year, but mainly in rains; Region A, Jan, Mar–Apr, Sept–Nov, season ill-defined; Region D, all months except Feb and Aug, peaking in long rains Apr–July.

INCUBATION: mainly by ♀. Period: 12–13 days. Mate scolds when observer near nest, but does not sing or seem overly flustered.

DEVELOPMENT AND CARE OF YOUNG: fledging period 17 days (n = 1).

BREEDING SUCCESS/SURVIVAL: in 1 nest, 3 young eaten by mongoose; observer then took 1 young from each of 2 nearby nests and placed them in the 1st nest; they were fed by their 2 foster parents and fledged (van Someren 1956).

References
Lynes, H. (1930).
van Someren, V. G. L. (1956).
Thorpe, W. H. (1972).
Todt, D. (1970).

Plate 9 (Opp. p. 160)

Cisticola nigriloris **Shelley. Black-lored Cisticola. Cisticole masquée.**

Cisticola nigriloris Shelley, 1897. Ibis, p. 536, pl. 12, fig. 2; Kombe, Misuku Hills, northern Nyasaland, 7000 ft.

Forms a superspecies with *C. hunteri* and *C. chubbi*.

Range and Status. Endemic resident. Highlands; S Tanzania on Ufipa Plateau and from Malaŵi border north to Mbeya range, Njombe highlands, Iringa and Mufindi districts, Udzungwe Mts (Ngwazi, Kihansi Falls (8° 32′S, 35° 46′E)), and S Kilosa (Uvidunda Mts); N Malaŵi south to N Viphya Plateau; extreme NE Zambia on Nyika Plateau and probably Mafinga Mts. Locally common to abundant.

Description. ADULT ♂: forehead to hindneck deep rich tawny; lores and adjacent feathers at edge of forehead, often extending around base of upper mandible, black; ear-coverts pale rufous-brown with faint pale shaft-streaks. Upperparts including scapulars unstreaked umber-brown, somewhat lighter on rump. Tail-feathers dark reddish brown, all except central pair with 3–5 mm black subterminal band and conspicuous 3–6 mm buffy tip; underside of tail much paler, with buffy greyish wash, black spots more contrasting. Underparts

Cisticola nigriloris

creamy buff with grey on sides of breast sometimes spreading onto centre of breast, flanks greyish brown, undertail-coverts light brown, thighs tawny. Wings dark brown, flight feathers narrowly edged tawny to light brown, undersides with broader buff margin to inner webs, axillaries and underwing-coverts reddish buff. Bill black, keel variably paler, inside of mouth black; eyes red-brown; legs and feet pale rufous-brown. Sexes alike. SIZE: (10 ♂♂, 10 ♀♀) wing, ♂ 67–71 (69·0), ♀ 63–71 (66·5); tail, ♂ 69–76 (72·0), ♀ 66–76 (71·4); bill to feathers, ♂ 12·5–14 (13·1), ♀ 12–14 (12·7), bill to skull, 2 ♂♂ 16, 16, 2 ♀♀ 16, 16·5; tarsus, ♂ (n = 12) 24–27 (25·7), ♀ (n = 12) 23–26 (24·6). WEIGHT: 2 ♂♂ 21·8, 21·8; 1 ♀ 17·7.

IMMATURE: 'crown paler and upperparts paler and redder than adult, so little contrast between head and back; most of forehead as well as lores black, forehead grading into crown with black feather bases . . . no yellow in plumage . . .' (Lynes 1930). However, immature in American Museum of Natural History is washed with sulphur above and below, top of head paler than adult, gingery red, lores and faint supercilium yellow, upperparts with rusty wash, tail greyer, less red, with black band very broad, *c*. 10 mm; chin, throat, centre of breast and belly sulphur-yellow, grey sides of breast with yellow wash, flanks and undertail-coverts cinnamon, thighs paler than adult. Bill of American Museum of Natural History specimen all yellowish white except culmen, gape ochre. Immatures collected by Vernon (1964) resemble Lynes's description, eyes grey-brown; eyes said to be pale by Willis (1986a).

NESTLING: has 2 black tongue spots.

Field Characters. A large, long-tailed cisticola confined to highlands of S Tanzania, N Malaŵi and NE Zambia. Combination of large size and group foraging gave observer impression of small babbler or brownbul *Phyllastrephus* until black tail-spots were seen (MacPherson 1966). Rusty crown contrasts with unstreaked brown upperparts; black lores prominent. Duetting behaviour easily distinguishes it from cisticolas in same habitat, which all have distinctive songs; in addition, Singing Cisticola *C. cantans* has unstreaked upperparts and head somewhat redder than back, but lighter upperparts and shorter, greyer tail, white lores, whiter underparts; Wailing Cisticola *C. lais*, Churring Cisticola *C. njombe* and Stout Cisticola *C. robustus* have rusty caps but heavily streaked upperparts.

Voice. Tape-recorded (45, 86, 87, F, KEI, LEM, MOY). Song a duet. 1st bird sings 3–4 clear, sweet, ringing notes, last often sibilant, typically see-sawing up and down the scale: 'hee-yo hay-ho', 'wah-tsee-tssay', 'hee-yo-see-tsssweep', 'hee-hoo-sit', 'wah-hee-hay', sometimes only ascending, 'wor-ter-pah-pee'; 2nd accompanies it with 1–3 short, nasal, rasping sounds, which have been likened to the creaking of taut leather or a squeaky cartwheel: 'dzidda', 'dziddle-wup', 'dzeep-dzup-dzup', 'dzip-dzip', or a single 'dzup'. 1st singer repeats a phrase several times, then switches to another. 2nd bird also varies both form and synchronization of its response; it usually sings at the same time, but start time variable; single-note accompaniment usually coincides with last note of phrase, e.g. 'hee-yo-hay-ho' has single 'dzup' synchronized with final 'ho', and response can also appear on its own between phrases. Although start time varies between phrases, it is constant for a given phrase, i.e. 2nd bird comes in at the same time until 1st bird changes the phrase. Alarm a loud 'chwee' or 'peat'. Snaps wings in display.

General Habits. Lives in brushy montane habitats: bracken-briar, bushes, weeds and rank growth of all kinds, along streams, in hollows and damp valleys, and beside patches of primary forest and regenerating second growth. Adapted to rough lawns and hedges of *Cotoneaster* and other exotics around rest camp (Nyika Plateau: MacPherson 1966). Altitudinal range in Tanzania 1100 m (Kihansi Falls) to 2400 m (Ufipa) (D. C. Moyer, pers. comm.); in Zambia/Malaŵi 1370–1740 m.

On Nyika, forages in parties of up to 9, hopping about in bushes and low undergrowth. Once 1 bird broke away from group and flew ahead, settled and started calling; others followed and joined in, song intensified and sometimes developed into a display but usually soon died down and birds resumed feeding; this group-singing was repeated at irregular intervals during the day; party of 7 studied for a morning sang every 15–30 min (Vernon 1964). Group of 6–8 spent much time on lawns around Nyika Rest House, often with Buff-shouldered Whydah *Euplectes psammocromius*. Family group of 5, including fledged young with short tails, followed raiding ants into dense vegetation. Pair duetting continues all year, even after young have left nest.

Regularly observed in close association with Common Stonechat *Saxicola torquata*, although advantages of this not clear (Harpum 1978b); whenever cisticolas duetted, stonechat was invariably visible close by on conspicuous bush or post.

Food. Insects.

Breeding Habits. Not territorial; nest not defended; some evidence of co-operative breeding. When a nest was found there was usually a party of birds nearby; 4 adults were seen carrying food to a nest, calling in alarm when disturbed. Display develops from joint song. 10 birds of 2 parties (4 and 6) assembled, 1 started calling, others joined in, and 3 or more were calling at once; sometimes first birds stopped as later ones started; calls were repeated faster as birds got excited; singing birds perched close to one another, bills pointed upwards, jerking bodies up and down, flitting wings and cocking tails. There was also countersinging by birds in adjacent bushes. Sometimes bird made short aerial flight, flying up $c.$ 1 m and tumbling back to perch, or flying down to ground, rapidly snapping wings against body. After display, birds resumed feeding (Vernon 1964). Displays on Nyika occurred all day at beginning of breeding season (Oct), but only at sunrise and sunset in Jan.

NEST: large, bulky ball, like small nest of *Centropus*, of broad blades of dry grass, mainly *Panicum*, also *Setaria*, *Digitaria*, *Urochloa* and *Imperata*, outside of nest wrapped mainly with grass blades, with only a trace of cobwebs; lined with entire seed-heads of grasses, once with ramenta of tree fern overlain with soft pappae of Compositae; 0.7–1.3 m above ground (once 2 m), attached to tall grass or wedged into bush, bracken or herb, once in exposed *Philippia* bush on bank of stream, once at top of pine sapling in windbreak; not concealed.

EGGS: 2–3, av. (12 clutches) 2·6. Ovate; smooth, with slight gloss; pale turquoise, turquoise-blue or pale green, finely and profusely marked all over with light brown, brown or red over purplish grey or lilac-grey, with tendency to form cap at large end. SIZE: (n = 5) 19·2–19·9 × 13·9–14·2 (av.?).

LAYING DATES: Tanzania, Nov–Feb, in rains; Zambia, Dec–Jan, Apr.

DEVELOPMENT AND CARE OF YOUNG: black tongue spots of nestling blur together and fade as it grows older. Recently fledged young has short tail and yellow gape flanges, fully grown young has yellow palate (black in adult).

Reference
Vernon, C. J. (1964).

Plate 9
(Opp. p. 160)

Cisticola aberrans (A. Smith). Rock-loving Cisticola; Lazy Cisticola. Cisticole paresseuse.

Drymoica aberrans A. Smith, 1843. Illustrated Zoology of South Africa, Aves, pl. 78; Port Natal (i.e. Durban), *errore* = Transvaal, restricted to Marico district by Clancey (1993).

Range and Status. Endemic resident, restricted to rocky hills from Mauritania to S Sudan, south to South Africa. Uncommon S Mauritania (Massif d'Assaba). Recorded Guinea (Konossou Mt at 1300 m), NE Sierra Leone. Uncommon Liberia, in montane areas north of forest zone, and once Monrovia (Gore 1994). Locally common on rocky hills in Ivory Coast. Locally common Ghana (N scarp of Volta basin, scarps in Mole, Bongo Hills, isolated populations on inselbergs on Accra Plains), N Togo. Uncommon and local in Mali south of $c.$ 15°N and in soudan zone of extreme SW Niger. Common Arli-Pendjari Nat. Parks (Benin-Burkina Faso). Locally frequent on inselbergs in guinea savanna of Nigeria, especially on Jos Plateau; in SW south to Idanre but not in SE. In Cameroon occurs on Adamawa and Benue Plateaux and Mandara District. Rare Bamingui-Bangoran Nat. Park, Central African Republic. Common Ouaddaï region of E Chad and probably occurs in other rocky areas of sahel and soudan zones, although absent Ennedi. Locally frequent E and S Sudan, 1600–2500 m. In Rwanda, NE Uganda, S Kenya, N Tanzania, local, sometimes common, on rocky ground at 600–2150 m. Common in mountains of Malaŵi west of Rift Valley above 1000 m, uncommon 700–1000 m; occurs E Malaŵi from Mulanje and Blantyre north and adjacent N Mozambique. Locally common E and S Zambia.

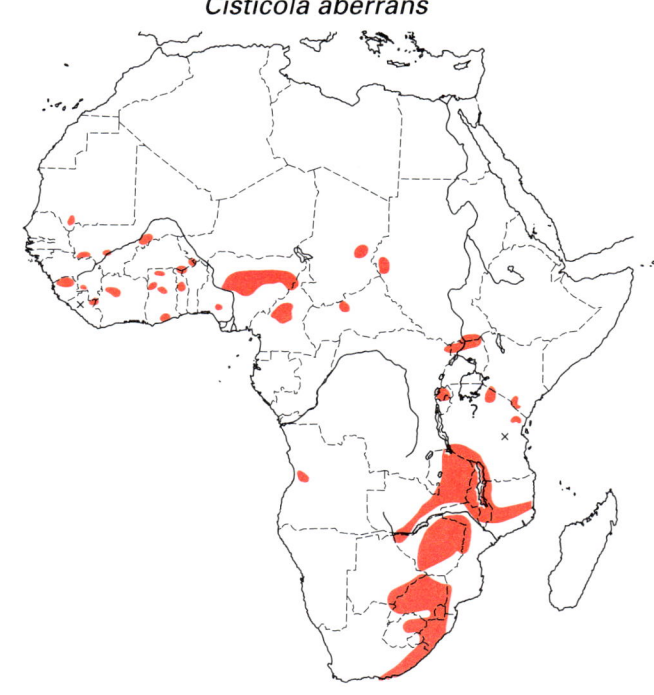

Cisticola aberrans

Common on Zimbabwe Plateau up to 1300 m and on adjacent Central Plateau of Mozambique; local W Zimbabwe. Rare to uncommon extreme SE and E Botswana. Locally common Transvaal, Natal, Lebombo Mts of S Mozambique and Swaziland southwest through Orange Free State to E Cape Province at Great Fish R. Central Angola, in highlands around Mt Moco.

Description. *C. a. petrophilus* (Alexander): N Nigeria, W Cameroon, SW Sudan, NE Zaïre. ADULT ♂ (breeding): forehead, crown and nape rufous-brown; lores and ear-coverts brownish buff; rest of upperparts dull brown; tail-feathers dull brown, all except central pair with subterminal dark brown bar and buff tip; underparts and underwing buff, richest on underwing and undertail-coverts, washed grey on flanks; primaries, secondaries, tertials and wing-coverts brown, with narrow paler fringes on outer webs, and with primaries and secondaries also having a broader, light brown fringe on inner webs. Upper mandible brown or dark grey, lower mandible paler grey; eyes reddish brown; feet pale pinkish brown. ADULT ♂ (non-breeding): head top richer rufous; underparts deep buff. Sexes alike. SIZE: wing, ♂ (n = 11) 58–64 (61·0), ♀ (n = 8) 50–58 (54·5); tail, ♂ breeding (n = 5) 52–56 (54·8), non-breeding (n = 6) 54–64 (58·7), ♀ breeding (n = 4) 47–53 (50·2), non-breeding (n = 4) 50–55 (52·5); bill, ♂ (n = 10) 15–17 (16·1), ♀ (n = 8) 14–16 (14·7); tarsus, ♂ (n = 11) 22–24 (23·1), ♀ (n = 8) 19–22 (21·2).

IMMATURE: like non-breeding adult.

NESTLING: unknown.

C. a. admiralis (Bates): Guinea, Sierra Leone, Mali, S Ghana. Like *petrophilus* but head top darker, dark brown, less rufous; rest of upperparts and tail darker (very dark brown); flanks and sides of breast washed olivaceous grey. Eye paler, yellow-brown. Smaller. SIZE: (1♂, 4♀♀) wing, ♂ 59, ♀ 50–52 (51·0); tail, ♂ 58, ♀ 47–53 (50·0); bill, ♂ 15, ♀ 13–14 (13·7); tarsus, ♂ 21, ♀ 20–21 (20·2).

C. a. emini Reichenow (including '*teitensis*', but see Zimmerman *et al*. 1996): N Tanzania, S Kenya. Like *petrophilus* except upperparts show slightly more prominent dark shaft-streaks; underparts paler buff. SIZE: (4♂♂, 2♀♀) wing, ♂ 60–62 (60·7), ♀ 55, 56; tail, ♂ 57–65 (59·2), ♀ 56, 60; bill, ♂ 14–15 (14·7), ♀ 14, 15; tarsus, ♂ 22–23 (22·7), ♀ 22, 22.

C. a. lurio (Vincent): Malawi east of Rift, adjacent NW Mozambique. Like *emini* except rufous of head less sharply cut off at nape and often darker; dark tail-spots often do not extend across outer webs; in fresh (non-breeding) plumage, upperparts more mottled (centres of feathers darker, more conspicuous). Smaller but longer-tailed. SIZE: (8♂♂, 8♀♀) wing, ♂ 56–61 (59·2), ♀ 50–54 (52·1); tail, ♂ breeding (n = 6) 55–65 (61·2), non-breeding (n = 2) 63, 71, ♀ breeding (n = 4) 55–58 (56·5), non-breeding (n = 4) 55–58 (56·2); bill, ♂ 14–16 (15·0), ♀ 14–15 (14·6); tarsus, ♂ 21–23 (22·2), ♀ 20–22 (21·0).

C. a. nyika Lynes: extreme SW Tanzania, Zambia east of line from Victoria Falls to Mpika and Abercorn, Zimbabwe, Malawi and central Mozambique west of Rift. Like *lurio* except dark tail-spots usually restricted to inner webs. In fresh (non-breeding) plumage, fringes on primaries, secondaries and tertials broader; upperparts average slightly lighter in colour; underparts deeper buff; differences slight and disappear with wear.

C. a. aberrans (A. Smith): Transvaal and adjacent Botswana, W Swaziland, highland Natal, E Orange Free State. Like *nyika* but crown slightly brighter rufous; tail-feathers unspotted (occasionally with trace of dark bar), without buff tips; underparts richer buff. Smaller and proportionately longer-tailed than *lurio*. SIZE: (6♂♂, 5♀♀) wing, ♂ 53–60 (56·2), ♀ 48–57 (50·8); tail, ♂ breeding (n = 2) 55, 56, non-breeding (n = 4) 68–76 (71·5), ♀ breeding (n = 2) 44, 51, non-breeding (n = 3) 52–71 (61·0); bill, ♂ 12–15 (14·2), ♀ 13–16 (14·2); tarsus, ♂ 20–21 (20·2), ♀ 19–21 (19·8).

WEIGHT (Southern African races): ♂ (n = 12) 13·1–16·9 (15·2), ♀ (n = 9) 12–15·9 (13·7), unsexed (n = 17) 10·5–14·1 (12·2).

C. a. minor Roberts: E Cape Province, lowland Natal, E Swaziland, S Mozambique (Lebombo Mts); intergrades with last in lowland E Transvaal. Like nominate but smaller. In fresh (non-breeding) plumage, upperparts less mottled, slightly paler; fringes on primaries, secondaries and tertials narrower; underparts paler. SIZE: (14♂♂, 7♀♀) wing, ♂ 50–55 (52·6), ♀ 47–49 (47·9); tail, ♂ breeding (n = 10) 55–70 (60·3), non-breeding (n = 4) 61–75 (70·0), ♀ breeding (n = 1) 52, non-breeding (n = 6) 51–64 (58·8); bill, ♂ 14–15 (14·2), ♀ 13–14 (13·7); tarsus, ♂ 19–22 (20·9), ♀ 19–21 (19·7).

C. a. bailunduensis Neumann: central Angola. Differs from preceding races in plain, unmottled upperparts; tail pattern like *lurio/nyika*. Smaller and proportionately longer-winged than *minor*. SIZE: wing, ♂ (n = 6) 50–55 (53·0), ♀ (n = 5) 48–50 (49·0); tail, ♂ breeding (n = 4) 44–52 (48·5), non-breeding (n = 1) 53, ♀ breeding (n = 5) 45–46 (45·4); bill, ♂ (n = 4) all 13, ♀ (n = 5) 13–14 (13·2); tarsus, ♂ (n = 6) 19–21 (19·8), ♀ (n = 5) all 20.

TAXONOMIC NOTE: *C. a. emini* is treated as a distinct species by Sibley and Monroe (1990) but as a race of *C. aberrans* by Watson *et al*. (1986), Short *et al*. (1990) and Dowsett and Forbes-Watson (1993). Morphological differences are slight, ecology is similar, and tape playback experiments suggest that the 2 are conspecific (Dowsett and Dowsett-Lemaire 1993b).

Field Characters. Medium-small, with plain brown upperparts, pale eyebrow and rusty cap; upperparts washed reddish in non-breeding season but still contrast with cap. Underparts vary, buffy in southern Africa, greyish sides in W and E Africa with belly and undertail-coverts rusty-buff to pale buff. No rufous in wing. Tail long and narrow in southern races, often partly cocked, giving prinia-like shape. Rocky habitat not shared by any other cisticola; in W Africa practically confined to savanna inselbergs, but elsewhere other habitats also used, especially in southern Africa. Voice distinctive. Neddicky *C. fulvicapillus* also has rusty cap and plain upperparts but different voice, shorter tail, no pale eyebrow, S and E races grey below.

Voice. Tape-recorded (45, 58, 87, 88, 91, B, F, ZIM). Song and calls vary both individually and geographically. Basic unit of song is a short (1–1·5 s) trill on one pitch; in W Africa it can be a dry rattle like that of Rattling Cisticola *C. chiniana* or a little more liquid, with the quality of Stout Cisticola *C. robustus*; in Kenya (*emini*) it is dry and short (D.A. Zimmerman, pers. comm.); in S-central Africa (race *nyika*) it is faster and higher-pitched; in South Africa it is dry and insect-like. The trill may be given on its own, or be preceded by a rasping 'chaaa', often repeated, which in W African birds can produce a form remarkably like that of Rattling Cisticola: 'chaaa-chaaa-chaaa-chaaa-trrrrrrr'. Trills are interspersed with a variety of chattering and clicking notes and harsh scolds, including a nasal, querulous bleat like that of Bleating Warbler *Camaroptera brachyura*, also a loud 'weep' or 'chweep' and a quieter 'choy-choy-choy', producing a medley of notes and trills without set form. Song becomes louder and faster during aggressive encounters (see Chappuis 1974). Contact calls of *emini* "a somewhat mournful, bisyllabic piping 'pee-u', repeated, and a squeaky 'squee-a' or 'squee-e-a'" (Zimmerman *et al.* 1996). Alarm, loud 'chu-ip' or 'tu-wheee', and harsh scolding notes; in Kenya, alarm call of *emini* "an insistent nasal 'tchaa' or 'zheea', often repeated and sometimes follwed by a rapid 'tsip-tsip-tsip-tsip' or a louder and more churring 'tsirrrrrrrr' to form a song: 'squee, chchchchchchchch' or 'skiew, skieee, chrrrrrrrrrr'" (Zimmerman *et al.* 1996).

General Habits. Inhabits rocky places, including rocky montane grassland (W Africa), outcrops, boulder screes, cliffs and inselbergs, often with scant vegetation or (in Serengeti) with lusher growth of 0·5–1 m-high grass, bushes and small trees between the rocks. In Serengeti, when this vegetation is temporarily destroyed by fire, may move into thickets of *Croton dichrogamus* and *Acacia brevispica* (Stronach 1990). Uses all vegetation strata, up to mature tree canopies, although spends most time in the grass layer and among bushes. In southern Africa (Malaŵi and Mozambique southwards) found in open *Brachystegia* woodland with short grass and low bushes, where it occurs in glades on moist ground or on rock outcrops or areas with boulders, or along ridges. In Zimbabwe often in association with *Brachystegia tamarindoides*. In Botswana often at foot of hills near streams. In South Africa found in a wider variety of habitats, including streamsides, but always with rank grass and scrub and usually with trees, on hillsides and in valleys; also suburban gardens; mostly absent from flat land.

Usually solitary or in pairs. Active but silent, flitting about rock outcrops, hopping and running mouse-like on rocks; wary, using rocks for cover from intruders; when flushed from grass, often flies fast and direct to rocks. Forages in crevices and small caves with surrounding vegetation. In southern Africa, stays on or near ground among short grass and low bushes. Normally carries tail at slight angle above back, with body leaning forward when perched. When alarmed, bobs by bending legs, flicks wings and jerks tail up, until vertical, while calling. Territorial defence may lead to fights.

In southern Africa, appears to undertake short-distance movements to find suitable habitat at different seasons.

Food. Insects, including caterpillars, grasshoppers and small beetles. Young in Zimbabwe given grasshoppers with hind legs removed.

Breeding Habits. Monogamous, territorial. Comparatively silent and no display noted. ♂ sings from outer part of bush or small tree, or from telegraph wire.

NEST: a ball with side entrance, lined with plant down, hidden low in bunch of grass surrounded by thorn scrub or in tangle of grass and herbs. One pair, Zimbabwe, built nest, then had 2–3 week delay before laying.

EGGS: Nigeria (2 clutches) 2–3; Zambia, W Malaŵi 2 (n = 2), 3 (n = 9), 4 (n = 1); Zimbabwe 2–4, usually 3, one laid per day; Natal 2 (n = 3), 3 (n = 6), 4 (n = 3). (*C. a. nyika*, *C. a. aberrans*, *C. a. lurio*) white to green or pale greenish blue, finely spotted with various shades of brown; (*C. a. minor*) white with fine pink spots in halo around big end. SIZE: (Malaŵi, *C. a. lurio*) 17·5–18·5 × 13·0–13·5; (South Africa, n = 3, *C. a. aberrans*) 17·3–17·7 × 12·4–12·8 (17·5 × 12·6); (n = 6, *C. a. minor*) 15·5–17·0 × 11·6–12·4.

LAYING DATES: Mauritania (carrying nest material Oct); Mali (carrying nest material Sept); Ghana, Oct; Nigeria, May, July; E Africa: Region C, Jan–Mar (mid-late rains); Malaŵi, Dec–Feb; Zambia, Nov; Mozambique, Oct–Nov (probably also later); Zimbabwe, Sept–Apr, peak Oct (77 records); Botswana, Dec; Transvaal, Oct–Dec, Mar; Natal, Oct–Feb, peak Dec.

INCUBATION: in Zimbabwe starts with penultimate egg; period 13–15 days.

DEVELOPMENT AND CARE OF YOUNG: nestlings brooded and fed by ♀; period 13–14 days. When young threatened, parents jump about and call agitatedly in nearby bush. Fledglings fed for at least 3 weeks, sometimes more than a month. Family may stay together all winter in Zimbabwe.

BREEDING SUCCESS/SURVIVAL: parasitized by Wahlberg's Honeybird *Prodotiscus regulus*.

References
Ginn, P. J. *et al.* (1989).
Lynes, H. (1930).
Stronach, N. (1990).
Vincent, J. (1935).

Plate 10 (Opp. p. 161) *Cisticola chiniana* (A. Smith). Rattling Cisticola. Cisticole grinçante.

Drymoica chiniana A. Smith, 1843. Illustrated Zoology of South Africa, Aves, pl. 79; north of Kurrichane (= near Zeerust), South Africa.

Range and Status. Endemic resident, E and southern Africa. Some records prior to 1974 in N Kenya, S Ethiopia, Eritrea and S Somalia may refer to Boran Cisticola *C. bodessa* (q.v.). Frequent in Sudan south of

Cisticola chiniana

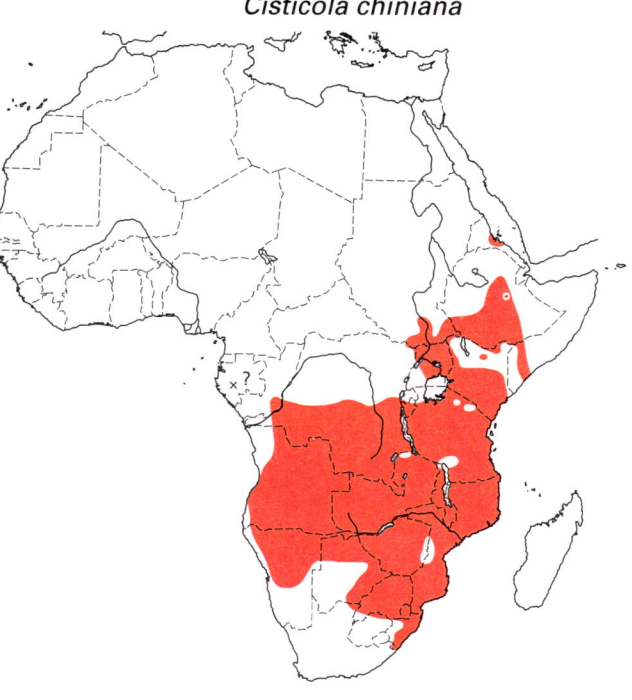

Sudd. Uncommon on E scarp of Eritrea at 1300–2000 m (Smith 1957). Locally common S Ethiopia. Recorded S Somalia (2 old records of *C. c. heterophrys*; others of *C. c. 'bodessa'* on Juba R. may be *C. bodessa*: Ash and Miskell 1983). Common N Uganda, S Kenya and Tanzania (including offshore islands), up to 2000 m; frequent Serengeti (16 birds per km^2: Folse 1982) but absent from highland Tanzania. Recorded Burundi. Recorded from Gabon without precise locality (Rand *et al.* 1959) and once seen singing near Mouila (Y.-M. de Viviès, pers. comm.). Local in central Congo. Locally common in Zaïre south of 3°S and in extreme NE. Locally common Angola, N Zambia (but absent between L. Tanganyika and L. Mweru), 300–1800 m. Common E and SW Zambia. Common Lilongwe District and abundant Shire Valley of Malaŵi, 1000–1500 m. Common Zimbabwe but absent eastern highlands and Chipinga Uplands. Uncommon N-central Mozambique (west of Malote), common further south, abundant south of Limpopo R., absent highlands. Locally abundant N and E Botswana. Common Transvaal, Natal from sea level to *c.* 1000 m (*c.* 1 pair per 4 ha in Transvaal acacia woodland). Recorded extreme N Cape Province, Swaziland. Uncommon N Orange Free State, rare in south. Occurs most of N Namibia, common north of *c.* 19°S.

Description. *C. c. simplex* (Heuglin): S Sudan to N Uganda, Adult ♂ (breeding): forehead, crown, nape, lores and ear-coverts buffy brown, streaked darker brown; rest of upperparts including scapulars and wing-coverts paler greyish brown, streaked very dark brown (feather centres dark brown), streaks fading out on rump, uppertail-coverts unstreaked; tail-feathers dull brown with subterminal dark brown bar and buff tip; entire underparts buff, paler (sometimes white) on chin, throat and belly, and washed grey on sides of breast and flanks; primaries, secondaries and tertials dull brown, edged buff; underwing buff. Bill very dark brown to black, lower mandible pale grey with darker terminal one-third; eyes reddish brown; feet pinkish brown. Adult ♂ (non-breeding): like breeding ♂ but head-top richer, more reddish; rest of upperparts and tail tawnier, dark brown streaks darker; underparts richer buff, especially on breast and flanks. Sexes alike. SIZE: wing, ♂ (n = 12) 62–70 (64·9), F (n = 8) 53–57 (54·7); tail, ♂ (n = 12) 52–62 (56·2), ♀ (n = 8) 46–51 (48·1); bill, ♂ (n = 10) 15–16 (15·7), ♀ (n = 8) 14–15 (14·5); tarsus, ♂ (n = 12) 22–24 (23·5), ♀ (n = 8) 19–22 (20·9). WEIGHT: (Uganda) 1♂ 25.

IMMATURE: like non-breeding adult but head-top paler, less rufous; rest of upperparts buffier (feather-edgings buff); underparts suffused with sulphur-yellow. Immatures are more strongly sulphur-tinged in north of range than in south.

NESTLING: unknown.

C. c. fricki (Mearns): S Ethiopia, N Kenya. Like corresponding plumages of *simplex* but head-top more prominently streaked with dark red-brown; rest of upperparts with richer, tawnier feather-edgings; breast and flanks richer buff. See *C. bodessa* for differences from that species.

C. c. humilis (Madarasz): W Kenya and NE Uganda highlands above 1800 m; Loliondo region of N Tanzania. Perennial dress like breeding plumage of *fricki*. Gape entirely black in dominant and single territorial ♂♂, partly pale in subordinate ♂♂, pale brown in ♀♀. SIZE: dominant ♂♂ larger than subordinates (Carlson 1986); wing, ♂ (n = 9) 67–72 (69·5), ♀ (n = 9) 56–59 (57·4); tail, ♂ (n = 9) 57–61 (59·4), ♀ (n = 8) 47–54 (51·2); bill, ♂ (n = 9) 15–17 (16·0), ♀ (n = 9) 14–16 (14·7); tarsus, ♂ (n = 9) 23–25 (24·1), ♀ (n = 9) 21–23 (22·0). WEIGHT: (Kenya) ♂ (n = 4) 16–17 (16·2), immature ♀ (n = 3) 9–14 (12·0).

C. c. ukamba Lynes: E Kenya uplands and NE Tanzania highlands. Like *humilis* but upperparts slightly brighter; redder on head-top; rest of upperparts more boldly marked, with greater contrast between darker, dark brown feather centres and paler, buffier edges; edgings to secondaries and tertials more rufous. SIZE: (10♂♂, 10♀♀) wing, ♂ 63–69 (65·7), ♀ 52–58 (55·2); tail, ♂ 53–62 (56·6), ♀ 44–54 (49·1); bill, ♂ 14–17 (15·5), ♀ 14–15 (14·5); tarsus, ♂ 22–25 (23·5), ♀ 20–22 (21·2). WEIGHT: (NE Tanzania) 2 ♂♂ 16·5, 17, 2 ♀♀ 12·5, 12·6.

C. c. victoria Lynes: Lake Victoria basin of SW Kenya and neighbouring Tanzania. Intermediate in size between *humilis* and *fischeri* (between whose ranges it lies) but dark streaks of upperparts broader than in either of these; upperpart colour more like *fischeri*.

C. c. fischeri (Reichenow): north-central Tanzania (south to Tabora Region). Like *humilis* but upperparts slightly colder, greyer, mottled with darker brown, rather than buff mottled with dark brown. SIZE: wing, ♂ (n = 15) 61–67 (64·1), ♀ (n = 7) 52–55 (53·9); tail, ♂ (n = 15) 49–59 (53·1), ♀ (n = 7) 44–47 (45·9); bill, ♂ (n = 15) 14–16 (15·1), ♀ (n = 7) 13–14 (13·9); tarsus, ♂ (n = 14) 21–24 (22·7), ♀ (n = 7) 20–22 (21·0).

C. c. keithi Parkes: S central Tanzania (Dodoma to Iringa); intergrades with *heterophrys* near Kilosa. Breeding dress indistinguishable from perennial plumage of *fischeri* but has distinctive non-breeding dress with head-top tawny buff, narrowly streaked with dark brown; rest of upperparts streaked buff and brown; underparts washed richer buff. SIZE: wing, ♂ (n = 15) 64–70 (67·1), ♀ (n = 11) 54–59 (55·5); tail, ♂ breeding (n = 10) 52–61 (55·8), non-breeding (n = 5) 54–62 (58·2), ♀ breeding (n = 7) 45–48 (46·4), non-breeding (n = 4) 48–51 (49·5); bill, ♂ (n = 15) 15–16 (15·4), ♀ (n = 11) 13–15 (14·2); tarsus ♂ (n = 15) 22–24 (22·9), ♀ (n = 10) 20–22 (21·1).

C. c. mbeya Parkes: Mbeya to Chimala, S Tanzania. Breeding dress like that of *keithi* but underparts less buff, central areas whiter, flanks and sides of breast darker, greyer. Non-breeding dress like *keithi*.

Plate 9

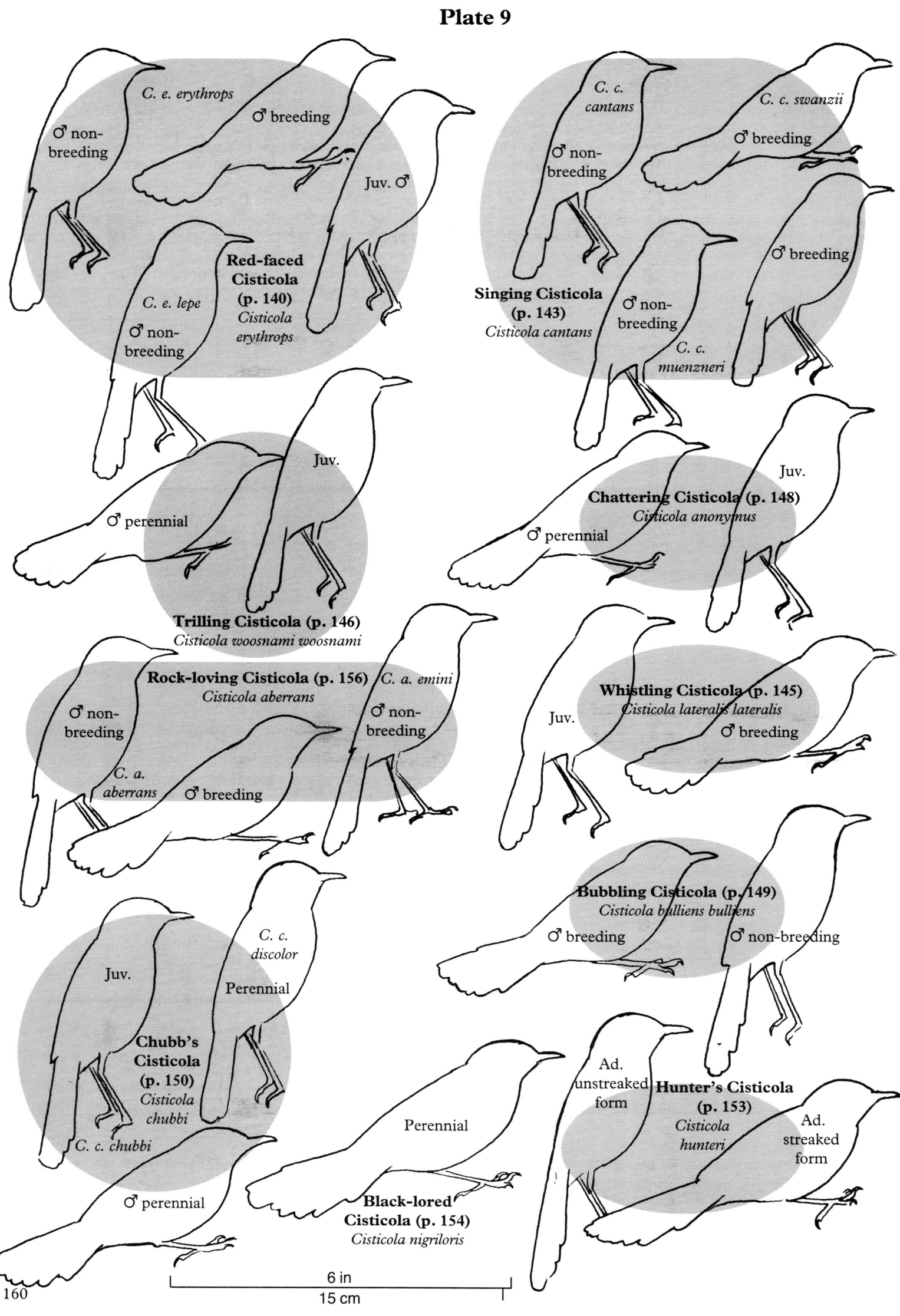

Plate 10

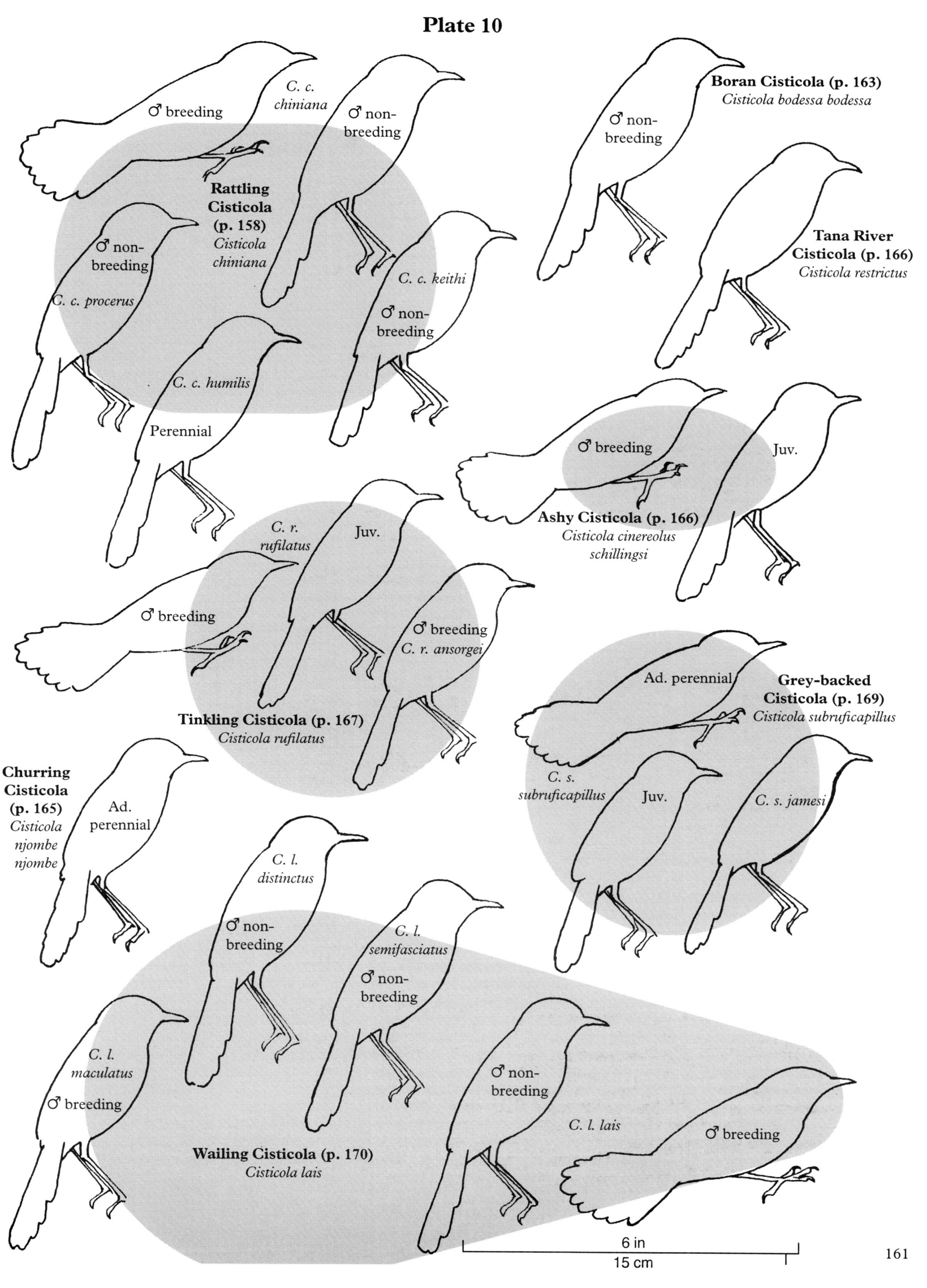

C. c. heterophrys (Oberholser): coastal Kenya and Tanzania, south to Dar-es-Salaam and inland to W Usambara. Perennial dress; upperparts plainer, less streaked than *ukamba*, *fischeri* and *keithi*; head-top more tawny, less rufous; rest of upperparts greyish brown, feathers with darker brown shaft-streaks; underparts more uniform buff; outer fringes of outer primaries tawny rufous. SIZE: wing, ♂ (n = 10) 58–65 (62·1), ♀ (n = 10) 50–55 (51·9); tail, ♂ (n = 10) 48–57 (52·2), ♀ (n = 10) 43–50 (45·4); bill, ♂ (n = 10) 15–17 (16·2), ♀ (n = 10) 14–15 (14·3); tarsus, ♂ (n = 9) 22–25 (23·3), ♀ (n = 10) 20–22 (21·0). WEIGHT: (Kenya) ♂ (n = 21) 8–18 (14·7), ♀ (n = 5) 6–20 (12·2); (Tanzania) 1 ♀ 12·6.

C. c. emendatus Vincent: SE Tanzania, N Mozambique, Malaŵi south to Blantyre, extreme E Zambia (Luangwa R. eastwards). Like *heterophrys* but in breeding plumage underparts whiter; in non-breeding dress richer rufous on upperparts. Larger than *heterophrys*, with greater difference in tail-length between breeding and non-breeding plumages. SIZE: (10 ♂♂, 10 ♀♀) wing, ♂ 63–69 (65·3), ♀ 51–55 (52·7); tail, ♂ breeding (n = 5) 52–60 (56·8), non-breeding (n = 5) 56–63 (59·0), ♀ breeding (n = 5) 43–47 (45·2), non-breeding (n = 5) 44–51 (48·0); bill, ♂ 14–17 (15·6), ♀ 14–15 (14·4); tarsus, ♂ 22–26 (23·8), ♀ 20–22 (21·0). WEIGHT: (Malaŵi) ♂ (n = 12) 17·1–20·0 (18·2), ♀ (n = 9) 10·1–14·0 (11·8).

C. c. procerus (Peters): Tete (Mozambique) and extreme S Malaŵi (Chiromo). Like *emendatus* but upperparts paler. In breeding dress, upperparts paler grey-brown; head-top more tawny, less rich brown. In non-breeding dress feather-edgings of upperparts paler buff, giving strikingly streaked appearance.

C. c. chiniana A. Smith (including '*vulpiniceps*' (Clancey 1992), whose distinguishing features not supported by material in Natural History Museum, London; no significant difference in size): Zimbabwe except extreme W, S Mozambique except S coast, Transvaal, SE Botswana. Breeding dress like *procerus* but more heavily streaked dark brown on upperparts. Non-breeding dress like *procerus* but slightly paler on upperparts (feather-edgings buffier), making dark streaks even more obvious; crown prominently streaked dark brown. WEIGHT: (Zimbabwe) ♂ (n = 23) 13·2–21·4 (18·1), ♀ (n = 8) 12·6–18·5 (14·4); (Botswana races) ♂ (n = 9) 9–20 (14·2), ♀ (n = 5) 11–19·5 (15·2).

C. c. campestris (Gould): Natal, Swaziland, S coastal Mozambique. Slightly darker above than nominate race, especially in non-breeding dress. Immature not yellow on underparts.

C. c. smithersi Hall (including '*huilensis*'): extreme W Zimbabwe, NW Botswana, Caprivi, extreme SW Zambia (S Barotseland), extreme N Namibia, S Angola (Cunene, NE Moçâmedes, central and N Huila districts). Paler and less heavily streaked on upperparts than corresponding plumages of nominate race: in breeding dress head-top mid-brown, rest of upperparts ash-grey; in non-breeding dress, upperparts more finely streaked than nominate, feather edges greyer, less buff.

C. c. frater (Reichenow): central Namibia. Like corresponding plumages of *smithersi* but colours richer, underparts much buffier. In breeding dress upperparts darker, more heavily streaked; in non-breeding dress upperparts with feather edges buffier, head-top brighter tawny. WEIGHT: (Namibia) ♂ (n = 5) 16·5–18 (17·5), 2 ♀♀ 12·5, 12·5.

C. c. fortis (Lynes): central and N Angola, Zambia except SW (south of 14°S) and extreme E (east of Luangwa R.), S Congo, Gabon, S Zaïre. Of adjacent races, most like *emendatus*, with which it intergrades in NE Zambia. In breeding plumage, darker: head-top dark rufous-brown; rest of upperparts less streaked, almost uniform dark greyish brown with dark shaft-streaks; underparts darker, with flanks and breast grey. Most characters clinal, upperparts becoming plainer and underparts darker further west, with NE Zambian birds (north of 12°S) intermediate in coloration between Angolan topotypical *fortis* and *emendatus*, but much larger than latter (size typical of largest *fortis*). In non-breeding dress, upperparts scarcely distinguishable from *emendatus* but tails slightly darker and underparts greyer (buffish grey). SIZE: birds in NW of range (N Angola, Kasai and Kinshasa): wing, ♂ (n = 5) 67–69 (68·0), ♀ (n 5) 53–56 (54·4); tail, ♂ breeding (n = 5) 55–59 (56·4), ♀ breeding (n = 5) 44–46 (44·8); bill, ♂ (n = 4) 15–16 (15·5), ♀ (n = 5) 13–15 (14·2); tarsus, ♂ (n = 5) 22–24 (23·0), ♀ (n = 5) 19–22 (20·6). Birds in SE (S Angola, SE Zaïre and Zambia): (8♂♂, 9♀♀): wing, ♂ 70–73 (70·9), ♀ 57–60 (58·1); tail, ♂ breeding (n = 6) 56–60 (58·0), non-breeding (n = 2) 63, 66, ♀ breeding (n = 6) 45–51 (48·2), non-breeding (n = 3) 51–52 (51·7): bill, ♂ 15–16 (15·4), ♀ 13–15 (14·2), tarsus, ♂ 22–25 (23·5), ♀ 20–22 (20·7). WEIGHT: (SE Zaïre) ♂ (n = 11) 17–21 (19·7), ♀ (n = 2) 13, 16, 1 unsexed 14; (Zambia) 1 ♂ 18·4, 2 ♀♀ 12·5, 13·6.

C. c. bensoni Traylor: S Zambia between ranges of *fortis* and *smithersi*. In nonbreeding dress upperparts greyer than *fortis*, browner than *smithersi* (intermediate between these two): dark streaks of upperparts broader than *smithersi* and more prominent than *fortis*; underparts like *fortis* (buffier than *smithersi*). Breeding dress like *fortis* but more streaked.

Field Characters. The common dry country (thornbush) cisticola of E and S-central Africa, common and noisy; its harsh scold and grating song 'jee-jee-jee-jee-tiddle-iddle' are a familiar sound in the bush. Medium large, with heavily streaked dull brown back, crown and wing-edgings with some rufous or reddish brown, varying with race and season, contrasting somewhat with back; upperparts and tail redder in non-breeding dress. Race *heterophrys* of coastal Kenya and Tanzania has much less pronounced streaking on upperparts, which can appear almost plain, more conspicuous red wing-patch and more varied habitat. For differences from Boran Cisticola, see that species.

Voice. Tape-recorded (9, 10, 21, 45, 87, 88, 91, B, C, F). Song extremely variable but always of 2 parts: 1–5 (typically 2–4) measured individual notes followed by faster ones which may be run into a trill. First notes usually harsh and scolding, like short version of alarm call; notes in second part usually dry: 'jee-jee-jee-jee-tutututu', 'jee-jee-jee-jee-jee-teetrroo', 'jee-jee-jee-tupiturree', 'jee-jee-trootroo', 'chur-chur-titicheeoo', 'widjiwidji-turrr', 'beeju-beeju-turrr', but sometimes almost liquid: 'tseee-tsi-wawawa', 'jee-jee-jee-doydoy', 'jee-jee-jee-joowijoowi'; occasionally first notes may be pure, 'peeyoo-peeyoo-tutututu', 'peeyoo-peeyoo-turrreee'. Alarm a loud, harsh, dry 'jaaa-jaaa' or 'jeee-jeee'.

General Habits. Inhabits scrub, light thickets and rank grass in open parkland, including *Acacia* and (in Burundi) *Parinari* savannas, Kalahari scrub, mixed dry woodland, coconut and cashew plantations, overgrown cultivation; often in the wetter areas, or near water. In S Ethiopia found in open mosaic of bush and grass, lacking trees over 5 m, where more common than Boran Cisticola, but also occurs in more closed bushland (R. Timmins, pers. comm.) and in towns. *C. c. fischeri* in Tanzania, Mozambique and nominate race in Transvaal live in thorn woodland, not *Brachystegia*, whereas *fortis* and other races in SE Zaïre, Angola, Zambia and Mozambique also inhabit the latter; other races in Zaïre inhabit plains near rivers and lakes; *heterophrys* in coastal Tanzania inhabits lusher, softer grass habitats than further inland. In Kenya and Uganda, *humilis* dwells in bushier country than other races there. In Mozambique, *procerus* associates more with clearings and forest edge, whereas *campestris* occurs in tangled secondary vegetation. In Namibia occurs in dense crowns of low-growing *Citrus*.

In non-breeding season, remains in social groups. In Kenya (Carlson 1986), these may include up to 4 apparently adult ♂♂ (father and sons or unrelated ♂♂) and 5 ♀♀; the ♂♂ defend an all-purpose territory; subordinate ♂♂ contribute substantially to defence and sing more than dominant ♂♂. Single ♂ joined by another ♂ increases its territory size; removal of subordinate ♂ results in territory shrinkage. Subordinate ♂♂ eventually replace single ♂. The major variable affecting group size is apparently bush density (more bushes, bigger groups), and territories with high bush density are also smaller.

Forages low in grass, bushes or on ground, where it hops. Flies to bushes when alarmed; singing ♂ drops down to bush. Approaches humans, calling. Noisy for much of the year, calling from bush tops or wire.

Food. Mainly insects: beetles (including Curculionidae), ants, termites, flies, grasshoppers, crickets up to 6 cm long, larvae including caterpillars up to 5 cm long; also snails, aloe nectar.

Breeding Habits. ♂ sings from high, thin branch on bush or low tree, jumping up and down and flitting tail up above back. Stops singing on appearance of intruder; gives warning call if danger persists or intensifies; then takes refuge in bush, becoming silent. ♀ leaves nest on hearing warning call and also seeks refuge in bush, silently. In courtship display, often given before sunrise, ♂ spreads wings, spreads and droops tail in front of ♀, while shivering and giving final part of song.

NEST: a sphere, pear shape, or broad oval, *c.* 10–13 cm high, 8 cm wide, with side or side-top entrance 3–4 cm in diam.; loosely built of dry grass, often with broader blades in the outermost layer, bound with spider web and lined with finer grass and seed-heads, plant down and web; attached by web, 5–100 cm above ground, usually in mixed dry/green grass or other herbage 30–70 cm high, in shelter of thorn bush, saplings or fallen branch.

EGGS: 2–5; Kenya (3 clutches) 4–5 (4·3); Tanzania (2 clutches) 3; Zambia, Zimbabwe, Malaŵi (81 clutches) 1 (n = 1), 2 (19), 3 (47) 4 (14); Malaŵi (another 10 clutches) 3; Transvaal (2 clutches) 3; Natal (53 clutches) 1–4 (3·4). Variable: white to bright blue, or pale greenish blue, much spotted or blotched with red-brown, pinkish brown, black, dark brown, maroon, dark purple or slate, occasionally evenly, more usually at big end, often in ring, where there may also be some larger spots of pale lilac or purplish grey; occasionally unspotted blue. SIZE: (n = 13, Kenya, *ukamba*) av. 16·3 × 13·0; (n = 2, Kenya, *humilis*) av. 17·1 × 13·0; (n = 3, Tanzania, *heterophrys*) av. 17·5 × 13·0; (n = 3, S Zaïre) 16·9–17·7 × 12·9–13 (17·5 × 13·0); (n = 4, S Zaïre, *fortis*) 16·5–17·6 × 12·9–13·0 (17·1 × 12·9); (n = 12, Malaŵi, *emendatus*) av. 17·2 × 12·5; (n = 34, Zimbabwe) 16·7–19·1 × 12·3–13·8 (18·2 × 12·9); (n = 6, Transvaal) av. 18·2 × 12·8.

LAYING DATES: Sudan, June–Sept, Dec; Ethiopia, Mar. E Africa: Kenya, Mar, May–Sept; Tanzania, Jan–May; Region A, Aug–Sept; Region B, Mar–July, Nov–Dec; Region C, Dec–Apr; Region D, Nov–Jan, Mar–June (peak Apr); rains. Burundi, ♂♂ singing Nov; S Zaïre/N Angola, Sept–Mar; S Angola, Feb; Zimbabwe, Sept–Apr, peak Nov–Jan (183 records); Malaŵi, Dec–Apr; W Zambia, Nov–Feb; Botswana, Oct–Apr; S Mozambique, Oct, Jan, probably double-brooded; Natal, Oct–Jan (peak Oct–Dec); Transvaal, Oct–Jan, Mar (peak Dec); Namibia (singing Oct).

INCUBATION: by ♀; readily deserts after nest inspections by humans.

DEVELOPMENT AND CARE OF YOUNG: young in nest *c.* 14 days. Both sexes give alarm notes near nest but become silent when real danger threatens.

BREEDING SUCCESS/SURVIVAL: one bird found dead with fowl-pox (contagious epithelioma) lesions and with ticks *Hyalomma rufipes* blocking ears (Steyn 1971).

References
Carlson, A. (1986).
Lynes, H. (1930).
Roberts, A. (1913).
Vincent, A. W. (1948a).

Cisticola bodessa Mearns. Boran Cisticola. Cisticole des Borans.

Plate 10
(Opp. p. 161)

Cisticola subruficapilla bodessa Mearns, 1913. Smithsonian Misc. Coll. 61 (11), pp. 2–3; Bodessa, S Ethiopia.

Range and Status. Endemic resident in extreme SE Sudan, S Ethiopia, N and W Kenya and probably Eritrea (Nefasit: Erard 1974b); perhaps also S Somalia (Ash and Miskell 1983) and N Uganda (Elliott 1972). Locally common to abundant, extreme SE Sudan and Kaffa, Sidamo, Harar, Shoa and northern Omo Provinces of Ethiopia; most common in Borana (part of Sidamo). In Kenya known from Moyale, Marsabit, Mt Kulal, Maralal, NE Elgon, Kito Pass, Laikipia Plateau, and locally common on the escarpments of Elgeyo, Iten and Kongelai, and on the N slopes of Mt Kenya (Lewa Downs, Timau-Isiolo). In Sudan up to 1200 m; elsewhere, usually above 1500 m and up to 2100 m; few records in Rift Valley at 900–1400 m (where Rattling Cisticola *C. chiniana* more common).

Cisticola bodessa

Description (plumage differences from *C. chiniana fricki* given in square brackets; many not so pronounced or consistent as suggested by Erard 1974b). *C. b. bodessa* Mearns: entire range except Gibe Gorge (Kaffa Province). ADULT ♂: forehead, crown, nape and hindneck rufous-brown streaked with darker brown; lores, ear-coverts and sides of neck mottled buff and rufous-brown; mantle, back, scapulars, rump and uppertail-coverts dark brown mottled with brownish buff (feather-edgings brownish buff), mottling less evident on rump and uppertail-coverts which are more uniform buffish brown; tail-feathers brown above, grey-brown below, with subterminal dark brown bar (less extensive on inner webs and very faint on central pair [no consistent difference from *chiniana*, contra Erard 1974b]), tipped brownish buff above and grey-brown (sometimes washed buff) below [*chiniana* has tip usually more strongly washed buff, but overlaps in this character]; chin, throat and belly creamy white, washed buff on centre of breast [*chiniana* averages deeper buff, but overlaps]; sides of breast and flanks buff, heavily washed olive-grey; undertail-coverts buff; primaries and secondaries dark brown, fringed rufous-brown; tertials and wing-coverts dark brown, fringed buffish brown or rufous-brown [primary and secondary fringes usually rather more rufous than tertial and covert fringes in *bodessa*; less contrast in *chiniana*]; alula fringed buff; underwing buff. Upper mandible dark brown, lower pale grey or horn; eyes reddish brown; feet pale brown. Sexes alike. SIZE (from specimens identified by voice: Benson's (1946b) series A): wing, ♂ (n = 10) 64–69 (66·8), ♀ (n = 1) 54; tail, ♂ (n = 10) 54–63 (58·5), ♀ (n = 1) 49; bill, ♂ (n = 10) 14–15 (14·7), ♀ (n = 1) 13; tarsus, ♂ (n = 10) 23–25 (24·0). WEIGHT: (Kenya) 1 ♂ 21, 1 ♀ 14.

IMMATURE AND NESTLING: unknown.

C. b. kaffensis Erard: Gibe Valley, Kaffa Province, NW Ethiopia. Darker. Upperparts duller, darker brown, lacking rufous, and with streaking poorly defined; crown more or less concolorous with mantle; margins of wing-coverts and primaries darker, redder brown; underparts richer buff; flanks washed darker brownish grey.

Field Characters. Very similar to Rattling Cisticola *C. chiniana*, with which it overlaps in Ethiopia and N Kenya; prefers somewhat thicker cover, but both can be heard singing at the same location; tends to use higher song perches than Rattling. Best distinction is very different song (see *Voice*). Plumage differences minor but may be useful at close range: crown browner, less rufous, but still contrasting somewhat with greyer back; back less heavily streaked, appearing almost plain in some birds, and overall rather darker; crown with indistinct streaks, less marked than in race *C. c. humilis* (Kenya) but about the same as that in *C. c. fricki* (Ethiopia). Underpart colour overlaps that of Rattling Cisticola, and wing-coverts and edgings of primaries and secondaries are also too variable to be useful characters.

Voice. Tape-recorded (10, 45, B, ERA, LEM, McVIC, ZIM). Song, loud, emphatic, short (1·5–2 s) phrases given at 2–3 s intervals, of 7–10 rather liquid notes with quality reminiscent of Whistling Cisticola *C. lateralis* but more rapid and run together; notes rise in pitch and intensity in middle of phrase and fall away at the end. Songs rendered (1) 'tchip-tchip-tchip-chipi ... CHUuuuuuuuuu'; (2) 'ti-ti-ti ... TIiiiiiiiiEW'; (3) ch-ch-ch-ch-ch ... CHUchichichi'; (4) 'chuchuch ... CHIchichichichichichichi-chichichiCHEW' (Zimmerman et al. 1996). Tikking calls precede phrases. Agitation call, a hard 'jep', 'chip' or 'tsip', often accelerating and leading into song.

General Habits. Occurs in dry highlands: dense, dry, tall-grass tree savanna, forest edge, open evergreen shrubland, secondary bush in clearings in juniper woodland (thicker, lusher habitat than the thorn-acacia bush preferred by Rattling Cisticola and often on steeper slopes than the latter). However, in Kenya and S Ethiopia, found on rocky, sparsely grassed hillsides with low thorn scrub. Occurs alongside Rattling Cisticola on Timau scarp, Kenya, and on rolling hills in Nechisar Nat. Park, Ethiopia, where local parapatry difficult to correlate with habitat selection; perhaps influenced by slope, with Rattling Cisticola on flatter ground (Safford *et al.* 1993). Sings at top of acacias at 3–5 m height (Rattling Cisticola sings low in bushes and from grasstops). Forages 1–3 m above ground (Rattling Cisticola 0–2 m). Moults Oct–Jan (Rattling Cisticola Sept–Dec) and perhaps occasionally also in spring.

Food. Insects.

Breeding Habits. NEST: a ball with side entrance, made of coarse grass or lichen and grass, lined with white seed-down. Placed *c.* 15–25 cm above ground in low herbs.

EGGS: (3 clutches) 4. Pale blue, immaculate, or pale green to white, with many pale, violet-red, indistinctly-edged flecks. SIZE: (n = 4, Ethiopia) 17·5–17·8 × 13 (17·6 × 13).

LAYING DATES: Ethiopia, Apr–June; probably also Sept–Nov (from state of skull ossification in juveniles).

BREEDING SUCCESS/SURVIVAL: one clutch of 4 hatched 3 young.

References
Ash, J. S. (1974).
Benson, C. W. (1946b).
Erard, C. (1974b).

Cisticola njombe Lynes. Churring Cisticola. Cisticole njombé.

Plate 10
(Opp. p. 161)

Cisticola aberrans njombe Lynes, 1933. Bull. Br. Orn. Club vol. 53, p. 170; Njombe, Tanganyika.

Range and Status. Endemic to highlands of E Africa. Common at 1830–3000 m in S Udzungwa Mts, and Njombe and S Mbeya highlands of S Tanzania, and above 1940 m on Nyika Plateau of adjacent Zambia and Malaŵi.

Description. *C. n. njombe* Lynes: southern highlands of Tanzania. ADULT ♂: forehead to nape rich reddish brown, obscurely mottled (almost plain) with darker brown; rest of upperparts and wing-coverts dull greyish brown, streaked with very dark brown, becoming more or less unstreaked on rump and uppertail-coverts; tail-feathers dull brown, fringed and tipped tawny and with very dark brown subterminal spot: chin and throat white; sides of neck and upper breast buff; rest of underparts pale olivaceous grey, palest (near-white) on centre of belly; primaries, secondaries and tertials dark brown, fringed with paler warm brown, with outer, basal part of fringes tinged rufous; underwing-coverts buff. Upper mandible dark brown to black, lower grey; eyes light brown; legs and feet brownish pink. Sexes alike. SIZE: wing, ♂ (n = 9) 52–57 (54·3), ♀ (n = 5) 49–51 (50·0); tail, ♂ (n = 8) 49–61 (54·7), ♀ (n = 4) 47–50 (48·2); bill, ♂ (n = 9) 11–13 (12·6), ♀ (n = 5) 12–13 (12·8); tarsus, ♂ (n = 9) 19–21 (19·9), ♀ (n = 5) 18–19 (18·6).

IMMATURE: like adult but head-top more prominently streaked; upperparts more tawny (feather-edgings more tawny); underparts washed with lemon-yellow.

NESTLING: unknown.

C. n. mariae Benson: Nyika Plateau of Malaŵi and NE Zambia. Like nominate race but underparts washed darker olivaceous grey. Larger. SIZE: wing, ♂ (n = 7) 55–59 (57·4), ♀ (n = 3) 52–54 (53·0); tail, ♂ (n = 6) 57–65 (61·3), ♀ (n = 3) 54–56 (55·0); bill, ♂ (n = 7) 13–14 (13·4) ♀ (n = 3) 12–14 (13·0); tarsus, ♂ (n = 7) 19–21 (20·3), ♀ (n = 3) 19–20 (19·7). WEIGHT: (Malaŵi) 1 ♂ 11·0.

Field Characters. Small-bodied and long-tailed; bright rufous cap contrasts with grey-brown upperparts; back lightly streaked, wings somewhat redder. Confined to short grass and bracken-briar in highlands of S Tanzania and on Nyika Plateau. Song dry and monotonous. Somewhat larger Wailing Cisticola *C. lais* is extremely similar but has distinctive clear trilling song and wailing calls and lives in rocky habitat. Black-lored and Singing Cisticolas *C. nigriloris* and *C. cantans* live in same habitat but have plain backs and ringing songs. Has been likened to Rattling Cisticola *C. chiniana*, a bulky bird of thornbush at lower levels, with rasping song, but they do not occur together.

Voice. Tape-recorded (45, 86, 87, B, F, LEM). Song, a dry, tuneless rattling note, 'chrreevit' or 'chrreeup', all on one pitch, given in series typically of 6–8, at rate of 3 notes per s.

General Habits. Occurs in montane grasslands, including short grass, rank grass along open valleys and bracken-briar, up to the borders of evergreen forest. Perches on termite mounds.

Food. Insects.

Breeding Habits.
NEST: undescribed.
EGGS: Tanzania (4 clutches) 2–3 (2·7); Zambia 2 (n = 2), 3 (3), 4 (1). Eggs 'like those of South African *C. aberrans*' (Lynes 1934).
LAYING DATES: E Africa: Tanzania, Dec, Feb; Region C, Dec–Feb (rains). Zambia, Dec–Jan.

166 SYLVIIDAE

Plate 10
(Opp. p. 161)

***Cisticola cinereolus* Salvadori. Ashy Cisticola. Cisticole cendrée.**

Cisticola cinereola Salvadori, 1888. Ann. Mus. Genova vol. 26, p. 254; Farré, Awash Valley, S Ethiopia.

Cisticola cinereolus

Range and Status. Endemic resident E Africa. One record (breeding) Sudan, at 4°N, 34°E on Kenya border. Frequent to common Ethiopia, in S, E and Rift Valley, at 900–1500m, occasionally to 1900 m; uncommon Somalia; scattered but locally common in Kenya on Huri Hills and E Kenya plateau at 300–1300m, south to NE Tanzania (Engaruka to N end of Pare Mts). 1–10 birds per 10 ha in Tsavo Nat. Park (Kenya).

Description. *C. c. schillingsi* (Reichenow): S Ethiopia and extreme SE Sudan to N Tanzania. ADULT ♂ (breeding): upperparts and sides of head ashy brown, streaked with dark brown; tail-feathers brown with dark brown subterminal band and buff tips; chin and throat white; breast, flanks and under-tail-coverts buff, washed olivaceous on flanks, centre of belly paler; primaries, secondaries, tertials and wing-coverts brown, fringed ashy buff; underwing-coverts buff. Bill dark brown, lower mandible pinkish or grey-brown; eyes orange-brown; legs and feet pinkish brown. Sexes alike. SIZE: (10 ♂♂, 10 ♀♀) wing, ♂ 60–66 (62·8), ♀ 52–54 (53·3); tail, ♂ 52–59 (54·2), ♀ 46–51 (47·7); bill, ♂ 13–15 (14·2), ♀ 12–14 (13·1), tarsus, ♂ 22–24 (23·1), ♀ 20–22 (21·0). WEIGHT: (Kenya) ♂ (n = 4) 15–15·5 (15·2), 1 ♀ 19.

IMMATURE: like adult but upperparts warmer, feathers edged richer buff.

JUVENILE: like immature but underparts sometimes with yellowish wash; bill yellow-ochre with brown culmen; eyes grey.

NESTLING: unknown.

C. c. cinereolus Salvadori: NE Ethiopia, N Somalia. Plumage seasonal. Upperparts slightly darker than *schillingsi* in breeding dress and paler in non-breeding; underparts like *schillingsi* in breeding dress but paler, less rich buff, in non-breeding. SIZE: (8♂♂, 2♀♀) wing, ♂ 59–65 (63·0), ♀ 54, 55; tail, ♂ 54–58 (55·9), ♀ 50, 50; bill, ♂ 14–16 (14·7), ♀ 14, 14; tarsus, ♂ 22–24 (23·0), ♀ 20, 22.

Field Characters. Rather slim and pale-looking; dull grey-brown, with no rufous on head, wings or tail; crown and upperparts heavily streaked. Warbling song unique in the genus. The only cisticolas sharing its habitat (dry lowland thornbush in E Africa) are Rattling *C. chiniana* (stout-bodied, some rufous on head and wings), Boran *C. bodessa* (plainer, darker upperparts contrast with rich brown crown), Desert *C. aridulus* (very small, rufous wash on back and especially on rump) and Tiny *C. nanus* (minute, unstreaked); all of these have distinctive songs.

Voice. Tape-recorded (10, 45, 87, B, C, GREG, KEI). Song a brief (1·5–2 s) bubbling warble, almost musical, preceded by 1–2 dry notes, 'jit-jit-widdly-widdly-woo'; may be ascending, 'ju-ju-toway-towee-towi' or descending, 'tsee-teedlee-dlaydlay-dloodloo'. Alarm given by parents with fledged young, liquid 'chi-weet-oo'.

General Habits. Found in arid open desert with annual grasses; semi-arid short-grass plains with scattered thorn bushes; acacia/short-grass savanna.

Keeps mostly to bushes and lower parts of small trees. ♂ sings from outer part of tree crown. Usually solitary; not shy.

Food. Insects, including caterpillars, dragonflies, mantids, crickets, small beetles.

Breeding Habits. Perhaps sequentially polygynous or cooperative breeder (Lynes 1930).

EGGS: Kenya (1 clutch) 4. Palest blue, immaculate or with a few dark brown spots near big end.

LAYING DATES: Sudan, Dec; Ethiopia, May–July. E Africa: Kenya, Apr (1 record); Region D, May (1 record).

Reference
Lynes, H. (1930).

Plate 10
(Opp. p. 161)

***Cisticola restrictus* Traylor. Tana River Cisticola. Cisticole du Tana.**

Cisticola restricta Traylor, 1967. Bull. Br. Orn. Club vol. 87, p. 45–48; Karawa, lower Tana R., Kenya.

Range and Status. Endemic to lower Tana R. basin, E Kenya, where recorded at five localities: Karawa, Garsen, Ijole, Mnazini and Sangole. Not seen since 1972.

Cisticola restrictus

Description. ADULT ♂ (breeding?): head-top and nape rusty brown, streaked dark brown; rest of upperparts greyish brown, streaked dark brown, plainer grey-brown on tail-coverts; tail-feathers brown above, brownish grey below, with subterminal dark brown band and buff tip; lores, sides of neck and entire underparts creamy white, paler on chin, throat and centre of belly, washed grey on flanks and sides of breast; primaries and secondaries brown, fringed buff; tertials and wing-coverts dark brown edged grey-brown; underwing-coverts buff; upper mandible dark grey or dark horn, lower pale horn; eyes tawny brown; legs and feet pinkish brown. Sexes alike. SIZE: (1♂, 2♀♀) wing, ♂ 60, ♀ 50, 52; tail, ♂ 55, ♀ 49, 54; bill, ♂ 14, ♀ 14, 14; tarsus, ♂ 20, ♀ 18, 20. WEIGHT: (Kenya) 2♂♂ 13, 15, 1♀ 12, 1 unsexed 13.

TAXONOMIC NOTE: the existence of this taxon is in doubt; specimens referred to it, collected in 1931–32, 1961–62, 1967 and 1972, may be aberrant Ashy Cisticolas *C. cinereolus* or hybrid *cinereolus/chiniana* (see *Voice*).

Field Characters. A rare bird of the lower Tana River, Kenya, almost unknown in life. Very similar to Ashy Cisticola *C. cinereolus* but paler and browner, with a touch of rusty on crown producing mild contrast with mantle (head and back concolorous in Ashy), streaking of upperparts narrower, grey wash on flanks and sides of breast, tail longer, browner, buff-tipped, subterminal black spots prominent. Looks like a pale, washed-out version of Wailing Cisticola *C. lais*, which is allopatric, with different habitat. Somewhat resembles the larger Rattling Cisticola *C. chiniana*, but tail proportionately longer; sympatric with *C. c. heterophrys*, which has unstreaked rufous crown and rufous wing-patch contrasting with indistinctly-streaked rather plain-looking back; more like *C. c. ukamba*, which it might possibly meet on middle Tana R., but crown streaked, not mottled, back streaking narrower.

Voice. Not tape-recorded. Song said to resemble that of Rattling Cisticola (A. D. Forbes-Watson, label data on topotype).

General Habits. Dwells in semi-arid bush.

Food. Grasshoppers.

Reference
Traylor, M. A. (1967b).

Cisticola rufilatus (Hartlaub). Tinkling Cisticola. Cisticole grise.

Plate 10
(Opp. p. 161)

Drymoica rufilata Hartlaub, 1870. *In* Finsch and Hartlaub, Vögel Ost-Afrikas (Decken, Reisen Ost-Afrika 4), p. 238; Damaraland (= Elephant Vlei), South West Africa.

Range and Status. Endemic resident. Plateau of Angola south of 8° 30′S, S Zaïre (Shaba), Zambia and N Malaŵi (from Tanzanian border) south through Zimbabwe, W Transvaal, Botswana and interior of Namibia to N Cape Province (Kuruman) and extreme N Orange Free State (Bloemhof dam: Colahan 1993). Common on Kalahari Sands, local to uncommon elsewhere. In Zimbabwe, 6 pairs/10 ha on stony hill top with scattered acacia bushes, reduced to 4 pairs/10 ha in subsequent season after area burnt over in dry season (Vernon 1987); in Transvaal (Nylsvlei), where patchily distributed, breeding density only *c.* 1 pair/12 ha in optimum habitat (Tarboton *et al.* 1987).

Description. *C. r. rufilatus* (Hartlaub): N Namibia and adjoining Angola, N Cape Province and Botswana to S and W Zimbabwe and adjacent S Zambia from Livingstone west and north to *c.* 15°S. ADULT ♂ (breeding): forehead, crown and nape rusty red with very pale buffy grey feather margins; lores, cheeks and supercilium reaching behind eye pale buff, streak through eye to ear-coverts rusty buff. Back with long, narrow red-brown streaks, feathers edged buff, rump unstreaked or almost so, sandy buff like uppertail-coverts. Tail bright rusty red, T1 plain or with obscure brown tip, T2 to T6 with black subterminal bar and buffy tip. Chin and throat whitish, breast very lightly suffused pale sandy buff, buff extending to flanks and thighs, centre of belly paler, undertail-coverts pale buff. Wings brown, primaries narrowly edged rusty buff, secondaries with broader buff edges; primary and secondary coverts

Cisticola rufilatus

with dull brown feather centres, broadly edged pale buff; inner webs of flight feathers pale rusty buff; shoulder, underwing-coverts and axillaries pale rusty buff. ADULT ♂ (non-breeding): streaking on mantle finer and redder, rusty brown, feather margins buffy mixed with pale silvery grey. Bill black, basal half of lower mandible pale grey, palate deep black; eyes bright hazel; legs and feet flesh-coloured. Sexes alike. SIZE: (34 ♂♂, 18 ♀♀) wing, ♂ 57–63 (60·0), ♀ 52–60 (55·0); tail, ♂ 53–70 (58·3), ♀ 46–58 (53·0); bill, ♂ 13·5–15 (14·0), ♀ 12·5–14 (13·4); tarsus, ♂ 18–22 (20·3), ♀ 18–21·5 (19·5). WEIGHT: ♂♀ (n = 23) 8–14·7 (11·3).

IMMATURE: like adult in non-breeding plumage.

NESTLING: hatches naked; gape yellow-ochre, interior of mouth yellowish with black lateral stripe on either side of lower mandible; palate black, tongue heavily spotted black on flanges; legs and feet pale yellowish flesh.

C. r. ansorgei Neumann (including '*venustula*'): Angola except SW, Zaïre, N Zambia, Malawi. Crown darker red, less clearly streaked; back darker brown; washed grey below.

C. r. vicinior Clancey: plateau of Zimbabwe, intergrading in S and W on Kalahari Sands with nominate *rufilatus*. Darker vinaceous above, feather margins browner; duller below, flanks more buffy tinged.

Field Characters. A bird of dry woodland and scrub in S-central Africa. Cap and stripe through eye bright rufous, separated by white supercilium. Back streaked dark red; tail long, bright rufous with black subterminal band and pale tip. Song diagnostic. Birds in higher rainfall areas have darker back streaks and might be confused with Rattling Cisticola *C. chiniana*, which also has rusty cap but is larger and has plain face with no white supercilium, brown tail and harsh calls. Range in W touches that of Grey-backed Cisticola *C. subruficapillus* which has same proportions and is same size but much darker and duller, with different voice.

Voice. Tape-recorded (45, 87, 88, 91, B, F, KEI). Song, 8–12 (9–10) regularly-spaced notes on one pitch, at rate of 3–4 per s, with intervals of 3–10 s between songs, 'wee-wee-wee-wee . . .'; clear, emphatic, far-carrying and slightly metallic, but hardly tinkling or bell-like as often described. Call (alarm?), a high-pitched, shivering trill, often preceded by thin, squeaky, mouse-like notes, or introduced by a single 'woy', e.g. 'woy-tisisisisisisisi'. Also snaps bill and wings (Maclean 1993).

General Habits. Habitat very variable; prefers semi-arid country dominated by Kalahari Sand, with scattered scrub and acacia bushes interspersed with grass; extends northwards on Kalahari Sand through W Zambia, where it occupies the woodland/grassland ecotone and straggling *Burkea* and *Terminalia* woodland. Elsewhere, from Transvaal north, inhabits thinner woodland types and edges of miombo woodland, often where sandy with patches of bare ground, scattered shrubs and sparse tufty grass; in this habitat it is highly localized and unevenly distributed although plentiful in apparently isolated pockets, and is absent from the major river valley systems (Limpopo, Zambezi and Luangwa); in Zambia, except in W, absent from extensive areas that appear suitable. Also frequents second growth and edges of cultivation. In N of range may be in competition with Rattling Cisticola on woodland edges.

Shy, elusive and very easily overlooked; spends most of time in long grass or bushes, creeping about on or near the ground and often slipping away when approached. In pairs or small family parties of 4–5 after breeding.

Food. Insects.

Breeding Habits. Territorial; territory probably maintained all year. In breeding season ♂ gives advertising song from top of small bush or tree; when disturbed ♂ responds with churring alarm call while ♀ slips away unseen, usually taking up a more prominent perch before moving off silently.

NEST: oval, with side-top entrance, loosely and untidily built of old, soft, greyish grass blades or occasionally fine rootlets, bound with a few cobwebs, surrounding a firmer structure of woolly plant down; ext. depth 100–150, entrance 22–60; placed and interlaced in clump of grass or shrub on firm base and lined with soft vegetable down.

EGGS: 3–4 (usually 3). White or pale blue, occasionally greenish, lightly to variably marked in shades of lilac and lilac-brown; others are marked with spots or blotches of pinkish brown or purplish grey, or streaky speckling, fairly evenly except where concentrated in broad ring or cap around the larger end. SIZE: (n = 70) 16·0–18·8 × 11·4 –13·2 (17·0 × 12·6).

LAYING DATES: Angola, Nov–Feb; Zambia, Jan; Malaŵi, Dec–Jan; Zimbabwe, Oct–Mar (Oct 2, Nov 6, Dec 12, Jan 14, Feb 5, Mar 2 clutches); Botswana, Oct, Jan; South Africa: Transvaal, Dec.

INCUBATION: by ♀ only.

DEVELOPMENT AND CARE OF YOUNG: both sexes feed young, both in nest and for 12–14 days after fledging. When approaching nest ♀ occasionally settles on ground several m away, then runs towards it.

BREEDING SUCCESS/SURVIVAL: probably parasitized by Wahlberg's Honeybird *Prodotiscus regulus*.

Reference
Vincent, A. W. (1948a).

Cisticola subruficapillus (A. Smith). Grey-backed Cisticola. Cisticole à dos gris.

Plate 10
(Opp. p. 161)

Drymoica subruficapilla A. Smith, 1843. Illustr. Zool. South Africa, Aves, pl. 76, fig. 2 (and text); western Cape Colony; restricted to Cape Town district, southwestern Cape.

Range and Status. Endemic resident. Common in South Africa from SW Cape and Karoo east to Port Elizabeth, and north through Cradock, Colesberg, Griqualand West and Upington; SW Orange Free State; N Cape and Bushmanland, lower Orange R. and S and W-central Namibia (absent from whole of NE) to borders of Namib Desert and adjacent SW Angola in Moçâmedes (Caraculo).

Description. *C. s. subruficapillus* (A. Smith): SW Cape east to Knysna and Oliphants R. ADULT ♂: forehead and crown streaked dark brown, feathers with paler, rusty brown edges; lores, cheeks and ear-coverts buffy, very finely streaked below eye. Nape, mantle and back streaked dark brown, with paler greyish brown feather margins; rump similar but more sparsely streaked. Tail warm brown, central (T1) feathers plain, remainder with broad (4–4·5 mm), subterminal blackish bar or spot and pale buffy tip (4–5 mm). Chin and throat pale buffy white; breast pale buffy white, lightly and variably streaked greyish brown, belly to undertail-coverts pale buffy, flanks somewhat darker. Wings dark brown, primaries margined reddish brown on outer webs; wing-coverts and scapulars streaked like back with greyish brown edges; inner margins of primaries and secondaries reddish brown; shoulder, underwing-coverts and axillaries buffy. Bill grey, culmen blackish brown; eyes brown; legs and feet pinkish flesh. Sexes alike, without seasonal plumage change. SIZE: (36 ♂♂, 30 ♀♀) wing, ♂ 53–60 (55·3), ♀ 47–52 (49·9); tail, ♂ 51–58 (54·5), ♀ 45–52 (49·5); bill, ♂ 12–13·5 (12·5), ♀ 11·5–13·5 (12·0); tarsus, ♂ 18·5–21·5 (19·7), ♀ 16·5–19 (17·6). WEIGHT: ♂ (n = 4) 10·8–14·7 (12·1); ♀ (n = 9) 8·0–12 (9·3).

IMMATURE: duller above, faintly washed with yellow below, particularly on face.

NESTLING: hatches naked.

C. s. namaqua Lynes: NW Cape between Oliphants R. and Orange R. and western Karoo. Streaking of upperparts reduced and paler; unstreaked below.

C. s. jamesi Lynes (including '*euroa*'): SE Cape between Port Elizabeth and Grahamstown, middle and E Karoo, Griqualand West and Orange Free State. Like nominate *subruficapillus* but buffier, less greyish below, without streaking.

C. s. karasensis (Roberts): NW Cape in Bushmanland, Gordonia and Kuruman districts and Orange R., and Namibia in Great Namaqualand east of the Fish R. and Rehoboth. Paler above than *jamesi*, less streaked, head not contrasting with back, whiter below.

C. s. windhoekensis (Roberts): central Namibia (Naukluft Mts north to Otjiwarongo and Waterberg Mts). Back washed fulvous, showing greater contrast with crown.

C. s. newtoni Rosa Pinto: Namibia (Kaokoveld) on edges of Namib Desert and adjacent SW Angola. Crown lacks streaking, back lightly streaked and conspicuously greyish, underparts cream (almost white), without streaking on flanks.

Field Characters. A common bird of grass, scrub and fynbos, endemic to SW Africa. Slim and lightly built, with quick, nervous movements and flirting of long tail. Rather drab, with brown head and tail, narrow dark streaks on grey back, greyish white underparts with a few streaks on breast. Northern birds have fewer, finer streaks, less grey back. Very similar to closely related but mainly allopatric Wailing Cisticola *C. lais*, which is larger and stouter, with broad streaks on grey-brown back and buffier underparts; their voices are similar but habitats different; they are ecologically separated where they meet in Cape and Orange Free State (see *General Habits*). Tinkling Cisticola *C. rufilatus* is much brighter, with conspicuous white supercilium below red cap, long rufous tail, different voice, more wooded habitat.

Voice. Tape-recorded (88, 91, F, LEM). Call, a brief (< 1 s) liquid trill, 'prrrrrip'; song consists of this trill followed by loud 'tee-tee-tee-tee . . .'. Alarm scold, a hard 'dzee'; also has squeaky conversational notes and a single, wailing 'weep'.

General Habits. Inhabits a wide variety of generally arid, rather open scrubby growth mixed with grass, principally fynbos, karoo and macchia; also drainage lines on mountain slopes. In Orange Free State Grey-backed Cisticola and Wailing Cisticola are found in the vicinity of Broom Karee *Rhus erosa* on rocky hill sides, but former favours karoo, fynbos and desert areas with sparser grass cover and lower hills and the 600 mm isohyet coincides with the limit of their ranges (du Plessis 1991); in the Cape over a large area from Piquetberg eastwards to Grahamstown and Barkly West, occurs alongside Wailing Cisticola, which is confined to the mountains and replaced by Grey-backed on the flats between; both also occur close together in the Suurberg Mts, Wailing on the grassland on south-facing slopes, Grey-backed in rain-shadow valleys in N (Vernon 1982); in Namibia found also in mountains (where Wailing absent).

Solitary or in pairs. Noisy; ♂ calls or scolds from conspicuous perch while ♀ keeps well out of sight. ♂ often flies over bushes flirting tail. Sings only in breeding season. Forages mostly within cover, in shrubs and grass tufts, occasionally on ground.

Food. Small insects.

Breeding Habits. Territorial. ♂ displays over ♀ in dancing flight 1–2 m above surrounding cover, hovering in air and flirting tail.

NEST: ball type, with side-top entrance, made of dry grass and cobwebs, in S Karoo, often seed-heads of *Galium tomentosum* (Rubiaceae) (Dean and Milton 1993); copiously lined with white plant down; diam. 80, depth 105; placed on or within 0·8 m of ground, often in *Helichrysum* bush; a few surrounding grasses are incorporated into roof and walls.

EGGS: 2–4, laid at 1 day intervals; av. (12 clutches) 2·8. Slightly glossy; white to pale turquoise blue, fairly well marked with red, brown or purple freckles and spots and a few bolder blotches. SIZE: (n = 78) 14·2–17·1 × 10·6–13 (15·9 × 11·9).

LAYING DATES: South Africa: Karoo, July–Dec (July 1, Aug 5, Sept 6, Oct 3, Nov 5, Dec 1 clutches); E Cape, Sept–Nov, Feb (Sept 1, Oct 4, Nov 4, Feb 1 clutches); SW Cape, July–Dec (July 17, Aug 119, Sept 117, Oct 45, Nov 7, Dec 6 clutches).

INCUBATION: period 15 days.

DEVELOPMENT AND CARE OF YOUNG: nestling period: 12 days.

References
du Plessis, D. (1991).
Vernon, C. J. (1983).
Vincent, A. W. (1948a).

Cisticola lais (Finsch and Hartlaub). Wailing Cisticola. Cisticole plaintive.

Plate 10 (Opp. p. 161)

Drymoica lais Finsch and Hartlaub, 1870. Baron Carl Claus von der Decken's Reisen, IV, Die Vögel Ost-Afrikas, p. 237; Natal (= Pinetown), South Africa.

Range and Status. Endemic resident. Locally common in highlands of Kenya and E Uganda at 1500–2750 m. Common on high plateaux of S Tanzania, above 1500 m. Frequent 1400–2600 m in Malaŵi. One old report SE Zaïre (Schouteden 1954–55), probably in error. Locally abundant Zambia, on Muchinga Mts and at 1800–2250 m on Nyika Plateau and other ranges in NE. Abundant in Zimbabwe at 1500–2600 m on Mashona Plateau east of Rusape and in E highlands, also down to 750 m on E slopes; in adjacent Mozambique on Chimanimani range, on massifs in N, mainly above 1800 m, occasionally down to 1500 m, and above 1700 m on Mt Gorongoza. Locally common Swaziland and Transvaal (Kruger Nat. Park and central and E highlands), extreme E Orange Free State to S Cape Province (one doubtful record SW Cape); common in mountains of Natal and Lesotho, at least up to 2400 m; abundant in hills (the 'Little Berg') below Drakensberg but uncommon near coast. Locally common in Angola from W Huila District to Huambo and S Cuanza Sul (in Huila, only in Serra de Chela and Planalto de Humpata).

Cisticola lais

Description. *C. l. semifasciatus* (Reichenow): Zambia, S Tanzania (Iringa Plateau), Malaŵi, N Mozambique. ADULT ♂ (breeding): forehead, crown and nape rufous-brown, mottled with dark brown; lores, supercilium and ear-coverts buff; rest of upperparts greyish brown, streaked with dark brown; tail tawny brown with faint subterminal darker band and tawny tip; entire underparts buff, palest on chin, throat and centre of belly, flanks and sides of breast washed olivaceous; primaries and secondaries brown, fringed rufous; tertials and wing-coverts dark brown, fringed tawny to grey-brown; underwing-coverts buff; axillaries olivaceous grey. Upper mandible dark brown, lower grey or pale brown; eyes red-brown; legs and feet brownish pink. ADULT ♂ (non-breeding): like breeding ♂ but head-top and upperparts appear streaked rather than mottled, due to broader fringes to feathers; fringes of feathers of mantle, back and scapulars slightly buffier brown, rather than grey-brown; underparts average richer buff. Sexes alike. SIZE: wing, ♂ (n = 10) 56–60 (58·1), ♂ (n = 10) 48–53 (50·8); tail, ♂ breeding (n = 5) 53–60 (57·2), non-breeding (n = 4) 59–64 (61·5), ♀ breeding (n = 6) 42–52 (47·3). non-breeding (n = 4) 50–53 (51·7); bill, ♂ (n = 9) 13–15 (14·1), ♀ (n = 10) 13–14 (13·3); tarsus, ♂ (n = 10) 20–22 (20·6), ♀ (n = 9) 19–20 (19·7).

IMMATURE: like non-breeding adult but head-top browner, less rufous; upperparts less grey, more rufous-brown, with narrower and fainter dark brown streaks; underparts rich yellow-buff, washed olivaceous on breast and flanks.

NESTLING: unknown.

C. l. distinctus Lynes ('Lynes's Cisticola'): Uganda, Kenya. Like breeding *semifasciatus* but head-top more heavily streaked with dark brown; feather-edgings of rest of upperparts slightly paler grey; tail darker, brown. Larger. SIZE: wing, ♂ (n = 9) 60–66 (63·1), ♀ (n = 7) 57–60 (58·4); tail, ♂ (n = 9) 58–63 (60·4), ♀ (n = 7) 53–60 (56·0); bill, ♂ (n = 8) 15–16 (15·7), ♀ (n = 7) 14–16 (15·3); tarsus, ♂ (n = 9) 23–26 (24·0), ♀ (n = 7) 21–24 (22·4). WEIGHT: (Uganda) ♂ (n = 2) 17, 17; ♀ (n = 3) 13–16 (15·0); juv ♀ (n = 2) 13, 16.

C. l. mashona Lynes: Zimbabwe, S Mozambique highlands, N Transvaal, Swaziland. Breeding adult like *semifasciatus* but head-top slightly redder; fringes of rest of upperparts buffier, less grey. Non-breeding adult differs from *semifasciatus* in having head-top more uniform and deeper red-brown (dark streaks fainter, less dark; edgings broader, darker); rest of upperparts brown, narrowly streaked dark brown (less grey than *semifasciatus*). Size like *semifasciatus*. WEIGHT: (Zimbabwe) ♂ (n = 3) 13·4–15·6 (14·3), 1 ♀ 10·8.

C. l. oreobates Irwin: Mt Gorongoza, Mozambique (*pace* Clancey 1980b, Irwin 1981, this race is not found in Zimbabwe, although birds from Chimanimani Mts are similarly saturated and may represent an undescribed taxon: M.P.S. Irwin, pers. comm.). Like *mashona* but in breeding dress crown and tail darker, less reddish; rest of upperparts darker (feather edges duller); underparts greyer, not buff.

C. l. monticola (Roberts): S Transvaal, west from Pretoria. Like corresponding plumages of *mashona* but in breeding dress head-top yellower, plainer; rest of upperparts slightly paler, with dark brown streaks lighter, less prominent. In non-breeding dress upperparts richer rufous-brown (feather-edgings of mantle, back, scapulars and wing more rufous or tawny). Larger. SIZE: wing, ♂ (n = 7) 60–65 (61·7), ♀ (n = 6) 54–56 (55·2); tail, ♂ breeding (n = 3) 52–58 (56·0), non-breeding (n = 4) 61–70 (65·5), ♀ breeding (n = 3) 51–52 (51·7), non-breeding (n = 3) 58–61 (59·3); bill, ♂ (n = 6) 14–15 (14·3), ♀ (n = 5) 12–14 (13·0); tarsus, ♂ (n = 7) 20–22 (20·7), ♀ (n = 6) 18–19 (18·5).

C. l. lais (Finsch and Hartlaub): Natal, SE Transvaal, E Orange Free State, Lesotho, E Cape Province. Like corresponding plumages of *monticola* but in breeding dress head-top prominently streaked with dark brown; rest of upperparts paler (feather-edgings broader). In non-breeding dress head-top and upperparts more prominently streaked dark brown. Some variation within the subspecies, with coastal (SE) birds slightly yellower on upperparts. Larger. SIZE: wing, ♂ (n = 10) 61–68 (63·6), ♀ (n = 10) 52–58 (54·6); tail, ♂ breeding (n = 5) 54–65 (58·0), non-breeding (n = 5) 60–71 (63·8), ♀ breeding (n = 5) 46–57 (51·0), non-breeding (n = 5) 52–64 (57·8); bill, ♂ (n = 10) 14–15 (14·5), ♀ (n = 9) 13–14 (13·2); tarsus, ♂ (n = 10) 20–23 (21·6), ♀ (n = 10) 19–20 (19·2). WEIGHT (South African races): ♂ (n = 3) 15·2–17·3 (16·0), ♀ (n = 3) 11·2–12·3 (11·7), unsexed (n = 8) 10–15·5 (12·2).

C. l. maculatus Lynes: S Cape Province east to Port Elizabeth. Like corresponding plumages of nominate race but underparts paler buff; breast spotted with dark brown (spots virtually absent in non-breeding dress); in non-breeding dress upperparts slightly greyer (feather edges greyer brown). Some intermediates in Grahamstown area show slight spotting on sides of breast. Smaller. SIZE: (10♂♂, 10♀♀) wing, ♂ 56–61 (59·5), ♀ 51–53 (52·1); tail, ♂ breeding (n = 5) 52–66 (59·0), non-breeding (n = 5) 55–63 (59·8), ♀ breeding (n = 5) 49–57 (54·0), non-breeding (n = 5) 52–59 (56·6); bill, ♂ 13–14 (13·6), ♀ 12–14 (12·9); tarsus, ♂ 19–22 (20·6), ♀ 18–20 (18·8).

C. l. namba Lynes: Angola. Most like *semifasciatus*. In breeding dress differs in having head-top plainer; underparts pale grey, not buff. In non-breeding dress head-top slightly redder and sharply cut off on nape from upper back, rather than colours merging; rest of upperparts slightly greyer, less yellow-tinged; underparts olivaceous buffy grey, rather than buff. SIZE: (8♂♂, 4♀♀) wing, ♂ 60–63 (61·0), ♀ 51–55 (53·7); tail, ♂ breeding (n = 7) 53–59 (56·7), non-breeding (n = 1) 60, ♀ breeding (n = 3) 48–51 (49·3), non-breeding (n = 1) 56; bill, ♂ 14–16 (14·6), ♀ 13–14 (13·7); tarsus, ♂ 20–23 (21·5), ♀ all 19.

TAXONOMIC NOTE: race *distinctus* sometimes regarded as separate species based on plumage, but vocal studies suggest it is conspecific with *lais* (Dowsett and Dowsett-Lemaire 1993b).

Field Characters. Medium-sized, with rufous crown and wings contrasting with grey-brown back; tail red-brown (darker and browner in race *distinctus*). Underparts buff (except in Angolan race *namba*), the main plumage difference from very similar Grey-backed Cisticola *C. subruficapillus*, which has off-white or pale grey underparts and streaked breast. Breast usually plain, but sometimes streaked in area of overlap with Grey-backed in S Cape. Plumage also differs from Grey-backed in some finer points: red of head paler and brighter in breeding plumge; back streaks bolder (broader) and grey between them lighter, producing greater contrast; wing-patch more rufous; breast markings include drop-shaped spots, and are bolder. Upperparts somewhat rustier in non-breeding season, and back stripes thinner, about same width as those of Grey-backed, but Grey-backed does not have rusty winter plumage. Appears more heavily built than Grey-backed; voices are similar but distinguishable; habitat different, steep rocky slopes rather than lowland scrub, although Grey-backed also found on grassy slopes. In E Africa (but not southern Africa) often occurs together with Rattling Cisticola *C. chiniana*, from which distinguished by smaller size, longer tail, rich buff underparts (extending to face), greyer back and wailing song. Occurs in rocky habitats and forages low on ground like Rock-loving Cisticola *C. aberrans*, but latter has plain upperparts, no red in wings or tail, dry trilling song. For differences from Churring Cisticola *C. njombe*, see that species.

Voice. Tape-recorded (45, 75, 87, 88, 91, B, C, F, KEI, McVIC). Varied vocabulary; distinction between songs and calls unclear. Bird in S Tanzania (KEI) consistently sang a liquid, shivering trill, quite short (1–1·5 s), increasing in intensity towards end, preceded by a few thin, high-pitched notes: 'tsip-tsip-tsi-swirrrrr'. Common 'call', often but not necessarily following 'song', a loud, upslurred 'pee-yip' or unslurred 'peeee', with somewhat wailing quality which gives the bird its name. 'Songs' and 'calls' frequently given independently of each other. Bird in SW Tanzania (87) gave similar liquid trill but filled in the gaps between trills with low, conversational raspy notes, 'wiji, rrij', also light 'psipsi' and thin 'pseee-pseee'. Song punctuated at intervals by loud 'WEEP' (from second bird?). Birds in South Africa seem to have switched trills with Grey-backed Cisticola; latter gives liquid trill very like that of Tanzanian Wailing, while local Wailing has a drier and shorter trill. In South Africa the short trill seems tied to the 'peee' calls and together they comprise the song, 'drrrreee-peee-peee-peee-peee-peee' (see sonagram in Maclean 1993). In Kenya (race *distinctus*), dry trill follows wailing notes to form a song, 'peeeee-peee-peee-peee-squirrrrrrr' (Zimmerman *et al*. 1996). Alarm, harsh 'jree-jree-jree' (southern Africa) or squeaky 'tspTSEEa' or 'tskSWEEA' (Kenya). Also has a buzzy scold, and in southern Africa a loud bubbling call note 'wititi-wititi' (Maclean 1993).

General Habits. Usually associated with dry, rocky ground at montane elevations. Inhabits sparse, rank grass and scrub, on short, open, often treeless grassland with low bushes and (in Mozambique) *Vellozia* tree-lilies or (South Africa) *Protea* trees, on rocky hill-tops and slopes, cliffs, kopjes and small rock outcrops and at forest edges; mainly on dry ground but often at base of hills or in valley bottoms, sometimes where wet; occasionally in grassy clearings in forest or on edge of evergreen forest (southern Africa); *C. l. oreobates* restricted to montane grassland.

Occurs in pairs and small family parties, including in non-breeding season. Restless, secretive; feeds in vegetation and on ground, flitting and scurrying among shrubs and boulders; runs on ground like a mouse, recalling Rock-loving Cisticola (Zimmerman *et al*. 1996). When alarmed, sidles up grass stem to watch intruder.

Food. Insects, including grasshoppers, small beetles, large caterpillars, other larvae.

Breeding Habits. Breeding ♂ perches and sings at top of bush or tall grass stem. Alarm calls warn ♀ of intruder.

NEST: a ball or upright oval, of dry, curled grass blades bound with spider web, lined with silky, usually white, plant down; placed up to 60 cm above ground in centre of tuft of green, broad-bladed grass or herbs, often among low shrubs, with tops of grass blades curled over and fixed with silk to nest; other blades left erect, concealing bent ones.

EGGS: Malaŵi 2 (n = 3), 3 (10), 4 (1) (once 5 juvs together but probably from 2 nests: Benson and Benson 1977); S Mozambique (3 clutches) 3; South Africa 2 (n = 1), 3 (3), 4 (2). Variable in colour: white to pale blue or turquoise, plain or finely speckled with dark reds, purples and browns. SIZE: (n = 2, Kenya, Uganda) 19·0–20·2 × 12·9–13·1; (n = 33, Malaŵi) av. 17·1 × 12·8; (South Africa) (n = 32, *C. l. lais*) 17·2–18·0 × 12·5–12·9 (17·5 × 12·4), (n = 7, *monticola*) av. 16·6 × 12·4.

LAYING DATES: E Africa: Tanzania, Nov–Dec, Feb; Region C, Nov–Dec, Feb; Region D, Apr; (all records in rains). Malaŵi, Oct–Jan (all Oct records from Mulanje, where possibly breeds early: Benson and Benson 1977); Zambia, Dec–Jan; Zimbabwe, Oct–Feb, peak Dec–Jan (32 records: Irwin 1981); S Mozambique, Oct–Mar; South Africa, Sept–Mar (peak Dec in Natal, Nov–Jan in Transvaal); Angola, Mar.

References
Lynes, H. (1930).
Roberts, A. (1913).

Cisticola galactotes (Temminck). Winding Cisticola; Black-backed Cisticola. Cisticole roussâtre.

Plate 11
(Opp. p. 176)

Malurus galactotes Temminck, 1823. *In* Temminck and Laugier, Nouveau recueil de planches col. d'oiseaux, livr. 11, pl. 65, fig. 1; 'New Holland', error for South Africa (Lynes, 1930, Ibis Suppl. p. 401), restricted to Durban by Clancey (1964, Birds of Natal and Zululand, p. 373).

Range and Status. Endemic, mainly resident with some local migration. Frequent in SW Mauritania (lower Senegal R. and adjacent lakes). Locally common resident in Gambia, Senegal and Sierra Leone. Recorded near Conakry, Guinea. Occurs throughout Niger inundation zone, Boucle du Baoulé and other major watercourses in Mali; perhaps undertakes local movements. Frequent S Burkina Faso. Abundant Ivory Coast, south of soudan savanna zone. Locally common resident S Ghana, rains visitor in north, apparently resident throughout Togo. Local (Nov, Mar–Apr) south of *c.* 14°N in Niger. Recorded S Benin. Common, locally abundant, resident in most of Nigeria; but in NE during rains, only frequent. Recorded in S Chad. Occurs throughout Cameroon but local in S. Frequent in S of Central African Republic, uncommon Bamingui-Bangoran Nat. Park. Common S Sudan, resident or local migrant in Nile valley south of *c.* 16°N and resident Darfur south of *c.* 14°N. Common resident Ethiopia, usually above 1500 m. Recorded in Eritrea at 600 m near Sudan border, and in small area about 15°N, 39°E at 1800–2500 m. Locally abundant in Somalia south of 3°N. Locally common up to 2300 m in W and S Kenya (abundant near coast, rare (rains) Tsavo), E and S Uganda, Tanzania except SE and highlands; mainly resident but may move locally. Abundant Burundi 1600–1700 m. Recorded Rwanda. Locally abundant in savannas of Gabon. Common near Brazzaville and elsewhere in S Congo; frequent Cabinda. Frequent in most of Zaïre, abundant in SE. Locally common resident in Zambia (mostly in isolated populations around lakes and along large rivers), Caprivi (along rivers), N Botswana and adjacent extreme W Zimbabwe (Zambezi Valley above Katombora Rapids). Local in S Malaŵi up to 1200 m (along Shire Valley and around Lakes Chilwa and Chiuta). Common S Mozambique, especially along coast and in lower Incomati Valley; less common north of Save R. Common in SE Zimbabwe. Occurs coastal Natal; 3 records (including breeding) Transvaal (Kruger Nat. Park).

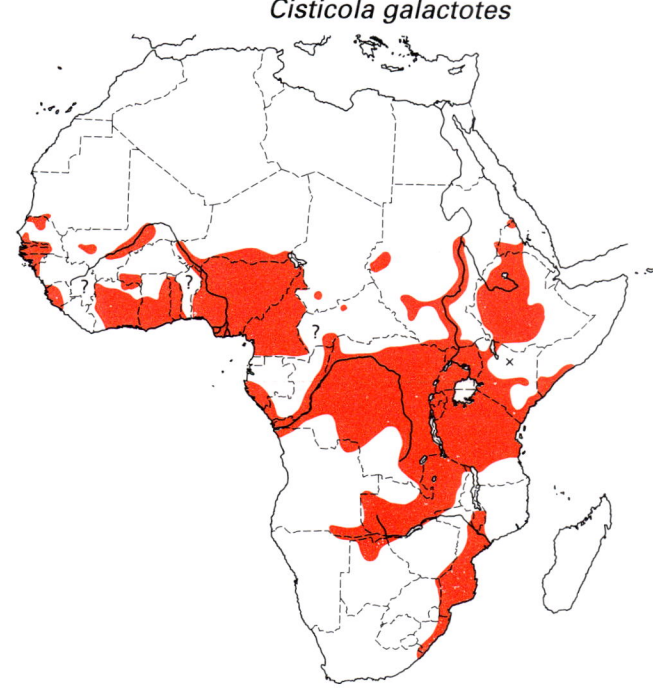

Cisticola galactotes

Description. *C. g. amphilectus* (Reichenow) (including '*griseus*'): Senegal to Ghana, coastal Nigeria, W Congo basin (specimens from E Zaïre, normally referred to this race, are closer to *nyansae* and are included thereunder). ADULT ♂ (breeding): forehead to nape brown to greyish brown, obscurely streaked with dark brown (feather centres darker); lores buff; ear-coverts pale brownish grey; mantle, back, scapulars and wing-coverts grey-brown, broadly streaked with very dark brown; rump and uppertail-coverts grey-brown, more or less unstreaked; tail-feathers dull brown, tipped buff, with subterminal dark brown band; underparts pale buff, whiter on throat and centre of belly, washed olivaceous grey on flanks; primaries, secondaries and tertials dark brown, most fringed buff, but rufous buff on outer fringes towards base of (especially) secondaries, giving a rufous patch on closed wing; underwing-coverts buff; axillaries olivaceous grey. Upper mandible dark brown, lower grey; eyes light brown; legs and feet pinkish brown. ADULT ♂ (non-breeding: rare plumage type – perennial dress usual over most of range): like breeding ♂ but entire upperparts and tail more rufous (mantle-feathers rufous-brown streaked with dark brown); underparts richer buff; fringes on flight feathers more rufous. Sexes alike. SIZE: (10 ♂♂, 10 ♀♀) wing, ♂ 59–65 (62·3), ♀ 53–60 (56·7); tail, ♂ 49–63 (54·2), ♀ 43–52 (47·8); bill, ♂ 16–18 (16·7), ♀ 15–17 (16·0); tarsus, ♂ 24–26 (24·8), ♀ 21–23 (22·2). WEIGHT: (Ghana) ♂ (n = 2) 17·5, 19·2, ♀ (n = 1) 18·5, unsexed (n = 4) 15–17 (16·2); (Congo) both sexes, 13–16.

IMMATURE: like non-breeding adult but head-top prominently streaked with dark brown.

JUVENILE: like immature but underparts white, washed with sulphur.

NESTLING: newly-hatched young pink with yellow gape.

C. g. zalingei Lynes: N Nigeria to Darfur (unclear whether race in Mali and Niger is this race or *amphilectus*). Seasonal plumage. Breeding dress like *amphilectus* but head-top browner, less grey; rest of upperparts darker (feather edges darker). Non-breeding dress like the rare non-breeding dress of *amphilectus* but head-top prominently streaked with dark brown; upperparts brighter, buffier brown. Smaller. SIZE: wing, ♂ (n = 9) 59–64 (61·8), ♀ (n = 10) 53–56 (54·6); tail, ♂ (breeding, n = 6) 48–57 (52·0), (non-breeding, n = 3) 66–70 (67·7), ♀ (breeding, n = 6) 46–53 (48·5), (non-breeding, n = 4) 53–60 (57·0); bill, ♂ (n = 9) 15–17 (16·0), ♀ (n = 8) 14–16 (15·4); tarsus, ♂ (n = 9) 22–24 (23·3), ♀ (n = 10) 20–22 (21·1).

C. g. marginatus (Heuglin): upper White Nile, S Sudan, N Uganda. Like *zalingei* but dark brown feather centres darker and broader, resulting in breeding dress upperparts appearing more prominently streaked and non-breeding upperparts appearing darker. Smaller still. SIZE: wing, ♂ (n = 6) 57–61 (57·8), ♀ (n = 6) 52–55 (52·8); tail, ♂ (breeding, n = 3) all 48, (non-breeding, n = 3) 46–62 (54·7), ♀ (breeding, n = 3) 42–49 (45·7), (non-breeding, n = 3) 47–57 (52·3); bill, ♂ (n = 6) 14–16 (15·2), ♀ (n = 6) 13–15 (14·0); tarsus, ♂ (n = 6) 21–23 (22·0), ♀ (n = 5) 20–21 (20·4).

C. g. lugubris (Rüppell): Ethiopia. Like *marginatus* but, in breeding dress, head-top browner; upperparts darker (feather centres broader); underparts slightly richer buff. In non-breeding dress, upperparts like *marginatus* but underparts richer buff. Larger than *marginatus*. SIZE: wing, ♂ (n = 6) 60–65 (62·8), ♀ (n = 6) 53–59 (55·2); tail, ♂ (breeding, n = 3) 51–54 (52·0), (non-breeding, n = 3) 57–64 (61·2), ♀ (breeding, n = 2) 46, 47, (non-breeding, n = 3) 53–57 (54·7); bill, ♂ (n = 5) 14–16 (15·0), ♀ (n = 6) 14–15 (14·3); tarsus, ♂ (n = 6) 22–23 (22·3), ♀ (n = 6) 20–21 (20·8). WEIGHT: ♂ (n = 2) 14, 14.

C. g. nyansae (Neumann): central and E Zaïre, Uganda except N, W Kenya; intergrades with *amphilectus* in Cameroon and Congo. Like breeding *amphilectus* but head-top brown, less greyish; rest of upperparts darker (feather edges darker), streaks less prominent. Some seasonal plumage change in Kenya: non-breeding dress like breeding but upperparts and tail more buffy (feather edges buffier brown); fringes on flight feathers brighter rufous. Slightly smaller than *amphilectus*. SIZE: wing, ♂ (n = 8) 60–64 (62·2), ♀ (n = 8) 53–59 (57·0); tail, ♂ (n = 8) 49–64 (54·1), ♀ (n = 7) 43–56 (50·1); bill, ♂ (n = 8) 14–18 (16·4), ♀ (n = 8) 14–16 (15·6); tarsus, ♂ (n = 8) 22–26 (24·1), ♀ (n = 8) 21–23 (22·1). WEIGHT: (Kenya, Uganda) ♂ (n = 5) 10–17 (14·6), ♀ (n = 2) 11, 12, unsexed (n = 31) 9·8–18 (12·8).

C. g. suahelicus (Neumann) (including '*isodactylus*'): central Tanzania, SE Zaïre, N Zambia (Mweru Marsh to L. Tanganyika), Malaŵi, Mozambique. Seasonal plumage. Breeding dress like *nyansae* except underparts creamier buff, less grey on flanks. In non-breeding dress, crown richer tawny; rest of upperparts brighter rufous-buff (feather edgings rufous-buff); underparts richer buff. Smaller than *nyansae*. SIZE: wing, ♂ (n = 8) 58–62 (60·6), ♀ (n = 8) 52–57 (53·9); tail, ♂ (breeding, n = 4) 46–52 (49·7), (non-breeding, n = 3) 55–61 (57·7), ♀ (breeding, n = 4) 46–49 (47·0), (non-breeding, n = 4) 46–51 (48·5); bill, ♂ (n = 8) 14–16 (15·4), ♀ (n = 8) 14–15 (14·7); tarsus, ♂ (n = 8) 22–24 (22·7), ♀ (n = 8) 19–22 (21·2). WEIGHT: (SE Zaïre) ♂ (n = 3) 10–14 (11·7), 1♀ 12, unsexed (n = 2) 12, 12, unsexed juveniles (n = 3) 10–14 (11·7).

C. g. haematocephalus (Cabanis): coastal S Somalia, Kenya and N Tanzania. Perennial dress, like *nyansae* or breeding *suahelicus* but head-top paler, less reddish; rest of upperparts slightly paler (feather edgings buffy grey rather than grey-brown); wing-patch paler, less rufous; underparts creamier buff, less grey. Smaller than *suahelicus*. SIZE: wing, ♂ (n = 6) 55–57 (56·5), ♀ (n = 6) 49–52 (50·5); tail, ♂ (n = 6) 50–55 (51·8), ♀ (n = 6) 48–52 (50·0); bill, ♂ (n = 5) 15–17 (15·8), ♀ (n = 6) 14–15 (14·8); tarsus, ♂ (n = 6) all 22, ♀ (n = 6) 20–22 (21·2). WEIGHT: (Kenya) 1♀, 10.

C. g. galactotes (Temmick): SE Zimbabwe, South Africa. Breeding dress like breeding *suahelicus*. Non-breeding like non-breeding *suahelicus* but head-top and feather edges of rest of upperparts richer rufous; underparts richer buff. Larger. SIZE: wing, ♂ (n = 4) 60–65 (63·0), ♀ (n = 2) 56, 56; tail, ♂ (breeding, n = 3) 51–58 (54·7), (non-breeding, n = 1) 70, ♀ (breeding, n = 1) 53, (non-breeding, n = 1) 63; bill, ♂ (n = 4) 17–18 (17·2), ♀ (n = 2) 16, 16; tarsus, ♂ (n = 4) 24–25 (24·2), ♀ (n = 1) 23.

C. g. luapula Lynes: only at L. Mweru and in L. Bangweulu basin of N Zambia. Breeding dress like breeding *suahelicus* but head-top and wing patch richer chestnut and darker; feather edges of rest of upperparts buffier, less grey. Non-breeding dress undescribed. Proportionately longer-winged. SIZE: (3♂♂, 3♀♀) wing, ♂ 60–66 (62·3), ♀ 54–56 (55·0); tail, ♂ 51–58 (55·0), ♀ 45–50 (47·7); bill, ♂ 14–15 (14·3), ♀ all 14; tarsus, ♂ 21–23 (22·0), ♀ all 21.

C. g. schoutedeni White: W Zambia, north of *c*. 15°S. Breeding dress like breeding *luapula* but crown more streaky. Non-breeding similar but feather edges of rest of upperparts richer buff. SIZE: (1 ♂) wing 67; tail 72; tarsus 24.

C. g. stagnans Clancey: Caprivi, N Botswana, adjacent S Zambia and W Zimbabwe. Breeding dress like *schoutedeni* but crown apparently unstreaked; non-breeding dress indistinguishable from *schoutedeni*. Size like *luapula*.

Field Characters. The most common and widespread of the marsh cisticolas, with rufous crown and bright reddish wing-patch contrasting with upperparts boldly streaked black on grey. Rump grey, tail longer in non-breeding plumage, grey with broad black subterminal band and whitish tip. In non-breeding plumage, crown streaked, upperparts more rufous, streaked black and reddish buff, rump tawny, tail-feathers with broad rufous edges. Races in N and W of range have duller crown, brown or olive, with touch of rufous on forehead, and back less strongly marked; in race *haematocephalus* (coastal E Africa), crown-feathers rufous when fresh, then abrade to dull brown. Normal song in W and E Africa a dry trill, but many other calls also given (see Voice). Very similar to Levaillant's Cisticola *C. tinniens* but seldom found with it; where their ranges meet, Levaillant's is in highlands, Winding in lowlands. In breeding plumage Levaillant's is somewhat brighter above, upperparts striped black and buff, not black and grey, tail feathers edged rufous, as in winter-plumage Winding; tail long at all seasons (tail of Winding shorter in breeding season). Best distinguished by calls. For differences from the 2 other more local marsh cisticolas, Carruthers's *C. carruthersi* and Chirping *C. pipiens*, see those species.

Voice. Tape-recorded (45, 87, 91, B, C, F). Wide range of vocalizations, with some geographic differences; division between songs and calls often unclear. Song of birds from W Africa to S Tanzania is a short (*c*. 1s), rapid, dry trill which has been likened to a grasshopper or to the winding of a watch (hence the name). The trill increases in intensity, trailing off slightly towards the end, and sometimes starts with a couple of short introductory notes, 'jit, ji-drrrrEEEEErr'. Birds from Zambia south apparently do not use this trill, at least not regularly, but instead give a series of dry notes, 'drit', 'beep', 'jit', sometimes almost a double note, 'd'rit', at rate of *c*. 2/s, sometimes increasing to 4/s. The 'drit' notes are given by displaying ♂ as he flies back and forth low over territory, both in E and southern Africa (Gibbon 1991, Zimmerman *et al.* 1996). Perched birds give a liquid 'chewy' which is identical in recordings by C. Chappuis from Chad and G. Gibbon from Natal. Other calls include a thin, buzzy 'dzweee' or 'bzee-bzee-bzee . . .' (Tanzania/Zambia), a liquid trill (E Zambia) and a loud, insistent chatter 'jididi . . .' (coastal Tanzania). In coastal Kenya race *haematocephalus* 'bleats rather than trills, giving an upslurred "brrrRRIP" from trees, shrubs or wires away from water, and also has an excited twittering "tic-titic-tic-tic-titic" call, sometimes given in dancing flight around bushes' (Zimmerman *et al.* 1996). For a fine collection of songs and calls from different localities, see Stjernstedt (87).

General Habits. Dwells in all types of marsh vegetation except dense papyrus, especially at edge of open water, including reed, sedge and *Echinochloa* swamps, rice paddies, rank grass in damp areas, river and lake edges, overgrown ditches, and adjoining bush, grassland and rainy-season flood-plains; also sometimes in dried-out channels and on cultivated land (including cotton fields, sugarcane and banana plantations) and low cover on dry ground. Not found along smaller forest rivers. In Ethiopia, as well as highland streams and marshes (including giant *Carex*), occupies highland grassland, broad-leaved tall-grass savanna, and *Acacia* short-grass savanna, although often associated with wet areas (Urban and Brown 1971). In coastal E Africa, inhabits bush, cultivation and dunes, as well as marshes, possibly an adaptation to recent clearance by humans (Pitman and Took 1973). Elsewhere in E Africa only in wet places, although up to 1 km from water. At L. Naivasha, most often on landward side of fringing swamp and usually within 1 m of ground (Tyler 1991). In some areas, sympatric with Carruthers's Cisticola *C. carruthersi* (Britton 1980), presence or absence of which did not appear to affect habitat use by Winding Cisticola (Tyler 1991). In E Zaïre occurs in elephant grass (Chapin 1953a).

Usually solitary or in pairs. In non-breeding season silent, stays hidden in dense vegetation. Straddles stems; may swing body vertical (head down) to pick prey (van Someren 1956). Forages mainly in undergrowth, at bases of trees and shrubs or on floating carpet of *Salvinia* (Tyler 1991). Sometimes roosts communally in Somalia, where 212 netted at one roost (Ash and Miskell 1988).

In Kenya, *C. g. nyansae* may show local movements between temporary breeding habitat and more permanent reed beds and riversides (van Someren 1956).

Food. Insects, including grasshoppers, crickets, small beetles, other winged insects, caterpillars; 'worms'; seeds in one stomach. Young given caterpillars, nymphs of mantids and grasshoppers, spiders, small snails, newly-emerged damselflies (van Someren 1956).

Breeding Habits. Territorial. Breeding territory size in marshy savanna < 0·1 ha (10–12 pairs per ha), more spaced-out in bushy savanna; pairs form in month following establishment of territory (Ruwet 1965). Noisy in breeding season. Breeding ♂ sings from bush, grass or top of low tree, occasionally flicking tail. Sometimes makes short, circular display flight, up to 15 m above perch or between perches, while singing and flicking tail, ending with a dive. Alarm calls given from grass-tops. Tail often raised and fanned when bird approaches nest.

A

NEST: a ball or vertical ellipse, *c.* 10 × 7–9 cm, with side-top entrance *c.* 3 cm in diam.; substantially constructed of dry vegetation around a frame of living blades, sometimes bound with spider web (**A**); lining absent or of plant down, often thick over lower half; sometimes decorated with cocoons, bark and coloured bits of plant; in Ethiopia occasionally a cup, with green blades bent over it (Cheesman and Sclater 1935); fixed among coarse grass or short reeds or weeds, 25–120 cm above ground, in marsh, sometimes over water; in drier areas in E Africa often placed in thorny bushes (Lynes 1930) or clump of *Citronella* (Jackson and Sclater 1938). Usually built by ♀; rarely by both sexes.

EGGS: Gambia 3–4; Ghana, S Nigeria, Gabon (*amphilectus*, 9 clutches) 2–4 (3·0); N Nigeria, Chad, Sudan (*zalingei*, 6 clutches) 3–5 (4·0); Ethiopia (*lugubris*, 5 clutches) 2–3 (2·6); Kenya (*haematocephalus*, 20 clutches) 2–5; Uganda (*nyansae*, 213 clutches) 2–4; Tanzania (*suahelicus*, 5 clutches) 3–5 (4·2); Zaïre/Zambia (2 clutches) 3; Malaŵi (6 clutches) 2–4 (3·3); Natal (18 clutches) 1–4 (2·7). Glossy; variable in colour but usually of a shade of red, generally deeper further south: in *amphilectus*, *zalingei*, *nyansae*, *haematocephalus*, round to ovate, smooth, white to pinkish cream, salmon pink or brick red, heavily spotted or blotched all over (or marks concentrated at big end, or sometimes nearly plain or clouded) with purple-brown to orange and underlying violet-greys; fade quickly. *C. g. lugubris*, white to pale brick red, boldly spotted and blotched darker brick red and violet, or immaculate white to dark pink. *C. g. suahelicus*, *schoutedeni*, creamy rufous to red, heavily stippled red-brown with underlying pale violet. *C. g. galactotes* and *luapula*, immaculate dark brick red; some with indistinct purplish marks at big end, or (*luapula*) some as previous race. SIZE: (n = 15, Darfur, N Nigeria: *zalingei*) 15·2–17·8 × 11·9–13·5 (16·7 × 12·4); (n = 18, Ethiopia: *lugubris*) 15·2–17·0 × 12·0–13·0 (16·9 × 12·6); (Gabon, Zaïre *amphilectus*)

Plate 11

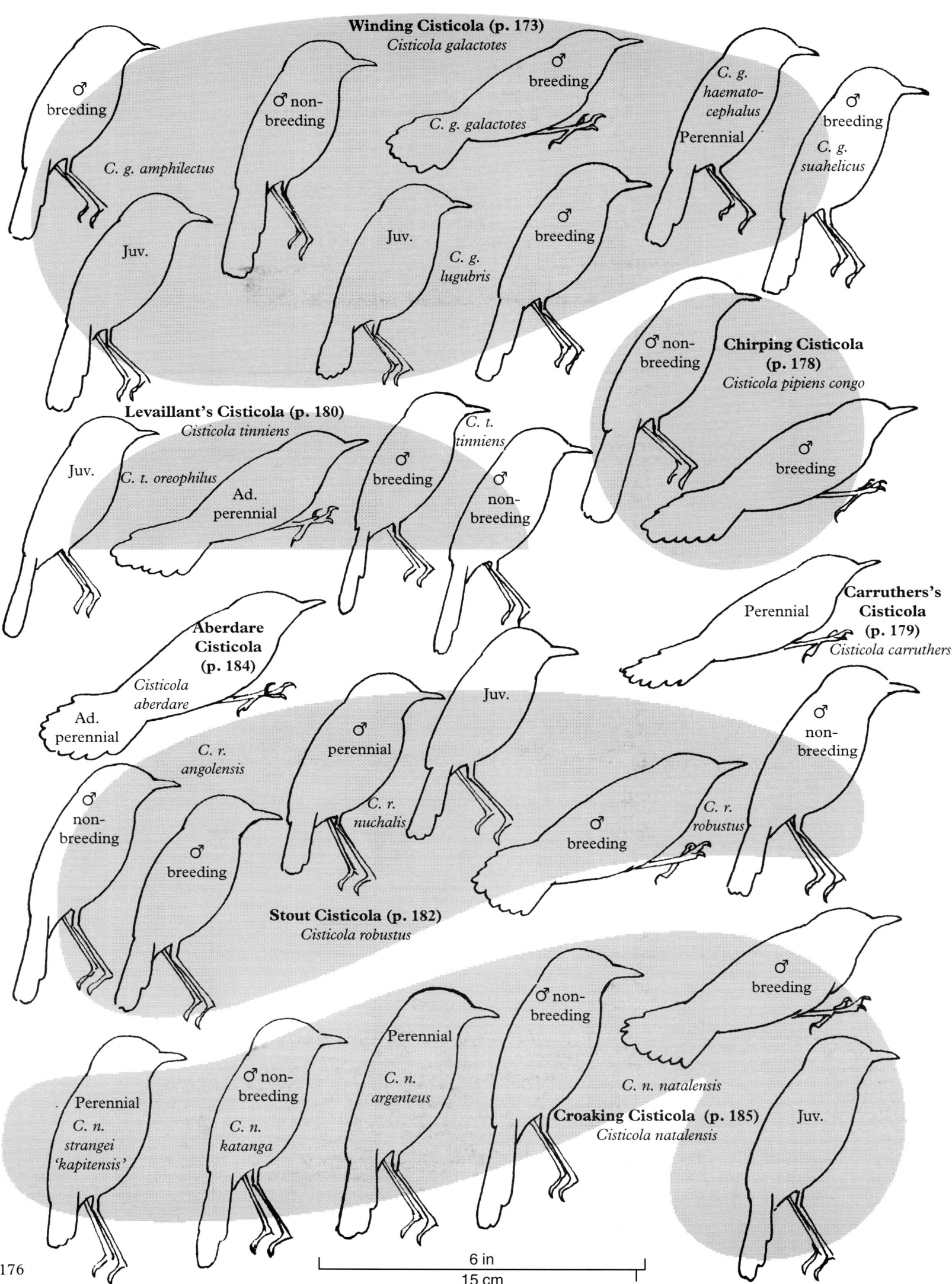

Plate 12

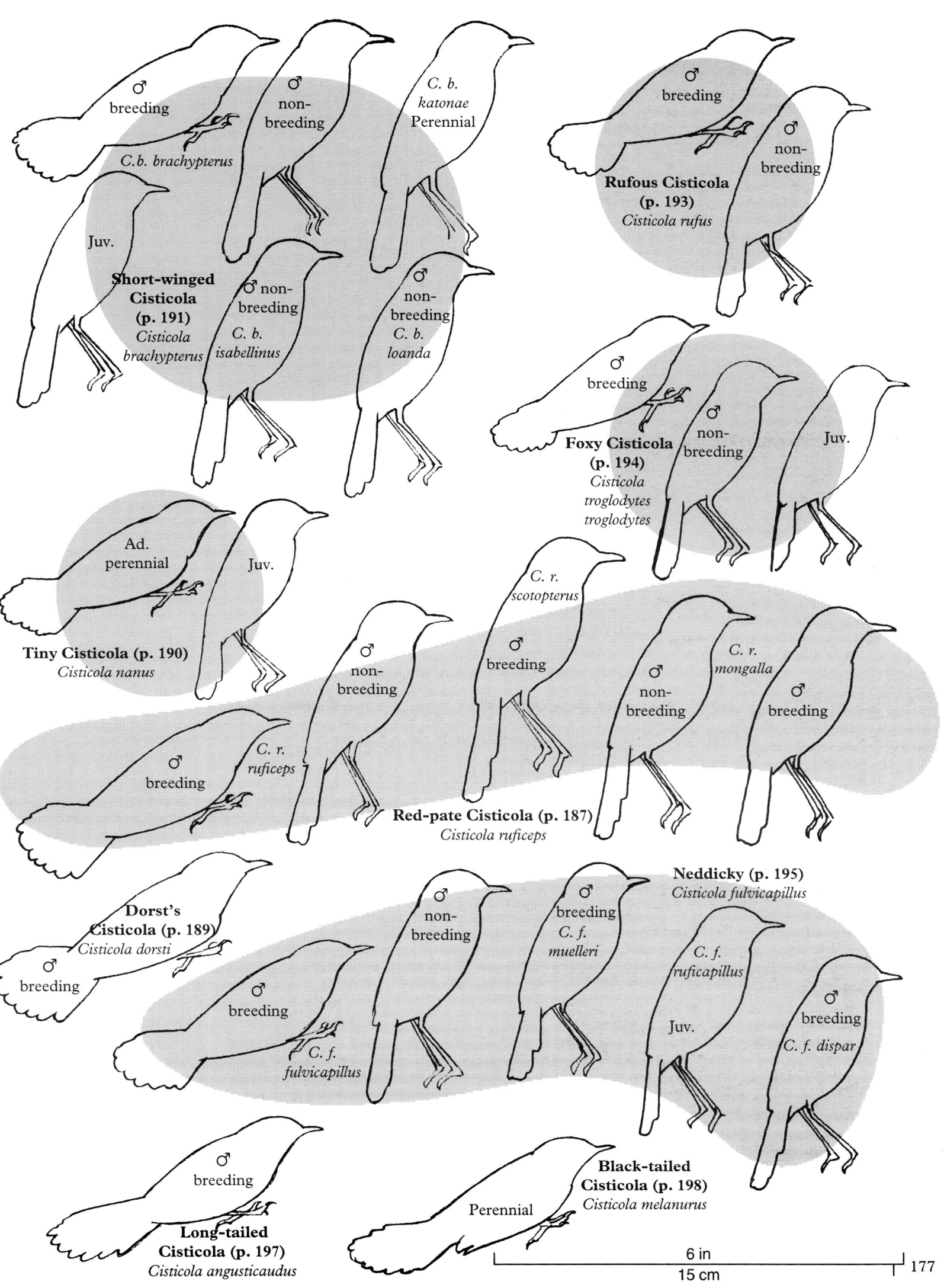

16·3–18·3 × 11·5–12·9; (n = 29, Kenya: *haematocephalus*) 14·6–17·3 × 11·5–13·0 (16·2 × 12·2); (n = 200, Uganda: *nyansae*) 14·9–18·3 × 11·3–14·0 (16·8 × 12·7); (n = 16, Tanzania: *suahelicus*) 16·2–17·9 × 12–13 (17·1 × 12·6); (n = 10, Malaŵi: *suahelicus*) av. 16·1 × 11·8; (n = 3, Zambia: *schoutedeni*) 15–15·8 × 12·3 (15·3 × 12·3); (n = 11, Zambia/S Zaïre: *luapula*) 16·0–17·2 × 11·8–13·5 (16·5 × 12·3); (n = 6, Natal) av. 16·5 × 12·5; (n = 1, South Africa) 17·0 × 12·6.

LAYING DATES: SW Mauritania, Sept–Nov; N Senegal, Aug–Nov; Gambia, Sept; S Senegal (in breeding dress July–Jan, singing Nov–Dec); Mali, Sept; Ghana, Apr, July–Aug, Nov; Togo, Apr–May, Sept; Nigeria, Mar–Dec; Gabon, July; Central African Republic (breeding condition Aug); Sudan, Feb, June–Sept; Ethiopia, June–Oct; E Zaïre (Uele) June–Nov (rains); S Zaïre/N Angola, Sept–Feb; SE Zaïre, Nov–Apr, occasionally May (double-brooded: first brood Dec–Feb, second Feb–Apr) (Ruwet 1965). E Africa: Kenya, mainly long rains, also (fewer individuals) short rains, often double-brooded; Kenya coast, Apr–June, Nov–Dec; Uganda (Entebbe), Mar–Aug, Oct; Tanzania, Jan–Mar; Region B, Feb–Aug, Oct–Nov (peak June); Region C, Jan–Apr (peak Feb); Region D, Oct–Jan, Apr–June, Aug; Region E, Apr–Aug; rains, but in Region C only late rains. Malaŵi, Jan–Apr (most records), June–Sept; W Zambia, Nov–Feb, probably to Apr; Zimbabwe, Jan (1 record); S Mozambique, Jan; Botswana, Nov, Apr; Transvaal, May (1 record); Natal, Oct–Feb, 1 record June.

INCUBATION: by both sexes but mostly by ♀; period 12–13 days (van Someren 1956) or 18–20 days (Priest 1948). Nest lining continues to be added during incubation; deserts eggs readily.

DEVELOPMENT AND CARE OF YOUNG: young in nest 14–17 days. Adults approach through vegetation. All food collected within 20 m of nest. Feathered nestlings preen.

BREEDING SUCCESS/SURVIVAL: nests lost to flooding, and trampling by domestic stock. Parasitized by Pintailed Whydah *Vidua macroura*, with whose chicks host's young may be reared (one nest contained 2 whydah chicks, 1 host chick) (Cheesman and Sclater 1935).

References
Chapin, J. P. (1953a).
Cheesman, R. E. and Sclater, W. L. (1935).
Jackson, F. J. and Sclater, W. L. (1938).
Lynes, H. (1930).
Pitman, C. R. S. and Took, J. M. E. (1973).
Ruwet, J.-C. (1965).
van Someren, V. G. L. (1956).

Plate 11
(Opp. p. 176)

Cisticola pipiens Lynes. Chirping Cisticola. Cisticole pépiante.

Cisticola pipiens Lynes, 1930. Ibis, *Cisticola* Suppl. p. 404; Huambo, Benguela, Angola.

Range and Status. Endemic resident. Plateau of W Angola (N Huila and Huambo to Cuanza Norte), E Angola and Zambia east to Muchinga Escarpment (except southeast); S Zaïre (Kasai to Shaba and Marungu Highlands), extreme SW Tanzania (Mumba R., very common); Burundi (Bujumbura and Rusizi R. Delta Nat. Park); N Namibia (Caprivi); N Botswana (common Okavango, sparse Chobe R.) and extreme NW Zimbabwe (Katombora Rapids). Common but localized. Density in SW Tanzania, 15 birds/10 ha (Moyer and Sikombe 1992).

Description. *C. p. congo* Lynes: E Angola, Zambia (except southwest), Zaïre, Tanzania and Burundi. ADULT ♂ (breeding): forehead and crown dark rusty red; lores, cheeks and ear-coverts rusty buff; nape greyish brown. Back heavily streaked black, feathers edged greyish brown, rump and uppertail-coverts greyish brown. Tail greyish brown, T1 and T2 with large blackish brown subterminal spot and greyish brown tip, T3 to T6 greyish brown with well-marked black subterminal bar (*c.* 8 mm) and broad (*c.* 6 mm) buffy white tips. Chin and throat to undertail-coverts buff, thighs rusty buff, centre of belly whiter. Wings dark brown, inner portion of outer webs of primaries and secondaries washed tawny-buff, scapulars blackish brown with paler buff margins, wing-coverts blackish brown, broadly edged buff on outer webs; shoulder, under-wing-coverts and axillaries buffy. ADULT ♂ (non-breeding):

Cisticola pipiens

margins of mantle feathers paler, washed buff; scapulars similar. Bill blackish with basal two-thirds of lower mandible grey, palate and interior of mouth black; eyes rich hazel; legs and feet brownish flesh. Sexes alike. SIZE: (44 ♂♂, 32 ♀♀) wing, ♂ 57–67 (61·8), ♀ 53–64 (56·1); tail, ♂ (breeding) 56–61 (58·3), (non-breeding) 59–67 (63·1), ♀ (breeding) 53–64 (56·3), (non-breeding) 57–65 (61·0); bill, ♂ 13–15 (14·5), ♀ 13·5–15 (14·2); tarsus, ♂ 22–25 (23·6), ♀ 21–24·5 (22·7). WEIGHT: ♂ (n = 5) 13·3–15·3 (14·1), ♀ (n = 6) 12·2–14·3 (12·9).

IMMATURE: feathers of upperparts with narrow blackish centres and tawny chestnut edges; crown noticeably streaked (similar streaking often retained in adult non-breeding plumage, perhaps not lost until year 2); bases and edges of tail-feathers redder. Very young birds (short-tailed) have sides of face and chin to belly white, buffy wash confined to flanks; scapulars and outer webs of primaries and secondaries rustier, tips of tail more strongly buff-washed.

C. p. pipiens Lynes: W Angola. Streaking above bolder, more heavily buff-washed below.

C. p. arundicola Clancey: SE Angola (Cuando Cubango), Zambian border on Kwando R., Namibia, Botswana and Zimbabwe. In breeding dress rusty red crown sharply demarcated from greyer upperparts; less buffy on underparts; seasonal plumage differences less well marked, mantle with blacker feather centres in non-breeding plumage.

Field Characters. The largest of the marsh cisticolas, with long, broad tail kept slightly spread and flirted. Confined to S-central Africa. Overlaps broadly with Winding Cisticola *C. galactotes*, less often with Levaillant's Cisticola *C. tinniens*; back of Chirping Cisticola looks browner, less black (dark feather centres more broadly edged brown or grey-brown), and underparts suffused, often heavily, with buff; best distinction is song.

Voice. Tape-recorded (35, 45, 87, 88, 91, B, F, KEI). Basic song/call, hard, buzzy 'dzit-dzidzeee' or 'dzit-dzit-dzidzeee'; second bird chimes in with 'wee-cherp-chyup-chyup-chyup', also calls 'chyup-chyup' alone. Another version, a longer 'chor-cher-ti-dee-dee-dee-dee-dee-dee', without buzzy element of basic buzz, and/or 5–6 purer, not burry, 'chyow' or 'chdyow' (chyow with a little hiccup in the middle). Also a louder, piping 'hweet-hweet-hweet...'. Sings from reed top or in flight.

General Habits. Frequents taller aquatic vegetation in permanent swamps where there are extensive reedbeds and papyrus, avoiding lower emergent growth although often flying over it; also seasonally inundated grassland and flood-plains. In Zambia occurs in tallest, lushest growth in dambos, frequently alongside Winding Cisticola *C. galactotes* (which tolerates drier ground), or on moist ground that has dried out; occasionally found with Levaillant's Cisticola in rather short vegetation on permanently wet ground. All 3 species can occur together (Aspinwall 1984) and may be in competition with one another; in lower and middle Kafue basin (Zambia) Levaillant's Cisticola appears to replace Chirping Cisticola entirely.

Usually in pairs or small family parties of 4–5, keeping within dense cover. Frequently climbs reed stem to call or sing; may perch on top of grass stem in early morning; skulks, but can move nimbly through rank cover; seldom far from swamp edge.

Food. Insects, especially green grasshoppers.

Breeding Habits. Territorial. ♂ constantly advertises territory by singing from exposed vantage point on reed; also sings and displays, making short, zigzag flights low over reeds and tall grass, in slow, hesitant fashion, with broad tail flopping from side to side; sings from top of grasses and bushes. 2 birds in excited courtship chase uttered piping 'hweet hweet hweet'.

NEST: pear-shaped ball of dry grass, narrow at top, broader at base, with largish side-top entrance, made of broad blades, coarse and reedy, frailest at top, bound on outside with a little cobweb and decorated with cocoons; well hidden, attached to surrounding grass stems; placed 45–90 cm up in dense growth, often flooded ground just above water.

EGGS: 3–4. Slightly glossed; cream or with rosy flush, closely and evenly freckled with brownish red or bright chestnut-brown, with underlying spotting and ashy blotches. SIZE: (n = 10) 15·7–19·1 × 12–13 (17·8 × 12·7).

LAYING DATES: Angola, Nov–Mar; Zaïre, Dec–Apr; Zambia, Oct–Jan, Apr; Botswana, Oct, Jan, Mar–Apr.

Reference
Vincent, A. W. (1948a).

Cisticola carruthersi Ogilvie-Grant. Carruthers's Cisticola. Cisticole de Carruthers.

Plate 11
(Opp. p. 176)

Cisticola carruthersi Ogilvie-Grant, 1909. Bull. Br. Orn. Club vol. 23, p.94; Mokia, near L. George, Uganda.

Range and Status. Endemic resident in swamps of upper Nile basin, at 1000 to 2200 m. Common to abundant Rwanda, Burundi, extreme E and NE Zaïre from Lendu Plateau to Kivu, S and W Uganda, Kenya from Kisumu west, NW Tanzania (Kagera basin). Population at Kadenge, L. Kanyaboli, Kenya, estimated at 30 per ha (Britton 1978).

Description. ADULT ♂ (breeding): forehead, crown, nape and ear-coverts dark reddish brown; chin and throat white, becoming buff on sides of neck; rest of upperparts streaked very dark brown and greyish brown (feather centres dark brown); tail dark brown, tipped buff; rest of underparts dirty white, washed buff, especially on breast and flanks; primaries, secondaries, tertials and wing-coverts very dark brown, edged brown (buffier on tertials and alula); underwing buff. Upper

Cisticola carruthersi

mandible black, lower grey; eyes pale brown; feet pale horn. Sexes alike. SIZE: wing, ♂ (n = 5) 57–58 (57·8), ♀ (n = 6) 52–58 (55·0); tail, ♂ (n = 5) 49–58 (54·8), ♀ (n = 6) 50–56 (53·2); bill, ♂ (n = 5) 14–16 (15·0), ♀ (n = 5) 14–16 (15·0); tarsus, ♂ (n = 5), 21–23 (22·0), ♀ (n = 6) 21–22 (21·2). WEIGHT: (Uganda) ♂ (n = 4) 11·9–12·1 (12·0), ♀ (n = 5) 10·7–13·9 (12·1), 1 juvenile ♂ 12, 1 unsexed 10; (Kenya) ♂ (n = 37) 10·1–14·4 (11·7), ♀ (n = 34) 9·5–12·4 (10·8).

IMMATURE: like adult but head-top less red; rest of upperparts browner, less streaked; throat and breast washed sulphur; bill paler; eyes greyer.

NESTLING: unknown.

Field Characters. A noisy bird of papyrus swamps from E Zaïre to W Kenya, easy to see. Has variety of loud calls, all very different from those of Winding Cisticola *C. galactotes*; sings in the air while flying from one papyrus stem to another. Somewhat smaller than Winding, with slimmer, longer bill; crown darker rufous, unstreaked, back streaked black and grey, wing-patch brown, not red; tail black (not grey with black band). Dark line from bill to eye and pale line above it; underparts whitish but flanks buff. Immature has lighter brown crown and back, latter lightly streaked dark brown, quite unlike bold pattern of adult. Winding Cisticola comes to edge of papyrus swamps but does not enter pure papyrus, where Carruthers's is the only cisticola; outside papyrus their habitats overlap.

Voice. Tape-recorded (BOUR, FISP, GREG, LEM, McVIC). Song, a series of sharp chattering notes run together into a 1–2 s trill, varying in speed and pitch, 'chi-chi-chi-chi . . .'; notes clearly separable to the ear, unlike 'winding' song of Winding Cisticola. Also described as "a chatter followed by a rapid series of high scratchy or squeaky notes, these run together and higher-pitched at the end: 'chchchchchchchch . . . tsik-tseeosisitseek' or 'tswi-squee-squee-squee . . . tisk-chee-oreeoo-tseek' (Zimmerman *et al.* 1996). Scolding calls include nasal 'cheeya' or 'nyaa', and protracted 'wick-tsyik-tsyik-tsyik . . .'.

General Habits. Mainly found in papyrus swamps; more or less restricted to them in some areas (e.g. Kenya, lower altitudes in E Rwanda and E Burundi), but in others found in all types of marsh, including coarse grass and rice (e.g. high altitudes in NW Rwanda). Usually parapatric with Winding Cisticola, whose presence seems responsible for the restriction of *carruthersi* to papyrus swamps in some areas; where *galactotes* is absent (high altitudes), *carruthersi* is found in a broader range of habitats (Vande weghe 1981).

Food. Insects.

Breeding Habits. NEST: rather loosely constructed oval with side-top entrance, of grass and vegetable down with cottony lining; supported by living grasses woven into structure; outside decorated with spider web and cocoons; placed in stunted shrub and grass.

EGGS: 2–4. Pale pink with brick red and liver-coloured spots. SIZE: (n = 5, Kenya, Uganda) 15·5–18 × 11·6–13 (16·7 × 12·2)

LAYING DATES: E Africa: Kenya, July; Uganda, Apr; Region B, Apr, June–July (rains).

References
Chapin, J. P. (1953a).
Lynes, H. (1930).

Plate 11
(Opp. p. 176)

Cisticola tinniens (Lichtenstein). Levaillant's Cisticola; Tinkling Cisticola. Cisticole à sonnette.

Malurus tinniens Lichtenstein, 1842. Verz. einer Sammlung von Säugetiere und Vögel Kaffernland, p. 13; 'Kaffirland' = Likwa (Vaal) R. at Valsch R. confluence, South Africa (Stresemann 1954, Ann. Mus. Roy. Congo Belge, vol. 1, p. 81).

Range and Status. Endemic, from Angola and Kenya southward; range discontinuous north of Zambezi R., where restricted to highlands; generally resident but may undertake local movements in Orange Free State (Earlé and Grobler 1987). Frequent at 2000–3200 m in W Kenya highlands. Occurs at 2800–3600 m in SW

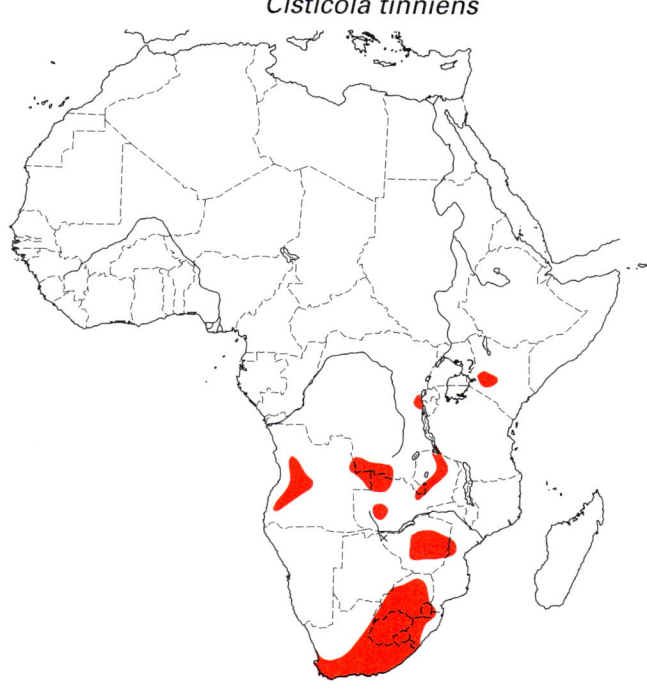

Cisticola tinniens

Tanzania, and extreme E Zaïre on border with Rwanda and Burundi between lakes Kivu and Tanganyika; also SE Zaïre (down to 1100 m in E Kasai). Uncommon and local N and W Zambia. Locally frequent Angola from N and W Huila to S Cuanza Norte, and in east, in S Lunda and N Moxico. Locally common Zimbabwe, usually above 1200 m, on plateau and E highlands, and in adjacent highland Mozambique in Manica e Sofala; in Bvumba highlands, only recorded Mozambique side. Botswana, once (near Lobatse, July). Common lowland Lesotho. Locally common Swaziland and South Africa up to 1900 m, but absent from arid N and NW Cape Province.

Description. *C. t. tinniens* (Lichtenstein): South Africa to Zimbabwe. ADULT ♂ (breeding): forehead to nape red-brown, obscurely streaked with very dark brown; indistinct supercilium buff; lores and ear-coverts pale red-brown; rest of upperparts streaked very dark brown (feather centres) and greyish buff; tail-feathers very dark brown, edged red-brown and tipped buff, with subterminal darker brown spot; entire underparts pale buff, sometimes streaked very dark brown on flanks; primaries, secondaries and tertials very dark brown, with outer fringes rufous, inner buff; wing-coverts very dark brown, fringed buff; underwing-coverts buff; axillaries grey. Upper mandible dark brown to black, lower grey; eyes red-brown; legs and feet brownish pink. ADULT ♂ (non-breeding): dark streaks on crown more prominent; rest of upperparts buffier, less grey (feather edges buffier); underparts darker, rich greyish buff. Sexes alike. SIZE: wing, ♂ (n = 10) 54–59 (55·6), ♀ (n = 7) 48–53 (50·3); tail, ♂ (breeding, n = 5) 50–61 (56·0), (non-breeding, n = 7) 57–68 (61·2), ♀ (breeding, n = 5) 45–56 (51·4), (non-breeding, n = 2) 57, 57; bill, ♂ (n = 9) 13–15 (13·6), ♀ (n = 7) 12–14 (13·0); tarsus, ♂ (n = 9) all 20, ♀ (n = 7) 17–19 (18·3). WEIGHT: (southern Africa) 1 ♂ 11·9, unsexed (n = 20) 8·0–18·5 (11·4).

IMMATURE: like non-breeding adult but upperparts duller; head less red; streaks on rest of upperparts less prominent (feather centres lighter brown, edges browner); underparts less buff (intermediate between breeding and non-breeding adult) and washed with very pale sulphur on breast.

NESTLING: at *c.* 5 days, feathers in pin dark grey; gape orange-yellow.

C. t. shiwae White: NE Zambia, adjacent extreme SE Zaïre and extreme SW Tanzania. Like nominate race but, in breeding dress, head-top darker brown, rest of upperparts darker (feather edges darker buff); in non-breeding dress black streaking above heavier.

C. t. perpullus Hartert: Angola, adjacent NW Zambia (east to *c.* 25°30'E) and S Zaïre (race of SW Zambian birds unknown). In breeding dress more like nominate than *shiwae*; upperparts paler than *shiwae*; head-top tawny brown, less red, obscurely streaked; rest of upperparts like nominate but feather edges duller grey-brown, less buff. Non-breeding dress like nominate but feather edges of mantle slightly duller and (at least in S Zaïre and E Angola) head-top more prominently streaked with very dark brown. Larger, and size clinal: central Angolan birds largest. SIZE: wing, ♂ (n = 7) 56–63 (59·0), ♀ (n = 6) 50–54 (52·2); tail, ♂ (breeding, n = 4) 55–60 (57·5), (non-breeding, n = 3) 64–66 (65·0), ♀ (breeding, n = 3) 50–54 (52·3), (non-breeding, n = 3) 54–58 (56·3); bill, ♂ (n = 7) 13–15 (14·0), F (n = 6) all 13; tarsus, ♂ (n = 7) 19–22 (20·6), ♀ (n = 5) 19–20 (19·8).

C. t. dyleffi Prigogine: mountains northwest of L. Tanganyika and west of Ruzizi Valley, over 2500 m. Like *perpullus* but tail-feathers and outer primaries darker.

C. t. oreophilus van Someren: highlands of central and W Kenya. Perennial plumage, like breeding dress of *shiwae* but head-top brighter red-brown (more like nominate race); underparts richer buff (richer still than nominate). SIZE: wing, ♂ (n = 6) 55–61 (58·2), ♀ (n = 5) 51–53 (52·0); tail, ♂ (n = 6) 56–65 (59·3), ♀ (n = 5) 50–56 (52·6); bill, ♂ (n = 5) 12–14 (13·2), ♀ (n = 5) 11–13 (12·2); tarsus, ♂ (n = 5) 19–21 (19·8), ♀ (n = 4) 19–20 (19·5). WEIGHT: ♂ (n = 2) 13, 14.

Field Characters. Rather long and slim, with proportionately long tail, giving prinia-like shape during jerky flitting flight low over tops of grass and reeds. Crown and hind-neck rufous, brighter than Winding Cisticola *C. galactotes*, and streaked black in non-breeding season; back heavily streaked black, appearing all black at a distance; reddish patch on wing. Rump streaked (plain grey in Winding). Tail-feathers broadly edged rufous, so that entire tail appears rufous (actually black and rufous), whereas tail of Winding is black and grey. Usually found at higher elevations than Winding, and voice very different. Allopatric with Carruthers's Cisticola *C. carruthersi*, but overlaps with Chirping Cisticola *C. pipiens*. Best distinguished by voice, but Chirping is also larger, with broader tail, browner or greyer (less black) back, buffier underparts.

Voice. Tape-recorded (45, 75, 87, 88, 91, B, C, F). Song brief (*c.* 1 s), 1–3 introductory notes followed by short rolling trill which ends with an emphatic note, 'tee-yoo, toodlyIP', 'whi-choo, poodlyEE', 'tip, plurillyROY'. Calls include a sharp 'wip-wip . . .' and a repeated scolding, buzzy 'zeee' or 'dzweee'.

General Habits. Inhabits rank weeds, sedge, grass and occasionally reeds, near water, including seasonal and perennial swamps and marshes, wet areas along seasonal river beds, montane grassland near streams or bogs; also open grassy places on mountain ridges, in bamboo. In W Zambia, occurs together with two other 'marsh' cisticolas, Winding and Chirping (Aspinwall

1984). In South Africa not shy; often found near buildings and in towns.

Usually in pairs, territorial. Forages low in vegetation but perches higher on tall stem when disturbed; jerks tail from side to side. Flight low, erratic and fluttering, flicking tail.

Food. Small insects, including beetles, grasshoppers and flies.

Breeding Habits. ♂ sings from grass top or low bush, occasionally giving short song flight, 2–3 m above vegetation level.

NEST: globular or upright oval c. 13 × 9 cm, with side-top entrance c. 3–4 cm diam.; occasionally a cup; thin outer structure of grass with spider web binding, lined (often thickly) with plant down; sometimes consists entirely of down, with no outer structure; outside usually covered with dark or rusty-coloured material, including rootlets, sometimes with green grass blades bent and fixed over top; usually slung between leaves and weed stalks, 10–100 cm above ground, in tall grass or low bushes, often on flooded ground or on stream bank; where robbing by humans frequent, often hidden close into bank under overhanging grass (Roberts 1913); one in young reed regrowth (50 cm high) after fire (Jackson and Sclater 1938).

EGGS: Kenya (1 clutch) 3; Zambia/Zimbabwe (50 clutches) 2–4 (3·2); Mozambique (1 clutch) 3; Natal (31 clutches) 2–5 (3·5); Angola (2 clutches) 3. Oval; moderately glossy; white to rich blue or pale green, occasionally pink, blotched or evenly freckled with reds and red-browns, or with spots of red, brown and purple-grey, sometimes marks concentrated over big half, or in ring there. SIZE: (n = 6, Kenya) 16·0–16·8 × 12·0–12·5 (16·3–12·2); (n = 13, Angola) av. 15·8 × 11·9 (n = 134, southern Africa) 13·5–17·6 × 10·9–13·3 (16·0 × 12·0).

LAYING DATES: E Africa: Kenya, June; Region A, Jan, June; Region D, Feb–July; Zambia, Sept, Jan; Zimbabwe, Sept–May, peak Dec–Feb (86 records: Irwin 1981); S Mozambique, Nov; Botswana, Oct, Jan; Natal, Nov–Apr, peak Dec–Feb; Transvaal, Sept–Apr, peak Nov–Jan; SW Cape Province, mainly Aug–Feb; Angola, Nov–Mar.

INCUBATION: period 14 days.

BREEDING SUCCESS/SURVIVAL: parasitized by Parasitic Weaver *Anomalospiza imberbis*.

References
Chapin, J. P. (1953a).
Lynes, H. (1930).
Roberts, A. (1913).
Vincent, A. W. (1948a).

Plate 11
(Opp. p. 176)

Cisticola robustus (Rüppell). **Stout Cisticola. Cisticole robuste.**

Drymoica robusta Rüppell, 1845. Systematische Uebersicht der Vögel Nord-Ost-Afrika's, p. 35; Shoa, Ethiopia.

Range and Status. Endemic resident. Common on Gangirwal and Chappal Waddi at 1900–2300 m, E Nigeria. In Cameroon, common in montane areas from Mt Manenguba to Bamenda Highlands, at 1800–2300 m. In Congo, only in Gamboma area. In Sudan, once at Shambe, White Nile. Frequent Ethiopian plateau, usually above 1200 m, commoner above 1900 m. In Eritrea only near Senafe (15°N, 39°E), where common. Common SW and central Kenya (east to Kitui and Simba) and N Tanzania at 1200–2500 m, mainly above 1600 m; also occurs at 2700 m on Kikuyu Escarpment, in L. Victoria basin of NW Tanzania to S and W Uganda, and in SW Tanzania. Abundant Rwanda, Burundi and adjacent Zaïre (1°N–4°S) (not in grasslands near L. Albert but abundant near L. Edward); also occurs in mountains southwest of L. Tanganyika (7–8°S, 29–30°E), and elsewhere in SE Zaïre (south of 8°S). In N Zambia, above 1300 m. In Angola, locally frequent on central plateau between N Huila and S Cuanza Sul, and in east.

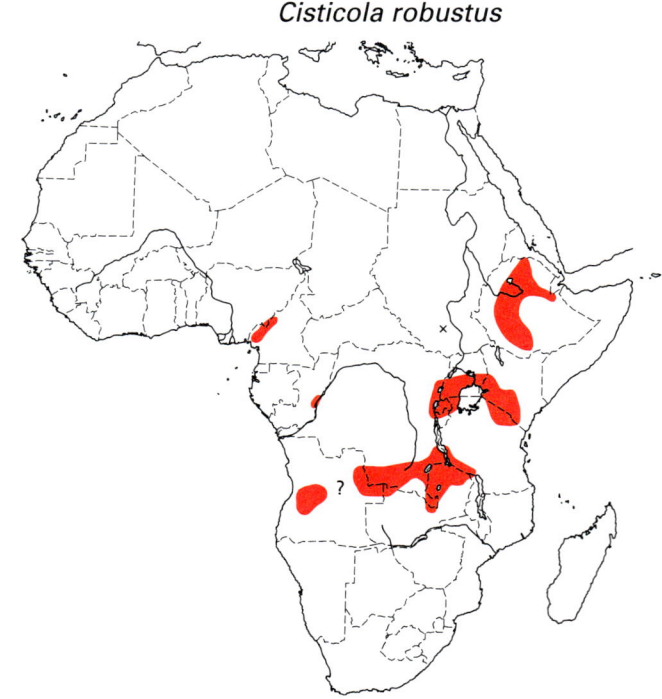

Cisticola robustus

Description. *C. r. robustus* (Rüppell): N Ethiopian plateau and Harar. ADULT ♂ (breeding): forehead to nape tawny brown, streaked with dark brown; lores and ear-coverts grey-brown; indistinct supercilium cream; rest of upperparts mottled very dark brown and buffy grey; tail-feathers dark brown, tipped and narrowly edged buff, with indistinct very dark brown subterminal band; entire underparts buff, richest on flanks and undertail-coverts, and washed olivaceous grey on sides of breast; primaries, secondaries and outer wing-coverts brown, fringed rufous-buff; tertials and inner wing-coverts very dark brown, fringed buff; underwing-coverts buff. Upper mandible dark brown to black, lower grey or horn to black; eyes orange-brown; legs and feet pinkish brown. ADULT ♂ (non-breeding): like breeding ♂ but darker, more rufous; head-top to mantle streaked rich rufous and black; rest of upperparts black, streaked with rufous-buff; tail fringed rufous; underparts richer buff. Sexes alike. SIZE (10 ♂ ♂, 10 ♀ ♀): wing, ♂ 73–81 (77·3), ♀ 62–66 (64·0); tail, ♂ (breeding, n = 5) 55–65 (60·8), (non-breeding, n = 5) 62–67 (63·2), ♀ (breeding, n = 5) 45–51 (47·2), (non-breeding, n = 5) 52–57 (54·6); bill, ♂ 16–18 (17·0), ♀ 14–17 (15·5); tarsus, ♂ 27–29 (27·9), ♀ 24–26 (24·8).

IMMATURE: like non-breeding adult but entire upperparts more or less uniformly streaked rufous and very dark brown; wing-feathers fringed rufous; entire underparts strongly washed with sulphur-yellow.

NESTLING: unknown.

C. r. schraderi Neumann: central Eritrea and adjacent N Ethiopia (Adigrat). Like nominate but, in non-breeding dress, upperparts paler: head-top streaked tawny and very dark brown; rest of upperparts very dark brown, streaked with sandy-buff; tail fringed tawny. Breeding dress undescribed.

C. r. omo Neumann and Lynes: S Ethiopian plateau, from Lekemte and Jimma south (Omo basin). Perennial dress, like non-breeding nominate but darker and less rufous: head-top black, lightly mottled with rufous; rest of upperparts black, lightly streaked with buff; tail-feathers fringed tawny; underparts intermediate between breeding and non-breeding plumages of nominate race.

C. r. nuchalis (Reichenow) (including '*ambigua*'): Kenya and N Tanzania to NE Zaïre; one record S Sudan; isolated population matching this race in Congo (Rand *et al.* 1959). Perennial dress like breeding nominate race but head-top more rufous; nape contrastingly tawny; rest of upperparts more narrowly streaked very dark brown (feather edges broader); buff tail fringes broader. Smaller, and size clinal: birds of Kenya and N Tanzania highlands ('*ambigua*') average larger. SIZE: (10 ♂ ♂, 10 ♀ ♀) wing, ♂ 64–70 (67·3), ♀ 57–64 (59·4); tail, ♂ 46–58 (50·7), ♀ 42–56 (47·1)); bill, ♂ 15–16 (15·4), ♀ 14–15 (14·6); tarsus, ♂ 24–27 (25·7), ♀ 21–24 (22·9). WEIGHT: (Kenya, Uganda) ♂ (n = 5) 16–22 (19·3), ♀ (n = 2) 14, 16·1.

C. r. awemba Lynes: SW Tanzania, SE Zaïre (E Shaba) and NE Zambia. Seasonal plumage. Breeding dress like *nuchalis* but averages slightly darker. In non-breeding dress, head-top darker, redder brown; feather edges of rest of upperparts browner. Larger. SIZE: wing, ♂ (n = 7) 69–73 (71·3), ♀ (n = 4) 57–61 (59·2); tail, ♂ (breeding, n = 3) 50–52 (50·7), (non-breeding, n = 3) 50–58 (53·7), ♀ (breeding, n = 3) 42–53 (48·3), (non-breeding, n = 1) 43; bill, ♂ (n = 7) 15–17 (16·0), ♀ (n = 4) 14–16 (15·0); tarsus, ♂ (n = 7) 25–27 (25·7), ♀ (n = 4) 22–23 (22·7).

C. r. angolensis (Bocage): S Zaïre (W Shaba) and NW Zambia (Mwinilunga) to Angola. Like *awemba* but, in breeding dress, upperparts darker (feather edges darker, brown); tail-fringes narrower. In non-breeding dress, head-top darker (feather edges darker red-brown) so appears less streaked; rest of upperparts duller (feather edges buffy brown, less rufous). Still larger. SIZE: (6 ♂ ♂, 4 ♀ ♀) wing, ♂ 72–76 (74·3), ♀ 60–65 (63·0); tail, ♂ (breeding, n = 3) 50–55 (51·7), (non-breeding, n = 3) 56–58 (57·0), ♀ (breeding, n = 3) 41–48 (45·0), (non-breeding, n = 1) 48; bill, ♂ 16–17 (16·8), ♀ 15–16 (15·2); tarsus, ♂ 25–29 (27·0), ♀ 23–25 (23·7). WEIGHT: (Zaïre) ♂ (n = 3) 21–25 (22·7), 1 ♀ 24, 1 unsexed 17.

C. r. santae Bates: highlands of W Cameroon and E Nigeria. Perennial dress, intermediate between breeding and non-breeding *angolensis*: head-top less red than non-breeding *angolensis*; upperparts more prominently streaked with paler buffy brown (feather edges broader than breeding *angolensis*, paler than non-breeding *angolensis*). Small. SIZE: (5 ♂ ♂, 5 ♀ ♀) wing, ♂ 65–69 (66·6), ♀ 55–59 (56·6); tail, ♂ 45–50 (47·6), ♀ 40–44 (42·4); bill, ♂ 15–16 (15·4), ♀ 13–15 (14·4); tarsus, ♂ 25–26 (25·4), ♀ 23–24 (23·2).

TAXONOMIC NOTE: race *angolensis* treated as a full species by Dowsett and Dowsett-Lemaire (1980) because of vocal differences. However, further studies of other races show that voice varies widely throughout the range and that *angolensis* cannot be maintained as a full species (Dowsett and Dowsett-Lemaire 1993b).

Field Characters. ♂ large, robust, with heavy bill; ♀ smaller. In race *nuchalis*, broad, unstreaked pale orange-rufous hind-collar separates darker orange-rufous crown from black-streaked grey back. Back streaks heavy, crown streaks lighter. Bright head and collar produce blaze of colour visible at some distance. Black bill and large eye contrast with pale face; tail rather short, grey but often appearing dark, allowing for confusion with Aberdare Cisticola *C. aberdare*. Immature sulphur-yellow below, hind collar streaked and less bright, merging with red-brown back. Mainly highlands; song, a dry or liquid trill. Equally large Croaking Cisticola *C. natalensis* lives in similar habitat at lower elevations; it has deeper bill and bull-necked appearance, no red crown or hind-collar, song a loud 'cheeee-WONK'. For differences from Aberdare Cisticola, see that species.

Voice. Tape-recorded (45, 87, B, F, KEI, LEM, MOY). Song in Kenya (race *nuchalis*), short trill preceded by 2 introductory notes, the second of which almost runs into the trill: a dry 'tri, tri-trrrrrrrrr' or a higher, sharper 'tsi, tsi-tsirrrrrrrr'; also said to have 'a long, descending, almost tremulo series of identical notes, "tew-tew-tew-tew-tew-tew" . . . rapidly run together' (Zimmerman *et al.* 1996). In S Tanzania and N Zambia (race *awemba*) has a trill without the introductory notes, slower, with individual notes audible, lasting 1 s or less, 'chititititi . . .' or more liquid 'tiwiwi-wiwiwiwi . . .'. Calls include a high, nasal 'bzeeee', and a sharp 'tsip-tsip-tsip . . .'.

General Habits. Often a highland bird but not restricted to montane habitats. Occupies moist, usually short, bushy and wooded grassland or damp areas in open woodland, including open mountain slopes, rocky hills, highland grassland, broad-leaved tall-grass savanna, *Acacia* short-grass savanna; also rank grass, bushes and aquatic vegetation by streams, marshes and lakes; cotton fields, banana plantations; in Ethiopia also grassland with evergreen forest (Benson 1946b). In Eritrea, often alongside Winding Cisticola *C. galactotes* (Smith 1957).

Perches and preens (♂ only?) on grass tops. Spreads tail.

Food. Insects, including small grasshoppers, small beetles, driver ants. Young given grasshopper nymphs,

A

crickets, moths, caterpillars, occasional chafer grubs, small weevils and glow-worms, and once a small scarab (van Someren 1956).

Breeding Habits. In Kenya, ♀♀ apparently commoner than ♂♂: perhaps polygynous (van Someren 1956); or may be a monogamous, cooperative breeder, with broods often attended by more than 2 adults (Lynes 1930). ♂ sings from grass top or low bush, sometimes high in tree. Occasionally sings in low display-flight between perches, with fluttering wingbeats.

NEST: a flimsy ball *c.* 12 (vertical) × 10 cm with side-top entrance *c.* 4 cm diam. (**A**); in Ethiopia, occasionally a cup; of grass, often with living grass blades bent down and woven with spiderweb into canopy over or enclosing nest; unlined or lined with seed-heads, grass fibres, down, feathers and wool, lining continuing to be added during incubation; placed up to 60 cm above ground among grass or in small shrub, sometimes on ground.

EGGS: Ethiopia (7 clutches) 2–4 (3·0); Uganda, SW Kenya 2–3; Kenya highlands 3–4; E Zaïre usually 3; Angola 2–3. Ovate; smooth, very slightly to moderately glossy; variable, most commonly white, spotted with variety of reds, or speckled and scrawled pale brown with ashy violet shell-marks; others plain blue, or blue to light greenish blue blotched and speckled, mainly at big end, with reddish, or brownish black or indistinct browns and ashy purple shell-marks; rarely pale pink. SIZE: (n = 22, Ethiopia) 18–20 × 13·5–14·2 (19·4 × 14·0); (n = 26, Kenya, Uganda, Rwanda) 16·6–19·8 × 12·3–14·4 (18·1 × 12·8); (n = 2, *awemba*) av. 18·1 × 13·4; (n = 6, Angola) av. 18·0 × 13·3.

LAYING DATES: Ethiopia, May–July, Sept–Oct; Eritrea, (building June); Rwanda, Mar–May, Nov, Dec; E Zaïre, probably in both rains; SE Zaïre, Nov–Jan; E Africa: Kenya, Nairobi, Apr–June, Dec–Jan; Nandi, May–July; Region A, Apr–July; Region B, Apr–May, July, Oct, Dec; Region D, Mar–July, Sept–Jan; mainly rains, especially long rains in Region D; Zambia, Oct–Feb; Angola, Nov–Mar (possibly to May).

INCUBATION: period *c.* 14 days.

DEVELOPMENT AND CARE OF YOUNG: adults approach nest through vegetation. Young in nest 14–17 days (less if much disturbed), leave when just able to fly; fledglings drop to ground when alarmed; first brood young may visit parents' second nest (van Someren 1956).

BREEDING SUCCESS/SURVIVAL: parasitized by Klaas's and African Emerald Cuckoos *Chrysococcyx klaas* and *C. cupreus*.

References
Chapin, J. P. (1953a).
Lynes, H. (1930).
van Someren, V. G. L. (1956).

Plate 11
(Opp. p. 176)

Cisticola aberdare Lynes. Aberdare Cisticola. Cisticole des Aberdares.

Cisticola robusta aberdare Lynes, 1930. Ibis Suppl., p. 426; Aberdare Range, Kenya.

Range and Status. Endemic to central Kenya. Resident between 2300 and 3700 m on both sides of Rift Valley, at Molo, Mau Narok and Aberdare Mts, where often the commonest cisticola.

Description. ADULT ♂: forehead, crown, nape and hindneck rusty, with thick black streaks; sides of face and neck paler rusty; mantle and back very dark brown with broad buff feather edgings giving mottled appearance; rump and uppertail-coverts streaked very dark brown and buff; tail very dark brown, fringed and broadly tipped buff. Chin and throat creamy white; breast and belly buff, richer on flanks and undertail-coverts; primaries, secondaries and primary-coverts very dark brown, fringed rusty-buff; tertials, scapulars and other wing-coverts very dark brown, fringed buff; underwing-coverts and axillaries rusty-buff. Bill grey or horn; eyes reddish brown; legs and feet pink. Sexes alike. SIZE: wing, ♂ (n = 10) 69–77 (72·5), ♀ (n = 10) 62–65 (63·1); tail, ♂ (n = 9) 55–61 (58·3), ♀ (n = 10) 51–56 (53·5); bill, ♂ (n = 9) 16–18 (17·1), ♀ (n = 10) 15–16 (15·4); tarsus, ♂ (n = 10) 26–28 (27·1), ♀ (n = 10) 21–24 (23·3). WEIGHT: 1 ♂ 24, 1 juv ♂ 16.

IMMATURE: like adult but entire underparts, underwing-coverts and lores washed with lemon-yellow.

NESTLING: unknown.

Cisticola aberdare

Field Characters. See *Range*; overlaps with closely-related Stout Cisticola *C. robustus* in several places at around 2300–2400 m. Larger, longer-tailed and darker than Stout Cisticola; tail black except for buff tips, *vs.* grey with black subterminal spots in Stout, although tail of Stout sometimes *appears* dark. Dull rufous of crown and hind-collar largely obscured by heavy streaking. Habitat similar but song different (see *Voice*).

Voice. Tape-recorded (45, B, C, NOR, ZIM). Song variable: (1) single trill, ringing and rather liquid, lasting 1·5–2 s, followed by a series of short notes, 'brrip'; (2) series of *c.* 8 trilled notes, 'trrreep, trrreep, trrreep . . .' (3) 'pieu-pieu-pieu-pieu-twirrrrr-chip-chipchip'; (4) 'pieu-pieu-pieu-pieu-tewtewtewtewtew-tschweep-tchweep-tchweep-chewchewchewchewchew'. Types (3) and (4), supplied by D.A. Zimmerman, incorporate the scold note. Scold, a loud, querulous 'pieu' or 'pee-pew-pew . . .', typically 2–3 notes, up to 8 or more if agitated. Also gives a tinny 'kek-kek-kek . . .'.

General Habits. Inhabits highland grassland.

Food. Insects, including 'flies', beetles.

Breeding Habits. Nest and eggs not described.
 EGGS: (1 clutch) 2.
 LAYING DATES: E Africa: Region A, Jan, Apr, May, Aug–Nov; not well defined.

Reference
Lynes, H. (1930).

Cisticola natalensis (A. Smith). Croaking Cisticola; Striped Cisticola. Cisticole striée.

Plate 11
(Opp. p. 176)

Drymoica natalensis A. Smith, 1843. Illustrated Zoology of South Africa, Aves, pl. 80: Port Natal (= Durban).

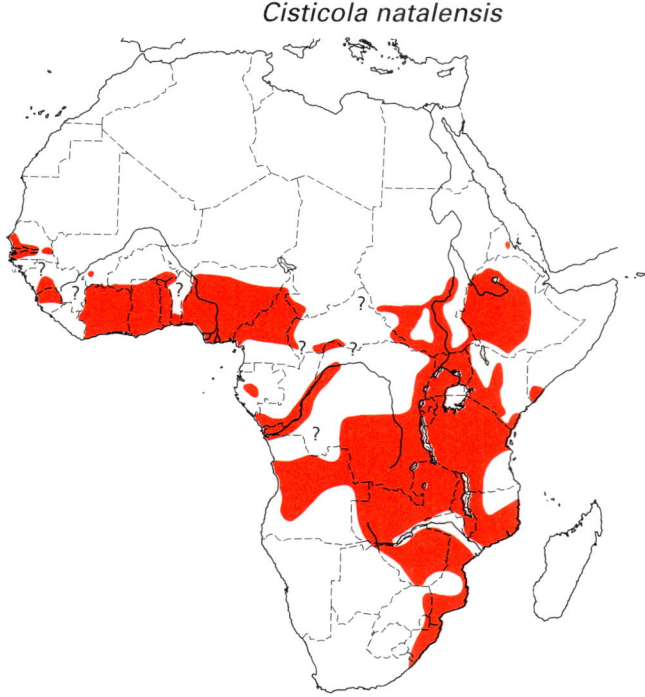

Range and Status. Endemic resident. Uncommon in Gambia along lower and middle river; few records in Senegal, near Cape Verde and near Gambian border, older records in east. Rare SW Mali. W Guinea (Conakry, Wassou, Kolenté, Labé). Frequent to common Sierra Leone. Rare Liberia (Mt Nimba). Abundant Ivory Coast. Common Ghana, Togo. Occurs S Burkina Faso and recorded Arli-Pendjari Nat. Parks (Benin-Burkina Faso) in Mar. Locally common S and central Nigeria north to *c.* 11°N. In Cameroon occurs at montane elevations, including Benue Plateau and Mandara Mts, south to forest edge. In Chad recorded near Moundou (June). In Central African Republic uncommon near Bangui and to the SW. Locally common Sudan south of 12°N. Uncommon to frequent W and central Ethiopia, 1300–2400 m. In Eritrea only at Kheren (16°N, 38°E). 3 old records in Juba, S Somalia. Locally common up to 2200 m in Uganda, S Kenya, Tanzania. Common in savannas of central Gabon (Y.-M. de Viviès, pers. comm.). Common S Congo, Cabinda and Zaïre along Congo R. from coast to 20°E. Common NE Zaïre up to 1900 m from 26°E to E border and south to 4°S and Rwanda and Burundi; and in

SE Zaïre south of 4°S, east of 20°E. Abundant in E Zambia, uncommon in W. Frequent 500–2100 m in Malaŵi but absent Nyika Plateau. Generally common NW and E Zimbabwe up to 1600 m, but sparse above *c.* 1000 m, and absent Zambezi Valley below Victoria Falls. In Mozambique, common Manica e Sofala, locally common Sul do Save (mostly south of Limpopo R.). Locally common breeder in extreme NE Botswana (Nochatsaa, Mpandamatenga), where perhaps seasonal visitor. One record E Caprivi (Namibia) (Branfield 1988). Locally common Swaziland, E Transvaal; perhaps only in breeding season in Kruger Nat. Park. Frequent to common coastal Transkei and Natal, mainly below 1000 m. In Angola recorded from N Huila District through N Bié to N Moxico (where frequent) and S Cuanza Norte, N Malange and central Lunda.

Description. *C. n. strangei* (Fraser) (including '*valida*', '*kapitensis*', '*littoralis*'): W Africa to Cabinda and W Zaïre, east to Bahr-el-Ghazal and upper White Nile (Sudan), NE Zaïre, Rwanda, Burundi, Uganda, central and S Kenya. Seasonality of plumage variable in E Africa, where some individuals have distinct breeding and non-breeding plumages, while others are intermediate between classic breeding and non-breeding (named as '*valida*', '*kapitensis*') and some populations have completely perennial dress (like non-breeding dress of seasonal individuals: '*littoralis*'); intergrades with nominate race in S Kenya and N Tanzania ('*littoralis*' and southern '*kapitensis*'). ADULT ♂ (breeding): entire upperparts, including wing-coverts, dull buffy brown, mottled with dark brown (feather centres dark brown); lores and ear-coverts buffy brown; supraloral streak buff; tail-feathers brown, with dark brown subterminal spot and buff tip; underparts buff, washed olivaceous grey on flanks and sides of breast; primaries and secondaries brown, fringed rufous; tertials dark brown, fringed buffy brown; underwing-coverts buff. Upper mandible dark brown to black, lower pink to grey; eyes yellow-brown; legs and feet pinkish brown. ADULT ♂ (non-breeding): like breeding ♂ but upperparts streaked very dark brown and rufous-buff; tail-feathers fringed rufous; underparts brighter buff, sides less grey. ADULT ♀ (breeding): like breeding ♂ but upperparts slightly brighter, buffier, less grey. ADULT ♀ (non-breeding): like non-breeding ♂. SIZE: (24 ♂♂, 18 ♀♀) wing, ♂ 63–73 (68·8), ♀ 55–61 (58·1); tail, ♂ (breeding, n = 6) 46–53 (49·0), (non-breeding, n = 5) 61–64 (62·0), (perennial '*kapitensis*', '*valida*', '*littoralis*' n = 13) 46–57 (50·0), ♀ (breeding, n = 5) 42–46 (44·0), (non-breeding, n = 4) 57–60 (58·0), (perennial '*kapitensis*', '*valida*', '*littoralis*' n = 9) 42–52 (45·8); bill, ♂ 15–17 (16·0), ♀ 13–15 (14·1); tarsus, ♂ 26–29 (26·8), ♀ 22–26 (24·1). WEIGHT: (Liberia) 2♂♂ 17, 22·8, 1♀ 16·1; (Ghana) 2 unsexed 20, 24; (Nigeria) ♂? (n = 5) 21·2–24·9 (23·1), ♀? (n = 5) 13·4–15·1 (14·2); (Central African Republic) ♂ (n = 2) 23, 26; (Kenya, Uganda) ♂ (n = 7) 16–23·0 (20·8), ♀ (n = 3) 10–15 (13·0), 1 juv. ♀ 15; (Tanzania) 2♂♂ 20, 25, 1♀ 14, 1 immature 20.
IMMATURE: like non-breeding adult but upperparts more uniform tawny brown (feather edges tawny, centres paler brown); underparts initially washed with rich lemon-yellow.
NESTLING: unknown.
C. n. tonga Lynes: White and Blue Niles, Sudan. Breeding dress unknown. Non-breeding like *strangei* but upperparts paler (feather edges sandy buff). Larger. SIZE: (5♂♂, 3♀♀) wing, ♂ 71–74 (72·2), ♀ 60–61 (60·3); tail (non-breeding), ♂ 60–70 (65·2), ♀ 56–60 (58·3); bill, ♂ 17–18 (17·4), ♀ 15–16 (15·7); tarsus, ♂ 26–28 (26·8), ♀ 23–25 (24·0).
C. n. inexpectatus Neumann: Ethiopian highlands. Plumages like *strangei* but bird larger. SIZE: wing, ♂ (n = 6) 72–77 (74·8), ♀ (n = 5) 58–63 (61·0); tail, ♂ (breeding, n = 3) 50–58 (53·0), (non-breeding, n = 3) 59–70 (64·0), ♀ (breeding, n = 2) 45, 47, (non-breeding, n = 2) 52, 62; bill, ♂ (n = 6) 15–17 (16·0), ♀ (n = 4) 14–16 (15·0); tarsus, ♂ (n = 6) 27–28 (27·9), ♀ (n = 5) 24–26 (25·0).
C. n. argenteus Reichenow: SE Somalia and S Ethiopia (Yavello) to N Kenya. Perennial dress like non-breeding *inexpectatus* or *strangei* but upperparts paler (feather centres less dark brown, edges buffy grey rather than rufous-buff); edges of primaries and secondaries pale brown, less rufous; underparts paler buff. Very large but short-tailed. SIZE: (4♂♂, 3♀♀) wing, ♂ 69–76 (73·2), ♀ 60–61 (60·7); tail, ♂ 55–56 (55·5), ♀ 49–51 (50·0); bill, ♂ 17–18 (17·5), ♀ all 15; tarsus, ♂ all 28, ♀ 24–25 (24·7).
C. n. natalensis (A. Smith) (including '*vigilax*': supposed distinguishing features not supported by material at Natural History Museum, London): interior and S Tanzania except extreme SW, SE Zambia along border from Zimbabwe to Malaŵi, S Malaŵi, Mozambique, E Zimbabwe, Swaziland, South Africa. Seasonal dress like *strangei* but, in breeding plumage, upperparts slightly darker (feather centres darker); in non-breeding dress upperparts more narrowly streaked with very dark brown, so appearing paler, more uniform. Larger. SIZE: (9♂♂, 7♀♀) wing, ♂ 70–75 (72·8), ♀ 57–62 (59·9); tail, ♂ (breeding, n = 6) 49–57 (51·3), (non-breeding, n = 3) 60–69 (63·3), ♀ (breeding, n = 4) 43–47 (45·7), (non-breeding, n =3) 57–61 (58·7); bill, ♂ 16–17 (16·2), ♀ 15–16 (15·1); tarsus, ♂ 24–30 (27·4), ♀ 24–26 (25·0). WEIGHT: (Malaŵi) ♂ (n = 4) 23·3–25·2 (24·3), ♀ (n = 3) 17·4–18·2 (17·7).
C. n. holubi (Pelzeln): extreme N Botswana, extreme W Zimbabwe and adjacent S Zambia north to Kalomo. Like nominate race but upperparts paler: in breeding dress, feather edges greyer, streaks more prominent; in non-breeding dress, feather edges sandy buff, streaks narrower.
C. n. katanga Lynes: SW Tanzania, adjacent N Malaŵi, NE Zambia, SE Zaïre and adjacent NE Angola (central Lunda, NE Moxico). Like nominate race but, in breeding dress, upperparts plainer (less contrast between feather centres and edges) and slightly richer brown; in non-breeding dress, upperparts slightly darker (feather-edges darker rufous). WEIGHT: (SE Zaïre) ♂ (n = 11) 20–29 (23·9), ♀ (n = 2) 15, 16, unsexed (n = 3) 17–24 (21·7), 1 unsexed juv. 23.
C. n. huambo Lynes: central and NW Angola. Like *katanga* but, in breeding dress, upperparts still plainer, slightly darker; head-top tawnier, contrasting more with back; in non-breeding dress upperparts slightly duller, darker, less prominently streaked (feather edges darker).

Field Characters. The bulkiest cisticola; large, with heavy body and thick neck; bill deep and culmen noticeably decurved; tail rather short, with broad black subterminal band conspicuous in flight; ♀ noticeably smaller. Overall rather dull brownish grey above, streaked on crown and hind-neck as well as back; no red hind-collar; crown concolorous with back or with just a hint of warmer colour; wing-patch warm brown to reddish. In non-breeding dress, upperparts striped blackish and rufous, underparts richer buff, tail longer. Very vocal, with wide repertoire of loud calls; typically in rank grass with scattered bushes. Stout Cisticola *C. robustus* is also large and striped, and can occur in same habitat, but has bright rufous crown and hind-collar, trilling song.

Voice. Tape-recorded (5, 45, 75, 87, 88, 91, B, C, F, KEI). Song from perched bird, series of loud, unmusical double notes with almost explosive delivery, 'dzurrr-WEEP', 'dzeee-CHOP', 'dzzeee-WURP, 'plong-KEE', 'jeee-ONK', 'cherrr-WIT', 'ti-PLOP', 't'werr-CHEE', or a frog-like croak, 'rrrrrodit'; also single notes repeated at rate of *c*. 2/s, 'grreee-grreee-grreee . . .' or prinia-like 'wip-wip-wip . . .'. Typical is a long buzzy note punctuated at beginning and end, 'chocheeeeezzzzWOP', 'd'zzzurrrrrrWOP'. Also duets, e.g. one bird goes 'whirrrrrrWIT' while the other gives single notes, 'jip-jip-jip . . .'. Notes of flight song similar but given faster (Chappuis 1974), 'plop-plop-plop-TEEplop-plop-plop-plopTEE. . . .', or guttural croaking 'kurrruk-krrk-krrk . . .'.

General Habits. Occurs in almost any open habitat; not inside forest, woodland or thick bush; most often in coarse, tall grass (0·5–1·5 m high) with scattered bushes or trees, on dry or moist ground or near water, including broad-leaved tall-grass savanna, reedbeds and edges of mangroves and other swamps; also, especially *strangei* in Kenya, in semi-arid, short, open grassland with or without scattered thorn bushes, *Acacia* short-grass savanna and open *Brachystegia* woodland, short grassland near evergreen forest, clearings and second-growth at forest edge, golf courses. In dry season in Zimbabwe, more restricted to swampy places and reedbeds (Irwin 1981). Often in valley bottoms in Natal, where enters cultivation in winter (Clancey 1964).

Singly or in small groups. Often feeds on ground. When disturbed perches on top of bush or tree, calling loudly. Mobs dogs (Jackson and Sclater 1938).

Food. Insects, mostly beetles and grasshoppers (including one so big as to fill stomach: Chapin 1953a), also alate termites, caterpillars, bugs, spiders, seeds.

Breeding Habits. Territorial: breeding territory *c*. 1 ha (Ruwet 1965). Sings in cruising flight up to 12 m above ground, with jerky wingbeats, or from tree, low bush or tall grass, flying frequently between perches. In display, circles high then plunges.

NEST: a ball 7–10 cm diam., or elliptical, with circular side entrance 25–35 mm diam.; of dry grass, with green blades woven over dome (constructed first), sometimes lined with layer of grass seed-heads and scanty inner lining of plant down over floor; placed 5–30 cm above ground in dense vegetation *c*. 30–60 cm tall, often in mixed dry and green grass on edge of patch of scrub (Roberts 1913).

EGGS: Gambia 2–3; Nigeria (4 clutches) 2–3 (2·5); Kenya, Uganda 3–4; Tanzania (3 clutches) 2–3 (2·7), one nest with a 4th egg, perhaps of cuckoo (Lynes 1934); NE Zaïre 2–3; S Zaïre/NE Angola (*katanga*) 2–3; W Angola (*huambo*, 1 clutch) 3; Zambia, Zimbabwe, Malaŵi 1 (n = 1), 2 (12), 3 (28), 4 (32), 5 (6); Mozambique (1 clutch) 3; Natal 2 (n = 3), 3 (13), 4 (16), 5 (1). Ovate or occasionally (Angola) long elliptical; smooth, matt or slightly glossy; variable in colour, white to pale blue, occasionally greenish blue; sometimes plain but usually with sparse spots or blotches of various shades of brown and ashy violet under-marks; or thinly or closely speckled with deep lilac and shades of brown; or with fine spots of lilac, purple-brown and grey, or black and grey (*katanga*); marks mostly at big end, often forming a ring. SIZE: (n = 1, Ghana) 17·5 × 12·7, (n = 5, Nigeria) 16·6–20 × 13·6–14·1 (18·7 × 13·8); (n = 1, Kenya, *strangei*) 18·0 × 13·6; (NE Zaïre) 17·5–18 × 12·5; (n = 27, S Zaïre) 17–19·8 × 12·6–14·2 (18·6 × 13·6); (n = 3, Angola, *huambo*) av. 20·3 × 12·8; (n = 33, Zimbabwe) 17·4–21 × 12·8–14·6 (18·9 × 13·8); (n = 105, southern Africa) 16–20·6 × 12·4–14·7 (18.7 × 13·7).

LAYING DATES: Gambia, Aug–Sept; Sierra Leone, July; Nigeria, May–Oct; Ethiopia, May–Oct; Rwanda, Mar–May, Oct–Dec; NE Zaïre, Apr–May, Oct; SE Zaïre, Jan–Apr (second half of rains), perhaps 1–2 months earlier in Kasai. E Africa: Uganda, Mar–early July, Sept–Dec; Tanzania, Dec–Feb; Region B, Apr–July, Sept–Oct; Region C, Dec–Feb; Region D, Mar, May; mostly confined to long rains. Malaŵi, Dec–Apr, peak Feb; Zambia, Dec–Feb (rains); Zimbabwe, Nov–Mar, peak Jan (162 records: Irwin 1981); Mozambique, Oct, Dec, Feb–Mar; Angola, Nov–Mar; Transvaal, Nov–Feb; Natal, Oct–Feb, peak Nov.

INCUBATION: by ♀.
DEVELOPMENT AND CARE OF YOUNG: young in nest *c*. 14 days.

References
Chapin, J. P. (1953a).
Lynes, H. (1930).
Vincent, A. W. (1948a).

Cisticola ruficeps (Cretzschmar). Red-pate Cisticola. Cisticole à tête rousse.

Plate 12
(Opp. p. 177)

Malurus ruficeps Cretzschmar, 1826. *In* Rüppell, Atlas zu der Reise, Vögel, p. 54, pl. 36a; Kordofan, Sudan.

Range and Status. Endemic resident. Rare in Gambia; locally common E Senegal. Widespread Mali south of 16°30′N. Uncommon in NE Ivory Coast, N Ghana and N Togo. Occurs in extreme SW Niger and Benin (no locality). Frequent to common in band across N Nigeria. In Cameroon, probably throughout N, south to Adamawa. In Chad, common in sahel zone, frequent soudan and guinea savanna zones. In Sudan, common south of *c*. 15°N, except in extreme SW and SE. Occurs NW and NE Uganda; sight record extreme NW Kenya (Lokichoggio). Locally common W Eritrea and NW Ethiopia below 900 m.

Cisticola ruficeps

Description. *C. r. guinea* Lynes: W Africa east to Adamawa (Cameroon) and N Nigeria (Maiduguri). ADULT ♂ (breeding): forehead to nape rufous brown; lores and ear-coverts paler brown; many individuals show a dark subloral spot, next to the eye; indistinct supercilium cream (occasionally absent); rest of upperparts grey-brown; tail-feathers brown, tipped buff (more broadly on outer feathers), with indistinct subterminal dark brown band, central pair rather paler, with less obvious spotting, outer margin of outermost feather buff; entire underparts creamy white, palest (near-white) on throat and centre of belly, richer buff on flanks and towards sides of breast, and washed olivaceous grey on sides of breast; primaries, secondaries and wing-coverts grey-brown, with outer fringes tawny, inner buff; tertials grey-brown, fringed buff; underwing-coverts buff. Outermost primary not notched ('blade-shaped'). Upper mandible dark brown to black, lower pinkish grey; eyes yellowish brown; legs and feet pinkish brown. ADULT ♂ (non-breeding): like breeding ♂ but head-top paler, indistinctly mottled with paler rufous; upperparts paler, indistinctly streaked with paler tawny; fringes to tail-feathers tawny; underparts deeper buff. ADULT ♀ (breeding): like breeding ♂ but upperparts average slightly paler. ADULT ♀ (non-breeding): like non-breeding ♂. SIZE: wing, ♂ (n = 9) 53–58 (55·9), ♀ (n = 7) 50–53 (51·6); tail, ♂ (breeding, n = 5) 40–41 (40·2), (non-breeding, n = 4) 43–49 (45·0), ♀ (breeding, n = 4) 33–38 (35·7), (non-breeding, n = 2) 40, 41; bill, ♂ (n = 9) 13–15 (14·0), ♀ (n = 7) 13–15 (13·6); tarsus, ♂ (n = 9) 19–20 (19·7), ♀ (n = 7) 18–20 (19·3). WEIGHT: (Ghana) unsexed (n = 4) 9·8–11 (10·2); (Nigeria) unsexed (n = 26) 8·3–12·0 (9·7); (Chad) unsexed (n = 25) 6–13 (10·0).

IMMATURE: like non-breeding adult but underparts at first lightly washed with sulphur.

NESTLING: unknown.

C. r. ruficeps (Cretzschmar): Chad to Kordofan and Bahr-el-Ghazal (Sudan). Intergrades with *guinea* in basin of L. Chad, including extreme NE Nigeria and extreme N Cameroon. Like *guinea* but in breeding dress upperparts slightly paler, sandier, in non-breeding dress upperparts more prominently streaked (feather edges paler, sandy buff).

C. r. scotopterus (Sundevall): White and Blue Nile Valleys of central and E Sudan, east to Eritrea. Like nominate but, in breeding dress, upperparts slightly greyer, less sandy; in non-breeding dress darker, head-top more rufous, rest of upperparts more heavily streaked (feather centres darker brown, edges greyer, less sandy).

C. r. mongalla Lynes: upper White Nile of S Sudan and N Uganda. Like *scotopterus* but, in breeding dress, upperparts darker; in non-breeding dress, head-top less rufous and streaked with dark brown, feather-edges of rest of upperparts duller, brownish grey, still less sandy, resulting in overall appearance duller, more streaked above.

TAXONOMIC NOTE: race *mongalla* considered a full species by Chappuis (1974), but vocalizations he ascribed to it were shown later to belong to newly-described *C. dorsti* (Chappuis and Erard 1991). Plumage and vocal differences from other races of *ruficeps* are only at the subspecies level.

Field Characters. A small cisticola of northern thorn-scrub and dry savanna, with 2 very different seasonal plumages. Crown and hindneck rufous at all times, brighter and paler when not breeding; in breeding season, back plain grey-brown, sometimes with a few faint streaks; wings and tail with dull rusty wash; in non-breeding season streaked grey-brown and pale buff above, wings and tail washed paler and brighter rufous. Outer margin of outermost tail-feathers buff, and buff tips to all except central pair. Other small cisticolas in same habitat are Rufous *C. rufus* (smaller, uniform plain rufous-brown above, no white in short tail), Foxy *C. troglodytes* (uniform bright red above), Short-winged *C. brachypterus* (brown above with dark mottling or streaking, head concolorous with back; prefers lusher habitats); and Desert *C. aridulus* (very small and short-tailed, streaked all year, no rufous cap, less white in tail).

Voice. Tape-recorded (45, CHA). Song of 2 main types, both very brief (less than 1 s): an introductory note followed by a little trill, 'tseee-tililililoo', or a short warble, ascending slightly in pitch, 'lululululu', with liquid quality reminiscent of Whistling Cisticola *C. lateralis* or Ashy Cisticola *C. cinereolus*, but higher-pitched. Calls include a sharp 'tsip' or 'tseep', and a buzzy 'dzeeeeip'.

General Habits. Inhabits dry, bushy and open wooded grasslands including scattered thornbush, acacia short-grass and semi-desert savannas, fallow land and rank grass, on plains and lower slopes of hills; also sometimes grassy swamps.

Moves through bushes and acacias and hops on ground.

Food. Insects, including Lepidoptera caterpillars.

Breeding Habits. ♂ sings from tall tree tops.

NEST: a ball with side entrance, of loosely assembled dry grass with live leaves stitched into structure, lined with plant down; slung from or woven to herb or low bush among vegetation (especially *Aristida* grass in sahel), 7–100 cm above ground.

EGGS: N Nigeria (1 clutch) 4; Chad 2; Sudan (7 clutches) 3–4 (3·9). Rounded-ovate; moderately glossy to almost matt; variable in colour, in Chad and Sudan white to turquoise-blue with variety of light or dark red marks; others (Nigeria) with reddish brown blotches and spots and greyish purple shell-marks, mainly in ring at big end. SIZE: (n = 1, Nigeria) 13 × 10·5; (n = 27, Sudan) av. 16·0 × 12·0.

LAYING DATES: Togo, Nigeria, Chad, Aug; Sudan, Aug–Sept.

DEVELOPMENT AND CARE OF YOUNG: if human approaches nest, parents fly about or circle 2–3 m above it.

Reference
Lynes, H. (1930).

Cisticola dorsti Chappuis and Erard. Dorst's Cisticola. Cisticole de Dorst.

Plate 12 (Opp. p. 177)

Cisticola dorsti Chappuis and Erard, 1991. Bull. Br. Orn. Club, vol. 111, pp. 59–70; Mokolo, Cameroon.

Range and Status. Presently known from NW Nigeria around Gusau, N Cameroon near Mokolo and S Chad around Bekao and Baïbokoum. Apparently sympatric with Red-pate Cisticola *C. ruficeps* but perhaps separated by altitude.

Description. (Provided by C. Erard.) ADULT ♂ (breeding): forehead, crown, nape and hind-neck cinnamon or rusty red-brown; lores pale buff; ear-coverts pale cinnamon, finely streaked buff; upper sides of neck pale cinnamon; lower sides of neck, mantle, back, rump and scapulars drab greyish red-brown, back somewhat more grey-brown; uppertail-coverts rusty red-brown; tail-feathers grey-brown above, greyer below, tipped greyish white, with subterminal black band which becomes narrower and incomplete on outer web of inner feathers; underparts pale buff, paler (creamy white) on chin, throat and centre of belly; sides of breast and flanks washed pale grey-brown; undertail-coverts deep rusty buff; primaries, secondaries and tertials grey-brown to dark brown, narrowly edged pale greyish buff along outer web, edges darker cinnamon on basal two-thirds of inner secondaries and tertials; wing-coverts greyish red-brown to greyish brown, narrowly edged cinnamon along outer web; axillaries, under-wing-coverts and edges of inner webs of flight feathers buff to rusty buff. Bill black or very dark brown, lower edge of lower mandible horn; eyes brown; legs and feet pinkish brown. ADULT ♂ (non-breeding) and ♀ undescribed. Field observations suggest sexes alike. SIZE: (3 ♂♂) wing, ♂ 56–58 (57·0); tail, ♂ 42–44 (43·0); bill, ♂ 13–14 (13·3); tarsus, ♂ 20–22 (21·3).

IMMATURE AND NESTLING: undescribed.

TAXONOMIC NOTE: a sibling species of *C. ruficeps*. Originally identified as *C. ruficeps mongalla* and given full species status (*C. mongalla*) by Chappuis (1974) because of vocal differences. Later found to differ from *mongalla*, and described as a new species by Chappuis and Erard (1991).

Field Characters. Very similar to Red-pate Cisticola, from which distinguished by buff vent and undertail-coverts, longer tail, and greyer, less white and narrower terminal band on underside of tail (Chappuis and Erard 1991). Habitat and song also different from Red-pate Cisticola.

Voice. Tape-recorded (45, CHA). Song a rapid, rather metallic trill lasting *c.* 1 s, sometimes with a single introductory note; quality somewhat similar to that of Tinkling Cisticola *C. rufilatus*. In areas where the bird is abundant, different individuals may suddenly all sing at the same time for a period of 20–40 seconds (Chappuis 1974). During aggressive encounters the trill becomes much slower, and is followed by 4–6 double notes, 'ti-shee, ti-shee, ti-shee'. The pitch is the same for both trills. Common call, given during aggressive encounters, a high, thin, down-slurred note.

General Habits. Inhabits grass steppe with clumps of thicket or a more continuous shrub layer; also cassava plantations and fallow. Perches 1–2 m above ground. Appears to inhabit lusher habitat and to perch lower when singing than Red-pate Cisticola.

Food and Breeding Habits. Unknown.

Reference
Chappuis, C. and Erard, C. (1991).

Cisticola dorsti

Cisticola nanus Fischer and Reichenow. Tiny Cisticola. Cisticole naine.

Cisticola nana Fischer and Reichenow, 1884. J. Orn. vol. 32, p. 260–261; Ngaruka, near Arusha, Tanganyika.

Range and Status. Endemic resident, E Africa. Once Sudan (Natoroputh Hills, 4°N, 34°E) near Kenya border (Cave and MacDonald 1955). Frequent to common in S Ethiopia at 900–1900 m. One old record Somalia, at 3°N, 41°E (Ash and Miskell 1983). Doubtful records NE Uganda (Britton 1980a). Uncommon to locally common central Kenya, NE Tanzania (south to Dar-es-Salaam and Selous Game Reserve), from sea level to 1500 m but mainly at 300–1300 m. Up to 5 birds per 10 ha in Tsavo Nat. Park (Kenya).

Cisticola nanus

Description. ADULT ♂ (breeding and non-breeding): forehead to nape rich reddish brown, fading and wearing to a paler tawny brown; ear-coverts paler brown; lores buff; rest of upperparts and wing-coverts greyish brown, broadly streaked with very dark brown (feather centres very dark brown), wearing to a more uniform dull dark brown as paler feather edgings are lost; tail-feathers dark brown, fringed and tipped buffish brown and with subterminal darker brown band; underparts buffish white, whitest on throat and centre of belly, often washed greyer on flanks; primaries, secondaries and tertials dark brown, fringed (except outer primaries) greyish brown; underwing-coverts buff and silvery white. Upper mandible dark brown, lower grey; eyes red-brown; legs and feet pinkish brown. Sexes alike. SIZE: (10♂♂, 10♀♀) wing, ♂ 45–48 (46·8), ♀ 42–45 (43·4); tail, ♂ 31–34 (32·2), ♀ 32–36 (33·7); bill, ♂ 10–11 (10·9), ♀ 11–12 (11·5); tarsus, ♂ 16–18 (16·9), ♀ 16–18 (17·0). WEIGHT: (Kenya) 1♀ 5.

IMMATURE: like adult but upperparts warmer brown (feather edges brown, not grey-brown); underparts richer buff, or washed yellowish.

NESTLING: unknown.

Field Characters. Tiny, round and stub-tailed, with bright red crown and hind-neck contrasting with grey-brown mantle; back looks plain except at close range, when some dark streaks can be seen, especially in fresh plumage; tail grey-brown with black subterminal spots, especially prominent from below; lores pale, underparts silky white; wings like back, without red. Young birds have head duller and back redder, with little contrast, and some are washed yellowish below. Gives clear, whistled song from exposed twig on treetop, then darts to top of next tree. In central Tanzania just meets Neddicky *C. fulvicapillus*, which also has rufous crown and grey-brown back, but Neddicky has narrow medium-length tail, grey wash on underparts, different voice. Overlaps with Short-winged Cisticola *C. brachypterus*, although usually occurs in drier habitat (low-altitude semi-arid thornscrub); adult Short-winged lacks red on crown, has longer tail, wispy voice; immatures have been confused, but immature Tiny is redder, with unstreaked crown and back, shorter tail.

Voice. Tape-recorded (87, B, F, GREG, McVIC). Song, a repeated series of measured notes all on one pitch; notes may be double, with first slightly down-slurred, 'chiu-wee, chiu-wee, chiu-wee . . .'; triple, 'tchew-tchew-eet' or 't'tchew-tchew-eet . . .'; or quadruple, 'tiditewi, tiditewi . . .'; numerous high 'tsick' or 'tseep' notes preceding or interspersed with song (D. A. Zimmerman, pers. comm.). Sounds like an *Apalis* except that notes are more liquid.

General Habits. Inhabits arid and semi-arid plains, dry thornbush, *Acacia* short-grass savanna, broad-leaved tall-grass savanna, tall bush and cultivation, isolated groups of large trees, open woodland, salt pans.

♂ sings from bush top or end of branch of tree up to 10 m high. Changes perch with buzzing flight. Display flight short, with quick swerves and side-slips at height of *c*. 10 m.

Food. Insects, including caterpillars and small red Orthoptera.

Breeding Habits. NEST: a ball *c*. 8 (vertical) × 7 cm, with side-top entrance *c*. 2·5 cm diam.; of dead grass blades bound with spider web to green growing blades, lined with plant down copiously at base, becoming thinner up sides; placed *c*. 15 cm above ground in a low clump of grass growing around and among base of low shrub or herbs. Lining augmented after laying.

EGGS: Kenya (5 clutches) 3–4 (3·8). Moderately glossy; pale greenish blue, dully freckled with pale purplish red, mostly near big end. SIZE: (n = 8, Kenya) av. 14·4 × 11·4.

LAYING DATES: Ethiopia, Apr–July. E Africa: Kenya, May; Region D, Jan, Apr, May.

Reference
Lynes, H. (1930).

Cisticola brachypterus (Sharpe). Short-winged Cisticola; Siffling Cisticola. Cisticole à ailes courtes.

Drymoeca brachyptera Sharpe, 1870. Ibis ser. 2, vol. 6, p. 476; Volta R., Gold Coast.

Plate 12
(Opp. p. 177)

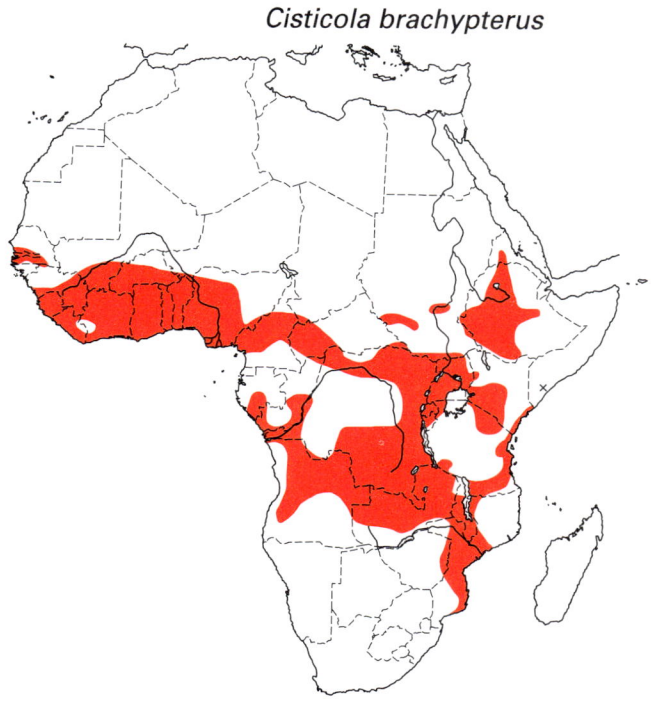

Cisticola brachypterus

Range and Status. Endemic resident or perhaps local migrant in places. Frequent Gambia (especially in Lower River) where may be a rains visitor; frequent S Senegal, Sierra Leone. Guinea in E (common Macenta) and widespread in SW, north at least to Boffa and Kindia (Demey 1995). Common resident coastal belt and guinea savanna of Liberia. Very abundant Ivory Coast. Locally common Ghana, Togo. Frequent (perhaps undertakes local movements) Mali south of 14°N. Recorded extreme SW Niger at Tapoa (July). Locally common resident S and central Nigeria. Common Cameroon north of forest zone, including highlands. W Central African Republic. In Sudan, locally frequent south of 10°N. In Ethiopia, frequent to common in highlands and S, rare in N (at 1500–2000 m). At least formerly in SW Eritrea. Two old and one modern records S Somalia, where probably resident (Douthwaite and Miskell 1991). Common S Gabon, S Congo, Cabinda, N and E Zaïre, also abundant Zaïre south of 4°S and east of 20°E. Reported in Rwanda, Burundi. Frequent to common N and central Angola, abundant in interior above 1000 m. Frequent Uganda, S Kenya, Tanzania up to 2200 m, but in Usambara Mts (NE Tanzania) mainly below 300 m, local up to 1200 m. Common central Mozambique and E Sul do Save, but sparse and local Manica e Sofala. Occurs E Zimbabwe, E and NW Zambia up to 2000 m. Common Malaŵi 500–2000 m (absent Nyika).

Description. *C. b. brachypterus* (Sharpe): W Africa to Central African Republic and Bahr-el-Ghazal (Sudan), south to N coastal Angola, N Zaïre; intergrades with *loanda* in central and S Congo. ADULT ♂ (breeding): entire upperparts dull greyish brown, slightly browner on crown, mottled with darker brown on back and scapulars, tinged rufous on rump and uppertail-coverts; lores, ear-coverts and sides of neck buffish brown; tail-feathers brown above, buffish grey below, marked with subterminal dark brown bar and buffish grey tip (marks less obvious on central pair). Underparts silvery white, washed buff on undertail-coverts and centre of breast, and brownish buff or greyish buff on sides of breast and flanks; primaries, secondaries, tertials and wing-coverts dark brown, fringed rufous-buff to rufous-brown; alula fringed buff; underwing-coverts and axillaries buff. Outermost primary notched, sharply pointed ('scimitar-shaped'). Upper mandible black, lower pale horn; eyes tawny; feet brownish pink. ADULT ♂ (non-breeding): upperparts usually more markedly mottled (feather centres darker, edges broader in fresh plumage); crown and wing-feather edgings usually more strongly rufous; breast and flanks usually deeper rusty buff. ADULT ♀: in both breeding and non-breeding plumages, upperparts often have more rufous tinge; underparts often richer buff. SIZE: (10♂♂, 9♀♀) wing, ♂ 48–52 (49.7), ♀ 42–46 (44.2); tail, ♂ breeding (n = 6) 32–37 (34.3), non-breeding (n = 4) 37–45 (40.2), ♀ breeding (n = 4) 34–42 (36.2), non-breeding (n = 5) 36–40 (38.0); bill, ♂ 12–13 (12.1), ♀ 12–13 (12.1); tarsus, ♂ 17–20 (17.9), ♀ 17–19 (18.0). WEIGHT: (Nimba, Liberia) ♂ (n = 7) 6.0–8.6 (7.5), ♀ (n = 3) 6.6–10.2 (7.8); (Nigeria) unsexed (n = 21) 6–8.8 (7.2); (Uganda) ♂ (n = 2) 8.5, 9.

IMMATURE: like non-breeding adult ♀ but upperparts often more rufous, or occasionally tinged greenish; less rufous-buff on breast, underparts washed with sulphur-yellow. Lower mandible brighter yellow; eyes dull olive-green.

NESTLING: unknown.

C. b. hypoxanthus (Hartlaub): N Uganda and adjacent NE Zaïre and SE Sudan. Like nominate but slightly less rufous on edges to greater coverts, tertials and secondaries; in non-breeding dress only, upperparts more heavily streaked. Larger. SIZE: (6 ♂♂, 5 ♀♀) wing, ♂ 50–54 (51.7), ♀ 44–48 (45.6); tail, ♂ breeding (n = 4) 34–36 (35.2), non-breeding (n = 2) 40, 45, ♀ breeding (n= 2) 33, 34, non-breeding (n = 3) 37–45 (41.3); bill, ♂ 11–13 (12.0), ♀ 11–12 (11.8); tarsus, ♂ 17–19 (18.5), ♀ 18–19 (18.6). WEIGHT: (Uganda) 1 ♂ 9.

C. b. zedlitzi (Reichenow): Eritrea and Ethiopian Plateau except in dry SE. Like *hypoxanthus* but, in breeding plumage, upperparts warmer brown, more obviously streaked; crown and wing-feather edgings more strongly rufous; underparts richer buff. In non-breeding dress upperparts darker (feather centres broader, blackish brown); underparts richer buff. Larger. SIZE: wing, ♂ (n = 8) 48–55 (52.9), ♀ (n = 5) 46–48 (46.8); tail, ♂ breeding (n = 4) 31–38 (35.2), non-breeding (n = 4) 46–49 (48.0), ♀ breeding (n = 3) 38–39 (38.7), non-breeding (n = 2) 46, 47; bill, ♂ (n = 6) 12–13 (12.7), ♀ (n = 5) 12–13 (12.4); tarsus, ♂ (n = 8) 18–20 (19.7), ♀ (n = 5) 19–20 (19.4).

C. b. katonae (Madarasz): interior of Kenya and of N Tanzania. Perennial dress like non-breeding *hypoxanthus* but slightly warmer brown above. Larger (even larger than *zedlitzi*). SIZE: wing, ♂ (n = 6) 52–57 (55.2), ♀ (n = 6) 46–51 (48.2); tail, ♂ (n = 5) 42–50 (44.4), ♀ (n = 5) 33–45 (41.2); bill, ♂ (n = 6) 12–13 (12.7), ♀ (n = 5) 11–13 (12.2); tarsus, ♂ (n = 6) 19–21 (20.2), ♀ (n = 6) 19–20 (19.5).

C. b. kericho Lynes: Kericho (SW Kenya). Like *katonae* but upperparts less strongly streaked.

C. b. reichenowi (Mearns): extreme S Somalia, coastal Kenya and N Tanzania. Perennial dress intermediate between breeding and non-breeding *hypoxanthus* in tone and degree of streaking, but crown averages slightly rustier brown. Small, size like nominate. SIZE: wing, ♂ (n = 5) 49–51 (50·0), ♀ (n = 4) 43–45 (43·7); tail, ♂ (n = 5) 33–39 (36·0), ♀ (n = 4) 33–37 (35·0); bill, ♂ (n = 5) all 12, ♀ (n = 3) 11–13 (12·0); tarsus, ♂ (n = 5) 18–19 (18·2), ♀ (n = 4) 17–18 (17·2). WEIGHT: (Kenya, Tanzania) ♂ (n = 4) 8·2–9·2 (8·6), 1 ♀ 7, immature unsexed (n = 2) 8, 8·1.

C. b. ankole Lynes: S Uganda, Rwanda, Burundi, adjacent E Zaïre and NW Tanzania (west of L. Victoria) to N end of L. Tanganyika. Perennial dress, intermediate between breeding and non-breeding *hypoxanthus* except that upperparts are more yellowish or rusty brown; streaks on upperparts slightly more pronounced than breeding *hypoxanthus*; colour of underparts like non-breeding *hypoxanthus*. Size like nominate race. SIZE: (5♂♂, 4♀♀) wing, ♂ 48–52 (49·4), ♀ 44–49 (45·5); tail, ♂ 37–39 (37·6), ♀ 33–37 (35·7); bill, ♂ 11–12 (11·4), ♀ 11–12 (11·7); tarsus, ♂ 17–18 (17·4), ♀ 17–19 (18·2).

C. b. isabellinus (Reichenow) (including '*tenebricosus*'): central Tanzania to Mozambique (south to Limpopo R.) and Zimbabwe, west to E Zambia. Non-breeding plumage like *ankole* but upperparts slightly yellower brown. Seasonal plumage change slightly more pronounced: less streaked on upperparts when breeding, and breeding ♂ duller, less yellow-brown (more like breeding *hypoxanthus*). Underparts like *ankole*, in ♂ with no obvious seasonal change, although breeding ♀ perhaps richer buff. SIZE: (6♂♂, 5♀♀) wing, ♂ 50–51 (50·5), ♀ 43–46 (44·8); tail, ♂ breeding (n = 3) 36–38 (36·7), non-breeding (n = 3) 40–43 (41·3), ♀ breeding (n = 2) 36, 38, non-breeding (n = 3) 35–42 (38·3); bill, ♂ 12–13 (12·3), ♀ 12–13 (12·2); tarsus, ♂ 17–20 (19·0), ♀ 17–19 (17·8). WEIGHT: (Malaŵi) unsexed (n = 6) 7·9–10·7 (9·2).

C. b. loanda Lynes: SE Zaïre, W Zambia, interior of Angola; intergrades with *isabellinus* in Zambia and with *ankole* in E Zaïre. Seasonal changes scarcely evident; like breeding *isabellinus* except yellower on upperparts than breeding ♂ of that race (like *ankole* or other plumages of *isabellinus*); less streaked on upperparts than *ankole* or non-breeding *isabellinus*; underparts like *isabellinus* but richer buff in non-breeding dress, and breeding ♀ noticeably brighter than *isabellinus*. WEIGHT: (SE Zaïre) ♂ (n = 10) 6–9 (7·4), ♀ (n = 2) 6, 7, unsexed (n = 2) 6, 8, juvenile ♂ (n = 2) 8, 8, juveniles unsexed (n = 2) 6, 7; (Zambia) ♂ (n = 1) 9·2, ♀ (n = 2) 8·3, 8·9, juveniles unsexed (n = 4) 8–8·5 (8·1).

Field Characters. A nondescript little bird, perhaps a candidate for the original 'little brown job'. Small, tail fairly short but not stumpy. Crown and upperparts generally plain brown in breeding season (some races with a little mottling), variably streaked dark in non-breeding season, but without the striking light-and-dark contrast of the cloudscrapers. Crown and wings virtually concolorous with rest of upperparts. Dark tail-spots present but indistinct; feather tips greyish. Underparts whitish buff. A tiny dot singing from the top of a bush or tree may well be this bird, but the feeble song gets lost if any other birds are singing. Has cruising display flight followed by swoop to earth like cloudscrapers but without vocalizations. Habitat not distinctive; occurs in all kinds of bushy and grassy country. In S-central and southern Africa might be confused with Neddicky *C. fulvicapillus*, but posture more hunched, tail shorter and appears broader, underparts buff-white, not greyish, no contrasting rufous crown; also, Neddicky has loud voice. In W and N-central Africa occurs with Rufous, Foxy and Tiny Cisticolas *C. rufus*, *C. troglodytes* and *C. nanus*; for differences, see those species.

Voice. Tape-recorded (45, 87, 91, B, C, F, KEI, PAY). Song, a series of notes, thin and high-pitched but clear and penetrating, 'tsip, tsup, tsu, tsik, tsep, tsup . . .'. Principal element is phrase of 3 descending notes, frequently repeated, but additional notes often added between phrases to form continuous song, which may be quite rapid. Quality of notes described by Lynes (1930) as like whistling through one's teeth or 'siffling', but while notes seem somewhat weak at a distance, at close range they sound quite sharp. Form of song varies throughout range, and variation may be clinal (Chappuis 1974). In Zambian dialect east of Luangwa valley, song ends with a trill (R. Stjernstedt, tape and notes). In Kenya (race *katonae*), a repeated 'su-SEET, su-SEET, su-SEET,' from low sapling (D. A. Zimmerman, pers. comm.). Call, 5 or more repeated notes, 'chip chip chip-ey chip', sometimes in display flight (Maclean 1993); alarm, given when guarding fledglings, a wispy squeak or 'chick'.

General Habits. Inhabits thin bush and scrub dotted with trees, well-wooded grassland and savanna woodland, palm scrub, cultivation and second-growth, short- or tall-grass flats, grassy clearings in woodland, thick scrub at forest edge, gallery woodland, land recently cleared and burnt for farms. In Usambara Mts (Tanzania) occurs on ridge tops with sparse trees and in sub-tropical savannas. In highland Uganda in marshes, among lobelias, tussocks of grass and sedge; also enters lake-side grassland and reed-beds in Angola and Zambia. Habitat usually contains elevated perches, e.g. dead trees, for song-posts, although also sings from high grass-stems.

Solitary, or in pairs or family parties of 3–4 birds. Unobtrusive; forages low in grass, but when disturbed, flies up into bush or tree.

In Zaria, Nigeria, 15 caught between 30 Jan and 11 Mar, only 6 in rest of year, but no other evidence of migration (Fry 1971).

Food. Insects, including termites, small grasshoppers, small beetles, pentatomid bugs. Young given moth caterpillars.

Breeding Habits. During breeding season, ♂ spends much of day on high, prominent perch in tree, bush or tall grass-head, from which he sings, occasionally rising with whirring wingbeats and momentary breaks and dips, to display. Cruises c. 30–60 m up, often in slow spiral or erratic and jerky; silent (or very feeble song?), tail slightly spread and flicked; ends with dive accompanied by swishing sound, followed by upswerve, repeated once or twice, with a final drop to grass or an upswerve to a tree-top. High cruise and sudden dives present in displays of *ankole* and *reichenowi* but, as far as known, not in *hypoxanthus* and *katonae* (Lynes 1937). Breeding ♀ keeps to cover.

NEST: oval (ext depth c. 80–100, ext diam. c. 60) or spherical, with 25 mm diam. side or side-top entrance; made of dry grass, lined with finer grass then plant down, especially on floor; secured to surrounding vegetation with gossamer; green blades often bent and secured over top. Hidden, up to 1 m above ground, in grass tussock or (Usambaras, Tanzania) 3 m up in stunted tree.

EGGS: Nigeria (5 clutches) 3; Ethiopia 2–3; interior Kenya, Tanzania (*katonae*) 3; coastal Kenya, Tanzania (*reichenowi*) 4; Malaŵi 2 (n = 6), 3 (11), 4 (1); Zaïre 2–3; Angola (4 clutches) 2–3; southern Africa 3–4. Ovate; slightly glossy; white or pinkish white to pale blue or greenish blue, marked with a variety of red or orange-brown spots, blotches or streaks, evenly distributed or denser at big end or forming a circle there, often over violet or grey shell-marks; or cream with a few scattered blotches of pale yellow-brown. SIZE: (n = 21, Nigeria) 14·2–15·9 × 11·0–11·7 (15·0 × 11·4); (n = 5, Ethiopia) 15·7–17 × 11·5–12·7 (16·3 × 12·2); (n = 5, interior Kenya, *katonae*) av. 16·2 × 11·7; (n = 4, coastal Kenya, *reichenowi*) av. 14·4 × 11·5, (n = 8, SW Zaïre) 15·2–16·3 × 11–12·1 (15·7 × 11·7); (n = 12, Malaŵi) av. 15·4 × 11·7; (n = 18, Angola) av. 15·7 × 11·6.

LAYING DATES: Gambia, July–Sept; S Senegal, Nov (♂ song July–Nov); Sierra Leone, July–Sept; Liberia, Aug–Sept; Ghana, Feb, July, Sept, Dec; Togo, Aug; Nigeria, May–Sept; Sudan, July; Ethiopia, Apr–Sept (possibly Oct); S Cameroon (gonads enlarged Aug–Nov); Congo, Nov; Rwanda, Sept, Dec, Jan, Mar–May; N Zaïre, May–Dec; SE Zaïre, Dec–May (rains); SW Zaïre, Angola, Oct–Apr (or May). E Africa Region A, May–June; Region D, Nov–June; Region E, July–Aug; mostly rains but in Region E at end of prolonged rainy season. Malaŵi, Dec–Apr (peak Jan–Feb); Zimbabwe, Nov–Mar; Mozambique, Jan, May.

INCUBATION: by ♀, which continues to add nest lining; period 14 days.

DEVELOPMENT AND CARE OF YOUNG: young in nest *c*. 17 days.

References
Chapin, J. P. (1953a).
Lynes, H. (1930, 1932).
Serle, W. (1940, 1943b).
van Someren, V. G. L. (1956).
Vincent, A. W. (1948a).

Cisticola rufus (Fraser). Rufous Cisticola. Cisticole rousse.

Plate 12
(Opp. p. 177)

Drymoica rufa Fraser, 1843. Proc. Zool. Soc. Lond. p. 17; Iddah, lower Niger R, Nigeria.

Forms a superspecies with *C. troglodytes*.

Range and Status. Endemic resident, W Africa. Rare Gambia, coastal and SE Senegal. Frequent south of 14°N in Niger inundation zone of Mali. Recorded in Burkina Faso. Frequent N Togo. Uncommon (perhaps often overlooked) in N Ghana and Nigeria, north to Maiduguri and east to Adamawa. In N Cameroon, from Ribao-Banyo to Benue Plateau and Mandara. In Chad, common in soudan savanna zone of south. Frequent in W Central African Republic south to Bozoum.

Description. ADULT ♂ (breeding): forehead to mantle warm brown; lores buff; ear-coverts buffy brown; rump and upper-tail-coverts rufous brown; tail-feathers dark brown, tipped buffy brown and with obscure (sometimes virtually absent) darker brown subterminal band; underparts buff, palest (near-white) on throat and centre of belly, richest rufous-buff on flanks, sides of breast and undertail-coverts; primaries, secondaries, tertials and wing-coverts brown, with outer fringes rufous, inner buff; underwing-coverts buff. Outermost primary not notched ('blade-shaped'). Upper mandible dark brown, lower pinkish brown to grey; eyes light brown; legs and feet pinkish brown. ADULT ♂ (non-breeding): upperparts average very slightly paler. Sexes alike. SIZE: wing, ♂ (n = 9) 49–52 (50·2), ♀ (n = 4) 43–46 (44·7); tail, ♂ (breeding, n = 5) 32–35 (33·6), (non-breeding, n = 4) 33–38 (36·2), ♀ (non-breeding, n = 2) 32, 35; bill, ♂ (n = 9) 12–13 (12·3), ♀ (n = 4) 11–13 (12·0); tarsus, ♂ (n = 9) 17–19 (18·0), ♀ (n = 4) 17–18 (17·7). WEIGHT: (Ghana) unsexed (n = 8) 7–8 (7·6).

IMMATURE. Like non-breeding adult but underparts washed with sulphur.

NESTLING: unknown.

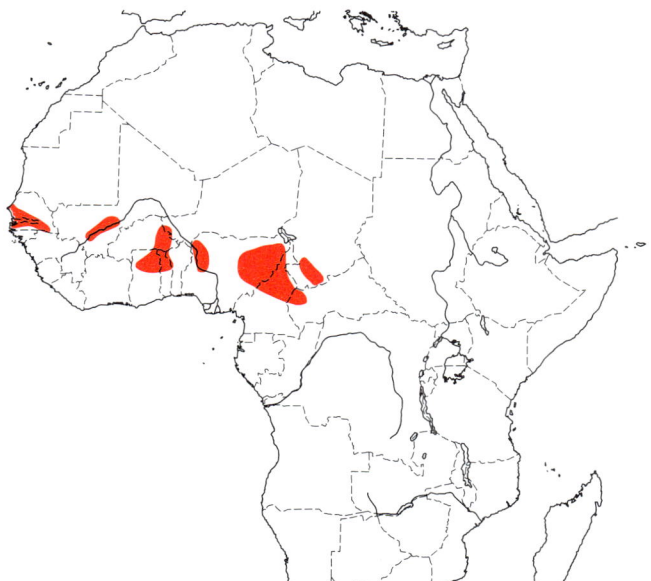

Cisticola rufus

Field Characters. A tiny, short-tailed inhabitant of W African Sahel and savanna. Plain reddish brown above, becoming brighter on rump; small reddish patch in centre of brown wing; white below, variably washed reddish buff, especially on breast and flanks. Barely meets (in Central African Republic/Chad) Foxy Cisticola *C. troglodytes*, which is bright red above but otherwise similar in size, voice and habitat. More likely to be confused with Short-winged Cisticola *C. brachypterus*, found in same habitat and with extremely similar song; distinguished from adult by paler, redder, unstreaked upperparts (Short-winged has some dark streaking in non-breeding plumage, a little mottling in breeding plumage); immature Short-winged is somewhat redder above than adult, especially on rump, but is still darker than Rufous, with some mottling, and underparts are washed with yellow. Red-pate Cisticola *C. ruficeps* has plain upperparts in breeding plumage but rufous cap contrasting with grey-brown back.

Voice. Tape-recorded (45, CHA, MOR). Song, both in the air and from perch, of 2 forms: (1) series of sharp 'tsik' notes on one pitch, intervals between notes varying slightly; (2) similar notes in rapid, rolling series, descending and ascending 2 or 3 notes, 'tsi-tse-tsu-tsi-tse-tsu-tse-tsi-tsu-tsi-tsu . . .'. Remarkably similar to the 2 song types of Short-winged Cisticola (see Chappuis 1974).

General Habits. Inhabits semi-arid and riparian bush, scrub, low trees and swamp, grassy dry watercourses.

Food. Insects.

Breeding Habits. ♂ sings from tree top.
NEST: of grass, lined with grass flower-heads; placed between 2 leaves sewn together.
EGGS: white with red-brown spots, or greenish blue with red-brown spots and pinkish grey shell-marks. SIZE: (n = 4) 15·0–16·4 × 11·5–12·3 (15·4 × 11·8).
LAYING DATES: Ghana, June, Oct.

Reference
Lynes, H. (1930).

Plate 12
(Opp. p. 177)

Cisticola troglodytes (Antinori). Foxy Cisticola; Sandy Cisticola. Cisticole russule.

Drymoica troglodytes Antinori, 1864. Cat. Desc. Coll. Uccelli, p. 38; Djur, Bahr-el-Ghazal, Sudan.

Forms a superspecies with *C. rufus*.

Range and Status. Endemic, usually resident, central and E Africa. One record (erroneous?) extreme E Mali, east of Azzawakh, Aug (Lamarche 1981). Recorded S Chad. Central African Republic: recorded in W, resident in Manovo-Gounda-St Floris Nat. Park. Locally common in S Sudan, probably resident, but recorded in S and central Darfur only Mar–Aug. Uncommon and local in W and S Ethiopia below 1800 m. Locally common N and NE Uganda, uncommon NW Kenya (1500–2200 m). In Zaïre recorded only along shore of L. Albert.

Cisticola troglodytes

Description. *C. t. troglodytes* (Antinori): Central African Republic to White Nile and W Kenya. ADULT ♂: entire upperparts, wing-coverts and tertials orange-brown to red-brown; lores and ear-coverts buffier brown; tail-feathers dark reddish brown above, greyer below, fringed and tipped red-brown and with obscure dark brown subterminal band (least marked on central pair); chin, throat and centre of belly pale buff, near-white; rest of underparts and underwing rich rufous-buff; primaries and secondaries dull dark brown, fringed orange-brown to red-brown. Outermost primary notched, sharply pointed ('scimitar-shaped'). Upper mandible brown, lower pinkish brown; eyes grey or light brown; legs and feet pinkish brown. Sexes alike. SIZE: wing, ♂ (n = 15) 45–48 (47·1), ♀ (n = 10) 43–46 (44·3); tail, ♂ (breeding, n = 10) 30–33 (31·6), (non-breeding, n = 5) 33–36 (34·4), ♀ (breeding, n = 5) 32–33 (32·2), (non-breeding, n = 5) 32–35 (33·8); bill, ♂ (n = 13) 10–12 (11·4), ♀ (n = 10) 11–12 (11·7); tarsus, ♂ (n = 15) 16–18 (16·9), ♀ (n = 8) 16–18 (17·4).

IMMATURE: like adult but underparts strongly washed with sulphur.

NESTLING: unknown.

C. t. ferrugineus (Heuglin): W Ethiopia and adjacent E Sudan (Blue Nile). Tail-feathers paler, concolorous or only slightly darker than rest of upperparts; underparts paler buff. Eyes red-brown. Larger. SIZE: (3 ♂♂, 1♀) wing, ♂ 48–51 (49·7), ♀ 45; tail, ♂ 35–40 (37·0), ♀ 33; bill, ♂ 12–13 (12·7), ♀ 12; tarsus, ♂ 17–19 (18·0), ♀ 18.

Field Characters. A tiny, short-tailed species of wooded grassland and thornbush north of equator from Chad to Ethiopia and N Kenya. Uniform bright foxy rufous above without streaking; tail similar (E Sudan, Ethiopia), or black and rufous (feathers black with broad rufous edges); below, buff with reddish sides of breast and flanks. Much brighter rufous than Rufous Cisticola *C. rufus*, which is only pale red-brown; just meets Rufous in Chad and replaces it to the east. Overlaps with Short-winged Cisticola *C. brachypterus* (darker and duller, brown with some dark mottling or streaking) and Red-pate Cisticola *C. ruficeps* (only crown red, contrasting with grey-brown back). Juv. Whistling Cisticola *C. lateralis* is reddish and has grey eyes like adult Foxy but is larger, with yellow-ochre bill (Zimmerman *et al.* 1996).

Voice. Tape-recorded (ZIM). Common call a soft 'tsit, tsit-tsit'; a rapid series of similar wispy notes is frequently delivered song-like from top of low savanna tree (e.g. *Combretum*, *Erythrina* on Mt Elgon) (D.A. Zimmerman, pers. comm.).

General Habits. Inhabits wooded grasslands, including broad-leaved tall-grass savanna with scattered trees (e.g. *Combretum*), *Acacia* short-grass savanna, thornbush and semi-desert savanna; also *Syzygium-Adina* riparian woodland and floodplains.

Food. Small black beetles.

Breeding Habits. ♂ sings from tree-top. Nest not described.

EGGS: white, cream, pale blue-grey, grey-green, bluish green or pale blue, spotted evenly, or in halo or generally more heavily near big end with red-brown or pink, violet or black.

LAYING DATES: Sudan, Apr, July, Sept; Ethiopia, probably June–Oct.

References
Lynes. H. (1930).
Urban, E. K. and Brown, L. H. (1971).

Cisticola fulvicapillus **(Vieillot). Neddicky; Piping Cisticola. Cisticole à couronne rousse.**

Plate 12
(Opp. p. 177)

Sylvia fulvicapilla Vieillot, 1817. Nouv. Dict. Hist. Nat., nouv. éd. 11, p. 217; based on 'Le Rousse-tête' of Levaillant, 1802, Hist. Nat. Ois. Afrique, 3, p. 69, pl. 124, figs. 1–2; Camdeboo ex Levaillant = Graaf-Reinet, eastern Cape Province, South Africa.

Forms a superspecies with *C. angusticaudus*.

Range and Status. Endemic resident. Zaïre (Kinshasa and Dilolo), and plateau of Angola south to NE Namibia; Zambia in W, S and NE to Malaŵi, S and E Tanzania from Matengo Highlands and Rovuma R. north to Kilosa, Soga and Dar es Salaam; south through Mozambique, Zimbabwe, N and E Botswana; South Africa to Transvaal, Natal and W Orange Free State to NE and S Cape Province. Common. Breeding density in Transvaal (Nylsvlei) 1 pair/5 ha in acacia, 1 pair/10 ha in broad-leaved woodland (Tarboton *et al.* 1987).

Description. *C. f. dexter* Clancey: plateau of Zimbabwe, Transvaal highveld to 26°S, and extreme E Botswana south to Kanye. ADULT ♂ (breeding): forehead, crown and nape chestnut; lores pale buff, cheeks and ear-coverts dusky buff, sides of neck washed greyish. Back to rump and uppertail-coverts uniform dark brown washed rufous. Tail graduated, outermost feather (T6) *c.* 10 shorter than T1, dark brown with indistinct blackish subterminal bar or spot (usually obsolete on T1) and pale buffy or brownish tips. Chin, throat and breast to undertail-coverts off-white, often with pale buffy or greyish wash extending to somewhat darker flanks; thighs pale tawny-buff. Wings earth-brown, inner primaries and secondaries with pale buff outer margins; inner webs of primaries and secondaries tawny-buff, scapulars washed rufous; shoulder, underwing-coverts and axillaries off-white, washed pale buff; outermost primary blade-shaped (not ending in acute point), exposed for *c.* 15–17 mm; P5 to P8 equal in length, P9 relatively broad and lacking emargination on outer web. ADULT ♂ (non-breeding): more reddish brown above, contrast between crown and upper back reduced. Bill dark brown to horn colour, lower mandible flesh, palate blackish; eyes pale brown or hazel; legs and feet flesh. Sexes alike. SIZE: (108 ♂♂, 48 ♀♀) wing, ♂ 47–54 (50·7), ♀ 45–49 (46·9); tail, ♂ 37–44 (41·0), ♀ 35–43 (37·9); bill, ♂ 11–12·5 (11·6), ♀ 11–12·5 (11·3); tarsus, ♂ 17–18·5 (17·9), ♀ 17–18 (17·5); hind claw, ♂♀ 4·5–6 (5·2). WEIGHT: ♂ (n = 26) 6–10·6 (8·6); ♀ (n = 16) 7–9·2 (7·9).

IMMATURE: more uniformly rusty red above and on tail, whiter below; grey-bellied forms are duller above, washed browner below.

NESTLING: hatches naked.

Cisticola fulvicapillus

C. f. dispar Sousa: W Zaïre, central plateau of Angola (except Huila) and NW Zambia (Mwinilunga and Balovale) south to Lungwebungu R. Darker above than *muelleri* or *dexter*, chestnut crown not so clearly demarcated from mantle, tail blacker (black subterminal spot usually obscured); colder below, greyer rather than buff; finer billed, ♂ (n = 14) 10·5–11 (10·8); immature with underparts olive-yellow.

C. f. muelleri Alexander: NW Zambia from Kabompo east to Malaŵi, E and S Tanzania, Mozambique south to Save R. and NE Zimbabwe. Paler above than *dispar*, nearer *dexter*, but crown darker, more chestnut; mantle and scapulars darker, more brownish olive, with rufous wash; below like *dexter*.

C. f. hallae Benson: S Angola (north to Huila), SW Zambia and NW Zimbabwe, N Botswana and N Namibia (Damaraland). Paler above than *muelleri* or *dexter*, crown lighter, more yellowish tawny, sharply demarcated from paler, greyer mantle; whiter, less washed with buff below.

C. f. ruficapillus (Smith): Transvaal highveld south of 26°S, W Orange Free State and N and NE Cape Province on Vaal and Orange rivers. Nearest to *dexter* but with slight blue-grey wash below.

C. f. lebombo (Roberts): Mozambique south of Save R., E Transvaal lowveld, N Zululand, Lebombo Mts, Swaziland and W Natal. Crown contrasting with back as in *dexter* and *fulvicapillus*; throat to flanks blue-grey, centre of belly white.

C. f. fulvicapillus (Vieillot): interior of E Cape Province east of Great Fish R., to Drakensberg escarpment, Natal Drakensberg and W Lesotho. Underparts blue-grey; subterminal tail-spots indistinct above, lacking below.

C. f. dumicola Clancey: W Zululand, Natal (except W), Griqualand East and coastal Transkei to S and E Cape Province as far as Knysna and George. Back darker, red crown not demarcated from rest of upperparts; like *fulvicapillus* below.

C. f. silberbaueri (Roberts): winter rainfall area of SW Cape Province. Darker above than *dumicola* and darker blue-grey below.

Field Characters. A small, slender grass-warbler of bush and woodland, common in southern Africa. When flushed from grass, flies up into trees with jerky flight and agitated calling. Breeding adult has bright rufous cap contrasting with plain brown back, underparts greyish in S of range, off-white to pale buff in N. Immature uniform reddish brown above including tail, underparts washed brown or yellowish buff. Small size and unstreaked upperparts distinguish it from all except Long-tailed Cisticola *C. angusticaudus* (q.v. for differences), and Short-winged Cisticola *C. brachypterus*, with which it occurs in miombo; Short-winged has uniform warm brown upperparts with no contrasting cap, shorter tail, different song.

Voice. Tape-recorded (11, 20, 45, 87, 91, B, C, F). Song monotonous, ventriloquial, not loud, a series of 10–30 single, pure, ringing or piping notes given at steady speed on same pitch; rate varies between but not within songs, from 2 to 3 notes per second. Notes sometimes more clipped, less ringing, sounding like an *Apalis*. Distinct song-dialects recognizable, especially in Zambia and S Tanzania. Call, hard tikking notes run into a trill, or followed by a burry, nasal 'chrreep'.

General Habits. Inhabits grass and scrub among bushes and trees, in N of range most typically in miombo woodland, especially on edges and in natural clearings; adapted to a very wide variety of woody vegetation including mopane and acacia; also frequents old cultivation and regenerating growth.

Usually in pairs or small family parties of 2–5; joins mixed woodland bird parties for brief periods but probably does not leave territory. Spends most of time hidden in long grass, foraging low down, probably largely on ground; climbs up and down rose bush gleaning off aphids. When disturbed flies up noisily, taking refuge in surrounding bushes and trees, flicking tail, moving jerkily about, giving alarm call. ♂ sings from top of bush or tree or similar vantage point, including telephone wire, keeping up monotonous little song throughout greater part of day; sings with little or no movement, appearing thin, with tail tightly closed.

Food. Insects: grasshoppers, moths, mantids, termites, aphids and caterpillars.

Breeding Habits. Territorial; territories vigorously advertised from song-post, but no aerial display.

NEST: a roughly oval or pear-shaped ball, with side-top entrance; of dry grass and some plant down with a few binding spider webs, with exterior of coarser grass blades, especially at base; cup neatly lined with downy plant material; open cup nest also known (Steyn 1966c); ext. depth 80–100, diam. 65, entrance diam. 35–40. Placed low down, rarely above 25 cm, sometimes almost at ground level, usually among scrub or second growth, in grass or thorny shrub or similar cover mixed with grasses. Apparently built by ♀, while ♂ sings nearby; lining continues to be added during incubation.

EGGS: 3–4 (rarely 5). Ground colour highly variable, from white or pinkish white to bluish, greenish, light turquoise blue or greenish blue; plain or marked with scattered freckles and spots of pinkish brown or rust-red and violet-grey, mainly around thick end, often forming a ring. SIZE: (n = 177) 13·3–16·9 × 10·5–12·6 (15·2 × 11·5).

INCUBATION: period 12–15 days (av. 13).

DEVELOPMENT AND CARE OF YOUNG: nestling period 12–14 days (av. 13). Alarm note uttered at nest only after 1st egg laid; threat display at nest at day 6 after young hatch, nearly always thereafter when intruder near.

LAYING DATES: Zaïre, Oct–Dec; Angola, Oct–Mar; Zambia, Oct, Dec–Feb; Malaŵi, Nov–Mar; Zimbabwe, Oct–Mar (Oct 5, Nov 33, Dec 63, Jan 40, Feb 23, Mar 12 clutches); Mozambique, Oct; Botswana, Oct–Jan; South Africa: Transvaal, Oct–Mar, May (Oct 9, Nov 24, Dec 27, Jan 16, Feb 8, Mar 2, May 1 clutches); Natal, Oct–Feb (Oct 16, Nov 23, Dec 8, Jan 5, Feb 1 clutches); Karoo, Sept–Nov, Feb; E Cape, July–Feb (July 2, Aug 1, Sept 19, Oct 43, Nov 25, Dec 14, Jan 7, Feb 2 clutches); SW Cape, Sept–Dec (Sept 15, Oct 9, Nov 3, Dec 4 clutches).

BREEDING SUCCESS/SURVIVAL: parasitized by Klaas's Cuckoo *Chrysococcyx klaas*, Wahlberg's Honeybird *Prodotiscus regulus* and Parasitic Weaver *Anomalospiza imberbis*.

Reference
Vincent, A. W. (1948a).

Cisticola angusticaudus Reichenow. Long-tailed Cisticola; Tabora Cisticola. Cisticole à queue fine.

Plate 12
(Opp. p. 177)

Cisticola augusticauda Reichenow, 1891. J. Orn. 39, p. 69 (corrected to *angusticauda* p. 440); Gonda (= Igonda, Ugunda), Tabora district, Tanganyika.

Forms a superspecies with *C. fulvicapillus*.

Range and Status. Endemic resident in miombo and similar woodlands. SW Kenya west of Rift (Muhoroni, Kendu Bay, Ruma Nat. Park east to Lolgorien and Mara Game Reserve); N and W Tanzania from Serengeti Nat. Park south to Iringa Highlands (L. Burigi, Tabora, Kondoa Irangi, Mpanda, Kitungulu and Katavi Plain); Rwanda (Akagera Nat. Park); SE Zaïre (Marungu to Lubumbashi) and N and central Zambia west of about 32°E to Ndola, Solwezi and Kasempa to E Mwinilunga (East Lunga R.), in places to c. 15°S. Common to local and uncommon.

Cisticola angusticaudus

Description. ADULT ♂ (breeding): forehead to hindcrown brick red, contrasting with grey-brown nape; cheeks, lores and ear-coverts buff. Back to rump and uppertail-coverts grey-brown. Tail-feathers long, narrow, strongly graduated, outermost (T6) c. 20 mm short of longest (T1); blackish brown with pale tips and distinct subterminal black spot, mostly on inner webs. Chin, throat and whole of underparts buff (warmer than Neddicky *C. fulvicapillus muelleri*). Wings earth-brown, rachis of primaries pale brown (as in Neddicky), outermost primary narrowed, acute (c. 12 mm exposed), P5 to P8 equal in length, P6 to P9 narrowed on inner webs, outer webs emarginated; shoulder, underwing-coverts and axillaries pale buff. ADULT ♂ (non-breeding): washed with tawny above, red of crown merging into back, closely resembling *C. f. muelleri*. Bill medium brown, lower mandible flesh (culmen horn in study skins), mouth flesh, palate blackish; eyes hazel-brown; legs and feet bright yellowish flesh. Sexes alike. SIZE: (50 ♂♂, 30 ♀♀) wing, ♂ 46–51 (48·1), ♀ 41–49 (45·4); tail, ♂ 48–61 (53·0), ♀ 43–53 (47·4); bill, ♂ 9·5–10·5 (10·0), ♀ 9–10·5 (9·9); tarsus, ♂ 15·5–17 (16·2), ♀ 16–17 (16·5). Tarsus less robust, feet more slender than Neddicky; hind claw ♂♀ 4–5·5 (4·5).

IMMATURE: crown and mantle uniform reddish brown; below washed yellowish olive; tail blacker.

NESTLING: hatches naked; black tongue-spots said to fade away early (Lynes 1936).

TAXONOMIC NOTE: the specific status of *C. angusticaudus* remains controversial. It is treated as conspecific with *C. fulvicapillus* by White (1962), Hall and Moreau (1970) and Dowsett and Forbes-Watson (1993), and as a distinct species by Traylor (*in* Watson *et al.* 1986). It hybridizes with *C. f. muelleri* in E and S Zambia, but in W where it meets *C. f. dispar/muelleri*, latter confined to Kalahari Sand while *C. angusticaudus* restricted to taller miombo. Their habitat, behaviour and ecology are similar, but there are differences in wing and tail structure, and *C. angusticaudus* has distinctive breeding dress. *C. fulvicapillus* is highly polytypic while *C. angusticaudus* is monotypic. They share song-types and both have distinct dialects, but in central and E Zambia their voices differ where they meet. For the moment we prefer to treat them as distinct species, pending further studies.

Field Characters. A small, slim bird of grass in miombo woodland, with long, slender, dark tail. Extremely similar to Neddicky C. *fulvicapillus*, with similar ecology and voice, and same habit of flushing from grass into trees and calling in agitated manner with cocked tail; differs in having longer and blacker tail, and in breeding plumage greyer back contrasts with brick-red crown. The 2 are largely allopatric, so identification is a problem only where their ranges meet in N-central Zambia.

Voice. Tape-recorded (86, 87, B). Song extremely similar to that of Neddicky, a series of ventriloquial ringing notes on 1 pitch; length of series varies from 15–20 to nearly 50 notes. Some songs faster than Neddicky: 2 from Zambia were at rate of 3 notes per s (47 notes/12 s and 47 notes/15 s). Local dialects differ in pitch and length of note and structure, some involving double notes; song differs from that of Neddicky where their ranges overlap. Purring alarm trill also very similar to that of Neddicky.

General Habits. Frequents small clearings and open places with long grass under canopy in well-developed miombo woodland; often more local and patchily distributed than Neddicky, perhaps absent in some areas; in Kenya, N Tanzania and Rwanda occurs in acacia country, creeping about low down in bushes or on the ground.

Occurs in pairs or small family parties of 3–4. Forages low down in grass or on ground; rises suddenly out of grass when disturbed, flying into surrounding bushes and small trees, calling noisily and cocking tail in agitated manner. Reported to have noisy, rattling flight, not seen by other observers (Böhm, quoted by Lynes 1936).

Food. Insects.

Breeding Habits. Territorial, probably all year.
NEST: ball type, with side-top entrance, composed of grasses and lined with plant down; placed low down at base of small shrub or in grass tuft; indistinguishable from that of Neddicky.
EGGS: 3–4. White, with small spots and freckles of Indian red.
LAYING DATES: Zaïre, Jan–Feb; Rwanda, Nov–Dec, Apr–May; Zambia, Dec–Mar.
BREEDING SUCCESS/SURVIVAL: parasitized by Wahlberg's Honeybird *Prodotiscus regulus*.

Plate 12
(Opp. p. 177)

Cisticola melanurus (Cabanis). Black-tailed Cisticola; Slender-tailed Cisticola. Cisticole à queue noire.

Dryodromas melanurus Cabanis, 1882. J. Orn. 30, p. 349; Angola. Restricted to Mona Quimbundo, Lunda Province.

Range and Status. Endemic resident in well-developed miombo woodland. NE Angola in Malanje and Lunda (Cafunfo, Cuango; Mona Quimbundo, 15 km southwest of Cacolo and Cazoa between Caxia and Rio Cassai) and sight record from Malanje (Kangandala Nat. Park: Dean *et al.* 1988); SE Zaïre at Gungu in Kwango (Pay Kikwanga) and Shaba (not uncommon upper Lufupa R. and Nasondoye; 7 specimens). Rare and extremely localized; only 14 specimens known.

Cisticola melanurus

Description. ADULT ♂: forehead and crown rich chestnut-red, merging without marked contrast into reddish brown back and sides of neck; lores, cheeks and ear-coverts buffish. Back and rump to uppertail-coverts reddish brown. Tail-feathers narrow, jet black, tipped grey below (except T1), without blackish subterminal spot of Long-tailed Cisticola *C. angusticaudus*, narrowly margined brown above, silvery grey below; outermost (T6) *c.* 24 mm shorter than T1, with more extensively grey tip and contrasting off-white outer web. Chin, throat and entire underparts colder and whiter, less buffy (bases of feather tracts blackish rather than brown-tinted), compared with Long-tailed and Neddicky *C. fulvicapillus dispar*. Wings markedly rounded, greyish brown, rachis of 5 outer primaries broadened and thickened, double the width of those of *C. angusticaudus* or *C. fulvicapillus*, glossy black (hard, strong and dense), outermost primary needle-shaped, acute, projects *c.* 6 mm beyond coverts, P7 shorter than P5 and P6;

shoulder, underwing-coverts and axillaries off-white. Sexes alike, but ♀ has outer web of outermost tail-feather (T6) uniform grey, and 5 outer primaries with rachis less broad and thick than in ♂. Non-breeding birds are not distinguishable, and there may be a single, post-breeding annual moult. ♂♂ from Angola may have greyer backs, ♀♀ redder, perhaps distinguishable from those from Shaba, Zaïre. Bill dark brown, lower mandible flesh (culmen blacker in skins), culmen more arched and robust than in *C. angusticaudus* or *C. fulvicapillus*, mouth flesh; eyes dull yellow-brown; legs and feet flesh. SIZE: (5♂♂, 6♀♀) wing, ♂ 48–49 (48·8), ♀ 43–49 (45·8); tail, ♂ 46–52 (49·4), ♀ 40–43 (41·8); bill, ♂ 10–11·5 (10·5), ♀ 10–11 (10·8); tarsus, ♂♀ 17–19 (18·0). WEIGHT: 1 ♂ 8·0; 1 ♀ 9·5.

Field Characters. A rare bird of miombo woodland in NE Angola and SE Zaïre. Feeds among leaves like an *Apalis*, in which genus it has often been placed. Small and slender, unstreaked, with grey-tipped glossy black tail (outer web of outer feather white in ♂); reddish crown does not contrast with red-brown back, unlike Long-tailed Cisticola, which in breeding plumage has distinct rufous crown-patch contrasting with grey-brown back. Long-tailed Cisticola otherwise very similar, with longer dark tail; both live in miombo but Long-tailed occupies grass under trees; their ranges come close in NW Zambia but are not known to meet. Overlaps in one locality with Neddicky *C. fulvicapillus*, which lives in grass and scrub below trees, and has dark brown tail, grey-brown back contrasting with rusty cap (although young birds have back same colour as head).

Voice. Not tape-recorded. Song not described but doubtless distinctive. Feeding birds keep in contact with continuous faint wispy squeak, very like that of Blue-breasted Cordon-bleu *Uraeginthus angolensis*. Makes purring sound with wings in breeding season, perhaps also at other times.

General Habits. Confined to open grassy places in well-developed miombo woodland. Very local and elusive. Behaves like an *Apalis*; travels in pairs, searching for insects in canopy, lower trees and bushes. Flies like clockwork toy, making loud clicking sound, making it very conspicuous (Neave 1910). When disturbed or excited makes purring sound with wings, for which thickened primary shafts may be adapted (Heinrich *in* Ripley and Heinrich 1960).

Food. Insects.

Breeding Habits. Unknown.
   LAYING DATES: Angola (♀ with egg in oviduct, Jan).

Reference
Irwin, M. P. S. (1991).

Cisticola juncidis (Rafinesque). Fan-tailed Cisticola; Zitting Cisticola. Cisticole des joncs.

Plate 13 (Opp. p. 240)

Sylvia juncidis Rafinesque, 1810. Caratteri Nuov. Gen. Nuov. Spec. Animali Planti Sicilia, p. 6; Roccella, Sicily.

Forms a superspecies with *C. haesitatus*.

Range and Status. Africa, SW Europe north to Netherlands coast, Cyprus and adjacent coasts, Yemen, Iraq; whole Indian subcontinent, SE and E Asia north to 35°N on mainland and to Japan (Honshu), Philippines, Java, Timor and coastal N Australia. Has spread dramatically north in Europe in last 80 years, also in Balkans. Vagrant Britain, Ireland, Madeira, Canary Is.

Resident, NW Africa, lower Nile, locally in W Africa and more or less widespread in much of E and southern Africa, mainly east of 20°E, except Somalia. Winter visitor in N Morocco from Europe.

Morocco, very common and widespread on plains and up to 750 m north of High Atlas Mts, locally abundant in winter, south to Sous R. and Merzouga; Algeria, very common throughout coastal plain, occurring south to Biskra and Laghouat; Tunisia, widespread but scarce, south to central plains, Gafsa, Sfax and (once) Achinina. Libya, old records at Wadi Kaam and Taguira; 20 near Tawarga Jan 1966, absent Apr. Egypt, abundant in Nile delta and along Nile south to Aswan, east to Suez Canal, west to Bahig, Faiyum, Siwa, Wadi el Natrun and Kharga; range expanding.

Range in W Africa and E Africa fragmented. Common in lower Senegal R. valley and delta, and

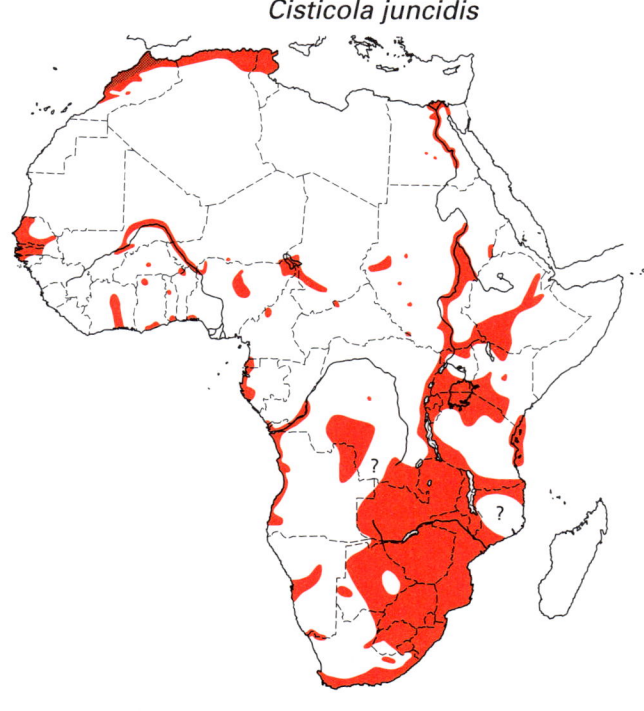

Cisticola juncidis

sparse in middle valley east to Kaedi and Keur Massène (Mauritania); common throughout Gambia; records in Senegal east to near Kidira, and in Guinea-Bissau. Mali, scarce on Dogon Plateau, common along Niger R. flood-plain from Mopti downstream through Niger to about Yelwa (Nigeria). Niger, quite common Niamey, recorded Maradi. Nigeria, locally frequent in NW (Niger R. above Yelwa; Sokoto, Zaria, Kano, whole of Jos Plateau) and probably in brackish marshes west of Lagos; in E, frequent in Gashaka-Gumti Res. and at Malamfatori (L. Chad). Occurs around south of L. Chad (Maiduguri, Nigeria; Mandara, Cameroon; and frequent near Ndjamena, Chad) and recorded in Chad along Chari R. south to Miltou. Sudan, sparse in Darfur and Kordofan, occurs on Nile between 18° and 20°N, and common on Nile and major tributaries south of Berbera; also near Kassala. Not known from Guinea, Sierra Leone, or Liberia. Ivory Coast, locally common from Abidjan to Ferkessédougou. Ghana, common from Cape Coast to Accra, rare in hinterland (e.g. Tumu). Burkina Faso, uncommon around Ouagadougou; frequent in Arli-Pendjari Nat. Parks (across border with Benin). Togo, frequent on coast (Lomé), once inland. Cameroon, only in extreme N (Mandara to L. Chad). Central African Republic, recorded Bamingui-Bangoran Nat. Park. Ethiopia, resident below 900 m in NE, E, Rift valley, SE, S and W; common at 1100 m in Nechisar Nat. Park (Safford et al. 1993). Absent from Djibouti and Somalia. Kenya, locally common in grassland mainly below 1500 m and mainly in subhumid areas in SW. Uganda, locally common except in N-central part. Tanzania, widespread in N and NW and up to 3000 m (Kitulo Plateau) in S; Dar es Salaam coast, also Pemba, Zanzibar and Mafia, where abundant. Humid Atlantic coastal regions in Rio Muni and Gabon and from Congo to SW Angola, and extends in grassy clearings up Congo R. from mouth to Bolobo. In Zaïre very common along W shore of L. Albert, in plains north of L. Edward, and south to Burton Bay (W L. Tanganyika); also throughout Kasai and adjacently in Lunda, Angola.

Common and widespread in SE Africa as mapped; in Lesotho common in lowlands, patchy in highlands; and occurs in SW Africa in Namibia (Sandwich Harbour to Waterberg), along lower Orange R., and from Table Bay eastward. Range contracts from drought areas and expands back into them when drought is broken (Botswana: E. H. Penry, pers. comm.).

Density of 10 nests in c. 24 ha, nearest nests < 30 m apart, South Africa (Pringle 1968), and of 65–86 singing ♂♂ per km^2 (Italy: Cramp 1992).

Description. *C. j. juncidis* (Rafinesque) (including 'cisticola'): North Africa, south to 24°N on Nile; also S Europe east to Turkey. ADULT ♂ (breeding): forehead blackish, crown heavily streaked black on dark brown, with narrow streaks of buff, streaks becoming broken and petering out on hindcrown; nape, hindneck and sides of neck pale brown, buffy or cinnamon-buff, nape with a few dark smudges; mantle and scapulars buff, heavily streaked with black or very dark brown, about 4 streaks each side, widest next to midline, progressively narrower outward; back pale rufous with a few small black streaks; rump uniform tawny or cinnamon; uppertail-coverts blackish with broad cinnamon fringes; tail short, graduated, dark brown with ochreous base and fringes above, narrowly white-tipped, and grey below, each feather with broad white tip and large black subterminal spot. Lores pale buff; black at side of crown ends abruptly, producing narrow but distinct pale buff superciliary stripe that merges with pale cream-buff sides of head and ear-coverts. Chin and throat white or pinkish cream, breast warm buff, merging into warm ochreous or cinnamon-washed buff on flanks and thighs; belly pinkish cream, vent and undertail-coverts buff. Primaries and secondaries dark brown, outer webs and tips narrowly fringed with warm buff, tertials blackish, widely fringed at base with cinnamon-buff; greater and median primary-coverts like primaries; alula uniform blackish; greater, median and lesser coverts black with broad buff fringes; underside of flight feathers greyish, underwing-coverts warm pink-buff. Bill with blackish culmen and yellowish horn cutting edges; base of lower mandible greyish; inside of mouth black; eyes yellow-brown, tawny or hazel; legs and feet flesh-pink. Breeding adult ♂ in worn plumage has many buff fringes abraded away; crown and mantle look blacker, and whole plumage less rufescent. ADULT ♂ and ♀ (non-breeding): black streaks on forehead, crown, mantle and scapulars much narrower than in breeding ♂; hindneck uniform but not very contrasting with crown or mantle; tail longer and narrower; underparts warmer buff; inside of mouth pink. ADULT ♀ (breeding): like breeding ♂ but black streaks not so broad; ground colour of upperparts less buffy but underparts buffier than in ♂; inside of mouth pink. SIZE: (Cramp 1992) wing, ♂ (n = 29) 49–56 (52·0), ♀ (n = 15) 45–50 (48·1); tail, non-breeding ♂ (n = 13) 38–43 (40·5), non-breeding ♀ (n = 14) 37–43 (39·8); tails of breeding adults av. 3–7 mm shorter; bill to skull, ♂ (n = 23) 11·4–13·2 (12·2), ♀ (n = 15) 11·6–13·2 (12·4); tarsus, ♂ (n = 17) 18·8–20·8 (19·8), ♀ (n = 12) 17·7–19·7 (18·6). WEIGHT: 1 ♂ Morocco 8·7; 2 ♀♀ Algeria 8·0, 8·5; unsexed (n = 69, Italy) 8·0–13·0, monthly averages from 8·9 (Apr–July) to 11·2 (Nov).

IMMATURE: like non-breeding adult ♂ but streaks narrower, browner (not black), and less well defined. Mouth pink, 2 black spots on tongue.

NESTLING: completely naked, skin orange-pink; gape and bill yellow; 2 black spots on tongue; legs pink.

C. j. uropygialis (Fraser) (including 'perennia'): N tropics and E Africa, from Senegambia to Ethiopia, north to 24°N in Sudan, south to S Nigeria, Rwanda, N Tanzania and Mafia I. Breeding plumage slightly paler than *juncidis*, colours a little less intense; underside of tail with cinnamon patch proximal to black subterminal spot. No non-breeding dress (i.e. breeding plumage perennial) in NE Zaïre and most of E Africa including Pemba, Zanzibar and Mafia I. On Pemba a melanoerythristic form is common, breeding freely with normal form (Archer and Turner 1993): mantle and back warm rusty brown with almost black streaks; throat grey, washed rusty; remaining underparts dark grey, heavily washed rusty. WEIGHT: (Kenya) 2 ♂♂ 7, 7, 1 ♀ 5.

C. j. terrestris (Smith): southern Africa, north to Rio Muni, Central Zaïre, Burundi, and S Tanzania (Mikindani, Iringa). Darker and browner than normal form of *uropygialis*.

Field Characters. A small (10 cm) streaky brown warbler of humid bushy grasslands, reedy ditches and adjacent cereal fields. The only cisticola in N Africa, where its black-striped crown and back, combined with tail conspicuously black-and-white-spotted below, distinguish it from all warblers except Graceful Warbler *Prinia gracilis*. Graceful has long, thin, graduated tail; Fan-tailed Cisticola has short round tail, jerkily spread during its protracted 'zitting' song-flights. South of Sahara the best known of the tiny streaked cisticolas, common in grassland at low and medium elevations, with broad habitat tolerance (long or short grass, wet or dry situations, also cultivation). Singing bird easily identified by bounding display flight accompanied by familiar sharp 'tsip' or 'zit'; does not snap wings. Cloud-scrapers have very different songs and displays and fly much higher; Desert Cisticola *C. aridulus* flies low but has ringing notes interspersed with wing snaps. When not singing easily confused with other species. Crown and back buff to tawny heavily streaked blackish, contrasting with light brown hind collar and plain tawny rump, but these features less clearly marked in non-breeding birds or birds in worn plumage; below buffy white. Tail often fanned in flight, medium-short, with 2-tone effect, light to dark brown with black subterminal band (greyer on underside and band more contrasting), and with the white tip common to the group; tails of other small species are uniform blackish, and shorter (except Desert Cisticola and non-breeding Black-backed Cisticola *C. eximius*, q.v. for differences).

Voice. Tape-recorded (45, 58, 87, 88, 91, B, C, F). Song, given by ♂ only, varies somewhat geographically (Chappuis 1976a) and is a short, sharp, decisive monosyllable, generally 'zit', 'tsip', 'dzeep', 'zeee' or 'klink', repeated regularly and persistently at 0·5–1·0 s intervals during song-flight and from perch; in undulating song-flight, one 'zit' note in rising part of each bound. Sometimes up to 12 rapid 'plick' notes mixed into song or given after it; sounds very like Red Crossbill *Loxia curvirostra* (Gibraltar: Elkins 1975). Courtship call by ♂, a weak 'pyr . . . pyr . . .' in cicada flight (see *Breeding Habits*). Contact call, 'zip' or 'chip'. Excitement call by territorial ♂ 'twud-twud-twud', occasionally accompanied by a wing-snap with every 'twud' note (Cramp 1992). Alarm, explosive 'tew', also series of toneless 'pt' or 'pitt' notes at 3 per s for 2 s, and wing-snapping (by both sexes). 5–7 day old nestlings make loud hissing when disturbed by person (Pringle 1968).

General Habits. 'The most remarkable small bird in the world' (Moreau 1966) because of its habitat adaptability. Inhabits wet and dry tropical grasslands, particularly where completely burnt in dry season, from sea level to 750 m (Morocco) and 1500 m (Kenya) (and able to endure night temperatures down to −8°C in Botswana and −10°C in Spain). Likes grass <1 m tall, on soft earth, but will use grasses and rushes >2 m tall on waterways and at edges of marsh; common in 1·15 m tussocky grass in Zambia, nesting in coarse grass *Stereochlaena cameroni* (Penry 1985). On Niger R. flood-plain in Mali inhabits *Vetiveria*, *Typha*, *Phragmites* and *Mimosa*. Also inhabits shrubby grassland next to woods (Zambia), saltmarshes with plenty of *Salicornia* (NW Africa), tamarisks, sedges, dry crops (sugar cane, *Eragrostis*, wheat, lucerne), wet crops (rice), seasonally-flooded areas without too many trees (that could obstruct song-flights) and, in South Africa, airfields and golf-courses.

Forages mainly on or near ground, where hops but can walk like a pipit *Anthus*, scuttling rapidly up into dense growth if disturbed, and there hopping, leaping and moving with legs often splayed, like an *Acrocephalus* warbler. Sometimes inquisitive; wary and secretive in breeding season. Occasionally makes sallies after insect flying past. In protracted breeding season ♂ spends much time aloft, patrolling territory in zitting song-flight. Sings from dawn to dusk; song-flights longest – up to 13 min – about 08.00 h, otherwise only 1 min or less. Also gives much shorter 'zit' songs from perch on grass head.

Solitary; ♂ is serially polygynous (see below) but does not consort very much with ♀ even in breeding season. Small group of juveniles may be harried by territorial ♂, flying high, disappearing to feeding grounds 2 km away (Malta, May, dawn: Cramp 1992). Noisy groups congregate before migrating (Gibraltar, July–Aug: Finlayson 1979) and cross to Africa in loose flocks of up to 30 birds. After breeding season, up to 80 birds may roost together (in samphire *Inula crithmoides*: Cramp 1992). Once 10 seen flocking with 4 Graceful Warblers, Egypt.

Moults in autumn (Malta); imm. ♀♀ take 75 days, imm. ♂♂ 81 days, adult ♀♀ 67 days, adult ♂♂ 92 days; many birds moult tertials twice, at beginning and end of moult period (Gauci and Sultana 1981).

Resident over most of African range, but juveniles disperse widely, and other reports of local movements. In NW Africa a conspicuous migrant, crossing from Gibraltar in Aug by day, abundance in Tingitane, Morocco, increasing from late July to early Oct (Pineau and Giraud-Audine 1979), and abundant on passage at Merja Zergla, Morocco, in late Oct (Thévenot *et al.* 1982).

Food. Insects, taken from ground or from vegetation just above it: larvae and beetles (Tunisia), 10–15 mm larvae (Zambia), mainly grasshoppers (southern Africa), dragonflies, also spiders (Malta) and larval tettigonid bush-crickets (Spain). 75 items (Japan) included 40 grasshoppers, 17 moths and caterpillars, 10 spiders, and fragments of snails. Young fed with caterpillars and small grasshoppers (Pemba); 1 grasshopper was 38 mm long.

Breeding Habits. Territorial, solitary nester. Mating system not studied in Africa, but polygynous (with successive ♀♀) in Japan, and evidently so in Malaysia, Australia, Malta, Spain, Portugal and France (McGregor *et al.* 1990). In Japan 50–70% of ♂♂ are polygamous (the rest 'monogamous' i.e. failed polygynists, or failed breeders); ♂ builds av. 6.5 nests during breeding season, with only 1 used for courting at any one time; when outer fabric completed, ♂ advertises the nest, leads ♀ to it with invitation flight (Motai 1973, Ueda 1984, 1986). One summer a ♂ built 18 nests and mated with 11 ♀♀. ♂ usually attracts 4 ♀♀. Occasionally ♂ is simultaneously polygynous, with up to 5 ♀♀. Pair-bond is weak and lasts only for single nesting.

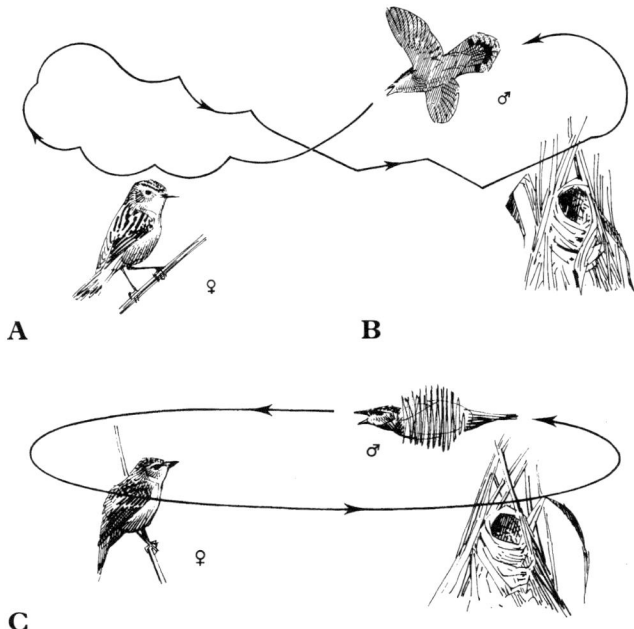

Territory size (n = 8, Portugal) 0·59–3·42 (1·44) ha (McGregor *et al.* 1990). Territory provides food, nesting site, perches, and song-flight area; its boundaries change daily, with undefended intrusions by neighbouring ♂ ♂ in song-flight, but area within a few m of nest is defended aggressively against conspecifics and other bird species (S France: Guichard 1959). Rival ♂ ♂ 1 m apart give fantail display in vegetation at edge of territories, turning away from each other, fanning and raising tails, but only rarely fighting thereafter. ♂ with large territory spends more time in song-flights, and gives more frequent 'zit' calls, than ♂ with small territory (Portugal: McGregor *et al.* 1990).

After ♂ builds outer fabric of nest, he attracts a ♀ with song-flights; ♂ flies down to ♀ in vegetation in a series of swoops, the pendulum flight (**A**), then shows her the nest location using zigzag flight (**B**). When ♀ approaches near to nest, ♂'s flight pattern changes abruptly to cicada flight, in a circle with whirring wings around perched ♀ (**C**). ♀ finds and inspects nest, ♂ perches nearby, ♀ invites copulation by wing-quivering, ♂ repeats pendulum flight, then they copulate. ♂ grips ♀'s back firmly with his feet and after copulation pair dangles upside down for 2 min, the ♂ still gripping ♀'s back, before falling and separating (K. Ueda *in* Cramp 1992).

NEST: beautifully-made bag (**D**), silvery white inside, in shape of pear or small upright bottle – the 'soda-bottle' nest unique amongst cisticolas – constructed by ♂ in a thick clump of living grass (occasionally dead grass) by attaching cobwebs to inner side of a ring of stems to make the neck, then binding downwards with cobwebs to a wider chamber and base. Cobweb (spiders' silk) adheres to grasses but nest is also sewn, some grass blades being pierced and silk passed through. 1 nest was made base-first: a small cup of cobweb pulled the grasses together and was lined with fluffy plant material; only later were grass stems drawn together above cup to form neck (E. H. Penry, pers. comm.). ♀ adds lining. Base of nest only 5–20 cm above ground. From outside to inside, base of finished nest consists of (1) cobweb, (2) a pad of fine dry grass, (3) plant down (e.g. *Asclepias fruticosa*), grass-heads, seeds and cobweb, (4) another pad of fine dry grass, (5) plant down, grass-heads and cobweb, and (6) an inner cobweb lining (Penry 1985, 1988). The cobweb lining extends up to narrow entrance which points upward (or is slightly bent over away from prevalent wind, i.e. facing north to northeast at aerodrome near Cape Town: Pringle 1968). Nest walls are tissue-like; some nests incorporate wool in addition to cobweb (Comins 1964). Nest bound to ring of 11–34 (av. 19) grass stems each 5 mm in diam., which with grass blades continue upwards and outwards above nest entrance as loose fronds, so that grass clump looks undisturbed (Penry 1985). Nest is well concealed and hard to find (Penry 1988). 8 nests in clumps of coarse grass *Stereochlaena cameroni* (Zambia) measured: int. depth 80–145 (103), entrance diam. 25–45 (35), ext. diam. of chamber 50–65 (57); height of entrance above ground 170–300 (226) mm. 5 nests (South Africa) measured: int. depth 104–124 (111), entrance diam. 34–53, ext. diam. of chamber 61–67. Nest in tuft of *Eragrostis curvula* was 110 high, ext. diam. of neck 30 and of base 50, entrance diam. 25. As nestlings grow, nest bulges and stretches, and the cobweb fabric may buttonhole (Penry 1988).

EGGS: 3–4 (tropics), 2–5 (southern Africa, av. 3·3), 4–6 (N Africa). Laid at daily intervals, from day after ♀ starts to line nest. Short ovals, glossy, colour very variable: white, cream, pale blue, pale turquoise or pinkish, sometimes immaculate but usually speckled with black or dark brown and sparsely or heavily blotched (mainly toward broad end) with purple, red-brown or dark brown. SIZE: (n = 181, southern Africa) 13·3–16·5 × 10·4–12·4 (15·1 × 11·4), (n = 10, Nigeria) 14·5–15·5 × 11·0–11·5 (15·2 × 11·3), (n = 100, Europe and N Africa) 14·8–17·1 × 10·2–13 (15·5–11·6). WEIGHT: av. 1·08.

LAYING DATES: Morocco, Mar–June; Egypt, Feb–June; Senegambia, July–Sept; Burkina Faso, Aug–Oct; Ghana (nest building May–July, Dec; nestlings Aug); Nigeria, June–Oct; Sudan, Aug–Mar; E Africa: Pemba, May–Mar; Region B, Apr, July–Aug; Region C, Jan, Mar, May–June; Region D, Jan, Feb, May–June, Oct–Nov; Angola, Nov–Mar; Zambia, Dec–June (mainly Dec–Jan: Anon. 1985); Malaŵi, Jan–Mar, May, Dec; Zimbabwe, Nov–Apr (of 131 clutches, Nov 4%, Dec 21%, Jan 32%, Feb 20%, Mar 17%, Apr 6%); Botswana, Dec–Jan, Mar–Apr; Transvaal, Oct–Mar (mainly Dec–Feb); Natal, Sept–Mar (mainly Nov–Jan); SW Cape, Aug–Jan.

INCUBATION: by ♀ only, starting with last egg. Period 13 days. ♀ continues to line nest throughout incubation period, and also interrupts incubation to forage.

DEVELOPMENT AND CARE OF YOUNG: skin remains orange, but crown skin black by day 5–6, when quills begin to sprout on edges of wings and midline of back; half feathered by day 7–8, wing quills 10 mm long, gape prominently yellow; general appearance blackish on day 9–10, crown-feathers sprouting, wing-feathers 20 mm; eyes open on day 11–12 when bird 75% of adult size, back plumage streaked black and orange-buff, tail quills appear; fully feathered by day 13–14, tail-feathers 6 mm, wing length 45 mm, gape less prominently yellow. Nestling period 11–15 (generally 13–14) days. Young normally cared for and fed before and after fledging by ♀ only, although both adults have been recorded feeding nestlings and removing faecal sacs (**D**) (Pringle 1968, Penry 1985). Young out of nest fed for 10–20 days by ♀ (who then starts another clutch, France, Japan).

BREEDING SUCCESS/SURVIVAL: heavily parasitized by Parasitic Weaver *Anomalospiza imberbis* (N Pemba I.). Breeding success better in dry than in wet grassland (Penry 1985). Overall success (eggs to fledglings) at least 67% (n = 27 eggs, Zambia); 75% (N France); only 30–33% (n = 570 eggs, Japan).

References
Cramp, S. (1992).
Guichard, G. (1959).
Penry, E. H. (1985, 1988).
Pringle, J. S. (1968).
Taillander, J. (1993).
Ueda, K. (1984, 1986).

*Cisticola haesitatus** (P. L. Sclater and Hartlaub). Socotra Cisticola. Cisticole de Socotra.

Plate 13
(Opp. p. 240)

Drymoeca haesitata Sclater and Hartlaub, 1881. Proc. Zool. Soc. Lond., p. 166; Socotra.

Forms a superspecies with *C. juncidis*.

Range and Status. Endemic to Socotra. Uncommon resident in two localities: *c.* 2·5 km west of Hadibu and at 900 m near Adho Dimellus (Ripley and Bond 1966).

Description. ADULT ♂ (breeding): entire upperparts streaked brown (feather-centres) and ash-grey (fringes); lores, sides of neck and ear-coverts ash-grey; tail-feathers brown with subterminal dark brown band and buff tip; chin, breast and flanks pale grey, washed buff on flanks; belly white; undertail-coverts buff; primaries, secondaries, tertials and wing-coverts brown, fringed ash-grey; underwing-coverts buff; axillaries ash-grey. Upper mandible grey to black, lower pinkish brown; eyes brown; legs and feet pinkish brown. ADULT ♂ (non-breeding): richer coloured than breeding dress. Upperparts streaked brown and buff; crown more prominently streaked; rump tawny buff; breast and flanks buffy grey. Sexes alike. SIZE: (3♂♂, 1♀) wing, ♂ 48–50 (48·7), ♀ 46; tail, ♂ 39–42 (40·7), ♀ 38; bill, ♂ 11–12 (11·7), ♀ 11; tarsus, ♂ 18–19 (18·3), ♀ 18. WEIGHT: (5♂♂, 1♀) 6–8.

IMMATURE: like non-breeding adult but upperparts paler (feather fringes broader, centres paler brown).

NESTLING: unknown.

Field Characters. A small warbler with a washed-out appearance. Crown and upperparts light grey-brown streaked darker; tail grey-brown with large black subterminal spots and narrower pale tips (on underside, pale tip and black spot are same width); underparts white. Has 'tsip' call in flight like that of Fan-tailed Cisticola

Cisticola haesitatus

* **Note added in proof.** The OSME Survey of Southern Yemen and Socotra in 1993 added much to our knowledge of this species. The results were published in *Sandgrouse* **17** (1996), too late to be included here.

C. juncidis, of which it is the island representative. Only other warbler on the island is larger Socotra Warbler *Incana incana*, which has completely plain upperparts, tail uniform except for white tips to underside; voice unknown but doubtless different.

Voice. Tape-recorded (OSME). Song/call, given in flight, single, repeated 'tsip'.

General Habits. Occupies stony ground with light scrub or almost entirely covered with thick, bush-like grass *c.* 60 cm high, or short-grass upland meadows, all with scattered larger bushes.

Usually in pairs, on territories of 0·4–0·8 ha; intruding ♂♂ chased off. Runs like a mouse. ♂ perches on grass or bush tops, ♀ usually in cover. Forages in bushes. ♂♂ most conspicuous in early morning, when singing from exposed perch such as bush top, or in cruising flight.

Food. Insects.

Breeding Habits. Family group of 2 adults with 3 young seen Apr. Otherwise unknown.

References
Lynes, H. (1930).
Ogilvie-Grant, W. R. and Forbes, H. O. (1903).
Ripley, S. D. and Bond, G. M. (1966).

Cisticola aridulus Witherby. Desert Cisticola. Cisticole du désert.

Plate 13 (Opp. p. 240)

Cisticola aridula Witherby, 1900. Bull. Br. Orn. Club 11, p. 13; Gerazi, White Nile, 60 miles south of Khartoum, Sudan.

Range and Status. Endemic, resident and perhaps local migrant in places. Uncommon S Mauritania, at least during rains (seasonal visitor?); uncommon N Senegal. Resident or perhaps local migrant Mali, in R. Niger inundation zone and Boucle du Baoulé; few records south of 14°30′N. Uncommon Niger south of *c.* 17°N; May–July in N of range, Feb–Mar in S, but in Aïr most of year. Uncommon resident most of N Nigeria but common in NE. Locally common L. Chad basin, Ouadi Rimé–Ouadi Achim Reserve and Ouaddaï (probably entire sahel zone and Ennedi) in Chad. In Sudan, common resident or local migrant, N Darfur but absent S Darfur, less common Kordofan, extending east to Nile; also on borders with Eritrea and Ethiopia in extreme NE and SE. In Ethiopia, locally common resident in E and NE, rare in S. Common in coastal plains of Eritrea below 150 m, from Sudan border, to *c.* 40°E. One record Djibouti, Dec. Fairly common resident NW Somalia west of 45°E. N Kenya (Huri Hills) common Nov–Dec; elsewhere in E Africa (N Tanzania, central and S Kenya) locally common, often abundant, up to 1800 m; uncommon resident Serengeti (Tanzania); in Tsavo (Kenya), 1 bird per 2 ha or less, mostly in rains. Uncommon W Sul do Save (Mozambique). Locally abundant Zambia. Widespread but local on central plateau of Zimbabwe, north to Harare; also shores of L. Kariba but not elsewhere in Zambezi Valley; may show seasonal movements. In Angola, extreme S Huila on lower Cunene R. and coastal plain from Congo R. south to Benguela. Locally common Namibia, including Caprivi. Common resident Botswana, especially in SW Kalahari and in SE of country. Locally common N and E Cape Province (south to Cradock), Transvaal (commonest in W and SW), Swaziland, Zululand, interior Natal; uncommon and local in Orange Free State and lowland Lesotho.

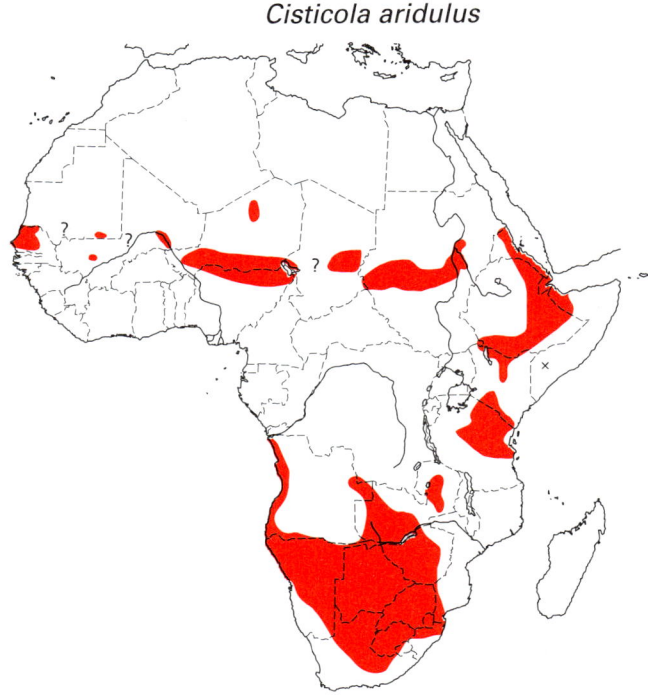

Cisticola aridulus

Description. *C. a. aridulus* Witherby: Mali, Niger, N Nigeria to S Sudan. ADULT ♂ (breeding): upperparts, including wing-coverts and tertials, streaked dark brown and sandy buff (feather centres dark brown, edges sandy); rump redder, less streaked; tail-feathers very dark brown (central pair rather paler brown), fringed buff and tipped white. Chin and throat white; breast, flanks and undertail-coverts sandy buff; belly white; primaries and secondaries pale brown, narrowly edged buff, and basal two-thirds of inner webs broadly fringed buff; underwing-coverts and axillaries pale buff. Upper mandible dark brown, lower pale grey-brown; eyes yellow-brown; feet pale pinkish brown. ADULT ♂ (non-breeding): upperparts paler, streaked dark brown and buff; underparts deeper buff. Sexes alike. SIZE: (10♂♂, 10♀♀) wing, ♂ 48–52 (50·4), ♀

43–50 (46·7); tail, ♂ 35–39 (37·4), ♀ 32–38 (35·3); bill, ♂ 11–13 (12·1), ♀ 11–12 (11·8); tarsus, ♂ 18–20 (18·7), ♀ 17–19 (18·1).

IMMATURE: like non-breeding adult.
NESTLING: unknown.

C. a. lavendulae (Ogilvie-Grant and Reid): S Ethiopia, coastal Eritrea, Somalia. Like nominate but upperparts greyer (feather edgings greyish white, not sandy); underparts paler buff.

C. a. tanganyika Lynes: Kenya, Tanzania; intergrades with *lavendulae* in N Kenya. Like *lavendulae* but upperparts darker (dark brown feather centres broader, edges light brown); underparts deeper buff (like nominate). SIZE: (10♂♂, 5♀♀) wing, ♂ 47–52 (49·9), ♀ 46–50 (48·0); tail, ♂ 34–39 (36·1), ♀ 32–37 (34·8); bill, ♂ 11–13 (11·8), ♀ 11–12 (11·8); tarsus, ♂ 17–19 (17·9), ♀ 17–19 (18·0). WEIGHT: (Kenya) 1♂ 8·5.

C. a. perplexus White: N Zambia (Bangweulu swamps). Like *tanganyika* but upperparts still darker (feather centres broader and blackish brown, edges darker brown).

C. a. traylori Benson and Irwin: E Angola and extreme W Zambia. Like *perplexus* but feathers of upperparts with paler centres (dark brown rather than blackish brown).

C. a. eremicus Clancey: S Angola, N Namibia (south to *c.* 20°S), S Zambia, Zimbabwe, N Botswana (south to *c.* 23°S in E). In breeding plumage like *tanganyika*, but upperparts paler; in non-breeding (fresh) plumage still paler, with broad unworn sandy buff feather edgings obscuring dark centres (appears more narrowly streaked dark). SIZE: (8♂♂, 5♀♀) wing, ♂ 50–53 (52·2), ♀ 44–50 (47·2); tail, ♂ 38–43 (40·5), ♀ 35–40 (38·6); bill, ♂ 11–12 (11·8), ♀ all 12; tarsus, ♂ 17–20 (18·6), ♀ 17–19 (18·4). WEIGHT: (Botswana, combining this race and *kalahari*) ♂ (n = 11) 5–11 (8·3), ♀ (n = 6) 8·8–13 (10·2).

C. a. kalahari (Ogilvie-Grant): central Namibia, S Botswana, South Africa east to *c.* 30°E (Barberton in Transvaal). Breeding plumage like *eremicus* but upperparts average slightly darker, browner, less sandy buff (feather edges darker brown); underparts richer buff. Non-breeding dress like *eremicus* but edges of feathers of upperparts more richly coloured, more tawny; underparts richer buff. WEIGHT: (Namibia) 1♀, Oct, 7·5; (South Africa) (combining this race and *caliginus*, unsexed) 7·8–10·1.

C. a. caliginus Clancey: Natal, Swaziland, S Mozambique and adjacent NE Transvaal. Breeding dress like *kalahari*. Non-breeding like *kalahari* but centres of feathers of upperparts broader, edges duller, darker; underparts still richer buff.

C. a. lobito Lynes: coastal Angola. Very like *eremicus* but duller on upperparts (difference slight and disappears in worn breeding dress). Smaller. SIZE: wing, ♂ (n = 10) 48–50 (49·0), ♀ (n = 6) 44–46 (45·0); tail, ♂ (n = 10) 35–39 (36·6), ♀ (n = 5) 32–36 (34·4); bill, ♂ (n = 8) 11–12 (11·6), ♀ (n = 5) all 12; tarsus, ♂ (n = 10) 18–19 (18·4), ♀ (n = 5) 18–19 (18·2).

Field Characters. Very small; tail longer than in other cloudscrapers, and display flight low over ground, not in the clouds. Upperparts rather sombre (streaked dark and light brown) all year in Kenya/Tanzania and in breeding season in southern Africa; otherwise considerably paler and more washed-out than other small species, rump plain pale reddish buff, not orange-red. Most resembles Fan-tailed Cisticola *C. juncidis* but has uniform blackish tail with pale tips (tail of Fan-tailed pale grey to pale brown with black subterminal band); also, tail of Fan-tailed *appears* longer even though about the same size. Calls and display very different from Fan-tailed (and all other species). Prefers more arid habitats than most small species, though often together with Fan-tailed, and generally not found over 1800 m.

Voice. Tape-recorded (38, 45, 87, 88, 91, B, F). Song, a series of high ringing notes, all on one pitch, given at rate of *c.* 3/s; also in display flight, clear 'hew', 'hew-hew' or 'ti-hew', alternating with wing-snaps, e.g. 'hew-hew/snap, hew-hew/snap'; latter song also given from perch. In between songs a variety of clicking and snapping noises, some possibly made by the bill (Chappuis 1974), and grating notes. Alarm, 'tuk . . . tuk . . .' accompanied by wing-snaps at 1-second intervals (Zimmerman *et al.* 1996). Nestlings give sharp hiss when nest touched (Maclean 1993).

General Habits. Inhabits arid and semi-arid open plains (not mountains), bushless or with scattered *Acacia* and *Balanites* trees or thorn bushes, and sparse grass (especially *Aristida* in sahel), often on sandy soil, including dunes, but also on fields of lava boulders. In Botswana *C. a. kalahari* habitat has bushes and this race also occurs in small glades in open woodland, while in adjacent W Zambia habitat of *C. a. perplexus* has no bushes. In Eritrea, Somalia and South Africa favours long grass, mainly *Panicum* steppe, and in southern Africa occasionally in taller moist grass; on shores of L. Kariba, in mats of *P. repens*. In Zimbabwe and South Africa occasionally found on old farmland on dry sandy soil, overgrown by low dense grass and herbage, with bare patches. Occurs in crops only where they border natural grassland.

Usually solitary or in pairs; in South Africa in small loose parties in non-breeding season. Not shy, although skulks in long grass and tufts of coarse vegetation; forages low in grass or on ground, where it runs; perches on grass and bush tops. Sings from grass stalk or in small bushes; when alarmed, darts about snapping wings and calling. Flies low, distances of 20–50 m when flushed, occasionally higher and further, when flight erratic and twisting. Forages in grass tufts; pecks at the soil like a lark.

Food. Small insects, including beetles, Orthoptera (grasshoppers, tree crickets); spiders.

Breeding Habits. Snaps wings in display. ♂ aerial display erratic, usually < 10 m above ground, over ♀ perched in grass: flies over her, swoops, rises and swoops again, up to 10 times.

NEST: a ball or pear-shaped structure (ext. depth 110, basal diam. 60, diam. at top 30), sometimes elongated (150 × 75), with large side-top entrance; of a thin, open construction above, more solid below; material bound around entrance; made of fine, soft, dry grass, mixed with and bound on outside by spider web, by which attached to growing grass stalks; lined with plant down, grass flower-heads and spider web. Well-hidden in a grass tuft, often in dead grass, *c.* 10–70 cm above ground. Takes at least 4 days from start of nest-building to laying first egg; ♀ brings lining material during incubation.

EGGS: 1–8 in one nest: Natal, Zambia, Zimbabwe, Tanzania 1 (n = 1), 2 (9), 3 (21), 4 (16), 5 (7), 8 (laid by 2 ♀♀?). *C. a. lavendulae* (Somalia) glossy; bluish white, minutely and sparsely speckled over whole surface with pale purple. *C. a. tanganyika* (Kenya) pale blue, sparsely marked with reds and red-browns. *C. a. lobito* white with purple-red and purple-brown flecks and spots. *C. a. kalahari* (Namibia, Zimbabwe) elongated oval; matt but shiny; white, with tiny red-brown and purple-grey spots and scrawls over whole surface but denser at big end, or occasionally unmarked, or greenish white covered with fine freckling of pale brownish lilac, denser and forming an irregular ring at big end. SIZE: (Kenya) 14·5–17 × 11·0–13; (n = 7, Zimbabwe) 14·3–15·8 × 11·1–11·9 (15·2 × 11·6); (n = 4, Namibia) 15·3–16 × 10·9–11·6 (15·6 × 11·3), av. wt. of shells 0·05; (n = 47, southern Africa) 13·9–15·9 × 10·8–12 (15·0 × 11·3); n = 38, *C. a. kalahari*) 14·0–16·6 × 10·9–11·6 (14·7 × 11·4).

LAYING DATES: S Mauritania, Aug–early Nov; N Senegal, Aug–Oct; Mali, June–Aug, Oct; Sudan, July–Oct; Ethiopia, Jan–Feb; Somalia, June. E Africa, mainly rains: Kenya, June; Tanzania, May; Region B, June; Region C, Dec; Region D, Jan–June (peak May). Zimbabwe, Nov–Apr, peak Dec–Jan (61 records); E Zambia, NW Botswana, Nov to at least Feb (rains); Namibia, Mar, Sept; coastal Angola, Feb–Mar; Botswana, Nov–Apr, July; Natal, Oct–Dec, Feb, Mar (peak Nov); Transvaal, Oct–Apr (peak Dec–Jan).

INCUBATION: probably by ♀ alone; period 14 days.

DEVELOPMENT AND CARE OF YOUNG: young in nest 18 days. ♀ snaps wings in alarm near nest or fledglings.

BREEDING SUCCESS/SURVIVAL: nests parasitized by Parasitic Weaver *Anomalospiza imberbis*.

References
Archer, G. and Godman, E. V. (1961).
Ginn, P. J. *et al.* (1989).
Hoesch, W. and Niethammer, G. (1940).
Lynes, H. (1930).
Vincent, A. W. (1948a).

Plate 13 (Opp. p. 240)

Cisticola textrix (Vieillot). Cloud Cisticola. Cisticole pinc-pinc.

Sylvia textrix Vieillot, 1817. Nouv. Dict. Nat., nouv. éd. 11, p. 208; based on 'Le Pinc-pinc' of Levaillant, 1802, Hist. Nat. Ois. Afrique, 3, p. 88, pl. 131; Cape Province ex Levaillant.

Range and Status. Endemic resident, locally common to uncommon in open grasslands from W Angolan plateau west of Kwanza R., and in east west to Vila Luso in Moxico and Missão de Luz, S Lunda, and NW Zambia in Kabompo and Balovale on Kansalya and Minyanya Plains; reappears in interior of South Africa in central Transvaal highveld, Orange Free State, Natal, W Lesotho and Swaziland; also S Mozambique to Maputo and W Zululand and Cape Province from Port Elizabeth westwards to SW Cape. Very common in W Lesotho.

Cisticola textrix

Description. *C. t. textrix* (Vieillot): S Cape Province from Port Elizabeth to Cape Town. ADULT ♂ (breeding): forehead and crown streaked black, feathers edged rufous; lores, cheeks and around eye to ear-coverts pale buff, feather centres with minute blackish spots. Hindneck, mantle and back streaked black with rufous feather edges, sides of back and scapulars with whitish feather margins; rump and uppertail-coverts tawny, narrowly streaked blackish. Tail dark brown, with narrow (c. 4 mm) off-white tips, mostly on inner webs. Chin and throat off-white; breast pale buff with longitudinal brown streaking; centre of belly to undertail-coverts pale buff to off-white, flanks washed tawny, streaked brown, thighs tawny. Primaries dark brown, edged buff, secondaries brown with whitish edges and pale greyish tips, wing-coverts black edged rufous; shoulder, underwing-coverts and axillaries tawny buff. ADULT ♂ (non-breeding): crown-feathers with paler edges, those of mantle and back with broader and brighter reddish margins. Bill dark horn, base of lower mandible yellowish pink; eyes light brown; legs and feet pinkish brown. Sexes alike. SIZE: (28 ♂♂, 26 ♀♀) wing, ♂ 51–56 (53·3), ♀ 47–51 (48·3); tail, ♂ 26–33 (27·6), ♀ 26–30 (27·2); bill, ♂ 10·5–12·5 (11·5), ♀ 10·5–12 (11·1); tarsus, ♂ 20–22·5 (20·8), ♀ 18–20·5 (19·5). WEIGHT: (South Africa) ♂ (n = 5) 11–12·5 (11·9), ♀ (n = 2) 8–10 (9·0).

IMMATURE: duller above, washed with bright lemon yellow below.

NESTLING: hatches naked; gape yellow.

C. t. major (Roberts): Transvaal, Orange Free State, W Natal, W Swaziland to E Cape (Grahamstown). In breeding ♂, streaking on underparts restricted to sides of breast, or sometimes very sparse streaking in centre. In non-breeding plumage much redder above, particularly on rump and tail; more reddish below, streaking absent or only very indistinct.

C. t. marleyi (Roberts): S Mozambique, NE Zululand, coastal Natal south to L. St Lucia. In breeding plumage resembles *major* but darker above, edges of mantle feathers and wing-coverts whiter; streaking on sides of breast sparse and indistinct; in non-breeding plumage somewhat brighter, whitish feather edges less distinct.

C. t. bulubulu Lynes: W Angolan highlands, intergrading with next race in NW Moxico at Munhango. In breeding plumage blacker above with strongly chestnut rump; streaking restricted to sides of breast, flanks tawnier. In non-breeding plumage very like *major*.

C. t. anselli White: E Angola west to Vila Luso in Moxico, Missão de Luz in S Lunda and NW Zambia. In breeding dress resembles *major* but slightly darker on crown; more distinct in non-breeding plumage, with mantle feathers edged whitish; paler below with sparse streaking at sides of throat.

Field Characters. A cloud-scraper confined to southern Africa, with very short tail. Very similar to Wing-snapping Cisticola *C. ayresii* but rather plumper, longer-legged and more tawny above; in S Cape and E Orange Free State distinguished by spots and streaks on underparts (not present in all birds). Lack of wing-snaps during song-flight helps distinguish it from Wing-snapping Cisticola and Dambo Cisticola *C. dambo*; Pectoral-patch Cisticola *C. brunnescens* also lacks wing-snaps but song is different and almost inaudible.

Voice. Tape-recorded (88, 91, F). Flight song in aerial display, 4–6 high-pitched, plaintive notes, first note often slightly lower, followed by a chatter, 'per-pee-pee-pee-chikchikchikchik...', repeated at 2–3 s intervals for several min at a time; varies regionally. Chatter and pure notes sometimes given at same time, suggesting a duet (see sonogram in Maclean 1993). Plaintive notes similar to Wing-snapping Cisticola but faster (4 notes per s). Rapid chatter also accompanies dive to ground after song-flight. Does not snap wings. Alarm, rapid 't-t-t-t-t-...' at rate of 5 't's per s.

General Habits. Favours areas with very short grass and bare ground between tufts; on coast of Cape Province estuarine marshes, in Transvaal, close-cropped grass with moderate basal cover on well-drained ground, often on slopes; co-exists with Wing-snapping Cisticola on *Cymbopogon-Themeda* grassland without obvious ecological differences, although it is larger than Wing-snapping, whereas Fan-tailed Cisticola *C. juncidis*, Pectoral-patch Cisticola and Desert Cisticola *C. aridulus* maintain distinct habitat preferences (Dean 1976). In Zambia restricted to dry watershed plains alongside Desert Cisticola. Spends most of time on ground; forages around grass tufts; remains extremely well concealed unless flushed.

Food. Small insects, especially grasshoppers.

Breeding Habits. Territorial in breeding season. Territory advertised by cruising flight: ♂ rises with whirring wings at angle of about 45°, often to height where it is no longer visible; begins singing during ascent at *c.* 7 m; cruises about for many min, maintaining steady flight without sudden swerves or stoops; during descent gives rapid chatter, checks at end of dive just above ground before landing and disappearing into grass.

NEST: ball of dry grass with 25–30 diam. side-top entrance, lined with plant down; ext. diam. 75, int. diam 55, int. depth 117–130; placed on or near ground in tuft of short grass, with fresh grass blades bent down and woven together over roof.

EGGS: 2–4, rarely 5, laid at 1 day intervals; av. (23 clutches) 3·1; larger (*c.* 3·6) in SW Cape. White, pale pink, turquoise blue or greenish, speckled and spotted with red-brown, mainly at large end. SIZE: (n = 58) 14–17·8 × 10·8–12·5 (15·7 × 11·6).

LAYING DATES: Angola, Feb, Dec; South Africa: Transvaal, Oct–Mar (Oct 1, Nov 6, Dec 8, Jan 9, Feb 5, Mar 3 clutches); Orange Free State, Oct–Feb (Oct 8, Nov 19, Dec 6, Jan 3, Feb 2 clutches); E Cape, Sept–Oct, Dec–Jan; SW Cape Aug–Nov (Aug 17, Sept 28, Oct 31, Nov 12 clutches).

INCUBATION: period 14 days.

DEVELOPMENT AND CARE OF YOUNG: down appears on day 3, eyes open on day 7, fully feathered by day 11 when only distinguishable from adult by yellow underparts and buffer upperparts; growth ceases at 11·5 days. Fledging period: 15–16 days.

BREEDING SUCCESS/SURVIVAL: parasitized by Parasitic Weaver *Anomalospiza imberbis*.

Reference
Vincent, A. W. (1948a).

Cisticola eximius (Heuglin). Black-backed Cisticola. Cisticole à dos noir.

Plate 13 (Opp. p. 240)

Drymoeca eximia Heuglin, 1869. Ibis ser. 2, vol. 5, p. 106–107; Upper Gazelle R. (Sudan).

Range and Status. Endemic resident. 2 records W Casamance (S Senegal) Jan; probably rare resident. Rare NE Ivory Coast and Mali (2 records Bamako, Oct–Nov). Locally common W Sierra Leone, and recently discovered in adjacent W Guinea (Demey 1995). S Ghana (Winneba–Accra Plains), central and S Nigeria. Uncommon guinea savanna zone of Togo. Recorded Chad (locality untraced). In Sudan, uncom-

Cisticola eximius

mon and local south of c 10°N. Frequent at 2100–2400 m in NW Ethiopia. Rare Eritrea. Locally common in Uganda at 900–1500 m. In Kenya, formerly frequent in SW at Mumias and Yala R. (Jackson and Sclater 1938) but no recent records. In Zaïre only in N (Uele savannas) at 20–27°E. Birds referred to this species common in seasonal wetlands at Odzala and between Owando and Oyo, Congo (F. Dowsett-Lemaire, pers. comm.).

Description. *C. e. eximius* (Heuglin): Ethiopia to extreme W Kenya, west to N Zaïre and S Central African Republic. ADULT ♂ (breeding): forehead, crown and nape rich reddish brown or paler, yellow-brown in some individuals (worn and faded?); supraloral streak cream; lores very dark brown; ear-coverts buffy brown; mantle, back, scapulars and uppertail-coverts very dark brown, almost black, streaked (feather edgings) rufous-buff; rump rufous; tail-feathers very dark brown, fringed rufous-buff; entire underparts and underwing buff, palest on chin, throat and centre of belly; flanks rich rufous-buff; primaries and secondaries brown, fringed brownish buff; tertials and wing-coverts very dark brown, fringed rufous-buff. Upper mandible dark brown, lower pinkish grey or horn; eyes light brown; legs and feet pinkish brown. ADULT ♂ (non-breeding): like breeding ♂ but head and back very dark brown, almost black, narrowly streaked buff, occasionally with some bright head colour remaining on nape; rump and uppertail-coverts paler rufous. ADULT ♀ (breeding): like breeding ♂ but head-top darker, streaked dark brown. ADULT ♀ (non-breeding): like non-breeding ♂. SIZE: (10 ♂♂, 10 ♀♀) wing, ♂ 47–53 (49·9), ♀ 43–49 (46·1); tail, ♂ 31–37 (34·6), ♀ 28–36 (33·3); bill, ♂ 11–12 (11·4), ♀ 11–12 (11·3); tarsus, ♂ 18–20 (18·6), ♀ 17–19 (18·0).
  IMMATURE: like breeding adult ♀ but upperparts (including head-top) more rufous, less dark, striped rufous and dark brown; underparts sulphur-yellow. First-winter like non-breeding adult.
  NESTLING: unknown.
  C. e. occidens Lynes: Casamance to Sierra Leone, S Mali, Nigeria. Like corresponding plumages of nominate race but paler, less rufous; underparts paler, chin, throat and centre of belly white; in breeding dress, head-top without red tinge.

C. e. winneba Lynes: Winneba, coast of Ghana. Like corresponding plumages of *occidens* but much paler. Breeding ♂ has head-top mottled buffy brown; rest of upperparts streaked buff and brown; rump and uppertail-coverts rufous; underparts white, washed buff on breast and richer buff on flanks. Breeding ♀ like ♂ but head-top streaked buff and brown, uniform with rest of upperparts. Non-breeding dress undescribed.

Field Characters. In E half of range, breeding ♂ stub-tailed and brightly coloured, with rich red-brown crown, plain or thickly-streaked back (appearing black), plain orange-rufous rump, reddish buff flanks. In W Africa duller, with buffish brown crown and paler, more obviously streaked back. Non-breeding ♂ has black crown with narrow white streaks, often separated from black back by rusty collar. Tail dark with white tip. Crown of ♀ streaked at all times. Voice and flight display distinctive. Occurs with Pectoral-patch Cisticola *C. brunnescens* in Ethiopia, but occupies water meadows while Pectoral-patch is on nearby dry ground, and similar habitat difference with Desert Cisticola *C. aridulus* in E Africa. Most closely resembles Wing-snapping Cisticola *C. ayresii* but always at lower elevations and has longer tail. Most likely to be confused in non-breeding season with Fan-tailed Cisticola *C. juncidis*, whose tail is about the same length but grey with contrasting black subterminal band; Fan-tailed also appears paler above, with much broader buff streaks on crown and back.

Voice. Tape-recorded (45, BRU). Birds in W Africa (Nigeria, Chad) sing irregular medley of notes from perch. Principal component is a thin, liquid, rather piercing 'trree', almost pure but with a slight rolling trill, repeated 3 to 10 times. The series may be preceded by 2–3 squeakier notes, 'ti-tsee-tsee-', and often ends with a lower-pitched, drier 'tsroy' or 'tsray'. Also included at intervals in the perch song are flight song notes and a pure note followed by a trill, 'peeee-tsrrrr', reminiscent of a lightweight Rattling Cisticola *C. chiniana*. Flight song, without wing-snaps, an irregular series of single or double short, burry, cricket-like notes, audible for some distance, which might be mistaken for an insect (Chappuis 1974). Alarm, a double 'troy-troy', like final notes of perch song. Voice in Guinea described as 'a distinctive, sharp, dissonant 'tchereet-tchereet'; a series of rather thin 'tsree-tsree-tsree-tsree . . .' uttered during display flight high above territory, sometimes accompanied by wing-snapping. The 'tchereet' call was occasionally uttered during the display flight' (Demey 1995). Song in Ethiopia apparently somewhat different. Displaying bird gives excited medley of notes during climb to cruising height, a loud 'chickle' during circling flight accompanied by wing-snaps, and whirring 'twee-twee' during dive to earth, audible for 50 m (Cheesman and Sclater 1935).

General Habits. Inhabits open, short-grass flats, laterite pans and plains on dry ground, but including areas subject to flood which become, in wet season, marshes and meadows with rank grass 1 m high; also (Nigeria) sandstone hills with poor, eroded soil; in Ethiopia occupies highland grassland and short, stiff, rush-like aquatic grassland which becomes marshy in rains.

Occasionally perches exposed on grass stalk, but usually keeps out of sight on ground. Runs away from human; when flushed, flies low *c.* 30 m, drops into grass, then runs away from the spot. Alarmed bird circles in flight *c.* 6 m high, with head high, tail low, calling (Cheesman and Sclater 1935).

Food. Ants, small homopteran bugs.

Breeding Habits. In display flight, *c.* 60 m high, describes circle *c.* 130 m diam. for *c.* 2–3 min while singing and snapping wings, then dives with wings closed, calling, almost to ground, then rises back up to *c.* 6 m once or twice, finally dropping into grass. Displaying sharply curtailed by heavy rain. Also sings from grass tops.

NEST: pear-shaped with side-top entrance; thin, of dry grass and spider web, lined with plant down; neighbouring green blades drawn down to form roof over nest; nest placed in tuft of grass or rushes, or, in wet habitat, often on grass tussock islet; may touch water surface.

EGGS: Nigeria (2 clutches) 3–4; Ethiopia (4 clutches) 2–3 (2·7). Medium gloss; white, immaculate or with purple-brown marks, small red-brown spots and occasional underlying greys, sometimes clustered at big end. SIZE: (n = 5, Ethiopia) av. 16·0 × 11·4.

LAYING DATES: Guinea (♀♀ carrying food, Sept); Sierra Leone, June (rains); Ghana, May; Nigeria, July–Aug, Nov; Ethiopia, July, Sept.

INCUBATION: nest lining added during incubation.

References
Chapin, J. P. (1953a).
Cheesman, R. E. and Sclater, W. L. (1935).
Lynes, H. (1930, 1932).

Cisticola dambo Lynes. Dambo Cisticola. Cisticole dambo.

Plate 13
(Opp. p. 240)

Cisticola dambo Lynes. 1931. Bull. Br. Orn. Club 52, p. 5; Nasondoye, southern Belgian Congo.

Range and Status. Endemic resident. NE Angola in S Lunda (Missão de Luz) and N Moxico (L. Cameia and L. Dilolo); S Zaïre in NW Kasai (near Banda) and Shaba (Dilolo and Kamina to Upemba and the Marungu Highlands) and NW Zambia in Mwinilunga and Balovale (Minyanya Plain). Locally common to uncommon within restricted range.

Description. *C. d. dambo* Lynes: range of species except NW Kasai. ADULT ♂ (breeding): forehead and crown blackish, finely and indistinctly streaked and flecked with buff; lores, cheeks and stripe over eye buff; nape buff, indistinctly streaked brown. Back streaked blackish, feathers edged rusty-buff, rump tawny, blackish streaked, uppertail-coverts tawny. Tail-feathers blackish margined tawny-buff, with paler, buffy-white tips. Chin and throat whitish, suffused with buff, sides of neck, breast and flanks to thighs tawny-buff, centre of belly white, undertail-coverts pale buff. Wings brown, outer webs of primaries buff-margined, scapulars and wing-coverts blackish, edged rusty-buff; shoulder, underwing-coverts and axillaries pale buff. ADULT ♂ (non-breeding): forehead and crown buffier, more uniform with back, mantle with broader buff feather margins; sides of breast and flanks tawnier. Adult ♀ in breeding plumage similar to ♂ but with forehead and crown more buff-streaked. Bill greyish flesh, culmen dusky, interior of mouth flesh; eyes brown; legs and feet flesh-coloured. SIZE: (37 ♂♂, 28 ♀♀) wing, ♂ 51–56 (54·7), ♀ 46–50 (47·8); tail, ♂ 40–46 (43·9), ♀ 35–41 (37·7); bill, ♂ 10·5–12 (11·1), ♀ 10·5–11·5 (10·9); tarsus, ♂ 19·5–21·5 (20·9), ♀ 18·5–21·5 (20·0). WEIGHT: ♂ (n = 9) 8–11 (9·3); ♀ (n = 4) 7–10 (8·7).

IMMATURE: like non-breeding adult above, chin to breast and upper flanks washed pale yellow, lower flanks washed reddish.

C. d. kasai Lynes: NW Kasai. Less heavily marked black above, broader reddish edges to secondaries and wing-coverts, underparts and tips to tail feathers rustier.

Field Characters. An uncommon and local cloud-scraper with a restricted range in central Africa (E Angola, SE Zaïre, NW Zambia); nevertheless overlaps with 5 other members of the group, and occupies similar short-grass habitat. Tail black and noticeably longer than in other cloud-scrapers; black and tawny-streaked upperparts similar to Pale-crowned Cisticola *C. cinnamomeus* and Wing-snapping Cisticola *C. ayresii*, but these have very short tails; other species appear paler. Best distinguished by voice and territorial behaviour, which includes wing-snaps.

Voice. Tape-recorded (86, 87, B, F, KEI). Clicking notes given during ascent into air for song-flight; circling birds give weak, short phrases: a burry 'wut-chichi', peevish 'chee-chee', a more pure 'woo-pee-pee', sometimes a longer 'cher-pee-pee-pee-pee'; all rather faint and insignificant. Wing-snapping accompanies aerial song and dives; hard 'tikki-tikki-tikki . . .' given during final descent. Song from perch, nasal 'jee-wit'.

General Habits. Frequents open, short, seasonally wet grasslands and watershed plains; in NW Zambia generally less marshy situations than Pale-crowned Cisticola and less dry ground than Wing-snapping Cisticola.

Occurs singly or in pairs during breeding season, apparently feeding on or near the ground in grass tufts, from which it can be flushed.

Food. Insects.

Breeding Habits. Territorial. In territorial display observed at Chitunta Plain, NW Zambia, several birds rose into air at once, gave sharp clicking notes similar to those made by local frog *Afrixalus* sp.; when *c.* 150 m in air they circled, calling (see *Voice*), for several min. At intervals they dived with snapping wings, then looped back up again and continued circling; in the end each bird plummeted to earth calling a hard 'tikki-tikki-tikki-tik'. Just before reaching the ground it bounced back up a short way, gave a few more 'chee' notes and wing-snaps, then dropped into grass (S. Keith, tape and notes).

NEST: of ball type, with side-top entrance, placed close to ground in short grass, built into living grass stems.

EGGS: 2 (occasionally 3); light to palest turquoise blue, variably marked with rather bold spotting to faint stippling, with shades of reds, browns and slate. SIZE: (n = 9) 14·9–15·6 × 11·5–11·7 (15·1 × 11·5).

LAYING DATES: Zaïre, Oct (♀ with egg in oviduct, Jan); Angola, Jan, Dec; Zambia (many immatures, late Apr).

Plate 13
(Opp. p. 240)

Cisticola brunnescens Heuglin. Pectoral-patch Cisticola. Cisticole brune.

Cisticola brunnescens Heuglin, 1862. J. Orn., vol. 10, p. 289; Gudofelassi, Serowi, Ethiopia, 6000ft.

Forms a superspecies with *C. cinnamomeus*.

Range and Status. Endemic resident. Locally common to abundant in montane areas of Cameroon, mostly above 1800 m. Common in savannas of Gabon (coast, Lopé and SE: P. Christy, pers. comm.) and Congo (Odzala Nat. Park, Léfini Reserve 3°20'S, 15°30'E: F. Dowsett-Lemaire, pers. comm.). Frequent to common highland Ethiopia 1400–2500 m, usually above 1800 m. Eritrea above 2300 m. Probable resident (no recent records) NW Somalia on border with Ethiopia. Locally common central Kenya to N Tanzania, 1400–2500 m. One record Uluguru Mts, central Tanzania, could be this species or Pale-crowned Cisticola *C. cinnamomeus* (Stuart and Jensen 1985).

Description. *C. b. brunnescens* Heuglin: NE, central, S and E Ethiopia, NW Somalia. ADULT ♂ (breeding): forehead, crown and nape mottled sandy brown and rusty brown; lores very dark brown; ear-coverts and sides of neck sandy buff; mantle, back and scapulars very dark brown, streaked sandy (feather centres very dark brown, fringes sandy); rump and uppertail-coverts streaked very dark brown and tawny rufous; tail-feathers very dark brown, fringed and tipped sandy buff. Entire underparts and underwing buff, palest on chin and centre of belly, richest on flanks, dark centres to feathers at sides of breast ('pectoral patches') usually hidden; primaries, secondaries, tertials and wing-coverts very dark brown, fringed tawny rufous. Upper mandible dark brown, lower horn; eyes pale reddish brown; feet pale brown. ADULT ♂ (non-breeding): like breeding ♂ but crown streaked, like back; underparts often richer buff (unfaded). ADULT ♀ (breeding and non-breeding): like non-breeding ♂ but underparts more richly coloured, rufous-buff when fresh. SIZE: wing, ♂ (n = 10) 52–59 (55·5), ♀ (n = 10) 46–52 (49·9); tail, ♂ (n = 10) 26–36 (31·2), ♀ (n = 9) 26–32 (29·0); bill, ♂ (n = 10) 11–13 (12·2), ♀ (n = 10) 11–13 (12·3); tarsus, ♂ (n = 10) 20–22 (20·7), ♀ (n = 10) 18–20 (19·0).

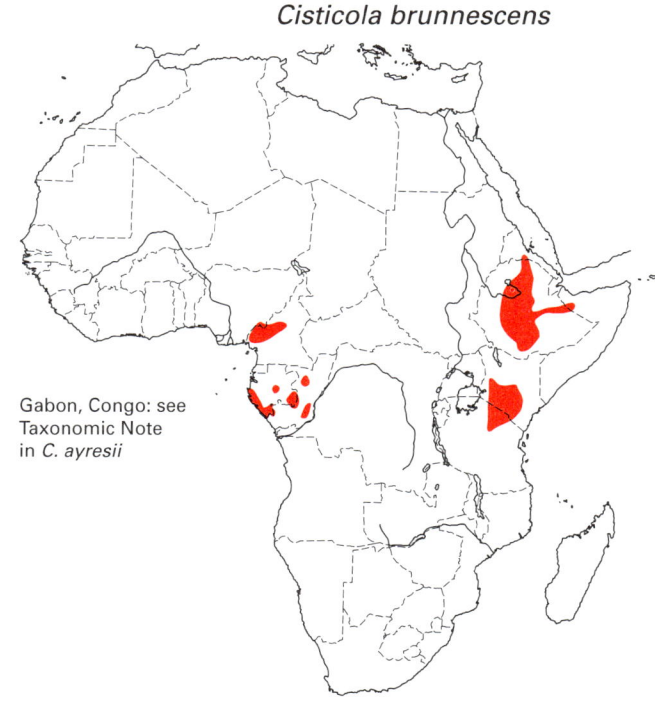

Cisticola brunnescens

Gabon, Congo: see Taxonomic Note in *C. ayresii*

IMMATURE: like adult ♀ but upperparts less heavily streaked dark (feather centres less broad); underparts washed sulphur-yellow.

NESTLING: unknown.

C. b. wambera Lynes: Wambera Plateau, NW Ethiopia. Darker than corresponding plumages of nominate: upperparts streaked blackish brown and rufous; breast, flanks and under-tail-coverts rufous-buff.

C. b. nakuruensis van Someren: Kenya and N Tanzania highlands, west of Rift Valley. Perennial dress: like breeding plumage of nominate but crown of ♂ often slightly darker; pale areas of upperparts paler, less rufous (especially in ♀). SIZE: (9 ♂♂, 3 ♀♀) wing, ♂ 52–55 (53·6), ♀ all 50; tail, ♂ 27–31 (29·4), ♀ 27–30 (29·0); bill, ♂ 12–13 (12·3), ♀ 12–13 (12·3); tarsus, ♂ 20–22 (20·4), ♀ all 19.

C. b. hindii (Sharpe): Kenya and N Tanzania highlands, east of Rift. Like *nakuruensis* but averages slightly paler on upperparts and underparts. WEIGHT: (Kenya) 1 ♂ 8, 1 immature ♂ 6·5.

C. b. lynesi (Bates): W Cameroon highlands. Perennial dress like breeding nominate but upperparts darker (crown, and feather-edgings of rest of upperparts, more rufous), especially in ♂; ♀ intermediate between breeding ♀ of *brunnescens* and ♂ *lynesi*. Eyes grey-brown. Smaller than other races. SIZE: wing, ♂ (n = 9) 52–54 (52·3), ♀ (n = 6) 46–49 (47·3); tail, ♂ (n = 9) 28–31 (29·2), ♀ (n = 6) 27–30 (28·5); bill, ♂ (n = 9) 11–12 (11·8), ♀ (n = 5) 11–12 (11·4); tarsus, ♂ (n = 9) 20–21 (20·4), ♀ (n = 5) 18–20 (19·0).

C. b. mbangensis Chappuis and Erard: Adamawa Plateau, N Cameroon. Like *lynesi* but upperparts paler, less streaky (streaks narrower and browner); crown rustier; rump bright, dark maroon. Smaller than *lynesi*.

TAXONOMIC NOTE: racial attribution of birds in Congo and Gabon undetermined (see Taxonomic Note under *C. ayresii*); their song matches that of Cameroon races (P. Christy, C. Chappuis and F. Dowsett-Lemaire, pers. comm.).

Field Characters. Tiny and short-tailed. ♂ in breeding plumage has top of head rusty, almost plain, contrasting with streaked nape and hindneck. Rump tawny but streaked blackish, contrasting little with back (rump brighter and redder in Wing-snapping Cisticola *C. ayresii*). Tail black with whitish tip. Blackish patches at sides of breast, obvious in museum skins, are hard to see in the field and of questionable utility; these become indistinct streaks in non-breeding plumage and in ♀. Very similar to Wing-snapping Cisticola but somewhat paler and more uniformly streaked above, without black on crown, tail longer in breeding season. Best distinguished by voice and display flight.

Voice. Tape-recorded (45, LEM). Flight song in Cameroon a long series of high-pitched, hard, dry notes, 'dzip-dzip-dzip . . .', at rate of 3 per s, with quality of stones being rubbed together, like Common Stonechat *Saxicola torquata*. At end of song notes change quality, 'dzeep-dzeep . . .' and accelerate as bird plunges to earth. Song accompanied by intermittent clacking or snapping sounds which might be taken for wing-snaps but they are also given by perched bird (Chappuis 1974). Birds in Kenya give 'monotonous repetition of sharp 'tsik' or 'tsip' notes, sometimes accelerating, and often (but not invariably) accompanied by wing-snaps given in flight' and end with faster, chattering 'tsiktsiktsiktsik . . .' as bird descends (Zimmerman *et al.* 1996).

General Habits. In Cameroon, inhabits highland meadows of very short grass on stony ground, with small patches of bushes (0·8–1 m high). Occupies similar habitat in Ethiopia and E Africa, mainly on dry ground, but also moister areas with tussock grass up to 1 m high and giant lobelia/*Alchemilla* moorland, giant heath moorland, *Rumex* 'heath', stunted heather-like scrub, other types of scrub and farmland. In W and central Kenya usually at lower elevations than *C. ayresii* and often on drier ground, but the 2 overlap. In Gabon, in coastal and inland short- or tall-grass savannas; in Congo, *Loudetia simplex* grassland at 300–700 m (F. Dowsett-Lemaire, pers. comm.). At Léfini, Congo, sympatric with Pale-crowned Cisticola, but separated by habitat (Pale-crowned in bogs).

Usually solitary or in pairs. Spends much time on ground, where it forages. When disturbed perches on thin stem; when flushed flies short distance then drops out of sight.

Food. Small insects, including grasshoppers, small black weevils.

Breeding Habits. ♂ performs high, steady or erratic, circular display flight, *c.* 400 m in diam., up to 100 m above ground, lasting *c.* 2 min.; ends in plunge with closed wings and (often) a final up-turn; sings, and sometimes snaps wings, during cruise and often calls during plunge. Also sings and calls from perch.

NEST: a sphere or oval, max. ext. diam. 100, with side-top entrance diam. 30; made of dry grass, strengthened with spider web; placed with opening away from prevailing wind, very near ground in grass tuft.

EGGS: NW Ethiopia 3–4; Kenya 2–4. Moderately glossy; very variable in colour, white, pale blue or pale greenish blue, freely freckled with rusty red, various of shades of brown and/or underlying purple-greys, mostly denser or in a ring at big end; in NW Ethiopia bluish white, marked all over with small violet spots, mostly at big end, or occasionally thickly spotted with dark purple. SIZE: (n = 3, NW Ethiopia) 17–18 × 12·3 (17·4 × 12·3); (n = 8, Kenya, *C. b. nakuruensis*) av. 15·1 × 11·6; (n = 4, Kenya, *C. b. hindii*) av. 16·0 × 12·0.

LAYING DATES: Cameroon, Feb–Apr; NW Ethiopia, Aug; Eritrea, June. E Africa mainly rains: Kenya, Feb, May–July, Sept, Nov–Dec; Region B, Apr; Region D, Jan, Mar–July (peak Apr–May).

INCUBATION: by both sexes (Cheesman and Sclater 1935). When disturbed, ♀ runs from nest.

BREEDING SUCCESS/SURVIVAL: in Ethiopia commonly parasitized by Pin-tailed Whydah *Vidua macroura*; sometimes 2 whydah eggs in one nest. Whydah young reared without host young, so host eggs or young probably all ejected.

References
Chappuis, C. and Erard, C. (1973).
Cheesman, R. E. and Sclater, W. L. (1935).
Jackson, F. J. and Sclater, W.L. (1938).
Lynes, H. (1930).

Plate 13
(Opp. p. 240)

Cisticola cinnamomeus Reichenow. Pale-crowned Cisticola. Cisticole châtain.

Cisticola cinnamomea Reichenow, 1904. Orn. Monatsber., vol. 12, p. 28; Ngomingi, Uhehe Dist., Tanzania.

Forms a superspecies with *C. brunnescens*.

Range and Status. Endemic resident. Common SE Gabon and Congo/Zaïre from about 1°N, 14°30'E (Odzala Nat. Park) to Gamboma, Kanungu and Léfini Reserve (3°20'S, 15°30'E), where sympatric with *C. brunnescens* (F. Dowsett-Lemaire, pers. comm.). Locally common S Tanzania (Iringa Plateau) above 1800 m; one record Uluguru Mts, central Tanzania, could be this species or *C. brunnescens* (Stuart and Jensen 1985). Local S Zaïre, Angola (Huambo, Mocussueze, extreme NE Moxico, Malange). Locally common Zambia: in N, west of Luangwa Valley but absent between L. Mweru and L. Tanganyika; in S but not Kafue; *c*. 20 pairs in 1 km² on dambo at Kitwe (Penry 1985). Recorded once on edge of Okavango Delta, N Botswana. Local on Mashona Plateau (Zimbabwe) above 1220 m and adjacent highland Mozambique. Common Limpopo floodplain in S Mozambique. Local in Swaziland and South Africa southeast to Transkei, but in Transvaal only in SE (where locally common); generally rare in Orange Free State, but common, at least in breeding season, near Harrismith; only recorded in summer in Orange Free State and Natal midlands.

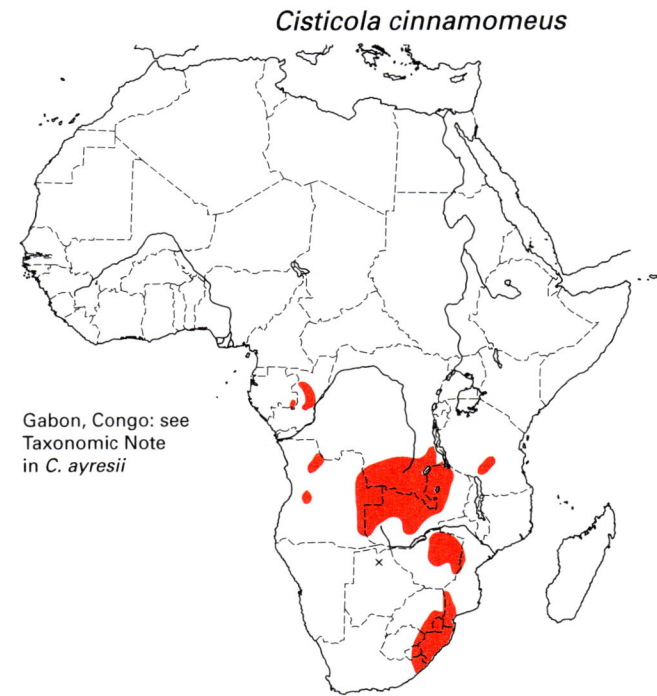

Cisticola cinnamomeus

Gabon, Congo: see Taxonomic Note in *C. ayresii*

Description. *C. c. cinnamomeus* Reichenow: S Tanzania, SE Zaïre, Zambia, Zimbabwe, Botswana, Angola. ADULT ♂ (breeding): forehead, crown and nape yellow-brown, mottled with dark brown; lores, supercilium and under eye black washed with yellow-brown; ear-coverts and sides of neck yellow-brown; mantle, back and scapulars very dark brown, virtually black, streaked with rufous-buff (feather centres black, edges rufous-buff); rump and uppertail-coverts dark rufous; tail-feathers very dark brown, edged rufous-buff; throat, breast, belly and undertail-coverts buff, flanks rufous; primaries, secondaries, tertials and wing-coverts very dark brown, darkest on tertials and wing-coverts, and edged rufous-buff; underwing-coverts and axillaries rich buff. Upper mandible dark brown, lower pinkish brown; eyes light olive-grey; legs and feet pinkish brown. ADULT ♂ (non-breeding): like breeding plumage but entire upperparts, including head-top and rump, streaked very dark brown and rufous; tail fringes rufous; ear-coverts and underparts darker rufous-buff. ADULT ♀ (breeding): upperparts like non-breeding ♂ but rump generally unstreaked rufous; underparts like breeding ♂. ADULT ♀ (non-breeding): like non-breeding ♂. SIZE: wing, ♂ (n = 10) 52–55 (53·6), ♀ (n = 10) 45–50 (47·8); tail, ♂ (breeding, n = 4) 32–34 (33·2), (non-breeding, n = 5) 37–42 (39·0), ♀ (breeding, n = 5) 32–33 (32·6), (non-breeding, n = 4) 32–40 (35·8); bill, ♂ (n = 10) 11–12 (11·5), ♀ (n = 10) 11–12 (11·4); tarsus, ♂ (n = 10) 19–22 (20·6), ♀ (n = 10) 18–20 (19·2).

IMMATURE: like non-breeding adult but upperparts more rufous (feather edgings broader); underparts washed pale sulphur.

NESTLING: unknown.

C. c. midcongo Lynes: SE Gabon (Teke Plateau), both banks of R. Congo at 2°S. Perennial dress, like breeding dress of nominate race but head-top plainer red, less mottled. Small. SIZE: (1 ♂) wing 50; tail 29; bill 12; tarsus 20.

C. c. egregius (Roberts) (including '*taciturnus*': colour differences quoted by Clancey (1992) not supported by material in Natural History Museum, London, inland specimens match '*taciturnus*' better than those from latter's coastal area; no significant size difference): S Mozambique to South Africa. Like *cinnamomeus* but breeding ♂ paler, duller on upperparts; other plumages very slightly duller on upperparts; breeding ♂ and ♀ paler on underparts. SIZE: (15 ♂♂, 5 ♀♀) wing, ♂ 51–55 (52·5) [including 6 '*taciturnus*' 51–54 (52·0)], ♀ 47–49 (47·8) [1 '*taciturnus*' 48]; tail, ♂ breeding (n = 14) 28–34 (31·5), non-breeding (n = 1) 39, ♀ breeding (n = 4) 29–33 (31·2), non-breeding (n = 1) 38; bill, ♂ 12–13 (12·3), ♀ all 12; tarsus, ♂ 20–22 (21·3), ♀ 19–20 (19·6).

Field Characters. Tiny and short-tailed. ♂ *egregius* in breeding plumage distinctive: top of head appears plain (some very light mottling at sides), dull greyish olive turning to whitish buff by end of summer (diagnostic: Maclean 1993); nape and hindneck variably mottled. Blackish area from base of bill to and part way around eye, separated by pale supercilium from narrow black line along side of crown. Breeding ♂ *cinnamomeus* has yellow-brown head top with dark mottling. Rump unstreaked and orange-rufous, conspicuous in flight. ♀ and non-breeding ♂ have streaked crown and are probably not separable in the field from other 'cloudscrapers'. Best distinguished by voice and display flight.

Voice. Tape-recorded (86, 87, 88, 91, B, F, MOY). Song, up to 7 very high, thin, measured notes with a dip at the end, 'eeeip' or 'eeely', at rate of 2 per s, followed by up to 13 somewhat lower notes with a slight tremolo, 'rreee, rreee. . . .' as bird dives, then 2 more 'rreee' as it swoops upward again; at end of flight display dives silently to grass without calls or wing-snaps (Maclean 1993). Alarm call of ♂ a nasal 'dee' reminiscent of Bleating Warbler *Camaroptera brachyura*; alarm call of ♀, quiet 'chitty chitty chitty' (Maclean 1993).

General Habits. Inhabits rank or fine, moist grassland and marshes. In Congo mainly in short, dry grassland, locally in moist habitat; where sympatric with Pectoral-patch Cisticola *C. brunnescens* (Léfini Reserve), separated by habitat, with Pectoral-patch on drier ground, Pale-crowned in bogs (F. Dowsett-Lemaire, pers. comm.). Feeds low in grass and on ground. When disturbed at nest, flies away and drops into cover.

Food. Small insects.

Breeding Habits. Perhaps a co-operative breeder: one adult ♀ in non-breeding dress was shot while consorting with a pair with fledglings (Lynes 1934). ♂ has circular display flight, ending in plunge; sings and snaps bill during cruise and descent. Also sings and snaps bill when perched.

NEST: a ball (diam. 60) or oval, with circular side-top entrance (diam. 40), loosely made of dry, soft grass with a little spider-web, lined thinly at sides, thickly (5 mm) at base, with plant down and web. More down added during incubation. Placed 2–15 cm above ground, hidden in dense tufty grass; often concealed by 'bower' of bent-over grass blades which are woven into nest; these bowers are built before nest and often several made, of which one chosen for nest. In marshes, may nest in drier patches.

EGGS: Zambia, Zimbabwe 2 (n = 2), 3 (14), 4 (7), 5 (1); Natal 1 (n = 2), 3 (6), 4 (2); Transvaal (3 clutches) 4–5 (4·7). Very variable, blue, pale greenish blue or white, with fine stippling of dull reds, purple-browns and/or purple-greys, often with zone or ring of slightly deeper shade, or denser speckling, at big end. SIZE: (n = 8, Tanzania) av. 14·9 × 11·0; (n = 11, S Zaïre) 14·3–16 × 10–11·8 (15·0 × 11·0); (n = 14, Transvaal) av. 15·5 × 11·6.

LAYING DATES: E Africa: Tanzania, Nov–Jan; Region C, Jan. Zaïre, late Oct–Jan; Zambia, Nov–Jan (peak Nov); Zimbabwe, Nov–May (peak Dec–Jan, 39 records); Angola, Nov–Feb; Mozambique, rains; Botswana, Oct; South Africa, Oct, Dec–Mar (peak Jan).

INCUBATION: period probably *c.* 11–13 days.

DEVELOPMENT AND CARE OF YOUNG: hatching spread over 2 days; young in nest *c.* 12–14 days.

References
Lynes, H. (1930).
Penry, E. H. (1985).
Vincent, A. W. (1948a).

Cisticola ayresii Hartlaub. Wing-snapping Cisticola; Ayres's Cisticola. Cisticole gratte-nuage.

Plate 13
(Opp. p. 240)

Cisticola ayresii Hartlaub, 1863. Ibis 5, p. 325; Natal.

Range and Status. Endemic resident, E and southern Africa from Congo and Sudan southwards. Common Imatong Mts above 2000 m (S Sudan) and W and central Kenya highlands 1700–3700 m (mostly above 2400 m); 1200–2000 m in NW Tanzania (frequent Serengeti) and S Uganda, and 1800–2400 m in Iringa Highlands (S Tanzania). Occurs in Kidepo Valley, NE Uganda. Abundant throughout Burundi above 1200 m; occurs Rwanda. Locally common to abundant in Zaïre, in highlands along E border and in SE; common at 1000 m on plains north of L. Edward, otherwise mainly above 1400 m. Status in Gabon and Congo unclear, due to confusion with *C. brunnescens*; the situation is currently under investigation (see Taxonomic Note). Locally abundant above 2000 m on Nyika Plateau and in N Mwinilunga District of Malaŵi. In Mozambique, only in W Manica e Sofala on border with Zimbabwe; locally common E highlands of Zimbabwe, generally above 1500 m (down to 1100 m in Chimanimani Mts), up to at least 2200 m. Common on highveld of Transvaal and Orange Free State; less common W Orange Free State; also in lowland Lesotho; in coastal areas of Natal and E Cape Province common from 300–2000 m, scarce lower. In Angola, common on interior plateau of N Huila District to Huambo and S Lunda.

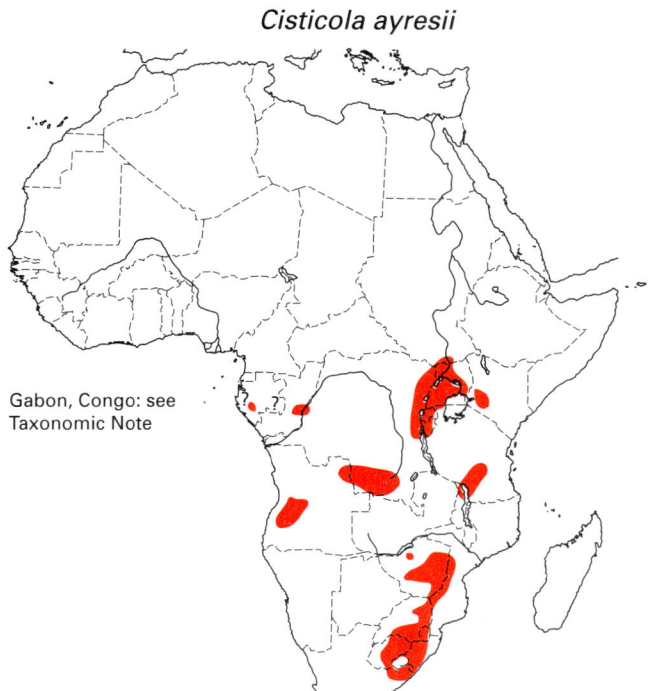

Gabon, Congo: see Taxonomic Note

Description. *C. a. ayresii* Hartlaub: South Africa, S Mozambique, E Zimbabwe, E Zambia, Malawi, S Tanzania, SE Zaïre (Shaba), isolated population in central Angola. ADULT ♂ (breeding): upperparts streaked blackish brown and sandy brown, more rusty on rump and uppertail-coverts and sometimes (faded?) on head; tail, primaries, secondaries, tertials and wing-coverts blackish brown, tipped and fringed buff. Face, underparts and underwing buff, palest on chin, throat and belly, deepest on flanks. Upper mandible very dark brown, lower pale grey or pale pinkish brown; eyes pale yellow-brown; feet pale pinkish brown. ADULT ♂ (non-breeding): pale streaks of upperparts richer rusty brown; underparts deeper buff; fringes on wing and tail broader. Sexes alike. SIZE: (10 ♂♂, 10 ♀♀) wing, ♂ 46–53 (48·7); ♀ 42–47 (45·2); tail ♂ breeding (n = 6) 22–30 (25·8), non-breeding (n = 4) 31–33 (32·0) ♀ breeding (n = 6) 25–28 (26·7), non-breeding (n = 4) 31–33 (32·2); bill, ♂ 11–13 (11·5), ♀ 9–12 (10·9); tarsus, ♂ 18–22 (19·4), ♀ 17–19 (18·1). WEIGHT: (SE Zaïre) ♂♂ (n = 3) 7–8 (7·3) unsexed (n = 3) 6–7 (6·7), 1 juvenile ♂ 8, 1 juvenile ♀ 8, 1 unsexed juvenile 8; (South Africa) 1 ♂ 16·4, 1 unsexed 9·4.

IMMATURE: like non-breeding adult but underparts washed lemon-yellow.

NESTLING: unknown.

C. a. itombwensis Prigogine: mts northwest of L. Tanganyika. Upperparts darker (blackish areas more extensive, paler areas duller rufous). Averages larger. SIZE: wing, ♂ (n = 9) 50–53 (50·9), F (n = 2) 47, 47.

C. a. mauensis (Someren): Kenya highlands. Like *ayresii* (breeding dress) but no obvious seasonal plumage change; crown slightly darker; upperparts with feather edges slightly paler, buffier, intermediate between breeding and non-breeding *ayresii*. SIZE: wing, ♂ (n = 5) 48–50 (49·0), ♀ (n = 6) 45–47 (46·0); tail, ♂ (n = 5) 23–25 (24·4), ♀ (n = 6) 23–29 (26·8); bill, ♂ (n = 5) 11–12 (11·2), ♀ (n = 5) 11–12 (11·6); tarsus, ♂ (n = 5) 19–20 (19·8), ♀ (n = 6) 18–19 (18·2).

C. a. entebbe Lynes: extreme E Zaïre from Kivu northward, Rwanda, Uganda, NW Tanzania, W Kenya. Like *mauensis* except underparts show similar seasonal change in colour as in nominate (but has less obvious change in upperparts than nominate). Smaller. SIZE: wing, ♂ (n = 6) 45–48 (46·5), ♀ (n = 6) 43–45 (43·8); tail, ♂ (n = 5) 21–26 (24·0), ♀ (n = 5) 23–27 (24·6); bill, ♂ (n = 5) 10–12 (10·8), ♀ (n = 6) 10–11 (10·7); tarsus, ♂ (n = 6) 18–19 (18·2), ♀ (n = 6) 18–19 (18·7).

C. a. gabun Lynes: Gabon, Congo, NW Zaïre. Like *entebbe* but dark streaks on upperparts narrower.

C. a. imatong Cave: Imatong Mts (S Sudan), possibly NE Uganda. Like *entebbe* but upperparts slightly darker; underparts much deeper rufous-buff. Seasonal change in plumage slight. Larger. SIZE: wing, ♂ (n = 6) 48–43 (51·7), ♀ (n = 6) 48–51 (49·5); tail, ♂ (n = 5) 28–34 (32·0), ♀ (n = 6) 28–35 (32·2); bill, ♂ (n = 5) 11–13 (12·4), ♀ (n = 6) 11–12 (11·8); tarsus, ♂ (n = 6) 19–21 (20·0), ♀ (n = 6) 19–21 (19·8).

TAXONOMIC NOTE: P. Christy (pers. comm.) reports that around Port-Gentil, Gabon, the type locality of *C. a. gabun*, and elsewhere in Gabon, he and P. Alexander-Marrack found *C. brunnescens* (identified by voice from a Chappuis tape) common, although it has not hitherto been reported from Gabon, while several years' observations revealed no sign of *C. ayresii*. A recording identified by Chappuis as *brunnescens* was also made in E Gabon (east of Franceville) by L. R. Macaulay. The possibility thus exists that *C. a. gabun*, both in Gabon and on the middle Congo R., may be referable to *C. brunnescens*. However, Y.-M. de Viviès (pers. comm.) reports *ayresii* common in savanna around Mouila (Gabon), while Lynes (1938b) reported co-occurrence of *C. 'a.' gabun* and *C. cinnamomeus midcongo* at Kunungu on the middle Congo R.; *gabun* differs noticeably in plumage from *midcongo*. Further, M. A. Traylor (pers. comm.) re-examined the series of *C. brunnescens* and *C. ayresii* in the Field Museum, Chicago, from Gamboma, Congo, and confirms that one of them is *ayresii*, the rest *brunnescens*. It is possible, therefore, that there exists widespread sympatry between *ayresii*, *brunnescens* and *cinnamomeus* in the region.

Field Characters. Neat and slim-looking, with extremely short tail; the smallest cloudscraper, although that is not apparent in the field except perhaps in the breeding season, when tail even shorter (tail of Pectoral-patch Cisticola *C. brunnescens* same length all year); legs shorter and slimmer than Cloud Cisticola *C. textrix*. In southern Africa breeding ♂ distinguished by plain rusty crown, mottled only on nape and hind-neck (crown of Pale-crowned Cisticola *C. cinnamomeus* plain but pale olive or whitish buff; crown of other species streaked). Black loral line present in breeding ♂ but narrower than in Pale-crowned, extending under eye but not over it, separated from crown by pale supercilium. In E Africa resembles Pectoral-patch but brighter and darker above; breeding ♂ has rusty crown streaked black, sometimes so heavily that whole top of head appears black, although forehead only lightly mottled, thus appearing paler; underparts strongly tinged rusty buff. Orange-red rump partly streaked, appearing more prominent than Pectoral-patch but less prominent than Pale-crowned. Tail black. Hind-collar paler than streaked crown and back, but this feature also present in some Pectoral-patch. Non-breeding ♂ and ♀ have crown completely streaked and pale lores; probably not safely distinguishable in the field from other cloudscrapers. In Kenya occurs at higher altitudes than Pectoral-patch, but partly sympatric with it elsewhere in E Africa, and in southern Africa sympatric with all other cloudscrapers. Best distinguished from all species by voice and wing-snapping display flight.

Voice. Tape-recorded (45, 87, 88, 91, B, C, F). Song, given only in air, a repeated phrase of 3–10 high, thin, lisping and piping notes: 'tsee hey-ho tsee-ha tsee hey-hoo tsee-ho' (Kenya); 'ter pee-pee-pee' (South Africa); also, e.g. 'chitik chitik tsi tsi tsi', 'chiki pee pee pee', 'ser SIT sue' (Maclean 1993). Songs vary considerably from one bird to another, but each bird tends to repeat its own song without variation. Cruising bird sometimes intersperses a medley of conversational notes between the set phrases. Song punctuated at intervals by single or double clicking notes and wing-snaps as bird dives: 'tik tik tik tik . . .' or 'pata pata pata pata . . .'; snaps often in loud, rapid volleys. Alarm, near nest or fledglings, rapid squeaky 'chiki chiki chiki' accompanied by loud wing-snaps (Maclean 1993).

General Habits. Found on short grasslands (grass usually < 45 cm high), including high-altitude meadows, open grassland of *Themeda triandra* and other spp.; on dry ground and damper areas with tussocks, often in areas with bushes or bracken; in SE Zaïre in marshes; in Natal near marshy ground in winter. Attracted to freshly burnt grass. Occurs on airfields.

Skulking, although occasionally perches on grass

tops. Solitary or in family parties. Flicks wings when disturbed, with sound like locust flight. Hops about in grass; also runs on ground. Gleans prey from grass stems and leaves.

Migrates locally away from burnt grassland in winter in southern Africa.

Food. Insects, mainly bugs, flies, grasshoppers, also small beetles; spiders and seeds.

Breeding Habits. A cloudscraper: sings in display flight at 100 m or more, cruising around for c. 5 min., broken at intervals by erratic swerves and swoops, pulling up and darting about. Snaps wings during this display, especially during swoops.

NEST: a ball with side entrance, loosely made of grass leaves, lined with plant down or fine seed-heads; hidden by growing grass blades bent and fixed over top; placed on or close to ground in grass tuft.

EGGS: 1–5. Angola 2; Zaïre 3–4; E Africa usually 3; Malaŵi, Zimbabwe 3 (n = 3), 4 (3); Natal 2 (n = 2), 3 (9), 4 (6), 5 (1). Very variable in colour, including white, light blue or pale green, plain or marked with rusty red, brown, slate or pinkish and purple spots and blotches. SIZE: (n = 2, Kenya) 15·3 × 11·2, 15·9 × 11·3; (n = 2, Angola) av. 14·7 × 10·6; (Zaïre) 15–15·5 × 11·4–11·9; (n = 33, southern Africa) 13·2–16·5 × 10·5–12·2 (15·2 × 11·5).

LAYING DATES: Sudan, Apr–June; Rwanda, Nov, Dec, Mar–May; E Zaïre (Itombwe,) probably Dec–May (non-breeding dress July, Sept–Oct); SE Zaïre, Jan–May. E Africa mainly in rains: Kenya and Uganda, Apr–July (long rains), possibly also Nov–Dec (short rains); Tanzania, Dec–Feb; Region A, Apr, May, Nov; Region B, Apr, May; Region C, Dec; Region D, Apr, May. Malaŵi, Nov, Dec; Angola, Nov–Feb; Gabon, Feb; Zambia, Jan; Zimbabwe, Oct–Mar; S Mozambique, Nov 'onward'; southern Africa, Sept–Mar (peak Dec–Feb in Natal, Dec–Jan in Transvaal).

INCUBATION: by ♀; period c. 14 days.

DEVELOPMENT AND CARE OF YOUNG: young in nest for c. 14 days.

BREEDING SUCCESS/SURVIVAL: in southern Africa, many nests predated by Barn Owl *Tyto alba* and Marsh Owl *Asio capensis*. Parasitized by Parasitic Weaver *Anomalospiza imberbis*, whose eggs are a close match.

References
Chapin, J. P. (1953a).
Jackson, F. J. and Sclater, W. L. (1938).
Lynes, H. (1930).
Maclean, G. L. (1993).

Genus *Incana* Lynes

Bill long, straight and compressed. Tail graduated, marked with white and black at tip as in *Cisticola*. Plumage plain brown and grey. Habitat grassy scrubland. Nest a neat ball with side entrance.

Monotypic, endemic to Socotra. Sometimes placed in *Cisticola*, e.g. by Watson *et al.* (1986) and Sibley and Monroe (1990), but in life it appears closer to *Prinia* (A. D. Forbes-Watson, *in* Hall and Moreau 1970), and we agree with Dowsett and Dowsett-Lemaire (1993b) that *Incana* should remain a monotypic genus, pending further studies.

*Incana incana** (Sclater and Hartlaub). Socotra Warbler. Fauvette de Socotra.

Plate 13
(Opp. p. 240)

Cisticola incana Sclater and Hartlaub, 1881. Proc. Zool. Soc. Lond., p. 166; Sokotra I.

Range and Status. Endemic resident, Socotra I. In scrub and on grassy plains near foothills up to 1400 m (Hagghier Mts), also on thickly covered plains at Ras Kharma and Kallansiya; widespread and abundant. Does not occur on the smaller island of Abd-el-Kuri.

Description. Texture of plumage loose and soft; feathers particularly dense and puffy on lower back and rump. ADULT ♂: entire head and nape dark cinnamon brown; back, rump and uppertail-coverts brownish olive. Tail (12 feathers) somewhat graduated, olive-brown above with white tips to all but T1, below similar but all feathers white-tipped, the white extensive, particularly on inner webs, and crossed by subterminal olive-brown bar. Underparts from chin to lower belly smoke grey, flanks darker, undertail-coverts whiter, thighs cinnamon-brown. Primaries, secondaries and upperwing-coverts olive-brown, primaries with thin, barely discernible, white edge to outer web; carpal joint and underwing-coverts silky white. Upper mandible blackish brown, lower mandible yellowish brown; eye-ring brown, eyes orange-brown or tawny; legs and feet pinkish to orange-flesh. ADULT ♀: like ♂, but underparts and flanks much paler, almost white in centre. SIZE: (18 ♂♂, 13 ♀♀) wing, ♂ 46–53 (50·3), ♀ 46–51 (48·4); tail, ♂ 43–50 (45·4), ♀ 41–47 (44·8); (6 ♂♂, 3♀♀) bill to feathers, ♂

* **Note added in proof.** The OSME Survey of Southern Yemen and Socotra in 1993 added much to our knowledge of this species. The results were published in *Sandgrouse* 17 (1996), too late to be included here.

216　SYLVIIDAE

Incana incana

11–12 (11·4), ♀ 11–12 (11·6); tarsus, ♂ 21–22 (21·8); ♀ 20–22 (21·1). WEIGHT: (n = ?) 9–12.

IMMATURE: similar to adult but browner above.
NESTLING: not described.

Field Characters. A small, lively, secretive, undistinguished warbler that runs mouse-like on ground, with dark cinnamon brown head, plain brown-olive upperparts and smoke-grey underparts. Tail graduated, underside spotted white and black at tip. Commoner than the only other warbler on the island, Socotra Cisticola *Cisticola haesitatus*, which has mottled upperparts, short tail, and 'tsip' call in flight, and occurs at lower altitudes. Where they overlap, Socotra Warbler tends to feed higher in vegetation (A. D. Forbes-Watson *in* Ripley and Bond 1966).

Voice. Tape-recorded (DAV). Possible contact call a series of scolding 'chip' notes often given 2–3 at a time: 'chip-chip, chip, chip-chip, chip-chip-chip, chip, chip-chip . . .'. Note is staccato with frequency range of 4–8 kHz; av. time between 'chip' notes 0·06 s (n = 12), av. time between chip sequences 0·44 s (n = 6). Call used constantly as family groups forage. Alarm, low-pitched churring trill, 'churrrrrrrr', made up of 20–59 (av. 41, n = 7) staccato notes with frequency range 2–5 kHz; duration 1·28–3·28 s (av. 2·29 s, n = 7).

General Habits. Occurs almost wherever low, dense scrub exists, even at high elevations. Noisy and active, moving from bush to bush searching for food, in pairs or family parties. Song not very often heard, but constantly chatters and scolds as one bird chases another up and down hill side. When disturbed, runs like mouse among thick bushes and rocks, disappearing in a moment (Ogilvie-Grant and Forbes 1903). Tail carried semi-erect when bird is on the move. Keeps within 2 m of ground (R. Porter, pers. comm.).

Food. Unknown.

Breeding Habits. Territorial, solitary nester.
NEST: only one described was a neat ball with side entrance, made from fine grass 'ornamented with patches of orange lichen' and placed *c.* 1 m high in thick bush. Both sexes evidently build.
EGGS: undescribed.
BREEDING INDICATIONS: 3 young nearly flying, early Jan; birds with nesting material end Jan; ♀ with well developed gonads end Jan; nest mid-Feb. Probably Dec–Feb or later.

References
Ogilvie-Grant, W. R. and Forbes, H. O. (1903).
Ripley, S. D. and Bond, G. M. (1966).

Genus *Scotocerca* Sundevall

Small, rather wren-like, mainly terrestrial, desert warbler. Sometimes merged with *Prinia* and probably close to it, but more robust, with broader head and shorter tail. Plumage soft and rather loose, variable in colour but usually buff to warm brown; tail darker, graduated, feathers tipped white; seemingly in constant motion. Liquid voice a further distinction from *Prinia* and *Cisticola*.

Plate 14
(Opp. p. 241)

Scotocerca inquieta (Cretzschmar). Streaked Scrub-Warbler. Dromoïque vif-argent.

Malurus inquietus Cretzschmar, 1827. *In* Rüppell, Atlas Reise Nordl. Afrika, Vögel (1826), p. 55, pl. 36, fig. b; Arabia Petraea.

Range and Status. Africa, Middle East and Arabian Peninsula east to Turkestan, Tadzhikistan, Afghanistan and W Pakistan.

Resident. Uncommon and local central Mauritania (Adrar); Western Sahara north from El Ainu, Semara and Saguia el Hamra into S Morocco at least as far as

Scotocerca inquieta

Taroudant, and single record north of breeding range at Essaouira; after an apparent gap reappears at Tenerhir and occurs in a broad band across SE Morocco, N-central Algeria and central and S Tunisia from the Atlas Mts and Hauts Plateaux south to N edge of desert; has recently extended breeding range in NE Morocco (Cramp 1992). Occurs in steppe country of N Libya but generally not near coast except around Gulf of Sirte and near Tobruk, and absent Jebel Akhdar. NE Egypt in mountains east and southeast of Cairo south from about Cairo–Suez road to *c.* 28° 30′N in Jebel el Galala; single record from Wadi Semna between Qena and Red Sea. Possibly occurs in NW near Libyan border. No documented records from Nile valley (Goodman and Meininger 1989), although sight records exist from there. Scarce to locally common.

Description. *S. i. saharae* (Loche) (including '*harterti*'): E Morocco to Libya. ADULT ♂: top of head and upperparts, including scapulars, tertials and upperwing-coverts, pale buffy brown, with pinkish tinge on lower back, rump and uppertail-coverts; forehead to hindneck with indistinct dark brown streaks; broad buffy white supercilium bordered above by narrow dark coronal stripe and below by dark line across lores and through and behind eye. Tail-feathers brown, outer pair with outer portion of outer web and broad (*c.* 5 mm) tip white, next pair with narrower white border to outer web and narrower (1–2 mm) white tip, rest edged and tipped pale buff. Chin, throat and breast white shading to pale buff on lower breast and belly and richer buff on flanks and undertail-coverts, a few indistinct streaks on sides of breast and flanks. Primaries and secondaries brown edged pale buff; axillaries, underwing-coverts, bend of wing and margin of inner webs of underside of primaries and secondaries pale buff. Birds from E Libya ('*harterti*') are slightly darker. Bill yellowish or pinkish brown, tip darker; eyes pale sulphur-yellow, yellowish grey, cream or hazel-brown; legs and feet bright red-brown, brownish or yellowish flesh, olive-brown or yellow-straw. Sexes alike. SIZE: (10 ♂♂, 9 ♀♀) wing, ♂ 43–48 (45·6), ♀ 41–46 (44·6); tail, ♂ 44–53 (47·6), ♀ 44–48 (46·0); bill, ♂ 10–12 (11·1), ♀ 9–11 (10·5); tarsus, ♂ 18–21 (19·3), ♀ 17·5–20 (18·7). WEIGHT: no data this race. Adults of Russian races 8·1–9·4 (Dementiev and Gladkov 1954). See also under *inquieta*.

IMMATURE: like adult but rather paler and less pink, head pattern only faint, underparts loosely feathered.

NESTLING: inside of mouth yellow, no spots on tongue; gape-flanges whitish yellow.

S. i. inquieta (Cretzschmar): NE Egypt (Sinai and S Israel to Persian Gulf). Browner above than *saharae*, crown-stripes blacker and more pronounced, tail blackish brown edged reddish buff; underparts more suffused with buff, some narrow dark streaks on throat and breast. WEIGHT: sex? (n = ?, Israel) 6–9.

S. i. theresae Meinertzhagen: S Morocco, presumably this race in Mauritania. Dark grey-brown above, much darker than *saharae* and *inquieta*, lores and front of cheeks cinnamon-rufous, belly and flanks deep pinkish brown; bill more slender than other races.

Field Characters. A small, rather large-headed and long-tailed warbler of desert scrub. Most widespread race (*saharae*, Morocco–Libya) is pale brown above and buffy white below, with variable narrow streaking on head and underparts, pronounced white supercilium and dark line through eye. Races in S Morocco and Egypt are darker above and buffer below, with blacker tails. Rather wren-like in shape and character, spending much time on ground, tail often cocked; restless, with excited, jerky movements. Warbling song distinctive. In Egypt often confused with the much commoner Graceful Prinia *Prinia gracilis*, which also has pale lores and long tail but a rather different shape, with slimmer head and proportionately longer tail, and stripes on back as well as top of head; Streaked Scrub-Warbler appears a lot fluffier (C. H. Fry, pers. comm.). Both birds frequently cock tails to show markings on underside; these markings are more conspicuous on the prinia (all except central tail-feathers have black and white spots, whereas Streaked Scrub-Warbler has dark tail with white tips to outer 3 pairs). Graceful Prinia is a noisy and conspicuous bird, with persistent chirping song; it frequents waterside vegetation and man-made habitats (agriculture, gardens, bushes around human habitation), while Streaked Scrub-Warbler avoids these areas, keeping to native scrub. Desert Warbler *Sylvia nana* is another pale bird of desert sands but rather larger, with shorter tail; it is pinker above and purer white below, with white-edged rufous tail and no streaks anywhere; it occurs in bushes rather than on ground; in Oman never seen in the same habitat as Streaked Scrub-Warbler (C. H. Fry, pers. comm.).

Voice. Tape-recorded (60, 68, 92, 93, B, HOL). Song of *saharae* a liquid trill 'lululululu . . .' lasting *c.* 1 s, sometimes repeated, on same pitch or descending; sometimes preceded by high-pitched, staccato 'zit' or 'tic', and frequently followed by 2–6 upslurred or downslurred notes, 'cheeyoo-chuwee . . .', less pure, with slightly husky quality; the latter may be the 'Whistle-call' described in Cramp (1992). Song of *theresae* said to be 'a tremulous musical "prrrriea"' (Cramp 1992). Various rattling, grating and rasping calls given in excitement or alarm. Songs and calls vary greatly throughout range, those of Asian birds differing from

African; in Oman, song, given freely, 'a soft, quiet, falling, fast trill, preceded by 2–4 very quiet, rapidly-uttered consonants: "s sp' la trrrrrrreee", the whole lasting <1 s' (C. H. Fry, pers. comm.). For complete catalogue of vocalizations, see Cramp (1992).

General Habits. In Mauritania inhabits sandy riverbeds with acacia and *Balanites* and areas with scattered trees, especially *Capparis decidua* and *Boscia senegalensis*. In North Africa occupies semi-desert and steppe with low shrubs, especially Salsolaceae, where it occurs with Spectacled Warbler *Sylvia conspicillata*; also brush mixed with euphorbias, vegetated wadis and rocky slopes, sandy plains with dry desert brush (Tunisia); in Egypt avoids the Nile and attendant cultivation and frequents areas of native vegetation.

Occurs singly or in pairs, sometimes groups of 3–5; travels in family parties after fledging. Briefly joins mixed-species flocks. Forages mainly on or near ground, even when in taller vegetation, rummaging in plant debris under bushes, sometimes entering cavities among stones, but also feeds in canopy of bushes and small trees (Cramp 1992). Sometimes hops across bare rock surface, evidently foraging on the way (Oman: C. H. Fry, pers. comm.). Bird of pair foraging *c.* 10 m apart gave bouts of tiny monosyllabic contact calls, 2–4 per s, for several s, then remained silent for 20 s while it caught an insect (Oman: C. H. Fry, pers. comm.). Runs mouse-like from bush to bush; restless and inquisitive, flicking wings and constantly jerking tail over head and from side to side; up to 17 jerks per 5 s noted in Oman (C. H. Fry, pers. comm.). Prefers to run rather than fly; flight short and straight. Often shy and secretive, skulking in bushes, but can also be tame and confiding. Once seen to feign injury (Algeria) by lying on side (**A**: see Cramp 1992).

Mainly sedentary in Africa, but recorded north of breeding range in Morocco (Essaouira) and some evidence of local movements in Algeria (Abadla).

Food. Mainly insects, but in winter seeds important. No detailed information for Africa. In Asia diet includes caterpillars, beetles and their larvae, small Orthoptera, Hymenoptera, ants and termites, also spiders and small snails. In Asian study, 253 food items brought to young 6–10 days old comprised (by number): 53% Lepidoptera (43·1% larvae, 9·9% adults), 15% bugs, 13·5% Diptera, 9·1% spiders, 2·8% cicadas, 1·6% Psocoptera, 1·6% Thysanura, 1·6% larval ant-lions, 1·2% lacewings (Chrysopidae) and 0·8% Hymenoptera (Cramp 1992).

Breeding Habits. Few data for Africa; information below mainly for Asian birds, from Cramp (1992). Monogamous, territorial. Pairs form in autumn, and remain on territory all year. Pair-bond long-term. 1 pair bred on the same territory in 2 successive years. Territory size 0·5–3 ha. Density dependent on vegetation type, very uneven, 1 pair/ha in best areas, elsewhere up to 1 km between nests. At territorial boundary, ♂♂ sing from bush tops; singing ♂ turns body from side to side, wags tail up, down and sideways.

NEST: ball-shaped, with side entrance (sometimes 2, 2nd used as exit), rather like nest of Common Chiffchaff *Phylloscopus collybita*; built of twigs, grass stems and other plant material, lined with feathers, fur and plant down; situated 0·2–1·5 m above ground in low scrub or dense thicket; built by both sexes in 6–9 days. New nest built for 2nd clutch.

EGGS: Algeria and Tunisia (n = 13) 2–5, mainly 3–4, 1 clutch of 8 probably laid by 2 ♀♀ (Heim de Balsac and Mayaud 1962); laid at 1 day intervals. Subelliptical; smooth, glossy; white or pink with very fine red-brown or purplish spots and speckles, usually in band around large end. SIZE: *S. i. inquieta* (n = 20) 13·8–16·0 × 10·7–12·4 (15·2 × 11·7), calculated weight 1·06; *S. i. saharae* (n = 40) 13·0–16·8 × 10·0–12·8 (14·8 × 11·2), calculated weight 0·96. Double-brooded.

LAYING DATES: Algeria and Tunisia, Feb–June, mainly Mar–Apr; Egypt, Mar–June.

INCUBATION: by both sexes; begins with last egg; all young hatch in 1 day. Period: 13–15 (14) days.

DEVELOPMENT AND CARE OF YOUNG: eyes open by day 5. Fed by both parents, who brought 1 g of food per h for a total of 230 g up to fledging (11–12 days) for brood of 5–6 (Asia: Cramp 1992). Parents eject eggshells from nest, and seen to feign injury when young fell out of nest. Fledging period 13–15 days, but young leave nest at 11–12 days if disturbed. Young unable to fly on leaving nest, hop from bush to bush, staying in cover; remain with parents for 1 month.

BREEDING SUCCESS/SURVIVAL: no data for Africa. Success rate (fledged young as percentage of eggs laid) in Asia (several studies), 41–60% (Cramp 1992); losses caused by late frost, strong winds, and predation by dogs, foxes, 'steppe cats', snakes and birds (Little Owl *Athene noctua*, Ground Jay *Podoces panderi*, Black-billed Magpie *Pica pica* and shrikes *Lanius*).

Reference
Cramp, S. (1992).

A

Genus *Phragmacia* Brooke and Dean

Resembles some species of *Prinia* in plumage (brown upperparts, whitish underparts, spotted breast) and number of tail feathers (10), but tail unpatterned and strongly graduated (**A**); bill very fine and slender, rictal bristles short (5 mm or less). Unlike *Prinia*, neither cocks nor swivels tail; nest a deep, unwoven cup made of dead vegetation, lined with feathers and vegetable down, placed near water; eggs blue, spotted and blotched with brown. Species confined to drainage lines where acacias grow amongst reeds.

Endemic. A single species in arid SW African Karoo. Evidence from morphology of tail, nest architecture, egg markings and vocalizations suggests generic separation from *Prinia*; for some putative relationships of this and other small-genera prinias, see Brooke and Dean (1990).

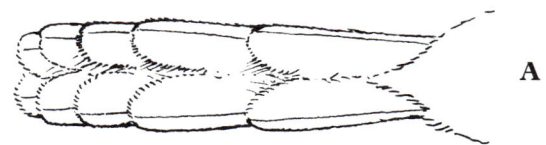

Phragmacia substriata (Smith). Namaqua Warbler. Prinia du Namaqua.

Plate 14
(Opp. p. 241)

Drymoica substriata A. Smith, 1842. Illus. Zool. South Africa, Aves pl. 72, fig. 1, and text; Oliphants River, = Clanwilliam, Cape Province, South Africa.

Range and Status. Endemic resident from interior of W Cape Province at Clanwilliam and Oliphants R. north to S Namibia (S Great Namaqualand and lower Orange R. and tributaries), east to upper reaches of Great Fish R., Colesberg and SW Orange Free State to about Bloemfontein. Locally common.

Description. *P. s. substriata* (Smith): Orange Free State, Karoo, S Little Namaqualand, Oliphants R. drainage and mid-Orange R. to Prieska and Great Fish R. valley, south to Oudtshoorn. ADULT ♂: forehead, crown, hindneck and nape rufous-brown; sides of neck ash-grey; ear-coverts and lores off-white, narrow white supercilium, fine spotting below eye made by dark feather tips, orbital ring of featherlets white; mantle and back to uppertail-coverts bright rufous-brown. Tail of 10 feathers, dark brown with pale tips; T1, T2 and T3 about equal in length, T4 *c.* 16 and T5 *c.* 28 mm shorter. Chin and throat white, breast off-white with narrow longitudinal blackish streaks extending from lower throat; belly and flanks off-white, thighs and undertail-coverts reddish buff. Wings brown, upperwing-coverts and edges of primaries and secondaries rusty brown; underwing-coverts and axillaries off-white. Bill and inside of mouth black; eyes grey; legs and feet pinkish flesh. Sexes alike. SIZE: (8 ♂♂, 6 ♀♀) wing, ♂ 53–58 (55·2), ♀ 50–56 (52·0); tail, ♂ 63–73 (68·2), ♀ 60–64 (62·6); bill, ♂ 12–14·5 (12·9), ♀ 11·5–12 (11·7); tarsus, ♂ 20–21 (20·5), ♀ 19·5–21 (20·3). WEIGHT: ♂ (n = 3) 11·5–13·2 (12·3), ♀ (n = 4) 9·6–10·8 (10·1).

IMMATURE: not known to differ from adult except for fluffier plumage.

NESTLING: hatches naked; skin dull orange pink, purplish over closed eyes; lacks tongue spots at hatching.

P. s. confinis (Clancey): arid lower reaches of Orange R. from confluence of Great Fish R. of Namibia and Richtersfeld, east to Bushmanland, Kenhardt district to Upington at 22°E. More fulvous brown above, streaking below finer. WEIGHT: ♂♀ (n = 3) 11·5–13·2 (12·3).

Field Characters. A long-tailed, prinia-like warbler confined to watercourses in the Karoo. Streaks on breast might cause confusion with Karoo Prinia *Prinia maculosa*, but distinguished by reddish upperparts, white underparts, tawny flanks, unspotted tail and trilling song; breast of Karoo Prinia is more heavily streaked and spotted, and underparts have yellow tinge.

Voice. Tape-recorded (88, 91, F, LEM). Song a dry descending trill preceded by 2–3 introductory notes, total time 1·5–2 s: 'tschip-tee-tee-trrrrrrrrrr', 'tschip-tee-trrrrrrrrrr' or 'tee-tee-trrrrrrrrrr'. Trill seems to run out of steam at the end. Second bird joins in with shorter, buzzier version, 'dzee-dzrrrrr', lasting *c.* 0·5 s. Contact call a sharp 'chit', repeated at intervals; alarm call, given by both sexes, repeated 'chewi-chewi-chewi . . .', continuing for 2 min or more (Maclean 1993). May change from alarm call to song on moving to prominent perch.

General Habits. Practically confined to drainage lines and immediately adjacent vegetation, with mixture of thorny acacia trees (usually *A. karroo*) often well ingrown with other plants and grasses; also reeds, normally *Phragmites australis*, sometimes bulrushes *Typha capensis*, and clumps of ovenbush *Conyza* spp. and patches of African thistle *Burkheya*. In arid NW Karoo, regularly occurs in drainage lines dominated by wild tamarisk *Tamarix usneoides* and *Salsola aphylla*. Sometimes forages in surrounding vegetation, e.g. *Lycium* thickets and clumps of aizoaceous plants 50 m from water, also *Prosopis* (but not thornless Australian spp.) and other vegetation close to water. Occurs in gardens.

Gleans branches, twigs and leaves in tangled riverside *Lycium* bushes, explores acacia thickets, flood debris and ground underneath, but almost never forages in open. Occurs in pairs or family parties after breeding; rarely singly. Movements quick and restless; keeps out of sight, creeping about on branches and on ground. ♂ sings from top of quite tall tree or bush, stretching neck and displaying black mouth-lining. Never cocks tail over back nor swivels it like typical prinia.

Food. Insects: strong predeliction for aculeate Hymenoptera; also beetles, including weevils, tortoise beetles (Cassidae), small hemipterans, myrmecine ants, caterpillars, insect eggs. Also fruit eaten whole; seeds and whole fruits of *Atriplex semibaccata*, *Lycium* seeds ingested with fruit.

Breeding Habits. Territorial; at Tierberg, Prince Albert, pairs spaced at 100 m intervals along 20–30 m wide strip of riverfront vegetation by permanent water.

NEST: deep, open cup, untidy on outside, composed of variety of small pieces of dead vegetation, shreds of *Typha* leaves, roots (often gathered from flood debris on river bank), grass, and leaves (some skeletonized). Inside of cup lined with feathers, fluffy seed-heads, hair, rarely wool; exterior decorated with bits of bark, dried leaves, lichens or small twigs. Ext. diam. 60–100, ext. depth 70–90; int. diam. 40, int. depth 40; nest size differs according to site – smaller and neater in upright bulrushes. Placed 0·3–2 m (usually <1 m) above ground, or often over water; frequently sited in flood debris on river bank, also in fallen *A. karroo*, green foliage of sweet thorn, centre of bulrush clump mixed with grasses, or in axils of large spinescent lower leaves of African thistles.

EGGS: 2–4, usually 3. Pale to deep bright blue, spotted and blotched with pale to dark reddish brown; spots and blotches usually round in shape. SIZE: (n = 9) 15·8–17·2 × 11·7–12·5 (16·6 × 12·0).

LAYING DATES: South Africa, Aug–Apr (Aug 5, Sept 7, Oct 5, Dec 1, Jan 2, Feb 2, Apr 1 clutches).

INCUBATION: period 16 days.

DEVELOPMENT AND CARE OF YOUNG: on day 5, hard, hair-like neossoptiles are visible in feather tracts on crown, back, wings, rump and thighs; chick well feathered by day 14; tongue spots (like *Prinia*) develop by day 6. Nestling period: 15 days.

Reference
Brooke, R. K. and Dean, W. R. J. (1990).

Genus *Prinia* Horsfield

Birds known in Africa as 'prinias' are heterogeneous and probably polyphyletic. We confirm the separation of *Heliolais*, *Urolais*, *Phragmacia*, *Oreophilais*, *Malcorus*, *Urorhipis* and *Schistolais* from *Prinia sens. lat.* on behavioural and morphological grounds. *Prinia sens. str.* is hard to diagnose morphologically: compact-bodied, wings short and rounded; tail long, strongly graduated, of 10 feathers, with dark subterminal bar or spot (sometimes vestigial) and pale tip. Distinguished behaviourally by diagnostic action of cocking tail over back and swivelling it from side to side, especially whilst calling agitatedly. Voice harsh, grating, unmusical, with high-pitched piping notes. Those species that we consider to be evolutionarily most advanced have plain backs and pale underparts with streaks or a breastband; most have well-marked breeding plumages and many have much shorter tails in breeding than in non-breeding plumage. Nests very characteristic, intricately woven and sewn, composed of fresh green material (mainly grasses), with side-top entrance; eggs hard-shelled, round, glossy ovals, variable in ground colour and markings. Well represented in Orient and probably Oriental in origin; prinias have evidently invaded Africa on several separate occasions.

6 African prinias occur in savanna and arid country: *P. subflava* (sometimes treated as conspecific with Oriental *P. inornata*), *P. somalica*, *P. fluviatilis*, *P. flavicans*, *P. maculosa* and the insular *P. molleri*. They often have complex local sympatry that precludes useful recognition of more than one limited superspecies (cf. Hall and Moreau 1970). Although fairly closely related, they appear to have a complex evolutionary history. *P. bairdii* and *P. gracilis* are atypical. *P. bairdii* is a large, forest-edge prinia with very long tail, strikingly barred underparts, open or semi-domed nest made of dry grass, and atypical eggs; it is sometimes placed in its own genus, *Herpystera*. *P. gracilis*, which ranges continuously from Egypt and N Somalia to Assam, occurs in the west of its range in desert scrub, is diminutive, has heavily streaked upperparts, a domed nest made of dry, unwoven materials, with a side-top entrance, and atypical eggs. It was placed in subgenus *Burnesia* by Traylor *in* Watson *et al.* (1986). Behaviourally, *P. bairdii* and *P. gracilis* are typical prinias.

21 species; 8 in Africa, 7 being endemic.

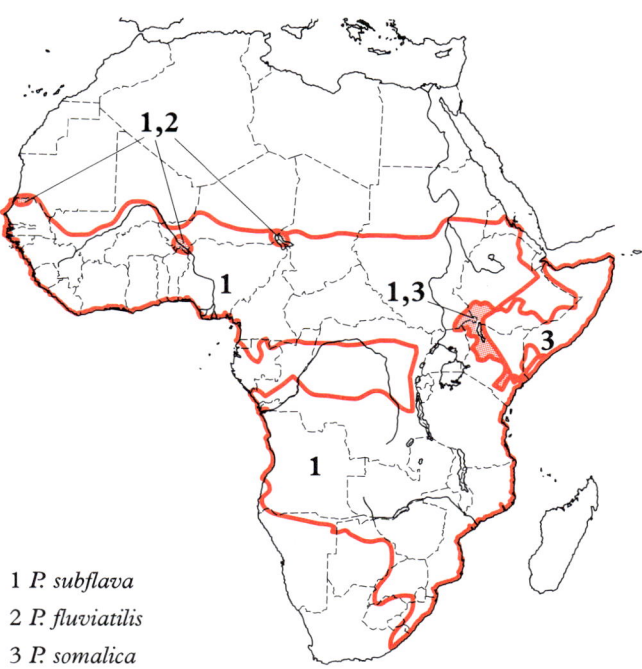

Prinia subflava superspecies

1 *P. subflava*
2 *P. fluviatilis*
3 *P. somalica*

Prinia subflava (Gmelin). Tawny-flanked Prinia. Prinia modeste.

Plate 14
(Opp. p. 241)

Motacilla subflava Gmelin, 1789. Syst. Nat. 1, p. 982; based on 'Figuier blond, du Sénégal' of Daubenton, 1765–81, Planches Enlum., pl. 584, fig. 2.

Forms a superspecies with *P. somalica* and *P. fluviatilis* (and in Asia with *P. inornata*).

Range and Status. Endemic resident, widespread throughout much of Africa and everywhere very common, north to southern borders of Sahara in savanna with long grass and bushes; scarce Orange Free State and Lesotho; absent from forested areas of Zaïre basin, arid NE and much of SW Africa.

Description. *P. s. affinis* (Smith): Zaïre (S Shaba), SW Tanzania and Zambia (except W) to E Botswana, Transvaal east to Lebombo Mts, and S Mozambique (Sul do Save).

ADULT ♂ (breeding): forehead, crown, nape and hindneck dark grey-brown. Cheeks off-white, lores and supercilium whitish buff, black pre-ocular line, ear-coverts dusky white, sides of neck dusky grey-brown, often suffused with buff. Mantle and upper back dark grey-brown, lower back, rump and uppertail-coverts russet-brown. Tail strongly graduated, outermost feather 30–35 mm shorter than longest (central) pair; brown, variably fringed tawny buff, paler rusty buff below with blackish subterminal bar or spot (except on central pair) and pale buff tip. Chin and throat whitish. Upper breast whitish, often suffused with buff, lower breast and belly washed buff, flanks, thighs and undertail-coverts rusty buff.

Prinia subflava

Wings dark brown, outer webs of primaries, secondaries and wing-coverts and inner webs of primaries and secondaries edged tawny; underwing-coverts and axillaries tawny. ADULT ♂ (non-breeding): paler on forehead and crown, mantle and back with variable tawny wash, rump tawny; breast and flanks rusty-buff; bill less intensely black, more horn-coloured. Bill black in both sexes (breeding), base of lower mandible blue-grey, in non-breeding ♀ horn-coloured, lower mandible flesh-coloured; eyes light brown to reddish brown, rim of eyelids reddish brown; legs and feet pinkish brown or flesh, toes light grey. Sexes alike except for seasonal colour changes in bill. SIZE: wing, ♂ (n = 80) 48–57 (51·7), ♀ (n = 60) 46–52 (49·3); tail, (breeding) ♂ (n = 40) 51–63 (53·0), ♀ (n = 40) 49–59 (54·2), (non-breeding) ♂ (n = 40) 61–73 (66·1), ♀ (n = 40) 60–76 (67·6); bill, ♂ (n = 80) 12·5–14 (13·0), ♀ (n = 60) 12–14 (12·9); tarsus, ♂♀ (n = 32) 19·5–21 (20·3). WEIGHT: Zimbabwe ♂ (n = 37) 6·7–11 (9·0), ♀ (n = 25) 7–9·6 (8·3); Kenya (Kisumu), ♂♀ (n = 31) 7·4–10·8 (9·1); Kenya (Mombasa), ♂♀ (n = 5) 7·5–8·7 (8·1); Malaŵi (Nchalo), ♂♀ (n = 90) 6·8–10·3 (8·2); South Africa (Johannesburg), ♂♀ (n = 30) 8–12 (9·4); South Africa (Donga Dell), ♂♀ (n = 23) 6·5–13 (10·0); Zaïre (Upemba), ♂ (n = 19) 7–12 (9·5), ♀ 9–11 (9·8).

IMMATURE: upperparts like adult in non-breeding plumage; sides of head, eye-stripe and underparts strongly washed yellow; bill brownish horn, light flesh below.

NESTLING: hatches naked; skin flesh-pink.

TAXONOMIC NOTE: variation strongly clinal, one form intergrading with another, so ranges of following populations are only approximate. Within equatorial rain belt, no marked seasonal plumage change occurs, but in drier savannas in N and S tropics, seasonal differences are accentuated: plumage becomes rufous-brown in dry season and tail becomes much longer.

P. s. pallescens Madarasz: sahelian savannas from Mali (Mopti) to Sudan (Darfur and Sennar), Ethiopia and NW Eritrea. Breeding plumage paler and greyer than *subflava*, less brown above; non-breeding plumage brighter, more tawny-buff, less brown.

P. s. subflava (Gmelin): soudanian savannas from Senegal to S Sudan and adjoining Uganda; S-central Ethiopia and S Eritrea. Like *melanorhyncha* but with marked seasonal plumage change; darker brown above in breeding dress than *pallescens*, non-breeding dress lighter brown above, tinged rufous. Tail, breeding ♂ 38–50, non-breeding ♂ 50–59.

P. s. melanorhyncha (Jardine and Fraser): guinean savannas and deforested zone from Sierra Leone to SW Nigeria, and Cameroon through N Zaïre to S Uganda, interior of Kenya (except where Pale Prinia *P. somalica* occurs) and NW Tanzania (to Kigoma). Like *subflava* but breeding plumage darker above; non-breeding plumage similar.

P. s. tenella (Cabanis): coastal E Africa from Somalia (Juba R.) to S Tanzania, inland to Usambara Mts and Iringa. Like *affinis* but with shorter wing: ♂♀ 44–50.

P. s. graueri Hartert (including '*canzelae*'): E Zaïre (Kivu) south to Mt Kabobo and west to Kasai and Angolan highlands; Rwanda. Browner above than *affinis*, whiter below; only slight difference between breeding and non-breeding plumage.

P. s. kasokae White: W Zambia (west of Zambezi R.) to 16° 30′S and adjoining E Angola. Darker brown above, less greyish in breeding plumage than *affinis*; in non-breeding plumage less tawny, more olive-brown above.

P. s. bechuanae Macdonald (including '*ovampensis*'): SW Angolan lowlands, N Namibia, N Botswana, extreme SW Zambia and W Zimbabwe. Paler and greyer than *affinis* in all plumages, whiter below, flanks with less tawny wash than *affinis*.

P. s. mutatrix Meise: interior of S Tanzania, Malaŵi, E Zambia, Mozambique north of Save R. and extreme E Zimbabwe. Breeding plumage darker and browner above than *affinis* and crown slate-grey; non-breeding plumage like *affinis*.

P. s. pondoensis Roberts: S Mozambique, Natal, E Swaziland to E Cape. Breeding plumage like *affinis* but browner; rump and edges of wing-feathers less reddish; in non-breeding plumage rump and tail less reddish.

Field Characters. The standard prinia throughout subsaharan Africa except in forest and desert; common and conspicuous in brushy habitats, calling and singing noisily with cocked and swivelled tail. Rather nondescript except for white supercilium; brown above, greyer on crown, with unmarked pale underparts and tawny flanks and rump; overall rustier in non-breeding plumage. Bill thin, black while breeding. Cisticolas have thicker bills and different voices, do not cock tails. Allopatric Pale Prinia *P. somalica* of NE Africa is paler and sandier overall; in soudanian and sahelian savannas sympatric with the little-known River Prinia *P. fluviatilis*, which is restricted to waterside habitats, is whiter below and has different voice. In SW Africa overlaps with Karoo Prinia *P. maculosa* (underparts streaked at all seasons) and Black-chested Prinia *P. flavicans*; latter has black breast-band in breeding season but loses it in non-breeding season, when distinguished by lemon-yellow underparts. Tawny wing-edgings form panel on wing which can be quite bright, often causing Tawny-flanked Prinia to be hopefully identified as the much rarer Red-winged Warbler *Heliolais erythroptera*; wing-patch of latter is much brighter red and more solid, rather than definable feather edgings, and in breeding season contrasts with brown back (upperparts washed tawny in non-breeding dress). Red-winged Warbler lacks white eye-stripe at all times, and has yellow-brown eye set in grey face.

Voice. Tape-recorded (22, 35, 45, 86, 88, 91, B, C, F). Song, used mainly to advertise presence, a tuneless, dry grating or clicking note repeated in short bursts of 5–10 or more often in lengthy series of 30–40 notes, at rates varying from 2·5/s to 7/s, latter sounding almost like a rattling trill; notes may be upslurred, 'zweet-zweet-zweet . . .'; downslurred, 'pseea-pseea-pseea . . .' or a slower, more drawling 'bzeea-bzeea-bzeea . . .'; on 1 pitch, 'psip-psip-psip . . .'; or run together 'pikpikpik . . .'; sometimes a double note, 'pitik-pitik-pitik . . .'. ♂ typically sings alone, but ♀ may join in with a rattling 'tuktuktuktuk . . .' to form synchronized duet. Calls include scolding 'chiky-chiky-dzhaaa-dzhaaa . . .' and a nasal, complaining 'bzeee, bzeee, bzeee . . .'.

General Habits. The commonest, most widely distributed and ecologically adaptable prinia; occurs throughout Afrotropical savannas but replaced in arid, scrubbier habitats by related species. Essentially arboreal, but always present where there is long grass or herbaceous cover or tangles; equally at home in forest clearings or in sahelian zone on borders of Sahara. Absent from forest interior and enclosed woodland such as miombo, but frequent on edges or where vegetation disturbed; also absent from open grass plains lacking woody vegetation and from montane areas, although it extends to 2750 m in Ethiopia, 2300 m in E Africa and 2000 m in Zimbabwe; hardly above 1700 m south of Zimbabwe. Occupies areas of scattered trees, bushes and long grass, edges of cultivation, regenerating growth, weeds and gardens; favours edges of swamps, reedbeds and similar cover along streams; often restricted to vicinity of water in drier country, although it extends into short grass savanna with scattered acacia trees and bushes so long as there is sufficient grass cover. Where sympatric with Pale Prinia in Kenya, found in lusher, moister situations, e.g. in Tsavo East Nat. Park, where *P. subflava* is in riverine vegetation and *P. somalica* in drier bush nearby. In South Africa (Transvaal) broadly sympatric with Black-chested and Karoo Prinias, former commonest in drier west, latter in hillier, mountainous country in moister east.

In pairs, or after breeding in small family parties. Noisy and conspicuous. ♂ calls from exposed position on top of small tree or bush in long grass or herbs; cocks tail over back, swivels it and fans it when excited or alarmed. Flies off into shelter of vegetation with low jerky flight, tail trailing; others often follow. Forages mainly within cover, gleaning insects from vegetation, occasionally flying out after one. Rather tame, fearless, popping out of cover into view at intervals, calling in open and allowing close approach.

Probably sedentary, though some dry season movement likely in drier areas; an immature, or adult in non-breeding plumage, at Makokou (Gabon) was far from known range (Brosset and Erard 1986).

Food. Insects: small beetles (including larvae), flies, caterpillars (including those of the mopane emperor moth *Imbrasia belina*: Styles 1995), moths (particularly noctuid and geometer larvae), and nymphs of acridid grasshoppers, tettigonid grasshoppers. Small spiders. Nectar (Maclean 1993).

Breeding Habits. Territorial; territory vigorously advertised by singing ♂, maintained throughout year. In broad-leaved woodland in Transvaal (Nylsvlei) breeding density of 1 pair per 15 ha (Tarboton *et al.* 1987), but in optimal, higher rainfall regions, density thought to be very much higher.

NEST: oval, with narrow side-top entrance, half domed or with pronounced porch over entrance (**A**); made with fine green grass blades, closely woven and interlaced; thin-walled and green at first, grass soon drying out to straw-colour and curling. Ext. depth 100–150, ext. diam. 50–70. Old nests often survive long after use, when surrounding vegetation has dried out. Bird occasionally weaves normal nest inside long-abandoned, durable nest of Red Bishop *Euplectes orix* (Steyn 1966b). Placed 0·6–1·8 m (rarely more than 2 m) above ground, usually in weedy herbage or coarse grass, attached to 2 or more supporting upright stems with twisted grass fibres; often incorporates surrounding leaves, sewing them into the structure, wrapping larger ones around nest or supports. To begin with, 2 supporting uprights are joined to form cradle, then bowl is added, followed by roof; later lined sparsely with fine grass or grass inflorescences (rarely seed down), mainly after eggs laid. 2 or more sites may be prospected; apparently ♂ selects site. Both sexes build.

EGGS: 1–6, usually 3–4; clutch av. 3·3. 67% of 593 clutches, Malawi, were of 3 eggs; in South Africa,

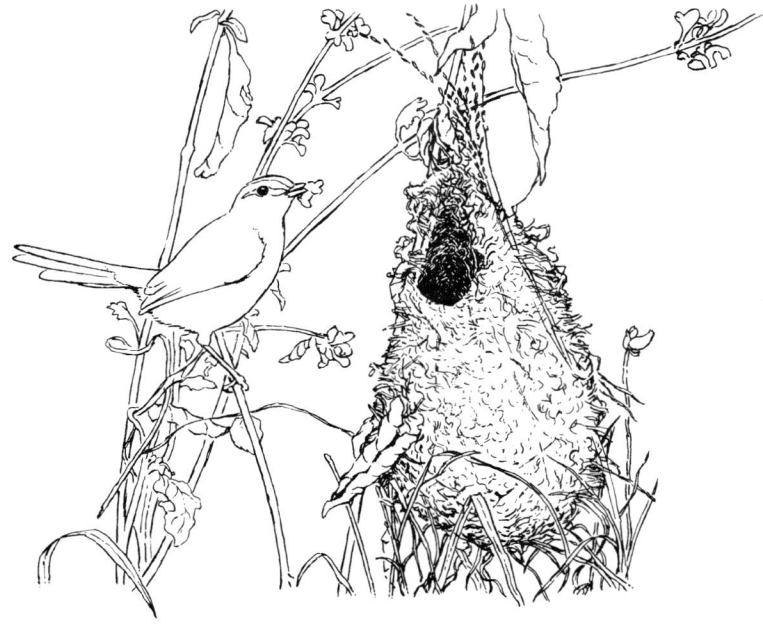

A

clutches of 3 and 4 are equally common. Eggs laid in early morning, at daily intervals. Oval, strong-shelled; glossy; highly variable, ground colour ranging from white, cream, greyish olive and pink to shades of blue; greenish may predominate; boldly scrawled or blotched with black, brown and red-brown, with irregular scroll marks, concentrated around larger end. Each ♀ lays one egg-type. SIZE: (n = 222) 14·1–17·9 × 10·5–12·8 (16·0 × 11·4). Double-brooded; possibly 3 broods raised in season.

LAYING DATES: Senegal, Aug–Dec; Gambia, Aug–Dec; Ghana, May, July (dependent young Oct–Nov); Togo, June; Nigeria, June–Nov; Ethiopia, May, July–Sept; Zaïre: Uele, Apr–Oct, Itombwe, Sept, Dec, Feb, May, Shaba and Kasai, Sept–Apr; Rwanda, Nov–Dec, Apr–May; Sudan, July–Oct; E Africa: Region A, June–Aug; Region B, Jan–Dec (mainly Mar–July); Region C, Jan–June, Dec; Region D, Jan–Aug (mainly Mar–June), Nov–Dec; Region E, Apr–May, Aug–Dec; breeds mainly in rains in all regions. Zambia, Jan–May, Sept–Dec (mainly Jan–Mar); Malaŵi, Jan–June (mainly Jan–Apr); Zimbabwe, Jan–Apr, Aug–Dec (Jan 164, Feb 138, Mar 81, Apr 15, Aug 2, Sept 2, Oct 10, Nov 60, Dec 184 clutches); Botswana, Jan, Mar–Apr, July–Aug, Dec; South Africa: Transvaal, Sept–Apr (Sept 2, Oct 3, Nov 28, Dec 32, Jan 21, Feb 12, Mar 9, Apr 1 clutches); Natal, Sept–Mar, June (Sept 2, Oct 17, Nov 41, Dec 50, Jan 19, Feb 2, Mar 1, June 1 clutches); Orange Free State, Apr; E Cape, Dec, Feb.

INCUBATION: mostly by ♀. Period: 13–15 (av. 14) days.

DEVELOPMENT AND CARE OF YOUNG: young fed by both sexes, brooded after feeding; if mate appears it passes food to brooding bird. As young develop, they are usually fed from upright nest support or similar perch, adult leaning forward with tail cocked and body held almost parallel to support. Most food obtained close to nest; feeding most frequent before 11.00 h and after 16.00 h. Adult may visit nests of other species in its territory, occasionally feeding their young (van Someren 1956). Nestling period 13–18 days (av. 14–15). Young remain with parents for 2–3 weeks after leaving the nest.

BREEDING SUCCESS/SURVIVAL: parasitized by Parasitic Weaver *Anomalospiza imberbis* (*P. subflava* is one of its principal hosts); also by African Emerald Cuckoo *Chrysococcyx cupreus* and Klaas's Cuckoo *C. klaas*. Longest period between ringing and recapture, 7·5 years (Malaŵi).

References
van Someren, V. G. L. (1956).
Vincent, A. W. (1948a).

Plate 14
(Opp. p. 241)

Prinia fluviatilis Chappuis. River Prinia. Prinia aquatique.

Prinia fluviatilis Chappuis, 1974. Alauda 42, p. 492; no type locality (type in Mus. Nat. Hist. Nat., Paris, from Fort Lamy, = N'Djamena, Chad).

Forms a superspecies with *P. subflava* and *P. somalica* (and in Asia with *P. inornata*).

Range and Status. Endemic resident. Locally common, perhaps discontinuously distributed in vicinity of water, from NW Senegal (L. Guier, 16° 25'N, 15° 45'W, Parc Nat. des Oiseaux de Djoudj and inundation zone of Senegal R. delta) to Mali (Niger R. between Tillabéri and Gao), Chad (N'Djamena, shore of L. Chad and lower Chari R.), SE Niger and N Cameroon. Guinea-Bissau, nests near Cacine thought to be of this species. Probably occurs along entire contact zone of sahelian and soudanian savannas; perhaps also to W Ethiopia (Gambela).

Description. ADULT ♂ (breeding): forehead, crown, nape and hindneck dark grey-brown; cheeks off-white; lores and supercilium whitish buff; black pre-ocular line; ear-coverts dusky white; sides of neck dusky grey-brown; lower back, rump and uppertail-coverts brown. Tail dark grey-brown, strongly graduated, outermost feather 30–35 mm shorter than longest (central) pair; brown, variably fringed tawny buff; paler rusty buff below with blackish subterminal bar or spot, except on central pair of feathers, and pale buff tip. Chin to upper belly silky white; lower belly, flanks, thighs and undertail-coverts warm buff. Wings dark grey-brown; underwing-

coverts and under margins to inner webs of primaries white, faintly washed tawny. ADULT ♂ (non-breeding): back greyish. Bill black, lower mandible pale at base; eyes light brown; legs and feet flesh-brown. Sexes alike. SIZE: (12 ♂♂) wing, 46–51 (48·5); tail, 44–52 (48·0); bill, 11·5–14 (13·0); tarsus, 17–20 (18·6).

Field Characters. Very like Tawny-flanked Prinia *P. subflava*; best distinguished by habitat and voice. Where they meet, River Prinia is always in aquatic vegetation, Tawny-flanked Prinia in dry areas. Appears smaller, paler and more slender-bodied than Tawny-flanked Prinia, with longer tail, somewhat longer bill, shorter legs and silky white underparts; overall rather greyer above.

Voice. Tape-recorded (45, CHA, MOR). Song, a rather shrill, high-pitched, down-slurred note repeated in variable-length series at rate of 2·5–3/s, 'pleeeu-pleeu-pleeeu . . .'. More pure, less grating than Tawny-flanked Prinia, with rather piercing quality; individual notes are longer, and song is uniform, with little variation in speed or rhythm. Each species has a strong reaction to playback of its own song but little or none to that of the other species.

General Habits. Inhabits swamp vegetation by permanent water, ponds and old cultivation. Where sympatric with Tawny-flanked Prinia, well-segregated ecologically, latter in drier situations, thorn scrub and wooded savanna. In Senegal occurs primarily in rather dense, thornless *Tamarix senegalensis* shrubs 3–4 m high on humid salt soils; more rarely in reeds, but avoids dense reedbeds; forages 1 m high in woody forb *Borreria verticillata*. In Niger frequents tall grass; in Chad, inundated reeds. Tends to be separated from Tawny-flanked Prinia by habitat, at least locally (e.g. L. Guier, Senegal).

Lives in pairs or small family parties. Forages from near ground to height of 3 m, hopping about in foliage, occasionally climbing up and down stems, gleaning twigs and especially flowers. Stretches out neck to obtain insect.

Food. Insects.

Breeding Habits. Territorial. Sings faster than usual to deter rivals in territorial disputes. Song posts well exposed, above or very close to water and bushes.

NEST: deep oval, with narrow side-top entrance, half-domed or with pronounced porch over entrance; strong and compact; made of closely-bound fine stems. Attached to extremity of hanging leaf of rush or drooping twig of *Tamarix* bush, over standing water; sited about half-way between water and top of reedbed, or sometimes near top of cluster of rush stems.

EGGS: 4 (2 clutches, Senegal, mid-Sept, when 3 other incomplete clutches of 1–2 eggs found). Immaculate, intense blue or blue-green, slightly glossy. SIZE: (n = 18) 13·75–16·0 × 10·1–13·8 (14·7 × 11·1).

LAYING DATES: Senegal, mid-Aug to late Oct.

References
Chappuis, C. *et al.* (1992).
de Naurois, R. and Morel, G. J. (1995).

Prinia somalica (Elliot). Pale Prinia. Prinia pâle.

Plate 14
(Opp. p. 241)

Burnesia somalica Elliot, 1897. Publ. Field Columbian Mus., Orn. Ser. 1, p. 45; Las Durban, Somaliland.

Forms a superspecies with *P. subflava* and *P. fluviatilis* (and in Asia with *P. inornata*).

Range and Status. Endemic resident, E Africa. In extreme SE Sudan (Lokororwa valley and Loelli) and extreme E Uganda (Mt Moroto); widespread in S Ethiopia, N and E Kenya below 1300 m (Kapenguria, L. Baringo, Isiolo, Tsavo, Taita Hills, Tana R.) and Somalia. In Somalia main race is widespread between 1°N and 8°N except in middle latitudes where it may be absent; another race occurs on and near the arid N coast, extending inland to Hargeisa and west of Odweina (9° 20′N). Common to uncommon, widespread or local.

Description. *P. s. erlangeri* Reichenow: Sudan, Ethiopia, Uganda, Kenya and Somalia between 1°N and 8°N. ADULT ♂: forehead, crown, nape and hindneck greyish brown. Lores dusky, whitish superciliary stripe extending behind eye, cheeks greyish brown, ear-coverts pale. Mantle and back to rump and uppertail-coverts greyish brown. Tail greyish brown above, greyish below, feathers with subterminal blackish brown spot, tips margined buffy white, central pair plain or almost so. Chin, throat and breast to belly and undertail-coverts creamy or buff white, flanks pale buff, thighs buff to creamy, variable but always contrasting with belly. Wings greyish brown, primaries and secondaries with paler greyish margins, wing-coverts margined off-white, underwing-coverts and axillaries off-white. No seasonal plumage change. Bill black, lower mandible light pinkish at base; eyes pale orange-brown; legs and feet pale pinkish flesh. Sexes alike. SIZE: (14 ♂♂, 10 ♀♀) wing, ♂ 46–52 (49·9), ♀ 47–54 (49·1); tail, ♂ 56–61 (57·8), ♀ 54–63 (58·0); bill, ♂ 11·5–13 (12·0), ♀ 10–13 (12·1); tarsus, ♂ 18·5–20 (19·3), ♀ 18·5–19·5 (18·8). WEIGHT: 1 ♂ 7·5.

IMMATURE: like adult but paler and fluffier.
NESTLING: unknown.

P. s. somalica (Elliot): Somalia south to about 9°N. Upperparts and wings paler and sandier; primaries margined sandy grey, margins of secondaries and wing-coverts whiter; underparts paler and creamier than in nominate race, thighs uniform with belly.

SYLVIIDAE

Prinia somalica

Field Characters. A pallid, washed-out version of the Tawny-flanked Prinia *P. subflava*, which it replaces in arid parts of E Africa; especially pallid in N of its range. Not known to differ from Tawny-flanked Prinia behaviourally. Graceful Prinia *P. gracilis* has brown upperparts with dark streaks.

Voice. Tape-recorded (B, F, STJ). Song weak, a single buzzy note on one pitch repeated at rate of *c*. 5/s, 'zzzi-zzzi-zzzi-...'. Also makes constant twittering and snapping sounds.

General Habits. Frequents grassy acacia savanna with stunted bushes, and thorn scrub in semi-desert; in Kenya, open, arid and semi-arid bushed country in areas with <500 mm rainfall; in coastal S Somalia, acacia. Plumage colour matches substrate of stones and dry vegetation. Where sympatric with Graceful Prinia in NW Somalia, latter is confined to narrow band of *Atriplex* bushes on beach (J. S. Ash, pers. comm.). Tame and confiding. Frequently fans and erects tail which almost touches back, and swivels it from side to side. Creeps through tangled cover and feeds in lower branches of thorn trees.

Food. Insects.

Breeding Habits.
NEST: purse-shaped, closely woven of fresh grass, with side-top entrance, fixed to upright grass stems or twigs in low bushes, lined with wool and vegetable down; 1 nest was 1·25 m high up in an *Acacia brevispica*.
EGGS: 4; oval; strong-shelled, glossy; pale green or rufous, spotted and blotched. SIZE: (n = 6) 15–16 × 10·5–11·1 (15·6 × 11·0).
LAYING DATES: Sudan, Dec; Ethiopia, Apr–May; Somalia, Apr–May, Nov.

Plate 14
(Opp. p. 241)

Prinia flavicans (Vieillot). Black-chested Prinia. Prinia à plastron.

Sylvia flavicans Vieillot, 1820. *In* Bonnaterre and Vieillot, Tabl. Encycl. Méthod. Trois Règnes Nat., Orn., livr. 89, p. 438; South Africa = Great Namaqualand, Namibia.

Range and Status. Endemic resident, SW Africa. Angola, coastal plain of Moçâmedes and Benguela and S Huila to N Bihe; common in acacia on Cunene R. and in Iona Nat. Park. Namibia, widespread except for driest coastal areas. Zambia, W and S Barotse Province, on Lungwebungu R., South Kasiji R., in Liuwa Plain, on South Lueti R. and Mashi R. east to Sesheke. Botswana, common and widespread south of 20°S but scarce and local north of it. Zimbabwe, only in SW, northwest to Main Camp, Hwange Nat. Park, and southeast to Shashe–Shashani confluence and Limpopo R. South Africa, common in most of Transvaal but scarce in E, east to Carolina, and vagrant in lowveld (N Kruger Nat. Park; Pongola), common throughout Orange Free State except for E and SE where uncommon and local, common in N Cape Province north of 30°S but south of it only from Olifantsvlei to about Colesburg; rare Lesotho. Density, in Zimbabwe 7 pairs/10 ha on stony hill top with scattered acacia (Vernon 1987); in Transvaal (Nylsvlei) breeding densities 1 pair/8 ha in acacia, 1 pair/21 ha in broad-leaved woodland (Tarboton *et al.* 1987).

Description. *P. f. nubilosa* Clancey: SW Zambia, E and NE Botswana, SW Zimbabwe and Transvaal north of 26°S. ADULT ♂ (breeding): forehead, crown, nape and hindneck pale earth-brown. Cheeks and ear-coverts greyish white, lores and narrow supercilium white, dark brown pre-ocular spot. Mantle and back to uppertail-coverts pale earth-brown. Tail strongly graduated, outermost feather 24–30 mm shorter than longest (central) pair; pale earth-brown, feathers with narrow, dusky brown subterminal bar or spot (often vestigial, and may abrade rapidly), tipped off-white. Chin and throat pure white; broad (7–8 mm) blackish brown breast-band extending to sides of neck; lower breast, belly and flanks to undertail-coverts bright lemon-yellow, fading rapidly with wear to pale whitish yellow; thighs buffy. Wings pale earth-brown, under-

Prinia flavicans

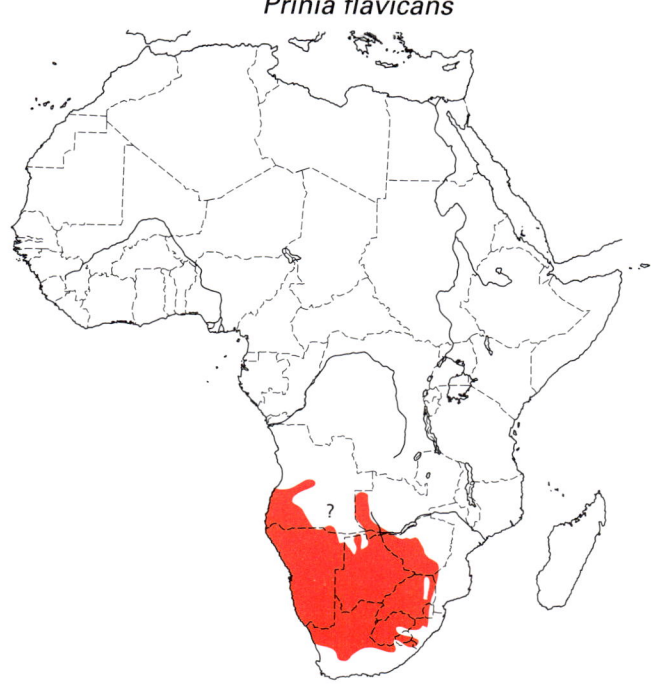

Field Characters. A common southern African prinia with white throat and yellow belly; easily told in breeding season by black breast-band, but later this changes to spots or may be entirely absent. Unspotted birds distinguished from Tawny-flanked Prinia *P. subflava* by yellowish belly contrasting with white throat and lack of tawny on flanks; voices and habitat also different. Karoo Prinia *P. maculosa* has spots or streaks on breast and flanks, and eastern race *hypoxantha* has yellow throat.

Voice. Tape-recorded (38, 75, 86, 88, 91, B, F). Song, series of rapidly-repeated grating or scraping notes with buzzy, insect-like quality, 'dzay-dzay-dzay-dzay . . .', 'zitzitzitzit . . .', or a more shrill 'dik-dik-dik-dik . . .'. Speed variable, from 4 notes/s to almost a trill, and length of series from brief phrase of 3–4 notes to 30 or more. Alarm a dry 'trrrrrrr', like a watch being wound; also a 'weeping alarm call' (Maclean 1993).

General Habits. Typically inhabits low scrub and thickets dominated by acacia, usually on Kalahari Sands, in arid interior of southern Africa. Broadly replaces Tawny-flanked Prinia in north and east and Karoo Prinia in south, all 3 species interacting socially in their extensive areas of overlap. Also extends northwards on Kalahari Sands in open areas of mesic savanna woodlands and edges of plains; range not wholly governed by rainfall. In coastal Angola occurs in sparse bush on rather arid plains; in N Cape (Kimberley) occupies mosaic of plant communities comprising low scrub dominated by *Rhizogum* bushes and taller, mixed *Tarchmanthus–Acacia* community; in NW Zimbabwe and SW Zambia associates with open savanna woodland, in former in low *Combretum-Bauhinia* scrub on edges and under canopy of fairly open teak (*Baikiaea*) woodland with *Terminalia sericea*, in latter in association with teak and clearings; race *bihe* occurs typically in *Burkea* savanna bordering plains and extends to edges of *Cryptosepalum* woodland and *Syzygium* forest.

Ecological and geographical replacement between *P. flavicans*, *P. subflava* and *P. maculosa* especially complex. *P. subflava* is always in moister places with trees and long grass, *maculosa* mainly in karoo vegetation. In Transvaal *P. flavicans* and *P. subflava* overlap widely but are segregated ecologically, but in E *subflava* and *P. maculosa hypoxantha* are found side by side. In NW Zambia (Minyanya Plain) *P. flavicans* occupies dry ground and nests on edges of riparian forest, while *P. subflava* occupies wet ground and nests in reeds and similar swamp growth (Aspinwall 1979). In N Cape (Kimberley) *P. flavicans* is broadly sympatric with Rufous-eared Warbler *Malcorus pectoralis* but they occupy different feeding niches (Dean 1976).

In pairs, or after breeding, in small family parties. Feeds mainly in low scrub and bushes out of sight, searching leaves and twigs for insects; occasionally forages on ground. Cocks tail over back, swivelling it from side to side, particularly when calling in breeding season; flies off just clearing low scrub, with tail trailing, before diving into cover.

wing-coverts and axillaries creamy white. ADULT ♂ (non-breeding): paler and sandier above, crown faintly streaked buff, tawnier on mantle and back, rump and uppertail-coverts; tail tawnier, often lacking discernible spotting; breast-band absent or reduced to variable spotting immediately below white throat; outer webs of primaries edged rusty-buff, inner secondaries broadly margined tawny-buff, wing-coverts with paler buffy edges. Bill black; eyes brown to yellowish brown; legs and feet brownish to pinkish flesh. ADULT ♀ (breeding): like breeding ♂ but breast-band usually less conspicuous. ADULT ♀ (non-breeding): like non-breeding ♂ but nearly always lacks breast spotting; bill dark horn. SIZE: (35 ♂♂, 20 ♀♀) wing, ♂ 52–57 (53·7), ♀ 50–54 (51·8); tail, ♂♀ (breeding) 57–62 (60·0), ♂♀ (non-breeding) (n = 25) 74–88 (82·0); bill, ♂♀ 11–13 (12·2); tarsus, ♂♀ 20–22 (21·0). WEIGHT: (Zimbabwe and Botswana), ♂ (n = 14) 8·6–10·9 (9·3), ♀ (n = 10) 6–9·8 (8·2).

IMMATURE: like adult ♀ in non-breeding plumage, but flanks more buffy-yellow, tail short.

NESTLING: unknown.

P. f. ansorgei Sclater: Angola (coastal plain: Benguela and Mossamedes) and Namibia (Namib Desert to Walvis Bay). ♂ in breeding plumage has black breast-band reduced to collar of broken spots; in non-breeding plumage paler above and buffy (rather than yellow) below.

P. f. bihe Boulton and Vincent: SW and central Angolan highlands to NW Huila; Zambia (Minyanya Plain south to Kalabo). ♂ in breeding plumage greenish brown above, yellow of underside washed with olive and breast-band narrower.

P. f. flavicans (Vieillot): Namibia, Botswana, South Africa (N and NW Cape Province). Like *nubilosa* but paler yellow below.

P. f. ortleppi (Tristram): South Africa (NE Cape Province, north of 31°S, west of 24°E; W Orange Free State, and extreme SW Transvaal); Lesotho lowlands. In non-breeding plumage darker and browner above, less grey; tail very dark with broad feathers; black collar greatly reduced or absent, belly sulphur-yellow, contrasting more sharply with white throat, flanks washed rusty-buff.

TAXONOMIC NOTE: hybridization with *P. maculosa* has been reported in Upper Karoo and Bushmanland (Rowan *in* Rowan and Broekhuysen 1962, Brooke 1993).

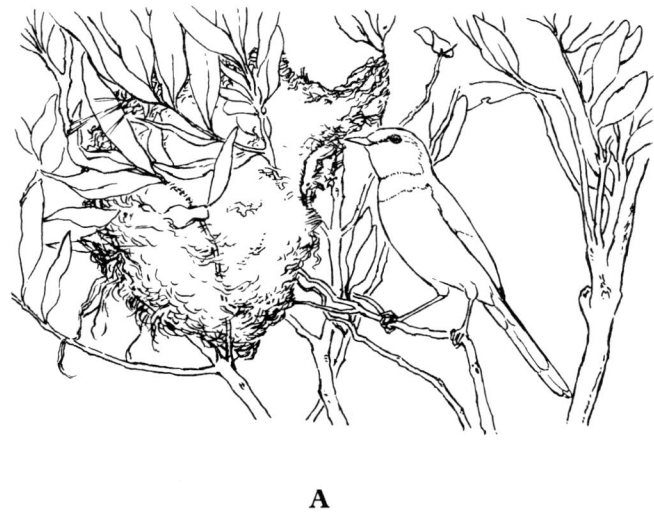

A

Sedentary, but some vagrancy into Transvaal from further west (Tarboton *et al.* 1987), and may move in N Cape.

Food. Insects: Hemiptera (Pentatomidae, Lygaeidae), Lepidoptera, Coleoptera (Coccinellidae), Hymenoptera (Chalcidoidea, Formicidae), aphids, moths and hairless caterpillars, also pseudoscorpions and spiders (Dean 1978) and once a small earthworm.

Breeding Habits. Territorial.
NEST: oval, with circular side-top entrance. Woven of fine green grass blades (often incorporating adjacent leaves, **A**) and lined (mostly after egg-laying) with plant down. Placed 0·9–1·5 m above ground in upright weeds, bush, shrub or long grass. Ext. depth *c.* 115, ext. diam. 65. Building takes up to 11 days.
EGGS: 1–5, mostly 3–4; laid at daily intervals. Oval; strong-shelled; pale blue, bluish white, turquoise blue, greenish or fawn, blotched, scrolled, clouded and marbled with black, browns, greys and mauve. SIZE: (n = 115) 13·7–17·8 × 10·5–12·4 (16·0 × 11·6).
LAYING DATES: Angola, Dec–Mar; Zambia, Oct; Zimbabwe, Sept–Mar (Sept 1, Oct 1, Nov 4, Dec 4, Jan 2, Feb 9, Mar 9 clutches); Namibia, Nov–Feb (nest-building July); Botswana, Oct–Apr (Oct 2, Nov 13, Dec 9, Jan 6, Feb 7, Mar 3, Apr 6 clutches); South Africa: Transvaal, Aug–May (Aug 5, Sept 3, Oct 26, Nov 119, Dec 53, Jan 75, Feb 56, Mar 19, Apr 3, May 1 clutches); Orange Free State, Nov–Feb; Karoo, Sept; W Cape, Nov–Apr. In arid areas breeds opportunistically in winter after rains, even without assuming breeding plumage (Maclean 1993).
INCUBATION: period 12–15 days (av. 14).
DEVELOPMENT AND CARE OF YOUNG: nestling period 11–13 days (av. 12); 2–3 broods may be raised in a season.
BREEDING SUCCESS/SURVIVAL: parasitized by Parasitic Weaver *Anomalospiza imberbis*.

Plate 14
(Opp. p. 241)

Prinia maculosa **(Boddaert). Karoo Prinia; Spotted Prinia. Prinia du Karoo.**

Motacilla maculosa Boddaert, 1783. Table Planches Enlum., p. 47; based on 'Fauvette tachetée, du Cap Bonne-Esperance' of Daubenton, 1765–1781, Planches Enlum., pl. 752, fig. 2; Cape of Good Hope = Swellendam, south-western Cape Province, South Africa.

Range and Status. Endemic resident, highveld and escarpment zones of N and E Transvaal to E and S Orange Free State, E Swaziland, interior and W Natal, W Zululand, Lesotho, Cape Province and S Namibia. Common.

Description. *P. m. maculosa* (Boddaert): South Africa from S Orange Free State and Cape Province in interior E Griqualand and Transkei to Algoa Bay and through Karoo north to S Namibia (Namaqualand). ADULT ♂ (breeding): top of head to uppertail-coverts dark olivaceous brown; lores, short supercilium and cheeks whitish. Tail strongly graduated, grey-brown, outermost feather *c.* 26 mm shorter than longest (central) pair, with dark subterminal bar (often indistinct) and pale off-white tips. Underparts lemon-yellow (subject to rapid fading) streaked blackish-brown, more sparsely on flanks; undertail-coverts buff, broadly streaked brown, thighs dark olivaceous brown. Wings dark olivaceous brown, outer webs of primaries and secondaries edged buff; underwing-coverts and axillaries buff. In subadult ♂ the crown often has distinct dark streaking. Bill black; eyes light brown; legs and feet pinkish brown.

ADULT ♂ and ♀ (non-breeding): deeper yellow below. SIZE: (20 ♂♂, 16 ♀♀) wing, ♂ 49–58 (53·0), ♀ 49–53 (50·5); tail, ♂ 64–77 (69·8), ♀ 58–68 (61·8); bill, ♂ 11–12·5 (11·9), ♀ 11·5–12·5 (12·0); tarsus, ♂ 21–22 (21·3), ♀ 19–21·5 (20·5). WEIGHT: ♂ (n = 9) 8·1–10·7 (9·1), ♀ (n = 4) 7·4–11·1 (8·7).
IMMATURE: upperparts tinged rufous; underparts washed bright yellow, with indistinct dusky streaking.
NESTLING: hatches naked; skin bright salmon-pink with a few threads of white down; bill and legs pinkish yellow, eyes prominent, dark grey, gape bright lemon-yellow, tongue with 2 lateral black streaks or spots.
P. m. hypoxantha (Sharpe): Transvaal, E Orange Free State, E Swaziland, interior Natal and W Zululand to E Cape Province in coastal Transkei as far as lower Great Fish R. and East London. Distinctive: overall reddish brown above; forehead and crown with faint shaft-streaking; cheeks and eye-stripe buff. Tail reddish brown with blackish subterminal spots and buffy tips; length, ♂ 62–82 (71·8), with outermost feather *c.* 48 mm shorter than longest (central) pair. Underparts, when freshly moulted, buffy yellow with finer brown streaking confined to lower throat and breast. Wings reddish brown, outer margins of primaries and secondaries bright tawny,

Prinia maculosa

underwing-coverts pale tawny or orange-buff. In non-breeding plumage more orange-buff below, particularly on sides of throat, breast and flanks. WEIGHT: ♂ (n = 8) 9·2–11·3 (10·7); ♀ (n = 4) 9·2–10·9 (10·1).

P. m. exultans Clancey: Lesotho and adjacent high elevations of NE Cape and W Natal (above 2360 m). Resembles *maculosa* but darker, colder greyish brown above; spotting of breast blacker, flanks washed grey; size like *maculosa* but at least one ♂ had longer tail (88 mm).

P. m. psammophila Clancey: W Cape Province (Berg R., Little Namaqualand northward); SW Namibia (Aus). Paler and greyer above than *maculosa*, whiter below.

TAXONOMIC NOTE: *P. m. hypoxantha* has been treated as a distinct species (e.g. by Clancey 1989, 1992b). It inhabits rich coastal vegetation in E Cape Province and is replaced by *maculosa* in kar31roid communities in interior. They are ecologically separated and have no known intermediates; but they do not differ in voice and we treat them as conspecific pending field studies. *P. maculosa* reportedly hybridizes with *P. flavicans* in Upper Karoo and Bushmanland (Rowan and Broekhuysen 1962, Brooke 1993).

Field Characters. A prinia confined mainly to South Africa, with a very long tail spotted at tip. Underparts vary from yellowish white in W (race *maculosa*), heavily spotted and streaked, including throat and flanks, to yellowish buff in E (race *hypoxantha*, Spotted Prinia), streaked on lower throat and breast only. Black-chested Prinia *P. flavicans* in non-breeding plumage may have some spots on breast but has white throat contrasting with pale yellow underparts. Namaqua Warbler *Phragmacia substriata* has some light streaks on breast but is more rufous above and white below, lacks tail spots, and is confined to streamside vegetation.

Voice. Tape-recorded (75, 88, 91, F, LEM, WALK). Song, a repeated staccato note, 'chik-chik-chik . . .' or faster 'chichichi . . .' or harder 'kikikiki . . .'; sometimes more sibilant, 'sweep-sweep-sweep . . .' or a 2-syllabled 'p'swik-p'swik-p'swik . . .'. Varies little in pitch but rate of delivery ranges from 2 to 8 (av. 5·4–6·5) notes per s. Song usually repeated at intervals of 2–22 s; occasionally sings for up to 1 min in breeding season. Has a variety of chattering and churring notes, including a buzzy 'trrrrr' like a watch being wound. Alarm calls wheezy and sibilant, including a series of 'skizz' notes. Contact calls in thick cover: one bird gives brief, varied clicking stutter, other answers with soft buzzing or 'skizz'.

General Habits. Frequents thick cover up to 1·3 m high in Cape macchia (fynbos), scrub in coastal dunes (including white dunes in arid W Cape) and rhenosterbosveld. Common in areas dominated by *Bobartia*; in Karoo in acacia thickets; in E Cape *maculosa* is confined to 'karoo' communities on mountain slopes in Drakensberg foothills and Transkei escarpment up to 3000 m, while *hypoxantha* occurs along coast and in immediate interior, so their ranges are mutually exclusive. In E Transvaal *hypoxantha* occupies hilly and mountainous country, frequenting forest edges and old cultivation, alongside Tawny-flanked Prinia *P. subflava*.

In pairs all year; in non-breeding season pair occupies fairly loosely defined territory. Occurs in family parties for a few weeks after nesting. Forages low down in vegetation, mates keeping in touch vocally; sometimes hawks insect in flight. When disturbed dives into cover, but at other times perches prominently on top of tall vegetation, even buildings; becomes tame near habitation. Cocks tail over back, swivelling it from side to side.

Food. Insects; nestlings at first fed very small items (mainly Diptera and Hymenoptera; spiders, caterpillars and soft-bodied acridid grasshoppers); later given larger spiders, Diptera and Hymenoptera, with fewer caterpillars; eventually Orthoptera predominate by bulk, followed by caterpillars, with longhorn grasshoppers less common than tougher-bodied shorthorns (Acrididae). Adults take ant-lions, mantids, cockroaches, Lepidoptera, Heteroptera and pentatomid bugs; one once took 15 cm long legless lizard, beating it vigorously against branch (Fraser 1987).

Breeding Habits. Territorial. Breeding territory c. 0·3 ha, vigorously advertised by singing ♂; boundaries appear to be set by owner and not adjusted between neighbours, neutral ground remaining unoccupied (Rowan and Broekhuysen 1962). Same territory may be used in consecutive years. If breeding cycle interrupted by nest loss, pairs relax territorial defence, only singing intermittently, and may feed outside boundaries; they withdraw to territory at start of new nesting cycle (Rowan and Broekhuysen 1962). Territory occasionally deserted suddenly. Pair's territory often distant from others so little interaction occurs. Territory may be defended in non-breeding season with prolonged calling, tail-flicking and display flights. Displaying bird flies over territory, advertising itself by dipping or bounding over bushes with thudding beat of wings, and by string of notes; display flights used in territorial encounters invoke aerial pursuit back and forth over disputed area.

During display, ♀ may half-fan and droop wings and flutter them rapidly, flying from perch to perch followed by ♂ behaving similarly; copulation may follow.

NEST: oval, with side-top entrance; begins as hammock constructed of thinly shredded green grass or occasionally sedge stems; dome completed 3–4 days after cup and walls finished, entire nest completed within 4 days (once 14 days). Lined with white vegetable down (rarely feathers or fur), which may be added 3–4 days after completion, in all averaging 9 days. Additional material may be added after incubation commences. Occasionally nests made entirely of sheep's wool; in S Karoo, seed-heads of *Galium tomentosum* frequently incorporated (Dean and Milton 1993). Ext. depth 90–130, ext. diam. 50–90 × 40–90, distance from entrance to back of interior wall 40–60. Placed low down in a variety of growing bushes, including exotics, slung from 3–4 stems; usually 0·3–0·9 m above ground, extremes from near ground up to 2·4 m or more; often overhangs stream or irrigation ditch; occasionally partially built inside nest of Yellow Bishop *Euplectes capensis* or Red Bishop *E. orix*. Both sexes build, travelling to and fro together every 1–2·5 min, sometimes only one bird carrying material, but if both do they enter nest by turns, and conduct all work inside; observations suggest that one bird may do most of the weaving but that both add lining equally.

EGGS: 1–5; av. 3·1–3·9 in W Cape over 8 year period (Rowan and Broekhuysen 1962) and 3·7 in E Cape. Laid in early morning at daily intervals, 2–6 days after nest complete (or up to 16 days after, in bad weather). Second clutch laid if first destroyed, but one record of being double brooded (Siegfried 1968b). Strong-shelled, oval, glossy; eggs of *maculosa* pale blue to greenish white, occasionally white or pale buff; spotted and blotched with reddish brown or lilac, more densely at broad end forming cap or ring; eggs of *hypoxantha* glossy pink, covered with large salmon-pink smears (W. R. J. Dean, pers. comm.). SIZE: (n = 135) 14·9–20 × 11–12·8 (16·5 × 11·8).

LAYING DATES: South Africa: Transvaal and Natal, Oct–Feb (Oct 3, Nov 9, Dec 18, Jan 16, Feb 3 clutches), E Cape, Aug–Apr (Aug 9, Sept 26, Oct 41, Nov 36, Dec 13, Jan 2, Feb 1, Mar 2, Apr 2 clutches), SW Cape, July–Jan (July 24, Aug 321, Sept 432, Oct 237, Nov 80, Dec 22, Jan 2 clutches) but peak a month later in Karoo (July 4, Aug 16, Sept 24, Oct 50, Nov 32, Dec 10 clutches); Namibia, Sept.

INCUBATION: by ♀ alone, period 12–17 (av. 14) days.

DEVELOPMENT AND CARE OF YOUNG: 6–24 h elapse between hatching of 1st and last chicks. Little change on days 1–2; day 3, head turns dark purplish brown, feather tracts visible on back and wings; days 4–5, sheaths of primaries and secondaries start breaking through; days 6–7, contour feathers appear; day 8, bird almost entirely covered by feathers in sheath; day 9, primaries start to open; day 12, bird well feathered, but bare patches persist ventrally for further 1–1·5 days. Eyes begin to open on days (4)–5–6, fully open by days 9–10. Bill darkens at this time. Chicks start squeaking on day 5 and rapidly become noisy. After day 12 chicks prone to leave nest if disturbed. Weight 0·7–1·2 at hatching, 4·2–5·3 on day 6, 8·2–9·6 on day 10, thereafter 8·0–10·5 (so post-fledging loss of *c.* 1–1·5 g).

Young brooded and fed by ♂ and ♀. Adult broods on days 1–4 in spells of 2–10 min, 2–3 times per h on warm days, brooding prolonged in cold or wet weather; brooding less regular on days 5–6 and ceases on day 7, although adult may continue to shade chicks until days 9–10. Up to 19 feeding visits to nest per h in first 3 h of daylight, then 11–17 visits per h. Young remain under parental care for 2–3 weeks and after that are still fed occasionally.

BREEDING SUCCESS/SURVIVAL: from av. clutch of 3·58 eggs, 2·96 chicks survive to fledging (83%). In another study, 11 out of 108 breeding attempts were abandoned; eggs laid in remaining 97 nests produced 42 fledglings (43%). Of 291 eggs, 44% hatched and 28% resulted in flying young.

Nest predators include several true shrikes and bushshrikes, ants *Iridomyrex humilis* (which kill nestlings) and striped mouse *Rhabdomys pumilio* (eating eggs). 5 cases of parasitism by Diederik Cuckoos *Chrysococcyx caprius*; parasitized also by Wahlberg's Honeybird *Prodotiscus regulus*. Nests appropriated or dismantled by Lesser Double-collared Sunbirds *Nectarinia chalybea*.

Reference
Rowan, M. K. and Broekhuysen, G. J. (1962).

Plate 14 (Opp. p. 241)

Prinia molleri Bocage. São Tomé Prinia. Prinia de São Tomé.

Prinia molleri, 1887. J. Sci. Math. Phys. nat. Lisboa, 44, p. 251; São Thomé.

Range and Status. Endemic resident, São Tomé I. Common to abundant in all edge habitats from sea-level to > 1000 m, in forest, cultivation and in São Tomé town. Breeds to altitude of 1400 m at least, and has occurred on summit of highest mountain, Pico (2145 m).

Description. ADULT ♂: forehead chestnut; crown and nape ochreous grey, grading into grey on hindneck; mantle, back and rump clear grey with olivaceous tinge; tail long and strongly graduated; T1 blackish, T2 to T4 grey with whitish tip and black subterminal spot showing far more clearly on underside than upperside, T5 short, pale greyish, unspotted; shafts black; underside of tail mainly greyish buff. Lores and

Prinia molleri

cheeks vivid chestnut or ochre; ear-coverts brown tinged with rufous; chin, throat, breast and belly white or dirty white, chin to upper breast flushed rich buffy-orange; flanks chestnut. Primaries and secondaries blackish or dark brown; tertials and greater wing-coverts brown with broad black shaft-streaks and white margins. Bill horn, with upper mandible darker than lower; eyes grey; legs and feet brownish flesh. ADULT ♀: like ♂ but duller and more uniformly coloured, lacking chestnut patches; mantle and back strongly olivaceous. SIZE: (6 ♂♂, 5 ♀♀) wing, ♂ 53–56 (54·5), ♀ 50–53 (51·5); tail, ♂ 72–80 (74), ♀ 60–72 (66); bill, ♂ 12–13 (12·7), ♀ 12–13 (12·6); tarsus, ♂ 21–24 (22·5), ♀ 21–25·5 (22·0). WEIGHT: 1 ♂ 8·5; 1 ♀ 9·0.

IMMATURE: forehead, crown, hindneck, mantle and back rufescent brown; cheeks and breast washed with rufous.

NESTLING: not described.

Field Characters. A small, long-tailed warbler, familiar in its island home, active in undergrowth, flying with clicking or snapping noise made by wings. ♂ distinctive, with chestnut forehead and face, grey back, long, graduated, blackish tail with black-and-white marks on underside, white underparts and chestnut thighs. ♀ duller, with olivaceous mantle and back. Juvenile also distinctive, with chestnut from forehead to back, and rufous cheeks and breast.

Voice. Tape-recorded (B, ALEX, GUL, JOPJ, TYE). Song a repeated single or double note, with many variations: 'tsee-tsee-tsee . . .', 'tsi-tsit, tsi-tsit . . .', 'tissy-tissy-tissy . . .', 'ti-whissy, ti-whissy . . .', 'tsip-willy, tsip-willy . . .', 'tseeup-tseeup . . .' or harsher 'jeeup-jeeup', and a loud 'chyap-chyap . . .'. A second bird often joins in with little 'tik's and 'tsip's and sometimes a high-pitched chatter. Contact call, a nasal 'dzik' or more continuous 'dzi-dzi-dzi-dzi . . .'.

General Habits. Occurs in all open, edge and disturbed habitats in São Tomé, in city, villages, gardens, grassland, coconut swamps, banana and cocoa plantations, clearings, and montane forest (but sparse in primary forest: Atkinson *et al.* 1991).

Extremely active; forages often in dense undergrowth, gleaning undersides of leaves, on ground and up to 5 m above it; but prefers well-lit vegetation with plenty of creepers and lianas and open airspace. Often feeds on ground, and hops about in streets of São Tomé city suburbs. Also feeds in isolated bush or tree, and particularly fond of crowns of palms 7–10 m high. Generally in pairs, but often in groups of 5–6 (once 17) birds, keeping in contact by frequent calling. Extremely vocal. On ground hops with tail held high. Clicks wings constantly, when perched and in flight. Tail is always flicked violently up and down when clicking occurs, and click appears to be made by upward movement of tail brushing past downwardly-held wings (Snow 1950) or by wing-tips striking tail when wings are raised.

In frequent display (context not studied – probably sexual), 2 birds perch facing each other on limb or horizontal palm midrib, and one repeatedly jumps *c.* 12 cm up, calling and wing-snapping. A more elaborate communal display is also common: up to 13 birds gather excitedly in group of bushes or forest clearing; one or several birds fly up and down through *c.* 1 m, like yo-yos, in open airspace 5–6 m above ground, wing-snapping constantly (**A**). Tail is turned up at top of each

A

flight. Calling is in unison, the 'tsip' notes being uttered simultaneously by 2 or more birds; the display is noisy and conspicuous and can be heard 100 m away.

Food. Winged insects including beetles; also caterpillars and some vegetable matter.

Breeding Habits. Solitary nester, evidently territorial and monogamous. Commonly-given displays (see above) are likely to be sociosexual.

NEST: closely-woven domed purse, with narrow entrance at one side towards top. Fixed into ends of mass of slender twigs, under fern frond, or into vegetation hanging down vertical bank; sited 1–3 m above ground. Built of fine grasses or creeper fibres, lined with moss, fabric of nest being pliant and semi-transparent but remarkably strong. SIZE: (n = 20), height 100–150, width 60–80, entrance 25–45 high and 22–30 wide; interior depth 48–70.

EGGS: 1–2 (7 C/1, > 10 C/2). Subspherical; brilliant blue, sparsely flecked with round reddish spots. SIZE: (n = 13), 17·6–22·0 × 13·7–15·8 (17·7 × 13·2).

LAYING DATES: (n = 40) Aug–Feb, mainly Oct–Nov.

BREEDING SUCCESS/SURVIVAL: some nests destroyed by rats.

References
de Naurois, R. (1984b).
Snow, D. W. (1950).

Plate 14
(Opp. p. 241)

Prinia bairdii (Cassin). **Banded Prinia. Prinia rayée.**

Drymoica Bairdii Cassin, 1855. Proc. Acad. Nat. Sci. Philadelphia 7, p. 327; Moonda (= Mondah) R., Western Africa = Gabon.

Range and Status. Endemic resident, from SE Nigeria to NW Angola and W Kenya. Nigeria, at 1200–1500 m on Obudu Plateau. Cameroon, widespread south of Mt Kupé and Rumpi Hills. Gabon, W Congo, NW Angola (Cabinda; Cuanza Norte at Ndala Tando, Canzele and Quiculungo). Central African Republic (Ubangi R.). Zaïre, in N east to Uele R., Ituri and Lindi R., in E in highlands mostly above 1500 m, along Albertine Rift from west of L. Albert (Lendu Plateau) to Rwenzori, Kivu and highlands northwest of L. Tanganyika (Itombwe). Uganda, in SW up to 2450 m (Bwamba Forest, Rwenzoris to Kibale and Bwindi-Impenetrable Forests and Ankole). Kenya, west of Rift at 1700–3000 m, from Mt Elgon and Kakamega Forest to Mau, Narok and Molo. Rwanda and Burundi, up to 2500 m. Common.

Prinia bairdii

Description. *P. b. bairdii* (Cassin): Nigeria to Cabinda, NE Zaïre and W Uganda; intergrades with *obscura* in Semliki valley and Bwamba lowlands, W Uganda. ADULT ♂: forehead and crown to back and uppertail-coverts, cheeks, lores, ear-coverts and sides of neck dark brown. Tail feathers narrow, steeply graduated, outermost c. 44 mm shorter than longest (central) pair, brown above, grey below, with blackish subterminal bar or spot and white tip. Chin and throat finely barred and breast to flanks and thighs heavily barred with black on white (each feather with multiple bars); undertail-coverts indistinctly barred black and white on brown, centre of belly creamy white. Wings dark brown, primaries and secondaries with paler buffy margins; innermost secondaries tipped white with subterminal blackish bar; primary and secondary-coverts with V-shaped white tips and blackish subterminal bar; bend of wing with slight dusky barring, underwing-coverts and axillaries white· Bill black; eyes yellow to yellow-ochre; legs and feet dark silvery grey. Sexes alike. SIZE: (16 ♂♂, 14 ♀♀) wing, ♂ 54–60 (56·8), ♀ 50–58 (54·4); tail, ♂ 61–80 (68·9), ♀ 57–65 (62·2); bill, ♂ 14–16 (15·4), ♀ 14–16 (15·4); tarsus, ♂ 21–23 (21·9), ♀ 20–22·5 (21·1). WEIGHT: Uganda, ♂ (n = 4) 10–16 (12·2), ♀ (n = 4) 10–14 (11·5); Kenya, unsexed (n = 15) av. 13·4 (Mann 1985).

IMMATURE: paler brown above than adult, buffier spots on wings; barring below browner, belly and flanks largely whitish; eyes light brown.

NESTLING: at hatching unknown; at day 10 like immature but earth-brown above, spotted buff on wings; below barred grey on buff, centre of belly white.

P. b. obscura (Neumann): highland areas of E Zaïre and W Uganda; Rwanda and Burundi; intergrades extensively with *bairdii* in Zaïre lowlands at Maboya; similar intermediates (or hybrids) occur around Kamituga, S Kivu without apparent contact with *bairdii*, which is unknown in adjacent E Zaïre lowlands (Prigogine 1979). Forehead, forecrown, face, sides of neck and upper throat black, barring commencing on lower throat. WEIGHT: (Uganda) ♂ (n = 8) 12–15 (13·4), ♀ (n = 13) 10–14 (11·8).

P. b. melanops (Reichenow and Neumann): Kenya. Black areas of head, sides of neck and throat sootier than *obscura*, less intensely black. WEIGHT: ♂ (n = 6) 10–13 (11·5); ♀ (n = 2) 10–11 (10·5).

P. b. heinrichi Meise: NW Angola in Cuanza Norte. Like *bairdii* but barring on underparts paler and narrower.

Field Characters. A robust, long-tailed prinia of forest-edge undergrowth with highly distinctive plumage: dark brown upperparts, black-and-white barred underparts, white-tipped tail and rows of white spots on wings. Only other forest warbler with barred underparts is Grauer's Warbler *Graueria vittata*, a little-known bird of mountains of Albertine Rift, but it is olive-green, without white in wings or tail, and has a quiet purring call, not a noisy prinia song.

Voice. Tape-recorded (32, 45, B, C, ERA, GREG, HOR, LEM). Song, a loud ringing note, 'bing-bing-bing-bing...', 'klui-klui-klui-klui...', or faster 'klikliklikli...', repeated at rate of *c.* 4/s to *c.* 8/s, in series lasting half a minute or more; notes slow down slightly toward end of series. ♂ may sing alone, or mate may join in, either with nasal 'gaaa' or 'geea', given at irregular rate, slower than ♂'s song, or with a high-pitched 'chip' synchronized note for note with ♂. Flying young give loud, insistent, nasal 'geeu'.

General Habits. Frequents lowland forest, also montane forest in E of range: edges, clearings, dense bushy growth, low tangles and semi-shaded moist places, open areas along rivers and streams; most abundant in second growth and regenerating cover.

Keeps within cover, skulking, scarcely showing itself even if alarmed, appearing every now and again. In pairs or family parties of 4–5. Small groups briefly followed ant-column in weedy, semi-open zone of stream valley (Willis 1986a).

Food. Insects: small beetles (frequently), large flies, insect larvae, caterpillars; small snails.

Breeding Habits. Territorial. Advertises territory by duetting, with vigorous disputes at territorial boundaries.

NEST: spherical, large for size of bird, ext. depth 130–140, ext. diam. 160; deep, unsewn bag with side-top entrance 40 mm in diam., one side raised to form partial roof over entrance; made of grass strips, interior decorated with plant down, lined with grass-heads. 1 rather elaborate nest had outer layer of green moss and tendrils, middle layer of green forbs and grass blades, leaves and some forb stems. Placed 0·5–1·5 m above ground in grass tangle, dead grass, herbaceous cover or bush, often near water; also in sedges at river edge.

EGGS: 2–3 (usually 3). Very variable, pale bluish green to greenish white or pinkish, with 2 distinct types of markings: either blotched or clouded with pale light red or grey, or minutely and densely freckled all over with pale light red forming zone around larger end. SIZE: (n = 4) 15–17·5 × 12·5–12·9 (16·0 × 12·7).

LAYING DATES: Cameroon, May, Aug; Gabon (Mar–May 3, June–Sept 10, Oct–Nov 4, Dec–Feb 6 clutches); Zaïre, apparently throughout year; Itombwe, Mar, Dec; Rwanda (nest-building Nov); E Africa: Region B, July, Sept–Oct.

INCUBATION: by ♀ alone. Period: 12 days.

DEVELOPMENT AND CARE OF YOUNG: both sexes feed young, sometimes assisted by non-breeding individuals.

BREEDING SUCCESS/SURVIVAL: at M'Passa (Gabon) 11 of 18 nests destroyed by predators; 2 nests were taken over by climbing mouse *Dendromus pumilio*. Ringed adult recaptured, Kenya, after 9 years (only 150 m from original ringing site).

References
Prigogine, A. (1953, 1979).

Prinia gracilis (Lichtenstein). Graceful Prinia. Prinia gracile.

Plate 14
(Opp. p. 241)

S(ylvia) gracilis Lichtenstein, 1823. Verz. Doubl. Zool. Mus. Berlin, p. 34; Nubia.

Range and Status. NE Africa; S Turkey to Sinai, Syria to Iraq, SW and SE Arabian Peninsula, Iran, Afghanistan, Pakistan, N India, Bangladesh and Assam. Several singing Crete, 1986. Vagrant Cyprus.

Resident, Egypt to Somalia. In Egypt, abundant in Nile delta and Wadi el Natrun, and common and locally abundant along whole Nile valley to Luxor; recorded at Aswan first in 1957, now spreading in low pioneering vegetation along shores of L. Nasser south to Abu Simbel, and colonized Wadi el Allaqi and Wadi el Rayan in 1980s; occurs along Suez Canal but absent from N Red Sea coast; specimens collected at Gebel Elba in 1928 but species does not now occur there (Goodman and Meininger 1989). In Sudan common along Nile to 15°N; extending range, with several recent records to 12°N in Kosti/Renk/Nuba Mts area and one near upper Rahad R.; occurs in Red Sea Hills south of Gebel Elba. In Ethiopia occurs in NE from coast up to 200 km inland in Awash valley. Djibouti, very common resident on coastal plain. In Somalia, common in NW and coastal areas of NE south to 10°N; also along coast between 4°N and 2°N.

Description. *P. g. carlo* Zedlitz: Sudan south of 12° 30′N, Ethiopia, Djibouti and Somalia. ADULT ♂: forehead, crown, nape, hindneck and back pale sandy brown with dark brown

Prinia gracilis

shaft-streaks, rump and uppertail-coverts sandy brown. Tail strongly graduated, T6 *c.* 28 mm shorter than T1; dark brown above with pale margins to feathers and narrow shadow barring along vanes; paler below with dark subterminal blackish bar and buffy white tips to all feathers except T1. Sides of head and ear-coverts ash-brown, lores cream, sides of neck ash-grey, chin and throat buffy white; breast to undertail-coverts buffy white, flanks washed brownish. Primaries and secondaries brown with pale buffy edges, upperwing-coverts streaked like back, underwing-coverts buffy white. Bill black, interior of mouth black; eyes light hazel; legs and feet flesh-colour. Sexes alike, but bill of ♀ flesh-brown with darker culmen. Size: (16 ♂♂, 10 ♀♀) wing, ♂ 45–49 (46·2), ♀ 44–48 (45·7); tail, 50–61 (55·8); ♀ 51–60 (54·6); bill, ♂ 10·5–12 (11·1), ♀ 11–11·5 (11·3); tarsus, ♂ 16–18 (17·2), ♀ 16·5–18·5 (17·3). Weight: of African races, unknown. *P. g. palaestinae* (see *deltae*, below): (Israel, Nov–Feb) 74 ♂♂ av. 7·2, 43 ♀♀ av. 7·0.

IMMATURE: very like adult, slightly buffier on belly and flanks; eyes very dark brown.

NESTLING: hatches naked; skin yellowish pink; bill pallid yellow, gape yellow, legs, feet and claws pallid yellow. Leaves nest fully feathered, but plumage brighter and bird shorter-tailed than adult.

P. g. gracilis (Lichtenstein): Nile valley from Cairo and El Faiyum (Egypt) to N Sudan at 12° 30′N. Upperparts slightly paler than in *carlo*, with broader and softer streaks. Shorter-winged and longer-tailed than *carlo*: wing, unsexed (n = 25) 41–46 (43·4); tail, unsexed (n = 24) 44–66 (57·4).

P. g. natronensis Nicoll: Wadi Natrun (Egypt). Like *gracilis* but streaks somewhat darker and sharper. Longer-tailed and long-billed: tail, unsexed (n = 11) 50–66 (59·0); bill, unsexed (n = 10) 12–13 (12·4).

P. g. deltae Reichenow: Nile delta (Egypt) east to Sinai and Israel where intergrades with paler *palaestinae* (even Suez Canal birds are slightly paler than Nile delta ones). Ground colour of upperparts distinctly darker and less sandy than in *gracilis*, with streaks broader, darker and sharper; underparts not so white. Size between *carlo* and *natronensis*: wing, unsexed (n = 31) 43–47 (45·2); tail, unsexed (n = 8) 54–59 (56·5).

Field Characters: a cisticola-like warbler of vegetated wadis, reeds and gardens, long-tailed but small-bodied (at 7 g, one of the Africa's lightest warblers). Active, not shy, conspicuous, generally in pairs, often on ground. Gracile but *not* graceful, short-winged, fine-billed, with body that can look slim or quite rotund, and – its best character – a long, steeply graduated tail, below blackish with white feather-tips, that is jerked, drooped, twitched sideways, sometimes spread and often cocked. Tail is narrow-based, the feathers also narrow, often dishevelled-looking; it looks particularly long and slender when bird flies. Flight straight, whirring and weak – bird constantly looks as though it might be blown away. Vocal; sings all year; distinctive song. Plumage not very distinctive: brown above with fine dark streaks on crown and softer ones, sometimes blotchy, on back; rump plain; tail above dark greyish brown with fine dark barring; lores pale; poorly marked eye-stripe (or none); underparts whitish or pale buff, warmer on flanks; legs pink-brown. Confusion species: Fan-tailed Cisticola *Cisticola juncidis* (Nile, S Red Sea coast) and Desert Cisticola *C. aridulus* (N Ethiopia) are short-tailed. Rattling Cisticola *C. chiniana* (N Ethiopia) has rufous crown, chunky shape, and different song; Tawny-flanked Prinia *Prinia subflava* (N Ethiopia), Pale Prinia *P. somalica* (NW Somalia) and Red-fronted Warbler *Urorhipis rufifrons* (Ethiopia and Somalia coast) are not streaked; Streaked Scrub-Warbler *Scotocerca inquieta* (Cairo, Suez) has well-marked eye-stripe, blackish eye-line and plain back; for further differences, see that species.

Voice. Tape-recorded (B, C, HOL, VIEL). Marked variation in tempo and structure of song, at least in Middle East (sonograms: Cramp 1992), evidently geographical, since no perceptible individual variation in Oman (C. H. Fry, pers. comm.). Song (Egypt: Simmons 1954) comprises repetition of a hard, thin, somewhat slurred disyllabic 'ze(r)wit' note, sometimes with slight tinkling quality, repeated 2–23 times (mostly 14–20) at an even tempo of *c.* 5 notes per 2 s. Notes run into each other so that song is a persistent, monotonous, quite loud and far-carrying 'ze(r)witze(r)witze(r)wit . . .'. Song repeated many times, with varying number of 'ze(r)wit's and with pauses between songs of *c.* 2–20 s, in singing bout of about 5 min, often trailing off in single notes. Notes can be more distinctly disyllabic: 'az-zik' or 'aslenk'. Contact call a drawn-out 'zeeet' (or 'breep' or 'tzeep'), somewhat variable. Alarm, a loud 'trrt', usually repeated and run together into a trilling scold 'trrt-trrt-trrt . . .' lasting 1–1·5 s; used at all seasons in response to person or cat. Agonistic call, in sexual and territorial encounters, a flat 'jit'. Bird defending territory snaps wings in flight, mechanically, making a crackling 'brrp-brrrp-brrrp' sound, audible up to 12 m. Nestlings utter thin, trilling food call.

General Habits. Inhabits vegetated areas of Nile delta, particularly cultivation and mixtures of thorn scrub and broad-leaved shrubs and small trees, also waterside *Typha*, reedbeds and rank growths of weeds. Commonly uses exotic flowering shrubs and young palms, foraging inside them, singing from near top, and feeding on shaded ground below them. Frequents loamy, sandy and stony soils, and commonly found on irrigated short turf lawns. Characteristic of dry wadis with thick tall grasses, acacias, tamarisks and oleanders at sides. In Ethiopia, tamarisk woodland. Occurs on off-shore islands in Red Sea and Gulf of Aden. In Somalia occupies young mangroves, stands of *Suaeda* in coastal flats behind dunes, wadis with tamarisk scrub and (SE coast) the narrow belt of *Atriplex* scrub on beach-head dunes (remaining in low, thick cover during monsoon winds). In N Somalia occurs alongside Pale Prinia, but in S confined to *Atriplex* and ecologically segregated.

Not shy; easy to approach and observe. Keeps in pairs year-round. When moving about in low vegetation, cocks tail loosely or twitches it from side to side with upward flicks, sometimes with much wing-snapping. (Like voice, wing-snapping varies geographically: it is rare in e.g. Oman.) Preens, scratches over wing, stretches, wipes bill and sometimes pecks at feet, much like other passerines. Sings from quite exposed perch at or near top of 3–5 m tree. Sings with bill open, showing black interior, turning head. In Egypt sings much less often in Oct–Jan than in Feb–Sept; sings all day. Regular song-posts are sited near edge of territory, though ♂ may sing anywhere in territory (and occasionally in flight).

Feeds mainly within 1–2 m of ground, rarely up to 4·75 m. Hops about energetically, making long spring at insect if necessary; slips smoothly through grass, wriggling forcibly through matted clumps; hops about in higher vegetation, clinging sideways to stems or slipping down them; may catch insect in flight or hover to pick one off leaf. Larger prey incapacitated before being eaten; bird flogs caterpillar against branch and beats grasshopper repeatedly on ground. When foraging on ground takes short hops, or flies as little as 2 m to join mate. In scrub, flights between bushes usually low, within 2 m of ground; occasionally flies 5–6 m high, between tree-top song stations or over a house. Flight whirring, rather slow and weak, direct and with little or no undulation; seldom flies >20 m. Pair stays together in territory all year (Egypt: Simmons 1954) or (Oman) up to 37 adult or subadult birds aggregate in hot weather after breeding season to forage on single 0·6 ha lawn.

Usually completely sedentary, but may move locally in SE Somalia where present Mar–Apr, absent earlier, and 2 birds once flew above beach at 6–10 m, heading into breeze, now and again dropping into *Atriplex* bushes before taking off and resuming flight (Ash 1982). Presumably moves between islands and mainland on Red Sea coast.

Food. Insects: flies, beetles, grubs, caterpillars and grasshoppers; also spiders. Small amounts of vegetable matter (including shoot of *Dahlbergia* and seeds). Green grasshoppers with wings removed commonly fed to young.

Breeding Habits. Monogamous; territorial for most or all of year (see above); in food-rich garden or by ploughed field breeding territory can be only *c.* 200 m^2, but 2000–3500 m^2 more usual; in vegetated wadi bank or in pond-side reedmace, territory is linear, 80–250 m long. At Fayid, Egypt, 5 pairs had territories in a 2·4 ha garden; territory size (n = 4) was 0·14–0·32 (0·23) ha (Simmons 1954). Pair remains in territory nearly all year and often for 2 or more successive years (Israel; Oman). Pair can build up to 7 nests in its territory per year and lay up to 5 clutches (Israel). Habitually 2–3 broods, with 10–18 days between successive ones. Replacement clutch laid within 4–7 days of loss of first. Juveniles disperse from natal area; others move back before onset of breeding, maintaining population at optimal level. Boundaries between territories clearly defined from mid-Feb (Egypt) by song-duels between adjacent ♂♂; duels continue without relaxation throughout breeding season. In more intense encounter, territorial ♂ threatens rival with wing-snapping, dancing and bowing. Wing-snaps are at perch or (usually) in flight; ♂ may also dance by shooting vertically up from tree top for 1–2 m, jerkily, with wings flailing and snapping, tail cocked, dropping to same or nearby perch. He may then chase rival in flight. In bowing-display ♂ cocks tail right over back and bows stiffly down, flicking wings rapidly and uttering 'jit' call. One confrontation lasted 2 min, with alternating wing-snapping, dancing, bowing and chasing (Simmons 1954). Rarely, rivals fight, coming to grips and falling to ground. ♂ display to ♀ only rarely seen; once, ♂ hopped forward with tail slightly elevated and wings extended horizontally almost at full stretch, shivering them slightly and continuously; another time ♂ raised tail and shivered wings in flight. ♀ becomes skulking and retires from contact with ♂ for up to 9 days before copulation begins.

NEST: domed oval, with entrance about two-thirds up the side (**A**); ext. depth 100–130, ext. diam. 25–40, int.

A

depth 25 and int. diam. 50–80. Constructed of dry grass (fresh green material not used), mixed with variety of material including vegetable down, small strips of leaves, scraps of bark, fibres, seed-cases, spider cocoons and cobwebs; lined with softer, downy materials (Egypt). Individual strands of grass are up to 30 cm long. In Somalia loose, untidy dome of rootlets and dry *Cymadocea*, firmly lined with thick cup of felted seed-heads of *Aerva* sp., placed 0·3–0·6 m up in *Atriplex* bush, sheltered from strong winds. One nest (Oman), placed 2 m up in large *Casuarina* tree, constructed entirely of *Casuarina* needles (and lined with downy seed-heads); another made entirely of fibres from fronds of palm *Washingtonia* mixed and lined with seed-head felt of *Aerva javanica*. Nest placed in fairly long grass or bushes, often near edge of territory at height above ground of 15–130 (av. 56) cm. exceptionally up to 3·5 m.

Cup and basal half of back of nest lightly constructed on day 1; on day 2 basic framework of wiry grass and entrance hole made; structure consolidated and almost completed by day 4; bird starts to line nest as soon as cup is formed. Once basic shell formed, all building is done from inside. Materials mostly collected within territory, often within 1–2 m of nest; material occasionally stolen from nest of neighbouring pair. Building is largely by ♂; ♀ assists later, mainly with lining. Material brought at intervals of 15–210 (av. 76) s. Lining is added to nest throughout incubation period, usually at change-over time.

EGGS: 3–5 (av. 3·8, Egypt); laid at daily intervals in early morning; larger clutches predominate in mid-season. Oval; thin-shelled; whitish or pinkish, spotted or washed with reddish, sometimes forming zone around larger end. SIZE: (n = 25) 14·0–15·2 × 10·5–11·6 (14·6 × 11·0).

LAYING DATES: Egypt, Mar–July (mainly May–June); Sudan, Feb–June; Ethiopia, Jan–May, Dec; Somalia, Jan, May, Dec.

INCUBATION: commences with penultimate egg; by both sexes. In hot weather (Egypt) eggs covered only 9% of the time, and change-overs frequent. Incubation spells short; in 135 min, ♂ had 14 spells totalling 52 min and ♀ 13 totalling 74 min. Period: 11–13 (av. 12) days.

DEVELOPMENT AND CARE OF YOUNG: chicks brooded for first few days, ♀ apparently taking greater share. Brooded for 22% of observed time; no brooding after day 4. In one brood of 2 (Oman), 48 h after 2nd bird hatched it weighed 1·6 g and the 1st, older bird weighed 3·1 g. Skin changes from yellowish to pink as chick grows; slits appear in eyelids at day 2–3; eyes not in full use until day 9–10. At 36 h feathers show below skin on spinal, humeral and ulnar tracts; by day 4 feather sheaths start to push through.

Food brought to nest once every 13·8 min on average; by both sexes. At first food items are minute; from day 6, main food at one nest was grasshoppers with wings removed, 11–24 (once 38) mm long. Adult feeds young perching as near nest as possible, sometimes partially or wholly on rim, leaning forward into nest; at each visit only 1 chick fed. Then parent waits for faecal sac, which is removed (mostly by ♀). Just after young hatch parent may eat faecal sac; later it is dropped on ground 1–100 m away; nest sanitation slackens at end of nestling period. Nestlings not noisy but utter intermittently thin, trilling call from day 8. Nestling period 13–14 days. On leaving nest young flutter some distance; they cannot fly well for about 12 days. They then take food independently but remain under parental care for at least a further week.

BREEDING SUCCESS/SURVIVAL: 28% of territorial adults survive to following year; 1 bird lived 5 years (Israel: Cramp 1992).

References
Ash, J. S. (1982).
Cramp, S. (1992).
Goodman, S. M. (1984).
Simmons, K. E. L. (1954).

Genus *Oreophilais* Clancey

Dull-plumaged, lacking salient pattern, brown above, off-white below with indistinct breast streaking; tail of 8 feathers, narrow, unpatterned, somewhat disintegrated, stepped rather than graduated, outermost feather (T4) *c.* 23 mm shorter than innermost (T1), and T2 and T3 somewhat clustered apically (**A**). No seasonal change in dress. Adult eye pale yellow. Bill black, long, fine; rictal bristles vestigial. Unique social display. Cocks tail over back but does not swivel it. Nest large, bulky, unwoven, domed, occasionally an almost open cup, made of dead grass. Eggs bright blue, boldly marked. Inhabits dense bushes and bracken-briar in forest. Differs from *Prinia* in bill and tail profiles and behaviourally – differences sufficiently great to require generic separation (Clancey 1991b, *contra* Dowsett and Dowsett-Lemaire 1993b).

Endemic, monotypic; in highlands of E Zimbabwe and W Mozambique.

Dowsett and Dowsett-Lemaire (1993b) do not accept *Oreophilais* and state that in the field 'it seems in all respects a typical *Prinia*'. However, we follow Clancey (1991b) and believe that the unwoven (unspecialized) nest and the presence of only 8 rectrices warrant recognition of a separate genus.

A

Oreophilais robertsi (Benson). Briar Warbler; Roberts's Prinia. Prinia de Roberts.

Prinia robertsi Benson, 1946. Bull. Br. Orn. Club 66, p. 52; Vumba (= Bvumba), near Umtali, Southern Rhodesia.

Plate 14
(Opp. p. 241)

Range and Status. Endemic resident in E Zimbabwe and adjacent Mozambique above 1250 m from Nyanga Highlands to Chimanimani Mts. Locally common.

Description. ADULT ♂: entire upperparts dark brownish grey, including central tail-feathers (T1). Lores and ear-coverts darkest, making blackish line through eye. Sometimes small whitish mark above lore. Sharp demarcation between dark lores/ear-coverts, and pale chin/throat. Tail of 8 narrow and somewhat disintegrated feathers, steeply graduated, T4 23 mm shorter than T1. Chin, throat and upper breast off-white, indistinctly streaked grey; lower breast to belly and undertail-coverts pale buffy white; flanks washed olive-brown, thighs more tawny. Wings brownish grey, underwing-coverts and axillaries pale buff. Bill black or very dark horn, interior of mouth blackish; eyes light yellow-ochre; legs and feet light brownish pink, soles whitish. Sexes alike. SIZE: (28 ♂♂, 36 ♀♀) wing, ♂ 49–57 (51·7), ♀ 48–54 (50·2); tail, ♂ 60–72 (61·8), ♀ 58–68 (61·8); bill, ♂ 13·5–15 (14·3), ♀ 14–15 (14·4); tarsus, ♂ 22–24 (22·8), ♀ 22–23·5 (22·6). WEIGHT: (n = 23) 8·7–11·5 (10·3); ♀ (n = 19) 9·0–11·4 (9·7).

IMMATURE: like adult but paler; eyes grey to brown; gape yellowish, interior of mouth pinkish with 2 lateral blackish tongue-spots.

NESTLING: undescribed.

Field Characters. A Zimbabwe/Mozambique endemic. Shaped like a prinia, with long tail, but lacks typical prinia song of repeated notes, having instead social chattering and harsh scolds. Plumage dingy, sooty grey-brown above, paler below, with pale eye set in dark face. Tail plain, without dark spots or pale tips. Occurs together with Tawny-flanked Prinia *Prinia subflava* but more of a skulker, and without any distinctive features except pale eye; Tawny-flanked Prinia has brown eye, pale eye-line, reddish wing-patch and spots on tail.

Voice. Tape-recorded (91, F, CHA, STJ, WALK). Song, a harsh, unmusical chattering, 'cha-cha-cha-cha . . .'; birds also give excited chorus of buzzy scolding notes, 'zizz-zizz-zizz . . .', 'zee-zee-zee-zee . . .', 'zoo-zoo-zoo-zoo . . .'; chorus starts suddenly, rises slightly in scale at first, then maintains constant pitch, and descends somewhat before ceasing.

Oreophilais robertsi

General Habits. Frequents forest edges and clearings, tangled montane vegetation and rank growth with scattered trees, tree heaths *Philippia*, bushes and streamside cover, sometimes ascending into canopy of flat-topped *Acacia abyssinica* and strips of gallery forest; adapts readily to exotic vegetation.

In pairs, or outside breeding season in fairly large, close-knit groups or family parties of 6 or more, foraging within cover. Tends to skulk and remain hidden, usually quite low down, never singing from exposed vantage point. In non-breeding season (Sept–Mar) there is very characteristic social display: members of feeding party come together, often all in same bush, and suddenly begin noisy, vociferous chattering (the 'zizz' call); each member shows great excitement, flitting about inside and out of cover for up to 1 min, then stopping suddenly as if at a signal. Cocks tail over back but does not swivel it. Sedentary, although 1 bird recovered 5 km from original ringing point.

Food. Insects: flies, small beetles, small green grasshoppers.

Breeding Habits. Territorial while breeding, and also has well-defined feeding area outside breeding season. Display not described; nests are defended with much calling.

NEST: bulky, dome-shaped structure, upright oval with large side-top entrance, often with porch. Built entirely of dry grass, very largely seedless inflorescences of multi-branched species, without complex construction and 'knitted' look of typical *Prinia* nests; materials are simply woven together and stuck into nest fabric. Flimsily attached to 5–6 stems of bushy weed, and bound with spider web-like fibres; spider webs used in strengthening lower rim of entrance and substantial weave immediately below it; a different type of grass used for the lining – smooth, hairlike strands *c*. 4 cm long, curled around bowl of internal chamber. Ext. depth 140–160, ext. diam. 70–80, depth of entrance (top to bottom) 60, diam. 40–50. Sometimes dome of nest lacking or vestigial, merely curving over top but not covering it, making structure little more than deep cup with marked depression at bottom where eggs lie.

EGGS: 2–3 (10 of 16 clutches, 2); bright turquoise blue, boldly marked with large round spots or blobs of chocolate and greyish lilac, rarely only with lilac undermarkings. SIZE: (n = 3) 17·5–17·6 × 12·7–13·0 (17·5 × 12·8).

LAYING DATES: Zimbabwe, Sept–Feb (Sept 2, Oct 9, Nov 5, Dec 4, Jan 1, Feb 1 clutches); Mozambique, Dec.

BREEDING SUCCESS/SURVIVAL: 1 ♂, ringed as adult, recovered after 7 years 51 days.

References
Clancey, P. A. (1991b).
Manson, A. J. and Manson, C. (1980).

Genus *Heliolais* Sharpe

Bill long and slightly decurved, rather broad at base, culmen high; rump feathers loose and long; tail *Prinia*-like, of 12 feathers, graduated, tipped white with subapical black spots. Breeding dress grey-brown, non-breeding dress tawny; underparts creamy to cinnamon, bright rufous patch on wing. Habitat woodland with ground cover of grass. Nest thin-walled, oval, with side entrance near top; not stitched.

Monotypic, endemic. Very close to *Prinia*, and often placed in it, e.g. by Watson *et al.* (1986), who did at least promote *Heliolais* to a subgenus. However, Hall and Moreau (1970) consider it sufficiently distinct to be kept in a monotypic genus, pointing out the bright rufous wings and the very different breeding and non-breeding plumages. We agree, and so do Dowsett and Forbes-Watson (1993).

Plate 14
(Opp. p. 241)

***Heliolais erythroptera* (Jardine). Red-winged Warbler. Prinia à ailes rousses.**

Drymoica erythroptera Jardine, 1849. Contrib. Orn. p. 15; Accra, Gold Coast.

Range and Status. Endemic resident; locally migratory in W Africa. Patchily distributed in N tropics and SE Africa. S Senegal (Nioro du Rip, south of 14°N), Gambia (lower and middle river), Guinea-Bissau and N and E Guinea (north of Fouta-Djalon). Local and uncommon S Mali (along Haut Bafing, Bakoye and Falémé rivers), S Burkina Faso (south of Ouagadougou). Widespread Ivory Coast, Ghana (throughout north and along coastal strip east of Cape Coast to Accra Plains), Togo, Benin, Nigeria (guinean zone north to Zaria, Kano and Jos Plateau and south to Ibadan and Afikpo Plains), N Cameroon (north of Adamawa Plateau) and central Cameroon (Galim, Djohong, Sala, Tibati plateau, Oku, Bamenda, Foumban, Obala). Rare S Niger (1 record between Maradi and Guidan Roumji). Occurs S Chad (Békao), Central African Republic (infrequent Lobaye Préfecture, Bangui area and frequent Bamingui-Bangoran Nat. Park) and east to S Sudan (where very local: most records along border from Li Rangu to Torit, Mt Korobe, also near Wau). Uncommon and local Ethiopia (Western Highlands and W Ethiopia, in valleys up to 1800 m), N Zaïre (Uele district), NW Uganda (West Nile, Arua, Murchison Falls Nat. Park), W Kenya (only 2 recent records, Ng'iya in May and Awasi, near Muhoroni, in Nov and Dec) and throughout coastal Tanzania from Bombo valley to Mikindani, inland to Tarangire Nat. Park, Mikumi Nat. Park,

Heliolais erythroptera

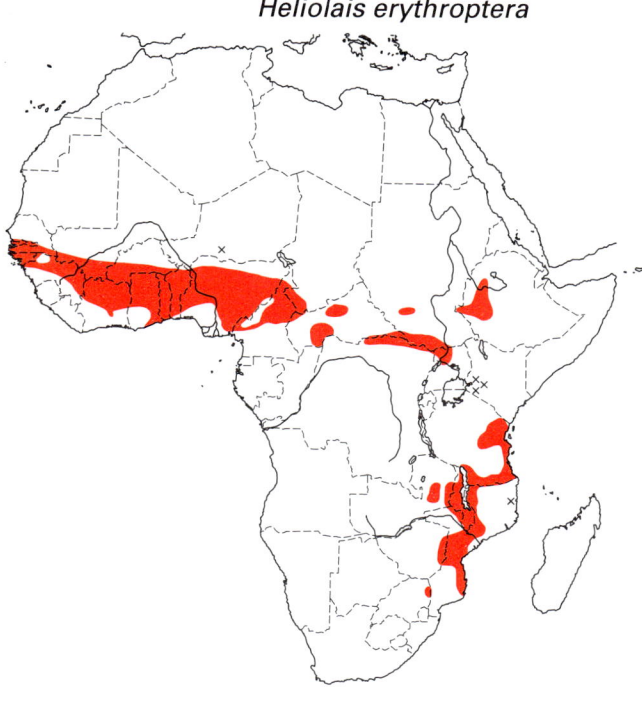

Kilosa region and L. Malaŵi. Fairly common throughout Malaŵi up to *c*. 1500 m (Dedza, Zomba, Tambo, Luchenza in Thyolo and Chiromo). Very localized and uncommon in E Zambia (Eastern Province at Chipata and Lundazi and eastern slopes of Muchinga Escarpment at Mpika and Musense). Generally uncommon E Zimbabwe (Honde valley, Vumba, Lusitu-Haroni area, Chipinga Uplands near Mt Selinda) and Mozambique (confined to east in N with an isolated record inland from Lurio), common and widespread north of Save R. (Inhaminga, Gorongoza Nat. Park, Dondo, Beira, Zinyumbo Hills), sparse south of Save R. and confined to coastal strip (Inhambane District). South Africa: isolated population in far north of Kruger Nat. Park at Punda Milia (Milstein and Milstein 1981).

Description. *H. e. erythroptera* (Jardine): Senegal to N Cameroon (north of Adamawa Plateau). ADULT ♂ (breeding): forehead, crown and nape dark grey with brown wash, lores and ear-coverts darker, shading into olive-brown on back; rump feathers similar but washed with cinnamon, particularly at tips. Uppertail-coverts tawny; tail strongly graduated (T6 27–34 shorter than T1 (n = 5): M. P. Adams, pers. comm.), same colour as mantle but washed tawny and all feathers tipped white with a black subterminal spot or bar; central feathers edged tawny. Cheeks, chin and throat white, merging into very pale cinnamon on breast, becoming deeper cinnamon on flanks and still more so on thighs; flank feathers long and loose. Middle of belly white. Undertail-coverts, axillaries and underwing-coverts same colour as thighs. Primaries and secondaries olive-brown with paler tips; outer webs edged chestnut; outer wing-coverts chestnut, inner ones like mantle. Bill and mouth cavity black; eyes yellowish brown or hazel; legs and feet pale yellow. ADULT: (non-breeding): like breeding ♂, but forehead to rump olive-brown strongly washed tawny, lores and ear-coverts darker; secondaries with pale chestnut wash; flight feathers darker, more blackish. Upper mandible rufous-brown, lower white (black in some specimens). Sexes alike. SIZE: (22 ♂♂, 16 ♀♀, breeding) wing, ♂ 53–60 (54·5), ♀ 48–60 (51·5); tail, ♂ 51–67 (58·8), ♀ 49–61 (53·9); bill to feathers, ♂ 13–17 (14·6), ♀ 13–16 (14·5); tarsus, ♂ 21–22 (21·4), ♀ 20–22 (21·0). WEIGHT: ♂♀ Nigeria (n = 12, Dec–Mar) 10·3–12·5 (11·5), (n = 10, Apr–June) 11·2–13·9 (12·3); Ghana (n = 9, July–Sept) 8–15 (12·5).

IMMATURE: ♀ resembles adult, but grey of crown and back intermediate between breeding and non-breeding plumages; tail-feathers without subterminal blackish markings.

NESTLING: not described.

H. e. jodoptera (Heuglin): central and S Cameroon to Sudan. Breeding dress more tawny and less brownish grey above; non-breeding dress darker and tawnier than nominate race. Wing longer than in nominate race; bill brownish even in breeding season, and slightly longer. WEIGHT: 1 ♀ (Cameroon, May) 16.

H. e. major (Blundell and Lovat): Ethiopia. Largest race: (4 ♂♀) wing 58–60 (59); tail 54–61 (56). Upperparts olive-brown washed tawny even in breeding dress; non-breeding dress tawnier than breeding dress.

H. e. rhodoptera (Shelley): Kenya to Mozambique. Darker than previous 2 subspecies; in non-breeding dress upperparts less tawny, in breeding dress more olive-brown, less brownish grey above. Bill slightly shorter than in nominate race. WEIGHT: ♀ (n = 2, Zimbabwe) 12·4, 13; ♀ (n = 3, Zimbabwe) 10·6–12·3 (11·6).

Field Characters. A prinia-like warbler of grassy woodland with long, graduated, tail with white tip and black subterminal spots. Distinguished at all seasons by solid red patch on wing and brownish yellow eye set in grey face. Upperparts grey-brown in breeding season, washed tawny in non-breeding dress and merging with rufous wings; bright cinnamon-buff flanks and belly contrast with creamy throat and breast. Most likely to be confused with the much commoner Tawny-flanked Prinia *Prinia subflava*, which has rusty wing edgings and tawny belly and flanks but quite different face pattern, with white eyebrow and dark eye set in dark eye-line. Red-winged Grey Warbler *Drymocichla incana* is a more robust, generally grey bird, with unspotted tail and red on wing confined to primaries. Briar Warbler *Oreophilais robertsi* has pale eye and unmarked grey face but is overall grey, with no red in wings.

Voice. Tape-recorded (7, 45, 86, 91, B, GRI, LEM, MOR, WALK). Song of ♂ a monotonous repetition of single loud slurred note 'tseep' or 'tseeu' (frequency falls from *c*. 5 kHz to 1 kHz in *c*. 0·1 s) which carries over a considerable distance. 2·7–4·0 'tseep's per s, and song lasts for 6–10 s or more. ♀ may sing in duet; her song, lasting for *c*. 5–6 s, is a high-pitched trill of *c*. 13–15 notes per s. Contact call (in family group moving through vegetation) a thin twittering 'tseek-tseek-tseek' repeated 2–6 times at rate of *c*. 2·5 notes per s; pitch and rate increases when birds are alarmed (up to *c*. 5 notes per s); another call a high-pitched rattle 'chirrrrrrr' lasting *c*. 2 s (Marchant 1942).

Plate 13

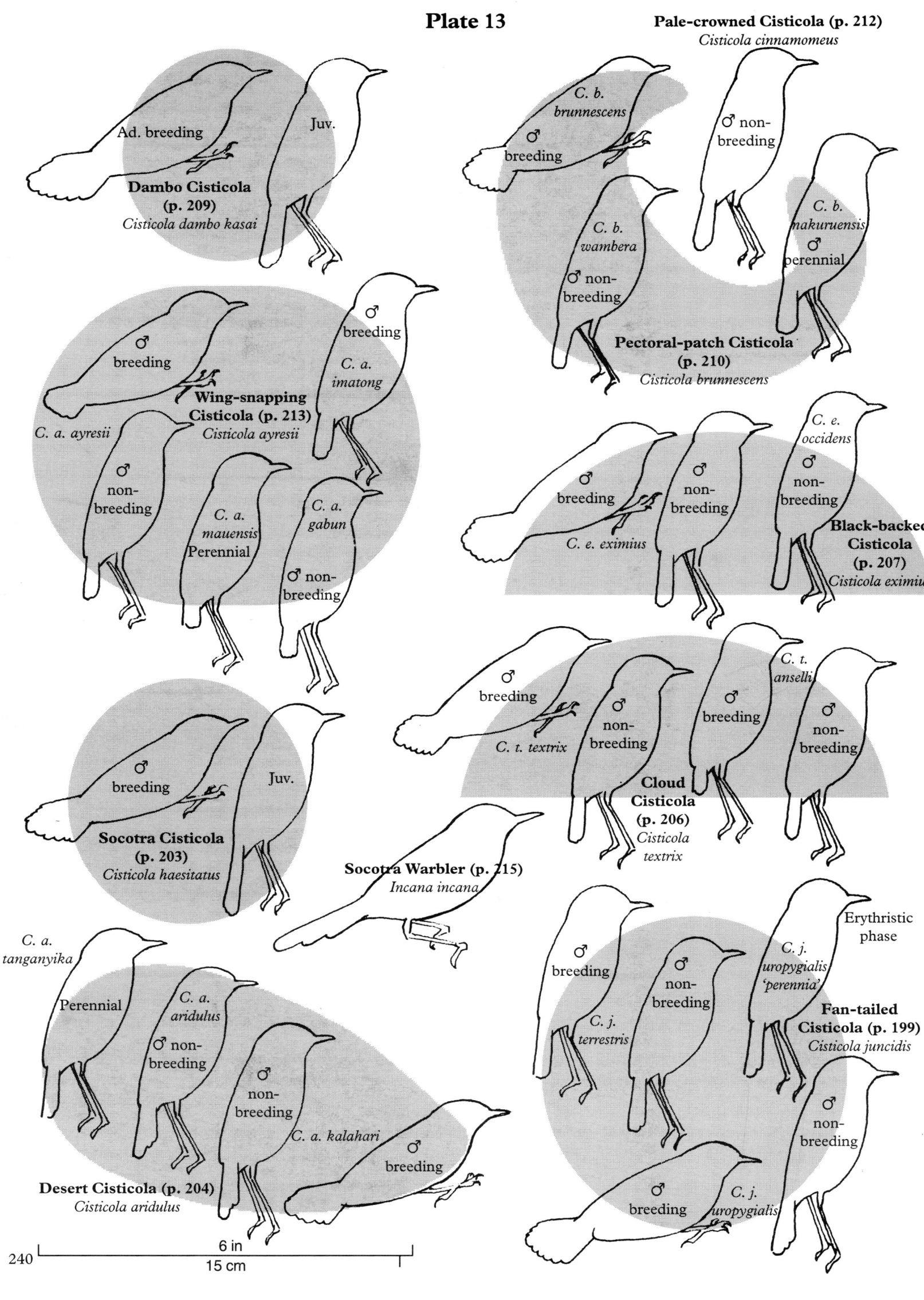

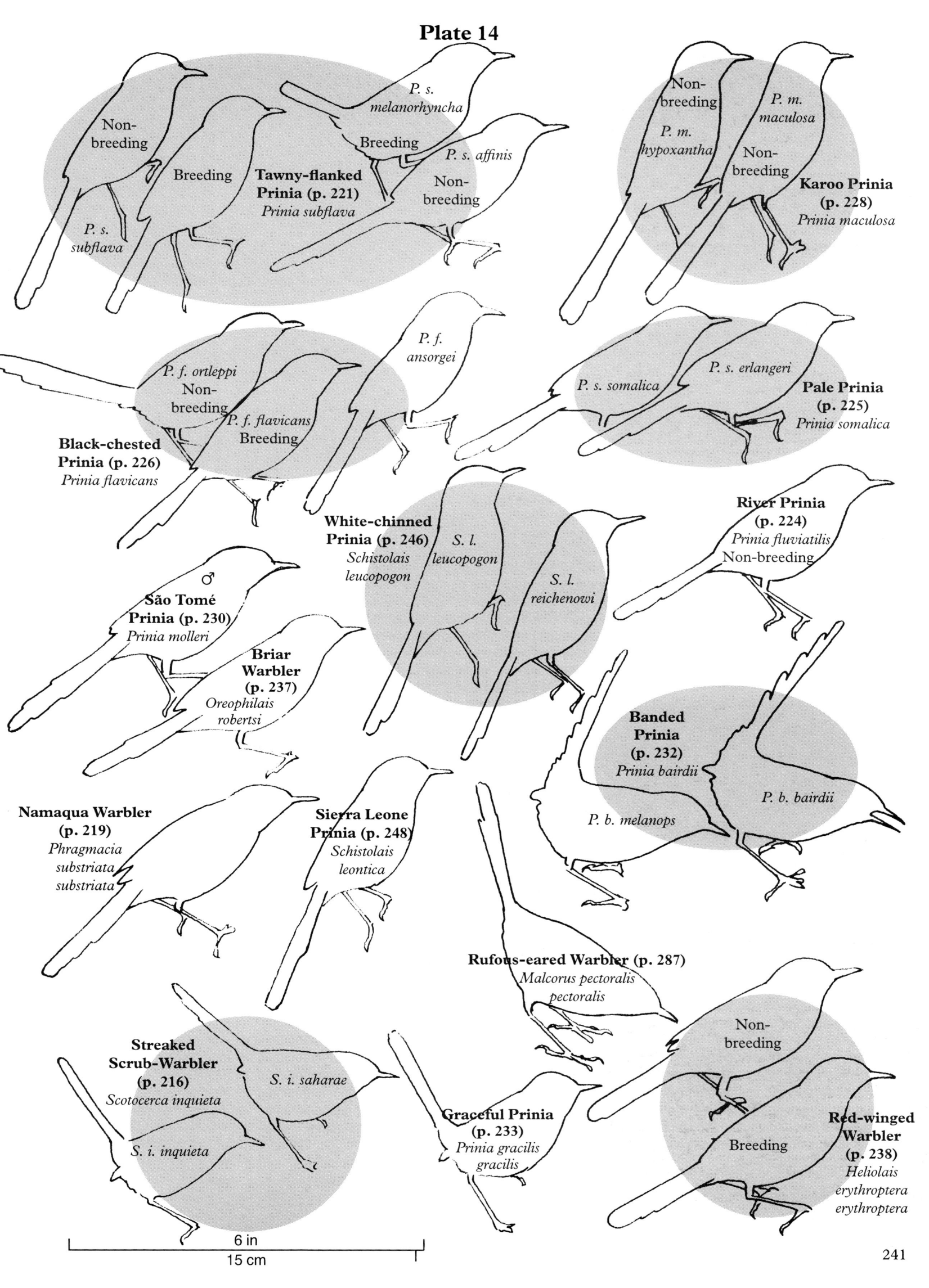

General Habits. Inhabits open woodland with a good ground cover of shrubs and long grass, also abandoned and overgrown cultivation (e.g. cassava). In E Africa occurs from sea level to 1500 m. North of Save R. in Mozambique occurs in miombo woodlands with tall grass understorey.

Restless and active, usually on the move. Lives in pairs or family parties of up to 6 birds which keep in audible contact with twittering calls. Occurs alongside Tawny-flanked Prinia throughout most of its range and may compete with it. Forages mainly in understorey, but will feed in canopy of small trees when grass layer is burned, sometimes with Green-capped Eremomela *Eremomela scotops*. When disturbed may fly up into canopy 12 m above ground. Singing bird perches in small tree or occasionally on telephone wire.

Probably resident throughout most of its range, but locally migratory within Mali, S Niger, S Chad and N Nigeria (Serti, S Sardauna Province, where a dry season visitor only), and possibly along Mozambique border with Zimbabwe (Muneni valley, present Oct–Jan only).

Food. Insects and spiders.

Breeding Habits. Solitary, territorial nester, breeding in the rains; moults into breeding dress at start of rainy season.

NEST: poorly known; one was an upright, thin-walled oval of dead grass, lined with feathery grass tops, with side entrance near top; bound with spider web and placed 0.45 m above ground in a *Tephrosia* shrub among rank grass and weeds. 2 nests in Malaŵi were similar to those of a Briar Warbler (Benson 1946a). Nest was formerly thought to be sewn, but that is now known to be incorrect (Plowes 1972). During nest-building ♂ and ♀ arrive near nest together and after some preliminary flitting about in nearby shrubbery one (the ♀?) enters the nest. ♂ often follows ♀ into nest or else remains flitting about in the shrubbery waiting for ♀ to emerge; both birds leave nest site together (Serle 1940).

EGGS: 2–3; (n = 4 clutches) av. 2·5 (southern Africa); Malaŵi 2–3 (n = 8 clutches) av. 2·4. Pale green with many fine streaks of pale rusty red concentrated at thick end (Plowes 1972). SIZE: (n = 9) 16.5–19 × 12–13 (17·5 × 12·5).

LAYING DATES: Senegal, June (1), July (1), Sept (2), Dec (2); Nigeria (breeding dress May–Sept, building nest early June, enlarged gonads May–Sept, fledglings early Nov, juvenile dress Sept–Oct); Zaïre (female with an enlarged ovary Nov); Tanzania (fledgling being fed mid-Jan); Zambia, (Dec 1, Jan 3, Feb 3, Mar 2 clutches); Malaŵi, (Jan 2, Feb 3, Mar 2, Nov 1 clutches); Zimbabwe, (Jan 1, Nov 2 clutches).

References
Milstein, P. le S. and Milstein, D. A. (1981).
Plowes, D. C. H. (1972).
Serle, W. (1957).

Genus *Urolais* Alexander

A medium-sized, slim warbler with thin bill and short, rounded wings. Tail long, narrow and graduated, with 10 feathers; T1 twice as long as T2, T5 very short. Upperparts olive-green, underparts white and grey with cinnamon wash on breast. Habitat montane forest. Loud clear song from canopy. Nest undescribed.

Monotypic, endemic to Bioko and Cameroon/Nigeria montane. Sometimes placed in *Prinia*, e.g. by White (1962), but more arboreal than *Prinia* and field experience indicates behaviour very different (Stuart 1986a, Dowsett and Dowsett-Lemaire 1993b). We follow Hall and Moreau (1970) and subsequent authors in regarding *Urolais* as a separate genus.

Urolais epichlora (Reichenow). Green Longtail. Prinia verte.

Burnesia epichlora Reichenow, 1892. J. Orn., p. 193; Buea, Mt Cameroon.

Range and Status. Endemic, resident and vertical migrant (on Mt Cameroon); confined to montane forest and cultivation in E Nigeria, W Cameroon and Bioko. Locally common in E Nigeria (Obudu Plateau); widespread and very common in W Cameroon and Bioko. At 1870 m on Mt Cameroon 3 singing ♂♂ heard simultaneously.

Urolais epichlora

Description. *U. e. epichlora* (Reichenow): Mt Cameroon, Muambong, Essosong, Mt Kupé, Mt Nlonako, Rumpi Hills (Dikome Balue), and Obudu Plateau (Nigeria). ADULT ♂: entire upperparts from forehead to uppertail-coverts yellowish olive-green; broad yellow stripe from base of bill to eye. Ear-coverts more yellowish than mantle. Tail dusky brown, feathers edged yellowish olive-green, tips paler. Chin and throat white tinged with pale cinnamon, rest of underparts similar but breast and flank feathers with smoke-grey wash, belly paler, almost white; undertail-coverts and thighs yellowish olive-green. Upperwing-coverts dusky brown at base fringed yellowish olive-green; primaries and secondaries dusky brown, all but outermost primaries edged yellowish olive-brown; secondaries have paler fringe on inner web. Axillaries and underwing-coverts pale yellow and white. Bill black; eyes light brown; feet and legs pale brown. ADULT ♀: like ♂ but some have pale lower mandible. SIZE: (10 ♂♂, 10 ♀♀) wing, ♂ 49–56 (53·0), ♀ 47–50 (48·7); tail, ♂ 60–93 (76·4), ♀ 55–63 (59·4) (inland populations may be longer-tailed); bill to feathers, ♂ 11–12 (11·5), ♀ 11–12 (11·4); tarsus, ♂ 21–22 (21·5), ♀ 21–22 (21·4). WEIGHT: ♂ (n = 5, Mt Cameroon) 11–12 (11·8); ♀ (n = 2, Mt Cameroon) 10, 11; ♂♀ (n = 13, Cameroon) 10–14 (11·3).

IMMATURE: small replica of the adult.
NESTLING: not described.

U. e. cinderella Bates: Mt Oku (from 2100 to 2500 m), Kumbo, Bamenda area, Ndu, Bafat-Ngemba Forest Reserve, Sabga Pass, between Mbengwi and Tinachong, and probably elsewhere in Bamenda Highlands; Mt Manenguba. Like *epichlora* but larger. SIZE: (6 ♂♂, 2 ♀♀) wing, ♂ 56–59 (57·4), ♀ 52, 56; tail, ♂ 78–92 (85·1), ♀ 69,79.

U. e. mariae Alexander: Bioko. Tails much longer, mantle brighter green and cheeks and neck yellower than on mainland. SIZE: (19 ♂♂, 8 ♀♀) wing, ♂ 50–55 (53·2), ♀ 49–52·5 (50·9); tail, ♂ 90–142 (121), ♀ 72–99 (84·3).

Field Characters. A medium sized, slim, long-tailed warbler with yellowish olive-green upperparts and undertail-coverts, white underparts washed with cinnamon on breast and grey on flanks; a broad yellow stripe from bill to eye. Usually in middle or upper strata or in canopy; not easily observed, but its montane habitat, song and contact calls give its presence away. Song has similarities to that of Common Chiffchaff *Phylloscopus collybita*. Its long tail is noticeable in flight (like Long-tailed Tit *Aegithalos caudatus*), especially on Bioko.

Voice. Tape-recorded (45, B, GRI, LEM). Territorial song a repeated thin, sharp note, 'chip-chip-chip-. . . .' or 'djib-djib-djib- . . .' (Eisentraut 1973), the 'chip' note given at a rate of 2·6 to 3·1 per s (n = 9), av. 2·8 per s; song lasts 3–18 s (av. of 20 songs, 6·8 s). Uttered invariably from canopy. Territorial song is not loud yet carries far, 60 m or more. Singing ♂ always responds to playback of his or another ♂'s song; without fail he can be brought to ground cover where he will continue to sing in response to playback (Grimes 1976c). Contact call a repeated 'seep-seep-seep-seep' or 'sis-sis-sis-sis', notes given at rate of *c.* 4 per s. Birds in canopy frequently use this call, often whilst a ♂ is singing. Duration of contact call, 1·1–3·7 s (av. of 29 calls, 2·2 s), call given every 2–8 s. In marked contrast to a singing ♂, birds using contact calls did not respond to playback of ♂ song. Alarm call (by adult with dependent young) is harsh, containing an harmonic component and repeated 3–4 times per s (sonograms of song and calls in Grimes 1976c).

General Habits. Inhabits montane forest on Mt Cameroon from 800 to 2100 m, Mt Manenguba from 1950 to 2200 m, Mt Nlonako from 1650 to 1750, Rumpi Hills from 1000 to 1700 m, Mt Oku from 2100 to 2500 m, Bamenda Highlands from 1700 to 2000 m; on Mt Cameroon descends to 500 m when not breeding. Also occurs in old secondary growth, grassy thickets in eucalyptus plantations above Buea on Mt Cameroon, and in isolated trees in savanna north of Mt Kupé. On Mt Kupé (and probably elsewhere) its abundance increases with altitude. In Bioko occurs from 1100 to 2800 m (Pérez del Val *et al.* 1994).

Occurs in family groups and in mixed flocks, often with Grey Apalis *Apalis cinerea* and sunbirds, foraging mainly among small twigs and leaves in middle and upper storeys but also in lower storey. Groups are continually on the move, travelling with dancing flight from tree to tree; birds keep in contact by calling, much like tits (Paridae). ♂ in ground cover, singing in response to playback, flicks wings audibly and at same time often cocks tail over its back. Its discovery at Saxenhof (500 m) at base of Mt Cameroon in the wet season suggests vertical migration of at least part of the population (Serle 1964).

Food. Mainly insects including caterpillars; once small seeds.

Breeding Habits. Probably a solitary, dry season breeder, Nov–Mar. Nest and eggs not described.

BREEDING INDICATIONS: Mt Cameroon (brood patches Dec, Mar; gonads developed Nov, Dec, Jan, Mar; oviduct egg Dec, adults feeding fledgling late Dec and early Jan); Obudu Plateau, Nigeria (fledgling Apr).

Reference
Grimes, L. G. (1976c).

244 SYLVIIDAE

Genus *Spiloptila* Sundevall

Culmen strongly arched; legs and feet weak. Feathers of crown and wing-coverts stiff. Tail of 12 feathers, strongly graduated, feathers broad, grey, each feather with broad white tip and subterminal black bar. Upperparts cinnamon, underparts buffy, wing-coverts with conspicuous white borders. Habitat desert edge. Nest oval, domed or partly domed, placed close to ground.

Monotypic, endemic to N tropical Sahel zone. Placed in *Prinia* by White (1962), but kept separate by later authors. '*Apalis*' *rufifrons* was placed in *Spiloptila* by Hall and Moreau (1970) and Dowsett and Forbes-Watson (1993), but is here considered a monotypic genus *Urorhipis*.

Plate 17
(Opp. p. 320)

Spiloptila clamans **(Temminck). Cricket Warbler; Scaly-fronted Warbler. Prinia à front écailleux.**

Malurus clamans Temminck, 1828. *In* Temminck and Laugier, Pl. col. livr. 78, pl. 466, fig. 2; Nubia.

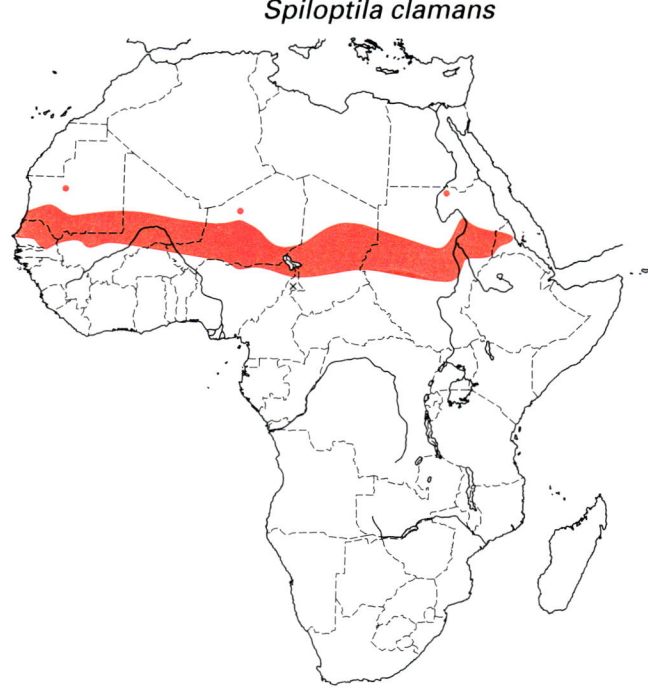

Spiloptila clamans

Range and Status. Endemic resident, Sahel zone. Locally common. N Senegal (common Richard-Toll, widespread north of 15°N); Mauritania (south of *c.* 18°N, rather common 50–80 km north of Rosso, Tidjikja, Tichit, also Atar in N); Mali (common throughout between 15°N and 18°N – possibly a partial migrant); Niger (Aïr south to Tillabéri and Niamey, Tem east to Kaadjia, Nguigmi and L. Chad); Nigeria (between Nguru and L. Chad, also Achilafia and Kazaure); sight record N Cameroon (north of Maroua: Robertson 1996); Chad (common throughout semi-desert zone south of Borkou-Ennedi region to latitude of about L. Chad; Marmarigna near Murdi Depression, Fada, widespread north of Arian and Ketrai dunes); Sudan (widely distributed, mainly between 12°N and 18°N, but reaches 21°N near Semna); common W Eritrea, below 1050 m.

Description. ADULT ♂: feathers of crown and forehead black edged with white; nape light grey shading into pale vinous cinnamon (salmon-pink) on mantle and back and becoming buff-yellow on rump and uppertail-coverts. Tail drab brown, strongly graduated (T6 18–28 shorter than T1 (n = 5): M. P. Adams, pers. comm.), feathers with broad white tip and dusky brown subterminal bar. Lores white, ear-coverts grey, thin white eyebrow; cheeks very pale buff, chin, throat and underparts white washed with pale buff (but much variation in tone through wear); undertail-coverts pale buff. Primaries drab dusky brown, darker on outer web; shafts of secondaries dusky brown, outer webs broadly edged salmon-pink; upperwing-coverts black, lesser wing-coverts with grey margins, others broadly edged white; underwing-coverts and axillaries white. Bill tip black, base of upper mandible pinkish brown, lower mandible darker; eyes yellowish brown; legs and feet yellowish flesh. ADULT ♀: similar to ♂ but feathers of crown and forehead less heavily marked: centres dusky brown, the dark area smaller than in ♂, giving head streaky look. Nape and hindneck pale vinous cinnamon (not light grey). Tail shorter than in ♂. SIZE: (10 ♂♂, 10 ♀♀) wing, ♂ 44–50 (47·3), ♀ 43–50 (45·3); tail, ♂ 53–62 (57·0), ♀ 43–53 (49·3); bill to feathers, ♂ 9–10 (9·5), ♀ 9–10 (9·4); tarsus, ♂ 17–18 (17·8), ♀ 17–18 (17·5). WEIGHT: ♂♀ (n = 3) 6–8 (7·0).

IMMATURE: head streaks dusky brown, less distinct than in adult. Inner secondaries broadly margined on both webs with pale vinous cinnamon, shading into dull white at tip and on inner web.

NESTLING: unknown.

Field Characters. Although very small, it is conspicuous in preferred habitat and easy to identify, with striking black and white feathers on forehead and wing-coverts, grey head, pale vinous cinnamon upperparts and pale cream underparts. Grey tail, about half the bird's length, is tipped white with a subterminal black bar or patch; ♂'s tail strongly graduated. Vocal, with rapid trill very like that of Tawny-flanked Prinia *Prinia subflava*, with which it occurs throughout most of its range; slower songs are reminiscent of tinkerbirds *Pogoniulus* (Gee 1984). Overlaps with Red-fronted Warbler *Urorhipis rufifrons* in Darfur, but latter occurs in hills and Cricket Warbler on the plains.

Voice. Tape-recorded (68, MOR). Several distinct songs in breeding season, each being repetition of a single note. Differences lie in duration of notes and their repetition rate. Common song is a lively, tinkling, rapid trill of *c.* 6–7 notes per s, much like a cricket; trill rate can be as low as 1·8 notes per s, each note lasting *c.* 0·4 s. Duets occur frequently, one bird singing a slow trill with some variations, the other a faster one; songs are uttered as birds move through vegetation, not pausing for a moment. Alarm, a hard 'zzt zzt'. Other calls include ringing whistles.

General Habits. Inhabits the Sahel, the dry thorn-scrub belt south of the Sahara. In Eritrea occurs in a wide variety of savannas including some with broad-leaved trees. Usually singly or in pairs, but also in family parties of 5–6. Active and restless; tail continually oscillated to and fro and up and down. Party easily observed as it moves through open vegetation searching for insects; also feeds on ground. Flight weak; when alarmed, escapes by running on ground rather than taking flight. Sings as it flits from branch to branch, never from a song perch. Resident throughout most of range, but may be partial migrant at its northern limit.

Food. Insects.

Breeding Habits. Solitary nester; territorial.
NEST: deep, egg-shaped, domed or partly domed cup, made of grass and dry herbage, lined with down, feathers or other soft material. In Senegal sited 0·5–2·0 m above ground in thorny, impenetrable *Balanites aegyptiaca* (G. J. Morel, pers. comm.). Also nests in small bushes close to ground and sometimes (Mali) amongst stems of tussock grass *Panicum turgidum*.
EGGS: 2–5, mainly 3–4 (N Senegal, n = 47) or 2–3 (Niger). Dull white with tiny red spots.
LAYING DATES: Mauritania, June–Aug (mainly July); N Senegal, Jan–Nov with peak in Sept (21%) and Oct (26%); Mali (fledgling late July, ♀ about to lay Sept, some with brood patches late Nov, immatures Nov, moult in primaries and tail-feathers June); Niger, Nov; Sudan, Jan–Apr, Aug; Chad (gonads in breeding condition late Jan to early Feb); N Darfur (tail-feathers in moult Jan and Feb); Eritrea (pair lining nest early July, reported nesting Aug).
BREEDING SUCCESS/SURVIVAL: in Senegal many nests are destroyed by grass fires (G. J. Morel, pers. comm.); nests also destroyed by browsing cattle.

Genus *Schistolais* Wolters

Warblers prinia-like in shape but large bodied, with plain, mainly grey, plumage (1 species with scaly forehead); strong bill; tail graduated, proportionately shorter than in *Prinia*, of 10 broad feathers with vestigial dark subterminal spots. May raise tail over back but neither cocks nor swivels it. Nest a deep pouch of strips of grass, sewn into broad-leaved *Afromomum* plant, opposing leaves of which are stitched together along edges. Eggs oval, white or greenish, sparingly marked, not glossy. Confined to edges of and clearings in central and W African rain forest.

Affinities unclear. Differs from other prinias in bill and tail profiles, lack of tail cocking/swivelling, and in plain grey plumage. Suspicion that it may be only distantly related to true prinias is strengthened by distinctive nest architecture (and social behaviour – that of 1 species is unknown); on the basis of nest architecture, relationship with Oriental genus *Franklinia* has been suggested (Brooke and Dean 1990).

Endemic. 2 species, comprising a superspecies.

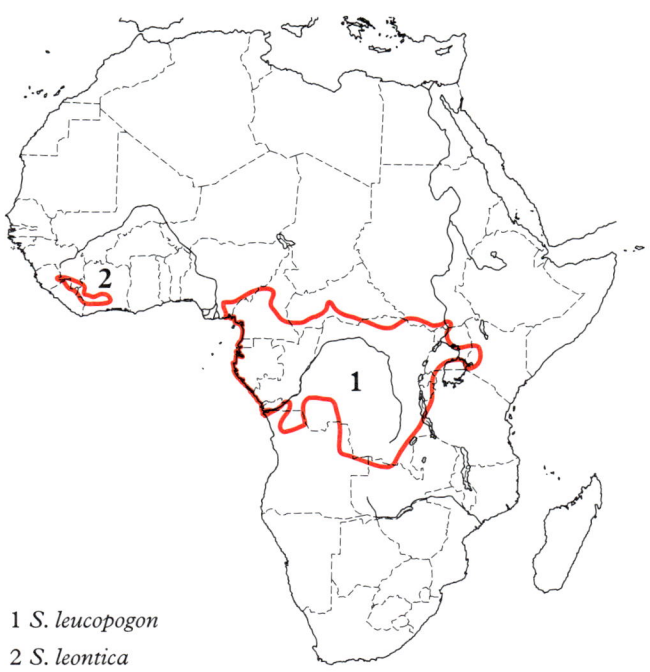

Schistolais leucopogon superspecies

1 *S. leucopogon*
2 *S. leontica*

Plate 14
(Opp. p. 241)

Schistolais leucopogon (Cabanis). White-chinned Prinia. Prinia à gorge blanche.

Drymoeca leucopogon Cabanis, 1875. J. Orn. 23, p. 235; Chinchoxo, Loango, Cabinda, Angola.

Forms a superspecies with *S. leontica*.

Range and Status. Endemic resident in forests from Nigeria to Kenya and Zambia. Nigeria, Calabar and Mambilla Plateau. Throughout Cameroon south of about 9°N. SW Central African Republic. S Sudan, Bengengai and Imatong Mts, up to 2500 m. Almost throughout Gabon, Congo and Zaïre. Uganda, north to Bunyoro, Acholi and Sezibwa R. W Kenya, west of Rift, up to 2400 m (Mt Elgon, S Nyanza south to Migori R. near Lolgorien, Kakamega Forest, Nandi, Mau). NW and W Tanzania (Bukoba to Kungwe Mahare Mts). Rwanda; Burundi. N Angola (Cabinda, N Cuanza Norte, N Malange, N and NE Lunda). N Zambia (Mwinilunga, Solwezi, Mporokoso). Common. Density in Gabon, 6–7 groups/km² of secondary forest.

Schistolais leucopogon

Description. *S. l. leucopogon* (Cabanis): range of species, east to middle Ubangi R. and to W side of L. Tanganyika. ADULT ♂: forehead to hindneck slate-grey, with blackish feather centres on forehead and forecrown giving scaly appearance; lores, area around eye and ear-coverts blackish grey, shade variable but usually darker than sides of crown. Mantle and back to uppertail-coverts slate-grey. Tail graduated, outermost feather *c*. 27 mm shorter than longest one (T1) (**A**), slate-grey, feathers with blackish subterminal spot, sometimes indistinct or absent; tips margined off-white or buffy. Chin and upper breast cream-white; lower throat, sides of neck and breast grey. Belly cream-white to off-white, flanks and undertail-coverts off-white washed with grey. Flight feathers and primary coverts brown, secondary coverts slate-grey, under-

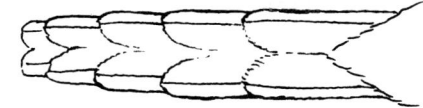

A

wing-coverts and axillaries white. Bill black; eyes vary from brown to reddish brown, red and scarlet (Bowen 1980a); legs and feet dull brownish pink. Sexes alike. SIZE: (22 ♂♂, 22 ♀♀) wing, ♂ 52–61 (58·4), ♀ 53–59 (56·0); tail, ♂ 56–66 (60·4), ♀ 52–58 (55·7); bill, ♂ 13–16 (14·8), ♀ 13·5–16 (14·8); tarsus, ♂ 19·5–22·5 (21·3), ♀ 20–22 (20·9). WEIGHT: Zaïre (Upemba), ♂ (n = 3) 12–13 (12·3); Zambia, ♂♀ (n = 12) 9·5–15 (11·7); Kenya, unsexed (n = 9) av. 13·5 (Mann 1985).

IMMATURE: resembles adult but has less speckling on forehead; chin and throat white; eyes brown; tip of bill horn-coloured, gape yellow.

NESTLING: hatches naked; mouth orange, 2 large oval black spots at base of tongue.

S. l. reichenowi (Hartlaub): Zaïre from middle Ubangi R. at Yakoma, eastward: Uganda, Kenya, Tanzania, Rwanda and Burundi. Grey on breast restricted; belly, flanks and undertail-coverts extensively buff; tail-feathers with paler tips, underwing-coverts and axillaries washed buff. WEIGHT: (Uganda) ♂ (n = 16) 10–15 (13·6), ♀ (n = 17) 10–15 (12·0).

Field Characters. A prinia-like warbler of forest edge, mainly grey with contrasting white throat, conspicuous in the field; belly white in western birds, buff in east (race *reichenowi*). Noisy, bubbling duet very different from the regular, repeated notes of Banded Prinia *Prinia bairdii*, another dark bird of forest edge; Banded Prinia has longer tail, white spots on wings and black-and-white barring on underparts.

Voice. Tape-recorded (32, 45, 86, B, C, ERA, HOR, LEM). Song always by a pair or family group, a bubbling mixture of harsh and pure notes, given in irregular bursts; the contribution of each individual is synchronized with those of the others. Totally different in form to the measured notes of individual birds in the typical *Prinia* song. Alarm, a harsh, 2-syllabled 'chipee-chipee'.

General Habits. Frequents forest edges (avoiding unbroken forest), gallery forest, edges of riparian forest, vicinity of rivers, dense bush, damp second growth and clearings, bracken-briar, rank grass and tangled cover. On Mambilla Plateau (Nigeria) common on edges of exotic gum plantations; in Ivindo basin (Gabon) favours areas of cleared forest and early stages of second growth (5–10 years old). Mainly in lowlands, but ascends up to 2400 m in E Zaïre, S Sudan, Uganda and W Kenya.

Sometimes in pairs but usually in family parties or social groups of 5–7, occasionally up to 10; birds of group remain together all year. Restless, constantly on the move; keeps within cover, seldom feeding higher than 5 m, but periodically emerges and perches in the open, calling in full view from branch or twig. Singing posts may be as high as 30 m (Chappuis 1974). Raises tail over back but does not cock or swivel it. In Nyungwe Forest, Rwanda, barely sings Oct–Dec, but much singing from Jan onward.

Food. Insects: beetles, eggs, caterpillars; spiders.

Breeding Habits. Territorial. Boundaries of territory are defended collectively by group of birds which call in synchrony, shaking wings and tail vigorously, keeping contact with incessant cries. Displaying ♂ raises tail over back.

NEST: deep, narrow pouch made of strips of large grass blades and similar material, sewn between 2 fairly broad, drooping leaves of *Aframomum* plant (**B**).

B

Opposing leaves are stitched together at the edges with tough spider web or cocoon silk, holes being punched in the margins; usually a small gap is left in front where edges are drawn together; entrance is at top rather than in side. Nest lined with hair-like fibres or shreds of grasses.

EGGS: 2, occasionally 3. Regular to long ovals, devoid of gloss; white to pale greenish blue, sparingly marked with rather large spots or blotches of reddish brown or chocolate and various underlying shades of lilac, grey or mauve. SIZE: (n = 20) 16–19·9 × 11·5–13 (17·6 × 12·2).

LAYING DATES: Cameroon, Apr–May, Sept; Zaïre, Apr–Sept, Itombwe, Jan–Feb (testes enlarged Sept–Oct; juvs Mar, May); Sudan, Apr; E Africa: Region B, Jan, Apr–Sept, Dec (prolonged breeding season in and after each rains, mainly Apr–June); Rwanda, Oct–Jan; Zambia, Nov–Dec, Mar.

Plate 14
(Opp. p. 241)

Schistolais leontica (Bates). Sierra Leone Prinia. Prinia du Sierra Leone.

Prinia leontica Bates, 1930. Bull. Br. Orn. Club 51, p. 51; Birwa Peak, Kono district, Sierra Leone.

Forms a superspecies with *S. leucopogon*.

Range and Status. Endemic resident. E Sierra Leone (Tingi Mts); Mt Nimba (Guinea/Liberia/Ivory Coast); Ivory Coast (Taï Forest, Man, Sipilou, Lamto). Very local and uncommon.

Description. ADULT ♂: forehead, crown, nape, hindneck, mantle and back grey; rump grey, tinged dark buff. Tail graduated, outermost feather *c.* 27 mm shorter than central one (T1); all feathers grey, narrowly margined off-white. Cheeks, lores, ear-coverts, chin, throat and sides of neck pale grey; breast pale grey, belly, flanks and undertail-coverts pale buff, centre of belly whiter. Wings grey-brown with pale margins to primaries, secondaries and wing-coverts; scapulars grey; underwing-coverts and axillaries pale buff. Bill black; eyes cream or silvery white; legs and feet dusky pink. Sexes alike. SIZE: (6 ♂♂, 4 ♀♀) wing, ♂ 53–55 (53·6), ♀ 51–53 (51·7); tail, ♂ 48–53 (51·0), ♀ 44–48 (45·5); bill, ♂ 13·5–14·5 (14·1), ♀ 13·5–14 (13·8); tarsus, ♂ 20–21·5 (20·7), ♀ 19·5–21 (20·1). WEIGHT: ♂ (n = 4) 11·9–13·6 (12·4), ♀ 12·2.

Field Characters. The only forest-edge prinia in its far-western range; a plain, entirely grey bird with one distinctive feature, a staring white eye.

Voice. Tape-recorded (KEI). Song a tuneless, unstructured, unsynchronized duet; one bird gives rapid, high-pitched 'sipsipsipsip . . .' and second one gives a lower, nasal, measured 'bur-bur-bur-bur . . .'. Other calls include 'psew-psew-psew . . .', 'pspi-psip-psip . . .' and 'bzee, bzee'.

General Habits. Inhabits forest edge, thick cover in mountain ravines, and thickets bordering streams; commonest in mountains between 580 and 1625 m. Found in parties, flitting about among vegetation, actively moving tail.

Food. Insects, including small black beetles.

Breeding Habits. Unknown. Birds with moderately enlarged gonads Liberia, Sept.

Schistolais leontica

Genus *Drymocichla* Hartlaub

Bill straight and black, no rictal bristles. Wing short and rounded, P9 broad and long, more than half length of P8; tail long and graduated, of 10 feathers, T5 much shorter than T4. Legs slender, feet small. Plumage grey with bright chestnut wing-patch, paler below, thighs buffy. Habitat dry shrubby savanna woodland near water. Nest unknown.

Monotypic, endemic to central N tropics. Closest relatives unknown, although Hall and Moreau (1970) saw some similarities with *Camaroptera*.

Drymocichla incana Hartlaub. Red-winged Grey Warbler. Prinia grise.

Drymocichla incana Hartlaub, 1881. Proc. Zool. Soc. Lond. (1880), p. 626; Magungo, northern Uganda.

Plate 17
(Opp. p. 320)

Range and Status. Endemic resident, E Nigeria to NW Uganda. In Nigeria known only from Gumti on Yim R., Gashaka-Gumti Game Reserve; Cameroon (W edge of Adamawa Plateau, Genderu Mts, Dodo, Tibati, and between Koncha and Tignere); uncommon in Central African Republic, occurring in Uam (= Uham = Ouham) district, Bozoum, Bamingui-Bangoran Nat. Park, Manovo-Gounda-Saint Floris Nat. Park and possibly Lobaye Préfecture (listed by Carroll 1988, but record thought uncertain by Germain 1992); widely distributed Sudan, south of 9°N (Kajo Kaji, Yei, Maridi, Lado, Wadelai, Wau and Boro R.); N Zaïre (Ubangi Chari, Faradje in Uele district where fairly common and 'where in every little patch of woods on swampy ground a pair could be expected' (Chapin 1953a), and W shore of L. Albert); uncommon in NW Uganda (600–1200 m), south to Murchison Falls Nat. Park, Masindi and NE shore of L. Albert, also on Semliki Flats near Bwamba Forest. 3 sightings in Guinea (Macenta Préfecture: Halleux 1994), and 1 old record from Chari R., Chad (Salvan 1967–69).

Drymocichla incana

Description. ADULT ♂: upperparts including crown, nape, back, rump, and uppertail-coverts medium grey; forehead paler than crown. Tail olive-brown washed grey, feathers strongly graduated (T6 21–26 shorter than T1 (n = 4): M. P. Adams, pers. comm.). Lores, feathers at base of bill and orbital ring white, ear-coverts grey. Chin, throat and central line of belly white, sides of breast and belly washed grey (paler than mantle). Undertail-coverts white, thighs white washed with buff. Primaries and most secondaries olive-brown, basal halves of P3–P9 chestnut, inner secondaries washed grey and merging into colour of mantle. Underwing-coverts white, rest of underwing brick red, axillaries pale grey; edge of wing near shoulder pale greyish white. Bill black; eyes pale greenish grey; legs and feet dull yellow. Sexes similar. SIZE: (10 ♂♂, 6 ♀♀) wing, ♂ 55–61 (58·4), ♀ 54–59 (56·1); tail, ♂ 51–57 (54·0), ♀ 49–56 (52·2); bill to feathers, ♂ 11–12·5 (11·7), ♀ 11·5–12·5 (11·8); tarsus, ♂ 20–21 (20·5), ♀ 20–21 (20·4). WEIGHT: (Uganda) 1 ♂ 10, 1 ♀ 10.

IMMATURE: like adult but tawny patch on wing paler, flanks and belly washed rufous, lower mandible and tip of upper mandible yellowish (not black).

NESTLING: not described.

Field Characters. A medium-sized warbler, about size of Red-winged Warbler *Heliolais erythroptera*, easily recognized by broad chestnut patch on primaries, conspicuous both at rest and in flight, contrasting with olive-brown wing and entirely grey upperparts; tail unpatterned, strongly graduated. Underparts are whitish, bill black. Grey eye and dull yellow legs and feet are also good field characters. The only warbler that it might be confused with is Red-winged Warbler, which is darker grey above, has chestnut (not grey) wing-coverts and a patterned tail, with pale tip and dark subterminal spots.

Voice. Tape-recorded (68, CHA). Song (of ♂?) unmusical and monotonous but not unpleasant; mainly a repetition of a double note (although sometimes only a single note is uttered); one 20 s sequence was 'three three-cheers three-cheers three-cheers-cheers three-cheers three three-cheers'. Second member of pair (♀?) repeats a single 'tchwee' note, 'tchwee tchwee tchwee tchwee tchwee tchwee', sometimes in duet with its mate. If sexes have been identified correctly, duetting is instigated by ♀, not ♂ (Chappuis 1974, 1979a).

General Habits. Inhabits tree savanna with a good shrub layer, often near water or in swampy areas in savanna woodland, and thick bushes fringing rivers in open grass savanna. Usually solitary or in pairs, but also in family parties and in mixed flocks when foraging (Green 1990). Typically forages by gleaning in shrubs or lower branches of trees. Moves constantly; flicks tail (Halleux 1994). Resident throughout its range.

Food. Small insects.

Breeding Habits. Unknown.

BREEDING INDICATIONS: Zaïre (Uele), adults with large gonads Apr, Aug, sexual activity subsiding Oct, Nov, ceased Dec (Chapin 1953a), pair nearly ready to breed early Sept, Kasenyi (L. Albert); Uganda (Murchison Falls), pair with 2 dependent young early Aug. Breeding evidently coincides approximately with rains.

Genus *Phyllolais* Hartlaub

A tiny warbler. Bill short, slender, compressed, with sharply ridged culmen; exposed bill shorter than hind toe plus claw. Wing short and rounded, like *Apalis*; tail strongly graduated. Upperparts olive, underparts yellow-buff, tail dark brown with all feathers but T1 having some white. Habitat open grassy woodland with flat-topped acacias, also forest edge. Nest purse-shaped with side entrance, material secured with cobweb and gummy acacia sap.

Monotypic, endemic to Africa north of 5°S, east of 10°E. Close to *Apalis* and considered congeneric by White (1960a), but Hall and Moreau (1970) admit that it is 'difficult to place systematically', having 'the appearance of an *Eremomela* with the tail of an *Apalis* or *Prinia*'; restricted to acacia, like *Eremomela*, but nest different. We follow them and subsequent authors in keeping *Phyllolais* separate.

Plate 17
(Opp. p. 320)

Phyllolais pulchella (Cretzschmar). Buff-bellied Warbler; Acacia Warbler. Phyllolaïs à ventre fauve.

Malurus pulchella Cretzschmar, 1827. In Rüppell, Atlas Reise Nördl. Afrika, Vögel (1826), p. 53, pl. 35, fig. a; Kordofan, central Sudan.

Phyllolais pulchella

Range and Status. Endemic resident. Local and rare in NE Nigeria (Kano, Potiskum, L. Chad, Maiduguri, Kauwa); Chad (Djimtillo, where common); N Cameroon (Gagadjé, Koza, Touroua, Mandara and Benue Plain), Central African Republic (Lobaye Préfecture); in S Sudan fairly common along Nile and to the east, but rare elsewhere (Maridi, Gogrial, Kordofan, Sennar-Darfur and further south). Common in NE Zaïre (Faradje, Nzoro, Aba in Upper Uele district, W shore of L. Albert, upper Semliki valley, S shore of L. Edward, foothills of Rwenzoris, Rutshuru Plain) but absent from Kivu highlands; NE Rwanda (Akagera Nat. Park). Uncommon in W and S Ethiopia, Western Highlands and Rift valley; widely distributed and locally common in W and central Kenya (Kora, Pokot, R. Kerio, Mt Elgon, L. Baringo, Kisumu, S Nyanza, Narok, L. Naivasha, R. Athi, Simba, Tsavo), rarer in arid NW (Turkana District, Ileret, Lodwar); widely distributed in Uganda except NE, south to Lango and Karamoja, and in W to Ankole; absent from moister areas west and north of L. Victoria; common in N Tanzania (Serengeti Nat. Park east to Arusha and south to Mt Hanang and Kondoa), 1 record S Tanzania (Usangu flats).

Description. ADULT ♂: whole of head, back, and rump olive-brown, uppertail-coverts olive. Tail strongly graduated (T6 12–17 shorter than T1 (n = 5): M. P. Adams, pers. comm.), feathers brownish olive, all except T1 broadly tipped white, T4 and outermost T5 with white outer web. Lores, eye-ring, and ear-coverts straw-yellow; entire underparts pale straw-yellow, including undertail-coverts, underwing-coverts and axillaries. Primaries and secondaries brownish olive, secondaries and upperwing-coverts pale-edged on both webs. Upper mandible brown, lower flesh-pink; eyes light brown; legs and feet dark flesh-pink. Sexes alike. SIZE: (10 ♂♂, 10 ♀♀) wing, ♂ 44–47 (45·8), ♀ 42–46 (44·1); tail, ♂ 40–47 (43·0), ♀ 39–43 (41·0); bill to feathers, ♂ 8–9 (8·5), ♀ 8–9 (8·5); tarsus, ♂ 15–16 (15·5), ♀ 15–17 (15·7). WEIGHT: (Uganda) 1 ♂ 6, 1 ♀ 6.

IMMATURE: like adult but more yellowish.
NESTLING: not described.

Field Characters. A rather undistinguished tiny warbler of arid acacia steppe, with pale olive-brown head and mantle, brown wings and tail, and pale yellow underparts. Best features are tail and voice. Strongly graduated tail is blackish with white tip and white sides; bird might be mistaken for a *Phylloscopus* or *Eremomela* warbler, but its tail is different and more like those of *Prinia* and *Apalis*. Difficult to spot in the canopy except when it moves among leaves; readily visible when feeding on insects that swarm on shoots of younger trees that are full of sap, and confiding when near nest. Its trilling call is strikingly like that of Chestnut-crowned Sparrow-Weaver *Plocepasser superciliosus*, with which it occurs in Ethiopia, Sudan, Uganda and Kenya. Immature similar to young Willow Warbler *Phylloscopus trochilus*.

Voice. Tape-recorded (68, B, F, McVIC, PAR, STJ). Song (or form of contact call?) a short, rapid, rather angry, rasping trill 'prrrreeeeet', lasting c. 0.5–0.7 s and repeated 9 times in 26 s, similar in quality to song of Winding Cisticola *Cisticola galactotes*, but shorter. Contact call when foraging, a 'zipzipzip', or 'cher cher cher' by ♂, and a higher 'tchit tchit tchit' by ♀ (van Someren 1956) (described as the song by Mackworth-Praed and Grant, 1973: a dry trill 'zit-zit-zit-char-char-chip', loud for such a small bird). Alarm, a single sharp, short, loud 'tchit' repeated 4–5 times per s, also an occasional low 'tcher' by ♂ (van Someren 1956).

General Habits. Inhabits open grassy woodland, especially flat-topped acacias with grass and scrub layer, also forest edge, and thornbush, at least in Ethiopia and some parts of Kenya. In E Africa occurs in acacia from 600 to 2000 m, most frequently above 1000 m and in areas with >500 mm annual rainfall. Occurs usually in groups of 3–5, actively and noisily foraging in canopy of trees. Occurs with the Collared Sunbird *Anthreptes collaris* at forest edge, but also commonly in mixed flocks foraging in acacia trees – particularly when food is short (Sinclair 1978). Resident throughout range, except possibly in Sudan where thought to move north in rains (Nikolaus 1987).

Food. Aphids and scale insects (van Someren 1956); insect larvae and spiders.

Breeding Habits. Territorial, solitary nester.
NEST: small purse-shaped bag with side entrance near top, made of closely felted vegetable down and scaly acacia bark worked in and secured with spiders' webs (**A**). Both birds of a pair were often observed collecting papery bark, carrying it to source of gummy sap exuding from broken branch of large acacia, passing bill over sap, then returning to nest and wiping material onto felted side of nest before working it in with its bill; bird then cleaned bill vigorously on a thorn (van Someren 1956). Nest built 1·2–9 m above ground, in thorny bush or acacia sapling with long, acutely sharp thorns that are criss-crossed and packed together in such a way that thorns around entrance point forward (**A**). Nest entrance so well protected that it is impossible to place a finger into it. Nest c. 70 mm deep, 50 wide at base, cup 37 deep; entrance 25 across.

EGGS: 2–4, usually 2–3. White or tinged light green or greenish blue with reddish brown, greyish brown or chocolate spots, mainly concentrated at large end. SIZE: c. 14·5 × 10.

LAYING DATES: Eritrea (nestlings mid-July); Ethiopia, June, July; Rwanda, Apr, May; Zaïre (birds with enlarged gonads May, late July, adults with full-grown young Oct); Uganda (birds with enlarged gonads May and Sept); Kenya (adults with fledglings late Dec 1988, pair active at nest mid-Sept, birds with enlarged gonads May, Sept); Tanzania (nest-building Aug, Nov, young in nest mid-June 1972). E Africa: Region A, May; Region B, Jan; Region D, Mar, Apr, May.

INCUBATION: period c. 12 days.

DEVELOPMENT AND CARE OF YOUNG: nestling period c. 16 days.

Reference
van Someren, V. G. L. (1956).

Genus *Apalis* Swainson

Small, distinctive, plain-plumaged or rather strikingly patterned warblers; sexes alike or considerably dimorphic (♂♂ more brightly coloured). Wings short and rounded; tail of 12 tail-feathers (10 in *pulchra* and *ruwenzorii*), quite long, strongly graduated, T6 half to one-third length of T1, pale tipped (plain in one species) or with outer feathers (T4–T6) extensively white. Bill varies from narrow and warbler-like to broad and flycatcher-like, but culmen and tomium nearly parallel, curving only at tip; nostrils with nasal operculum; rictal bristles poorly developed.

Occur principally in evergreen forest (especially montane), a few species in savanna (only one widespread there); most frequently in forest canopy and mid-stratum, edges and thickets, rarely in undergrowth. Ecologically *Apalis* spp. are strictly segregated although 8 can occur in a single forest. Up to 6 are regularly sympatric, with much interspecific competition and counter-singing. Songs simple, much alike; all species duet. All species are sedentary (or wander very locally). Many have disjunct ranges. Sociable, forming family parties outside breeding season. Nests oval or bag-shaped with side-top entrance; eggs highly variable.

Endemic; 21 species. Several clear-cut phyletic and ecological lines are recognizable, but genus is very uniform, its species immediately recognizable as such. All appear to be closely related and to be in process of continuing differentiation. They form 3 groups, thought to be related as follows: (1) *thoracica/pulchra/ruwenzorii*; *ruddi*; *flavida*; *binotata/personata*; *jacksoni*; *chariessa*; *nigriceps*; (2) *melanocephala/chirindensis*; *porphyrolaema/chapini*; *sharpii*; *rufogularis/karamojae*; *bamendae*; *goslingi*; (3) *cinerea/alticola*. Generic relationships uncertain; displays of a few *Apalis* species are similar to those of *Camaroptera*. *Apalis* has often been placed next to *Prinia*, but a close relationship between them remains to be demonstrated.

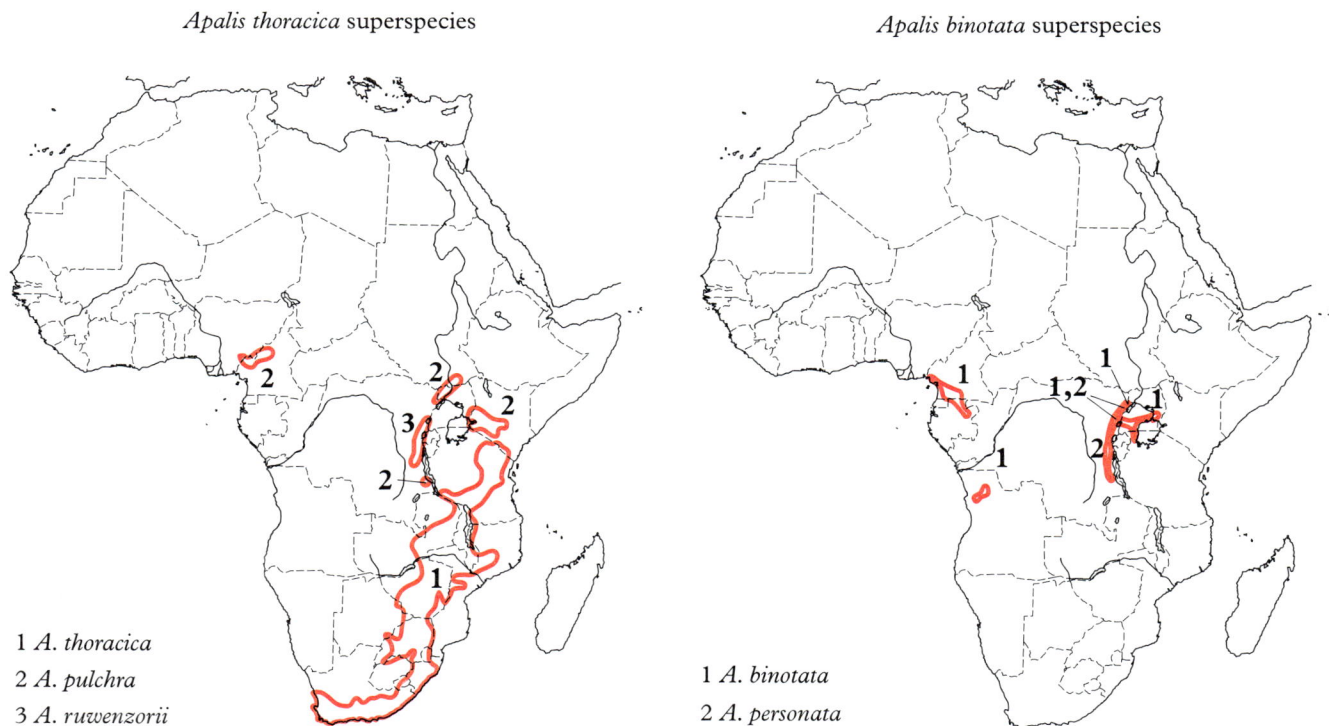

Apalis thoracica superspecies

1 *A. thoracica*
2 *A. pulchra*
3 *A. ruwenzorii*

Apalis binotata superspecies

1 *A. binotata*
2 *A. personata*

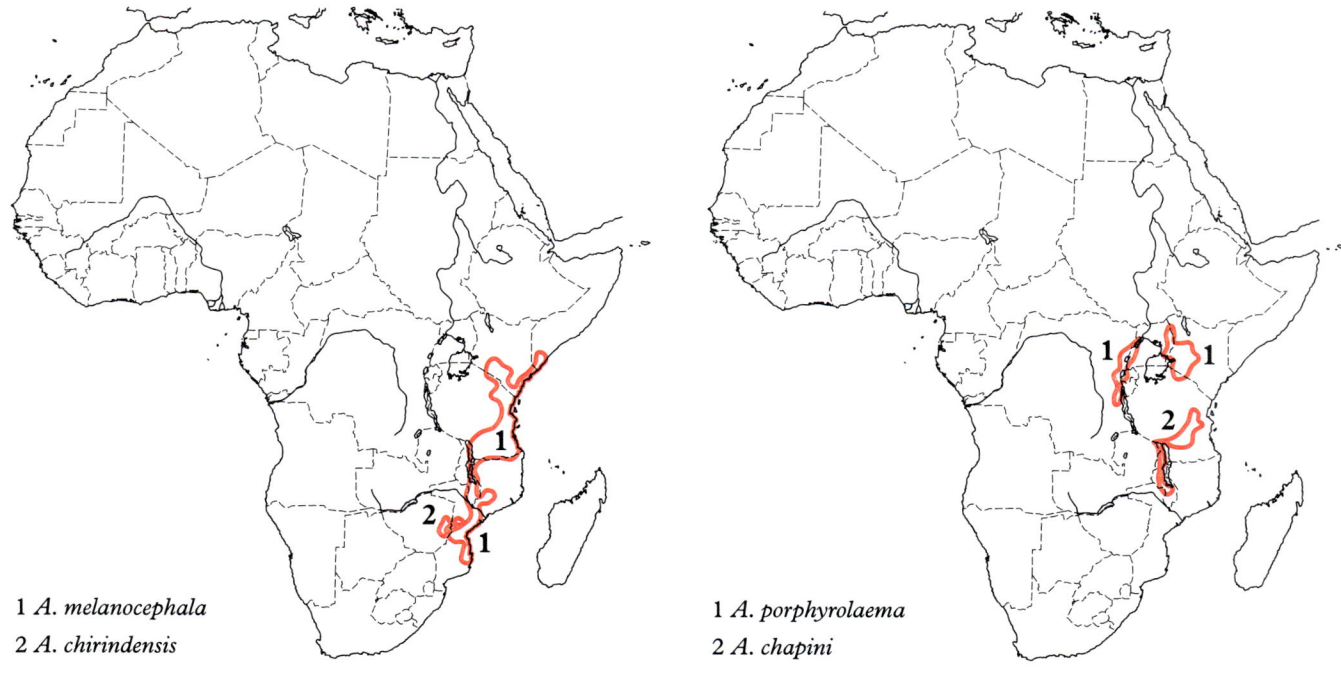

Plate 15
(Opp. p. 256)

Apalis thoracica (Shaw). Bar-throated Apalis. Apalis à collier.

Motacilla thoracica Shaw, 1811. *In* Shaw and Nodder, Nat. Misc. 22, pl. 969; Grahamstown, Cape Province, South Africa.

Forms a superspecies with *A. pulchra* and *A. ruwenzorii*.

Apalis thoracica

Range and Status. Endemic resident. Common in forest and thickets: SE Kenya, highlands of central and E Tanzania, E Zambia, Malaŵi, interior of Mozambique, E and SW Zimbabwe, extreme E Botswana, and South Africa from Transvaal and Natal to SW Cape Province. Densities of 8–10 pairs/5 ha (Malaŵi: Dedza, 2150 m, Mlunduni, 1800–2000 m) and >10 pairs in 40 ha of lowland forest (Malaŵi: Thambani, 1100–1200 m) (Dowsett-Lemaire 1989). Density of 31 pairs or 45 singletons per 100 ha in secondary forest, Udzungwa Mts, Tanzania (Moyer 1993). On Nyika Plateau, Malaŵi, pairs occupy forest patches of 0·12–0·15 ha; circular patches av. 1 pair/0·6 ha (once, 17 pairs in 7 round patches totalling 10·2 ha); long, narrow patches av. 1 pair/0·4 ha (once, 9 pairs in 3·8 ha) (Dowsett-Lemaire 1983). In Zimbabwe linear territories in wooded ravines and scattered savanna thickets are much larger (M. P. S. Irwin, pers. obs.).

Description. *A. t. arnoldi* Roberts: E Zimbabwe and adjoining Mozambique frontier from central Nyanga Highlands to Mt Selinda. ADULT ♂: forehead, crown, cheeks and ear-coverts sooty brown; dark loral streak continuing through eye. Nape, mantle and back to uppertail-coverts dark moss green. Tail blackish, feathers edged dark grey, T6, T5 and T4 with extensive white tips, T6 almost wholly white. Chin and throat off-white; narrow, blackish, crescent-shaped breast-band reaching to sides of neck. Breast, belly and flanks to undertail-coverts white, strongly washed yellow; thighs greyish olive. Wing feathers blackish, margined dark grey; underwing-coverts and axillaries white. Varies geographically within narrow limits in intensity of green above and yellow below (populations 30 km apart readily separable). Bill black; eyes creamy white; legs and feet pinkish flesh. Sexes alike, but ♀ usually has narrower breast-band. SIZE: (60 ♂♂, 40 ♀♀) wing, ♂ 51–58 (55·4), ♀ 49–56 (53·8); tail, ♂ 48–58 (52·8), ♀ 47–53 (50·9); bill, ♂ 13·5–16 (14·7), ♀ 13–15, (14·2); tarsus, ♂ 20·5–22 (21·6), ♀ 20–22 (20·9). WEIGHT ♂ (n = 70, Zimbabwe) 9·3–13·1 (11·0); ♀ (n = 60, Zimbabwe) 8·3–12·5 (10·0).

IMMATURE: like adult but with duller, less well-defined and narrower breast-band.

NESTLING: hatches naked; skin dusky flesh-pink; bill yellowish horn, gape cream, inside of mouth and tongue yellow with 2 black marks at base.

A. t. fuscigularis Moreau: SE Kenya (Taita Hills). Top and sides of head earth-brown, ear-coverts paler; back and wings blackish grey with dark greenish wash on rump and uppertail-coverts; chin and throat blackish, merging with dark breast-band; belly, flanks and underwing-coverts off-white to sooty grey (due to extensive blackish feather bases); T5–T6 white with dusky margins, T4 white-tipped.

A. t. griseiceps Reichenow (including '*iringae*'): SE Kenya (Chyulu Hills); mountains of N Tanzania (Mt Kilimanjaro to 3200 m, Mt Meru and from Ketumbeine, Monduli and Longido to the Crater Highlands: Oldeani, Nou, Mbulu, Mt Hanang and Ngorongoro), E Tanzania at Kiboriani (Mpwapwa), and in S Highlands (Iringa, Uzungwa Mts). Top and sides of head brown, back green, tail grey-black, T4–T6 white except for dusky outer web of T4; chin, throat and immediately below breast-band white, belly and flanks bright yellow. WEIGHT: unsexed (n = 12, Udzungwa Mts, Tanzania) 10.5–12.1 (11.1) (J. O. Svendsen, pers. comm.).

A. t. pareensis Ripley and Heinrich: South Pare Mts. Dark grey above; belly and flanks almost white, yellow restricted to vent.

A. t. murina Reichenow: NE Tanzania (W and E Usambara Mts), E Tanzania (Nguru and Ukaguru Mts) and S Highlands (Mbeya and Rungwe Mts, Poroto Mts to Tukuyu and Njombe); N Malaŵi in Mafinga Mts (and adjacent Zambia) and Misuku Hills (Wilindi and Matipa). Crown to nape brown, back grey (sometimes tinged olive), rump grey-green; yellow restricted to lower belly, flanks strongly washed olive-green.

A. t. uluguru Neumann: NE Tanzania (Uluguru Mts). Forehead, crown and sides of head dark sooty brown, remainder of upperparts green; tail feathers blackish grey, white restricted to outer half of outer webs and to tips of T5–T6: chin and throat immaculate white (but with grey feather bases), in strong contrast to bright yellow belly and flanks.

A. t. youngi Kinnear: SW Tanzania (Ufipa Plateau at Sumbawanga), NE Malaŵi littoral (Livingstonia, Nyankhowa and Uzumara), Nyika Plateau (extending into Zambia) and N and S Viphya Mts. Crown brown, back to rump and uppertail-coverts clear slate-grey (olive washed); belly to undertail-coverts and flanks clear white.

A. t. whitei Grant and Mackworth-Praed: E Zambia (west of Muchinga Escarpment to *c.* 14°20′S); S Malaŵi above 1000 m west of Nyasa-Shire Rift (Dzalanyama to Kirk Mts and Ntchisi), and adjacent Mozambique (Zobue). Zambian and Malaŵian populations isolated and differ slightly from one another. Back green, contrasting with brownish crown; belly washed yellow.

A. t. flavigularis Shelley: SE Malaŵi east of Nyasa-Shire Rift (Mulanje, Zomba, Malosa) and adjacent Mozambique; Chiperoni Mt (this race?). ♂ has forehead, crown and sides of head black with greenish wash, ear-coverts dark grey; mantle and back bright green; tail black, outer webs of feathers margined green; wings green, underwing-coverts yellowish; underparts (except for black breast-band) bright yellow, flanks washed olive. ♀ has sooty brown crown and blackish loral streak. Immatures are individually variable, more uniformly green above.

A. t. lynesi Vincent: N Mozambique (Namuli Mt). Like *flavigularis* but ♂ has forehead, crown, nape and sides of head sooty grey, lores deep black; chin and throat to upper breast deep black (obscuring breast-band), belly olive-yellow, brighter in centre, flanks and undertail-coverts deep olive. ♀ has lores, chin, throat and upper breast sooty rather than lustrous black, centre of breast invaded by dark grey (a vestigial breast-band), washed green at sides; flanks with marked olive wash. Tail short: (n = 9) 46–51 (48·8), T5 and T6 with white only on tips and outer webs.

A. t. quarta Irwin: NE Zimbabwe (Mt Nyangani) and Mozambique (Gorongoza Mt). Forehead and crown darker than contiguous *arnoldi*, mantle and back darker green than *arnoldi*; belly and flanks yellow, strongly washed olive.

A. t. rhodesiae Gunning and Roberts: Zimbabwe plateau from western rain-shadow of Nyanga Highlands, Rusape and west of Save R. valley to Zambezi Escarpment at 30°E and south to Matobo Hills, Shashani R., Plumtree and NE Botswana (north of Francistown). Crown paler and greyer than *arnoldi*, back greyish, faintly washed pale green; chin and throat cream-white, belly and flanks washed pale yellow.

A. t. spelonkensis Gunning and Roberts: South Africa (E and NE Transvaal from Soutpansberg, Louis Trichardt and Woodbush to Lydenberg and Pilgrim's Rest). Forehead and crown greyish brown, contrasting with greenish back; lower throat and underparts bright cream-yellow.

A. t. flaviventris Gunning and Roberts: South Africa (N and W Transvaal east to Pretoria and north to Blouberg) and SE Botswana (Gaborone, Kanye and Lobatsi). Forehead and crown grey, contrasting with more olive-green upperparts; duller yellow below than *spelonkensis*.

A. t. lebomboensis Roberts: NE Zululand (Lebombo Mts), E Swaziland and S Mozambique. Like *venusta* but forehead and crown entirely green, more yellowish green on back.

A. t. drakensbergensis Roberts: South Africa (SE Transvaal in Drakensberg Mts, interior of Natal and E Orange Free State); W Swaziland. Forehead and crown green as in *lebomboensis*, but throat and underparts paler yellow.

A. t. venusta Gunning and Roberts (including '*darglensis*'): Zululand, Durban and interior parts of Natal, E Griqualand west to Great Kei R. Forehead and crown grey, contrasting with olive-green mantle and back; belly paler yellow.

A. t. thoracica (Shaw): South Africa (SE Cape Province from Great Kei and Gamtoos R. to Umtata). Forehead and crown light grey, back olive-green like *venusta*; throat white, contrasting with primrose yellow belly.

A. t. claudei Sclater: South Africa (S Cape from Knysna to Humansdorp and north to Beaufort West). Whole of upperparts deep rusty olive-brown; throat pale rusty buff, belly white, flanks tinged olive-brown.

A. t. capensis Roberts: South Africa (S and SW Cape from Paarl to Oudtshoorn and Mossel Bay). Crown and back grey; white below with greyish flanks.

A. t. griseopyga Lawson: South Africa (coastal W Cape from Lamberts Bay south to Cape Town). Forehead, crown and back pale blue-grey; below like *capensis*.

TAXONOMIC NOTE: geographical variation is striking; highly distinctive races occur in forest on particular mountains, e.g. in E Africa and Malaŵi (*fuscigularis, pareensis, uluguru, lynesi, flavigularis, whytei*). In Tanzania, disjunct populations of 2 widespread races (*griseiceps, murina*) replace one another irregularly with little or no variation between isolates; both occur in S highlands, where they are apparently mutually exclusive on particular mountains. From Zambia and Malaŵi southwards variation is more broadly clinal, except for *flavigularis* and *lynesi* east of Rift. Populations appear not to differ biologically, except that *fuscigularis* does not react to taped call of neighbouring races (D. A. Turner, pers. comm.). There may be some incipient speciation, and further studies are needed.

Field Characters. A widespread and highly variable apalis, with different combinations of brown, grey or sometimes black head, grey or green back, and white or yellow underparts (throat black in 2 isolated races). All races have black band separating throat from breast. Darker and duller above than Yellow-breasted Apalis *A. flavida* and Rudd's Apalis *A. ruddi*, back less green; best seperated from them by white eye and extensive white in outertail, conspicuous in flight. Colour of underparts highly variable in Bar-throated Apalis but constant in Yellow-breasted Apalis (broad yellow breast-band with black, if present, *below* the yellow) and Rudd's Apalis (narrow olive-yellow breast-band below the black bar, olive-yellow flanks, white belly). Voices distinctive.

Voice. Tape-recorded (14, 20, 86, 88, 91, B, F, HAN, SVEN). Song a duet. First bird (♂?) gives a monotonous, dry, rapidly repeated single or double note, 'tlip, tlip, tlip, tlip . . .', 'pilly, pilly, pilly, pilly . . .', 'tjil, tjil, tjil, tjil . . .', with regional variations; notes are grouped into phrases of variable length, sometimes only 2–4 notes per phrase, often 8–12 or even more. Constant pitch and tempo are maintained. The sound has a penetrating quality and can carry for a considerable distance. Second bird responds with thin, high-pitched notes, either at about same rate as partner's, 'ti, ti, ti, ti . . .', or run together into a trill, 'titititiiti . . .'. Sometimes several birds call together. Alarm calls, a single, abrupt 'deg' or 'dew' and a rapidly repeated 'tik-tik-tik-tik-tik-tik', often given by several birds at once; also snaps bill.

General Habits. Highly adaptable, frequenting wide range of cover from wet evergreen forest to thickets in semi-arid savanna, coastal dune scrub at sea level in South Africa, and coastal forest in Natal. In Zimbabwe occurs in forest, scrub in ravines, semi-deciduous thickets, and on hillsides with vegetated gullies; absent from all low-lying river valley systems; has also adapted to parks and gardens in towns; in Zambia and Malaŵi in riparian fringing forest, but always most numerous in evergreen forest. Also occurs in second growth and thickets nearby, and wanders along riparian fringes; occupies exotic vegetation with abundant cover. In E Africa becomes more montane, with many isolated populations occurring between 1300 and 2600 m (and up to 3200 m on Kilimanjaro). In forest principally on edges and in lower mid-stratum; only occasionally in canopy, although more frequently so at higher altitudes where forest is stunted.

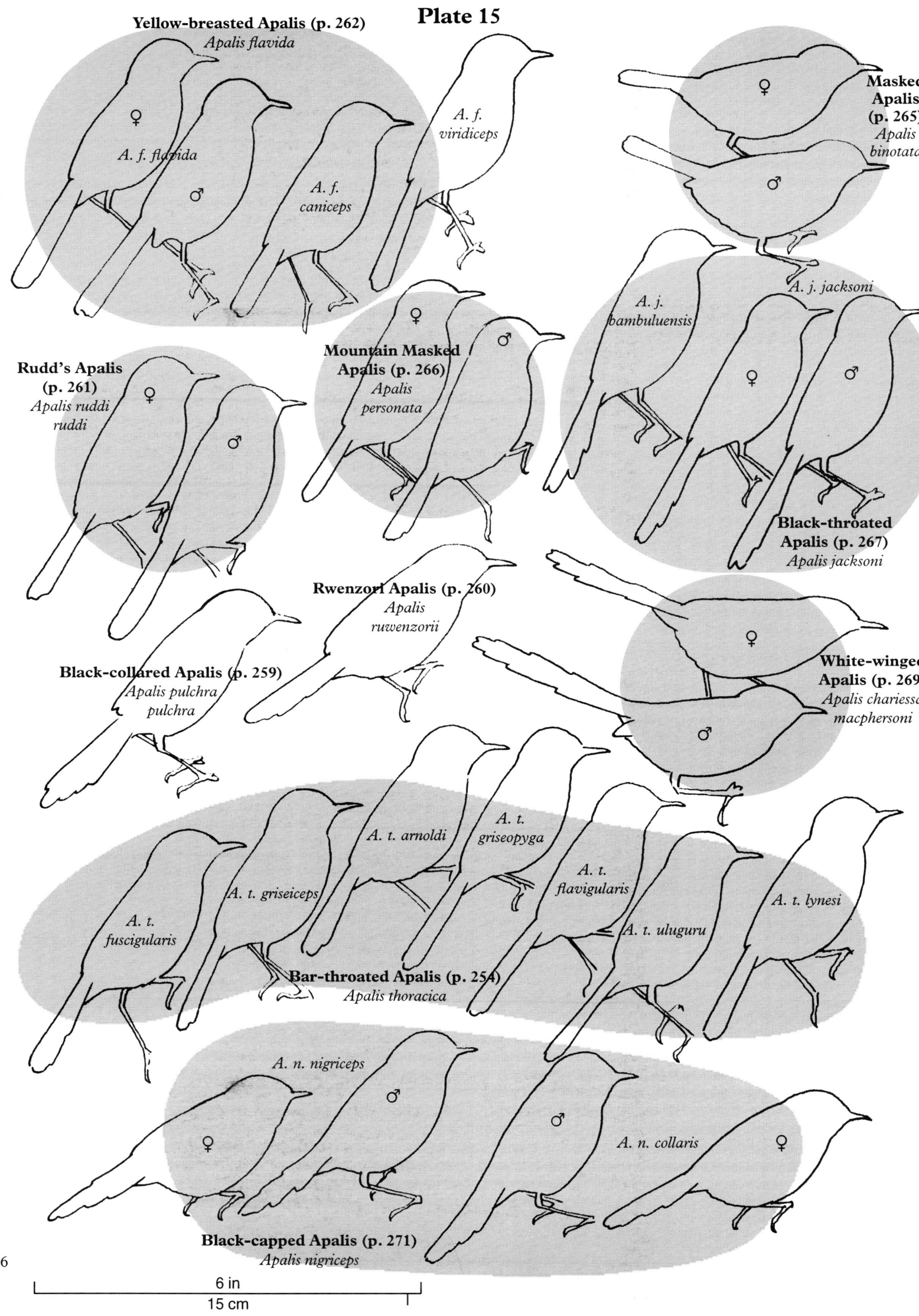

Plate 16

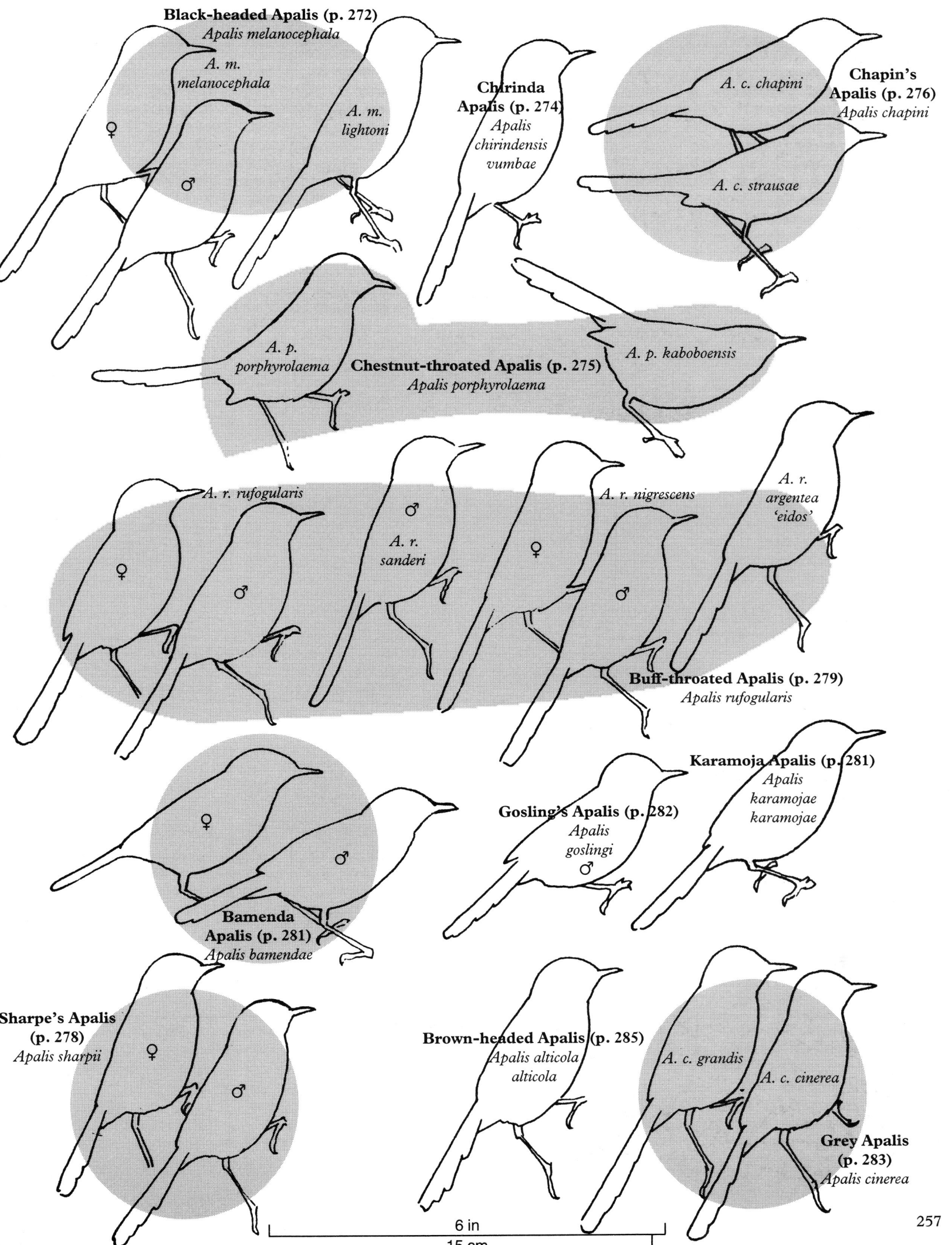

Highly sociable, moving around in pairs or small family parties of 4–6 (occasionally more), searching leaves, twigs, branches and tree trunks; often descends to ground; flycatches frequently. In Transvaal forest reported to forage mainly from horizontal trunks and twigs of c. 20 mm diam., gleaning 6 times as often as snatch-and-hovering (Earlé 1983, 1989); regional feeding differences seem probable. Occasionally found at ant-columns, probably attracted by other birds rather than ants, when pecks at branches and stems of vegetation rather than leaves, 0·3–3·0 m above ground (Willis 1986a). On Nyika Plateau (Malaŵi) feeds at higher levels (in canopy and on forest edge) in Dec–Jan than at other times of year (Dowsett and Dowsett-Lemaire 1984).

Ecology and relationships with other species studied on Nyika Plateau (Dowsett-Lemaire 1989). Bar-throated Apalis inhabits understorey and mid-stratum of tall forest (16–18 m) but all levels in lower, scrubby montane forest (8–15 m), as in Zimbabwe. In Malaŵi often countersings with Chapin's Apalis *A. chapini* wherever they overlap, but competes more obviously with Brown-headed Apalis *A. alticola* (on SW Nyika where they co-exist between 1950 and 2150 m, Brown-headed Apalis keeps to canopy). In Njombe (Tanzania) *A. thoracica* inhabits forest and *A. alticola* replaces it in riparian tree growth (Lynes 1934). In Malaŵi density and altitudinal distribution vary with race; *murina* widespread though not numerous in Wilindi-Matipa (Misuku Hills, and neighbouring Tanzania) but absent 8 km away in Mugesse Forest; no records from Mafinga Mts apart from specimen collected in 1937; also absent from neighbouring Jembya and Musisi (where *A. alticola* occurs); *youngi* numerous throughout, with c. 10 pairs per 5 ha of fragmented forest at 1600–2450 m. Some unmated ♂ ♂ wander a few km away into riparian plantations (Nyika) and small pockets of forest in miombo woodland (Viphya). *A. t. whitei* thrives in mid-altitude rain and riparian forest (Dowsett-Lemaire 1989).

Most populations are sedentary except for some wandering by young birds; but some move locally, particularly downward for the winter.

Food. Insects: caterpillars, Diptera, small grasshoppers, beetles and their larvae, moth eggs; spiders. Occasionally berries and seeds.

Breeding Habits. Territorial; most pairs apparently maintain year-round territory. A bird reacted to its reflection in window, dancing up and down, flicking wings, snapping bill and calling (Williams 1991).

NEST: bulky structure of moss, strengthened with grass, rootlets and other material (occasionally almost wholly grass and tendrils); domed, with side-top entrance (sometimes with 2 entrances), lined with vegetable down, hair and rootlets (**A**: see Benson and Benson, *Ibis* 1947, 285 and Fig. 3). The dome is also lined. Ext. depth 100, ext. diam 50–80, int. depth 70, diam. of entrance 40–50, depth of cup 50. Sited near ground, often in bank or at side of path; in forest up to 1 m or more above ground, in more open habitats up to at least 3 m. Lightly attached to twigs, or bound onto stem, or in bank recess.

A

EGGS: 2–4 (usually 3 in South Africa and 2 further north). Oval: very variable ground colour, pale blue, greenish or white, with fine to heavy spotting of reddish brown, rufous or slate, uniformly distributed or concentrated at larger end, but rarely forming ring. SIZE: (n = 106) 15·5–19·1 × 11·2–13·5 (17·2 × 12·2). Often double-brooded.

LAYING DATES: E Africa: Region C (in pre-rains and rains), Oct, Dec; Region D, Feb, Dec; Malaŵi, Sept–Feb (Sept 3, Oct 23, Nov 23, Dec 6, Feb 1 clutches): Zambia, Jan, Oct, Dec; Zimbabwe, Aug–Mar (Aug 7, Sept 23, Oct 40, Nov 48, Dec 42, Jan 19, Feb 3, Mar 3); South Africa: Transvaal, Sept–Jan (Sept 1, Oct 8, Nov 4, Dec 8, Jan 4); Karoo, Aug–Nov; Natal, Aug–Jan (Aug 1, Sept 5, Oct 16, Nov 19, Dec 19, Jan 2); E Cape, Aug–Jan (Aug 3, Sept 14, Oct 34, Nov 14, Dec 7, Jan 1); SW Cape, July–Jan (July 6, Aug 20, Sept 33, Oct 16, Nov 4, Dec 1, Jan 1); Mozambique, Oct–Mar; Botswana, Aug.

INCUBATION: both sexes incubate. Eggs laid at daily intervals, usually in early morning; some ♀ ♀ may dump eggs in nests of conspecifics (Hustler 1990). Period: 14–17 (av. 15) days.

DEVELOPMENT AND CARE OF YOUNG: both sexes feed young. They often forage close to nest, approach it through bush at entrance level, and leave from point just above nest. At first adults feed young through nest entrance; later, they perch on nest and bend down to feed young. Food brought at av. interval of 3·4 min; less often just before fledging (Winterbottom 1954). Faeces removed at irregular intervals. Fledging period: 13–18 (av. 16) days.

BREEDING SUCCESS/SURVIVAL: on Nyika Plateau (Zambia) evidence points to surplus non-breeding population: ringed mates are replaced within a day (Dowsett-Lemaire 1985). Adult recaptured after 6 years (Zimbabwe). Parasitized by Klaas's Cuckoo *Chrysococcyx klaas* (n = 13, South Africa).

References
Dowsett-Lemaire, F. (1989).
Vincent, A. W. (1948a).

Apalis pulchra Sharpe. Black-collared Apalis. Apalis à col noir.

Apalis pulchra Sharpe, 1891. Ibis, p. 119; Mt Elgon.

Forms a superspecies with *A. thoracica* and *A. ruwenzorii*.

Plate 15
(Opp. p. 256)

Range and Status. Endemic resident in montane forest undergrowth in SE Nigeria and Cameroon and E Africa. Common to local in Nigeria at 1550–2300 m on Mambilla Plateau and Gotel Mts, possible also Obudu Plateau; in SW Cameroon at 1700–2900 m in Bamenda Highlands north to Tchabal Mbada, but absent from forest patches between Mbengwi (6° 1′N, 10° 1′E) and Tinachong (6° 2′N, 9° 53′E) and all mountains south of Bamenda Highlands, including Mt Kupé and Mt Cameroon. In NE Zaïre in highlands of Albertine Rift northwest of L. Albert (Lendu Plateau) and adjacent highlands in E Ituri; SE Sudan (Imatong Mts, above 1800 m); SE Uganda (Mt Elgon, around 2325 m); and W and central Kenya at 1550–2400 m (Mt Elgon, Kakamega Forest, Kericho, Mau Forest and east to Mt Kenya, south to Nyanza, Nairobi and Lolgorien); SE Zaïre (Marungu Plateau above 1550 m).

Apalis pulchra

Description. *A. p. pulchra* Sharpe: range of species except Marungu Plateau. ADULT ♂: forehead, crown, nape, hindneck, cheeks, lores, ear-coverts, mantle and back to uppertail-coverts dark slate grey. Tail blackish, of 10 feathers, T5 and T4 extensively white, T3 and T2 with 10–15 mm white tips, central pair (T1) narrowly tipped white. Chin and throat creamy white with faint buffish tinge. Black crescent-shaped breast-band 6–8 mm deep; breast to belly and undertail-coverts off-white, flanks and thighs deep reddish chestnut. Wings dark slate-grey; underwing-coverts and axillaries white. Bill black; eyes warm light brown, tinged reddish in adult ♂; legs and feet dusky grey, toes lighter. Sexes alike. SIZE: (24 ♂♂, 18 ♀♀) wing, ♂ 53–61 (56·8), ♀ 51–57 (53·3); tail, ♂ 52–64 (58·0), ♀ 49–62 (53·4); bill, ♂ 15–17 (15·7), ♀ 14·5–16 (14·7); tarsus, ♂ 22–24 (23·0), ♀ 21–23 (21·3). WEIGHT: Cameroon, ♂♀ (n = 11) 8–11; Kenya, ♂ (n = 16) 7–12 (9·1), ♀ (n = 8) 6–10 (7·7).

IMMATURE: paler and duller above, breast-band slate-grey, flanks paler chestnut.

NESTLING: undescribed.

A. p. murphyi Chapin: Marungu Plateau, Zaïre. Lighter grey above, flanks pale cinnamon-buff, thighs grey, faintly washed buff, tail more extensively white (except T1), T5 wholly white.

Field Characters. A fairly large apalis, usually first encountered as a voice from the undergrowth. Uniform dark grey above, white below, with black breast-band and rusty flanks; race *murphyi* overall paler. Does not resemble any other apalis in its range, and has different habits: small groups feed low down in thick cover, while most congeners join mixed-bird parties at higher levels.

Voice. Tape-recorded (68, B, F, GREG, HOR, KEI, STJ). Typical song, 4 loud, insistent, slightly burry, downslurred notes with a complaining quality, 'dzeew, dzeew, dzeew, dzeew', given at rate of 4/s; responds to playback with agitated song at twice the speed. In other versions, notes are more clear and ringing, 'tuit, tuit, tuit, tuit' or 'peet, peet, peet, peet'. Often duets. Alarm scold, a drawn-out, nasal, buzzy 'djeeeaaa'.

General Habits. Frequents forest undergrowth, bamboo and thickly tangled creepers and bushes, especially along streams; also at forest edges, in clearings and second growth. On Marungu Plateau inhabits riparian growth and relict forest patches.

In pairs or small family parties of 3–6; tame, restless, very active, often calling to keep in touch with one another. As it moves about, holds head into body with bill straight forward, contour feathers loose, longish rump feathers overlapping wings; tail held high and well forward, usually highly mobile, moved up and down, jerked from side to side; bird sidles up and down stem in undergrowth.

Gleans leaves, including underside; probes into plants; pokes head into clusters of dead vegetation and flowers; darts forward to catch insect. In Cameroon usually feeds near ground, occasionally 10 m up; in gallery forest along streams, within 1–3 m of ground.

Food. Insects: mantid nymphs, small nymphal grasshoppers, green caterpillars; spiders.

Breeding Habits. Territorial in breeding season and probably throughout year; territory *c.* 0·5 ha. Apparent social display by birds showing generally excited behaviour, with exaggerated tail-wagging, tail held over back, several members of party flitting about in short jerky flight, making little clicking noises and excited calls (Serle 1950a).

A

NEST: domed, with side-top entrance, made almost entirely of long strands of moss bound with a few cobwebs and plentifully lined with feathers (**A**). Placed in bush or sapling, 1–14 m or more above ground. Often uses domed and woven nests of other species (warblers, sunbirds, weavers), reconditioning them, reducing entrance size and adding feathers as lining. Same nest used in consecutive seasons; may build only 1 nest in 7 attempts (van Someren 1956).

EGGS: 2–3, usually 2; elongate, almost long ovals; very pale greenish white or occasionally pinkish green, with bold spots of greyish or lilac, overlain by evenly but sparsely distributed bolder spots and blotches of red-brown and maroon. Some eggs are more pointed and lack spots at narrow end. SIZE: 17·5–18 × 11–12. Sometimes double-brooded.

LAYING DATES: Cameroon, Mar–Apr; Sudan, Jan–Apr; E Africa: Region A, Dec, Region B, Dec, Region D, May–June, Nov–Dec, breeds mainly in long rains, also in short rains (May).

INCUBATION: by both sexes; ♂ brings food to incubating ♀ which leaves nest to take it. Period: 13–14 days.

DEVELOPMENT AND CARE OF YOUNG: ♂ continues to bring food to incubating ♀ when eggs hatch; ♀ brings food to brooding ♂. At one nest, in 2½ h at midday, ♂ fed young 14 times, ♀ 20 times. Both sexes brood and remove faecal sacs. Fledging period: 15–16 days.

BREEDING SUCCESS/SURVIVAL: mortality high, 70% of nests unsuccessful (van Someren 1956).

Reference
van Someren, V. G. L. (1956).

Plate 15
(Opp. p. 256)

Apalis ruwenzorii Jackson. Rwenzori Apalis. Apalis du Ruwenzori.

Apalis ruwenzorii Jackson, 1904. Bull. Br. Orn. Club 15, p. 11; Ruwenzori.

Forms a superspecies with *A. pulchra* and *A. thoracica*.

Range and Status. Endemic resident. Common in montane forest and lower transition zone in highlands of Albertine Rift in E Zaïre from Rwenzori Mts and highlands west of L. Edward at 1550–2100 m to Itombwe and Mt Kabobo up to 2450 m; SW Uganda at 1550–3100 m (Rwenzori, Kibale and Bwindi-Impenetrable Forests); Rwanda (Nyungwe Forest) and Burundi (Teza, Ijenda and Bururi Forests). Density of 1–2 pairs/ha (Nyungwe Forest: Dowsett-Lemaire 1990).

Description. ADULT ♂: forehead, crown, nape, cheeks, lores, ear-coverts, hindneck, mantle and back to uppertail-coverts dull sooty grey. Tail sooty grey, of 10 feathers, unpatterned. Chin and throat reddish buff, lower throat off-white or faintly buffy; a sooty grey breast-band 4–7 mm deep in midline; sides of breast, flanks, thighs and undertail-coverts buffish chestnut, centre of belly white, the white sometimes reaching up to breast-band. Wings sooty grey; underwing-coverts and axillaries white, sometimes washed buff. Bill black; eyes bright orange-brown; legs and feet light pinkish brown, often washed grey, claws dark grey. Sexes alike except that ♀ has less well-developed breast-band and more rufescent eyes. SIZE: (11 ♂♂, 9 ♀♀) wing, ♂ 47–53 (50·5), ♀ 48–50 (49·0); tail, ♂ 42–47 (44·6), ♀ 38–44 (40·6); bill, ♂ 14·5–16 (15·5), ♀ 14·5–15 (14·8); tarsus, ♂ 19–20 (19·3), ♀ 18·5–20 (19·2). WEIGHT: Uganda, ♂ (n = 16) 8·5–13 (10·0), ♀ (n = 11) 7–12 (9·7).

Apalis ruwenzorii

IMMATURE: paler and duller above, breast-band paler, less well-marked; lower mandible horn-coloured.

NESTLING: unknown.

TAXONOMIC NOTE: White (1962) treated *A. ruwenzorii* as conspecific with *A. pulchra*, a view supported by Dowsett-Lemaire and Dowsett (1990) because of strong vocal similarities. Traylor *in* Watson *et al*. (1986) included both in a superspecies with *A. thoracica* and they are surely each other's closest relatives. But *ruwenzorii* and *pulchra* have tails of 10 feathers, *thoracica* has 12; *ruwenzorii* is smaller than either and differs from *pulchra* in plumage, including unpatterned tail. It occupies a central position along the Albertine Rift with populations of *pulchra* at either end (and in W Africa) indicating independent speciation. It is treated here as specifically distinct, following Traylor *in* Watson *et al*. (1986).

Field Characters. A small, slim version of Black-collared Apalis *A. pulchra*, which it replaces in mountains of central Albertine Rift. Lacks white in tail and has reddish underparts; unlike any other apalis in its range. Another voice from the undergrowth; sounds very like Black-collared Apalis but somewhat higher-pitched.

Voice. Tape-recorded (32, C, LEM). Song very like Bar-throated Apalis *A. thoracica*, a series of unmusical, rather squeaky notes, 'dzeew, dzwee, dzeew, dzeew . . .' or 'peek, peek, peek, peek . . .', repeated 4–8 times, sometimes in longer series; delivered at different speeds, from 2·5 to 8 notes/s; in faster songs, notes are shorter, 'bee-bee-bee-bee . . .'. Duets; ♀ gives same song at slightly higher pitch and a little faster. Alarm call like notes of song but squeakier and given at varying intervals.

General Habits. Frequents undergrowth in montane forest and transition zone, with tangled creepers, dense bushes and herbaceous cover, in some places also midstratum and second growth; in Nyungwe Forest (Rwanda) occurs in dense understorey in deep shade, remaining in cover, usually below 4 m; in Bwindi-Impenetrable Forest (Uganda) feeds principally by sally-gleaning in low undergrowth; replaced by Mountain Masked Apalis *A. personata* in taller undergrowth and lower canopy (Bennun 1986).

In pairs or small family parties of 3–4, searching leaves and twigs, tail often slightly raised. Pairs in undergrowth briefly followed ant-columns but normally fed far from them (Uganda, Kibale: Willis 1986a); in Rwanda (Nyungwe), frequently attracted to ant swarms, picking prey from twigs and saplings just above them (Dowsett-Lemaire 1990). When singing or agitated opens wings and raises tail which may be flicked from side to side.

Food. Insects and their larvae, including ants and small caterpillars; spiders.

Breeding Habits. Territorial, solitary breeder.

NEST: dome-shaped, with side-top entrance; loosely constructed of soft green moss; unexpectedly large – ext. depth 220, ext. diam. 120, diam. of entrance 35; lining of entrance and probably of nest interior fine whitish lichen. Attached to sapling 0·4–1·2 m above ground, suspended from top, well concealed by herbs. Adult carried nest material from less than 3 m away.

EGGS: 2, pure white, with very light to dark pinkish brown spots and blotches, mostly at larger end forming ring. SIZE: (n = 3) 16–17·4 × 12·2–13 (16·4 × 12·7).

LAYING DATES: Zaïre (Itombwe) (breeding condition Mar–May); Rwanda (feeding fledged young Oct, Dec); E Africa: Region B, Apr, Sept, Dec (fledgling Jan).

Apalis ruddi Grant. Rudd's Apalis. Apalis de Rudd.

Apalis ruddi Grant, 1908. Bull. Br. Orn. Club 21, p. 93; Coguno, Inhambane District, Mozambique.

Plate 15 (Opp. p. 256)

Range and Status. Endemic resident. Thickets and coastal forest from SE Mozambique (north to about Save R.) to extreme SE Transvaal (Lebombo Range), E Swaziland (Lubuli), Zululand and Natal to 28° 45′S; a small, isolated population in Malaŵi, in lower Shire valley (Nchalo) and Lengwe Nat. Park (Hanmer 1977). Density in optimal habitat (Zululand) of 1 pair/2 ha (Beven 1944). Rare to locally common.

Description. *A. r. ruddi* Grant: coastal areas of Mozambique from about the Save R. and Vilanculos south to lower Incomati R. ADULT ♂: forehead, crown, nape and cheeks to ear-coverts dull slate-grey, with dark grey lores and pale supraloral line. Mantle and back to uppertail-coverts yellowish green. Tail green, T3–T6 with 4 mm yellow tips, T2 narrowly tipped yellow, T1 plain. Chin and throat creamy buff, black breast-band 5 mm deep in midline, sides of neck and breast green, centre of belly off-white, flanks and thighs olive-green, undertail-coverts yellow. Primaries and secondaries dusky brown, outer webs margined green, upperwing-coverts yellowish green; underwing-coverts and axillaries yellow. Bill black; eyes light brown; legs and feet brownish flesh. Sexes alike, but ♀ with narrower breast-band, c. 3·5 mm deep in midline. SIZE: (8 ♂♂, 6 ♀♀) wing, ♂ 48–52 (49·3), ♀ 46–50 (47·3); tail, ♂ 50–56 (53·0), ♀ 45–50 (46·8); bill, ♂♀ 11·3–13 (12·6); tarsus, ♂♀ 18·5–22·5 (20·4).

IMMATURE: like adult but paler, breast-band less distinct.

NESTLING: hatched naked.

A. r. caniviridis Hanmer: S Malaŵi. Nearest to *ruddi* but crown slightly more bluish grey, rest of upperparts greyer, less yellow-green; ♂ eye deep red, ♀ browner red. WEIGHT: ♂ (n = 3) 9·8–10·5 (10·1), ♀ (n = 3) 9·6–10 (9·6).

A. r. fumosa Clancey: Mozambique (Maputo district) to coastal Natal. Crown and nape darker slate-grey than *ruddi*, back darker green, less yellowish; tail darker green with an ill-defined blackish olive subterminal area (not forming clearly defined spot), tipped greenish white. Breast-band slightly broader. WEIGHT: ♂ (n = 8) 9·2–10·5 (9·9), ♀ (n = 4) 9·3–9·8 (9·5).

Apalis ruddi

Field Characters. Small, with restricted range in lowlands of SE Africa. Shares black throat-bar with Bar-throated Apalis *A. thoracica*, but differs in greener back, contrasting with grey head, dark eye, white line above dark lores, green tail with narrow yellow tips (tail of Bar-throated dark grey with extensive white). Similar above to Yellow-breasted Apalis *A. flavida* but underparts different: has narrow yellow breast-band with complete black bar always present *above* it, whereas Yellow-breasted has broad yellow breast-band with incomplete black spot or bar, if present, *below* the yellow. Voices of all 3 distinctive.

Voice. Tape-recorded (75, 88, 91, F, ROC). Song a dry rattling or clicking trill, given in bursts of 5–15 notes at rate of *c*. 6/s to *c*. 9/s, 'klikliklikli . . .'; fast, but notes distinguishable to the ear; sounds like rapid tapping of stones. Often duets, second bird answering with a faster, more liquid trill, reminiscent of tinkerbird *Pogoniulus*. Contact call, low 'churg-churg' (Maclean 1993); also snaps bill.

General Habits. Occurs principally in low-canopied coastal forest and dense thickets, patches of *Strelitzia* thicket (Zululand) and adjoining bushy areas; also woodland patches, often with low dense undergrowth.

Gleans twigs and foliage, picking insects off leaves, making short flights one after another. ♂ sometimes calls from top of dead tree or other conspicuous vantage point and is answered by 2–3 others of either sex.

Food. Insects and their larvae, caterpillars; minute flower buds.

Breeding Habits. Territorial, probably throughout year. Displaying ♂ follows ♀ restlessly; ♀ pauses, ♂ jumps about excitedly, flirting tail and uttering soft 'chook' call, and ♀ turns body quickly from side to side.

NEST: oval ball, taller than wide, with side entrance near top; very loose transparent framework of silk taken from caterpillar colonies, decorated outside with bits of moss, tree lichen or vegetable down; interior lined with plant down. Sited in low bush or occasionally in fork, from near ground up to 3 m high; leaves may be sewn into sides, and low nests often incorporate plant stems or grass growing from below. Built by ♂, which can be seen working inside.

EGGS: 1–3, usually 2; ovate; bright turquoise blue or light greenish blue, with variable-sized large rusty spots, mostly at blunt end. SIZE: (n = 5) 16·0–17·8 × 11·6–12·7 (16·9 × 12·2).

LAYING DATES: Malaŵi, 3 juvs June (from eggs laid Mar), ♀♀ with brood patches Jan and Apr; Mozambique (breeding condition Oct); South Africa (Zululand), Jan, Sept–Nov.

Reference
Beven, G. (1944).

Plate 15
(Opp. p. 256)

Apalis flavida (Strickland). Yellow-breasted Apalis. Apalis à gorge jaune.

Drymoeca flavida Strickland, 1852. *In* Jardine (ed.), Contrib. Orn., p. 148; Damaraland, South West Africa (error for Ngamiland, Bechuanaland).

Range and Status. Endemic resident. Widespread in mesic savanna woodlands and forest edges in N tropics mainly south of 10°N, and in S tropics mainly north of 20°S except that it is locally common in E South Africa. Absent from interior of SW Africa, forested W Africa and Congo Basin. Common to local and uncommon.

W and central Gambia (over 50 records: Wacher 1993), Guinea-Bissau, Sierra Leone (Loma Mts), uncommon in Liberia and Ivory Coast, extreme SW Burkina Faso (Ouagadougou), Ghana north to Du on White Volta, Togo north to 9°31′N and Benin to 9°54′N, Nigeria north to Yankari and Zaria, Cameroon, Gabon, N Angola (Cabinda and Uige), Chad (S. Chari R.), Central African Republic; Zaïre on northern borders of forest zone and in E, south to Rutshuru; SW Sudan (Tambura), W Uganda (Murchison Falls Nat. Park, Rwenzori, Mbarara) and from N end of L. Victoria to W Kenya (Kisumu, Siaya) (*A. f. caniceps*);

Apalis flavida

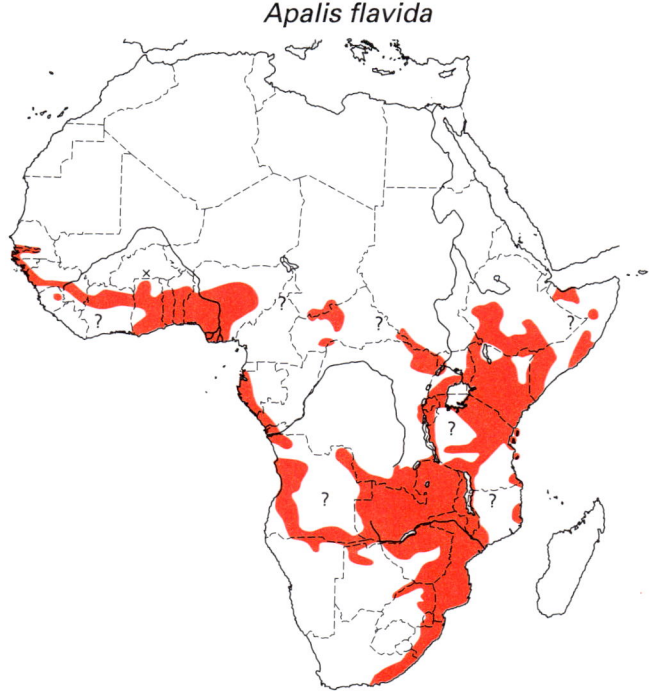

SE Sudan, S Ethiopia, N and SW Somalia, Kenya (but absent from or rare in parts of N and E), Rwanda and Tanzania; *neglecta* in SE Kenya (Mombasa), SW Tanzania and coastal lowlands in E including Pemba, Zanzibar and Mafia Is, S Zaïre (Baraka to Shaba), E Angola (Malanje and Lunda), Zambia, Malaŵi, Zimbabwe, Mozambique, E Transvaal, Swaziland and N Natal; other races in W Angolan escarpment zone south of Cuanza R., N Namibia, N Botswana (Okavango and Chobe R.), and South Africa in Natal and Cape Province west to Knysna.

Description. *A. f. neglecta* (Alexander) (including '*renata*', '*lucidigula*', '*niassae*' (= '*canora*') and '*tenerrima*'): E Angola to SE Kenya and N Natal (see above); inland in Tanzania to Nguru Mts, Morogoro, Luipa R. and Mahenge. ADULT ♂: forehead, crown, lores, cheeks and ear-coverts pale grey, hind-crown washed green (grey in NE of range); nape, mantle, back, rump and uppertail-coverts green; tail green, feathers tipped yellow, T6 and T5 all yellow. Chin and throat white, breast and sides of neck yellow, sometimes washed olive, bordered below by variable black patch in centre, pointing upward in the midline and often extending sideways to form an incomplete breast-band; belly, flanks, thighs and undertail-coverts white. Primaries and secondaries blackish with outer webs margined green, upperwing-coverts green; underwing-coverts and axillaries yellow. Bill black, cutting edges of maxilla and mandible greyish white; eyes bright hazel, rim of eyelids brick red; legs and feet pinkish buff, claws dark grey. ADULT ♀: like ♂ but lacks black breast spot. SIZE: (50 ♂♂, 50 ♀♀) wing, ♂ 46–55 (51·0), ♀ 45–51 (48·8); tail, ♂ 40–56 (49·3), ♀ 41–53 (44·7); bill, ♂ 13–15 (13·7), ♀ 12·5–15 (13·5); tarsus, ♂♀ 18–20 (19·1). WEIGHT: (Zimbabwe) ♂ (n = 22) 6·8–9·5 (8·5), ♀ (n = 14) 7·1–9·0 (7·9).

IMMATURE: wholly dull greenish above, washed pale yellow below; subadult ♂ that has not yet acquired black breast spot resembles adult ♀.

NESTLING: hatches naked; skin pink.

A. f. caniceps (Cassin): Gambia to W Kenya and N Angola (see above). Crown wholly grey, breast yellow, belly and flanks white, washed greyish; tail-feathers deep olive-green with yellow or whitish tips. Lacks black breast spot. Intergrades with *golzi* in W Uganda. Tail, ♂ 35–40 (36·7). Immature has yellow throat and breast, contrasting with white belly.

A. f. abyssinica Erard: SW Ethiopian Highlands (Wollega, Illubabor and N Kaffa provinces; Agaro, Dembi and Jimma). Above more olive than *caniceps*, less yellowish; yellow of breast with olive suffusion; some ♂♂ may have black breast spot; bill noticeably short (c. 12 mm) and thin; tail, ♂ 36–39 (37.0).

A. f. flavocincta (Sharpe) (including '*malensis*'): SE Sudan (Yoelli), N Uganda (Mt Moroto and Kidepo Valley Nat. Park), S Ethiopian lowlands, Somalia south of 2°N and West of Shebelle R., E and NE Kenya below 1400 m (including Lamu and Manda Is.), NE Tanzania (Tanga). Forehead grey, crown and nape green; breast less heavily washed olive than *viridiceps* (more so in ♂♂), and both sexes lack black breast spot; tail brown but T6 and T5 yellowish; tail, ♂ 53–62 (55·2).

A. f. viridiceps Hawker: N Somalia west of 48°E, and adjoining Ethiopia. Crown almost entirely dull green with little or no grey on forehead; mantle and back duller green; yellow of breast heavily suffused olive-green, yellower at sides; ♂♂ show incipient dusky olive breast-band or spot; tail brownish, very faintly margined with green, T6 and T5 off-white, remainder tipped yellow. Tail longest of any race: ♂ 55–62 (59·2).

A. f. pugnax Lawson: Kenya (highlands at 1460–2200 m and Chyulu Hills). Forehead grey, crown wholly green; breast clearer yellow than in *flavocincta*, undertail-coverts yellow. ♂ and ♀ with black breast spot (sometimes forming bar); tail green. Intergrades with *golzi* in S Kenya and parts of N Tanzania. Tail, ♂ 46–56 (52·0).

A. f. golzi (Fischer and Reichenow): SE Kenya (Taita Hills), interior of Tanzania (except extreme SW and coastal areas occupied by *neglecta*); Rwanda (Kagera valley). Like *pugnax* but top of head and nape entirely grey; undertail-coverts yellow. ♂ and ♀ with black breast spot. Tail, ♂ 44–55 (48·9).

A. f. flavida (Strickland): W Angola, N Namibia, N Botswana and SW Zambia. Hindcrown grey mixed with green; chin white, throat yellow. Prominent black breast spot in ♂ only; undertail-coverts yellow. Tail, ♂ 43–55 (49·8).

A. f. florisuga Reichenow: South Africa (central Natal southwards). Chin and throat white, breast markedly yellow; both sexes lack black breast spot. Tail, ♂ 46–59 (51·6).

TAXONOMIC NOTE: potentially or apparently reproductively-isolated populations occur in several places, but may intergrade elsewhere; it precludes recognition of more species than a single, highly polytypic one. Differences are reinforced vocally, by habitat, and visually by the presence or absence of a black breast spot. In Tanzania *neglecta* of the moist coastal lowlands meets *golzi* of the dry interior without apparent intergradation (Moreau and Moreau 1939). In Kenya *pugnax* and *flavocincta* occur in wet and dry country respectively without intergrading and may be locally sympatric (Lewis 1982b, 1989); they are distinct vocally. In S Ethiopia (Kaffa), *flavocincta* occupies thornbush in the south, but only 250 km away *abyssinica* in north is found in moist forest and thicket (Erard 1971). However, intergrades between *flavocincta* and *neglecta*, *golzi* and *pugnax*, and *caniceps* and *golzi* have been reported (Lawson 1968), and reproductive isolation may be incomplete.

Field Characters. Small and slender; green above with variable amount of grey on head, white below with broad yellow breast-band, with or without a black spot or bar in centre below the yellow. Found in wide variety of bushy and wooded habitats but usually not in forest except at edge, and so ecologically separated from most congeners. In southern Africa overlaps with Rudd's Apalis *A. ruddi* (similar above but with white line in front of eye and crescent-shaped black bar always present *above narrow* yellow breast-band), and with Bar-throated Apalis *A. thoracica* (larger, darker above, with white eye, dark tail with much white in outer feathers, complete black breast-band always present, often no yellow on breast). Voices of all 3 distinctive.

Voice. Tape-recorded (58, 68, 75, 86, 88, 91, B, C, F, KEI). Song typically a duet, although ♂ may sing on his own. Song of ♂ has considerable geographical and local variation. In Kenya highlands ♂ repeats single, double or sometimes triple note, 'clip-clip-clip-clip . . .', 'turrit-turrit-turrit-turrit . . .', 'cli-cli-click, cli-cli-click . . .', at rate of 3–4 per s, sometimes in short bursts, often in lengthy series of 25 or more, which may continue after brief pause. Longer series often have rolling or undulating quality due to variations in pitch. ♂ starts to click and 2–10 s later ♀ joins in excitedly with high-pitched, strident, grating 'jee-jee-jee-jee . . .' which sounds more like a querulous scold than a song, especially as notes are variable in speed, without regular beat of ♂. Duet may go through several rising and falling series; exceptionally ♀ may complete 4–5 series of up to 15 notes each. Dry country birds in Kenya often lack duet; ♂ gives rapid 'crit-crit-crit . . .', which may be initiated by a rasping, 2-s 'terrrrrsk'. W African birds also tend to click, but in southern Africa ♂'s voice often has a buzzy quality, 'jeeu-jeeu-jeeu . . .' or a shorter 'jib-jib-jib . . .', sometimes a 2-note 'jidit-jidit . . .'. Alarm, buzzy churr or sharp 'krit-krit' (Maclean 1993); also snaps bill.

General Habits. The most widespread and ecologically adaptable apalis, found in a broad range of savanna, riparian and forest edge habitats. In southern Africa, where absent from dry acacia savannas, local in coastal forest, thick bush, scrub and thickets, and particularly riparian fringing forest and acacia canopy (especially *A. albida* and *A. xanthophloea* on alluvial soils) in major river valleys; also edges and canopy of lowland and mid-altitude forest. In the miombo belt from Angola to Mozambique, frequents forest edges and mushitus, riparian cover and thickets; also wanders into surrounding woodland. In E Africa has broader habitat range: occurs in moist coastal areas and dry interior acacia savanna and scrub, particularly crowns of flat-topped acacias. In Kenya *pugnax* occurs in moist, higher-altitude environments with >500 mm rainfall; *flavocincta* abruptly replaces it in arid and semi-arid areas; the 2 are possibly sympatric on Mt Maralal, where former mainly at higher altitudes than latter. On Zanzibar occurs in coral rag bush and scrub; in Somalia and S Ethiopia in arid acacia scrub, and in latter also in forest and moist thickets; in Sudan in low bushed grassland and dry stony acacia country. In W Africa generally confined to moister savannas where localized; in Nigeria also in coastal mangroves and gallery forest, in Ghana on forest edges, coastal thickets and riverine bush; an isolated forest population in Sierra Leone (Loma Mts) (replaced 45 km away in Tingi Mts by Sharpe's Apalis *A. sharpii* and Black-capped Apalis *A. nigriceps*: Field 1974).

Forages restlessly in pairs or small family parties of 4–5, gleaning insects from twigs and leaves, constantly on the move, now and again duetting; searches thicker vegetation and particularly fine-leaved crowns of trees, also clumps of parasitic *Viscum* and *Loranthus*. Snaps bill and makes snapping sound with wings in alarm. In dry Tsavo East Nat. Park, of 70 food items, 81% were taken from leaves, 14% from twigs and 5% from ground; of these 70% were from inside tree or bush, 20% inside edge and 10% at edge, with mean feeding height 2·4 m (Lack 1985); but situation must vary greatly with habitat.

Pair silent when birds can see one another; otherwise they keep in contact by duetting. ♂ flies off, immediately starts to give 3–5 'galloping' phrases, increasing in volume and duration until ♀ replies or joins him. ♀ adopts listening posture when ♂ is calling far off: she stiffens, motionless, head raised, bill pointing up at angle of 45°, gives 2–3 muted squeaks, and continues feeding when ♂ stops calling (Lewis 1982a). When listening, prominent white throat is displayed, probably important in individual recognition. Duetting is probably pair-specific (otherwise unseen bird would duet with others in neighbouring territories); individual differences in ♂♂ apparent even to human ear. Duets are not used territorially; counter duetting between 2 pairs never occurs (Lewis 1982b).

Resident and mainly sedentary, but in Tsavo East Nat. Park occurs only May–Oct (Lack 1985), although habitat much altered there. Once at Ouagadougou (Burkina Faso) far from normal range, Mar (Thonnérieux *et al.* 1989).

Food. Insects: Coleoptera (small weevils, cetoniids), small nymphal grasshoppers and mantids, ants and ant pupae, scale insects, caterpillars and other insect larvae; spiders, plentifully; fruit and nectar (Maclean 1993).

Breeding Habits. Territorial; feeding territory maintained throughout year. During courtship tail carried high, often expanded spasmodically, wings held down; ♂ chases ♀. In threat display ♂ and ♀ react to reflection in window pane, striking it and fluttering up and down, and sometimes move about agitatedly, with rapid turns and jerks. In stationary threat display by ♂ only, bird produced very rapid upwards and sideways sweep of head, neck and upper body, maintaining stance for 3–10 s; faced reflection at right angles, stretched to full length with neck fully extended, pointing bill up at 60° above horizontal, cocked tail vertically over back, and repeatedly uttered low, grating 2–3 s 'churrr', sometimes followed by 4–6 low monosyllabic clicks (Lewis 1982a). White throat was very prominent during display, and muscular movements were apparent in producing call and fanning throat feathers (Lewis 1982b).

NEST: purse-shaped bag, almost pendant, domed or semi-domed, with side-top entrance (**A**); variable, top sometimes open but covered by thick vegetation, leaving deep cup. Made almost entirely of small bits of lichen, bound together with spider webs, often with spider cocoons attached. Bits of vegetable down are worked in; usually scantily lined with seed-pod down (*Marsdenia, Ceropegia* or asclepiad); bowl is felted but

A

not sides or top. Int. diam. 40, int. depth 50–80, ext. depth 90–100. Placed between 1·0 and 1·6 m above ground, in low bush or small sapling, mostly in leafy situations, particularly in bush with close-set leaves and twiggy branchlets; suspended from end of branch, securely bound to supporting twigs with spider webs, strands of down or lichen. Uses old nests of ploceid weavers and estrildines (e.g. *Ploceus, Uraeginthus*), lining bowl with vegetable down and adding fibrous coating of seed or flower heads when nest has a spout (Benson 1944). ♂ brings nest material.

EGGS: 2–4 (av. 2·5); laying interval 1 day; elongate or rounded oval; slightly glossy; pale bluish green with small but distinct red-brown spots, fairly evenly distributed, or concentrated to form ring towards thick end, with underlying marks of ashy violet or ochreous grey. SIZE: (n = 35) 14·2–17·7 × 10–12 (15·8 × 11·3).

LAYING DATES: Ivory Coast (breeding condition Feb, Aug); Ghana (fledgling fed by parents Oct); Nigeria, Aug; Ethiopia, Apr; Somalia, June; Zaïre (Shaba), Oct; Rwanda, Sept–Jan, Mar–June; E Africa: Region A, May, Region C, Jan–Apr, Dec, Region D, Mar–July, Aug, Nov; all Regions A and C records are in rains, most of Region D ones in long rains in Apr–June and some dry months before short rains in Sept–Oct; all of the last are from Arusha (Tanzania) with long rains peak May–June, Nairobi (Kenya) long rains Mar–June; Angola, Nov; Zambia, Mar, Sept–Oct; Malaŵi, May, Sept (nest-building Mar); Zimbabwe, Sept–Apr (Jan 1, Feb 1, Mar 1, Apr 1, Sept 1, Oct 4, Nov 7, Dec 1 clutches); Botswana, Sept; South Africa: Transvaal, Oct–Nov; Natal, Sept–Jan; Cape, Jan–Feb, June, Nov.

INCUBATION: by both sexes, mostly ♀; bird sits closely, with head towards entrance, point of bill visible. Period: 12–14 days (usually 12–13).

DEVELOPMENT AND CARE OF YOUNG: ♀ broods for first 3 days, fed by ♂. Young always fed first but large items eaten by brooding bird. Once 8 visits in *c*. 10 min, when young became satiated. Adults at nest fearless, feeding young close to observer. Fledges by day 15, sometimes by day 17.

BREEDING SUCCESS/SURVIVAL: percentage egg-loss very high; principal predators (Kenya) are *Lanius* shrikes, arboreal rodents, tree snakes; parasitized by Klaas's Cuckoo *Chrysococcyx klaas* (Kenya). Longest-lived birds 9·5 and 10 years (Malaŵi).

References
Lewis, A. D. (1982b, 1989).
van Someren, V. G. L. (1956).

Apalis binotata Reichenow. Masked Apalis. Apalis masquée.

Plate 15
(Opp. p. 256)

Apalis binotata Reichenow, 1895. Orn. Monatsber. 3, p. 113; Jaunde (= Yaoundé), Cameroon.

Forms a superspecies with *A. personata*.

Range and Status. Endemic resident. 3 populations: (1) SW Cameroon, up to 1000 m (Rumpi Hills, Mt Manenguba and Mt Kupé) and in lowlands (Dja R., Yaoundé and Efulen), and NE Gabon (Ivindo R.); (2) N Angola in Cuanza Norte (Camabatela, Canzele and Ndala Tando); and (3) E Zaïre around 1000 m, at Burondo and Kasebere on Beni-Butembo ridge, W and E Uganda at 1200–1500 m (Itwara, Kibale, Malibigambo and Sango Bay Forests), base of Mt Elgon to 1800 m, and NW Tanzania (Bukoba and Minziro). Density in Gabon 3 pairs/35 ha in regenerating growth (Brosset and Erard 1986). Common to local and uncommon.

Description. ADULT ♂: forehead, crown, lores, cheeks and ear-coverts slate-grey; nape green, well demarcated from hindcrown. Mantle and back to uppertail-coverts green. Tail green, with yellowish tips to all but central feathers. White line down side of throat from chin to below and behind ear-coverts; chin, throat and upper breast black, clearly demarcated from belly; sides of breast and flanks washed green; centre of belly white, undertail-coverts yellow. Wings with outer webs of flight feathers margined green, upper wing-coverts green; underwing-coverts and axillaries whitish with some greenish suffusion. Bill black, tomia whitish; eyes chestnut-brown, eyelids pinkish brown; legs and feet brownish pink or flesh-coloured. ADULT ♀: like ♂ but black on breast less extensive, more restricted to throat, and bordered with yellow; white line down sides of throat broader and more conspicu-

Apalis binotata

ous. SIZE: (12 ♂♂, 8 ♀♀) wing, ♂ 47–52 (49·0), ♀ 45–48 (46·0); tail, ♂ 38–45 (40·5), ♀ 33–37 (35·0); bill, ♂ 12–13·5 (12·5), ♀ 12–13·5 (12·6); tarsus, ♂ 17·5–19 (18·2), ♀ 17–18·5 (17·6). WEIGHT: Uganda, ♂ (n = 8) 6·0–10·0 (8·2), 1 ♀ 8·5.

IMMATURE: uniform green above from crown to uppertail-coverts; throat washed pale yellow, becoming paler on breast, belly whitish or with yellow wash.

NESTLING: undescribed.

Field Characters. A dumpy, short-tailed bird of undergrowth and mid-levels, with dark grey head, green upperparts, green and white underparts and black bib on throat and breast, separated from dark head by conspicuous white stripe. Likely to be confused only with closely-related Mountain Masked Apalis *A. personata*, which it meets in E Zaïre and W Uganda; differs mainly in white neck stripe, lacking in Mountain Masked Apalis, which has all-black head and white patch behind ear-coverts. Black-throated Apalis *A. jacksoni* has similar head pattern (white stripe separating dark throat and head) but longer tail with much white, yellow underparts, very different song, and forages at higher levels.

The tiny Black-capped Apalis *A. nigriceps* is another canopy bird, distinguished by white throat, black breast patch, restless behaviour.

Voice. Tape-recorded (32, 68, C, F, ERA, STJ). Song of ♂, single sharp burry note repeated at regular rate of *c.* 4 per s, with slight upward or downward bias at end, 'crreew-crreew-crreew-crreew . . .', 'brreep-brreep-brreep-brreep . . .', 'trreee-trreee-trreee-trreee . . .', sometimes sounding like a double note, 'turrrip-turrrip-turrrip-turrrip'. ♀ often joins with hard, rapid 'ta-ta-ta-ta . . .'; in synchronized duet, these notes are given at exactly twice speed of ♂'s; at other times they vary, may speed up and slow down within same song. Foraging birds give a low conversational churr (contact?).

General Habits. Inhabits old clearings and second growth (Cameroon), forest patches and edges, regenerating growth and bushes around cultivation (Gabon), gallery forest (Angola), and dense undergrowth in dry forest (Malibigambo Forest, Uganda). Mainly in lowlands but extends into intermediate forest in some places.

In pairs or small family parties of 3–4, moving actively through vegetation, gleaning from foliage, hopping and flitting from twig to twig.

Food. Insects: small beetles, small grasshoppers, and caterpillars 3–4 cm long.

Breeding Habits. Territorial; individual territories vigorously defended against conspecifics with much fighting on boundaries.

NEST: loosely woven pocket with side-top entrance, made of *Usnea* lichen with some cobwebs running through it; suspended from twig quite low down, attached with gossamer threads; lined with fine grass or with grass heads woven into interior.

EGGS: 1–2; long narrow ovals, blunt at thinner end, very slightly glossed; dull greenish blue, washed rufous towards large end, marked with very small light red spots with smeared edges.

LAYING DATES: Cameroon, Mar, July (fledged nestlings July, Sept); Angola (juv. May); Gabon, apparently throughout year.

Plate 15 (Opp. p. 256)

Apalis personata **Sharpe. Mountain Masked Apalis. Apalis à face noire.**

Apalis personata Sharpe, 1902. Bull. Br. Orn. Club 13, p. 9; Ruwenzori, Uganda.

Forms a superspecies with *A. binotata*.

Range and Status. Endemic resident. Highlands of E Zaïre, mainly above 1500 m, along Albertine Rift from west of L. Albert (Lendu Plateau) and Rwenzori Mts to west of L. Edward, Beni-Butembo ridge at 1400 m, west of L. Kivu, northwest of L. Tanganyika (Itombwe and Itombwe ridge down to 1270 m) and Mt Kabobo at 1980–2480 m; SW Uganda at 1500–2800 m (Bwindi-Impenetrable Forest and Rwenzoris); Rwanda (Nyungwe Forest); Burundi (Teza and Ijenda Forests) and SE Zaïre (Marungu Plateau above 1800 m). Common.

Apalis personata

Description. *A. p. personata* Sharpe: range of species except Marungu Plateau. ADULT ♂: forehead, crown, lores and cheeks blackish; white patch behind ear-coverts, bordered posteriorly with greenish yellow; nape green, not well demarcated from dark crown; mantle to back and uppertail-coverts green. Tail green, unpatterned, occasionally with pale yellowish tips to outer feathers. Chin, throat and breast black, belly off-white, washed greyish and invaded in centre by black of breast; flanks pale olive, thighs dark olivaceous brown. Wings dark grey-brown, outer webs of flight feathers green, upperwing-coverts green; underwing-coverts and axillaries pale yellow. Bill black, narrowly edged greyish along cutting edge of upper mandible, gape yellowish; eyes light brown tending to rufous, rim of eyelids pinkish brown; tarsus pinkish grey, feet brownish pink. Sexes alike. SIZE: (10 ♂♂, 8 ♀♀) wing, ♂ 55–58 (56·1), ♀ 53–56 (54·3); tail, ♂ 42–48 (46·0), ♀ 41–45 (42·8); bill, ♂ 12·5–14·5 (13·2), ♀ 12·5–13·5 (13·0); tarsus, ♂ 18–19 (18·6), ♀ 17·5–19 (18·1). WEIGHT: Uganda, ♂ (n = 9) 9–12 (11·0), ♀ (n = 8) 9–12 (10·9).

IMMATURE: head and back uniform green; chin and throat cream-white, rest of underparts silvery grey; subadult has crown washed green; in intermediate plumage develops black first on middle and lower throat.

NESTLING: unknown.

A. p. marungensis Chapin: SE Zaïre (Marungu Plateau). Face greyish black; black of foreneck bordered grey; sides of breast and flanks greener than nominate race.

TAXONOMIC NOTE: previously considered a highland race of *A. binotata*, but treated as specifically distinct following Irwin (1988) and Louette (1988). Differs in facial pattern and other plumage details, and is larger and heavier. There is limited sympatry, but the 2 are largely separated altitudinally.

Field Characters. The montane representative of the Masked Apalis *A. binotata*, with proportionately longer tail. Differs mainly by having white and green patch behind ear-coverts instead of white stripe on neck; black on underparts extends further down, onto centre of belly. Voices very similar.

Voice. Tape-recorded (LEM). Song very similar to that of Masked Apalis, q.v., with the same hard quality. Note of ♂ slightly longer, sounding trisyllabic, 'pritritru'; ♀ has similar rattle, 'tuc-tuc-tuc . . .' or 'tec-tec-tec . . .'. For a comparison of sonograms of the 2 songs, see Dowsett-Lemaire and Dowsett (1990). Playback of a tape of Masked Apalis produced mixed results.

General Habits. Inhabits montane forest, including gallery forest, edges and second growth. Feeds in lower canopy and at mid-levels, also in undergrowth. In Bwindi-Impenetrable Forest, Uganda, often in parties with other apalises; usually in the upper undergrowth, lower canopy or dense hanging vines; replaced in low undergrowth by Rwenzori Apalis *A. ruwenzorii* (Bennun 1986). In Nyungwe Forest (Rwanda), most numerous of 6 *Apalis* spp.; present throughout, with no altitudinal preferences, gleaning from foliage at all levels but spending most or time at mid-levels, especially in lianas around isolated trees; also feeds in second growth, clearings and edges. Here too the only overlap is with Rwenzori Apalis (which often responds to its song: Dowsett-Lemaire 1990). On Marungu Plateau (Zaïre) inhabits canopy and forest edges, 'mushitu' forest and forest galleries, extending into dense undergrowth nearby. Forages in pairs or family parties of 3–4, moving actively through foliage, gleaning insects from twigs and leaves.

Food. Insects and their larvae, including small beetles and caterpillars.

Breeding Habits. Almost unknown. Territorial, probably throughout year. Nest and eggs undescribed.

BREEDING INDICATIONS: Zaïre, juvs Itombwe Mar, enlarged gonads Mar–Apr, June, Aug; Uganda, adult feeding fledglings Apr, breeding condition Nov; Rwanda, independent juvs Oct, still with adults, indicating active breeding in dry season; 2 pairs still feeding young from estimated Aug clutches. E Africa: Region B, fledglings with adults Apr.

***Apalis jacksoni* Sharpe. Black-throated Apalis. Apalis à gorge noire.**

Apalis jacksoni Sharpe, 1891. Ibis p. 119; Mt Elgon.

Plate 15
(Opp. p. 256)

Range and Status. Endemic resident. SE Nigeria at 1500–1650 m on Mambilla Plateau at Ngel Nyaki and Gotel Mts; local and uncommon SW Cameroon at 1050–2250 m (Mt Cameroon, Rumpi Hills, Mt

Apalis jacksoni

Nlonako and Bamenda Highlands in scattered localities north to Mt Oku), also in lowlands (Dja R.); Gabon (1 km west of Bokaboka: Robertson 1995); Congo (Odzala Nat. Park: F. Dowsett-Lemaire, pers. comm.); locally in N Angola in Cuanza Norte and W Malanje (Canzele, Lubanda, Quela, Ndala Tando and Gabela), Cuanza Sul (Quitondo) and S Lunda (Cacolo); reappears in E Zaïre lowlands at Angu (Uele) and Ituri, and probably elsewhere on N borders of forest; common in highlands along Albertine Rift from northwest of L. Albert (Lendu Plateau); Rwenzori Mts; mountains west of L. Edward at 1000–1870 m and northwest of L. Tanganyika at 1400–2000 m (Itombwe), and Mt Kabobo at 1250–2100 m; S Sudan (Imatong and Didinga Mts above 1800 m); Rwanda (Nyungwe Forest); Burundi (Ijenda, Teza and Bururi Forests); S and W Uganda at 1200–2400 m from Budongo to Kibale and Bwindi-Impenetrable Forests and east to Malibigambo Forest, Sango Bay and Jinja; highlands of W and central Kenya at 1700–2400 m (Mt Elgon and Cherangani Hills to Kakamega and Mau Forests, and from Aberdare Mts and Mt Kenya to near Nairobi and Nguruman Hills); NW Tanzania (Bukoba and Minziro). Uncommon to common.

Description. *A. j. jacksoni* Sharpe: Sudan, Kenya, Uganda, Zaïre and Angola. ADULT ♂: forehead, crown and nape sooty grey, bordered behind by yellow band. Lores, cheeks and ear-coverts jet black; creamy white malar streak from base of bill to below ear-coverts. Chin and throat jet black, black area broadening on lower throat. Mantle and back rich olive-green, rump olive-green, uppertail-coverts olive-green tipped grey. Tail dark grey, strongly graduated, T5 and T6 broadly tipped and margined with white on outer webs, T2–T4 with 6–10 mm white tips, T1 with white margin. Breast to belly bright yellow, flanks washed green, thighs greyish white, undertail-coverts white. Flight feathers blackish, secondaries with narrow white margins on outer webs, tending to form wing-bar; primary and secondary coverts dark grey, lesser wing-coverts greenish; underside of outer webs of primaries and secondaries margined white; axillaries white, underwing-coverts white with some yellow edges. Bill black; eyes grey-brown to dark brown; legs and feet brown to dark flesh. ADULT ♀: like ♂ but forehead, crown, nape, cheeks, lores and ear-coverts to chin and throat grey, feathers on upper breast bordered olive. Remainder of underparts yellow, less olive-washed on flanks than ♂; tail-feathers with narrower white tips. Primaries and outer secondaries narrowly margined grey, inner secondaries narrowly margined white on outer webs, some ♀♀ with as much white as ♂♂; underwing-coverts and axillaries white washed yellowish. SIZE: (15 ♂♂, 12 ♀♀) wing, ♂ 52–56 (54·2), ♀ 49–54 (50·8); tail, ♂ 55–62 (58·1), ♀ 45–55 (48·5); bill, ♂ 11·5–12·5 (12·0), ♀ 11·5–12·5 (11·9); tarsus, ♂ 17·5–18·5 (17·6), ♀ 17·5–18·5 (17·6). WEIGHT: Uganda, ♂ (n = 14) 7–10 (8·8), ♀ (n = 9) 6–10 (8·5).

IMMATURE: ♂ has crown olive-green like back but with greyish suffusion, grey extending onto throat; streak on sides of neck washed yellow. ♀ has head and lower throat washed yellow, like back, less grey than in ♂.

NESTLING: unknown.

A. j. bambuluensis Serle: Nigeria and Cameroon highlands. ♂ has deep black crown, concolorous with sides of face; ♀ has chin and throat sooty black; both sexes less golden green on back than nominate *jacksoni*.

A. j. minor Ogilvie-Grant: lowland Cameroon and Zaïre (Dja R., Uele R. and Ituri). Like *jacksoni* but smaller: ♂♀ wing 45–51, tail 44–52.

TAXONOMIC NOTE: *A. j. albimentalis* Meise (N Angola) may prove to be distinguishable from *minor* and nominate *jacksoni*, being brighter yellow below, with throat patch extending further down onto breast (especially in ♀); variation may be oversimplified (Parkes 1987) but this has been questioned (Louette 1989).

Field Characters. Long-tailed and brightly coloured, with green back, yellow underparts and much white in tail, pale patch on wing. Head and throat blackish (♂) or grey (♀) with conspicuous white stripe across face. Inhabits upper levels in forest, where it may occur alongside Black-capped Apalis *A. nigriceps*; seen from below, Black-capped Apalis shows shorter tail, white throat, black bar across breast, and has more restless actions and twittering calls. Masked Apalis *A. binotata* has dark head and white neck stripe but dull whitish underparts, dumpy shape, shorter tail and grating song, and generally feeds at lower levels.

Voice. Tape-recorded (32, 68, B, C, GREG, HOR, LEM, MOY). Song a duet. Both birds give a plaintive 'piu' or 'pu', soft in quality yet far-carrying (up to 200 m: Chappuis 1979); repeated in variable length series, at rates of 2·5–4/s in E of range (Kenya/Uganda/Rwanda), slower in Cameroon (2/s or less). One bird invariably sings 3 semitones higher than its partner. Notes are pure, with an almost musical quality, quite unlike the hard dry songs of most apalises. Duet typically asynchronous, one bird singing slightly faster than the other, so that at one moment the notes coincide and the next they are out of sync. again (Keith and Gunn 1971), but synchronized duet recorded in Cameroon (Chappuis 1979a).

General Habits. Inhabits montane forest, in mid-stratum and canopy; avoids interior of lowland forest but occurs in clearings; also in second growth, coffee forest and tree canopy in woodland nearby. In Cameroon in forest canopy and on edges where uncommon and thinly distributed. In Sudan prefers tops of large forest trees, with marked preference for *Albizia* spp. In Nyungwe Forest (Rwanda) one of 4 canopy-frequenting apalises, found throughout below 2500 m but in smaller numbers than its congeners; above 2150–2200 m outnumbered by Chestnut-throated Apalis *A. porphyrolaema* and below 2000 or 2100 m by Grey Apalis *A. cinerea* and Buff-throated Apalis *A. rufogularis*; feeds predominantly in canopy, hopping along branches, pecking undersides of leaves; has special preference for light-foliaged trees such as *Albizia*, *Dombeya*, *Newtonia*, *Neobutonia* and *Macaranga*; also in tangles and lianas at mid-levels, and occasionally smaller trees in clearings (Dowsett-Lemaire 1990). Common in Bwindi-Impenetrable Forest, Uganda, and confined to canopy, feeding mainly by sally-gleaning; has strong preference for valley forest (Chestnut-throated Apalis replacing it in hillside forest, the 2 species never actually seen together: Bennun 1986).

Forages in pairs (rarely singly), or small family parties of 3–4 (sometimes 6); hops along branches, searches small twigs, pecks from underside of leaves; sally gleans, makes short flight after insect; occasionally joins mixed bird parties.

Food. Insects and their larvae: beetles, flies, small hemipterans, moths and caterpillars; spiders.

Breeding Habits. Territorial, probably throughout year.
NEST: only 1 known (Jackson and Sclater 1938) was a deep cup or purse made of fragments of *Usnea* lichen and thick spider webs; frail, net-like, almost transparent. Placed in straggling bush, securely attached at sides to several upright twigs forming fork.
EGGS: 2; light bluish green with small, evenly distributed reddish spots.
LAYING DATES: Cameroon (breeding condition Aug, adult feeding juv. Sept); Zaïre (Itombwe) (breeding condition Feb, Apr–May); Angola (juv. Apr); E Africa: Region B, Apr, June, adult feeding young Sept.

Apalis chariessa Reichenow. White-winged Apalis. Apalis à ailes blanches.

Apalis chariessa Reichenow, 1879. Orn. Centralb. 4, p. 114; Mitole, lower Tana R., Kenya.

Plate 15
(Opp. p. 256)

Range and Status. Endemic resident. SE Kenya in gallery forest below 100 m on lower Tana R. at Mitole and 37 km north of Garsen (now probably extinct: only 3 specimens obtained, 1877 and 1961, and the forest has now been destroyed); E Tanzania, Uluguru Mts at 1250–1400 m in Kinole Forest, Udzungwa Scarp forests at 1000–1500 m (Mwanihana, Magombera and Chita Forests; Ndundulu and Nyumbanitu Mts 1350–1660 m, occasionally to 2000 m); N Mozambique (Chiperoni Mt) and SE Malaŵi east of Shire Rift at 500–1500 m (Chikala Hill; Shire Highlands: Zomba, Lisau Mt, Ndirande Mt, Bangwe and Malabvi Hills, Thyolo Mt, Kongeni Escarpment; Mulanje Mt; once west of Rift at Dzonze, Kirk Range). In Malaŵi 15–18 pairs/150 ha on Lisau Mt, 6–7 pairs/100 ha on Thyolo Mt, lowest density on Mulanje (2 isolated pairs) in 1983–84; total Malaŵi population cannot be much above 100 pairs (Dowsett-Lemaire 1989). Density of 17 pairs or 51 singletons per 100 ha, secondary forest in Udzungwa Mts, Tanzania (Moyer 1993). Locally common to rare. 'Near-threatened' status (Collar and Stuart 1985).

Description. *A. c. macphersoni* Vincent: range of species except Kenya. ADULT ♂: forehead, crown, nape, lores, cheeks and ear-coverts glossy blue-black. Mantle and back to upper-tail-coverts glossy blue-black. Tail blue-black, very strongly graduated, T6 *c.* 56 mm shorter than T1; T6 white, next 3 pairs extensively tipped white and with white outer webs, T2 with white area of *c.* 4 mm and T1 with vestigial white on tip. Chin and throat white, white extending to sides of neck. Lower throat and breast with broad black band *c.* 10–12 mm deep in midline, bordered below by orange band; belly and flanks bright yellow; undertail-coverts white. Wing feathers blue-black, inner primaries with white outer margins, secondaries more extensively white on outer webs, forming conspicuous white patch; underwing-coverts and axillaries white. Bill black; eyes grey-brown; legs and feet light brown. ADULT ♀: forehead, crown and sides of head greyish; chin, upper throat and sides of neck off-white (duller than ♂). Mantle and back to uppertail-coverts olive-green. Tail dark grey with dull white tips to feathers (including central pair) and white outer webs. Sooty grey band on lower throat and upper breast, bordered below with orange-buff; belly and flanks yellowish. Wings olive-green except for grey primary and secondary coverts, pale patch less well marked. SIZE: (5 ♂♂, 2 ♀♀) wing, ♂ 50–54 (52·6), ♀ 49–52 (50·5); tail, ♂ 66–94 (77·8), ♀ 59–62 (60·5); bill, ♂ 12–13 (12·5), ♀ 11–12 (11·5); tarsus, ♂ 18–19 (18·4), ♀ 15–17 (16·0).

IMMATURE: imperfectly known; ♂ with unossified skull differed in having the mantle partially olive-green.

NESTLING: unknown.

A. c. chariessa Reichenow: lower Tana R. (Kenya). Chestnut wash below breast-band forms well-defined zone. ♂ small: (n = 3) wing, 45–48 (46·6); tail, 55–67 (62·3).

Field Characters. A brightly-coloured apalis of the canopy, with very long, graduated, white-tipped tail, longest of any apalis. ♂ has head, breast-band, upperparts and wings black, with white throat and wing-patch, yellow belly; ♀ similar but smaller, with grey head, green back, smaller wing-patch. Quite unlike any other bird in its range.

Voice. Tape-recorded (B,F, LEM). Song a lively piping duet of 4 notes of melodious quality, repeated in quick succession at rate of 6–7 notes/s, 'tee-lu dee-lu tee-lu dee-lu . . .'; more rarely utters 2 notes, 'tee-lu . . . te-lu', leaving gap for ♀'s part.

General Habits. In Malaŵi inhabits lowland and mid-altitude forest canopy, edges, or riparian forest nearby. In Mwanihana Forest (Tanzania), moderately common at intermediate elevations, in luxuriant growth, in pairs or mixed-species flocks, usually in canopy of larger-crowned trees, moving very actively through foliage of 30 m canopy, occasionally in mid-stratum; in Ndundulu and Nyumbanitu Forest almost exclusively in canopy, but twice seen down to 3–4 m in sunlit patches. In Malaŵi, densities highest in relatively dry, mid-altitude *Albizia*-dominated forest of Lisau Mt, and lowest on Mulanje in wet *Newtonia* forest and at Chikala. On Mt Thyolo mostly on slopes below 1300 m, and lowland *Albizia–Khaya* forest on nearby tea estates at 1000–1100 m. On Blantyre Hills it is more noticeable in secondary forest on slopes than in mature forest (as on top of Mt Soche); in riverine forest at Zomba 1000–1100 m, and Khonjeni 500–600 m, SE slope of Shire Plateau (Dowsett-Lemaire 1989). In Udzungwa Mts (Tanzania) regularly associates with more numerous Black-headed Apalis *A. melanocephala* (Stuart *et al.* 1981) but where they occur together in these forests Black-headed Apalis occurs more regularly on edges and in mid-stratum (Jensen and Brøgger-Jensen 1992).

Forages in pairs or family parties of 3–4, searching leaves and twigs for insects. In Malaŵi gleans leaves and twigs in canopy and at edges, preferring lighter-foliaged, wide-spreading crowns, especially *Albizia* spp.

Resident; but in Malaŵi a record of singing ♂ in 12 ha forest at Dzonze, separated by 80 km of inhospitable country from nearest regular site (Zomba), is remarkable example of exploratory movement, across Rift. Unfortunately little forest remains at Dzonze for species to become established there (Dowsett-Lemaire 1989).

Food. Insects.

Breeding Habits. Territorial, evidently throughout year.

NEST: bag-shaped, with side entrance near top; made of *Usnea* lichen, lined with white silken seed pappi, probably from asclepiads. Placed 8–16 m up in tree on forest edge or clearing, not concealed but easily overlooked, suspended from bare branch (**A**: see Benson and Benson, *Ibis* 1947, 286 and Fig. 3). Built by both sexes. One was made of material already *in situ* and shaped out, comparatively little brought from elsewhere; birds added lining, working inside for up to 2 min at a time, although ♂ once remained inside nest for 10 min. Once they were inside together, shaping inner bag; they used 2 entrances at opposite sides until lining

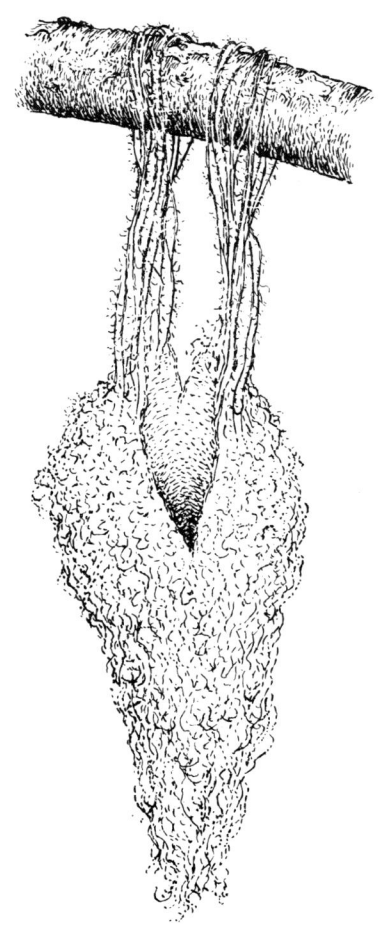

A

was completed, then used normal entrance to gain access. ♂ once brought material and passed it through one of the holes to ♀. Building period in one instance exceeded 24 days without eggs being laid (Benson and Benson 1947).

EGGS: 2–3. Narrow, elongate; dull-surfaced; pale green with faint underlying slate, spotted with light brown and light yellow-brown, generally but not very thickly distributed; in 1 egg spots formed a definite cap at thick end. SIZE: (n = 3) 16·5–17·1 × 11·3–11·5 (16·8 × 11·3).

LAYING DATES: Malaŵi, Oct–Nov, Jan.
INCUBATION: by ♀.

Reference
Dowsett-Lemaire, F. (1989).

Apalis nigriceps (Shelley). Black-capped Apalis. Apalis à calotte noire.

Plate 15 (Opp. p. 256)

Dryodromas nigriceps Shelley, 1873. Ibis, p. 139; Abouri (= Aburi), Aguapim (= Akwapim), Gold Coast.

Range and Status. Endemic resident. NE Sierra Leone at 1300–1400 m (Tingi Plateau and forested slopes immediately below); Guinea (common on Ziama Massif above 700 m); Liberia (Mt Nimba); Ivory Coast (Yapo and Tai Forests, Mt Tonku, Comoé, Ferkessédougou, Lamto and Maraoué); S Ghana north to 6° 30′N; SE Nigeria (Sapoba); Bioko; S Cameroon (Mt Kupé at 900–950 m, Korup Nat. Park, Lolodorf, Dja R. and Yaoundé); Equatorial Guinea; NE Gabon (Makokou, M'Passa and Bélinga); SW Central African Republic (between Nola and Mbaiki). Absent W and central Zaïre, but occurs in NE, on borders of forest (Ituri R.), also in Kivu and Itombwe (Kamituga); S and W Uganda at 1000–1400 m (Budongo and Bugoma Forests, Entebbe, Mabira and Mubende Forests and L. Nabugabo). Common to uncommon.

Description. *A. n. nigriceps* (Shelley): Sierra Leone to Gabon and Central African Republic. ADULT ♂: forehead to hindcrown, cheeks, lores and ear-coverts glossy black. Hindneck golden yellow; mantle and back to rump and uppertail-coverts yellowish green. Tail strongly graduated, slate grey, T2 tipped white, extent of white increasing on T3 to T6. Chin and throat white. Broad, glossy black throat circlet *c.* 6 mm wide extending to sides of neck; sides of breast golden yellow, belly whitish with greyish wash, especially on flanks which are suffused with yellow; undertail-coverts white. Wings blackish, primaries narrowly fringed white along outer webs, secondaries fringed yellowish green, upperwing-coverts yellowish green; axillaries and underwing-coverts white. Bill black, lower mandible greyish; eyes light reddish brown; legs and feet pinkish buff. ADULT ♀: like ♂ but forehead to hindcrown, lores, cheeks and ear-coverts slate grey, washed with dark olive-green on nape. Chin and throat cream-white; band around throat only 4 mm wide, and sooty blackish grey. SIZE: (17 ♂♂, 8 ♀♀) wing, ♂ 45–50 (47·3), ♀ 45–48 (46·2); tail, ♂ 40–48 (44·3), ♀ 37–42 (39·0); bill, ♂ 10–12 (11·0), ♀ 10·5–12 (11·1); tarsus, ♂ 16–17·5 (16·6), ♀ 15·5–17·5 (16·3). WEIGHT: Liberia, ♂ (n = 8) mean 8·4, ♀ (n = 3) mean 8·3; Cameroon, ♂♀ (n = 7), 7–9 (8·2).

IMMATURE: crown olive-green like back; underparts duller, lacking throat band. Newly fledged young has grey eye, blackish bill, yellowish base of lower mandible and mouth flanges, greyish white feet.

NESTLING: unknown.

A. n. collaris van Someren: E Zaïre and SW Uganda. Outer 3 tail-feathers wholly white. WEIGHT: Uganda, ♂ (n = 10) 7·0–9·0 (8·0), ♀ (n = 6) 8·0–10·2 (9·0).

TAXONOMIC NOTE: it has been suggested (Irwin 1987) that this species may not belong to the genus *Apalis* and may be closer to *Eremomela*, but pending further evidence it is retained here.

Field Characters. A tiny bird which moves in restless, twittering flocks through the canopy. From below appears white with black band across breast and white-tipped, graduated tail; could be confused with Rufous-crowned Eremomela *Eremomela badiceps*, with which it often travels; latter has similar black breast-band on pale underparts but shorter, ungraduated tail without white tips. From side view, Black-capped Apalis shows yellow-green back contrasting with black crown, rather than brown crown and grey back. Overlaps in some places with Black-throated Apalis *A. jacksoni*, which has black throat, yellow underparts and longer tail, and a ringing duet.

Voice. Tape-recorded (68, C, KEI). Song a steady trill on one pitch lasting *c.* 2 s, not unlike that of Chestnut-throated Apalis *A. porphyrolaema* but drier and harder. ♀ often joins in with similar trill. Twittering contact calls similar to those of Rufous-crowned Eremomela but song of latter very different, a jumble of tuneful notes.

General Habits. Inhabits primary or mature secondary forest, mostly in lowlands, occasionally higher. Habitat evidently varies geographically: in Sierra Leone occupies gallery forest around 1200 m; in Ivory Coast, undisturbed lowland and secondary forest in wet and dry zones, where practically confined to crowns and higher canopy of large *Piptadeniastrum africanum* trees (L. D. C. Fishpool, pers. comm.); in Ghana, lower canopy of mature and secondary forest, also shade trees above cocoa plantations. In Gabon confined to crowns of emergent trees in primary and mature secondary forest; in Zaïre found in crowns of mature forest trees, but also occurs widely in clearings and creepers at forest edges, and in mid-stratum and second growth.

In pairs or small family parties, usually of 3–4 but frequently of 6 or more. Gleans foliage. Members of party call to one another when feeding, flitting about, cocking tail, ruffling feathers. Frequently found with other apalis species and in mixed-bird parties. In Sierra Leone calls continuously in wet season, less so in heat of day; in Gabon calls from Oct to Mar.

Food. Insects: mostly beetles; ants, caterpillars. Spiders.

Breeding Habits. Little known. Nest and eggs not described but pair seen entering nest in hanging moss high in tree on Mt Nimba, Liberia, Nov (Gore 1994). OTHER BREEDING INDICATIONS: Cameroon (about to lay, Oct); Gabon (young being fed Mar); Zaïre: Itombwe (♂ with enlarged testes Apr, ♀ about to lay, Feb).

Plate 16
(Opp. p. 257)

Apalis melanocephala (Fischer and Reichenow). Black-headed Apalis. Apalis à tête noire.

Burnesia melanocephala Fischer and Reichenow, 1884. J. Orn. 32, p. 56; Pangani, coastal Tanganyika.

Forms a superspecies with *A. chirindensis*.

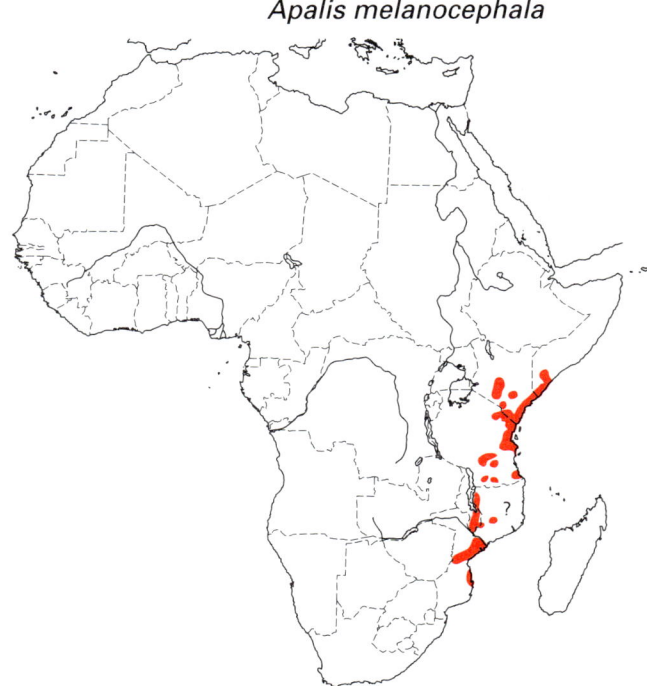

Apalis melanocephala

Range and Status. Endemic resident, E African coastal, mid-altitude and montane forests from S Somalia to S Mozambique. Somalia, lower Juba R. and in extreme south. Kenya, in dense woodlands throughout coastal lowlands north to lower Tana R.; on Mt Endau, Taita Hills and Mt Kasigau, and in central highlands up to 2000 m from Meru southwest to Kiambu and Nairobi. Tanzania, coast at Tanga and Pugu Forest, and inland on Mts Meru, Kilimanjaro, Ngurus, Ukagurus, Ulugurus, Handeni, Kidugallo, West Usambaras, Udzungwas, Mahenge (Muhulu Forest), and in SE corner east to Songea. Malaŵi, hills in SE. Mozambique, in N on Njesi Plateau, Unango, Mt Chiperoni and Mt Namuli; south of Zambezi ranges from Inhamitanga to Dondo, and in coastal forests from Vilanculos to Inhambane; also occurs inland along upper Buzi R. and its tributaries. Also adjacently in SE Zimbabwe, in Lusitu-Haroni Forest. Common, at least locally. Density, up to 1 pair or family per ha (Malaŵi, S Mulanje at 1000 m and Thyolo at 1000–1400 m: Dowsett-Lemaire 1989), and 30 pairs or 90 singletons per 100 ha in secondary forest, Udzungwa Mts, Tanzania (Moyer 1993).

Description. *A. m. lightoni* Roberts: S Mozambique from Inhamitanga and east of Cheringoma Plateau south to Beira (Dondo) and inland to SE Zimbabwe (Lusitu-Haroni Forest). ADULT ♂: forehead, crown, nape, cheeks, lores and ear-coverts blackish slate; hindneck, mantle and back to uppertail-coverts blackish grey. Tail blackish, long and strongly graduated, all but central feathers tipped off-white. Chin and throat creamy white. Breast, belly, flanks and undertail-coverts creamy white. Wings blackish grey, underwing-coverts and axillaries cream-white. Bill black, interior of mouth black; eyes yellowish brown to grey-brown; legs and feet pinkish flesh. Sexes alike. SIZE: (20 ♂♂, 12 ♀♀) wing, ♂ 45–52 (49·0), ♀ 45–52 (47·5); tail, ♂ 57–66 (61·9), ♀ 47–53 (49·5); bill, ♂ 13·5–15·0 (14·1), ♀ 13·5–14·5 (14·0); tarsus, ♂ 16·5–18·5

(18·0), ♀ 16·5–18·5 (17·6). WEIGHT: ♂ (n = 15) 7·7–10·1 (9·0), ♀ (n = 8) 7·5–9·8 (8·4).

IMMATURE: olive green above; throat yellow, belly and flanks dull yellowish; bill horn-coloured, lower mandible yellowish.

NESTLING: undescribed.

A. m. melanocephala (Fischer and Reichenow): S Somalia, coastal Kenya, coastal NE Tanzania, E Usambara Mts and Pangani R. to Pugu Forest, Dar es Salaam. ♂ with blacker crown and face, blackish brown back, olive-tinged rump; whiter below than *lightoni*. ♀ paler above, more olive-tinged, underparts whiter than *lightoni*.

A. m. nigrodorsalis Granvik: Kenya highlands including Mts Endau and Kasigau. ♂ sooty blackish brown above, shade intermediate between *moschi* and *melanocephala*; ♀ tinged olive above.

A. m. moschi van Someren: Taita Hills, Kenya, and highlands of E Tanzania from Meru and Kilimanjaro to W Usambaras and Udzungwa Mts. ♂ paler and grey above, including crown; face blackish; heavier olive wash on lower back and rump than *nigridorsalis*; secondaries edged green; less buff on belly and flanks. ♀ slightly more olive above. This and following interior races longer-tailed: tail, ♂ (n = 18) 58–77 (67·3), ♀ 49–58 (53·4).

A. m. muhuluensis Grant and Mackworth-Praed: SE Tanzania at Mahenge (Muhulu Forest) and Songea (Luwiri-Kitessa Forest). Blacker above and lighter below than next race.

A. m. tenebricosa Vincent: N Mozambique (Njesi Plateau, Unangu; Mts Chiperoni and Namuli). Sooty black above; breast and flanks washed dusky, contrasting with cream-coloured throat. Tail, ♂ (n = 8) 61–81 (69·0).

A. m. fuliginosa Vincent: SE Malaŵi (Mulanje and Thyolo Mts). Less sooty black above than *tenebricosa*; ♂ washed greyish olive on back, similar to *tenebricosa* below. ♀ like *tenebricosa* with contrast between sexes less marked. Tail, ♂ (n = 6) 60–69 (64·0).

A. m. adjacens Clancey: SE Malaŵi (Namizimu Hills, Mangochi, Chikala; Shire Highlands: Zomba, Chiradzulu (Lisau), Ndirande, Soche, Bangwi, Malabvi). ♂ like *lightoni* but darker, more sooty, less leaden grey above; more creamy white below than *fuliginosa* and *tenebricosa*; ♀ more olive on back than these races. Tail, ♂ (n = 5) 68–72 (69·9).

A. m. addenda Clancey: S Mozambique (south of Save R.; E Inhambane from Vilanculos to Massinga). Both sexes sootier grey above than *lightoni*, throat brighter buff, belly whiter. Tail-length similar to *lightoni*.

Field Characters. Distinctly two-toned, with blackish upperparts and creamy white underparts. Tail long, tipped white. Overlaps marginally in Zimbabwe with closely-related Chirinda Apalis *A. chirindensis*, which has hissing quality to song and more uniform appearance, grey above and pale grey below. In central Kenya just meets Grey Apalis *A. cinerea* which is white below but has grey-brown crown, grey back. Chapin's Apalis *A. chapini* and Chestnut-throated Apalis *A. porphyrolaema* have grey backs and rufous on throat or breast.

Voice. Tape-recorded (32, 91, B, F, CHA, GREG, HOR). Song geographically variable, evidently increasing in complexity from coast to interior, unrelated to geographic variation (Dowsett-Lemaire 1986). In coastal Kenya a single repeated note, more drawn-out, less clipped and hard, than other apalises, e.g. Grey Apalis *A. cinerea*, 'wee-wee-wee-wee . . .'; in Zimbabwe a more sibilant 'sweep-sweep-sweep-sweep . . .'; given at rate of c. 3/s in often lengthy series. ♀ joins in with similar note. This song is never heard in Malaŵi. In S Malaŵi most frequent motif consists of 3 (occasionally 4) clear, piping notes repeated in series for a few s without a break, 'titipu, titipu . . .', 'pititi, pititi . . .'. Calls include a piping 'pee-pee-pee-pee', short trills, a rolled 'prru' and an alarm note 'puit'. In Mwanihana Forest (Tanzania), song quite different: varied, di- or trisyllabic, some with accent on first note. Another frequent song is a piping, 3-syllabled 'chiririt-chiririt-chiririt . . .', (sometimes 'chiri' or just 'chi'), answered by second bird. Duetting often incessant. Also gives a very low trill audible only at close range. Call, a quiet 'seet' (Maclean 1993). Soft warble given during display.

General Habits. Lives in forest canopy, usually in tallest trees, coming lower down on edges and in clearings. Favours tangles of creepers and vines festooning trees; in coastal forest wanders into woodland nearby, particularly taller miombo (as in Dondo Forest, Mozambique, where habitats merge or adjoin). Also occupies dry forest on sand substrate with scattered baobabs *Adansonia digitata* at Inhamitanga (Mozambique), stunted secondary miombo and coastal scrub (Massinga, Mozambique); at Lusitu-Haroni (Zimbabwe) and elsewhere frequent in riparian forest.

In SE Zimbabwe marginally sympatric with Chirinda Apalis at around 350–450 m; in E Kenya highlands segregated altitudinally from Chestnut-throated Apalis (latter in higher level forest); in E Tanzania virtually parapatric with Chapin's Apalis (latter at higher altitudes) in Uluguru and Nguru Mts; in Udzungwa Mts Black-headed Apalis ascends to 1700 m, Chapin's down to 1100 m but uncommon below 1500 m (Jensen and Brøgger-Jensen 1992).

Gleans foliage, moving through canopy in pairs or family parties of 4–6, often joining mixed-species parties. Assiduously searches leaves and twigs; flies out to catch insect; restless, may stop momentarily to sing or call but then moves away through branches. Flits from one tree crown to another with somewhat jerky flight, when tail can look almost transparent against sunlight. In Malaŵi noisy all year round, with full song commonest Aug-Nov.

In Malaŵi vagrant up to 2200 m on Mulanje Plateau; regularly occurs as low as 800 m in July and even down to 600 m on Ruo R.; similar altitudinal movements may take place elsewhere.

Food. Insects, particularly grasshopper nymphs, caterpillars and small flies; once a seed.

Breeding Habits. Territorial; pair or family group probably maintains feeding territory throughout year. 2 displaying birds face one another, adopting upright posture with bodies extended and tails raised vertically, warbling softly. Once a very long-tailed (♂?) bird leapt vertically up for 1 m with head, body and tail in rigid straight line; it curved over sharply at zenith and dropped back almost vertically to perch; flight was repeated 3 times with warbled accompaniment (Moreau and Moreau 1939).

A

NEST: domed, with side-top entrance, composed of very fine foliaceous lichen, mixed with small amounts of *Usnea*; walls thick (*c.* 14 mm), densely lined with white, wool-like asclepiad seed pappi (**A:** see Benson and Benson, *Ibis* 1947, 286 and Fig. 3). Placed 4–7 m up, extremely well hidden by leaves, difficult to locate; bound to adjacent branches with strands of spider web.

EGGS: 2–3. Pale greenish, boldly marked all over with spots of bright chestnut (occasionally dull brown), with underlying blotches and specks of faint purplish slate, sometimes almost imperceptible. SIZE: (n = 2) 19 × 13.

LAYING DATES: E Africa: Region D, Apr; Region E, July; Malaŵi, Oct–Nov; Zimbabwe, Oct–Nov.

Plate 16
(Opp. p. 257)

Apalis chirindensis Shelley. Chirinda Apalis. Apalis de Chirinda.

Apalis chirindensis Shelley, 1906. Bull. Br. Orn. Club 16, p. 126; Chirinda Forest (= Mt Selinda), Gazaland.

Forms a superspecies with *A. melanocephala*.

Range and Status. Endemic to E Zimbabwe and adjacent Mozambique from Nyanga Highlands to Chimanimani Mts and Mt Selinda, with isolated population on Mt Gorongoza. Common resident at 350–2200 m.

Description. *A. c. chirindensis* Shelley: Mozambique (Mt Gorongoza), Zimbabwe (Banti Forest Reserve to Mt Selinda). ADULT ♂: forehead, crown and hindneck pale grey; cheeks, lores and ear-coverts pale grey; mantle and back to uppertail-coverts pale grey, lighter on rump. Tail pale grey, strongly graduated, all but 2 central pairs of feathers tipped white. Chin and throat greyish white; breast dull greyish white, belly to undertail-coverts clearer white, thighs grey, flanks with greyish wash. Wings dark grey with paler grey edges to secondaries and upperwing-coverts; underwing-coverts and axillaries off-white. Bill black, base of lower mandible pale flesh in ♀, inside of mouth pinkish flesh, palate darker; eyes medium brown to orange-brown; legs and feet light flesh-brown. Sexes alike. SIZE: (30 ♂♂, 20 ♀♀) wing, ♂ 48–52 (50·4), ♀ 46–50 (48·0); tail, ♂ 52–65 (61·3), ♀ 47–54 (49·6); bill, ♂ 13–15 (14·6), ♀ 13–15 (13·7); tarsus, ♂♀ 18–20 (19·1). WEIGHT: ♂ (n = 32) 7·3–9·7 (8·7), ♀ (n = 16) 7·5–10·6 (8·4).

IMMATURE: upperparts tinged greenish; chin, throat and centre of belly washed yellowish, flanks fawn-coloured; eyes grey, minute ring of white feathers around eyelid; slightly older birds become whitish grey below.

NESTLING: undescribed.

A. c. vumbae Roberts: Zimbabwe (Nyanga Highlands to Bvumba Mts). Slightly paler grey above, tail with more extensive white on tips; bill dusky horn, lower mandible flesh-coloured.

Apalis chirindensis

Field Characters. Plain grey, paler below but without the strong contrast between upper- and underparts shown by Black-headed Apalis *A. melanocephala*.

Sibilant song also distinctive. Similar-sized White-tailed Crested Flycatcher *Elminia albonotata* occurs in same forests and is also grey but much darker, especially on head, and has different feeding habits and quite different jizz, with peaked head and fan-shaped, white-edged tail which it constantly flirts. Vocal and active, continually on the move through canopy.

Voice. Tape-recorded (32, 91, F, CHA). Song a loud, strident, sibilant note, 'swik' or 'swink', with a slightly ringing quality, repeated at rate of 2–3/s in variable length series, sometimes 20 or more times, with no variation in pitch or speed. ♀ responds with lower-pitched 'chip'; song becomes faster and more excited during duets, especially during breeding season, when ♀ intervenes with variety of 'peep' and 'chip' notes. Calls include a sharp chatter in monotone, and in alarm, low quivering trill interspersed with quiet clicking notes.

General Habits. Inhabits montane and mid-altitude forest at 350–2200 m. Occurs in pairs or small family parties of 3–5, moving actively through canopy, coming lower down on edges and in clearings and sunlit places. Gleans foliage, searches leaves and twigs; occasionally snaps up insect in flight. Birds constantly call to one another in party; calls throughout year, with marked fall-off in activity in cold weather (Apr–July).

Sedentary, maintaining feeding territory throughout year, although some birds move downward to lower forests in winter; perhaps dependent on continuous forest (now often destroyed). Where sympatric with Black-headed Apalis in Lusitu-Haroni Forest, some may occur only in colder months but others present year round.

Food. Insects: small beetles, flies and caterpillars.

Breeding Habits. Territorial all year.

NEST: domed, with side-top entrance below point of attachment, with false entrance higher up in the one nest adequately described (perhaps due to earlier damage: Lorber *et al.* 1983). Composed of epiphytic liverwort *Frullaria*, skeletonized leaves, achenes and pappi of composite plant, and bits of *Asparagus* fern; seed cases and *Usnea* lichen incorporated into exterior; lined with plant down from forest creepers, rest of material gathered from shrub layer or ground stratum. Placed *c.* 20 m up in lower canopy, suspended from small branch at a point where always heavily covered with moss and well-concealed. During 1 h of observation, presumed ♂ made 17 visits, carrying vegetable down, landing on twig, looking around before entering nest, with same behaviour when leaving. The process was accompanied by singing at intervals; ♀ answered.

EGGS: undescribed.

LAYING DATES: Zimbabwe, Oct–Dec, Feb (Oct 1, Nov 4, Dec 2, Feb 1 clutches); Mozambique (enlarged gonads Nov).

INCUBATION: presumed ♂ approaches nest quickly and silently; incubating bird given caterpillar.

DEVELOPMENT AND CARE OF YOUNG: young fed by both adults, at intervals of 5–8 min. When chicks stopped demanding food, one adult would settle into nest and brood them for up to 40 min (usually less); every 10–15 min its mate brought food, which sitting bird either consumed or gave to chicks. When chicks became restless, sitting bird left nest and both birds fed them (Ginn 1996).

BREEDING SUCCESS/SURVIVAL: 1 bird recaptured after 3 years, 137 days.

Apalis porphyrolaema **Reichenow and Neumann. Chestnut-throated Apalis. Apalis à gorge marron.**

Plate 16 (Opp. p. 257)

Apalis porphyrolaema Reichenow and Neumann, 1895. Orn. Monatsb. 3, p. 75; Eldoma (= Eldama), Mau, Kenya.

Forms a superspecies with *A. chapini*.

Range and Status. Endemic resident, E African montane forest. E Zaïre along Albertine Rift at 2000–3220 m (mountains west of L. Edward, Kivu, Itombwe and Mt Kabobo at 1600–2480 m); Rwanda (Nyungwe Forest); Burundi (Teza, Ijenda and Bururi Forests); Uganda in SW (Kigezi, Bwindi-Impenetrable Forest and Rwenzori Mts) and E (Mts Moroto and Elgon); W and central Kenya Highlands at 1700–3400 m (rarely lower), Mt Elgon, Kakamega Forest, Cherangani Hills, Aberdares and Maralal south to Londiani, Sotik, Kikuyu and Mt Kenya; N Tanzania (Mt Loliondo). Common to fairly common.

Description. *A. p. porphyrolaema* Reichenow and Neumann: range of species except SW Uganda and E Zaïre. ADULT ♂: forehead, crown, cheeks, lores, ear-coverts, nape, sides of neck, hindneck, mantle, back and rump to uppertail-coverts slate-grey. Tail dark slate-grey, strongly graduated, all but central pair of feathers tipped off-white. Chin and throat chestnut. Breast to undertail-coverts pale grey, lighter on centre of belly; thighs grey. Wings slate-grey; underwing-coverts and axillaries white, variably washed chestnut. Bill black; eyes light ochreous brown, rim of eyelids brownish pink; legs and feet light pinkish buff, claws dark grey. Sexes alike. SIZE: (10 ♂♂, 9 ♀♀) wing, ♂ 50–54 (53·2), ♀ 49–54 (52·0); tail, ♂ 57–61 (59·8), ♀ 48–60 (52·4); bill, ♂ 12–13 (12·5), ♀ 12–13·5 (12·5); tarsus, ♂ 17–18·5 (17·6), ♀ 17–19 (17·7). WEIGHT: Uganda, ♂ (n = 7) 7–11 (8·8), ♀ (n = 4) 7–10 (8·0).

IMMATURE: like adult but overall duller grey, upperparts washed greenish; chin and throat creamy buff.

NESTLING: unknown.

A. p. affinis Ogilvie-Grant: SW Uganda. Like nominate *por*-

Apalis porphyrolaema

phyrolaema but underparts paler whitish grey, contrasting with slightly darker chestnut throat.

A. p. kaboboensis Prigogine: E Zaïre (Mt Kabobo). Like nominate race but chin, throat and upper breast dark grey; underwing-coverts and axillaries pure white. Immature resembles *porphyrolaema* with creamy buff chin and throat; back and wings with greenish wash, trace of yellow on centre of belly.

TAXONOMIC NOTE: *A. p. kaboboensis* has sometimes been treated as a full species (Prigogine 1985a) but is now widely regarded as a well-marked race of *porphyrolaema*; it differs materially only in its grey, not chestnut throat.

Field Characters. A slim, grey, long-tailed apalis of montane forest canopy; tail blackish, graduated, tipped white. Chestnut throat separates it from Black-headed Apalis *A. melanocephala*, which it barely meets in central Kenya, and from broadly sympatric Grey Apalis *A. cinerea*, which has outer half of tail pure white. ♀ Buff-throated Apalis *A. rufogularis* has reddish buff throat and breast but blackish upperparts and shorter, mainly white tail. Chestnut-throated Apalis easily distinguished from all 3 by trilling song, and in areas of overlap it is also segregated ecologically.

Voice. Tape-recorded (32, 68, B, C, F, LEM, ZIM). Song, 1 long or 2 shorter trills with a few introductory tiks or chips, 'ti-titi-trrrreeeee, trrrreeeee'; 'chip-chipa-trrrrreeeeeeeee'; 'cha-rrrreeeee, cha-rrrreeeee'; 'cha-churrrrreeeeeeeee'; total time, 1·5–2 s. High-pitched and rather insect-like but with ringing quality and fuller tone than the dry, grasshopper-like Black-capped Apalis *A. nigriceps*. Sometimes duets. Also gives a low, churring rattle like fingernail run along teeth of comb; when one bird rattles others often join in.

General Habits. Inhabits montane forest, second growth, gallery forest and edges; prefers canopy and crowns of taller trees, also medium-sized trees and liana tangles. In Kenya highlands replaces Black-headed Apalis west of about Nairobi, where they are believed to overlap (although never seen together: L. Bennun, pers. comm.). Chestnut-throated Apalis occurs mainly at higher altitudes. In Bwindi-Impenetrable Forest, Uganda, inhabits mainly hillside forest, thus reducing competition with Black-throated Apalis *A. jacksoni*, which is mainly in valleys; also more flexible in its feeding preferences, descending even to ground although rarely below 2 m (Bennun 1986). In Nyungwe Forest (Rwanda) occurs mainly above 2000 m, where along an altitudinal gradient its density is inversely related to those of Buff-throated Apalis and Grey Apalis (Dowsett-Lemaire 1990).

In pairs or small family parties, occasionally of 6–8; constantly on the move, searching twigs and leaves; commonly sally-gleans, catches insect in flight. Pair often duet from tree top where they forage, and may countersing with another pair 50 m away. Foraging bird flits from twig to twig, bowing and swaying body sideways, with drooping, partly opened wings, and tail erect, fanned, and flicked from side to side (Jackson and Sclater 1938).

No evidence of vertical migration, and resident even at 3000 m at Mau-Narok (Kenya).

Food. Insects: beetles, Hymenoptera, Diptera and caterpillars; also spiders.

Breeding Habits. Territorial. Nest and eggs not described.

BREEDING INDICATIONS: Zaïre: Itombwe (breeding condition Mar–June, Aug); Rwanda (many pairs accompanied by flying young Oct, which infers dry season breeding, e.g. Aug–Dec); E Africa: Region A, Dec; Region B, fledgling fed by adult Apr; Region C, Jan, Dec.

Plate 16
(Opp. p. 257)

Apalis chapini Friedmann. Chapin's Apalis. Apalis de Chapin.

Apalis chapini Friedmann, 1928. Proc. New Engl. Zool. Club 10, p. 47; Nyingwa, Uluguru Mts, Tanganyika.

Forms a superspecies with *A. porphyrolaema*.

Range and Status. Endemic resident. Montane forest, in E Tanzania above 1500 m on Nguru, Ukaguru, Uluguru and Udzungwa Mts, and in south in highlands of Mt Rungwe, Poroto Mts and Umalila to Njombe; NE

Apalis chapini

Zambia on Nyika Plateau, Mafinga and Mukutu Mts; and Malaŵi, west of Rift, at 1100–2300 m north of 14°S and 1550–2000 further south, from Misuku Hills, Mafinga Mts and Nyika Plateau to N and S Viphya Plateaux, W shore of L. Malaŵi (Kalwe, Nkuwadzi and Mzuma), Kuwilwe Hill, Chipata and Ntchisi, Dzalanyama Range, Chongoni and Chirobwe Mts. In Malaŵi pairs occupy small forest patches of 0·5–1 ha on SW Nyika, but overall densities of 3 pairs/10 ha occur in more continuous forest at 2000–2200 m (Dowsett-Lemaire 1989). Common. Density of 16 pairs or 45 singletons per 100 ha in secondary forest, Udzungwa Mts, Tanzania (Moyer 1993).

Description. *A. c. strausae* Boulton: Tanzania south of 8°30′S, Zambia and Malaŵi. ADULT ♂: forehead and crown grey, washed cinnamon-brown; cheeks, lores and ear-coverts deep chestnut-brown. Nape, mantle and back uniform slate-grey; uppertail-coverts washed rusty buff. Tail slate-grey, all but central pair of feathers tipped very pale whitish buff. Chin, throat and breast cinnamon-chestnut, not as dark as face; lower breast and belly to undertail-coverts pale buffy, flanks washed greyish, thighs chestnut. Wings slate-grey, underwing-coverts and axillaries white. Bill black; eyes dark red-brown; legs and feet pinkish flesh. Sexes alike. SIZE: (14 ♂♂, 12 ♀♀) wing, ♂ 50–54 (51·5); ♀ 49–53 (50·8); tail, ♂ 53–63 (57·3); ♀ 47–53 (50·6); bill, ♂ 12–13·5 (12·9), ♀ 12–13·5 (12·5); tarsus, ♂ 17·5–18·5 (17·8), ♀ 17·5–18·5 (18·0). WEIGHT: unsexed (n = 4) 8·8.

IMMATURE: head, cheeks, mantle and back dull grey, chin and throat buff, uniform with underparts.

NESTLING: undescribed.

A. c. chapini Friedmann: E Tanzania north of 8°30′S. Forehead and crown to nape grey with variable rufescent wash. Chin white, upper throat creamy buff; lower throat and breast cinnamon-chestnut, merging with buff of belly and flanks.

TAXONOMIC NOTE: formerly treated as conspecific with *A. porphyrolaema* but head and breast pattern and voice markedly different. Here considered specifically distinct, following Dowsett and Dowsett-Lemaire (1980).

Field Characters. The only apalis in its range with chestnut face, throat and breast; does not overlap with Chestnut-throated Apalis *A. porphyrolaema*, and both Black-headed and Brown-headed Apalises *A. melanocephala* and *A. alticola* have white underparts. Can be confused with African Tailorbird *Artisornis metopias*, which has chestnut face and throat but brown upperparts and shorter tail and feeds in undergrowth, typically below 2·5 m but occasionally up to 6 m; Chapin's Apalis generally forages above 8 m but sometimes down to 2–3 m so overlap possible; leg colour is also useful, blackish in African Tailorbird, yellowish red in Chapin's Apalis (L. Svendsen and J. Hansen, pers. comm.). Songs are different but some calls are similar.

Voice. Tape-recorded (86, F, CHA, HAN, LEM, SVEN). Song geographically variable. In Zambia and Malaŵi, ♂ gives an initial 'dzee' followed by a series of 10–30 regularly repeated single or double tuneless notes, 'dzee-tsi-tsi-tsi . . .' or 'dzee-tsili-tsili-tsili . . .', at rate of 4 notes/s. ♀ accompanies with faster tikking notes, often ending with a few squeaky ones. In Tanzania ♂ gives high-pitched, sibilant 'psi-psi-psi-psi . . .' at rate of 5 notes/s or double 'tsizi-tsizi-tsizi . . .' at rate of 3/s, or a lower-pitched 'pui-pui-pui . . .'; ♀ may join in with irregular squeaky notes.

General Habits. Inhabits montane forest; mainly in canopy, but often comes down to 2–3 m above ground at forest edges in early morning or late afternoon, particularly when sun is out; also mid-stratum cover exposed to midday sun. Sometimes wanders into nearby riparian forest, and in central Malaŵi occasionally extends into secondary forest.

Quite strongly segregated vertically where sympatric with Bar-throated Apalis *A. thoracica* (in Malaŵi) where aggression and countersinging between them extremely frequent; in Malaŵi co-exists with Brown-headed Apalis which it dominates in forest canopy but in Misuku Hills excludes it with altitudinal segregation above 1700 m. Outnumbered by Brown-headed Apalis at 1600–1800 m in Mafinga Mts, and they frequently countersing; in SW Nyika they overlap at 1950–2150 m; above 2050 m Chapin's Apalis dominates, and at 2000 m some pairs keep separate territories, countersinging and chasing one another. At Misuku, Mafinga and Musisi, where Brown-headed Apalis occurs, Chapin's Apalis is not found below 1700 m, but it descends to 1100 m in Kaningina Hills and Viphya where Brown-headed Apalis is absent (Dowsett-Lemaire 1989). In Udzungwa Mts (Tanzania) uncommon below 1500 m, showing hardly any altitudinal overlap with White-winged Apalis *A. chariessa* and Black-headed Apalis *A. melanocephala* (Jensen and Brøgger-Jensen 1992).

A foliage gleaner, moving about in pairs or small family parties of 3–4 birds, searching leaves and twigs. Very vocal; frequently stops to sing. In Malaŵi sings throughout year, most active Aug–Jan.

Food. Insects.

278 SYLVIIDAE

Breeding Habits. Territorial; in SW Nyika (Malaŵi) pairs occupy forest patches of 0·5–1·0 ha. All patches of closed forest greater than 0·75 ha are occupied by a pair.

NEST: vertically elongated bag with single point of attachment at top, and narrow side entrance 3 cm from top and beneath conspicuously protruding perching platform; composed almost entirely of fibre-lichen which straggles out untidily below; lined with a little vegetable wool (probably seed-coating). Ext. depth 115, ext. diam 50, diam. of entrance 25. Sited 16 m above ground on edge of evergreen forest.

EGGS: undescribed.

LAYING DATES: E Africa: Region C, Dec–Jan; Malaŵi, Oct–Jan.

Plate 16
(Opp. p. 257)

Apalis sharpii Shelley. Sharpe's Apalis. Apalis de Sharpe.

Apalis sharpii Shelley, 1884. Ibis, p. 45; Gold Coast.

Range and Status. Endemic resident. Very common wherever tall forest survives, in Sierra Leone (Freetown to Tingi Hills and Gola Forest); Guinea (Ziama and Kounounkan Massifs: Halleux 1994, Hayman *et al.* 1995); Mt Nimba (S Guinea/Liberia); Ivory Coast (widespread in S, north to Mt Nimba, Bouaké and Comoé Nat. Park) and Ghana (east to scarp of Volta basin and 6° 40′N).

Apalis sharpii

Description. ADULT ♂: forehead, crown, nape, cheeks, ear-coverts, neck, mantle, back, rump, uppertail-coverts and tail sooty grey. Tail-feathers with pale buffish tips to all except 2 central pairs. Chin, throat and upper breast sooty grey, flanks paler grey, thighs sooty grey, centre of belly to undertail-coverts off-white. Wings sooty grey; underwing-coverts and axillaries white. Bill black, lower mandible greyish, gape pink; eyes brown to red-brown; legs and feet dark reddish flesh. ADULT ♀: like ♂ above and on sides of head, but washed with deep sooty olive. Chin and throat light brick-red; breast and flanks grey, centre of belly white. SIZE: (16 ♂♂, 8 ♀♀) wing, ♂ 48–52 (49·1), ♀ 45–50 (47·5); tail, ♂ 45–50 (47·5), ♀ 41–46 (42·6); bill, ♂ 12·5–14·5 (13·5), ♀ 12·5–14 (12·9); tarsus, ♂ 16·5–18 (17·3), ♀ 16–18 (16·9). WEIGHT: Liberia, ♂ (n = 14) av. 9·1, ♀ (n = 7) av. 8·3.

IMMATURE: ♂ upperparts including sides of head and wings olive-brown. Chin and throat pale yellow, darkening to olive on breast; flanks olivaceous, centre of belly washed yellow. Lower mandible pinkish horn, gape pink. Before adult plumage attained, bird becomes darker grey below and develops pale buff patch on chin and throat. ♀ at first like ♂ but greener above except for grey crown; chin and throat pale buffy, breast less greyish and with less yellow wash on centre of belly.

NESTLING: unknown.

TAXONOMIC NOTE: Chappuis (1980), on the basis of vocal similarities, treated plain-tailed *sharpii* as either conspecific or forming a superspecies with *A. rufogularis*, which has extensively white outer tail-feathers. They are best treated as independent species, following Traylor *in* Watson *et al.* (1986); both are in turn closely related to *A. bamendae* and *A. goslingi* (the latter widely sympatric with *rufogularis*); at this stage the recognition of any superspecies is inadvisable.

Field Characters. A dull grey, rather dumpy, short-tailed bird of the canopy. ♂ fairly uniform; ♀ has reddish throat. Confined to forest west of the Dahomey Gap; no similar warbler within its range.

Voice. Tape-recorded (68, B, CHA, KEI, MAC). Song of ♂ a sharp, dry, tuneless note on one pitch repeated at rate of 3/s, in series usually of 5–25 notes, 'chivi-chivi-chivi . . .'. In duet, ♂ gives series of faster, ringing notes, 'pi-pi-pi-pi- . . .' at rate of 7/s, ♀ joins in with slower, drier chip notes. Also gives dry rattle at rate of 10 notes/s. Song very like that of Buff-throated Apalis *A. rufogularis* but less modulated, with frequency of 3000–3500 Hz vs 3000–4500 Hz in *rufogularis* (Chappuis 1979a, 1980b); to the ear the note of Sharpe's Apalis sounds more uniform, less broken into individual elements. Fledged young give continuous rapid nasal twitter.

General Habits. Inhabits lowland forest, second growth and gallery forest; in Ghana moist evergreen forest, moist semi-deciduous and dry deciduous forests. Mainly in canopy and lower canopy (5–25 m) in Ghana; descends on edges where canopy is discontinuous. Lives in pairs or family parties of 4–5, gleaning insects from fine-leaved trees. In Yapo Forest, Ivory Coast, mostly in mixed-species flocks in canopy, where favours leguminous trees with finely divided bipinnate leaves, e.g. *Piptadeniastrum africanum*, whose leaf structure (very small leaflets with midribs providing perches) possibly favours gleaning activities (Demey and Fishpool 1994). Calls throughout day, including in midday heat.

Food. Insects: small beetles, grasshoppers, larval mantids and caterpillars.

Breeding Habits. Territorial. Displaying ♂ cocks tail, bounces up and down, either on branch, like Bleating Warbler *Camaroptera brachyura*, sometimes pointing bill skywards, stretching neck to full extent, or in jerky flight, with tail fanned (Field 1974).

NEST: 3 described, all inaccessible, placed near top of slender tree in sub-canopy. A bag made apparently of moss, lined with white plant down. Built by both sexes.

EGGS: unknown; clutch probably 3 (3 juvs fed on 5 different occasions).

BREEDING INDICATIONS: Liberia, probably May–June; Sierra Leone, mostly Feb–Mar (height of dry season), but observed nest-building or feeding newly fledged young in all months except July–Oct (height of rains); Ghana, (pair feeding fledglings Apr).

Apalis rufogularis (Fraser). Buff-throated Apalis. Apalis à gorge rousse.

Plate 16
(Opp. p. 257)

Drymoica rufogularis Fraser, 1843. Proc. Zool. Soc. Lond., p. 17; Clarence (= Malabo), Fernando Po.

Forms a superspecies with *A. karamojae*.

Range and Status. Endemic resident, Nigeria to Kenya, Angola and Zambia. Nigeria, rain forest zone west to Lagos and north to Nindam Forest, Kagoro. Bioko; S Cameroon; common and widespread in Gabon. SW Central African Republic, Mbaiki to east of Ubangi R. Angola, Cuanza Norte (Canzele, Canhoca, Ndala Tando, Golungo Alto, Bolongongo), Cuanza Sul (Calulo, Quitondo, Roca Congulu, Gabela), Malange (Quela, Pungo Andongo) and E Lunda (Luachimo R., Cassai Post). SW Sudan in Bangangai Forest and on Aloma Plateau. Zaïre from Tshuapa and Lukolela to N edge of forest zone, Kivu, Mt Kabobo and SW Shaba. NW Zambia in Mwinilunga. S and W Uganda in forest at 700–1800 m in Bwindi, Malabigambo, Kalinzu, Budongo, Mubende, Sezibwa R., Kifu and Mabira, to Mt Elgon. Adjacently in Kenya west of Rift at 1700–2400 m from Elgon and Mumias to N Nandi and Kakamega Forests; and in NW Tanzania (Bukoba, Minziro). Rwanda (Nyungwe Forest); Burundi (Bururi Forest). W Tanzania (Lukolansala R. at 1350 m, Mt Kungwe at 1800–2200 m, Katuma R. near Mpanda). Common to local and uncommon. Density (Gabon, M'Passa) 6–8 pairs/km² (Brosset and Erard 1986) and (Rwanda, Nyungwe) *c*. 2–3 pairs/ha of forest excluding clearings (Dowsett-Lemaire 1990).

Description. *A. r. rufogularis* (Fraser): Nigeria east of Niger R., Bioko, Cameroon, Gabon and Central African Republic. ADULT ♂: forehead, crown, lores, cheeks, ear-coverts, chin, throat and breast and entire upperparts to uppertail-coverts sooty grey-black. Tail with 2 central pairs of feathers uniform blackish brown, 4 outer pairs white (T3 and T4 sometimes with narrow dusky margins near tip and on outer webs). Breast strongly demarcated, more intensely black next to white belly; belly to undertail-coverts creamy white, flanks washed greyish, thighs dirty white. Wings sooty black, underwing-coverts and axillaries white. Bill black, lower mandible grey to greyish white; eyes light brown to brick-red; legs and feet light reddish brown, toes greyer. ADULT ♀: duller above; forehead and crown dark greyish olive, cheeks, ear-coverts and neck grey, occasionally a buff loral spot; mantle and back to uppertail-coverts with dark olive-green wash. Chin, throat and breast chestnut-buff, centre of belly off-white, suffused pale buff, flanks washed greyish, undertail-coverts white. Wings dark olive-green. SIZE: (14 ♂♂, 12 ♀♀) wing, ♂ 45–51 (47·5), ♀ 45–48 (45·6); tail, ♂ 44–54 (48·0), ♀ 39–45 (41·1); bill, ♂ 12–14 (12·6), ♀ 12–14 (12·6); tarsus, ♂ 15·5–17·5 (16·4), ♀ 15–17 (15·7).

IMMATURE: ♂ paler above, washed greenish olive; yellow below with olive wash on sides of breast. ♀ has underparts whitish, tinged pale lemon-yellow, washed grey at sides of breast.

NESTLING: undescribed.

A. r. sanderi Serle: SW Nigeria (Lagos to Ife and Niger R.). ♂ has blackish crown and face and black throat. ♀ like nominate ♀.

A. r. angolensis Bannerman: NW Angola (Cuanza Norte, Cuanza Sul from Calulo to Quitondo, Malange). Both sexes have dark grey-brown crown, contrasting with greyer back and rump; ♂ has whitish chin and throat, breast and whole of underparts creamy white except for grey wash at sides of breast; ♀ has pale reddish buff throat and breast.

A. r. brauni Stresemann: W Angola (escarpment zone in Cuanza Sul: Roca Congulu and Gabela). ♂ like *angolensis* but lighter, more uniform grey on crown and back; chin and throat clearer white; ♀ like ♀ *angolensis* but slightly paler.

A. r. nigrescens Jackson: SW Sudan, Zaïre, NW Zambia, NE Angola (E Lunda), Uganda except Kigezi, NW Tanzania. ♂ has upperparts blackish olivaceous brown; underparts creamy white; ♀ upperparts dusky olive-brown, chin, throat and breast light rufous. Immature washed olive-brown above. ♂♂ in SW parts of range are lighter, more grey-brown. WEIGHT: Uganda, ♂ (n = 26) 6-10 (8·7), ♀ (n = 8) 7-9 (8·0).

A. r. kigezi Keith, Twomey and Friedmann: SW Uganda at 1100–1550 m in Kigezi (Bwindi-Impenetrable Forest). ♂ greyer, less olive-brown above than *nigrescens*. ♀ has grey upperparts; rufous on throat only just reaches breast.

A. r. argentea Moreau (including '*eidos*'): E Zaïre along Albertine Rift at 1500–2350 m: Idjwi Is. in L. Kivu, Rwanda (Nyungwe Forest), Burundi (Bururi Forest) and W Tanzania (Lukolansala R. at 1350 m, Mt Kungwe at 1800–2200 m and Katuma R. near Mpanda). ♂ grey above, whitish below; ♀ variably washed with green on back and wings, uniform whitish below; immature washed green above. Tail 49–55.

TAXONOMIC NOTE: *argentea* (with *eidos*) has been treated as a distinct species (e.g. Hall and Moreau 1970, Collar and Stuart 1985). Dowsett and Dowsett-Lemaire (1993b) state that *argentea* and *eidos* are conspecific with *A. rufogularis*, 'a decision reached independently by Traylor (1986) and Short et al. (1990)'. Only Hall and Moreau (1970) give reasons; nonetheless we follow majority opinion in treating the assemblage as a single species.

Field Characters. Small and short-tailed, with monotonous, dry, grating song often submerged by louder voices around it. Upperparts blackish, outer two-thirds of tail white. In W of range ♂ distinctive, with black throat and breast, but from Zaïre east, underparts entirely white. Could be confused with Grey Apalis *A. cinerea*, with which it occurs in same flocks; whiter below than Grey Apalis, tail shorter and more extensively white; spreads tail (displaying white) more readily than Grey Apalis, e.g. during short flights or chases and when changing direction (Dowsett-Lemaire 1990). ♂ likely to be travelling with rufous-throated ♀ (sexes alike in Grey Apalis). In SW of range overlaps with Brown-headed Apalis *A. alticola*, which differs in dark tail with white only at tips. Rufous-throated ♀ resembles sympatric Gosling's Apalis *A. goslingi*, but latter lacks white in tail. Also overlaps with Chestnut-throated Apalis *A. porphyrolaema*, but latter has red only on throat (not extending to breast), longer tail with white only at tip, trilling song.

Voice. Tape-recorded (68, 86, B, F, ASP, CART, GREG, KEI, LEM). Song of ♂, a dry, burry, grating, tuneless note repeated at rate of 3/s in often lengthy series of up to 30 or more, typically 12–20, but variable; in pauses between longer series may give only a few notes or even just a single one. Elements of each note audible to human ear, giving sense of brief rattle, 'drrrrit-drrrrit-drrrrit . . .'. Sometimes notes have sharper, less burry quality, 'trrreee-trrreee-trrreee . . .', and may sound upslurred, 'trrueee-trrueee . . .' or downslurred, 'beeeu-beeeu . . .'. Song of ♂ typically unaccompanied, but ♀ may join in with higher-pitched, thin, downslurred 'tsyu' or 'tsip', synchronized with and immediately following ♂'s 'brrrr', to produce double-sounding 'brrrr-tsyu, brrrr-tsyu . . .'. ♀ may also accompany ♂ with faster, sharper, higher-pitched 'tee-ti-ti-ti- . . .', ending with buzzy 'dzee-dzee'.

General Habits. Inhabits tall, mature forest, from sea level to 2400 m. Occurs in undergrowth, mid-levels and canopy; also in secondary growth at forest edge. As well as forest, in W Tanzania inhabits bamboo and riverine strips of dry woods where it occurs particularly in canopy of fringing *Albizia*. In Nyungwe Forest, Rwanda, widespread below 2100 m but ascends to 2350 m on drier W slopes. In 12 m tall dry ridge forest there, feeds in canopy foliage and liana tangles with small gaps at mid-levels, commonly in company of Grey Apalis; but unlike Grey Apalis, Buff-throated Apalis does not readily descend along forest edges nor forage below 8 m (Dowsett-Lemaire 1990).

In pairs or family parties of 3–5; flocks of 12–20 seen at Idjwi. Gleans foliage and small twigs; flies out to catch insect; regularly associates with mixed-species parties. Regularly displays white outer tail-feathers, as contact signal between members of group. In Gabon most vocal Apr–Sept.

Food. Insects and their larvae: small beetles, bugs, 3-cm mantis nymphs, ants, small caterpillars; also spiders and seeds.

Breeding Habits. Territorial. Disputes frequent along boundaries, with much calling and display of white outer tail-feathers. At least 3 birds called and displayed in canopy and flew about with much wing-snapping, like Bleating Warbler *Camaroptera brachyura* (W. R. J. Dean, pers. comm.).

NEST: one, described as shallow cup of twigs (Gray 1972), probably belonged to some other species; another, located 3·5 m up in a dense-foliaged tree with many lichens, could not be examined (Vande weghe and Loiselle 1987).

EGGS: undescribed.

BREEDING INDICATIONS: Cameroon, eggs reported Feb, Dec, immatures with adults June; Gabon, young under parental care Mar, June, Sept; Angola, young Mar, breeding condition May–Oct; Zaïre (Itombwe), juvs Jan–Feb, Apr–May, Aug–Sept; Ituri, enlarged gonads Jan, Sept–Oct; Idjwi, probably breeds, June–July; Burundi, young being fed late Mar; E Africa: Region B, Feb–May, July–Aug; Zambia, juv. Sept.

Apalis karamojae (van Someren). Karamoja Apalis. Apalis du Karamoja.

Plate 16
(Opp. p. 257)

Eupirnoides (sic) *karamojae* van Someren, 1921. J. East Afr. Uganda Nat. Hist. Soc. 16, p. 25; Mount Kamalinga, Karamoja, Uganda.

Forms a superspecies with *A. rufogularis*.

Range and Status. Endemic resident. Extreme NE Uganda at Kanatorok (Kidepo Valley Nat. Park), south to Mt Moroto, Mt Kamalinga and Mt Kadam (Debasien); and 720 km to the south in N-central Tanzania (Nzega District) from Nzega to Igunga and Wembere Steppe at Ngongoro and Itumba to Ndala; probably also in NE at Ndutu (Serengeti Nat. Park). Rare or uncommon to locally common; range in Tanzania threatened by habitat clearance.

Description. *A. k. karamojae* (van Someren): Uganda. ADULT ♂: forehead, crown, nape and cheeks ashy grey, ear-coverts darker grey; white stripe from nares to above eye, pre-orbital blackish loral spot. Mantle and back to uppertail-coverts ashy grey. Tail with central feathers (T1) plain blackish, T6 and T5 wholly white except at base, T4 white along outer half, T3 with distal half of inner web white, T2 white-tipped. Underparts off-white, thighs off-white mottled blackish. Wings blackish brown, 4 inner secondaries with outer webs broadly edged white, forming wing flash; underwing-coverts and axillaries white. Bill black; eyes brown; legs and feet black. Sexes alike. SIZE: (2 ♂♂, 1 ♀) wing, 49–50 (49·7); tail, 43–47 (45·0); bill, 15·8–16·8 (16·4); tarsus, 17–19·1 (17·8). WEIGHT: 1 ♀ 9.

IMMATURE AND NESTLING: unknown.

A. k. stronachi Stuart and Collar: Tanzania (Nzega district). ♂ has upperparts darker brownish grey; underparts creamy white, mottled grey, paler on throat and belly, breast and flanks darker grey; ♀ paler than nominate, cream-white below, lightly and variably suffused with grey. SIZE: ♂ ♀ (n = 4) wing, 52–54 (53·3); tail, 47–50 (48·3); bill, 15·3–16·7 (16·2); tarsus, 17·8–19·1 (18·5) (Stuart and Collar 1985).

TAXONOMIC NOTE: we place *A. karamojae* in the *A. rufogularis* superspecies on account of striking similarities in size, proportions, bill form and tail pattern between them (particularly *A. r. argentea*).

Field Characters. Rare and little known, with very restricted range in NE Uganda and N Tanzania. Plain grey with much white in tail and white wing-flash. Not likely to be confused with any warbler, but in N Uganda might just meet Grey Tit-Flycatcher *Myioparus plumbeus*, another grey and white bird with white in tail and with warbler-like feeding habits. Grey Tit-Flycatcher lacks wing-flash and has white in tail confined to outermost feather; it also has characteristic habit of raising and fanning tail continuously while feeding.

Apalis karamojae

Voice. Unknown.

General Habits. In Kidepo Nat. Park, Uganda, inhabits stands of dwarf *Acacia drepanolobium*; elsewhere in Uganda occurs in small trees lining seasonal watercourses. In Tanzania inhabits seasonally flooded ground along Wembere R. with acacias *A. kirkii* and *A. drepanolobium*, also *A. mellifera* mixed with *Commiphora ugogoensis*, and bushy growth, especially on hard, imperfectly drained sodic soils. Such suitable habitat appears to stretch in narrow belt for 175 km along Wembere River.

Reported to move about in small foraging parties.

Food. Unknown.

Breeding Habits. Unknown.

References
Collar, N. J. and Stuart, S. N. (1985).
Stuart, S. N. and Collar, N. J. (1985).

Apalis bamendae Bannerman. Bamenda Apalis. Apalis du Bamenda.

Plate 16
(Opp. p. 257)

Apalis bamendae Bannerman, 1922. Bull. Br. Orn. Club 42, p. 131; between Bemba (= Bamenda) and Chang (= Dschang), Cameroon.

Range and Status. Endemic resident. Known only from very few specimens from SW and central Cameroon, in scattered localities in Bamenda Highlands (Banso, Bamali, near Bamenda and

Apalis bamendae

Dschang, Bali and Mt Nkogham; apparently absent from Bafut-Ngemba Forest Reserve and Mt Oku); Adamawa Plateau (Tello and Ngaoundaba). Uncommon or rare.

Description. ADULT ♂: forehead brown, washed pale rufous; cheeks, lores and ear-coverts greyish with rufous wash, area surrounding eye pale rufous. Crown, nape, mantle and back to uppertail-coverts grey with olive-brown wash. Tail dark brown. Chin and throat pale buffy rufous; breast to belly grey, washed fulvous, paler in centre, flanks darker buff, thighs chestnut, undertail-coverts white. Wings grey-brown, secondaries and upperwing-coverts with paler margins; underwing-coverts and axillaries white, washed buff. Bill black; eyes pale yellowish brown; legs and feet flesh. Sexes alike or almost so. SIZE: (2 ♂♂, 2 ♀♀) wing, ♂ 54, 55, ♀ 50; tail, ♂ 44, 46, ♀ 36, 37; bill, ♂ 12·5, 13·5, ♀ 12·5, 14; tarsus, ♂ 17, ♀ 16·5, 18.
　IMMATURE AND NESTLING: unknown.
　TAXONOMIC NOTE: the specific status of *bamendae* has been controversial. Mackworth-Praed and Grant (1960) thought it conspecific with the long-tailed *A. chapini* of E Africa, but similarities appear convergent. White (1962) and Hall and Moreau (1970) treated it as conspecific with short-tailed *A. sharpii* and *A. goslingi* which, with *A. rufogularis*, appear to be its closest relatives. Traylor *in* Watson *et al.* (1986) retained it as a species and this is the course adopted here, supported by Chappuis (1979a) on vocal evidence.

Field Characters. Grey above, tail short and dark, with no white. The only apalis with rufous face and throat in its very limited range in Cameroon highlands. ♀ Buff-throated Apalis *A. rufogularis* has reddish throat but shows much white in tail, and is a lowland bird in Cameroon. Much commoner Grey Apalis *A. cinerea* has grey-brown face and pale buff underparts, white outer tail, different voice.

Voice. Tape-recorded (68, CHA, STJ). Song, variable repeated phrases which almost always begin with a single downslurred note: 'chyee-pipi, chyee-pipi, chyee-pipi'; 'tyu-wee-wee-wee-wee-wee'; or a faster, rattling 'chyu-titititititititititititi'. Notes are sharp and tuneless. ♂ may sing alone, or ♀ may join in with irregular squeaky notes, or in the case of the rattling song, with a similar rattle. For sonograms and discussion of the songs of this and related species, see Chappuis (1979a, 1980b).

General Habits. Inhabits forest, second growth and isolated trees near forest, at intermediate elevations, 1100–1600 m; absent from higher altitude montane forests. Frequents forest patches, gallery forest, riverine thickets and forest relicts in farmland; also recorded in very narrow belt of 10–15 m high trees in savanna country, so dependence on forest is uncertain (Stuart 1986a). Forages high up, moving around in small parties; occasionally solitary.

Food. Insects.

Breeding Habits. Unknown.

Reference
Stuart, S. N. (1986a).

Plate 16
(Opp. p. 257)

Apalis goslingi Alexander. Gosling's Apalis. Apalis de Gosling.

Apalis goslingi Alexander, 1908. Bull. Br. Orn. Club 21, p. 89; Guruba (= Gurba) R., Upper Uele, Belgian Congo.

Range and Status. Endemic resident. Common to uncommon in lowland forest from S Cameroon (Dja R.) and NE Gabon (Makokou, Ivindo R.) to Central African Republic (Manovo-Gounda-Saint Floris Nat. Park and Lobaye Préfecture), NE Angola in N Lunda (Luachimo R.) and Zaïre (Equateur and Kinshasa to Uele R. and Ituri; Kasai at Chikapa).

Description. ADULT ♂: forehead, crown, sides of neck and mantle to back and uppertail-coverts dull brownish grey, somewhat paler and more pure grey on rump. Small blackish grey patch on lores and around and behind eye. Tail-feathers greyish brown, with narrow (less than 1 mm) buffy or whitish tip variably present on T5–T6. Chin and throat whitish to pale buff; rest of underparts light grey, almost whitish in centre of breast, belly and undertail-coverts. Wings greyish brown, basal

Apalis goslingi

half of underside of inner webs of inner primaries and secondaries edged white; axillaries and underwing-coverts white. Bill black; eyes light yellowish brown or red, eyelids with slight reddish tinge; legs and feet pinkish buff. ADULT ♀: like ♂ but lacks dark face patch. SIZE: (3 ♂♂, 4 ♀♀) wing, ♂ 47–48 (47·7), ♀ 43·5–45 (44·2); tail, ♂ 40–42 (41·7), ♀ 31–33 (32·5); bill, ♂ 12·5–13·5 (12·8), ♀ 11·5–13 (12·2); tarsus, ♂ 16–17·5 (16·8), ♀ 15·5–16·5 (15·9).

IMMATURE: upperparts with greenish wash; throat pale yellow, rest of underparts washed greenish yellow, especially on centre of belly and undertail-coverts. Iris and rim of eyelids dull greenish grey; bill dusky greenish, corners of mouth yellow; legs and feet pale buff.

NESTLING: unknown.

Field Characters. A short-tailed, nondescript apalis largely confined to the vicinity of forested rivers. Mainly dull grey, paler below, ♂ with black face patch; tail dark except for narrow pale tips to outer feathers. Overlaps broadly with Buff-throated Apalis *A. rufogularis*, which has white underparts (throat black or white in ♂, reddish in ♀) and large amount of white in tail. Voices very different.

Voice. Tape-recorded (68, ERA). Song rapid, a single sharp note repeated at rate of 7/s to 10/s, the latter almost a rattle. For detailed discussion see Chappuis (1979a, 1980b). Very different from the slow, dry notes of Buff-throated Apalis.

General Habits. Occurs principally along rivers and streams in lowland forest; in Gabon each pair occupies 400–800 m of river frontage; also forest edge, secondary forest and regenerating growth. In pairs or small family parties, gleaning insects from leaves and twigs; searches clumps of dead vegetation lodged 10–15 m up in trees. In Gabon calls most actively Apr–Aug.

Food. Insects: beetles, 15–20 mm orthopterans, small mantids, caterpillars, eggs; spiders.

Breeding Habits. Territorial; territories in Gabon often marked by boundaries along streams; 60–70 m of river frontage vigorously defended by pair. Nest and eggs undescribed.

BREEDING INDICATIONS: Cameroon, testes enlarged Jan; Gabon, probably Apr–Aug; Zaïre, breeding condition Apr, Aug; young Aug, Oct.

Reference
Brosset, A. and Erard, C. (1986).

Apalis cinerea (Sharpe). Grey Apalis. Apalis cendrée.

Plate 16
(Opp. p. 257)

Euprinodes cinereus Sharpe, 1891. Ibis, p. 120; Mt Elgon.

Forms a superspecies with *A. alticola*.

Range and Status. Endemic resident. 3 populations: (1) SE Nigeria at 1300–2250 m (Obudu and Mambilla plateaux and Gotel Mts); SW Cameroon at 850–2900 m (Mt Tchabal Mbaba, Bamenda Highlands, Mt Oku, Rumpi Hills, Mt Kupé, Mt Manenguba and Mt Cameroon); Bioko; and at about 1000 m in NE Gabon (Belinga, M'Passa). In Cameroon, the most abundant and widespread apalis. (2) W Angola in Cuanza Sul, Huambo, SE Benguela (Mts Soque and Moco, Mucuio) and N Huila (Chingoroi, Sá da Bandeira, Serra de Chela and Lubango); and (3) E Zaïre along Albertine Rift from mountains northwest of L. Albert (Lendu Plateau up to 1800 m) to highlands west of L. Edward (Litunguru and Bitakongo up to 1800 m), Itombwe and Mt Kabobo; apparently absent from Rwenzori and from Kivu; S Sudan above 1830 m (Imatong, Didinga and Dongotona Mts); NE Uganda (Mt Lonyili); S and W Uganda at 1200–2400 m (Budongo to Kibale, Kibirau, Mpumu and Bwindi-Impenetrable Forests); Kenya highlands at 1700–3000 m from Mt Elgon, Kakamega, Mau Forest and Kericho to Meru, Nairobi and Nguruman Hills, with isolated populations on Ndotos, Mt Marsabit and Mathews Range; N and NW Tanzania at 1200–2200 m (Loliondo and Kahambwe R.); Rwanda (Nyungwe and Cyamudonga Forests); and Burundi (Bururi, Teza and

Apalis cinerea

Ijenda Forests). Common. In Nyungwe Forest (Rwanda) density av. 2–3 pairs/ha, not counting clearings (Dowsett-Lemaire 1990).

Description. *A. c. funebris* Bannerman: Nigeria and Cameroon (except Mt Cameroon). ADULT ♂: forehead, crown and nape dark grey, cheeks and ear-coverts sooty grey-brown. Mantle and back to uppertail-coverts sooty grey-brown. Central tail-feathers dusky brown, T2 tipped white, T3 tipped white and white along shaft, 3 outermost feathers (T4-T6) wholly white. Chin, throat, breast and belly pale buff, centre of belly whiter, undertail-coverts off-white, flanks washed greyish, thighs sooty brown. Wings sooty grey-brown, underwing-coverts and axillaries white. Bill black; eyes rufous-brown, rim of eyelids dull pinkish brown; legs and feet pinkish brown, claws dusky brown. Sexes alike. SIZE: (20 ♂♂, 20 ♀♀): wing, ♂ 52–60 (56·1), ♀ 50–55 (53·0); tail, ♂ 52–65 (60·3), ♀ 48–56 (51·6); bill, ♂ 13·5–15·5 (14·3), ♀ 13–14 (13·4); tarsus, ♂ 18–20·5 (19·2), ♀ 18–19·5 (18·5).

IMMATURE: washed dark olive above; secondaries and wing-coverts olive-brown; underparts washed yellow, brightest on chin and throat.

NESTLING: hatches naked; skin flesh-pink.

A. c. sclateri (Alexander): Mt Cameroon and Bioko. Crown grey-brown; underparts creamy buff, lacking dusky wash.

A. c. grandis Boulton: W Angola. Crown lighter than *funebris* and *sclateri*, bird otherwise like *cinerea* but sooty grey above and whiter below. Larger: (9 ♂♂, 9 ♀♀) wing, ♂ 60–63 (61·9), ♀ 55–60 (57·8); tail, ♂ 63–73 (70·0), ♀ 56–59 (57·6).

A. c. cinerea (Sharpe): Zaïre eastwards. Crown grey-brown, contrasting with sooty grey back; whiter below; most closely resembles *grandis* but similar in size to *funebris*.

WEIGHT: Kenya and Uganda, ♂ (n = 10) 8–12·5 (10·5), ♀ (n = 5) 7–11 (9·3).

Field Characters. A common bird of montane forest whose dull plumage would cause it to be overlooked but for its insistent song. Plain grey or grey-brown above, whitish below, but less contrastingly 2-toned than either Black-headed Apalis *A. melanocephala* or ♂ Buff-throated Apalis *A. rufogularis*, which are neatly black-and-white. Has much white in tail like Buff-throated Apalis but tail longer; Black-headed Apalis has equally long tail but white only at tips; voices of all 3 distinctive. For differences from Brown-headed Apalis *A. alticola*, see that species.

Voice. Tape-recorded (32, 68, B, C, F, LEM). Song of ♂ a dry 'bek-bek-bek-bek . . .' or a sharper 'bik-bik-bik-bik . . .' repeated monotonously up to 50 times, more typically 10–20, at rate of 4/s, sometimes 5/s. ♂ often sings alone, but ♀ may accompany him in latter part of song; they also sing in asynchronous duet, which either bird may start; while ♂ gives 'bek-bek-bek . . .', ♀ joins in either with some irregular tikking notes or with a high-pitched trill followed by a few sunbird-like 'tsip's, 'tirrrrrrrrrr-tsip-tsip-tsip'. This trill is somewhat reminiscent of Chestnut-throated Apalis *A. porphyrolaema*.

General Habits. Inhabits montane forest, usually canopy, also mid-stratum; liana tangles, edges (to within a few m of ground) and second growth; occasionally thickets, forest galleries, riparian fringes, and canopy of nearby woodland. Favours acacia trees on forest edge in Sudan, montane forest along escarpment and in highlands in Angola.

In Nyungwe Forest (Rwanda) the most numerous canopy apalis, up to 2100 m, very aggressive and noisy, even when moulting in Oct–Nov, attacking 'intruder' in reaction to tape-recording with noisy wing-beats and bill clipping, with tail spread and slightly raised, as it hops around calling loudly. Countersings with other species, especially Black-throated Apalis *A. jacksoni* and interacts with Buff-throated Apalis (Dowsett-Lemaire 1990). Occurs in pairs, but usually in family parties of 4 to 6 or 8; rarely singly. Moves restlessly through vegetation, usually at considerable height. Movements agile, graceful, with much flicking and fanning of tail. Calls from highest branches, usually in early mornings, late afternoons and after showers. Searches leaves and small twigs like *Phylloscopus*, picking off insects; also makes short sallies, catching insects in flight. Joins mixed bird parties; in Cameroon often with the Green Longtail *Urolais epichlora*. Sedentary.

Food. Insects (beetles, small caterpillars) and small spiders.

Breeding Habits. Territorial throughout year.

NEST: a rather bulky, elongated ball with side-top entrance, built almost entirely of *Usnea* lichen and bits of tree moss, bound together with spider webs and fine bark fibres; thickly lined with feathers and some vegetable down. Ext. diam. 180, ext. depth 100 from lower edge of entrance, int. diam. 90. Nests are placed high (6–10 m) up on outer aspect of tree, with entrance facing outward. Both sexes build, taking about 14 days, but construction does not go on throughout the day.

EGGS: 3. Elongate, pale blue, finely speckled with discrete red-brown spots, with tendency for heavier markings at broad end. SIZE: (n = ?) av. 15 × 11.5 mm.

LAYING DATES: Cameroon, Feb–Mar (juvs Apr, June); Sudan, Feb, July; E Africa: Region B (adult feeding fledgling June); Region C, Jan–Feb; Region D, Apr (also Mar–July, Dec); Zaïre (Itombwe) (breeding condition Mar, Sept, juvs July–Aug); Rwanda, Sept–Jan; Angola, commences Sept.

INCUBATION: almost entirely by ♀. Period: 13–14 days.

DEVELOPMENT AND CARE OF YOUNG: young fed from early morning until 11.00 h when rate slackens; another peak at 16.30–17.00 h, when adult visits every 5 min. When young hatch, encapsulated faeces are swallowed; faeces carried off later as young develop. Nestling period: 15–16 days.

BREEDING SUCCESS/SURVIVAL: of 8 nests closely observed 5 were attacked by predators and only 3 were successful.

Reference
van Someren, V. G. L. (1956).

Apalis alticola (Shelley). Brown-headed Apalis. Apalis à tête brune.

Plate 16 (Opp. p. 257)

Cisticola alticola Shelley, 1899. Bull. Br. Orn. Club 8, p. 35; 'Nyasaland', i.e. Fife (= Isoka), Zambia.

Forms a superspecies with *A. cinerea*.

Range and Status. Endemic resident. SW Kenya (Nguruman Hills); Tanzania at 1200–2200 m in north (Oldeani, Loliondo (Salempo 1994), Crater Highlands and Mbulu), re-appearing in S Highlands from Ufipa Plateau (Sumbawanga), Mbeya Mt, Umalila, Poroto and Rungwe Mts, Njombe, Iringa and Udzungwa Mts south to Matengo Highlands; N Malaŵi at 1600–2050 m (Misuku Hills, Mafinga Mts, Nyika Plateau, Jembya Plateau and Musisi Hill); N Zambia in Mafinga and Makutu Mts, Nyika and west of Luangwa Rift south to Danger Hill, west through Solwezi and Kasempa to Mwinilunga, and into adjacent S Zaïre in Shaba, Marungu Plateau and Upemba; central and N Angola in Cuanza Norte, NW Malange (Duque de Bragança and Cuanza R.), S Lunda (Cacolo and Cassai Post) and N Bié. Common to local and uncommon. In Malaŵi on SW Nyika Plateau, pairs occupy small forest patches of 0·5 ha at 1925 m, with densities of *c.* 3 pairs/10 ha at *c.* 2000 m (Dowsett-Lemaire 1989).

Description. *A. a. alticola* (Shelley): range of species except for *dowsetti*. ADULT ♂: forehead, crown, nape, cheeks, lores and ear-coverts dark brown. Hindneck, mantle and back to uppertail-coverts grey-brown. Tail grey-brown, all feathers but central pair tipped white. Underparts creamy white, flanks washed grey. Wings grey-brown, underwing-coverts and axillaries white. Bill blackish, interior of mouth black; eyes pale orange to raw sienna; legs and feet pinkish brown. Sexes alike. SIZE: (34 ♂♂, 30 ♀♀) wing, ♂ 55–64 (58·2), ♀ 52–58 (54·7); tail, ♂ 55–62 (58·4), ♀ 46–54 (48·5); bill, ♂ 13–15·5 (14·4), ♀ 13–14·5 (13·6); tarsus, ♂ 19–22 (20·6), ♀ 18–20 (18·8). WEIGHT: Angola, ♂ (n = 3) 11·5–13 (12·5), ♀ (n = 2) 10, 11·5; Zaïre, ♂ (n = 9) 10–15 (12·2), ♀ (n = 4) 10–15 (11·8); Malaŵi, ♂♀ (n = 10) av. 10·4.

IMMATURE: upperparts and wings washed olive-green, crown duller, not contrasting with back (but head darkens rapidly); a slight yellowish supercilium; chin to belly and undertail-coverts yellowish.

NESTLING: undescribed.

A. a. dowsetti Prigogine: Marungu Plateau. ♂ has outermost tail-feather white and T5 with 20–24 mm apical white area; ♀ like nominate ♀.

TAXONOMIC NOTE: *A. alticola* was treated as conspecific with *A. cinerea* by Dowsett and Dowsett-Lemaire (1980), Short *et al.* (1990) and Dowsett and Forbes-Watson (1993), primarily because of vocal similarity, and as a separate species by Traylor (*in* Watson *et al.* 1986) and Sibley and Monroe (1990). Recent discovery that the 2 are sympatric in S Kenya and N Tanzania reinforces our belief that they should be considered different species.

Field Characters. The southern representative of the Grey Apalis *A. cinerea*, which it closely resembles in both plumage and voice. The 2 overlap in S Kenya and N Tanzania. Top of head dark brown, rather than grey-brown in Grey Apalis, but main difference is in tail, dark with white tip in Brown-headed, outer third pure white in Grey. Race of Brown-headed on Marungu Plateau, Zaïre, has outermost tail-feather white, but Grey Apalis not present there. Black-and-white ♂ Buff-throated Apalis *A. rufogularis* (sympatric in SE Zaïre/NW Zambia) has shorter tail with even more white than Grey Apalis, different voice.

Voice. Tape-recorded (86, B, F, CHA, KEI, LEM, PAR). Extremely similar to voice of Grey Apalis, q.v. Song of ♂ a sharp, percussive, dry note monotonously repeated at rate of 3–4/s. ♂ may sing alone, or ♀ may join in with hurried tikking notes and brief trills.

General Habits. Inhabits montane and gallery forest. Habitat preference varies regionally, strongly influenced by congeners. In Kenya (Nguruman Hills) occurs on forest edge (Grey Apalis in interior), and is in apparent vocal competition there with Chestnut-throated Apalis *A. porphyrolaema* (Turner 1992). In N Tanzania co-exists widely with Bar-throated Apalis *A. thoracica* in forest in Crater Highlands and Oldeani, whereas in S Tanzania (Njombe) lives on forest edge while Bar-throated Apalis is in the interior. In Malaŵi, where thoroughly studied, occupies lower forest storey at Musisi, Jembya and Mafinga and forest regrowth a few m high in Misukus; on SW Nyika at 1925 m occupies small patches of forest of 0·5 ha with some territories defended against Chapin's Apalis *A. chapini*; rare above 2070 m. Countersings with Chapin's Apalis at Nyika, Jembya and Musisi; in Misuku Hills almost totally excluded from forest by congeners. In Mugesse Forest (where Bar-throated Apalis absent) occurs below 1700 m in all forest strata and secondary growth; in Wilindi-Matipa (where Bar-throated present) it is confined to secondary growth and (locally) forest edges. Clearly outnumbers Chapin's Apalis in Mafinga Mts between 1600 and 1800 m, and in riparian forest at 1450 m at foot of Mafinga Mts and Misukus. In N Zambia, S Zaïre and Angola occupies mid-stratum of gallery forest, also rich dense thickets; in N Mwinilunga (Zambia) Buff-throated Apalis may locally exclude it.

Gleans foliage, searching twigs and leaves, and catches insects on wing. In pairs, or small family parties after breeding.

Food. Insects (small beetles, caterpillars); also spiders.

Breeding Habits. Territorial throughout the year. On Nyika Plateau (Malaŵi) occupies forest patch as small as 0·5 ha. Interspecifically territorial with Chapin's Apalis when each exclusively occupied adjacent forest patches of 0·9 and 1·2 ha separated by 40 m of grassland; both countersang from their respective patches. On two occasions one pair crossed into the other's territory, which resulted in vigorous singing and interspecific chases in canopy (Dowsett-Lemaire 1989).

Nest and eggs undescribed.

BREEDING INDICATIONS: Malaŵi, probably Mar; Zambia, gonads enlarged Sept, Dec, recently fledged young Mar, Dec.

Reference
Dowsett-Lemaire, F. (1989).

Genus *Malcorus* A. Smith

Plumage distinctive: upperparts streaked, face reddish chestnut, and a narrow blackish breast-band. Tail of 10 feathers, long, very strongly graduated, outermost pair extremely short (occasionally almost vestigial), unpatterned (**A**). Bill fine, as in *Prinia*, black, with vestigial rictal bristles. Lifts tail when running about on ground but neither cocks it over back nor swivels it. Nest a bulky, unwoven structure with side-top entrance, made of dry grass and lined before egg-laying; eggs immaculate, very pale blue.

Endemic, monotypic, confined to arid W southern Africa in dry scrubby vegetation. Often placed in genus *Prinia* but may not be at all closely related, differing in plumage, largely terrestrial habits, and in nest.

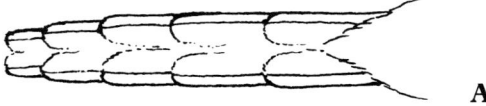

A

Malcorus pectoralis Smith. Rufous-eared Warbler. Prinia à joues rousses.

Plate 14
(Opp. p. 241)

Malcorus pectoralis A. Smith, 1829. South Afr. Commercial Advertiser 4 (27 June); Karroo country, north of Oliphants R. = Bitterfontein, Cape Province, South Africa.

Range and Status. Endemic resident from W Cape Province to W Orange Free State, SW Transvaal, SW Botswana, and northwards in Namibia to Ovamboland. Absent from Namib Desert. Common in arid scrubby vegetation. Very numerous in Karoo vegetation. 1 record W Lesotho (P. Osborne and B. Tigar, pers. comm.).

Description. *M. p. pectoralis* Smith: interior W Cape from about Mamre to upper Great Fish R., Griqualand West and SW Orange Free State. ADULT ♂: forehead, crown, nape and hindneck greyish brown, streaked dark reddish brown, feathers margined buffy white. Lores, feathers around eyes, cheeks and ear-coverts bright rusty-chestnut. Mantle and back to uppertail-coverts greyish brown, heavily streaked blackish brown, feathers margined buff. Tail brown, very strongly graduated, outermost feather very small, c. 40–56 mm shorter than longest (central) pair, outer 3 feathers margined buff along outer webs. Chin and throat white, breast greyish white, blackish breast-band 4 mm wide, extending to sides of neck; centre of belly and flanks pale buff; undertail-coverts buff with dark centre streaks, thighs brownish buff. Primaries and secondaries dark brown, edged rusty buff on outer webs, wing-coverts streaked and edged buff; underwing-coverts and axillaries pale buff. ADULT ♀: like ♂ but face duller rusty-red; breast-band narrower. Bill black; eyes reddish hazel; legs and feet pinkish flesh. SIZE: (22 ♂♂, 15 ♀♀) wing, ♂ 49–53 (51·4), ♀ 48–51 (49·4); tail, ♂ 66–79 (74·0), ♀ 62–77 (67·0); bill, ♂ 11–13 (11·8), ♀ 11–12·5 (11·6); tarsus, ♂ 19·5–21·5 (20·1), ♀ 19·5–21 (20·0). WEIGHT: ♂ (n = 9) 10·3–11·8 (10·7), ♀ (n = 4) 9·2–10·5 (9·6).

IMMATURE: resembles adult but face paler chestnut, throat more clearly white; breast-band paler and narrower (or almost absent).

NESTLING: hatches naked; skin pink, mouth lining deep yellow, tongue with 2 lateral black spots.

M. p. ocularius (Smith): N Cape in Gordonia, Kuruman and Bechuanaland districts, NW Orange Free State, extreme SW Transvaal, SW Botswana and Namibia north to S Damaraland (Erongo). Much paler; crown to nape rusty chestnut, white feather margins with silvery appearance, back paler, narrowly streaked, feathers extensively edged sandy buff; 2 outermost tail-feathers predominantly buff; face lighter tawny, underparts whiter than *pectoralis*, less contrast between throat and belly, flanks more clearly washed pale buff. Breast-band narrower, in freshly moulted post-breeding plumage much reduced, sometimes vestigial or restricted to broken circle of tawny spots. Immature has back more heavily streaked.

M. p. etoshae (Winterbottom): Namibia: N Damaraland (Windhoek) north to Etosha Pan. Even paler than *ocularius*, less reddish above, feather edges more yellowish-tinged, wider grey margins to crown-feathers; whiter still below, sometimes with pale rufous wash.

Field Characters. Prinia-shaped, with long tail, but upperparts streaked, unlike all southern prinias; reddish face and black breast-band complete distinctive plumage. Black-chested Prinia *Prinia flavicans* has black breast-band in breeding plumage but no red on face, upperparts unstreaked, underparts yellowish. Feeds mainly on the ground, running from one patch of cover to another, with tail cocked but not swivelled.

Voice. Tape-recorded (88, 91, F). Song, prinia-like series of single notes, downslurred, usually with buzzy quality, 'deea-deea-deea-deea . . .', sometimes more pure, 'seeu-seeu-seeu-seeu . . .'; repeated at rates of 4/s to 6/s, occasionally 2/s. Alarm call near nest, plaintive 'peeeee'; also gives a quiet 'chit'. Begging call of chicks, wheezy 'swee-swee'.

General Habits. Frequents low, scrubby bushes in arid, sandy and semi-arid shrubland; bushy hill sides in E of range, scrub cover on Namib Desert edge, particularly along drainage lines in more open places. In Kalahari, dune troughs covered with *Rhizogum* scrub and *Monechma incana* < 1 m tall, also dune crests with scrub cover; at Okauejo (Namibia), open grassland with scattered low bushes.

In pairs or small family parties after breeding; occasionally solitary. Forages low down; 72% of food items taken on ground and 26% in bushes (Dean 1976); gleans from twigs and small leaves. Cocks tail over back but never swivels it; bounds quickly along ground from one patch of cover to another, tail raised. Calls from vantage point, like typical prinia.

Food. Insects: shield bugs (Pentatomidae), plant-hoppers (Cicadellidae), Coleoptera (Curculionidae, Cassidae, small ground beetles: Brachininae or Lebiinae), Isoptera (particularly Hodotermitidae), Hymenoptera (Formicidae), Lepidoptera and caterpillars; also spiders. Fruits, plant material and seeds: 1 stomach contained 16 *Lycium* seeds (W. R. J. Dean, pers. comm.).

Breeding Habits. Territorial. ♂ advertises territory with penetrating calls from top of bush. Simple apparent display consists of ♂ perched a little above ♀, jerking body and tail, fluttering wings, ♀ responding with wing-flicking.

NEST: in Kalahari, untidy oval, with *c.* 34 mm diam. side-top entrance, lacking porch, built of dry, unwoven greyish grass stems and leaves; lined before egg-laying with soft vegetable down. Placed in wide variety of low-growing shrubs, in Kalahari often in *Rhizogum* bushes, 0·2–1·22 (av. 0.54) m above ground. Nest-entrance often faces E (9 of 27 nests). In S Karoo nest often built of dry strips of grey bark of *Asclepias buchenavianus*, lined with fluffy seeds of *Eriocephalus* spp.; often neat, (resembling that of *Apalis*), sited mostly in woody vegetation, shrubs of *Pteronia pallens, Osteospermum sinuatum, Galenia fruticosa* and *Rosinia humilis*, also in succulents such as *Malephora lutea*.

EGGS: 1–7, usually 3–5 (av. 3·82). Very pale blue (almost white), unmarked. SIZE: (n = 53) 14·2–17·1 × 10·5–12·3 (15·5 × 11·4).

LAYING DATES: Botswana, Jan–Mar, July, Oct; South Africa, breeds opportunistically after rains at any season, e.g. Transvaal, Oct; Orange Free State, Mar; N Cape, July, Sept–Mar, May; Karoo, July–Aug, Oct–Nov, Jan–Feb; E Cape, Sept; SW Cape, Aug–Sept.

INCUBATION: role of sexes undetermined. When disturbed, incubating bird drops straight to ground, hopping away rapidly. Period: 12–13 days.

DEVELOPMENT AND CARE OF YOUNG: pin feathers well formed by day 6, bird fully feathered and ready to leave nest at day 11 when it resembles adult except for short tail and indistinct (or absent) breast-band. Young fed by both sexes, mainly on grasshoppers and caterpillars. Period: 11–13 days. Fledged young remain motionless in shrub when approached by person to within 1 m. Period of dependency on adults unknown.

BREEDING SUCCESS/SURVIVAL: eggshells once found below abandoned nest appeared to have been regurgitated by egg-eating snake *Dasypeltis*.

Reference
Maclean, G. L. (1974).

Genus *Urorhipis* Heuglin

Bill like that of *Apalis*. Tail long, black, outer feathers tipped white; often held vertically and waved from side to side. Upperparts grey-brown with reddish forehead, underparts buffy white. Habitat semi-arid scrub and woodland. Song a repeated chirping note; often several birds join in to produce a noisy chorus. Nest a neat, long pocket, open or domed, with side entrance near top, placed in low bush.

Endemic, monotypic. Has been placed in *Apalis* (Watson *et al.* 1986) and *Spiloptila* (Hall and Moreau 1970, Short *et al.* 1990, Dowsett and Forbes-Watson 1993), which may be closer to *Prinia* than *Apalis* (Short *et al.* 1990). Differs from *Spiloptila* in feather pattern and structure and bill shape, and we follow Irwin (1989) in placing it in its own genus.

Urorhipis rufifrons Rüppell. Red-fronted Warbler. Apalis à front roux.

Prinia rufifrons Rüppell, 1840. N. Wirbelt. Vög.; p. 110; Eritrea.

Range and Status. Endemic resident. Rare E Chad (recorded 20 km east and 10 km south of Abéché); in Sudan uncommon between 13°N and 20°N (Darfur, Kordofan, Khartoum, Port Sudan and Red Sea Province), rare in extreme SE on Natoporoputh Hills and along Kenya border in dry acacia scrub; locally common Ethiopia (NE, S, SW, SE and Rift valley) and Djibouti (northern side of Gulf of Tadjoura mainly on coastal plain, also Goula, Tôha, Obock and Arta); common and widespread in Somalia (south to *c.* 1°N); rare in NE Uganda (between Iriri and Kang'ole, S Karamoja); locally common in N Kenya (Ileret, Turkana region, Marsabit, Wajir south to Kerio valley, Wamba, Isiolo, Garissa, Kozibiri Hill and L. Baringo, east to Lorian Swamp, also Kora Nat. Park at Mwitimba and Mansumbi) and SE Kenya (Tana R. to Tsavo region); a disjunct population in S Kenya and NE Tanzania from Olorgesailie to L. Magadi, L. Natron, foothills of North Pare Mts, Mkomasi Game Reserve and Mombo; also recorded Jangalo southeast of Kondoa, Tanzania, but absent from Serengeti Nat. Park. Absent from semi-arid coastal strip in Kenya.

Urorhipis rufifrons

Description. *U. r. rufifrons* Rüppell: Chad, N Sudan, NE Ethiopia, Djibouti and NW Somalia. ADULT ♂: forehead pale red-brown, crown to rump smoke-grey, uppertail-coverts darker. Tail strongly graduated and fan shaped (ratio of length of T1 to T6 (n = 5) 1·66–2·05 (1·80): M. P. Adams, pers. comm.), feathers dusky brown, darker underneath, with white tips to all but inner pairs (T1 and T2, in some T2 has white tip), amount of white depending on wear and extending along webs of other feathers, but extensive on outer web of T6. Lores and orbital feathers minutely flecked white on black in some individuals, others have thin light brown orbital ring of bare skin; cheeks and ear-coverts smoke-grey. Chin, throat, breast and central belly silky buff-white, lower belly and sides washed with buff; undertail-coverts white; thighs pale tawny. Many SE Kenya birds have thin band of buff feathers, each with dark centres, across breast (van Someren 1922, Traylor 1967b). Primaries, secondaries and wing-coverts pale brown, margins of outer web of primaries paler, outer web of inner secondaries and wing-coverts thinly edged grey-white, more so in some populations than others; underwing greyer, underwing-coverts similar but mottled with brown, axillaries white. Bill black in most birds but lower mandible paler with darker tip in some; mouth greyish black; eyes light brown, yellowish brown or red-brown; legs and feet dull yellow, yellowish brown or light brown, soles white. Sexes alike, except mouth of ♀ pale flesh. SIZE: (20 ♂♂, 20 ♀♀) wing, ♂ 44–48·5 (46·2), ♀ 39–46 (43·7); tail, ♂ 53–61 (56·5), ♀ 48–56 (52·4); bill to feathers, ♂ 10–11 (10·6), ♀ 10–11 (10·5); tarsus, ♂ 19–21 (19·8), ♀ 19–20 (19·7). WEIGHT: unsexed (n = 8, Kenya) 5–7·5 (6·7).

IMMATURE: like adult but generally duller, forehead washed with pale chestnut, less white on tail.

NESTLING: not described.

U. r. smithi (Sharpe): S Sudan, SW and SE Ethiopia, Somalia (except NW border), Uganda, Kenya (except Tsavo), Tanzania. Darker grey above, chestnut on head extends over crown, white edges to inner secondaries and wing-coverts broader than in *rufifrons*.

U. r. rufidorsalis (Sharpe): SE Kenya (Tsavo). Darker grey above, rufous of crown extends onto mantle, and breast feathers have dark centres (often obscured in ♀, and may be absent in young birds: van Someren 1922, Traylor 1967b).

Field Characters. A small, long-tailed warbler with pale red-brown forehead and plain grey-brown upperparts contrasting with blackish brown tail, which has white tip to all but inner 2 pairs of feathers and outer web of outer pair white; narrow, greyish white wing-bar; creamy white underparts. Long, graduated, fan-shaped tail, half total length of bird, and white wing-bar are diagnostic. Characteristically wags tail, and flits from bush to bush much like tits (Paridae). Calls not unlike Tawny-flanked Prinia *Prinia subflava*. Unlike any other warbler found in similar habitat. In Darfur occurs alongside Cricket Warbler *Spiloptila clamans*, which has mottled crown and wing-coverts and cinnamon upperparts, but Red-fronted Warbler is mainly at higher elevations (Lynes 1925).

Voice. Tape-recorded (B, GREG, McVIC, PAY, STJ). Song a clear chirping repetition of 'tick': 'tick tick tick tick tick . . .'. Contact call a single thin note 'tseep' lasting *c*. 0·1–0·2 s and repeated 15 times in 18 s (Somalia): 'tseep-tseep-tseep . . .'. Of 3 birds moving excitedly in tops of 3 m high thorn trees (Kenya: in territorial dispute?), 1 uttered 2–3 staccato notes 'tick-tick' in quick succession at irregular intervals, which sometimes merged into one continuous noisy racket as the others joined in with a series of 'tcheepe' notes: 'tcheepe-tcheepe-tcheepe . . .' repeated *c*. 15 times in 13 s, the whole lasting up to 60 s or more. The calls and tempo are like those used in territorial disputes by Chubb's Cisticola *Cisticola chubbi* on Mt Cameroon. Alarm call a loud 'seep-seep', a ringing 'sippe-sippe' and a bubbling trill 'spi-spi-hee-hee-hee'.

General Habits. Inhabits thorn scrub, acacia and sometimes broad-leaved trees in arid and semi-arid zones, occurring from sea level to 1800 m in Somalia (Mt Wagar) and from 300 to 1300 m in E Africa, being commonest in areas with annual rainfall <500 mm (at least in Kenya).

In non-breeding season, usually in small parties of 12–15, secretive but actively and systematically searching shrub and grass layer for food. Whilst on the move, bird often raises tail almost vertically and waves it from side to side; tail wagging occurs when birds interact in apparent territorial dispute (R. McVicker, pers. comm.). Bird sometimes gives short burst of audible wing-flapping.

Resident throughout its range, except perhaps Chad where only recorded twice (mid-Jan and end July).

Food. Small winged insects, ants.

Breeding Habits. Solitary, territorial nester.

NEST: neat long pocket, open or domed, with side-top entrance without porch, made loosely of grass and fibres, lined with softer material such as kapok; cocoons and/or plant wool placed on outside of nest. Placed in low bush and slung between twigs some 15–60 cm above ground.

EGGS: 3–4; cream, white or greenish white, speckled all over with reddish spots and flecks on pale undermarkings and forming zone at large end. SIZE: (n = 3) 15·5–16 × 11·5. Probably double-brooded (Archer and Godman 1961).

LAYING DATES: N Sudan, probably June–Aug; S Sudan, Dec; Eritrea, Mar (nestling June); Ethiopia, Mar–June; NW Somalia (nest-building, Erigavo, late May, fledglings, Hargeisa, late June); Somalia, late Jan, late Mar, June; Kenya/Sudan border (gonads well developed Dec); Tanzania (nest-building mid-Dec); E Africa: Region D, Feb, Apr, June.

Genus *Artisornis* Friedmann

Very small, sombre-coloured greyish warblers, unpatterned, with reddish face and rust-coloured thighs. Wings rounded, tail short, graduated, of 10 feathers. Rictal bristles vestigial. Bill of *A. moreaui* long, straight, somewhat flattened (**A**); that of *A. metopias* short, with robust base (**B**). Tarsus long. Nest of tailorbird-type, a deep bag of fine grass strips, sewn between leaves of a plant. Inhabit dense undergrowth in wet montane forest; with restricted range in E Africa.

One species formerly placed in genus *Apalis* (e.g. Hall and Moreau 1970), but overall similarities in plumage and nest architecture suggest close relationship to Oriental genus *Orthotomus* (Fry 1976), and the African tailorbirds have been included in that genus (Traylor *in* Watson *et al.* 1986). *Orthotomus* has 12 tail-feathers, central ones seasonally elongated in some species. For that reason we keep *Artisornis* separate, but recognize that when more is known about its species it might be better treated as congeneric with *Orthotomus*.

Endemic. 2 sympatric species, 1 very rare.

Plate 17
(Opp. p. 320)

Artisornis moreaui (Sclater). **Moreau's Tailorbird. Couturière de Moreau.**

Apalis moreaui Sclater, 1931. Bull. Br. Orn. Club 51, p. 109; near Amani, Usambara district, Tanganyika.

Range and Status. Endemic resident. NE Tanzania (E Usambara Mts at 900–1050 m); and NW Mozambique (Njesi Plateau, Unangu, at 1650 m). These 2 populations are 1000 km apart. Highly local and uncommon. Endangered, threatened by forest clearance; in E Usambara only about 110 km² of suitable habitat remains; known from several localities but distribution may now be discontinuous, with low density; at Njesi (where status unknown) *c.* 25 km² of suitable habitat may remain (Collar and Stuart 1985).

Description. *A. m. moreaui* (Sclater): Tanzania. ADULT ♂: forehead olive-grey, faintly washed with rusty red, ear-coverts browner, crown, nape, lores and cheeks olive-grey. Mantle and back to uppertail-coverts olive-grey. Tail olive-grey, strongly graduated, outermost feather (T5) *c.* 25 mm shorter than T1. Entire underparts greyish white, thighs rusty chestnut. Wings olive-grey, underwing-coverts and axillaries white. Bill black, whitish at tip, lower mandible greyish or whitish horn; eyes pale brown; legs and feet pale flesh-brown. Sexes alike. SIZE: (2 ♂♂, 3 ♀♀) wing, ♂ 48, 48, ♀ 43–45 (44·0); tail, ♂ 48, 48, ♀ 39–41 (40·0); bill, ♂ 18, 18, ♀ 17–18 (17·5); tarsus, ♂ 20·5, 21, ♀ 20–21·5 (20·8). WEIGHT: ♂ (n = 5) 8·3–10 (9·1), ♀ (n = 3) 8·3–10 (9·0).

IMMATURE: lacks rusty wash on forehead, more uniformly grey below.

NESTLING: unknown.

A. m. sousae (Benson): Mozambique. Forehead, lores and ear-coverts washed chestnut; darker olive-brown above, darker grey below.

Field Characters. An inconspicuous, greyish warbler of forest undergrowth, with chestnut wash on forehead (Tanzania) and also on face (Mozambique). Tail relatively short, bill very long and thin. Might be confused with African Tailorbird *A. metopias* but latter is darker and browner, with extensive red on head and throat, shorter bill.

Voice. Tape-recorded (STJ). Song apalis-like, a repeated, percussive double note, given in groups of 4–5, 'peedoo-peedoo-peedoo-peedoo', at rate of about 3 notes/s.

Occasionally quite far from surviving forest; originally probably in natural clearings with fallen trees, now increased by felling. At Njesi occurs in montane forest gullies; said to be in canopy there, with African Tailorbird in understorey (Benson 1946c).

Very secretive, extremely inconspicuous, shy. Feeds no higher than 10 m, searching foliage for insects, avoiding sunlit places; joins mixed-bird parties. Sings infrequently, from vines just above ground level. One caught could not fly more than *c.* 3·3 m, and species may be unable to cross open areas, although it moves through vegetation with ease (Stuart 1981b).

At Usambara similar in ecology to African Tailorbird and Bleating Warbler *Camaroptera brachyura*, with former dominant in the zone above it and the latter below it, both perhaps restricting its altitudinal range (Stuart 1981b).

Food. Insects.

Breeding Habits. Apparently territorial.

NEST: 1 found below song-post and attributed to this species: cup of rootlets, moss and hair, strung from sapling, with about 5 leaves bound around it (R. Stjernstedt, pers. comm.).

General Habits. At Usambara inhabits forest edge and clearings with dense undergrowth, especially in small valleys, and areas of cultivation with matted vines.

Artisornis metopias (Reichenow). African Tailorbird. Couturière d'Afrique.

Prinia metopias Reichenow, 1907. Orn. Monatsb. 15, p. 30; Usambara, Tanganyika.

Plate 17
(Opp. p. 320)

Range and Status. Endemic resident in E Tanzania, at 1000–2500 m in E and W Usambaras, Nguru, Uluguru, Ukaguru and Udzungwa Mts, Iringa Highlands (Itanga and Mdanda Forests), Matengo Highlands and Songea (Luwiri-Kitessa), and in NW Mozambique (Njesi Plateau, Unangu at 1560 m). Generally common.

Description. *A. m. metopias* (Reichenow) (including '*pallidus*'): range of species except Uluguru Mts. ADULT ♂: forehead and crown deep reddish chestnut (brightest on forehead), merging into dark olive-brown on hindneck; cheeks, lores, ear-coverts, chin and throat chestnut-brown (off-white in centre of throat). Mantle and back to uppertail-coverts dark olive-brown. Tail blackish grey, moderately graduated, unpatterned, with dull reddish wash. Breast dull whitish, with greyish suffusion, belly dull whitish, flanks washed buffy red, undertail-coverts pale buff, thighs dark chestnut. Wings dark olive-green, underwing-coverts and axillaries white. Bill black; eyes brown; legs and feet brown. Sexes alike. SIZE: (10 ♂♂, 10 ♀♀) wing, ♂ 48–52 (49·8), ♀ 46–49 (47·4); tail, ♂ 36–42 (38·2), ♀ 32–38 (35·5); bill, ♂ 14·5–16 (15·0), ♀ 14–15·5 (14·8); tarsus, ♂ 22–24 (22·8), ♀ 21·5–23 (22·3). WEIGHT: Tanzania, ♂ (n = 11) 8·3–10 (9·1), ♀ 8–10 (9·1); ♂♀ (n = 27) 7·5–9·3 (8·4).

IMMATURE: belly slightly washed with yellow, lower mandible yellowish horn.

NESTLING: hatchling unknown; bird about to fledge had upperparts grey, head and neck warmer, underparts olive-yellow; culmen grey, edges of upper mandible and entire lower mandible yellow, gape yellow, legs purple-brown.

A. m. altus (Friedmann): Tanzania (Uluguru Mts). Chestnut on sides of throat and neck darker, extending further onto throat; upper breast and flanks darker.

Field Characters. An olive-brown warbler of the forest understorey with reddish head and throat; tail short and uniformly dark. Chapin's Apalis *Apalis chapini* has chestnut face, throat and breast but longer, pale-tipped tail and is usually in the canopy; for further differences, see that species. Moreau's Tailorbird *A. moreaui* is mainly grey, with just a wash of reddish on forehead and face only, and has a very long bill.

Voice. Tape-recorded (B, LEM, STJ). Song usually a duet. First bird (♂?) repeats single insistent note on one pitch, either downslurred, 'tseeu-tseeu-tseeu-tseeu . . .' or upslurred, 'tsui-tsui-tsui-tsui . . .'; may sing alone, but often second bird joins in at intervals with short, buzzy note, 'dzz-dzz . . . dzz . . . dzz-dzz-dzz . . .'.

General Habits. Inhabits low dense ground vegetation in forest; in Chita Forest (Tanzania) especially in wet forest with dense undergrowth and many tree ferns *Cyathea* spp. Very rare in E Usambaras (not seen there for 50 years), perhaps due to presence of Moreau's Tailorbird (Stuart 1981b); but on Njesi Plateau, Mozambique, they are sympatric and apparently common (perhaps ecologically segregated), with Moreau's Tailorbird in low (8–9 m) canopy of gully forest (Benson 1946c). Gleans insects from leaves. Follows ant-columns (Willis 1986a), though normally occurs in pairs or small groups hopping through low *Begonias* and other bushes and herbs at forest edge or on landslide areas, far from driver ants; birds benefiting from presence of ants were mainly in cut-over areas, on weedy edge of pine plantation and in similar dense, disturbed places. 1 bird hopped about in vegetation 0·2–6·0 m above ground, pecking tiny prey items from leaves (16 times) and stems (twice), now and then with long leg stretches; once pecked item from spider web (Willis 1986a).

Food. Insects.

Breeding Habits.
NEST: delicate cup or pouch of fine materials bound together with spider web, sewn between drooping leaves of *Solanum* plant; leaves pierced along edges with long threads of yellow-green, moss-like silk; nest lined with scraps of forked grey lichen and a fungus resembling horsehair (probably *Marasmius*). Int. diam. 40 and depth 26, ext. diam. 65 and depth 76.
EGGS: 2; oval, very slightly narrowed towards one end; white with dark red-brown marks evenly spread over entire surface, with some greyish marks between.
LAYING DATES: E Africa: Region D, Dec–Feb (Dec 1, Jan 1, Feb 2 clutches), (nest-building mid-Nov).

Genus *Poliolais* Alexander

Bill quite long, almost straight, upper mandible overlapping lower but not hooked; tail of 10 feathers, very short, slightly rounded, mainly white; legs long, outstretched feet extending considerably beyond tail. Plumages of ♂, ♀ and immature all distinct: ♂ slate-grey; ♀ olive-brown and grey with chestnut cheeks and crown; immature olive-brown above and yellowish below. Habitat montane forest undergrowth. Nest unlike *Camaroptera*, a suspended ball with side entrance and porch, not sewn to leaves.

Monotypic; endemic to Bioko and Cameroon/Nigeria montane. Hall and Moreau (1970) noted resemblances to the tailorbird '*Orthotomus*' (= *Artisornis*) *metopias*, but admitted that 'nothing can be resolved until its nest is discovered'. Placed by White (1962) in *Camaroptera*, which has a tailorbird-like nest, and field observations indicate that in behaviour it is very like other *Camaroptera* spp. (Stuart 1986a). Linked to *Camaroptera* via *C. chloronota*, resembling it in colour, sexual dimorphism (although much more marked in *Poliolais*), and cicada-like song. However, we now know the nest of *Poliolais* is not stitched, so we prefer to retain it in its own genus.

Poliolais lopesi (Alexander). White-tailed Warbler. Poliolaïs à queue blanche.

Apalis lopezi Alexander, 1903. Bull. Br. Orn. Club. 13, p. 35; Bakoki, Fernando Po.

Range and Status. Endemic resident, confined to Nigeria, Cameroon and Bioko. Locally fairly common in SE Nigeria (Obudu plateau only), and W Cameroon (Bonenza on S slopes of Mt Cameroon, Mt Manenguba, Foto near Dschang, Mt Nlonako, Mt Kupé, and Dikome Balue in Rumpi Hills); quite common throughout Bioko about 800 m.

Poliolais lopesi

Description. *P. l. manengubae* Serle: Nigeria, Cameroon except Mt Cameroon. ADULT ♂: entire upperparts, from forehead to tail, also lores, ear-coverts and upperwing-coverts, dark blackish grey, with chestnut flecks on some feathers of crown, forehead and lores. Central pair of tail-feathers dark brown-grey, T2 white edged with brown on both webs, remainder white. Chin greyish white becoming more slate-grey on throat, breast and belly, thighs olive-brown, flanks pale brown; undertail-coverts tinged pale rufous. Primaries and secondaries dark olive-brown, outer webs of secondaries with olive wash; underwing-coverts and axillaries white shading into pale olive near margins of wing. Bill dark brown; eyes brown; legs and feet brown. ADULT ♀: forehead, crown and sides of breast dark chestnut-brown, lores, feathers around eye, ear-coverts and cheeks bright rufous-chestnut. Remaining upperparts, primaries and secondaries dark olive-brown; outer webs of secondaries with olive wash; underwing-coverts and axillaries white becoming pale yellow along margins of wing. Tail like ♂ but central feathers brown. Chin and throat off white merging into chestnut on cheeks and into dusky olive on breast; centre of breast and belly pale yellow, sides and flanks greenish olive-brown with faint rufous wash. Undertail-coverts buff with chestnut wash at tips. SIZE: (10 ♂♂, 10 ♀♀) wing, ♂ 52–56 (53·7), ♀ 46–51 (49·1); tail, ♂ 32·5–38 (34·2), ♀ 25–29 (27·3); bill to feathers, ♂ 12–13·5 (12·7), ♀ 12–13 (12·6); tarsus, ♂ 22–23·5 (22·8), ♀ 22–23 (22·6). WEIGHT: ♂ (n = 3) 10·8–13·4 (12·1); 1 ♀ 12·2.

IMMATURE: ♂ has entire upperparts dark olive-brown; underparts olive-green washed with olive-brown on sides and becoming yellower on belly and central line of throat and breast; flank feathers long and loose, olive-brown and tipped with rufous-brown. ♀ like ♂ but crown and nape with rufous-brown wash and ear-coverts washed chestnut; breast lighter and markings on flanks more distinctive than in ♂.

NESTLING: not described.

P. l. alexanderi Bannerman: Mt Cameroon. Adult ♂ like *manengubae* but crown and rest of upperparts dark olive-green; edges of outer webs of primaries and secondaries tinged olive-green, broader on secondaries; underparts olive except for grey on throat, centre of breast and belly, thighs olive-brown washed with rufous, undertail-coverts tinged pale rufous. Adult ♀ much like *manengubae*, but chestnut does not extend onto sides of breast; upperparts with greener wash and underparts with paler yellow wash. WEIGHT: ♂ (n = 4) 12–13 (12·2); ♀ (n = 3) 11–13·4 (12·2).

P. l. lopesi Alexander: Bioko. Adult ♂ has upperparts lighter than *manengubae* and more brownish grey, but underparts darker and flanks dark sooty grey, slightly paler on belly; thighs dark brown. Adult ♀ similar to *manengubae* but chin and throat olive-grey becoming sooty grey on breast and belly (much like ♂ *lopesi*), more olive-brown on flanks; thighs brown, undertail-coverts grey. Rufous of crown more clearly demarcated from brown of mantle than in other races. WEIGHT: ♂♀ (*lopesi* and *alexanderi*, n = 9) 10·5–15 (12·6).

Field Characters. A tiny warbler restricted to montane forests of Nigeria, Cameroon and Bioko. Best identified by very short, white tail. ♂ and ♀ very different (once thought to be different species) and immature different again. ♂ mainly slate-grey, ♀ olive-brown above and grey below with chestnut cheeks and crown and white throat, immature dark olive-brown above and olive-green below. Flight weak and brief, with whirring wings producing a distinctive buzz. Tail sometimes fanned and wings flicked; frequently gives bursts of insect-like song. In shape and size much like a camaroptera, especially Olive-green Camaroptera *Camaroptera chloronota*, which is olive-green above and pale brownish below and has a dark tail and ringing, ventriloquial song; it is usually found at lower elevations but they can overlap.

Voice. Tape-recorded (68, GRI, LEM). Song high-pitched and thin but pure; one form, often lasting several min, is a regular repetition of 2-note phrase 'pee-eep peuuu' at rate of *c.* 7–10 phrases per 10 s: 'pee-eep peuuu, pee-eep peuuu, pee-eep peuuu, pee-eep peuuu . . .'. Pitch slightly rises in the first note and decreases in the second. Another form is insect-like and almost 'electronic' in quality, a continuous, prolonged repetition of the 'pee-eep' note at rate of 15–20 notes per 10 s: 'pee-eep pee-eep pee-eep . . .'. Contact call a fizzing noise; alarm a series of high-pitched chirps 'peep peep peep . . .', rate *c.* 1 per s.

General Habits. Inhabits thick bush and tangled undergrowth of primary and secondary montane forest, forest edge and forest clearings. Ranges on Mt Cameroon from 800 to 2200 m (but does not occur at highest elevations of montane forest); Mt Kupé from 900 to 1050 m; Mt Manenguba from 1950 to 2200 m; Mt Nlonako from 1250 to 1650 m; Rumpi Hills from 1100 to 1700 m. Usually in pairs or small parties, actively searching leaves and twigs for food, usually very close to ground. Not shy and responds to 'squeaks' but difficult to observe in gloom of the undergrowth; does not join mixed-species flocks. On Bioko occurs up to 1600 m (Pérez del Val *et al.* 1994).

Food. Insects.

Breeding Habits. Territorial, solitary nester.

NEST: 2 known (Eisentraut 1963, Stuart 1986a); thick-walled bottle-shaped bag or ball of moss, lined with thick plant down and other soft plant material; no arthropod silk used in structure; side entrance near top, with a porch. Nest suspended 1–1·5 m above ground from fern frond or leaf canopy of tall herb.

EGGS: 1–2; immaculate white. Size not recorded.

LAYING DATES: Mt Cameroon, Jan (brood patch developing mid-Jan, enlarged gonads Nov, Jan, Feb, ovary slightly enlarged July); Foto (fluffy juv. Mar); Mt Manenguba (well-developed brood patch mid-Feb); Mt Kupé (enlarged gonads Nov, Jan, Feb, primary moult beginning early Feb); Bioko, Nov, Jan. Breeding appears to begin at end of rains and continue into the dry season. On Mt Cameroon breeds in dry season only (Tye 1991).

BREEDING SUCCESS/SURVIVAL: one of the 2 nests found held a warbler egg and an egg of either a honeyguide (Indicatoridae) or a Pin-tailed Whydah *Vidua macroura*.

Reference
Stuart, S. N. (1986a).

Genus *Camaroptera* Sundevall

Small, leggy warblers of forest and bush. Bill long, not hooked at tip; nares lidded. Wing rounded and short, P9 more than half length of P8. Tail of 12 feathers, short, often cocked over back. Plumage greenish above and greyish below; thighs contrastingly coloured; rump-feathers soft and abundant; one species (*C. chloronota*) sexually dimorphic; some races of another (*C. brachyura*) have different breeding and non-breeding plumages. *C. superciliaris* has naked blue skin at side of neck which is conspicuous when bird sings. Nest built inside a covering of large, living leaves stitched together with silk rivets. Displays elaborate. Songs vary widely between species but distinctive cat-like bleating call is common to all.

Endemic. 3 species. Closely related to *Calamonastes* and to Asian *Orthotomus* (see Fry 1976 and Watson *et al.* 1986); there are good grounds, in fact, for treating all 3 groups congenerically. *C. brachyura* contains grey-backed forms (*brevicaudata* group), green-backed forms (*brachyura* group) and a green-tailed form (*harterti*) which have been treated as 3 species. In some respects grey- and green-backed populations behave as separate species in e.g. Transvaal (see Taxonomic Note). With some misgivings we follow Dowsett and Dowsett-Lemaire (1993b) in keeping them a single species.

Plate 17
(Opp. p. 320)

Camaroptera brachyura (Vieillot). Bleating Warbler. Camaroptère à tête grise.

Sylvia brachyura Vieillot, 1820. Encycl. Meth. II, 459; Knysna.

Range and Status. Endemic resident; common to abundant almost throughout subsaharan Africa south of *c.* 17°N, except driest parts of NE and SW; in South Africa only in Transvaal and along SE coast. Mauritania, common in Senegal R. valley and up to 50 km north of Rosso, Assaba and Tagant hills, and Sahel zone. Mali, common in Boucle du Baoulé Nat. Park, and widespread south of *c.* 17°N; Niger, widespread in Niger R. valley, records from Parc du W, Makalondi, Niamey, Dosso, Tahoua, Agadès, Takoukout and Kaadjia; Chad, common in soudanian and Sahel regions, north to Oum Chalouba (15°47′N in Ouadi Rime-Ouadi Achim Faunal Res.), absent Ennedi; Sudan, very common north to Khartoum and in E to Red Sea Hills; Ethiopia, widespread in W and S, absent from lowlands in NE (E Tigré, E Wollo) and SE (Harrar, Bale Provs.); Djibouti, uncommon and local (Forêt du Day, Mabla Mts, Bankoualé, Simoneau), not on coastal plain; common in extreme NW and SE Somalia, as mapped; Kenya, absent from arid E, scarce in NW except near L. Turkana. Tanzania, common in W, N and E; status in centre not known, but probably widespread; common Zanzibar and Mafia, absent Pemba. Absent desert areas of extreme SW Angola. Namibia, rare to common in N (Ovambo area, Namutoni, Twee Koppies and Otjivasanda in Etosha Game Res.) and south to Okandukaseibe Farm, Warmquelle, Spitzkoppe and Brandberg Mts, also Sturmfield in Sandveld Kalahari (Winterbottom 1969). Common to very common N and E Botswana, sparse to common in SE along Molopo R. to Bray and to west of Khakhea (Penry 1994). In South Africa, common and

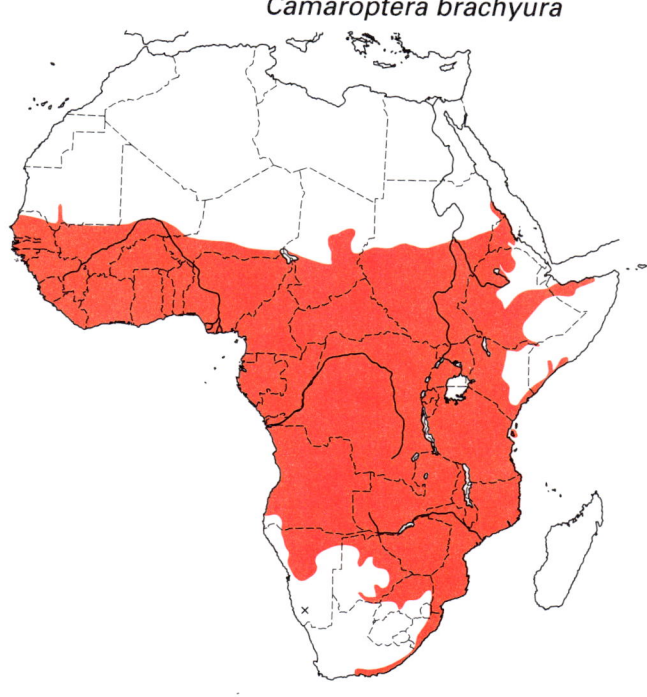
Camaroptera brachyura

widespread throughout Transvaal bushveld and lowveld, vagrant in highveld at Barberspan and Potchefstroom; common Zululand, coastal Natal and Transkei, scarce inland, west to high forests at Tonti, Qebe, Ingeli and Weza (but not Sulenkama); coastal Cape Province west to Knysna and George, inland to Grahamstown.

Density, Liberia, 4 pairs per ha in coffee plantations and abandoned farms, up to 12 pairs per ha in burned montane forest (W. Gatter, pers. comm.); Uganda, 3–15 birds per ha (15 where cover thickest); c. 5000 birds in 10 km² strip of Kazinga Channel (Fogden and Fogden 1979). At M'Passa, Gabon, 3–4 birds per ha (Brosset and Erard 1986).

Description. *C. b. brevicaudata* (Cretzschmar): northern savannas from Senegal and Sierra Leone to central Sudan and lowlands of NW Ethiopia. ADULT ♂ (breeding): forehead, crown, nape and neck greyish brown, lores, ear-coverts, cheeks, mantle, back and rump greyer, tips of rump-feathers lighter; uppertail-coverts brown. Tail brown, darker terminally (new feathers with white tips which soon wear off). Chin to upper breast grey, mottled with some buff feathers, lower breast, belly and flanks buffy white, sides of belly and flanks washed grey, thighs orange-buff, undertail-coverts whitish. Primaries and secondaries dark brown, inner edges of primaries paler, outer edges of P1–P8 and edges of secondaries yellowish olive. Scapulars and upperwing-coverts yellowish olive with brown centres; marginal upperwing-coverts yellow. Marginal and lesser underwing-coverts bright lemon-yellow, mottled olive; axillaries and remaining underwing-coverts white. Bill greenish black, black or dark horn, mouth black; eyes brown, hazel-brown to pale buff, or neutral orange; legs and feet fleshy-pink, light brown, or pale pinkish brown, soles yellowish. Sexes alike. ADULT ♂ (non-breeding): upperparts from forehead to rump light brown, feather bases grey, rump washed grey (feathers pale at tips). Tail like mantle but darker brown terminally, feathers tipped greyish white when fresh, uppertail-coverts brown. Chin and upper breast pale grey, mottled whitish, lower breast and upper belly washed light brown, sides of breast browner, lower belly whitish washed buff, flanks and sides of belly darker buff. Primaries and secondaries paler than in breeding dress. Sexes alike. SIZE: (11 ♂♂, 9 ♀♀) wing, ♂ 50–61 (54·6), ♀ 48–52 (49·8); tail, ♂ 31–40 (34·3), ♀ 30–36 (32·0); bill to feathers, ♂ 11–13 (12·4), ♀ 11–13 (12·0); tarsus, ♂ 18–23 (21·5), ♀ 19–22 (19·5). WEIGHT: unsexed (n = 7, Mole, N Ghana, July–Sept) 9·0–12·5 (10·9); (Chad, n = 37) 8–12 (10·0); (n = 29, Zaria, Nigeria, Oct–June) 9·5–11·7 (10·5).

IMMATURE: upperparts olive-brown, feather bases grey, wings as in non-breeding adult, underparts mostly lemon-yellow, sides and breast washed olive, extent of olive wash increasing with age; bill dark grey with dull yellow tip; eyes light greyish brown; feet light brown.

NESTLING: blind, naked, without down, skin pale orange, alar tracts forming dark line down each wing; gape yellow; tongue yellow with 2 triangular black spots at base; body dark pink by day 4 and legs and bill yellow.

C. b. tincta (Cassin): forested areas of W Africa and Congo Basin from Liberia to W Kenya, south to embrace range of *harterti* in NW Angola, and southeast to extreme NW Zambia (Mwinilunga) along border with Zaïre; W Tanzania (Kigoma and Nkungwe–Mahare Mts) also W Uganda, Rwanda, Burundi, border areas of W Kenya (south of Mt Elgon), northern limits abut with *brevicaudata*. No non-breeding dress. Like *brevicaudata* but upperparts darker and browner grey, but washed brown, rump darker grey; chin to lower breast rather darker grey mottled with light grey, underparts whiter, sides of belly and flanks darker grey (mottled buff, feathers with pale tips), wings and tail more olive-brown (4 ♂♂, Cameroon, have dark yellowish olive-green tails: Serle 1954). Intergrades with *aschani* in Uganda. WEIGHT: (Liberia) ♂ (n = 11) 10·3–12 (10·7), ♀ (n = 3) 8·1–9·1 (8·7); (Cameroon) ♂ (n = 6) 11–13 (11·1) (Eisentraut 1963); (Central African Republic) ♂ (n = 3) 11–12 (11·7); (Kenya) ♂ (n = 2) 10, 11; (Uganda) ♂ (n = 23) 6–22 (11·6), ♀ (n = 15) 5–21 (10·7) (the 6 g and 5 g weighings erroneous?).

C. b. harterti Zedlitz: NW Angola escarpment zone (southern Cuanza Norte and Malange, to Luanda and Vila Salazar in west, and south along escarpment to Gabela, Cuanza Sul). Like *tincta* but tail green and underparts even whiter; ♀ with lower breast and upper belly buffier than ♂. Immature like adult but upperparts from forehead to tail olive-yellow, chin to lower belly with lemon-yellow wash. SIZE: (5 ♂♂, 4 ♀♀) wing, ♂ 52–57 (54·8), ♀ 50–58 (53·0); tail, ♂ 36–38 (36·6), ♀ 33–35 (34·3); bill to feathers, ♂ 11–13 (12·4), ♀ 11–12 (11·8); tarsus, ♂ 20–22 (20·8), ♀ 19–21 (19·8).

C. b. aschani Granvik: highlands of Kenya (Ngong to Mt Elgon), extreme SW Uganda and Kivu, Zaïre. Underparts almost entirely dark grey. Intergrades with *tincta* in Uganda. Large: wing 52–61.

C. b. abessinica Zedlitz: S Sudan, NE Zaïre (Uele, Mahagi Port), Ethiopia except NW and Ghere region, N Uganda, N Kenya, Djibouti, N and (just) W Somalia. Breeding dress like *brevicaudata* but upperparts darker grey and underparts much more uniform olive-grey; non-breeding dress like breeding dress but upperparts washed brown. NE Zaïre population lacks seasonal change (Chapin 1953a). WEIGHT: (Uganda) ♂ (n = 9) 10·1–12·8 (11·7), ♀ (n = 3) 10–12·5 (11·2); (Nchalo, Kenya) ♂♀ (n = 51) 8·0–11·5 (10·19); (L. Shala, Ethiopia) 1 unsexed 12·0.

C. b. insulata Desfayes: Ethiopia, rain forest margins in Ghere region and probably Gore, Kaffa Prov. Like *abessinica* but head and back dark, throat, breast and underparts dark grey, wings olive-green. Darker than *tincta*, especially below, with greener, less yellow wings.

C. b. erlangeri Reichenow (including '*albiventris*'): S Somalia, coastal Kenya, NE Tanzania west to Amani, south to Handeni and Mpwapwa. Like *abessinica* but in breeding and non-breeding dress much whiter below, sides of belly sometimes with buffish wash in non-breeding dress. WEIGHT: (Somalia) unsexed (n = 31) 7·1–10·3 (8·7) (Wood 1989); (Kenya) ♂ (n = 54) 4–20 (9·2), ♀ (n = 24) 5–15 (7·8).

C. b. griseigula Sharpe: W half of Kenya, except N and W borders, E Uganda (Mts Elgon and Moroto, N Tanzania (Ngorongoro Crater highlands, Kilimanjaro and Kibaya) east to Tsavo Region and Taru. Adult like *abessinica* but upperparts more brownish grey and underparts more olive-grey, flanks brown; belly white with olive wash. Like *tincta* but paler above. Immature has upperparts washed greenish, breast and upper belly washed and obscurely mottled greenish grey or olive. Intergrades with *erlangeri*. WEIGHT: (Riakanau, Kenya) ♂ (n = 40) 5–13 (9·6), ♀ (n = 18) 5–11 (7·4); (Mt Nyiru, Kenya) ♂♀ (n = 11) 10–15; (Nyanza, Kenya) unsexed (n = 74) 9·7–13·0 (11·45 ± 0·85); (Tanzania) ♂ (n = 4) 10·7–11·5 (10·9), ♀ (n = 3) 8·6–10·5 (9·8).

C. b. intercalata White: Zaïre (Shaba), W Tanzania (east to Iringa Region, Hanang and north to L. Victoria), W, central and E Angola (between *tincta* to north and *sharpei* in south), NW, N and E Zambia (west of Mankoya). Much whiter below than *tincta*. In breeding and non-breeding dress similar to non-breeding *sharpei*, but a little browner in its non-breeding dress; chin to upper breast and sides of body more uniform olive-grey than *sharpei*, tawny wash confined to lower belly and flanks, upperparts more grey-brown; smaller (Clancey 1970b, 1974b, 1980a). Intergrades with *sharpei* in Zambia and *fugglescouchmani* at Isoka, Zambia. WEIGHT: (Kazungula, Zimbabwe) unsexed (n = 11) 9·5–12.

C. b. sharpei Zedlitz (including '*noomei*'): S Angola, Zambia south of range of *intercalata*, Malawi (Shire Valley south of *c.* 13°S, also Mzimba (L. Kasuni) and Rumphi (Nkhamango) districts), Namibia, N and NE Botswana, SW Zimbabwe (W Matabeleland) and W Transvaal. Breeding and non-breeding plumages different. Like *tincta* but breeding dress paler grey above, non-breeding dress browner on back and washed buff below. Upperparts (breeding dress) paler than in *transitiva*. Intergrades with *intercalata* in Zambia and with *bororensis* in S Malawi. SIZE: (10♂♂, 10♀♀) wing, ♂ 55–60 (56·8), ♀ 50·5–54·5 (52·9); tail, ♂ 42–46 (43·7), ♀ 36·5–40·5 (38·6) (Clancey 1974b). WEIGHT: (Namibia) unsexed (n = 8) 11–12 (11·5) (Friedmann and Northern 1975); (S Malawi) unsexed (n = ?) 8–11·5 (10·2).

C. b. transitiva Clancey: Zimbabwe except SW and extreme E, Transvaal, SE Botswana. In breeding dress like *intercalata*, in non-breeding dress like *intercalata* but upperparts darker, underparts buffier, wings greener. Intergrades widely with *constans* in SE Zimbabwe. SIZE: (10 ♂♂, 10♀♀) wing, ♂ 50–55 (53·3), ♀ 48–52·5 (50·6); tail, ♂ 35–40·5 (38·5), ♀ 32·5–36·5 (35·2) (Clancey 1974b).

C. b. beirensis Roberts (including '*marleyi*'): Mozambique, north of Save R., and probably E Zimbabwe (Umtali highlands; Nyamkwarara and Honde valleys) reaching south through core of *constans* to reach NE Zululand (Clancey 1993). Breeding and non-breeding plumages similar; upperparts, especially crown and nape, greyer than in *transitiva* (less so in non-breeding dress), back washed green, wings greener, underparts generally whiter than in *transitiva*. Hybridizes with *constans* in Mozambique (P. A. Clancey, pers. comm.). WEIGHT: ♂ (n = 14) 10·5–13 (11·7), ♀ (n = 5) 10–11·5 (10·7).

C. b. brachyura (Vieillot): South Africa from George to Natal and W Zululand. ADULT ♂: forehead to crown greyish olive, hind crown to rump yellowish olive-green, feathers of lower back loose; uppertail-coverts greenish olive-brown. Tail brownish olive with green wash. Ear-coverts grey faintly streaked whitish, cheeks paler (colour merging into chin); chin, throat, breast and mid-line of underparts whitish grey, breast washed buffish grey, sides of body and flanks dark olive-grey, thighs buffish with yellowish wash, darker on upper part, undertail-coverts white. Primaries and secondaries brown, outer webs edged yellowish olive-green, inner webs edged pale brown, upperwing-coverts brown, broadly edged yellow-green; underwing-coverts white, marginal coverts washed yellow, leading edge of wing yellow, axillaries olive-brown; flight feathers silver-grey below. Bill black or blackish horn; eyes brownish yellow, umber brown, or light reddish brown; feet pinkish flesh. No non-breeding dress. Sexes alike. SIZE: (10 ♂♂, 8 ♀♀) wing, ♂ 50–57 (53·8), ♀ 48–55 (51·0); tail, ♂ 31–37 (35·9), ♀ 31–37 (34·3); bill to feathers, ♂ 11–13 (12·1), ♀ 12–13 (12·6); culmen (unsexed, n = 21) 11·2–15 (13·5); tarsus, ♂ 18–21 (19·5), ♀ 18–21 (18·9). WEIGHT: (southern African races) ♂ (n = 20) 9·2–12·1 (10·8), ♀ (n = 15) 9–11·5 (10·1), unsexed (n = 66) 5·1–13·9 (10·8). Immature has entire upperparts yellowish olive-green; chin, breast and central line of belly washed yellow, rest of underparts whiter than in adult; tail brown without greenish wash of adult. Outer edge of primaries whitish becoming olive yellow on inner section; thighs buff.

C. b. constans Clancey: E Zululand, except where *beirensis* penetrates from north (Clancey 1993), E Swaziland, Transvaal, SE Zimbabwe, Mozambique south of Save R. Like *brachyura* but forehead and crown paler, mantle, wings and tail paler and more greyish yellow-green, throat and underparts whiter, and flanks distinctly paler. E and N Transvaal Escarpment forest populations more like nominate form, with upperparts darker and fresher green. Intergrades with *beirensis* north of Save R. and with *transitiva* in E Zimbabwe and NW Mozambique (P. A. Clancey, pers. comm.).

C. b. bororensis Gunning and Roberts: S Tanzania (Songea), S Malawi, N Mozambique. Upperparts more yellow-green than *constans* and grey of forehead extends further back (Lawson 1963a). Intergrades with *sharpei* in S Malawi (Mangochi and Nsanje districts). WEIGHT: (Mopeia, Mozambique) (n = 11) 10·1–13·1 (11·2).

C. b. pileata Reichenow: S Kenya to lowland SE Tanzania, inland to the E Usambaras; Zanzibar and Mafia islands. Similar to *bororensis* but underparts greyer, less white. WEIGHT: (coastal Kenya) unsexed (n = 65) 7·8–10·5 (9·29 ± 0·75).

C. b. fugglescouchmani Moreau: E Tanzania from Ulugurus to Mahenge and Udzungwa Mts; Zambia; Malawi south to 13°S. Like *pileata* but throat darker grey and flanks olive. Intergrades with *intercalata* at Isoka, Zambia.

TAXONOMIC NOTE: within *C. brachyura* there are green-backed forms (*brachyura* group), grey-backed forms (*brevicaudata* group) and one race with green tail (*C. b. harterti*). Hall and Moreau (1970), Watson *et al.* (1986), Sibley and Monroe (1990) and Clancey (1992a, 1993) treated green-backed and grey-backed forms as separate species; Short *et al.* (1990), Dowsett and Dowsett-Lemaire (1993b) and several field workers treat them conspecifically. Hall (1960b) showed that *harterti* is rather distinctive though later she made it a race of '*C. brevicaudata*' (Hall and Moreau 1970). Except for *harterti*, which is almost unknown in life, the voice and behaviour of green-backed and grey-backed forms seem to be identical. In a few places they are apparently parapatric and ecologically separated, but elsewhere they hybridize; in Transvaal they 'are readily identifiable, frequent different habitats and overlap broadly in their respective ranges' (Tarboton *et al.* 1987). Clancey (1992a, 1993) points out that the grey-backed *beirensis* interposits between 2 green-backed populations (*bororensis*, *constans*) and also occurs without hybridization in centre of range of *constans* (Malamala, E Transvaal, Big Bens, Swaziland and Mkuzi R., Zululand). He treats grey- and green-backed birds as separate species, *C. brevicaudata* and *C. brachyura*.

Field Characters. A small, compact warbler with short tail often cocked over back. Head grey, underparts greyish white, wings yellow green, eye dark; back green in *brachyura* group ('Green-backed Camaroptera', coastal E and S Africa) and grey elsewhere (*brevicaudata* group, 'Grey-backed Camaroptera'), tail green in race *harterti* (Angola), otherwise brown. Monotonous dry clicking song recalls prinia, but bleating calls sufficient to identify it even in depths of thicket. Common but a skulker. In extreme SW, grey-backed form almost meets Karoo Eremomela *Eremomela gregalis*, which is distinguished by pale eye, green back, whiter underparts, yellow flanks and undertail-coverts, and high, thin voice.

Voice. Tape-recorded (9, 32, 58, 68, 75, 86, 88, 91, B, C, F, LUTG, WAT). Song a series of loud, sharp, penetrating notes, either single or double, 'chup-chup-chup-...', 'chit-chit-chit-...', 'pitchu-pitchu-pitchu-...', 'pitip-pitip-pitip-...', or something in between, 'ch'wit-ch'wit-ch'wit-... 'ch'yup-ch'yup-ch'yup-...'. Notes evenly spaced; number varies, 4–12, usually 5–8; song sometimes preceded by one or two 'warm-up' notes, 'tirip... tirip-tirip... tirip-tirip-tirip-tirip-tirip'. During aggressive encounters song becomes drier, lower and harder, 'drit-drit-drit-...'. Length of song (n = 24) 1·2–3 s; singing bout sometimes lasts up to 7 min. 3–4 bill snaps by ♂ may precede, accompany or follow song, but whether song itself is by ♂ or ♀ is uncertain (Ginn *et al.* 1989). During bill-snapping notes are faster (*c.* 11 in 2 s). ♂ may add rattle to song which sounds like bill-snapping but may be vocal (D. Watts, pers. comm.). During courtship flight ♂ gives short 'bzeea' or 'beejip' at rate of *c.* 3 per s for 8 s or more, each note followed by a rattle and wing-flapping. Call a low, rather hollow, slightly buzzy 'kwer', 'ger', 'berp' or 'puw'; alarm call similar but higher-pitched and more nasal, like bleating lamb or mewing cat, 'baaaa', 'meeeuw', 'bzeeee'.

General Habits. Inhabits all manner of woody growth: mist-belt evergreen forest, riparian and coastal forests, thickly wooded valleys, mixed bush, riverine woods, deciduous leafy thickets, forest clearings and edges, savanna woodland, dry water-courses, less arid parts of thornveld, exotic vegetation and gardens, and desert oases. Often near man; on farms, in hedgerows, and town and village gardens. In Botswana commonest in fringing forest, common in *Acacia* and mopane, but absent from *Baikiaea* woodland. Keeps near ground, seldom over 3 m high, in heavy shade and thick leaves of herbaceous and woody vegetation. Where grey-backed and green-backed forms are sympatric, e.g. in Transvaal, former occurs in thickets in dry areas, deciduous woodland and patches of tangled thickets in savanna, and latter restricted to edges of evergreen forest and in wetter areas with evergreen growth (Tarboton *et al.* 1987). Quarter of range in Kenya is arid (annual rainfall av. of < 250 mm), but bird common in interior of Sokoke forest on coast. Not dependent on water, and does not drink even in the most severe drought (Ethiopia: Safford *et al.* 1993). From coastal lowlands up to 1650 m in Liberia (W. Gatter, pers. comm.) and 2200 m in E Africa.

Solitary, in pairs, or family parties of up to 4 birds. Secretive, but sometimes inquisitive. Vocal and territorial, concealed ♂ singing from same spot for much of year. Territorial ♂ commonly gives rapidly repeated 'bzeep's for 8 s or more, each 'bzeep' followed by a rattle of 3–4 notes and by wing 'fripping'. Moves restlessly but inconspicuously through undergrowth and low foliage in search for insects; also forages on ground, prying under fallen leaves and twigs and around buttress roots of large trees. Habitually cocks tail high over back. Holds wings loosely. Flies without undulation, with quick whirring beats; flies only short distances. Snaps bill loudly and often flies with wings 'fripping': 'frrrp-frrrp-frrrp-frrrp'. Sings from concealed perch low down, sometimes 2–3 m high, rarely in canopy; sings all year in e.g. Gabon (Brosset and Erard 1986), or only seasonally. Associates little with other species, but in Zambia once seen feeding in canopy with Southern Hyliota *Hyliota australis*, Chinspot Batis *Batis molitor* and Collared Sunbird *Anthreptes collaris*.

In Uganda (where it breeds twice a year, but also whenever conditions are suitable) most birds moult in Apr–May, but rate varies, and moult interrupted if bird starts to breed. During ovulation fat reserves of ♀ decline markedly and she becomes relatively inactive (Fogden and Fogden 1979).

Sedentary throughout most of range, possibly a partial migrant at northern edge of range and Namibia. Present all year in S Mauritania but in Senegal R. valley commoner July–Jan than Feb–June, and occurs north of Rosso in dry season. Recorded throughout year in 'W' Park, Niger, but elsewhere records are Feb–Aug only. In N Nigeria, 4 times as many birds caught in spring as in winter (dry season), suggesting a small-scale shift in centre of gravity of population (Fry 1971). Only occurs Bamingui-Bangoran Nat. Park, Central African Republic, Aug and Sept. Recorded Chaub Farm, N Namibia, Oct–Nov.

Food. Moth larvae, nymphal mantids, termite larvae, other small insects, spiders; occasionally fruits (small seeds in stomach); chironomid midges an important fat source.

Breeding Habits. Solitary nester, strongly territorial, presumably monogamous. In courtship display, which may last up to 10 s or more, ♂ jumps up and down in tight loop from perch whirring his wings and giving an intense call in time with the jumps (Bannerman 1939, Ginn *et al.* 1989); jumps are 30–40 cm high and are continuous at rate of 1 per s, as though bird at end of a rubber band. ♂ also described as moving up and down, drooping wings, cocking tail well over back, and dashing with wings snapping towards ♀; ♀ flies off, lands on ground to face ♂ then slips into cover when he arrives, loudly snapping her wings; ♂ follows, and performance is repeated in ground cover or sometimes in trees (van Someren 1956). During display, a second ♂ may arrive and the 2 fight on ground by pecking each other. When disturbed near nest or when fledglings are present, bird uses anxiety call incessantly (van Someren 1956).

NEST: hollow ball built among growing leaves, usually broad leaves of undergrowth, also among broad blades

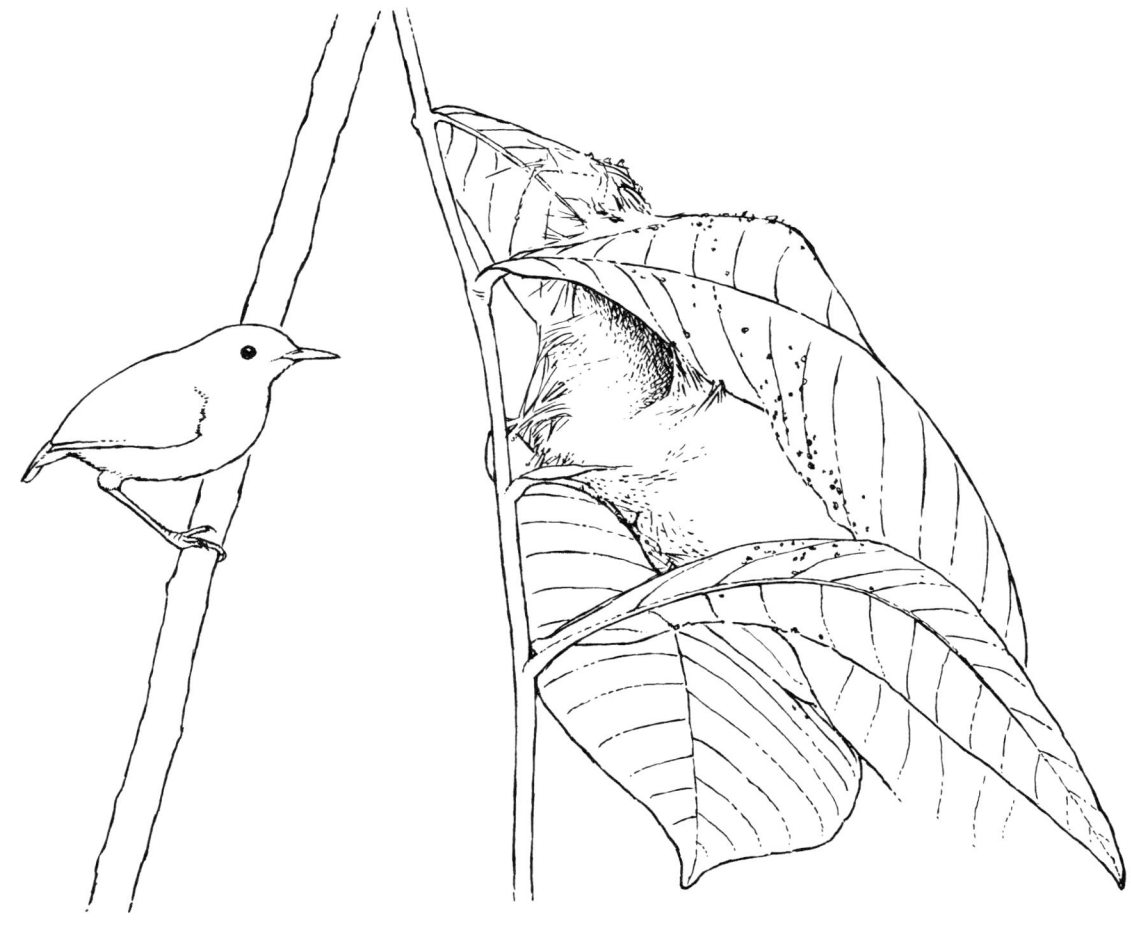

A

of *Panicum maximum* grass and even in needle-leaved *Asparagus* shrub. Leaves are sewn together with spider web poked through 10–20 tiny perforations mostly towards margins of leaf (**A**). Bird uses bill to push short masses of spider web or fine plant fibres through 2 holes in adjacent leaves to draw them together. Ends of the rivet are then teased to prevent them slipping out (Steyn 1970). Sometimes only a single leaf is used, its sides drawn together by sewing or riveting to form a vertical cylinder and its apex turned down as a roof (Brosset and Erard 1986). Bird also rivets leaf edge to nest proper (not always to another leaf), and occasionally brings a separate leaf and stitches it on top of nest to form dome or roof (van Someren 1956). Inner shell is of stiff dry grass, built inside joined leaves and lined with soft plant down. Entrance to pouch is at side, generally facing vertical stem of plant; inside strengthened by spider web. Lower part of pocket is filled with strands of bark, fine fibres, tendrils and very thin pieces of grass laid down in circles; if silky or woolly down of Compositae or Asclepiadaceae is available it is invariably added to lining. Lining forms loosely-constructed or quite compact and felty open cup; sometimes lining is stitched upward onto covering leaf so cup faces outward as well as upward. Covering leaf may form porch, drooping over and obscuring entrance to nest. Nests not completely covered by leaves are decorated on the outside with cocoons of Lepidoptera (of 5 such nests, 4 had < 15 cocoons, the other 27). Lining usually shows, but is sometimes quite concealed by leaves, making nest cryptic and hard to find. Nest takes 5–10 days to build, the lining a further 4–7 days. At one nest, one bird (ringed) of a pair did most of the building. Bird chooses plant with large leaves that do not readily drop off, typically a sturdy herb or sapling, sometimes a tree, vine or broad-leaved grass; leaves generally 1–4 times as long as bird. Nesting material collected up to 200 m from nest (Earlé 1980). Nest frail, not as strong as *Calamonastes* or *Cisticola* nests. Sited usually in undergrowth, 0·15–1·1 m above ground, sometimes up to 2·5 m, exceptionally up to 6 m; in *Solanum, Trema, Vitis, Passiflora, Lantana*, or plant of families Marantaceae, Compositae, Labiatae or Acanthaceae. Height above ground depends on height of vegetation; nests occasionally placed on ground, but later in season when vegetation has grown may be 1–1·5m above ground (Earlé 1980).

EGGS: 2–4, usually 3; av. 2·7, (26 clutches), also 2·7 (44 clutches Natal), and 3·1 (7 clutches Pietermaritzburg, Natal). Oval or long oval; somewhat glossy; white, very pale pink or pale blue-green, immaculate, or blotches and spots of red-brown and lilac sparingly scattered all over. SIZE: (n = 9) 15·5–18 × 11–12·5 (16·6 × 11·7); (n = 12) 17·0–18·8 × 12·4–13·0 (17·6 × 12·7); (n = 4, Kano, Nigeria) 15·5–16·5 × 11·5–12 (16·2 × 11·75).

LAYING DATES: Senegal and Gambia, June (2

clutches), July (2), Aug (2), Sept (5), Oct (2), Nov (2); Liberia, Nov; Nigeria, Apr–Sept, Nov–Jan (probably double brooded); W Cameroon, Oct (1), Dec (1), Jan (1), Mar (3), Apr (1), June (3) (Serle 1981); Chad, July–Sept; S Sudan, Feb, July, Oct, Nov; Ethiopia, Feb, July, Aug; Gabon, Oct (1), Dec (5), Jan (1), Feb (1), Mar (1) (mainly dry season: Brosset and Erard 1986); N Zaïre, May–Dec, Itombwe, Aug–June, Kivu, Aug–June; Rwanda, Nov (1), Dec (1), Feb (1), Apr (1), May (4); E Africa: Region A, Sept; Region B, Jan–Aug, Region C, Jan–Feb, Dec; Region D, Jan, Mar–June, Aug–Sept, Dec; Region E, May – in Uganda breeds twice a year but erratically, according to timing and duration of rainfall (Fogden and Fogden 1979); Angola, (juveniles June, July); Malaŵi, Sept (1), Oct (2), Jan (1), Feb (1), Mar (3), Apr (2), May (1); Zambia, Oct (1), Dec (4), Jan (3), Feb (3), May (1); Zimbabwe, Oct (2), Nov (19), Dec (22), Jan (9), Feb (3), Mar (1); South Africa, Transvaal, Nov (4), Dec (2); Natal and Zululand, Sept–Feb (Sept 1, Oct 7, Nov 11, Dec 12, Jan 3, Feb 3 clutches: Dean 1971).

INCUBATION: 12–15 days; mostly by ♂. In a clutch of 3, incubation started after laying of the second egg. Down often added to nest during and after egg-laying.

DEVELOPMENT AND CARE OF YOUNG: day 4, quills of primaries 3 mm long, tracts on underparts clearly defined; day 6, eyes open, primary quills 16 mm long, tracts on upperparts and underparts well developed, some feathers breaking out, dorsal ones olive-green, ventral ones whitish; day 8, primaries dull grey-brown, unfurled 4 mm from quills, tail-quills 4 mm long with feathers just beginning to appear, all body feathers breaking from quills, feathers of underparts fluffing out; day 10, bill slate-black, tip yellowish, legs dark pink, primaries 9 mm out of quills, rest of body covered except for centre of breast and belly; days 10–12, feather growth very rapid; day 12, fully feathered: crown and mantle grey, back and wings yellowish green, throat olive, breast and belly whitish, flanks buffy, wing-feathers 21 mm long, tail-feathers brownish, eyes brown; utters mewing 'pee-u', and if handled plaintive 'tweep tweep' of alarm (Steyn 1966a). At one nest adults fed young 25 times in an hour. Faecal sacs removed by both parents and dropped well away from nest. Nestling period 14–15 days; young leave nest before they can fly and do not roost in nest thereafter. If young threatened, ♂ draws attention to himself by calling and snapping wings. Young fed by both parents.

BREEDING SUCCESS/SURVIVAL: mortality during severe dry seasons probably due to lack of fat reserves (Fogden and Fogden 1979). From 22 eggs 7 young fledged (Earlé 1980). Parasitized by Wahlberg's Honeybird *Prodotiscus regulus* (Vernon 1970a), and African Emerald Cuckoo *Chrysococcyx cupreus*, Klaas's Cuckoo *C. klaas* and Diederik Cuckoo *C. caprius*. Estimated longevity (Malaŵi), 7–7·5 years (4 ♂♂), 8–8·5 years (one bird), 9–9·5 years (one ♂) (Hanmer 1989). In Zimbabwe one adult at least 7 years old.

References
Brosset, A. and Erard, C. (1986).
Clancey, P. A. (1970b, 1974, 1980a, 1992a, 1993).
Earlé, R. A. (1980).
Fogden, M. P. L. and Fogden, M. P. (1979).
Ginn, P. J. *et al.* (1989).
van Someren, V. G. L. (1956).
Steyn, P. (1966a, 1970).

Camaroptera superciliaris (Fraser). Yellow-browed Camaroptera. Camaroptère à sourcils jaunes.

Plate 17
(Opp. p. 320)

Sylvicola superciliaris Fraser, 1843. Ann. Mag. Nat. Hist. XII, 440; Clarence, Fernando Po.

Range and Status. Endemic resident. Sierra Leone, uncommon; records from Yonnibanna, Mafwe, Buedu, Sefadu and Nimmini Mts where not uncommon, absent from Njala, and Outamba and Kilimi area in NW. E Guinea, N'Zerekore, frequent Ziama massif. Liberia, locally fairly common from coast to N highlands. Ivory Coast, widespread from coast to guinea savanna zone. Ghana, widespread and not uncommon within forest zone; records from Ejura, Mampong Ashanti, Bia Nat. Park, Tarkwa, Prahsu, Akim Oda, Akropong Akwapim. Togo, Ahoué-houé, Atakpamé, Ebeva. Nigeria, uncommon in SW north to Ibadan and Ife, in SE at Calabar, Umuagwu, Ikpan Block, Oban West, Oban hills, Boshi-Okwango Forest. Cameroon, widespread in forest zone. Widespread Bioko (Clarence, Basakate, Bilelipi, Basakate and Apu rivers), Equatorial Guinea. Central African Republic, occurs Lobaye Préfecture at Ngoundji, Ndélé, Bangui, Botambi and west of Baroua, but status uncertain. Gabon, common and widespread. Congo, Mayombe, Brazzaville where the commonest camaroptera. Zaïre, probably throughout forest zone; also Shaba (Kasaji). Locally common in isolated forests of W Uganda from Budongo Forest south through Ankole to Bwindi-Impenetrable Forest, and eastwards in S to Sezibwa R., Kifu and Mabira forests. Frequent in N Angola (Cuanza Norte), recorded Canzele, Canhoca, Quicolungo and Ndala Tando. Record from Kombo, SE Sudan requires confirmation (Nikolaus 1989). Density (M'Passa, Gabon) 10 pairs per km^2 (Brosset and Erard 1986).

Camaroptera superciliaris

Description. ADULT ♂: forehead to rump dark yellowish olive-green, uppertail-coverts green in midline, yellow at sides; tail brown, each feather edged yellowish green. Ear-coverts and sides of neck yellow, lores black, superciliary stripe broad and yellow, extending beyond eye. Underparts whitish, sometimes (breast always) faintly tinged buff in centre and grey on sides; bare patch of pale or dark blue skin at side of throat; small yellowish olive patch on sides of lower breast; thighs green, variably tinged with yellow; undertail-coverts bright yellow. Primaries and secondaries brown, outer webs edged yellowish green, inner webs edged whitish, tips of secondaries yellowish green; upperwing-coverts brown broadly edged yellowish green. Underwing-coverts mostly white, marginal coverts yellow. Bill black with tip yellow; eyes variable, dark brown, brown or light brown, sometimes with blue-grey flecks; feet and legs variable, brownish grey or pale flesh-brown or dusky pink. Sexes alike. SIZE: (10 ♂♂, 10 ♀♀) wing, ♂ 47–52 (48·9), ♀ 43–47 (46·0); tail, ♂ 21–29 (25·7), ♀ 24–27 (25·8); bill to feathers, ♂ 12–14 (12·9), ♀ 11–13 (11·9); tarsus, ♂ 17–20 (18·4), ♀ 16–19 (17·3). WEIGHT: (Liberia, Mt Nimba) ♂ (n = 7) 7–10·8 (9·2), ♀ (n = 4) 7·5–9·5 (8·3); (Cameroon) 1 ♂ 10·5, 1 ♀ 9·5; (Bioko) unsexed (n = 6) 10–12 (10·8); (Central African Republic) 1 ♂ 10·5; 1 juv. 8·2; (Angola) 1 juv. ♂ 11·1; (Uganda) unsexed (n = 6) 9–11.

IMMATURE: upperparts like adult but with more greenish yellow wash, cheeks duller yellow, eye-stripe less conspicuous; throat and breast yellow becoming paler on sides, flanks yellow, thighs washed yellow, belly white, undertail-coverts yellow. Bare patch of skin present on throat but partly concealed by yellow feathers. In subadults yellow confined to chin; sides of breast and throat olive-yellow.

TAXONOMIC NOTE: Uganda and Angola birds are alike but brighter and yellower than Zaïre and West African ones; Bioko birds are lighter above and have more extensive yellow on cheeks (Eisentraut 1973). Several races have been recognized, but we follow White (1962) in regarding differences as too slight to warrant recognition.

Field Characters. A small, short-tailed warbler, bright yellow-green above, whitish below, with prominent yellow superciliary stripe, black lores, and yellow on face, sides of neck and under tail. Characteristic far-carrying nasal song. Bleating Warbler *C. brachyura* has bright yellow-green wings but grey head and back, no eye-stripe, clicking song; Olive-green Camaroptera *C. chloronota* is dingy green, with no eye-stripe and ringing song.

Voice. Tape-recorded (68, BRU, ERA, GUL, MAC). Song of ♂, a series of monotonously repeated low-pitched notes with a tuneless, nasal, rather gravelly quality; either a 2-syllable note is given twice, followed by a pause, 'jerwa-jerwa, jerwa-jerwa, jerwa jerwa . . .'; sometimes slightly higher-pitched, 'jeewa jeewa . . .', or a 3-syllable note, rising and falling in the middle, is given once, followed by a pause 'jerwya, jerwya, jerwya . . .'. ♀ may join in with a nasal mewing 'maaaa', like that of Bleating Warbler but more muted, less sharp. Series lasts *c.* 12–15 s and is repeated at *c.* 3 s intervals. Song carries over considerable distance. Scolding call note (Chapin 1953a).

General Habits. Inhabits undergrowth of secondary lowland forest about 7–10 years old, overgrown cultivation within forest, forest edge and gallery forest. Favours vine-covered trees and tangles of creepers; also coffee plantations (Angola) and open bush, cocoa plantations and cultivated areas outside forest (Bioko). Occurs up to 700 m in Liberia (Mt Nimba), 900 m in Bioko and 1200 m in E Zaïre, and at 700–1550 m in W Uganda.

Lives in pairs or family groups; does not join mixed parties. Forages actively inside undergrowth and foliage in mid-strata of forest and at its edges, making it difficult to see, although not afraid of man. Gleans insects from vegetation, dislodging them (notably caterpillars) by flapping or fluttering wings. Wing-fluttering, and frequent calling, also serve to keep pair in contact. Often attracted down to lower levels by feeding Red-cheeked Wattle-eyes *Dyaphorophyia blissetti*, and the 2 species often interact (Brosset and Erard 1986). Occurs in same forests as Bleating Warbler but forages higher (Dean *et al.* 1988). Sings often, from within cover of vines and creepers; singing ♂ lifts throat feathers, making bright blue skin on sides of throat highly conspicuous.

Food. Insects, including Orthoptera, termites, small moths and caterpillars.

Breeding Habits. Territorial and monogamous. During territorial disputes, ♂♂ perch in full view of each other at tops of bushes; each cocks tail, moves wings forward and lowers tips, pulls head into shoulders, puffs out breast then immediately calls 'kwa-kwa' whilst spreading out undertail-coverts and turning them towards the other bird. The birds change perches continually and flap their wings violently and audibly.

NEST: constructed from 3 leaves at end of a branch, stitched together with spider web, and lined inside with grass fibres and plant down. Nest built in bush at edge of clearing *c.* 2·5 m above ground.

EGGS: 3. Very pale blue, spotted and speckled with dark brown.

LAYING DATES: Ivory Coast, Dec–Mar; Angola, Mar (Traylor 1963); Zaïre, in some areas probably breeds all year (Lippens and Wille 1976), Itombwe, Oct–Feb (enlarged gonads Apr–May, Aug); Gabon, Feb (fledglings, Mar).

Reference
Brosset, A. and Erard, C. (1986).

Camaroptera chloronota Reichenow. Olive-green Camaroptera. Camaroptère à dos vert.

Plate 17 (Opp. p. 320)

Camaroptera chloronota Reichenow, 1885. Orn. Mber., p. 96; Misahöhe, Togoland.

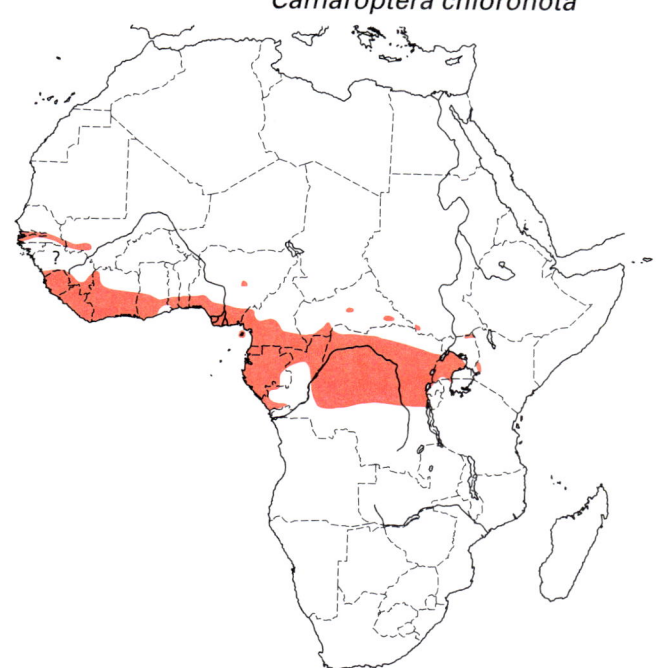

Camaroptera chloronota

Range and Status. Endemic resident, Senegal to Kenya. Senegal (2 records Badi), Gambia, rare, Abuko and Pirang Forest. S Mali, Bafing-Makana, Kangaba. Abundant SE Guinea along borders with Sierra Leone and Liberia. Frequent or common in forests of Sierra Leone, Liberia (from coast to N highlands), Ivory Coast (commoner in W from Taï to Mt Nimba and Odienné than from Abidjan to Ferkessédougou), Ghana, and Togo (Ounabé, Ebeva, Dzogbegan, Misahohé). Locally not uncommon S Benin (from coast to at least Abomey, Arian and Kétou and probably further north) and Nigeria (mainly in SW, also Kagoro (Jos Plateau) and in SE at Enugu, Boshi-Okwango Forest, Oban Hills, Leinde Fadali, Ngel Nyaki, Ikpan and Mamu Forest, rare in Gashaka-Gumti Game Reserve). Common Cameroon within forest zone, and in Bioko. Central African Republic, in Botambi, Bambari, Maboké and Baroua. Sudan, Bangangai Forest only. Equatorial Guinea, Gabon, S Congo from Kouilou Basin and Mayombe at least to Brazzaville and Gamakala, probably further; across Zaïre in forest zone, south to Kasai and Itombwe (rare), north to Ubangi and Ituri districts and across Semliki valley into Uganda at Bwamba. Widespread and common in forests of W and S Uganda from Budongo south to Bwindi-Impenetrable Forest, east to Kifu, Mabira and Sango Bay, and south into extreme NW Tanzania at Minziro Forest; also in NE Uganda on Mt Lonyili. Kenya from Mt Elgon to Kakamega, Kaimosi and North Nandi Forest. Occurs Rwanda (Dowsett and Forbes-Watson 1993).

Description. *C. c. chloronota* Reichenow: Togo to S Cameroon and perhaps this race in Gabon, Equatorial Guinea and Congo. ADULT ♂: forehead to uppertail-coverts greenish olive, crown slightly tinged brown; tail greyish brown, strongly washed with greenish olive, particularly at edges of feathers. Lores and ear-coverts grey; chin and upper throat pale grey, merging into slate-grey on lower throat; breast, belly and flanks slate-grey, upper breast washed greenish olive; thighs brownish olive-green, undertail-coverts greyish white. Primaries and secondaries brown, outer webs edged greenish olive, more markedly on secondaries, margins of inner webs light brown, paler underneath; upperwing-coverts brown edged greenish olive. Underwing-coverts pale yellow, yellow extending onto leading edge of wing, axillaries white tinged yellow. Upper mandible black, lower grey; eyes light brown, tawny or pinkish brown; bare skin at front corner of eye yellow; legs and feet yellowish brown, orange-brown, or pale brown, soles of feet and toes yellowish. ADULT ♀: like ♂ but green wash on upper breast paler, belly white, flanks grey. SIZE: (10 ♂♂, 9 ♀♀) wing, ♂ 50–57 (53·6), ♀ 46–53 (49·0); tail, ♂ 18–25 (22·0), ♀ 16–20 (18·6); bill to feathers, ♂ 11–12 (11·9), ♀ 10–12 (11·2); tarsus, ♂ 19–22 (20·0), ♀ 17–20 (18·9). WEIGHT: (Nigeria) ♂ (n = 7) 10–13, ♀ (n = 5) 9–10·5, unsexed (n = 10) (10·9); (Cameroon) ♂ (n = 10) 10·5–13 (11·8), ♀ (n = 6) 8·5–10 (9·4).

IMMATURE: upperparts brown with greenish olive wash, underparts greyish yellow in midline, washed grey at sides, tail brown tinged with greenish olive. Subadult much like adult ♀ with throat and midline of belly silky white, breast faintly washed pale yellow or buffy, flanks and breast pale grey. Bill blackish, pale at tip; eyes brown; legs and feet brownish flesh.

C. c. kelsalli Sclater: Senegal to Ghana. Ear-coverts and region around eye rufous-buff; upper breast of ♂ grey not olive. WEIGHT: (Liberia) ♂ (n = 10) 9·2–12.3 (10·0), ♀ (n = 7) 9·0–12·7 (10·6); unsexed (n = 13) 9·5–12·3 (11·2); (Mt Nimba, Liberia) ♂ (n = 16) (10·7 ± 1·2), ♀ (n = 9) (10·3 ± 1·2), juv. ♂ (n = 11) (10·1 ± 0·9); (Ghana) unsexed (n = 3) 8·0–12·3 (10·1).

C. c. granti Alexander: Bioko. Like nominate race but underparts deeper grey. Larger: (4 ♀♀) wing 48–54 (50·8), tail 23–26 (24·3), bill to feathers 14–15. WEIGHT: (Bioko) ♂ (n = 2) 12, 13, ♀ (n = 4) 10–12 (11).

C. c. toroensis (Jackson): SE Central African Republic, SW Sudan, Zaïre (middle Congo R., N Kasai north and east to Uele R., Ituri and L. Kivu), Uganda, W Kenya and NW Tanzania. Like nominate race but forehead, lores, around eye, cheeks, chin and throat are more brown or rufous-buff, and tail and upperparts not quite so green. WEIGHT: (Central African Republic) ♂ (n = 3) 10·5–12 (11·2); (W Kenya) unsexed (n = 21) (10·4 ± 1·1); (Sango Bay Forest, Uganda) ♂ (n = 7) 10–13, ♀ (n = 5) 9–10·5; (Kalinzu Forest, Uganda) unsexed (n = 8) 8–13; (Bwamba, Uganda) unsexed (n = ?) 9–12, 1 bird 18.

C. c. kamitugaensis Prigogine: Zaïre in Itombwe area. Very like *toroensis* but forehead, side of head, breast and flanks less buffy, breast-band darker and belly whiter. SIZE: (10 ♂♂, 4 ♀♀) wing, ♂ 52·5–58 (55), ♀ 48–50·5 (49·6); tail, ♂ 31·5–38 (34·2), ♀ 43·5–52·5 (50·9).

Field Characters. A small, compact warbler with dingy plumage matching its undergrowth habitat, dull green above, greyish below, some races with buffy or rusty face. Tail short, often cocked, legs quite long. Easily overlooked unless its ventriloquial ringing song is heard. Face lacks yellow eye-stripe of Yellow-browed Camaroptera *C. superciliaris*; sympatric forms of Bleating Warbler *C. brachyura* have grey back and clicking song and inhabit forest edge and bush.

Voice. Tape-recorded (32, 68, B, F, ASP, CART, GRI, MOY, ZIM). Song a clear ringing note, 'ee' or 'hee', repeated at a rate of 4–5/s for lengthy periods, sometimes for only a minute but can continue for up to 4 or 5 min (Chappuis 1979a). Notes are all more or less on one pitch and of the same intensity, although as song progresses they sound as if they are descending slightly and starting to fade away. Sometimes the 'ee' has a slightly vibrant quality, or may be lower in pitch, 'ü'. Song may start up abruptly or be introduced by a few faster notes which merge into song, either ringing, like song, 'pipipipi-ee-ee-ee-ee- . . .', or a sharp chatter, 'chichichichichichipa-ee-ee-ee-ee- . . .'. Individual notes are pure, drawn-out, starting and ending gradually, and are centred around 2·0 kHz (Chappuis 1971). Ventriloquial; song seems alternatively to become louder and softer, and to come from different directions, but singer remains completely still, concealed among leaves (Keith and Gunn 1971). One singer later gave slower (2 per s) series of different notes, upslurred, drier, not ringing, 'toowi-toowi-toowi . . .' or preceded by brief note, 'ti-toowi-toowi . . .', but it is uncertain whether this was a song variation or a repeated call (Keith and Gunn 1971). Also has nasal call typical of the genus, 'jeewee' or 'juwee'.

General Habits. Occurs mainly in dense undergrowth of lowland and montane primary and secondary forest, in forest clearings and beside forest tracks, also in thickets and cocoa plantations near forest villages; in Guinea, all wooded habitats, including coffee plantations (Halleux 1994). Confined in Kenya to humid region with > 1000 mm annual rainfall. Occurs up to 2000 m in Uganda (Mt Lonyili), 1800 m in Kenya, 1300 m in Bioko, 1600 m in Nigeria (Ngel Nyaki), 1400 m in Cameroon (Mt Nlonako).

Occurs singly, in pairs, or family parties, and joins mixed-bird parties when foraging. Skulks in lower stratum of forest and dense undergrowth when feeding, usually below 2 m but as high as 10 m, jumping smartly from twig to twig and flying through foliage; also feeds on ground, where it hops; catches insects disturbed by red ants. ♂ sings concealed either within low thicket or else in a tangle of vines, 4–5 m above ground, and shakes or bobs body up and down (Good 1953). In Gabon sings all year, but mainly in Sept–Mar. Slightly larger than Yellow-browed Camaroptera, but the 2 are evidently not segregated ecologically.

Food. Lepidoptera caterpillars and eggs; small beetles; spiders.

Breeding Habits. Monogamous and territorial. Defends territory with flight display involving rhythmical wing-flaps.

NEST: edges of a large horizontal leaf are bent down and sewn together with vegetable fibres to form a tube or pouch (Brosset and Erard 1986), or a spray of leaves is sewn together to make a pouch, with one or more leaves forming a roof; pouch is filled with white kapok or down, bound with tough spider web (Bates 1911). Placed c. 30 cm from ground, well hidden and sheltered from rain.

EGGS: 2 (n = 2, Gabon). White or pale greenish blue, finely spotted with pale yellowish, pale brown or lilac. SIZE: 17·5 × 12·5 (2 eggs, Cameroon).

LAYING DATES: Liberia (enlarged gonads Mt. Nimba, Feb, Apr, Sept, Oct; recently laid, Sept); Nigeria (enlarged gonads, Jan–Oct); Cameroon, May, Nov (juvs, Dec, May), a wet season breeder in lowland forest, but on Mt Cameroon breeds in dry season (Tye 1991); Bioko (enlarged gonads, Dec); Sudan, Feb, Mar; Gabon (enlarged gonads, Dec–Feb (dry season)); Zaïre, Itombwe, all months except Feb, July and Oct (juvs Jan, Mar, June–Aug); E Africa: Region B, May, Dec; Uganda (enlarged gonads, Feb).

BREEDING SUCCESS/SURVIVAL: longevity records 4·1 and 8·5 years (Kakamega, Kenya).

Reference
Brosset, A. and Erard, C. (1986).

Genus *Calamonastes* Sharpe

Warblers of dry woodland and thornbush. Bill long and slender with gently arched culmen; wings rounded and short, P5–P6 longest, P9 more than half length of P8. Tail of 12 feathers, rounded, and faintly barred throughout. Tarsus long, strongly scutellated, rough to the touch. Plumage grey or brown, with barred underparts and spotted wing-coverts. Monomorphic, except in juveniles of 2 races of *C. undosus*. Nest similar to *Camaroptera*, delicate, cradled in living leaves which are stitched or rivetted together.

Endemic; 3 species. Close to *Camaroptera*, sometimes united with it (Hall and Moreau 1970, Dowsett and Forbes-Watson 1993), more often not (Watson *et al.* 1986, Sibley and Monroe 1990, Short *et al.* 1990, Maclean 1993). Tail longer than *Camaroptera* and often flirted; barring on underparts lacking in *Camaroptera*; songs very different, but this may not be significant since songs within *Camaroptera* vary considerably. One clear link to *Camaroptera* is the nasal 'maaaa' call. *Calamonastes* may be the product of an earlier incursion into dry country by forest-based *Camaroptera*; today the *C. brachyura* group is making another incursion, although still present in forest as well.

Plumage and voice are so similar within *Calamonastes* that all forms might be considered well-marked races of a single species were it not that *C. simplex* and *C. undosus* meet without intergradation in the north and *C. undosus* and *C. fasciolatus* do the same in the south. Northern races of *C. undosus* are greyer with partly barred underparts and song similar to *C. simplex*; southern races are browner with wholly barred underparts and different voice, and have been considered a separate species, *C. stierlingi*, but since the 2 groups hybridize in a number of places and in Zambia hybrids even produce a hybrid song, we consider *undosus/stierlingi* all one species.

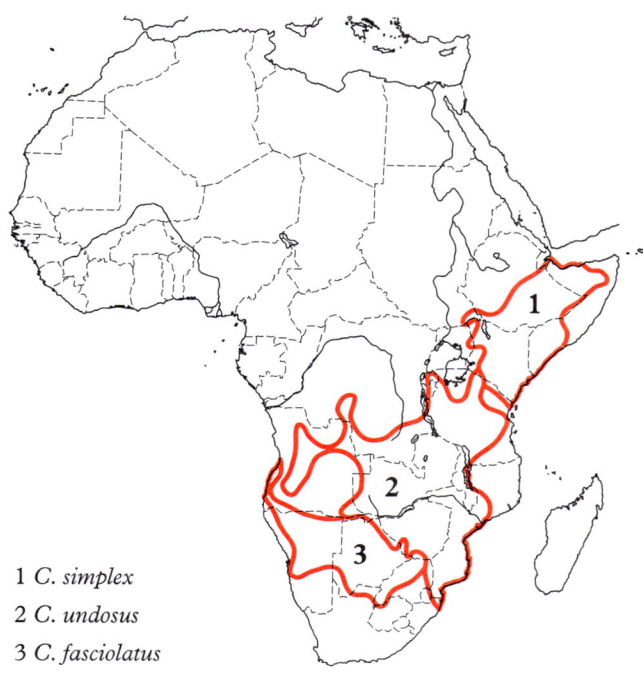

Calamonastes simplex superspecies

1 *C. simplex*
2 *C. undosus*
3 *C. fasciolatus*

Calamonastes simplex (Cabanis). Grey Wren-Warbler. Camaroptère modeste.

Plate 17 (Opp. p. 320)

Thamnobia simplex Cabanis, 1878. J. Orn., pp. 205, 221; Ndi, Teita District, Kenya.

Forms a superspecies with *C. undosus* and *C. fasciolatus*.

Range and Status. Endemic resident NE Africa, common and widespread. Somalia, fairly common and widespread in N (upper Guban, Golis range, across plateau to Burao, Oadweina, Hargeisa and Gebileh) and in S, south of *c.* 4°N; absent elsewhere. Ethiopia, common and; widespread in NE, E, SE, Rift and in S (Arero, Yavello, Mega, Békaoulé, Omo-Raté). Sudan, fairly common in extreme SE. Uganda, only Mt Moroto

Calamonastes simplex

area. Kenya, wide-ranging in drier parts, up to 1300 m, but absent from moist W and from very arid areas in E and east of L. Turkana. NE Tanzania along Kenya border from Loliondo and L. Natron to Pangani R.

Description. ADULT ♂: upperparts dark brown, lower back and rump with greyish wash, feathers of lower back loose; uppertail-coverts brown. Tail blackish brown, each feather with faint lighter bands at 2–3 mm intervals; T1–T3 tipped buff, T4–T6 tipped white, old feathers fade and tips abrade. Lores, ear-coverts and cheeks brown, mottled with white shaft streaks. Chin and upper throat narrowly barred brown and greyish white, breast, belly, sides and flanks dark greyish olive-brown, sometimes faintly barred greyish white, especially on centre of belly; undertail-coverts dark brown with white tips, thighs brown mottled with white or pale brown. Primaries, secondaries and upperwing-coverts dark brown, inner webs of primaries edged buffish white, edges of wing-coverts paler brown especially at tips, marginal coverts on bend of wing mottled brown and white. Underwing-coverts white, each with subterminal brown bar making them mottled; axillaries dark brown. Bill black; eyes red, reddish brown, hazel or brown; legs and feet pale brown, red-brown, brown or dusky flesh. Sexes similar. SIZE: (10 ♂♂, 10 ♀♀) wing, ♂ 55–60 (58·2), ♀ 53–61 (56·8); tail, ♂ 50–58 (53·0), ♀ 50–56 (53·2); bill to feathers, ♂ 12–14 (12·7), ♀ 12–13 (12·5); tarsus, ♂ 19–21 (20·0), ♀ 19–21 (19·5).

IMMATURE: like adult but upperwing-coverts margined with whitish buff, in older birds only tips whitish buff. Underparts paler grey with white bars across centre of belly, hardly discernible in some birds, clear in others; flanks buffish white, barred grey-brown. Bill horn.

TAXONOMIC NOTE: here considered monotypic. Races formerly listed under this species, e.g. by Watson *et al.* (1986), are now considered part of *C. undosus*.

Field Characters. The only dry country warbler in E Africa that is entirely slate-grey (sometimes a little pale barring on belly). Long legs; tail rather long, constantly raised and lowered. More often heard than seen – song a distinctive, loud 'CHUP'. Meets Miombo Wren-Warbler *C. undosus* in SW Kenya and NE Tanzania; Miombo is distinctly barred below; it has similar song but note slurred, less dry, 'piewk'.

Voice. Tape-recorded (68, B, F, CART, NOR, STJ). Song a repeated, sharp, loud and far-carrying 'tchup', 'chok', 'tewp', 'click', 'click-clack' or 'chip-chip' emphatic and percussive, like sound of 2 pebbles or 2 thin pieces of hard dry wood struck together; repeated at steady, even tempo at intervals of just under 1s. Note lasts only *c.* 0·04 s; pitch falls from 4·0 to 1·5 kHz. Intervals between single notes 0·8–0·92 s (n = 50, Kenya, 3 localities). In double-note song, time between 'click' and 'clack' ranges from 0·16–0·19 to 1·0–1·1 s (n = 15, Maungu, Kenya). Bird may switch from 1-note to 2-note song during single bout of singing. Alarm, a soft percussive 'tchup', usually 4–5 together, sometimes more, at rate of 7 per s. Call, a soft 'oo-tu tu tu tu', audible only at close quarters (Moreau and Moreau 1939).

General Habits. A bird of *Acacia-Commiphora* thornbush and dry *Acacia–Chrysopogon* wilderness. Keeps to undergrowth except when singing and skulks in thick thorn scrub. Ranges from 760 to 1520 m in Somalia and Ethiopia (higher on Mt Wagar) and from 100 to 1300 m in Kenya. Often drops to ground for insect prey. Tail generally slightly fanned, and constantly and deliberately wagged up and down through about 45°. Sings from topmost branch of tree. In 'click-clack' song bird bobs up on the 'click' and comes down sharply on the 'clack'. At every note bill snaps open and shuts and 'whole bird shakes from stem to stern' (Moreau and Moreau 1939).

Food. Insects.

Breeding Habits. Monogamous and territorial.
NEST: 2 described: remarkably strong globular structure with side entrance, made of exceedingly fine fibres like untreated silkworm cocoon, supported by 2 stems in its sides, covered by leaves drawn together with stitches, and with fine fibre cup added inside. Ext. diam. 150, entrance diam. 50. Sited 30 cm above ground in leafy shrub. 'One bird, presumably the female, did the work; the male called frequently, greeted the hen whenever she brought material, and with outspread tail described an aerial arc over the nest' (Moreau and Moreau 1939).

EGGS: 3–4. Greyish white (or light blue: oviduct egg), thickly marked with minute dark or pale brown dots slightly concentrated into a belt around large end. SIZE: (n = 3, E Africa) 18·8–19·3 × 12·5–12·5 (19·0 × 12·5) (Jackson and Sclater 1938, Moreau and Moreau 1939).

LAYING DATES: Ethiopia, April (and gonads enlarged Feb, May); Somalia (nest-building Jan, Apr); E Africa: Region D, May, Nov–Jan (fledglings Kibwezi, Kenya, late May).

Reference
Moreau, R. E. and Moreau, W. M. (1939).

Calamonastes undosus (Reichenow). Miombo Wren-Warbler. Camaroptère du miombo.

Plate 17 (Opp. p. 320)

Drymoica undosa Reichenow, 1882. J. Orn., p. 211; Kakoma, Tabora.

Forms a superspecies with *C. simplex* and *C. fasciolatus*.

Range and Status. Endemic resident in woodlands of S-central and SE Africa; possible migrant N Botswana. Angola in N (Cabinda: Landana, Malange to S Lunda), on central plateau (central Huila to S Cuanza Sul, Huambo and N Bié) and in NE on borders of Zaïre and NW Zambia, also in SE on border of Namibia. Namibia, common in W Caprivi but vagrant in E (Branfield 1990, Brown 1990). Zaïre, occurs Kinshasa, status in rest of lower Congo R. area uncertain; widespread in S and SE (Kasai, Shaba); Rwanda (Kibungu and along Akanyaru R.); most of Tanzania except L. Victoria basin, E highlands and SE lowlands, north to Ngara, Biharamulo, N Tabora and Serengeti Nat. Park (Stronach 1990), into Kenya at Loita Hills and Mara R., east to foothills of highlands, and E lowlands from Bagamoyo to Songea. Zambia, throughout except SW, Malaŵi, fairly common throughout except N, from lake littoral to 1450 m, north to Mzimba and Nkhata Bay. Mozambique, widespread and quite common south of Save R., less common north of Save R. (Lacerdonia, Inhamitanga, Inhaminga, Vila Paive de Andrade, Gorongoza Nat. Park, Muanza, Chiniziua, Garuso, Revue, Dondo and Beira); few records in N Mozambique (Tete in Zambezi R. valley and W Niassa). Zimbabwe, absent from parts of N and from area around Plumtree and southwest of Nata R., but widespread over rest of country, from L. Kariba across Kalahari Sands on central plateau to Chipinga uplands, Malilangwe hills and lower Limpopo valley; absent from Sabi valley and E highlands. Botswana, common resident in NE and sparse west to E end of Moremi Wildlife Reserve; also patchily in E near Nata and south to N Tuli region. NE and E Transvaal in lowveld, including Kruger Nat. Park, north to area north of Soutpansberg and west to Messina. E Swaziland and NE Zululand, south to L. St Lucia. At Mosope, Botswana, 6 ♂♂ singing in an area of 20 ha (Herremans and Herremans 1992).

Description. *C. u. undosus* (Reichenow): Rwanda, Tanzania (except E) and Kenya. ADULT ♂: forehead to upper back uniform brown, slightly darker on head, lower back and rump greyish brown, feathers of rump sometimes with buffish tips; uppertail-coverts brown. Tail brown, each feather with faint darker bands at regular intervals of 2–3 mm throughout; tip of T6 and T5 buffy. Lores, ear-coverts and cheeks brown, feathers of lores and cheeks with buffish white tips giving mottled effect. Chin mottled brown and white, or distinctly barred, throat and upper breast pale olive-brown, lower breast and belly buff or buffy white, faintly barred with pale grey-brown, sides and flanks darker with cinnamon wash and barred more distinctly grey-brown; undertail-coverts whitish buff, broadly barred brown; thighs cinnamon, mottled brown in some birds. Primaries, secondaries and upperwing-coverts brown, primaries with pale outer edges, tips of upperwing-coverts pale brown. Underwing-coverts buff, sometimes mottled dark brown; marginal coverts mottled brown and white; axillaries whitish brown; underside of primaries and secondaries buffish white. Bill black; eyes light brown, or red-brown; legs brown-

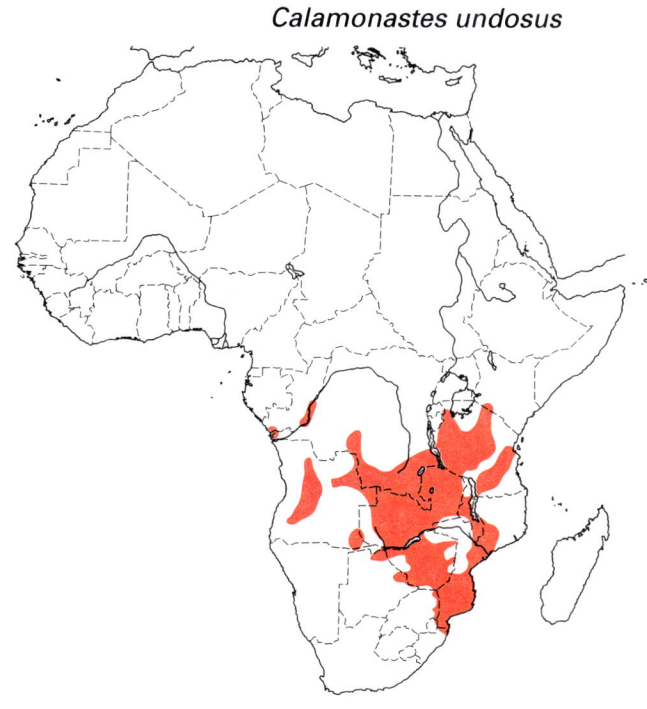

Calamonastes undosus

ish flesh. Sexes alike. SIZE: (10 ♂♂, 8 ♀♀) wing, ♂ 60–65 (62·8), ♀ 55–68 (60·1); tail, ♂ 45–52 (48·6), ♀ 43–52 (48·1); bill, ♂ 14–16 (14·7), ♀ 13–15 (14·4); tarsus, ♂ 20–23 (21·7), ♀ 21–22 (21·5). WEIGHT: (S Tanzania) 1 ♂ 12·5, 1 ♀ 12.

IMMATURE: like adult but edges of secondaries, wing-coverts and primaries paler brown, undertail-coverts cinnamon-brown, chin greyish white, breast buffish and barred (each feather with subterminal dark brown bar), belly with cinnamon wash and flanks more distinctly cinnamon, belly sometimes with yellow wash.

NESTLING: unknown.

C. u. cinereus Reichenow: Congo, W Zaïre, NW and NE Angola, NW Zambia (N Mwinilunga and Ndola districts). Upperparts greyer and lighter than *undosus*, throat and breast pale grey, any barring on underside faint and confined to throat and belly, belly paler grey, flanks whitish. Immature dimorphic, either like immature *undosus*, or (uncommon) dull buffy or greyish green, paler on underparts (Traylor 1967a).

C. u. katangae (Neave): Zaïre (Shaba), N Zambia (Northern Luapula and North-Western Provinces, but not Ndola and Mwinilunga districts). Like *undosus* but more greyish brown above and lighter below, lower breast and belly whitish grey or pale buff, barring below absent or virtually so; sides of belly buff, flanks whitish grey. SIZE: wing, ♂ (n = 15) 53–60 (56·5), ♀ (n = 6) 50–64 (61). WEIGHT: (Upemba, Zaïre) unsexed (n = 13) 11–16 (13·0).

C. u. huilae (Meise): W-central Angola. Similar to *cinereus* but larger, and thighs brown. Immature possibly with same dimorphism as *cinereus* (Traylor 1967a). SIZE: wing, ♂ (n = 22) 63–68 (65·5), ♀ (n = 7) 59–63 (60·4).

C. u. stierlingi (Reichenow) (including '*buttoni*', '*neglectus*'): SE Angola, NE Namibia, S Zambia east to Eastern Province, E Tanzania (except coast), S Malaŵi east of Shire R. and N Mozambique. Similar to *undosus* but forehead to hindneck dark olive-brown with rufous wash, back and rump brown, tail-feathers edged rufous, lores mottled black and white, throat and breast evenly barred blackish brown and white, undertail-coverts rufous, primaries and secondaries olive-brown, medium wing-coverts tipped with large white blobs, less marked in greater wing-coverts. IMMATURE: like adult but upperparts more reddish brown, breast with yellowish wash, thighs barred pale cinnamon and brown; upper mandible dark reddish brown with black tip, lower mandible paler; eyes dark brown; legs and feet flesh-pink; greater and median upper-wing-coverts edged pale brown, not blotched as in adult; underwing-coverts whitish with horeshoe-shaped blotches (not bars), median underwing-coverts (carpal joint) pale yellow. SIZE: (6 ♂♂, 3 ♀♀) wing, ♂ 59–65 (62·7), ♀ 56–60 (58·0) (also, wing, ♂ (n = 12) 63·5–68 (65·3), ♀ (n = 12) 57–62·5 (59·6): Clancey 1984); tail, ♂ 43–51 (47·3), ♀ 40–46 (43·0); bill to feathers, ♂ 12–14 (12·9), ♀ 11–12 (11·8); tarsus, ♂ 21–24 (22·3), ♀ 20–21 (20·7). WEIGHT: (E Tanzania) ♂ (n = 4) 12–14 (13·0), ♀ (n = 2) 10·5, 13.

C. u. olivascens (Clancey): coastal Mozambique, north of Limpopo R., probably this race in lowland S Malaŵi west of Shire R. and in coastal E Tanzania. Like *stierlingi* but rather darker and more olivaceous above. Smaller: wing, ♂ (n = 13) 58·5–62 (60·9), ♀ (n = 6) 51–56·5 (54·2).

C. u. irwini (Smithers and Paterson): Botswana, Zimbabwe, extreme E Zambia, Malaŵi (except S lowlands), interior of Mozambique between Zambezi and Limpopo. Like *stierlingi* in size and plumage but upperparts including wings and tail more reddish brown, underparts white with blacker bars. WEIGHT: (Umtali, Zimbabwe) ♂ (n = 3) 12·3, 13·4, 14·0, 1 ♀ 13·1

C. u. pintoi (Irwin): NE South Africa, Swaziland, Mozambique south of R. Limpopo. Upperparts like *irwini*, underparts whiter, transverse barring dark olive-brown rather than blackish, barring on flanks generally indistinct. Slightly larger than *olivascens*: wing, ♂ (n = 3) 61·5–63·5 (62·3), ♀ (n = 5) 54–57·5 (56·3).

Field Characters. A bird of miombo and other dry woodlands, ecologically separated from its close relatives Grey Wren-Warbler *C. simplex* and Barred Wren-Warbler *C. fasciolatus*, which live in thorn scrub and acacia savanna. Upperparts dark grey-brown in N to brown or olive-brown in S; underparts paler, partly barred in N, wholly barred in S. Northern birds sing a single 'piewk' at 1-s intervals, similar to the dry 'chup' of Grey Wren-Warbler; latter best distinguished by its all-dark plumage. Southern birds have a trill with distinguishable notes, 'tiriririp', while song of Barred Wren-Warbler has same quality but is a single long note, 'preep' or 'kirruk'. Miombo told from Barred Wren-Warbler by whiter underparts boldly barred black from chin to undertail; Barred is richer brown above (no olive wash) and has buffy underparts, less heavily marked. Miombo has shorter tail, and at close range shows brighter eye (orange to red-brown) and blacker bill.

Voice. Tape-recorded (72, 86, 88, 91, B, C, F, CART, CHA, HA, HOR, HRS, MOY). Song of *C. u. undosus* and *C. u. cinereus* similar in rhythm to that of *C. simplex*, a single loud note repeated at intervals of *c*. 1 s, less dry, more ringing than *simplex* and often slurred, 'piewk . . . piewk . . . piewk . . .'; variable, sometimes almost 2-syllabled, 'pwee', 'weeyi', 'tchewup'. Note typically lasts 0·1–0·14 s (n = 14) and pitch falls from 4·0 kHz to 2 kHz; interval between notes 0·9–1·3 s (av. 1·05 s, n = 11, Zambia) but sometimes 1·5–2·0 s (Dowsett and Dowsett-Lemaire 1980). Another song of *cinereus* was 'chi, chee, chi-chi-chi-chi, cheee' lasting 3 s and repeated every 3–4 s (C. Carter, pers. comm.). Song of *stierlingi* group a repeated, 3–4 syllable 'tiririp' or 'tiriririp', high-pitched, metallic and ringing; sometimes 2 syllables, 'tirrip', sometimes accelerated into cricket-like trill, 'tirrrrrrrip'. Number of calls varies from 4 to 50 (once 90, Harare); tempo almost twice as fast as *simplex*; varies regionally from 1·8 to 3·4 per s: Zambia (Lusaka, n = 5) 1·9–2·05 (2·0); Zimbabwe (Harare, n = 8) 2·3–2·9 (2·4); Transvaal (n = 12) 1·8–2·1 (1·8); Natal (Sodhwana Bay, n = 4) 3·3–3·6 (3·4). A hybrid *C. u. cinereus* × *C. 's'. stierlingi* recorded in Zambia by R. Stjernstedt (tape 86) sang a high 'pee-uu', 'pleeyi' or 'plee' at rate of 1·6 times per s; when this song was played back hybrid gave 'tiririp' song. Calls include a muffled, telephone-like double 'prree-prree' or 'kew-kew' and a bleating, buzzy 'maaaa' like that of *Camaroptera* spp.; alarm, sharp 'tsik tsik'.

General Habits. Throughout range inhabits miombo (*Brachystegia*) woodland, living in small thickets often on and around termite mounds. In Zambia occurs also in shrub layer of mutemwa (*Baikiaea*) woodland and occasionally in mopane (*Colophospermum mopane*), and in Zimbabwe in understorey of gusu (*Baikiaea plurijuga*) woodland on Kalahari Sand. In S and SW of range inhabits *Acacia*, *Baikiaea*, mopane and other woodlands (Clancey 1984). In Tanzania inhabits woodland with grassy, partially bare ground; in Natal mixed woodlands on littoral plain; in Transvaal restricted to lowveld. Occurs from sea level to 610–760 m (Malaŵi) and 1700 m (Tanzania).

Singly, in pairs, and small family parties in the dry season. Occasionally joins bird parties. Forages mostly on or near the ground, often among bramble-like scrub on floor of tall miombo woods, but also at tips of canopy in small *Acacias* (da Rosa Pinto and Lamm 1953). Ascends into canopy when disturbed. When bird flies up from ground, wings make an audible rattling (Vincent 1935). Sings from within canopy, stops singing when observer approaches: bird either flies further away to sing or, more often, dives into ground cover and remains silent. Responds vigorously to playback of its song.

Usually absent from Botswana June–Sept in dry summers, but present in July when earlier rainfall is prolonged.

Food. Insects.

Breeding Habits. Solitary and monogamous. Territorial, defending area often centred on termite hill thicket. In display, flies fast, dodging between trees, high then low but always below the canopy; sings in flight (Winterbottom 1938). In courtship flight flutters like a butterfly. ♂ at Harare, Zimbabwe, sang from 2 m high perch, then flew upwards uttering faint 'peep peep'; from height of *c*. 3 m above perch it dropped vertically into cover (J. Cook, pers. comm.).

NEST: loosely made, almost entirely of felted vegetable down, sometimes with a little grass, bark or rootlets; 15 mm thick at base; built into a cradle of leaves, usually a hanging cluster, stitched or rivetted together by spider web, which is plugged through tiny holes made in adjacent leaves, each hole 7–10 mm from leaf edge. Shape varies with leaf shape and size: nest spherical (diam. 65), oval, or long oval, with entrance (diam. 25–30) at side and near top. Entrance usually concealed by overhanging leaves stitched together to form a porch. Placed 0·5–4·5 m high, well within broad-leaved tree such as *Brachystegia spiciformis*, *Julbernardia globiflora*, *Ziziphus mucronata*, *Combretum mollis* (a favourite in Zambia), *Faurea speciosa* and *Terminalia sericea* (Masterson 1972). Nest is readily deserted.

EGGS: 2–3 (av. of 4 clutches 2·5). White or pale blue, sometimes rich turquoise blue, marked all over with minute brown or grey spots or speckles. SIZE: (n = 10, southern Africa) 15·5–18·3 × 11·5–13·1 (17·2 × 12·3); (n = 5, Zimbabwe) 17·1–18·3 × 13·5–14·0 (17·7 × 13·7) (Priest 1935).

LAYING DATES: Angola, Oct and Nov; S Zaïre, Sept–Mar; E Africa: Region C, Oct–May (fledglings early June, Serengeti Nat. Park: Stronach 1990); Zambia, Oct (1 clutch), Nov (11), Dec (5), Jan (9), Feb (3), Mar (1); Zimbabwe, Sept (3 clutches), Oct (17), Nov (17), Dec (14), Jan (4), Feb (2), Mar (4); Malaŵi: Oct (2 clutches); Transvaal, Nov.

References
Clancey, P. A. (1970a, 1984).
Dowsett, R. J. and Dowsett-Lemaire, F. (1980).
Herremans, M. and Herremans, D. (1992).
Irwin, M. P. S. (1960).

Calamonastes fasciolatus (Smith). Barred Wren-Warbler. Camaroptère barrée.

Plate 17
(Opp. p. 320)

Drymoica fasciolata A. Smith, 1847. Illus. Zool. South Africa, Aves, Pl. 111, fig.2; northeast of Latakoo.

Forms a superspecies with *C. undosus* and *C. simplex*.

Range and Status. Endemic resident. Angola, fairly common in SW (Lobito Bay and arid coastal plain of Benguela) and south into Namibia, where widespread but local in N and E, from Ovamboland and Etosha Pan south in W (present Spitzkoppe but absent Brandberg) to Naukluft Mts and east into Botswana; also Caprivi (common in W, uncommon in E along Kwando/Mashi R.). Sparse to common throughout Botswana except SW duneland. Zimbabwe, widespread but local and uncommon in SE along Botswana border and along Transvaal border east to Beit Bridge. Transvaal, common in bushveld of NW, east to Mutale R.; N Cape Province from Botswana border south through Kuruman and just into NW Orange Free State near Hoopstad (de Swardt 1991). Common; density in Transvaal (Nylsvlei) 1 pair /4 ha.

Description. *C. f. fasciolatus* (Smith): central Namibia, N Cape Province., Botswana (except SE) and Zimbabwe (W Matabeleland). ADULT ♂ (breeding), Nov–Apr: forehead to back cinnamon-brown, rump and uppertail-coverts cinnamon. Tail-feathers cinnamon-brown, darker distally, tips whitish, with tawny outer margins, faint dark bands uniformly spaced along each feather. Cheeks, lores, and chin white narrowly barred black, throat and upper breast brown, sometimes flecked with buffy white at sides of breast, more broken up and flecked with buff in birds moulting into non-breeding dress; lower breast to upper belly whitish, broadly barred brown-black, lower belly and flanks similar but strongly washed tawny, thighs tawny, undertail-coverts tawny with subterminal brown V-shaped blotches. Primaries and secondaries dark brown with tawny outer margins to webs, broadest in secondaries; tips of secondaries pale, margins of inner webs pale. Greater upperwing-coverts dark brown, edges of webs tawny, tips with whitish blobs; median upperwing-coverts and marginal coverts dark brown tipped white. Underwing-coverts white, washed with pale tawny, each with subterminal brown blotch giving barred effect, axillaries brown tipped whitish. Bill black; eyes bright reddish hazel or pale light brown; legs and feet reddish brown or dark brown. Breeding plumage when worn is similar, but tawny wash of underparts paler, whitish tips of tail and secondaries usually abraded and absent. ADULT ♂ (non-breeding), May–Oct: like breeding ♂, but throat and upper breast white, narrowly barred black. ADULT ♀ (breeding and non-breeding): like non-breeding ♂ barring on throat and belly sometimes darker. In freshly moulted non-breeding dress, ♀ has upper breast barred brownish, sides of body and flanks not barred (or vestigially barred), rest of underparts plain; throat unmarked (in 50% of birds: Clancey 1970a). Bill black to blackish brown; eyes light brown or grey; legs and feet dark brown or pinkish brown. SIZE: wing, ♂ (n = 39) 57–64 (61·1), ♀ (n = 23) 54–62 (58·1); tail, ♂ (n = 11) 52–57·5 (av. of 7, 52·0), ♀ (n = 9) 47–52 (av. of 3, 47·3); bill to feathers, ♂ (n = 7) 12–14 (12·9), ♀ (n = 3) 12–13 (12·3); tarsus, ♂ (n = 7) 21–24 (22·3), ♀ (n = 3) 21–23 (22·0), ♂♀ (n = 20) 20–23. WEIGHT: no data; see under *C. f. europhilus*.

IMMATURE: very like non-breeding adult, but upperparts more rufous-brown and breast transiently washed with yellow, barring of underparts duller, heavier and less clearly defined.
NESTLING: unknown.

C. f. europhilus (Clancey): Transvaal, Zimbabwe (S Matabeleland) and SE Botswana. Like nominate race in non-breeding dress but darker above; throat buffish white mottled blackish; barring on underparts blacker, more distinct and more extensively developed on sides, flanks and upper belly. In breeding dress (Sept–Mar) not always separable from nominate race, many birds more heavily barred below, flanks and undertail-coverts washed deeper buff. Intergrades with nominate race in SE Botswana (Kanye) and Zimbabwe (Sentinel Ranch) (Clancey 1970a). WEIGHT: (Transvaal) unsexed (n = 19) 11·1–14·3 (13·1).

C. f. pallidior Hartert: Angola. Differs from nominate subspecies in having breast, belly and lower flanks unbarred. Slightly longer-tailed: ♂ (n = 4) 49–55 (52·3), 1 ♀ 46.

Field Characters. A warbler of thornbush and acacia savanna in SW Africa, with longish tail often cocked and fanned over back as it skulks in low cover. Rich brown above; underparts buffy, barred except that ♂ in breeding season has breast plain chocolate brown. Can only be confused with Miombo Wren-Warbler *Calamonastes undosus*, which it meets in NE of range. Barring less distinct than on Miombo Wren-Warbler and lacking on undertail; Miombo has clear white underparts with blackish barring throughout. At close range Barred Wren-Warbler shows dull brownish eye (not orange) and browner bill. Songs are similar in tone but differ in structure, see *Voice*.

Voice. Tape-recorded (38, 58, 72, 88, 91, B, F, LUT, WAT). Song of ♂ 2–7 (up to 12) sharp, high-pitched, vibrant notes like pea-whistle or cricket, 'prreep-prreep . . .', 'kirruk-kirruk . . .', 'kerwuck-kerwuck' . . ., 'kreep-kreep . . .'. Notes are *c.* 0.3 s in length and are repeated about every half second; intervals between notes: Transvaal (n = 27) 0·7–1·1 s (0·85 s), Namibia (n = 8) 0·35–0·49 s. In display, a long cricket-like trill, 'keeeer-rrrrrr', repeated every 2·0–2·3 s. Calls include chattering 'dzheee, dudududududu' and a buzzy, scolding 'dzee-zee-zee . . .'. Snaps wings.

General Habits. Inhabits semi-arid savanna with *Acacia*, *Combretum* and *Commiphora*, thick bush with small-leaved thorny bushes and patchy ground cover, deciduous woodland and *Acacia* thornveld savanna.

Usually in pairs or small groups, often in bird parties; forages by working up through bush or tree, tail often cocked and fanned over back, then flying to foot of next bush. Pair keeps in contact by calling. Where it occurs alongside Miombo Wren-Warbler, it is found in scrub while Miombo occurs in taller woodland, e.g. in Zimbabwe, where it avoids *Commiphora mollis* which is occupied by Miombo; forages more in interior of cover than Miombo and feeds less on ground. Resident, but in winter appears to move from *Brachystegia* into mixed woodland (Vernon 1977).

Food. Insects, caterpillars.

Breeding Habits. Monogamous and territorial. In display flight, ♂ (usually with ♀ in attendance) flies vertically into air and swoops downward with purring wings.

NEST: envelope of growing leaves of herb, shrub or tree which are riveted together by pushing spider web though small holes pierced in them; envelope lined or half-filled with plant down, grass and rootlets, which are attached to the leaves; nest has side-top entrance, usually concealed by leaves. Placed 1–3 m above ground.

EGGS: 1–4 (av. 2·6, n = 25, Zimbabwe) and 2–4 (av. 2·8, n = 6, South Africa). Bluish white or cream, evenly speckled with lilac and purplish brown. SIZE: (n = 16, South Africa) 15·2–18·5 × 11·4–13·5 (16·8 × 12·4).

LAYING DATES: Transvaal, Nov–Mar; Botswana, Nov–Dec, Feb–Mar; Namibia, Dec–Jan (nest-building Dec, juv. Apr).

References
Clancey, P. A. (1986b).
Irwin, M. P. S. (1969).
de Swardt, D. H. (1991).

Genus *Euryptila* Sharpe

A small warbler with 12 tail-feathers, unlidded nares and distinctive brown and cinnamon plumage; confined to specialized rocky habitat in extreme SW Africa. Indistinct ventral barring suggests relationship to *Calamonastes*, with which it was allied by Irwin (1969) and thought congeneric by Hall and Moreau (1970), but it differs from *Calamonastes* and *Camaroptera* in 2 important respects: the nares are small, oval and unlidded, and the nest is not stitched (Fry 1976). We follow Fry (1976) in maintaining *Euryptila* as a separate genus.

Endemic; monotypic.

Euryptila subcinnamomea (Smith). Cinnamon-breasted Warbler. Camaroptère cannelle.

Drymoica subcinnamomea A. Smith, 1847. Illus. Zool. South Africa, Aves, Pl. 111, fig. 1, and text; Kamiesberg Mts, Little Namaqualand.

Plate 20
(Opp. p. 337)

Range and Status. Resident, endemic to Namibia and South Africa. Mountains and plateaux in Namibia, north to Naukluft Mts, where only vagrant (Boyer and Bridgeford 1988); W Cape Province south to Eendekuil (60 km south of Clanwilliam) and Touws R. and east to Murraysburg and to Verwoerd Dam on Orange R. (R. Martin, pers. comm.). Reports from Orange Free State need confirmation (Earlé and Grobler 1987). Uncommon to locally fairly common.

Description. *E. s. subcinnamomea* (Smith): Cape Province except extreme NW. ADULT ♂: forehead reddish chestnut, shading to cinnamon-brown on top of head to rump and scapulars; uppertail-coverts reddish chestnut. Tail-feathers blackish, sometimes with narrow light reddish tips, basal portion of outer webs edged chestnut. Lores, around eye and cheeks speckled black and pale grey; ear-coverts chestnut-brown; chin, throat and breast pale grey with black bases to feathers sometimes showing through to produce banded effect; scattered spots on breast, broad band across upper belly, and undertail-coverts reddish chestnut; flanks and thighs cinnamon-brown; centre of lower belly grey-black. Wings brown, coverts edged cinnamon-brown, axillaries and underwing-coverts cinnamon-brown. Bill black or blackish horn, paler at base; eyes brown, brownish grey, olive or greenish; legs and feet dark horn to flesh-brown. Sexes alike. SIZE: wing, ♂ (n = 6) 51–55 (52·3), ♀ (n = 3) 52–53 (52·7), ♂♀ (n = 6) 52–56 (53·6); tail, ♂ (n = 6) 54–57 (55·8), ♀ (n = 3) 53–60 (56·0), ♂♀ (n = 6) 56–61 (58·0); bill to skull ♂ (n = 2) 16, 17, ♂♀ (n = 6) 17–17·5, bill to feathers ♂ (n = 3) 11·5–12 (11·8), ♀ (n = 3) all 12; tarsus, ♂ (n = 6) 18–21 (19·4), ♀ (n = 3) 18–19·5 (18·8). WEIGHT: 2 unsexed, 10·5, 12·5.

IMMATURE: like adult but upperparts more rufous, rusty band across underparts paler, extending up onto breast.

NESTLING: unknown.

E. s. petrophila Clancey: Namibia and extreme NW Cape. Upperparts paler and more ochreous or tawny, chin to breast vinaceous-buff showing little black on feather bases, belly and flanks paler and more buffish brown, tail-feathers browner, sharply tipped with hazel. Tail somewhat longer, especially in ♀.

Field Characters. An uncommon warbler with specialized habitat and restricted range in SW Africa. Brown and cinnamon, with grey throat and breast, and black tail frequently held cocked. Few birds share its rocky habitat, and none resemble it. Rufous-eared Warbler *Malcorus pectoralis* has red face and shares habit of hopping along ground with tail raised, but has streaked upperparts, white underparts with black breast-band, piping song; Karoo Eremomela *Eremomela gregalis* is a green bird with white underparts, high-pitched song.

Voice. Tape-recorded (91). Pair call (sing ?) together with very different voices: one bird gives a scolding 'chwee-chwee-chwee . . .', while the second answers it with a long, upslurred whistle, 'pooweeee'; the latter bears no resemblance to the call of Diederik Cuckoo *Chrysococcyx caprius*, as has been claimed (Clancey 1985, Maclean 1993), being richer and more powerful, and a single note, not 5–7 separable notes. Call, plaintive 'eeeeee' lasting *c.* 2 s (R. Martin, pers. comm.).

General Habits. Inhabits bushes and small trees on mountain slopes, bases of kopjes, jumbles of boulders, rocky outcrops, scree slope in old quarry; scrub in steep ravines, sometimes rocks and scrub near seasonal watercourses. Occurs in pairs during breeding season, at other times either singly or in small, loose groups of 3–4 birds (R. Martin, pers. comm.). Active but skulking; runs rapidly from bush to bush with tail cocked, like North American Sage Sparrow *Amphispiza belli* or Australian grasswren *Amytornis*; also flicks tail sideways and sometimes fans it. Inspects crevices among rocks with jerky movements, hopping quickly from rock to rock; clings to rock face; pauses from time to time to perch on top of bush or rock. Presence of birds often first detected by 'eeeeee' call, given from top of prominent rock, usually near summit of kopje or rocky hill side (R. Martin, pers. comm.). Sedentary.

Food. Insects, including grasshoppers; spiders; small land molluscs.

Breeding Habits. Little known.

NEST: one described by Martin and Martin (1965) was thick-walled, oval, with wide entrance near top, built of 'round-stemmed grasses thickly padded and interwoven with a mixture of the down of the plant Kapokbossie *Eriocephalus africanus* and a little sheepswool, liberally held together with cobwebs'; outside covered with spider web and a little down, general colouring grey; bottom of nest so heavily padded with down that nest contents were raised to near the entrance and could easily be seen without touching the nest; ext. diam. 100–115, int. diam. 50, ext. depth 105, int. depth 65, entrance 30 × 35; placed in hollow in dense matting of bush *Restio gaudochaundianus*, *c*. 1 m above ground, sheltered by large boulders.

EGGS: 2–4, usually 2–3; white or pale bluish white spotted with brown and lilac-grey, mainly at large end. SIZE: (n = 2) 18·4–19 × 13·1–13·8.

LAYING DATES: South Africa, May–Oct.

References
Clancey, P. A. (1985, 1990).
Martin, J. and Martin, R. (1965).

Genus *Graueria* Hartert

Similar to *Macrosphenus*, but bill stronger and shorter, not quite as long as head, less hooked at tip; rictal bristles weak; tail longer, nearly as long as wing, tail-feathers wider; body plumage firmer, without long, lax feathers on flanks and rump; underparts barred.

Considered close to *Macrosphenus* by Hartert. The strong, hooked bill caused it to be placed in babblers Timaliidae, e.g by Sclater (1930), but Chapin (1953a) placed it in Sylviidae because of its colour; this has been followed by Hall and Moreau (1970) and subsequent authors. Barred underparts resemble pattern found in *Calamonostes fasciolatus* and *C. simplex*, and except for heavy bill the colour, pattern and build suggest closer relationship to *Camaroptera* than any other genus (Hall and Moreau 1970).

Endemic; monotypic.

Plate 20
(Opp. p. 337)

Graueria vittata Hartert. Grauer's Warbler. Grauérie striée.

Graueria vittata Hartert, 1908. Bull. Br. Orn. Club 23, p. 8; forest 90 km west of L. Edward, Belgian Congo, 1600 m.

Range and Status. Endemic resident. E Zaïre from highlands west of L. Edward south to Itombwe; SW Uganda (Bwindi-Impenetrable Forest); W Rwanda (Nyungwe Forest). Scarce to locally common; density hard to assess because song unknown, but observations in Rwanda suggest it may be reasonably common there (Dowsett-Lemaire 1990).

Description. ADULT ♂: top of head dark greyish olive, feathers of forehead with buff bases, producing scaly effect; rest of upperparts including scapulars dull greenish olive, somewhat paler on rump and uppertail-coverts; tail olive-brown. Lores, around eye, cheeks, ear-coverts, chin and throat barred buff and black, breast similar but with olive wash, centre of belly barred dark olive and whitish, flanks and undertail-coverts greenish olive. Wing-feathers dark brown, outer webs mainly olive-green, underside of inner webs of primaries and secondaries margined greenish white, axillaries and underwing-coverts yellowish white barred dark olive. Upper mandible blackish, lower mandible bluish grey tipped pale grey, gape green, inside of mouth yellow; eyes red-brown, rim of eyelids dull greenish; legs and feet bluish grey or dull greenish blue, toes tinged green, claws dark grey-brown. Sexes alike. SIZE: (12 ♂♂, 7 ♀♀) wing, ♂ 57–64 (60·4), ♀ 56–64 (60·3); tail, ♂ 54–59 (56·9), ♀ 54–59 (56·1); bill, ♂ 15–17·5 (16·3), ♀

Graueria vittata

16–17 (16·3); tarsus, ♂ 18–20 (19·3), ♀ 18·5–19·5 (19·1).
WEIGHT: ♂ (n = 6) 14·5–18 (15·8), ♀ (n = 4) 14–17 (15·3).
 IMMATURE: crown greenish like back; barring less distinct, more restricted to throat.
 NESTLING: unknown.

Field Characters. A little-known warbler of mountain forests of Albertine Rift; skulking and secretive, but once its purring call is known it can be seen by watching and waiting. Olive-green, with black and buff barring on throat and upper breast and dark scalloping on pale olive belly. Distinguished from all warblers except Banded Prinia *Prinia bairdii* by barred underparts; Banded Prinia has longer tail, dark brown upperparts, white spots on wings and tail, and is heavily barred black and white below. From behind might be mistaken for Grey Longbill *Macrosphenus concolor*, but rather darker, and top of head greyish; Grey Longbill lacks barring on underparts.

Voice. Tape-recorded (V.DAE). Song unknown. Contact (?) call a brief, frog-like, rolling purr, fairly high-pitched and vibrant, lasting 1·25–2 s, repeated every 3–4 s, once given 17 times at very short intervals (Chapin 1978). Alarm, soft, descending 'feuu' (Dowsett-Lemaire 1990).

General Habits. Inhabits montane primary and gallery forest, at 1500–2080 m in Zaïre, 1600–2300 m in Uganda, 1700–2500 m in Rwanda. Occurs in dense vegetation at all levels, from shrubs and vine tangles in understorey to creepers (especially *Sericostachys*) cloaking tall trees. Usually in pairs, remaining out of sight in foliage, but becomes more active and easy to see when it joins mixed-species flocks moving through its territory. Forages by gleaning small branches and leaves, with long tail slightly raised.

Food. Insects, including eggs and larvae: beetles, weevils, small ants, caterpillars, nymphs of Hemiptera; tiny spiders; small seeds.

Breeding Habits. Almost unknown. In Zaïre, territories appear widely spaced; calls all months except Sept–Oct, especially Mar–May, and probably breeds throughout most of year (Chapin 1978); in Uganda, birds in breeding condition Mar.

Reference
Chapin, R. T. (1978).

Genus *Macrosphenus* Cassin

Bill long, slender and straight, upper mandible with tiny hook at tip curving down over tip of lower mandible; rictal bristles present but not well developed; wing short and rounded, P4 usually longest; tail short, about two-thirds length of wing, but longer than wing in *kretschmeri*; tarsus scutellated anteriorly. Feathers of lower back, rump and flanks long and silky, erectile.

Generally considered close to *Graueria*, which in turn may be part of the *Camaroptera* group (see Hall and Moreau 1970), but *M. kretschmeri* has been considered a bulbul, and any placement of *Macrosphenus* remains speculative.

Endemic; 5 species, 2 superspecies (*M. kempi/M. flavicans* and *M. concolor/M. pulitzeri*). *M. kempi* and *M. flavicans* have been considered conspecific (White 1960a, and suggested by Chapin 1953a), and their songs are quite similar, but they have recently been found to occur together in Cameroon (Korup Nat. Park), and we follow Hall and Moreau (1970) and Traylor (*in* Watson *et al.* 1986) in considering them members of a superspecies. *M. pulitzeri* appears to be the geographical representative of *M. concolor* in Angola, and we tentatively place the two in a superspecies, as did Hall and Moreau (1970) and Traylor (*in* Watson *et al.* 1986), but its voice and habits are unknown and it may prove to be less closely related. We do not follow the above authors in including *M. kretschmeri* in this same superspecies. *M. kretschmeri* differs from other species in having a long tail; it was originally thought to be a bulbul and placed in the monotypic genus *Suaheliornis*. Its voice bears no resemblance to that of *M. concolor*, and its relationships remain unclear.

312 SYLVIIDAE

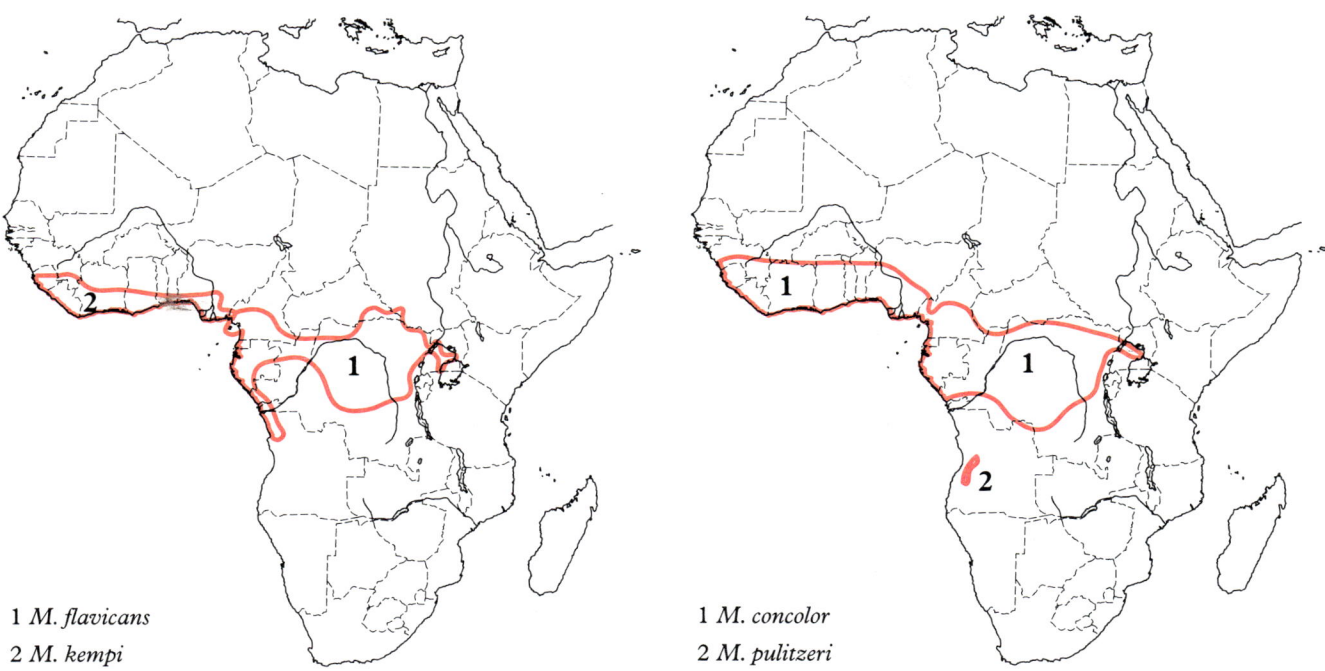

Macrosphenus flavicans superspecies

1 *M. flavicans*
2 *M. kempi*

Macrosphenus concolor superspecies

1 *M. concolor*
2 *M. pulitzeri*

Plate 20
(Opp. p. 337)

***Macrosphenus flavicans* Cassin. Yellow Longbill. Nasique jaune.**

Macrosphenus flavicans Cassin, 1859. Proc. Acad. Nat. Sci. Philadelphia, p. 42; Camma R., Western Africa (= Sette Cama, Gabon).

Forms a superspecies with *M. kempi*.

Range and Status. Endemic resident. Bioko; W Cameroon in lowlands and slopes of highlands to 1250 m (Mt Cameroon north to Korup Nat. Park and Mts Manenguba, Kupe and Nlonako), and widespread in S Cameroon south of *c.* 5° 00′N; Gabon (widespread and fairly common); S Congo (Kouilou basin, Mayombe, common); Angola in Cabinda and Cuanza Norte; SE Central African Republic; Zaïre, probably throughout lowland forest zone, common to very common; SW Sudan in Bengengai Forest, perhaps also in Imatong Mts (Talanga Forest, calling bird: Nikolaus 1987); S and W Uganda in Budongo, Bugoma, Bwamba, Kifu, Mabira, Lugalambo and Malabigambo forests (common in forest undergrowth) and NW Tanzania at Minziro Forest. Density in NE Gabon, 18–20 pairs/km².

Macrosphenus flavicans

Description. *M. f. hypochondriacus* (Reichenow): E Zaïre, Uganda, Central African Republic, Sudan. ADULT ♂: top of head to hindneck, lores, cheeks and ear-coverts grey-brown; rest of upperparts and scapulars olive-green, with variable amount of yellow on rump and uppertail-coverts. Tail-feathers grey-brown edged olive-green. Chin and throat pale grey becoming a shade darker on lower throat and upper breast;

breast and belly yellow mixed with some grey in centre; flanks and sides of lower breast yellow-orange, undertail-coverts rich yellow. Wing-feathers grey-brown edged pale olive-green, axillaries, underwing-coverts and bend of wing whitish, underside of inner webs of primaries and secondaries margined greyish white. Feathers of back, rump and flanks long and silky. Upper mandible black with white cutting edges, lower mandible pale grey, keel paler; eyes yellow; legs and feet grey or blue-grey. Sexes alike, ♂♂ somewhat larger. SIZE: (10 ♂♂, 5 ♀♀) wing, ♂ 58–64 (61·0), ♀ 53–57 (55·6); tail, ♂ 40–47 (43·9), ♀ 35–39 (36·2); bill, ♂ 18–19·5 (18·8), ♀ 18–18·5 (18·1); tarsus, ♂ 19–21·5 (20·6), ♀ 18–20 (19·1). WEIGHT: ♂ (n = 18) 12–16 (14·3); ♀ (n = 6) 11–14 (12·5).

IMMATURE: top of head and upperparts including scapulars brown, slightly yellower on rump and uppertail-coverts; cheeks, ear-coverts and sides of neck grey with greenish wash. Chin, throat and upper breast grey shading to brown on lower breast and pale yellow on belly; sides of breast and flanks brown, latter with a touch of orange, lower flanks and undertail-coverts greenish yellow. Wings like adult except inner secondaries, tertials and coverts edged brown, longest alula feather with small white spot at tip, axillaries and underwing-coverts tinged yellow. Bill like adult except cutting edge yellow, gape yellow, inside of mouth yellow with 2 black spots at back of tongue; eyes grey. This plumage apparently lasts for only a short time (Chapin 1953a). SUBADULT: crown olive-green like back, throat washed yellow-green, centre of breast and upper belly pale grey, no orange on flanks.

NESTLING: unknown.

M. f. flavicans Cassin: Cameroon to Angola and W Zaïre. Underparts less richly coloured, greenish yellow, with no orange on flanks.

Field Characters. Slightly larger than sympatric Grey Longbill *M. concolor*, with longer and slimmer bill; upperparts very similar; greyish crown contrasts somewhat with greenish back but this is only apparent in good light (head and back concolorous in Grey Longbill). Underparts yellow with orange flanks, but even this can be hard to see in undergrowth. The 2 are instantly identified by their songs; Yellow has measured notes descending in semitones, Grey has a bubbling duet. Overlaps in W Cameroon with Kemp's Longbill *M. kempi*, which has similar voice but greyish underparts with reddish chestnut flanks.

Voice. Tape-recorded (68, C, ERA, GRI, KEI, LEM). Song, a series of rather mournful downslurred whistles descending the scale in semitones, typically 9–10, sometimes 12, sometimes shorter series of 5–7, even 3–4. Notes are evenly spaced, given at rate of 2 per s; pause between songs *c*. 5 s. Song in Cameroon said to resemble that of Brown-chested Alethe *Alethe poliocephala* (Rodewald *et al.* 1994). Call, high-pitched 'tsiss'.

General Habits. Inhabits primary and mature secondary forest, also logged forest in Cameroon, chiefly in lowlands but also transition and lower montane forest, up to 1820 m in E Zaïre, and at 1000–1400 m in Uganda. Within forest occurs at all levels, from dense undergrowth to tangled masses of vines on tree trunks and in canopy, also thickets at edges of clearings. In Gabon descends to lower levels most often during rains and in short dry season, frequenting canopy during long dry season (Brosset and Erard 1986). In pairs or family parties; often joins mixed-species flocks. Hops about on twigs and branches in a leisurely manner. Usually sings from high perch.

Sedentary, at least in NE Gabon.

Food. Insects, especially moths and caterpillars; Orthoptera, Coleoptera, leafhoppers; also spiders and some fruit.

Breeding Habits. Territorial. Nest and eggs unknown. Birds in breeding condition, Cameroon, May, July, carrying nest material Dec; in Gabon, singing increases Sept–Mar; breeding condition Angola, Sept; NE Zaïre, breeding condition Apr–June, Aug, immature Nov; Itombwe, breeding condition Nov–Apr, juvs Sept.

Macrosphenus kempi (Sharpe). Kemp's Longbill. Nasique de Kemp.

Amaurocichla kempi Sharpe, 1905. Bull. Br. Orn. Club 15, p. 38; Bo, Sierra Leone.

Forms a superspecies with *M. flavicans*.

Plate 20
(Opp. p. 337)

Range and Status. Endemic resident. Sierra Leone, SE Guinea, Liberia; Ivory Coast from Sassandra and Abidjan north to Nimba, Béoumi and Abengourou; S Ghana north at least to Bia Nat. Park, Akropong and Atewa Forest Reserve; S Nigeria from Lagos and Ife east to *c*. 7°30′E; W Cameroon (Korup Nat. Park: Rodewald and Bowden 1995). Uncommon to locally common; Red Data Book candidate.

Description. *M. k. kempi* (Sharpe): Sierra Leone to SW Nigeria. ADULT: ♂ top and sides of head, upperparts, including scapulars and wing-coverts, and tail dark olive-brown; long silky feathers of lower back and rump grey. Chin and throat whitish grey to grey; upper breast olive-grey, long silky feathers of sides of breast and flanks orange-chestnut, sometimes with olive wash on breast and olive-buff wash on flanks; centre of breast to belly grey-white, undertail-coverts buffy white tinged olive, thighs olive-brown. Flight feathers dark brown edged olive-brown, inner margins greyish white, upperwing-coverts dark olive-brown, outermost with narrow pale edge; axillaries, underwing-coverts and bend of wing

Macrosphenus kempi

greyish white with olive wash. Upper mandible black with pale tip, lower mandible pinkish horn or whitish; eyes yellow and pale grey; legs and feet dark horn, plumbeous or slate-blue, soles yellowish. Sexes alike. SIZE: wing, ♂ (n = 16) 59·5 ± 1·5, ♀ (n = 5) 55·5 ± 2·4; tail, ♂ (n = 16) 33·6 ± 3·0, ♀ (n = 5) 29·8 ± 1·0; bill to skull, ♂ (n = 16) 21 ± 0, ♀ (n = 5) 20·7 ± 0·5, to feathers, ♂ (n = 3) 17–19 (18·0), 1 ♀ 18; tarsus, ♂ (n = 3) 20–22 (20·7), 1 ♀ 20. WEIGHT: ♂ (n = 15) 10–14·4 (13·4), ♀ (n = 4) 9·8–14·8 (11·9).

IMMATURE: upperparts olive-green, underparts without any chestnut, pale yellowish green with olive wash across breast, thighs yellowish; gape dull yellow; eyes grey-brown.

NESTLING: unknown.

M. k. flammeus Marchant: SE Nigeria and W Cameroon. Chestnut on underparts richer and redder, extending from sides of breast onto centre but not quite meeting. SIZE: (6 ♂ ♂, 1 ♀) wing, ♂ 57–61 (58·7), ♀ 52; tail, ♂ 33–39 (36·0), ♀ 30; bill to feathers, ♂ 16–18 (17·3), ♀ 18; tarsus, ♂ 20–22 (20·8), ♀ 20.

Field Characters. Replaces Yellow Longbill *M. flavicans* from Nigeria west, but sympatric with it in W Cameroon. Brownish above, grey below, with sides orange-chestnut (Sierra Leone to W Nigeria) or reddish chestnut (east of Niger R.). Main song similar to Yellow Longbill, secondary song different; both help to distinguish it from sympatric Grey Longbill *M. concolor*, which is olive-green above, greyish below, without any bright colour on flanks.

Voice. Tape-recorded (DEM, KEI, MAC). Song in Liberia of 2 types. Type (a) is very similar to song of Yellow Longbill *M. flavicans*, q.v., a series of 5–6 downslurred, mellow, rather mournful whistles descending (sometimes ascending) in semitones, first note a little higher, second sometimes lower, with an optional unslurred single note at the end higher in pitch than the others, 'hee, here-here-here-here-heep' or 'hee, hoo, here-here-here-here'. Song in Cameroon, apparently different from that of local Yellow Longbill, described as 'a series of about eight clear whistles rising slightly in pitch' (Rodewald *et al.* 1994). Type (b), not shared with Yellow Longbill, is a series of 8–12 tuneless notes, faster at first and becoming slower as they descend the scale, 'didididi-der-der-der-der-dar-dar-daar-daar'; final notes have a husky, nasal quality. In between songs maintains subdued 'conversation' like that of Grey Longbill *M. concolor*, but many of the notes are less scratchy, almost mellow; also included are bulbul-like churrs, hard trills, sharp 'tik-tik', and a quiet, low-pitched version of song type (a), almost a subsong. Alarm, squeaky notes.

General Habits. Inhabits primary and old secondary lowland forest, chiefly in undergrowth and thickets and at edges. Usually in pairs, sometimes in mixed-species flocks. Creeps about like nuthatch *Sitta* in low, dense vegetation; stretches neck out to pry into small dead branches for ants and other insects; once seen to attend army ants. When alarmed fluffs out feathers of back and flanks and gives squeaky notes. Neck appears longer in the field than it does in skins.

Food. Insects, including beetles and ants.

Breeding Habits. Nest and eggs unknown. Birds in breeding condition, Liberia, May, June, Aug, Dec, juvs Apr, July, Sept–Oct, i.e. probably breeds most of year; birds in breeding condition Nigeria, Aug.

Plate 20 (Opp. p. 337)

Macrosphenus concolor (Hartlaub). Grey Longbill. Nasique grise.

Camaroptera concolor Hartlaub, 1857. Syst. Ornith. Westafrika's, p. 62; Guinea.

Forms a superspecies with *M. pulitzeri*.

Range and Status. Endemic resident. S Guinea, north at least to Kounounkan Massif (Hayman *et al.* 1995), Conakry and nearby coastal islands, and in SE along borders of Sierra Leone and Liberia; Sierra Leone, Liberia (common), Ivory Coast north to Odiemé, Korhogo and Comoé Nat. Park, common; S Ghana, Togo, Benin, S Nigeria (common and widespread); Bioko; common throughout Cameroon lowlands,

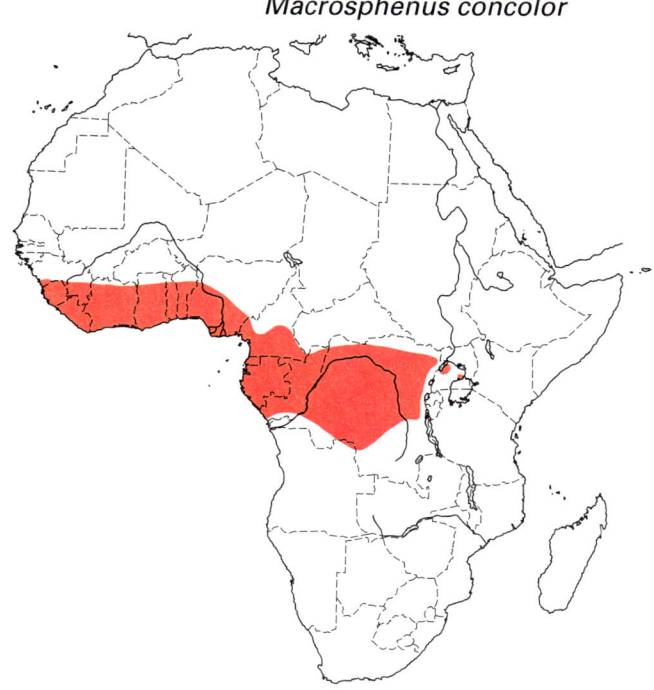
Macrosphenus concolor

ascending lower slopes of S highlands; common and widespread Gabon, common Congo and throughout forest zone of Zaïre; Angola in NE Lunda; S and W Uganda in Budongo, Bugoma and Mabira forests, scarce. Density in NE Gabon, 7–10 pairs/km^2.

Description. ADULT ♂: top and sides of head and neck and entire upperparts olive-green, variably yellower on rump and uppertail-coverts (feathers with yellowish tips). Tail-feathers olive-brown edged olive-green. Underparts light greyish olive, palest on throat, variably washed greenish or yellowish, somewhat darker on sides of breast and flanks; feathers of lower back, rump and flanks long and silky with yellowish tips. Wing-feathers dark brown edged olive-green, axillaries, underwing-coverts and margins of underside of inner webs of primaries and secondaries whitish, becoming yellow along bend of wing. Upper mandible black or dark brown, lower mandible grey, sometimes tinged pinkish; inner portion of iris whitish or greenish grey, outer portion light brown or orange-buff (Chapin 1953a), rim of eyelids pinkish to purplish pink; legs and feet light brown, brownish pink or flesh. Sexes alike. SIZE: (mainland) (8 ♂♂, 5 ♀♀) wing, ♂ 55–59 (57·2), ♀ 54–57 (55·6); tail, ♂ 37–43 (40·4), ♀ 36–39 (37·6); bill, ♂ 15·5–18 (16·6), ♀ 16–17 (16·3); tarsus, ♂ 18·5–20·5 (19·8), ♀ 18–20 (19·0). Birds from Bioko somewhat larger: (3 ♂♂) wing 58–63 (61·0), tail 42–44 (43·0), bill 17–18 (17·3), tarsus 20·5–21·5 (20·8). WEIGHT: (Liberia) ♂ (n = 9) 13·9 ± 0·9, ♀ (n = 6) 13·9 ± 0·8.

IMMATURE: paler overall, underparts dirty white washed pale olive, breast tinged yellow.

NESTLING: unknown.

Field Characters. A nondescript olive warbler with long bill, short tail, and habit of fluffing out back and flank feathers. Bill shorter and broader than sympatric Yellow Longbill *M. flavicans* and Kemp's Longbill *M. kempi*, neck shorter, eye brown, not yellow; Yellow has yellow underparts and orange flanks, Kemp's has grey underparts and reddish or chestnut flanks. Might be confused with 2 other dingy forest warblers which creep among leaves, the smaller Bleating Warbler *Camaroptera brachyura* (grey with contrasting green wings, nasal call note and chirping song), or much smaller Olive-green Camaroptera *Camaroptera chloronota* (dull brownish olive, song a repeated ringing note), but Grey Longbill easily distinguished from all of the above by bubbling song (see *Voice*). Also easy to confuse with Fraser's Sunbird *Anthreptes fraseri*, which is the same size and colour and usually does not show the scarlet tufts; Grey Longbill lacks head markings and has pale flesh-coloured legs, while Scarlet-tufted Sunbird has conspicuous eye-ring and dark greyish black or brownish legs; these marks are easier to see than bill differences (the sunbird has a slightly curved bill, with dark upper mandible gradually tapering to a point, pale horn and almost straight lower mandible (Demey and Fishpool 1994)).

Voice. Tape-recorded (68, B, C, ERA, KEI, LEM, MAC). Song a hurried, bubbling, lilting warble; typically a 2-part phrase repeated 6–20 times without pause, 'tarradiddle-worradiddle, tarradiddle-worradiddle . . .', increasing in power but not accelerating; variation almost endless, often suggesting human words: 'wichita-weetatata, wichita-weetatata . . .', 'video rental, video rental . . .', 'widicheepa-wakeepa . . .', 'get it cheaper in duka . . .', 'wertily-fiddly, wertily fiddly . . .', 'we-keepy-chirpy . . .', 'the cat and the fiddle . . .'; song may end in middle of phrase, 'widicheepa-wakeepa, widichi', or with a decisive note, 'tarradiddle-chop'. Singing bouts last for several min and are followed by period of silence. One bird may sing alone, or each member of a pair may sing its own song, either separately or at the same time, the second chiming in after first bird starts; their songs are different, e.g. the bird singing opposite 'we-keepy-chirpy' sang 'tuwee-tuwer, tuwee-tuwer . . .', or gave a loud series of downslurred notes 'fee-tew-tew-tew-tew'. Singing is not antiphonal; entry point for second bird is random. In one variant, bubbly phrase is replaced by 3–4 distinct notes going up the scale, like weak version of song of White-browed Robin-Chat *Cossypha heuglini*. Foraging pair often keeps up a conversation of soft chips, sunbird-like squeaks, grating notes, and little trills like beginning of song of Ruby-crowned Kinglet *Regulus calendula*; these often lead into sweet notes of song, in manner of Garden Warbler *Sylvia borin*. A quite different type of song, less frequent and perhaps with some sexual significance, noted in Gabon (Brosset and Erard 1986; see also Chappuis 1979a); it consists of numerous imitations of other species present in the forest, given rapidly; species imitated include African Pied Hornbill *Tockus fasciatus*, Rufous Flycatcher-Thrush *Neocossyphus fraseri*, Gosling's Apalis *Apalis goslingi*, Grey-throated Tit-Flycatcher *Myioparus griseigularis*, Black-and-White Flycatcher *Bias musicus*, Blue-headed Crested Flycatcher *Trochocercus nitens* and Bates's Paradise-Flycatcher *Terpsiphone batesi*.

General Habits. Inhabits primary and (more often) old secondary forest, forest edge and riverine forest;

mainly in lowlands, but up to 1220 m in Cameroon, 1460 m in Zaïre, and at 1000–1400 m in Uganda. Occurs at all levels from undergrowth to lower canopy, with a distinct preference for creeper-clad tree trunks and hanging masses of lianas. Travels in pairs or small family parties; very often in mixed-species flocks, where it sings a lot; buries itself among dense leaf clusters while it searches for insects and can be very hard to see, even when singing.

Sedentary, at least in NE Gabon.

Food. Insects and their eggs, including small caterpillars, moths and small Orthoptera.

Breeding Habits. Nest not described but one was stated to be '2·5 m high in a liana' (Halleux 1994). Eggs unknown. Probably territorial. Nest in Guinea, Mar; birds in breeding condition Liberia, Oct, Ghana, Dec, Nigeria Nov–Jan, Angola, June; Cameroon, immatures Apr; Gabon, singing increases Sept–Feb; Zaïre, breeding condition Aug–Feb, immatures Sept–Dec.

Plate 20
(Opp. p. 337)

Macrosphenus pulitzeri **Boulton. Pulitzer's Longbill. Nasique de Pulitzer.**

Macrosphenus pulitzeri Boulton, 1931. Ann. Carnegie Mus. 21, p. 50; Chingoroi, Benguela district, Angola, 2200 feet.

Forms a superspecies with *M. concolor*.

Range and Status. Endemic resident. Known from 5 localities on the escarpment of W Angola from Vila Nova do Seles to the Chingoroi area. Status given as indeterminate in the IUCN/ICBP Red Data Book. May possibly occur in other patches of relict forest on same escarpment, but these are unlikely to total more than a few hundred square miles (Hall and Moreau 1962); certainly threatened by forest clearance, although it might possibly be able to survive in coffee plantations (A. A. da Rosa Pinto, *in* Collar and Stuart 1985).

Description (based on notes on the type by R. K. Panza and K. C. Parkes). ADULT ♂: top of head, back, rump and scapulars olivaceous brown, ear-coverts somewhat paler; lores and spot in front of eye charcoal-grey; uppertail-coverts cinnamon-brown; tail-feathers olivaceous brown, shafts dark reddish brown. Chin and throat dirty pale yellow with slight greenish cast; breast, belly and flanks olive, darker on flanks, with dull brownish yellow central line, 10 mm wide on lower breast, tapering to 5 mm on lower belly; undertail-coverts and thighs cinnamon-brown, bases of body-feathers slate. Primaries and secondaries dark greyish olive, tertials and coverts paler, all feathers edged olivaceous brown, primaries narrowly and secondaries more broadly; underside of inner webs of primaries and secondaries edged greyish white, axillaries and underwing-coverts dirty yellowish white. SIZE: (1 ♀) wing (flat) 63, (chord) 60; tail, 50·5; bill to skull, 20·1; tarsus 22.

General and Breeding Habits. Voice and field appearance unknown, but it should look like a large version of Grey Longbill *M. concolor*. Inhabits dry evergreen forest and secondary forest. Searches restlessly for insects low down (often near ground) in forest; hard to observe, often hidden in dense foliage. Birds in breeding condition Sept, Dec.

Reference
Collar, N. J. and Stuart, S. N. (1985).

Macrosphenus kretschmeri (Reichenow and Neumann). Kretschmer's Longbill. Nasique de Kretschmer.

Phyllostrephus kretschmeri Reichenow and Neumann, 1895. Orn. Monatsber. 3, p. 75; Kibosho, Kilimanjaro, c. 2500 m.

Plate 20 (Opp. p. 337)

Macrosphenus kretschmeri

Range and Status. Endemic resident. Extreme SE Kenya at Kitovu Forest (1 record, possibly a wanderer: Lewis and Pomeroy 1989); E Tanzania from Mt Kilimanjaro, Moshi and Usambaras south in highlands to Ngurus, Ulugurus and Udzungwas (Mwanihana and Magombera forests), and on coast in Kiono, Pugu, Kazimzumbwe and Kiwengoma forests, and in SE at Mikindani; N Mozambique at Netia (common). Uncommon to locally common.

Description. *M. k. kretschmeri* (Reichenow and Neumann): Kenya to central Tanzania. ADULT ♂: top of head, lores, cheeks and ear-coverts dull greyish olive; rest of upperparts brownish olive-green, paler on rump and uppertail-coverts. Tail-feathers olive-brown edged greener, shafts maroon above, straw below. Underparts dull olive-yellow, greyer on chin and throat, yellower on centre of breast and belly, thighs and undertail-coverts greenish brown. Primaries and secondaries dark brown edged olive-green, tertials, scapulars and wing-coverts olive-brown; inner webs of underside of primaries and secondaries margined yellowish white, axillaries, underwing-coverts and bend of wing greenish yellow. Upper mandible black to horn-grey, lower mandible whitish to flesh, pale horn or purplish horn; eyes brown, red, yellowish or creamy, eyelids pink; legs and feet purplish flesh or pinkish to grey, bluish or red-brown. SIZE: wing, ♂ (n = 10) 63–68 (65·1), ♀ (n = 8) 57–70 (63·9); tail, ♂ (n = 10) 61–69 (63·8), ♀ (n = 8) 55–69 (61·3); bill to skull, ♂ (n = 2) 19, 20, ♀ (n = 3) 17–19 (18·3), to feathers, ♂ (n = 7) 15–17 (16·3), ♀ (n = 4) 15–16 (15·3); tarsus, ♂ (n = 10) 18–23 (19·7), ♀ (n = 8) 18–24·5 (21·6). WEIGHT: 2 ♂♂ 20·5, 22; 1 ♀ 17·5.

IMMATURE: like adult but top of head greener, eyes pale brown.

NESTLING: unknown.

M. k. griseiceps Grote: SE Tanzania and Mozambique. Paler, top of head greyer, less olive.

Field Characters. A large longbill, the only one in its range. Normal-length (i.e. not short) tail gives it the appearance of a small bulbul, which it was orginally thought to be. Plumage undistinguished, dull, olive-green above and olive-yellow below, greyer on top of head; eye variable, usually reddish or yellowish. Skulks in thickets, presence often detected only by distinctive 4-note call. Similar-looking Tiny Greenbul *Phyllastrephus debilis* is same size and occurs in same forests, but has greyer head, brighter green back, paler yellow underparts, shorter bill; eye also variable but usually pale. More important, Tiny Greenbul is not shy and often travels in groups, keeping up a constant chatter.

Voice. Tape-recorded (32, STJ). Song a quick 4-note phrase of loud pure notes lasting c. 1 s, first note loudest and highest, followed by a slight pause, next 3 shorter, third slurred upwards, 'weet, chu-moi-check'; last note sometimes 'chu', producing, by a slight stretch of the imagination, 'pleased to meet you'; emphatic, unmusical, monotonous, sometimes continued for long periods; occasionally followed by jumble of sweet musical sound (Sclater and Moreau 1932). 'Throaty gurgling warble' sometimes given by party feeding on ground (Vincent 1935); alarm, low 'charrr'.

General Habits. Inhabits lowland and intermediate forest, up to 1800 m. Frequents dense, tangled cover in forest interior under tall trees, from undergrowth to mid-stratum, also forest edge and coastal thickets. Buries itself in thick vegetation; elusive and hard to see, even though it sings constantly. Occurs singly or in pairs, or parties of 5–6; scratches about among dead leaves on forest floor. Joins in bird parties with bulbuls, flycatchers, puffbacks and drongos, usually towards rear of flock (Mozambique: Vincent 1935).

Food. Insects.

Breeding Habits. Unknown; nest and eggs not described; birds in breeding condition, Tanzania, Feb, Apr.

Reference
Vincent, J. (1935).

318　SYLVIIDAE

Genus *Eremomela* Sundevall

Small, active warblers, predominantly grey or green on upperparts and grey, green or yellow on underparts. Difficult to diagnose morphologically: intermediate in character between *Sylvietta* and *Apalis*. Some species rather short-tailed, but tail always extends beyond the wing-tip and bird never as short-tailed as *Sylvietta*; some species with medium-length rounded tails like *Apalis*. Tail is always plain and unpatterned, never strongly graduated. Bill and feet weaker than *Sylvietta* and similar to *Apalis*. Wings fairly long, with P6, P7 and P8 more or less equal and longest and P10 about half length of P9. P6 is emarginated and there may also be some emargination on P5 and/or P7.

Insectivorous birds, most of wooded country, foraging mainly in canopy; they inhabit almost the whole of lowland Africa south of 20°N. Usually seen in parties of up to about 6; most breed co-operatively and keep in family parties year-round. Nest a neat cup, unlike *Sylvietta* or *Apalis*.

Endemic. 11 species, comprising 2 superspecies and a species group. Some at least are close to *Sylvietta*. One superspecies, with very short tails and in the driest country, contains a widespread species *E. icteropygialis* and 2 with limited ranges, *E. flavicrissalis* and *E. salvadorii*; there is no indication of intergrading or interbreeding between the first two species. Some birds in NW Zambia, named *E. i. lundae* by Grant and Mackworth-Praed (1941), are said to be intermediate between *salvadorii* and *E. i. polioxantha*, but there appears to be no interbreeding where *salvadorii* meets *icteropygialis* in E Angola. A second superspecies, with rather longer rounded tails, contains 3 wide-ranging species of moister savannas, *E. pusilla*, *E. canescens* and *E. scotops*, and 1 with a restricted range, *E. gregalis*. The remaining 4 species, all with longer tails and in wetter savannas, woodland and secondary forest, are independent: *E. badiceps* is widespread in forests in W Africa, *E. turneri* is restricted to a small area of forest in Zaïre and Uganda; *E. atricollis* occurs mainly in *Brachystegia* in Zambia and Angola, and *E. usticollis* in *Acacia* in southern Africa.

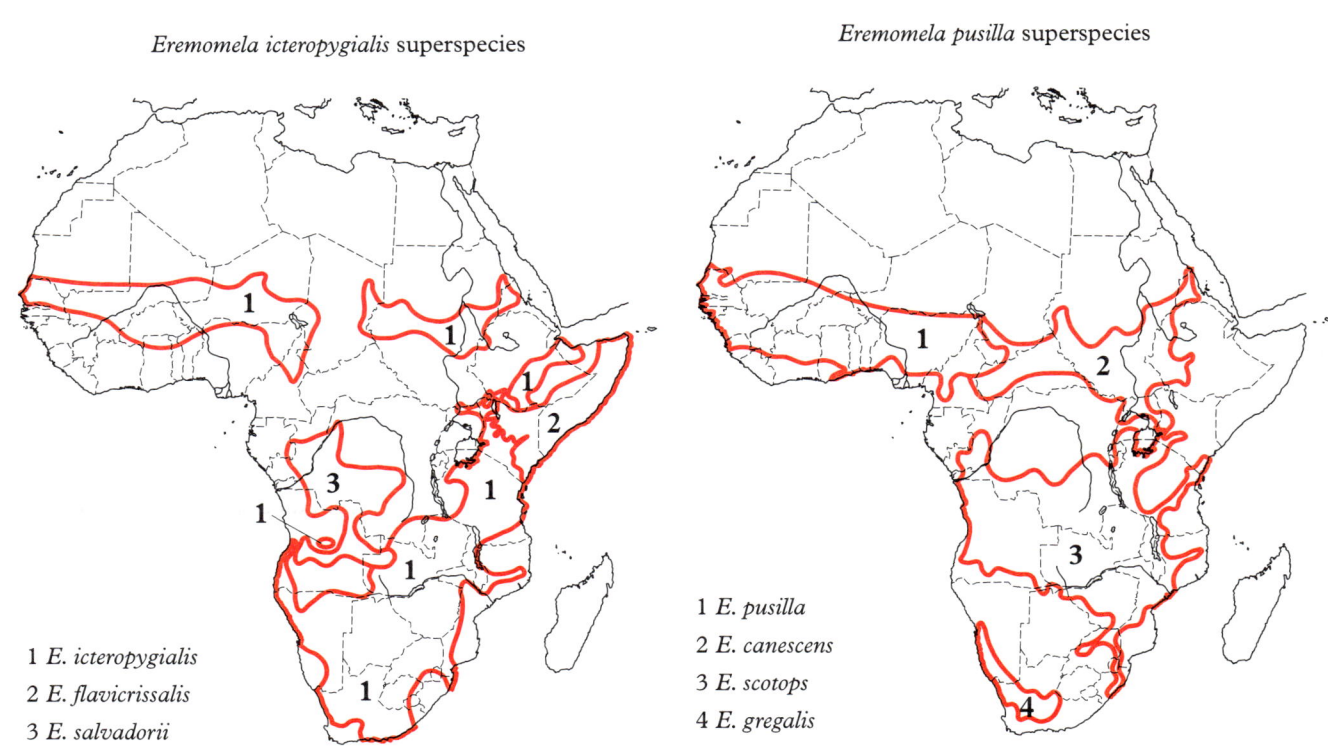

Plate 18
(Opp. p. 321)

Eremomela icteropygialis (Lafresnaye). Yellow-bellied Eremomela. Érémomèle à croupion jaune.

Sylvietta icteropygialis Lafresnaye, 1839. Rev. Zool., p. 258; Orange River, South West Africa.

Forms a superspecies with *E. flavicrissalis* and *E. salvadorii*.

Range and Status. Endemic resident, widespread in much of non-forested Africa south of Sahara. Occurs in Sahel zone and soudanian savanna zone: N Senegal (where it has moved farther south as acacias have

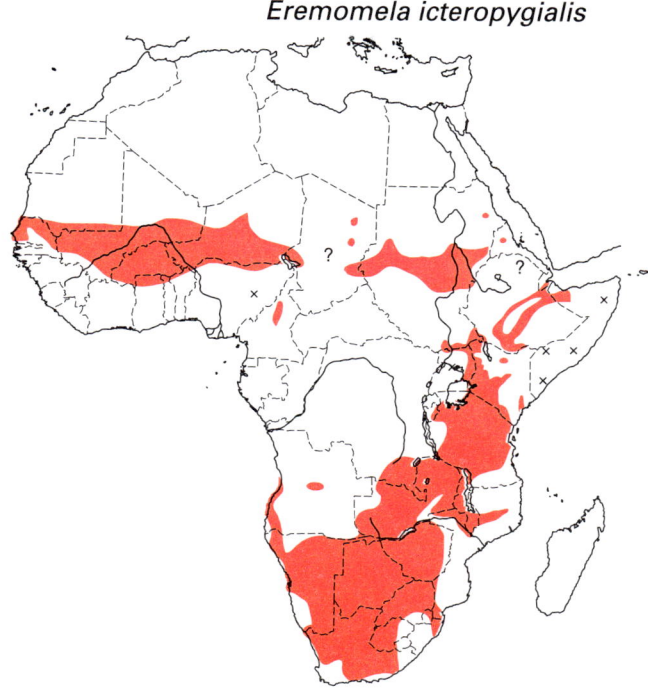

Eremomela icteropygialis

extended south) and S Mauritania (north in rains to 18°N, although resident in Senegal R. valley); fairly common in Niger R. delta and woody savanna in Mali but apparently absent south of 13°N; widespread in Niger except in NE; uncommon in N half of Ghana south to Mole Nat. Park; status in Burkina Faso unknown. Common in N Benin, and across N Nigeria from Sokoto to L. Chad and recorded in Yankari Game Reserve; in extreme N Cameroon, south to Ngaoundéré.

Status in Chad unclear; perhaps a break in range; occurs around L. Chad but no other records between there and E Chad where recorded at Fada, Koboué and Bodhoué in Ennedi, and is recorded as a migrant from mid-July to mid-Sept at Ouaddai. Uncommon across central Sudan and to coast in S Red Sea Province; occurs in N and central lowland Ethiopia and in northern part of W Highlands and in SE Highlands; also in NW Somalia. 2 records in S Somalia and 1 in NE. Sparse in S Sudan; in Uganda occurs in Karamoja in NE and a sight record at Nakaongola in centre. Seldom common in Kenya where absent from semi-arid and arid regions of N and E, though present on coast near Tana R. Through Tanzania in broad belt from east of L. Victoria to S end of L. Tanganyika and the coast opposite Zanzibar to near N end L. Malaŵi. 3 records in NW Tanzania west of L. Victoria. Throughout Zambia except Luangwa and Middle Zambezi valleys, between Lakes Mweru and Tanganyika and in the southwest where it may not reach the Mashi/Kwando R. on the Angolan border or the Namibian border. Through SE Zaïre, extending to S Huila and Cuanza Sul in Angola. Throughout Malaŵi, Zimbabwe, Botswana and Namibia (where uncommon along watercourses in east); extends north on coastal plain of Angola to Namibe (Moçâmedes) and Benguela. In S Mozambique uncommon and restricted to S and W Sul do Save and extreme SW; 2 records in NE Mozambique. Throughout South Africa except for extreme S and E Cape Province, much of Natal and most of the highveld. 1 record W Lesotho (P. Osborne and B. Tigar, pers. comm.).

Ranges up to 1900 m, but occurs mainly below 1000 m, where common. Sparse in dry areas where also often migratory or nomadic. Density in Transvaal of 4 birds in 100 ha of *Burkea* woodland (and none in *Acacia*: Tarboton 1980); in karroid broken veld 8 birds in 838 ha (*c.* 1 per km^2: Winterbottom 1968); in Zimbabwe 1 pair per km^2 in mixed woodland (Vernon 1985a); in Ethiopia up to 6 seen per day in Sermale bushland, and in Nechisar Plains 4 per day (Safford *et al.* 1993).

Description. *E. i. alexanderi* Sclater and Mackworth-Praed: Senegambia to Sudan (Darfur and Kordofan). ADULT ♂: forehead grey with variable but usually very little white; crown to uppertail-coverts and scapulars grey with slight brownish wash; tail brownish grey becoming paler on outer feathers. Lores and faint stripe over and behind eye dirty white; ear-coverts greyish. Chin and throat white; breast greyish white, suffused with brown; upper belly dirty white; lower belly and undertail-coverts pale yellow. Wings brownish grey, primaries with outer edges whitish, secondaries and tertials with outer edges pale brown and tips pale; upperwing-coverts like primaries but with broad whitish tips, making bar on closed wing; all tips and edges of wing-feathers and wing-coverts with variable amount of whitish. Upper mandible dark brown or black, lower mandible paler; eyes red-brown or dark brown; legs and feet black. Sexes alike although yellow on ♀ usually paler. Most variation is clinal, with races in dry SW Africa similar to those in the dry N tropics, and paler and less yellow than races in the moister centre of range. SIZE: (15 ♂ ♂, 15 ♀ ♀) wing, ♂ 50–57 (53·8), ♀ 49–56 (52·5); tail, ♂ 31–41 (35·7), ♀ 29–38 (33·6); bill, ♂ 9·9–12·2 (11·0), ♀ 9·9–11·7 (11·0); tarsus, ♂ 17·0–20·5 (18·6), ♀ 16·2–19·4 (18·1). WEIGHT: 1 unsexed (Niger, June) 6·7.

IMMATURE: yellow on underparts duller; edges of secondaries, tertials and wing-coverts much browner.

NESTLING: naked at first, skin ochre but darkening rapidly; gape and mouth yellow, latter becoming more orange and with 3 black spots.

E. i. griseoflava Heuglin (including '*karamojensis*' and '*crawfurdi*'): Sudan (except Darfur and Kordofan), Ethiopia, Somalia, E Uganda, Rwanda, W Kenya and Mwanza District of Tanzania. Upperparts darker grey than in *alexanderi*, breast paler and whiter, and yellow brighter. Rather variable. WEIGHT: unsexed (n = 12, Ethiopia) 6·9–8·2 (7·4), (n = 5, Kenya/Uganda) 5–6 (5·3).

E. i. abdominalis Reichenow: Kenya (except W) and N Tanzania. Lacks white eye-stripe; yellow on underparts very bright; sides of breast greyer than in *alexanderi*.

E. i. polioxantha Sharpe: S Zaïre and S Tanzania to SW Zimbabwe, E Transvaal and Mozambique. Yellow belly brighter than in *alexanderi* and extending to lower breast where it meets grey of breast; upperparts darker than in *alexanderi*; larger and longer-billed. SIZE: wing, ♂ (n = 3) 60–62 (61·0), ♀ (n = 4) 58–61 (59·8); bill, ♂ (n = 3) 11·4–11·5 (11·4), ♀ (n = 3) 10·5–11·5 (11·0). WEIGHT: ♂ (n = 10, Zimbabwe) 6·0–8·1 (7·3), (n = 1, Mozambique) 7·8; ♀ (n = 7, Zimbabwe) 7·0–9·0 (7·8), (n = 2, Zaïre, Feb, June) 5, 7·5, (n = 1, Mozambique) 8·4.

E. i. helenorae Alexander (including '*viriditincta*'): Caprivi Strip (Namibia), Zimbabwe (except SW), and SW Zambia to W Mozambique. Yellow more restricted and less pure than in

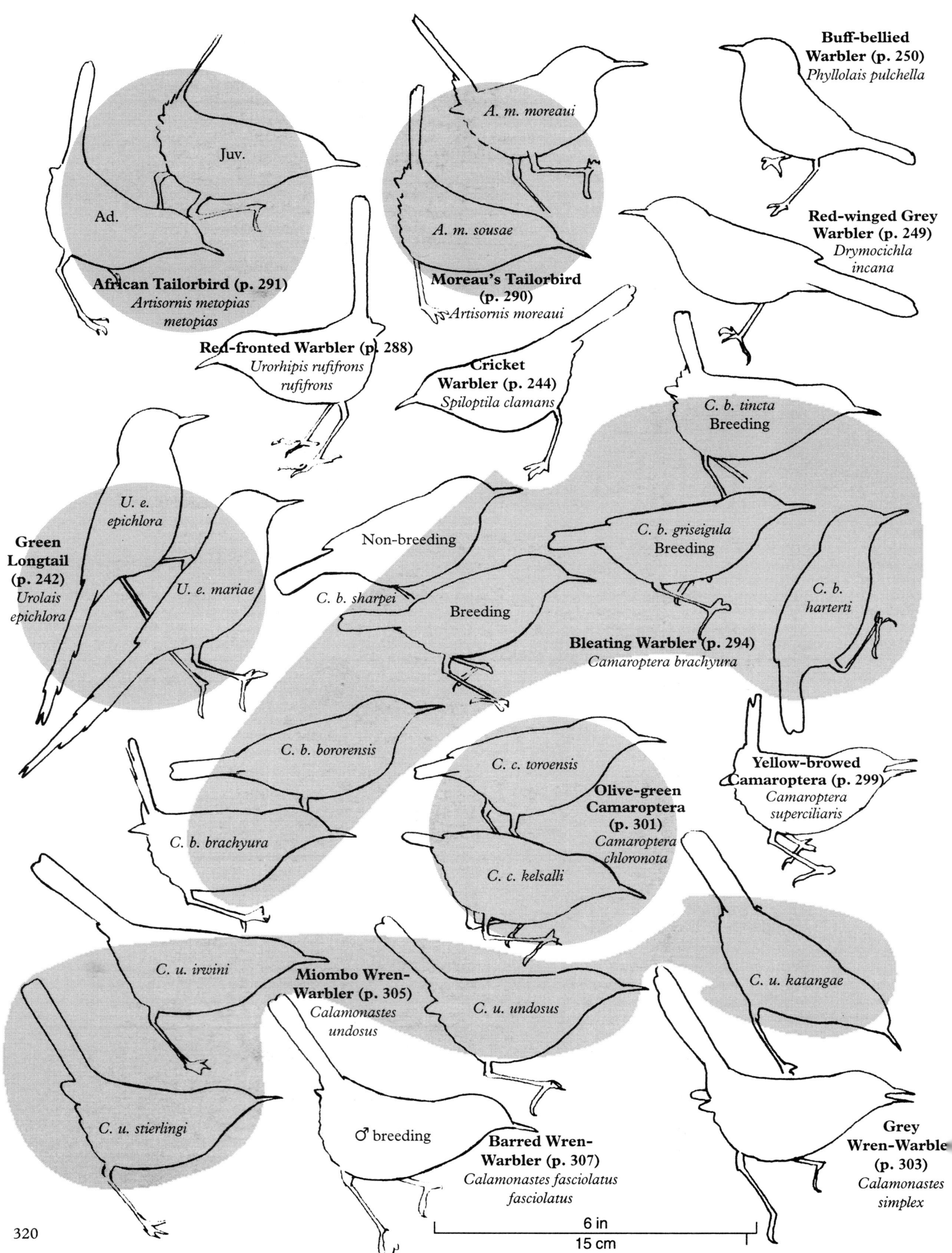

Plate 17

Plate 18

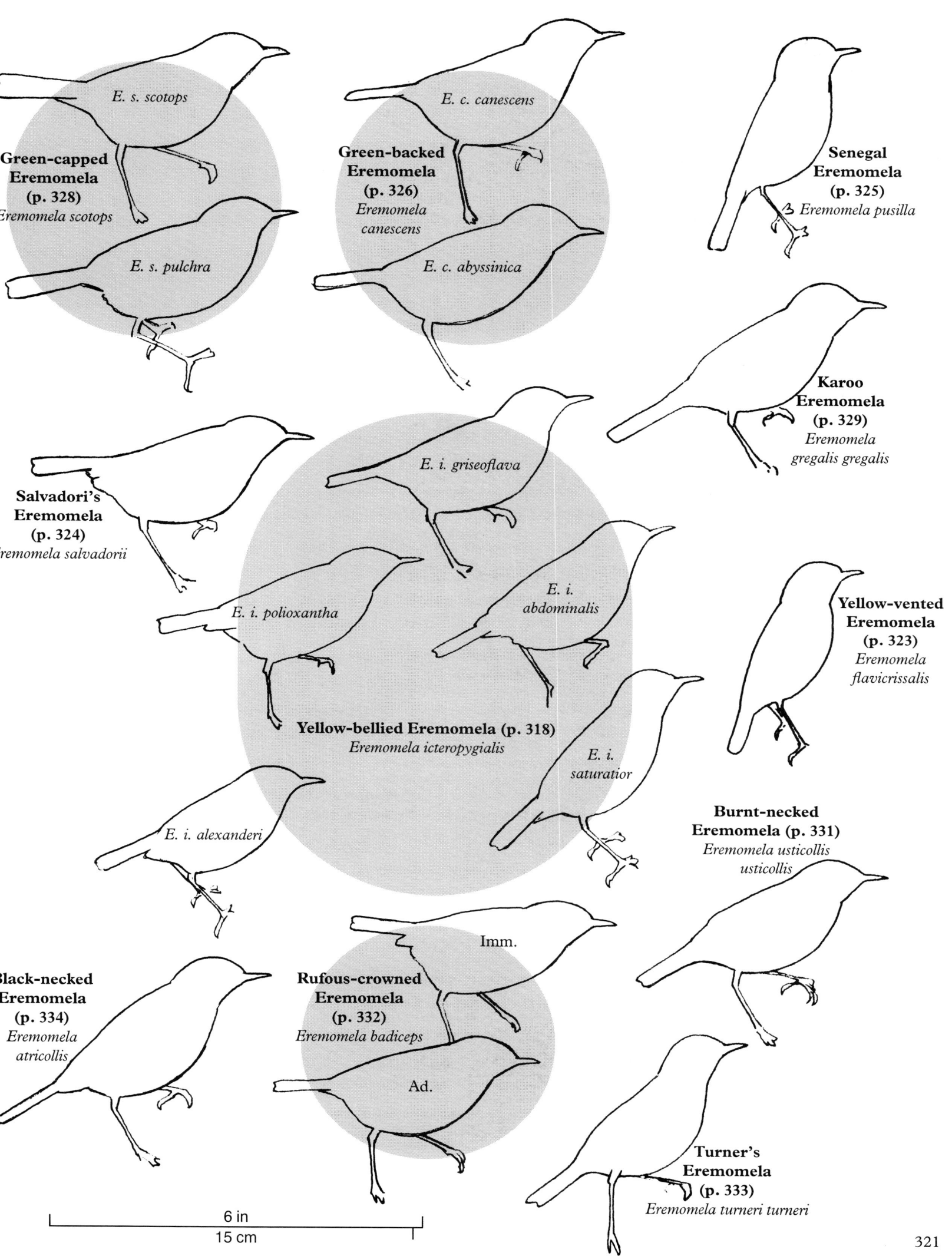

polioxantha giving belly a greenish tinge; upperparts and breast darker than in *alexanderi*. WEIGHT: ♂ (n = 8, Botswana) 7·0–8·8 (7·7); ♀ (n = 6, Botswana) 7·0–8·6 (7·9).

E. i. icteropygialis (Lafresnaye): Namibia (except Caprivi Strip) and W Botswana. Rather whiter on breast and upper belly than *alexanderi*.

E. i. puellula Grote: SW Angola. Upperparts slightly darker and yellow less extensive and paler than in *icteropygialis*. Smaller: wing, ♂ (n = 7) 49–55 (51·9), ♀ (n = 4) 48–52 (50·5).

E. i. perimacha Oberholser: NW Cape Province, W Transvaal and S Botswana. Upperparts slightly browner, breast slightly greyer and yellow slightly brighter than in *icteropygialis*.

E. i. saturatior Grant (including '*sharpei*'): South Africa (Cape Province (except NW) to Orange Free State). Upperparts browner, throat, breast and upper belly more brownish grey, and yellow on belly darker than in *icteropygialis*.

Field Characters. A rather dumpy short-tailed warbler with grey upperparts, whitish breast and yellow belly. Resembles crombecs *Sylvietta* spp., penduline tits *Anthoscopus* spp. and white-eyes *Zosterops* spp. in behaviour and plumage; all 4 genera associate. Told from penduline tits by restriction of the yellow to belly, longer bill and whitish eye-stripe; and from white-eyes by grey (not olive) back. Overlaps most congeners at some point in range, but other grey and yellow ones differ as follows: in Yellow-vented Eremomela *E. flavicrissalis* the yellow is confined to small area around vent and is paler; Salvadori's Eremomela *E. salvadorii* has olive-green rump and uppertail-coverts; Green-capped Eremomela *E. scotops* has a pale green crown; Karoo Eremomela *E. gregalis* has wholly olive-green upperparts; and Green-backed Eremomela *E. canescens* and Senegal Eremomela *E. pusilla* have yellowish green mantle and rump, and more yellow in underside, extending onto breast. Yellow-bellied is easily confused with other eremomelas, especially with Green-backed in W Africa, and so it has been overlooked; in Ethiopia and Somalia it can also be confused with Philippa's Crombec *Sylvietta philippae* which, however, has pale legs and a face mask.

Voice. Tape-recorded (68, 86, 88, 91, B, F, HOR). Typical song a quite loud 7–8 note 'te-te-tu, tetetu-u' with middle part slightly higher pitched and with the last note elided, repeated several times. Very similar to Northern Crombec *Sylvietta brachyura* (suggesting interspecific vocal convergence: Chappuis 1979a). Heard to imitate 5 other species on 1 day at start of breeding season (Cape Province, South Africa: Martin 1990). Alarm a plaintive whistle, rising in pitch. Has a harsh 'dzzz' before flying off. Begging young have a quiet persistent 'cheep'.

Sings most of year. When wandering in non-breeding season, at times sings a quiet subsong consisting largely of imitations (Natal, Zimbabwe: Vernon 1973).

General Habits. A widespread bird in almost any woodland or scrub, although in the more arid parts of southern Africa largely restricted to riparian habitats. In Zambia and Zimbabwe inhabits *Brachystegia* (especially), acacia and other primary and secondary woodlands, but not mopane, and where it overlaps with Burnt-necked Eremomela *Eremomela usticollis* it is rare in acacia. Commoner in drier areas. Also occurs in gardens, cassava cultivation and on the edge of forest. Reported feeding in an irrigated lucerne field (South Africa: Martin and Pepler 1995). In E Africa mainly in lower and medium rainfall areas but replaced by Yellow-vented Eremomela in driest parts of N and E Kenya and Somalia; where they co-exist in Kenya and Somalia, Yellow-bellied Eremomela is in lusher, greener areas (Lewis and Pomeroy 1989). In Sudan, Ethiopia and W Africa it occurs in drier savanna scrub and especially in acacias.

Forages in canopy and lower down in trees; also in bushes and occasionally on ground; the least specialized member of the genus. In Zimbabwe prefers soft foliage and trees with fewer thorns. Restlessly works through a bush or tree from bottom to top searching mainly the leaves – in 35 observations 46% of birds were in leaves, 11% in twigs and 43% in bushes (Vernon 1980). Foraging behaviour very similar to Cape Penduline Tit *Anthoscopus minutus*.

Fairly tame and confiding, although inconspicuous and rather quiet, usually in pairs or parties of up to 8 birds. Joins mixed parties; seen in 12 out of 60 parties in *Brachystegia* woodland in Zimbabwe (L. Kyle) with 1–2 eremomelas (av. 1·1) in each party, and in 13 out of 50 sightings of a party in mixed woodland (av. 1·3 birds: Vernon 1980). Forages almost entirely very early or very late in day. Keeps to shade for much of day.

In W Africa resident in southern part of range, but migratory in northernmost areas, where present only in rains, especially July–Sept. At L. Kyle, Zimbabwe wanders seasonally into *Brachystegia* woodland from adjacent habitats in winter (Vernon 1980); and some migration to lower elevations into Middle Zambezi valley (Zimbabwe) in June–July.

Food. Tiny insects: Homoptera, larval Lepidoptera, beetles, and larvae. A butterfly once taken to nest. 1 apparently fed on nectar of *Acacia mellifera* (Duckworth *et al.* 1992).

Breeding Habits. Monogamous and territorial as far as known.

NEST: purse-shaped, deep cup; a compact structure of plant down and cobwebs, lined with thin layer of fine grass; rather thin-walled (**A**). Ext. diam. 50, int. diam. 35–40, ext. depth 45–55, int. depth 40–50. Slung in fork of bush or small tree, normally 1–3 m from ground, often at edge in open, but may be hidden in foliage.

EGGS: 1–3; southern Africa av. (n = 35 clutches) 2·3. Oval; white with no gloss, lightly spotted at large end with dark red or sepia spots, and greenish brown and light grey undermarkings. SIZE: (n = 133, southern Africa) 13·0–18·0 × 10·4–12·2 (15·8 × 11·4).

LAYING DATES: Senegambia, all year; Mauritania, June–Aug; Mali (breeding condition June–July); Chad

A

(breeding condition Feb); Niger, June; Nigeria, Feb; Zaïre, Sept; Sudan, Feb, June; Ethiopia, Feb–June, Aug–Dec; Somalia, May–June; E Africa: Region C, Mar–June, Oct, Nov; Region D, Feb–Apr; Region E, June; Tanzania, Feb; Zambia, June–Oct (46 of 50 clutches Sept–Oct); Malaŵi, Oct; Zimbabwe, Aug–Apr (especially Sept–Nov); Angola, Jan; Botswana, Oct–Nov, Mar; South Africa, Sept–Mar; Namibia, May (nest-building July).

INCUBATION: ♂ once incubated (Ethiopia). 2 eggs laid on consecutive days and bird sat during day on first egg (Namibia). Period: 13–14 days.

DEVELOPMENT AND CARE OF YOUNG: young fed by both parents and prone to leave nest prematurely if disturbed. Parents continue feeding young >14 days after they fledge; once 2 juveniles seen with parents 4 weeks after fledging (Jensen and Clinning 1974). Nestling period 15–16 days.

BREEDING SUCCESS/SURVIVAL: 3 out of 12 nests parasitized by Klaas's Cuckoo *Chrysococcyx klaas* in Namibia (Jensen and Clinning 1974). A bird ringed in Bulawayo was retrapped 2 years and 23 days later.

References
Jensen, R. A. C. and Clinning, C. F. (1974).
Vernon, C. J. (1980).

Eremomela flavicrissalis Sharpe. Yellow-vented Eremomela. Érémomèle à ventre jaune.

Plate 18
(Opp. p. 321)

Eremomela flavicrissalis Sharpe, 1895. Proc. Zool. Soc., Lond., p. 48; Webi Shebeli, southern Abyssinia.

Forms a superspecies with *E. icteropygialis* and *E. salvadorii*.

Range and Status. Endemic resident, confined to semi-arid areas of E Africa. 1 record NE Uganda (Moroto); uncommon through much of N and E Kenya from L. Turkana and Baringo south to Tsavo East Nat. Park and the coast. Also in S and SE Ethiopia and in Somalia except NW and NE and south of 3°N.

Uncommon in Kenya and common in Somalia, up to 1200 m; in Kenya, range is evidently contracting from west as all records along SW edge of range are pre-1970 (Lewis and Pomeroy 1989).

Description. ADULT ♂: upperparts uniformly grey with brownish tinge but uppertail-coverts tipped very pale yellow; tail grey with brownish tinge, feathers paler around edges and especially on tips. Cheeks somewhat paler than crown but darker than underparts; chin to belly dirty white, sides of breast greyish; a small patch around vent very pale yellow; undertail-coverts white with a yellowish wash. Primaries like upperparts but slightly darker, with pale outer edges forming pale panel on closed wing; secondaries and tertials a little paler than primaries and with pale outer edges; upper wing-coverts similar; primary coverts with pale tips. Bill dark; eyes light brown; legs and feet grey or black. Sexes alike. SIZE: (3 ♂♂, 2 ♀♀) wing, ♂ 48–52 (50·3), ♀ 49, 49; tail, ♂ 26 (all

Eremomela flavicrissalis

Dashed line shows pre-1970 range

26), ♀ 23, 27; bill, ♂ 9·4–10·6 (10·0), ♀ 9·6, 9·9; tarsus, ♂ 16·0–17·4 (16·7), ♀ 16·3, 16·4. WEIGHT: 1 ♀ (Kenya) 4·5.

IMMATURE: lower belly white or with only faint yellowish tinge.

NESTLING: unknown.

Field Characters. A fairly small, dry country eremomela, similar to Yellow-bellied Eremomela *E. icteropygialis*, but with yellow confined to just around vent (sympatric race of Yellow-bellied has belly entirely yellow). Yellow-vented further distinguished from Yellow-bellied by being smaller with a much more slender bill. Similar to sympatric Philippa's Crombec *Sylvietta philippae*, which is slightly larger and has reddish legs, not grey or black; it also has a darker face mask and a faint pale eye-stripe.

Voice. Not tape-recorded. Has a weak 'tssp' call.

General Habits. Fairly widespread in wooded and bushed habitats in semi-arid and arid areas (80% of squares in which recorded have less than 500 mm annual rainfall: Lewis and Pomeroy 1989). In Tsavo East Nat. Park occurs mainly in the more open parts of *Commiphora* woodland (1 seen every 10 h) and occasionally in other savanna habitats with trees (Lack 1985). Always solitary in Tsavo, but in Somalia 50% of observations were of 1 bird, 44% of 2 birds and 6% of 3 birds (J. S. Ash, pers. comm.).

Forages in bushes and trees, busily taking insects from leaves and twigs up to 10 m from the ground in canopy. Gleans leaves 3 times as much as twigs. Seen twice to drop to ground to take insect from grass; once seen to take insect in flight (P. C. Lack, pers. obs.).

Food. Small insects.

Breeding Habits.
NEST: small cup of grasses or cobwebs, sometimes with a few small twigs; ext. diam. *c.* 50; sited 1·5–2 m from ground in bushes or low trees; one nest in top of almost bare bush 2 m from ground was slung in fork between several twigs. Built by both sexes.

EGGS: 1–2; white with a few dark brown or sepia spots and fine speckling. SIZE: (n = 4, Somalia and Kenya) 14–15 × 10·1–11·8 (14·2 × 11·0).

LAYING DATES: E Africa: Region D, Jan; Ethiopia, May; Somalia, nest-building Apr, May, Dec.

Plate 18
(Opp. p. 321)

Eremomela salvadorii Reichenow. Salvadori's Eremomela. Érémomèle de Salvadori.

Eremomela salvadorii Reichenow, 1891. J. Orn., p. 64; Leopoldville, western Congo.

Forms a superspecies with *E. icteropygialis* and *E. flavicrissalis*.

Range and Status. Endemic resident, restricted mainly to area each side of Zaïre/Angola border. Fairly common in Lekoni area, SE Gabon. In Zaïre, from Kinshasa up Congo R. to just north of equator, east to E Kasai and west across river into Congo; fairly common in Angola on the central plateau from central and N Huila to S Cuanza Sul; in Zambia restricted to Zambezi District.

At least 10 recorded daily in open *Brachystegia* woodland in Gabon (D. Sargeant, pers. comm.).

Description. ADULT ♂: head and nape grey except for white stripe over eye and dark streak through it; mantle and scapulars olive-green, becoming more yellowish green towards rump and bright yellowish green on uppertail-coverts; tail brown. Chin and throat white; lores and cheeks pale grey; breast white with grey sides; lower breast to undertail-coverts and flanks bright yellow. Wings brown with paler edges to basal half of primaries and to all of secondaries and tertials; upper wing-coverts same brown as wings; underwing white washed with yellow. Bill blackish; eyes brown or black; legs and feet black. Sexes alike. SIZE: (4 ♂♂, 2 ♀♀) wing, ♂ 54–64 (57·8), ♀ 50, 57; tail, ♂ 33–37 (34·3), ♀ 29, 31; bill, ♂ 11·7–12·1 (11·9), ♀ 10·9, 11·2; tarsus, ♂ 17·0–19·5 (18·2), ♀ 16·7, 17·2.

IMMATURE and NESTLING: unknown.

Field Characters. A large eremomela with greyish green back and yellow belly. Rump and uppertail-coverts green not grey, yellow in underparts considerably brighter than in Yellow-bellied Eremomela *Eremomela icteropygialis*, and sharply divided from white on breast. For other distinctions see that species. Of other sympatric congeners, Black-necked Eremomela *E. atricollis* has orange around head, yellow chin and throat and black breast-band, and Burnt-necked Eremomela *E. usticollis* has rusty breast-band, grey not greenish upperparts and no yellow on underparts.

Voice. Unknown.

General Habits. Inhabits bushes at edge of *Brachystegia*, acacia and other woodland. In Gabon in open grassy plains with well-spaced *Brachystegia* (D. Sargeant, pers. comm.). Habitats like those of Yellow-bellied Eremomela but Salvadori's is usually in more open and probably in lusher country. In S Zaïre and Angola it is a bird of high escarpment woodlands (with the Yellow-bellied in dry lowlands). In Zambia once recorded in mavunda (*Cryptosepalum*) woodland (D. Aspinwall, pers. comm.).

Usually in pairs and almost always in mixed-species parties, often with sunbirds and Green-capped Eremomela *Eremomela scotops*. Usually in small trees of up to 8 m in which it feeds 2–6 m up (D. Sargeant, pers. comm.).

Food. Insects: small beetles found in 1 stomach.

Breeding Habits. Unknown. Zaïre 'not breeding condition' July.

Eremomela pusilla Hartlaub. Senegal Eremomela. Érémomèle à dos vert.

Plate 18
(Opp. p. 321)

Eremomela pusilla Hartlaub, 1857. Orn. West Afr., p. 59; Senegal.

Forms a superspecies with *E. canescens*, *E. scotops* and *E. gregalis*.

Range and Status. Endemic resident; the commonest eremomela in most non-forested areas of W Africa. Common in guinea savannas in Senegambia, although rare in lower Senegal; frequent in S Sahel region of Mauritania but rare in Senegal R. delta; widespread in Guinea and Sierra Leone, not yet recorded in Liberia, frequent or common across W Africa to Cameroon, between about 15°N and N edge of forest belt (e.g. south to Accra and Keta plains in Ghana, to Ibadan, Abeokuta, Okigwi, Ofikpo and Obudu in Nigeria, and Benue plain and Yaoundé in Cameroon). Eastern boundary, about 15°E, uncertain due to field confusion and some hybridization with Green-backed Eremomela *E. canescens*. Only records of *E. pusilla* in Chad are at Ndjamena, Fort Archambault (Sarh) and Pala. Occurs in NW Central African Republic (Manovo-Gounda St Floris Nat. Park); specimens from Koum and Tello on E Cameroon/Central African Republic border appear to be hybrids.

Densities in Ivory Coast up to 10 birds in 50 ha (July–Aug, in guinea savanna: Thiollay 1973); in Guinea–Bissau 28 birds per 10 ha in open *Avicennia* mangrove 1·5–3 m high and in dense *Rhizophora* mangrove 5 m high, but only 8 per 10 ha in dense *Rhizophora* 3–8 m tall (Altenburg and van Spanje 1989).

Eremomela pusilla

Description. ADULT ♂: forehead to nape, cheeks and ear-coverts pale grey with sandy brown tinge; faint whitish eye-stripe; mantle and scapulars pale grey with outer edges of feathers yellowish green giving appearance of grey with many diffuse green streaks; rump more yellowish green with no grey; uppertail-coverts green with less yellow than rump. Tail brown suffused with green, edges to feathers greener, the green becoming paler and yellower towards outer feathers; pale tips on outer feathers larger than on inner ones. Chin and throat white; upper breast whitish with pale buffy tinge; lower breast to undertail-coverts, flanks and underwing bright yellow. Wings brown (darker than tail) with outer edges of primaries pale greenish yellow grading to those of tertials yellowish green; upper wing-coverts a little paler than wings with variable amount of pale green on tips and edges. Upper mandible dark brown, lower mandible much paler, often dull yellow; eyes grey, sometimes with greenish or brownish tint; feet brown often tinged olive-green. Sexes alike. SIZE: (10 ♂♂, 10 ♀♀) wing, ♂ 46–51 (49·2), ♀ 45–52 (47·9); tail, ♂ 38–44 (42·3), ♀ 33–42 (39·6); bill, ♂ 11·7–12·9 (12·5), ♀ 10·4–13·0 (11·8); tarsus, ♂ 14·1–17·7 (16·3), ♀ 14·8–17·4 (15·9). WEIGHT: unsexed (n = 38, Ghana, July–Sept) 5·0–8·4 (6·2), (n = 5, Ghana, Nov–Dec) 6·0–6·4 (6·1), (n = 4, Nigeria, Nov–June) 6·9–7·5 (7·1).

IMMATURE: like adult except that upper mandible has yellow tip.

NESTLING: unknown.

Field Characters. A small arboreal warbler characterized by grey head, green mantle, white throat and yellow underparts, with no dark streak through eye. Distinguished from Green-backed Eremomela by lack of dark streak through eye, lack of white eye-stripe and of buffy area on breast between the white throat and yellow belly. Easy to confuse with Yellow-bellied Eremomela *E. icteropygialis*, which has grey-brown mantle and less yellow on underparts.

Voice. Tape-recorded (68, B, GRI, MOR). Song, sung only at dawn, a monotonous chipping, a short double note lasting 0·1 s and repeated evenly and monotonously about 139 times per min for 10–15 min: 'chip, chip, chip, chip . . .' (sonogram in Grimes 1975). Song-like contact or foraging call (formerly described as song) is a 'sharp-toned chittering noise as if the bird is shivering with the cold' (Bates 1930): 3–4 harsh notes followed by a high-pitched, clear, rising then falling trill; one trill of 19 notes lasted 1·6 s (Grimes 1975). This can be uttered at any time of day (C. H. Fry, pers. comm.). Also has a 3-note cricket-like rattle and a small 'tsep'.

General Habits. Inhabits all types of wooded savannas, cultivated areas with small trees, acacia grassland, degraded savanna, orchard bush and gardens; also mangrove. Occurs along forest edges, especially in moister areas, but does not penetrate forest itself. In gallery forests and small patches, keeping mainly to more open parts. In Senegal, one of the few passerines common in leafless dry-season woodland (but favours thicker vegetation than Yellow-bellied Eremomela in the same area).

Occurs in small groups, typically of 4–6 birds, and a frequent member of mixed-bird parties. Forages actively mainly in treetops, calling frequently. Feeds largely in twiggy foliage; sometimes jumps from twig up to leaves above, hovering momentarily; rarely, takes flying insect in air (V. Salewski, pers. comm.). Continually flicks tail while flitting from twig to twig. Flights short and undulating.

Finishes primary moult, Ivory Coast, in late Nov/early Dec (V. Salewski, pers. comm.).

Food. Insects: mainly ants, beetles and caterpillars and other larvae; also blattids, flies, hymenopterans, lepidopterans, orthopterans, mantids, bugs, termites; spiders. At Lamto (Ivory Coast) 70% of food was adult arthropods and 30% larvae; also a few fruits recorded (Thiollay 1973).

Breeding Habits. Breeds co-operatively. Song occurs only prior to and during breeding season, Jan-early Sept (Ghana: Grimes 1975). Double-brooded (Nigeria).

NEST: tiny well-made cup of small pieces of leaf and bark, bound together with silken threads of spiders or vegetation. Ext. diam. 53, int. diam. 35, int. depth 20. Usually suspended from branch or fork in bush, normally 1–2 m up.

EGGS: 2; greenish blue with a zone or wreath of brown or lilac markings near larger end. SIZE: (n = 4) 14–15 × 10·5–10·9 (14·5 × 10·6).

LAYING DATES: Senegambia, Apr; Burkina Faso, May; Ivory Coast, Jan–Oct; Niger, May; Ghana (fledglings, Apr–Aug, Oct); Nigeria, Feb, May (and fledglings Oct).

DEVELOPMENT AND CARE OF YOUNG: fledglings usually fed by more than 2 adults.

BREEDING SUCCESS/SURVIVAL: in Ivory Coast, 2-egg clutches produced av. 1·2 flying young (no sample size given) (Thiollay 1971).

Reference
Grimes, L. G. (1975).

Plate 18
(Opp. p. 321)

Eremomela canescens Antinori. **Green-backed Eremomela.** Érémomèle grisonnante.

Eremomela canescens Antinori, 1864. Cat. Coll. Ucc., p. 38; Djiu R., Bahr-el-Ghazal, south-west Sudan.

Forms a superspecies with *E. pusilla*, *E. scotops* and *E. gregalis*.

Range and Status. Endemic resident, N tropical Africa from Chad to Ethiopia and Kenya. Recorded near Garoua in Cameroon. Occurs in S Chad and Central African Republic where it meets range of

Eremomela canescens

Senegal Eremomela *E. pusilla* at about 15°E; they hybridize along Cameroon/Central African Republic border (Koum and Tello). All Central African Republic except W; in N Zaïre south to Mondjo, Uele R. and L. Albert; fairly common in Sudan south of *c.* 14°N; uncommon in central and W Eritrea, frequent on higher ground of central and W Ethiopia to just east of Rift valley at 41°30'E, and south to Yavello and 50 km south of Adola; in Uganda south to L. Albert, just to southwest of Ntoroko and L. Kioga, and W Kenya to Mt Elgon, Kericho, Sotik and Kongelai escarpment.

Frequent to common, mainly between 500 m and 2000 m, but sometimes up to 2200 m; in Ethiopia and Eritrea only up to 1800 m. In Kenya mainly a highland bird and only in areas with more than 1000 mm rain annually.

Description. *E. c. canescens* Antinori: Central African Republic and Chad to Sudan (Mongalle), Uganda and Kenya. ADULT ♂: forehead to nape ash-grey; cheeks and broad streak through eye black; variable white stripe from base of bill over eye; mantle light olive-green sharply defined from nape and becoming more yellowish on rump and upper-tail-coverts. Tail brown, paler at tip, with outer feathers paler and outer edges of all feathers yellowish green; pale tip to T6 1–2 mm, breadth decreasing to little or nothing on inner feathers. Chin to upper breast white; lower breast to under-tail-coverts, and underwing, bright yellow. Wings dark brown, with outer edges of feathers yellowish green; scapulars bright olive-green; upperwing-coverts dark brownish grey with green tips and outer edges. Bill black; eyes yellowish brown; feet light brown or flesh. Sexes alike. SIZE: (10 ♂♂, 7 ♀♀) wing, ♂ 52–57 (54.5), ♀ 51–55 (53.4); tail, ♂ 38–45 (41.8), ♀ (n = 6) 32–43 (38.7); bill, ♂ 12.5–14.0 (13.0), ♀ 12.2–13.4 (12.9); tarsus, ♂ 16–18.3 (17.2), ♀ 16.2–17.5 (16.7). WEIGHT: 1 ♂ (Kenya) 6; 1 ♀ (Uganda) 8.

IMMATURE: mantle tinged olivaceous.
NESTLING: unknown

E. c. elegans Heuglin: Sudan: Darfur and Kordofan to Sennar. A poorly defined race with upperparts slightly paler grey and less bright green and underparts paler than *canescens*.

E. c. abyssinica Bannerman: Ethiopia, Eritrea, and Sobat and Ayod in Sudan. Darker generally than *canescens* and with no eye-stripe.

E. c. elgonensis van Someren: Mt Elgon to S Nandi. Larger and more brightly coloured than *canescens* and ear-coverts blacker.

Field Characters. Grey head contrasting with green back, and white throat sharply demarcated from yellow belly, together with black streak through the eye, distinguish it from all congeners. The black eye-streak, white supercilium, clear grey head and green mantle, lack of a buffy zone between white throat and yellow belly, distinguish it from the very closely related Senegal Eremomela. Yellow-bellied Eremomela *E. icteropygialis* has a grey brown mantle and less extensive yellow on underparts; Yellow-vented Eremomela *E. flavicrissalis* has even less yellow on underparts and no green on back; Green-capped Eremomela *E. scotops* has a yellowish grey head and ash-grey back and mantle.

Voice. Tape-recorded (68). Unmusical but joyful continuous chittering song, less melodious and more complex than that of Senegal Eremomela (Chappuis 1979a). Call a sharp, harsh, double note, often repeated.

General Habits. Locally common in open wooded savannas, mainly on higher ground especially in acacia bush, all *Combretum* woodland and thick riparian bushes and scrub. In Ethiopia very much a bird of *Combretum/Terminalia* woodland; also in juniper and other evergreen scrub. In dry zones of N Zaïre inhabits woodland edges and cultivation.

A shy bird occurring in parties of 4–6; often joins mixed-species flocks. Forages very actively, searching for insects in crowns of trees.

Food. Insects. Small caterpillars and hard sclerites found in stomachs.

Breeding Habits. Not known to differ from Senegal Eremomela.

NEST: cup of leaves, twigs and stems, suspended from fork or branch by threads.

EGGS: 1–2, always 1 in Zaïre; bright bluish green with brown and lilac spots on larger end. SIZE: *c.* 14 × 10.5.

LAYING DATES: Zaïre, Feb–Apr; Sudan, Feb, Mar; Ethiopia, Feb; E Africa: Region B, May, June.

Plate 18
(Opp. p. 321)

Eremomela scotops **Sundevall. Green-capped Eremomela. Érémomèle à calotte verte.**

Eremomela scotops Sundevall, 1850. Oefr. K. Vet-Akad. Forh. 7, p. 103; Mohapoani, Witfontein Mts, western Transvaal.

Forms a superspecies with *E. pusilla*, *E. canescens* and *E. gregalis*.

Range and Status. Endemic resident, S-central Africa. In Lekoni area of SE Gabon where a few seen daily; SE Congo at Brazzaville, Ossale (Kwamouth) and Djambala, and throughout Angola except for extreme SW. Zaïre south of a line from about Mbandaka near Congo border to L. Tanganyika opposite Kigoma; Burundi, Rwanda and SW Uganda on Kagera R. Around S side of L. Victoria, extending into SW Kenya through Mara Game Reserve, to Nairobi and Thika and including Kedong valley, east to 20 km east of Embu. Throughout Tanzania except for the more arid central areas and east of Loita Hills, although it occurs all along the coast inland to the foothills of the E Usambara Mts, and north into Sokoke Forest on Kenya coast (although this population perhaps isolated). Throughout Zambia, except for Luangwa and Middle Zambezi valleys and largely absent where there is no miombo such as the area between Lakes Mweru and Tanganyika, Bangweulu and the Kafue Flats. In Namibia only in Caprivi Strip; throughout Zimbabwe, Malaŵi and Mozambique, where local and uncommon in Sul do Save but relatively common north of Save R.; N and NE Botswana, and South Africa in central and E Transvaal (scarce) and N coastal Natal.

Occurs from sea level to 1800 m, being commoner above 1500 m. Common in most of range, becoming rarer in N and E; scarce and local in South Africa and S Mozambique and the more arid parts of Zimbabwe and Kenya. Density of 5 pairs per 100 ha in *Brachystegia* in Zimbabwe (Vernon 1985a).

Description. *E. s. scotops* Sundevall (including '*chlorochlamys*' and '*occipitalis*'): E Kenya, E Tanzania, E Malaŵi, E Botswana, Zimbabwe, Mozambique, Transvaal and Natal. ADULT ♂: forehead, forecrown and sides of face yellowish green with a greyish wash; hindcrown green and grey mixed; streak from lores to eye rather variable, dusky or grey; nape to uppertail-coverts ash-grey with a greenish wash; tail grey with faint brown wash and outer feathers a little paler. Underparts pale yellow, with throat and breast brighter yellow than chin, belly and undertail-coverts. The yellow is more extensive in E and NE birds, and is restricted to breast in central and S populations. Primaries brownish grey grading to greyish brown tertials; outer edges of all flight feathers whitish, in primaries forming a pale panel in closed wing; broad whitish stripe along inner edge of all flight feathers; scapulars and upperwing-coverts grey; underwing white. Bill black; eyes pale yellow; legs and feet reddish or pinkish brown. Sexes alike. SIZE: (10 ♂♂, 10♀♀) wing, ♂ 53–61 (58·1), ♀ 53–59 (56·5); tail, ♂ 45–49 (47·1), ♀ 42–48 (46·4); bill, ♂ 13·2–14·6 (13·9), ♀ 12·6–14·7 (13·5); tarsus, ♂ 17·8–21·2 (19·2), ♀ 18·0–20·9 (19·1). WEIGHT: ♂ (n = 21, Zaïre) 6–11·0 (8·8), (n = 20, Zimbabwe) 8–10·4 (9·0), (n = 7, Mozambique) 7·5–9·1 (8·4); ♀ (n = 12, Zaïre) 8–11 (9·2), (n = 11, Zimbabwe) 8·3–11·0 (9·7), (n = 4, Mozambique) 7·6–8·5 (8·1), (n = 2, Kenya) 5, 5; unsexed (n = 9, Zaïre) 7–10 (8·3).

IMMATURE: upperparts grey with green tinge; edges of flight feathers and of wing-coverts green; belly white; yellow restricted to throat and breast.

NESTLING: unknown.

E. s. kikuyuensis van Someren: central highlands of Kenya. Belly whiter, with only slight yellowish wash; olive-green on head paler than in *scotops* and extends less onto nape.

E. s. pulchra (Bocage) (including '*extrema*'): S and central Angola, Zaïre, Zambia, W Malaŵi. Underparts white; yellow only on throat and upper breast, sharply demarcated from the white; eye-stripe yellowish green. Intergrades with *congensis* in Angola.

E. s. citriniceps (Reichenow): W Tanzania, Uganda and W Kenya. Like *pulchra* but head more greenish yellow especially on cheeks and eye-stripe.

E. s. congensis Reichenow (including '*angolensis*'): Congo, NW Zaïre and N Angola. Like *scotops* but underparts duller yellow, upperparts washed with olive-green. Intergrades with *pulchra* in Angola.

Field Characters. A small arboreal warbler of *Brachystegia* woodland where it is the commonest eremomela, living in small tight flocks all year, with all individuals tending to act in concert (Vernon and Vernon 1978). Characterized by whitish eye in dusky streak, yellow breast and white belly. Much yellower on head than any other eremomela; yellowish head and ash-grey back distinguish it from *Phylloscopus* warblers. Grey back and lack of eye-ring distinguish it from white-eyes *Zosterops* spp. Continuous chattering call attracts the attention and is a good character.

Voice. Tape-recorded (86, 88, 91, B, C, F). Song a loud, rapid, incessant, and rather variable 'piou, piou, piou', the first note actually being double but not distin-

Eremomela scotops

guishable as such by ear (see sonogram in Maclean 1993). Each note lasts 0·17 s; notes are repeated at rate of 144 per min (Grimes 1975). Also a more liquid 5–6 note song. Contact call a continuous chatter of rather variable buzzes, 'squeeechee' or 'chiaou', often of quality of *Acrocephalus* warbler. Alarm, a loud rasping 'churrr'.

General Habits. Primarily a bird of *Brachystegia* woodland but occurs widely in many types of open woodland, including *Baikiaea*. Keeps mainly in canopy. In SE Zimbabwe especially common in tall riparian acacia woodland, and in S Mozambique almost confined to mopane. Also occurs in open woodland in moister areas, riparian forest, evergreen forest edges and clearings and gardens. Prefers thicker habitats than Greenbacked Eremomela *E. canescens*, where they overlap in Kenya. In Zimbabwe, where *E. scotops* occurs with 2 other eremomelas, it inhabits only tall trees.

Tame. Occurs in pairs or family parties; in Zimbabwe flocks in breeding season were of 2–7 birds (av. 3.3, n = 40), and in rest of year 2–8 birds (av. 4.1, n = 35) (Vernon and Vernon 1978). Activities of flock are communal; all birds of a group forage, preen and bathe together; they sometimes allopreen, and sunbathe; bathing is in dew or rain on leaves in canopy (Vernon and Vernon 1978). The species is a frequent, often a core, member of mixed-bird parties. In *Brachystegia* woodland at L. Kyle (Zimbabwe) it was one of 14 main species, occurring in 39 out of 60 bird parties, with 2–6 birds (av. 3·8) in each party (Vernon 1980). Recorded in 60% of parties in E Zambia and in 40% in W Zambia (Winterbottom 1943). At all times of year when groups meet there is great excitement and much continual chasing around and above tree tops. Interactions usually end abruptly after up to 5 min (Vernon and Vernon 1978). Very active and noisy; makes a continual flipping sound with wings. Forages on small insects, taken mainly from leaves (58%), twigs (22%), and bushes (19%) with 1% from the air (119 observations: Vernon 1980). Contrarily, said to catch insects in air more often than other eremomelas, often with audible snap of bill (Irwin 1953). Flights between trees rapid and slightly undulating. Flocks return to same tree 2 or 3 times in a morning (Vernon and Vernon 1978). Groups roost together in a row, often returning to same site on successive nights (Vernon and Vernon 1978).

2 complete moults per year in Zaïre, Nov–Feb and Apr–June, and has partial post-juvenile moult.

Food. Small insects.

Breeding Habits. Co-operative breeder; groups of 2–7 (av. 3·3) defend territory together all year and jointly participate in all breeding duties (Vernon and Vernon 1978). Sings, usually from top of tree, from before dawn to just after sunrise, then ceases abruptly and rejoins flock. Sings only just prior to and during breeding season. 3 singing birds once noted *c*. 40 m apart, but normally they are more widely spaced. Displays include bill-snaps, wing-flicks and much chasing.

NEST: delicate, thin-walled cup of plant fibres, small bits of leaf and sepals, bound with silk thread; ext. diam. 50, int. diam. 38, ext. depth 55, int. depth 35. Slung from fork or at end of small branch attached at 3–4 points on rim, usually concealed among leaves, often 6–7 m from ground. Built co-operatively by up to 5 birds; bird arrives alone, builds, and then sits on nest until next one arrives; 2 arriving together each build in turn; some individuals come to nest more often than others (Vernon and Vernon 1978).

EGGS: 2–3, usually 2. Pale blue, lightly spotted with rufous and lilac markings especially at larger end. SIZE: (n = 9) 15–17 × 11–12.5 (16·3 × 11·8).

LAYING DATES: Zaïre, July–Nov; Zambia, Sept–Nov (and active gonads Aug, Apr); Zimbabwe (Aug 2, Sept 18, Oct 9, Nov 6, Dec 1, Jan 1, Feb 1 clutches); Malaŵi, Oct–Dec, Feb; Tanzania, Feb (and feeding young early Dec); Mozambique (about to nest, Sept); South Africa, July–Feb (mainly Sept–Nov).

INCUBATION: by up to 5 birds. In 20 h of observation over 11 days, 3 ringed birds came 8, 13 and 19 times and 2 unringed birds came 51 times in all (Vernon and Vernon 1978).

DEVELOPMENT AND CARE OF YOUNG: once 3 birds fed young in nest (Ginn 1986). At least 3 birds (members of 2 different flocks of 4 and 5 birds) fed fledglings, and all 4 of another group did so (Vernon and Vernon 1978).

BREEDING SUCCESS/SURVIVAL: in Zimbabwe, in 1971–74, 14 flocks had 1–2 (av. 1·5) fledglings. In one year 2 of 5 flocks in 100 ha reared fledglings and in following year 3 of the 5 did (Vernon and Vernon 1978). Fledglings may remain in flock and help the following season.

References
Vernon, C. J. (1980)
Vernon, F. J. and Vernon, C. J. (1978).

Eremomela gregalis (A. Smith). Karoo Eremomela. Érémomèle du Karroo.

Plate 18 (Opp. p. 321)

Malcorus gregalis A. Smith, 1829. S. Afr. Commercial Advertiser, 27 June 1829; northern Little Namaqualand.

Forms a superspecies with *E. pusilla*, *E. canescens* and *E. scotops*.

Range and Status. Endemic resident, confined to small area of South Africa and Namibia. In Namibia ranges from Oösop on Swakop R., extending along coast and about 100 km inland, into NW Cape Province south to about 33°S, and east to Oudtshoorn and Colesburg.

Eremomela gregalis

Common in drier habitats. Numbers averaged 89 birds in 838 ha (4·2 per 40 ha) in karoo broken veld (Winterbottom 1968).

Description. *E. g. gregalis* (A. Smith) (including '*albigularis*'): range of species except Oösop area in N Namibia. ADULT ♂: head and upperparts greyish olive-green, greyer on forehead and nape and greener on hindneck and mantle, and greener still on rump; uppertail-coverts olive-green. Tail brown with olive tinge, becoming green with slight yellow tinge on outer edges of feathers and especially so on the outermost 2 pairs. Cheeks and ear-coverts grey; lores dark; chin white; rest of underparts whitish grey with breast and flanks greyest and centre of belly whitest; undertail-coverts pale yellow. Wings and upperwing-coverts brown, primaries with whitish outer edges especially on basal half, secondaries and tertials with edges green not whitish; scapulars like mantle but greener on outer edges; underwing white. Bill black, palate black with gullet white; eyes yellow; feet brown. Sexes alike, except that ♀ has pink palate and dull white gullet. SIZE: (6 ♂♂, 9 ♀♀) wing, ♂ 48–56 (52·3), ♀ 48–55 (50·9); tail, ♂ 48–53 (51·3), ♀ 43–55 (48·4); bill, ♂ 10·8–12·1 (11·5), ♀ 10·5–11·8 (11·3); tarsus, ♂ 17·5–19·8 (18·5), ♀ 16·5–19·0 (17·7).

IMMATURE: upperparts much browner and underparts whiter than adult.

NESTLING: unknown.

E. g. damarensis Wahlberg: Namibia (Oösop on Swakop R.). Generally paler and rump more yellowish.

Field Characters. An uncommon and local bird of the karoo and semi-desert scrub, living in drier habitats than the few similar-looking birds in its range; often shy and hard to find. Best character is yellow eye set in plain grey face. Yellow-bellied Eremomela *E. icteropygialis* lacks green on upperparts and has white eye-stripe and dark eye, and yellow belly; Bleating Warbler *Camaroptera brachyura* has chunkier shape, dark eye, chirping song and characteristic mewing call. Immature Burnt-necked Eremomela *E. usticollis*, which lacks rusty face and throat-bar, has pale eye but grey upperparts, pale yellow underparts.

Voice. Tape-recorded (91, F). Song a continuous repeated far-carrying (up to 400 m) high-pitched 'seep'. Each burst continues for up to 2 min at *c*. 2 notes per s; slight tonal variation between individual birds. Song may start 1 h before sunrise and stop at sunrise, but it often becomes difficult to distinguish and locate singing birds for the last 15 min when other birds start (Frost and Vernon 1978). Group rallying call after an interaction or disturbance 'zii-zii-zii'. Call a persistent sharp 'twink'. Alarm 'chwit'.

General Habits. Occurs widely in karoo, semi-desert and desert scrub, especially along watercourses and among scattered bushes at base of and on rocky hill sides. Where sympatric with Yellow-bellied Eremomela it prefers drier habitats and more stunted bush growth. Common in karoo broken veld (Winterbottom 1968).

Occurs singly, in pairs or groups of up to 7 birds. In Cape Province (plains of Doring R.) group sizes were: Nov (n = 5) 2–6 (4·0), Dec (n = 4) 3–5 (3·8), Feb (n = 4) 2–7 (4·8), and in E Namib Desert and N Richtersveld in Dec (n = 8) 2–6 (3·8). All these, except perhaps the 7, could be families, as they were just after the breeding season; in a group, the birds extra to the pair all resembled ♀♀ and begged for and got food (Frost and Vernon 1978). Secretive and hard to locate; forages in branches and twigs of low bushes, and often on ground.

Roosts sometimes in nest. Apparently some nests are not used for breeding; 2 out of 6 nests in Oct–Nov had faeces around them whereas 2 from which young had fledged and 2 others not used for breeding were clear of faeces (Frost and Vernon 1978). Many droppings were found below a *Euphorbia mauretanica* bush in which a bird was singing before dawn, suggesting a regular roost (Frost and Vernon 1978).

Food. Small insects.

Breeding Habits. Territorial. Possibly a co-operative breeder like some others of genus. Singing birds *c*. 200 m apart. Most interactions and territorial behaviour vocal, with persistent alarm calls and some chasing. Once ♂ seen to fly above bushes in fluttering display flight. One pair foraged in *c*. 5 ha when they were feeding 3 1-day-old nestlings (Frost and Vernon 1978).

NEST: thick-walled cup of thin grass, twigs and plant down, thickly lined with soft downy seeds; not as tightly woven as by other eremomelas but more insulated; ext. diam. 80–120, int. diam. 40, ext. depth 70, int. depth 52, wall thickness 19. Supported from below in fork of small bush under canopy, usually shaded, with clear horizontal view, *c*. 30–40 cm from ground.

EGGS: 3–4. Oval; pale blue spotted with reddish brown and pale lilac especially at large end. SIZE: (n = 4, 1 clutch, South Africa) 15·5–16·3 × 11·0–11·5 (16·0 × 11·3).

LAYING DATES: southern Africa, Aug–Dec (mainly Oct–Nov).

Reference
Frost, P. G. H. and Vernon, C. J. (1978).

Eremomela usticollis Sundevall. Burnt-necked Eremomela. Érémomèle à cou roux.

Eremomela usticollis Sundevall, 1850. Oefr. K. Vet-Akad. Forh. 7, p. 102; Leroma, Pillaansburg, western Transvaal.

Plate 18
(Opp. p. 321)

Range and Status. Endemic resident, S-central Africa. Occurs commonly in acacias from Angola (S Huila) and E half of Namibia across NW, NE and E Botswana (but seems to be absent from central and SW Kalahari although it occurs on Molopo R. south of Kakia), Zambia north to 16°S west of Zambezi and to 14°S in Kafue basin, and also in Luangwa valley between 10°S and 14°S. In Malaŵi north to Benga on L. Malaŵi; common throughout Zimbabwe although absent from Inyanga, highlands east of Sabi valley and a small area along Botswana border in west; S and central Mozambique, but absent or very rare in moister parts between the Tropic of Capricorn and just to the south of Zambezi R. Common through Transvaal south to *c.* 25°S, becoming rarer farther south along Natal coast.

Density of 1 pair per 17 ha at Nylsvlei, Transvaal (Tarboton *et al.* 1987); 11·4 birds per 100 ha in acacia woodland in Transvaal (Tarboton 1980).

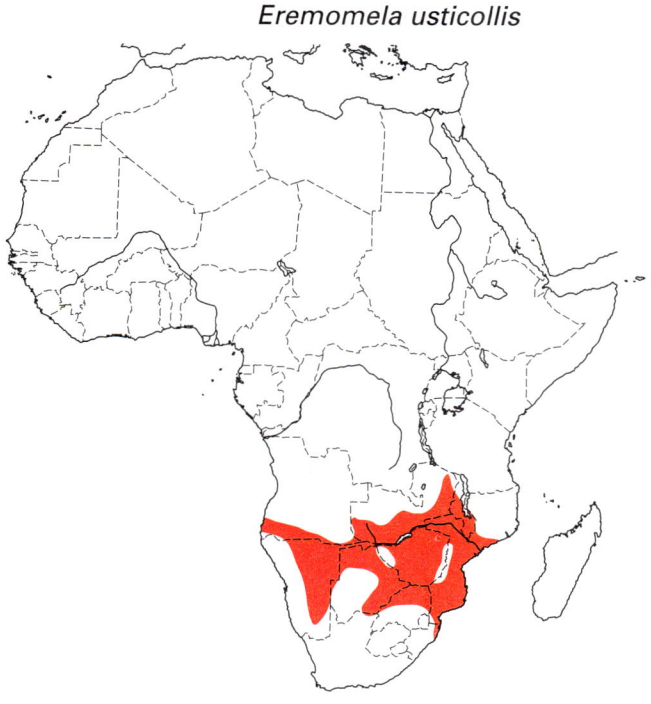

Eremomela usticollis

Description. *E. u. usticollis* Sundevall: (including '*baumgarti*'): Angola, Namibia, S and SW Zimbabwe, Botswana, Mozambique south of Save R. and South Africa. ADULT ♂: upperparts, sides of head, scapulars, wing-coverts and tail ash-grey. Chin and throat whitish; usually a variable pale rusty to chestnut bar across lower throat and a paler one between white throat and grey cheek; rest of underparts pale or very pale buff. Wings somewhat darker and with paler outer edges to feathers; underwing white. Bill horn; eyes pale yellow; legs and feet horn or dark orange, more powerful than in rest of genus. Sexes alike. SIZE: (12 ♂♂, 6 ♀♀) wing, ♂ 53–58 (56·0), ♀ 52–56 (54·5); tail, ♂ 40–47 (43·4), ♀ 42–45 (42·8); bill, ♂ 11·4–13·3 (12·2), ♀ 10·8–13·9 (12·1); tarsus, ♂ 19·1–21·4 (20·2), ♀ 19·0–20·6 (19·8). WEIGHT: ♂ (n = 23, Zimbabwe) 7·7–10 (8·6), (n = 2, Botswana) 8·4, 9; ♀ (n = 16, Zimbabwe) 6·5–9·6 (8·2); unsexed (n = 1, Namibia, Nov) 8; (n = 5, southern Africa) 7·3–9·3 (8·3).

IMMATURE: lacks chestnut around head and throat; bill black; eyes brown; feet yellowish.

NESTLING: unknown.

E. u. rensi Benson: S Zambia, S Malaŵi, Mozambique north of Save R., Zimbabwe. Upperparts darker; smaller: wing, ♂ (n = 12) 51–55 (52·8), ♀ (n = 4) 50–53 (51·3).

Field Characters. The smallest eremomela; an extremely plain bird, grey above and buffy white below, adult distinguished by its pale yellow eye and (when present) narrow rusty band under throat. Easily overlooked except for distinctive far-carrying trilling call. Green-capped Eremomela *E. scotops* has pale eye but greenish upperparts, yellow underparts; Yellow-bellied Eremomela *E. icteropygialis* has dark eye, pale supercilium, white breast and yellow belly. Where it overlaps with Yellow-bellied Eremomela it remains strictly in canopy of acacia trees while Yellow-bellied is usually in bushes.

Voice. Tape-recorded (86, 88, 91, B, F, CHA, LUTG, PAY). Song a high-pitched 'teeup-ti-ti-ti-ti' at a fairly rapid but variable speed usually on a rising scale, and often followed by a short trill on a lower note. Introductory notes before trill may be a repeated 'tu-ti-ti tu-ti-ti'. Call a thin musical 'di-di-di-di-di' or 'dyup-dyup-dyup-dyup'; also a rather harsher 'tee-up, tee-up, tee-up . . .' or often 'too-ee-up, too-ee-up, too-ee-up . . .' with the middle syllable on a slightly higher pitch. The 'too-ee-up' call has been mistaken for the song.

General Habits. Inhabits mainly *Acacia* woodland, especially edges of riparian forest, in tall trees, e.g. *A. albida*; also mixed woodland, and dry thornbush, even patches of < 0·5 ha, but not *Burkea* woodland. In miombo belt it is highly localized and patchy in distribution and occurs only along main river valleys (M. P. S. Irwin, pers. comm.).

Occurs in pairs or parties of up to 5; joins mixed-species flocks of other warblers, penduline tits *Anthoscopus* and white-eyes *Zosterops*. Very active; forages largely in canopy, gleaning leaves, twigs and thorns. Scratches head indirectly – foot up over wing from behind (Simmons 1961).

Food. Small insects and larvae; spiders.

Breeding Habits. Territorial, solitary nester. May breed co-operatively, since birds extra to the pair have been seen at a nest (Grimes 1976b).

NEST: small thin-walled cup of soft vegetable down, adorned with small pieces of mantis egg capsule (pale green), bound with fine silky web from moth cocoons; ext. diam. 50–55; int. diam. 30–36; ext. depth 30–41; int. depth 25–35. Bound onto or slung from twigs or forks in tops of low acacia trees usually 3–6 m up and attached at 3–4 points.

EGGS: 1–4; southern Africa (n = 6 clutches) 2–4 (2·6). Oval; almost uniform, white, very pale green or blue, with small, bold, nearly uniformly scattered spots of light brown. SIZE: (n = 12) 13·0–17·3 × 10·5–12·6 (16·0 × 11·8).

LAYING DATES: Zambia, Mar (breeding condition, Oct, Nov, Feb); Zimbabwe, Sept–Apr; Malaŵi, Mar; Botswana, Oct–Nov; Transvaal, Oct–Apr.

INCUBATION: by both sexes. Once an incubating bird sang in response to mate nearby.

Reference
Vernon, C. J. (1963).

Plate 18 (Opp. p. 321)

Eremomela badiceps (Fraser). Rufous-crowned Eremomela. Érémomèle à tête brune.

Sylvia badiceps Fraser, 1842. Proc. Zool. Soc., Lond., p. 144; Clarence, Fernando Po.

Range and Status. Endemic resident, lowland forests of W and central Africa. Inhabits high forest canopy, so probably widely overlooked and likely to be commoner than following regional remarks suggest. Occurs in Tingi Hills and logged parts of Gola Forest (in small numbers) in E Sierra Leone, Guinea (Macenta, common, also Sereonin and Mt Nimba (Guinea/Liberia), uncommon; frequent S Ivory Coast from Taï to Sipilou, Lamto and Bouaké; in Ghana uncommon, north to Bia Nat. Park, Kade and Begoro; in Togo at Djodji; not yet recorded in Benin. In Nigeria sparse between Ilaro and Owerri, north to Ife and near Ado-Ekiti; occurs on Bioko and throughout forested part of Cameroon, into extreme SW Central African Republic; frequent in Gabon and Congo; in N Angola in Cuanza Norte forest and probably Cabinda; across Zaïre in broad belt, between about 3°N and a line from N end of L. Tanganyika to S Kasai. In W Uganda in Budongo (where common), Bugoma, Bwamba and Kasyoha-Kitomi Forests and fairly common in the Imatong Mts and Talonga Forest (S Sudan).

Common in forest from sea level to 1400 m.

Eremomela badiceps

Description. *E. b. badiceps* (Fraser): Bioko, Nigeria, Congo Basin to Angola and Uganda. ADULT ♂: forehead and crown reddish chestnut; nape to uppertail-coverts grey; tail blackish brown with outer feathers rather paler than inner ones. Ear-coverts grey; streak through and below eye black; chin and throat white usually with creamy or pale yellow area where white abuts black eye-streak; broad black band across upper breast, widest in centre; rest of underparts grey at sides, paler than back, and creamy white, or rarely pale yellow, in centre. Wings brown; scapulars grey; upper wing-coverts brown with grey edges and tips; underwing white. Bill black; eyes dark brown; legs yellow or pale brown. Sexes alike. SIZE: (19 ♂♂, 17 ♀♀) wing, ♂ 49–56 (52·6), ♀ 53–58 (54·6); tail, ♂ 31–40 (38·0), ♀ 34–41 (38·5); bill, ♂ 12·1–13·8 (13·1), ♀ 12·5–14·5 (13·4); tarsus, ♂ 15·8–18·3 (17·3), ♀ 16·2–19·0 (17·9). ♀♀ significantly larger than ♂♂ statistically. WEIGHT: ♂ (n = 10, Uganda) 9–11 (10·3); ♀ (n = 4, Uganda) 10–11·5 (10·9); unsexed (n = 8, Liberia) 8·8–10·6 (9·7); 2 ♂♂, 1 ♀ (Bioko) 12–14.

IMMATURE: upperparts including crown olive-green with cinnamon wash on crown; streak through eye, and breast-band very poorly defined or absent; underparts pale yellow, somewhat brighter on chin and throat. ♀ larger than ♂: wing, ♂ (n = 4) 48–51 (49·8), ♀ (n = 8) 51–53 (52·1).

NESTLING: unknown.

E. b. fantiensis Macdonald: Sierra Leone to W Nigeria. Like *badiceps* but belly creamy or yellowish rather than white.

E. b. latukae Hall: S Sudan. Very like *badiceps* but chestnut of crown rather duller and extends less far onto nape.

Field Characters. The only eremomela mainly in the equatorial rain forest. The chestnut crown, black streak through eye and black band across breast distinguish it from others of genus except Turner's Eremomela *E. turneri* (which is very restricted in range, has the chestnut paler and restricted to forehead, more slender bill and much weaker legs and feet). Rufous-crowned Eremomela has a very similar song and call and is similar behaviourally to Black-capped Apalis *Apalis nigriceps*, which is sympatric over a wide area and occurs in same bird parties, but has black (not chestnut) head and greenish yellow (not grey) back; from below, Black-capped Apalis best told by longer, graduated tail with white tips. Song also similar to sympatric Green

Crombec *Sylvietta virens*, which has greenish upperparts, no throat-bar, and lives in undergrowth.

Voice. Tape-recorded (68, C, CHA, ERA, KEI, MAC). Song an irregular series of sharp, high-pitched notes interspersed with a few lower, more musical ones, sometimes run together into brief trills. Contact call, high-pitched twittering with some sharp notes and a few buzzy scolds.

General Habits. In much of range primarily a bird of lowland forest canopy. In Gabon occurs mainly in tree tops at forest edge (D. Sargeant, pers. comm.) and in secondary forest, also in the environs of villages, traditional-agricultural areas, plantations and the first stages of regrowth after agriculture is abandoned; in later stages of regrowth it remains only in clearings (Brosset and Erard 1986). Elsewhere frequents clearings, secondary growth and gallery forests as well as primary forest; in open woods in Liberia.

Rarely descends below 20 m, although singing bird recorded at 1–2 m (Chappuis 1979a). Occurs in noisy groups of 2–7 (mainly 4–5) birds, probably family parties. Calls incessantly when feeding. Forages for insects by searching leaves and twigs; also seen at ant and termite swarms. In Gabon a core member of mixed species parties (Brosset and Erard 1986), but in Cameroon only once found associating with other species (sunbirds: Serle 1965), and in only single-species parties in Zaïre (Chapin 1953a). Where it overlaps with Turner's Eremomela, Rufous-crowned seems to be slightly lower in the trees (Prigogine 1958).

Primary moult completed in Sept (2 ♂♂, Liberia, Mt Nimba).

Food. Insects: caterpillars, eggs of Lepidoptera, small beetles. 29 stomachs contained only insects (Cameroon: Serle 1965); 15 others all contained insects including caterpillars in 6 of them (Uganda). Spiders, berries and seeds also recorded.

Breeding Habits. Territorial. Probably a co-operative breeder like some congeners: territory defended by song and communal display, birds shaking tail, shivering wings and hopping about and making many noisy short flights. More than 2 adults may accompany fledglings (A. Brosset *in* Grimes 1976b).

NEST and EGGS: unknown.

LAYING DATES: Liberia (♀ with enlarged ovary and yolk, Sept; 4 ♂♂ with enlarged testes, Sept–Oct); Ivory Coast, Mar–Nov; Ghana (adults with dependent young late Sept and late Dec); Nigeria (nest with young Oct); Cameroon (fluffy juv. collected Dec); Gabon (flying young being fed by parent, Oct, Feb, Mar); N Zaïre (breeding condition Nov–Apr); Kivu, Zaïre (breeding condition, Mar–July).

Reference
Brosset, A. and Erard, C. (1986).

Eremomela turneri van Someren. Turner's Eremomela. Érémomèle de Turner.

Eremomela badiceps turneri V.G.L. van Someren, 1920. Bull. Br. Orn. Club 40, p. 92; Yala River, Kavirondo District, western Kenya Colony.

Plate 18
(Opp. p. 321)

Range and Status. Endemic resident, known only from a few scattered localities: E-central Zaïre (Abyaloze, Kailo, Kalima, Makayobo and Mazali) and once in NE at 50 km north of Beni; extreme SW Uganda (Nyondo Forest); and W Kenya (Mt Elgon, Kakamega Forest, and S Nandi Forest (type specimen, and sight record of a party in 1982)). Locally common Kakamega, rare elsewhere.

Description. *E. t. turneri* van Someren: W Kenya. ADULT ♂: forehead and front of crown reddish chestnut; rest of upperparts slate grey; tail dark slate, almost black. Chin and upper throat white but sides cream; slate grey streak through eye; black bar across lower throat, broad in centre; rest of underparts greyish, paler in centre. Wings, upperwing-coverts dark slate, almost black. Bill black; eyes brown; feet pinkish flesh. Sexes alike. SIZE: wing, ♂ (n = 7) 45·5–52 (47·2), 2 ♀♀ 43, 47·5; tail, ♂ (n = 7) 28–35 (30·5), ♀♀ 28, 31·5; bill, ♂♀ (n = 5) 11·1–12·9 (11·6); tarsus, 2 ♀♀ 15·8, 16·8. WEIGHT: 1 ♂ (Kenya) 6, 1 ♀ (Kenya, Mar) 9.

IMMATURE: upperparts greenish brown; no chestnut on head; underparts pale yellow; no black band across lower throat at first.

NESTLING: unknown.

E. t. kalindei Prigogine: Zaïre and Uganda. Paler and distinctly browner on head and mantle, wings and tail darker.

Field Characters. Very similar to Rufous-crowned Eremomela *E. badiceps*, with a chestnut head, slate grey upperparts and paler grey underparts, but the chestnut is confined to forehead and front of crown (in the hand its shorter wing, more slender bill and smaller and weaker feet are evident). Where the 2 species overlap in Zaïre, Turner's Eremomela feeds higher in the vegetation.

Voice. Tape-recorded (C, KEI, McVIC). Call a weak chippering.

334　SYLVIIDAE

Eremomela turneri

General Habits. Inhabits lowland and intermediate forest edges and clearings, secondary growth and the environs of plantations, especially along streams. Occurs in primary forest at times. 3 birds recorded on a 20 ha plot of closed forest in Kakamega Forest in 1963, 2 on the same plot in 1965 and 3 in 1966 (Zimmerman 1972), but the bird is commoner in more open parts near the edge of Kakamega Forest (A. W. Diamond *in* Collar and Stuart 1985).

Remains in tree tops, especially where it overlaps Rufous-crowned Eremomela in Zaïre. Works through the canopy in parties: in Kakamega of 3–6 birds and in Zaïre of up to 10–15. Commonly feeds alongside Buff-throated Apalis *Apalis rufogularis* (Kakamega).

Food. Insects, including caterpillars.

Breeding Habits. Zaïre, birds in breeding condition Feb and Aug (Prigogine 1958).

References
Collar, N. J. and Stuart, S. N. (1985).
Prigogine, A. (1958).

Plate 18
(Opp. p. 321)

Eremomela atricollis Bocage. **Black-necked Eremomela. Érémomèle à cou noir.**

Eremomela atricollis Bocage, 1894. Jorn. Ac. Real. Sci. Lisboa (2), 3, pp. 153 and 162; Galanga, Benguella, Angola.

Range and Status. Endemic resident, in miombo woodland across S-central Africa. In Angola, frequent to common in highlands of N Huila, Huambo, N Bié and E Benguela, and across SE Zaïre to near S end of L. Tanganyika. Frequent to common in Zambia east from the Zambezi R. in Zambezi District to Muchinga Escarpment, south to at least 15°S in Mongu and Kaoma Districts, Chikana in Kabwe Rural and to nearly 14°S in Mkushi District; in the NE, range does not quite reach Isoka or Mbala and bird is largely absent from low-lying non-miombo woodlands between Lakes Mweru and Tanganyika.

Frequent to common but mostly rather local, from less than 500 m to at least 1500 m.

Eremomela atricollis

Description. ADULT ♂: forehead, forecrown and eye-stripe orange-yellow; crown and nape ash-grey with olive tint; lores and ear-coverts black, forming mask from bill to hindneck; rest of upperparts ash-grey; tail grey with a slight brownish tint and rather paler towards outer feathers. Chin and throat bright yellow with black band across lower throat, broader in centre; breast to undertail-coverts white with greenish or yellowish wash. Wings brown with paler outer edges to feathers especially of secondaries and tertials and, in fresh plumage, a trace of yellowish green on outermost edges; scapulars and upper wing-coverts brownish grey; underwing white. Bill black; eyes dark brown; legs slate grey with feet paler and contrasting. Sexes alike. SIZE: (3 ♂♂, 4 ♀♀) wing, ♂ 58–60 (59·3), ♀ 54–61 (58·3); tail, ♂ 45–50 (47·0), ♀ 47–51 (49·3); bill, ♂ 12·9–14·5 (13·9), ♀ 13·7–14·7 (14·3); tarsus, ♂ 18·7–19·7 (19·3), ♀ 18·3–19·6 (19·2). WEIGHT: (Zaïre) ♂ (n = 4, Sept) 11–12 (11·3), ♀ (n = 3, Sept) 9–11 (9·7), (n = 1, Feb) 10.

IMMATURE: like adult but duller and with less contrast, especially on sides of head and throat; breast-band and eye-stripe rather indistinct.

NESTLING: unknown.

Field Characters. A miombo woodland bird, whose black face mask set in yellow face and black throat-bar easily distinguish it from all other birds in its range.

Voice. Tape-recorded (86, B, F, CART, STJ). Flocking call a fairly strident rasp or rattle 'zwut, zwut, zwut' often given when feeding. Indulges frequently in 'squabbling parties', with the call having more of a buzzing quality than that of Green-capped Eremomela *Eremomela scotops* (D. R. Aspinwall, pers. comm.).

General Habits. Largely confined to canopy of miombo (*Brachystegia*) woodland, especially the richer parts of it. Also in bushy savannas in Angola and Zaïre.

Usually in small parties and is an active member of mixed-bird parties moving through the canopy of *Brachystegia*; mingles with Green-capped Eremomela without apparent ecological segregation. Very active when feeding; often feeds hanging upside down like tit *Parus*.

Moult in Zaïre follows breeding Nov–Dec, and a second moult is completed during dry season.

Food. Insects.

Breeding Habits. Extra birds ('helpers') occur at nest and with fledglings (A. Brosset *in* Grimes 1976a).

NEST: cup of fibres and cobwebs, with many adornments, hung from flexible branch well concealed in foliage, usually 6–7 m up.

EGGS: 2; white, evenly spotted with reddish brown and grey. SIZE: *c*. 15·5 × 11·0.

LAYING DATES: Zaïre, Sept–Dec; Zambia, Aug, Oct–Dec. Breeds at end of dry season and start of rains.

Genus *Sylvietta* Lafresnaye

Small, very short-tailed warblers, predominantly grey on upperparts and buff, yellow or reddish on underparts and around face. Similar to *Eremomela* but with stronger feet and bill (related to habit of foraging by probing stems, twigs, branches or trunks rather than gleaning foliage). Tail hardly extends beyond rather long uppertail-coverts, and closed wing-tips often extend beyond tail. Wings rounded with P6, P7 and P8 more or less equal and longest and P10 more than half length of P9. Bill fairly long, slender and slightly downcurved. Nest is usually a bag-like structure hung from thin twig at edge of bush, built of grass and fibres bound together with cobwebs, and ornamented outside with cocoons, seed-heads, bits of bark and small flowers.

Endemic; almost entirely sedentary. 9 species, together occupying the whole of subsaharan Africa. Most are birds of dry wooded country. There is little range overlap between species; when there is, one is usually in the canopy and the other in the undergrowth. Relationships within the genus not entirely clear but 2 superspecies occupy dry savanna woodland: the *S. rufescens*/*S. isabellina* superspecies with very long bills; the *S. whytii*/*S. brachyura*/*S. philippae* superspecies with rather short bills. A hybrid between *S. whytii* and *S. brachyura* has been found in S Ethiopia (Benson 1946b). The other 4 species are independent. *S. virens* and *S. denti* occupy lowland rainforests of W and central Africa, *S. leucophrys* is in a rather small area of E African montane forest, and *S. ruficapilla* inhabits *Brachystegia* woodland in the southern tropics.

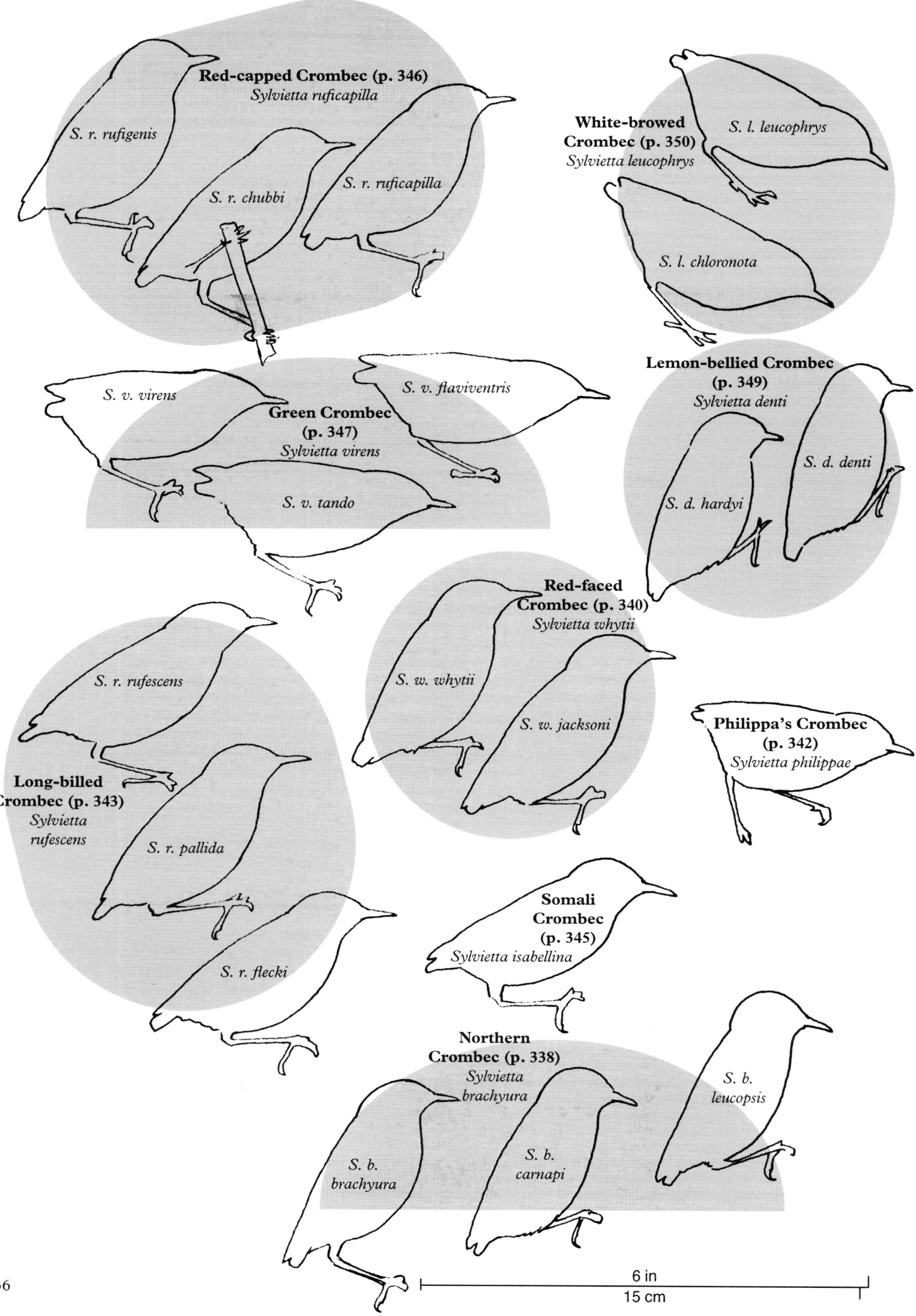

Plate 19

Plate 20

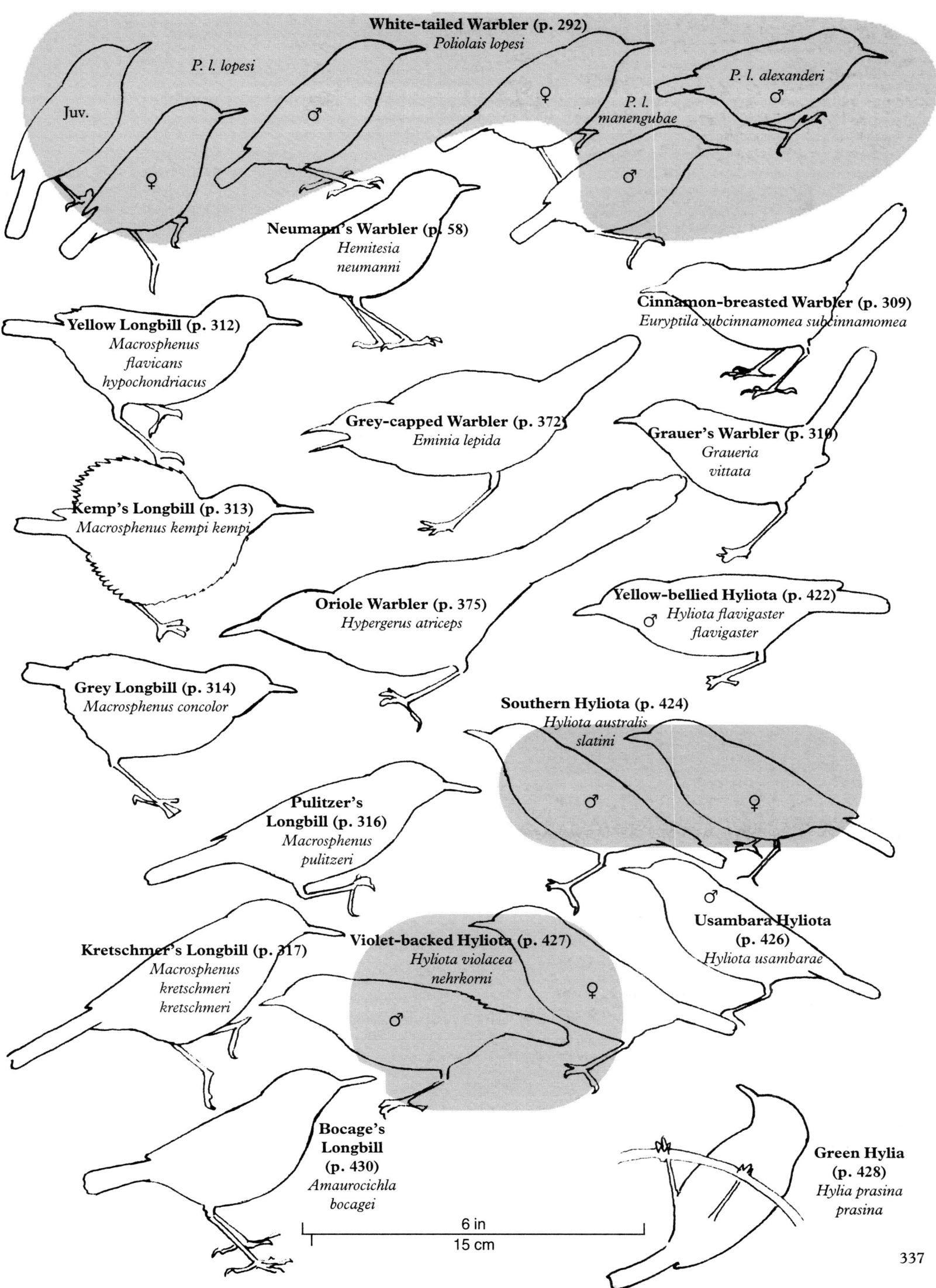

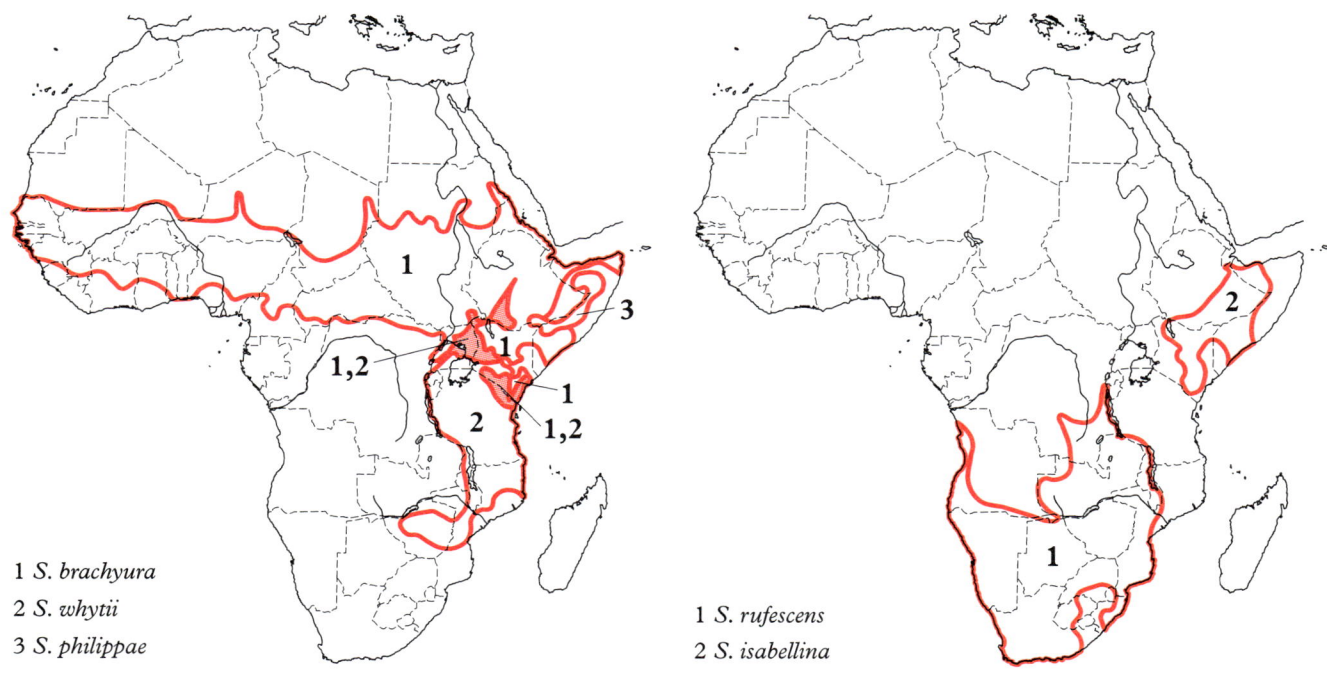

Plate 19
(Opp. p. 336)

Sylvietta brachyura Lafresnaye. Northern Crombec. Crombec sittelle.

Sylvietta brachyura Lafresnaye, 1839. Rev. Zool., p. 258; Senegambia.

Forms a superspecies with *S. philippae* and *S. whytii*.

Range and Status. Endemic resident and partial intra-African migrant in savannas of N tropics, from Senegambia (never very common, less so than elsewhere), S Mauritania and Sierra Leone (fairly common in interior and in Freetown peninsula mangroves) to Red Sea coast of Sudan, Ethiopia and NW Somalia. Northern limit is about 18°N; in W Africa it reaches that latitude only in wet season. Southern boundary in W is N edge of forest belt at about 7°N, in Ivory Coast and W Ghana; reaches coast in E Ghana, Togo and Benin (where no forest). Common and widespread in savannas of Nigeria north of Niger and Benue rivers, extending south to Ibadan, Mekko, Enugu, Obudu and Serti. In Cameroon south to Yaoundé and Batouri, Central African Republic except extreme SW, Zaïre north of 4°N; Ethiopia except E Ogaden; scarce resident in Djibouti, although quite common in Mabla Mts; NW and S Somalia west of 46°E, and NE Somalia; Uganda south to Toro and Queen Elizabeth Nat. Park and L. Victoria, uncommon in Kampala area; across all of Kenya except SW between L. Victoria and Amboseli Nat. Park, and perhaps absent in highlands and humid west. Common in Mkomazi Game Res., Tanzania, south to N end of Usambara Mts.

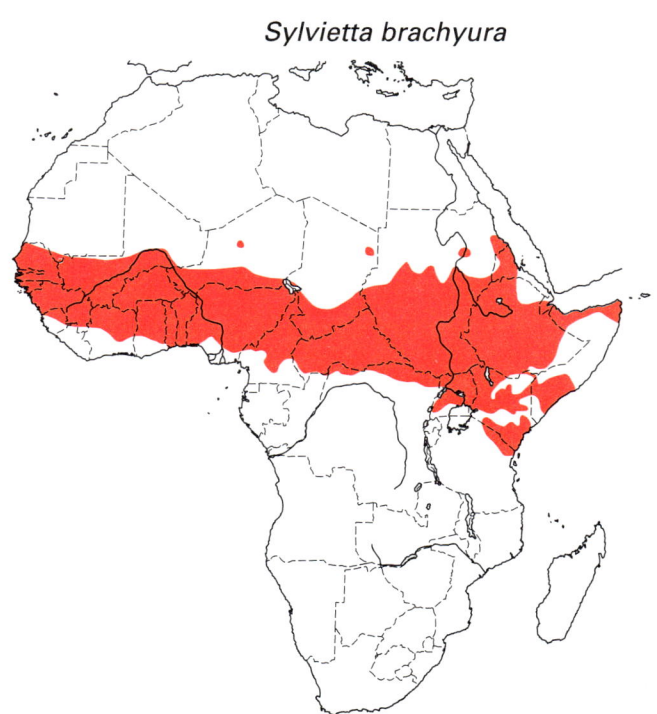

Common and widespread up to 1500 m; locally common up to 2000 m in Kenya, and reaches 2400 m in Sudan. In Nechisar Nat. Park, Ethiopia, scarce (max. 1 seen per day) in plains and common (5–6 seen per day) in bush (Safford *et al.* 1993). 9 per day (Goula) and 5 (Bankouate) in Djibouti (Welch *et al.* 1989). In mangroves in Guinea–Bissau netted at rate of 1 bird per 100 m mist net per 10 h (Altenberg and van Spanje 1989).

Description. *S. b. brachyura* Lafresnaye: Senegambia and Sierra Leone to Sudan and N Eritrea. ADULT ♂: upperparts pale grey with a slight tawny wash; stripe from base of bill over and behind eye tawny or buffy cream, and dark brownish grey streak through eye; tail greyish brown becoming buffish at tips of feathers. Chin and centre of belly nearly white; cheeks, ear-coverts and throat to undertail including flanks reddish tawny, this forming broad band on breast, between chin and belly. Wings browner than upperparts and with pale outer edges to feathers; upperwing-coverts browner with grey edges; underwing reddish tawny. Bill dark horn, lower mandible paler; eyes orange-brown; legs and feet light brown or flesh. Sexes alike. SIZE: wing, ♂ (n = 46) 49–61 (56·1), ♀ (n = 35) 50–57 (53·1); (10 ♂♂, 10 ♀♀) tail, ♂ 20–26 (23·5), ♀ 18–24 (20·8); bill, ♂ 11·6–13·2 (12·3), ♀ 11·6–13·3 (12·4); tarsus, ♂ 16·5–19·4 (17·8), ♀ 16·0–17·7 (17·0). WEIGHT: ♂ (n = 3, Kenya) 5–6 (5·3); ♀ (n = 3, Kenya) 5·5–8 (6·8); unsexed (n = 8, Ghana, July–Sept) 7·0–8·5 (7·9), (n = 6, Chad) 7–9 (8·0), (n = 9, Nigeria, Nov–Feb) 7·6–9·0 (8·2), (n = 8, Nigeria, Apr–June) 8·1–9·5 (8·5), (n = 85, Ethiopia) 6·5–9·6 (8·1).

IMMATURE: like adult but wing-coverts have tawny tips.
NESTLING: unknown.

S. b. carnapi Bannerman (including '*dilutior*'): Cameroon to Uganda and W Kenya. Underparts deeper chestnut and upperparts rather darker than *brachyura*.

S. b. leucopsis Reichenow: S Eritrea, Ethiopia, SE Sudan and Somalia to Tanzania. Eye-stripe, chin and throat white, not buffy cream, and rump paler than *brachyura*.

Field Characters. A familiar little warbler, common throughout dry and leafy northern tropical woodlands, readily observable. Immediately told as a crombec by its stumpy tail, only about 5 mm longer than wing-tip. The uniform grey upperparts, rufous face and underparts preclude confusion with all other warbler genera, even eremomelas; its colour, short bill and well-defined eye-stripe serve to distinguish it from all other crombecs. The pale eye-stripe and dark streak through the eye are especially prominent in areas of overlap with Red-faced Crombec *S. whytii* (SW Ethiopia, Uganda, Kenya). Northern Crombec also has much shorter bill, is slightly smaller, and has richer buff underparts more mixed with white than the Red-faced Crombec. Philippa's Crombec *S. philippae* is yellow below, not tawny, and Somali Crombec *S. isabellina* is larger and has longer bill and considerably paler underparts.

Voice. Tape-recorded (68, B, GREG, KÖN, LEM, MOR, PAY). Song, a sweet, shivering jumble of 4–6 notes with 2 short, clear notes at end 'te-tu', sometimes with a light 'churr' at start. Notes descend scale, reminiscent of Willow Warbler *Phylloscopus trochilus* but shorter. Song is very similar to that of Yellow-bellied Eremomela *Eremomela icteropygialis*, with which it is sympatric over a wide area, suggesting interspecific vocal convergence (Chappuis 1979a). Alarm, a clicking 2-note 'chit-chit'; another call closely resembles the 'chack' of a Common Whitethroat *Sylvia communis*.

General Habits. Occurs in almost all types of woody vegetation in savannas from semi-arid thorn scrub to acacia woodland, especially in drier areas; also around human habitations and suburban areas where trees are present. Less frequent in riparian woodland, forest edges and at high altitudes.

Lives singly or in pairs, also family parties. When foraging, busily searches twigs and leaves and crevices in larger branches; usually fairly low down in vegetation. Took 78% of food items (n = 80) from twigs and branches; prefers larger twigs and branches, and seems not to use the smallest twigs (Lack 1985). Moves almost continuously and calls frequently; systematically visits each tree and shrub, walking or running along branches and twigs, especially horizontal ones, and will hang upside down to feed. Seems to 'do the rounds' 2–3 times each day, when moving between trees and bushes with a looping flight. Joins mixed-bird parties, especially parties of other warblers. 1 pair roosted together 2 m high at edge of bush on 3 successive nights, perching head to tail; they crouched low on nearly vertical twig which bent under their weight, to be well clear of other vegetation (Irvine and Irvine 1991).

Mainly sedentary but in W Africa only occurs in northern parts in rains and in southern parts in dry season, e.g. in Mauritania present only in June–Dec.

Food. Small invertebrates, including Lepidoptera larvae and eggs, winged termites, and spiders. 10 stomachs all contained only invertebrates.

Breeding Habits. Monogamous and territorial.
NEST: deep, hanging pocket, somewhat sunbird-like but not roofed over and without a long 'tail' underneath; more stretched lengthwise than nest of Red-faced Crombec (Moreau and Moreau 1939). Cavity more or less spherical, material drawn out at top where suspended by its higher side, the entrance more a large gap than a small hole, facing sideways and upwards; base of nest conical, *c.* 25 deep. Strongly made of fine plant stems, fibres and grasses, bound with spider's web, lined with fine rootlets and grass heads; decorated outside with cocoons or small brown seeds; suspended mainly by spider's web. Overall height *c.* 150; diam. of subspherical part of nest *c.* 50. Hung from twig near outside of thorn tree, fig or bush, nest often right out in open, hanging free at extreme end of leafless branch.

EGGS: 2; white with scattered olive-brown, rufous or grey spots. SIZE: (n = 7, Gambia, Somalia) 16·2–18·5 × 11·4–12·5 (17·1 × 11·9).

LAYING DATES: Senegambia, Mar–Dec; Sierra Leone, Apr; Mauritania, June–Dec; Mali, June–July; Niger, June–July; Ghana, Oct, Feb; Nigeria, Mar, Apr, June; Zaïre, Jan; Sudan, Jan–June, Sept, Oct, Dec; Ethiopia, Apr (nest-building July); E Africa: Region B, Feb, Mar, June; Region D, Jan, May, Oct; Uganda, June, Sept; Somalia, Mar.

340   SYLVIIDAE

INCUBATION: bird once seen incubating with back to entrance of nest (Jackson and Sclater 1938).

DEVELOPMENT AND CARE OF YOUNG: young sit with backs to entrance, and would have to turn heads over their backs to be fed; adults bring food by climbing down nest attachment from above, not by perching below (Moreau and Moreau 1939).

BREEDING SUCCESS/SURVIVAL: several birds seen mobbing black mamba *Dendroaspis polylepis* in tree and 1 was caught by it (Short and Horne 1987).

Plate 19
(Opp. p. 336)

Sylvietta whytii (Shelley). Red-faced Crombec. Crombec à face rousse.

Sylviella whytii Shelley, 1894. Ibis, p. 13; Zomba, southern Nyasaland.

Forms a superspecies with *S. brachyura* and *S. philippae*.

Sylvietta whytii

Range and Status. Endemic resident, eastern Africa from Ethiopia to Zimbabwe and Mozambique. Fairly common in SW Ethiopia; rather uncommon in extreme SE Sudan and Uganda south and east of Kidepo valley to Rwanda border, but uncommon in Kampala area; Kenya east to N Uaso Nyiro R. and in highlands south to Mt Kilimanjaro. Many old records in coastal strip of Kenya north to Manda I., but almost none there since 1970. Through Rwanda; very abundant in savannas in Burundi. Widespread and locally common throughout Tanzania. E half of Malaŵi, including W side of lake; reaches lip of escarpment of Middle Zambezi valley (replaced on other side by Red-capped Crombec *S. ruficapilla*). Locally common in Mozambique south to Save R. In Zimbabwe, sparse on Mashonaland Plateau, north to Zambezi escarpment; commoner in Kalahari sand woodland, west to Hwange Nat. Park (Main Camp), south to Umgusa Forest Reserve; widespread to east, common along Mozambique border.

Locally fairly common up to 2000 m, especially above 1000 m. Density of 3 pairs per 100 ha in *Brachystegia* in Zimbabwe (but none in mixed woodland: Vernon 1985a); 0·14 birds per ha at Masalani in S Kenya (Pomeroy and Tengecho 1982); and 3–5 seen per day in bush of Nechisar Nat. Park in Ethiopia (Safford *et al.* 1993).

Description. *S. w. whytii* (Shelley) (including '*nemorivaga*'): coastal S Tanzania, Mozambique, Zimbabwe, S Malaŵi. ADULT ♂: upperparts and tail grey, sometimes slightly tinged tawny on nape. Chin and throat buff, with grey subterminal spots to feathers giving a speckled appearance; lores, ear-coverts, cheeks, and breast to undertail-coverts, including flanks and underwing, tawny buff, centre of belly rather paler. Wings rather darker grey than upperparts, but outer edges to primaries and secondaries rather paler. Upper mandible dark brown, culmen brownish sepia, lower mandible horn; eyes yellow to pale reddish brown; legs and feet reddish or flesh. Sexes alike. SIZE: (9 ♂♂, 5 ♀♀) wing, ♂ 55–60 (58·1), ♀ 54–57 (55·6); tail, ♂ 18–27 (24·0), ♀ 18–23 (20·0); bill, ♂ 12·5–13·3 (12·9), ♀ 11·8–12·9 (12·3); tarsus, ♂ 17·5–20·1 (18·5), ♀ 16·5–18·8 (17·6); a clinal size decrease from west to east (Irwin 1968). WEIGHT: (Tanzania, May–Sept) ♂ (n = 5) 10·5–12·5 (11·1), ♀ (n = 6) 9·5–11·2 (10·7); (Zimbabwe) ♂ (n = 17) 8·0–11·3 (10·0), ♀ (n = 14) 8–11·5 (9·8); (Mozambique) ♂ (n = 8) 8·9–11·0 (10·0), ♀ (n = 6) 9·1–10·3 (9·7).

IMMATURE: similar to adult but upperparts have tawny tint, wing-coverts have slight tawny tips, and underparts are darker tawny, especially on throat, flanks and undertail.

NESTLING: unknown.

S. w. jacksoni Sharpe: S and E Uganda, SW Kenya, W Tanzania, N Malaŵi. Upperparts darker grey and underparts much more tawny than *whytii*. WEIGHT: (Kenya) ♂ (n = 10) 7–12 (9·7), ♀ (n = 8) 5–9 (7·0).

S. w. loringi Mearns (including '*abayensis*'): Sudan and Ethiopia to N Uganda, N and W Kenya, NE Tanzania. Upperparts more olive-brown, less pure grey, and underparts much paler, whiter on mid-belly, than *jacksoni*.

S. w. minima Ogilvie-Grant: coastal Kenya, E Tanzania. Generally paler than *whytii*, sometimes with olive tinge on back.

Field Characters. A crombec with light grey upperparts and wholly rufous underparts, including face. Overlaps with the 2 other species with rufous underparts, Northern Crombec *S. brachyura* in the north and Long-billed Crombec *S. rufescens* in the south; told from both by more solid rufous of underparts, without any white areas, the rufous extending up over face, to above eye, and being especially bright on the face; both Northern and Long-billed have pale supercilium, dark eye-line and pale chin and throat. Northern also has paler rufous underparts and white belly. Long-billed is somewhat larger, with longer bill; where they overlap, Red-faced occurs mainly in tree canopy and Long-billed in undergrowth; song and calls distinct in areas of overlap but similar where the species are allopatric.

Voice. Tape-recorded (91, B, F, GREG, HOR, McVIC, STJ). In southern Africa a far-carrying trill like Red-capped Crombec; also a phrase of 3–7 whistled syllables on same pitch 'see-sisi-seee'. Alarm call a sharp 'tip' which becomes a penetrating rattle when uttered continuously. In Ethiopia and Kenya, and Tanzania south to Rukwa valley, song indistinguishable from that of Long-billed Crombec, i.e. 'tui-tu-tuti' or 'wheet-chu' repeated 6–8 times. In Kenya also a melodic, whistling, warbled 'witch-ee, witch-ee, witch-ee-eeee', sometimes longer; and a more melodic complex warble, occasionally a trill, but quieter and less prolonged than above, this last also noted in Malaŵi. Call in Ethiopia and Kenya highlands 'jing-a-jing' but in Malaŵi this call used by Long-billed but not Red-faced Crombec. In Zimbabwe, in August and September its usual call is a rapid 'tee-tee-tee-tee-tee-tee . . .' becoming louder at the end, but when the chicks have hatched call is a ticking sound (Parkes 1993).

General Habits. Typically inhabits fairly well-developed woodland and edges of forest areas including acacia, *Brachystegia* and *Baikiaea*. Usually in moister areas but occurs in fairly arid thorn scrub and low bushes in Sudan, and in dry acacia areas in Tanzania.

Generally in pairs but often also in mixed bird parties where 1–4 (av. 2·2) per party. Seen in 44 of 60 parties in *Brachystegia* with no particular associations with other species (Zimbabwe: Vernon 1980); however, associated with but ecologically separated from Yellow White-eye *Zosterops senegalensis* in Zimbabwe (Irwin 1959). Once seen to supplant a Yellow-breasted Apalis *Apalis flavida*, one of only 4 within-party interactions seen which did not involve Fork-tailed Drongo *Dicrurus adsimilis* (Vernon 1980). Seen in parties 8 times and on its own 4 times in S Kenya (Pomeroy and Tengecho 1982).

Forages mainly in canopy of trees, only rarely coming within 3–4 m of ground (especially where it overlaps with Long-billed Crombec). Occasionally takes flying insects. Searches systematically all around branches, twigs and seed-pods, looking into lichens, crevices and curled leaves. In 65 observations of Red-faced Crombecs in bird parties, 57% of food items were taken from twigs, 31% from branches, 2% from trunks, 5% bushes and 5% from leaves (Vernon 1980).

Posture hunched with head drawn into shoulders. Roosting behaviour of one bird was studied in Kenya (Horne and Short 1986). It foraged actively in tree in evening, suddenly ceased moving and simply perched; 1 h later its plumage was fluffed out, concealing tail and

feet, making bird into a 'headless' ball; this sleeping posture was maintained all night; bird woke at 06·00 h, depressed its plumage, sat for 2 min, then started feeding voraciously. Next night bird returned to exactly the same spot at the same time; it did not return the following one which was windy, but did so again for the next 3, and again a month later on 4 nights. In morning it woke up, 'unfluffed', stretched, and left after 2–7 min.

Sedentary in most of range, but in Kenya may be rare in or absent from areas near northern limit for several months in dry weather.

Food. Insects, including scale insects and caterpillars; also spiders and small worms. 9 stomachs from Kenya all contained only insects. Dried and bleached 'rings' from dead millipedes fed to chicks (Ginn 1989).

Breeding Habits. Monogamous and territorial.

NEST: oval pouch, with side-top entrance (**A**), often resembling old insect nest or collection of odd bits of vegetation; made of spider or caterpillar webs of silk, and strips of *Combretum* bark, and decorated with rotten wood chips or lumps of caterpillar droppings and occasionally a leaf or seed-pod; usually in fairly open site, mostly leafless or very sparsely leafed trees, hanging from a fork of a twig 0·5–5·5 m from ground. Most nests thick-walled, total length 93 (**B**), top to lip 63, width across mouth 37, mouth to bottom 42 (Parkes 1993). Birds in captivity take a long time to build nest (Kleefisch 1985). One branch of *Anacardium* was used for 3 consecutive seasons and the same tree in a 4th season (Kenya: van Someren 1947).

EGGS: 1–3, usually 2; southern Africa (n = 7 clutches) av. 1·7; Zimbabwe (n = 7) av. 2·3. Oval; white with many fine reddish brown speckles especially at larger end, and with grey or sepia undermarkings. SIZE: (n = 13) 16·6–20·8 × 11·7–13·7 (17·7 × 12·4).

LAYING DATES: Sudan, Dec, Feb; Ethiopia, Mar; E Africa: Region B, Mar, June; Region C, Feb, Apr, May, Sept–Dec; Region D, Mar–June, Sept–Jan. In E Africa peaks ill-defined, but mainly just before rains or in early

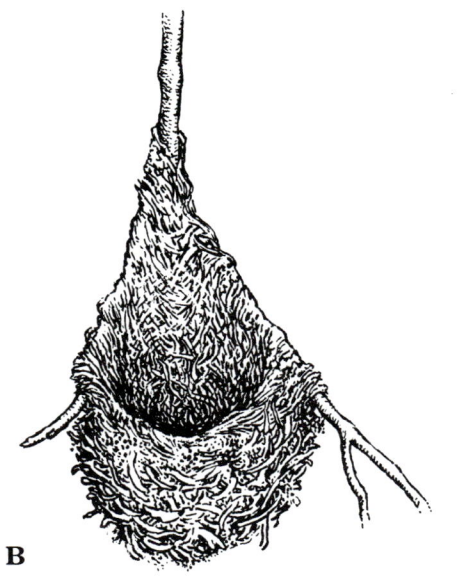

B

rains, and in both rainy seasons where 2 occur (Region D). Malaŵi, Sept–Dec; Mozambique, Sept–Nov; Zimbabwe, Aug–Dec (mainly Sept–Nov); southern Africa, Aug–Dec (especially Sept–Nov).

INCUBATION: mainly by ♀; incubates with tail towards nest entrance (Moreau and Moreau 1939); by both sexes in captivity (Kleefisch 1985). Period: at least 14 days (13–14 days in captivity).

DEVELOPMENT AND CARE OF YOUNG: young fed by both parents in wild. Nestling period at least 17 days. In captivity, one chick left nest after 14 days and was fed by both parents; ♂ mated again after 10 days; ♀ continued to feed young for 3 weeks after fledging until she was separated from it. Changes to adult plumage after 4 months. Incubation and fledging period 28·5 days (± 1·5) (Kleefisch 1985).

BREEDING SUCCESS/SURVIVAL: 1 adult found dead, caught in spider's web (Vernon 1976).

References
Donnelly, B. G. and Irwin, M. P. S. (1969).
Irwin, M. P. S. (1968).
Parkes, D.A. (1993).
Vernon, C. J. (1980).

Plate 19
(Opp. p. 336)

Sylvietta philippae Williams. **Philippa's Crombec. Crombec de Somalie.**

Sylvietta philippae Williams, 1955. Ibis, p. 582; Galkayu (= Rocco Littoria), central Italian Somaliland.

Forms a superspecies with *S. brachyura* and *S. whytii*.

Range and Status. Endemic resident, known only from NW and W Somalia and adjoining parts of Ethiopia. In Somalia known from Burao and Daba Dalol (in NW) south to Iesomme (central Somalia) and around Wajit and Lugh to the west. 2 records at each of Bogol Mayo and Gheraro in adjoining areas of Ethiopia. Suspected to be more widespread in Ethiopia than is currently known (Ash 1982).

Fairly common, mainly at about 300 m, but ranges up to 900 m.

Description. ADULT ♂: head and upperparts ashy grey becoming paler towards uppertail-coverts. Tail greyish brown. Thin white stripe over eye from base of bill; streak through and below eye to ear-coverts dark greyish brown. Chin and throat white; breast, belly and undertail-coverts pale lemon-yellow, becoming browner on flanks. Wings greyish brown with pale outer edges to primaries and secondaries; upperwing-coverts brown with grey edges; underwing pale brownish grey. Bill lavender grey with pinkish base to lower mandible; eyes brown; legs brownish red. Sexes alike. SIZE: wing, ♂ (n = 5) 52–55 (54·1), ♀ (n = 7) 50–54 (52·1); tail, 1 ♂ 21, 1 ♀ 24; bill 1 ♂ 11·1 ♀ 11·3; tarsus, 1 ♂ 16·6, 1 ♀ 16·4. WEIGHT: (Somalia, Apr)1 ♂ 7·4, 1 ♀ 7·9; (Somalia, Nov) 1 ♂ 10, 1 ♀ 9.

IMMATURE AND NESTLING: unknown.

Field Characters. A short-billed crombec confined to a restricted area in NE Africa, with pale yellow underparts; the yellow is visible on sides of uppertail-coverts

Sylvietta philippae

when bird is at rest. Underpart colour separates it from Northern Crombec *S. brachyura* (rufous) and Somali Crombec *S. isabellina* (pale buff); latter also has longer bill. Bears a striking resemblance to Yellow-vented

Eremomela *Eremomela flavicrissalis* but is stouter, with reddish brown, not grey, legs, dark face mask and white eye-stripe; the eremomela also has yellow confined to vent and longer bill.

Voice. Not tape-recorded. Song an often repeated sequence of 3 notes 'ti-churr-cheesis', the middle one ascending and the last subdued. Call 'churr' as in the song; also a loud metallic 'chink' and a rarely uttered quiet 'tip'.

General Habits. Frequents fairly dense thickets in semi-desert acacia and *Commiphora* scrub, preferring areas of rocky ground and red sandy soil.

Usually tame; in pairs or groups of 3–4 birds. Often joins mixed parties with other warblers. Searches branches, foliage and flowers, especially of various acacias.

Food. Minute beetles and other insects. 1 stomach full of small green caterpillars.

Breeding Habits. Almost unknown. Somalia: ♀ shot off nest with eggs by Archer, 22 June (specimen in American Museum of Natural History) but neither nest nor eggs were described (J. E. Miskell, pers. comm.); fully developed gonads in Apr; pair feeding fledgling late May.

Reference
Ash, J. S. (1982).

Sylvietta rufescens (Vieillot). Long-billed Crombec. Crombec à long bec.

Plate 19
(Opp. p. 336)

Dicaeum rufescens Vieillot, 1817. N. Dict. d'Hist. Nat. 9, p. 407; Oliphants R., western Cape Province, South Africa.

Forms a superspecies with *S. isabellina*.

Range and Status. Endemic resident, southern Africa. In narrow strip all across SW and W Huila Province of S Angola, reaching edge of escarpment at Leba and extending north up coastal plain to Luanda. Zambia, in west occurs north to 12°S but rare/sparse north to 14°S, in NE sparse or absent above 1200 m; elsewhere common. Extends north in extreme E Zaïre to Ruzizi Plain in Burundi; may be more widespread in SE Zaïre than map shows. Tanzania, only in SW, around Tatanda and Kasanga. Common throughout drier parts of Malaŵi and in Mozambique in S Sul do Save, south of Limpopo R., but very local north of river; also in Tete District near Zambezi R. Throughout Zimbabwe except for area along Mozambique border, Botswana, Namibia and South Africa except for parts of E Cape, E Orange Free State, S Transvaal, S Natal and Transkei; very common in Kruger Nat. Park. 3 records in Lesotho, in SW.

Uncommon and rather local in north of range but common in south; occurs mainly at lower altitudes but up to 2000 m in Zaïre. Much overlooked unless call is known. Density of 40 birds per 100 ha in *Acacia* woodland and 16·7 per 100 ha in *Burkea* (Transvaal: Tarboton 1980), 3 pairs per 100 ha in mixed woodland, Zimbabwe (and none in *Brachystegia*: Vernon 1985a); 3 birds seen in 838 ha (0·4 per 100 ha) in karroid broken veld (Winterbottom 1968). It may have benefitted from destruction of *Brachystegia* woodland in Zimbabwe (Irwin 1981).

Description. *S. r. rufescens* (Vieillot): W and NW Cape Province, SW Transvaal, S Botswana. ADULT ♂: entire upperparts brownish grey; tail greyish brown. Face, chin and throat buffy white mottled with grey; cheeks rather greyer; faint supercilium over eye and back towards nape buffy white,

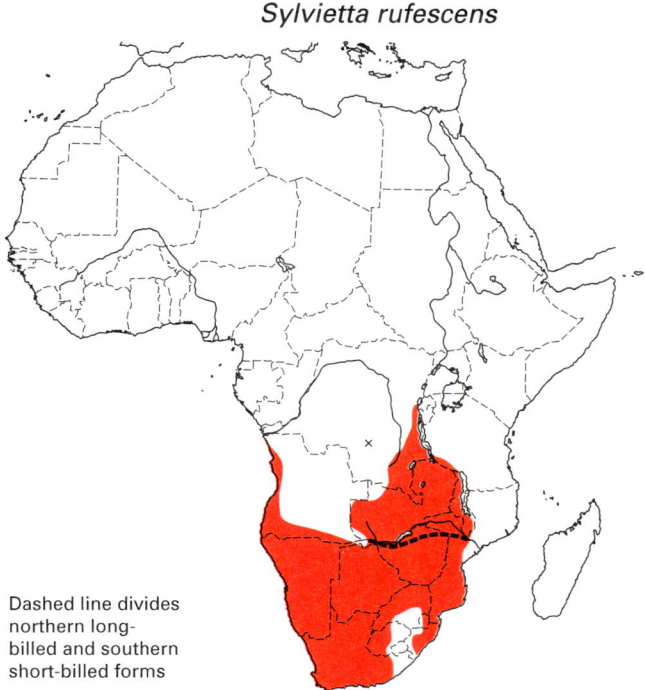

Sylvietta rufescens

Dashed line divides northern long-billed and southern short-billed forms

darker streak through eye; breast to undertail-coverts including flanks and underwing rufous-buff. Wings and upperwing-coverts greyish brown, primaries and secondaries with paler edges to feathers. Bill brown or blackish brown; eyes light brown; feet reddish brown or reddish flesh. Sexes alike. SIZE: (11 ♂♂, 10 ♀♀) wing, ♂ 56–64 (61·3), ♀ 57–63 (59·6); tail, ♂ 26–31 (28·4), ♀ 23–28 (25·1); bill, ♂ 14·3–17·6 (15·8), ♀ 14·1–16·8 (15·4); tarsus, ♂ 17·7–19·4 (18·8), ♀ 17·7–20·0 (19·2). WEIGHT: (southern Africa) ♂ (n = 6) 10·1–12·7 (11·3), ♀ (n = 3) 9·2–12·4 (10·8), unsexed (n = 76) 8–23 (11·6).

IMMATURE: not described; immature *S. r. pallida* has underparts a little paler than adult.

NESTLING: initially naked, with dark pinkish olive skin which turns blackish; yellowish gape.

S. r. diverga Clancey: SW to E Cape Province, Orange Free State, S Transvaal. Upperparts darker and underparts richer cinnamon than *rufescens*.

S. r. resurga Clancey: Natal. Underparts like *diverga* but upperparts paler and more bluish grey.

S. r. flecki Reichenow (including '*mossamedes*' and '*ochrocara*'): S Angola, E Namibia, N and E Botswana, SW Zambia, uplands Zimbabwe. Upperparts like *resurga*, and clearer grey and less brownish than *rufescens*; underparts much paler than *resurga*; short bill. WEIGHT: (Botswana) ♂ (n = 25) 10·0–14·9 (12·1), ♀ (n = 11) 7·0–12 (9·8); (Zimbabwe) ♂ (n = 27) 10·0–13·3 (11·7), ♀ (n = 26) 9·2–14·1 (11·3); (Namibia), 3 ♂♂ 11, 11, 13, 1 ♀ 14.

S. r. pallida Alexander: N Zululand, N and E Transvaal, Mozambique, E Zimbabwe, SE Zambia, S Malaŵi. Like *flecki* but face whiter and underparts generally paler. Bill much smaller: ♂ (n = 10) 13–14·5 (13·4), ♀ (n = 8) 12·1–14·2 (13·1). WEIGHT: (Middle Zambezi valley) ♂ (n = 8) 7–10 (8·5), ♀ (n = 3) 9–11 (9·8); (Transvaal) adult (n = 6) 9·9–12·9 (11·9), 2 imm. 11·1, 11·4.

S. r. adelphe Grote: Zaïre, much of Zambia and N Malaŵi. Somewhat smaller and breast and belly brighter rufous than *pallida*. WEIGHT: (Zaïre) 1 ♂ (Feb) 10, 1 unsexed (Aug) 9;

S. r. ansorgei Hartert: coastal Angola (Benguela to Luanda). Upperparts browner and less grey than *flecki*, eye-stripe white, not buff, throat whiter, and wing shorter: ♂ (n = 6) 56–60 (57·8).

Field Characters. Tailless appearance and reddish underparts easily distinguish it from all other birds except Red-faced Crombec *S. whytii*. Bill longer than Red-faced and appearing slightly decurved, but the 2 are best separated by face pattern: Red-faced has whole facial area rufous, continuous with underparts, while Long-billed has white eyebrow, dark line through eye and pale throat. Upperparts of Red-faced are a paler, ashy grey and underparts richer rufous. Voices also different.

Voice. Tape-recorded (22, 58, 75, 86, 88, 91, B, C, F). Song a fairly high-pitched, rather variable and slightly bubbled, 'tip-tip, tee-teeo' or 'tee-i, ti-tu' repeated several times. Almost identical to song of Red-faced Crombec in Kenya and Tanzania (q.v.). Also has a tripping 'chirrit, chirrit, titrr, trrt, chip, chip, tria, tria, tria, tria, chiploi, chiploi, chiploi . . .'. Call is more varied and louder than that of Red-faced and punctuated by short snatches of song. Also a short buzzing 'trrp, trrp'.

General Habits. Inhabits mainly thick bushes in drier areas although often in dongas and near water. Acacia scrub, open mopane woodland, secondary and mixed woodland of many types, edges of more well-developed woodland and thorn scrub are preferred habitats, but the species also occurs in very arid areas in Namibia. Avoids forests and interior of such woodland as mature *Brachystegia*. A rare visitor to a rural garden in Stellenbosch (Siegfried 1968a) but quite common in gardens in Namibia (Winterbottom 1971b).

Occurs mainly in undergrowth, and only in undergrowth wherever the canopy-dwelling Red-faced and Red-capped *S. ruficapilla* Crombecs are present. Where they are absent, Long-billed is more adaptable and occurs in tree canopy; it favours canopy in parts of Limpopo and Sabi valleys and in SE Zimbabwe. It uses leafless trees and aloes if there are no bushes.

Fairly tame; lives solitarily or in pairs or family parties. Readily joins mixed-species parties; of 60 different bird parties in Zimbabwe, Long-billed Crombecs (av. 1·2 birds) occurred in 9. One party was observed 50 times in an area containing 3 Long-billed Crombecs, and an av. 2·0 Long-billeds joined it 16 times (Vernon 1980). Territory of the resident pair of Long-billed Crombecs was larger than the area used by this party and individual birds sometimes changed to adjacent parties after interactions between parties (Zimbabwe: Vernon 1980). Long-billed Crombecs found in 52% of parties in Western Province and 17% in Eastern Province, Zambia (Winterbottom 1943). When Long-billed and Red-faced Crombecs are both in a party, their foraging horizons differ strictly, with Red-faced in canopy and Long-billed in undergrowth (Irwin 1981).

Forages restlessly, methodically searching bushes and sometimes trees, from bottom to top (usually up to 3 m), before flying with bouncy undulating flight a short distance to the next bush. Forages from twigs and leaves, and only rarely from branches, trunks or the ground: of 43 observations 30% involved twigs, 7% leaves and 63% bushes (Zimbabwe: Vernon 1980). Calls frequently whilst foraging.

Complete moult Jan–Mar (Zaïre).

Largely sedentary, but moves locally, especially at edges of range.

Food. Insects, including mantids and beetles and caterpillars of the mopane emperor moth *Imbrasia belina* (Styles 1995); also ticks, and grass seeds.

A

Breeding Habits. Monogamous and territorial. Calls frequently in breeding season.

NEST: purse-shaped, quite large and bulky; strongly made of fibres and dry grass (**A**), with less spider and caterpillar silk than that of Red-faced Crombec, ornamented with bits of grass, leaves, lumps of spider's web, leaf vanes and some wood chips; some lined with sheep's wool. Ext. diam. up to 60, and depth up to 120 (**B**); int. diam. 32 and cup depth at least 45–50. Sited nearly always within 1 m of ground, but recorded up to 4.3 m, hanging from end of one of lower branches of a tree, often in space on inside, screened by leaves; often on an acacia. Uses same site year after year (Jensen and Clinning 1974). In one case 1st egg laid 7 days from nest completion (Jensen and Clinning 1974).

EGGS: 1–3, usually 2; southern Africa (n = 72 clutches) and Malaŵi (n = 8) av. 1·8; Zimbabwe (n = 21) av. 2·1; Namibia (n = 11) av. 2·3. Laid on consecutive days. Elongated oval; white, slightly glossed, thinly and evenly spotted and blotched with reddish and brownish, and with lilac or grey undermarkings except at larger end where densely marked. Occasionally (2 of 8 clutches) heavily speckled and blotched all over. SIZE: (n = 65) 17·1–20·9 × 12·0–13·3 (18·6 × 12·6); 2 eggs in 1 clutch elongated: 22·4 × 12·5 and 20·1 × 12·5 (Vincent 1935).

LAYING DATES: Zaïre, Jan; Zambia, Aug–Mar (54 of 58 clutches in Sept–Nov); Zimbabwe, Aug–Mar (95% of 278 clutches in Sept–Dec); Malaŵi, Apr, Sept–Feb; Mozambique, Mar, Nov; Botswana, Sept–Feb (Sept 4, Oct 4, Nov 7, Dec 3, Jan 5, Feb 3 clutches); Namibia, Oct, Nov; Cape Province, Aug–Dec; Natal, Oct–Jan; Transvaal, Sept–Mar (88% of 113 clutches in Oct–Jan); Lesotho (Dec, fledgling).

INCUBATION: usually starts with 2nd egg but sometimes with 1st. Period: 14 days.

DEVELOPMENT AND CARE OF YOUNG: nestlings fed by both parents; fledglings fed by both parents for at least 2 weeks. Young prone to leave nest prematurely if disturbed. Period 14 days.

BREEDING SUCCESS/SURVIVAL: one of main hosts of Klaas's Cuckoo *Chrysococcyx klaas* (see Vol. III, p. 87), in Namibia (Jensen and Clinning 1974), Transvaal and Zimbabwe (Jensen and Jensen 1969) and Malaŵi (Hanmer 1982b). In Namibia 1 nest parasitized out of 34 found.

Reference
Jensen, R. A. C. and Clinning, C. F. (1974).

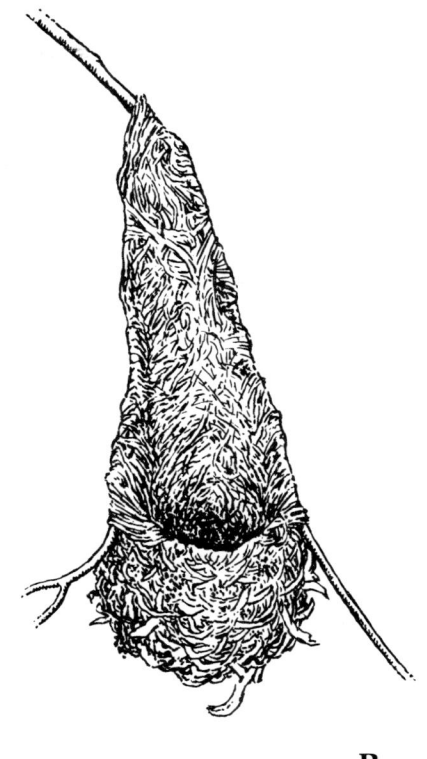

B

Sylvietta isabellina Elliott. Somali Crombec. Crombec isabelle.

Sylvietta isabellina Elliott, 1897. Field-Columbian Mus. Publ. Orn. ser. 1, p. 44; Le Gud, Ogaden, eastern Abyssinia. (Le Gud is actually in Somalia.)

Plate 19
(Opp. p. 336)

Forms a superspecies with *S. rufescens*.

Range and Status. Endemic resident, restricted to dry thornbush country in NE Africa. Locally common up to 1000 m in NW Somalia, although not near border with Djibouti; in S and SE Ethiopia, north to Awash R. In Kenya mainly below 500 m south to Lokitaung (west of L. Turkana), Horr valley, Samburu Game Reserve, Meru Nat. Park, Garissa, near Wamba, Kitui and Tsavo East and West Nat. Parks. In S Somalia only west of 46°E.

Description. ADULT ♂: top of head, nape and upperparts pale grey; stripe from bill over eye to top of ear-coverts white; cheeks and ear-coverts paler grey than upperparts; tail greyish brown with paler edges to feathers. Underparts pale buff, throat with grey bases to feathers, with pale tawny wash on flanks and around legs. Wings and upperwing-coverts greyish brown with paler edges to feathers. Bill sepia or black, very long; eyes light reddish brown; legs and feet pale brown. Sexes alike but ♀ distinctly smaller. SIZE: (9 ♂♂, 7♀♀) wing, ♂ 55–66 (58·9), ♀ 52–58 (54·9); tail, ♂ 23–29 (26·2), ♀ 21–24 (22·6); bill, ♂ (n = 8) 15·1–16·2 (15·6), ♀ 13·7–16·5 (14·9); tarsus, ♂ 18·5–19·7 (19·0), ♀ 17·3–19·2 (18·3). WEIGHT: (Kenya) 1 ♂ 10, 1 ♀ 10.

IMMATURE: like adult but duller, with throat slightly mottled.

NESTLING: unknown.

Sylvietta isabellina

Crombec *S. philippae*, which, besides much shorter bill, has pale yellow underparts. Also overlaps with similar-sized Yellow-vented Eremomela *Eremomela flavicrissalis*, which has white underparts and yellow vent.

Voice. Undescribed.

General Habits. A poorly known bird of thorn scrub in arid and semi-arid areas. Prefers acacia bushes or taller trees, and does not skulk in undergrowth, as does Northern Crombec where they are sympatric. Also occurs along banks of dry watercourses. Appears to be absent in *Commiphora* scrub where there is no *Acacia*. Tame and easily approached.

Food. 2 stomachs contained spiders, 1 also insect larvae and eggs.

Breeding Habits.
NEST: bottle-shaped and decorated with spider's web and plant down; hangs from twig.
EGGS: 2–3, usually 2; white with much olive-brown speckling and grey undermarkings. SIZE: (n = 2, Somalia) 17·5 × 11·9, 17·8 × 12·0.
LAYING DATES: E Africa: Region D, June; Kenya (breeding condition) Mar; Somalia, Apr; Ethiopia, Mar.

Field Characters. A long-billed crombec of dry NE Africa with pale buff underparts. Both Northern *S. brachyura* and Red-faced Crombec *S. whytii* have rufous underparts; more likely to be confused with Philippa's

Plate 19
(Opp. p. 336)

Sylvietta ruficapilla Bocage. Red-capped Crombec. Crombec à calotte rousse.

Sylvietta ruficapilla Bocage, 1870. J. Acad. Real. Sci. Lisboa 6, p. 160; Caconda, Benguella, Angola.

Range and Status. Endemic resident, S central Africa. Ranges from coast in S Congo (Manyanga, Mah), Zaïre (Kinshasa) and N Angola, extending up Zaïre R. to about 3°S, and across S half of Zaïre to L. Tanganyika and into SW Tanzania near Tatanda. Also occurs in a belt across central Angola between c. 10°S and 15°S (N Huila and Huambo to S Lunda and N Moxico) and Zambia except SW, Luangwa and Middle Zambezi valleys, L. Mweru to L. Tanganyika, Bangweulu and Kafue Flats. 1 record in Zimbabwe, at Nampini Ranch, west of Victoria Falls. Occurs in W Malawi to edge of high plateaux, and in Tete Province of NW Mozambique east to Kirk Mts and south to Furancungo.

Locally fairly common; mainly at 700–1500 m.

Sylvietta ruficapilla

Description. *S. r. ruficapilla* Bocage: central Angola to SW Shaba. ADULT ♂: crown and nape tawny chestnut, nape feathers with pale buffish bases; sides of face and ear-coverts similar but a little darker; rest of upperparts and scapulars grey; tail somewhat browner. Chin and throat grey with white tips to feathers; pale chestnut band across upper breast; rest of underparts and flanks pale grey, palest on undertail-coverts. Primaries grey brown with pale outer edges; secondaries and tertials greyer; upperwing-coverts dark brown with grey edges; underwing white with yellowish or buff tinge. Bill dark horn with pale lower mandible; eyes yellow to orange-red; legs and feet reddish flesh to brown. Sexes alike. SIZE: (8 ♂♂, 8 ♀♀) wing, ♂ 62–70 (66·5), ♀ 62–68 (65·0); tail, ♂ 25–31 (28·0), ♀ 24–29 (25·6); bill, ♂ 12·2–14·2 (13·3), ♀ 12·1–13·4 (13·0); tarsus, ♂ 18·0–20·9 (19·7), ♀ 19·1–21·4 (20·1). WEIGHT: (Zaïre) ♂♂ (Feb) 12, (May) 10, 12, (June) 12; ♀♀ (May) 10·5, 11·5, (July) 10, (Nov) 11. (There is slight clinal variation, with larger birds in W: Irwin 1968.)

IMMATURE: like adult but whiter below; throat, underwing and flanks have distinctly rufous tint; tips of primaries rounded.

NESTLING: unknown.

S. r. gephyra White: W Shaba to Zambia and Zimbabwe. Crown more sandy and paler than the dark rufous ear-coverts, as compared with *ruficapilla*. WEIGHT: (Zimbabwe) 1 ♂ 12·8.

S. r. chubbi Ogilvie-Grant: SE Shaba to Malaŵi and N Mozambique. Crown and nape greyer and uniform with mantle, which is rather browner than *ruficapilla*; belly often with yellowish wash.

S. r. makayii White: interior of N Angola. Like *ruficapilla* but upperparts washed with olive and primaries edged with yellow.

S. r. rufigenis Reichenow: lower Congo inland to Kasai. Crown and nape as *gephyra*, but upperparts washed with olive, underparts washed with yellow, and primaries have yellow edges.

S. r. schoutedeni White: E Zaïre from L. Tanganyika to Marungu and Mt Kabobo. Crown and back as *chubbi*, but upperparts more olive and a strong yellow tint to underparts.

Field Characters. Largest crombec, with moderately long bill, found mainly in *Brachystegia* woodland. Combination of reddish crown (most races), cheeks, ear-coverts and breast-band, plain face without eye-stripes and pale underparts often washed with yellow distinguish it from other crombecs. Both Red-faced and Long-billed Crombecs *S. whytii* and *S. rufescens* have entire underparts rufous; song of Red-capped is different from these 2 where they overlap but similar elsewhere.

Voice. Tape-recorded (86, 91, B, F, CART, CHA, LEM). Song a loud and ringing 'tichi-tichi-tee-teee-oo' with emphasis on the 'oo', repeated several times. Slight variations of emphasis, e.g. sometimes more like 'titi, tee-ti, cheeoo', and is sometimes a rather longer phrase. West of Luangwa valley in Zambia, song is the same as song of Red-faced Crombec in Zimbabwe and Malaŵi: a far-carrying trill; also a shrill whistled phrase. Occasionally duets. Feeding call a high-pitched, rather quiet 'tsssp'.

General Habits. Very similar ecologically and behaviourally to Red-faced Crombec q.v., but the 2 are almost completely allopatric. Red-capped occurs mainly in wooded savanna, especially *Brachystegia* woodland, and occasionally in thicker secondary woodland or riverine thickets, in woodland edges and in bushes over termite mounds. Almost exclusively in canopy of trees (like Red-faced, but unlike Long-billed Crombec with which Red-capped overlaps extensively). Where Long-billed is absent, Red-capped uses undergrowth as well as canopy.

Occurs singly or in family parties and is an active member of mixed-bird parties. Recorded several times in same party as Long-billed (Britton 1970a, D. R. Aspinwall, pers. comm.). Searches leaves.

Complete annual moult occurs in Feb–Apr (Malaŵi) and Jan–Apr (Zaïre).

Food. Insects.

Breeding Habits. Monogamous and territorial as far as known. Sings all year.

NEST: purse-shaped structure of grass and fibres ornamented on outside; more rounded and compact, less elongated, and with thicker walls than nest of Long-billed Crombec. Int. diam. 35, depth 25–35. Placed usually 3–4 m up in bush, screened by branches, foliage or weedy growth but sometimes with no particular attempt at concealment (except for ornamentation). Sited in fork attached all around, or hung from end of lower branch.

EGGS: 2; white, smooth and dull-surfaced, well marked with sepia and grey markings especially at larger end. SIZE: Zambia (n = 4) 17·7–18·5 × 11·0–12·2 (18·3 × 11·8); Mozambique (n = 6) 18·6–21·4 × 11·9–13·3 (19·8 × 12·6).

LAYING DATES: Zaïre, Sept–Apr; Zambia, Sept–Nov; Malaŵi, Oct; Mozambique, Sept, Nov. Breeds mainly in rainy seasons or at end of dry season.

INCUBATION: ♂ has been collected off a nest containing eggs.

Sylvietta virens Cassin. Green Crombec. Crombec vert.

Plate 19
(Opp. p. 336)

Sylvietta virens Cassin, 1859. Proc. Nat. Acad. Sci., Philadelphia, 11, p. 39; Camma River, Gabon.

Range and Status. Endemic resident in forests of W and central Africa. Fairly common in forest patches in coastal Senegambia although surprisingly not recorded in Abuko Forest (Wacher 1993). Fairly common in farmbush of Gola Forest area in Sierra Leone, less so in cocoa and forestry plantations. W Guinea in Kindia area (Demey 1995) and Sougueta (T. Aversa, pers. comm.) and in SE in Macenta (Halleux 1994); abundant Mt Nimba (Guinea/Liberia). Occurs in forest belt in S Ivory Coast, Ghana, Togo, Benin and Nigeria, and extends into gallery forests in the guinea savannas of Ivory Coast, Ghana and Togo. In Nigeria occurs north to Ibadan, Akure and Enugu and also on Chappal Waddi at 1400 m. Mali, 1 collected near Kangaba 'reported as this species' (Lamarche 1981). S Cameroon, SW Central African Republic, Gabon (fairly common throughout), Congo and NW Angola south to Ndala Tando, Canhoca and Roca Congulu,

Sylvietta virens

and at Dundo in N Lunda. The commonest crombec in Zaïre, where it occurs south to Kasai District; apparently absent from Shaba Province in SE. In S Sudan rather uncommon in Lotti Forest, Imatong Mts and Bengengai; in Uganda ranges widely in W and S, from Acholi and Murchison Falls Nat. Park to Bwindi-Impenetrable, Malabigambo and Namlala Forests, east to Busoga and Mt Elgon (Britton 1980a) and south to Bukoba in NW Tanzania; the commonest crombec in dense undergrowth around Kampala; status in Rwanda and Burundi uncertain. In Kenya at Mungatzi in Busia District (Turner *et al.* 1990; these authors place renewed doubt on the occurrence of this species in Kakamega despite Gerhart and Paxton (1980) and subsequent acceptance of their record).

Common and widespread from sea level to at least 1400 m, but in Uganda and Kenya rarely if ever above 1200 m. Density of 3 pairs in 15 ha in liana-strewn forest at M'Passa, Gabon (Brosset and Erard 1986).

Description. *S. v. virens* Cassin: SE Nigeria, Cameroon and Gabon to mid-Zaïre (Kunungu and great bend of Ubangi R.). ADULT ♂: crown and upper nape dark chocolate brown; very thin brown stripe from immediately over eye to hindcrown with a thin pale brown stripe above it; rest of upperparts dark brownish olive-green; tail brown with olive-green edges to all feathers. Lores, cheeks, ear-coverts, throat and breast brown with rufous tinge; chin paler; lower breast to undertail-coverts dirty grey with whiter centre of belly and somewhat darker flanks. Some Nigerian birds have a yellowish tint in centre of upper belly. Primaries brown with pale yellowish brown outer edges; secondaries and tertials rather paler and edges greener and less yellowish; scapulars and upperwing-coverts dark brown with olive-green edges; underwing yellow. Bill dark brown with paler lower mandible; eyes light brown or orange-brown; legs and feet flesh brown. Sexes alike. SIZE: (11 ♂♂, 10 ♀♀) wing, ♂ 48–53 (50·3), ♀ 46–49 (47·6); tail, ♂ 17–19 (18·1), ♀ (n = 7), 15–18 (16·9); bill, ♂ 11·2–12·6 (11·8), ♀ 11·3–12·5 (11·7); tarsus, ♂ 16·5–18·9 (17·4), ♀ 16·1–18·0 (17·0). WEIGHT: (Mt Nimba) ♂ (n = 5) 7·3–9·0 (7·8), ♀ (n = 8) 7·0–9·4 (7·9); (Ghana, n = 3, unsexed) 7·6–8·3 (7·9); (Cameroon, Apr) 2 ♂♂ 9, 10, 2 ♀♀ 8·5, 9; (Angola, Feb, May) ♂ (n = 8) 9–10·5 (9·7), ♀ (n = 5) 8·5–9·5 (8·9); (Uganda) ♂ (n = 11) 6–10 (8·3), ♀ (n = 12) 5–9·5 (8·5), unsexed (n = 5) 7·5–9 (8·7); (Congo, n = 3) 7·8–8·6 (8·2).

IMMATURE: upperparts generally greener and underparts yellower than adult.

NESTLING: unknown.

S. v. flaviventris (Sharpe): Sierra Leone to SW Nigeria. Mantle to uppertail-coverts olive-green; chin white; throat and upper breast pale rufous-brown merging to yellow on lower breast and upper belly; lower belly white. Immature has belly almost wholly yellow.

S. v. baraka Sharpe: east of *virens* to Kenya. Duller brown than *virens*; throat and breast greyer with less rufous tinge.

S. v. tando W. L. Sclater (including '*meridionalis*'): Congo, Zaïre south of *virens*, Angola. Chin white, sharply demarcated from rather orange-brown throat and breast; rest of underparts white, with yellow tint in centre of belly; flanks rather greyer and upperparts and tail more olive green than *virens*.

Field Characters. A dingy green, brown and grey crombec of forest and second growth. Only other lowland forest crombec is Lemon-bellied *S. denti*, with which it overlaps extensively, but Green lives mainly in undergrowth, Lemon-bellied in canopy. Most races of Green distinguished by lack of yellow in underparts; yellow-bellied race *flaviventris* told from Lemon-bellied by eye-stripes and larger bill. Barely overlaps (in E Zaïre) with the montane forest White-browed Crombec *S. leucophrys*, which has brown cap, greener upperparts, broad white eyebrow, white face and throat and grey underparts; their songs are similar but Green's has slightly different intonation and is higher-pitched, important as birds often difficult to see. Bleating Warbler *Camaroptera brachyura* is larger, with a normal-length tail and very different voice.

Voice. Tape-recorded (32, 68, B, C, ERA, LEM, MOY). Song a rather thin, rapid, high-pitched series of 3 notes followed by a trill then 2–3 clear notes descending the scale; in Gambia typically the last 1–2 notes go back up the scale (Wacher 1993); repeated about 12 times in a song burst. Call a sharp 'pririt-pririt'.

General Habits. Occurs primarily in thick undergrowth of primary forest, within 300 m of the edge; also in secondary forest, thickets (including of *Lantana*) and gallery forests. Mainly at low altitude: up to 1390 m in Zaïre. Fairly common at edges of forest and clearings, in thickets in savannas, plantations, and even copses, villages and gardens and abandoned cultivation if there is plenty of cover. Occurs in drier areas in coastal Angola but always stays in thick cover.

Singly, in pairs or family parties; does not occur in mixed-species flocks. More often heard than seen. Feeds largely from stems, especially lianas and other creepers, mainly within 3 m of ground, often at top of bush layer, though will occasionally move into canopy. In Nigeria 74% of observations were in scrub below 1.5 m and 26% from 1·5 to 3 m; of 141 foraging sites 6% were on ground, 72% were stems and twigs, 18% were leaves and 3% were seed heads; bird did not use air, branches, trunks, berries or flowers (Edington and

Edington 1983). Always active as it forages, continually opening and closing wings and moving tail up and down. When in canopy crombecs were chased off by sunbirds but not by Bleating Warblers (Edington and Edington 1983).

At least 2 moulting periods during year on Mt Nimba: primaries growing in Feb–Mar and Sept–Oct.

Food. Small arthropods; black ants, hemipterans, beetles, orthopterans, mantids, flies, hymenopterans, lepidopterans, blattids, moth eggs, caterpillars and other larvae, spiders and small helicoid land snails. Ova, fruits and grass seeds also recorded. 9 stomachs in Cameroon and 23 in Uganda contained only insects. In Ivory Coast diet is 60% adult arthropods, 20% larvae, and 20% fruits (Thiollay 1973).

Breeding Habits. Monogamous and territorial; a record of 2 young in nest being fed by 3 adults (Brosset and Erard 1986).

NEST: neat, pear-shaped structure of grasses, plant fibres, bark and spider's webs, decorated with small bits of bark, spider's web and wood dust; hung from extremity of twig among lianas or foliage; 1–3 m above ground, normally 1–2 m.

EGGS: 2–3, usually 2; Ivory Coast av. (n = 11 clutches) 2. White or very pale blue with some faint rufous markings. SIZE: 16·5 × 11.

LAYING DATES: Liberia (Mt Nimba, enlarged ovaries Feb, Mar, June, Oct); Ivory Coast, June–Aug; Ghana (fledglings Jan, Aug); Nigeria, May (feeding young, Oct); Gabon, Oct–Feb; Congo, Oct; Zaïre, Jan–Apr, July; Angola (breeding condition, Mar); E Africa: Region B, Feb–June; Sudan, Oct.

BREEDING SUCCESS/SURVIVAL: from av. 2 eggs, av. 1·2 young fledged (sample size unknown, Ivory Coast: Thiollay 1971).

References
Brosset, A. and Erard, C. (1986).
Edington, J. M. and Edington, M. A. (1983).

Sylvietta denti (Ogilvie-Grant). Lemon-bellied Crombec. Crombec à gorge tachetée.

Plate 19
(Opp. p. 336)

Sylviella denti Ogilvie-Grant, 1906. Bull. Br. Orn. Club 19, p. 25; 10 miles northwest of Fort Beni, eastern Congo.

Range and Status. Endemic resident in forests of W and central Africa. Gambia, 3 isolated records, near Tanji, Brufut and Jakhaly. Uncommon Sierra Leone and Liberia; Guinea in W at Kindia (Demey 1995) and SE in Macenta (common: Halleux 1994); S Ivory Coast (Abidjan and Tai to Sipilou and Bouaké) and Ghana (4 localities in Ashanti, Tafo, Bia Nat. Park); in Nigeria recorded only at Sapoba, Odo-Akure and in the Ikpan forest block. S Cameroon and extreme SW Central African Republic south to N Gabon where widespread and dense, and scattered records all across Zaïre, perhaps continuous, in narrow belt (c. 4°S to 4°N) to Budongo Forest in W Uganda. Frequent in logged forest at Koubotchi near coast of Congo, seen at Moukalaba, SW Gabon (D. Sargeant, pers. comm.) and should be expected in Mayombe area on coast of Gabon (Dowsett-Lemaire and Dowsett 1991). Occurs S Zaïre in gallery forest along Luachimo R., and in adjacent NE Angola (N Lunda).

Occurs up to at least 1200 m in Zaïre. Uncommon and local, and probably underrecorded, especially if call is not known. In Yapo Forest, Ivory Coast, was noted on 24 visits out of 52 (L. Fishpool, pers. comm.). Occurs in many of the same localities as Green Crombec *S. virens*, but is much less common. Density of 8–10 pairs per 100 ha in primary and old secondary forest in Gabon (Brosset and Erard 1986).

Description. *S. d. denti* (Grant). Cameroon to Zaïre. ADULT ♂: crown and nape grey with olive wash; rest of upperparts olive-green; tail brown. Chin and cheeks dull white, with dark tips to feathers; ear-coverts pale grey; throat pale rufous merging to pale olive-green on breast; belly to undertail-coverts and flanks pale yellow. Wings dark brown, primaries with

Sylvietta denti

white outer edges washed with yellow for basal three-quarters. Bill black; eyes light brown; legs and feet greyish horn or brown. ADULT ♀: like ♂ but slightly duller, especially on belly. SIZE: (including *S. d. hardyi*) (5 ♂♂, 11 ♀♀) wing, ♂ 47–51 (49·0) ♀ 45–48 (46·2); tail, ♂ 17–19 (18·4), ♀ 15–18 (16·9); bill, ♂ 9·2–10·5 (9·9), ♀ 9·2–10·4 (9·8); tarsus, ♂ 14·6–15·5 (15·1), ♀ 14·5–15·9 (15·4). WEIGHT: (Mt Nimba) ♂ (Oct) 8·1, ♀ (Sept) 8·7.

IMMATURE: like adult except for broad pale yellowish terminal bars on feathers of mantle, crown, secondaries, and especially obvious (2 mm) on greater coverts.

NESTLING: unknown.

S. d. hardyi Bannerman: Sierra Leone to Ghana; race in Gambia and Nigeria unknown. Each block of colour more clearly demarcated than in *denti*; less mottling on cheeks, belly brighter yellow, throat brighter rufous and breast more olive.

Field Characters. A small crombec of secondary forest with very small bill. Yellow underparts distinguish it from Northern Crombec *S. brachyura*, and from eastern forms of Green Crombec which have white belly. Western race of Green, *S. v. flaviventris*, has yellow belly but also dark brown crown, pale eyebrow and brown upper and pale lower mandible; Lemon-bellied has grey crown, no eyebrow (or a very indistinct one), all-black bill, and rufous on throat of western race is fairly conspicuous in the field. Lemon-bellied occurs in canopy, but Green in undergrowth.

Voice. Tape-recorded (68, CHA, ERA). Song a short, high-pitched jumble of notes, followed by 7-11 clear, rather piercing, notes on one pitch or slightly descending. The clear single notes may be quick pairs of notes 'te-tu' with first higher-pitched than second. Third type is slower series of elided notes 'teeo, tui, teeo, tui . . .', the first note, 'teeo' rising then falling and the second, 'tui', falling then rising. Has subsong like Green Crombec (Chappuis 1979a); it is given from dense thicket, but territorial song is from tree top.

General Habits. Inhabits secondary forest, especially in forest gaps; also some primary and gallery forests, forest edge undergrowth and thickets in abandoned villages.

Stays mainly in crowns of large trees, where acrobatically and actively searches among leaves, branch tips, twigs and particularly lianas. Very difficult to spot.

Mainly seen in pairs, often joins mixed-bird parties.

Food. Insects, including caterpillars and beetles.

Breeding Habits. Monogamous and territorial.

NEST: thin-walled, compact but solid, flexible structure bound loosely with gossamer threads, with little or no lining, and made of small stems, strips of bark, flowers and the 'nests' of caterpillars found locally; and ornamented especially with pupae filled with droppings and scraps of rotting bark. Ext. diam. 80, ext. depth 90. Hung at edge of bush or tree over open ground, 3–4 m above ground, sometimes up to 12 m. Built by ♀ while ♂ accompanies her singing.

EGGS: usually only 1; Cameroon (n = 5) always 1. Yellowish clay colour mottled all over with brown and grey. SIZE: (Cameroon) $16·5 \times 11$, 19×12.

LAYING DATES: Mt Nimba, Liberia, Sept–Oct; Ivory Coast, Jan–Apr; Cameroon, Mar, Apr; Gabon, Sept, Jan, Mar.

INCUBATION: incubating parent was aggressive at approach of observer.

Reference
Brosset, A. and Erard, C. (1986).

Plate 19 (Opp. p. 336) *Sylvietta leucophrys* (Sharpe). White-browed Crombec. Crombec à sourcils blancs.

Sylviella leucophrys Sharpe, 1891. Ibis, p. 120; Mt Elgon, Uganda–Kenya border.

Range and Status. Endemic resident in highland forests of central Africa. 4 disjunct populations: (1) Lendu Plateau, NE Zaïre (probably extinct); (2) Mountains either side of Albertine Rift from Rwenzoris and west of L. Edward through E Zaïre to Mt Kabobo and through SW Uganda (Kigezi and Ankole Districts) and W Rwanda to forests of W Burundi (Bururi, Rwegura, Teza); (3) Mahari Mts of W Tanzania; (4) Kenya highlands from Mt Elgon to Nandi, Mau, Aberdares, Mt Kenya, Limuru and Ngong Hills; rare in Kakamega.

Locally common, between 1550–2600 m, rarely somewhat lower, and up to 3000 m in bamboo in Zaïre.

Description. *S. l. leucophrys* (Sharpe): Kenya and Uganda (Kibale Forest, Rwenzori). ADULT ♂: head, nape and ear-coverts chocolate brown; broad white stripe from bill over eye with a thinner extension farther back; mantle and back olive-brown or brown with grey bases to feathers; rump and upper-tail-coverts similar but rather greener; tail olive-brown with greener edges to feathers. Chin, throat and lores white; breast to belly grey becoming paler on lower belly and browner grey on sides of breast; undertail-coverts bright yellow. Wings dark brown with bright green outer edges of feathers; edges of primaries more yellowish green; scapulars and upperwing-coverts green becoming more yellowish on outer wing; underwing yellow. Bill light brown or flesh; eyes dark red or chestnut; legs and feet light brown. Sexes alike. SIZE: (Rwenzori, 5 ♂♂, 6 ♀♀) wing, ♂ 62–66 (64·2), ♀ 57–61 (58·3); tail, ♂ 22–26 (24·4), ♀ 20–24 (22·3); bill, ♂ 10·8–12·6 (11·8), ♀ 11·5–12·6 (12·1); tarsus, ♂ 19·6–22·4 (21·5), ♀ 20·7–22·1 (21·4); (Mt Elgon) wing, ♂ 59, ♀ 55; tail, ♂ 22, ♀ 23; bill, ♂ 11·5, ♀ 10·6; tarsus, ♂ 20·2, ♀ 20·6. WEIGHT: (Kakamega, Kenya, n = 4) av. 11·2; (Uganda) ♂ (n = 16) 10–13 (11·7), ♀ (n = 12) 8–13·5 (11·0); (central Africa, n = 7) av. 10·4 (SD 0·88); (Burundi, n = 5) av. 10·9 (SD 0·48); (Rwanda, n = 2) av. 9·3.

IMMATURE: like adult but sides of face browner; chin, throat and underparts brownish grey with centre of belly whitish; rump with some light barring.

NESTLING: unknown.

S. l. chloronota Hartert (including '*arileuca*' (Parkes 1987), best considered as an intermediate between *leucophrys* and *chloronota* (Louette 1989)): Kigezi, SW Uganda, E Zaïre and W Tanzania. Reddish brown behind eye extends farther onto cheeks and ear-coverts than in *leucophrys*; upperparts rather greener, and less brown-tinged. Immature lacks white eye-stripe of adult or has only slight indication; mantle heavily washed rusty; underparts much darker than adult. SIZE: wing, 2 ♂♂ 58, 59, 3 ♀♀ 55–59 (56·7).

S. l. chapini Schouteden: Lendu plateau, Zaïre. Chestnut cap extends over face; no white eye-stripe; belly has yellow-buff wash. May be extinct as forest on Lendu plateau has been destroyed (Louette 1989).

Field Characters. The only montane forest crombec, barely meeting Green Crombec *S. virens* and Red-capped Crombec *S. ruficapilla* in E Zaïre; distinguished from both by broad white eyebrow, face and throat, grey underparts and yellow undertail-coverts.

Voice. Tape-recorded (C, CHA, GREG, HOR, LEM, McVIC, ZIM). Song a rapid 'teeo, tititi', with the first note emphasized and dropping in pitch slightly, and the 'tititi' 3 very rapid short notes slightly rising; all repeated 3–7 times in a song burst. Also a slightly variable series of high, clear phrases often with a trill, and sometimes repeated. A less stereotyped series of notes, with the tone of a white-eye *Zosterops*, is apparently not a territorial song (Dowsett-Lemaire 1990). Contact call a low trill.

General Habits. Occurs only in thick undergrowth inside and at edge of montane forest and in the bamboo zone, especially under a broken canopy; almost entirely between 1500 and 2500 m. Favours thickets with many vines and bushes, and low trees growing among bamboo. There is no geographical and almost no altitudinal overlap with Green Crombec in Zaïre; White-browed occurs down to 1290 m but 80% of records are over 1580 m; and the Green occurs only up to 1390 m (Prigogine 1980a).

Feeds in cover, in open or moving through epiphytes on vertical trunks and lianes, mostly within 6 m of ground but up to 12 m. Occurs in mixed parties, and feeds close to Masked Apalis *Apalis binotata* and Black-collared Apalis *A. pulchra* in low thickets.

Moult lasts about 37 days (Mann 1985); in Rwanda it occurs Sept–Oct.

A

Food. Small insects: beetles, caterpillars, eggs; also spiders. 36 stomachs all contained insects and 1 also contained a small snail.

Breeding Habits. Sings occasionally Oct–Dec, with much territorial activity and countersinging also in Jan (Nyungwe Forest, Rwanda: Dowsett-Lemaire 1990).

NEST: larger than nests of many other crombecs, purse- or bucket-shaped (**A**), consisting almost entirely of moss and grass and lined with a few feathers or tendrils; slung from a fork, twig or grasses, usually about 1 m from ground. In 2 consecutive years nest was built at end of same twig, but in 2nd year new nest was built for 2nd brood in nearby tree (van Someren 1947).

EGGS: 2–3, usually 2. Glossy; white or rose-pink, sparsely spotted and blotched with reddish or maroon, especially at larger end. SIZE: (n = 2) 14·9 × 11·8, 15·3 × 12·0.

LAYING DATES: E Africa: Region A, Mar, Apr; Region D, Dec, Jan, Mar–July; Zaïre, Jan–June; Burundi, Feb; Rwanda, Aug, Sept (large young being fed Oct).

References
Dowsett-Lemaire, F. (1990).
Prigogine, A. (1980a).

Genus *Phylloscopus* Boie

Small warblers with small pointed bills and slender legs. Rictal bristles present, sometimes very small; tail medium length, square or slightly forked with 12 rectrices; wing rounded or bluntly pointed. Most have yellow or green tinged plumage and a conspicuous pale supercilium; some have pale wing-bars. Sexes alike. Arboreal; mostly species of forest or woodland, often of high canopy. Build domed nests, usually on or near ground. Produce high, clear songs.

38 species, all in Old World. 12 in Africa, of which 6 are endemic residents south of Sahara, 4 are Palearctic migrants (2 also breeding in NW Africa) and 2 are Palearctic vagrants.

The Afrotropical residents, all forest species, were formerly placed in *Seicercus*. They have short rounded wings with P10 almost half as long as P9, rather broad bills and well-developed rictal bristles. Most are grey or grey-green above and greyish below, with contrasting coloured heads and throats. The Palearctic species have longer, more pointed wings with very small P10, and slender bills; they are plain olive or greenish above and whitish or pale yellow below. Most inhabit open woodland.

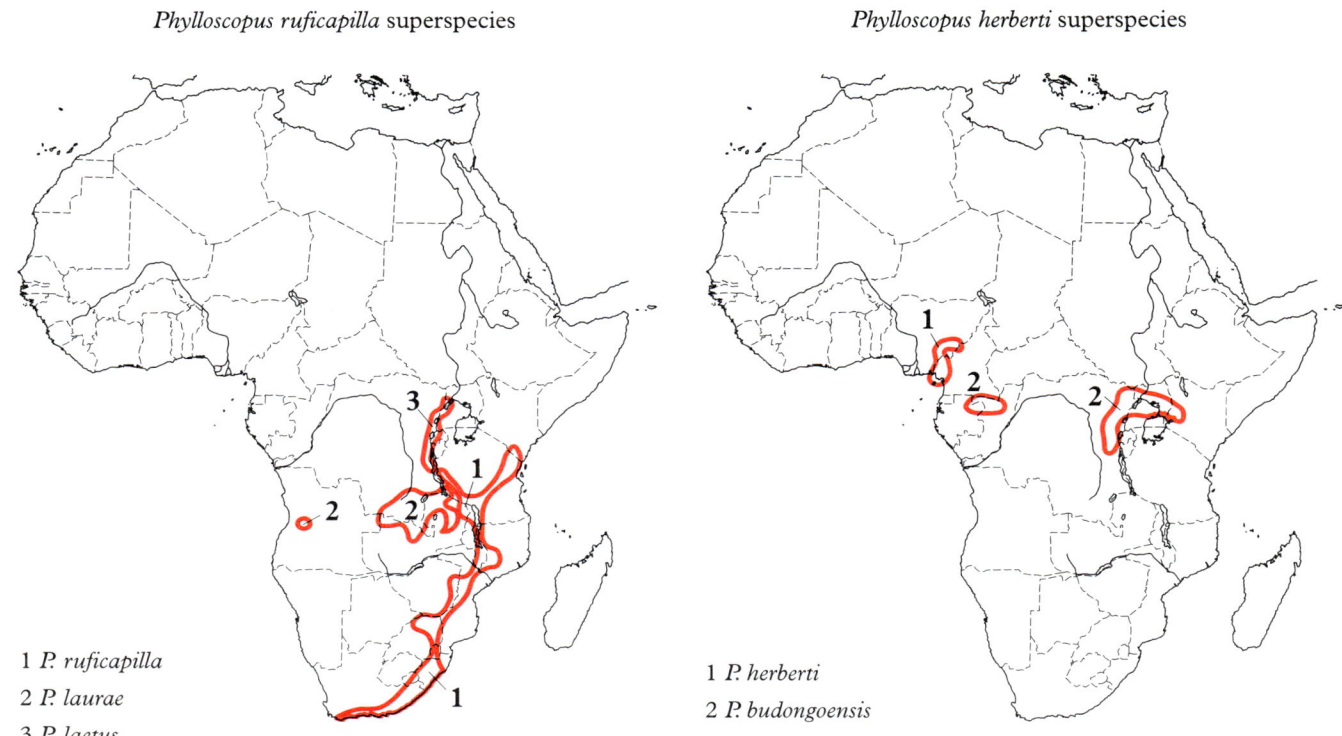

Phylloscopus ruficapilla superspecies

1 *P. ruficapilla*
2 *P. laurae*
3 *P. laetus*

Phylloscopus herberti superspecies

1 *P. herberti*
2 *P. budongoensis*

Plate 21 *Phylloscopus trochilus* (Linnaeus). Willow Warbler. Pouillot fitis.
(Opp. p. 400)
Motacilla trochilus Linnaeus, 1758. Syst. Nat. (10th ed.), p. 188; England.

Range and Status. Europe from *c*. 45° N north to the Arctic, and east through Siberia to *c*. 175° E; winters Africa.

Palearctic migrant. Winters in wooded savanna and forest edge from Gambia, S Senegal, S Mali (north to Niger delta), S Burkina Faso (north to Ouagadougou), SW Niger (Park du W), Nigeria (except extreme N), Cameroon (except N), Central African Republic, Zaïre, central and S Uganda, and W, central and SE Kenya, south to the Cape, except arid W Namibia and W South Africa; mainly south of 11–12° N in W Africa and 1–2° N in E Africa; a few also W and central Ethiopia (north to L. Tana) and SE Somalia (J. S. Ash, pers. comm.); mainly common to abundant. Occasional Dec–Jan records north of Sahara in Tunisia, Algeria and Egypt.

Phylloscopus trochilus

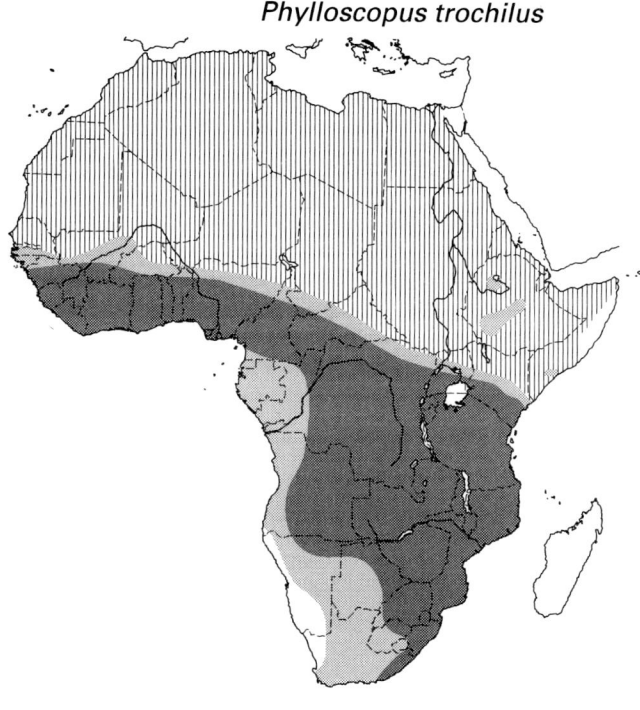

A

On passage in N Africa, common to abundant in autumn in Morocco and Egypt, frequent to common Algeria to Libya; abundant throughout in spring, though with fewer in Tunisia and Libya. Common to abundant both seasons in Saharan oases. South of Sahara, locally abundant to very abundant on southward passage Mauritania, Senegal, Mali, Burkina Faso, N Nigeria, Chad, Sudan and W and central Ethiopia; also Uganda, W and central Kenya, Zambia and Malaŵi. Often abundant to very abundant on northward migration in Uganda, Kenya, Ethiopia, Eritrea, Djibouti and N Somalia; and in W African soudanian and sahelian savanna belts. Commonly oversummers Mali in Niger delta and Sahel to north of 18° N.

Description. *P. t. trochilus* (Linnaeus): Europe north to S Sweden, east to S Poland and Romania; winters W Africa east to Cameroon. ADULT ♂: entire upperparts olive-brown, tinged yellowish green. Tail-feathers dark brown, edged olive-green. Supercilium yellow; lores and indistinct eye streak dusky brown; cheeks and ear-coverts brownish, streaked pale yellow. Chin to belly whitish streaked with varying amount of yellow, especially on throat and breast, often with pale brownish streaks and wash at sides of breast; flanks pale yellowish buff; undertail-coverts pale yellow. Primaries and secondaries dark brown, outer webs edged olive-green; tertials and upper-wing-coverts brown with broad olive-green fringes and tips. Bend of wing, axillaries and underwing-coverts yellow. Wing-tip bluntly pointed; P7–P8 longest, P6 1–3 shorter, P5 5–8; P1 13–18; P9 4–7, usually falls between P5 and P6; P10 very small; P6–P8 emarginated (**A**). Plumage colour variable; yellow and green pigment is suppressed in some individuals and populations, and streaking on underparts reduced. Bill brown, base of lower mandible pale; eyes hazel; legs and feet pale brown to blackish brown. Sexes alike. SIZE: (10 ♂♂, 10 ♀♀, W Europe) wing, ♂ 62–70 (66·3), ♀ 60–65 (62·4); tail, ♂ 49–53 (50·7), ♀ 46–51 (48·1); bill, ♂ 12–13 (12·9), ♀ 12–13 (12·5); tarsus, ♂ 19–21 (20·3), ♀ 19–20 (19·5). WEIGHT: unsexed: (n = 13, Algerian Sahara at *c.* 30° N, autumn) 8·2–12 (9·0), (n = 119, Morocco, Mar–May) 6–10·3 (7·4), (n = 21, Ghana, Nov–Dec) 6·5–10·8 (7·8); N Nigeria, presumed this race and *acredula* (n = 127, autumn) 6·5–10 (8·0), (n = 142, spring) 7–12·8 (9·7).

2 complete moults per year, one on breeding grounds July–Aug, the other in Africa Dec–Feb.

IMMATURE: 1st winter like adult but with almost uniform canary yellow underparts.

P. t. acredula (Linnaeus): Norway, N and central Sweden, NE Europe and Russia east in Siberia to R. Yenisei; winters E and southern Africa, west to Cameroon, W Zaïre and rarely Mauritania. Highly variable; many impossible to distinguish individually from birds of nominate race. Typically paler and browner above, often with greyish streaking; whiter below with yellow streaking reduced or almost absent, but some have bright greenish streaking above and yellow-streaked breast. Slightly larger than nominate: wing (10 ♂♂, N Scandinavia and Russia) 67–71 (68·7). WEIGHT: unsexed: (n = 43, Egypt, Aug–Sept, presumed mostly this race) 7·1–11·1 (9·2); Ethiopia, mostly this race (n = 127, Jan–Mar) 6·4–15·1 (8·9), (n = 125, Apr–May) 6·3–14·3 (9·9); Kenya and Uganda (n = 244, Sept–Feb) 6·3–11·3 (8·1), (n = 500, Mar–May) 6·5–14·6 (8·6); (n = 34, Zambia, Feb) 8·2–12.

P. t. yakutensis Ticehurst: E Siberia from Taimyr peninsula east to Anadyr; winters E and southern Africa west to Zaïre. Differs from *acredula* in having upperparts grey-brown with only a trace of olive-green on rump and edges of wing and tail-feathers; underparts dull white, greyer on breast; supercilium white, more pronounced than in nominate race; no yellow except at bend of wing and on axillaries, underwing-coverts and tibia. 1st winter bird lacks yellow below. The largest race: wing (18 ♂♂) 66–74 (70·1).

Field Characters. A slim, fairly long-winged leaf warbler, olive or brownish above, and tinged yellow below, especially on breast. However, some Siberian birds are greyish, with whiter underparts. Pale supercilium fairly distinct. Distinguished from similar Common Chiffchaff *P. collybita* by diagnostic song, paler (usually brown rather than blackish) legs, and more noticeable pale base to lower mandible. Typically less buffy than Common Chiffchaff and more yellowish below, especially young birds in autumn. In hand, separated by longer and more pointed wing and lack of emargination on P5 (**A**). More lightly built than Wood Warbler *P. sibilatrix*, with more slender legs and finer bill; browner, less brightly coloured, without sharp colour contrast between breast and belly; wing-tip less pointed (**A**).

Voice. Tape-recorded (11, 22, 57, 86, 88, 89, 91). Song, a sweet liquid descending cadence; rather faint at start, becoming louder, notes slower and more confident, then finishing with a quiet flourish: 'se-se-se-se-see-see-su-su-sut-sut-sueet-sueetoo'; lasting *c.* 3 s. Contact-alarm call, a plaintive 'whooeet', similar to that of Common Chiffchaff but more disyllabic.

General Habits. Winters in a wide range of habitats; chiefly woodland, wooded savanna, secondary growth, gardens, disturbed lowland forest, edges of montane forest and cultivated areas; swamp vegetation and tall grass country in Sierra Leone. Partial to acacia and *Brachystegia*, and occupies upper layers of miombo woodland. Typically occurs up to 2200 m, but occasionally reaches tree heath zone above 3000 m. On passage, occurs also in more arid bush, desert scrub, lake edge vegetation and mangroves. Often solitary and dispersed in winter quarters, but more typically in small loose groups, sometimes 20–30 in a single tree; sometimes gathers in hundreds during migration. Commonly joins mixed-species flocks. At Naivasha, Kenya, *c.* 15 birds seen per ha in each of 3 different habitats (Rabøl 1987). Some birds remain attached to one site for weeks, but little aggressive or territorial behaviour noted. The relatively low retrap rates suggest considerable local movement during winter. Easily observed foraging in tree canopy and tops of bushes; also in low scrub and at times among grass stems and on ground. Gleans from foliage, and darts out to catch flying insects. In Tsavo, Kenya, 81% of prey taken from leaves, 9% from air, 3% from twigs; mean foraging height 5·2 m (Lack 1985). Carriage typically horizontal in tree canopy. Flicks wings and regularly moves tail up and down. Flight quick, light and agile. Sings commonly in Africa from sunrise to middle of day, in Oct–Nov and again in Feb–Mar after moult.

Moults completely in Africa as well as on Palearctic breeding grounds; typically Dec–Feb W Africa, Jan–Mar E Africa, late Dec–Feb southern Africa. Winter moult relatively slow, duration *c.* 68 days Uganda, *c.* 52 days S-central Africa and *c.* 49 days South Africa (Underhill *et al.* 1992).

Nominate birds from W and central Europe and S Scandinavia migrate southwest in autumn to enter Africa mainly through Morocco, while N Scandinavian and Finnish *acredula* head south-southeast towards Egypt (Hedenström and Pettersson 1987). Birds from further east (Russian *acredula* and *yakutensis*) require progressively more westerly heading to reach NE Africa. Main passage in Morocco, mid-Sept to mid-Oct; Egypt and N Sudan, late Aug to late Sept with stragglers to Nov. Birds grounded in stony desert in N Egypt were very fat (Biebach *et al.* 1986). Heavy passage occurs through sahelian and soudanian savanna belts: Senegal in early Sept to mid-Oct, Mali from late Aug, Chad in mid-Sept to mid-Oct; also through Ethiopia early Sept to mid-Nov. Arrives Ghana and Nigeria from mid-Sept, Cameroon and Gabon from Oct. Some move south quickly, reaching Uganda, Zambia and Zimbabwe by mid-late Sept; 1 bird ringed Finland already in South Africa 30 Sept. However, main passage in E Africa is in early Oct to mid-Dec, Zambia and Malaŵi mid-Oct to mid-Dec. Northward movement begins southern Africa from late Feb, Kenya and Uganda from mid-Mar. Main E African passage late Mar to late Apr, but a few birds still present southern and E Africa to early May. Large numbers pass through Kenya, Ethiopia, Eritrea and N Somalia, fewer through Uganda and Sudan; migration apparently centred further east than in autumn. W African wintering areas vacated by early April. Passage in Senegal begins mid-Mar and peaks late Mar–Apr; present Mali to late Apr to early May. Spring passage in N Africa takes place on a broad front mid-Mar to mid-May, with last birds early June. Ringing suggests Fennoscandian birds tend to use more easterly route across or around Mediterranean than in autumn (Hedenström and Pettersson 1987).

Many birds ringed in Europe recovered N Africa: over 40 from Britain and W Europe to Morocco (spring and autumn), 11 from Britain, France, Holland and Denmark to Algeria, and 3 from Belgium, Denmark and Sweden to Tunisia; also over 30 from Finland, 3 from Sweden and 1 from Denmark to Egypt (mainly autumn). 1 ringed in Algeria recovered Denmark, 1 ringed Tunisia recovered Sweden. South of Sahara, recoveries from Britain to Mauritania (3), Senegal (2), Mali (1), Ivory Coast (4) and Ghana (1); France to Mauritania (1), Burkina Faso (1) and Nigeria (1), Spain to Nigeria (1); Holland to Ghana (1); Germany to Niger (1) and Malaŵi (1); Norway to Ivory Coast (12) and Togo (1); Sweden to Ghana (1), Togo (1), Niger (1), Nigeria (2), Cameroon (1), Congo (1), Zaïre (4), Central African Republic (1), Uganda (1), Zambia (1) and Zimbabwe (1); Finland to Zaïre (11), Uganda (2), Zimbabwe (3) and South Africa (2). 1 ringed Nigeria (Nov) recovered in Belgium, and 1 ringed Kenya (Apr) recovered Finland.

Ringing evidence thus suggests nominate race winters mainly or entirely in W Africa; it is also reported as common in southern Africa (e.g. Clancey 1970c) but yellow-streaked individuals are difficult to assign. Birds examined from Zaïre, Rwanda and Burundi were all considered *acredula* (Prigogine 1982). Finnish birds (*acredula*) were recovered in Zaïre basin mostly during migration season, and this population probably winters mainly in southern Africa. Russian *acredula* presumably all migrate to E, central and southern Africa. E Siberian *yakutensis* comprise *c.* 5% of birds in Kenya and Uganda and perhaps over 10% in southern Africa (Clancey 1970c), and range west to Zaïre and Namibia.

Food. Insects and their larvae; spiders; small worms. Few details for Africa.

References
Cramp, S. (1992).
Hedenström, A. and Pettersson, J. (1987).
Underhill, L. G. *et al.* (1992).

Phylloscopus collybita (Vieillot)*. Common Chiffchaff. Pouillot véloce.

Sylvia collybita Vieillot, 1817. Nouv. Dict. d'Hist. Nat. II, p. 235; France.

Plate 21 (Opp. p. 400)

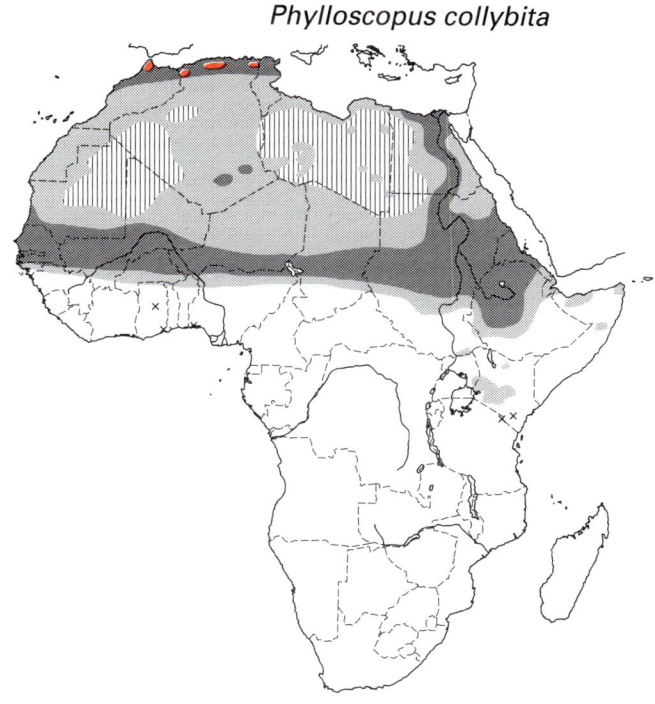

Phylloscopus collybita

Range and Status. Europe and NW Africa, Russia north to the Arctic and east through Siberia to *c*. 150° E, and N Kazakhstan; also N Turkey, Caucasus and N Iran. Mainly migratory, wintering Africa, S Europe and SW and S Asia.

Palearctic migrant. Also scarce local breeder, probably resident, highlands of Morocco (Tangier peninsula) and N Algeria (near Tlemcen, near Blida, probably Grand Kabylie and near El Kala; singing June at Babors and near Oran). Winters south to 11–12° N in W and to equator in E. Common to abundant north of Sahara; in Saharan oases and along Nile; and in narrow belt south of Sahara, to Senegal, Gambia, Mali, Burkina Faso, Niger, N Nigeria (south to Zaria), N Cameroon (Waza), central Chad, central Sudan, W and central Ethiopia and Eritrea; locally also Djibouti and N Somalia (north of 7° N). Rare Ghana (Yegi, Accra, perhaps overlooked in N) and once S Nigeria (Lagos). Locally frequent to common in E African highlands, in SE Sudan, E Uganda (Elgon) and N, W and central Kenya; once N Tanzania (Kilimanjaro). Occasionally at lower altitudes in Kenya and Uganda in Dec and Mar, south to Nyanza and Ngulia, presumably on passage. Pronounced autumn and spring passage movements in N Africa and Saharan oases. A few birds oversummer in Mali.

Description. *P. c. collybita* (Vieillot): Europe from Pyrenees north to Denmark and east to Poland and Romania; occurs throughout N and W African range, and reported Sudan, Eritrea and Ethiopia. ADULT ♂: entire upperparts brownish olive, tinged greener on rump. Tail-feathers dark brown, edged olive-green. Supercilium yellowish white, not well pronounced; narrow eye-ring creamy white; lores dusky; cheeks and ear-coverts mottled buff and olive. Below, chin to breast and flanks tinged buff and variably streaked yellow, belly whitish, undertail-coverts yellowish white. Primaries and secondaries dark brown, narrowly edged olive-green; tertials and upperwing-coverts dark brown, edged and tipped olive-green; bend of wing, axillaries and underwing-coverts pale yellow. Wing-tip rounded; P6–P8 longest, P5 1–3 shorter, P4 3·5–6, P1 9–12; P9 5–8, usually falls between P5 and P6; P10 very small, 4–9 longer than primary coverts; P5–P8 emarginated (**A**). Bill dark horn, paler at base of lower mandible; eyes dark brown; legs and feet dark brown to blackish. Sexes alike. SIZE: (10 ♂♂, 10 ♀♀, Europe) wing, ♂ 58–63 (59·7), ♀ 54–61 (56·3); tail, ♂ 47–51 (48·1), ♀ 44–51 (46·6); bill, ♂ 11–13 (11·9), ♀ 11–13 (12·0); tarsus, ♂ 19–20 (19·8), ♀ 19–20 (19·6). WEIGHT: unsexed: Algerian Sahara, presumed mostly this race (n = 20, autumn) 6·7–10·9 (8·3), (n = 54, winter) 6·5–9·5 (7·8); (n = 17, Morocco, Mar–May) 5–7·9 (6·3); Senegal (n = 132, Oct–Nov) 5·6–9 (7·0), (n = 259, Dec–Feb) 5·5–10·5 (7·0).

Complete moult on breeding grounds; pre-nuptial body moult in Africa.

IMMATURE: juvenile browner above than adult, more yellow below. 1st winter like adult.

NESTLING: short sparse down on upperparts only, dark grey. Mouth dull yellow, no tongue-spots; gape-flanges pale yellow.

P. c. abietinus (Nilsson): Norway and central Sweden east to Urals, and Caucasus to N Iran; winters SE Europe, SW Asia and throughout African range, reaching 3° S in E. Slightly greyer, cooler-toned above than nominate, whiter (less yellow and buff) below; slightly larger, wing of 10 ♂♂ 59–66 (62·9). WEIGHT: unsexed: Ethiopia, presumed mostly this race (n = 50, Oct–Feb) 5·7–10·5 (7·2), (n = 84, Mar–Apr) 5·5–10·3 (7·5).

P. c. tristis Blyth: Russia from Urals eastwards, and NE Iran; recorded once Sudan. Paler, greyish olive above; white below with pale greyish olive sides to breast and flanks.

P. c. brehmii (van Homeyer): SW France, Iberia, N Morocco, N Algeria. Like nominate race but upperparts tinged more yellowish olive; underparts with more yellow streaking, belly whiter; undertail-coverts deeper yellow. Voice distinct.

Field Characters. A slim brownish leaf warbler with pale underparts. Very similar to Willow Warbler *P. trochilus*, but slightly more compact with shorter wing. Best separated by darker, usually blackish, legs, darker

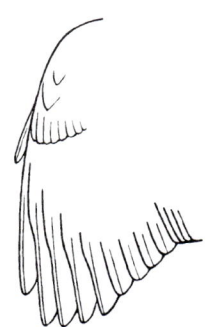

A

---

*TAXONOMIC NOTE added in press: Helbig *et al.* (1996) have found genetic and bioacoustic evidence that '*P. collybita*' comprises several species. Those in Africa are *P. c. collybita/abietinus*, *P. brehmii* and a semispecies *P. (?c.) tristis*.

base to bill and buffier underparts; crown appears slightly rounder, eye-ring more noticeable. Song diagnostic, but call similar to Willow-Warbler. In hand, P9 shorter than in Willow Warbler and P5 emarginated (see *P. trochilus*, **A**). Brown Woodland-Warbler *P. umbrovirens* is richer brown on breast and flanks, with strongly contrasting green wing- and tail-edging and much shorter wings.

Voice. Tape-recorded (62, 73, 89, 92, 93, B, C, F, CHA). Song (of *collybita* and *abietinus*) a steady repetition of 2 notes in irregular sequence, 'chif-chaf-chif-chif-chaf-chif-chaf . . .', delivered at *c.* 2 notes per s for up to *c.* 15 s; often preceded by quiet churring notes. Song of *brehmii* less pure-toned and structurally different, with introductory notes and a middle section accelerating into a short trill, 'tit-tit-tit-tit-tswee-tswee-chit-it-it-it-it', lasting *c.* 4 s. Contact-alarm call, a soft plaintive 'hweet', like similar note of Willow Warbler but more monosyllabic; also a shorter shriller down-slurred 'psiu', more characteristic of *brehmii*.

General Habits. Breeds in Algeria in high oak forests. Winters in a variety of habitats: bushes and scrub, oases and vegetated wadis, large trees near water, patches of *Acacia nilotica*, willow thickets and even mangroves; in N Africa, palm gardens and stream edges; from sea level to 3700 m. In E Africa, *abietinus* winters in montane forest and bushy forest edge, at higher altitudes than Willow Warbler, typically at 2200–3500 m, up to tree heath zone.

Often solitary, but commonly also in parties, especially on migration. In N Sudan, occurs in mixed groups with Lesser Whitethroat *Sylvia curruca* and Olivaceous Warbler *Hippolais pallida* in acacia trees (Hogg *et al.* 1984). At Tamanrasset oasis, Algeria, *c.* 2000 were present during Feb in 2 km², foraging in loose parties of 5–10 (Gaston 1970). Wintering birds may remain localized, but territorial behaviour not pronounced. Population apparently dispersed and stable in N Senegal in Nov–Feb (Morel and Roux 1966). Fidelity to wintering site documented by ringing in Morocco and Senegal.

Forages actively in canopy, moving from tree to tree, often in mixed-species parties; also in bushes, edges of low scrub and thickets and at times on ground. Picks prey from foliage and twigs, sometimes while hovering, and darts out to take flying insects. Often quite tame. Carriage usually horizontal. Frequently flicks wings and lowers tail and twitches it sideways (Cramp 1992). Flight light and agile, rather more fluttery than Willow Warbler (Cramp 1992). Sings throughout winter.

Birds from W and central Europe migrate in autumn towards NW Africa, those from E Europe towards E Mediterranean. In N Scandinavia (all *abietinus*) there is a clear migratory divide, with Swedish and Norwegian birds moving southwest, Finnish birds southeast (Zink 1973). Main passage and arrival N Africa in late Sept to early Nov. First birds reach Mali and Nigeria late Sept. Main movement through Senegal in Oct (earliest 20 Sept), and wintering population stabilizes early Nov (Morel and Roux 1966). Main passage and arrival in N Sudan mid-Oct to early Nov. Present in Ethiopia in late Oct, but not until Dec in highlands of S Sudan, Uganda and Kenya. Spring movement begins Feb, but is mainly Mar to early Apr. Birds leave E African highlands in early to mid-Mar, Senegal by early Apr, Mali and N Sudan by Apr, Ethiopia by early May. In SE Morocco, local wintering population leaves in Apr, after passage of most birds from further south. Migrates through Egypt mid-Feb to late Apr, with peak in Mar, latest 18 May.

Ringing recoveries in N Africa demonstrate south-southwesterly migration direction of most W European birds: over 100 in Morocco (mainly from Britain, France, Belgium, Holland and Germany) and 50 in Algeria (mainly from France and Germany, 1 from Sweden and 1 Denmark), but only 3 in Tunisia (from France and Germany) and 3 in Egypt (from Germany, Poland and Sweden). Afrotropical recoveries are mostly in extreme W: Britain to Mauritania (1), Senegal (22), Gambia (3) and Mali (1); France to Senegal (4); Belgium to Mauritania (1); but 1 ringed Finland was recovered in Uganda. *P. c. collybita* winter as main race in Libya and Egypt, and these are presumably mostly from SE Europe. Most birds wintering in Sudan and Ethiopia and all in E African highlands are *abietinus*, presumably from E Baltic and W Russia.

Food. Mainly insects and their larvae; small spiders; occasionally small molluscs, seeds and berries. Autumn birds in Senegal fed on beetles and flies (Morel 1968).

Breeding Habits. Monogamous; occasionally polygynous. Solitary and territorial. Territory 0·1–1 ha Europe; used for courtship and nesting, but birds often feed outside it. ♂ typically sings from high canopy while moving about territory. ♀ remains low and feeds in herb layer. During courtship ♂ chases ♀, giving excited song variant and snapping bill; may descend to ♀ in floating butterfly flight, with wings spread and flapping slowly.

NEST: roughly spherical, domed, with wide side entrance; of grass, dead leaves and moss, lined with finer grasses and feathers; on or close to ground in tall herbaceous growth or low shrub; ext. depth 110–140, ext. diam. 80–100; site chosen and nest built in 4–10 days by ♀, accompanied and guarded by ♂.

EGGS: usually 5–6 Europe; glossy white, spotted and speckled dark purple, purplish brown and black. SIZE: (n = 318 Europe) 13–18 × 11–14.

LAYING DATES: Algeria, early May. 2 broods common in Europe.

INCUBATION: starts when clutch complete; by ♀. Period: 13–14 days.

DEVELOPMENT AND CARE OF YOUNG: feeding and nest sanitation mainly by ♀; young remain in nest 14–16 days; after fledging brood sometimes divided between parents; young independent after 10–19 days.

BREEDING SUCCESS/SURVIVAL: of 235 eggs Switzerland, 44% produced fledged young (Cramp 1992).

Reference
Cramp, S. (1992).

Phylloscopus sibilatrix (Bechstein). Wood Warbler. Pouillot siffleur.

Plate 21
(Opp. p. 400)

Motacilla sibilatrix Bechstein, 1793. Naturforscher, Halle, 27, p. 47; Thuringia, Germany.

Range and Status. Europe, from Britain, France, Italy and Yugoslavia north to the Baltic, and east through S Russia to *c*. 90°E; winters Africa.

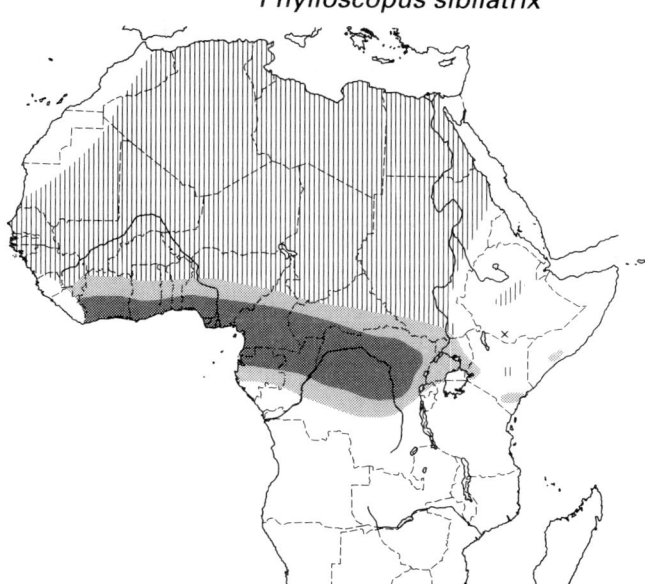

Palearctic migrant. Winters W and central Africa from 9–10°N to *c*. 5°S; from Sierra Leone and S Guinea to Ivory Coast, central and S Ghana, S Nigeria, S Cameroon, Gabon, Central African Republic, Zaïre (south to Kasai and Kivu), SW Sudan (Bangangai, Imatong Mts), Uganda, Rwanda and W Kenya (Kakamega, Mara); locally common E Liberia (Nimba) to Zaïre, but uncommon Sierra Leone, and scarce Sudan, Uganda, Rwanda and Kenya. Presumed wintering birds north of usual range S Senegal (Basse Casamance, Feb) and S Mali (Bamako, Feb). Apparently isolated populations S Somalia (Webi Shebele, frequent) and SE Kenya (Mtito Andei, Mida, uncommon). On passage N Africa, scarce in autumn, mainly Libya and Egypt, common in spring from E Morocco to Egypt; frequent Saharan oases, mainly in spring. South of Sahara, rare on autumn passage in Mauritania and Senegal, and uncommon Mali and Niger; common S Burkina Faso, N Nigeria and central Chad, and locally abundant E Chad (Ennedi) and W Sudan (Darfur); uncommon Sudan coast, rare central Ethiopia, once in S near Yavello (29 Nov 1994: Robertson 1995). Occasional Nov–Dec at Ngulia, SE Kenya, apparently still on passage. On northward migration, rare in Senegal, Mali, Burkina Faso and NW Liberia (Wonigizi Mts, single bird 21 Mar: Dickerman *et al.* 1994), but frequent to common E Liberia, N Ghana, N Nigeria and central Chad; once central Kenya (Mt Kenya, Apr) and once Ethiopia (May).

Description. ADULT ♂: entire upperparts yellowish olive-green. Tail-feathers dark brown edged yellowish green. Supercilium yellow and conspicuous; lores and eye-streak dark olive; cheeks and ear-coverts greenish yellow. Below, throat and upper breast yellow, sharply demarcated from white lower breast, belly and undertail-coverts; tibial feathers pale yellow. Primaries and secondaries dark brown edged yellowish green; tertials and upperwing-coverts dark brown, tipped and edged yellowish green; axillaries and underwing-coverts bright yellow, mixed with white. Wing-tip bluntly pointed; P8 longest, P7 0·5–2 shorter, P6 4–6, P5 8–11·5, P1 20–22; P9 1·5–5 shorter, falls between P6 and P8; P10 minute, usually shorter than primary coverts; P6–P8 emarginated (**A**). Bill dark brown above, yellowish flesh below; eyes blackish brown; legs and feet usually pale yellowish brown. Sexes alike. SIZE: (10 ♂♂, 10 ♀♀, Europe) wing, ♂ 74–80 (77·4), ♀ 70–74 (72·0); tail, ♂ 51–53 (52·2), ♀ 45–50 (47·4); bill, ♂ 13–15 (13·9), ♀ 13–14 (13·8); tarsus, ♂ 18–20 (19·1), ♀ 18–19 (18·4). WEIGHT: unsexed: (n = 43, Morocco, Mar–May) 6·8–10·2 (8·4); N and central Nigeria (n = 93, autumn) 6·4–10·5 (8·4), (n = 19, spring) 8·1–15 (11·3); (n = 8, Kenya, Nov–Jan) 8–9·8 (9·0).

Complete moult in Africa Dec–Mar; post-nuptial body moult on breeding grounds.

IMMATURE: 1st winter like adult, but wing and tail-feathers fresh and brightly edged until Oct.

Field Characters. A compact leaf warbler, greenish above, with a prominent yellow supercilium, and yellow throat and breast contrasting with white belly. More brightly coloured than Willow Warbler *P. trochilus*, with longer wings, shorter tail, stronger bill, thicker, pale coloured legs, and different voice. In hand, P9 longer than in Willow Warbler, P10 shorter (see *P. trochilus*, **A**). Yellow-throated Woodland-Warbler *P. ruficapilla* superficially similar in colour, but smaller and dumpier with much shorter wing and brown crown.

Voice. Tape-recorded (57–62, 73, 93, B, C, F). 2 songs; one a high-pitched staccato introduction 'vit-vit-vit-vit . . .' accelerating into a shivering descending trill; lasts 2–3 s. Other one a series of *c*. 5–15 melancholy piping notes, increasingly drawn out, 'püü-püü- . . . püüü-püüü', given less frequently than trilling song. Contact-alarm call, a single plaintive 'püüü'.

General Habits. Winters in humid lowland habitat near the equator; forest, especially edges and clearings, and more scattered trees with open spreading foliage; also tall secondary growth, moist thickets, gardens,

A

plantations, riverine *Ficus* canopy in Somalia, and even coastal mangroves; generally below 1400 m. May occur on passage in more open savanna, and as high as 2000 m in E Africa.

Usually solitary in winter quarters, but sometimes in small groups. Locally numerous W Africa, e.g. tens per ha Mt Nimba (Brosset 1984). No territorial behaviour noted. Quiet and unobtrusive, more confined to broadleaved canopy than other Palearctic *Phylloscopus* spp. Gleans actively; hovers to pick food from underside of leaves; flies out to take aerial prey (Cramp 1992). Moves through foliage more slowly than Willow Warbler, carriage markedly horizontal, long wings often drooping (Cramp 1992). Does not flick wings or tail. Song heard in Ghana from Nov onwards, but is irregular in most wintering areas and mainly in Feb–Mar.

Main autumn migration from Europe is across central and E Mediterranean. Passage N and NE Africa mid-Aug to early Oct, but sparse, suggesting most birds fly direct to tropics. Passage south of Sahara Sept–Nov more substantial, from Burkina Faso to W Sudan, but few records extreme W Africa or E Africa. Arrives main wintering area Ivory Coast–Zaïre Oct–Nov; Somalia from late Nov. Vacates winter quarters late Mar–Apr. Northward passage noted N Ghana, Nigeria and Chad, but most birds apparently take off south of soudanian–sahelian belt. N African passage much heavier than in autumn, extending west to Morocco, mainly early Apr to mid-May.

Birds ringed Britain (1), Belgium (1) and Russia (1, St Petersburg) were recovered in Algeria (Apr–May), and 1 ringed Germany in Morocco (Apr). 1 ringed Tunisia (Apr) was recovered in France. South of Sahara, recoveries from France to Cameroon (1, Nov), Germany to Burkina Faso (1, Apr) and Yugoslavia to Zaïre (1, Nov). 1 ringed N Nigeria (Aug) recovered in Greece.

Food. Insects and other invertebrates; no information for Africa.

References
Cramp, S. (1992).
Fouarge, J. G. (1968).

Plate 21
(Opp. p. 400)

Phylloscopus bonelli (Vieillot). Bonelli's Warbler. Pouillot de Bonelli.

Sylvia bonelli Vieillot, 1819. Nouv. Dict. d'Hist. Nat., nouv. éd., 28, p. 91; 'Piémont'.

Range and Status. NW Africa, W and central Europe north to S Germany, Balkans, W Turkey and Levant; winters Africa.

Palearctic migrant. A common breeding visitor to massifs and highlands of N and W Morocco (Rif, Middle Atlas, High Atlas) and N Algeria (Atlas Tellien from Mts de Tlemcen east to Djurdjura, Babor and Constantinois; l'Aurès; Djelfa; Monts des Ksours), and frequent in NW Tunisia (Kroumirie, east to Sedjenane and south to Le Kef). On passage N Africa, frequent in autumn in Morocco and Egypt, rare Algeria; common throughout in spring, occurring along coast and Nile, and in small numbers in Saharan oases.

Winters south of Sahara in soudanian–sahelian belt. Locally common S Mauritania (Senegal R.), Senegal, Gambia, S Mali (mainly in Sahel), Burkina Faso (south to 11°N), extreme N Ghana (Navrongo), S Niger (near Niamey), N Nigeria (Sokoto to L. Chad and south once to Zaria) and N Cameroon (Waza); apparently scarce central and NE Sudan (Darfur to Nile and Red Sea coast, south to c. 13°N) and rare Eritrea (once Massawa) and S Sudan (once Malakal). On southward passage, uncommon Sudan coast, frequent to common S Niger (near Zinder) and central Chad (Ouadi Rime). Northward passage noted Chad and Sudan. Occasional birds oversummer in Mali and Egypt.

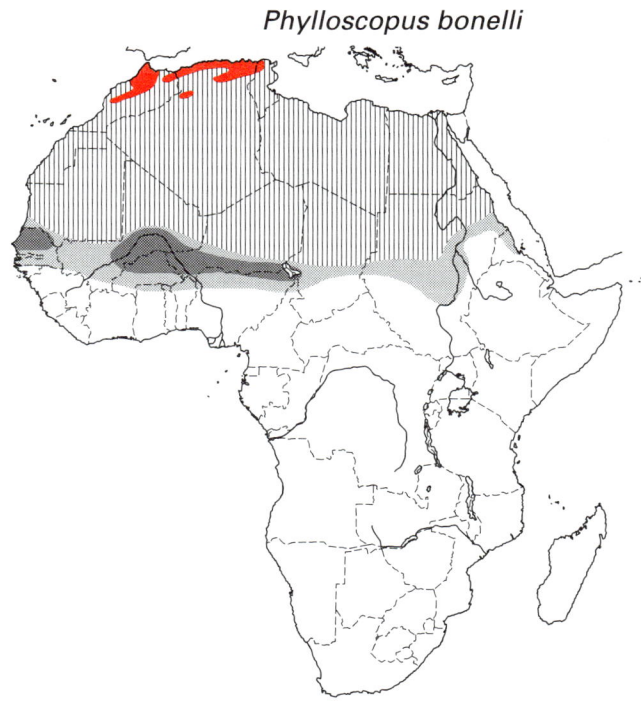

Phylloscopus bonelli

Description. *P. b. bonelli* (Vieillot): W and central Europe, winters Senegal and Gambia to L. Chad. ADULT ♂: upperparts brown, often with slight greenish streaking, rump bright greenish yellow, contrasting with mantle. Tail-feathers dark brown, edged greenish yellow. Supercilium creamy white; lores dusky; cheeks and ear-coverts pale brown. Underparts white, greyer on sides of breast and flanks. Primaries, secondaries and upperwing-coverts dark brown, edged light greenish yellow, forming greenish or golden-brown panel on closed wing; tertials dark brown, fringed and tipped whitish; axillaries, underwing-coverts and bend of wing bright yellow. Wing-

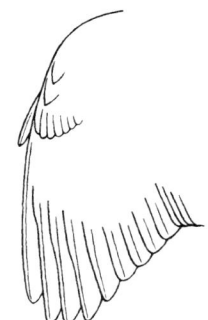

A

tip rather rounded; P7–P8 longest, P6 0·5–3 shorter, P5 3–6, P1 13–17; P9 5–7 shorter, falls between P4 and P5 in length; P10 very small, 3–8 longer than primary coverts; P6–P8 (and sometimes P5) emarginated (**A**). Bill horn above, pinkish below; eyes dark brown; legs and feet dull brown. Sexes alike. SIZE: (10 ♂♂, 10 ♀♀, SW Europe and NW Africa) wing, ♂ 59–66 (63·4), ♀ 57–66 (62·1); tail, ♂ 48–52 (49·4), ♀ 46–52 (47·8); bill, ♂ 12–14 (13·0), ♀ 12–14 (13·1); tarsus, ♂ 19–20 (19·5), ♀ 19–20 (19·1). WEIGHT: unsexed: (n = 31, Morocco, Mar–May) 5·5–8·4 (7·2); (n = 1, N Nigeria, Dec) 8.

Complete moult in Africa in Oct–Dec.

IMMATURE: juvenile darker with brown suffusion above. 1st winter bird like adult but rump less bright and contrasting.

NESTLING: covered with long, sparse grey down above; mouth orange; gape flanges yellow.

P. b. orientalis (C. L. Brehm): Balkans, Turkey and Levant; winters in Sudan (and probably further west). Greyer above than nominate race, underwing-coverts and axillaries paler yellow; averages larger, wing of 10 ♂♂ 64–70 (68·2). WEIGHT: unsexed: (n = 50, Cyprus, spring) 6·5–10·2 (8·3); (n = 4, NE Sudan, Aug) 5·5–6 (5·8).

TAXONOMIC NOTE: recent DNA sequence studies conclude that *bonelli* and *orientalis* should be treated as separate species (Helbig *et al.* 1995).

Field Characters. Similar in size to Willow Warbler *P. trochilus* and Common Chiffchaff *P. collybita*, but rather greyer above and whiter below, supercilium less distinct and sides of head paler. Bright yellow-green patch on folded wing distinctive at close range. Pale yellowish rump often hard to see. Bill longer, less fine than in Common Chiffchaff, more pinkish below. Song distinctive.

Voice. Tape-recorded (57, MOR – 62, 73, 93, B). Song, a loose sibilant trill, recalling that of Wood Warbler *P. sibilatrix*, but shorter (*c.* 1 s), slower and lower-pitched, without acceleration; often repeated several times per min. Call note of nominate *bonelli*, 'hooeet', resembling that of Willow Warbler but more distinctly disyllabic, and a trisyllabic variant 'hooeeoo'; of *orientalis*, an abrupt monosyllabic, rather metallic 'chip' or 'chup'.

General Habits. Breeds NW Africa in a variety of deciduous, pine, cork oak and cedar woodland with ground cover, typically on slopes at 500–2200 m. In Morocco, favours dry sunlit pine woods (Thévenot 1982). Winters in tropics at low altitude in open steppe, bushy savanna with acacia and *Balanites*, dry woodland, dry riverine and lakeside thickets, and in oases. Usually solitary, but spring migrants in Egypt occur in small loose parties. In E Burkina Faso, where abundant in Feb, almost all small trees are defended as territories, birds 50–80 m apart (Chappuis 1978). In Mauritania, 2·6 birds per km^2 in dry steppe (Browne 1982). In Senegal, 20 birds per km^2 in bushy savanna, perhaps more in *Acacia tortilis* woodland, but avoids *A. nilotica* groves used by other migrants (Morel and Roux 1966). Active and restless, foraging mainly in tree crowns and tops of bushes. Gleans from outer foliage, often hovering, and dives to catch flying insects. Sings frequently in Senegal, and at all times of day in N Nigeria in Mar (Elgood *et al.* 1966).

Nominate *bonelli* appears to head west of S initially in autumn, and *orientalis* east of S. Passage concentrated through NW Africa (early Sept to early Nov) and Egypt (mid-Aug to late Sept), but low numbers suggest overflying; thin passage also on Sudan coast in late Aug to early Sept. Reaches tropics early: Mali from late Aug, Senegal from early Sept, Niger from mid-Sept, Nigeria early Oct; rapid passage Chad Sept to early Oct. Western birds winter Senegal to L. Chad, but migrations of *orientalis* poorly known; a few winter in Sudan but others perhaps further west, since this race occurs in spring in Tunisia, W Libya and Malta. Spring migration begins late Feb to early Mar, but some birds remain in most wintering areas until early to mid-Apr. Widespread passage over Sahara in mid-Mar to mid-May. Marked spring passage in N Africa, on wider front than in autumn; in N Morocco late Mar to mid-May (peak late Apr), Egypt early Mar to mid-May (peak late Mar to early Apr). Arrives Moroccan breeding grounds April. A bird ringed Spain was recovered in Morocco in spring; 1 ringed Israel in spring recovered Egypt in autumn.

Food. Mainly insects, spiders and small snails. Entirely insects in Senegal (Morel 1968). Stomachs of birds in Mali Oct–Nov contained only small adult and larval insects (Bates 1934).

Breeding Habits. Monogamous; solitary and territorial. Territories vary greatly in size; often >100 pairs per km^2 S Europe; 8–32 pairs per km^2 Morocco in various woodland types (Thévenot 1982). ♂ sings nearly all day to advertise and defend territory, typically while moving about in tree crown; may fly with vibrating wings from one tree to the next. Neighbouring ♂♂ perform song-duels, and threaten in crouched posture with wings drooped and tail fanned; this may lead to fights. ♀ usually enters territory inconspicuously, keeping to low cover; chooses ♂ and nest-site simultaneously (Cramp 1992). Copulation takes place in vegetation; perched ♀ solicits in crouched posture, with wings drooped and fanned tail quivering (Cramp 1992).

NEST: domed, with flattened oval side entrance; of dried grasses, leaves and moss, lined with finer grasses and rootlets; often camouflaged with materials such as pine needles; usually on ground in bank or depression; ext. depth *c.* 60, ext. diam. *c.* 110, entrance hole 35 ×

25, depth of cup below hole 30. Built by ♀, accompanied by ♂, in *c.* 5 days; material gathered within 150 m (Cramp 1992).

EGGS: NW Africa 3–5; mean (n = 21 clutches) 4·43. In Europe 4–6. Laid daily, starting 1–4 days after nest completed. White, heavily spotted reddish or purple-brown. SIZE: (n = 64, Europe) 14–17 × 11–13 (18·8 × 13·1).

LAYING DATES: NW Africa, mid May–early July; single-brooded.

INCUBATION: by ♀ alone, beginning with last egg. Period: 12–15 days.

DEVELOPMENT AND CARE OF YOUNG: young fed by both parents; brooded by ♀ for 7–8 days; leave nest at 12–13 days. After fledging, brood may be divided between parents.

Reference
Cramp, S. (1992).

Not illustrated

Phylloscopus fuscatus (Blyth). Dusky Warbler. Pouillot brun.

Phillopneuste fuscata Blyth, 1842. Jour. Asiat. Soc. Bengal, 11, p. 113; Calcutta.

Range and Status. Siberia to Mongolia and Manchuria, and through central China to the E Himalayas; winters S and SE Asia; autumn vagrant Europe and SW Asia.

Palearctic vagrant Morocco (1 trapped near Settat, 11 Feb 1974; Thouy 1978a) and (just outside Africa as here defined) Sinai (1, Sharm el Sheik, Oct 1988; Baha el Din 1996).

Description. *P. f. fuscatus* (Blyth): Siberia to Mongolia and N and NE China; only race likely in Africa. ADULT ♂: upperparts brown to olive-brown, tinged rufous. Tail-feathers dark brown, edged olive. Supercilium long, rather narrow, whitish in front of eye, tinged warm brown above and behind. Dusky brown stripe through lore and behind eye; cheeks and ear-coverts mottled rufous and brown. Underparts creamy white, tinged greyish; sides of breast and undertail-coverts washed tawny-buff, flanks more strongly so. Wing feathers dark brown, edged pale olive-brown. Axillaries and underwing-coverts buffish white. Wing-tip rounded; P6–P7 longest, P5 1–1·5 shorter, P4 4–6, P1 9–11; P8 1–3 shorter, P9 7–10, falling between P1 and P3; P10 small (*c.* 30% P7); P5–P8 emarginated. Eyes brown to dark brown; bill dark brown, base of lower mandible pale yellowish; legs and feet brownish flesh to greenish brown. Sexes alike. SIZE: (10 ♂♂, 10 ♀♀) wing, ♂ 59–65 (62·8), ♀ 56–64 (58·6); tail, ♂ 49–55 (52·8), ♀ 45–55 (49·1); bill, ♂ 12–13 (12·5), ♀ all 12; tarsus, ♂ 22–24 (22·9), ♀ 20–22 (21·3). WEIGHT: (n=8, unsexed, W Europe, autumn) 8·1–12·5 (9·2).

Complete moult near breeding grounds.

IMMATURE: first-winter bird like adult, but upperparts slightly more olive-brown, underparts tinged more yellowish.

Field Characters. A small warbler, brown above and creamy white below with tawny flanks. Long whitish supercilium contrasts with dark eye-stripe, and brown legs are rather long and slender. Lacks the greenish tones on upperparts and wing-edgings of Common Chiffchaff *P. collybita*. Bill slightly longer and more pointed than in Common Chiffchaff with pale base to lower mandible; wings relatively shorter and rounder.

Voice. Tape-recorded (93–96, 98). Call-note of migrants a hard, sharp 'chek' or 'tchk', often given twice.

Phylloscopus fuscatus

General Habits. Well known in Asia. Only marginally arboreal. Migrants frequent scrub, tall grass, low bushes and hedges, and only occasionally enter low canopy (Cramp 1992); often associate with wet habitat. Secretive and skulking, preferring to creep away low rather than fly when disturbed. Feeds mainly on or near ground. Flight light and flitting. May dart after insects but does not hover persistently. Carriage horizontal. Constantly flicks wings and tail (Cramp 1992).

Food. Mainly insects.

Reference
Cramp, S. (1992).

Phylloscopus proregulus (Pallas). Pallas's Warbler. Pouillot de Pallas.

Not illustated

Motacilla Proregulus Pallas, 1827. Zoographia Rosso-Asiat., I, p. 499; Ingoda River, southern Transbaicalia.

Range and Status. Asia from Russian Altai east to Sakhalin, south to N Mongolia; also central China and Himalayas to borders of Afghanistan; winters S and SE Asia; widespread autumn vagrant in W Europe, where the nominate race has occurred annually in Oct–Nov since the 1960s, with marked influxes in some years. Very rare in spring.

Palearctic vagrant, Morocco: 1 at Dulma, near Agadir, 7 Dec 1985 (C. Hjort, *per* M. Thévenot, pers. comm.).

Phylloscopus proregulus

Description. *P. p. proregulus* (Pallas): Russian Altai east to N Mongolia and Sakhalin; only race likely to occur in Africa. ADULT ♂: upperparts bright olive-green; broad lemon-yellow band across rump. Tail-feathers dark brown, edged bright green. Central crown-stripe yellow or yellowish white, bordered by dusky olive lateral stripe. Supercilia long and broad, meeting above base of bill and extending back to side of hindneck, bright yellow in front of eye, paler behind; dark olive line through lores extending back and broadening on side of neck; cheeks and ear-coverts yellow, mottled olive. Underparts dull white, tinged yellow on flanks; chin and undertail-coverts pale yellow. Primaries and secondaries blackish brown, narrowly edged bright green; tertials dark brown with broad yellowish white edges and tips; larger wing-coverts dark brown edged green, greater and median coverts broadly tipped yellow or yellowish white to form conspicuous double wing-bar; lesser coverts bright olive-green; axillaries and underwing-coverts pale yellow. Wing-tip bluntly pointed; P6–P7 longest, P5 1–2 shorter, P4 3–5, P1 7·5–10; P8 0–1 shorter, P9 5–7, falling between P1 and P4; P10 very small; P5–P8 emarginated. Bill blackish brown, base of lower mandible yellowish; eyes dark brown; legs and feet dark olive-brown or blackish brown. Sexes alike. SIZE: (10 ♂♂, 10 ♀♀) wing, ♂ 50–54 (53·3), ♀ 47–53 (49·9); tail, ♂ 37–40 (38·7), ♀ 35–38 (36·3); bill to skull, ♂ 10–12 (11·0), ♀ 10–11 (10·2); tarsus, ♂ 16–18 (17·1), ♀ 16–17 (16·3). WEIGHT: (n = 12, unsexed, W Europe, autumn) 4·5–6·1 (4·9).

IMMATURE: 1st winter like adult.

Field Characters. A very small, restless leaf warbler, mainly green above and white below, with prominent double yellowish wing-bar, strikingly patterned head, and yellow rump. Distinguished from Yellow-browed Warbler *P. inornatus* by yellowish crown-stripe and broad yellow rump patch which may show conspicuously when bird flicks wings or hovers. Hovering to glean insects from leaves, and flitting flight, may suggest a kinglet *Regulus* rather than a *Phylloscopus*.

Voice. Tape-recorded (62, 94–97). Call-note of migrants, a rather soft, usually monosyllabic 'dwee' or 'tweet', less shrill than call of Yellow-browed Warbler.

General Habits. Well known in Asia and as autumn vagrant in Europe. On migration, occurs mainly in woodland and gardens; often in mixed-species parties. Forages in tree tops, also in scrub and bushes. Not shy, but highly active. Picks prey from foliage, often while hovering outside bough or bush-top. Flicks wings and tail when excited.

Food. Mainly insects.

Reference
Cramp, S. (1992).

Phylloscopus inornatus (Blyth). Yellow-browed Warbler; Inornate Warbler. Pouillot à grands sourcils.

Plate 21
(Opp. p. 400)

Regulus inornatus Blyth, 1842. J. Asiat. Soc. Bengal, II, p. 101; near Calcutta.

Range and Status. N Urals and through Siberia to *c.* 170°E, south to N Manchuria, N Mongolia, Tien Shan Mts and W Himalayas; also central China; winters S and SE Asia; widespread autumn vagrant Europe.

Palearctic vagrant N Africa, recorded Egypt (twice Bahig, Oct, once Giza, Mar), Libya (once Serir, Nov), Algeria (twice near El Goléa, and once 60 km south of Ghardaïa, all Oct 1985) and Morocco (once Tamrakht

Phylloscopus inornatus

valley near Immouzer Ida-ou-Tanane, Nov 1988). Vagrant Senegal (1 ringed Podor, Sept 1987; J. Betlem, pers. comm.).

Description. *P. i. inornatus* (Blyth): Siberia south to L. Baikal and N Manchuria; only race in Africa. ADULT ♂: upperparts brownish green, back and rump brighter and more yellowish. Tail-feathers blackish brown, edged green. Crown often with indistinct pale line down centre; supercilium broad, yellowish white, extending to back of crown; narrow dark line through lores to behind eye; ear-coverts mottled greenish brown and yellow. Underparts dull white, tinged creamy, flanks washed brownish green. Primaries and secondaries blackish brown, narrowly edged yellowish green; tertials dark brown, broadly edged creamy white; larger wing-coverts dark brown edged green, greater and median coverts broadly tipped creamy white to produce conspicuous double wing-bar; lesser coverts greenish brown; axillaries and underwing-coverts white tinged yellowish green. Wing-tip bluntly pointed; P6–P8 longest, usually equal, P5 1·5–3 shorter, P4 4·5–6·5, P1 9–12; P9 5–7 shorter, equals P4 or falls between P3 and P4; P10 very small; P5–P8 emarginated. Bill dark brown, base of lower mandible yellowish; eyes dark brown; legs and feet olive-brown or greyish brown. Sexes alike. SIZE: (10 ♂♂, 10 ♀♀) wing, ♂ 54–57 (56·2), ♀ 52–55 (52·9); tail, ♂ 40–44 (41·9), ♀ 36–43 (39·2); bill, ♂ 11–12 (11·3), ♀ 11–12 (11·3); tarsus, ♂ 18–19 (18·3), ♀ 17–18 (17·7). WEIGHT: unsexed: (n = 1, Senegal, Sept) 5·5; (n = 23, Britain, autumn) 5·4–7 (6·4).

Complete moult on breeding grounds.
IMMATURE: 1st winter like adult.

Field Characters. A small, highly active, rather short-tailed leaf warbler, greenish above and dull white below, with conspicuous yellowish white supercilium. Easily distinguished from Willow Warbler *P. trochilus* and Common Chiffchaff *P. collybita* by small size, prominent whitish double wing-bar and creamy fringes to tertials. Legs brown or olive.

Voice. Tape-recorded (62, 73, 94, B, CHA, HAZ). Call note, a fairly loud 'weest', sharper and much shriller than call of Willow Warbler; recalls Coal Tit *Parus ater*.

General Habits. Well known in Asia; also as autumn vagrant in Europe. In winter and on migration occurs in woodland, groves and gardens. Often in mixed-species groups, sometimes with other leaf warblers. Carriage and behaviour typical of the genus. Forages in tree canopy, mainly at middle height among outer foliage, but also in scrub and bushes on migration. Active and restless; picks prey from leaves and twigs, constantly fluttering and flicking wings and tail; darts out to take flying insects, but rarely hovers.

Regularly reaches Europe and the Near East in mid-Sept to early Nov (mainly nominate race), with pronounced influxes some years. This involves mainly 1st year birds, and probably results from displacement in anticyclonic conditions coupled with reverse migration. Exceptionally overwinters in Europe. Rare in spring.

Food. Mainly insects; spiders; other small invertebrates.

Reference
Cramp, S. (1992).

Plate 21
(Opp. p. 400)

Phylloscopus umbrovirens (Rüppell). **Brown Woodland-Warbler. Pouillot ombré.**

Sylvia umbrovirens Rüppell, 1840. N. Wirbelt. Fauna Abyss. Vög., p. 112; Semien, N Abyssinia.

Range and Status. Breeds NE and E Africa and SW Arabia. Resident. Locally common to abundant in highlands: in Eritrea (north to Asmara); Ethiopia (W Highlands, SE Highlands east to Harar, south to Mega); N Somalia (near Sheikh, and Erigavo district); S Sudan (Imatong, Dongatona and Didinga Mts); N and E Uganda (Mts Morongole, Kadam and Elgon); Kenya (W and central highlands, forest islands north to Mts Nyiru, Kulal and Marsabit, and the Chyulus); N and E Tanzania (N highlands south to Mt Hanang, east to N Pares; Uluguru Mts and in W in Gombe Nat. Park (Stanford and Msuya 1995); E Zaïre, SW Uganda and W Rwanda (Rwenzoris, Volcano highlands, Kivu, Nyungwe and Itombwe).

Phylloscopus umbrovirens

Description. *P. u. mackensianus* (Sharpe): S Sudan, E Uganda and W, N and central Kenya. ADULT ♂: upperparts dark, rich olive-brown. Tail-feathers blackish brown, narrowly edged green. Supercilium pale greyish, narrow and indistinct; lores and streak behind eye dark brown; ear-coverts and cheeks paler, warmer brown. Below, chin to breast greyish with faint streaks; belly and undertail-coverts whitish; flanks warm brown. Flight feathers and larger upperwing-coverts blackish brown, broadly edged green; lesser coverts brown; bend of wing greenish yellow; underwing-coverts and axillaries white, tinged yellowish. Upper mandible blackish, lower mandible yellowish brown; eyes dark brown; legs and feet bluish horn or slaty grey. Sexes alike. SIZE: (10 ♂♂, 10 ♀♀) wing, ♂ 57–63 (61·1), ♀ 54–59 (57·1); tail, ♂ 47–50 (48·6), ♀ 40–47 (45·3); bill, ♂ 12–13 (12·7), ♀ 12–13 (12·8); tarsus, ♂ 20–22 (21·0), ♀ 20–22 (20·7). WEIGHT: unsexed (n = 5) 8·5–10 (9·5).

IMMATURE: washed yellowish below.
NESTLING: unknown.

P. u. dorcadichrous (Reichenow and Neumann): SE Kenya (Chyulus); N and NE Tanzania. As *mackensianus* but washed brown on throat and breast; smaller: wing, ♂ (n = 5) 53–62 (57·8).

P. u. umbrovirens (Rüppell): Eritrea, N and central Ethiopia, NW Somalia. Slightly paler above than *mackensianus*, more olivaceous; more generally washed brownish below, but throat and centre of belly whitish; size similar. WEIGHT: unsexed (n = 8) 8·6–9·7 (9·0).

P. u. omoensis (Neumann): W and S Ethiopia. Like nominate, but more greenish brown above.

P. u. williamsi Clancey: N Somalia (Erigavo district). Rather richer and brighter than nominate, with whiter underparts; lacks green edges to upperwing-coverts.

P. u. alpinus (O. Grant): Rwenzori Mts. Darker and greyer above than *mackensianus*, greyer on flanks; generally pale brown below, chin to breast with deeper russet tinge. Larger: wing, ♂ (n = 6) 60–66 (63·7), ♀ (n = 4) 58–63 (60·3). WEIGHT: unsexed (n = 3) 10·5–11.

P. u. wilhelmi (Gyldenstolpe): Zaïre, Rwanda, SW Uganda (Kivu Volcanoes). Like *alpinus* but duller; greyer above and lacks warm tone below, belly whiter; size smaller.

P. u. fugglescouchmani (Moreau): E Tanzania (Uluguru Mts). Similar to *mackensianus* but chin to breast and undertail-coverts rich brown as well as flanks.

Field Characters. A predominantly brown leaf warbler with green-edged wing and tail-feathers. Slightly larger than other resident African *Phylloscopus* warblers but slimmer, with a finer bill. May suggest Common Chiffchaff *P. collybita* or Willow Warbler *P. trochilus*, but has much shorter wings, and is darker and more warmly coloured, with only a faint supercilium. Voice distinctive.

Voice. Tape-recorded (B, C, GREG, KEI, LEM, PEA, ZIM). Song highly musical and rather melancholy; more like that of Palearctic congeners than songs of other African *Phylloscopus*. Variable phrases, usually 2–3 s long. Typically a rapid succession of loud, clear notes, mainly in descending sequences; an alternating 'chicher-chi-cher-chi-cher-chee-chew-chee-chew . . .', a descending glissando recalling structure of Willow Warbler song, though faster in tempo, and more rapid trills recalling Wood Warbler *P. sibilatrix*. Sometimes delivers longer sequences (up to 12–14 s) with continuous cascades of contrasting high and low notes and rising and falling glissandi (Dowsett 1990). Call, 'tee, teewe' (Williams and Arlott 1980); alarm, a sharp 'teewik'.

General Habits. Inhabits montane forest, woodland and scrub, typically at 2100–3600 m, down to 1900 m on N Pares (Cordeiro 1995), especially open *Hagenia* woodland and tree heath; on Rwenzoris frequently in more open *Senecio*, up to 4300 m; in N Somalia in juniper forest at 1500–2000 m. Very common in Rwanda on dry ridges of high mountains dominated by heath, and in *Erica*-dominated marshes (Dowsett 1990). In areas of sympatry with congeners, typically separated at higher altitudes, e.g. in E Zaïre, where Red-faced Woodland-Warbler *P. laetus* also occurs, found mainly above 2500 m, and in the Ulugurus, where it overlaps with Yellow-throated *P. ruficapilla*, found above 2200 m. Solitary or in pairs. Usually located by song, but not difficult to see. Forages mainly in tree canopy and among hanging lichens, but also in tall bushes. Gleans foliage and branches and also darts out to take flying prey. On Mt Kilimanjaro joined mixed flocks of white-eyes and flycatchers (Cordeiro 1994). Movements and horizontal carriage recall Willow Warbler.

Food. Insects, spiders; stomach of 1 bird in Kenya contained diptera, spiders and Lepidoptera larvae.

Breeding Habits. Monogamous; territorial.
NEST: thick-walled, domed structure of dried grass, moss and leaves with large side opening, lined with feathers and plant down; on bank or ground among

roots and dry leaves, attached to stem of small sapling, or in hanging lichens about 1 m above ground.

EGGS: 2–3; white or pinkish, spotted reddish or lavender grey. SIZE: (n = 4, Kenya) 17–19 × 12–13 (17·6 × 12·6).

LAYING DATES: E Ethiopia, Mar–Apr; central Kenya, Jan–Mar, July, Sept, Nov; N Tanzania, Oct–Dec; Rwenzoris, Dec–Jan. Mainly in dry season.

Reference
Dowsett, R. J. (1990).

Plate 21
(Opp. p. 400)

Phylloscopus ruficapilla (Sundevall). Yellow-throated Woodland-Warbler. Pouillot à gorge jaune.

Pogonocichla ruficapilla Sundevall 1850. Oefv. K. Sv. Vet.-Akad. Forhandl., vol. vii, p. 105; 'Caffraria Inf.' = Durban.

Forms a superspecies with *P. laurae* and *P. laetus*.

Range and Status. Endemic, resident. Range fragmented, but locally common from E Africa south to Cape. SE Kenya (Taita Hills) and E Tanzania (S Pare, Usambara, Nguru, Uluguru and Udzungwa Mts); W Tanzania (Kungwe-Mahale Mts); S Tanzania (Mt Rungwe) to N Malaŵi and adjacent NE Zambia (Misuku and Mafinga Mts south to E Nyika and N Viphya plateaux, Ntchisi Mt); highlands of SE Malaŵi (Mangochi Mt, Chikala Hill, Shire highlands, Mulanje Mt) and adjacent NW Mozambique (east to Mt Namuli); highlands along Zimbabwe–Mozambique border; Mt Gorongoza; W Swaziland and South Africa (Drakensberg escarpment to Natal and coastal districts of SE and S Cape Province).

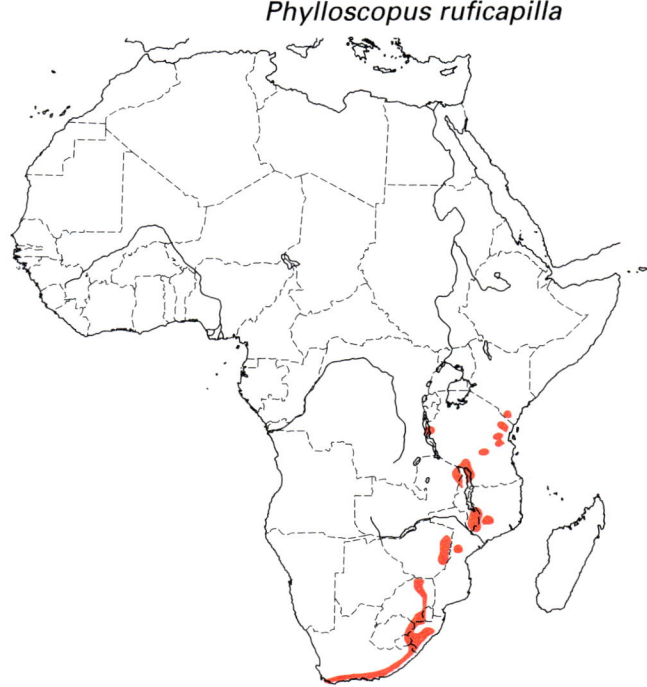

Phylloscopus ruficapilla

Description. *P. r. ruficapilla* (Sundevall) (including '*ochraceiceps*'): Natal and E Transvaal. ADULT ♂: forehead to hindneck bright cinnamon-brown; rest of upperparts, including sides of neck and scapulars, olive-brown. Tail-feathers dark brown, edged bright green. Supercilium bright yellow; lores and streak behind eye dusky; posterior ear-coverts bronzy. Cheeks and chin to breast bright yellow; flanks dull grey; belly creamy; undertail-coverts yellow. Flight feathers dark brown, edged bright green; upperwing-coverts bright green; underwing-coverts and axillaries bright yellowish green. Upper mandible dusky, lower mandible pale yellow; eyes dark brown; legs and feet greenish grey. Sexes alike. SIZE: (10 ♂♂, 10 ♀♀) wing, ♂ 52–56 (54·3), ♀ 52–55 (54·4); tail, ♂ 40–45 (42·4), ♀ 41–43 (42·2); bill, ♂ 12–13 (12·8), ♀ 12–13 (12·8); tarsus, ♂ 20–21 (20·4), ♀ 20–21 (20·6). WEIGHT: (southern African races; unsexed, n = 94) 6·2–9·5 (7·7) (MacLean 1993).

IMMATURE: greener on breast than adult.
NESTLING: downy upon hatching; no black tongue spots.

P. r. voelckeri (Roberts): S and E Cape Province. Differs from nominate race in having yellow extending to lower breast and flanks.

P. r. alacris (Clancey): E Zimbabwe and W Mozambique (Mt Gorongoza). Differs from nominate race in having yellow practically confined to chin and throat.

P. r. johnstoni (W. Sclater): Malaŵi, NW Mozambique, NE Zambia and S Tanzania. Like *alacris* but back washed greenish rather than olive-brown. WEIGHT: unsexed (n = 22, Malaŵi) 7–10·2 (8·5).

P. r. quelimanensis (Vincent): N Mozambique (Mt Namuli). Greyer on back than *johnstoni*, almost without olive; throat paler yellow. Slightly smaller.

P. r. minullus (Reichenow): SE Kenya and E Tanzania. Differs from nominate in having cap olive-brown; upperparts washed olive-green; below, yellow confined to chin and throat, breast grey. WEIGHT: unsexed: (n = 4, Kenya) mean 7·8; (n = 6, Tanzania) 7·7–8·9 (8·3).

P. r. ochrogularis (Moreau): W Tanzania (Kungwe-Mahale Mts). Cap and upperparts as in *johnstoni*, but supercilium, cheeks and chin to upper breast saffron, not yellow.

Field Characters. A small, dumpy warbler with a bright yellow supercilium and yellow throat contrasting with whitish underparts (breast also yellow in southern populations). Upperparts olive, wings and tail tinged bright green. Differs from other green-and-yellow leaf warblers in having a brown or red-brown cap. Allopatric Laura's Woodland-Warbler *P. laurae*, of lower elevations, has green crown and is greyer below. Palearctic Wood Warbler *P. sibilatrix*, with colour of face and underparts superficially similar, is larger and slimmer, with much longer wings and greenish top to head.

Voice. Tape-recorded (32, 58, 75, 86, 88, 91, B, F, HAN, SVEN). Song, clear and high-pitched: phrases of alternating higher and lower notes lasting *c.* 1·5–2 s, separated by short pauses: 'cheeper-cheeper-cheeper-cheeper-chee, chuichi-chuichi-chuichi-chui, chuchi-chichew-chuchichichew-chuchichichew'. Much higher-pitched than song of Red-faced Woodland-Warbler *P. laetus*; note-pattern simpler than in either Red-faced or Laura's Woodland-Warblers, at times recalling Coal Tit *Parus ater*. More rollicking song from Taita Hills (Kenya) not unlike that of some North American wood warblers (Keith and Gunn 1971). Call notes, a plaintive repeated 'tieuu', and a high, wispy 'zit-zit'; alarm, 'peee-trrrrrr'.

General Habits. Inhabits moist evergreen forest, forest edge and woodland. In E Africa, ranges up to 2300 m, including bamboo zone, but occurs as low as 900 m in Ulugurus and E Usambaras and 800 m in Udzungwes. In Malaŵi, occurs at 1200–2350 m in N and 1200–1950 in S, but down to 900 m on S Mulanje; quite common in some submontane forests, e.g. as many as 10 singing ♂♂ per 10 ha, Misuku, less so in drier, more secondary types, e.g. Mangoche (Dowsett-Lemaire 1989). In South Africa, ranges to over 1500 m, but also occurs near sea level in Natal and Cape Province. Usually in pairs or small groups; often in mixed-species parties. Forages actively, mostly in middle to upper strata (at 8–30 m high in Malaŵi), moving from tree to tree; occasionally lower down in shrubby undergrowth. Gleans from foliage, sometimes hovering or fluttering out to take insects in flight (Dowsett-Lemaire 1989).

Some altitudinal and intermontane movements occur. In Malaŵi, wanders down to 600–700 m at foot of Mulanje during Jan–Sept non-breeding season, and to lowland forest near Thyolo (1000–1100 m) in Feb–Aug; has occurred up to 80 km from known breeding forests (Dowsett-Lemaire 1989).

Food. Insects, spiders; stomachs of 2 birds in Tanzania contained Diptera and Lepidoptera larvae.

Breeding Habits. Monogamous.
NEST: domed with entrance hole in side; usually of moss with some plant fibre and cobwebs, sometimes lined with grass and a few feathers; usually in niche in mossy bank, or low down in herbage or bush, often near forest track; one in South Africa was in ground cover of clubmoss *Selaginella*, fern *Adiantum* and grass *Oplismenus*; one Taita Hills, Kenya, *c.* 2 m high in mass of dead pine needles caught on branch, and built of needles.
EGGS: 2–4; white or pinkish, with a few reddish or brown spots. SIZE: (n = 8, southern Africa) 15–17 × 11–13 (16·2 × 12·0).
LAYING DATES: Kenya, Dec; Tanzania, Sept, Nov; Malaŵi, Sept–Oct; Mozambique (juv. Feb); southern Africa, Oct–Dec; apparently mainly in dry season.
INCUBATION: period 17 days (Maclean 1993).
DEVELOPMENT AND CARE OF YOUNG: young remain in nest *c.* 16 days.
BREEDING SUCCESS/SURVIVAL: parasitized by African Emerald Cuckoo *Chrysococcyx cupreus*. Bird ringed as adult recaptured after 4 years 326 days.

Reference
Dowsett-Lemaire, F. (1989).

Phylloscopus laurae (Boulton). **Laura's Woodland-Warbler. Pouillot de Laura.**

Plate 21
(Opp. p. 400)

Seicercus laurae Boulton, 1931. Ann. Carnegie Mus., 21, p. 52; Mt Moco, Angola.

Forms a superspecies with *P. ruficapilla* and *P. laetus*.

Range and Status. Endemic to S-central Africa; resident. Restricted to Mt Moco, W Angola, and to mid-altitude forest patches in SE Zaïre (Upper Shaba), N and NW Zambia (Mwinilunga, Solwezi, Western and Luapala Provinces, Northern Province east to the Muchinga escarpment), and extreme SW Tanzania (Kitungula Forest); locally common.

Description. *P. l. eustacei* (Benson): SE Zaïre, N and NW Zambia and SW Tanzania. ADULT ♂: entire upperparts bright green. Tail-feathers blackish brown, broadly fringed green. Broad yellow supercilium from base of bill to well behind eye; lores and streak behind eye dark green. Ear-coverts, cheeks and chin to upper breast bright greenish yellow; lower breast and flanks greyish, belly whitish; undertail-coverts bright greenish yellow. Flight feathers and upperwing-coverts blackish brown, broadly edged green; underwing-coverts and axillaries whitish, washed greenish yellow. Upper mandible dark horn, lower mandible dull yellowish; eyes dark brown; legs and feet greyish flesh. Sexes alike. SIZE: wing, ♂ (n = 9) 55–60 (58·7), ♀ (n = 12) 53–61 (56·0); tail, 1 ♂ 42; bill, 1 ♂ 13; tarsus, 1 ♂ 19. WEIGHT: unsexed (n = 16, Zambia) 7·7–9·5 (8·5).
IMMATURE: duller above than adult, with brownish tinge; throat dull greenish yellow.
NESTLING: unknown.

P. l. laurae (Boulton): W Angola (Mt Moco). Yellow of face, chin to breast and undertail-coverts less bright than in *eustacei*, green of upperparts duller, belly darker grey. Size similar: 1 ♂, wing 57, tail 47, bill 13, tarsus 20; 1 ♀, wing 55, tail 40, bill 12, tarsus 19.

Phylloscopus laurae

Field Characters. A small, brightly coloured, arboreal warbler. Plain green above, including top of head, with broad yellow supercilium. Below, yellow throat to upper breast and yellow undertail-coverts contrast with pale greyish belly. Similar but allopatric Yellow-throated Woodland-Warbler *P. ruficapilla* of montane forest is duller above with brownish cap, and whiter below; Zambian race has yellow confined to chin and throat.

Voice. Tape-recorded (86, B, F, CART). Song clear and high-pitched; a series of somewhat crescendo phrases, each lasting 1·5–2 s, separated by short pauses; each phrase includes 4–6 repetitions of same motif, 'chirri-prichirri-prichirri-prichirri-prichee', 'titiwee-titiwee-titiwee- . . .', 'wichity-wichity-wichity . . .', with only slight variation. Similar in form and pitch to song of Yellow-throated Woodland-Warbler but faster, phrases rather longer and note pattern more complex. May sing non-stop in Zambia for up to 30 min (C. Carter, pers. comm.).

General Habits. In Zambia and Zaïre, inhabits swamp forest, riparian strips and dry evergreen forest patches in miombo woodland at 1100–1400 m. On Mt Moco, montane forest at 2200 m. Usually occurs singly or in pairs, or in small, loosely-associated groups. Sometimes in mixed-species bird parties. Forages typically in mid-stratum, also in canopy and at forest edge; sometimes nearer ground. Gleans from leaves and also flycatches. Sings frequently and vigorously (D. R. Aspinwall, pers. comm.).

Food. Insects, including caterpillars.

Breeding Habits. Unknown. Nest and eggs not described. Birds in breeding condition Zambia, July, Sept, Oct; Zaïre, Sept; Angola, Aug.

Plate 21
(Opp. p. 400)

Phylloscopus laetus (Sharpe). Red-faced Woodland-Warbler. Pouillot à face rousse.

Cryptolopha laeta Sharpe, 1902. Bull. Br. Orn. Club 13, p. 9; Ruwenzori Mts, W Uganda.

Forms a superspecies with *P. ruficapilla* and *P. laurae*.

Range and Status. Endemic to highlands on either side of the Albertine Rift; resident. NE and E Zaïre (Lendu, Kahuzi, Regege, Itombwe, Kabobo); SW Uganda (Rwenzoris, Bwindi-Impenetrable Forest, Mafuga Plantation); Volcano Highlands (Uganda/Rwanda); and W Rwanda to W Burundi (south to Bururi Forest). Common and widespread.

Description. *P. l. laetus* (Sharpe): Lendu plateau and Rwenzoris south to Burundi and Itombwe. ADULT ♂: entire upperparts green, head tinged slightly brownish. Tail-feathers dark brown, edged green. Lores and streak behind eye greenish; supercilium, ear-coverts, cheeks and chin warm buff-brown. Underparts whitish, grey on flanks, paler buff-brown on throat and breast. Flight feathers and upperwing-coverts dark brown, broadly edged green; underwing-coverts and axillaries white, washed yellowish. Upper mandible brown, lower mandible yellowish; eyes dark brown; legs and feet olive-green to greyish green. Sexes alike. SIZE: (9 ♂♂, 6 ♀♀) wing, ♂ 55–59 (58·4), ♀ 51–56 (53·8); tail, ♂ 40–46 (43·5), ♀ 39–43 (41·2); bill, ♂ 12–13 (12·6), ♀ 12–13 (12·8); tarsus, ♂ 19–21 (20·2), ♀ 19–21 (19·8). WEIGHT: ♂ (n = 17) 9–11 (9·9), ♀ (n = 11) 8–10·5 (9·2).

IMMATURE: lacks reddish colour on face; cheeks washed yellowish green; chin and throat whitish.

NESTLING: unknown.

P. l. schoutedeni (Prigogine): E Zaïre (Mt Kabobo). Richer orange-brown on cheeks, chin and throat; breast grey with faint brownish tinge.

Field Characters. A small dumpy warbler, plain green above and whitish below, with a reddish brown face, chin and supercilium. Wings short and rounded, tail rather short. Brown Woodland-Warbler *P. umbrovirens* has brown upperparts and flanks and a pale face; Uganda Woodland-Warbler *P. budongoensis*, usually at lower altitudes, has greenish upperparts but white face and supercilium; songs of all 3 distinctive.

Phylloscopus laetus

General Habits. Inhabits primary and secondary forest and gallery forest, mainly between 1400 and 2600 m (limits 1100–3120 m). Common in bamboo zone, ranging to near its upper limit. Occurs lower than Brown Woodland-Warbler, however, and largely segregated by altitude in areas of broad sympatry; higher in E Zaïre than Uganda Woodland-Warbler, which it meets only in Itombwe. Usually in pairs or small groups, and often in mixed-species parties. An active gleaner which forages at all levels, but mainly lower down than Brown Woodland-Warbler; in closed canopy, more open canopy, and secondary growth at forest edges.

Food. Stomachs in Zaïre contained spiders, hemipterans and beetles.

Breeding Habits.
NEST: dome of moss, dead leaves and herbs, with side entrance; in shrubbery *c*. 0·5 m above ground; also recorded in tangled mass of moss-covered epiphyte roots *c*. 10 m high in tree.
EGGS: 2; cream with a band of orange-brown speckles.
LAYING DATES: Zaïre (Kivu), mainly Jan–Aug; Uganda, Sept; Rwanda, Aug.
DEVELOPMENT AND CARE OF YOUNG: young remain with parents for some weeks after fledging.

Voice. Tape-recorded (C, F, KEI, LEM). Song, clear and pleasant, but poorly structured; 2–3 short bubbling phrases (*c*. 0·5 s) are delivered in quick succession followed by a short pause. Phrases varied, but consist of a few quick repetitions of disyllabic notes: 'putreeputriputri', 'pitjupitju . . .' (Dowsett 1990). Lower-pitched than more simply structured songs of Yellow-throated Woodland-Warbler *P. ruficapilla* and Laura's Woodland-Warbler *P. laurae*. Call note, a piercing 'sip-sip'.

Reference
Dowsett, R. J. (1990).

Phylloscopus herberti (Alexander). Black-capped Woodland-Warbler. Pouillot à tête noire.

Plate 21
(Opp. p. 400)

Cryptolopha herberti Alexander, 1903. Bull. Br. Orn. Club 13, p. 35; Bakaki, Fernando Po.

Forms a superspecies with *P. budongoensis*.

Range and Status. Endemic resident. Restricted to highlands of W Cameroon (Mt Cameroon, Mt Nlonako, Mt Manenguba, Rumpi Hills, Mt Kupé) and SE Nigeria (Obudu plateau, Gangirwal, Ngel Nyaki Forest), and Bioko. Locally frequent to common; not immediately threatened (Stuart 1986a).

Description. *P. h. camerunensis* (Ogilvie-Grant): Cameroon and Nigeria. ADULT ♂: top of head blackish brown, rest of upperparts bright golden green. Tail-feathers blackish, broadly edged bright green. Supercilium pale buffish, broad, extending back to sides of nape; lores and broad band behind eye blackish brown; lower ear-coverts and cheeks grey. Chin and throat buffish white; breast dull grey, tinged buffish in centre; flanks grey; belly and undertail-coverts pale greyish white. Flight feathers blackish, broadly edged bright green; upperwing-coverts golden green; underwing-coverts and axillaries white, tinged yellowish green. Upper mandible blackish, lower mandible yellowish; eyes brown; legs and feet grey. Sexes alike. SIZE: (10 ♂♂, 10 ♀♀) wing, ♂ 53–57 (55·4), ♀ 48–51 (49·5); tail, ♂ 37–42 (39·5), ♀ 35–39 (37·4); bill, ♂ 12–14 (13·0), ♀ 11–12 (11·9); tarsus, ♂ 18–20 (18·9), ♀ 17–19 (17·9). WEIGHT: 1 ♀ 9.
IMMATURE: green of upperparts duller than in adult; breast paler, chin and throat whiter, belly and undertail-coverts tinged yellow. Supercilium yellowish green.
NESTLING: unknown.
P. h. herberti (Alexander): Bioko. Supercilium, cheeks, chin and throat deep buff. Slightly brighter green above than *camerunensis*. Smaller: wing, ♂ (n = 4) 52–55 (54·1). WEIGHT: ♂♀ (n = 8) 8·5–9·5 (8·9).

368 SYLVIIDAE

Phylloscopus herberti

Field Characters. A small arboreal warbler with short wings, shortish tail and a short, fine bill. Bright golden green above and grey below with a whitish belly. Black crown and black streak through eye contrast with buff supercilium and greyish or buff sides to face.

Voice. Tape-recorded (57, B, STJ). Song, a short, variable sequence of high ringing notes, e.g. 'chirchircherchewee', lasting *c.* 1 s and ending with a flourish, either upslurred or downslurred; repeated at intervals of a few s. Like song of Uganda Woodland-Warbler *P. budongoensis*, but slightly lower-pitched, individual notes delivered more rapidly and often run together. Call, a sibilant 'see-eer', prolonged into a more urgent 'see-err-err' when alarmed (Bannerman 1939).

General Habits. Confined to wetter, mature montane forest (moss forest in Bioko) at altitudes of 900–2200 m in Cameroon (down to 700 m on Mt Cameroon); in Bioko at 1100–1800 m, with a single bird at 50 m within undisturbed lowland forest (Perez del Val *et al.* 1994). Occurs singly, or in pairs or small parties, sometimes together with sunbirds. Often in high canopy, but also forages in mid-stratum and in undershrubs; in Nigeria in understorey and liana tangles at forest edges (Ash *et al.* 1989a). Very active, searching and gleaning food from leaves and twigs.

Food. Insects.

Breeding Habits. Nest and eggs undescribed. Birds in breeding condition, Nigeria, Jan–Feb, ♀ with egg in oviduct Dec.

Reference
Stuart, S. N. (1986a).

Plate 21
(Opp. p. 400)

Phylloscopus budongoensis (Seth-Smith). Uganda Woodland-Warbler. Pouillot de l'Ouganda.

Cryptolopha budongoensis Seth-Smith, 1907. Bull. Br. Orn. Club 21, p. 12; Budongo Forest, W Uganda.

Forms a superspecies with *P. herberti*.

Range and Status. Endemic resident. E Zaïre (Bondo Mabe, Semliki valley and E Ituri Forest south to Itombwe); Uganda (Budongo, Mpanga and Kifu Forests and Mt Elgon); and W Kenya (Kakamega and N and S Nandi Forests). Common at Kakamega but uncommon to frequent elsewhere. Recent records from N Congo (Odzala Nat. Park, not rare) and E Gabon (Bélinga) extend the range considerably to the west (F. Dowsett-Lemaire, pers. comm.).

Description. ADULT ♂: top of head and hindneck olive-brown; rest of upperparts olive-green. Tail-feathers blackish brown, broadly fringed green. Supercilium broad, creamy, from base of bill to well behind eye; lores and streak behind eye blackish brown; ear-coverts and cheeks creamy, flecked dusky. Underparts creamy white, washed grey on sides of breast and flanks; undertail-coverts pale yellow. Flight feathers blackish brown, broadly edged green; upperwing-coverts green; underwing-coverts and axillaries creamy, washed greenish yellow. Upper mandible black, lower mandible black with yellowish tip; eyes dark brown; legs and feet bluish grey. Sexes alike. SIZE: (5 ♂♂, 3 ♀♀) wing, ♂ 52–56 (55·0), ♀ 52–56 (53·7); tail, ♂ 35–39 (36·6), ♀ 32–35 (34·3); bill, ♂ 12–13 (12·2), ♀ 11–12 (11·3); tarsus, ♂ 18–19 (18·8), ♀ 18–19 (18·3). WEIGHT: unsexed (n = 9, Kenya) 7–9 (8·2).
IMMATURE: duller than adult; breast washed with olive.
NESTLING: unknown.

Field Characters. A very small arboreal warbler, olive-green above and creamy below, with a prominent pale supercilium and dark streak through eye extending to side of nape. Smaller than Willow Warbler *P. trochilus* or

Phylloscopus budongoensis

Zaïre, occurs mainly between 900 and 1200 m, generally allopatric with Red-faced Woodland-Warbler *P. laetus* which occupies higher altitudes; it overlaps locally with *laetus* in Itombwe, however, where it occurs up to 1800 m (Prigogine 1980b). Ranges from 1200–1900 m in Uganda and Kenya. In N Congo occurs in cool forests on rivers at *c.* 400 m (F. Dowsett-Lemaire, pers. comm.). Usually singly or in pairs near clearings, occasionally in mixed-species parties. Forages by gleaning, mainly in high canopy over 20 m from ground; also on branches extending down to lower canopy and in tops of mid-level trees, but not in undergrowth (Zimmerman 1972). Unobtrusive but quite vocal. In Kakamega, frequents thinly foliaged trees and thus not difficult to observe (Zimmerman 1972).

Food. Insects.

Breeding Habits.
NEST: well-concealed domed structure built of moss, deep cup thickly lined with fibres and plant down; ext. depth *c.* 190, ext. diam. *c.* 120; rounded entrance hole *c.* 45 mm diam. towards top of outer facing side (Holyoak 1990); in small cleft in tree, near ground; or within hanging mat of moss in slight recess between buttresses at base of tree, *c.* 50 cm above ground (Holyoak 1990).
EGGS: (2 clutches, Kenya) 3; white with spots and small blotches of light chestnut, more concentrated round larger end. SIZE: (n = 2, Kenya) both 16 × 12.
LAYING DATES: Kenya, Aug, Dec; Zaïre (Kivu), (breeding condition Mar–July).

Common Chiffchaff *P. collybita*, with shorter wings and tail, bolder head markings, greener wings and upperparts and greyer legs. Voice distinctive. White-browed Crombec *Sylvietta leucophrys* has similar pale supercilium and dark eye-line but very short tail, yellow-green undertail-coverts and short trilling song.

Voice. Tape-recorded (B, F, CHA, GREG, HOR, KEI, PEA, STJ, ZIM). Song, a short high-pitched sequence of 5–6 notes, lasting *c.* 1·2 s, 'chi-cher-chi-chewi'; varied slightly and repeated at intervals of a few s.

General Habits. Inhabits interior of tall, shady primary or secondary forests at medium elevations. In E

References
Holyoak, D. T. (1990).
Prigogine, A. (1980b).
Zimmerman, D. A. (1972).

Genus *Regulus* Cuvier

Tiny, warbler-like birds of temperate coniferous and broad-leaved woods; aspects of feeding, physiology and nesting determined by their minute size. Greenish; crowns of adults brilliant ('goldcrest', 'firecrest', 'flamecrest', 'rubycrown'), differing sexually. Nostrils with an operculum, partly covered by single, stiff feather. Rictal bristles soft. Wings fairly long, rounded, P10 up to half length of P9; P7–P5 longest. Tail-feathers pointed at tip. Front of tarsus booted. Nest semi-pensile, in twigs.

5 species, including 1 superspecies; 3 Palearctic, 2 Nearctic, 2 reach N Africa as migrants, 1 also breeding there.

Plate 21
(Opp. p. 400)

Regulus ignicapillus (Temminck). Firecrest. Roitelet à triple bandeau.

Sylvia ignicapilla Temminck, 1820. Man. d'Orn., 2nd ed., i, p. 231; France.

Range and Status. W Palearctic from Madeira and NW Africa north to S England, S Denmark, S Latvia and east to E Belarus and NW Turkey, with a few small populations near E end of Black Sea. Summer visitor in NE of range, resident in centre, winter visitor in S. Also dispersive, and an altitudinal migrant.

Resident and winter visitor, NW Africa. Morocco, widespread and frequent breeding resident in Middle Atlas, Jebel Ayachi, Rif up to 2160 m, and High Atlas between 1000 m and 2600 m; elsewhere, a scarce winter visitor (Thévenot *et al.* 1982). Algeria, abundant around Aurès; common and widespread in cedar forests up to 2000 m (Babor, Djurdjura) and 2100 m (Jebel Chélia); in places down to sea level (e.g. Tonga). Tunisia, uncommon breeding resident at Kroumirie; regular but scarce winter visitor in N, along E coast, south to Gabès. Libya, scarce in winter on NW coast and Jebel Nafusa. Egypt, 1 in NW, Nov 1916, 1 Bahig, Mar 1973.

Breeding density of *c.* 132 pairs per km² in humid conifer forest, N Morocco.

Regulus ignicapillus

Description. *R. i. balearicus* Jordans: NW Africa and Balearic Is. ADULT ♂: forehead white, confluent with superciliary stripe, narrow above lores and very broad above and behind eye; crown bright orange, in long median stripe, bordered by black stripes which meet in front, above white forehead; nape, hindneck and sides of neck greyish olive; mantle, scapulars, back, rump and uppertail-coverts olive-green. Tail blackish, feathers with olive-green edges. Lores and small mark behind eye black; small arc just under eye white, ear-coverts grey; short, dusky moustachial stripe. Chin, throat and breast greyish cream-white; 'shoulders' – area between breast and mantle – bronze or olivaceous orange; belly and undertail-coverts whitish, flanks white washed brown-grey. Primaries and secondaries brown-black, with narrow yellowish outer edges, white at base of outer primaries; tertials blackish with outer vanes broadly tipped white; beyond tips of greater coverts, edges of inner primaries, all secondaries, and outer tertials black for 4–5 mm, forming black rectangle in closed wing. Primary coverts and greater coverts with 2–3 mm white tips, forming a small wing-bar; median coverts with 1–2 mm white tips, forming a smaller wing-bar, lesser coverts broadly tipped yellowish green. Underwing-coverts and axillaries white, tinged yellow. Bill black; eyes dark brown; legs and feet brown. ADULT ♀: like ♂ but centre of crown bright lemon-yellow. SIZE: (*R. i. ignicapillus*, Netherlands) wing, ♂ (n = 34) 51–56 (53·3), ♀ (n = 14) 48–53 (50·2); tail, ♂ (n = 31) 39–43 (40·7), ♀ (n = 13) 37–43 (39·2); bill to skull, ♂ (n = 12) 10·6–12·0 (11·2), ♀ (n = 14) 10·1–11·9 (10·8); tarsus, ♂ (n = 11) 16·0–17·4 (16·9), ♀ (n = 10) 15·5–17·5 (16·4). WEIGHT: (*R. i. ignicapillus*, Netherlands) ♂ (n = 31) 4·2–6·5 (5·8), ♀ (n = 14) 4·8–6·5 (5·4).

IMMATURE: crown lacks black-sided yellow/orange stripe; upperparts greyish olive-brown, sides of crown may be mottled with black in subadult birds. Superciliary stripe indistinct, narrow, pale olive. Underparts pale buff.

NESTLING: hatchling naked except for short down on crown.

R. i. ignicapillus Temminck: Europe (south to S Spain, Sicily, Cyprus); scarce winter visitor Morocco, Tunisia, probably Algeria; vagrant Egypt. Like *balearicus* but warmer colour, tinged brown on upperparts and underparts.

Field Characters. A tiny warbler- or tit-like bird of heath and oak scrub, at 5·5–6 g the smallest breeding bird in NW Africa. Plump, rather short-tailed, vocal, approachable. Adult olive-green above, white below, with double wing-stripe and variegated wing, instantly told from all birds other than Goldcrest *R. regulus* by bright orange (♂) or yellow (♀) median crown-stripe, edged with black, with well-defined white superciliary stripe. Juvenile has plain brown-olive crown and indistinct eye-stripe, but has adult's variegated wing. Adult distinguished from Goldcrest by black-and-white forehead (grey in Goldcrest), broad white supercilium, black stripe through eye and large bronzy patch on 'shoulder'.

Voice. Tape-recorded (CHA – 62, 73, 89, 93, B). ♂ song a succession of 10–15 very high-pitched 'see' notes, lasting 1·5–2·5 s, increasing in volume towards vibrant notes at end. ♂ and ♀ have soft, warbling subsong. Many calls described (Cramp 1992), all thin and high-pitched, e.g. contact and flight call 'ssii' or 'sisisi', aggressive call 'zick', display call 'ssri'.

General Habits. In Morocco, breeds up to 1000 m in cork oak *Quercus suber* forests, also in rich holm oak *Q. ilex* forest; above 1600 m breeds in pure cedar *Cedrus atlantica* forests, and inhabits junipers and thorny xerophytes up to 2600 m; uncommon in maritime Aleppo pine *Pinus halepensis* (Snow 1952, Thévenot *et al.* 1982). In S Spain common in cork oak woods and stands of ivy-covered alder *Alnus*. On Porquerolle Is

(Mediterranean France), ecologically separated from Goldcrest by using scrub mainly of tree heath *Erica arborea*, holm oak and strawberry tree *Arbutus unedo*, up to 3 m high (Cramp 1992).

Usually in pairs; family parties in summer and small loose groups in winter. Forages almost continuously in daylight hours; so small, that in captivity 1 hour's starvation was fatal. Roosts in thick foliage, 2 or more birds in contact. Feeds by creeping, climbing, hanging by feet under pine-needle spray, and hovering, often taking short flights between branches; forages low down in vegetation, in cold weather sometimes on ground. Beats prey against a branch; chisels at fixed insect eggs or cocoons with strong blows of bill. Gleans broad leaves, needles, twigs and trunks.

Resident in Morocco, but some descend from mountains in winter and occupy coastal plain scrub habitats. Some immigration from Europe in autumn, to Morocco, Algeria, Tunisia and NW Libya. Well-known as autumn migrant in Gibraltar and Malta; in spring, passage at Gibraltar Feb–Apr and at Cap Bon, Tunisia, Mar–Apr. Bird ringed Belgium recovered NW Morocco.

Food. Large variety of small arthropods, especially springtails, spiders and aphids (Austria) and hymenopterans (Spain).

Breeding Habits. Territorial, solitary breeder. Territory established and defended by ♂ singing, 5–7 songs per min at dawn, 4–6 per min throughout day, with up to 15 songs per min when reacting with rival. In display flight ♂ silently dives from tree top nearly to ground then rises up again. For details of ruffling display, forward display, combat, and appeasement behaviour, see Cramp (1992).

NEST: subspherical, almost elastic, cup with small entrance at top, made in 3 layers: outer layer of moss, lichen and cobweb, middle layer of moss, inner layer (lining) of up to 3000 feathers, and hair. Nest bound to surrounding twigs and leaves with cobwebs; suspended near end of branch of conifer or oak. Ext. height 60, ext. diam 80×90, int. diam. 40, int. depth 40. Built by ♀, in 20 days.

EGGS: 4–7 (Morocco), 7–12 (Europe). Subelliptical, not very glossy, uniform buff or with very fine (or occasionally coarse) brown spots. SIZE: (n = 140) 12·5–15·1 × 9·9–11·5 (13·6 × 10·6). WEIGHT: (n = 45) av. 0·692 g.

LAYING DATES: Morocco, May–July.

INCUBATION: by ♀, starting before clutch complete. Period: 14·5–16·5 days.

DEVELOPMENT AND CARE OF YOUNG: young brooded by ♀ for at least 7 days and 8 nights; fed by ♂ and ♀. Nestling period: *c.* 22–24 days.

BREEDING SUCCESS/SURVIVAL: nest success rate is low – predated by Black-billed Magpie *Pica pica*, Eurasian Jay *Garrulus glandarius* and dormouse *Dryomys nitedula*.

Reference
Cramp, S. (1992).

Regulus regulus (Linnaeus). Goldcrest. Roitelet huppé.

Plate 21
(Opp. p. 400)

Motacilla Regulus Linnaeus, 1758. Syst. Nat. (10th ed.) i, p. 188; Sweden.

Range and Status. Palearctic, from Ireland and N Spain to Japan. Partial migrant within Palearctic, south (in west) to Africa, Sicily and Cyprus. Endemic races in Azores and Canary Is.

Vagrant and scarce winter visitor to N Africa. Morocco, 1 Tanger Nov 1975 (and only 3 records, Gibraltar); Algeria, small numbers winter regularly in extreme north, south to 35° 30′N (Batna); Tunisia, irregular, late Dec to early Feb, in extreme north, perhaps regular in Kroumirie; Egypt, several Giza, Nov 1910–Feb 1911, 1 Bahig Oct 1968.

Description. *R. r. regulus* (Linnaeus): Eurasia (only subspecies in Africa). ADULT ♂: forehead grey, crown in midline bright yellow in front and orange-red behind, broadly edged with black; nape grey, hindneck and sides of neck olive-grey; mantle, scapulars, back, rump and uppertail-coverts olive-green. Tail-feathers blackish with olive-green edges. Area around eye white, merging into pale grey at sides of crown, lores and cheeks; ear-coverts olive-grey. Short, thin, blackish moustachial stripe. Chin pale grey or whitish, throat and breast olivaceous buff, belly and undertail-coverts whitish in midline, olive-buff at sides. Primaries and secondaries blackish with narrow olive-yellow outer edges, white at base of outer primaries; tertials black with outer vanes broadly tipped white; beyond tips of greater coverts, edges of inner primaries, all secondaries, and outer tertials black for 5–8 mm, forming conspicuous black rectangle in closed wing. Primary coverts and greater coverts with yellow-green edges; greater coverts with 2–3 mm white tips, forming a wing-bar; median coverts with 1–2 mm white tips, forming a smaller wing-bar; lesser coverts broadly tipped yellow-green. Underwing-coverts and axillaries white, tinged yellow. Bill black; eyes dark brown; legs and feet brown-grey. ADULT ♀: like ♂ but centre of crown bright yellow behind as well as in front. SIZE: (Netherlands) wing, ♂ (n = 42) 51–57 (54·4), ♀ (n = 32) 50–54 (52·4); tail, ♂ (n = 17) 39–44 (41·3), ♀ (n = 19) 39–42 (39·9); bill to skull, ♂ (n = 40) 9·5–11·5 (10·3), ♀ (n = 30) 9·2–11·6 (10·6); tarsus, ♂ (n = 25) 15·8–17·7 (16·8), ♀ (n = 18) 15·3–17·2 (16·4). WEIGHT: (Sweden) ♂ (n = 491) 4·6–7·1 (5·6), ♀ (n = 334) 4·6–7·0 (5·5).

IMMATURE: crown lacks black-sided yellow/orange stripe; entire upperparts greyish olive-brown. Sides of crown may be mottled with black in subadult birds. Underparts greyer than in adult.

Field Characters. A tiny, greenish, thin-voiced bird of pine woods, with brilliant crown-stipe (orange-scarlet in ♂♂, lemon-yellow in ♀♀) accentuated by broad black

372 SYLVIIDAE

Regulus regulus

lines at side, 2 white wing-bars, and black, white and yellow areas in flight feathers of closed wing. Similar to Firecrest *R. ignicapillus*, but face has 'blank' look, without black-and-white stripes or grey ear-coverts, and there is no bronzy patch between breast and mantle. Juvenile lacks crown-stripes and is plain olive except for adult wing pattern; told from Yellow-browed Warbler *Phylloscopus inornatus* (rare vagrant to N Africa) by lack of long yellowish superciliary stripe.

Voice. Tape-recorded (62, 73, 89, 93, B). Song of ♂, very high-pitched, sweet, tinkling 'pit*eeti*ly' or 'eedle' given 3 or 4 times without break, lasting 2·0–3·5 s. Other calls, 'siii', 'zik', 'zee', and 'ssri', are all very thin and high-pitched.

General Habits. Strictly arboreal; mainly in spruce *Picea abies*, Douglas fir *Pseudotsuga taxifolia*, silver fir *Abies alba*, mountain pine *Pinus mugo*, Scots pine *P. sylvestris*, Aleppo pine *P. halepensis*, maritime pine *P. pinaster*, and larch *Larix*. In Tunisia winters in unspecified conifers. Like Firecrest a tiny bird, at 5·5 g the smallest in NW Africa and W Palearctic, like a short-tailed warbler in appearance but rather like Coal Tit *Parus ater* in habits; forages in pairs or family parties in trees, from 2 m above ground to the tops, contact-calling often, moving restlessly, hopping, creeping and making short flights within tree and between trees; gleans twigs and pine-needle clumps by standing and pecking, fluttering, climbing about, and hovering.

Food. Huge variety of small insects; also spiders, harvestmen, mites, snails, and spruce and pine seeds.

Reference
Cramp, S. (1992).

Genus *Eminia* Hartlaub

Large and heavy-set, with fairly short, rounded tail; feathers of rump and flanks loose, extending to tip of tail. Plumage patterned, grey, black and green with chestnut throat patch and bend of wing. Habitat undergrowth and thickets, often near water. Nest globular, large, untidy, with side entrance, and suspended, much like a sunbird's. Song a series of powerful, varied trills, some of which resemble some trills of *Hypergerus*, but we agree with Chappuis (1978) that the songs are not closely related.

Monotypic, endemic to E Africa. Chapin (1953a) noted 'not a few points of similarity' between *Hypergerus* and *Eminia*, without stating what these were, but then went on to say that the bill, wings and tail of *Hypergerus* are all much more elongated. *Eminia* was merged with *Hypergerus* by Watson *et al.* (1986), and their nests are similar, but plumage is totally different and proportions and song bear little resemblance; we prefer to retain *Eminia* in its own genus, while keeping it next to *Hypergerus* in the family sequence.

Plate 20
(Opp. p. 337)

Eminia lepida (Hartlaub). Grey-capped Warbler. Éminie à calotte grise.

Eminia lepida Hartlaub, 1881. Proc. Zool. Soc. Lond. (1880), p. 625; Magungo, north end of L. Albert.

Range and Status. Endemic resident. Common in S Sudan near border (Gilo, Talanga-Wald, Wald, Imatong Mts); local and uncommon in N Zaïre (Uele district), becoming common in E (Lendu Plateau and

Eminia lepida

W shore of L. Albert south to L. Kivu), but less so in Rwanda and Burundi (Bukeye, Bugarama, Kayero). Locally common in Uganda north to Wadelai in NW, Kenya (W and central highlands from Mt Elgon and Mt Kenya south to Kisumu, Kericho and Mara region), and Tanzania (NW and N in moist areas around L. Victoria, south in W to Kasulu, reaches Crater Highlands in NE; rare in Serengeti Nat. Park). Record of singing ♂ at Tihon Mariam, Ethiopia, 1400 km from known range, identified only by comparing taped song with recordings of M. E. W. North, needs confirmation (Desfayes 1975).

Description. ADULT ♂: forehead grey, crown darker; black band around crown from lores through eye to nape. Hindneck greyish, rest of upperparts including wings and tail olive-green. Cheeks, ear-coverts, and underparts smoke-grey, becoming paler on belly and shading into olive-green on undertail-coverts and flanks; chestnut patch in centre of chin and throat, and most birds have chestnut patch on vent. Primaries dusky brown, outer webs edged olive-green, secondaries the same but inner webs also edged olive-green; marginal and lesser coverts chestnut, underwing-coverts chestnut (near angle of wing) and buff. Bill and mouth cavity black; eyes reddish brown; legs and feet reddish brown. Sexes alike. SIZE: (10 ♂♂, 10 ♀♀) wing, ♂ 67–73 (69·6), ♀ 64–73 (66·7); tail, ♂ 60–67 (63·5), ♀ 54–67 (58·0); bill to feathers, ♂ 15–17 (16·4), ♀ 15–16·5 (15·8); tarsus, ♂ 25–26 (25·4), ♀ 25–26 (25·6). WEIGHT: ♂ (n = 3, Kenya) 16–24 (20·0); (n = 1, Zaïre) 23; imm. ♂ (n = 2) 22·5, 23; ♀ (n = 4, Kenya) 17·5–20 (18·4), (n = 4, Zaïre) 21–22·5 (21·5).

IMMATURE: a dull form of the adult; chestnut patch on throat smaller and paler, eye of ♂ bright brown.

NESTLING: naked and without down when hatched, gape and inside of mouth yellow, tongue with 2 distinct black dots on surface near base; at *c.* 8 days old eyes open, green feathers on back emerging from sheaths, tail short and sheathed (Chapin 1978).

Field Characters. A large, rather heavy-set warbler of dense undergrowth near water or on damp hill sides. The loud, arresting song is usually heard long before the bird is seen. The green upperparts and pale grey underparts are shared by many species but the head pattern (broad black band around grey crown, chestnut throat patch) is unique. Black mouth noticeable when bird is singing. Feathers of rump and flanks are loose and extend to tip of wing, accentuating thinness of the rounded tail and giving the bird a characteristic shape, e.g. when balancing momentarily on a liana (van Someren 1956).

Voice. Tape-recorded (5, 30, 34, 38, 66, B, C, F, LEM). Song of ♂ a series of loud, powerful trills and runs lasting 6–20 s or more, like those of a Nightingale *Luscinia megarhynchos* but longer, or a domestic canary turned up to double volume, e.g. 'peurrrrrrr, cheo cheo cheo che che che che che, prit prit prit prit, pri-r-r-riii . . .' and can include 'chree-e-eero', 'chi chi chi', 'p-eurr-r-r-r' and 'chweri-chi' phrases (van Someren 1956); 'one of our most spectacular singers' (North 1958). Songs are sustained and variable; one of North's (1958) recordings includes imitations of Cinnamon Bracken Warbler *Bradypterus cinnamomeus*, another singer with a loud trill. Song of ♀ a long dry trill, of up to 18 notes in 2·5 s, given intermittently while ♂ is singing (Grimes 1974). In one 7-min recording (Gituru, Kenya), which was not continuous, ♂ gave 15 bursts of song, all except one made up of a repeated phrase, and ♀ trilled in duet 8 times. Each of ♂'s 13 phrases differs in number of notes (range 7–20) and duration (range 1·0–1·7s). In most phrases there is only one repeated note, uttered either at even tempo with or without initial longer whistle, or at increasing rate. All are about the same pitch. Intervals between phrases are 1–4 s or more. Song of another ♂ (Nakuru, Kenya) contained one 2 s phrase of 8 notes, which was repeated 12 times, and one 2·5 s phrase of 3 long and 18 short notes, repeated 7 times.

General Habits. Inhabits damp, dark and dense undergrowth at forest edge and tangles of bushes and creepers along river banks, also suburban gardens, edges of swamps and lakes, and bushes on hill sides 'where permanent springs come trickling down in deep water-worn cuttings' (van Someren 1956). It does not occur in isolated forest islands (Lewis and Pomeroy 1989), and avoids centres of dense reedbeds and papyrus swamps. In Sudan occurs only above 1600 m and in SW Uganda reaches 2040 m. In Kenya ranges from 800 to 2500 m and is commonest above 1500 m and in areas with > 500 mm of annual rainfall. Reaches *c.* 1700 m on W but 2100 m on E slopes of Rwenzori.

Shy and retiring, usually preferring the half light of a stream bed, but not averse to more open areas on hill sides; easily overlooked except for its call. ♂ sings from exposed perch, regularly in breeding season and occasionally in non-breeding season, particularly during morning and evening. When duetting ♂ and ♀ sit together or hop around each other, with bodies erect and plumage sleeked but throat patch expanded and vibrating (**A**). Grey crown framed by broad black line,

A

and chestnut throat, are prominently displayed, as are chestnut marginal and lesser wing-coverts; tail is cocked vertically and fanned (Kunkel 1974). When duetting ceases, ♂ bends forward, puffs out back, rump and flank feathers so that he resembles a feathery ball (much like puff-back shrike *Dryoscopus*), and turns to right and left, his whole body quivering; as he relaxes, he closes his feathers and trills loudly (van Someren 1956). Usually in pairs. ♂ and ♀ clump together when roosting; allopreening has not been observed (Kunkel 1974). Forages actively, searching every crack and cranny in bark, curled-up leaves and clumps of suspended debris; prods any rotten holes on branches with thin, sharp bill; few insects survive its searching.

Sedentary.

Food. Insects (eggs, mantids, geometer moth larvae and other caterpillars, weevils, earwigs, small grasshoppers, very occasionally moths); also spiders and millipedes.

Breeding Habits. Monogamous, territorial, solitary nester, probably pairs for life.

NEST: domed; side entrance with platform (25 mm long) and porch (80 long), like a sunbird's nest but larger and very untidy (**B**). Made of dry grass-like fibres, tendrils, leaves, some feathers, spider web, strips of moss loosely interlocked, with closely felted lining of fibre, rootlets, vegetable down, and the occasional feather. Suspended, usually freely, from a slender branch of tree or creeper within shade of creepers or thick vegetation (often *Carissa*, *Gymnosporia* or *Scutia*); one attached at its sides to 2 upright saplings, one to 4 twigs of a *Bougainvillea*. In *Raphia* palm or *Dracaena* sometimes placed as high as 5 m, usually only 1–3 m. When over water, nest looks like flotsam deposited by flood in overhanging vegetation. One nest took 2 weeks to complete; both birds build, but ♂ often trills while ♀ builds. Old nests are sometimes reconditioned and occupied for 2 seasons. Once a weaver's nest was refurbished; the entrance was rebuilt and long dangling strips of fibre were added to outside of nest as camouflage.

EGGS: 2–3. Long and ovate; smooth and slightly glossy; white, pink or pale blue; most are immaculate, others have variable markings, some sparsely spotted purplish brown all over, others with spots concentrated at large end. SIZE: (n = 3) 17·8–21·0 × 11·8–13·8 (18·0 × 12·5); (n = ?, Kenya) 21·0–22·0 × 13·0–14·0; (n = 6, Zaïre) 19·5–20·1 × 13·7–14·5 (19·8 × 14·0). Double-brooded.

LAYING DATES: mainly in the rains throughout its range. Sudan (reported breeding Nov); Zaïre end Oct, mid-Mar, mid-June (gonads enlarged Apr–Aug); Uganda (nests Apr and May); Kenya (nest May, Nandi, nestling late May, Nairobi and Aug, Elgeyu, fledgling out of nest mid-Nov, Kiambu, reported nesting Nov–Jan, Meru district); E Africa: Region A, Apr–July (n = 3); Region B, Apr–Oct (n = 21); Region D, Mar–Dec (n = 19).

INCUBATION: by both sexes, mainly by ♀. Period: 12–13 days.

DEVELOPMENT AND CARE OF YOUNG: fledgling has soft and downy feathering, short tail, black crown, rusty-tinged wings, and throat only tinged with chestnut; yellow of gape very pronounced. When perched huddled together with feathers fluffed they look like puff-balls (van Someren 1956). At one nest, on 2 consecutive afternoons, parents brought food 6 times and 10 times in 1 h; ♂ often passes food to ♀ who takes it to nestlings. Faecal sacs removed by both parents, but apparently not swallowed by them. ♀ approaches nest with stealth, with long pauses between movements (Serle 1943b); at other times flies onto platform of entrance hole, the impact causing nest to swing 15 cm from the vertical. Quietly attentive at nest. Noisy if alarmed when young are on the wing. Nestling period *c.* 16 days.

BREEDING SUCCESS/SURVIVAL: mortality rate high, only 4 broods raised by 3 pairs in 5 years (van Someren 1956). Nests often washed away by flood water, others destroyed by wandering cattle.

References
Grimes, L. G. (1974).
van Someren, V. G. L. (1956).

B

Genus *Hypergerus* Reichenbach

Large, with long rounded tail. Plumage striking, with black head and yellow body giving oriole-like appearance; quite unlike any other African warbler. Habitat edges of gallery forest near water. Nest globular, large and untidy, with side entrance, and suspended, more like a sunbird's than a warbler's nest. Song a rich, mellow undulating warble; pairs frequently duet.

Monotypic, endemic to W Africa. The taxonomic position of *Hypergerus* is uncertain. It has been considered a babbler because its shape resembles e.g. *Turdoides*, while the combination of black head and yellow and green body give it a superficial resemblance to the Asiatic bulbuls *Pycnonotus atriceps* and *P. melanicterus*; however, its fine melodious song is unlike that of a bulbul or babbler, and Chapin's (1953a) impression was that *Hypergerus* was an 'overgrown warbler'. It is best left *incertae sedis*, along with its presumed closest relative *Eminia*.

Hypergerus atriceps (Lesson). Oriole Warbler; Moho. Noircap loriot.

Plate 20
(Opp. p. 337)

Moho atriceps Lesson, 1831. Traité d'Orn., p. 646; Gold Coast.

Hypergerus atriceps

Range and Status. Endemic resident and possible partial migrant. Common S Senegal, Gambia (local, not uncommon behind coastal sand dunes at Kotu and Tanji, also occurs middle and upper river to Bansang), Guinea-Bissau (Gunnal); uncommon Guinea (coastal scrub near Sangaria Bay, Dubreka, Conakry, Labe and north of Fouta-Djalon Mts); common W Sierra Leone; uncommon and local S Mali (Yanfolila, Sikasso, Boucle du Baoulé Nat. Park); S and central Burkina Faso; rare SW Niger ('W' Nat. Park, Nov–Mar, also Mekrou); common N Ivory Coast (gallery forests in guinea zone, uncommon as far south as Lamto but occurs near coast in Azagny Nat. Park, 90 km west of Abidjan); Ghana (throughout northern savanna, south through Volta Region to Accra Plains and westwards along coast to Elmina); Togo and Benin (widespread inland and coast); Nigeria (SW coastal area north to Danbagudu, Anara Forest Reserve, Dunbi Woods, Gubuchi and Zaria (Fry 1975), Jos Plateau, Yankari, Aliya and north of Benue R. except for Mambilla Plateau at 1500–2000 m); Cameroon (forest/savanna montane areas and Adamawa Plateau, also Benue reserve and within forest zone at Ebolowa in S); S Chad (Baibokoum, and S border with Central African Republic); western half of Central African Republic and N Zaïre (near great bend of R. Ubangi).

Description. ADULT ♂: head, neck, throat and upper breast black, feathers of crown, nape, cheeks, and neck bordered with silvery grey. Rest of upperparts olive-yellow, including wing-coverts and tail, yellowest on uppertail-coverts. Tail strongly graduated (T6 41–51 shorter than T1 (n = 5): M. P. Adams, pers. comm.). Lower breast and belly bright yellow, shading into olive-yellow on flanks and thighs; undertail-coverts paler yellow and tinged rufous. Primaries dusky brown, outer webs edged olive-yellow, secondaries similar but yellow-olive wash also on inner webs; underwing-coverts and margins of wings yellow. Bill black; eyes olive-brown; legs and feet brown. Sexes alike. SIZE: (10 ♂♂, 10 ♀♀) wing, ♂ 79–90 (84·7); ♀ 76–81 (79·2); tail, ♂ 85–98·5 (93·0), ♀ 79–94 (86·2); bill to feathers, ♂ 20–21 (20·4), ♀ 20–21 (20·6); tarsus, ♂ 28–30 (28·6), ♀ 28–29 (28·4). WEIGHT: ♂ (n = 2, Cameroon, June, July) 28, 28; ♂♀ (n = 7, Ghana, July–Dec) 26·4–31·5 (28·9).

IMMATURE: like adult but black of head more greenish, upperparts more olive, edges of wings, middle of breast and belly duller yellow, eyes grey.

NESTLING: at *c.* 6 days old downy feathers well developed (Lang 1969).

Field Characters. A large and brightly coloured warbler, long-tailed and slim; head black, throat and upper breast mottled white, upperparts, including closed wings and fan-shaped tail, olive-yellow, and underparts bright yellow; bill black, long and slender. Colour and pattern more akin to a bright forest bulbul (e.g. Green-tailed Bristle-bill *Bleda eximia*) or an oriole (e.g. Western Black-headed Oriole *Oriolus brachyrhynchus*). If sighted briefly in poor light, might even be mistaken for a weaver (e.g. Yellow-backed Weaver *Ploceus melanocephalus*).

Voice. Tape-recorded (57, B, C, GRI, MEES, MOR). ♂ song (Ivory Coast) a phrase of cheerful, bold, undulating whistles with a full, rich tone, 'here, tea-purry'; 'woy-toy-tea-tay-purry', repeated without pause 4 times in 3·5 s. Phrase varies considerably; another ♂ (Ivory Coast) sang 'wheera ti-ti-t-t whirr ti-ti-ti', duration 2·0 s, repeated 6 times without audible pause in 12 s; 'wheera' and 'whirr' notes rounded, musical, not high-pitched, some of the 'ti' and 't' notes high-pitched and a bit squeaky. Pairs regularly duet. Presumed ♂ usually sings from an exposed perch and dominates the duet; his song is a combination of phrases, some of which are incomplete forms of others; favoured phrases can be repeated in unbroken sequence. Intervals between notes within a phrase are 0·04–0·15 s long; intervals between phrases 0·09–0·17 s; melodious song sounds regular and flowing and lasts up to 2 min or more. Song of presumed ♀ is trilled repetition of one note; increasing in speed from 10 to 12 notes per s, and remaining constant or decelerating as duet finishes (further details in Grimes 1974).

General Habits. Inhabits thick undergrowth of forest edge, dry thicket on inselbergs, gallery forest, oil palm plantations, and mangrove; often occurs near water. Shy and secretive, occurring usually singly or in pairs, conspicuous only while it duets, when ♂ invariably uses exposed perch. Moves slowly through vegetation while foraging; also feeds on ground.

Sedentary throughout most of its range, but probably locally migratory in Mali (occurring Yanfolila in Mar and Sikasso in Apr and May) and elsewhere at N edge of range.

Food. Insects (grasshoppers).

Breeding Habits. Territorial, monogamous, solitary nester.

NEST: globular, like a large untidy sunbird's nest or nest of African Broadbill *Smithornis capensis* (R. K. Brooke, pers. comm.), with side entrance *c*. 50 diam. near top and slightly overhanging porch; made of grasses, lined with finer grass, sometimes with pieces of dry plants woven into main structure. Ext. diam. 200 at its widest, chamber *c*. 150 deep; below the chamber dangles an untidy mass of ragged grass ends *c*. 200 long. Suspended from palm frond, twig or tendril, 1·8–3·6 m above ground, usually over water; woven attachment to tendril *c*. 70 thick. Replacement nest was built within 8 days at end of same tendril that had supported the first. Old nests may be renovated, new material added for strength, new porch built above entrance, and used again (suggesting that the species is double-brooded: Lang 1969). Same palm tree once used for at least 3 consecutive years (Gambia: Cawkell and Moreau 1963).

EGGS: 2–3, usually 3 in Gambia. Pale blue with violet or reddish brown spots and blotches. SIZE: (n = 1) 23 × 15·5 (Lang 1969).

LAYING DATES: Gambia, June–Nov, mainly Oct–Nov (3 feathered young in nest early Nov); Guinea-Bissau, mid-July; Sierra Leone (gonads enlarged May–June); Ghana (nest-building July, Aug); Burkina Faso (nest-building late July); Nigeria, July, Aug, Oct (breeding condition June–Oct).

INCUBATION: period *c*. 14 days.

DEVELOPMENT AND CARE OF YOUNG: nestling period *c*. 12 days.

References
Fry, C. H. (1975).
Grimes, L. G. (1974).
Lang, J. R. (1969).

Genus *Sylvia* Scopoli

W Palearctic warblers centred on the Mediterranean. Robust, sleek birds, bill not compressed and with culmen more or less curved. Rictal bristles not well developed. Shape of P10 variable; wing usually long. Many species capped, with contrasting malar stripe; upperparts bluish grey or rufous-brown, underparts off-white, buff or in several species reddish. Tail moderately long, square, round or graduated, usually white-sided. Sexes different; alike in a few species. Songs mainly scratchy, some melodious. Food insects, also small soft fruits. Inhabit low vegetation, understorey of woods, scrub, macchia, acacias. Nest open, eggs spotted. Mostly migratory, many intercontinentally.

17 species (or 19 if *minula* and *althaea* separated from *S. curruca*), including 5 species in 2 superspecies. W Palearctic, 4 species ranging east to Siberia. All occur in Africa; 1 endemic to NW Africa (*S. deserticola*); 3 endemic to W Mediterranean (*S. sarda*, *S. undata*, *S. cantillans*), two breeding in NW Africa; 2 endemic to whole Mediterranean (*S. conspicillata*, *S. melanocephala*), breeding in NW Africa; *S. melanocephala* comprises a superspecies with the Middle Eastern *S. mystacea*, which winters in NE Africa, and with an E Mediterranean endemic, *S. melanothorax* of Cyprus; the latter and another E Mediterranean endemic, *S. rueppelli*, do not breed in Africa but some winter there; *S. hortensis*, endemic to Mediterranean and Middle East and breeding in NW Africa, is very similar to *S. leucomelaena*, breeding in SW Arabia and adjacently in Africa, and we propose that they are of 'immediate common descent' and so form a superspecies; 1 species, *S. nana*, breeds in central Asia and also W Sahara; and 5 range widely in the Palearctic (*S. atricapilla* in W Palearctic), all wintering in subsaharan Africa, and *S. atricapilla* and *S. communis* breeding in NW Africa. In all, 10 species breed in Africa; of Eurasian migrant species, 6–7 winter in Africa north and 10 south of Sahara.

Sylvia hortensis superspecies

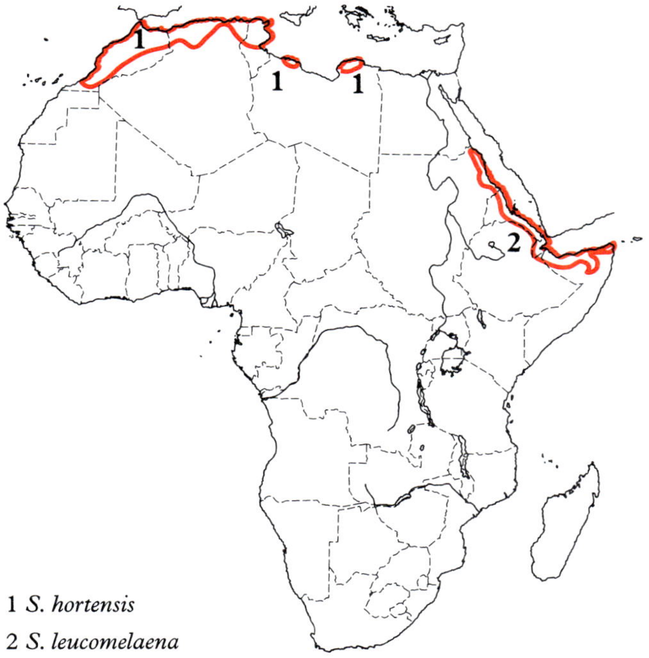

1 *S. hortensis*
2 *S. leucomelaena*

Sylvia nisoria (Bechstein). Barred Warbler. Fauvette épervière.

Motacilla nisoria Bechstein, 1795. Gemeinn. Naturgesch. Deutschlands, 4, p. 580, pl. 17; central and northern Germany.

Plate 23
(Opp. p. 416)

Range and Status. Central and E Europe to S Urals and Caucasus; also N and E Kazakhstan to W Mongolia and Tien Shan; winters Africa.

Palearctic migrant. Winters in Kenya north and east of the highlands (L. Turkana and Turkwell and Kerio Rivers south to Baringo and Samburu; and Isiolo and Meru districts south to Tsavo and Taita-Taveta, west to Nairobi and Kajiado and east to Lower Tana R.; common to locally abundant); also in NE and central Tanzania (Mkomazi south to Dodoma; frequent to locally common); and probably SW Ethiopia border areas. Occasionally winters S Uganda (Entebbe, Kampala); rare W Kenya (Nyanza). On passage, rare to uncommon in Egypt (13 autumn and spring records in last decade, mainly N coast and Red Sea, once Nile valley); locally common to abundant both migrations in E Sudan (especially Red Sea coast), Eritrea, Ethiopia (mainly Rift valley and W areas) and NW, central and

Sylvia nisoria

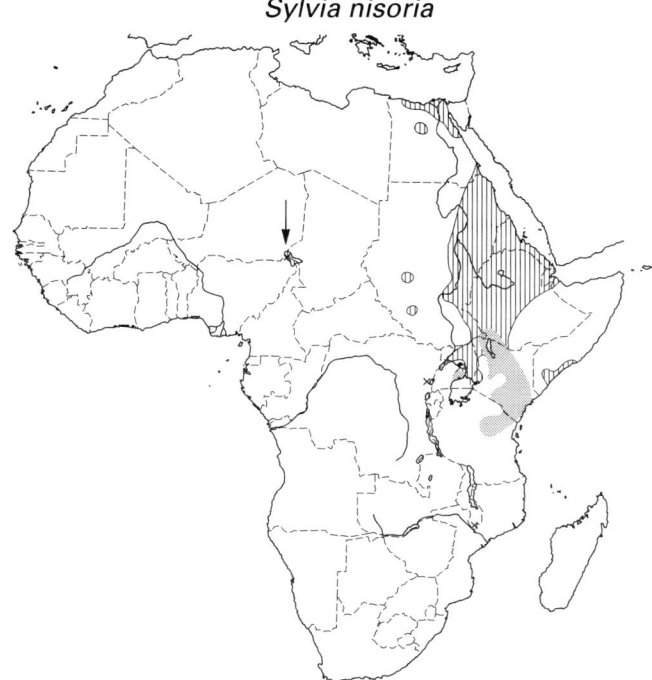

SE Kenya; occasional in SW Sudan and E and N Uganda (west to Butiaba and Entebbe); rare Djibouti and NW Somalia (spring); occasional S Somalia (autumn and spring).

Vagrant to Senegal (once, Djoudj, Nov), Nigeria (once, Malamfatori, Oct), Zaïre (once, L. Edward, Apr) and Malaŵi (once, Nchalo, Jan).

Wintering density up to *c.* 3–4 birds per ha in thick bush SE Kenya (Lindström *et al.* 1993); more than 10 per ha in riverine *Salvadora* in N Kenya.

Description. ADULT ♂: upperparts brownish grey, feather tips with white fringes bordered by dark grey subterminal bands, giving scaly appearance. Lores and ear-coverts grey flecked with white. Tail grey-brown, feathers with white fringe along tip; T4 (sometimes also T3) with broader white area at tip of inner web; T5 and T6 with distal 10–30 mm of inner web white except along shaft, T6 also with narrow white outer edge. Underparts white, tinged creamy from chin to breast and on flanks; chin and throat closely barred with grey; breast and flanks more coarsely scalloped dark grey; side of belly and often centre of belly and vent with short grey bars; undertail-coverts with dark grey subterminal chevrons. Primaries, secondaries and primary coverts dark grey, narrowly edged greyish white; tertials and rest of upperwing-coverts brownish grey with contrasting creamy white fringes bordered subterminally by blackish bars, tips of greater coverts forming distinct whitish wing-bar; axillaries and underwing-coverts creamy white barred grey. Wing-tip bluntly pointed; P8 longest (occasionally = P9), P7 0·5–3 shorter, P6 3–7, P5 5–10, P10 16–22; P9 0–2·5, between P7 and P8 (sometimes P6 and P7); P10 minute, 5–10 shorter than primary coverts; P7–P8 emarginated, sometimes also tip of P6. Bill blackish brown, base of lower mandible yellowish; eyes golden yellow; legs and feet yellowish grey. ADULT ♀: like adult ♂, but on average tinged browner, less strongly barred above; barring on underparts often more restricted, belly more extensively white. Eyes dirty yellow or yellow-green. SIZE: (10 ♂♂, 10 ♀♀, Europe and Asia) wing, 87–92 (89·0), ♀ 86–90 (87·8); tail, ♂ 67–73 (70·0), ♀ 66–72 (68·3); bill, ♂ 16–18 (17·1), ♀ 17–18 (17·6); tarsus, ♂ 23–25 (24·4), ♀ 23–25 (24·3).

WEIGHT: unsexed: NE Sudan (n = 64, Aug–Sept) 17–27 (21·5), (n = 20, Mar–Apr) 21·5–39·5 (28·8); Ethiopia (n = 64, Oct–Feb) 19·6–28 (23·7), (n = 70, Mar–Apr) 18·7–27 (23·1); N Kenya (n = 15, Mar–Apr) 21–34·2 (27·1); SE Kenya (n = 618, Nov–Dec) 18–28·8 (22·5), (n = 79, Jan–Feb) 19·8–27 (23·5), (n = 46, Mar–Apr) 21·4–34·9 (26·3).

Moults on breeding grounds, July–Aug, complete except for secondaries and most tail-feathers; also in Africa, Nov–Jan, complete except for primaries and primary coverts.

IMMATURE: 1st winter bird almost completely unbarred; browner above than adult, with noticeable pale supercilium; underparts creamy white, tinged buff on breast and flanks, faint grey barring, if any, confined to lower flanks and undertail-coverts; tail-feathers tipped buff, not white; fringes of tertials and wing coverts pale buff, those on greater coverts forming distinct wing-bar; axillaries and underwing-coverts creamy white; eyes dark brown, greyish or pale brown; legs and feet grey. 1st summer ♂ like adult ♀; thus browner, less barred than adult ♂; eyes yellow or yellowish brown. 1st summer ♀ often with little barring on underparts and underwing-coverts; eyes sepia or greenish brown.

Field Characters. A large grey or grey-brown *Sylvia* with a strong bill, strong legs and feet and a rather long and full tail. Shape and habits often suggest a large Common Whitethroat *S. communis*. Well-marked adults are highly distinctive, with barred grey upperparts, conspicuously dark-barred creamy underparts, bright yellow eye, whitish wing-bar and white corners and outer edge to tail. Dull ♀♀ are browner grey above, less barred, and have a greenish eye. 1st winter birds are browner and practically unbarred, and the rather plain face and dark eye may suggest Garden Warbler *S. borin*. Distinguished, however, by heavier and longer-tailed appearance and less stubby bill, and often by noticeable buff wing-bar and tertial edges. Young Orphean Warbler *S. hortensis* is also large, but shows contrast between white throat and dark ear-coverts and has broad white sides to tail.

Voice. Tape-recorded (62, 73, 93, B, PEA). Song in Europe is a vigorous rich warbling. Short variable phrases, usually *c.* 3–5 s long, are separated by intervals of a few s, at times interspersed with rattle calls (see below). Recalls song of Garden Warbler though less throaty and less evenly delivered. Song in Africa quieter, with bouts of warbling commonly 10–20 s long; very like subdued song of Garden Warbler. Contact-alarm calls, a hard rattling 'trrrt' or 'trrrrt-trrt' and a harsh 'tschurrrr', often a more extended rattle lasting *c.* 3 s and slowing towards end, 'trrrt-t-t-et-et'; occasionally a soft 'tsek' (Cramp 1992).

General Habits. Occurs mainly in semi-arid and arid country, inhabiting bush, thickets, secondary growth and dry open woodland with a bushy underlayer. In E Kenya, typically in bush or woodland dominated by *Acacia* and *Commiphora*, with thickets of e.g. *Grewia*, *Combretum* and *Premna*; in arid N and NW Kenya, mainly in *Salvadora persica* thickets along dry watercourses. On passage in NE Sudan, in dry riverine acacia thickets. Solitary in winter quarters, or in small loose groups. Sometimes 20 or more birds together at passage times. Secretive and usually detected by calls. Forages mainly within bushy green thickets, or in low tree canopy, taking berries or picking insects from foliage. Rarely attempts to catch flying prey. Occasionally feeds on ground near cover. In Kenya (Tsavo), 96% of items taken from leaves, 4% from twigs or stems; usually fed well inside bushes or canopy; av. foraging height 3·0 m (Lack 1985). Carriage usually horizontal, with long tail slightly raised. Moves with rather heavy hopping gait. Flies directly and often quite low with tail slightly spread, usually disappearing quickly back into leafy cover. Rattle call given commonly in Africa, especially at dawn when several birds may answer in chorus. Sings commonly from low thickets in E Africa during Feb–Mar; typically a prolonged warble, at first quiet and subdued, but becoming stronger towards spring departure.

Moves south from breeding grounds in Aug, with passage of all populations concentrated through Levant and Middle East. No N African records west of Egypt. Main entry into Africa is across Red Sea to NE Sudan and Eritrea during late Aug–Sept. Occurs on passage and stopover in Ethiopia from late Sept to mid-Dec. Arrives in Kenya from early Nov; locally abundant some northern areas by mid-Nov, but main build up in southeastern wintering sites not until Dec–Jan. Passage at Ngulia is protracted, from early Nov to early Jan. Adults reach Africa with new primaries, but with most secondaries and tail-feathers unmoulted (Nikolaus and Pearson 1991). Adult and 1st winter birds moult secondaries, tail, tertials and body feathers in Kenya or Tanzania during Nov–Jan (Lindström *et al.* 1993). Birds leave Kenya wintering sites in late Mar to early Apr. Passage through Kenya and E Uganda occurs mid-Mar to mid-Apr (last record 28 Apr). Some birds apparently move north earlier. Thus, some fat birds in SE Sudan in mid-Feb, and passage begins in Ethiopia in early Feb, lasting to early Apr. Strong passage on Eritrean and Sudan coasts, lasting to early May, but all populations apparently again skirt E Mediterranean, and very few spring records Egypt. Migration through Middle East occurs mainly Apr to early May. A bird ringed Poland (June) was recovered Sudan (*c.* 13°N, 32°E) the following Oct; 1 ringed Finland recovered Sudan; and 1 ringed Kenya (Nov) was on passage Saudi Arabia (*c.* 26°N, 44°E) the following Sept.

Food. Invertebrates, especially beetles, Hymenoptera, Hemiptera and Orthoptera. Also fruit in autumn and winter: Kenya wintering birds fed on berries of *Premna* and *Grewia*; a vagrant in Senegal took *Bridelia* berries.

References
Cramp, S. (1992).
Lindström, Å. *et al.* (1993).
Pearson, D. J. (1978).

Sylvia hortensis (Gmelin). Orphean Warbler. Fauvette orphée.

Plate 23
(Opp. p. 416)

Motacilla hortensis Gmelin, 1789. Syst. Nat. I, p. 955, ex Brisson, Buffon, Daubenton, pl. 579, fig. 1; France and Italy, restricted to France.

Forms a superspecies with *S. leucomelaena*.

Range and Status. N Africa, S Europe, Turkey, Levant, Caucasus, and Iran to W Pakistan and W Tien Shan; winters in Africa, and from Arabia to India.

Palearctic migrant. Breeding summer visitor N Africa south to Sahara borders. Locally frequent to common in N and W Morocco (south to High Atlas and Anti-Atlas), N Algeria (south to Atlas Tellien, Biskra and Laghouat), Tunisia (south to Gafsa and Gabès), NW Libya (Tripoli, Jebel Nafusa) and NE Libya (Jebel Akhdar).

Winters south of Sahara, mainly in Sahel between 14°N and 17°N; in SW Mauritania (to *c.* 18°N) and N Senegal, and occasionally south to coastal Gambia; and from N Burkina Faso, Niger (north to Aïr) and extreme N Nigeria (south to Sokoto and Maiduguri) to central Chad (north to Fada, south to Ouadi Haddat), W Sudan (Darfur), central Sudan (south to Melut), NE Sudan, Eritrea and N Ethiopia (south in Rift to about 11°S); locally frequent to common. Occasional midwinter records in and north of Sahara, in Morocco (Agadir) and Algeria (Algiers, Beni Mellal, Djanet). On passage, rare Morocco–Tunisia in autumn, frequent in spring

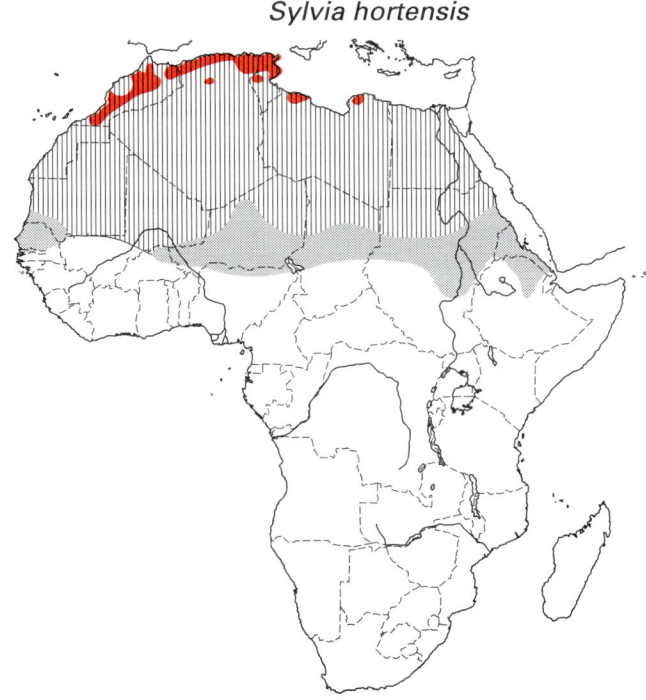

when present in Saharan oases; rare Libya (mainly spring); rare Egypt (mainly the east, mainly spring); locally common both seasons in NE Sudan (Red Sea coast).

Description. *S. h. hortensis* (Gmelin): Morocco to NW Libya, and SW Europe; winters S Mauritania and N Senegal to W Sudan. Top and sides of head down to ear-coverts dull black, merging into grey-brown on hindcrown and sides of neck; rest of upperparts uniform brownish grey. Tail dull black, T2–T3 with narrow white fringe at tip, T4 with small white spot at tip of inner web, T5 with small white wedge along shaft of inner web, T6 white with broad dusky border along inner web. Underparts white, lower throat, breast, flanks and undertail-coverts suffused creamy pink or creamy buff. Primaries, secondaries and primary coverts blackish grey with narrow grey-white fringes; tertials and greater coverts dark brownish grey, fringed paler grey; remainder of upperwing brownish grey like rest of upperparts. Underwing-coverts and axillaries pale grey or whitish. Wing-tip bluntly pointed: P7–P8 longest; P6 1–3 shorter, P5 3–7, P1 14–19; P9 2–7 shorter, between P5 and P6; P10 minute, 2–8 longer than primary coverts; P6–P8 emarginated. Bill blackish grey above, blue-grey below with blackish tip; eyes pale brown with broad whitish or yellowish white ring; legs slate-grey to greyish brown. ADULT ♀: like adult ♂, but forehead and crown dark grey rather than blackish; rest of upperparts slightly browner than in ♂, underparts suffused more buffish, less pink; eyes pale brown with narrow broken whitish ring or spotting. SIZE: (10 ♂♂, 10 ♀♀) wing, ♂ 77–84 (79·3), ♀ 73–84 (78·2); tail, ♂ 60–67 (64·3), ♀ 61–64 (62·2); bill, ♂ 17–19 (18·5), ♀ 17–19 (17·9); tarsus, ♂ 22–24 (22·9), ♀ 21–23 (22·4). WEIGHT: N Senegal, sexes combined (n = 20, Oct–Nov) 18·5–22 (20·3), (n = 9, Dec–Feb) 17·6–22·8 (20·1), (n = 64, Mar–Apr) 17·8–33·8 (24·0).

IMMATURE: juvenile has top of head grey-brown, uniform with rest of upperparts; lores and ear-coverts darker than crown, but contrasting less with throat than in adult; sometimes a faint pale supraloral stripe; sides of breast and flanks more buffish; tail without white tips to T2–T4, and with white marking on T5–T6 suffused greyish or buff; fringes of tertials and greater coverts more olive-brown, less grey; eyes dark olive-brown. 1st winter like adult ♀, but upperparts browner; retains juvenile tail, tertials and greater coverts.

NESTLING: naked; pale grey down develops at 2–3 days.

S. h. crassirostris (Cretzschmar): NE Libya, SE Europe to Levant, Turkey and Caucasus; winters central and NE Sudan, Eritrea and N Ethiopia. Underparts slightly whiter, extent of pinkish or buffish wash reduced; bill slightly longer: ♂ (n = 10) 18–20 (19·5). WEIGHT: NE Sudan, ♂♀ (n = 44, Aug–Sept) 16·5–25 (19·8), (n = 6, Apr) 21·5–35 (30·6).

Field Characters. A rather large warbler with strong bill and heavy head. Uniform grey-brown upperparts and dark grey to black top and sides of head contrast with whitish underparts, especially with white chin and upper throat. Legs dark greyish. Eye of adult usually shows prominent whitish ring, especially in western birds. Typically steep forehead and broad chest add to impression of bulk about head and forebody (Cramp 1992). Might be confused with Sardinian Warbler *S. melanocephala*, Ménétries's Warbler *S. mystacea* and ♀ Rüppell's Warbler *S. rueppelli*, but they are all much smaller and slimmer looking, with reddish eye and pale legs. Lesser Whitethroat *S. curruca*, also smaller, has shorter bill and more contrast between dark ear-coverts and grey crown. Near Red Sea coast, Orphean occurs with resident Red Sea Warbler *S. leucomelaena*, similar in size and plumage, but with blacker cap, whiter underparts, shorter bill, shorter wing, and longer, blacker tail with only narrow white outer edge.

Voice. Tape-recorded (CHA – 57, 62, 73, 93, B). Song melodious and far-carrying; that of nominate *hortensis* consists of short, unhurried phrases of *c.* 1·5 s duration given about 6–7 times per min, each phrase composed of several repetitions of a mellow, thrush-like, usually paired unit, 'weecha-weecha-weecha-weecha, weecha-wrytty-wrytty-wrytty-uh' (Cramp 1992); song of *crassirostris* fuller, with longer, more continuous and more varied phrases, including mimicry; a loud, vigorous thrush-like warble, without harsh notes, lacking monotony of nominate *hortensis* (Beven 1971). Contact-alarm calls, a single 'tak', recalling Lesser Whitethroat *S. curruca* or Blackcap *S. atricapilla*, or a rapid series of 'tek-tek-' notes; also a rattling 'trrr', recalling Sardinian Warbler but less hard and prolonged (Cramp 1992).

General Habits. Breeds in open woodland, citrus and olive groves, orchards and gardens, in lowlands as well as on hill sides and mountain foothills (Cramp 1992). In NW Africa, especially in sparse evergreen oak woods with little undergrowth; ranges up to 1600 m in the Atlas and 1700 m on Djebel Amour, but rarely at higher levels. Winters in Sahelian tree-steppe, and acacia thickets and woodland. In Oman, mainly frequents small-leaved thorn trees (*Acacia, Ziziphus*).

Solitary in winter quarters or in small groups; in NE Sudan associates on autumn migration with other *Sylvia* warblers. In Senegal, *c.* 10 per km^2 in woodland savanna (Morel and Roux 1966). In S Mauritania, 1·3 per km^2 in open steppe (Browne 1982). Not shy. Forages in tree canopy and tops of bushes with rather slow deliberate movements, picking insects from leaves and branches. Descends to ground less often than other *Sylvia* warblers (Cramp 1992). Gait hopping, but powerful leaps allow fast progress among foliage. Carriage generally above horizontal. Raises crown-feathers when excited. Sometimes cocks tail. Flight fluent and confident. Bird often flies some distance when disturbed, heading for tree canopy. Makes heavy entry into cover with tail slightly spread (Cramp 1992). Sings occasionally in winter quarters, e.g. in Eritrea in early Mar (Smith 1957).

Moults near breeding grounds July–Aug, but moult usually suspended so that many secondaries (sometimes also primaries) retained for autumn migration.

European nominate *hortensis* enters and leaves Africa from Morocco east to Tunisia. Autumn passage Sept to early Oct, but inconspicuous, suggesting long unbroken flights to tropics. Leaves NW African breeding areas in Sept. Reaches winter quarters mainly Sept–Oct.

Passage occurs through N Senegal in mid Sept–Oct. Northward migration begins late Feb to early Mar and continues to May. Present Niger until Apr, sometimes May, N Senegal to early May. Spring passage noticeable in Senegal, Mauritania, and Algerian and Moroccan Sahara. More conspicuous near NW African coast than in autumn, mainly late Mar to mid-May.

Race *crassirostris* is scarce in autumn in Egypt; apparently enters NE Africa mainly across Sudan coast, where main passage occurs late Aug to early Oct. Spring migration begins in Feb, but most birds remain on wintering grounds in Sudan and Eritrea until Mar. Main route apparently as in autumn, with only a few late Mar to late Apr records in Egypt.

Food. Mainly invertebrates; also fruit. In breeding areas, figs, fruit and berries important in diet in Aug–Sept. In N Senegal, Mar, takes *Salvadora* berries; in Niger, May, *Salvadora* and insects; in Mali, Oct–Nov, large insects and fruit (Cramp 1992). In Arabia, ate large locusts (Meinertzhagen 1954); in Oman, eats small insects.

Breeding Habits. (Best known for extralimital *S. h. jerdoni*: e.g. Kovshar' and Rukina 1968.) Monogamous; solitary and territorial; territory large for a *Sylvia*, up to 2·4 ha; used for courtship, nesting and raising young. In Morocco, 18–23 pairs per km² in one valley in NE (Cramp 1992); 1 pair per km² in maquis (Thévenot 1982). ♂ uses habitual song-posts, sometimes up to 200 m from nest. Usually sings from cover of small trees or bush tops 2–8 m above ground; occasionally in horizontal song-flight *c.* 10 m above ground between perches 20–30 m apart (Cramp 1992).

NEST: well-built cup of grass, plant stems and usually a few twigs, bound with moss, cobwebs and plant down and lined with finer grasses and fibres; in outer branches of small tree or bush, usually 1–3·5 m above ground; outer diam. *c.* 110, inner diam *c.* 68, cup depth *c.* 45; built by both sexes in up to 4–5 days. Surplus nests occur, probably built by ♂.

EGGS: 3–5 in NW Africa, occasionally 6 (60 clutches in Algeria and Tunisia, av. 4·2). White or faintly tinged bluish, sparsely spotted and blotched brown, olive and grey, mainly at broad end. SIZE: (n = 120 SW Europe) 18–22 × 13–16 (19·2 × 14·5). Single-brooded.

LAYING DATES: NW Africa, mid-Apr to early June.

INCUBATION: by both sexes but mainly by ♀. Period: 12–13 days.

DEVELOPMENT AND CARE OF YOUNG: young cared for and fed by both parents, brooded for 4–5 days; remain in nest 12–13 days; fed by parents for 5–6 days after fledging.

BREEDING SUCCESS/SURVIVAL: in 29 clutches in W Tien Shan (*S. h. jerdoni*), 56% of eggs hatched and 61% of these fledged: overall success 34%, with av. 1·4 young fledged per nest (Kovshar' and Rukina 1968).

References
Beven, G. (1971).
Cramp, S. (1992).
Kovshar', A. F. and Rukina, A. K. (1968).

Sylvia leucomelaena (Hemprich and Ehrenberg). Red Sea Warbler; Arabian Warbler. Fauvette d'Arabie.

Plate 23
(Opp. p. 416)

Curruca leucomelaena Hemprich and Ehrenberg, 1833. Symb. Phys. Avium, fasc. cc, footnote 7; Arabia.

Forms a superspecies with *S. hortensis*.

Range and Status. Resident on both sides of Red Sea and Gulf of Aden, north to Arava valley (Israel) and east to Dhofar (Oman) and Cape Guardafui (Somalia), ranging from coast to about 100 km inland.

Breeding resident from Egypt to Somalia. Egypt, common in well-vegetated wadis in foothills of Gebel Elba, on Sudan border; may possibly occur further north, in central Red Sea Mts (Goodman and Meininger 1989). Sudan, fairly common in acacia woodland in Red Sea hills. Ethiopia, uncommon in hills of S Eritrea. Djibouti, widespread, commonest on coastal plain (10 seen in a day, between Obock and Waddi). Somalia, frequent in acacias on maritime plain and in adjacent mountains and on plateau to south, as mapped.

Description. *S. l. blanfordi* Seebohm: SE Egypt, E Sudan. ADULT ♂: forehead, crown, lores and ear-coverts matt black with dark brown tinge; conspicuous orbital ring of white featherlets (variable in extent, often broken); nape, hindneck and sides of neck dark greyish brown; mantle, scapulars, back, rump and uppertail-coverts greyish brown. Tail black, T6 with white edge to outer vane, T5 and T6 (and sometimes T4) with 2–3 mm white tip. Underparts white, breast tinged pink in centre, flanks pale buffy grey tinged pink, undertail-coverts tinged buff. Thighs buffy. Wings blackish brown, tertials broadly and upperwing-coverts narrowly fringed ash-grey. Underwing-coverts and axillaries white, tinged with buff-pink. Bill slaty black, with proximal half of lower mandible bluish grey; eyes brown (Shirihai 1988b, 1989; not with the 'clear whitish grey ring' of Cramp 1992); legs and feet dark grey. ADULT ♀: like ♂ but cap (particularly hindcrown) browner, less blackish, white eye-ring less conspicuous, always

Sylvia leucomelaena

broken, remaining upperparts slightly browner than in ♂, flanks slightly buffier, underparts with less of a pink tinge. SIZE: wing, ♂ (n = 13) 65–70 (67·5), ♀ (n = 10) 65–67 (65·8); tail, ♂ (n = 12) 59–67 (64·7), ♀ (n = 8) 56–63 (61·1); bill to skull, ♂ (n = 11) 11·5–16·7 (14·7), ♀ (n = 9) 12–16·6 (15·0); tarsus, ♂ (n = 13) 20–23·4 (21·9), ♀ (n = 10) 20·5–24 (22·2). WEIGHT: (Egypt) ♂ (n = 5) 11·2–13·2 (12·3), 1 ♀ 12·5, unsexed (n = 29) 12·5–16 (13·9).

IMMATURE: like adult ♀ but rather browner above, small whitish patch above lores, iris dark brown, no white orbital ring; wing-coverts with rusty buff fringes; underparts buff, without pink tinge; white in tail is buffy.

NESTLING: not described; gape light orange.

S. l. somaliensis Sclater and Mackworth-Praed: Eritrea, Djibouti, Somalia. Cap of ♂ less blackish and upperparts slightly browner than in ♂ *blanfordi*; far more white on T6.

TAXONOMIC NOTE: there is little current support for placing this species any longer in *Parisoma* (see diagnosis of that genus). In the field we judge it a typical *Sylvia*. Making it an allospecies of *S. hortensis* is novel, although Shirihai (1988a, 1989) presents abundant evidence for that interpretation. We propose also that *S. leucomelaena* is closely linked with the ill-known Arabian warbler *Parisoma (Sylvia) buryi*.

Field Characters. A rather large, short-winged *Sylvia* of semi-arid woodland near Red Sea and Gulf of Aden coasts and hills, black-capped with contrasting white throat; dark greyish brown upperparts, blackish tail (narrow white edge usually difficult to see), white or pale buffy underparts. Soft parts blackish, but adult, particularly ♂, has narrow white eye-ring. Juvenile browner, but lores and ear-coverts blackish, making it look like *Lanius* shrike. Warbling, *Turdus* thrush-like song. Distinctive habit of flopping tail downwards a few times after each time bird moves. In plumage quite like Ménétries's Warbler *S. mystacea*, which told by its clearly white-sided tail which is constantly wagged, and by red eye of adult ♂, but more like Orphean Warbler *S. hortensis*. Latter does not droop its tail; and usually has yellow eye. Juveniles of Red Sea and Orphean are alike, but former has 5–12 mm projection of primaries beyond tips of tertials in closed wing, and latter 16–20 mm projection (Shirihai 1989).

Voice. Tape-recorded (92, HOL, SHI). Rather silent outside breeding season. ♂ song a loud warble lasting 5–15 s, melodious but with some harsh, scratchy and gurgling notes, strongly repetitive. Audible several hundred m away. Recalls songs of thrush *Turdus*, Upcher's Warbler *Hippolais languida*, Blackcap *Sylvia atricapilla* and Common Whitethroat *S. communis*. ♀ sings occasionally. Call, a quiet, soft 'chack' or 'chack-chack-chack' or 'tscha-tscha-tscha'; also a soft rattle, sometimes used between songs.

General Habits. Inhabits thick woodland of *Acacia tortilis* and *A. raddiana* at about 550 m (Gebel Elba, Egypt/Sudan), steep gullies with sparse *Acacia* cover below 320 m (Eritrea), open *Acacia/Commiphora* bushland (Yemen) and thick or open woods with or without acacias, on coastal plain and foothills, often near a stream (Oman). Ascends higher outside Africa: resident up to 1900 m in Yemen, where 'winters' up to 2800 m. In Israel inhabits wadis with dense stands of *A. tortilis* and *A. raddiana*, tamarisks and date palms; also with the shrubs *Ochradenus*, *Zygophyllum*, *Calligosum*, *Lycium*, *Gymnocarpus* and *Nitraria*. Life is centred upon dense stands of *Acacia* (Hebrew name is Acacia Warbler: Afik and Pinshow 1984).

Habits barely known in Africa. Elsewhere, occurs solitarily or in pairs, not shy, easily to be seen foraging and preening: feeds in typical *Sylvia* manner, by moving in short (sometimes long) hops through thicker parts of trees and bushes, keeping mainly at 2–3 m above ground, gleaning twigs, leaves, thorns and small branches for invertebrates, and – the commonest method of insect capture – picking caterpillars out of acacia tree bark. Hawks flying insects, including dragonflies; sometimes forages on ground under acacias by digging into surface of soil with bill; small insects eaten whole, large ones pounded against branch (Afik and Pinshow 1984). Characteristically droops tail 1–3 times after each hop. ♂ sings conspicuously from bush top; at other times rather silent.

Mainly resident; disperses altitudinally in Yemen.

Food. Soft-bodied insects, beetles, and small, pulpy fruits; also seeds (Egypt). In Israel most insect prey is caterpillars of pyralid moths; fruits include *Lycium shawii*, *Nitraria retusa* and *Ochradenus baccatus*.

Breeding Habits. Monogamous, solitary, territorial breeder (Israel). Pair defends territory of 20–70 h, by flying together from tree to tree along boundary, flight undulating and purposeful, ♂ singing at each stop. Boundary patrol may be hundreds of m from nest tree, but when nestlings are being reared adults forage within a few tens of m of nest. Aggressive conflict with conspecifics is uncommon.

NEST: cup made of stems and fine branches of dry annual plants, interwoven with fibres of *Erodium hirtum* and other plants (Israel) and lined with hairy *E. hirtum* seeds; some nests fairly robust, others thin and transparent. Height 70, ext. diam. (n = 15) av. 85, int. diam. (n = 15) av. 55, cup 30 deep. Sited in periphery of *Acacia* tree canopy, 0·8–3·0 m above ground, av. (n = 15) 1·6 m. Built by ♂ and ♀, in 4–6 days.

EGGS: 2–4, av. (n = 10, Israel) 2·7. Laid before dawn at 24 h intervals. Glossy; smooth; subelliptical; cream-white, speckled with brown and grey. Size unknown. 2 broods, occasionally 3.

LAYING DATES: none known, Africa. Israel, Mar–May.

INCUBATION: starts on completion of clutch; by ♀ and ♂, in spells of about 20 min, once 4 h. Period: 15–16 days.

DEVELOPMENT AND CARE OF YOUNG: chicks brooded constantly for 2 days after hatching; parents ate chicks' faecal sacs, but after day 3 carried them away and dropped them. Young call loudly from day 6. Wing-feathers sprout on day 6; primaries 20 mm long on day 10. Nestling period 14–17 days. Young of 1st brood tended for 21, 22 and 40 days and of 2nd brood for 14 and 32 days.

BREEDING SUCCESS/SURVIVAL: 27 eggs in 10 nests produced 10 fledglings (Israel).

References
Afik, D. and Pinshow, B. (1984).
Cramp, S. (1992).
Goodman, S. M. (1988).
Shirihai, H. (1989).

Sylvia borin (Boddaert). Garden Warbler. Fauvette des jardins.

Motacilla borin Boddaert, 1783. Table Planches Enlum., p. 35; France.

Plate 22
(Opp. p. 401)

Range and Status. Europe north to the Arctic, and through Russia to W Siberia and NW Kazakhstan; also N Turkey and Caucasus; winters Africa.

Palearctic migrant. Winters from Gambia, extreme S Mali, S Burkina Faso, central Nigeria, S Central African Republic, central Uganda and W and S Kenya, south to Angola, Botswana, Namibia (south to Okahandja), W and S Mozambique and South Africa (Transvaal, N Cape Province and Natal). Vagrant further south: Durbanville (Dec), Grahamstown (Feb) and Tygerberg Nat. Res. (4, Feb) (Underhill 1992, Allan *et al.* 1995). Widespread and common to abundant in W Africa in more humid savannas from *c.* 10°N southwards, and in E and southern Africa in green bush and woodland from *c.* 2°N, mainly below 1800 m. The most common Palearctic warbler in much of Zaïre, Rwanda and Uganda. In N Africa, locally common to abundant both passages, and frequent to common in Saharan oases. South of Sahara, frequent to locally abundant on passage in SW Mauritania, Senegal, Mali (to *c.* 17°N), N and central Nigeria, central and S Chad (mainly autumn), central, E and S Sudan (mainly autumn), Eritrea and Djibouti (autumn), and Ethiopia. Uncommon Niger (mainly autumn); rare N Somalia (spring). Wintering numbers augmented by double passage in Gabon, Zaïre, Rwanda, Uganda (especially autumn), Kenya, Zambia and Malâwi. Old reports suggest breeding N Tunisia (young bird caught July 1952) but no recent records.

Description. *S. b. borin* (Boddaert): W and central Europe and Scandinavia, east to S Urals and Caucasus; winters throughout African range. ADULT ♂: upperparts brown with olive tinge, often slightly greyer on crown and hindneck. Lores and inconspicuous supercilium pale greyish; narrow eye-ring creamy white, usually distinct; ear-coverts and side of neck

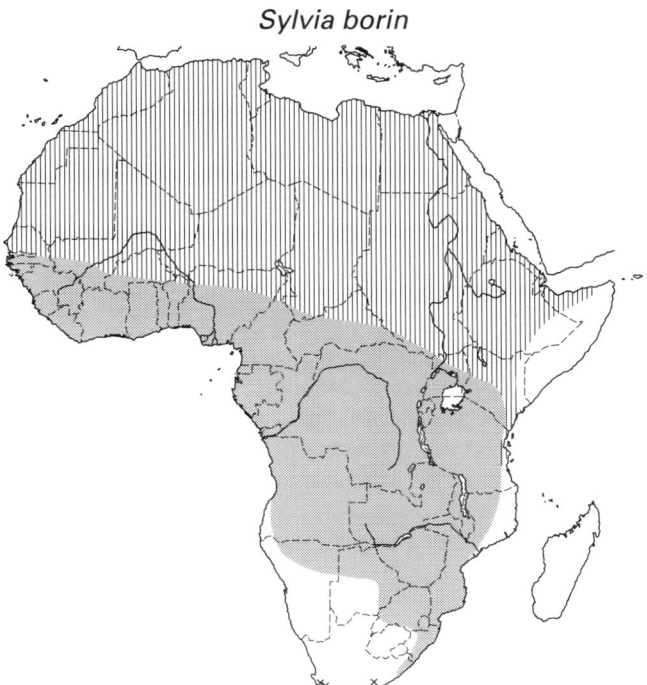

light brown, the latter usually with pale grey suffusion. Tail drab-brown tinged olive-brown along outer webs. Underparts whitish, tinged brownish buff on upper breast and flanks; centres of undertail-coverts pale brownish grey; lower throat sometimes faintly streaked with grey. Primaries, secondaries and primary coverts dark greyish brown, narrowly edged olive-brown; tertials and rest of upperwing-coverts olive-brown; axillaries and underwing-coverts pale tawny-buff. Wing-tip bluntly pointed; P8 longest (occasionally = P9), P7 2–4 shorter, P6 5–9, P5 8–13, P1 17–23; P9 0–2, between P3 and P4; P10 minute, 2–6 shorter than primary coverts; P8 emarginated, often also tip of P7. Bill dark brown, base of lower mandible pale flesh or pale horn; eyes dark brown; legs and feet dark horn, brownish grey or slate-grey. Sexes alike. SIZE: (10 ♂♂, 10 ♀♀, Europe) wing, ♂ 77–83 (79·8), ♀ 75–82 (78·5); tail, ♂ 52–56 (53·6), ♀ 52–55 (53·8); bill, ♂ 14–16 (14·8), ♀ 14–15 (14·7); tarsus, ♂ 20–22 (20·9), ♀ 20–22 (21·0). W European birds smaller than those from N and central Europe: wing, (10 ♂♂, Britain) 75–79 (76·9), (18 ♂♂, Poland and W Russia) 79–85 (81·2). WEIGHT: unsexed: (n = 23, Morocco, autumn) 15·5–25 (21·4); (n = 43, NW Morocco, spring) 14·4–22·7 (17·1); (n = 17, SE Morocco, spring) 13·3–20·8 (16·3); (n = 377, N Algeria, autumn) 15–29·9 (19·6); (n = 129, central Algeria, autumn) 11·7–29·6 (23·5), (n = 15, S Algeria, autumn) 16·9–23 (19·3); (n = 75, Egypt, autumn) 13·3–32·4 (21·1); N Senegal (n = 71, Oct–Nov) 12·7–23·3 (15·8), (n = 63, Apr) 13·9–28 (24·1); L. Chad, Nigeria (n = 103, autumn) 14–22·6 (17·4), (n = 302, spring) 14·5–35·5 (22·2); Zaria, N Nigeria (n = 14, autumn) 15–20·5 (17·8), (n = 11, spring) 15·4–33 (21·5); Vom, central Nigeria (n = 85, autumn) 13·8–23 (17·3), (n = 71, spring) 17–32·5 (21·4); S Nigeria (n = 20, Nov–Mar) 15·0–22 (17·8), (n = 8, Apr) 18–37 (25·2); (n = 78, NE Sudan, Aug–Oct) 13·0–24·5 (16·0); Ethiopia (n = 130, Sept–Mar) 12·7–22·2 (17·1), (n = 202, Apr–May) 14–34·3 (20·0); Uganda (n = 237, Oct–Dec) 15–24·5 (19·2), (n = 72, Jan–Feb) 15·6–21·6 (19·3), (n = 143, Mar–Apr) 16·6–28·2 (20·2); Kenya (n = 518, Oct–Dec) 13·5–24·7 (18·2), (n = 170, Jan–Feb) 14·8–23·3 (18·9), (n = 275, Mar–Apr) 16·1–28·7 (20·8); (n = 40, Zambia, Oct–early Apr) 17–26·1 (20·3); Malaŵi (n = 179, Nov–Feb) 16–22·5 (19·2), (n = 9, Mar–Apr) 17·4–26 (20·0); (n = 22, N Mozambique, Nov–Jan) 16–20·3 (18·0); South Africa (Transvaal), (n = 79, Dec–Feb) 15·4–21·8 (19·3), (n = 12, Mar) 19·7–27·7 (22·8). (Data for Sudan, Ethiopia, Uganda, Kenya and southern Africa refer to this race and *woodwardi*.)

Moults completely in Africa, Dec–Mar.

IMMATURE: 1st winter bird similar to adult.

S. b. woodwardi (Sharpe): N European Russia and W Siberia; winters E and southern Africa, west to Zaïre. Colder olive above than nominate race, slightly greyer on crown and rump; paler below, with throat and belly more extensively white. Wing, ♂ (n = 8) 77–84 (80·0).

Field Characters. A rather plump medium-sized warbler, plain dull brown or greyish brown above and buffish white below, with a short stubby bill, domed head and greyish brown legs. Face rather plain, but eye dark, emphasized by pale fore-supercilium and eye-ring. Wings relatively long, but square tail quite short. When glimpsed briefly, can look like some other *Sylvia* spp. ♀ Blackcap *S. atricapilla*, similar in size and proportions, is distinguished by reddish brown cap; 1st winter Barred Warbler *S. nisoria* is larger and slightly greyer, with longer pale-edged tail, less stubby bill and distinct pale wing-bar. Can also be confused with grey-brown *Hippolais* spp. and brown variants of Icterine Warbler *H. icterina*, but these are less compact-looking, with relatively long bills and flat foreheads.

Voice. Tape-recorded (57, 86, 88, 91, B, C, F). Song, a vigorous, mellow warbling; even-flowing phrases of 3–6 s (sometimes up to 20 s) are separated by pauses of c. 3 s. More sustained and more contralto than song of Blackcap, and lacks the variety and distinct phrasing of that species; slightly throatier and more even than song of Barred Warbler. In Africa, phrases typically long, commonly 10–30 s with only brief pauses. Subsong, a subdued fast sweet warbling, sometimes continuous for over 30 s. Contact-alarm call, a hard 'vik' or 'vik-vik'; also a low grating 'tchurrr' like Common Whitethroat *S. communis*, occasionally used in threat.

General Habits. Frequents densely foliaged habitats: forest edges, forest/grass mosaic, woodland with abundant thickets and undergrowth, tall moist bush, secondary growth and gardens; also drier bush and acacia thickets on migration. Mainly at 1000–1800 m in E Africa, but commonly above 2000 m on passage; up to 2400 m in Zambia and 1800 m in Zimbabwe. Sometimes solitary but usually in small groups, although flocking less marked during winter moult. Gathers to feed at fruiting bushes and trees; commonly remains attached to an area of a few ha for weeks in autumn, and for up to 4 months in winter. Same birds retrapped in successive years at transit sites in Ethiopia and Uganda, and at wintering sites in Uganda, Malaŵi and Zimbabwe. Several recaught at Nchalo, Malaŵi, up to 5 years after being ringed (Hanmer 1986). Easily overlooked; forages unobtrusively in leafy bushes or low tree canopy. Picks insects from leaves and twigs and sometimes hovers or flies out to catch airborne prey. Moves with hopping or sidling gait, carriage typically rather horizontal (Cramp 1992). Flight fast, usually direct from bush to bush. Low-intensity song or subsong given throughout stay in Africa, but mainly during autumn (Sept–Nov), and then increasingly in winter quarters from Feb up to spring departure, when volume may be full or almost so.

Birds head generally south to southwest from breeding areas in autumn. Most from W Europe migrate through Iberia, those from the Baltic area through Italy. N Africa crossed on a broad front, with strong passage near Mediterranean and Red Sea coasts. Seems usually to cross Sahara in unbroken flights, but some make landfall in desert, and refuel in Saharan oases. High average weight values reported from N Sahara, but in Egyptian oasis c. 280 km south of Mediterranean coast birds were much less heavy than those grounded in nearby 'desert' site (Bairlein 1991). Passage through Morocco and Algeria lasts from late Aug to early Nov (mainly Sept–Oct), through Mauritania and Senegal mainly late Sept to early Nov. Migration is earlier in E than in W, with main movement through Egypt mid-Aug to late Sept, central Chad in Sept, and central and E Sudan and Eritrea from late Aug to early Oct. Present in Ethiopia from Sept to late Nov. Common in Nigeria from mid-Sept, and a few reach wintering areas in Zimbabwe by late Sept. However, main arrival in Liberia and Ghana is from mid-late Oct, and in parts of S Ghana and S Nigeria not until Dec. In Uganda and Rwanda, main influx occurs in late Oct to mid-Nov, with strong passage continuing to mid-Dec. Birds commonly remain several weeks in Uganda transit sites and gain up to 4–5 g in weight before moving on. Main arrival Zambia and Malaŵi from mid-Oct, with strong passage Nov, continuing through Dec. In Kenya, main influx in highlands is late Oct–Nov, but passage continues through SE bush areas to early Jan.

Moults completely in winter quarters, during mid-Dec to late Mar in W Africa, Uganda and Kenya, but often earlier (from late Nov) in southern Africa (Williamson 1968b). Wing-moult duration in individual birds in Uganda *c.* 71 days (Pearson 1973).

Departs from Transvaal and Zimbabwe in late Feb to mid-Mar, but remains later in Zambia and Malaŵi with local influxes to end of March and last birds mid-Apr (latest 19 Apr). Passage begins Rwanda late Feb, but main movement in NE Zaïre, Uganda and Kenya is late Mar to late Apr (last birds early May). Passage in Ethiopia from early Mar, but mainly mid-Apr to early May (latest 27 May); recorded coastal Sudan early May. In W Africa, leaves Gabon by mid-Mar and Liberia early Apr, but many remain in Ivory Coast, Ghana and Nigeria to late Apr to early May; passage peaks second half Apr in central Nigeria, and lasts from mid-Apr to late May in Senegal. Generally small numbers in Sahel indicate that main migration commences from further south; very high spring weights S and central Nigeria suggest capacity for direct flights to Mediterranean from those latitudes. Passes through N Africa on broad front in Apr–May (peak late Apr, latest early June); common near coast and frequent in oases.

Over 50 birds ringed Britain and W Europe recovered autumn and spring in Morocco. Recoveries from Germany, Denmark and Finland to Algeria (5), from France and Germany to Tunisia (2), and from France and Holland to Egypt (2). Spring birds ringed Morocco to France (1) and Germany (1), from Algerian Sahara to France (1), and from Tunisia to Italy (4). South of Sahara, recoveries from Britain to Ghana (5) and Nigeria (1); France, Belgium and Holland to Guinea (2), Guinea-Bissau (1), Ghana (1), Cameroon (1) and Zaïre (2); Germany to Liberia (1), Mali (1), Ivory Coast (1), Ghana (1), Nigeria (1), Congo (1), Zaïre (2) and Central African Republic (2); Slovakia to Central African Republic (1); Denmark and Sweden to Mauritania (1), Togo (1) and Zaïre (2); Finland to Ghana (1), Zaïre (6) and Zambia (1); and Estonia to Uganda (1); most of these recoveries in winter (Dec–Feb). Also recoveries from Nigeria to Italy (1), Germany (1) and Finland (1). Birds breeding further west in Europe thus tend to winter further west in Africa; some from Holland, Germany and further east cross the equator. 1, ringed Nigeria in Oct and found Zaïre the following Feb, implies reorientation south-eastwards after crossing Sahara. Ringing recovery data lacking from breeding grounds further east, but birds wintering southern and E Africa are mainly from Russia; the majority are eastern *woodwardi* but nominate *borin* also occur. *S. b. woodwardi* is also common among passage migrants in Sudan and Egypt and occurs west to Tunisia. A bird ringed Kenya in Feb was recovered in Jordan the following Apr.

Food. Mainly berries in autumn and winter; also invertebrates, especially beetles, caterpillars, Hymenoptera, Diptera and spiders. In tropical Africa, invertebrate food includes winged termites; fruit includes that of *Rubus*, *Solanum torvum*, mulberry, fig, *Lantana*, *Salvadora*, *Trema*, *Hoslundia*, *Premna*, *Securingea*, *Phyllanthus*, *Canthium*, and *Parinari* (Cramp 1992). In Uganda, eats mainly insects in late winter when *Lantana* berries exhausted (Pearson 1972).

References
Bairlein, F. (1991).
Cramp, S. (1992).
Pearson, D. J. (1973).

Sylvia atricapilla (Linnaeus). Blackcap. Fauvette à tête noire.

Motacilla atricapilla Linnaeus, 1758. Syst. Nat. (10th ed.), p. 187; Sweden.

Plate 22 (Opp. p. 401)

Range and Status. Europe (including W Mediterranean islands) to W Siberia; Black Sea to Caucasus and N Iran; NW Africa; Madeira, Canary Is, Cape Verde Is and Azores. Northern and eastern populations wholly migratory, wintering in Africa, Mediterranean basin and W Europe; southwestern populations partially resident, and those of Mediterranean and Atlantic islands presumed wholly so.

Palearctic migrant; also resident NW Africa. Breeds NW Morocco (Tangiers and Ouezzane south to Middle Atlas), NE Morocco (Gorges du Zegzel), N Algeria (throughout the Tell) and probably N Tunisia (Kroumirie); locally common; these birds presumed mainly resident. Migrants winter commonly to abundantly south to N edge of Sahara in Morocco, Algeria, Tunisia and coastal NW Libya; scarce and local NE Libya and NE Egypt. South of Sahara, occupies separate wintering areas in savannas of W Africa and in highlands of NE and E Africa. In W Africa, winters from W Mauritania, Senegal, S Mali (south of 15°N) and S Niger (Parc du W) south to Sierra Leone, Liberia (Mt Nimba), Ivory Coast, Ghana (rare), Nigeria

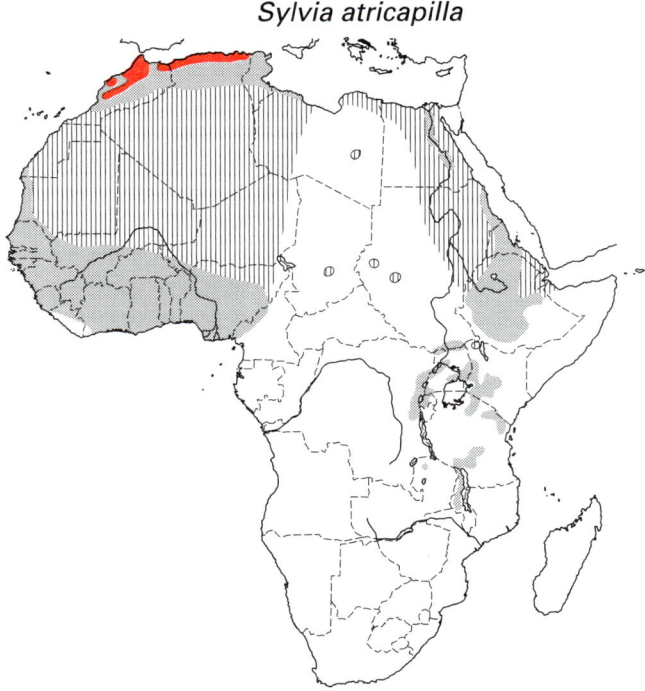

Sylvia atricapilla

(mainly the south) and Cameroon (rare, once Yagoua); mainly in more humid savannas and forest edge; frequent to common locally, but numbers variable. Annual fluctuations in Senegal controlled by productivity of fruiting bushes such as *Balanites*, *Ziziphus* and *Salvadora*. In the east, a few winter coastal Sudan and Eritrea, but common to abundant in highlands of S Sudan, Ethiopia, Kenya, N Tanzania and E Uganda; less common in highlands of W and SW Uganda, Rwanda, Burundi and E Zaïre (mainly Rwenzori, also Kivu). Small numbers reach S Tanzanian highlands, mountains of Malaŵi south to *c.* 14° 30′S, and NE Zambia on Nyika plateau, and at least 3 birds in Njalamimba mushitu, Mporokoso District, at *c.* 1600 m (Aspinwall 1994). Vagrant South Africa (Transvaal; once, near Johannesburg, Dec). Passage through NW Africa partly masked by wintering but evident W Morocco and W Mauritania and in spring from coastal Morocco to Tunisia. Mainly a passage migrant in Libya and Egypt, scarce in autumn but common in spring with frequent records Saharan oases. South of Sahara, passage noted in Senegal, Sierra Leone, Mali, Burkina Faso and N Nigeria, but rarely in E Chad and W Sudan. Common to abundant on both migrations in E Sudan (especially Red Sea coast); uncommon Djibouti and N Somalia. A few oversummer in Senegal and Mali; and recorded in June in Ethiopia.

Description. *S. a. atricapilla* (Linnaeus): Europe, Siberia and (probably this race) NW Africa; occurs throughout African wintering range. ADULT ♂: forehead and crown black; hindneck and side of head and neck grey; rest of upperparts brownish olive. Tail dark grey brown, feathers edged olive. Underparts whitish, tinged light grey on breast and olive-grey on flanks; undertail-coverts olive-grey with white fringes. Primaries, secondaries and primary coverts blackish grey, narrowly edged olive-grey. Tertials and rest of upperwing-coverts brownish olive. Axillaries and underwing-coverts creamy white. Wing-tip bluntly pointed; P8 longest (sometimes = P7); P7 0–1 shorter, P6 2–4, P5 5–10, P1 14–20; P9 4–7, usually between P5 and P6; P10 very small, 0–6 longer than primary coverts; P6–P8 emarginated. Bill blackish, base of lower mandible paler grey; eyes brown; legs and feet blackish or slate-grey. ADULT ♀: forehead and crown rufous-brown; grey of hindneck and side of head and neck with olive tinge; rest of upperparts slightly browner than in adult ♂, and breast and flanks tinged more olive. SIZE: (10 ♂♂, 10 ♀♀, Europe) wing, ♂ 73–79 (76·1), ♀ 70–76 (73·2); tail, ♂ 55–60 (58·0), ♀ 50–58 (55·2); bill, ♂ 14–16 (14·5), ♀ 14–15 (14·6); tarsus, ♂ 20–22 (21·1), ♀ 20–22 (21·1). WEIGHT: SE Morocco, ♂ (n = 93, spring) 11–19 (14·1), ♀ (n = 87, spring) 11–19 (14·1). Sexes combined: (n = 26, NW Morocco, Apr–early May) 14–23 (17·9); N Senegal (n = 148, Oct–Nov) 12–25·1 (15·9), (n = 117, Dec–Feb) 14·6–23·5 (17·7), (n = 156, Mar–Apr) 11·1–30 (20·4); Vom, Nigeria (n = 5, winter) 14–17 (15·3), (n = 7, Mar–Apr) 15–27 (20·4); NE Sudan (n = 71, Oct–Nov) 13–19·5 (15·4), (n = 140 Mar–Apr) 15–30 (20·4); Ethiopia (n = 266, Sept–Feb) 12·1–22·8 (17·1), (n = 371, Mar–May) 14·1–30·5 (19·7); Kenya (mainly *dammholzi*), (n = 143, Nov–Feb), 14·3–23·4 (17·9), (n = 8, Mar–Apr) 16·1–22·4 (19·1).

Complete moult on breeding grounds, partial moult in winter quarters, Jan–Feb.

IMMATURE: juvenile like adult ♀ with rufous forehead and crown, but underparts more buffish and tertials and wing-coverts often with rufous-brown fringes. 1st winter ♂ like adult ♂ but forehead often tinged brown, and crown-feathers commonly with some rufous fringes. 1st winter ♀ like adult ♀.

NESTLING: naked; bare skin including bill and legs flesh pink; mouth flesh red, 2 faint tongue-spots; gape flanges pale yellow to ivory.

S. a. dammholzi Stresemann: Caucasus and N Iran; winters Ethiopia, Kenya and Uganda. Upperparts of ♂ paler than in nominate race, olive-grey rather than olive-brown; underparts whiter. ♀ generally paler, cap paler rufous. Rather large: wing (n = 10 ♂♂) 75–80 (78·1).

Field Characters. A compact, medium-sized warbler, plain brownish or olive above and pale grey below, with a short bill, longish wings and a rather short plain tail. Size and silhouette like Garden Warbler *S. borin*, but readily distinguished by short cap, black in male, rufous-brown in female and juvenile. Other warbler species with dark cap have contrasting white throat and white outer tail-feathers.

Voice. Tape-recorded (57, 86, 89, 91, B, C, F). Song rich and musical; composed of warbling phrases *c.* 3–7 s long, which usually begin with a chattering section and end with a louder section of clear fluty notes; less rapidly delivered and more varied than song of Garden Warbler, with longer pauses between phrases; often includes mimicry. Subsong, a quiet, prolonged rambling warble, like that of Garden Warbler but with high-pitched discordant notes, audible only at a few m. Song in Kenya in late winter also with long continuous phrases, but louder than subsong, and with typical structure and fluty sections discernible. Contact-alarm call, a hard 'tacc', often repeated persistently.

General Habits. Breeds in woodland with well developed broad-leaved trees and a good shrub layer; also in parkland and suburbs; in NW Africa mainly on lower and middle mountain slopes up to above 1500 m. Widespread in NW Africa in winter, from the coast and oases on N Sahara fringes up to high cedar forest in Middle Atlas. In W Africa, winters in acacia steppe, dry savanna with bushes and scattered trees, moist secondary bush and scrub, edges of lowland and montane forest, mangroves and gardens; up to 1200 m on Mt Nimba. Near Red Sea coast frequents scrub and riverine acacia. In E and southern Africa, winters mainly between 1600 and 2400 m, in forest edge and secondary growth, gallery forest, *Lantana* scrub, dense thickets by permanent swamps, and gardens. Sometimes associates with Garden Warbler *S. borin*, especially on passage, but often separated from it at higher altitude; recorded up to 3600 m on Mt Kenya. Sometimes solitary outside breeding season, but more typically in loose groups, often of 20 or more birds at fruit sources. Occurs in mixed parties, sometimes with other *Sylvia* spp. In Kenya, birds often appear territorial in late winter, with regular song-posts (Pearson 1978). Forages in low shrubs and bushes, also commonly in tree canopy up to 20 m high (Cramp 1992). In breeding season picks insects from leaves and twigs, and may hover or fly out to take slow-moving flying prey. In autumn and winter takes mainly berries, usually eaten whole (Cramp 1992). In winter in N Africa, also exploits food from human domestic sources. Often skulking, but bolder and more inquisitive than Garden Warbler, typically with more upright stance, although movements and flight are similar. Raises crown-feathers in excitement or alarm, and flicks wings and tail. Gives subsong throughout stay in winter quarters, usually hidden in tall bush or tree canopy. This strengthens to full song, e.g. from late Jan in Kenya.

Birds from northern breeding areas leave earlier, migrate faster and tend to winter further south than southern ones. Many from Mediterranean region remain in breeding area. European ringing shows clear migratory divide (Klein *et al.* 1973, Zink 1973). Most birds from west of *c.* 12°E head southwest in autumn to Iberia, S France and NW Africa. Some continue on to winter in W Africa. Birds from Europe east of 12°E, and most birds from Scandinavia, funnel towards E Mediterranean and Levant. They change migratory direction, and together with populations from Russia and SW Asia continue on to reach NE and E Africa, mainly via the Red Sea coast.

Autumn passage in NW Africa occurs late Sept to Nov, but is masked by arrival of wintering birds. Earliest Senegal record 15 Oct, and first birds reach Sierra Leone and Ivory Coast late Oct. Main arrival in W Africa occurs from Nov. Numbers fluctuate locally during same winter and from year to year in response to fruit availability. Many birds make midwinter movements within Mediterranean area. Most recoveries NW Africa not until Jan, suggesting late movement southward from Iberia. Highest numbers in W Morocco and N Tunisia in Jan–Mar when fruit is ripe. Annual variation in numbers in W Africa suggests that some birds winter north of Sahara in one year and south of it in another (Cramp 1992).

Southward migration somewhat earlier in NE Africa than in NW. Passage rather light through Egypt, mainly mid-Sept to late Oct (earliest 23 Aug), but heavy through coastal Sudan, early Sept to mid-Nov (earliest 18 Aug), and Eritrea (earliest 26 Aug). Birds reach Kenya and Uganda from late Oct, but mainly in Nov, and Malaŵi from mid-Nov (earliest 9 Nov).

Departs from Zambia and Malaŵi after short stay; few after early Jan, latest 21 Feb. Leaves Kenya and Uganda abruptly in late Mar (latest 14 Apr), and Ethiopia mainly Apr (latest 25 May). Strong passage across Red Sea coasts Mar–Apr, and protracted movement on a broad front through Egypt in mid-Feb to mid-May (mainly Mar–Apr). Migration begins in W Africa in Feb; birds leave most wintering areas Mar to early Apr, though in some years a few remain in Sahel until June. In N Senegal, passage is mainly late Mar to mid-Apr, when many birds are very fat; in W Mauritania Feb–Mar, N Nigeria Mar–Apr. Marked spring migration through N Africa; mainly Feb–Mar in Algeria, Mar to early May in Libya.

Over 500 European ringed birds recovered in N Africa, mainly in winter and spring: from Britain and W Europe most to Morocco, many to Algeria, a few to Tunisia; from Germany, most to Algeria, many to Morocco, others to Tunisia (4), Libya (2) and Egypt (1). Further returns from Denmark to Morocco (1), Algeria (1) and Tunisia (1); from Czech Republic to Tunisia (1); and from Austria and Poland to Libya (2). 7 birds ringed in Morocco found in Britain and France, and 1 ringed Tunisia found in Italy. Recoveries south of Sahara as follows: Spain to Mauritania (1); France to Senegal (2); Britain to Mauritania (1), Senegal (3), Gambia (1), Guinea-Bissau (1) and Guinea (1); Belgium to Senegal (2) and Nigeria (1); and Germany to Senegal (2). Also from Sudan coast to Cyprus (2, Sept, Nov); from S Sudan to Sweden (1) and Lebanon (1, May); from Ethiopia to Lebanon (3, Apr–May) and Syria (1, Sept); and from Kenya to Iran (1, Apr). 1 ringed Egypt (Sept) found 2 years later on Sudan coast (Sept).

Food. Mainly insects in the breeding season, but fruit, especially berries, at other times. Experiments with captive birds showed preference for fruit (rather than insects) during post-juvenile moult, autumn migration and winter controlled by endogenous rhythm (Berthold 1976). Invertebrates taken in breeding season are mainly beetles, bugs, Hymenoptera, Diptera, and caterpillars. Invertebrates taken in Senegal in winter include bugs and small grasshoppers (Morel 1968). Main fruits eaten in Burkina Faso, Mali and Senegal are, successively, *Balanites*, *Ziziphus* and *Salvadora*. In Sierra Leone, eats fruits of *Parinari*, *Trema guineensis*, *Phyllanthus*, and *Canthium*; in Kenya, mainly fruit, including *Lantana*, mulberry and indigenous forest trees such as *Pauridiantha* (Cramp 1992).

Breeding Habits. Monogamous. Solitary and territorial; territories of 0·2–1·2 ha in Europe. In Morocco, up to 46 pairs per km² in subhumid/humid zones (Thévenot 1982). Territory is used for courtship, mating and nesting, and supplies most of food. Migrant ♂♂ reach northern breeding areas together with or only shortly ahead of ♀♀. ♂ occupies territory, and sings from cover, usually 6–10 m high; typically from a preferred song-post, but when excited also in song-flight. ♂ defends territory aggressively by threat and sometimes fighting; often builds one or more 'cock nests'. Pairs formed quickly; ♂ chases ♀, often with excited, continuous chattering song. Displaying ♂ sometimes flies towards ♀ with slow wing-beats; then at perch may raise crown-feathers, ruffle back, droop or gently beat wings, and flirt, raise or lower tail. ♀ inspects cock nests, and if she accepts one both continue building; sometimes ♀ chooses a new site. ♂ sings mainly in early morning, with lesser peak at dusk. Lengthy song-period in Europe lasts to July.

NEST: small, neat cup of dry grass, herb stems and rootlets, and sometimes moss, wool and spider webs, lined with finer rootlets and grasses or hair; outer diam. *c.* 100, inner diam. *c.* 63, depth of cup *c.* 40; usually in bush or herbage within 1 m of ground, but sometimes 2–3 m high in shrub or on tree branch. Usually started by ♂ and completed by both birds; cock nest takes 1–2 days to build, completion of nest by pair another 2–5 days (Cramp 1992).

EGGS: usually 4–5 in Europe, 4 in NW Africa; white or pale buff or olive, with brown, reddish or purplish markings. SIZE: (n = 412) 17–23 × 13–16 (19·7–14·7). Sometimes double-brooded.

LAYING DATES: NW Africa, mid-Apr to early June.

INCUBATION: begins with penultimate egg. Period: 10–16 days; in Britain, mean 11 days (n = 68).

DEVELOPMENT AND CARE OF YOUNG: young cared for and fed by both parents; brooded by both parents for 7–8 days; fledging period 10–14 days (in Britain, mean 11 days, n = 45), young usually still unable to fly and remain as noisy group near nest for 4–5 days; family bonds maintained for 2–3 weeks after fledging.

BREEDING SUCCESS/SURVIVAL: of 2484 eggs laid in Britain, 76% hatched and 80% of these fledged: overall success 60%. Mean of 2·6 young fledged from 446 nests (Mason 1976).

References
Cramp, S. (1992).
Klein, H. *et al.* (1973).
Mason, C. F. (1976).

Plate 22
(Opp. p. 401)

***Sylvia communis* Latham. Common Whitethroat. Fauvette grisette.**

Sylvia communis Latham, 1787. Gen. Synopsis Birds, suppl. 1, p. 287; England.

Range and Status. Europe and NW Africa, and through S Russia and N Kazakhstan east to Altai; also Turkey, Near East and Caucasus to N Iran, N Afghanistan, Tien Shan and NW China; winters Africa.

Palearctic migrant. Breeding summer visitor N Morocco (frequent to common Loukkos plains, Djabala, Rif and Middle Atlas, and locally in High Atlas foothills near Marrakesh), extreme N Algeria (locally frequent between Sidi-Bel-Abbès and Reghaïa, perhaps breeds near Annaba, and in the Aurès), and probably N Tunisia; mainly in highlands.

Winters south of Sahara, across W African sahelian and soudanian belts and through E Africa south to Zimbabwe and Botswana. In W Africa, mainly in drier savannas north of *c.* 11°N, from SW Mauritania, Senegal and Gambia and probably N Guinea to S Mali (north to 17°N), N Ivory Coast (Comoé Nat. Park), Burkina Faso, N Ghana (south to Mole Game Reserve), S Niger (north to Takoukout), N Nigeria (south to Borgu Game Reserve and Jos), N Cameroon, and central and S Chad; frequent to common. Rare further south, in S Ivory Coast (Lamto), S Ghana (Accra), S Nigeria (Ibadan, Ife), S Cameroon, NE Gabon

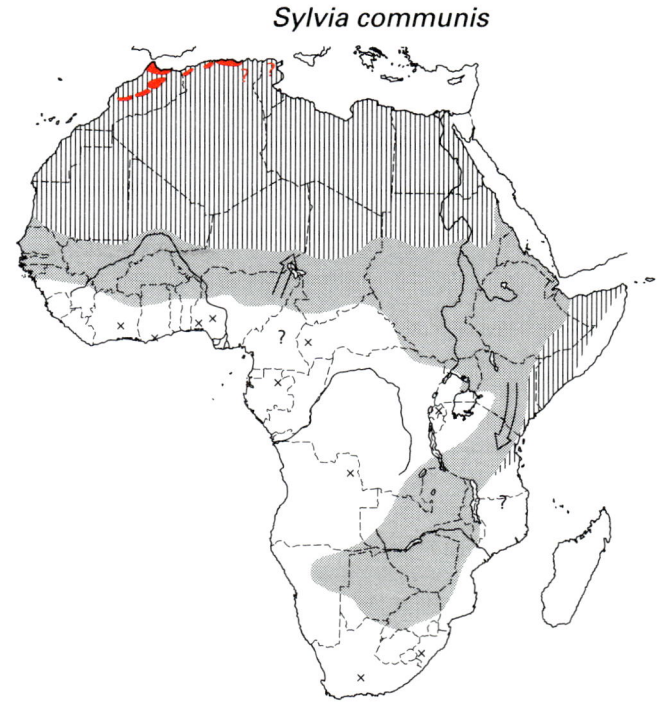

Sylvia communis

(M'Passa), and E Central African Republic (Sarki). In E and southern Africa, winters mainly in drier areas, where frequent to locally abundant: from W, central and NE Sudan to Eritrea, Ethiopia, S Sudan, NE Zaïre, Uganda (mainly in E), Kenya (mainly centre and SE), Tanzania (mainly NE, centre and SW), SE Zaïre, Zambia (scarce in NW), Malaŵi (scarce), Zimbabwe (mainly in W and S), Botswana (except SW), NE Namibia (south to Okahandja) and N South Africa (Transvaal, N Cape Province and rarely to c. 33°S). Rare or occasional in 'winter' also in NW Somalia, Rwanda, and Angola (once, Saurimo, Lunda), and probably reaches W Mozambique. Rare winter records on N edge of Algerian Sahara, Dec and Jan.

On autumn passage, scarce from Morocco to Libya, but common to abundant in Egypt. In spring, common to abundant across whole of N Africa, including Saharan oases. Pronounced migrations south of Sahara: in Senegal (locally abundant both passages), N Nigeria (very abundant in spring Malamfatori, L. Chad), central and E Chad (common to abundant both passages), Sudan (locally abundant in autumn, Darfur, Nile valley system and Red Sea coast), Eritrea and central Ethiopia (locally abundant in spring), Somalia (common in spring in north), central and SE Kenya (abundant to very abundant both passages), and S Zambia (locally abundant both passages).

A decline of 50–90% in various parts of W and central Europe from 1968 to about 1980 (partially reversed in some areas since) was considered due to drought and environmental deterioration in the African Sahel zone (Winstanley *et al.* 1974).

Description. *S. c. communis* Latham (including '*cinerea*' and '*jordonsi*'): NW Africa and W Europe, east to Scandinavia, W Poland and Yugoslavia; winters in N tropics east to Sudan. ADULT ♂: crown and nape ash-grey, feathers tipped brown, more extensively towards rear; rest of upperparts brown, uppertail-coverts greyer. Lores grey; narrow eye-ring whitish; greyish supercilium, noticeable in fresh plumage; ear-coverts and side of neck grey-brown. Tail dark brown, feathers narrowly edged grey-buff; T5 (often also T4) with narrow white fringe on tip, sometimes extending to white spot or wedge at tip of inner web on T5; T6 white with grey base to inner web extending as wedge along shaft to near tip. Underparts whitish, breast suffused pinkish buff, sides of breast and flanks deeper buff. Primaries, secondaries and primary coverts blackish brown, primaries narrowly edged and primary coverts edged and tipped buffish white, secondaries more broadly edged buff or rufous-buff; tertials and greater coverts dark brown, broadly fringed rufous-cinnamon; rest of upperwing coverts tipped greyish; axillaries and underwing-coverts greyish white. Wing-tip bluntly pointed; P8 longest (often = P7); P7 0–1 shorter, P6 1–3, P5 3–7, P1 12–15; P9 0·5–3 shorter, between P7 and P8; P10 minute, 0·5–6 shorter than primary coverts; P7–P8 emarginated, sometimes tip of P6. Bill dark horn or greyish horn, base of lower mandible flesh; eyes light brown, yellow-brown or red-brown; legs and feet pale brown or flesh-brown. ADULT ♀: like adult ♂, but crown brown like rest of upperparts, chest without pinkish tinge, median and lesser coverts browner, eyes olive-grey, olive-brown or light brown. SIZE: (10 ♂♂, 10 ♀♀, W Europe) wing, ♂ 68–74 (71·9), ♀ 68–73 (71·0); tail, ♂ 56–62 (59·9), ♀ 54–61 (57·4); bill, ♂ 13–15 (14·0), ♀ 14–15 (14·5); tarsus, ♂ 21–22 (21·6), ♀ 20–22 (21·3). WEIGHT: (Morocco) ♂ (n = 35, spring) 11·6–15·5 (13·7), ♀ (n = 22, spring) 11·7–15 (13·7). Sexes combined: N Senegal, (n = 168, Nov–Feb) 10·4–18·5 (13·5), (n = 211, Mar–Apr) 11·1–26·5 (15·2); (n = 55, Zaria, N Nigeria, Nov–Apr) 11·3–16·6 (14·0); L. Chad, Nigeria (this race and/or *volgensis*), (n = 93, autumn) 11·2–17 (14·3), (n = 1776, spring) 10·9–24·6 (15·1).

Complete moult near breeding grounds, July–Aug; partial moult in Africa in late winter.

IMMATURE: juvenile buff-brown or grey-brown above; off-white below, with sides to breast buff-brown, breast, flanks and undertail-coverts pale buff; tail as adult but lacks pure white, T6 pale brown with dark shaft and greyish inner web; dark centres to tertial and greater coverts narrower, less sharply defined. 1st winter like adult ♀ but tail, tertials and outer greater coverts still juvenile; eyes dark brown or dull grey brown.

NESTLING: naked, bare skin pink at first, darkening to blackish within a few days; mouth flesh-pink; 2 blackish spots at base of tongue and 2 grey marks near tip, fading rapidly; gape flanges pale yellow; legs flesh.

S. c. volgensis Domanieski: E Europe and W Siberia to foothills of Altai; winters N central, E and southern Africa. Similar to nominate race but slightly paler drab brown above, paler, less buff below, breast of ♂ paler pink. Slightly larger: wing, (11 ♂♀, Volga valley) 73–77 (74·5), (16 ♂♂, W Siberia) 72–79 (Cramp 1992).

Birds from Volga basin eastwards have complete moult in Africa.

S. c. icterops Ménétries: Turkey, Near East and Caucasus to W and N Iran and W Turkmenistan; winters E and probably southern Africa. Darker and greyer above than nominate race, crown of adult ♂ darker grey, mantle and scapulars grey-brown; paler below, flanks less buff, chest of ♂ less pink; fringes of tertials and greater coverts paler cinnamon or pale buff.

Complete moult in Africa.

S. c. rubicola Stresemann: E Iran, N Afghanistan and Tien Shan to NW China and W Mongolia; winters E and southern Africa. Paler than *icterops*, upperparts slightly greyer. Larger: wing, (41 ♂♀) 72–80 (76·1) (Cramp 1992). WEIGHT: sexes combined (this race, *icterops* and *volgensis*): (n = 31, Egypt, autumn) 9·9–20 (15·8); (n = 99, NE Sudan, Aug–Oct) 9·5–16 (12·2); Ethiopia (n = 329, Aug–Mar) 10·9–17·8 (13·7), (n = 29, Apr–May) 11·3–22·1 (15·8); Kenya (n = 571, Nov–Dec) 11·8–20·6 (14·7), (n = 114, Jan–early Mar) 12·8–18 (14·8), (n = 198, Apr) 12·1–22·3 (15·2); South Africa (Transvaal), (n = 14, Dec–Jan) 13–15·4 (14·5), (n = 28, Feb–Mar) 13·5–21 (15·7).

Complete moult in Africa, either (in Ethiopia) Oct–Nov or (further south) in Dec–Mar.

TAXONOMIC NOTE: the racial treatment adopted here follows Cramp (1992).

Field Characters. A slim, medium-sized warbler with a rather long tail. Brown or grey-brown above and whitish below, with a distinctive rufous or sandy-buff wing-panel, white or whitish outer tail-feathers and pale brown legs. Crown of ♂ grey or brownish grey, contrasting with pure white throat; of ♀ and immature, brown. Skulks in bushes with horizontal posture, tail raised and crown often peaked. Frequent scolding 'tcharrr' call is characteristic. May be confused with much smaller Subalpine Warbler *S. cantillans* and Spectacled Warbler *S. conspicillata*, especially ♀♀ and juveniles, but Subalpine has a reddish eye, and Spectacled has buffier underparts, a more uniform bright chestnut wing-panel and more prominent white eye-ring. Lesser Whitethroat *S. curruca* lacks wing-panel of Common Whitethroat, has darker ear-coverts, dark legs and different song and call-note.

Voice. Tape-recorded (86, 88, 91, B, C, F, PEA). Song from perch, a short, brisk, rather scratchy warble, lasting *c*. 1·5 s and delivered about 10 times per min. In song-flight, song more protracted, up to 10 s long. Subsong, a quiet, usually prolonged warble, audible up to *c*. 20 m; more varied and more musical than full song. Main contact-alarm call, a harsh 'tcharrr'; occasionally a typical *Sylvia* 'tacc'. Also gives a characteristic, drawn-out, nasal 'whuiid', often as a series, 'whuiid-whuiid-whid-whid-whid', when excited or as a prelude to song.

General Habits. Breeds in low patchy scrub or herbaceous cover, along hedgerows and woodland edges, and among young trees; in Morocco, ranges from coastal plains to highlands, including bare summits in Middle Atlas above 2000 m. Winters in N tropics in thorny tree steppe, scattered scrub and thickets, bushes at oases and along wadis, open woodland with good shrub cover, and gardens; in Sahel typically in scattered acacias, *Balanites*, *Ziziphus*, and other evergreen or semi-evergreen trees and shrubs (Cramp 1992); in Niger delta transition zone (Mali) inhabits patches of *Guarea* (Curry and Sayer 1979); in Sudan and Eritrea mainly in bushy acacia country. In Kenya, winters in scrub and thickets in arid or semi-arid areas, especially acacia dominated bush or *Commiphora* woodland at 700–1200 m, but mainly where leaf and undergrowth are retained through the Feb–Mar dry season; often together with Barred Warbler *S. nisoria*, but tends to prefer cooler, less arid areas than that species, and lower cover. On migration, habitat varies from sparse bare scrub to dense coastal bushes, hedges, *Lantana* thickets and gardens in highlands up to 2000 m. In Zambia and Zimbabwe, mainly in drier areas, in acacia scrub and thicket, and secondary vegetation including *Lantana*.

Solitary in winter quarters or in loose groups, commonly with mixed-species parties. Often in large concentrations on migration, and many hundreds can occur locally, e.g. at Malamfatori (Nigeria), Mar–Apr, and Ngulia (Kenya), Nov–Dec. In Kenya, individuals remain attached to local stopover or wintering sites for weeks, sometimes up to 3 months, but birds vacate bush areas when conditions become too dry. Several ringed birds returned to same sites in Ethiopia and Kenya in successive years. Sometimes skulking, but less secretive, more inquisitive than most *Sylvia* spp. Forages in bushes, low tree canopy or herb layer, searching twigs and foliage and picking off berries (Cramp 1992). In Kenya (Tsavo) foraging height is 1·7 m (n = 4), 85% of items taken from leaves, 13% from twigs or stems (Lack 1985). Occasionally takes insects in flight or from ground. Carriage typically horizontal with head held up and tail slightly raised. Raises and lowers crown-feathers when excited. Flight usually low, darting and rather undulating, the long tail frequently fanned and flirted (Cramp 1992). Prolonged subsong often given in winter quarters, usually with bird well hidden within bush or low tree canopy.

Ringing shows a migratory divide, with birds from NW Europe moving initially to Iberia, those from central Europe and Fennoscandia towards the central and E Mediterranean. E European birds also enter Africa through E Mediterranean. Autumn passage in NW Africa (Morocco to Libya) lasts from late Aug to early Nov, but is sparse, suggesting most birds leave S Europe and cross Mediterranean and Sahara without landfall. However, movement through Egypt is strong and protracted, from late July to early Nov (mainly mid-Aug to late Oct). First birds reach Sahel in late Aug, with main arrival late Sept to early Oct; common in Nigeria from early Oct, and in N Ghana from Nov. Passage movements through N Senegal (Oct–Nov) and Darfur, W Sudan (late Sept to late Oct), imply substantial wintering further south; passage E Chad, late Dec–Jan, probably of birds moving south from Sahel (Salvan 1968–69).

Heavy passage through central and E Sudan, Aug–Oct (peak late Aug to late Sept), and through Eritrea, involves eastern races migrating to Africa via the Middle East. In Ethiopia, first arrives late Aug, with main passage/stopover mid-Sept to late Oct. Earliest birds in Kenya usually mid-Oct; very heavy passage Nov to mid-Dec, confined to narrow front east of the highlands. Large numbers attracted regularly to lights at Ngulia, Tsavo, where over 35,000 ringed 1969–91 (Backhurst and Pearson 1993). Movement continues through SE Kenya until mid-Jan. Birds reach Zambia from late Oct, but mainly from mid-Nov, with passage to mid-Dec. In Zimbabwe, arrives from mid-Nov, but mainly in Dec with passage to end of the month. Present in Botswana and Transvaal from Dec.

European birds wintering in N tropics (nominate *communis* and western elements of *volgensis*) arrive with fresh flight feathers and have only a partial moult, during Jan–Feb. Asian birds (*icterops*, *rubicola* and eastern *volgensis*) reach Africa with old flight feathers and moult there completely (Stresemann and Stresemann 1968), either in NE Africa in Sept–Nov, or in E/southern Africa in Dec–Mar. Some birds arrive in Kenya in Nov–Dec after recent full moult, presumably completed in Ethiopia (Pearson and Backhurst 1976).

Northward migration begins southern Africa in Mar, with passage through Zimbabwe mid-Mar to mid-Apr (latest 24 Apr) and heavy movement Zambia and Malawi Mar to mid-Apr (latest 29 Apr). Locally wintering birds leave Kenya late Mar to early Apr, but strong passage occurs through central and E areas throughout Apr (latest 15 May); large scale overflying east of highlands is revealed by occasional huge 'falls'. Movement is heavy through Ethiopia (peak late Mar to early Apr) and Eritrea (to early May), and noticeable in May in N Somalia. This contrasts with low numbers in central and E Sudan, and suggests many birds follow a more easterly route through Arabia in spring than in autumn. W African wintering sites are vacated mainly in Mar, but late birds remain in Sahel until May. Strong passage occurs in N Senegal, late Mar to mid-Apr, and during same period at L. Chad (Nigeria) where over 700 ringed in one season; also in E Chad, mid-Mar to early Apr. Spring passage N Africa is heavier and on a broader front than in autumn, with more records from Sahara. Occurs mainly late Mar to mid-May in Morocco and Algeria, early Apr to mid-May in N Tunisia. Begins mid-Feb in Egypt, with main movement mid-March to mid-May.

European ringed birds recovered in N Africa in autumn and spring, as follows: Britain to Morocco (13); Belgium to Morocco (2) and Algeria (1); Germany to Morocco (7) and Egypt (1); France to Morocco (2), Tunisia (2), Libya (1) and Egypt (1); Czech Republic to Tunisia (1); Italy to Tunisia (1); Malta to Egypt (1); Denmark to Morocco (2); Sweden to Egypt (6); and Finland to Algeria (1), Libya (1) and Egypt (4). Birds ringed in Tunisia (spring) recovered in Italy (2), Germany (1) and Poland (1), and in subsequent migrations in Libya (1) and Egypt (1); ringed in Egypt (autumn) recovered in Finland (1). South of Sahara, recoveries from Britain to Senegal (3); Belgium to Senegal (1); Germany to Ghana (1); Sweden to Chad (1); and Finland to W Sudan (1). Birds ringed Nigeria (Apr) found in Libya (2, Apr) and Egypt (1, Sept); birds ringed Kenya (Nov–Dec) found Syria (1, Aug), Saudi Arabia (1, Sept), Yemen (1, May), Oman (1, May) and Russia (2, Tatar and Kuybyshev).

Thus, W European birds winter mainly in W Africa; N, central and E European ones mainly in N central or NE Africa. No evidence from ringing or moult schedules that nominate *communis* reach southern Africa (but see Clancey 1974a). Asian birds (*icterops*, *rubicola* and eastern *volgensis*) winter in E and southern Africa, but details of their separate distributions not established.

Food. In breeding season mainly insects, especially beetles, caterpillars, bugs (Hemiptera) and Hymenoptera; on autumn migration and in winter, largely berries. In Senegal, takes *Salvadora* and *Maerua* berries; in Niger (May) *Salvadora* berries and adult and larval insects. In Libya eats mulberries and grasshoppers in spring (Cramp 1992). At L. Chad, Nigeria, in Mar–Apr, gizzards of 97 birds contained, by dry weight, 80% *Salvadora* fruits and flowers, 15% midges *Tanytarsus* and 5% other insects (Fry *et al.* 1970).

Breeding Habits. Essentially monogamous, but ♂♂ occasionally bigamous. Pair-bond lasts only for breeding season. Solitary and territorial; territories separated, contiguous or sometimes overlapping; typically 0·3–1·0 ha in W Europe (Cramp 1992); used for courtship, nesting and collection of most of food for young. ♂ arrives first and establishes territory; moves about singing and may build flimsy 'cock nests'. ♂ has several song-posts within territory, typically in low cover; unpaired ♂ often uses exposed perch. In song-flight ♂ rises to 10 m or more, sings near top of ascent with crown raised and tail raised and spread, then descends in a series of jerky swoops (Cramp 1992). ♂ defends territory by song-display, and threat-display in which crown and back feathers are raised and tail spread and lowered and flicked up and down. He sometimes chases rival ♂. ♀ arrives about a week after ♂; ♂ then becomes more continuously active and vocal. He displays by crouching above ♀ with plumage sleeked but crown-feathers raised, then dashes at her, swerving away just before contact, and darts back and forth singing all the time. ♀ usually responds by fanning tail and spreading and quivering wings (Cramp 1992). ♀ may select a cock nest and add to it, or choose her own site. Bigamous ♂ has ♀♀ in 2 territories, usually simultaneously; typically pairs with second ♀ after first has completed clutch.

NEST: deep cup of dry grasses and leaves with some roots, plant down and cobwebs, lined with fine grasses, rootlets, hair and wool; usually situated in scrub within 30 cm of ground, supported by twigs or plant stems; outer diam. *c*. 105, inner diam. *c*. 65, cup depth *c*. 48; started by ♂ and completed by ♀, or built together by ♂ and ♀.

EGGS: 4–5 in NW Africa, occasionally 6; usually pale blue or green with fine ochre or dark grey markings, sometimes with larger greyish blotches. SIZE: (n = 396 Europe) 16–21 × 12–15 (18·2 × 13·8).

LAYING DATES: NW Africa, mainly early May to mid-June. Normally 1 brood in N and W Europe, commonly 2 in south of range.

INCUBATION: starts usually with penultimate egg; by both sexes. Period: 9–14 days (mean in Britain 11 days, n = 56).

DEVELOPMENT AND CARE OF YOUNG: young cared for, fed and brooded by both parents; remain in nest 8–15 (usually 10–12) days; become independent 15–20 days after fledging.

BREEDING SUCCESS/SURVIVAL: in 869 British clutches, 69% hatched and 85% of these fledged: overall success 59%, with av. 2·7 young fledged per nest (Mason 1976).

References
Cramp, S. (1992).
Fry, C. H. *et al.* (1970).
Mason, C. F. (1976).
Pearson, D. J. and Backhurst, G. C. (1976).
Winstanley, D. *et al.* (1974).

Sylvia curruca (Linnaeus). Lesser Whitethroat. Fauvette babillarde.

Motacilla curruca Linnaeus, 1758. Syst. Nat. (10th ed.), p. 184; Sweden.

Range and Status. NW, central and E Europe; Turkey and Caucasus; Levant; and Asia east to Yakutia and Transbaykalia, south to Iran, the Himalayas and NW China. Winters in Africa, Arabia and S Asia.

Palearctic migrant. Winters mainly south of Sahara in central and eastern part of N tropical belt. Locally common in Egypt south of 30°N (Nile valley, Fayoum, Western Desert oases). Locally common to very abundant Chad (north to Fada, south to soudanian zone, mainly east of 19°E), Sudan (from Wadi Halfa south to *c.* 10°N, once Juba), Eritrea, and N and central Ethiopia (south in Rift Valley to *c.* 6°N). Further west, smaller numbers in N Cameroon (Yagoua, Sir, Waza), N Nigeria (locally common Nguru to L. Chad, occasional west to Sokoto) and S Niger (west to Maradi); uncommon S Mali (Niger delta north of 15°N) and N Senegal (Djoudj). Rare NW Somalia (once, Mar). On passage, common E Libya in spring but scarce in autumn; widespread and common to abundant both seasons in Egypt and N Sudan; frequent to common in Eritrea, mainly in spring; uncommon Djibouti. Uncommon Morocco (*c.* 25 records, mainly spring); rare Algeria (once Tamanrasset, Mar) and Tunisia (once Zembra, Sept).

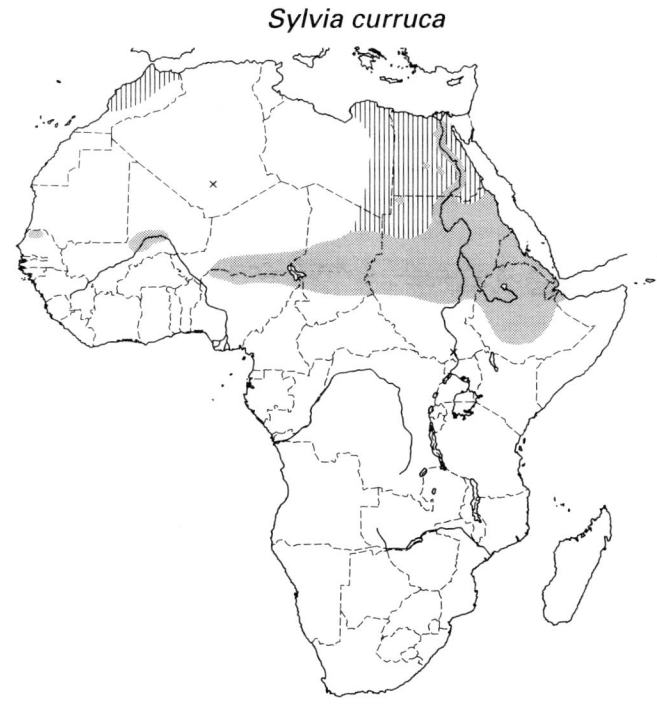

Sylvia curruca

Description. *S. c. curruca* (Linnaeus) (including '*blythi*'): NW, central and E Europe and W Siberia to *c.* 70°E; winters throughout African range. ADULT ♂: forehead, crown and hindneck slate-grey; rest of upperparts grey-brown, slightly paler and greyer on rump and uppertail-coverts. Dark grey or blackish face mask from lores below eye to ear-coverts. Tail dark grey, T1–T5 with narrow whitish outer edge, and narrow white tip, broader towards T5; tip of T5 inner web often with whitish spot or small wedge along shaft; T6 white apart from grey basal wedge on edge of inner web and sometimes grey mark along shaft at tip. Breast and flanks greyish with pinkish buff tinge; rest of underparts white. Primaries and secondaries blackish, narrowly edged greyish white; primary coverts blackish with narrow greyish buff edges and tips; tertials and greater coverts blackish brown with buff fringes; rest of upperwing-coverts grey-brown; axillaries and underwing-coverts white, tinged creamy buff. Wing rather short, tip bluntly pointed; P8 longest (sometimes also P7); P7 0–2 shorter, P6 1–3, P5 3–6, P1 11–14; P9 1·5–5 shorter, usually between P5 and P6 (rarely shorter than P5); P10 very small, 0–4 longer than primary coverts; P6–P8 emarginated. Bill dark horn to slate-black, base of lower mandible paler; eyes brownish grey to dark brown; legs and feet greyish brown, slate-grey or black. Sexes alike. SIZE: (10 ♂♂, 10 ♀♀, Europe) wing, ♂ 65–68 (66·6), ♀ 61–66 (64·2); tail, ♂ 51–55 (52·6), ♀ 48–54 (50·2); bill, ♂ 12–14 (13·0), ♀ 12–14 (12·6); tarsus, ♂ 20–21 (20·3), ♀ 19–21 (20·0). WEIGHT: unsexed: (n = 48, NW Egypt, autumn) 9·5–15·5 (12·6); N Senegal (n = 18, Nov–Feb) 9·5–13·7 (11·6), (n = 3, Mar) 11·9, 12·5, 15·4; NE Nigeria (n = 29, autumn) 10·1–12·9 (11·4), (n = 29, spring) 11–16·6 (13·8); Wadi Halfa, N Sudan (n = 22, Feb) 9·7–11 (10·0), (n = 138, Mar–Apr) 8·2–14·8 (11·0); NE Sudan (n = 19, Sept–Oct) 8–13 (10·5), (n = 190, Mar–Apr) 10–18·5 (13·5); Ethiopia (n = 173, Sept–Feb) 9·1–15·9 (11·3), (n = 69, Mar–Apr) 9·1–18·3 (12·6).

Complete moult on breeding grounds; partial moult in Africa Feb–Mar.

IMMATURE: 1st winter like adult, except that pale tips to tail-feathers are poorly defined, and T6 pale grey with buffish white confined to outer web and distal fringe of inner web.

S. c. caucasica Ognev and Banjkovski: mountains of SE Europe and Middle East; probably winters Arabia and tropical NE Africa. Crown darker grey than in nominate race, rest of upperparts greyer, underparts more buff.

TAXONOMIC NOTE: the small pale forms of the central Asian deserts, and the large, dark, grey race of the W central Asian mountains, are often treated as separate species, *S. minula* and *S. althaea* respectively. However, a zone of intergradation links the desert forms with *S. c. curruca* and *S. c. blythi* to the north, and *S. c. caucasica* has characters intermediate between nominate *curruca* and *althaea*. We therefore prefer to regard all these forms as races of *S. curruca*. Only nominate *curruca* and *caucasica* are known to visit Africa, but *althaea*, common in Oman in winter, could well reach Somalia.

Field Characters. A rather small, slim *Sylvia*, dull grey-brown above and whitish below, with white outer tail-feathers. Grey crown contrasts slightly with mantle, and with darker face mask formed by blackish lores and ear-coverts. Some ♂ Common Whitethroats *S. communis* have similar head and mantle coloration, but Lesser is distinguished by its darker ear-coverts, lack of sandy or rufous panel on wing, and dark legs. Smaller than Common Whitethroat, with shorter wings and shorter tail. Sardinian Warbler *S. melanocephala*, Subalpine Warbler *S. cantillans* and Ménétries's Warbler *S. mystacea* are of similar size to Lesser Whitethroat; all have grey upperparts and white outer tail-feathers and can show contrast between dark ear-coverts and pale throat. However, all have a reddish eye or eye-ring, and pale legs.

Voice. Tape-recorded (57, 62, 73, 93, B, F). Song, a dry rattling trill, 'rutututututu', usually preceded by a short, quiet varied warble; the whole lasts *c.* 1.5–2 s and is repeated about 6 times per min; rattle audible up to 200 m. Subsong, a continuous quiet warbling sequence, including thin squeaky notes, similar to that of Common Whitethroat. Contact-alarm call, a hard 'tuc', often repeated in a loose series; occasionally also a harsh 'charr'.

General Habits. Winters in dry savannas; in acacias and semi-open woodland, steppe with sparse bushes and trees, palm groves, and thickets by wadis and pools, including *Salvadora persica*; also in gardens and moister grassy places; mainly at low altitude. Sometimes solitary but usually a few together, often in parties with other warblers. Loose groups of 20 or more common on migration. In N Sudan, no firm winter territories, birds moving within a wide but delimited area (Mathiasson 1971), but spring passage birds in Egypt remain within small areas for several days (Simmons 1954). Mutual chasing of conspecifics occurs in N Sudan, and is most frequent where birds are densely distributed. Site fidelity shown by 5 birds ringed in Ethiopia and retrapped 2 years later (Ash 1981), and by 5 birds ringed on migration at Bahig, Egypt, and retrapped there in a subsequent season. More shy and secretive than Common Whitethroat and often remains hidden within cover. Forages within bushes and acacia canopy, taking prey from leaves, twigs and bark; commonly searches small acacia flowers in N Sudan (Mathiasson 1971). Hops and sidles with typically horizontal carriage. Flight fast and direct, similar to Common Whitethroat, but bird appears more compact with shorter narrower tail (Cramp 1992). Gives persistent 'tuc' call when alerted. Frequent subsong heard in Mar in Egypt, Oct in Eritrea, and throughout winter in N Sudan; one bird gave full song in Sudan from Oct to Mar (Cramp 1992).

Most or all birds of races *curruca* and *caucasica* enter Africa through the Middle East. Populations breeding west of *c.* 30°E head southeast in autumn to E Mediterranean, while those from SW Asia and from Russia east to W Siberia head south or southwest. Scarce in autumn in E Libya, but a strong passage in NE Egypt and N Sudan from late Aug to early Nov (peak mid-Sept to early Oct); fewer in Eritrea and Djibouti. Common in Egyptian Sahara; in oasis *c.* 280 km south of Mediterranean coast most birds were lean and some stayed up to 17 days, while at nearby 'desert' site most grounded birds were fat and none stayed more than a day (Biebach *et al.* 1986). Arrivals in central Ethiopia are from mid-Sept (but mainly late Oct), in S Sudan from end of Sept, and Darfur (W Sudan) from mid-Oct. Birds reach central Chad mainly in mid-late Oct, with main passage at Abéché in early Nov. Small numbers reach N Nigeria, Niger, Cameroon, Mali and Senegal from Oct–Nov, presumably by westward extension of migration from main winter quarters.

Northward migration begins early and lasts several weeks. Birds leave S Sudan in late Feb to early Mar, Nigeria and Chad in Mar to early Apr, and central Ethiopia mainly by mid-Mar. Passage occurs through Egypt mid-Feb to mid-May, extending regularly to E Libya (Cyrenaica and Serir oasis). Peak movement occurs late Mar to early Apr, with last birds in Egypt late May. ♂♂ tend to migrate a few days earlier than ♀♀ (Horner 1980). Stopovers of up to 19 days occur in Suez corridor (Simmons 1954). Passage through Eritrea is much stronger in spring than autumn, lasting from early Apr to early May.

Over 40 birds ringed W and N Europe (Britain, France, Belgium, Holland, Germany, Denmark, Sweden and Finland) recovered in autumn in NE Egypt. By contrast, spring recoveries are all from Levant, suggesting a more easterly return migration for these populations. 1 Egyptian spring bird recovered Germany, but most Egyptian spring migrants are probably bound for breeding areas east of *c.* 30°E (Cramp 1992). Birds ringed Sweden (1) and Finland (1) recovered Chad; 1 ringed N Sudan (Sept) recovered Hungary (June).

Food. Mainly arthropods, especially Hymenoptera, caterpillars, Hemiptera, Coleoptera and spiders; also berries in autumn and in winter quarters. In Sudan, takes fruit of tamarind *Tamarindus indica*, and once seen eating large dragonfly (Mathiasson 1971).

References
Cramp, S. (1992).
Mathiasson, S. (1971).

Sylvia nana (Hemprich and Ehrenberg). Desert Warbler. Fauvette naine.

Curruca nana Hemprich and Ehrenberg, 1833. Symb. Phys. Avium, fol. cc.; Sinai.

Plate 22
(Opp. p. 401)

Range and Status. Resident or partial migrant in NW Sahara and breeding summer visitor to central Asia, wintering in NE Africa, Arabia and east to Pakistan. NW African race vagrant to Italy, Malta and Atlantic islands, other races to Britain, Netherlands, Germany, Sweden and Finland.

Sylvia nana

In NW Sahara resident, but occurs only or mainly as non-breeding visitor in Aïr Massif, N Niger, and Awana region, Mali. Morocco, known from Merzouga, Tarda, Oum-Jerane, Erfoud and Tamlelt. Algeria, known from a limited number of localities (Heim de Balsac and Mayaud 1962) including Erg Admer, Oued In Djerane, Amguid, Timimoun, the Arak gorges, Béchar, El Oued, El Abiodh Sidi Cheikh, and south of Ghardaia (Ledant *et al.* 1981). Tunisia, 3 records: Douz, Moulares, and 'Bou Suad' (probably near Remada: Thomsen and Jacobsen 1979). Mali, locally quite common in central Awana (Aklé Awana) and Erigat in Nov–Jan, and seen carrying nest material in the area; once Arawan, Feb. 1 bird of NW African race obtained near Salum, Libya, in 1920 (Goodman and Meininger 1989).

Asiatic race *nana* is winter visitor of NE Africa and may breed in SW Fezzan, Libya (Bundy 1976). Scarce in Libya; records from south of Sirte, Ghadames, Al Adem, Wadi Tanezzuft, Bir Tahala, Ghat, Tin Abunda, Edeyin and Serir. Egypt, once west of Nile (between Bahariya and Cairo), rare passage migrant and local winter visitor in Eastern Desert, and recorded numerously in Red Sea mountains, where frequent south to Gebel Elba. Sudan, uncommon to common on Red Sea coast, rare inland. Eritrea, frequent on entire length of coast. Djibouti, 7 recent records of 1–4 birds, mainly from coast (Welch and Welch 1989). Somalia, fairly common on coast west of 48°E.

Breeding density in Africa unknown, in Asia high, ♂♂ only 50 m apart, av. 10·4 pairs in 10 km, and 13 pairs in 56-km transect in saxaul *Halozylon* desert (Cramp 1992).

Description. *S. n. deserti* (Loche): NW Africa, east to W and possibly E Libya. ADULT ♂: forehead, crown, nape, hindneck, sides of neck, mantle, scapulars and back bright yellow-buff with pinkish tinge; rump sandy cinnamon, uppertail-coverts rufous-cinnamon. Tail rufous-cinnamon with white sides: T1 bright rufous, T2–T5 rufous with blackish inner vanes, T4 and sometimes T3 with small white spot at tip, T5 with mainly white outer vane and large white wedge on inner vane, and T6 white. Lores pale grey; eye-ring, broad above and narrow below eye, pinkish off-white. Underparts rather uniformly pinkish cream-buff, paling to white or buffy white on belly and undertail-coverts. Wings much like mantle; primaries, secondaries and greater primary-coverts dark grey-brown with narrow pale rufous edge to outer vanes; tertials and greater coverts dark grey-brown with broad pale rufous edges and tips; median and lesser coverts bright yellow-buff. Underwing-coverts and axillaries cream-pink. Bill horn-yellow with brown culmen, gonys and tip of upper mandible; mouth yellow; eyes bright lemon-yellow, edge of eyelids black; legs and feet pale yellow. Sexes alike. SIZE: (12 ♂♂, 12 ♀♀) wing, ♂ 56–59 (57·5), ♀ 54–60 (56·6); tail, ♂ 46–51 (47·9), ♀ 43–49 (46·1); bill to skull, ♂ 12–13·4 (12·6), ♀ 11·6–13·5 (12·2); tarsus, ♂ 18·8–20·4 (19·6), ♀ 18·7–20·7 (19·8). WEIGHT: (S Algeria) ♂ (n = 3) 9, 9, 9, one ♀ 10·6.

IMMATURE: very like adult but slightly paler above and below; tail (T2–T5) browner, the white marks tinged with buff; eyes lead-grey, clouded yellow, or yellow with thin greenish orange ring (Shirihai 1988b).

NESTLING: hatches naked.

S. n. nana Hemprich and Ehrenberg: breeds from Caspian Sea to Mongolia. Upperparts darker, head tinged grey, uppertail-coverts and T1 bright foxy red, contrasting with rump and back; bill darker; T1 with black shaft, T2–T5 black-brown with rufous fringe to outer vane and tip, T6 white with dusky base to inner vane. Fractionally larger. WEIGHT: (Sudan) 2 unsexed 7·0, 7·5; (central Asia) ♂ (n = 26) 7–10·5 (9·0), ♀ (n = 4) 7–10 (9·0).

Field Characters. 11·5 cm. One of the smallest *Sylvias*, and the only one that is unmistakable in all plumages. Sandy or yellowish buff, with red-brown or bright foxy red tail and uppertail-coverts, tail white-sided, all very conspicuous as bird flies away low down. Eye bright yellow (greyish in juvenile). In low desert shrubs, spending much time on ground.

Voice. Tape-recorded (57, 60, 73, 93, B). Song of *deserti* a jaunty warble, duration 1.5 s, starting with harsh, emphatic rattle 'krrrrr' and ending with rising whistle; repeated every 5–6 s; recalls song of Common Whitethroat *S. communis* in duration and pitch and Blackcap *S. atricapilla* in timbre. Once, Algeria, song was a varied babble, including alarm calls, rather like Barn Swallow *Hirundo rustica* (Cramp 1992). Song of nominate *nana*, occasionally heard in winter, has starting rattle and finishing whistle, but middle is monotonous 'dididididididididi'. Subsong (Oman) a harsh, repeated 'chur-ur-ur-zick'. Contact and other calls 'krrrrr' (as in song) or a weak, dry 'drrrrrrr', 'tirrr', 'djjj-djjj-djjj', a high-pitched 'chee-chee-chee-chee' and a harsh 'zee-zee-zee'. Alarm, a staccato rattle of 25 notes in 2·0 s.

General Habits. Breeding habitat in NW Sahara is sandy or clayey plains, sometimes stony, with patches of low shrubs or tufts of herbage such as *Nucularia*; occurs also in thorn and tamarisk scrub on dunes and stony foothills (Cramp 1992). In Mali winters (and breeds?) in pasture of *Aristida acutiflora*, *A. pungens* and *Cornulaca monacantha*. Winter habitat in N Libya is *Thala* woodland (Bundy 1976) and on African Red Sea coast is open sandy plains with sparse cover of *Acacia* and *Suaeda*, and scrub on coastal sand dunes and salt flats. In Oman (where wintering habitat probably same as in NE Africa) occurs sparsely in vegetated mountains and wadis but commonly within 2 km of sea, on salty, sandy flats dominated by dense, knee-high patches of *Limonium nov. sp.* ('*axillare*'), *Suaeda maritima*, *S. fruticosa*, *Halopeplis perfoliata* and *Sphaerocoma aucheri*, with plenty of bare soil between clumps, with or without thin woodland of acacia or tamarisk.

Solitary, or 2–3 birds forage in same bush or within a few m of each other on ground. Wary, but open nature of habitat makes bird easy to watch. On observer's approach at slow walk, bird hidden in dense low growth flies up at distance of 5–8 m, sometimes only 2 m; it flies low for 10–15 m into thorn tree where it can eye observer, or drops quickly back into herbage. Nominate subspecies said to use voice frequently throughout year (Cramp 1992), but wintering birds in Oman (and Africa?) call less frequently and more quietly than other *Sylvia* spp.; 3 birds feeding in thick 0·5 m high *Limonium* seem to keep in contact without calling. Sings in Algeria in Dec–Apr. May raise crown-feathers when giving alarm call. Excited bird (Arabia) once sharply bobbed rear of body and flicked tail up and down; may also cock tail high and droop and shiver wings. Forages about equally on bare soil and within 0·3 m of ground in twigs and thin branches of leafless thorn tree. Bird foraging solitarily (or sometimes 2 together) often associates with larger bird perching on bush top above it and uses it as sentinel. On ground keeps beneath or within 0·5–1 m of shrub; gait a short, creeping hop, tail held up a little but not cocked high; also has brief hopping run on ground. Freely uses open, microphyllous thorn trees, perching on top at 2–3 m high to preen in early-morning sun, or if nervous to view a person, but seldom forages higher than 0·5–1 m.

Desert Warbler associated with sentinel in 28% of 104 observations (Oman); it forages briskly low in shrub below sentinel or, rather more frequently, on ground under shrub, and almost invariably flies a few s after sentinel flies, perhaps some distance, to another shrub top. Commonest sentinel species in Oman are 5 spp. of wheatears *Oenanthe*; Little Green Bee-eaters *Merops orientalis*, Arabian Babblers *Turdoides squamiceps*, Red-tailed Shrike *Lanius isabellinus* and Great Grey Shrike *L. excubitor* are often used. Ignores or at least does not obviously follow other *Sylvia* spp., redstarts *Phoenicurus*, Spotted Flycatcher *Muscicapa striata*, sunbirds, sparrows or bulbuls. Often follows its sentinel repeatedly; one once followed a Red-tailed Wheatear *Oenanthe xanthoprymna* as it changed its hunting perch 10 times in 7 min in area of *c.* 1 ha; each time warbler flew a few s after the wheatear and alighted low in the same bush or in tussock or on ground within 5 m of it. Sentinel-use in Africa not yet reported.

Resident in NW Africa; winter visitor Nov–Jan to N Mali where may breed, Nov–Dec to Aïr (Niger), Sept–Apr to Egyptian Red Sea hills (mainly Sept–Mar; once late July), Nov–Mar to Eritrea, Oct–Dec to Djibouti and Jan–Feb to Somalia. In Asia ♀♀ and young seem to emigrate earlier than ♂♂.

Food. Mainly small insects: ants, bugs and caterpillars. In Saudi Arabia seeds, spiders, bugs, Lepidoptera, ants, beetles and *Lycium* berries.

Breeding Habits. Monogamous, solitary, territorial. Pair closely bonded. In Asia ♂ sings from top of bush, also sometimes on ground; favourite song-posts include nest-bush; singing ♂ may spread and half-raise tail; uncommonly, rises from bush to 10 m or more in air, starts to sing before peak of ascent, and sings in parachuting descent with bill and tail raised. Sings from sunrise, less at midday, with resurgence in evening; may sing after sunset. ♂ patrols territory using regular sequence of perch sites (Asia: Cramp 1992); rival ♂♂ confront each other at territorial boundary and may fight in air. ♂ both tolerates and chases other warblers. ♂ builds cock nest, evidently prerequisite for attracting and courting ♀. Courtship includes chasing in air, through scrub, and on ground; ♂ raises and fans tail and droops wings; tail said to be folded over back (Algeria).

NEST: substantial cup of twigs, leaves and grass stems, with plant down and spider web, lined with down, fine grasses and fibres (Cramp 1992). Size (Asia, n = 10), ext. diam. 100–120 (102), ext. depth 84–130 (99), int. diam. 44–62 (50), int. depth 52–70 (60). Sited within 1·1 m of ground, in shrub.

EGGS: 2–5 (8 clutches, Algeria, av. 3·0). Subelliptical; smooth; glossy; white or palest blue, finely specked and scrawled with brown and grey, particularly at large end. SIZE: (*deserti*, n = 10), 14·5–16·8 × 11·0–12·6 (15·9 × 12·4). WEIGHT: *c.* 1·26. 1 brood in north of African range, often 2 in south.

LAYING DATES: Morocco, Mar, May; Algeria, Jan–May.

INCUBATION: by ♀ (mainly) and ♂ (significantly); both have brood patches. Period unknown.

DEVELOPMENT AND CARE OF YOUNG: young cared for by ♂ and ♀. Nestling period unknown. Parents very secretive when feeding young.

BREEDING SUCCESS/SURVIVAL: overall success rates of 59% (4 clutches, Uzbekistan) and 31% (14 clutches, Turkmeniya); most nests lost to ground predators (Cramp 1992).

Reference
Cramp, S. (1992).

Plate 23
(Opp. p. 416)

Sylvia rueppelli Temminck. Rüppell's Warbler. Fauvette de Rüppell.

Sylvia ruppeli [sic] Temminck, 1823. *In* Temminck and Laugier, Planches col., livr. 41. pl. 245, fig. 1; Crete.

Sylvia rueppelli

Range and Status. Greece, Turkey, Crete and Aegean Is; winters wholly in Africa. Overflies E Mediterranean in autumn, making landfall in NE Egypt; in spring some birds return via NW Saudi Arabia. Vagrant Malta, Italy, Romania, Russia, Algeria, France, Britain, Faeroes and Finland.

Winter visitor, Sudan and Chad and much more rarely further west. Passage migrant in Egypt in autumn and Libya and Egypt in spring. In midwinter, fairly common across Sudan mainly between 17°N and 12°N (Hogg *et al.* 1984, Nikolaus 1987), recorded south to 11°N, north to Egyptian border and once at Aswan (Egypt), and east to Red Sea hills. Common in winter in NE Chad: Tibesti, Ellala, Ouadi Souala, Zouar, Yatroum, Fada and Ouadi Rime-Ouadi Achim Faunal Reserve south to Ouadi Enné (13° 30′N, 19°45′E); in the Faunal Reserve, the second commonest wintering *Sylvia* (after Lesser Whitethroat *S. curruca*: Newby 1980). Winter range in northern halves of Chad and Sudan probably much wider than shown. Niger, rare: 2 records Ngourti (15° 19′N, 13° 12′E) and a few from N Aïr, Feb and Mar. Mali, rare: records at Tessalit, Tin Karr and Kara, Sept–Oct. Eritrea, vagrant, 1 Massala, Feb. Algeria, vagrant, 1 Reghaïa, Mar.

On autumn passage fairly common in NE Egypt, Sept, recorded mainly along coast from El Alamein and Bahig to Suez and Hurghada; in 3 years 356 birds noted, mostly near Bahig, in late Aug to late Sept; inland, 25 once caught at Bahariya Oasis, late Aug to mid-Sept.

In spring common in NE Egypt, early Mar to late Apr (occasional Feb, May) and found in Western Desert at Gebel Uweinat, Dakhla, Kharga, Farafra, Siwa, Wadi Natroun and Bahariya (Goodman *et al.* 1986). Once (spring?) hundreds in Giza gardens (Meinertzhagen 1954). In Libya, common in Cyrenaica in Mar–Apr with peak passage late Mar to early Apr, and scarce but regular on Tripoli coast. Recorded in desert interior at Jaghbub, Jalo, Kufra and Serir; once 50 at Serir, early Apr.

Description. ADULT ♂ (breeding): forehead, crown, superciliary area, lores and cheeks black; ear-coverts, sides of neck, nape, hindneck and mantle bluish grey, quite sharply demarcated from crown; scapulars, back, rump and uppertail-coverts dark grey with slight brownish tinge. Tail black, T1 and outer webs of T2–T5 tinged with grey, T6 white, T5 with white wedge near tip of inner web 17–28 mm long, T4 with similar white mark 6–11 long, T3 sometimes with small white spot at tip. Wings brownish black, primaries, secondaries and greater primary coverts with narrow but well-defined grey fringes on outer webs, tertials and greater coverts with conspicuous, broad pale grey or pinkish outer margins; lesser and median coverts dark blue-grey. Sharp white moustachial stripe; chin and throat black. Breast, flanks and thighs pale grey with faint pink wash; belly, vent and undertail-coverts white (concealed feather bases grey). Undersides of tail and wings dark grey, underwing-coverts and axillaries pale grey. Bill dark horn to black, with basal half of lower mandible pinkish or horn-brown; eyes orange to hazel-brown or chestnut-brown; orbital ring of bare skin orange-red; legs and feet light brown or reddish brown, soles yellowish. ADULT ♂ (non-breeding): like ♀ (see below) but forehead and crown black, or black with grey feather fringes, and chin and throat white with slight or heavy black mottling. ADULT ♀: forehead and crown dark grey-brown, speckled blackish; lores, cheeks and ear-coverts brown; remainder of upperparts brown, with greyish tinge on hindneck; wings and tail dark brown with same white marks in tail and pale feather-edgings in wing as ♂, although wing-feather margins are pale buff-grey rather than pale grey. Sometimes a broken ring of minute whitish featherlets around eye. Chin and throat white with small blackish speckles; moustachial stripe white, remaining underparts dark buff where ♂ grey and pale buff where ♂ white; undertail-coverts sometimes patterned with dark brown. Soft parts like ♂. SIZE: wing, ♂ (n = 70) 67–74 (70·0), ♀ (n = 19) 65–72 (68·8); tail, ♂ (n = 40) 56–64 (59·8), ♀ (n = 14) 54–62 (57·8); bill to skull, ♂ (n = 38) 14·2–16·1 (15·0), ♀ (n = 13) 14·5–15·7 (15·0); tarsus, ♂ (n = 38) 20·1–22·1 (21·2), ♀ (n = 14) 20·5–22·2 (21·1). WEIGHT: (Cyprus, spring) ♂ (n = 19) 9–16 (12·6), ♀ (n = 6) 11·5–14·9 (13·2).

IMMATURE: like adult ♀ but edges of wing-feathers (especially tertials, also secondaries and greater coverts) rufescent brown rather than pale buff-grey, iris and orbital ring dark brown or red-brown, no white moustachial streak (hence dark cap not as well defined as in adult ♀), and chin and throat buff-white, usually without dark speckles.

Field Characters. 14 cm. A typical *Sylvia*, wintering in central and E Sahara, skulking in low vegetation in wadis and oases, particularly the shrub *Capparis aphylla*; size and shape of Common Whitethroat *S. communis*. Adult ♂ readily identified by conspicuous white moustache separating black face from black throat; red eye and eye-ring, white sides to tail, white fringes to tertials and red-brown legs. Acquires this plumage by moult in

Jan–Apr; before then, ♂ is browner overall but can be told by its black or grey-black forehead and forecrown, white moustache, and dark-speckled white throat. ♀ is rather plain brown above and buffy below, with same red eye, white-sided tail, red legs and greyish-edged tertials as ♂; can be distinguished if white chin and throat are dark-speckled, producing pale moustachial stripe, from all except ♀ Cyprus Warbler *S. melanothorax*, but if they are not speckled ♀ almost impossible to distinguish from juvenile and winter ♀ Sardinian Warbler *S. melanocephala*. Rüppell's larger than Sardinian, with squarer and proportionately shorter tail. Very like ♀ Ménétries's Warbler *S. mystacea*, but does not depress/wag tail as latter regularly does. ♀ without speckled throat, and juvenile, also very like Subalpine Warbler *S. cantillans* but larger (Subalpine 12 cm) (Svensson 1992). Can also be confused with Lesser Whitethroat *S. curruca*, which is smaller, with black legs, dark mask, white throat, and no red in eye; and Orphean Warbler *S. hortensis*, which is larger, with whitish eyes and blackish legs.

Voice. Tape-recorded (62, 73, 93, B). ♂ song a brief, rapid, jerky chatter containing pure and harsh notes: 'trr-cht-titchrtrr . . .', lasting 2 s, repeated up to 16 times per min. Alarm a hard 'tak', also 'tak-tak'. Also gives rattling 'tkt-tkt-tkt-tkt . . .' and blunt 'chrr'.

General Habits. On wintering grounds associated in particular with leafless, many-stemmed, non-aromatic, 4-m-tall shrub *Capparis decidua* (= *C. aphylla*), and probes its profuse red flowers; *C. decidua* grows on stony foothills, near farmed land, in wadis and on dunes. Also occurs in trees and scrub on stony slopes, at oases, and on passage in Egypt in hedges, canebrakes and gardens. In Crete nests mainly in broom *Calycotome*. In Europe occurs where July isotherm is 25–30°C, a warmer climate than tolerated by congeners; in dry thorny scrub on rocky slopes with ravines and fissures, and undergrowth in oak and cypress woods.

Forages by hopping low down in tree or shrub; movements rather slow; frequently flicks wings and tail; sometimes half cocks tail. Gleans insects acrobatically from leaves; sometimes comes to ground to feed. Skulks, but is approachable when deep in bush. Usually solitary in Africa, but sometimes 2–4 together in clump of desert shrubs, and concentrates with its own species and with Common Whitethroat, Lesser Whitethroat, Subalpine Warbler and other *Sylvia* spp. on passage. Spring migrant in Egypt may remain in same small area for some time; ♂ once defended 150 m^2 of fenceline in late Mar (Cramp 1992).

Present in winter quarters from Oct (Chad) and Nov (Sudan) until about Mar. For passage times in Libya and Sudan, see *Range and Status*.

Food. Poorly known. Mainly insects, including caterpillars (Crete). Probes *Capparis* flowers presumably for insects; eats *Capparis* berries in Saudi Arabia.

Reference
Cramp, S. (1992).

Sylvia melanothorax Tristram. Cyprus Warbler. Fauvette de Chypre.

Plate 23
(Opp. p. 416)

Sylvia melanothorax Tristram, 1872. Ibis, p. 296; En-Gedi, Palestine.

Forms a superspecies with *S. melanocephala* and *S. mystacea*.

Range and Status. Cyprus: common resident and partial migrant. Many emigrate, but wintering grounds not known. Regular on passage in Israel, Oct and (Eilat) Feb–Mar, and a few winter there (Jericho, Arava valley, Eilat). Vagrant Lebanon, W Saudi Arabia, Palestine, Egypt and Sudan. Main wintering area may be Sinai and Egyptian eastern desert (Cramp 1992).

Vagrant, NE Africa: Egypt, 4 records (Wadi Hof, Mar 1910, Wadi Umm Rus and Gebel Elba, Dec 1938, Sharm el Sheikh [Sinai], Jan 1982); Sudan, 3 records (all Erkowit, Nov 1980: Nikolaus 1981).

Description. ADULT ♂: forehead to hindcrown, lores, cheeks, ear-coverts and sides of neck glossy black; nape to rump and scapulars dark bluish grey or slate-grey; uppertail-coverts almost black. Tail black with white sides: T6 with outer web white and 12–16 mm white tip to inner web, T5 with black outer web, inner web with white edge and large white spot at tip, T4 (and sometimes T3) with small white spot on inner web. Eyes light brown; bare reddish eye-ring with minute white featherlets above and below it. Malar stripe white. Chin and throat barred black and white, breast dusky black with white scallops, upper belly and sides whitish with large blackish triangles or chevrons (all of these feathers are black, with white fringes). Lower belly, flanks, thighs and vent grey; undertail-coverts dark grey with broad white fringes. Primaries and secondaries dull black, narrowly edged grey; tertials, greater primary coverts, alula and median coverts dull black, very narrowly edged white. Bill bluish black, upper mandible with silvery or yellowish cutting edges, lower mandible yellowish or pale pinkish, with blackish tip; bare, papillate orbital ring pink to dark red; eyes brown, red-brown or dark yellow; legs and feet flesh brown or yellowish brown. ADULT ♀: like ♂ but upperparts much less blackish and underparts less spotted; orbital ring skin brown, sometimes with white featherlets. Forehead, crown, mantle to uppertail-coverts, scapulars, ear-coverts, tertials and wings, dark grey-brown. Chin to belly and flanks buff or buffy grey, with small dusky spots on chin and throat, larger ones on breast, none on upper belly. SIZE: wing, ♂ (n = 13) 59–63 (60·5), ♀ (n = 5)

Sylvia melanothorax

60–63 (61·3); tail, ♂ (n = 7) 53–57 (55·3), ♀ (n = 5) 50–51 (50·5); bill to skull, ♂ (n = 12) 13·0–14·2 (13·7), ♀ (n = 15) 12·8–14·0 (13·4); tarsus, ♂ (n = 12) 19·0–20·4 (19·6), ♀ (n = 15) 18·7–20·2 (19·4). WEIGHT: (Cyprus) ♂ (n = 54) 9·4–13·7 (11·3), ♀ (n = 39) 9·5–15·0 (11·5).

IMMATURE: like adult ♀ but browns warmer, less greyish; chin, throat and breast indistinctly speckled; white parts of tail suffused with buff.

Field Characters. Adult ♂ readily distinguished by black hood, reddish eye and eye-ring, barred black and white chin, and heavily black-mottled throat and breast. Adult ♀ has dusky-speckled throat and cannot be distinguished reliably from ♀ Rüppell's Warbler *S. rueppelli* in field; in the hand, main distinction is greater amount of white in tail of Rüppell's (Svensson 1992). Barred Warbler *S. nisoria* has finely barred underparts, grey head, yellow eye, and white in corner of tail, not side. Unspotted juvenile Cyprus Warbler not safely distinguishable in field from young Rüppell's or Sardinian *S. melanocephala*.

Voice. Tape-recorded (BERG, GORD). ♂ song a rapid mixture of dry rattles and tonal notes, lasting 2·5–3·5 s; like Sardinian Warbler but more wooden; also like Common Whitethroat *S. communis*. Subsong a long rambling warble, rich in mimicry. Contact calls a staccato 'tchek' or 'tzigg' and a dry, rasping 'zrik'. Alarms, lasting 3–5 s, a ticking rattle and a churring rattle (Cramp 1992).

General Habits. In Cyprus inhabits coastal maquis scrub: *Pistacia, Rhamnus, Cistus, Salvia*; also gorse and broom, edges of woods with pines, oaks, juniper, myrtle and carob; up to 1400 m. Limits closely follow 340–350 mm isohyets around edges of central Cyprus plain (Mesaoria) (Flint and Stewart 1983). In pairs. Feeds inside bushes and trees by moving along branches and through twigs to glean insects; sometimes darts out to take butterfly from herb layer. Sings Jan–June, mainly Mar–June; uses elevated song-post; occasionally sings in display flight.

Food. Insects: midges, flies, pierid and other butterflies, possibly aphids; also spiders.

Reference
Cramp, S. (1992).

Plate 23
(Opp. p. 416)

Sylvia melanocephala (Gmelin). Sardinian Warbler. Fauvette mélanocéphale.

Motacilla melanocephala Gmelin, 1789. Syst. Nat., i, ii, p. 970; Sardinia.

Forms a superspecies with *S. mystacea* and *S. melanothorax*.

Range and Status. Endemic resident and migrant. Mediterranean basin in S Europe (Portugal to Turkey), Levant, Canary Is, and NW Africa (Morocco to Libya). Breeding summer visitor Yugoslav republics, Bulgaria, E Greece and NW Turkey; partial migrant in rest of breeding range. Winter visitor to S Turkey, Cyprus, Sinai and much of Sahara south to about 16°N. Vagrant north to Sweden and Finland, east to Kuwait.

Widespread resident, passage migrant and winter visitor to Maghreb. Morocco, isolated population breeds near coast at 24°N (Cramp 1992), and species widespread and locally common north of 27°N, occurring in Atlas Mts up to 2400 m; autumn migrants common in W Moroccan lowlands; small numbers winter around Errachidia and south to Boudenib, Rissani, Alnif, Tazzarine and Zagora. Algeria, common and widespread in coastal regions and very common in the Tell, less common on Hauts-Plateaux; scarce in Aurès; autumn migrants common in E Algerian oases; winters south to Tassili and Hoggar. Tunisia, common or very common in warm lowlands south to Sfax and Djerba I.; absent from high plateaus, rare and local in Kroumirie and Kairouan-Sbeitla area; after breeding disperses to semi-desert and oases. Libya, breeds in Jebel Akhdar and coastal areas east to about Derna; in winter widespread in N Cyrenaica in some years, localized in oth-

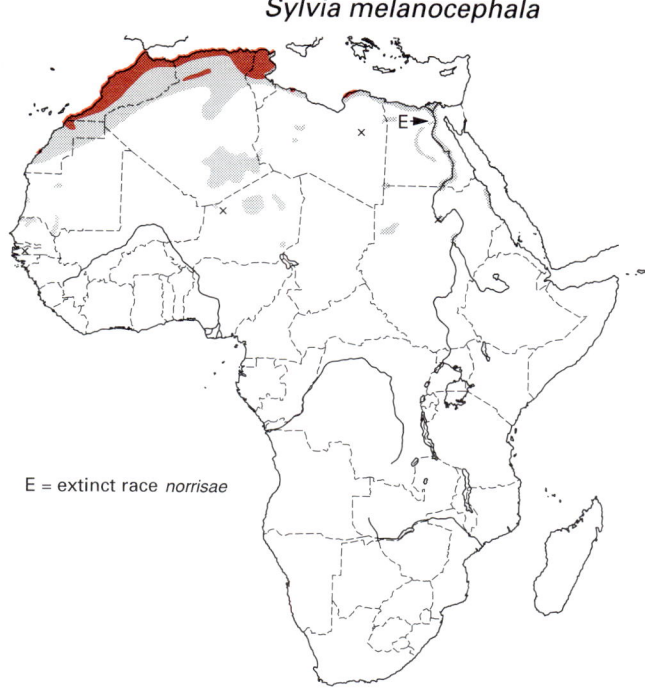

Sylvia melanocephala

E = extinct race *norrisae*

ers, and sometimes common in N Tripoli where a few breed near Tripoli (Cramp 1992); in Fezzan, recorded from Jofra area, Sebha, and commonly in SW in Ghat-Serdeles region; once Jaghbub and Serir. Egypt, common passage migrant and winter visitor to Mediterranean coast, Nile delta and valley, and Western Desert oases (Siwa, Natrun, Bahariya, Farafra, Dakhla, Kharga) and Gebel Uweinat. Until about the 1920s an endemic subspecies, *S. m. norrisae*, nested numerously in late Mar at L. Qarun, which became saline and is now without vegetation; *norrisae* last recorded in 1939 and is now evidently extinct.

Mauritania, frequent in Nouakchott, lower Senegal R. valley, Guidimaka, and Assaba, Tagant and Adrar massifs; once Tanoudert; quite frequent and widespread west of 12°W, more so than shown on map; in SW, 60 birds encountered in 120 h (Browne 1982). Senegal, winters regularly in NW on L. Guier; records at Richard-Toll, Ndioum and in SW at Bignona. Not yet recorded in Mali or Chad, but in Niger several records from Ngourti (north of L. Chad), Assamaka and Arlit, common at Yebjeyeba and very common in Kaour region and Aïr-Ténéré Nat. Nature Reserve. Sudan, a few recorded in Wadi Howar-Jebel Meidob area in W, in Nile valley south to near Khartoum, and on Red Sea coast.

Breeding densities of *c*. 40 pairs per km² in Moroccan maquis, *c*. 50–60 in semi-arid plains and high valleys with plentiful shrubs, and 8 pairs per km² in 'good' habitat (Cramp 1992). Wintering densities of 50 birds in 15 ha in oasis at Tamanrasset, Algeria, Dec, and 100 per km² there, Feb (Niethammer and Laenen 1954).

Description. *S. m. melanocephala* Gmelin: range of species except Canary Is, Faiyum (Egypt), Sinai and Levant. ADULT ♂: forehead, crown, nape, lores, cheeks and ear-coverts black, slightly glossy, sharply demarcated; hindneck, sides of neck, mantle, scapulars and back dark bluish grey, rump and uppertail-coverts the same but a purer tone. Tail-feathers black, fringed grey at tip and along outer webs, but T6 largely white; white spot at tip of inner web of T4 and sometimes of T3; T5 with white mark at end of inner web 5–15 long; T6 with outer web and distal half of inner web white; tail well rounded, T6 7–15 shorter than T1. Chin and throat white, breast grey, darker at sides, grey sometimes forming distinct band; belly white or greyish white, vent white; flanks, thighs and undertail-coverts grey, the last sometimes brownish. Primaries and secondaries dull greyish black, narrowly edged and tipped with pale grey, tertials and greater primary-coverts blackish with dark grey edges, remaining upperwing-coverts blackish with indistinct greyish edges; underwing-coverts and axillaries pale grey. Bill dark slate-grey with base of lower mandible pale grey or yellowish; eyes light reddish brown, bare orbital ring red-brown to crimson; legs and feet yellowish or pinkish olive-brown. ADULT ♂ (1st winter): variable, often like adult ♀, cap brownish and not well demarcated from hindneck and sides of neck, iris and eye-ring dull mauve. ADULT ♀: forehead, forecrown, lores, cheeks and ear-coverts blackish grey, hindcrown and sides of neck bluish grey, nape, hindneck, mantle to uppertail-coverts and wings dark greyish brown, tail dark greyish brown with same white sides as ♂, chin and throat buffy white in centre and off-white at sides, contrasting sharply with cheeks, remaining underparts buff. Eyes like ♂ but bare orbital ring, if apparant at all, thin, brown and inconspicuous. SIZE: wing, ♂ (n = 39) 57–64 (60·4), ♀ (n = 25) 57–62 (59·2); tail, ♂ (n = 28) 56–68 (60·8), ♀ (n = 23) 55–63 (58·3); bill to skull, ♂ (n = 39) 13·4–15·2 (14·2), ♀ (n = 21) 13·3–15·4 (14·2); tarsus, ♂ (n = 24) 19·8–22·4 (20·8), ♀ (n = 24) 19·9–21·9 (20·9). WEIGHT: (Morocco, Algeria, Niger) ♂ (n = 13) 9·4–14·0 (11·3), ♀ (n = 5) 10·9–13·0 (11·8), unsexed (n = 485) 9·6–20·0 (11·7).

IMMATURE: upperparts dull brown, rather uniform, but greyer around eyes; tail dark grey, feathers fringed brown; underparts cream-white, buffy on sides of breast, flanks, thighs and undertail-coverts; throat weakly contrasted with cheeks; eyes grey-brown, no bare orbital ring but sometimes narrow ring of whitish featherlets.

NESTLING: hatches naked.

S. m. norrisae Nicoll: Faiyum, Egypt. Probably extinct. Hindneck, mantle, back, and edges of tertials paler than in nominate race, sandy grey rather than blue-grey; underparts warm pinkish buff where nominate race grey. Smaller: wing, ♂ (n = 8) 56–59 (58·1), 1 ♀ 56·5; tail, ♂ (n = 8) 55–57 (56·5), 1 ♀ 54·5.

S. m. momus (Hemprich and Ehrenberg): Levant, and presumably this race which breeds in Sinai. Winter records in Egypt from Arminna and Bahig; probably winters quite numerously in NE Africa, since a common spring migrant at Eilat, Israel. Like nominate race but greys of upperparts paler and underparts more extensively white. Smaller: (26 ♂♂, 8 ♀♀) wing, ♂ 55–60 (57·6), ♀ 55–58 (56·5), tail, ♂ 51–57 (54·1), ♀ 50–54 (51·8).

Field Characters. 13·5 cm. A slim, rather long, medium-sized warbler, the commonest *Sylvia* throughout much of its breeding range. Bold, vocal, curious, and angry-looking (red eye in dark face; forecrown often peaked with forehead steeply raised). Adult ♂ distinguished from all except Ménétries's Warbler *S. mystacea* by combination of black head, red eye and eye-ring, white throat, dark grey upperparts, pale grey underparts, and quite long, rounded or graduated tail with white sides and corners. ♂ Ménétries's waves tail and has pinkish throat and white moustache. ♀ Sardinian is dark greyish brown above, distinctly hooded, with well-demarcated white throat, buff flanks

Plate 21

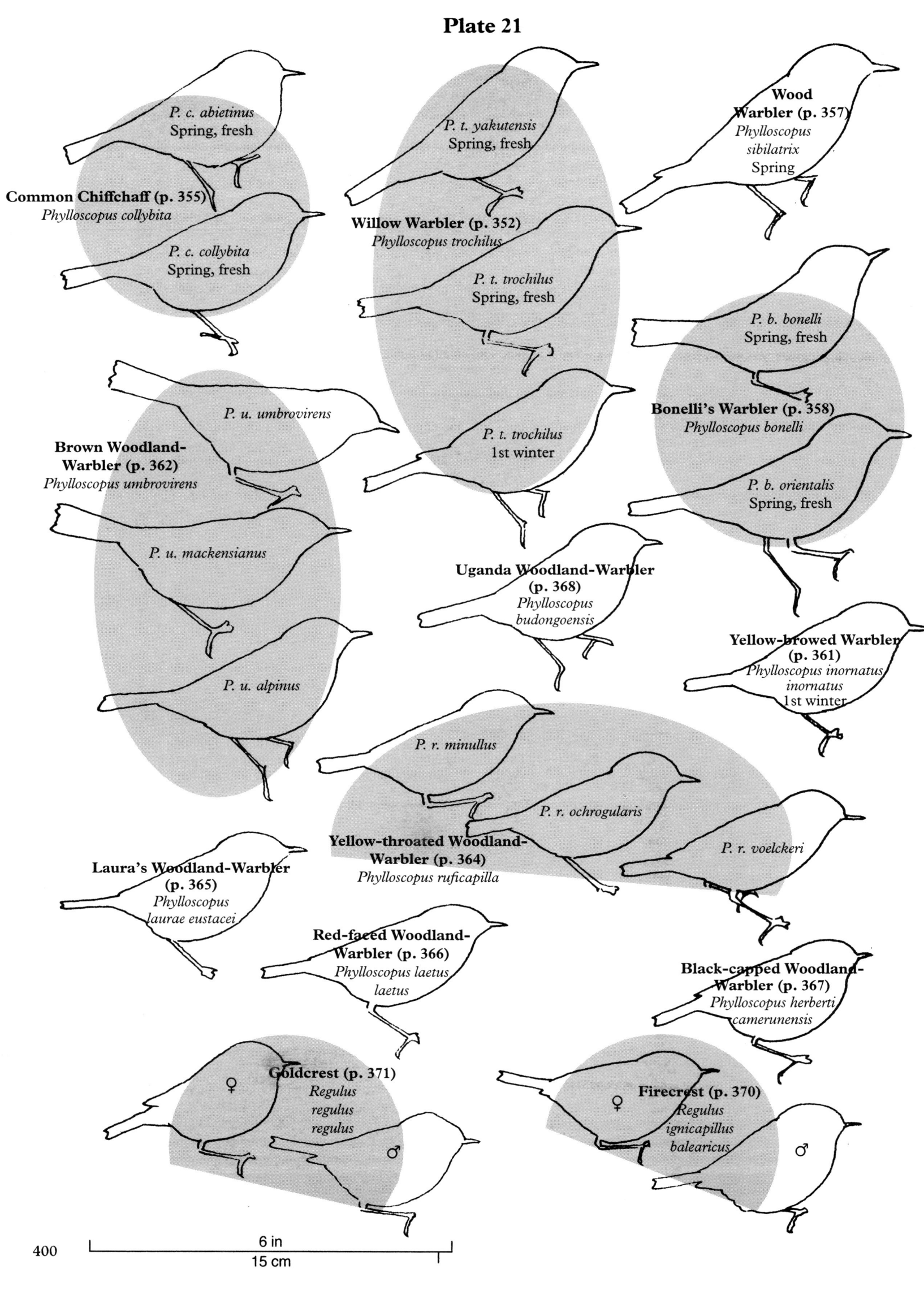

400

Plate 22

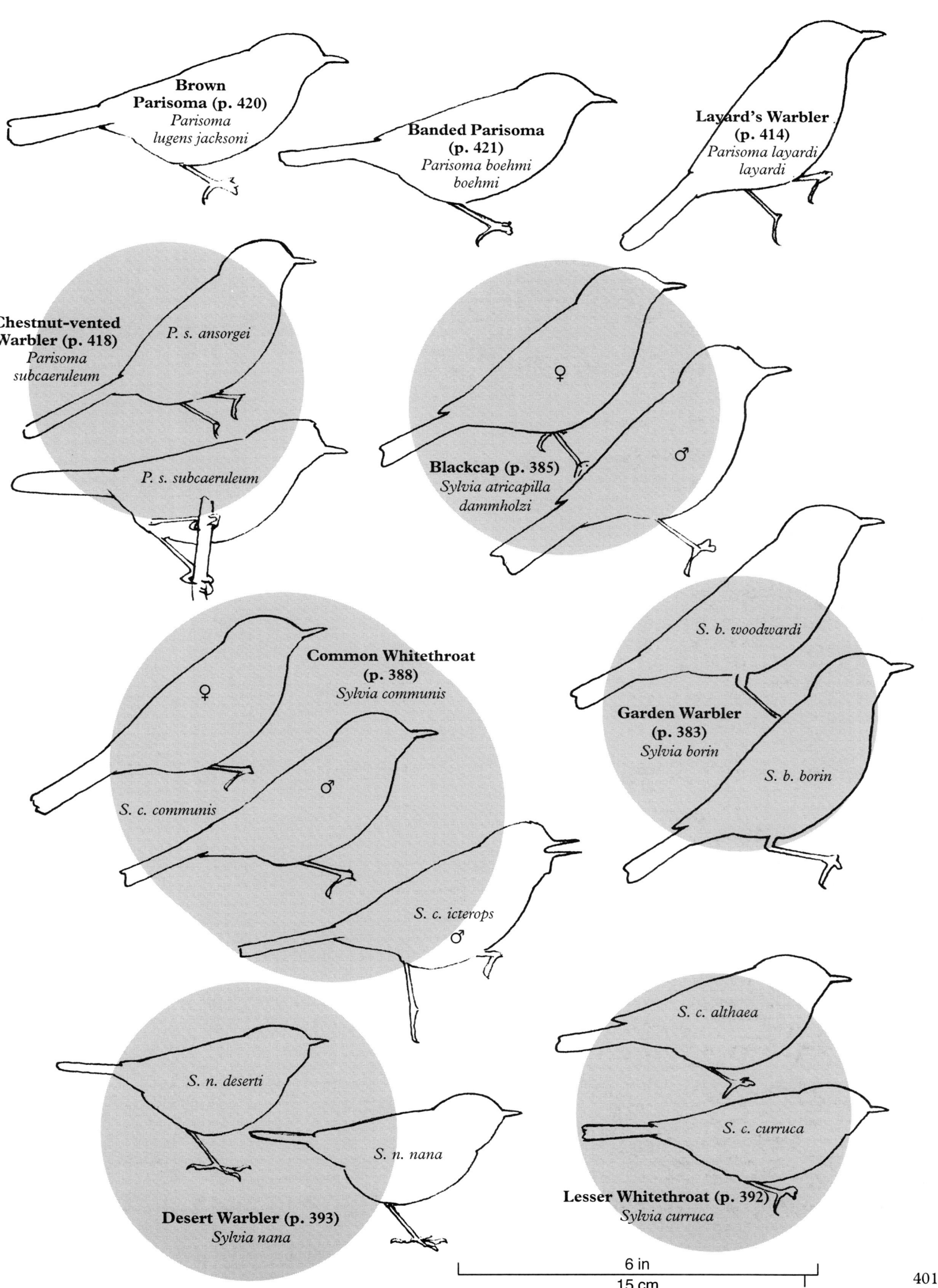

and red-brown eyes – similar to many other ♀ or juvenile *Sylvia* spp., but lack of rufous in wings eliminates several species. ♀ Ménétries's (overlaps Sardinian in E Sudan) has pale, not dark, lores, cheeks and ear-coverts, wags tail, and has distinctive 'mig' call. ♀ Cyprus Warbler *S. melanothorax* and ♀ Rüppell's Warbler *S. rueppelli* have dark-speckled throats. ♀ and juvenile Subalpine Warbler *S. cantillans* are paler, sandier, cleaner-looking birds than ♀ Sardinian, with pale orange or pinkish throats.

Voice. Tape-recorded (57 – 62, 73, 93, B). ♂ song from top of bush a staccato but melodious warble interspersed with many loud rattles; duration av. 3·3 s; ♂ song in flight similar but much longer (at least 22 s), a phrase of 2–2·5 s being repeated without pause 8 times or more, giving song a cyclical, flowing quality. Song sometimes preceded by soft 'chuck' or 'kuick' note repeated 2–3 times. Main call, given all year, is loud, far-carrying rattle 1–1·5 s long, beginning and ending abruptly, like wooden football-rattle, or machine-gun. Various alarm and other calls described (Cramp 1992).

General Habits. In NW Africa breeds in poor stands of holm oak *Quercus ilex* dominated by other shrubs, and in juniper scrub and light woodland dominated by lentisc *Pistacia lentiscus*, also in exotic acacias and seaside *Salicornia* scrub (Morocco). Avoids cork oak *Q. suber* and zen oak *Q. faginea*. Forages in wide variety of woody habitats, particularly crowns of Aleppo pine *Pinus halapensis* and holm oak, shrubby maquis undergrowth and garigue, tamarisk, olive groves, vineyards, brambles, roadside hedges and 'wild' gardens. Indifferent to proximity of people and habitations; freely uses artifacts for perching, at least where within *c*. 3 m of ground and if plenty of cover nearby. In winter occurs in herb layer and (at least on passage) thin, low desert shrubs, in palmeries with cultivation (Errachidia, Morocco), gardens, plantations and around market-gardens (Mauritania), tamarisk stands and edges of mangrove (Senegal), and arid acacia grassland (Sudan).

Outside breeding season solitary or in pairs. Bold, curious, noisy and restless; when breeding, quarrelsome and highly territorial. Often comes briefly into view at edge of vegetation, but spends most of time concealed inside brambly growth, low shrubs or crown of close-growing trees. Moves in open from one patch of bushes to another with rapid, whirring flight, spreading tail as it alights; when creeping through bush usually holds tail tightly closed. Often cocks tail and steeply raises forecrown and forehead feathers. Usually raises tail somewhat before flying. Spends much time on ground where hops, sidles and at times runs, with body kept low. In Gibraltar spends *c*. 5% of time feeding on ground in summer and up to 21% in winter; 47·5% of foraging is <1 m high, 8·5% at >3 m high and rest in between (Cramp 1992); av. foraging height 1.8 m (Sardinia). Feeds rather slowly and methodically, mainly in centre of bushes. Once seen flycatching for insects; searches flowers of aloe for nectar and of *Antholysa* and *Mimosa* for insects; once seen taking scraps from bin (Algeria).

When singing, ♂ perches rather upright with plumage loose and wings drooping, white throat-feathers contrasting strongly with black head. Sings late Feb, peaking in Apr, with song-flights occasionally from Dec but mainly from late Mar (Gibraltar); sings in winter in Libya and S Algeria. Occupies temporary territories in winter, chasing other warblers out (Egypt).

Mainly resident in NW Africa. Much commoner on Atlantic coast of Morocco in autumn than in winter, indicating migration then; late autumn occurrence of fat birds in Gibraltar suggests trans-Saharan flights, evidently to W Mauritania and N Senegal. Present on wintering grounds in SE Morocco from late Oct or early Nov. Winters in Mauritania Sept–Apr, mainly Oct–Mar, and in Senegal Nov–Feb. In Algeria passage occurs across Tell coast (Reghaïa) in Sept–Oct and Mar; migrants arrive in Saharan oases mainly in Nov, some in Sept. Winters in Tassili and Hoggar ergs and N Niger in Dec–Mar. Strong passage once noted near Tunis, Tunisia, mid-Oct. Bird ringed in NW Tunisia (Mahibens) in July recovered next Feb in E Tunisia (Sousse). Very fat birds occur in Tunisia in late Mar. In Libya present in SW Fezzan from Sept. In Egypt passage and wintering are from early Sept to mid (or late) Apr.

Food. Insects, and fruit in winter. Earwigs, termites, ants, caterpillars, beetles, small hymenopterans (Niger); beetles, small snails, and *Myoporum* and other berries (Tunisia). Wide range of invertebrates known as prey in Europe, including tiny springtails, grasshoppers, moths, flies, spiders, ticks, pseudoscorpions, centipedes, millipedes and snails, and once a lizard. European plant food includes fruits of lentisc *Pistacia*, olive *Olea*, myrtle *Myrtus*, dogwood *Cornus*, fig *Ficus*, honeysuckle *Lonicera*, mulberry *Morus*, buckthorn *Rhamnus*, bramble *Rubus* and many others; also some seeds, and nectar of *Aloe* which a bird sometimes defends. 84% of annual crop of honeysuckle berries eaten in Israel (Cramp 1992). Olive trees visited more than other plants, Gibraltar; small olive fruits eaten whole, but only flesh of large ones (Rodriguez de los Santos *et al*. 1986). In SE Spain, 2884 animal food items in winter were Hymenoptera (mainly ants) 34·6%, bugs 30·3%, beetles 19·5%, snails 4%, caterpillars 3%, flies, crickets, spiders, roaches and lacewings 9%; nearly all of the birds sampled had eaten olives. Also in Spain, estimated to eat 52·5% of fruit by volume and 47·5% animal matter.

Breeding Habits. Monogamous, solitary, territorial nester. Pair occupies territory all year (Malta); winter (and all year?) territories held in Libya (Willcox and Willcox 1978). In Gibraltar and Egypt, bird may defend winter territory for several weeks, but most are solitary and nomadic then. Breeding territory size in Gibraltar 0·12 ha in maquis and 0·31 ha in garigue and in Sardinia (n = 11) 0·23–0·64 (0·38) ha (Cramp 1992). Nests sometimes only 2 m apart (Majorca, Spain). Territory defined and defended by song-flights. No courtship behaviour known.

NEST: compact cup of grass stems, other stalks, grass leaves, vegetable down and spider web, lined with fine grass and rootlets. In 12 nests, ext. diam. 83–99 (91), ext. depth 50–85 (62), int. diam. 53–65 (58) and int. depth 30–50 (40) (Cramp 1992). Sited in clump of tall grasses or brambles, or low shrub, 0·6–3·5 m above ground, usually 0·75–1·35 m. Built by ♂ and ♀; ♂ also builds cock nests.

EGGS: 3–6, usually 4–5, av. 4·2 (53 clutches, NW Africa). Laid at daily intervals, within 2 h of dawn. Subelliptical, glossy, smooth; greenish, pinkish or (usually) buffy white, or buff, speckled with minute red-brown dots, or blotched mainly at large end with pale grey, olive or brown. SIZE: (n = 125) 15·2–20·0 × 13·1–14·5 (18·0 × 13·6). WEIGHT: 1·73. 2, occasionally 3, broods (Malta), next clutch started av. 16 days after previous brood fledged.

LAYING DATES: NW Africa, Apr–June; (formerly Egypt, Mar.).

INCUBATION: begins with penultimate egg laid; by ♀ and ♂; ♀ incubates twice as long as ♂ during day; only ♀ incubates at night. Daytime spells of 2–46 (18) min by ♀ and 2–32 (10) min by ♂. Period: 12–15 (13) days.

DEVELOPMENT AND CARE OF YOUNG: young cared for by both parents. Fledging period 12–13 days; young independent 2–3 weeks later.

BREEDING SUCCESS/SURVIVAL: overall success rate c. 45% (Malta); annual adult mortality, Gibraltar, 45% (Cramp 1992).

Reference
Cramp, S. (1992).

Sylvia mystacea Ménétries. Ménétries's Warbler. Fauvette de Ménétries.

Plate 23
(Opp. p. 416)

Sylvia mystacea Ménétries, 1832. Cat. Raisonné Objets Zool. Rec. Voy. Caucase Perse, p. 34; Saliane (= Salyany), lower Kura River, Azerbaijan.

Forms a superspecies with *S. melanocephala* and *S. melanothorax*.

Range and Status. Breeds from E Turkey north along W Caspian to Volga R. delta, east through N Iran to N Afghanistan, Uzbekistan and W Tadzhikistan, and southeast through Mesopotamia to SW Iran. May breed in Lebanon. Winters in S Iran, Arabia and NE Africa. Vagrant to Portugal, Kuwait, Israel, Jordan and Nigeria.

Common winter visitor, coastal plains from Sudan to Somalia; uncommon inland to Nile. Sudan, common on Red Sea coast, uncommon (6 records) in interior, southwest to west of Nile near Umm Ruwaba. Ethiopia, common, Eritrean coastal plain below about 300 m. Djibouti, 2 records, Bankoualé and As Eyla. Somalia, scarce, 10 records from Saba Wanak, Laskhoreh, Suksode, Ber, Warabod, Gumbowerin, Arowein, Biyo Dai and Gedka Debta, all below 1000 m. Egypt, vagrant, 2 old and 5 recent records, Cairo, Hurghada, Qena, Gebel Elba and Wadi Hagul (Suez). Nigeria, 1 netted Gaya, Kano State, Apr (Best 1975).

Description. *S. m. mystacea* Ménétries: NE Turkey, Transcaucasia. ADULT ♂: forehead, forecrown, lores and around eye dull greyish black; hindcrown, nape, ear-coverts and sides of neck blackish grey; nape to scapulars and uppertail-coverts dark olivaceous grey. Tail black, T1 washed grey, all feathers grey-edged; T5 and sometimes T4 with narrow white tips which may be white triangles 5 mm long; T6 with white outer vane and long white wedge (12–16 mm) at end of inner vane. Eye surrounded by naked, thick papillate red orbital skin. Well-defined white malar stripe. Chin white, throat and upper breast dark pink, lower breast, belly, flanks, vent and undertail-coverts greyish off-white, tinged pink. Thighs brown. All wing-feathers dull black, narrowly edged grey or bluish grey; underwing-coverts and axillaries pale

Sylvia mystacea

pink-grey. Bill dark brown, cutting edges and lower mandible proximally yellow-horn; eyes variably yellow, light brown, red-brown or orange-brown; legs and feet yellowish flesh-brown. ADULT ♀: forehead rufous-buff, lores greyish buff, crown brownish grey, neck, ear-coverts, mantle, back, rump and scapulars sandy greyish brown, uppertail-coverts dark grey; tail like ♂; iris like ♂ but orbital ring skin yellow or orange-brown, sometimes with white featherlets; all wing-feathers paler grey than in ♂ and fringed isabelline brown; no white malar stripe; chin and throat buffy white, tinged pink, breast and flanks buffy greyish white, belly white in midline, undertail-coverts pale grey. SIZE: (*M. m. rubescens*, 14 ♂♂, 10 ♀♀) wing, ♂ 58–63 (60·3), ♀ 58–62 (59·8); tail, ♂ 51–58 (53·0), ♀ 50–56 (52·6); bill to skull, ♂ 12·3–13·4 (12·9), ♀ 12·3–13·6 (13·0); tarsus, ♂ 18·4–19·9 (19·1), ♀ 18·3–19·8 (19·0). WEIGHT: (various races) unsexed (n = 1, Nigeria) 9·3, (n = 34, Ethiopia, Oct–Feb) 9·5–10·4 (9·8); (N Iran, Mar–Apr) ♂ (n = 6) 9–11 (10·3), unsexed (n = 30) 8–11·5 (9·9); (Turkey) unsexed (n = 27, Aug) 7–11 (9·7), (n = 33, Sept) av. 10·9.

IMMATURE: like adult ♀ but buffier, less greyish; iris olive-brown; whites in tail are uniform pale buff.

S. m. rubescens Blanford: Lebanon and SE Turkey to SW Iran. Like nominate race but in ♂ black cap slightly greyer, remaining upperparts slightly paler, underparts silky white, sides very pale grey and throat and upper breast very pale pink. Slightly larger than nominate race (which has wing, n = 43, 56–59).

S. m. turcmenica Zarudny and Bilkevich: E Iran, Afghanistan, Uzbekistan and Tadzhikistan. ♂ like *rubescens* but breast pinker.

TAXONOMIC NOTE: formerly treated as race of Sardinian Warbler *S. melanocephala*. Birds wintering in Africa have been called *S. melanocephala mystacea* to discriminate them from wintering *S. melanocephala momus*, but they have not been discriminated as between *S. mystacea mystacea*, *rubescens* and *turcmenica*. All 3 races of *S. mystacea* are detailed above, since all are likely to occur in Africa.

Field Characters. ♂ very like Sardinian Warbler, with blackish cap and face offset by white malar stripe, red iris and bare eye-ring, grey upperparts and white-sided black tail, but differs in strong pink tinge to underparts. ♀ and immature very similar to Sardinian and Subalpine Warblers *S. cantillans*; best diagnostic feature for the species is constant tail-waving, side-to-side and up-and-down, every few s when foraging undisturbed (but beware tail-drooping Red Sea Warbler *S. leucomelaena*); quiet 'mig' call also characteristic.

Voice. Tape-recorded (73, LOS). ♂ song, frequently heard in Jan (Eritrea), is melodious, conversational warble, interspersed with harsh notes and rattles. ♂ subsong (Eritrea), a subdued harsh warbling, like Common Whitethroat *S. communis*. Much the commonest call, probably contact call, is quiet monosyllable 'mig' or 'meeg' where *m* sound is sometimes nearer *j* or *p*, not far-carrying, but uttered frequently by 1 or both birds of pair in winter territory (Oman). Other calls, a rattle 'trrrrt' or 'tschrrrrt', a repeated 'tchk' or 'tac-tac', and a 'zsch' of alarm.

General Habits. In winter occupies low vegetation near coast, feeding readily in such tall trees as occur (5 m high *Ziziphus spinachristi* and *Acacia tortilis*) but keeping mainly to bushes, especially acacias, and low vegetation. Spends much time in knee-high sea-lavender *Limonium axillare* (coast of Oman), evidently in preference to acacia scrub 50 m away; throughout range, is associated more with low bushes than trees. Also inhabits tamarisks, low-growing thorn bushes, desert scrub, palm groves and hedgerows. Marked preference for low scrub on sandy coastal plains. In breeding range, nests mainly in tamarisk.

On winter grounds mainly in pairs, ♂ and ♀ often in same bush or only 20 m apart. Evidently territorial in winter, ♂♂ chasing each other through bushes and airspace for 500 m (Oman). Site fidelity shown by bird trapped in Eritrea where ringed 3 winters earlier (Ash 1981). Not as restless as sometimes claimed, though like most *Sylvias* is wary, its foraging readily disturbed by people. Pair forages up to 4 m high on opposite side of large open tree, or 1–3 m high in adjacent small bushes, flitting from twig to twig, making short flights from branch to branch, gleaning insects from thorns, twigs and leaves, sometimes chasing an insect in flight, often staying put in one clump of twigs for 20–30 s at a time. Uses full height of tree, but seldom perches on top for more than a few s (to scan around before flying low into next tree) and seldom comes to ground. Whilst foraging waves tail up-and-down or side-to-side or with circular motion, 2 or 3 times per min; wave is about 45° in horizontal and about 30° in vertical plane. Frequently calls '(m)eeg'. Occasionally occurs in small flocks (Somalia). Occurs up to 2400 m on passage (Yemen).

Migrants arrive in Sudan and Eritrea in Sept; most depart in Mar, some not until mid-Apr. Nigerian vagrant was latest African record (17 Apr). Egyptian records mainly in Mar, suggesting passage. Distribution on passage through Middle East and Arabia indicates entry into Africa via Arabia (Cramp 1992).

Food. Insects: grasshoppers, bugs, moths, butterflies, flies, wasps, ants, beetles, larvae and eggs. Seeds and *Lycium* berries (Saudi Arabia).

Reference
Cramp, S. (1992).

Plate 24
(Opp. p. 417)

Sylvia cantillans (Pallas). **Subalpine Warbler. Fauvette passerinette.**

Motacilla cantillans Pallas, 1764. *In* Kroeg's Cat. Verzam. Vogelen, Adambratiunculae, p. 4; Italy.

Range and Status. NW Africa, SW and S Europe east to Aegean and W Turkey; winters Africa.

Palearctic migrant. Breeding summer visitor N and W Morocco (Debdou-Tlemcen ridge, Rif, Middle and

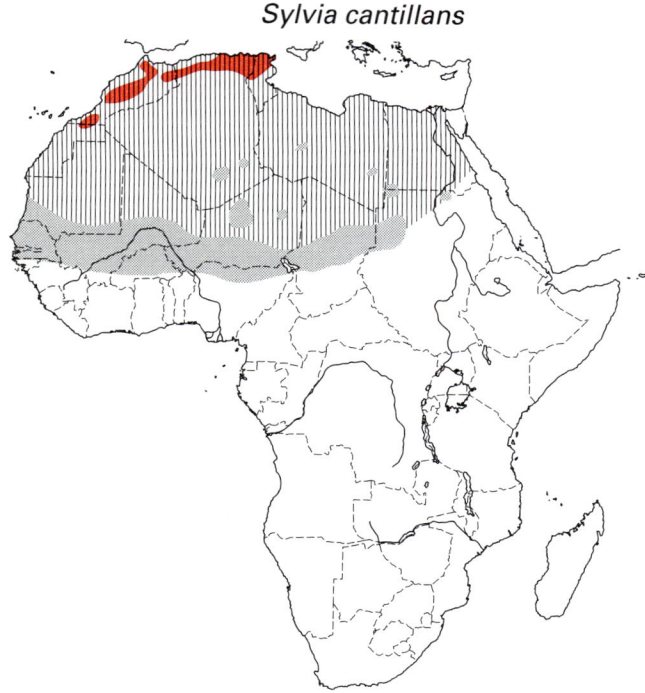

Sylvia cantillans

High Atlas, and near SW coast), N Algeria (Tell, Aurès Mts), NW Tunisia and probably NW Libya (near Tripoli); mainly in highlands; locally common.

Winters mainly south of Sahara, in W and central parts of Sahel belt: in S Mauritania (north to *c.* 18°N), Senegal, Gambia, S Mali (north to *c.* 17°N), Burkina Faso (south to *c.* 12°N), Niger (north to Aïr and Arrigui), N Nigeria (Sokoto to L. Chad, south to Zaria), central Chad, NW Sudan and Egypt/Libya border (Djebel Uweinat); locally common to abundant, but scarce in Gambia and E Chad–W Sudan. Occasional in midwinter in Saharan oases of Algeria (Laghaout south to Tamanrasset) and Libya (Sebha, Brak, Kufra), and once, early Mar, in Nile valley (Wadi Halfa, N Sudan). Passage migrant through N Africa and Saharan oases: in autumn common Morocco–Algeria, frequent Tunisia–W Libya, scarce E Libya–Egypt; in spring common and widespread from Morocco to Egypt (least numerous E Egypt).

Description. *S. c. cantillans* (Pallas): SW and S Europe, east to Italy; winters from Senegal to Chad. ADULT ♂: upperparts and sides of head and neck blue-grey, tinged brown when fresh, especially on mantle; lores and ear-coverts slightly darker. Usually a conspicuous narrow white moustachial stripe from base of bill to below middle of ear-coverts. Tail dark grey, outer webs (both webs of T1) fringed pale grey; T3–T4 with narrow white tip; T5 with larger white patch at tip, extending as wedge along shaft on inner web; T6 white apart from dark grey basal area on inner web. Chin, throat, breast, flanks and sides of belly deep vinous pink or vinous red; centre of belly, vent and undertail-coverts white, tinged brownish pink. Primaries, secondaries and primary coverts blackish brown, narrowly edged pale pinkish buff; tertials and greater coverts dark grey-brown, fringed pinkish buff; median and lesser coverts grey with darker centres; axillaries and underwing-coverts grey with pinkish brown suffusion. Wing rather short, tip rounded: P7–P8 longest; P6 0–2·5 shorter, P5 2–5, P1 8–13; P9 1–3·5 shorter, between P4 and P6 (rarely longer than P6); P10 minute, 2·5 shorter to 3·5 longer than primary coverts; P6–P8 emarginated. Bill dark horn, base of lower mandible pale flesh or yellow; eyes pale brown or yellowish; bare ring around eye orange-red or bright red; legs and feet light brown or yellowish brown. ADULT ♀: upperparts rather paler and browner than in adult ♂, lores and ear-coverts distinctly paler than crown; narrow eye-ring pale buff or whitish; narrow whitish moustachial stripe contrasts with ear-coverts, but often poorly demarcated from throat. Underparts much paler and whiter than in adult ♂, throat, breast and flanks typically pinkish buff or creamy buff, but variable and sometimes vinous pink as in less saturated ♂♂. SIZE: (10 ♂♂, 8 ♀♀) wing, ♂ 56–63 (59·9), ♀ 56–62 (58·4); tail, ♂ 45–51 (48·7), ♀ 46–50 (48·3); bill, ♂ 12–14 (13·0), ♀ 13–14 (13·4); tarsus, ♂ 18–20 (19·2), ♀ 18–19 (18·9). WEIGHT: (SE Morocco, spring) ♂ (n = 11) 7·4–9·2 (8·2), ♀ (n = 2) 8·0, 9·2; unsexed (n = 9, NW Morocco, Mar) 8·2–10·3 (9·3), (n = 60, N Algeria, autumn) 8·7–14·1 (10·8); unsexed (N Senegal) (n = 328, Nov) 7–13·2 (9·6), (n = 474, Dec–Feb) 7·1–12·2 (9·2), (n = 135, Mar–Apr) 7·6–16·1 (11·0); unsexed (N Nigeria) (n = 31, autumn) 7·6–11·4 (9·4), (n = 6, Nov–Feb) 8–9 (8·5), (n = 58, spring) 8·8–14·8 (11·3).

IMMATURE: juvenile buffish brown above; buffish white below with whiter chin, and bright buff flanks contrasting with whitish underbody; shows buffish white eye-ring, but usually lacks sharp contrast between ear-coverts and lower cheek; wing and tail-feathers browner than in adult, pale tail areas brownish white and confined to T6 and tip of T5; secondaries, tertials and greater coverts with more distinct sandy buff edges. First winter like adult ♀, but browner above with grey confined mainly to sides of head, underparts usually without pink tinge; flight feathers and tail paler and browner, tail retaining juvenile pattern; eyes olive-brown or dull yellowish.

NESTLING: naked.

S. c. inornata Tschusi: Morocco to Tunisia and perhaps NW Libya; winters Senegal to W Niger. Like nominate race, but underparts of ♂ brick-red rather than pinkish red, flanks deep chestnut; ♀ rather pale below.

S. c. albistriata (C. L. Brehm): SE Europe and W Turkey; winters from E Mali to SW Egypt and W Sudan. Chin, throat and breast of ♂ darker brick-red than in nominate race, contrasting with pale rufous flanks and extensively white belly; moustachial stripes slightly broader, feathers of throat and breast often flecked with white. ♀ and immature slightly greyer on upperparts, whiter on underparts than nominate race. Larger: wing, ♂ (n = 10) 61–65 (63·5); P9 between P4 and P7 (commonly longer than P6).

Field Characters. A rather small, slim warbler with a medium length tail, short bill and pale legs. ♂ blue-grey above and dark pinkish below, with white moustachial stripe, white outer tail-feathers and bare red ring around bright reddish or yellowish eye. ♀ more sombre, grey-brown above and creamy or pinkish buff below, but has contrast between dark head sides and pale throat, and often a clear white moustachial stripe as in ♂. Immature browner than ♀ with sandy buff wing-edges (recalling the larger, longer-tailed Common Whitethroat *S. communis*), and buff rather than white outer tail-feathers. ♂ Subalpine might be confused with Tristam's Warbler *S. deserticola* or ♂ Ménétries's Warbler *S. mystacea*; distinguished from former by moustachial stripe, red eye-ring and lack of broad rufous wing-edging, from latter by deep red underpart colour and paler grey tail with different white patterning. ♀ and immature Subalpine easily confused with several other small *Sylvia* spp. Separated from all plumages of Tristam's Warbler and

Spectacled Warbler *S. conspicillata* by lack of rufous wing-panel; from ♀ and immature Ménétries's by strong cheek contrast, more prominent wing-feather edgings, and browner, less graduated tail; from Lesser Whitethroat *S. curruca* by absence of darker grey ear-coverts, by warm pinkish or buffish tones to underparts and wing-edgings, and by pale rather than dark grey legs.

Voice. Tape-recorded (57 – 62, 73, 93, B). Song, a clear lively chattering warble, phrases lasting *c*. 3 s and delivered about 12 times per min. Recalls Common Whitethroat, but more sustained and musical. Similar to songs of Tristam's Warbler and Sardinian Warbler *S. melanocephala*, but lacks hard rattle sounds of those species (Cramp 1992). Contact-alarm call, a sharp 'tec', often given in series in alarm or excitement, 'tec, tec, tec-tec'.

General Habits. Breeds in dense, often xerophytic, scrub of hillsides and ravines, or in woodland undergrowth. In NW Africa, inhabits the rich bush layer of open woodlands, especially of cork oak *Quercus suber* and cedar *Cedrus atlantica*; mainly in the highlands, from 600 m up to 2400 m (Cramp 1992). Winters in Sahelian bush- and tree-steppes, living in *Tamarix* thickets and acacia scrub as well as in more continuous acacia woodland; also in suburban gardens. In the Niger delta (Mali) foraged in *Diospyros*, and reported in large concentrations in lake shore reeds (Lamarche 1981). In Senegal recorded in mangroves.

Solitary outside breeding season; or in loose groups often with other *Sylvia* spp. Usually skulks in thick cover. Forages in low scrub and thicket, but also in tree foliage. Hunts *Acacia*-flower insects in canopy. Constantly active when feeding, hopping and sidling through thick cover with noticeably horizontal carriage, and picking prey from leaves and flowers. Raises crown-feathers and tail in general excitement and flicks wings frequently (Cramp 1992). Flight light and fluent like Common Whitethroat. Sings occasionally in winter quarters, Dec–Mar.

Complete moult near breeding ground July–Aug, sometimes suspended; partial moult in Africa Jan/Feb–Mar.

Autumn passage of nominate race through NW Africa begins late Aug, main migration in Sept–Oct. Birds are then widespread and numerous in E Morocco and Algeria, but scarce in NW and W Morocco, suggesting this area skirted or overflown. High numbers occur in wadis and oases of Algerian and W Libyan Sahara until Nov. Nominate *cantillans* passes through breeding areas of *inornata*, whose movements are thus poorly known. Further east, *albistriata* occurs only in small numbers on passage through Egypt and Libya, mainly early to mid-Sept. Its scarcity on coast and in Nile delta indicates that main landfall is well inland. Birds arrive in Senegal from mid- to late Aug, but mainly in Sept, Chad in Oct, and N Nigeria mainly from mid-Oct. Present in Niger inundation zone (Mali) from Sept, but largest numbers Jan–Mar. Most birds leave tropics in Mar, but a few remain in Chad until early to mid-Apr and in N Senegal until late Apr and occasionally May. Some birds reach high pre-migratory weights in N Nigeria and N Senegal. Birds collected in May in S Mali and S Niger were *inornata*, suggesting that this race tends to delay spring migration. Returns to NW African breeding grounds late Mar–early May. The nominate race is widespread and numerous on spring passage from W Morocco to W Libya, with many birds in Sahara; movement lasts from Mar (occasionally late Feb) to early May. Heavy spring passage of *albistriata* occurs through Egypt and Libya (especially coast and Nile delta) from late Feb to mid-May (mainly mid-Mar to early Apr).

Food. Mainly insects; also much fruit, especially in late summer and early autumn. Young fed entirely on invertebrates. In winter, in W Africa, eats adult and larval insects; berries, especially *Salicornia*, also important.

Breeding Habits. Monogamous. Solitary and territorial; territory 0·1–1·7 ha Europe (Cramp 1992); used for courtship, nesting and raising young. ♂ sings from bushes or lower tree branches, typically within cover 1·5–3 m above ground. Also has song flight like Common Whitethroat; flies up and sings during dancing, fluttering descent, or while fluttering in curve from one perch to another nearby (Cramp 1992). Soliciting ♂ cocks and fans tail over back, turning it to show pattern; droops and shivers wings and ruffles breast-feathers (Cramp 1992). ♂ builds cock nests not used for laying.

NEST: deep cup of grasses and other leaves and stems, with some roots, plant down and cobwebs, lined with finer grasses, rootlets and hair; usually in low bushes 30–60 cm from ground, occasionally in low trees; outer diam. *c*. 85, inner diam. *c*. 55; cup depth *c*. 35; built by both sexes (Cramp 1992).

EGGS: 3–5; Algeria av. (34 clutches) 4·2; whitish, sometimes with greenish or pinkish tinge, faintly spotted or blotched red-brown, olive or buff. SIZE: (n = 75 SW Europe) 15–19 × 12–14 (16·5–12·9). Normally double-brooded.

LAYING DATES: NW Africa, early May to mid-June. Eggs laid daily.

INCUBATION: by both sexes (Beven 1967). Period: 11–12 days.

DEVELOPMENT AND CARE OF YOUNG: young cared for and fed by both parents; remain in nest 11–12 days.

References
Beven, G. (1967).
Cramp, S. (1992).

Sylvia conspicillata Temminck. Spectacled Warbler. Fauvette à lunettes.

Sylvia conspicillata Temminck, 1820. Man. Orn., ed. 2, 1, p. 210; Sardinia.

Range and Status. Breeding summer visitor in W Mediterranean (Spain, S France, Corsica, Sardinia, Italy, Sicily, Menorca, barely Portugal), resident in Cape Verde Is, Madeira, Canary Is and NW Africa and winter visitor south to Senegal and Niger; also resident Crete and Levant and winter visitor to NE Egypt. Vagrant to Netherlands, Germany, Switzerland, SE Europe, Iraq and Kuwait.

Morocco, widespread resident from N Atlas foothills to N and W coasts, common nester in pre-Saharan steppes (Boumalne du Dadès, Tinerhir, Tinjdad, Goulmina, Alnif, Errachidia, Erfoud, Rissani). In winter vacates coastal NE Morocco, less common than in summer between Casablanca and Essaouira and Tafilalet, frequent further south. Algeria, resident in NW (Macta, Tlemcen Mts, Ksour Mts – very common at Djenien Bou Rezg – steppes near Abadla) and NE (Kabylie, Hodna Mts, Arris, Bou Lhilet, and Jebel Chelia at 1700 m); Djenien population augmented in winter, when frequent in wadis and oases south to Tassili N'Ajjer and Hoggar. Autumn passage noted on Tell coast (Reghaïa) and spring passage in Hoggar. Tunisia, rare or scarce breeder from L. Tunis and Zaghouan to Gabès, and in W from Chott Djerid to Tozeur and Nefta, and scarce winter visitor in same area. In Libya winters commonly on Tripoli coast and may breed near Tunisian border; 3 records in Cyrenaica, and once 10 migrants at Serir. Mauritania, winter visitor and passage migrant; common to very common near coast from Senegal R. to 18° 30′N; occasional records inland east to Wadan, Adrar Mts. Senegal, frequent winter visitor to NW, from Djoudj Nat. Park to south of Saint-Louis. Niger, recorded in most winter months in Aïr and Ténéré Nat. Nature Reserve (Newby *et al.* 1987), and Dec records from Tazolé and Arrigui. Egypt, uncommon to locally frequent winter visitor to N, mainly El Khanka dunes near Cairo (18 in 2 days), northern Nile delta, and wadis near El Amarna; in E, uncommon in area between Qena, Luxor, Ras Gemsa and Quseir. Records from Sidi Barrani, Suez, Abu Simbel and Gebel Elba.

No African data on population density. On Canary Is and in Europe, breeding densities of 0·3, 1·2, 4·6, 6·5 and 35–40 pairs per km^2 (Cramp 1992).

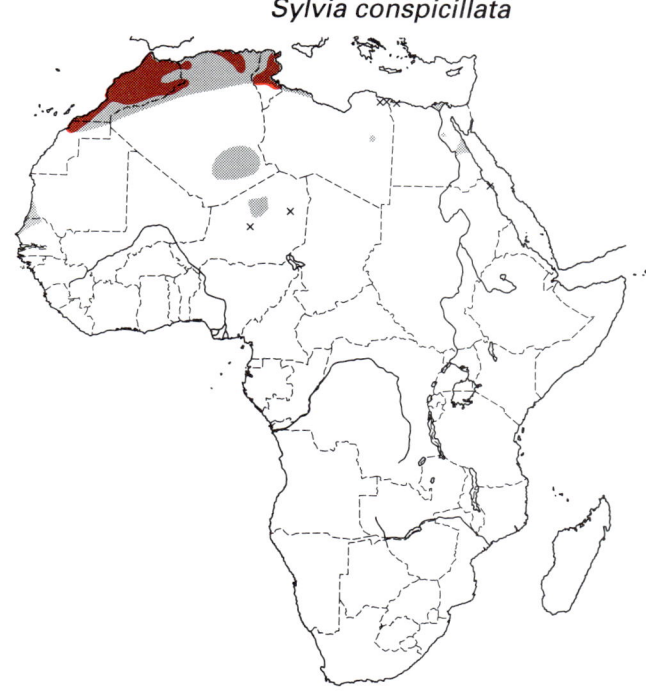
Sylvia conspicillata

Description. *S. c. conspicillata* Temminck (only race in Africa): range of species except Canary Is, Madeira and Cape Verde Is. ADULT ♂ (breeding): forehead, crown, nape and most of ear-coverts grey; orbital ring of featherlets white, ring broken in front of and behind eye; lores, cheeks and area around orbital ring blackish grey; hindneck, sides of neck, mantle, scapulars and back warm brown tinged olive or rufescent olive; rump and uppertail-coverts olivaceous grey. Tail blackish with white sides: T1 dark slaty grey, edged pale grey; T2–T5 black with outer vanes edged grey; T4 and T5 (and sometimes T3) with white fringe at tip, T6 white. Chin, upper throat and sides of throat (moustachial stripe) white, in sharp contrast with blackish cheeks; lower throat grey, merging to vinous pink on breast, buffy pink on flanks and pinkish white on belly, vent and undertail-coverts; thighs whitish. Primaries black with pale buff-grey outer edges, secondaries and tertials black with broad foxy rufous outer edges (secondaries) or tips and outer vanes (tertials); greater primary-coverts like primaries; alula mainly black, greater coverts like secondaries, median and lesser coverts rufescent brown; underside of flight feathers slaty, underwing-coverts and axillaries grey-pink or vinous. Bill dark brown, cutting edges yellowish, basal two-thirds of lower mandible yellowish flesh; eyes red-brown, rich dark brown, or red-orange, with narrow bare black orbital ring, surrounded by broken white ring of featherlets broader above than below eye; legs and feet yellowish flesh. ADULT ♂ (non-breeding): like breeding ♂ but crown buffy grey and mantle brighter rufous, underparts slightly paler. ADULT ♀: wings and tail like ♂ but forehead, crown, ear-coverts, rump and uppertail-coverts warm olivaceous brown (like back), lores, cheeks and around eye pale grey; underparts cream or buff-cream, hardly washed with pink at all; no grey on throat; lacks white moustache. SIZE: (16 ♂♂, 13 ♀♀) wing, ♂ 53–59 (55·7), ♀ 53–57 (55·3); tail, ♂ 47–53 (49·5), ♀ 45–53 (48·9); bill to skull, ♂ 11·8–13·2 (12·4), ♀ 11·6–12·8 (12·2); tarsus, ♂ 17·6–19·3 (18·4), ♀ 17·7–19·4 (18·6). WEIGHT: (Algeria, Niger) 1 ♂ 9·0, 1 ♀ 7·3; unsexed (France, n = 6) 8·8–11 (10·1), (Cyprus, n = 10) 8·7–10 (9·4).

IMMATURE: like adult ♀: head and upperparts light greyish brown, lores paler; orbital ring of featherlets buffy, not contrasting with rest of head; upperparts off-white, washed with pale greyish brown on breast and with buff on flanks and undertail-coverts.

NESTLING: hatches naked; skin black and pink; gape flanges pale yellow, mouth bright orange.

Field Characters. 12.5 cm. A small NW African and W Mediterranean *Sylvia* warbler of low scrub on saltings, in all plumages with rufous wings like Common Whitethroat *S. communis* and Tristram's Warbler *S. deserticola*. Breeding ♂ closely resembles ♂ Common Whitethroat also in its grey head, white throat, red-brown eye, brown back, white-edged tail and yellow-flesh legs; differs mainly in smaller size, shorter tail, blacker face; white restricted to upper throat (often looking like white moustachial stripe), with grey lower throat, pink underparts, and narrow but distinct white eye-ring. ♂ Tristram's Warbler distinguished by grey back and brick-red throat and underparts. ♀, juvenile and winter ♂ Spectacled Warblers are cleaner-looking than Common Whitethroat and Desert Warbler *S. nana*, with closed wing solidly bright rufous contrasting with black marks on tertials, wing-tip and alula, warm brown upperparts and buff underparts; adult ♀ Tristram's is grey-backed, underparts with at least flanks brick-red, and juvenile is overall greyer and drabber than Spectacled. ♀ and juvenile Common Whitethroat hard to distinguish from Spectacled, best told by size, habitat, voice, and only vestigial whitish eye-ring. Juveniles and adult ♀♀ (in some plumage) of several *Sylvia* spp. have rufescent brown wings, and such stages of Subalpine Warbler *S. cantillans*, Ménétries's Warbler *S. mystacea* and Sardinian Warbler *S. melanocephala* cannot be separated from Spectacled, even in the hand, without measurement and careful appraisal of tone (Svensson 1992).

Voice. Tape-recorded (CHA – 62, 73, B, C, HAZ, HOL). Song of ♂ varied, a sweet, pleasing, quite melodious warble usually < 2 s in duration but up to 4.5 s, consisting of random mixture of twitters, pure high-pitched whistles, and short rattles; quite like song of better-known Sardinian Warbler but less raucous. Song commonly repetitious, with variation, after pauses of a few s. Contact call a high-pitched 'tsee'. Alarm call: 'churr' which at high intensity becomes a loud, metallic, rasping 'kirrrrr' or harsher 'skerrrrrr', often prolonged and rolling, like alarm of Winter Wren *Troglodytes troglodytes*.

General Habits. Breeding habitat is saltflats and semi-desert in plains, lowlands and hills, with low, coarse, shrubby vegetation (garigue). In Tunisia, mainly in coastal plains with lagoons and salinas, and salty plains with dense, low scrub near lakes and wadis, and in Algeria, hamada desert and *Nucularia* steppe. In Canary Is, cultivated areas with tree-heath *Erica* and *Cytisus* scrub on dissected, rocky land; in Cape Verde Is occurs nearer human habitation, in *Acacia* and *Lantana* scrub; in Europe also in *Salicornia* on dry mudflats, in bramble *Rubus*, also *Cistus*, *Phillyrea* and *Pistacia*. Wintering birds in Mauritania and Senegal inhabit littoral halophytic steppe of *Suaeda*, *Arthrocnemum*, *Euphorbia balsamifera*, tamarisk, *Nitraria retusa*, *Salsola* and *Zygophyllum*.

Forages in low bushes and short vegetation by hopping and seeking out arthropods; sometimes feeds among roots and on ground (Algeria). Skulks. Likes crevices in dry-stone walls. Often quite approachable; usually shy. Takes berries in spring and autumn; takes insects on the wing (Malta); follows wheatear from bush to bush, once for 30 min, foraging actively below the perched wheatear, much like Desert Warbler; seen following Desert Wheatear *Oenanthe deserti* (Egypt: Short and Horne 1981) and Finsch's Wheatear *O. finschii* (Israel: Arnies 1990). Gait on ground a bouncy but sidling hop.

Solitary or (usually) in pairs; not gregarious, although small aggregations can form on migration. Wintering ♂♂ territorial (Libya: Willcox and Willcox 1978). Rarely sings in winter (once Egypt, Nov); breeding-territory singing starts in late Jan.

Passage migration across African coasts never conspicuous; no concentrations of migrants are known. Regular Aug–Oct in Gibraltar. 1 recovery, from W Spain to W Morocco (Safi), 1 (2?) year later. Weak passage through Morocco in Sept (also Oct–Nov), quite strong movement through Nouakchott from mid-Sept (♂♂ later than ♀♀: Gee 1984). Quits Mauritania in Feb–Mar, and N Niger and SE Algeria in Mar. Passage in Morocco and N Algeria from mid-Mar; passage stronger than in autumn. Main migration at Tafilalet (Morocco) mid-Mar to mid-Apr. Still some movement through Gibraltar in early May. Winters Libya late Sept/Oct to early Apr, with some passage in late Apr. Winters Egypt early Sept to mid-Apr.

Food. Mainly insects. In Tunisia and S Algeria, orthopterans, small beetles, flies, hymenopterans, and berries of *Myoporum*. Also eats caterpillars, insect eggs, spiders, blackberries and mulberries.

Breeding Habits. Monogamous, territorial, solitary breeder. Nests usually *c.* 230 m apart (Cyprus), sometimes only 15 m apart. ♂ sings from top of small bush at edge of territory, or at times from top of 6 m tree, with throat puffed out and head thrown back. ♂ also sings in flight, rising quickly to 8–9 m then singing whilst planing down on rapidly beating wings, with tail fanned. Tolerates congeners but not conspecies in its territory; several Spectacled Warblers, some carrying food, often chase in circles, periodically perching on bush or ground and uttering alarm call or snatch of song (Cramp 1992). Apparent pair-formation (Nov, Libya) consists of chasing in wheeling flight, with tails slightly fanned, for over 2 h (Willcox and Willcox 1978).

NEST: neat, loosely-made cup of dry grass stems, leaves and rootlets, with spider web, wool, rag and paper, lined with soft plant down, and some fine roots and hair. Ext. diam. (n = 12) 82–100 (93), ext. depth 55–80 (65), int. diam. 52–63 (56), int. depth 35–50 (43). Sited in tussock, matted grass and thistles, or shrub, usually < 0·7 m above ground but up to 2 m. Site chosen by ♂. Nest built by ♂ and ♀ equally at first and mainly by ♀ at later stages; building takes up to 2 weeks (Cramp 1992).

EGGS: 3–5, usually 5. Laid daily. Subelliptical, smooth, glossy; very pale green or buff-white, either very finely peppered with brown or olive (sometimes forming dusky zone at broad end), or with fewer brown and grey speckles mainly at broad end, or almost immaculate. SIZE: (n = 100) 14·0–18.6 × 11.5–13.7 (16.3 × 12.8). WEIGHT: 1.37. 2 (occasionally 3) broods.

LAYING DATES: SW Morocco, from, mid-Feb; N Morocco, Apr; NW Africa, Mar–June. (Cape Verde Is, Sept–May.)

INCUBATION: by ♂ and ♀. Period: 12–13 days.

DEVELOPMENT AND CARE OF YOUNG: young at first brooded by ♀. ♂ and ♀ participate equally in feeding young and removing faecal sacs. Fledging period: 11–12 days. Young out of nest fed by both parents for 10–20 days.

BREEDING SUCCESS/SURVIVAL: 20% of 95 nests predated by *Coluber* snake.

Reference
Cramp, S. (1992).

Sylvia deserticola Tristram. Tristram's Warbler. Fauvette de l'Atlas.

Sylvia deserticola Tristram, 1859. Ibis, p. 58; Southern Algerian Sahara.

Plate 24
(Opp. p. 417)

Range and Status. Endemic resident NW Africa, and partial migrant well into W Sahara. Breeds N Morocco, N Algeria and NW Tunisia; winters from there to W Mauritania, SE Algeria, SW and central Libya.

Morocco, breeding birds widespread from 1400 to 2500 m, frequent in High Atlas, less so in Middle and Anti-Atlas Mts, from Tanalt in SW to Debdou and near Oudja in NE; in winter occupies southern fringe of breeding range, at lower altitudes: Ouarzazarte (Todra), Jebel Sarho (Mecissa to Alnif), Tafilalet (Errachidia, Erfoud, Merzouga, Erg Chebbi), and extends to coast (Agadir, Goulimine) and to oases in northern fringes of Sahara, where very common in some winters (e.g. at Tafilalet). Algeria, breeds from Oudja east to Tlemcen, in Monts-des-Ksours and abundantly in Saharan Atlas Mts east to Djelfa, also in Aurès Mts east to Tunisia; a partial migrant, some remaining in Monts-des-Ksours in snowy winters, but many moving to N Saharan oases and to central Sahara: Tindouf (27° 50'N, 08° 04'W), Plateau du Tadmaït (Heim de Balzac 1926) and highlands of Tassili n'Ajjer and Hoggar Mts. In some winters abundant at oases, e.g. Beni Abbès, and locally common in S Algeria. Tunisia, scarce breeder above 1000 m in mountains from Bou Chebka to Maktar; frequent on Jebel Chambi; occurs, and may breed, south to Gafsa; winters southeast to coast at Ben Gardane and Libyan border at Dehibat. Winter visitor to Mauritania, sight records from Nouakchott, Oct, and 5 localities in Adrar region, Nov–Feb; and Libya, common at Jofra oases Jan–Mar and common in several wadis at Ghat (E end of Tassili n'Ajjer hills, SW Libya) Oct–Mar; Jan–Feb records at Sebha and Jebel Soda.

Description. *S. d. deserticola* Tristram: Algeria (except Tlemcen area) and Tunisia. ADULT ♂ (breeding): forehead to scapulars, rump and uppertail coverts slate-grey or bluish grey. Tail black with white sides; T1 greyish black, T2–T5 black, the outer webs edged pale grey, T4–T5 narrowly white-tipped, T5 with white wedge up to 8 mm long at end of inner vane, T6 white (with blackish base). Lores dark slate-grey; circle of minute white featherlets around eye; cheeks, ear-coverts and sides of neck same grey as back, sharply demarcated from rufous throat. Chin-feathers rufous with white tips; throat, breast and flanks vinous rufous-brown; mid-belly white; vent white, thighs brown; undertail-coverts mixed pink and white. Wing-feathers blackish, primaries and primary coverts narrowly edged pale buff, secondaries and greater coverts edged pale rufous, and tertials broadly edged cinnamon-rufous.

Sylvia deserticola

Underwing-coverts and axillaries pale pinkish brown. Upper mandible dark brown, lower mandible yellowish with dark tip; eyes light brown to bright yellow; legs and feet yellowish flesh-brown. ADULT ♂ (non-breeding): upperparts slightly buffier; underparts slightly paler and pinker than in breeding ♂. ADULT ♀ (breeding): like breeding ♂ but upperparts brown-grey not slate-grey and underparts much less solidly rufous. Chin white; throat-feathers rufous with broad white tips; centre of breast whitish and sides rufous and white mixed; belly white; flank-feathers rufous with white fringes. ADULT ♀ (non-breeding): mantle, scapulars, back and rump sandy buff, tinged with cinnamon; underparts creamy buff, tinged with pink. SIZE: wing, ♂ (n = 19) 53–58 (55·4), ♀ (n = 15) 52–58 (55·0); tail, ♂ (n = 15) 50–57 (52·9), ♀ (n = 13) 50–56 (53·1); bill to skull, ♂ (n = 17) 11·7–12·6 (12·2), ♀ (n = 13) 11·9–13·1 (12·5); tarsus, ♂ (n = 13) 18·0–19·8 (18·9), ♀ (n = 13) 18·2–19·6 (18·9). WEIGHT: ♂ (n = 5) 8·5–10 (9·1). ♀ (n = 3) 7·7, 8·0, 8·5.

IMMATURE: like non-breeding adult ♀, but tail brown with T6 not pure white, and underparts not tinged pink.

NESTLING: not described.

S. d. maroccana Hartert: Morocco, except Ouarzazarte area and eastward. Upperparts slightly darker than *deserticola*; rufous fringes of tertials and secondaries much narrower; T6 blackish with white outer vane. Wing and tail av. 2 mm shorter than in *deserticola*.

S. d. ticehursti Meinertzhagen: Morocco (Ouarzazarte area and perhaps to east: migrants at Beni Abbès). Said to be distinctive, generally isabelline (Etchécopar and Hue 1967).

Field Characters. A small *Sylvia* warbler, breeding high in NW African mountains and wintering in Sahara, with high crown, small bill and slender tail. Adult has blue-grey upperparts, rufous wing-patch, and rufous breast and flanks. Eye brown to yellowish; lacks coloured bare orbital ring but has very narrow ring of white feathers. In breeding plumage looks like a cross between Dartford Warbler *S. undata* and Spectacled Warbler *S. conspicillata*; Dartford told by its long tail, red eye and orbital ring and grey wing and Spectacled by its brown back, broad white malar stripe and pink underparts. Non-breeding ♀ and immature much browner above, very like ♀ and immature Spectacled Warbler but ♀ pinker below, rather than buff, and often with some rufous feathers on throat. Immature Spectacled is browner above and has more rufous in wing. Song like Dartford and very like Subalpine Warbler *S. cantillans* songs.

Voice. Tape-recorded (60, 93, B, CHA). ♂ song in song-flight, lasting *c*. 7·5 s, a hurried chattering with squeaky quality, containing hard 'tack' or 'chit' notes, 2–4 short rattles, and ending with hard 'tsuk' notes. Song of perched ♂ similar, but shorter and with more melodious middle part with notes recalling Great Tit *Parus major*. Calls, a harsh stuttering rattle 'cht-cht-chchch-chit...' like Winter Wren *Troglodytes troglodytes*, a sharp 'chit' or 'chir-it', a 'tscherr' by ♂ chasing another, and a high-pitched, descending 'eeeeeee' (Cramp 1992).

General Habits. Breeding habitat is mountain sides and hills with bushes and dense scrub 1–3 m tall of junipers, *Buxus*, *Pistacia* and *Cistus*, with open spaces and sometimes with well-spaced trees (holm oak *Quercus ilex*, Atlas Cedar *Cedrus atlantica*). Intolerant of habitat disturbance by people. Winters at low elevation, in scrubby foothills, oases and saline wadis in Sahara, especially where there are tamarisks and palm scrub. Foraging habits like Dartford Warbler; keeps low down in bushes, half-cocks tail, hops from twig to twig, feeds among clumps of grass. Skulking but not shy. Sings Feb–Mar, sometimes Jan, from dawn to midday. Sings from bush top, sometimes hovering over bush (with nest?), occasionally in song-flight.

Dispersal and migrations, see above.

Food. Includes ants and beetles; young fed with beetles.

Breeding Habits. Solitary, territorial nester. ♂ in song-flight ascends a few m, then descends, singing, with head and tail raised. ♂ ♂ chase each other in circles, through bushes and flying low over ground. Perched ♂ threatens with feathers ruffled, wings shivering, tail cocked up and down, then chases rival away from its song-post (Cramp 1992).

NEST: deep cup made of coarse grasses, lined with finer grass and sometimes horsehair, placed 1–1·5 m above ground in bush (Cramp 1992). Ext. diam. 100, diam. of cup 50.

EGGS: 3–5, av. (n = 23 clutches) 3·9. Subelliptical; glossy; pale green, finely spotted with dark brown and grey, especially at broad end. SIZE: (n = 28) 15·0–17·3 × 12·0–13·0 (15·6 × 12·5). WEIGHT: (calculated) 1·25.

LAYING DATES: Morocco, Mar–July; Algeria, Apr–May (and ♂, out of normal breeding range, repeatedly carrying nest material, Dec); Tunisia, Apr–May.

Reference
Cramp, S. (1992).

Plate 24
(Opp. p. 417)

Sylvia undata (Boddaert). Dartford Warbler. Fauvette pitchou.

Motacilla undata Boddaert, 1783. Table Planches Enlum., p. 40, based on Daubenton, 1765–81, Planches Enlum., pl. 655, fig. 1; Provence, France.

Range and Status. Resident and partial migrant, SW Europe and NW Africa; breeds north to S England, east to S Italy. Vagrant north to Sweden, east to Greece.

Resident in mainly coastal regions and foothills of Morocco and Algeria; rare in Tunisia, where may occasionally breed; winters south to 28–29°N. Morocco, locally frequent in Rif (Jebel Tisirène, Taounate, Jebel Masgout) and abundant (once 'extremely abundant') in coastal Maroc Oriental (Cap des Trois Fourches, Mechra Klila, Mechra Homadi), mainly in the Tétouan-Chaouen-Ouezzane-Bab Berred sector; in winter occurs sparsely, locally frequently, south to Marrakesh, Oued Sous, Ouarzazate, Tafilalet and Erfoud, and rare south to Cap Dra; on Mediterranean, occurs at mouth of Moulouya in winter but not summer. Algeria, breeds on N slopes of Atlas Mts of Tell coastal area, quite commonly east to Ouarsenis and on coast between Algiers and Annaba; locally abundant on Dahras foothills near Ténès; in winter occurs sparsely south to Biskra, Laghouat, El Oued, Ghardaïa, El Goléa and Timimoun; spring passage noted at Reghaïa. Tunisia, rare records in N from Cap Bon, Forêt Feidja

Sylvia undata

and Zembra; breeding likely but not proven, and perhaps some confusion with Marmora's Warbler *S. sarda*; in winter a regular visitor, most frequent on desert edge between Nefta, Jebel Tebaga and Gabès. Libya, singing ♂ near Homs and 1 near Wadi Turghat, both in spring.

Winter density of 68 birds per km² in scrub and 5.6 birds per km² in cork oak *Quercus suber* (Gibraltar: Arroyo and Tellería 1984). Breeding density (England and Corsica) *c.* 10–20 pairs per km² and (Sardinia) up to 90 pairs per km².

Description. *S. u. toni* Hartert: NW Africa, S Spain and S Portugal. ADULT ♂: upperparts almost uniformly dark brownish grey, wing and tail-feathers mostly black with brownish grey fringes; bare orbital ring and surrounding narrow ring of featherlets red; outer tail-feather (T6) with sharply-defined outer edge and tip, T3–T5 usually with small white mark at tip. Underparts somewhat variable in tone, usually rich dark brick-red or vinous red, feathers of chin and throat with small white triangle at tip; mid-belly and vent whitish, flanks and undertail-coverts grey-brown, thighs vinous-grey. Alula black, larger feathers narrowly tipped white. Underwing-coverts and axillaries grey; outer marginal coverts white. Upper mandible dark grey, culmen and tip black, cutting edges and lower mandible yellowish or orange-flesh; eyes red; legs and feet yellowish flesh or brownish orange. ADULT ♀: often indistinguishable from ♂ but worn plumage browner above and paler and more uniform below. Eyes perhaps paler. SIZE: wing, ♂ (n = 14) 52–56 (53·5), ♀ (n = 11) 51–55 (52·5); tail, ♂ (n = 5) 61–64 (62·5), ♀ (n = 11) 56–65 (60·0); bill to skull, ♂ (n = 13) 11·9–12·8 (12·4), ♀ (n = 10) 11·9–13·6 (12·6); tarsus, ♂ (n = 6) 18·6–19·8 (19·0), ♀ (n = 10) 18·7–20·0 (19·4). WEIGHT: (Algeria, Nov–Dec), 3 ♂♂ 9, 9·5, 9·6, 1 ♀ 8·5; (Spain) 1 ♂ 9·2, 3 ♀♀ 9·7, 9·8, 10·5.

IMMATURE: upperparts uniform earth-brown; underparts buffy or faintly rufescent brown, grading to dirty whitish on belly, flanks buffier; T6 buff; eyes dark brown or red-brown.

NESTLING: hatches naked. Not described, but has black tongue-spots.

(*S. u. undata* (Boddaert): SW Europe, north of *toni*. Long suspected to winter in NW Africa although proof still lacking. Upperparts slate-grey, underparts possibly brighter than in *toni*. Slightly larger: (8 ♂♂, 8 ♀♀, Italy) wing, ♂ 53–58 (54·8), ♀ 51–55 (53·0), tail, ♂ 61–70 (65·1), ♀ 59–67 (63·1).)

Field Characters. 12·5 cm. One of the smallest-bodied *Sylvia* warblers and in breeding plumage easy to identify: upperparts slate-grey, underparts rich wine- or brick-red, eye and its surround scarlet, white speckles on throat, long tail with white edge, and whitish belly. Cocks tail high, even over back. Characteristic flight (see below), and high-crowned profile. Very similar to Marmora's Warbler *S. sarda* in silhouette, easy to confuse in poor light, but Marmora's has grey not red underparts; over most of range the 2 are mutually exclusive but they seem to co-exist in N Tunisia. ♂ Tristram's Warbler *S. deserticola* has vinous red underparts but short tail, rufous wing-patch and white eye-ring. ♂ Subalpine Warbler *S. cantillans* has shorter tail and white moustachial streak.

Voice. Tape-recorded (CHA – 62, 73, 93, B). ♂ song is 'whimsical, wistful and spirited' medley of sweet piping notes with metallic rattles (Cramp 1992); duration 1·5–2·0 s; repeated after pauses of 2 s or more. Quite like song of Common Whitethroat *S. communis* but more melodious, less jerky. Also described as cheerful, rambling, abrupt, tinny, hurried, and grating. Subsong sweeter, like distant Skylark *Alauda arvensis*. Commonest call a grating 'tchirr', ♀'s said to be lower-pitched than ♂'s; sometimes followed by liquid note: 'tchirr-chiwee'. Several other calls described (Cramp 1992).

General Habits. In NW Africa confined almost entirely to scrub in coastal foothills with kermes oak *Quercus coccifera*. Elsewhere in W Mediterranean basin, frequents open garigue, low *Halimium* thicket, *Cistus*, myrtle, bramble, herbs, grasses and grubbed-up vineyards, and strongly associated with gorse *Ulex*, up to 1500 m altitude. Breeds at av. July temperature of 30° C; cannot tolerate winter freezing conditions.

In territorial pairs in and out of breeding season, or solitary out of it; occurs in small aggregations on migration; different families of young at times join together, with a post-breeding adult (Cramp 1992), and family may associate with one of Marmora's Warblers (Lévêque 1976). 2 birds of pair keep close together most of time. Forages entirely below 2 m high in shrubs and ground layer, av. 0·98 m high (27 observations, Corsica and Sicily). Moves through vegetation slowly and carefully, by hopping and sideways shuffling on stems. Often comes to ground, where usually creeps but can run very fast. Secretive and shy, at least in inclement weather, but if sunny shows itself openly at times. When excited, perches on highest spray of bush and calls with throat and crown-feathers ruffled. When deep in cover tends to make sudden break from it and makes off low down; dives straight into thick bush.

Habitually cocks tail high, about vertically, and often flicks it nervously. Often follows Common Stonechat *Saxicola torquata* from bush to bush in winter, the chat perching on bush top and the warbler well below it (England); similarly follows Black Wheatear *Oenanthe leucura* (Spain) (significance unclear, but see Desert Warbler *S. nana*). Roosts communally in England, Dec–Feb, but roosting behaviour in Africa unknown. ♂ has a few favoured song-perches, one generally very close to nest; sometimes sings from tall post or telegraph wire. Not a persistent singer, and can stay silent for hours (Cramp 1992). In France sings Apr–June but more in July, silent in late summer, then song or subsong is renewed in autumn; subsong given from within cover, with bill closed. Sings sporadically in winter.

Regular migrant across Strait of Gibraltar and on Tingitane peninsula (Morocco), commoner in autumn (mainly Oct) than in spring. European birds seem to enter Africa between Tangier and Algiers. Winters in Tafilalet, Morocco, early Nov to mid-Feb. Most records in Moroccan wintering zone are in Dec–Mar (in Tizi-n-Test to mid-Apr in one year). Winters Tunisia Dec–Feb, occasionally Mar. Spring migration at Strait of Gibraltar in Mar–Apr; NE Moroccan wintering grounds vacated in Mar.

Food. No African studies. In Europe mainly arthropods (damselflies, grasshoppers, moths, butterflies, caterpillars, flies, small wasps, bugs, spiders, harvestmen, millipedes); also snails, and berries (*Rubus*, *Daphne*, *Myrtus*, *Pistacia*, *Vaccinium*, *Phillyrea*, *Phytolacca*).

Breeding Habits. Habits poorly known in Africa, well known in Europe. Monogamous, solitary nester; territorial activity starts or is renewed in autumn. Territories overlap those of larger congeners, without aggression. Not a very aggressive species, but sings vigorously to defend its territory and chases off conspecific intruders; quarrels with adjacently-nesting Marmora's Warblers (Lynes 1912). In territorial song-flight, ♂ hovers and flutters 6–7 m above bushes, singing, and may fly in high arc before descending to different perch. ♂ attracts ♀ by song. In courtship, ♂ and ♀ in bush extend and flap wings, fan and flirt tail, and ♂ erects slaty cheek feathers. ♂ sometimes makes flimsy, unlined cock nest (Cramp 1992).

NEST: quite neat, compact cup made of grass leaves and fine twigs and rootlets, with some vegetable down, spider web and sometimes feathers; lined with hair and finer rootlets. Ext. diam. (12 nests) 80–98 (88), ext. depth 40–60 (51), int. diam. 49–60 (55), int. depth 25–50 (36). Built by ♂ and ♀, collecting material up to 100 m away. For 2nd or repeat clutch, ♀ sometimes selects a cock nest and completes its construction and lining by herself. Sited inside thick, low shrub, in England almost always heather with canopy height of only 0·5 m.

EGGS: 3–5, av. 4·0 (England). Subelliptical; smooth; slightly glossy but can also look matt and chalky. Buff, sometimes whitish, blotched with grey mainly around large end, with overlying speckles and blotches of olive-brown or red-brown, evenly over whole surface or mainly at large end. SIZE: (n = 11, *S. u. toni*) 16·3–18·0 × 12·5–13·4 (17·1 × 13·0). WEIGHT: 1·52. 2–3 broods (W Europe), or only 2 (Sardinia).

LAYING DATES: Algeria, Apr–June.

INCUBATION: by ♀, mainly, and ♂. Nest-relief extremely unobtrusive; with indirect approach and departure (Cramp 1992). Period: 12–14 days.

DEVELOPMENT AND CARE OF YOUNG: young brooded for up to 8 days, by both parents, and fed by both parents. ♀ once perched on rim and energetically rummaged with bill in floor of nest, re-arranging it (Bibby 1979). Fledging period c. 12 days. Young leave nest before they can fly, staying within 5 m of nest for 1–2 days or more. Young independent of parents 10–15 days after leaving nest.

BREEDING SUCCESS/SURVIVAL: no African data. Breeding success in England only moderately high, and population crashes in hard winters (Cramp 1992).

References
Bibby, C. J. (1979).
Cramp, S. (1992).

Plate 24
(Opp. p. 417)

Sylvia sarda Temminck. Marmora's Warbler. Fauvette sarde.

Sylvia sarda Temminck, 1820. Man. Ornith., ed. 2, 1, p. 204; Sardinia.

Range and Status. W Mediterranean islands: Balearic Is, Corsica, Sardinia, Pantelleria. Recently displaced from Menorca (Balearic Is) by Dartford Warbler *S. undata*. Resident and migrant. Winter visitor to Sicily (where recent marked decrease) and NW Africa. Rare in Gibraltar, on passage and in winter. Vagrant Britain, France, Greece, Malta, Egypt.

Winter visitor to N Africa. Scarce in N Morocco, from Tingitane peninsula eastward (Cramp 1992), once in SW at Taroudant. Algeria, uncommon, Oct–Mar, in NE and oases on fringe of Sahara (Batna, El Kantara, Biskra); recorded on NW coast at Mostaganem, also deep in Sahara, from Ghardaïa and Tamanrasset (Mar 1971). Tunisia, scarce but regular

Sylvia sarda

winter visitor, Oct–Feb (rare, Mar), south to Nefta, Sfax, Kerkenna Is and Zarzis; possibly bred at Zembra where seen July–Aug 1962. (Pantelleria I. (Italy), where Marmora's Warbler is a breeding summer visitor, is < 80 km from Tunisia.) Libya, scarce to common in NW, Nov to late Mar, from Tunisian border and Jefara to Jebel Nafusa and near Homs; also twice Serir, Apr. Egypt, 1 Salum Jan 1928, 2 claimed Wadi Hibran, 1870.

Description. *S. s. sarda* Temminck: Corsica, Sardinia, Pantelleria I. ADULT ♂: upperparts from forehead to uppertail-coverts slaty or bluish grey, darkest on forehead, blackish on lores and around eye, slightly paler on rump and uppertail-coverts. Tail long, graduated, blackish grey, feather edges a trifle paler, T6 with very narrow white outer edge and tip. Chin whitish; throat grey; breast and belly grey with pink tinge, paler in the midline; flanks, thighs and vent grey; undertail coverts slaty grey. Wing flight feathers and upperwing-coverts all slaty grey, narrowly fringed mid-grey; tertials blackish, fringed brown-grey; alula feathers black, with outer vanes narrowly white-edged. Underside of tail and wings slate-grey. Bill dark brown, cutting edges and proximal half of lower mandible pale pink-yellow; inside of mouth yellow-ochre; orbital ring wine-red; eyes blood-red or light brown; legs and feet yellowish or pinkish brown. ADULT ♀: very like ♂ but slightly paler and browner, above and below; mid-belly whiter; lores and area around eye brown-grey, not blackish. SIZE: wing, ♂ (n = 20) 53–59 (55·7), ♀ (n = 13) 51–56 (53·5); tail, ♂ (n = 21) 54–62 (58·5), ♀ (n = 13) 53–60 (57·8); bill to skull, ♂ (n = 20) 11·5–13·2 (12·3), ♀ (n = 13) 11·5–13·1 (12·2); tarsus, ♂ (n = 17) 19·0–20·8 (19·8), ♀ (n = 10) 18·9–20·3 (19·8). WEIGHT: (N Algeria) 2 unsexed 8·5, 8·9, ♂ ♀ (n = 6, Sardinia) av. *c.* 10·5.

IMMATURE: like adult ♀ but mantle, back, rump and scapulars browner; wing flight feathers and coverts edged warm brown, tertials edged almost rufescent brown; underparts buffy, particularly flanks. Eyes olive-brown; bare orbital ring yellow-brown or pale orange.

(*S. s. balearica* von Jordans: Balearic Is. Not certainly identified in Africa, but Moroccan and some Algerian birds likely to belong to this race. Slightly paler than *sarda*; forehead, lores and around eye black; pinker tinge on underparts. Smaller: wing, ♂ (n = 18) 48–54 (50·8), ♀ (n = 7) 48–51 (49·9), tail, ♂ (n = 12) 52–58 (54·4), ♀ (n = 5) 51–57 (53·8); but bill long: 13 ♂♂ av. 12·8, 6 ♀♀ av. 12·8. Weight: unsexed (n = 6) 7·5–13·5 (9·9)).

Field Characters. A small *Sylvia* warbler with long, slim, blackish tail; bluish grey, paler below, with base of lower mandible yellowish, legs and feet yellow-brown, reddish eye and bare red eye-ring. ♀ browner; juvenile much browner (and with brown eye). In poor light adult looks blackish. Cocks tail straight up; high, peaked crown; short wings; forages near ground; tame, curious; song a short, fast warble. In most of these features very similar to Dartford Warbler, which differs in its red underparts and white-speckled throat. Juvenile nondescript, not safely distinguishable from juvenile Dartford – even in hand, best to compare live bird with series of skins (Svensson 1992).

Voice. Tape-recorded (62, 73, 93, B, CHA). Sometimes sings in winter. ♂ song (Corsica) is weak high-pitched warble '(se-)WEE-se(-se)' followed by whirring tremolo, stuttering, and slurred trills. Gives 9–12 phrases in 50–60 s, with 3–4 s pauses; timbre recalls European Robin *Erithacus rubecula* or Hedge Accentor *Prunella modularis*. In Balearic Is, song in 3 parts, 1st and 3rd like a soft Sardinian Warbler *S. melanocephala*, the 2nd like a soft Lesser Whitethroat *S. curruca* (Cramp 1992). Various short, explosive, rasping calls.

General Habits. Inhabits islets, coastal slopes, cliff-top heaths, and hill sides, with uniform dense cover *c.* 1 m tall of garigue, i.e. heath and tree-heath *Erica*, palmetto *Chamaerops*, live or dead *Cistus*, *Genista*, *Pistacia* and *Arbutus*, sometimes with a few spaced trees (junipers, pines). In Libya, well-vegetated flat steppe. Forages in low vegetation, spending a third of daylight hours on ground, where it can hop, without flying, for up to 200 m. Climbs sideways up twigs, gleaning insects; flycatches.

Resident and partial migrant (see above).

Food. Caterpillars, grasshoppers, many spiders.

References
Cramp, S. (1992).
Garcia, J.C. (1994).

Genus *Parisoma* Swainson

An assemblage of warblers, 4 endemic to Africa and 1 to Yemen, closely related to *Sylvia* and recently merged with that genus (Sibley and Monroe 1990, Sibley and Ahlquist 1990, Dowsett and Forbes-Watson 1993). DNA analysis of southern African *P. subcaeruleum* suggests that it should be congeneric with *Sylvia* (Dowsett and Dowsett-Lemaire 1993b). The 2 genera are linked by *P. buryi* (the Arabian endemic) and *S. leucomelaena* (Israel to Somalia). *P. buryi* has been regarded as a systematic oddity (Brooks 1987). *S. leucomelaena* has been placed in *Parisoma* (Sclater and Mackworth-Praed 1918, Meinertzhagen 1949, Afik 1984); but its closest relative is now recognized to be *S. hortensis*, which is in fact an allospecies of it (see Shirihai 1988a, 1989). In our view the closest relative of *buryi* is *leucomelaena*: plumages, songs and nest form are all similar.

'*Parisoma*' is probably a radiation of *Sylvia* warblers into Afrotropical acacia habitats. All *Parisoma* spp. remain poorly known; they seem to embrace considerable variation in morphology and behaviour, making the genus, if valid monophyletically, difficult to diagnose and discriminate from *Sylvia*, which is itself quite diversified. We retain *Parisoma* as a separate entity but recognize that at least 3, the African *P. lugens*, *P. layardi* and *P. subcaeruleum*, are probably *Sylvia*s.

Plumages, proportions and songs like *Sylvia*; grey-brown warblers mainly with relatively short, stout bill, pale iris, dark legs, speckled throat, poorly-demarcated cap, white-sided tails, short wings, and little sexual dimorphism. 2 species with rufous crissum, 1 with pectoral band. Associated with *Acacia*, mostly in highlands.

5 species: 1 in Arabia, 2 endemic to E and 2 to southern Africa. No superspecies.

Plate 22
(Opp. p. 401)

Parisoma layardi Hartlaub. Layard's Warbler; Layard's Tit-babbler. Parisome de Layard.

Parisoma layardi Hartlaub, 1862. Ibis, p. 147; Malmesbury, South Africa.

Range and Status. Endemic resident, Namibia, Lesotho and South Africa. Occurs in semi-arid montane scrub from W Namibia at about 22°S to SW Cape (Blouberg) and E end of Little Karoo, and inland east to Lesotho. Fairly common in SW Cape, in N coastal lowlands and along borders of Karoo; common in Worcester-Robertson valley. Fairly common in S Orange Free State but rare in N (north to Bloemhof and Kroonstad). 3 records Natal (Mkuzi Game Reserve, Mtunzini, Karkloof). Uncommon, Lesotho in uplands and high plateaux. Quite common in W Namibia, e.g. around Kleinkaras and between Witputs and Brandberg Mts.

Description *P. l. layardi* Hartlaub: extreme SW Cape Province, north to Port Nolloth. ADULT ♂: upperparts from forehead to uppertail-coverts, also ear-coverts, cheeks, sides of neck, scapulars and upperwing-coverts, uniform slate-grey (occasionally brown), except that forehead is tinged with olive. Tail glossy black, graduated, with white sides: T6 12 mm shorter than T1, white except for edge of inner vane; T5 6 shorter than T1, black with white end 11 long; T4 with white tip 6 long at shaft; T3 with 1 mm white tip; T2 with < 1 mm grey tip; T1 all black. Chin off-white, usually with yellowish or buffy wash; throat and upper breast similar but softly and generally quite heavily streaked with blackish grey; lower breast clear grey, pale in centre, becoming dark at sides; belly white, becoming grey on flanks; thighs grey; undertail-coverts grey-white. Primaries, secondaries and tertials blackish brown, outer primaries fringed white proximal to place of emargination, inner primaries and secondaries broadly fringed grey, tertials with broad grey outer fringes and narrow white tips and inner fringes. Outermost primary covert edged white; alula feathers broadly edged white, and small coverts underlying them white, making conspicuous white patch near wrist. Underside of primaries glossy grey, with whitish inner edges; underwing-coverts clear grey, silky white near primaries. Lores matt blackish. Bill black; eyes pale yellow or cream-white; legs and feet black. ADULT ♀: like ♂ but upperparts browner (difference apparent only in series); iris same yellow, sometimes grey-tinged. SIZE: wing, ♂ (n = 5) 63–67 (64·8), ♀ (n = 3) 64–66·5; tail, ♂ ♀ (n = 10) 61–64; bill, ♂ (n = 5) 14–15 (14·2); tarsus, ♂ ♀ (n = 10) 18–21. WEIGHT: (3 ♂ ♂) 14, 14·9, 15·5, (2 unsexed, Great Karoo) 13, 16.

IMMATURE: a brown rather than grey bird; upperparts like adult but warm brown, lacking any grey tones; patch near wrist, and edges of outer primaries buffy grey (not white). Chin and throat greyish, unstreaked; breast warm greyish buff, unstreaked, darker at sides; belly whitish in centre; flanks and undertail-coverts buffy grey; iris yellow, grey-tinged.

NESTLING: pre-feathered pullus not described. Nestling like immature, but iris grey and, at stage when tail only just sprouting, white marks at edge of wing near wrist are conspicuous.

P. l. aridicola Winterbottom: NW Cape. Upperparts paler, slightly more olive than in *layardi*. Larger: wing, ♂ ♀ (n = 19) 65–68 (66·5); bill, ♂ ♀ (n = 19) 13–17 (15·1).

P. l. barnesi Vincent: Lesotho. Upperparts greyer than in *aridicola* and underparts whiter. Same size as *aridicola*.

P. l. subsolana Clancey: Cape in Great Karoo and Little Karoo. Like *aridicola* but crown darker, and mantle and rump darker and browner.

Field Characters. A neat, solidly-built, rather retiring warbler of arid mountainside scrub. Rather nondescript, grey, with stubby black bill, yellow-white eyes; grey-brown back, black tail with white sides and corners, whitish chin and softly white-and-dark-grey-striped throat and upper breast. Stripey throat conspicuous (adult); juvenile has plain throat. Very like Chestnut-vented Warbler *P. subcaeruleum*, which differs by its rufous undertail-coverts. Chestnut-vented also has conspicuous black-and-white alula, and more white in tail. The 2 occupy same habitat in W Cape coastal lowlands, but in karroid areas Layard's inhabits scrub on rocky hill sides and Chestnut-vented occurs mainly in riverine acacias; in E Cape (Cradock) Layard's found where *Acacia karroo* penetrates kloofs, both species in hillside scrub, Chestnut-vented also in acacia veld, riverside bush and large gardens (Collett 1982). Melodious song, like Chestnut-vented Warbler's, but more metallic and tinkling, less throaty and churring.

Voice. Tape-recorded (75, 88, 91, F, LEM, ROC). Song a hurried mixture of varied notes, defined into phrases by brief pauses, without marked changes in pitch within or between phrases. Most readily identified component is a short, melodious 'chewi-chewi-chewi-chewi' or 'whee-whee-we-we'; interspersed irregularly between such phrases are clear fluty notes, scratchy ones, gritty churs, and very thin, high 'sees' notes like alarm call of many woodland birds. Song often preceded by 1 or 2 contact calls. One song (Gibbon 1991) lasted *c*. 20 s, divided by 5 pauses into 6 phrases all differing in duration and types of notes included. Contact call, a short, explosive 'ptchrrrrrr'. Song and call both strongly reminiscent of Mediterranean *Sylvia* warblers.

General Habits. Inhabits arid and semi-arid mountain scrub, including *Olea* bushes, see above. In N-central Namibia occurs in small shrubs, along watercourses, in valleys and around mountains. Occurs at quite high altitudes on mountains of W Cape (Hockey *et al.* 1989)

A

and up to 2750 m in exposed situations in Lesotho (Osborne and Tigar 1990). Solitary and in pairs; also small family parties (Macdonald 1957). Spends much time skulking in thick vegetation, where forages low down, searching leaves and twigs for insects. Inquisitive, responds to person 'spishing'. In breeding season more conspicuous, singing from perch on bush top and in display flight. Flies high between bushes, in curved (not straight) line, sometimes snapping wings (Maclean 1993).

Fewer records in SW Cape in summer than in winter, perhaps due merely to seasonal differences in conspicuousness. Probably resident.

Food. Insects; fruits including *Ficus cordata*. Nestlings given small caterpillars.

Breeding Habits. Solitary, territorial breeder. Territory defended by singing from perch or in short aerial display flight; bird bounds in air, singing at each bound, then descends steeply on stiff wings with head up, lark-like (Maclean 1993).

NEST: quite neat and thin-walled cup made of thin plant fibres and thin green grass blades, woven with cobweb (**A**). Built low down in dense, matted bush.

EGGS: 3 (n = 4 clutches). White, evenly spotted with brown and grey. SIZE: (n = 12) 17·2–19 × 12·7–14 (17·8 × 13·3).

LAYING DATES: South Africa, Oct–Nov; Lesotho, Dec.

References
Hockey, P. A. R. *et al.* (1989).
Macdonald, J. D. (1957).
Maclean, G. L. (1993).

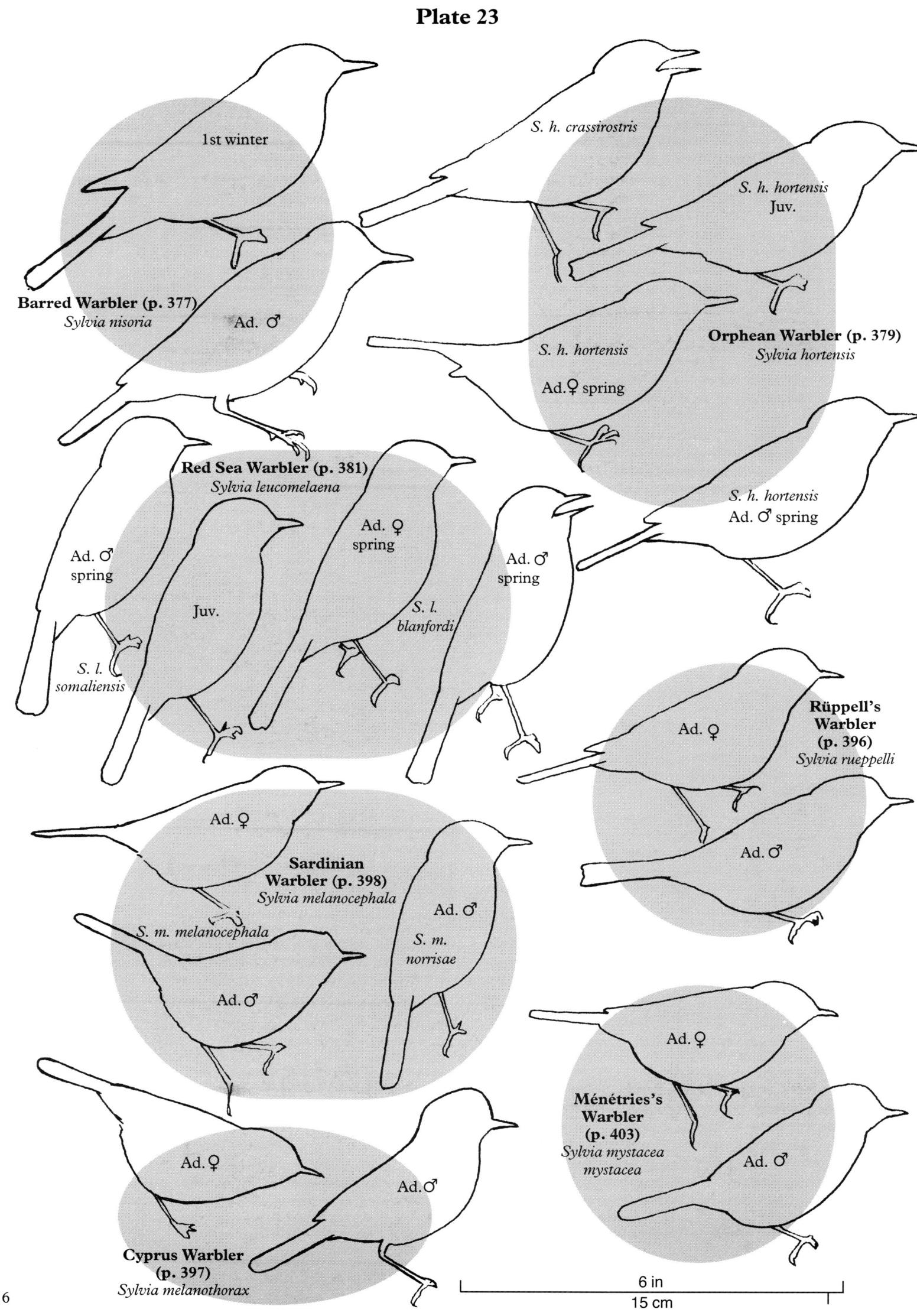

Plate 24

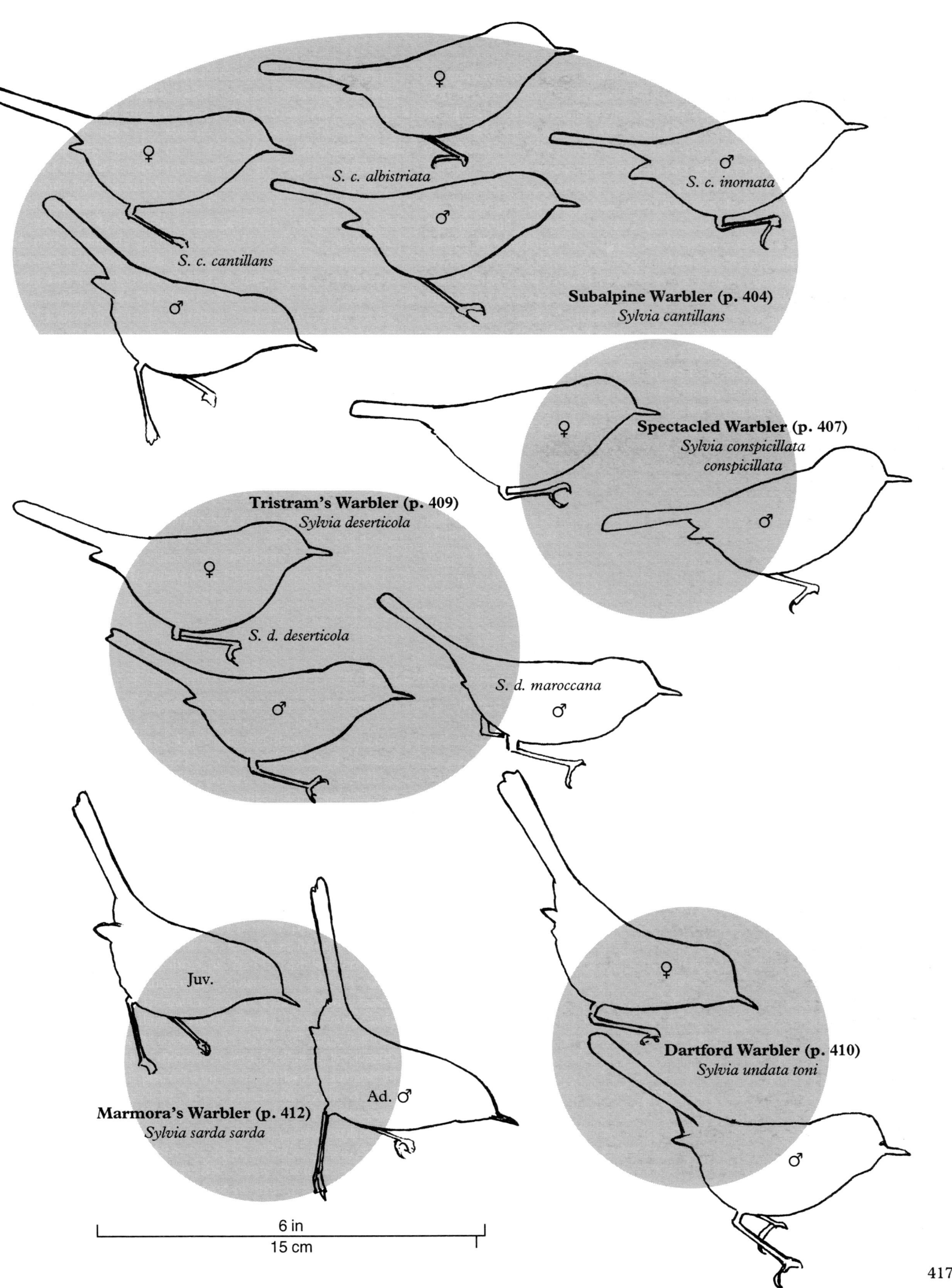

Plate 22
(Opp. p. 401)

Parisoma subcaeruleum (Vieillot). Chestnut-vented Warbler; Tit-babbler. Parisome grignette.

Sylvia subcaerulea Vieillot, 1817. Nouv. Dict. d'Hist. Nat., 11, p. 188; Gouritz River, Cape Province.

Range and Status. Endemic resident, southern Africa. SW Angola in coastal plains of Moçâmedes and Benguela, and hinterland in Huila from Humpata to Humbe; very common in Iona Nat. Park. Zambia, local and uncommon in SW Barotse Province from Nangweshi, Shangombo and Sinjembele to Chunga Pools and Sesheke, also Southern Province (Mambova, Livingstone, Mubi Pools). Namibia, widespread in W, sparse in E. Zimbabwe, widespread in acacia savanna on central plateau, northwest to Wankie Nat. Park (Main Camp) but absent from NE and from W Matetsi; north to Beatrice, east to Rusape, Fort Victoria and Sentinel Ranch. N and SE Botswana. Transvaal, common west of line from Tuli Block (Botswana) to Lydenburg to Vereeniging; common in acacia in west half of Orange Free State and in Cape Province west of Grahamstown; an outlying population in Natal. 2 records in Lesotho (Jacot-Guillarmod 1963, Osborne and Tigar 1990), and single record in S Mozambique (Hall and Moreau 1970).

Description. *P. s. subcaeruleum* Vieillot: Cape, SW Orange Free State. ADULT ♂: forehead brownish grey, washed olive; crown to uppertail-coverts, scapulars and upperwing-coverts brownish dark grey; cheeks, ear-coverts and sides of neck the same, lores paler, area immediately around eye darker, ear-coverts minutely shaft-streaked with white. Tail blackish, with white corners: T1 and T2 black, T3 with greyish tip visible only below, T4 with terminal white patch 11 mm deep, T5 with white end 19 long on inner web and 22 on outer, T6 with distal 22 mm of inner web white and outer web all white except near base. Tail rounded: T4 3 mm, T5 5 and T6 8 shorter than T1–T3. Chin mottled grey-brown and white. Throat and upper breast smoke-grey with long, soft blackish streaks; lower breast and flanks clear grey becoming cream-white in centre of belly; thighs grey; whole of vent and undertail-coverts deep cinnamon-orange, some feathers overlying thighs the same. Primaries, secondaries and tertials dark brown, edged greyish; outer webs of outer primaries edged pale grey proximal to point of emargination; P9 blackish, with distinct white line along distal 8 mm of outer web. Inner secondaries very narrowly tipped grey white. Primary coverts blackish, fringed grey and narrowly tipped whitish; alula very conspicuous: feathers black, with broad white chevron tips, overlying white featherlets. Underside of primaries grey-brown, inner edges of webs whitish; axillaries grey-white. Bill black; eyes white, greyish white, cream or pale yellow; legs and feet black or dark grey. SIZE: wing, ♂ (n = 29) 62–71 (66·9), ♀ (n = 17) 62–69 (64·9); tail, ♂♀ (n = 46) 62–74; bill, ♂♀ (n = 46) 10–13·5; tarsus, ♂♀ (n = 46) 19–23. WEIGHT: ♂ (n = 6, Namibia) 13–16 (14·1), ♀ (n = 7, Namibia) 10–16 (14·3), unsexed (n = 84, southern Africa) 11·6–18·6 (14·6).

IMMATURE: like adult but chin, throat and breast grey, plain or slightly mottled dusky on chin; breast not streaked. Feathers around thighs, vent, and undertail-coverts pale rufous.

NESTLING: hatchling not described. One-third-grown nestling, eyes just opening, has feathered crown and large pale yellow gape flanges; two-thirds-grown bird like adult but eyes dark brown, undertail-coverts paler rufous than adult, inside mouth orange, gape flanges cream-coloured, thin (photos *in* Ginn *et al.* 1989).

Parisoma subcaeruleum

P. s. cinerascens Reichenow: Namibia. Slightly paler; larger area of white on belly. Intergrades with *subcaeruleum* and *ansorgei*.

P. s. ansorgei Zedlitz: Angola. Fractionally paler above, less sooty-looking than nominate race; paler below, more whitish; streaks in chin, throat and upper breast shorter, narrower and better separated, lower breast and flanks silky grey-white, breast and belly washed very pale yellow, vent and undertail-coverts not so deep an orange.

P. s. orpheanum Clancey: Zimbabwe, Transvaal, Natal, W Zululand, Orange Free State except SW, lowland Lesotho. More bluish violet-grey above than nominate, with little or no sooty; streaks in throat broader and more intensely black; bill longer, ♂♀ (n = ?) 14–16.

Field Characters. A neat, solidly-built, rather retiring warbler of acacia thornveld, behaving like a Palearctic *Sylvia* warbler. Grey-brown above, greyer below, with staring white or yellow eye, chestnut undertail-coverts and white-tipped black tail, conspicuous in flight and as bird alights. White-edged black alula is a further good field mark. Chin and throat pale grey, speckled dark brown. Inconspicuous; best detected by 'jeriktik' call and by melodious, imitative song. Layard's Warbler *P. layardi* is very similar but has white vent.

Voice. Tape-recorded (11, 22, 42, 75, 88, 91, F). Song a cheerful series of bold, varied phrases, loud and emphatic, mainly variations on 2 notes: a clear, ringing 'see', and a slow, falling, scolding rattle 'chiur-r-r-r'. Song often preceded by 'cheriktiktik' contact call. One song lasted *c.* 22 s, divided by 5 pauses into 6 distinct phrases. Typical phrase: 'er-see, chiur-r-r-r, chiur-r-r r, see-er, see-sir-chk, chirur-r r', but no 2 successive phrases alike. Song often answered by another bird. At times highly imitative; includes voices of Diederik Cuckoo *Chrysococcyx caprius*, Greater Honeyguide *Indicator indicator*, European Bee-eater *Merops apiaster*, White-browed Scrub-Robin *Cercotrichas leucophrys*, Karoo Prinia *Prinia maculosa*, Long-billed Crombec *Sylvietta rufescens*, Cape Batis *Batis capensis*, Pririt Batis *B. pririt*, Southern Grey Tit *Parus afer*, Cape White-eye *Zosterops pallidus*, Fiscal Shrike *Lanius collaris*, Bokmakierie *Telophorus zeylonus*, White-throated Canary *Serinus albogularis* and Cape Bunting *Emberiza capensis* (Steyn 1969, Maclean 1993). Contact call the 'Tjeriktik' of the Afrikaans name: an abrupt, explosive rattle '(j)err(s)ik, tk tk tik tick', with emphasis on second and last syllables. Call usually answered by mate. This call, and non-mimetic song, strongly reminiscent of Mediterranean *Sylvia* warblers. Alarm, a rattle (like Grey-backed Cisticola *Cisticola subruficapillus*).

General Habits. Inhabits semi-arid areas of acacia and thick thorny scrub, thornveld, thickets, bushes and scrub by streams and on hill sides, and savanna woodlands. In W Cape, largely restricted along borders of Karoo and Namaqualand to acacia woods near watercourses, and on W coast to strandveld (where Layard's Warbler also occurs). In E Iona Nat. Park, Angola, typically occurs in *Colophospermum/Commiphora* scrub on rocky outcrops; along Cunene R. in acacia woodland. A vagrant was found in miombo woodland in cold spell in Zimbabwe, far from its usual habitat.

Singly or in pairs; sometimes joins mixed-species foraging party. Forages inconspicuously inside bush or canopy of small thorn tree by creeping and hopping, picking insects from bare twigs. Inconspicuous unless singing; generally keeps within dense bush although, if undisturbed, often shows itself to the patient observer; sings from exposed perch; calls frequently and readily detected by 'jeriktik' call. Inquisitive. Slips out of bush suddenly and flies typically low-down to next one, or dives into clump. Aggressive to conspecifics, at least in breeding season.

Resident; sedentary in some gardens; but above-mentioned Zimbabwe bird, at Malimasindi, was *c.* 100 km from normal range. Status in E Transvaal lowveld uncertain.

Food. Small arthropods; caterpillars, a mantis; small soft fruits (Ginn *et al.* 1989); spiders; feeds on ticks hiding under bark of trees (Earlé 1993).

Breeding Habits. Solitary, territorial breeder. Sings from perch at top of bush and in brief, shallow songflight.

NEST: neat, thin-walled cup of dry grasses and rootlets, often with tendrils of *Clematis* and thin twigs (**A**); base is thin and open, made of woolly plant down overlain with tendrils, rootlets and sometimes horsehair; or lining may be only of very fine dry grass. In S Karoo, seedheads of *Galium tomentosum* also incorporated into nest structure (Dean and Milton 1993). Int. diam. *c.* 47 mm, int. depth *c.* 27 mm. Nest lightly attached by cobwebs to surrounding twigs. Sited 1·5–3·5 m above ground in thick foliage usually in outer part of low acacia or other bush, on a cradle of twigs, often in clump of mistletoe *Viscum*, sometimes in centre of bush where the stem forks; 1 in Botswana in *Dodonaea viscosa* hedge (Skinner 1993).

A

EGGS: 2–4, usually 2, av. 2.5. Laid at 1 day intervals. White, spotted mainly at broad end with brown, olive and blue-grey. SIZE: (n = 155) 16·3–20·3 × 12·6–14·8 (18·2 × 13·8).

LAYING DATES: Angola, Nov; Zimbabwe, Sept–Mar (mainly Oct–Nov, also Sept; uncommon in Dec–Mar); Botswana, Sept–Feb; Transvaal, Aug–Mar (63% Oct–Nov); Natal, Oct–Jan; N Cape (SW Kalahari), Dec–Apr (somewhat opportunistically, after rain: Maclean 1993); SW Cape, July–Nov.

INCUBATION: if disturbed by person, incubating bird leaves nest at last moment and stays almost in reach; sometimes it does not leave until touched. At one nest incubating ♀ allowed person to touch her but still stayed put. Period (n = 2): 13, 15 days.

DEVELOPMENT AND CARE OF YOUNG: nestling period (n = 2) 14, 15 days. At one nest ♀ brooded young apparently for whole of first 9 days after they hatched, and ♂ fed young and ♀ in nest; later, ♀ helped feed young (Herremans and Herremans-Tonnoeyr 1993).

BREEDING SUCCESS/SURVIVAL: parasitized by Jacobin Cuckoo *Oxylophus jacobinus*.

References
Ginn, P. J. *et al.* (1989).
Herremans, M. and Herremans-Tonnoeyr, D. (1993).
Maclean, G. L. (1993).
Vincent, A. W. (1947).

Parisoma lugens (Rüppell). Brown Parisoma. Parisome brune.

Sylvia lugens Rüppell, 1840. N. Wirbelt., p. 113; Simen, Abyssinia.

Parisoma lugens

Range and Status. Endemic resident, with fragmented distribution in highlands of E Africa from Begemdir Province, Ethiopia, to Mwanza District, Malaŵi. In Ethiopia frequent in olive–juniper–*Podocarpus* forest between 1600 m and 2100 m in Begemdir (north to Gondar), Gojjam, Wollega, Illubabor, Kaffa, Shoa, Arussi, Sidamo, Bale and Harrar Provinces; a second subspecies occurs on Mendebo-Areanna Mts, Bale, at 3500–3700 m, surrounded by the first one much lower down. Uncommon to rare in Imatong and Didinga Mts, SE Sudan, and Mt Morongole, NE Uganda. In Kenya 1 record, Kakamega, and locally common at 1400–2500 m from Mt Elgon, Elgeyu, Laikipia and Mt Kenya to Kericho, Loita and Nguruman Hills, and adjacently in N Tanzania in Longido and Crater Highlands, all within 500 mm rainfall areas. 2 isolated populations in E Zaïre, in Itombwe (rare, 2620–3220 m) and Marungu (uncommon, 1840–2020 m). Also in S Tanzania (Matengo Highlands), N Malaŵi/NE Zambia (W side of Nyika Plateau, very local, between 1400 and 1850 m), and S Malaŵi (uncommon, Chongoni south to Tambo).

Description. *P. l. lugens* Rüppell: Ethiopia, except Bale Mts. ADULT ♂: entire upperparts very dark brown, crown fractionally darker than mantle to rump and upperwing-coverts, and tail and outer wing darker still. Plumage rather soft and lax: fluffy grey feather-bases sometimes show through in mantle, back and, particularly, rump, which then appear greyer. Outer tail-feather has outer vane entirely white or greyish white, and white triangle 12 mm long near tip of inner vane; the next feathers have progressively smaller greyish white tips to both vanes, from 7 mm long on T5 to 1 mm on T2; T1 blackish brown, sometimes with greyish tip < 1 mm long; tail rounded, T6 9 mm shorter than T1. Lores very dark brown; ear-coverts and cheeks dark brown, cheeks greyer, with minute white streaks. Chin and throat silky grey-white or buffy white, finely and softly mottled with dark brown. Breast pale grey with brownish wash, particularly towards sides of neck, sides of breast and flanks; breast sometimes faintly streaked. Belly creamy white in centre, greyish buff or warm ochreous or pinkish buff at sides; thighs warm buff; undertail-coverts buffy. Grey feather-bases show through, making underparts patchy. Primaries, primary coverts, secondaries and tertials blackish brown, very narrowly fringed grey; alula black-brown, leading edge of wrist cream-white. Underside of primaries and secondaries grey-brown, inner edges white; underwing-coverts pinkish buff. Bill black; eyes dark brown or red-brown; legs and feet plumbeous grey. Sexes alike. SIZE: (3 ♂♂, 8 ♀♀) wing, ♂ 61–63·5 (62·0), ♀ 60·5–64 (62·0); tail, ♂ 56·5–58 (57·2), ♀ 57–61 (58·4); bill to skull, ♂ 11·5–12 (11·7), ♀ 11·5–12·5 (11·9). WEIGHT: ♂ (n = 3, Ethiopia) 14–16 (14·7), (n = 1, Kenya) 14; ♀ (n = 8, Ethiopia) 14–18 (15·7), (n = 3, Kenya) 12–15 (13·3); unsexed (n = 2, Ethiopia) 14, 15·5, (n = 1, Kenya) 15.

IMMATURE: 1, still with yellow gape-flanges, like adult but upperparts with rufous wash; wing-coverts and flight feathers rufous-edged; rufous wash on flanks, lower belly and undertail-coverts. 1 (Kenya, Mt Elgon), also with head rufous.

NESTLING: unknown.

P. l. griseiventris Erard: Bale Mts at 3500–3700 m, Ethiopia. Upperparts darker than *lugens*, blackish sepia, crown and mantle concolorous; chin and throat darker than in *lugens*, breast and belly uniform pale grey, throat to belly indistinctly streaked grey-brown; no buff on undertail-coverts or sides of belly (Erard 1978). Larger than *lugens* but lighter in weight. SIZE: (3 ♂♂, 1 ♀) wing, ♂ 64–66 (65·3), ♀ 63·5; tail ♂ 61–64·5 (62·5), ♀ 62; bill from skull ♂ 12–12·5 (12·3), ♀ 12. WEIGHT: ♂ (n = 3) 13–14 (13·7), 1 ♀ 13.

P. l. jacksoni Sharpe: Sudan and Uganda (Didinga and Imatong Mts), Kenya, N Tanzania, Zaïre (Marungu Highlands), Malaŵi. Like *lugens* but mottling more extensive on throat; and belly, flanks, thighs and undertail-coverts grey-buff rather than warm pinkish buff. Larger: ♂ (n = 5, Mt Elgon) wing 64–67, tail 58–65, bill 10–11·5, tarsus 20–21·5; 1 ♀ (Mt Elgon) wing 64, tail 58, bill 11, tarsus 20·5.

P. l. clara Meise: Tanzania (Matengo Highlands). Like *jacksoni* but chin and throat duskier, much less obviously streaked, becoming clear grey on breast. Belly and undertail-coverts lack buff tinge.

P. l. prigoginei Schouteden: Zaïre (Itombwe). Like *jacksoni* but upperparts darker, with olive-brown wash; crown markedly darker than mantle; chin blackish, throat and belly whitish, breast contrastingly dark. Larger. SIZE: ♂♀ (n = 8) wing 66–72, tail 62–67.

Field Characters. A plain dark grey-brown warbler, like a *Sylvia* warbler in posture, characteristic of highland acacias; tail with white sides and whitish tip; throat faintly mottled and streaked; underparts brown-grey, belly whitish in centre, buffy at sides. Eyes dark; legs lead-grey. Chin blackish (Zaïre: Itombwe); cap blackish (Ethiopia: Bale); undertail-coverts buff (Ethiopia). Juvenile tinged with rufous. Appears dull, dark and featureless until one glimpses the white sides and tip of tail. *Hippolais* warblers are much paler brown above and whitish below.

Voice. Tape-recorded (86, B, HOR). A short song of *c.* 7 notes, duration *c.* 0·7 s, starting and stopping abruptly, the first 4 notes almost as 2 pairs, delivered rapidly, song ending with 3 more discrete notes: 'parisoma-tuk-tuk-tuk', or better 'p'r's'z-choo-too-took' with emphasis on *r* and *took* (Naivasha, Kenya, Mar).

General Habits. In Ethiopia inhabits light savanna woodland with shrubs and acacias; gallery forest with acacias and *Ficus*; and woody moorlands dominated by *Erica*, *Euphorbia* and *Hypericum*. In Kenya it occurs especially in *Acacia abyssinica* woodland; also wooded grassland, forest edges, watercourses and gardens. Altitude of the Itombwe, Zaïre, population suggests that it inhabits subalpine zone, with bamboos, heaths, *Helichrysum*, *Alchemilla* and *Senecio* (Prigogine 1975, Erard 1978). In Dedza, Malaŵi, confined to *Acacia woodii* woodland.

Forages in canopy of flat-topped acacia, and in heaths, much like a *Sylvia* warbler (Erard 1978).

Food. Insects, including caterpillars 10–20 mm long and aphids; also spiders, small seeds and berries.

Breeding Habits. Barely known.
NEST: light cup of moss and rootlets, lined with grass and bark-fibre. Sited on a stump or in thin fork of tree (Mackworth-Praed and Grant 1960).
EGGS: 2. Whitish olive or cream, speckled and spotted with fine reddish brown, liver and grey marks. SIZE: *c.* 18 × 14.
LAYING DATES: E Africa: Region D, Apr, May, Dec; Zaïre (Itombwe, enlarged gonads Apr); Malaŵi, Oct–Nov; Zambia (Nyika, enlarged gonads May).

Reference
Erard, C. (1978).

Parisoma boehmi Reichenow. Banded Parisoma. Parisome sanglée.

Parisoma böhmi Reichenow, 1882. J. Orn., p. 209; Seke, Ugogo, Tanganyika.

Plate 22
(Opp. p. 401)

Range and Status. Endemic resident, dry acacia country up to 1700 m in E Africa. Occurs from NW Somalia to central Tanzania, but range fragmented. Fairly common in NW Somalia and adjacently in Ethiopia. Occurs in S Ethiopia, NE Kenya (1 record), and central Kenya from Mt Nyiru and Marsabit to Isiolo and Meru. Main range is S Kenya and NE Tanzania, where quite common, from E shores of L. Victoria to Nairobi, Simba, Tsavo, Taru, south to Serengeti (uncommon), NE Tabora, Singida, Dodoma, Kibedya, N Iringa (uncommon), Usangu Flats and near Rukwa.

Description. *P. b. somalicum* Friedmann: Ethiopia, Somalia.
ADULT ♂: forehead to uppertail-coverts, sides of neck and scapulars brownish grey. Tail, graduated, black with white sides: T1 and T2 black with buffy tips 0·5 mm deep; T3 with white tip 1 mm deep; T4 with 2 mm white tip, deeper on outer web; T5, inner web with white tip, outer web with distal half white; T6, inner web with white tip 12 deep, outer web entirely white. T5 5 shorter and T6 9·5 shorter than T1. Lores and sometimes featherlets around base of upper mandible and front half of eye buffy white. Ear-coverts and cheeks grey, mottled, with brownish cast – general effect like crown, but mottled or even barred. Chin buffy white; throat greyish white, spotted and blotched with blackish brown, coalescing to form blackish band across upper breast (feathers brown-black with grey-buff tips). Breast-band sharply demarcated from silky white lower breast and belly; lower breast washed buff, grey towards sides; flanks and thighs warm buff, light orange, or rich rufous-orange; undertail-coverts deep rufous-orange. Primaries, secondaries and tertials blackish brown, the tertials with outer webs broadly fringed white, making conspicuous white line in closed wing; outer primaries narrowly edged greyish, proximal to point of emargination; greater coverts and median coverts blackish brown, broadly tipped white, forming 2 white bars in closed wing; primary coverts black-brown; alula feathers black-brown, with white outer edges. Undersides of primaries grey-brown, inner edges of webs whitish; axillaries pale orange, remaining underwing-coverts buffy white, with patch of dark grey at base of P10. Upper mandible black-brown or slate-grey, lower mandible pale horn or whitish with dark tip; eyes pale yellow or yellow-white; legs and feet grey or grey-brown, soles pale grey or whitish. Sexes alike. SIZE: (5 ♂♂, 5 ♀♀) wing, ♂ 59–62·5 (60·9), ♀ 58–63 (60·5); tail, ♂ 54–61 (58·1), ♀ 55–60 (57·2); bill, ♂ 13–14 (13·3), ♀ 12–14·3 (13·2); tarsus, ♂ 20–21 (20·7), ♀ 20–23 (21·3).

IMMATURE: like adult, but upperparts rather browner, rump fulvous, and uppertail-coverts rufescent dark brown (not grey); underparts lack throat speckles and blackish breast-band; feathers of chin, throat, breast and belly soft and fluffy, white, heavily dappled with grey caused by feather-bases showing through; flanks pale orange, undertail-coverts rich orange.

NESTLING: not described.

P. b. boehmi Reichenow: S Kenya, Tanzania. Upperparts slightly darker than *somalicum*.

P. b. marsabit van Someren: N-central Kenya. Flanks and undertail-coverts paler than *somalicum*.

Field Characters. Rather a plump bird, *Parus* tit-like in appearance and habits. In small flocks. Readily distinguished by combination of yellow eye, speckled throat, blackish breast-band, buff-rufous flanks and deep orange undertail-coverts. Tail black, narrowly white-sided; tertials conspicuously white-fringed.

Voice. Tape-recorded (10, B, GREG, HOR, LEM, McVIC, STJ). Cheerful, rollicking song a medley of piping notes, rattles and musical rolling trills blending into one another; often more than one bird takes part (North and McChesney 1964). Call, a squeaky 'chicky-wurrah-chick-wurr'.

General Habits. Inhabits woodland, wooded grassland and bushland, mainly of acacias, on arid and semi-arid plateaux with 250–1000 mm annual rainfall, between 500 and 1700 m. In Somalia, acacias bordering dry watercourses and pans. A sprightly bird, in parties of 3–4 and up to 7, constantly on the move, searching leaves and hopping about branches inside canopy of flat-topped acacia, or flying one after another between trees. Rather silent. Thought to feed 'on the yellow shoots of the mimosas and on seeds and small berries; it certainly does not hunt flies and insects in the habitual manner of the flycather' (Archer and Godman 1961).

Mainly resident. Wanders to higher latitudes in Kajiado distinct, Kenya.

Food. See above.

Breeding Habits.

NEST: small neat cup of grass and rootlets, lined with grass heads. Sited in thornbush, c. 1 m above ground.

EGGS: 2–3. Ivory white, freckled and blotched with light olive-brown, with irregular band of violet-grey marks around broad end. SIZE: 16·5 × 14, and about 17 × 12·5.

LAYING DATES: Ethiopia (possibly Nov); Somalia (juvs Jan, June); E Africa: Region C, Jan; Region D, Jan, Mar, Apr.

Genus *Hyliota* Swainson

Small birds of woodland canopy with rather slender and flat bill, exposed nostrils, rather long tarsus and pointed wings. Forage like a warbler (e.g. *Eremomela*) but have flycatcher-like plumage with platysteirid pattern. Nestling has mouth spots; juvenile plumage not spotted; adult ♂ plumage matt or glossy blue-black above, creamy to rusty below, with white panel in wing; thighs black; rump feathers rather fluffy with white spots at base. Sexes dissimilar except in *usambarae*. Travel in family parties and may prove to be co-operative breeders; often in mixed-species foraging flocks. Nest a neat, lichen-decorated cup, rather like *Batis* nest.

Status uncertain. Hyliotas have been treated as warblers (Traylor 1970b, Traylor *in* Watson *et al.* 1986, Dowsett and Dowsett-Lemaire 1980, 1993b, Dowsett and Forbes-Watson 1993), muscicapid or monarch flycatchers, and platysteirids (Erard 1987). Mouth spot, unspotted juveniles and foraging behaviour are warbler characters. Adult plumage pattern, colour and quality are very like some *Ficedula* flycatchers, but even more like Platysteiridae. Nidification is platysteirid. *Hyliota* might even turn out to be a diminutive malaconotid shrike: DNA studies would be most revealing.

Endemic; 4 species (or 3 if *usambarae* and *australis* prove to be conspecific).

Plate 20
(Opp. p. 337)

Hyliota flavigaster Swainson. Yellow-bellied Hyliota; Yellow-breasted Hyliota. Hyliote à ventre jaune.

Hyliota flavigaster Swainson, 1937. Birds W. Afr., 2, p. 47; Senegal.

Range and Status. Endemic resident, tropics. Widely but rather sparsely distributed in mature and little-disturbed miombo woodlands (*Isoberlinia* in N tropics, *Brachystegia* in S tropics) from Senegal to Ethiopia and south to Angola, Zambia and S Mozambique. Not very common (or easily overlooked in its habitat of the canopy of tall trees), and these woodlands are being increasingly fragmented, making distribution difficult to

Hyliota flavigaster

plot. Gambia and S Senegal, moderately common. From there, south through Guinea to N Sierra Leone, and east in moist guinean wooded savannas in band 5° wide but nowhere extending to Atlantic coast, to Bamingui-Bangoran Nat. Park, Central African Republic. Also in Mbomou district, E Central African Republic. Widespread in SW Sudan, fairly common southwest of Gogrial-Yei-Imatongs and around Boma. Uncommon in W Ethiopia, single record in NW. In E Africa uncommon, widespread in Uganda except W and NE; in Kenya confined to W; in Tanzania from Ufipa to Gombe Stream Game Reserve, Kibondo and Biharamulo, and from Rufiji R. to Songea. Burundi, Kigamba area only. Zaïre: in NE, W, probably SW, and widespread and moderately common in SE but evidently absent from Mweru basin (and on Zambian side). Angola, central plateau from Zaïre border in NE through Huambo and Malanje to N Huila. Zambia, throughout, except for Mweru, Luangwa valley, Middle Zambezi valley, and Senanga and Sesheke Districts in SW. Absent from Zimbabwe. Malaŵi, not uncommon in miombo in wooded hills in W, reaching W lake shore between Bana and Nkhotakota, and from Namwera south to N Blantyre District. Mozambique, as mapped but probably more widespread north (and south?) of Zambezi; in S sight record near Vila Paiva de Andrada and 3 specimens from Macia ('confined to Mozambique littoral': Maclean 1993). Namibia, E Caprivi (Maclean 1993).

Yellow-bellied Hyliota was common around Enugu, SE Nigeria, in early 1950s (Serle 1957) but was not found there 20 years later (Cowper 1977), probably due to destruction of mature woodland.

Description. *H. f. flavigaster* Swainson: N tropics, south to L. Victoria. ADULT ♂: upperparts from forehead to tail uniform glossy blue-black; lores matt black. Underparts rather uniform creamy pale orange or rich orange-buff, darkest on centre and sides of breast, palest on undertail-coverts. Primaries, secondaries, and tertials glossy blue-black; greater primary-coverts glossy blue-black, median primary-coverts white, greater and median coverts entirely white, lesser coverts bluish black. Underside of flight feathers silvery dark grey; underwing-coverts and axillaries white. Upper mandible black or blackish, basal two-thirds of lower mandible pale bluish grey or pale flesh-horn; inside mouth yellow or flesh; eyes dark sepia-brown; legs and feet bluish slate or dark steely grey. ADULT ♀: like ♂, but forehead to mantle, lores, cheeks and ear-coverts dark grey-brown, slightly glossed blue on crown; mantle, scapulars and back dark grey-brown, heavily glossed blue; rump grey with white feather-bases showing through; uppertail-coverts glossy blue-black; tail brownish black. Underparts like ♂, but a less distinct division between black cap and buffy throat. SIZE: (5 ♂♂, 5 ♀♀) wing, ♂ 68·5–76 (70·1), ♀ 64–68·5 (66·0); tail, ♂ 43–45·5 (43·8), ♀ 38·5–42·5 (40·7); bill to skull, ♂ 16–17 (16·3), ♀ 15–17 (16·1); tarsus, ♂ 16–18 (16·2), ♀ 16·5–17·5 (17·0). WEIGHT: (Central African Republic) 2 ♂♂ 13·0, 15·0, 1 ♀ 12·5; (Angola) 1 unsexed 11·7; (Ghana) 1 ♀ 12·2, 1 juv. 11·4.

IMMATURE: like adult ♀, but upperparts without any gloss, feathers grey-brown (cap) or dark brown (mantle, back) with narrow buff fringes: ear-coverts and nape look buffy rather than blackish. Tail black. White in wings like adult; primaries, secondaries and tertials narrowly edged and tipped with whitish. Underparts creamy white with buffy wash, greyish feather bases showing through patchily.

NESTLING: black spots on tongue. Late nestling (*H. f. barbozae*) like immature, but white in wing has buff wash. No distinct demarcation between brown head and buffy throat. Tail projects only 2 mm beyond wing tips.

H. f. barbozae Hartlaub (including '*marginalis*'): S tropics, north to L. Victoria. ♂ like ♂ *flavigaster* but breast rather more orange, and outer vanes of innermost secondaries with long white wedges up to 4 mm wide. Most ♀♀ dark grey where ♂ is glossy blue-black, but some are ♂-plumaged, although black in upperparts sootier, feathers duller and only slightly glossy, not violaceous. SIZE: (Zambia): ♂ (n = 10) wing 70·5–78 (73·4), tail 45–52·5 (48·2), ♀ (n = 10) wing 68–74·5 (71·6), tail 42–49 (46·7). *H. f. barbozae* is larger than nominate subspecies; but Mozambique population ('*marginalis*') is smaller than nominate: ♂ (n = 9) wing 65·5–70·5 (68·6), tail 42–47 (44·7). WEIGHT: (NE Angola) ♂ (n = 8) 11·5–14·0 (12·7), ♀ (n = 3) 11·5–13·0 (12·1), 1 juv. 12; (Mozambique) 1 unsexed 11·7.

Field Characters. Length 12·5 cm. A bird of miombo savanna woodland tree tops, usually seen in family parties and mixed-species foraging flocks; restlessly gleans leaves, like eremomela. Weak voice. Distinctive appearance: ♂ glossy black above with large white patch or stripe in wing, and cream-yellow throat and breast; ♀ like ♂ but dark grey or dark brown where ♂ is black; juvenile like ♀ but paler throughout. White wing-panel suggests *Ficedula* flycatcher, but flycatchers have no yellow, and perch upright – hyliotas keep body horizontal. Very like Southern Hyliota *H. australis*, and their ranges overlap in Angola, Zambia, Malaŵi and S Zaïre; Southern told by paler and yellower underparts and white in wing-panel on greater coverts only; southern (not northern) race of Yellow-bellied has longer white wing-stripe formed by greater coverts and white edges to tertials. ♂ Yellow-bellied has glossy, not dull, upperparts; ♀ darker and greyer, less brown, than ♀ Southern.

Voice. Tape-recorded (86, 91, B, F, CART, ERA). Song a rapid tumble of notes, 'tsweet, sweeti, sweetiswi' lasting *c*. 1 s, babbled and tuneless but not unpleasant to the ear. Call, perhaps foraging contact call, an irregularly-repeated monosyllable, nearly a bisyllable, 'p'lic', of quality of flight-call of Eurasian Nuthatch *Sitta europaea*: 'plic-plic, splic plic . . .' with about 8 notes in 3 s; sometimes interspersed with it (and given separately?), a tiny sound 'tsi'. Contact call also described as a slight, bisyllabic whistle.

General Habits. Inhabits canopy of middle-sized and tall trees in *Isoberlinia*, *Brachystegia* and (Zambia) *Cryptosepalum* woodland. In pairs, often in family parties, and spends much time in mixed-species foraging flocks, which may even include Southern Hyliota. In Ghana often occurs with Senegal Batis *Batis senegalensis*. Members of family party keep in close touch.

Systematically explores upper foliage and bark, gleaning them for insects; sometimes hangs momentarily upside-down and examines undersides of branches. Occasionally snatches insect in flight. Restless; flock moves ever onward through canopy – 'one has to walk quite quickly in order to keep in touch with them, although the birds are feeding all the time' (Vincent 1935).

Food. Insects.

Breeding Habits. Little known. Solitary, territorial breeder, evidently monogamous.

NEST: small, neat, open cup made of plant stems, moss and lichens; somewhat resembles batis nest. Sited in tree 'at almost any height above ground' (Maclean 1993).

EGGS: 2. White, dull, with brown and lilac spots and squiggles around broad end (Maclean 1993). SIZE: *c*. 17 × 13.

LAYING DATES: Guinea (dependent young, Oct); Ghana (large gonads, May); Nigeria (juveniles in family parties, July–Sept); E Africa: Region B, Dec (1 record); Angola (enlarged gonads Sept–Nov); Zambia and Malaŵi, Oct–Dec (9 records, mainly Oct) (gonads active Sept).

Reference
Lawson, W. J. (1964).

Plate 20
(Opp. p. 337)

Hyliota australis Shelley. Southern Hyliota; Mashona Hyliota. Hyliote australe.

Hyliota australis Shelley, 1882. Ibis, p. 258; Umvuli River, Southern Rhodesia.

Range and Status. Endemic resident, southern tropics, with 2 outlying populations north of Equator, in Zaïre, Uganda and Kenya; also Cameroon (see below). Zaïre, sparse, Semliki and from west of L. Albert to west of L. Edward (also across Uganda border, in Bwamba Forest); widespread in S Shaba. Kenya, scarce, in Kakamega Forest and Nandi Forest only. Main range from Angola to Mozambique. Angola, poorly known: local in N Moxico, central Huila (Cului R.), Cuanza Sul (Gabela), Cacolo, and Luanda Province. Zambia, sparse and local, confined to miombo woodland; widespread on the Copperbelt and in Mumbwa and somewhat commoner than elsewhere in Southern Province; in NW occurs south to Mayau, Kabompo and Balovale, with 1 sight record west of Zambezi R.; in SW (Barotse Province) known only from Mongu and Senanga; absent from E Zambia as mapped, except for Danger Hill, Mpika, Chimpili Plateau, upper Lubansenshi R., Kawambwa, and in S only on Chiromwe Hill (Feira). Malaŵi, quite common in extreme S of Lilongwe and Dedza Districts and extreme W of Mwanza District (along Mozambique border), uncommon on Mpingwe Hill, Blantyre; 1 old record in N Zimbabwe, not uncommon on Mashonaland plateau, from edge of Zambezi Escarpment to 21°S in upper Nuanetsi valley and to E highlands where ranges locally from Inyanga to Chimanimani Mts, on Makurupini R. and in Chipinga Uplands; 5 other populations probably isolated from main range: Kapsuku Hills (Kenyemba) and Naodsa R. (Kariba basin) (both overspills of Zambian populations), Matopos Hills in SW, in Kalahari Sand miombo woodland from Charama Plateau to Chirisa, Chizarira Nat. Park, Kana R. and Nkai, and in SE at Sabi R.–Lundi R. confluence, Malilangwe Hills and Gonarezhou Nat. Park; ranges down to 350 m and may be only seasonal visitor to lowveld (Irwin 1981). Mozambique, records from 10 localities as mapped, but fairly common in miombo and so probably more widespread in fact. Sight record and nest-building reported NW Kruger Nat. Park, South Africa, where there may be a very small resident population (Hockey 1995).

Description. *H. a slatini* Sassi: NE Zaïre, W Uganda, W Kenya. ADULT ♂: forehead to back velvety black, matt or only faintly glossy; rump greyish: feathers of lower back and rump dense and fluffy, black-tipped, silky whitish or pale grey subterminally which often shows through as whitish or greyish patches; uppertail-coverts glossy black. Tail glossy black, T1–T4 with very narrow whitish edge to outer web, T5 with 0.5 mm white edge, and T6 with all outer web white and indistinct whitish spot near tip of inner web. Underparts very pale yellow-orange or creamy buff, feathers white, soft, rather fluffy and silky, with dark grey bases and cream-orange tips; rather uniform, but breast and sides of throat tend to be a stronger colour and belly and undertail-coverts weakest – the last sometimes greyish white. Inner thigh white, outer thigh blackish. Primaries, secondaries and tertials black, slightly glossy; bases of secondaries white below the greater coverts; greater and median coverts white, all other upperwing-coverts black. Underside of flight feathers silvery dark grey, underwing-coverts and axillaries white; underside of tail black, less glossy than upperside. Upper mandible black, lower mandible black with blue-grey base; eyes dark brown; legs and feet slategrey, slender and rather weak. ADULT ♀: forehead to uppertail-coverts velvety dark brown or very dark brown sometimes with faint rufescent or greyish tinge; lores, cheeks and ear-coverts with a few paler brown feathers. Tail glossy black, T1–T4 with very narrow whitish edge to outer webs, T5 with 0.5 mm white edge and T6 with all outer web white and indistinct whitish spot near tip of inner web. Underparts like ♂ but slightly paler, sometimes with greyish tinge. Wings brownish black, all flight feathers, tertials and greater primary coverts very narrowly edged and tipped with pale buff, secondaries with concealed white bases like ♂, greater and median coverts white, all other upperwing-coverts black. SIZE: (5 ♂♂, 5 ♀♀) wing, ♂ 67.5–72 (69.1), ♀ 61–69 (66.3); tail, ♂ 43–46 (44.3), ♀ 36–44 (40.0); bill to skull, ♂ 12·0–13.3 (12.8), ♀ 13·0–14.4 (13.9); tarsus, ♂ 19–20 (19.4), ♀ 18–20 (18.8). WEIGHT: (*H. a. australis*, NE Angola) ♂ (n = 4) 9·5–11·7 (10·4), ♀ (n = 5) 10·0–12.1 (10.7), 1 juv. 9.

IMMATURE: like adult ♀ but upperparts grey-brown, without gloss, feathers with narrow buff fringes; primaries, secondaries and tertials narrowly edged and tipped with whitish or pale buff; underparts creamy white with buffy wash.

NESTLING: black spots on tongue.

H. a. australis Shelley (including '*rhodesiae*'): Zimbabwe, Mozambique. ♂ like ♂ *slatini* but rump greyer and may even look whitish in life, and tail with white sides: T5 with white patch 10 mm long and 1 mm wide on outer web, 18 mm from tip, and T6 with outer web white except for distal 14 mm (which is black) and inner web narrowly edged white. ♀ like ♀ *slatini*. Larger. SIZE: ♂ (n = 6) wing 72.5–76.5 (74.1), tail 47–50.5 (48.6); ♀ (n = 4) wing 70.5–74 (71.5), tail 44–49 (47.4).

H. a. inornata Vincent (includes '*pallidipectus*'): Angola, Shaba, Zambia, Malawi and NW Mozambique. Intergrades with *slatini*. ♂ blacker above, slightly less buff-orange below than nominate race. Much white in tail: T3 with narrow white outer edge, T4 with outer web mainly white except near tip and inner edge white near tip, T6 with both webs white, except for tip of outer web and base and tip of inner web. ♀ like ♀ *slatini*, but T3 with narrow white outer edge, T4 with broader white line along outer edge and around tip, and T6 white, base of inner web black, and small dusky mark near tip of both webs; tertials and greater primary coverts broadly edged white. WEIGHT: (2 ♂♂, Mozambique) 10.5, 10.8.

H. a. ? slatini: a ♀-plumaged bird was collected at 1385 m at Dikume Balue (04° 55′N, 09° 15′E) in W Cameroon, and could not be sexed (Serle 1965). Although over 2000 km from the known range of *slatini* it is exactly like a ♀ *slatini*, right down to the white patterns in T5 and T6. Wing 65, tail 40, bill 11, tarsus 16.

Field Characters. Length 11.5–13 cm. A bird of miombo savanna woodland tree tops, usually seen in family parties and mixed-species foraging flocks; restlessly gleans leaves, like eremomela. Weak voice. Distinctive appearance: ♂ matt black above with large white panel in wing, and pale creamy yellow-buff below; ♀ like ♂ but dark brown where ♂ is black; juvenile like ♀ but paler throughout. Somewhat resembles *Ficedula* flycatcher but flycatchers have no yellow, and perch upright – hyliotas keep body horizontal. Very like Yellow-bellied Hyliota *H. flavigaster* and their ranges overlap in Angola, Zambia, Malawi and S Zaïre; Southern told by paler and yellower underparts and white in wing a panel on greater coverts only; southern race of Yellow-bellied has long white wing-stripe. ♂ Southern has matt, not glossy, upperparts, and ♀ is browner than ♀ Yellow-bellied.

Voice. Tape-recorded (86, 91, B, F). Song a sustained, tuneless babble in a monotone, with irregular short pauses (< 0.5 s) and barely-separable 'syllables': 'tswi-t-swiswi-s-st-w-brr-st . . .'. Can last up to 15 s. Contact call a twitter; also utters a thin, sharp, high-pitched 'tsi' (3–6 kHz, duration 0.1 s or less), like that of Yellow-bellied Hyliota. Song also described as pleasant, stuttered, chippering whistles followed by sweet trilling warble (Vincent 1935) and as repeated squeaking whistles followed by warbling trill (Maclean 1993).

General Habits. Inhabits canopy of *Brachystegia* and other miombo woodland trees; commonest in thick undisturbed woodland; in Zimbabwe also associates with Mountain 'Acacia' *Brachystegia glaucescens* and to some extent with riparian *A. albida* and riparian fringing forest. Occurs alongside ecologically very similar Yellow-bellied Hyliota in several localities; at Mayau, Zambia, Southern Hyliota keeps to *Cryptosepalum* forest (mavunda) and Yellow-bellied to *Brachystegia* (miombo). Habits like those of the more familiar eremomelas *Eremomela* spp.: forages in tops of large savanna trees (hence easy to overlook) by gleaning for insects; singly or in pairs or family parties, which in winter almost invariably keep in mixed-species bird flock, for instance with Long-billed Crombec *Sylvietta rufescens*, Green-capped Eremomela *E. scotops*, Spotted Creeper *Salpornis spilonotus*, Brubru *Nilaus afer*, weavers and small woodpeckers. Foraging habits of Yellow-bellied Hyliota (q.v.) 'equally applicable to Southern Hyliota' (Vincent 1935). Young stay with parents until next year's breeding season; family party maintains permanent territory; party joins mixed-bird flock (or probably acts as nucleus for flock) at daybreak; flock moves slowly through woodland with little interaction until it reaches boundary of hyliotas' territory, where excited squabbling, chasing and calling occurs between adjacent hyliota families as mixed-bird flock moves into new hyliota territory (C. J. Vernon *in* Ginn *et al.* 1989).

Mainly resident; but likely to enter Zimbabwe lowveld (Kariba basin; Sabi–Lundi area) only for the winter months (Irwin 1981).

Food. Insect fragments in stomachs.

Breeding Habits. Solitary breeder, strongly territorial; evidently monogamous but may prove to be co-operative breeder. Rival ♂♂ interact at territorial boundary, facing each other in silence, body held horizontally, head feathers ruffled, tail cocked, bird bowing all the time; when joined by ♀, ♂♂ displayed more vigorously and both flitted about ♀, singing (C. J. Vernon *in* Ginn *et al.* 1989). ♂♂ can pursue each other through topmost foliage of trees, with tails semi-erect and fanned, and wings half-opened with drooping tips, exactly like White-tailed Blue Flycatcher *Elminia albicauda* (Vincent 1935).

NEST: neat, round, deep, thick-walled cup made of somewhat loosely-woven fine rootlet-like plant material, with pieces of whitish lichen up to 2 cm long worked into outer few mm of wall and adhering to exterior of nest. Nest like *Batis* nest but not quite so compact, and not bound with spider web. Sited on stout lichen-covered fork in topmost branches of tree, 6–7.5 m above ground.

EGGS: 2–4, usually 3. Cream or pinkish white, freckled with minute spots of brown and grey, concentrated in wreath around broad end. SIZE: (n = 10) 15·0–18·9 × 11.5–14.6 (17.2 × 13.2).

LAYING DATES: Zambia, July–Aug, Oct (and gonads active, Sept); Malaŵi, Oct; Zimbabwe, Aug–Jan (Aug 1, Sept 22, Oct 9, Nov 9, Dec 1, Jan 1 clutches).

DEVELOPMENT AND CARE OF YOUNG: nestling period details unknown; young stay with parents in territory nearly all year.

References
Ginn, P. J. *et al.* (1989).
Lawson, W. J. (1964).
Vincent, J. (1935).

Hyliota usambarae Sclater. Usambara Hyliota. Hyliote des Usambaras.

Plate 20 (Opp. p. 337)

Hyliota australis usambarae Sclater, 1932. Bull. Br. Orn. Club 52, p. 104; Amani, Tanganyika.

Range and Status. Endemic resident, NE Tanzania from Ruvu R. to E and W Usambara Mts. Known from 3 specimens at Amani, NE Tanzania; pairs seen at 300 m (Kihuhwi-Sigi Forest Reserve, E Usambara) in Feb 1980 and Mar 1981 and at 1000 m (Dindira, W Usambara) in May 1981; a 1920s record from Rubu R. near Dar es Salaam (S. N. Stuart, pers. comm.); searched for in Usambara in early 1990s by A. Tye and D. A. Turner, but not found; however, 1 at Manga and pairs commonly seen at Kisiwani (about 400 m, E Usambara) in Aug 1994 to Feb 1995 (T. Evans, pers. comm.).

Description. ADULT ♂: all upperparts from forehead to uppertail-coverts and tail glossy blue-black; underside of tail black, less glossy than upperside; feathers of lower back and rump dense and fluffy, black tipped, silky whitish or pale grey subterminally, which can show through as greyish patches. Chin and throat orange, breast rich rufous-orange, rest of underparts like throat, feathers white, soft, rather fluffy and silky, with dark grey bases and rufous tips. All of thighs black. Tertials glossy black, broadly edged white; primaries and secondaries black, narrowly edged white; bases of secondaries white below the greater coverts; greater and median coverts white, all other upperwing-coverts black. Underside of flight feathers silvery dark grey, underwing-coverts and axillaries white. Upper mandible black, lower mandible slate-blue with dark brown tip; eyes black; legs and feet slate-grey. Sexes

Hyliota usambarae

alike. SIZE: 1 ♂, wing 64, tail 39, bill to feathers 10, bill to skull 14, tarsus 17·7; 1 ♀ wing 64, tail 35, bill to feathers 10, bill to skull 13·7, tarsus 17. WEIGHT: ♂ (n = 3) 10·2–11·2 (10·7); 1 ♀ 9·5.

TAXONOMIC NOTE: we raise this taxon, originally described as a race of *H. australis*, to the rank of full species on morphological evidence that is admittedly slender. In so doing we bring to attention a population that is rare and possibly endangered. The 3 adult skins of *usambarae* examined by us are alike, but labelled as 2 ♂♂ and 1 ♀, and at least 10 birds seen often in pairs were black and none brown (T. Evans, pers. comm.). Thus this small, forest hyliota appears to be monomorphic, with ♂-plumaged ♀. The other hyliotas are dimorphic.

Field Characters. Length 10 cm. A glossy blue-black warbler-like bird of forest canopy with bright orange underparts, particularly the breast. The only hyliota in its range. Both sexes very similar to ♂ Southern Hyliota *H. australis*, but smaller and with all-black thighs.

Voice. Not tape-recorded. A thin squeak 'mbambasana' (Sclater and Moreau 1933).

General Habits. Inhabits forest canopy and coffee plantations, at 300–1000 m. In E Usambara, pair seen twice from large rock that overlooked forest canopy (S. N. Stuart, pers. comm.), and at Kisiwara pairs forage conspicuously in crowns of tallest trees, >15–20 m, often trees bare of leaves, birds keeping to small branches and twigs, and seldom calling (T. Evans, pers. comm.). No evidence of breeding in Mar or Sept (Sclater and Moreau 1933).

Food. Insects.

Breeding Habits. Unknown.

References
Lawson, W. J. (1964).
Sclater, W. L. and Moreau, R. E. (1933).

Hyliota violacea Verreaux. Violet-backed Hyliota. Hyliote à dos violet.

Plate 20
(Opp. p. 337)

Hyliota violacea Verreaux, 1851. Rev. Mag. Zool., p. 308; Gabon.

Range and Status. Endemic, resident. Lowland forest of E Guinea (frequent), Liberia (common, Mt Nimba), Ivory Coast (rare to frequent, Taï Forest to Mt Nimba, Gagnoa, Lamto and Yapo Forest), Ghana (frequent or common, Central and Kwahu regions, reserves inland from Takoradi and Cape Coast, also Accra and Prashu), Togo (Misahohe), Nigeria (rare or uncommon, Erin-Ijesha, Gonesesso, Bashu Okpambe), Cameroon (north to Korup Nat. Park, Mt Kupé, Dikome Balue, Rumpi hills, Yaoundé, Efulen, Bitye, Dja R., also Ngaoundéré), Rio Muni, Gabon (rare or uncommon, Woleu N'Tem and Ogooué-Ivindo, southward including coast), Congo (Mayombe and Kouilou Basin), Central African Republic (reports from Gounda-St Floris in soudanian–sahelian zone, and Lobaye require confirmation), Zaïre (Mayombe, Salonga Nat. Park, and Tshuapa, Manyema and Itombwe, Kivu). Rare to frequent.

Description. *H. v. violacea* Verreaux: Nigeria and Cameroon to Zaïre. ADULT ♂: upperparts, lores, ear-coverts, sides of neck, scapulars, wings, uppertail-coverts and tail black glossed with deep purple (bluer on flight feathers and tail-feathers); inner greater upperwing-coverts white; rump feathers whitish grey with broad glossy purple-black tips. Entire underparts white with uniform pale buff wash; flanks greyish; thighs black. Underwing-coverts and axillaries glossy purple-black; underside of flight feathers and tail dull black. Bill black, base of lower mandible bluish; eyes brown; legs and feet black, soles greenish yellow. ADULT ♀: like ♂ but inner median upperwing-coverts black with deep purple gloss or with concealed white basal spot; underparts rich rusty or ferruginous orange, lower belly much paler, undertail-coverts white. SIZE: (6 ♂♂, 5 ♀♀) wing, ♂ 77–78·5 (77·8), ♀ 72–77·5 (74·1); tail, ♂ 41–44 (42·7), ♀ 40–42 (40·9); bill, ♂ 15–16 (15·4), ♀ 14·5–16 (15·3); tarsus, ♂ 18–19 (18·4), ♀ 17·5–19 (18·3). WEIGHT: (Liberia) ♂ (n = 4) 14–17 (15·7), ♀ (n = 5) 15·8 ± 1·1.

IMMATURE: like adult but upperparts duller, more sooty black, less glossy purple-blue; underparts of ♂ irregularly mottled rusty orange, particularly at throat and breast; bill greyer at base of lower mandible.

JUVENILE: like immature ♀ but upperparts, wing and tail-feathers dull sooty brown, almost without purple gloss; underparts duller and browner, less orange.

NESTLING: unknown.

H. v. nehrkorni Hartlaub: Liberia to Ghana and Togo. Inner median upperwing-coverts of ♂ glossy black, not white.

Field Characters. Active foliage-gleaner with stocky body, thin bill, long legs and short tail. Upperparts and sides of head black (purple gloss seen only in good light); underparts pale buff (♂) or rusty orange (♀). Forest habitat helps separate it from Yellow-bellied Hyliota *H. flavigaster*, which is a woodland bird. Yellow-bellied Hyliota has a broader and more extensive white wing-bar (all median coverts), and underparts in both sexes are orange-buff, intermediate between ♂ and ♀ Violet-backed Hyliota (deeper than ♂, paler than ♀); ♀ Yellow-bellied Hyliota has grey upperparts.

Voice. No known tape-recording. Song unknown. Contact call, metallic 'pic-pic' or 'psi-psi'; gives dry 'tic-tic-tic-tic-tic' or rapid 'tsec-tsec-tsec-tsec' on take-off.

General Habits. Inhabits mature and secondary lowland forest, also woodland; occurs along roadsides and is especially fond of old plantations and cultivated clearings such as cocoa farms, e.g. in Ghana; up to 1100 m in Itombwe, Zaïre, 2000 m in Rwanda. Near Ngaoundéré, Cameroon, a pair once seen in clump of trees in grassland.

In pairs or small family groups of 3–4 individuals (up to 6 in Itombwe, Zaïre). Lives in canopy, but also recorded in middle storey in Ghana. Usually gleans foliage but often seen on dead or temporarily leafless trees where it searches bark; often in or around flowering or fruiting trees in Ivory Coast. Rather active; hops along branches and from twig to twig; moves quickly, gleaning and snatching prey from bark or under leaves; sometimes hovers. Frequently joins mixed-species flocks; in Gabon associates particularly with foliage gleaners but also with woodpeckers and other bark gleaners (C. Erard, pers. obs.; P. Christy, pers. comm.).

Food. Insects, including beetles, grasshoppers, moths and caterpillars; also spiders.

Breeding Habits. Monogamous, probably territorial but behaviour undocumented. Nest and eggs unknown. In Gabon, ♀ observed collecting lichens (P. Christy, pers. comm.).

BREEDING INDICATIONS: Gabon (♀ building nest Nov); Zaïre (♂ ♂ with enlarged testes Mar).

DEVELOPMENT AND CARE OF YOUNG: young stay with adults for long periods; in NE Gabon, 1 or 2 fledged young observed with parents for at least 3 months.

References
Brosset, A. and Erard, C. (1986).
Erard, C. (1987).

Genus *Hylia* Cassin

Bill shorter than head, fairly strong but slender and narrow with slightly decurved culmen; nostrils slit-like with well-developed nasal operculum, rictal bristles minute; tongue with frayed tip, hyoid apparatus with broad, flattened ends as in sunbirds Nectariniidae but not so long; wing fairly long but rounded, outer primary about half the length of second, and P5 and P6 the longest; tail not graduated; plumage firm. Eggs white, unmarked; nest globular, with side-top entrance, not suspended, i.e. not like a sunbird's nest.

Relationships uncertain. At times it has been placed in warblers Sylviidae, tits Paridae and weavers Ploceidae. Often considered a sunbird because of its peculiar tongue; it was placed by Bates (1930) in its own family, Hyliidae, next to sunbirds. Chapin (1953a) considered it to have little in common with sunbirds except the tongue, and returned it to Sylviidae, although he admitted that its firm plumage, rather strong bill and well-developed nasal operculum were rather exceptional in warblers. In foraging behaviour it has resemblances to both sunbirds and warblers; its voice is unique, unlike either sunbirds or warblers; the nest is not sunbird-like. We follow Hall and Moreau (1970) in keeping it *incertae sedis*, and tentatively place it at the end of the Sylviidae.

Endemic; monotypic.

Plate 20
(Opp. p. 337)

Hylia prasina (Cassin). Green Hylia. Hylia verte.

Sylvia prasina Cassin, 1855. Proc. Acad. Nat. Sci. Philadelphia 7, p. 325; Moonda (= Mondah) River, Western Africa (= Gabon).

Range and Status. Endemic resident. W Gambia south of R. Gambia in coastal remnant forest, once inland near Brumen Bridge (Wacher 1993); SW Senegal (Casamance), Guinea-Bissau, Guinea, extreme SW Mali (Falea, Sagabari, Kangaba, uncommon); Sierra Leone, Liberia (abundant), Ivory Coast (almost

Hylia prasina

throughout, abundant), S half of Ghana (common and widespread), S Togo. Throughout S Nigeria (common); Bioko; widespread and common in lowlands of S Cameroon, north in highlands to Bamenda; SW and SE Central African Republic; Gabon, Congo, Angola in Cabinda and locally in W from Cuanza Norte to Cuanza Sul, also N Lunda. Zaïre, common to very common throughout forest zone and out into forest/savanna mosaic, south in E to Itombwe; extreme S Sudan (fairly common); widespread and common in Uganda, north to Bwamba and Budongo, east to Mabira, south to Bukoba, NW Tanzania; Mt Elgon (Kenya/Uganda) and W Kenya in Kakamega and Nandi forests. Density in Gabon, 15–18 pairs/km^2.

Description. *H. p. prasina* (Cassin): range of species except Bioko. ADULT ♂ (breeding): top of head dark olive-green becoming blackish toward base of bill; rest of upperparts, including scapulars, upperwing-coverts and tail, olive-green, brighter than head; tail-feathers with very narrow pale tips (variable), shafts dark brown above, whitish below. Long, bright yellow-white stripe across lores and over and behind eye, below which is black line from base of bill through eye and broadening behind it; cheeks and ear-coverts bright olive-yellow. Chin and upper throat buffy white grading to light grey on rest of underparts, flanks and undertail-coverts washed olive. Flight feathers dark brown edged olive-green, axillaries and underwing-coverts white, bend of wing pale yellow, underside of flight feathers pale grey on inner webs. Bill black, tip sometimes paler, some birds with gape and basal half of tomium of lower mandible yellow; eyes brown to dark brown; legs and feet brownish olive to olive-green. ADULT ♂ (non-breeding): like breeding ♂ but yellowish tone of supercilium, cheeks and ear-coverts much less bright. Sexes alike but ♀ smaller. SIZE: (5 ♂♂, 4 ♀♀) wing, ♂ 61–66 (64·2), ♀ 55–59 (57·4); tail, ♂ 40–48 (45·2), ♀ 39–43 (40·5); bill, ♂ 13–14 (13·7), ♀ 12·5–13 (12·9); tarsus, ♂ 18·5–20 (19·1), ♀ 18–19 (18·8). WEIGHT: Liberia, ♂ (n = 10) 12·7 ± 1·4, ♀ (n = 11) 10·5 ± 1·3; Uganda ♂ (n = 48) 10–16·5 (14·1), ♀ (n = 30) 10–15 (12·1). Diurnal weight changes (n = 2) of 13·7% and 14·7% recorded Kenya (Mann 1985).

IMMATURE: very young birds have top and sides of head and nape dark grey, crown and nape washed olive, supercilium continued along sides of occiput pale grey, throat washed pale grey; bill, legs and feet yellow. At a later stage, crown becomes olive, brighter than adult, supercilium pale greenish white, black eye-stripe extends to part of cheeks and ear-coverts; underparts darker and more olive than adult. 1st year birds retain yellow bill and feet for some time after moulting into adult plumage.

NESTLING: unknown.

H. p. poensis Alexander: Bioko. Same size as mainland birds but throat purer white; not a well-marked race.

Field Characters. A common forest bird that stays out of sight most of the time in dense foliage; the first sign of its presence is usually a loud, sharp 'peee-peee' from the undergrowth. Plumage of subdued forest colours (green and grey) is broken only by long whitish supercilium and black eye-line. Other dingy forest warblers (*Sylvietta, Camaroptera, Macrosphenus*) lack white supercilium except for White-browed Crombec *Sylvietta leucophrys*, which has a rusty cap and very short tail, and is often seen in the open. In montane forests of Albertine Rift might be confused with the rare Neumann's Warbler *Hemitesia neumanni*; for differences see that species.

Voice. Tape-recorded (C, F, CHA, ERA, GRI, KEI, LEM, MOR). Call, a loud, high-pitched double note, 'peee-peee'; pure, and not varying in pitch; not repeated; given at irregular intervals. This may constitute the 'song', since no other song has been heard from this common species, or it may be for contact. Brief purring contact call noted in Gabon (Brosset and Erard 1986). Alarm, a brief, hard dry chatter, not repeated and not connected to the 'peee-peee' call.

General Habits. Inhabits primary and secondary lowland, transitional and lower montane forest, up to 2100 m in Cameroon, 1830 m in Zaïre, and at 700–1800 m in E Africa. Occupies all stages of regenerating forest and spreads outside closed forest into forest/grassland mosaic, swamp forest, relict forest patches in cleared areas, even thick high-bush savanna (Nigeria) and plantations. Typically found in thick forest undergrowth and tangled vegetation in clearings, along tracks and at forest edge, but also ascends to middle levels and sometimes canopy on creeper-clad tree trunks and liana tangles.

Occurs singly, in pairs, or in family parties. Often joins mixed-species flocks, where it tends to keep to the periphery, catching flushed insects, especially butterflies. Gleans foliage in fluttering flight, and sometimes makes short aerial sallies; searches for small ant nests plastered to underside of leaves; when following in path of Fraser's Sunbird *Anthreptes fraseri* gleans leaf petioles and the point where they branch from the main stem (Gabon: Brosset and Erard 1986). Occasionally feeds on ground. Once noted visiting raiding ants *Dorylus wilverthi* as they passed fallen tree in forest; bird darted from one shrub to another pecking here and there, with sunbird-like actions; also seen to open tunnel webs (perhaps embiid), on branch (Gabon: Willis 1986a). Acrobatic foraging actions recall tit *Parus*, tipping over head first and turning in circles as it clings to underside of leaf or twig; at intervals flicks wings and jerks tail from side to side.

Duration of primary moult of 2 birds in Kakamega, Kenya, 82 and 116 days (Mann 1985).

Tendency to be sedentary in fixed home range noted in Uganda, where measured range distance, as represented by greatest distance between trapping points, was 260 m (Zika Forest: Okia 1976).

Food. Insects, including beetles, grasshoppers, ants (including arboreal *Oecophylla*) and scale insects, also scale wax.

Breeding Habits. Courtship and terriorial behaviour not known.

NEST: oval, domed, with side-top opening *c*. 20 diam, or open at the top; loosely-built, rather fragile structure, with outer layer of dry leaves, leaf skeletons and a few twigs, middle layer of fibres, and thick inner lining of plant down; in one nest the walls included stems of Compositae and the lining was of white flowers of Compositae (Prigogine 1984a); ext. diam. 60–90, ext. depth 120–170; placed 0·4–1·3 m above ground in forest understorey in vertical fork in shrub, bush, lone sapling, young *Raphia* palm or bramble, also outside forest in pineapple plant or on low branch of cacao or kola tree; not well hidden. ♀ builds many nests; last one built is used as roost; only 2 out of 'scores' of nests contained eggs (Ghana: Holman *in* Bannerman 1948).

EGGS: usually 1, sometimes 2. Regular oval; smooth; not glossy; slender-shelled; immaculate white. SIZE: (n = 4) 19–20 × 13·6–14·5 (19·5 × 13·9).

LAYING DATES: Liberia (breeding condition Sept); Ivory Coast (juvs May, Aug); Ghana (breeding condition June); Nigeria (juvs Dec); Cameroon, Mar–Apr, July, Sept, Dec (brood patch Jan); Gabon, Nov–Apr (breeding condition Oct, nest-building July, Sept); Angola (breeding condition June, Aug–Sept); Central African Republic (breeding condition June); Sudan, Jan; Zaïre: Ituri (breeding condition June, Sept, imm. Nov, probably breeds throughout rains); Itombwe, Oct–June, especially Jan–Feb; Uganda (breeding condition Jan–Feb, June–Aug); Kenya, June.

References
Bannerman, D. A. (1948).
Brosset, A. and Erard, C. (1986).

Genus *Amaurocichla* Sharpe

Endemic. Monotypic. A long-billed, long-legged, round-winged, short-tailed brown bird endemic to São Tomé island; anatomy never investigated and affinities uncertain. Eyes large. Bill slender, straight, not hook-tipped; no rictal bristles. Wing with 9 primaries, P8 longest, P9 nearly as long. Tail short, with 10 tail feathers; some specimens with rachis-tip extending beyond vane (but vanes may be worn). Flank-feathers loose and fluffy. External morphology suggests affinity with warblers Sylviidae, babblers Timaliidae or Neotropical groups such as woodhewers Dendrocolaptidae, White-throated Treerunner *Pygarrhichas albogularis* (Furnariidae) or gnatwrens *Ramphocaenus* (Sylviidae).

Plate 20
(Opp. p. 337)

***Amaurocichla bocagei* Sharpe. Bocage's Longbill. Nasique de Bocage.**

Amaurocichla bocagei Sharpe, 1892, Proc. Zool. Soc. London (1892), p. 228; São Miguel, São Thomé.

Range and Status. Endemic resident, São Tomé Island. Formerly known from 6 specimens: 1 collected at Binda (0°13′N, 6°28′E), 1 at Juliana de Sousa (0°12′N, 6°28′E) and 1 at São Miguel (0°08′N, 6°30′E) all in 1890–1891, and 3 on Rio Quija (very near São Miguel) in 1928. All localities are close together, near W coast. Sight record, Rio Caué near E coast in 1987 (Eccles 1988) was probably a misidentification (Atkinson *et al.* 1991). 2 populations discovered in 1990, of 4·1–6·3 pairs per km of Rio Xufexufe and about 5·6 pairs per km of Rio Ana Chaves (SW and S-central São Tomé). In 1992 searched for but not found on Rio Quija: forest there has been disturbed (Atkinson *et al.* 1991). Not immediately threatened by forest destruction, since areas in which bird recently found are inaccessible.

Amaurocichla bocagei

Description. ADULT ♂: forehead, crown, nape, hindneck, mantle, back, rump, uppertail- and wing-coverts uniform brown; tail and wings slightly darker brown. Tail short, with 10 feathers; some specimens 'needle-tailed', with rachis extending beyond vane (**A**). Lores and cheeks dark brown, ear-coverts and sides of neck rufescent brown; narrow russet malar stripe from base of bill to below hind cheek. Chin and throat dirty white; breast, belly and underwing-coverts isabelline; indistinct brownish bar between breast and belly; posterior flank-feathers fluffy. Colour of soft parts not known; in skins, bill and legs blackish brown. Sexes alike. SIZE: (2 ♂♂, 2 ♀♀) wing, ♂ 65, 68, ♀ 64, 65; tail, ♂ 36, 39, ♀ 35, 36; bill, ♂ 16, 16·5, ♀ 16, 16; tarsus, ♂ 24·5, 25, ♀ 24, 25.
IMMATURE AND NESTLING: unknown.

A

Field Characters. A small, short-tailed, long-billed, long-legged ground bird with undistinguished plumage: dark brown above, buffier below with whitish chin and throat. Restricted to forest streams in S São Tomé, where no other bird with similar habits occurs.

Voice. Not tape-recorded. Call a loud, high-pitched, forced 'tsee', by both sexes. Song of ♂, at night, 'a longer version of the call' (Atkinson et al. 1991).

General Habits. Inhabits borders of rivers and streams in lowland primary forest. Generally keeps within 10 m of running water with overhanging vegetation and mossy rocks, and spends much time on ground; several also found some 100 m from water within primary forest, where it occurs on fallen logs and low branches within 2 m of the forest floor (Sargeant 1994). Forages on gravel at edge of stream, flicking aside gravel and fallen leaves; also on mossy boulders, by probing and picking. Seen also on sand in middle of creek, on muddy ground and dry sticks on a stony stream bed.

Solitary or in pairs, once 3 seen together. Elusive, but easily located by call. Territorial, pairs holding linear territories along waterways. Often calls from a large boulder. ♂ sings at night, from branch *c*. 3 m high, close to stream. Reluctant to fly: when alarmed tends to run off rather than fly. Flight very weak, not at all rapid, mainly within 3 m of ground; bird usually flies for only a few m before alighting on ground again (Atkinson et al. 1991). Gait very pipit-like (Sargeant 1994).

Its discoverer, F. Newton, noted that bird climbs up long tree-trunks like a treecreeper *Certhia*. That would account for fact that several specimens have abraded tail tips with shafts projecting like needles (photo in Atkinson et al. 1991), but in light of recent observations treecreeping needs confirmation. Said by Correia in 1928 to be rail-like on ground.

Food. Small arthropods.

Breeding Habits. Unknown.

References
Atkinson, P. et al. (1991).
Collar, N. J. and Stuart, S.N. (1985).
de Naurois, R. (1982).

Family MUSCICAPIDAE: flycatchers

Small to medium-sized birds, often sexually monomorphic. Bill variously shaped (adaptive structure linked to foraging habits) but mainly dorsoventrally flattened and relatively wide at base with well to moderately developed rictal bristles; tarsometatarsus usually scutellate, rather spindly, weak and short; claws short, usually strongly curved and sharp. Plumage soft and rather full; grey or brown though some species are black and white, or are patterned yellow, blue or red; usually replaced by a single moult. True flycatchers and thrushes share a derived condition of the syrinx, the 'turdine thumb' (Ames 1975, Olson 1989), the 'passeroid' carpometacarpal condition (Pocock 1966) as well as a double humeral fossa, similar jaw musculature and young with spotted plumage. Main hunting behaviour includes many sallies from preferred perch. Territorial; monogamous (some polygamous). Nest rather bulky, often placed in hole or open cavity. Arboreal; in every habitat with trees from dry thorn scrub to primary evergreen forest.

Represented in Africa by 37 species (32 endemic) in 7 genera: *Fraseria, Melaenornis, Empidornis, Muscicapa, Myioparus, Stenostira* and *Ficedula*, all of which are endemic except *Muscicapa* and *Ficedula*. Various treatments have been proposed for this family: (1) a huge group including thrushes, Old World flycatchers, warblers, Australo–Papuan 'warblers', babblers and their allies (Mayr and Amadon 1951, Mayr and Greenway 1956); (2) an assemblage consisting of Muscicapinae, Platysteirinae, Monarchinae, Rhipidurinae and Pachycephalinae (Wetmore 1960); and (3) a family limited to the former Muscicapinae, with more than 100 species distributed in Africa, Eurasia and SE Asia but none in the Americas; this is the position adopted here, see also Voous (1977). In his phylogenetic classification Cracraft (1981) placed old Muscicapinae along with Turdinae in the family Muscicapidae. Recent DNA analyses conducted by Sibley and Ahlquist (1990) conclude that the family Muscicapidae consists of 2 subfamilies, Turdinae (typical thrushes) and Muscicapinae, the latter including 2 tribes, Muscicapini (true flycatchers) and Saxicolini (chats and robins); the Muscicapidae are the sister group of the Sturnidae.

Genus *Fraseria* Bonaparte

Large-eyed, robust flycatchers with shrike-like head and bill, rather thick tarsus, well scutellated anteriorly, with large, strongly curved and sharp claws; *F. cinerascens* has bill less hooked and more ridged, tarsus more slender and claws weaker than *F. ocreata*. Bill well feathered at base with large bristles over nostrils. Tail slightly rounded. Slate to blackish above, white below with lower throat and breast scalloped. Sexes alike. Juveniles spotted and/or scalloped. Nest rather bulky, usually in open cavity or niche. Inhabit forested areas.

Included by Watson *et al.* (1986) and Short *et al.* (1990) in *Melaenornis*; but like Vaurie (1953), Sibley and Monroe (1990), Dowsett and Dowsett-Lemaire (1993b) and Dowsett and Forbes-Watson (1993), we prefer to keep it separate, particularly on ecological and behavioural grounds.

Endemic. 2 species.

Fraseria ocreata (Strickland). Fraser's Forest-Flycatcher. Gobemouche forestier.

Tephrodornis ocreatus Strickland, 1844. Proc. Zool. Soc. London, p. 102; Fernando Po.

Range and Status. Endemic resident. Forests of Sierra Leone, Liberia (uncommon to common), Ivory Coast (from Taï and Yapo forests to Comoé) and Ghana (widespread, not uncommon, coast north to Bia Nat. Park, Fumsu and Atewa Forest Reserve), Bioko and S Nigeria (coast north to Erin-Ijesha, Benin and Ikom, also Nindam Forest Reserve, Kagoro), S Cameroon (north to Korup Nat. Park, Kumba, Eseka, Avélé and Yokadouma), Central African Republic (Dzanga Reserves, Ile Bokassi, upper Kemo R.), Rio Muni, Gabon (widespread, frequent to common), Congo (common at least in Kouilou), N Angola (Dundo and N Malanje) and most of Zaïre (Uele and Ituri south to Itombwe and Kasai); W Uganda (Budongo and Bugoma to Ankole). Frequent to common. In NE Gabon, av. density of 2·1 groups (16 individuals)/km^2.

Fraseria ocreata

Description. *F. o. ocreata* (Strickland): Bioko and Nigeria to Uganda and Angola. ADULT ♂: upperparts dark slate-grey, blacker on forehead and crown. Tail-feathers black, outer webs fringed slate-grey. Lores and ear-coverts black; sides of neck slate-grey; underparts white, more or less densely scalloped with blackish crescents on lower throat, breast, flanks and upper belly (each feather blackish, broadly edged white and narrowly fringed blackish grey); chin, lower belly and undertail-coverts white with blackish base to feathers; thighs blackish grey barred white. Primaries, secondaries and tertials black, edged slate-grey on outer web; upperwing-coverts brownish black, edged slate-grey, especially on outer web; scapulars slate-grey; axillaries slate-grey tipped white, underwing-coverts blackish broadly edged white. Bill black; eyes olive-brown; legs and feet blackish grey. Sexes alike except that ♀ has lores and ear-coverts browner, less black. SIZE: (6 ♂♂, 6 ♀♀) wing, ♂ 92–98 (94·9), ♀ 89–92 (91·0); tail, ♂ 67–74 (71·4), ♀ 66–70 (67·9); bill, ♂ 19–21 (20·4), ♀ 19–21 (20·0); tarsus, ♂ 22–24 (23·2), ♀ 23–24 (23·4). WEIGHT: (Gabon) ♂♀ (n = 8) 31–36 (34·0); (Uganda) ♂ (n = 7) 34–42 (37·7), ♀ (n = 7) 28–38 (35·1).

IMMATURE: like adult ♀, i.e. lores and ear-coverts browner than ♂, and forehead and crown less black; apical spots on lesser and median upperwing-coverts russet to rufous.

JUVENILE: upperparts blackish smoke-brown; sparse russet spots on forehead, crown, hindneck and mantle, denser on lesser and median upperwing-coverts; underparts white, irregularly and very narrowly barred blackish on sides of throat and upper breast.

NESTLING: unknown.

F. o. prosphora (Oberholser): Liberia to Ghana. Crown grey like mantle. WEIGHT: (Liberia, Jan, Mar, June–Oct, Dec) ♂ (n = 9) av. 32·2, ♀ (n = 8) av. 30·3.

F. o. kelsalli (Bannerman): Sierra Leone. Paler above than other subspecies.

Field Characters. A noisy, active, gregarious bird; gleans foliage in small chattering groups, with characteristic tail-swinging and partly spread wings. Stout and big-headed; dark slate above, white below with dark scalloping, densest on breast. White-browed Forest-Flycatcher *F. cinerascens* looks similar but has very different habits and voice. At a distance, horizontal stance prevents confusion with ♂ Shrike-Flycatcher *Megabyas flammulatus* which perches upright and has pure white underparts and rump; ♂ puffback shrikes *Dryoscopus* are black-and-white leaf gleaners but smaller, with unmarked underparts, no tail movements and different voice.

Voice. Tape-recorded (C, CHA, ERA, GRI, MAC). 3 song types: (a) most frequent, a short series of buzzy notes introduced by and interspersed with purer whistles, used by group and subgroup leaders; (b) semi-continuous and longer: groups of whistles and some harsh grating notes organized into themes or motifs, used by group leader either to maintain its rank or during territorial encounters with other groups; (c) a continuous and melodious rather turdine song, i.e. series of phrases of 2–7 distinct notes, lasting more than 1 min, highly repetitive, including imitations of other species; used for territorial advertising by group leader. Anxiety call, a chatter of 5–10 rasping notes increasing in length and volume; alarm call, similar but louder, with notes nearly all of same length and volume. Pre-flight call, usually by subgroup leader, 2 raspy, nasal notes, second much shorter than first. Individual returning to group, or brooding ♀ joining group, gives monosyllabic 'tchlip' like wagtail *Motacilla*. Subordinate reacts to approaching dominant with slow, prolonged, rasping 'kchee-kchee-kchee-kchee'. Dominants give chatter of 5–12 notes, 'tcheew-chuck-chuck-chuck . . . chuck-chuck-chuck', similar to alarm call but first note higher. Adults in hand emit harsh nasal notes. Nestlings give rhythmic shrill buzzes; in hand they use either panic calls, irregular series of short and long high-pitched nasal notes, or distress calls, rhythmic prolonged piercing wheezes.

General Habits. Inhabits primary and secondary forest, in lowlands and up to 1600 m. Prefers dense continuous mature forest, generally along watercourses although not confined to water; avoids clearings and liana-rich areas; favours forest with closed, thick and high canopy. In second growth, occupies riparian forest and final stages of regenerating forest on old fields and abandoned plantations, where vegetation layers are numerous and ill-defined. Occurs in all vegetation levels but especially in canopy and emergents.

Social organization is of groups of 3–20 individuals; in NE Gabon, av. 7·2 (2 = 12) adults and immatures (excl. juveniles). Group occupies territory of c. 30 ha and is surprisingly stable. Groups consists of 2–4 pairs or trios of adults accompanied permanently by young from previous and current years. There are 4 times as many pairs as trios; a trio consists of a pair (♂, ♀) and an adult 'helper'; the 3 stay together all year, with other units in the group; it is likely that, as in numerous other African co-operative breeding species, the 'helper' participates in all nest and territorial duties. Leader of each subgroup uses first song type, group leader uses all 3 song types. Each pair or trio has one bird (the ♂ or oldest ♂?) which dominates all group members in that part

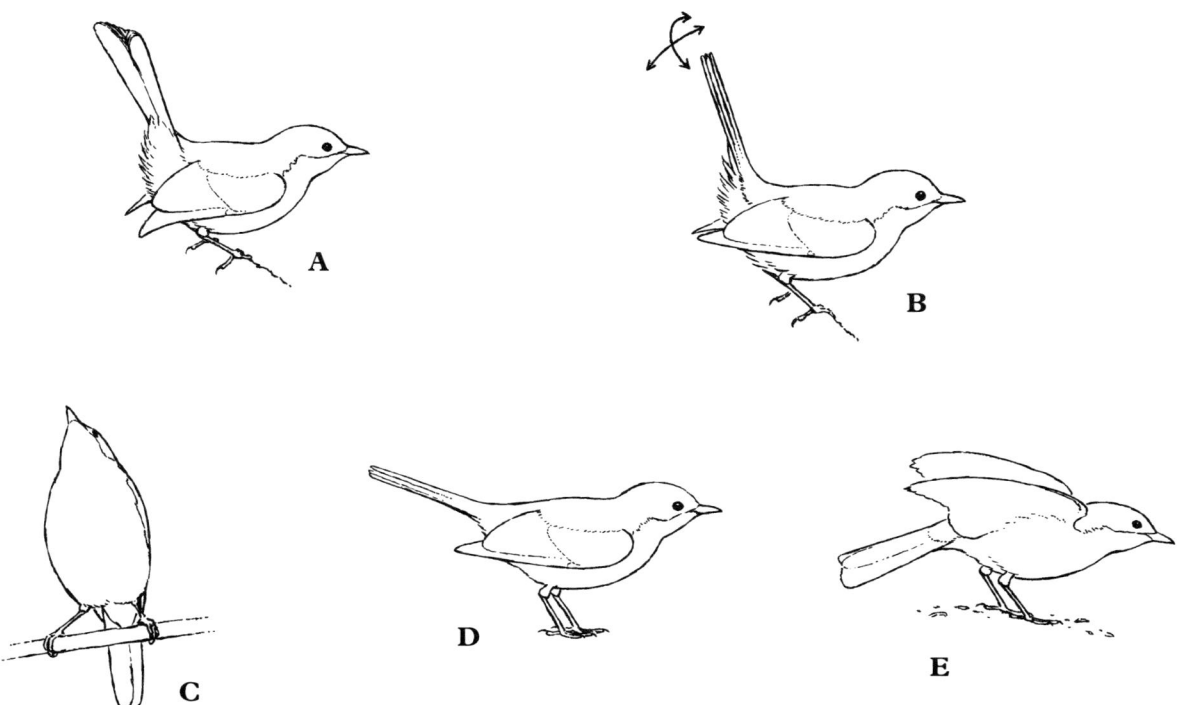

of the group territory which 'belongs' to the pair or trio (presumably the part it nests in). Group leadership changes as the group moves through its entire territory (which is visited every day), according to which pair's or trio's part of the territory the group is presently in. Pairs or trios with their families bond together more closely during long dry season; at other seasons they remain more or less separate but maintain regular vocal (first song type) and/or visual contact. Dominant birds usually display with horizontal body, spread and drooping wings, breast and throat prominent, and swinging of closed tail; subordinates sit in semi-upright stance with sleeked plumage and face away with bill pointing upward, or crouch and slowly wag tail vertically or lower wings and spread tail, or keep body horizontal, stretch legs, lift tail, and extend head and neck forward and upward.

Foliage-gleans in crown of trees and shrubs, also leafy liana tangles; hops along large branches, particularly those covered with moss, lichens or epiphytes; in typical foraging behaviour crouches with drooping or half-spread wings (**A**), swings hind part of body and makes lateral sweeping movements with closed and raised tail (**B**). Other typical hunting postures are figured in **C, D**, and **E**. May forage like true flycatcher, mostly when feeding on swarming winged termites and ants. Particularly skilful at unrolling leaves containing pupating caterpillars. Descends to ground to follow raiding army-ants. Often joins mixed flocks, less during breeding season. Activity most intense in the morning, less in the afternoon, resumes in the evening. When active, group proceeds steadily, usually 200–250 m in 40–50 min, but sometimes intensively searches small area (0·25–0·5 ha), then moves off some distance.

Group members react collectively to danger. Anxiety expressed by calls, crouching, jerky lateral body movements and slow tail-swinging. Alarm includes calls, spasmodic wing-flicking, and irregular jerking of tail from side to side with broad sweep. Group also chases flying raptors (e.g. African Goshawk *Accipiter tachiro toussenellii*) and mobs resting raptors (e.g. Lizard Buzzard *Kaupifalco monogrammicus*).

Sedentary, at least in NE Gabon.

Food. Mostly insects: moths, caterpillars (hairy and smooth), Coleoptera, ants, Orthoptera, winged termites, Hemiptera, Odonata, Dictyoptera, Mantodea, Diptera, Heteroptera, Dermaptera; also spiders and small fruit, once a newly hatched small snake. Prey size, 5–55 mm, mostly 15–45.

Breeding Habits. Monogamous or polygamous; shows every evidence of being a co-operative breeder with helpers at the nest. Group territory av. 30 ha. Dominant ♂ gives advertising song in upright posture, and flaps both wings when dominant of other group responds. All members participate in territorial defence, except young in spotted plumage, which follow adults and give alarm calls. In territorial defence gives second song type, adopts warning posture (upright stance with head forward), gives threat display (body horizontal, legs stretched, wings drooped, carpal joints away from body, tail wagging, open bill pointing forward), and fights with body horizontal, wings held away from body and open bill pointing forward. After a serious conflict, group dominant often gives third song type, and sub-group members allopreen.

In pre-copulatory behaviour, ♀ solicits ♂, crouches,

spreads tail to one side, droops wings, ruffles rump-feathers, rapidly opens and closes wings, then quivers them rapidly as bowing ♂ approaches; he holds head horizontal, half extends and droops wings, fluffs out vent feathers and undertail-coverts, and keeps closed tail upright.

NEST: a bulky, coarse base of dead leaves (mostly shafts), rootlets and tendrils of epiphytes or lianas, supporting a neat cup lined with small pieces of decaying leaf blades and torn bark, often bound with *Marasmius*; ext. diam. 90–100, int. diam. and depth 55–75. Placed 0·5–22 m above usually dry ground, in NE Gabon, mean 5·9 m high (10 nests); in open cavity, e.g. hole in tree trunk or dead stump, opened-up old woodpecker or barbet nest, slit under epiphyte clump, crack at end of broken branch, or wedged between tree trunk and piece of exfoliating bark (Rodewald *et al.* 1994); also among vegetable detritus left by floods on branches of riparian trees or shrubs hanging low above water or dry land, and once in opening in reinforced concrete mast of high-voltage line. Built by several subgroup members; group leader once seen displaying at very beginning of nest construction, pivoting in horizontal crouched posture with fully open and drooping wings, half-spread tail and bill full of nest material.

EGGS: 2–3, laid at 1 day intervals; NE Gabon mean (8 clutches) 2·8. Oval; glossy; pale olive-green or yellow-ochre, heavily streaked and spotted dark brown, red-brown and dark grey, especially at larger end. SIZE: *c.* 21 × 17.

LAYING DATES: Liberia, Jan, Nov (♀♀ with moderately enlarged ovaries June–Aug; ♂♂ with enlarged testes Mar, June, Aug–Sept); Ivory Coast (juvs Aug, Dec); Nigeria, Sept, Oct; Cameroon, Sept, Dec (nest-building Mar); Gabon, Oct–Mar; Zaïre (♂♀ with enlarged gonads Dec–June).

INCUBATION: begins with last egg; by ♀. Period: 17 days. ♀ sits for 65–110 min at a time, leaves for 20–40 min. Group members with singing leader come and fetch ♀ from nest, she leaves with 'tchlip' wagtail call, all go foraging; ♀ fed by group members, although she also catches prey by herself; they then escort her back to nest.

DEVELOPMENT AND CARE OF YOUNG: young remain in nest 15–16 days; fed by group members, with relatively large prey, about every 40–45 min during the first 4–5 days, thereafter every 30 min. Weight increases steadily during development, from 12–13 g at 4 days to 26–27 g at 13–16 days. Fledged young fed by group members, at least for first 3 weeks.

BREEDING SUCCESS/SURVIVAL: only 3 out of 7 nests with clutches under observation fledged young in NE Gabon; 1 clutch destroyed by storm. Lives > 12 years: 3 adults making a trio were ringed in Nov 1973 and were still together in May 1985.

References
Brosset, A. and Erard, C. (1986).
Erard, C. (1987, 1990a).

Fraseria cinerascens Hartlaub. White-browed Forest-Flycatcher. Gobemouche à sourcils blancs.

Plate 26
(Opp. p. 481)

Fraseria cinerascens Hartlaub, 1857. Syst. Ornith. Westafrica's, p. 102; Ashanti, Gold Coast.

Range and Status. Endemic, resident. Forests of Senegal and Gambia (Gambia and Basse Casamance), Guinea-Bissau, Guinea (Kindia), Sierra Leone, Liberia (rare), Ivory Coast (common, north to Sipilou and Comoé), S Ghana (uncommon, at least type locality and Akwapim hills). S Nigeria (not uncommon, forests of coast, from west of Lagos east to Benin, also gallery forests north to Kainji, Pandam and Kagoro, and in SW at Mberubu); Cameroon north to Korup Nat. Park, Sanaga, Mieri and Batouri; Central African Republic (Bangui, Bangassou, R. Ouossi, NW Zemio); Gabon (widespread, common), Congo, Mayombe and Angola (Cabinda), and Zaïre east to Upper Uele, Ituri and Kasai. Uncommon to abundant. In NE Gabon on av. 7 pairs/km of linear river bank, but 70 pairs/km² on periodically flooded forested islands.

Description. *F. c. ruthae* Dickerman: S Nigeria and Cameroon east to Zaïre. ADULT ♂: upperparts plain slate-grey, forehead and forecrown black. Tail brownish black, feathers narrowly edged slate-grey along outer web. Lores and anterior ear-coverts black, contrasting with pure white supraloral spot extending backward over eye; posterior ear-coverts and sides of neck slate-grey; underparts white (base of

feathers blackish), heavily scalloped with broad blackish crescentic marks, narrow on throat, wider on breast and upper flanks; chin, belly and undertail-coverts white, thighs black barred white. Primaries and secondaries brownish black, edged slate-grey on outer webs; tertials brownish black suffused with slate-grey; scapulars slate-grey; upperwing-coverts blackish fringed slate-grey; axillaries grey edged white; underwing-coverts blackish edged white. Bill black; eyes olive-brown; legs and feet black. ADULT ♀: like ♂ but face greyer, less black. SIZE: (7 ♂♂, 4 ♀♀) wing, ♂ 78–83 (80·8), ♀ 74–82 (78·5); tail, ♂ 61–68 (64·3), ♀ 58–64 (60·7); bill, ♂ 16–18 (17·1), ♀ 16–19 (17·3); tarsus, ♂ 18–20 (19·5), ♀ 19–20 (20·0). WEIGHT: (Liberia, July, Nov, Dec) ♂ ♀ (n = 4) 13–21 (17·5); (Gabon) ♂ ♀ (n = 9) 21–24 (23·1); (Congo) ♂ ♀ (n = 2) 20·8, 23; (Central African Republic, May–June) ♂ (n = 5) 21·5–24·5 (23·0), ♀ (n = 4) 18–22 (19·7).

IMMATURE: like adult but all grey parts heavily suffused with olive; apical spots on upperwing-coverts, tertials and tail-feathers rufous; rusty ochre streaks sparse on upper mantle, numerous on lores and ear-coverts; white replaced by ochre in centre of breast, by russet on sides of breast and flanks; supraloral stripe ochre, not white, and small.

JUVENILE: entire upperparts and upperwing-coverts dark smoky brown densely covered with large spots (stripes on crown) of buff or rusty (head-feathers dark brown with narrow buff median stripe, mantle feathers with triangular buff mark near tip narrowly edged with blackish brown); flight feathers and tail dark brown, tertials tipped and edged buff on outer web; underparts whitish, washed buff and scalloped and freckled blackish on breast and flanks.

NESTLING: at hatching young have dark brown skin and long, sparse, dusky brown down which changes to pinkish brown with age, matching colour of nest rim, thus increasing camouflage.

F. c. cinerascens Hartlaub: Senegal and Gambia to Ghana. Upperparts paler, head not darker than back; breast weakly edged with sooty black.

Field Characters. A quiet, solitary bird that hunts from perches in shady riverine vegetation. Smaller size, thrush-like shape, upright stance and lack of tail movements further distinguish it from its noisy cousin Fraser's Forest-Flycatcher *F. ocreata*. Blackish slate above, with broad white mark in front of and over eye, conspicuous in the field; eye large; crescentic marks on underparts more diffuse, less clear-cut.

Voice. Tape-recorded (CHA, ERA, VIEL). Song, a series of lengthy (av. 0·5 s), high-pitched, thin, trilled, almost insect-like notes separated by intervals of av. 0·7 s, lasting 30–50 s, sometimes more, with many repetitions; 70% of notes have overtones (i.e. 2 independent frequencies emitted simultaneously); songs highly individual, but because of their high pitch (8 kHz), only a few can be distinguished by the human ear in the field. Adults keep in contact with high-pitched, prolonged 'tseeee', and use same type of call when coming to nest. Anxiety call, harsh 'kchee'; alarm calls when danger nearby, a long and loud, high-pitched whistle from ♂, a harsh 'shhhh' note from ♀. Young in nest give rapid, short, high-pitched trills which become longer as birds grow older. Fledged young give shrill, slightly descending whistles.

General Habits. Inhabits lowland riparian evergreen forest, including river banks in seasonally flooded forest; always near water, never farther than 300 m from river, although occupies entire surface of periodically flooded forested islets; not found along small forest streams, but occurs at their junction with wider watercourses; also along closed meanders of rivers. Prefers dark shady places, usually under dense overhanging foliage, also tight stands of trees with high aerial roots (e.g. *Uapaca* sp.); needs decaying stumps, old trunks and fallen dead trees emerging from water. Typical habitat is damp ground covered with scattered seedlings and saplings up to 3–5 m high, underneath dense overhanging foliage in shade of trees.

Solitary or in pairs. Partners often move separately, using contact calls when out of sight; on meeting, ♂ approaches ♀, sits on exposed perch in horizontal stance with drooping wings and raised tail, displaying conspicuous white eye-patches. Sits upright, but shape more like European Robin *Erithacus rubecula* than true flycatcher *Muscicapa*. 4 types of foraging behaviour: (1) true flycatching, i.e. sits on exposed perch (low branch, loop of hanging liana, stump) in shady place under dense foliage mixed with tangles of leafy lianas, and sallies in looping or hovering flight to pick up prey on leaf or trunk; (2) foliage-gleaning, i.e. hops and flies up and down among branches, then often sits under area it has searched and flycatches as in (1); (3) fallen-trunk gleaning, i.e. hops around decaying stumps and along trunks covered with moss and lichens that have fallen along river banks or across streams, or searches huge horizontal branches covered with epiphytes low over water; (4) dropping from a very low perch (less than 1 m above ground) to catch prey on ground, where it may hop. When foraging, often chases species of similar stance or behaviour.

Forages actively all day, usually in bouts of 40–60 min; less active in the afternoon, when it bathes in shallow water, either half submerged, with jerky movements of spread tail and wing feathers, or sprinkling water on its back with much flapping; sunbathes with half-open drooping wings and fluffed out body plumage. When anxious, stops motionless in more or less crouched posture; when alarmed, jerks closed tail up and down and droops wing-tips.

Food. Mostly arthropods; various insects: Coleoptera, Hymenoptera (wasps and especially ants), Diptera, Lepidoptera (caterpillars (**A**) and moths), Heteroptera, Hemiptera, Orthoptera, Mantodea; many spiders; also small millipedes, worms, and small fruits (e.g. *Ficus natalensis*). Prey size: 5–65 mm, mostly 10–40.

Breeding Habits. Monogamous; probably pairs for life; in NE Gabon, partners still together 12 years after being ringed as adults. Territory size av. 1·5 ha, usually in linear shape along river banks, but may occupy entire surface of small island. Territorial all year; patrols entire territory daily. ♂ sings in upright stance; when he hears rival song, he sits on exposed perches and crouches, droops wing-tips, raises tail, inflates breast and fully expands white eye-patches; sometimes he moves about in dense low vegetation. When other ♂ approaches, territory owner flies rapidly back and forth with noisy wing-flaps and bill-snapping, displaying all white parts of his body toward opponent. In threat display, holds wings half open and drooped, inflates breast, flicks out carpal joints and jerks half-spread tail up; may chase intruder in flapping flight with vigorous bill-snapping. During pair-formation, ♂ crouches facing away from ♀, quivering drooping wings, holding head between hunched shoulders, fluffs out crown-feathers, expands white eye-patches, throws out breast, and jerks tail up; ♀ remains upright, facing away, with fluffed out body plumage, and at intervals picks up some moss or thin rootlets.

NEST: a bulky coarse base of moss, decaying twigs, dead leaves, lichens, epiphyte rootlets, and vegetable fibres, supporting a neat and well-shaped cup made of gauze-like thin leaf skeletons, lined with some moss, rootlets or *Marasmius*. Ext. diam. 90–100, ext. depth 90; int. diam. 50–70, int. depth 30–40; nest varies in size according to situation; e.g. decreases with size of cavity but larger in open cavity of tree trunk (**A**) than among tangles of roots. Placed less than 3 m above ground, usually 1–2 m; in NE Gabon, 72% of nests placed above dry land, 28% above water; land nests up to 110 m from river bank (on island), usually less than 20 m from water; nests above water less than 20 m from bank; in open cavity or fracture in dead stump, tree trunk, large liana stem or low-hanging branch, or among thick tangle of roots and rootlets of periodically flooded trees, in recess in steep river bank, or in termite mound. Nest site may be re-used in the same or successive years. Nest built by ♀ in 6–8 days; ♂ sings and guards her.

EGGS: 1–2, laid at 1 day interval, beginning 3–5 days after completion of nest. NE Gabon mean (29 clutches) 1·8. Oval; glossy; cream or whitish more or less tinged green, heavily washed with blotches and spots of different shades of brown (reddish, rusty, chocolate and violaceous), forming darker and more continuous cap at large end. SIZE: *c.* 22–24 × 14·5–15. A few pairs lay replacement clutches; some breed twice in the same cycle, sometimes in the same nest.

LAYING DATES: Senegambia, May, Sept; Ghana, Jan, Apr, May; Liberia (♀♀ with enlarged ovaries July, Dec); Nigeria, Feb, Apr (♂ with enlarged testes Nov); Cameroon, Jan, Feb (young Mar); Gabon, Sept–Feb, mostly Dec–Feb during short dry season (moulting young June); Central African Republic (imm. Apr, ♂ with enlarged testes May); Zaïre (young Oct, ♂ in breeding condition May).

INCUBATION: begins with last egg; by ♀. She sits for periods of 1–33 min (av. 15·5), leaves for 1–27 min (av. 11·8). Period: 14 days. Sits deep down in nest with white eye-patches concealed. ♂ guards and occasionally feeds ♀. He reacts to approaching danger with alarm behaviour, and displays his white underparts toward nest in a rapid zigzag flight.

A

DEVELOPMENT AND CARE OF YOUNG: weight increases steadily from 2 g at hatching to 18–19 g at 10 days. Nestling period: 11–12 days.

Young fed by both parents. Prey size increases and food delivery rate decreases with age of young. ♂ actively defends nest surroundings, chasing every bird coming near. Both ♂ and ♀ give alarm and resist threatening predator, but ♀ keeps in shadow in the background whilst ♂ mobs and attacks it with much wing-flapping and bill-snapping, also drooping wings, raising tail and pointing head forward.

Fledged young remain together in a small area for 3–4 days, then move along river bank with adults, each parent caring for a particular young. Young fed for 1–2 months after fledging, sometimes for 70 days; stay on parental territory until next breeding cycle.

BREEDING SUCCESS/SURVIVAL: in NE Gabon, out of 30 clutches (54 eggs), 13 (21 eggs) hatched and 10 (17 young) fledged. Lives > 12 years.

References
Brosset, A. and Erard, C. (1986).
Erard, C. (1987, 1990a).

Genus *Melaenornis* Gray

Flycatchers like *Muscicapa* but more chat-like and generally larger, with longer and thicker scutellate legs and usually more robust and narrower thrush-like bills. Wing rounded. Upperparts brown, grey or glossy black, underparts the same or white; plumage not striped (except crown of 1–2 species); 1 species with white in wing and tail, 1–2 with conspicuous eye-ring; sexes alike; juveniles like adults but spotted or scaly. Small rictal bristles. Cup nest. Inhabit forest and savanna woodland, mainly the latter. Forage mainly by pouncing to ground, also by fly-catching sallies, leaf- and trunk-gleaning, and (*M. annamarulae*) by dashing along boughs.

Endemic. We recognize 13 species which have been distributed amongst several genera in the past: the black *Melaenornis*, brown *Bradornis*, spectacled *Dioptrornis* and pied *Sigelus*. Taxonomic opinion remains divided on how many genera to recognize, with various treatments by Vaurie (1953), Watson *et al.* (1986), Short *et al.* (1990), Sibley and Monroe (1990), Dowsett and Dowsett Lemaire (1993b) and Dowsett and Forbes-Watson (1993). We consider division into 4 genera to be essentially a matter of convenience, based on differences in colour; habitat separation is not clear-cut. We follow Watson *et al.* (1986) in recognizing only one genus, and agree with most authorities that it is closely related to *Empidornis* and *Fraseria*. *Melaenornis* flycatchers exhibit few, if any, significant biological differences – fewer even than within African *Muscicapa* species. 2 superspecies: *M. brunneus*/*M. chocolatinus*/*M. fischeri* and *M. pammelaina*/*M. edolioides*.

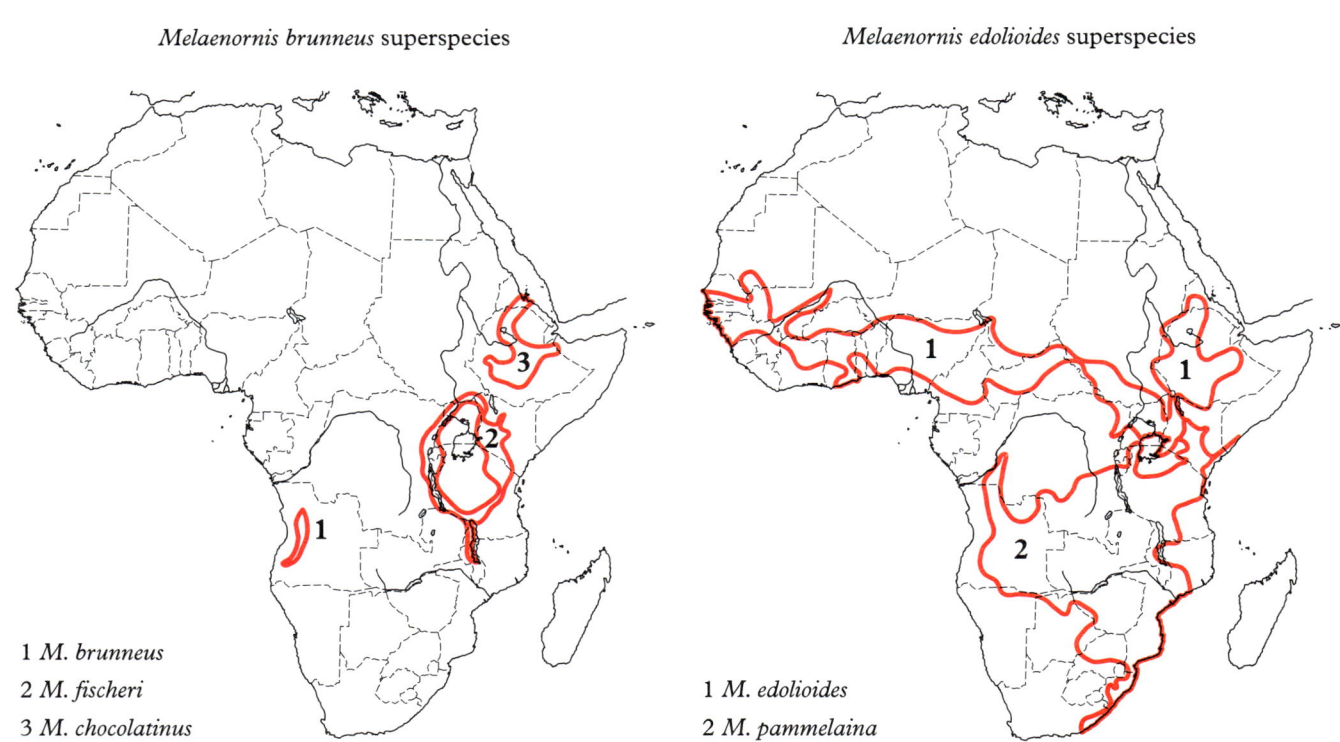

Melaenornis brunneus superspecies

1 *M. brunneus*
2 *M. fischeri*
3 *M. chocolatinus*

Melaenornis edolioides superspecies

1 *M. edolioides*
2 *M. pammelaina*

Plate 25 *Melaenornis brunneus* (Cabanis). Angola Slaty Flycatcher. Gobemouche de l'Angola.
(Opp. p. 480)

Dioptrornis brunnea Cabanis, 1886. J. Orn., p. 92; Pungo Andongo, Angola.

Forms a superspecies with *M. fischeri* and *M. chocolatinus*.

Range and Status. Endemic resident, Angola. W Malanje, S Cuanza Sul, Huambo, S Benguela and on Serra da Chela, Huila; ranges from 2000 to 2200 m.

Description. *M. b. brunneus* Cabanis: N end of W Angola escarpment. ADULT ♂: forehead warm dark brown, changing to pale buffy grey within 3 mm of bill; lores, chin and 2 mm

Melaenornis brunneus

white. Wings warm dark brown above and below. Bill slate-grey or blue-grey with black-brown tip; eyes dark brown to light grey; legs and feet dark grey to black. Sexes alike. SIZE: (1 ♂, 2 ♀♀) wing, ♂ 84, ♀ 77·5, 87; tail, ♂ 69, ♀ 62, 67; bill to skull, ♂ 15, ♀ 14, 14·5; tarsus, ♂ 20, ♀ 20, 22·5.

IMMATURE AND NESTLING: not known.

M. b. bailunduensis Neumann: Mt Moco and central Angolan highlands. Like *brunneus* but upperparts darker and less warm in tone.

Field Characters. A poorly-known flycatcher, dark brown, with pale grey about eye and base of bill, and white belly. Pale Flycatcher *M. pallidus* is a pale grey-brown bird of low woodlands. May just overlap Sooty Flycatcher *Muscicapa infuscata* which also forages in forest canopy; it is long-winged, short-tailed and dark-bellied.

Voice. Not known.

General Habits. Inhabits patches of montane forest and isolated trees among quartzite outcrops on rocky hillsides, also woodland along hill streams and forest patches in mountain valleys. Forages by hawking for insects from crown of tall tree.

Food. Insects.

Breeding Habits. Not known. Birds have gonads enlarged in Aug–Nov.

wide zone around eye grey; crown, nape, hindneck, sides of neck, mantle, scapulars, back, rump and uppertail-coverts warm dark brown. Tail dark brown, nearly square-ended: T6 5–7 shorter than T1; underside of tail grey-brown. Cheeks and ear-coverts warm dark brown, feathers with pale shafts. Throat and breast warm dark brown, upper belly buffy white, quite sharply demarcated from brown breast, lower belly and flanks white, thighs and vent light brown, undertail-coverts

Melaenornis fischeri (Reichenow). White-eyed Slaty Flycatcher. Gobemouche de Fischer.

Plate 25 (Opp. p. 480)

Dioptrornis fischeri Reichenow, 1884. J. Orn., p. 53; Mt Meru, Tanganyika.

Forms a superspecies with *M. brunneus* and *M. chocolatinus*.

Range and Status. Endemic resident, E African highlands from S Sudan to Malaŵi. Sudan, common above 1600 m in Imatong, Dongotona and Didinga Mts. Uganda, common on Lonyili and Morongole Mts in N, Moroto and Elgon Mts in E, and in Ankole, Kigezi and Rwenzori Mts in SW. Kenya, common and widespread at 1350–3000 m in SW; present in forest islands of Loima, Kulal, Nyiru, Ndoto and Mathews Mts; density around Nairobi increased greatly with suburbanization in 1940s and 1950s (van Someren 1956). Zaïre, 3 populations: highlands west of L. Albert; from northwest of L. Edward and L. Kivu (including Idjwi I.) to Itombwe where fairly common south to Baraka (4°06′S); and Marungu Highlands where widespread at about 1860 m. Rwanda and Burundi, quite common in highlands. Tanzania, Kilimanjaro, Meru, Monduli and N Pare Mts, Crater Highlands and Mbulu; Ulugurus and Ukagurus to Njombe, Mt Rungwe, Isoko and Ufipa Plateau. Zambia, west sides of Nyika Plateau and Mafinga Mts only. Malaŵi, widespread above 1700 m from Misuku Hills and Nyika Plateau to Dedza; down to 1475 m on Mt Mafinga.

Description. *M. f. fischeri* Reichenow: Sudan, NE and E Uganda, Kenya, and NE Tanzania. ADULT ♂: forehead, crown, nape, hindneck, sides of neck, ear-coverts, mantle, scapulars, back, rump and uppertail-coverts dark bluish grey. Lores and broad ring around eye whitish grey: ring *c.* 3 mm wide in front of, below and behind eye. Chin, throat and breast bluish grey (much paler than upperparts), colour paling to white in centre of belly, with pale cream tinge; flanks, vent and undertail-coverts same pale grey as lower breast, thighs somewhat darker. Feathers of wings and tail blackish, glossy, most of those parts exposed when feathers are folded being dark bluish grey. Undersides of flight feathers and tail-feathers silvery grey; underwing-coverts and axillaries whitish grey. Bill black at tip merging to bluish at base; eyes dark brown; legs and feet black. Sexes alike. SIZE: (5 ♂♂, 3 ♀♀) wing, ♂ 90–97 (92·1), ♀ 89–96 (93·2); tail, ♂ 69–77 (72·8), ♀ 75·5–79 (77·4); bill to skull, ♂ 15·0–17·2 (16·1), ♀ 15·2–16·3 (15·7); tarsus, ♂ 21·5–22·5 (22·0), ♀ 22·5–23·5 (22·8). WEIGHT: ♂ (n = 15) 19–33 (24·1), ♀ (n = 13) 16–30 (23·5).

IMMATURE: mantle blue-grey, rest of upperparts sootier, forehead, crown, hindneck and mantle heavily spotted with pale buff (feathers have blackish fringe, pale buff subterminal spot and grey base); back, scapulars and upperwing-coverts similarly spotted but more sparsely; rump and uppertail-coverts diffusely marked with pale buff crescents or bars; ter-

Melaenornis fischeri

tials and primary coverts with whitish tips. Plumage of underparts fluffy; chin and throat grey, breast pale buff heavily marked with dark brown crescents, belly and flanks silky off-white heavily marked with dark brown crescents and bars, vent and undertail-coverts whitish, almost unmarked. Eye-ring barely shows in very young bird.

NESTLING: hatches with skin dark flesh-brown; mouth rich yellow; swollen gape yellow.

M. f. toruensis Hartert: highlands of SW Uganda and E Zaïre from Kivu to Itombwe; Rwanda, Burundi. Like *fischeri* but eye-ring narrow and inconspicuous: only a little paler than crown and confluent with pale lore. WEIGHT: ♂ (n = 12) 18–24 (21·8), ♀ (n = 15) 20–26 (23·3).

M. f. nyikensis Shelley (including '*ufipae*'): Zaïre (Marungu Highlands), Tanzania (Ufipa to Crater Highlands and Ulugurus) and Malaŵi. Upperparts slightly darker than in *fischeri*; narrow but distinct white eye-ring, 1 mm wide, broken behind eye. Bill black. Some birds in SW Tanzania and Shaba ('*ufipae*') have whiter throat and belly.

M. f. semicinctus Hartert: highlands west of L. Albert, Zaïre. Like *fischeri* but white eye-ring interrupted in front of eye and to smaller extent behind it.

Field Characters. A medium-sized flycatcher (length about 15 cm), dark grey above, pale grey below becoming almost white on belly, recognized by its white eye-ring. Eye-ring wide and unbroken from Sudan to NE Tanzania, narrow and broken behind eye in rest of Tanzania and Marungu area (Zaïre), wide and broken before and behind eye west of L. Albert (Zaïre), and distinct but inconspicuously narrow in Kivu (Zaïre). Ashy Flycatcher *Muscicapa caerulescens* is smaller and shorter-tailed, with quite different face pattern: dark line through eye, white streak above lore and eye and tiny white crescent below eye, i.e. no real eye-ring; it is also a canopy bird.

Voice. Tape-recorded (B, F, CHA, GREG, McVIC, PAR, STJ). Song a high-pitched, rather thin but sweet, simple phrase lasting 1–1·5 s: 'tsp, swee-oo-sweetoo', varying only in the initial, tiny 'tsp' notes. Call a harsh, nasal 'screescreet', quite loud, like calls of monarchine flycatchers (*Trochocercus, Terpsiphone*). Alarm a single or double 'tchuiskere cheeweet'.

General Habits. A bird of dense montane forest, less of its interior than of borders. Perches low along forest edges, on bare boughs, saplings and bushes, enters forest where canopy is broken, and flycatches in gaps around tall trees; disappears into forest if alarmed. Also inhabits woodland, trees in villages, and *Hagenia* woods of central Kivu volcanoes at about 3380 m (but altitudinal range mainly about 1400–2500 m). Once seen foraging inside forest, attracted in from nearby clearing by an ant swarm (Dowsett-Lemaire 1990). In Kenya occurs at edges of many highland forest types – fringing forest, dry forest, open woodlands and scattered trees in open areas with short grass; often in gardens; freely uses telephone wires and electricity cables as foraging lookouts.

Forages about equally by making swift sallies after passing insects, and shrike-like, by scanning ground for insect movement, flying down to snatch prey in bill and returning with it to same or nearby perch. Attracted to flying swarms of ants and termites, and to insects flushed by ants swarming on tree-trunks: party once watched feeding inside forest at height of 6–10 m, by sallying and pecking on branches. On returning to its perch taps prey sharply against it and eats it whole. However, honeybees *Apis* sp. treated differently (Gwinner 1986); bee is caught in air and taken to thick horizontal branch, where its abdomen is repeatedly squeezed until sting emerges. Sting either sticks into branch, or remains attached to bee's abdomen. In the latter event, bird lays bee down, grasps stinging apparatus in its bill and flings it away with sudden movement of its head. It swallows the bee then wipes its beak vigorously. Bee de-stinging behaviour of immature flycatcher is clumsy. On ground, insects taken from bare earth, newly dug soil and leaf debris; if moth takes refuge in leaves, bird flicks leaves over or frightens it out by fluttering over them. Stance on ground quite thrush-like: stands high on legs, wings depressed, tail slightly up, head a little to one side.

Wary but bold; will dash down to take insect almost at person's feet, but shies off at once if observer pays attention to it. Usually in pairs; a dozen seen around a bird-bath; 2 families sometimes feed together. Forages actively until almost dark – one of the last birds to retire in Nairobi gardens. Fond of water-bathing, often long after sunset (van Someren 1956).

Food. Grasshoppers, mantids, scarabaeid beetles (only small black beetles in 16 stomachs), noctuid moths, ants, honeybees, termites and flies. Nestlings given mainly grasshopper nymphs, crickets, mantids and small chafer beetles; also a few moth larvae, wireworms, a very small skink, 2 parts of a tree-frog, a small millipede and occasional small berries (van Someren 1956).

Breeding Habits. Solitary, monogamous breeder; territorial. 3 adults once attended nest (Beesley 1972b).

NEST: cup made mainly of moss, usually mixed with lichen and spider web; foundation thick, depending on amount of space to be filled in between multiple underlying twigs (**A**); cup neatly finished with fine rootlets, fragments of lichen and a few very fine tendrils, and lined profusely with hairs and feathers. Exterior rough and irregular, making nest resemble collection of debris caught up in twigs. Int. of cup 50 wide and 30 deep. Nest sited in lichen- and mistletoe-infested tree, around Nairobi (Kenya) commonly old *Schrebera* and olive trees and tops of *Maba* trees. Built by ♂ and ♀, taking nearly 2 weeks to complete, once 4 weeks. Pair often nests in successive seasons in one particular tree; new nest sometimes built on top of old one. Less frequently, nest reconditioned and used straight away for second clutch.

EGGS: 2–4, usually 2–3 (about equally: Kenya). Pale greenish, with blue-grey blotches, regular overlying umber-brown flecks, and fine red-brown freckles often forming zone around widest diameter or cap at large end. SIZE: about 22 × 15·5. Usually double-brooded, in Kenya in long rains.

LAYING DATES: E Africa: Region A, Nov–Dec (7 records) and Feb; Region B, Jan; Region C, Dec, Apr–May; Region D, all months, mainly Sept–Jan (51 records) and Apr–May (26 records) (16 records in remaining months); breeds in short rains and long rains, but in Region A, above 2100 m, in dry season; Zaïre, (various indications, Feb, May–June, Aug); Rwanda, (several fledglings being fed, from Aug clutches); Zambia, Sept–Oct. Malaŵi, Oct.

INCUBATION: by ♀ (mainly) and ♂. ♀ fed on nest by ♂. Clutch left for spells of up to 10 min in late evening (Nairobi), when ♂ and ♀ may bathe together. Period:

A

12–13 days. Feathers often added to nest lining before and even after young hatch.

DEVELOPMENT AND CARE OF YOUNG: young brooded closely for first 1–2 days, thereafter fed actively by both parents except for brooding spells by ♀ at midday. Young regurgitate pellets of insect sclerites. After each feed parent waits until faeces voided, when sac is removed (not eaten). Nestling period 16 days. Fledglings sit in nest-clump or thick foliage nearby for about a day before leaving. They remain with parents for at least 4 weeks, unless adults start another nest.

References
Gwinner, E. (1986).
van Someren, V. G. L. (1956).

Melaenornis chocolatinus (Rüppell). **Abyssinian Slaty Flycatcher. Gobemouche chocolat.**

Plate 25
(Opp. p. 480)

Muscicapa chocolatina (Rüppell), 1840. N. Wirbelt. Vög., p. 107; Simen, Ethiopia.

Forms a superspecies with *M. fischeri* and *M. brunneus*.

Range and Status. Endemic resident, highlands of Ethiopia and Eritrea. Uncommon in Eritrea, in wet woodland above 2150 m in southern mountains, from Adi Caieh to Senafe. Frequent to common in Ethiopia in W and SE Highlands, and in Rift Valley above 1800 m; common at Alghe, in Kwera Hills, in Ghera region and in Balaghes valley bottom. Occurs up to 2500 m.

Description. *M. c. chocolatinus* Rüppell: Eritrea and Ethiopia except W. ADULT ♂: forehead, crown, nape, hindneck, cheeks, ear-coverts, sides of neck, mantle, scapulars, back, rump and uppertail-coverts dark brown with greyish or olivaceous tinge. Tail dark brown to blackish brown, nearly square-ended: T6 3–7 shorter than T1; underside of tail dark grey. Underparts smoky grey with buffy wash on breast; belly markedly paler than breast but colours not sharply demarcated; flanks like belly; vent and undertail-coverts buffy grey-white. Upperwing-coverts dark brown like back, primaries, secondaries and tertials very dark brown, slightly glossy; underwing-coverts and axillaries like breast; undersides of flight feathers of wing and tail shiny dark grey-brown. Bill pale grey or bluish grey with dark culmen and tip; eyes pale yellow or very pale straw; legs and feet black, soles grey. SIZE: (5 ♂♂, 5 ♀♀) wing, ♂ 81–87 (85·2), ♀ 82–86 (85·1); tail, ♂ 63·5–70 (68·3), ♀ 69–74 (70·2); bill to skull, ♂ 15·4–17·2 (16·5), ♀ 14·0–16·6 (15·4); tarsus, ♂ 22·5–24 (23·4), ♀ 20·5–23·5 (22·8). WEIGHT: 2 ♂♂ 20, 23, 2 ♀♀ 25, 25.

IMMATURE: upperparts dark brown, with numerous small buff spots on forehead and crown, and numerous large, more diffuse, buff spots on mantle, back and upperwing-coverts and tips of tertials. Tail and flight feathers dark brown, unspotted. Underparts pale buff or greyish white, heavily scaled with dark brown on chin, throat, breast and upper belly. Subadult bird-like adult but with a few scattered buff spots on upperparts and a few dark brown crescents and irregular marks on under-

Melaenornis chocolatinus

parts – mainly sides of breast and upper belly. Eyes brown; bill yellowish (in skins).

M. c. reichenowi (Neumann): W Ethiopia (Wallegha to Gimirra). Weakly differentiated; throat and breast greyer and slightly darker than nominate subspecies; upper belly with suggestion of dark stripes in some specimens.

Field Characters. The common flycatcher of upland forest and woods in Ethiopia: slim, moderately large, quite long-tailed, dark brown above and grey below, colour of underparts paling to buffy grey-white on undertail-coverts. Coloration, combined with forest habitat, makes confusion with other flycatchers unlikely. Northern Black Flycatcher *M. edolioides* is all black. African Dusky Flycatcher *Muscicapa adusta* is very small and has buff or white throat.

Voice. Tape-recorded (C). Very like voice of White-eyed Slaty Flycatcher *M. fischeri* (Benson 1946b).

General Habits. Inhabits heavy woodland, not-too-dense forest, evergreen forest edges, juniper woods and coffee plantations, above 1800 m; often near streams. Habits very like those of White-eyed Slaty Flycatcher (Benson 1946b); hunts 'in typical fly-catcher manner' in canopy and from lower branches, usually > 3 m above ground, and sometimes darts to ground for an insect (Desfayes 1975). Occasionally wags tail up and down.

Food. Not known; presumably insects.

Breeding Habits.
NEST: neat cup of moss and rootlets, lined with rootlets; placed on horizontal bough (Mackworth-Praed and Grant 1960).
EGGS: 3(?). Pale green or turquoise-blue, blotched blue-grey, with overlying clay-coloured flecks and spots. SIZE: about 22×15.5.
LAYING DATES: Ethiopia, suspected to breed in Jan–Feb and Mar–June; gonads enlarged, June, Dec and Mar–May.
BREEDING SUCCESS/SURVIVAL: adult shot 6 years after it was ringed, Ethiopia (Tyler 1987a).

Plate 25
(Opp. p. 480)

Melaenornis annamarulae **Forbes-Watson. Nimba Flycatcher. Gobemouche noir du Nimba.**

Melaenornis annamarulae Forbes-Watson, 1970. Bull. Br. Orn. Club, 90, p. 145–149; Grassfield, Mt Nimba, Liberia.

Range and Status. Endemic resident Guinea (Macenta, Ziama Massif: Wilson 1990, Halleux 1994); foot of Mt Nimba, Liberia, and northern half of Taï Nat. Park, SW Ivory Coast. Scarce, but easily overlooked; > 30 seen on Mt Nimba in 4 years, and pairs or trios seen in Taï Nat. Park 5 times in 53 days (Gartshore 1989). Gravely at risk in Liberia from forest destruction (timber extraction, mining); even Taï Forest has been suffering from illegal encroachment (Collar and Stuart 1985).

Description. ADULT ♂: uniform dark blue-grey except that throat, belly and flanks slightly paler; primaries and secondaries black with shiny blackish grey undersides; tail with T1 and T2 blackish next to shaft, T3–T6 with blackish inner webs; underside of tail black, a little glossy. Tail long, square-ended or slightly rounded; shortest feather 4 mm shorter than longest. Bill quite broad at base, stubby, hooked at tip; c. 5 thin, weak rictal bristles; bill black, eyes dark brown, rim black; legs and feet black, soles dull cream. SIZE: (7 ♂♂, 4♀♀) wing, ♂ 101–109 (106), ♀ 102–104 (103); tail ♂ 76.5–81 (78.5), ♀ 76–81 (77.75); bill, ♂ 16.5–19.5 (17.5), ♀ 17–17 (17.0); tarsus, ♂ 22–23 (22.7), ♀ 21–24 (22.75). WEIGHT: ♂ (n = 7) 37.4–42.0 (40.0), ♀ (n = 4) 37.0–42.3 (39.7).
IMMATURE AND NESTLING: not known.

Field Characters. A large, drab, dark flycatcher, thick-bodied, with quite long tail and small bill. Bluish grey in the hand but in poor forest light looks blackish. Same size and shape as Square-tailed Drongo *Dicrurus ludwigii*, but only superficially similar – 'bears only a passing resemblance to a drongo' (Balchin 1990). At close range told from drongo by dark brown eyes (drongo has red eyes). Feeds high in tall trees in forest or a clearing, by flycatching and running along branches. Some habits and calls very like those of Fraser's Forest-Flycatcher *Fraseria ocreata*.

Melaenornis annamarulae

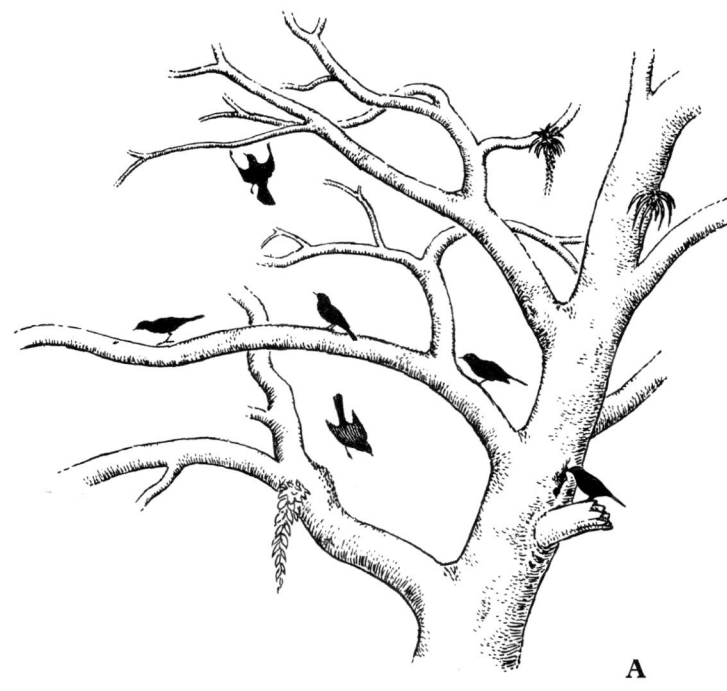

A

Voice. Tape-recorded (CHA, GAR, KEI). Song, 5–7 loud, ringing notes, with many variations; the following recorded at Mt Nimba by S. Keith: (a) 3–4 long, pure, downslurred, measured whistles followed by faster lower notes, 'peeew-peeew-peeew-peeew-toytoytoy', (b) 5 upslurred notes, 'tooey-tooey-tooey-tooey-tooey'; (c) 3 long downslurred notes preceded and followed by shorter ones, 'per-peeew-peeew-peeew-doy'; (d) 'pee-pee-pee-purpurpur' or 'pee-pee-purpurpurpur'. Intervals between songs *c.* 3 s; usually does not repeat the same song, prefers to mix them up. Calls described as loud and strident, like certain calls of drongos and Fraser's Forest-Flycatcher; also said to give a repeated thin, soft 'wheep-wheep' and a faint churring.

General Habits. Frequents highest parts of tall trees, in primary lowland rain forest in Liberia (never on slopes) and in middle of clearings planted with maize or often in large leafless trees on inselbergs in Ivory Coast. In continuous forest commonly perches conspicuously above closed canopy, at 30–50 m above ground, sometimes in leafless emergent tree. Seldom occurs below 20 m. Lives in pairs or groups of 3–6, staying in one vicinity for several days in succession. Inactive; perches upright and spends long periods doing nothing; group moves slowly through treetops, feeding as it goes. Hunts in rather infrequent, short aerial flights; also inspects crevices and moss of larger limbs of trees; and often hops or runs for about half a metre along a large limb, with head held down close to the branch (**A**). Seizes a food item during the run, in manner of malimbe *Malimbus* or cuckoo-shrike *Campephaga* (Balchin 1990, Demey and Fishpool 1991). Flycatching sallies are upward to glean insect from undersides of overhead branches, or downward to pounce on prey below perch (Gartshore 1989). Longer flights between trees slightly undulating, with tail tightly closed.

Food. Insects, especially large black alate hymenopterans, also small black beetles, small metallic beetles and a hairless caterpillar.

Breeding Habits. Not known. Birds with enlarged ovaries, Liberia, July–Aug.

Reference
Forbes-Watson, A. D. (1970).

Melaenornis ardesiacus Berlioz. Yellow-eyed Black Flycatcher. Gobemouche de Berlioz.

Plate 25
(Opp. p. 480)

Meloenornis ardesiaca Berlioz, 1936. Bull. Mus. Hist. Nat. Paris sér. 2, 8, p. 329; Mbwahi, west of L. Kivu.

Range and Status. Endemic resident, mountain forests of Albertine Rift from SW Uganda to N end of L. Tanganyika. Uganda, fairly common in Bwindi-Impenetrable Forest at 1550–2150 m. Zaïre, frequent and locally common from west of L. Edward to Ruzizi area and Itombwe highlands at 1300–2000 m. 8 birds

Melaenornis ardesiacus

seen at several places near Mohange (0°25′S, 28°55′E) in 2 days (Chapin 1953a); numerous near Tshibati (2°14′S) at 1785–2155 m (Chapin 1978). Rwanda, only west of Crête Zaïre-Nil, including Ruzizi-Gisovu area, at 1700–2450 m (Dowsett-Lemaire 1990).

Description. ADULT ♂: uniform dark blue-grey except that belly and flanks slightly paler and chin, lores and forehead blackish; primaries and secondaries black with shiny blackish grey undersides; tail long, well rounded, with T1 and T2 blackish next to shaft, T3–T6 with blackish inner webs; underside of tail black, a little glossy. Bill quite broad at base, stubby, hooked at tip; 6 quite strong rictal bristles; bill black, eyes bright pale yellow; legs and feet black, soles dull cream. Sexes alike. SIZE: (7♂♂, 6♀♀) wing, ♂ 85–90 (87), ♀ 80–86 (84); tail, ♂ 83–89 (85), ♀ 76–85 (82); bill, culmen ♂ 11–13 (12), ♀ 11–13 (12, to skull 18·2–20·0); tarsus ♂ 20–22 (21), ♀ 21–22 (21). WEIGHT: (Uganda) ♂ (n = 10) 26–35 (29·1), ♀ (n = 6) 27–34 (28·9), (Zaïre) 1 ♂ 33·7, 1 ♀ 29·7.

IMMATURE: like adult but fine whitish spots on breast, particularly at its sides, and less distinct ones on belly and undertail-coverts. Tail narrow at base, with rounded or wedge-shaped tip (photo in Prigogine 1971). Eyes dark, or greyish yellow.

NESTLING: undescribed.

Field Characters. A quite large, slim flycatcher of Rift mountain forests, dark bluish grey (but can look slaty black in poor light); adult distinguished from Northern Black Flycatcher *M. edolioides* by forest habitat and pale yellow eyes, and from grey *Muscicapa* flycatchers by larger size and longer tail. Immatures have dark eyes but white spots on underparts.

Voice. Tape-recorded (LEM). Call a rasping 'raap, raap' and harsh 'tch-tchec', also described as a slow 'tchiip, tchiip' followed by a short 'tche' repeated several times rapidly.

General Habits. Quite a common flycatcher of edges of and clearings in primary and secondary forest; also in bottoms and slopes of open bushy valleys, using scattered shrubs, small trees and columns of lianas. Sometimes occurs quite high in tall trees at sides of forest roads. Solitary or in pairs or family groups of 3–4, perching conspicuously 1–12 m above ground, on projecting branches of bare or leafy bushes and on tall flower spikes of *Lobelia giberroa*. Constantly raises and lowers tail. Does not join mixed-species foraging flocks. Restless. Flycatches from open perch, and searches foliage like a warbler (Williams 1959), also pounces and hops on bare ground and roads (Dowsett-Lemaire 1990). When disturbed, disappears quickly into nearest thick growth. Silent.

Food. Insects, including caterpillars and beetles.

Breeding Habits. Solitary nester, evidently monogamous; territorial. Territory size 5–10 ha; pairs often not in contact with others (Dowsett-Lemaire 1990).

NEST: cup made of 2 types of soft green moss, loosely built, lined with fine stems and brown fibres; placed on forks 4–12 m (mainly 4–5) above ground in small trees. One nest on topmost fork of 12 m tall *Alangium chinense* tree. Ext. diam. 180–200; int. diam. 60, int. depth 40–50.

EGGS: 2. Brownish white, with numerous brown spots. SIZE: (n = 2?) c. 23 × 15; (n = 1), 22·0 × 16·9.

LAYING DATES: Zaïre, Jan–Apr.

DEVELOPMENT AND CARE OF YOUNG: at c. 12 days young have long bluish grey down on head, median tract of bluish grey feathers on nape, back and throat, and tracts of same colour on sides of breast and flanks; wings feathered; feathers appearing on rump and thighs. At c. 15 days young well feathered, blue-grey, feathers of neck, breast, belly and tail pale-spotted at end; bill yellow, dusky on culmen, palate and gape yellow-orange. At c. 19 days wing 45–47 and tail 12–17 long and at 25 days wing 60–65, tail 25–30 (hand-reared birds).

References
Chapin, R. T. (1978).
Dowsett-Lemaire, F. (1990).
Prigogine, A. (1971).
Williams, J. G. (1959).

Melaenornis edolioides (Swainson). Northern Black Flycatcher. Gobemouche drongo.

Melasoma edolioides Swainson, 1837. Birds Western Africa, 1 (Jardine, Naturalist's Library, 17, Ornith., 7), p. 257, pl. 29; Senegal.

Forms a superspecies with *M. pammelaina*.

Range and Status. Endemic, resident. S Mauritania (between Assaba and Senegal R.), Senegambia (south of 14°N and along coast north to Dakar), Guinea-Bissau, Guinea in N and from Futa Djalon to Conakry, Sierra Leone (Kilimi area), not certainly recorded from Liberia, Ivory Coast (northern guinea savanna woodland south to Maroué and Toumodi), Mali (uncommon, south of 14°N, along Niger valley from Mopti to Bamako), Burkina Faso (around Ouagadougou, and Bobo Djoulasso to Dangouadougou, also Arli-Pendjari Nat. Parks), Ghana (rare or uncommon, Shai hills, Cape Coast, Weija, Achimota, Akropong, Accra plains and Kete Kratchi north to Mole Nat. Park, Tumu and Gambaga), Togo (Bafilo, Lama-Kara, Landa-Pozanda), Benin (Arli-Pendjari Nat. Parks to Cotonou), Niger ('W' Nat. Park and Korogoungou), Nigeria (common and widespread, Badagri to Lagos and soudanian zone north to Sokoto, Kano, Kagoro and Gashaka-Gumti G. R.), Cameroon (south to Korup Nat. Park (occasional), Bamali, Mbankuop, Kounden, Lepopomo, Bertoua and Kombetiko). Congo records from Brazzaville, Kwamouth and Gamakala (Malbrant and Maclatchy 1949, Salvan 1972) require confirmation: confusion with Southern Black Flycatcher *M. pammelaina*? S Chad (Baibokoum and Moundou north to Mao, Bongor, Massenya, Sarh and Moissala), NW Central African Republic (south to Bozoum, Bangui and Haut Kemo), NE Zaïre (relatively common, Dika and Mauda, upper Uele to Rwindi, Molindi R. and L. Edward), Sudan (southern Darfur to Didinga hills, also Boma and Blue Nile), S Eritrea, Ethiopia (uncommon or frequent, widespread W Tigre, Gondar, central Godjam, E Wollega and W Illubabor to NW Harar, N and W Bale and SE Sidamo), Kenya (Moyale and from west shore of L. Turkana to S Nyanza and Mara R., including Nakuru and Uraguess), Uganda (except parts of SW and NE), NW Tanzania (Mwanza). Common to frequent.

Description. *M. e. edolioides* (Swainson): Senegambia to W Cameroon. ADULT ♂: entire plumage very dark slate or bluish grey, almost black, wings and tail slightly darker, more brownish black; crown, mantle, scapulars, tertials, upperwing-coverts and upperside of tail-feathers glossed greenish blue-grey. Tail slightly rounded (T6 15–20 mm shorter than T1); tail-feathers and tertials faintly barred. Median third of outer web of inner primaries narrowly edged blue-grey; underside of flight feathers with greenish bronze sheen contrasting with black underwing-coverts and axillaries. Bill black; eyes olive-brown to dark brown; legs and feet black. ADULT ♀: like ♂ but duller, less black; upperparts blackish sooty grey, more slate-grey on rump and uppertail-coverts; underparts slightly paler and more blackish ash-grey. SIZE: (7♂♂, 8♀♀) wing, ♂ 92·5–104·5 (98·9), ♀ 91–102·5 (97·5); tail, ♂ 89–112 (101·7), ♀ 92–106·5 (100·3); bill, ♂ 17·5–20·5 (19·0), ♀ 17·5–20·5 (19·2); tarsus, ♂ 23–24

Melaenornis edolioides

(23·6), ♀ 22–24·5 (23·4). WEIGHT: (Senegal) 2 ♀♀ 30, 30; (Nigeria) unsexed (n = 3) 28–36 (32·0).

IMMATURE: like adult ♀ but with ochraceous spots at tip of median and greater upperwing-coverts, wings and tail browner.

JUVENILE: entirely sooty black, upperparts and underparts including scapulars, axillaries, underwing-coverts, upper- and undertail-coverts densely spotted rusty ochre (each feather with dark grey base and blackish tip, with wide crescent of rusty ochre in centre); face, forehead, crown and nape streaked, not spotted. Wings sooty brown-black, upperwing-coverts with broad subterminal rusty ochre crescent, primaries with median third of outer web greyish buff. Tail sooty black. Bill, legs and feet brown.

NESTLING: undescribed.

M. e. lugubris (Müller): E Cameroon to Eritrea and W Ethiopia (Kaffa and Gemu Gofa north to upper Mareb and Mai Aini, Bahar Dar and Gondar, Tigre, east to SW Sidamo), W Kenya (L. Turkana to South Nyanza and Mara R.) and Tanzania. Duller, more sooty black, base of inner web of flight feathers pale grey, ♂ without blue-grey gloss. WEIGHT: (Uganda, May, Dec) ♂ (n = 3) 27–36 (30·7), 1 ♀ 36; (Kenya, June, Sept) ♂ (n = 3) 35–42 (39·3); (Ethiopia, May–July, Sept, Oct) ♂ (n = 5) 30–34 (31·6), ♀ (n = 9) 27–36 (31·8).

M. e. schistaceus (Sharpe): N and E Ethiopia (SE Tigre to Sidamo, Harrar and W Ogaden) and N Kenya (Moyale). Much less black, greyer. WEIGHT: (Ethiopia, Feb, May, Oct) ♂ (n = 2) 30, 32, ♀ (n = 4) 28–34 (31·2).

Field Characters. A large, all-black, rather slim and long-tailed open-country flycatcher. Similar but allopatric Southern Black Flycatcher *M. pammelaina* is blacker, more glossy steel-blue and has a square or very slightly forked, not rounded tail. Tameness, unobtrusive behaviour, thin bill, brown eye, lack of gloss in plumage and rounded tail separate it from drongos *Dicrurus* spp. Foraging behaviour, upright stance, slender shape, thin bill and legs differentiate it from all-black boubous *Laniarius* spp.

Voice. Tape-recorded (C, CHA, ERA, GREG, GRI, MOR). Territorial song, a lengthy series of short phrases delivered in leisurely manner, with gaps between phrases; most notes pure but with a buzzy overtone, some thin and very high-pitched, others rather scolding, 'tsee-u-weet', 'tuwee-teewiteewi', 'yewee-teedudit' or 'yeweet-weeliew', 'ptee-yay', 'tsee-ew', 'psee-tewee' or 'tsieu-tee-teeh'; given by ♂ or in duet by ♂ and ♀; during aggressive encounters song more continuous, sustained, with more rasping notes. Flying young give similar but simpler version of song (C. Chappuis, pers. comm.). Contact call, a long, thin prolonged 'tseeeu' or short clicking 'tsic'. Warning call, a long, deep, rasping 'kchchew'.

General Habits. Inhabits moist woodland, acacia woodland, edges of riverine forest in open woodland (in Chad, gallery forest dominated by *Khaya senegalensis*), bushland, orchard-bush savanna, woodland savanna, inselbergs, cultivation, even banana plantations and gardens; also occurs in almost arid environments (e.g. in Turkana). Up to at least 1300 m in Adamawa, Cameroon, and 1800 m in Kenya.

In pairs or family groups of 3–4. Joins mixed-species flocks. Forages mostly on ground, either darting from low perch, where it sits upright with body and tail at the same angle, or hopping on ground to snatch prey; also sallies to snap up prey on leaf or bark, particularly when exploiting swarming ants or termites. When alarmed, crouches on perch with body horizontal and legs bent, slowly raises and lowers tail, and emits warning call.

Mobs snakes, even cobras, with aggressive fluttering flight, calling and clapping bill.

Moves seasonally with the rains, at least in dry northern Kenya.

Food. Arthropods, mainly insects: beetles, mantids, grasshoppers, cicadas, ants, caterpillars; also small millipedes and spiders.

Breeding Habits. Monogamous. Territorial; perches on top of bush or small tree up to 11 m above ground and gives territorial song; responds vehemently to playback of its song, displaying underside of spread and raised wings and giving nervous movements of spread tail. ♂ and ♀ greet each other with song duet; ♂ leans forward, flicks closed wings without showing underside and wags tail, first drooping tail and raising wings, then the reverse; ♀ quivers wing-tips in upright posture with tail depressed vertically.

NEST: coarse, shallow cup, made of twigs, dry grass, plant stems and pieces of bark, lined with fine rootlets and fibres; placed in concealed situations, e.g. cavity at top of dead stub, 2–6 m above ground or even in bunch of bananas; sometimes uses old nest of other species.

EGGS: 2–3. Oval; glossy; white, greenish white or very pale rufous, heavily spotted and blotched rufous, with some pale purple or lilac markings. SIZE: (n = 2) $21-21 \cdot 3 \times 15 \cdot 3-15 \cdot 7$.

LAYING DATES: Senegal (young July); Ghana, May–July; Nigeria, May, June; Cameroon, Oct (♂ with very enlarged testes Mar); Central African Republic (young Oct); Zaïre, May, Nov, Dec (♂♀ with very enlarged gonads Feb–June); E Africa: Region A, Mar–May; Region B, Oct, Dec–June; Region C, Dec; breeds mainly in the rains; Ethiopia, Apr, June.

INCUBATION: by ♀; period unknown.

DEVELOPMENT AND CARE OF YOUNG: young fed by both parents; nestling period unknown.

Plate 25
(Opp. p. 480)

Melaenornis pammelaina (Stanley). **Southern Black Flycatcher. Gobemouche sud-africain.**

Sylvia pammelaina Stanley, 1814. In Salt, Voy. Abyss. App., p. 47; Mozambique.

Forms a superspecies with *M. edolioides*.

Range and Status. Endemic resident, southern Africa. Somalia, 2 old records from Juba forests. Kenya, uncommon or locally common east of Rift Valley, from sea-level to 1800 m, but scarce or absent in much of E plateau and lowlands; makes close approach to range of Northern Black Flycatcher *M. edolioides* and may overlap it (the 2 difficult to separate in the field) on Mt Nyiru and Ndoto Mts and between L. Baringo and Nakuru; present on lower Tana R.; a few records in arid Tsavo East Nat. Park, but species unlikely to breed there. Tanzania, in NE from Mbulu to Kilimanjaro; widespread but uncommon in W and SW, from Uganda and Rwanda borders to Mozambique border, east to Tabora and through highlands from Ufipa and Iringa to Ngurus and Usambara. Zaïre, SE Kivu, Shaba, S Kasai Occidental. S Burundi. Angola, widespread throughout, except N and coastal lowlands. Zambia, throughout. Malaŵi, throughout, but sparse above 1500 m and at lowest altitudes and absent from Nkhata Bay District. Namibia, Ovamboland near Angolan border, and Caprivi strip. Botswana, uncommon to fairly common in N; along E border to Limpopo R.; in E Kalahari

Melaenornis pammelaina

woodland west of Serowe; in SE west to Moshaneng. Zimbabwe, locally quite common, but sparse in open savanna and absent between Ramaquabane R. and Diti Tribal Trust Land (Limpopo valley) and from suitable-looking woodland in E highlands on moist, E-facing slopes; sparse in Chipinga Uplands. Mozambique, widespread and quite common; may be more widespread in N than shown. South Africa: Transvaal, scarce in semi-arid in W and frequent to common in E, with breeding density of 1 pair per 30 ha (Nylsvley); E Swaziland; Natal and E Cape west to Bushman's R.

Description. *M. p. ater* (Sundevall): SW Malaŵi, lowland S and E Zimbabwe, Mozambique (Sul do Save), Swaziland, South Africa and SE Botswana. ADULT ♂: uniformly black; with steel-blue gloss on head and body feathers and exposed parts of tertials and tail-feathers; concealed parts of tail-feathers only slightly glossy; primaries and secondaries mainly brownish black, not glossy. Underside of flight feathers and tail-feathers glossy brownish black. Tail square-ended or slightly forked: T1 3–9 shorter than T6. 2 main rictal bristles on each side, rather weak, and *c.* 10 thinner bristles curling forward under chin. Bill black; eyes dark brown; legs and feet black. Sexes alike. Albinos known (Marshall 1978, Edwards 1988). SIZE: (10 ♂♂, 10 ♀♀) wing, ♂ 105–116·5 (110), ♀ 97–114 (105); tail, ♂ 87–100 (93·7), ♀ 81–97 (90·4); bill to skull, ♂ 17–19·5 (18·1), ♀ 17–19·5 (18·4); tarsus, ♂ 20·5–22·5 (22·3), ♀ 21–24·5 (22·4). WEIGHT: (Kenya) ♂ (n = 5) 22–30 (27·0), ♀ (n = 4) 21–30 (26·3); (southern Africa) unsexed (n = 8) 24·2–32·4 (29·8).

IMMATURE: very dark brown, plumage rather fluffy, not glossy, with many small buff spots on head and neck, sparser ones on mantle, and sparser and larger ones – short bars rather than round spots – on upperwing-coverts, back, rump and uppertail-coverts; underparts heavily marked with buffy crescents. Primaries, secondaries, tertials and tail unmarked brown-black. Eyes dark grey. In older bird, upperparts glossy blue-black with a few small buff spots mainly on mantle and upperwing-coverts, underparts less glossy, with a few buff bars mainly on chin and lower breast, and still profuse whitish buff bars on lower belly and vent. Subadult bird has upperparts like adult but tiny buff spots at tips of tertials and tail-feathers and buff fringes to greater wing-coverts; underparts matt greyish black, with whitish buff tips to vent feathers and undertail-coverts.

NESTLING: pinkish grey, sparsely covered with grey down (Ade 1976).

M. p. pammelaina (Stanley): Mozambique (north of Sul do Save), SE Malaŵi, S and E Tanzania. Like *ater* but smaller. SIZE: ♂ (n = 6) wing 100–107 (103), tail 82·5–89·5 (85·6); ♀ (n = 2) wing 101, 103, tail 85, 85.

M. p. diabolicus (Sharpe): NW Botswana, N Namibia (not Caprivi Strip), S Angola. Like *ater* but larger. SIZE: 1 ♂, wing 117·5, tail 99; ♀ (n = 3) wing 107–114 (110), tail 95–101 (98·3).

M. p. tropicalis (Cabanis): Zaïre (except S?) and E Africa north of *pammelaina* and *poliogyna*. Size of *ater* and like it except that upperparts are glossed with violet.

M. p. poliogyna Lawson: Zimbabwe (centre, W and N), Namibia (Caprivi Strip), Zambia, Angola (not S), S Zaïre?, NW Malaŵi and SW Tanzania. Size of *ater* and like it except that ♀ has throat and breast washed grey.

TAXONOMIC NOTE: races are poorly defined, clinal and often not recognized; see Clancey (1967).

Field Characters. 19–22 cm. A silent and retiring but quite conspicuous woodland flycatcher with glossy all-black plumage and black soft parts; in flight, bases of primaries and secondaries look pale and translucent. Juvenile blackish, buff-spotted on head, browner below with black crescents. In Kenya range abuts that of Northern Black Flycatcher, which is greyer and lacks gloss, and has a rounded tail. Closely resembles Square-tailed Drongo *Dicrurus ludwigii* in size, shape, colour and behaviour; they can occur in same habitat but drongo is usually in forest, and has notched tail, deep red eye and harsh voice. Also resembles Fork-tailed Drongo *D. adsimilis*, which is larger (22–25 cm), noisy, with deeply forked fishtail-shaped tail, wide-based bill, red eye and flat, sloping forehead; forehead of Southern Black Flycatcher is high and rounded. Another look-alike species is ♂ Black Cuckoo-Shrike *Campephaga flava* without yellow on wing; it differs in having conspicuous yellow gape, rounded tail, different foraging habits.

Voice. Tape-recorded (9, 86, 88, 91, B, F, MOY, PAY). Song in Zambia somewhat variable, duration 2–5·5 s, a warble of measured notes uttered at rate of 18 notes in 5 s; a pleasant-sounding medley, if a little thin and high-pitched. In South Africa song is sweet, bold, almost thrushlike series of phrases separated by 1-s pauses. Most phrases are 3 clear whistles, 'tseep-tsoo-tsoo', with slight variants, the 'tseep' higher-pitched than the 'tsoo's. Some phrases are a fast, musical 'whee-wheewheewhee'. One song, of 18 phrases at regular intervals in 25 s, was 'tseep-tsoo-tsoo, tseep-tsoo-tsoo, tseep-cheeu, tseep-tsoo-tsoo, wheewheewhee, tseep-tsoo-tsi, tseep-tsoo-tsoo . . .'.

General Habits. Prefers open woodland, bushland, thornveld, parks, cultivation and forest clearings; typical of miombo woodland with scant undergrowth; also *Baikiaea* woods, tall riparian and alluvial acacia woods, well-developed mopane and mixed broadleaved woodland. Readily uses burnt ground. Occurs in coastal dune forest (Natal), plantations and gardens. Often found alongside Fork-tailed Drongo and Pale Flycatcher *M. pallidus* and sometimes with Mariqua Flycatcher *M. mariquensis*. Pale Flycatcher favours cover in understorey and Southern Black more open places below woodland canopy, perching in places with unrestricted view of ground.

Solitary or in pairs and family parties; in winter several associate loosely as members of mixed-species foraging party (Natal). Undemonstrative; largely silent. Feeds mainly on ground (locally, almost entirely on ground). Perches on low twig, fence or wire and patiently scans ground, shrike-like; when prey seen, bird drops to ground and quickly hops forward to capture it. Less commonly, perches high in tree on projecting branch and makes short dashes after insects passing in flight. Of 228 forays, 72% were pounces to ground, 24% flycatching sallies and 4% hovering to glean trunks and leaves (Fraser 1983); 96% of forays were outside canopy and 4% within. Prey is taken back to perch to be eaten; occasionally bird eats prey on ground then hops to catch another insect nearby. It may spend some time on ground catching ants near their holes.

Resident, but breeding summer visitor in extreme S of range (E Cape).

Food. Insects and their larvae; ants; beetles, spiders, centipedes and worms (Ginn *et al.* 1989); some prey items quite large, e.g. a locust (Vincent 1947). Also consumes aloe nectar and small fruits including *Solanum nigrum*.

Breeding Habits. Solitary breeder, evidently monogamous, not markedly aggressive in defence of territory; chases away Fork-tailed Drongos (but sometimes tolerates them near nest). Resemblance to drongos often cited as instance of 'aggressive mimicry' (Ginn *et al.* 1989).

NEST: thin, loosely-made, shallow cup, little more than a thin platform of twiglets and petioles overlain by neat lining of wiry, yellowish rootlets and tendrils (**A**). Bird uses side of crevice in which nest placed to form nest rim. Sited in thin woodland, parkland or open scrub growth; built in hollow in top of short stump or end of broken branch, often fire-blackened; also in fork where peeling bark provides support, between jagged edges of torn wood, and in deserted nests of Arrow-marked Babbler *Turdoides jardineii*, Kurrichane Thrush *Turdus libonyanus* and Red-headed Weaver *Anaplectes rubriceps* (Beasley 1985); old doves' and thrushes' nests frequently used in Zaïre. In the pendant weaver nest, flycatcher used a torn side opening and made its own nest inside, a rudimentary hemisphere, unlined, 50 mm wide. Tree-stumps much the commonest site, 0·5–5 m above ground, occasionally up to 9 m; av. of 71 nests, 2·6 m above ground. Nest generally in shade, sometimes in full sun. Other sites: bunch of bananas, aloe leaf bases, hollow in wall, top of drainpipe and rear light of abandoned tractor (bare electrical wires incorporated

A

into base of nest). Nest built by ?♀, accompanied by ?♂ when collecting material (S. K. Frost *in* Ginn *et al.* 1989).

EGGS: 2–4. Av. of 66 southern African clutches 2·6. In Malawi, 2–3, av. of 36 clutches 2·8. Once, clutch of 5 (Masterson 1981). Cream, beige or pale green, with ash-grey blotches heavily overlain with dense, heavy red-brown freckles, or lightly flecked with dark brown spots which coalesce to form ring around broad end. SIZE: (n = 160) 19·5–24·3 × 14·7–17·2 (21·6 × 15·9). Weight of shell (n = 26) 0·123–0·189 (0·158) (Masterson 1981).

LAYING DATES: E Africa: Region C, Oct; Region D, Mar, Nov; Angola, Aug–Oct; Zambia, Aug–Dec (mainly Sept–Nov); Malawi, Sept–Dec (mainly Oct); Botswana, Jan; Zimbabwe, Aug–Jan (Aug 10, Sept 170, Oct 159, Nov 71, Dec 20, Jan 4 clutches); South Africa: Transvaal, Sept–Feb (mainly Oct–Nov; 50 out of 69 clutches); Natal, Sept–Jan (mainly Oct–Nov).

INCUBATION: period 13–14 days. Incubating bird well disguised against fire-blackened wood background, and can be very difficult to detect (Ade 1976).

DEVELOPMENT AND CARE OF YOUNG: brood fed by ♂ and ♀; typically one parent arrives with food item on treetop near nest, calls quietly, and when mate appears carrying food, both birds fly straight to nest, land on it together and feed the young in turn.

BREEDING SUCCESS/SURVIVAL: adult retrapped 2·2 years after being ringed (Zimbabwe).

References
Lawson, W. J. (1964b).
Maclean, G. L. (1993).
Vincent, A. W. (1947).

Melaenornis pallidus (Müller). Pale Flycatcher. Gobemouche pâle.

Musicapa (sic) *pallida* J. W. von Müller, 1851. Naumannia, 1, (4), p. 28; Abyssinia and Kordofan, Sudan; restricted to Kordofan by Rothschild, 1913, Bull. Br. Orn. Club, 33, p. 65.

Plate 25
(Opp. p. 480)

Range and Status. Endemic, resident (some populations moving locally with rains?). Savannas of S Mauritania (records from Guidimaka and upper Senegal valley), S Senegambia (south and west of Nioro du Rip to Kédougou), Guinea (Doko and Fouta Djalon to Kindia east to borders of Mali and Ivory Coast), Sierra Leone (Karina district, Outemba area); not certainly recorded from Liberia; Mali (rare or uncommon, S Mts Malingués, Upper Falémé and Bafing R. and generally south of 12°N but north to 15°30′N at Ban Markala in central delta of Niger R.), Burkina Faso (south of Ouagadougou to Arli-Pendjari Nat. Parks), Niger ('W' Nat. Park), Ivory Coast (widespread and not uncommon throughout guinea savanna woodland, Lamto to 10°N, also locally in forest and coastal zones), Ghana (frequent, coast at Winneba and Cape Coast, Akwapim hills and Accra plains north to woodland savanna of e.g. Mole Nat. Park), Togo (south to Tomegbe and Fazao), Benin (south to Bassila), Nigeria (common and widespread, guinea savannas south to coast at Lagos, also soudanian zone north to Sokoto and Ririwai, Yankari Reserve, Falgore Game Reserve, Gashaka-Gumti Game Reserve), S Chad (Bessao and Moundou areas to N'Djamena), N Cameroon (south to Meiganga, Tibati, Nkolngem, Bafia and Bamali, once Korup Nat. Park: Rodewald *et al.* 1994), Mbini (mention by Fa 1990 requires confirmation), S Gabon (L. Onangué, Nyanga, Haut-Ogooué and Batéké Plateau), Congo (coast and sublittoral north of Pointe Noire, Brazzaville to Djambala and Batéké Plateau), SW Central African Republic (Bozoum to Bangui and Gribingui R.), Zaïre (lower Congo R. up to Kwamouth, upper Uele to L. Albert, Semliki R. to lakes Kivu and Tanganyika then to Kasai southward), Sudan (south of 13°N, Darfur, Kordofan and Bahr al Ghazal, Didinga Mts), Ethiopia and Eritrea (west of Rift Valley, Keren and W Tigre and Simen to Matama and NE Shoa, west and south to W Illubabor, S Gemu Gofa and Sidamo; status in Ogaden unclear; recorded in E Haud), Somalia (coast north to Mogadishu, Juba and Shebelle R., and north to E Haud along Ethiopian border and near El Hamurre), Kenya (except north and east of W and Central highlands), Uganda, Rwanda, Burundi, most of Tanzania, Mozambique (Cabo Delgado, Nampula and Zambezia, southward), Malaŵi (common on slopes of Nyika Plateau, sparse Nkhata Bay and Mzimba districts, lake littoral near Salima and Bana, also Mangoshi, Thyolo and Nsanje districts), Zambia (common and widespread, except Luangwa and Middle Zambezi valleys), Zimbabwe (widespread although uncommon, mostly on central plateau and SE lowveld; sparse, e.g. Middle Zambezi and Limpopo valley (absent Sabi valley), in major river valley systems), NE Namibia (Ovamboland and Waterberg), Botswana (fairly common in NE, uncommon in N and NW and in E near Francistown), South Africa (common, N and

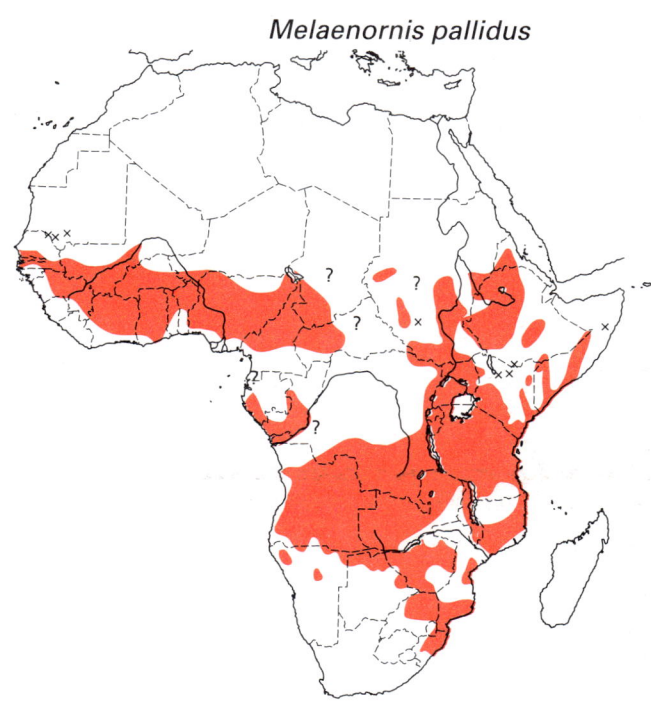

Melaenornis pallidus

central Transvaal and Zululand), E Swaziland and Angola (Cabinda and central plateau, south to central Huila, north to Luanda and W Malanje). Frequent to common.

Description. *M. p. parvus* (Reichenow): SW Ethiopia (west of lakes, north to Gibbe R.), Sudan (borders of Baro R.), Zaïre (possibly Uele district) and NW Uganda (south to Masindi). ADULT ♂: entire upperparts including sides of head, sides of neck, scapulars and uppertail-coverts dark grey-brown; lores and narrow eye-ring rusty buff. Chin, throat, foreneck and belly creamy white washed pale pinkish brown-buff; breast suffused pinkish buff to pinkish brown-buff; flanks, thighs, vent and undertail-coverts pinkish brown-buff. Wings and tail blackish brown; flight feathers edged rusty buff along outer web, secondaries and tertials also along tip of inner web; upperwing-coverts with faint rusty edges; tail-feathers with pale rusty tips (particularly in fresh plumage), T6 with narrow buffy edge along outer web; underside of flight feathers blackish brown with silver gloss, feathers broadly edged pink brown-buff along inner web; underwing-coverts and axillaries pinkish brown-buff. Bill blackish, base of lower mandible horn; eyes olive to chestnut-brown; legs and feet blackish. Sexes alike. SIZE: (5 ♂♂, 5 ♀♀) wing, ♂ 82–88 (85·6), ♀ 81–89 (86·0); tail, ♂ 72–75 (73·2), ♀ 65–78 (71·8); bill, ♂ 16–18 (16·8), ♀ 16–18 (17·0); tarsus, ♂ 21–22 (21·6), ♀ 19–21·5 (20·4). WEIGHT: (Ethiopia, June, Sept–Nov) ♂ (n = 10) 21–27 (23·8), ♀ (n = 10) 20–28 (23·4); (Ethiopia, May) 1 ♂ 26, 1 ♀ 26.

IMMATURE: like adult but duller and darker brown; median and greater upperwing-coverts, alula, secondaries and tertials irregularly but broadly edged and tipped buff or buffy white; tail-feathers narrower and more pointed, with faint (abraded) tips buff.

JUVENILE: upperparts, face, scapulars and uppertail-coverts

dark brown, densely streaked creamy white or pale buff (each feather with grey base, wide arrow-shaped creamy white spot in centre and brownish black edges, spots larger and more pearl-shaped on mantle, narrower on head, tear-shaped on scapulars). Wings and tail blackish brown; primaries narrowly edged pale buff along outer web, secondaries and tertials broadly edged buff-white on both webs, lesser and median wing-coverts tipped buff-white, greaters edged and tipped rusty buff to creamy white; tail-feathers narrowly fringed and tipped pale rusty buff. Underparts creamy white washed pale buff; breast, upper belly and flanks densely scalloped blackish brown (irregular edges to feathers).

NESTLING: undescribed.

M. p. pallidus (Müller): semi-arid regions from Senegambia to N Ghana and N Nigeria, east to Sudan and W Ethiopia (Baro R.), possibly also Zaïre (Uele District). Much paler, with a warm buff tone. The exact distribution of this subspecies and the following one is still badly documented.

M. p. modestus (Shelley) (including '*nigeriae*'): south of *pallidus*, from Guinea, SE Mali, to Central African Republic. Underparts colder than *parvus*, less buffy, greyer, particularly on breast and flanks. Larger but shorter tailed: (7 ♂♂, 4 ♀♀) wing, ♂ 85–91 (87·4), ♀ 82–86 (84·3); tail, ♂ 66–73 (69·5), ♀ 65–67 (66·0). WEIGHT: (Nigeria) unsexed (n = 14) 16·5–24·0 (20·7); (Nigeria) unsexed (n = 13, Apr–May) 18·2–23·6 (20·7), (n = 12, Nov–Feb) 17·9–23·5 (21·3).

M. p. bowdleri (Collin and Hartert): Eritrea and central Ethiopia (south to Burji, L. Abaya, Arussi Plateau). Browns greyer than *parvus*. Larger: wing, ♂ (n = 8) 85–95 (90·3), ♀ (n = 5) 85–88 (86·4). WEIGHT: (Ethiopia) ♂ (n = 4) 24–28 (26·2), 1 ♀ 27.

M. p. duyerali (Traylor): NE Ethiopia (Duyer Ali) and Central Somalia (El Bur). Paler than *parvus*, with grey upperparts and whitish underwing-coverts.

M. p. erlangeri (Reichenow): S Somalia (Bardera and Serenli to Hanole). Like *pallidus* but more sandy, less buffy brown, almost without breast-band, flanks and crissum washed buffy; smaller, with striking sexual size dimorphism: wing, ♂ (n = 6) 79–84 (81·7), ♀ (n = 13) 73–78 (76·0).

M. p. subalaris (Sharpe): coastal and E Kenya (lower Tana R. and Lamu to Bura, Lali, Samburu) and Tanzania (coast to Moa, inland to Amani). Like *erlangeri* but without sexual size dimorphism: wing, ♂ (n = 40) 77–86 (82·5), ♀ (n = 21) 76–85 (81·2). WEIGHT: (Kenya, Feb, Mar) ♂ (n = 5) 20–22 (21·2), 1 ♀ 20.

M. p. bafirawari (Bannerman): S Ethiopia (Kambata, Bishoftu and Negelle, Sidamo), NE Kenya (Garissa and Wajir to Djiroko). Breast-band and upperparts greyish brown, underwing-coverts white. Small: wing, 2 ♂♂ 73, 78, 1 ♀ 74. WEIGHT: (Ethiopia) 2 ♂♂ 22, 22.

M. p. murinus (Hartlaub and Finsch): W and S Kenya, N Tanzania, Uganda (except NW) and Sudan (Didinga Mts) to Gabon, Congo and Angola through Zaïre and to NE Namibia, N Botswana, W and S Zambia (Southern and Central Provinces, Barotse Province northwards into Mankoya and Balovale) and Zimbabwe (except SE). Paler, more grey-brown and much larger than *parvus*: (70 ♂♂, 41 ♀♀) wing, ♂ 89–105 (98·9), ♀ 87–102 (94·8). WEIGHT: (Kenya, June, Sept, Nov) ♂ (n = 4) 19–28 (23·8); (Zambia, May) ♀ (n = 2) 21·6, 24·5; (Zimbabwe) ♂♀ (n = 3) 21·1–23·2 (22·2).

M. p. aquaemontis (Stresemann): central Namibia (Waterberg Plateau). Very pale grey, palest subspecies, almost without breast-band.

M. p. griseus (Reichenow): extreme SE Kenya and Central Tanzania (Taveta, Dar es Salaam and N Tabora south to Iringa, possibly south-east to Mikindani) to N and E Zambia, and N and Central Malaŵi. Upperparts more brownish than *murinus*; smaller: (5 ♂♂, 7 ♀♀) wing, ♂ 90–101 (94·0), ♀ 84–94 (90·0).

M. p. divisus (Lawson): SE Zambia, S Malaŵi (Nsanje district), SE Zimbabwe, Mozambique (south to Delagoa Bay), and Swaziland. Much greyer than *griseus*; smaller: (4 ♂♂, 5 ♀♀) wing, ♂ 89–94 (91·8), ♀ 84–92 (88·0).

M. p. sibilans (Clancey): Mozambique (south of Sul do Save) and South Africa (Natal). Less ochraceous brown than *griseus*; upperparts darker and browner than *divisus*. Large: (6 ♂♂, 10 ♀♀) wing, ♂ 92–96 (93·8), ♀ 86–91 (87·7). WEIGHT: (Mozambique, May, June) 2 ♂♂ 23·2, 23·5, 1 ♀ 21·1.

TAXONOMIC NOTE: in *M. pallidus*, subspecies *bafirawari*, *bowdleri*, *duyerali*, *pallidus* and *parvus* have a long tail representing 82–95% of wing whereas the other subspecies have a tail only 73–80% of wing, like *M. microrhynchus*. What does such a structural difference mean? Are there 2 species? Furthermore, *M. p. subalaris* does not intergrade with *griseus* in SE Kenya; together with *erlangeri*, it might represent a separate species (see Traylor 1970a). The geographical variation of both African Grey and Pale Flycatchers has been studied only in museum skins; more field work on ecology and behaviour is required to understand the *M. pallidus/M. microrhynchus* complex and the taxa involved.

Field Characters. A nondescript, chat-like, round-headed flycatcher which sits on a low perch and sallies to ground for prey. Plumage varies with locality, upperparts darker or lighter brown or grey-brown, underparts paler, grey-buff or brown-buff but not white except on throat and centre of belly. Overlaps in southern Africa with similar-sized Mariqua Flycatcher *M. mariquensis*, which has pure white underparts, browner upperparts (no grey tone) and pale panel on wing. More difficult is distinction from African Grey Flycatcher *M. microrhynchus* in eastern Africa. Pale Flycatcher is larger and browner, shows less contrast between dark breast and pale throat, has horn base to lower mandible, lacks streaks on head and has less definite pale wing-edgings.

Voice. Tape-recorded (72, 86, 88, 91, B, F, CHA). Song an irregular jumbled series of raspy notes, including repetitions of motifs, e.g. 'turru-tee-tweet', 'ptiew-tew-teeweet' or 'piew-tee-ttwit'; described in southern Africa as 'treeky-treeky, trip-tricky, treeky chip chippy-chee, witti-wttittee, witty witty witty . . .' (Maclean 1993). In reaction to playback song is faster, more continuous and varied, and louder. Contact call, a soft 'churr', also a persistent 'chrr' before roosting in evening; alarm calls, a thin mammal-like squeak 'see-see', a rasping 'tsek' or a rasping, rolled 'krrrr'; young emit a thin, high-pitched hiss, 'tchchcheee', also a low 'djjjèèè' or a petulant 'squeer'.

General Habits. Inhabits miombo (*Brachystegia*), gusu (*Baikiaea*) and mopane woodland, moist open woodland, wooded savanna, orchard-bush savanna, both thorny and soft-foliaged (e.g. stands of *Nauclea* and *Hymenocardia* in S Gabon) bush country, shrubs (e.g. *Butyrospermum*), leleshwa scrub (growth of *Tarconanthus* with scattered trees and undergrowth of short grass), wooded steppes, grass steppes with thicker clumps of bushes, inselbergs, coastal thickets, and a wide variety of open or secondary habitats, including clearings, cultivation and gardens, with well-developed understorey or shrub layer. Up to at least 1300 m in

Adamawa, Cameroon; not above 1600 m in Zimbabwe but up to 1750 m in Zambia and 2000 m in Kenya. In southern Africa occurs mainly in broad-leaved woodland and savanna but enters acacia savanna where Mariqua Flycatcher absent; on Waterberg plateau, Namibia, inhabits low sand dunes and sandy landscapes; the same ecological exclusion exists also in Zimbabwe and Zambia where the ranges of the two species interdigitate and form a mosaic. In Tanzania and Kenya occurs in moister habitats than African Grey Flycatcher; in low rainfall areas restricted to edges of richer woodland and forest along rivers, although the subspecies *bafirawari* ('Wajheir Grey Flycatcher') inhabits < 250 mm rainfall areas below 500 m.

In pairs or small family groups (exceptionally up to 12 birds). Somewhat chat-like; gently flicks tail on settling; sometimes slightly spreads and closes wings, particularly when alerted or on returning to perch. Tame, unobtrusive and rather retiring and silent. Perches on low outer branch of small tree or shrub at edge of clearing or along earthen track, drops to or even hops on ground to catch prey; also sallies and makes swooping or hovering flight to snap up prey on or under leaf or on bark; sometimes hawks insects in flight, mainly when exploiting swarming ants and termites.

Migratory in some countries, e.g. in Sudan where subspecies *murinus* is probably a rains visitor, nominate *pallidus* migrates north with the rains while *parvus* probably moves to Uganda; but these movements are poorly documented.

Food. Insects, especially beetles, flies, hymenopterans, Lepidoptera (including caterpillars), bugs, Orthoptera, mantises, swarming winged termites; also small fruits, e.g. *Rhus pyroides*.

Breeding Habits. Monogamous; territorial, but behaviour undocumented except that birds (usually pair members but up to 3 individuals) aggressive to play-back of their territorial song.

NEST: a ragged but strongly built, thin-walled bowl or cup-shaped structure of fibres, roots, grass, creeper stems, flower-tops, twigs, sometimes including dead leaves or lichens, lined with fine rootlets and tendrils, sometimes so flimsy that eggs can be seen through bottom; ext. diam. 88, ext. depth 54, int. diam. 60, int. depth 44; placed 0·5–6 m above ground (mean of 35 southern Africa nests 2·5), usually in fork against trunk or far out on branch of densely foliaged tree, shrub or sapling; thornless trees are mostly preferred.

EGGS: 2–3; Ivory Coast av. 2, Malaŵi mean (16 clutches) 2·6, Zimbabwe mean (21 clutches) 2·6, southern Africa mean (77 clutches) 2·5. Oval; slightly glossy; greenish white or pale bluish green, densely speckled, streaked or spotted all over with burnt umber, pale red-brown or yellow-brown, with underlying spots of ash, violet-grey or pale lilac, streaks thicker at larger end where they may concentrate to form ring. SIZE: (n = 13, Nigeria) 17·9–20·4 × 13·3–14·7 (19·2 × 14·1); (n = 36, Zimbabwe) 18·0–22·7 × 12·9–15·9 (19·9 × 14·8); (n = 169, southern Africa) 17·8–23·4 × 12·9–16·5 (19·8 × 14·8).

LAYING DATES: Gambia, June; Ivory Coast, Apr–July; Ghana, Apr–Aug; Nigeria, Feb–Sept, mainly May–July; Gabon (young out of nest but still dependent on parents Jan, Sept, Nov); Central African Republic (♂♂ with testes enlarged Jan); Sudan, Apr–June; Ethiopia, Apr–June; Somalia, Apr (dependent young Sept); E Africa: Region A, Mar, Aug; Region B, Mar–May; Region C, Mar; Region D, Oct–May; Region E, Aug, Sept, Apr–June; breeding mainly in the rains; Rwanda, Nov, Jan, Mar; Zaïre (♂♀ in breeding condition Sept–Oct at beginning of rains in lower Congo, juvs in Mar–Apr in Upper Uele); Malaŵi, Sept–Dec, mainly Oct–Nov; Zambia, Sept–Dec, mainly Sept–Oct; Zimbabwe, Aug–Jan, mainly Sept–Nov; Mozambique (fledged young being fed Jan); Transvaal, Aug–Feb (mainly Oct); Natal, Sept–Jan (mainly Oct–Nov); Angola, Aug–Sept.

INCUBATION: by ♀. Period: 14 days.

DEVELOPMENT AND CARE OF YOUNG: nestlings fed by both parents; period: 17 days. Young remain with parents for several months after leaving nest.

BREEDING SUCCESS/SURVIVAL: in Ivory Coast mean brood 1·6 fledglings (Thiollay 1971). In Zimbabwe, length of a complete nesting cycle 35 days, 37% of nests with eggs and 37% of eggs laid produced one or more fledglings; pairs made mean of 0·6 breeding attempts per season and raised 0·5 young per year; successful pairs produced mean of 2·3 fledged young (Vernon 1984). Bird ringed as adult in Zimbabwe was retrapped at same locality 4 years and 66 days later.

References
Thiollay, J.-M. (1971).
Traylor, M. A. (1970a).
Vernon, C. J. (1984).

Melaenornis infuscatus (Smith). Chat Flycatcher. Gobemouche traquet.

Saxicola infuscata Smith, 1839. Ill. Zool. S. Afr. Aves, pl. 28; Nieuwerust.

Plate 25
(Opp. p. 480)

Range and Status. Endemic resident, SW Africa. Angola, arid coastal plain of Moçâmedes and Benguela. Namibia, throughout but rare or absent in N and W; possibly much more widespread than shown. Botswana, fairly common and locally very common, north to 20°S and east to 26–27°E; in Makgadikgadi basin and north to Nxai Pan where locally common; uncommon in Xaudum. Common and widespread in

Melaenornis infuscatus

W South Africa. In Cape Province absent south of Nuweveldberge range but fairly common, if localized, resident west of it in SW Cape, where confined to low-lying areas north of Berg R., although until 1950s it occurred south to between Malmesbury and Moorreesburg (where suitable habitat now destroyed); common north of Elands Bay. In SW Transvaal fairly common in Christiana–Bloemhof area; records from Lichtenburg, Koster and Magaliesburg. In Orange Free State fairly common but very local, mainly between Christiana, Herzogville and Boshof and west of Luckhoff; record from Parys. Absent from desert where av. rainfall < 100 mm.

Description. *M. i. infuscatus* Smith: SW Namibia (Witputs, Great Namaqualand) and South Africa (lower Orange R. to SW Cape). ADULT ♂: forehead, lores, crown, nape, hindneck, sides of neck, mantle, scapulars and back greyish mid brown; rump and uppertail-coverts the same or warmer brown. Tail square-ended, dark brown, all feathers narrowly edged greyish or buffy. Ear-coverts and cheeks mid brown. Narrow buff-white orbital ring of featherlets, joining short buffy eye-stripe above lore. Chin and throat buffy off-white; breast greyish buff, merging into pinkish grey-buff on flanks (flanks sometimes brown) and into cream-white in middle of belly; belly of some birds with faint shaft-streaking; thighs brown; vent and undertail-coverts pale pinkish buff. Primaries dark brown with narrow pale buff outer edges and tips, secondaries dark brown with broad (1 mm) pale buff outer edges and tips, tertials dark brown with 2-mm wide rufescent or grey-buff outer edges and 1 mm pale tips; upperwing-coverts dark brown, fringed buff; underside of flight feathers and tail shiny dark brown; underwing-coverts and axillaries pale orange. Bill black; eyes dark brown; legs and feet black. Sexes alike. SIZE: (4 ♂♂, 8♀♀) wing, ♂ 117–124·5 (120·5), ♀ 110–121·5 (114); tail, ♂ 87–93 (88·6), ♀ 82·5–90 (86·5); bill to skull, ♂ 24·5–28 (25·7), ♀ 24–26·5 (24·7); tarsus (7♂♀) 25–28 (25·8). Much the largest subspecies; noticeably large-billed. WEIGHT: 1, unsexed, 37.

IMMATURE: forehead to back, lores, ear-coverts, neck, scapulars and upperwing-coverts warm, slightly rufescent dark brown, with numerous small pale buff spots on head and large ones on back. Primaries and secondaries edged pale rufous and tipped pale buff; tertials with broad pale buff tips; greater coverts and greater primary coverts broadly fringed pale rufous. Tail-feathers tipped pale rufous-buff, and outer ones edged with same colour. Feathers of underparts fluffy and silky; chin, throat, lower belly and undertail-coverts dirty white, with a few dark marks on chin and throat; breast and upper belly whitish with irregular dark brown or rufous-brown connecting streaks or mottles.

NESTLING: at *c*. 10 days skin dark red-brown, inside mouth deep orange, gape pale yellow, short grey down on head, pin feathers on crown, hindneck, sides of neck, scapulars and wings blackish with pale buff vanes breaking out (photo in Ginn *et al.* 1989).

M. i. seimundi (Ogilvie-Grant): South Africa (S-central and E Cape to upper Great Kei R. drainage, W Griqualand and SW Orange Free State). Upperparts warm brown, paler and less grey than in *infuscatus*; belly buffier than in *infuscatus* – underparts darker, more uniform, and tinged vinaceous. Intergrades with previous and following race. Smaller: (18 ♂♂, 6 ♀♀) wing, ♂ 112–119 (116), ♀ 102·5–111 (107·5); tail, ♂ 83·5–88·5 (85·6), ♀ 78–85 (81·6); bill to skull, ♂ 20·5–23·5 (21·8), ♀ 21–23 (21·9); bill slender.

M. i. namaquensis (Macdonald): Namibia except NW and SW. Browns of upperparts paler and redder than in *infuscatus*; lores and ear-coverts paler than crown, forming indistinct superciliary stripe; ear-coverts rufescent; throat whiter than in *infuscatus*; underparts vinaceous buff. Same size as *seimundi* except for tail. Tail, ♂ (n = 8) 79–82·5 (81·1), ♀ (n = 6) 77–80·5 (78·8). WEIGHT: 1 ♂ 39, 1 unsexed 37.

M. i. placidus Clancey: Botswana except extreme W and SW, N Cape (east of *namaquensis*), NW Orange Free State, W Transvaal. Like *namaquensis*, but slightly darker and grey above, less vinaceous above and below, and breast and flanks greyer. Smaller: (6 ♂♂) wing 102·5–110·5 (108), tail 74–82 (78·5); bill slighter.

M. i. benguellensis Sousa: Angola and NW Namibia (Kaokoveld). Like *namaquensis* but upperparts paler, ear-coverts more rufous, and upperwing-coverts and flight feathers more broadly edged pale rufous and buff; rump pinkish brown; underparts pale – creamy white with grey-buff wash, strongest on breast and flanks. Same size as *placidus* except that bill av. 1 mm longer.

Field Characters. A robust, rather chat-like, plain brown flycatcher with thrush-like bill (**A**) and quite strong, black legs. Secondaries and tertials pale-fringed; pale buff mark above lores, extending backwards as a short superciliary stripe. Smaller in north of range (19 cm) than south (21 cm). In pairs in open, arid country silently perching on low bush-top and dropping to ground for prey. Immature strongly marked, with buff spots and black streaks above and irregular dark V-marks below, and rufescent wings; family groups are

A

often seen. Juvenile blackish, could be mistaken for Dusky Lark *Pinarocorys nigricans* were it not for the plain, chat-like face (Hobson 1995). Pale Flycatcher *M. pallidus* is smaller and greyer; their ranges do not quite meet, although in Botswana they occur within < 50 km of each other west of Tsodilo Hill, at head of Selinda Spillway east of Seronga, and around Nata. Sympatric Mariqua Flycatcher *M. mariquensis* has white underparts.

Voice. Tape-recorded (88, 91, F). Song a quiet, conversational chirruping, 'chirr chirr cheep cheerrr chip chirrr krrt chirr cheep . . .', duration 2–5 s. 3–4 birds counter-singing together for 25 s, Transvaal, uttered confused babble of harsh 'cht' notes and liquid chirrups, sounding like cross between a caged Budgerigar *Melopsittacus undulatus* and a subsinging Great Reed-Warbler *Acrocephalus arundinaceus*. Anxiety call, thin 'tsee tsee'. Alarm, harsh buzzing 'krzzzt, kzzt kzzt peep' and near nest, loud 'chek chek', 'chrrt' or 'chrr'. Contact call a soft but carrying 'irrioo', 3–4 times. Adult soliciting mate utters soft 'chee-kreeow-kreeow'. Parents tending fledglings give continuous chattering or chippering.

General Habits. A bird of acacia savanna and open grass plains in semi-arid bushland with scattered shrubs and low trees. In SW Cape, slightly undulating, sandy, burnt-over ground; thorn scrub with sparse grass cover and widely spaced trees.

Singly; or in pairs, birds keeping close; often in family party of 4–5. Rather silent and sluggish; spends much time sitting conspicuously on top of small bush, looking for prey on ground. Also uses fences and telephone wires, but tends not to perch on low branches of shrubs and trees. Obtains food largely on ground; drops onto ground heavily or half-heartedly and hops toward prey with wings partly opened. Flies with prey back to same perch or new perch; sometimes eats prey on ground. Catches a fair amount of food on the wing; adept as a fly-catcher, if rather clumsy. Hovers readily. Longer flights between trees quite powerful, straight and low for short distance, or if higher, only slightly undulating, with final swoop.

Food. Insects including ants and grasshoppers, and small reptiles including blind-worms *Typhlops*.

Breeding Habits. Solitary nester, evidently monogamous. Highly secretive. Adult once begged from its mate, fluttering wings and raising and lowering its tail.

NEST: untidy cup of coarse grass, rootlets and thin stems built on base of twigs, lined with rootlets, whitish cottony bolls of kapok-karoobush and soft, furry leaves of *Helichrysum* (resembles nest of Karoo Chat *Cercomela schlegelii*). Int. diam. 66, int. depth 45. Sited in fork in *Rhigozum*, *Acacia*, *Boscia*, *Ziziphus*, scrub-mimosa or similar bushy shrub or tree, at height of (n = 30) 0·5–2·5 (1·2) m above ground. Nest not well concealed.

EGGS: 2–3, av. of 44 clutches 2·6. Clear bright greenish blue, boldly marked with black squiggles or red-brown and grey spots, concentrated at broad end. SIZE: (n = 83) 20·0–25·9 × 16·2–18·2 (23·6 × 17·2).

LAYING DATES: Botswana and Transvaal, Oct–Dec. In semi-arid areas further west, all months, according to spasmodic rainfall (Maclean 1993).

INCUBATION: period 14 days (n = 2).

DEVELOPMENT AND CARE OF YOUNG: eyes open on day 4, feathers appear on day 7. Young fed by both parents.

References
Blignaut, J. (1959).
Clancey, P. A. (1958).
Ginn, P. J. *et al.* (1989).
Maclean, G. L. (1993).
Vincent, A. W. (1947).

Melaenornis microrhynchus (Reichenow). African Grey Flycatcher. Gobemouche à petit bec.

Plate 25
(Opp. p. 480)

Bradyornis microrhynchus Reichenow, 1887. J. Orn., 35, p. 62; Irangi (= Kondoa Irangi) district, Tanganyika.

Range and Status. Endemic, resident. Somalia (Djibouti border and Bender Merhagno to Gardo, Eil and Obbia, and Ogaden border to Meregli and Mogadishu, south along coast and Juba basin) and central Ethiopia (N Somalian border to L. Zwai and S Shoa, thence south to Wabi Shebeli and SE Ogaden in E and S Kaffa in W, recorded in E Illubabor), SE Sudan (west to Mongalla), NE Uganda (Moroto and Kidepo Valley Nat. Park), Kenya (throughout except damp highlands, S coastal strip and driest parts of NE) and Tanzania (east to Pare Mts, south to Dodoma, Iringa, L. Nyasa and southern L. Tanganyika). Frequent.

Melaenornis microrhynchus

Description. *M. m. neumanni* (Hilgert): Sudan, S Ethiopia (north to Arussi), central and S Somalia, Uganda, N Kenya (south to Kapenguria, Murang'a and Wajir). ADULT ♂: entire upperparts including face, sides of neck, scapulars and uppertail-coverts grey-brown, lores, forehead and narrow eye-ring pale buff; forehead, crown and nape narrowly streaked blackish brown. Tail-feathers blackish brown, greyer on underside, edged buff along outer web and tipped buff when fresh. Foreneck, throat and chin white; breast pale greyish brown, darker on sides, rest of underparts including undertail-coverts buffy white; flanks and thighs mouse-brown. Flight feathers blackish brown, primaries and secondaries narrowly edged pale greyish buff along outer web, tertials edged and tipped pale greyish white-buff to brownish buff; upperwing-coverts dark greyish brown, lessers and medians edged pale greyish brown-buff, greaters fringed pale greyish white-buff along outer web and tip of inner web; underwing-coverts dark brown, edged fawn to rusty buff; axillaries mouse-brown, underside of flight feathers blackish brown with silver gloss, edged creamy to buffy white along inner web. Bill black; eyes olive-brown; legs and feet black. ADULT ♀: like ♂ but chin, throat and foreneck less white, more washed pale buffy brown. SIZE: (65 ♂♂, 34 ♀♀): wing, ♂ 72–86 (78·5), ♀ 73–85 (77·5); tail, ♂ 55–68 (61·6), ♀ 58–67 (61·2); bill (**A**), ♂ 15–18 (16·1), ♀ 15–17 (16·0); tarsus, ♂ 18·5–22 (20·2), ♀ 19–22 (20·7). WEIGHT: (Ethiopia, Nov, Dec, Feb) ♂ (n = 12) 17–22 (19·5), ♀ (n = 6) 17–20 (18·5); (Kenya, June, Oct) 1 ♂ 14, 1 ♀ 17; (Uganda, May) ♂ (n = 6) 19–22 (20·3).

IMMATURE: like adult but upperparts darker, browner, and underparts whiter; uppertail-coverts, tail-feathers, upperwing-coverts, secondaries and tertials with broad creamy white edges narrowing with wear.

A

JUVENILE: upperparts, face, scapulars and uppertail-coverts dark grey-brown densely streaked with creamy white (each feather with pale grey base, wide arrow-shaped creamy white spot in centre and brownish black edge, spots larger and round on mantle, narrower on head, tear-shaped on scapulars). Wings and tail brownish black; primaries and secondaries narrowly edged buff along outer web and creamy white at tip of inner web, tertials broadly edged buff-white along both webs, lesser and particularly median wing-coverts with broad buff-white spot at tip, greaters edged along outer web and with large pear-shaped spot at tip pale rusty buff to creamy white; tail-feathers tipped pale buff. Underparts creamy white; breast, upper belly and flanks densely scalloped (irregular blackish edges to feathers).

NESTLING: hatches naked, dark brown above, pinkish on belly.

M. m. pumilus (Sharpe): Ethiopia (Rift from L. Shalla and L. Zwai to Addis Ababa and to Dire Daua and Harar), and N Somalia. Darker, browner (brownish grey wash extends over belly, not confined to breast) and larger than *neumanni*: wing, ♂ (n = 34) 76–87 (82·1), ♀ (n = 17) 79–84 (81·2). WEIGHT: (Ethiopia, Mar) ♂ (n = 4) 20–22 (20·5).

M. m. burae (Traylor): Somalia (Juba R.) and E Kenya (Garba Tula and Garissa, south to Ijara and Lali hills). Paler, greyer and smaller than *neumanni*: wing, ♂ (n = 10) 72–77 (74·9), ♀ (n = 8) 71–75 (72·8). WEIGHT: (Kenya, Mar, Apr) ♂ (n = 5) 14–20 (15·8), ♀ (n = 4) 11–17 (14·8), 1 unsexed 16.

M. m. taruensis (van Someren): SE Kenya (Mbuyuni to Voi and Taru). Darker, browner and slightly smaller than *neumanni*: wing, ♂ (n = 11) 75–80 (77·6), ♀ (n = 5) 75–77 (75·8).

M. m. microrhynchus (Reichenow): SW Kenya (west of *burae* and *taruensis*, north to Kisumu, Simba and Thika) and Tanzania (Serengeti, Arusha and Pare Mts to Dodoma, N Iringa, S Mbeya and SW Tabora). Much greyer and larger than *neumanni*: wing, ♂ (n = 39) 78–92 (85·4), ♀ (n = 16) 82–91 (86·6). WEIGHT: (Kenya, Jan, Mar, Apr) ♂ (n = 5) 16–21 (18·2), ♀ (n = 6) 15–24 (21·2).

TAXONOMIC NOTE: *M. microrhynchus* has a less rounded wing than *M. pallidus*: P10 is 40–44% of length of P9 in *M. microrhynchus*, 41–51% of P9 in *pallidus*. In both species, wing-length is correlated negatively with the ratio of P10 to P9. Graphically the species separate clearly: at any wing-length the races of *microrhynchus* show a lower ratio than races of *pallidus*, e.g. at wing-length of 75, 42–44% for *microrhynchus* and 48–51% for *pallidus* and at wing-length of 85, 40–42% for *microrhynchus* and 45–48% for *pallidus* (for details see Traylor (1970a) who gives tail/wing ratios and other characters for separating specimens from the same localities).

The subspecies *pumilus* has sometimes been considered a full species, e.g. by Sibley and Monroe (1990) because it has been claimed that *microrhynchus* and *pumilus* would be sympatric though ecologically separated, but we follow White (1963), Traylor (1970a), Britton (1980a), Short et al. (1990) and Dowsett and Dowsett-Lemaire (1993b) and consider that specific status is unjustified.

Field Characters. A nondescript, chat-like flycatcher, brownish grey above and whitish below; 'generally resembles a plump Spotted Flycatcher *Muscicapa striata*' (Williams and Arlott 1980), but without the breast streaks. Separated from very similar Pale Flycatcher *M. pallidus* by drier habitat, smaller size, greater contrast between grey breast-band and flanks and white throat and belly, conspicuous pale wing-edgings, all black bill, faint streaks on forehead and crown, pale forehead, lores and eye-ring; generally looks paler and greyer above, whiter below.

Voice. Tape-recorded (CART, GREG, McVIC, PAR). Call, a thin, low 'tweety'.

General Habits. Inhabits open thorn bush and scrub of rather xerophytic deciduous type, orchard country, bushland, wooded grassland and woodland, particularly clearings and secondary growth; up to 2000 m in Kenya. Ecologically separated from Pale Flycatcher by much drier habitat although interdigitating habitats allow them to live almost side by side in S and W Kenya and Tanzania.

In pairs or family groups. Sits upright and silent, with head down between shoulders, on exposed perch; sporadically flicks wings and wags tail up and down, scans ground by turning head from side to side, darts down to pick up prey on ground and returns to perch; prey is rubbed, hammered and dismembered before being swallowed; makes sallies to catch flying insects; sometimes hovers to pick up insect on leaf, trunk or flower. Joins mixed-species flocks at least temporarily, following foliage-gleaners and capturing fleeing arthropods (S Ethiopia: C. Erard, pers. obs.).

Food. Caterpillars and other insect larvae, moths, grasshoppers, melolonthid beetles, nymphal mantids, winged ants and termites; also spiders.

Breeding Habits. Monogamous; territorial. ♂ feeds ♀ while she is begging and quivering wings.

NEST: small shallow cup, of very fine twiglets and rootlets rather loosely interlaced, lined with finer grass and bark fibre, sometimes also down or feathers; placed mostly 1·5–5 m above ground, sometimes up to 6 or 7 m, usually in thorn tree, in fork of horizontal branch.

EGGS: 1–4, laid at 1–2 day intervals. Oval; glossy; pale greenish or bluish green, densely freckled pale brownish; often liberally smeared with white excrement (Belcher 1942). SIZE: (n = 2, Kenya) 19 × 14·7, 19·2 × 15. In Ukamba country, Kenya, attempts 3 broods a year (van Someren 1956).

LAYING DATES: Ethiopia, Apr, Aug–Oct (young June, Nov); Somalia, May; E Africa: Region A, Mar, May; Region B, June; Region C, Aug–June with peak Sept–Oct during hot pre-rains; Region D, Aug, Dec, Feb–June, with 2 peaks, in short rains in Dec and in long rains in Apr–June.

INCUBATION: by ♀, regularly fed by ♂. Period: *c.* 12 days.

DEVELOPMENT AND CARE OF YOUNG: young closely brooded by ♀ during first few days, all being fed by ♂ who brings food, passes it to ♀ who gives some to young and eats the rest; later both parents feed young. Nestling period 15 days. Fledglings, hardly able to fly, hide up in leafy thorn tree for 1–2 days, then follow adults, but soon learn to feed for themselves.

References
van Someren, V. G. L. (1956).
Traylor, M. A. (1970a).

Melaenornis mariquensis (Smith). Mariqua Flycatcher; Marico Flycatcher. Gobemouche du Marico.

Plate 25
(Opp. p. 480)

Bradornis mariquensis Smith, 1847. Ill. Zool. S. Afr., Aves, pl. 13; Marico River.

Range and Status. Endemic resident, Kalahari. Angola, acacia woodlands of central and S Huila and Cubango, fairly common along Cunene R. Namibia, widespread in interior, but absent from parts of Caprivi strip. Botswana, common to very common almost throughout, but sparse or absent in heavily wooded parts of N and NE. Zambia, SW Barotse Prov. in Senanga and Sesheke, from Lupuka and Nangweshi to Namibian border; east to Sesheke, and in Southern Prov. to Mulanga. Zimbabwe, common in SW Matabeleland south to main miombo belt; sparse and local on central plateau northeast to Harare and Rusape, and in Limpopo valley to Sinyoni. South Africa, common in NW Transvaal in bushveld and thornveld, also in SW Transvaal, vagrant to central Highveld, and very local resident in E (Bangu and Pafuri areas, Kruger Nat. Park and Hoedspruit). In Orange Free State uncommon to locally quite common in Hoopstad district; also west of Boshof, near Vredefort, and in Bloemfontein and Lindley areas. N Cape, widespread north of Orange R. and Modder R. Density of 1 pair per 36 ha (Nylsvley, Transvaal).

Description. *M. m. mariquensis* (Smith): Zambia, Zimbabwe, E Caprivi Strip of Namibia, Botswana except SW, South Africa except western N Cape. ADULT ♂: lores pale grey or brown; forehead, crown, nape, hindneck, sides of neck, ear-coverts, mantle, scapulars, back, rump and uppertail-coverts warm mid brown. Very thin ring of minute buff feathers around eye. Tail dark brown, nearly square-ended (T6 4–8 shorter than T1), each feather edged and quite broadly tipped with buff. Underparts white, grey feather bases sometimes showing through patchily. Wings dark brown, primaries narrowly and secondaries more broadly edged with buffy rufous; most upperwing-coverts fringed buff-rufous. Underwing-coverts and axillaries white. Bill black; inside of mouth black; eyes dark brown; legs and feet black. Sexes alike. SIZE: (12 ♂♂, 12 ♀♀) wing, ♂ 81–87·5 (84·7), ♀ 82–87 (84·7); tail, ♂ 71–75·7 (73·1), ♀ 69–74 (72·6); bill to skull, ♂ 16–18 (17·0), ♀ 16·5–18 (17·1); tarsus, ♂♀ (n = 32) 19–24. WEIGHT: (all subspecies) ♂ (n = 11) 23–27 (24·3), ♀ (n = 9) 22–36 (26·1).

IMMATURE: upperparts dark brown, mottled and spotted with buff; rump rufescent; underparts dark fawn, breast boldly dark-streaked.

NESTLING: not described.

M. m. acaciae (Irwin): Angola, Namibia except Caprivi Strip, SW Botswana (south of line Ghanzi–Damara Pan–Molepolole), South Africa (Gordonia and Kuruman dis-

Melaenornis mariquensis

tricts). Slightly paler and more ochraceous on mantle; rump rufescent. Larger: wing, ♂ (n = 12) 85–91 (88·5), ♀ (n = 12) 85·5–90 (87·9).

M. m. territinctus Clancey: NW Botswana and NE Namibia (Okavango R.); intergrades with adjacent subspecies. Very like nominate subspecies but upperparts greyer and underparts whiter.

Field Characters. A thornveld bird. Readily distinguished from dingy brown and grey sympatric congeners Chat Flycatcher *M. infuscatus* and Pale Flycatcher *M. pallidus* by pure white underparts. Upperparts warm brown; buff wing-edgings form paler panel on closed wing. Immature more heavily streaked below than Chat Flycatcher. Spotted Flycatcher *Muscicapa striata* is smaller, with streaked breast and crown.

Voice. Tape-recorded (88, 91, F, WALK). Song a protracted series of variable but monotonous and unmusical chirps of 1–3 syllables, coarse and somewhat sparrowlike. Each chirp lasts *c*. 0·2 s. In one 27 s song, 9 chirps uttered at rather regular intervals in 16 s: 'chitic'p, chitik, chreep, chirrup, chirrup, tkitik, screet, chichu, chitikip', followed by 13 monosyllabic chirps at irregular intervals in 11 s. Alarm a thin 'tsee, tsee'. Anxiety call near nest a sharp 'chiu, *chik*ichiu', also a buzzing, scolding 'chuurr' which brings round other small birds in the vicinity. Contact call on approaching brood, 'cheww'. Nestlings have continous high-pitched begging call.

General Habits. Restricted to acacia savanna, with tall thorn bushes on otherwise bare ground or with sparse grass cover. Rarely in wide grassy plains with a few small acacias; often near edge of large dry pan. Absent from miombo, mopane and *Baikiaea* woodlands except where acacias are mixed in. Somewhat overlaps range of the more northern and eastern Pale Flycatcher *Melaenornis pallidus*, but ecologically exclusive with it (Pale inhabits miombo and *Baikiaea*); borders of ranges of 2 species are a mosaic, with marginal contact where habitats adjoin (Irwin 1981, Frost 1987, 1990).

Forages singly, in pairs or trios, and family parties of 4–5. Adults normally occur in groups of 3, occupying home range of a few ha. Bold and aggressive. Rather silent in presence of casual observer, but members of group regularly call to keep contact whilst foraging. Perches low on outer leafless branch of tree, and often on telegraph wires, fence wires and fenceposts (only 8 out of 91 perch-positions were on top of or inside tree: Fraser 1983). One-fifth of forays are in pursuit of insect inside canopy seen by bird perching on edge of canopy; otherwise flycatcher feeds entirely by scanning ground from perch *c*. 2 m high and flying down swiftly to land near prey, then hopping quickly to catch it (Fraser 1983). Prey generally eaten on ground, then bird flies up, nearly always to a different perch. When ants or harvester termites are hatching, bird stays hopping on ground and catches them as they emerge. Will also pounce into grass. Habitually flicks tail on settling. On ground, posture more upright than at perch in tree, tail held well clear of ground and bird more lively. When nesting, group forages in only small area around nest site.

Normally sedentary, but occurrence of at least 6 individuals in Kruger Nat. Park where it is rare, 30 Sept 1995, possibly regular event in times of drought (Terblanche 1996).

Food. Insects; including termites, caterpillars, and large black ants; mostly small grasshoppers (Vincent 1947). Small fruits of *Rhus pyroides* and *Euclea undulata*. > 80% of food items given to nestlings are minute (< 6 mm); larger prey included 10 caterpillars, 5 beetles, 5 'worms', a grasshopper, a spider and a slug (Frost 1990).

Breeding Habits. Solitary nester; co-operative breeder; territorial, aggressively excluding at least Pale Flycatchers. Brood raised by 3–4 adults or by 2 adults helped by full-grown young from nesting 2 months earlier (Brooke and Borrett 1972, Frost 1990). No courtship behaviour so far observed. A Mariqua Flycatcher once attended nestlings of Chestnut-vented Warbler *Parisoma subcaeruleum* (Steyn 1969).

NEST: a rather ragged, shallow, open framework of coarse rootlets, dry stems, soft grass, thin creepers and yellowish or reddish tendrils, sometimes with bits of string, wool or rag, around a cup of soft, chaffed dry grass, lined with fine plant fibres, animal hairs and feathers, generally large soft white ones. Feathers always used, often profusely, and can curl inwards from nest rim. One nest made entirely of pale wood shavings, lined with feathers. Most nests quite strong; some flimsy, slight and loosely made. Ext. diam. 70–77, ext. depth 37–45, int. diam. 45–60 (usually 45–50), int. depth 20–38 (usually 20–25). Sited in fork against trunk of thorn tree or more often on horizontal fork supported by side twigs in outer canopy, at height of (n = 71) 1·5–5 (2·6) and (n = 38) 1–14 (3·0) m. At Nylsvley, Transvaal, uses 5 species of *Acacia* (n = 20), *Dichrostachys cinerea* (n = 13), *Boscia albitrunca* and *Euclea undulata*. Nest tree is generally bare or just coming into leaf. Some nests placed in clump of mistletoe-like epiphyte.

EGGS: 2–4, av. (n = 18 clutches, Nylsvley) 3·0. Rough-textured; pale olive with greenish or warm brownish tinge, often 2 eggs in clutch with very faint pink band around broad end (fades to greyish in blown egg). SIZE: (n = 80) 18·0–22·0 × 13·2–15·7 (20·2 × 14·5). Second clutch started *c*. 4 weeks after first brood fledges. Replacement clutch laid 8 days after previous clutch destroyed.

LAYING DATES: Angola, Nov; Botswana, Aug–Apr, mainly Oct–Dec; Zambia (gonads active Aug, Sept); Zimbabwe, July, Aug–Jan and Mar, mainly Sept–Nov (91 out of 111 records); Transvaal, Nylsvley, Sept–Feb (Sept 5, Oct 13, Nov 7, Dec 3, Jan 2, Feb 1 clutches); elsewhere, Aug–Jan (n = 42), mainly Oct–Nov (n = 25). In drier regions breeds opportunistically after rain (Maclean 1993).

INCUBATION: at nest attended by 3 adults, only 1 bird incubated (in 4 h of observation) and was fed large insects 3 times by others; recipient on nest made begging calls like fledgings (Frost 1990). Period unknown.

DEVELOPMENT AND CARE OF YOUNG: young fed by both parents and by helpers. Parents brought 31 feeds in 3 h to 7-day brood of 3 chicks, and a parent brooded them after half of the feeds. Fledglings fed by 1 adult; they forage for themselves 1 month after leaving nest. Same 5 birds, parents and young, brought food to second brood 7 weeks later av. 15·3 times per h; adults made 9 times as many visits as juveniles. Both adults and juvenile helpers removed faecal sacs and chased away White-browed Sparrow-Weavers *Plocepasser mahali*. One adult shaded brood at midday by standing with wings outstretched over nest, for periods of 1–27 (10) min; others continued to bring food whilst chicks were being shaded, giving food to shading bird which quivered its wings and passed food to chicks.

BREEDING SUCCESS/SURVIVAL: one breeding group (2 parents, 3 juvenile helpers) attempted to nest 5 times in 3 months, but all 5 nests robbed of eggs (by small mammal?). Parasitized by Diederik Cuckoo *Chrysococcyx caprius* (Brooke and Borrett 1972, Gargett and Webb 1973).

References
Clancey, P. A. (1976, 1979).
Fraser, W. (1983).
Frost, S. K. (1987, 1990).
Vincent, A. W. (1947).

Melaenornis silens (Shaw). Fiscal Flycatcher. Gobemouche fiscal.

Lanius silens Shaw, 1809. Gen. Zool. 7, p. 330; Knysna, South Africa.

Plate 26
(Opp. p. 481)

Range and Status. Endemic resident and partial migrant, almost confined to South Africa. In Botswana a small, resident, breeding population in SE, north to Kanye and Molepolole, and in winter fairly widespread in SE as non-breeding visitor from South Africa, in some years reaching as far north as Orapa and Francistown (H. Penry, pers. comm.). In South Africa a common resident with much local movement in winter; sparse or absent in large parts of NW Cape, SE Transvaal and W Swaziland; but no evidence of movement in SW Cape, where bird is widespread resident, commoner in S and SE of the area than in arid N; rare in (largely treeless) mountains. In N Cape known from Asbestos Mts, Kuruman, Gordonia district and Molopo valley on Botswana border. In W Cape, north to Little Namaqualand (Viool's Drift). Transvaal, fairly common breeding resident in SW, W and central Highveld, Kalahari, thornveld and S edge of bushveld, east to Ohrigstad; scarce winter visitor to central and N bushveld and lowveld. Rare visitor to lowland NW Lesotho. Winter visitor to S Mozambique, E Swaziland and lowland Natal. One seen and photographed at Main Camp, Zimbabwe 18°44′S, 26°57′E (Hustler and Irwin 1995). Density in SW Cape thought to have increased with agricultural tree-planting.

Description. *M. s. silens* (Shaw): range of species except that of *lawsoni*. ADULT ♂: forehead, lores, ear-coverts, cheeks, crown, nape, hindneck, sides of neck, mantle, scapulars, back and uppertail-coverts black with bluish gloss, rump a trifle greyer. Tail brownish black, slightly glossy, T1 black, T2 black (sometimes with white crescent in middle of inner web), T3–T6 black-ended, with proximal two-thirds of inner webs and proximal half of outer webs white; black tips of outer and inner web respectively about 30, 20 mm (T3), 30, 20 (T4), 35–40, 25 (T5) and 40–50, 25–30 (T6); tail rounded, T5 2–3 and T6 10–14 shorter than T1. Chin white with yellowish wash and with many thin black whiskers. 4 rictal bristles on each side. Throat off-white; breast pale grey, merging to white on belly; flanks pale grey; thighs black and white, vent and undertail-coverts white. Wings black but: marginals white; P7–P1 with proximal part of inner webs white, P6–P1 with small white patch near base of outer web, S3–S6 with broad white wedge along most of outer web, narrowly sharply towards tip; inner secondaries narrowly tipped grey-white. Underside of flight feathers dark grey; underwing-coverts dark grey with white fringes, axillaries white. Bill black; eyes dark brown; legs and feet black. ADULT ♀: greyish very dark brown above where ♂ is black; underparts like ♂. SIZE: wing, ♂ (n = 24) 91–98 (95·6), ♀ (n = 8) 87–93 (89·8); tail, ♂ (n = 10) 81–83 (81·8), ♀ (n = 10) 74–79 (76·7), ♂♀ (n = 32) 73–89; bill to skull, ♂ (n = 10) 19–21 (19·5), ♀ (n = 10) 18–20·5 (19·3), to feathers, ♂♀ (n = 32) 13·5–17·5; tarsus, ♂♀ (n = 32) 21–24·5. WEIGHT: (Orange Free State) ♂ (n = 12) 23–31 (25·6), ♀ (n = 10) 23·6–37 (26·9) (Herholdt 1988), unsexed (n = 163) 21–34·7 (26·2).

IMMATURE: when newly fledged, upperparts dark brown, each feather of forehead, crown, nape, lores, cheeks, hindneck, mantle and scapulars with roughly diamond-shaped buff spot; ear-coverts and upperwing-coverts similarly spotted; rump and uppertail-coverts rufescent dark brown with indistinct buffy spots; white in secondaries becomes buff or warm light brown towards tips; underparts fluffy, dappled evenly and softly with dark brown and off-white, undertail-coverts fluffy grey-white. Later, underparts dappled unevenly: breast darker than chin, throat, belly and undertail-coverts.

NESTLING: skin dark reddish brown at hatching, interior of mouth bright orange, swollen gape pale yellow; thin, short, pale grey down on crown and back.

M. s. lawsoni Clancey: N Cape from Asbestos Mts and Gordonia to SE Botswana, NW Orange Free State and dry W Transvaal. Poorly differentiated, with throat and sides of breast paler than in nominate subspecies and tinged with buff, particularly in ♂. Wing, tail and bill *c.* 1 mm shorter.

Field Characters. 17–20 cm. Adult ♂ is a boldly pied, tree-perching but ground feeding flycatcher that is unlikely to be confused with any other bird except Fiscal Shrike *Lanius collaris*. Jet black above, greyish white below; tail with black centre and tip and rectangular white sides; wings with broad white stripe conspicuous in flight, appearing at rest as small patch at base of primaries and long panel in secondaries. ♀ very dark grey rather than black above, with same white marks as ♂. Juvenile dark brown with rufescent rump, mottled with buff above and whitish below, best told from other juvenile flycatchers by accompanying adults. Adult plumage thought to have evolved in mimicry of Fiscal Shrike; but shrike can be distinguished by its robust, hooked bill, V-shaped white mark across back, lack of white wing-stripe, greyish rump, tail long and graduated with all-white sides, and (♀ shrike only) rufous flanks (sometimes concealed).

Voice. Tape-recorded (88, 91, F, WALK). Most vocalizations are feeble, barely audible even from close by. Song a series of indefinite duration (at least 25 s) of phrases of thin, reedy, high-pitched, clear, sibilant 'tswee' and 'sooo' notes interspersed with a few 'trp's. Each phrase lasts 0·5–1·0 s. In one 25-s song, 9 phrases of 4–5 types uttered at about equal intervals, with 2-s pauses: 'swee't'swee, tp'see-seoo, swee't'swee, seeoo-seeoo, trp-see't'see-see-seoo, trp-see't'see-seoo', trp-see't'see-seoo, trp-see't'see-seoo, trp-see't'see-seoo'. Notes vary from 3 to 7 kHz. Song also rendered 'tsip tsip chuk tweeu-kik-rrr, tswippy tswip trree-up-up, tsippy tsip twee-up ...'. Sometimes mimics other birds, including Cape Robin-Chat *Cossypha caffra*, when robin-song dominated short snatches of flycatcher's own song (Craig 1977). Alarm, a sharp 'skisk' and 'kirr-kirr-kirr'.

General Habits. Inhabits open country with scattered trees, keeping mainly to small, dense clumps of large trees and thickets; in parks and large, suburban gardens, along paths and highways with woods and outlying native or exotic trees, on tree-lined watercourses in farmland; bushveld, grassland with a few trees, coastal macchia, karoo, fynbos and thornveld with semi-arid scrub. In Botswana occurs in small patches of woodland and clumps of mixed deciduous trees near open grassland, but mainly in *Acacia*. In South Africa commonly uses wattles, eucalypts and alien acacias.

Singly or in pairs; up to 10 congregate at fruiting trees in winter. Conspicuous and confiding. Rather silent, ♀ particularly mute. Forages by keeping watch at perch mainly (58% of 196 observations) on top of bush or small tree, also on its lower sides, and flying down to ground to settle and snatch prey. Readily uses telegraph wires, fences and rooftop aerials. Also hawks for insects in flight; of 61 observed forays 45 were to ground and 16 in air (Fraser 1983). Does not take insects inside canopy; how it takes fruit is not on record. Insect prey eaten on ground or taken back to perch; bird returns to a new perch 8 times as often as to original perch.

Mainly resident, but a non-breeding visitor to SE Botswana in May–Sept, central Botswana in June–Aug, and N and E Transvaal, S Mozambique and coastal Natal in Apr–Aug.

Food. Mainly insects including ants; small fruits of *Halleria lucida*, *Chrysanthemoides monilifera*, *Cotoneaster* and *Euonymus*, and nectar of aloe (Maclean 1993); 65–70 mm earthworm fed to nestling; once, ate porridge from dogbowl.

Breeding Habits. Solitary nester, monogamous, not strongly territorial although it chases other bird species away from home range. One once fought with Fiscal Shrike near nest, feathers flying. Displaying pair flies excitedly from tree-top to tree-top, ♂ at intervals fanning tail and uttering low wheezing calls and 'chat-chat' notes.

NEST: bulky bowl, usually shallow, sometimes deep, composed of twiglets, rootlets, some dry grass, green or dry weed-stems, grey everlasting plant stems and at times pieces of string, with smooth, compact, quite thick, felt-like lining of plant-down and fine grasses with a few feathers within it, sticking out at the rim. In S Karoo, seedheads of *Galium tomentosum* (Rubiaceae) incorporated into 5 nests (Dean and Milton 1993). Small pieces of lichen applied to outsides of some nests. Nest neat or rather untidy, made using long, fine and pliant materials and strongly bound onto supporting twigs. Ext. diam 110–130, ext. depth 65, int. diam. 65, int. depth *c.* 37. Sited on fork well within crown of tree or near end of branch, supported by side twigs; usually rather well concealed in dense foliage; also at base of aloe leaf and in vine clump. Nest placed at 1–7 m above ground in shrub or tree; wattle-trees particularly favoured.

EGGS: 2–4, usually 3, av. (n = 31 clutches) 2·8. Laid at daily intervals. Pale green-blue or olive-green, closely and evenly freckled all over with dull olive or light warm brown; speckles sometimes concentrated at wide end. SIZE: (n = 92) 18·5–24·2 × 14·4–17·0 (21·4 × 15·9).

LAYING DATES: Transvaal, Aug–Jan (Aug 2, Sept 27, Oct 37, Nov 17, Dec 12, Jan 2 clutches); SW Cape, Aug–Jan; Natal, Oct–Dec.

INCUBATION: at one nest by ♀, sitting mainly in early morning and evening, with long spells off nest. Period 13–15 days.

DEVELOPMENT AND CARE OF YOUNG: at one nest the 2 eggs hatched within a few h. Eggshells are removed. Chicks fed by ♀ only, ♂ present but taking little interest. Both chicks left nest on day 16 (Haigh 1986).

References
Haigh, H. (1986).
Lawson, W. J. (1962a).
Maclean, G. L. (1993).
Vincent, A. W. (1947).

Genus *Empidornis* Reichenow

An endemic genus whose single species is like *Melaenornis* morphologically but is distinguished by silver-grey and russet plumage, also only 2 main rictal bristles; builds domed nests; thrush-like song.

Empidornis semipartitus (Rüppell). Silverbird. Gobemouche argenté.

Muscicapa semipartita Rüppell, 1840. N. Wirbelt. Vög., p. 107; Gondar, Abyssinia.

Plate 25
(Opp. p. 480)

Range and Status. Endemic resident and wanderer, Sudan to Tanzania. Sudan, widespread and locally common in Darfur and from Bahr el Arab to Kajo Kaji and Malakal; also in Sobat R. tributaries on Ethiopian border; rare north to NE Blue Nile Province. Ethiopia, uncommon to locally abundant in W; scarce, north to Gondar. Uganda, locally common in NE, south to Bunyoro, Buruli and Sebei. Kenya, locally common at 1000–1500 m in and west of Rift Valley, less common down to 400 m; in north known from Lokichokio and S Turkana, and a record on central island, L. Turkana; south to Mara Nat. Res. and Loita Hills and east to Isiolo; records at 1800 m in highlands near Nairobi and 2300 m on Mt Elgon. Tanzania, from Mwanza to Shinyanga and central Tabora, east to W Arusha and N Dodoma.

Description. ADULT ♂: upperparts rather uniform french grey or pale bluish grey, sharply delineated from underparts which are uniform bright rufous. Lores, cheeks, ear-coverts and sides of neck french grey; sometimes an olive-tinged round spot on side of forehead; grey on side of neck extends forward in a blunt point between throat and side of breast. Tail silvery grey above and below; graduated, with T5 3–8 and T6 17–28 shorter than T1; feathers sometimes with very narrow whitish line around all edges. Wings grey above, orange below, including axillaries; primaries and secondaries dark brownish grey, edged silvery grey; tertials uniformly silver-grey. Bill black; eyes dark brown to pale sepia; legs and feet black. Sexes alike. SIZE: (5 ♂♂, 5 ♀♀, Uganda) wing, ♂ 90·5–95 (93·2), ♀ 87–95 (91·6); tail, ♂ 77–85 (81·3), ♀ 74–82 (79·1); bill to skull, ♂ 11·2–16·4 (15·8), ♀ 15–16 (15·6); tarsus, ♂ 20–22 (21·0), ♀ 20–22 (21·0). A cline of increasing wing length, from 82–95 in Sudan to 95–101 in E Africa, on which character 2–3 races have sometimes been recognized. WEIGHT: 2 ♂♂, Uganda, 22, 23.

Empidornis semipartitus

IMMATURE: all feathers of upperparts except primaries and secondaries french grey with neat black tip and black shafts (tertials) or black fringe (crown), and with large buff or pale rufous subterminal spot, well demarcated on mantle, upper-wing-coverts and tertials, the last having rudiments of a second pale spot. Chin to belly buff mottled and scalloped with dark brown; undertail-coverts uniform pale rufous; under-wing-coverts orange. Subadult birds the same but with orange feathers growing patchily in underparts.

Field Characters. About 18 cm. An easily-identified, slim, quite long-tailed flycatcher of acacia bush, silvery grey above, rufous below, the colours sharply demarcated. Juvenile looks quite different, although equally as beautiful as adult: upperparts grey, mottled with buff on mantle and wings, and finely streaked and spotted with black from head to rump; underparts buff, scalloped with dark brown.

Voice. Tape-recorded (GREG, McVIC, PAR). Song described as a soft warbling, also sweet and rich, like thrush or robin-chat, mainly a repetition of 3 compound notes.

General Habits. Inhabits acacia grassland and semi-arid bushland with scattered large trees; essentially arboreal, perching on top of acacia bush or on outer and inner branches of open, larger tree, but also uses fences and other man-made objects. Solitary or in pairs; sometimes shy and wary. Hawks insects from a perch, but frequently also flies to ground to seize insect there, returning immediately to its perch. Described as hopping along a post-and-wire fence (Bednall 1990), running about on the ground (Mackworth-Praed and Grant 1960); and sitting still then hopping sedately from branch to branch (Edwards 1973). Attracted to fruiting fig trees, evidently to catch insects at the fruits.

Resident, but several extralimital and out-of-habitat records suggest considerable dispersion and wandering (Lewis and Pomeroy 1989).

Food. Insects, mostly small and wingless, including many ants. Captive birds ate locusts and mealworms.

Breeding Habits. Solitary nester, apparently monogamous.

NEST: substantial dome made of dry grass and small thorn twigs (Mackworth-Praed and Grant 1960), lined with fine grass-heads; commonly uses old weaver nests, including those of White-headed Buffalo-Weaver *Dinemellia dinemelli*, Red-headed Weaver *Anaplectes rubriceps*, White-browed Sparrow-Weaver *Plocepasser mahali*, also Rufous Sparrow *Passer motitensis*. Appropriated weaver nests are at extremities of thorn-tree branches (or once in a pepper-tree *Schinus molle*), 3–6 m above ground. Flycatcher takes a week or more to repair domed weaver nest and line it with fine grass stems (Durand 1949). Captive pair, England, made well-finished, domed nest with top entrance, using teased string and raffia (Scamell 1973).

EGGS: 2 (or 3 in captivity, England). Rather long, slightly glossy. Pale olive-green, with ill-defined brown streaks forming brown cap at large end, or finely speckled with red-brown. SIZE: (n = 4) 21·5–23·0 × 15·0–15·5.

INCUBATION: in captivity, England, same ♀ laid several clutches one summer in quick succession; incubation, mainly or entirely by ♀, began with third egg. Period: 13–15 days.

LAYING DATES: Sudan, Darfur (breeds May–Aug); E Africa: Region A, Feb, May; Region B, Dec; Region C, Nov–Feb; Region D, Apr–May, Sept. Breeds in both wet and dry seasons; all 6 Region C records in rains.

DEVELOPMENT AND CARE OF YOUNG: in captivity, ♀ flew away with faecal sacs, nestling period 19–21 days (3 broods), juvenile fed for itself within 14 days of leaving nest, and juveniles started to moult and to lose spotted plumage at 2 and 4 months.

BREEDING SUCCESS/SURVIVAL: 1 bird survived 7 years in captivity.

References
Durand, M. (1949).
Scamell, K. M. (1973).

Genus *Muscicapa* Brisson

Small brownish flycatchers. Bill wide at base, tapering gradually, ratio of width to length 0·54–0·96, gape-tip-gape angle of 30°–50°, dorsoventrally flat and shallow to quite robust with well-curved culmen, or deep at base, sharp and with nearly straight culmen – almost *Parus*-like in *M. boehmi*. Upper mandible weakly hook-tipped in most species. Up to 7 rictal bristles, strong, nearly straight, reaching forward to level of nostrils or well beyond them. Nostrils partly overlain with short feathers and fine bristles. Eyes quite large in a few species. Tarsus short and slender, often not scutellated, heel joint feathered in front but bare at back, feet weak. Plumage mid-brown, bluish grey-brown or sooty brown in African species, partly rufescent in some Oriental ones; plain, but many species with narrow pale eye-ring, white chin and throat, whitish belly, and pale-edged dark wing feathers. Underparts uniform or softly streaked; breast sharply arrow-marked in *M. boehmi*, or softly mottled and forecrown streaked in a few others. Plumage soft and full, making most species rather large-headed. Wings quite long, with conspicuous primary projection. P10 shorter or slightly longer than primary coverts or up to 6 mm longer (*M. dauurica*); P7–8 generally longest, P9 and sometimes P6 nearly as long. Wing shape varies from round to pointed, Afrotropical species, particularly forest ones, having a rounder wing. 12 rectrices; tail square, or slightly rounded, or shallowly forked in *M. ussheri*. Sexes alike; juveniles spotted and scalloped. Typical flycatchers, stand on exposed vantage point from which they sally forth to snap up passing insect, returning to same perch. Rather silent, song usually poor, vocalizations high-pitched.

24 species (Sibley and Monroe 1990), in Africa, Palearctic, S and E Asia to Philippines and Lesser Sundas; most widespread genus of Muscicapini. High latitude ones, particularly Palearctic and Himalayan species, migratory. 15 in Africa, all but one endemic, mainly sedentary or subject to local movements. They embrace 3 superspecies: *M. gambagae* and the Palearctic *M. striata*, *M.lendu/M. olivascens*, and *M. infuscata/M. ussheri*. The African *M. adusta* is very like the temperate-zone Asiatic *M. dauurica* in plumage. The latter is a long-distance migrant with a tropical population, presumably derivative, breeding 'close' to Africa in India's Western Ghats, and it also has 2 tropical allospecies in SE Asia, *M. randi* and *M. segregata*. All inhabit forest at about 1000–2000 m (*M. adusta* also higher) and are altitudinal migrants. We propose that *M. adusta* is recently derivative of *M. dauurica* and forms a superspecies with it and *M. randi* and *M. segregata*. African species more diverse morphologically than Asiatic ones, and have been assigned to many genera (*Alseonax*, *Apatema*, *Hypodes*, *Pedilorhynchus*, *Artomyias*, *Myopornis*) in the past.

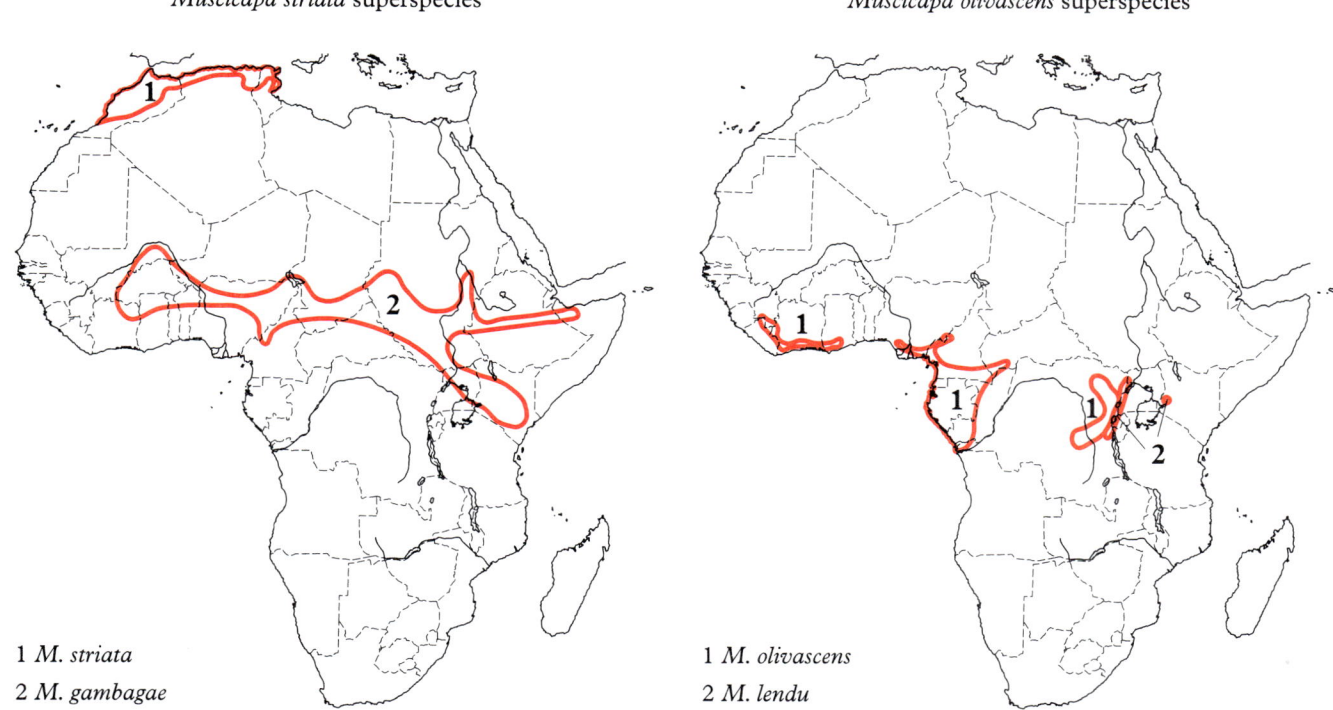

Muscicapa striata superspecies

1 *M. striata*
2 *M. gambagae*

Muscicapa olivascens superspecies

1 *M. olivascens*
2 *M. lendu*

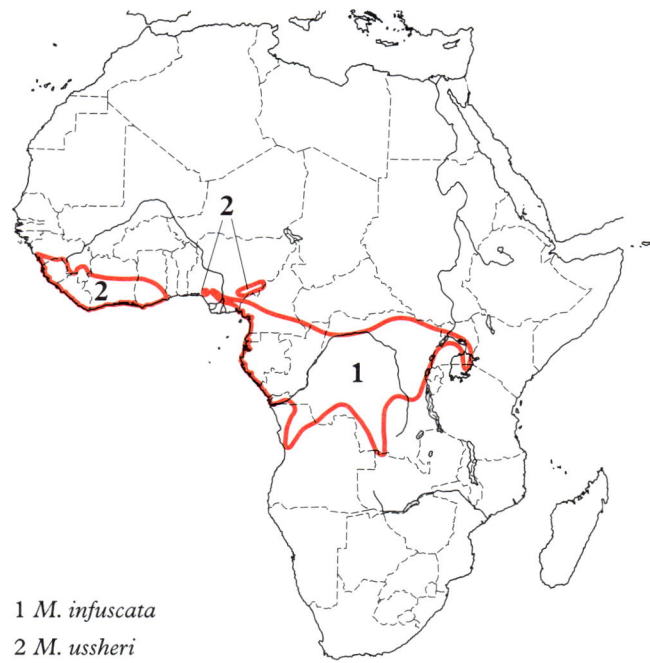

Muscicapa infuscata superspecies

1 *M. infuscata*
2 *M. ussheri*

Plate 26
(Opp. p. 481)

Muscicapa striata (Pallas). Spotted Flycatcher. Gobemouche gris.

Motacilla striata Pallas, 1764. In Vroeg, Cat. raisonné Coll. Oiseaux, Adumbr., p. 3; Holland.

Forms a superspecies with *M. gambagae*.

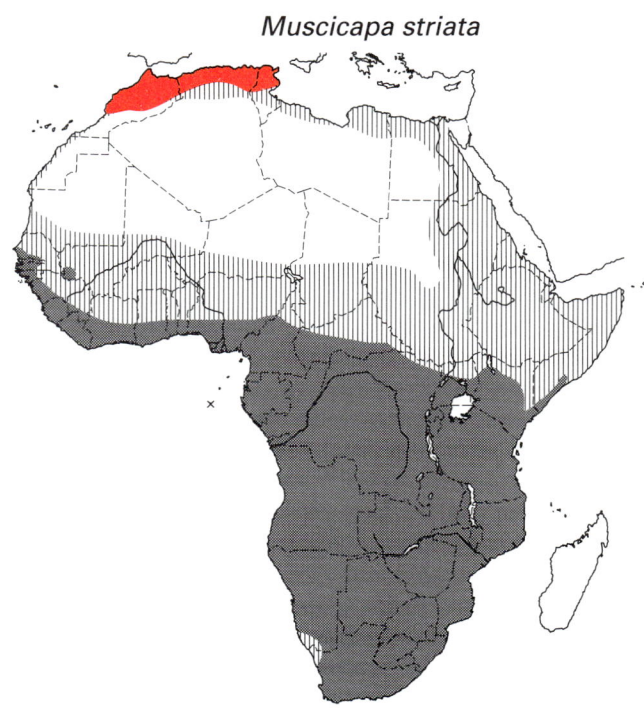

Muscicapa striata

Range and Status. Breeds N Africa, British Isles and Scandinavia east to Altai, China (Sinkiang), Mongolia and SE Transbaikalia, through Europe, N Turkey, Transcaucasia, W, N and E Iran and Siberia, with isolated populations in Crete, some Aegean islands, Cyprus and Levant; winters mainly in Afrotropics.

Breeds in Maghreb: Morocco (Sous, southern edge of High Atlas, Errachidia, up to 2400 m), Algeria (Tell-Atlas, Aurès, possibly Ksour Mts, up to 2200 m in Djebel Chelia) and Tunisia (particularly in N, south to Gafsa and Gabès). Local populations present until Oct–Nov, even early Dec; once Feb, Morocco. Frequent to abundant. Av. density in Moroccan woodland, 12 pairs/km^2 (Thévenot 1982); 2·5 to 10 pairs/km^2 in oak bushland or woodlands in Algeria (Bellatrèche 1993).

Palearctic migrant wintering south of Sahara. Libya (passage Sept–Oct, Apr–May). Egypt (both nominate *striata* and *neumanni*, 20 Aug–28 Oct, chiefly Sept, and 30 Mar–28 May, with some occasional mid-winter records); Mauritania, Senegal and Gambia (passage early Sept–end Oct and 27 Apr–16 May; rare in winter (Nov–Apr), from Dakar and Saloum to Casamance and throughout Gambia); Guinea-Bissau; Mali (passage Sept–Oct and Mar–Apr, widespread but uncommon everywhere though locally common in Mts Mandingues where some may overwinter); Guinea (Macenta and Beyla, Oct–Nov); Sierra Leone (recorded Dec); Liberia; Ivory Coast (common everywhere outside forest Sept–Apr; mainly *striata*, also *balearica*); Burkina Faso (Ouagadougou, Oct and May); Ghana (not uncommon, regular on passage Sept–Oct in N, but records up to Apr at Kumasi, Tafo, Accra, Achimota, Legon and Akropong in Akwapim, *striata*, but *balearica* suspected); Togo (regular on passage, present Atakmé 7 Sept–15 Mar); Benin and Niger (Sept–Dec, mostly Oct–Nov, and Mar–May); Nigeria (common throughout as passage migrant Sept–Nov and Mar–May, some winter locally in SW – Dec records, Lagos, Kagoro); Chad (passage migrant mainly Sept–Oct and Apr–May); Cameroon (throughout, passage migrant, winters in south and Bénoué Plain, Oct–May); Rio Muni, Gabon (early Oct to end Apr); Congo (e.g. Djambala, Oct–Dec); Central African Republic (Sept–Apr, winters Dzanga reserves, Bouar and Bozoum east to Bangui and Bangassou); Sudan (nominate *striata*, and *neumanni* east of Nile; throughout but particularly Darfur, Nile valley, Sudd swamps and flood plains; common Sept–Nov, uncommon Mar–May); Ethiopia (double passage throughout, e.g. Eritrea mid-Aug to end Oct and end Apr to mid-May); Somalia (common Sept–Oct and Apr–May, a few winter Dec–Mar south of 3°N); Uganda, Kenya and Tanzania, mainly *neumanni*, also *striata*, winter visitor rather thinly distributed, but common on coast and locally in E plateau, and passage migrant, late Sept–Nov and late Mar to mid–Apr; Rwanda and Burundi, common end Sept to mid-Apr, commonest Oct to early Nov; Zaïre, throughout, present Upemba Nov–Mar, Itombwe late Sept to mid-Apr); Angola, common throughout, end Sept to early Apr; Zambia (throughout, mid-Oct to Mar with a few June–Sept, very common Nov–Mar); Zimbabwe (throughout, particularly in dry west, extreme dates 10 Oct and 27 Apr). In Zambia and Zimbabwe, *neumanni* 1·5 times commoner than *striata*. Malaŵi, present throughout, Nov–Mar, occasional Oct and Apr, *striata* commoner than *neumanni*?); Mozambique, throughout, Oct–Mar; and southern Africa south to Cape Town, including Swaziland, lowlands of W Lesotho and Botswana, but absent SW Namibia and NW Cape Province (common early Oct to Mar, sometimes to mid-Apr).

Density, 25–27 birds/100 ha in uncleared bush, 8 in old cleared bush in Kenya (Pomeroy and Muringo 1984).

Description. *M. s. striata* (Pallas): N Africa and Europe to W Siberia (Irtysh R.); winters in whole of Africa. ADULT ♂: upperparts plain mouse grey-brown, forehead and forecrown paler; forehead to nape streaked brown-black (feathers grey-brown with arrow-shaped brown-black spot in centre and narrow edges off-white). Tail dark brown (feathers with whitish tips, more conspicuous in fresh plumage); uppertail-coverts grey-brown with pale grey edges. Lores whitish; ear-coverts and sides of neck mouse grey-brown; underparts dull white, flanks, thighs and breast washed pale greyish buff, breast, sides of chin and throat and upper flanks irregularly streaked mouse grey-brown. Alula, outer greater upperwing-coverts, primaries and secondaries dark brown; scapulars, median and lesser upperwing-coverts dark brown narrowly edged pale mouse grey-brown; tertial and inner secondaries dark brown edged whitish along outer web and tip of inner web; inner greater upperwing-coverts dark brown narrowly fringed pale brown-grey along outer web. Underside of flight feathers brown with grey sheen (feathers edged pale greyish buff along inner web); underwing-coverts grey-brown broadly fringed pale buff, axillaries pale brownish buff. Bill brown-black, base of lower mandible pinkish to fleshy horn; eyes dark olive-brown; legs and feet brown-black, soles dull yellow. Sexes alike. SIZE: (28 ♂ ♂, 25 ♀ ♀) wing, ♂ 79–89 (85·6), ♀ 82·5–90 (86·1); tail, ♂ 54–65 (60·4), ♀ 58–62 (60·6); bill, ♂ 14·5–17·5 (15·5), ♀ 14·5–17 (15·7); tarsus, ♂ 13·5–16 (14·5), ♀ 13·5–16 (14·7). WEIGHT: unsexed (n = 13, El-Mohammadia, Algeria, Sept–Oct) 13·3–21·1 (16·7); (n = 13, Oran, Algeria, Sept) 11·2–18·4 (13·8); (n = 7, Tunisia) 13·6–15·7 (14·2); (n = 11, Libya, Apr–May) 11·5–17·5 (14·8); (n = 14, Sahara) 15·2–20·5 (18·0); (n = 16, Nigeria, Apr, May) 13·6–21·9 (17·9); (n = 11, L. Chad, autumn) 12·1–16·0 (14·0); (n = 7, L. Chad, spring) 21·2–25·3 (22·7); (n = 47, southern Africa) 12–18·6 (15·2); (Upemba, Zaïre) ♂♀ (n = 14) 13–16 (14·8); (Uganda, Nov) 1 ♀ 15; (Kenya, Mar, Oct) 2 ♂ ♂ 14, 15, 2 ♀ ♀ 15, 16, 1 unsexed 13.

IMMATURE: like adult but uppertail-coverts tipped rufous-buff, greater upperwing-coverts with buffy-white spot at tip.

JUVENILE: upperparts, face and scapulars blackish brown very densely spotted and scaled ochre-buff (each feather with grey-brown base, wide ochre-buff crescent in centre and brownish black edge, spots larger and scalier on mantle, narrower on head, tear-shaped on scapulars). Tail and wings blackish brown; upperwing-coverts grey brown with dark brown subterminal spot and broad rufous-buff tip; outer greater upperwing-coverts, primaries and inner secondaries faintly edged pale rusty along outer web, inner secondaries and tertials broadly edged ochre-buff along outer web and tip of inner web, lesser and median coverts broadly tipped ochre-buff, greaters fringed on outer web and tip of inner web with ochre-buff; tail-feathers narrowly fringed and tipped pale rusty buff. Underparts creamy white washed pale buff; chin and throat faintly spotted, breast, upper belly and flanks densely scalloped blackish brown (feathers unevenly edged). Bill brown, base of lower mandible pinkish; eyes dark brown; legs and feet grey-brown.

NESTLING: hatches naked. Bill pale straw-yellow, gape-flanges whitish yellow.

M. s. balearica Jordans: Balearic islands; winter records from Ivory Coast, Cameroon, Pagalu and Namibia. Paler and sandier, less streaked on crown and breast, smaller: (10 ♂ ♂, 9 ♀ ♀) wing, ♂ 80–83 (81·3), ♀ 78·5–82 (79·7).

M. s. tyrrhenica Schiebel: Corsica and Sardinia, migrations unknown but winter quarters most probably African. Upperparts warmer brown, breast less densely marked, with spots rather than streaks.

M. s. neumanni Poche: Crete, Cyprus, Levant, Turkey, N and W Iran, and Siberia, east of Irtysh R., to Altai and China; migrates through E Africa, winters Sudan to Somalia, Tanzania, Malaŵi, Zambia, Zimbabwe and South Africa. Upperparts paler, forehead and underparts whiter. WEIGHT: (Egypt) unsexed (n = 39) 12·4–20·2 (16·5); (Kenya, Feb, Mar, Nov) 2 ♂ ♂ 16, 20, 1 ♂ 13; (Zimbabwe) ♂♀ (n = 7) 13·6–17·7 (14·9); (Transvaal, may include nominate *striata*, Nov–Apr) unsexed (n = 41) 12·0–18·6 (15·0); (Namibia, Nov) 2 ♀ ♀ 16, 17, 1 unsexed 16.

Field Characters. A medium-sized, slim, active, typical flycatcher, with short legs, broad flat bill (**A**), large flat head, grey-brown upperparts, white underparts,

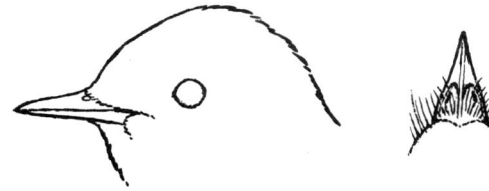

A

streaked on head and breast, pale wing edgings, particularly on tertials. Similar Gambaga Flycatcher *M. gambagae* is smaller, with lower mandible horn with blackish tip; forehead and crown dark, these and breast only faintly streaked; breast streaks mouse grey-brown, not blackish brown, and extending onto throat; underparts suffused with pale greyish brown; wings broadly edged ochre-buff to whitish, particularly on secondaries, median and greater upperwing-coverts, producing two wing-bars and a wing-panel. African Grey Flycatcher *Melaenornis microrhynchus* is chat-like with long legs, stout bill, whitish lores and eye-ring, and unstreaked breast.

Voice. Tape-recorded (86, 88, 91, B, C, F – 62). Song, unobtrusive series of short, squeaky, very high-pitched single or disyllabic notes, interspersed with some trills, without definite phrases, 'sip-sip-sree . . . sree-ti-sree-sip . . .' or a thin, quick 'sip-sip-see-sitti-see-see'. Usual call, a thin sibilant 'seep' or harsh 'chirrt'. Threat call, a repeated 'dset'. Alarm call, 'tek-tek', 'tsee-tec', 'see-tic', 'tsee-tay-tay', or a rattling 'tsee-tictictic'. Mobbing or chatter-call, a rattling 'ch-r-r-r-r-rer'. Call of young, a vibrant 'sree' or buzzy 'zeet'. For further details see Cramp and Perrins (1993) and Glutz von Blotzheim and Bauer (1993).

General Habits. In N Africa inhabits deciduous and evergreen broad-leaved and coniferous forests and woodlands; in Kabylie des Babors, Algeria, more abundant in *Quercus coccifera* woodland than bushland (10 vs. 5 pairs/100ha) or than in *Quercus suber* woodland or forest (5 pairs/100ha), or in mixed *Quercus suber* and *Q. faginea* forest (2·5 pairs/100ha), and is absent above 900 m even in oak forests (Bellatréche 1993). Elsewhere occurs in forest edges, second growth, open woodland of any type (e.g. mopane, miombo), wooded and bushed grassland, mosaic of high-grass savanna and small woodlands, plantations, dry thornscrub, orchards, gardens, parks and exotic plantations around and in villages and towns. Dependent on raised perches; avoids open areas devoid of trees and bushes, also dense forest. In NE Gabon, occurs in open habitats which structurally resemble those where it breeds in temperate regions, but it does not enter continuous forest; also frequents ecotones around villages: gardens, hedges and regrowth with scattered trees and bushes, and recent clearings, although its preferred habitat is managed plantations (cacao, coffee, citrus, cassava, avocado) overgrown with regular layers of vegetation up to 25 m, recalling managed temperate forests. Although it does not enter continuous primary or secondary forest, it does occur in openings along roads or wide earth tracks. Similarly in Sierra Leone said to winter irregularly in forest, entering along logging trails. Found at all levels, up to 2340 m in Itombwe, Zaïre, and 3000 m in E Africa, though mainly below 2200 m.

Usually solitary, but in small parties during migration (e.g. in NE Gabon up to 12 birds on 1 ha). Generally quiet and unobtrusive; returns to same set of perches day after day and may spend all winter in same small area. Active all day; in NE Gabon practically up to nightfall, particularly during migration. Forages at any height from ground level to upper canopy of trees along forest edges; near habitation perches fairly low on branch, wire or fence, sallying out to catch insects in flight; on returning to perch flicks wings more frequently than other flycatchers. Most usual foraging tactics: sits motionless or flicking one wing on exposed perch and makes sallies to pick prey from leaf or bark or to snap it up in the air, usually after a pursuit including some loops or butterfly-flights; returns to perch to swallow largest items. May also dart to ground to catch prey and carry it back to perch. Gleans leaves and trunks, jumping or fluttering from twig to twig. In NE Gabon, regularly seen at sunrise or in the very early morning hopping on ground near houses or fluttering along or against walls to pick up insects attracted there by lights during the night; seen being chased away by African Pied Wagtail *Motacilla aguimp*. Sometimes evicted from perches by African Paradise-Flycatcher *Terpsiphone viridis*. Usually does not join mixed flocks but, like many other species, attracted by swarming flying insects. Regurgitates indigestible prey parts as pellets measuring *c.* 10 × 5 mm. Sings Mar–Apr in Transvaal and Zimbabwe.

Sometimes same individuals occur in same locality in successive years or seasons; bird ringed near Bulawayo, Zimbabwe, was recaptured at same place 1 year and 37 days later and again after a further 2 years 87 days. Said to be territorial in winter in Zambia, but it is not generally so in Gabon, although individuals may hold successive territories. Temporary loose territoriality occurs during migration stop-overs in N Africa and S Europe.

During migration, perched birds frequently flick both wings briefly over back in V. In W Palearctic, ringing recoveries indicate migratory divide at *c.* 12°E: eastern birds head south to southeast and W European birds head southwest to south-southwest, then change direction to south or south-southeast in Iberian Peninsula and NW Africa. However it probably crosses entire Mediterranean and N Africa on broad front at both seasons. Movements in NE and E Africa also on broad front. Little evidence of non-stop flights; rather, birds progress gradually, apparently in relation to rains. In W Palearctic, autumn migration begins in Aug, though some movements detected late July; birds enter Africa mid-Aug to mid-Nov, particularly Sept–Oct in Mediterranean regions and E Africa. Some move quickly (e.g. early Sept records in Gambia, L. Chad, Ethiopia, and mid- to late Sept records in Kenya, Uganda, Tanzania, Rwanda and Angola); but most move gradually. Passage continues in W Africa in Oct–Nov; most birds reach E Africa in Oct, and South Africa in late Oct to Dec (mean arrival date Bloemfontein 21 Nov, 1950–1986). At Ngulia, SE Kenya, night migration takes place during last 10 days of Oct, diminishes during Nov and almost ceases by early Dec (Backhurst and Pearson 1984).

Winters south of 10°N, mainly south of equator. Retraps of ringed birds indicate site fidelity, at least in Zambia, Zimbabwe and Transvaal. Spring migration starts late Feb; departs from Orange Free State 17–26 Mar; departure date influenced mainly by day-length (Kok *et al.* 1991). In Transvaal, latest 14 Apr; extreme dates Botswana 4 Oct, 5 May; passage in Gabon mainly first half Apr, in Zaïre basin extreme dates 26 Mar and 9 May, in Kenya mid- to late Apr, in Sierra Leone until late Apr, in Nigeria until mid-May. Passage through N Africa and Mediterranean region chiefly mid-Apr to end of May, with stragglers early June. Spring migration is more north to northeast in W Africa, so birds less common in spring than in autumn west of Nigeria. As evidence of loop-migration, a bird ringed in autumn in SW France was recovered the following spring in Tunisia.

4 birds ringed in Britain and 1 each in Belgium, Germany and Switzerland recovered in migration in Nigeria, and a bird ringed there in Oct recovered in June in SW England; 1 bird ringed Britain recovered Senegal and another in Ghana; 1 ringed Holland recovered Mauritania, 1 ringed Germany recovered Burkina Faso. 2 birds ringed in Britain and 1 each in Belgium, Sweden and Malta recovered 17 May and 20 Sept–10 Oct in coastal Congo, from Djambala north to Kellé (Dowsett-Lemaire and Dowsett 1991), 1 from Germany and 1 from Poland also recovered in Congo (Dowsett *et al.* 1988, Glutz von Blotzheim and Bauer 1993). 1 ringed Sweden and 1 ringed Ireland recovered Jan in Angola. In Kasai and Lualaba, Zaïre, 8 autumn and 27 spring recoveries in Aug, Nov and Mar–Apr of birds from Finland (22), Sweden, Britain, France, Belgium, Italy and Poland. In South Africa, birds ringed in Finland recovered Dec–Feb in Johannesburg, Swaziland and Natal; 1 bird ringed Wales recovered Mar at Kei Road (E Cape); bird ringed Sweden recovered from 7 km west of Cullinan, near Pretoria (Underhill and Oatley 1994). Thus Finnish birds spend 3 months in breeding area, 4 months on wintering grounds (av. stay 121 days at Bloemfontein) and 5 months along migration routes.

Food. Small insects: Diptera, bees, wasps, winged ants, Lepidoptera (mainly moths, including caterpillars of emperor moth *Imbrasia belina*: Styles 1995), Homoptera, Trichoptera, Neuroptera, Odonata, small beetles, bugs, Mantodea, winged termites; also spiders, worms, kernels of sunflower seeds (stolen from White-throated Canary *Serinus albogularis*), and small fruits, e.g. *Ochna pretoriensis* and *Trema orientalis*.

Breeding Habits. Data mainly from Europe. Monogamous, though some cases of bigamy recorded. Pair-bond known to be maintained for 2nd brood but no other information on mate fidelity. Pairing may begin before arrival on nesting grounds; some courting may occur in Africa (courtship flight in Nov and singing in Mar observed on Pemba (Pakenham 1943) although this could have been merely territorial behaviour). During pair formation, ♂ sings actively, demonstrates suitable nest-site in crouched posture, leaning body forward with ruffled crown and throat feathers, bows and moves head up and down and side to side, flicking tail, shuffling to and fro. Songs mostly sexual rather than territorial; emitted from arrival to nest-building, rarely after egg-laying though again at hatching and when young become independent. Territorial. Territory size variable, 0·24 to 0·7 or 1·0 ha in high-density areas; typically includes 6–10 favoured hunting perches, usually 10–20 m apart. Territorial defence includes much bill-snapping, wing-flicking, chasing with rattling calls, and return to perch with 'tchick' calls and tail-wagging. ♂ feeds ♀ quite frequently up to and including hatching period (see Haartman 1969, Davies 1976). ♀ solicits ♂ in crouched posture with lowered tail, ruffled uppertail-coverts and shivering wings. Copulation may occur without any preliminary display though often ♂ has food in bill, and ♂ and ♀ solicit each other with fluttering flight, crouched postures, wing-shivering and tail-raising or flicking; sometimes ♂ waves wings raised in V in 'wind-mill display' with loose-jointed movements (see Cramp and Perrins 1993, Glutz von Blotzheim and Bauer 1993).

NEST: more or less bulky cup of loosely piled fine twigs, rootlets, dead leaves, pieces of decaying bark, moss, dry grass, lichens and vegetable fibres, bound with hair, lined inside with hair, feathers and finer material. Ext. diam 80–100, height 45–85, int. diam. 45–60, int. depth 30–35; size highly variable according to location. Placed 1–15 m above ground, mainly 2–5 m, on natural or artificial ledge, in niche, at base of basket-shaped tuft of twigs against tree-trunk, in small or large hole in tree trunk or stump, in creeper against trunk or wall, or on top of flat branch or in shallow open cavity in broken thick branch; may use old nests of other species and open-fronted nest-boxes. Built soon after arrival, mostly by ♀ in 3 to 7 days; often ♂ displays with material in bill but does not build.

EGGS: 2–7 but mainly 3–4 in Maghreb; laid at 1-day intervals (interval affected by temperature). Oval; smooth but not glossy; pale blue, blue-green, buff or dirty creamy white, variably mottled and blotched red-brown and purple-grey, often as cap at larger end. SIZE: (Europe, n = 494) 16·5–21·0 × 12·5–15·0 (18·7 × 14·0); fresh weight 1·3–2·2 g. Replacement clutches laid after egg loss; 2 broods regular in some areas at least in Europe.

LAYING DATES: NW Africa: Apr–July, mainly mid-May to early June; some second clutches early July in Algeria.

INCUBATION: begins with last egg; by ♀ but sometimes also by ♂. Period 10–17 days (13·2) in Britain. ♂ frequently feeds ♀ on and out of nest, guards and defends nest surroundings, particularly when mate absent. ♀ often leaves nest for brief spells, dominates and displaces ♂ at feeding perches. Feeding range, 50–200 m around nest.

DEVELOPMENT AND CARE OF YOUNG: in Europe, young brooded more or less continuously by ♀ during first 6–7 days; fed by both parents at delivery rate of 7–57 visits/h; for first 2–3 days food mostly brought by ♂ to ♀ which passes it to young; young given fewer Hymenoptera but more large Diptera in England, where size of prey increases during nestling period but decreases after fledging. Weight of young increases steadily from *c.* 4 g at 24 h to 15·5 g at 9 days. Nestling period 10–17 days, mainly 12–16 in Britain.

After leaving nest, young still fed by both parents for 12–32 days, longer during warm weather with good food supply. Young of first brood may stay near nest of second brood until it fledges (Epprecht 1985) and may help parents to feed fledglings (Erard 1991). Breed at 1 year.

BREEDING SUCCESS/SURVIVAL: no statistics in N Africa. In Britain, 77·9% of eggs produced hatchlings of which 81·3% fledged; overall success 61% in May, 63% in June and 67% in July; predators account for 10% of egg-losses (Cramp and Perrins 1993). In Switzerland, predators took 6% of eggs and further 5% of nestlings; over an 18-year period 66% of first broods and 88% of second broods fledged young. In Germany for 14 years, production rate was 1·13–4·45 (2·67) young/pair/year. Longevity records of ringed birds, 5 years for ♀, 8–9 years for ♂; mean annual mortality rate, 0·45 for ♂, 0·65 for ♀ (Glutz von Blotzheim and Bauer 1993).

References
Cramp, S. and Perrins, C. M. (1993).
Epprecht, W. (1985).
Erard, C. (1987, 1990a, 1991).
Glutz von Blotzheim, U. N. and Bauer, K. M. (1993).
Kok, O. B. *et al.* (1991).
Summers-Smith, D. (1952).

Plate 26
(Opp. p. 481)

Muscicapa gambagae (Alexander). Gambaga Flycatcher. Gobemouche de Gambaga.

Alseonax gambagae Alexander, 1901. Bull. Br. Orn. Club 12, p. 11; Gambaga, Gold Coast.

Forms a superspecies with *M. striata*.

Range and Status. SW Arabia and N tropics of Africa.

Resident and migrant, breeding about 10°–15°N from Mali to Somalia and wintering a few degrees further south in W Africa and to at least 3°S in E Africa. Rare or uncommon; possibly confused with Spotted Flycatcher *M. striata* and so widely overlooked. Mali, occurs sparsely in S Gourma and on E edge of Dogon Plateau (Sourou, Kassa). Ivory Coast, widespread but scarce in N or at least NE (Comoé Nat. Park: Balchin 1988, Demey and Fishpool 1991). Ghana and Togo, uncommon resident, records from Gambaga, Nangodi Bridge, Pong Tamale, ?Mole, Fosse aux Lions and Pewa. Nigeria, known only from 3 areas: northeast of Yola, Jos Plateau (Kigom Hills, Vom) and Potiskum. Cameroon, sparse, from Adamawa Plateau (Garua, Rei Buba) to L. Chad, once in W, at Foumban (05°43′N). Chad, thinly distributed in S (L. Chad, Chari R., Sarh, Zakouma Nat. Park). Sudan, scarce in W basin of Darfur and low zone of Jebel Marra, either resident (Lynes *in* Bannerman 1936) or breeding visitor (Nikolaus 1987); once Sennar, twice in S Blue Nile Prov.; a few records in S (Lado, Mvolo). Uganda, rare, only in N (Budongo, Gulu, Labwor). Zaïre, frequent on hills in extreme NE (Aba, Dramba). Kenya, records from Mt Elgon, Ololokwe, Lerata, Archer's Post, Isiolo, Meru Nat. Park (Ura Gate), Garissa, Galana Ranch, Wamba, Baragoi and Ngulia. (All S Sudan, Zaïre, Uganda and Kenya records are in Nov–Apr: see *General Habits*). Ethiopia, Bonga area and near Welkite (Erard 1974a), and 2 further east at Arsi Neghelli, 10 Nov 1995 (Robertson 1996). Somalia, only in forest at 1850–2000 m on Mt Wagar, where apparently fairly common in June–Sept.

Description. ADULT ♂: forehead and crown brown, faintly dark-streaked; nape to rump and uppertail-coverts uniform brown. Tail brown, all feathers very narrowly fringed buff-grey. Lores, eye-ring and featherlets at base of upper mandible creamy; ear-coverts brown. Chin and throat whitish, breast soft pale brown in centre, sometimes with faint streaks, more solidly brown at sides; flanks brownish; vent and undertail-coverts silky white, thighs brown. Primaries and secondaries brown, narrowly edged buff, tertials brown broadly margined with buff, greater and median primary-coverts and alula darker brown; lesser wing-coverts brown; greater and median coverts brown fringed with buff, often forming 2 bars in folded wing. Underside of flight feathers shiny grey, underwing-coverts and axillaries whitish. Upper mandible black, lower mandible black with pinkish base; eyes dark brown; legs and feet black. SIZE: wing ♂ (n = 5) 73–75 (74·2), ♀ (n = 5) 73–76 (75·5); tail, ♂ (n = 5) 51·5–55 (53·6), ♀ (n = 5) 46–59 (54·0); bill to feathers, ♂ (n = 7) 9–11, ♀ (n = 2) 9, 10; tarsus, ♂ (n = 7) 15–16, ♀ (n = 2) 15, 16. WEIGHT: 1 ♂ Kenya, 12; 1 ♀, Ethiopia, 14, and juvenile there 12.

Muscicapa gambagae

Intra-tropical migrant: see text

IMMATURE: like adult but all upperparts with profuse, large pale buff spots, tertials broadly margined pale buff, tips of tail feathers with small buff spot; breast streakier than in adult.

NESTLING: not known.

Field Characters. A small, plain grey-brown flycatcher with trace of soft streaking on breast. Like Spotted Flycatcher but smaller (12–13 cm; Spotted 15–16 cm), and has much fainter breast streaking and little or no spotting or streaking on forehead. Wings shorter than in Spotted, tips reaching only to base of tail. Forecrown less peaked and head more rounded than in Spotted Flycatcher, and carriage more horizontal. In fresh plumage in spring, a prominent pale panel in dark wing, formed by edges of closed secondaries and broad pale fringes of tertials (Pearson *et al.* 1980).

Voice. Tape-recorded (McVIC). Call an arresting, repeated 'chik', sharper than that of Spotted Flycatcher (Pearson *et al.* 1980).

General Habits. Inhabits dry woodland, *Acacia* grassland, bushland, and thick woodland with open grassy patches, shrubby savanna, and dry montane forest (Mt Wagar); *Combretum–Terminalia–Acacia* savanna in Ethiopia, with seasonal *Hyparrhenia* grasses; recorded mainly in valleys of large rivers or on hills and plateaux at 1000–1500 m; breeds up to about 2000 m in Somalia and in winter can occur below 1000 m in Kenya. Almost nothing known about its social or foraging behaviours, which are presumably much like those of Spotted Flycatcher (the 2 species are often regarded as conspecific).

Partially migratory: 3 juveniles netted at night at Ngulia lights (S Kenya) in Nov. All E African records south of 70°N are in Nov–Apr, although species may prove to be resident north and northeast of Mt Kenya (Pearson *et al.* 1980). In north of range it is a late-rains and early dry-season visitor to Mali, Aug–Dec, presumably non-breeding (breeds further south in Mar–Apr), but appears to be breeding visitor in Darfur (Sudan) and Somalia, latter in June–Sept.

Food. Insects.

Breeding Habits. Breeds solitarily.

NEST: one a cup made of fine grass laid in a circle, lined with cocoon silk and a little seed-down, held in place with spider web; sited in hollow in end of small dead tree with top broken off 4–4·5 m above ground; another, a neat cup of fine grasses and spider web, sited 1 m above ground in middle of small acacia bush.

EGGS: 2 (2 clutches). Slightly glossy; greenish white, covered all over with very faint grey and less faint yellowish brown freckles. SIZE: (n = 2) 15·5–17·5 × 13.

LAYING DATES: NE Nigeria or N Cameroon, Mar (Bates 1927, and breeding condition Apr). Sudan, Darfur (spotted young July). Kenya, Kerio Valley, Mar; Garissa (spotted young with adult, Nov).

References
Bates, G. L. (1927).
Pearson, D. J. *et al.* (1980).
Richards, D. (1992).

Muscicapa caerulescens (Hartlaub). Ashy Flycatcher; Blue-grey Flycatcher. Gobemouche à lunettes.

Plate 27
(Opp. p. 496)

Butalis caerulescens Hartlaub, 1865. *In* Gurney, Ibis, p. 267; Natal.

Range and Status. Endemic, resident, possibly intra-African migrant. Widespread in forested areas of SE Guinea (Macenta area), Sierra Leone, Liberia, Ivory Coast (not uncommon, Danané to Tiassalé, southward), S Ghana (rare to uncommon, at least Eastern Region and up to Kete Kratchi), Togo (Misahöhe), Benin (Pendjari Nat. Park), S Nigeria (uncommon, mainly west of lower Niger R.), S Cameroon (north at least to Korup Nat. Park, Rumpi Hills, Mt Kupé, Abong Mbang and Dja R.), Rio Muni, Gabon (widespread but uncommon), Congo, Zaïre (frequent to common, widespread, Mbandaka, Uele R., southward), S Sudan (S Equatoria); S Somalia (lower Webi Shebelli and Juba rivers), W, E and S Kenya (Kakamega and eastern edge of central highlands to Taita, also on coast), S and W Uganda (Bwindi–Impenetrable and Malabigambo Forests to Budongo Forest and Jinja), Tanzania, Mozambique, Malaŵi, Zambia, Zimbabwe, south to E Cape Province (King William's Town), Transkei, Natal, Swaziland, NE Transvaal (scarce or uncommon, Lowveld and Escarpment regions), NE Botswana (fairly common to very common, Okavango, NE and E border), N Namibia (south to N Damaraland) and Angola (north of S Huíla and Benguella). Uncommon to common; av. density in NE Gabon, 3 pairs/km^2.

Description. *M. c. brevicauda* O. Grant: Nigeria, Cameroon and Gabon to Sudan, E Zaïre, Rwanda, Uganda and W Kenya (Kakamega), south to NW Angola (Cuanza Norte and Cuanza Sul) and Kasai. ADULT ♂: upperparts plain slate-grey, forehead and crown sparsely and faintly streaked. Tail-feathers brownish black, outer webs fringed slate-grey, tiny white apical spot on T3 to T6. Narrow black line through lores and eyes, bordered above by narrow white line extending above and behind eyes; narrow but well-marked eye-ring white; ear-coverts and sides of neck slate-grey, short black and white

Muscicapa caerulescens

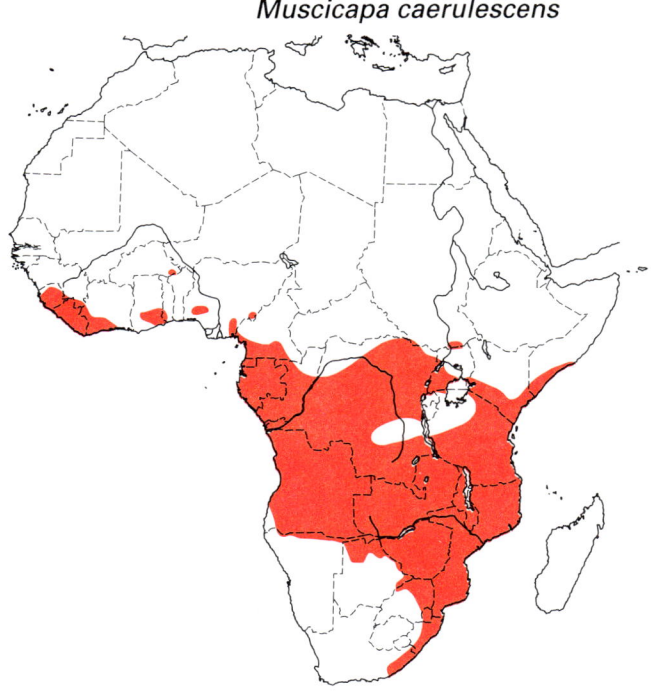

moustachial stripe; chin and throat white suffused with pale slate-grey; breast and flanks light slate-grey, centre of belly and undertail-coverts white, thighs dark slate-grey. Primaries, secondaries and tertials brownish black, outer webs of inner secondaries and tertials greyer, outer webs and tips of tertials edged whitish. Scapulars and upperwing-coverts brownish black edged slate-grey; axillaries and underwing-coverts white, the latter with brownish black base. Bill (**A**) brownish black, lower mandible mostly bluish horn, lighter at base; eyes olive-brown; legs and feet dark grey to brownish black. Sexes alike except that ♀ is duller than ♂. SIZE: (5 ♂♂, 5 ♀♀) wing, ♂ 67–74 (70·4), ♀ 67–69 (68·6); tail, ♂ 48–51 (49·6), ♀ 46–50 (48·7); bill, ♂ 14–15 (14·9), ♀ 13–14 (13·9); tarsus, ♂ 16–17 (16·2), ♀ 15–16 (15·3). WEIGHT: (Uganda) ♂ (n = 6) 10–17 (13·9), ♀ (n = 10) 11–18 (14·7), unsexed (n = 3) 15–19 (16·8); (Upemba, Zaïre, Oct) ♂ (n = 11) 16–20 (17·9), ♀ (n = 9) 16–19 (17·2).

IMMATURE: like adult but with buff tips to greater upperwing-coverts and tertials, the latter edged cream (not white) on outer web; mottled undertail-coverts sometimes retained.

JUVENILE: like adult but upperparts from forehead to uppertail- and upperwing-coverts densely spotted creamy to russet (feathers grey with broad subterminal creamy or russet band, and narrow smoky brown-black tip); underparts white, heavily mottled smoky blackish brown, especially on breast, flanks and undertail-coverts (these feathers are white, tipped with dark crescent), edges and apical spots on tertials and tail-feathers creamy white to ochre.

NESTLING: unknown.

M. c. caerulescens (Hartlaub): S Mozambique and E Swaziland south to E Cape Province. Somewhat larger and paler above and especially below, with whiter throat.

M. c. vulturna Clancey: Mozambique (south of Zambézia), S Malaŵi, E and S Zimbabwe, E Transvaal and N Swaziland. Larger and much paler than *brevicauda*, less bluish slate above, underparts much less washed with grey.

M. c. impavida Clancey: SW Angola (S Huila north to S Benguella and east to central Lunda) and Namibia to S Zaïre (Shaba), Tanzania (from Songea and Njombe to Ufipa), Malaŵi, N Mozambique, Zimbabwe (Middle Zambezi R. system) and W Transvaal. Largest and palest race, especially below, underparts almost white, with very faint pale greyish tinge on breast and flanks. WEIGHT: (Mozambique, May, June) ♂ (n = 3) 16–18 (17·1), ♀ (n = 3) 16–18 (16·7); (Zimbabwe) ♂ (n = 9) 15·2–18·4 (17·3), ♀ (n = 4) 13·8–16·4 (15·7); (Zambia, June) unsexed (n = 4) 15–21 (16·6).

M. c. cinereola Hartlaub and Finsch: Somalia, Kenya east of the Rift and Tanzania (coastal lowlands, inland to Iringa, Kilosa, Kondoa and Mt Kilimanjaro). Underparts paler than *brevicauda* but darker than *impavida*, size intermediate between them. WEIGHT: (Kenya) ♂ (n = 16) 14–20 (16·0), ♀ (n = 14) 14–18 (15·6); (Somalia) unsexed (n = 10) 15·6–17·2 (16·4).

M. c. nigrorum Collin and Hartert: Guinea to Togo. Slightly paler than *brevicauda* but more mouse-grey above, more uniformly suffused with pale mouse-grey below.

Field Characters. Widespread in woodland and forest edge, almost throughout Afrotropics. Does not enter forest, so cannot be confused with grey flycatchers of the forest interior. Grey above, whitish below shading into pale grey on breast; face marked by black line through eye bordered by white line, incomplete white eye-ring, indistinct short dark moustachial stripe, no white in wings or tail. Usually in middle and upper levels, where it hawks for insects like typical flycatcher; this distinguishes it from the tit-flycatchers *Myioparus* which feed like warblers.

Voice. Tape-recorded (72, 86, 88, 91, B, F, ERA, KEI, WALK, ZIM). Advertising song in Gabon, a series of shrill chattering notes or, in more intense territorial battles, a brief, piercing, buzzy, peevish 'teep-tup-tyup-tyeep'; repeated several times. Advertising song recorded in coastal Kenya (Gedi), 3–4, occasionally 5, high-pitched, thin but pure notes, 'tyee-too-way-tee', 'tyee-per-tee', 'tyee-tutu', 'tyee-tutu-way-tee'; sometimes a short phrase is lengthened into a pleasant medley of 6–8 notes ending in a little chatter. When this song was played back on an apparent territorial boundary, 2 birds responded and engaged in a vigorous song battle, a speeded-up, continuous version of the song, interspersed with buzzy chattering and high-pitched trills; they also fought briefly in the air (S. Keith, tape and notes). Courtship song, a rather soft twittering 'tip-tip-tip-tyuteerew' including jingling and rolling notes; also a variable, quick sibilant stuttering 'tsip-tsip-tsip-tsip-tseep-tslipsipsipsip' (Maclean 1993). Contact call, short, high-pitched buzzy chirp or peep (♂) or short, high-pitched wheeze (♀); also a husky 'zay' and a short series of notes starting high and dying away as they descend. Alarm call, prolonged piercing descending hiss. Birds in hand utter shrill, drawn-out buzzing notes. Begging call of young out of nest, shrill, short, rattling squeak.

General Habits. Inhabits open wooded areas, canopy of riverine woodland, thickets in moister savanna; mostly under 1500 m, up to 1800 m in E Africa. Does not enter virgin evergreen forest, but penetrates forest that has been logged or where roads or wide tracks have been opened; remains in secondary growth, old and recent cultivation, e.g. cacao, coffee and other plantations requiring tree canopy, peanut and cassava fields with scattered tall trees and borders of shrubs and bushes.

Solitary or in pairs. Usually forages in upper levels of vegetation, between treetops and under canopy; sits upright on exposed twig and makes short circular flights to catch flying insects, or hovers to snatch prey from foliage; often flicks wings on return to perch. Prefers more open canopy; occasionally gleans foliage like *Phylloscopus* warbler. In Angola, as in NE Gabon, where both Ashy and Dusky-blue Flycatchers *M. comitata* are present, e.g. in coffee forest, Ashy forages at higher levels (Dean *et al.* 1988). Rarely joins mixed-species flocks. Anxiety and alarm behaviours very similar to those of Cassin's Flycatcher *M. cassini* (adopts upright and horizontal postures when anxious; flicks tail and wing with rapid and jerky movements when alarmed). Active, moves about a lot, frequently traversing entire home range; able to adapt to rapid changes in habitat.

Food. Mostly insects, e.g. beetles, Diptera, Orthoptera, moths, winged ants and termites, caterpillars; sometimes small fruit. Prey size 5–35 mm, mostly 15–20.

Breeding Habits. Monogamous, territorial; pairs maintain rather large territory (*c.* 20 ha in NE Gabon) but breeding bird is confined to restricted areas (1–4 ha). Unpaired ♂ patrols territory, giving brief courtship songs while fanning tail and rapidly opening and closing drooping wing-tips. Territorial advertising behaviour similar although song more far-carrying, with bill pointing upward and plumage more fluffed out. ♂ sings to ♀ with sleeked plumage, extended neck, almost vertically held bill, drooped wings and flexed tarsi. Territorial defence by ♂ very similar to that of Cassin's and Little Grey Flycatcher *M. epulata* (when contestants stay at a distance, sings with body horizontal, droops wing-tips and jerks tail up and down; at closer range, threatens opponent with upright body, tail bent forward, carpal joints of folded wings away from body and open bill).

NEST: bulky cup made of moss, dry grass, matted vegetable fibres and rootlets; internal cup lined with finer material; usually placed 3–10 m above ground in crevice in bark, in shallow open cavity in trunk or thick branch, or in narrow fork. (Records of breeding in old weaver nests must refer to Dusky-blue Flycatcher.) Nest in Zimbabwe was built on top of remains of previous year's nest and consisted mainly of plant fibres, usually thin strips of bark with a few bits of spider web on the underside; cup lined with fine twiglets or similar material; placed *c.* 3 m above ground on verandah of house, on top of metal pipe supporting roof; built mostly in early morning by both birds, who were fairly noisy (Oxford 1994).

EGGS: 2–3; South Africa mean (17 clutches) 2·8; Malaŵi mean (4 clutches) 2·8. Oval; glossy; creamy white to light buff, finely speckled yellowish brown or reddish. SIZE: av. 19 × 14·5.

LAYING DATES: Nigeria, Dec; Gabon, Oct; Angola, Oct (spotted young Jan, Nov); Zaïre, Nov (birds with enlarged gonads Jan–Aug); Rwanda, Jan, Nov; E Africa: Uganda, (young Oct); Region B, Feb, May, June, Aug, Sept; Region E, May; Malaŵi, Sept–Nov (♂♂ with enlarged testes Dec); Zambia, Sept–Dec; Botswana, Sept–Dec; Zimbabwe, Sept–Jan; Transvaal and Natal, Oct–Dec.

INCUBATION: period estimated at 14 days (Oxford 1994).

DEVELOPMENT AND CARE OF YOUNG: young fed by both parents.

References
Brosset, A. and Erard, C. (1986).
Erard, C. (1987, 1990a).
Oxford, B. (1994).

Muscicapa aquatica Heuglin. Swamp Flycatcher. Gobemouche des marais.

Plate 26
(Opp. p. 481)

Muscicapa aquatica Heuglin, 1864. J. Orn., p. 256; Wau.

Range and Status. Endemic resident. Senegal to S Sudan, and Uganda to Zambia. Senegal, frequent on Taoué R. in N, common and widespread in S and E. Mali, common in Boucle du Baoulé Rés. Biosphère in W; frequent in Delta (L. Débo, Diaka, Bani valley); probably widespread in S. Burkina Faso, but does not occur as far north as Ouagadougou; abundant in Arli Nat. Park in SE. Ivory Coast, sight records from swamps on Kolonkoko R. (Comoé Nat. Park) and Korhogo. Ghana, uncommon; regular at Mole and Sugu; old record L. Volta area. Benin, Beterou, and abundant in Pendjari Nat. Park. Niger, frequent in 'W' Nat. Park and on lower Niger R. Nigeria, frequent to uncommon, north to Sokoto, Kano, Kirikasama and Malamfatori, south to Upper Ogun Res. and Benue R. (Makurdi, Ibi). Cameroon, sparse, Adamawa Plateau (Bini) to L. Chad. Chad, frequent along Logone R. and Chari R. to L. Chad, and in marshes at confluence of Sounta and Bahr Azoum. Central African Republic, occurs in Bamingui-Bangoran Nat. Park and Manova-Gounda-Saint Floris Nat. Park and south to about 07°30′N. Sudan, uncommon to frequent in Equatoria, north to E Bahr el Arab, L. No and Malakal, east to R. Nile. Throughout Uganda except for Karamoja, at

Muscicapa aquatica

700–1400 m; also Rwanda and W Burundi. Kenya, common in W, mainly at 1130–1400 m from Mt Elgon and Kitale–Kapenguria area to Keekorok, Mara. Tanzania, occurs in Mwanza district between L. Victoria and Rwanda and along shores of L. Victoria; also in W in Kigoma–Ujiji area; in SW a record in W Poroto region and species may occur at N end of L. Malaŵi. Zaïre, locally frequent in NE, common along W shore of L. Albert; from Luofu, SW L. Edward, to Kivu Highlands; on Lualaba R. above Kasongo to L. Kisale and L. Upemba where abundant in papyrus; also on Zaïre side of L. Mweru. Zambia, frequent on L. Mweru, in Mweru Marsh, Pambashye Swamp, along lower Luapula R. and in L. Bangweulu area; isolated population (birds very uncommon) at Suye Lake and Lukanga Swamp, Kabwe.

Description. *M. a. aquatica* Heuglin: Gambia to SW Sudan (east to about 28°E) and N Zaïre (Ubangui). ADULT ♂: entire upperparts uniform greyish or faintly olivaceous brown; tail, primaries, secondaries and tertials darker and glossy; greater coverts, secondaries and tertials with narrow pale brown fringes. Lores dark with small ochreous-buff spot above gape. Underparts white, breast and flanks washed with buff-brown or grey-brown. Underwing-coverts brownish white, axillaries white. Upper mandible black, lower mandible black with pale brown base; eyes dark brown; legs and feet brownish black. Sexes alike. SIZE: (12 ♂♂, 7 ♀♀) wing, ♂ 67–71 (67·8), ♀ 67–70 (67·6); tail, ♂ 52–56 (53·3), ♀ 49–54 (51·8); bill to feathers, ♂ 11–12 (11·4), ♀ 10–12 (11·3); tarsus, ♂ 15–16 (15·6), ♀ 15–16 (15·5). WEIGHT: (E Africa) ♂ (n = 3) 10·0–12·5 (11·5).

IMMATURE: like adult but forehead, crown, hindneck, mantle, scapulars, smaller upperwing-coverts, back and rump covered with diffuse buff spots; tertials sharply bordered with buff; breast and sides of throat with indistinct dark streaks.

NESTLING (*infulata*): skin flesh-brown, with tufts of long down on head and short down on base of wings, back and pelvis; gape yellow, inside mouth darker yellow (van Someren 1956).

M. a. infulata Hartlaub (including '*ruandae*'): Sudan (east of 28°E, north to 10°N) to N Zambia; E Zaïre, including Lualaba R. from Kasongo to about Mwanza. Upperparts dark blackish brown, mantle slightly rufescent in some specimens; throat and belly white, quite well delineated from breast, which is darker than in nominate race; flanks brown. Wing 64–72 but in highlands from Kigezi to Kivu and Rwanda 68–76 ('*ruandae*').

M. a. lualabae Chapin: upper Lualaba lakes, Shaba, Zaïre; perhaps this race also on W shores of L. Mweru, in which case it may extend from Mweru to Lualaba. Shade of upperparts and prominence of breast band intermediate between *aquatica* and *infulata*. Small; wing 58–65.

M. a. grimwoodi Chapin: S Zambia (Kabwe district: Suye L., Lukanga Swamp). Like *infulata* but crown to mantle somewhat greyer; breast band like *aquatica*. Large; wing 68–71.

Field Characters. A rather nondescript, small, grey-brown or dark brown flycatcher, with white throat, mid-belly and undertail-coverts and brownish breast and flanks. Best told by its habitat of swamps and lakesides. Particularly nondescript west of Sudan, although generally looks white-throated; could be confused there only with Gambaga Flycatcher *M. gambagae*, which is not obviously white-throated, has slightly streaky breast and inhabits dry wooded grassland. In E Africa Swamp Flycatcher is dark above, and white throat contrasts quite strongly with brown breast and flanks: bird can appear to have white underparts with a poorly-demarcated brown breast-band. Highland birds near Rift Valley larger than lowland ones. Zambian birds are noticeably white-throated. Easily confused with African Dusky Flycatcher *M. adusta*, which is slatier and inhabits woods and gardens. Smaller and darker than Pale Flycatcher *Melaenornis pallidus*, a bird of woody savanna scrub. Juvenile Swamp Flycatcher is spotted with buff above and mottled below and looks very different from adult.

Voice. Tape-recorded (GREG, MOR, McVIC). Song, 2 high-pitched and thin but quite sweet notes, the first downslurred, the second upslurred, followed by a trill, 'tseeuw-tsweee-titititi'; notes of trill may be audibly separate, as in this transcription, or run together, 'tssrrrr'. Trill sometimes given on its own, 'tsitsitsitsi'. In duet, presumed ♂ sings variations on the above theme while partner has an unsynchronized accompaniment, an irregular medley of little chattering and whistled notes, 'tsik-tsuk', 'charrr', or a longer 'wee-tsik-tsuk-tsway, trrrr', sometimes lasting longer than ♂'s song. Call a sharp 'pzitt'.

General Habits. Inhabits papyrus swamps, reeds and dead branches on lakeshores and large streams; woody and grassy vegetation around small stagnant pools; on L. Chad occurs in papyrus and reeds interspersed with ambatch trees *Aeschynomene elaphroxylon* and bare bushes in shallow water. Sometimes a few hundred m away from water, perching near tips of elephant grass. Solitary or in pairs. Forages by making short sallies after passing insects from a twig or papyrus-stem or reed-head; in flight takes small green grasshopper nymphs from grasstops (van Someren 1956). 'Has also been noticed running about on lily-pads and picking insects off them like a wagtail' (Mackworth-Praed and Grant 1960).

Resident, but in Niger recorded only in Jan–Oct, and may be forced to quit marshes in seasonally drying-out wadis in north of range.

Food. Insects, including flies (Diptera) and grasshopper nymphs (see above).

Breeding Habits. Solitary nester, apparently monogamous, and if territorial not obviously so. Nests inside old weavers' nests.

NEST: a scant lining of feathers placed inside the abandoned roofed nest of Yellow-backed Weaver *Ploceus melanocephalus*, Golden-backed Weaver *P. jacksoni*, Northern Brown-throated Weaver *P. castanops* and Slender-billed Weaver *P. pelzelni*. Said to nest also in trees, in crevices in rough bark, and to line nest with fine grass as well as feathers (Mackworth-Praed and Grant 1960); an unseen nest near L. Chad, Nigeria, was in cavity below thatched roof of house (J. H. Elgood, pers. comm.).

EGGS: 2–3. Bluish green or pale blue, freckled and spotted with dark-brown or reddish brown, mainly towards large end. SIZE: (n = 3) 16·0–17·5 × 12·0 (16·8 × 12·0).

LAYING DATES: Senegal, Mar (and juvs being fed, Nov); Nigeria (inaccessible nest probably with young, Mar); E Africa: Region B, 13 clutches Mar–June, 5 in Aug–Oct, 7 in Nov; Region C, June; Zaïre: Ituri Aug; Shaba Jan (in Zaïre likely breeding season July–Oct in NE, Jan–Mar and July–Oct in SE); Zambia, Oct.

INCUBATION: by both sexes, mainly ♀. Period: 12 days.

DEVELOPMENT AND CARE OF YOUNG: nestling period about 15 days.

References
Chapin, J. P. (1953a).
van Someren, V. G. L. (1956).

Muscicapa cassini Heine. Cassin's Flycatcher. Gobemouche de Cassin.

Plate 27
(Opp. p. 496)

Muscicapa cassini Heine, 1859. J. Orn., 7, p. 428; Sette Cama, Gabon.

Range and Status. Endemic resident. Widespread along rivers and streams in lowland forested areas: Sierra Leone (Kilimi), Liberia (common), Ivory Coast (south of 9°30′N), Ghana (uncommon or frequent, Ankobra R., Central Region, Kwahu, Cape Coast); Togo (Misahöhe, Tasso), Benin (Bétérou), Bioko, S Nigeria (common, coast north to Asejire and Iowo); S Cameroon (widespread in forest south from Korup Nat. Park, Kumba, Obala, Akonolinga); Gabon (frequent to locally abundant, widespread), Congo, N Angola (common in Cabinda, less so Cuanza Norte to Gabela); S Central African Republic (Sangha, Lobaye, Bangui, Bangassou); most of Zaïre (Mbomu R., Kibali R. and Itombwe to S Kasai); Rwanda (Koko R., Nyungwe), forests of W Uganda from Budongo to Bwamba, Kibale, Ankole and Impenetrable; NW Tanzania at Minziro Forest; NE and NW Zambia (Kalungwisi R. and Mwinilunga). Sight records away from forest in Mali (Boucle du Baoulé) and Niger ('W' Nat. Park) require confirmation. Uncommon to common. Density in NE Gabon, at least 1 pair/km on large rivers, 1 pair/500 m on smaller ones; in Cameroon, up to 12 pairs/2·5 km.

Description. ADULT ♂: upperparts plain bluish slate-grey, slightly streaked on forehead and crown. Tail-feathers brownish black with narrow whitish tips, more conspicuous in fresh plumage. Lores whitish; ear-coverts and sides of neck slate-grey; underparts white, flanks and breast washed grey, the latter indistinctly streaked. Primaries and secondaries brownish black; scapulars and lesser coverts brownish black, broadly edged slate-grey; tertials and inner secondaries brownish black narrowly fringed whitish on outer web and tip of inner web; median and greater coverts brownish black, narrowly edged grey in fresh plumage. Underwing-coverts brownish, broadly fringed white, axillaries white. Bill (**A**) black, lower mandible paler at base; eyes olive-brown; legs and feet blackish brown. ADULT ♀: like ♂ but upperparts less bluish, with a brownish tinge. SIZE: (5♂♂, 6♀♀) wing, ♂ 72–74 (73·1), ♀ 69–73 (71·1); tail, ♂ 54–57 (55·6), ♀ 54–56 (55·2); bill, ♂

×× = sight records

A

16–17 (16·8), ♀ 16–17 (16·5); tarsus, ♂ 14–15 (14·4), ♀ 13–14 (14·0). WEIGHT: (Upemba, Zaïre, Oct) 1 ♂ 16; (Uganda, June, July) 1 ♂ 19, 1 ♀ 17.

IMMATURE: like adult but wings and tail browner, upperwing-coverts and secondaries edged pale rufous.

JUVENILE: upperparts and sides of head dark brown, densely spotted light rusty to pale ochre (each feather with greyish base, wide crescent of rusty ochre in centre and narrowly edged brownish black). Wings and tail brownish black, upperwing-coverts tipped rusty ochre; outer web of inner secondaries edged creamy white; tail-feathers tipped whitish. Underparts creamy white, buffier on flanks; breast scalloped (irregular narrow blackish edges to feathers).

NESTLING: hatches naked, skin pinkish, with short sparse grey down on sides of crown, nape, wings and thighs; inside of bill yellow, flanges creamy white, bill purplish brown.

Field Characters. A small bluish grey flycatcher, with contrasting black wings and tail, faint streaks on crown, white underparts with pale grey flanks and breast-band, the last slightly streaked. Always near water: this is the only small grey flycatcher that habitually feeds along forest rivers. Similar-sized Olivaceous Flycatcher *Muscicapa olivascens* is brown, and feeds mainly in canopy; Grey-throated Tit-Flycatcher *Myioparus griseigularis* is also grey with dark wings and tail but is uniform grey below, and gleans foliage rather than hawking insects from a perch. Dusky-blue Flycatcher *Muscicapa comitata* and Tessmann's Flycatcher *M. tessmanni* are much darker, bluish slate, with dark breasts and contrasting white throats. Cassin's has a more hunched appearance than Ashy Flycatcher *M. caerulescens*, to which it is more similar in plumage; latter is paler overall, with uniform whitish underparts, white lores and eye-ring, and yellow lower mandible, and has different habitat (canopy of forest edge and second growth).

Voice. Tape-recorded (CHA, ERA, VIEL). Song by ♂ during territorial contests and courtship, a pleasing medley of high-pitched chirps, whistles, liquid trills, chips and buzzes, in phrases lasting 3–4 s, forming singing bouts of 0·5–3 min; notes often repeated singly or grouped; 21% have overtones. Contact call between partners, a single high-pitched buzzy note. Alarm, 5–8 buzzy notes, the first note longer, 'bzeee-bz-bz-bz-bz'. Food begging ♀ during courtship or incubation gives rasping calls.

General Habits. Confined to forest vegetation along watercourses, mainly in lowlands, but up to 1200 m in Cameroon, 1300 m in Zaïre (Itombwe) and 1800 m in Uganda; prefers open, seasonally flooded forest and junctions of rivers, also forested riverbanks, open old arms of rivers with still water, and flooded reed beds; uses branches overhanging water, and especially stumps and dead trees standing in water, for watching-posts and breeding sites; also perches on stones in river. Regularly excluded from all except waterside habitats and from denser parts of riparian vegetation by White-browed Forest-Flycatcher *Fraseria cinerascens*.

Solitary or in pairs. Sits upright, silent and motionless on exposed perch; makes sallies to catch aerial prey; sometimes hovers to pick up insect resting on or under leaf or caught in colonial spider web; also takes prey on ground; typically catches flying insects low over water, mostly less than 6 m from perch, sometimes up to 20 m. Uses as vantage points small rocky outcrops, stumps and dead trunks projecting from water, exposed branches and twigs, and hanging lianas. Very active, always on the move; works its way along riverbanks often for several hundred m. Swallows smaller prey in flight, kills larger items on perch, picking them apart with jerky head movements.

Like other flycatchers, scratches head indirectly (i.e. foot over wing). Bathes during hottest parts of day, usually in the afternoon, dipping to water surface like small kingfisher but immersing only lower body; back on perch it preens with fluffed out plumage, quivering wings. When anxious, e.g. about approaching animal or human observer, first adopts more upright posture, then pre-flight horizontal posture. When alarmed, flicks tail and closed wing with rapid, jerky movements; as alarm increases gives alarm calls while drooping wing-tips and giving nervous up and down movements of closed tail; finally, in panic stage, spreads tail and flaps wings while swaying from side to side, pointing open bill towards nearby source of danger (e.g. threatening snake).

Food. Mostly flying insects, especially Diptera, Odonata (dragonflies) and Lepidoptera (in NE Gabon mainly butterflies, also some moths), also Ephemeroptera, Neuroptera, Hymenoptera (small bees and wasps, winged ants), Isoptera (winged termites), caterpillars (for young) and Coleoptera (beetles), occasionally small millipedes. Prey size mainly 5–20 mm, sometimes up to 70 mm (av. 15 mm, n = 123, NE Gabon).

Breeding Habits. Monogamous, territorial. Pairs establish linear territories along rivers, or broader ones of *c.* 3 ha in open flooded forest. Pair-formation behaviour unknown. Exhibits territorial behaviour all year, although not frequently; usually contestants stay at a distance from each other but within sight or hearing, singing vehemently with body horizontal and wing-tips drooped, jerking tail up and down. At closer range adopts more elaborate and threatening postures, with upright body, spread wings, open bill and depressed tail directed towards opponent; chases and harasses rival in flight, vigorously snapping wings and bill. Courtship feeding common, begins as early as nest building, before copulation. During precopulatory behaviour, partners face each other in horizontal posture, with wings drooped and tail raised, and ♂ feeds ♀.

NEST: compact open cup; mainly of dry grasses, fibres

and pieces of bark, sometimes small twigs, bound together with thin rootlets and spider webs and plastered outside with moss; lined with strands of vegetable matter, thin fibres from shredded bark, and even down from hairy wind-blown seeds; ext. diam. 90–100, ext. depth 70–80, int. diam. av. 45; usually < 2 m above ground, on stump a few m high or dead tree fallen in river bed, also on bare branch sticking out of the water covered with vegetable detritus left by floods, in hanging clumps of leaves, or in tuft of grass on small rock in the water; nests on stumps or trunks are often placed in crevice or open cavity, or in old nest of White-throated Blue Swallow *Hirundo nigrita*; on branches usually camouflaged among vegetable detritus, but sometimes fully exposed; rarely above dry ground, mostly above water, 5–20 m from bank; often excluded from closer sites by White-browed Forest-Flycatcher. Built by ♀, escorted and guarded by ♂.

EGGS: 1–3, laid at 1 day intervals; in NE Gabon 22 clutches all of 2; Cameroon mean (10 clutches) 1·7. Oval; glossy; light green, finely and densely speckled reddish brown or light rufous-brown and ashy purple, often as a cap at larger end. SIZE: (n = 16, Cameroon) 19·3–20·1 × 13·7–14·4 (18·7 × 13·6). Bird occasionally produces second clutch (30–35 days after first brood fledged) or replacement clutch (18–27 days after destruction of previous one).

LAYING DATES: Liberia, Mar, Nov; Nigeria, Jan, Apr–May; Cameroon, July, Oct–Mar (mainly during dry season; spotted young May, Aug); Gabon, Nov–Mar (particularly during short dry season); Congo (Mayombe), Sept; Zaïre, Sept, Dec–Jan, March (moulting spotted young Jan, ♂♂ with enlarged testes Aug, Sept); E Africa: Uganda, May–July, Nov; Zambia, July–Aug; Angola, Sept.

INCUBATION: begins with last egg; by ♀, perhaps also by ♂ (Cameroon). Period: 14–15 days. By day ♀ sits 80% of time, in sessions of 1–68 (13·8) min, with absences of 1–12 (3·8) min; ♂ guards environs of nest; ♀ receives 17–36% of her food from ♂, who gives her 20–25% of the prey he catches. Sitting bird on nest on wooden stump in Zambezi R., NW Zambia, left nest to eat butterfly brought to it by mate; later, it begged for food when its mate appeared on the nest (Beel 1995).

DEVELOPMENT AND CARE OF YOUNG: young fed by both parents, although often ♂ passes food to ♀ which gives it to young; intervals between feedings longer when prey larger; prey size increases with age of brood. ♀ broods young for first 3–4 days, sporadically up to 7 days, and during heavy rains. Nestling period 11–13 days. After leaving nest, young are usually fed for at least a month by both parents (although parents may take care of one young each). Young begin to moult when 1·5–2 months old.

BREEDING SUCCESS/SURVIVAL: in NE Gabon, out of 17 nests in which eggs were laid, 7 (41%) survived until young fledged; 36·7% of eggs produced young leaving nest. ♂ ringed as adult was recaptured at the same place 3 years later. Nest predators include monitor lizard *Varanus niloticus* and snake *Boulengerina annulata*.

References
Brosset, A. and Erard, C. (1986).
Erard, C. (1987, 1990a).
Hoi, H. (1987).

Muscicapa olivascens (Cassin). Olivaceous Flycatcher. Gobemouche olivâtre.

Plate 26
(Opp. p. 481)

Parisoma olivascens Cassin, 1859. Proc. Acad. Nat. Sci. Philadelphia, p. 52; Camma R., Western Africa = Sette Cama, Gabon.

Forms a superspecies with *M. lendu*.

Range and Status. Endemic, resident. Distribution disjunct: SE Guinea (Macenta area), Liberia (rare Mt Nimba), Ivory Coast (rare to uncommon, Mt Nimba, Taï Nat. Park, Gagnoa, Lamto and Yapo Forest); Ghana (rare, Fanti); S Nigeria (Benin, Owerri Province, Oban Hills, Mambilla Plateau), S Cameroon (south from Korup Nat. Park, Mt Kupé, Victoria, Bityé on Dja R. and Assobam on Boumba R.), Gabon (frequent to common in Woleu N'Tem and Ogooué-Ivindo, rare to uncommon elsewhere), S Central African Republic (Dzanga Reserves, possibly Lobaye) and Congo (locally abundant in Kouilou basin and Mayombe); and E Zaïre (N Ituri to Semliki Valley, Itombwe and Kasai). Uncommon to locally abundant; in NE Gabon, av. density, 2·1 pairs/km^2; in Congo, locally up to 12 pairs/15 ha (Koubotchi), less common in Mayombe (3 pairs/50 ha) (Dowsett-Lemaire and Dowsett 1991).

Description. *M. o. olivascens* (Cassin): Ghana and Nigeria to Zaïre. ADULT ♂: upperparts plain olive-brown, browner on forehead and crown. Tail dark olive-brown. Lores whitish, ear-coverts and sides of neck greyish olive-brown; chin and throat white, producing an ill-defined bib; breast and flanks pale greyish olive-brown, belly and undertail-coverts white, thighs dark mouse-brown. Primaries, secondaries and tertials dark olive-brown, with paler edges, almost whitish at tips of inner tertials; scapulars and upperwing-coverts olive-brown (greater coverts darker on inner web); axillaries, underwing-coverts and fringes to undersurface of inner primaries and secondaries white. Upper mandible brownish black, lower mandible yellow to whitish; eyes olive-brown; legs and feet horn to light brown. Sexes alike. SIZE: (4 ♂♂, 2 ♀♀) wing, ♂ 69–80 (75·2), ♀ 70–74 (72·2); tail, ♂ 50–62 (56·7), ♀ 52–57 (55·0); bill, ♂ 13–14 (14·1), ♀ 13–14 (14·0); tarsus, ♂ 16–18 (17·4), ♀ 15–16 (15·7). WEIGHT: (Gabon, June) unsexed (n = 2) 15·7, 16.

IMMATURE: like adult, but browner, with tawny or buff apical spots on greater and median upperwing-coverts and tertials.

Muscicapa olivascens

JUVENILE: cheeks and ear-coverts with blackish scales; feathers of upperparts, upperwing-coverts and tertials warm brownish olive broadly edged tawny, pale rufous or rufescent buff, giving spotted appearance (drop-shaped spots, each narrowly outlined with black-brown, small on sides of crown); underparts white freckled dark grey to blackish on sides of breast and centre of belly.

NESTLING: unknown.

M. o. nimbae Colston and Curry-Lindahl: Liberia and Ivory Coast (Mt Nimba). Upperparts and wings warmer rufous brown, underparts whiter. WEIGHT: (Liberia, Apr–Oct) (8 ♂♂) av. 17·6, (3 ♀♀) av. 13·5.

Field Characters. A small, almost featureless brown flycatcher of the canopy of tall lowland primary forest. Uniform brown above, including wings and tail; paler brown below with whitish throat and centre of belly; lower mandible yellowish white. Most sympatric small flycatchers are grey and do not live in canopy; Ashy Flycatcher *M. caerulescens* frequents canopy but is grey above and white below, and occurs in primary forest only at edges and in clearings. Chapin's Flycatcher *M. lendu* and African Dusky Flycatcher *M. adusta* are brown but inhabit montane forest; migrant Spotted Flycatcher *M. striata* is brown and visits forest clearings but has white underparts with streaks on breast and head; Sooty Flycatcher *M. infuscata* and Ussher's Flycatcher *M. ussheri* are uniform dark brown and live in clearings.

Voice. Tape-recorded (CHA, ERA, LEM). Courtship songs, a brief high-pitched nasal twitter and a longer, varied series of grating and squeaky notes audible only at close quarters. Territorial song, in Gabon a short phrase of 2–3 shrill notes, the last note longest, buzzy and loud; in Congo (Mayombe) a phrase of 6–10 regularly spaced high-pitched whistles ('tsi, tsiu, tsic, tsup, tsic, tsi') lasting 2·5–3·5 s. Pair maintains contact with high-pitched nasal notes; ♂ greets ♀ with short, feeble twitter. Anxiety call, a slightly descending high-pitched hiss; alarm note similar but longer. Fleeing bird gives brief, high-pitched chatter. Fledged young repeatedly utter short shrill squeaks.

General Habits. Inhabits lowland primary forest; in Mayombe also flooded forest. Favours tall forest with almost continuous canopy and many emergents, and rather open understorey without well-defined strata. Frequents canopy, sometimes also upper layers of understorey; avoids recent tree falls and openings, also liana-rich areas. Excludes Yellow-footed Flycatcher *Muscicapa sethsmithi* from canopy and upper levels of understorey.

Solitary or in pairs. Active while foraging; covers a distance of 100–150 m in less than 15 min, moving from tree to tree. Covers entire territory daily but has favoured smaller areas (1–4 ha); sits on exposed horizontal perch (leafy twig, dead branch, loop of liana) and makes sallies to catch insects passing between tree crowns or hovers to pick them from leaves; also searches for prey in leaf clumps and on bark. Pairs may work through foliage together to dislodge insects. Takes insects stuck in spider webs, particularly in colonies of social spiders. Once observed dropping to ground like a falling leaf in pursuit of a small butterfly; it caught the insect with a few short hops, then returned to the tree before swallowing it (Erard 1987). Pairs remain in visual or vocal contact, and greet each other with calls and crouching postures. General behaviour otherwise very like that of Cassin's Flycatcher *M. cassini* (for instance, upright and horizontal postures when anxious; jerky tail and wing flicks when alarmed; spreads tail and flaps wings when mobbing). Very occasionally follows mixed-species flocks.

Food. Various insects: beetles, small grasshoppers, large flies, moths, winged termites and ants; also caterpillars for fledged young. Prey size, 5–30 mm, mostly 5–15.

Breeding Habits. Monogamous. Pair maintains large territory (*c.* 20 ha), although during the breeding season ♂ and ♀ restrict their activity to *c.* 2–4 ha. Followed by ♀, ♂ patrols territory and defends it all year with singing. Territorial encounters between adjacent pairs or between established pairs and occasional intruders are rare, because of low population density. Playback of territorial song induced ♂ to react with threat behaviour similar to that of Little Grey Flycatcher *M. epulata*, q.v. (faces opponent in upright posture with bill pointing upward, or adopts a humped posture, droops wing-tips, bends tail forward and points head and bill toward opponent). ♂ once hopped from branch to branch around ♀, trying to copulate, after courtship feeding.

NEST: undescribed. Twice in Gabon ♀ was observed building nest among dead leaves piled in webs of colonies of social spiders hanging from leaf clumps, 12–18 m above ground. ♂ followed ♀ around, sometimes feeding her, and occasionally visiting nest site.

LAYING DATES: Liberia (♀ with moderately enlarged ovary Aug, juv. Oct); Ivory Coast (pair with juv. Feb); Nigeria, Sept; Cameroon (juv. Feb, oviduct egg Oct), Dec; NE Gabon, Jan, Feb, June, Aug; Zaïre (♂ in breeding condition May).

References
Brosset, A. and Erard, C. (1986).
Erard, C. (1987, 1990a).

Muscicapa lendu (Chapin). Chapin's Flycatcher. Gobemouche de Chapin.

Alseonax lendu Chapin, 1932. Amer. Mus. Novit., no. 570, p. 11; Djuga, west of L. Albert.

Plate 26 (Opp. p. 481)

Forms a superspecies with *M. olivascens*.

Range and Status. Endemic resident, E Zaïre, SW Uganda and W Kenya. Rare. Zaïre, Lendu Plateau, where known only by the type specimen collected in 1926, at 1700 m at Djuga (1°55′N, 30°30′E), and from Itombwe Mts at Ibachilo, Milanga and Butokolo, where 13 examples found in late 1950s: 11 at Ibachilo (1750–1820 m, 3°41′S, 28°31′E) and one each at Milanga (1700 m, 3°23′S, 28°25′E) and Butokolo (1470 m, 2°43′S, 28°17′E). Also 2 seen Bunyole (NW L. Kivu, 2100 m) in 1957 (Chapin 1978). Uganda, one Bwindi–Impenetrable Forest at 1500 m (Keith and Twomey 1968). Kenya, Kakamega Forest at 1600–2150 m and N Nandi Forest at 2130 m; in the first, 4 birds collected in 1963 and 1965 and a few subsequent sight records; in the second, one record, in 1978.

This species, already rare, is threatened by forest clearance in Bwindi, Kakamega and N Nandi Forests, on Lendu Plateau and in Itombwe Mts, near the mining centre of Kamituga (Collar and Stuart 1985).

Description. *M. l. itombwensis* Prigogine. Itombwe Mts, Zaïre. ADULT ♂: upperparts darker than in *M. olivascens*, warm olive-brown, slightly darker on forehead and crown. Tail dark olive-brown, T2–T6 with warm brown outer webs and dull blackish inner webs. Lores and eye-ring greyish, ear-coverts warm brown, a little streaky, sides of neck greyish olive-brown. Chin and throat off-white, producing an ill-defined bib, grading into greyish on cheeks and into grey-brown on breast; flanks pale greyish olive-brown, belly silky white, undertail-coverts white or pale greyish brown, thighs dark mouse-brown. Primaries, secondaries and tertials dark olive-brown edged rufescent pale brown, almost whitish at tips of inner tertials; scapulars and upperwing-coverts olive-brown (greater coverts darker on inner web); axillaries, under-wing-coverts and fringes to undersurface of inner primaries and secondaries white or greyish white. Upper mandible brownish black, lower mandible black; eyes olive-brown; legs and feet grey, horn, or light brown. Sexes alike. SIZE: (3 ♂♂, 3 ♀♀) wing, ♂ 78–79 (78·7), ♀ 70–75 (73·0); tail, ♂ 57–63 (60·0), ♀ 52–56 (54·7); bill to skull, ♂ 13–14, ♀ 13–14; tarsus, ♂ 15–16·5 (15·5), ♀ 15·5–16 (15·7). WEIGHT: unknown.

IMMATURE: not described.

NESTLING: hatchling has pale yellow-brown down.

M. l. lendu (Chapin): Lendu Plateau, Zaïre; SW Uganda; W Kenya. Like *itombwensis* but upperparts slightly paler, breast paler and in less contrast with throat, flanks and belly; bill slighter, wing less pointed.

TAXONOMIC NOTE: *itombwensis* has a rather more pointed wing than *lendu* and so they may be separate species (Vaurie 1953, Keith and Twomey 1968); but differences in bill size and shape are insignificant (Prigogine 1957), and we follow Short *et al.* (1990) and Dowsett and Forbes-Watson (1993) in treating the 2 birds conspecifically.

Field Characters. A small brown flycatcher of the canopy of highland forest. Olive-brown above with rufescent wings, paler brown below with whitish throat and white belly and undertail-coverts; bill black, eyes and legs brown. Very rare, and sight records require the fullest documentation. Combination of brown (not grey) colour and montane forest habitat eliminates most congeners. Similar-looking brown Olivaceous Flycatcher *M. olivascens* lives in lowland forest and has yellow lower mandible. Sooty Flycatcher *M. infuscata* is a very different bird, uniform dark brown, long-winged and fork-tailed, hawking insects in clearings in sailing flight. African Dusky Flycatcher *M. adusta* is the only

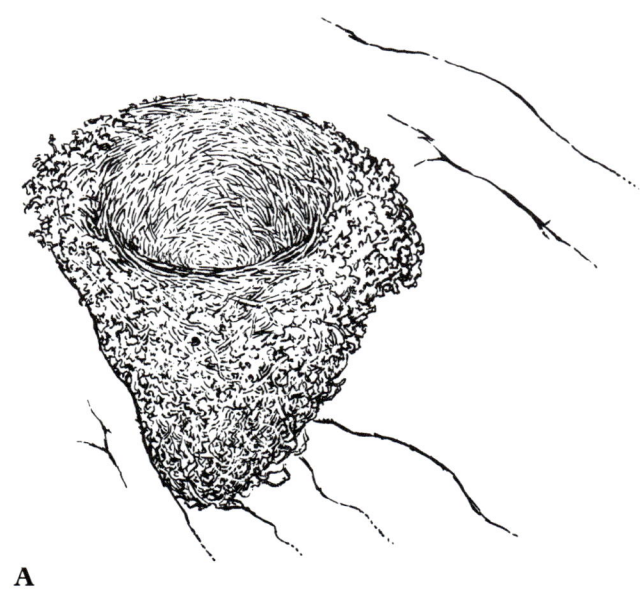

A

real contender; it is smaller, with smudgy streaks on breast and sides of throat; sympatric race *pumila* has chin and belly warm buff, breast and flanks grey-buff.

Voice. Not known.

General Habits. Inhabits canopy of dense, hilltop montane forest, keeping mainly to bare or nearly leafless branches high in trees. Occurs in pairs or flocks of 3–4; joins mixed-species flocks. A nest, almost certainly of this species, was sited over a pond.

Food. Insects fed by bird to its brooding mate and to hatchlings.

Breeding Habits.
NEST: the only one found (see above) was a bulky deep cup (**A**) made of fibres, smoothly lined with finer fibres and contour-feathers of African Black Duck *Anas sparsa*; ext. diam. *c.* 100, int. diam. *c.* 50, ext. depth *c.* 75; built in fork of dead liana hanging against trunk of small tree in middle of large artificial pond in forest.
EGGS: 2 (i.e., nest contained 2 newly-hatched young).
LAYING DATES: NE Zaïre, June. In E Africa, Region B, recently-fledged young begged from adult, Jan.
DEVELOPMENT AND CARE OF YOUNG: newly-hatched young brooded by one parent which was visited and usually fed by mate about every 10 min; mate once fed nestlings.

References
Chapin, R. T. (1978).
Collar, N. J. and Stuart, S. N. (1985).
Keith, S. and Twomey, A. (1968).
Prigogine, A. (1957).

Plate 26
(Opp. p. 481)

Muscicapa adusta (Boie). African Dusky Flycatcher. Gobemouche sombre.

Butalis adusta Boie, 1828. Isis von Oken, col. 318; Auteniquoi (= George), S Cape.

Range and Status. Endemic resident, in Cameroon and Bioko I., and more or less continuously from Eritrea to Swaziland and South Africa with an arm west from Shaba and N Zambia to SW Angola. Some migration within southern Africa.

Montane forests of Bioko and Cameroon from Mt Cameroon to Kumba, Oku, Bamenda, Tibati, Mt Genderu and Adamawa Plateau. Ethiopia, rare in Eritrea, frequent to common in all highlands. S Sudan, very common on Imatong, Dongotona and Didinga Mts, above 1800 m. Uganda, Mts Lonyili and Morongole in NE, and widespread down to 900 m in S Uganda south to Masindi and Mabira. Kenya, Mt Nyiru, Mt Marsabit, Ndotos, common in highlands in W and SW from Elgon to Mara, east to Nanyuki and Thika; Chyulus and Taita Hills. Rwanda, Burundi. Zaïre, highlands of E; Shaba; S Kasai; numerous on central Kivu volcanoes at 3400 m, frequent at 2150 m, probably ascends to 3700 m. Tanzania: between Rwanda and L. Victoria; Mt Ketumbeine; montane forests in NE – Usambara, Pare, Kilimanjaro, Meru, Essimingor, Lolkissale, Endulen, Ngaruka, Monduli and Crater Highlands; Ngurus, Ukagurus, Ulugurus, Kilosa, Kungwe, Iringa to Rungwe; Mt Mahari;

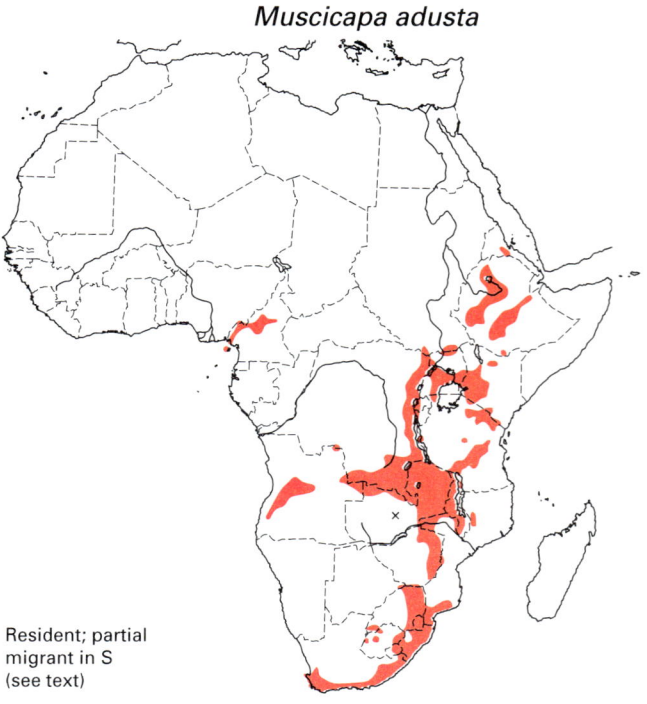

Muscicapa adusta

Resident; partial migrant in S (see text)

Matengo highlands. Zambia, scarce in Northern Prov. and Luangwa Valley, commoner on Muchinga Escarpment north to Mpika, occurs in Eastern Prov. on Nyika Plateau, Chipata and Chiromwe Hill; from Copperbelt at Mwekera west to Solwezi and N Mwinilunga south to Mundwiji Plain and N Kabompo (where rare); sight record on Kafue R. at 14°37′S. Malaŵi, throughout, breeding at 900–2150 m, moving down to Lake level afterwards. Angola, throughout central plateau from W and N Huila to S Cuanza Sul; S Lunda and N Moxico. Mozambique, in NW, W (Mt Gorongoza, Chimanimani Mts, Vumba, Mt Garuso) and S (to coast in winter). Zimbabwe, widespread in E up to 1200 m and less frequently to 1600 m, and from Mashonaland Plateau northwest to 30°40′E and to Zambezi Escarpment. Swaziland. South Africa: common in escarpment region of E Transvaal, including Soutpansberg and Blouberg, and Lowveld; W, central and E Orange Free State in several isolated populations; Natal, from Drakensbergs to coast; Cape Province, locally common in coastal areas west to Cedarburg Mts, Saldanha Bay and Cape Peninsula.

Description. *M. a. adusta* (Boie): South Africa east to Umzimvubu R. and Griqualand East. ADULT ♂: upperparts including wings and tail uniform dark brown except for narrow buffish crescent under eye and pale buff or whitish mark between nostril and top-front of eye. Crown sometimes slightly dark-streaked. Chin and throat whitish, with broad, diffuse, buffy olive-brown and whitish streaks at sides, merging into brown of cheeks and sides of neck. Breast brown, a little paler than upperparts, not forming a breast-band, but brown grading into white of throat and belly; the brown not uniform, but with soft, diffuse, broad streaks or blotches. Belly buffy off-white, flanks buffy light brown, vent and undertail-coverts white. Thighs brown. Underwing-coverts and axillaries buff; underside of flight feathers and tail shiny dark brown. Upper mandible black or dark brown, lower mandible the same with dull yellow base; eyes dark brown or brown-black; legs and feet black or brown-black, pinkish brown or greyish. Sexes alike. SIZE: (10 ♂♂, 10 ♀♀) wing, ♂ 66·5–72 (68·9), ♀ 65–68 (66·5); tail, ♂ 53–58 (55·1), ♀ 50·5–55·5 (52·7); bill to skull, ♂ 14–19 (15·2), ♀ 13·5–15·5 (14·5); tarsus, ♂♀ 14·5–17·0 (15·5). WEIGHT: (South Africa, 1 ♀ and 4 unsexed) 9·6–12·9 (11·2).

IMMATURE: recently-fledged birds have upperparts blackish brown with small buff spots on head, larger drop-shaped buff spots on upperwing-coverts and tips of tertials, and still larger (but less discrete) buff marks on mantle, back and rump. Tail with small buff spots at tip. Primaries and secondaries narrowly and tertials broadly edged with warm buff, sometimes rufescent. Feathers of underparts silky and fluffy, buff, with dusky marks, mostly crescentic, on breast and flanks; belly white.

NESTLING: skin at hatching orange-yellow, inside mouth yellow; scantily covered with white down (Skead 1963) or rather dark down (Skead 1966). When bird feathered, gape pale yellow, inside mouth orange-yellow.

M. a. fuscula Sundevall: South Africa from Umzimvubu R. and Griqualand East north to N Transvaal; Swaziland; winter visitor to S Mozambique (Brooke 1984). Like *adusta* but underparts less buffy; streaks on lower throat and breast grey not buffy olive and better defined, belly white.

M. a. subadusta (Shelley) (including '*angolensis*'): E highlands of Zimbabwe, Zambia, Malaŵi, Mozambique north of Limpopo R., Angola, Zaïre (S Kasai, Shaba), Tanzania (Matengo highlands). Like *adusta* but upperparts paler and greyer brown; breast greyer, slightly paler, and more uniform than in *adusta*; bill finer; tail shorter, ♂ (n = 10) 48–51 (49·3), ♀ (n = 10) 45–51 (47·7); '*angolensis*' has colour shades intermediate between *adusta* and *subadusta* and is often regarded as a valid subspecies.

M. a. mesica Clancey: Zimbabwe except for E highlands. Upperparts, particularly crown, slightly paler and less greyish than in *subadusta*, rump and uppertail-coverts tinged isabelline, flight feathers much less blackish.

M. a. fuelleborni Reichenow: Tanzanian highlands from Poroto Mts and Njombe to Ngurus, also Mahari Mts. Like *adusta* but upperparts and breast a cooler grey.

M. a. murina Fischer and Reichenow (including '*roehli*'): NE Tanzanian highlands north of 5°30′S (except Mt Ketumbeine) and S Kenya (Taita Hills). Upperparts like *fuelleborni*, underparts browner, slightly olivaceous. WEIGHT: ♂ (n = 3) 6–7 (6·7).

M. a. minima Heuglin: Ethiopia, Eritrea. Upperparts slightly warmer brown than in *adusta* and tinged olivaceous; underparts deeper buff than in *adusta* with larger areas of cream-white on throat and belly. Small: wing, ♂♀ (n = 6) 59–64·5 (61·3).

M. a. marsabit (van Someren): Kenya (Mt Marsabit, Moyale). Upperparts slightly warmer brown than in *adusta*; chin warm buff, breast and flanks grey-buff with rufous tinge, belly varies from cream-white to rufescent buff. Small: wing, ♂♀ (n = 5) 61–63·5 (62·6). WEIGHT: ♂ (n = 3) 5–11 (8·7), ♀ (n = 3) 5–10 (7·7).

M. a. pumila (Reichenow) (including '*grotei*', '*subtilis*', '*interposita*' and '*chyulu*'): NW Tanzania, Burundi, Rwanda, E Zaïre north of Shaba, Uganda, Kenya (except Mt Marsabit and Taita Hills), N Tanzania (Mt Ketumbeine), S Sudan, Cameroon (Adamawa Plateau). Upperparts like *adusta*; chin and belly warm buff (paler than in *marsabit*), breast and flanks grey-buff; plumage darkens with increasing altitude (Zaïre: '*subtilis*'). Immature birds more rufescent than in those of *adusta*, *subadusta*, *fuelleborni* and *murina*. WEIGHT: (Kenya) ♂ (n = 14) 5–11 (8·2), ♀ (n = 9) 5–12 (8·1), unsexed, 10, 10, 10.

M. a. obscura (Sjöstedt) (including '*poensis*', '*albiventris*', '*kumboensis*' and '*okuensis*'): Bioko I.; Cameroon northeast to Tibati and Mt Genderu. Upperparts like *subadusta*, underparts like *murina*. WEIGHT: (Buea, Cameroon) 2 ♂♂, 10, 11, 1 ♀ 10, 1 unsexed 11.

TAXONOMIC NOTE: 20 subspecies have been recognized in the past, but *M. adusta* has considerable intrapopulation and clinal variability, as well as variation in the appearance of skins arising from different taxidermal procedures; we recognize only 10 subspecies. The African representative of a widespread Asiatic superspecies (*M. dauurica* supersp.).

Field Characters. The common small brown flycatcher of forest edge and rich woodlands in the eastern half of Africa, mainly in highlands in tropics. Often the only member of the genus present in its preferred niche. Rather plump and short-tailed; some races greyer above; smudgy breast contrasts with pale throat. In southern (but not eastern) Africa has narrow pale broken eye-ring. Generally silent. Range overlaps with Ashy Flycatcher *M. caerulescens* which is more of a lowland bird, larger and blue-grey rather than brown, whitish below, with well-marked face (black lores and white supra-loral line) and distinctive song, and typically feeds in upper levels and canopy. Migrant Spotted Flycatcher *M. striata* is larger and longer-tailed, with streaked breast and crown, and usually in more open habitats; regularly flicks wings. Unlikely to meet Gambaga Flycatcher *M. gambagae*, a bird of dry lowland habitats, although their ranges overlap; Gambaga is like a small, washed-out Spotted Flycatcher, with indistinct breast and crown streaks and pale underparts (local races of Dusky are brown and buff below). Swamp Flycatcher *M. aquatica* is not a contender since it lives in papyrus.

Voice. Tape-recorded (58, 75, 86, 88, 91, B, C, F). Song a hurried, high-pitched jumble of thin notes including 1–2 more emphatic short trills; duration 5 s (Zambia). Also a measured rendition of 7 short sharp notes, monosyllabic or nearly so, no two quite alike, in 9 s (South Africa). Call a thin, very high-pitched 'tseeeee' or 'seeeuu' lasting 1–1.3 s, pitch falling from 8 to 7 kHz, repeated a few times, like 'seeea' whistle of Grey Cuckoo-Shrike *Coracina caesia*. Another call, quite different, is short (0.3–0.5 s) phrase of *c*. 10 rapid, scolding, clicking notes. Also a repeated, sharp 'tsip-tsip-tsirrrrt' or 'tsi-rit-tsirit-tsirit'.

General Habits. Inhabits glades in and edges of evergreen and riverine forests, forested hillside gullies, dense woodland, parks, well-wooded gardens, orchards and plantations; often near water. In Zimbabwe, Zambia and Malaŵi occurs in rich stands of *Brachystegia* woodland, particularly where *Uapaca* trees are plentiful; also on edges of *Cryptosepalum* forest. In SW Cape Prov. wooded ravines on hillsides, particularly near boulder-strewn perennial streams in deep shade of oak groves. On Kivu volcanoes, open *Hagenia* woods. A highland bird throughout most of tropics, from 900 to at least 3400 m, mainly above 1500 m in E Africa although widespread in lowlands north and west of L. Victoria; but in southern Africa occurs from 400 to 1200 in Zimbabwe (rarely up to 1600 m) and commonly down to sea level in South Africa and Mozambique.

Solitary or in pairs. Quiet, unobtrusive, tame and confiding but can also be secretive; keeps itself in shade, often over wooded stream. Perches upright, occasionally flicks wings. Sometimes joins mixed bird parties (Zimbabwe: Beasley 1995). Forages by hawking out for 2–3 m from low projecting branch, 1–3 m above ground, with clear airspace below or around; also hovers briefly to pluck prey from tree-trunk or foliage and picks prey from trunk in full flight; occasionally lands on ground to snatch small insect, and hunts from small rock in a stream-bed, making swift sallies after passing insects, each time returning to same perch or one close by. One bird made numerous flights upwards for 1–2 m from rock in stream, returning to it every time 'as if flicked back on a length of elastic' (Vincent 1947). Insect caught with audible snap of bill. On Mt Kilimanjaro usually forages in canopy and mid-stratum, occasionally in understorey (Cordeiro 1994); in Zambia feeds in woodland canopy. Once attracted to spray of water from garden hose-pipe (Cape). Seen attending ant swarms, where it dived for insects flushed by ants (Kilimanjaro: Cordeiro 1994).

Resident, but south of 10°S also partly migratory. Evidently a dry-season visitor, May–Aug, to Kasaji (10°20′S, 23°25′E, Zaïre: Aspinwall 1985). In Zambia some birds move into Luangwa Valley from surrounding plateaux in July, returning in December (Aspinwall 1985). In Malaŵi breeds mainly at 900–2150 m and moves below 1000 m (and down to 50 m in Shire valley) in Apr–Aug. In Zimbabwe occurs down to 400 m in July and probably moves from Zambezi Escarpment into Zambezi Valley then. An altitudinal migrant in parts of South Africa, moving from montane forests to escarpment and coastal ones and, in winter, into closed forest; winters in gardens and blue gum plantations around Tzaneen, lowveld NE Transvaal (Ginn *et al.* 1989). Resident and partial migrant in SW Cape, many birds moving after breeding to winter in Natal and S Mozambique, although recent records from Newlands (SW Cape) indicate it is more common there in winter than in summer (Silbernagl and Silbernagl 1995, Barnes 1996). In Mozambique a winter visitor only, Apr–Sept (Clancey 1971).

Food. Winged insects including small beetles, syrphid wasps, midges, many noctuid moths and their caterpillars, and a few (supposedly distasteful) *Acraea* butterflies (van Someren 1956); most food items very small; also small fruits including the alkaloid-rich *Vepris undulata* (Rutaceae) (Maclean 1993) and mulberries (Ginn *et al.* 1989).

Breeding Habits. Solitary breeder, monogamous, not markedly territorial but pair once seen to chase Southern Boubou *Laniarius ferrugineus* out of nest tree. Territory size 1–2 ha (Dowsett-Lemaire 1989). Courtship behaviour not known. Pair nests in same site, such as hanging spray of a particular olive or *Schrebera* tree, year after year (once for 5 years: van Someren 1956), but builds fresh nest each year.

NEST: quite robust, thick-walled cup, sometimes neat but usually rather untidy; made of green moss, lichen, fine rootlets, pine needles, *Clematis* seeds, dry grass, with some old cobweb and bits of bark fibre, lined with fine fibres and feathers (of doves, hens, neck feathers from captive Grey Crowned Crane *Balearica regulorum*). Feather lining compact, often thick, sometimes curling over lip of the cup, once entirely hiding the incubating bird. Some nests are made within growing moss of cobwebby leaf debris caught in twiggy branches, such that nest structure blends with the mass – nest more a lined depression in the surrounding vegetation than a discrete entity. Sited 1–15 (mainly 5–12) m above ground, in deep cleft in tree trunk or behind wood splinters or peeling bark on trunk, in hanging twigs with dead-leaf debris or much lichen, hanging bunch of ferns, multi-stemmed hanging vine, old *Lobelia* flower-spike, clump of 'fern-lichen' growing upright on a branch (van Someren 1956), mistletoe clump, crotch of lichen-clad fork of thick branch, old nest of Baglafecht Weaver *Ploceus baglafecht* (2 records), on mossy rock ledge (2 records) and in niche in wall. Nest generally well concealed, in shade, but one in hanging vine was in very open situation. Ext. diam. 90–115, ext. depth 70, int. diam. 50–55, int. depth 36–39. Built by ♂ and ♀.

EGGS: 2–4, usually 2–3, av. (n = 40 clutches, southern Africa) 2·7. Greenish white or pale green, finely freckled with dull reddish brown or light rust-brown, either closely and evenly all over or concentrated mainly at large end. SIZE: (n = 35) 16·4–19·3 × 12·8–15·0 (18·2 × 13·4).

LAYING DATES: Cameroon, Jan–Apr; E Africa: Region A, Jan–Aug, mainly Apr–May; Region B, Apr–May (9 clutches), July–Aug (4), Nov–Feb (4); Region C, Oct–Nov; Region D, Sept–June (37 clutches), mainly Jan (8), Feb (6), Mar (6) and Apr (4) (in Regions A and B breeds mainly in rains); E Zaïre, July, Oct–Dec; Angola, Sept–Oct; Zambia, Oct–Dec; Malaŵi, Sept–Dec; Zimbabwe, Sept–Jan (Sept 2, Oct 7, Nov 10, Dec 6, Jan 1 clutches); Mozambique, Oct–Nov; South Africa: Transvaal Oct–Dec; Natal Oct–Jan; SW Cape Sept–Jan.

INCUBATION: almost entirely by one parent (presumed ♀), which sits very close; occasionally relieved by the other. At one nest bird incubated for 74% of daytime, leaving nest every 15 min on average; incubating spells were 3–23·5 min, clutch unattended for 0·25–9·5 min. Mate fed incubating bird 5 times in 8·5 h: twice by perching on rim of nest, and 3 times sitting bird flew several m away from nest to accept food from mate. Period: 11–12 days (Kenya) and 14–15 days (Cape). At one nest the 2 eggs hatched several h apart; eggshells soon removed.

DEVELOPMENT AND CARE OF YOUNG: only 1 parent (presumed ♀) broods young; at first young brooded for 68% of daytime, in spells of 0·75–10·75 min, and unattended for spells of 0·5–6·0 min. Eyes half open at day 5; quills visible under skin on days 3–4, and emerge on back and wings by days 5–6; feather vanes break out of quills about day 7. On day one 2 young were fed 35 times in 260 min period: 16 times by '♀', 11 by '♂', and 8 times by the '♂' passing food to brooding '♀' who fed the chicks; greatest time between feeding visits, 23 min. Fledging period unknown.

References
Clancey, P. A. (1974).
Lawson, W. J. (1963c).
Skead, D. M. (1966).
van Someren, V. G. L. (1956).
Vincent, A. W. (1947).

Muscicapa epulata (Cassin). Little Grey Flycatcher. Gobemouche cendré.

Plate 27 (Opp. p. 496)

Butalis epulatus Cassin, 1855. Proc. Acad. Nat. Sci. Philadelphia, 7, p. 326; Moonda (= Mondah) R., Western Africa = Gabon.

Range and Status. Endemic, resident. Disjunct distribution in lowland forest. SE Guinea, Liberia (Mt Nimba), Ivory Coast (Taï, Lamto and Yapo Forest), Ghana (uncommon, Aburi and Tarkwa, north to Ejura and Tafo); S Nigeria (Nindam Forest Reserve and possibly near Erin-Ijesha), S Cameroon (north to Sanaga, Kribi, Lolodorf and Dja R., also Korup Nat. Park; juv. recorded near Tibati, misidentification?); Rio Muni, Gabon (frequent at least north of Ogooué R.), possibly S Central African Republic (Lobaye), and Congo (Mayombe); E Zaïre (Uele and Ituri districts, from Titule to Semliki). Uncommon to frequent; in NE Gabon, density 4–6 pairs/km².

Muscicapa epulata

Description. ADULT ♂: upperparts, including scapulars, plain mouse-grey, indistinctly streaked on forehead and crown. Tail brownish black. Lores whitish; ear-coverts and sides of neck mouse-grey, the former washed brown; chin and throat white mottled mouse-grey, forming poorly defined white bib; breast and flanks grey; belly white, laterally mottled mouse-grey; undertail-coverts white. Primaries brownish black, wing-coverts and secondaries browner, narrowly fringed brownish grey, edges of inner secondaries and of tertials almost whitish; axillaries and underwing-coverts white. Bill brownish black, lower mandible yellow at base; eyes olive-brown; legs and feet black. Sexes alike. SIZE: (6 ♂♂, 7 ♀♀) wing, ♂ 54–59 (56·7); ♀ 54–59 (56·5); tail, ♂ 38–42 (39·7), ♀ 36–41 (38·6); bill, ♂ 11–12 (11·4), ♀ 10–12 (11·5); tarsus, ♂ 11–12 (11·7), ♀ 11–12 (11·6). WEIGHT: (Liberia) ♂ (Jan, n = 2) 9·3, 10·7, ♀ (Aug, Sept, n = 2) 8·8, 9·5.

IMMATURE: like adult but wings browner; with russet apical spots on greater upperwing-coverts, tertials, upper- and undertail-coverts.

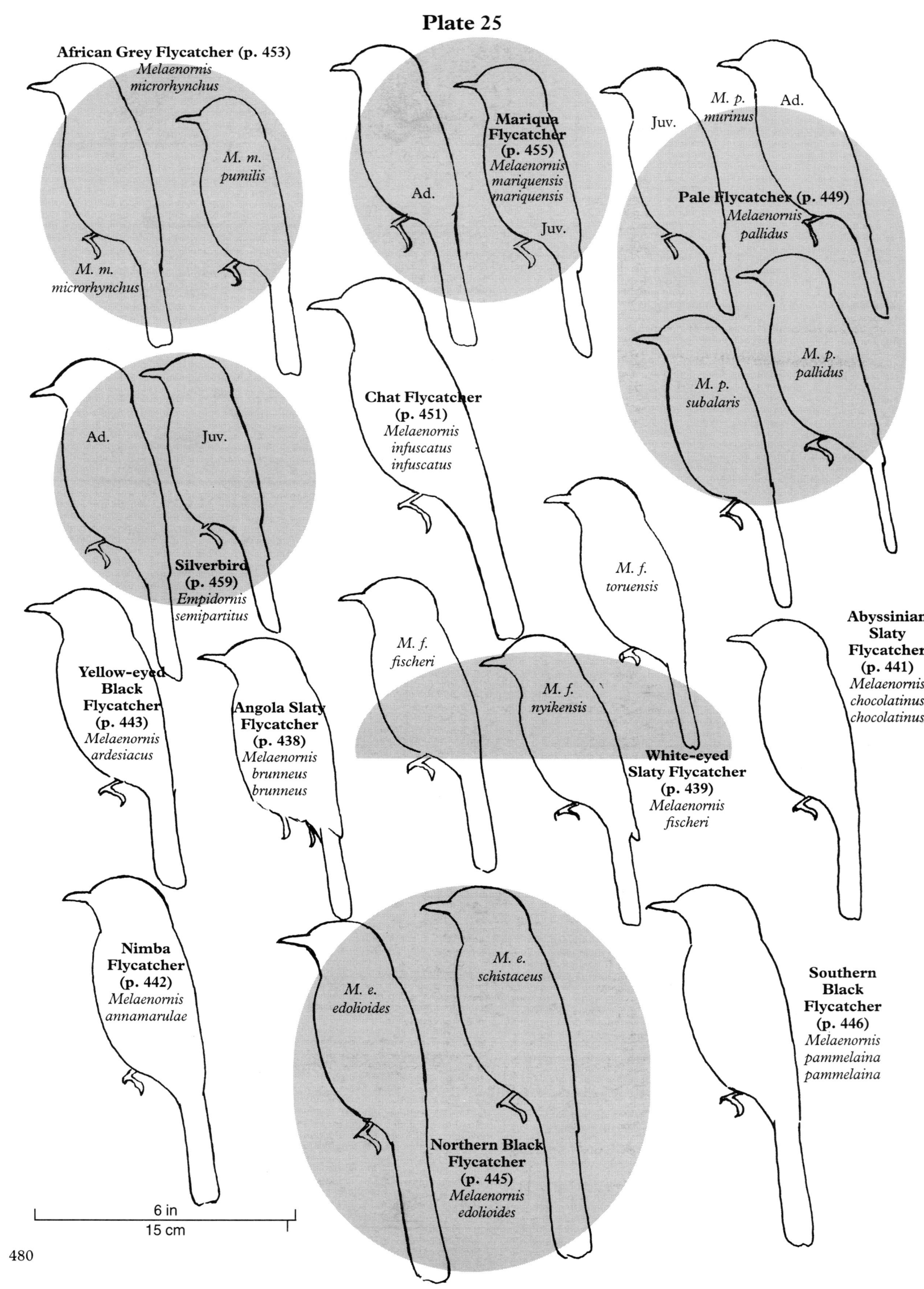

Plate 25

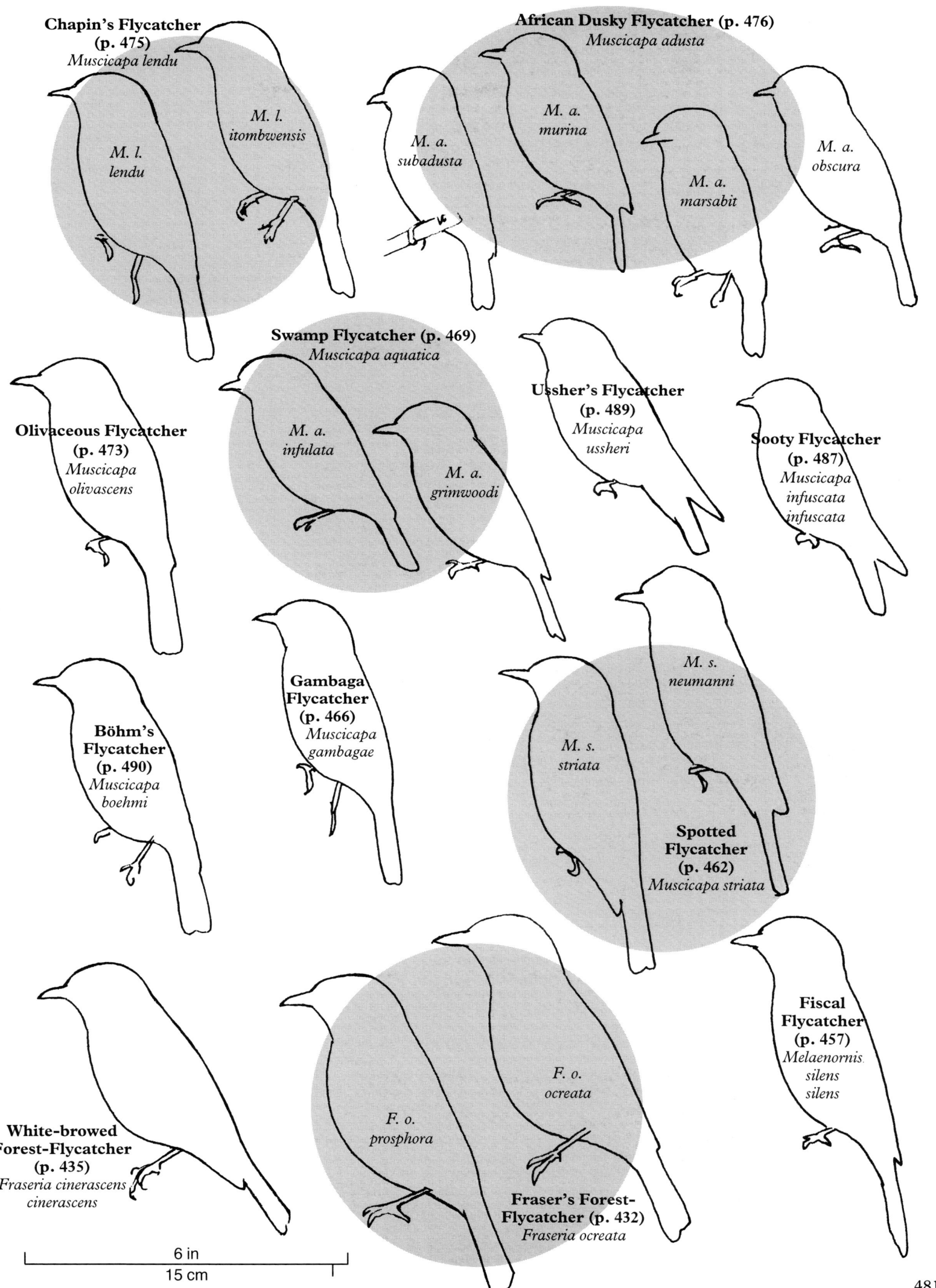

JUVENILE: upperparts grey, densely scalloped with black and pale ochre or buff (each feather with grey base, V-shaped broad pale ochre subterminal band and narrow blackish edge); wings and tail dark greyish brown, upperwing-coverts and tertials broadly edged pale ochre; underparts white, throat, breast and flanks heavily freckled and scalloped with V-shaped blackish marks.

NESTLING: unknown.

Field Characters. Tiny (9·5 cm), even smaller than Yellow-footed Flycatcher *M. sethsmithi* (10·5 cm). Plain grey above, rather blotchy below, whitish with grey mottling on breast, flanks and sides of throat. Differs from darker and bluer Yellow-footed Flycatcher by grey plumage, ill-defined white throat, black legs and feet and lower mandible yellow only at base. Diminutive size distinguishes it from all other grey flycatchers.

Voice. Tape-recorded (ERA). Song, lasting 4–12 s, a rapid medley of thin, high-pitched notes, some buzzy, some pure, interspersed with little rattling calls. Contact call, high-pitched, short, descending squeaky notes. Flight call, a high-pitched, whistled 'tsee-see'. Fledged young give short, down-slurred, high-pitched whistles, higher and faster when begging.

General Habits. Inhabits lowland forest, mainly second growth, although on Mt Nimba, Liberia, occurs in small relict stands of primary forest. Typically found in old overgrown cultivation where vegetation is not too uniform or dense; also cultivated land with trees and shrubs. Occurs at all levels but mainly in middle strata, 10–20 m in medium-sized trees.

Solitary or in pairs. Active, moving about in tree and frequently changing trees; works through 50–200 m of forest in less than 1 hour. Sits upright on exposed twig or branch; makes circular flights to catch insects passing 5–10 m from perch, usually between or under crowns of trees but often in openings in dense foliage. Frequently hovers to snap up prey on or under a leaf, or flutters among leaves to flush out insects it captures on the wing. Does not usually join mixed-species foraging flocks, although it may follow them for a short distance. Behaviour otherwise very similar to that of Yellow-footed Flycatcher.

Food. Small insects (mostly 5–12 mm, up to 16 mm): beetles, grasshoppers, moths, swarming winged ants and termites. Also small berries.

Breeding Habits. Monogamous. Defends territory of 6–7 ha throughout year; territory usually unstable because of rather fast-changing habitat (in second growth, plants grow so fast that contrary to mature forest, birds cannot find the same habitat in the same place for two successive annual cycles). Territorial behaviour similar to that of Cassin's Flycatcher *M. cassini*. In threat display, faces opponent in upright posture with fluffed-out plumage, wings held out from body and bill pointing upward; in more intense version, sleeks plumage and droops wing-tips in humped posture, with closed tail bent slightly forward, and head and bill pointing toward opponent. Overall breeding behaviour very similar to that of Yellow-footed Flycatcher (for instance ♂ courtship-feeds ♀; only ♂♂ engage in contests).

NEST: rather voluminous for such a small bird; bulky cup of dry moss, dead leaves, small twigs and lichens, loosely built but bound with spider webs; internally, the small cup has a compact, even lining of very thin vegetable fibres; in Cameroon said to be placed often in colony-nests of social spiders, among clumps of dead leaves in the webs (Bates 1936); in NE Gabon one nest was found in old processionary caterpillars' nest hanging at end of leafy liana 18 m above ground.

EGGS: unknown; one broken oviduct egg very similar in colour and size to that of Yellow-footed Flycatcher. Clutch-size probably 2.

LAYING DATES: Liberia (♀ with enlarged ovary Jan); Cameroon, Mar, Apr, July (spotted young Mar, Aug, Sept); NE Gabon, Jan, Feb, Oct.

INCUBATION: by ♀; ♂ guards nest and feeds ♀; period unknown.

DEVELOPMENT AND CARE OF YOUNG: fledged young fed by both parents; ♂ still feeding ♀.

References
Bates, G. L. (1936).
Brosset, A. and Erard, C. (1986).
Erard, C. (1987, 1990a).

Plate 27
(Opp. p. 496)

Muscicapa sethsmithi (van Someren). Yellow-footed Flycatcher. Gobemouche à pattes jaunes.

Pedilorhynchus epulatus seth-smithi van Someren, 1922. Novit. Zool., 29, p. 96; Budongo Forest.

Range and Status. Endemic, resident; disjunct distribution in lowland forest. (1) E Nigeria (not uncommon, Beebuo-Boshi and Obudu Plateau), S Cameroon (common, southern montane district north to Mt Kupé, Mt Nlonako, Rumpi Hills and Foto, and from coast at least to Bitye, on Dja R.), Bioko, Rio Muni, Gabon (frequent to common and widespread, except savannas), Congo (Kouilou, Mayombe) and Central African Republic (Dzanga Reserves, possibly Lobaye); (2) E Zaïre (common, Uele to Rift Valley lakes, south to Itombwe and Kasai) and W Uganda (Budongo Forest). Uncommon to frequent or common. In NE Gabon, av. density is 8 pairs/km^2.

Muscicapa sethsmithi

A

Description. ADULT ♂: top of head and upperparts, including scapulars, plain dark slate-grey, very indistinctly streaked on forehead and crown. Tail brownish black. Lores whitish; ear-coverts and sides of neck slate-grey; chin and throat white, sharply delineated from slate-grey breast and flanks; belly mottled white and grey, whiter in the centre; undertail-coverts white. Primaries and secondaries brownish black; lesser, median and inner greater coverts, tertials and inner secondaries brownish black fringed with slate-grey. Underwing-coverts and axillaries grey fringed white. Upper mandible brownish black, lower mandible and gape lemon yellow; eyes olive-brown; legs and feet straw yellow. ADULT ♀: like ♂ but slate-grey parts more brownish, less blue. SIZE: (10 ♂♂, 7 ♀♀) wing, ♂ 55–59 (57.4), ♀ 55–57 (55.5); tail, ♂ 34–39 (36.7), ♀ 34–36 (35.1); bill, ♂ 11–12 (11.6), ♀ 10–12 (11.1); tarsus, ♂ 10–12 (11.2), ♀ 11–12 (11.5). WEIGHT: (Cameroon, Dec, Feb) unsexed (n = 3) 9.2–9.6 (9.4); (Gabon, June) 1 ♂ 8.8, 1 ♀ 7.9; (Uganda, Apr–May) ♂ (n = 5) 8–9 (8.8), ♀ (n = 4) 8–10 (9.6).

IMMATURE: like adult ♀ but wings browner, with russet apical spots on greater upperwing-coverts and tertials, also on upper- and undertail-coverts.

JUVENILE: entire upperparts dark slate-grey heavily spotted rufous (mantle feathers with grey base, narrow blackish subterminal band and rufous edges); wings and tail dark greyish brown, upperwing-coverts and tertials broadly edged rufous; face and underparts whitish with greyish breast band, all except belly densely speckled grey and rufous.

NESTLING: hatches naked, skin yellowish to pinkish, with sparse tufts of brown down which changes to pale pinkish brown with age, increasing camouflage; yellowish white gape-flanges.

Field Characters. A very small (10.5 cm), plump flycatcher; hunts low down in open forest and edges of clearings. Dark bluish slate with contrasting white throat and whitish belly, bright yellow legs; bill short and broad (**A**), lower mandible and gape yellow. Similar-looking Dusky-blue Flycatcher *M. comitata* is larger (13 cm) and greyer, with black bill and legs; Grey-throated Tit-Flycatcher *Myioparus griseigularis* is uniform grey with narrow, mainly dark bill and dark legs, and feeds like a warbler. Little Grey Flycatcher *Muscicapa epulata* is even smaller (9.5 cm) and greyer, with mottled breast, poorly defined white throat, black legs, and only base of lower mandible yellow.

Voice. Tape-recorded (CHA, ERA). Voice thin and high-pitched, notes often long, buzzy, insect-like. During territorial advertising and contact between partners not in view, piercing, short, rising whistles, often trilled. Greeting song of ♂ highly individual, a series of 3–5 high-pitched, buzzy notes with overtones. Courtship song of ♂ a continuous, subdued, high-pitched chatter. Contact calls of ♂ single high-pitched, short, buzzy notes (like ♀) or shrill double notes. Contact calls of ♀, high-pitched chirps or short buzzes with overtones. On seeing motionless raptor or ground predator, ♂ and ♀ give shrill, 3–5-note chatter introduced by brief whistle; at sight of raptor moving through forest understorey, both ♂ and ♀ utter long, high-pitched, descending whistles. Young emit short high-pitched notes and also hiss in nest, buzzy trills after leaving it.

General Habits. Widespread in virgin evergreen lowland forest and old second growth, from sea-level to 1550 m in Cameroon, 2100 m in E Zaïre (Itombwe). Typically in new openings and clearings, also in open understorey or along tracks in forest. Needs exposed hunting posts, particularly on loop or arch of hanging liana. Usually in lowest strata, rarely in canopy where excluded by Olivaceous Flycatcher *M. olivascens*.

In pairs, rarely solitary. Hunts actively from exposed perches; makes 35–50 sallies/h; sits upright while watching for prey; makes short circular flights to catch small passing insects, usually within 5 m of perch. Sometimes hovers to snap up insect on or under leaf, but never gleans leaves like a warbler, whilst perched. Seldom joins mixed-species flocks, but occasionally attends swarming army ants. Remains for long periods at same location or within very small area. Partners are always in close contact by both sight and calls. Anxiety, alarm and warning behaviours very similar to those of Cassin's Flycatcher *M. cassini*.

Food. Very small flying insects, < 10 mm in length (mostly *c*. 5 mm); some large Diptera (e.g. *Chrysops*) and Microlepidoptera (white moths up to 35 mm); regularly takes swarming winged termites and ants.

Breeding Habits. Monogamous; apparently pairs for life. Territorial. Throughout year defends territory of 7–8 ha. ♂ patrols it several times a day, giving piercing

B

advertising calls audible for 150–200 m, even 250 m under good conditions. Calls are accentuated by visual signal of bright yellow gape. Conflicts between neighbours generally rare. Territorial warning and threat displays very like those of Cassin's Flycatcher. Usually only ♂♂ engage in contests but when pairs come into close contact ♀♀ may also act aggressively to each other.

♂ begins to courtship-feed ♀ shortly before start of nest-building. Precopulatory behaviour similar to that of Cassin's Flycatcher, including courtship feeding; after copulation, ♂ sometimes 'dances' near ♀, jumping from side to side in horizontal posture, with wings drooped and tail raised.

NEST: a bulky structure, rather large for the size of the bird; an open cup, mainly of moss, loosely built on a foundation of dead leaves, rootlets, pieces of bark and small decaying twigs (**B**); inner cup lined with thin fragments of vegetation such as leaf veins; ext. diam. 90–120, ext. depth 70–80; int. diam. 40–50, int. depth 32–40. Placed often in first fork in trunk of shrubby tree in understorey, also at end of leafy hanging liana, in leaves at top of small tree, among large leaves of epiphyte, or at top of rotting stump; sited 1·5–10 m above ground, mainly 2–6 m. Built by ♀ in 6–10 days; ♂ helps construct base and sometimes brings material to ♀.

EGGS: 2, laid at 1-day intervals; in NE Gabon 21 clutches all of 2. Elongated oval; not glossy; greenish white to light green, indistinctly spotted all over with reddish brown or reddish grey. SIZE: *c.* 17 × 13·5. No record of second brood in NE Gabon, but sometimes lays replacement clutch.

LAYING DATES: Nigeria, Oct (breeding condition July–Aug); Cameroon, Dec–Mar, June; Gabon, every month except July and Apr, mainly Aug–Sept (end of long dry season) and Dec–Mar (short dry season and beginning of rains); Congo (Mayombe), Aug–Sept; Zaïre, Apr, July (♂♂ with enlarged testes Jan, Feb, July, Sept, Dec; juvs Feb, Apr–May).

INCUBATION: begins with last egg; by ♀. Period: 14–16 (15) days. By day ♀ incubates 60% of time, for periods of 1–35 (12·1) min., with absences of 1–20 (7·6) min.; ♂ guards nest and defends it against intruders, e.g. sunbirds *Anthreptes* or greenbuls *Andropadus* and *Phyllastrephus*. Av. number of sallies for food per h: 70 for ♂, 25 for ♀. ♀ receives at least 16–41% of her food from ♂, which gives ♀ at least 10–19% of what he catches. ♂ uses the most productive perches while ♀ is incubating but surrenders them to her when she leaves nest.

DEVELOPMENT AND CARE OF YOUNG: weight of young at hatching 1·5, at fledging 8·5. Young fed equally by both adults; brooded by ♀ for first 3 days for 30–50% of daylight time. ♂ gives at least 3·7% of food caught to ♀; av. number of feeding sallies per h: 94 for ♂, 70 for ♀; rate of food delivery increases with age of brood. Nestling period: 11–13 days.

After leaving nest, young are fed for up to 64 days by both parents; each parent feeds both young without preference. When independent, young at first remain on parental territory but disperse when they moult from juvenile spotted plumage into grey immature plumage, i.e. when 2–3 months old.

BREEDING SUCCESS/SURVIVAL: in NE Gabon, of 19 clutches, 11 (58%) produced fledged young, 6 were destroyed and in 2 cases young were taken by predator; 52% of eggs laid produced fledglings. Heavy rains destroyed 2 nests and stopped construction of 8 others. 1 ♂ and 1 ♀ ringed as adults were still paired on the same territory 5 years later.

References
Brosset, A. and Erard, C. (1986).
Erard, C. (1987, 1990a).

Muscicapa comitata (Cassin). Dusky-blue Flycatcher. Gobemouche ardoisé.

Butalis comitatus Cassin, 1857. Proc. Acad. Nat. Sci. Philadelphia, p. 35; Muni R., West Africa = Gabon.

Plate 27
(Opp. p. 496)

Range and Status. Endemic, resident. Forested areas of Sierra Leone, SE Guinea, Liberia, Ivory Coast (north to Sipilou, Boron and Yapo Forest), Ghana (not uncommon, widespread north to Bia Nat. Park, Goaso, Kumasi and Tafo), Nigeria (uncommon, Ibadan to Calabar, also Pandam), S Cameroon (north to Korup Nat. Park, Rumpi Hills, Nkolngem, Dimako, Yokadouma), Central African Republic (Dzanga Reserves, possibly Lobaye), Rio Muni, Gabon (widespread, frequent to common), Congo, NW Angola (frequent in Cabinda, numerous in Cuanza Norte), widespread but nowhere numerous eastward through Zaïre (Uele and Ituri, south to Itombwe and W Kasai) to Uganda (Ankole and Kigezi north to Bwamba and Budongo forests, east to Entebbe, Kampala, Kifu and Mabira forests) and SW Sudan (Aloma Plateau). Uncommon to common; av. density in NE Gabon 12·5 pairs/km^2.

Description. *M. c. comitata* (Cassin) (including '*stuhlmanni*'): Cameroon and Angola to Sudan and Uganda. ADULT ♂: upperparts plain dark slate-grey. Tail brownish black. Lores black bordered above by narrow whitish line extending above eye. Ear-coverts mottled grey and white; sides of neck dark slate-grey; chin and throat white. Breast and flanks dark slate-grey; belly whitish, washed creamy ochre; thighs mouse-brown; undertail-coverts ochre. Primaries, secondaries and tertials brownish black; scapulars and lesser upperwing-coverts dark slate-grey; median and greater upperwing-coverts brownish black edged slate-grey; axillaries and underwing-coverts mouse-grey broadly bordered whitish. Bill black; eyes olive; legs and feet brownish black. ADULT ♀: less slaty, paler and browner than ♂. SIZE: (7 ♂♂, 4 ♀♀) wing, ♂ 64–67 (65·3), ♀ 62–66 (63·7); tail, ♂ 50–55 (53·0), ♀ 47–53 (51·1); bill, ♂ 14–15 (14·2), ♀ 13–14 (13·5); tarsus, ♂ 15–16 (15·7), ♀ 15–16 (15·5). WEIGHT: (Gabon) ♂♀ (n = 4) 13–15 (14·3); (Uganda) ♂ (n = 7) 12–15 (14·1), ♀ (n = 7) 13–16 (14·1).

IMMATURE: like adult ♀ but browner grey.
JUVENILE: unspotted dark greyish blue, underparts paler with clear-cut white throat, wholly yellowish bill.
NESTLING: unknown.

M. c. aximensis (Sclater): Sierra Leone to Nigeria. Bluer slate than *comitata*, white areas of throat and belly washed with buffish.

M. c. camerunensis (Reichenow): Mt Cameroon. Differs from nominate subspecies in fulvous tinge to throat and from *aximensis* in having less white on throat and belly.

Field Characters. Stocky and stout-billed (**A**). More dusky than blue, dark slate-grey with bluish wash; white throat contrasts sharply with dark breast; narrow loral line white; bill and legs black. Sits upright and motionless on low perch; rather quiet and inconspicuous. The closely related Tessmann's Flycatcher *M. tessmanni* is larger and somewhat paler, with light grey breast and much more white on belly, and has no white line on lores. Yellow-footed Flycatcher *M. sethsmithi* has similar pattern of dark breast and contrasting white throat, but is smaller and bluer, with yellow legs and lower mandible.

A

Voice. Tape-recorded (C, ERA). Courtship song, short and rapid; advertising song shriller and longer; both songs recall those of Ashy Flycatcher *M. caerulescens*. Contact calls, short rapid series of high-pitched staccato notes. Alarm, long series of grating churrs (♂) and prolonged high-pitched buzzing notes (♀). Birds in hand give harsh notes.

General Habits. Inhabits clearings and other open areas in forest. Readily adapts to man-made habitats, e.g. in Gabon, where typically found in areas recently logged for agriculture, in open cultivated fields of, e.g. peanut and cassava, and in well-maintained plantations of cacao, coffee, oil-palm and banana. Also occurs in early stages of regenerating forest, and in forest edge and riverine brush close to villages; disappears when shrubby vegetation becomes too dense. Inhabits 'gallery forest along rivers' and occurs 'along dry water courses, frequenting the coffee trees and low bushes' in Angola (Dean *et al.* 1988). Lowlands, up to 1400 m in Zaïre (Itombwe), 1600 m in Uganda and 2100 m on Rwenzori Mts.

Solitary or in pairs. Pair members usually remain some distance apart; they maintain more or less regular vocal or sight contact, calling in upright posture with slow and jerky raising and lowering of closed tail. Moves around frequently; may stay in one spot for half an hour or more, but more often travels 200–300 m in a relatively short time; covers entire territory daily. Forages like typical flycatcher: sits upright and motionless on exposed perch, then sallies to catch flying insects, or swoops down or hovers to pick prey from leaf. Usually perches low down, on dead branch or stump, or on exposed leafy twig at top of shrub or just under crown of small tree. Forages usually below 10 m, very occasionally above 20 m. Occasionally joins slowly-moving mixed-species flocks, but never stays with them for long.

General behaviour similar to that of Ashy Flycatcher. Alarm calls and displays similar to those of Cassin's Flycatcher *M. cassini*, i.e. alternates upright and pre-flight horizontal postures with rapid and jerky movements of wings and tail, but with flicking movements of both unfolded wings and spread tail.

Food. Insects: mostly beetles, Hymenoptera (bees, wasps and winged ants, also Ichneumonoidea), Orthoptera, Diptera, Lepidoptera (caterpillars, moths and sometimes butterflies), Heteroptera (Lygaeidae), winged termites; also spiders and occasionally small fruit. Prey size, 5–30 mm, mostly 5–15.

Breeding Habits. Monogamous, territorial. Territory size in NE Gabon *c.* 7 ha. Birds advertise territory by patrolling it with loud rattling calls, sitting on very exposed perches and making irregular fluttering flights along territorial boundaries. During contests, ♂ chases intruder with vigorous wing flapping and territorial song, using postures much like those of Cassin's or Little Grey Flycatchers *M. epulata* (upright, wings held out from body, tail bent forward and bill open and pointing toward opponent).

NEST: uses old weaver nest (e.g. of *Ploceus nigricollis*, *P. cucullatus*, *P. melanogaster* or *Malimbus* spp.), 2–5 m above ground, at extremity of overhanging branch. Cup placed inside bag built by weavers; bird adds rudimentary base of dry grass supporting a loose cup lined with thin grass stems.

EGGS: 1–2. Oval; glossy; olive-green or greenish beige, completely covered with very small, indistinct flecks and spots of rusty brown, yellowish brown or pale reddish brown, sometimes more dense at large end, forming cap. SIZE: *c.* 20–21 × 13–14.

LAYING DATES: Ghana, Jan, Mar, May–July; Nigeria, Apr; Cameroon, Feb–Apr, June–Oct, Dec; Gabon, Jan, Feb, Aug; Angola, Mar, Sept; Zaïre, Mar, Sept, Oct (birds with enlarged gonads July–Feb); Uganda, July, Oct.

INCUBATION: by ♀; duration unknown. ♀ sits for periods of 5–40 min, leaves to feed for 7–14 min; ♂ guards nest and chases other birds with threatening flights and vigorous bill clapping. ♂ sings more or less regularly when ♀ on nest, facing nest entrance with body upright, plumage fluffed out, head up, and tail spread; he also sings when ♀ comes back to nest. ♂ regularly feeds ♀.

DEVELOPMENT AND CARE OF YOUNG: both ♂ and ♀ feed young, coming to nest every 6–7 min; they bring rather large prey. Adult seen feeding fledgling with yellow gape; both still in same location 20 days later (T. M. Butynski, pers. comm.).

References
Brosset, A. and Erard, C. (1986).
Erard, C. (1987, 1990a).

Muscicapa tessmanni (Reichenow). Tessmann's Flycatcher. Gobemouche de Tessmann.

Pedilorhynchus tessmanni, Reichenow 1907. Orn. Monatsb., p. 147; Alén, Spanish Guinea.

Range and Status. Endemic resident, Congo Basin and W African rain forest. Ivory Coast, Nimba (but not known from other flanks of Mt Nimba in Liberia and Guinea) and in SW fairly common around San Pedro, in Taï Nat. Park, Mopri Forest Reserve near Tiassele (M. Gartshore, pers. comm) and east to Lamto. Ghana, rare; collected Prahsu and Fumsu 90 years ago and Goasa 60 years ago; twice seen recently Cape Coast. Nigeria, 2 old records: no locality (Bates 1930) and Shonga (Bannerman 1936). Cameroon, rare, in S, east

Muscicapa tessmanni

of Sanaga R. Rio Muni, 2 old records in N on or near Cameroon border. Congo, known from Mayombe. Zaïre, occurs in Ituri from Panga to Avakubi.

Description. ADULT ♂: forehead to uppertail-coverts, lesser wing-coverts, lores, cheeks, ear-coverts and sides of neck, dark bluish slate, uniform except that: forehead and crown faintly dark-streaked, ear-coverts and in some specimens broad ring of feathers around eye paler, and an indistinct pale grey curving line of feathers between nostril and top-front of eye. Tail and wings blackish brown, rather glossy; tail rounded, T6 5–7 shorter than T1, projecting 25–30 mm beyond wing-tips. Tertials, inner secondaries, greater and median coverts edged dark bluish grey. Chin and throat cream-white in centre merging to grey at sides; breast and flanks grey (much paler than upperparts); centre of belly cream-white in a rather well-demarcated patch; vent and undertail-coverts cream-grey. Underwing-coverts and axillaries greyish white; undersides of remiges and tail dull black. Bill black, robust (A); eyes dark brown; legs and feet dark grey. Sexes alike. SIZE: (3 ♂♂, 2 ♀♀, W Africa) wing, ♂ 70–79 (74.0), ♀ 73, 74; tail, ♂ 54–63 (58.3), ♀ 53, 56; bill to skull, ♂ 14.8–17.7 (16.2), ♀ 14.5, 15.0; tarsus, ♂ 17–18 (17.5), ♀ 16, 17; (2 ♂♂, 1 ♀, NE Zaïre) wing 70, 73, 75.5, tail 53, 55, 57, bill to feathers 15, 16, 16.

IMMATURE: unknown. One subadult specimen (Dec) has narrow rufous-buff tips to greater upperwing-coverts, tertials, and undertail-coverts.

NESTLING: not known.

Field Characters. Very similar to closely-related Dusky-blue Flycatcher *M. comitata*, and occurs together with it, but has a more restricted range and is much less common. Larger and paler than Dusky-blue, with light grey breast-band and much more white on belly, and lacks white loral line; bill stouter and deeper.

 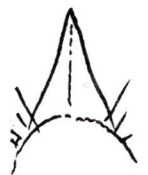

Yellow-footed Flycatcher *M. sethsmithi* is smaller, with yellow legs and lower mandible; Little Grey Flycatcher *M. epulata* is smaller still and greyer, with mottled breast and poorly-defined white throat.

Voice. Tape-recorded (GAR). Song a medley of weak but quite sweet notes interspersed with a few buzzy ones, usually starting with a few thin, high-pitched squeaks; notes are brief, and high and low ones alternate, producing a rather jerky effect. Length of songs 3–10 s, with intervals of a few s between them. Song, ♂, loud (M. Gartshore, pers. comm.).

General Habits. Frequents tops of secondary tangle (7–20 m high) of failed sipo (*Endraphragma utile*)/acaju (*Khaya ivorensis*) plantations and farm bush (Mopri, Ivory Coast). ♂ flies from singing post to singing post, conspicuous and boisterous, with ♀ sitting quietly nearby (M. Gartshore, pers. comm.). Almost completely unknown elsewhere. In Zaïre 3 'were all in pairs [*sic*] in second growth near roads and clearings' (Chapin 1953a). Foraging behaviour unknown; bill much deeper and more robust than in any congener, and bird may be more a leaf-gleaner (since it eats caterpillars) than a fly-catcher.

Food. Insects, including 2 caterpillars (in 3 stomachs).

Breeding Habits. Unknown.

Muscicapa infuscata (Cassin). Sooty Flycatcher. Gobemouche enfumé.

Butalis infuscatus Cassin, 1855. Proc. Acad. Nat. Sci. Philadelphia, 7, p. 326; Moonda (= Mondah) River, Western Africa = Gabon.

Forms a superspecies with *M. ussheri*.

Range and Status. Endemic resident. Forests of S Nigeria (uncommon to locally common in coastal areas from Gambari to Calabar, north to Bashu), S Cameroon (common and widespread, north to Korup

Muscicapa infuscata

A

Nat. Park, Rumpi Hills, Mt Kupé, Nkolngem, Djaposten and Yokadouma); Central African Republic (uncommon, Dzanga reserves, Lobaye); Gabon (widespread, uncommon to locally common), S Congo (locally common in N Mayombe, Sibiti), W Angola (Cabinda and N Malanje south to Gabela, Cuanza Sul, but not reported recently from last locality: W. R. J. Dean and M. A. Huntley, pers. comm.); Zaïre (Mbandaka and Uele, south to Manyema and Kasai); SW Sudan (2 old records from Aloma Plateau), Uganda (locally common, Budongo, Bugoma, Bwamba, Kalinzu and Maramagambo forests, east to Kifu and Mabira) and N Tanzania (old record from Ukerewe I., SE L. Victoria); NW Zambia (N Mwinilunga, Salujinga, Zambezi rapids and Kalene Hill). Uncommon to abundant. In NE Gabon, av. density 2·1 pairs/km² in man-modified forests.

Description. *M. i. infuscata* (Cassin) (including '*chapini*'): Nigeria to Angola and Zaïre (except NE). ADULT ♂: upperparts, sides of head (except blackish lores), wings and tail uniformly ferruginous sooty brown, hindneck and mantle slightly scaly (fringes of feathers less dark than centres); primaries, secondaries and tail-feathers browner and darker, slightly glossy, especially in fresh plumage. Underparts paler, heavily mottled light ferruginous brown or ochre and sooty brown, with indistinct narrow brown streaks (feathers ferruginous ochre or brown, whitish on throat, more tawny or rufescent on flanks, with irregular sooty brown shaft streak narrowing towards tip). Undertail-coverts pale rusty brown with median and subterminal bands dark sooty brown, narrowly edged whitish or buffish. Axillaries and underwing-coverts pale ferruginous sooty brown. Bill black, very short, quite deep, like that of *M. ussheri* (**A**); eyes dark olive-brown; legs and feet black. Sexes alike. SIZE: (8 ♂♂, 4 ♀♀) wing, ♂ 78–86 (83·0), ♀ 78–83 (81·5); tail, ♂ 44–50 (48·0), ♀ 42–47 (46·2); bill, ♂ 10–12 (11·0), ♀ 10–11 (10·7); tarsus, ♂ 13–14 (13·2), ♀ 12–13 (12·7). WEIGHT: (Gabon, Mar, Oct) ♂♀ (n = 2) 20·0; (Uganda, Nov, Dec, Apr, May), ♂ (n = 6) 15–19 (17·6), ♀ (n = 7) 15–20 (17·9), unsexed (n = 3) 17–18 (17·5).

IMMATURE: like adult but underparts rustier and more distinctly and boldly streaked; narrow edges to wing-coverts (forming 2 narrow wing bars), tail-feathers and inner secondaries pale rusty buff or whitish; tertials broadly edged and tipped buff.

JUVENILE: upperparts much darker than adult; chin buff, breast more distinctly and boldly streaked, blotched or scalloped sooty brown with diffuse buff spots, belly mainly buff; feathers of nape and lower mantle, upper tail-coverts, tail-feathers, lesser wing-coverts, alula, primaries and outer secondaries with tiny white or buff-white tips; inner secondaries, median and greater wing-coverts edged creamy or buffy white.

NESTLING: undescribed, except near to fledging when in almost full juv. plumage.

M. i. minuscula Grote: Zaïre (E Congo basin from Bolobo on middle Congo R. to Uele and Semliki); birds in Sudan and Uganda probably of this subspecies. Like nominate *infuscata* but underparts more rufescent and more uniform, with pale buff streaks less conspicuous. Wing, (NE Zaïre, n = 20) ♂♀ 78–84 (80·9).

TAXONOMIC NOTE: merging the long-winged flycatchers, often still separated as *Artomyias*, with *Muscicapa* necessitates change of specific epithet of this species from *fuliginosa* to *infuscata*. Many authors regard *M. infuscata* as monotypic, but NE Zaïre specimens that we have examined are subspecifically separable.

Field Characters. A plump, rather short-tailed dark brown flycatcher with characteristic habit of hunting from exposed perch in tall dead tree. Wags tail; sails out after insects in swallow-like flight. Other small flycatchers are light brown or grey, and use more protected perches. At close range, indistinct streaks on underparts give mottled appearance.

Voice. Tape-recorded (C, CHA, ERA). Territorial advertising call is a shrill descending whistle, shorter when given singly, longer and uttered at rate of *c.* 1 per s when other bird responds. Courtship song is a rapid medley of low-intensity, rolling, clicking and nasal notes. Contact call, a short harsh, nasal descending 'tsiew'. Excitement call an irregular 'tchick-tchick-tchictchictchic'.

General Habits. Inhabits clearings and other openings in forest. Readily adapts to man-made habitats, such as recently logged areas and agricultural land with scattered tall trees both live and dead. Regularly occurs around villages, in plantations and in early stages of regenerating forest on old cultivation, as long as vegetation is stratified. Also frequents forest edge and riparian woodland, where it keeps to the tallest emergent trees; in Gabon usually only transient in these habitats. Lowlands, up to 1150 m in E Zaïre (Itombwe), 1600 m in Uganda.

In pairs or small parties. When foraging sits on very exposed perch, typically on top of tall dead tree, and sallies to catch flying insects, mainly 3–15 m from hunting post, but may pursue prey up to 45 m in swallow-like circular and zigzag flight. Often begins sallies with ascending, flapping flight like lark or pipit. Mainly an aerial feeder but approaches tops of trees in fluttering flight; uses perches from 5 m above ground to tops of tallest trees, mostly above 20 m. Remains in area of 0·25–0·75 ha for period of < 30 min to several hours, then moves to another location; each day visits 8–10 such locations scattered throughout its territory. Diurnal activity pattern constant throughout year: forages all day, up to sunset, with some decrease in the afternoon (9–15 sallies per h vs 90–120).

Contact between partners mostly visual, including slow raising and rapid lowering of tail; calls used at close range. When anxious, gives Excitement calls (see *Voice*), while rapidly wagging tail up and down. Harasses potential predator in flight with measured Excitement call. Members of pairs or trios often approach each other closely and may allopreen. In NE Gabon, 50% of social units are pairs, and 50% are trios (i.e. a pair plus another adult which may be a breeding adult, a non-breeding adult or a 'young' of previous brood more than 2 years old); all include young from the current and/or previous year (Erard 1990a).

Food. Insects: mostly Coleoptera (beetles), Hymenoptera (bees and winged ants), Diptera, Lepidoptera (small butterflies); regularly feeds on swarming winged termites; may take small berries. Prey size up to 20 mm.

Breeding Habits. Monogamous, co-operative breeder (trios of adults) – see above and below. Defends large territory (36–45 ha) throughout the year. Advertises territory by calling from high exposed perch and wagging tail rapidly up and down. In threat display adopts upright posture with breast puffed out and wings drooping, or crouches with bill open and head directed forward. During breeding season ♂ often greets ♀ with 1–3 notes of his individually-specific courtship song, makes fluttering circular flight, then perches and bows with spread tail while giving courtship song. ♂ sings and defends nest site against other species.

NEST: rather bulky and untidy shallow cup of moss, vegetable fibres, grasses and rootlets, bound together with spider webs. Placed in fork near end of small branch or in outer part of main branch of tall tree; once in hole in metal girder on top of concrete pylon supporting high voltage line. 6–35 m above ground; built by ♀ alone (pairs) or by ♀ and one other bird (trios).

EGGS: not described. Clutch possibly 2–3, since 2–3 young in nest.

LAYING DATES: Nigeria, May, Sept–Oct; Cameroon, Feb, (♀ with recent ovarian scars May, juvs June, Nov, Dec); Gabon, Feb, Mar, Sept, Oct; Zaïre (♂♂ with enlarged testes Mar, May; juvs Jan, May, Aug, Dec); Angola, Mar; Zambia, Feb (♂♀ with enlarged gonads Sept, Nov).

INCUBATION: by ♀; she sits for periods of 10–20 min, sometimes up to 45 min, with absences of 5–10 min (when she feeds). ♂ guards and defends nest against other species. Period: at least 14 days.

DEVELOPMENT AND CARE OF YOUNG: young remain in nest for 13 days; fed by members of social unit, i.e. adults (pair or trio members) plus, at least in one case, 3 immatures of previous brood (Erard 1990a); av. of 1 food delivery every 5·4 min at nest (of 2 young fed by 2 adults).

Fledged young fed for at least 6 weeks after leaving nest; they begin to forage by themselves after 3 weeks. They remain with parents until following breeding season, some probably longer, i.e. some trios could be formed of 2 adults and their offspring < 2 years old. Immatures help adults to build nest and feed young.

BREEDING SUCCESS/SURVIVAL: 1 died in spider web (Bannerman 1936: 244).

References
Brosset, A. and Erard, C. (1986).
Erard, C. (1987, 1990a).

Muscicapa ussheri (Sharpe). Ussher's Flycatcher. Gobemouche d'Ussher.

Plate 26
(Opp. p. 481)

Artomyias ussheri Sharpe, 1871. Ibis, p. 416; Abrokonko, Gold Coast.

Forms a superspecies with *M. infuscata*.

Range and Status. Endemic resident, W African rain forests from Sierra Leone to Nigeria. Sight record Senegal (Basse-Casamance Nat. Park, Géroudet 1983) doubted (Morel and Morel 1990). Sierra Leone, widespread and frequent but not yet reported from NW; records from Kwendu, Kailahun, Sefadu, Tembikunda and Rotifunk. Guinea, records from Kakoulima (Richards 1982) and Macenta, and sight record Salla R. (Fouta Djalon) (T. Aversa, pers. comm.). Liberia, abundant throughout. Ivory Coast, widespread from coast north to Sipilou and Béoumi; one of the commonest flycatchers in Taï Nat. Park (Gartshore 1989). Ghana, frequent east to Aburi and north to Mampong and near Wenchi; probably easily overlooked in closed forest, but seen commonly in logged forest and on farms in SW (Dutson and Branscombe 1990). Nigeria, 3–4 sight records from mouth of Benin R., also at Utange and Serti (1 netted, Hall 1977c); this species, or Sooty Flycatcher *M. infuscata*, seen also near Benin R. at Erin Ijesha.

Muscicapa ussheri

Description. ADULT ♂: upperparts very dark brown or blackish brown, rather uniform, the slightly lighter feather margins giving forehead and crown barely perceptible scaly look; wings and tail even darker, and glossy. Wings long; tips fall only *c*. 14 mm short of tail tip. Circumorbital featherlets pale in some birds. Chin whitish; lores, cheeks, ear-coverts, throat and rest of underparts including underwing-coverts and axillaries dark greyish brown, vent greyer, feathers of belly obscurely tipped whitish, undertail-coverts long, blackish, with 1-mm-deep white tips. Underside of wings and tail brownish black. Bill black, very short, rather deep; inside of mouth blackish; eyes brown or brownish black, legs and feet dark brown or brownish black. Sexes alike. SIZE: wing, ♂ (n = 9) 81–89 (84·3), ♀ (n = 8) 81–90 (84·0); tail, ♂ (n = 9) 43–47 (44·0), ♀ (n = 8) 40–51 (43·75); bill to skull, ♂ (n = 3) and ♀ (n = 3), all 12·0; tarsus, ♂♀ (n = 11) 13–15 (14·0). WEIGHT: (Mt Nimba, Liberia) ♂ (n = 3) 16·3–19·0 (17·5), ♀ (n = 3) 16·8–19·8 (18·2), 1 unsexed 18·0.

IMMATURE: like adult but greater coverts, greater primary-coverts, primaries, secondaries and tail-feathers have small white tips, and breast and belly feathers have minute white tips.

NESTLING: not known.

Field Characters. Replaces closely-related and very similar-looking Sooty Flycatcher *M. infuscata* from Ghana westwards. Generally dark brown or blackish, but somewhat greyer below than Sooty, and without mottling. Has same chunky shape and habits, hawking from exposed perch high up in dead tree in forest clearing. Long pointed wings, short forked tail and sailing flight make it look like a brown martin in the air.

Voice. Tape-recorded (CHA). Call (?) an irregular series of buzzy, chattering notes. Also said to have 'a quiet squeak' (Smith 1966a).

General Habits. Inhabits clearings in and edges of rain forest, also gallery forests in derived savanna, tall mangrove, logged forest and adjacent farms; usually keeps high up; requires open areas with several dead standing tall trees, or leafless boughs emergent from the canopy. At Serti (Nigeria), open derived savanna near edge of riverine woodland. Occurs usually in groups of 3–5 birds, perching near to one another on dead branch 15 m high or more; forages by flying out after passing insect, snapping it up and sailing or flying back to the same perch. Feeds actively after a rain shower passes. Occasionally flycatches from dead branch only 3–5 m above cleared ground. This species or Sooty Flycatcher seen in mixed foraging flocks of warblers, batises, flycatchers and tinkerbirds (Greig-Smith 1977).

Food. Insects.

Breeding Habits. Almost unknown. In Taï Nat. Park, Ivory Coast, adult on nest (18 m high in dead tree at edge of main clearing), Oct, juvs seen mostly July–Aug, also Apr (Demey and Fishpool 1994).

Plate 26
(Opp. p. 481)

Muscicapa boehmi (Reichenow). Böhm's Flycatcher. Gobemouche de Böhm.

Bradyornis boehmi Reichenow, 1884. J. Orn., p. 253; Tabora, Tanganyika.

Range and Status. Endemic resident in *Brachystegia* woodland, from central plateau of Angola to W Malaŵi, north to W Tanzania. Angola, common in N Huila, Huambo, N Bihe and NW Moxico districts; once collected at Xá-Cassan, Lunda. Zambia, sparse and local; Zambezi valley (e.g. Balovale: rare); throughout North-Western Prov., frequent at Mwinilunga, with records south to east of Mankoya, Chunga and once Ngoma in Kafue Nat. Park, and Kafue valley in N Mumbwa; Western Prov., south to Kabwe; Luapula Prov. south and east of Mansa; Northern Prov. east to Chambeshi R. and Chinsali, but largely absent north of 10°S; on Muchinga Escarpment from Kolala to Kanona, but absent from Luangwa Valley; Eastern Prov. from Lundazi to Chadiza and Chipata. Malaŵi, 1000–1500 m from Rumphi (and Chitipa?) districts south to Mzimba, Kasungu and Mchinji (Kapiriuta). Zaïre, known from S Kasai (Tshisika) and Shaba (Lubumbashi where very uncommon, Marungu Highlands where quite common around Lubenga at 750–1750 m, Mwanza). Tanzania, Tabora area, southwest to near Ugalla Game Res., also near Tukuyu northwest of L. Malaŵi.

Muscicapa boehmi

Description. ADULT ♂: crown, nape, hindneck, sides of neck, ear-coverts and middle of forehead warm brown, finely streaked with black; mantle, scapulars and back warm brown, rufescent in some birds, diffusely and sparsely streaked with blackish; rump and uppertail-coverts light rufous-brown, sometimes grey-brown. Tail-feathers mid- to dark brown, fringed buff distally and pale or bright rufous proximally. Lores and often sides of forehead whitish; well-pronounced, if narrow, eye-ring of cream feathers. Cheeks brown, merging to white on moustachial area; well-defined narrow black malar stripe; chin and throat white with a few small black triangles; breast white, heavily marked with precise black triangles; flanks buffy, with diffuse brown streaks; belly, vent and under-tail-coverts white with buffy wash. Primaries brown, all but P10 with buff outer edges, secondaries dark brown with quite broad, light rufous-brown outer edges, tertials and all upper-wing-coverts dark brown, broadly fringed rufous-buff (fringe of greater and median coverts 2 mm wide). Underwing-coverts and axillaries white, mottled buff-brown; underside of remiges and of tail shiny grey. Bill quite deep (**A**), blackish brown or black, greyer at base; eyes dark brown; legs and feet dark grey, soles pale. SIZE: (5 ♂♂, 5 ♀♀) wing, ♂ 75–81 (78·8), ♀ 76–80·5 (79·0); tail, ♂ 47–53 (49·1), ♀ 47–51 (49·0); bill to skull, ♂ 13·6–15·3 (13·9), ♀ 13·0–16·2 (15·1); tarsus, ♂ 17·5–18·5 (17·9), ♀ 17·5–20·0 (18·4).

IMMATURE: upperparts dark brown, liberally spotted on head, mantle and wing-coverts with large rufous or buff marks (feathers rufous-buff, with blackish margins). Rump rufescent, dark-banded. Tail-feathers, flight feathers and tertials more broadly edged and tipped with rufous-brown than in adult. Underparts white or greyish white, heavily marked (mainly on breast, not on lower belly) with scaly blackish crescents. Legs and feet pale grey.

NESTLING: not known.

A

Field Characters. In most behavioural respects a typical *Muscicapa* flycatcher, arboreal, perching upright, sallying after insects, silent; but readily distinguished from all other flycatchers by its rufescent brown plumage and boldly marked breast – black arrowheads on a white ground. Spotted Flycatcher *M. striata* has softly streaked grey breast. Juveniles of several *Melaenornis* flycatchers have scaly underparts. In fact Böhm's Flycatcher is more like some small, upright-perching pipit, although Tree Pipit *Anthus trivialis* (for instance) quickly told by its streaky back, white outer tail-feathers, walking and flight call.

Voice. Tape-recorded (STJ). Song variable, lasts *c.* 1·5 s, repeated somewhat differently about 5 times in 20 s; main element, starting and usually ending song, is 't'chee', 'tch-tee' or 'sh-shee', which may be given on its own, 't'chee-t'chee-t'chee', or include a brief (0·3–0·7 s) cheerful warble. Otherwise a very silent bird; an 'indeterminate chattering' once heard from a family party.

General Habits. Inhabits *Brachystegia* woodland. Occurs solitarily, in pairs and family parties. Forages by making aerial flycatching sallies from tree, but much less active than most *Muscicapa* species. Perches upright; between flycatching forays sits almost motionless on branch for considerable periods. Retiring, probably easily overlooked, but not shy and can easily be approached. 'Judging from crop-contents, much of its food is composed of black tree-ants, which can be caught without flying' (White 1947).

Food. Insects including black ants.

Breeding Habits. Breeds solitarily, using old nests of weavers.

NEST: made inside old roofed nest of weavers, including Olive-headed Weaver *Ploceus olivaceiceps*, Chestnut-mantled Sparrow-Weaver *Plocepasser rufoscapulatus* and Red-headed Weaver *Anaplectes rubriceps*. The only one described was in old nest of the last; weaver nest was the usual short-funnelled retort shape, made of soft intermingled leaves and twiglets, with entrance underneath; flycatcher had added pad of yellowish brown vegetable down on which eggs rested; nest sited at end of down-hanging branch in large *Brachystegia* tree by huge termite mound in open glade.

EGGS: 3–4, usually 4. Dull pale green, covered with very faint, close, pale pinkish brown freckles. SIZE: (n = 1) 17·6 × 13·4.

LAYING DATES: Zaïre (Lubumbashi), Sept; Zambia, Sept–Nov; Malaŵi, Oct; Angola (spotted young indicate eggs early in rains, about Oct–Nov).

References
Vincent, A. W. (1947).
White, C. M. N. (1947).

MUSCICAPIDAE

Genus *Myioparus* Roberts

Small and slender warbler-like flycatchers. Bill slender, attenuated and compressed laterally, most of lower mandible whitish horn. Nostrils partly feathered, rictal bristles present but short and weak. Tarsometatarsus relatively robust though slender, claws stongly curved and sharp. Tail-feathers slightly graduated, outer white in *plumbeus*. Plumage grey with white or whitish belly. Sexes alike. Juveniles spotted. Glean leaves, acting like warblers. Vocalizations distinctive, with contact calls made of short, weak, repeated plaintive notes, and song consisting of a phrase of mournful, quavering notes. Nest rather bulky, in tree hole. Arboreal; forest and woodland.

Endemic, 2 species. Formerly considered to be warblers of the genus *Parisoma* until, on the basis of juvenile plumage, Vaurie (1957) made *plumbeum* a flycatcher and suggested that *griseigularis* might be one also. Hall and Moreau (1970) placed them both in *Myioparus*, but Watson *et al.* (1986) kept them generically apart. We follow Dowsett and Dowsett-Lemaire (1980) and Erard (1987).

Plate 27
(Opp. p. 496)

Myioparus griseigularis (Jackson). Grey-throated Tit-Flycatcher. Gobemouche à gorge grise.

Alseonax griseigularis Jackson, 1906. Bull. Br. Orn. Club, 19, p. 19; Kibiran, Toro, Uganda.

Range and Status. Endemic, resident. Lowland forests of Liberia (at least Mt Nimba), Ivory Coast (Nimba, Taï, Dueké, Abidjan, Lamto, Yapo Forest), Ghana (rare, Tafo); SE Nigeria (rare, Ahoada, Umuagwu and Ebe R. near Ikpan) and S Cameroon (north to Korup Nat. Park, Lolodorf, Bitye, Yokadouma); Gabon (widespread, frequent to locally abundant); N and central Congo; NW Angola (Canzele, N Cuanza Norte); S-central and E Zaïre (north to Bolobo in west, Kasai, Itombwe and Kivu north to Ituri, Kisangani and Uele), S Central African Republic (Ouossi R., NW Zemio, La Maboké, possibly Lobaye), forests of S and W Uganda north to Budongo and east to Mabira and Sango Bay, and into NW Tanzania (Bukoba). Possibly Rwanda (Nyungwe). Uncommon to abundant. In NE Gabon, av. density 5·8 pairs/km^2.

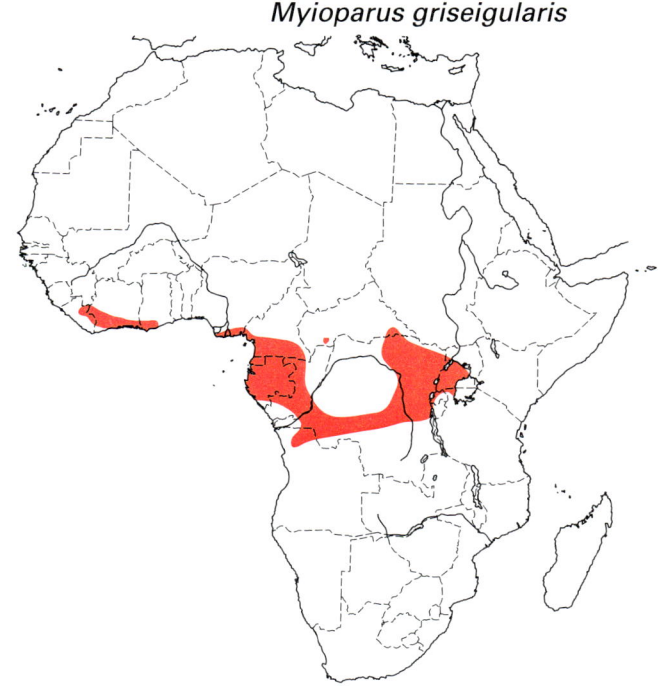

Myioparus griseigularis

Description. *M. g. griseigularis* (Jackson): Nigeria to L. Victoria and Angola. ADULT ♂: entire upperparts, sides of head and neck, scapulars and upperwing-coverts plain dark bluish slate-grey. Tail blackish slate-grey, 2 outer feathers shorter and brownish slate-grey. Lores blackish slate. Underparts pale bluish slate-grey, centre of belly and under-tail-coverts white. Primaries and secondaries blackish slate-brown, edged pale slate-grey on outer webs, tertials with bluish grey wash, more pronounced in fresh plumage. Underwing-coverts, axillaries and fringes to inner webs of underside of flight feathers white. Bill black or blackish brown, lower mandible mostly pale bluish horn; eyes olive-brown; legs and feet black. Sexes alike except that ♀ is duller. SIZE: (4 ♂♂, 5 ♀♀) wing, ♂ 60–64 (63·1), ♀ 59–63 (61·5); tail, ♂ 47–51 (49·5), ♀ 47–50 (49·3); bill, ♂ 14–16 (15·1), ♀ 14–15 (14·7); tarsus, ♂ 16–17 (16·5), ♀ 16–17 (16·9). WEIGHT: (Gabon) unsexed (n = 7) 12–17 (14·2); (Uganda) ♂ (n = 29) 9–15 (12·6), ♀ (n = 14) 10–15 (12·5).

IMMATURE: like adult but wings and tail browner, with narrow rusty or ochre edges and a few spots on upperwing-coverts, sometimes also on inner primaries, secondaries, tertials and tail-coverts.

JUVENILE: entire upperparts and upperwing-coverts dark greyish brown densely covered with triangular rusty buff or pale rufous spots (feathers with grey-brown base, tear-shaped pale rufous spot and narrow blackish edge); wings and tail dark greyish brown, feathers edged buff or pale rufous; underparts greyish densely covered with pale rufous spots narrowly edged blackish.

NESTLING: unknown.

M. g. parelii Traylor: Liberia to Ghana. More bluish slate; lower mandible black with a trace of horn at base; wing shorter (54–56), tail longer (52–53). WEIGHT: (Liberia, Aug, Sept) ♂ ♀ (n = 6) 11–14 (11·6).

Field Characters. A small grey flycatcher which looks and acts more like a warbler. Does not hunt from perch but gleans leaves with thin warbler-like bill; on branch has horizontal rather than upright stance; wags tail. Rather featureless, with no face markings or contrasting pale throat, which is same grey as breast. Replaces Grey Tit-Flycatcher *M. plumbeus* in forest; if they should overlap at the forest edge, Grey Tit-Flycatcher easily told by white in tail, paler colour, white lores and eye-ring. Dark brown wings and tail contrasting with grey upperparts give superficial resemblance to Cassin's Flycatcher *Muscicapa cassini*, but latter lives along streams and hawks for insects in typical flycatcher manner, also has pale throat and belly separated by grey breast-band.

Voice. Tape-recorded (32, B, CHA, ERA, STJ). Advertising song a phrase of mournful, quavering notes very similar in quality to those of closely-related Grey Tit-Flycatcher. In Uganda usually 2 notes, 'peee-pyurrr', second note lower; in Gabon, usually 3 notes, 'eeeee-yieee-tieeew', occasionally 4 or 2; form varies somewhat individually. In aggressive encounter once heard to give warbling song with a few short buzzy notes, reminiscent of Garden Warbler *Sylvia borin* (Erard 1990a). Contact calls very different, a series of phrases of 3–4 weak, short, plaintive notes without quavering quality of song, repeated several times without pause; second note lower than first, third note low but with higher harmonic, fourth note like first: 'pip-pu-puee-pip' (Uganda) or 'tee-tyiew-ew' (Gabon). Alarm, a harsh 'tseee-tye'.

General Habits. Inhabits primary lowland rain forest, also old second-growth, particularly in later stages of forest regeneration; also enters young second-growth, uncleared or disused plantations bordering forest, and edges of riparian forest. Avoids clearings or logged areas; fond of places rich in lianas. Up to 1800 m in Zaïre (Itombwe) and Uganda.

Solitary or in pairs. Occupies all vegetation levels from 2 m above ground to canopy, mainly 10–25 m. Forages by gleaning foliage rather than in typical flycatcher manner; searches among leaves in outer branches of trees and shrubs, also in dense clumps of leafy lianas. Perches horizontally or at a slight angle; before entering leaves spreads and droops wings and raises tail a little (**A**); may cock and wag tail and flirt outer tail feathers (**B**); moves noisily among leaves with rapidly beating wings, twisting tail from side to side; picks prey from leaves or stems, or snaps it up in the air after circular descending flight; may also forage like a tit, clinging to underside of leaf. Moves around a lot; covers entire territory each day. Pair members maintain contact by continuous calling, especially when foraging in thick vegetation. Frequently joins mixed-species flocks, in particular during non-breeding season. Actively feeds on swarming winged ants and termites. When anxious adopts upright but somewhat hunched position, nervously swinging tail; when alarmed perches upright and jerks wings.

Food. Mostly insects: Coleoptera, Hymenoptera (mainly ants), Hemiptera, Lepidoptera (moths and especially caterpillars), Orthoptera, termite alates; also spiders and small fruits. Prey size 8–15 mm; up to 30 mm for caterpillars.

Breeding Habits. Monogamous; remains paired for long periods (birds ringed as adults still together 5 years later: NE Gabon). Territorial throughout the year; defends large territory (17–21 ha in NE Gabon); advertises it by singing for lengthy periods in upright position. Low-intensity threat posture upright, with fluffed-out plumage, open bill pointing toward opponent; high-intensity threat and fighting postures very similar to those of Cassin's and Little Grey Flycatchers *Muscicapa cassini* and *M. epulata* (upright, wings spread, head and neck extended toward opponent with open bill, depressed tail bent forward).

NEST: a loose bulky structure of moss, rootlets, dry leaves and vegetable fibres, supporting poorly-built small cup lined with thin materials, sometimes just a few tendrils; placed in hole in dead tree trunk, in open, unfinished woodpecker hole, or in opened-up old nest hole of cavity-nesting species; once even in a large old hanging nest of Blue-throated Brown Sunbird *Nectarinia cyanolaema*. Sited 1·7–12 m above ground. Built by ♀, accompanied by ♂ which calls and sings.

EGGS: 2. Oval; glossy; buff or greenish beige heavily covered with small spots and blotches of dark brown, rusty brown and violaceous grey-brown, more numerous at larger end, where indistinct grey blotches sometimes present. SIZE: *c.* 20 × 14.

LAYING DATES: Liberia, Sept (♀ with enlarged ovary Aug); Ghana, (adult with young Feb); Gabon, Jan, Feb, Oct, (♀ fed by ♂ Mar, prospecting nest cavity Dec,

moulting young Aug); Zaïre, May (moulting young Sept; breeding condition Dec–May).

INCUBATION: by ♀; sits continuously for 1–3 h, leaves at irregular intervals for feeding bouts of 10–20 min; ♂ feeds brooding ♀ about every 15 min with large caterpillars; at other times guards nest surroundings. Period: 12 days.

DEVELOPMENT AND CARE OF YOUNG: young remain in nest for 13 days; fed by both parents, on av. every 20 min; ♀ may visit nest more than ♂, but ♂ often gives prey to ♀.

BREEDING SUCCESS/SURVIVAL: lives > 5 years (see above).

References
Brosset, A. and Erard, C. (1986).
Erard, C. (1987, 1990a).

Plate 27
(Opp. p. 496)

Myioparus plumbeus (Hartlaub). Grey Tit-Flycatcher; Fan-tailed Flycatcher. Gobemouche mésange.

Stenostira plumbea Hartlaub, 1858. J. Orn., 6, p. 41; Casamance River, Senegal.

Range and Status. Endemic, resident. Senegambia (south of Gambia R.), Mali (rather common, upper Bafing R., Plateau Mandingue, upper Baoulé R.), Guinea-Bissau (?), Guinea (Kouratongo, Mambia), at least coastal Sierra Leone and Liberia, Ivory Coast (sparse, from Taï Nat. Park, Lamto and Yapo Forest to northern border, mainly north of Séguéla, Sipilou and Bouaké), Ghana (rare, from Cape Coast and Kete Kratchi to Mole, Tumu and Gambaga), Togo, Benin, Nigeria (widespread but uncommon, Lagos to Borgu, Leraba Gare, Kano and Gashaka-Gumti, mostly Borgu to Yankari, Aliya and Potiskum), Cameroon (frequent to common, south of Benue Plain), Rio Muni, Gabon (widespread, unfrequent to locally common), W and SW Congo, S Chad (upper Logone and Chari), W and SE Central African Republic (Dzanga Reserves to Bangui, Kémo and Gounda St Floris, and NW Zémio), Zaïre (Bolobo and lower Congo R. to Luebo, Ubundu, Kisangani and Upper Uele, south to Shaba), S Sudan (north to Upper Nile and Bahr el Ghazal), SW Ethiopia, W Uganda (west of Acholi and Victoria Nile), W Kenya (west of Kerio Valley), and in coastal lowlands inland to Kitui and Kibwezi; Tanzania except N and central interior, but recently discovered in Kahe II Forest Reserve, 20 km south of Moshi (Cordeiro *et al.* 1995). Malawi (widespread but uncommon), Zimbabwe (except Kalahari Sand woodlands), Zambia (nowhere common), W Mozambique, N Natal, Transvaal except S and SW, Botswana (sparse to uncommon, NW and SE), Namibia (Ovamboland) and Angola (south to central Huila and northern Moxico). Uncommon to common. In NE Gabon, av. density 1·8 pairs/km².

Description. *M. p. plumbeus* (Hartlaub): Senegambia to Ethiopia, Uganda, NW Tanzania (south to Biharamulo and Kibondo) and Zaïre (Congo R., Kasai and Kivu). ADULT ♂ entire upperparts, sides of head and neck, and scapulars slate-grey. Uppertail-coverts blackish grey. Tail mainly black, graduated (T6 shortest); T6 almost entirely white; T5 has broad tip to both webs and narrow fringe on outer web white; T4 has narrow white tip. Lores blackish slate, supraloral line and eye-ring white. Chin, throat, breast and flanks pale slate-grey. Rest

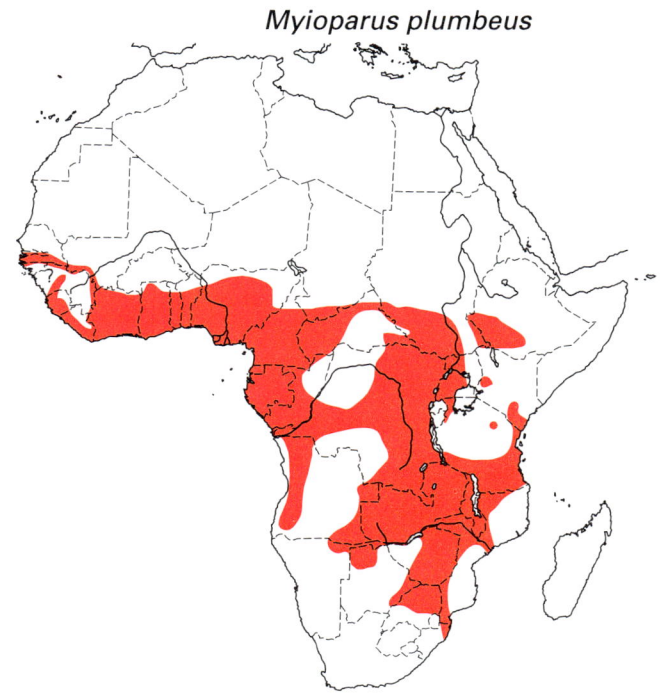

Myioparus plumbeus

of underparts, underwing-coverts, axillaries and fringes to inner webs of underside of flight feathers white. Undertail-coverts white suffused with buff. Wings dark to blackish grey-brown, primaries and secondaries very narrowly edged greyish white, upperwing-coverts fringed pale slate-grey except greaters and tertials which have distinct white edge to outer webs. Bill dark brown to black, base of lower mandible bluish horn; eyes olive-brown; legs and feet grey-brown to black. Sexes alike. SIZE: (10 ♂♂, 3 ♀♀) wing, ♂ 66–70 (67·5), ♀ 62–65 (64·0); tail, ♂ 55–64 (59·2), ♀ 55–59 (57·5); bill, ♂ 14–15 (14·7), ♀ 14–15 (14·2); tarsus, ♂ 17–18 (17·9), ♀ 17–18 (17·5). WEIGHT: (Nigeria) 1 unsexed 14; (Gabon, July) 1 unsexed 14; (Central African Republic, June) 1 ♂ 13; (Upemba, Zaïre, Oct) 7 ♂♂ 11–15 (13·3), 7 ♀♀ 11–15 (12·4), 2 unsexed 16, 16; (Uganda, Apr, May) 2 ♀♀ 13, 14·5.

IMMATURE: similar to adult but browner grey; underparts washed buff; upperwing-coverts, inner primaries and secondaries edged tawny to russet; bill brown, lower mandible almost entirely horn.

JUVENILE: upperparts and upperwing-coverts spotted (feathers with ochre to tawny apical spot and blackish terminal band); underparts scalloped (feathers narrowly tipped black); wing margins pale buff, not white; white reduced on T4; bill paler than immature.

NESTLING: unknown.

M. p. orientalis (Reichenow and Neumann): E Kenya (south of Kitui-Malindi, east of Kibwezi-Taita) and E Tanzania (east of Amani-Morogoro-Songea) to S Malaŵi, Zimbabwe (NW and Middle Zambezi Valley), E Transvaal and Natal (Zululand). Paler and more bluish grey than nominate race, undertail-coverts white. WEIGHT: (Kenya) 1 ♂ 10.

M. p. catoleucum (Reichenow) (including '*grandior*'): SE Tanzania (Sumbawanga to Iringa and Dodoma) and N Malaŵi to Angola, SE Zaïre (Shaba), Namibia, Botswana, S and E Zimbabwe, Transvaal and Natal. More ashy and much whiter below than other races.

Field Characters. A small, warbler-like flycatcher always on the move through foliage. Frequently displays white outer tail-feathers by raising and fanning tail over back, also flirting it in flight. Bluish grey above, paler below, with white belly, white lores and eye-ring, black and white tail, white edges to tertials and wing-coverts. Its forest relative, Grey-throated Tit-Flycatcher *M. griseigularis*, has very similar song but lacks white in tail and wings and on face. Ashy Flycatcher *M. caerulescens* is very similar in general coloration, including pale lores and eye-ring, but has no white in tail or pale wing edgings, and hunts from perch like typical flycatcher.

Voice. Tape-recorded (86, 88, 91, B, F, CHA, ERA, PAY, WALK). Advertising song a 3-note phrase of rather mournful, quavering notes, 'pee-lee-peeerr' or 'peely-peeerr'. Second note follows first very closely; gap between second and third notes is more perceptible. In Gabon, notes ascend the scale, whereas in southern Africa last note is lowest. Also gives a single, drawn out 'peeerrrp' (Maclean 1993). Contact call between members of pair, a series of 3 down-slurred whistles, similar to those of Grey-throated Tit-Flycatcher.

General Habits. Inhabits primary forest edges, secondary forest, clearings, light woodland, bushveld, *Brachystegia* woodland, riverine forest. Man-made habitats include cultivated fields with scattered tall trees, plantations (e.g. cacao) with tall tree cover and various stages of regenerating forest. Lowlands, and up to 2000 m.

Solitary or in pairs. Gleans foliage like Grey-throated Tit-Flycatcher: moves like a warbler among leaves, raising and lowering tail and fanning it, flashing white feathers. Frequents tops of tallest trees in NE Gabon, rarely below 20 m, but regularly at middle levels in southern Africa. Moves about constantly; covers entire territory daily. Regularly joins mixed-species flocks, particularly during lean dry season.

Food. Insects, mainly Coleoptera, also moths, caterpillars, Orthoptera, ants and termite alates; spiders. Prey size 8–20 mm.

Breeding Habits. Monogamous. Pair members maintain close contact mainly with visual cues, spreading tail to expose white feathers. Territorial all year; defends large territories (30–40 ha in Gabon) by singing for lengthy periods while sitting upright on perch. During territorial defence adopts threatening postures with fluffed-out plumage, carpal joints of folded wings away from body, tail bent forward, and open bill pointing toward opponent. Chases and harasses intruder in flight, vigorously snapping wings and bill. Interspecifically territorial with Grey-throated Tit-Flycatcher.

NEST: untidy, fairy sizeable mass of grass, fine straws, rootlets, shredded bark, small feathers and lichens, stuffed into cavity, e.g. old nest-hole of barbet or woodpecker, or natural hole in branch; placed 5·5–10 m above ground. Nest in Kenya was in former Nubian Woodpecker *Campethera nubica* nest hole in telegraph pole; diam. of hole 60, diam. of cup 60, depth 25; scantily built of fine interwoven grass, lined with dried bougainvillea flowers and small feathers (Wilson and Wilson 1994b). Built by both sexes.

EGGS: 2. Oval; glossy; dirty white to greenish white, densely speckled and blotched olive-brown and grey-brown or lavender. 2 eggs in Kenya were pale sea-green, heavily marked all over with dark olive-green streaks and spots, more concentrated at large end (Wilson and Wilson 1994b). SIZE: (n = 4, South Africa) 17·0–17·5 × 12·5–13·3 (17·2 × 13·0); (n = 2, Kenya) 17 × 12·5.

LAYING DATES: Ghana, (juv. July); Nigeria, Mar, Dec; Cameroon, Feb–Aug; E Africa: Kenya, Mar (nest-building Apr–May); Region B, Apr; Rwanda, Aug; Zaïre (breeding condition Jan–Apr); Angola, Aug, Oct; Malaŵi, May, Oct, Nov; Zambia, Sept, Oct (breeding condition Aug); Botswana, Oct–Nov; Transvaal, Natal and Zimbabwe, Oct–Dec.

INCUBATION: by both sexes; continuously except during afternoon heat.

BREEDING SUCCESS/SURVIVAL: adult taken by Little Sparrowhawk *Accipiter minullus* (Wilson and Wilson 1994b).

References
Brosset, A. and Erard, C. (1986).
Erard, C. (1987, 1990a).
Fraser, W. (1983).
Wilson, N. and Wilson, V. G. (1994b).

Plate 27

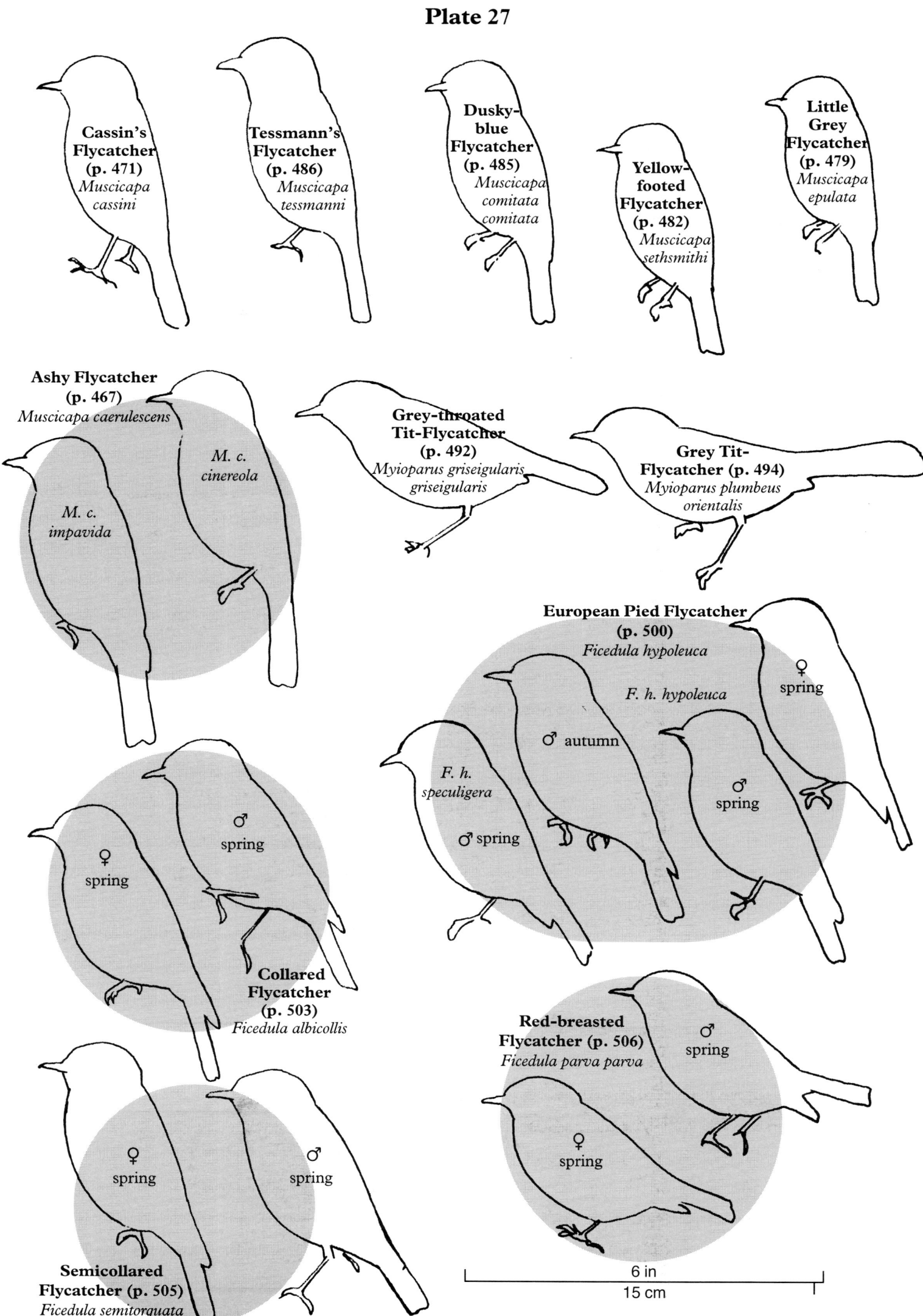

Plate 28

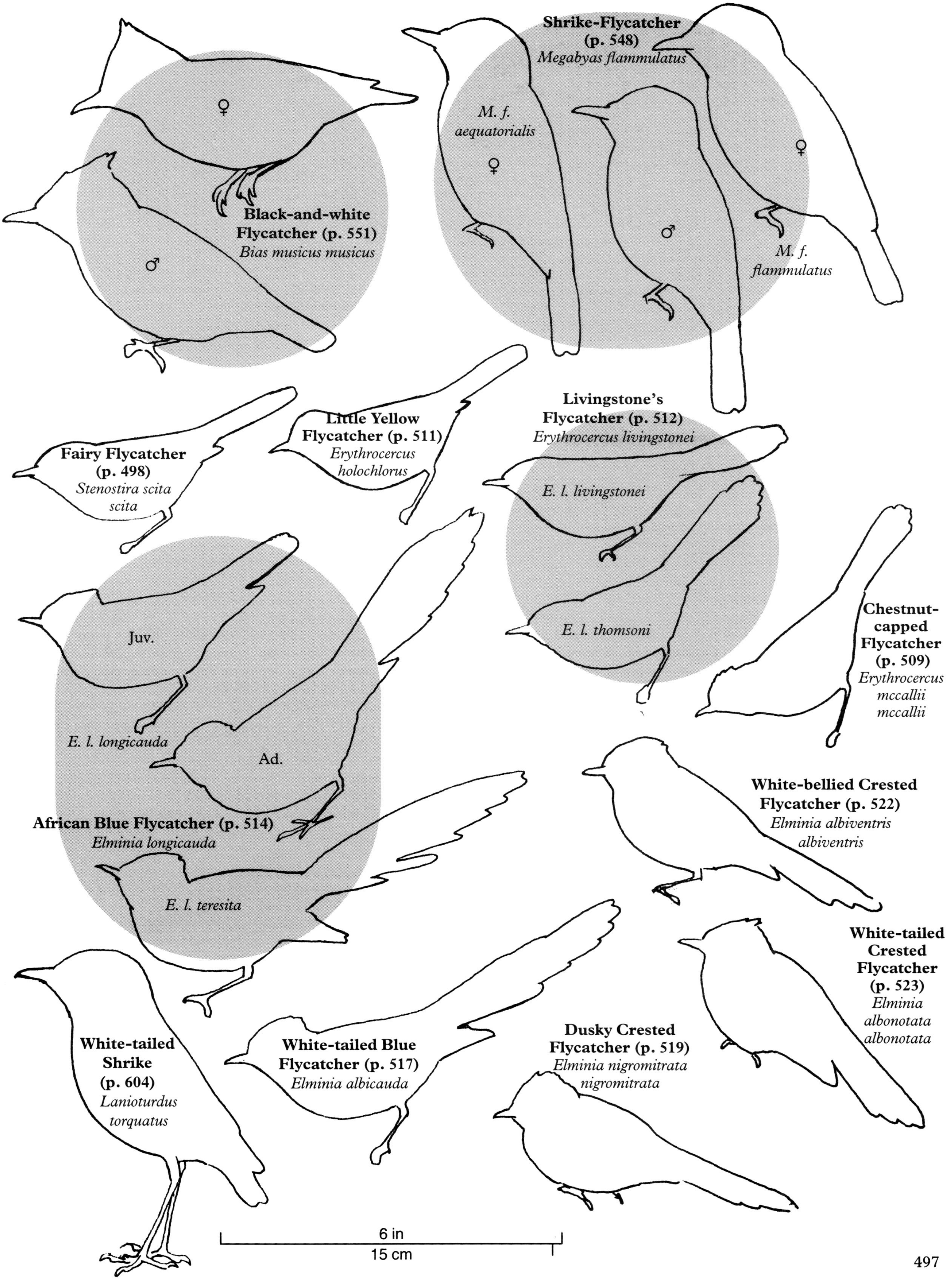

Black-and-white Flycatcher (p. 551)
Bias musicus musicus
♀
♂

Shrike-Flycatcher (p. 548)
Megabyas flammulatus
M. f. aequatorialis ♀
♂
M. f. flammulatus ♀

Fairy Flycatcher (p. 498)
Stenostira scita scita

Little Yellow Flycatcher (p. 511)
Erythrocercus holochlorus

Livingstone's Flycatcher (p. 512)
Erythrocercus livingstonei
E. l. livingstonei
E. l. thomsoni

Chestnut-capped Flycatcher (p. 509)
Erythrocercus mccallii mccallii

African Blue Flycatcher (p. 514)
Elminia longicauda
Juv.
Ad.
E. l. longicauda
E. l. teresita

White-bellied Crested Flycatcher (p. 522)
Elminia albiventris albiventris

White-tailed Crested Flycatcher (p. 523)
Elminia albonotata albonotata

White-tailed Shrike (p. 604)
Lanioturdus torquatus

White-tailed Blue Flycatcher (p. 517)
Elminia albicauda

Dusky Crested Flycatcher (p. 519)
Elminia nigromitrata nigromitrata

6 in
15 cm

Genus *Stenostira* Cabanis and Bonaparte

Monotypic; endemic to southern (practically, South) Africa. Tiny (6 g) and slim; weak, slender, warbler-like bill, rictal bristles, legs and feet; plumage grey, with *Batis*-like black mask accentuated by white borders, and black and white wings and tail; throat and belly feathers white with pink bases; tail moderately long and graduated. Sexes similar. Nestling with neossoptiles. Juvenile not spotted. Nest a cup; courtship-feeds; eggs plain. Thin sweet song; forages restlessly in leafy tree; droops wings, fans and cocks tail.

Often regarded as a muscicapid flycatcher (e.g. Hall and Moreau 1970, Ginn *et al.* 1989, Maclean 1993), but treated as a monarchid flycatcher by Watson *et al.* (1986), Dowsett and Dowsett-Lemaire (1993b) and Dowsett and Forbes-Watson (1993), and as a sylviid warbler by Traylor (1970b), Clancey (1980b) and Sibley and Monroe (1990). Plumage and behaviour recall *Myioparus* (Muscicapidae, especially *M. plumbeus*), *Apalis* (Sylviidae, e.g. *A. karamojae*, *A. melanocephala*) and *Eliminia* (Monarchidae, e.g. *E. albonotata*). Vernon (1985) showed that it is not a warbler (because it has neossoptiles and lacks tongue spots) and found more monarch than muscicapid characters; but possibility of *Stenostira* being a monarch flycatcher is discounted by its voice, mouth colour, lack of eye-ring and shape of its carpometacarpus (Pocock 1966).

We consider *Stenostira scita* to be quite close to *Myioparus plumbeus*; in fact they should probably be made congeneric (in *Stenostira*, which pre-dates *Myioparus*), even though *scita* is much smaller and with a different song and proportionately smaller, needlelike bill. Lack of spotted juvenile plumage need not debar *S. scita* from Muscicapidae: nestlings of *M. plumbeus* have spots only on back (not on breast) and lose them very quickly. The 2 species are parapatric. Voices are not very similar, but nests are very alike though *M. plumbeus* uses tree holes; *Stenostira* courtship-feeds, *Myioparus* does not – a character probably correlated with their nest-site differences.

Plate 28 (Opp. p. 497)

Stenostira scita (Vieillot). Fairy Flycatcher; Fairy Warbler. Mignard enchanteur.

Muscicapa scita Vieillot, 1818. Nouv. Dict. d'Hist. Nat. 21, p. 474; Lower Orange River, Cape Province.

Range and Status. Endemic to Africa south of Tropic of Capricorn; migrant within that region. Breeds from SW Cape (north to at least Olifants R.) to Lesotho, with a few summer records on S border of Orange Free State and in SW Drakensberg Mts; fairly common, less so in Lesotho. Winters throughout most of breeding range but withdraws about 100 km from southwestern extremity (Hockey *et al.* 1989), and partially migrates north and northeastwards to winter north to Orange R. valley, Orange Free State (where patchily distributed, uncommon to locally common), W Natal and Transvaal north to Vaalwater and Potgietersrus (once Vivo); widely but thinly distributed in Transvaal Highveld and central Bushveld (in Kruger Nat. Park 1 old record and 1 recent, in Shawu). 1 record S Mozambique, 2 in S Zimbabwe (Sentinel Ranch, July, Bulawayo, Aug, both 1964), 20 at 13 localities in Botswana, all in SE except 2 at Khakhea (24°42'S, 23°30'E) and Martin's Drift (21°45'S, 21°44'E) (Penry 1988), and sparse records in S Namibia, north to Otjiwarongo and west of Etosha.

Description. *S. s. scita* (Vieillot): South Africa: W and NW Cape, wintering north to S Namibia. ADULT ♂: forehead to rump uniform dark, soft, bluish grey with olive tinge; long silky white flank-feathers can displace rump-feathers, making rump look white. Uppertail-coverts bluish black. Tail glossy black, graduated, with white sides: T1–T3 black, T4 4 mm shorter with white end 8 mm long on inner web and 15 on outer, T5 9 shorter and T6 14 shorter than T1, both white, except for blackish near base and inner web of T5. Lores, very narrow line linking lores across base of upper mandible, area under eye, cheeks and ear-coverts, glossy black. Very narrow white line above the black across base of bill, confluent with long, narrow, well-defined superciliary stripe; lower eyelid bordered by well-defined, very narrow white line 3 mm long; distinct long white malar stripe, the feathers being long and silky. Chin white, upper throat white with salmon-pink feather bases showing through strongly, at least in centre; lower throat, sides of neck and breast same grey as back; grey extends onto forepart of flanks, becoming paler; rear flank feathers long, loose, silky, white; belly, vent and undertail-coverts white, centre of belly creamy white with feather bases bright pink, the colour showing through patchily, sometimes barely visible; thighs black. Primaries, secondaries and tertials black, primaries and secondaries very narrowly edged glossy grey, and 3–4 largest tertials with outer web widely fringed white proximally and very narrowly fringed white distally; primary coverts and alula black, greater and median coverts broadly tipped white, lesser coverts brownish black, some tipped with grey; underside of flight feathers dark grey, inner edges of feathers pale grey or whitish; underwing-coverts and axillaries white. Bill black, eyes dark sepia-brown; legs and feet black, slender. Sexes similar, but ♀ less clear blue-grey above and breast paler grey. SIZE: wing, ♂ (n = 2) 46, 51, ♀ (n = 2) 48, 51·5, ♂♀ (n = 9) 47–50 (49·3); tail, ♂ (n = 4) 45–51 (49·0), ♀ (n = 2) 48·5, 52, ♂♀ (n = 9) 49–55 (50·6); bill, ♂ (n = 3) 12·0–13·2 (12·7), ♀ (n = 2) 10·3, 13·0; tarsus, ♂ (n = 3) 18·0–20·5 (18·6), ♀ (n = 2) 19·3, 19·4, ♂♀ (n = 23) 16–19. WEIGHT: ♂♀ unsexed (n = 48) 4–8 (5·9).

IMMATURE: (Cape Province: Natural History Museum, London): like adult but upperparts browner, breast grey washed very pale yellow, more white on chin; upper throat and belly without pink, belly white or grey-white washed very pale yellow, grey of breast does not extend onto flanks, which are like belly. (Natal?, described by Clancey 1964): rusty brown above and on wings, buffy below, the breast darker.

NESTLING: not described, but hatches naked with a few neossoptiles (down) (Vernon 1985).

S. s. rudebecki Clancey: Lesotho, wintering in Transvaal. Greys slightly darker than in *saturatior*. Larger: wing, ♂♀ (n = 14) 50–56 (52·6), tail, ♂♀ (n = 14) 51·5–57.

S. s. saturatior Lawson: South Africa: Great and Little Karoos and eastward, wintering north to W Orange Free State. A poor race, a trifle darker than *scita*.

Field Characters. A vivacious, delicate, very small, grey, warbler-like flycatcher in leafy trees in Africa south of Limpopo R., readily told by combination in all plumages of quite long, graduated tail, black with white sides, and black mask bordered above and below by narrow white lines. Closed wing black with conspicuous white stripe for its whole length. Chin and belly pinkish white. Narrow black and white lines across forehead; tiny white mark below eye. Usually in pairs. Often fans tail and droops wings. From below, tail looks white with a black tip. Completely unlike any other bird in its range.

Voice. Tape-recorded (88, 91). Song a structured but quite variable, thin, wispy or squeaky 'tsee-tsi-zee-tseepy-tsweeu' or 'tisee-tchee-tchee' or 'tsee-tsi-tsippy-tseeu-ts-twee-zzzz', lasting 1·5–2·0 s, falling slightly, repeated several times in a few min., recalling a Lesser Double-collared Sunbird *Nectarinia chalybea*. Calls (functions?) a short sibilant trill 'kisskisskisskiss', sometimes only 2 or 3 syllables; and 3 descending discords 'cher, cher, cher' or 'zrrt, zrrt, zrrt'. ♀ on nest gives characteristic buzzing sound when she sees ♂ coming to courtship-feed her (C. J. Vernon *in* Ginn *et al.* 1989).

General Habits. Breeding habitat is dry karoo bush and scrub, fynbos, bushy hillsides, thorn thickets and dry, scrub-filled mountain kloofs and ravines. Shares breeding mountain habitat with Layard's Warbler *Parisoma layardi* and Mountain Wheatear *Oenanthe monticola*. In winter moves into acacia savanna woodlands, montane scrub, camelthorn, plantations and gardens. Often near rivers. Active and restless, usually in pairs, sometimes singly, in family party, or in small mixed-species foraging party; flits about inside leafy bush or among small branches in outer canopy of tree, usually within 7 m of ground, quickly darting at small prey from leaf or twig and moving on with constant creeping hops (hopping stride is short despite its long legs, which are half-concealed under feathers for much of the time) or short flights within canopy; moves freely between tree top and small bushes below but does not come close to ground; makes short, fluttery sally after flying insect or a spider moving 50 cm away. Rather a silent bird. Fans and bobs tail and raises fanned tail much of the time when foraging; folded wings then hang loosely and droop.

Except for lowlands north of Great Berg R., breeds mainly above 300 m altitude. Moves to lower elevations and many migrate well to north and northeast after breeding, withdrawing completely from only a small part of range (e.g. from Worcester–Darling–Picketberg–upper Dweka R. area of SW Cape) in Nov–Apr. Occurs year-round in rest of breeding range; winters in SE Botswana May–Sept; in Orange Free State in all months but most sparsely in Dec–Jan; in Transvaal Apr–Sept (rare Oct), numbers varying from year to year; and in Natal Apr–Sept (rare Mar, Oct, Nov).

Food. Not studied; apparently small insects and spiders.

Breeding Habits. Solitary nester.

NEST: a small, deep cup, neatly made of fine materials; walls thick and compact, round and relatively smooth, composed of dead leaves, dead grass and thin bark fibres, bound with spider silk, sometimes camouflaged with lichen, thin shreds of bark, and bits of old dry leaves; thickly lined with feathers, wool, hair and plant floss. Int. depth 30 mm, int. diam. 38 (James 1922). Sited deep in almost impenetrable vegetation: a thick hedge, flood debris, mass of dead twigs around thorn-tree trunk; hence very hard to find. Height, 0·2–2·3 m above ground. Built by ♀ only, taking *c.* 4 days (Maclean 1993).

EGGS: 2–3, av. (10 clutches) 2·3. Broad ovals, glossy; thin-shelled, ground colour cream-buff to pale greenish buff, sometimes covered with small pale greenish brown speckles, with indistinct zone of darker buff around broad end or with distinct zone of confluent brown blotches around middle. SIZE: (n = 21) 14·5–16·9 × 10·8–12·7 (15·2 × 11·5).

LAYING DATES: South Africa, Sept–Dec; Lesotho (Jan, courtship feeding).

INCUBATION: incubating bird quits nest well before person approaches. Period: 17–18 days.

References
Clancey, P. A. (1955).
James, H. W. (1922).
Maclean, G. L. (1993).
Penry, H. (1988).
Wyndham, C. (1939).

Genus *Ficedula* Brisson

Small slim-bodied flycatchers with rather rounded heads and short bills. Males strikingly patterned, many blackish above with white wing-bar and white sides to tail, some rufous or yellow below. Females and immatures dull and brownish, often difficult to identify. Usually unobtrusive. Inhabit forest trees and undergrowth, wooded areas and thickets.

27 species, most in Asia; 4 in Africa, all Palearctic migrants but 1 also breeding in NW Africa.

Ficedula hypoleuca (Pallas). European Pied Flycatcher. Gobemouche noir.

Motacilla hypoleuca Pallas, 1764. *In* Vroeg's Cat. Adambratiunculae, p. 3; Holland.

Range and Status. NW Africa, W, central and N Europe, and through Russia at *c.* 53°–63° N, east to Ob R.; winters Africa.

Palearctic migrant. Locally common migrant breeder highlands of N Morocco (south to High Atlas) and N Algeria (Atlas Tellien, Aurès Mts); uncommon N Tunisia (Kroumirie). On passage N Africa, common to abundant in autumn W Morocco, frequent to common central and E Morocco and Algeria, scarce Libya and Egypt; in spring, common to abundant Morocco to Libya, frequent Egypt, often common in oases.

Winters W Africa: scarce Gambia and S Senegal; locally common Guinea, S Mali (Bamako) and Burkina Faso (south of 11°N) to Sierra Leone, N Liberia, Ivory Coast, Ghana, Nigeria (south of *c.* 9°N), Cameroon and Central African Republic; uncommon Gabon and N Zaïre (Ubangi R. east to Ituri and Mahagi); rare N Angola (once Cabinda). Some mid-winter reports NW Africa, but probably does not winter there regularly (Zink 1985). On southward passage, common to abundant in SW Mauritania and W Senegal, locally common S Mali (south of 15° N), S Burkina Faso and N Nigeria, uncommon Chad. On northward passage, frequent to common Senegal, Mali, N Nigeria and Saharan oases east to Chad (Tibesti). No satisfactory records for E Africa, where confusion has occurred with Semicollared Flycatcher *F. semitorquata* (Britton 1980b, Pearson 1981), nor Ethiopia (Tyler 1987b) nor for southern Africa. Rare Sudan (2 trapped Red Sea coast, Aug).

Ficedula hypoleuca

Description. *F. h. hypoleuca* (Pallas): Europe (except Iberia) east to Urals, winters throughout African range. ADULT ♂ (breeding): upperparts blackish (W Europe) or dark brown to grey-brown (central Europe eastwards), with small white patch on forehead. Tail-feathers blackish; T6 with basal ¾ of outer web white, sometimes extending onto base of inner web, T5 often with white subdistal patch on outer edge, T4 occasionally also with small white mark. Underparts white, tinged cream. Primaries and secondaries blackish brown; basal part of outer web of P1–5 white, extending *c.* 3 mm beyond primary coverts; base of outer web of secondaries white, extending *c.* 4 mm beyond greater coverts. Inner tertial black with wholly white outer web or broad white outer margin; middle tertial white apart from black distal or subterminal part of inner web; outer tertial black with white base and outer edge to outer web. Primary coverts dark grey-brown. Inner greater coverts black basally, white distally; outer greater coverts dark brown with narrow creamy tips. Median and lesser coverts blackish. White parts of tertials, inner greater coverts and secondary bases together form prominent white patch, connected to white wing-bar. Exposed white at primary bases forms additional small mark. Underwing-coverts grey-brown; axillaries creamy with grey-brown bases. Wing rather long, tip bluntly pointed; P8 longest; P7 1–3 shorter (occasionally equal), P6 3–6, P5 8–13, P1 19–24; P9 4–7, usually shorter than P6; P10 minute; P6–P8 emarginated. Bill blackish brown; eyes dark brown; legs and feet black. ADULT ♂ (non-breeding): upperparts brown with black confined to uppertail-coverts, forehead almost uniform with crown. Underparts tinged buffish on throat, breast and flanks. Tertials dark brown, with narrow whitish outer margin extending around tip, middle and outer feathers with wholly white base to outer web. Greater coverts all dark brown with narrow whitish tips. White tertial patch and wing-bar thus narrower than in breeding adult ♂. Median and lesser coverts dark brown with paler tips. Otherwise like breeding adult ♂. ADULT ♀ (breeding): similar to non-breeding adult ♂. Upperparts uniform brown, though feathers of forehead often slightly creamy and longest uppertail-coverts dark brown or blackish. Tail feathers dark brown, T6 with basal ¾ of outer web whitish, T5 with whitish subdistal mark on outer edge, T4 sometimes with small whitish mark on outer edge. Underparts whitish, tinged pale brown on breast and flanks. Primaries and secondaries dark brown, white bases to outer webs of inner primaries extending only up to 1 mm beyond primary coverts, white bases to secondaries extending *c.* 2 mm beyond greater coverts. Tertials, upperwing-coverts, underwing-coverts and axillaries as non-breeding adult ♂. In general, wing markings less prominent than in non-breeding ♂. Bare parts like adult ♂. ADULT ♀ (non-breeding): like breeding adult ♀, but forehead browner, underparts more buffish. SIZE: (10 ♂♂, 10 ♀♀, W Europe) wing, ♂ 76–81 (78·1); ♀ 75–79 (77·0); tail, ♂ 50–54 (51·3); ♀ 48–51 (49·6); bill, ♂ 11–13 (12·5), ♀ 12–13 (12·2); tarsus, ♂ 16–18 (16·8), ♀ 16–18 (17·0). WEIGHT: unsexed: (n = 60, Morocco, spring) 9·7–14·3 (11·6); (n = 15, N Algeria, autumn) 10·7–16·6 (13·4); N Nigeria (n = 20, autumn) 10·3–13·8 (11·6), (n = 2, spring) 16·7, 16·8; central Nigeria (n = 75, autumn) 9·8–12·8 (9·6), (n = 58, spring) 11·4–22·2 (16·0).

Complete moult in breeding area; partial moult in Africa, Jan–Feb.

IMMATURE: juvenile dark brown above, spotted pale buff; buffish white below with blackish feather tips; wing and tail as first winter. First winter ♂ and ♀ like non-breeding adult ♀, except that whitish margins on outer webs of tertials broader, *c.* 2–3 mm near shaft at tip, and contrasting with narrow (*c.* 0·5–1 mm) margins on inner web tips. First summer ♂ like breeding adult ♂, but upperparts usually browner, and primaries and tail feathers like adult ♀.

NESTLING: covered with scanty grey down, mainly on upperparts; mouth orange-yellow; gape flanges pale yellow; no tongue spots.

F. h. tomensis (Johansen) (equals '*sibirica*'): Siberia east from Urals; probably winters throughout African range. Upperparts of breeding ♂ usually grey-brown; slightly larger than nominate race: wing of 21 ♂♂ 79–84 (81·2) (Vaurie 1959).

F. h. iberiae (Witherby): Iberia; winters W Africa. Breeding ♂ intermediate between nominate form and *speculigera*: white forehead patch usually larger than in nominate, inner webs of tertials whiter, white at bases of primaries and secondaries more extensive; less white in outer tail-feathers.

F. h. speculigera (Bonaparte): N Morocco to N Tunisia; recorded Senegal and Ivory Coast. Breeding ♂ blacker above than nominate, white forehead and white area on wing larger. Inner webs of tertials whiter, inner greater coverts wholly white and median coverts usually tipped white; white at bases of primaries (often P1–8) and secondaries more extensive. White in tail usually restricted to T6. Often shows partial, occasionally almost complete, white collar; rump sometimes whitish. ♀ has more prominent white wing markings than in nominate race.

Field Characters. A compact flycatcher with a rather long wing and stubby bill. Breeding ♂ blackish above and white below, with conspicuous white tertial patch extending as bold bar across wing, and broad white sides to tail. N European birds separated from Collared Flycatcher *F. albicollis* and Semicollared Flycatcher *F. semitorquata* by lack of even a partial collar, by blackish rump, less extensive white on wings, with only a small mark at the base of primaries and smaller white forehead patch. ♀ and first winter and non-breeding ♂ are brown above and brownish white below, with narrow white wingbar and tertial edges, and white sides to dark tail; they are very difficult to distinguish from corresponding plumages of Collared and Semicollared Flycatchers but white wing markings are typically less conspicuous with 'flash' at base of primaries very small or not visible. In the hand, Pied usually has a shorter P9 than Collared or Semicollared (**A**) and, except in N African race, lacks white at base of P6 and P7. ♀, non-breeding ♂ and first winter distinguished from Collared by lack of whitish band on feathers of hindneck (see *F. albicollis* **A**). Voice distinctive. N African and Iberian ♂♂, with more white in wing, and sometimes partial collar and pale rump, are more easily mistaken for Collared or Semicollared, but can be separated by shorter P9, and by call-note and song.

Pied   **A**   Collared

Voice. Tape-recorded (62, 73, 89, 93, B, C). Song, a short cheerful phrase of about 15 notes, typically alternating in pitch and often ending in an abrupt trill recalling Common Redstart *Phoenicurus phoenicurus*, 'si-tsu-tsi-tsu-tsi-tsu . . . tsrr-tsirr-tsirr'; lasts about 3 s, and is delivered up to 10 times per min; sometimes preceded or followed by an ascending series of whistled sounds, 'hy-hi-hi' (Cramp and Perrins 1993). Contact-alarm call, typical of migrants, a sharp, loud 'vit', often repeated insistently about twice per s; also a clicking 'tck', sometimes combined when excited, e.g. 'vit-tk-tk'; and a soft 'tsrr' used in pair-contact.

General Habits. Breeds NW Africa in cedar, oak and pine woods at c. 1200–1800 m. In winter, inhabits edges of lowland forest, gallery forest, savanna woodland, gardens and cultivated areas with large trees; also forest on mountain slopes, up to 2000 m on Mt Nimba. Often in drier, bushier areas on migration. Apparently solitary and often sedentary; 1 trapped 6 times at same site Nigeria (Smith 1966). Highly territorial on autumn migration in Portugal (Bibby and Green 1980). Rather shy; often remains silent and inconspicuous within thick canopy. Forages at all levels in shady as well as more open trees and in tall bushes. Searches and gleans from foliage and branches. Frequently darts out to take flying insects, constantly changing perch. Often feeds on or near ground. Perches less upright than Spotted Flycatcher *Muscicapa striata*, and tends to hold tail slightly raised. Flicks tail up and down when excited. Gives repeated 'vit' call when alarmed, accompanied by tail and wing flicking. ♂♂ often sing on spring passage in Egypt (Moreau and Moreau 1928).

All races are believed to winter in W Africa. Birds from Europe, including Russia east to the Urals, migrate southwest to Iberia (some via N Italy). Onward passage is concentrated along NW coast of Africa, but many presumably then turn southeast to reach wintering grounds between Ivory Coast and Nigeria. Autumn records from Algerian Sahara and Chad, however, suggest some birds use more easterly route from S Europe to W central Africa. Passage occurs Morocco late Aug–early Nov, mainly Sept–early Oct; central Nigeria from mid–Sept, peak mid-Oct; Chad mainly late Sept–early Oct. Birds reach wintering areas quite early: Sierra Leone from Oct, Ivory Coast from Sept, S Nigeria and Ghana from Oct. Northward migration begins Mar, but most birds leave wintering areas in Apr, some not until early May. Passage peaks central Nigeria third week Apr, latest 23 May. Many spring weights in central Nigeria are 50–90% above typical autumn weights, suggesting that birds embark on long flights from c. 10°N. Most birds appear to have a loop migration, with the route across NW Africa and the Mediterranean further east in spring: birds scarcer in spring than autumn in W Morocco, but far commoner in spring than autumn along N African coast, with passage extending to Egypt and Levant. Peak migration occurs in Mediterranean area in mid-Apr–early May. Birds return to NW African breeding grounds late Apr–early May.

A loop migration is also indicated by N African passage recoveries of birds ringed in W and N Europe (Zink 1985). Of c. 90 in autumn, most are near W Moroccan coast (from Britain, France, Holland, Germany and Scandinavia, and 1 from Russia at 38°E), with few in central and E Morocco, and only 2 in Algeria (from Finland). Of c. 250 in spring, fewer are from W than from central and E Morocco, with c. 50 in Algeria and Tunisia, 12 in Libya (ringed France, Germany, Finland, Latvia and Russia at 41°E) and 1 in Egypt (ringed S France Oct). South of Sahara, birds ringed in Britain recovered Guinea (1), Ivory Coast (1), Ghana (1, Apr) and Central African Republic (1, Feb); ringed Germany recovered Ghana (1, Sept); ringed Norway recovered Mauritania (1) and Guinea (1); and ringed Finland recovered Guinea (1, Feb), Sierra Leone (1, Mar) and Liberia (1). One ringed in N Nigeria (Oct) recovered Cyprus (spring).

Food. Flying and non-flying insects, especially Hymenoptera, Diptera, Coleoptera, and adult and larval Lepidoptera; also spiders; occasionally fruit or seeds. Took *Lantana* berries in winter in Sierra Leone. Birds in Nigeria Nov and Jan contained only insects (Cramp and Perrins 1993).

Breeding Habits. One of the best-studied European passerines (see e.g. Lundberg and Alatalo 1992). Essentially monogamous, but successive polygyny (with 2 ♀♀ or occasionally 3) is frequent. Rate of polygyny as high as 39% in Sweden; generally higher where breeding density low (Alatalo and Lundberg 1984). Solitary and territorial. Territory small, c. 400–2000 m² Europe; used for pair formation and nesting, and pair forages outside defended area. Often polyterritorial; ♂ with primary mate already laying may establish secondary territory and seek to attract another ♀. In Morocco, 9·5 pairs per km² in woodland, up to 38 pairs per km² in humid forest (Thévenot 1982). In Europe, breeding density up to 400 pairs per km²; higher in deciduous than coniferous forest and greatly increased by provision of nest-boxes (Cramp and Perrins 1993).

♂♂ return to breeding grounds about 1 week before ♀♀ (older birds first), select prospective nest-holes and defend territory within c. 20–30 m radius. Resident ♂ may threaten intruder near hole, ruffling plumage, flicking wings and tail, and calling (Cramp and Perrins 1993). ♀ apparently attracted by quality and size of ♂'s territory. Enticed to hole by ♂'s conspicuous flight and song-display, and if satisfied soon begins nest-building. Copulation usually takes place on branch near nest. On arrival in territory ♂ sings vigorously from high in tree or from wire or building, but in presence of ♀, song focused on nest-site, and briefer and less frequent after pairing (Cramp and Perrins 1993).

NEST: loosely built cup of dead leaves, plant stems and moss, lined with rootlets, fine grasses, hair and sometimes feathers; in hole in tree, wall or building; in W Europe mainly in nest-boxes; ext. diam. 130, int. diam. 55, cup depth 45; built by ♀ in 4–11 days.

EGGS: 4–8 (3–10), laid at daily intervals; NW Africa (20 clutches) 4–7, mean 5·15; mean clutch (many European studies) 5·7–7·5 (Cramp and Perrins 1993). Smooth and slightly glossy; very pale blue, rarely with fine reddish speckling. SIZE: (n = 30, NW Africa) mean 18 × 13·5. Occasionally double-brooded in Europe.

LAYING DATES: NW Africa, late Apr–late May.

INCUBATION: by ♀ only; period usually 13–15 days; increased food availability shortens incubation time (Sanz 1996).

DEVELOPMENT AND CARE OF YOUNG: young usually fed by both parents; brooded for 3–9 days by ♀ only; remain in nest c. 14–18 days; leave territory within a few hours of fledging, but still fed by parents for at least another week.

BREEDING SUCCESS/SURVIVAL: in Germany, of 3724 eggs in 606 clutches, 71% hatched, 62% produced fledged young (Creutz 1955).

References
Alatalo, R. V. and Lundberg, A. (1984).
Cramp, S. and Perrins, C. M. (1993).
Creutz, G. (1955).
Curio, E. (1959a).
Lundberg, A. and Alatalo, R. V. (1992).
Zink, G. (1985).

Ficedula albicollis (Temminck). Collared Flycatcher. Gobemouche à collier.

Plate 27 (Opp. p. 496)

Muscicapa albicollis Temminck, 1815. Manuel d'Orn., ed. 1, p. 100.

Range and Status. Central and SE Europe north to the S Baltic and east in S Russia to c. 45°E; winters Africa.

Palearctic migrant. Winters southern Africa, generally south and southwest of Semi-collared Flycatcher *F. semitorquata*, but limits poorly known. Widespread and frequent Zambia (east of 25°E) and Malaŵi to Zimbabwe; presumably also adjacent W Mozambique. Rare Angola (2 records, Huila, NE border), N Namibia (1 record) and N South Africa (2 records, Transvaal). Probably winters S Zaïre, but records almost all from passage months. On southward migration, rare N Africa, Sahara and Sahel: SE Algeria (Djanet), Libya (once, Tripoli), Egypt, Niger (once, Aïr), N Nigeria (Kano), Chad (Ennedi, Abéché) and N and W Sudan. Uncommon Rwanda and E Zaïre, but widespread and frequent central and S Zaïre; rare W Kenya (once, Nyanza) and few records W Tanzania. On northward migration, occasional central and S Zaïre, and 1 W Tanzania (Kibondo) late Feb presumed on passage. Uncommon NE Nigeria. More numerous N Africa in spring than autumn: uncommon E Algeria and Tunisia, and recorded SE Morocco (Defilia) and NW Algeria (Beni Abbès); frequent to common coastal Libya (especially Cyrenaica) and Egypt; frequent in Sahara south to Djanet and Tibesti (N Chad). Status in W Africa uncertain due to apparent confusion with *F. hypoleuca*, notably with the NW African race *speculigera*. Claimed specimens from Ghana are all misidentified *F. hypoleuca* (L. Svensson, pers. comm.).

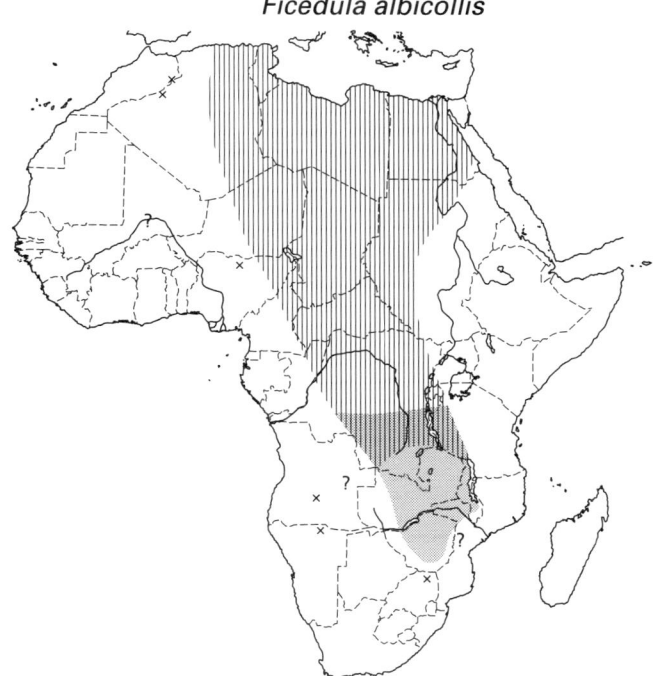

Ficedula albicollis

Description. ADULT ♂ (breeding): top of head, including lores, ear-coverts and around eye, glossy black; large patch on forehead, broad collar around hindneck, and lower back and rump white; uppertail-coverts black. Tail-feathers black, outer web of T6 sometimes with white subdistal mark, of T4–T5 occasionally with small white mark. Underparts white. Primaries and secondaries blackish brown, base of P1–8 (sometimes 1–7) white, extending 7–12 mm beyond primary coverts (much further than in *F. hypoleuca*); base of secondaries white, extending beyond greater coverts. Inner tertial with blackish brown inner web and white outer web, middle and outer tertials white except for black near tip of inner web. New inner greater coverts black with broad white tips, older outer greater coverts dark brown with narrow whitish tips. New inner median and lesser coverts black, older outer ones brown. White parts of tertials, greater coverts and secondary bases form large patch on closed wing; a smaller patch is formed by exposed white primary bases. Primary coverts and alula blackish brown. Underwing-coverts brown to grey-brown; axillaries white with blackish bases. Wing rather long, tip bluntly pointed; P8 longest; P7 0–2 shorter, P6 4–7, P5 10–14, P1 23–28; P9 3–6, usually longer than P6 (cf. *F. hypoleuca*); P10 minute; P6–P8 emarginated. Bill black; eyes dark brown; legs and feet black. ADULT ♂ (non-breeding): upperparts greyish brown with black confined to uppertail-coverts; pale areas less distinct. Feathers of forehead with greyish tips and white bases; feathers of hindneck greyish with white subterminal band. Tail feathers dark brown, T4–T6 with white outer marking as in adult ♀. Underparts tinged creamy buff. Tertials dark brown with narrow white outer margins extending around tip, middle and outer feathers with base of outer web also white. Inner greater coverts dark brown, outer feathers blackish, all edged buffish white; inner median and lesser coverts brown, outer feathers blackish. White tertial patch and wingbar thus less pronounced than in breeding ♂. Outer upperwing-coverts, together with primary coverts and alula form black patch on closed wing. Otherwise like breeding adult ♂. ADULT ♀ (breeding): upperparts greyish brown, forehead slightly creamy. Pale collar usually visible; feathers of hindneck with brown tip and white subterminal band. Paler rump usually noticeable; feathers with white bases. Large uppertail-coverts blackish brown. Tail-feathers dark brown, T5–T6 with white base tapering into white distal margin, T4 often with white mark on outer edge. Underparts whitish. Primaries and secondaries dark brown, white bases of outer webs of P1–8 (sometimes 1–7 or 1–6) extending 2–5 mm beyond primary coverts, and forming a small mark at base of closed primaries. Primary coverts and alula dark brown. Tertials like non-breeding adult ♂. Greater coverts dark brown with narrow whitish tips which, together with exposed white secondary bases form narrow wing-bar (narrower than in non-breeding ♂). Median and lesser coverts dark brown with paler margins. Underwing-coverts, axillaries and bare parts like adult ♂. ADULT ♀ (non-breeding): tinged more buffy below than breeding adult ♀. SIZE: (10 ♂♂, 10 ♀♀, Europe) wing, ♂ 80–85 (82·6), ♀ 79–83 (80·3); tail, ♂ 48–51 (49·4), ♀ 47–50 (47·9); bill, ♂ 11–12 (11·6), ♀ 11–13 (11·9); tarsus, ♂ 16–18 (17·0), ♀ 16–18 (17·2). WEIGHT: unsexed: Nigeria (n = 7, Oct) 10·9–13·5 (12·2), (n = 2, Feb–Mar) 10·5, 13; (n = 2, Zambia, Oct) 11·4, 11·8.

Complete moult in breeding area; partial moult in Africa Jan–Feb.

IMMATURE: first winter ♂ and ♀ like adult ♀ but usually browner; white margins of outer webs of tertials broader, contrasting on tips with much narrower margin of inner webs. First summer ♂ like breeding adult ♂, but upperparts often less black with some brown or grey feathers admixed; collar more greyish white; white mark at primary bases small, as in ♀; tail as in adult ♂.

Field Characters. Breeding ♂ strikingly pied, with broad white neck collar. Blacker above than European Pied Flycatcher *F. hypoleuca*, with larger white patches on closed wing, but lacks extra white bar on median coverts shown by many ♂ Semicollared Flycatchers *F. semitorquata*. Differs from Pied and Semicollared in having full white collar, larger white patch on forehead, white bar across rump and often all-black tail. Non-breeding ♂ greyish above, but paler areas usually still show, and also large white patch on closed primaries. ♀ and first winter very similar to Pied Flycatcher. Tail pattern like ♀ Pied, but white wing markings generally bolder, white primary bases showing as a distinct small patch. Adult ♀ rather greyer brown above than in Pied, usually with paler rump, and many also show paler collar. In the hand, ♀ and first winter distinguished from Pied Flycatcher by white base to P6 and usually also P7, and typically by longer P2 (see *F. hypoleuca* **A**). Separated from ♀ and first winter Pied and Semicollared Flycatchers by whitish subterminal band

 Collared  Pied

A

on hindneck feathers (**A**). Voice differs from Pied Flycatcher (see below).

Voice. Tape-recorded (62, 73, 91, 93, B, F). Song, a short series of high, thin, drawn out notes, recalling European Robin *Erithacus rubecula*; delivery slower and notes higher pitched than in Pied Flycatcher; phrases typically 3–4 s, often separated by long interval or interspersed with 'seep' calls. Contact-alarm, a thin, clear, drawn out 'sieb' or 'seep'.

General Habits. Frequents moist woodland and forest edge, also more open country with scattered trees. In Zambia, usually the edges and interior of *Brachystegia* woodland at 800–1600 m (R. J. Dowsett, pers. comm.), but sometimes in acacia. Rather shy; perches inconspicuously among foliage. Usually occurs singly, although sometimes 2–3 together on migration. Commonly in mixed-species parties in Zambia and Zimbabwe. ♂♂ defended territories during spring stopover in Egypt (Simmons 1954). Perches in upper and lower canopy, usually in enclosed, shaded part of tree (D. R. Aspinwall, pers. comm.). Darts out to take flying insects, returning to different perch; also picks prey from leaves and twigs. In Zambia, feeds on ground, flying to small shrubs or termite mound thickets when disturbed (D. R. Aspinwall, pers. comm.). Carriage and movements like Pied Flycatcher. When excited, gives 'seep' call, and flicks tail and wings.

Migration routes and wintering range rather poorly known, due to problems of separation from other *Ficedula* species. Migrates further in Africa than Pied and Semicollared Flycatchers, all probably wintering south of the equator. Heads south to south southeast from Europe, and lack of autumn records in N Africa and Middle East suggests non-stop flights across Mediterranean and Sahara. Passes in small numbers through Egypt late Aug–late Sept, and more commonly Zaïre mid-Sept–mid-Oct (earliest 7 Sept). Reaches Zambia mainly from late Oct (earliest 5 Oct), with small passage Nov; Malaŵi and Zimbabwe from Nov. Leaves wintering grounds early. Latest record Malaŵi 5 Mar, Zimbabwe 16 Mar; few Zambia after early Mar (last 29 Mar), but recorded in W Zaïre to Apr. Few reports of spring migration in tropics, but a scattering of records from Saharan oases and marked passage N Africa late Mar–mid-May.

Birds ringed Sweden recovered W Zaïre (Kwango, Apr) and NE Angola (mid-Oct); 1 ringed Hungary also recovered W Zaïre (Apr).

Food. Flying insects and other arthropods. Birds collected Libya and Tunisia in Apr had eaten ants and beetles.

Reference
Cramp, S. and Perrins, C. M. (1993).

Ficedula semitorquata (Von Homeyer). Semicollared Flycatcher. Gobemouche à demi-collier.

Plate 27 (Opp. p. 496)

Muscicapa semitorquata Von Homeyer, 1885. Zeitschr. gesamm. Ornith., 2, p. 185, pl. 10; Caucasus.

Ficedula semitorquata

Range and Status. Balkans, Turkey, Caucasus and N and W Iran; winters Africa.

Palearctic migrant. Winters in central E Africa, but range poorly known: E Zaïre (Kivu), Rwanda, Burundi (probably this species), W Tanzania (Tabora, Iringa, Njombe), SW Uganda (Toro) and W Kenya (Kakamega, Mara); uncommon to locally frequent. On passage, rare to uncommon Sudan, Ethiopia, NW Somalia (2 records, autumn), Uganda, and W and central Kenya; frequent in autumn in Eritrea, and frequent to common Rwanda, Burundi, E Zaïre (W shore of L. Albert and Kivu) and NW Tanzania (Kibondo). Rare autumn migrant Egypt, but frequent in spring. Reports of breeding in Algeria and Morocco are probably due to confusion with white-necked variants of *F. hypoleuca speculigera*; rare spring migrant Tunisia.

Description. ADULT ♂ (breeding): mantle to top of head, including lores, ear-coverts and around eye, glossy black; patch on forehead white; sides of neck behind ear-coverts white, forming incomplete collar; back and rump grey, feathers with whitish bases; uppertail-coverts black. Tail-feathers mainly black; T6 usually with wholly white outer web, black inner web with white base and margin; T5 with outer web white with black tip, inner web all black or with white base; T4 black with white subdistal patch on outer web. Underparts white. Primaries and secondaries blackish brown, base of P1–8 with white extending 4–7 mm beyond primary coverts (thus intermediate between *F. hypoleuca* and *F. albicollis*); base of secondaries white, extending beyond greater coverts. Inner tertial black, outer web with broad white margin; middle and outer tertials white except for black near tip of inner web. New inner greater coverts black with broad white tips, old outer greater coverts dark brown with narrow creamy tips; new inner median coverts black with broad white tips, old outer feathers brown with creamy tips; lesser coverts black. White parts of tertials, greater coverts and secondary bases thereby form large white patch on closed wing, while white primary bases form small patch; tips of median coverts typically form an additional narrow white wing-bar, but this may be absent in birds which retain old outer feathers. Primary coverts and alula blackish brown. Underwing-coverts grey-brown; axillaries whitish with black bases. Wing rather long, tip bluntly pointed; P8 longest; P7 0–1·5 shorter, P6 4–6, P5 9–13, P1 22–27; P9 2–3, usually longer than P6 (cf. *F. hypoleuca*); P10 minute; P6–P8 emarginated. Bill black; eyes dark brown or hazel; legs and feet black. ADULT ♂ (non-breeding): upperparts grey-brown, paler on rump; forehead almost uniform with crown; uppertail-coverts black. Underparts tinged creamy buff. Tertials dark brown with narrow whitish outer margins extending around tip, middle and outer feathers with white base to outer web. Inner greater coverts dark brown, outer feathers black, all narrowly tipped white; inner median coverts dark brown, tipped creamy, outers black, narrowly tipped white; inner lesser coverts brown, outer feathers black. Primary coverts and alula black, and with outer upperwing-coverts form black patch on closed wing. Tertial patch and wing-bars less bold than in adult ♂ breeding. Otherwise like breeding adult ♂. ADULT ♀ (breeding): upperparts greyish brown, forehead slightly creamy; rump usually paler, feathers with whitish bases. Tail-feathers dark brown, T5–T6 usually with white middle part to outer web, sometimes reaching tip on T6; T4 sometimes with whitish outer mark. Underparts whitish, tinged pale brown on breast and flanks. Primaries and secondaries dark brown, base of outer web on P1–6 (sometimes also P7 and P8) white, extending only 1–4 mm beyond greater coverts. Tertials as in non-breeding adult ♂. Greater coverts brown with whitish tips, which together with exposed white secondary bases form narrow wing-bar (narrower than in non-breeding ♂); median coverts brown, usually with creamy tips forming additional wing-bar; lesser coverts brown. Primary coverts and alula dark brown. Underwing and bare parts as adult ♂. ADULT ♀ (non-breeding): as breeding adult ♀, but forehead browner, underparts more buffy brown. SIZE: (9 ♂♂, 5 ♀♀) wing, ♂ 79–85 (82·4), ♀ 79–83 (80·6); tail, ♂ 49–51 (50·0), ♀ 49–50 (49·4); bill, ♂ 11–12 (11·9), ♀ 12–13 (12·2); tarsus, ♂ 16–17 (16·9), ♀ 17–18 (17·2). WEIGHT: ♂♀ (n = 7, Turkey and Iran, Mar–Apr) 13–15 (14·1); (n = 7, unsexed, NE Sudan, Aug) 8–12 (9·6).

Complete moult in breeding area; partial moult in Africa Jan–Feb.

IMMATURE: first winter ♂ and ♀ like adult ♀ but browner; tertial fringes differ as in first winter *F. hypoleuca* (q.v.) and *F. albicollis*. First summer ♂ like breeding adult ♂, but upperparts often less black; white patch at primary bases very small, as in ♀.

Field Characters. Breeding ♂ differs from Collared Flycatcher in having partial collar, extending to sides of neck only, and grey (not white) rump. Blacker above than breeding ♂ European Pied Flycatcher with more extensive white in the wing, and usually distinguished from both Collared and Pied by additional narrow white bar along median coverts. White forehead patch conspicuous (though smaller than in Collared). White sides to tail more extensive than in either Collared or typical Pied. Non-breeding ♂ greyish above, with indications of partial collar and pale buffish median covert bar usually visible. ♀ and first winter very difficult to separate in field from Pied and Collared Flycatchers. Usually show pale median covert bar, but this is occasionally indicated in Collared. Wing markings bolder than in Pied with more conspicuous small whitish patch at base of primaries. Adult ♀ (but not first winter) greyer above than in Pied, often with paler rump. In the hand, longer P2 and white at base of P6 (often also P7) are good distinctions from ♀ and first winter Pied. Separation from Collared depends on pattern on hindneck feathers (see *F. albicollis* **A**).

Voice. Tape-recorded (62, 73, CHA). Song, a short phrase of 2–3 s, comprising a few thin, high-pitched notes like those of Collared Flycatcher, but tempo faster, more like that of Pied. Contact-alarm calls, a dry, hard 'tec', and a clear 'sjieb', rather sharper than similar call of Collared.

General Habits. In Africa, frequents forest edge, gallery forest, light woodland, gardens and more open country with leafy trees; occurs in acacias on passage; winters mainly in miombo woodland in W Tanzania. Usually in ones or twos, often with mixed parties. Behaviour like Collared Flycatcher. Feeds from tree canopy or sometimes from bushes, darting out frequently to take flying insects. Feeds less from ground than Pied Flycatcher (Cramp and Perrins 1993). Flicks wings and tail, and often gives 'sjieb' call.

Due to identification difficulties, and common treatment of Semicollared as race of Collared, migrations and wintering range are poorly known. In autumn, scarcity in E Mediterranean and Middle East, including Egypt, suggests direct flights to tropics. Recorded sparsely in NE Africa early Aug–mid-Oct, from Sudan, Chad, Ethiopia, Eritrea (where sometimes more common) and NW Somalia. Arrives early in E and central Africa, with thin passage Uganda and W Kenya mid-Sept–mid-Nov, and more pronounced passage Rwanda, Burundi and NW Tanzania mid-Sept–early Nov. The few definite records Dec–early Feb are from E Zaïre (Kivu), Rwanda, SW Uganda, W Kenya and W Tanzania. Begins to move north early, and scattered late Feb–Mar records in NW Tanzania, Rwanda, S and W Uganda, W Kenya and S Sudan probably refer mostly to passage birds. Recorded in early Apr in Ethiopia, and up to mid-Apr in Kenya and Sudan. Regular and fairly common on spring passage in Egypt, mid-Mar–mid-May.

Food. Mainly flying insects. No information for Africa.

References
Cramp, S. and Perrins, C. M. (1993).
Curio, E. (1959b).
Pearson, D. J. and Turner, D. A. (1986).
Tyler, S. (1987b).

Plate 27
(Opp. p. 496)

Ficedula parva (Bechstein). Red-breasted Flycatcher. Gobemouche nain.

Muscicapa parva Bechstein, 1794. Latham's Allg. Uebersicht der Vögel, 2, p. 356; Thüringerwald.

Range and Status. Central Europe, S and E Baltic, and east through Russia to Anadyr, Kamchatka and Mongolia; also Caucasus to N Iran; winters S and SE Asia; regular migrant NE Africa and autumn vagrant W Europe.

Palearctic visitor. Scarce passage migrant N Libya and N Egypt, south in Sahara to Serir and Bahariya; regular and locally frequent in autumn, more sporadic in spring. Rare Tunisia (4 records autumn, once Jan), Algeria (Sahara: 4 trapped Oct, Hassi Marraket, once Nov, Djanet) and Morocco (7 records Oct–Nov, once Dec). Rare or vagrant Senegal (twice Nov, Djoudj) and N Sudan (once Nov, Red Sea Hills).

Description. *F. p. parva* (Bechstein): Europe east to W Siberia and Caucasus to N Iran; only race in Africa. ADULT ♂ (breeding): upperparts brown, tinged grey on forehead and crown; uppertail-coverts blackish brown tipped grey. Tail-feathers blackish brown, T3 with middle third of outer web white, often extending onto middle of inner web, T4–T6 with basal two-thirds of both webs white. Lores, ear-coverts and sides of neck ashy grey, sides of breast bordering bend of wing purer grey; narrow eye-ring off-white. Chin, throat and centre of breast orange-red; rest of underparts white, tinged creamy buff on flanks and sides of lower breast. Flight feathers and upperwing-coverts brown; underwing-coverts and axillaries creamy buff. Wing-tip bluntly pointed: P7 and P8 longest; P6 1–2 shorter, P5 5–7, P1 15–19; P9 5–8; P10 minute; P6–P8 emarginated. Bill horn, lower mandible yellowish; eyes brown; legs and feet dark brown. ADULT ♂ (non-breeding): like breeding adult ♂, but orange-red on chin and throat less extensive, flanks more buffish. ADULT ♀: upperparts brown, blackish on uppertail-coverts. Lores pale brown; ear-coverts brown; narrow eye-ring buffish. Chin, throat, sides of breast and flanks creamy buff, rest of underparts white. Wings, tail and bare parts as adult ♂. SIZE: (10 ♂♂, 10 ♀♀) wing, ♂ 68–72 (69·8), ♀ 64–70 (66·4); tail, ♂ 47–52 (49·3), ♀ 47–50

Ficedula parva

(48·8); bill, ♂ 11–13 (12·1), ♀ 11–13 (12·0); tarsus, ♂ 16–18 (17·1), ♀ 16–18 (17·3). WEIGHT: ♀ (n = 1, Sudan coast, Nov) 7·5; unsexed: (n = 4, Algerian Sahara, Oct) 7–9 (8·1); (n = 4, Sinai, autumn) av. 8·5.

Complete moult in breeding area; partial moult in winter quarters.

IMMATURE: first winter ♂ and ♀ like adult ♀, but inner secondaries narrowly edged, and tertials, greater coverts and outer median coverts edged and tipped rufous-buff.

Field Characters. A small, compact flycatcher with a small bill; rather plain brown above and whitish below. Tail pattern distinctive, with dark centre and terminal band and broad white patches on sides. ♂ has striking orange-red throat and breast centre bordered by greyish face and greyish sides to neck and breast. ♀ and first winter have creamy or buffish throat and breast and lack grey tinge on face and neck. Distinguished from other *Ficedula* flycatchers in Africa by lack of white bars or patches on wing, although in first winter bird may show narrow buffish wing-bar and buff tertial edges; also smaller with proportionately longer legs and tail. Call distinctive.

Voice. Tape-recorded (62, 73, 93, B, CHA). Call a short, high-pitched 'dzik', and sometimes a loose ticking series, 'tk tk . . . tk tk tk'. In alarm, a rattling 'zrrrt' and a plaintive 'hveet'.

General Habits. On migration occurs in woodland, plantations and gardens, often in large trees, but also in taller bushes and thickets in drier country, and in Saharan oases. Usually solitary. Not shy, but unobtrusive, often skulking in canopy or within bushes. Feeds mainly from lower twigs and leaves, hopping and creeping among foliage and sometimes hovering like a *Phylloscopus* warbler. Makes short looping sallies to take flying prey, and frequently feeds from ground (Cramp and Perrins 1993). Flight agile, with rapid wing-beats; usually darts back quickly into cover when disturbed. Frequently flicks wings, and cocks and flirts tail to display distinctive pattern (**A**).

A

Most European birds migrate southeast to winter in India, but a few, presumably from western populations, pass regularly across the E Mediterranean and NE Africa, early Sept–mid-Nov. Also occurs each autumn west and southwest of the breeding grounds, with Oct–Jan records south to Morocco, Algeria, Tunisia and N Senegal. A few may winter regularly south of Sahara. Occasional spring birds E Libya and Egypt mid-Apr–early May.

Food. Mainly insects and other invertebrates.

Reference
Cramp, S. and Perrins, C. M. (1993).

Family MONARCHIDAE: paradise-flycatchers and monarchs

Small to medium-sized flycatchers, sexually mono- or dimorphic. Typical members of this family display an amphirhinal condition of ossification of the nostril and a corvine humerus with single pneumatic tricipital fossa; *Erythrocercus* and *Elminia* lack this pattern of nasal ossification and have non-pneumatic humeri with a double fossa (Olson 1989). Differ from true flycatchers (Muscicapidae) in jaw musculature (Beecher 1953) and lack 'turdine thumb' condition of the syrinx of Muscicapidae (Ames 1975). At least *Terpsiphone* shows 'lanioid' carpometacarpal process (Pocock 1966). Bill dorsoventrally flat, relatively enlarged at base, culmen with well-marked ridge, upper mandible often notched or hooked at tip; well developed rictal bristles; nostrils often overlain with short feathers. Gape and inside mouth often bright yellow or greenish yellow, conspicuous when bird calls. Some species have narrow circumorbital skin wattle. Tarsometatarsus short, claws small but strongly curved and sharp. Tail-feathers graduated, in some species (e.g. *Terpsiphone*) middle pair particularly elongated. Often crested. Plumage soft and full; coloration often bright with various patterns of blue, orange and/or rufous associated with grey, black and/or white. Juveniles with neossoptiles but with plumage unspotted and without tongue spots. Nest usually a neat, tight and relatively small cup, with fibres, moss, lichens and spider webs (though *Erythrocercus* makes deep ovoid pouches). Usually vocally and visually obtrusive. Social organization in pairs or family groups, but some live in more organized groups. Woodlands and forested areas.

2 subfamilies: Monarchinae (16 genera, 87 species) and Rhipidurinae (1 genus, 37 species) plus 3 genera (5 species) *incertae sedis*. Widespread in Africa and neighbouring islands, India and Asia to Japan, south to Tasmania, New Zealand and Chatham I., east to Society, Marquesas and Hawaiian Is (Watson *et al.* 1986). 4 genera (16 species) in Africa: *Erythrocercus*, *Elminia*, *Trochocercus* and *Terpsiphone*, all endemic except the last. Formerly regarded as true flycatchers and placed in the Muscicapidae, usually in a subfamily either with the present Platysteiridae (Mayr and Greenway 1956) or alone (Traylor 1970b). Grouped with Platysteirinae, Sylviinae, Timaliinae, Rhipidurinae, Pachycephalinae and Orthonychinae in the family Sylviidae by Cracraft (1981); in his phylogenetic classification true flycatchers and thrushes constitute the family Muscicapidae and jointly form the infraorder Muscicapi, and are thus separated from Lanii. Ranked at the family level by Watson *et al.* (1986). DNA comparisons (Sibley and Ahlquist 1990) suggest placing monarchs and Australian magpie-larks (*Grallina*) together in the tribe Monarchini, which with Dicrurini (drongos) and Rhipidurini (fantails) form the subfamily Dicrurinae of Corvidae. Anatomical characters (see above, Olson 1989) show *Elminia* and *Erythrocercus* to be rather different from the others and relationships, especially with oriental *Rhipidura*, need further study. Note also that the correct name of the family should be Myiagridae (Boles 1981, Olson 1989) but we follow Mayr's more conservative argument (footnote in Watson *et al.* 1986: 464).

Genus *Erythrocercus* Hartlaub

Small; body slim; bill quite short and broad; well developed, long, strong rictal bristles; tarsometatarsus short though relatively longer than in other monarchine genera; claws short, strongly curved and sharp. Plumage soft and full. General coloration grey, olive, rufous, buff or yellow. No clear sexual dimorphism. Wings rounded, short erectile crest; fan-shaped tail relatively shorter than in other monarchine genera and with median feathers not particularly elongated; flicked and spread continuously. Nest ball-shaped with side entrance, unusual for a monarchine. Considered by Hall and Moreau (1970) to be similar in many features of structure and habits to *Erannornis* (= *Elminia*) and *Trochocercus*, but resemblances are superficial. Has also been judged to be close to Oriental *Culicicapa* (Traylor 1970b). As in *Elminia*, lack of turdine 'thumb' on the syrinx, of corvine configuration of humerus and of monarchine nostril ossification (Ames 1975, Olson 1989) raises questions about the real affinities of this genus.

Endemic; 3 species in a superspecies.

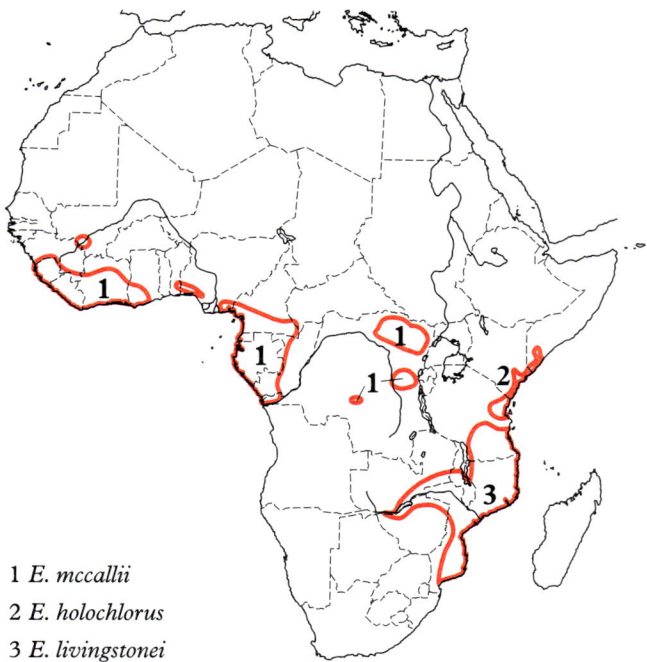

Erythrocercus mccallii superspecies

1 *E. mccallii*
2 *E. holochlorus*
3 *E. livingstonei*

Erythrocercus mccallii (Cassin). Chestnut-capped Flycatcher. Érythrocerque à tête rousse.

Plate 28
(Opp. p. 497)

Pycnosphrys mccallii Cassin, 1855. Proc. Acad. Nat. Sci. Philadelphia, 7, p. 326; Moonda R., Gabon.

Forms a superspecies with *E. holochlorus* and *E. livingstonei*.

Range and Status. Endemic resident. Lowland forest of Sierra Leone, Guinea (Macenta district, common to abundant), SW Mali (eastern Mts Mandingues, Kourémalé, Kangaba, uncommon), Liberia, Ivory Coast (north to Mt Nimba, Korhogo and Comoé, common), Ghana (north to Bia Nat. Park, Goaso, Mampong, Kwahu, Bogoro, not uncommon), Benin (Pobé), S Nigeria (not uncommon, Lagos to Sapele, north to Ibadan and Ife, also recorded Umuagwu), Cameroon (Korup Nat. Park, Mt Cameroon, Rumpi Hills and Mt Kupé to Yaoundé, Dja R. and Messea, southward), Central African Republic (Dzanga reserves, also reported from Lobaye), Rio Muni, Gabon (Libreville, Woleu N'Tem and Ogooué-Ivindo south to Franceville and Mouila, common to abundant), Congo (Loango coast and Mayombe, probably also northwestern corner), Angola (Cabinda), Zaïre (lower Congo R., then Uele to Ituri and Kivu, Kasongo, Manyema and Luebo, Kasai) and W Uganda (Budongo Forest). Uncommon to abundant. Density in NE Gabon, av. 4·7–6·1 groups/km^2 or a mean of 26·4–29·6 individuals/km^2 (Erard 1987).

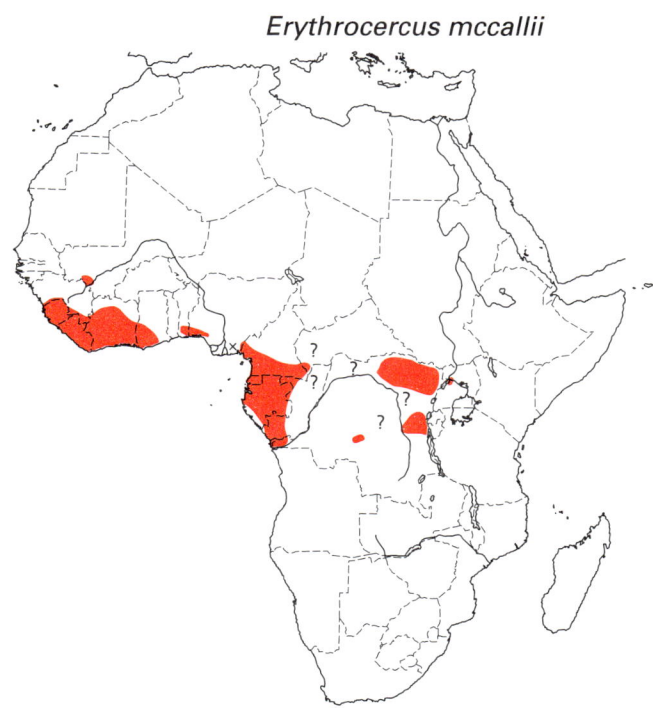

Erythrocercus mccallii

Description. *E. m. mccallii* (Cassin): SE Nigeria (Umuagwu) and Cameroon to Zaïre (Mayombe). ADULT ♂: forehead and forecrown (including eye-ring and forepart of ear-coverts) dark cinnamon with narrow buff streaks (each feather cinnamon with buff shaft); hindcrown, hindpart of ear-coverts, sides of neck, hindneck, scapulars, mantle, back and rump mouse-grey, the latter washed cinnamon; uppertail-coverts and tail dark cinnamon; lores, chin, throat, foreneck, upper breast and undertail-coverts buff, rest of underparts creamy white, flanks washed pale grey; primaries and secondaries dark brown, narrowly edged greyish buff along outer web; tertials dark brown with rusty ochre edges, broader along inner web; upperwing-coverts dark brown narrowly edged rusty buff; axillaries creamy white, underwing-coverts and edges of inner webs of flight feathers buff. Upper mandible dark brown edged horn, lower horn; eyes hazel- or red-brown or even almost red; legs brown. Sexes alike. SIZE: (5 ♂♂, 6 ♀♀) wing, ♂ 47–49 (48·3), ♀ 47–49 (47·9); tail, ♂ 42–47 (44·5), ♀ 43–47 (44·4); bill, ♂ 10–12 (10·8), ♀ 10–11 (10·5); tarsus, ♂ 15–16 (15·4), ♀ 15–16 (15·4). WEIGHT: (Cameroon) unsexed (n = 6) 7–8 (7·2).

IMMATURE: like adult but upperparts and edges of upperwing-coverts and secondaries more olive-rufous.

JUVENILE: unspotted; like immature but forehead and crown olive-grey tinged brownish, without cinnamon; throat very pale brownish.

NESTLING: inside of mouth and tongue bright orange, margin of gape yellowish white.

E. m. nigeriae Bannerman: Sierra Leone to SW Nigeria (east to Sapele). More olive-brown above, throat paler buffy. WEIGHT: (Liberia) ♂ (n = 6) 7·3 ± 1, 1 ♀ 7·1.

E. m. congicus Ogilvie-Grant: E and S Zaïre (Uele to Kivu and Kasai) and Uganda. More olive-brown above, crown and throat darker. WEIGHT: (Uganda, May) ♂ (n = 6) 6–7·5 (7·0), ♀ (n = 3) 7–8 (7·4).

Field Characters. Tiny, active, noisy, gregarious little bird, constantly on the move. Bright cinnamon or chestnut cap and tail contrast with mouse-grey upperparts, buff or fawn face and breast, and white belly; tail often flashes because of sunlight shining through feathers. 'Swinging' foraging behaviour with tail widely fanned and wings drooped and half-spread also characteristic. Cannot be confused with any other forest species.

Voice. Tape-recorded (C, CHA, ERA, KEI). Territorial song, a rapid, rather high-pitched warble or twitter, beginning with some short individual notes and running into a trill, 'pit . . . pit . . . pit-pit-pit . . . pit-ptit-ptilililili' or 'ptit . . . ptit-ptit-ptit . . . ptilulululu'. In territorial defence gives long series of songs without pause 'ptit-ptit-ptililili-ptit-tilililili-ptit-pit-tililili . . .'. Foraging flock gives continuous series of little chips and squeaky notes; at frequent but irregular intervals chips break into brief (1–1·5 s), tuneless, sibilant little song, which makes it difficult to separate calls from songs and understand their significance.

General Habits. Inhabits lowland forest, both primary and secondary. Favours forests with closed canopy and areas rich in small-leaved trees such as *Penthacletra eetveldeana* (Mimosaceae) and particularly *Scorodophloeus zenkeri* (Caesalpiniaceae). Occurs also in flooded forests and forested islands of large rivers, in old uncleared or disused plantations on edges of forest where vegetation layers blur, and in riverine forest through cultivated land; even enters small forest remnants around villages. Up to 1250 m in Cameroon and 1450 m in Itombwe, Zaïre. In Budongo forest, Uganda, frequents undergrowth, mid-stratum and canopy of forest.

Usually in small groups of 3–12 birds (mean 5·1 in NE Gabon), but may be seen in pairs or even alone. Larger groups sometimes, particularly during rains, subdivide in subgroups, pairs or trios of adults with young. Each group or subgroup led by an adult which is much more active than the others; it calls and sings frequently, keeps crest almost always erect, moves jerkily from side to side, flaps wing-tips and swings raised and fanned tail. Groups largest during dry season. Gleans foliage in crown of trees, also leafy parts of dense liana tangles; particularly attracted by small-leaved trees. In NE Gabon, forages at all levels of vegetation from *c*. 2 m to upper canopy, though seasonal variation important: more in canopy during long dry season (June–Aug), with 76% of records above 20 m including 29% above 30 m; during the following rainy season (Sept–Nov) is localized more in the understory with 42% of records under 20 m, including 11% below 10 m; during short dry season (Dec–Jan), mostly in upper layers of understorey and canopy, with 82% of records 10–30 m including 53% 20–30 m; during following rains (Feb–May) occurs in intermediate levels with 85% of records 10–30 m including only 7·5% both above 30 m and under 10 m (Erard 1987). Always rapidly on the move; makes jerky or even acrobatic short flights or restless short forward jumps, pivoting from side to side in horizontal posture among and along branches; flicks wings and makes wide sweeps with fanned tail, giving contact calls or brief songs. This dislodges insects which it snaps up with a quick jump or more often a short sally or fluttering spiral flight. Does not hunt like a true flycatcher, even though it regularly feeds on swarming ants and termites. Usually progresses upward to crown of tree from lower branches; moves outward along branch, then in fluttering flight makes a wide loop to reach basal part of higher branch which it again works outward, repeating the process until it reaches the crown, then in fluttering flight swoops down to next tree. Regularly joins or even 'catalyses' mixed-species flocks, where it is one of the core species, at least in NE Gabon: 98·2% of records were in mixed flocks, without any marked seasonal variation. Often associates with Fraser's Sunbird *Anthreptes fraseri*.

Group members react collectively to danger; express anxiety with rapid clicking calls, crouch and nervously pivot from side to side, rapidly open and close tail. When alarmed or when mobbing, they intensify calling, erect crest, flick wings and make broad sweeping movements with fanned tail.

Sedentary, at least in NE Gabon.

Food. Small arthropods, mainly insects: grasshoppers, cicadas, beetles, small bees, ants, alates of termites, caterpillars, moths; also spiders. Prey size 5–30 mm, mainly 5–15.

Breeding Habits. Monogamous or polygamous; communal breeder; territorial. Territory defended by group members all year. Group patrols entire territory daily. Territory size 10–19 ha (av. 13·8, n = 9) in NE Gabon, size correlated to that of group (Erard 1987). Leader sings actively, particularly when another mixed-species flock approaches or when near territory boundary. During territorial contests, several birds of the group may participate against members of the other group. They move about rapidly with erect crest, inflated breast, flicked-out carpal joints of folded and drooped wings, and raised and fully fanned tail; make swooping display flights and flutter about singing vehemently. Confrontations are rather rare because groups indicate their presence by calls and songs and so avoid coming into physical contact, each keeping on its own territory.

NEST: 2 observed in NE Gabon correspond to Bates's (1911) description of a nest in Cameroon: much like a leafy cisticola or wren nest, i.e. a deep ovoid pouch with subapical side entrance, made of fibres, pieces of dry grass and green leaves, small pieces of bark and small twigs and vegetable stems, bound together with spider web and densely covered outside with small green leaves (e.g. *Scorodophloeus zenkeri*) held on with spider webs, inside lined with thin fibres and *Marasmius*, also vegetable down; outer diam. 75, height 145; inner diam. 44, depth 33; diam. of entrance hole 43. Placed 8–13 m above ground toward end of leafy branch, tightly fastened between 2 small twigs with spider web; one of Gabonese nests placed against stem of fallen, dry, almost skeleton-like leaf of *Musanga* tree hanging near end of leafy branch of small tree. 1 nest was built by all 4 members of social group.

EGGS: not described. Clutch-size unknown; 1 nest in Cameroon contained 2 young and 1 in NE Gabon 3 young.

LAYING DATES: Ivory Coast (adults with young Feb, May); Cameroon, Feb (♂♂ with enlarged testes May, June, Sept, Dec, ♀ with small ova in ovary July); Gabon, Feb, Mar; Zaïre (♂♀ with gonads enlarged Jan–Apr, July, Aug).

INCUBATION: by ♀. Period unknown. Brief observations at one nest indicate bird sits for long periods (60–120 min), is fed at nest by group members, or leaves nest for 40–70 min when group comes to fetch it, with leader singing.

DEVELOPMENT AND CARE OF YOUNG: nestling period about 12 days. In NE Gabon, at one nest 3 young fed communally by at least 6 group members; one bird sat in nest for periods of 40–60 min, with absences of 25–30 min for first few days; frequency of food delivery increased from av. 1 every 10–11 min on days 3–4 to 1 every 5–6 min on days 7–8 at nest containing 3 young. Of the 2 Gabonese nests, one fledged young, the other was partly destroyed by nocturnal predator during incubation.

References
Brosset, A. and Erard, C. (1986).
Erard, C. (1987).

Erythrocercus holochlorus Erlanger. Little Yellow Flycatcher. Érythrocerque jaune.

Plate 28 (Opp. p. 497)

Erythrocercus holochlorus Erlanger, 1901. Orn. Monatsber. 9, p. 181; Salole, Juba River, Italian Somaliland.

Forms a superspecies with *E. livingstonei* and *E. mccallii*.

Range and Status. Endemic resident. Somalia in forests along Juba R. and south along coast to Kenya, uncommon; Kenya in coastal lowlands and inland along lower Tana R. and to Shimba Hills, common; coastal Tanzania south to Dar-es-Salaam region and Kazimzumbwe Forest Reserve, inland to foothills of E and W Usambaras, Ngurus and Ulugurus, common. Status in S Tanzania unclear; recorded once in Rondo Forest (Holsten *et al.* 1991).

Description. ADULT ♂: top of head, ear-coverts and upperparts olive-yellow, yellower on rump and uppertail-coverts; short crest. Tail-feathers brown edged olive-yellow, subterminal dark smudge variably present. Indistinct pale loral line variably present; eye-ring yellowish white; entire underparts golden yellow, slight olive wash on sides of breast and flanks. Primaries and secondaries dark brown edged olive-yellow, upperwing-coverts brown broadly edged and tipped olive-yellow, underside of inner webs of primaries and secondaries margined yellowish white, axillaries, underwing-coverts and bend of wing yellow. Upper mandible dusky, lower mandible pale horn, rictal bristles black; eyes brown to pale brown; legs and feet pale brown, sometimes grey-brown or flesh. Sexes alike. SIZE: (6 ♂♂, 6 ♀♀) wing, ♂ 46–49 (47·0), ♀ 42–47 (44·5); tail, ♂ 40–44 (41·8), ♀ 38–40 (38·8); bill, ♂ 9–11 (9·9), ♀ 9–10 (9·7); tarsus, ♂ 15–16 (15·7), ♀ 15–16 (15·6). WEIGHT: ♂ (n = 7) 5–6·7 (5·8), ♀ (n = 5) 5–7 (6·2).

IMMATURE: slightly paler and yellower above, with more fluffy appearance.

NESTLING: olive-green above, lighter greenish below, bill and eye-ring yellow, gape pink.

Field Characters. A tiny flycatcher of E African coastal lowlands, with green upperparts and tail and bright yellow underparts. Restless feeding actions recall Livingstone's Flycatcher *E. livingstonei*, which it just overlaps in S Tanzania. Livingstone's also has yellow underparts and green back, but has rufous tail with black subterminal band, and weaker, less tuneful song.

Erythrocercus holochlorus

Voice. Tape-recorded (B, GREG, HOR, KEI). Song a brief (1·5–2 s), rather hurried, high-pitched sweet warble, quite penetrating at close range, introduced by some little chips and chattery notes; these chip notes are given irregularly but frequently by foraging birds, and at intervals lead into the song.

General Habits. Inhabits primary and secondary forest, moist thickets and riverine forest, sea level to 500 m. In Sokoke Forest (Kenya), common in all 4 habitat types (Britton and Zimmerman 1979).

Usually travels in small parties, moving in dainty and lively manner. Gleans insects from leaves like *Phylloscopus* warbler, only occasionally taking them in the air.

Food. Tiny spiders and insects, including Lepidoptera larvae.

Breeding Habits.
NEST: hanging ball type with side entrance; only 2 described (Moyer *et al.* 1992); (1) made of dead leaves of *Drypetes reticulata* and at least 2 other tree species, bound together and secured to twigs of supporting tree with spider webs; webs not bound around outside of leaf blades, thereby preserving camouflage; surrounding green foliage incorporated into outer layer; interior cup made of spider webs and plant down; ext. diam. 55, int. diam. (cup) 40, ext. depth 105; superficially similar to nests of arboreal forest ants *Oecophylla*; on one side near entrance, 2 leaves had daubs of mud in hexagonal pattern, most probably formed by mud-dauber wasps *Synagris*; situated 4·5 m above ground in dense foliage near top of understorey tree in old tree-fall gap; (2) untidy ball of green leaves bound together from within with spider web; leaf edges on outside of nest not bound together, making nest difficult to distinguish from surrounding foliage; living foliage from surrounding leaf cluster appeared to be incorporated into nest walls; situated 6 m above ground and 2 m below canopy in tree at edge of gap in forest; built by 2 birds from group of 3, apparently all adults, not certainly always by same 2 birds; bird carrying nest material (probably spider webs) remained in nest 30 s to 2 min, while second bird remained close, giving constant excited twittering call, moving actively about in foliage and probing around outside of nest; in between periods of building activity they joined the third bird to forage; third bird remained nearby, often close to nest.
EGGS: not described.
LAYING DATES: Kenya, Nov; Tanzania (nest-building Dec).
INCUBATION: unknown.
DEVELOPMENT AND CARE OF YOUNG: nestlings fed by one parent (♀?) while the other (♂?) sings repeatedly nearby.

Reference
Moyer, D. C. *et al.* (1992).

Plate 28
(Opp. p. 497)

***Erythrocercus livingstonei* Gray. Livingstone's Flycatcher. Érythrocerque de Livingstone.**

Erythrocercus Livingstonei Gray, 1870. *In* Finsch and Hartlaub, Vögel Ost-Afrikas (Decken, Reisen Ost-Afrika, 4), p. 303; Zambezi (restricted to Zumbo, Zambia–Mozambique border, by Irwin 1957, Bull. Br. Orn. Club 77, 119).

Forms a superspecies with *E. holochlorus* and *E. mccallii*.

Range and Status. Endemic resident. S Tanzania north to W Songea, Udzungwa Mts, Luwegu R., Liwale and Kingupira; Zambia in Zambezi valley and tributaries below 800 m, west to about Victoria Falls, and Luangwa valley and tributaries south of *c.* 14°S; common S Malaŵi north to Benga, below 600 m in thicket canopy, up to 760 m in riparian evergreen forest in SW Mangochi. Mozambique, probably almost throughout except montane areas, south to Macia, just south of Limpopo R.; N fringes of Zimbabwe in Zambezi valley and tributaries (Gairezi, Mazoe, Ruenya), west to about Victoria Falls, once on plateau at Nyamasaka (17°25′S, 28°51′E: Tree 1987). Uncommon to locally common.

Erythrocercus livingstonei

Description. *E. l. francisi* Sclater: Malaŵi in lower Shire valley south from Liwonde; Mozambique from below Tete to Netia and Monapo R. and south to Macia. Top of head to hindneck and ear-coverts brownish grey, lores and area around base of upper mandible sometimes paler; short crest; mantle to upper rump and scapulars dull yellowish green, upper rump slightly tinged sulphur-yellow, lower rump and uppertail-coverts tawny. Tail tawny with broad (10–12 mm) black bar across all feathers. Chin and throat white, shading to pale yellow on breast and bright yellow on rest of underparts, flanks and undertail-coverts with slight tawny wash. Wings dark brown, flight feathers narrowly edged greenish yellow, coverts edged and tipped dull yellowish green, underside of inner webs of primaries and secondaries narrowly edged pale brown, axillaries, underwing-coverts and bend of wing pale yellow. Upper mandible dark brown, lower mandible pale brown to whitish; eyes brown or dark brown; legs and feet pinkish, brownish or black. Sexes alike. SIZE: (9 ♂♂, 5 ♀♀) wing, ♂ 48–53 (50·2), ♀ 46–48·5 (47·5); tail, ♂ 45–50 (47·2), ♀ 43–47 (45·4); bill to feathers, ♂ (n = 7) 7–8·5 (8·0), ♀ (n = 4) 8–8·5 (8·1), to skull, 2 ♂♂ 10·5, 11, 1 ♀ 10·5; tarsus, ♂ (n = 7) 14–16 (14·9), ♀ (n = 4) 13–15 (14·0). WEIGHT: 1 ♀ 5·6, 2 unsexed, both 6.

IMMATURE: like adult but yellowish tinge of upperparts extends to back of head, sides of head and immediately behind eye yellowish; tail lacks black band except for a trace of black on outer margins of some feathers; inner edge of inner greater coverts tinged chestnut, not whitish grey as in adult.

NESTLING: unknown.

E. l. livingstonei Gray: Zambia, Zimbabwe, NW Mozambique to about Tete. Overall slightly paler above, back with more yellowish tinge, more pronounced sulphur-yellow wash on upper rump; tail paler tawny, with discrete black spot on each web (not continuous bar as in *francisi*), broadest (but only *c.* 7 mm) on central pair, becoming smaller outwardly, sometimes lacking on outer feathers.

E. l. thomsoni Shelley: Tanzania; Malaŵi north of range of *francisi*, intergrading with it at Liwonde in Lisungwe valley; Mozambique south to Lurio R. Like *francisi* but top of head and ear-coverts greenish yellow, concolorous with back, slightly darker (more olive) on crown, cheeks pale yellow, lores whitish; outer tail feathers (T5, T6) lack black spots, T3 and T4 have separate spots as in *livingstonei*, 2 central pairs have spots run together, as in *francisi*. Below, chin (only) white, rest of underparts from throat to flanks, thighs and undertail-coverts bright yellow, more golden than *francisi*, especially on breast.

Field Characters. Tiny, slim bird of treetops, with greenish upperparts, green or grey top of head, yellow underparts with diagnostic longish rufous tail with black subterminal band. Constantly in motion, darting about in foliage and making quick, fluttering flights, all the while flitting and fanning tail. Yellow-bellied Eremomela *Eremomela icteropygialis* has similar restless manner of feeding but has short grey tail which is not fanned, yellow confined to belly, grey upperparts. Overlaps with closely-related Little Yellow Flycatcher *E. holochlorus* in S Tanzania; for differences see that species.

Voice. Tape-recorded (86, 91, B, PAY). Song, a brief (1–2 s) sweet warble of 4–6 notes, sometimes on its own, more often introduced by a single 'chup' or a little chatter; 2 songs may be run together with a liquid twitter. Call, an *Apalis*-like double note, second note slurred, 'tsi-tsreea', often repeated 2–3 times; in flight a short, sunbird-like 'sweet'; also, sharp 'zert' and 'chip' (alarm?) (Maclean 1993).

General Habits. Inhabits damp woodlands, riparian forest, stream margins in drier country, bushland with thickets and creeper tangles; rich *Brachystegia*, *Uapaca* and mopane woodland; often but not always near water. Frequents taller trees with thickest foliage, usually in canopy. Sea level to 800 m.

Travels in pairs, or small family parties after breeding; in cool weather sometimes joins mixed-species flocks. Forages with rapid movements, like *Trochocercus* and *Elminia*, hopping quickly about, hover-gleaning insects from leaves, making brief flights to catch insects in air with loud snap of bill. Indefatigable; seldom still; constantly fans tail.

Mainly sedentary, although some post-breeding wandering recorded in Zimbabwe.

Food. Insects.

Breeding Habits. Little known.

NEST: ball-shaped with large side entrance (diam. *c.* 50), made of leaves bound together with spider web; deeply lined with soft grass fibres; (n = 1) ext. diam. 60, ext. depth 108, distance from base of nest to lower edge of entrance 48; situated 4–5 m above ground in dense scrub.

EGGS: 2 (3 clutches) or 3 (1 clutch). White, with thick overlay of tiny spots of bright chestnut, red-brown and rufous over lilac, with irregular hair-like scrawls around the top. SIZE: (n = 1) 13·2 × 10·6.

LAYING DATES: Zambia, Dec–Jan; Malaŵi, Jan–Mar (fresh nest Dec); Mozambique, Feb–Mar.

Reference
Vincent, J. (1935).

MONARCHIDAE

Genus *Elminia* Bonaparte

Small fan-tailed flycatchers. Bill dorsoventrally flat with wide base and narrow tip, culmen ridged and hooked at tip; nostrils partly covered with small bristles and short feathers; well developed rictal bristles, stiff and long; tarsometatarsus short and thin; claws short, curved and sharp; long and strongly graduated tail-feathers, with white pattern in 2 species; often crested or at least with erectile crown-feathers. Plumage soft, predominantly slate, grey-blue or blue, with white belly (except *E. nigromitrata*). Tail-fanning behaviour typical, usually with drooped wings and body pivoting from side to side. Juvenile plumage unspotted, though young of *E. longicauda* and *E. albicauda* have narrow pale bars on cap, mantle and upperwing-coverts. Song a melodious warble of short phrases with chattery trills and jingles. Nest a compact, neat cup of tightly woven material, often moss and lichens, bound with cobwebs. Arboreal; forest.

Endemic, 5 species. 3 were formerly placed in the genus *Trochocercus* but removed on the basis of habits, shape and proportions (Dowsett and Sternstedt 1973, Dowsett and Dowsett-Lemaire 1980) and also skull anatomy (Erard 1987). Lack of 'turdine thumb' in syrinx, of monarchine nasal ossification and of humerus condition of typical corvine configuration, raises questions about the true affinities of this genus (Olson 1989). External morphology and habits recall Oriental *Rhipidura* (see also Mayr and Moynihan 1946, Dowsett and Dowsett-Lemaire 1980). Sometimes *E. nigromitrata*/*E. albiventris*/*E. albonotata* considered as constituting a superspecies (Watson *et al.* 1986) but allopatry of *albiventris* is altitudinal not geographical.

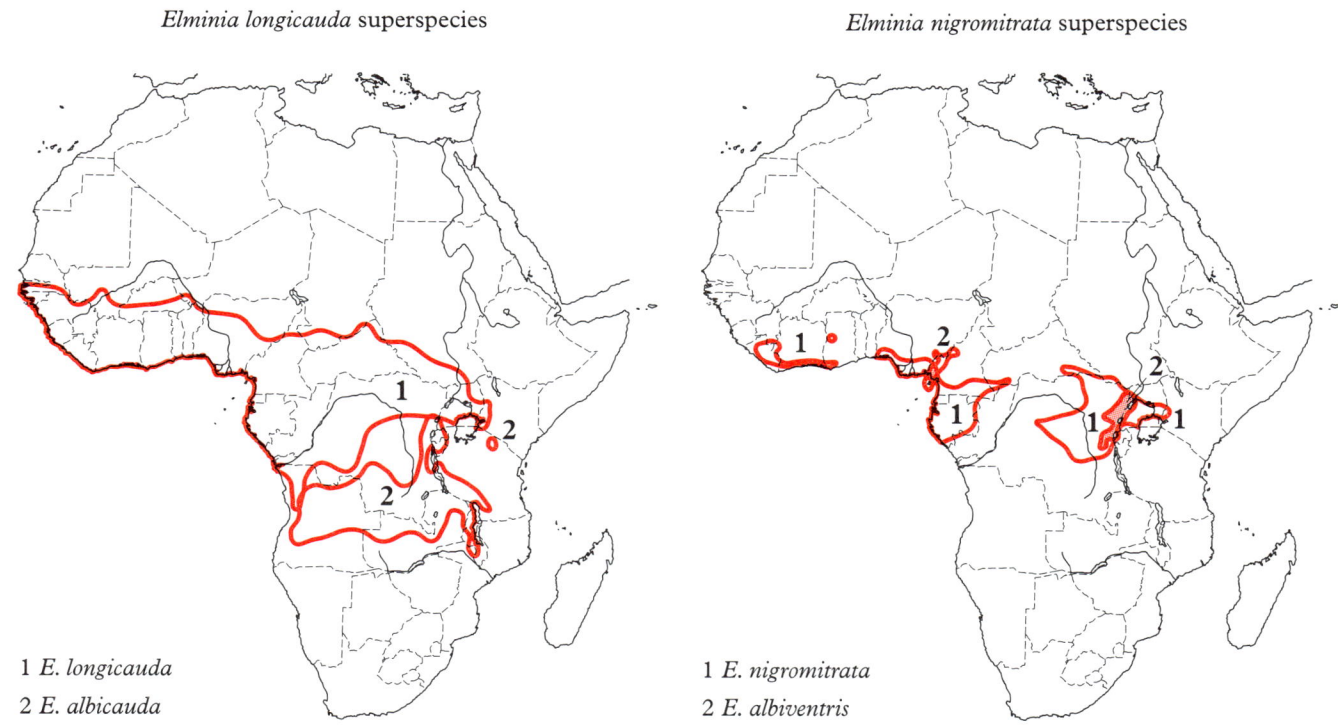

Elminia longicauda superspecies

1 *E. longicauda*
2 *E. albicauda*

Elminia nigromitrata superspecies

1 *E. nigromitrata*
2 *E. albiventris*

Plate 28 *Elminia longicauda* (Swainson). African Blue Flycatcher. Tchitrec bleu.
(Opp. p. 497)

Myiagra longicauda Swainson, 1838. Flycatchers (Jardine, ed., Naturalist's Library, 21, Ornith. 10), p. 210, pl. 25; New Holland; error for Senegal *fide* Hartlaub, 1857, Syst. Ornith. Westafrika's, p. 93.

Forms a superspecies with *E. albicauda*.

Range and Status. Endemic, resident. Lowland woodlands of Senegal and Gambia (south of 14°N); Guinea–Bissau, Mali (uncommon, Kulikoro, Bamako, eastern Mts Mandingues); Guinea (coast and N, including Futa Djalon); Sierra Leone (Freetown to Makeni, Tingi Mts and E boundary); Liberia (sus-

Elminia longicauda

pected but not confirmed); Ivory Coast (throughout guinea zone and also along coast); Burkina Faso (upper Comoé R., Ouagadougou and Arli and Pendjari Nat. Parks); Ghana (not uncommon, Mole Nat. Park and N wooded savanna, south to inselbergs on Accra plains); Togo (Paio); Benin (Pendjari and Arli Nat. Parks, Bembereke area); Nigeria (Lagos to Badagri, and Ilori, Serti and Gashaka-Gumti Game Res., north to Zaria, Nindam Forest Res., Kagoro and Jos plateau, also Ofemiri, Obudu plateau and Potiskum); Cameroon (Korup Nat. Park, Mt Cameroon, Rumpi hills and Bamenda to Mt Genderu, Poli, Koum and Bénoué, south to Garoua-Moulai, M'Bakaou and Dja R.); Chad (Dagbao, Mts Lam, Moundou and Logone area to Maro); Central African Republic (Birao to Bouar, south to Bangui and probably Lobaye, also Zemio); Rio Muni, Gabon, Congo (Mayombe to Djambala and Makoua); NW Angola (Cabinda south to Cuanza Norte and along escarpment to Gabela, common to locally abundant); SW to NE Zaïre (Mayombe to Kasai and to Eola and Salonga Nat. Park, east to Uele and Ituri, south to L. Edward); S Sudan (north to Ferti, Wau and Torit); Uganda (except Teso and Karamoja); NW Tanzania (Bukoba and Musoma) and W Kenya (Mt Elgon, Saiwa and Bungoma to Nandi, Kakamega, Yala, Ng'iya, Kibigori, Mau and Mara R.). Uncommon to common. Density in NE Gabon, generally 3·6 groups/km^2 in second growth but 6·2 groups/km^2 in patches of more suitable habitat (Erard 1987); in Kenya (Kakamega Forest, Zimmerman 1972) 2 adults in 8 ha census tract.

Description. *E. l. teresita* (Antinori): Cameroon to Angola, east to Sudan, Uganda and Kenya. ADULT ♂: forehead cobalt-blue mottled black, crown, ear-coverts, sides of neck, hind-neck, scapulars, mantle, back and rump cobalt-blue (feathers with pale grey base, blue edges; feathers of crown elongated, with iridescent whitish blue stripe along shaft, giving a streaked appearance); lores black. Chin, throat and foreneck grey with intense wash of cobalt-blue. Uppertail-coverts and tail greenish blue, tail-feathers very graduated (T1 longest, projecting 10–12 mm, greenish blue, T2–T6 brown-black, with broad edge along outer web greenish blue; underside of shaft white). Breast pale blue-grey, with centre whiter; axillaries, belly and undertail-coverts white; flanks pale grey washed blue. Primaries, secondaries, tertials and upperwing-coverts black with broad edges cobalt-blue, particularly along outer web; underwing-coverts grey broadly edged white; underside of flight feathers brown-black, margins of inner web white. Bill black; eyes brown; legs and feet brown-black to black. ADULT ♀: like ♂ but more greenish, less cobalt-blue; T1 projecting only 8–10 mm; throat and breast greyer, less blue. SIZE: (6 ♂♂, 6 ♀♀) wing, ♂ 64–67 (65·7), ♀ 61–65 (62·9); tail, ♂ 76–85 (79·9), ♀ 70–84 (77·6); extension of T1, ♂ 9–13 (11·1), ♀ 5–12 (9·0); bill, ♂ 12–14 (13·2), ♀ 12–13 (12·5); tarsus, ♂ 15–16 (15·8), ♀ 15–16 (15·6). WEIGHT: (Nigeria) unsexed (n = 5) 7–10 (8·4); (Cameroon) ♂ (n = 3) 8–12 (10·0), 2 ♀♀ 10, 11; (Gabon) 1 unsexed 10; (Kenya, May, Nov) 1 ♂ 7, 1 unsexed 6; (Uganda, May) 2 ♂♂, 7, 12, 2 ♀♀ 9, 9.

IMMATURE: like adult ♀ but much greyer, cobalt-blue reduced to edges of wing and tail-feathers; line at base of forehead, lores and ear-coverts sooty brown, top of head dark bluish grey-brown, each feather with narrow edge at tip pale brown, feathers of back with indistinct narrow barring buff and dark grey; secondaries and tertials with narrow whitish tip; throat, breast and flanks mouse-grey; bill brown.

JUVENILE: like immature but still duller, feathers of head, mantle and wing-coverts edged pale brown to sandy buff, primaries with dirty white; underparts greyish but belly and undertail-coverts whitish, no blue on throat and breast.

NESTLING: see (**C**).

E. l. longicauda (Swainson): Senegal and Gambia to Nigeria. Darker, underparts much less white, greyer.

Field Characters. A small, active, noisy bird quite at home in man-made habitats; gleans foliage in small chattering groups, with characteristic tail fanning and partly spread and drooped wings (**A**). Blue above and on throat, paler and greyer below, with short crest, long tail and black lores. Differs from very similar White-tailed Blue Flycatcher *E. albicauda* in lack of white on outer tail-feathers and much bluer coloration, particularly on head.

Voice. Tape-recorded (C, CHA, ERA, KEI, LEM, MOR, STJ). Calls and songs grade into one another, making it hard to define where one ends and the other

A

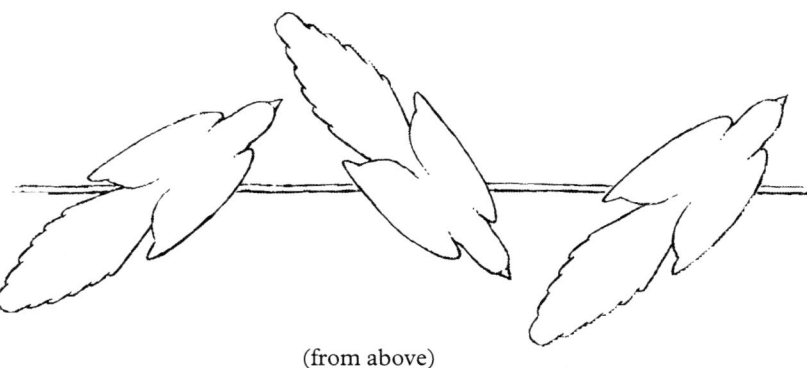

(from above)

B

begins; this is linked to social organization. Call notes thin and high-pitched, 'zip', 'tsup', 'dzwip', 'ptit', 'pit-pit', 'ptiay-twi-ti'. 'Song' is a casual, unstructured, tuneless medley or jumble of notes strung together; they may speed up in the middle, 'tsip, chip, . . . tsee-tsu-tsee-tsair . . . chair, tsee, chup, chup, tsee'; faster section in middle may be repeated, '. . . tsee-tsoo-tsee-dzidzidzi . . . tsee-tsoo-tsee-dzidzidzi . . .'; or they may lack fast section, 'ti, tiay, pti, ti, wi, tut, tit . . .'. Quality rather sunbird-like but more varied and rich.

General Habits. Inhabits wooded savanna, riverine or swampy forest patches, inselbergs, moist woodland; in forested areas, clearings and edges, open secondary forest, edges of riverine forest in man-made habitats, regenerating forest around villages, cultivated land with trees, plantations of cocoa, rubber, cypress and *Podocarpus*, gardens in villages and towns with their shrubs and hedges; occurs also in coastal mangrove and along montane forest edges. In Futa Djalon, Guinea, occurs in poorly wooded savannas with thick patches of bushes on termite mounds (G. Jarry, pers. comm.). In NE Gabon found only within 500 m of human habitation, in open areas with scattered trees, both cleared and disused plantations (cocoa, avocado, mango, citrus) dominated by trees with fairly open foliage 20–25 m above ground, and in open thickets in and around villages. Up to at least 1350 m in Nigeria and 2000 m in Cameroon, and at 800–2400 m in E Africa. Separated altitudinally from White-tailed Blue Flycatcher, which occurs at higher elevations; however in Angola, near Vila Salazar, both species occur almost side by side, Blue Flycatcher in gardens and recent plantations, White-tailed in old abandoned plantations and forest patches (C. Erard, pers. obs.).

In pairs or more often in small groups of 3–7 individuals, sometimes up to a dozen, including several adults; occasionally solitary. Larger groups sometimes subdivided in subgroups, pairs or trios of adults with immatures and young; each group or subgroup has a leader who sings frequently and is much more active than the others, almost always with crest erect, flicking wings and swinging raised and fanned tail. Groups are larger during dry season.

Gleans foliage in crowns of trees and shrubs, also leafy liana tangles; particularly attracted to Leguminosae. Usually forages at higher levels, but in Freetown, Sierra Leone, birds observed in gardens around houses finding food among dead leaves on ground, flushing winged insects which they quickly caught (Thompson 1925), and similar activities noted in NE Gabon in an enclosure with 2 captive western bush-pigs. In Gabon, searches all levels of vegetation from near ground to upper canopy, though 44% of records are between 14 and 22 m (Erard 1987). Always on the move, making quick little flights or restlessly moving among branches, flicking wings and fanning tail, giving sharp calls or quick disjointed songs. In typical foraging action makes short forward jumps, pivoting from side to side (**B**) with each jump in crouched posture, flicking drooped or half-spread wings and sweeping raised tail from side to side while fanning and closing it; this dislodges prey which it snaps up with a quick pounce or brief sally, or in fluttering spiral flight. Sometimes sallies from fixed perch like true flycatcher to catch flying insects around leaves or swarming in air, but waits in crouched not upright posture and flight more weaving, less straight. Often feeds on swarming ants and termites. Regularly joins mixed-species flocks, and is a core species in NE Gabon, where 77·4% of records were in mixed flocks.

Group members react collectively to danger. Anxiety expressed by clicking calls, crouching, pivoting body and opening and closing tail; sometimes stands and freezes in upright posture. When alarmed or when mobbing, intensifies calling, erects crest, spasmodically flicks wings and irregularly swings fanned tail from side to side in broad sweeps.

Sedentary, at least in NE Gabon.

Food. Very small insects, e.g. bee-like Hymenoptera, flies, beetles, moths. Prey size mainly 3–6 mm, up to 10–20 mm (NE Gabon).

Breeding Habits. Monogamous or polygamous; sometimes communal breeder. Territorial. Defends group territory of *c.* 12 ha all year in NE Gabon. Group patrols entire territory each day, led by dominant ♂ who sings repeatedly when entering new area. Usually group forages in a small patch of vegetation and then moves to another one 100–300 m away; ♂ sings before leaving first place and when arriving at the next. Density low, so that contacts between groups rather rare, particularly because around many small native villages there is no more than one group. During territorial defence leader and other adult ♂ ♂ face intruders, sing actively in long bouts and crouch with erect crest, fluffed-out breast plumage, raised fanned tail and wings drooped and held away from body, and display with bill pointing up or bill open. During sexual interactions between ♂ and ♀, birds sit parallel, erect crest, raise and tilt fanned tail toward partner and quiver drooping wing-tips or gently flap wings.

NEST: beautiful small, compact, neat and well-shaped open cup (**C**); mainly of moss, fine grasses, thin plant stems, pieces of bark, thick vegetable down, gossamer and lichens; externally covered with fragments of pale green lichen bound with gossamer so that it looks silver-green and is rather well camouflaged, resembling a large bump on a branch and matching colour of foliage; internally lined with fine fibres and grass-heads. Ext. diam. 52–56, height 46; int. diam. 44, int. depth 23–25 (2 nests, Gabon, Cameroon). Placed 2–15 m above ground, in a fork toward end of horizontal leafy branch of small tree, e.g. avocado.

EGGS: 1–2; Uganda mean (4 clutches) 1·8. Oval with pointed end truncated; almost no gloss; creamy white to creamy grey with dense mottling of fine spots and speckles, ashy or greyish or greenish grey and yellow-brown to mauve or lilac-grey, concentrated in a wide ring slightly away from centre toward larger end. SIZE: (Cameroon, n = 4) 15·3–17·0 × 11·8–12·5 (16·1 × 12·2).

LAYING DATES: Ghana, June; Nigeria, Mar, June–Aug (immature June); Cameroon, Feb–Aug; Gabon, Jan, Apr, June, Sept, Nov, Dec; Angola, Aug; Zaïre, Apr, June, Aug; E Africa: Uganda, Feb–Apr; Region A, Apr, July; Region B, during long rains, Mar–Aug, Oct, Dec.

INCUBATION: by ♀, for long sessions (40–60 min), interrupted by absences of 20–30 min. Period unknown. ♀, both in and out of nest, fed by group partners. Not well studied (brief observations at one nest).

DEVELOPMENT AND CARE OF YOUNG: young stay in nest at least 14 days. Fed by both parents or by group members. At one nest in NE Gabon 4 individuals regularly fed nestlings together. Food delivery rate slower (av. every 15 min) during first half of morning and in late afternoon than during hottest part of the day, when there are feeding bouts of 15–30 min at a rate of one every 1–2 min, separated by pauses of 3–7 min. After leaving nest, young fed by group members for at least 3 weeks; they soon imitate foraging movements of adults and move with drooped wings and fanned tail.

BREEDING SUCCESS/SURVIVAL: of 6 nests in NE Gabon, 5 produced fledglings: 2 fledged one, 3 two.

C

References
Brosset, A. and Erard, C. (1986).
Erard, C. (1987).
Hubbard, J. P. and Hubbard, C. L. (1970).

Elminia albicauda Barbosa du Bocage. White-tailed Blue Flycatcher. Tchitrec à queue blanche.

Plate 28
(Opp. p. 497)

Elminia albicauda Barbosa du Bocage, 1877. J. Sci. Math. Phys. Nat., Lisbon, 6, p. 159; Caconda, Angola.

Forms a superspecies with *E. longicauda*.

Range and Status. Endemic, resident. Woodlands of SW Uganda (Kigezi and S Ankole), Rwanda (Nyungwe Forest), Burundi, Tanzania (in NW at Kifunzo and Bugarama, further south at Mahari Mt and from Ufipa Plateau to Mt Rungwe and Songea, also in NE from Crater Highlands and Ngorongoro to Mbulu); Malaŵi

Elminia albicauda

west of Rift (Mwanza and Ntcheu districts north to Chitipa and Karonga districts), N Mozambique (Tete district), N Zambia (local though often common, Kawambwa and Northern Province from Mbala and Mporokoso to Eastern Province on Malaŵi border, Muchinga escarpment west of Luangwa valley and Mkushi R., also Copperbelt and Northwestern Province from Solwezi and Mwinilunga to Kasempa, Mankoya, Kabompo and Balovale), S and E Zaïre (north to Kasai, Luluabourg and Merode, L. Edward and former Parc Nat. Albert), and Angola (Benguela and Huila to Malanje, eastward on central plateau). Uncommon to frequent.

Description. ADULT ♂: forehead, crown, nape, hindneck and sides of neck grey-blue streaked whitish blue (feathers dark slate-grey with grey-blue margins and narrow shaft streak black on median third, broader and whitish blue on distal third); narrow line at base of forehead black; feathers of crown lanceolated and elongated, longest on hindcrown, forming crest; lores and ring around eyes velvet-black, ear-coverts velvet-black tipped grey-blue and irregularly streaked whitish blue. Mantle, back, rump and uppertail-coverts grey-blue (feathers with grey base and grey-blue tip). Tail graduated (longest T1 extending 3–12 beyond T2, shortest T6 15–25 shorter than T2), blackish brown-grey washed grey-blue, 4 pairs of outer feathers with white tip increasing in breadth outwardly; pattern variable, but usually T1 washed grey-blue on both webs, T2 washed and edged grey-blue along outer web, T3 edged grey-blue along outer web and with broad tip and distal half of shaft white, T4 like T3 but entire shaft white and white tip wider, extending more on inner web, T5 with shaft and distal half of both webs white, T6 entirely white. Underparts white, chin, upper throat, sides of breast, flanks and thighs heavily washed pale slate-grey. Primaries, secondaries, tertials and greater upperwing-coverts blackish brown-grey, all except outer primaries edged grey-blue along outer web; bend of wing, lesser and median upperwing-coverts blackish brown-grey with broad grey-blue edges on both webs and faint whitish blue shaft streaks. Axillaries white washed pale slate-grey, underwing-coverts dark brown-grey edged pale slate-grey, underside of flight feathers dark brownish grey, inner webs edged whitish grey. Bill black; eyes brown to dark sepia; legs and feet black to brown-black. ADULT ♀ like ♂ but duller, upperparts greyer, but underparts whiter, grey wash less pronounced, more restricted to breast; crest shorter; T1 extending less beyond T2. SIZE: (10 ♂♂, 8 ♀♀) wing, ♂ 66–74 (68·8), ♀ 62·5–69 (66·1); tail, ♂ 78–86·5 (82·1), ♀ 76–84·5 (79·5); extension of T1 beyond T2, ♂ 3–12 (6·7); ♀ 3–6 (4·7); bill, ♂ 11·5–13 (12·2), ♀ 11·5–13 (11·9); tarsus, ♂ 14·5–16 (15·3), ♀ 14–15·5 (15·0). WEIGHT: (Upemba, Zaïre) unsexed (n = 16) 6–9 (7·8).

IMMATURE: like adult ♀ but upperparts greyer, less blue; cheeks dark slate-grey faintly streaked whitish; crest small; upperwing-coverts, particularly greaters, blackish brown with only faint blue wash and small whitish tip; inner primaries, secondaries and tertials with narrow white tip; tail like adult except T2 with whitish tip.

JUVENILE: like immature but still greyer; crestless; blue restricted to tail and wings, almost none on head and upperparts, the latter finely barred pale buff; lower rump whitish; upperwing-coverts with pale buff edges, whiter and broader on inner medians and greaters; inner secondaries and tertials with whitish tip; breast and flanks washed pale buff.

NESTLING: undescribed.

Field Characters. An active, noisy bird which travels in small groups, constantly fanning its tail and flicking its wings. Grey-blue above with small crest and black line from bill to eye, whitish below with grey sides, tail long and graduated, with much white in sides. Closely related African Blue Flycatcher *E. longicauda* lacks white in tail and is cerulean blue above, with no grey tone, blue extending onto throat and sides of breast. White-tailed Crested Flycatcher *E. albonotata* has white in tail but is otherwise very different, mainly dark grey, with white belly; other crested flycatchers in *Elminia* and *Trochocercus* are similarly dark, and have no white in tail.

Voice. Tape-recorded (LEM, MOY). Song, a short pretty warble beginning and ending with 'tip-tip' notes; described as a serin-like jingle (Moreau and Sclater 1938). Contact call a low, sharp, double 'tip-tip', alternating with whistled chirrups 'teereet' or 'chiri'.

General Habits. Inhabits open lowland and mid-altitude evergreen forest and its openings and edges, open second-growth in forested areas, riparian forest, bamboo, taller and denser *Brachystegia* and *Isoberlinia* woodlands (not particularly in rock-strewn areas as sometimes suggested), rich miombo and other woodlands (though not mavunda in Zambia), savanna woods, stream banks, trees in shrub country, roadsides and clearings with scattered trees; up to 2000 m in Itombwe, Zaïre, 2100 m in Burundi and Malaŵi, 2500 m in Rwanda. In W Malanje, Angola, in area of contact with African Blue Flycatcher, inhabits edges of primary and old unstratified secondary forest, including old coffee plantations, and is more of a canopy bird, while Blue Flycatcher occurs in recent clearings, gardens and well-stratified second growth (C. Erard, pers. obs.). In E Africa, usually at higher elevations than Blue Flycatcher, 1200–2450 m, seldom below 1600 m; also segregated ecologically from White-tailed Crested Flycatcher, keeping to edge of forest, but locally extends into interior when competitor absent.

In pairs or small parties. Constantly on the move. Forages at mid-levels and in open canopy; zigzags and hops hither and thither among and along branches, turning and twisting, flitting quickly from twig to twig, drooping half-open wings and brushing foliage with tail fanned, catching disturbed insects in short acrobatic flights. Very often joins mixed-species flocks, associating with or followed by hyliotas, eremomelas and crombecs. Exploits swarming ants and termites.

Some local off-season downward movements suspected in Malaŵi.

Food. Small insects: flies, beetles, moths, ants and termites; also small spiders.

Breeding Habits. Monogamous, territorial; behaviour undocumented.

NEST: neat, compact little cup made of fine wood fibres, sometimes lined with feathers, and thickly plastered outside with spider web, giving a smooth silver-grey appearance; ext. diam. 50, ext. depth 50; int. diam. 33, int. depth 21; placed in fork of twigs at end of projecting branch 9–12 m above ground, firmly fixed on support by spider webs.

EGGS: 2. Malaŵi mean (3 clutches) 2·0. Oval; glossy; white or buffy, densely speckled grey-green with underlying smoky grey clouding with a few olive spots in broad band near larger end, much less densely speckled elsewhere. SIZE: (n = 2) 17·2 × 11·9 and 16·9 × 11·9.

LAYING DATES: Uganda, Jan, Aug; Malaŵi, Oct–Dec; Zambia, Nov (♂♀ with gonads active Sept); Zaïre, Oct–Dec; Angola (juv. Dec).

INCUBATION: by ♀; pair spend much time in or in immediate vicinity of nest tree. Period unknown.

DEVELOPMENT AND CARE OF YOUNG: young fed by both parents. Nestling and dependent periods unknown.

Reference
Vincent, A. W. (1947).

Elminia nigromitrata (Reichenow). Dusky Crested Flycatcher. Tchitrec à tête noire.

Plate 28
(Opp. p. 497)

Terpsiphone nigromitratus Reichenow, 1874. J. Orn., 22, p. 110; Cameroon R., Cameroon.

Forms a superspecies with *E. albiventris*.

Range and Status. Endemic, resident. Disjunct distribution in lowland forests: Liberia (forest north to Mt Nimba), Guinea (Macenta district, frequent), Ivory Coast (Yapo – not uncommon, and possibly Taï), Ghana (not uncommon, Atewa and Kakum forest reserves, Bia Nat. Park), S Nigeria (uncommon, Lagos to Oban, also north to Gambari and Benin, and Okwango Forest Res.), S Cameroon (north to Korup Nat. Park (common), Nlonako, Efulen, Dja R. and Dimako), Rio Muni, Gabon (uncommon to frequent in all forested areas from coast eastward, south to Tchibanga, Mimongo, M'Bigou and Franceville), NW Congo (Odzala), Central African Republic (Dzanga reserves and region of Baroua, Ouossi R., possibly also Lobaye), S Sudan (Bengengai), E Zaïre (Salonga Nat. Park to upper Uele and Ituri, south to Itombwe and Kasongo), S and W Uganda (Budongo, Bugoma and Bwamba to Ankole, Kigezi and Sango Bay, east to Kifu, Mabira and Mt Elgon), W Kenya (Kakamega forest, also recorded Chemoni forest) and NW Tanzania (Bukoba). Uncommon to frequent. Density in NE Gabon, 3·2 to 5·8 pairs/km² (Erard 1987); in Kakamega forest, Kenya, 8–10 adults and 1 immature in 8 ha census tract (Zimmerman 1972).

Description. *E. n. nigromitrata* (Reichenow): from Cameroon eastward. ADULT ♂: forehead, lores and crown black, crown-feathers somewhat elongated. Ear-coverts, sides of neck, hindneck, scapulars, mantle, back, rump and uppertail-coverts dark bluish slate-grey. Entire underparts slate-

grey, belly, flanks, axillaries, underwing-coverts and undertail-coverts paler. Wings and tail brown-black, tail-feathers narrowly edged bluish grey on outer web. Bill black; eyes dark brown; legs black. ADULT ♀: like ♂ but less blue, more slate-grey. SIZE: (7 ♂♂, 8 ♀♀) wing, ♂ 61–65 (62·3), ♀ 57–61 (59·3); tail, ♂ 59–61 (60·2), ♀ 53–60 (56·3); bill, ♂ 14–16 (15·2), ♀ 14–16 (14·7); tarsus, ♂ 15–17 (16·1), ♀ 15–17 (15·6). WEIGHT: (Liberia, Mt Nimba) ♂ (n = 12) 9·6 ± 0·9, ♀ (n = 12) 8·5 ± 0·9; (Cameroon) ♂♀ (n = 7) 10–12 (11·2); (Gabon) ♂♀ (n = 3) 10·7–11 (10·8); (Kenya, June, July, Oct, Nov) ♂ (n = 5) 6–9 (7·4), ♀ (n = 4) 7–9·5 (8·1); Uganda (Jan, Feb) ♂ (n = 10) 8–10·5 (9·5), ♀ (n = 3) 10–10·5 (10·2), (Apr, May) ♂ (n = 12) 7–11 (9·2), ♀ (n = 8) 6–10 (8·2), (June, July) ♂ (n = 5) 10–11 (10·4), 2 ♀♀ 9, 10, (Oct, Nov) 2 ♂♂ 10·5, 11·5.

IMMATURE: like adult ♀ but more uniform sooty brown-grey.

JUVENILE: like immature but duller, without bluish tinge, head dull slaty black.

NESTLING: hatches naked; later, brownish pink skin, sparse dark brown down tufts on nape, middle of back and wings.

E. n. colstoni Dickerman: Liberia to Nigeria. Slightly paler and brighter blue, less dull, flat dark grey; distinct tendency to have white-tinged collar on hindneck.

Field Characters. Small, blackish (bluish in good light), active foliage-gleaner which keeps low in forest undergrowth. Identified by uniform dark slate-blue plumage except for black cap, foraging behaviour with fanned tail and drooping partly spread wings, and distinctive song. Might be confused with ♀ Blue-headed Crested Flycatcher *Trochocercus nitens* but more uniform in colour, blacker and more slate-coloured rather than ash-grey, sides of head and entire underparts much darker. Separated from very similar White-bellied Crested Flycatcher *E. albiventris* of montane forests by uniform dark underparts. White-tailed Crested Flycatcher *E. albonotata* has black head, white belly and broad white ends to tail-feathers.

Voice. Tape-recorded (C, ERA, HOR, MOY, STJ). Song highly variable, a continuous series (1–2 min) of phrases, each lasting 10–20 s and composed of groups of whistles and liquid trills interspersed with much chattering and thin high notes 'weee-tee-tewrwit, weeee-tee-tewrweet, tweet-tweet-tweet, tew-weet, tew-weet-tew-weet-tee-tee-tee-tee, tew-weet-tew-weet-teww-weet, tweet-tweet-tweet, teeteeteetee . . . weet-tee-tew, weet-tee-tew, ew-tic-tew, twee-tic-tew, twee-tee-tic-tweet-tew, ptew-tee-tee-tee-tee-tew . . .'. Often, at least in Gabon, song includes imitations of calls and songs of other species, sometimes up to 13 species in one song. Species mimicked include *Andropadus ansorgei*, *Baeopogon indicator*, *Pycnonotus barbatus*, *Stiphrornis erythrothorax*, *Neocossyphus fraseri*, *Sylvietta denti*, *Macrosphenus concolor*, *Fraseria ocreata*, *Myioparus griseigularis*, *Erythrocercus mccallii*, *Terpsiphone batesi*, *Illadopsis cleaveri*, *Dicrurus atripennis*, *Nectarinia olivacea* and *Spermophaga haematina*. Call a persistent 'chip-chip-chip-chip . . .' noted in Cameroon; contact call given as a dry 'tsep' or feeble 'tsip' or 'tsièp-tsièp' in Gabon (P. Christy, pers. comm.) but more thorough study in NE Gabon (Erard, pers. ob.) gives usual contact call between pair members a soft 'ptee-tyiew-tyiew' or 'ptee-dew-dew' easily mistaken for that of Grey-throated Tit-Flycatcher *Myioparus griseigularis*. Around nest, ♂ guarding or greeting or accompanying ♀ gives either slow 'tlee-dyew-tee' or rapid 'teedyewtew', and other varied calls: 'pweetit-tee-tyew', 'teedew-tee-dewdew', 'teedewtu-teedewtu', 'teeee-dew-tew', 'teedew-tewlulu', 'tyew-tu-tee-tit', 'pteedee-tew-teedew', usually emitted in phrases, each repeated 3 to 7 times with varying intervals. Fight call, a high-pitched squeaky 'tseeee'; anxiety call, dry harsh 'ptayc' or 'tchayc'; alarm call, sharp scratchy 'ptric' or 'tsic', or 'ptic-ptic-ptaydec' or quick clinking 'ptic-ptic-ptic'; mobbing call, a rapid tit-like 'ptit-ptit-ptit . . .'; call of ♀ on nestlings, a soft rapid 'pseetseetyew'. Bird in hand emitted soft 'ptiyew-ptiyew' and a low rasping.

General Habits. Inhabits mature and secondary lowland forest, up to 800 m in Cameroon, 900 m in Gabon, 1490 m in Itombwe (Zaïre) and 2150 m in Uganda. Mainly in primary rain forest, where it favours vicinity of tree falls and moister areas such as borders of rivers and streams, swamp forest or flooded *Gilbertiodendron* forest. Fond of shady places with dense undergrowth, rich in thick *Alchornea* stands and clumps and tangles of lianas and tall herbs (e.g. Marantaceae, broad-leaved Graminaceae, *Aframomum*); at highest altitudes occurs where understorey is rich in mosses, ferns, lichens and epiphytes. Occurs in dense secondary forest, at every stage of regenerating forest on old cultivation, especially in later stages, also uncleared disused plantations; approaches native villages in riverine forest or where large patches or corridors of natural forest remain; once in a tall tree in savanna at M'Bigou, Gabon (Rand *et al.* 1959).

Solitary or in pairs or small family groups. Very active, always on the move, especially in the morning and evening, less so in the afternoon. In Kenya, exclusively an undergrowth bird, seldom seen more than 1 m above ground, even when in mixed-species flocks. Likewise in Gabon never seen above 6 m; 79% of records fall below 2 m; during long dry and short rainy seasons (i.e. June to Nov) keeps to lowest layers (av. height 1·2 m above ground) but during rest of year forages also in higher layers (av. 1·9 m above ground) (Erard 1987). Searches dense vegetation of saplings and shrubs, tangles of fallen leafy branches and trees, leafy lianas, or aerial roots and rootlets; usually avoids sunny areas and gaps caused by recent tree-falls, moving quickly from one dense part of forest understorey to another. This restriction to low, impenetrable forest undergrowth, combined with its dark colour, high mobility and rather infrequent vocalizations make it quite inconspicuous and might lead one to consider it rare, but mist-netting has shown it to be common.

Foraging behaviour reminiscent of African Blue Flycatcher *Elminia longicauda* but less elaborate. Gleans foliage, making brief sallies or hopping rapidly along and among branches; flicks wings, fans tail and nervously pivots from side to side in crouched posture; sometimes droops wings and swings fanned and raised tail from side to side. Catches dislodged insects with quick pounce or sally or pursues them in fluttering spiral descent. When in mixed-species flocks or around army-ants or in food-rich places, may sally from post like a true flycatcher to catch flying insects among leaves, on bark or swarming in the air, but posture on perch crouched not upright, and flight erratic. May even hop on the ground and work its way into low vegetation covered with dead and decaying leaves and branches. Regularly feeds on swarming ants and termites. Regularly joins mixed-species flocks, e.g. 90% of records in NE Gabon, varying from 100% during long dry season to 81% during rains. On several occasions seen being attacked by ♂ or ♀ Blue-headed Crested Flycatcher in intimidation posture giving 'castanets' call and harsh songs with bill snapping; this occurred when Dusky Crested Flycatcher foraged higher than 2–3 m in mixed-species flocks where Blue-headed were present.

When alarmed gives clicking calls, crouches and bends forward or stands in semi-erect, sometimes hunched posture with fluffed out crown-feathers, and pivots body from side to side, rapidly opening and closing wing-tips and tail-feathers; in more intense and aggressive response, flicks one or both wings nervously and sings. When encountered it often behaves like this, appearing on an open branch or liana, then dropping down into cover.

Food. Small arthropods, mainly insects, moths, caterpillars, grasshoppers, cockroaches and flies; also spiders. Prey size mostly 5–10 mm, up to 30.

Breeding Habits. Monogamous; pairs for a long time, probably for life. Pair formation behaviour unknown. Solitary birds in mixed-species flocks giving lengthy songs with much mimicry of other species may have been trying to attract partners, since mimicry does not appear to be territorial but sexual: birds do not react vehemently to play-back of these songs. Territorial. Patrols entire territory of 13–15 ha every day in NE Gabon. Contact call appears to be used also as a territorial signal, as well as the highly variable song; on the other hand, a recording of these songs from Uganda provoked no reaction from Gabonese birds which had responded aggressively to their own songs.

NEST: well made, neat cup of very tightly woven dark green moss and fibres, sometimes with greenish or grey-green lichen bound with spider web, internally lined with thin material, e.g. rootlets and black *Marasmius*, or fine brownish seed heads of, probably, *Panicum* sp. In Uganda, one nest typically 'funnel-shaped', i.e. an elongated cone; elliptical exterior 90 × 60, ext. depth 110, int. diam. 40, int. depth 30; in NE Gabon, 2 nests, ext. diam. 60–65, height 60–65, int. diam. 43–49, depth 25–27. Uganda nest placed *c.* 4·5 m above ground in tree *c.* 5·5 m high and *c.* 6 cm thick in open montane forest with dense herb layer and scattered medium sized trees (A. Forbes-Watson and T. Butynski, pers. comm.), another one in Uganda placed *c.* 3·5 m up in fork of sapling in secondary growth (Chapin 1953a). Nest in Cameroon was placed in fork of shrub in understorey. The 2 Gabonese nests were placed in forks of low saplings, 0·6 and 0·7 m above ground. Built by both sexes though mainly by ♀, accompanied by ♂ who gives varied calls and songs, particularly imitative songs.

EGGS: 2, laid at 1 day intervals, 2 days after end of nest construction. Elongated oval; almost no gloss; creamy white with slight bluish tinge, with conspicuous broad ring of dense small dark to rufous-brown spots toward larger end (though in one clutch near middle of shell, which is irregularly spotted dark to slate-brown; smaller end unmarked). This description fits Pitman's from Uganda given by Chapin (1953a) but not Bates's (1930) from Cameroon, who stated eggs resemble those of *Terpsiphone*, probably confusing Blue-headed with Dusky Crested Flycatcher. SIZE: *c.* 18 × 13.

LAYING DATES: Liberia (♀♀ with moderately enlarged ovaries June, July); Ivory Coast (carrying nest material Jan, adults with dependent young Mar, June); Nigeria, Jan; Cameroon, Feb–May (breeding condition July, Dec); Gabon, Aug, Sept (young fed by adult Mar); Zaïre, Jan, Oct, Nov (♂♂ with enlarged testes Dec, Mar); E Africa: Uganda, Feb; Region B, Sept.

INCUBATION: begins with last egg; by ♀. Period: 16 days (n = 2). By day ♀ sits for 40–50 min with absences of 15–30 min. ♂ guards nest surroundings; calls and sings regularly, facing nest with fluffed-out crest, half-spread tail and slow flicking of half-open wings. Greets ♀ when she leaves nest and accompanies her with varied slow or rapid calls.

DEVELOPMENT AND CARE OF YOUNG: young fed by both parents. Often ♂ passes food to ♀ which gives it to young, particularly during first 2 days while she is brooding them (68% of time on day 1, 35% on day 2). Intervals between food deliveries decrease as young grow: av. 12·8 min on day 1, 7·4 on day 3, 4·7 on day 5, 3·3 after day 6. Weight of young increases steadily from 1·6–1·8 g at birth to 8·8–9·1 g at 10 days. Nestling period: 12–13 days. ♂ and ♀ defend nest against intruders with threat posture (crouched, breast and crown plumage fluffed out, raised carpal joints of folded and drooped wings, and spread tail-feathers slightly raised) and give harsh scolding calls, and may hit observer handling calling young.

BREEDING SUCCESS/SURVIVAL: both nests observed in NE Gabon fledged 2 young; ringed birds were still present on the same territory 3–5 years later. In Kakamega Forest, Kenya, 3 ringed birds were recaptured 6 to 16 months later at the same spot (Zimmerman 1972).

References
Brosset, A. and Erard, C. (1986).
Erard, C. (1987).

Plate 28
(Opp. p. 497)

Elminia albiventris (Sjöstedt). **White-bellied Crested Flycatcher. Tchitrec à ventre blanc.**

Trochocercus albiventris Sjöstedt, 1893. Orn. Monats., 1, p. 43; Cameroon Mt.

Forms a superspecies with *E. nigromitrata*.

Range and Status. Endemic, resident. Disjunct range, in 2 widely separated areas: (1) montane forests of SE Nigeria (Obudu Plateau, Gotel Mts, Ngel Nyaki Forest Res. and Gashaka-Gumti Forest Res.), Cameroon (Mt Cameroon, Mt Kupé, Mt Manenguba, Rumpi Hills and Bamenda highlands north to Mt Oku and Mt Lefo) and Bioko; and (2) highlands of E Zaïre (Albertine Rift, W of L. Albert and L. Edward from Lendu Plateau and E Ituri to Itombwe), Rwanda (Nyungwe Forest) and SW Uganda (Kibirau and Kibale in Toro and Bwindi–Impenetrable Forest, Kigezi). Frequent to abundant. In Cameroon, 3 to over 15 birds recorded per observer per day at Mts Cameroon, Kupé, Manenguba and Rumpi Hills, scarcer elsewhere (Stuart and Jensen 1986).

Elminia albiventris

Description. *E. a. albiventris* Sjöstedt: Nigeria, Cameroon and Bioko. ADULT ♂: forehead, lores and crown black, hindcrown feathers slightly elongated; ear-coverts, sides of neck, hindneck, scapulars, mantle, back, rump and uppertail-coverts dark sooty slate-grey with bluish tinge. Entire underparts sooty black except for white belly; flanks, axillaries, underwing-coverts and undertail-coverts pale slate-grey; wings and tail brown-black, tail-feathers narrowly edged bluish grey on outer web. Bill black; eyes dark brown; legs black. ADULT ♀: like ♂ but duller slate-grey, less blue. SIZE: (5 ♂♂, 6 ♀♀) wing, ♂ 60–64·5 (61·8), ♀ 57·5–61 (59·3); tail, ♂ 61·5–63 (62·3), ♀ 58·5–61·5 (60·5); bill, ♂ 12·5–14·5 (13·4), ♀ 12–13 (12·7); tarsus, ♂ 17–18 (17·5), ♀ 16–17·5 (16·5). WEIGHT: (Cameroon, Oct–Apr) unsexed (n = 29) 7–10 (8·9).
    IMMATURE: like adult ♀ but duller slate-grey, wings and tail sooty brown-grey.
    JUVENILE: like immature, still duller and with more sooty brown wings and tail.
    E. a. toroensis Jackson: Zaïre, Rwanda and Uganda. Upperparts more bluish grey, throat greyer, less sooty. WEIGHT: (Uganda, Mar, May, June) ♂ (n = 3) 9–10 (9·5), ♀ (n = 5) 8–10 (8·6), unsexed (n = 3) all 8.

Field Characters. The montane forest counterpart of closely-related Dusky Crested Flycatcher *E. nigromitrata*, with similar wing and tail movements and same undergrowth habitat. Dark sooty plumage with black cap and contrasting white belly. Dusky Crested Flycatcher of lowland forests is bluer grey and has uniform dark underparts. ♀ Blue-headed Crested Flycatcher *Trochocercus nitens*, another lowland forest bird, is paler with ash-grey underparts. The other montane forest congener, White-tailed Crested Flycatcher *E. albonotata* is much paler and has conspicuous broad white tips and edges to tail-feathers.

Voice. Tape-recorded (CHA, LEM). Song in Cameroon said to be a soft melodious, rather hesitant warble (Stuart and Jensen 1986) but usually consists of either an irregular series of separate notes, 'psip . . . psip . . . psip . . . psip-trr . . . plee-plee-plee . . . tsee . . . tsee-tsi . . . tlee-lee-lee . . .' followed by a short twitter including some fluted notes, or in a more sustained and repetitive medley of high-pitched notes with a metallic tone, 'ti-ti-ti-ti . . . ti-ti-tew . . . plee-plee . . . tu-tu-tut . . . tee-tweet . . . piup-piup . . . peeu-peeu . . . ti-ti-ti-wee-wit . . . tiu-weee-weee-weee . . .'. In Rwanda, song is a high-pitched twitter 'twilit, tchri-ri-ri-rit . . .', almost identical to that of White-tailed Crested Flycatcher *E. albonotata*; one bird imitated 2 song types of the local Neumann's Warbler *Hemitesia neumanni* (Dowsett-Lemaire 1990). Contact call a striking, sharp 'pink' or 'slip' in Cameroon, and dry 'frrit, frrit' in Nigeria.

General Habits. Inhabits primary and secondary montane forests, including more open areas; particularly common in fairly closed to semi-open understorey with small gaps, avoiding large clearings (Dowsett-Lemaire 1990); above 1350 m, mainly at 1700–2200 m in Nigeria, between 900 and 2500 m in Cameroon, between 1210–1300 and 1980 m in Itombwe, Zaïre, usually below 2200 m, locally to 2350 m in Rwanda and E Africa. Said to be sympatric with Dusky Crested Flycatcher at least between 1300 and 1400 m in Itombwe (Prigogine 1971, 1978a, 1984a), though collecting localities differ, suggesting checkerboard distribution.

Single or, usually, in pairs or family parties. Rather tame and inquisitive. General behaviour apparently much like that of Dusky Crested Flycatcher but detailed study highly desirable. Behaves and moves more like a tit or warbler than a true flycatcher. Forages actively, gleaning foliage like a tit or warbler, hopping along branches and from twig to twig, twisting body with open wings and sideways movements of partly fanned tail; keeps to low vegetation of forest understorey, where it picks detected prey from under a leaf or snaps it up in the air after short, often descending flight. Frequently joins mixed-species flocks.

Food. Insects, including very small beetles and Hemiptera.

Breeding Habits. Monogamous. Territorial, but behaviour still undocumented.

NEST: only two described; small, compact, neat and well-formed bowl-shaped cup; thick wall mainly of moss bound with spider web and covered outside with lichens, particularly around rim; inside of cup lined with plant down and fine vegetable fibres or grass-stems; fixed to support by gossamer; ext. diam. 65–70, ext. depth 51–80; int. diam. 38–40, int. depth 30–36. Placed 0·5–1 m above ground in vertical fork of small sapling.

EGGS: 2 (only 3 clutches known). Blunt ovate; little or no gloss; shell thin; yellowish white with a solid ring of dull chocolate-, yellow-, tawny- and purplish brown spots round broadest part of shell, near centre. SIZE: $c.$ 17×12 (Toro, Uganda); $16·8 \times 12$ and $16·9 \times 12·2$ (Mt Cameroon).

LAYING DATES: Nigeria (♀ with ovary enlarged Mar, ♂ with testes enlarged Dec); Cameroon, Nov, Feb, during dry season; Zaïre (♂♂ with testes enlarged Apr); Rwanda, Aug, Sept; E Africa: Uganda, Sept; Region B, Apr.

Nothing further known.

References
Stuart, S. N. and Jensen, F. P. (1986).
Dowsett-Lemaire, F. (1990).

Elminia albonotata (Sharpe). White-tailed Crested Flycatcher. Tchitrec à queue frangée.

Plate 28
(Opp. p. 497)

Trochocercus albonotatus Sharpe, 1891. Ibis, 6 (3), p. 121; Mt Elgon, Uganda/Kenya.

Range and Status. Endemic, resident. Montane and highland forests in SW Kenya (Mt Elgon, Cheranganis and Elgeyu south to Kakamega Forest, Eldama ravine and Mau Forest, east to Nyambenis, Mt Kenya and Limuru, also Taita Hills), SW Uganda (Rwenzoris and Bwindi–Impenetrable Forest), E Zaïre (Rwenzoris and L. Edward to Mt Kabobo), Rwanda (Nyungwe Forest), Burundi (Bukeye, Kibira Nat. Park, Rwegura, Teza), Tanzania (Ufipa plateau and E and S highlands (not Mts Kilimanjaro or Meru), south to Mt Rungwe, Umalila, Njombe and Songea), NE Zambia (Nyika Plateau and Northern Province at Muzyatama, Danger hills, Kasama, Mbala and Mporokoso), Malaŵi (absent only on Dzalanyamas, Ntchisi and Chipata Mts), Mozambique (Milange and Namuli Mt to Cholo Mt, also Gorongoza Mt and neighbouring highlands) and Zimbabwe (common, Inyanga and Chipinga highlands and Chimanimani Mts). Uncommon to common. In Malaŵi, density 5–10 pairs/10 ha (Dowsett-Lemaire 1989); smallest forest patch with 1 pair 0·3 ha, and with 2 pairs 1·7 ha (Dowsett-Lemaire and Dowsett 1984). 67 pairs or 201 individuals/100 ha in secondary forest in Udzungwa Mts, Tanzania (Moyer 1993).

Description. *E. a. albonotata* (Sharpe): Zaïre, Rwanda, Burundi, Uganda, Kenya, SW Tanzania (Ufipa Plateau), N Malaŵi (Viphya Mts northward), and Zambia. ADULT ♂: forehead, crown, lores, ear-coverts, chin, throat and neck black, more velvety on forehead and crown which have elongated feathers, forming crest; scapulars, mantle, back and rump dark blue-grey. Uppertail-coverts and tail brown-black, tail-feathers graduated, T3–T6 with broad white tip, broader on outer web, underside of shafts white. Breast and flanks dark slate-grey, centre of lower breast, belly and undertail-coverts white. Primaries, secondaries, tertials and upperwing-coverts brown-black, edged grey along outer web. Axillaries and underwing-coverts white (coverts with grey base and broad white tip); underside of flight feathers dark brown edged whitish along inner web. Bill sooty black; gape deep pink to yellow; eyes dark sepia or brown; legs dark slate to black, soles greyish yellow. ADULT ♀: like ♂ but more slate-grey, less blue; throat less black; tail-feathers less graduated. SIZE: (13 ♂♂, 12 ♀♀) wing, ♂ 61·5–71 (65·8), ♀ 60–65 (63·0); tail, ♂ 68·5–79·5 (73·1), ♀ 66·5–73·5 (70·3); bill, ♂ 12–13·5 (12·8), ♀ 11·5–13 (12·4); tarsus, ♂ 16·5–18·5 (17·2), ♀ 15·5–18 (16·9). WEIGHT: (Uganda, Mar, May, June) ♂ (n = 3) 8·5–9 (8·8), ♀ (n = 4) 8–10 (9·2); (Kenya, Sept) ♂

(n = 8) 5–10 (7·4), ♀ (n = 3) 6–8 (6·7), 1 unsexed 7; (Tanzania) 2 ♂♂ 8·3, 8·7, 2 ♀♀ 7·7, 7·8; (Zambia, Mar, June) 1 ♂ 9·1, 2 ♀♀ 7·8, 8·2.

IMMATURE: like adult ♀ but duller; head greyer, particularly throat and upper breast; gape yellowish grey.

JUVENILE: like immature but still duller, more sooty with browner head and wings; with swollen, bright yellow skin at gape; legs and feet pale slate with soles markedly yellow; rictal bristles very fine or absent.

NESTLING: hatches naked, bill pale yellow, later covered with scanty short grey down.

E. a. subcaerulea (Grote): E and S Tanzania, central and S Malaŵi (north to Dedza and Chongoni Mts) and Mozambique north of Zambezi. Upperparts and breast paler than nominate subspecies, less sooty grey. WEIGHT: (Tanzania, May, Nov) 2 ♂♂ 6, 7, 2 ♀♀ 8, 10; (Udzungwa Mts, Tanzania) unsexed (n = 17) 6·6–9·3 (8·2).

E. a. swynnertoni (Neumann): E Zimbabwe and Mozambique south of Zambezi. Somewhat smaller than nominate *albonotata*, cheeks and throat grey, back almost slate-black, less white on tail. WEIGHT: (Zimbabwe) ♂ (n = 3) 8·2–9·1 (8·5), ♀ (n = 3) 7·0–8·2 (7·7), unsexed (n = 3) 7·1–7·9 (7·4); (Mozambique) 1 ♂ 7·3, 1 ♀ 8·7.

Field Characters. Small and dull-coloured, almost uniform dark grey (blacker on head) except for pale centre of belly. Distinguished from all *Elminia* and *Trochocercus* flycatchers except White-tailed Blue Flycatcher *E. albicauda* by white in tail, conspicuous when tail is fanned; further told from Blue-mantled Crested Flycatcher *T. cyanomelas* by lack of white in wing. White-tailed Blue Flycatcher is much paler, blue-grey above and whiter below, with black line from bill to eye.

Voice. Tape-recorded (86, 88, 91, B, C, F, KEI, LEM). Song weak, a formless jumble of notes, some sharp, some sweet, some buzzy, interspersed with short chattery trills: 'swit-chip-sweep-tak-weechit-bzeet-weedrowee-trrrr . . .'; song in Zimbabwe, a rich 'tilip-tulweet-tulweet-taytaytaytay-tiplu-tiplu-pli-piptuweetiti wee-taktak-uipteewee-uitteeweet . . .'; described in South Africa as 'a stuttering series of thin sharp twittering whistled notes', 'chrit-it-it-it-it' or 'tait-chrit-itit-itit-chreep' (Maclean 1993), and as a high-pitched twitter 'twilit, tchri-ri-ri-rit . . .' (Dowsett-Lemaire 1990). Contact call, single 'zitt', or a 2-syllabled note, also a fairly rapid 'weet-weet-weet' when feeding, and a repeated rapid trill, 'ptitititititi'. Warning call, 3–4 rapid 'zitt's followed by a chitter; alarm, sharp 'zitt'; nestling call, rapid 'ti-ti-ti-ti'.

General Habits. Inhabits Afromontane and transitional vegetation types (e.g. on Gorongoza Mt in Mozambique), montane and mid-altitude evergreen forests, even when degraded, and adjacent bracken or scrub, also riparian forest patches along rivers; up to 1800 m in E Tanzania and 2050 m in Burundi; confined to altitudes above 2250 m in Nyungwe Forest, Rwanda, by its competitor, White-bellied Crested Flycatcher *E. albiventris*; occurs from 1850 to 2770 m in Itombwe, Zaïre, up to 2300 m in Zimbabwe and 2700 m in Uganda, but down to 1200 m and even 1050 m in Malaŵi where White-bellied Crested Flycatcher absent.

In Malaŵi, Zambia and Zimbabwe, segregated ecologically from slightly larger Blue-mantled Crested Flycatcher, which lives either at lower altitudes or in different habitat, e.g. fringing thickets and lower vegetation levels.

In pairs or family parties. Frequents understorey and lower canopy. Capture/recapture study gave estimate of home range, Zimbabwe, of *c.* 150 cm^2 (Manson 1992), but that estimate seems far too low. Forages restlessly with raised crest among foliage, at all levels up to canopy, although usually 1–8 m above ground, where Blue-mantled Crested Flycatcher absent; hops and flits rapidly along branches in zig-zag fashion, dislodges insects with constant weaving movements of body with fanned tail and drooping wings; makes frequent sallies after prey in looping flight, picking it from leaves or bark, or in small open space in understorey. Flight jerky. Highly vocal. Regularly joins mixed-species flocks.

In Malaŵi, some movement between forest patches: 39 ringed birds moved 0–2450 m (mean 251 ± 441 m). Long-term site fidelity followed for 38 ringed birds: 20 were present at same site for one or more of the following seasons, 14 for 2 years, 3 for 3 years, 1 for 4 years and 2 for 7 years (Dowsett 1985). Translocation experiments showed that empty territories may take up to 3 years to be refilled. Some post-breeding downward movement recorded in Malaŵi in Mar, July and Dec, when birds occur at altitudes below 1200 m. Similar local movements also recorded in Zimbabwe in May–Aug.

Food. Small arthropods, mainly insects: flies, ants, small grubs and moths (even distasteful ones); also spiders.

Breeding Habits. Monogamous, territorial. Territory size in Malaŵi (n = 23) 1·5 ha (i.e. av. diam. 140 m: Dowsett 1985). ♀ begs and is fed by ♂ for only a few days at onset of breeding. ♂ seen displaying around ♀ with tail raised vertically and wings drooped below feet (Sclater and Moreau 1933).

NEST: small, neat, tight V-shaped or goblet-shaped cup of closely woven green moss bound with spider webs, sometimes also with greyish white lichens; inner cup lined with thin materials, e.g. pine needles; ext. diam. 70, ext. depth 95; int. diam. 40, int. depth 30. Usually placed in vertical fork near top of sapling, weed or shrub, 1–2 m above ground, sometimes higher, up to 6·5 m. In Zimbabwe, most favoured trees are *Peddiea africana* and *Xymalos monospora* (Manson 1992). Built by ♀ in *c.* 12 days, accompanied by ♂ which calls and sings.

EGGS: 1–3; Malaŵi mean (3 clutches) 2·0. Oval to pyriform; glossy; white or buffy cream, blotched and spotted brown, olive and lilac-grey, sometimes in zone around larger end. SIZE: (n = 3, southern Africa) 14·2–16·3 × 11·7–13·1 (15·5 × 12·4).

LAYING DATES: Zaïre (♂ in breeding condition May, Sept, Nov, Dec); Burundi (♂♂ with enlarged testes Aug); E Africa: Region D, Oct, before short rains; Malaŵi, Sept–Dec, with peak in Sept–Oct; Zambia (birds with gonads active Nov); Zimbabwe, Oct–Jan, mainly Nov–Dec; Mozambique, Jan.

INCUBATION: begins with last egg; by ♀. Period: probably 15–16 days. ♀ sits at night and for more than two-thirds of day; ♂ feeds her at nest; ♀ leaves nest frequently, often escorted by ♂. Sitting ♀ may join ♂ and defend territory against intruding neighbouring pair. ♂ and sitting ♀ call frequently during early stages of incubation but later become more secretive.

DEVELOPMENT AND CARE OF YOUNG: both adults forage in vicinity of nest and feed nestlings in quick and silent visits, using set approach routes. Faecal sacs frequently removed. When human observer nears recent fledglings, ♂ gives agitated displacement display, threatening and calling with drooped wings and fanned tail.

BREEDING SUCCESS/SURVIVAL: of 13 nests in Zimbabwe 5 were successful, 1 probably so, 5 unsuccessful and outcome unknown in remaining 2 (Manson 1992). During a 10-year ringing study in Zimbabwe, 19 juvs were caught, none was recaptured after 8 months; of 22 immatures, 9 were last recaptured between 1 and 4 years; of 73 adults, 8 were retrapped 1–6·5 years later (Manson 1992). One bird ringed as immature recaptured after 6 years, 117 days (Zimbabwe: Harwin *et al.* 1994). On Nyika Plateau, Malaŵi, av. 36% of pairs (60%, 8% and 46% in 3 successive years) had fledglings, av. 1·0 per successful pair but 0·4 per established pair (0·6, 0·1 and 0·5 per established pair in the same 3 successive years) (Dowsett-Lemaire 1985). In the same area, over a period of 10 years, ringing data gave a mortality rate of 40% for ♂♂ (n = 11) and 61% for ♀♀ (n = 13), with an average further life expectancy of 1·1–2·0 years; minimum age of oldest individual, 7·4 years (Dowsett 1985).

References
Dowsett, R. J. (1985).
Dowsett-Lemaire, F. (1985, 1989).
Manson, A. J. (1992).
Vincent, A. W. (1947).

Genus *Trochocercus* Cabanis

On anatomical grounds considered by Erard (1987) and Olson (1989) as a synonym of *Terpsiphone*, of which it might be a subgenus; full genus status adopted here is conservative. Very like *Terpsiphone* but smaller, bill flatter and a bit broader, crest longer and more pointed. Sexually dimorphic. Glossy black head and breast, grey or glossy black back contrasting with white or pale grey belly; white in wing in one species (*cyanomelas*). Foraging behaviour characteristic, with sweeping fanned tail, drooped wings and rapid pivoting movements of body from side to side. Though calls similar to *Terpsiphone*, song is a prolonged ringing trill or tremolo. Nest, a neat, tight open cup, more rounded, less V-shaped than *Terpsiphone*, with more vegetable fibres. Arboreal; forest.

Endemic. 2 species.

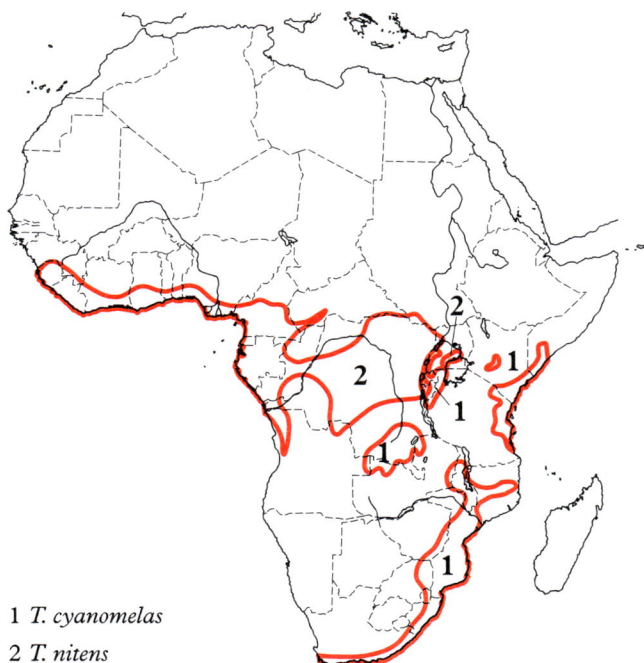

Trochocercus cyanomelas superspecies

1 *T. cyanomelas*
2 *T. nitens*

Plate 29
(Opp. p. 544)

Trochocercus cyanomelas (Vieillot). Blue-mantled Crested Flycatcher. Tchitrec du Cap.

Muscicapa cyanomelas Vieillot, 1818. Nouv. Dict. Hist. Nat., nouv. éd., 21, p. 473; Auteniquoi, Knysna district, Cape Province, South Africa.

Forms a superspecies with *T. nitens*.

Range and Status. Endemic, resident. Distribution patchy and discontinuous. Coastal and lowland forests of Somalia (lower Juba R. to Kenya border), Kenya east of Rift, on coast north to lower Tana R. and inland to Tsavo, and separate population in highlands from Ngong, Nairobi and Mt Endau to Mt Kenya, Meru and Mt Uraguess; most highland records are old and range may be contracting, but pair in Meru Forest, 1991 (B. W. Finch *in* Backhurst 1993); Tanzania (Zanzibar and coast, inland to Mt Kilimanjaro, Arusha, Kidugallo, Ulugurus, Ngurus, E Usambara and Luwegu R., and in W from Bukoba, Karagwe, Kibondo, Gombe Stream Game Reserve to Mt Mahari and Karema); Uganda (Bwindi–Impenetrable Forest, old records from Lugalambo, Kibirau, Kibale, Mubende and Bagaya I.), Rwanda (uncommon, Nyungwe forest), Burundi, Zaïre (Shaba), N and W Zambia (uncommon, Luapula Province at Kawambwa and Samfya, and Northwestern Province from Solwezi and Mundwiji plain to Sakeji Stream and source of Zambezi, also Salujinga, Mayau and Kasempa), Malaŵi (south of Nkhata Bay but numerous only in lower Shire valley and lake littoral at Nkhata Bay district); Mozambique (north to Zambezi and locally to Cabo Delgado, but apparently absent from most of N); E Zimbabwe (fairly common Sabi valley, sparse and local Turgwe R., Mutema, Honde valley and Chipinga uplands to Mt Selinda; has bred Bvumba Highlands (Harwin *et al.* 1994); one record Victoria Falls); E Transvaal (scarce to fairly common, Soutpansberg and Escarpment region), Swaziland, Zululand, and coastal areas from Natal and Transkei to SW Cape Province. Uncommon to frequent. Density in Malaŵi, at least 10 pairs per 10 ha (Dowsett-Lemaire 1989).

Trochocercus cyanomelas

Description. *T. c. bivittatus* Reichenow: Somalia, Kenya, E Tanzania and Zanzibar. ADULT ♂: forehead, crown, lores, chin, throat, foreneck and upper breast glossy blue-black; feathers of crown elongated, forming long pointed crest; ear-coverts and hindneck dark sooty grey. Mantle, back, rump and scapulars dark mouse-grey to dark blue-grey, slightly glossed bluish. Uppertail-coverts dark slate-grey; graduated tail-feathers brown-black, edged mouse-grey along outer web, underside of shafts white. Lower breast and flanks white mottled dark sooty grey; belly and undertail-coverts buffy white (feathers of belly white with dark grey base). Primaries dark brown; secondaries brown-black narrowly edged mouse-grey along outer web; tertials white; upperwing-coverts brown-black, lesser coverts edged slate-grey. Axillaries white; underwing-coverts dark mouse-grey, underside of flight feathers dark brown with buff edge to inner web. Bill black with blue sheen or bluish slate-grey with black tip and blackish around nostrils; gape yellow; eyes brown to dark sepia; legs and feet slate-grey to black. ADULT ♀: like ♂ but lores whitish, ear-coverts grey, chin, throat, foreneck, sides of neck and upper breast dull black with greenish gloss, mottled with white drop-shaped spots, sometimes giving striped effect (each feather blackish with shaft and broad subterminal spot white and narrow bluish grey margin which becomes broader and glossy black at point where breast meets white underparts); tertials brown-black edged slate-grey along outer web; median and greater upperwing-coverts brown-black, edged slate-grey on outer web and tipped white. Bill grey-blue; eyes brown; legs blue-grey to slate-blue. SIZE: (9 ♂♂, 14 ♀♀) wing, ♂ 62·5–72 (68·6), ♀ 62·5–69·5 (65·5); tail, ♂ 68·5–77 (73·1), ♀ 64–76 (70·0); bill, ♂ 14–15 (14·4), ♀ 13·5–15 (14·3); tarsus, ♂ 16–17·5 (17·1), ♀ 15–17·5 (16·4). WEIGHT: (Somalia) ♂ (n = 13) 9·6–12·1 (10·2), ♀ (n = 9) 9·3–11·2 (10·1); (Kenya, Jan–May) ♂ (n = 11) 7–12 (9·7), ♀ (n = 6) 6–12 (9·7), 2 unsexed 9, 10.

IMMATURE: like adult ♀ but duller; crest shorter and greyer, much less glossy; upperparts washed with olive-brown; greater upperwing-coverts with small rufous tips; bill brownish horn, yellowish white basally.

JUVENILE: like immature but more grey-brown; without metallic gloss on head and mantle; crest very short, grey; upperwing-coverts tipped rusty, spots smaller; flight feathers much browner.

NESTLING: undescribed.

T. c. cyanomelas (Vieillot): South Africa: W Transkei to about Swellendam in SW Cape Province. Whiter on upperwing-coverts, ♀ less grey, ♂ with shorter crest.

T. c. vivax Neave: Uganda and W Tanzania to Zaïre and Zambia, possibly also N Zimbabwe (Victoria Falls). Upperwing-coverts not white, only tipped grey. WEIGHT: (Upemba, Zaïre, July) 1 ♂ 14.

T. c. megalolophus Swynnerton: Malaŵi and N Mozambique to Zimbabwe and E Zululand (north from Mtunzini and Richard's Bay). Crest much longer than *bivittatus*.

T. c. segregus Clancey: E Transvaal to Natal and W Zululand (north to Lebombo). Smaller, tail shorter, upperparts paler blue-grey, upperside of tail greyer, less blackish, and underparts whiter than nominate subspecies.

Field Characters. A noisy little flycatcher which seems most at home in coastal lowlands. ♂ has glossy blue-black head and breast, grey upperparts, black tail and white belly; ♀ similar but greyer, with mottled breast and white lores. In eastern races both sexes have conspicuous white wing-patch which, together with lack of white in tail, distinguishes them from White-tailed Crested Flycatcher *E. albonotata.* Western race *vivax* lacks white patch but has 2 narrow pale wingbars. Song and calls very like those of paradise-flycatchers *Terpsiphone*, very different from the weak little voices of other crested flycatchers (*Elminia* spp.). White-bellied Crested Flycatcher *E. albiventris*, which it overlaps in W, lacks crest and is mainly blackish above and grey below, with white only on belly.

Voice. Tape-recorded (9, 14, 58, 86, 88, 91, KEI, B, F). Song loud and ringing, introductory notes followed by liquid trill, 'zweep-tyew-too-tutututututu . . .', lasting *c.* 2 s, or mellower and slower 'zweet-zweet-zweet-kew-ew-ew-ew-ew-ew-ew-ew-ew-ew'; often interspersed with harsh 'zweet' or followed by a series of loud clicks; variable; other versions include high nasal, 'tzeet-zerdt-zerdt-wee-wee-wee-wee', or more liquid 'psî-psî-ti-ti-trree-u-u-u-u-u-u-u-u' or 'tsee-wu-u-u-u-u-u-u-u'; all strongly reminiscent of paradise-flycatcher. Less frequently, a quick *Prinia*-like 'tip-tipp'ing (Vincent 1935). Fighting or mobbing call, a harsh guttural 'tchrray-tchrray-tchrray' or 'tchrray-tchrray-rrrrwayway'. When excited, gives rapid 'pikpikpik'. Contact call, a rasping, *Terpsiphone*-like double or triple 'zi, za', 'zweet-zwa', 'tyi-tyay', 'tyi-tay', 'tchi-tchayt-tchayt', 'tzeet-zerdt-zerdt', first note highest in pitch, also 'tsi-tyay-tyay' or 'zweet-zweet-zwa', interspersed with twittering 'titititi'.

General Habits. Inhabits dense moist thickets of coastal, lowland and mid-altitude evergreen forest, mavunda (*Cryptosepalum*) forest (in Zambia), also riparian forest, and muhulus in Shaba (Prigogine 1969b); up to 1100 m in E Tanzania, 1900 m on Mt Kilimanjaro, 1400 m in Malaŵi, 1800 m in Kenya, 2300 m in Rwanda and Burundi. In Nyungwe Forest, Rwanda, occurs especially on dry ridges with thin canopy and understorey thickets, e.g. of *Mimulopsis*. Occupies even very small forest patches. Usually in understorey and lower and middle strata of forest, usually below 8 m, typically in thick undergrowth and liana tangles; often patchily distributed because confined to densest parts of undergrowth.

Usually solitary or in pairs. Highly vocal. Rather restless and shy; forages actively among foliage; constantly flits and hops about, swinging body from side to side and often fanning tail and drooping wings; not as demonstrative and movements not as rapid as *Elminia* spp. Gleans prey from leaves and twigs in short looping flights or picks it from branch. Sometimes joins mixed-species flocks.

Some local seasonal movements may occur in southern Africa, in June–July, also in Dec in Zimbabwe.

Food. Small insects.

Breeding Habits. Monogamous, territorial. Territory advertised by song; ♂ first responds to rival ♂ with contact and excitement calls while raising and opening crest, drooping wings and fanning tail, then sings, and finally gives fight calls if intruder enters territory or when song played back.

NEST: neat, thick cup of bark fibres, fine grass and green moss, with silvery lichens on outside, bound together with spider webs. Usually placed 0·5–8 m above ground in leafy fork of tree or bush.

EGGS: 2–3; southern Africa mean (9 clutches) 2·2. Oval; glossy; cream or whitish, tinged pink, blotched and speckled red-brown, greenish brown and brownish and purplish grey, often in ring at larger end. SIZE: (n = 10, southern Africa) 15·8–22·9 × 12·2–15·1 (18·8 × 13·8).

LAYING DATES: Zaïre, Sept; Rwanda, Sept; Uganda, Apr; Zambia (birds with gonads active Sept); Malaŵi, Feb, Dec; Zimbabwe, Oct–Jan, mainly Nov–Dec; Transvaal, Dec; Natal, Oct–Dec; SW Cape Province, Oct–Jan.

References
Dowsett, R. J. (1989).
Maclean, G. L. (1993).

Trochocercus nitens Cassin. Blue-headed Crested Flycatcher. Tchitrec noir.

Plate 29
(Opp. p. 544)

Trochocercus nitens Cassin, 1859. Proc. Acad. Nat. Sci. Philadelphia, 11, p. 50; Camma R., Gabon.

Forms a superspecies with *T. cyanomelas*.

Range and Status. Endemic resident. Lowland forests of Sierra Leone, Guinea (Macenta district, and sight record, Sougueta (T. Aversa, pers. comm.)), Liberia, Ivory Coast (Taï and Fresco to Mt Nimba, Gagnoa, Lamto and Yapo forest), Ghana (known from Fanti, Prashu, Mampong in Ashanti and Akropong in Akwapim), S Nigeria (Lagos to Sapele, Ibadan and Ife, also recorded Umuagwu and Yankari), W and S Cameroon (from Korup Nat. Park, Mt Kupé and Mt Nlonako, south to upper Sanaga R., Dja, Boumba and Sangha R.), Central African Republic (Dzanga reserves, Botambi and Bangui area, also possibly Lobaye), Rio

Trochocercus nitens

Muni, Gabon, Congo (coast and Mayombe, probably also area between NE Gabon and SE Cameroon), Angola (Cabinda, Uige, Canzele and Gabela), Zaïre (coast and Mayombe, and from middle Congo R. east to Uele and Ituri, south to Itombwe and Kasai), Rwanda, Burundi, Uganda (Bwamba, Fort Portal, Lugalambo, Malabigambo and Mabira), S Sudan (Bengengai). Uncommon to common. Density in NE Gabon av. 22 pairs/km^2 but varies between 29·6 and 21·4 pairs/km^2 depending on abundance of dense clumps of lianas (Erard 1987).

Description. *T. n. nitens* Cassin: Nigeria and Cameroon to Angola and Uganda. ADULT ♂: entirely black with bright blue gloss, except: lower breast, belly, flanks and undertail-coverts ash-grey; narrow whitish line across breast separating black upper breast from grey lower breast; primaries and secondaries matt, not glossy; underside of flight feathers blackish grey. Crown feathers elongated. Tail-feathers graduated (T1 longest, T6 one fourth shorter). Bill blue-grey, upper mandible blacker; gape yellow; eyes brown; legs and feet blue-grey to blackish brown. ADULT ♀: forehead and crown black with bright bottle-green gloss, crown feathers elongated but shorter than ♂; lores, ear-coverts and sides of neck pale grey mottled blackish grey. Nape, hindneck, mantle, back and rump dark slate-grey, with a slight greenish gloss. Uppertail-coverts and tail dark brown-grey with slight green gloss, underside of shaft white. Entire underparts ash-grey. Primaries, secondaries and tertials dark brown-grey with slight greenish grey gloss on outer web; upperwing-coverts dark brown-grey, broadly edged slate-grey with greenish gloss. Axillaries and underwing-coverts whitish. Bill dark slate-grey, upper mandible blacker; gape greenish yellow; eyes brown; legs dark brown. SIZE: (7 ♂♂, 4 ♀♀) wing, ♂ 63–67 (64·7), ♀ 62–64 (63·4); tail, ♂ 69–74 (70·9), ♀ 66–71 (67·9); bill, ♂ 14–17 (15·2), ♀ 14–16 (14·7); tarsus, ♂ 16–17 (16·3), ♀ 16–17 (16·2). WEIGHT: (Cameroon) ♂♀ (n = 7) 9·8–13·5 (11·8); (Gabon) ♂♀ (n = 6) 11–13 (11·7); (Congo) 1 ♀ 10·8; (Uganda, June, July, Nov) ♂ (n = 7) 10–13 (11·8), ♀ (n = 4) 10–15 (11·9).

IMMATURE: like adult ♀ but top of head without gloss; upperparts and underparts browner.
JUVENILE: like immature but upperparts dark sooty brown, underparts pale grey-brown, darker on breast and greyer on throat and sides of head; bill dark brown, most of lower mandible horn; legs pale brown-grey.
NESTLING: hatches naked, skin pink, later sparse down sooty brown, then nestling plumage like juvenile, unspotted.
T. n. reichenowi Sharpe: Sierra Leone to Togo. Underparts darker grey. WEIGHT: (Liberia) ♂ (n = 6) 10·9 ± 1·4.

Field Characters. A small, dark, crested and fan-tailed foliage-gleaner, foraging actively among tangles with widely spread tail and wings. ♂ glossy blue-black except for grey breast and belly; ♀ glossed only on top of head and crest, otherwise dark grey above, pale grey below, with spotted face and neck. Sympatric Dusky Crested Flycatcher *Elminia nigromitrata* is mainly sooty black, and montane White-bellied Crested Flycatchers *E. albiventris* is like it but with white belly. Distinguished from Blue-mantled Crested Flycatcher *T. cyanomelas* and montane White-tailed Crested Flycatcher *E. albonotata* by lack of white on tail or wing. Overlaps in E Zaïre with larger Bedford's Paradise-Flycatcher *Terpsiphone bedfordi*, which differs in being uniform dark blue-grey with black head.

Voice. Tape-recorded (C, CAR, CHA, ERA, LEM). Territorial song, a rapid series of short whistles, a long tremolo 'ou-ou-ou-ou-ou-ou-ou-ou-ou- . . .'; not particularly loud, rather low-pitched for such a small bird, but far-carrying; often introduced, especially in conflicts or high intensity territorial defence, by a harsh, guttural 'tchict-krr-krr-kiwéé wéé wéé' which may be used as fight call, and also by an excited rattle, 'tictictic-tictic . . .' or 'kwickwickwickwick . . .' ('castanets' call). Contact and alarm calls harsh, buzzy and squeaky, e.g. 'tchiit-tchict' or 'tchitchiitchweet', which sound like distorted and higher-pitched calls of paradise-flycatchers *Terpsiphone*; ♀ call when building nest, a soft 'tchtchtchtchtch'.

General Habits. Inhabits lowland primary and secondary forest, up to 900 m in Uganda, 1100 m in Itombwe (Zaïre) and 1200 m in Cameroon. Preferred habitat is large, dense tangles and curtains of vines and lianas. Occurs in wettest parts of forest, in dense tangles along rivers and streams, and in low-lying disturbed areas clogged with vine tangles which make understorey almost impenetrable; also around old tree-falls, in old second growth, old uncleared and disused plantations where forest is growing back, riverine forest, and in tall stands of Marantaceae and *Aframomum* herbs among cultivation.

Solitary or in pairs. Forages actively most of day, though less during hottest hours of afternoon. Usually pair members not far from each other, maintain acoustic or visual contact. Gleans foliage and tangles of leafy lianas or tightly packed stems of vines, from near ground to canopy; in primary forest in NE Gabon, 66% of records at 14–20 m during long dry season (June–Aug) and long rains (Feb–May) but at 6–12 m

during short rainy (Sept–Nov) and dry (Dec–Jan) seasons (Erard 1987). Moves actively, always in shadow of dense tangles and leaves; searches vegetation with fully fanned tail, spread and drooped wings (**A**), body moving up and down and twisting from side to side; rapidly beats wings and twists tail from side to side to dislodge prey; snaps up insects in air in short sally or circular descending flight. Uncommonly, moves among leaves, jumping from branch to branch in upright posture, gleaning undersides of leaves or hovering. Sometimes, particularly when feeding on swarming ants or termites, forages from perch like true flycatcher, assuming upright stance (**B**) and making quick sallies after flying insects, including those dislodged by other gleaners, e.g. sunbirds *Anthreptes* and bulbuls *Phyllastrephus*. Works underside of leaves in undulating flight, brushing them with its tail-feathers and snapping up prey with a quick dart. Often stretches wings wide (**C**) before pushing itself into mass of dead leaves or other very dense vegetation. Works its way inward and upward along branches and hanging leafy vines. Repertoire of foraging tactics thus similar to that of true *Terpsiphone* species but differs in relative use of each method. Regularly joins mixed-species flocks, e.g. 42·9% of records in NE Gabon, more frequently during long rainy and dry seasons (48·6 and 58·1% respectively) than during short rainy and dry seasons (34·7 and 33·5% respectively). When alarmed, e.g. in response to hawk flying nearby, freezes on branch with sleeked plumage, closed crest, wings and tail. Usual alert posture is crouched body with closed crest, wings and tail, head hunched between shoulders; when alarmed stands more upright with crest erect and calls while pivoting from side to side, often flapping wings spasmodically.

Sedentary, at least in NE Gabon.

Food. Arthropods, mostly insects: moths, Orthoptera, Coleoptera, Hymenoptera (mainly ants), cockroaches, also winged termites and spiders. Prey size av. 12·6 mm (n = 58) in NE Gabon, items eaten by adults may be larger than those given to nestlings, but further study required.

Breeding Habits. Monogamous; apparently pair-bond in NE Gabon lengthy, perhaps for life. During pair formation, courting ♂ approaches ♀ in upright posture, with sleeked plumage, swings partly fanned tail, shivers wing-tips and gives soft guttural calls, displaying inside of mouth, then droops, quivers wings and crouches with closed crest straight and erect and tail fanned and raised, shivers and gently flaps wings and sings; responsive ♀ crouches with raised spread crest and fanned tail, and quivers drooped wing-tips; ♂ then makes a bow with rapid wing-quivering, while ♀ crouches with tail spread and quivers drooped wing-tips; ♂ then sings in more upright posture and continues courting in alternately crouched and upright positions. Territorial. Territory size av. 3·3 ha (n = 13) in NE Gabon; ♂ defends it all year and patrols it every day, singing from dense vegetation. ♂ approaches intruder in upright posture, giving guttural and rattling calls with wide open bill, then attacks and chases opponent with 'castanets' calls (see *Voice*). In fight posture ♂ holds body upright, with sleeked plumage but breast puffed out, straight erect crest, bill pointing up, wings held away from body with drooped tips, swings half spread tail and spasmodically flaps one or both wings, and gives 'castanets' call. Then with breast still inflated and closed crest raised like a spur, fully fans tail and crouches, facing intruder, with open bill, tail raised, wing-tips drooped, carpal joints held well away from body; plumage flattened except sides puffed out. During precopulatory behaviour, pair members approach each other in crouched posture with erect crest and fanned and raised tail and quiver drooped wing-tips; ♂ displays coloured inside of mouth and gives guttural rattle and songs. During courtship feeding, ♀ crouches and behaves like a young bird.

NEST: neat compact cup of vegetable fibres tightly interwoven with some dead leaves at base and bound with spider web, lined internally with finer material; outer diam. 60, height 60, inner diam. 45, int. depth 33. Placed 2–15 m above ground; in NE Gabon av. height 10·2 m (n = 7); typically 2–3 m underneath a very thick dome of dense leafy lianas, attached near end of thin hanging dead vine, either suspended as a basket between 2 thin (diam. *c.* 5 mm) parallel stems or placed in fork. Built by both sexes, though ♀ works more than ♂. ♂ chooses nest-site, where he crouches, extends head and neck horizontally toward ♀, raises and quivers fanned tail, opens bill wide and softly calls and twitters, also sings.

EGGS: 2, laid at 1-day intervals. Very similar to those of *Terpsiphone* spp. Oval; glossy; creamy white, finely speckled red-brown to bright ochre-rufous, often as a cap or crown at larger end. SIZE: *c.* 20 × 13.

LAYING DATES: Liberia (♂ with enlarged testes Dec); Ghana, (♂ with enlarged testes Feb); Cameroon (young June–Aug, Oct); Gabon, Sept–Feb (adults in breeding condition Aug); Congo, Sept; Central African Republic, (♀ in breeding condition July); Zaïre, Feb, Aug, Sept, Nov (♂♂ with enlarged testes Feb, Sept–Nov); E Africa: Region B, Apr.

INCUBATION: begins with last egg; by both sexes. Period unknown. ♂ and ♀ equally share incubation time by day, each sitting for 30–60 min; ♀ sits by night. ♂ guards nest surroundings and may sing from nest. Nest relief rapid and silent; ♂ often sings as he approaches, but becomes silent as he reaches nest. Adults do not attack stuffed owl placed near nest containing eggs: they stay hidden in dense vegetation in alarm.

DEVELOPMENT AND CARE OF YOUNG: nestling period *c.* 12 days. Young fed by both parents; mean interval between food deliveries 11·6 min (2–45 min; n = 39). Parents brood nestlings for first 4 days, sporadically later, and during heavy rains. At this stage both adults but particularly ♂ very aggressive toward stuffed owl placed near nest.

After leaving nest, young fed by both parents for more than one month; each parent may take care of a particular young. As soon as young able to follow their parents, they search leaves or move along vine stems with tail fanned and more or less spread wings.

BREEDING SUCCESS/SURVIVAL: of 6 nests observed in NE Gabon, only one fledged young, 3 were destroyed by storm, one was torn apart by an unknown animal before eggs were laid, and fate of last one unknown. During construction, one nest was visited several times by Dusky Long-tailed Cuckoo *Cercococcyx mechowi*.

References
Brosset, A. and Erard, C. (1986).
Erard, C. (1987).

Genus *Terpsiphone* Gloger

Medium-sized flycatchers, rather large-headed. Bill broad, more or less flattened, culmen usually ridged, cutting edges sharp; nostrils mostly overlain with short feathers and bristles; rictal bristles strong; gape luminous yellow or yellowish green; tarsometatarsus short but not weak; claws short, curved and sharp. Tail-feathers often rufous and strongly graduated; in some species median pair very long, black, rufous or white. Head often entirely or partly glossy black and crested. Plumage soft and full; general coloration rufous, grey, black or white; upperparts often contrast with underparts; in some species ♂ highly variable in colour and pattern (e.g. *T. viridis*, *T. paradisi*, *T. mutata*, see Salomonsen 1933a,b,c, Chapin 1948, Erard 1987). Sexually dimorphic. Juveniles unspotted. Very demonstrative birds, with rasping calls and loud, liquid and rippling songs (a character shared with the closely related oriental *Hypothymis azurea*). Nest a rather compact, neat and well formed V-shaped cup, well secured to support and bound with spider web or *Marasmius*, often covered with moss and/or lichens. Sometimes forage with fanned tail, but this behaviour is not characteristic. Arboreal; forest and savanna woodland.

13 species worldwide: in Africa, Malagasy Region, India and E Asia to Japan, Philippines and Lesser Sundas. We recognize 6 species in Africa, all but one endemic; but further work on ecological and behavioural segregation is needed to clarify species limits better. African species, smaller, with flatter bills and less curved culmens, were long separated in genus *Tchitrea* from Oriental species. Mainly sedentary, though some populations of *T. viridis* make long-distance migrations. One superspecies: *T. viridis*/*T. atrochalybeia* (also extralimital *T. mutata*/*T. corvina*/*T. bourbonnensis* (Malagasy) and Oriental *T. paradisi*/*T. atrocaudata*).

Terpsiphone viridis superspecies

1 *T. viridis*
2 *T. atrochalybeia*

Terpsiphone viridis (Müller). African Paradise-Flycatcher. Tchitrec d'Afrique.

Plate 29
(Opp. p. 544)

Muscicapa viridis P. L. S. Müller, 1776. Linné Natursyst. Suppl., p. 171; Senegal.

Forms a superspecies with *T. atrochalybeia* (and with *T. mutata* (Madagascar and Comoro Is), *T. corvina* (Seychelles), *T. bourbonnensis* (Mascarene Is), and *T. paradisi* and *T. atrocaudata* (Asia)).

Range and Status. Africa and Arabia (SW Saudi Arabia, Yemen and S Oman).

Resident and intra-African migrant. Mauritania (migrant, Senegal valley and Guidimaka), Senegal and Gambia (north to 15°N, occasionally to Richard-Toll); Guinea-Bissau, Guinea, Sierra Leone, Liberia (rare to uncommon, possibly migrant); Ivory Coast (uncommon over entire forest and guinea zones, but absent inside forest); Mali (north to Niafounké, Tombouctou and L. Faguibine, present all year only south of 13°N); Burkina Faso (north at least to Ouagadougou); Ghana (common in northernmost areas May–June, resident but less common elsewhere to Volta region, occasional on inselbergs of Accra plains, Legon and Weija, absent from coastal thicket zone); Togo, Benin, Nigeria (breeds south to about 7°N; migrates to forest areas in dry season, to dry N, e.g. Sokoto and Maiduguri, in rains); S Niger ('W' Nat. Park, Gaya and Gangara); S Chad (west and south of Sarh (Fort Archambault) to Bokoro and N'Djamena, common); Cameroon north to Koza and Waza; Rio Muni, Gabon, Congo; Angola (throughout except NW and extreme SW); Zaïre; Central African Republic (Dzanga reserves and Bouar, north and east to Birao, Ippy and Zemio); Sudan north

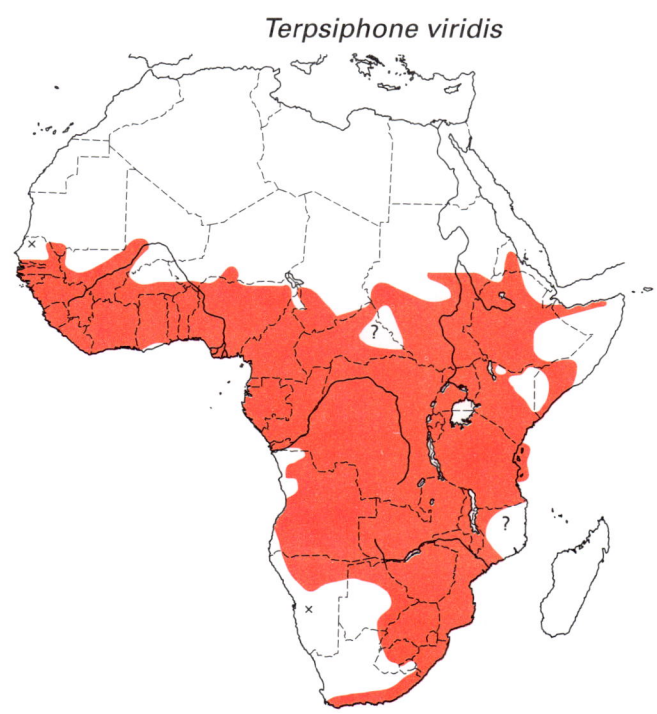

Terpsiphone viridis

to 14°N in Darfur, Kordofan, El Gezira and Dinder Nat. Park; Ethiopia except driest parts of Eritrea, Tigre, Wele, Harerge and Bale); S and N Somalia (north to 4°N on Shebelle and Guban to Carear Mts); Kenya except dry NE, breeding mostly in highlands and making off-season movements to arid and semi-arid country; Uganda, Rwanda, Burundi, Tanzania, Zambia, Malaŵi; Mozambique (except Cabo Delgado, E Niassa and N Zambézia?); Zimbabwe; South Africa (Zululand, Natal, Transvaal, Orange Free State, to Cape Town), Swaziland; Lesotho, 4 records, Maseru (Bonde 1993); N and E Botswana and N Namibia; vagrant Naukluft Mts (Boyer and Bridgeford 1988). Uncommon to abundant. Mean density in 2 localities in NE Gabon, respectively 15·5 and 17·9 pairs/km^2 (Erard 1987); 21 pairs or 63 individuals/km^2 in secondary forest in Udzungwa Mts, Tanzania (Moyer 1993).

Description. *T. v. speciosa* (Cassin): S Cameroon (east and south of Kumba and Bamenda) to Sudan, Zaïre (Mayombe east to Kasai and Manyema) and NE Angola (Kasai R.). ADULT ♂: highly variable in colour of upperparts, wings and tail; authors often speak of colour phases and call populations polymorphic, but at any given place there is a wide variety of colour types with all intermediates, so this is not true polymorphism but plumage variability (cf. Erard 1987). *Colour type 1*: entire head and neck glossy greenish blue-black, crown-feathers elongated. Scapulars, mantle, back, rump, uppertail-coverts and tail rufous, tail-feathers graduated (longest T1, extending 10–20 beyond T2, shortest T6, 8–15 shorter than T2); breast, belly and thighs slate-grey, former darker, more blackish slate with slight greenish blue sheen, flanks slate-grey washed brownish, undertail-coverts rusty to rufous. Alula and greater primary coverts brownish grey, upperwing-coverts and tertials rufous, primaries and secondaries dark brown-grey (inner primaries and secondaries edged rufous along outer web, latter also along inner web and tip); underside of wing dark slate-grey, flight feathers edged rusty along inner web, axillaries white. *Colour type 2*: like type 1 but breast darker, like head, upperwing-coverts mostly white (lessers white with base and narrow streak along shaft black, medians and greaters black with broad white edge along outer web). *Colour type 3*: like type 2 but darker, crest longer. Upper mantle like head, lower mantle, back and rump chestnut. Uppertail-coverts and tail black, tail-feathers strongly graduated (T1 extending 20–40 beyond T2) and with irregular rufous marks on inner web (T6 to T4) and along distal part of shaft (T3 to T1). Scapulars mottled dark grey and chestnut; upperwing-coverts black, medians with irregular tips, greaters with white edges along outer web; primaries, secondaries and tertials black, inner primaries with narrow pale grey to white edge along outer web, secondaries and outer tertials narrowly edged rufous along outer web. Axillaries pale grey; undertail-coverts slate-grey. *Colour type 4*: like type 3 but blacker. Entire upperparts, uppertail-coverts and tail black except long T1 (extending 40–60 beyond T2) white with small triangular tip and most of shaft black. Entire underparts glossy black, except that lower belly and undertail-coverts are dark slate-grey to sooty black without gloss, latter irregularly mottled white. Scapulars black; upperwing-coverts white, greaters with irregular black marks on inner web and along shaft; primaries, secondaries and tertials black, the last 2 with more or less narrow white edges along outer web. Axillaries black with whitish tip. *Colour type 5*: like type 4 but with much white. Entire upperparts, upperwing-coverts and broad edges along outer web of secondaries and tertials white. Tail black mottled white with very long white streamers (T1 entirely white, extending 50–120 beyond T2; T2–T6 with irregular white marks on both webs, particularly along shaft). Belly, flanks and undertail-coverts dark slate-grey; thighs, axillaries and underwing-coverts white. Every kind of intermediate exists between these colour types. Bill bright cobalt blue to greenish blue with black tip, inside of mouth greenish yellow; narrow wattle around eye bright cobalt-blue; eyes brown; legs and feet bluish slate. ADULT ♀: like ♂ of colour type *1* but crest shorter, face, throat and breast less glossy, underparts paler grey, rufous parts of plumage browner, particularly on wings and mantle. Bill blacker, less blue; eye-wattle much smaller, almost lacking. SIZE: (24 ♂♂, 8♀♀) wing, ♂ 73·5–84 (78·7), ♀ 73–78 (75·8); tail (from base to tip of T2), ♂ 71·5–93 (83·9), ♀ 72–79·5 (75·6); extension of T1 beyond T2, ♂ 5–144 (50·2), ♀ 2–7 (4·5); bill, ♂ 17–19·5 (18·6), ♀ 17·5–20 (18·7); tarsus, ♂ 14·5–16 (15·2), ♀ 14–16 (15·3). WEIGHT: (Cameroon) 1 ♂ 15, ♀ (n = 4) 13–15 (14·5); (Gabon) ♂♀ (n = 7) 13–16·5 (14·5). In NE Gabon, from colour types *1* to *5*, i.e. from rufous to white plumage, ♂♂ get larger, with proportionately longer and more graduated tail, with longer streamers; variation not age-dependent (Erard 1987).

IMMATURE: like adult ♀ but almost crestless, upperparts still browner, less rufous, head greyer with gloss more limited to crown. Bill black, legs and feet dark brown.

JUVENILE: unspotted, like immature but still duller, head dark brown, upperparts rufous-brown, underparts greyish tawny to very pale brown-rufous.

NESTLING: hatches naked with some sparse brownish down, skin pink. Later, body downy (**A**).

T. v. viridis (P. L. S. Müller): Senegal and Gambia to Sierra Leone. ♂ with upperparts deep rufous, white edges on wing,

A

glossy blue-black of throat extending onto breast, tail and undertail-coverts rufous.

T. v. ferreti (Guérin-Méneville): Mali and Ivory Coast to N Cameroon, Chad, N Central African Republic, NE Zaïre and Sudan, east to Somalia, S Kenya (E coast of L. Victoria to Mombasa and Taita, absent from central highlands), NE Tanzania (Mwanza and Mt Kilimanjaro to Mpanda and Dodoma). ♂♂ as variable as *speciosa*, but breast greyer, undertail-coverts dull grey to greyish buff, tail feathers other than T1 usually rufous. WEIGHT: extremes, ♂ 10, 20, ♀ 9·9, 22. Best data from Kenya – overall, ♂ (n = 142) 10–20 (13·8), ♀ (n = 29) 10–16 (12·7) (averages, Feb–Mar, 112 ♂♂ 13·8, 10 ♀♀ 12·8; May–June, 12 ♂♂ 14·4, 6 ♀♀ 13·0; July–Sept, 13 ♂♂ 13·6, 7 ♀♀ 12·7; Oct–Dec, 5 ♂♂ 12·8, 6 ♀♀ 12·3). Other averages: Uganda (May–July) 3 ♂♂ 16·0, 3 ♀♀ 12·7, (Nov–Dec) 7 ♂♂ 15·4, 8 ♀♀ 13·5; Ethiopia (Apr–June) 7 ♂♂ 14·8, ♀ 15, (Nov–Mar) 7 ♂♂ 13·2, 4 ♀♀ 12·3; Somalia, 6 ♂♂ 13·1, 8 ♀♀ 11·3; Sudan (Feb–Mar) 4 ♂♂ 15·3; Nigeria, 2 ♂♂ 15·4, 3 ♀♀ 15·5.

T. v. suahelica Reichenow: W Kenya and N Tanzania (south to Usandawe region). Like *ferreti* but undertail-coverts white to rusty-brown and ♂♂ never white-tailed.

T. v. kivuensis Salomonsen: SW Uganda, Zaïre (Kivu), Rwanda and Burundi to NW Tanzania. Head and throat as in *ferreti*, upperparts and tail rufous, no white in tail and usually none on wing, undertail-coverts bright rufous. WEIGHT: (Uganda, Mar–July) ♂ (n = 12) 13–17 (15·4), ♀ (n = 7) 13–18 (15·1), 2 unsexed 14·5, 16.

T. v. restricta (Salomonsen): Uganda (Nkose and Sese Is, and along N shore of L. Victoria). Like *viridis* but upperparts darker rufous, glossy blue-black of throat extending onto breast, long median tail-feathers sometimes white.

T. v. ungujaensis (Grant and Mackworth-Praed): Pemba, Zanzibar, Mafia and E Tanzania (Usambaras and Dar-es-Salaam to Amani, Kilosa, Njombe and Rovuma R.). Centre of belly and undertail-coverts white, usually no white on wing, only rufous edges to wing feathers. Only rufous-backed and rufous-tailed colour types.

T. v. plumbeiceps Reichenow (including '*violacea*' and '*subrufa*'): Angola (north to Cuanza Norte and Lunda), Zambia (throughout; mainly migratory, present Sept–Apr, some all year in Luangwa and Middle Zambezi valleys, and in N Mwinilunga), S Zaïre (Shaba), SW Tanzania (occurs north to Kigoma but breeding restricted to southernmost regions, e.g. Songea), Malaŵi (throughout below 1500 m; mostly migratory, though a few remain in dry season), Mozambique, Zimbabwe (throughout; migratory, present Sept–May, a few all year from Middle Zambezi and Sabi valleys eastward), South Africa (Transvaal, Cape Province and Natal), Botswana and Namibia; migrates to Cameroon, Gabon, SE Zaïre (Upemba), Burundi, SE Kenya (Vanga to Mombasa, Rabai and Sokoke forest), Pemba and Mafia; one record from Sudan. Upperparts and tail always rufous, head crested dark grey, undertail-coverts usually white or pale rusty white. WEIGHT: (Upemba, Zaïre) ♂ (n = 19) 11–17 (14·8), ♀ (n = 11) 12–16 (14·2), unsexed (n = 4) 11–16 (12·8); (Zimbabwe) ♂ (n = 11) 11·1–14·7 (13·3), ♀ (n = 3) 11·6–13·6 (12·6); (Namibia, Oct, Nov) 2 ♂♂ 15, 12, 1 ♀ 13.

T. v. granti (Roberts): Natal to SW Cape Province, including most of Zululand; migrates to Zimbabwe (June–Sept; at least in Sabi valley and Lusitu–Haroni area), Zambia (June–Aug; Luangwa valley, Mpika, Chipata, Mumbwa, Kasama and Museshya), Malaŵi (May–Sept; Nsanje, Zomba and Mangochi districts) and S Tanzania (June–Aug). Like *ungujaensis* but head glossy green not blue.

Field Characters. The common paradise-flycatcher of subsaharan Africa, found almost everywhere except arid zones, and the only one present in most of E and S Africa. In forest zone of W and central Africa overlaps with 3 other species, but partially ecologically segregated, preferring open wooded habitats of all types and usually keeping to edges and clearings in forest, while the others are true forest birds. Plumage extremely variable, but consistent characters are glossy dark crested head, tail with long streamers, and greyish underparts; upperparts, wings and tail rufous with variable admixture of white or black. Red-bellied Paradise-Flycatcher *T. rufiventer* has rufous underparts. Bates's Paradise-Flycatcher *T. batesi* has blue-grey head without crest, brighter, more orange tone to rufous upperparts and tail, and tail without long streamers except in a few populations (e.g. Angola). Rufous-vented Paradise-Flycatcher *T. rufocinerea* very similar to African, with glossy black crested head, rufous upperparts, wings and long tail, grey underparts and rufous undertail-coverts; never has white in wing or grey undertail-coverts, but sometimes African also lacks white and has rufous undertail-coverts, and some individuals may not be distinguishable in the field.

Voice. Tape-recorded (10, 49, 86, 88, 91, B, C, F, ERA, KEI). Territorial song of ♂ very varied, individually and geographically. Typically a loud, rippling phrase of 5–6 liquid notes, normally repeated, almost without break; in Tanzania has falling cadence, 'wheeo-whee-wheeo-whit-whit'; in South Africa, rendered 'tzzeee switty-tsweep-sweepy-taweep' (Maclean 1993). Variations include a 2-second 'tzeee swit-ty-swee-ty-tsweep-swee-py-ta-weep'; a high-pitched, rapid 'zweet-tweetoo, tweetoo, tweetoo-tweeto' (rasping note followed by short pause, then by rapid warble, then pause and final note with last syllable accentuated), a 4-syllabled 'squee, squee, woo, woo' with notes slowly and deliberately articulated, and a 'tweeoo-wheet-wheet' repeated 2 or more times. In NE Gabon, a raspy 'tyewtee-trik' followed by piping 'twit-twee-tee-tee-tyew-tee-tyewt', 'twit-twee . . . tyew-tee-tyew-tee-tyew-twee-tit' or 'hweet-tweet . . . tee-twee-tee-twee-ti'; full advertising song, 'ptee-tee-twit tcheetdiay . . . tu-tu-tee-twee-tweet . . . tu-tu-tee-twee-tweet' (with accent on 'twee'); when more aggressive, interspersed with rapid, raspy 'tetewit-tchitchay-ta-tee-tiwet-tatateeteetiwet'; high intensity courting song of ♂, a high-pitched guttural warble 'blibliblibliblibli . . .'; greeting call, 'tii tweet tit-ti tii tweet'; contact call, 'ti-twit tee-twit' or 'twee-tweet'; call of ♀ 'chui . . . tiu-ti-chuit'. In South Africa, contact call a simple 'zeet, zwayt', 'zeet,er' or 'zwayt,er', in Tanzania, a soft 'zi'zk'zk'. Excited call of nest-building ♀, a high-pitched tripping 'chirree, chirree, chirree' or 'chizzareereet'. Alarm call, 'zwayt' or 'zwayt-er', shorter, higher-pitched, raspier and jauntier than contact call, also a loud, raspy 'tchee-tchiay', 'zweet-zwayt' or 'zweet-zweet-zwayt' or 'zwä-i-zwei'. When mobbing snake, a husky 'skizzit'.

General Habits. Inhabits savanna woodland, park-like country, open forest and plantations, avoiding dry thornscrub and heavy forest; upland savannas, open areas beside woods, bushland, open woodland of any

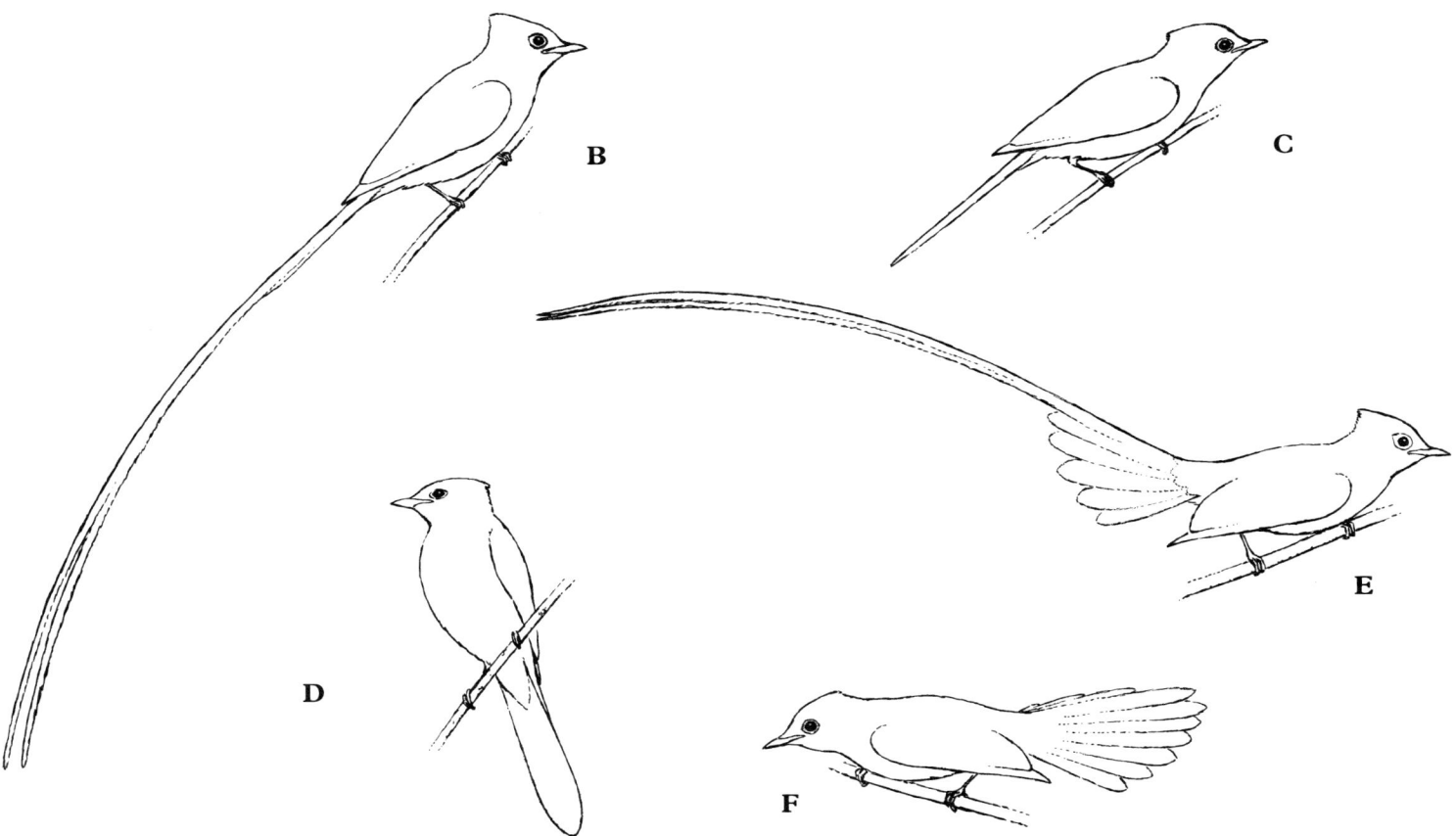

type (commonest in miombo *Brachystegia* woodland in Zambia and Zimbabwe); dense thickets, coastal and riparian fringing forest, edges and large clearings of moist evergreen forest, sometimes entering forest, e.g. on Ntchisi Mt, Malaŵi, also in Zambia and Zimbabwe; on Mt Kilimanjaro occurs up to *Podocarpus*/*Erica* forest, especially where canopy is thin or broken (Cordeiro 1994); secondary forest (e.g. *Aucoumea* forest in Congo and in Gabon), vegetation around ponds in savanna country, exotic plantations (e.g. *Eucalyptus* with shrubby undergrowth in Malaŵi), old and recently cultivated land, particularly on forest-savanna edges; particularly attracted to well-established gardens, orchards, native cultivation (e.g. cassava), and large shady trees in and around villages and towns. Up to 1450 m in Itombwe, Zaïre, 1800 m in Zimbabwe and E Tanzania, 2150 m in Malaŵi, 2300 m in Burundi, 2500 m in E Africa. In Malaŵi some ecological exclusion seems to exist between African Paradise-Flycatcher and White-tailed Crested Flycatcher *Elminia albonotata*.

Solitary or in pairs. Quite vocal; tame but unobtrusive. Almost always on the move, either flitting about in foliage or dashing through open spaces with undulating and graceful flight; ♂ may shoot forward like white or rufous rocket, or fly more slowly with long floating tail streamers waving up and down. Forages among leaves, though avoids dense clumps. Foraging methods are: (a) like a true flycatcher, sits on perch (**B**) looking out for insects passing nearby or dislodged by another animal, and hawks them in quick sallies from perchor in hovering flight; this method often used at edges of thickets or in small or large gaps in understorey and middle levels;

(b) moves continuously among foliage (**C**), hopping from branch to branch in upright posture (**D**), snatching prey from under leaves or in short looping or hovering flight; (c) flits among leaves like a butterfly, twisting around to catch flushed insects; (d) like an *Elminia*, forages in horizontal posture with partly open and drooping wings, and tail-feathers more or less fanned and moved both sideways and up and down (**E**, **F**), then swoops down onto dislodged insect. In South Africa, used combination hawking-gleaning technique (flying out to capture stationary prey) in 52% of observations (58% of which included hovering), versus 32% for hawking insect in air and 12% for gleaning; also more inside than outside canopy (93% *vs* 7%) (Fraser 1983). Prey usually swallowed only after having been mandibulated in the bill; larger items are dismembered on perch, sometimes largest (e.g. cicadas) or harmful ones are grasped with the foot and torn apart with the bill. Frequently joins mixed-species flocks, and follows large mammals (e.g. bushbuck and other antelope) or even game birds; in NE Gabon does this more during long rainy and dry seasons (Mar–Aug) than during short rainy and dry seasons (Sept–Feb). In NE Gabon, forages somewhat differently in flocks than when alone: foraging method (a) accounts for 91·4% (n = 70) of prey captures as against 77·5% (n = 129) when alone; methods (c) and (d) account only for 5·8% of prey-captures as against 20·9% when alone. Bathes by fluttering or plunge-diving into water; later preens on perch. Bird in South Africa bathed in swimming pool while people were swimming (Oschadleus 1996). Actively mobs cuckoos and shrikes, also arboreal and terrestrial snakes and terrestrial mammals, harassing them in fluttering hovering flight with erect crest and swooping attacks with bill clapping, giving loud, squeaky and rasping calls. Seen attacking other sallying birds like Spotted Flycatcher *Muscicapa striata* and immature Black-and-white Flycatcher *Bias musicus*. When alarmed, gives raspy calls, shivering wings over back.

Many populations migrate or at least make local movements but these movements are still not well understood and are poorly documented. See subspecies, above. *T. v. plumbeiceps* is a breeding visitor present in Tanzania only from Oct to Feb, with stragglers until 20 Apr; in southern Africa it is widespread except in highest and drier parts, and though present in all months in Zululand, Mozambique and locally in N Zimbabwe, it is only a migrant in N Namibia, NE Botswana, most of Zimbabwe, Transvaal, Swaziland, Orange Free State, and Natal to Cape Town (common Sept to Mar or May, rare to June); a few birds stay in austral winter (Skead 1967). Bird ringed Pietermaritzburg recovered Vila Junqueiro, Mozambique. Populations of this subspecies are known for their long distance migration, occurring as far north as Tanzania, with stragglers to E Zaïre, Gabon and Cameroon. Populations of sahelian and even soudanian zones, in Mauritania, Senegal and Gambia, Mali, Niger, Ivory Coast, Nigeria, Chad and Sudan, also migrate, leaving during dry season and returning to breed during the rains when insects are plentiful; in Mauritania present May–Nov, a few remaining in Dec and Jan; in Nigeria it is a savanna woodland bird which enters dense secondary forest for part of the year, and from Zaria northward occurs only during the rains.

Food. Arthropods, mainly insects: eggs, larvae (particularly caterpillars), nymphs and imagos, Coleoptera, Hymenoptera, Diptera, Orthoptera (grasshoppers), Neuroptera, Archiptera, cicadas, moths, butterflies (particularly Pieridae, e.g. *Terias senegalensis* and Lycaenidae, e.g. *Oberonia bueronica*), mantids, cockroaches, flying ants and termites; also spiders and small berries. Prey size (n = 51) 2–30 mm (10·9) for adults but (n = 77) 2–20 mm (8·3) for nestlings in NE Gabon (Erard 1987).

Breeding Habits. Monogamous, territorial. Pair-bond does not last long, at least in NE Gabon, where pair members change after each breeding season and even in the course of one season. During pair formation, ♂ approaches and courts any ♀, even brooding ones at nest. Courting ♂ approaches ♀ in undulating flight, displaying upperparts and long tail-streamers; next he perches upright with sleeked plumage and erect crest, swings tail, shivers wing-tips and calls with bill wide open, and sometimes dances on perch; then he crouches with drooping wings and fanned tail raised, shivers wings and opens bill wide while twittering softly; responsive ♀ does the same, and they lean toward each other. Territory defended by ♂ who patrols it every day, advertising it with loud calls and territorial songs, particularly at sunset, and in the morning and late afternoon. Singing increases with approach of breeding; in NE Gabon, sings all year, though less frequently during long rainy season; at Amani, Tanzania, sings from late Oct to mid-Mar. Mean territory size 5·1 ha in NE Gabon, but size increases according to colour type of ♂: 4·3 ha for types 1–3, 5·2 ha for type 4 and 5·6 ha for type 5 (Erard 1987). Territoriality may be confined to vicinity of nest: some nests placed no more than 10 yards apart (Moreau 1949, Skead 1967) (pairs did not interact territorially, and both appeared to defend both nests when person approached). In South Africa, sings during establishment of territories and during breeding, then sings less, with irregular bursts until departure. ♂ defending territory against intruder calls and sings in rapid direct flight; on approaching opponent makes short undulating flights interrupted by sudden pauses, when he perches upright and calls with wide-open bill; then attacks and chases opponent with squeaky and guttural songs, or faces it in crouched horizontal posture with erect crest, sides of breast fluffed out, flicking out carpal joints while opening and closing tail and drooping wings slightly.

NEST: neat, well-formed, tight, shallow cup (**A, G, H**) of vegetable fibres, small pieces of bark and dried leaves, fine rootlets, tendrils, vines and grass-stems, bound and externally covered with spider web, often also decorated outside with small whitish, greenish and/or grey lichens, often with 'tail' consisting of some dead leaves and spider web dangling below; lined with horsehair, fine rootlets, fibres or dry grass; usually with stiff rim of strong fine fibres laid parallel and sewn together with spider web; sometimes twig or branch embedded in structure (**G**); firmly fixed with cobweb to support. In

G

NE Gabon, moss is rarely if ever used; in South Africa, nests built mainly of green moss that grows on trees in kloofs, bound with lichens, although nests vary according to the availability of materials nearby (Skead 1967); similarly, nests in Tanzania were made of green moss (Moreau 1949). Shape variable; ext. diam. 65–70, height 40–90; int. diam. 45–50, depth 30–40. Placed up to 10 m above ground, mostly 1·5–4·5 m, in open site, not hidden, mostly on exposed fork at or toward end of slender horizontal or down-sloping branch, or in cluster of dead twigs or tendrils, in bush or small tree (e.g. in NE Gabon, *Solanum*, cassava, avocado, bamboo, *Trema*), or in fork of leafy creeper hanging from large tree; at least once attached to 2 herb stems; in South Africa and Tanzania, often over water or dry streambed and usually in shade. ♂ chooses site, flies around and perches on it, squats down and turns around, with much chattering and twittering, and displays most highly coloured parts of its plumage, bends and waves its long tail-feathers up and down. Built in 2–9 days, equally by both sexes, though in some cases mainly by ♀. Same nest may be restored and used for successive broods, up to 3 times in Kenya and South Africa, but this was never noted in NE Gabon.

EGGS: 1–5, NE Gabon 1–3, mean (20 clutches) 1·9; Malaŵi 1–5, mainly 2–3, mean (42 clutches) 2·6; Zambia 1–3, mean (21 clutches) 2·2; Zimbabwe 1–3, mean (44 clutches) 2·4; South Africa 1–4, usually 2–3, mean (197 clutches) 2·4. Laid at 1-day intervals, 1–7 days after nest completion. Oval to rounded but abruptly pointed at one end; glossy; creamy or pinkish white (pink when fresh), sparingly spotted with reddish or umber-brown, salmon-red or rusty, particularly in ring or cap around large end, sometimes with some underlying mauve or slate spots. SIZE: 17–20 × 12·5–15 in Cameroon; 18·0–19·4 × 13·4–14·2 in Zaïre; 21–23 × 13·5–14 in Kenya; (n = 32, Zimbabwe) 17·2–20·0 × 13·6–14·6 (18·8 × 13·9); (n = 111, southern Africa) 17·3–21·5 × 13·0–15·1 (18·8 × 14·1). Produces second clutches, mostly replacement clutches.

LAYING DATES: Senegal and Gambia, May, Aug; Burkina Faso, Aug; Ghana, (juv. Aug); Nigeria, May–Aug, in rains; Cameroon, Apr, July, Sept, probably all year (Bates 1911); Gabon, every month but little breeding mid-Apr to mid-Aug and peak mid-Dec to mid-Feb; Central African Republic, June, July, during rains; Zaïre, Mar–July, Sept, Nov, Dec; Rwanda, Aug–Oct, Dec, Mar, Apr; Ethiopia, Mar–June (♂♂ with enlarged testes and young Aug); E Africa: Uganda, Aug; Region A, Mar, Apr, June, July; Region B, Mar–July; Region C, Jan, Mar, Apr, Nov, Dec; Region D, every month; breeds in rains in regions A–C, in short rains and between rains in parts of Region D, and in long rains in other parts; Malaŵi, Oct–Mar (mainly Nov–Jan); Mozambique, Sept–Mar; Zambia, Aug–Feb (mainly Oct–Dec); Zimbabwe, Sept–Mar (mainly Oct–Dec); Botswana, Oct–Jan, July, mainly Oct–Dec; Transvaal, Oct–Mar (mainly Oct–Jan); Natal, Sept–Mar (mainly Oct–Dec); Cape, Oct–Mar; Namibia, Nov–Feb; Angola, Sept (with active gonads Sept–Nov).

INCUBATION: begins with last egg in NE Gabon, but

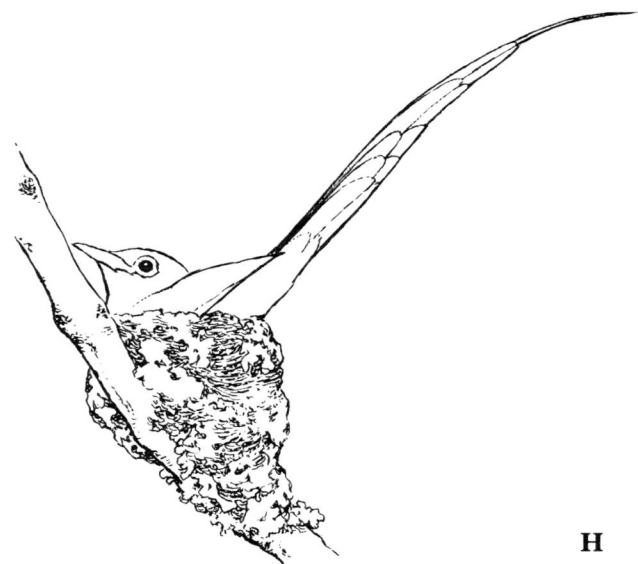

H

with first egg in South Africa (Skead 1967); by both sexes, with frequent nest-reliefs; in Tanzania, eggs brooded for > 90% of time; usually brooding bird does not leave nest until partner arrives; individual spells on nest mostly 30–40 min but up to 120 (Moreau 1949). In NE Gabon, ♀ incubates at night, both parents by day. At 3 nests observed for 1082, 834 and 438 min, adults spent together 98·7, 99 and 93·4% of time on nest; ♂ brooded for 61·5, 56·9 and 58·2% of time; mean duration of sessions 39·7 min (14–64, n = 13), 38·4 min (7–89, n = 12) and 42·5 min (15–84, n = 6) whereas ♀ brooded for 37·2, 28·5 and 35·2% of time; mean duration of sessions 23·7 min (10–36, n = 13), 28·5 min (14–46, n = 11) and 15·5 min (11–33, n = 4). Period: 13 days in Gabon (n = 6); 14–15 days in Zimbabwe; in South Africa, 12–15, rarely up to 17 days, mean 13·7 (n = 10). During incubation birds more silent, territorial songs of ♂ mostly in early morning and during nest-reliefs when ♂ comes to nest, though he may sing on nest. ♀ (and ♂) may be very tame when sitting (**H**). When ♂ arrives, ♀ often raises and quivers wings over back and erects crest, and ♂ may then hover like displaying ♂ Pin-tailed Whydah *Vidua macroura*, with erect crest and feeble trill; at one nest, in NE Gabon, ♀ regularly sang on nest in response to approaching ♂. Both ♂ and ♀ mob intruder or potential predator but stay at a distance. Defends nest against almost every bird coming within a 5 m radius; most aggressive towards cuckoos, shrikes, hornbills and drongos, although in NE Gabon ignores *Chrysococcyx* cuckoos.

DEVELOPMENT AND CARE OF YOUNG: some down on tracts on 1st day; eyes open and sheaths of primaries and secondaries emerge on 3rd day; sheaths of upper breast-feathers emerge on 4th day; feathers of back and wings out of sheaths on 8th day; chicks well feathered on 9th day. A few days before leaving nest, young perch on rim. Weight increases steadily from 2 g at hatching to 12 g at 8 days. Nestling period: 11 days in NE Gabon (n = 5); 14–16 days in Kenya (Ngong) but shorter periods recorded on coast; 10–12, rarely up to 16 days, in South Africa.

Young regularly fed by both parents. For first 5 days brooding time by both sexes about same as during incubation but for much smaller spells. In Tanzania, feeding rates (number of feeds per nestling per hour) 2·6–5·4 for broods of two vs 1·4–2·5 for broods of three (Moreau 1949); in NE Gabon, feeding rates of 5·8 on first day for brood of two and from 2·1 on second day to 7·8 on 7th day for brood of one. Food may be broken and given bit by bit to small young. Nest well sanitised; egg-shells carried away and dropped at hatching; faeces swallowed during first week, later carried away. ♂ and ♀ pugnacious towards potential predators. In South Africa seen harassing Cape Glossy Starling *Lamprotornis nitens* and Sombre Bulbul *Andropadus importunus*; ♀ seen making clicking sounds and mobbing Hadada *Bostrychia hagedash*; in Tanzania recorded attacking Half-collared Kingfisher *Alcedo semitorquata*, Little Grebe *Tachybaptus ruficollis*, Dwarf Bittern *Ixobrychus sturmii*, Red-chested Cuckoo *Cuculus solitarius* and Fork-tailed Drongo *Dicrurus adsimilis*.

Young dependent on parents for at least 1 week after leaving nest.

BREEDING SUCCESS/SURVIVAL: in NE Gabon, out of 35 nests in which eggs were laid, only 10 produced 1 or more fledglings. In South Africa, out of 15 nests built, 11 had eggs, 4 had hatchlings and 3 produced fledglings; 9 out of 27 eggs hatched, only 6 producing fledglings (Skead 1967). In W and E Cape, Transvaal and Natal, hatching success 74·6% (94/126 eggs), fledging success 53·7% (66/123 eggs) and nest success 58·7% (27/46 nests produced 1 or more young); in Zimbabwe, hatching success 77·5% (111 eggs), fledging success 52·3% (88 eggs) and nest success 50% (42 nests); in Zambia, hatching and fledging success both 72·5% (40 eggs) and nest success 76·2% (21 nests) (Winterbottom 1967).

References

Brosset, A. and Erard, C. (1986).
Chapin, J. P. (1948).
Erard, C. (1987).
Little, J. de V. (1964).
Moreau, R. E. (1949).
Skead, C. J. (1967).
van Someren, V. G. L. (1956).

Terpsiphone atrochalybeia (Thomson). São Tomé Paradise-Flycatcher. Tchitrec du São Tomé.   Plate 29

Muscipeta atrochalybeia Thomson, 1842. Ann. Mag. Nat. Hist., 10, p. 204; São Tomé.   (Opp. p. 544)

Forms a superspecies with *T. viridis* (and with *T. mutata* (Madagascar and Comoro Is), *T. corvina* (Seychelles), *T. bourbonnensis* (Mascarene Is), and *T. paradisi* and *T. atrocaudata* (Asia)).

Range and Status. Endemic resident, São Tomé Island. Common and widespread. May have been adversely affected by pesticide spraying in plantations (de Naurois 1984a), but now appears to have recovered to pre-1971 abundance (Jones and Tye 1988). Greatest density, c. 125 birds per km^2 (Atkinson *et al.* 1991).

Description. ADULT ♂ (breeding): uniformly glossy blue-black, with brightest blue sheen on crest, mantle, wing-coverts and tail. Short erectile crest. Central tail-feathers (T1) greatly elongated and flexible. Lores and around eye bare: skin dark blue. Bill shiny blue-black, eyes deep blue, legs and nails bluish. ADULT ♂ (non-breeding): same, but tail not elongated. ADULT ♀: head like ♂ but without crest; mantle greyish, in form of backward-pointing triangle, back and rump dull ochre, tail-feathers including underside dull ochre with blackish tips; underparts from chin to vent whitish, but on breast blackish feather-bases show through to give mottled appearance, undertail-coverts rusty, underside of wing dark grey. SIZE: (5 ♂♂, 5 ♀♀) wing, ♂ 79–84 (82), ♀ 75–80 (77); tail, ♂ 150–195, ♀ 80–87 (83); bill, ♂ 9·0–10·3, ♀ 9·0–10·3. WEIGHT: 1 ♂ 12·2.

IMMATURE: ♀ like adult ♀; ♂ at first like adult ♀, then crest turns blue-black, and ochre on back, tail, secondaries and belly replaced patchily by glossy blue-black feathers.

Terpsiphone atrochalybeia

Field Characters. ♂ in blue-black breeding plumage, with long flowing tail, is unmistakable. ♀ and immature are mainly ochre, not long-tailed. Harsh calls, often given, are useful character; bird often heard but not seen. The only flycatcher on São Tomé, where there are no other flycatcher-like birds.

Voice. Tape-recorded (B, ALEX, GUL, JOPJ, TYE). Typical calls, given when flycatching, are low, grinding and rasping, 'jaaajaaajaaa . . .' or faster chatter, 'jejejejeje . . .'. Sometimes gives a higher-pitched 'dzeek' or 'dzik' with a more *Terpsiphone*-like quality, but no-one seems to have heard it give the normal *Terpsiphone* rolling song.

General Habits. Inhabits primary and secondary forest, plantations with *Erythrina* or other shade trees, clearings, dry woodland, wooded rivers, maize fields and grassy areas; uncommon in gardens; in north of São Tomé occurs sparsely in copses in open, cleared country; from sea-level up to 1400–1600 m. Solitary, in pairs (♂♂, ♀♀ and ♂♀) or groups of 3–6 birds. Often tame and curious, approaching observer closely (Jones and Tye 1988). Restless and noisy. Forages solitarily or in group, catching insects on the wing between trees or within a tree, mainly at height of 2–6 m. One bird followed a São Tomé Weaver *Ploceus sanctithomae* along a branch, apparently picking up insects disturbed by it (Eccles 1988). Foraging flock is very noisy; rasping and grinding cries may serve to frighten insects into the open. Often flycatches from exposed perch by river, and flies down to pick insect from water (Atkinson *et al.* 1991). Probes into ends of broken twigs.

Food. Insects; mainly beetles (Nitidulidae, Curculionidae), also Hemiptera, Diptera and Lepidoptera; also banana nectar (Keulemans 1866, *in* Atkinson *et al.* 1991).

Breeding Habits. Territorial, evidently monogamous; solitary nester.
NEST: neat, compact cup, attached to drooping branches and twigs. Fabric of 3 layers: central layer is made of dry, more-or-less skeletonized leaves built vertically against one another at sides and stacked at bottom; outer layer is of moss, fern-blades and fibres, ornamented with lichen, bound with spider web; inner layer (lining) is of long fibres wound horizontally against middle layer. Ext. diam. (3 nests) 68–75, ext. height 70–75, int. depth 30–45. Sited in descending branch of cocoa tree or bamboo, or on branch of forest tree; at height of 2–5 (exceptionally 8) m.
EGGS: 1–2, (n = 14 clutches) mean 1·7. Rounded ovals; pale ochre-orange, with neat, round, reddish spots. SIZE: (n = ?) 17·6–22·0 × 13·6–15·8 (20·2 × 14·15).
LAYING DATES: July–Jan, mainly (33 out of 45 clutches) Nov–Dec.

Reference
de Naurois, R. (1984a).

Plate 29
(Opp. p. 544)

Terpsiphone rufocinerea **Cabanis. Rufous-vented Paradise-Flycatcher. Tchitrec du Congo.**

Terpsiphone rufocinerea Cabanis, 1875. J. Orn., 23, p. 236; Tschintschoscho (= Chinchoxo), Portuguese Congo (= Cabinda).

Range and Status. Endemic, resident. SE Nigeria (Calabar). Said to occur in W and S Cameroon north to Korup Nat. Park, Mt Cameroon, Kumba, Efulen, Yaoundé and Bitye on Dja R., but specimens examined suggest a more restricted distribution, north only to Eseka, Yaoundé, and Sangmélima. Rio Muni (along coast), Gabon (along coast, inland to Ogooué and Mouila; further east similar birds interbreed freely with and behave like colour types of *T. viridis*), Congo (sublittoral, Mayombe, Brazzaville), Angola (Cabinda, south to Loanda, N Cuanza Norte and N Lunda) and Zaïre (Mayombe inland to Kwamouth on lower Congo R.; single records from Ngombi on middle Congo R. and Kamiembi in Kasai). Common to frequent. Density in Mouila, Gabon, estimated at *c.* 2 pairs/km on line-transect (Y.-M. de Viviès, pers. comm.).

Description. ADULT ♂: entire head, neck and upper breast glossy blue-black (though in Angola and Cameroon some birds have chin to upper breast bluish slate-grey), nape feathers slightly elongated into a short crest. Scapulars, mantle, back, rump, uppertail-coverts and tail rufous-chestnut (though in lower Zaïre and Angola upperparts of some birds have more orange-rufous tone), tail-feathers graduated (longest T1 extending 50–100 beyond T2; T6 shortest, *c.* 20 shorter than T2). Lower breast, flanks, and belly slate-grey; vent and undertail-coverts rufous-chestnut to orange-rufous. Primaries and secondaries dark brown-grey, primaries narrowly edged slate-grey along outer web, secondaries edged rufous-chestnut along outer web; tertials mostly rufous-chestnut (outer ones with dark grey wash along shaft and on inner web); alula and greater primary coverts dark brown-grey, rest of upperwing-coverts chestnut-rufous. Axillaries orange-buff to chestnut-rufous, underwing-coverts slate-grey, underside of flight feathers grey, edged pale chestnut-rufous to buff

Terpsiphone rufocinerea

along inner web. Bill bright blue, gape greenish yellow; eyes dark brown, rim of eyelid blue; legs and feet blue. ADULT ♀: like ♂ but duller, glossy blue-black of head restricted to forehead to nape, crest much shorter; upperparts less bright, more cinnamon-rufous than chestnut-rufous; median tail-feathers less projecting; underparts paler, particularly belly which is washed with rusty. SIZE: (11 ♂♂, 7 ♀♀) wing, ♂ 73·5–83·5 (79·7), ♀ 72–77 (75·7); tail (from base to tip of T2), ♂ 72–94·5 (81·9), ♀ 73·5–81 (75·7); extension of T1 beyond T2, ♂ 3–113 (41·8), ♀ 2–10 (5·2); bill, ♂ 18·5–20·5 (19·3), ♀ 17–20 (18·5); tarsus, ♂ 14·5–16 (15·4), ♀ 14·5–15·5 (14·9). WEIGHT: (Congo) 1 ♀ 15·4.

IMMATURE: like adult ♀ but duller, top of head less glossy; upperwing-coverts brown-black, edged rufous-brown; tertials with pronounced dark brown area along shaft; bill blacker, less blue.

JUVENILE: like immature but with much paler underparts, browner head and browner rufous parts of plumage; bill brown-black; legs greyer, less blue.

NESTLING: not described.

TAXONOMIC NOTE: we follow recent authors (e.g. Watson *et al.* 1986, Sibley and Monroe 1990, Dowsett and Forbes-Watson 1993), in considering this as a full species, though field studies have shown that it hybridizes with and replaces *T. viridis speciosa* in Gabon (Brosset and Erard 1977, 1986, Erard 1987), and with *T. v. plumbeiceps* in Angola, and thus has been treated as just a subspecies of *T. viridis*, e.g. by White (1963). It is usually linked with *batesi* and *bannermani*, which are treated here under *T. batesi*. Records of *rufocinerea* outside the above range, e.g. in W Africa (Mali: Dekeyser and Derivot 1968), Central African Republic (Bamingui-Bangoran Nat. Park: Green 1984) or E Zaïre are probably colour morphs of *T. viridis* or *T. batesi*.

Field Characters. Restricted to lowland forests of W Congo basin from S Cameroon to W Zaïre and N Angola. Basically very similar to African Paradise-Flycatcher *T. viridis*, with glossy black crested head, narrow blue wattled eye-ring, rufous upperparts, wings and long graduated tail, grey underparts and rufous undertail-coverts. Some colour types of African Paradise-Flycatcher (q.v.) have white in wing and grey undertail-coverts, but others lack white and have rufous undertail-coverts, and some individuals not distinguishable in the field. Bates's Paradise-Flycatcher *T. batesi* has blue-grey head without crest, more orange-rufous upperparts, and tail lacking long streamers except in Angola. Red-bellied Paradise-Flycatcher *T. rufiventer* has rufous underparts.

Voice. Tape-recorded (CHA, ERA, LEM). Songs and calls typical of *Terpsiphone*, e.g. *T. viridis*, i.e. territorial song, a mellow 'lulululululu' or more sibilant 'thui-huit-tui-hui-hui-huit-huit-huit', little twittering trills, and the rasping contact call, 'gzwe' or 'zweet'. Birds in coastal Gabon have a different dialect from those in NE, a rather slower song, 'tyew-teevoureevouee-vouee-vouee-vouee'. In Zaïre song said to be a slow repeated 'zee, zee, zee', not whistled (Chapin 1953a).

General Habits. In Kouilou Basin, Congo, inhabits low riverine thickets in open marshes and more locally understorey of flooded forest, once even in secondary dry-land forest; replaced by Red-bellied Paradise-Flycatcher in undisturbed and secondary forest. In coastal Gabon found in *Rhizophora* and *Avicennia* mangroves, in flooded or coastal forest with *Chrysobanus icaco*, *Syzigium guineense*, *Manilkara lacera* and *Anthostemma aubryanum*, also papyrus stands with scattered trees and bushes (e.g. *Nauclea*, *Anthostemma*, *Anthocleista*) (P. Christy, pers. comm. and C. Erard, per. obs.); meets African Paradise-Flycatcher along edges of these habitats and replaces it in rubber and cocoa plantations and in and around villages. In Angola occurs in primary and mainly secondary woods, scattered shorter trees in coffee plantations and in gallery forest (Heinrich 1958, and C. Erard, pers. obs.).

Solitary or in pairs. Usually quite vocal and rather tame. Behaviour very like that of African Paradise-Flycatcher (q.v.). Almost always on the move. Forages among foliage; hunts like a true flycatcher, perching upright and hawking insects passing nearby or dislodged by another animal with rapid looping sallies; gleans leaves, working through them with upright body, snatching prey from under leaves or in short sally or hovering; occasionally searches leaves with horizontal body and partly open and drooping wings, swinging fanned tail-feathers, then swoops down to catch dislodged prey. Moves in all levels of vegetation but mainly at low and medium ones; sometimes comes to ground and picks up insects among dead leaves. Frequently joins mixed-species flocks and exploits swarms of flying ants and termites.

Food. Insects: moths, grasshoppers, cicadas, beetles, also winged ants and termites.

Breeding Habits. Monogamous, territorial. Behaviour similar to that of African Paradise-Flycatcher (q.v.): ♂ regularly patrols territory, perching in upright posture, singing and giving sharp loud calls. Some observations in coastal Gabon suggest territory size 4–6 ha.

NEST: neat regular open cup of moss scantily lined with dry grass and an inner layer of fine fibres, bound to support by thin spider web; placed in fork of dead branch hanging down or caught up in a trailing vine, up to 6 m above ground in small clearing; ext. diam. 70, ext. depth 55; int. diam. 50, int. depth 45.

EGGS: 1–2; 3 clutches of 2, Cameroon. Like those of African Paradise-Flycatcher, i.e. oval; slightly glossy; creamy or pinkish white (pink when fresh), sparingly spotted with reddish, salmon-red or rusty, mainly in ring or cap around larger end. SIZE: c. 19 × 14.

LAYING DATES: Cameroon, Jan; Gabon, Jan, Feb, Aug, Nov; Zaïre (♂♂ with enlarged testes Jan, young Jan, June, Dec); Angola (♂♀ with enlarged gonads May, Oct).

INCUBATION: by both sexes. Period unknown; at least 10 days.

DEVELOPMENT AND CARE OF YOUNG: young fed by both sexes. Nestling period unknown.

Plate 29 (Opp. p. 544)

Terpsiphone batesi Chapin. Bates's Paradise-Flycatcher. Tchitrec de Bates.

Terpsiphone batesi Chapin, 1921. Amer. Mus. Novit., no. 7, p. 6, fig. 3; Medje, northern Ituri district, Belgian Congo.

Range and Status. Endemic, resident. Widespread in lowland rain forest of S Cameroon (north to Mt Cameroon, Mt Kupé, Nkongsamba, southeast of Bafia, Yaoundé, Mbalmayo, Sangmelima, Bitye and Assobam), NE Rio Muni, Gabon (Woleu N'Tem and Ogooué–Ivindo, south to Lastoursville), SW Central African Republic (Dzanga to Bangui), Congo (Odzala Nat. Park, Téké Plateau (R. J. Dowsett, pers. comm.) and probably entire area between Gabon, Central African Republic and Zaïre, but precise data still lacking), Angola (Cabinda and Cuanza Norte to Gabela) and Zaïre (lower Congo R., north to Eola, Bolafa and Uele, south to Kwango and Kasai, east to Ituri, Kivu and N Itombwe). Rare to common. Av. density in NE Gabon, 14·0 pairs/km².

Description. *T. b. batesi* Chapin: S Cameroon, Rio Muni and Gabon to E Zaïre. ADULT ♂: entire head, neck, upper mantle and underparts blue-grey. Thighs, lower belly, undertail-coverts and rest of upperparts including scapulars, upper-wing-coverts and tertials, bright deep orange-rufous; some white on lower belly; rump brighter than back (base of feathers orange). Tail orange-rufous (feathers graduated, T6 8–19 shorter than T2). Primaries and secondaries blackish brown, edged dark orange-rufous along outer web. Axillaries and edges of inner webs of underside of flight feathers pale orange-rufous; underwing-coverts orange-rufous. Bill cobalt-blue, tip pearly; gape bright yellow; narrow eyelid-wattle cobalt-blue; eyes brown; legs and feet brown-black. ADULT ♀: like ♂ but less orange, more rufous; bill less bright; eyelid-wattle reduced and duller; median tail-feathers shorter. SIZE: (6 ♂♂, 4 ♀♀) wing, ♂ 74–78 (76·1), ♀ 71–75 (72·9); tail (from base to tip of T2), ♂ 87–114 (95·6), ♀ 70–80 (74·9); extension of T1 beyond T2, ♂ 8–33 (16·5), ♀ 3–5 (3·7); bill, ♂ 18–21 (19·7), ♀ 18–20 (18·7); tarsus, ♂ 14–16 (15·0), ♀ 14–15 (14·7). WEIGHT: (Cameroon) ♂♀ (n = 6) 16–17 (16·5); (Gabon) unsexed (n = 12) 13–17 (15·2).

IMMATURE: like adult ♀ but duller, upperparts browner.

JUVENILE: like immature but paler, head and underparts brown-grey, upperparts rufous-brown; unspotted.

NESTLING: hatches naked, skin pink; later covered with sparse long brownish down.

T. b. bannermani: Angola, Zaïre (lower Congo R.) and Congo. Like *batesi* but lower belly whiter and median tail-feathers much longer, projecting 35–105.

TAXONOMIC NOTE: both *batesi* and *bannermani* are often considered as subspecies of *T. rufocinerea* but field studies in Gabon (Brosset and Erard 1977, 1986, Erard 1987) lead to the conclusion that *batesi* is specifically distinct from *rufocinerea*. Populations named *bannermani* are here included in the species *batesi* only on morphological grounds, based on examination of specimens. Further comparative eco-ethological studies are required.

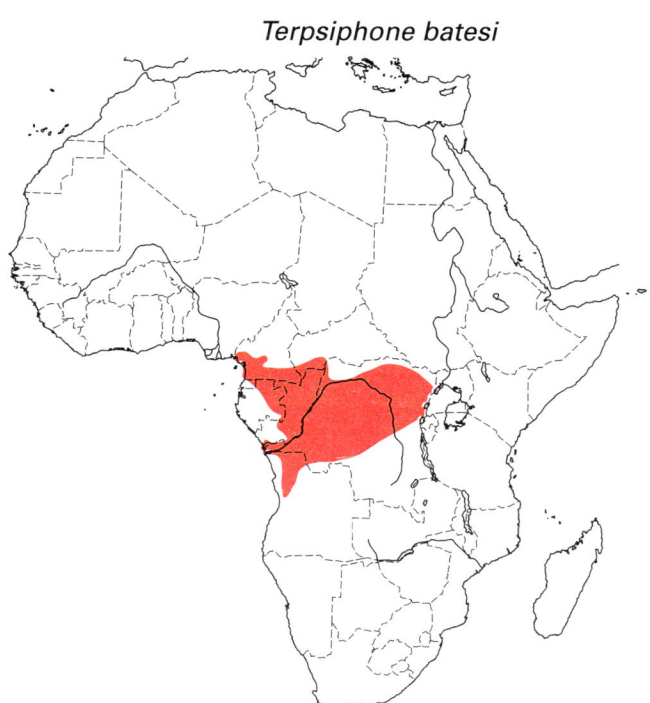

Terpsiphone batesi

Field Characters. Distinguished from sympatric paradise-flycatchers by rounded blue-grey head without crest and brighter, more orange tone to rufous upperparts, wings, tail and lower belly. Tail rather long and graduated but central feathers not extending far except in some populations in Cameroon and Angola. Red-bellied Paradise-Flycatcher *T. rufiventer* has rufous underparts, African and Rufous-vented Paradise-Flycatchers *T. viridis* and *T. rufocinerea* have blackish, crested head, tail with long streamers, and darker, more chestnut tone to rufous areas, and former often has white in wing.

Voice. Tape-recorded (B, CHA, ERA, LEM). Songs and calls very similar to those of other paradise-flycatchers. Territorial song of ♂, typically a loud, short, 3- or 4-note, rippling phrase, often repeated, almost without a break, 'weeweeweewee-weeweeweewee' or a high-pitched, repeated 'tuwi-wi-wi-wi-wi', sometimes introduced by a raspy 'tchee-tchyay'; high intensity courting song or fight song of ♂, a high-pitched, guttural 'bliblibliblibli . . .' or 'tchitchitchitchi . . .'; greeting call, a rapid tit-like 'tidudududu'; contact call, a shrill 'tchyee' or 'tchyee-tchyay' or 'tchyee-tchyay-tchyay'; alarm call, loud, raspy 'tchee' and 'tchee-tchiay'; call of young, a thin sharp 'tsii', also a rapid thin 'titititi'.

General Habits. Inhabits mainly primary but also secondary forest, including swampy and flooded forests; lowlands and up to 1300 m in Cameroon. Usually in mature forest with relatively unstratified understorey and with foliage evenly distributed up to 20 m. Occurs in regenerating forest, particularly old second growth where forest structure is almost back to its original form, i.e. where vegetation layers are ill-defined. Tends to avoid dense liana-rich forest. Occurs in old uncleared and disused plantations, e.g. of cacao and coffee, also in riverine forest in cultivated areas. In NE Gabon and S Cameroon, replaces African Paradise-Flycatcher in virgin forest and in the final stages of regenerating forest. In Odzala Nat. Park, occurs locally in moister forest with Marantaceae, also along edges of swamp forest, between habitats of African and Red-bellied Paradise-Flycatchers (F. Dowsett-Lemaire, pers. comm.).

In pairs or family parties; in all vegetation layers from eye level to canopy and even emergent trees, but mainly in understorey, avoiding tangles and 'curtains' of lianas. In NE Gabon, like many other forest species, shows seasonal preference for various levels: remains between upper understorey and lower canopy during long dry season, and mostly under 12 m during short rains (Erard 1987). Partners maintain regular, almost permanent visual or vocal contact; when not breeding they move and search foliage often together. Partners greet one another with rapid twitter in crouched posture: with body horizontal they flex legs, keep head up, quiver wing-tips and half-spread tail. Active all day; covers entire home range daily. Foraging behaviour much like that of African Paradise-Flycatcher (q.v.), i.e. forages among leaves but not in dense clumps, likewise among individual hanging lianas but not dense tangles. Forages either: (a) like a true flycatcher, hawking insects in quick sallies from perch or in hovering flight, or (b) moves continuously among foliage, jumping from twig to twig in upright posture, snatching prey from under leaves or with a brief flight or hover, or (c) flies erratically among leaves, and twists around to catch dislodged prey, or (d) searches leaves in horizontal posture, drooping partly open wings and swinging partly fanned tail, then swooping down to snatch dislodged insect in air. Food usually rapidly swallowed after some manipulation in the bill; larger items are beaten to death on perch, very large or dangerous ones are grabbed with the foot and torn apart with the bill, just like a drongo or a small raptor. Frequently joins mixed-species flocks, particularly during non-breeding season. In NE Gabon, when foraging solitarily (n = 110 observations), the above tactics (a, b, c, d) account respectively for 32·7, 20·0, 14·6 and 32·7% of prey captures, but when foraging in flocks (n = 383) they account for 86·7, 3·9, 1·0 and 8·4%. Sometimes pirates other species, harrassing them and stealing the prey they are pursuing. Regularly feeds on swarming ants and termites. Regularly also bathes in small rivers and forest pools, plunging repeatedly from overhanging bare dead branch 1–2 m high; flies about among wet leaves and, when drenched, perches and preens. When alarmed, flexes legs, stands in upright posture, flicks tail rapidly, jerks head from side to side and gives shrill alarm calls, and often retreats into dense foliage. Usually one of the 'sentinels' of mixed-species flocks.

Sedentary, at least in NE Gabon.

Food. Arthropods, mainly insects, e.g. beetles, grasshoppers, bees, wasps, ants, cockroaches, moths, dragonflies, Neuroptera, flies, alates of termites, also spiders and small snails. Mean length of 103 items in NE Gabon, 21·1 mm.

Breeding Habits. Monogamous; pair-bond lengthy, probably mates for life in NE Gabon. Territorial. Defends territory all year; mean territory size 6·1 ha in NE Gabon. Behaviour very like that of African Paradise-Flycatcher. During pair formation, courting ♂ approaches ♀ in undulating flight and displays bright orange-rufous upperparts and tail, then stands in upright posture with sleeked plumage, swings tail, shivers wing-tips and gives feeble guttural twitter with wide-open bill; then droops and quivers wings and crouches with fanned tail raised, opens bill widely and twitters rapidly; responsive ♀ does the same and also flaps wings; ♂ then makes a bow while rapidly quivering wings, ♀ crouches with tail spread and quivers drooping wings; ♂ then stands upright and sings, and continues courting in alternately crouched and upright positions. ♂ patrols territory every day, advertising it with loud calls and territorial songs. When defending territory ♂ calls and sings in rapid darting direct flight; approaches opponent in short undulating flight with intense bill-snapping, alternating with sudden pauses in upright posture; gives scolding calls and loud twitters with wide open bill, then attacks and chases opponent with guttural twitters. In fight posture ♂ holds body upright (though not as upright as African Paradise-Flycatcher), with sleeked plumage, and calls loudly with wide-open bill showing bright yellow interior, then inflates breast, ruffles crown-feathers, spreads tail-feathers and crouches, facing intruder, with open bill, fully spread tail raised, wing-tips drooped and carpal joints held well out from body. Courtship feeding not rare: ♀ may crouch and act like a young bird. During precopulatory behaviour, partners approach each other in crouched posture with spread tail and quiver and droop wing-tips; ♂ may have wide-open bill and give guttural twitter. ♂ chooses nest-site, where he crouches, extends head and neck horizontally toward ♀, raises and quivers half-spread tail and opens bill widely and flutters throat as if giving an inaudible twitter.

NEST: rather loose-looking cup made of dark green moss bound with spider web, tightly lined inside with pieces of bark and particularly blackish fibres, mostly *Marasmius*; in NE Gabon and apparently also in Cameroon bottle-green colour, use of moss, looser structure and more elongated conical shape differentiate Bates's from African Paradise-Flycatcher's nest; for 5 nests in NE Gabon, outer diam. 55–70 (63), height 65–80 (70), inner diam. 45–55 (49·6), depth 22–35 (29·6). Placed in fork near top of sapling or on branch of isolated small tree in forest understorey, often also on broken dead branch hanging from lianas, or in dense foliage in open space between vegetation layers; 1·5 to 20 m above ground; in NE Gabon (n = 51) mean 4·8 m (1·5–18). Built by both sexes.

EGGS: 1–3, NE Gabon mean (23 clutches) 1·9. Laid at 1-day intervals. Oval; glossy; creamy to pinkish white, with fine spots of red-brown or salmon-red and lilac-grey forming ring or cap at larger end. SIZE: *c.* 19 × 14. Produces second clutches, mostly replacement clutches.

LAYING DATES: Cameroon, Jan, Mar, June, Sept (♀ with small ovary Oct, ♂♂ with enlarged testes Jan, Feb, Sept, Oct); Gabon, every month but mainly at beginning of rains in Sept–Oct (10/60 records) and in short dry season Jan–Mar (35/60 records); Zaïre probably all year (birds in breeding condition Jan, Mar, May, June, Sept, Dec); Angola, Oct–Dec.

INCUBATION: begins with last egg; at night by ♀, during day by both sexes. Eggs brooded more than 90% of time: brooding bird does not usually leave nest until partner arrives; individual spells on nest mostly 40–60 min but up to 130. Period: 13 days. Birds usually fairly silent; frequency of territorial songs of ♂ reduced; both sexes may sing during nest-reliefs, but ♂ sings mainly when coming to nest, occasionally when on nest. Usually very tame when sitting, particularly ♀. Both sexes (but mainly ♂) mob intruder or potential predator but stay at a distance. Defends nest against almost every animal coming within a 5 m radius.

DEVELOPMENT AND CARE OF YOUNG: young fed equally by both parents; brooded for first 3–4 days, sometimes more, and during heavy rains; fed relatively large items whose size increases with age of young; usual delivery rate 4–5/h, but may be reduced to 2/h, particularly during hottest hours of the afternoon. A few days before leaving nest, fledglings perch on rim. Nestling period: 10–11 days. Nest kept very clear; both ♂ and ♀ actively defend its surroundings.

After leaving nest young usually fed for at least a month by both parents; each parent may take care of a particular young. When independent, young stay on parental territory for several months, almost until next nesting cycle.

BREEDING SUCCESS/SURVIVAL: in NE Gabon, out of 23 nests with eggs only 5 fledged young; 16 of these 23 nests were observed from beginning of construction: 15 received a full clutch but only 5 clutches hatched and young fledged from 4 nests. Several clutches were destroyed by elephants moving through the understorey and trampling the saplings supporting the nests. Several birds ringed there as adults were still present 5 and even 7 years later.

References
Brosset, A. and Erard, C. (1986).
Erard, C. (1987).

Plate 29
(Opp. p. 544)

Terpsiphone rufiventer (Swainson). **Red-bellied Paradise-Flycatcher. Tchitrec à ventre roux.**

Muscipeta rufiventer Swainson, 1837. Birds Western Africa, 2 (Jardine, ed., Naturalist's Library, 19, Ornith., 8), p. 53, pl. 4; west coast of Africa, restricted to Senegal by Meise, 1968, Zool. Beitr., N. F., 14, p. 14.

Range and Status. Endemic, resident. Widespread in lowland rain forest of Senegal and Gambia (south of 14°N, particularly Casamance, Gambia River), Guinea–Bissau, Guinea (locally abundant), Mali (southern Mts Mandingues, Kourémalé to Gourbassi), Sierra Leone, Liberia, Ivory Coast (Taï to Abidjan, Yapo and Abengourou, north to Odienné, Korhogo and Comoé, common to locally abundant), Ghana (north to Bia Nat. Park, Kete Kratchi and Akropong), Togo (north to Djodji), Benin (north to Kétou and Begon), Nigeria (Ilaro to Oshogbo, east to Port Harcourt, Mamu forest, Enugu and Obudu plateau), Cameroon (Korup Nat. Park and Mt Nlonako to Nsimalen, Sangmelima and Dja R.), Central African Republic (Dzanga reserves and Lobaye to Bangui, also Ouossi R., NW Zemio; recorded Bamingui–Bangoran and Gounda-St Floris), Rio Muni (Nkolentangan), Bioko, Pagalu (Annobon) I. where common; Gabon (common to locally abundant, Bitam and Oyem to Libreville and Mayumba, east to Booué, Lastourville), Congo (sublittoral and Mayombe to Lukolela, also Odzala Nat. Park), NW Angola (Cabinda and Cuanza Norte, Malanje), Zambia (northern Mwinilunga at source of Zambezi and on Lisombo stream), Zaïre (Eala and Jyonba to Bolafa and lower Uele south to Kivu and Lualaba, Kasai), Rwanda, Burundi, NW Tanzania (lower Kagera R. and Bukoba), S and W Uganda (north to Budongo and 2°N) and W Kenya (Kakamega, Kaimosi). Uncommon to abundant; in NE Gabon, density of *c.* 10–14 pairs/km^2.

Terpsiphone rufiventer

Description. *T. r. neumanni* Stresemann: SE Nigeria (east of lower Niger), Cameroon (except SE), Gabon, S Congo, coastal Zaïre and Cabinda, Angola. ADULT ♂: entire head, neck and upper mantle velvety blue-black; rest of upperparts including uppertail-coverts and scapulars slate-grey (sometimes rump with sparse orange-rufous feathers). Tail brown-black, washed slate-grey (T1 on both webs, T2–T6 along outer web), tail-feathers graduated (longest T1 extending *c.* 10 mm beyond T2; T6 shortest, *c.* 12 mm shorter than T2). Breast, belly, flanks and undertail-coverts bright orange-rufous, upper breast separated from throat by narrow pale orange-rufous line (formed by short throat-feathers exposing bases of longer breast-feathers). Flight feathers brown-black, primaries and secondaries edged slate-grey along outer web, tertials along both webs; upperwing-coverts brown-black edged slate-grey along outer web. Axillaries, lesser and median underwing-coverts orange-rufous; greater underwing-coverts grey-brown. Bill cobalt-blue, with black borders; gape bright sulphur yellow; eyes chestnut-brown; narrow eyelid-wattle cobalt-blue; legs black to dark blue-grey. ADULT ♀: like ♂ but throat greyer, median tail-feathers shorter, less projecting; bill less blue; eyelid-wattles duller. SIZE: wing, ♂ (n = 31) 71–83 (76·9), ♀ (n = 14) 71–78 (74·6); tail, ♂ (n = 29) 71–102 (87·0), ♀ (n = 13) 67–81 (75·3); extension of T1 beyond T2, ♂ (n = 29) 1–26 (10·6), ♀ (n = 13) 1·5–5·5 (2·7); bill, ♂ (n = 31) 18·5–21·5 (19·7), ♀ (n = 14) 18·5–20 (19·2); tarsus, ♂ (n = 10) 15–16 (15·4), ♀ (n = 14) 14·5–16 (15·2). WEIGHT: unsexed: (n = 44, Cameroon) 14–23 (16·5), (n = 13, Congo) 14·3–20·7 (16·5).

IMMATURE: like adult ♀ but still duller, browner; upperparts less grey, breast more smoky, less clearly delineated from rest of underparts which are less orange-rufous; bill brown, lower mandible horn; legs brown-black.

JUVENILE: like immature but much greyer; almost completely grey, paler on breast, lower underparts pale rufous buff.

T. r. tricolor Neumann: Bioko. Similar to *neumanni*, slightly larger. SIZE: (♂♂) wing, (n = 22) 80–87 (83·0); tail, (n = 15) 85–105 (92·6); extension of median tail-feathers, (n = 15) 6–23 (10·3); bill, (n = 22) 19·5–22 (20·7). WEIGHT: ♂♀ (n = 12) 16–19 (17·6).

T. r. rufiventer (Swainson): Senegal and Gambia, Guinea-Bissau and W Guinea (south and west from Futa Djalon and Kindia). ADULT ♂: mainly orange-rufous; head black glossed with steel-blue, slightly crested; wings black (except scapulars, inner upperwing-coverts and tertials orange-rufous), median and greater coverts and some secondaries broadly edged white along outer web; median tail-feathers greatly elongated. ADULT ♀: like ♂ but without white in wing, head duller.

T. r. nigriceps (Hartlaub): Senegal and Gambia (reported from lower Casamance and Gambia, as a migrant?), Sierra Leone and Guinea (east of Futa Djalon) to Togo and SW Benin (Abomey). Like *rufiventer* but darker; head velvet-black with purple gloss; no crest; tertials brown-black edged dark orange-rufous along outer web; median tail-feathers projecting 25–70. WEIGHT: (Liberia) ♂ (n = 25) 14·4 ± 1·4, ♀ (n = 11) 13·7 ± 2·1.

T. r. smithii (Fraser): Pagalu. Like *nigriceps* but head more glossy blue, flight feathers, greater upperwing-coverts and tail-feathers grey, median tail-feathers projecting 5–10.

T. r. ignea (Reichenow): E Central African Republic, Zaïre (Evenaar, Kisangani and Uele to Ituri, Kivu, Kasai and Lualaba), NE Angola and NW Zambia. Like *nigriceps* but smaller; velvet-black of head less bright but extends onto upper mantle and upper breast; tail browner, median tail-feathers shorter, projecting 5–25. WEIGHT: (Angola) 1 ♂ 16, 1 ♀ 15; (Central African Republic, June) ♂ (n = 3) 15–15·5 (15·2), 2 ♀♀ 15·5, 16.

T. r. fagani (Bannerman): Benin and SW Nigeria (west of lower Niger). Like *ignea* but head black with purple-blue gloss; mantle and upperwing-coverts rufous olive-brown; tail olive-brown tinged rufous, median tail-feathers projecting 5–10.

T. r. schubotzi (Reichenow): SE Cameroon, SW Central African Republic; recorded from Zaïre (Kunun and NW Kasai) but this might refer to *mayombe*. Like *ignea* but darker, less reddish orange, head dark grey-blue; tail and wings blue-grey, upperwing-coverts and tertials margined rufous, median tail-feathers projecting 5–30.

T. r. mayombe (Chapin): Congo and W Zaïre (Lukolela, Mayombe, Ubangi, recorded from Yangambi). Like *ignea* but darker and more reddish orange, tail and wings grey; median tail-feathers projecting 15–25.

T. r. emini Reichenow: Tanzania, Kenya, SE Uganda (Entebbe, Toro). Variable, paler than *ignea*, more orange, less red, median tail-feathers much longer, projecting 80–150. WEIGHT: (Kenya) 2 ♂♂ 14, 15; ♀ (n = 3) 14–15 (14·3).

T. r. somereni Chapin: Uganda (Bwamba, Bugoma, Ankole, Rwenzori, Budongo, Kibale, Sango Bay, Kyagwe, Mabira). Like *nigriceps* but median tail-feathers much shorter, projecting only 5–25; inner secondaries more orange-rufous. WEIGHT: (Uganda, Oct–Feb), ♂ (n = 23) 10–17 (14·3), ♀ (n = 14) 10–19 (13·4), unsexed (n = 3) 14–16 (15·2); (Uganda, Apr–July), ♂ (n = 21) 12–19·5 (15·8), ♀ (n = 13) 12·5–16 (14·8), 2 unsexed 14, 15.

TAXONOMIC NOTE: examination of museum skins shows local hybridization with African Paradise-Flycatcher *T. viridis* in Cameroon, Gabon, Uganda and Kenya, with Bedford's Paradise-Flycatcher *T. bedfordi* in E Zaïre, and with Bates's Paradise-Flycatcher *T. batesi* in lower Congo R. and in Cameroon. Field studies would be most useful.

Field Characters. The only paradise-flycatcher with rufous underparts. Head rather big and flat, black, without crest; blue bill, grey or rufous to orange upperparts and wings; tail graduated but usually not with long streamers. Rufous-vented Paradise-Flycatcher *T. rufocinerea* and African Paradise-Flycatcher have grey underparts and long streamers, the last also having white in wing. Bates's Paradise-Flycatcher and African Paradise-Flycatcher *T. v. plumbeiceps* have blue-grey

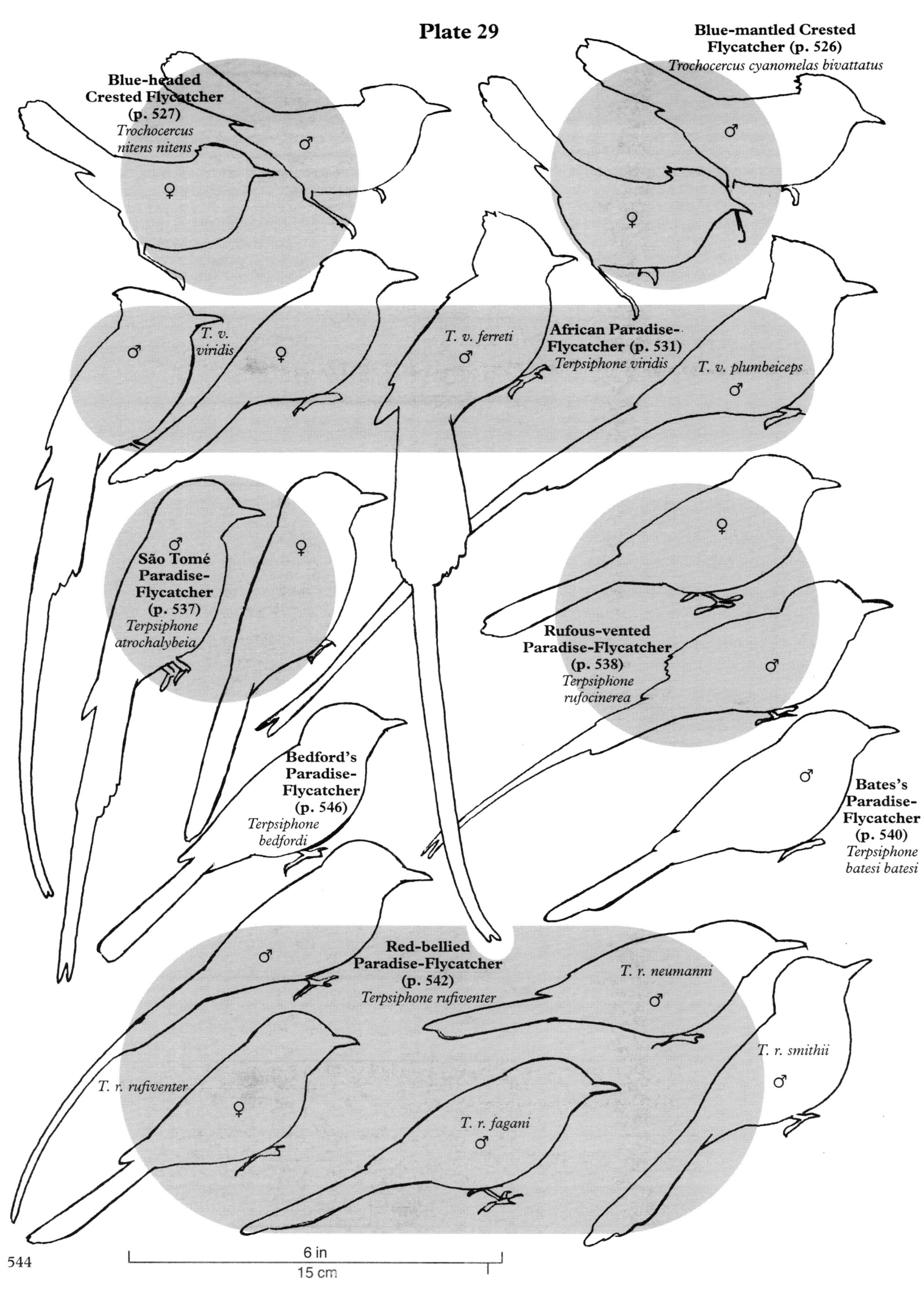

Plate 30

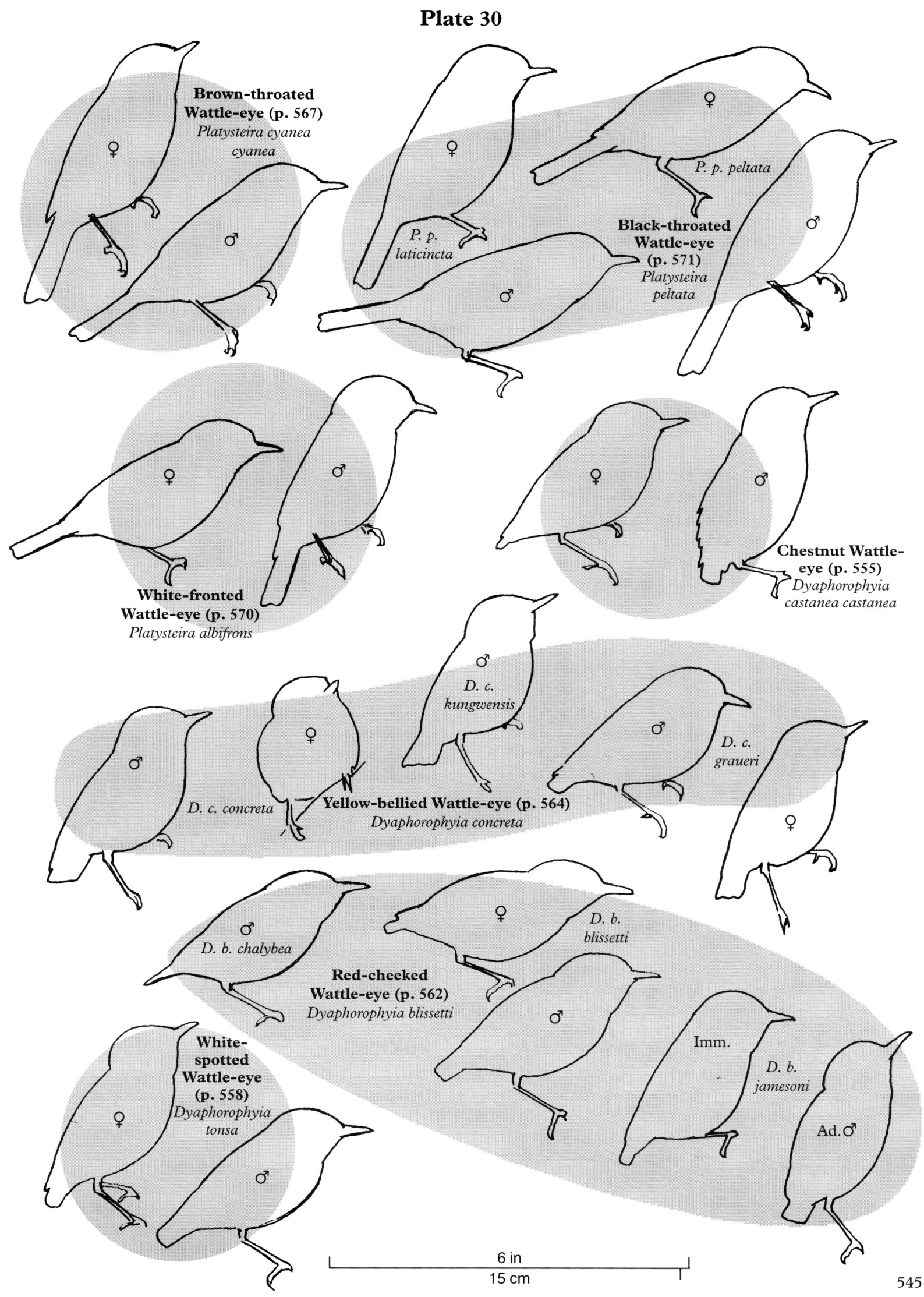

head and grey underparts, the last also having long tail streamers.

Voice. Tape-recorded (C, F, BRU, CHA, GRI, HOR, KEI, LEM). Very like that of other paradise-flycatchers, with rasping calls, twittering trills and liquid songs. In Lopé Nat. Park, Gabon and Odzala Nat. Park, Congo, voice very similar if not identical to that of Bates's Paradise-Flycatcher (C. Erard, pers. obs., and F. Dowsett-Lemaire, pers. comm.). Both sexes sing. In Nigeria, phrase of 3 silvery notes, repeated twice 'tsisisi-tsisisi' (Marchant 1953). In Gabon, a soft 'wee-wee-wee-wee-wee' with the last two notes higher pitched (P. Christy, pers. comm.). On Pagalu, call of both sexes described as a loud, rasping 'scleep, scree-scree' very similar to that of Bioko birds (Fry 1961). In Gabon, calls are vibrant 'gzwéi' or 'gzwè-zwà-gzwà' or 'gzwè-gzwei' (P. Christy, pers. comm.).

General Habits. Inhabits primary and secondary lowland and montane forest, also forest-grassland mosaic, oil-palm groves in forest clearings, gallery forest and thickets in savanna, cocoa plantations and large gardens; may also enter mangroves (e.g. N Angola). In Congo it is the common forest *Terpsiphone* species in both secondary and undisturbed forest in the Kouilou basin; in Odzala Nat. Park it is the commonest one in marshy forests, replaced on the edges and in moister areas of Marantaceae forest by Bates's Paradise-Flycatcher (F. Dowsett-Lemaire, pers. comm.). In Gabon, its main distribution matches that of *Aucoumea kleineana* (Burseraceae) (P. Christy, pers. comm.). In Cameroon, also occurs in rubber plantations, alone or together with Bates's Paradise-Flycatcher (Young 1946). Up to 1340 m in Zaïre (Itombwe), 1650 m in Cameroon and 1800 m in Uganda.

Solitary, in pairs or family groups. Rather vocal and tame. Behaviour very similar to that of African and Bates's Paradise-Flycatchers (q.v.). Almost always on the move in undergrowth and middle layers of forest, usually between 5 and 15 m above ground; in Nigeria, most feeding records between 2 and 12 m (Johnson 1984). Mostly active in the morning. Forages either like a true flycatcher, perching upright and looking for insects passing nearby or flushed by an animal, hawking them in quick looping sallies from perch; or gleans foliage, searching with horizontal body and partly open and drooping wings, swinging body from side to side and fanning tail, then swooping down to snap up dislodged prey; also works steadily through leaves with upright body, snatching prey from under leaves or in brief sally or hovering. Regularly joins mixed-species flocks, and feeds on swarms of flying ants and termites. Regularly bathes in small rivers and forest pools, plunging repeatedly from overhanging bare dead branch 1–2 m high.

Food. Mostly insects: moths, grasshoppers, beetles, cicadas, mantids, small dragonflies, caterpillars, Diptera, Neuroptera, Hemiptera, winged ants and termites; also Arachnida and a few unidentified seeds of small fruits.

Breeding Habits. Monogamous, territorial. Behaviour like that of African and particularly Bates's Paradise-Flycatchers. Defends territory of at least c. 2·5 ha (see fig. in Johnson 1984); in Gabon at La Lopé Nat. Park, distribution of pairs in forest very similar to that of Bates's Paradise-Flycatcher in Makokou, i.e. territories of c. 6 ha. Courtship includes bowing and tail-fanning.

NEST: strong and neat little cup, made of grasses, moss, lichens and other thin fibres bound externally with spider web, lined with finer material, including vegetable down and even feathers. Ext. diam. c. 60, int. depth c. 40. Placed in fork of small tree or sapling, often along stream or near water; usually low, up to a few m high; also at fork of bare thin drooping branch. Built by both sexes.

EGGS: usual clutch 2 in Gambia, 3 in Ghana. Oval; glossy; translucent when fresh; cream to creamy white with solid belt of red-brown to carmine-red spots and blotches at larger end. SIZE: (n = 5, Ghana and Zaïre) 18–19 × 13–15·6 (18·4 × 14·0).

LAYING DATES: Gambia, May–July; Guinea, July; Liberia, Mar, May, June (birds with enlarged gonads Jan–Apr, June; many young June); Ivory Coast, Feb–Sept; Ghana, Mar, Apr, June (♂ with enlarged testes Feb); Togo, Feb; Nigeria, Apr, July, Nov (♀♀ with active ovaries Sept); Cameroon, Dec–May, July, Oct; Bioko, July, Aug; Gabon (♂♀ in breeding conditions Jan, Aug; pairs with dependent young Feb, Mar, May–July, Sept); Congo, Sept, Oct; Central African Republic (♂♀ with gonads active Jan, Mar, June, Sept, Nov); Zaïre, Jan–Mar (♂♀ in breeding condition June–Feb); E Africa: Region B, Oct–Jan, Mar–July, mainly during rains, particularly long rains Apr–June.

INCUBATION: by ♂ and ♀; rather shy when potential predator present, flees, does not defend nest.

DEVELOPMENT AND CARE OF YOUNG: nestlings fed by both parents; parents vehemently defend nestlings and fledglings against intruders and predators.

Reference
Johnson, D. N. (1984).

Terpsiphone bedfordi (Ogilvie-Grant). Bedford's Paradise-Flycatcher. Tchitrec de Bedford.

Trochocercus bedfordi Ogilvie-Grant, 1907. Bull. Br. Orn. Club, 19, p. 40; Mawambi, Ituri, Zaïre.

Range and Status. Endemic, resident. Lowland forest of E Zaïre, in 2 areas: (1) NE Ituri, from Mawambi to c. 50 km north of Irumu (1°31′N, 29°49′E), north to near Arebi (2°46′N, 29°34′E), and (2) west of Itombwe and

Terpsiphone bedfordi

Kahuzi Mts (2°13′S, 28°42′E) from at least Kabunga (1°40′S, 28°10′E) south to southern limit of lowland forest. Rare to frequent.

Description. ADULT ♂: forehead, crown, nape, hindneck, sides of neck, ear-coverts, lores and chin deep velvet-black to glossy steel blue; throat and foreneck variable, dark blue-grey to deep velvet-black or glossy steel blue. Rest of upper- and underparts, including scapulars, axillaries, upper- and undertail-coverts slate-blue grey; steel blue-black of head sometimes extends over upper mantle. Tail brown-black, washed blue-grey (T1 on both webs, T2–T6 on outer web), feathers graduated, longest (T1) extending 5–10 beyond T2, shortest (T6) 7–17 shorter than T2. Primaries, secondaries and tertials brown-black, outer webs margined blue-grey, particularly from inner primaries inwards; upperwing-coverts brown-black deeply washed blue-grey, particularly on outer web; underwing-coverts sooty slate-grey. Bill dark cobalt-blue, gape bright yellow; eyes brown, narrow eyelid-wattle cobalt-blue; legs and feet brown-black. ADULT ♀: like ♂ but duller, neck and throat less black, plumage greyer, less blue, median tail-feathers shorter, eyelid duller. SIZE: (27 ♂♂, 26 ♀♀) wing, ♂ 74–81 (78·4), ♀ 71–77 (74·8); tail, ♂ 74–90 (80·9), ♀ 66–78 (73·0); extension of T1 beyond T2, ♂ 2–25 (10·8), ♀ 1–8 (4·6); bill, ♂ 18·5–20 (19·4), ♀ 18–19 (18·6); tarsus, ♂ 15–16 (15·5), ♀ 14–15 (14·6). WEIGHT: no data.

IMMATURE: like adult ♀ but duller, greyer, particularly wing and tail-feathers, head less black, eyelid-wattle absent and bill browner, less blue.

JUVENILE: like immature but still duller, almost without blue, wings and tail more grey-brown.

NESTLING: undescribed.

TAXONOMIC NOTE: often considered a subspecies of *T. rufiventer*, but hybridization with parapatric, locally sympatric, *T. r. ignea* is not general and is confined to a very narrow belt which is best considered an 'overlap and hybridization zone', where 2 species *T. rufiventer* and *T. bedfordi* overlap with a low hybridization rate, not a 'hybrid zone' where 2 subspecies meet and mix (see Prigogine 1976).

Field Characters. A short-tailed paradise-flycatcher with a very restricted range in E Zaïre; uniform bluish grey except for head which is black with some gloss. Some of the so-called mutants of African and Red-bellied Paradise-Flycatchers *T. viridis* and *T. rufiventer* (probably in fact hybrids between these 2 or between Red-bellied and Bates's Paradise-Flycatcher *T. batesi*) can resemble Bedford's but usually show some rufous or brown, at least on lower belly and undertail-coverts. Larger than both Dusky Crested Flycatcher *Elminia nigromitrata*, which has greyer, less blue underparts and black cap without gloss, and Blue-headed Crested Flycatcher *Trochocercus nitens*, which has glossy blue-black head, pronounced crest and light grey belly; also does not share their habit of continuously fanning tail and drooping wings.

Voice. No known tape-recording. Song and calls (e.g. 'zre-zre') of same type as other *Terpsiphone* spp.

General Habits. Inhabits lowland primary evergreen forest and transition forest, occasionally deciduous forest; between 900 and 1900 m in Itombwe.

In pairs or family parties. Foraging behaviour similar to that of other *Terpsiphone* spp.: in forest understorey, gleans foliage in upright posture, often calling, with frequent hops from twig to twig or branch to branch; sallies to catch dislodged or passing insects in looping flight. Frequently joins mixed-species flocks. Usually not found in same habitat as Red-bellied Paradise-Flycatcher *T. rufiventer ignea*: prefers transition to lowland forest, occurs only in primary forest, not in both primary and secondary forest, also lower in understorey, not in upper storey and canopy, and the 2 species 'do not join in a common flock' (Prigogine 1976, 1980d).

Food. Insects: Hemiptera, Lepidoptera (moths) and grasshoppers.

Breeding Habits. Monogamous, territorial. When range overlaps with Red-bellied Paradise Flycatcher, mixed pairs occur and there is some hybridization but the 2 usually remain separate when they meet, even in mixed-species flocks; in one verified case ♂ was *T. bedfordi* and ♀ *T. rufiventer ignea* (Prigogine 1980d).

NEST: deep, well-made open cup made of moss, lined internally with fine grasses; ext. diam. 65, ext. depth 60–70; int. diam. 50, int. depth 45.

EGGS: 2. Oval; glossy; creamy white with small red-brown speckles forming zone at larger end. SIZE: c. 19·0 × 14·1.

LAYING DATES: Zaïre, all year in Itombwe, although the only nest found was in Mar; ♀♀ with active ovaries Feb and Oct, Nov, ♂♂ with enlarged testes Apr–Sept.

INCUBATION: by both sexes.

DEVELOPMENT AND CARE OF YOUNG: no data.

References
Prigogine, A. (1971, 1976, 1980d).

Family PLATYSTEIRIDAE: shrike-flycatchers, wattle-eyes and batises

Sexually dimorphic, small to medium-sized, flycatcher-like birds, larger species somewhat shrike-like. Bill dorsoventrally flat, moderately wide, upper mandible notched or hooked at tip; well-developed rictal bristles; iris yellow or red, or circumorbital skin wattle variously coloured and shaped; short-tailed; tarsometatarsus short and stout or moderately long and spindly; claws short, strongly curved and sharp. ♂♂ mainly white and glossy black, ♀♀ usually rufous below where ♂ is black; rump-feathers often fluffy, white or yellow, or with concealed white spots. No neossoptiles, tongue spots or spotted juvenile plumage. Jaw musculature not like true flycatchers (Muscicapinae) (Beecher 1953); 'lanioid' carpometacarpal process (Pocock 1966). Nest a rather small, compact, neat and well-formed open cup, well secured to support, densely covered outside with spider webs and/or lichens. Social organization in family groups with tight pair-bond. Arboreal (*Lanioturdus* largely terrestrial) in forest and savanna woodland.

7 genera: African *Megabyas*, *Bias*, *Lanioturdus*, *Batis*, *Platysteira* and *Dyaphorophyia*, and Madagascan *Pseudobias*. Formerly regarded as flycatchers Muscicapidae, which family was divided into Platysteiridae, Monarchidae and Muscicapidae *sens. str.* by Traylor (1986) and Watson *et al.* (1986). Wolters (1975–82) placed them in subfamily Platysteirinae in family Laniidae. In his phylogenetic classification, Cracraft (1981) placed these genera in subfamily Platysteirinae (with Monarchinae, Sylviinae, Timaliinae, Rhipidurinae, Pachycephalinae and Orthonychinae) in family Sylviidae, separating them from Muscicapinae which he associated with Turdinae in family Muscicapidae. He included both families in the infraorder Muscicapi, different from Lanii (Laniidae and Vangidae). On the basis of DNA studies (Sibley and Ahlquist 1990, Sibley and Monroe 1990), *Batis*, *Platysteira* and *Dyaphorophyia* were placed in family Corvidae: Malaconotinae: Vangini (with *Prionops*, Malagasy vangas and Oriental *Philentoma* and *Tephrodornis*). Close relationship between African platysteirines and malaconotine shrikes also claimed by Harris and Arnott (1988) after analysis of > 90 characters (unpublished); he places platysteirines in subfamily Malaconotinae (family Laniidae, with prionopine and laniine shrikes).

Genus *Megabyas* J. and E. Verreaux

Head large, rounded; no crest; bill broad with ridge on culmen; nostrils feathered; tail short; wings pointed; legs short. ♂ black and white, patterned like *Dryoscopus*, ♀ rufous above, streaked below. Inhabits forest canopy. Sometimes placed in *Bias* (e.g. Traylor 1986) and their skulls are similar (Erard 1987), but bill is hooked and shrike-like and not so flat, and we consider that flight and general behaviour and vocalizations are generically different.

Endemic. Monotypic.

Plate 28
(Opp. p. 497)

Megabyas flammulatus **Verreaux. Shrike-Flycatcher. Bias écorcheur.**

Megabyas flammulata J. and E. Verreaux, 1855. Rev. Mag. Zool., (2), 7, p. 348; Muni R., Gabon.

Range and Status. Endemic, resident and local migrant. Forested regions of Senegambia (one record Abuko Nat. Res., Gambia), Mali (recorded from Bafing–Makana), Guinea (Macenta district, frequent), Sierra Leone, Liberia, Ivory Coast (Taï Nat. Park, Fresco, Azagny Nat. Park, Abidjan and Yapo Forest north to Oumé and Maraoué Nat. Park), Ghana (frequent, north to Kintampo, Ejura, Begoro and Amedzofe), Togo (recorded from Misahöhe), S Nigeria (uncommon, Ibadan to Enugu, though mostly west of lower Niger R., also recorded Nindam Forest Reserve, Kagoro), Bioko, Cameroon (north to Korup Nat. Park, Mt Kupé, Efulen and Dja R.), Central African Republic (Haute Kémo and possibly Lobaye), Rio Muni, N and E Gabon (rare to uncommon, Mts de Cristal, Woleu N'Tem and Ogooué–Ivindo to Lopé and coast), Congo (Brazzaville, Kai Bumba, Odzala), Angola (Cuanza Norte), Zaïre (Uele and Ituri south to Itombwe and Kasai, west to Salonga Nat. Park and lower Congo R., also S Shaba), S and W Uganda (uncommon,

Megabyas flammulatus

Budongo, Bugoma, Bwamba, Bwindi–Impenetrable and Malabigambo Forests east to Entebbe, Kifu and Mabira), NW Tanzania (Minziro Forest), W Kenya (uncommon, Kakamega and Nandi Forests) and S Sudan (Bengengai and Lotti Forest). Uncommon to frequent; density in Kenya (Kakamega Forest) 2–5 adults and 1 immature in 8 ha census tract (Zimmerman 1972).

Description. *M. f. flammulatus* Verreaux: Sierra Leone to W Zaïre. ADULT ♂: black and white. Forehead, lores, crown, ear-coverts, nape, hindneck, mantle, wings, scapulars, uppertail-coverts and tail black glossed with steel-blue. Rump-feathers white with black bases; entire underparts including axillaries, underwing- and undertail-coverts white; flanks mottled black and white; thighs black narrowly barred white. Bill black; eyes red-orange to reddish hazel-brown; legs brown to reddish purple. ADULT ♀: forehead, crown, nape, sides of neck, hindneck and upper mantle grey-brown; lores whitish, ear-coverts mottled whitish and grey-brown. Mantle, back and scapulars dark cinnamon; rump paler (each feather tawny with broad blackish base and cinnamon tip). Uppertail-coverts tawny, washed cinnamon; tail dark cinnamon (feathers with blackish shaft and shaft-streak broadening toward tip, largest on T1, decreasing outwardly). Chin, throat, foreneck, breast, flanks and upper belly white with broad blackish streaks (each feather white, with broad edges, but not tip, blackish), belly white washed buff, with sparse rusty crescentic marks on sides; thighs blackish barred pale buff; undertail-coverts white washed buff. Primaries and secondaries blackish brown, the latter broadly edged cinnamon along outer web, at tip and along distal part of inner web; tertials, lesser and median upperwing-coverts cinnamon with blackish centre; greater upperwing-coverts blackish brown with very broad cinnamon edges along outer web. Axillaries dark grey with cinnamon edges, underwing blackish brown, inner webs of flight feathers edged tawny. Bill black; eyes hazel-brown; legs reddish purple. SIZE: (5 ♂♂, 5 ♀♀) wing, ♂ 82–90·5 (87·3), ♀ 87–91 (88·7); tail, ♂ 57–62·5 (60·1), ♀ 60·5–64·5 (63·1); bill, ♂ 21·5–23 (22·2), ♀ 23–24 (23·4); tarsus, ♂ 15·5–16·5 (15·9), ♀ 15·5–16·5 (16·0). WEIGHT: (Liberia) ♂ (n = 7) 27·6 ± 1·7, ♀ (n = 5) 25·7 ± 2·2; (Kenya) 1 ♂ 28, 1 ♀ 25.

IMMATURE: like adult ♀ but duller, head and neck browner, upperwing- and uppertail-coverts edged and tipped whitish.
NESTLING: undescribed.

M. f. aequatorialis Jackson (including '*carolathi*'): Angola, Zaïre (Equateur and Kasai eastward), Kenya and Uganda. ♀ with tail largely blackish. WEIGHT: (Uganda) ♂ (n = 12) 29–34 (30·7), ♀ (n = 12) 22–32 (28·1).

Field Characters. A rather quiet flycatcher with very upright posture (**A**); often remains motionless except for characteristic habit of swinging tail slowly from side to side. ♂ is blue-black above, with white rump and underparts; it superficially resembles a puffback, e.g. Red-eyed Puffback *Dryoscopus senegalensis* (in secondary forest) and Sabine's Puffback *D. sabini* (in primary forest), but distinguished from them by shorter legs, lack of red eye, and by shape, behaviour and voice. Puffbacks have a more horizontal stance, move continually while foraging, and are often noisy, with quite different calls. ♂ Black-and-white Flycatcher *Bias musicus* is noisy and active and has crest, chunky, short-tailed shape, black breast, white wing-patch and yellow eye. ♀ and immature have cinnamon upperparts but are distinguished from ♀ Black-and-white Flycatcher by paler head and rump, brown eyes, dark streaks on whitish underparts and direct not butterfly flight, in addition to shape and behaviour. Streaked underparts might suggest African Broadbill *Smithornis capensis*, which has different shape, with big head, short tail and stout bill, also black cap, pale wing-edgings and white patch on back in flight.

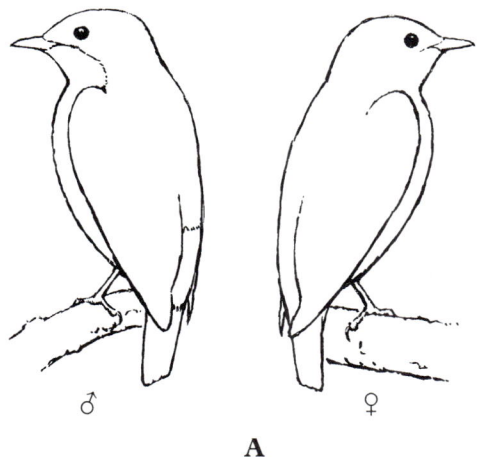

A

Voice. Tape-recorded (32, F, CAR, ERA, LEM, MAC). Song by ♂ only, sustained and highly repetitive, brief phrases of jingling notes separated by silences of same length: 'titiwit titiwit titiwit piwit piwee tit tit pityew pityew piwit piwit tutuwit tutuvit pirrit tutuwit tuwit pirrit . . .'. In another version, a series of phrases of dry, tuneless upslurred notes works its way up the scale, the notes becoming shorter and higher-pitched: 'chewy-chewy-chewy . . . chewy-chewy-chewy . . . chui-chui . . . teesy-teesy . . . ticlik-ticlik'. The metallic clicking notes which often end the series do not sound vocal and might be made by the bill (Keith and Gunn 1971).

Contact call of both sexes, a thin, very high-pitched whistle 'tseee' or 'pseet', also a harsh nasal 'tsit'; pre-flight call, a short 'psit'. Flight call of ♂ and ♀ in NE Gabon, a rolling 'tsrrrt', which may be the same as the insect-like rasping trill 'prrrt' near nest (Ryall and Stoorvogel 1995); another call, a melodious 'tuwick' with a double whistle (Marchant 1942) is probably same as the 2-syllable 'chuick-chuick' used by both sexes; a brittle 'chip-chip' call noted from ♀ near nest; harsh grating screams given during mobbing (Walker 1939).

General Habits. Inhabits both primary lowland rain forest and second growth, including old clearings with tall *Albizzia* and *Piptadeniastrum* trees. In NE Gabon, found in forest in areas disturbed by storms or mining, also around clearings and other openings in canopy and tallest emergent trees. Occurs both in old and recent second growth, uncleared or disused plantations, also in riparian woodland; enters man-made habitats such as recently logged areas and agricultural land with scattered tall trees and regrowth, also around villages. Up to 1600 m in Itombwe, Zaïre, and 2150 m in Uganda.

In pairs, family parties or small groups; in NE Gabon 4 groups included 2 adult ♂♂, once by themselves, once with 1 young ♂, and twice with 1 adult ♀; all were most probably migrants (Erard 1987). Elsewhere in Gabon, often in pairs although sometimes a solitary ♀ or ♀ with young (P. Christy, pers comm.). In Sierra Leone, Ghana and Kenya, frequents higher undergrowth and middle storey of both mature and secondary forest; in Gabon, keeps to higher levels of forest vegetation, i.e. 15–40 m, in second-growth down to 10 m above ground. In Ivory Coast, often noted near flowering and fruiting trees.

Pairs or group members forage together, keeping quiet, calling only on the move or when approaching one another closely. Flight strong and powerful. Sits upright like a shrike, slowly swings or twitches closed tail from side to side almost continuously; inspects nearby foliage with lateral movements of head. Perches in lowest part of tree crown, mostly on rather leafless limbs or thick branches; avoids twigs and thinner branches. Usually hops and flies up and down among branches and then flycatches, sallying in direct, fluttering, looping or hovering flight to pick prey from leaf; may also sit on exposed perch and sally to catch insects on leaves of nearby tree, 12–20 m from hunting post; sometimes crouches and hops along branch. Occasionally joins mixed-species flocks, but remains high, avoiding undergrowth. Regularly feeds on swarming ants and termites.

Mainly resident but may move locally (e.g. recorded in NE Gabon only Nov–Mar; and in Kakamega Forest, Kenya, thought to be only post-breeding visitor).

Food. Insects: mostly beetles, also moths, grasshoppers, cockroaches, bugs and cicadas; some winged ants and termites. Prey size 15–25 mm.

Breeding Habits. Monogamous, but probably with helpers; territorial. Defends territory against congeners and also hornbills, e.g. African Pied Hornbill *Tockus fasciatus*. Territorial behaviour undocumented. In Ivory Coast a second ♂ in full adult plumage noted at nest several times (in incubation period, although it did not incubate: Ryall and Stoorvogel 1995). Both ♂ and ♀ are said to have slow flight with rapid wing-beats in aerial display, ♂ singing during up-and-down 'pipit-like' courtship flight (Mackworth-Praed and Grant 1960); this requires confirmation.

NEST: very similar to that of Black-and-white Flycatcher; a small, neat, compact, well-formed, open, smooth, pale grey cup, mainly of thin vegetable fibres and moss, also rotten wood, flakes of bark and lichen, tightly bound outside with spider webs and well secured to support; ext. diam. *c.* 40, height *c.* 15. A nest in Guinea resembled a bump on the forked branch where it was placed, *c.* 15 m high in an *Albizzia sassa* (Halleux 1994). Another one in Taï Nat. Park, Ivory Coast, was built by ♂ and ♀ in fork of thick leafless branch near top of *Klainedoxa* tree, 35–40 m above ground in forest canopy (Ryall and Stoorvogel 1995).

EGGS: not known for certain. Described by Mackworth-Praed and Grant (1960: 196) as 'three, bluish or greenish grey with small spots and blotches of umber brown and lilac grey tending to form a zone round the middle of the egg; about 21 × 16 mm', but they later (1973: 126) say their data may refer to Black-and-white Flycatcher.

LAYING DATES: Liberia, (♂♂ and ♀♀ with enlarged gonads Oct); Guinea, Mar; Ivory Coast, Dec (juvs Mar, May); Ghana (♂ in breeding condition Feb, ♀♀ with juvs Mar); Nigeria (♀ with active ovary Apr, pair with enlarged gonads June); Cameroon, May; Gabon, Jan; Zaïre (specimens with enlarged gonads Mar–Nov, mostly June–Sept); E Africa: Region B, Apr, June.

INCUBATION: by ♂ and ♀ (both in Guinea and in Ivory Coast). Period unknown; at least 16 days in Ivory Coast where birds coming to nest displayed with much tail-wagging and 2-syllable calls, and also with dipping or nodding (head and upper body briefly tilted downwards) directed toward partner or nest; ♂ and ♀ replaced each other roughly every 30 min but intervals varied from 3 to > 95 min; nest was sometimes left unattended for periods of 15–60 min (Ryall and Stoorvogel 1995).

DEVELOPMENT AND CARE OF YOUNG: young fed by both ♂ and ♀ at short intervals. Nestling period unknown. Young apparently stay for a long time with their parents, like those of Black-and-white Flycatcher. Dependent young (being fed by ♀) in Gabon already shows typical tail-swinging.

References
Brosset, A. and Erard, C. (1986).
Erard, C. (1987).
Ryall, C. and Stoorvogel, J.J. (1995).

Genus *Bias* Lesson

Head large, crown flat, bill broad, flat, with sharply ridged culmen; nostrils feathered; iris yellow; hindcrown crest; tail short; wings rounded; legs very short, stout and yellow. ♂ black and white, ♀ rufous and whitish. Noisy and demonstrative. Many behaviours *Batis*-like. Sings in flight; song includes grating and rasping notes; flight fluttering; territorial flight erratic, butterfly-like, with rump feathers fluffed. Nest compact, cryptic; ♀ presses head down onto rim to compact the nest. Forested areas.

Endemic. Monotypic.

Bias musicus (Vieillot). Black-and-white Flycatcher; Vanga Flycatcher. Bias musicien.

Plate 28 (Opp. p. 497)

Bias musicus Vieillot, 1818. Nouv. Dict. Hist. Nat. 27, p. 15; Malimbe (= Malembo), Cabinda.

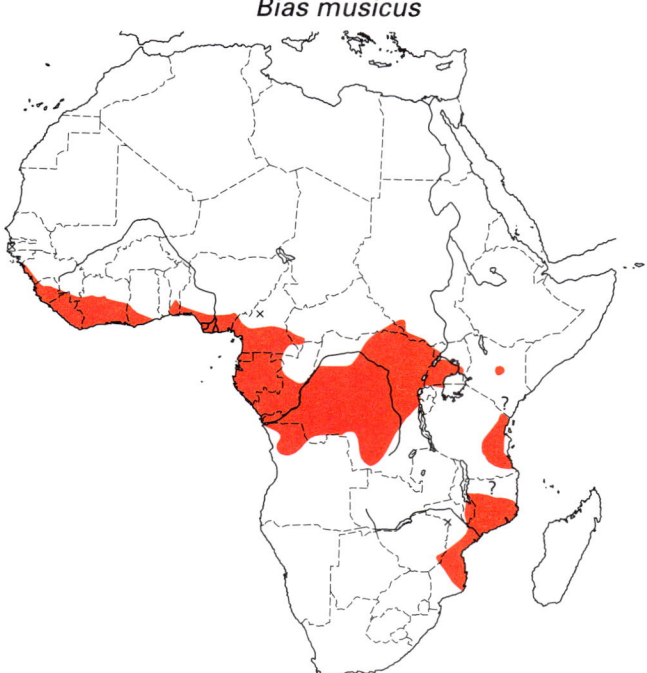

Bias musicus

Range and Status. Endemic, resident. Forested areas of Senegambia (one record, lower Gambia R.), Guinea-Bissau, Sierra Leone, Guinea (coast and Macenta Préfecture, frequent), Liberia, Ivory Coast (Abidjan and San Pedro to Béoumi-Bouaké and Gouessesso), Ghana (not uncommon, Accra Plains, Akwapim hills and Jukwa, north to Mampong, Ashanti and Kumasi), Benin (Pobé, Kétou), Nigeria (not uncommon, Badagri and Calabar to Ibadan and Obudu Plateau), S Cameroon (north to Korup Nat. Park, Rumpi Hills, Nyassosso, Sakbayeme, Nachtigal, Lomié and Kombetiko; once near Tibati), Rio Muni, Gabon, Congo (at least north to Gamboma), Central African Republic (Dzanga; Lobaye?), Zaïre (Kunungu and Lukolela to Uele and Ituri, south to Itombwe, Shaba and Kasai), Angola (rare to locally frequent, N Lunda, Malanje, Cuanza Norte and Cabinda), S Sudan (Bengengai), S and W Uganda (common, Bwamba, Rwenzori and Kibale to Bwindi-Impenetrable Forest, Mbarara and Sango Bay east to S Busoga), Kenya (isolated population in Nyambenis, Luazomela R., Meru and Ngaia Forest; old records on coast), Tanzania (in NW (Minziro Forest, Bukoba) and in E from Usambaras and Kilosa to Selous, Nachingwea and Mikindani), S Malawî (lower Shire valley and Mitongwe), Mozambique (Limpopo R. and Inhambane northward, but northern limits obscure), Zimbabwe (Lusitu-Haroni confluence and Mazoe R.). Uncommon to common. Density in NE Gabon, 2·15 pairs/km².

Description. *B. m. musicus* (Vieillot) (including '*feminina*' and '*pallidiventris*'): Senegambia and Sierra Leone to Uganda, NW Tanzania (Bukoba), Zaïre (Kasai and SW Shaba) and Angola. ADULT ♂: black and white. Entire head, neck, upperparts (including scapulars and uppertail-coverts), tail-feathers and upper breast glossy bottle-green to steely blue-black. Feathers of hind crown long and lanceolated. Concealed spots on rump white (feathers grey with wide subterminal white crescent and velvet-black tip, edged glossy blue or green). Lower breast, belly and undertail-coverts white (base of feathers blackish); flanks white scalloped with broad, semi-concealed black crescents; thighs black. Entire wing black with green to blue gloss except for white patch at base of both webs of P1–P8 and of outer web of P9, decreasing from basal two-thirds of P1 to less than basal quarter of P9. Bill black, gape lemon-yellow; eyes golden yellow; legs and feet straw to greenish yellow. ADULT ♀: top of head and ear-coverts brownish black, lores buff, feathers of hind crown long and lanceolated; back and sides of neck mottled dark cinnamon-rufous and brownish black. Rest of upperparts cinnamon-rufous with indistinct dark streaks on upper mantle and concealed whitish spots on rump. Uppertail-coverts and tail-feathers dark cinnamon-rufous. Underparts creamy white, breast and undertail-coverts washed cinnamon; flanks cinnamon-rufous, feathers with indistinct whitish spots and blackish edges. Entire wing dark cinnamon-rufous except for brown-black distal portion of primaries, increasing from tip on P8 to distal half on P1. Bill, eyes and legs like ♂. SIZE: (6 ♂♂, 6 ♀♀) wing, ♂ 85–90 (87·7), ♀ 84–90 (86·9); tail, ♂ 47–51 (49·3), ♀ 46–54 (49·7); bill, ♂ 21–24 (23·0), ♀ 21–24 (22·0); tarsus, ♂ 13–15 (14·0), ♀ 13–15 (13·9). WEIGHT: (Gabon, Nov) 1 ♂ 24, 1 ♀ 21; (Uganda) ♂ (n = 10) 20–25 (22·1), ♀ (n = 3) 21–22 (21·3).

IMMATURE: like adult ♀ but duller, head and neck browner, upper mantle more marked with ill-defined streaks.

JUVENILE: like immature but breast scalloped with narrow dark crescents; more thickly spotted with buff over crown and back.

NESTLING: hatches naked; later, skin greenish yellow with sparse greyish down.

B. m. changamwensis van Someren: Kenya and E Tanzania. Smaller, ♀ paler than nominate *musicus*, without blackish centres to feathers of mantle.

B. m. clarens Clancey: Mozambique, Malaŵi, Zimbabwe. Similar to *changamwensis* but ♀ and young ♂ paler, whiter below, with crown matt black, not sooty brown, and malar streaks with only vestigial rusty overlay.

Field Characters. A medium-sized flycatcher of second-growth, noisy and conspicuous, often around villages. Chunky, with big, flat, crested head, short tail, broad bill and yellow eyes and legs; ♂ black except for white belly and white wing patch conspicuous in flight, ♀ with dark head, smaller crest, chestnut upperparts, white or buffy underparts. ♂ frequently gives aerial displays, flapping slowly on broad, rounded wings and calling loudly. ♀ Black-throated Wattle-eye *Platysteira peltata* has colour pattern of ♂ Black-and-white Flycatcher but is different shape, with slimmer, crestless head, smaller bill, longer tail, red wattle over eye, very different voice. Similar-sized Shrike-Flycatcher *Megabyas flammulatus* is a crestless and much less active forest species with different flight, longer tail and dark eyes; black and white ♂ has entirely white underparts, conspicuous white rump, no white in wing; brown ♀ has broad brown streaks on underparts.

Voice. Tape-recorded (32, 91, B, C, F, CHA, ERA). Territorial song of ♂, usually given in flight, very varied, both individually and geographically; typically 3–4 loud, measured, tuneless notes, 'tiew-tyip-tiewp', 'tiew-tyip-tiewp-tiewp', 'wee-tew-tew'; sometimes with clicking notes at beginning or end, 'tik-chee-chew', 'weep-weep-weep-titiku'; sometimes only 2 notes, or a single, long, down-slurred 'cheeeyuw'. Higher intensity song is longer, accelerating and rising in pitch 'twee-tyee-tyee-tyee-tyee-tyee-tyee-tyee-tyew', last note down-slurred, or 'wup-weep-weep-cheew-cheew-cheew-cheew', last 4 notes all the same, with an agitated tone. Bird in Uganda (Bwamba) kept up running conversation of low chips and nasal notes while on perch, warming up into louder 'chup's followed by a low 'cher-weeer', then flew up and gave intense flight song, 'wee-chi-chi-chi-chi-weeer, wee-chi-chi-chi-chi-cher-weeer', ending with slow, deliberate 'ta-chwee-chu' repeated 3 times (S. Keith, pers. comm.). In more aggressive situations, ends song with separate lower-pitched 'tyew-tyiew', first note sharply down-slurred, and when fighting repeatedly gives long, whistled 'tyiew'. In aggressive encounters both ♂ and ♀ give 3–4 brief guttural notes followed by a descending grating call. Normal call of ♀ a long, rasping 'kchchew'. Calls of fledglings and immatures like those of ♀; begging nestlings give long nasal calls.

General Habits. Inhabits large clearings and other openings in forest, tree tops on edges of primary lowland and montane rain forest, second-growth, edges of riparian woodland; also enters open *Brachystegia* woodland. Adapts well to man-made habitats, including villages, gardens, farms, cultivated fields with scattered tall trees both live and dead, plantations (e.g. cacao) with tall tree cover, recently logged areas and early stages of regenerating forest on old cultivation. Lowlands, up to 1100 m in E Zaïre (Itombwe), 1300 m in E Tanzania, and 1700 m in Uganda.

Solitary, in pairs or family groups. Occurs at all vegetation levels but mainly in crown of tallest trees, 20–40 m. Active, moves about in tree and frequently changes trees. Covers entire home range daily. Forages around outside but not interior of tree crowns; sits upright (**A**), typically in hunched posture (**B**), and slowly wags head from side to side; perches at end of leafy branches and makes short, descending butterfly-flights or hovers to snatch prey from foliage; flutters along clumps of leafy lianas, or among crowns of flowering trees to dislodge insects which it snaps up in the air; when prey abundant, i.e. swarming winged ants or termites, behaves in typical flycatcher manner, making short circular flights from an exposed post to catch passing insects. Active all day from about an hour after sunrise to an hour before sunset. Roosting behaviour conspicuous: ♂ leads, sings high in tallest trees, then makes steep spiral glides above roosting site, a bush or small tree (e.g. *Solanum*, *Trema*, avocado), perches at end of pliant leafy branch or vine and gives loud, scolding territorial songs; ♀ then joins him, followed by young, which roost a little apart from adults. Roosting birds fluff out plumage and tuck bill

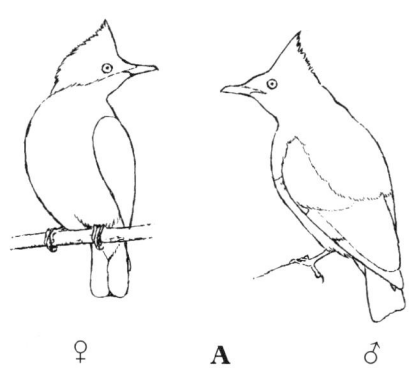

♀   **A**   ♂

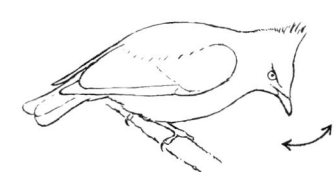

B

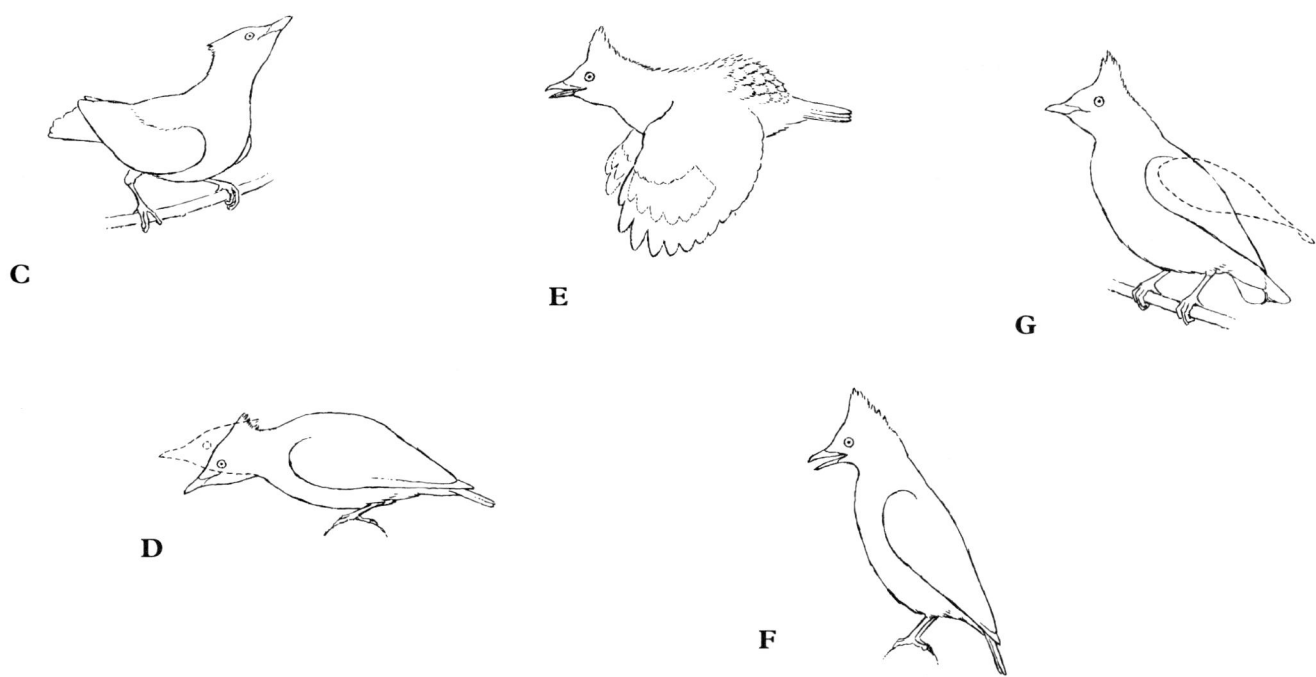

into scapulars; in this position white spots on rump of ♂ are exposed, giving him speckled appearance.

Pair members usually remain close together but may be 100–300 m apart, particularly in non-breeding season when they maintain looser contact and when each may be followed by a free-flying young. ♂ on the move gives territorial song, especially when pair traverses large open area. When pair members reunite, ♂ greets ♀ with territorial song and acrobatic flights, displaying crest and white wing and rump spots; ♀ perches on exposed branch in upright posture, with bill pointing upward, flicks wing-tips and wags tail, then crouches (**C**), fans and closes outer-tail feathers in jerky manner and bobs head up and down (**D**). Perched birds maintain minimum distance of 30–60 cm; if one comes too close, the other erects crest, rapidly fans outer tail-feathers and flicks wing-tips, and gives hoarse fight calls; sometimes also points bill at partner and makes biting intention movements. When anxious, crouches in hunched posture and sleeks plumage, lowers flattened head and extends neck forward; when alarmed, stretches legs and body upward, erects crest, jerkily flicks both wings, nervously fans outer tail-feathers and gives territorial song (♂) or rasping call (♀).

Perhaps migratory in Mozambique, whence Zimbabwe birds may have come (Irwin 1981).

Food. Mainly arthropods: spiders and various insects including moths, caterpillars, Diptera (particularly Syrphidae), Orthoptera, wasps, bees, winged ants, Odonata, alate termites, Coleoptera, Dictyoptera, Heteroptera; also small lizards. Mean prey size: 20·0 (n = 186) in NE Gabon.

Breeding Habits. Monogamous; pairs remain together for lengthy periods (3–5 years in NE Gabon). Territorial; av. territory size 42·2 ha (n = 4), in NE Gabon. ♂ defends it all year; patrols it regularly and sings, flying erratically like a large black-and-white butterfly, with many acrobatics; he erects crest, partially fluffs out rump feathers, and displays white wing patches in shallow flapping flight with rounded wings fully extended (**E**). When intruder is detected, ♂ perches on very exposed post, adopts upright posture with fully erect crest and neck extended (**F**), flicks wings, wags head from side to side (**G**) and gives loud territorial song, then makes series of territorial flights; in more intense defensive action, he gives high-intensity territorial song (see *Voice*) and dives at trespasser giving fight calls and snapping bill vigorously; sometimes ♀ fights other ♀, diving, clapping bill and giving harsh calls.

Details of pair formation unknown, except that once (NE Gabon) unmated ♂ was seen to court paired ♀ in flight; he gave intense territorial songs and circled and swooped around her, exposing all white parts of plumage, as in territorial display, but with much slower wing beats; at close range he frequently displayed white underparts and ruffled rump. As breeding season approaches, ♂ and ♀ spend more and more time close together; ♂ makes more frequent butterfly-flights around ♀, giving various territorial songs, particularly when she moves from tree to tree; ♀ spends more time prospecting for nest sites in isolated bushes and exposed branches and forks of trees, and makes nest-building movements; ♂ chooses nest site, where he crouches, stretches neck and head horizontally, conceals crest, spreads forward and quivers wings, and ruffles rump feathers, exposing white spots.

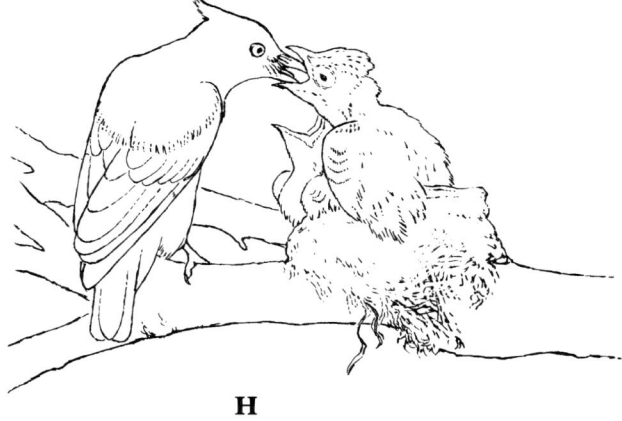

H

NEST: small, compact, neat and well-formed open shallow cup (**H**); mainly of thin stems and rootlets, leaf stalks and pieces of leaf skeletons, also pieces of bark, vegetable fibres and small bits of decaying wood; covered outside with spider webs and lichens, so it appears silver-grey, resembling a protuberance on a branch; well secured to support; ext. diam. 80–90, ext. depth 40–45; int. diam. 60–80, int. depth 30–35. Placed 1·2–30 m above ground; 11 out of 14 nests in NE Gabon < 10 m; in bamboo, bush or tall tree, at extremity of branch, sometimes completely uncovered; usually in a crotch on stout horizontal bough where 2 or 3 twigs grow upright. Built by ♀, in about a week, mainly in late morning and early afternoon; she tramples interior of nest with her feet and compacts wall by leaning over and pressing it with bill (Tarboton 1987); ♂ guards nest site, making display flights with territorial songs towards congeners, also when greeting ♀ and accompanying her on her flights to and from nest; defends site against African Pied Hornbill *Tockus fasciatus*, Purple-throated Cuckoo-Shrike *Campephaga quiscalina*, Red-backed Shrike *Lanius collurio* and Brown-headed Tchagra *Tchagra australis* in NE Gabon, and against Samango monkeys *Cercopithecus albogularis* and Narina's Trogon *Apaloderma narina* in Zimbabwe.

EGGS: 2–3, laid at 1-day intervals, beginning 4–8 days after completion of nest. NE Gabon mean (9 clutches) 2·0. Oval; glossy; dirty white to very pale blue-green, heavily blotched and spotted with dark brown, rufous-brown, grey and/or lilac, forming continuous ring or cap at large end. SIZE: *c.* 21 × 16. In NE Gabon, no regular second clutch observed, but some pairs lay replacement clutches; once a pair whose 2 young were taken by a predator a few days before leaving nest laid 2 eggs 14 days later.

LAYING DATES: Liberia, Apr, May; Ghana (♀ with ovary egg Mar); Nigeria (♀ with enlarged ovary Apr); Cameroon, Feb, Mar, June; Gabon, Sept–Mar, also May, during rains and short dry season; Central African Republic, Mar; Zaïre, Apr–Aug, Dec (♀ in breeding condition Sept, fledglings Nov, Dec); E Africa: Region B, Mar–May, July, Sept; Region C, Nov; Region D, Jan; Region E, June; Angola, Nov–Jan; Zimbabwe, Oct.

INCUBATION: begins with last egg; by both sexes. Period: 18–19 days. During the day ♂ and ♀ share incubation time equally; spells last 10–71 min (av. 28·9); timing may vary, particularly in places much frequented by man. During nest relief, ♂ much noisier than ♀, giving long series of loud territorial songs and making butterfly-flights. ♂ guards nest even when brooding, sometimes singing from nest but more often leaving it to give territorial display and to drive other birds away; harasses raptors, coucals, cuckoos and hornbills. At night, ♀ incubates, ♂ roosts 50–120 m from nest.

DEVELOPMENT AND CARE OF YOUNG: young fed by both parents, at mean rate of once every 5·3 min, each adult coming about every 10 min; usual pattern is a series of deliveries at short intervals followed by another one at longer intervals. Adults remove faeces, and eat pieces of feather sheaths or even give them to young. Any animal approaching within 1 m of brood is repelled; cuckoos, shrikes and raptors within 15 m of nest are vigorously harassed: ♂ and ♀ dive, snap bill and give fight calls. For first 5–6 days young call and beg only when adult arrives at nest; from 5 to 10 days they respond to calls of approaching adult; after that they call more and more, almost continuously during last week in nest. Nestling period: 18–23 days.

Fledged young remain near nest site for 3–4 days, then follow adults around; each parent cares for a particular young. Frequently at least one young dies in the first 15 days; other(s) are fed for 1–2 (sometimes 3) months. Young remain with parents until next breeding season.

BREEDING SUCCESS/SURVIVAL: in NE Gabon, out of 11 nests in which eggs were laid, 10 produced 1 or more fledglings, to give av. of 1·64 young/nest; 13/14 eggs hatched; 17/21 hatchlings fledged.

References
Brosset, A. and Erard, C. (1986).
Erard, C. (1987).
Field, G. D. (1971).

Genus *Dyaphorophyia* Bonaparte

Small, plump, flycatcher-like birds with a large head. Bill black, broad, more or less flat, with ridged culmen; nostrils unfeathered; tail without white margins; wings rounded, without any white patch; very short tail and short legs; grey, blue or green eye wattles; ♂ underparts white or yellow with pectoral band or gorget; ♀ upperparts rufous, grey or green, and underparts white or yellow or rufous. Songs simple, also grinding noises, bill snapping and wing flapping. Inhabit forests.

Endemic. 4 species, though some authors separate *chalybea* and *jamesoni* from *blissetti*. Sometimes made congeneric with *Platysteira*. 2 species-pairs: *concreta/blissetti* and *castanea/tonsa*.

Dyaphorophyia castanea (Fraser). Chestnut Wattle-eye. Pririt châtain.

Plate 30 (Opp. p. 545)

Platysteira castanea Fraser, 1843. Proc. Zool. Soc. London (1842), p. 141; Clarence (= Malabo), Fernando Po (= Bioko).

Dyaphorophyia castanea

Range and Status. Endemic, resident. Lowland forests of Sierra Leone, Guinea (Ziama Massif, Macenta district, abundant), Liberia, Ivory Coast (common, north to Korhogo and Comoé), Ghana (frequent, north to Bia Nat. Park, Kumasi, Mampong in Ashanti, Aduamoah, Atewa Forest Reserve and Akropong), Togo (Bismarckburg, Misahöhe), Benin (Uyere), S Nigeria (common, north to Ibadan, Enugu and Obudu Plateau, also Oban hills), S Cameroon (frequent or common, north to Korup Nat. Park, Rumpi Hills, Nkongsamba, Obala, Dimako, and Mieri), Bioko, Rio Muni, Gabon (common, widespread), Congo (coast, sublittoral and Mayombe to Djambala), S Central African Republic (Dzanga Reserves, Lobaye, Bangui, Bambari, NW Zemio), extreme S Sudan, Uganda (common, north to Budongo and Mabira), Kenya (Kakamega and S Nandi, old records from Mt Elgon and Mumias), NW Tanzania (Bukoba), Zaïre (common, Uele to Ituri and Kivu, south to Itombwe, Shaba and Kasai, west to Gemena, Mbandaka and Mayombe) and Angola (common, Cabinda to Gabela, Cuanza Sul and Canzele, Malanje). Uncommon to abundant. Density in NE Gabon, 42·3 pairs/km^2 in primary forest and 20–25 pairs/km^2 in regrowth and disused plantations (Erard 1987); in Kenya (Kakamega Forest) 8–11 adults and 2 immatures in 8 ha census tract (Zimmerman 1972).

Description. *D. c. castanea* (Fraser): Benin and Nigeria south to Angola, east to Kenya and Tanzania. ADULT ♂: black and white; crown, sides of head, lores, hind neck, mantle, back, wings, uppertail-coverts and tail black with steel-blue gloss, particularly on head; small white spot above eye obscured by wattle; scapulars black, edged white; rump white (long white hairy feathers with blackish base). Entire underparts, including sides of neck, white except for broad pectoral band black glossed blue; thighs black barred white. Undersurface of primaries and secondaries dark grey with narrow white edge along inner web (P9 falciform, with inner web sharply narrowed along distal half); axillaries white. Bill black; fleshy wattle above eye purplish blue-grey, dark liver or purplish black; eyes chestnut to purple-brown; legs reddish blue to dark purple. ADULT ♀: forehead, crown, nape, upper hindneck, lores and ear-coverts dark slate grey; small spot above eye white and chestnut, concealed by wattle. Chin white, throat, sides of neck, foreneck, chest, lower hindneck, mantle, scapulars and upperwing-coverts bright chestnut, greater primary coverts dark brown; rump mottled grey, white and chestnut (long hairy feathers white, tipped grey or chestnut, with dark base). Uppertail-coverts brown-black; tail black (feathers with very narrow whitish tips in fresh plumage). Belly and undertail-coverts white; flanks dark grey; thighs brownish grey. Primaries and secondaries blackish brown, narrowly edged cinnamon along outer web (P9 falciform and narrowed along outer web), tertials chestnut with brownish black streak on shaft and greyish wash on inner web. Underwing grey with white edges on underwing-coverts and along inner web of flight feathers; axillaries white. Bill black; fleshy wattle above eye purple-grey to purple-black; eyes chestnut to purple-brown; legs reddish to dark purple. SIZE: (15 ♂♂, 10 ♀♀) wing, ♂ 56–61 (58·7), ♀ 55·5–60·5 (58·8); tail, ♂ 24·5–27 (25·6), ♀ 24·5–26·5 (25·6); bill, ♂ 14–16·5 (15·0), ♀ 14–15 (14·7); tarsus, ♂ 15–16 (15·4), ♀ 15–16

(15·4). WEIGHT: (Bioko) ♂♀ (n = 10) 15–17 (15·6); (Cameroon) ♂♀ (n = 7) 13–16 (14·4); (Gabon) ♂♀ (n = 27) 13–17 (14·7); (Congo) unsexed (n = 3) 12·2–14 (13·2); (Central African Republic) ♂♀ (n = 3) 13–16 (14·3); (Uganda) ♂ (n = 29) 12–16 (14·3), ♀ (n = 18) 11–15 (13·9), unsexed (n = 5) 12–14 (13·2).

IMMATURE: like adult ♀ but head browner; throat and breast chestnut mottled white, with more or less distinct blackish vermiculations; upperparts and wings browner, not bright chestnut; upperwing-coverts and tertials dark brown with chestnut edges, not uniformly chestnut; eye-wattle small and greyer; eyes brown-grey; legs dark brown.

JUVENILE: like immature but head grey-brown, mottled white on cheeks; grey-brown extends to upper mantle; chin and throat dirty white; breast pale grey-brown with dense blackish vermiculations; eye-wattle barely visible.

NESTLING: hatches naked, skin blackish, soon with sparse brownish down.

D. c. hormophora Reichenow: Sierra Leone to Togo. ♂ with white collar. Small; (15 ♂♂, 16 ♀♀): wing, ♂ 56·8 ± 1·3, ♀ 55·8 ± 1·6; tail, ♂ 19·6 ± 1·2, ♀ 19·4 ± 1·9; bill, ♂ 15·1 ± 0·3, ♀ 15 ± 0. WEIGHT: (Mt Nimba) ♂ (n = 15) 11·7 ± 1·3, ♀ (n = 16) 11·4 ± 1.

A

Field Characters. Wattle-eyes in the genus *Dyaphorophyia* are small and plump with big, round heads and short wings, and appear almost tailless (**A**). This species has a grey eye-wattle; ♂ is black and white, ♀ is chestnut with grey head and white belly. Very similar to White-spotted Wattle-eye *D. tonsa* from which distinguished by larger eye-wattle extending further backward, lack of conspicuous white or buffy spot behind eye, and by habitat preference within forest (understorey and middle levels rather than canopy); also ♂ lacks white collar (except in Upper Guinea) and ♀ has top of head grey, not black, i.e. cheeks and crown same colour, not contrasting. The other 2 members of the genus have large bright wattle around (not just above) eye; Red-cheeked *D. blissetti* has black throat in both sexes (although throat tawny in immature), Yellow-bellied *D. concreta* has underparts entirely yellow or chestnut.

Voice. Tape-recorded (34, B, C, BRU, ERA, GREG, KEI, ZIM). Territorial song in NE Gabon a long, regular, monotonous series of 'tuck' notes (95–110/min), so similar to song of Red-rumped Tinkerbird *Pogoniulus atroflavus* (70–80/min) that some playback experiments attracted also the barbet; in Ivory Coast, a regular series of 'pti', 'kak' or a ringing 'pink' (72/min), to which Gabonese birds do not react. Ritualized territorial fight song, rhythmic series of 'ptik-kwow' or 'p'qwonk' (1 per s), a guttural note followed by an explosive one sounding like a frog. Guttural territorial song, a monotonous series of regularly emitted rough 'ptock' (*c.* 80/min); in territorial defence, gives frog-like 'kwow' or 'twonk'. Courtship call, an explosive, hiccough-like 'kwow' or 'p'kwup'. Nasal chattering frequent. Contact calls, soft, irregular 'wa', 'tchwa', 'wo' or 'wop'. Anxiety call, nasal 'kowo' or 'koway', or 'kowet-kowet-kowet . . .'; alarm, rapid 'pop-pop-pop . . .', then nasal 'pop-pway-pop-pway . . .'. In appeasement gives irregular series of soft notes, 'u-u-u . . .'. Also gives grunting 'gä-gä-gä . . .', and in excitement, e.g. during nest-building, thin 'tsstsstss' or crackling 'titititit'. Begging calls of young and ♀, harsh 'kchay-kchay-kchay . . .'.

General Habits. Inhabits primary and secondary forest, including swampy and flooded forest, in lowlands and up to 1150 m in Cameroon, 1600 m in Zaïre (Itombwe) and 1800 m in Uganda. Usually in mature forests with relatively unstratified understorey (foliage uniformly distributed between 8 and 20 m above ground). Frequents liana-strewn openings and clearings but avoids too open understorey. Occurs in regenerating forest, particularly in final stages, also in stands of parasol-trees *Musanga cecropioides* with undergrowth of 'gorilla herbs' (*Aframomum*), and in uncleared and disused plantations (e.g. cacao) where vegetation layers are ill-defined. Also found in forest-grassland mosaics, isolated tree clumps and even gardens (e.g. at Mt Nimba).

In pairs or family parties; partners (i.e. family members) maintain permanent and close visual and vocal contact, non-breeding pair members move always together. Searches foliage from about 2 m to lower canopy (up to 25 m above ground in NE Gabon, 18 m in Nigeria) but seasonal variability marked, at least in NE Gabon: during the rains and short dry season, i.e. during breeding period, birds feed at lower levels. Where White-spotted Wattle-eye is present in canopy and Yellow-bellied Wattle-eye in lowest strata, feeding niche narrows to 8–22 m. Regular territorial fights observed between White-spotted and Chestnut Wattle-eyes (Erard 1987). Active all day; covers entire home range daily. Forages energetically, mostly at extremity or on distal third of branches, also in tangles of leafy lianas; hops from twig to twig in upright or crouched posture; catches prey mostly under leaves like typical hover-gleaner, with short upward sally to snap up prey in brief hovering flight or swoop. Pairs dislodge insects from leaves in noisy flapping flight. Noxious and large prey (e.g. bees, wasps, caterpillars) are smashed on perch, while beetles, grasshoppers and cicadas, are dismembered. When foraging, sometimes aggressive toward other insectivores, e.g. Grey-throated Tit-Flycatcher *Myioparus griseigularis*, Blue-headed Crested Flycatcher *Trochocercus nitens* and Bates's Paradise-Flycatcher *Terpsiphone batesi*. Often joins mixed-species flocks, particularly during non-breeding season. Bathes in holes in trees, or more often flutters among wet leaves until drenched, after which it perches and preens.

Individuals roost under leaf at extremity of exposed thin twig. Anxiety expressed by calls, rapid head movements and plumage sleeking in crouched posture. When alarmed, erects crown feathers and calls while flying from branch to branch with measured noisy wing flaps. Mobbing bird swoops toward intruder with erected crest and expanded eye-wattles, and vigorously flaps wings and snaps bill.

Sedentary, at least in NE Gabon.

Food. Arthropods. In NE Gabon, analysis of 53 stomach contents showed Coleoptera in 50 of the stomachs, caterpillars in 10, spiders in 5, moths in 4, ants in 3, mantises in 3, grasshoppers in 2, cockroaches in 2, insect eggs in 2 and scorpions in 1 (Erard 1987); 166 observed prey were: unidentified insects (30), caterpillars (71), moths (39), beetles (11), grasshoppers (5), bees and wasps (5), cicadas (4), spiders (3), termite (1), fly (1). Young were given unidentified insects (41), caterpillars (38), moths (16), grasshoppers (7), beetles (3), mantis (1), ants (1), spider (1) and millipede (1). Other insects (e.g. Hemiptera, Neuroptera) and even small fruits sometimes eaten. Prey size (mm): $18 \cdot 3 \pm 10 \cdot 1$ (n = 166) for adults, $13 \cdot 2 \pm 8 \cdot 9$ (n = 105) for young.

Breeding Habits. Monogamous, apparently pairs for life. Helpers at some nests. Territorial. Pair defends territory throughout year; mean territory size $2 \cdot 4 \pm 0 \cdot 3$ ha (n = 64) in NE Gabon. ♂ advertises territory by singing, usually in canopy and upper levels of understorey, mostly in morning and early afternoon, during pauses in feeding activity. Both ♂ and ♀ defend territory against individuals of same sex. When intruders seem ready to enter territory, owners emit irregular series of alarm calls and move towards opponents with noisy wing flaps; then, perched, give territorial defence song in upright posture, expose throat and breast patterns, fluff out rump and erect eye-wattles. In next stage they fly noisily with intense grunting calls and bill snapping, flying with measured wing flaps, exposing rump, erecting crown-feathers and eye-wattles, and raising front of body; usually at this stage eye-wattles are swollen with blood and bright red, and protude like horns on sides of head. When opponent gets closer (< 2 m) territory owner becomes more aggressive, approaches and perches less than 1 m from it, usually above it, in threat posture (crouches, points bill towards opponent, ruffles flank feathers, erects crest and bright red eye-wattles) and emits scolding territorial fight song: 'ptick' note while facing opponent and 'kwow' with head turned to one side; when perched at level of opponent, owner adopts bill-up posture, presenting fluffed breast. When pairs meet at boundary of their territories, ♂♂ and ♀♀ engage in ritualized fight: birds of same sex face each other in threat posture and duet with fight song, turning heads from side to side so that when one says 'ptick' the other says 'kwow' (human ear hears only one rhythmic song). According to sex of intruder, ritualized fights can be by ♂♂ only, ♀♀ only, or by ♂♂ and ♀♀ but independently of each other; they may be joined by 2–3 pairs holding territories and several unpaired ♂♂ and/or ♀♀. When only ♂♂ are engaged in the ritualized fight chorus, unpaired ♂♂ court ♀♀ whether paired or not. Usually, after such fights, ♂ which owns territory gives series of guttural territorial songs, then rejoins family group with appeasement song. In courtship display, ♂ crouches and approaches ♀, often faces away, and gives measured contact calls and appeasement songs; hops about in branches and then passes behind ♀, perching less than 1 m above her, crouches, bends body forward and downward, fluffs out rump, droops wings, expands blood-swollen red eye-wattles; then swoops down close in front of ♀ while giving a single loud call (one ritualized fight song note) and perches below her. This display is repeated several times and for several days until pair-bond is definitely established. Courtship feeding is common and begins just before nest-building; observation of ♂ feeding ♀ is a good breeding record, meaning that ♀ will lay in the very near future (less than 2 weeks) or is already incubating. To find the nest, follow ♀ not ♂.

NEST: small tight structure; an open cup, made of vegetable fibres, dried grasses, small stems, fine rootlets, pieces of bark, lichens, bound with spider web, inner cup lined with thinner material, particularly fungus *Marasmius*, sometimes with vegetable down, rim lined with spider webs; ext. diam. 55–65, ext. depth 25–35; int. diam. 41–42, int. depth 21–25; usually placed in fork near end of small leafy branch, sometimes at top of young sapling, always with a leaf (at least 6 cm wide) very close above it, acting as a roof. The leaf is bent down and secured by spider web or *Marasmius* fixed on the rim, and is adjusted by the bird during nest-building. Placed at various heights, e.g. a nest in Nigeria only 0·3 m above ground, but 37 nests in NE Gabon 1·45–16 m, mostly 3–8 (mean 5·7). Built in 14–16 days by both sexes, often helped by young of a previous brood. During nest-building and even incubation, birds are very particular about the distance which separates the nest from its leafy roof, so observers must never cut or displace spider web or *Marasmius* which holds leaf in the correct position.

EGGS: 1–2, laid at 1 day intervals, beginning 1–4 days after completion of nest; NE Gabon mean (13 clutches) 1·9. Oval; glossy; green or blue-green or bluish white, finely spotted dark brown and grey, particularly as a cap at broad end. SIZE: 16–18 × 12–13·5. In NE Gabon neither second nor replacement clutch observed although birds breeding unsuccessfully in Sept–Oct may rebuild in Dec–Feb.

LAYING DATES: Liberia, (♀♀ with active ovary Apr, July, ♂♂ with enlarged testes Jan–July); Ivory Coast, Jan, July–Nov (juvs Feb–Nov); Ghana, Dec, Jan; Nigeria, May, Dec; Cameroon, May, Oct (♂♀ in breeding condition Jan, Feb, Mar, Aug, Dec); Bioko, (♂♂ with gonads strongly enlarged Sept–Nov, Jan; ♀ with very active ovary Sept); Gabon, July–Apr, particularly Sept–Oct and Dec–Feb; Congo, Aug, Sept; Zaïre, Feb, Mar, Sept, Oct (♂♂ with enlarged testes Feb–Nov); Central African Republic (imm. May, ♀♀ in breeding condition Apr–May); E Africa: Uganda,

Mar; Region B, May; Angola (♀ with active ovary Feb, juvs Mar, Nov, Dec).

INCUBATION: begins with last egg; by ♀. From morning to early afternoon she sits for periods of 60–90 min and leaves for 60–75 min, then sits continuously until late afternoon when she leaves for 60–75 min. ♂ patrols territory while ♀ sits, giving territorial song. ♂ feeds ♀ regularly and intensively, both on and out of nest, though she may occasionally feed by herself; from nest-building to hatching, ♀ receives most of her food from ♂. When at times ♂ comes near nest without bringing food, he sings in appeasement! Period: 17 days.

DEVELOPMENT AND CARE OF YOUNG: young remain in nest 14–16 days; brooded by ♀ for first 7–8 days for 50–70% of daylight; fed by both parents, often helped by young of a previous brood. For first 4–7 days ♂ passes food to ♀ which feeds herself or gives it to young. Each young fed once every 30 min in first 2 days, once every 10 min at 8 days, but food delivery rate is irregular, feeds being more frequent in the morning and late afternoon; the usual pattern is a series of visits at short intervals (1–15 min) separated by pauses of 30–60 min. Young increases weight steadily from 2·7 g at 2 days old to 10 g at 8 days. Fledged young fed by both parents and often also by one or two young of a previous brood, at least for 2 months. Young stay in family group for almost 2 years and help adults to build nest and feed young. They attain full adult plumage at the end of their second year and even then may still help parents.

BREEDING SUCCESS/SURVIVAL: in NE Gabon, of 27 nests observed from the building stage, eggs were laid in only 13, young hatched in 6 and fledged in 4. 13 nests were followed from laying to fledging: 25 eggs were laid, 11 hatched and 8 produced fledglings. In Kakamega Forest, Kenya, bird ringed as juv. was still present in same spot 4 years later.

References
Brosset, A. and Erard, C. (1986).
Erard, C. (1987).
Johnson, D. N. (1984).

Plate 30 (Opp. p. 545)

Dyaphorophyia tonsa Bates. White-spotted Wattle-eye. Pririt à taches blanches.

Diaphorophyia tonsa Bates, 1911. Bull. Br. Orn. Club, 27, p. 86; Bitye, Ja R., S Cameroon.

Range and Status. Endemic, resident. Distribution disjunct: (1) SE Nigeria (rare and local, Umuagwu, Omanelu, Owerri, Oban-Ipkan forests), W and S Cameroon (north to Korup Nat. Park (fairly common), Mamfe, Kumba, Ototomo, Bitye, Ja R.), Rio Muni and Gabon (widespread, frequent to common, Woleu N'Tem and Ogooué-Ivindo south at least to Franceville, Mouila and Tchibanga); NW Congo (Odzala). (2) Central and E Zaïre (Salonga Nat. Park, and lower Uele to Ituri and Itombwe). Reports from Ivory Coast (Abidjan, San Pedro and Lamto) require confirmation since confusion with Chestnut Wattle-eye *D. c. hormophora* is possible. Probably more widespread, particularly in Congo basin, but more readily detected by voice than by sight. Uncommon to common. Density in NE Gabon, 13·7 pairs/km².

Dyaphorophyia tonsa

Description. ADULT ♂: black and white; crown, sides of head, wide breast-band, mantle, wings, uppertail-coverts and tail black with steel-blue gloss; scapulars black broadly edged white. Supercilium (extending behind eye-wattle), hindneck, rump (long silky white hairy feathers with dark bases), chin, throat, foreneck, sides of neck, belly and undertail-coverts white; thighs black barred white. Underwing-coverts black edged white, axillaries white, underside of primaries and secondaries dark grey edged white along inner web. Bill black; wattle above but not extending behind eye purplish; eyes chestnut; legs black or dark purplish grey. ADULT ♀: forehead, crown and nape velvet-black; lores and ear-coverts slate-grey; streak above eye and extending behind wattle buffy white. Chin white; neck, throat, breast, mantle, back, scapulars, lesser and median upperwing-coverts bright chestnut; lower rump greyish (long hairy feathers pale grey more or less edged white). Uppertail-coverts and tail black. Belly and undertail-coverts white, flanks dark grey. Primaries and secondaries blackish brown edged tawny (primaries) and chestnut (secondaries), tertials chestnut washed dark grey, greater primary coverts dark brown-grey, narrowly edged chestnut on outer web, greater upperwing-coverts almost entirely chestnut (outer feathers with dark brown-grey inner web). Underwing-coverts grey edged pale chestnut, axillaries white, underside of flight feathers dark grey with very narrow whitish edges along inner web of secondaries and inner primaries. Bill black; wattle, above eye but not behind it, grey; eyes chestnut; legs blackish grey. SIZE: (10 ♂♂, 5 ♀♀) wing, ♂ 52·5–57·5 (54·2), ♀ 52·5–55·5 (54·5); tail, ♂ 20·5–24 (22·1), ♀ 22–23·5 (22·6); bill, ♂ 14–15 (14·3), ♀ 13–15 (14·0); tarsus, ♂ 14–15·5 (14·9), ♀ 14·5–15·5 (15·0). WEIGHT: (Gabon) 1 ♂ 11·5.

IMMATURE: like adult ♀ but head more rusty brown, scarcely black on crown; throat and breast mottled greyish chestnut-brown, white and rufous, with dark vermiculations; chestnut of back and wings browner; eye-wattle small.

JUVENILE: like immature but more grey-brown, not so chestnut, particularly head and barred breast which are almost dark grey.

NESTLING: unknown.

A

Field Characters. Overlaps with similar-looking Chestnut Wattle-eye *D. castanea*, q.v., but is restricted to Congo basin and adjacent regions. Less noisy than Chestnut Wattle-eye and found higher up within forest, mainly in canopy; has eye-wattle same grey colour but different shape, ending above eye rather than extending behind it, and instead having a white or buff postocular spot (**A**). ♀ further distinguished by black top of head, ♂ by white collar. For distinction from other congeners, see Chestnut Wattle-eye.

Voice. Tape-recorded (C, CAR, CHA, ERA). Low intensity territorial claim song of ♂ a regular whistle, 'ou', repeated at rate of 60–70/min; at medium intensity (by ♂ or often in duet by ♂ and ♀), a double 'ou-u' (second note shorter and higher-pitched), at rate of 65–70/min, with triple 'ou-u-u' regularly interspersed (interval between second and third note shorter than that between first and second); at high intensity (by ♂ or ♀), a rhythmic 'u' at rate of 80–85/min which at highest intensity accelerates into a higher-pitched 'u-u-u-u . . .' (165–170/min). Territorial defence song, series of 'ptoc-ou' or more intense 'ptoc-u', followed by more aggressive, scolding 'ku-lee' or 'ku-li', sometimes 'ku-lili', all at a rate of 80–85/min. Fight call, same as courtship call, a repeated harsh nasal 'ptooee'. In excitement and aggression, a harsh 'prrrt' or scolding 'ptec' or 'ptedetec', irregularly repeated, followed by a prolonged grunt 'gägägägä' or 'krrkrrkrrrk'. In anxiety and alarm, a guttural, repeated 'ptek', 'ptick' or 'ptck' (♂), or a harsh 'krrr' (♀). Contact call, irregular soft 'pec', 'prrek', 'piup' or 'tuctucpiup'. Young give harsh nasal 'kwet'.

General Habits. Inhabits rain forest in lowlands and up to 1500 m in Zaïre (Itombwe). Occurs particularly in primary forest but also in overgrown plantations beside riverine forest, in regenerating forest, and in old second-growth where vegetation is tall and almost unstratified; avoids clearings and open forest, also man-made habitats; frequents forests with thick and continuous high canopy with tall emergents, even when understorey vegetation is not dense, i.e. in clear tall forest on hill-tops and slopes.

In pairs or family parties; partners maintain constant visual and vocal contact; covers entire territory daily. Although sometimes found in understorey, essentially a bird of the canopy and emergents, i.e. more than 20 m above ground. Very often joins mixed-species flocks, more in dry than in rainy season. Forages actively among foliage of tree-crowns and leafy liana tangles; foraging behaviour very like Chestnut Wattle-eye, i.e. hops along branches and from branch to branch, in upright or half-crouched posture, and sallies horizontally or upward, in direct whirring flight, for 2–5 m from perch, to catch prey on or near leaf. Sometimes pair or group members fly through dense foliage to dislodge insects, but they usually follow other gleaners. Noxious prey are rubbed or crushed on perch. Chestnut and White-spotted Wattle-eyes exclude each other among foliage, maintaining a distance of at least 5 m between them with territorial songs and calls; they flap wings and snap bill before engaging in territorial fight and pursuit. When the 2 are together in mixed-species flocks, as often happens, White-spotted confines its foraging activity to above 22 m; it may descend to 14 m when Chestnut is not present (Erard 1987).

Comfort behaviour very similar to that of Chestnut Wattle-eye, i.e. bathes in holes in trees and by flying among wet foliage, then preens vigorously. When anxious, calls and rapidly turns head, and may crouch; when alarmed, raises forecrown-feathers, calls, and flies with rapid, continuous wing-beats (i.e. less rhythmic than Chestnut Wattle-eye).

Sedentary, at least in NE Gabon.

Food. Arthropods, mostly insects: beetles, moths, caterpillars, cicadas, bees, wasps, winged ants and termites, flies, Neuroptera; also spiders. Most prey 7–15 mm long, sometimes up to 25–30.

Breeding Habits. Monogamous; young of previous brood(s) help parents, at least to build nest and feed older fledglings; territorial. Pair patrols territory daily and defends it throughout year; mean size 6·4 ha (n = 22) in NE Gabon. Advertises territory during the day with low- or medium-intensity songs. When congeners or even just a mixed-species flock advance toward them in canopy, ♂ and ♀ become excited, raise crest and expand eye-wattles, move rapidly and noisily under or in upper canopy, emitting grunts, highest-intensity advertising and territorial defence songs. ♂ and ♀ approach trespassers with grunts, guttural calls and rhythmic fight calls; at close range they engage in duets of fight calls with erect and blood-swollen eye-wattles

Plate 31

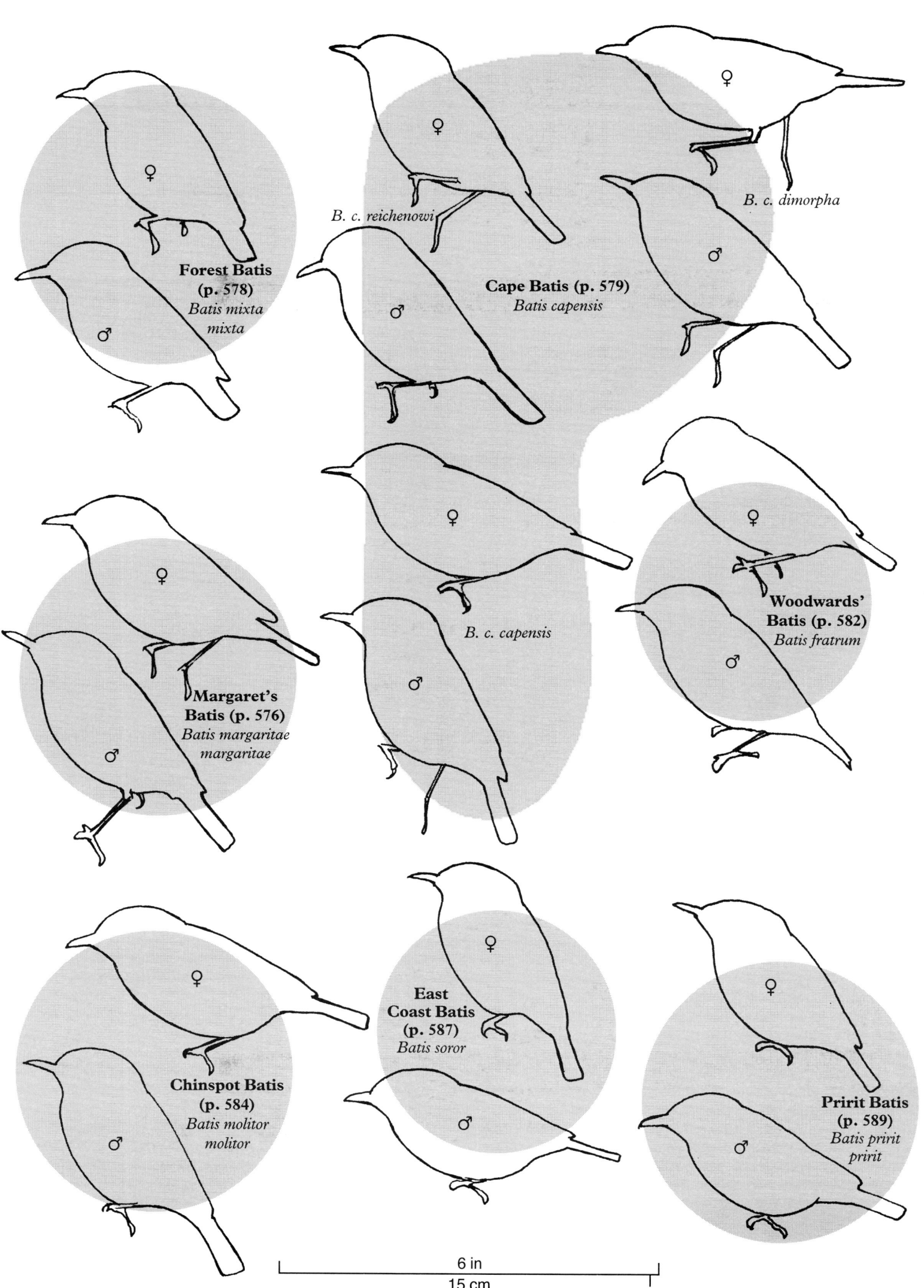

560

6 in / 15 cm

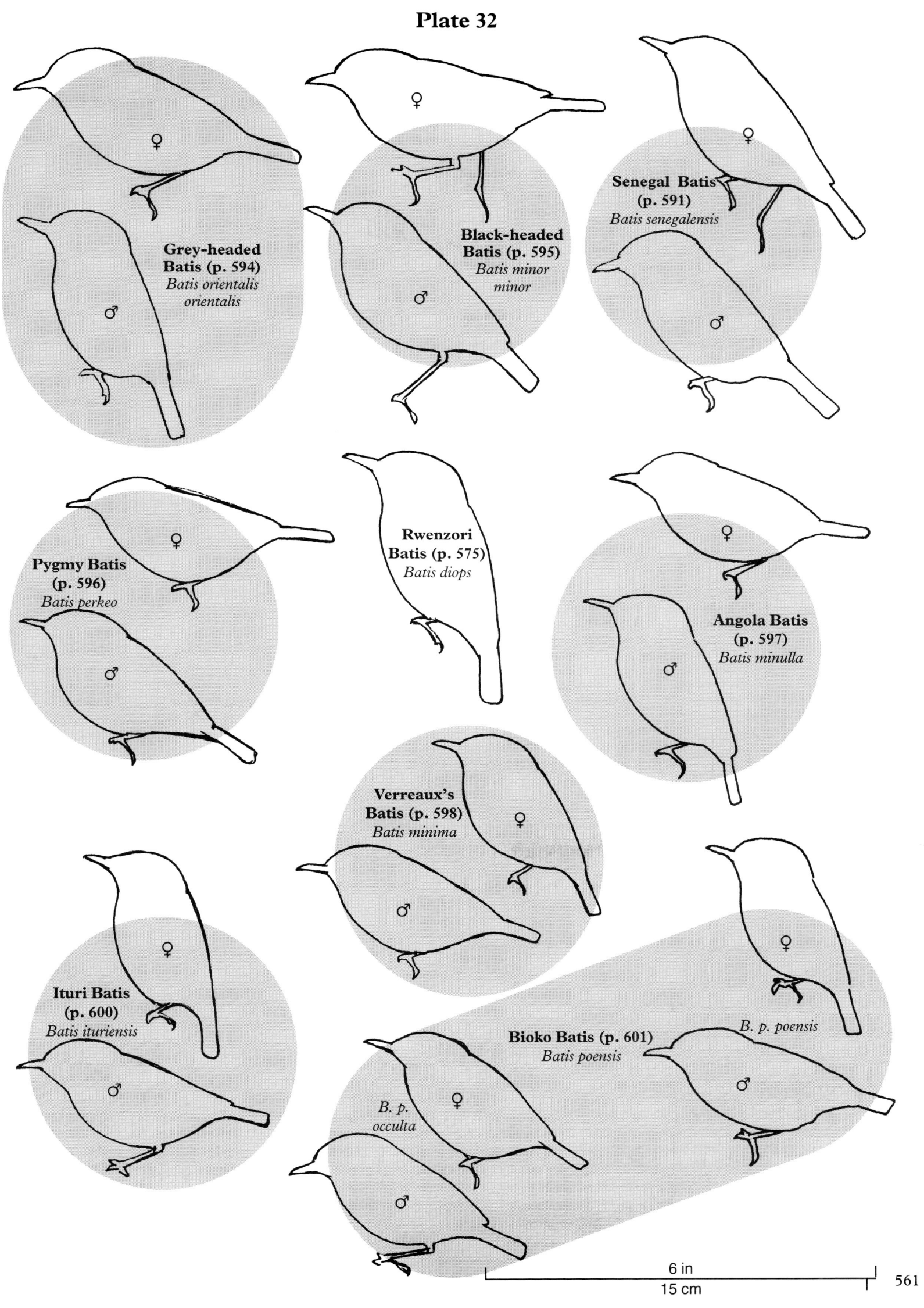

but these fights are much less formal than those of Chestnut Wattle-eye (antagonists do not face each other so closely, nor do they time so precisely their calls and postures). ♂ and ♀ pursue fleeing trespassers with whirring flight and rapid bill-snaps.

NEST: very similar to that of Chestnut Wattle-eye: small open cup, made of small fine stems or rootlets and vegetable fibres, bound with spider web and fungus *Marasmius*. Built by ♂ and ♀ helped by young of previous brood. The only nest found was placed 15 m above ground at end of horizontal leafy branch of understorey tree in forest.

EGGS: unknown.

LAYING DATES: Nigeria, Mar (♂♂ with testes enlarged Oct, Nov); Cameroon, Apr–May (♂♂ with testes enlarged Mar, Sept, ♀♀ with ovary slightly enlarged Dec); Gabon, Dec–Feb, Apr, Sept–Oct, probably similar to Chestnut Wattle-eye, q.v.; Zaïre (♂♀ with active gonads Feb, Mar, May, Sept, Nov).

INCUBATION: by ♀; period unknown. Incubation behaviour similar to that of Chestnut Wattle-eye, i.e. ♀ sits for long periods and is fed by ♂ in and out of nest.

DEVELOPMENT AND CARE OF YOUNG: nestling period unknown. Fledglings fed by both adults, helped by young of previous brood, for at least 2 months, and remain with family group almost 2 years.

References
Brosset, A. and Erard, C. (1986).
Erard, C. (1987).

Plate 30 (Opp. p. 545)

Dyaphorophyia blissetti Sharpe. Red-cheeked Wattle-eye. Pririt de Blissett.

Diaphorophyia Blissetti Sharpe, 1872. Ann. Mag. Nat. Hist., (4) 10, p. 451; Wassaw, Gold Coast.

Range and Status. Endemic resident. Distribution disjunct and patchy: E Guinea (Macenta district, uncommon, also Kindia area), Sierra Leone, Liberia, Ivory Coast (locally common, Taï to Abidjan, north to Sipilou and Béoumi) and Ghana (common and widespread, north to Aburi and Ejura); S Nigeria (rare to frequent, Lagos to Sapele, north to Ibadan and Ife, also Kagoro and Enugu), S Cameroon (frequent or common, Mt Cameroon, Rumpi Hills and Kumba, east and south to Nkolngem, Bitye and Dja R.), Bioko, Rio Muni and Gabon (frequent or common, Woleu N'Tem and Ogooué-Ivindo south to Franceville and Mouila); Angola (Gabela); E Zaïre (Uele to Ituri, Kivu and Itombwe), S Sudan (southernmost Western Equatoria and Bari), S and W Uganda (common, Budongo, Bugoma, Bwamba, Kibale, Bwindi–Impenetrable and Malabigambo Forests east to Kifu, Mabira and Mt Elgon), W Kenya (Kakamega and Nandi Forests) and NW Tanzania (Bukoba). Uncommon to common. Density in NE Gabon, 13·5 pairs/km² in early stages of regenerating forest, 7·1 pairs/km² in cultivated areas (Erard 1987); in Kenya (Kakamega Forest), 10–11 adults and 3 immatures in 8 ha census tract (Zimmerman 1972).

Dyaphorophyia blissetti

Description. *D. b. chalybea* Reichenow ('Black-necked Wattle-eye'): Bioko, W and S Cameroon (Rumpi Hills and Mt Kupé, eastward and southward) to Gabon and Angola. ADULT ♂: entirely black with bottle-green sheen except that: lower rump pure white, lower breast, belly, axillaries and undertail-coverts silky white with well-marked uniform yellow tinge; upperside of tail metallic bottle-green, underside of tail and wings grey-black; flanks blackish grey; thighs black. Bill black; large flat-topped wattle around eye bright turquoise-blue; eyes chestnut-brown; legs grey-blue or purple; claws white. ADULT ♀: like ♂ but duller, black of plumage less glossed, eye-wattles much smaller. SIZE: (10 ♂♂, 3 ♀♀) wing, ♂ 51–54 (52·3), ♀ 52–54 (51·8); tail, ♂ 22–25 (23·5), ♀ 23–24 (23·8); bill, ♂ 14–15·5 (14·8), ♀ 14–15 (14·5); tarsus, ♂ 17–18 (17·5), ♀ 17–18 (17·3). WEIGHT: (Bioko) ♂♀ (n = ?) 9–12·5 (11·1); (Cameroon) 2 unsexed 11·2, 11·4; (Gabon) ♂♀ (n = 10) 11–13 (11·7).

IMMATURE: like adult ♀ but no yellow wash on underparts, chin white, throat, foreneck and upper breast pale tawny with sparse black feathers, encircled by chestnut and black (moustachial stripe extends onto sides of neck and joins narrow breast-band). Bill blackish, tipped horn; vestigial eye-wattle greenish blue; eyes greyish brown; legs and feet brown-grey to greyish purple.

JUVENILE: entire upperparts sooty black, entire underparts dirty white.

NESTLING: undescribed.

D. b. blissetti Sharpe: Guinea and Sierra Leone to Nigeria and W Cameroon (Kumba, Mt Cameroon, Victoria, Saxenhof). Sides of neck and of throat bright dark chestnut; black throat extending less far onto upper breast; immature with chin and throat tawny, sides of neck and breast-band cinnamon, with black feathers on sides of throat and sides of breast-band. WEIGHT: (Liberia) ♂ (n = 14) 11 ± 1, ♀ (n = 8) 11 ± 0·9; (Cameroon, Jan) 1 ♂ 11, 1 ♀ 11·5.

D. b. jamesoni Sharpe ('Jameson's Wattle-eye'): Zaïre to Sudan, Kenya and Tanzania. Like nominate *blissetti* but chestnut restricted to sides of neck. WEIGHT: (Uganda) ♂ (n = 29) 10–13 (11·5), ♀ (n = 28) 9–15 (11·6), unsexed (n = 6) 9–15 (11·2); (Kenya) ♂ (n = 8) 10–12 (11·1), ♀ (n = 12) 9–12 (10·2).

TAXONOMIC NOTE: because both *D. b. blissetti* and *D. b. chalybea* occur in W Cameroon (although altitudinally separated: Serle 1954, Eisentraut 1973), many authors rank *chalybea* as a full species. Recent authors (Traylor 1986, Sibley and Monroe 1990, Dowsett and Dowsett-Lemaire 1993b) consider *blissetti*, *chalybea* and *jamesoni* to be allospecies of a superspecies. Pending a detailed study of the biology of these forms, including geographical analysis of behaviour and vocalizations, we think it best to retain them in the same species.

Field Characters. Small, active and noisy (songs, calls and wing snapping); plump, with short wings and tail, and broad flat bill; hunts low down in understorey. Black above except for white rump, breast to belly silky white or pale yellow, throat black, conspicuous eye-wattle blue-green. Diagnostic chestnut neck-patch from ear-coverts to sides of throat and neck in nominate *blissetti*, and smaller patch in *jamesoni*; these areas entirely black in *chalybea*. Adults of all races distinguished from congeners by colour of eye-wattle, black throat, and lack of chestnut on underparts (♀ just a duller version of ♂); immature has tawny throat and upper breast but told from ♀♀ of Chestnut Wattle-eye *D. castanea* and White-spotted Wattle-eye *D. tonsa* by lack of chestnut on upperparts and wings.

Voice. Tape-recorded (C, ERA, GRI, HOR, KEI, LEM, MAC, McVIC, ZIM). Advertising song of *jamesoni* in Kenya, long series, 'ptee-dee-deee …', accelerating from 20 notes in 25 s to 20 in 17 s, mixed with excited 'ptik' and wing-flaps; that of *chalybea* very different, 4–6 descending sweet notes, reminiscent of Brown-throated Wattle-eye *Platysteira cyanea*, preceded by excited 'ptiucteedee' or 'ptiuc-ti-di tititit'. Territorial defence song of all races, long series of ringing notes on one pitch at rate of *c*. 3 notes per s, like weak version of Olive-green Camaroptera *Camaroptera chloronota*; in *jamesoni*, introduced by hoarse 'trr-trr' and followed by nasal 'hee'. Fight call of *chalybea*, 'kweck-kweck- …' accelerating to 'ptedecptedecptedec …', or in flight, grunting, nasal 'gehgehgehgeh …'. In excitement, repeated high-pitched 'ptick'; when anxious or alarmed, rapid 'pwit-pwit-pwit…', faster 'pwickpwickpwick…' or hard 'pteckpteckpteck…'

General Habits. Inhabits lowland forest up to 700 m (*blissetti*), at 1050–1950 m in Cameroon (*chalybea*), up to 1420 m in Zaïre (Itombwe) and 2300 m in Uganda (*jamesoni*). In Ghana, Ivory Coast and Cameroon, in places where Yellow-bellied Wattle-eye *D. concreta* is absent, nominate *blissetti* frequents dense undergrowth of both mature and secondary forest, riverine forest, forest edges and plantations (e.g. cocoa); found also in coastal thickets and on some inselbergs; but where Yellow-bellied Wattle-eye is present, typically inhabits dense understorey of secondary forests. The same is also true for *chalybea* in Cameroon. In NE Gabon, where Yellow-bellied Wattle-eye is common, *chalybea* is typically a bird of secondary forest, but unlike Yellow-bellied Wattle-eye, it occupies islands of Ivindo R. in regenerated forest of a low, flooded and liana-rich type. Prefers early stages of regenerating forest, i.e. dense beds of 4 m-high *Aframomum* herbs underneath continuous stands of parasol trees *Musanga cecropioides*; also occurs around villages in overgrown cultivation and plantations (coffee, cocoa, cassava) and in fallow ground completely covered with dense low vegetation, e.g. thickets of *Solanum* and other shrubs. *D. b. jamesoni* inhabits mostly secondary forest, but in Uganda and Kenya also understorey of primary forest, together with both Yellow-bellied and Chestnut Wattle-eyes; the ecological relationships of the 3 species are not clear, although their niches may be stratified, with Red-cheeked Wattle-eye foraging at lowest level.

In pairs or family parties; partners maintain regular and close visual or vocal contact, ♂ singing and both giving contact calls, also flapping wings. Always in lowest strata of vegetation, in NE Gabon and Kenya, < 4 m above ground, usually < 2 m. Forages among dense foliage of low shrubs, broad-leaved herbs and tangles of leafy lianas; hops along fallen and decaying trunks and large stems in thickets; searches piles of branches with dead leaves; hops from twig to twig in upright or crouched posture, and carefully inspects the surrounding vegetation with nervous lateral movements of body or head; often flies with whirring wings among tangles and thickets to dislodge insects, then, from perch below the worked area, makes short upward sallies or hovers to snatch prey, or stretches legs and body to pick insects off leaves. Frequently joins mixed-species flocks, particularly during long rains and dry seasons, i.e. mostly when not breeding.

Food. Insects: beetles, moths, caterpillars, ants, large flies, winged termites; also spiders. Prey length 10–15 mm, up to 25–30.

Breeding Habits. Monogamous; territorial. In NE Gabon, because of rapid habitat change, pairs are rarely together or present on the same spot for more than one breeding season and we do not know how long the pair-bond lasts. Defends territory throughout year; mean territory size 8·6 ± 1·4 ha (n = 7) in NE Gabon. ♂ regularly patrols entire territory, advertising it with much singing. He sings in the more open areas above the undergrowth, flying from one perch to another with whirring wings and fluffed-out rump-feathers; sings in upright posture with bill pointing up and swollen eye-wattles bright green. When potential intruder detected (e.g. approaching mixed-species flock or singing neighbour), ♂ gives lengthy advertising and territorial defence songs, and flies back and forth with whirring wings along territory boundary, making himself conspicuous with fluffed-out rump feathers. Intruder crossing territorial boundary is harassed in whirring flight with fight calls and vigorous bill snapping until it leaves territory. During breeding season, ♂ keeps close to ♀ and feeds her.

NEST: small neat open cup of vegetable fibres, small stems, rootlets, pieces of bark and dead leaves, with lichens and pieces of decaying leaves outside, bound with spider web; inner cup lined with finer elements; ext. diam. 45, ext. depth 45; int. diam. 40, int. depth. 30; placed in fork of sapling or at point where thin, low large-leaved vines cross, under leaf (but not so close to leaf as in *D. castanea*); 0·4–1 m above ground.

EGGS: 2 (only 2 clutches known). Oval; glossy; pale greenish or greenish white with a cap of brown to brown-grey spots at larger end. SIZE: (n = 2) 17·5–18 × 12.

LAYING DATES: Liberia (♀♀ with enlarged ovaries Apr, moderately enlarged Mar, Apr, Aug, Nov; ♂♂ with enlarged testes Apr, Aug, Sept); Ghana (♂ with enlarged testes Feb); Nigeria, Dec–Feb; Bioko (♂♀ with gonads enlarged Oct, Nov); Cameroon, Feb–Apr, Sept (♂♀ with gonads active Jan); Gabon, Jan, July; Zaïre (♂ with enlarged testes May, Sept); Angola (♂ with much enlarged testes Aug); E Africa: Region B, June.

INCUBATION: by ♀; period unknown (at least 14 days). ♀ sits for long periods; ♂ regularly feeds her in nest and out of it; when ♀ at nest, ♂ stands guard and sings nearby.

DEVELOPMENT AND CARE OF YOUNG: at 1 nest, young stayed in nest for 15 days, regularly fed by both adults and by a young from previous brood. Young remain with parents for several months, probably until next breeding season. In Kakamega Forest, Kenya, 2 birds ringed as juvs were retrapped 2 years and 7 months later (Zimmerman 1972).

References
Brosset, A. and Erard, C. (1986).
Erard, C. (1987).

Plate 30
(Opp. p. 545)

Dyaphorophyia concreta (Hartlaub). Yellow-bellied Wattle-eye. Pririt à ventre doré.

Platystira concreta Hartlaub, 1855. J. Orn., 3, p. 360; Guinea, but restricted to Ghana by Serle, 1952, Ibis, 94, p. 686.

Range and Status. Endemic, resident. Distribution disjunct: (1) E Guinea (Macenta district, common), Sierra Leone (Loma Mts), Liberia (north to Mt Nimba), Ivory Coast (rare or uncommon, Taï Nat. Park to Mt Nimba, Sipilou, Lamto and Yapo Forest) and Ghana (2 records: unknown type locality, and Lateh); (2) SE Nigeria (rare or uncommon, Umuagwu to Calabar and Oban, also SW escarpment of Obudu Plateau), W and S Cameroon (north to Korup Nat. Park, Rumpi Hills, Mt Kupé, Nlonako, Yaoundé and Dja R.), Central African Republic (Dzanga Reserves), Rio Muni, Gabon (uncommon to common, Woleu N'Tem and Ogooué-Ivindo south to Tchibanga, Mouila and Franceville), Congo (Mayombe, Sibiti) and W Zaïre (Mayombe); (3) W Angola (escarpment from Cuanza Norte to Chingoroi, northern Huila); (4) E Zaïre (Ituri south to Itombwe), Rwanda (Nyungwe Forest), Burundi, W Uganda (Kalinzu, Kasyoha–Kitomi, Maramagambo and Bwamba forests) and Tanzania (Mt Nkungwe, Mahari Mt); and (5) W Kenya (Kakamega and Nandi forests). Uncommon to common. Mean density in NE Gabon, 5·1 pairs/km², but varies from 2·6 to 6·4 pairs/km² (Erard 1987); in Kakamega Forest, Kenya, 2 adults in 8 ha census tract (Zimmerman 1972).

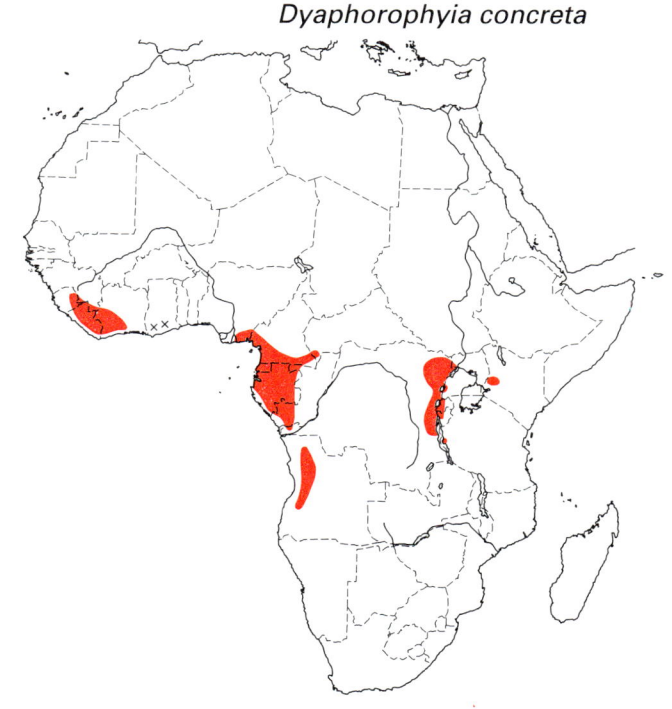

Description. *D. c. graueri* Hartert (including '*kumbaensis*', '*harterti*' and '*silvae*'): Nigeria to Gabon, Zaïre and Kenya. ADULT ♂: entire upperparts, lores, ear-coverts and scapulars greenish blue-grey, with dark green-blue gloss on head; very narrow line between base of upper mandible and forehead and stripe above lores golden yellow; lower rump-feathers broadly tipped bright golden yellow. Uppertail-coverts and upperside of tail-feathers metallic steel-blue, underside dark grey. Entire underparts and axillaries bright golden yellow (but in E Zaïre c. 20% of ♂♂ have chin yellow and rest of underparts chestnut, Prigogine 1969a); flanks greenish grey. Primaries, secondaries and tertials blackish, edged greenish blue-grey along outer web; upper- and underwing-coverts blackish broadly edged metallic green-blue. Bill black; large flat-topped wattle around eye, wider in front, bright emerald or apple-green; eyes red-brown to purplish brown with inner ring pale bluish or lavender; legs purplish grey, claws whitish. ADULT ♀: like adult ♂ but entire upperparts and wings greyer, less green, with less metallic gloss; rump grey; narrow forehead and supraloral stripe, chin and narrow moustachial stripe whitish; throat, breast and upper sides of belly chestnut; rest of underparts yellowish washed pale chestnut; eye-wattle small. SIZE: (4 ♂♂, 5 ♀♀) wing, ♂ 59–61 (59·8), ♀ 58–61 (59·1); tail, ♂ 26·5–27·5 (27·0), ♀ 26·5–28 (27·1); bill, ♂ 14–15 (14·6), ♀ 14–15·5 (14·6); tarsus, ♂ 16–17 (16·9), ♀ 16–17 (16·7). WEIGHT: (Cameroon) ♂♀ (n = 37) 10·9–14·8 (12·6); (Gabon) ♂♀ (n = 26) 12–15 (12·7); (Congo) unsexed (n = 3) 12·5–13·1 (12·7); (Uganda, Nov) ♂♀ 12–15 (13·7); (Kenya, June, Aug) 1 ♂ 9, 1 ♀ 9·8.

IMMATURE: like adult ♂ but much duller; upperparts green-grey, greyer on head, greener on mantle, underparts dull pale golden yellow, with grey-green wash on breast; throat and upper breast with sparse chestnut feathers in ♀, belly brighter golden yellow in ♂; some narrow tawny tips on greater upper-wing-coverts and tertials; very small eye-wattle grey; bill blackish tipped horn; legs dark brown, claws white.

JUVENILE: like imm. but duller, upperparts greyer, speckled brown with sparse pale olive edgings; underparts whiter, much less yellow and more washed olive-grey particularly on breast; upperwing-coverts and tertials with more tawny or buff tips; no eye-wattle; bill browner.

NESTLING: hatches naked, skin pale pinkish brown, with sparse pale brown down.

D. c. concreta (Hartlaub): Sierra Leone to Ghana. Upperparts darker, underparts of ♂ rich chestnut (feathers with golden yellow bases) except that chin and upper throat are golden yellow. WEIGHT: (Liberia) ♂ (n = 23) 11·7 ± 1·0, ♀ (n = 22) 11·1 ± 1·2.

D. c. ansorgei (Hartert): Angola. Poorly differentiated, paler and greyer above than *graueri*. Some ♂♂ show a black pear-shaped patch on throat and breast (Traylor 1960). WEIGHT: (Angola, Aug), ♂ (n = 7) 9–12·5 (11·1), 2 ♀♀ 10·5, 10·5.

D. c. kungwensis Moreau: Tanzania. Paler below, more slaty grey above than *graueri*, eye-wattle china blue.

Field Characters. Small, plump, active hover-gleaner of forest understorey, with short wings and tail. Easily distinguished from congeners by having entire underparts golden yellow and/or bright chestnut. Greenish or greyish above, lower rump yellow, large eye-wattles green. Some congeners have chestnut breast but belly always white; ♀♀ of Chestnut and White-spotted Wattle-eyes *D. castanea* and *D. tonsa* have chestnut upperparts and wings.

Voice. Tape-recorded (32, ERA, GREG, HOR, McVIC, SAR, ZIM). Advertising song in Angola, rapid 3-syllabled 'tututu', third syllable sometimes strongly accentuated (Heinrich 1958), or a whistled 'phee, phee, pheat' and a longer sweet 'pih, pih, puh, puh, puh, puhh' preceded by a call not unlike alarm of Common Bulbul *Pycnonotus barbatus* (Hall 1960a). In Uganda, song is 4-syllable descending melodious whistle, 'ti-ti-ti-tyew' or 'ptit-ti-ti-tyew', last note longest. In Kenya, 3 short notes, 'kikiki', followed by 2 whistled ones, each a third lower than the last, and a final loud, grating, upslurred one, 'kikiki-pee-pur-GWICK' (Keith and Gunn 1971). In Gabon, a very different two-note high-pitched whistle, 'eee-eee', very often followed by some loud scolding 'houit' or 'whick'; Gabonese birds did not react to playback of song from Uganda. When ♂ very excited, this song, 'eee-eee-houit-houit', is often introduced by some buzzy notes, e.g. 'trr-trr-trr ptipptip eee-eee-houit-houit'. The difference in pitch, volume and character between 'eee' and 'houit' produces a somewhat ventriloquial effect, which has caused observers to think they are hearing a duet between ♂ and ♀. Territorial defence song in Gabon, a long, regular series of loud 'houit' or 'pwick' notes; song in Uganda very similar, but Gabon birds showed only slight reactions to it during playback. At least in Gabon, all these territorial songs are emitted only by ♂. Aggressive flight call, a long, rapid, series of regular, hoarse 'gehgehgehgeh . . .' notes followed by rapid 'pwick-pwick-pwick . . .'; contact calls by ♂ and ♀, irregular conversational medley of various soft notes, 'hit', 'ptic', 'pwik', 'peep', 'pwayt', 'plup', 'pilup' or 'prec', mixed with hard, nasal 'kcht', 'treck', 'tchek' or 'terwit'. Anxiety and alarm calls, irregular series of rapid, grating 'pweck' or 'taytayk', mixed with softer, buzzy 'twaytwi', 'turrit' or 'prirrit'. Call of young, 'tictic' or 'pwick'. In hand, adults give nasal 'kwee-kway-kwee' and 'coïk', young softer 'pwee-pway-pwee'.

General Habits. Inhabits primary and secondary lowland forest, up to 1800 m in Zaïre (Itombwe), 1950 m in Cameroon (Mt Kupé), and 2500 m in Uganda. In Liberia, around Mt Nimba, inhabits forest-grassland mosaic as well as pure forest. In Angola, occurs in taller bushes and low trees in dense and liana-tangled patches of secondary thickets in hills; once noted in coffee plantation. In Gabon, inhabits primary forest and is quite exceptional in second growth, where replaced by Red-cheeked Wattle-eye *D. blissetti chalybea*. Avoids clearings and areas where forest has been disturbed by storms, also areas of low and liana-rich forest; particularly common where understorey is relatively open without well-defined vegetation layers, and in areas with many small understorey trees, e.g. *Alchornea floribunda* (Euphorbiaceae) which provide good nesting sites. In Kenya, occurs in deep, shady forest with well-developed layer of low trees and little ground cover. On Mahari Mt, Tanzania, inhabits forest and bamboo at 1800–2500 m.

In pairs or family parties; partners always move together, maintaining regular but not close or continuous vocal or visual contact, except when breeding; in NE Gabon, besides calls, ♂ gives 'eee-eee' song. Covers entire territory daily; active all day but mostly in morning and late afternoon. Searches foliage of lower strata of forest understorey, usually < 4 m above ground, mostly < 2 m, although during dry season may forage in higher strata, at 5–15 m, or even up to 20 m. A foliage-gleaner; hops in upright or crouched posture from twig to twig in saplings and small trees, or among dense forest herbs or thick tangles of low leafy lianas, also among dry leaves and dead or decaying twigs; snatches prey from vegetation, either stretching upward from perch or hovering briefly in flight to catch it in the air or from under leaf; sometimes also works through dense vegetation with whirring flight to dislodge insects. Frequently joins mixed-species flocks particularly during dry season and when not breeding. When anxious crouches with sleeked plumage and makes rapid lateral movements with head and body; when alarmed, ♂ stretches legs, grunts and flies off rapidly, while ♀ usually remains crouching among leaves.

Sedentary, at least in NE Gabon.

Food. Mostly insects: beetles, wasps, bees, ants, moths, caterpillars, grasshoppers, large flies, winged termites, insect eggs and pupae; also spiders; fruits once in Angola. Prey 8–35 mm long, mostly 12–18.

Breeding Habits. Monogamous, with long pair-bond; apparently pairs for life in NE Gabon. Territorial; defends territory throughout year; mean territory size 17·0 ± 3·7 ha (n = 7) in NE Gabon (Erard 1987). ♂ proclaims territory by rapidly patrolling it while singing advertising and defence songs, mostly while foraging in morning and early afternoon, particularly when in mixed-species flock; sings in open understorey, 2–4 m above ground. When another pair approaches, territory owner rapidly repeats 'houit' note and makes lengthy, fast, whirring flights with fluffed out rump and aggressive flight calls, then perches in open, exposing bright yellow underparts and fully extended emerald eye-wattles, and gives advertising song with bill tilted upward. When another ♂ is seen in territory, owner harasses it, flying rapidly with fluffed rump and snapping bill until it leaves. ♀♀ are usually passive during these territorial encounters, although ♀ may chase off intruding ♀.

NEST: small and neat, tight open cup of vegetable fibres, small stems, fine rootlets, pieces of bark, covered outside with pieces of dry leaves bound together with spider webs, particularly around rim in order to make it level with the small pile of dead leaves on which it is placed; inner cup lined with thinner material, particularly blackish *Marasmius* fungus; ext. diam. 55–75, ext. depth 40–55; int. diam. 46–48, int. depth 25–36; in NE Gabon, usually placed 1–3 m above ground (mean 1·9, n = 17) in middle of leaf of *Alchornea floribunda* (Euphorbiaceae), a common understorey plant whose leaves form receptacles for dead and decaying vegetable material which provide camouflaged nest sites, especially for this species. Built by ♀; ♂ follows her and sings, often feeds her.

EGGS: 1–2, laid at 1-day interval; NE Gabon mean (8 clutches) 1·9. Oval; glossy; cream, slightly pinkish, or pale green, with ring of fine dark brown-grey and rusty brown spots at larger end (once near centre), and wash of smaller brown-grey to lavender-grey speckles. SIZE: (n = 5, Gabon) 18–21 × 12–14 (19·3 × 12·9). In NE Gabon, neither second nor replacement clutch observed but birds laying in Aug–Oct, if unsuccessful, may lay again in Jan–Feb.

LAYING DATES: Sierra Leone (enlarged gonads Apr); Liberia (juv. Feb, ♂♂ and ♀♀ with moderately enlarged gonads Jan–Aug); Nigeria (♂♀ with slightly enlarged gonads Apr); Cameroon, Nov–Feb, Apr, June; Gabon, all months except Dec, mostly Aug–Oct and Jan–Feb; Congo, Oct; Angola (juvs Mar); Zaïre (♂♀ with enlarged gonads Feb–June, juvs Aug, Dec); E Africa: Region B, Nov; Angola, Feb (♂♂ with enlarged testes Sept–Nov).

INCUBATION: begins with last egg; by ♀; period 16 days. ♀ sits for long periods (60–95 min), regularly fed by ♂; ♂ patrols territory and sings. ♀ leaves nest for periods of 40–60 but sometimes up to 120 min; ♂ always stays with her and feeds her regularly, although she also feeds herself; both keep up an almost continuous babble of squeaking notes as they forage low down, almost on the forest floor.

DEVELOPMENT AND CARE OF YOUNG: young remain in nest 12–13 days. ♀ broods chicks for first 3–4 days; ♂ brings food to her which she gives to young; later both feed young. For first 3–4 days young fed several times in quick succession, with intervals of 1–8 min, then pauses of 30–40 min before they are fed again; later fed every 5–10 min. Weight of young increases steadily from 1·5–2·2 g at hatching to 9–9·5 g at 11 days. Young stay with parents for 3–5 months, but are fed only for first 60–70 days; usually ♂ and ♀ each care for a particular young. Independent young usually remain on their family territory, but some move to other parts of the forest; there, because of their immature plumage, they are only threatened, not attacked, by territory owners.

BREEDING SUCCESS/SURVIVAL: in NE Gabon, of 12 nests observed from inception, 8 contained total of 15 eggs, and only 2 produced fledglings, for a total of 4 young. In NE Gabon, most birds ringed as adults in 1976 were still present in 1981 and even in 1985, and had the same mates.

References
Brosset, A. and Erard, C. (1986).
Erard, C. (1987).

Genus *Platysteira* Jardine and Selly

Small, plump, big-headed birds, intermediate between *Dyaphorophyia* and *Batis* (Erard 1987, Dowsett and Dowsett-Lemaire 1993b). Bill broad and flat with ridged culmen; nostrils exposed; wings more or less pointed; tail short with white margins; legs rather thin and long; red wattle above eye; upperparts grey or black, some white feather-edges in wing; underparts white, ♂♂ with black breast-band, ♀♀ with rufous or black gorget (underparts all white in *albifrons*). 'Frip' wings in flight. Forests, woodlands and mangrove.

Endemic. 3 species, though some authors separate Mt Cameroon *laticincta* from *peltata*; they form a superspecies.

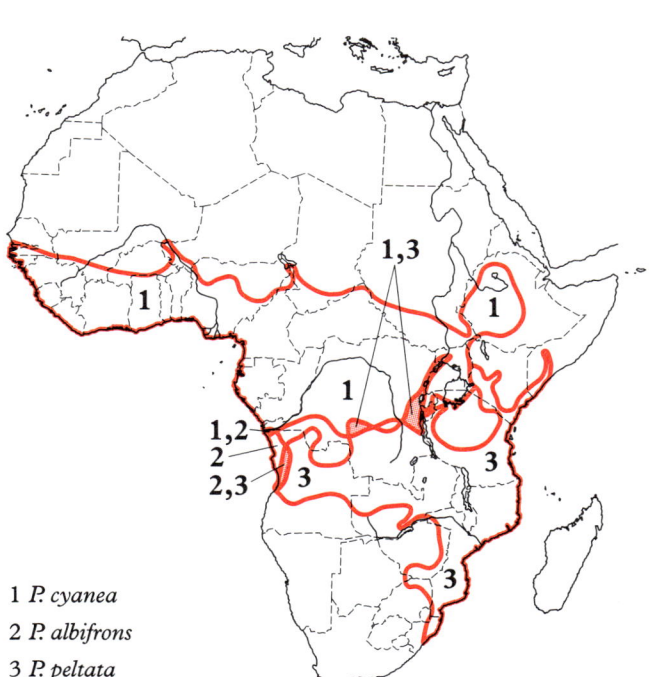

Platysteira cyanea superspecies

1 *P. cyanea*
2 *P. albifrons*
3 *P. peltata*

Platysteira cyanea (Müller). Brown-throated Wattle-eye. Pririt à collier.

Muscicapa cyanea P. L. S. Müller, 1776. Linné Natursyst. Suppl., p. 170; Sénégal.

Forms a superspecies with *P. albifrons* and *P. peltata*.

Plate 30
(Opp. p. 545)

Range and Status. Endemic resident. Savannas and forest edges of Senegal and Gambia (south of 14°N), Guinea Bissau, Mali (Baoulé, Kangaba, Bafin–Makana, Bougouni), Guinea (throughout, common), Sierra Leone, Liberia, Ivory Coast (guinean zone and locally on coast), Ghana (common in savanna woodland and coastal mangrove and thickets, occasional in forest clearings), Togo, Benin, SE Burkina Faso (Arli and Pendjari Nat. Parks), Niger (along Niger, Mékrou and Tapoa rivers), Nigeria (north to 10°N, occasionally to Kaduna and Aliya, also in Kagoro, Gashaka-Gumti Game Res. and Obudu plateau), Cameroon (north to Benoué R., Sidderi and Garoua), S Chad (north to Léré, Moundou, Laï and Sarh, also along upper Logone and Chari R.), W Central African Republic (Dzanga and Sarki northeast to Birao, Bambari and Rafai), Rio Muni, Gabon, Congo (at least coast, sublittoral, Mayombe to Djambala, situation elsewhere unclear), Angola (Cabinda, Noqui), Zaïre (lower Congo R. to Mbandaka, east to Bumba and Uele to Ituri, south to Kivu and west to S Kasai), Rwanda (Nyungwe), Burundi (Cyohoha L., Musigaki, Bujumbura), NW Tanzania (West Lake and Mwanza), Uganda, W Kenya (Elgon and Nandi to Nyanza and Mara Game Reserve), S Sudan (north to Fertit, Wau, Mandari and Boma) and Ethiopia (west of Rift, north to Tigre). Uncommon to common. Density in NE Gabon, 12·5 to 21·9 pairs/km^2 around villages, but absent from some villages (Erard 1987); in Kakamega forest, Kenya, 1 adult in an 8 ha census tract (Zimmerman 1972).

Platysteira cyanea

Description. *P. c. cyanea* (Müller): Senegal and Gambia to W Central African Republic (east to Fafa R.) and Zaïre (lower Congo R.) to Angola. ADULT ♂: forehead, crown, lores, ear-coverts, hindneck, sides of neck, mantle, back and scapulars black with purplish blue to steel-blue sheen; rump slate-grey becoming white on lower rump (feathers grey edged white). Uppertail-coverts and tail glossy blue-black; T6 with narrow edge along outer web and tip white, other tail-feathers with narrow tip white, becoming smaller from T5 to T1. Entire underparts white except for breast-band, 10–15 mm wide and glossy blue-black; flanks grey; thighs black. Primaries, secondaries and tertials black, innermost secondaries and outer tertials narrowly edged white along outer web; upperwing-coverts glossy blue-black, most median and innermost greater coverts with broad edge on outer web and tip white. Axillaries and underwing-coverts white with black base, except lessers and wing margin black, underside of flight feathers blackish grey, narrowly edged white along inner web. Bill black; eyes pale blue-grey; wattle above eye red; legs purplish black. ADULT ♀: like ♂ but upperparts dark slate-grey, rump paler; chin and narrow malar stripe white; throat, sides of neck and foreneck deep chestnut, bordered black on upper breast; eye-wattles smaller. SIZE: (8 ♂♂, 8 ♀♀) wing, ♂ 62–66 (64·0), ♀ 63–66 (64·4); tail, ♂ 47–49 (47·6), ♀ 48–52 (50·2); bill, ♂ 16–18 (16·9), ♀ 16–18·5 (17·2); tarsus, ♂ 18–20 (19·4), ♀ 18–20 (19·2). WEIGHT: (Liberia) ♀ (n = 3) 15–15·1 (15·0); (Nigeria) ♂ (n = 3) 13–14 (13·5), 2 ♀♀ 13·5, 15; (Cameroon, Nov–Apr) ♂♀ (n = 21) 12–17 (14·6); (Cameroon) ♂ (n = 4) 14–16 (15·0), ♀ (n = 6) 14–16 (15·3); (Gabon) 3 ♂♀ 15–16·5 (15·5); (Congo) unsexed (n = 7) 14·7–16 (15·5).

IMMATURE: like adult ♀ but upperparts browner, particularly on back and upperwing-coverts; flight feathers narrowly edged tawny on outer web; white marks on wing heavily washed cinnamon; greater upperwing-coverts tipped dark tawny or cinnamon; throat and foreneck whitish, washed cinnamon or pale chestnut, sides of throat and neck chestnut; narrow breast-band greyish chestnut; eye-wattle greyish yellow, much reduced; eye olive to grey-brown; bill dark grey, paler at tip; legs brown.

JUVENILE: like immature but browner upperparts with small dark ochre spots particularly on crown, nape, sides of head, axillaries and upperwing-coverts; flanks greyer; broader white margins on tail-feathers.

P. c. nyansae Neumann: Central African Republic (Bangui and Fafa R., eastward) and Zaïre (except lower Congo R.) to Sudan (west of upper Kangen R.) and Kenya. Similar but with a narrow white frontal band extending above lores. WEIGHT: (Uganda, Feb–June) ♂ (n = 5) 14–17 (15·7), ♀ (n = 5) 14–17 (15·5).

P. c. aethiopica Neumann: SE Sudan (Boma) and Ethiopia. Like *nyansae* but greyer above and smaller. WEIGHT: (Ethiopia, Apr, May, Nov) ♂ (n = 4) 10–14 (12·2), ♀ (n = 7) 10–16 (12·3).

Field Characters. A small, active foliage-gleaner whose presence is often established by the plaintive descending whistles of a duetting pair; the duet is given all day, even during the hottest hours, and all year. Rather large head, conspicuous red wattle above pale eye, plump body, blackish upperparts; underparts white with black breast-band in ♂ and chestnut throat and breast in ♀ (**A**); broken white line on wing formed by edges of greater upperwing-coverts and inner secondaries; white-edged tail. ♀ separated from Black-throated and White-fronted Wattle-eyes *P. peltata* and *P. albifrons* by deep chestnut throat and breast; ♂ distinguished from Black-throated by white on wing and broad black breast-band, from White-fronted by entirely black (not grey) upperparts and narrow white edges of outer tail-feathers. Separated from all sympatric *Batis* species by red eye-wattle and larger size, and by chestnut throat of ♀. Other sympatric wattle-eyes are smaller, short-tailed and lack white on wing.

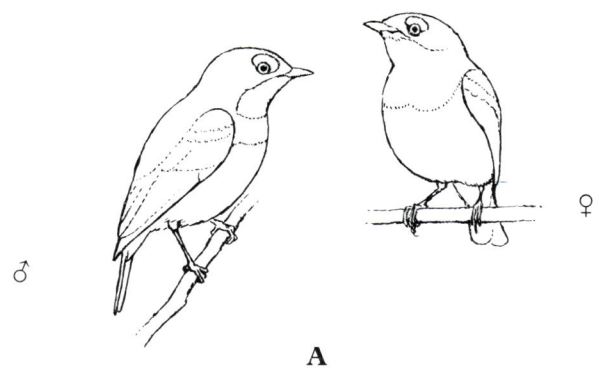

A

Voice. Tape-recorded (10, 32, 36, 40, B, C, ERA, GRI, MEES, ZIM). Territorial song given from perch, highly variable both geographically and individually; usual pattern, a short series of sweet, plaintive, descending whistles, 'teep-tyee-tyee', or 'eee-yew-yeew', or 'huit-eee-yew-yeew', or 'tee-tee . . . ptyup-tudu-tu-heet', or 'ptee-tu-ptee-tuu-trrr', or often introduced by a short medley of hoarse, nasal, raspy and whistled notes, 'zurr-zurr . . . heepu-heepu-heep', or 'huip-tip-tu-turru . . . tip tee-tu-tuu', or 'ptick-ptick-tick-tawaa . . . tsee-tsuu-tee-tuu', or 'ptip-tip-ti-ouai-ouai . . . hee-hee-hu-hu'. Usual pattern is ♀ combining its notes with those of ♂ in both parts, although first part sometimes given by ♂ and second part by ♀; in all cases with almost no break, so that listener hears only one song, e.g. 'ptip-tip-tee-way-way . . . hee-hee-hu-hu' or 'ptick-ptick-tick-tawa-psee-tsic-tsuc-tee-tuu', or 'tup-tu-tu-rouayee-hee-hu'. Territorial defence flight song a descending 'hee-hee-hu-hu', or 'huip-tu . . . ee-teee', or 'huip-tu-ee-teee . . . pu-ee', or 'pi-tyew . . . hee-tee', or rhythmic 'ptiup . . . touay-touay' changing in 'ptyup . . . toway-toway . . . ptip-tyay'; aggressive call, a raspy nasal 'toc-roway-roway' or 'ptoc-rouay-oway'. Excitement call, a harsh rasping 'cray-cray-cray-cray-cray-cray . . .'. Call of begging young, a rapid nasal 'tzrray-tzrray'.

General Habits. Inhabits lowland forest clearings and edges, regenerating forest, forest-grassland mosaics, lowland open woodlands, riverine forest in wooded savanna, moist guinea savanna, wooded grassland, bushland (though avoids bush savannas unless there are clumps of trees sufficiently large to harbour at least one pair), inselbergs, coastal woods and thickets, and mangroves (e.g. *Avicennia* and *Rhizophora* mangrove in coastal Gambia, Guinea, Ghana, Nigeria, Cameroon and Gabon); gardens and cultivated land, plantations; in and around villages, e.g. in forested areas of NE Gabon, where it occurs only in villages or within 300 m of human dwellings. Only occasional in montane forest clearings (e.g. Mt Nimba in Liberia or Mt Cameroon and Mt Manenguba in Cameroon), although in Ethiopia occurs in *Juniperus* and *Podocarpus* forests, and also in moist subtropical forests. Up to 1300 m in Burundi, 1450 m in Zaïre (Itombwe), 2000 m in Nigeria, 2200 m in Cameroon and Uganda. In Bamenda highlands, Cameroon, replaced above 1500 m by *P. peltata laticincta*.

Occurs singly, in pairs or family parties. Flight not rapid, rather hesitant, producing audible flaps. Partners maintain close contact all year either by sight or by duetting. Active all day; covers entire territory daily; forages among leaves, mostly between 8 and 18 m above ground, sometimes lower, rarely higher; moves among branches, frequently working through bunches of leaves in erratic and whirring flight to dislodge prey; hops from twig to twig and sallies or jumps to snatch insect in flight or from under leaves, sometimes after a brief hovering flight or swoop. Prey usually taken 0·5–1·5 m from perch, sometimes up to 4–5 m. Large, venomous, or hard-bodied insects are rubbed and pounded on perch and taken apart before being swallowed. Often joins mixed-species flocks, but not for long, because flocks soon move beyond its limited territory; in NE Gabon only 35·6% of birds recorded were in mixed flocks. Regularly feeds on swarming alate termites and ants. Foraging postures: (**B**).

In coastal Gabon, pair observed sunning at noon, landing on beach along an old mangrove and lying several min on hot sand, with ruffled plumage and wide open bill (P. Christy, pers. comm.). When anxious, calls and crouches with sleeked plumage, jerking head rapidly from side to side and nervously wagging tail. When more alarmed, erects crown-feathers and calls while flying from branch to branch with measured noisy wing flaps and tail flicks. Mobbing bird swoops toward predator with erected crest and exposed eye-wattles, and vigorously flaps wings and snaps bill, making rattling noise.

Sedentary, at least in NE Gabon.

Food. Mostly insects, imagos, larvae, chrysalises and pupae: mainly beetles, caterpillars, moths, bees, wasps, ants, grasshoppers, flies, cicadas, mantises, cockroaches; readily catches swarming winged ants and termites; also eats small snails. Prey size 5–40 mm, mostly 10–20.

Breeding Habits. Monogamous; territorial. Apparently pairs for life, at least in NE Gabon, where pair defends territory throughout year; mean territory size 4·2 ha (n = 4); unmated ♂ defended 2·5 ha (Erard 1987). ♂ advertises territory with territorial claim song, usually from tree canopy, top of bush or from understory; sings all day but mostly in morning and early afternoon, when not feeding; particularly demonstrative at sunrise and sunset. Sings all year. Territory defended by both ♂ and ♀, which sing and duet with fluffed out plumage and stiff postures. When conspecific intruder seems ready to enter territory, owners emit irregular series of grunts and move towards opponent with (particularly the ♂) noisy wing flaps and bill snapping, ruffled rump-feathers and rasping song; then ♂ gives territorial defence song from perch in upright posture, exposing throat and breast. When opponent gets closer, territory owner becomes more aggressive, approaches antagonist, usually above it, in threat posture (crouches, points bill towards opponent, ruffles crown, breast and flank-feathers and flicks tail and wings) and gives scolding territorial fight song, usually accompanied by ♀. Courtship feeding common, begins just before nest building: observation of ♂ feeding ♀ is thus a good breeding record. Harasses cuckoos, coucals and shrikes.

NEST: small, neat, tight open cup made of vegetable fibres, lined with fine grass stems, covered with flat grey-green lichens on outside, bound with grey silky spider web; outer diam. 65–75, height 55–65; int. diam. 45–50, int. depth 25–30; placed in small fork near extremity of thin leafy or hanging branch of small tree, e.g. avocado, at 4–8 m above ground. Built by both sexes.

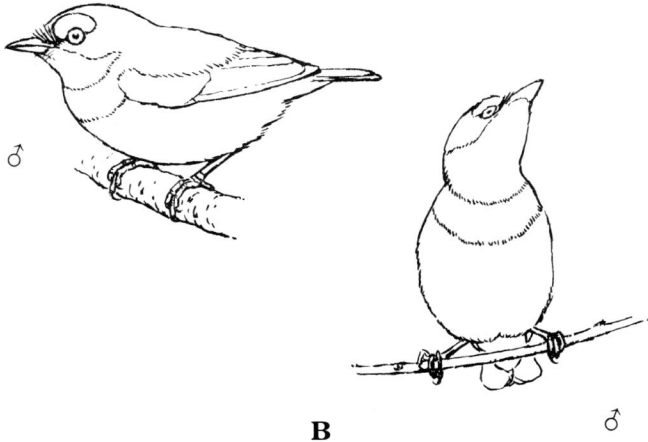

B ♂

EGGS: 2, laid at one day intervals. Oval; glossy; pale greenish or brownish white, with dark brown and violaceous brown spots, dense towards larger end, sparse elsewhere. SIZE: *c.* 18–19 × 14.

LAYING DATES: Gambia, Jan, July, Nov, Dec; Sierra Leone, Nov, Dec (♂♂ with enlarged testes, Apr); Liberia (♀♀ in breeding condition July, Oct); Ivory Coast, Apr–Aug; Ghana (dependent young/juvs June, Aug, Nov); Nigeria, Dec–Aug, possibly all year; Cameroon, Feb, Mar, May, Dec (♂♂ with enlarged testes Mar, Nov; juvs Mar, May, Aug); Gabon, Feb, Aug (♂♂ with enlarged testes Oct–Feb; young Feb–Apr, June, Oct); Congo, Aug, Sept; Rwanda, Aug, Sept, Dec; Central African Republic (♀♀ with active ovaries Mar, July, Aug); Zaïre, Mar (♂♀ with enlarged gonads Jan, Mar, July); E Africa: Region B, Feb, Apr–June, Aug, Nov, Dec, i.e. mainly during long rains.

INCUBATION: by ♀; ♂ guards with scolding notes, bill snapping and wing flapping; regularly feeds ♀. Period unknown.

DEVELOPMENT AND CARE OF YOUNG: young fed by both parents; ♂ brings food to ♀ which passes it to young while she is brooding them during first week after hatching. Nestling period unknown. After leaving nest young fed by both parents at least for one month, stay on territory almost to next breeding season.

References
Brosset, A. and Erard, C. (1986).
Erard, C. (1987).

Plate 30 (Opp. p. 545)

Platysteira albifrons Sharpe. White-fronted Wattle-eye. Pririt à front blanc.

Platystira albifrons Sharpe, 1873. Ibis, p. 159; Loge R., Angola.

Forms a superspecies with *P. cyanea* and *P. peltata*.

Range and Status. Endemic, resident. Known only from W Angola, from Congo mouth to Benguela, inland to Canhoca and Dondo, i.e. mostly along escarpment. May occur in lower Zaïre (e.g. Kwango R.; records from Ngombe Lutete still unconfirmed) and Congo. At Congo mouth, sympatric with Brown-throated Wattle-eye *Platysteira cyanea*. Frequent at least at Kishama Nat. Park and Barra do Cuenza (Dean *et al.* 1988).

Description. ADULT ♂: narrow line at base of forehead and over lores white; forehead, crown, nape and hindneck dark ash-grey, hindcrown mottled black, nape and hindneck mottled black and white; lores, ear-coverts and sides of hindneck black slightly glossed with steel-blue. Mantle, back and rump slate-grey (feathers of lower back and rump long and fluffy, grey with partly concealed broad subterminal spot white). Uppertail-coverts and tail deep glossy blue-black, T6 with outer web and tip white, other tail-feathers with narrow white tip decreasing in size from T5 to T1. Entire underparts, including sides of neck, from chin to undertail-coverts white, except for wide glossy blue-black breast-band, greyish flanks and black thighs, narrowly barred white. Scapulars, primaries, secondaries and tertials dull black, primaries and secondaries narrowly edged pale greyish white to pale pearl-grey along outer web, outer tertials edged white along outer web; upper-wing-coverts glossy blue-black, medians with broad white tip, innermost greaters with broad margin or entire outer web and tip white; axillaries white, bend of wing and underwing-coverts black, greaters broadly tipped white; underside of flight feathers blackish grey narrowly fringed white on inner web. Bill black; wattle above eye scarlet-vermilion; eyes pale greyish blue with narrow inner ring white; lower eye-lid olive-green; legs and feet dark purple. ADULT ♀: like ♂ but entire upperparts dark pearl- to slate-grey without black, except for white line at base of forehead and over lores; entire underparts white without breast-band or gorget. SIZE: (4 ♂♂, 5 ♀♀) wing, ♂ 59·5–62 (60·7), ♀ 59·5–62·5 (61·0); tail, ♂ 46–49 (47·5), ♀ 47–51 (49·0); bill, ♂ 16·5–18 (17·5), ♀ 16–17 (16·5); tarsus, ♂ 18–19·5 (18·6), ♀ 18·5–19·5 (19·0). WEIGHT: no data.

Platysteira albifrons

IMMATURE: like adult ♀ but buffier below, with irregular pale rusty tips and edges to upperwing-coverts and inner secondaries.

JUVENILE: like immature but pale tawny from chin to breast, upperwing-coverts and innermost secondaries with edge of outer web and tip pale tawny.

NESTLING: unknown.

Field Characters. Small, plump, active, big-headed and broad-billed foliage-gleaner; with dark grey upperparts, white underparts and red eye-wattles. Similar to Brown-throated Wattle-eye *P. cyanea*, with white in wing and ♂ with broad breast-band, but upperparts grey, not black; ♀ has top of head grey, ♂ has grey forehead and forecrown contrasting with black lores and sides of head; both have nape and hindneck mottled with white and broad white edge to outer tail-feathers. ♀ distinguished from ♀♀ of Brown-throated and Black-throated Wattle-eye *P. peltata* by entirely white throat and breast; ♂ Black-throated Wattle-eye has very narrow breast-band and both sexes lack white in wing.

Voice. Unknown.

General Habits. Very poorly known. Inhabits thickets and gallery forest in lowland wooded savannas, also mangrove forest. In pairs or family groups; habits evidently very similar to those of Brown-throated Wattle-eye; actively gleans foliage, hopping from twig to twig among branches, making short sallies to catch prey on or under leaf, sometimes hovering. Joins mixed-species flocks.

Food. Small arthropods, particularly insects.

Breeding Habits. Monogamous and territorial but behaviour not documented. Nest and eggs unknown. Juveniles found in Angola in Nov.

Platysteira peltata Sundevall. Black-throated Wattle-eye; Wattle-eyed Flycatcher. Pririt à gorge noire.

Plate 30
(Opp. p. 545)

Platystira peltata Sundevall, 1850. Ofversigt K. Vetenskaps-Akad. Förhandlingar, Stockholm, 7, p. 105; Caffraria inferiore; type from Umlalazi R., Zululand, Natal, *fide* Gyldenstolpe, 1934, Ibis, 13 (4), p. 291.

Forms a superspecies with *P. cyanea* and *P. albifrons*.

Range and Status. Endemic, resident. Isolated population in W Cameroon (Bamenda highlands at Mt Oku, Bafut–Ngemba Forest Reserve, Sagba Pass, Ndu, Mt Lefo). Main range, S Somalia (Juba R. south of 3°N), S Kenya (coast and SW north to Mt Elgon and Nairobi), S and W Uganda (Nkose Is., Budongo, Kibale and Murchison Falls Nat. Park), Rwanda (Nyungwe Forest), Burundi, Tanzania (locally in coastal regions, inland to Poroto Mts, Crater Highlands and Mbulu, also Mt Mahari and from Tukuyu to Kibondo and Mwanza, also Mafia I.); S Zaïre (north to Kasanga, Kwango, Luluabourg, Manyema, Itombwe and L. Kivu), Angola (west to Malanje and Cuanza Norte, south to Huila), Zambia south to Kalabo, Mongu, Mumbwa, Chunga, Namwala, Mapungu and Chirundu; Malaŵi (Chitipa and Nkhotakota districts, Ntchisi and Mangoshi Mts, Zomba, Thyolo and Chikwawa districts), Mozambique except north of Lurio R., Zimbabwe (locally common, Middle Zambezi valley, Mazoe-Ruenya R., Mashonaland plateau (once Harare: Tree 1993), Sabi valley, eastern highlands in Honde and Nyamkwarara valleys, Lusiti-Haroni area, Buzi R. in Chipinga uplands, also Chirinda Forest); extreme E Botswana (Shashe R.); fairly common NE and E Transvaal in lowveld and Limpopo Valley, and Natal, wandering south to East London. Frequent to abundant. Still common in Cameroon, but endangered because restricted to montane forest patches that are being rapidly cleared (Stuart and Jensen 1986).

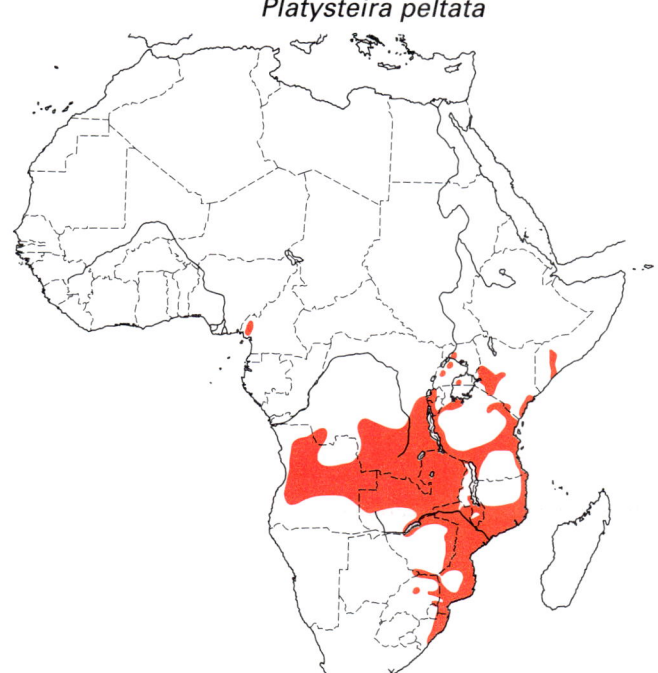

Platysteira peltata

Description. *P. p. peltata* Sundevall: Zambia (Eastern Province Plateau, Middle Zambezi and Luangwa valleys to Kondolilo Falls and Kanona), S Malaŵi, Mozambique (south of Zambezi R.), E Zimbabwe and South Africa. ADULT ♂: forehead, lores, crown, nape, ear-coverts, sides of neck and hindneck deep glossy greenish blue to oily green-black, latter with base of feathers white. Mantle and back glossy dark greenish slate-grey; rump slate-grey (hairy feathers pale grey edged dark slate-grey), lowest rump-feathers whitish (base pale grey, wide edge white washed slate-grey). Uppertail-coverts and tail glossy green-black, T6 with narrow white edge along outer web, T5 with faint white fringe on outer web in fresh plumage. Underparts from chin to undertail-coverts white, except for narrow (2–3 mm in centre, 5–6 on sides) breast-band glossy greenish blue-black, flanks greyish white (feathers grey edged white) and thighs black faintly barred white. Scapulars, upperwing-coverts and tertials glossy greenish blue-black, primaries and secondaries unglossed black, outer webs narrowly edged whitish grey in fresh plumage; axillaries white with base dark grey, bend of wing, lesser and median underwing-coverts black, greaters slate-grey edged white; underside of flight feathers glossy grey-black, narrowly

edged white along inner web. Bill black; wattle above and in front of eye bright scarlet-vermilion; eyes brown, outer ring purplish or grey, iris white; legs and feet blackish slate-grey. ADULT ♀: like ♂ but chin white, throat and breast glossy greenish blue-black. SIZE: (4 ♂♂, 3 ♀♀) wing, ♂ 66–69 (67·6), ♀ 65–70 (67·8); tail, ♂ 54–56 (55·2), ♀ 52·5–55·5 (54·3); bill, ♂ 17–18·5 (17·6), ♀ 16–17 (16·7); tarsus, ♂ 19–20 (19·7), ♀ 18–19 (18·5). WEIGHT: no data.
 IMMATURE: upperparts bluish grey, wings and tail blackish brown, upperwing-coverts, tertial and inner secondaries edged tawny to rusty; underparts creamy white, throat and sides of breast mottled tawny and buff, sometimes in the form of an irregular band; eye-wattle small, dull red.
 JUVENILE: upperparts pale brown and rufous with pale patches on wing-coverts, underparts creamy white; eye red, wattle small, orange-vermilion.
 NESTLING: hatches blind and naked; skin pale pink to dark brown, gape yellow.
 P. p. laticincta Bates: Cameroon. Upperparts glossy blue-black, ♂ with broader breast-band than nominate *peltata*. WEIGHT: (L. Oku, Cameroon, Mar) ♂♀ (n = 9) 13·2–15·2 (14·3).
 P. p. mentalis Barbosa du Bocage: Kenya west of Rift (Nguruman hills, Naivasha and Sotik to Kakamega Forest, Elgeyu, Saiwa and Mt Elgon), W Tanzania (Mt Mahari and Tukuyu to Mwanza), Uganda, Rwanda, Zaïre, Zambia (Isoka, Mpika and Serenje to Mongu), Angola. Upperparts, breastband of ♂ and throat and breast of ♀ glossy blue-black. WEIGHT: (Upemba, Zaïre) ♂ (n = 12) 11–17 (15·0), ♀ (n = 17) 10–17 (15·0); (Kenya, Mar, Sept, Dec) 2 ♂♂ 11, 14, ♀ (n = 4) 11–14 (12·0); (Uganda, May) 2 ♂♂ 15, 15.
 P. p. cryptoleuca Oberholser: Somalia, Kenya east of Rift (Meru and Nairobi to Tsavo and Lamu), E Tanzania (locally in Poroto Mts, Crater Highlands and Mbulu east to Mafia I.), Mozambique (north of Zambezi R.), Malaŵi, Zambia (south and east of Luangwa Valley), Zimbabwe (E highlands). Mantle darker and blacker than nominate *peltata*, strongly glossed green. WEIGHT: (Somalia) 2 ♂♂ 12·8, 13·2, ♀ (n = 6) 11·2–12·8 (11·9); (Kenya, Feb, Mar, May) 2 ♂♂ 11·5, 12, ♀ (n = 3) 10–13 (11·2).
 TAXONOMIC NOTE: *P. p. laticincta* is sometimes considered a separate species (cf. Stuart and Jensen 1986, Traylor 1986), but morphological and ecological differences are small and we agree with Dowsett and Dowsett-Lemaire (1993b) that it should be kept as a subspecies, pending further investigation.

Field Characters. Large-headed and broad-billed, with conspicuous red eye-wattle. Upperparts and wings blackish, ♂ white below with narrow black breast-band, ♀ with black throat and breast. Larger and longer-tailed than forest wattle-eyes in genus *Dyaphorophyia*, none of which has red eye-wattle. Separated from both Brown-throated and White-fronted Wattle-eyes *P. cyanea* and *P. albifrons* by lack of white on wings; ♂ has much narrower breast-band than either of them, and ♀ distinguished by black throat (throat brown in ♀ Brown-throated, white in ♀ White-fronted). Sympatric batises *Batis* spp. lack eye-wattle and have white supercilium and white in wing.

Voice. Tape-recorded (9, 86, 88, 91, B, C, F). Advertising song in southern Africa, harsh, rasping and unmusical, introductory single notes followed by double notes, 'djip-djip-djip-djip-zipweet, zipweet, zipweet, zipweet . . .', and a similar song described from Angola by Heinrich (1958), 'tek-tek-tek-tedaah-tedaah'. In southern Africa also gives a tinkling 'er-er-fee, er-er-fee-fu' (Maclean 1993). In Zambia, short, clucking introductory notes followed by a single drawn-out harsh one, 'chik-cluck-cluck-jeeery' or longer 'cluck-cluk-clik-cluk-cluk-jeeery, clo-jeeery', with variations; ♀ sometimes interjects harsh notes. Flight song in territorial defence, rapidly repeated 'ptec' followed by long unbroken series of harsh sawing notes 'jutzy-jutzy-jutzy-jutzy . . .' or 'dzitta-dzitta-dzitta-dzitta . . .', accompanied by wing flaps. Contact call a loud, repeated 'zing', 'jing' or 'zee', 'like beating of a hammer on an anvil' (Benson 1940); also several short bursts of harsh chattering like African Dusky Flycatcher *Muscicapa adusta* (Vincent 1935). Alarm, harsh 'tsit-tsit'. Mobbing call, sharp, staccato 'chit chit chit' like 2 bits of metal tapped together, often with 2 additional notes run together at the end (van Someren 1956).

General Habits. Inhabits montane forest between 1700 and 2450 m in Cameroon, up to 2000 m in Itombwe, Zaïre, 2220 m in Burundi, 3000 m in E Africa; in the last, as well as in Zambia, Zimbabwe, Malaŵi, Mozambique and South Africa, also occurs in lowland evergreen forest patches and edges, with a preference for gallery forest and riverine evergreen scrub near or even over water. Other habitats include *Sizygium* bush, evergreen and bushland thickets, coffee forests and even gardens; in South Africa also mangrove swamps, usually low down near water. In Zambia replaced by Cape Batis *Batis capensis* in moist montane forest on Nyika Plateau, and in Zimbabwe and Malaŵi also there is evidence that the 2 species are mutually exclusive.
 Usually in pairs or family groups, occasionally single. Rather silent, less vocal than batises, though it makes conspicuous wing-snapping flights across open spaces between thickets. Pair members maintain contact by duets. In Cameroon montane forest, forages at any height but usually in undershrubs. Continuously on the move; flits about in foliage, hopping along branches or from twig to twig, sometimes using its wings to flush prey; darts upward or hovers to snap up prey on or under leaf or branch; also sallies to catch flying insects. Frequently joins mixed-species flocks. Aggressively mobs arboreal snakes with sharp staccato calls and snapping wings. Maintenance behaviour includes indirect scratching of head.
 Sedentary, at least in Zimbabwe, but off-season downward movement suspected in Malaŵi and in South Africa: presence in East London attributed to local migration (Vernon 1989).

Food. Insects, especially moths, caterpillars and flies, also grasshoppers and crickets; small green caterpillars given to young nestlings.

Breeding Habits. Monogamous, territorial. Helpers suspected. Advertises and defends small territory (said to be < 1 ha but probably underestimated) with flight song, bill snapping and aggressive calls; ♂ in open perches in upright posture with tail slightly fanned,

expands throat and sways head from side to side. During territorial interactions, opponents bow, swing body from side to side and make short zigzag flights with measured wing flaps. Often ♀ joins ♂ in duet. During courtship makes wing-snapping display flight with bill clapping, tail fanning and fluffed-out rump-feathers. When nesting, particularly aggressive towards other species. Courtship feeding is very common and begins before nest-building.

NEST: small shallow neat cup of fine plant material, grass and fibres, and small pieces of bark, bound with spider web (**A**), frequently covered outside with lichens, lined with beard-lichen and bark fibre, sometimes plant down and small feathers though may be unlined; ext. diam. *c.* 50, int. depth *c.* 25. Placed 0·3–9 m above ground in crotch of small fork of tree, or on drooping branch below thick foliage, sometimes above water. Built by both sexes though major part by ♀. Strong fidelity to nest site (Scott 1984).

EGGS: 1–2. Zimbabwe mean (25 clutches) 1·9. Oval; glossy; pale grey-green, green or blue, or greenish white, densely marked with spots and blotches of red-brown and various shades of brown or grey-brown, with pale purple, slate, lilac and grey-blue undermarkings, particularly around larger end, elsewhere mostly finely speckled pale brown. SIZE: (n = 8, southern Africa) 16·5–18·5 × 13–14·3 (18·0 × 13·7).

LAYING DATES: Cameroon, Oct–May, mainly Dec–Feb; Zaïre, Sept; E Africa: Region A, Dec; Region D, Apr–June, Aug–Nov; Malaŵi, May–Dec; Mozambique, Aug–Mar; Zambia, Sept–Dec, Apr, May; Zimbabwe, Sept–Feb (mainly Sept–Dec); Natal, Oct–Dec; NE Transvaal, July–Mar (mainly Sept–Nov); Angola, Sept, Oct (♂ with enlarged testes Aug).

INCUBATION: period: 16–18 days; mainly by ♀ but partly by ♂. ♀ fed regularly and intensively both on and out of nest by ♂; from nest building to hatching she receives most of her food from ♂.

DEVELOPMENT AND CARE OF YOUNG: within a week wing and tail quills start to emerge and thereafter young feather quickly. Between 25 and 40 days, pale orange eye-wattles start to develop, and faint pale brown smudges appear on breast. At 55 days upperparts greyer, edges of wing-coverts pale golden brown. Sexual dimorphism begins to appear at *c.* 95 days; eye-wattles become dark red at *c.* 115 days.

After hatching both parents take turn in brooding and throughout nestling period young closely and vigorously guarded and defended by one adult. Young fed by both parents. At nest containing young honeyguide, 6 feedings per half hour reduced to 3 per hour during hottest hours of day (van Someren 1956). Food items picked from nearby foliage and twigs by ♀, brought from further away by ♂ which usually passes it to ♀; she solicits with quivering and drooping wings, then gives food to young or eats it. Prey size increases with age of brood. ♀ shelters young from hot sun. Nestling period 14–16 days. Young dependent on parents for some weeks after leaving nest and remain with them for *c.* 6 months. May breed in the year following their birth.

BREEDING SUCCESS/SURVIVAL: in Zimbabwe, in 25 nests 31 of 48 eggs produced fledglings. Identified predators were Tropical Boubou *Laniarius ferrugineus* and Grey-headed Bush-shrike *Malaconotus blanchoti*; also monkeys considered as important nest predators. Parasitized by African Emerald Cuckoo *Chrysococcyx cupreus*, possibly also by Klaas's Cuckoo *C. klaas*, Eastern Honeyguide *Indicator meliphilus* and Cassin's Honeybird *Prodotiscus insignis*. In Zimbabwe, bird ringed as adult was retrapped at same locality 3 years 45 days later.

References
Brooke, R. K. and Manson, A. J. (1979).
Hanmer, D. (1979).
Harris, T. and Arnott, G. (1988).
van Someren, V. G. L. (1956).
Scott, J. A. (1984).
Stuart, S. N. and Jensen, F. P. (1986).

Genus *Batis* Boie

Small flycatcher-like and shrike-like birds, grey above, with white or pale in upperwing-coverts and inner secondaries, white underparts, usually with distinct breast-band (sexually dimorphic) and mask. Bill short, wide and flat, culmen ridged; palate black; nostrils small, oval, exposed; rictal bristles long and strong. Wing rounded, P10 more than half length of P9; P9 shorter than secondaries. Tarsus slender and quite long. Iris bright yellow. Rump plumage soft and erectile. Forage by flycatching; 'frip' wings in flight. Voice includes grinding noises and bill-snapping. Nest a tiny, neat, compact open cup, moulded onto branch, disguised with lichen.

Formerly treated as flycatchers Muscicapidae, and recently as monarch flycatchers Monarchidae. Placed by Harris and Arnott (1988) in subfamily Malaconotinae (with Prionopinae and Laniinae in family Laniidae); by Sibley and Monroe (1990) with vangas, helmet-shrikes, and wattle-eyes in family Corvidae (Vangini: Malaconotinae); and by Taylor (1986) and Dowsett and Forbes-Watson (1993) in family Platysteiridae, placed between Muscicapidae and monarchs.

Endemic. 16 species, mainly allopatric or parapatric. Where 2 or more occur together they are segregated by habitat. Opinions about species limits and superspecific affiliations have varied greatly (White 1963, Hall and Moreau 1970, Lawson 1986 and 1987, Erard 1987, Harris and Arnott 1988, Short *et al.* 1990, Sibley and Monroe 1990, Dowsett and Dowsett-Lemaire 1993b, Dowsett and Forbes-Watson 1993). Forest-dwelling southern African populations are generally regarded as comprising a single superspecies, and we concur: *B. capensis*, *B. fratrum*, *B. margaritae*, *B. mixta* and *B. diops*. 4 W and central African forms inhabit forest, forest edge, and rich mesic woodlands, and we regard them as 2 independent species, *B. poensis* and *B. minulla*, and a superspecies, *B. minima*/*B. ituriensis*. 7 savanna woodland and steppe species were treated as a single superspecies by Hall and Moreau (1970) and as 3 independent species (*B. pririt*, *B. perkeo*, *B. minor*) and 2 superspecies (*B. molitor*/*B. soror*, *B. orientalis*/*B. senegalensis*) by Dowsett and Forbes-Watson (1993). Short *et al.* (1990) thought that *molitor* and *soror* are a superspecies, probably with *pririt*, and that *orientalis* forms another either with *senegalensis* or with *minor*. Lawson (1987) concluded from UPGMA analysis (Unweighted Pair-Group Method using Arithmetic averages) that *soror* and *pririt* are very close to each other and close to *molitor*, that *orientalis* and *minor* are close, and that *perkeo* and *senegalensis* are independently distant. Influenced as much by parapatry as by plumage comparisons, we propose that the geographical series *B. senegalensis*/*B. orientalis*/*B. molitor*/*B. soror*/*B. pririt* forms a superspecies; we leave *B. minor* and *B. perkeo* as independent species, nearly parapatric with each other (and with *B. molitor*) but widely sympatric with *B. orientalis*. The final say will be had by students of vocal and other social communication characteristics of the various batis populations in the field.

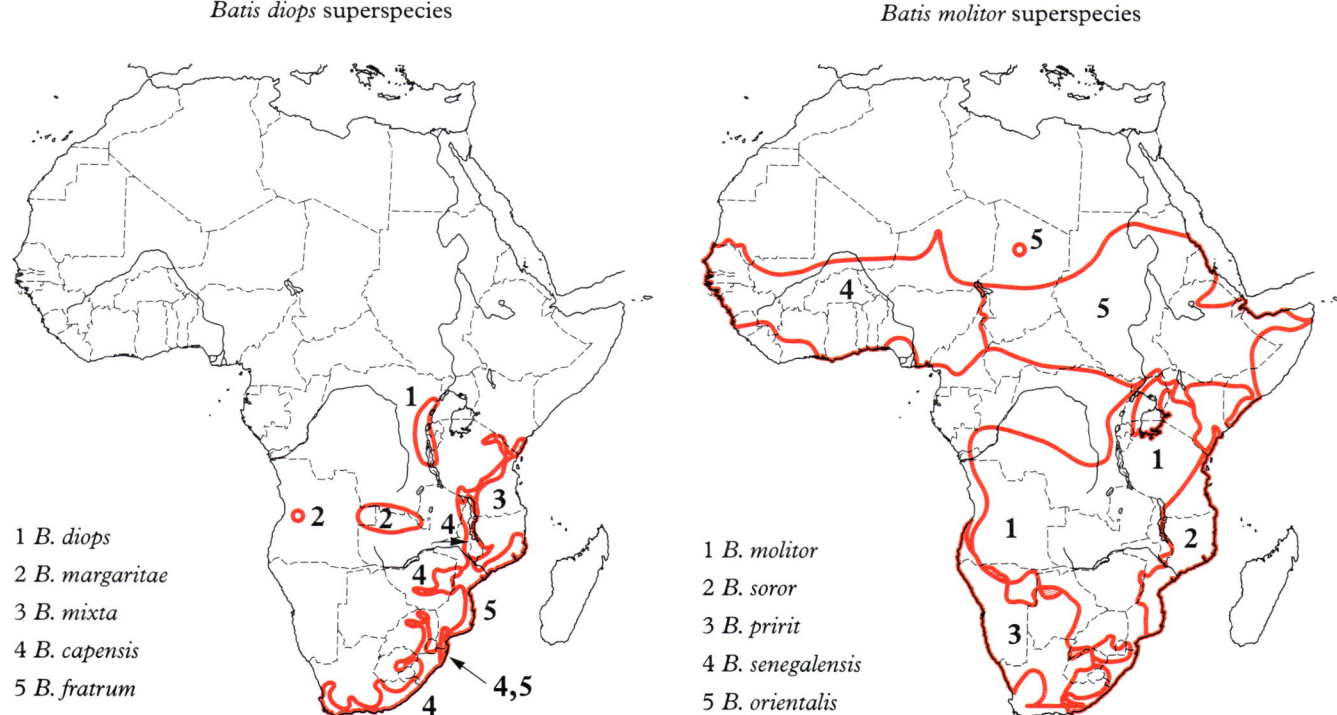

Batis diops superspecies

1 *B. diops*
2 *B. margaritae*
3 *B. mixta*
4 *B. capensis*
5 *B. fratrum*

Batis molitor superspecies

1 *B. molitor*
2 *B. soror*
3 *B. pririt*
4 *B. senegalensis*
5 *B. orientalis*

Batis minima superspecies

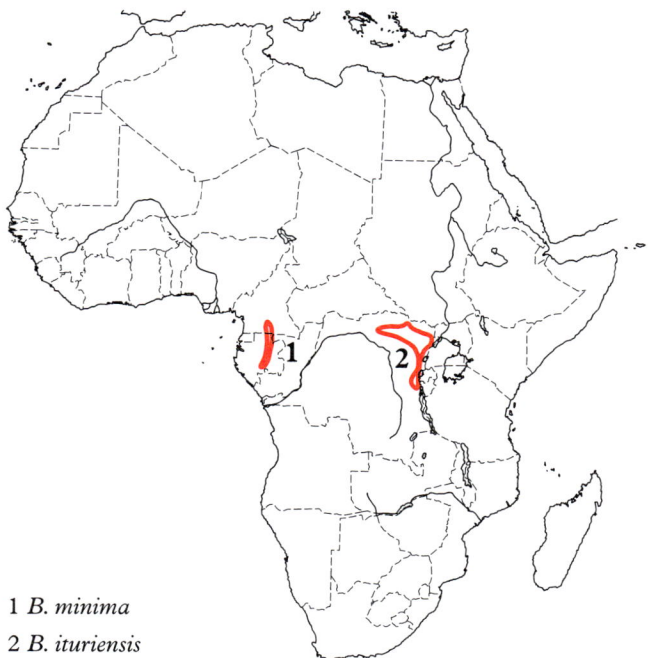

1 *B. minima*
2 *B. ituriensis*

Batis diops Jackson. Rwenzori Batis. Pririt du Ruwenzori.

Plate 32
(Opp. p. 561)

Batis diops Jackson, 1905. Bull. Br. Orn. Club, 15, p. 38; Ruwenzori.

Forms a superspecies with *B. capensis*, *B. fratrum*, *B. margaritae* and *B. mixta*.

Range and Status. Endemic resident, Rift montane evergreen forests, from Rwenzori to Itombwe and NE Shaba. Uganda, common at 1600–2000 m in Rwenzoris and Bwindi–Impenetrable Forest. Rwanda, common in W half. (Tanzania, a specimen labelled Kagera, more likely to be from Rwanda.) Burundi, common in W, south to about Makamba. Zaïre, W Rwenzoris about Beni, through Kivu where very common in forest at 1480–2140 m, less common towards altitudinal limits of 1340–2780 m (Itombwe), to mountains west of L. Tanganyika, south to Kabobo. Upper altitudinal limit 3050 m or even 3300 m (Lippens and Wille 1976). Density of 2–3 pairs per 10 ha in optimum habitat.

Description. ADULT ♂: forehead glossy bluish black with white spot on each side (hence 'diops' or 2-eyes); forecrown, sides of crown, lores, cheeks and ear-coverts glossy bluish black, including wide area below lore and above eye, the black patch extending backwards to become a narrow black line around lower hindneck; hindcrown and nape dark bluish grey; mantle dark grey intermixed with glossy bluish black feathers; scapulars black; rump and uppertail-coverts dark grey. Tail black, T6 with white edge to outer vane and narrow white tip, T2–T5 with narrow white tips. Chin white with small black spot at top; throat white, the white extending far back, forming broken white collar when bird sits with neck withdrawn; breast glossy bluish black, in well-defined band *c*. 15 mm deep; belly, flanks, vent and undertail-coverts white, sides

Batis diops

(concealed beneath folded wing) grey, thighs black. Primaries black, slightly glossed bluish; secondaries black, narrowly tipped white, tertials black; tertials and a few inner secondaries with broad white fringe to outer vane; greater and median primary coverts and alula black; outer and inner greater coverts black, middle ones white, median coverts white, lesser coverts black; underwing-coverts white, but black near leading edge of wing. Bill black; eyes golden yellow, with blackish brown outer edge; legs and feet black or dark slate-grey. Sexes alike except that ♀ has red or orange eyes.
SIZE: (23 ♂♂, 16 ♀♀) wing, ♂ 60–66·5 (63·1), ♀ 61–66 (63·25); tail, ♂ 38–43 (40·9), ♀ 40–43 (41·6); bill, ♂ 14·5–16 (15·5), ♀ 14–16 (15·4); tarsus, ♂ (n = 4) 18–19 (18·5), ♀ (n = 2) 19, 19. WEIGHT: (Uganda) ♂ (n = 19) 11–15·5 (12·8), ♀ (n = 6) 8–15 (12·4).

IMMATURE: none examined; said to be like adult but spotted with buff-brown on crown and mantle.

NESTLING: not known.

TAXONOMIC NOTE: treated as a member of the *B. capensis* superspecies by Hall and Moreau (1970), Traylor (1986) and Short *et al.* (1990), but as an independent species by Dowsett and Forbes-Watson (1993).

Field Characters. A small, dumpy, active 'flycatcher', grey-backed but in poor light appearing entirely black and white but for yellow or orange eyes. Upperparts dark grey, with white spot on side of forehead and white line in folded wing, underparts white with sharply-defined black breast-band. Similar to ♂♂ of 4 sympatric wattle-eyes; ♂ Black-throated Wattle-eye *Platysteira peltata* has very narrow breast band, ♂ Brown-throated *P. cyanea* has red wattle and pale grey rump; and ♂ Chestnut and ♂ White-spotted Wattle-eyes *Dyaphorophyia castanea* and *D. tonsa* have purple wattles and white rumps. Even more similar to 3 sympatric congeners: Black-headed Batis *B. minor*, Chinspot *B. molitor* and Ituri *B. ituriensis*. ♀ Black-headed and Chinspot readily told by their rufous breast-bands, and ♂♂ by their long thin white superciliary stripes. Ituri Batis (sexes alike or nearly so) is almost identical to Rwenzori Batis in plumage but is smaller; ♀ sometimes has thin white superciliary stripe; overlap unlikely since Ituri Batis is confined to lowland forest.

Voice. Tape-recorded (C, LEM). Song a series of low-pitched, ventriloqual, eerie whistles, 'fuu' or 'heeoooo', recalling Grey-headed Bush-Shrike *Malaconotus blanchoti* (Bennun 1986). Each 'fuu' note repeated at intervals of 2 s or more; 'fuu' lasts 0·25 s and in sonagram registers at 2·7 kHz and at 7·2 kHz; 'fuu' followed by 'chik' in excited or aggressive bird (Dowsett-Lemaire 1990). Also a chuntering or compressed churring 'ch-k-oick-k-k-', used in alarm and aggression. Frips wings in flight; snaps bill.

General Habits. At its most abundant in low, closed, bamboo forest on ridges, up to about 2800 m, but also inhabits mixed forest and scrub country, e.g. on lower slopes, down to 1500 m, of Virunga volcanoes. Keeps low down in dense undergrowth and low trees, singly or in pairs, foraging at low and middle levels of understorey, often at 2–6 m above ground, but feeds also in middle growth of high trees and their tops. Takes small insects from bark and undersides of leaves by hopping and snatching, and in mid-air in short flights; comes to edge of thickets but usually stays under cover (Dowsett-Lemaire 1990).

Food. Insects, evidently mainly beetles. Contents of 24 stomachs given as 'insects' or 'small insects' (12 stomachs), 'beetles' (4), 'small' or 'very small' beetles (5) and 'small flies' or 'small flying insects' (3). Once a 50 mm caterpillar.

Breeding Habits. Neighbouring pairs dispute territorial boundaries (Dowsett-Lemaire 1990). Several nests have been found, but neither nest nor contents described. Breeding ♀ regularly fed by ♂.

BREEDING INDICATIONS: Zaïre, Itombwe, Mar–Apr, June (and juvs Aug, Oct); Rwanda, Nyungwe, Aug–Jan (1 clutch each month); Rwenzori (♀ had recently laid, mid-Nov); Uganda (juv. Aug).

References
Chapin, J. P. (1953a).
Dowsett-Lemaire, F. (1990).
Gyldenstolpe, N. (1924).

Plate 31
(Opp. p. 560)

Batis margaritae **Boulton. Margaret's Batis. Pririt de Boulton.**

Batis margaritae Boulton, 1934. Proc. Biol. Soc. Wash., 47, p. 47; Mt Moco.

Forms a superspecies with *B. capensis*, *B. mixta*, *B. diops* and *B. fratrum*.

Range and Status. Endemic resident, south-central Africa: NW Zambia, Shaba (Zaïre) and W Angola. In Angola known only from forested upper slopes of Mt Moco and Manje Mts, where common, but almost certainly occurs in the extensive *Cryptosepalum* forests adjacent to NW Zambia (Bowen 1979). Zaïre, 1 specimen from Shilatembo. Not uncommon, and locally very common, in NW Zambia, between Mwinilunga, Solwezi and about Kabompo, recorded mainly along west Lunga R. and Kabompo R. and their tributaries. Single records, Kasempa, Ndola, and west of Luanshya (Benson and Irwin 1965).

Batis margaritae

Description. *B. m. kathleenae* White: Zambia, Zaïre. ADULT ♂: forehead, crown, nape, mantle, back, rump and uppertail-coverts grey; scapulars dark grey. Tail black, T6 with outer vane white, T2–T6 with narrow white tips. Lores grey, cheeks and ear-coverts black, forming mask that extends onto hindneck: almost a collar. Chin and throat white, in patch extending back below mask nearly to hindneck; breast black, in discrete band *c.* 15 mm deep, belly white, flanks white tinged grey, thighs grey, vent and undertail-coverts white. Primaries and secondaries black with narrow white outer edges, tertials black with broad white outer edges and narrow white tips; greater and median primary-coverts and alula black; outer and inner greater coverts black, middle ones white, median coverts white, lesser coverts black; underside of flight feathers dark grey, underwing-coverts white. Bill black; eyes golden yellow, tinged orange, legs and feet black. ADULT ♀: like ♂ but non-black feathers in wing rufous not white; lores speckled with white and above them an indistinct whitish line in some specimens; black breast-band only 12 mm deep, and a small rusty buff patch below it on foreflank; flanks with slight buffy wash; T6 tipped buff; eyes claret-red. SIZE: (16 ♂♂, 16 ♀♀) wing, ♂ 61–65 (62·7), ♀ 59–64 (61·7); tail, ♂ 40–45 (42·7), ♀ 39–44 (41·5); bill, ♂ 14–17 (15·3), ♀ 14–16 (15·0). WEIGHT: ♂ (n = 4) 11·5–13 (12·3), ♀ (n = 4) 12·5–15·5 (14·1), 1 imm. ♀ 10·5.

IMMATURE: like adult ♀, but flight feathers edged buff not white, whitish superciliary stripe longer, extending over eye, mantle and back washed with olive; eyes brown.

NESTLING: not known.

B. m. margaritae Boulton: W Angola. ♂ with upperparts darker than ♂ *kathleenae*, ♀ with wing-coverts edged rich rufous-brown. Larger: (5 ♂♂, 5 ♀♀) wing, ♂ 65–68 (66·4), ♀ 66–69 (67·4), tail, ♂ 45–46·5 (46·0), ♀ 45·5–47·5 (46·6), bill, ♂ 16–17 (16·25), ♀ 15·5–16 (15·9).

Field Characters. Overlaps only with Chinspot Batis *B. molitor*. Margaret's is decidedly bulkier and dumpier. ♀♀ of the 2 species readily identified, Chinspot having rufous chin spot and breast band and Margaret's white chin and throat and black breast band. ♂♂ almost exactly alike, but ♂ Chinspot has narrow breast-band nicked in the centre, and ♂ Margaret's has noticeably broader breast-band, not narrowed or nicked in the centre. Despite some variation in ♂ Chinspot, this difference is constant (Bowen 1979). ♂♂ differ also in eye-stripe: Margaret's Batis has vestigial one above lores, not visible in field; Chinspot has long, very narrow white eye-stripe. Songs also differ.

Voice. Tape-recorded (86, B). Song a soft, uninflected whistle repeated about 10 times, in 5 s (range, 3–26 times). First whistle slightly higher pitched than following ones, which are all on same pitch. Song carries some distance, although notes often thin and weakly delivered (Bowen 1979). Call, a harsh, scolding rattle.

General Habits. Primary habitat is *Cryptosepalum* forest: a dark, dense growth of dry evergreen forest on Kalahari sand dominated by *C. pseudotaxus*, usually low but attaining 18 m, with understorey of *Canthium malacocarpum*, *Diospyros undabunda*, and lianas including *Carpodinus*, with springy mosses predominating on the ground, to the exclusion of grasses. Margaret's Batis inhabits mid-stratum, in extensive tracts of forest, e.g. in S Mwinilunga Province (Zambia), and in small patches with burnt and farmed edges, e.g. in N Mwinilunga. It occurs also in dry-ground riparian forest and anthropogenic thickets, and has been recorded from wet forest (Bowen 1979). Chinspot Batis *B. molitor* also inhabits Zambian *Cryptosepalum* forest, but tends to forage above Margaret's, in the canopy.

Solitary or in pairs; very confiding; much less active than Chinspot, often remaining motionless for 1 min or more. Margaret's Batis has been observed by many ornithologists; they have compared it ecologically with Chinspot in the same forests but have published little about its foraging and social habits, which are therefore inferred to be so similar to those of Chinspot Batis, q.v., as not to have excited comment.

Probably resident in riparian forest, but appears to vacate *Cryptosepalum* forest from Nov (or Oct) to Apr, i.e. the wet season (Bowen 1979).

Food. Not known.

Breeding Habits. Not known. In Zambia, gonads active Sept.

References
Benson, C. W. and Irwin, M. P. S. (1965) (see Appendix).
Bowen, P. St J. (1979).

Batis mixta (Shelley). Forest Batis. Pririt à queue courte.

Plate 31
(Opp. p. 560)

Pachyprora mixta Shelley, 1889. Proc. Zool. Soc., Lond., p. 359; Kilimanjaro.

Forms a superspecies with *B. capensis*, *B. margaritae*, *B. diops* and *B. fratrum*.

Range and Status. Endemic resident, SE Kenya to N Malaŵi. Kenya, common in forest in SE coastal lowlands (Sokoke Forest, Shimba Hills, Shimoni, Mrimi, Jadini, Ganda, Rabai); also Taveta. Tanzania, from Mt Kilimanjaro, Manyara and Arusha Nat. Park to Pugu Hills, Kazimzumbwe For. Res., Usambara Mts, Uluguru Mts, Kiono Forest (06°10′S, 38°35′E), Udzungwa Mts, Ukinga Mts, Mt Rungwe, Poroto Mts, Isoko on Malaŵi border, also in Songea (perhaps a separate population). Malaŵi, extreme north in Masuku Mts. Ascends to 2300 m on Mt Kilimanjaro. Common throughout most of range.

Batis mixta

Description. *B. m. mixta* (Shelley): N Malaŵi (Masuku Mts), Tanzania and Taveta (Kenya). ADULT ♂: very like ♂ *B. capensis dimorpha*: forehead, crown, nape, hindneck, mantle, scapulars, back, rump and uppertail-coverts dark bluish grey; concealed white spots on rump. Tail short, black, with white sides, T6 with white outer vane, T2–T6 narrowly tipped white. Lores white, area above and below eye, and ear-coverts, black, forming a well-defined black patch that extends backwards towards nape. Chin and throat white, the white extending backwards onto sides of neck; breast black in a well-demarcated band 14–23 (19·5) deep. Flanks and belly white; thighs blackish; vent and undertail-coverts white. Primaries and secondaries blackish, narrowly edged white; tertials blackish, broadly edged white; greater and median primary coverts and alula blackish; outer greater coverts dark grey, inner ones white; median coverts broadly tipped white, lesser coverts dark grey; underside of flight feathers dark grey, underwing-coverts and axillaries white. Bill black; eyes red; legs and feet black. ADULT ♀: upperparts from forehead to rump olivaceous dark grey, scapulars blackish olive-brown; wing feathers olive-brown with russet outer webs; throat patch, breast-band and forepart of flanks pale rufous, the feathers white-tipped; throat patch practically merges with breast-band; rest of underparts white. SIZE: (33 ♂♂, 30 ♀♀) wing, ♂ 59–64·5 (62·6), ♀ 58–66 (61·7); tail, ♂ 31·5–38 (35·6), ♀ 31–38 (35·0); bill, ♂ 14·5–17·5 (15·6), ♀ 14·0–16·5 (15·3); tarsus, ♂ 17·9–19·4 (18·6), ♀ 18·0–19·9 (18·7). Tail length varies regionally, from 30–35 in Usambaras to 35–37 in Ulugurus and Kilimanjaro (Moreau 1940). WEIGHT: (Kenya) ♂ (n = 9) 7–10 (8·8), ♀ (n = 7) 6–12 (9·4), in same data set another ♂ given as 5·0 and another ♀ as only 4·0 (Los Angeles County Museum); (Amani, Tanzania) ♂ (n = 4) 12·0–14·2 (12·85), ♀ (n = 5) 11·0–13·5 (12·4); (Uluguru, Tanzania) 1 ♀ 12; (Udzungwa Mts, Tanzania) ♂ (n = 14) 11·7–13·5 (12·5), ♀ (n = 24) 10·5–13·6 (12·1) (J. O. Svendsen, pers. comm.).

IMMATURE: like adult ♀ but upperparts more olivaceous and dark patch on side of head not well demarcated from crown or throat. Soft parts paler.

NESTLING: not known.

B. m. ultima Lawson: SE Kenya. ♂ like ♂ *B. m. mixta*, but ill-defined, narrow white superciliary stripe, and breast-band narrower: (n = 14) 8–16 (11·3). ♀ like ♀ *mixta* but demarcation between grey crown and olive mantle less clearly defined, white eyestripe longer, throat patch pale and ill-defined, rufous breast feathers with more white on tip; forepart of flanks barely invaded by rufous. Smaller; wing, ♂ (n = 14) 58–63 (60·2), ♀ (n = 13) 55·5–60 (58·35); tail, ♂ (n = 14) 30·5–34·5 (32·0), ♀ (n = 13) 29–33 (31·0).

Field Characters. Very like closely-related Cape Batis *B. capensis*, but ♂ lacks rufous on flanks and in wings; their ranges come close in N Malaŵi but do not overlap. ♂ has very broad black breast-band and no white line above eye; ♀ has rufous on wings, flanks and breast, continuing without gap onto throat; tail especially short. Range overlaps 3 other batises, all of woodland rather than forest; they differ from Forest Batis as follows: ♂ Chinspot Batis *B. molitor* has white line over eye, narrower breast-band, ♀ has small rufous chin spot separated by white from rufous breast-band, no rufous on wings or flanks; characteristic song; East Coast Batis *B. soror* is a smaller and paler version of Chinspot Batis; both sexes of Black-headed Batis *B. minor* have blackish top of head and upperparts and long white supercilium, ♀ has no rufous chinspot on white throat.

Voice. Tape-recorded (KEI). Call (song?) of ♂, a single, low whistle with a ringing, metallic overtone, with quality of bush-shrike *Malaconotus*, repeated at rate of 1 every 2 seconds *ad lib.*, with great regularity.

General Habits. Inhabits undergrowth, middle storey and sometimes canopy of forest, from sea-level to 2300 m. In Kenya, near-coastal forest and thickets. On Mt Kilimanjaro, forages at all levels of the forest but mostly below the mid-stratum. Pair seen once feeding in canopy with 1 Grey Cuckoo-Shrike *Coracina caesia* and 2 Black-headed Apalis *Apalis melanocephala* (Cordeiro 1994). When singing, raises head and puffs out throat feathers with each whistle (Sclater and Moreau 1933).

Food. Insects. Contents of 5 stomachs simply 'insects' and of one, termites.

Breeding Habits. Solitary, territorial breeder. Details unknown. Breeding indications: Tanzania, Dec (Fuggles-Couchman 1986); Kenya, Sokoke, (oviduct egg early May); E Africa: Region E, May–June (2 records). SW Tanzania, Njombe and Dabaga (evidence of recent breeding, Dec, and post-breeding moult, Feb: Lynes 1934).

Batis capensis (Linnaeus). Cape Batis. Pririt du Cap.

Muscicapa capensis Linnaeus, 1766. Syst. Nat. (12th ed.), 1, p. 327; Cape of Good Hope.

Plate 31 (Opp. p. 560)

Forms a superspecies with *B. mixta*, *B. margaritae*, *B. diops* and *B. fratrum*.

Range and Status. Endemic resident, SW Cape to N Malaŵi. Locally common in S Cape Province, South Africa, but uncommon in arid northern districts of SW Cape. Orange Free State, sparse resident in tall *Leucosidea* scrub above 1800 m in E at Golden Gate (Colahan 1989). Lesotho, population near Ficksburg in NW (Glass 1986). Natal, common and widespread but absent north of St Lucia estuary, although it occurs in S Mozambique. Transvaal, very common in wooded parts of Escarpment region, including Blouberg and Soutpansberg; common in larger forests in valleys in SE Highveld, west to Amersfoort (Magaliesberg) and Kransberg (Waterberg). Separate populations in Zimbabwe, along eastern border from Inyanga Highlands (ranging from 900 to 2250 m) to Chipinga Uplands; at Marandellas; on Mt Wedza; and from Mt Mwenji to Bikita and west to Mt Buhwa, Mt Emberengwa and Matobo Hills. These isolates are in forests on the moist, southeast-facing mountain slopes; the species is widespread in Matobo Hills and on eastern border, also adjacently in Mozambique. Malaŵi, frequent in S (and adjacently in Mozambique) north to Ntchisi; another population in NW, south to Mwantjati (12°16′S), north to Nyika Plateau, where ranges from 1150 to 2450 m. Zambia, W Nyika Plateau; record from Chiri R. (Lawson 1986). A distant and distinct population between Lindi and Mikindani, Tanzania.

Density of 10 pairs per 10 ha in most localities on Nyika Plateau, Malaŵi.

Description. *B. c. capensis* (Linnaeus): South Africa, east to Orange Free State and Natal. ADULT ♂: forehead, crown, nape and hindneck dark bluish grey; mantle, scapulars, back, rump and uppertail-coverts warm olive-brown. Tail black, with white sides, T2–T6 narrowly tipped white. Lores, area above and below eye, and ear-coverts, black, forming a well-defined black patch that extends backwards towards nape. Chin and throat white, the white extending backwards onto sides of neck; breast black in a well-demarcated band; belly white; flanks russet-brown or chestnut; thighs blackish; vent and undertail-coverts white. Primaries and secondaries blackish, narrowly edged white; tertials blackish, broadly edged rufous; greater and median primary-coverts and alula blackish; outer greater coverts dark grey, inner ones rufous; median coverts rufous, lesser coverts dark grey. Underside of flight

feathers and tail dark grey, underwing-coverts and axillaries white. Bill black; palate black; eyes lemon-yellow or orange-yellow, said to be sometimes red in breeding season; legs and feet black. ADULT ♀: differs from ♂ in having: very narrow whitish superciliary stripe; chin and upper throat rufous in middle, white at sides; breast band rufous; eyes red in breeding season (photos of pairs at nest, R. E. Viljoen, N. Myburgh). SIZE: (113 ♂♂, 97 ♀♀) wing, ♂ 56·5–66·5 (61·6), ♀ 56·5–67 (60·5); tail, ♂ 40·5–49·5 (45·1), ♀ 40–48 (44·3); bill, ♂ 14–18 (16·6), ♀ 14·5–17·5 (16·1); tarsus, ♂♀ (n = 38) 18–22 (18·8). WEIGHT: (subspecies?: Maclean 1993) ♂ (n = 101) 9·3–14·5 (11·8), ♀ (n = 200) 8·9–13·9 (11·5).

IMMATURE: like adult ♀, but upperparts more olivaceous, dark grey, mantle spotted with light orange-buff, greater and median coverts broadly tipped orange-buff forming a double wing-bar; black mask absent, throat buffy, eyes brown.

NESTLING: hatches naked. No down. At 4 days pteryla capitis indistinct, pteryla spinalis distinct but short and restricted; pterylae humeralis, alaris and femoralis all distinct

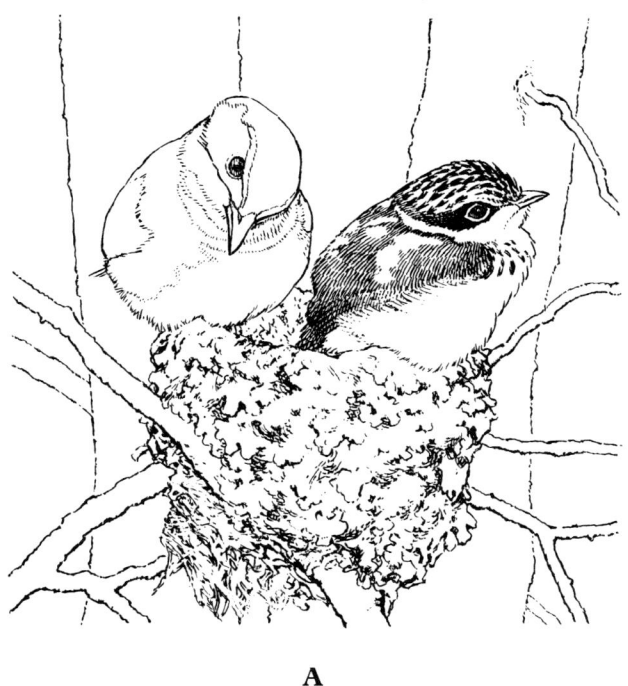

A

(Broekhuysen 1958). Later, bill purplish grey, gape creamy, tongue and palate yellow (Serle 1955) (**A**).

B. c. hollidayi Clancey: South Africa: Zululand, Swaziland, Mozambique (Lebombo Range). Like nominate subspecies but crown clearer blue-grey, mantle to rump greyish olive, rufous of flanks less intense.

B. c. erythrophthalma Swynnerton: E Zimbabwe and adjacently in Mozambique. Like *capensis* but mantle grey; not much contrast between blue-grey crown and grey mantle; black breast-band wider. ♀ has mantle warmer olive-brown than in ♀ *capensis*; breast-band wider and darker; sides of belly rufous – only the middle white. Iris of ♂ with red outer ring and narrower orange inner one; iris of ♀ dusky carmine-red, with fine silver inner ring around pupil (Swynnerton 1908). WEIGHT: 1 ♂ 12·1, 1 ♀ 11·1.

B. c. kennedyi Smithers and Paterson: SW and central Zimbabwe. Like *capensis* but rufous/chestnut parts are paler. Larger; wing, ♂ (n = 9) 62·5–67 (64·2), ♀ (n = 12) 62–68 (63·5), tail, ♂ (n = 9) 45·5–48·5 (46·9), ♀ (n = 12) 45–49·5 (47·0). WEIGHT: 1 ♂ 12·3, 1 ♀ 12.

B. c. dimorpha (Shelley): S Malaŵi and adjacent Mozambique. ♂ with flanks white, ♀ with scapulars blackish olive-brown.

B. c. sola Lawson: N Malaŵi (Mwantjati to Nyika). ♂ like ♂ *dimorpha*; ♀ has greyer mantle than ♀ *dimorpha*, scapulars broadly black laterally; lesser coverts black not brown; tertials fringed white not rufous.

B. c. reichenowi Grote: SE Tanzania (Mikindani to Lindi). ♂ with upperparts pale grey, wing-feathers edged grey not rufous; black mask does not reach so far towards hindneck; flanks white. ♀ with tinge of rufous on throat; rufous breast-band admixed with grey at sides. Size of *B. c. capensis* except for very short tail: ♂ (n = 3) 31–32 (31·5).

TAXONOMIC NOTE: *reichenowi* has characters of both *B. capensis* and *B. mixta*, and has been treated as a race of the latter (e.g. Britton 1980b) and as a separate species *B. reichenowi* (Lawson 1986). We follow Dowsett-Lemaire (1989) in placing it as a 'vocal' race of *B. capensis*.

Field Characters. Batises are approachable, plump, large-headed, short-tailed flycatcher-like birds of forest and savanna. Cape Batis likely to be confused only with congeners. It overlaps with Pririt Batis *B. pririt* (S Cape), Chinspot Batis *B. molitor* (E Cape to Malaŵi) and East Coast Batis *B. soror* (E highlands of Zimbabwe, S Malaŵi and adjacent Mozambique); and it closely approaches range of Forest Batis *B. mixta* in northernmost Malaŵi and Woodwards' Batis *B. fratrum* in Natal. ♂ Cape Batis told from others in its range by combination of broad black breast-band, rufous in wing, rufous flanks, and lack of white eye-stripe. ♀ has narrow white stripe reaching to but not behind eye, and rufous wing-patch and flanks; only other ♀ with rufous rather than white in wing is Woodwards' Batis, which is smaller, with distinct white eye-stripe extending behind eye to sides of neck, and diffuse underpart colour without defined throat patch. ♀ Chinspot Batis has white flanks and darker and browner breast-band and throat-patch; latter is smaller and more neatly defined. ♀ East Coast Batis has diffuse throat-patch but is very much smaller and lacks rufous on wings and flanks.

Voice. Tape-recorded (11, 58, 75, 86, 88, 91, B, F). Commonest call, often by ♂ and ♀ in concert, a loud, plaintive, piping, pure whistle 'whew', 'hui', 'tu' or 'tee', with quality of call of Grey-headed Bush-Shrike *Malaconotus blanchoti*, given on one pitch 3 times in 1 s or 4 times in 1·5 s: 'whew-whew-whew'; also used by ♂ as a song, in advertising and territorial defence, when note may be longer, 'hueet', sometimes disyllabic, 'hu-wit', and repeated about 12 times in 5 s: 'tu tu tu tu . . . teee teee', with the last 2–3 notes, 'teee', higher pitched. ♂♂ counter-sing, often using different-pitched whistles and producing a seesaw effect: 'hee', 'huu', 'hee', 'huu. . .' (Harris and Arnott 1988). A very different, variable call is sometimes given by ♀ when ♂ sings and is used by both sexes in territorial threat (**B**), often accompanied by wing fripping; call impossible to transliterate satisfactorily, mainly a low-pitched 'jirrowirro wirr . . .' or 'jk-jk-jjwrjjwrjjwr . . .' containing dry stone-grinding or stone-rubbing notes, hoarse or harsh rolling 'chi chi chch chi ri ri ri' phrases and softer 'jjwee' or 'jjizjjiz' notes, which ♀ can use alone as a buzzing 'zizz zizz' call. Contact call by ♂ and ♀, 'chik . . .' or 'jk . . .' and churring. Alarm, harsh 'chikchik' and bill-snapping. ♀ begs from courtship-feeding ♂ using chattering 'kish-kish . . .'.

B

General Habits. Inhabits lower mid-stratum of forest, from sea-level up to 2150 m, wandering freely into adjacent woodland, secondary growth, wattle and other plantations, and gardens. In Chimanimani Mts, Zimbabwe, moves to *Brachystegia/Uapaca* woodland in non-breeding season, when it may join mixed bird parties (Beasley 1995). In South Africa mainly a bird of shady kloofs, forest edges, clearings, wooded gorges, stream thickets and dense bush; sometimes uses bracken (Harris and Arnott 1988) and dense acacias on hillside dominated by spekboom *Portulacaria afra* (Pocock 1963). Restricted in E Orange Free State to tall stands of *Leucosidea* scrub on mountain slopes and adjacent *Podocarpus* forest but near Excelsior occurs in indigenous ravine-forest (Colahan 1989). In Lesotho inhabits well-treed areas with pines and some deciduous trees. Requires forest patch of at least 0·30–0·35 ha (in one locality, only 0·25 ha) (Nyika, Malaŵi, Dowsett-Lemaire 1983). Lives in pairs, territorially active for much of year, and family parties. In weeks before breeding, often congregates in flocks of 10 (rarely, 30), sometimes all ♀♀, known as 'batis parliaments', when much chasing, wing-fripping and excited calling. Inquisitive, little afraid of man; vocal. Pair forages together, actively hopping about, gleaning leaves, making short flights of 1–2 m to pick insect off branch or from under leaf with audible snap of bill; flycatches; pecks on bark whilst perched and takes small insects fleeing up trunk above ants. Makes light, bouncing flights from bush to bush; sometimes joins mixed-species foraging flock in winter. Feeds at all levels of understorey up to canopy of large trees at 20–25 m (Malaŵi). When alarmed makes fripping sound with wings, and snaps bill.

Post-nuptial moult starts soon after breeding; on Nyika Plateau, Malaŵi, most birds breed Sept–Dec and most moult Dec–Mar (< 10%, Nov, Apr: Dowsett and Dowsett-Lemaire 1984).

Mainly sedentary; but in Zimbabwe some move into low country (Honde valley; Lusitu-Haroni area, at 350 m) in May–Sept. Similar altitudinal downward shift outside breeding season in Malaŵi. On Nyika Plateau > 60 birds retrapped had moved horizontally for up to 750 m (Dowsett 1985), once 2·6 km.

Food. Insects: flies, beetle larvae, caterpillars; also spiders. One large beetle grub, swallowed entire, filled stomach. Food brought to young: flies, small cockroaches, moths, a phasmid.

Breeding Habits. Solitary breeder, monogamous, territorial. A second ♀ sometimes present in breeding territory, perhaps immature; species may therefore be a facultative cooperative breeder. The species shows strong fidelity to territory; of 63 adults at Nyika, Malaŵi, 21 were in same territory 3 years later and 1 10 years later (Dowsett 1985). Territory size 0·3–1·0 ha. ♂ advertises and defends territory of *c*. 1 ha by singing 'hui' whistle; singing posture is squat, body and head inclined, white throat feathers expanded and conspicuous, tail slightly fanned. Territorial defence by both sexes also involves bouncing and zigzag flights with tail fanned, wing-fripping, bill-snapping, and wide variety of voices including grating 'stone-rubbing' calls. ♂ fluffs out rump-feathers. Courtship is similar; ♂ flies about constantly, with much wing-fripping; display flight by ♂ and ♀ is undulating with rolling or churring calls; birds perch, both raise wings above back, fan tails, fluff rumps and copulation may follow, ♀ crouching with bill pointing up. Pair builds nest in same area each year. 1–2 broods.

NEST: beautifully made, small, neat cup, thick-walled, robust; sides convex – nest is subglobular with deep cup in centre (**C**). Made of variety of fine, often fibrous, dry plant material, profusely or scantily lined with fine branching stems of herbs, bound and strengthened with spider web; material extends along and partly around small branch on which nest built, moulding nest onto the branch. Built by ♀ and ♂ (or ♀ closely attended by ♂); both apply spider web with side-to-side weaving or wiping movement of head (Harris and Arnott 1988). When nest nears completion, with each beakful of material brought ♀ squats down in cup and revolves slowly to shape it. Outside of nest is finished with covering of green moss, and heavily decorated with concealing large flakes of pale greenish lichen, or filamentous lichen (*Usnea*), bound in place with a little spider web. Ext. diam. (n = 1) 60 × 70, height 40; int. diam. 50, depth 30. Sited on stout horizontal branch or against tree trunk on small branch or in simple or multiple fork, not hidden among foliage, but placed in poor light under canopy of upper foliage of leafy bush or tree, mainly indigenous evergreen but also for example oak, *Royena* and *Pittisporum*, 0·8–9·5 m above ground (n = 27, mean 4·1 m, and n = 138, mean *c*. 3 m).

C

EGGS: 1–3 (n = 65 clutches) av. 2·1. 2nd egg (n = 1) laid 24–28 h after 1st. Eggs of 2 distinct types: (a) very pale green with dark brown spots and blotches; (b) white or very pale pink, spotted and blotched with purple-grey and red-brown; both have the markings in a well-defined belt around broad end (Swynnerton 1909, MacLeod and Hallack 1956). SIZE: (n = 40) 16·0–21·0 × 12·6–15·2 (18 × 13·9).

INCUBATION: by ♀ only, beginning with last egg laid. Incubating ♀ courtship-fed by ♂; she is sometimes timid, but can often be actually touched on nest by person. In one 12-h period, ♀ incubated for 77% of time and left nest 25 times at regular intervals for 2–12 (6·6) min (Broekhuysen 1958). Incubating ♀ often stands up 2–4 times in quick succession, to turn eggs with bill. Period (n = 4) 17–21 days (18 days).

LAYING DATES: South Africa (all parts), Sept–Jan, mainly Oct–Nov; Mozambique, Oct; Zimbabwe, Sept–Feb, mainly Oct–Dec; Malaŵi, Sept–Feb (Nyika, n = 168 clutches, 31 Sept, 49 Oct, 37 Nov, 40 Dec, 8 Jan, 3 Feb: Dowsett and Dowsett-Lemaire 1984); Botswana (fledglings being fed, Mar, 2 years).

DEVELOPMENT AND CARE OF YOUNG: rectrices in pin sheath stage on day 7; day 9, flight feathers break open, ventral pterylae feathery; day 12, young feathered, flight feathers up to 7 mm long; day 13, well feathered, crown steaky, dark grey and light buff; ear-coverts dull blackish with white line below them, mantle and scapulars streaky (photos, Broekhuysen 1958); day 15, fully feathered; legs pearl-grey. Eyes are narrow slits on day 5; on day 7 eyes closed most of the time but open when chick begs for food; eyes wide open by day 12. Weight increases from 1·5 g on day 1 to plateau of 11·5–12·5 g on days 11–16. Tarsus grows at uniform rate from 6 mm on day 1 to 20 mm on day 15. Nestlings fed 1–6 times per h by ♀, once every few h by ♂. ♀ broods young for c. 80% of daytime on day 5 and 45% on day 16; incubating ♀ fed 0–5 times per h by ♂ (she is also fed by ♂ when she is not incubating). Chick faeces not eaten by ♀ but removed by her from nest. Period: 16 days (n = 2). Young sometimes stay with parents for at least a year.

BREEDING SUCCESS/SURVIVAL: parasitized by Klaas's Cuckoo *Chrysococcyx klaas*; in one study 3 nests parasitized out of 5. Nestlings eaten by boomslang *Dispholidus*; adult killed by Fork-tailed Drongo *Dicrurus adsimilis*. One ringed bird survived 5 years, another 6 years 28 days, and a third 8 years 17 days.

References
Broekhuysen, G. J. (1958).
Colahan, B. D. (1989).
Harris, T. and Arnott, G. (1988).
Lawson, W. J. (1964a).
Swynnerton, C. F. M. (1908).
Vincent, A. W. (1947).

Plate 31
(Opp. p. 560)

Batis fratrum Shelley. Woodward's Batis. Pririt de Woodward.

Pachyprora fratrum Shelley, 1900. Ibis, 8th ser., Vol. VI, p. 522; Zululand.

Forms a superspecies with *B. capensis*, *B. margaritae*, *B. mixta* and *B. diops*.

Range and Status. Endemic resident, lowland Mozambique, just entering S Malaŵi, E Zimbabwe, and Natal (Zululand). From Richards Bay (Natal) to Sul do Save coast, at many localities from Maputo to Beira; inland west and north of Beira, to Lusitu–Haroni area of easternmost Zimbabwe where species is at altitude of 350–450 m, and Inhaminga, Gorongoza Mt, and lower Shire R. valley into Malaŵi, ranging north to Malaŵi Hill and Lengwe Nat. Park. Ascends to summit of Malaŵi Hill, c. 900 m. Also in N coastal lowlands of Mozambique at Netia and Mirrote (where 'tolerably abundant' 60 years ago: Vincent 1935); likely to range continuously from there to Zambezi mouth and Beira. Locally not uncommon. Malaŵi Hills forests hold >100 pairs, with greatest density of 10 pairs per 10 ha.

Description. ADULT ♂: forehead blue-grey, with short, narrow white streak at sides; crown and nape dark bluish grey; hindneck black; mantle and back dark bluish grey, scapulars edged bluish black, rump and uppertail-coverts blue-grey; rump-feathers with concealed penultimate white spot. Tail black, outer web of T6 white, tips of T2–T6 white. Lores, superciliary area, cheeks and ear-coverts black, forming mask that reaches back practically to hindneck. Chin, throat and sides of neck white, breast pale rufous, the feathers tipped

Batis fratrum

buffy white; belly, flanks, vent and undertail-coverts white. Thighs dark grey. Primaries and secondaries black with narrow white edges, tertials black with broad white edges. Inner and outer greater coverts slaty, but middle ones largely white; median coverts white, lesser coverts and other upperwing-coverts slaty; underwing-coverts black, axillaries white. Bill black; eyes yellow, orange or scarlet; legs and feet black or very dark slate-grey. ADULT ♀: head like ♂, but a narrow white superciliary stripe on sides of forehead, forecrown and hindcrown; chin and throat pale rufous; mantle, back and scapulars olive-grey; wings pale cinnamon where ♂ is white; belly and flanks buffish off-white. SIZE: (17 ♂♂, 19 ♀♀) wing, ♂ 58·5–64 (61·2), ♀ 57·5–62·5 (60·0); tail, ♂ 33–38 (34·8), ♀ 32–35 (33·2); bill, ♂ 15·5–17·5 (16·5), ♀ 15–16 (15·4); tarsus, ♂♀ (n = 8) 17·5–19. WEIGHT: ♂ (n = 43) 10·4–13·8 (12·2), ♀ (n = 23) 10·3–13·6 (11·7).

IMMATURE: like adult ♀, but superciliary stripe buffy; mask buffy brown, not black; broad rufous-buff margins to wing-coverts and flight feathers; legs dusky flesh-colour.

NESTLING: not described. Fledgling has grey-brown eyes, and upper-parts spotted with buff and brown.

TAXONOMIC NOTE: treated as a member of *B. capensis* superspecies by Hall and Moreau (1970), Traylor (1986) and Short *et al.* (1990) but as an independent species by Dowsett and Forbes-Watson (1993). White (1963) recognized 2 subspecies, nominate *fratrum* and *ultima*; but we treat the latter as a subspecies of *Batis mixta*. *B. fratrum* includes '*sheppardi*', the darker population north of Save R., too weakly differentiated to recognize taxonomically (Clancey 1969).

Field Characters. A medium-sized batis with restricted range in coastal lowlands of SE Africa. Sexes similar, both with diffuse pale tawny breast-band which in ♀ continues somewhat onto throat and flanks. ♀ also has pale supercilium and tawny in wing, lacking in ♂. ♂ easily distinguished from ♂♂ of other batises by lack of breast-band; both told from sympatric ♀♀ by paler and more diffuse colouring of underparts, with no clear-cut breast-band or throat-patch; these areas are darker, more chestnut on ♀♀ of Chinspot and East Coast Batises *B. molitor* and *B. soror*, which also lack tawny in wing. ♀ Cape Batis *B. capensis* has tawny in wing but also on flanks and throat, and is larger.

Voice. Tape-recorded (75, 88, 91, C, F). Song of ♂ rather slow and deliberate, a clear, ringing, low-pitched whistle 'whhhh' or 'huu' or 'poo', repeated up to 10 times, at rate of one note every 1·1–2·3 s. Quality of call is very similar to Grey-headed Bush-Shrike *Malaconotus blanchoti*, pure, mournful, slightly quavering, spooky, pitched at 1·3 kHz, but Woodwards' call is much shorter: 0·20–0·25 s. Song duration, up to 23 s, is determined by variation in length of pause, not of notes. Notes in any 1 song delivered at unvarying pitch and frequency, yet they can sound a little hesitant and indefinite; sometimes note has 'echo' at end: 'whhhh-wh' or 'hui'. ♀ song, often sung in alternating duet with ♂, is quite different, a very dry, discordant note of c. 0·3 s duration, 'krjjer', repeated about 6 times at same frequency as ♂ song, each ♀ note evoking immediate 'whhhh' response from ♂. On occasion ♀'s contribution to song-duet is limited to a few notes at random intervals (Harris and Arnott 1988). ♂ song can be more rapid and much higher pitched, when it sounds squeaky. 2 ♂♂ sometimes counter-sing using 'whhhh' notes which, if at rather different pitches, produce a seesaw effect. Wide variety of other calls, by ♂ and ♀, including indefinite 'chik', and chuntering, a nasal, mechanical 'jnananaee' and 'jneeeu', bill-snapping and wing-fripping. Alarm, rasping, jerky 'prer-rer-rert' (Vincent 1935).

General Habits. Inhabits low and mid-levels of evergreen lowland forest, evergreen thickets in deciduous dry forest, riparian forest, thick low bushes in dune forest and dense stands of acacia.

Occurs usually in pairs, also family parties, and in vocally-interacting social 'parliaments' of up to 12 adults. Retiring; and less vocal than some batises. Restless, spending all day foraging; seldom still for more than a few s. Arboreal, insectivorous, making numerous short sallies to snap up insects from twigs, branches and leaves, often hovering momentarily before seizing prey with audible snap of bill. Often feeds within a few cm of ground (Harris and Arnott 1988). Joins mixed-species foraging parties, particularly out of breeding season. Readily mobs a predator, with rapid bill-snapping and harsh 'chikchik' calling. Socially highly interactive. ♂ defends territory by singing, perching at edge of tree in crouched posture and uttering 'whhhh' whistles with head inclined up, neck stretched, neck feathers forming a stiff-looking ruff and stumpy tail fanned. Singing posture of ♀ similar but less pronounced. 'Parliaments' not studied and function unknown; several birds make full repertoire of calls, with short flights and much wing-fripping and bill-snapping.

Food. Insects.

Breeding Habits. Solitary, monogamous, territorial. Territory size c. 0·5 ha. Territory defended by ♂, also perhaps by ♀, singing (see above). High-intensity interactions with adjacent pair involve much zigzag flying with wing-fripping, 'chik' calls, 'jnananaee' and 'jneeeeu' calls, and bill-snapping; ♂'s white-spotted rump-feathers are fluffed out as he lands and wings a little drooped (Harris and Arnott 1988).

NEST: small, neat cup of fine grass, bark and moss, bound with spider web, sometimes decorated outside with lichen. Well hidden; sited c. 2 m above ground in fork or on horizontal branch of bush or tree. One nest seemed to have been built on top of nest of another species (Harris and Arnott 1988).

EGGS: 1–3, av. 2·4. Creamy white or bluish white, spotted and blotched with red, brown and grey, mainly around broad end. SIZE: (n = 8) 16·3–19·3 × 12·4–14·1 (18·1 × 13·1); WEIGHT (n = 2) 1·8.

LAYING DATES: Natal, Oct–Dec.

Nothing further known, except that ♀ seen incubating and brooding, and fledglings known to be attended by both parents.

References
Harris, T. and Arnott, G. (1988).
Lawson, W. J. (1962b).

Plate 31
(Opp. p. 560)

Batis molitor (Hahn and Kuster). Chinspot Batis. Pririt molitor.

Muscicapa molitor Hahn and Kuster, 1850. Vög. aus Asien, Lief 20, pl. 2; Baviaans River, Cape Province.

Forms a superspecies with *B. senegalensis*, *B. orientalis*, *B. soror* and *B. pririt*.

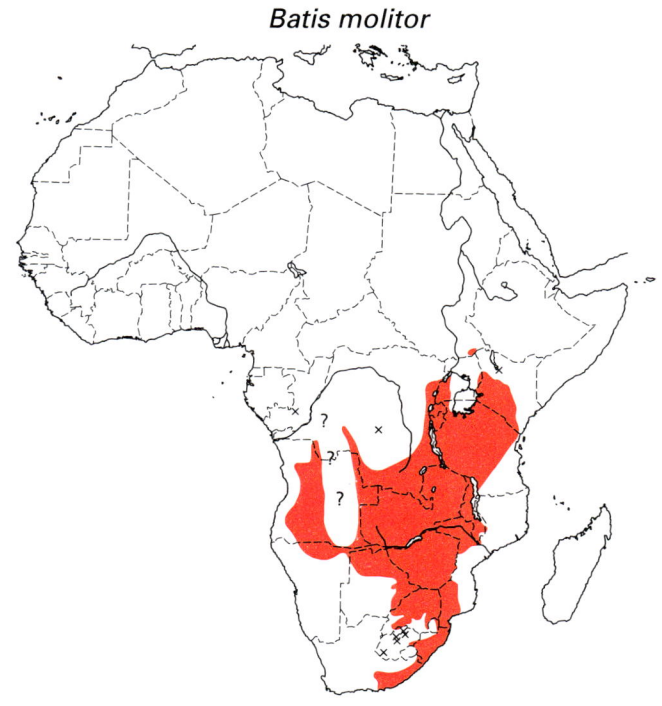

Batis molitor

Range and Status. Endemic resident, southern African savanna woodland. Sudan and N Uganda, fairly common in Didinga Mts. E Africa, common in SW and SE Uganda, W and S Kenya, Rwanda, Burundi, and throughout Tanzania except for coast and SE, from 500 m to 3000 m. Zaïre, widespread in E and SE, commoner below than above 1400 m but occurring up to 2500 m in Rwenzoris and on the volcanoes; a few records in Kasai and 1 in Congo, where species may be more widespread than records suggest. South of Kenya, eastern limits of range are defined by range of East Coast Batis *B. soror*, with which Chinspot is precisely parapatric south to Limpopo R. mouth; and from SW Angola to SE Cape western limits of range of Chinspot are defined, a little less precisely, by range of Pririt Batis *B. pririt*. Angola, widespread except for coastal regions, probably also throughout the corridor shown on map; fairly common also in N Namibia. Common throughout Zambia and Malaŵi up to 2060 m on Nyika Plateau, but in SE Malaŵi absent from Zomba, Blantyre, Thyolo, Chiradzulu and Mulanje Districts where replaced by *B. soror*. Zimbabwe, very common, absent only from area occupied by *B. soror*, i.e. lower levels in E highlands, where Chinspot lives in stunted miombo woodland up to 1680 m (Chimanimani Mts) and 1800 m (Inyanga Highlands). Mozambique, common in N Manica e Sofala Province, Limpopo valley and south of Limpopo; also common on Zimbabwe border at confluence of Sabi R. and Lundi R. (replaced by *B. soror* only 40 km downstream). Botswana, frequent in N and E. South Africa, absent west of 24°E; Transvaal, common and widespread in bushveld and lowveld woods, extending onto highveld only locally; Orange Free State, widespread but uncommon in acacia savanna in NW, sparse or rare in Bloemfontein and Bethulie districts, and a small population near Harrismith; Natal, widely distributed in thornveld and small patches of dry woodland from sea level to 900 m, sparser and more local up to 1200 m; E Cape, common west to Great Fish River, Mountain Zebra Nat. Park and Bloemfonteinberge; N Lesotho, rare.

Breeding density of 1 pair per 5 ha in acacia veld and 1 pair per 15 ha in broadleaved woodland (Nylsvley, Transvaal).

Description. *B. m. molitor* Hahn and Kuster: SE South Africa and S Mozambique, from Port Elizabeth to mouth of Limpopo R. ADULT ♂: forehead, crown and nape blue-grey, darker and slightly olivaceous on nape; sides of forehead and forecrown white, wide in front, the white tailing away over eye; small whitish patch on hindneck. Mask (lores, above and below eye, ear-coverts, sides of hindneck) glossy black; mantle, scapulars and back blue-grey, rump and uppertail-coverts grey: rump-feathers dark grey with white tips. Tail black, T6 with broad white edge to outer web and white tip to inner web. Chin, throat and side of neck below ear-coverts white; breast glossy black in well-defined band; flanks, vent and undertail-coverts white, thighs dark grey. Wings black, with long white stripe in folded wing formed by broad white ends to median and inner greater coverts and broad white edges to inner secondaries. Underside of flight feathers and tail shiny dark grey; underwing-coverts black and axillaries white. Bill black; palate black; eyes bright lemon-yellow (sometimes iris is yellow only in narrow ring immediately around pupil); legs and feet black. ADULT ♀: like ♂ but breast-band chestnut and a distinct triangular patch of the same colour in centre of chin and upper throat; white eye-stripe longer than ♂'s and can extend to hindneck, but variable and often no larger than in ♂. SIZE: (94 ♂♂, 68 ♀♀) wing, ♂ 55·5–66·5 (62·0), ♀ 54–66 (60·8); tail, ♂ 41–51·5 (46·9), ♀ 41·5–52·5 (46·7); bill, ♂ 14·5–18·0 (16·1), ♀ 14·5–17·0 (15·6); tarsus, ♂♀ (n = 124) 16–20. WEIGHT: (Kenya) ♂ (n = 11) 5–12 (9·6), ♀ (n = 7) 5–11 (7·8), 2 unsexed 9, 11·5; (Uganda) ♂ (n = 3) 11–12 (11·3), ♀ (n = 4) 11–12 (11·25); (southern Africa: Harris and Arnott 1988) ♂ (n = 78) 8–14 (11·2), ♀ (n = 75) 8–14 (12·4).

IMMATURE: like adult ♀ but crown and mantle mottled with rufous-buff, and white parts of upperwing buffy; mask less clear-cut than in adult; eye-stripe and sides of neck slightly buffy; ♂ can have chestnut breast-band looking tawny, with a few black feathers. Eye paler yellow than in adult.

NESTLING: hatches naked, blind, without mouth spots.

B. m. palliditergum Clancey: S Angola, S Zambia, middle Zambezi valley, Zimbabwe, Botswana, N Namibia, Transvaal, Orange Free State. Crown and mantle slightly paler; tail averages 2 mm shorter.

B. m. pintoi Lawson: Angola (except S), NW Zambia. ♂ with shade of crown and mantle between those of *molitor* and *palliditergum*; crown suffused with glossy blue-black; breast-band tinged glossy green not blue. ♀ with crown and mantle darker than in *B. molitor*. Larger. SIZE: (37 ♂♂, 23 ♀♀) wing, ♂ 61–69 (65·2), ♀ 58–67·4 (64·1); tail, ♂ 41–47 (44·3), ♀ 41·5–46·5 (44·1); bill, ♂ 15–17 (15·7), ♀ 14·5–17 (15·2).

B. m. puella Reichenow: Zaïre, E Africa. Like *molitor* but mantle more glossy blue-black, white on edges of primaries and secondaries narrower (sometimes none in ♂♂); breast-band and throat spot of ♀ dark maroon-chestnut. Wing same length as in *molitor* but tail averages 4 mm shorter.

Field Characters. The best-known batis, common in all types of woodland. 12–13 cm. An alert, lively, arboreal 'flycatcher', usually in pairs; ♂ strikingly pied, ♀ with well-defined chestnut breast-band and throat patch. Yellow eye. Noisy, like all batises; ♂ with characteristic 'three-blind-mice' song, both sexes with variety of calls, some froglike, others like grinding dry stones together, and with bill-snapping and, in flight, noisy wing-fripping. ♂ very like ♂ Pririt Batis *B. pririt* and ♂ East Coast Batis *B. soror* but has: (a) 3-note song (Pririt has long series of 'hee' notes, East Coast has > 3 'p'lk' or 'tlop' notes, often a long series), (b) white flanks (dusky speckles in Pririt), and (c) broad breast-band (narrow in East Coast Batis). ♀, with sharp 'chinspot', readily told from ♀ Pririt (throat and breast uniformly buff) but can be confused with ♀ East Coast Batis (diffuse 'chinspot', narrow breast-band, rufous on sides behind breast-band).

Voice. Tape-recorded (10, 75, 86, 88, 91, B, C, F). ♂ song a pure, 2–3-note whistle, falling by about a quarter-tone, lasting *c.* 1·25 s, repeated about 4 times in 15 s; 'whee-whew-whew', 'he-ho-hu', 'kreep-kroop-chowp', 'weep-woop-chuk' or 'three blind mice' (the last a useful mnemonic rather than an accurate transliteration). In N Botswana territorial ♂ readily imitates song of Pririt Batis, giving song usually of 5–7 whistles (max. 15) (Herremans 1992). Another variant is 'whee-whew-hlut', where 'hlut' not a whistle. ♀ often joins ♂ in duet, giving strident 'wik' notes, the pair producing 'he-ho-wik-hu-wik'. Out of breeding season, ♂ has single whistle, 'he', or double 'he ho'. In Nyungwe Forest, Rwanda, and in Kenya, 2 song types: (a) 'three-blind-mice' or sometimes 'three-blind', and (b) up to 9 monotonous whistles at rate of 2 per s, preceded by a higher one (very like song of Black-headed Batis *B. minor*: Dowsett-Lemaire 1990). Other notes: an exited, emphatic 'chikchik-chik . . .' before roosting, and in flight in winter 'parliament' assemblages; frog-like 'grrruu-grrruu' sounds and a rolling whistle 'prrreeo' by ♂ establishing or defending territory; 'chik-prrreo' by ♂ and 'weeeo-tik' by ♀ during courtship (Harris and Arnott 1988), former also described as 'querk-querk-querk' (van Someren 1956); 'wikwikwik' of low-intensity alarm; buzzing or chattering 'ksshkisshkissh . . .' by begging nestling; and bill-snapping and wing-fripping by adults, in several contexts.

General Habits. Inhabits savanna woodlands, deciduous woods, gardens, farmland with trees, riparian thickets, bushveld and edges of forest, open scattered scrub, orchard bush and conifer plantations in Malaŵi although avoids plantations in Transvaal. In central Africa ascends mountains to about 3000 m, inhabiting open types of montane forest, avoiding deep shade and keeping to crowns of medium and tall trees where canopy is discontinuous (Dowsett-Lemaire 1990).

Forages inside open tree but mainly at edge of leafy tree, at all levels up to *c.* 10 m, and sometimes flies down to seize insect on ground. Extremely active; at perch moves head constantly to search for prey, twisting and jerking body and flicking tail, cocking head at all angles, to scrutinize vegetation and airspace. Hops along and between branches. Forages by making long hops and short flights; sights prey, then flies swiftly upwards and outwards to seize it. Momentarily hovers before seizing prey, or grabs it in unbroken flight as bird passes, or lands next to it then takes it; seldom returns to same perch. Of 112 feeds, 79% were inside and 21% at outer edge of canopy; when perched outside canopy, bird looks inward rather than outward. 24% of the sallies were hawking, 63% hawking-gleaning and 13% gleaning; bird feeds from leaves 4 times as often as from trunks, branches and twigs (Fraser 1983). Often hangs head downwards when prying into crevice in bark (Vincent 1935). Large insect is beaten against branch, and held down under one foot to be dismembered by tearing with bill (Harris and Arnott 1988). In Tsavo, Kenya, 49 insects were captured one-by-one by bird in flight, bird returning to perch each time; 69% were gleaned from leaves, 14% from twigs, 14% airborne; 81% were taken deep inside vegetation, 13% just inside, 6% at edge; av. height 4·8 m; of insects gleaned from leaves, nearly all were from underside, only 1 or 2 from upperside of leaf (Lack 1985). Rarely forages on ground. In Botswana bird spent 35 min and another 10 min feeding actively on ground among very short herbs, taking prey from ground (never on wing) and using lookouts such as stones, dung, earth and sticks; did not use low hanging branches even when passing under tree (Herremans and Herremans-Tonnoeyr 1994).

Lives in pairs, maintaining contact visually and with frequent quiet 'chik' calls. One or both birds commonly join mixed-species foraging party – in Kenya seen in mixed party 4 times as often as on its own. In months before breeding, 12 birds or more (sometimes of same sex) congregate in 'parliament', chattering loudly and excitedly with 'chikchikchik' calls, fripping wings in bouncing flight, and snapping bills continuously. Family party roosts < 1 m from ground, birds all separated from each other in same bush; sleeps in hunched-up posture. Before going to roost, family flies about excitedly, wing-fripping and calling.

Strongly territorial, nearly all year. ♂ advertises and defends territory of *c.* 5 ha with loud, repeated song; when singing, ♂ inclines body upward, stretches neck, raises bill, and slightly fans tail. ♀ may join in with 'wik' notes in duet. Sings from prominent perch on small tree. Aggressive, particularly to boubou shrikes *Laniarius* spp. One of the first birds to locate a roosting owl; which is then mobbed strenuously. Vigorously chases away *Chrysococcyx* cuckoos. Territorially aggressive ♂ has 'jinksing' display, with short circular flight above tree tops, head held high, white-spotted grey rump-feathers fluffed out, tail fanned, wings fripping, with vocal 'grrruu' and 'prrreeo' calls. ♀ may join in with 'wik' and 'chik' calls; perched bird sometimes sways head slowly from side to side (Harris and Arnott

1988); also bends forward with rump-feathers standing on end to make a soft powder-puff (batises were formerly called 'puff-backed flycatchers').

Almost exactly parapatric with East Coast Batis (q.v. for overlap areas). Parapatric with Pririt Batis but overlaps its range in (a) N Namibia, (b) N Botswana, (c) SW Transvaal between Wolmaranstaad and Christiana and adjacently in NW Orange Free State between Boshof and Viljoenskroon, also central Orange Free State between Virginia, Ventersburg and Winburg, and (d) Cradock District of S Cape (Collet 1982). Details of (a) and (b) under *B. pririt*.

Food. Insects including flies and many beetles. Small noctuid caterpillars, moths and spiders fed to nestlings.

Breeding Habits. Monogamous, territorial, solitary nester. There may also be low incidence of co-operative breeding; additional bird sometimes present in breeding territory, the offspring in an earlier breeding season of the territorial pair; but such third birds usually chased out eventually by breeding ♀ (Harris and Arnott 1988). ♂ courtship display like territorial defence (above) except that flights are lower down and in zigzags from branch to branch rather than in circle above tree top. ♂ flicks tail, fluffs out rump, flies in bouncing flight with wings fripping at each bounce, alights in front of ♀ with his wings quivering, calls 'grrruu'; ♀ gives harsh begging 'kisshkisshkiss' call in crouched position, and flutters wings; copulation follows, accompanied by much vibrating of both birds' wings. Copulation can occur without prelude other than courtship-feeding of ♀ by ♂. One pair remained mated for 3+ years. Single-brooded; occasionally a second brood (southern Africa).

NEST: small cup, not as deep as nest of Cape Batis *B. capensis*, made of flakes of lichen and bark with fine grass and fibres, strengthened with spider web, lined with very fine grass, rootlets, wood fibres, thin petioles and hair, and finished outside with bits of lichen, bark and dry leaves so that, on a lichen-covered branch, nest is extremely well camouflaged (**A**). Some nests built entirely of lichen and spider web, with a few flakes of lichen lying loose in cup. Sited (n = 263) 0·45–12 m up open tree with only sparse foliage, av. *c.* 3 m up, in middle of large horizontal branch or in inclined or vertical fork. Thorny *Acacia* and *Ziziphus* tree often used, but bird also builds in wide variety of other trees; in Zaïre often uses down-hanging branch in small tree on side of *Macrotermes* termite mound (Vincent 1947). Site chosen by pair after investigating several places (process described by van Someren 1956). Pair first works spider web onto chosen site, then binds silky tendrils and bits of lichen on, using spider web; nest built by both birds but mainly ♀, taking 10–14 days. Materials are brought from trees other than nest tree. Bird alights on edge of nest making tiny croaking noises, and presses newly-brought material down with closed bill; then, with bill open, bird packs down edge of nest around its entire circumference; then bird enters nest and wriggles vigorously so as to shape it; and finally rolls its head and neck around the edges to smooth them. At one nest ♀ did 5

A

times as much nest building and forming as ♂ (Vincent 1935). Partly-built nest sometimes abandoned in favour of another site. Completed nest can remain empty for up to 10 days before eggs are laid. SIZE: ext. height 25–40 and diam. 55, int. diam. 40 and depth 20–25.

EGGS: 1–3 (nearly always 2; occasional clutches of 4 thought to be product of 2 ♀♀). Laid on consecutive days. Pale grey-green or green-blue, evenly freckled with dark brown or red-brown on underlying grey spots and blotches, forming irregular and ill-defined band around large end. SIZE: (n = 92, southern Africa) 15·4–19·0 × 11·9–14·2 (17·3 × 13·0).

INCUBATION: by ♀, sitting so tightly that she can sometimes be touched by person. ♂ once recorded relieving ♀ for short periods (van Someren 1956). ♀ sits low in nest, back lower than rim, only bill and tail above rim, pointing upward. ♀ fed on and off nest by ♂, who announces arrival with 'chik-chik' call from a distance. ♀ generally begs loudly and flutters wings, and may leave nest to receive food from ♂ (Harris and Arnott 1988). ♀ also leaves clutch to forage for herself. Period: *c.* 17 days.

LAYING DATES: E Africa: Region A, Apr–May; Region B, Jan, Oct; Region C, Oct–Jan, Mar, June (mainly Nov–Jan); Region D, Oct, Dec–Feb, Apr–June (mainly Oct, Dec) (breeds in Region C mainly in pre-rains and early rains). Further Kenya data (nestlings Feb, fledglings Feb, May, Aug, Nov). Zaïre, Rwenzori, Dec–Jan; Lubumbashi, Sept–Oct; Angola, Aug–Nov; Zambia, Aug–Nov, mainly Oct; Malaŵi, Oct–Nov; Mozambique, Mar–Apr; Zimbabwe, Aug–Feb (Aug 5%, Sept 30%, Oct 40%, Nov 20%, Dec 4%; n = 260); Botswana, Sept–Feb, May; Transvaal, Aug–Feb (83% of 98 clutches in Oct–Dec); Natal, Oct–Dec; E Cape, Sept–Oct.

DEVELOPMENT AND CARE OF YOUNG: eyes open at day 5 and are dark brown, but turn whitish at fledging time. At fledging resembles immature but duller, more mottled, feathers of forehead, crown, nape, mantle and back with small buffy tips, breast-band and throat patch indistinct. At first, young fed by ♀ who gives quiet 'whoit-whoit' call; later, young fed by ♂ and ♀, mainly with caterpillars. Parent eats faecal sac or carries it away. Nestling period 16–18 days. Fledglings depend on parents for about 8 weeks, but can remain with them until start of next breeding season. Mottled fledgling plumage retained for 2 months, when young bird becomes dull, pale-eyed version of adult. Young ♂ can give adult whistles at 4 months.

BREEDING SUCCESS/SURVIVAL: parasitized by Klaas's Cuckoo *Chrysococcyx klaas* in E and southern Africa. Longevity: a ♀ retrapped after nearly 5 years.

References
Harris, T. and Arnott, G. (1988).
Herremans, M. (1992).
van Someren, V. G. L. (1956).
Vincent, A. W. (1947).
Vincent, J. (1935).

Batis soror Reichenow. East Coast Batis; Mozambique Batis. Pririt pâle.

Plate 31 (Opp. p. 560)

Batis puella soror Reichenow, 1903. Vög. Afr., 2, p. 485; Quelimane.

Forms a superspecies with *B. senegalensis*, *B. orientalis*, *B. molitor* and *B. pririt*.

Range and Status. Endemic resident, woodland in SE African coastal lowlands; S Kenya to S Mozambique. Kenya, common from Lamu and lower Tana R. southward within 50 km of coast. Tanzania, common in miombo and other low-lying woodlands along entire coastline; Zanzibar and Mafia; inland to Amani, Morogoro, Selous Game Res., Songea and S shores of L. Malaŵi; extends up to 1100 m. Malaŵi, uncommon or only locally at all common: Namwera (14°22′S, 35°30′E); and SE Malaŵi in Zomba, Blantyre, Chiradzulu, Mulanje and Thyolo Districts. Zimbabwe, localized in extreme E: E-facing slopes in Pungwe, Honde and Nyamkwarara valleys, Banti, along Musapa R., Lusitu-Haroni area, Chimanimani Mts, and confluence of Sabi R. and Lundi R.; ascends to 750 m in Nyamkwarara, 1500 m in Chimanimanis, and 1550 m at Banti and above Musapa and Haroni valleys. Mozambique, common and widespread, but absent from Tete Province, NW Manica e Sofala Province, Gaza Province, and W Inhambane Province (west of about Covane, Mabate, Funhaloura, Panda, and Va de João Belo).

Description. ADULT ♂: forehead white near nostrils, blue-grey in centre; crown and nape blue-grey, darker and slightly olivaceous on nape; from side of forehead a narrow white superciliary stripe extends above black mask, almost to hindneck (supraloral spot and eye-stripe sometimes vestigial). Small whitish patch on hindneck. Mask (lores, above and below eye, ear-coverts, sides of hindneck) glossy black. Mantle, scapulars and back blue-grey, faintly dappled whitish; rump and uppertail-coverts grey. Tail black, T6 with broad white edge to outer web and white tip to inner web. Chin, throat and side of neck below ear-coverts white; breast glossy black in well-defined band 6–14 mm wide in midline, av. 9·8 (n = 46); belly, flanks, vent and undertail-coverts white, thighs grey. Wings black, with long white stripe in folded wing formed by broad white ends to median and inner greater coverts and broad white edges to inner secondaries. Underside of flight feathers and tail shiny dark grey; underwing-coverts and axillaries white. Bill black; palate black; eyes bright lemon-yellow; legs and feet black. ADULT ♀: like ♂ but a diffuse pale rufous spot on chin and upper throat, and breast-band rufous, slightly narrower in midline than in ♂: width 6–13 (8·8) mm (n = 56). SIZE: a cline of increasing wing and tail lengths from N to S (Irwin 1962). E Africa: wing, ♂ (n = 5) 52–59 (55·4), ♀ (n = 9) 51–56 (53·5); tail, ♂ (n = 5) 36–37 (36·4), ♀ (n = 9) 34–37 (35·8). S Mozambique: wing, ♂ (n = 15) 55–58 (56·2), ♀ (n = 20) 54–57 (54·9); tail, ♂ (n = 15) 36–40 (38·5); ♀ (n = 20) 37–43 (39·1); bill, ♂♀ (n = ?) 12–14; tarsus, ♂♀ (n = ?) 17·5–19. WEIGHT: (Zimbabwe, Mozambique) ♂ (n = 36) 8·8–11·0 (9·6), ♀ (n = 38) 8·0–13·1 (9·5); (Tanzania) 1 ♂ 10·5, ♀ (n = 4) 10–10·5 (10·3).

IMMATURE: ♂ like adult ♀ but breast-band can look tawnier when black feathers appear after first complete body moult; ♀ like adult ♀ but white nape-patch, white sides of neck and supraloral spot and white eye-stripe are suffused with rufous.

NESTLING: not described.

TAXONOMIC NOTE: formerly treated as a subspecies of *B. molitor*; taxonomic status reviewed by Lamm (1953), Rand (1953), Irwin (1962), Lawson (1987) and others, with widely different conclusions. Recent opinion is unanimous in regarding *soror* and *molitor* as separate species.

Field Characters. A small version of Chinspot Batis *B. molitor*, with which it is parapatric, overlapping only at a few places in extreme E Zimbabwe. Length only 11 cm – smallest batis in its range. Both sexes have narrower and less clearly defined breast-band than Chinspot, ♂ has white dappling on back and sometimes black spots on flanks, ♀ has pale tawny rather than chestnut breast-band and throat-patch, and latter is diffuse, not clear-cut. Songs are very different; East Coast has a monotonous, repeated, rather dry 'tlop-tlop-tlop . . .', without any of the tuneful or ringing quality of Chinspot. Sympatric with medium-sized Black-headed Batis *B. minor* (blackish head and upperparts, ♀ with chestnut breast-band, no throat-patch) and Woodward's Batis *B. fratrum* (diffuse tawny underparts in both sexes, ♀ with tawny in wing), and with much larger Cape Batis *B. capensis* and Forest Batis *B. mixta* (♂ with no white supercilium, very broad breast-band, ♀ with much tawny on wing and underparts).

Voice. Tape-recorded (88, 91, F). In Malaŵi, ♂ song is 'p'lk' repeated at constant rate for at least 31 s. Rate varies between songs (i.e. between individuals?), from 6 to 9 'p'lk's in 5 s. Sings also in short sequences, characteristically of 6 notes descending the scale (Benson and Benson 1977). ♂ song in Zimbabwe described as deliberate piping of 4–12 notes, 'tlop-tlop-tlop . . .' or 'hu-hu-hu . . .', on same pitch except that first note slightly higher. ♀ joins ♂ in duet, giving strident 'wik' notes. On Pungwe R. (Zimbabwe) ♂ song is a slow, mournful 'pook, pook, pook' (C. J. Vernon *in* Ginn *et al.* 1989). Alarm and threat: repeated 'wik' notes, with bill-snapping and wing-fripping; threat also signalled by 'gruuk-gruuk' calls with wing-fripping in 'jinksing' flight (Harris and Arnott 1988). Courtship voices of ♂: a strident 'weeek-weeek-weeek', a muted 'prrrup-prrrup', a deliberate slow froglike croaking 'grrruk-grrruk' and a call 'klop-klop', and of ♀ an excited, repeated 'wikwik', and (begging) a chattering 'kisshkissh'. Contact call a soft 'wik'. In Tanzania makes insect-like buzzing and noise like 'windy squeak sometimes made when bicycle is pumped up' (Sclater and Moreau 1933).

General Habits. Inhabits lowland *Brachystegia* woodland (miombo), and commonest below 500 m where such woodland is little disturbed and forms mosaics with thickets and forest. Inland, along Mozambique/Zimbabwe border, occurs in hot valleys and ascends to 1550 m, in tall dense stands of *Uapaca kirkiana* trees in high-rainfall areas, also in *Philippia* thickets and mountain acacia *Brachystegia glaucescens* woodland (Beasley 1995).

Lives in pairs, but in months before nesting commences occurs also in mainly single-sex gatherings or 'parliaments'. Active and alert; arboreal. Forages by flycatching: makes short sallies to glean an insect, barely checking in flight, from leaves or branch or trunk, with audible snap of bill. Feeds mainly in mid-canopy of small-leaved trees. Hops along branches. Flicks tail before flying, after landing, and whenever it moves at perch. Flights between trees, generally of only 5–15 m, are undulating, even bouncing, with a few half-second bouts of wing-fripping. In 'parliaments', birds fly about excitedly, calling, fripping wings and snapping bills (Harris and Arnott 1988). In 'winter' pair frequently joins mixed-species foraging party. At night roosts low down in dense bushes, hunched up, plumage fluffed out, head drawn in.

Exactly parapatric with Pygmy Batis *B. perkeo* in SE Kenya (Lewis and Pomeroy 1989). Parapatric with Chinspot Batis, overlapping it within 10 km of Sabi/Lundi confluence, Zimbabwe (Irwin 1962), and – probably with altitudinal separation – from Chimanimanis to Nyamkwarara (Irwin 1981). Broadly sympatric with Woodwards' Batis *B. fratrum* in Mozambique; and in certain regions overlaps Cape Batis *B. capensis* and Forest Batis *B. mixta*.

Food. Insects.

Breeding Habits. Evidently monogamous, territorial, solitary breeder. 2 ♀♀ once seen apparently competing for a ♂. Pair tolerates a third bird near nest, 'probably a first-year bird from the previous season' (Harris and Arnott 1988). Pair-territory probably < 3 ha. Both sexes chase intruders out of territory. Courting ♂ flies in rapid zigzag flights with tail fanned, white-spotted grey rump-feathers fluffed out, loud wing-frips, 'weeek-weeek-weeek' call and other calls (see above) (Harris and Arnott 1988).

NEST: cup made of plant fibres and bark, bound with spider web and decorated outside with bits of lichen. Placed on fork, 4–15 (av. 8) m (n = 4) high in tree. One nest, exposed to weather in open tree, survived whole year and was then used again (Amani, Tanzania).

EGGS: 2 (7 clutches, Malaŵi, Zimbabwe), once 3 (E Africa). Eggs not described.

INCUBATION: by ♀ alone; ♀ returning to nest is sometimes escorted to it by ♂ (Harris and Arnott 1988). Period unknown.

LAYING DATES: E Africa: Region D, Jan; Region E, Jan, May; Malaŵi, Sept–Oct; Zimbabwe, Oct–Nov; Mozambique (Vumba), Nov.

DEVELOPMENT AND CARE OF YOUNG: at first, young fed by ♀ with food supplied by ♂; later, young attended by both parents.

References
Harris, T. and Arnott, G. (1988).
Irwin, M. P. S. (1962).

Batis pririt (Vieillot). Pririt Batis. Pririt de Vieillot.

Muscicapa pririt Vieillot, 1818. Nouv. Dict. d'Hist. Nat., 21, p. 486, Somerset East, Cape Province.

Plate 31
(Opp. p. 560)

Forms a superspecies with *B. soror, B. molitor, B. orientalis* and *B. senegalensis*.

Range and Status. Endemic resident, SW Africa. Widespread and locally common in acacia savannas and watercourse scrub, from Angola coastal lowlands at Benguela (common in parts of Iona Nat. Park), through all Namibia except NE and the most-barren coastal desert, north in Botswana to about 20°S and east to 25°E, and South Africa. Occupies most of western Cape Province south to Olifants R. (about Birdfield), upper Bree R. (about Robertson) and to near Uitenhage; also W Orange Free State and extreme SW Transvaal. Common in Namibia. Uncommon and local in SW Cape, where confined to fringes of karoo and Namaqualand. Uncommon resident in Orange Free State, east to 28°E; in Transvaal occurs in Kalahari thornveld northeast to Schweizer Reneke and Leeudoringstadt. Once in W Lesotho (P. Osborne and B. Tigar, pers. comm.).

Description. *B. p. pririt* (Vieillot): Cape Province south of Orange R., Griqualand West, Orange Free State and Transvaal. ADULT ♂: forehead, crown and nape dark grey, bordered by distinct but narrow, long, white superciliary stripe; hindneck, mantle, scapulars, back, rump and upper-tail-coverts grey. Tail black, T6 with white outer web and tip, T2–T5 with narrow white crescents at outer corners. Lores, 1–2 mm area above and below eye, cheeks, ear-coverts, and sides of neck behind ear-coverts, black, forming a shrikelike mask, extending back to form an almost complete collar. Chin and throat white, in patch extending backwards across lower sides of neck; breast black, belly white, flanks white with indistinct dark grey flecks or bars, thighs grey, vent and undertail-coverts white. Primaries and secondaries black, narrowly edged white, tertials black, broadly edged white, greater and median primary-coverts and alula black, greater coverts black except for 2 near middle that are broadly white-fringed, median coverts white, lesser coverts black, underside of flight feathers dark grey, underwing-coverts and axillaries white. Bill black; inside of mouth black; eyes pale yellow; legs and feet black. ADULT ♀: upperparts like ♂, but crown with olivaceous tinge and white eyebrow less distinct; chin, throat, lower sides of neck, and breast, light orange, shading to white on belly and greyish white on flanks. SIZE: (10 ♂♂, 10♀♀) wing, ♂ 56–58·5 (56·7), ♀ 55–57 (55·8); tail, ♂ 44–46 (44·7), ♀ 42·5–46 (44·1); bill, ♂ 14·0–15·5 (14·9), ♀ 14·0–15·0 (14·5). WEIGHT: (Namibia) ♂ (n = 3) 9–10 (9·7), 1 ♀ 9·7, (South Africa) ♂ (n = 10) 8–10 (8·9), ♀ (n = 14) 8–14 (9·5).

IMMATURE: like adult ♀, but breast buffy orange, mottled with black, crown mottled with white, and eyes very pale yellow.

NESTLING: hatches naked and blind.

B. p. affinis (Wahlberg): Angola, Namibia, W Botswana, NW Cape Province east to Kuruman. Upperparts paler; black mask does not form a complete collar; secondaries broadly edged white. Slightly larger: (10 ♂♂, 10 ♀♀) wing, ♂ av. 57·4, ♀ av. 57·7, tail, ♂ av. 44·2, ♀ av. 45·0, bill, ♂ av. 15·6, ♀ av. 14·8.

Field Characters. The batis of the SW arid zone, and the only one in most of its range. Overlaps narrowly with Chinspot Batis *B. molitor* around periphery of range. ♂ ♂ are extremely similar, main difference being black flecks on flanks of Pririt (flanks of Chinspot pure white); ♀ Pririt has chin to breast uniform pale orange instead of sharply defined chestnut breast-band and throat-patch. Songs have similar ringing quality but different form, that of Pririt being lengthy repetition of a single note, slightly descending the scale, rather than the well-known 'three blind mice' of Chinspot. Range barely overlaps with Cape Batis *B. capensis* in S Cape but doubtful if they occur together since their habitats are so different; Cape Batis is larger, and both sexes of local race *capensis* have tawny on flanks and wings.

Voice. Tape-recorded (38, 88, 91, B, F, LUTG). ♂ song a long series of whistles, up to 51, '(w)eee't' or 'hee', repeated in nearly but not quite perfectly even tempo, somewhat crescendo to start with, then rising and falling irregularly and only just perceptibly. Duration of whistle 0·15 s. Delivered at rate of 30 whistles in 21 s; rate varies somewhat. Whistles start as 'hee', descent slowly and evenly, changing gradually to 'ho' then to a mournful 'huu'; at end of song the whistles have a slight rolling quality: 'trrop'. ♀ song, a series of about 10 strident notes, 'wik', given at same tempo as ♂'s song; they sing in duet, ♀'s 'wik' 0·1 s after ♂'s whistle, ♀'s song lasting 7 s and coinciding with middle or end of ♂'s song. Contact call by ♂ and ♀, 'chik', repeated a few times. Threat, mainly by ♂ at edge of territory, high-pitched whistles 'he', also, in display flight, 'preet-preet-preet' and a deep 'gruu-gruu' accompanied by wing-fripping; ♀ joins in with 'wik', chattering 'chikchik-chik', and bill-snapping (Harris and Arnott 1988).

General Habits. Inhabits savanna grassland with scattered acacia trees, open woodland, scrub, and in particular dense stands of tamarisk *Tamarix usneoides* growing along dry watercourses in desert, where also in groves of bushes near wells.

Arboreal, very active, inquisitive, foraging in mid- and low horizons of trees, but often using also the canopy of large acacias. Foraging techniques very similar to those of Chinspot Batis, q.v. (Fraser 1983); perches within canopy, looking around and upwards, and searches a section of canopy foliage systematically before moving on to another. Hovers before seizing prey, more than Chinspot does. Of 75 feeds, bird hawked in 33%, hawkgleaned in 60% and gleaned in 7%. Makes short flights within tree as it darts to an insect seen on a branch 1–2 m away or after one disturbed into flight; hops along branches, gleans insects amongst branches, twigs and leaves, and sometimes on ground. Prey caught with audible snap of bird's bill. Constantly flicks tail. Holds large prey down with one foot, like shrike (Harris and Arnott 1988).

Occurs singly or (generally) in pairs, also in family parties and in winter occasionally in 'parliaments'; in the last, up to 12 birds gather, of one sex or dominated by one sex, and fly about excitedly in small clump of trees, making chattering and dry chuntering noises and 'chikchik-chik' calls, fripping wings and snapping bills. Flight undulating, usually frips wings about twice per s in flights between trees. Joins mixed-species foraging flocks. Mobs some predators, e.g. Pearl-spotted Owlet *Glaucidium perlatum*. At all times of day and year vocal, singing, contact-calling, and with variety of other noises. Singing ♂ perches with body crouching and inclined, bill raised, neck stretched, throat-feathers expanded, and tail fanned.

Resident, with some evidence of movement into SW Cape in winter; in Transvaal an erratic visitor to Barberspan.

Pririt overlaps ranges of only 2 congeners: Cape Batis and Chinspot Batis. Pririt lives alongside Cape Batis only in South Africa, in Oudtshoorn District (where both are rare: Pocock 1963) and perhaps in W Cradock District (Collett 1982). Pririt and Chinspot are sympatric in 100-km-wide band in N Namibia east of 15°E (Komen 1987), in N Botswana between Tsau, Bushman Pits and L. Dow, in W Orange Free State, SW Transvaal and a small area of adjacent N Cape Province, and in Cradock District (S Cape).

Food. Insects in 2 stomachs; no qualitative information.

Breeding Habits. Solitary, strongly territorial, monogamous. ♂ advertises and defends large territory by singing from vantage points on side of tree; the same song is used all year but may serve in territorial defence only in breeding season; courtship behaviour unstudied:

A

evidently like other batises (Harris and Arnott 1988). ♂ courtship-feeds ♀, who assumes submissive, crouched posture and flutters wings. Threatens another bird by using steeply-undulating, bobbing or 'jinksing' flight with 'preet' and 'gruu' calling and wing-fripping; rump-feathers are fluffed out.

NEST: a small, neat cup made of plant fibres, fine grasses and rootlets, bound with spider web and cocoon silk; outside decorated with small pieces of bark, leaves or (2 out of 24 nests) lichen, which camouflage nest so effectively that it looks like a gall or knot. Ext. height 60 and diam. 30; int. diam. 40. Built mainly but not exclusively by ♀; sited (n = 60) 0·7–8·0 m high, usually 1·5–3 m, on slender horizontal branch of *Acacia* or *Tamarix* bush; in Namibia on dead branches or twigs, usually in shade, sometimes in the open.

EGGS: 1–4, usually 2 (n = 35). White, cream or greenish white, spotted and blotched with shades of brown overlain with spots and blotches of black-brown. SIZE: (n = 24) 15·7–18·1 × 12·0–13·3 (16·5 × 12·7).

INCUBATION: only by ♀, fed at nest by ♂. Period: (n = 1) 17 days.

LAYING DATES: Namibia, Sept–Mar, mainly Oct–Feb; Botswana, Aug–Dec; South Africa, Transvaal (fledglings late Jan); southern Africa (breeding season July–Mar, mainly Oct–Dec).

DEVELOPMENT AND CARE OF YOUNG: young fed by both parents. At one nest fully exposed to the sun (Namibia), ♀ shaded young by crouching over them with head and body feathers ruffled and wing-tips constantly quivering (**A**), interpreted as fanning (Nöller 1975). Period: (n = 1) 17 days. Fledglings dependent for about 6 weeks.

BREEDING SUCCESS/SURVIVAL: 7% of 76 Namibian clutches parasitized by Klaas's Cuckoo *Chrysococcyx klaas* (in one locality, inland of Walvis Bay, 8% of 48 clutches) (Jensen and Clinning 1974).

Reference
Harris, T. and Arnott, G. (1988).

Batis senegalensis (Linnaeus). Senegal Batis; Senegal Puff-back Flycatcher. Pririt du Sénégal.

Plate 32
(Opp. p. 561)

Muscicapa Senegalensis Linnaeus, 1766. Syst. Nat. (12th ed.), 1, p. 327; Senegal.

Forms a superspecies with *B. orientalis*, *B. molitor*, *B. soror* and *B. pririt*.

Range and Status. Endemic resident, W Africa from Mauritania to Cameroon. Mauritania, common in Senegal R. valley, north to Aftout es Saheli, resident south of 16°N. Senegal, common and widespread. Gambia, uncommon resident on lower river, rare on upper. Mali, quite common and widespread in Sahel and in better-wooded river valleys, north to 17°N. Niger, uncommon (< 30 records), east to 10°E and north to Agadès, Dabaga, Bagzans Mts and Timia (18°07′N) in Aïr. Guinea, known from Koundara and Gaoual districts and Mt Nimba, and in W from Mambia and Koba (Demey 1995). Sierra Leone, frequent in Kilimi district, records Waterloo, Bintumane, Sassa, Sefadu and elsewhere; probably widespread in N. Liberia, vagrant (Gatter 1988). Ivory Coast and Ghana, widespread and quite common north of forest; in south, records from Lamto, Abidjan and inselbergs on Accra Plain. Togo, Benin and Nigeria, widespread and common, south in Benin to near coast and in Nigeria to Lagos, Enugu, and once on coast east of Port Harcourt; commoner in guinean savannas than in drier woodlands further north; records (specimens) in NE Nigeria near L. Chad (Arege, Malamfatori). Cameroon, north of 9°N: Touroua, Gagadjé, Yagoua, Mokolo and Kosa, also 2 records in E, east of Meiyganga and near Koum, and 2 in S near Bafia and Bikoué (last 3 records possibly in error for *B. minor*, known from the same localities).

Thought to be decreasing in Gambia, due to loss of habitat. In Senegal, in dry sahelo-soudanian acacia

woodland at Bandia, near M'Bour, 1 pair every 500 m of line transect, but in dry soudanian woodland in Diambour, near Koumpenntoum, 1 pair every 400 m.

Description. ADULT ♂: forehead and crown blackish slate-grey bordered by broad white supercilium almost joining white patch of centre of nape and hindneck; lores, face, upper sides of neck and sides of nape deep glossy blue-black; chin, throat and lower sides of neck white. Mantle and upper back slate-grey with more or less pronounced brownish wash; lower back and rump mottled grey, white and black (feathers grey with wide subterminal white spot and narrow black tip). Uppertail-coverts glossy blue-black. Tail velvety black; T6 outer web and broad tip on inner web white, T5 with outer web narrowly edged white, T4 with faint narrow fringe on outer web white. Breast-band deep; glossy blue-black; flanks white mottled blackish grey (feathers white with blackish grey tip); thighs black faintly barred white; rest of underparts and undertail-coverts white. Primaries and secondaries brownish black, outer web narrowly edged greyish brown-buff; tertials velvet black, outer web and tip broadly edged white. Scapulars slate-grey tipped blackish; wing-edge white, lesser and most of inner greater upperwing-coverts velvet black, outer greater coverts brownish black narrowly edged greyish buff, median and innermost greater coverts black with broad tip white (white margin on wing-coverts increasing from the outside in, from about half the feather on outer to almost the whole feather on inner coverts). Axillaries and underwing-coverts velvety black, the latter edged white; underside of wing dull black with bronzy sheen, flight feathers fringed white on inner web. Bill black; eyes golden yellow; legs and feet black. ADULT ♀: forehead and crown sooty grey-brown bordered by broad buff to orange-ochre supercilium joining buff patch of centre of nape and hindneck; lores, face, upper sides of neck and sides of nape sooty black; chin, throat and lower sides of neck white, chin and upper throat suffused with pale orange-buff. Mantle and upper back greyish orange-brown mottled white (feathers grey with white spot in centre and greyish orange-brown tip), upper mantle suffused with orange-buff; lower back and rump mottled white, grey and black (feathers grey with wide subterminal spot and narrow black tip; feathers of lower rump with almost entirely white tip). Uppertail-coverts velvety black. Tail sooty black; T6 with outer web and broad tip on inner web white, T5 with outer web narrowly edged white, T4 with very faint narrow fringe on outer web white. Breast-band deep orange-buff, twice as wide on sides as in centre; flanks white mottled blackish grey (feathers white with blackish grey tip), with slight ochre wash; thighs black faintly barred white; rest of underparts and undertail-coverts white. Primaries and secondaries brownish black, outer web narrowly edged pale greyish brown-buff; tertials brownish black, outer webs and tips broadly edged pale buff. Scapulars grey-brown; wing-edge pale buff, browner narrowly edged pale greyish brown, innermost greater median coverts blackish brown with broad tip buff to buffy white (buff margin on wing-coverts increasing from the outside in, from about half the feather on outer coverts to almost the whole feather on inner coverts). Axillaries and underwing-coverts sooty black, the latter edged white; underside of wing dull blackish with bronzy sheen, wing-quills narrowly fringed white on inner web. Bill black; eyes golden yellow; legs and feet black. SIZE: (11 ♂♂, 7 ♀♀) wing, ♂ 56·5–61·5 (58·0), ♀ 54–58·5 (55·8); tail, ♂ 40–45·5 (42·3), ♀ 39–43 (40·9); bill, ♂ 15–16·5 (15·7), ♀ 15–17 (16·1); tarsus, ♂♀ (n = 36) 16–17. WEIGHT: (Senegal, May) ♂ (n = 4) 9–11 (9·9), 1 ♀ 9·5; (Nigeria) ♂ (n = 3) 9–9 (9·0), ♀ (n = 5) 8–11 (9·1); (Nigeria, Nov–June) ♂ (n = 9) 9·8–10·5 (10·0), ♀ (n = 9) 9–10 (9·3); (Ghana) ♂ (n = 9) 8·6–10·2 (9·3), ♀ (n = 13) 8–11·4 (9·2).

IMMATURE: like adult ♀ but browner; lesser and greater upperwing-coverts with sparse pale rusty buff tips; white of outer tail-feathers suffused with pale buff.

JUVENILE: like immature but head and upperwing-coverts more marked with pale rusty buff tips, white parts of head, back and wings buff; underparts white mottled grey-brown at breast.

NESTLING: hatches naked, skin dark purple-grey, with sparse down pinkish brown.

Field Characters. Length 10 cm. A small, arboreal 'flycatcher' of humid and dry savanna woodlands throughout W Africa. Yellow eye is conspicuous; rather short-tailed and spindly-legged. Readily identified: ♂ with black head, long, broad white eye-stripe, dark grey back, pied wings and tail, and white underparts with black breast-band; accompanying ♀ resembles ♂ except that eyebrow, hindneck collar, wing-stripe and breast-band are rufous. The only batis in most of its range. In Cameroon meets Grey-headed Batis *B. orientalis* a few km west of Logone R., in far north and possibly in E. Very like Grey-headed, which has upperparts darker and crown blackish. Meets Black-headed Batis *B. minor* near S Cameroon/Nigeria border and possibly in S-central Cameroon; ♂ and ♀ Black-headed told by much narrower white eye-stripe and less white in wing. Black-headed has very different, ringing song. Makes close approach to range of forest-dwelling Bioko Batis *B. poensis* (q.v.) on Mt Nimba and near coast from Ivory Coast to SW Nigeria.

Voice. Tape-recorded (CHA, ERA, MOR). Song brief and feeble, a long buzzy note preceded by 1 or 2 whistled ones, 'klop-zheet', 'cherp-chup-zheet'; sometimes reversed, 'peet-jit'; responds to playback with agitated 'zitit-YEEEER'. Territorial defence song, a measured scolding series of 3–8 clear 'tlut' notes (2·5 per s), the series repeated 5–10 times. Threat, to repel intruder from territory boundary, a harsh downslurred 'ptiuurrr' repeated in more or less long, measured series, or, to keep intruder away, a high-pitched upslurred scold, 'puuiiit' or 'peweeeet' given in rather regular rhythm. Territorial song a series of disyllabic notes, 'pew-pweeet'; territory advertising, 'pewt-tic-pew-pwee-pew-pweet' or less intensely 'pew-pwit', 'pew-pweet' or 'ju-veet'. Contact call, soft 'pui' or 'piew'; aggressive or irritated call, a chat-like 'tec-tec-tec-tec'; mobbing or harassing call, a harsh, buzzy, downslurred 'ptchyèèè', usually with rhythmic flaps of wings and claps of bill. Other calls include 'jwit-jwit-jwit', 'p(h)eerk', 'tsit-tsit-tsit', and a variety of dry, froglike notes; also makes mechanical noises – bill-snapping and wing-fripping in flight.

General Habits. Inhabits lowland dry thornscrub, grasslands with sparse trees and bushes, woody savannas rich in *Grewia*, *Ziziphus*, *Feretia*, *Combretum*, *Balanites*, *Gueira*, *Pilostigma* and other high bushes, open acacia and baobab woodland, dry woodlands with *Sterculia*, *Bombax* and *Pterocarpus*, orchard bush, smallholdings, farms, cassava fields, *Daniella* groves, and areas of bushes, small trees and spaced large trees with thick growth of grass 1–2 m tall.

Solitarily or, usually, in pairs or family parties. Pair, and to lesser extent family, keeps in close contact by sight and sound. Quite active; easily found by calls; mobile from sunrise to sunset but less so during hottest hours. A typical foliage-gleaner; searches outer parts of branches and sometimes around trunks and stems; entirely arboreal and not known to come to ground; forages mainly at 1–5 m above ground. Moves over entire territory each day but uses a core area and rotates over the rest in 3–5 days. Hops from twig to twig among branches in bushes and crown of trees, and makes sallies or jumps (usually horizontal or upward, 0·5–2 m from perch) to pick up prey under leaves or on bark, sometimes after a brief hover; works actively through foliage to dislodge prey that it catches in a sudden swoop. May behave like a true flycatcher when feeding on swarming ants and termites, but does not return to same perch. Dangerous prey items are rubbed or hit on perch and dismembered. Kills large caterpillar (over twice length of bill) by impaling it on acacia thorn, then tugging with bill (Senegal). Kills large, chitinous item at perch, picking it apart with jerky head movements and excited calls; when doing so sometimes pirated by Fork-tailed Drongo *Dicrurus adsimilis*. Makes short flights between trees; flight dipping; sometimes makes puffball of rump-feathers. Often joins mixed flocks, associating with Senegal Eremomela *Eremomela pusilla*, also with Northern Crombec *Sylvietta brachyura* or West African Penduline Tit *Anthoscopus parvulus*. During hottest hours of day, joins restless mixed flocks of Red-cheeked Cordon-bleus *Uraeginthus bengalus*, Grey-headed Sparrows *Passer griseus*, Melba Finches *Pytilia melba*, Scaly-fronted Weavers *Sporopipes frontalis* and others.

Anxious bird quivers tail, like Black Redstart *Phoenicurus ochruros*; when alarmed, crouches and swings body jerkily from side to side, with tail swinging amply. Harasses shrikes, particularly Brubru *Nilaus afer* (which from above resembles a large Senegal Batis) by snapping bill and wings; also mobs hornbills, cuckoos and Pearl-spotted Owlet *Glaucidium perlatum*. Facing the last, batis crouches, raises foreparts, inclines bill and flicks tail jerkily up and down. Vehemently harasses tree snakes, hovering and making rattling calls and bill-snapping. Spends night usually on isolated bush or small tree; roosting bird fluffs out plumage and tucks bill into scapulars; it then loses clear-cut black-and-white pattern and looks speckled.

Food. Small arthropods, mainly insects: beetles, cicadas (in Senegal, particularly *Soudaniella laticeps*), caterpillars, moths, young mantids, flies, small hymenopterans, grasshoppers, bugs, lacewings, termites; also spiders. Prey size, 5–55, mostly 20–30 mm. Regularly feeds on swarming ants and termites.

Breeding Habits. Monogamous; long pair-bond (in Senegal pair stayed together for 2 years). Territorial, all year. Solitary breeder. Defends territory of from 4 ha (open woodland) to 16 ha (dense acacia woods). ♂ patrols entire territory daily, giving advertising songs and often sitting on high exposed perch. When a rival ♂ is detected, ♂ gives spacing songs, moves towards territory boundary where it gives defence songs at perch or (more often) in bouncing flight with bill up, crown ruffled, rump-feathers fluffed out, tail half spread, and wings fripping noisily. Agonistic behaviour includes upright posture with bill up, breast and crown plumage fluffed out and wrists exposed; bird swings rear end, makes slow, jerky, sweeping movements of tail up and down and from side to side, the tail half spread and spasmodically opened and closed. In full threat, crouches with ruffled crown, exposed wrists, slightly hunched back, half-spread tail, and bill pointing at opponent. ♀ may approach and make some rattling flights and calls.

NEST: tiny, neat, compact cup of small pieces of dried grass and leaves, and small bits of lichen and bark, bound with spider web, well secured to support; ext. diam. 57, ext. depth 41; int. diam. 43 and depth 35. Placed in fork of shrub (once on bamboo stalk) 0·6–1·5 m above ground (Nigeria: Jourdain and Shuel 1935) or (n = 6) 1–4 m (av. 2·0) high in fork or on thick stem or branch of small tree or bush up to 6 m tall (Senegal). Nest is exposed but well camouflaged, resembling knob on branch (particularly in *Grewia* or *Ziziphus* tree fork, favourite nesting trees in Senegal). Built by both sexes. In Senegal one pair began nest-building less than 24 h after first shower of season.

EGGS: 2 (Senegal, n = 4 clutches); laid at one-day interval. Oval; glossy; pale greenish or blue-green, distinctly spotted with red-brown and violet almost all over, but toward larger end spots larger, more violet and thicker, forming a cap or ring on underlying small rusty or lilac dots and blotches. SIZE: (n = 2, Nigeria) 15·1 × 12·1, 15·2 × 11·8. Second clutches may be laid; 2 pairs began new nests 2 and 3 days after uncompleted first nest was destroyed (Senegal).

LAYING DATES: Senegal/Gambia, Mar, May–July; Mali, Feb–June; Niger, Apr, May; Ivory Coast, Jan–June; Ghana, May, July, Aug (♂♂ with testes enlarged Feb); Nigeria, Feb–June.

INCUBATION: by ♀; begins when clutch complete (once began with first egg). Period: 15 days. ♀ sits by night and during 80–90% of day. By day sits for long spells with only short breaks. During hottest hours, eggs shaded rather than covered. ♂ guards nest surroundings and feeds ♀ about once every 30 min, both in and out of nest. All other birds near nest are attacked. Harasses tree squirrel *Heliosciurus gambianus* but not ground squirrel *Xerus erythropus*.

DEVELOPMENT AND CARE OF YOUNG: young fed by both parents. Period: 18 days. Young brooded by ♀ for 60–78% of day time in first 2 days, and 38% of 6th day. She stays for up to 33 min, usually 7–12 min, leaving brood for av. 3–6 min on first 2 days and 14 min on 6th day. For first week ♂ provides nearly all food for young and for ♀; ♂ gives item to ♀ who passes it to young. Food delivered once every 9 min at nest with 2 young, once every 17 min at nest with 1 young. Parent repels other birds within 20 m of nest (once even a pair of Fork-tailed Drongos feeding nestlings 25 m away).

After leaving nest, young still fed by parents (period undetermined). Young stay on parental territory until next breeding cycle when young ♂♂ disperse or, if they stay on parental territory, they avoid parents, who repel them by day but allow them to sleep nearby at night, usually near end of thin leafy branch of small bush.

BREEDING SUCCESS/SURVIVAL: in Ivory Coast pairs produce av. 1·8 fledglings (Thiollay 1971); in Senegal, of 6 nests (4 pairs of birds) followed from start, only 4 received eggs and 2 fledged young.

Plate 32
(Opp. p. 561)

Batis orientalis (Heuglin). **Grey-headed Batis. Pririt à tête grise.**

Platystira orientalis Heuglin, 1871. Orn. Nordost Afr., 1, p. 449; Modat Valley, Eritrea.

Forms a superspecies with *B. senegalensis*, *B. molitor*, *B. soror* and *B. pririt*.

Range and Status. Endemic resident, frequent to common in wooded grasslands from L. Chad to Somalia. Nigeria, between Arege and Malamfatori, W shore of L. Chad (also 1 record from Mambilla Plateau, a netted bird, more likely to have been *B. minor*). Chad, rare in Ennedi, quite common in sahelian and soudanian zones in S. Cameroon, confined to NE: records from Logone Birni, Tchéré, Koum and Rei Bouba. Central African Republic, a few records in W, frequent in N in Manovo–Gounda–Saint Floris Nat. Park. Sudan, widespread except in SW and N. Ethiopia, almost throughout, mainly below 2100 m. Djibouti, frequent, commoner in Mabla Mts and Forêt du Day; max. of 9 seen in a day. Somalia, common north of 8°N and in SE, mainly in lower Juba R. and Shebelle R. valleys. Zaïre, 1–2 records in NE (Lawson 1987). Uganda, rare in NW and sparse at about 1200 m on Mt Moroto in NE. Kenya, sparse in N from 1200 m (Moyale) down to 400 m (L. Turkana); and 1 record on coast, near Dar Es Salam (Somali border).

Batis orientalis

Description. *B. o. bella* (Elliot) (including '*somaliensis*'): dry acacia woodland in E, central and S Ethiopia, from S Shoa to E Harrar; Djibouti, N Somalia, N Eritrea and N Kenya. ADULT ♂: forehead, crown and nape bluish grey; small white supraloral spot, confluent with moderately broad, diffuse-edged, white superciliary stripe that runs at side of crown backwards to hindneck; hindneck with small white patch; lores, area above eye, cheeks, ear-coverts and broad line down side of neck glossy blue-black; mantle and back dark grey, tinged glossy bluish; scapulars blackish; rump whitish, the feathers grey-tipped; uppertail-coverts black. Tail black, T6 with edge of outer web and tip of inner web white, T5 white-tipped. Chin and throat white; breast glossy black, forming well-demarcated band; belly white, flanks greyish, thighs grey, vent and undertail-coverts white. Wings black, except for line formed in closed wing by white median coverts and white outer webs of inner greater coverts and of inner secondaries and tertials; primaries and outer secondaries narrowly edged white. Underwing-coverts blackish, axillaries white. Bill black; eyes yellow; legs and feet black. ADULT ♀: like ♂ but breast-band russet. SIZE: (30♂♂, 22 ♀♀) wing, ♂ 56–60 (57·9), ♀ 55–60 (56·8); tail, ♂ 37–45·5 (40·4), ♀ 36·5–41·5 (40·0); bill, ♂ 14–17 (15·3), ♀ 14–15·5 (15·0); tarsus, ♂♀ 15–17·5. WEIGHT: (Kenya) 2 ♀♀ 9, 10; (Uganda) 1 ♂ 9.

IMMATURE: like adult ♀ but mantle with brown tinge and median wing-coverts tawny.

NESTLING: hatchling not known. Feathered nestling like adult but pattern not so clear-cut, crown and mantle speckled with white and breast-band blackish.

B. o. orientalis (Heuglin): very arid acacia scrub in N Ethiopia, from interior of Eritrea to N Shoa. Like *bella* but crown and mantle darker grey, and larger. SIZE: (2♂♂, 2♀♀) wing, ♂ 63·5, 64·5, ♀ 60·5, 61·5; tail, ♂ 48, 48·5, ♀ 45, 45; bill, 4 ♂♀ 16–16·5.

B. o. chadensis Alexander: extreme NE Nigeria and Zaïre, Chad, Central African Republic, Sudan except Red Sea Province, extreme W Ethiopia. ♂ like ♂ *bella* but crown and mantle darker grey: identical to ♂ *orientalis* although smaller. ♀ like ♀ *bella* but crown and mantle washed with brown, white superciliary stripe wider, flight feathers brown not black, and breast-band narrower. Same size as *bella*.

B. o. lynesi Grant and Mackworth-Praed: E Red Sea Province, Sudan, in subdesert steppe. ♂ same as ♂ *bella*; ♀ like ♀ *bella* but breast-band paler, with yellow tinge. Wing av. 1 mm shorter and tail 0·5 mm shorter than *bella*.

Field Characters. Length, ♂ 11 cm, ♀ 10 cm. A poorly-known batis, extremely similar in plumage to Black-headed Batis *B. minor* and not known to differ substantially from other batises behaviourally. Differs from Black-headed Batis in having grey crown and mantle and somewhat broader breast-band, which is darker in ♀. Distinguished from Pygmy Batis *B. perkeo*, which overlaps with it in S Ethiopia and occupies the same habitats (Benson 1946b), by having full-length rather than half-length white eye-stripe and (♀) darker breast-band and no yellowish wash on chin and throat.

Voice. Not tape-recorded. ♂ song described as 'weet, weet, weet, seeerr', the first 'weet' more emphatic than the next 2, and 'seeerr' less loud (Benson 1946b); alternatively 'a penetrating, metallic clicking . . . tink tink tink tink' (von Heuglin *in* Archer and Godman 1961); or 'bell-like, usually of about four syllables' (Mackworth-Praed and Grant 1973).

General Habits. Inhabits acacia woodland and subdesert steppe, from sea level to 1700 m in N Somalia and 2100 m in Ethiopia. Keeps to low-growing *Acacia* bushes and *Ziziphus* and other thorn trees, spending much time in shade in centre of tree, where it evidently forages. In pairs, or 3 together; very active, seldom still for more than a few s; forages in short sallying flights inside tree, gleaning vegetation. Sings mainly Dec–Apr (N Somalia). Pair always to be found in same patch of bushes.

Food. Insects; numerous ants in 2 stomachs (Mt Wagar, Somalia).

Breeding Habits. Solitary breeder, evidently monogamous and territorial.

NEST: a small, neat, perfectly circular cup, made of fine shreds of fibre and dead bark of twigs, bound with spider web, lined with fine grass, and with large strips of bark applied to exterior. Nest sited on central fork of thorn tree at height of about 2 m, and very well camouflaged.

EGGS: 2–3. Very pale blue, finely spotted with purple-brown in zone around large end, or flecked with black-brown on underlying lilac blotches at large pole. SIZE: (n = 4) av. 17·3 × 13.

LAYING DATES: Sudan, Apr; Ethiopia, July (and some evidence for Feb–Apr); N Somalia, June.

References
Archer, G. and Godman, E. M. (1961).
Lawson, W. J. (1987).

Batis minor Erlanger. Black-headed Batis. Pririt à joues noires.

Plate 32
(Opp. p. 561)

Batis orientalis minor Erlanger, 1901. Orn. Monatsb., p. 181; Salole, Juba River.

Range and Status. Endemic resident, tropics east of 10°E and north of 10°S. Range fragmented, and species is thought to be nearly absent from savanna lacunae as mapped, although perhaps more widespread in parts of E Central African Republic, W Ethiopia and N Angola. Main range in 3 parts: (a) northern tropics from Mambilla Plateau, Nigeria (1 record) and Cameroon to Ethiopia and from Khartoum, Sudan, to Burundi; (b) southern tropics from S Gabon and Cabinda to Kasai Oriental, Zaïre, and N Lunda Province, Angola; and (c) coastal lowlands of E Africa from S Somalia SE Kenya, inland to Tsavo West Nat. Park, and Tanzania inland to Moshi, Kilosa and Rufiji, south to Kilwa and Mikindani. Outlying populations in Darfur highlands and Gebel Elba (Sudan), N Eritrea, N Somalia, and evergreen forests of Cuanza Norte and central Malanje Provinces, Angola. Frequent to common throughout.

Description. *B. m. erlangeri* Neumann (including '*congoensis*', '*nyansae*' and '*batesi*'): range of species except E African coastal lowlands. ADULT ♂: forehead, crown and nape blackish grey, tinged glossy bluish; small white supraloral spot, confluent with moderately broad, diffuse-edged, white superciliary stripe that runs at side of crown backwards to hindneck; hindneck with small white patch; lores, area above eye, cheeks, ear-coverts and broad line down side of neck glossy blue-black; mantle and back dark grey, tinged glossy bluish; scapulars blackish; rump whitish, the feathers grey-tipped; uppertail-coverts black; tail black, T6 with edge of outer web and tip of inner web white, T5 white-tipped; chin and throat white; breast glossy black, forming well-demarcated band; belly white, flanks greyish, thighs grey, vent and undertail-coverts white. Wings black, except for line formed in closed wing by white median coverts and white outer vanes of inner greater coverts and of inner secondaries and tertials; primaries and outer secondaries narrowly edged white. Underwing-coverts blackish, axillaries white. Bill black; eyes yellow; legs and feet black. ADULT ♀: like ♂ but mantle blackish and breast-band russet-maroon. SIZE: (66 ♂♂, 49 ♀♀) wing, ♂ 57–67·5 (61·8), ♀ 57–64·5 (61·0); tail, ♂ 38–51 (44·2), ♀ 40–50 (44·1); bill, ♂ 14–17·5 (16·5), ♀ 15–17·5 (16·0); tarsus, ♂ (n = 20) 16–17, ♀ (n = 13) 15–16.

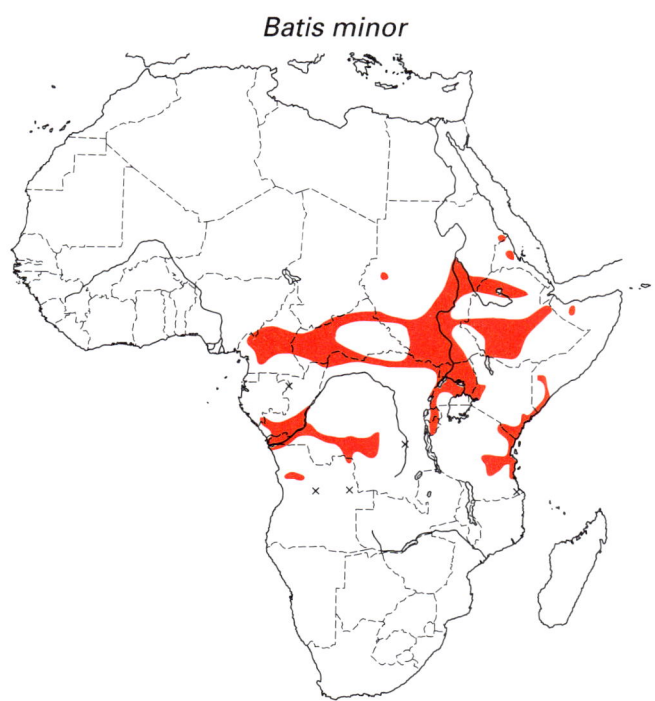

Batis minor

IMMATURE: like adult ♀, but buffy and blackish brown on upperparts where adult is white and black-grey respectively, each feather with buff tip; young birds have dusky cinnamon breast-band, later turning russet; immature ♂ retains russet breast-band for 6–7 months, then moults to black one.

NESTLING: hatches completely naked, without any natal down.

B. m. minor Erlanger: S Somalia. ♂ with glossier, greyer crown and less blackish mantle, ♀ with mantle less olive than in *erlangeri*. Smaller: (4 ♂♂, 5 ♀♀) wing, ♂ 54–56 (55·25), ♀ 52–53 (52·5); tail, ♂ 34·5–39 (36·25), ♀ 34–36·5 (35·5); bill, ♂ av. 15·2, ♀ av. 14·8.

B. m. suahelicus Neumann: Kenya and Tanzania. Like *minor*, but ♀ with narrower breast-band. SIZE: (24 ♂♂, 28 ♀♀) wing, ♂ 54–59 (56·75), ♀ 53–59 (56·4); tail, ♂ 36–41·5 (38·5), ♀ 36·5–43 (39·35); bill, ♂ av. 15·8, ♀ av. 15·4.

Field Characters. Widespread in woodlands in tropics. Distinguished from sympatric congeners by blackish crown and upperparts. White supercilium extends well behind eye; ♀ has maroon breast-band and white throat (no chin-patch). ♀ Chinspot *B. molitor* and East Coast Batises *B. soror* have chin-patch, and patch and breast-band of ♀ East Coast is paler, more tawny. Tiny Pygmy Batis *B. perkeo* has short supercilium ending above eye, ♀ has narrow pale tawny breast-band. Most resembles Grey-headed Batis *B. orientalis*, which has grey crown and mantle, somewhat broader breast-band which in ♀ is darker chestnut. Forest Batis *B. mixta* is larger, with very broad breast-band, supercilium lacking or vestigial, and different habitat.

Voice. Tape-recorded (B, BRU, CHA, ERA, GREG, McVIC, ROC, STJ). Song varies regionally, but consists of the ringing whistled notes characteristic of the genus. In Nigeria and Congo, 8 monotonous whistles, at rate of 2 per s, preceded by higher one; in Cameroon, similar monotonous whistles on same pitch but repeated 20 or more times, at rates of 2–2·5 per s. In Kenya, slower, longer and sweeter notes, usually in groups of 2 or 3, first note often lower and upslurred: 'sweet-sweet', 'dooey-sweet-sweet'; sometimes 3–5 'sweets' on same pitch. In Usambara Mts, Tanzania, described as clear, ringing 'thwi-thwee' or triple 'thwi-thwe-thwee', probably same song as in Kenya. Alarm, a buzzing 'zz-zz-zz'.

General Habits. Inhabits wooded grasslands including woodland edges, bushland, gardens, tall grass and second growth, from sea-level to 1600 m in E Africa; in Cameroon 'orchard-bush' savanna at 1100–1230 m; common in juniper woods (Yavello, Ethiopia). In pairs and small parties. Arboreal, not known ever to come to ground. Forages by making short flycatching sallies; in Tsavo, Kenya, 40 insects were captured one-by-one by bird in flight, bird returning to perch each time, and 5 taken at perch; 53% of them were gleaned from leaves, 18% from twigs, 24% airborne; 29% were taken deep inside vegetation, 53% just inside, 18% at edge; av. height 9·8 m (Lack 1985). In Chad, searches for minute insects in foliage, particularly of *Acacia raddiana*. Both sexes make loud fripping noises with wings in flight, which is bouncily undulating between trees. Flirts tail upwards with every change of position at perch. Snaps bill.

Food. Insects.

Breeding Habits. Solitary breeder, evidently monogamous and territorial.

NEST: small, neat cup of fine vegetable material and spider web, with flat pieces of dark bark fixed to exterior. Built in bush or small tree, on branch or fork, usually only 1 m above ground; sometimes sited on top of stump; once in fork of dead acacia sapling 50 cm high, fully exposed to sun (in this instance termites had coated entire sapling and outside of nest up to rim with thin crust of mud, but nestlings were not affected: Moreau and Moreau 1937). Ext. diam. *c.* 50 mm.

EGGS: 2. Pale greenish-grey or whitish, thickly covered with small dark grey or reddish spots with some dark brown blotches; spots form a wreath around large end. SIZE: (n = 3) 15·3–16·2 × 12·0–12·8 (15·7 × 12·5).

INCUBATION: by ♀, sitting tightly and allowing person to approach to within 1 m. Period unknown.

LAYING DATES: Cameroon (newly fledged, Apr); NE Zaïre, Mar–May; Ethiopia, Mar, May, Aug–Sept; E Africa: Region A, Mar; Region B, Apr–May; Region D, May; Region E, July, Oct; Tanzania (Mkomasi), June.

DEVELOPMENT AND CARE OF YOUNG: young fed and nest cleaned of droppings by both parents. Parents once brought food 12 times in 50 min. At one nest fully exposed to sun, ♀ stood motionless over young twice in an hour, for 7 min each time, shading them. At this nest the 2 half-grown young filled the nest, crouching flat, their upperparts, with irregular whitish blotches, concealing them effectively: 'they looked like scabby lichen or droppings on a grey branch' (Moreau and Moreau 1937). Dependent fledglings cared for by both sexes.

References
Chapin, J. P. (1953a).
Lawson, W. J. (1987).
Moreau, R. E. and Moreau, W. M. (1937).

Plate 32
(Opp. p. 561)

Batis perkeo Neumann. Pygmy Batis. Pririt pygmée.

Batis perkeo Neumann, 1907. J. Orn., p. 352; Darassam, southeast Ethiopia.

Range and Status. Endemic resident, dry E African savannas. Ethiopia, frequent to common in S. Sudan, frequent in dry acacia scrub in Natoporoputh Hills and Ilemi Triangle. Somalia, records in N at Eildab and Las Anod and to southeast, and common in bushland between 4°N and Equator. Kenya, fairly common below 1200 m in W and central highlands and adjacent plateaux, uncommon to rare around L. Turkana, absent from most of NE corner, and south of Equator present only in Kitui-Galana R.-Tsavo-Taita Hills area; uncommon in Tsavo Nat. Park. Uganda, in extreme E only, east of Mt Moroto and north of Mt Elgon. Tanzania, in extreme NE in lowlands east of Mt Kilimanjaro, with records south to Dar-es-Salaam.

Batis perkeo

Description. ADULT ♂: forehead, crown and nape bluish grey; white supraloral stripe, extending to above eye; small whitish patch on hindneck; mask (lores, above and below eye, ear-coverts, sides of hindneck) glossy black; mantle, scapulars, back, rump and uppertail-coverts grey, faintly dappled whitish. Tail black, T6 with broad white edge to outer web and white tip to inner web. Chin, throat and side of neck below ear-coverts white; breast glossy black in narrow, well-defined band; belly, flanks, vent and undertail-coverts white; thighs grey. Wings black, with long white stripe in folded wing formed by broad white ends to median and inner greater coverts and broad white edges to inner secondaries. Underside of flight feathers and tail shiny dark grey; underwing-coverts and axillaries white. Bill black; palate black; eyes bright lemon-yellow; legs and feet black. ADULT ♀: like ♂ but front of superciliary stripe with rusty tinge; hindneck patch small, with buffy tinge; suggestion of a buffy hindneck collar; throat white in midline, pale buff at sides; breast-band pale tawny-buff. SIZE: (25 ♂♂, 13 ♀♀) wing, ♂ 48·5–55 (52·0), ♀ 49–53·5 (50·7); tail, ♂ 30–37·5 (33·9), ♀ 30–35·5 (32·7); bill, ♂ 14–16·5 (15·3), ♀ 13·5–15·5 (14·5); tarsus, 1 ♂ 16·5, 1 ♀ 15. WEIGHT: (Kenya) ♂ (n = 7) 5–8 (6·2), ♀ (n = 3), 7–8 (7·7); (Uganda) 2 ♂♂ 8, 8, 1 ♀ 6.

IMMATURE AND NESTLING: not known.

TAXONOMIC NOTE: White (1963) considered that *perkeo* is nearest within the genus to *soror*. *B. perkeo* marginally overlaps the *soror* superspecies (*B. molitor*, in Kenya); it appears to overlap *B. minor* in 5 areas; we therefore treat it as an independent species, not a paraspecies.

Field Characters. Length 8–9 cm; the smallest E African batis. ♂ further distinguished by short white supercilium ending above eye from ♂♂ of Chinspot Batis *B. molitor*, East Coast Batis *B. soror* and Grey-headed Batis *B. orientalis*; from ♂ Black-headed Batis *B. minor* by crown colour; and from all by voice. ♀ has white chin and throat and narrow, tawny breast-band (♀ Chinspot and East Coast have rufous patch on chin and throat, ♀ Grey-headed and Black-headed have white throat and deep chestnut breast-band).

Voice. Tape-recorded (B, CAR, GREG, McVIC). Harsh churrs and weak peeps.

General Habits. Inhabits acacia trees and thorn scrub in arid and semi-arid regions mainly with 250–500 mm annual rainfall; in Tsavo Nat. Park wooded and bushy grassland, and woodland, but not riverine woods. Occurs in pairs. Strictly arboreal; flycatches from a perch of av. height 3·4 m, making short sallies to glean insects from leaves, twigs and the air, returning to perch each time. In Tsavo, of 49 insects caught, 55% were taken from leaves, 31% from twigs and 14% airborne; 50% were caught deep inside foliage, 21% just inside and 29% at edge (Lack 1985). Habits have not been reported upon in any additional detail; evidently social and other behaviours do not differ markedly from those of congeners. 1 bird once mist-netted with migrants at Ngulia, Kenya, in early Dec; but otherwise species appears sedentary.

Overlaps Chinspot Batis in 2 areas: S Kenya between Laitokitok, Voi, Galana R. at E border of Tsavo Nat. Park, and Galana R. east of Darajani; and zigzag corridor *c.* 100 km wide in W-central Kenya, centred on Kacheliba, Baringo, Barsalinga and Meru Game Res. Widely sympatric with Grey-headed Batis; parapatric with East Coast Batis in S Kenya; marginally overlaps range of Black-headed Batis in 5 areas: Ethiopia south of Batu Mts; S Ethiopia from lower Omo R. valley east to about 39°E; Juba R. valley, Somalia; S Kenya from Laitokitok in Tsavo R. and Galana R. valleys east to about 39°30′E; and NE Tanzania south to Dar-es-Salaam.

Food. Insects, including small beetles.

Breeding Habits. Breeding indications: Ethiopia, Feb; SE Kenya (soft egg in dead bird, late Mar). Nothing further known.

Batis minulla (Bocage). Angola Batis. Pririt de l'Angola.

Plate 32
(Opp. p. 561)

Platystira minulla Bocage, 1874. J. Lisboa 5, p. 37; Biballa.

Range and Status. Endemic resident, dry forests in W Angola, Congo and W Zaïre. Angola, occurs in Cabinda, Cuanza Norte, W Malanje, south along escarpment to N Moçâmedes and Capangombe, and in Quissama Nat. Park, Luanda. Congo, widespread in S; also in NW in Odzala Nat. Park (F. Dowsett-Lemaire, pers. comm.). Zaïre, widespread north of Congo R. near its mouth, including Shiloango R. valley; also in

Batis minulla

Kwango Province and W Kasai Occidental Province; probably much more widespread than in map.

Description. ADULT ♂: forehead glossy black with small white supraloral spot at side, crown bluish grey with thin white superciliary stripe at side, arising from the supraloral spot; mask (lores, cheeks, ear-coverts, and line above eye and below superciliary stripe) glossy black, reaching backwards to nape and running down side of neck as a black stripe; hindneck white; mantle grey; scapulars blackish; back, rump and uppertail-coverts blackish grey with blotched white spots. Tail short, black with white sides, outer webs of T5 and T6 having white edges. Chin and throat white, in patch extending back to black neck stripe; breast black, in band c. 10 mm deep, sharply demarcated; belly, flanks, vent and undertail-coverts white; thighs dark. Primaries black, secondaries black, narrowly edged on both webs with white, tertials white with proximal halves of inner webs black; greater and median primary-coverts black, greater ones narrowly edged white; alula black; greater and median coverts white-tipped, lesser coverts black. Underside of flight feathers and tail blackish grey; underwing-coverts white, marginals blackish; axillaries white. Bill black; eyes bright yellow or lemon-yellow; legs and feet black. ADULT ♀: like ♂ but breast-band chestnut, superciliary stripe extends to eye, white hindneck patch smaller and mantle has rusty tinge. SIZE: (17 ♂♂, 19 ♀♀) wing, ♂ 53–57·5 (55·1), ♀ 50·5–56 (54·0); tail, ♂ 35·5–39·5 (38·0), ♀ 36–40·5 (38·0); bill, ♂ 14–16 (15·25), ♀ 14–15 (14·6); tarsus, (5 ♂♀) 16–17.

IMMATURE AND NESTLING: unknown.

Field Characters. A tiny batis restricted to SW Congo basin and Angola, in forest patches in rich woodland bordering forest; length 10 cm. White supercilium, thin and inconspicuous. North of 10°S, extensively sympatric with Black-headed Batis *B. minor*, which is larger, with less white in wing, especially tertials, more white at tip of tail, a longer and more conspicuous white eye-stripe and smaller supraloral spot, and dusky (not white) underwing-coverts, and inhabits bushy grassland. Marginally sympatric with Chinspot Batis *B. molitor*, a considerably larger bird with more conspicuous white eye-stripe and small white corners to tail, ♀ with rufous chin spot, and savanna woodland habitat; probably does not overlap Verreaux's Batis *B. minima* (tiny; ♀ grey-breasted; forest canopy habitat).

Voice. Not known.

General Habits. Inhabits vestigial patches of forest, gallery forest and thick woodland, sometimes dry, below Angolan escarpment, along rivers and near southern borders of rain forest in SW Congo basin. Its habits have not been reported, but behaviour and voice are said to be like those of Chinspot Batis *B. molitor*.

Food. Insects.

Breeding Habits.
NEST: one was a small cup composed of soft bark and spider web, placed at head-height above ground on a fork in a small tree killed by fire; nest built by ♂ and ♀ (Chapin 1953a). Another, a small but deep cup strongly reinforced with spider web, 2 m high on a fork, also in a recently-burnt tree (Lippens and Wille 1976).
EGGS: 2 (only clutch found; eggs not described).
LAYING DATES: Zaïre (Kwango) July, (Msuata), (nest-building July).

Plate 32
(Opp. p. 561)

Batis minima (Verreaux). Verreaux's Batis. Pririt de Verreaux.

Platystira minima J. and E. Verreaux, 1855. Rev. Mag. Zool., Ser. 2, 7, p. 219; Gabon.

Forms a superspecies with *B. ituriensis*.

Range and Status. Endemic resident, lowland rain forest in Gabon and adjacently in S Cameroon. Rare. 3 collected in 'Gabon' (no localities given) in 1850s, and 4 at Bitye, Ja R., 3°10'N, 12°20'E in Cameroon in 1907, 1914 and 1924 (Erard and Colston 1988). Known from S Gabon ('found in pairs amid the bushes bordering the heavy forest': Chapin 1953a); and from N Gabon (Ogooué–Ivindo Prov. south to M'Bigou, also on coast between Libreville and Cocobeach). Density in Makokou forest 3 pairs per 2 km^2 in old plantations or only 1 pair per km^2 in younger growth; pair (or family party) requires home range of 18–20 ha.

Batis minima

Description. ADULT ♂: forehead and forecrown velvety black, supraloral spot white; lores, cheeks, ear-coverts and upper sides of neck velvety black; hindcrown blackish grey; narrow whitish superciliary line; hindneck and centre of nape white (seen from behind making a white V); mantle and upper back velvety black more or less mottled with dark slate-grey; lower back and upper rump mottled dark slate-grey, white and black (feathers grey with wide subterminal spot white and narrow black tip); lower rump white (feathers grey with white tip). Uppertail-coverts and tail velvety black; T6 has faint narrow white border on outer web. Chin and throat white, in patch extending back nearly to hindneck; breast glossy blue-black, in well-demarcated band 10 mm wide; flanks white mottled blackish grey (feathers blackish grey with white tip); thighs black faintly barred white; rest of underparts and undertail-coverts white. Primaries, secondaries and tertials velvet black, outer webs and tips of tertials broadly edged white. Axillaries and underwing-coverts white, the latter with base black; underside of wing dull black with bronzy sheen, flight feathers fringed white on inner web. Bill black; eyes golden yellow; legs and feet black. ADULT ♀: like ♂ but breast-band dark grey and only 8 mm wide, and blacks duller – forehead and sides of head matt black, not glossy; mantle less speckled with black; white supraloral spot smaller. One specimen has inner median coverts buffy. SIZE: 4 ♂♂, 4 ♀♀, 7 ♂♀) wing, ♂ 49–50 (49.5), ♀ 47.5–49 (48.0), ♂♀ av. 47.9; tail, ♂ 31–33 (32.1), ♀ 29–33 (31.75), ♂♀ av. 30.5; bill, ♂ 11.5–13.5 (12.1), ♀ 11.5–12.0 (11.75), ♂♀ av. 12.0; tarsus, ♂ 11.5–13.5 (12.5), ♀ 12–13 (12.5). WEIGHT: (Gabon, Mar, Aug) ♂♀ (n = 5) 8–12 (10.0).

IMMATURE: like adult ♀ but lower throat, lower breast and edges of median upperwing-coverts suffused with pale rusty buff, outer greater upperwing-coverts tipped buffy white; outer tail-feathers fringed buffy white, T6 with distal half and edge of basal half of outer web and tip of inner web buffy white, T5 with faint edge and broad tip on outer web and narrow edge along tip of inner web buffy white.

JUVENILE: nondescript; moulting specimen suggests white of patches on wings reduced and replaced by buff or ochre, fine whitish speckles on head, back and upperwing-coverts, breast-band indistinct if present.

NESTLING: unknown.

Field Characters. A rare batis of secondary forest canopy; a tiny, black-and-white 'flycatcher' that might draw attention to itself by its short upward sallies after insects underneath middle-stratum or canopy leaf-clusters. Upperparts black, with white spot above lore, narrow white eye-stripe, broad white wing stripe; tail short, black with white sides; underparts white with black (♂) or grey (♀) breast-band and black thighs. With good view, likely to be confused only with sympatric Bioko Batis *B. poensis*, ♀ of which has chestnut breast-band.

Voice. Tape-recorded (C, ERA). Territorial song a long series of thin, tuneless, short notes on one pitch: 'pti-pti-pti . . .', at rate of 2 per s, sometimes up to 3; resembles song of Bioko Batis but not introduced by buzz or trembling notes, 0.7–1.0 kHz lower-pitched, and usually slower (intervals between notes 0.5 not 0.35 s). Similarity complicated by the fact that they seem to copy one another when they meet. Contact calls a variety of soft, clucking, buzzing, downslurred or up-slurred notes emitted by both ♂ and ♀.

General Habits. Inhabits lowland forested areas mainly below 800 m, but not primary forest except within 500 m of edge. In NE Gabon favours secondary but well regrown forest with ill-defined vegetation layers, a dense low understorey and a dense but discontinuous canopy. Inhabits uncleared old cacao and coffee plantations where tree layers are beginning to disappear. Avoids open cultivated land even with scattered tall bushes and trees, avoids neighbourhood of villages and other man-made habitats, and may avoid forest 'saturated' with wattle-eyes, particularly White-spotted Wattle-eye *Dyaphorophyia tonsa*. The 2 species countersing in forest interior about 500 m from edge (Erard 1987).

In pairs or family parties. Active but unobtrusive; moves over almost entire territory daily. Pair maintains constant close contact, mostly by sight. Forages in foliage 5–40 m from ground, mainly at 14–24 m (57% of records); moves about in highest part of tree canopy but does not frequently change trees. Alert foraging postures shown in **A**. Favours small-leaved trees with foliage not too dense; much attracted by flowering trees.

A

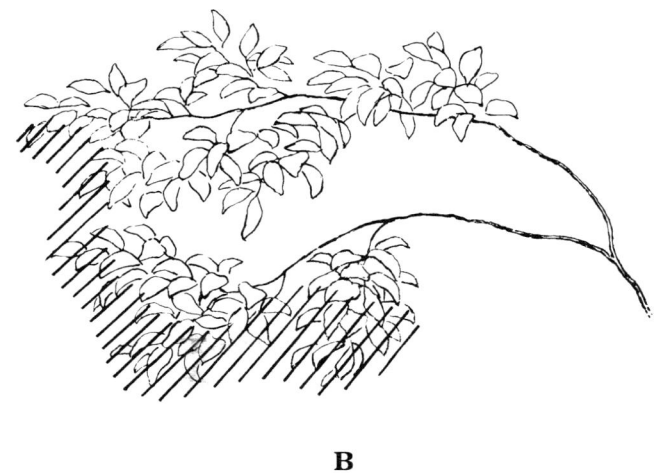

B

Typically hover-gleans; moves among branches, hops from twig to twig and makes short upward sallies or jumps (usually 0·5–2 m from perch) to snatch prey in flight or pick it from under leaves (**B**, feeding station hatched; compare with **B** in Bioko Batis), usually after a brief hover. Also makes short circular whirring flights and snaps up moving prey, and works actively through foliage to dislodge prey that is caught after a rapid swoop. May behave like a *Muscicapa* flycatcher when feeding on swarming ants and termites. Large, hard prey items are rubbed and hit on perch and dismembered before ingestion. Often joins mixed flocks, and associates with Yellow-fronted Penduline Tit *Anthoscopus flavifrons*, Lemon-bellied Crombec *Sylvietta denti* and Rufous-crowned Eremomela *Eremomela badiceps*, which it follows from underneath to snap up insects that they disturb. Regularly attacked and repelled by Bioko Batis; reacts to it by ruffling crown feathers and, in upright or crouched posture, moving body from side to side, with jerky tail- and sometimes also wing-flicks. Sunbathes during hottest hours of day with back feathers fluffed out and wings half spread and drooping.

Sedentary, at least in Gabon.

Food. Arthropods: mainly beetles, also hymenopterans and spiders. Field observation suggests most prey items are 5–15 mm long, including termites captured on the wing.

Breeding Habits. Monogamous, territorial. Defends territory, of 18–21 ha, all year (NE Gabon: Erard 1987). Pair patrols territory, the ♂ leading and irregularly giving short territorial song, particularly in morning and late afternoon.

NEST: unknown.

EGGS: unknown.

LAYING DATES: Gabon (probably Sept–Feb: young observed during rainy season following short dry season).

INCUBATION: unknown.

DEVELOPMENT AND CARE OF YOUNG: young stay with parents almost to next breeding season, though some solitary, probably dispersing immatures seen in July–Aug during the long dry season.

References
Brosset, A. and Erard, C. (1986).
Erard, C. (1975, 1987).
Erard, C. and Colston, P. R. (1988).

Plate 32
(Opp. p. 561)

Batis ituriensis **Chapin. Ituri Batis. Pririt de l'Ituri.**

Batis ituriensis Chapin, 1921. Amer. Mus. Novit., 7, p. 5; Gamangui, Nepoko R.

Forms a superspecies with *B. minima*.

Range and Status. Endemic resident, lowland forests of eastern Congo basin: Zaïre and (just) Uganda. Ranges from lower Uele R. (Buta, 2°48′N, 24°47′E) to Avakubi (1°24′N, 27°40′E) and Gamangui (2°10′N, 27°15′E), and from about Mongbwalu (2°N, 30°E) through Ituri to Irumu and Beni (0°28′N, 29°28′E), and in Kivu from Ibanga (2°47′S, 28°25′E), Kilobozi (3°03′S, 28°09′E) and nearby Kamituga to Kakanda (3°14′S, 28°20′E); also in Budongo Forest, W Uganda. Ranges altitudinally from about 900 m up to 1300 m. Rare; 22 examples were collected in S Kivu (Itombwe) in 1950–67 and bird described as rare there (Prigogine 1971); during a long sojourn at Avakubi Chapin (1953a) saw only one, and in 8 days' march from Irumu to Beni in 1926 he saw 3.

Description. ADULT ♂: forehead glossy bluish black with white supraloral spot; forecrown, sides of crown, lores, cheeks and ear-coverts, glossy bluish black, including wide area below lore and above eye, the black patch extending well backwards; hindcrown and nape dark bluish grey; hindneck dark grey; mantle glossy bluish black or dark grey; scapulars blackish; rump and uppertail-coverts dark grey, rump fluffy, with indistinct white spots. Tail black, T6 with white edge to outer vane and sometimes with narrow white tip, T5 all black or in some birds like T6, T2–T4 black or with narrow white tips. Chin and throat white, the white extending far back; breast glossy bluish black, in sharply demarcated band; belly, flanks, vent and undertail-coverts white, thighs velvety black. Primaries black, slightly glossed bluish; secondaries black, narrowly tipped white, tertials black; tertials and a few inner secondaries with broad white fringe to outer vane; greater and median primary coverts and alula black; greater coverts tipped

Batis ituriensis

white, outer median coverts white, lesser coverts black; underwing-coverts white, but black near leading edge of wing. Bill black; eyes yellow; legs and feet black. ADULT ♀: sexes very similar, but ♀ has thin whitish superciliary line, and one has T6 with whole outer web and tip of inner web white. SIZE: (8 ♂♂, 4 ♀♀) wing, ♂ 47–52 (49·1), ♀ 47–51·5 (49·6); tail, ♂ 29–32·5 (30·5), ♀ 27–33 (30·2); bill, ♂ 14·0–14·5 (14·1), ♀ 13·5–14·5 (14·0); tarsus, ♂ (n = 3) 13·5–14 (13·7), ♀ (n = 4) 12·5–14·5 (13·2).

IMMATURE: like adult ♀ but duller; sides of throat and neck and edges of median and greater upperwing-coverts buff to ochre; alula and greater primary coverts with tiny buff spot at tip; T6 with white edge on outer web broader. Juvenile has head and upperparts dark greyish, densely spotted with creamy white on head; ear-coverts and sides of nape blackish; underparts white, sides of breast dusky greyish; tail and wings blackish; edges of outer webs of inner secondaries, median and inner greater upperwing-coverts creamy buff.

NESTLING: one bird (late nestling or recently fledged: Gyldenstolpe 1924) had feathers of whole head greyish, each with triangular cream spot at tip; remaining upperparts similar although spots sparser; sides of neck blackish, unspotted; underparts white, except for grey sides of breast; creamy buff in wing where adult white. Iris dark.

TAXONOMIC NOTE: formerly treated as conspecific with *Batis minima*, partly because of diminutive size of *ituriensis*. Resembles *B. diops* in plumage, and in andromorphic ♀.

Regarded as an allospecies of *B. poensis* by Dowsett and Forbes-Watson (1993); but Lawson's (1986) analyses show *ituriensis* to be very closely allied only to *minima*.

Field Characters. The only batis in E Zaïre lowland forest. Upperparts black, with white spot at side of forehead, white wing stripe, and white tail-edge; eye yellow; underparts white with black breast-band in both sexes and black thighs. Separated from Rwenzori Batis *B. diops* by smaller size, black not grey upperparts and less pronounced supraloral white spots.

Voice. No tape-recording known though F. Dowsett-Lemaire (pers. comm. to C. Erard) while tape-recording 'Jameson's Wattle-eye' *Dyaphorophyia blissetti jamesoni* in Uganda recorded a bird the song of which is very reminiscent of that of *B. poensis*: some rhythmic vibrated notes followed by whistled ones, e.g. 'pirrit-pirrit-pirrit pti pti pti . . .' (last notes *c.* 4/sec.); the bird remains unidentified but could have been Ituri Batis.

General Habits. Inhabits degraded lowland forest, second growth and cultivated areas with scattered high trees; does not enter primary forest. Up to 1300 m (Itombwe). In pairs or family groups. Forages in tree tops but sometimes descends to 5–6 m above ground; hops from twig to twig to glean foliage. Frequently joins mixed species flocks, particularly where primary and secondary forests meet. 'They were always members of an "upper-level" bird party which kept well up in the higher boughs and might include *Scoptelus castaneiceps, Mesopicos xantholophus, Apalis nigriceps, Hyliota australis, Parus funereus, Anthreptes tephrolaema, Malimbus erythrogaster*, and a variety of other birds' (Chapin 1953a). Nothing known further.

Food. Insects including termites.

Breeding Habits. ♂ seen to feed ♀ (Uganda). Nest and eggs unknown.

LAYING DATES: Zaïre, Gamangui, ('breeding' Feb), Wambutti, Ituri (nestling, June), Itombwe (enlarged gonads Mar, Apr, Aug); Uganda (♂ fed ♀, Mar).

References
Ash, J. S. *et al.* (1991).
Chapin, J. P. (1953a).

Batis poensis Alexander. Bioko Batis; Lawson's Batis. Pririt de Lawson.

Plate 32
(Opp. p. 561)

Batis poensis Alexander, 1903. Bull. Br. Orn. Club, 13, p. 34; Bakaki, Fernando Po.

Range and Status. Endemic resident, forested areas of W Africa from Liberia to Bioko and Gabon. Not uncommon on Mt Nimba (Liberia/Ivory Coast) and Bioko (e.g. San Carlos), widespread and frequent in N Gabon, elsewhere rare. Liberia, Mt Nimba only (presumably occurs on Guinea flank of Mt Nimba too);

Batis poensis

Ivory Coast, known from Taï and Yapo forests, Mt Nimba, Kpapekou, Soubré, Gagnoa and Lamto; Ghana, old records at Sekondi and Aburi, recent sight record at Jukwa; Benin, Begon; Nigeria, Iju (Lagos), Gambari, Akure, Ondo, and sight records Bashu, and Gashaka and Waddi in Gashaka–Gumti Game Res. (Green 1990). Bioko; Cameroon, Mt Oku, Kribi to Ebolowa and Dja R. and to south (early records from Bitye R. are referable to *B. minima*: Erard and Colston 1988); Gabon, widespread in Woleu N'Tem and Ivindo Basin; Congo, Odzala Nat. Park, and song heard at Goumina, Kouilou Basin, could only be this batis (Dowsett-Lemaire and Dowsett 1991). Density of about 32 pairs per 10 km² of old secondary growth around primary rain forest; home range (n = 3 pairs) av. 27 ha (Erard 1987).

Description. *B. p. occulta* Lawson: Liberia to Cameroon. ADULT ♂: forehead glossy bluish black with white supraloral spot at side; crown, lores, cheeks, ear-coverts and sides of neck glossy bluish black, with thin white superciliary stripe and square white patch on nape and hindneck; mantle to rump slate-grey, back and rump mottled with black and with partly concealed white spots; uppertail-coverts black. Tail short, black, with white sides (outer webs of T5 and T6 with white edges). Chin and throat white, breast glossy bluish black, 11 mm deep, sharply demarcated; belly, flanks, vent and undertail-coverts white; sides (beneath folded wings) with dusky marks; thighs black. Scapulars and wings black, except for long white stripe in folded wing formed by white tips to median coverts and broad white outer edges to inner greater coverts and broad white edges to tertials and inner secondaries. Underside of flight feathers and tail black, underwing-coverts black tipped white, marginal ones black, axillaries white. Bill black; eyes lemon-yellow or golden yellow, legs and feet black. ADULT ♀: like ♂ but crown duller, not so glossy black, mantle with slight brownish wash, rump with more white speckles, and breast-band dark vinous chestnut. SIZE: (9 ♂♂, 4 ♀♀) wing, ♂ 47–52 (49·8), ♀ 48–51 (49·5); tail, ♂ 27–31·5 (30·1), ♀ 28·5–31 (29·8); bill, ♂ 13·5–15·5 (14·3), ♀ 13–15 (14·5); tarsus, (7 ♂♀) 13–14. WEIGHT: (Mt Nimba, Liberia) 2 ♂♂ 8·8, 9·4, 2 ♀♀ 8·7, 9·0, 1 imm. ♂ 9·2 (av. of the 5, 9·0).

IMMATURE: like adult, but sides of throat pale rufous, and white line in closed wing rufous-tinged.

NESTLING: not described.

B. p. poensis Alexander. Bioko, ♂ differs from ♂ *occulta* as follows: supraloral spot small; no white superciliary streak; mantle blackish; white edges of tertials and outer tail-feathers narrower, breast-band broader. ♀ like ♀ *occulta* but hind border of chestnut breast not well-defined – upper belly and forepart of flanks suffused with chestnut. Immature like adult ♀ but back washed olive-ochre, white marks of wing washed ochre, dots at tip of primary upperwing-coverts creamy buff, breast-band paler, more orange-chestnut. Larger. SIZE: (9 ♂♂, 4 ♀♀) wing, ♂ 50–57 (53·4), ♀ 51·5–54 (52·3); tail, ♂ 29–32 (30·3), ♀ 29–33 (31·1); bill, ♂ 15–17 (15·7), ♀ 14–15·5 (15·0); tarsus, ♂ 14–15 (14·6), ♀ 14–15 (14·3). WEIGHT: (Bioko) ♂♀ (n = 11) 9–10 (9·6).

TAXONOMIC NOTE: Lawson (1984, 1986) argued that *poensis* and *occulta* (*'occultus'*) should be treated as 2 different species. But Erard and Colston (1988), in correcting the identity of S Cameroon (Bitye R.) batises, show that differences between *poensis* and *occulta* are not entirely clear-cut. As they remark, further study is required, particularly with regard to vocalizations and mating systems.

Field Characters. A forest batis that forages in canopy of tallest trees and is therefore seen mainly from below. Underparts white, breast-band black in ♂, chestnut in ♀. Head black with yellow eye and square white patch on hindneck; wing black with long white stripe when wing folded; tail short, black with white sides; remaining upperparts grey. The only batis in its limited range, except in Gabon; there it overlaps with Verreaux's Batis *B. minima*, which is very similar in plumage but smaller, with more conspicuous supraloral spot and less conspicuous white hindneck patch; ♀ has grey, not chestnut, breast-band. Verreaux's Batis feeds lower down in forest but when they are together in tree crown, this species forages inside foliage (**B**) and Verreaux's at periphery. Range may prove to meet ranges of Black-headed Batis *B. minor* in SW Cameroon and Senegal Batis *B. senegalensis* in secondary forest edges on Mt Nimba and near coast from Ivory Coast to Nigeria; for distinctions, see those species. Might be confused with Rufous-crowned Eremomela *Eremomela badiceps* but always in pairs or small family parties, not in groups, not so fast-moving in foliage, has more upright stance, bigger head, longer tail, yellow eyes, stout bill, black and white plumage with wide breast-band, and lacks rufous cap.

Voice. Tape-recorded (CHA, COR, ERA). Song, typically a dry, grating trill followed by one or more plaintive whistles: 'jurrrrweep' or a more rolling 'juririweep'; 'churr-weee-pitpit' or just 'weee-pitpit'; churr followed by several 'weeps' – 'juririri-weep-weep-weep-weep'. Sometimes repeats single 'jirrreeet' on one pitch over 40 times at rate of 2·5–3 per s, and may also repeat single 'peet' in series.

General Habits. Inhabits overgrown cultivation in rain forest, where vegetation is not uniform or dense,

and large clearings with plantations of, e.g. cacao or cassava, and with scattered tall trees (often flat-topped) and shrubs around villages. Much less frequent along primary forest edges and in old second growth. In W, tall trees in gallery forest and sometimes in degraded forest; but in NE Gabon and S Cameroon it is a treetop bird of man-made, managed habitats, particularly roadsides in farms around villages. Occurs in dense leafy canopy of very tall trees; forages mainly at 35–40 m above ground, only rarely moving below 25–30 m (once down to 12 m) and uncommonly above 42 m (once, 45 m). Foraging horizon is amongst the highest of all 'flycatchers', and similar to that of Black-and-white Flycatcher *Bias musicus*. Lowland, up to 1350 m in E Nigeria and possibly 2500 m in Cameroon. On Bioko inhabits woodlands up to lower limit of montane forest, keeping to tree tops. Where it overlaps with Verreaux's Batis, Bioko Batis occurs inside foliage near tops of tallest trees and Verreaux's in small trees in middle strata.

In pairs or family groups. Strongly territorial but very mobile within territory; 2 birds of pair keep constantly in contact, moving about in basal and middle thirds of canopy and often changing trees, covering entire territory every day. Flight not rapid, rather hesitant, wingflaps audible at close range; but rapidly crosses open spaces of 200–300 m. A foliage gleaner; moves among branches, works actively through foliage to dislodge prey, hops from twig to twig, makes short inclined sallies or jumps to snatch prey in flight or to pick at it under leaves, mainly after a brief hover or swoop; also makes short irregular zigzagging, whirring flights and snaps up dislodged prey. Prey usually taken 0·5–3 m from perch, but from up to 4–5 m away; large, hard insects are rubbed and hit on perch and dismembered. Perches with bill inclined, seeming to look upwards (**A**), and a bird feeds from a horizontal branch by flying up to leaves above it, returning to different places (**B**, feeding station hatched). Joins mixed-species foraging parties, when prone to follow Rufous-crowned Eremomela *Eremomela badiceps*, catching flushed insects on the wing. Sedentary.

Food. Insects 5–20 mm long, including alate ants, small beetles, moths, caterpillars (one 35 mm long) and cicadas.

Breeding Habits. Monogamous, territorial. Defends territory all year. Av. territory size (n = 3) 27·1 ha in NE Gabon (Erard 1987). Pair (but mainly ♂) patrols territory, frequently advertising with short songs even during hottest hours. When it detects another pair or ♂, bird makes circular flight, diam. 10–20 m, rather bouncing, just above or well above treetops, singing, with lower back and rump plumage fluffed out, crown feathers erected, wings snapping, and often bill snapping also.

NEST: 2 found. One was 30 m above ground, on fork near end of branch in canopy of tall forest tree; seen from ground it appeared to be a typical batis nest – a small cup of plant fibres decorated with bits of bark making nest resemble branch. Other was a neat, tightly-built, well-camouflaged, small cup made of fibres, covered outside with pieces of bark; placed 6 m up in small fork.

EGGS: not known.

LAYING DATES: Gabon, ♀ incubating Jan (family parties Nov–Mar, mainly Feb–Mar); Nigeria (large gonads Jan, June); Bioko (large gonads Sept–Jan).

INCUBATION: by ♀, fed on and out of nest by ♂.

DEVELOPMENT AND CARE OF YOUNG: young appear to stay with parents in territory almost until next breeding season, when they disperse. Dispersal seems important, as species readily colonizes new clearings and other degraded forested habitats in NE Gabon.

References
Brosset, A. and Erard, C. (1986).
Erard, C. (1987).

Genus *Lanioturdus* Waterhouse

Endemic, monotypic. A semi-terrestrial insectivore, chat-like on ground but with strong plumage and behavioural resemblances to *Batis*. Thought to be a malaconotine shrike by Harris and Arnott (1988). Sexes alike; plumage grey, black and white; eye yellow; bill moderately robust, somewhat hook-tipped, with well-developed rictal bristles; quite long wings and legs; very short tail. Differs from genus *Batis* in large size (2–4 times weight of *Batis*), white tail, and in having the sexes alike; resembles *Batis* in short tail, hindneck patch and general plumage features, food and feeding strategies (although *Lanioturdus* much more terrestrial); voice, wing-'fripping', aerial display with bouncing flight and puffed-out rump-feathers; cryptic cup-nest bound with spider web; and other breeding habits. Eggs like those of helmet-shrikes *Prionops*.

Plate 28
(Opp. p. 497)

Lanioturdus torquatus Waterhouse. White-tailed Shrike. Lanielle à queue blanche.

Lanioturdus torquatus Waterhouse, 1838. In Alexander, Exped. Int. Afr. 2, App., p. 264; Bull's Mouth Pass, Naukluft Mts, inland of Walvis Bay, Damaraland.

Range and Status. Endemic, resident and partial migrant, SW Africa. SW Angola, common but local in Huila; from Mossamedes (Capangombe) north along coast and Benguela escarpment to about Lobito and Chipepe; common in E Iona Nat. Park. NW Namibia, common from Cunene R. through mountains and interior south to Windhoek, Tsondab and Sossus R.

Description. ADULT ♂: forehead white, crown, nape, lores and ear-coverts slightly glossy black, hindneck black with long, narrow, white patch in midline, mantle, scapulars, back, rump and uppertail-coverts blue-grey, dappled with white. Tail white, with long tear-drop black mark on distal half of T1. Chin, throat and sides of neck white, sharply demarcated from all adjacent areas; upper breast black, slightly glossy, in a band *c*. 7 mm deep in midline, narrowing to points at sides; lower breast blue-grey at sides paling to nearly white in midline; forepart of flanks blue-grey, hindpart of flanks, thighs, vent and undertail-coverts white. Wings black, except for white bases to primaries, small white tips to primaries, secondaries and tertials, and white greater and median coverts and greater primary-coverts; alula feathers usually have small white tips. Bill black; palate black; eyes yellow; legs and feet black. SIZE: wing ♂ (n = 24) 80–90 (86·7), ♀ (n = 21) 84–91 (86·0); tail, ♂♀ (n = 44) 39·8–40·0 (44·2); bill, ♂♀ (n = 39) 20·6–24·0 (21·8); tarsus, ♂♀ (n = 44) 25·6–32·0 (29·5). Angolan birds ('*mesicus*') slightly larger and darker than others. WEIGHT: unsexed (n = 26) 22·5–45.

IMMATURE: like adult, but eyes brown, hindneck mottled, and breast-band narrower.

NESTLING: hatches blind and naked; later, skin purplish black; no mouth spots. Fledgling has dark eyes.

Field Characters. A distinctive and unmistakable bird in trees and bushes but mainly on ground, where hops in long bounds. Large-headed and short-tailed. Looks and behaves like a large batis: restless and vocal; shows much white in wing in flight, and stumpy tail almost entirely white; head black with white forehead, glaring yellow eye, and white throat; wings strongly pied; rest of plumage mainly blue-grey, with conspicuous black breast-band in front view and quite conspicuous white

hindneck band in rear view (**A**). Can look plump when perching in bush, but on ground usually attenuated, tail pointing down, body upright, head held high (**B**). Seldom flies far; flight direct, heavy and ungainly, with rapid, shallow wing-beats, accompanied by penetrating whistles and churring.

Voice. Tape-recorded (74, 91, F.). Song of ♂ a mixture of 2 notes: a loud, clear, high-pitched crescendo whistle 'eeee', 'oo-eee' or 'hooou' at 2·5 kHz and lasting 0·5 s, repeated up to 5 times, and an abrupt, short, liquid, *Lamprotornis*-starling-like 'tewp' or 'chiu' in lower register than the whistle. One 25 s recording contained: 3 whistles, 'tewp', whistle, 10 'tewp's, then 6 whistles, with varying-length pauses between successive notes. Some whistles are double, the second note following after the first without pause, but lower-pitched and quieter. Song of ♀ a number of low-pitched discords 'tchzrr' or 'kirrrr' and 'ktchurr', given at irregular intervals in duet with ♂; in the above recording 5 'tchzrr's, interpolated only with 'tewp' notes (not with 'eeee' whistles). Contact call a single, hollow, metallic 'chink'. Other calls: a ratchet-like rattle 'ch-ch-kkkkkk . . .' used by ♂♂ in high-intensity territorial interaction, lasting > 2 s, with 20 'k's in 1·7 s; discordant, nasal 'kzzzzp, kzzzp', by ♂ and 'ktchzrrr' by ♀, also during territorial interactions; discordant 'chuk' and 'tok' notes on going to roost; slow 'chuk-chuk' calls signalling apprehension, a nasal stuttering 'chk-chk-chk-chk' of alarm, and discordant 'wrwr-wawawa' and explosive 'skzeer, skzeer' of high-intensity alarm (Harris and Arnott 1988).

General Habits. In N Huila (Angola) inhabits climax and degraded *Brachystegia* woodland but in remainder of range occurs in *Colophospermum mopane* woods and scrub and in thornbush savanna, keeping mainly to denser patches of woody vegetation on rocky hillsides and in watercourses. Occurs in pairs, also singly and (in winter) in flocks of up to 12. Shy but inquisitive, and active; generally keeps to dense vegetation. Forages mainly in low bushes and on ground below them, also commonly in trees, up to their crowns. Gleans food from trunks, branches and leaves; occasionally flies out to catch insect in flight; on ground hops about restlessly, often in long bounds, and seizes prey on soil surface or jumps up to seize insect at end of grass stem. Sings from exposed perch, mainly about July–Oct, both sexes with neck stretched and bill raised. Perches on top of bush or tree when alarmed. Birds roost separately in thick foliage of large tree, sometimes nest tree. Scratches indirectly.

Mainly resident, but some local movement in Namibia in winter, and in N Huila (Angola) is noticeably commoner in Apr–Aug than in Sept–Mar.

Food. Invertebrates, mainly insects, including beetles, moths, butterflies, caterpillars, mantids and grasshoppers.

Breeding Habits. Monogamous, territorial, solitary breeder. Pair occupies territory of *c.* 5 ha. ♂ advertises and defends territory using loud, ringing, tonal 'eeee' or 'heeeeu' whistles uttered at exposed perches; ♂ also sings, with ♀ joining in antiphonally. Sometimes makes batis-like 'fripping' noise with wings in flight. Courting ♂ displays in flight, rising steeply with rapid wing-beats then descending in bouncing glide, with whistling, and rump feathers fluffed out; he pursues ♀ by hopping rapidly along ground and climbing up through low bush; they sometimes hop in tandem; ♂ and ♀ countersing 'heeee-tchzurr-heeee . . .'; copulation occurs without any further obvious display (Harris and Arnott 1988). Pair nests in same vicinity in successive years.

NEST: deep, neat cup, round and smooth, made of fine dry grasses and bark strips bound and plastered with spider web, and scantily lined with fine grass stems. Ext. diam. *c.* 75. Sited about 3 m above ground (range 0·4–8·0 m, n = 18), usually in canopy of acacia or other small thorn tree, on horizontal branch or in fork, often quite exposed, particularly when tree is leafless. Built by both sexes, mainly ♀, in about 6 days.

EGGS: 1–3, av. (16 clutches) 2·4. Laid at 1-day intervals. Oval, glossy, whitish or pale greenish, sparingly spotted, mainly around broad end, with red-brown and blue-grey. SIZE: (n = 9) 20·2–20·9 × 15·4–16·8 (20·4 × 15·8). Sometimes double-brooded.

LAYING DATES: Angola, Nov–Dec; Namibia, Sept–May, mainly Oct–Mar.

INCUBATION: by ♀; incubating ♀ very confiding; ♂ feeds her on nest and off it. Incubating ♀ sometimes sings antiphonally when ♂ sings. ♂ arrives on nearby tree with food item (often large green caterpillar) and gives trilling call; ♀ immediately leaves nest, joins ♂, takes offered food, hops about with him for a few minutes, returns to nest-tree where approaches nest cautiously and indirectly, then resumes incubating; well concealed on nest (Dedekind 1987). ♀ may solicit food with weak wing-quivering. ♂ appears to be attentive; one stayed near nest after ♀ left it and called 'heeeu' repeatedly when she remained away too long. ♂ known sometimes to roost in nest-tree. Period: (once) 15 days (Joerges 1972).

DEVELOPMENT AND CARE OF YOUNG: young fed at first by ♀, with food items given by ♂; after a few days young fed by both parents. Beg using high-pitched 'seee-seee' calls. Young feathered in adult plumage at c. 12 days. Nestling period (n = 4) 19–21 (20) days. Fledglings dependent for at least 2 weeks and remain in parental territory for at least 19 weeks (Harris and Arnott 1988).

References
Harris, T. (1986).
Harris, T. and Arnott, G. (1988).

BIBLIOGRAPHY

The bibliography is in three parts: (1) general and regional references, (2) family references and (3) acoustic references. These lists comprise all of the works cited in text and all other significant works consulted, in particular the sections relevant to species covered in the current volume of *The Birds of Africa*.

If a reference cited in the text does not appear in the appropriate family list (2), it should be looked for under 'General and Regional References' (1). Further, 'References' at the end of a family or generic account list significant works consulted but not necessarily cited during the preparation of the text.

We have divided the bibliography of the large Family Sylviidae into three sections: 1) *Cisticola*; 2) *Incana* to *Macrosphenus* (including *Scotocerca, Phragmacia, Prinia, Oreophilais, Heliolais, Urolais, Spiloptila, Schistolais, Drymocichla, Phyllolais, Apalis, Malcorus, Urorhipis, Artisornis, Poliolais, Camaroptera, Calamonastes, Eurypilla* and *Graueria*); 3) all other warblers.

Conversely, we have combined the often similar bibliographies of the families Muscicapidae, Monarchidae and Platysteiridae into a single section.

The citation of 'Peters Check-list of the Birds of the World' varies among references including the four previous volumes of this work. Some use the names of the editors (e.g. Mayr and Cottrell 1986) or even 'Peters 1986'. We cite the authors who prepared the specific families, that is, Ripley (1964), Traylor (1986), and Watson *et al.* (1986).

1. General and Regional References

Ali, S. and Ripley, S.D. (1973). 'Handbook of the Birds of India and Pakistan', Vol. 8. Oxford University Press, Bombay.

Allport, G., Ausden, M., Hayman, P.V., Robertson, P. and Wood, P. (1989). 'The Conservation of the Birds of Gola Forest, Sierra Leone'. *ICBP Study Report* No. 38, International Council for Bird Preservation, Cambridge.

Amadon, D. (1953). Avian systematics and evolution in the Gulf of Guinea. *Bulletin of the American Museum of Natural History* **100**(3), 393–452.

Anonymous (1991). 'Protected Areas of the World: a Review of National Systems', Vol. 3. International Union for Conservation of Nature and Natural Resources, Gland, Switzerland.

Archer, G. and Godman, E.M. (1961). 'The Birds of British Somaliland and the Gulf of Aden', Vol. 4. Oliver and Boyd, Edinburgh and London.

Ash, J.S. (1969). Spring weights of trans-Saharan migrants in Morocco. *Ibis* **111**, 1–10.

Ash, J.S. (1980). Migrational status of Palaearctic birds in Ethiopia. *Proceedings of the Fourth Pan-African Ornithological Congress* pp. 199–208. Southern African Ornithological Society, Johannesburg.

Ash, J.S. (1983). Over fifty additions to the Somali list including two hybrids, together with notes from Ethiopia and Kenya. *Scopus* **7**, 54–79.

Ash, J.S. (1985). Midwinter observations from Djibouti. *Scopus* **9**, 43–49.

Ash, J.S. (1990). Additions to the avifauna of Nigeria, with notes on distributional changes and breeding. *Malimbus* **11**, 104–116.

Ash, J.S. and Miskell, J.E. (1983). Birds of Somalia, their habitat, status and distribution. *Scopus* Special Supplement No. 1, 1–97.

Ash, J.S. and Miskell, J.E. (1988). Observations on birds in Somalia in 1978–1982, together with a bibliography of recent literature. *Scopus* **11**, 57–78.

Ash, J.S., Dowsett, R.J. and Dowsett-Lemaire, F. (1989a). New ornithological distribution records from eastern Nigeria. *Tauraco Research Report* No. 1, 13–27.

Aspinwall, D.R. (1979). Bird notes from the Zambezi district, North-western Province. *Zambian Ornithological Society Occasional Paper* No. 2, 1–60.

Atkinson, P., Peet N, and Alexander, J. (1991). The status and conservation of the endemic bird species of São Tomé and Príncipe, West Africa. *Bird Conservation International* **1**, 255–282.

Backhurst, G. (Ed.) (1993). East African Bird Report 1991. *Scopus* **15**(3).

Backhurst, G.C., Britton, P.L. and Mann, C.F. (1973). The less common Palaearctic migrant birds of Kenya and Tanzania. *Journal of the East Africa Natural History Society and National Museum* **140**, 1–38.

Baker, N.E. and Hirslund, P. (1987). Minziro Forest Reserve: an ornithological note including seven additions to the Tanzanian list. *Scopus* **11**, 9–12.

Bannerman, D.A. (1922). The birds of southern Nigeria. *Revue Zoologique Africaine* **9**, 254–426.

Bannerman, D.A. (1936, 1939, 1951). 'The Birds of Tropical West Africa', Vols 4, 5, and 8. The Crown Agents for the Colonies, London.

Barnes, J. and Bushell, B. (1989). 'Birds of the Gaborone Area'. Botswana Bird Club, Gaborone.

Barreau, D., Bergier, P. and Lesne, L. (1987). L'avifaune de l'Oukaimeden, 2200–3600 m (Haut Atlas, Maroc). *L'Oiseau et la Revue Française d'Ornithologie* **57**(4), 307–367.

Basilio, A. (1963). 'Aves de la isla de Fernando Poo'. Editorial Coculsa, Madrid.

Bates, G.L. (1911). Further notes on birds of southern Cameroon. *Ibis* **1911**, 479–545, 581–631.

Bates, G.L. (1930). 'Handbook of the Birds of West Africa'. John Bale, Sons and Danielsson Ltd, London.

Bates, G.L. (1934). Birds of the southern Sahara and adjoining countries in French West Africa. *Ibis* **1934**, 439–466.

Beasley, A.J. (1995). The birds of the Chimanimani Mountains. *Honeyguide* **41**, Supplement No.1, 1–57.

Beesley, J.S. (1972a). Birds of the Arusha National Park, Tanzania. *Journal of the East Africa Natural History Society and Natural Museum* **132**, 1–30.

Beesley, J.S.S. and Irving, N.S. (1976). The status of the birds of Gaborone and its surroundings. *Botswana Notes and Records* **8**, 231–261.

Benson, C.W. (1937). Miscellaneous notes on Nyasaland birds. *Ibis* **1937**, 551–582.

Benson, C.W. (1940, 1941). Further notes on Nyasaland

Benson, C.W. (1942). Additional notes on Nyasaland birds. *Ibis* **1942**, 299–337.
Benson, C.W. (1944). Notes from Nyasaland. *Ibis* **86**, 445–480.
Benson, C.W. (1946a). A visit to the Vumba Highlands, Southern Rhodesia. *Ostrich* **17**, 280–296.
Benson, C.W. (1946b, 1948). Notes on the birds of southern Abyssinia. *Ibis* **88**, 180–205; **90**, 325–327.
Benson, C.W. (1947). Observations from the Kota-kota district of Nyasaland. *Ibis* **89**, 553–566.
Benson, C.W. and Benson, F.M. (1947). Some breeding and other records from Nyasaland. *Ibis* **89**, 279–290.
Benson, C.W. and Benson, F.M. (1949). Notes on birds from northern Nyasaland and adjacent Tanganyika Territory. *Annals of the Transvaal Museum* **21**(2), 155–177.
Benson, C.W. and Benson, F.M. (1977). 'The Birds of Malaŵi'. Montfort Press, Limbe, Malaŵi.
Benson, C.W., Brooke, R.K., Dowsett, R.J. and Irwin, M.P.S. (1971). 'The Birds of Zambia'. Collins, London.
Betts, F.N. (1966). Notes on some resident breeding birds of southwest Kenya. *Ibis* **108**, 513–530.
Bibby, C.J., Collar, N.J., Crosby, M.J., Heath, M.F., Imboden, C., Johnson, T.H., Long, A.J., Stattersfield, A.J. and Thirgood, S.J. (1992). 'Putting Biodiversity on the Map: Priority Areas for Global Conservation'. International Council for Bird Preservation, Cambridge.
de Bie, S. and Morgan, N. (1989). Les oiseaux de la Réserve de la Biosphère "Boucle du Baoulé", Mali. *Malimbus* **11**, 41–60.
Black, J.G., Loiselle, B.A. and Vande weghe, J.-P. (1990). Weights and measurements of some central African birds. *Le Gerfaut* **80**, 3–11.
Bonde, K. (1993). 'Birds of Lesotho'. University of Natal Press, Pietermaritzberg.
Bowden, C.G.R. (1986). The use of nest-netting for studying forest birds in Cameroon. In 'Conservation of Cameroon Montane Forests'. (Stuart, S.N., Ed), pp. 130–174. International Council for Bird Protection, Cambridge.
Boyer, H.J. and Bridgeford, P.A. (1988). Birds of the Naukluft Mountains: an annotated checklist. *Madoqua* **15**(4), 295–314.
Bretagnolle, F. (1993). An annotated checklist of north-eastern Central African Republic. *Malimbus* **15**, 6–16.
Brewster, C. (1991). Birds of the Gumare area, northwest Botswana. *Babbler* Nos. 21 & 22, 12–61.
Britton, P.L. (1970a). Birds of the Balovale District of Zambia. *Ostrich* **41**, 145–190.
Britton, P.L. (Ed.) (1980a). 'Birds of East Africa'. East Africa Natural History Society, Nairobi.
Britton, P.L. and Dowsett, R.J. (1969). More bird weights from Zambia. *Ostrich* **40**, 55–60.
Britton, P.L. and Zimmerman, D.A. (1979). The avifauna of Sokoke Forest, Kenya. *Journal of the East Africa Natural History Society and National Museum* **169**, 1–16.
Brooke, R.K. (1966). Distribution and breeding notes on the birds of the central frontier of Rhodesia and Mozambique. *Annals of the Natal Museum* **18**(2), 429–453.
Brosset, A. (1961). Écologie des oiseaux du Maroc oriental. *Travaux de l'Institut Scientifique Chérifien, Série Zoologie* No. 22, 1–155. Université Mohammed V, Rabat.
Brosset, A. (1984). Oiseaux migrateurs européens hivernant dans la partie guinéene du Mont Nimba. *Alauda* **52**(2), 81–101.
Brosset, A. and Erard, C. (1986). 'Les Oiseaux des Régions Forestières du Nord-Est du Gabon', Vol. 1. Société Nationale de Protection de la Nature, Paris.
Brown, C.J. (1990). Birds of the West Caprivi Strip, Namibia. *Lanioturdus* **25**, 22–37.
Brown, C.J. (1993). The birds of Owambo, Namibia. *Madoqua* **18**(2), 147–161.
Brown, L.H. and Britton, P.L. (1980). 'The Breeding Seasons of East African Birds'. East Africa Natural History Society, Nairobi.
Browne, P.W.P. (1982). Palaearctic birds wintering in south-west Mauritania: species, distributions and population estimates. *Malimbus* **4**, 69–92, 104.
Brunel, J. (1958). Observations sur les oiseaux du Bas-Dahomey. *L'Oiseau et la Revue Française d'Ornithologie* **28**, 1–38.
Buckley, P. and McNeilage, A. (1989). An ornithological survey of Kasyoha-Kitomi and Itwara forests, Uganda. *Scopus* **13**, 97–108.
Bundy, G. (1976). The birds of Libya. BOU Check-list No. 1. British Ornithologists' Union, London.
Burgess, N.D., Huxham, M.R., Mlingwa, C.O.F., Davies, S.G.F. and Cutts, C.J. (1991). Preliminary assessment of forest birds in Kiono, Pande, Kisiju and Kiwengoma coastal forests, Tanzania. *Scopus* **14**, 97–106.
Campbell, B. and Lack, E. (1985). 'A Dictionary of Birds'. T. and A.D. Poyser, Calton.
Carroll, R.W. (1988). Birds of the Central African Republic. *Malimbus* **10**, 177–200.
Carswell, M. (1986). Birds of the Kampala area. *Scopus* Special Supplement No. 2, 1–89.
Cave, F.O. and Macdonald, J.D. (1955). 'Birds of the Sudan'. Oliver and Boyd, Edinburgh and London.
Cawkell, E.M. and Moreau, R.E. (1963). Notes on birds in The Gambia. *Ibis* **105**, 156–178.
Chapin, J.P. (1953a, 1954). The birds of the Belgian Congo, Vols 3 and 4. *Bulletin of the American Museum of Natural History* **75A** and **75B**.
Chapin, R.T. (1978). Brief accounts of some central African birds, based upon the journals of James Chapin. *Revue de Zoologie Africaine* **92**, 805–836.
Chappuis, C. (1974, 1975, 1978, 1979a). Illustration sonore de problèmes bioacoustiques posés par les oiseaux de la zone éthiopienne. *Alauda* **42**(4), 467–500; **43**(4), 427–474; **46**(4), 327–355; **47**(3), 195–212.
Chappuis, C. (1976a). Origine et évolution des vocalisations des certains oiseaux de Corse et des Baléares. *Alauda* **44**(4), 475–495.
Chappuis, C. (1979b). Emissions vocales nocturnes des oiseaux d'Europe. *Alauda* **47**(4), 277–299.
Chappuis, C. (1980a). List of sound-recorded Ethiopian birds. *Malimbus* **2**(1), 1–15, 82–98.
Chappuis, C. (1986). Revised list of sound-recorded Afrotropical birds. *Malimbus* **8**(1), 25–39, 79–88.
Cheesman, R.E. and Sclater, W.L. (1935). On a collection of birds from north-western Abyssinia. *Ibis* **1935**, 594–622.
Cheke, R.A. (1982). More bird records from the Republic of Togo. *Malimbus* **4**, 55–63.
Cheke, R.A. and Walsh, J.F. (1980). Bird records from the Republic of Togo. *Malimbus* **2**, 112–120.
Cheke, R.A. and Walsh, J.F. (1984). Further bird records from the Republic of Togo. *Malimbus* **6**, 15–22.
Cheke, R.A., Walsh, J.F. and Sowah, S.A. (1986). Records of birds seen in the Republic of Togo during 1984–1986. *Malimbus* **8**, 51–72.
Cheke, R.A. and Walsh, J.F. (1996). The birds of Togo. BOU Check-list No. 14. British Ornithologists' Union, Tring.
Christy, P. and Clark, W. (1994). 'Guide des oiseaux de la Réserve de la Lopé'. ECOFAC, Gabon, Libreville.

Claffey, P.M. (1995). Notes on the avifauna of the Bétérou area, Borgou Province, Republic of Benin. *Malimbus* **17**, 63–84.

Clancey, P.A. (1964). 'The Birds of Natal and Zululand'. Oliver and Boyd, Edinburgh and London.

Clancey, P.A. (1971). A handlist of the birds of southern Moçambique. *Memórias do Instituto de Investigação Cientifica de Moçambique*, Série A, **11**, 1–167.

Clancey, P.A. (1980a). On birds from the mid-Okavango Valley on the South West Africa/Angola border. *Durban Museum Novitates* **XII**(9), 87–127.

Clancey, P.A. (Ed.) (1980b). 'S.A.O.S. Checklist of Southern African Birds'. Southern African Ornithological Society, Johannesburg.

Clancey, P.A. (1985). 'The Rare Birds of Southern Africa'. Winchester Press, Saxonwold.

Clancey, P.A. (Ed.) (1987). S.A.O.S. Checklist of Southern African Birds; first updating report. Southern African Ornithological Society, Johannesburg.

Clark, G. (1985). Bird observations from northwest Somalia. *Scopus* **9**, 24–42.

Collar, N.J. and Stuart, S.N. (1985). 'Threatened Birds of Africa and Related Islands'. ICBP/IUCN Red Data Book, part 1, 3rd edition. International Council for Bird Preservation, Cambridge.

Collar, N.J. and Stuart, S.N. (1988). Key forests for threatened birds in Africa. *ICBP Monograph* No. 3, 102 pp. International Council for Bird Preservation, Cambridge.

Colston, P.R. and Curry-Lindahl, K. (1986). 'The Birds of Mount Nimba, Liberia'. British Museum (Natural History), London.

Cordeiro, N.J. (1994). Forest birds on Mt. Kilimanjaro, Tanzania. *Scopus* **17**, 65–112.

Cracraft, J. (1981). Toward a phylogenetic classification of the recent birds of the world (Class Aves). *Auk* **98**, 681–714.

Craig, A.J.F.K. (1983). Moult in southern African passerine birds: a review. *Ostrich* **54**, 220–237.

Cramp, S. (Ed.) (1988, 1992). 'The Birds of the Western Palearctic', Vols V, VI. Oxford University Press, Oxford.

Cramp, S. and Perrins, C.M. (Eds) (1993, 1994). 'The Birds of the Western Palearctic', Vols VII; VIII and IX. Oxford University Press, Oxford.

Cyrus, D. and Robson, N. (1980). 'Bird Atlas of Natal'. University of Natal Press, Pietermaritzburg.

Dean, W.R.J. (1971). Breeding data for the birds of Natal and Zululand. *Durban Museum Novitates* **IX**(6), 59–91.

Dean, W.R.J. (1974). Breeding and distributional notes on some Angolan birds. *Durban Museum Novitates* **X**(8), 109–125.

Dean, W.R.J. and Huntley, M.A. (1984). An updated list of the birds of Angola. Unpublished manuscript (available Tierberg Karoo Research Centre, Prince Albert, South Africa).

Dean, W.R.J. and Milton, S.J. (1993). The use of *Galium tomentosum* (Rubiaceae) as nest material by birds in the southern Karoo. *Ostrich* **64**, 187–189.

Dean, W.R.J., Huntley, M.A., Huntley, B.J. and Vernon, C.J. (1988). Notes on some birds of Angola. *Durban Museum Novitates* **XIV**(4), 43–92.

Dejaifve, P.-A. (1994). Contribution à l'étude de l'avifaune de la savane guinéenne du Nord-Ubangui, Zaïre. *Le Gerfaut* **84**, 63–71.

Dekeyser, P.L. (1951). Mission A. Villers au Togo et au Dahomey (1950). III. Oiseaux. *Études Dahoméennes* **5**, 47–90.

Demey, R. (1995). Notes on the birds of the coastal and Kindia areas, Guinea. *Malimbus* **17**, 85–99.

Demey, R. and Fishpool, L.D.C. (1991). Additions and annotations to the avifauna of Côte d'Ivoire. *Malimbus* **12**, 61–86.

Demy, R. and Fishpool, L.D.C. (1994). The birds of Yapo Forest, Ivory Coast. *Malimbus* **16**, 100–122.

Desfayes, M. (1975). Birds from Ethiopia. *Revue de Zoologie Africaine* **89**(3), 505–535.

Devillers, P. and Ouellet, H. (1993). 'Noms Français des Oiseaux du Monde'. MultiMondes Inc., St.-Foy, Québec.

Donnelly, B.G. (1985). The birds of the Matobo (formerly Matopos) National Park, Zimbabwe. *Honeyguide* **31**, 11–23.

Douthwaite, R.J. and Miskell, J.E. (1991). Additions to *Birds of Somalia, their habitat, status and distribution* (Ash and Miskell 1983). *Scopus* **14**, 37–60.

Dowsett, R.J. (1971). The avifauna of the Makutu Plateau, Zambia. *Revue de Zoologie et de Botanique Africaines* **84**, 312–333.

Dowsett, R.J. (1985). Site-fidelity and survival rates of some montane forest birds in Malaŵi, south-central Africa. *Biotropica* **17**(2), 145–154.

Dowsett, R.J. (Ed.) (1989). A preliminary natural history survey of Mambilla Plateau and some lowland forests of Eastern Nigeria. *Tauraco Research Report* No. 1.

Dowsett, R.J. (Ed.) (1990). Enquête faunistique et floristique dans la Forêt de Nyungwe, Rwanda. *Tauraco Research Report* No. 3.

Dowsett, R.J. (1991). Gazetteer of zoological localities in Congo. *Tauraco Research Report* No. 4, 335–340.

Dowsett, R.J. (1993). Afrotropical avifaunas: annotated country checklists. *Tauraco Research Report* No. 5, 1–322.

Dowsett, R.J. and Dowsett-Lemaire, F. (1980). The systematic status of some Zambian birds. *Le Gerfaut* **70**, 151–199.

Dowsett, R.J. and Dowsett-Lemaire, F. (1984). Breeding and moult cycles of some montane forest birds in south-central Africa. *Revue d'Écologie (La Terre et La Vie)* **39**, 89–111.

Dowsett, R.J. and Dowsett-Lemaire, F. (1989a). Liste préliminaire des oiseaux du Congo. *Tauraco Research Report* No. 2, 29–51.

Dowsett, R.J. and Dowsett-Lemaire, F. (Eds) (1991). Flore et faune du bassin du Kouilou (Congo) et leur exploitation. *Tauraco Research Report* No. 4.

Dowsett, R.J. and Dowsett-Lemaire, F. (Eds) (1993a). A contribution to the distribution and taxonomy of Afrotropical and Malagasy birds. *Tauraco Research Report* No. 5.

Dowsett, R.J. and Dowsett-Lemaire, F. (1993b). Comments on the taxonomy of some Afrotropical bird species. *Tauraco Research Report* No. 5, 323–389.

Dowsett, R.J. and Forbes-Watson, A.D. (1993). 'Checklist of Birds of the Afrotropical and Malagasy Regions'. Tauraco Press, Liège, Belgium.

Dowsett, R.J. and Prigogine, A. (1974). The avifauna of the Marungu Highlands. *Hydrobiological Survey of the Lake Bangweulu Luapula River Basin* **XIX**, 1–67.

Dowsett, R.J. and Stjernstedt, R. (1973). The birds of the Mafinga Mountains. *Puku* **7**, 107–123.

Dowsett, R.J., Backhurst, G.C. and Oatley, T.B (1988). Afrotropical ringing recoveries of Palaearctic migrants. 1. Passerines (Turdidae to Oriolidae). *Tauraco* **1**, 29–63.

Dowsett-Lemaire, F. (1983). Ecological and territorial requirements of montane forest birds on the Nyika Plateau, south-central Africa. *Le Gerfaut* **73**, 345–378.

Dowsett-Lemaire, F. (1985). Breeding productivity and the non-breeding element in some montane forest birds in Malaŵi, south-central Africa. *Biotropica* **17**(2), 137–144.

Dowsett-Lemaire, F. (1988). Fruit choice and seed dissemination by birds and mammals in the evergreen forest of upland Malaŵi. *Revue d'Écologie (La Terre et La Vie)* **43**, 251–285.

Dowsett-Lemaire, F. (1989). Ecological and biogeographical aspects of forest bird communities in Malaŵi. *Scopus* **13**, 1–80.

Dowsett-Lemaire, F. (1990). Eco-ethology, distribution and status of Nyungwe Forest birds (Rwanda). *Tauraco Research Report* No. 3, 31–85.

Dowsett-Lemaire, F. and Dowsett, R.J. (Eds) (1989a). Enquête faunistique dans la forêt du Mayombe, et check-liste des oiseaux et des mammifères du Congo. *Tauraco Research Report* No. 2.

Dowsett-Lemaire, F. and Dowsett, R.J. (1989b). Liste commentée des oiseaux de la forêt du Mayombe (Congo). *Tauraco Research Report* No. 2, 5–16.

Dowsett-Lemaire, F. and Dowsett, R.J. (1989c). Zoogeography and taxonomic relationships of the forest birds of the Cameroon Afromontane region. *Tauraco Research Report* No. 1, 48–56.

Dowsett-Lemaire, F. and Dowsett, R.J. (1990). Zoogeography and taxonomic relationships of the forest birds of the Albertine Rift Afromontane region. *Tauraco Research Report* No. 3, 87–109.

Dowsett-Lemaire, F. and Dowsett, R.J. (1991). The avifauna of the Kouilou basin in Congo. *Tauraco Research Report* No. 4, 189–239.

Dowsett-Lemaire, F., Dowsett, R.J. and Bulens, P. (1993). Additions and corrections to the avifauna of Congo. *Malimbus* **15**, 68–80.

Duckworth, J.W., Evans, M.I., Safford, R.J., Telfer, M.G., Timmins, R.J. and Chemere Zewdie (1992). A survey of Nechisar National Park, Ethiopia. *ICBP Study Report* No. 50. International Council for Bird Preservation, Cambridge.

Duhart, F. and Descamps, M. (1963). Notes sur l'avifaune du delta central nigérien et régions avoisinantes. *L'Oiseau et la Revue Française d'Ornithologie* **33**, No. spécial, 1–107.

Dyer, M., Gartshore, M.E. and Sharland, R.E. (1986). The birds of Nindam Forest Reserve, Kagoro, Nigeria. *Malimbus* **8**, 2–20.

Earlé, R. and Grobler, N.J. (1987). 'First Atlas of Bird Distribution in the Orange Free State'. National Museum, Bloemfontein.

Edington, J.M. and Edington, M.A. (1983). Habitat partitioning and antagonistic behaviour amongst the birds of a West African scrub and plantation plot. *Ibis* **125**, 74–89.

Eisentraut, M. (1963). 'Die Wirbeltiere des Kamerungebirges'. Paul Parey, Hamburg and Berlin, 353 pp.

Eisentraut, M. (1973). Die Wirbeltierfauna von Fernando Poo und Westkamerun. *Bonner Zoologische Monographien* **3**, 1–427.

Elgood, J.H., Sharland, R.E. and Ward, P. (1966). Palaearctic migrants in Nigeria. *Ibis* **108**, 84–116.

Elgood, J.H., Fry, C.H. and Dowsett, R.J. (1973). African migrants in Nigeria. *Ibis* **115**, 375–409.

Elgood, J.H., Heigham, J.B., Moore, A.M., Nason, A.M., Sharland, R.E. and Skinner, N.J. (1994). The birds of Nigeria, 2nd edition. BOU Check-list No. 4. British Ornithologists' Union, Tring.

Elliot, C.C.H. (1972). An ornithological survey of the Kidepo National Park, northern Uganda. *Journal of the East Africa Natural History Society and National Museum* No. 129, 1–31.

van den Elzen, R. and König, C. (1983). Vögel des (Süd-)Sudan: taxonomische und tiergeographische Bemerkungen. *Bonner Zoologische Beiträge* **34**, 149–196.

Erard, C. (1974a). Notes faunistiques et systématiques sur quelques oiseaux d'Ethiopie. *Bonner Zoologische Beiträge* **25**, 76–86.

Erard, C. (1987,1990a). Écologie et comportement des gobemouches (Aves: Muscicapinae, Platysteirinae, Monarchinae) du Nord-Est du Gabon. Vols 1 and 2. *Mémoires du Muséum National d'Histoire Naturelle, Série A, Zoologie* **138**, 1–256; **146**, 1–234.

Etchécopar, R.D. and Hüe, F. (1967). 'The Birds of North Africa'. Oliver and Boyd, Edinburgh and London.

Fairon, J. (1975). Contribution à l'ornithologie de l'Aïr (Niger). *Le Gerfaut* **65**, 107–134.

Feather, P.J. (1986). The Bulawayo garden bird survey 1973–1982. *Honeyguide* **32**, 13–33.

Finlayson, C. (1992). 'Birds of the Strait of Gibralter'. T. and A.D. Poyser, London.

Frade, F. (1951). Aves coligidas pela Missão zoológica de Moçambique. *Anais da Junta das Missões Geográficas e de Investigações Coloniais, Lisboa*, Vol. VI, Tomo IV, Fasc. III, 220 pp.

Frade, F. (1951). Catálogo das Aves de Moçambique. *Anais da Junta das Missões Geográficas e de Investigações Coloniais, Lisboa*, Vol. VI, Tomo IV, Fasc. IV, 294 pp.

Frade, F. and Bacelar, A. (1959). Catálogo das aves da Guiné portuguesa, II. *Memórias da Junta Investigações do Ultramar* **7**, 116 pp.

Frandsen, J. (1982). 'Birds of the South Western Cape'. Sable Publishers, Sloane Park, South Africa.

Friedmann, H. (1937). Birds collected by the Childs Frick Expedition to Ethiopia and Kenya Colony, Part 2. Passeres. *Bulletin of the United States National Museum* **153**, 1–506.

Friedmann, H. (1962). The Machris Expedition to Tchad, Africa: birds. *Los Angeles County Museum Contributions in Science* No. 59, 1–26.

Friedmann, H. (1966). A contribution to the ornithology of Uganda. *Bulletin of the Los Angeles County Museum of Natural History* No. 3, 1–55.

Friedmann, H. (1978). Results of the Lathrop Central African Republic Expedition 1976, Ornithology. *Los Angeles County Museum Contributions in Science* No. 287, 1–22.

Friedmann, H. and Northern, J.R. (1975). Results of the Taylor South West Africa Expedition 1972, Ornithology. *Los Angeles County Museum Contributions in Science* No. 266, 1–39.

Friedmann, H. and Stager, K.E. (1964). Results of the 1964 Cheney Tanganyikan Expedition, Ornithology. *Los Angeles County Museum Contributions in Science* No. 84, 1–50.

Friedmann, H. and Stager, K.E. (1967). Results of the 1966 Cheney Expedition to the Samburu District, Kenya. Ornithology. *Los Angeles County Museum Contributions in Science* No. 130, 1–34.

Friedmann, H. and Stager, K.E. (1969). Results of the 1968 Avil Expedition to Mt. Nyiru, Samburu District, Kenya. Ornithology. *Los Angeles County Museum Contributions in Science* No. 174, 1–30.

Friedmann, H. and Williams, J.G. (1969). The birds of the Sango Bay Forests, Buddu County, Masaka District, Uganda. *Los Angeles County Museum Contributions in Science* No. 162, 1–48.

Friedmann, H. and Williams, J.G. (1970a). The birds of the Kalinzu Forest, southeastern Ankole, Uganda. *Los Angeles County Museum Contributions in Science* No. 195, 1–27.

Friedmann, H. and Williams, J.G. (1971). The birds of the lowlands of Bwamba, Toro Province, Uganda. *Los Angeles County Museum Contributions in Science* No. 211, 1–70.

Fry, C.H. (1971). Migration, moult and weights of birds in northern guinea savanna in Nigeria and Ghana. *Proceedings of the Third Pan-African Ornithological Congress, Ostrich Supplement* **8**, 239–263.

Fry, C.H., Ash, J.S. and Ferguson-Lees, I.J. (1970). Spring weights of some Palaearctic migrants at Lake Chad. *Ibis* **112**, 58–82.

Fuggles-Couchman, N.R. (1953). The ornithology of Mt. Hanang, in northern-central Tanganika Territory. *Ibis* **95**, 468–482.

Gatter, W. (1987). Zugverhalten und Überwinterung von palaärktischen Vögeln in Liberia (Westafrika). *Verhandlungen der Ornithologischen Gesellschaft in Bayern* **24**, 479–508.

Gatter, W. (1988). The birds of Liberia (West Africa). *Verhandlungen der Ornithologischen Gesellschaft in Bayern* **24**, 689–723.

Gaugris, Y., Prigogine, A. and Vande weghe, J.-P. (1981). Additions et corrections à l'avifaune du Burundi. *Le Gerfaut* **71**, 3–39.

Gee, J.P. (1984). The birds of Mauritania. *Malimbus* **6**, 31–66.

Germain, M. (1992). Sur quelques données erronées concernant l'avifaune de la Lobaye, République Centrafricaine. *Malimbus* **14**(1), 1–6.

Germain, M. and Cornet, J.-P. (1994). Oiseaux nouveaux pour la République Centrafricaine ou dont les notifications de ce pays sont peu nombreuses. *Malimbus* **16**(1), 30–51.

Germain, M., Dragesco, J., Roux, F. and Garcin, H. (1973). Contribution à l'ornithologie du Sud-Cameroun. II. Passeriformes. *L'Oiseau et la Revue Française d'Ornithologie* **43**(3), 212–259.

Gibbon, G. (1991). 'Southern African Bird Sounds'. Southern African Birding cc., Durban.

Gillet, H. (1960). Observations sur l'avifaune du massif de l'Ennedi (Tchad). *L'Oiseau et la Revue Française d'Ornithologie* **30**, 99–134.

Ginn, P.J. (1976) Birds of Makgadigadi: a preliminary report. *Wagtail* No. 15, 21–96

Ginn, P.J. (1979). 'Birds of Botswana'. Chris van Rensburg, Johannesburg.

Ginn, P.J., McIlleron, W.G. and Milstein, P.le S. (1989). 'The Complete Book of Southern African Birds'. Struik Winchester, Cape Town.

Giraudoux, P., Degauquier, R., Jones, P.J., Weigel, J. and Isenmann, P. (1988). Avifaune du Niger: état des connaissances en 1986. *Malimbus* **10**(1), 1–140.

Good, A.-I. (1953). The Birds of French Cameroon, Part II. *Mémoires de l'Institut Français d'Afrique Noire, Série Sciences Naturelles* **3**, 1–269. Centre du Cameroun, Douala.

Goodman, S.M. and Meininger, P.L. (Eds) (1989). 'The Birds of Egypt'. Oxford University Press, Oxford.

Gore, M.E.J. (1990). Birds of The Gambia, 2nd revised edition. BOU Check-list No. 3. British Ornithologists' Union, Tring.

Green, A.A. (1983). The birds of Bamingui-Bangoran National Park, Central African Republic. *Malimbus* **5**, 17–30.

Green, A.A. (1984). Additional bird records from Bamingui-Bangoran National Park, Central African Republic. *Malimbus* **6**, 70–72.

Green, A.A. (1989). Avifauna of Yankari Reserve, Nigeria: new records and observations. *Malimbus* **11**, 61–72.

Green, A.A. (1990). The avifauna of the southern sector of the Gashaka-Gumti Game Reserve, Nigeria. *Malimbus* **12**, 31–51.

Green, A.A. and Carroll, R.W. (1991). The avifauna of Dzanga-Ndoki National Park and Dzanga-Sangha Rainforest Reserve, Central African Republic. *Malimbus* **13**, 49–66.

Green, A.A. and Sayer, J.A. (1979). The birds of Pendjari and Arli National Parks (Benin and Upper Volta). *Malimbus* **1**, 14–28.

Greig-Smith, P.W. and Davidson, N.C. (1977). Weights of West African savanna birds. *Bulletin of the British Ornithologists' Club* **97**, 96–99.

Grimes, L.G. (1987). The birds of Ghana. BOU Check-list No. 9. British Ornithologists' Union, London.

Grote, H. (1928). Uebersicht über die Vogelfauna des Tschadgebiets. *Journal für Ornithologie* **76**, 739–783.

Günther, R. and Feiler, A. (1986). Zur Phänologie, Ökologie und Morphologie angolanischer Vögel (Aves), II. Passeriformes. *Faunistische Abhandlungen, Staatliches Museum für Tierkunde in Dresden* **14**(1), 1–29.

Gyldenstolpe, N. (1924). Zoological results of the Swedish expedition to Central Africa 1921. *Kungliga Svenska Vetenskapsakademiens Handlingar*, Third series, **1**(3), 326 pp.

Hall, B.P. (1960a). The ecology and taxonomy of some Angola birds. *Bulletin of the British Museum of Natural History (Zoology)* **6**(7), 367–462.

Hall, B.P. and Moreau, R.E. (1962). A study of the rare birds of Africa. *Bulletin of the British Museum of Natural History (Zoology)* **8**, 313–378.

Hall, B.P. and Moreau, R.E. (1970). 'An Atlas of Speciation in African Passerine Birds'. British Museum (Natural History), London.

Hall, D.G. (1983). Birds of Mataffin, Eastern Transvaal. *Southern Birds* **10**, 1–55.

Hall, P. (1977a). Birds of the Chad basin boreholes. *Bulletin of the Nigerian Ornithologists' Society* **13**(43), 37–42.

Hall, P. (1977b). The birds of Maiduguri. *Bulletin of the Nigerian Ornithologists' Society* **13**(43), 15–36.

Hall, P. (1977c). The birds of Serti. *Bulletin of the Nigerian Ornithologists' Society* **13**, 66–79.

Halleux, D. (1994). Annotated bird list of Macenta Prefecture, Guinea. *Malimbus* **16**, 10–29.

Hanmer, D.B. (1976). Birds of the lower Zambezi (Mozambique). *Southern Birds* **2**, 1–66.

Harding, D.P. and Harding, R.S.O. (1982). A preliminary checklist of birds in the Kilimi area of northwest Sierra Leone. *Malimbus* **4**(2), 64–68.

Harris, T. and Arnott, G. (1988). 'Shrikes of Southern Africa'. Struik Winchester, Cape Town.

Harrison, C. (1982). 'An Atlas of the Birds of the Western Palaearctic'. Collins, London.

Harrison, J. (1995). The atlas of Southern African birds. *Birding in Southern Africa* **47**(2), 42–45.

Harvey, W.G. and Howell, K.M. (1987). Birds of the Dar es Salaam area, Tanzania. *Le Gerfaut* **77**, 205–258.

Harwin, R.M., Manson, A.J., Manson, C. and Mwadziwana, P. (1994). The birds of the Bvumba Highlands. *Honeyguide* **40** Supplement No. 1, 1–51.

Hayman, P.V., Prangley, M., Barnett, A. and Diawara, D. (1995). The birds of the Kounounkan Massif, Guinea. *Malimbus* **17**, 53–62.

Heaton, A.M. and Heaton, A.E. (1980). The birds of Obudu, Cross River State, Nigeria. *Malimbus* **2**, 16–24.

Heim de Balsac, H. and Mayaud, N. (1962). 'Les Oiseaux du Nord-Ouest de l'Afrique'. Paul Lechevalier, Paris.

Heinrich, G. (1958). Zur Verbreitung und Lebensweise der Vögel von Angola. *Journal für Ornithologie* **99**, 322–362, 399–421.

Hines, C.J.H. (1985–1987). The birds of eastern Kavango, SWA/Namibia. *Journal of the South West Africa Scientific Society* **40–41**, 115–147.

Hockey, P.A.R. (1995). Rare birds in South Africa, 1991–1992. *Birding in Southern Africa* **47**(1), 14–19.

Hockey, P.A.R., Underhill, L.G., Neatherway, M. and Ryan, P G. (1989). 'Atlas of the Birds of the Southwestern Cape'. Cape Bird Club, Cape Town.

Hoesch, W. and Niethammer, G. (1940). Die Vogelwelt Deutsch-Südwestafrikas. *Journal für Ornithologie* **88** Sonderheft, 1–390.

Hogg, P., Dare, P.J. and Rintoul, J.V. (1984). Palaearctic migrants in the central Sudan. *Ibis* **126**, 307–331.

Holman, F.C. (1947). Birds of the Gold Coast. *Ibis* **89**, 623–650.

Holyoak, D.T. and Seddon, M. B. (1989). Distributional notes on the birds of Burkina Faso. *Bulletin of the British Ornithologists' Club* **109**, 205–216.

Howells, W.W. (1985). The birds of the Dande Communal Lands, Middle Zambezi Valley, Zimbabwe. *Honeyguide* **31**, 26–48.

Irwin, M.P.S. (1981). 'The Birds of Zimbabwe'. Quest Publishing, Harare.

Irwin, M.P.S., Niven, P.N.F. and Winterbottom, J.M. (1969). Some birds of the lower Chobe River area, Botswana. *Arnoldia (Rhodesia)* **4**(21), 1–40.

Jackson, F.J. and Sclater, W.L. (1938). 'The Birds of Kenya Colony and the Uganda Protectorate'. Vol. II. Gurney and Jackson, London.

Jackson, H.D. (1989). Weights of birds collected in the Mutare Municipal Area, Zimbabwe. *Bulletin of the British Ornithologists' Club* **109**, 100–106.

Jensen, F.P. and Brøgger-Jensen, S. (1992). The forest avifauna of the Uzungwa Mountains, Tanzania. *Scopus* **15**, 65–83.

Jensen, J.V. and Kirkeby, J. (1980). 'The Birds of Gambia'. Aros Nature Guides, Århus, Denmark.

Johnston-Stewart, N.G.B. (1984). Evergreen forest birds in the southern third of Malaŵi. *Nyala* **10**, 99–119.

Johnston-Stewart, N.G.B. and Heigham, J.B. (1982). 'Bridging the Bird Gap'. Mountford Press, Limbe.

Keith, S. and Gunn, W.W.H. (1971). 'Birds of the African Rain Forests'. Sounds of Nature #9, Federation of Ontario Naturalists, Don Mills, Ontario and the American Museum of Natural History, New York.

Keith, S., Twomey, A., Friedmann, H. and Williams, J. (1969). The avifauna of the Impenetrable Forest, Uganda. *American Museum Novitates* No. 2389.

Kemp, A.C. (1976). The distribution and status of the birds of the Kruger National Park. *Koedoe Monograph* **2**, 1–130.

Kettle, R. and Boswall, J. (1988). Palearctic bird Sound Recordings 1981–1987. *Bioacoustics* **1**, 209–239.

Lack, P.C. (1985). The ecology of the land-birds of Tsavo East National Park, Kenya. *Scopus* **9**, 2–23, 57–96.

Lamarche, B. (1981). Liste commentée des oiseaux du Mali. Passereaux. *Malimbus* **3**, 73–102.

Lamarche, B. (1988). Liste commentée des oiseaux de Mauritanie. *Études Sahariennes et Ouest-Africaines* **1**(4), 1–162. Private printing, Nouakchott and Paris.

Laurent, A. (1990). 'Catalogue commenté des oiseaux de Djibouti'. Ministère du Commerce, des Transports et du Tourisme, Djibouti.

Lawson, P.C. and Edwards, J.A. (1983). Birds of Kangwane (Mswati District). *Southern Birds* **11**, 1–84.

Lawson, W.J. (1963a). A contribution to the ornithology of Sul do Save, southern Moçambique. *Durban Museum Novitates* **VII**(4), 73–124.

Ledant, J.-P., Jacob, J.-P., Jacobs, P., Malher, F., Ochando, B. and Roché, J. (1981). Mise à jour de l'avifaune algérienne. *Le Gerfaut* **71**, 295–398.

Lewis, A. and Pomeroy, D. (1989). 'A Bird Atlas of Kenya'. A.A. Balkema, Rotterdam.

Lippens, L. and Wille, H. (1976). 'Les Oiseaux du Zaïre'. Lannoo à Tielt, Belgique.

Liversidge, R. (1991). 'The Birds Around Us: Birds of the Southern African Region'. Fontein, Parklands, South Africa.

Louette, M. (1981). 'The Birds of Cameroon: an Annotated Check-List'. Paleis der Academiën, Brussel.

Louette, M. and Prévost, J. (1987). Passereaux collectés par J. Prévost au Cameroun. *Malimbus* **9**, 83–96.

Lynes, H. (1925). On the birds of north and central Darfur, with notes on the west-central Kordofan and north Nuba Provinces of British Sudan. *Ibis* **1925**, 71–131.

Lynes, H. (1934). Contribution to the ornithology of southern Tanganyika Territory. Birds of the Ubene-Uhehe highlands and Iringa uplands. *Journal für Ornithologie Supplement* **82**, 1–147.

Lynes, H. (1938a). Contribution to the ornithology of the southern Congo basin. Lynes-Vincent tour of 1933–34. *Revue de Zoologie et de Botanique Africaines* **31**, 1–129.

Macdonald, J.D. (1957). 'A contribution to the ornithology of western South Africa'. Trustees, British Museum, London.

Mackworth-Praed, C.W. and Grant, C.H.B. (1960). 'Birds of Eastern and North Eastern Africa', Vol. 2. Longmans, London.

Mackworth-Praed, C.W. and Grant, C.H.B. (1963). 'Birds of the Southern Third of Africa', Vol. 2. Longmans, London.

Mackworth-Praed, C.W. and Grant, C.H.B. (1973). 'Birds of West Central and Western Africa', Vol. 2. Longman Group Ltd, London.

Maclean, G.L. (1993). 'Roberts' Birds of Southern Africa', 6th edition. Trustees of the John Voelcker Bird Book Fund, Cape Town.

Malbrant, R. (1954). Contribution à l'étude des oiseaux du Borkou-Ennedi-Tibesti. *L'Oiseau et la Revue Française d'Ornithologie* **24**, 1–47.

Malbrant, R. and Maclatchy, A. (1949, 1952). 'Faune de l'équateur africain français'. I and II. Oiseaux. Paul Lechevalier, Paris.

Malzy, P. (1962). La faune avienne du Mali. *L'Oiseau et la Revue Français d'Ornithologie* **32**, No. spécial, 1–81.

Mann, C.F. (1985). An avifaunal study in Kakamega Forest, Kenya, with particular reference to species diversity, weight and moult. *Ostrich* **56**, 236–262.

Marchant, S. (1942). Some birds of the Owerri Province, S. Nigeria. *Ibis* **1942**, 137–196.

Mayaud, N. (1988, 1990). Les oiseaux du nord-ouest de l'Afrique. Notes complémentaires. *Alauda* **56**, 113–125; **58**, 135–140, 143–148, 187–194.

Mayr, E. and Amadon, D. (1951). A classification of recent birds. *American Museum Novitates* No. 1496.

Mees, G.F. (1970). Birds of the Inyanga National Park, Rhodesia. *Zoologische Verhandelingen* **109**, 3–74.

Mlingwa, C.O.F., Huxham, M.R. and Burgess, N.D. (1993). The avifauna of Kazimzumbwe Forest Reserve, Tanzania: initial findings. *Scopus* **16**, 81–88.

Moreau, R.E. (1936a). Breeding seasons of birds in East African evergreen forest. *Proceedings of the Zoological Society of London* **1936**, 631–653.

Moreau, R.E. (1938). The avifauna of the mountains along the Rift Valley in north central Tanganyika Territory (Mbulu District). Part II. *Ibis* **1938**, 1–32.

Moreau, R.E. (1940). Contributions to the ornithology of the East African islands. *Ibis* **1940**, 48–91.

Moreau, R.E. (1944). Clutch-size: a comparative study, with special reference to African birds. *Ibis* **86**, 286–347.

Moreau, R.E. (1950). The breeding seasons of African birds. *Ibis* **92**, 223–267, 419–433.

Moreau, R.E. (1972). 'The Palaearctic-African Bird Migration Systems'. Academic Press, London.

Moreau, R.E. and Moreau, W.M. (1939). Observations on some East African birds. *Ibis* **1939**, 296–323.

Moreau, R.E. and Moreau, W.M. (1940). Incubation and fledging periods of African birds. *Auk* **57**, 313–325.

Morel, G.J. and Morel, M.-Y. (1982). Dates de reproduction des oiseaux du Sénégambie. *Bonner Zoologische Beiträge* **33**(2–4), 249–268.

Morel, G.J. and Morel, M.-Y. (1988). Liste des oiseaux de Guinée. *Malimbus* **10**, 143–176.

Morel, G.J. and Morel, M.-Y. (1990). 'Les oiseaux de Sénégambie (Notices et cartes de distribution)'. Centre de l'Office de la Recherche Scientifique et Technique Outre-Mer, Paris.

Morel, G. and Roux, F. (1966). Les migrateurs paléarctiques au Sénégal. *La Terre et La Vie* **113**, 143–176.

Moyer, D.C. (1993). A preliminary trial of territory mapping for estimating bird densities in Afromontane forest. *Proceedings of the Eighth Pan-African Ornithological Congress, Annales Musée Royal de l'Afrique Centrale (Zoologie)* **268**, 302–311.

Moyer, D.C., Lovett, J.C. and deLeyser, E.A. (1990). Birds of Ngwazi, Mufindi District, Tanzania. *Scopus* **14**, 6–13.

de Naurois, R. (1969). Peuplements et cycles de reproduction des oiseaux de la côte occidentale d'Afrique. *Mémoires du Muséum National d'Histoire Naturelle, Série A, Zoologie* **56**, 9–312.

Newby, J.E. (1980). The birds of the Ouadi Rime – Ouadi Achim Faunal Reserve: a contribution to the study of the Chadian avifauna. *Malimbus* **2**, 29–50.

Newby, J.E., Grettenberger, J. and Watkins, J. (1987). The birds of the northern Aïr, Niger. *Malimbus* **9**, 4–16.

Newman, K. (1989). 'Newman's Birds of Botswana'. Southern Book Publishers, Cape Town.

Newman, K. (1991). 'Newman's Birds of Southern Africa', update edition. Halfway House: Southern Book Publishers, Cape Town.

Newman, K., Johnston-Stewart, N. and Medland, B. (1992). 'Birds of Malawi'. Southern Book Publishers, Cape Town.

Niethammer, G. (1955). Zur Vogelwelt der Ennedi-Gebirges (Französisch Äquatorial-Afrika). *Bonner Zoologische Beiträge* **6**, 29–80.

Nikolaus, G. (1979a). Notes on some birds new to south Sudan. *Scopus* **3**, 68–73.

Nikolaus, G. (1987). Distribution atlas of Sudan's birds with notes on habitat and status. *Bonner Zoologische Monographien* **25**, 1–322.

Nikolaus, G. (1989). Birds of south Sudan. *Scopus* Special Supplement No. 3, 1–124.

North, M.E.W. and McChesney, D.S. (1964). More Voices of African Birds. Houghton Mifflin Co., Boston.

Oatley, T.B. (1969). Bird ecology in the evergreen forests of north western Zambia. *Puku* **5**, 141–180.

Ogilvie-Grant, W.R. and Forbes H.O. (1903). Aves. *In* 'The Natural History of Sokotra and Abd-el-Kuri' (Forbes, H.O., Ed.). H.R. Porter, London.

Olson, S.L. (1989). Preliminary systematic notes on some Old World passerines. *Rivista Italiana di Ornitologia* **59**, 183–195.

Osborne, P.E. and Tigar, B.J. (1990). The status and distribution of birds in Lesotho. Unpublished manuscript, Nature Conservation Bureau, Newbury, UK, 336 pp.

Pakenham, R.H.W. (1979). The birds of Zanzibar and Pemba. BOU Check-list No. 2. British Ornithologists' Union, London.

Pearson, D.J. and Turner, D.A. (1986). The less common Palaearctic migrant birds of Uganda. *Scopus* **10**, 61–82.

Penry, H. (1994). 'Bird Atlas of Botswana'. University of Natal Press, Pietermaritzburg.

Perez del Val, J., Fa, J.E., Castroviejo, J. and Purroy, F.J. (1994). Species richness and endemism of birds in Bioko. *Biodiversity and Conservation* **3**, 868–892.

Pineau, J. and Giraud-Audine, M. (1979). Les oiseaux de la Péninsule Tingitane. *Travaux de l'Institut Scientifique, Série Zoologie*, **38**, 1–147. Université Mohammed V, Rabat.

Priest, C.D. (1935). 'The Birds of Southern Rhodesia', Vol. 3. William Clowes and Sons, London and Beccles.

Prigogine, A. (1953). Contribution à l'étude de la Faune ornithologique de la région à l'ouest du lac Edouard. *Annales du Musée Royal du Congo Belge, série in-8vo, Sciences zoologiques* **24**, 1–114.

Prigogine, A. (1960). La faune ornithologique du Massif du Mont Kabobo. *Annales du Musée Royal du Congo Belge, série in-8vo, Sciences zoologiques* **85**, 1–46.

Prigogine, A. (1967). La faune ornithologique de l'île Idjwi. *Revue de Zoologie et de Botanique Africaines* **75**, 249–274.

Prigogine, A. (1971, 1978a, 1984a). Les oiseaux de l'Itombwe et de son hinterland, Vols I, II, III. *Annales du Musée Royal de l'Afrique Centrale, série in-8vo, Sciences zoologiques* **185, 223, 243**.

Prigogine, A. (1980a). The altitudinal distribution of the avifauna in the Itombwe Forest (Zaïre). *Proceedings of the Fourth Pan-African Ornithological Congress* pp. 169–184. Southern African Ornithological Society, Johannesburg.

Prigogine, A. (1980b). Etude de quelques contacts secondaires au Zaïre oriental. *Le Gerfaut* **70**, 305–384.

Prigogine, A. (1985a). Recently recognised bird species in the Afrotropical region – a critical review. *Proceedings of the International Symposium on African Vertebrates, Bonn 1985*, pp. 91–114.

Prigogine, A. (1987). Disjunctions of montane forest birds in the Afrotropical region. *Bonner Zoologische Beiträge* **38**, 195–207.

Quickelberge, C.D. (1989). 'Birds of the Transkei'. Durban Natural History Museum, Durban.

Raijmakes, J.M.H. and Raijmakes, J.H.F.A. (1994–1995). Distribution, size and moult of migrant warblers in the southern Transvaal. Parts One; Two. *Safring News* **23**, 65–71; **24**, 3–12.

Rand, A.L. (1951). Birds from Liberia. *Fieldiana: Zoology* **32**(9), 561–653.

Rand, A.L., Friedman, H. and Traylor, M.A. (1959). Birds from Gabon and Moyen Congo. *Fieldiana: Zoology* **41**(2) 221–411.

Reichenow, A. (1902–1905). 'Die Vögel Afrikas'. Band II, III. J. Neumann, Neudamm.

Reichenow, A. (1910). Über eine Vogelsammlung vom Rio Benito im Spanischen Guinea. *Mitteilungen aus dem Zoologischen Museum in Berlin* **5**, 71–87.

Richards, D.K. (1982). The birds of Conakry and Kakulima, Democratic Republic of Guinea. *Malimbus* **4**, 93–103.

Ripley, S.D. (1952). The thrushes. *Postilla* **13**, 1–48.

Ripley, S.D. (1964). Subfamily Turdinae. *In* 'Check-list of Birds of the World', Vol. X (Mayr, E. and Paynter, R.A., Jr., Eds), pp. 13–227. Museum of Comparative Zoology, Cambridge, Massachusetts.

Ripley, S.D. and Bond, G.M. (1966). The birds of Socotra and Abd-el-Kuri. *Smithsonian Miscellaneous Collections* **151**(7), 1–37.

Ripley, S.D. and Heinrich, G.H. (1960, 1966a). Additions to the avifauna of northern Angola, I and II. *Postilla* **47**, 1–7; **95**, 1–29.

Ripley, S.D. and Heinrich, G.H. (1966b). Comments on the avifauna of Tanzania I. *Postilla* **96**, 1–45.

Roberts, A. (1935). Scientific results of the Vernay-Lang Kalahari Expedition, March to September, 1930. *Annals of the Transvaal Museum* **16**, 1–185.

Rodewald, P.G., Dejaifve, P.-A. and Green, A.A. (1994). The

birds of Korup National Park and Korup Project area, Southwest Province, Cameroon. *Bird Conservation International* **4**, 1–68.
de Roo, A.E.M. (1970). Contribution à l'ornithologie de la République du Togo. 2. *Revue de Zoologie et de Botanique Africaines* **81**, 163–172.
de Roo, A., de Vree, F. and Verheyen, W. (1969). Contribution à l'ornithologie de la République du Togo. *Revue de Zoologie et de Botanique Africaines* **79**, 309–322.
de Roo, A., Hulselmans, J. and Verheyen, W. (1971). Contribution à l'ornithologie de la République du Togo, 3. *Revue de Zoologie et de Botanique Africaines* **83**, 84–94.
de Roo, A., de Vree, F. and van der Streten, E. (1972). Contribution à l'ornithologie de la République du Togo. 4. *Revue de Zoologie et de Botanique Africaines* **86**, 374–384.
da Rosa Pinto, A.A. (1970). Um catálogo das aves do Distrito da Huila (Angola). *Memórias e Trabalhos do Instituto de Investigação Científica de Angola* **6**, 5–160.
da Rosa Pinto, A.A. (1972). Contribuição para o estudo da avifauna do Distrito de Cabinda (Angola). *Memórias e Trabalhos do Instituto de Investigação Científica de Angola* **10**, 9–90.
da Rosa Pinto, A.A. (1973). Aves da Colecção do Museu do Dundo. *Publicações Culturais da Companhia de Diamantes de Angola* No. 87, 129–178.
da Rosa Pinto, A.A. (1983). 'Ornithologie de Angola'. Instituto de Investigação Científica Tropical, Lisbon.
da Rosa Pinto, A.A. and Lamm, D.W. (1953). Contribution to the study of the ornithology of Sul do Save (Mozambique). *Memórias do Museu Dr. Álvaro de Castro* nr. **2**, 65–85, **3**, 125–159, **4**, 107–167.
Ryall, C. (1991). Avifauna of Nguuni near Mombassa, Kenya, between September 1984 and October 1987: Part I- Afrotropical species. *Scopus* **15**, 1–23.
Safford, R.J., Duckworth, J.W., Evans, M.I., Telfer, M.G., Timmins, R.J. and Zewdie, C. (1993). The birds of Nechisar National Park, Ethiopia. *Scopus* **16**, 61–80.
Salvan, J. (1968, 1969). Contribution à l'étude des oiseaux du Tchad. *L'Oiseau et la Revue Française d'Ornithologie* **38**, 249–273; **39**, 38–69.
Sargeant, D.E. (1994). Recent ornithological observations from São Tomé and Príncipe Islands. *Bulletin of the African Bird Club* **1**, 96–102.
Sayer, J., Harcourt, C.S. and Collins, N.M. (1992). 'The Conservation Atlas of Tropical Forests: Africa'. Simon and Schuster, New York.
Schmidl, D. (1982). The Birds of the Serengeti National Park, Tanzania. BOU Check-list No. 5. British Ornithologists' Union, London.
Schönwetter, M. and Meese, W. (1972, 1974–1977). 'Handbuch der Oologie', Lieferung 20–26. Akademie-Verlag, Berlin.
Schouteden, H. (1938). Oiseaux. *In* 'Exploration du Parc National Albert. I. Mission G.F. de Witte (1933–1935)', Fascicule 9. Institut des Parcs Nationaux du Congo belge, Bruxelles.
Schouteden, H. (1954–1955). De vogels van belgisch Congo en van Ruanda-Urundi, III. *Annales du Musée Royale du Congo Belge, série in-4to, Zoologie* **IV**, fasc. 1–2.
Schouteden, H. (1957). Faune du Congo belge et du Ruanda-Urundi, IV. Oiseaux Passereaux (1). *Annales du Musée Royal du Congo Belge, série in-8vo, Sciences zoologiques* **57**.
Schouteden, H. (1961). La faune ornithologique des districts de la Tshaupa et de l'equateur. *Documentation Zoologique, Musée Royal de l'Afrique Centrale* No. 1.
Schouteden, H. (1962). La faune ornithologique du territoire de Mushie. *Documentation Zoologique, Musée Royal de l'Afrique Centrale* No. 2.
Schouteden, H. (1966). La faune ornithologique du Rwanda. *Documentation Zoologique, Musée Royal de l'Afrique Centrale* No. 10.
Sclater, W.L. (1930). 'Systema Avium Aethiopicarum', pt. 2. British Ornithologists' Union, London.
Sclater, W.L. and Moreau, R.E. (1933). Taxonomic and field notes on some birds of north-eastern Tanganyika Territory. *Ibis* **1933**, 1–33, 399–439.
Seebohm, H. (1881). Catalogue of the Passeriformes or Perching Birds, Vol. 5. *In* 'The Collection of the British Museum'. Trustees, British Museum, London.
Serle, W. (1940). Field observations on some northern Nigerian birds, part II. *Ibis* **1940**, 1–47.
Serle, W. (1943a). Further field observations on northern Nigerian birds. *Ibis* **85**, 413–437.
Serle, W. (1949a). Notes on the birds of Sierra Leone (Part III). *Ostrich* **20**(2), 70–85.
Serle, W. (1950a). A contribution to the ornithology of the British Cameroons. *Ibis* **92**, 602–638.
Serle, W. (1954). A second contribution to the ornithology of the British Cameroons. *Ibis* **96**, 47–80.
Serle, W. (1955). Miscellaneous notes on the birds of the eastern highlands of Southern Rhodesia. *Ostrich* **26**, 115–127.
Serle, W. (1957). A contribution to the ornithology of the Eastern Region of Nigeria. *Ibis* **99**, 628–685.
Serle, W. (1964). The lower altitudinal limit of montane forest birds of the Cameroon Mountain, West Africa. *Bulletin of the British Ornithologists' Club* **84**, 87–91.
Serle, W. (1965). A third contribution to the ornithology of the British Cameroons. *Ibis* **107**, 60–94.
Serle, W. (1981). The breeding seasons of birds of the lowland rainforest and in the montane forest of west Cameroon. *Ibis* **123**, 62–74.
Sharland, R.E. and Wilkinson, R. (1981). The birds of Kano State, Nigeria. *Malimbus* **3**, 7–30.
Short, L.L., Horne, J.F.M. and Muringo-Gichuki, C. (1990). Annotated check-list of the birds of East Africa. *Proceedings of the Western Foundation of Vertebrate Zoology* **4**(3), 61–246.
Sibley, C.G. and Ahlquist, J.E. (1990). 'Phylogeny and Classification of the Birds of the World'. Yale University Press, New Haven.
Sibley, C.G. and Monroe, B.L., Jr. (1990). 'Distribution and Taxonomy of Birds of the World'. Yale University Press, New Haven.
Siegfried, W.R. (1968a). Ecological composition of the avifaunal community in a Stellenbosch suburb. *Ostrich* **39**, 105–129.
Siegfried, W.R. (1983). Trophic structure of some communities of Fynbos birds. *Journal of South African Botany* **49**, 1–43.
Sinclair, J.C. (1984). 'Field Guide to the Birds of Southern Africa'. C. Struik, Cape Town.
Sinclair, J.C., Hockey, P. and Tarboton, W. (1993). 'Birds of Southern Africa'. C. Struik, Cape Town.
Skead, D.M. (1974). Bird weights from the Central Transvaal bushveld. *Ostrich* **45**, 189–192.
Skinner, N.J. (1995). The breeding seasons of birds in Botswana. 1: passerine families. *Babbler* Nos. 29–30, 9–23.
Smith, K.D. (1955a). Recent records from Eritrea. *Ibis* **97**, 65–80.
Smith, K.D. (1957). An annotated check list of the birds of Eritrea. *Ibis* **99**, 307–337.
Smith, K.D. (1965). On the birds of Morocco. *Ibis* **107**, 493–526.
Smith, K.D. (1968). Spring migration through southeast Morocco. *Ibis* **110**, 452–492.

Smithers, R.H.N. (1964). 'A Check List of the Birds of the Bechuanaland Protectorate and the Caprivi Strip'. Trustees of the National Museums of Southern Rhodesia, Salisbury.

van Someren, V.G.L. (1916). A list of birds collected in Uganda and British East Africa, with notes on their nesting and other habits. Part II. *Ibis* **1916**, 373–472.

van Someren, V.G.L. (1922, 1932). Notes on the birds of East Africa. *Novitates Zoologicae* **29**, 1–246; **37**, 252–380.

van Someren, V.G.L. (1939). Report on the Coryndon Museum expedition to the Chyulu Hills, Part 2. The Birds of the Chyulu Hills. *Journal of the East Africa and Uganda Natural History Society* **XIV**, Nos. 1–2, 15–129.

van Someren, V.G.L. (1956). Days with birds. *Fieldiana: Zoology* **38**, 1–520.

van Someren, V.G.L. and van Someren, G.R.C. (1949). The birds of Bwamba. *Uganda Journal* **13**, Special Supplement, 111 pp.

Stuart, S.N. (1981a). A comparison of the avifaunas of seven east African forest islands. *African Journal of Ecology* **19**, 133–151.

Stuart, S.N. (Ed.) (1986a). 'Conservation of Cameroon Montane Forests'. International Council for Bird Preservation, Cambridge.

Stuart, S.N. (1986b). Records of other species of birds from western Cameroon. *In* 'Conservation of Cameroon Montane Forests' (Stuart, S.N., Ed.), pp. 106–129. International Council for Bird Preservation, Cambridge.

Stuart, S.N. and Adams, R.J. (1990). Biodiversity in Sub-Saharan Africa and its islands. *Occasional Papers IUNC Species Survival Commission* No. 6, 242 pp. International Union for Conservation of Nature and Natural Resources, Gland, Switzerland.

Stuart, S.N. and Hutton, J.M. (Eds) (1977). The avifauna of the East Usambara Mountains, Tanzania. *Report on the Cambridge Ornithological Expedition to East Africa* pp. 1–90.

Stuart, S.N. and Jensen, F.P. (1981). Further range extensions and other notable records of forest birds from Tanzania. *Scopus* **5**, 106–115.

Stuart, S.N. and Jensen, F.P. (1985). The avifauna of the Uluguru Mountains, Tanzania. *Le Gerfaut* **75**, 155–197.

Stuart, S.N. and Jensen, F.P. (1986). The status and ecology of montane forest bird species in western Cameroon, pp. 38–105. *In* 'Conservation of Cameroon Montane Forests' (Stuart, S.N., Ed.). International Council for Bird Preservation, Cambridge.

Stuart, S.N., Jensen, F.P. and Brøgger-Jensen, S. (1987). Altitudinal zonation of the avifauna in the Mwanihana and Magombera forests, eastern Tanzania. *Le Gerfaut* **77**, 165–186.

Styles, C. (1995). Notes on the bird species observed feeding on mopane worms. *Birding in Southern Africa* **47**(2), 53–54.

Svensson, L. (1992). 'Identification Guide to European Passerines'. British Trust for Ornithology, Thetford.

Tarboton, W.R., Kemp, M.I. and Kemp, A.C. (1987). 'Birds of the Transvaal'. Transvaal Museum, Pretoria.

Taylor, P.B. and Taylor, C.A. (1988). The status, movements and breeding of some birds in the Kikuyu Escarpment Forest, central Kenya highlands. *Tauraco* **1**(1), 72–89.

Thévenot, M. (1982). Contribution à l'étude écologique des Passereaux forestiers du Plateau Central et de la corniche du Moyen Atlas (Maroc). *L'Oiseau et la Revue Française d'Ornithologie* **52**(1), 21–86, 97–152.

Thévenot, M., Bergier, P. and Beaubrun, P. (1980). Compte-rendu d'ornithologie marocaine, année 1979. *Documents de l'Institut Scientifique, Université Mohammed V, Rabat* No. 5.

Thévenot, M., Bergier, P. and Beaubrun, P. (1981). Compte-rendu d'ornithologie marocaine, année 1980. *Documents de l'Institut Scientifique, Université Mohammed V, Rabat* No. 6.

Thévenot, M., Beaubrun, P., Baouab, R.E. and Bergier, P. (1982). Compte-rendu d'ornithologie marocaine, année 1981. *Documents de l'Institut Scientifique, Université Mohammed V, Rabat* No. 7.

Thiollay, J.-M. (1971). L'avifaune de la région de Lamto (moyenne Côte d'Ivoire). *Annales, Université d'Abidjan: série E: Ecologie* **4**(1), 5–132.

Thiollay, J.-M. (1973). Place des oiseaux dans les chaînes trophiques d'une zone préforestière en Côte d'Ivoire. *Alauda* **41**, 273–300.

Thiollay, J.-M. (1985). The birds of the Ivory Coast: status and distribution. *Malimbus* **7**, 1–59.

Thomsen, P. and Jacobsen, P. (1979). 'The Birds of Tunisia'. Nature-Travels, Copenhagen.

Thonnérieux, Y. (1988a). Commentaires sur la distribution de quelques migrateurs paléarctiques au Burkina Faso. *Le Gerfaut* **78**, 317–362.

Thonnérieux, Y. (1988b). Etat des connaissances sur la reproduction de l'avifaune du Burkina Faso (ex Haute-Volta). *L'Oiseau et la Revue Française d'Ornithologie* **58**(2), 120–146.

Thonnérieux, Y., Walsh, J.F. and Bortoli, L. (1989). L'avifaune de la ville de Ouagadougou et ses environs (Burkina Faso). *Malimbus* **11**, 7–40.

Traylor, M.A. (1962). 'Notes on the birds of Angola, Passeres'. *Publicações Culturais da Companhia de Diamantes de Angola* No. 58.

Traylor, M.A. (1963). 'Check-list of Angolan Birds'. *Publicações Culturais da Companhia de Diamantes de Angola* No. 61.

Traylor, M.A. (1965a). A collection of birds from Barotseland and Bechuanaland. *Ibis* **107**, 357–384.

Traylor, M.A., Jr. (1986). Family Platysteiridae. *In* 'Checklist of Birds of the World', Vol. XI (Mayr, E. and Cottrell, G.W, Eds), pp. 376–390. Museum of Comparative Zoology, Cambridge, Massachusetts.

Traylor, M.A., Jr. and Archer, A.L. (1982). Some results of the Field Museum 1977 Expedition to south Sudan. *Scopus* **6**, 5–12.

Urban, E.K. and Brown, L.H. (1971). 'A Checklist of the Birds of Ethiopia'. Haile Sellassie I University Press, Addis Ababa.

Vande weghe, J.-P. (1973). Les périodes de nidification des oiseaux du Parc National de l'Akagera au Rwanda. *Le Gerfaut* **63**, 235–255.

Vande weghe, J.-P. (1979). The wintering and migration of Palaearctic passerines in Rwanda. *Le Gerfaut* **69**, 29–43.

Vande weghe, J.-P. (1981). L'avifaune des papyraies au Rwanda et au Burundi. *Le Gerfaut* **71**, 489–536.

Vande weghe, J.-P. and Loiselle, B.A. (1987). The bird fauna of the Bururi Forest, Burundi. *Le Gerfaut* **77**, 147–164.

Vaurie, C. (1959). 'The Birds of the Palearctic Fauna', Order Passeriformes. H.F. & G. Witherby, London.

Verheyen, R. (1953). Oiseaux. *In* 'Exploration du Parc National de l'Upemba. I. Mission G.F. de Witte (1946–1949)', Fascicule 19. Institut des Parcs Nationaux du Congo belge, Bruxelles, 687 pp.

Vernon, C.J. (1977). Birds of the Zimbabwe Ruins area. *Southern Birds* **4**.

Vieillard, J. (1971–1972). Données biogéographiques sur l'avifaune de l'Afrique Centrale. *Alauda* **39**, 227–248; **40**, 63–92.

Vincent, A.W. (1947, 1948a). On the breeding habits of some African birds. *Ibis* **89**, 163–204; **90**, 284–312.

Vincent, J. (1934, 1935, 1936). The birds of northern Portuguese East Africa. *Ibis* **1934**, 126–160; **1935**, 355–397, 485–529, 707–762; **1936**, 48–125.

Voous, K.H. (1960). 'Atlas of European Birds'. Nelson, London.
Voous, K.H. (1977). List of recent Holarctic bird species. Passerines. *Ibis* **119**, 223–250.
Wacher, T. (1993). Some new observations of forest birds in The Gambia. *Malimbus* **15**, 24–37.
Walsh, J.F., Cheke, R.A. and Sowah, S.A. (1990). Additional species and breeding records of birds in the Republic of Togo. *Malimbus* **12**, 2–18.
Watson, G.E., Traylor, M.A., Jr. and Mayr, E. (1986). Families Sylviidae, Muscicapidae, and Monarchidae. In 'Check-list of Birds of the World', Vol. XI (Mayr, E. and Cottrell, G.W., Eds), pp. 3–294, 295–375, 464–556. Museum of Comparative Zoology, Cambridge, Massachusetts.
Welch, G.R. and Welch, H.J. (1984a). Birds seen on an expedition to Djibouti. *Sandgrouse* **6**, 1–23.
Welch, G.R. and Welch, H.J. (1984b). 'Djibouti expedition March 1984'. Published by the authors (21a East Delph, Whittlesey, Cambridgeshire PE7 1RH, UK).
Welch, G.R. and Welch, H.J. (1992). Djibouti III. Published by the authors (Minsmere Reserve, Westleton, Saxmundham, Suffolk, IP17 3BY, UK).
Welch, G.R., Welch, H.J., Coghlan, S.M. and Denton, M.L. (1989). Djibouti II – autumn '85. Published by the authors (21a East Delph, Whittlesey, Cambridgeshire PE7 1RH, UK).
Wetmore, A. (1960). A classification for the birds of the world. *Smithsonian Miscellaneous Collections* **139**(11), 1–37.
White, C.M.N. (1960a, 1962). A Check List of the Ethiopian Muscicapidae (Sylviinae). *Occasional Papers of the National Museums of Southern Rhodesia* **24B**, 399–430; **26B**, 653–738.
White, C.M.N. (1963). 'A Revised Check List of African Flycatchers, Tits, Tree Creepers, Sunbirds, White-eyes, Honey Eaters, Buntings, Finches, Weavers and Waxbills.' Government Printer, Lusaka, 218 pp.
Williams, J.G. and Arlott, N. (1980). 'A Field Guide to the Birds of East Africa'. Collins, London.
Wilson, J.D. (1987). The status and conservation of the montane forest avifauna of Mount Oku, Cameroon in 1985. International Council for Bird Preservation, Cambridge.
Winterbottom, J.M. (1964). Results of the Percy Fitzpatrick Institute-Windhoek State Museum joint ornithological expeditions: Report on the birds of Game Reserve No. 2. *Cimbebasia* No. 9, 1–75.
Winterbottom, J.M. (1966). Results of the Percy Fitzpatrick Institute-Windhoek State Museum joint ornithological expeditions: 5. Report on the birds of the Kaokoveld and Kunene River. *Cimbebasia* No. 19, 1–71.
Winterbottom, J.M. (1971a). 'A Preliminary Check List of the Birds of South West Africa'. South West Africa Scientific Society, Windhoek.
Witherby, H.F., Jourdain, F.C.R., Ticehurst, N.F. and Tucker, B.W. (1938). 'The Handbook of British Birds', Vols 1–2. H.F. and G. Witherby Ltd., London.
Wolters, H.E. (1979–1980). 'Die Vögelarten der Erde', Lieferung 4–6. Paul Parey, Hamburg and Berlin.
Zedlitz, O.G. (1911). Meine ornithologische Ausbeute in Nordost-Afrika. *Journal für Ornithologie* **59**, 1–92.
Zimmerman, D.A. (1972). The avifauna of the Kakamega Forest, western Kenya, including a bird population study. *Bulletin of the American Museum of Natural History* **149**, Article 3.
Zink, G. (1973, 1975, 1981, 1985). 'Der Zug europäischer Singvögel', Vols. 1–4. Vogelzug-Verlag, Möggingen.

2. Family References

Family TURDIDAE: *Monticola, Zoothera, Psophocichla* and *Turdus*

Anonymous (1971). Groundscraper Thrushes at the nest. *Honeyguide* No. 67, 20–21.
Anonymous (1993). Conserving the Spotted Thrush. *Birding in Southern Africa* **45**, 40.
Ashmole, M.J. (1962). The migration of European thrushes: a comparative study based on ringing recoveries. *Ibis* **104**, 314–346, 522–559.
Aspinwall, D.R. (1977). European Rockthrush at Mbala. *Bulletin of the Zambian Ornithological Society* **9**(1), 25–26.
Aspinwall, D.R. (1981). Comments on Groundscraper Thrush (October 1976 – September 1978, January 1979 – January 1981). *Zambian Ornithological Society Newsletter* **11**(12), 163–165.
Backhurst, G.C. (1987). Behaviour of Rock Thrush in winter quarters. *British Birds* **80**, 77–78.
Backhurst, G.C. (Ed.) (1992). East African Bird Report 1990. *Scopus* **14**(3).
Baker, N.E. and Baker, E.M. (1992). Four Afrotropical migrants on the East African coast: evidence for a common origin. *Scopus* **15**, 122–124.
Beals, E.W. (1970). Birds of a *Euphorbia-Acacia* woodland in Ethiopia: habitat and seasonal changes. *Journal of Animal Ecology* **39**, 277–297.
Beasley, A. (1993). Foliage-bathing by Black-collared Barbets and Kurrichane Thrushes. *Honeyguide* **39**(1), 30–31.
Bednall, D.K. (1958). Heller's Thrush on Kilimanjaro. *Journal of the East Africa Natural History Society* **XXIII**, No. 2 (99), 17.
Belcher, C.F. (1930). 'The Birds of Nyasaland'. Technical Press, London.
Bennun, L.A. (1985). The Spotted Ground Thrush *Turdus fischeri fischeri* at Gede in coastal Kenya. *Scopus* **9**, 97–107.
Bennun, L.A. (1987). Ringing and recapture of Spotted Forest Thrushes *Turdus fischeri fischeri* at Gede, Kenya coast; indications of site fidelity and population size stability. *Scopus* **11**, 1–5.
Bennun, L.A. (1989). E.A.N.H.S. nest record scheme: 1985–1989. *Scopus* **13**(3), 165–180.
Benson, C.W. (1950). Miscellaneous taxonomic notes (b) *Geokichla gurneyi*. *Ostrich* **21**, 28–30.

Benson, C.W. (1950). Some notes on the Spotted Forest Thrush *Turdus fischeri*. *Ostrich* **21**, 58–61.

Benson, C.W. (1951). The Spotted Forest Thrush *Turdus fischeri*. *Ostrich* **22**, 121.

Benson, C.W. (1952). A further note on the Spotted Forest Thrush *Turdus fischeri*. *Ostrich* **23**, 48.

Benson, C.W. (1954). The status of *Turdus fischeri belcheri* Benson, *Ostrich*, 1950, p. 58. *Bulletin of the British Ornithologists' Club* **74**, 88–90.

Benson, C.W. (1969). The relationship of *Turdus pelios bocagei* (Cabanis) and *Turdus pelios stormsi* Hartlaub. *Bulletin of the British Ornithologists' Club* **89**(5), 133–134.

Beven, G. (1968). Studies of less familiar birds 148. Blue Rock Thrush. *British Birds* **61**, 303–307.

Beven, G. and England, M.D. (1969). Studies of less familiar birds 152. Rock Thrush. *British Birds* **62**, 23–25.

Brieschke, H. (1989). 578 Spotted Thrush *Turdus fischeri*. In 'An annotated list of rare and poorly documented birds recently recorded in the eastern Cape Province.' *Bee-eater* **40**, 42.

Britton, P.L. and Rathbun, G.B. (1978). Two migratory thrushes and the African Pitta in coastal Kenya. *Scopus* **2**, 11–17.

Broadley, D.G. (1974). Predation by birds on reptiles and amphibians in south-eastern Africa. *Honeyguide* No. 78, 11–18.

Broekhuysen, G.J. (1941). Some observations on the diet of a young Cape Rock Thrush (*Monticola rupestris*). *Ostrich* **12**(2), 71–74.

Broekhuysen, G.J. (1965). Nesting behaviour of the Sentinel Thrush *Monticola explorator* (Vieillot). *Ostrich* **36**(1), 41–42.

Brooke, R.K. (1986). Visitors to a fruiting *Ficus* at Karoi. *Honeyguide* **32**, 93.

Brooks, D.J. (1987). Feeding observations of birds in North Yemen. *Sandgrouse* **9**, 115–120.

Brooks, D.J., Evans, M.I., Martins, R.P. and Porter, R.F. (1987). The status of the birds in North Yemen and the records of the OSME Expedition in autumn 1985. *Sandgrouse* **9**, 4–66.

Brosset, A. and Erard, C. (1976). Première description de la nidification de quatre espèces de la forêt gabonaise. *Alauda* **44**, 205–235.

Brosset, A. and Erard, C. (1977). New faunistic records from Gabon. *Bulletin of the British Ornithologists' Club* **97**, 125–132.

Brown, C.J. and Riekert, B.R. (1984). Range extensions to the distribution of some birds occurring in South West Africa/Namibia. *Lanioturdus* **20**(6), 3–12.

Burrell, J.H., Abel, R. and Abel, Mrs. (1976). A not so 'extinct' thrush on the Kenya coast. *East Africa Natural History Society Bulletin* **1976**, 32–33.

Campbell, N.A. (1973). Letter to the Editor. *Honeyguide* No. 74, 40.

Carlson, A. (1986). A comparison of birds inhabiting pine plantations and indigenous forest patches in a tropical mountain area. *Biological Conservation* **35**, 195–204.

Carlson, A. and Moreno, J. (1986). Foraging behaviour and parental care in the Fieldfare. *Ardea* **74**, 79–90.

Carpentier, C.-J. (1983). Contribution à l'étude de l'ornithologie marocaine les oiseaux du pays Zaïan. *Bulletin de la Société des Sciences Naturelles et Physiques du Maroc* **13**, 23–68.

Chiazzari, W.L. (1952). Some observations on the Natal Spotted Forest Thrush *Turdus fischeri natalicus*. *Ostrich* **23**, 50.

Chittenden, H. (1982). Kurrichane Thrush builds seven nests in one season. *Bokmakierie* **34**(3), 67–68.

Clancey, P.A. (1952). Miscellaneous taxonomic notes on African birds (4). A new race of Sentinel Rock-Thrush *Monticola explorator* (Vieillot). *Durban Museum Novitates* **IV**(1), 13–14.

Clancey, P.A. (1952). Three new races of South African birds. *Bonner Zoologische Beiträge*, **1–2**(3), 17–22.

Clancey, P.A. (1955a). Further as to the present status of *Turdus fischeri natalicus* Grote. *Ostrich* **26**, 164–165.

Clancey, P.A. (1955b). Geographical variation in the Orange Thrush *Turdus gurneyi* Hartlaub of eastern and south-eastern Africa. *Bulletin of the British Ornithologists' Club* **75**, 70–78.

Clancey, P.A. (1956). The South African races of the Ground-scraper Thrush *Turdus litsipsirupa* (Smith). *Durban Museum Novitates* **IV**(17), 288–291.

Clancey, P.A. (1957a). Further records of the Spotted Thrush being killed on migration. *Ostrich* **28**, 126–127.

Clancey, P.A. (1957b). Some further records of the Spotted Thrush on migration. *Ostrich* **28**, 37.

Clancey, P.A. (1965a). Miscellaneous taxonomic notes on African birds XXIII. 2: Racial variation in the rock-thrush *Monticola angolensis* Sousa. *Durban Museum Novitates* **VIII**(1), 9–17.

Clancey, P.A. (1965b). The characters and ranges of the subspecies of the Kurrichane Thrush. *Arnoldia (Rhodesia)* **2**(2), 7 pp.

Clancey, P.A. (1968). The status of *Monticola pretoriae* Gunning and Roberts, 1911. *Bulletin of the British Ornithologists' Club* **88**, 126–128.

Clancey, P.A. (1972). Miscellaneous taxonomic notes on African birds XXXIV. New races of a rockthrush and a warbler from Angola. *Durban Museum Novitates* **IX**(11), 146–152.

Clancey, P.A. (1982). The Olive Thrush *Turdus olivaceus* Linnaeus in southern Africa. *Durban Museum Novitates* **XIII**(7), 65–70.

Clancey, P.A. (1991). The generic status of the Spotted Thrush. *Bee-eater* **42**, 23.

Clancey, P.A. (1992–1993). The ground thrush *Zoothera guttata* (Vigors 1831) in the southern Afrotropics. *Le Gerfaut* **82–83**, 45–50.

Clark, G.A. (1983). An additional method of foraging in litter by species of *Turdus* thrushes. *Wilson Bulletin* **95**(1), 155–157.

Cody, M.L. (1983). Bird diversity and density in South Africa forests. *Oecologia* **59**, 201–215.

Cole, D. (1984). The specific epithet of *Turdus litsitsirupa* (Smith). *Bokmakierie* **36**(1), 11–12.

Cooper, R. (1945). Redwing, *Turdus musicus* Linnaeus, in Morocco. *Ibis* **89**(4), 566–567.

Cornwallis, L. and Porter, R.F. (1982). Spring observations on the birds of North Yemen. *Sandgrouse* **4**, 1–36.

Cunningham-van Someren, G.R. and Schifter, H. (1981). New races of montane birds from Kenya and southern Sudan. *Bulletin of the British Ornithologists' Club* **101**(4), 355–370.

Dean, W.R.J. (1987). Birds of the upper Limpopo River Valley. *Southern Birds* **14**, 1–76.

Dean, W.R.J. and Macdonald, I.A.W. (1981). A review of African birds feeding in association with mammals. *Ostrich* **52**, 135–155.

Diamond, A.W. and Keith, S. (1980). Avifaunas of Kenya forest islands, I – Mount Kulal. *Scopus* **4**, 49–55.

Dickerman, R.W., Cane, W.P., Carter, M.F., Chapman, A. and Schmitt, C.G. (1994). Report on three collections of birds from Liberia. *Bulletin of the British Ornithologists' Club* **114**, 267–274.

Dorst, J. (1950). Considérations systématiques sur les grives du genre *Turdus* L. *L'Oiseau et la Revue Française d'Ornithologie* **20**, 212–248.

Douglas, R. (1995). Utilization of an urban fish pond by the Olive Thrush *Turdus olivaceus*. *Mirafra* **12**(1&2), 7–8.
Dranzoa, C. (1994). Lyre-tailed Honeyguide *Melichneutes robustus* and Grey Ground Thrush *Zoothera princei batesi*: new records for Uganda. *Scopus* **18**, 128–130.
Earlé, R.A. and Oatley, T.B. (1983). Population, ecology and breeding of the Orange Thrush at two sites in eastern South Africa. *Ostrich* **54**, 205–212.
Eisentraut, M. (1970). Die vertikale Rassendifferenzierung und Rassen-integration bei *Turdus olivaceus* von West-Kamerun und Fernando Poo. *Bonner Zoologische Beiträge* **21**(1–2), 119–132.
Elliott, C.C.H. and Jarvis, M.J.F. (1972, 1973). Fifteenth ringing report. *Ostrich* **43**(4), 236–295; **44**(1), 34–78.
Erard, C. (1990b). Aide au nourrissage chez le Merle noir *Turdus merula*. *L'Oiseau et la Revue Française d'Ornithologie* **60**(1), 56–57.
Etchécopar, R.-D. (1950). Contributions oologiques à l'étude systématique du genre *Turdus*. *L'Oiseau et la Revue Française d'Ornithologie* **20**(3/4), 249–262.
Evans, D.J. and Dijkstra, K.D. (1993). The birds of Gezira, Egypt. *Ornithological Society of the Middle East Bulletin* **30**, 20–26.
Farkas, T. (1962a). Zur Biologie und Ethologie der südafrikanischen Arten der Gattung *Monticola* (Boie). II. Teil: *Monticola rupestris* L. *Die Vogelwelt* **83**(6), 161–173.
Farkas, T. (1962b). Zur Biologie und Ethologie der südafrikanischen Arten der Gattung *Monticola* (Boie). *Die Vogelwelt* **83**(4), 97–116.
Farkas, T. (1963). Zur Biologie und Ethologie der südafrikanischen Arten der Gattung *Monticola* (Boie). III. Teil: *Monticola explorator* (Vieillot). *Die Vogelwelt* **84**(1), 11–22.
Farkas, T. (1966). Zur systematischen Stellung des Pretoria-Steinrötels *Monticola pretoriae* Gunning & Roberts. *Die Vogelwelt* **87**(2), 33–48.
Farkas, T. (1968). Zur Biologie und Ethologie der Angola-Waldmerle, *Monticola angolensis* Sousa. *Revue de Zoologie et de Botanique Africaines* **77**(1/2), 162–189.
Farkas, T. (1979). A further note on the status of *Monticola pretoriae* Gunning and Roberts, 1911. *Bulletin of the British Ornithologists' Club* **99**, 20–21.
Field, G.D. (1973). Ortolan and Blue Rock Thrush in Sierra Leone. *Bulletin of the British Ornithologists' Club* **93**, 81–82.
Friedmann, H. and Williams, J.G. (1968). Notable records of rare and little-known birds from western Uganda. *Revue de Zoologie et de Botanique Africaines* **77**, 11–36.
Fry, C.H. (1975). The northern limits of fringing forest birds in North Central State, Nigeria. *Bulletin of the Nigerian Ornithologists' Society* **11**(40), 56–64.
Fuggles-Couchman, N.R. and Elliot, H.F.I. (1946). Some records and field-notes from north-eastern Tanganyika territory. *Ibis* **88**, 327–347.
Ginn, P.J. (1982). Letter to the editor. *Safring News* **11**(2), 72–73.
Goodman, J.D. and Goodman, S.M. (1985). Song mimicry, duetting and antiphonal song in African birds. *East Africa Natural History Society Bulletin* **1985**, 75–77.
Goodman, S.M. and Watson, G.E. (1984). Records of Palearctic thrushes (*Turdus* spp.) in Egypt and north-eastern Africa. *Le Gerfaut* **74**, 145–161.
Goodman, S.M., Meininger, P.L. and Mullié, W.C. (1986). The birds of the Egyptian Western Desert. *Miscellaneous Publications Museum of Zoology, University of Michigan*, No. 172.
Grimes, L.G. (1972). The successive use of the same nest by the Laughing Dove *Streptopelia senegalensis* and its subsequent use by the Kurrichane Thrush *Turdus libonyanus*. *Bulletin of the Nigerian Ornithologists' Society* **9**(35), 57.
Heim de Balsac, H. (1952). Rythme sexuel et fécondité chez les oiseaux du nord-ouest de l'Afrique. *Alauda* **20**(4), 213–242.
Hezekia, G. (1987). Helpers at nest of Groundscraper Thrush. *Honeyguide* **37**(1), 18.
Hillman, J.C. and Hillman, S.M. (1986). Notes on some unusual birds of the Bangangai area, south west Sudan. *Scopus* **10**, 29–32.
Holsten, B., Braunlich, A. and Huxham, W. (1991). Rondo Forest Reserve, Tanzania: an ornithological note including new records of the East Coast Akalat *Sheppardia gunningi*, the Spotted Ground Thrush *Turdus fischeri*, and the Rondo Green Barbet *Stactolaema olivacea woodwardi*. *Scopus* **14**, 125–128.
Howell, K.M. and Msuya, C.A. (1979). Bare-eyed Thrush *Turdus tephronotus* from the Selous Game Reserve. *Scopus* **3**, 29.
Ince, S.A. and Sclater, P.J.B. (1985). Versatility and continuity in the songs of thrushes *Turdus* spp. *Ibis* **127**, 355–364.
Irvine, G. and Irvine, D. (1988). Down at Diani – October 1987. *East Africa Natural History Society Bulletin* **18** (1), 10–11.
Irwin, M.P.S. (1960). A history of the supposed occurrence of the Olive Thrush *Turdus olivaceus* in Ngamiland. *Bulletin of the British Ornithologists' Club* **80**(6), 95–98.
Irwin, M.P.S. (1983). The status of the Groundscraper Thrush in the middle Zambezi Valley. *Honeyguide* No. 114/115, 55.
Irwin, M.P.S. (1984). The genera of African thrushes and the systematic position of the Groundscraper Thrush. *Honeyguide* **30**(1), 13–20.
Irwin, M.P.S. (1992). The Spotted Thrush and Spotted-winged Thrush: an Afrotropical-Oriental connection. *Honeyguide* **38**, 142–144.
James, G.L. (1951). Nest of the Angola Rock Thrush *Monticola angolensis*. *Ostrich* **22**(1), 47–48.
Jensen, J.V. and Kirkeby, J. (1987). Records of Rock Thrush *Monticola saxatilis* in The Gambia. *Malimbus* **9**, 123–124.
Jones, P.J. and Tye, A. (1988). A survey of the avifauna of São Tomé and Príncipe. *ICBP Study Report* No. 24, 1–64. International Council for Bird Protection, Cambridge.
Jonsson, L. (1993). 'Birds of Europe with North Africa and the Middle East'. Princeton University Press, Princeton.
Keith, S. (1968). Notes on birds of East Africa, including additions to the avifauna. *American Museum Novitates* No. 2321, 1–15.
Keith, S. and Garrett, K.L. (1994). Oberländer's Ground-Thrush *Zoothera oberlaenderi* in the Impenetrable Forest, Uganda. *Scopus* **17**, 141–142.
Keith, S. and Twomey, A. (1968). New distributional records for some East African birds. *Ibis* **110**, 537–548.
Keith, S. and Urban, E.K. (1992). A summary of present knowledge of the status of thrushes in the *Turdus olivaceus* species complex. *Proceedings of the Seventh Pan-African Ornithological Congress* pp. 249–260. PAOC Committee, Nairobi.
Kelsey, M.G. and Langton, T.E.S. (1984). The conservation of the Arabuko-Sokoke Forest, Kenya. *ICBP Study Report* No. 4. International Council for Bird Protection, Cambridge.
Kirkpatrick, C.W.M. (1993). Miombo Rock Thrush incubation period. *Honeyguide* **39**, 25–26.
Koen, J.H. (1988). Birds of the Eastern Caprivi. *Southern Birds* **15**, 73 pp.

LaCroix, T. (1989). Inter-generic aggression in Turdidae. *Nyala* **14**(1), 39–52.

Liversidge, R. (1957). A note on the Natal Thrush, *Turdus fischeri*. *Ostrich* **28**, 179–180.

Liversidge, R. (1985). Alien bird species introduced into southern Africa. *Proceedings of the Symposium on Birds and Man, Johannesburg 1983*, pp. 31–44.

Lorber, P. (1973). Multiple brooding in the Kurrichane Thrush *Turdus libonyanus*. *Ostrich* **44**(1), 84.

Lübcke, W. and Furrer, R.K. (1985). 'Der Washolderdrossel *Turdus pilaris*'. A. Ziemsen, Wittenberg, Lutherstadt.

Macdonald, I.A.W. and Birkenstock, P.J. (1980). Birds of the Hluhluwe-Umfolozi Game Reserve complex. *Lammergeyer* **29**, 1–56.

Macdonald, J.D. (1948). A new race of Orange Ground Thrush from Uganda. *Bulletin of the British Ornithologists' Club* **69**, 12.

Mearns, E.A. (1913). Descriptions of four new African thrushes of the genera *Planesticus* and *Geocichla*. *Smithsonian Miscellaneous Collections* **61**(10), 1–5.

Meinertzhagen, R. (1932). Occurrence of *Turdus philomelos clarkei* in Algeria. *Ibis* **1932**, 349.

Meinertzhagen, R. (1951). Some relationships between African, Oriental and Palaearctic genera and species, with a review of the genus *Monticola*. *Ibis* **93**(3), 443–459.

Melde, F. and Melde M. (1991). 'Die Singdrossel'. Ziemsen, Wittenberg, Lutherstadt.

Milstein, P. le S. (1968). Affinity of *Turdus litsipsirupa*. *Bulletin of the British Ornithologists' Club* **88**(1), 1.

Moreau, R.E. (1936b). A contribution to the ornithology of Kilimanjaro and Mount Meru. *Proceedings of the Zoological Society of London (1935)* **1936**, 843–891.

Moreau, W.M. (1943). Common Rock-Thrush feeding in winter quarters. *Ibis* **85**(1), 103.

Morel, G.J., Monnet, C. and Rouchouse, C. (1983). Données nouvelles sur *Monticola solitaria* et *Monticola saxatilis*. *Malimbus* **5**, 1–4.

Moyer, D.C. and Lovett, J.C. Community structure, ecology, and conservation status of montane forest birds in the southern Udzungwa Mountains, Tanzania. *Scopus*, in press.

Munro, R.C. (1969). Fledgling period of Gurney's Thrush. *Ostrich* **40**, 214.

de Naurois, R. (1984c). Les *Turdus* des Iles de São Tomé et Príncipe *T. o. olivaceofuscus* (Hartlaub) et *T. olivaceofuscus xanthorhynchus* Salvadori (Aves, Turdinae). *Revue de Zoologie Africaine* **93**, 403–423.

Ng'weno, F. (1986). Happy birthday, Olive Thrush. *East Africa Natural History Society Bulletin* **16**(2), 17–18.

Niethammer, G. (1963). Zur Vogelwelt der Hoggar-Gebirges (Zentrale Sahara). *Bonner Zoologische Beiträge* **14**, 129–150.

Nikolaus, G. (1982). A new race of the Spotted Ground Thrush *Turdus fischeri* from south Sudan. *Bulletin of the British Ornithologists' Club* **102**, 45–47.

Ogilvie-Grant, W.R. (1910). 'Aves' in 'Ruwenzori Expedition Reports'. *Transactions of the Zoological Society of London* **XIX** (19), Pt. IV, 253–481.

Parnell, G.W. (1974). Treble-brooding of Kurrichane Thrush. *Honeyguide* No. 79, 37.

Parnell, G.W. (1975). Further note on Kurrichane Thrush breeding behaviour. *Honeyguide* No. 82, 44–45.

Payne, R.B. (1981). Review of "A new ground-thrush from Africa" by A. Prigogine. *Journal of Field Ornithology* **51**, 199.

Pérez-Gonzalez, J.A. and Soler, M. (1990). Le régime alimentaire en automne-hiver de la grive draine *Turdus viscivorus* dans le sud-est de l'Espagne. *Alauda* **58**(3), 195–202.

Phillips, N.R. (1982). Observations on the birds of North Yemen in 1979. *Sandgrouse* **4**, 37–59.

Pikula, J. and Beklova, M. (1983). Nidobiology of *Turdus merula*. *Acta Scientiarum Naturalium Academiae Scientiarum Bohemoslovacae, Brno* **17**(7), 1–46.

Pitman, C.R.S. (1961). The Kurrichane Thrush *Turdus libonyanus tropicalis* Peters as host of the Red-chested Cuckoo *Cuculus solitarius* Stephens in Southern Rhodesia. *Bulletin of the British Ornithologists' Club* **81**(3), 48–49.

Prigogine, A. (1965). Notes sur les *Geokichla* de la République du Congo. *Revue de Zoologie et de Botanique Africaines* **71**, 230–244.

Prigogine, A. (1977). The Orange Ground Thrush *Turdus tanganjicae* (Sassi) is a valid species. *Bulletin of the British Ornithologists' Club* **97**, 10–15.

Prigogine, A. (1978b). A new ground-thrush from Africa. *Le Gerfaut* **68**, 482–492.

Prigogine, A. (1984b). Les populations occidentales de la Grive terrestre d'Abyssinie, *Zoothera piaggiae*, et description d'une nouvelle sous-espèce du Ruwenzori. *Le Gerfaut* **74**, 383–389.

Prigogine, A. (1985b). Révision des espèces africaines appartenant au genre *Zoothera*. *Le Gerfaut* **75**, 285–315.

Prigogine, A. and Louette, M. (1984). A new race of the Spotted Ground Thrush, *Zoothera guttata*, from Upemba, Zaïre. *Le Gerfaut* **74**, 185–186.

Quickelberge, C.D. (1966). The status of Gurney's Thrush *Turdus g. gurneyi* (Hartlaub) in the eastern Cape. *Ostrich* **37**, 230–231.

Quickelberge, C.D. (1969). Notes on the Spotted Thrush *Turdus fischeri*. *Ostrich* **40**, 133–134.

Rensch, B. (1923). Die Formenkreise *Turdus libonyanus* (A. Sm.) und *Turdus olivaceus* L. *Journal für Ornithologie* **71**, 95–104.

Rouchouse, C. (1985). Sédenterisation de *Monticola solitarius* au Cap de Naze, Sénégal. *Malimbus* **7**, 91–94.

de los Santos, M.R., Cuadrado, M. and Castellano, A. (1987). Prey-catching techniques of Blue Rock Thrush. *British Birds* **80**, 578.

Schmidt, E. and Farkas, T. (1974). 'Des Steinrötel'. Wittenburg, Lutherstadt.

Serle, W. (1962). Remarks on the taxonomy of *Turdus nigrilorum* Reichenow and *Turdus saturatus* (Cabanis) in Southern British Cameroon. *Bulletin of the British Ornithologists' Club* **82**(7), 124–126.

Sessions, P.H.B. (1966). Notes on the birds of Lengetia Farm, Mau Narok. *Journal of the East Africa Natural History Society and National Museum* **26**, 18–48.

Simms, E. (1978). 'British Thrushes'. Collins, London.

Sinclair, A.R.E. (1978). Factors affecting the food supply and breeding season of resident birds and movements of Palaearctic migrants in tropical African savannah. *Ibis* **120**, 480–497.

Smith, J.N. and Duckworth, G.M. (1994). The influence of rainfall on the timing of laying for some seed, nectar and insect-eating birds at Falcon College, Esigodini. *Honeyguide* **40**(2), 87–91.

Smith, K.D. (1960). Passage of Palaearctic migrants through Eritrea. *Ibis* **102**, 536–544.

Snow, D.W. (1950). The birds of São Tomé and Príncipe in the Gulf of Guinea. *Ibis* **92**, 579–595.

Snow, D.W. (1987). 'The Blackbird'. Shire Natural History, Shire Publications, Aylesbury.

Stephen, B. (1985). 'Die Ansel'. Die Neue Brehm-Bucherei, Gräfenhainichen.

Steyn, P. (1965). Persistant Kurrichane Thrush. *Bokmakierie* **17**(2), 38.

Steyn, P. (1985). Olive Thrush mimicking other birds. *Promerops* No. 169, 12.

Steyn, P. and Brooke, R. K. (1973). Kurrichane Thrush nests close together. *Ostrich* **44**(3&4), 266.
Sueur, F. (1985). Les Mollusques dans le régime alimentaire de la Grive musicienne *Turdus philomelos*. Dép. de la Somme (France). *Nos Oiseaux* **38**, 77–79.
Tarboton, W.R. and Clinning, C.F. (1977). Nesting associations between Groundscraper Thrush *Turdus litsipsirupa* and Fork-tailed Drongo *Dicrurus adsimilis*. *Madoqua* **10**(1), 87–89.
Tree, A.J. (1967). *Turdus olivaceus* as host to *Cuculus solitarius*. *Ostrich* **37**(4), 288.
Tree, A.J. (1994). Recent reports. *Honeyguide* **40**, 259.
Tree, A.J. (1995b). Recent reports. *Honeyguide* **41**, 186.
Triplet, P. and Yésou, P. (1993). Observation de *Sphenoeacus* (=*Melocichla*) *mentalis* en Mauritanie. *L'Oiseau et la Revue Française d'Ornithologie* **63**, 222.
Tye, A. (1981). Ground-feeding methods and niche separation in thrushes. *Wilson Bulletin* **93**(1), 112–114.
Tyler, S. (1987). How long do garden birds in Addis live? *Ethiopian Wildlife and Natural History Society Newsletter* **217**, 3–4.
Tyrväinen, H. (1969). The breeding biology of the Redwing (*Turdus iliacus* L.). *Annales Zoologici Fennici* **6**, 1–46.
Urban, E.K. (1975). Weights and longevity of some birds from Addis Ababa, Ethiopia. *Bulletin of the British Ornithologists' Club* **95**(3), 96–98.
Vernon, C.J. (1968). A year's census of Marandellas, Rhodesia. *Ostrich* **39**, 12–24.
Vernon, C.J. (1973). Vocal imitation by southern African birds. *Ostrich* **44**(1), 23–30.
Vittery, A. (1978). Wahlberg's Eagle nesting in Ethiopia. *Scopus* **2**, 18–19.
Vuilleumier, F. and Mayr, E. (1987). New species of birds described from 1976 to 1980. *Journal für Ornithologie* **128**, 137–150.
Walsh, J.F. and Walsh, B. (1983). Possible thrush 'anvils' in Upper Volta. *Malimbus* **5**, 54–55.
White, C.M.N. (1961). Notes on African species of *Turdus*. *Bulletin of the British Ornithologists' Club* **81**(9), 164–166.
Whitelaw, D. (1983). Yet more on longevity. *Safring News* **12**(1), 24–26.
Whyte, I. (1993). Bill malformation in a Kurrichane Thrush. *Birding in Southern Africa* **45**(2), 42.
Willis, E.O. (1985). East African Turdidae as safari ant followers. *Le Gerfaut* **75**, 140–153.
Willis, E.O. (1986b). West African thrushes as safari ant followers. *Le Gerfaut* **76**, 95–108.
Wilson, J. (1978). Letter to the Editor. *Honeyguide* No. 96, 46.
Winterbottom, J.M. (1960). Range of Gurney's Thrush. *Ostrich* **31**, 27.
Winterbottom, M.G. (1966). A study of the Cape Thrush *Turdus olivaceus* L. *Ostrich* **37**(1), 17–22.
Zamora, R. (1990). The fruit diet of Ring-Ouzels (*Turdus torquatus*) wintering in the Sierra Nevada (South-East Spain). *Alauda* **58**(1), 67–70.

Family SYLVIIDAE: *Cisticola*

Anonymous (1985). Breeding data – Fan-tailed Cisticola (*Cisticola juncidis*). *Zambian Ornithological Society Newsletter* **15**, 161–162.
Archer, A.L. and Turner, D.A. (1993). Notes on the endemic species and some additional new birds occurring on Pemba Island, Tanzania. *Scopus* **16**, 94–98.
Ash, J.S. (1974). The Boran Cisticola in Ethiopia. *Bulletin of the British Ornithologists' Club* **94**, 24–26.
Aspinwall, D. (1984). A record of three march (sic) Cisticola species *Cisticola galactotes*, *C. pipiens* and *C. tinniens* together in Western Province. *Bulletin of the Zambian Ornithological Society* **16**, 15–16.
Bates, G.L. (1927). Notes on some birds of Cameroon and the Lake Chad region: their status and breeding times. *Ibis* **1927**, 1–64.
Bennun, L.A. (1989). E.A.N.H.S. Nest Record Card Scheme: 1985–1989. *Scopus* **13**, 165–180.
Benson, C.W. and Irwin, M.P.S. (1966). A new subspecies of Desert Cisticola, *Cisticola aridula* Witherby. *Arnoldia (Rhodesia)* **2**(27), 1–3.
Benson, C.W. and White, C.M.N. (1957). 'Check-list of the Birds of Northern Rhodesia'. Government Printer, Lusaka.
Benson, C.W., Brooke, R.K. and Vernon, C.J. (1964). Bird breeding data for the Rhodesias and Nyasaland. *Occasional Papers of the National Museums of Southern Rhodesia* **27B**, 30–105.
Bowen, W.W. (1926–1931). 'Catalogue of Sudan Birds'. Sudan Govenment Museum (Natural History), Khartoum.
Branfield, A. (1988). New bird records for the East Caprivi, Namibia. *Lanioturdus* **25**, 4–21.
Britton, P.L. (1970b). The immature plumages of two African warblers. *Bulletin of the British Ornithologists' Club* **90**, 26–28.
Britton, P.L. (1978). Seasonality, density and diversity of birds of a papyrus swamp in western Kenya. *Ibis* **120**, 450–466.
Britton, P.L. and Sugg, M.St.J. (1973). Birds recorded on the Kimilili Track, Mt Elgon, Kenya. *Journal of the East Africa Natural History Society and National Museum* **143**, 1–7.
Butler, A.L. (1905). A contribution to the ornithology of the Egyptian Soudan. *Ibis* **1905**, 301–401.
Butler, A.L. (1908). A second contribution to the ornithology of the Egyptian Soudan. *Ibis* **1908**, 205–263.
Carlson, A. (1986). Group territoriality in the Rattling Cisticola, *Cisticola chiniana*. *Oikos* **47**, 181–189.
Chapin, J.P. (1932). Fourteen new birds from tropical Africa. *American Museum Novitates* No. 570, 1–18.
Chapin, J.P. (1953b). A new race of *Cisticola lateralis* (Fraser). *Bulletin of the British Ornithologists' Club* **73**, 83–84.
Chappuis, C. and Erard, C. (1973). A new race of Pectoral-patch Cisticola from Cameroun. *Bulletin of the British Ornithologists' Club* **93**, 143–144.
Chappuis, C. and Erard, C. (1991). A new Cisticola from West-central Africa. *Bulletin of the British Ornithologists' Club* **111**, 59–70.
Clancey, P.A. (1992). Taxonomic comment on southeastern representatives of two wide-ranging African cisticolas. *Bulletin of the British Ornithologists' Club* **112**, 218–225.
Clancey, P.A. (1993). The status of the *Cisticola aberrans* subspecies *C. a. nyika* Lynes, 1930. *Bulletin of the British Ornithologists' Club* **113**, 18–21.

Colahan, B.D. (1993). Bird notes from Free State Nature Reserves: January to July 1993. *Mirafra* **10**(4), 74–79.

Comins, D.M. (1964). Nesting material used by *Cisticola juncidis*. *Bulletin of the British Ornithologists' Club* **84**, 141–142.

Dean, W.R.J. (1976). Sympatric *Cisticola* spp. and the competition exclusion principle. *Bulletin of the British Ornithologists' Club* **96**, 38–39.

Eisentraut, M. (1956). Notizen über einige Vögel des Kamerungebirges. *Journal für Ornithologie* **97**, 291–300.

Elkins, N. (1975). Voice of the Fan-tailed Warbler. *British Birds* **68**, 45.

Elliot, H.F.I. (1947). A new race of *Cisticola* from Tanganyika Territory. *Bulletin of the British Ornithologists' Club* **68**, 10–11.

Erard, C. (1974b). The problem of the Boran Cisticola. *Bulletin of the British Ornithologists' Club* **94**, 26–38.

Ey, A. (1977). Notes on the streaked grass warbler. *Sunbird* **8**, 20–21.

Field, G.D. (1974). 'Birds of Freetown Peninsula'. Fourah Bay College Bookshop, Freetown.

Finlayson, J.C. (1979). Movements of the Fan-tailed Warbler *Cisticola juncidis* at Gibraltar. *Ibis* **121**, 487–489.

Folse, L.J. (1982). An analysis of avifauna-resource relationships on the Serengeti plains. *Ecological Monographs* **52**, 111–127.

Fry, C.H. (1965). The birds of Zaria II, IV. *Bulletin of the Nigerian Ornithologists' Society* **2**(6), 35–44; (8), 91–102.

Gauci, C. and Sultana, J. (1981). The moult of the Fan-tailed Warbler. *Bird Study* **28**, 77–86.

Gore, M.E.J. (1994). Bird records from Liberia. *Malimbus* **16**, 74–87.

Gowthorpe, P. (1977). Territorialité et structures sociales d'une population de *Cisticola anonyma* (Muller) au Gabon: un nouveau cas d'aide au nourrissage. *L'Oiseau et la Revue Française d'Ornithologie* **47**, 243–252.

Gowthorpe, P. (1978). Notes sur la reproduction de *Cisticola anonyma* (Muller) au Gabon. *L'Oiseau et la Revue Française d'Ornithologie* **48**, 37–43.

Greig-Smith, P.W. and Davidson, N.C. (1977). Weights of West African savanna birds. *Bulletin of the British Ornithologists' Club* **97**, 96–99.

Grimes, L.G. (1976a). The duets of *Laniarius atroflavus*, *Cisticola discolor* and *Bradypterus barratti*. *Bulletin of the British Ornithologists' Club* **96**, 113–120.

Guichard, G. (1959). Notes sur la biologie du cisticole des joncs (*Cisticola juncidis* Temm.). *L'Oiseau et la Revue Française d'Ornithologie* **29**, 88–95.

Harpum, J. (1978b). Species-pair association of Stonechat and Black-lored Cisticola in southwest Tanzania. *Scopus* **2**, 99–101.

Irwin, M.P.S. (1990). On *Cisticola melanura* (Cabanis) and its type locality. *Honeyguide* **36**, 54.

Irwin, M.P.S. (1991). The specific characters of the Slender-tailed Cisticola *Cisticola melanura* (Cabanis). *Bulletin of the British Ornithologists' Club* **111**, 228–236.

Irwin, M.P.S. (1993). Further remarks on the grasswarblers *Cisticola melanura*, *anguisticauda* and *fulvicapilla*. *Honeyguide* **39**, 36–38.

Irwin, M.P.S. and Benson, C.W. (1967). Notes on the birds of Zambia, Part 3. *Arnoldia (Rhodesia)* **3**(4), 1–30.

Keith, S. (1993). On the supposed occurrence of Chubb's Cisticola *Cisticola chubbi* at Bukoba, Tanzania. *Scopus* **16**, 114.

King, D.G. (1973). The birds of the Shira Plateau and west slope of Kibo, Kilimanjaro. *Bulletin of the British Ornithologists' Club* **93**, 64–71.

Lack, P.C., Leuthold, W. and Smeenk, C. (1980). Check-list of the birds of Tsavo East National Park, Kenya. *Journal of the East Africa Natural History Society and National Museum* **170**, 1–25.

Lewis, A.D. (1981–1982). Field identification of the genus *Cisticola* (Aves) in Kenya, Parts 1, 2, 3. *East Africa Natural History Society Bulletin* Nov/Dec **1981**, 108–114; Jan/Feb **1982**, 2–9; Mar/Apr **1982**, 28–37.

Lewis, A.D. (1982). Further notes on Cisticolas. *East Africa Natural History Society Bulletin* Nov/Dec **1982**, 104–105.

Lynes, H. (1930). Review of the genus *Cisticola*. *Ibis* **1930**, Cisticola Supplement.

Lynes, H. (1932). Cisticolas. *Sierra Leone Studies* **17**, 65–89.

Lynes, H. (1936). *Cisticola angusticauda* not *Apalis angusticauda*. *Bulletin of the British Ornithologists' Club* **56**, 112–113.

Lynes, H. (1937). Tips for Cisticolas. Eastern Africa. Private Publication (available Edward Grey Institute of Field Ornithology, Oxford University).

Lynes, H. (1938b). Announcement of a new race of *Cisticola brunnescens*. *Revue de Zoologie et de Botanique Africaines* **31**, 182–184.

Lynes, H. and Sclater, W.L. (1933–1934). Lynes-Vincent tour in central and West Africa in 1930–31. *Ibis* **1933**, 694–729; **1934**, 1–51.

MacPherson, D.W.K. (1966). A note on the Black-lored Cisticola *Cisticola nigriloris* (Shelley). *Ostrich* **37**, 58.

Madden, J.F. (1935). Notes on the birds of southern Darfur, Part II – Passerine birds. *Sudan Notes and Records* **18**, 103–118.

Manson, A.J. (1985, 1986). Results of a ringing programme at Muvuwati Farm, Mazowe. *Honeyguide* **31**, 203–211; **32**, 34–41.

Masterson, A.N.B. (1992). Alarm calls of the Red-faced and Singing Cisticolas. *Honeyguide* **38**, 187.

McGregor, P.K., Clayton, H.S., Kolb, U., Stockley, P. and Young, R.J. (1990). Individual differences in the displays of Fan-tailed Warblers *Cisticola juncidis*: associations with territory and mate quality. *Ibis* **132**, 111–118.

Moreau, R.E. (1966). 'The Bird Faunas of Africa and its Islands'. Academic Press, London.

Motai, T. (1973). Male behaviour and polygamy in *Cisticola juncidis*. *Journal of the Yamashina Institute for Ornithology* **7**, 87–103.

Moyer, D.C. and Schulenberg, T.S. (1994). First record of Chirping Cisticola *Cisticola pipiens* from Burundi. *Bulletin of the British Ornithologists' Club* **114**(1), 63–64.

Moyer, D.C. and Sikombe, S.W. (1992). Notes on birds from southwest Tanzania including an addition to the East African avifauna. *Scopus* **16**(1), 55–56.

Neave, S.A. (1910). On the birds of Northern Rhodesia and Katanga District of Congoland. *Ibis* **1910**, 78–155.

Parkes, K.C. (1987). Taxonomic notes on some African warblers (Aves:Sylviinae). *Annals of the Carnegie Museum* **56**, 231–243.

Peirce, M.A. (1984). Haematozoa of Zambian birds V. Redescription of *Haemoproteus wenyoni* from *Cisticola erythrops* (Sylviidae). *Journal of Natural History* **18**, 555–557.

Penry, E.H. (1985). Notes on breeding of *Cisticola brunnescens* and *C. juncidis* in Zambia. *Ostrich* **56**, 229–235.

Penry, E.H. (1988). Short notes on a Fantailed Cisticola nest. *Mirafra* **5**, 33–35.

Pitman, R.S. and Took, J.M.E. (1973). The eggs of the African Marsh Grass-Warbler, *Cisticola galactotes* (Temminck). *Arnoldia (Rhodesia)* **6**(24), 1–12.

du Plessis, D. (1991). Distribution and habitat of the Grey-backed and Wailing Cisticolas in the Orange Free State. *Mirafra* **8**, 48–51.

Priest, C.D. (1948). 'Eggs of Birds Breeding in Southern Africa'. Glasgow University Press, Glasgow.

Prigogine, A. (1957). Un nouveau *Cisticola* de l'Itombwe (Congo belge). *Revue de Zoologie et de Botanique Africaines* **55**, 33–38.

Pringle, J.S. (1968). The Common Fantail Cisticola. *Bokmakierie* **20**, 45–46.

Robert, J.-C. and Bellard, J. (1975). La nidification de la Cisticole des joncs *Cisticola juncidis* en baie de Somme. *Alauda* **43**, 475–485.

Roberts, A. (1913). The grass warblers of southern Africa (*Cisticola* and *Hemipteryx*). *Annals of the Transvaal Museum* **3**, 227–266.

da Rosa Pinto, A.A. (1965). Contribuição para o Conhocimiento da Avifauna da Região Nordeste, do Distrito do Moxico, Angola. *Boletim do Instituto de Investigação Científica de Angola (Luanda)* **1**, 153–249.

da Rosa Pinto, A.A. (1967). Geographical variation of the African Cisticola *Cisticola subruficapilla* (A. Smith), and description of a new subspecies from Angola. *Boletim do Instituto de Investigação Científica de Angola (Luanda)* **4**, 7–14.

Ruwet, J.-C. (1965). 'Les oiseaux des plaines et du lac-barrage de la Lufira supérieure (Katanga Méridional)'. Fondation de l'Université de Liège pour les Recherches Scientifiques en Afrique centrale, Liège.

Sala, A. (1983). Inventaire de l'avifaune du département d'Oussoye et particulièrement du Parc national de Basse Casamance (Sénégal). *Bulletin de l'Institut Fondamental d'Afrique Noire, Dakar*, Série A, **45**, 342–366.

Schmidt, R.K. (1965). Abnormal nest of Le Vaillant's Cisticola *Cisticola tinniens* (Lichtenstein). *Ostrich* **36**, 135.

Serle, W. (1943b). Notes on East African birds. *Ibis* **85**, 55–82.

Serle, W. (1949b). The Chattering Grass-Warbler *Cisticola anonyma*, v. Müller. *Nigerian Field* **14**, 28–29.

Sessions, P.H.B. (1966). Notes on the birds of Lengetia Farm, Mau Narok. *Journal of the East Africa Natural History Society and National Museum* **26**, 18–48.

Steyn, P. (1966c). Abnormal nest of *Cisticola fulvicapilla* (Vieillot). *Ostrich* **37**, 195.

Steyn, P. (1971). Rattling Cisticola with fowl-pox lesions. *Ostrich* **42**, 74.

Stockley, P. (1988). Territoriality in the Fan-tailed Warbler *Cisticola juncidis*. *University of Nottingham Terrestrial Zoology Field Course Report 1988*, 171–198.

Stronach, N. (1990). Habitat and distribution of the Rock-loving Cisticola *Cisticola aberrans* in Serengeti National Park, Tanzania. *Bulletin of the British Ornithologists' Club* **110**, 32–34.

Sultana, J. and Gauci, C. (1976). Polygamy in *Cisticola juncidis*. *Il Merill* **17**, 28–29.

Taillandier, J. (1993). Reproduction de la Cisticole des joncs, *Cisticola juncidis* dans les prairies d'un marais salant (Guérande, Loire-Atlantique). *Alauda* **61**, 39–51.

Thorpe, W.H. (1972). 'Duetting and Antiphonal Song in Birds – its extent and significance'. *Behaviour* Supplement **XVIII**. E.J. Brill, Leiden.

Todt, D. (1970). Die antiphonen Paargesänge des ostafrikanischen Grassängers *Cisticola hunteri prinioides* Neumann. *Journal für Ornithologie* **111**, 332–356.

Traylor, M.A. (1964). Three new birds from Africa. *Bulletin of the British Ornithologists' Club* **84**, 81–84.

Traylor, M.A. (1965b). Winter dress of *Cisticola chiniana bensoni*. *Bulletin of the British Ornithologists' Club* **85**, 135–136.

Traylor, M.A. (1967a). A new race of *Cisticola galactotes*. *Bulletin of the British Ornithologists' Club* **87**, 58.

Traylor, M.A. (1967b). A new species of Cisticola. *Bulletin of the British Ornithologists' Club* **87**, 45–48.

Turner, D.A. and Pearson, D.J. (1991). Species report. *Scopus* **13**, 138–164.

Tyler, S.J. (1991). Birds of Lake Naivasha 2: foraging niches and relationships between migrant and resident warblers in papyrus swamps. *Scopus* **14**, 117–124.

Ueda, K. (1984). Successive nest building and polygyny of Fan-tailed Warblers *Cisticola juncidis*. *Ibis* **126**, 221–229.

Ueda, K. (1986). A polygamous social system of the fan-tailed warbler *Cisticola juncidis*. *Ethology* **71**, 43–55.

Ueda, K. (1989). Re-use of courtship nests for quick remating in the polygynous Fan-tailed Warbler *Cisticola juncidis*. *Ibis* **131**, 257–262.

Vernon, C.J. (1964). Observations on *Cisticola njombe* and *nigriloris*. *Bulletin of the British Ornithologists' Club* **84**, 124–131.

Vernon, C.J. (1982). Notes on the identification of the Wailing and Greybacked Cisticolas. *Bee-eater* **33**, 29–30.

Vernon, C.J. (1987). Comments on some birds of Zimbabwe. *Honeyguide* **33**, 54–57.

Walsh, J.F. (1987). Records of birds seen in north-eastern Guinea in 1984–1985. *Malimbus* **9**, 105–122.

White, C.M.N. (1960b). Notes on some savanna species of the genus *Cisticola*. *Bulletin of the British Ornithologists' Club* **80**, 128–130.

Willis, E.O. (1986a). Vireos, wood warblers and warblers as ant followers. *Le Gerfaut* **76**, 177–186.

Wright, B. (1987). Fan-tailed Cisticola feeding on open ground. *Honeyguide* **33**, 103.

Young, C.G. (1946). Notes on some birds of the Cameroon Mountain District. *Ibis* **88**, 348–382.

Zimmerman, D.A., Turner, D.A. and Pearson, D.J. (1996). 'The Birds of Kenya and Northern Tanzania'. Christopher Helm and A. & C. Black Ltd, London.

Family SYLVIIDAE: *Incana* to *Macrosphenus* (*Incana, Scotocerca, Phragmacia, Prinia, Oreophilais, Heliolais, Urolais, Spiloptila, Schistolais, Drymocichla, Phyllolais, Apalis, Malcorus, Urorhipis, Artisornis, Poliolais, Camaroptera, Calamonastes, Euryptila, Graueria, Macrosphenus*)

Ash, J.S. (1982). A major extension in distribution of the Stripe-backed Prinia *Prinia gracilis* in Somalia. *Bulletin of the British Ornithologists' Club* **102**, 2–5.

Aspinwall, D.R. (1992). The status of Stierling's Barred Warbler *Camaroptera stierlingi* in northern Botswana. *Babbler* No. 24, 25.

Bennun, L.A. (1986). Montane birds of the Bwindi (Impenetrable) forest. *Scopus* **10**, 89–91.

Bennun, L.A. (1991b). E.A.N.H.S. nest record scheme: 1985–1989. *Scopus* **13**, 165–180.

Benson, C.W. (1946c). A collection from near Unangu, Portuguese East Africa. *Ibis* **88**, 240–241.

Benson, C.W. (1956). A contribution to the ornithology of Northern Rhodesia. *Occasional Papers of the National Museums of Southern Rhodesia* **21B**, 1–51.

Benson, C.W. and Benson, F.M. (1975). Studies of some Malaŵi birds. *Arnoldia (Rhodesia)* **7**(32), 1–27.

Benson, C.W. and Irwin, M.P.S. (1964a). Some additions and corrections to 'A Check List of the Birds of Northern Rhodesia'. *Occasional Papers of the National Museums of Southern Rhodesia* **27B**, 106–127.

Berruti, A., Harrison, J.A. and Navarro, R.A. (1994). Seasonal migration of terrestrial birds along the southern and eastern coasts of southern Africa. *Ostrich* **65**, 54–65.

Beven, G. (1944). Rudd's Bush Warbler (*Apalis ruddi*). *Ostrich* **15**, 178–187.

Blignaut, J. (1958). Some points of interest about the nesting of the Namaqua Prinia *Burnesia substriata* at Graaff-Reinet. *Ostrich* **29**, 47.

Boulton, R. and Rand, A.L. (1952). A collection of birds from Mount Cameroun. *Fieldiana Zoology* **34**, 35–64.

Bowen, P. St. J. (1980a). Notes on the juvenile plumage of the Brown-headed Apalis *Apalis cinerea*. *Bulletin of the Zambian Ornithological Society* **12**, 47–48.

Bowen, P. St. J. (1983). The White-chinned Prinia *Prinia leucopogon* in Mwinilunga district. *Bulletin of the Zambian Ornithological Society* **13/15**, 16–22.

Branfield, A. (1990). New bird records for the east Caprivi, Namibia. *Lanioturdus* **25**, 4–21.

Brooke, R.K. (1965). Ornithological notes in the Furancungo District of Mozambique. *Arnoldia (Rhodesia)* **10**(2), 1–13.

Brooke, R.K. (1993). A case of apparent hybridization between Spotted and Blackchested Prinias *Prinia maculosa* and *flavicans*. *Ostrich* **64**, 137.

Brooke, R.K. and Dean, W.R.J. (1990). On the biology and taxonomic position of *Drymoica substriata* Smith, the so-called Namaqua Prinia. *Ostrich* **61**, 50–55.

Brown, C. (1991). Birds of the Brandberg and Spitzkoppe. *Lanioturdus* **26**(1), 25–29.

Butynski, T.M. (1993). Nest record of the Collared Apalis *Apalis ruwenzorii* and description of a partial albino. *Scopus* **16**, 116–118.

Chappuis, C. (1971). Un example de l'influence du milieu sur les émissions vocales des oiseaux: l'évolution des chants en forêt équatoriale. *La Terre et La Vie* **XXV**, 183–202.

Chappuis, C. (1980b). Study and analysis of certain vocalisations as an aid in classifying African Sylviidae. *Proceedings of the Fourth Pan-African Ornithological Congress* pp. 57–63. Southern African Ornithological Society, Johannesburg.

Chappuis, C., Erard, C. and Morel, G.J. (1988). Type specimens of *Prinia subflava* (Gmelin) and *Prinia fluviatilis* Chappuis. *Bulletin of the British Ornithologists' Club* **109**, 108–110.

Chappuis, C., Erard, C. and Morel, G.J. (1992). Morphology, habitat, vocalisations and distribution of the River Prinia *Prinia fluviatilis* Chappuis. *Proceedings of the Seventh Pan-African Ornithological Congress* pp. 481–488. PAOC Committee, Nairobi.

Clancey, P.A. (1957). Are *Prinia maculosa* and *Prinia hypoxantha* conspecific? *Ibis* **99**, 513–516.

Clancey, P.A. (1960). On the races of *Prinia pectoralis* (Smith). *Bulletin of the British Ornithologists' Club* **80**, 15–16.

Clancey, P.A. (1969a). An undescribed race of the Black-headed Apalis. *Bulletin of the British Ornithologists' Club* **89**, 92–94.

Clancey, P.A. (1969b). Miscellaneous taxonomic notes on African birds XXVII. *Durban Museum Novitates* **VIII**(15), 227–274.

Clancey, P.A. (1970a). An overlooked race of Barred Bush Warbler from the Transvaal. *Durban Museum Novitates* **VIII**(17), 337–339.

Clancey, P.A. (1970b). Comments on some anomalous Bleating Bush Warblers from Eastern Rhodesia. *Durban Museum Novitates* **VIII**(17), 335–337.

Clancey, P.A. (1974b). Subspeciation studies in some Rhodesian birds. *Arnoldia (Rhodesia)* **6**(28), 1–43.

Clancey, P.A. (1982). Namibian Ornithological Miscellanea. *Durban Museum Novitates* **VIII**, 62–63.

Clancey, P.A. (1984). Subspeciation in the Eastern Barred Bush Warbler of the south-eastern Afrotropics. *Honeyguide* **30**, 21–23.

Clancey, P.A. (1986a). Additional mensural data for a subspecies of the Eastern Barred Bush Warbler. *Honeyguide* **32**, 44.

Clancey, P.A. (1986b). Endemicity in the southern African avifauna. *Durban Museum Novitates* **XIII**(20), 245–284.

Clancey, P.A. (1989). Four additional species of southern African endemic birds. *Durban Museum Novitates* **XIV**, 140–152.

Clancey, P.A. (1990). Variation in the Cinnamon-breasted Warbler of the South West arid zone of the Afrotropics. *Bonner Zoologische Beiträge* **41**, 109–112.

Clancey, P.A. (1991a). On the generic status and geographical variation of the Namaqua Prinia. *Bulletin of the British Ornithologists' Club* **111**, 101–104.

Clancey, P.A. (1991b). The generic status of Roberts' Prinia of the south-eastern Afrotropics. *Bulletin of the British Ornithologists' Club* **111**, 217–222.

Clancey, P.A. (1992a). Subspeciation, clines and contact zones in the southern Afrotropical avifauna. In 'Avian Systematics and Taxonomy' (Monk, J.F., Ed.). *Bulletin of the British Ornithologists' Club* **112A**.

Clancey, P.A. (1992b). The eastern Cape contact zone between the Karoo and Saffronbreasted Prinias. *Bee-eater* **42**, 67–69.

Clancey, P.A. (1993). The background to "new" species of birds on the southern African list. *Birding in Southern Africa* **45**(4), 99–103.

Dean, W.R.J. (1976). Niche occupation of Rufouseared Warbler and Blackchested Prinia. *Ostrich* **47**, 67.

Dean, W.R.J. (1978). An analysis of avian stomach contents from southern Africa. *Bulletin of the British Ornithologists' Club* **98**, 10–13.

Dementiev, G.P. and Gladkov, N.A. (1954). 'Birds of the Soviet Union'. State Publishing, Soviet Sciences, Moscow.

Dowsett, R.J. and Dowsett-Lemaire, F. (1989c). Natural history notes. *Nyala* **14**, 39.

Dowsett-Lemaire, F. (1986). Vocal variations in two forest apalises of Eastern Africa, *Apalis (porphyrolaema) chapini* and *A. melanocephala*. *Scopus* **10**, 92–98.

Earlé, R.A. (1980). Notes on the breeding of the Bleating Bush Warbler. *Ostrich* **51**, 128.

Earlé, R.A. (1983). Foraging overlap and morphological similarity among some insectivorous arboreal birds in an eastern Transvaal forest. *Ostrich* **54**, 36–42.

Earlé, R.A. (1989). Aspects of the foraging behaviour of some insectivorous birds in three Transvaal indigenous forests. *Ostrich Supplement* **14**, 71–74.

Erard, C. (1971). *Apalis flavida caniceps* (Cassin) in Ethiopia. *Bulletin of the British Ornithologists' Club* **91**, 84–88.

Every, B. (1990). Prinia distributions in the eastern Cape. *Bee-eater* **41**, 20–23.

Field, G.D. (1974). The distribution and behaviour of Apalis warblers in Sierra Leone. *Ostrich* **45**, 258–260.

Fogden, M.P.L. and Fogden, P.M. (1979). The role of fat and protein reserves in the annual cycle of the Grey-backed Camaroptera in Uganda (Aves: Sylviidae). *Journal of Zoology, London* **189**, 233–258.

Fraser, M.W. (1987). Spotted Prinia catching legless lizard. *Promerops* No. 180, 17.

Friedmann, H. (1948). 'The Parasitic Cuckoos of Africa'. Washington Academy of Sciences.

Fry, C.H. (1976). On the systematics of African and Asian tailor-birds (Sylviinae). *Arnoldia (Rhodesia)* **8**(6), 1–15.

Ginn, P. (1996). First photographs of the elusive Chirinda Apalis. *Zimbabwe Wildlife* **83**, 9.

Goodman, S.M. (1984). The validity and relationships of *Prinia gracilis natronensis* (Aves: Sylviidae). *Proceedings of the Biological Society of Washington* **97**, 1–2.

Gore, M.E.J. (1994). Bird records from Liberia. *Malimbus* **16**, 74–87.

Grant, C.H.B. and Mackworth-Praed, C.W. (1941). Notes on eastern African birds: on the races of the Black-breasted Bush Warbler. *Bulletin of the British Ornithologists' Club* **61**, 41–45.

Gray, H.H. (1972). The nests of three forest birds. *Bulletin of the Nigerian Ornithologists' Society* **9**, 24–25.

Grimes, L.G. (1976c). The vocalizations of the Green Longtail *Urolais epichlora*. *Bulletin of the British Ornithologists' Club* **96**, 99–101.

Guichard, K.M. (1947). Birds of the inundation zone of the River Niger, French Soudan. *Ibis* **89**, 450–489.

Hald-Mortensen, P. (1971). A collection of birds from Liberia and Guinea. *Steenstrupia* **1**, 115–125.

Hall, B.P. (1960b). The faunistic importance of the scarp of Angola. *Ibis* **102**, 420–442.

Hanmer, D.B. (1977). An undescribed subspecies of Rudd's Apalis *Apalis ruddi* from southern Malaŵi. *Bulletin of the British Ornithologists' Club* **99**, 27–28.

Hanmer, D.B. (1978). Incubation and nestling periods in the Tawnyflanked Prinia. *Ostrich* **49**, 205.

Hanmer, D.B. (1989). The end of an era – final longevity figures for Nchalo. *Safring News* **18**(1&2), 19–30.

Harkrider, J.R. (1993). Garden and farm-bush birds of Njala, Sierra Leone. *Malimbus* **15**(1), 38–46.

Heppel, R.E. (1985). Birds of Outamba area, northwest Sierra Leone. *Malimbus* **7**(2), 101–102.

Herremans, M. and Herremans, D. (1992). The status of Stierling's Barred Warbler *Camaroptera stierlingi* in Botswana. *Babbler* No. 23, 39–40.

Hustler, H. (1990). Variable egg colour in nest of Bar-throated Apalis. *Honeyguide* **30**, 41–42.

Irwin, M.P.S. (1959). Some remarks on *Prinia flavicans* and its allies. *Bulletin of the British Ornithologists' Club* **79**, 127–128.

Irwin, M.P.S. (1960). Relationships within the *Camaroptera fascioloata-stierlingi-simplex* complex of warblers. *Durban Museum Novitates* **VI**(3), 43–60.

Irwin, M.P.S. (1966). Remarks on the *Apalis thoracica-rhodesiae-arnoldi-whitei* group of races of the Bar-throated Apalis, with the description of a new subspecies from Gorongoza Mountain, Moçambique. *Durban Museum Novitates* **VIII**, 47–52.

Irwin, M.P.S. (1969). *Camaroptera stigmosus* (Reichenow), the male breeding dress of *Camaroptera fasciolata* (Smith). *Bulletin of the British Ornithologists' Club* **89**, 44–48.

Irwin, M.P.S. (1979). The Zimbabwe Rhodesian and Moçambique highland avian endemics: their evolution and origins. *Honeyguide* No. 99, 5–11.

Irwin, M.P.S. (1987). What are the affinities of the Black-capped Apalis *Apalis nigriceps*? The need for field studies. *Malimbus* **9**, 130–131.

Irwin, M.P.S. (1988). The relationships of the African warblers *Apalis binotata* and *A. (b.) personata*. *Bulletin of the British Ornithologists' Club* **108**, 58–61.

Irwin, M.P.S. (1989). The number of rectrices in some Afrotropical warblers (Sylviidae). *Honeyguide* **35**, 179–180.

Irwin, M.P.S. and Jackson, H.D. (1971). Geographical variation and relationships in *Apalis chirindensis* Shelley and *Apalis melanocephala* (Fischer and Reichenow) in Rhodesia and southern Moçambique. *Bulletin of the British Ornithologists' Club* **91**, 49–56.

James, H.W. and Brooke, R.K. (1972). Breeding data on the Karoo Prinia in Great Fish basin. *Ostrich* **43**, 137–138.

Jones, P.J. and Tye, A. (1988). A survey of the avifauna of São Tomé and Príncipe. *ICBP Study Report* No. 24, 1–64. International Council for Bird Protection, Cambridge.

Kunkel, P. (1974). Mating systems of tropical birds: the effects of weakness or absence of external reproduction-timing factors, with special reference to prolonged pair bonds. *Zeitschrift für Tierpsychologie* **34**, 265–307.

Lawson, W.J. (1965). The geographical races of the Bar-throated Apalis *Apalis thoracica* (Shaw and Nodder), occurring in southern Africa. *Ostrich* **36**, 3–8.

Lawson, W.J. (1968). Geographical variation in the Yellow-breasted Apalis *Apalis flavida* of Africa. *Durban Museum Novitates* **VIII**, 199–226.

Lewis, A.D. (1982a). Form and function of duetting of the Yellow-breasted Apalis *Apalis flavida*. *Scopus* **6**, 95–100.

Lewis, A.D. (1982b). Threat display of the Yellow-breasted Apalis *Apalis flavida*. *Scopus* **6**, 73–74.

Lewis, A.D. (1983) Why doesn't the Yellow-breasted Apalis duet at Kenyatta University College? *East Africa Natural History Society Bulletin* **1983**, 2–4.

Lewis, A.D. (1989). Behavioural, physical and environmental differences between races of the Yellow-breasted Apalis *Apalis flavida* in Kenya. *Scopus* **12**, 83–86.

Lorber, P., Manson, A.J. and Sharp, C. (1983). The undescribed nest of the Chirinda Apalis. *Honeyguide* No. 114/115, 69–71.

Louette, M. (1988). Additions and corrections to the avifauna of Zaïre (2). *Bulletin of the British Ornithologists' Club* **108**, 43–50.

Louette, M. (1989). Additions and corrections to the avifauna of Zaïre (4). *Bulletin of the British Ornithologists' Club* **109**, 217–225.

Lowe, P.R. (1937). Report on the Lowe-Waldron expeditions to the Ashanti forests and northern territories of the Gold Coast, Part II. *Ibis* **1937**, 635–690.

Lynes, H. (1936). *Cisticola angusticauda* not *Apalis angusticauda*. *Bulletin of the British Ornithologists' Club* **56**, 112–113.

Macdonald, J.D. and Hall, B.P. (1957). Ornithological results of the Bernard Carp/Transvaal Museum expedition to the Kaokoveld, 1951. *Annals of the Transvaal Museum* **23**, 1–39.

Macdonald, M.A. and Taylor, I.R. (1977). Notes on some uncommon forest birds in Ghana. *Bulletin of the British Ornithologists' Club* **97**, 116–120.

Mackworth-Praed, C.W. and Grant, C.H.B. (1938). On the exact type locality of *Apalis thoracica* (Shaw and Nodder) and its races. *Ibis* **1938**, 528–533.

Mackworth-Praed, C.W. and Grant, C.H.B. (1942). On the status of *Motacilla subflava* Gmelin and the races of the Tawny-flanked Long-tail. *Ibis* **1942**, 265–269.

Maclean, G.L. (1974). The breeding biology of the Rufous-eared Warbler and its bearing on the genus *Prinia*. *Ostrich* **45**, 9–14.

Manson, C. and Manson, A.J. (1980). Some notes and a preliminary analysis of ringing data on the Forest Prinia. *Honeyguide* No. 102, 12–15.

Martin, J. and Martin, R. (1965). The nest of Cinnamon-breasted Warbler *Euryptila subcinnamomea* (A. Smith). *Ostrich* **36**, 136–137.

Martin, J. and Martin R. (1975). Cinnamon-breasted Warbler *Euryptila subcinnamomea*. *Ostrich* **46**, 177.

Masterson, A.N.B. (1972). The nest and eggs of Rhodesian "Taylor-birds". *Honeyguide* No. 69, 21–25.

Milstein, P.le S. and Milstein, D.A. (1981). Occurrence of Red-wing Warbler *Heliolais erythroptera* and some other observations from the northern Kruger National Park. *Koedoe* 24, 109–117.

Morel, G.J. and Bourlière, F. (1962). Relations écologiques des avifaunes sédentaires et migratrices dans une savane sahélienne du Bas Sénégal. *La Terre et la Vie* 16(4), 371–393.

de Naurois, R. (1984b). *Prinia molleri* Bocage, 1887, endemique de l'île de São Tomé. *Rivista Italiana di Ornitologia* 54, 191–206.

de Naurois, R. and Morel, G.J. (1995). Description des oeufs et du nid de la Prinia aquatique *Prinia fluviatilis*. *Malimbus* 17, 28–31.

Parkes, K.C. (1987). Taxonomic notes on some African Warblers (Aves: Sylviinae). *Annals of the Carnegie Museum* 56, 231–243.

Penry, E.H. (1976). White-chinned Prinia (*Prinia leucopogon*) on East Tunga river. *Bulletin of the Zambian Ornithological Society* 8, 26–27.

Plowes, D.C.H. (1972). The nest of the Redwing Warbler (*Heliolais erythroptera rhodoptera* (Shelley)). *Honeyguide* No. 69, 34–35.

Pompert, W. (1988). Graceful Warbler in Cyprus October 1987. *Dutch Birding* 10, 29.

Priest, C.D. (1948). 'Eggs of birds breeding in southern Africa'. R. MacLehose and Co. Ltd., Glasgow.

Prigogine, A. (1961). Une nouvelle forme de *Camaroptera chloronota* Reichenow du Congo. *Revue de Zoologie et de Botanique Africaines* 63, 142–144.

Prigogine, A. (1972). Une nouvelle race de *Apalis alticola* de l'est de la République du Zaïre. *Revue de Zoologie et de Botanique Africaines* 86, 173–178.

Prigogine, A. (1973). Nouveau nom pour *Apalis alticola marungensis* Prigogine (Aves Sylviidae). *Revue de Zoologie et de Botanique Africaines* 87, 456.

Prigogine, A. (1979). Relation entre les Prinias Rayées *Prinia bairdii obscura* et *Prinia bairdii bairdii*. *Le Gerfaut* 69, 305–318.

Randall, R.D. (1993). Rare birds to look for in the Kasane/north Chobe area. *Babbler* No. 25, 23–26.

Robertson, I. (1995). Black-throated Apalis *Apalis jacksoni*, a new bird for Gabon. *Malimbus* 17, 28.

Rodewald, P.G. and Bowden, C.G.R. (1995). First record of Kemp's Longbill *Macrosphenus kempi* in Cameroon. *Bulletin of the British Ornithologists' Club* 115, 66–68.

Rowan, M.K. and Broekhuysen, G.J. (1962). A study of the Karoo Prinia. *Ostrich* 33, 6–30.

Salempo, E. (1994). Birds recorded from the Loliondo area of northern Tanzania. *Scopus* 17, 124–128.

Salvan, J. (1972). Notes ornithologiques du Congo-Brazzaville. *L'Oiseau et la Revue Française d'Ornithologie* 42, 241–252.

Serle, W. (1949c). New races of a warbler, a flycatcher, and an owl from West Africa. *Bulletin of the British Ornithologists' Club* 69, 74–76.

Shuel, R. (1938). Further notes on the eggs and nesting habits of birds in northern Nigeria (Kano Province). *Ibis* 1938, 463–480.

Siegfried, W.R. (1968b). Note on repeat-breeding in the Karoo Prinia. *Ostrich* 39, 199.

Simmons, K.E.L. (1954). The behaviour and general biology of the Graceful Warbler *Prinia gracilis*. *Ibis* 96, 262–292.

Sinclair, A.R.E. (1978). Factors affecting the food supply and breeding season of resident birds and movements of Palaearctic migrants in a tropical African savannah. *Ibis* 120, 480–497.

Snow, D.W. (1950). The birds of São Tomé and Príncipe in the Gulf of Guinea. *Ibis* 92, 579–595.

Steyn, P. (1966a). A note on the nesting of *Camaroptera brevicaudata* (Cretzchmar). *Ostrich* 37, 60–61.

Steyn, P. (1966b). Tawny-flanked Prinias, *Prinia subflava* (Gmelin), utilising Red Bishop, *Euplectes orix* (Linnaeus), nests. *Ostrich* 37, 195.

Steyn, P. (1970). An exquisite nest. *African Wildlife* 24, 337–340.

Stronach, N. (1990). New information on birds in Serengeti National Park, Tanzania. *Bulletin of the British Ornithologists' Club* 110, 198–202.

Stuart, S.N. (1981b). An explanation for the disjunct distribution of *Modulatrix orostruthus* and *Apalis* (or *Orthotomus*) *moreaui*. *Scopus* 5, 1–4.

Stuart, S.N. and Collar, N.J. (1985). Subspeciation in the Karamoja Apalis *Apalis karamojae*. *Bulletin of the British Ornithologists' Club* 105, 86–89.

Stuart, S.N. and Turner, D.A. (1980). Some range extensions from eastern and northeastern Tanzania. *Scopus* 4, 36–41.

Stuart, S.N., Howell, K.M., van der Willigen, T.A. and Geertsema, A.A. (1981). Some additions to the forest avifauna of the Uzungwa Mountains, Tanzania. *Scopus* 5, 46–50.

de Swardt, D.H. (1991). Barred Warbler now in the south. *Mirafra* 8(1), 26–27.

Traylor, M.A. (1967a). A case of dimorphic juvenal plumage. *Bulletin of the British Ornithologists' Club* 87, 58–60.

Traylor, M.A. (1967b). A spotted breast band in *Apalis rufifrons*. *Bulletin of the British Ornithologists' Club* 87, 112.

Traylor, M.A. (1967c). Notes on *Apalis cinerea* and *Apalis chariessa*. *Bulletin of the British Ornithologists' Club* 87, 95–96.

Turner, D.A. (1977). Status and distribution of East African endemic species. *Scopus* 1, 2–11.

Turner, D.A. (1992). Brown-headed Apalis *Apalis alticola* occurring alongside Grey Apalis *A. cinerea* in the Ngurumans, southwestern Kenya. *Scopus* 16, 57–58.

Tye, H. (1991). Reversal of breeding season by lowland birds at high altitudes in western Cameroon. *Ibis* 134, 154–163.

Ulfstrand, S. (1960). The juvenile plumage of *Apalis argentea* Moreau 1941 and a note on the habitat of the species. *Bulletin of the British Ornithologists' Club* 80, 2–3.

Vernon, C.J. (1970a). New hosts for three cuckoos. *Ostrich* 41, 258–259.

Vernon, C.J. (1978). Breeding seasons of birds in deciduous woodland at Zimbabwe, Rhodesia from 1970 to 1974. *Ostrich* 49, 102–115.

Vernon, C.J. (1987). Comments on some birds of Zimbabwe. *Honeyguide* 33, 54–57.

Williams, J. (1991). Odd behaviour by a Bar-throated Apalis. *Honeyguide* 37, 16.

Willis, E.O. (1986a). Vireos, wood warblers and warblers as ant followers. *Le Gerfaut* 76, 177–186.

Winterbottom, J.M. (1938). Further notes on some Northern Rhodesian birds. *Ibis* 1938, 269–277.

Winterbottom, J.M. (1954). A note on the feeding rate of the Bar-throated Apalis, *Apalis thoracica*. *Ostrich* 25, 99–101.

Winterbottom, J.M. (1957). On the races of *Prinia pectoralis* A.Smith. *Bulletin of the British Ornithologists' Club* 77, 155–156.

Winterbottom, J.M. (1969). On the birds of the Sandveld Kalahari of South West Africa. *Ostrich* 40, 182–204.

Wood, B. (1989). Biometrics, iris and bill coloration, and moult of some Somali forest birds. *Bulletin of the British Ornithologists' Club* 109, 11–22.

Family SYLVIIDAE: all other warblers *(Hemitesia, Cettia, Bradypterus, Bathmocercus, Melocichla, Sphenoeacus, Schoenicola, Locustella, Acrocephalus, Chloropeta, Hippolais, Eremomela, Sylvietta, Phylloscopus, Regulus, Eminia, Hypergerus, Sylvia, Parisoma, Hyliota, Hylia, Amaurocichla)*

Afik, D. (1984). [Observations on the breeding biology of the Arabian Warbler.] (In Hebrew.) *Tzufit* **2**, 57–67.

Afik (Aizik), D. and Pinshow, B. (1984). Notes on the breeding biology of the Arabian Warbler *Sylvia leucomelaena* in the Arava Valley (Rift Valley), Israel. *Ibis* **126**, 82–89.

Aidley, D. and Wilkinson, R. (1987a). Moult of some Palaearctic warblers in northern Nigeria. *Bird Study* **34**, 219–225.

Aidley, D. and Wilkinson, R. (1987b). The annual cycle of six *Acrocephalus* warblers in a Nigerian reedbed. *Bird Study* **34**, 226–234.

Allan, D.G., Tarboton, W.R., Filmer, R.J. and Bassi, J. (1988). Breeding of the Broadtailed Warbler in South Africa. *Ostrich* **59**, 137.

Allan, D., Fraser, M., Graham, J. and Martin, R. (1995). Rare birds in the Western Cape — the second report of the Cape Bird Club Rare Birds Committee. *Promerops* No. 221, 7–9.

Altenburg, W. and van Spanje, T. (1989). Utilisation of mangroves by birds in Guinea-Bissau. *Ardea* **77**, 57–74.

Amadon, D. (1954). A new race of *Chloropeta gracilirostris* Ogilvie-Grant. *Ostrich* **25**, 140–141.

Amies, P.A. (1990). Spectacled Warblers feeding in association with Finsch's Wheatear. *British Birds* **83**, 72–73.

Anonymous (1995). African round-up. *Bulletin of the African Bird Club* **2**(1), 1–7.

Arroyo, B. and Tellería, J.L. (1984). La invernada de las aves en el area do Gibraltar. *Ardeola* **30**, 23–31.

Ash, J. S. (1973). Six species of birds new to Ethiopia. *Bulletin of the British Ornithologists' Club* **93**, 3–6.

Ash, J.S. (1977). Four species new to Ethiopia and other notes. *Bulletin of the British Ornithologists' Club* **97**, 4–9.

Ash, J. S. (1978). A Basra Reed Warbler *Acrocephalus griseldis* in Moçambique. *Bulletin of the British Ornithologists' Club* **98**, 29–30.

Ash, J. S. (1978). Ethiopia as a presumed wintering area for the eastern Grasshopper Warbler *Locustella naevia straminea*. *Bulletin of the British Ornithologists' Club* **98**, 22–24.

Ash, J.S. (1979). A new species of serin from Ethiopia. *Ibis* **121**, 1–7.

Ash, J.S. (1981). Bird-ringing results and ringed bird recoveries in Ethiopia. *Scopus* **5**, 85–101.

Ash, J.S. (1982). The Somali Short-billed Crombec *Sylvietta philippae* in Somalia and Ethiopia. *Bulletin of the British Ornithologists' Club* **102**, 89–92.

Ash, J.S. (1994). Weights of migrant Palaearctic *Sylvia* warblers in Ethiopia. *Scopus* **17**, 119–123.

Ash, J.S. and Miskell, J.E. (1981). Basra Reed Warblers *Acrocephalus griseldis* overwintering in Somalia. *Scopus* **5**, 81–82.

Ash, J.S. and Watson, G.E. (1974). *Locustella naevia* in Ethiopia. *Bulletin of the British Ornithologists' Club* **94**, 39–40.

Ash, J.S., Pearson, D.J., Nikolaus, G. and Colston, P.R. (1989b). The mangrove reed warbler of the Red Sea coasts. Description of a new subspecies of the African Reed Warbler *Acrocephalus baeticatus*. *Bulletin of the British Ornithologists' Club* **109**, 36–43.

Ash, J.S., Coverdale, M.A.C. and Gullick, T.M. (1991). Comments on status and distribution of birds in western Uganda. *Scopus* **15**, 24–29.

Aspinwall, D.R. (1994). Recent reports: Zambia. *Bulletin of the African Bird Club* **1**, 107.

Backhurst, G.C. and Pearson, D.J. (1976). Savi's Warbler in Tsavo, an addition to the Kenya list. *East Africa Natural History Society Bulletin* **1976**, 21–22.

Backhurst, G.C. and Pearson, D.J. (1984). The timing of the southward night migration of Palaearctic birds over Ngulia, southeast Kenya. *Proceedings of the Fifth Pan-African Ornithological Congress* pp. 361–369. Southern African Ornithological Society, Johannesburg.

Backhurst, G.C. and Pearson, D.J. (1993). Ringing and migration at Ngulia, Tsavo, autumn 1991. *Scopus* **15**, 172–177.

Baha el Din, M. (1996). The first Dusky Warbler *Phylloscopus fuscatus* in Egypt. *Sandgrouse* **18**(1), 69.

Bairlein, F. (1988). Herbstlicher Durchug, Körpergewichte und Fettdeposition von Zugvögeln in einem Rastgebiet in Nordalgerien. *Die Vogelwarte* **34**, 237–248.

Bairlein, F. (1991). Body mass of Garden Warblers (*Sylvia borin*) on migration: a review of field data. *Die Vogelwarte* **36**, 48–61.

Bairlein, F., Lehnart, J., Schule, H., Stadelmeier, H. and Wildski, M. (1988). First record of the Yellow-browed Warbler *Phylloscopus inornatus* for Algeria. *Alauda* **56**, 178–179.

Beasley, A. (1988). Moustached Warbler in Chimanimani Mountains. *Honeyguide* **34**, 128–129.

Becker, P. and Lutgens, H. (1976). Sumpfrohrsänger (*Acrocephalus palustris*) in Sudwest Afrika. *Madoqua* **9**, 41–44.

Beier, J. (1981). Untersuchungen an Drossel- und Teichrohrsänger (*Acrocephalus arundinaceus*, *A. scirpaceus*): Bestandsentwicklung, Brutbiologie, Ökologie. *Journal für Ornithologie* **122**, 209–230.

Belcher, C.F. (1948). Two African rarities. *Oologists' Record* **22**, 56–57.

Bennun, L.A. (1986a). A nest record for *Phylloscopus laetus* the Red-faced Woodland Warbler. *Scopus* **10**, 113–114.

Bennun, L.A. (1986b). Montane birds of the Bwindi (Impenetrable) Forest. *Scopus* **10**, 87–91.

Bennun, L.A. (1991a). An avifaunal survey of the Trans-Mara Forest, Kenya. *Scopus* **14**, 61–72.

Bensch, S. and Hasselquist, D. (1991). Territorial infidelity in the polygynous Great Reed Warbler *Acrocephalus arundinaceus*: the effect of variation in territory attractiveness. *Journal of Animal Ecology* **60**, 857–871.

Bensch, S., Hasselquist, D., Hedenström A. and Ottoson, U. (1991). Rapid moult among Palaearctic passerines in West Africa – an adaptation to the oncoming dry season? *Ibis* **133**, 47–52.

Benson, C.W. (1939). On the status of the genus *Bradypterus* in Nyasaland, including a new race, *Bradypterus usambarae granti*. *Bulletin of the British Ornithologists' Club* **59**, 108–113.

Benson, C.W. (1954). A new race of warbler from Northern Rhodesia. *Bulletin of the British Ornithologists' Club* **74**, 77–79.

Benson, C. (1958). Notes from Northern Rhodesia. *Bulletin of the British Ornithologists' Club* **78**, 90–93.

Benson, C.W. (1961). Some notes from Northern Rhodesia. *Bulletin of the British Ornithologists' Club* **81**, 145–147.

Benson, C.W. and Irwin, M.P.S. (1964b). Some birds from

the North Western Province, Zambia. *Arnoldia (Rhodesia)* **29**, 1–11.
Benson, C.W. and Irwin, M.P.S. (1965). The birds of Marquesia thickets in Northern Mwinilunga District, Zambia. *Arnoldia (Rhodesia)* **30**, 1–4.
Benson, C.W. and Pitman, C.R.S. (1966). Further breeding records from Zambia (formerly Northern Rhodesia) (No. 5). *Bulletin of the British Ornithologists' Club* **86**, 21–33.
Berruti, A., Taylor, P.J. and Vernon, C.J. (1993). Morphometrics and distribution of the Knysna Warbler *Bradypterus sylvaticus* Sundevall and Barratt's Warbler *B. barratti* Sharpe. *Durban Museum Novitates* **18**, 29–36.
Berruti, A., Harrison, J.A. and Navarro, R.A. (1994). Seasonal migration of terrestrial birds along the southern and eastern coasts of southern Africa. *Ostrich* **65**, 54–65.
Berthold, P. (1976). Animalische und vegetabilische Ernahrung omnivorer Singvogelarten: Nahrungsbevorzugung, Jaresperiodik der Nahrungswahl, physiologische und ökologische Bedeutung. *Journal für Ornithologie* **117**, 145–209.
Best, D. (1975). Menetries' Warbler *Sylvia mystacea* new to Nigeria and West Africa. *Bulletin of the Nigerian Ornithologists' Society* **11**, 85–86.
Beven, G. (1967). Studies of less familiar birds. *143* Subalpine Warbler. *British Birds* **60**, 123–129.
Beven, G. (1971). Studies of less familiar birds. *163* Orphean Warbler. *British Birds* **64**, 68–74.
Bibby, C.J. (1979). Breeding biology of the Dartford Warbler *Sylvia undata* in England. *Ibis* **121**, 41–52.
Bibby, C.J. and Green, R.E. (1981). Autumn migration strategies of Reed and Sedge Warblers (*Acrocephalus scirpaceus* and *schoenobaenus*). *Ornis Scandinavica* **12**, 1–12.
Bibby, C.J., Green, R.E., Pepler, G.R.M. and Pepler, P.A. (1976). Sedge Warbler migration and reed aphids. *British Birds* **69**, 384–399.
Biebach, H., Friedrich, W. and Heine, G. (1986). Interaction of bodymass, fat, foraging and stopover period in trans-sahara migrating passerine birds. *Oecologia* **69**, 370–379.
Blanchet, A. (1957). Oiseaux de Tunisie. *Mémoires de la Société de Sciences Naturelles de Tunisie* **1**, 93–216.
Bowen, P. St.J. (1983b). Some notes on the African Yellow Warbler *Chloropeta natalensis* at Mwinilunga. *Bulletin of the Zambian Ornithological Society* **13–15**, 1.
Bowen, W.W. (1931). East African birds collected during the Gray African Expedition 1929. *Proceedings of the Academy of Natural Sciences of Philadelphia* **83**, 11–79.
van den Brink, B. and Loske, K.-H. (1990). Botswana and Namibia as regular wintering quarters for European Reed Warblers. *Ostrich* **61**, 146–147.
Britton, P.L. (1970b). The immature plumages of two African warblers. *Bulletin of the British Ornithologists' Club* **90**, 26–28.
Britton, P.L. (1978a). The Andersen collection from Tanzania. *Scopus* **2**, 24–25.
Britton, P.L. (1978b). Seasonality, density and diversity of birds of a papyrus swamp in western Kenya. *Ibis* **120**, 450–466.
Britton, P.L. (1981). Notes on the Andersen collection and other specimens from Tanzania housed in some West German museums. *Scopus* **5**, 14–21.
Britton, P.L. and Britton, H.A. (1977). An April fall of Palaearctic migrants at Ngulia. *Scopus* **1**, 109–111.
Britton, P.L. and Harper, J.F. (1969). Some new distributional records for Kenya. *Bulletin of the British Ornithologists' Club* **89**, 162–165.
Britton, P.L. and Sugg, M. St. J. (1973). Birds recorded on Kimilili track, Mt. Elgon, Kenya. *Journal of the East Africa Natural History Society* **143**, 1–7.
Brooke, R.K. (1954). A note on the Yellow Flycatcher. *Babbler* No. 11, 11–12.
Brooks, D.J. (1987). The Yemen Warbler in North Yemen. *Sandgrouse* **9**, 90–93.
Brosset, A. (1956). Les Oiseaux du Maroc oriental de la Méditerranée à Berguent. *Alauda* **24**, 161–205.
Brosset, A. (1971). Territorialisme et défense du territoire chez des migrateurs paléarctiques hivernant au Gabon. *Alauda* **31**, 127–131.
Brown, P.E. and Davis, M.G. (1949). 'Reed Warblers'. Foy Publications, Inc., East Molesey.
Castan, R. (1963). Notes de Tunisie (région de Gabès). *Alauda* **31**, 294–303.
Catchpole, C.K. (1974). Habitat selection and breeding success in the Reed Warbler (*Acrocephalus scirpaceus*). *Journal of Animal Ecology* **43**, 363–380.
Chapin, J.P. (1916). Four new birds from the Belgian Congo. *Bulletin of the American Museum of Natural History* **35**, 23–29.
Chapin, J.P. (1948a). A new genus of Sylviidae from the vicinity of Lake Tanganyika. *Auk* **65**, 291–292.
Chapin, J.P. (1948b). Two new passerine birds from Angola. *Annals of the Carnegie Museum* **31**, 1–3.
Chapin, R.T. (1973). Observations on *Bradypterus carpalis* and *Bradypterus graueri*. *Bulletin of the British Ornithologists' Club* **93**, 167–170.
Chappuis, C. (1976b). Note sur la genre *Bathmocercus* Reichenow, discrimination acoustique de *B. rufus* et *B. cerviniventris*. *Revue de Zoologie Africaine* **90**, 1028–1031.
Christy, P. (1994). La "redécouverte" de la fauvette du Dja au Gabon. *Canopée* **2**, 7.
Clancey, P.A. (1954). Comments on geographical variation in the Tit-Babbler *Parisoma subcaeruleum* (Vieillot) and the description of a new race from the high interior of Natal, South Africa. *Bulletin of the British Ornithologists' Club* **74**, 30–33.
Clancey, P.A. (1955a). Comments on geographical variation in the Knysna Scrub-Warbler *Bradypterus sylvaticus* Sundevall of South Africa. *Bulletin of the British Ornithologists' Club* **75**, 26–28.
Clancey, P.A. (1955b). A review of the races of Barratt's Scrub-Warbler *Bradypterus barratti* Sharpe of South Africa. *Bulletin of the British Ornithologists' Club* **75**, 38–44.
Clancey, P.A. (1961). The Reed Warbler *Acrocephalus scirpaceus* in Natal, a species new to the South African list. *Ostrich* **32**, 143–144.
Clancey, P.A. (1963). Miscellaneous taxonomic notes on African birds XX. *Durban Museum Novitates* **VI**(19), 231–264.
Clancey, P.A. (1969b). Miscellaneous taxonomic notes on African birds XXVII. *Durban Museum Novitates* **VIII**(15), 227–274.
Clancey, P.A. (1970c). On the races of *Phylloscopus trochilus* (Linnaeus) occurring in southern Africa, with special reference to the status of *P. t. yakutensis* Ticehurst 1935. *Durban Museum Novitates* **VIII**, 331–334.
Clancey, P.A. (1973). Subspeciation in the Grassbird *Sphenoeacus afer* (Gmelin) (Aves: Sylviidae). *Arnoldia (Rhodesia)* **6**(5), 1–6.
Clancey P.A. (1974a). The races of the Whitethroat *Sylvia communis* Latham reaching the South African Sub-Region. *Durban Museum Novitates* **XI**, 117–120.
Clancey, P.A. (1975a). On the species limits of *Acrocephalus baeticatus* (Vieillot) (Aves: Sylviidae) of Ethiopian Africa. *Arnoldia (Rhodesia)* **7**, 1–14.
Clancey, P.A. (1975b). The Great Reed Warbler *Acrocephalus*

arundinaceus (Linnaeus) in the South African subregion. *Durban Museum Novitates* **X**, 232–235.

Clancey, P.A. (1975c). Variation in *Acrocephalus palustris* (Bechstein). *Durban Museum Novitates* **X**, 235–238.

Clancey, P.A. (1976). The austral African races of the Rush Warbler *Bradypterus baboecala* (Vieillot). *Durban Museum Novitates* **X**, 120–128.

Clancey, P.A. (1994). Further comment on *Acrocephalus baeticatus* and *A. cinnamomeus* of the Afrotropics. *Honeyguide* **40**, 262–267.

Collett, J. (1982). Birds of the Cradock District. *Southern Birds* **9**, 1–65.

Colston, P.R. and Morel, G.J. (1984). A new subspecies of the African Reed Warbler from Senegal. *Bulletin of the British Ornithologists' Club* **104**, 3–5.

Cordeiro, N.J. and Kiure, J. (1995). An investigation of the forest avifauna in the North Pare Mountains, Tanzania. *Scopus* **19**, 9–26.

Cowper, S.G. (1977). Dry season birds at Enugu and Nsukka. *Bulletin of the Nigerian Ornithologists' Society* **13**(43), 57–63.

Curry, P.J. and Sayer, J.A. (1979). The inundation zone of the Niger as an environment for palearctic migrants. *Ibis* **121**, 20–40.

Davies, N.B. and Green, R.E. (1976). The development and ecological significance of feeding techniques in the Reed Warbler (*Acrocephalus scirpaceus*). *Animal Behaviour* **24**, 213–229.

Dean, W.R.J. (1978). An analysis of avian stomach contents from southern Africa. *Bulletin of the British Ornithologists' Club* **98**, 10–13.

Depuy, A. and Johnson, E.D.H. (1967). Capture d'une locustelle fluviatile (*Locustella fluviatilis*) au Sahara algérien. *L'Oiseau et la Revue Française d'Ornithologie.* **37**, 143.

Devillers, P. and Dowsett-Lemaire, F. (1978). African Reed Warblers (*Acrocephalus baeticatus*) in Khouar (Niger). *Le Gerfaut* **68**, 211–213.

Diamond, A.W. and Keith, G.S. (1980). Avifaunas of Kenya forest islands. 1–Mt Kukal. *Scopus* **4**, 49–55.

Dickerman, R.W., Cane, W.P., Carter, M.F., Chapman, A. and Schmitt, C.G. (1994). Report on three collections of birds from Liberia. *Bulletin of the British Ornithologists' Club* **114**, 267–274.

Diesselhorst, G. (1959). Die geographische Variabilität von *Melocichla mentalis* (Fraser). *Opuscula Zoologica* No. 36, 1–12.

Donnelly, B.G. and Irwin, M.P.S. (1969). Some adaptations in the crombecs *Sylvietta rufescens* (Vieillot) and *Sylvietta whytii* Shelley as reflected by their osteology. *Arnoldia (Rhodesia)* **4**(16), 1–15.

Dowsett, R.J. (1968). Migrants at Malamfatori, Lake Chad, spring 1968. *Bulletin of the Nigerian Ornithologists' Society* **5**, 53–56.

Dowsett, R.J. (1969). B.O.U. supported research at Lake Chad in 1968. *Ibis* **111**, 449–452.

Dowsett, R.J. (1969). Breeding biology of the Olivaceous Warbler *Hippolais pallida laeneni*. *Bulletin of the Nigerian Ornithologists' Society* **6**, 107–108.

Dowsett, R.J. (1969). Migrants at Malamfatori, Lake Chad, autumn 1968. *Bulletin of the Nigerian Ornithologists' Society* **6**, 39–45.

Dowsett, R.J. (1970). A collection of birds from the Nyika Plateau, Zambia. *Bulletin of the British Ornithologists' Club* **90**, 49–53.

Dowsett, R.J. (1972). The River Warbler *Locustella fluviatilis* in Africa. *Zambia Museums Journal* **3**, 69–79.

Dowsett, R.J. (1975). Sight record of an Olive-tree Warbler near Mambova. *Bulletin of the Zambian Ornithological Society* **7**, 19–20.

Dowsett, R.J. (1979). The status of the Broad-tailed Warbler (*Schoenicola platyura*) in Zambia. *Bulletin of the Zambian Ornithological Society* **11**(2), 11–13.

Dowsett, R.J. and Fry, C.H. (1969). Weight losses of trans-Saharan migrants. *Ibis* **113**, 531–533.

Dowsett, R.J. and Hopson, A.J. (1969). Additions and amendments to the list of birds of Malamfatori, Lake Chad. *Bulletin of the Nigerian Ornithologists' Society* **6**, 53–55.

Dowsett, R.J. and Lemaire, F. (1976). The problem of the African Reed Warbler (*Acrocephalus baeticatus*) in Zambia. *Bulletin of the Zambian Ornithological Society* **8**, 62–63.

Dowsett, R.J. and Stjernstedt, R. (1979). The *Bradypterus cinnamomeus-mariae* complex in Central Africa. *Bulletin of the British Ornithologists' Club* **99**, 86–94.

Dowsett-Lemaire, F. (1979). The imitative range of the song of the Marsh Warbler *Acrocephalus palustris* with special reference to imitations of African birds. *Ibis* **121**, 453–468.

Dowsett-Lemaire, F. (1981). The transition period from juvenile to adult song in the European Marsh Warbler. *Ostrich* **52**, 253–255.

Dowsett-Lemaire, F. and Dowsett, R.J. (1985). Breeding biology of the Arabian Warbler *Sylvia leucomelaena* in Israel: comments and suggestions for further research. *Ibis* **127**, 567.

Dowsett-Lemaire, F. and Dowsett, R.J. (1987a). European Reed and Marsh Warblers in Africa: migration patterns, moult and habitat. *Ostrich* **58**, 65–85.

Dowsett-Lemaire, F. and Dowsett, R.J. (1987b). European and African Reed Warblers, *Acrocephalus scirpaceus* and *A. baeticatus*: vocal and other evidence for a single species. *Bulletin of the British Ornithologists' Club* **107**, 74–85.

Dyrcz, A. (1981). Breeding ecology of great reed warbler *Acrocephalus arundinaceus* and reed warbler *A. scirpaceus* at fish-ponds in SW Poland and lakes in NW Switzerland. *Acta Ornithologica* **18**, 307–334.

Earlé, R.A. (1993). Bird blood parasites – a new dimension to bird ringing. *Safring News* **22**, 5–9.

Erard, C. (1978). A new race of *Parisoma lugens* from the highlands of Bale, Ethiopia. *Bulletin of the British Ornithologists' Club* **98**, 43–49.

Fairon, J. (1972). Analyse de contenus stomacaux d'oiseaux provenant de Kaouar (Niger). *Le Gerfaut* **62**, 325–330.

Fayad, V.C. and Fayad, C.C. (1977). A Grasshopper Warbler *Locustella naevia* from Kenya. *Scopus* **1**, 84–85.

Field, G.D. (1973). Subalpine and Grasshopper Warblers in Sierra Leone. *Bulletin of the British Ornithologists' Club* **93**, 101–103.

Flint, P.R. and Stewart, P.F. (1992). The birds of Cyprus, 2nd edition. BOU Check-list No. 6. British Ornithologists' Union, Tring.

Fogden, M.P.L. (1972). Premigratory dehydration in the Reed Warbler (*Acrocephalus scirpaceus*) and water as a factor limiting migratory range. *Ibis* **114**, 548–552.

Forbes-Watson, A.D. (1994). Bird surveys of three reserves in northwestern Uganda. *Scopus* **17**, 128–137.

Fouarge, J.G. (1968). Le pouillot siffleur. *Le Gerfaut* **58**, 177–368.

Fraser, M.W. (1986). Icterine Warbler and Redbacked Shrikes in the Cape of Good Hope Nature Reserve. *Promerops* No. 172, 11.

Fraser, M. and McMahon, L. (1995). European Marsh Warblers on the Cape Peninsula. *Safring News* **24**, 75–76.

Friedmann, H. and Williams, J.G. (1968). Notable records of rare or little-known birds from western Uganda. *Revue de Zoologie et de Botanique Africaines* **77**, 11–36.

Friedmann, H. and Williams, J.G. (1970b). Additions to the known avifauna of the Bugoma, Kibale, and Impenetrable forests, west Uganda. *Los Angeles County Museum Contributions in Science* No. 198, 1–20.

Frost, P.G.H. and Vernon, C.J. (1978). Notes on the Green Eremomela. *Ostrich* 49, 86–88.

Fry, C.H. (1970). Migration, moult and weights of birds in northern Guinea savanna in Nigeria and Ghana. *Proceedings of the Third Pan-African Ornithological Congress, Ostrich Supplement* 8, 239–263.

Fry, C.H. (1975). The northern limits of fringing forest birds in North Central State, Nigeria. *Bulletin of the Nigerian Ornithologists' Society* 11(40), 56–64.

Fry, C.H. (1976). On the systematics of African and Asian tailor-birds (Sylviinae). *Arnoldia (Rhodesia)* 8(6), 1–15.

Fry, C.H. (1977). Taxonomy of the *Acrocephalus baeticatus* complex of African Marsh Warblers. *Nigerian Field* 42, 134–137.

Fry, C.H. (1990). Foraging behaviour and identification of Upcher's Warbler. *British Birds* 83, 217–221.

Fry, C.H., Ferguson-Lees, I.S. and Ash, J.S. (1969). Mite lesions in Sedge Warblers and bee-eaters in Africa. *Ibis* 111, 611–612.

Fry, C.H., Williamson, K. and Ferguson-Lees, I.J. (1974). A new subspecies of *Acrocephalus baeticatus* from Lake Chad and a taxonomic reappraisal of *Acrocephalus dumetorum*. *Ibis* 116, 340–346.

Fuggles-Couchman, N.R. (1986). Breeding records of some Tanzanian birds. *Scopus* 10, 20–26.

Gaston, A.J. (1970). Birds in the central Sahara in winter. *Bulletin of the British Ornithologists' Club* 90, 61–66.

Gaugris, Y. (1976). Additions à l'inventaire des oiseaux du Burundi (décembre 1971 – décembre 1975). *L'Oiseau et la Revue Française d'Ornithologie* 46, 273–289.

Gerhart, J.D. and Paxton, R.O. (1980). The Green Crombec *Sylvietta virens* in Kenya. *Scopus* 4, 47.

Geroudet, P. (1950). La fauvette orphée aux environs de Genève. *Nos Oiseaux* 20, 221–232.

Ginn, P. (1986). Birds using helpers at the nest. *Honeyguide* 32, 45.

Ginn, P. (1989). What birds feed their chicks. *Honeyguide* 35, 170–171.

Ginn, P. (1993). More on the nests of the Long-billed and Red-faced Crombecs. *Honeyguide* 39, 197–198.

Gladwin, T.W. (1963). Increases in weight of *Acrocephali*. *Bird Migration* 2, 319–324.

Goodman, S.M. (1988). Patterns of geographic variation in the Arabian Warbler *Sylvia leucomelaena* (Aves: Sylviidae). *Proceedings of the Biological Society of Washington* 101, 898–911.

Goodman, S.M., Meininger, P.L. and Mullié, W.C. (1986). The birds of the Egyptian Western Desert. *Miscellaneous Publications of the Museum of Zoology, University of Michigan* 172, 1–91.

Grant, C.H.B. and Mackworth-Praed, C.W. (1940). A new genus of African swamp warbler. *Bulletin of the British Ornithologists' Club* 60, 91–92.

Grant, C.H.B. and Mackworth-Praed, C.W. (1941). A new Alethe from Tanganyika Territory; a new race of Sparrow Lark from the Sudan; and a new race of Yellow-bellied Eremomela from Angola. *Bulletin of the British Ornithologists' Club* 61, 61–63.

Green, R.E. and Davies, N.B. (1972). Feeding ecology of Reed and Sedge Warblers. *Wicken Fen Group Report* No. 4, 8–14.

Grieve, A. (1992). First record of Thick-billed Warbler *Acrocephalus aedon* in Egypt. *Sandgrouse* 14, 123–124.

Grimes, L.G. (1974). Duetting in *Hypergerus atriceps* and its taxonomic relationship to *Eminia lepida*. *Bulletin of the British Ornithologists' Club* 94, 89–96.

Grimes, L.G. (1975). The dawn song of the Grey-backed Eremomela *Eremomela pusilla*. *Bulletin of the British Ornithologists' Club* 95, 92–93.

Grimes, L.G. (1976a). The duets of *Laniarius atroflavus*, *Cisticola discolor* and *Bradypterus barratti*. *Bulletin of the British Ornithologists' Club* 96, 113–120.

Grimes, L.G. (1976b). The occurrence of cooperative breeding behaviour in African birds. *Ostrich* 47, 1–15.

Grimes, L.G. (1978). Weights of some Ghanaian birds and distribution data for two species. *Bulletin of the Nigerian Ornithologists' Society* 14, 84–85.

Gunn, W.W.H. and Gulledge, J.L. (1977). 'Beautiful Bird Songs of the World'. National Audubon Society and Cornell Laboratory of Ornithology, Ithaca, New York.

Hall, B.P. (1960b). The faunistic importance of the scarp of Angola. *Ibis* 102, 420–442.

Hanmer, D.B. (1979). A trapping study of Palaearctic birds at Nchalo, southern Malaŵi. *Scopus* 3, 81–92.

Hanmer, D.B. (1982a). Fidelity to winter quarters by Palaearctic passerines. *Safring News* 11, 41–43.

Hanmer, D.B. (1982b). Klaas's Cuckoo parasitizing Whitebellied Sunbird. *Ostrich* 53, 58.

Hanmer, D.B. (1986). Migrant Palaearctic passerines at Nchalo, Malaŵi. *Safring News* 15, 19–28.

Harpum, J. (1978a). Olive-tree Warbler at Dodoma, central Tanzania. *Scopus* 2, 24–25.

Hasselquist, D., Hedenström, A., Lindström, Å. and Bensch, S. (1988). The seasonally divided flight feather moult of the Barred Warbler *Sylvia nisoria* – A new moult pattern for European passerines. *Ornis Scandinavica* 19, 280–286.

Hedenström, A. and Petterson, J. (1984). The migration of Willow Warblers *Phylloscopus trochilus* at Ottenby. *Vår Fågelvärld* 43, 217–228.

Hedenström, A. and Petterson, J. (1987). Migration routes and wintering areas of Willow Warblers *Phylloscopus trochilus* (L.) ringed in Fennoscandia. *Ornis Fennica* 64, 137–143.

Hedenström, A., Bensch, S., Hasselquist, D. and Ottosson, U. (1990). Observations of Palaearctic migrants rare to Ghana. *Bulletin of the British Ornithologists' Club* 110, 194–197.

Hedenström, A., Bensch, S., Hasselquist, D., Lockwood, M. and Ottosson, U. (1993). Migration, stopover and moult of the Great Reed Warbler *Acrocephalus arundinaceus* in Ghana, West Africa. *Ibis* 135, 177–180.

Heim de Balzac, H. (1926). Contributions à l'ornithologie du Sahara central et du sud-Algérien. *Mémoires de la Société d'Histoire Naturelle de l'Afrique du Nord, Alger* 1, 1–127.

Helbig, A.J., Seibold, I., Martens, J. and Wink, M. (1995). Genetic differentiation and phylogenetic relationships of Bonelli's Warbler *Phylloscopus bonelli* and Green Warbler *P. nitens*. *Journal of Avian Biology* 26, 139–153.

Helbig, A.J., Martens, J., Seibold, I., Henning, F., Schottler, B. and Wink, M. (1996). Phylogeny and species limits in the Palaearctic chiffchaff *Phylloscopus collybita* complex: mitochondrial genetic differentiation and bioacoustic evidence. *Ibis* 138(4), 650–666.

Herremans, M. (1994). Major concentration of River Warblers *Locustella fluviatilis* wintering in northern Botswana. *Bulletin of the British Ornithologists' Club* 114, 24–26.

Herremans, M. and Herremans-Tonnoeyr, D. (1993). Tenacious nest attendance by female Titbabbler *Parisoma subcaeruleum*. *Babbler* No. 25, 38.

Hofmeyer, J.H., Hofmeyer, P.K., Broekhuysen, G.J. and Stanford, W. (1961). The nest of the Knysna Scrub

Warbler (*Bradypterus sylvaticus*) and some notes on parental behaviour. *Ostrich* **32**, 177–180.

Holyoak, D.T. (1990). The nest and eggs of *Phylloscopus budongoensis*. *Bulletin of the British Ornithologists' Club* **110**, 159–160.

Horne, J.F.M. and Short, L.L. (1986). Roosting behaviour of Red-faced Crombec. *Scopus* **10**, 49–51.

Horner, K.O. (1980). Spring migration of *Sylvia* species on the north coast of the arab republic of Egypt. *Proceedings of the Fourth Pan-African Ornithological Congress* pp. 215–216. Southern African Ornithological Society, Johannesburg.

Hustler, K. (1995). First breeding record, incubation period and density of the Greater Swamp Warbler in Zimbabwe. *Honeyguide* **41**(3), 161–163.

Irvine, G. and Irvine, D. (1991). Crombec roosting, Ndara Ranch. *East Africa Natural History Society Bulletin* **21**, 34–35.

Irwin, M.P.S. (1953). Notes on some birds of Mashonaland, Southern Rhodesia. *Ostrich* **24**, 37–49.

Irwin, M.P.S. (1959). The siblings *Sylvietta rufescens* and *Sylvietta whytii*. *Occasional Papers of the National Museums of Southern Rhodesia* **23B**, 286–294.

Irwin, M.P.S. (1968). The relationships of the crombecs *Sylvietta ruficapilla* Bocage and *Sylvietta whytii* Shelley. *Bonner Zoologische Beiträge* **19**, 249–256.

Jacot-Guillarmod, C. (1963). Catalogue of the birds of Basutoland. *The South African Avifauna Series* **8**, 111 pp. Percy Fitzpatrick Institute of African Ornithology, Cape Town.

Jany, E. (1960). An Brutplatzen des Lannerfalken. *Proceedings of the Twelfth International Ornithological Congress* pp. 343–352, Helsinki.

Jarry, G. and Larigauderie, F. (1974). Notes faunistique sur quelques oiseaux de Sénégal. *L'Oiseau et la Revue Française d'Ornithologie* **44**, 62–71.

Jensen, R.A.C. and Clinning, C.F. (1974). Breeding biology of two cuckoos and their hosts in South West Africa. *Living Bird* **13**, 5–50.

Jensen, R.A.C. and Jensen, M.K. (1969). On the breeding biology of southern African cuckoos. *Ostrich* **40**, 163–181.

Johnson, D.N. and Maclean, G.L. (1994). Altitudinal migration in Natal. *Ostrich* **65**, 86–94.

Jones, P.J. and Tye, A. (1988). A survey of the avifauna of São Tomé and Príncipe. *ICBP Study Report* No. 24, 1–64. International Council for Bird Protection, Cambridge.

Karr, J.R. (1976). Weights of African birds. *Bulletin of the British Ornithologists' Club* **96**, 92–96.

Keith, S. and Vernon, C. (1966). Notes on African warblers of the genus *Chloropeta* Smith. *Bulletin of the British Ornithologists' Club* **86**, 115–120.

Keith, G. S. and Vernon, C. J. (1969). Bird notes from northern and eastern Zambia. *Puku* No. 5, 131–139.

Kelsey, M.G. (1989). A comparison of the song and territorial behaviour of a long-distance migrant, the Marsh Warbler *Acrocephalus palustris*, in summer and winter. *Ibis* **131**, 403–414.

Kleefisch, T. (1985). Some insectivorous softbills bred in 1984. *Avicultural Magazine* **91**, 204–207.

Klein, H., Berthold, P. and Gwinner, E. (1973). Der Zug europäischer Garten- und Mönchsgrasmucken (*Sylvia borin* und *S. atricapilla*). *Die Vogelwarte* **27**, 73–134.

Komen, J. and Myer, E. (1988). European Reed Warblers in Namibia. *Ostrich* **59**, 142–143.

Kovshar', A.F. and Rukina, A.K. (1968). Breeding biology of Orphean Warbler in the Western Tien Shan (in Russian). *Trudy Instituta Zoologii Akademiya Nauk Kazakhskoy SSR* **29**, 58–63.

Kunkel, P. (1974). Mating systems of tropical birds: the effects of weakness or absence of external reproduction-timing factors, with special reference to prolonged pair bonds. *Zeitschrift für Tierpsychologie* **34**, 265–307.

Lang, J.R. (1969). The nest and eggs of the Oriole Babbler or Moho (*Hypergerus atriceps*). *Bulletin of the Nigerian Ornithologists' Society* **6**, 127–128.

Lawson, W.J. (1964). Systematic notes on African birds II. *Durban Museum Novitates* **VII**(6), 141–155.

Leisler, B. (1972). Artmerkmale am Fuss adulter Teich- und Sumpfrohrsänger und ihre Funktion. *Journal für Ornithologie* **113**, 366–372.

Lévêque, R. (1976). Observations ornithologiques en Sardaigne. *Alauda* **44**, 190–192.

Lindström, Å., Pearson, D.J., Hasselquist, D., Hedenström, A., Bensch, S. and Åkesson, S. (1993). The moult of Barred Warblers *Sylvia nisoria* in Kenya – evidence for a split wing-moult pattern initiated during the birds' first winter. *Ibis* **135**, 403–409.

Louette, M. (1976). Notes on the genus *Bathmocercus* Reichenow (Aves: Sylviidae). Part 1. The different plumages of *B. cerviniventris* and its relationship to *B. rufus*. *Revue de Zoologie Africaine* **90**, 1021–1027.

Louette, M. (1988). Additions and corrections to the avifauna of Zaïre (2). *Bulletin of the British Ornithologists' Club* **108**, 43–50.

Louette, M. (1989). Additions and corrections to the avifauna of Zaïre (4). *Bulletin of the British Ornithologists' Club* **109**, 217–225.

Lynes, H. (1912). Field-notes on a collection of birds from the Mediterranean. *Ibis* **1912**, 121–187.

Macdonald, J.D. (1948). Breeding of Olivaceous Warbler in the Sudan. *Bulletin of the British Ornithologists' Club* **69**, 17.

Maclean, G.L. (1967). The breeding biology and behaviour of the Double-banded Courser *Rhinoptilus africanus* (Temminck). *Ibis* **109**, 556–569.

Maclean, G.L. and Vernon, C.J. (1976). Mouthspots of passerine nestlings. *Ostrich* **47**, 95–98.

MacLeod, J.G.R. (1946). *Cryptillus victorini*. *Ostrich* **7**, 202–203.

MacLeod, J.G.R. and Broekhuysen, G.J. (1951). The nest and eggs of the Victorin's Scrub-Warbler. *Ostrich* **22**, 44.

MacLeod, J.G.R. and Hallack, M. (1956). Some notes on the breeding of Klaas's Cuckoo. *Ostrich* **27**, 5.

MacLeod, J.G.R., Stanford, N.R. and Broekhuysen, G.J. (1958). Notes on the parental behaviour of the Victorin's Warbler *Bradypterus victorini* Sundevall. *Ostrich* **29**, 71–73.

Madge, S. G. (1972). African Yellow Warbler feeding in trees. *Bulletin of the Zambian Ornithological Society* **4**, 60.

Martin, R. (1986). Suspected Garden Warbler in CBC area. *Promerops* No. 173, 11.

Martin, R. (1990). Yellowbellied Eremomela mimicking other birds. *Promerops* No. 195, 12.

Martin, R. and Martin E. (1993). European Marsh Warbler in the southwestern Cape. *Promerops* No. 208, 12.

Martin, R. and Pepler, D. (1995). Birds feeding in lucerne. *Promerops* No. 219, 11.

Martin, R., Martin, J., Martin, E., Neatherway, P., Neatherway, M. and Tyler, D. (1982). A note on the distribution of the Knysna Scrub-Warbler in the South Western Cape Province. *Bokmakierie* **34**, 13.

Mason, C.F. (1976). Breeding biology of the *Sylvia* warblers. *Bird Study* **23**, 213–232.

Masterson, A.N.B. (1981). Notes from the Rwenzori mountains, including description of the nest and eggs of Archer's Ground Robin *Dryocichloides archeri*. *Scopus* **5**, 33–34.

Mathiasson, S. (1971). Untersuchungen an Klappergrasmucken (*Sylvia curruca*) im Niltal in Sudan. *Die Vogelwarte* **26**, 221–227.

Meinertzhagen, R. (1949). On the status of *Parisoma leucomelaena* (Hemp. & Ehr.). *Bulletin of the British Ornithologists' Club* **69**, 109–110.

Meinertzhagen, R. (1954). 'Birds of Arabia'. Oliver and Boyd, Edinburgh and London.

Meininger, P.L., Sorensen, U.G. and Atta, G.A.M. (1986). Breeding birds on the lakes in the Nile Delta, Egypt. *Sandgrouse* **7**, 1–20.

Moreau, R.E. (1940). Distributional notes on East African birds. *Ibis* **1940**, 454–463.

Moreau, R.E. (1946). The adult of Mrs. Moreau's Warbler. *Bulletin of the British Ornithologists' Club* **66**, 44.

Moreau, R.E. and Dolp, R.M. (1970). Fat, water, weights and winglengths of autumn migrants in transit on the northwest coast of Egypt. *Ibis* **112**, 209–228.

Morel, G.J. (1968). Contribution à la synécologie des oiseaux du Sahel sénégalais. *Mémoires de l'Office de la Recherche Scientifique et Technique Outre Mer* **29**.

Moyer, D.C. and Stjernstedt, R. (1986). A new bird for East Africa and extensions of range for some species in southwest Tanzania. *Scopus* **10**, 99–102.

de Naurois, R. (1982). Une énigme ornithologique: *Amaurocichla bocagei*, 1892. *Bulletin de l'Institut Fondamental d'Afrique Noire, Dakar, série A*, **44**, 200–212.

de Naurois, R. (1985). Sur la reproduction de la Rousserolle *Calamocichla rufescens* ssp. dans la région des Niayes (Sénégal Nord-Occidental). *Alauda* **53**, 182–185.

Niethammer, G. and Laenen, J. (1954). Hivernage au Sahara. *Alauda* **22**, 25–31.

Nikolaus, G. (1979b). The first record of the Basra Reed Warbler in the South Sudan. *Scopus* **3**, 103.

Nikolaus, G. (1981). Palaearctic migrants new to north Sudan. *Scopus* **5**, 121–124.

Nikolaus, G. (1982). Further notes on some birds new to south Sudan. *Scopus* **6**, 1–4.

Nikolaus, G. (1983). An important passerine ringing site near the Sudan Red Sea coast. *Scopus* **7**, 15–18.

Nikolaus, G. (1984). Further notes on birds new or little known in the Sudan. *Scopus* **8**, 38–42.

Nikolaus, G. and Pearson, D.J. (1982). Autumn passage of Marsh Warblers *Acrocephalus palustris* and Sprossers *Luscinia luscinia* on the Sudan Red Sea coast. *Scopus* **6**, 17–19.

Nikolaus, G. and Pearson, D.J. (1991). The seasonal separation of primary and secondary moult in palearctic migrants on the Sudan coast. *Ringing and Migration* **12**, 46–47.

North, M.E.W. (1958). 'Voices of African Birds'. Cornell University Press, Ithaca.

Okia, N.O. (1976). Birds of the understorey of lake-shore forests on the Entebbe Peninsula, Uganda. *Ibis* **118**, 1–13.

Olson, S.L. (1984). Syringeal morphology and relationships of *Chaetops* (Timaliidae) and certain South African Muscicapidae. *Ostrich* **55**, 30–32.

Olson, S.L. (1990). Preliminary systematic notes on some Old World passerines. *Rivista Italiana di Ornitologia* **59**(3–4), 183–195.

Parker, S.A. and Harrison, C.J.O. (1963). The validity of the genus *Lusciniola* Grey. *Bulletin of the British Ornithologists' Club* **83**, 65–69.

Parkes, D.A. (1993). Nests of the Red-faced and Long-billed Crombecs. *Honeyguide* **39**, 91–93.

Parkes, D.A. (1994). Further on the differences between the nests of the Red-faced and Long-billed Crombecs. *Honeyguide* **40**, 99–100.

Parkes, K.C. (1987). Taxonomic notes on some African warblers. (Aves: Sylviinae). *Annals of the Carnegie Museum* **56**, Art. 13, 231–243.

Peach, W., Baillie, S. and Underhill, L. (1991). Survival of British Sedge Warblers *Acrocephalus schoenobaenus* in relation to west African rainfall. *Ibis* **133**, 300–305.

Pearson, D.J. (1971). Weights of some Palaearctic migrants in southern Uganda. *Ibis* **113**, 173–184.

Pearson, D.J. (1972). The wintering and migration of Palaearctic passerines at Kampala, southern Uganda. *Ibis* **114**, 43–60.

Pearson, D.J. (1973). Moult of some Palaearctic warblers wintering in southern Uganda. *Bird Study* **20**, 24–36.

Pearson, D.J. (1975). The timing of complete moult in the Great Reed Warbler *Acrocephalus arundinaceus*. *Ibis* **117**, 506–509.

Pearson, D.J. (1978). The genus *Sylvia* in Kenya and Uganda. *Scopus* **2**, 63–71.

Pearson, D.J. (1980). Northward passage of Palaearctic passerines across Tsavo. *Scopus* **4**, 25–28.

Pearson, D.J. (1982). The migration and wintering of Palaearctic *Acrocephalus* warblers in Kenya and Uganda. *Scopus* **6**, 49–59.

Pearson, D.J. (1989). Palaearctic migrants in the Middle and Lower Jubba valley, southern Somalia. *Scopus* **12**, 53–60.

Pearson, D.J. (1989). The separation of Reed Warblers *Acrocephalus scirpaceus* and Marsh Warblers *A. palustris* in eastern Africa. *Scopus* **13**, 81–89.

Pearson, D.J. (1992). Northward passage of Palearctic songbirds through Kenya. *Proceedings of the Seventh Pan-African Ornithological Congress* pp 113–124. PAOC Committee, Nairobi.

Pearson, D.J. and Backhurst, G.C. (1976). The southward migration of Palaearctic birds over Ngulia, Kenya. *Ibis* **118**, 78–105.

Pearson, D.J. and Backhurst, G.C. (1983). Moult in the River Warbler *Locustella fluviatilis*. *Ringing and Migration* **4**, 227–230.

Pearson, D.J. and Backhurst, G.C. (1988). The characters and taxonomic position of the Basra Reed Warbler. *British Birds* **81**, 171–178.

Pearson, D.J., Britton, P.L. and Britton, H.A. (1978). Substantial wintering populations of the Basra Reed Warbler *Acrocephalus griseldis* in eastern Kenya. *Scopus* **2**, 33–35.

Pearson, D.J., Backhurst, G.C. and Backhurst, D.E.G. (1979). Spring weights and passage of Sedge Warblers *Acrocephalus schoenobaenus* in central Kenya. *Ibis* **121**, 8–19.

Pearson, D.J., Finch, B.W. and Backhurst, D.E.G. (1988). A second Savi's Warbler *Locustella luscinioides* at Ngulia. *Scopus* **12**, 94.

Pikulski, A. (1986). Breeding biology and ecology of Savi's Warbler *Locustella luscinioides* at Milicz fishponds. *Ptaki Slacska* **4**, 2–39.

Pomeroy, D.E. and Tengecho, B. (1982). Studies of birds in a semi-arid area of Kenya. II. Bird parties in two woodland areas. *Scopus* **6**, 25–32.

Prigogine, A. (1958). The status of *Eremomela turneri* van Someren and the description of a new race from the Belgian Congo. *Bulletin of the British Ornithologists' Club* **78**, 146–148.

Prigogine, A. (1972). Nids et oeufs recoltés au Kivu, 2. *Revue de Zoologie et de Botanique Africaines* **85**, 203–224.

Prigogine, A. (1975). Contribution à l'étude de la distribution verticale des oiseaux orophiles. *Le Gerfaut* **64**, 75–88.

Prigogine, A. (1980c). *Bradypterus alfredi kungwensis* au Zaïre. *Le Gerfaut* **70**, 279–280.

Prigogine, A. (1982). Les sous-espèces du pouillot fitis *Phylloscopus trochilus* au Zaïre, Rwanda et Burundi. *Le Gerfaut* **72**, 55–72.

Pringle, J.S. (1977). Breeding of the Knysna Scrub Warbler. *Ostrich* **48**, 112–114.

Rabøl, J. (1987). Coexistence and competition between overwintering Willow Warbler *Phylloscopus trochilus* and local warblers at Lake Naivasha, Kenya. *Ornis Scandinavica* **18**, 101–121.

Raijmakes, J.M.H. (1995). Record of a European Reed Warbler at Vanderbijlpark. *Birding in Southern Africa* **47**(3), 95.

Raymer, H. (1993). The elusive River Warbler. *Witwatersrand Bird Club News* **161**, 19.

Robertson, I. (1995). Notes on birds in Ethiopia. *Scopus* **19**, 61–62.

Rodriguez de los Santos, M., Cuadrado, M. and Arjona, S. (1986). Variation in the abundance of Blackcaps (*Sylvia atricapilla*) wintering in an Olive (*Olea europea*) orchard in southern Spain. *Bird Study* **33**, 81–86.

de Roo, A. and Deheegher, J. (1969). Ecology of the Reed Warbler *Acrocephalus arundinaceus* (Linné) wintering in the southern Congo savanna. *Le Gerfaut* **59**, 260–275.

Saurola, P. (1994). Saharan ylittäjät: Mitä rengaslöydöt kertovat. *Linnut* **29**(3), 8–14.

Schmidt, R.K. (1965). Incubation period of African Marsh Warbler *Acrocephalus baeticatus*. *Ostrich* **36**, 34.

Schmitt, M.B. (1975). New distributional data, 6. *Ostrich* **46**, 177.

Schulze-Hagen, K. (1989). Bekanntes und weniger Bekanntes vom Seggenrohrsänger *Acrocephalus paludicola*. *Limicola* **3**, 229–246.

Schulze-Hagen, K. and Barthel, P.H. (1993). Die Bestimmung der europäischen ungestreiften Rohrsänger *Acrocephalus*. *Limicola* **7**, 1–34.

Schulze-Hagen, K., Flinks, H. and Drycz, A. (1989). Brutzeitliche Beutewahl beim Seggenrohrsänger *Acrocephalus paludicola*. *Journal für Ornithologie* **130**, 251–255.

Sclater, W.L. (1932). *Hyliota australis usambarae*, subsp. nov. *Bulletin of the British Ornithologists' Club* **52**, 104.

Sclater, W.L. and Mackworth-Praed, C. (1918). A list of the birds of the Anglo-Egyptian Sudan. Pt. 2. *Ibis* **1918**, 602–721.

Serle, W. (1943b). Notes on East African birds. *Ibis* **85**, 55–58.

Shirihai, H. (1987). Identification of Upcher's Warbler. *British Birds* **80**, 473–482.

Shirihai, H. (1988a). A new subspecies of Arabian Warbler *Sylvia leucomelaena* from Israel. *Bulletin of the British Ornithologists' Club* **108**, 64–68.

Shirihai, H. (1988b). Iris color of *Sylvia* warblers. *British Birds* **81**, 325–328.

Shirihai, H. (1989). Identification of Arabian Warbler. *British Birds* **82**, 97–113.

Short, L.L. and Horne, J.F.M. (1981). Bird observations along the Egyptian Nile. *Sandgrouse* **3**, 43–61.

Short, L.L. and Horne, J.F.M. (1987). Black mamba takes Northern Crombec from mobbing bird group. *Scopus* **11**, 53–54.

Siegfried, W.R. (1992). Conservation status of the South African endemic avifauna. *South African Journal of Wildlife Research* **22**, 61–64.

Simmons, K.E.L. (1952). Some observations of the Olivaceous Warbler *Hippolais pallidus* in Egypt. *Ibis* **94**, 203–209.

Simmons, K.E.L. (1954). Field notes on the behaviour of some passerines migrating through Egypt. *Ardea* **42**, 140–151.

Simmons, K.E.L. (1961). Problems of head-scratching in birds. *Ibis* **103a**, 37–49.

Simms, E. (1985). 'British Warblers'. Collins, London.

Sinclair, J.C. (1976). Identification of Olive-tree Warbler. *Bokmakierie* **28**, 19.

Skinner, N.J. (1993). The nest record card scheme. *Babbler* No. 25, 43–46.

Smith, K.D. (1953). *Locustella luscinioides* in Eritrea. *Ibis* **95**, 698–699.

Smith, K.D. (1961). On the Clamorous Reed Warbler *Acrocephalus stentoreus* (Hemprich and Ehrenberg) in Eritrea. *Bulletin of the British Ornithologists' Club* **81**, 28–29.

Smith, K.D. (1964). *Acrocephalus dumetorum* in Africa. *Bulletin of the British Ornithologists' Club* **84**, 172.

Smith, V.W. (1966). Autumn and spring weights of some Palaearctic migrants in central Nigeria. *Ibis* **108**, 492–572.

Snow, D. (1952). A contribution to the ornithology of North-West Africa. *Ibis* **94**, 473–498.

van Someren, V.G.L. (1919). New forms from Africa. *Bulletin of the British Ornithologists' Club* **40**, 20–29.

van Someren, V.G.L. (1947). Onset of sexual activity. *Ibis* **89**, 51–56.

Squelch, P. and Safe-Squelch, W. (1994). Didric Cuckoo being fed by Little Rush Warblers. *East Africa Natural History Society Bulletin* **24**(3), 39–40.

Staav, R. (1993). 1000 tropikaterfynd. *Vår Fågelvärld* **52**(8), 6–15.

Stam, D. and Voous, K.H. (1963). African record of Aquatic Warbler. *Ardea* **51**, 74.

Stanford, C.B. and Msuya, P. (1995). An annotated list of the birds of Gombe National Park, Tanzania. *Scopus* **19**, 38–46.

Steyn, P. (1969). Marico Flycatcher tending tit-babbler nestlings. *Ostrich* **40**, 51–54.

Steyn, P. (1994). Knysna Warbler in Newlands. *Promerops* No. 212, 10.

Stjernstedt, R. and Moyer, D.C. (1982). Some new birds and extentions of range for southwest Tanzania. *Scopus* **6**, 36–37.

Stresemann, E. and Stresemann, V. (1968). Winterquartier und Mauser der Dorngrasmucke, *Sylvia communis*. *Journal für Ornithologie* **109**, 303–314.

Stronach, N. (1990). New information on birds in Serengeti National Park, Tanzania. *Bulletin of the British Ornithologists' Club* **110**, 198–202.

Swynnerton, C.F.M. (1908). Further notes on the birds of Gazaland. *Ibis* **1908**, 1–107.

Tarboton, W.R. (1970). Nest and eggs of Burnt-necked Eremomela. *Ostrich* **41**, 212.

Tarboton, W.R. (1980). Avian populations in Transvaal savanna. *Proceedings of the Fourth Pan-African Ornithological Congress* pp. 113–124. Southern African Ornithological Society, Johannesburg.

Thomas, D.K. (1977). Wing moult in the Savi's Warbler. *Ringing and Migration* **1**, 125–130.

Thouy, P. (1978a). Nouvelles captures et observations hivernales au Maroc. *Alauda* **46**, 87–93.

Thouy, P. (1978b). Première capture au Maroc de l'hypolais ictèrine *Hippolais icterina*. *Alauda* **46**, 98.

Took, J.M.E. (1959). Breeding of *Schoenicola brevirostris* in Southern Rhodesia. *Ostrich* **30**, 138–139.

Traylor, M.A. (1966). The race of *Acrocephalus rufescens* in Zambia. *Bulletin of the British Ornithologists' Club* **86**, 161–162.

Traylor, M.A. (1970b). Notes on African Muscicapidae. *Ibis* **112**, 395–397.

Tree, A.J. (1991). Recent reports. *Honeyguide* **37**, 134.

Tree, A. J. (1995a). Recent reports. *Honeyguide* **41**, 30–40, 117–130.

Tucker, J.J. (1978). A River Warbler *Locustella fluviatilis* wintering and moulting in Zambia. *Bulletin of the British Ornithologists' Club* **98**, 2–4.

Turner, D.A. (1977). Status and distribution of the East African endemic species. *Scopus* **1**, 1–11.

Turner, D.A., Pearson, D.J. and Finch, B.W. (1990). Green Crombecs *Sylvietta virens* in Busia District, Western Kenya. *Scopus* **14**, 31–32.

Tyler, S.J. (1991). Birds of Lake Naivasha 2. Foraging niches and relationships between migrant and resident warblers in papyrus swamp. *Scopus* **14**(2), 117–124.

Underhill, G.D. (1992). Garden Warbler *Sylvia borin* in the southwestern Cape Province. *Safring News* **21**, 60.

Underhill, L.G., Prŷs-Jones, R.P., Dowsett, R.J., Herroelen, P., Johnson, D.N., Lawn, M.R., Norman, S.C., Pearson, D.J. and Tree, A.J. (1992). The biannual primary moult of Willow Warblers *Phylloscopus trochilus* in Europe and Africa. *Ibis* **134**, 286–297.

Vande weghe, J.-P. (1983). Sympatric occurrence of the White-winged Warbler *Bradypterus carpalis* and Grauer's Rush Warbler *Bradypterus graueri* in Rwanda. *Scopus* **7**, 85–88.

Vande weghe, J.-P. (1992). New records for Uganda and Tanzania along the Rwandan and Burundian borders. *Scopus* **16**, 59–60.

Vernet, R. (1973). Présence de l'hypolais ictérine et mortalité de migrateurs dans le Sahara nord-occidental. *Alauda* **41**, 425–426.

Vernon, C.J. (1962). Variation in the nest of *Chloropeta natalensis*. *Ostrich* **33**, 52–53.

Vernon, C.J. (1963). Notes on the breeding of the Burnt-neck Eremomela *Eremomela usticollis* in Southern Rhodesia. *Ostrich* **34**, 175.

Vernon, C.J. (1968). A year's census of Marandellas, Rhodesia. *Ostrich* **39**, 12–24.

Vernon, C.J. (1970b). Palaearctic warblers in the Transvaal. *Ostrich* **41**, 218.

Vernon, C.J. (1973). Vocal imitation by southern African birds. *Ostrich* **44**, 23–30.

Vernon, C.J. (1976). Red-faced Crombec caught in spider's web. *Honeyguide* No. 85, 41.

Vernon, C.J. (1980). Bird parties in central and south Africa. *Proceedings of the Fourth Pan-African Ornithological Congress* pp. 313–325. Southern African Ornithological Society, Johannesburg.

Vernon, C.J. (1983). A specimen of the African Yellow Warbler from Amanzi. *Bee-eater* **34**, 24.

Vernon, C.J. (1985a). Bird populations in two woodlands near Lake Kyle, Zimbabwe. *Honeyguide* **31**, 148–161.

Vernon, C.J. (1985b). Striped Pipit at Sapkamma and Yellow Warbler at Amanzi. *Bee-eater* **36**, 23.

Vernon, C.J. (1989). Observations on the forest birds around East London. *Ostrich Supplement* **14**, 75–84.

Vernon, C.J. (1993). Yellow Warbler in the Eastern Cape. *Bee-eater* **44**, 11.

Vernon, C.J., Macdonald, I.A.W. and Dean, W.R.J. (1989). Birds of an isolated tropical lowland rainforest in eastern Zimbabwe. *Ostrich Supplement* **14**, 111–122.

Vernon, F.J. and Vernon, C.J. (1978). Notes on the social behaviour of the Duskyfaced Warbler. *Ostrich* **49**, 92–93.

Vincent, J. (1948b). New races of a tit-babbler and a lark from the Basutoland Mountains. *Bulletin of the British Ornithologists' Club* **68**, 145–146.

Walker, G.R. (1939). Notes on the birds of Sierra Leone. *Ibis* **1939**, 401–450.

Wallace, D.I.M. (1969). Palearctic migrants in West Lagos: November 1968 to May 1969. *Bulletin of the Nigerian Ornithologists' Society* **6**, 45–49.

Walsh, J.F. and Grimes, L.G. (1981). Observations on some Palaearctic land birds in Ghana. *Bulletin of the British Ornithologists' Club* **101**, 327–334.

White, C.M.N. (1952). The status of the genus *Calamocichla*. *Ibis* **94**, 685–686.

Wilkinson, R. and Aidley, D. (1982). The status of Savi's Warbler in Nigeria. *Malimbus* **4**, 48.

Wilkinson, R. and Aidley, D. (1983). African Reed Warbler in northern Nigeria; morphometrics and the taxonomic debate. *Bulletin of the British Ornithologists' Club* **103**, 135–138.

Willcox, D.R.C. and Willcox, B. (1978). Observations of birds in Tripolitania, Libya. *Ibis* **120**, 329–333.

Williams, J.G. (1951). Notes on *Scepomycter winifredae* and *Cinnyris loveridgei*. *Ibis* **93**, 469–470.

Williams, J.G. (1955). A new species of *Sylvietta* from Italian Somalia. *Ibis* **97**, 582–583.

Williamson, K. (1967). Identification for ringers. The genus *Phylloscopus*. B.T.O. identification guide no. 2. British Trust for Ornithology, Tring.

Williamson, K. (1968a). Identification for ringers. The genera *Cettia*, *Locustella*, *Acrocephalus* and *Hippolais*. B.T.O. identification guide no.1. British Trust for Ornithology, Tring.

Williamson, K. (1968b). Identification for ringers. The genus *Sylvia*. B.T.O. identification guide no. 3. British Trust for Ornithology, Tring.

Willis, E.O. (1986a). Vireos, wood warblers and warblers as ant followers. *Le Gerfaut* **76**, 177–186.

Wink, M. (1981). On the diets of warblers, weavers and other Ghanaian birds. *Malimbus* **3**, 114–115.

Winstanley, D., Spencer, R. and Williamson, K. (1974). Where have all the Whitethroats gone? *Bird Study* **21**, 1–14.

Winterbottom, J.M. (1943). On woodland parties in Northern Rhodesia. *Ibis* **85**, 437–442.

Winterbottom, J.M. (1958). A new subspecies of *Parisoma layardi* Hartlaub. *Bulletin of the British Ornithologists' Club* **78**, 148–149.

Winterbottom, J.M. (1968). The bird population of karroid broken veld in the Klaarstroom area, Prince Albert Division. *Ostrich* **39**, 85–90.

Winterbottom, J.M. (1971b). Results of the garden bird counts organised by the Percy Fitzpatrick Institute of African Ornithology. *Ostrich* **42**, 110–118.

Families MUSCICAPIDAE (*Fraseria, Melaenornis, Empidornis, Muscicapa, Myioparus, Stenostira, Ficedula*); MONARCHIDAE (*Erythrocercus, Elminia, Trochocercus, Terpsiphone*); PLATYSTEIRIDAE (*Megabyas, Bias, Diaphorophyia, Platysteira, Batis, Lanioturdus*)

Ade, B. (1976). Notes on breeding of the Black Flycatcher. *Honeyguide* No. 85, 39.

Alatalo, R.V. and Lundberg, A. (1984). Polyterritorial polygyny in the pied flycatcher *Ficedula hypoleuca* – Evidence for the deception hypothesis. *Annales Zoologica Fennici* **21**, 217–228.

Alexander-Marrack, P.D., Aaronson, M.J., Farmer, R., Houston, W.H. and Mills, T.R. (1985). Some changes in the bird fauna of Lagos, Nigeria. *Malimbus* **7**, 121–127.

Altenburg, W. and van der Kamp, J. (1991). Ornithological importance of coastal wetlands in Guinea. *ICBP Study Report* No. 47. International Council for Bird Preservation, Cambridge.

Ames, P.L. (1975). The application of syringeal morphology to the classification of the Old World insect eaters

(Muscicapidae). *Bonner Zoologische Beiträge* **26**, 107–134.
Ash, J.S., Dowsett, R.J. and Dowsett-Lemaire, F. (1991). Additions to the East African avifauna. *Scopus* **14**, 73–75.
Aspinwall, D.R. (1985). Movement analysis charts: comments on Dusky Flycatcher. *Zambian Ornithological Society Newsletter* **15**, 37–39.
Backhurst, G.C. and Pearson, D.J. (1984). The timing of the southward night migration of Palaearctic birds over Ngulia, southeast Kenya. *Proceedings of the Fifth Pan-African Ornithological Congress* pp. 361–369. Southern African Ornithological Society, Johannesburg.
Bairlein, F. (1985). Red-breasted Flycatcher (*Ficedula parva*) in the central Sahara. *Le Gerfaut* **75**, 101–103.
Bairlein, F. (1988). Herbstlicher Durchzug, Körpergewichte und Fettdeposition von Zugvögeln in einem Rastgebiet in Nordalgerien. *Die Vogelwarte* **34**, 237–248.
Balchin, C.S. (1988). Recent observations of birds from the Ivory Coast. *Malimbus* **10**, 201–206.
Balchin, C.S. (1990). Further observations of birds from the Ivory Coast. *Malimbus* **12**, 52–53.
Bannerman, D.A. (1932). Account of the birds collected (i) by Mr G.L. Bates on behalf of the British Museum in Sierra Leone and French Guinea; (ii) by Lt-Col. G.J. Houghton, R.A.M.C., in Sierra Leone, recently acquired by the British Museum. Part II. *Ibis* **1932**, 1–33.
Barnes, E. (1996). Dusky Flycatcher in winter. *Promerops* No. 222, 15.
Bates, G.L. (1905). Field-notes on the birds of Efulen in the West-African colony of Kamerun. *Ibis* **1905**, 89–98.
Bates, G.L. (1908). Observations regarding the breeding-seasons of the birds in southern Kamerun. *Ibis* **1908**, 558–570.
Bates, G.L. (1909). Field-notes on the birds of southern Kamerun, West Africa. *Ibis* **1909**, 1–74.
Bates, G.L. (1927). Notes on some birds of Cameroon and the Lake Chad Region: their status and breeding-times. *Ibis* **1927**, 1–64.
Bates, G.L. (1936). A bird-spider nesting association. *Ibis* **1936**, 817–818.
Beasley, A.J. (1985). Black Flycatcher nesting in old Red-headed Weaver's nest. *Honeyguide* **31**, 175–177.
Bednall, D.K. (1990). Silverbird near Mt. Elgon. *East Africa Natural History Society Bulletin* **20**, 26.
Beecher, W.J. (1953). A phylogeny of the oscines. *Auk* **70**, 273–333.
Beel, C. (1995). Species notes for August 1995. Cassin's Grey Flycatcher. *Zambian Ornithological Society Newsletter* **25**, 104.
Beesley, J.S.S. (1972b). Possible polygamy amongst two species of flycatcher. *East Africa Natural History Society Bulletin* **1972**, 38.
Beesley, J.S.S. (1973). The breeding seasons of birds in the Arusha National Park, Tanzania. *Bulletin of the British Ornithologists' Club* **93**, 10–20.
Belcher, C.F. (1925). Birds on the Luchenya Plateau, Mlanje, Nyasaland. *Ibis* **1925**, 797–814.
Belcher, C.F. (1942). Field notes from Kenya. *Ibis* **1942**, 91–93.
Bellatrèche, M. (1993). Ecologie et biogéographie de l'avifaune forestière nicheuse de la Kabylie des Babors (Algérie). Thesis, Université de Bourgogne, Dijon.
Bennun, L.A. (1986). Montane birds of the Bwindi (Impenetrable) Forest. *Scopus* **10**, 87–91.
Benson, C.W. and Irwin, M.P.S. (1965). The birds of *Cryptosepalum* forests, Zambia. *Arnoldia (Rhodesia)* **1**(28), 1–12.
Bibby, C.J. and Green, R.E. (1980). Foraging behaviour of migrant pied flycatchers, *Ficedula hypoleuca*, on temporary territories. *Journal of Animal Ecology* **49**, 507–521.
Blignaut, J. (1959). Chat Flycatchers *Bradornis infuscatus* at Mount Stewart. *Ostrich* **30**, 140–141.
Boles, W.E. (1981). The subfamily name of the monarch flycatchers. *Emu* **81**, 50.
de Bournonville, D. (1967). Notes d'ornithologie guinéenne. *Le Gerfaut* **57**, 145–158.
Bowen, P. St.J (1979). Some notes on Margaret's Batis, *Batis margaritae* in Zambia. *Bulletin of the Zambian Ornithological Society* **11**, 1–10.
Britton, P.L. (1980b). *Ficedula* flycatchers in East Africa. *Scopus* **4**, 21–22.
Broekhuysen, G.J. (1958). Notes on the breeding behaviour of the Cape Flycatcher *Batis capensis*. *Ostrich* **29**, 143–152.
Brooke, R.K. (1958). Incubation and nestling periods revealed by Rhodesian nest-record cards. *Ostrich* **29**, 133–136.
Brooke, R.K. (1962). Further incubation and nestling periods revealed by Rhodesian nest-record cards. *Ostrich* **33**, 23–25.
Brooke, R.K. (1984). An overlooked source of Moçambican bird records, and its bearing on the local races of the Dusky Flycatcher. *Ostrich* **55**, 170–171.
Brooke, R.K. and Borrett, R.P. (1972). A new biological host of the Didric Cuckoo. *Ostrich* **43**, 235.
Brooke, R.K. and Manson, A.J. (1979). Towards a natural history of the wattle-eyed flycatcher, particularly in Rhodesia. *Honeyguide* **97**, 15–17.
Brosset, A. and Erard, C. (1977). Faunistic records new for Gabon. *Bulletin of the British Ornithologists' Club* **97**, 125–132.
Brown, L.H. (1948). Notes on birds of the Kabba, Ilorin and N. Benin Provinces of Nigeria. *Ibis* **90**, 525–538.
Brunel, J. and Thiollay, J.-M. (1969–1970). Liste préliminaire des oiseaux de Côte d'Ivoire. *Alauda* **37**, 230–254, 315–337; **38**, 72–73.
Butynski, T.M. and Kalina, J. (1993). Further additions to the known avifauna of the Impenetrable (Bwindi) Forest, southwestern Uganda (1989–91). *Scopus* **17**, 1–7.
Byrom, D.A. (1961). Notes on the breeding behaviour of the Paradise Flycatcher, *Terpsiphone viridis plumbeiceps*. *Ostrich* **32**, 174–176.
Cawkell, E.M. (1965). Notes on Gambian birds. *Ibis* **107**, 535–540.
Chapin, J.P. (1948). Variation and hybridization among the paradise flycatchers of Africa. *Evolution* **2**, 111–126.
Chapin, J.P. (1963). The supposed "grey mutants" of *Terpsiphone viridis*. *Ibis* **105**, 198–202.
Cheke, R.A., Walsh, J.F. and Fishpool, L.D.L. (1985). Bird records from the Republic of Niger. *Malimbus* **7**, 73–90.
Clancey, P.A. (1955). Geographical variation in the Fairy Flycatcher *Stenostira scita* (Vieillot) of South Africa. *Bulletin of the British Ornithologists' Club* **75**, 1–3.
Clancey, P.A. (1957). The systematic position of *Parisoma plumbeum*. *Ibis* **99**, 512–513.
Clancey, P.A. (1958). Miscellaneous taxonomic notes on African birds XI. *Durban Museum Novitates* **V**(10), 117–142.
Clancey, P.A. (1959). The South African races of the Cape Batis *Batis capensis* (Linnaeus). *Bulletin of the British Ornithologists' Club* **79**, 57–60.
Clancey, P.A. (1966). A catalogue of birds of the South African sub-region (part IV: Families Sylviidae – Prionopidae). *Durban Museum Novitates* **VII**(12), 465–544.
Clancey, P.A. (1966). The avian superspecies of the South

African avifauna. *Proceedings of the Second Pan-African Ornithological Congress, Ostrich Supplement* **6**, 13–39.

Clancey, P.A. (1967). On the South African races of the Black Flycatcher *Melaenornis pammelaina* (Stanley). *Ostrich* **38**, 50–51.

Clancey, P.A. (1969). Miscellaneous taxonomic notes on African birds XXVII. *Durban Museum Novitates* **VIII**(15), 227–274.

Clancey, P.A. (1970). Miscellaneous taxonomic notes on African birds: two new subspecies of passerine birds from western Angola. *Durban Museum Novitates* **IX**, 8–11.

Clancey, P.A. (1974). Subspeciation studies in some Rhodesian birds. *Arnoldia (Rhodesia)* **6**(28), 1–43.

Clancey, P.A. (1975). On the endemic birds of the montane evergreen forest biome of the Transvaal. *Durban Museum Novitates* **X**(12), 151–180.

Clancey, P.A. (1976). Subspeciation in the Marico Flycatcher *Melaenornis mariquensis* (Smith) of south-western Africa. *Bulletin of the British Ornithologists' Club* **96**, 53–57.

Clancey, P.A. (1979). Miscellaneous taxonomic notes on African birds LV. *Durban Museum Novitates* **XII**(5), 47–61.

Colahan, B.D. (1989). The distribution of the Cape Batis in the Orange Free State, South Africa. *Mirafra* **6**, 104–106.

Collett, J. (1982). Birds of the Cradock District. *Southern Birds* **9**, 1–65.

Cordeiro, N.J., Lehmberg, T. and Kiure, J. (1995). A preliminary account of the avifauna of Kahe II Forest Reserve, Tanzania. *Scopus* **19**, 1–8.

Craig, A. (1977). Fiscal Flycatcher mimics Cape Robin. *Ostrich* **48**, 42–43.

Creutz, G. (1955). Der Trauerschnäpper (*Muscicapa hypoleuca* Pall.). Eine Populationsstudie. *Journal für Ornithologie* **96**, 241–326.

Curio, E. (1959a). Verhaltensstudien am Trauerschnäpper. *Journal für Ornithologie* **100**, 176–209.

Curio, E. (1959b). Beobachtungen am Halbingschnäpper *Ficedula semitorquata* in mazedonischen Brutgebiet. *Zeitschrift für Tierpsychologie* Supplement 3.

Davies, N.B. (1976). Parental care and the transition to independent feeding in the young Spotted Flycatcher. *Behaviour* **59**, 280–295.

Davies, N.B. (1977). Prey selection and the search strategy of the Spotted Flycatcher: a field study on optimal foraging. *Animal Behaviour* **25**, 1016–1033.

Decoux, J.-P. and Erard, C. (1992). *Tauraco macrorhynchus* and *T. persa* in north-eastern Gabon: behavioural and ecological aspects of their coexistence. *Proceedings of the Seventh Pan-African Ornithological Congress* pp. 369–380. PAOC Committee, Nairobi.

Dedekind, H. (1987). Whitetailed Shrike nesting in the Naukluft Park. *Lanioturdus* **22**, 82–83.

Dejonghe, J.F. and Cornuet, J.F. (1982). La migration du Gobemouche noir en France et dans le Maghreb: une analyse des reprises. *L'Oiseau et la Revue Française d'Ornithologie* **52**, 259–288.

Dekeyser, P.L and Derivot, J.H. (1968). Les oiseaux de l'Ouest Africain, III. *Initiations et Etudes Africaines* **XIX**. Institut Fondamental d'Afrique Noire, Dakar.

Dickerman, R.W. (1994). Notes on birds from Africa with descriptions of three new subspecies. *Bulletin of the British Ornithologists' Club* **114**, 274–278.

Dickerman, R.W., Cane, W.P., Carter, M.F., Chapman, A. and Schmitt, C.G. (1994). Report on three collections of birds from Liberia. *Bulletin of the British Ornithologists' Club* **114**, 267–274.

Dowsett, R.J. and Dowsett-Lemaire, F. (1989b). Avifaune du Congo: additions et corrections. *Tauraco Research Report* No. 2, 17–19.

Dowsett, R.J. and Fry, C.H. (1971). Weight loss of trans-Saharan migrants. *Ibis* **113**, 531–533.

Dowsett-Lemaire, F. and Dowsett, R.J. (1984). The effects of forest size on montane bird populations. *Proceedings of the Fifth Pan-African Ornithological Congress* pp. 237–248. Southern African Ornithological Society, Johannesburg.

Durand, M. (1949). A note on the eggs of an African Flycatcher. *Oologists' Record* **23**, 10.

Dutson, G. and Branscombe, D. (1990). Rainforest birds in south-west Ghana. *ICBP Study Report* No. 46, 1–70. International Council for Bird Protection, Cambridge.

Dyer, M., Gartshore, M.E. and Sharland, R.E. (1986). The birds of Nindam Forest Reserve, Kagoro, Nigeria. *Malimbus* **8**, 2–20.

Eccles, S.D. (1988). The birds of São Tomé – record of a visit, April 1987 with notes on the rediscovery of Bocage's Longbill. *Malimbus* **10**, 207–217.

Edwards, E.A. (1988). Albinism in Black Flycatcher. *Honeyguide* **34**, 129.

Edwards, J.K. (1973). Silverbird *Empidornis semipartitus* at Karen. *East Africa Natural History Society Bulletin* **1973**, 87.

Epprecht, W. (1985). 21 Jahre Grauerschnäpper-bruten am gleichen Nestort. *Der Ornithologische Beobachter* **82**, 169–184.

Erard, C. (1975). Affinités de *Batis minima* (J. et E. Verreaux) et de *B. ituriensis* Chapin. *L'Oiseau et la Revue Française d'Ornithologie* **45**, 235–240.

Erard, C. (1991). Aide au nourrissage chez le Gobe-mouche gris *Muscicapa striata*. *L'Oiseau et la Revue Française d' Ornithologie* **61**, 154–155.

Erard, C. and Colston, P.R. (1988). *Batis minima* (Verreaux) new for Cameroon. *Bulletin of the British Ornithologists' Club* **108**, 182–184.

Erard, C. and Etchécopar, R.D. (1970). Some notes on the birds of Angola. *Bulletin of the British Ornithologists' Club* **90**, 158–161.

Erard, C. and Larigauderie, F. (1972). Observations sur la migration pré-nuptiale dans l'ouest de la Libye (Tripolitaine et plus particulièrement Fezzan). *L'Oiseau et la Revue Française d'Ornithologie* **42**, 81–169, 253–284.

Fa, J.E. (1990). 'La conservation des écosystèmes de la Guinée équatoriale'. International Union for the Conservation of Nature, Gland.

Field, G.D. (1971). Breeding behaviour of a pair of Black and White Flycatchers *Bias musicus* at Freetown. *Bulletin of the Nigerian Ornithologists' Society* **8**(29), 3–6.

Forbes-Watson, A.D. (1970). A new species of *Melaenornis* (Muscicapinae) from Liberia. *Bulletin of the British Ornithologists' Club* **90**, 145–148.

Fotso, R.C. (1993). Breeding of Bannerman's Turaco *Tauraco bannermani* and Banded Wattle-eye *Platysteira laticincta* in Cameroon. *Proceedings of the Eighth Pan-African Ornithological Congress* p. 438. PAOC Committee, Nairobi.

Fraser, W. (1983). Foraging patterns of some South African flycatchers. *Ostrich* **54**, 150–155.

Frost, S.K. (1987). Factors affecting habitat separation in the Pallid Flycatcher *Melaenornis pallidus* and Marico Flycatcher *Melaenornis mariquensis*. M.Sc. thesis, Universiy of Cape Town.

Frost, S.K. (1990). Notes on the breeding behaviour of Marico and Pallid Flycatchers in the central Transvaal, South Africa. *Ostrich* **61**, 111–116.

Fry, C.H. (1961). Notes on the birds of Annobon and other islands in the Gulf of Guinea. *Ibis* **103a**, 267–276.

Fuggles-Couchman, N.R. (1939). Notes on some birds of the Eastern Province of Tanganyika Territory. *Ibis* **1939**, 76–106.

Fuggles-Couchman, N.R. (1986). Breeding records of some Tanzanian birds. *Scopus* **10**, 20–26.

Gargett, V. and Webb, D.G. (1973). A Marico Flycatcher feeds a Didric Cuckoo. *Ostrich* **44**, 79.

Gartshore, M.E. (1989). An avifaunal survey of Taï National Park, Ivory Coast. *ICBP Study Report* No. 39, 1–51. International Council for Bird Protection, Cambridge.

Géroudet, P. (1983). Engoulevent musicien et Gobemouche d'Ussher en Basse-Casamance (Sénégal). *L'Oiseau et la Revue Française d'Ornithologie* **53**, 84–85.

Glass, D.R. (1986). Cape Batis in N. W. Lesotho. *Mirafra* **3**, 5.

Glutz von Blotzheim, U.N. and Bauer, K. (1993). 'Handbuch der Vögel Mitteleuropas'. Band 13/I. Passeriformes. AULA-Verlag, Wiesbaden.

Gore, M.E.J. (1994). Bird records from Liberia. *Malimbus* **16**, 74–87.

Gray, H.H. (1972). The nests of three forest birds. *Bulletin of the Nigerian Ornithologists' Society* **9**(34), 24–25.

Greig-Smith, P.W. (1977). Mixed-species flocking of Nigerian forest birds. *Bulletin of the Nigerian Ornithologists' Society* **13**, 53–56.

Grimes, L.G. (1976b). The occurrence of cooperative breeding behaviour in African birds. *Ostrich* **47**, 1–15.

Guichard, K.M. (1947). Birds of the Inundation Zone of the River Niger, French Soudan. *Ibis* **89**, 450–489.

Gwinner, E. (1986). Sting-removal from bees by White-eyed Slaty Flycatchers *Melaenornis chocolatina*. *Scopus* **10**, 51–52.

von Haartman, L. (1956). Territory in the Pied Flycatcher *Muscicapa hypoleuca*. *Ibis* **98**, 460–475.

von Haartman, L. (1969). The nesting habits of Finnish birds. I. Passeriformes. *Commentationes Biologicae, Societas Scientiarum Fennica* **32**, 3–187.

Haigh, H. (1986). Some notes on the breeding biology of the Fiscal Flycatcher in the SW Cape. *Promerops* No. 174, 13.

Hall, B.P. (1960b). The faunistic importance of the scarp of Angola. *Ibis* **102**, 420–442.

Hanmer, D. (1979). Some data on the Wattle-eyed Flycatcher in Malaŵi. *Honeyguide* No. 100, 33–35.

Happel, R.E. (1985). Birds of Outamba area, north-west Sierra Leone. *Malimbus* **7**, 101–102.

Harris, T. (1986). Is the Whitetailed Shrike a terrestrial batis? *Lanioturdus* **22**, 47–51.

Harrison, H.J.S. (1990). A recent survey of the birds of Pagalu (Annobon). *Malimbus* **11**, 135–143.

Herholdt, J.J. (1988). Bird weights from the Orange Free State (Part II: passerines). *Safring News* **17**, 43–58.

Herremans, M. (1992). Interspecific song of Chinspot Batis *Batis molitor* to Pririt Batis *Batis pririt*. *Babbler* No. 23, 40–41.

Herremans, M. and Herremans-Tonnoeyr, D. (1994). Persistent foraging on the ground by Chinspot Batises *Batis molitor*. *Babbler* Nos. 26–27: 18–19.

Herroelen, P. (1955). Notes sur quelques nids et oeufs inconnus d'oiseaux africains observés au Congo belge. *Revue de Zoologie et de Botanique Africaines* **52**, 185–192.

Hobson, T. (1995). Identification of juvenile Chat Flycatchers. *Bee-eater* **46**, 5.

Hoesch, W. (1938). Von Paradiesfliegenschnäpper *Terpsiphone viridis plumbeiceps*. *Journal für Ornithologie* **86**, 328–329.

Hoi, H. (1987). Brutaufteilung und Habitatnutzung beim Cassinschnäpper (*Muscicapa cassini*). *Journal für Ornithologie* **128**, 338–342.

Holstein, B., Braunlich, A. and Huxham, M. (1991). Rondo Forest Reserve, Tanzania: an ornithological note including new records of the East Coast Akalat *Sheppardia gunningi*, the Spotted Ground Thrush *Turdus fischeri* and the Rondo Green Barbet *Stactolaema olivacea woodwardi*. *Scopus* **14**, 125–128.

Holyoak, D.T. and Seddon M.B. (1990). Distributional notes on the birds of Benin. *Malimbus* **11**, 128–134.

Hubbard, J.P. and Hubbard, C.L. (1970). Foraging behavior in the Blue Flycatcher. *Auk* **87**, 154–156.

Hustler, K. and Irwin, M.P.S. (1995). Fifth report of the OAZ rarities committee. *Honeyguide* **41**, 103–106.

Hustler, K., Allan, D., Cassidy, R. and Salinger, T. (1985). Collared Flycatchers in north-west Matabeleland. *Honeyguide* **31**, 172–173.

Irwin, M.P.S. (1957a). A new race of Marico Flycatcher *Bradornis mariquensis* from South-West Africa. *Bulletin of the British Ornithologists' Club* **77**, 118.

Irwin, M.P.S. (1957b). The southern grey-headed races of Livingstone's Flycatcher *Erythrocercus livingstonei* C.R. Gray. *Bulletin of the British Ornithologists' Club* **77**, 118–119.

Irwin, M.P.S. (1962). The specific status of *Batis soror* and its relationship to *Batis molitor*. *Ostrich* **33**(3), 17–28.

Isenmann, P. (1991). Le Gobemouche à collier *Ficedula albicollis* au Maghreb. *Alauda* **59**, 177.

Jackson, H.D. (1972,1976). Avifaunal survey of the Umtali Municipal Areas I – II. *Arnoldia* **6**(1), 1–10; **8**(5),1–10.

Jackson, H.D. (1986,1987). Avifaunal survey of the Mutare Municipal Areas III – V. *Arnoldia* **9**(25), 325–332; **9**(28), 353–360; **9**(29), 361–367.

James, H.W. (1922). Notes on the nest and eggs of *Stenostira scita* (Vieill.). *Ibis* **1922**, 254–256.

Jensen, R.A.C. and Clinning, C.F. (1974). Breeding biology of two cuckoos and their hosts in South West Africa. *Living Bird* **13**, 5–50.

Joerges, B. (1972). Der Drosselwürger – *Lanioturdus torquatus*. Eine Verhaltenbeobachtung. *Mitteilungen der Ornithologischen Arbeitsgruppe der Südwest Afrika Wissenschaftliche Gesellschaft* **8**(1/2), 6–8.

Johnson, D.N. (1984). The coexistence of the Red-bellied Paradise Flycatcher *Tchitrea rufiventer* and the Chestnut Wattle-eye *Dyaphorophyia castanea*. *Proceedings of the Fifth Pan-African Ornithological Congress* pp. 263–274. Southern African Ornithological Society, Johannesburg.

Jones, P.J. and Tye, A. (1988). A survey of the avifauna of São Tomé and Príncipe. *ICBP Study Report* No. 24, 1–64. International Council for Bird Protection, Cambridge.

Jourdain, F.C.R. and Shuel, R. (1935). Notes on a collection of eggs and breeding-habits of birds near Lokoja, Nigeria. *Ibis* **1935**, 623–663.

Keith, S. and Twomey, A. (1968). New distributional records of some East African birds. *Ibis* **110**, 537–548.

Kok, O.B., van Ee, C.A. and Nel, D.G. (1991). Daylength determines departure date of the Spotted Flycatcher *Muscicapa striata* from its winter quarters. *Ardea* **79**, 63–65.

Komen, J. (1987). Batis meets batis in Namibia. *Lanioturdus* **22**, 79–81.

Lamm, D.W. (1953). Taxonomic status of *Batis molitor soror*. *Ostrich* **24**, 171–173.

Lawson, W.J. (1962a). Geographical variation in the Fiscal Flycatcher *Melaenornis silens* (Shaw) of South Africa. *Bulletin of the British Ornithologists' Club* **82**, 135–137.

Lawson, W.J. (1962b). Geographical variation in Woodwards' Batis *Batis fratrum* Shelley. *Durban Museum Novitates* **VI**, 220–224.

Lawson, W.J. (1963b). Geographical variation in *Batis pririt* (Vieillot). *Bulletin of the British Ornithologists' Club* **83**, 29–32.

Lawson, W.J. (1963c). Geographical variation in the southern African populations of the Dusky Flycatcher *Muscicapa adusta* (Boie). *Bulletin of the British Ornithologists' Club* **83**, 4–7.

Lawson, W.J. (1963d). On the geographical variation of the Wattle-eye Flycatcher *Platysteira peltata* (Sundevall). *Bulletin of the British Ornithologists' Club* **83**, 114–116.

Lawson, W.J. (1964a). Geographical variation in the Cape Batis *Batis capensis* (Linnaeus). *Durban Museum Novitates* **VII**(8), 189–200.

Lawson, W.J. (1964b). Systematic notes on African birds II. *Durban Museum Novitates* **VII**(6), 141–155.

Lawson, W.J. (1984). The West African mainland forest dwelling population of *Batis*; a new species. *Bulletin of the British Ornithologists' Club* **104**, 144–146.

Lawson, W.J. (1986). Speciation in the forest-dwelling populations of the avian genus *Batis*. *Durban Museum Novitates* **XIII**(21), 285–304.

Lawson, W.J. (1987). Systematics and evolution in the savanna species of the genus *Batis* (Aves) in Africa. *Bonner Zoologische Beiträge* **38**, 19–45.

Little, J. de V. (1964). Notes on the breeding behaviour of the Paradise Flycatcher. *Ostrich* **35**, 32–41.

Louette, M. (1974, 1975). Contribution to the ornithology of Liberia. *Revue de Zoologie Africaine* **88**, 741–748; **92**, 639–643.

Lundberg, A. and Alatalo, R.V. (1992). 'The Pied Flycatcher'. T. and A.D. Poyser, London.

Macdonald, J.D. and Usher, H.B. (1952). The relationship of *Dyaphorophyia concreta* and *ansorgei*. *Ibis* **94**, 356–358.

MacLeod, J.G.R. and Hallack, M. (1956). Some notes on the breeding of Klaas's Cuckoo. *Ostrich* **27**, 5.

Mahé, E. (1988). Contribution à la liste des oiseaux du Parc National de la Bénoué, Nord Cameroun. *Malimbus* **10**, 218–221.

Mann, C.F. (1973). Two new bird breeding records for Kenya. *East Africa Natural History Society Bulletin* **1973**, 39.

Manson, A.J. (1992). Biology of the White-tailed Crested Flycatcher. *Honeyguide* **38**, 6–11.

Marchant, S. (1953). Notes on the birds of south-eastern Nigeria. *Ibis* **95**, 38–69.

Marchant, S. (1966). (Letter). *Bulletin of the Nigerian Ornithologists' Society* **3**, 75.

Marshall, B. (1978). A partially albino Black Flycatcher. *Honeyguide* No. 94, 42.

Masterson, A.N.B. (1981). Full house. *Honeyguide* No. 106, 14–15.

Mayr, E. and Greenway, J.C., Jr. (1956). Sequences of passerine families (Aves). *Breviora* **58**, 1–11.

Mayr, E. and Moynihan, M. (1946). Evolution in the *Rhipidura rufifrons* group. *American Museum Novitates* No. 1321, 1–21.

Meise, W. (1968). Zur Speciation afrikanischer, besonders angolischer Singvögel der Gattungen *Terpsiphone*, *Dicrurus* und *Malaconotus*. *Zoologische Beiträge* **14**, 1–60.

Mild, K. (1994). Field identification of Pied, Collared and Semi-collared Flycatchers. *Birding World* **7**, 139–151, 231–240, 325–334.

Moali, A., Samraoui, B. and Benyacoub, S. (1991) Première nidification du Gobemouche à collier *Ficedula albicollis* cf. *semitorquata* en Algérie. *Alauda* **59**, 51–52.

Moreau, R.E. (1949). The breeding of a paradise flycatcher. *Ibis* **91**, 256–279.

Moreau, R.E. (1961). Problems of Mediterranean-Saharan migration. *Ibis* **103a**, 373–427, 580–623.

Moreau, R.E. and Moreau, W.M. (1928). Some notes on the habits of Palaearctic migrants while in Egypt. *Ibis* **1928**, 233–252.

Moreau, R.E. and Moreau, W.M. (1937). Biological and other notes on some East African birds. *Ibis* **1937**, 321–345.

Moreau, R.E. and Sclater, W.L. (1938). The avifauna of the mountains along the Rift Valley in North Central Tanganyika Territory (Mbulu District). *Ibis* **1938**, 1–32.

Moyer, D.C., Short, L.L., Horne, J.F.M. and Wachira, J. (1992). First description of the nest of Yellow Flycatcher *Erythrocercus holochlorus*. *Scopus* **15**, 129–132.

de Naurois, R. (1984a). Le Moucherolle endemique de l'île de São Tomé, *Terpsiphone atrochalybeia* (Thomson, 1842). *Alauda* **52**, 31–44.

Nikolaus, G. (1981). Palearctic migrants new to the North Sudan. *Scopus* **5**, 121–124.

Nöller, R. (1975). Pririt Flycatcher fanning young. *Bokmakierie* **27**, 16.

Oatley, T.B. and Tinley, K.L. (1989). The forest avifauna of Gorongosa Mountain, Mozambique. *Ostrich Supplement* **14**, 57–61.

Oschadleus, D. (1996). Paradise Flycatcher bathing. *Laniarius* No. 60, 13.

Oxford, B. (1994). Incubation period of the Blue-grey Flycatcher. *Honeyguide* **40**, 182.

Paget-Wilkes, A.H. (1931). Birds of the region south of Lake Nyasa. Part II. Passerine birds. *Ibis* **1931**, 475–490.

Pakenham, R.H.W. (1943). Field notes on the birds of Zanzibar and Pemba. *Ibis* **85**, 165–189.

Pearson, D.J. (1981). The identity of two *Ficedula* flycatchers recently collected in Kenya. *Scopus* **5**, 59–60.

Pearson, D.J., Lewis, A.D. and Turner, D.A. (1980). The Gambaga Flycatcher *Muscicapa gambagae* in northern Kenya. *Scopus* **4**, 96.

Penry, H. (1988). A review of the Fairy Flycatcher in Botswana. *Babbler* No. 15, 21–25.

Pocock, T.N. (1963). Occurrence of Cape and Pririt Flycatcher *Batis capensis* and *B. pririt* respectively together in the Oudtshoorn District. *Ostrich* **34**, 174–175.

Pocock, T.N. (1966). Contributions to the osteology of African birds. *Proceedings of the Second Pan-African Ornithological Congress, Ostrich Supplement* **6**, 83–94.

Pomeroy, D. and Muringo, G. (1984). Studies of birds in a semi-arid region of Kenya. *Proceedings of the Fifth Pan-African Ornithological Congress* pp. 179–199. Southern African Ornithological Society, Johannesburg.

Prigogine, A. (1957). La redécouverte de *Muscicapa lendu* (Chapin). *Revue de Zoologie et de Botanique Africaines* **55**, 3–4.

Prigogine, A. (1969a). Polymorphism of the Chestnut-bellied Wattle-eye *Dyaphorophyia concreta*. *Ibis* **111**, 95–97.

Prigogine, A. (1969b). Trois oiseaux nouveaux du Katanga, république démocratique du Congo. *Revue de Zoologie et de Botanique Africaines* **79**, 110–115.

Prigogine, A. (1976). Relation entre les gobe-mouches de paradis *Terpsiphone rufiventer ignea* et *Terpsiphone bedfordi* et statut de ce dernier. *Le Gerfaut* **66**, 171–205.

Prigogine, A. (1980d). Hybridization between the paradise flycatchers *Terpsiphone rufiventer* and *Terpsiphone bedfordi*. *Proceedings of the Fourth Pan-African Ornithological Congress* pp. 17–22. Southern African Ornithological Society, Johannesburg.

Rand, A.L. (1953). Notes on flycatchers of the genus *Batis*. *Fieldiana: Zoology* **34**, 133–149.

Reichenow, A. (1911). Die ornithologischen Sammlungen der Zoologie-Botanik. Kamerun-Expedition 1908 und 1909. *Mitteilungen aus dem Zoologischen Museum in Berlin* **5**, 205–258.

Richards, D. (1992). First breeding record for Gambaga Flycatcher *Muscicapa gambagae* in East Africa. *Scopus* **15**, 137–138.

Riddiford, N. (1990). Collared Flycatcher in Senegal. *Malimbus* **11**, 149–150.

Riddiford, N. (1991). A field character for identification of Collared Flycatcher in female and non-breeding plumage. *British Birds* **84**, 19–23.

Robertson, I. (1996). Recent reports. *Bulletin of the African Bird Club* **3**, 60–63.

da Rosa Pinto, A.A. (1962). As observações de maior destaque das expedições ornitologicas do Instituto de Investigação científica de Angola. *Boletim do Instituto de Investigação Científica de Angola (Luanda)* **1**, 21–38.

Rougeot, P.C. (1957). Note sur la biologie de quelques Muscicapidés du Gabon. *L'Oiseau et la Revue Française d'Ornithologie* **27**, 277–283.

Ryall, C. and Stoorvogel, J.J. (1995). Observations on nesting and associated behaviour of the Shrike Flycatcher *Megabyas flammulatus* in Tai National Park, Ivory Coast. *Malimbus* **17**, 19–24.

Salomonsen, F. (1933a). Les gobe-mouches de paradis de la région Malagache. *L'Oiseau et la Revue Française d'Ornithologie* **3**, 603–614.

Salomonsen, F. (1933b). Remarks on the Madagascar paradise flycatchers. *Bulletin of the British Ornithologists' Club* **53**, 119–124.

Salomonsen, F. (1933c). Revision of the group *Tchitrea affinis* Blyth. *Ibis* **1933**, 730–745.

Salvan, J. (1972). Notes ornithologiques du Congo-Brazzaville. *L'Oiseau et la Revue Française d'Ornithologie* **42**, 241–252.

Sanz, J.J. (1996). Effect of food availability on incubation period in the Pied Flycatcher (*Ficedula hypoleuca*). *Auk* **113**, 249–253.

Scamell, K.M. (1973). Breeding the Silver Bird *Empidornis semipartitus*. *Avicultural Magazine* **79**, 183–189.

Scott, J.A. (1984). Breeding behaviour of the Black-throated Wattle-eye. *Honeyguide* **30**, 72–74.

Serle, W. (1943b). Notes on East African birds. *Ibis* **85**, 55–82.

Serle, W. (1950b). Notes on the birds of southwestern Nigeria. *Ibis* **92**, 84–94.

Seth-Smith, L.M. (1913). Notes on birds around Mpumu, Uganda. *Ibis* **1913**, 485–508.

Sharpe, R.B. (1873). On the genus *Platysteira* and its allies. *Ibis* **1873**, 156–177.

Sharpe, R.B. and Bates, G.L. (1907). On further collections of birds from the Efulen district of Cameroon, West Africa. Part IV. *Ibis* **1907**, 416–464.

Silbernagl, P. and Silbernagl, S. (1995). Dusky Flycatcher in Newlands in winter. *Promerops* No. 221, 15.

Simmons, K.E.L. (1954). Field notes on the behaviour of some passerines migrating through Egypt. *Ardea* **42**, 140–151.

Skead, D.M. (1963). Dusky Flycatchers, (*Muscicapa adusta* (Boie)) using the same nest over two seasons. *Ostrich* **34**, 253.

Skead, D.M. (1966). The Dusky Flycatcher, *Muscicapa adusta* (Boie), in the southern Cape. *Ostrich* **37**, 143–145.

Skead, C.J. (1967). A study of the Paradise Flycatcher *Terpsiphone viridis* (Muller). *Ostrich* **38**, 123–132.

Skead, D.M. (1977). Weights of birds handled at Barberspan. *Ostrich Supplement* **12**, 117–131.

Smith, K.D. (1955b). The winter breeding season of landbirds in eastern Eritrea. *Ibis* **97**, 480–507.

Smith, P.A. (1966a). Ussher's Flycatcher: a new bird for Nigeria. *Bulletin of the Nigerian Ornithologists' Society* **3**, 48.

Smith, T.B. and McNiven, D. (1993). Preliminary survey of the avifauna of Mt Tchabal Mbabo, west-central Cameroon. *Bird Conservation International* **3**, 13–19.

Smith, V.W. (1966b). Autumn and spring weights of some Palaearctic migrants in central Nigeria. *Ibis* **108**, 492–512.

Steyn, P. (1969). Marico Flycatcher tending tit-babbler nestlings. *Ostrich* **40**, 51–54.

Stoneham, H.F. (1928). Field notes on a collection of birds from Uganda. Part II. *Ibis* **1928**, 252–285.

Summers-Smith, D. (1952). Breeding biology of the Spotted Flycatcher. *British Birds* **45**, 153–167.

Swynnerton, C.F.M. (1908). Further notes on the birds of Gazaland. *Ibis* **1908**, 1–107.

Tarboton, W. (1987). Breeding of the Vanga or Black and White Flycatcher. *Honeyguide* **33**, 63–64.

Taylor, P.B. (1978). The Collared Flycatcher (*Muscicapa albicollis*) in Copperbelt Province. *Bulletin of the Zambian Ornithological Society* **10**, 40–41.

Terblanche, S. (1996). Rarities and unusual observations: Marico Flycatcher. *Laniarius* No. 60, 17.

Thomas, J. (1991). Birds of Korup National Park, Cameroon. *Malimbus* **13**, 11–23.

Thomson, W.R. (1925). Field notes on the birds of Sierra Leone. *Ibis* **1925**, 47–70.

Traylor, M.A. (1960). Mutation in an African flycatcher, *Dyaphorophyia concreta*. *Auk* **77**, 80–82.

Traylor, M.A. (1970a). East African *Bradornis*. *Ibis* **112**, 513–531.

Traylor, M.A. (1970b). Notes on African Muscicapidae. *Ibis* **112**, 395–397.

Tree, A.J. (1987). Recent reports. *Honeyguide* **33**, 65–69.

Tree, A.J. (1993). Recent reports. *Honeyguide* **39**, 207.

Tyler, S. (1987a). How long do garden birds in Addis live? *Ethiopian Wildlife and Natural History Society Newsletter* No. 217, 3–4.

Tyler, S.J. (1987b). *Ficedula* species in Ethiopia. *Scopus* **11**, 54–55.

Underhill, L. and Oatley, T. (1994). Spotted Flycatcher from Sweden. *Birding in Southern Africa* **46**, 96.

Vande weghe, J.-P. (1974). Additions et corrections à l'avifaune du Rwanda. *Revue Zoologique Africaine* **88**, 81–98.

Vaurie, C. (1952). Geographical variation in the Chat Flycatchers (*Bradornis infuscatus*). *American Museum Novitates* No. 1599, 1–9.

Vaurie, C. (1953). A generic revision of flycatchers of the tribe Muscicapini. *Bulletin of the American Museum of Natural History* **100**, 453–538.

Vaurie, C. (1957). Notes on the genus *Parisoma* and on the juvenal plumage and systematic position of *Parisoma plumbeum*. *Ibis* **99**, 120–122.

Vernon, C.J. (1984). Population dynamics of birds in Brachystegia woodland. *Proceedings of the Fifth Pan-African Ornithological Congress* pp. 201–216. Southern African Ornithological Society, Johannesburg.

Vernon, C.J. (1985). The systematics of the southern African flycatchers, Muscicapidae. *Honeyguide* **31**, 93–98.

Vernon, C.J. (1989). Observations on the forest birds around East London. *Ostrich Supplement* **14**, 75–84.

Vernon, C.J., Macdonald, I.A.W. and Dean, W.R.J. (1989). Birds of an isolated tropical lowland rainforest in eastern Zimbabwe. *Ostrich Supplement* **14**, 111–122.

Vernon, C.J., Macdonald, I.A.W. and Dean, W.R.J. (1990). The birds of the Haroni-Lisitu. *Honeyguide* **36**, 14–35.

Walker, G.R. (1939). Notes on the birds of Sierra Leone. *Ibis* **1939**, 401–450.

Walsh, J.F. (1986). Notes on the birds of Ivory Coast. *Malimbus* **8**, 89–93.

Walsh, J.F. (1987). Records of birds seen in north-eastern Guinea in 1984–85. *Malimbus* **9**, 105–122.

Wells, D.R. (1967). Possible range extension of the Dusky Flycatcher. *Bulletin of the Nigerian Ornithologists' Society* **4**(13/14), 40.

White, C.M.N. (1943). Field notes on some birds of Mwinilunga, Northern Rhodesia. *Ibis* **85**, 128–131.
White, C.M.N. (1946). The ornithology of the Kaonde-Lunda Province, Northern Rhodesia. *Ibis* **88**, 68–103.
White, C.M.N. (1947). Field notes on some birds of Mwinilinga, Northern Rhodesia. *Ibis* **85**, 127–131.
White, C.M.N. (1948). Weights of some Northern Rhodesia birds. *Ibis* **90**, 137–138.
White, C.M.N. (1951). Systematic notes on African birds. (7) *Bradornis infuscatus* in Angola. *Ibis* **93**(3), 464–465.
Wilkinson, R. and Beecroft, R. (1985). Birds in Falgore Game Reserve, Nigeria. *Malimbus* **7**, 63–72.
Williams, J.G. (1959). *Melaenornis ardesiaca* in East Africa. *Bulletin of the British Ornithologists' Club* **79**, 51.
Wilson, J.D. (1989). Range extensions of some bird species of Cameroon. *Bulletin of the British Ornithologists' Club* **109**, 110–115.
Wilson, N. and Wilson, V.G. (1994a). Avifauna of the southern Kerio Valley with emphasis on the area around the Kenya Fluorspar Mine site, August 1989 – July 1993. *Scopus* **18**, 65–115.
Wilson, N. and Wilson, V.G. (1994b). Description of the nest and eggs of the Lead-coloured Flycatcher *Myioparus plumbeus* from the Kerio Valley, Kenya. *Scopus* **18**, 60–61.
Wilson, R. (1990). Annotated bird list for the Forêt Classée de Ziama and immediate environs. Unpublished report, International Union for the Conservation of Nature and Natural Resources.
Winterbottom, J.M. (1936). Distributional and other notes on some Northern Rhodesian birds. *Ibis* **1936**, 763–791.
Winterbottom, J.M. (1939). Miscellaneous notes on some birds of Northern Rhodesia. *Ibis* **1939**, 712–734.
Winterbottom, J.M. (1942). A contribution to the ornithology of Barotseland. *Ibis* **1942**, 337–389.
Winterbottom, J.M. (1958). Systematic notes on birds of the Cape Province. Part VI – *Bradornis infuscatus* (Smith). *Ostrich* **29**, 157–159.
Winterbottom, J.M. (1967). Footnote on breeding of Paradise Flycatcher. *Ostrich* **38**, 133–134.
Wood, B. (1989). Biometrics, iris and bill coloration, and moult of some Somali forest birds. *Bulletin of the British Ornithologists' Club* **109**, 11–22.
Wyndham, C. (1939). The Fairy Flycatcher. *Ostrich* **10**, 47–50.
Young, C.G. (1946). Notes on some birds of the Cameroon Mountain District. *Ibis* **88**, 348–382.

3. Acoustic References

Section A: Discs and Cassettes

5. North, M. E. W. (1958). Voices of African Birds. Cornell University Press. 159 Sapsucker Woods Road, Ithaca, N.Y. 14850. One 12-inch, 33⅓ r.p.m. disc. 42 species. The first African record concerned mainly with identification. Species are presented in systematic order, grouped on separate bands, and details given of circumstances, place and date of recording.
7. Haagner, C. H. (1961). Birds of the Kruger National Park. International Library of African Music, P.O. Box 138, Roodeport, near Johannesburg, South Africa. Two 7-inch, 45 r.p.m. discs, Nos. XTR 17044 and XTR 27045. 31 species in systematic order following Roberts (1957. 'Birds of South Africa', Trustees of the John Voelcker Bird Book Fund, Cape Town.) and with the Roberts number; each on a separate band.
9. Haagner, C. H. (1964). Birds of Zululand. Same publisher as No. 7. Two 7-inch, 45 r.p.m. discs, Nos. XTR 4 7094 and XTR 5 7095, 27 species.
10. North, M. E. W. and McChesney, D. S. (1964). More Voices of African Birds. Houghton Mifflin Co., Boston, USA. One 12-inch, 33⅓ r.p.m. disc. 90 species. Details of recordings are given in an accompanying booklet. These 2 discs (Nos. 5 and 10) together contain the voices of 132 species, and are the first major reference work for African bird voices.
11. Stannard, J. and Niven, P. (1966). Bird Songs of Amanzi. Percy FitzPatrick Institute of African Ornithology, University of Capetown, Rondebosch 7700, South Africa. One 12-inch 33⅓ r.p.m. disc, No. ACP 524; No 1 in 'Bird Song Series', 37 species. On one side the birds are heard in their natural surroundings, the emphasis being on atmosphere or ambience; on the other side they are singled out and identified.
12. Pooley, A. C. (1966). Wildlife Calls of Africa. Percy FitzPatrick Institute, University of Cape Town, Rondebosch 7700, South Africa. One 12-inch, 33⅓ r.p.m. disc. 24 species.
13. Hayes, C. and Hayes, J. (1966). East African Birdsong; No. 2 in *Heartbeat of Africa*, Series 1. Sapra Studios, Box 5882, Kimathi and York Streets, Nairobi, Kenya. One 7-inch, 45 r.p.m. disc. 25 species.
14. Stannard, J. and Niven, P. (1967). Bird Song of the Forest. Percy FitzPatrick Institute (address under No. 12). One 12-inch 33⅓ r.p.m. disc. GALP 1559. 32 species.
20. Henley, A. and Pooley, A. C. (1970). Birds of the Drakensberg. Published by the authors and obtainable from Wildlife Society of South Africa. One 12-inch 33⅓ r.p.m. stereo disc, BD 100. 41 species.
21. Reucassel, R. and Adendroff, A. (1970). Nature's Melody. Published by the authors; obtainable from Wildlife Society of South Africa. One 12-inch 33⅓ r.p.m. stereo disc, SWL 3. 53 species.
22. Walker, A. (1970). Garden Birds of Southern Africa. Gallo (Africa) Ltd., Johannesburg; obtainable from Wildlife Society of South Africa. One 12-inch 33⅓ r.p.m. disc, SGALP 1598. 40 species.
32. Keith, G. S. and Gunn, W. W. H. (1971). Birds of the African Rain Forests. *Sounds of Nature* No. 9. Federation of Ontario Naturalists, 1262 Don Mills Road, Don Mills, Ontario M3B 2WB, Canada, and American Museum of Natural History, New York. Two 12-inch, 33⅓ r.p.m. discs. 95 species. The most important reference work

since the records of North (Nos. 5 and 10) and the first specializing in forest birds, many of which are here published for the first time. Most species are from East Africa, some central Africa. Species are arranged in systematic order and grouped in bands; a simple announcement of the name accompanies each species, but a lot of information is provided on the jacket.

33. Stannard, J. (1971). Bird Sounds and Songs. FitzPatrick Institute (address under No 12). Issued in conjunction with *Ostrich* Supplement 9. One 7-inch, 45 r.p.m. disc, NV1. 20 species.

34. Chappuis, C. (1971). Ambiances des plaines et savanes d'Afrique orientale. *Afrique sauvage* No. 1. One 12-inch $33\frac{1}{3}$ r.p.m. disc, JAC9. Edition *Jacana*, 32 rue St Marc, 75002 Paris. 44 species.

35. Martin, R. B. (1971). Journey Across Africa. Parlophone PCSJ (D) 12.79. Obtainable from Wildlife Society of South Africa. One 12-inch $33\frac{1}{3}$ r.p.m. disc. 34 species.

36. Ker, A. (1972). Satan 99. Equator Sound Studios Ltd., 30068, Nairobi, Kenya. One 12-inch $33\frac{1}{3}$ r.p.m. disc, ESS 1001. 63 species.

38. Keibel, W. D. (1972). Wildlife of West South Africa. Wildlife Society of South Africa, P.O. Box 3508, Windhoek, Namibia. 1 cassette, 48 species.

40. Roche, J.-C. (1973). Birds of West Africa – Senegal. Birds and Wild Beasts of Africa No. 3. L'Oiseau Musicien, France. One 12-inch, $33\frac{1}{3}$ r.p.m. disc, SP 002, and 16 colour slides. 16 species.

42. Worman, D. (1974). African Birds. Soundpics Enterprises (Pty) Ltd, P.O. Box 61055, Marshalltown 2107, South Africa. One 10-inch $33\frac{1}{3}$ r.p.m. disc, SP 002, and 16 colour slides. 16 species.

45. Chappuis, C. (1974). *Les Oiseaux de l'Ouest Africain*, Sound Supplement to *Alauda*. Disc 2: Sylviidae 1 (*Cisticola*); 32 species. Disc 3: Bucerotidae, Alcedinidae, Meropidae, Coraciidae and Sylviidae 2; 48 species. Two 12-inch, $33\frac{1}{3}$ r.p.m. mono discs, ALA 3 and 4, ALA 5 and 6; commentary in *Alauda* 42, no. 4, 467–500. For details, see No. 53.

46. Anon. (1966). A Night at Treetops. Sapra Studios (address under No. 13). *Heartbeat of Africa*, Series 1, No. 3. One 17-cm 45 r.p.m. disc. About 10 species.

49. Reucassel, D. (1975). Calls of the Wild. Published by the author and available from Wildlife Society of South Africa. One 30-cm $33\frac{1}{3}$ r.p.m. disc, AVI; also available as a cassette. Comes with 32 colour slides. 17 species.

53. Chappuis, C. (1975). *Les Oiseaux de l'Ouest Africain*. Disc 5: Timaliidae, Pycnonotidae (first part), 32 species. Disc 6: Pycnonotidae (end), Turdidae (first part), 44 species. *Alauda*, Sound Supplement accompanying commentary in *Alauda* 43, 450–474, M.N.H.N., Laboratoire d'Ecologie, 4 Avenue du Petit Château, 91.800 Brunoy, France. Two 12-inch, $33\frac{1}{3}$ r.p.m. discs. These records are a part of a series whose aim is to present all known recordings for species of a particular region, including different forms of songs and calls and geographical variation. Details of the recordings are provided in the accompanying article in *Alauda*, of which reprints may be requested when ordering the records.

These records represent a landmark in the history of African voice-recording. This is a lengthy series covering large numbers of species in great detail, and the accompanying commentaries in *Alauda* are of considerable scientific value.

56. Gunn, W. W. H. and Gulledge, J. L. (1977). *Beautiful Birds Songs of the World*. Cornell Laboratory of Ornithology, Sapsucker Woods, Ithaca, New York 14850. Two 12-inch, $33\frac{1}{3}$ r.p.m. discs, 13 species in Vol. IV.

57. Chappuis, C. (1978). *Les Oiseaux de l'Ouest Africain*, Sound Supplement to *Alauda*. Disc 8: Turdidae (end), Sylviidae 3; 35 species. One 12-inch, $33\frac{1}{3}$ r.p.m. mono disc, ALA 15 and 16; commentary in *Alauda* 46, no. 4, 327–335. For details, see No. 53.

58. Natal Bird Club (*c.* 1978). Bird Calls, Vols 1 and 2. Natal Bird Club, P.O. Box 10909, Marine Parade, Durban 4056, South Africa. 2 cassettes. 136 species presented in random order. Lengthy and numerous cuts are provided for each species.

60. Roché, J.-C. (1968). *Guide sonore des oiseaux d'Europe*, Tome II: Maghreb. Edwards Records, 58, Rue du Docteur Calmette, 59320 Sequedin, France. Five 17-cm, $33\frac{1}{3}$ r.p.m. discs.

62. Palmer, S. and Boswall, J. (1969–1972). A Field Guide to the Bird Songs of Britain and Europe. SR Records, Swedish Broadcasting Corp., 105 10 Stockholm, Sweden. Twelve 12-inch, $33\frac{1}{3}$ r.p.m. discs, RFLP 5001–5012. 530 species, nesting or accidental in Europe, mostly wintering in Africa. Presented in systematic order, on separate bands, announced by scientific name. The most important reference work for Palearctic birds wintering in Africa.

63. Palmer, S. and Boswall, J. (1973). A sequel to No. 62. 2 discs, RFLP 5013 and 5014. Includes 23 African species.

66. Kabaya, T. (1978). Birds of the World. I: Africa. King Records Co., Japan. One 30-cm, $33\frac{1}{3}$ r.p.m. stereo disc, King Records SKS (II) 2007. 20 species.

68. Chappuis, C. (1979). *Les oiseaux de l'Ouest africain*. Sound supplement to *Alauda*. Disc 10; Sylviidae (end), Paridae, 37 species. One 12-inch $33\frac{1}{3}$ r.p.m. mono disc, ALA 19 and 20. Commentary in *Alauda* 47, No. 3, 195–212. For details, see no. 53.

72. Audio Three (1981). Bird Calls. See No. 58. 3 cassettes, of which the first 2 are the same as those of No. 58; the third contains additional species.

73. Palmer, S. and Boswall, J. (1981). A Field Guide to the Bird Songs of Britain and Europe. 16 cassettes, RFLP 5021–5036. An updated edition of Nos. 62, 63 and 70. 612 species, in boxes of 4 cassettes with commentary and list of species in each box. A first class reference collection.

74. Gillard, L. and Gibbon, G. (1982). A Field Guide to the Bird Calls of southern Africa. Published by the authors; P.O. Box 394, Greenside 2034, Johannesburg, or P.O. Box 10123, Ashwood, 3600 Pinetown, South Africa. 2 cassettes, about 420 species. The large number of species makes this one of the most important and comprehensive collections of African bird voices so far published. Several types of songs or calls are often given per species. Species are grouped by environment, and in systematic order within the environment.

75. Audio Three, Bird Calls: Bird Families, Vol IV. 2 cassettes, 171 species. Many of these species already appear on No. 72, but here all are in systematic order.

76. Chappuis, C. (1984). *Oiseaux de France: Migrateurs et Hivernants*, parts I and II. Obtainable from the author, 10 Vallon du Fer à Cheval, 76530 La Bouille, France. Two cassettes with booklets. Present mainly flight and contact calls, not full songs, of Palearctic birds; useful because these are the vocalizations typically made in Africa by migrants. 147 species.

86. Stjernstedt, R. (1986–1990); last updating in 1993. *Bird Songs of Zambia*. Obtainable from the author, Tan y Coed, Derwenlas, Machynlleth, Powys SY 20 8PZ, UK. Three mono cassettes; No. 1: Non-passerines, Eurylaemidae, Pittidae, 164 species. No. 2: Alaudidae to Sylviidae, 108 species, No. 3: Sylviidae (end) to Fringillidae, 140 species. One of the major collections of African bird voices. The

species are presented in systematic order, often with several types of vocalization per species.

87. Stjernstedt, R. (1988). *Cisticolas of Africa*. One mono cassette; 25 species and 11 subspecies, mainly from southern Africa. For further details, see No. 86
88. Gillard, L. (1987), *Southern African Bird Calls*. Gillard Bird Cassettes, P.O. Box 72059, Parkview 2122, Johannesburg, South Africa. Revised and enlarged edition of Nos. 77 and 79 (see 'The Birds of Africa', Vol. III). Three cassettes of 90 min., 540 species presented in systematic order, often with several types of song and call per species. The large number of species makes this one of the most important and comprehensive collections of African bird voices so far published.
89. Chappuis, C. (1990). *Sounds of Migrant and Wintering Birds*, Western Europe. Obtainable from F. Franklin, 13 Carden Hill, Hollingbury, Brighton BN1 8AA, UK, or from C. Chappuis (see No. 76). Two mono cassettes; English (and revised) version of No. 76.
91. Gibbon, G. (1991). *Southern African Bird Sounds*. Southern African Birding cc. P.O. Box 24106, Hillary 4024, South Africa. Six 90-min. mono cassettes in a box with booklet. 880 species presented in systematic order. The most important collection of African bird voices yet published. The quality of sound reproduction is excellent. A major reference work, even though the large number of species makes the time devoted to each one relatively short.
92. Mild, K. (1990). *Birds Songs of Israel and the Middle East*. Obtainable from the author, Kopparvägen 23, S-175 72 Järfälla, Sweden. 2 cassettes with booklets. 114 species presented in systematic order. Many species were recorded in Africa (north of the Sahara).
93. Roché, J.-C. and Couzens, D. (1985). *The Bird-Walker*. L'Oiseau Musicien, Chateaubois, Mayres-Savel 38350, La Mure, France. 3 cassettes devoted to birds of Europe, with a few species from North Africa; 406 species.
94. Veprintsev, B. N. (1987). *Birds of the Soviet Union, a sound guide: Thrushes and Rock Thrushes, Leaf Warblers and Hippolais Warblers*. Two 30-cm 33⅓ r.p.m. discs. Melodiya C 90, 25473001 and 25475001.
95. Schubert, M. (1982). *Stimmen Vögel der Zentral Asien*. Two 30-cm 33⅓ r.p.m. mono discs. Eterna 822 575–6.
96. Svenson, L. (1984). *Soviet Birds*. Mono cassette, LSKB 1.
97. Pukinsky, Y. and Ilinsky, L. (1985). *Bird Sounds of the Ussuri Taiga*. 17-cm mono 33⅓ r.p.m. disc. Melodiya M 92, 46559009.
98. Mild, K. (1987). *Soviet Bird Songs*. 2 cassettes with booklet. Obtainable from the author (for address, see under #92).

Section B: Most Important Discs and Cassettes by Region

East Africa: Nos 5, 10, 32, 36
West Africa: Nos 53, 57, 68

Southern Africa: Nos 58, 75, 86, 88, 91
Palearctic migrants: Nos 62, 63, 73, 76, 89

Section C: Institutions with Sound Libraries

A. Audio Three, 6, Larch Road, Durban, South Africa.
B. British Library of Wildlife Sounds (BLOWS). The British Library, National Sound Archive, 29 Exhibition Road, London SW7 2AS, UK.
C. Cornell University, Library of Natural Sounds, Laboratory of Ornithology, 159 Sapsucker Woods Road, Ithaca NY 14850, USA.
E. Fonoteca Zoologica, Museo de Zoologia, Parc de la Ciutadella, 08003 Barcelona, Spain.
F. FitzPatrick Bird Communication Library, Bird Department, Transvaal Museum, P.O. Box 413, Pretoria 0001, South Africa.
N. Natal Bird Club, P.O. Box 10909, Marine Parade, Durban 4056, South Africa.
S. South African Broadcasting Corporation. Library of Wildlife Sounds, P.O. Box 4559, Johannesburg 2000, South Africa.

Section D: Individual Recordists

ADE	Adendorff, G.
ALEX	Alexander-Marrack, P. D., BLOWS
ANO	Anon.
ASP	Aspinwall, D. R.
BERG	Bergman, H. H.
BOUR	Bourguignon, C.
BRU	Brunel, J.
CART	Carter, C., FITZ
CHA	Chappuis, C.
CHR	Christy, P.
DAV	Davidson, P.
DEM	Demey, R.
DYE	Dyer, M.
ERA	Erard, C.
FAR	Farkas, T.
FISP	Fishpool, L. D. C.
FOR	Forbes-Watson, A. D.
GAR	Gartshore, M. E.
GIB	Gibbon, G.
GIL	Gillard, L.
GORD	Gordon, J.
GREG	Gregory, A. R.
GRI	Grimes, L., BLOWS
GUL	Gullick, T., BLOWS
HA	Haagner, C. H., FITZ
HAN	Hansen, L.
HRS	Harris, T., FITZ
HAY C	Hayes, C.
HAY J	Hayes, J., BLOWS, BBC
HAZ	Hazevoet, C. J.
HEN	Henley, T.
HOL	Hollom, P. A. D.

HOR	Horne, J. F. M.	PAR	Parker, V., FITZ
JOJ	Jones, J.	PAY	Payne, R. B.
JO PJ	Jones, P. J.	PEA	Pearson, D. J.
KAB	Kabaya, T.	POO	Pooley, T.
KEIB	Keibel, W. D., FITZ	PRIN	Pringle, J. S.
KEI	Keith, S., Cornell	REU	Reucassel, D.
KER	Ker, A.	ROC	Roché, J.-C.
KÖN	König, C.	SAR	Sargeant, D. E.
LEM	Dowsett-Lemaire, F.	SHI	Shirihai, H.
LOS	Loskot, V. M.	SIN	Sinclair, J. C.
LUT	Lutgens, H., BLOWS	STA	Stannard, J., FITZ
MAC	Macaulay, L. R., Cornell	STJ	Stjernstedt, R., BLOWS
McVIC	McVicker, R.	SVEN	Svendsen, J. O.
MANN	Mannery, D.	TYE	Tye, A., BLOWS
MAR	Martin, R. B.	V.DAE	Van Daele, P.
MEES	Mees, V., BLOWS	VEP	Veprintsev, B. N.
MIL	Mild, K.	VIEL	Vieilliard, J.
MOR	Morel, G.	WALK	Walker, A.
MOY	Moyer, D.	WAT	Watts, D. E.
NOR	North, M. E. W., Cornell	WOR	Worman, D.
OSME	OSME expedition to Socotra	ZIM	Zimmerman, D. and M.

ERRATA TO PREVIOUS VOLUMES

The following errors and omissions have been brought to our attention by reviewers and others.

Volume I

p. 414, Verreaux's Eagle, Description, line 2 should read 'back, rump, upper tail-coverts . . .'.

Volume II

p. 22, Barbary Partridge, *A. b. barbata*, not *A. b. barbara*.

p. 24, generic diagnosis, *Francolinus*, line 2, Pliocene, 5–2 million, not 26–17 million years ago.

p. 66, Djibouti Francolin, line 9, dense cover, not dense over.

p. 286 and 314, Sanderling and Spotted Redshank, Vroeg's Cat., not Kroeg's Cat.

p. 299, Subfamily GALLINAGINAE, not GALLINAGININAE.

p. 309, Black-tailed Godwit, Range and Status, line 10, Kenya, not Keyna.

p. 330, Red-necked Phalarope, Zimbabwean distribution under this heading applies to Grey Phalarope, *Phalaropus fulicarius*, p. 331.

p. 447, Rock Dove, in southern Africa occurs in most towns.

Volume III

p. 82, Barred Long-tailed Cuckoo, Breeding Habits, line 11, for Vimba Forest, read Vumba.

p. 84, African Emerald Cuckoo, egg size, $19 \cdot 1 \times 14 \cdot 1$, not $19 \cdot 1 \times 18 \cdot 0$.

Plate 11, opp. p. 176, Brown Nightjar does not have the robust rictal bristles depicted.

p. 240, opposite Plate 13, Scarce Swift, for *myioptilus*, read *myoptilus*.

p. 250, Speckled Mousebird, *C. s. rhodesiae*, not *C. s. rhodesia*.

p. 260, Narina's Trogon, General Habits, subject to extensive seasonal or migratory movement in Zimbabwe.

p. 394, African Pied Hornbill, map, delete arrows.

p. 406, Silvery-cheeked Hornbill, in original description replace Rüppell 1835 with Friedmann 1929.

p. 430, Green Tinkerbird, Range and Status, south of the Zambezi in Mozambique known only from a single specimen from Chicomo at 24°33′S, 34°11′E (Funhalouro district), and not throughout coastal area as shown on map.

p. 440, Yellow-fronted Tinkerbird, map, occurs throughout Zimbabwe.

p. 490, Green-backed Honeybird, eggs also blue in colour.

p. 496, generic diagnosis, *Indicator*, line 1, delete 'Contains largest and smallest members of family'.

p. 523, Bennett's Woodpecker, Breeding Habits, lines 9 and 10, for *Berlinia* read *Isoberlinia*.

Volume IV

p. viii, add *Ammomanes cincturus* to C.H. Fry's authorship.

p. 22, Rufous-naped Lark, M.P.S. Irwin (pers. comm.) notes that *Mirafra africana irwini* described from Cuando Cubango, Angola, by A.A. da Rosa Pinto in *Bonner Zoologisches Beitrage* heft 3/4, 1968, 280–288, has been overlooked (the same applies to *M. angolensis niethammeri* and *Macronyx grimwoodi cuandocubangensis* and also *Mirafra africana anchietae* which is included in *Bol. Inst. Invest. cient. Angola* 4(2), 1967, 29–31).

p. 28, Angola Lark, *Mirafra angolensis niethammeri*, see above under p. 22.

p. 54, Red Lark, French name is Alouette ferrugineuse.

pp. 65 and 594, specific epithet is *clotbey*, not *clot-bey*.

p. 194, Common House Martin, Range and Status, line 10, sentence beginning 'Widespread' should read 'Widespread at least over higher ground, but less often encountered than might be expected – flies very high and easily overlooked.'

p. 262, Grimwood's Longclaw, *Macronyx grimwoodi cuandocubangensis*, see above under p. 22.

p. 313, Yellow-necked Greenbul, Range and Status, line 4, for Bamingui read Bangui.

Plate 22, opp. p. 353, White-starred Robin, the adult illustrated is *P. s. ruwenzorii*, not *P. s. stellata*.

p. 392, Forest Robin, title line, delete Robin after Forest Robin.

p. 448, opposite Plate 16, White-crowned Black Wheatear, bird labelled ♀ is immature, bird labelled ♂ is an adult (sexes alike).

p. 494, generic diagnosis, *Saxicola*, insert Clancey 1990 before Tye 1989.

p. 582, add Clancey, P.A. (1990). The generic status of the Buff-streaked Chat of the southern Afrotropics. *Gerfaut* **80**, 179–191.

INDEXES

Bold page numbers indicate the main account of an entry in the index; *italic*: the relevant plate illustration.

Scientific Names

A

abayensis, Sylvietta w. 340
abdominalis, Eremomela i. **319**, *321*
aberdare, Cisticola 138, *176*, 183, **184**
   Cisticola r. **184**
aberrans, Cisticola 138, **156**, *160*, 165, 172
   Cisticola a. **157**, *160*
   Drymoica 156
abessinica, Camaroptera b. 295
abietinus, Phylloscopus c. **355**, *400*
abyssinica, Apalis f. **263**
   Eremomela c. *321*, **327**
abyssinicus, Bradypterus b. **64**
   Turdus o. **35**, *48*
acaciae, Melaenornis m. **455**
Accipiter badius 43
   minullus 495
   tachiro 434
Achaetops 58, 85
   pycnopygius xii, **58**, *97*
acredula, Phylloscopus t. **353**
Acrocephalinae 58, 61
Acrocephalus 57, 60, 92, **98**, 123, 124, 127, 128, 129, 201, 329
   aedon **122**
   arundinaceus 98, *112*, **114**, 117, 118, 120, 121, 134, 453
   baeticatus 64, 99, **104**, *106*, 107, 110, *112*, 128
   cinnamomeus 108
   dumetorum 108
   gracilirostris 64, 99, 106, 108, *112*, **120**, 124, 128
   griseldis 98, *112*, 114, **116**, 118, 121
   melanopogon 99, **100**, 103, *112*
   paludicola **101**, 103, *112*
   palustris 103, 105, **109**, *112*, 117, 130, 137
   rufescens 64, 99, 108, *112*, **119**, 121, 128
   schoenobaenus 92, 100, **102**, 106, 111, *112*, 125
   scirpaceus 86, 95, 99, 103, **104**, 107, 108, 110, *112*, 115, 117, 130
   stentoreus 98, *112*, 115, **118**
adamauae, Cisticola c. **143**
   Melocichla m. 86
adametzi, Cisticola c. **151**
addenda, Apalis m. **273**
adelphe, Sylvietta r. **344**
adjacens, Apalis m. **273**
admiralis, Cisticola a. **157**
adsimilis, Dicrurus 30, 341, 447, 537, 582, 593
adusta, Butalis 476
   Muscicapa 442, 461, 470, 474, 475, **476**, *481*, 572
   Muscicapa a. **477**
Aëdon, Muscicapa 122
aedon, Acrocephalus **122**
Aegithalos caudatus 243
aequatorialis, Megabyas f. **497**, 549
aethiopica, Platysteira c. **568**
afer, Nilaus 425, 593
   Parus 419
   Sphenoeacus 62, 79, 86, **87**, *97*
   Sphenoeacus a. **88**, *97*
affinis, Apalis p. **275**
   Batis p. **589**
   Prinia s. **221**, *241*
afra, Muscicapa 97
aguimp, Motacilla 464
alacris, Phylloscopus r. **364**
Alauda arvensis 411
alba, Tyto 215
albicauda, Elminia 426, *497*, 514, 515, **517**, 524
albicollis, Ficedula 496, 501, **503**, 505
   Muscicapa 503
albifrons, Platysteira 545, 567, 568, **570**, 571, 572
   Platystira 570
albigularis, Eremomela g. **330**
albimentalis, Apalis j. **268**
albistriata, Sylvia c. **405**, *417*
albiventris, Camaroptera b. 295
   Elminia *497*, 514, 519, 520, 522, 524, 527, 528
   Elminia a. **497**, **522**
   Muscicapa a. **477**
   Trochocercus 522
albogularis, Pygarrhichas 430
   Serinus 419, 465
albonotata, Elminia 275, **497**, 498, 514, 518, 520, 522, 523, 527, 528, 534
   Elminia a. **497**, **523**
albonotatus, Trochocercus 523
albus, Corvus 43
Alcedo semitorquata 537
Alethe 1
   poliocephala 18, 313
alexanderi, Eremomela i. **319**, *321*
   Poliolais l. **293**, *337*
alexinae, Schoenicola p. **90**, *97*
alfredi, Bradypterus 61, **69**, *72*, 96
   Bradypterus a. **70**, *96*
alpestris, Turdus t. 56
alpinus, Phylloscopus u. **363**, *400*
Alseonax 461, 466, 475, 492
althaea, Sylvia 377, 392
   Sylvia c. **392**, *401*
alticola, Apalis 252, 257, 258, 277, 280, 283, 284, **285**
   Apalis a. **257**, **285**
   Cisticola 285
altumi, Bradypterus l. 71
altus, Artisornis m. **291**
Amaurocichla 313, 430
   bocagei *337*, **430**
amaurourus, Melocichla m. **86**
ambigua, Cisticola r. 183
Amphilais 89
amphilectus, Cisticola g. **173**, *176*
Amphispiza belli 309
Amytornis 309
Anaplectes rubriceps 448, 460, 491
Anas sparsa 476
Andropadus 484
   ansorgei 520
   importunus 537
angolensis, Apalis r. **280**
   Cisticola 183
   Cisticola r. *176*, **183**
   Eremomela s. 328
   Monticola 1, 5, **6**, 8, 10, *32*
   Monticola a. **6**
   Muscicapa a. **477**
   Uraeginthus 199

angusticaudus, Cisticola 138, *177*, 195, 196, **197**, 198
ankole, Cisticola b. **192**
annamarulae, Melaenornis 438, **442**, *480*
Anomalospiza imberbis 142, 182, 197, 203, 206, 207, 215, 224, 228
anonyma, Drymoeca 148
anonymus, Cisticola 138, 146, **148**, 149, 150, 151, *160*
anselli, Cisticola t. **207**, *240*
ansorgei, Acrocephalus r. **112**, **119**
  Andropadus 520
  Cisticola r. *161*, **168**
  Dyaphorophyia c. **565**
  Parisoma s. *401*, **418**
  Prinia f. **227**, *241*
  Sylvietta r. **344**
Anthoscopus 322, 331
  flavifrons 600
  minutus 322
  parvulus 593
Anthreptes 484, 529
  collaris 251, 297
  fraseri 315, 429, 510
  tephrolaema 601
Anthus 201
  trivialis 491
antinorii, Cisticola l. **145**
Apalis 190, 196, 199, 250, **252**, 288, 290, 318, 498, 513
  alticola 252, *257*, 258, 277, 280, 283, 284, **285**
  bamendae 252, *257*, 278, **281**
  binotata 83, 252, *256*, **265**, 266, 268, 351
  chapini 252, *257*, 258, 273, 275, **276**, 282, 286, 292
  chariessa 252, *256*, **269**, 277
  chirindensis 252, *257*, 272, 273, **274**
  cinerea 243, 252, *257*, 269, 273, 276, 280, 282, **283**, 285
  flavida 252, *255*, *256*, **262**, 341
  goslingi 252, *257*, 278, 280, **282**, 315
  jacksoni 252, *256*, 266, **267**, 271, 276, 284
  karamojae 252, *257*, 279, **281**, 498
  melanocephala 252, *257*, 270, 272, 274, 276, 277, 284, 498, 578
  nigriceps 252, *256*, 264, 266, 268, **271**, 276, 332, 601
  personata 83, 252, *256*, 261, 265, **266**
  porphyrolaema 252, *257*, 269, 272, 273, **275**, 276, 280, 284, 286
  pulchra 252, *254*, *256*, **259**, 260, 261, 351
  ruddi 252, *255*, *256*, **261**, 264
  rufifrons 244
  rufogularis 252, *257*, 269, 276, 278, **279**, 281, 282, 283, 284, 286, 334
  ruwenzorii 252, *254*, *256*, 259, **260**, 267
  sharpii 252, *257*, 264, **278**, 282
  thoracica 252, **254**, *256*, 259, 260, 261, 262, 264, 277, 286
Apaloderma narina 554
Apatema 461
apiaster, Merops 419
aquaemontis, Melaenornis p. **450**
aquatica, Muscicapa **469**, 478, *481*
  Muscicapa a. **470**
Aquila wahlbergi 38
arcana, Cisticola e. **141**
ardesiaca, Meloenornis 443
ardesiacus, Melaenornis **443**, *480*
argentea, Apalis r. *257*, **280**, 281
argenteus, Cisticola n. *176*, **186**
aridicola, Parisoma l. **415**
aridulus, Cisticola 138, 166, 188, 201, **204**, 207, 208, 234, *240*
  Cisticola a. **204**, *240*
arileuca, Sylvietta l. 351
arnoldi, Apalis t. **254**, *256*
Artisornis 58, 81, **290**
  metopias 82, 277, 290, **291**, 292, *320*
  moreaui 290, 292, *320*
Artomyias 461, 488
arundicola, Cisticola p. **179**
arundinaceus, Acrocephalus 98, *112*, **114**, 117, 118, 120, 121, 134, 453
  Acrocephalus a. *112*, **114**
  Turdus 114
arvensis, Alauda 411
aschani, Camaroptera b. **295**
Asio capensis 215
ater, Melaenornis p. **447**
  Parus 362, 365, 372
aterrimus, Turdus m. **54**
Athene noctua 218
atricapilla, Motacilla 385
  Sylvia 377, 380, 382, 384, 385, 394, *401*
  Sylvia a. **386**
atricauda, Melocichla m. 86
atriceps, Hypergerus 84, *337*, **375**
  Moho 375

Pycnonotus 375
atricollis, Eremomela 318, *321*, 325, **334**
atripennis, Dicrurus 520
atrocaudata, Terpsiphone 530, 531, **537**
atrochalybeia, Muscipeta 537
  Terpsiphone 530, 531, **537**, *544*
atroflavus, Pogoniulus 556
augusticauda, Cisticola 197
australis, Hyliota 297, *337*, 422, **424**, 426, 427, 601
  Hyliota a. **425**
  Tchagra 554
avicenniae, Acrocephalus b. **108**, *112*
awemba, Cisticola r. **183**
aximensis, Muscicapa c. **485**
ayresii, Cisticola 138, 207, 208, 209, 210, 212, **213**, *240*
  Cisticola a. **214**, *240*
azurea, Hypothymis 530

B

baboecala, Bradypterus 61, **63**, 66, 67, 69, 76, 78, *96*, 128
  Bradypterus b. **63**
  Sylvia 63
badiceps, Eremomela 271, 318, *321*, **332**, 333, 600, 602
  Eremomela b. **332**
  Sylvia 332
badius, Accipiter 43
Baeopogon indicator 520
baeticatus, Acrocephalus 64, 99, 104, 106, **107**, 110, *112*, 128
  Acrocephalus b. **108**, *112*
  Sylvia 107
bafirawari, Melaenornis p. **450**, 451
baglafecht, Ploceus 478
bailunduensis, Cisticola a. **157**
  Melaenornis b. **439**
bairdii, Drymoica 232
  Prinia 221, **232**, *241*, 247, 311
  Prinia b. **232**, *241*
Balearica regulorum 478
balearica, Muscicapa s. **463**
  Sylvia s. 413
balearicus, Regulus i. **370**, *400*
bambuluensis, Apalis j. *256*, **268**
bambusicola, Turdus o. **35**
bamendae, Apalis 252, *257*, 278, **281**
bangwaensis, Bradypterus c. 62
  Bradypterus l. 62, **72**
bannermani, Terpsiphone b. 539, **540**

Terpsiphone r. 540
baraka, *Sylvietta v.* **348**
 Turdus o. 35, 48
barakae, *Bradypterus l.* **72**, **96**
barbatus, *Pycnonotus* 125, 520, 565
barbozae, *Hyliota f.* **423**
barnesi, *Parisoma l.* **415**
barratti, *Bradypterus* 62, 64, 72, 73, 74, **75**, 78, **96**
 Bradypterus b. **76**, **96**
batesi, *Batis m.* **595**
 Chloropeta n. 113, **124**
 Terpsiphone 315, 520, 533, 539, **540**, 543, *544*, 547, 556
 Terpsiphone b. **540**, *544*
 Terpsiphone r. 540
 Zoothera p. 18
Bathmocercus 58, **81**
 cerviniventris 81, 82, **84**, **97**
 rufus 81, **82**, 84, **97**
 winifredae **81**, **97**
Batis xii, 422, 426, 498, 548, 551, 567, 568, 572, **574**, 604
 capensis 78, 419, 560, 572, 574, 575, 576, 578, **579**, 582, 583, 586, 588, 590
 diops 561, 574, **575**, 576, 578, 579, 582, 601
 fratrum 560, 574, 575, 576, 578, 579, 580, **582**, 583, 588
 ituriensis 561, 574, 576, 598, **600**
 margaritae 560, 574, 575, **576**, 578, 579, 582
 minima 561, 574, **598**, 600, 601, 602
 minor 561, 574, 576, 578, 585, 588, 592, 594, **595**, 597, 598, 602
 minulla 561, 574, **597**
 mixta 560, 574, 575, 576, **578**, 579, 580, 582, 583, 588, 596
 molitor 297, 560, 574, 576, 577, 578, 580, 583, **584**, 587, 588, 589, 590, 591, 594, 596, 597, 598
 occultus 602
 orientalis 561, 574, 584, 587, 589, 591, 592, **594**, 595, 596, 597
 perkeo 561, 574, 588, 594, **596**
 poensis 561, 574, 592, 599, **601**
 pririt 419, 560, 574, 580, 584, 587, **589**, 591, 594
 puella 587

 reichenowi 580
 senegalensis 424, *561*, 574, 584, 587, 589, **591**, 594, 602
 soror 560, 574, 578, 580, 583, 584, **587**, 589, 591, 594, 596, 597
baumgarti, *Eremomela u.* 331
bechuanae, *Prinia s.* **222**
bedfordi, *Bradypterus b.* 64
 Terpsiphone 528, 543, **544**, 546
 Terpsiphone r. 547
 Trochocercus 546
beirensis, *Camaroptera b.* **296**
belcheri, *Zoothera g.* **26**
bella, *Batis o.* **594**
belli, *Amphispiza* 309
 Cisticola c. **143**
bengalus, *Uraeginthus* 593
benguellensis, *Bradypterus b.* 64
 Melaenornis i. **452**
bensoni, *Chloropeta g.* 113, 123, **128**
 Cisticola c. **162**
 Monticola 1
bewsheri, *Turdus* 30
Bias 548, **551**
 musicus 315, 497, 535, 549, **551**, 603
bihe, *Prinia f.* **227**
binotata, *Apalis* 83, 252, 256, **265**, 266, 268, 351
bivittatus, *Trochocercus c.* **526**, 544
blanchoti, *Malaconotus* 82, 573, 576, 580, 583
blandfordi, *Sylvia l.* 381, **416**
Bleda exima 376
blissetti, *Dyaphorophyia* 300, 545, 555, 556, **562**, 565, 601
 Dyaphorophyia b. 545, **563**
blythi, *Sylvia c.* **392**
bocagei, *Amaurocichla* 337, **430**
 Turdus p. 39, 48
bodessa, *Cisticola* 138, 158, *161*, **163**, 166
 Cisticola b. *161*, **164**
 Cisticola c. 159
 Cisticola s. 163
boehmi, *Bradyornis* 490
 Muscicapa 461, *481*, **490**
 Parisoma 401, **421**
 Parisoma b. 401, **422**
böhmi, *Parisoma* 421
bonelli, *Phylloscopus* **358**, **400**
 Phylloscopus b. **358**, **400**
 Sylvia 358
borin, *Motacilla* 383
 Sylvia 315, 378, **383**, 386, 401, 493
 Sylvia b. **383**, **401**

bororensis, *Camaroptera b.* **296**, 320
Bostrychia hagedash 537
boultoni, *Bradypterus l.* **72**
bourbonnensis, *Terpsiphone* 530, 531, 537
bowdleri, *Melaenornis p.* **450**
brachyptera, *Drymoeca* 191
brachypterus, *Cisticola* 138, *177*, 188, 190, **191**, 194, 195, 196
 Cisticola b. *177*, **191**
brachyrhynchus, *Oriolus* 376
brachyura, *Camaroptera* 157, 212, 279, 280, 291, **294**, 300, 302, 303, 315, *320*, 330, 348
 Camaroptera b. **296**, *320*
 Sylvia 294
 Sylvietta 322, 335, *336*, **338**, 340, 342, 346, 350, 593
 Sylvietta b. *336*, **339**
Bradornis 438, 455
Bradyornis 453, 490
Bradypterus **61**
 alfredi 61, **69**, 72, **96**
 baboecala 61, **63**, 66, 67, 69, 76, 78, **96**, 128
 barratti 62, 64, 72, 73, 74, **75**, 78, **96**
 carpalis 61, 64, **65**, 68, **96**
 cinnamomeus 61, 70, 72, 73, 75, **96**, 373
 grandis 61, 67, **68**, **96**
 graueri 61, 64, 66, **67**, 68, **96**
 lopezi 61, **70**, 72, 74, **96**
 mariae 72
 sylvaticus 61, 76, **77**, **96**
 victorini 62, 74, **79**, **96**
brauni, *Apalis r.* **280**
brehmii, *Phylloscopus* 355
 Phylloscopus c. **355**
brevicauda, *Muscicapa c.* **467**
brevicaudata, *Camaroptera* 296
 Camaroptera b. 294, **295**
brevipes, *Monticola* 1, 2, 3, **4**, 7, *32*
 Monticola b. **4**, *32*
 Petrocincla 4
brevirostris, *Schoenicola* 89
 Schoenicola p. 86, **90**, **97**
brunnea, *Dioptrornis* 438
brunnescens, *Acrocephalus s.* **118**
 Cisticola 138, 207, 208, **210**, 212, 213, 240
 Cisticola b. **210**, 240
brunneus, *Melaenornis* **438**, 439, 441, *480*
 Melaenornis b. **438**, *480*
budongoensis, *Cryptolopha* 368
 Phylloscopus 352, 366, 367, **368**, **400**

bulubulu, Cisticola t. 207
bulliens, Cisticola 138, 146, 148, **149**, *160*
   Cisticola b. **150**, *160*
burae, Melaenornis m. **454**
Burnesia 221, 225, 242, 272
buryi, Parisoma 382, 414
   Sylvia 382
Butalis 467, 476, 479, 485, 487
buttoni, Calamonastes u. 305

C

caerulescens, Butalis 467
   Muscicapa 18, 440, **467**, 472, 474, 477, 486, 495, *496*
   Muscicapa c. **468**
caesia, Coracina 478, 578
caffra, Cossypha 78, 458
Calamocaetor 123
Calamocichla 99, 119, 123
Calamoherpe 116, 120
Calamonastes 58, 294, 298, **303**, 308
   fasciolatus 303, 305, 306, **307**, 310, *320*
   simplex **303**, 305, 306, 307, 310, *320*
   stierlingi 303
   undosus 303, 304, **305**, 307, 308, *320*
Calamonastides 123
calendula, Regulus 315
caliginus, Cisticola a. **205**
Camaroptera 58, 248, 252, 292, **294**, 303, 306, 308, 310, 311, 314, 429
   brachyura 157, 212, 279, 280, 291, **294**, 300, 302, 303, 315, *320*, 330, 348
   brevicaudata 296
   chloronota 292, 293, 294, 300, **301**, 315, *320*, 563
   superciliaris 294, **299**, 302, *320*
cameronensis, Geocichla 16
   Zoothera **16**, 20, *33*
   Zoothera c. **16**, *33*
camerunensis, Bradypterus l. **71**
   Muscicapa c. **485**
   Phylloscopus h. **367**, *400*
Campephaga 443
   flava 447
   quiscalina 554
campestris, Cisticola c. **162**
Campethera nubica 495
canescens, Eremomela 318, *321*, 322, 325, **326**, 328, 329
   Eremomela c. *321*, **327**
caniceps, Apalis f. **256**, **263**
caniviridis, Apalis r. **261**
canora, Apalis f. 263

cantans, Cisticola 138, 141, **143**, 151, 155, *160*, 165
   Cisticola c. **143**, *160*
   Drymoeca 143
cantillans, Motacilla 404
   Sylvia 377, 389, 392, 397, 402, **404**, 408, 410, 411, *417*
   Sylvia c. **405**, *417*
canzelae, Prinia s. 222
capensis, Apalis t. **255**
   Asio 215
   Batis 78, 419, **560**, 572, 574, 575, 576, 578, **579**, 582, 583, 586, 588, 590
   Batis c. 560, **579**, 590
   Emberiza 419
   Euplectes 154, 230
   Muscicapa 579
   Smithornis 376, 549
caprius, Chrysococcyx 65, 230, 299, 309, 419, 457
carlo, Prinia g. **233**
carnapi, Sylvietta b. *336*, **339**
carolathi, Megabyas f. 549
carpalis, Bradypterus 61, 64, **65**, 68, *96*
carruthersi, Cisticola 138, 174, 176, **179**, 181
cassini, Muscicapa 469, **471**, 474, 482, 483, 486, 493, *496*
castanea, Dyaphorophyia 545, 555, 558, 563, 565, 576
   Dyaphorophyia c. 545, **555**
   Platysteira 555
castaneiceps, Scoptelus 601
castaneus, Bradypterus l. **72**
castanops, Ploceus 471
Catharus 1
   guttatus 20
   ustulatus 20
cathkinensis, Bradypterus b. **76**
catoleucum, Myioparus p. **495**
caucasica, Sylvia c. **392**
caudatus, Aegithalos 243
cavei, Bradypterus c. **74**
centralis, Bradypterus b. **64**, *96*
   Turdus p. 36, **39**, *48*
Centropus 90, 156
   superciliosus 38
Cercococcyx mechowi 530
Cercomela schleglii 453
   sordida 154
Cercotrichas leucophrys 419
   leucosticta 18
   signata 22, 27
Certhia 431
certhiola, Locustella 91
cerviniventris, Apalis 84
   Bathmocercus 81, 82, **84**, *97*

Cettia 59, 61
   cetti 59, **60**, *96*
cetti, Cettia 59, **60**, *96*
   Cettia c. **60**, *96*
   Sylvia 60
chadensis, Acrocephalus r. **119**
   Batis o. **594**
   Bradypterus b. **64**
Chaetops frenatus 80
chalybea, Dyaphorophyia 555, 563
   Dyaphorophyia b. 545, **562**, 565
   Nectarinia 230, 499
Chamaetylas 18
changamwensis, Bias m. **552**
chapini, Apalis 252, 257, 273, 275, **276**, 282, 286, 292
   Apalis c. **257**, **277**
   Muscicapa i. **488**
   Sylvietta l. **351**
chariessa, Apalis 252, 256, **269**, 277
   Apalis c. **270**
cherina, Cisticola 138
chiguancoides, Turdus p. **39**, *48*
chiniana, Cisticola 138, 146, 148, 157, **158**, *161*, 163, 165, 166, 167, 168, 172, 208, 234
   Cisticola c. *161*, **162**
   Drymoica 158
chirindensis, Apalis 252, 257, 272, 273, **274**
   Apalis c. **274**
chlorochlamys, Eremomela s. **328**
chloronota, Camaroptera 292, 293, 294, 300, **301**, 315, *320*, 563
   Camaroptera c. **301**
   Sylvietta l. *336*, **351**
Chloropeta 123
   gracilirostris *113*, 123, **127**
   natalensis *113*, **123**, 127
   similis *113*, 123, 124, **126**
chobiensis, Turdus l. 42
chocolatina, Muscicapa 441
chocolatinus, Melaenornis 438, 439, **441**, 480
   Melaenornis c. **441**, *480*
Chrysococcyx 536, 585
   caprius 65, 230, 299, 309, 419, 457
   cupreus 184, 224, 299, 365, 573
   klaas 145, 184, 197, 224, 258, 265, 299, 323, 345, 573, 582, 587, 591
chubbi, Cisticola 138, 148, **150**, 153, 154, *160*, 289
   Cisticola c. **151**, *160*

Sylvietta r. 336, **347**
chuka, Zoothera g. **21**, 24, *33*
chyulu, Melocichla m. 86
   Muscicapa a. 477
   Zoothera g. 21
chyuluensis, Bradypterus c. 74
cinclorhynchus, Monticola 1
cinderella, Urolais e. **243**
cinerascens, Fraseria 18, 432, 433, **435**, 472, *481*
   Fraseria c. **436**, *481*
   Parisoma s. **418**
cinerea, Apalis 243, 252, *257*, 269, 273, 276, 280, 282, **283**, 285
   Apalis c. *257*, **284**
   Sylvia c. 389
cinereola, Muscicapa c. **468**, *496*
cinereolus, Cisticola 138, *161*, **166**, 167, 188
   Cisticola c. **166**
cinereus, Calamonastes u. 305
   Euprinodes 283
cinnamomea, Salicaria 73
cinnamomeus, Acrocephalus 108
   Acrocephalus b. **108**, *112*
   Bradypterus 61, 70, 72, **73**, 75, 96, 373
   Bradypterus c. **74**, *96*
   Cisticola 138, 209, 210, **212**, 214, *240*
   Cisticola c. **212**
   Sylvia 73
cirlus, Emberiza 13
Cisticola 57, 92, **138**, 215, 216, 298
   aberdare 138, *176*, 183, **184**
   aberrans 138, **156**, *160*, 165, 172
   angolensis 183
   angusticaudus 138, *177*, 195, 196, **197**, 198
   anonymus 138, 146, **148**, 149, 150, 151, *160*
   aridulus 138, 166, 188, 201, **204**, 207, 208, 234, *240*
   augusticauda 197
   ayresii 138, 207, 208, 209, 210, 212, **213**, *240*
   bodessa 138, 158, *161*, **163**, 166
   brachypterus 138, *177*, 188, 190, **191**, 194, 195, 196
   brunnescens 138, 207, 208, **210**, 212, 213, *240*
   bulliens 138, 146, 148, **149**, *160*
   cantans 138, 141, **143**, 151, 155, *160*, 165
   carruthersi 138, 174, *176*, **179**, 181
   cherina 138
   chiniana 138, 146, 148, 157, **158**, *161*, 163, 165, 166, 167, 168, 172, 208, 234
   chubbi 138, 148, **150**, 153, 154, *160*, 289
   cinereolus 138, *161*, **166**, 167, 188
   cinnamomeus 138, 209, 210, **212**, 214, *240*
   dambo 138, 207, **209**, *240*
   discolor 151
   distinctus 171
   dorsti 138, *177*, 188, **189**
   erythrops 138, **140**, 144, *160*
   exilis 138
   eximius 138, 201, **207**, *240*
   fulvicapillus 138, 157, *177*, 190, 192, **195**, 197, 198
   galactotes 138, *173*, *176*, 179, 180, 181, 183, 250
   haesitatus 138, 199, **203**, 216, *240*
   hunteri 138, 150, 151, **153**, 154, *160*
   juncidis 138, **199**, 203, 204, 205, 207, 208, 234, *240*
   lais 138, 155, *161*, 165, 167, 169, **170**
   lateralis 138, **145**, 147, 148, 151, *160*, 164, 188, 195
   lepe 141
   melanurus 138, *177*, **198**
   mongalla 189
   nanus 138, 166, *177*, **190**, 192
   natalensis 138, *176*, 183, **185**
   nigriloris 138, 150, 153, **154**, *160*, 165
   njombe 138, 155, *161*, **165**, 172
   pipiens 138, 174, *176*, **178**, 181
   restrictus 138, *161*, **166**
   robustus 61, 138, 155, 157, *176*, **182**, 185, 186
   ruficeps 138, *177*, **187**, 189, 194, 195
   rufilatus 138, *161*, **167**, 170, 189
   rufus 138, *177*, 188, 192, **193**, 194, 195
   subruficapilla 163
   subruficapillus 138, *161*, 168, **169**, 172, 419
   textrix 138, **206**, 214, *240*
   tinniens 138, 174, *176*, 179, **180**
   troglodytes 138, *177*, 188, 192, 193, **194**
   woosnami 138, **146**, 148, 149, 151, *160*

cisticola, Cisticola j. 200
Cisticolidae 58
Cisticolinae 58
citriniceps, Eremomela s. **328**
clamans, Malurus 244
   Spiloptila **244**, 289, *320*
clara, Parisoma l. **420**
clarens, Bias m. **552**
clarkei, Turdus p. 45
claudei, Apalis t. **255**
cleaveri, Illadopsis 520
coelebs, Fringilla 10
collaris, Anthreptes 251, 297
   Apalis n. 256, **271**
   Lanius 43, 80, 419, 458
collurio, Lanius 134, 554
collybita, Phylloscopus 130, 218, 243, 353, **355**, 359, 360, 362, 363, 369, *400*
   Phylloscopus c. **355**, *400*
   Sylvia 355
colstoni, Elminia n. **520**
comitata, Muscicapa 469, 472, 483, **485**, 487, *496*
   Muscicapa c. **485**, *496*
comitatus, Butalis 485
communis, Sylvia 339, 377, 378, 382, 384, **388**, 392, 394, 396, 398, *401*, 404, 405, 407, 411
   Sylvia c. **389**, *401*
concolor, Camaroptera 314
   Cisticola c. **143**
   Macrosphenus 311, **314**, 316, *337*, 520
concreta, Dyaphorophyia 545, 555, 556, 563, **564**
   Dyaphorophyia c. 545, **565**
   Platystira 564
confinis, Phragmacia s. **219**
congensis, Eremomela s. **328**
congicus, Erythrocercus m. **510**
congo, Cisticola p. *176*, **178**
congoensis, Batis m. **595**
conspicillata, Sylvia 218, 377, 389, 406, **407**, 410, *417*
   Sylvia c. **407**, *417*
constans, Camaroptera b. **296**
Coracina caesia 478, 578
Corvidae 508, 548, 574
corvina, Terpsiphone 530, 531, 537
Corvus albus 43
Cossypha caffra 78, 458
   heuglini 315
   natalensis 27
   semirufa 36
crassirostris, Sylvia h. **380**, *416*
crawfurdi, Eremomela i. **319**
crossleyi, Turdus 19
   Zoothera 15, **19**, 20, *33*
   Zoothera c. **19**, *33*

cryptoleuca, Platysteira p. **572**
Cryptolopha 366, 367, 368
cucullatus, Ploceus 486
Cuculus solitarius 2, 38, 43, 537
Culicicapa 508
culminans, Turdus o. 35
cunenensis, Acrocephalus g. **121**
cupreus, Chrysococcyx 184, 224, 299, 365, 573
Curruca 118, 129, 131, 381, 393
curruca, Motacilla 392
   Sylvia 130, 356, 377, 380, 390, **392**, 396, *401*, 406, 413
   Sylvia c. **392**, *401*
curvirostra, Loxia 201
cyanea, Muscicapa 567
   Platysteira 545, 563, **567**, 568, 570, 571, 572, 576
   Platysteira c. 545, **568**
cyanolaema, Nectarinia 493
cyanomelas, Muscicapa 526
   Trochocercus 524, 525, **526**, 527, 528, *544*
   Trochocercus c. **526**

D

damarensis, Eremomela g. **330**
dambo, Cisticola 138, 207, **209**, *240*
   Cisticola d. **209**
dammholzi, Sylvia a. **386**
darglensis, Apalis t. 255
dauurica, Muscicapa 461, 477
debilis, Phyllastrephus 317
deckeni, Turdus o. 35, *48*
deichleri, Turdus v. **49**, *50*
deltae, Prinia g. **234**
Dendrocolaptidae 430
denti, Sylviella 349
   Sylvietta 335, *336*, 348, **349**, 520, *600*
   Sylvietta d. *336*, **349**
deserti, Oenanthe 408
   Sylvia n. **394**, *401*
deserticola, Sylvia 377, 405, 407, **409**, 411, *417*
   Sylvia d. **409**, *417*
dexter, Cisticola f. **195**
diabolicus, Melaenornis p. **447**
Dicaeum 343
Dicrurinae 508
Dicrurini 508
Dicrurus 446
   adsimilis 30, 341, 447, 537, 582, 593
   atripennis 520
   ludwigii 442, 447
dilutior, Sylvietta b. **339**
dimorpha, Batis c. *560*, 578, **580**

Dinemellia dinemelli 460
dinemelli, Dinemellia 460
diops, Batis 561, 574, **575**, 576, 578, 579, 582, 601
Dioptrornis 438, 439
discolor, Cisticola 151
   Cisticola c. 148, **151**, *160*
dispar, Cisticola f. *177*, **196**, 197, 198
distinctus, Cisticola 171
   Cisticola l. **161**, **171**
disruptans, Zoothera g. **21**
diverga, Sylvietta r. **344**
divisus, Melaenornis p. **450**
domesticus, Passer 135
dorcadichrous, Phylloscopus u. **363**
dorsti, Cisticola 138, *177*, 188, **189**
dowsetti, Apalis a. **285**
drakensbergensis, Apalis t. **255**
Drymocichla 58, **248**
   incana 239, **249**, *320*
Drymoeca 140, 143, 148, 191, 203, 207, 246, 262
Drymoica 85, 145, 156, 158, 167, 169, 170, 182, 185, 193, 194, 219, 232, 238, 279, 305, 307, 309
Dryodromas 198, 271
Dryoscopus 374, 433, 548
   sabini 549
   senegalensis 549
dumetorum, Acrocephalus 108
dumicola, Cisticola f. **196**
duyerali, Melaenornis p. **450**
Dyaphorophyia 548, **555**, 567, 572
   blissetti *300*, 545, 555, 556, **562**, 565, 601
   castanea 545, **555**, 558, 563, 565, 576
   chalybea 555, 563
   concreta 545, 555, 556, 563, **564**
   jamesoni 555, 563
   tonsa 545, 555, 556, **558**, 563, 565, 576, 599
dyleffi, Cisticola t. **181**

E

edolioides, Melaenornis 438, 442, 444, **445**, 446, *480*
   Melaenornis e. **445**, *480*
   Melasoma 445
egregius, Cisticola c. **212**
eidos, Apalis r. **257**, *280*
elaeica, Hippolais p. *113*, **130**, 132
elegans, Eremomela c. **327**
elgonensis, Bradypterus b. **64**, *96*
   Eremomela c. **327**

Elminia 498, 508, 513, **514**, 527
   albicauda 426, *497*, 514, 515, **517**, 524
   albiventris *497*, 514, 519, 520, **522**, 524, 527, 528
   albonotata *275*, *497*, 498, 514, 518, 520, 522, **523**, 527, 528, 534
   longicauda *497*, 514, **517**, 518, 520
   nigromitrata *497*, 514, **519**, 522, 528, 547
elusa, Cisticola e. **141**
Emberiza capensis 419
   cirlus 13
emendatus, Cisticola c. **162**
emini, Cisticola a. **157**, *160*
   Terpsiphone r. **543**
Eminia 372, *375*
   lepida 337, **372**
Empidornis 432, 438, **459**
   semipartitus **459**, *480*
entebbe, Cisticola a. **214**
epichlora, Burnesia 242
   Urolais **242**, *284*, *320*
   Urolais e. **243**, *320*
epulata, Muscicapa 469, 474, **479**, 483, 486, 487, 493, 496
epulatus, Butalis 479
   Pedilorhynchus 482
Erannornis 508
eremicus, Cisticola a. **205**
Eremomela 58, 250, 271, **318**, 335, 422, 425
   atricollis 318, *321*, 325, **334**
   badiceps 271, 318, *321*, **332**, 333, 600, 602
   canescens 318, *321*, 322, 325, **326**, 328, 329
   flavicrissalis 318, *321*, 322, **323**, 324, 327, 343, 346
   gregalis 297, 309, 318, *321*, 322, 325, 326, 328, **329**
   icteropygialis **318**, *321*, 323, 324, 325, 326, 327, 330, 331, 339, 513
   pusilla 318, *321*, 322, **325**, 326, 328, 329, 593
   salvadorii 318, *321*, 322, 323, **324**
   scotops 242, 318, *321*, 322, 325, 326, 327, **328**, 329, 331, 335, 425
   turneri 318, *321*, 332, **333**
   usticollis 318, *321*, 322, 325, 330, **331**
Erithacus rubecula 413, 436, 503
erlangeri, Batis m. **595**
   Camaroptera b. **295**
   Melaenornis p. **450**
   Prinia s. **225**, *241*

Erythrocercus **508**
  holochlorus 497, 509, **511**, 512, 513
  livingstonei 497, 509, 511, **512**
  mccallii 497, **509**, 511, 512, 520
erythrogaster, Malimbus 601
erythrophthalma, Batis c. **580**
erythrops, Cisticola 138, **140**, 144, *160*
  Cisticola e. **141**, *160*
  Drymoeca 140
erythroptera, Drymoica 283
  Heliolais 222, **238**, *241*, 249
  Heliolais e. **239**, *241*
erythrothorax, Stiphrornis 520
etoshae, Malcorus p. **287**
Eupirnoides 281
Euprinodes 283
Euplectes 94
  capensis 154, 230
  orix 223, 230
  psammocromius 155
euroa, Cisticola s. 169
europaea, Sitta 424
europhilus, Calamonastes f. **308**
Euryptila 58, **308**
  subcinnamomea **309**, *337*
eustacei, Phylloscopus l. **365**, *400*
excisus, Sphenoeacus a. **88**
excubitor, Lanius 395
exilis, Cisticola 138
exima, Bleda 376
  Drymoeca 207
eximius, Cisticola 138, 201, **207**, *240*
  Cisticola e. **208**, *240*
explorator, Monticola 1, 2, **3**, 5, 6, 32
  Monticola e. **3**, *32*
  Turdus 3
extrema, Eremomela s. 328
exultans, Prinia m. **229**

F

fagani, Terpsiphone r. **543**, *544*
fantiensis, Eremomela b. **332**
fasciatus, Tockus 41, 315, 550, 554
fasciolata, Drymoica 307
fasciolatus, Calamonastes 303, 305, 306, **307**, 310, *320*
  Calamonastes f. **307**, *320*
feminina, Bias m. 551
ferreti, Terpsiphone v. **533**, *544*
ferrugineus, Cisticola t. **195**
  Laniarius 78, 478, 573
Ficedula 422, 423, 425, 432, **500**
  albicollis 496, 501, **503**, 505
  hypoleuca 496, **500**, 503, 505

parva 496, **506**
semitorquata 496, 500, 503, **505**
finschii, Oenanthe 408
fischeri, Cisticola c. **159**
  Dioptrornis 439
  Melaenornis 438, **439**, 441, 442, *480*
  Melaenornis f. **439**, *480*
  Zoothera g. **26**
flammeus, Macrosphenus k. **314**
flammulatus, Megabyas 433, 497, **548**, 552
  Megabyas f. **497**, **549**
flava, Campephaga 447
flavicans, Macrosphenus 311, 312, 313, 314, 315, *337*
  Macrosphenus f. **313**
  Prinia 221, 222, **226**, 229, *241*, 287
  Prinia f. **227**, *241*
  Sylvia 226
flavicrissalis, Eremomela 318, 321, 322, **323**, 324, 327, 343, 346
flavida, Apalis 252, 255, *256*, **262**, 341
  Apalis f. **256**, *263*
  Drymoeca 262
flavifrons, Anthoscopus 600
flavigaster, Hyliota *337*, **422**, 425, 428
  Hyliota f. **423**
flavigularis, Apalis t. **255**, *256*
flaviventris, Apalis t. **255**
  Sylvietta v. *336*, **348**, 350
flavocincta, Apalis f. **263**
flecki, Sylvietta r. *336*, **344**
florisuga, Apalis f. **263**
fluviatilis, Locustella 93, 95, 97
  Prinia 221, 222, **224**, 225, *241*
  Sylvia 93
fortis, Cisticola c. **162**
foxi, Acrocephalus r. 119
francisi, Erythrocercus l. **513**
Franklinia 245
fraseri, Anthreptes 315, 429, 510
  Neocossyphus 315, 520
Fraseria **432**, 438
  cinerascens 18, 432, 433, **435**, 472, *481*
  ocreata **432**, 436, 442, *481*, 520
frater, Cisticola c. **162**
fraterculus, Acrocephalus b. 108
fratrum, Batis 560, 574, 575, 576, 578, 579, 580, **582**, 583, 588
  Pachyprora 582
frenatus, Chaetops 80
fricki, Cisticola c. **159**, 164

Fringilla coelebs 10
frontalis, Sporopipes 593
fuelleborni, Muscicapa a. **477**
fugglescouchmani, Camaroptera b. **296**
  Phylloscopus u. **363**
fuliginosa, Apalis m. **273**
  Artomyias 488
fulvicapilla, Sylvia 195
fulvicapillus, Cisticola 138, 157, 177, 190, 192, **195**, 197, 198
  Cisticola f. **177**, **196**
fumosa, Apalis r. **261**
funebris, Apalis c. **284**
funereus, Parus 601
Furnariidae 430
fusca, Locustella l. **95**, *97*
fuscata, Phillopneuste 360
fuscatus, Phylloscopus **360**
  Phylloscopus f. **360**
  Turdus o. 35
fuscigularis, Apalis t. **254**, *256*
fuscula, Muscicapa a. **477**
fuscus, Acrocephalus s. **105**, 110, 112

G

gabun, Cisticola a. **214**, *240*
galactotes, Cisticola 138, **173**, 176, 179, 180, 181, 183, 250
  Cisticola g. **174**, *176*
  Malurus 173
gambagae, Alseonax 466
  Muscicapa 461, 462, 464, **466**, 470, 477, *481*
Garrulacinae 58
Garrulus glandarius 371
Geocichla 15, 16, 28, 30
gephyra, Sylvietta r. **347**
glandarius, Garrulus 371
Glaucidium perlatum 590, 593
godfreyi, Bradypterus b. **76**, *96*
golzi, Apalis f. **263**
goslingi, Apalis 252, *257*, 278, 280, **282**, 315
gracilirostris, Acrocephalus 64, 99, 106, 108, *112*, **120**, 124, 128
  Acrocephalus g. **112**, **121**
  Calamoherpe 120
  Chloropeta 113, 123, **127**
  Chloropeta g. 113, 123, **128**
gracilis, Prinia 200, 217, 221, 226, **233**, *241*
  Prinia g. **234**, *241*
  Sylvia 233
Grallina 508
grandior, Myioparus p. **495**

grandis, Apalis c. 257, **284**
   Bradypterus 61, 67, **68**, *96*
   Melocichla m. 86
granti, Bradypterus l. 71
   Camaroptera c. 301
   Terpsiphone v. 533
granviki, Melocichla m. 86
Graueria 310, 311
   vittata 233, **310**, *337*
graueri, Bradypterus 61, 64, 66, **67**, 68, *96*
   Dyaphorophyia c. 545, **565**
   Prinia s. 222
   Turdus p. 39
   Zoothera c. 17, *33*
gregalis, Eremomela 297, 309, 318, *321*, 322, 325, 326, 328, **329**
   Eremomela g. *321*, **330**
   Malcorus 329
grimwoodi, Muscicapa a. **470**, *481*
griseiceps, Apalis t. **254**, *256*
   Macrosphenus k. 317
griseigula, Camaroptera b. **296**, *320*
griseigularis, Alseonax 492
   Myioparus 315, 472, 483, **492**, 495, 496, 520, 556
   Myioparus g. **492**, *496*
griseiventris, Parisoma l. 420
griseldis, Acrocephalus 98, *112*, 114, **116**, 118, 121
   Calamoherpe 116
griseoflava, Eremomela i. **319**, *321*
griseopyga, Apalis t. **255**, *256*
griseus, Cisticola g. 173
   Melaenornis p. 450
   Passer 593
grotei, Muscicapa a. 477
guiersi, Acrocephalus b. **108**
guinea, Cisticola r. **188**
gularis, Monticola 1
gurneyi, Geocichla 15
   Turdus 20
   Zoothera 15, 16, **20**, 24, 26, *33*, 36
   Zoothera g. **21**, *33*
guttata, Zoothera 15, **25**, 28, 29, *33*
   Zoothera g. **26**, *33*
guttatus, Catharus 20
   Turdus 25

H

hadii, Zoothera p. 24
haematina, Spermophaga 520
haematocephalus, Cisticola g. **174**, *176*
haesitata, Drymoica 203
haesitatus, Cisticola 138, 199, 203, 216, *240*
hagedash, Bostrychia 537
hallae, Acrocephalus b. **108**
   Cisticola f. **196**
hardyi, Sylvietta d. 336, **350**
harterti, Camaroptera b. **294**, 295, *320*
   Dyaphorophyia c. 565
   Scotocerca i. 217
hebridensis, Turdus p. 45
heinrichi, Prinia b. 233
helenorae, Eremomela i. 319
Heliolais 58, 220, **238**
   erythroptera 222, **238**, *241*, 249
helleri, Turdus o. **36**, *48*
Hemitesia 58
   neumanni 58, *337*, **429**, 522
herberti, Cryptolopha 367
   Phylloscopus 352, **367**, 368, 400
   Phylloscopus h. **367**
Herpystera 221
heterophrys, Cisticola c. 159, **162**, 167
heuglini, Cossypha 315
hindii, Cisticola b. **211**
Hippolais 111, 123, **129**, 384, 420
   icterina *113*, 124, 127, 129, 135, **136**, 384
   languida *113*, 129, 130, **131**, 134, *382*
   olivetorum *113*, 129, 132, **133**, 137
   pallida *113*, **129**, 132, 135, 137, *356*
   polyglotta *113*, 130, **134**, 137
Hirundo nigrita 473
   rustica 394
hollidayi, Batis c. **580**
holochlorus, Erythrocercus 497, 509, **511**, 512, 513
holubi, Cisticola n. **186**
hopsoni, Acrocephalus b. 108
hormophora, Dyaphorophyia c. **556**, 558
hortensis, Motacilla 379
   Sylvia 377, 378, **379**, 381, 382, 397, 414, *416*
   Sylvia h. **380**, *416*
huambo, Cisticola n. **186**
huilae, Calamonastes u. **305**
huilensis, Cisticola c. 162
humilis, Cisticola c. **159**, *161*, 164
hunteri, Cisticola 138, 150, 151, **153**, 154, *160*
   Cisticola h. 153
Hylia 428
prasina 59, *337*, **428**
Hyliidae 428
Hyliota **422**
   australis 297, *337*, 422, **424**, 426, 427, 601
   flavigaster *337*, 422, 425, **428**
   usambarae *337*, 422, **426**
   violacea *337*, **427**
hylophila, Monticola a. **7**, *32*
Hypergerus 372, **375**
   atriceps 84, *337*, **375**
hypernephela, Cisticola h. 153
hypochondriacus, Macrosphenus f. **312**, *337*
Hypocoliidae 58
Hypocolius 58
Hypodes 461
hypoleuca, Ficedula 496, **500**, 503, 505
   Ficedula h. 496, **501**
   Motacilla 500
Hypothymis azurea 530
hypoxantha, Prinia 229
   Prinia m. 227, **228**, *241*
hypoxanthus, Cisticola b. 191

I

iberiae, Ficedula h. **501**
icterina, Hippolais *113*, 124, 127, 129, 135, **136**, 384
   Sylvia 136
icterops, Sylvia c. **389**, *401*
icteropygialis, Eremomela 318, *321*, 323, 324, 325, 326, 327, 330, 331, 339, 513
   Eremomela i. **322**
   Sylvietta 318
ignea, Terpsiphone r. **543**, 547
ignicapilla, Sylvia 370
ignicapillus, Regulus **370**, 372, 400
   Regulus i. 370
iliacus, Turdus 45, **46**, *49*, 51, 52, 56
   Turdus i. **46**, *49*
Illadopsis cleaveri 520
imatong, Cisticola a. **214**, *240*
imberbis, Anomalospiza 142, 182, 197, 203, 206, 207, 215, 224, 228
imerinus, Monticola 1
impavida, Muscicapa c. **468**, *496*
importunus, Andropadus 537
Incana 215
   incana 204, **215**, *240*
incana, Cisticola 215
   Drymocichla **239**, *249*, *320*
   Incana 204, **215**, *240*
incanus, Melocichla m. 86
Indicator meliphilus 573

indicator 419
indicator, Baeopogon 520
Indicator 419
Indicatoridae 294
inexpectatus, Cisticola n. **186**
infulata, Muscicapa a. **470**, *481*
infuscata, Muscicapa 439, 461, 474, 475, *481*, **487**, 489
 Muscicapa i. *481*, **488**
 Saxicola 451
infuscatus, Butalis 487
 Melaenornis 451, 456, *480*
 Melaenornis i. **452**, *480*
inornata, Hyliota a. **425**
 Prinia 221, 224, 225
 Sylvia c. **405**, *417*
inornatus, Phylloscopus 361, 372, *400*
 Phylloscopus i. **362**, *400*
 Regulus 361
inquieta, Scotocerca 216, 234, *241*
 Scotocerca i. **217**, *241*
inquietus, Malurus 216
insignis, Prodotiscus 573
insulata, Camaroptera b. **295**
intercalata, Camaroptera b. **296**
intermedius, Sphenoeacus a. **88**
interposita, Muscicapa a. 477
iringae, Apalis t. 254
irwini, Calamonastes u. **306**, *320*
isabellina, Sylvietta 335, 336, 339, 342, 343, **345**
isabellinus, Cisticola b. *177*, **192**
 Lanius 395
isodactylus, Cisticola g. 174
itombwensis, Cisticola a. **214**
 Muscicapa 475
 Muscicapa l. **475**, *481*
ituriensis, Batis 561, 574, 576, 598, **600**
Ixobrychus sturmii 537

J

jacksoni, Acrocephalus g. **121**
 Apalis 252, 256, 266, **267**, 271, 276, 284
 Apalis j. 256, **268**
 Bathmocercus r. 82
 Parisoma l. **401**, **420**
 Ploceus 471
 Sylvietta w. **336**, **340**
jacobinus, Oxylophus 419
jamesi, Cisticola s. *161*, **169**
jamesoni, Dyaphorophyia 555, 563
 Dyaphorophyia b. **545**, **563**, 601
jardineii, Turdoides 448
jerdoni, Sylvia h. 381

jodoptera, Heliolais e. **239**
johnstoni, Phylloscopus r. **364**
jordonsi, Sylvia c. 389
juncidis, Cisticola 138, **199**, 203, 204, 205, 207, 208, 234, 240
 Cisticola j. **200**
 Sylvia 199
Jynx torquilla 6

K

kaboboensis, Apalis p. *257*, **276**
kaffensis, Cisticola b. **164**
kalahari, Cisticola a. **205**, *240*
kalindei, Eremomela t. **333**
kamitugaensis, Camaroptera c. **302**
kaokensis, Monticola b. 4
kapitensis, Cisticola n. *176*, 186
karamojae, Apalis 252, *257*, 279, **281**, 498
 Apalis k. *257*, **281**
 Eupirnoides 281
karamojensis, Eremomela i. 319
karasensis, Cisticola s. **169**
kasai, Cisticola d. **209**, *240*
kasokae, Prinia s. **222**
katanga, Cisticola n. *176*, **186**
katangae, Calamonastes u. **305**, *320*
kathleenae, Batis m. **577**
katonae, Cisticola b. *177*, **191**
Kaupifalco monogrammicus 434
keithi, Cisticola c. **159**, *161*
kelsalli, Camaroptera c. **301**, *320*
 Fraseria o. **433**
kempi, Amaurocichla 313
 Macrosphenus 311, 312, **313**, 315, *337*
 Macropheus k. **313**, *337*
kennedyi, Batis c. **580**
kericho, Cisticola b. **191**
kibalensis, Zoothera 15, 17
 Zoothera c. **16**, *17*, *33*
kigezi, Apalis r. **280**
kikuyuensis, Eremomela s. **328**
kilimensis, Zoothera p. **23**, *33*
kivuensis, Terpsiphone v. **533**
klaas, Chrysococcyx 145, 184, 197, 258, 265, 299, 323, 345, 573, 582, 587, 591
kretschmeri, Macrosphenus 311, **317**, *337*
 Macrosphenus k. **317**, *337*
 Phyllostrephus 317
kumbaensis, Dyaphorophyia c. 565
kumboensis, Muscicapa a. 477
kungwensis, Bradypterus a. **70**
 Dyaphorophyia c. **545**, **565**

L

laeneni, Hippolais p. **130**
laeta, Cryptolopha 366
laetus, Phylloscopus 363, 364, 365, **366**, 369, *400*
 Phylloscopus l. **366**, *400*
lais, Cisticola 138, 155, *161*, 165, 167, 169, **170**
 Cisticola l. *161*, **171**
 Drymoica 170
Lamprotornis 605
 nitens 537
languida, Curruca 131
 Hippolais 113, 129, 130, **131**, 134, 382
Laniarius 446, 585
 ferrugineus 78, 478, 573
Lanii 508, 548
Laniidae 548, 574
Laniinae 574
Lanioturdus xii, 548, **604**
 torquatus 497, **604**
Lanius 218, 265, 382
 collaris 43, 80, 419, 458
 collurio 134, 554
 excubitor 395
 isabellinus 395
lateralis, Cisticola 138, **145**, 147, 148, 151, *160*, 164, 188, 195
 Cisticola l. **145**, *160*
 Drymoica 145
laticincta, Platysteira 567, 572
 Platysteira p. **545**, **569**, **572**
latukae, Eremomela b. **332**
laurae, Phylloscopus 352, 364, **365**, 366, 367, *400*
 Phylloscopus l. **365**
 Seicercus 365
lavendulae, Cisticola a. **205**
lawsoni, Melaenornis s. **458**
layardi, Parisoma **401**, **414**, 418, 499
 Parisoma l. **401**, **414**
lebombo, Cisticola f. **196**
lebomboensis, Apalis t. **255**
lendu, Alseonax 475
 Muscicapa 461, 473, 474, **475**, *481*
 Muscicapa l. **475**, *481*
leontica, Prinia 248
 Schistolais 241, 246, **248**
lepe, Cisticola 141
 Cisticola e. **141**, *160*
lepida, Eminia 337, **372**
leptorhynchus, Acrocephalus g. *112*, **121**
leucocapilla, Monticola b. 4
leucomelaena, Curruca 381
 Sylvia 132, 377, 379, 380, **381**, *404*, *414*, *416*

leucophrys, Cercotrichas 419
   Sylvietta 335, *336*, 348, **350**, 369, 429
   Sylvietta l. *336*, **350**
leucopogon, Drymoeca 246
   Schistolais 241, **246**, 248
   Schistolais l. 241, **246**
leucopsis, Sylvietta b. *336*, **339**
leucosticta, Cercotrichas 18
leucura, Oenanthe 412
libonyanus, Merula 41
   Turdus 7, 29, 30, 34, 36, 38, 41, 44, *48*, 448
   Turdus l. 42, *48*
lightoni, Apalis m. 257, **272**
lippensi, Zoothera g. **26**
litsitsirupa, Merula 28
   Psophocichla 26, **28**, *33*, 45
   Psophocichla l. 29, *33*
littoralis, Cisticola n. 186
livingstonei, Erythrocercus 497, 509, 511, **512**
   Erythrocercus l. 497, **513**
loanda, Cisticola b. 177, **192**
lobito, Cisticola a. 205
Locustella 57, **91**
   certhiola 91
   fluviatilis **93**, 95, 97
   luscinioides 60, 92, **94**, 97
   naevia **92**, 95, 97
Lonchura 41
longicauda, Elminia 497, **514**, 517, 518, 520
   Elminia l. 497, **515**
   Myiagra 514
longirostris, Monticola s. 13
lopesi, Poliolais **292**, *337*
   Poliolais l. **293**, *337*
lopezi, Apalis 292
   Bradypterus 61, 70, **72**, 74, 96
   Bradypterus l. **71**, 96
   Phlexis 70
loringi, Sylvietta w. **340**
Loxia curvirostra 201
lualabae, Muscicapa a. 470
luangwae, Melocichla m. **86**
luapula, Cisticola g. **174**
lucidigula, Apalis f. 263
ludoviciae, Turdus o. 36, *48*
ludwigii, Dicrurus 442, **447**
lufira, Cisticola w. **147**
lugens, Parisoma 401, 414, **420**
   Parisoma l. **420**
   Sylvia 420
lugubris, Cisticola g. **174**, *176*
   Melaenornis e. **445**
lundae, Eremomela i. 318
lurio, Cisticola a. **157**
Luscinia luscinia 60
   megarhynchos 60, 373

luscinia, Luscinia 60
luscinioides, Locustella 60, 92, **94**, 97
   Locustella l. **95**, 97
   Sylvia 94
Lusciniola melanopogon 99
lynesi, Apalis t. **255**, *256*
   Batis o. **594**
   Cisticola b. **211**
lysis, Bradypterus b. 76

M

macdonaldi, Bradypterus c. 74
mackensianus, Phylloscopus u. **363**, *400*
macphersoni, Apalis c. *256*, **269**
Macrosphenus 58, 310, **311**, 429
   concolor 311, 313, **314**, 316, *337*, 520
   flavicans 311, **312**, 313, 314, 315, *337*
   kempi 311, 312, **313**, 315, *337*
   kretschmeri 311, **317**, *337*
   pulitzeri 311, 314, **316**, *337*
macroura, Vidua 144, 178, 211, 294, 536
maculatus, Cisticola l. *161*, **171**
maculosa, Motacilla 228
   Prinia 219, 221, 222, 227, **228**, *241*, 419
   Prinia m. **228**, *241*
mahali, Plocepasser 457, 460
   Prinia m. **228**
major, Bradypterus b. 76
   Chloropeta n. 113, **124**, 127
   Cisticola t. **206**
   Heliolais e. **239**
   Parus 410
makayii, Sylvietta r. 347
Malaconotinae 548, 574
Malaconotus 578
   blanchoti 82, 573, 576, 580, 583
Malcorus 220, **286**, 329
   pectoralis 227, *241*, **287**, 309
malensis, Apalis f. 263
Malimbus 443, 486
   erythrogaster 601
Malurus 173, 180, 187, 216, 244, 250
manengubae, Bradypterus l. 72
   Poliolais l. **293**, *337*
margaritae, Batis 560, 574, 575, **576**, 578, 579, 582
   Batis m. 560, **577**
marginalis, Hyliota f. 423
marginatus, Cisticola g. **173**
mariae, Bradypterus 72
   Bradypterus l. **71**, 96
   Cisticola n. **165**

Urolais e. **243**, *320*
mariquensis, Bradornis 455
   Melaenornis 448, 450, 453, **455**, *480*
   Melaenornis m. **455**, *480*
marleyi, Camaroptera b. **296**
   Cisticola t. **207**
maroccana, Sylvia d. **409**, *417*
marsabit, Muscicapa a. **477**, *481*
   Parisoma b. **422**
marungensis, Apalis p. **267**
   Cisticola c. **151**
masaba, Cisticola h. **153**
mashona, Cisticla l. **171**
massaica, Chloropeta n. 113, **124**
mauensis, Cisticola a. **214**, *240*
mauritanicus, Turdus m. 49, *54*
maxis, Zoothera g. **26**
mayombe, Terpsiphone r. **543**
mbangensis, Cisticola b. **211**
mbeya, Cisticola c. **159**
mccallii, Erythrocercus 497, **509**, 511, 512, 520
   Erythrocercus m. 497, **509**
   Pycnosphrys 509
mechowi, Cercococcyx 530
Megabyas 548
   flammulatus 433, 497, **548**, 552
megalolophus, Trochocercus c. **526**
Megalurinae 58, 89
Megalurus 89
megarhynchos, Luscinia 60, 373
Melaenornis 432, **438**, 459, 491
   annamarulae 438, **442**, *480*
   ardesiacus **443**, *480*
   brunneus **438**, 439, 441, *480*
   chocolatinus 438, 439, **441**, *480*
   edolioides 438, 442, 444, **445**, 446, *480*
   fischeri 438, **439**, 441, 442, *480*
   infuscatus **451**, 456, *480*
   mariquensis 448, 450, 453, **455**, *480*
   microrhynchus 450, **453**, 464, *480*
   pallidus 439, 448, **449**, 453, 454, 456, 470, *480*
   pammelaina 438, 445, **446**, *480*
   pumilus 454
   silens **457**, *481*
melanicterus, Pycnonotus 375
melanocephala, Apalis 252, 257, 270, **272**, 274, 276, 277, 284, 498, 578
   Apalis m. 257, **273**
   Burnesia 272
   Motacilla 398

Sylvia 377, 380, 392, 397, **398**, 403, 404, 406, 408, 413, *416*
Sylvia m. **399**, *416*
melanocephalus, *Ploceus* 376, 471
melanogaster, *Ploceus* 486
melanopogon, *Acrocephalus* 99, 100, 103, *112*
   Acrocephalus m. **100**, *112*
   Luciniola 99
   Sylvia 100
melanops, *Prinia b.* **233**, *241*
melanorhyncha, *Prinia s.* **222**, *241*
melanothorax, *Sylvia* 377, **397**, 398, 402, 403, *416*
melanurus, *Cisticola* 138, *177*, **198**
   Dryodromas 198
   Passer 80
Melasoma 445
melba, *Pytilia* 593
meliphilus, *Indicator* 573
Melocichla **85**, 87
   mentalis **85**, 88, *97*
Meloenornis 443
Melopsittacus undulatus 453
menachensis, *Turdus* 30, 34, 38, 41, 44
mentalis, *Drymoica* 85
   Melocichla **85**, 88, *97*
   Melocichla m. **86**, *97*
   Platysteira p. **572**
meridionalis, *Melocichla m.* 86
   Sylvietta v. 348
Merops apiaster 419
   orientalis 395
Merula 28, 41
merula, *Turdus* 13, 16, 25, 30, 34, 36, 44, 47, 49, 50, 52, **53**, 56
   Turdus m. **49**, **54**
mesica, *Muscicapa a.* **477**
mesicus, *Lanioturdus t.* 604
Mesopicos xantholophus 601
metopias, *Artisornis* 82, 277, 290, **291**, 292, *320*
   Artisornis m. **291**, *320*
   Orthotomus 292
   Prinia 291
microrhynchus, *Bradyornis* 453
   Melaenornis 450, **453**, 464, *480*
   Melaenornis m. **454**, *480*
midcongo, *Cisticola c.* **212**, 214
milanjensis, *Turdus o.* 35
mildbreadi, *Bradypterus c.* **74**, *96*
minima, *Batis* 561, 574, **598**, 600, 601, 602
   Muscicapa a. **477**
   Platystira 598

Sylvietta w. **340**
minor, *Apalis j.* **268**
   Batis 561, 574, 576, 578, 585, 588, 592, 594, **595**, 597, 598, 602
   Batis m. **561**, **595**
   Batis o. 595
   Cisticola a. **157**
minula, *Sylvia* 377, 392
minulla, *Batis* 561, 574, **597**
   Platystira 597
minullus, *Accipiter* 495
   Phylloscopus r. **364**, *400*
minuscula, *Muscicapa i.* **488**
minutus, *Anthoscopus* 322
mitoni, *Bradypterus l.* **71**
mixta, *Batis* 560, 574, 575, 576, **578**, 579, 580, 582, 583, 588, 596
   Batis m. **560**, **578**
   Pachyprora 578
modesta, *Cisticola l.* **145**
modestus, *Melaenornis p.* **450**
modularis, *Prunella* 61, 413
Moho 375
molitor, *Batis* 297, 560, 574, 576, 577, 578, 580, 583, **584**, 587, 588, 589, 590, 591, 594, 596, 597, 598
   Batis m. **560**, **584**
   Muscicapa 584
molleri, *Prinia* 221, **230**, *241*
momus, *Sylvia m.* **399**, 404
Monarchidae xii, 498, **508**, 548, 574
Monarchinae 432, 508, 548
Monarchini 508
mongalla, *Cisticola* 189
   Cisticola r. *177*, **188**, 189
monogrammicus, *Kaupifalco* 434
Monticola xii, **1**
   angolensis 1, 5, **6**, 8, 10, *32*
   bensoni 1
   brevipes 1, 2, 3, **4**, 7, *32*
   cinclorhynchus 1
   explorator 1, 2, **3**, 5, 6, *32*
   gularis 1
   imerinus 1
   rufiventris 1
   rufocinereus 1, **7**, 10, *32*
   rupestris **1**, 3, 5, *32*
   saxatilis 1, 7, 8, **9**, 13, *32*
   sharpei 1
   solitarius 1, 11, **12**, *32*
monticola, *Cisticola l.* **171**
   Oenanthe 499
moreaui, *Apalis* 290
   Artisornis **290**, 292, *320*
   Artisornis m. **290**, *320*
   Bradypterus b. 64
moschi, *Apalis m.* **273**

mossamedes, *Sylvietta r.* **344**
Motacilla 92, 102, 221, 228, 254, 462, 500
   aguimp 464
motitensis, *Passer* 460
msiri, *Bradypterus b.* **64**
muelleri, *Cisticola f.* *177*, **196**, 197
muenzneri, *Cisticola c.* **143**, *160*
muhuluensis, *Apalis m.* **273**
murina, *Apalis t.* **254**
   Muscicapa a. **477**, **481**
murinus, *Melaenornis p.* **450**, *480*
murphyi, *Apalis p.* **259**
Muscicapa 432, 436, 438, 444, **461**
   adusta 442, 461, 470, 474, 475, **476**, *481*, 572, 600
   aquatica **469**, 478, *481*
   boehmi 461, *481*, **490**
   caerulescens 18, 440, **467**, 472, 474, 477, 486, 495, *496*
   cassini 469, **471**, 474, 482, 483, 486, 493, *496*
   comitata 469, 472, 483, **485**, 487, *496*
   dauurica 461, 477
   epulata 469, 474, **479**, 483, 486, 487, 493, *496*
   gambagae 461, 462, 464, **466**, 470, 477, *481*
   infuscata 439, 461, 474, 475, *481*, **487**, 489
   itombwensis 475
   lendu 461, 473, 474, **475**, *481*
   olivascens 461, 472, **473**, 475, *481*, 483
   randi 461
   segregata 461
   sethsmithi 474, **482**, 485, 487, *496*
   striata 137, 395, 454, 456, 461, **462**, 466, 474, 477, *481*, 491, 502, 535
   tessmanni 472, 485, **486**, *496*
   ussheri 461, 474, *481*, 487, **488**, **489**
Muscicapi 508, 548
Muscicapidae xii, 1, 57, **432**, 498, 508, 548, 574
Muscicapinae 432, 548
Muscicapini 432
Muscipeta 537, 542
Musicapa 449
musicus, *Bias* 315, 497, 535, 549, 551, 603
   Bias m. **497**, **551**
mutata, *Terpsiphone* 530, 531, 537
mutatrix, *Prinia s.* **222**
mwaki, *Turdus o.* 35

Myadestes 1
Myiagra 514
Myiagridae 508
Myioparus 432, 468, **492**, 498
   griseigularis 315, 472, 483, **492**, *495*, *496*, *520*, *556*
   plumbeus 281, *492*, *493*, **494**, *496*, 498
Myiophonus 1
Myopornis 461
mystacea, Sylvia 132, 377, 380, 382, 392, 397, 398, 399, **403**, 405, 408, *416*
   Sylvia m. **403**, *416*

N

naevia, Locustella 92, 95, 97
   Locustella n. **92**, *97*
   Motacilla 92
nakuruensis, Cisticola b. **211**, 240
namaqua, Cisticola s. **169**
namaquensis, Melaenornis i. **452**
namba, Cisticola l. **171**
nana, Curruca 393
   Sylvia 217, 377, **393**, *401*, 412
   Sylvia n. **394**, *401*
nanus, Cisticola 138, 166, *177*, **190**, 192
narina, Apaloderma 554
natalensis, Chloropeta 113, **123**, 127
   Chloropeta n. *113*, **124**
   Cisticola 138, *176*, 183, **185**
   Cisticola n. *176*, **186**
   Cossypha 27
   Drymoica 185
   Sphenoeacus a. **88**, *97*
natronensis, Prinia g. **234**
Nectarinia chalybea 230, 499
   cyanolaema 493
   olivacea 520
   violacea 36
Nectariniidae 428
neglecta, Apalis f. **263**
neglectus, Acrocephalus g. **121**
   Calamonastes u. **305**
nehrkorni, Hyliota v. *337*, **428**
nemorivaga, Sylvietta w. 340
Neocossyphus 1
   fraseri 315, 520
neumanni, Hemitesia **58**, *337*, 429, 522
   Melaenornis m. **454**
   Muscicapa s. **463**, *481*
   Sylvietta 58
   Terpsiphone r. **543**, *544*
newtoni, Cisticola s. **169**
niassae, Apalis f. **263**
   Monticola a. 6
nigeriae, Erythrocercus m. **510**

Melaenornis p. 450
nigrescens, Apalis r. 257, **280**
nigricans, Pinarocorys 453
nigriceps, Apalis 252, 256, 264, 266, 268, **271**, 276, 332, 601
   Apalis n. 256, **271**
   Dryodromas 271
   Terpsiphone r. **543**
nigricollis, Ploceus 486
nigriloris, Cisticola 138, 150, 153, **154**, *160*, 165
nigrilorum, Turdus p. 36, **39**, 48
nigrita, Hirundo 473
nigrodorsalis, Apalis m. **273**
nigromitrata, Elminia 497, 514, 519, 522, 528, 547
   Elminia n. *497*, **519**
nigromitratus, Terpsiphone 519
nigrorum, Muscicapa c. **468**
Nilaus afer 425, 593
niloticus, Acrocephalus r. 119
   Cisticola e. **141**
nimbae, Muscicapa o. **474**
nisoria, Motacilla 377
   Sylvia 134, 377, **384**, 390, 398, *416*
nitens, Lamprotornis 537
   Trochocercus 315, 520, 522, 525, 526, **527**, *544*, 547, 556
   Trochocercus n. **528**, *544*
njombe, Cisticola 138, 155, *161*, 165, 172
   Cisticola a. 165
   Cisticola n. *161*, **165**
noctua, Athene 218
noomei, Camaroptera b. **296**
norrisae, Sylvia m. **399**, *416*
nubica, Campethera 495
nubilosa, Prinia f. **226**
nuchalis, Cisticola r. *176*, **183**
nyansae, Batis m. 595
   Cisticola g. **174**
   Platysteira c. **568**
nyasa, Cisticola e. **141**
nyassae, Bradypterus c. **74**, *96*
nyika, Cisticola a. **157**
nyikae, Turdus o. 35
nyikensis, Melaenornis f. **440**, *480*

O

oberlaenderi, Geocichla g. 15
   Zoothera **15**, 17, 20, 26, *33*
obscura, Muscicapa a. **477**, *481*
   Prinia b. **232**
obscurior, Locustella n. **92**
occidens, Cisticola e. **208**, *240*
occipitalis, Eremomela s. **328**
occulta, Batis p. *561*, **602**

occultus, Batis 602
ochraceiceps, Phylloscopus r. 364
ochrocara, Sylvietta r. 344
ochrogularis, Phylloscopus r. **364**, 400
ochruros, Phoenicurus 8, 593
ocreata, Fraseria **432**, 436, 442, *481*, 520
   Fraseria o. **433**, *481*
ocreatus, Tephrodornis 432
ocularius, Malcorus p. **287**
Oenanthe 13, 395
   deserti 408
   finschii 408
   leucura 412
   monticola 499
   oenanthe 11
   xanthoprymna 395
oenanthe, Oenanthe 11
okuensis, Muscicapa a. **477**
oldeani, Turdus o. **36**, *48*
olivacea, Nectarinia 520
olivaceiceps, Ploceus 491
olivaceofuscus, Turdus 30, **31**, *49*
   Turdus o. **31**, *49*
olivaceus, Turdus 14, 21, 25, 26, 29, 30, 31, **34**, 38, 41, 44, *48*, 78
   Turdus o. **35**, *48*
olivascens, Calamonastes u. **306**
   Muscicapa 461, 472, **473**, 475, *481*, 483
   Muscicapa o. **473**
   Parisoma 473
olivetorum, Hippolais 113, 129, 132, **133**, 137
   Salicaria 133
omo, Cisticola r. **183**
omoensis, Phylloscopus u. **363**
opaca, Hippolais p. 113, **130**, 137
oreobates, Cisticola l. **171**
Oreophilais 58, 220, **236**
   robertsi 144, **237**, 239, *241*
oreophilus, Cisticola t. *176*, **181**
orientalis, Batis 561, 574, 584, 587, 589, 591, 592, **594**, 595, 596
   Batis o. *561*, **594**
   Cettia c. 60
   Melocichla m. **86**, *97*
   Merops 395
   Myioparus p. **495**, *496*
   Phylloscopus 359
   Phylloscopus b. **359**, *400*
   Platystira 594
Oriolus brachyrhynchus 376
orix, Euplectes 223, 230
orpheanum, Parisoma s. **418**
Orthonychinae 508, 548
Orthotomus 290, 294
   metopias 292

ortleppi, Prinia f. 227, *241*
otomitra, Zoothera g. 21
ovampensis, Prinia s. 222
Oxylophus jacobinus 419

P

Pachycephalinae 432, 508, 548
Pachyprora 578, 582
pallescens, Prinia s. 222
pallida, Curruca 129
    Hippolais 113, **129**, 132, 135, 137, 356
    Hippolais p. **113**, **130**
    Musicapa 449
    Sylvietta r. 336, **344**
pallidior, Bradypterus c. 74
    Calamonastes f. **308**
pallidipectus, Hyliota a. 425
palliditergum, Batis m. **584**
pallidiventris, Bias m. 551
pallidus, Artisornis m. 291
    Melaenornis 439, 448, **449**, 453, 454, 456, 470, *480*
    Melaenornis p. **450**, *480*
    Zosterops 36, 419
paludicola, Acrocephalus **101**, 103, *112*
    Sylvia 101
palustris, Acrocephalus 103, 105, **109**, *112*, 117, 130, 137
    Sylvia 109
pammelaina, Melaenornis 438, 445, **446**, *480*
    Melaenornis p. **447**, *480*
    Sylvia 446
panderi, Podoces 218
paradisi, Terpsiphone 530, 531, 537
pareensis, Apalis t. 254
parelii, Myioparus g. **492**
Paridae 428
Parisoma 123, 382, **414**, 492
    boehmi 401, **421**
    buryi 382, 414
    layardi 401, **414**, 418, 499
    lugens 401, 414, **420**
    plumbeum 492
    subcaeruleum 401, 414, 415, **418**, 456
Parus 422, 430, 461
    afer 419
    ater 362, 365, 372
    funereus 601
    major 410
parva, Ficedula 496, **506**
    Ficedula p. **496**, **506**
    Muscicapa 506
parvulus, Anthoscopus 593
parvus, Acrocephalus g. *112*, **121**
    Melaenornis p. **449**

Passer domesticus 135
    griseus 593
    melanurus 80
    motitensis 460
pauciguttatus, Psophocichla l. **29**
pectoralis, Malcorus 227, *241*, **287**, 309
    Malcorus p. **241**, **287**
Pedilorhynchus 461, 482, 486
pelios, Turdus 18, 30, 34, 36, **38**, 41, 44, *48*
    Turdus p. **39**
peltata, Platysteira 545, 552, 567, 568, 570, **571**, 576
    Platysteira p. **545**, **571**
    Platystira 571
pelzelni, Ploceus 471
perennia, Cisticola j. 200, 240
perimacha, Eremomela i. **322**
peripheris, Turdus l. **42**
perkeo, Batis 561, 574, 588, 594, **596**
perlatum, Glaucidium 590, 593
perplexus, Cisticola a. **205**
perpullus, Cisticola t. **181**
personata, Apalis 83, 252, 256, 261, 265, **266**
    Apalis p. **267**
Petrocincla 4
petrophila Euryptila s. **309**
petrophilus, Cisticola a. **157**
Philentoma 548
philippae, Sylvietta 322, 324, 335, *336*, 338, 339, 340, **342**, 346
Phillopneuste 360
philomelos, Turdus 26, 28, 29, 40, **45**, 47, *49*, 50, 52, 54, 56
    Turdus p. **45**, *49*
Phlexis 70
phoenicuroides, Phoenicurus p. 8
Phoenicurus 1, 395
    ochruros 8, 593
    phoenicurus 8, 11, 70, 502
phoenicurus, Phoenicurus 8, 11, 70, 502
Phragmacia **219**, 220
    substriata **219**, 229, *241*
Phyllastrephus 155, 484, 529
    debilis 317
Phyllolais 250
    pulchella **250**, *320*
Phylloscopus 123, 125, 127, 137, 250, 284, 328, **352**, 469, 507, 512
    bonelli **358**, *400*
    budongoensis 352, 366, 367, **368**, *400*
    collybita 130, 218, 243, 353, **355**, 359, 360, 362, 363, 369, *400*

    fuscatus **360**
    herberti 352, **367**, 368, *400*
    inornatus **361**, 372, *400*
    laetus 352, 363, 364, 365, **366**, 369, *400*
    laurae 352, 364, **365**, 366, 367, *400*
    orientalis 359
    proregulus **361**
    ruficapilla 352, 357, 363, **364**, 365, 366, 367, *400*
    sibilatrix 62, 78, 353, **357**, 359, 363, 364, *400*
    trochilus 13, 124, 127, 250, 339, **352**, 355, 356, 357, 359, 362, 363, 368, *400*
    umbrovirens 356, **362**, 366, *400*
Phyllostrephus 317
piaggiae, Turdus 23
    Zoothera 16, 20, 21, **23**, *33*
    Zoothera p. **23**, *33*
Pica pica 218, 371
pica, Pica 218, 371
pictipennis, Cisticola c. **143**
pilaris, Turdus 47, 49, 50, 52
pileata, Camaroptera b. **296**
pilettei, Zoothera c. **20**, *33*
Pinarocorys nigricans 453
pintoi, Batis m. **584**
    Calamonastes u. **306**
pipiens, Cisticola 138, 174, *176*, **178**, 181
    Cisticola p. **179**
placidus, Melaenornis i. **452**
Platysteira 548, 555, **567**
    albifrons 545, 567, 568, **570**, 571, 572
    cyanea 545, 563, **567**, 570, 571, 572, 576
    laticincta 567, 572
    peltata 545, 552, 567, 568, 570, **571**, 576
Platysteiridae xii, 422, 508, **548**, 574
Platysteirinae 432, 508, 548
Platystira 564, 570, 571, 594, 597, 598
platyura, Schoenicola 86, **89**, *97*
    Thimalia 89
Ploceidae 428
Plocepasser mahali 457, 460
    rufoscapulatus 491
    superciliosus 250
Ploceus 84, 120, 265
    baglafecht 478
    castanops 471
    cucullatus 486
    jacksoni 471
    melanocephalus 376, 471
    melanogaster 486

Ploceus (continued)
  nigricollis 486
  olivaceiceps 491
  pelzelni 471
  sanctithomae 538
plumbea, Stenostira 494
plumbeiceps, Terpsiphone v. 533, 539, 543, *544*
plumbeum, Parisoma 492
plumbeus, Myioparus 281, 492, 493, **494**, 496, 498
  Myioparus p. **494**
Podoces panderi 218
poensis, Batis 561, 574, 592, 599, **601**
  Batis p. 561, **602**
  Hylia p. **429**
  Muscicapa a. 477
  Turdus p. 36, **39**
Pogoniulus 245, 262
  atroflavus 556
Pogonocichla 364
poliocephala, Alethe 18, 313
poliogyna, Melaenornis p. 447
Poliolais 58, **292**
  lopesi **292**, *337*
polioxantha, Eremomela i. 318, 319, *321*
Polyboroides typus 27
polyglotta, Hippolais *113*, 130, **134**, 137
  Sylvia 134
pondoensis, Bradypterus s. **78**
  Prinia s. **222**
  Turdus o. 35
porphyrolaema, Apalis 252, *257*, 269, 272, 273, **275**, 276, 280, 284, 286
  Apalis p. *257*, **275**
prasina, Hylia 59, *337*, **428**
  Hylia p. *337*, **429**
  Sylvia 428
pretoriae, Monticola b. **5**, *32*
priesti, Bradypterus b. **76**, *96*
prigoginei, Parisoma l. **420**
princei, Chamaetylas 18
  Zoothera 16, 17, **18**, 20, *33*, 40
  Zoothera p. **18**, *33*
  Prinia 57, 145, 215, 216, 219, **220**, 236, 238, 242, 244, 245, 247, 250, 252, 286, 288, 527
  bairdii 221, **232**, *241*, 247, 311
  flavicans 221, 222, **226**, 229, *241*, 287
  fluviatilis 221, 222, **224**, 225, *241*
  hypoxantha 229
  gracilis 200, 217, 221, 226, **233**, *241*

  inornata 221, 224, 225
  maculosa 219, 221, 222, 227, **228**, *241*, 419
  molleri 221, **230**, *241*
  somalica 221, 222, 224, **225**, 234, *241*
  subflava 141, **221**, 224, 225, 227, 229, 234, 237, 239, *241*, 245, 289
prinioides, Cisticola h. 153
Prionopinae 574
Prionops 548, 604
pririt, Batis 419, *560*, 574, 580, 584, 587, **589**, 591, 594
  Batis p. *560*, **589**
  Muscicapa 589
procerus, Cisticola c. 161, **162**
Prodotiscus insignis 573
  regulus 158, 169, 197, 198, 230, 299
proregulus, Motacilla 361
  Phylloscopus 361
  Phylloscopus p. 361
prosphora, Fraseria o. **433**, *481*
Prunella modularis 61, 413
psammocromius, Euplectes 155
psammophila, Prinia m. 229
Pseudobias 548
Psophocichla xii, **28**
  litsitsirupa 26, **28**, *33*, 45
puella, Batis 587
  Batis m. **584**
puellula, Eremomela i. **322**
pugnax, Apalis f. **263**
pulchella, Malurus 250
  Phyllolais **250**, *320*
pulchra, Apalis 252, 254, **256**, **259**, 260, 261, 351
  Apalis p. **256**, **259**
  Eremomela s. *321*, **328**
pulitzeri, Macrosphenus 311, 314, **316**, *337*
pumila, Muscicapa a. 476, **477**
pumilus, Melaenornis 454
  Melaenornis m. **454**, *480*
pusilla, Eremomela 318, *321*, 322, **325**, 326, 328, 329, 593
Pycnonotidae 58
Pycnonotus 13
  atriceps 375
  barbatus 125, 520, 565
  melanicterus 375
pycnopygius, Achaetops xii, 58, **97**
Pycnosphrys 509
Pygarrhichas albogularis 430
pyrrhomitrus, Cisticola e. **141**
Pyrrhula pyrrhula 10
pyrrhula, Pyrrhula 10
Pytilia melba 593

Q

quarta, Apalis t. **255**
quelimanensis, Phylloscopus r. **364**
quiscalina, Campephaga 554

R

raineyi, Zoothera g. **21**
Ramphocaenus 430
randi, Muscicapa 461
Regulidae 58
regulorum, Balearica 478
Regulus 361, **369**
  calendula 315
  ignicapillus **370**, **372**, *400*
  regulus 370, **371**, *400*
regulus, Motacilla 371
  Prodotiscus 158, 169, 197, 198, 230, 299
  Regulus 370, **371**, *400*
  Regulus r. **371**, *400*
reichenowi, Batis 580
  Batis c. *560*, **580**
  Batis m. 580
  Cisticola b. **191**
  Melaenornis c. **442**
  Schistolais l. *241*, **247**
  Trochocercus n. **528**
reiseri, Hippolais p. **130**
renata, Apalis f. **263**
rensi, Eremomela u. **331**
restricta, Terpsiphne v. **533**
restrictus, Cisticola 138, *161*, **166**
resurga, Sylvietta r. **344**
Rhipidura 508, 514
Rhipidurinae 432, 508, 548
Rhipidurini 508
rhodesiae, Apalis t. **255**
  Hyliota a. 425
rhodoptera, Heliolais e. **239**
robertsi, Oreophilais 144, 237, 239, *241*
  Prinia 237
robusta, Cisticola 184
  Drymoica 182
robustus, Cisticola 61, 138, 155, 157, *176*, **182**, 185, 186
  Cisticola r. *176*, **182**
roehli, Bradypterus l. **72**
  Muscicapa a. 477
  Turdus o. **35**, 48
rowei, Zoothera p. **24**
ruandae, Muscicapa a. 470
rubecula, Erithacus 413, 436, 503
rubescens, Sylvia m. **404**
rubicola, Sylvia c. **389**
rubriceps, Anaplectes 448, 460, 491
ruddi, Apalis 252, 255, 256, **261**, 264

Apalis r. 256, **261**
rudebecki, Stenostira s. **499**
rueppelli, Sylvia 377, 380, **396**, 398, 402, *416*
rufa, Drymoica 193
rufescens, Acrocephalus 64, 99, 108, *112*, **119**, 121, 128
   Acrocephalus r. *112*, **119**
   Calamocichla 119
   Dicaeum 343
   Sylvietta 125, 335, *336*, 340, 343, 345, 347, *419*, 425
   Sylvietta r. *336*, **343**
ruficapilla, Phylloscopus 352, 357, 363, **364**, 365, 366, 367, *400*
   Phylloscopus r. **364**
   Pogonocichla 364
   Sylvietta 335, *336*, 340, **346**, 351
   Sylvietta r. *336*, **346**
ruficapillus, Cisticola f. *177*, **196**
ruficeps, Cisticola 138, *177*, **187**, 189, 194, 195
   Cisticola r. *177*, **188**
ruficollis, Tachybaptus 537
   Malurus 187
rufidorsalis, Urorhipis r. **289**
rufifrons, Apalis 244
   Prinia 288
   Urorhipis 234, 244, 245, **288**, *320*
   Urorhipis r. **289**, *320*
rufigenis, Sylvietta r. *336*, **347**
rufilata, Drymoica 167
rufilatus, Cisticola 138, *161*, **167**, 170, 189
   Cisticola r. *161*, **167**
rufiventer, Muscipeta 542
   Terpsiphone 533, 539, 540, **542**, *544*, 546, 547
   Terpsiphone r. **543**, *544*
rufiventris, Monticola 1
rufocinerea, Terpsiphone 533, **538**, 540, 543, *544*
   Terpsiphone v. 539
rufocinereus, Monticola 1, 7, 10, *32*
   Monticola r. **8**, *32*
   Saxicola 7
rufoflavidus, Bradypterus c. 74
rufogularis, Apalis 252, 257, 269, 276, 278, **279**, 281, 282, 283, 284, 286, 334
   Apalis r. 257, **279**
   Drymoica 279
rufoscapulatus, Plocepasser 491
rufus, Bathmocercus 81, **82**, 84, *97*
   Bathmocercus r. **83**
   Cisticola 138, *177*, 188, 192, **193**, 194, 195

rupestris, Monticola 1, 3, 5, *32*
   Turdus 1
ruppeli, Sylvia 396
rustica, Hirundo 394
ruthae, Fraseria c. **435**
ruwenzorii, Apalis 252, 254, 256, 259, **260**, 267
   Zoothera p. **24**, *33*

S

sabini, Dryoscopus 549
saharae, Scotocerca i. **217**, *241*
Salicaria 73, 133
Salpornis spilonotus 425
salvadorii, Bradypterus c. 74
   Eremomela 318, *321*, 322, 323, **324**
sanctithomae, Ploceus 538
sanderi, Apalis r. 257, **279**
santae, Cisticola r. **183**
sarda, Sylvia 377, 411, **412**, *417*
   Sylvia s. **413**, *417*
sarmatica, Locustella l. **95**
saturatior, Eremomela i. *321*, **322**
   Stenostira s. **499**
saturatus, Turdus p. 18, 36, **39**, 48
saxatilis, Monticola 1, 7, 8, **9**, 13, *32*
   Turdus 9
Saxicola 7, 451
   torquata 155, 211, 412
Saxicolini 432
Scepomycter winifredae 81
schillingsi, Cisticola c. *161*, **166**
schistaceus, Melaenornis e. **445**, *480*
Schistolais 220, **245**
   leontica *241*, **246**, 248
   leucopogon *241*, **246**, 248
schleglii, Cercomela 453
Schoenicola **89**
   brevirostris 89
   platyura 86, **89**, *97*
schoenobaenus, Acrocephalus 92, 100, **102**, 106, 110, *112*, 125
   Motacilla 102
schoutedeni, Cisticola g. **174**
   Phylloscopus l. **366**
   Sylvietta r. **347**
schraderi, Cisticola r. **183**
schubotzi, Terpsiphone r. **543**
scirpaceus, Acrocephalus 86, 95, 99, 103, **104**, 107, 108, 110, *112*, 115, 117, 130
   Acrocephalus s. **105**, 110, *112*
   Turdus 104
scita, Muscicapa 498
   Stenostira 497, **498**
   Stenostira s. 497, **498**

sclateri, Apalis c. **284**
   Monticola r. 8
Scoptelus castaneiceps 601
Scotocerca **216**
   inquieta **216**, 234, *241*
scotops, Eremomela 242, 318, *321*, 322, 325, 326, 327, **328**, 329, 331, 335, 425
   Eremomela s. *321*, **328**
scotopterus, Cisticola r. *177*, **188**
segregata, Muscicapa 461
segregus, Trochocercus c. **526**
Seicercus 123, 352, 365
seimundi, Melaenornis i. **452**
semicinctus, Melaenornis f. **440**
semifasciatus, Cisticola l. *161*, **171**
semipartita, Muscicapa 459
semipartitus, Empidornis **459**, *480*
semirufa, Cossypha 36
semitorquata, Alcedo 537
   Ficedula 496, 500, 503, **505**
   Muscicapa 505
senegalensis, Acrocephalus r. **119**
   Batis 424, 561, 574, 584, 587, 589, **591**, 594, 602
   Dryoscopus 549
   Muscicapa 591
   Streptopelia 38, 41, 43
   Zosterops 341
septentrionalis, Cisticola b. **149**
Serinus albogularis 419, 465
sethsmithi, Muscicapa 474, **482**, 485, 487, 496
seth-smithi, Pedilorhynchus e. 482
sharpei, Camaroptera b. **296**, *320*
   Eremomela i. 322
   Monticola 1
sharpii, Apalis 252, 257, 264, **278**, 282
sheppardi, Batis f. 583
shiwae, Cisticola t. **181**
Sialia 1
sibilans, Melaenornis p. **450**
sibilatrix, Motacilla 357
   Phylloscopus 62, 78, 353, **357**, 359, 363, 364, *400*
sibirica, Ficedula h. 501
Sigelus 438
signata, Cercotrichas 22, 27
silberbaueri, Cisticola f. **196**
silens, Lanius 457
   Melaenornis **457**, *481*
   Melaenornis s. **458**, *481*
silvae, Dyaphorophyia c. 565
simensis, Psophocichla l. **29**, *33*
similis, Chloropeta 113, 123, 124, **126**
simplex, Calamonastes **303**, 305, 306, 307, 310, *320*
   Cisticola c. **159**
   Thamnobia 303

Sitta 314
   europaea 424
sjöstedti, Bradypterus l. 71
slatini, Hyliota a. *337*, **425**
smithersi, Cisticola c. **162**
smithi, Turdus o. **35**, *48*
   Urorhipis r. **289**
smithii, Terpsiphone r. **543**, *544*
Smithornis capensis 376, 549
sola, Batis c. **580**
solitarius, Cuculus 2, 38, 43, 537
   Monticola 1, 11, **12**, *32*
   Monticola s. **13**, *32*
   Turdus 12
somalica, Burnesia 225
   Prinia 221, 222, 224, **225**, 234, *241*
   Prinia s. **225**, *241*
somalicum, Parisoma b. **421**
somaliensis, Batis o. **594**
   Sylvia l. **382**, *416*
somereni, Terpsiphone r. **543**
sordida, Cercomela 154
soror, Batis 560, 574, 578, 580, 583, 584, **587**, 589, 591, 594, 596, 597
   Batis m. 588
   Batis p. 587
sousae, Artisornis m. **290**, *320*
sparsa, Anas 476
speciosa, Terpsiphone v. **532**, 539
speculigera, Ficedula h. **496**, **501**, 503, 505
spelonkensis, Apalis t. **255**
Spermophaga haematina 520
Sphenoeacus 85, **87**
   afer 62, 79, 86, **87**, *97*
spilonotus, Salpornis 425
spiloptera, Zoothera 15
Spiloptila **244**, 288
   clamans **244**, 289, *320*
Sporopipes frontalis 593
squamiceps, Turdoides 395
stagnans, Cisticola g. **174**
Stenostira 58, 432, 494, **498**
   scita *497*, **498**
stentorea, Curruca 118
stentoreus, Acrocephalus 98, *112*, **115**, **118**
   Acrocephalus s. *112*, **118**
stierlingi, Calamonastes 303
   Calamonastes s. **306**
   Calamonastes u. **305**, *320*
   Psophocichla l. **29**
Stiphrornis erythrothorax 520
stormsi, Turdus p. **39**, 42, *48*
straminea, Locustella n. **92**
strangei, Cisticola n. **176**, **186**
strausae, Apalis c. **257**, **277**
Streptopelia senegalensis 38, 41, 43

striata, Motacilla 462
   Muscicapa 137, 395, 454, 456, 461, **462**, 466, 474, 477, *481*, 491, 502, 535
   Muscicapa s. **463**, *481*
stronachi, Apalis k. **281**
stuhlmanni, Muscicapa c. 485
sturmii, Ixobrychus 537
Sturnidae 432
Sturnus vulgaris 13, 47
suahelica, Terpsiphone v. **533**
suahelicus, Acrocephalus b. **108**
   Batis m. **595**
   Cisticola g. **174**, *176*
Suaheliornis 311
subadusta, Muscicapa a. **477**, *481*
subalaris, Melaenornis p. **450**, *480*
subcaerulea, Elminia a. **524**
   Sylvia 418
subcaeruleum, Parisoma 401, 414, 415, **418**, 456
   Parisoma s. **401**, **418**
subcinnamomea, Drymoica 309
   Euryptila **309**, *337*
   Euryptila s. **309**, *337*
subflava, Motacilla 221
   Prinia 141, **221**, 224, 225, 227, 229, 234, 237, 239, *241*, 245, 289
   Prinia s. **222**, *241*
subrufa, Terpsiphone v. 533
subruficapilla, Cisticola 163
   Drymoica 169
subruficapillus, Cisticola 138, 161, 168, **169**, 172, 419
   Cisticola s. **161**, **169**
subsolana, Parisoma l. **415**
substriata, Drymoica 219
   Phragmacia **219**, 229, *241*
   Phragmacia s. **219**, *241*
subtilis, Muscicapa a. 477
sudanensis, Bradypterus b. **64**
superciliaris, Camaroptera **294**, **299**, 302, *320*
   Sylvicola 299
superciliosus, Centropus 38
   Plocepasser 250
swanzii, Cisticola c. **143**, *160*
swynnertoni, Elminia a. **524**
   Turdus o. **35**
sylvaticus, Bradypterus 61, 76, 77, *96*
   Bradypterus s. **78**, *96*
Sylvia 58, 123, 132, **376**, 414, 415, 418, 420
   althaea 377, 392
   atricapilla 377, 380, 382, 384, **385**, 394, *401*
   borin 315, 378, **383**, 386, *401*, 493
   buryi 382

   cantillans 377, 389, 392, 397, 402, **404**, 408, 410, 411, *417*
   communis 339, 377, 378, 382, 384, **388**, 392, 394, 396, 398, *401*, 404, 405, 407, 411
   conspicillata 218, 377, 389, 406, **407**, 410, *417*
   curruca 130, 356, 377, 380, 390, **392**, 396, *401*, 406, 413
   deserticola 377, 405, 407, **409**, 411, *417*
   hortensis 377, 378, **379**, 381, 382, 397, 414, *416*
   leucomelaena 132, 377, 379, 380, **381**, 404, 414, *416*
   melanocephala 377, 380, 392, 397, **398**, 403, 404, 406, 408, 413, *416*
   melanothorax 377, **397**, 398, 402, 403, *416*
   minula 377, 392
   mystacea 132, 377, 380, 382, 392, 397, 398, 399, **403**, 405, 408, *416*
   nana 217, 377, **393**, *401*, 408, 412
   nisoria 134, 377, 384, 390, 398, *416*
   rueppelli 377, 380, **396**, 398, 402, *416*
   sarda 377, 411, **412**, *417*
   undata 377, **410**, 412, *417*
sylvia, Cisticola e. **141**
Sylvicola 299
Sylviella 349, 350
Sylvietta 58, 318, 322, **335**, 429
   brachyura 322, 335, *336*, **338**, 340, 342, 346, 350, 593
   denti 335, *336*, 348, **349**, 520, 600
   isabellina 335, *336*, 339, 342, 343, **345**
   leucophrys 335, *336*, 348, **350**, 369, 429
   philippae 322, 324, 335, *336*, 338, 339, 340, **342**, 346
   rufescens 125, 335, *336*, 340, **343**, 345, 347, 419, 425
   ruficapilla 335, *336*, 340, 344, **346**, 351
   virens 333, 335, *336*, **347**, 349, 351
   whytii 335, *336*, 338, 339, **340**, 342, 344, 346, 347
Sylviinae 58, 508, 548
Sylviidae xii, 57, 310, 428, 430, 498, 508, 548
syriacus, Turdus m. **54**

T

Tachybaptus ruficollis 537
tachiro, Accipiter 434
taciturnus, Cisticola c. 212
tando, Sylvietta v. 336, **348**
tanganjicae, Zoothera 15
   Zoothera p. 16, **24**, *33*
tanganyika, Cisticola a. **205**, 240
taruensis, Melaenornis m. **454**
Tchagra australis 554
Tchitrea 530
teitensis, Cisticola a. 157
Telophorus zeylonus 419
tenebricosa, Apalis m. **273**
tenebricosus, Cisticola b. 192
tenebriformis, Monticola e. **3**
tenella, Prinia s. **222**
tenerrima, Apalis f. 263
Tephrodornis 432, 548
tephrolaema, Anthreptes 601
tephronotus, Turdus 30, 34, 36, 38, 41, **44**, *49*
teresita, Elminia l. *497*, **515**
Terpsiphone 124, 440, 508, 520, 525, 527, 528, **530**
   atrocaudata 530, 531, 537
   atrochalybeia 530, 531, **537**, *544*
   batesi 315, 520, 533, 539, **540**, 543, *544*, 547, 556
   bedfordi 528, 543, **544**, **546**
   bourbonnensis 530, 531, 537
   corvina 530, 531, 537
   mutata 530, 531, 537
   paradisi 530, 531, 537
   rufiventer 533, 539, 540, **542**, *544*, 547
   rufocinerea 533, **538**, 540, 543, *544*
   viridis 84, 464, 530, **531**, 537, 538, 540, 543, *544*, 547
terrestris, Cisticola j. **200**, 240
territinctus, Melaenornis m. **456**
Tesia 58
tessmanni, Muscicapa 472, 485, **486**, *496*
   Pedilorhynchus 486
textrix, Cisticola 138, **206**, 214, 240
   Cisticola t. **206**, 240
   Sylvia 206
Thamnobia 303
theresae, Scotocerca i. **217**
Thimalia 89
thomsoni, Erythrocercus l. *497*, **513**
thoracica, Apalis 252, **254**, **256**, 259, 260, 261, 262, 264, 277, 286
   Apalis t. **255**

Motacilla 254
ticehursti, Sylvia d. **410**
Timaliidae xii, 57, 310, 430
Timaliinae 508, 548
tincta, Camaroptera b. **295**, *320*
tinniens, Cisticola 138, 174, *176*, 179, **180**
   Cisticola t. *176*, **181**
   Malurus 180
Tockus fasciatus 41, 315, 550, 554
tomensis, Ficedula h. **501**
tonga, Cisticola n. **186**
tongensis, Bradypterus b. **64**
toni, Sylvia u. **411**, *417*
tonsa, Dyaphorophyia 545, **555**, 556, **558**, *563*, 565, 576, 599
toroensis, Camaroptera c. **302**, *320*
   Elminia a. **522**
torquata, Saxicola 155, 211, 412
torquatus, Lanioturdus *497*, **604**
   Turdus 30, 47, 49, 51, 52, **55**
   Turdus t. 49, **56**
torquilla, Jynx 6
toruensis, Melaenornis f. **440**, *480*
toussenellii, Accipiter t. 434
transitiva, Camaroptera b. **296**
transvaalensis, Bradypterus b. **64**
   Turdus o. 35
traylori, Cisticola a. **205**
tricolor, Terpsiphone r. **543**
tristis, Phylloscopus c. **355**
trivialis, Anthus 491
trochilus, Motacilla 352
   Phylloscopus 13, 124, 127, 250, 339, **352**, 355, 357, 359, 362, 363, 368, *400*
   Phylloscopus t. **353**, *400*
Trochocercus 440, 498, 508, 513, 514, 518, 524, **525**
   cyanomelas 524, 525, **526**, 527, 528, *544*
   nitens 315, 520, 522, 525, 526, **527**, *544*, 547, 556
Troglodytes troglodytes 60, 408, 410
troglodytes, Cisticola 138, *177*, 188, 192, 193, **194**
   Cisticola t. *177*, **194**
   Drymoica 194
   Troglodytes 60, 408, 410
tropicalis, Melaenornis p. **447**
   Turdus l. **42**
tsanae, Acrocephalus g. **121**
turcmenica, Sylvia m. **404**
Turdidae xii, 57
Turdinae 1, 432, 548
Turdoides 375
   jardineii 448

squamiceps 395
Turdus xii, 1, 3, 14, 28, **30**, 382
   bewsheri 30
   guttatus 25
   iliacus 45, **46**, *49*, 51, 52, 56
   libonyanus 7, 29, 30, 34, 36, 38, **41**, 44, *48*, 448
   menachensis 30, 34, 38, 41, 44
   merula 13, 16, 25, 30, 34, 36, 44, 47, *49*, 50, 52, **53**, 56
   olivaceofuscus 30, **31**, *49*
   olivaceus 14, 21, 25, 26, 29, 30, 31, **34**, 38, 41, 44, *48*, 78
   pelios 18, 30, 34, 36, **38**, 41, 44, *48*
   philomelos 26, 28, 29, 40, **45**, 47, *49*, 50, 52, 54, 56
   pilaris 47, *49*, 50, **52**
   tephronotus 30, 34, 36, 38, 41, **44**, *49*
   torquatus 30, 47, 49, 51, 52, **55**
   viscivorus 28, 29, 30, 45, 47, *49*, 50, 52, 54, 56
turneri, Eremomela, 318, *321*, 332, **333**
   Eremomela b. 333
   Eremomela t. *321*, **333**
typus, Polyboroides 27
tyrrhenica, Muscicapa s. **463**
Tyto alba 215

U

ufipae, Bradypterus l. **72**
   Melaenornis f. 440
ukamba, Cisticola c. **159**, 167
ultima, Batis f. **583**
   Batis m. **578**
uluguru, Apalis t. **254**, *256*
umbrovirens, Phylloscopus 356, **362**, *366*, *400*
   Phylloscopus u. **363**, *400*
   Sylvia 362
undata, Motacilla 410
   Sylvia 377, **410**, 412, *417*
   Sylvia u. 411
undosa, Drymoica 305
undosus, Calamonastes 303, 304, **305**, 307, 308, *320*
   Calamonastes u. **305**, *320*
undulatus, Melopsittacus 453
ungujaensis, Tersiphone v. **533**
Uraeginthus 265
   angolensis 199
   bengalus 593
Urolais 220, **242**
   epichlora **242**, *284*, *320*
uropygialis, Cisticola j. **200**, *240*
Urorhipis 220, 244, **288**
   rufifrons 234, 244, 245, **288**, *320*

usambarae, Bradypterus l. 72
   Hyliota 337, 422, **426**
   Hyliota a. 426
ussheri, Artomyias 489
   Muscicapa 461, 474, *481*, 487, 488, **489**
usticollis, Eremomela 318, *321*, 322, 325, 330, **331**
   Eremomela u. *321*, **331**
ustulatus, Catharus 20

V

valida, Cisticola n. 186
Vangidae 548
Vangini 548, 574
venusta, Apalis t. 255
venustula, Cisticola r. 168
verreauxi, Turdus l. 42
vicinior, Cisticola r. 168
victoria, Cisticola c. 159
victorini, Bradypterus 62, 74, **79**, 96
Vidua macroura 144, 178, 211, 294, 536
vigilax, Cisticola n. 186
vincenti, Cisticola l. 145
violacea, Hyliota 337, **427**
   Hyliota v. **427**
   Nectarinia 36
   Terpsipone v. 533
virens, Sylvietta 333, 335, *336*, 347, 349, 351
   Sylvietta v. *336*, **348**
viridiceps, Apalis f. 256, **263**
viridis, Muscicapa 531
   Terpsiphone 84, 464, 530, **531**, 537, 538, 540, 543, *544*, 547

Terpsiphone v. **532**, *544*
viriditincta, Eremomela i. 319
viscivorus, Turdus 28, 29, 30, 45, 47, *49*, **50**, 52, 54, 56
   Turdus v. **50**
vittata, Graueria 233, **310**, *337*
vivax, Trochocercus c. **526**
voelckeri, Phylloscopus r. **364**, *400*
volgensis, Sylvia c. **389**
vulgaris, Sturnis 13, 47
vulpinus, Bathmocercus r. **82**, 97
vulpiniceps, Cisticola c. 162
vulturna, Muscicapa c. **468**
vumbae, Apalis c. 257, **274**

W

wahlbergi, Aquila 38
wambera, Cisticola b. **211**, 240
whitei, Apalis t. 254
whytii, Sylvietta 335, *336*, 338, 339, 340, 342, 344, 346, 347
   Sylvietta w. *336*, **340**
wilhelmi, Phylloscopus u. 363
williamsi, Phylloscopus u. 363
wilsoni, Bradypterus b. 76
windhoekensis, Cisticola s. **169**
winifredae, Artisornis 81
   Bathmocercus **81**, 97
   Scepomycter 81
winneba, Cisticola e. **208**
winterbottomi, Acrocephalus g. **121**
woodwardi, Sylvia b. **384**, *401*
woosnami, Cisticola 138, **146**, 148, 149, 151, *160*
   Cisticola w. **147**, *160*

X

xantholophus, Mesopicos 601
xanthoprymna, Oenanthe 395
xanthorhynchus, Turdus o. **31**, *49*

Y

yakutensis, Phylloscopus t. **353**, 400
youngi, Apalis t. 254
   Bradypterus l. 71

Z

zalingei, Cisticola g. **173**
zarudnyi, Acrocephalus a. **112**, 114
zedlitzi, Cisticola b. **191**
zeylonus, Telophorus 419
Zoothera xii, 1, **14**, 28, 30
   cameronensis **16**, 20, *33*
   crossleyi 15, **19**, 20, *33*
   gurneyi 15, 16, 19, **20**, 24, 26, *33*, 36
   guttata 15, **25**, 28, 29, *33*
   kibalensis 15, 17
   oberlaenderi **15**, 17, 20, 26, *33*
   piaggiae 16, 20, 21, **23**, *33*
   princei 16, 17, **18**, 20, *33*, 40
   spiloptera 15
   tanganjicae 15
Zosteropidae 58
Zosterops 322, 328, 331, 351
   pallidus 36, 419
   senegalensis 341
zuluensis, Acrocephalus g. 121

English Names

A

Accentor, Hedge 60, 413
Alethe, Brown-chested 18, 313
Apalis, Bamenda 257, **281**
   Bar-throated **254**, *256*, 261, 262, 264, 277, 286
   Black-capped *256*, 264, 266, 268, **271**, 276, 332
   Black-collared *256*, **259**, 261, 351
   Black-headed 257, 270, **272**, 274, 276, 277, 284, 578
   Black-throated *256*, 266, **267**, 271, 276, 284
   Brown-headed 257, 258, 277, 280, 284, **285**

Buff-throated *257*, 269, 276, 278, **279**, 282, 283, 284, 286, 334
Chapin's *257*, 258, 273, **276**, 286, 292
Chestnut-throated *257*, 269, 272, 273, **275**, 277, 280, 284, 286
Chirinda *257*, 273, **274**
Gosling's *257*, 280, **282**, 315
Grey 243, *257*, 269, 273, 276, 280, 282, **283**, 285
Karamoja *257*, **281**
Masked *256*, **265**, 267, 268, 351
Mountain Masked *256*, 261, **266**

Rudd's 255, *256*, **261**, 263
Rwenzori *256*, **260**, 267
Sharpe's *257*, **264**, *278*
Yellow-breasted 255, *256*, **262**, 341
White-winged *256*, **269**, 277

B

Babbler, Arabian 395
   Arrow-marked 448
Batis, Angola *561*, **597**
   Bioko *561*, 592, 599, **601**
   Black-headed *561*, 576, 578, 585, 588, 592, 594, **595**, 597, 598, 602

Cape 78, 419, *560*, 572, 578, **579**, 583, 586, 588, 590
Chinspot 297, *560*, 576, 577, 578, 580, 583, **584**, 588, 590, 596, 597, 598
East Coast *560*, 578, 580, 583, 584, **587**, 596, 597
Forest *560*, **578**, 580, 588, 596
Grey-headed *561*, 592, **594**, 596, 597
Ituri *561*, 576, **600**
Lawson's 601
Margaret's *560*, **576**
Mozambique 587
Pririt 419, *560*, 580, 584, **589**
Pygmy *561*, 588, 594, **596**
Rwenzori *561*, 575, 601
Senegal 424, *561*, **591**, 602
Verreaux's *561*, **598**, 602
Woodwards' *560*, 580, **582**, 588
Bee-eater, European 419
Little Green 395
Bishop, Cape 154
Red 223, 230
Yellow 230
Bittern, Dwarf 537
Blackbird, Eurasian 13, 16, 25, 34, 36, 44, 47, *49*, 50, 52, **53**, 56
Somali 36
Blackcap 380, 382, 384, **385**, 394, *401*
Bokmakierie 419
Boubou, Southern 78, 478
Tropical 573
Bristle-bill, Green-tailed 376
Broadbill, African 376, 549
Brubru 425, 593
Budgerigar 453
Buffalo-Weaver, White-headed 460
Bulbul, Common 125, 565
Sombre 537
Bullfinch, Eurasian 10
Bunting, Cape 419
Cirl 13
Bush-Shrike, Grey-headed 82, 573, 576, 580, 583
Buzard, Lizard 434

C

Camaroptera, Green-backed 296
Grey-backed 297
Olive-green 293, 300, **301**, 315, *320*, 563
Yellow-browed **299**, 302, *320*

Canary, White-throated 419, 465
Chaffinch 10
Chat, Karoo 453
Moorland 154
Robin- [see Robin-Chat]
Chiffchaff, Common 130, 218, 243, 353, **355**, 359, 360, 362, 363, 369, *400*
Cisticola, Aberdare *176*, 183, **184**
Ashy *161*, **166**, 167, 188
Ayres's 213
Black-backed *173*, 201, **207**, *240*
Black-lored 154, *160*, 165
Black-tailed *177*, **198**
Brown-backed 148
Boran 158, *161*, **163**, 166
Bubbling **149**, *160*
Carruthers's 174, *176*, **179**, 181
Chattering 148, 150, 151, *160*
Chirping 174, *176*, **178**, 181
Chubb's **150**, 153, *160*, 289
Churring 155, *161*, **165**, 172
Cloud **206**, 214, *240*
Croaking *176*, 183, **185**
Dambo 207, **209**, *240*
Desert 166, 188, 201, **204**, 207, 208, 234, *240*
Dorst's *177*, **189**
Fan-tailed **199**, 203, 205, 207, 208, 234, *240*
Foxy *177*, 188, 192, **194**
Grey-backed *161*, 168, **169**, 172, 419
Hunter's 151, **153**, *160*
Lazy 156
Levaillant's 174, *176*, 179, **180**
Long-tailed *177*, 196, **197**, 198
Pale-crowned 209, 210, **212**, 214, *240*
Pectoral-patch 207, 208, **210**, 213, 214, *240*
Piping 195
Rattling 146, 148, 157, **158**, *161*, 163, 165, 166, 167, 168, 172, 208, 234
Red-faced **140**, 144, *160*
Red-pate *177*, **187**, 189, 194, 195
Rock-loving **156**, *160*, 172
Rufous *177*, 188, 192, **193**, 195
Sandy 194
Short-winged 188, *177*, 190, **191**, 194, 195, 196
Siffling 191
Singing 141, **143**, 151, 155, *160*, 165
Slender-tailed 198
Socotra **203**, 216, *240*

Stout 155, 157, *176*, **182**, 185, 186
Striped 185
Tabora 197
Tana River *161*, **166**
Tinkling *161*, **167**, 170, 180, 189
Tiny 166, *177*, **190**, 192
Trilling **146**, 151, *160*
Wailing 155, *161*, 165, 167, 169, **170**
Whistling **145**, 147, 148, 151, *160*, 164, 188, 195
Winding **173**, *176*, 179, 180, 181, 183, 251
Wing-snapping 207, 208, 209, 211, **213**, *240*
Zitting 199
Cordonbleu, Blue-breasted 199
Red-cheeked 593
Coucal, White-browed 38
Crane, Grey-crowned 478
Creeper, Spotted 425
Crombec, Green 332, *336*, **347**, 349, 351
Lemon-bellied *336*, 348, **349**, 600
Long-billed 125, *336*, 340, **343**, 347, 419, 425
Northern 322, *336*, **338**, 340, 342, 346, 350, 593
Philippa's 322, 324, *336*, 339, **342**, 346
Red-capped *336*, 340, 344, **346**, 351
Red-faced *336*, 339, **340**, 344, 346, 347
Somali *336*, 339, 342, **345**
White-browed *336*, 348, **350**, 369, 429
Crossbill, Red 201
Crow, Pied 43
Cuckoo, African Emerald 184, 224, 299, 365, 573
Diederik 65, 230, 299, 309, 419, 457
Dusky Long-tailed 530
Jacobin 419
Klaas's 145, 184, 197, 224, 258, 265, 299, 323, 345, 573, 582, 587, 591
Red-chested 2, 38, 43, 537
Cuckoo-Shrike, Black 447
Grey 478, 578
Purple-throated 554

D

Dove, Laughing 38, 41, 43
Drongo, Fork-tailed 30, 341, 447, 537, 582, 593

Drongo (*continued*)
Square-tailed 442, 447
Duck, African Black 476

E

Eagle, Wahlberg's 38
Eremomela, Black-necked *321*, 325, **334**
Burnt-necked *321*, 322, 325, 330, **331**
Green-backed *321*, 322, 325, **326**, 329
Green-capped 242, *321*, 322, 325, 327, **328**, 331, 335, 425
Karoo 297, 309, *321*, 322, **329**
Rufous-crowned 271, *321*, **332**, 333, 600, 602
Salvadori's *321*, 322, **324**
Senegal *321*, 322, **325**, 327, 593
Turner's *321*, 332, **333**
Yellow-bellied **318**, *321*, 324, 325, 326, 327, 330, 331, 339, 513
Yellow-vented *321*, 322, **323**, 327, 342, 346

F

Fieldfare 47, *49*, 50, **52**
Finch, Melba 593
Firecrest **370**, 372, *400*
Flycatcher, Abyssinian Slaty **441**, *480*
African Blue *497*, **514**, 518, 520
African Dusky 442, 470, 474, 475, **476**, *481*, 572
African Grey 450, **453**, 464, *480*
Angola Slaty **438**, *480*
Ashy 18, 440, **467**, 472, 474, 477, 486, 495, *496*
Black-and-white 315, *497*, 535, 549, **551**, 603
Blue-grey 467
Blue-headed Crested 315, 520, 522, **527**, *544*, 547, 556
Blue-mantled Crested 524, **526**, 528, *544*
Böhm's *481*, **490**
Cassin's **469**, **471**, 474, 482, 483, 486, 493, *496*
Chapin's 474, **475**, *481*
Chat **451**, 456, *480*
Chestnut-capped *497*, **509**
Collared *496*, 501, **503**, 505
Dusky-blue 469, 472, 483, **485**, 487, *496*
Dusky Crested *497*, **519**, 522, 528, 547
European Pied *496*, **500**, 503, 505
Fan-tailed Flycatcher 494
Fairy *497*, **498**
Fiscal **457**, *481*
Forest- [see Forest-Flycatcher]
Gambaga 464, **466**, 470, 477, *481*
Little Grey 469, 474, **479**, 483, 486, 487, 493, *496*
Little Yellow *497*, **511**, 513
Livingstone's *497*, 511, **512**
Marico 455
Mariqua 448, 450, 453, **455**, *480*
Nimba **442**, *480*
Northern Black 442, 444, **445**, 446, *480*
Olivaceous 472, **473**, 475, *481*, 483
Pale 439, 448, **449**, 453, 454, 456, 470, *480*
Paradise- [see Paradise-Flycatcher]
Red-breasted *496*, **506**
Semi-collared *496*, 500, 503, **505**
Senegal Puff-back 591
Shrike- [see Shrike-Flycatcher]
Sooty 439, 474, 475, *481*, **487**, 489
Southern Black 445, **446**, *480*
Spotted 137, 395, 454, 456, **462**, 466, 474, 477, *481*, 491, 502, 535
Swamp 469, **478**, *481*
Tessmann's 472, 485, **486**, *496*
Tit- [see Tit-Flycatcher]
Ussher's 474, *481*, **489**
Vanga 551
Wajheir Grey 451
Wattle-eyed 571
White-bellied Crested *497*, 520, **522**, 524, 527, 528
White-eyed Slaty **439**, 442, *480*
White-tailed Blue 426, *497*, 515, **517**, 524
White-tailed Crested 275, *497*, 518, 520, 522, **523**, 527, 528, 534
Yellow-eyed Black **443**, *480*
Yellow-footed 474, **482**, 485, 487, *496*
Flycatcher-Thrush, Rufous 315
Forest-Flycatcher, Fraser's **432**, 436, 442, *481*
White-browed 18, **433**, **435**, 472, *481*

G

Goldcrest 370, **371**, *400*
Goshawk, African 434
Grassbird 87
Fan-tailed 89
Grass-Warbler, Cape 62, 79, 86, **87**, *97*
Moustached **85**, 88, *97*
Grebe, Little 537
Greenbul, Tiny 317
Ground-Thrush, Abyssinian 16, 20, 21, **23**, *33*
Black-eared **16**, 20, *33*
Crossley's **19**, *33*
Grey 16, 17, **18**, 20, *33*, 40
Oberländer's **15**, 17, 20, 26, *33*
Orange 16, **20**, 24, 26, *33*, 36
Spotted **25**, 29, *33*

H

Hadada 537
Harrier-Hawk, African 27
Hawk, Harrier- [see Harrier-Hawk]
Honeybird, Cassin's 573
Wahlberg's 158, 169, 197, 198, 230, 299
Honeyguide, Eastern 573
Greater 419
Hornbill, African Pied 41, 315, 550, 554
Hylia, Green 59, *337*, **428**
Hyliota, Mashona 424
Southern 297, *337*, **424**, 427
Usambara *337*, **426**
Violet-backed *337*, **427**
Yellow-bellied *337*, **422**, 425, 428
Yellow-breasted 422

J

Jay, Eurasian 371
Ground 218

K

Kingfisher, Half-collared 537
Kinglet, Ruby-crowned 315

L

Lark, Dusky 453
Longbill, Bocage's *337*, **430**
Grey 311, 313, **314**, 316, *337*
Kemp's **313**, 315, *337*

Kretschmer's **317**, *337*
Pulitzer's **316**, *337*
Yellow **312**, 314, 315, *337*
Longtail, Green **242**, 284, *320*

M

Magpie, Black-billed 218, 371
Moho 375

N

Neddicky 157, *177*, 190, 192, **195**, 197, 198
Nightingale 60, 373
Nuthatch, Eurasian 424

O

Oriole, Western Black-headed 376
Ouzel, Ring 47, *49*, 51, 52, **55**
Owl, Barn 215
   Little 218
   Marsh 215
Owlet, Pearl-spotted 590, 593

P

Paradise-Flycatcher, African 84, 464, **531**, 539, 540, 543, *544*, 547
   Bates's 315, 533, 539, **540**, 543, *544*, 547, 556
   Bedford's 528, 543, *544*, **546**
   Red-bellied 533, 539, 540, **542**, *544*, 547
   Rufous-vented 533, **538**, 540, 543, *544*
   São Tomé **537**, *544*
Parisoma, Banded *401*, **421**
   Brown *401*, **420**
Pipit, Tree 491
Prinia, Banded **232**, *241*, 247, 311
   Black-chested 222, **226**, 229, *241*, 287
   Graceful 217, 226, **233**, *241*
   Karoo 219, 222, 227, **228**, *241*, 419
   Pale 222, **225**, 234, *241*
   River 222, **224**, *241*
   Roberts's 144, 237
   São Tomé **230**, *241*
   Sierra Leone *241*, **248**
   Spotted 228, 229
   Tawny-flanked 141, **221**, 225, 226, 227, 229, 234, 237, 239, *241*, 245, 289
   White-chinned *241*, **246**
Puffback, Red-eyed 549
   Sabine's 549

R

Redstart, Black 8, 593
   Common 8, 11, 70, 502
Redwing 45, **46**, *49*, 51, 52, 56
Reed-Warbler, African 64, 106, **107**, 110, *112*, 128
   Basra *112*, 114, **116**, 118, 121
   Clamorous *112*, 115, **118**
   Eurasian 95, 103, **104**, 108, 110, *112*, 115, 117, 130
   Great *112*, **114**, 117, 118, 120, 121, 134, 453
   Robin, European 413, 436, 503
   Scrub- [see Scrub-Robin]
Robin-Chat, Cape 78, 458
   Red-capped 27
   Rüppell's 36
   White-browed 315
Rockjumper, Cape 80
Rockrunner *97*
Rock-Thrush, Blue 11, **12**, *32*
   Cape **1**, 3, 5, *32*
   Little **7**, 10, *32*
   Miombo 5, **6**, 8, 10, *32*
   Mountain 7, 8, **9**, 13, *32*
   Sentinel 2, **3**, 5, *32*
   Short-toed 2, 3, **4**, 7, *32*
Rush-Warbler, Grauer's 67
   Little **63**, 66, 67, 69, 76, 78, *96*, 128

S

Scrub-Robin, Brown 22, 27
   Forest 18
   White-browed 419
Scrub-Warbler, Ja River **68**, *96*
   Streaked **216**, 234, *241*
Shikra 43
Shrike, Bush- [see Bush-Shrike]
   Cuckoo- [see Cuckoo-Shrike]
   Fiscal 43, 80, 419, 458
   Great Grey 395
   Red-backed 134, 554
   Red-tailed 395
   White-tailed *497*, **604**
Shrike-Flycatcher 433, *497*, **548**, 552
Silverbird **459**, *480*
Skylark 411
Sparrow, Cape 80
   Grey-headed 593
   House 135
   Rufous 460
   Sage 309
Sparrowhawk, Little 495
Sparrow-Weaver, Chestnut-crowned 250
   Chestnut-mantled 491
   White-browed 457, 460

Sprosser 60
Starling, Cape Glossy 537
   Common 13, 47
Stonechat, Common 155, 211, 412
Sunbird, Blue-throated Brown 493
   Collared 251, 297
   Fraser's 315, 429, 510
   Lesser Double-collared 230, 499
   Orange-breasted 36
Swallow, Barn 394
   White-throated Blue 473
Swamp-Warbler, Grauer's 64, 66, **67**, *96*
   Greater 64, 108, *112*, **119**, 121, 128
   Lesser 64, 106, 108, *112*, **120**, 124, 128
   White-winged 64, **65**, 68, *96*

T

Tailorbird, African 82, 277, **291**, *320*
   Moreau's **290**, 292, *320*
Tchagra, Brown-headed 554
Thrush, African 18, 34, 36, **38**, 42, 44, *48*
   Bare-eyed 36, 40, 42, **44**, *49*
   Flycatcher- [see Flycatcher-Thrush]
   Ground- [see Ground-Thrush]
   Groundscraper 26, **28**, *33*, 45
   Gulf of Guinea **31**, *49*
   Hermit 20
   Kurrichane 7, 29, 36, 39, **41**, 44, *48*, 448
   Mistle 29, 45, 47, *49*, **50**, 52, 54, 56
   Olive 21, 25, 26, 29, **34**, 39, 42, 44, *48*, 78
   Rock 9
   Rock- [see Rock-Thrush]
   Song 26, 28, 29, 40, **45**, 47, *49*, 50, 52, 54, 56
   Swainson's 20
Tinkerbird, Red-rumped 556
Tit, Cape Penduline 322
   Coal 362, 365, 372
   Great 410
   Long-tailed 243
   Southern Grey 419
   West African Penduline 593
   Yellow-fronted Penduline 600
Tit-babbler 418
   Layard's 414

Tit-Flycatcher, Grey 281, 493, **494**, *496*
  Grey-throated 315, 472, 483, **492**, 495, *496*, 520, 556
Treerunner, White-throated 430
Trogon, Narina's 554

W

Wagtail, African Pied 464
Warbler, Acacia 250, 382
  African Sedge 63
  African Yellow *113*, **123**, 127
  Aquatic **101**, 103, *112*
  Arabian 132, 381
  Bamboo **69**, 72, *96*
  Barratt's 64, 72, 75, 78, *96*
  Barred 134, **377**, 384, 390, 398, *416*
  Black-faced Rufous **82**, *97*
  Black-headed Rufous 82, **84**, *97*
  Bleating 157, 212, 279, 280, 291, **294**, 300, 302, 315, *320*, 330, 348
  Bonelli's **358**, *400*
  Briar 144, **237**, 239, *241*
  Broad-tailed 86, **89**, *97*
  Buff-bellied **250**, *320*
  Cetti's **60**, *96*
  Chestnut-vented *401*, 415, **418**, 456
  Cinnamon Bracken 70, 72, **73**, *96*, 373
  Cinnamon-breasted **309**, *337*
  Cricket **244**, 289, *320*
  Cyprus **397**, 402, *416*
  Dartford **410**, 412, *417*
  Desert 217, **393**, *401*, 408, 412
  Dusky **360**
  Eurasian River **93**, 95, *97*
  Evergreen-Forest **70**, 74, *96*
  Fairy **498**
  Garden 315, 378, **383**, 386, *401*, 493
  Graceful 200
  Grass- [see Grass-Warbler]
  Grasshopper **92**, 95, *97*
  Grauer's 233, **310**, *337*
  Grey-capped *337*, **372**
  Icterine *113*, 124, 127, 135, **136**, 384
  Inornate 361
  Knysna 76, **77**, *96*
  Layard's *401*, **414**, 418, 499
  Marmora's 411, **412**, *417*
  Marsh 103, 105, **109**, *112*, 117, 130, 137
  Melodious *113*, 130, **134**, 137
  Ménétries's 132, 380, 382, 392, 397, 399, **403**, 405, 408, *416*
  Mountain Yellow *113*, 124, **126**
  Moustached **100**, 103, *112*
  Mrs Moreau's **81**, *97*
  Namaqua **219**, 229, *241*
  Neumann's 58, *337*, 429, 522
  Olivaceous *113*, **129**, 132, 135, 137, 356
  Olive-tree *113*, 132, **133**, 137
  Oriole 84, *337*, **375**
  Orphean 378, **379**, 382, 397, *416*
  Pallas's **361**
  Papyrus Yellow *113*, **127**
  Red-fronted 234, 245, **288**, *320*
  Red Sea 380, **381**, 404, *416*
  Red-winged 222, **238**, *241*, 249
  Red-winged Grey 239, **249**, *320*
  Reed- [see Reed-Warbler]
  Rufous-eared 227, *241*, **287**, 309
  Rüppell's 380, **396**, 398, 402, *416*
  Rush- [see Rush-Warbler]
  Sardinian 380, 392, 397, **398**, 404, 406, 408, 413, *416*
  Savi's 60, 92, **94**, *97*
  Scaly-fronted 244
  Scrub- [see Scrub-Warbler]
  Sedge 86, 92, 100, **102**, 106, 111, *112*, 125
  Socotra 204, **215**, *240*
  Spectacled 218, 389, 406, **407**, 410, *417*
  Subalpine 389, 392, 397, 402, **404**, 408, 410, 411, *417*
  Swamp- [see Swamp-Warbler]
  Thick-billed **122**
  Tristram's 405, 407, **409**, 411, *417*
  Upcher's *113*, 130, **131**, 134, 382
  Victorin's **79**, *96*
  White-tailed **292**, *337*
  White-winged 65
  Willow 13, 124, 127, 250, 339, **352**, 355, 357, 359, 362, 363, 368, *400*
  Wood 62, 78, 353, **357**, 359, 363, 364, *400*
  Woodland- [see Woodland-Warbler]
  Wren- [see Wren-Warbler]
  Yellow-browed **361**, 372, *400*
Wattle-eye, Black-throated **545**, 552, 568, **571**, 576
  Brown-throated *545*, 563, **567**, 570, 572, 576
  Chestnut *545*, **555**, 558, 563, 565, 576
  Jameson's 563, 601
  Red-cheeked 300, *545*, 556, **562**, 565
  Yellow-bellied *545*, 556, 563, **564**
  White-fronted *545*, 568, **570**, 572
  White-spotted *545*, 556, **558**, 563, 565, 576, 599
Weaver, Baglafecht **478**
  Buffalo- [see Bufffalo-Weaver]
  Golden-backed **471**
  Northern Brown-throated 471
  Olive-headed **491**
  Parasitic 142, 182, 197, 203, 206, 207, 215, 224, 228
  Red-headed 448, 460, 491
  São Tomé 538
  Scaly-fronted 593
  Slender-billed 471
  Sparrow- [see Sparrow-Weaver]
  Yellow-backed 376, 471
Wheatear, Black 412
  Desert 408
  Finsch's 408
  Mountain 499
  Northern 11
  Red-tailed 395
White-eye, Cape 36, 419
  Yellow 341
Whitethroat, Common 339, 378, 382, 384, **388**, 392, 394, 396, 398, *401*, 404, 405, 407, 411
  Lesser 130, 356, 380, 390, **392**, 396, *401*, 406, 413
Whydah, Buff-shouldered 155
  Pintailed 144, 178, 211, 294, 536
Woodland-Warbler, Black-capped **367**, *400*
  Brown 356, **362**, 366, *400*
  Laura's 364, **365**, 367, *400*
  Red-faced 363, 365, **366**, 369, *400*
  Uganda 366, **368**, *400*
  Yellow-throated 357, 363, **364**, 366, 367, *400*
Woodpecker, Nubian 495
Wren, Winter 60, 408, 410
Wren-Warbler, Barred 306, **307**, *320*
  Grey **303**, 306, *320*
  Miombo 304, **305**, 308, *320*
Wryneck, Northern 6

French Names

A

Apalis à ailes blanches 256, **269**
  du Bamenda 257, **281**
  à calotte noire 256, **271**
  cendrée 257, **283**
  de Chapin 257, **276**
  de Chirinda 257, **274**
  à col noir 256, **259**
  à collier 254, **256**
  à face noire 256, **266**
  à front roux 288, **320**
  à gorge jaune 256, **262**
  à gorge marron 257, **275**
  à gorge noire 256, **267**
  à gorge rousse 257, **279**
  de Gosling 257, **282**
  du Karamoja 257, **281**
  masquée 256, **265**
  de Rudd 256, **261**
  du Ruwenzori 256, **260**
  de Sharpe 257, **278**
  à tête brune 257, **285**
  à tête noire 257, **272**

B

Bathmocerque à capuchon 84, **97**
  à face noire 82, **97**
  de Winifred 81, **97**
Bias écorcheur 497, **548**
  musicien 497, **551**
Bouscarle à ailes blanches 65, **96**
  des bambous 69, **96**
  cannelle 73, **96**
  caqueteuse 63, **96**
  de Cetti 60, **96**
  des fourrés 75, **96**
  géante 68, **96**
  de Grauer 67, **96**
  de Knysna 77, **96**
  de Lopes 70, **96**
  de Victorin 79, **96**

C

Camaroptère barrée 307, **320**
  cannelle 309, **337**
  à dos vert 301, **320**
  du miombo 305, **320**
  modeste 303, **320**
  à sourcils jaunes 299, **320**
  à tête grise 294, **320**
Chloropète aquatique 113, **127**
  jaune 113, **123**
  de montagne 113, **126**
Cisticole des Aberdares 176, **184**
  à ailes courtes 177, **191**
  babillarde 148, *160*
  des Borans 161, **163**
  brune 210, *240*
  de Carruthers 176, **179**
  cendrée 161, **166**
  chanteuse 143, *160*
  châtain 212, *240*
  de Chubb 150, *160*
  à couronne rousse 177, **195**
  dambo 209, *240*
  du désert 204, *240*
  de Dorst 177, **189**
  à dos gris 161, **169**
  à dos noir 207, *240*
  à face rousse 140, *160*
  gratte-nuage 213, *240*
  grinçante 158, *161*
  grise 161, **167**
  de Hunter 153, *160*
  des joncs 199, *240*
  masquée 154, *160*
  murmure 149, *160*
  naine 177, **190**
  njombé 161, **165**
  paresseuse 156, *160*
  pépiante 176, **178**
  pinc-pinc 206, *240*
  plaintive 161, **170**
  à queue fine 177, **197**
  à queue noire 177, **198**
  robuste 176, **182**
  roussâtre 173, **176**
  rousse 177, **193**
  russule 177, **194**
  siffleuse 145, *160*
  de Socotra 203, *240*
  à sonnette 176, **180**
  striée 176, **185**
  du Tana 161, **166**
  à tête rousse 177, **187**
  de Woosnam 146, *160*
Couturière d'Afrique 291, **320**
  de Moreau 290, **320**
Crombec à calotte rousse 336, **346**
  à face rousse 336, **340**
  à gorge tachetée 336, **349**
  isabelle 336, **345**
  à long bec 336, **343**
  de Neumann 58, **337**
  sittelle 336, **338**
  de Somalie 336, **342**
  à sourcils blancs 336, **350**
  vert 336, **347**

D

Dromoïque vif-argent 216, *241*

E

Éminie à calotte grise 337, **372**
Érémomèle à calotte verte 321, **328**
  à cou noir 321, **334**
  à cou roux 321, **331**
  à croupion jaune 318, *321*
  à dos vert 321, **325**
  grisonnante 321, **326**
  du Karroo 321, **329**
  de Salvadori 321, **324**
  à tête brune 321, **332**
  de Turner 321, **333**
  à ventre jaune 321, **323**
Érythrocerque jaune 497, **511**
  de Livingstone 497, **512**
  à tête rousse 497, **509**

F

Fauvette d'Arabie 381, *416*
  de l'Atlas 409, *417*
  babillarde 392, *401*
  de Chypre 397, *416*
  épervière 377, *416*
  grisette 388, *401*
  des jardins 383, *401*
  à lunettes 407, *417*
  mélanocéphale 398, *416*
  de Ménétries 403, *416*
  naine 393, *401*
  orphée 379, *416*
  passerinette 404, *417*
  pitchou 410, *417*
  de Rüppell 396, *416*
  sarde 412, *417*
  de Socotra 215, *240*
  à tête noire 385, *401*

G

Gobemouche de l'Angola 438, *480*
  ardoisé 485, **496**
  argenté 459, *480*
  de Berlioz 443, *480*
  de Böhm 481, **490**
  de Cassin 471, **496**
  cendré 479, **496**
  de Chapin 475, *481*
  chocolat 441, *480*
  à collier 496, **503**
  à demi-collier 496, **505**
  drongo 445, *480*
  enfumé 481, **487**

Gobemouche (*continued*)
    fiscal 457, *481*
    de Fischer 439, *480*
    forestier 432, *481*
    de Gambaga 466, *481*
    à gorge grise 492, *496*
    gris 462, *481*
    à lunettes 467, *496*
    des marais 469, *481*
    du Marico 455, *480*
    mésange 494, *496*
    nain *496*, 506
    noir *496*, 500
    noir du Nimba 442, *480*
    olivâtre 473, *481*
    pâle 449, *480*
    à pattes jaunes 482, *496*
    à petit bec 453, *480*
    sombre 476, *481*
    à sourcils blancs 435, *481*
    sud-africain 446, *480*
    de Tessmann 486, *496*
    traquet 451, *480*
    d'Ussher *481*, 489
Graminicole à queue large 89, *97*
Grauérie striée 310, *337*
Grive du Cameroun 16, *33*
    de Crossley 19, *33*
    draine *49*, 50
    de Gurney 20, *33*
    litorne *49*, 52
    mauvis 46, *49*
    musicienne 45, *49*
    d'Oberlaender 15, *33*
    olivâtre 18, *33*
    de Piaggia 23, *33*
    tachetée 25, *33*

H

Hylia verte *337*, 428
Hyliote australe *337*, 424
    à dos violet *337*, 427
    des Usambaras *337*, 426
    à ventre jaune *337*, 422
Hypolaïs ictérine *113*, 136
    des oliviers *113*, 133
    pâle *113*, 129
    polyglotte *113*, 134
    d'Upcher *113*, 131

L

Lanielle à queue blanche *497*, 604
Locustelle fluviatile 93, *97*
    luscinioïde 94, *97*
    tachetée 92, *97*
Lusciniole à moustaches 100, *112*

M

Mélocichle à moustaches 85, *97*
Merle africain 38, *48*
    cendré 44, *49*
    kurrichane 41, *48*
    litsitsirupa 28, *33*
    noir *49*, 53
    olivâtre 34, *48*
    à plastron *49*, 55
    de São Tomé 31, *49*
Mignard enchanteur *497*, 498
Monticole angolais 6, *32*
    à doigts courts 4, *32*
    espion 3, *32*
    merle-bleu 12, *32*
    merle-de-roche 9, *32*
    rocar 1, *32*
    rougequeue 7, *32*

N

Nasique de Bocage *337*, 430
    grise 314, *337*
    jaune 312, *337*
    de Kemp 313, *337*
    de Kretschmer 317, *337*
    de Pulitzer 316, *337*
Noircap loriot *337*, 375

P

Parisome brune *401*, 420
    grignette *401*, 418
    de Layard *401*, 414
    sanglée *401*, 421
Phyllolaïs à ventre fauve 250, *320*
Phragmite aquatique 101, *112*
    des joncs 102, *112*
Poliolaïs à queue blanche 292, *337*
Pouillot de Bonelli 358, *400*
    brun 360
    à face rousse 366, *400*
    fitis 352, *400*
    à gorge jaune 364, *400*
    à grands sourcils 361, *400*
    de Laura 365, *400*
    ombré 362, *400*
    de l'Ouganda 368, *400*
    de Pallas 361
    siffleur 357, *400*
    à tête noire 367, *400*
    véloce 355, *400*

Prinia à ailes rousses 238, *241*
    aquatique 224, *241*
    à front écailleux 244, *320*
    à gorge blanche *241*, 246
    gracile 233, *241*
    grise 249, *320*
    à joues rousses *241*, 287
    du Karoo 228, *241*
    modeste 221, *241*
    du Namaqua 219, *241*
    pâle 225, *241*
    à plastron 226, *241*
    rayée 232, *241*
    de Roberts 237, *241*
    de São Tomé 230, *241*
    du Sierra Leone *241*, 248
    verte 242, *320*
Pririt de l'Angola 561, *597*
    de Blissett *545*, 562
    de Boulton *560*, 576
    du Cap *560*, 579
    châtain *545*, 555
    à collier *545*, 567
    à front blanc *545*, 570
    à gorge noire *545*, 571
    de l'Ituri *561*, 600
    à joues noires *561*, 595
    de Lawson *561*, 601
    molitor *560*, 584
    pâle *560*, 587
    pygmée *561*, 596
    à queue courte *560*, 578
    du Ruwenzori *561*, 575
    du Sénégal *561*, 591
    à taches blanches *545*, 558
    à tête grise *561*, 594
    à ventre doré *545*, 564
    de Verreaux *561*, 598
    de Vieillot *560*, 589
    de Woodward *560*, 582

R

Roitelet huppé 371, *400*
    à triple bandeau 370, *400*
Rousserolle africaine 107, *112*
    à bec fin *112*, 120
    des cannes *112*, 119
    effarvate 104, *112*
    à gros bec 122
    d'Irak *112*, 116
    stentor *112*, 118
    turdoïde *112*, 114
    verderolle 109, *112*

S

Sphénoèque du Cap 87, *97*

T

Tchitrec d'Afrique **531**, *544*
  de Bates **540**, *544*
  de Bedford *544*, **546**
  bleu *497*, **514**
  du Cap **526**, *544*
  du Congo **538**, *544*
  noir **527**, *544*
  à queue blanche *497*, **517**
  à queue frangée *497*, **523**
  du São Tomé **537**, *544*
  à tête noire *497*, **519**
  à ventre blanc *497*, **522**
  à ventre roux **542**, *544*